Chemistry

ADDISON-WESLEY

TEACHER'S EDITION

ANTONY C. WILBRAHAM

DENNIS D. STALEY

MICHAEL S. MATTA

EDWARD L. WATERMAN

Needham, Massachusetts
Upper Saddle River, New Jersey
Glenview, Illinois

Program Components

Student Edition
Teacher's Edition
Teaching Resources
- Chapter Review Modules
- Solutions Manual for Chapter Reviews
- Laboratory Practicals
- Laboratory Recordsheets
- Graphing Calculator Problems
- Overhead Transparency Preview
- Spanish Supplement
- Block Scheduling Guide
- How to Assess Student Work

Overhead Transparency Package
Laboratory Manual, ATE
Laboratory Manual
Small-Scale Chemistry Laboratory Manual, ATE
Small-Scale Chemistry Laboratory Manual
Probeware Lab Manual for Computers and Calculators
Guided Reading and Study Workbook, ATE
Guided Reading and Study Workbook
Prentice Hall Assessment System
Color Transparencies
Chem ASAP! CD-ROM
ActivChemistry CD-ROM
Resource Pro CD-ROM
Assessment Resources CD-ROM
Web site at www.phschool.com
SCIENCE NEWS at www.phschool.com
The Chemistry Place Internet Site
Chemistry Alive! Videodisc
Small-Scale Lab Video and Videodisc
CHEMedia™ Videodiscs
Problem Pro CD-ROM

Cover photographs: Clockwise from top left: Test tube with zinc in acid, Richard Megma, Fundamental Photographs; Molecular structure of buckyball, Ken Eward, Photo Researchers; Bunsen burner flame, and flask containing precipitate of lead iodide, Richard Megma, Fundamental Photographs; Scanning tunneling microscope image, Fran Heyl Associates.

About the cover: The collage of images on the cover illustrates the beauty and excitement of chemistry: the old and the new; the macro view and the micro view; the applied and the theoretical. Cover designed by Amanda Kavanagh/Ark Design, New York

ISBN 0-13-054847-2
4 5 6 7 8 9 10 06 05 04 03 02

Brief contents

TEACHER'S EDITION

STUDENT EDITION

Chemistry

ADDISON-WESLEY

Its content supports your teaching, its approach inspires your students

Addison-Wesley Chemistry is the nation's #1 chemistry program, providing more students with a core foundation in chemistry than any other program available. Comprehensive content coverage and an inviting, student-friendly approach have made this program the favorite of teachers and students across the country.

The new edition takes this winning combination a step further—providing even more resources to make chemistry accessible to all students. Still rich in content and rigor, the new edition of *Addison-Wesley Chemistry* provides the Chem/ASAP! student CD-ROM, dozens of small-scale labs, and a host of other new features.

PRENTICE HALL EXCLUSIVE!

Chem ASAP! CD-ROM

HELP YOUR STUDENTS MASTER CONCEPTS & IMPROVE THEIR PROBLEM-SOLVING SKILLS

ANIMATIONS illustrate important concepts, chemical processes, and applications to technology.

SIMULATIONS allow students to manipulate variables, observe results, and draw conclusions.

ASSESSMENT OPPORTUNITIES are plentiful, with more than 500 varied and interactive assessment questions from every section in the text.

PROBLEM-SOLVING TUTORIALS provide 100 detailed solutions to practice problems in the text. Each tutorial follows the established three-step problem-solving approach and concludes with a related practice problem.

See page xxi for screen grabs from this exciting product!

SMALL-SCALE LAB

MEASURING MASS AS A MEANS OF COUNTING

SAFETY

Wear your safety glasses and follow standard safety procedures as outlined on page 18.

PURPOSE

To determine the mass of several chemical compound samples and use the data to count atoms.

MATERIALS

- pencil
- paper
- plastic spoon
- chemicals shown in Figure A
- balance
- ruler

PROCEDURE

Measure the mass of one level teaspoon of sodium chloride (NaCl), water (H_2O), and calcium carbonate ($CaCO_3$). Make a table similar to Figure A to record your measured and calculated data.

	$H_2O(l)$	$NaCl(s)$	$CaCO_3(s)$
Mass (grams)			
Molar Mass (g/mol)			
Moles of each compound			
Moles of each element			
Atoms of each element			

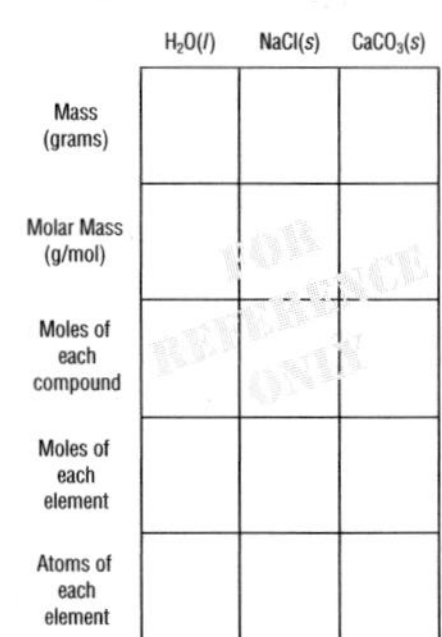

Figure A

ANALYSIS

Using your data, record the answers to the following questions in or below your data table.

1. Calculate the moles of NaCl contained in one level teaspoon and record the result in your table.

$$\text{moles of NaCl} = ?\text{ g NaCl} \times \frac{1\text{ mol NaCl}}{58.5\text{ g}}$$

2. Repeat Step 1 for the other compounds in Figure A. Use the periodic table if necessary to calculate the molar mass of water and calcium carbonate.
3. Calculate the moles of each element present in the teaspoon-sized sample of H_2O.

$$\text{moles of H} = ?\text{ mol } H_2O \times \frac{2\text{ mol H}}{1\text{ mol } H_2O}$$

Repeat for all the other compounds in your table.

4. Calculate the number of atoms of each element present in the teaspoon-sized sample of H_2O.

$$\text{atoms of H} = ?\text{ mol H} \times \frac{6.02 \times 10^{23}\text{ atoms H}}{1\text{ mol } H_2O}$$

Repeat for all the other compounds in your table.

5. Which of the three teaspoon-sized samples contains the greatest number of moles?
6. Which of the three compounds contains the most atoms?

YOU'RE THE CHEMIST!

The following small-scale activities allow you to develop your own procedures and analyze the results.

1. Design It! Can you use the technique of measuring volume as a means of counting? Design and carry out an experiment to do it!
2. Design It! Design an experiment that will determine the number of atoms of calcium, carbon, and oxygen it takes to write your name on the chalkboard with a piece of chalk. Assume chalk is 100 percent calcium carbonate, $CaCO_3$.

Chemical Quantities 187

Small-Scale Labs

A PRENTICE HALL EXCLUSIVE!

SMALL-SCALE LABS ALLOW YOU TO REGULARLY BRING CHEMISTRY CONTENT INTO THE LAB. With small-scale labs, students can do and learn more chemistry than through conventional labs. Students learn to be creative and inventive with chemistry, to be good problem solvers, and to communicate their findings.

SMALL-SCALE LABS ARE TIME-EFFICIENT. Setup is made easy for the teacher and the student because the experimental format does not rely on complicated instructions. Experimental results are immediate and obvious, and the time students take to set up, manipulate, and clean equipment is minimal.

SMALL-SCALE LABS USE SAFE, INEXPENSIVE EQUIPMENT. Nontraditional, student-designed and built equipment draws on everyday materials—materials that are safe for the classroom and friendly for the environment. Cleanup with a paper towel takes seconds, and most materials can be easily disposed of in the wastebasket.

New student edition tools for inspiring today's students

Discover It!
Quick kick-off activity at the start of every chapter allows students to explore and discover concepts before they read about them in the chapter.

Section Openers
Motivating bits of information that capture student interest—include historical nuggets, unusual occurrences, and everyday applications of chemistry.

Small-Scale Labs
Abundant, hands-on labs utilize inexpensive, readily available materials. Now your students can carry out sophisticated and diverse investigations on a small scale!

Mini Labs
Quick, effective hands-on labs allow students to apply chemistry concepts and skills—one per chapter.

CHEMath
Full-page, point-of-use math refresher includes step-by-step instructions and practice problems.

Chemistry Serving
Innovative new feature illustrates the important role chemistry plays in serving the environment, industry, the consumer, and society.

Chemistry in Careers
Highlighted on each Chemistry Serving page and featured on pages 868–881, this feature discusses career opportunities that require an understanding of chemistry.

Link to...
Sidebar feature illustrates how concepts in chemistry relate to both science and nonscience disciplines.

Standardized Test Prep
At the end of each chapter, there is a practice test with a variety of question formats to prepare your students for state and national exams.

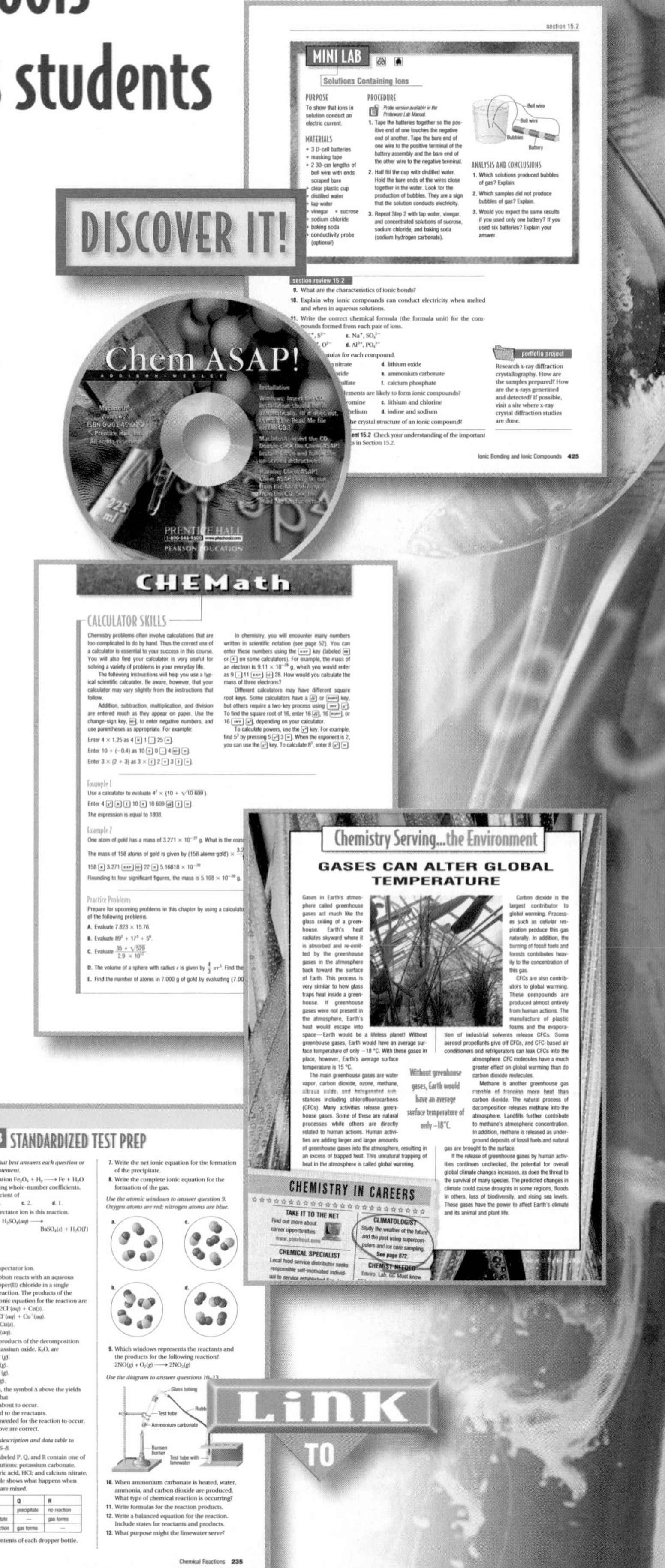

Technology tools designed to inspire today's chemistry students

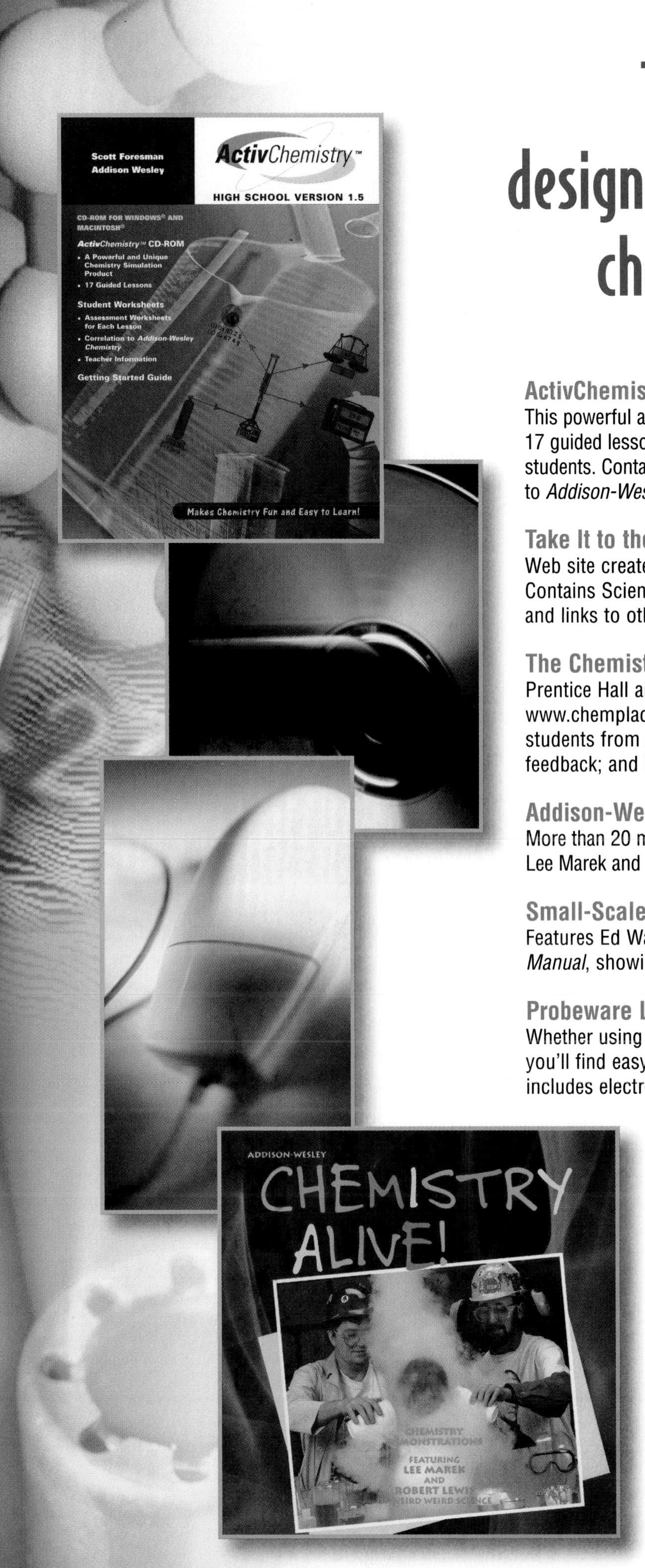

ActivChemistry™ CD-ROM
This powerful and unique chemistry simulation product contains 17 guided lessons that make tough-to-visualize concepts come alive for students. Contains assessment worksheets for each lesson, correlations to *Addison-Wesley Chemistry*, and teacher support materials.

Take It to the Net
Web site created exclusively for use with *Addison-Wesley Chemistry*. Contains Science News® updates, activities for every chapter, self-tests, and links to other sites. **www.phschool.com**

The Chemistry Place™ Web Site
Prentice Hall and Peregrine Publishers have partnered to create www.chemplace.com, containing investigative projects involving students from around the world; review exercises with instant feedback; and interactive tutorials.

Addison-Wesley Chemistry Alive!
More than 20 motivating and dramatic chemistry demonstrations from Lee Marek and Robert Lewis, high school teachers and gifted presenters.

Small-Scale Chemistry Lab Videodisc or Videotape
Features Ed Waterman, author of the *Small-Scale Chemistry Laboratory Manual*, showing you and your students small-scale techniques.

Probeware Lab Manual
Whether using Vernier, Texas Instruments or PASCO® probes and sensors, you'll find easy-to-use labs with step-by-step instructions. CD-ROM includes electronic labs for quick start up and technical support.

Problem Pro CD-ROM
Problem/Solution generator lets you customize and generate an endless number of practice problems. Helps create worksheets, quizzes, and tests.

CHEMedia™
Videodiscs add real-world connections to chemistry topics through demos, field trips, animations, and explorations.

Science News®
Keep current with Science News—weekly science content updates on the Web! Organized by chapter in the Student Edition.

Comprehensive Teacher Resources provide a wealth of classroom tools

Laboratory Manual with Teacher's Edition
52 laboratory investigations to extend learning; quantitative problems for developing students' critical thinking and inquiry skills; and a wide variety of teaching strategies.

Small-Scale Chemistry Laboratory Manual with Teacher's Edition
40 additional laboratory investigations using inexpensive materials. Each lab has a "Now It's Your Turn" feature that can be used for performance-based assessment; ideal for block scheduling.

Laboratory Practicals
30 lab practicals help you assess your students' lab skills and provide alternative assessment opportunities; structured for testing students individually or in pairs.

Laboratory Recordsheets
Reproducible recordsheets give students a structured opportunity to answer all of the questions in the mini labs and small-scale labs and allow you to quickly assess performance.

Chapter Review Modules
Seven chapter modules, covering four chapters each. Chapter modules contain planning guides, section reviews, practice problems, interpreting graphics, vocabulary reviews, chapter tests, and quizzes.

Solutions Manual for Chapter Reviews
Detailed solutions to all of the problems in the student text. You can have your students use the manual to assess their own work.

Spanish Supplement
Translation of all Chapter Review concept summaries, plus definitions of all key terms from the chapter.

Block Scheduling Guide
Comprehensive, daily outlines—written by block-scheduling practitioner Ed Waterman—for a complete chemistry course organized by chapter and topic. Contains suggested labs, mini-labs, other activities, guided practice, and homework assignments.

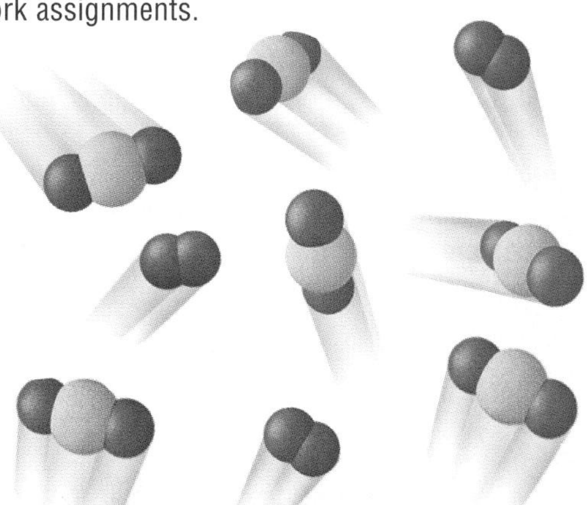

Additional support materials allow you to more effectively manage instruction & motivate students

Teacher's Edition
Three-step lesson cycle—Engage, Teach, and Assess—provides simple, effective structure for presenting content. Numerous activities and teacher demos provide a wealth of opportunities for hands-on learning. Convenient aids for course organization allow you to customize lessons for conceptual, standard, and honors courses. Extra practice problems provided at point-of-use.

Overhead Transparency Package
Full-color transparencies—80 in all—for key concepts in the program, accompanied by a booklet of strategies for using them.

Assessment Resources CD-ROM
User-friendly software for creating quizzes, tests, and exams.

Resource Pro® CD-ROM
Contains all of the teaching resources on one CD-ROM—view, customize, and print resources with the click of a button! Lesson planning software allows you to create lesson plans for the day, month, and year, as well as import local objectives.

Probeware Lab Manual
Whether using Vernier, Texas Instruments or PASCO® probes and sensors, you'll find easy-to-use labs with step-by-step instructions. CD-ROM includes electronic labs for quick start up.

Prentice Hall Assessment System
Make sure your students have the content and skills they need for high school exit exams. The System includes practice tests, test-taking strategies, and options for content remediation.

Guided Reading and Study Workbook
Promotes and enhances study skills using innovative questioning strategies and exercises. Can build a record of student work to use as a study aid.

PROGRAM PHILOSOPHY

Chemistry for All Students

Chemistry is a fascinating, dynamic, and exciting area of study. A successful high school chemistry course must reflect these characteristics. It must sufficiently prepare those students who want to learn the basics of chemistry and those who are preparing for a more rigorous course of study at the post-secondary level. For too many high school students, the study of chemistry is little more than hundreds of unrelated details and vocabulary words. *Addison-Wesley Chemistry* offers a coherent approach that connects these seemingly unrelated details into broad patterns and applies these connections to the observable world.

The following goals guided the efforts of all those involved in the program.

- **To motivate, instruct, and excite students about the science of chemistry**
- **To provide comprehensive scope and in-depth coverage of major topics appropriate for a high school chemistry course**
- **To address the spirit and specifics of the National Science Standards in a clear and comprehensive manner**
- **To emphasize process and higher-order thinking skills in ways that engage and reward students**
- **To use real, familiar, and frequent examples of chemistry concepts that involve students in learning and ensure their success**
- **To provide a unique and attractive design that enhances content presentation, concept development, and visual-learning opportunities**
- **To support heterogeneous student populations and diverse teaching styles with a wide range of instructional tools**
- **To incorporate a full-spectrum technology program into the product**

"Doing" Science

Studies have shown that comprehension and retention increase dramatically when the instructional focus is shifted from the role of the student as a passive receptor of facts to that of an active participant in the discovery process. Toward this goal, *Addison-Wesley Chemistry* provides a variety of opportunities to develop and practice the art of doing science. In each chapter, Mini Labs and Small-Scale Labs allow students to experience chemistry as a process. The *Laboratory Manual* and *Small-Scale Chemistry Laboratory Manual* extend the hands-on experience by providing more than ninety opportunities for active learning. The *Probeware Lab Manual for Computers and Calculators* offers Vernier, PASCO®, and Texas Instruments versions of selected Student Edition labs. A CD-ROM is included, which provides User's Guides and electronic versions of the labs.

Skills Development

Addison-Wesley Chemistry provides ample opportunities for the student to develop science process skills, such as observing, predicting, and measuring, as well as writing and communication skills. Numerous and varied suggestions for activities expand the range of skill-building by encouraging students to conduct independent research, deliver oral presentations, design visual displays, express themselves in a range of written modes, and delve deeply into specific subject areas.

Interest and Readability

No textbook, no matter how well planned, can effectively reach its audience if it is not written in a manner that is friendly, inviting, readable, and understandable. The goal of *Addison-Wesley Chemistry* is to engage the interest of all students, and encourage them to learn chemistry because it is exciting, interesting, useful, relevant, and *fun!* Current and compelling stories, enlightening introductions, informative openers, and connections among content blocks help draw the student into the subject. Features expand the scope of the text, making chemistry fascinating and real.

THE LESSON CYCLE

Learning is an ongoing process that occurs in cycles. Similarly, each lesson should follow a cycle that encourages and enhances the learning process. In the Teacher's Edition, the lesson cycle of engage, teach, and assess helps make learning fun and meaningful.

1 Engage

In the first step of the lesson cycle, the teacher engages the student's interest and inspires curiosity. Students are encouraged to activate prior knowledge and ask what they can learn about the new topic. During this stage of the lesson, misconceptions are explored and the teacher can identify target areas on which to focus. The *Addison-Wesley Chemistry Teacher's Edition* employs the following techniques to engage the students in the first step of the lesson cycle.

Use the Visual provides suggestions for launching a discussion using the photograph and text that introduces the section.

Check Prior Knowledge, which appears in many sections, helps teachers determine students' previous knowledge of the subject to be studied. This feature allows the teacher to assess the students' background and unveil misconceptions.

Motivate and Relate provides suggestions for connecting content and visuals to real-world situations and applications, with a focus on students' relevant experiences.

Teacher Demo or Activity serves to pique students' interest with a relevant, active experience. The activity inspires the students to seek ways to answer "how" and "why" questions, rather than to simply acquire information.

2 Teach

During the second step of the lesson cycle, the teacher serves as a guide while the students explore phenomena, attempt to answer questions raised during the first step of the cycle, and begin to construct an understanding of a concept through hands-on participation. The teacher helps students to develop understanding through communication, construction of models, and critical thinking. Connections are made to other disciplines as well as to social and community issues.

Discuss presents topics that relate to or reinforce the concepts in the student text and provides suggested topics for discussion among students.

Use the Visual provides suggestions for incorporating a photograph, diagram, or illustration into the lesson, which helps to engage the visual, spatial, or nonverbal learner in concept development.

Teacher Demo and Activity provide an opportunity for students to witness or experience the concepts described in the text. All students should participate in class activities, which may include the use of manipulatives or role playing.

Build Writing Skills helps teachers to encourage writing as an integral part of learning. Writing activities may include research, letters to the media, and creative writing projects.

Think Critically encourages students to think beyond what is written on the page. This feature may present a problem or provide a series of questions to help guide students.

Cooperative Learning gives practical suggestions for group work in situations such as problem-solving and hands-on activities.

Connections to other sciences and disciplines, which appear in many sections, extend the learning cycle. These strategies give teachers an opportunity to integrate a variety of sciences and other disciplines into their curriculum.

3 Assess

During the assessment step of the lesson cycle, students demonstrate their understanding and mastery of concepts and skills by answering questions based on their observations and experiences during the lesson. This activity allows teachers to identify areas needing further clarification, and allows the students to evaluate their own success in achieving learning goals.

Evaluate Understanding uses questioning, student participation, written explanations, and student generated materials to demonstrate learning.

Reteach presents the teacher with an alternative strategy to reinforce section content and concepts.

COOPERATIVE LEARNING

Cooperative learning is an organizational scheme for the learning environment that requires students to work collaboratively to carry out structured activities. Rather than stressing individual or competitive accountability in which students work alone or against one another, cooperative learning emphasizes group responsibility, mastery, respect, and success. Students work in small groups to solve a problem or create a product to which each group member contributes and for which each group member is held equally accountable. Interdependence is essential for the group to be successful in achieving its designated goal.

Cooperative Learning Groups

Cooperative learning groups should range in number from two to six members and should be heterogeneous with respect to gender, ethnicity, and ability. Once a group is established, it should remain together until the activity is completed. Do not dissolve a group that is having difficulty working together socially or staying on task. Keeping the group intact helps students learn the social interaction skills needed to solve problems and complete tasks through cooperation and collaboration.

Collaborative/Social Skills

The development and enrichment of social skills are basic to successful collaboration and cooperation. For each cooperative learning activity, you should assign a specific social skill. Social skills include listening carefully, taking turns, sharing resources, encouraging participation, treating others with respect, providing constructive feedback, resolving conflict, reaching consensus, and explaining and helping without simply giving answers.

Self-Evaluation

In cooperative learning environments, students should be encouraged to become involved in the evaluation process. You can accomplish this by giving them adequate time to reflect on the activity. You should remind students that for any project all members of a group should be prepared to explain what the group has done.

Suggested Roles in Cooperative Groups

Members of a cooperative learning group must assume a variety of roles for the group to collaborate effectively. The roles will be determined primarily by the nature of the project. Some suggested roles are

- **Principal Investigator** **The Principal Investigator manages the tasks within the activity and insures that all members understand activity goals and content. The Principal Investigator serves as a liaison between the teacher and the group, reads instructions, checks results, and facilitates group discussions.**
- **Primary Researcher** **The Primary Researcher assumes the role of group manager for activities involving research. The Primary Researcher should explain tasks and identify resources.**
- **Materials Manager** **The Materials Manager assembles and distributes the materials and equipment. The Materials Manager is responsible for operating equipment, checking the results of the activity, and ensuring that all equipment is returned clean and in good working order. The Materials Manager also reports damaged or unsafe equipment.**
- **Data Collector** **The Data Collector is responsible for gathering, recording, analyzing, and organizing the group's data into the presentation format.**
- **Data Organizer** **The Data Organizer coordinates the group data and certifies each group member's contributions to the activity.**
- **Presenter** **The Presenter reports the results of an activity. The Presenter may field questions posed by classmates about the group's activity or delegate specific types of questions to other members of the group.**
- **Timekeeper** **The Timekeeper is responsible for keeping track of time, monitoring noise level, checking the results of an activity for accuracy, and making sure proper safety precautions are observed.**

CONCEPT MAPPING

Concept mapping helps students learn in a more meaningful way by making key concepts and abstract information more understandable and useful. A concept map links together main ideas in a format that emphasizes interactions, interrelationships, pathways, and hierarchies. Concept mapping relies on a basic learning principle: People learn best by incorporating new information into what they already know. Using concept maps, students can assimilate new information, grasp new relationships, and perceive new connections.

Concept mapping helps students to

- **Link new ideas with previously learned material**
- **Organize information and establish relationships between ideas**

Concept mapping helps teachers to

- **Introduce new concepts to students in an interesting way**
- **Determine student misconceptions**

Most important, concept mapping can give both students and teachers new insights into chemistry.

Constructing a Concept Map

Here are some suggestions for creating concept maps.

- **Place the main concept in an oval at the top of the map.**
- **Draw branches from the main concept to subordinate concepts also in ovals.**
- **Draw branches from the subordinate concepts to specific examples.**
- **Use each concept only once in the map.**
- **Write the concepts as nouns or short phrases.**
- **Link the nouns by verbs, adverbs, and prepositions.**
- **Link every concept to at least one other concept.**
- **Do not let linking lines intersect.**
- **Make sure linking lines between concepts run in different directions on the map.**
- **Make certain that any two concepts and their linking words form a complete thought.**

Teaching Concept Mapping Skills

The best way to teach students about concept mapping is to provide plenty of practice. Every chapter of *Addison-Wesley Chemistry* has a concept mapping exercise in the Student Study Guide. Students use the terms provided to construct a concept map that organizes major ideas of the chapter.

To encourage students, show them examples of good maps. Good maps link concepts in a logical arrangement and do not contain misconceptions. However, remind students that concept maps are intended to be flexible tools for learning: There is no single correct way to do a concept map because there are many different ways to represent the same ideas.

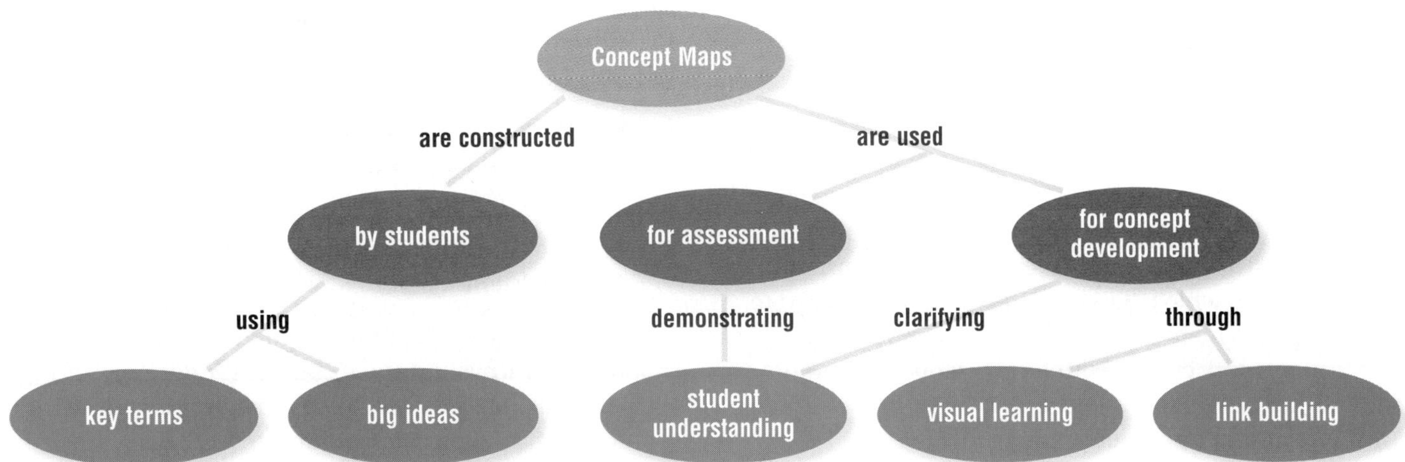

STUDENTS WITH INDIVIDUAL NEEDS

Today's science teachers must structure their teaching to accommodate a student population with diverse abilities, interests, and learning styles that span a broader range than ever before. A science classroom should be a positive learning environment, with optimum opportunities for all students to achieve success. Challenges such as visual, physical, or hearing impairments, limited English proficiency (LEP), and different levels of learning abilities can be addressed through specific teaching strategies. General strategies that will help you prepare for the challenge of teaching diverse student populations are presented here. Specific tips for applying these strategies appear throughout the Teacher's Edition in boxes titled Meeting Diverse Needs.

Learning Styles

Even among students with similar abilities and skills, there is a variety of cognitive and behavior patterns with which students approach the task of learning. Some students learn by visualizing spatial and physical relationships. Others learn through understanding patterns, rhythms, or motion. Some students prefer to work alone, while others are more successful in groups or cooperative situations.

To accommodate the wide variety of learning styles that are likely to comprise the population of any science classroom, it is essential that the instructional program be flexible and varied. A mixture of techniques should be combined within each lesson to give all students an equal opportunity to succeed. Class time should include a blend of lecture, demonstrations, and hands-on and minds-on inquiry, as well as the use of text and multimedia resources.

Physical, Visual, or Hearing Impairments

Students with a wide variety of impairments are now able to share in the process of learning science alongside their nonimpaired classmates. Teachers are faced with the task of ensuring that these students participate in the learning activities and have an equal opportunity to succeed. The nature and severity of the impairment will determine which classroom techniques will need to be adjusted to include all students. It is essential that the teacher work closely with the students, the parents, and the school guidance counselor to determine how best to address the students' individual needs.

- **Encourage independence and help nonimpaired students and adults to understand the students' special needs.**
- **Make sure the physical arrangement of classroom furniture and other objects can accommodate wheelchairs, guide dogs, or other equipment. Materials should be within reach and accessible to all students.**
- **If appropriate, enlist the aid of one or two students to assist the challenged student as necessary, such as reading aloud, acting as a sight peer during lab work, or providing physical assistance.**
- **During physical activities, provide physically challenged students with roles in which they can participate in a nonphysical way, such as Timekeeper or Materials Manager.**
- **Seat students with limited hearing or visual problems near the front of the classroom. Face the class when speaking.**

Students "At Risk"

Certain students, due to their ability levels, behavioral or emotional challenges, or previous educational experiences, are at greater risk than others of failing to achieve success. To maintain a positive learning environment, emphasis should be placed on reinforcing success and minimizing failure.

Concepts should be presented in different ways, using relevant, familiar examples. Learning goals should be presented frequently, at a pace that enables success to be achieved. *Addison-Wesley Chemistry* provides course planning options on pages T16 and T17, and in the Time Saver features at the beginning of each chapter, to help you construct an appropriate curriculum. Other general strategies for teaching at-risk students include

- **Provide clear, specific instruction, assignments, and learning goals that students can achieve.**
- **Use hands-on activities and demonstrations to reinforce concepts. Provide opportunities to repeat and practice new skills.**
- **Check for understanding frequently. Provide students with a variety of choices for demonstrating their learning, including speaking, writing, drawing, and multimedia presentations.**

- **Use Section Review Worksheets to build and reinforce comprehension; use Vocabulary Worksheets to build confidence in using challenging scientific terms. Both items can be found in the *Review Modules* in the Teacher's Resource Package.**
- **Use the worksheets in the *Guided Reading and Study Workbook* to provide students with an organized approach to concept mastery.**

Limited English Proficiency (LEP) Students

Students who speak English as a second language face a variety of challenges in the science classroom. In addition to the challenge of achieving mastery in science, students must overcome a language barrier and discrepancies in educational background, as well as the social concerns resulting from cultural differences. It is difficult to grasp complex and abstract concepts when vocabulary and pronunciation provide barriers. Strategies for teaching LEP students include

- **Speak clearly, using body language and gestures for emphasis.**
- **Have students create their own "dictionary" of key terms and unfamiliar words that includes a term in English, the term in their native language, and a brief definition.**
- **Use illustrations and other visual learning resources to help students grasp concepts. Provide labeling and vocabulary exercises to help build power.**
- **Provide newspapers and magazines to make connections between the science content and cultural issues. Incorporate examples from various cultures into discussions and encourage student participation.**
- **Encourage students to make use of English-language dictionaries.**
- **Provide audiocassettes of lectures and/or text materials to reinforce classroom learning and to provide model language patterns.**
- **Have picture dictionaries, wall charts, and other visual learning resources available.**
- **Use the worksheets in the *Guided Reading and Study Workbook* to provide students with an organized approach to concept mastery.**

Gifted Students

Providing resources and enriched learning opportunities for the gifted student can be as important as accommodating impaired students. When learning is no longer challenging and stimulating, educational opportunities can easily be missed. Although gifted students can sometimes provide assistance in developing other students' comprehension, the activities of the gifted student must focus on maximizing his or her own educational opportunities.

- **Provide students with the opportunity to work on independent projects of interest to the student. If possible, make arrangements for the student to accelerate his or her progress in selected subjects.**
- **Encourage students to use public resources, such as materials provided by government agencies, and to become involved in community activities.**
- **Emphasize the use of high-level cognitive skills, such as developing analogies, synthesizing new concepts, and analyzing relationships within and beyond the science disciplines. Encourage students to take intellectual "risks."**

SKILLS IN CHEMISTRY

Like all sciences, chemistry involves more than just acquiring a knowledge foundation. Instead, the study of chemistry can offer students opportunities to learn, to practice, and to master skills that are relevant to their everyday world and to their professional goals.

Chemistry is more than a collection of information to be read, discussed, tested, and assessed for a numerical grade. Content mastery can no longer be evaluated by students' abilities to recall vocabulary terms, memorize unrelated facts, or repeat a theory without understanding how the content and concepts are interconnected. Inquiry, discovery, and investigation must also be a part of each student's science experience. Students need to develop process skills and thinking skills.

Throughout the *Addison-Wesley Chemistry* program, there are many opportunities for students to learn, practice, and master a variety of skills. These skills include thinking skills, science process skills, and communication skills—all essential to academic, personal, and professional success.

Thinking Skills

Critical thinking is a process that uses a variety of skills including analyzing and solving problems, making decisions, and evaluating reasoning. Throughout this program, students are provided with many opportunities to develop reasonable conclusions, evaluate information, make decisions, and assess issues with facts and reasoned judgment. Critical thinking and process skills can be found in the following features of the *Addison-Wesley Chemistry* program: *Section Reviews, Sample Problems, Mini Labs, Small-Scale Labs, Chapter Reviews,* and *Chem ASAP! CD-ROM.*

Process Skills

Process skills are those skills usually associated with gathering, recording, organizing, interpreting, and analyzing information about the natural world. While the process skills are often identified as *science* skills, they are also used in many other settings to make decisions and to assess situations. As students use process skills, they will also use thinking skills—and a pattern emerges linking the two types of skills. Students have ample opportunities to learn and practice process skills throughout the *Addison-Wesley Chemistry* program. A wide variety of activities helps the students to develop the following skills: classifying; measuring; predicting; observing; estimating; inferring; modeling; hypothesizing; collecting and recording data; organizing and analyzing data.

Communications Skills

The ability to communicate effectively is of paramount importance in today's world. Communication skills such as writing, reading, speaking, and listening are prerequisites to acquiring, analyzing, and applying information. In *Addison-Wesley Chemistry* skills are an integral part of the Student Edition and other components.

Reading In a society that transmits an ever-increasing amount of information, reading is an essential life skill. Students must be effective readers who can read with understanding. Because the number of new terms in chemistry may be formidable, reading and comprehending the content and concepts can be difficult for many students. However, it is essential that students develop a working knowledge of the content-specific terms.

Writing Writing is a powerful tool for learning. Competency in writing is necessary for success in the classroom and through life. Writing uses a number of process and critical thinking skills. It involves collecting, recording, and organizing information, as well as sequencing, establishing relationships among ideas, ordering and organizing paragraphs, and revising written materials until a final polished draft has been completed.

To help all students develop effective reading and writing skills, numerous and varied pedagogical tools have been incorporated into *Addison-Wesley Chemistry.* They include: *Section Objectives; Key Terms; Concept Maps; Discover It! Activities; Mini Labs; Small-Scale Labs; Chemistry Serving* and *Link to* features; *Concept Challenges;* and *Portfolio Projects.*

COURSE PLANNING OPTIONS

Most chemistry teachers are faced with the challenge of educating students of varied academic abilities. It takes careful planning to map out a method by which students can best be exposed to the major topics in the high school chemistry curriculum.

Addison-Wesley Chemistry is organized in a way that can meet the needs of a variety of students and teachers. The textbook is comprehensive enough to satisfy the most comprehensive high school chemistry curriculum, while remaining readable, interesting, and accessible to less-academically motivated students. In addition, the division of chapters into manageable numbered sections provides today's teachers with the flexibility needed to teach classes comprised of students with diverse learning levels.

Pages T16 and T17 contain section planning options to assist teachers in designing lessons that meet the needs of all students. Symbols are used as follows.

The Conceptual Curriculum

In recent years, more students have been required to take chemistry, either in a mainstream setting or in a conceptual chemistry course. Thus, classroom populations have more students with below-average academic abilities in reading comprehension and math skills. The conceptual curriculum allows for a more concrete approach to ensure these students achieve their maximum potential. Course options are designated with symbols as follows.

● **Concrete concepts**—These sections typically describe fundamental (concrete) chemical concepts without relying on highly theoretical (abstract) or mathematical principles. These topics will help students focus on qualitative aspects of chemistry.

○ **More abstract concepts or math/problem-solving**—These sections offer a combination of concrete topics plus theoretically abstract topics and/or mathematically challenging problem solving. In these sections, the emphasis on theoretical aspects or the degree of problem solving can be determined by the teacher.

Some sections have no symbol because they are beyond the scope of a typical conceptual curriculum and do not include concepts that are prerequisites for later chapters.

The Standard Curriculum

The standard curriculum is designed for students considered average in ability and achievement levels. These students should have little difficulty reading and understanding most of the material presented, and they will have sufficient math skills for tackling the problems in the textbook. Course options are designated with symbols as follows.

■ **Core content**—These sections are suitable for all students in a standard curriculum. A course that includes most of these sections will prepare students for other science courses and for becoming scientifically literate citizens.

□ **Extension topics**—These sections may be covered as time allows or according to the interest of the teacher and the needs and/or problem-solving skills of the students.

Some sections have no symbol because a teacher would not have sufficient time to cover the topic well, or because the material is too challenging or abstract, or is not usually required in state or national curricula. Assume if there is no symbol that the section is not a prerequisite for later chapters.

The Honors Curriculum

The standard curriculum is designed to be used for students considered above average in their reading comprehension and general scholastic abilities. These students should be able to master most of the content in the textbook with less time and effort. Course options are designated with symbols as follows.

◆ **Core honors content**—These sections are suitable for all students in an honors curriculum. A course including these sections will prepare students for advanced science courses and for college coursework.

◇ **Options to accelerate**—These sections may be covered more rapidly using homework assignments. Students may only need to review this material because they have studied it in a previous course or a previous chapter.

COURSE PLANNING OPTIONS

		C	S	H
1	**Introduction to Chemistry**			
1.1	Chemistry	●	■	◇
1.2	Chemistry Far and Wide	●	■	◇
1.3	Thinking Like a Scientist	●	■	◇
1.4	How to Study for Chemistry	●	■	◇
2	**Matter and Change**			
2.1	Matter	●	■	◇
2.2	Mixtures	●	■	◆
2.3	Elements and Compounds	●	■	◆
2.4	Chemical Reactions	●	■	◆
3	**Scientific Measurement**			
3.1	The Importance of Measurement	○	■	◆
3.2	Uncertainty in Measurements	○	■	◆
3.3	International System of Units	●	■	◇
3.4	Density	●	■	◆
3.5	Temperature	●	■	◆
4	**Problem Solving in Chemistry**			
4.1	What Do I Do Now?	●	■	◇
4.2	Simple Conversion Problems	○	■	◆
4.3	More-Complex Problems		□	◆
5	**Atomic Structure and the Periodic Table**			
5.1	Atoms	●	■	◇
5.2	Structure of the Nuclear Atom	●	■	◆
5.3	Distinguishing Between Atoms	○	■	◆
5.4	The Periodic Table: Organizing the Elements	●	■	◇
6	**Chemical Names and Formulas**			
6.1	Introduction to Chemical Bonding	●	■	◆
6.2	Representing Chemical Compounds	●	■	◆
6.3	Ionic Charges	●	■	◆
6.4	Ionic Compounds	●	■	◆
6.5	Molecular Compounds and Acids	●	■	◆
6.6	Summary of Naming and Formula Writing	●	■	◆
7	**Chemical Quantities**			
7.1	The Mole: A Measurement of Matter	○	■	◆
7.2	Mole–Mass and Mole–Volume Relationships	○	■	◆
7.3	Percent Composition and Chemical Formulas		■	◆
8	**Chemical Reactions**			
8.1	Describing Chemical Change	●	■	◆
8.2	Types of Chemical Reactions	●	■	◆
8.3	Reactions in Aqueous Solutions	○	■	◆
9	**Stoichiometry**			
9.1	The Arithmetic of Equations	●	■	◇
9.2	Chemical Calculations		■	◆
9.3	Limiting Reagent and Percent Yield		□	◆
10	**States of Matter**			
10.1	The Nature of Gases	○	■	◆
10.2	The Nature of Liquids	●	■	◆
10.3	The Nature of Solids	○	■	◆
10.4	Changes of State		□	◆
11	**Thermochemistry—Heat and Chemical Change**			
11.1	The Flow of Energy—Heat	○	■	◆
11.2	Measuring and Expressing Heat Changes	○	■	◆
11.3	Heat in Changes of State	○	□	◆
11.4	Calculating Heat Changes			◆
12	**The Behavior of Gases**			
12.1	The Properties of Gases	●	■	◇
12.2	Factors Affecting Gas Pressure	●	■	◇
12.3	The Gas Laws	○	■	◆
12.4	Ideal Gases		■	◆
12.5	Gas Molecules: Mixtures and Movements	○	■	◆
13	**Electrons in Atoms**			
13.1	Models of the Atom	●	■	◆
13.2	Electron Arrangement in Atoms	○	■	◆
13.3	Physics and the Quantum Mechanical Model	○	□	◆

Conceptual Curriculum	Standard Curriculum	Honors Curriculum
● Concrete concepts	■ Core content	◆ Core honors content
○ More abstract concepts or math/problem-solving	□ Extension topics	◇ Options to accelerate

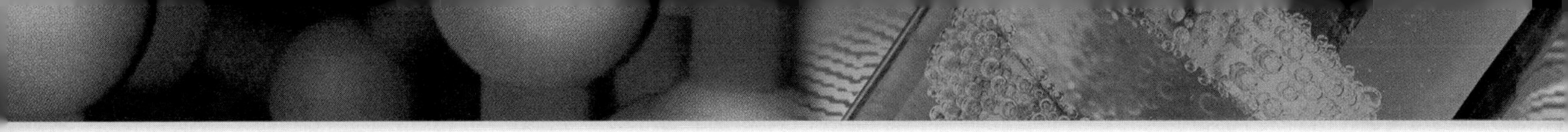

14 Chemical Periodicity

Section	Title	C	S	H
14.1	Classification of the Elements	●	■	◇
14.2	Periodic Trends	●	■	◆

15 Ionic Bonding and Ionic Compounds

Section	Title	C	S	H
15.1	Electron Configuration in Ionic Bonding	●	■	◆
15.2	Ionic Bonds	●	■	◆
15.3	Bonding in Metals	○	□	◆

16 Covalent Bonding

Section	Title	C	S	H
16.1	The Nature of Covalent Bonding	●	■	◆
16.2	Bonding Theories		□	◆
16.3	Polar Bonds and Molecules	●	■	◆

17 Water and Aqueous Systems

Section	Title	C	S	H
17.1	Liquid Water and Its Properties	●	■	◆
17.2	Water Vapor and Ice	●	■	◆
17.3	Aqueous Solutions	●	■	◆
17.4	Heterogenous Aqueous Systems		□	◇

18 Solutions

Section	Title	C	S	H
18.1	Properties of Solutions	●	■	◆
18.2	Concentrations of Solutions	○	■	◆
18.3	Colligative Properties of Solutions		■	◆
18.4	Calculations Involving Colligative Properties		□	◆

19 Reaction Rates and Equilibrium

Section	Title	C	S	H
19.1	Rates of Reaction	●	■	◆
19.2	Reversible Reactions and Equilibrium	○	□	◆
19.3	Determining Whether a Reaction Will Occur		□	◆
19.4	Calculating Entropy and Free Energy			◆
19.5	The Progress of Chemical Reactions			◆

20 Acids and Bases

Section	Title	C	S	H
20.1	Describing Acids and Bases	●	■	◇
20.2	Hydrogen Ions and Acidity	○	■	◆
20.3	Acid–Base Theories	○	■	◆
20.4	Strengths of Acids and Bases		□	◆

21 Neutralization

Section	Title	C	S	H
21.1	Neutralization Reactions	●	■	◆
21.2	Salts in Solution		□	◆

22 Oxidation-Reduction Reactions

Section	Title	C	S	H
22.1	The Meaning of Oxidation and Reduction	●	■	◆
22.2	Oxidation Numbers		■	◆
22.3	Balancing Redox Equations		□	◆

23 Electrochemistry

Section	Title	C	S	H
23.1	Electrochemical Cells	●	■	◆
23.2	Half-Cells and Cell Potentials	○	□	◆
23.3	Electrolytic Cells	○	■	◆

24 The Chemistry of Metals and Nonmetals

Section	Title	C	S	H
24.1	The *s*-Block Elements: Active Metals	●	□	◇
24.2	The *p*-Block Elements: Metals and Nonmetals	●	□	◇
24.3	The *d*- and *f*-Block Elements: Transition and Inner Transition Metals	●	□	◇
24.4	Hydrogen and Noble Gases	●	□	◇

25 Hydrocarbon Compounds

Section	Title	C	S	H
25.1	Hydrocarbons	●	■	◆
25.2	Unsaturated Hydrocarbons	○	■	◆
25.3	Isomerism		□	◆
25.4	Hydrocarbon Rings	○	□	◆
25.5	Hydrocarbons from the Earth	●	■	◇

26 Functional Groups and Organic Reactions

Section	Title	C	S	H
26.1	Introduction to Functional Groups	○	■	◆
26.2	Alcohols and Ethers		□	◆
26.3	Carbonyl Compounds		□	◆
26.4	Polymerization		□	◆

27 The Chemistry Of Life

Section	Title	C	S	H
27.1	A Strategy for Life	●	■	◇
27.2	Carbohydrates		□	◆
27.3	Amino Acids and Their Polymers		□	◆
27.4	Lipids		□	◆
27.5	Nucleic Acids		□	◆
27.6	Metabolism		□	◆

28 Nuclear Chemistry

Section	Title	C	S	H
28.1	Nuclear Radiation	●	■	◆
28.2	Nuclear Transformations	○	■	◆
28.3	Fission and Fusion of Atomic Nuclei	●	■	◆
28.4	Radiation in Your Life	●	■	◇

ASSESSING STUDENT PROGRESS

Learning Styles

Learning occurs at a variety of levels. Similarly, students process new information using different learning styles. The learning style of a student refers to the way in which the student is best able to learn and retain information. Some students learn best by reading; others learn by listening or seeing a visual presentation. Still other students learn best by touching and manipulating objects as in a laboratory environment. To ensure student success, it is important to provide opportunities for use of each of these learning styles.

Assessment Options

Just as there are different styles of learning, there are also different methods for assessing student progress and understanding. These methods, which can be formal or informal, include the following: questions for student self-assessment; traditional paper-and-pencil tests; classroom discussions; performance assessment; presentations and discussions; portfolios; journals; interviews; projects; investigations. The following assessment devices that are part of the *Addison-Wesley Chemistry* program can be used informally or formally.

- **Section Review** Each section concludes with questions that can be used as a self-check for students, a class-discussion exercise, or a homework assignment. Many sections have a suggested Portfolio Project. Often, these projects require students to prepare a written report to help them develop their writing skills.
- **Chapter Review** This review provides students with an opportunity to check their understanding of key vocabulary terms, content, and concepts. Each Chapter Review also includes a concept mapping activity, critical thinking questions, and problem-solving challenges.
- **Concept Mapping** Concept maps provide students with a visual study tool that shows relationships among ideas. Concept mapping is an open-ended assessment tool. Because different students learn and relate ideas in different ways, there is no single correct way to organize a concept map.
- **Portfolios** Portfolios are collections of representative work done by students during the year. Portfolios require a significant amount of time for planning and managing. Many kinds of work can be included in portfolios: photographs, artwork, videotapes, audiotapes, computer programs, results of research projects, lab reports, journals, experimental designs, and stages of a "work in progress." Students can select what they want to place in their portfolios within guidelines set by the teacher or the student. Periodic portfolio reviews help to identify a student's strengths and weaknesses, assess a student's progress, and adjust instruction to meet individual needs.

Standardized Test Prep

- **Student Edition** Each chapter has a practice test to prepare students for state and national exams. Formats include multiple-choice, classification, interpretation of tables and graphs, visualization questions, essays, and true/false questions that ask students to decide whether a pair of true statements have a causal relationship.
- **Prentice Hall Assessment System** Prepare students for high school exit exams with test-taking strategies, practice tests, and options for remediation.

Teacher's Resource Package

- **Section and Vocabulary Reviews** The Review Modules in the Teacher's Resource Package include Section Review worksheets for every section, plus a Vocabulary Review and an Interpreting Graphics exercise for each chapter. The worksheets can be used by students for informal review or by teachers as quizzes.
- **Tests and Quizzes** The Review Modules contain a comprehensive testing program with chapter quizzes and two different tests for each chapter. The tests offer a variety of testing formats, including multiple-choice and essay questions, matching, completion, problem solving, and true/false. The Teacher's Resource Package also includes Laboratory Practicals that can be used to test students' laboratory skills.

Assessment Technology

On the *Chem ASAP!* CD-ROM, you will find Assessments to supplement the Section Reviews in the textbook. *Chem ASAP!* also allows students to create electronic concept maps for each chapter. The *Assessment Resources* CD-ROM is a comprehensive test generator with 2200 questions. The companion Web site at www.phschool.com has a Self-Test for every chapter.

LABORATORY SAFETY

Chemistry teachers are no longer exempt from successful prosecution for accidents that occur in their classrooms. That threat coupled with our increased awareness of the insidious effects of chemical exposure explains the demand for safety education, particularly in science classrooms.

When it comes to laboratory chemistry, the safety of students and teachers is the primary concern of Prentice Hall. To that end, both the traditional *Laboratory Manual* and the *Small-Scale Chemistry Laboratory Manual* give detailed safety information for the student and the teacher at the beginning of each lab and in the front of each laboratory manual. In the Student Edition, the labs display appropriate safety icons, which are defined for the students on page 19 as part of the introduction to laboratory safety.

Safety in the Chemistry Laboratory

The following information about safety issues in the high school chemistry laboratory is intended to be used by classroom chemistry teachers. This reference section is intended to be accessible to the teacher for planning demonstrations and preparing for lab. The material included here is by no means comprehensive. Prentice Hall takes no responsibility for the completeness of this information or for the implementation of these suggestions.

The safety of the students in a high school laboratory is traditionally assigned to the teacher. But the teacher teaches safety awareness by holding each student accountable for the safety of all other participants in the lab, including the teacher. Have students sign a safety contract at the beginning of the year. There is a sample contract in each laboratory manual. Of course, the teacher will most effectively teach safety awareness by modeling safe practices and by making a discussion of safety conerns the first priority for all lab activities. The following suggestions are intended to help you establish a program of safety awareness for staff and students.

Education

- **Put together a safety committee for your school. Include faculty, students, and administrators.**
- **Schedule regular meetings of the safety committee to discuss issues, accident reports, and the results of lab inspections.**
- **Implement a safety education program for staff and students.**
- **Require students to know the lab safety rules, assess this by giving a safety quiz, and have the students sign a safety contract, saying that they know the rules and will comply with them. Send the contract home with a set of rules and require that one parent or guardian sign the contract, having read the rules.**
- **Give a safety grade with each lab, based on each student's behavior during the lab. Penalize unsafe behavior by removing students from the lab.**
- **Encourage school staff to become certified in CPR and first aid.**
- **Require that science staff read any of several publications on safety in the high school science classroom.**

Management

- **Report all accidents to the safety committee.**
- **Periodically inspect laboratories and report findings to the safety committee.**
- **Establish an emergency plan and review it periodically with students and staff.**
- **Keep phone numbers for the fire department, police department, local ambulance, and poison control center next to every phone in a lab or stockroom.**
- **Display warning signs and safety posters.**
- **Prohibit student access to stockrooms.**
- **Require students to keep aisles clear and counters clean. There should be two exits in a laboratory or stockroom, and they should be kept free of obstacles.**
- **Keep a current chemical inventory. Update the chemical inventory often. Include a Material Safety Data Sheet (MSDS) for each chemical.**
- **Label all chemicals with the date and possible hazards. Establish a policy for disposing of and buying chemicals.**
- **In prelab discussions, stress safety issues relevant to each specific lab.**

Lab Techniques

- Do not allow any student or staff member to work alone in the laboratory.
- Do not leave any experiment unattended.
- Do not eat, smoke, or drink in the lab.
- Do not store food in chemical refrigerators. Store chemicals only in chemical refrigerators.
- Never taste a chemical in the laboratory.
- Always add acids to water, never water to acids.
- When pipetting, use a suction bulb. Never use your mouth.
- When heating the contents of a test tube, direct the mouth of the test tube away from yourself and away from others.

Personal Safety

- Always wear goggles in lab. Require students and anyone entering the lab to wear goggles.
- Provide lab coats and aprons for all participants in the lab.
- Take into account student health problems such as allergies or asthma.
- Do not allow untied long hair, loose clothing, bare feet, open-toed or -heeled shoes, or dangling jewelry to be worn in the lab.
- Strongly discourage the wearing of contact lenses in lab. In the event of a chemical splash into the eye, the lenses may hold chemicals in contact with the eye, potentially causing greater injury.

Equipment

- Make available the following items in all labs and demonstration areas: an ABC fire extinguisher, a smoke alarm, a safety shower, an eye-wash faucet, a fume hood.
- Maintain and check all safety equipment regularly.
- Keep a first-aid kit stocked and available.
- Keep flammable chemicals in an approved fireproof cabinet.
- Attach compressed gas cylinders to counters or walls. Move them on dollies on which they are secured.
- Use safety shields when there is a possibility of explosion, implosion, chemical splashing, or fire.
- Use hot plates in lieu of open flames as often as possible.
- Use dispensing devices such as syringes, spigots, or droppers, rather than having students pour chemicals from their containers.
- Be prepared for spills, especially acid spills. Spill-control devices are available from suppliers. To avoid mercury spills, use alcohol thermometers.

Storing Chemicals

- Store chemicals in stockrooms, not in the lab.
- Do not store chemicals alphabetically. Use an approved system of chemical storage, and post legends in the stockroom.
- Store only the chemicals you will use during a school year.

Disposal

- Dispose of broken glass in a dedicated container kept out of the way and clearly labeled.
- Dispose of chemicals in the following ways:

  Neutralize acids and bases and flush down the drain.

  Flush light metal salts and their solutions down the drain with excess water.

  Precipitate and filter heavy metals. Evaporate heavy metal solutions down to salt. Keep heavy metals salts in labeled containers and dispose of them by following local regulations.

  When in doubt, check with local authorities about local regulations.
- Have a disposal plan for each lab, and include directions for disposal in your prelab discussion.

SMALL-SCALE CHEMISTRY

Small-scale chemistry is an innovative, low cost, safe, time efficient, conservation-based teaching model for chemistry. It is not a curriculum, but a highly flexible approach to teaching chemistry at all levels. Small-scale science is actually a misnomer—it describes an attitude more than it does a size. (It is **not** microscale!) The goal of small-scale science is to have every student involved in hands-on learning become more efficient in process and more knowledgeable in content while using this highly effective discovery approach to learning.

Small-scale chemistry is more than just a methodology. It is a systemic medium that creates a learning environment that fosters creativity, invention, and scientific problem-solving. As you and your students perform the Small-Scale Labs included in each chapter of the Student Edition, you will see that small-scale chemistry does not use traditional equipment like Bunsen burners, beakers, and test tubes. Instead, small-scale chemistry employs non-traditional, inexpensive, student-designed and built laboratory equipment made from readily available materials. The *Small-Scale Chemistry Laboratory Manual* has instructions for building balances, spectroscopes, pH meters, electrolysis devices, and digital burets.

Small-scale chemistry is not designed to simply do better what traditional laboratory methodologies try to do. (It does not, for example, miniaturize traditional experiments.) Small-scale science creates an environment where students and teachers participate together in the interplay of logic and experiment. It promotes the interchange of information among students and between students and teachers.

This approach is different from other student-centered approaches in an important way: it emphasizes that asking good questions is more important than giving right answers. It is a unique and effective blend of process and content. Each time a teacher supervises a lesson, the lesson might be different. In each class, students often contribute new insights, useful innovations, and fresh ideas.

Why Use Small Scale Chemistry?

Small-scale chemistry is relatively safe. Hazardous reagents and procedures were eliminated before experiments were designed. Small-scale builds in safety at the design stage of an experiment rather than consigning teachers to try to control hazards at the execution stage. For example, solution concentrations are usually about 0.2*M* or less, and concentrations never exceed 1.0*M*. Plastic thin-stemmed pipets, originally designed for biomedical research, serve both as liquid-storage vessels and transfer devices. Each pipet delivers only about 20 microliters of dilute solution. The potential for major accidents in traditional labs is replaced by minor inconveniences in small-scale labs.

Small-scale chemistry is time-efficient. Setup is made easy for the teacher and the student because the experimental format does not rely on lengthy written instructions. Instead, the directions are easy to read and follow. Experimental results are immediate and obvious, and the time students take to set up, manipulate, and clean equipment is minimal. Teachers are available to spend their valuable time and energy interacting with students during the laboratory. The simplicity of the experiments also enhances learning outcomes. The simple organization under the reaction surface helps students accurately record, compare, and correlate their data. Students focus on what's happening rather than on what to do. Best of all, the teacher decides and chooses which learning outcomes are most practical for each class.

Small-scale chemistry labs have less environmental impact than do conventional chemistry labs. Clean-up with a paper towel takes seconds, and the minute quantities of chemicals can usually be disposed of safely, usually in the waste basket. As many school districts trim budgets and tighten regulations on the use and disposal of school-laboratory chemicals, small-scale offers immediate solutions. The small amounts of common, relatively safe chemicals employed never become a waste-management problem.

Small-scale chemistry is a lab-based program that can be used as the central focus of a high school chemistry course. Teachers can build the conventional content around lab where it seems to fit. For example, students write and balance only those chemical equations for reactions that they have carried out in the lab. Quantitative concepts, such as stoichiometry, are well suited for lab-centered teaching. Thus teachers can bring the content of chemistry into the laboratory. Through small-scale, students can and do learn much more chemistry than they can through conventional labs. In addition, they learn to be creative and inventive with chemistry; to be good problem solvers; and to communicate effectively with the teacher, with other students, and with themselves.

How to Set Up and Run a Small-Scale Chemistry Laboratory

Efficient organization on the part of the teacher is a key to running a successful small-scale chemistry lab program. Here are some suggestions for the setup and day-to-day operations of a small-scale laboratory.

1. Assemble the following list of basic equipment. The quantities here assume you plan to convert your laboratory to small-scale chemistry entirely and further, that you will teach five to six sections of chemistry per day, with 24 students per section. You will need to adjust the numbers if you plan to use fewer small-scale labs your first year to augment your traditional program.
   - **4 boxes of thin-stemmed plastic disposable pipets (500 per box), to store and deliver liquids**
   - **50 plastic reagent bottles (250 mL), to store prepared solutions**
   - **10 plastic reagent bottles (1.0 L), to store prepared solutions**
   - **60 plastic screw-capped specimen jars (60 mL), to fill pipets**
   - **60 plastic cups (12-ounce), to store sets of pipets.**
   - **120 plastic pill vials (30–60 mL), to store and dispense solid reagents**
   - **1 plastic "egg crate" light panel (2 feet × 4 feet, cut) to hold pipets for easy access**
   - **48 small-scale reaction surfaces (file protectors or overhead projection transparencies), to mix chemicals qualitatively**
   - **24 well plates (1× 8), to calibrate pipets**
   - **12 pairs of scissors, to construct instruments**
   - **12 office punches (1/4-inch) to construct instruments**
   - **6 standard staplers, to construct instruments**
   - **150 plastic cups (1-ounce), for titrations and balance pans**
   - **150 plastic cups (3 1/2-ounce), for titrations and chromatography**
2. Prepare the reagent solutions described in the Master Solutions List on pages T26–27. Store each reagent solution in an appropriately sized plastic reagent bottle. Label each bottle with the concentration, the chemical formula of the contents, the number of grams of solute per volume of solution, and the date of preparation (for example:*1M NaCl, 14.6 g per 250 mL, 01/20/02)*. When a reagent bottle is empty, simply make a new solution from the labeled directions and change the date. You do not have to make all the solutions at once; prepare the necessary solutions only for the first lab. Then as the year progresses, you can add solutions as you need them. By the end of your first year of small-scale chemistry, you will have a complete set that will need only occasional maintenance. Once a small-scale lab is set up, a teacher in charge of 5 sections of 24 students each should count on replenishing about 10 to 15 solutions a year.
3. Store each reagent bottle in a designated place. The storage area should be secure from students but easily accessible by a teacher during a laboratory period. The prepared solutions and solids in this book require approximately 36 feet of shelf space.
4. For a classroom of 24 students, prepare and label 15 small-scale pipets and one 60-mL screw-capped plastic specimen jar (reagent-filling container) for each of the solutions listed in the Master Solutions List on pages T26–27. Fill each set of pipets and store them stem-up in a labeled 12-ounce plastic cup on the shelf along with the reagent bottle and plastic specimen jar. To save time, you may want to spend a few minutes in class having students pull pipets and tape labels on them. Use a computer to make labels or print the number of labels you need before you cut them out and tape them to the pipets.
5. Cut the egg-crate plastic light panel into small sections to use as pipet racks. Wear your safety glasses and use a hacksaw. This is a job you have to do only once. It's convenient to cut sections that are 3 squares × 10 squares, which will hold as many as 30 pipets.
6. When it is time to set up a lab, set out a 12-ounce storage cup containing 15 pipets of each solution needed in the lab. Students can pick up one pipet from each cup and quickly assemble a complete set of solutions needed for the lab. You may wish to have

two students share a set of reagents, even if they are doing the lab individually. At the end of the lab, students can return the pipets to their proper cups, ensuring that everything is ready for the next class.

7. Fill empty pipets as students call your attention to them. Because students often need only one or two drops of each reagent, many pipets will not need filling until after several sections of students have done the lab. For labs that require large quantities of reagents and, thus, frequent refills of pipets, you may want to teach students how to prepare pipets for some reagents, as described in "Safe and Efficient Techniques for Using Small-Scale Equipment" in the *Small-Scale Chemistry Laboratory Manual.*

8. When the lab is finished, return the pipet storage cups to the storage compartment. You may want to leave one or two sets out for another day or two to accommodate students who were absent during the lab.

9. For a classroom of 24 students, prepare and label plastic pill vials (30–60 mL) for each solid on the Master Solids list on page T28. Punch small holes in the lids and cover them with tape. To use the vials, students can remove the tape and carefully shake out the desired quantity. Store the solids in the plastic pill vials with the openings taped. You can store each set in a large plastic cup or in a plastic ice-cube tray.

Evaluation of Student Learning Through Small-Scale Laboratory

The following list provides some suggestions for how you could evaluate what your students learn from doing small-scale chemistry laboratories.

1. **Lab Notebook** As an alternative to using the Laboratory Recordsheets, you may want to have students keep a laboratory notebook. The lab notebook is a real-time, permanent, hand-written record of a student's entire laboratory experiences, including dates, titles, experimental procedures, data, conclusions, and further research and investigations. Pick up the lab notebooks periodically, and grade them for completeness rather than for specific details. With only a quick glance you can verify that a student has kept his or her work current. You may be less interested in format than in how well students can use lab records to solve problems. In this case, look for signs of creativity, invention, problem-solving, and good thought processes. Be sure to tell students your expectations. You may want to give extra credit for evidence of creativity or invention. For example, students might invent interesting new questions or novel ways to solve problems. Or a student might choose to write the lab procedure in cartoon format.

2. **Open-lab Notebook Tests** To check for details and understanding, you may want to give students regular lab tests and allow them to use their own handwritten notes to answer the questions. Try to balance the range of questions from knowledge-level questions to those that evaluate comprehension, analysis, synthesis, evaluation, and creativity. Students soon learn that to score well on these exams, they must have a well-written lab book and show that they can use it to interpret their results.

3. **Closed-lab Notebook Tests** Some skills learned in lab need to be internalized by any student of chemistry. Use closed-book tests to evaluate how well students have learned these skills. For example, test students' abilities to write and balance equations.

4. **Performance-based Assessment** Small-scale chemistry lends itself well to the direct hands-on evaluation of problem-solving using both qualitative and quantitative analysis. Identification of unknowns are always a good source of performance-based assessment as are quantitative titrations of unknowns. The Student Edition small-scale labs have activities in the "You're the Chemist" sections that you can use to assess students' reasoning skills. The 30 practicals in *Laboratory Practicals* can be used to test both laboratory skills and comprehension.

5. **Writing Essays** Consider having students write short essays on their conclusions for various labs. Try to assign essays that encourage creativity and original thought—ones that prompt students to write something that would be interesting for you to read.

SUPPLIES FOR SMALL-SCALE LABS

Master Solutions List

The following list gives directions for preparing all the chemical solutions necessary for doing all the small-scale labs in *Addison-Wesley Chemistry*. The most efficient way to organize your small-scale lab is to make up the appropriate aqueous solutions, then use and replace them as needed. Store each solution in a plastic reagent bottle. Depending on the labs you do, some solutions will have to be replaced several times during the year; others will last the entire year. Most solutions are 0.1*M* or 0.2*M*. In no case is any solution of greater concentration than 1.0*M*. Except for sodium chloride, the labs do not require you to prepare multiple solutions for the same chemical. For example, all the experiments that use sodium hydroxide use 0.5*M* NaOH.

Label each reagent bottle with the chemical formula, the concentration, the number of grams of solute per volume of solution, and the date. When the bottle is empty, simply replenish it by following the directions on the label, and change the date. When making dilute acid solutions from concentrated acids, always add acid to water slowly and carefully. The Teacher's Edition of the *Small-Scale Chemistry Laboratory Manual* has a combined list of solutions for the labs in the Student Edition and the *Small-Scale Chemistry Laboratory Manual.*

Solution	Preparation	Experiments
alizarin yellow R, 0.02%	50 mg AYR in 250 mL	606
aluminum chloride, 0.2*M*	12.1 g $AlCl_3 \cdot 6H_2O$ in 250 mL	371, 711
ammonia, 1.0*M*	67 mL of 15*M* NH_3 in 1.0 L **Caution! Using proper ventilation, add ammonia to water carefully and slowly!**	467, 489
ammonium chloride, 1.0*M*	13.4 g NH_4Cl in 250 mL	273, 426
borax, 4%	10 g $Na_2B_4O_7 \cdot 10H_2O$ in 250 mL	801
bromcresol green*, 0.04%	100 mg BCG in 14.3 mL 0.01*M* NaOH; dilute to 250 mL	606
bromphenol blue*, 0.04%	100 mg BPB in 14.9 mL 0.01*M* NaOH; dilute to 250 mL	606
bromthymol blue*, 0.04%	100 mg BTB in 16.0 mL 0.01*M* NaOH; dilute to 250 mL	273, 606, 698
calcium chloride, 0.5*M*	13.9 g $CaCl_2$ in 250 mL	157, 229, 371, 711
citric acid, 0.2*M*	9.6 g in 250 mL	329
copper sulfate, 0.2*M*	12.5 g $CuSO_4 \cdot 5H_2O$ in 250 mL	44, 157, 371, 698, 820
ethanoic acid, 0.8*M* CH_3COOH	undiluted white vinegar	329, 467, 489, 625, 670
hydrochloric acid, 1.0*M*	82 mL of 12*M* HCl in 1.0 L **Caution! Always add acid to water carefully and slowly!**	251, 273, 329, 426, 489, 625, 670, 820
hydrogen peroxide, 3% H_2O_2	undiluted household supply	44
iron(III) chloride, 0.1*M*	6.8 g $FeCl_3 \cdot 6H_2O$ in 25 mL of 1.0*M* NaCl; dilute to 250 mL	44, 157, 329, 371, 426, 711
lead(II) nitrate, 0.2*M*	16.6 g $Pb(NO_3)_2$ in 250 mL	157, 229, 397, 426
magnesium sulfate, 0.2*M*	6.0 g $MgSO_4$ in 250 mL	157, 371
metacresol purple*, 0.04%	100 mg MCP in 26.2 mL 0.01*M* NaOH; dilute to 250 mL	606
methyl orange, 0.05%	125 mg MO in 250 mL	606

Master Solutions List continued

Solution	Preparation	Experiments
nickel(II) sulfate, 0.2*M*	2.94 g $NiSO_4$ in 250 mL	371
nitric acid, 1.0*M*	63 mL of 15.8*M* HNO_3 in 1.0 L **Caution! Always add acid to water carefully and slowly!**	397, 426. 489, 625, 670, 711
pH buffer, Solution A	14.9 g boric acid and 12.6 g citric acid in 1.2 L	606
pH buffer, Solution B	45.6 g $Na_3PO_4 \cdot 12H_2O$ in 1.2 L	606
phenolphthalein, 0.1%	250 mg phenolphthalein in 250 mL of 70% 2-propanol	606, 625
potassium bromide, 0.2*M*	6.0 g KBr in 250 mL	397, 698
potassium chloride, 0.4*M*	7.5 g KCl in 250 mL	397
potassium fluoride, 0.4*M*	5.8 g KF in 250 mL	397
potassium hydroxide, 0.2*M*	2.8 g KOH in 250 mL	711
potassium iodate, 0.1*M*	5.35 g KIO_3 in 250 mL	44
potassium iodide, 0.1*M*	4.2 g KI in 250 mL	44, 73, 387, 426, 698
potassium permanganate, 0.02*M*	0.8 g $KMnO_4$ in 250 mL	44
potassium thiosulfate, 0.1*M*	2.4 g KSCN in 250 mL	426
2-propanol (70% isopropanol)	undiluted rubbing alcohol	467, 489
silver nitrate, 0.05*M*	2.1 g $AgNO_3$ in 250 mL	157, 229, 371, 397, 426
sodium carbonate, 1.0*M*	26.5 g Na_2CO_3 in 250 mL	157, 229, 329, 371
sodium chloride, 0.1%	1.0 g NaCl in 1.0 L	467
sodium chloride, 1.0*M*	14.6 g NaCl in 250 mL	229, 371, 698
sodium hydrogen carbonate, 1.0*M*	21 g $NaHCO_3$ in 250 mL	329
sodium hydrogen sulfite, 1.0*M*	26 g $NaHSO_3$ in 250 mL	273
sodium hydroxide, 0.5*M*	20.0 g NaOH in 1.0 L	157, 229, 251, 273, 371, 426, 489, 625, 711, 820
sodium hypochlorite, 1%	50 mL liquid bleach (NaOCl) in 200 mL	44, 397
sodium nitrite, 0.4*M*	6.9 g $NaNO_2$ in 250 mL	44, 273
sodium phosphate, 0.1*M*	9.5 g $Na_3PO_4 \cdot 12H_2O$ in 250 mL	157, 229, 426
sodium sulfate, 0.2*M*	7.1 g Na_2SO_4 in 250 mL	157, 229, 426, 698
starch, 20%	50 mL liquid starch in 200 mL	44, 698
sulfuric acid, 1.0*M*	56 mL of 18*M* H_2SO_4 in 1.0 L **Caution! Always add acid to water carefully and slowly!**	489, 625, 670
thymol blue*, 0.04%	100 mg TB in 21.5 mL 0.01*M* NaOH; dilute to 250 mL	251, 606
zinc chloride, 0.2*M*	6.8 g $ZnCl_2$ in 250 mL	371, 711

*If the indicator is available as the sodium salt, dissolve directly in 250 mL of deionized or distilled water.

SUPPLIES FOR SMALL-SCALE LABS

Master Solids List

For each solid that can be poured, fill six 30- to 60-mL plastic pill vials. Use a 1/4-inch office punch to make a hole in each lid. Cover the holes with plastic tape. Label and store each set in a plastic ice cube tray.

Name	Experiments
ammonium chloride, NH_4Cl	557
calcium carbonate, $CaCO_3$	187
calcium chloride, $CaCl_2$	557
guar gum	801
iron, Fe (staples)	426, 670
limestone	329
magnesium, Mg	670
magnesium sulfate, $MgSO_4$ (Epsom salts)	489
potassium chloride, KCl	489, 557
potassium iodide, KI	489
sodium carbonate, Na_2CO_3 (washing soda)	489, 557
sodium chloride, NaCl (table salt)	44, 187, 489, 557
sodium chloride, NaCl (lab grade)	516
sodium hydrogen carbonate, $NaHCO_3$ (baking soda)	251, 489, 557
sodium phosphate, Na_3PO_4	329, 557
sucrose (table sugar)	489, 516
zinc, Zn (galvanized nails)	670

Master Equipment List

Name	Experiments
balance	73, 96, 122, 187, 251, 319, 516, 820
chromatography paper	467
conductivity tester	489
electrolysis device	698
pipet	44, 96, 251, 426, 516, 625, 801
volumetric bottle, 50-mL plastic	516
well strip, 8-well	96, 625

Master Household Products List

Store liquids in their original containers.

Name	Experiments
alcohol thermometer	557
aluminum cans	96
aluminum foil	319
antacid tablets	44, 329, 489
baking powder	251, 329, 489
beverages, assorted	489
blocks, toy	73
candles, votive	319
candy (colored)	122, 467
cereal	44
coffee filters	467
construction paper, colored	44
cornstarch	489, 801
cotton swabs	273, 426
cups, plastic	96, 122, 251, 273, 467, 557, 625, 801
detergents (laundry and dish)	44, 329
dice	96, 852
eggs	820
fertilizer, solid	426
food coloring	467
glue, white	801
ice, crushed	557
medicine droppers	273, 426
milk, powdered	820
modeling clay	758
pennies (pre- and post-1982)	96, 670, 852
pickle brine	489
powdered soft drinks	329, 467
reaction surface	44, 157, 229, 273, 329, 371, 397, 426, 489, 606, 670, 698, 711
sea shells	329
spoons, plastic	187, 557, 801
starch, liquid laundry	801
straws	801
toothpicks	467, 758

MINI LAB AND DISCOVER IT! MATERIALS LIST

Material	Mini Lab	Discover It!
air freshener, solid	286	
aluminum foil	697	50
ammonia, household	593	
ammonia, 6*M* 100mL 15*M* NH_3 in 250 mL **Caution! Using proper ventilation, add ammonia to water carefully and slowly!**	794	
antacid tablets, effervescent, varied brands	346, 615	532
baking powder		28
baking soda ($NaHCO_3$)	425, 593	28, 412, 576
balance, centigram	195, 259	170
ball-and-stick molecular model kit	757	
balloons, round, different colors	259, 346	326, 436
barium chloride, 0.1*M* 5.2 g $BaCl_2$ in 250 mL	383	
batteries, D-cell	425	
beakers, 100-mL	224, 383	50
bell wire	425	
bleach, powder	669	
borax	593	772
bowl		2
box	112	
Bunsen burner	195, 383 556	704
calcium chloride ($CaCl_2 \cdot 2H_2O$)	195	
calcium chloride, 0.1*M* 2.8 g $CaCl_2$ in 250 mL	383	
calculator, scientific		840
can, 6-oz and 12-oz	448	
candle	735	742
carbonated beverage	593	
cardboard, thick, cut in strips	286	
cardboard tube from paper towel roll	826	
chalk		50
clock or watch	346	532, 612
cloth, clean white	593	
coat hanger	448	
colored fabric	669	

Material	Mini Lab	Discover It!
copper(II) chloride, 0.1*M* 3.4 g $CuCl_2$ in 250 mL	383	
copper strip	224	676
copper(II) sulfate, 0.05*M* 3.1 g $CuSO_4 \cdot 5H_2O$ in 250 mL	224	
copper(II) sulfate ($CuSO_4 \cdot 5H_2O$)	195	
copper wire		644
cornstarch	508	2
cup, 8 oz plastic, clear	17, 35, 286, 425, 508, 593, 615	28, 412, 474, 532, 612, 704
cup, plastic foam	308	772
dark room		360
dish or other shallow container	286, 493, 697	644, 742
dishwashing liquid	17, 493, 593	474
distilled water	425, 508	412
dropper bottles, plastic	163	
duct tape	858	
egg		612
Epsom salts ($MgSO_4 \cdot 7H_2O$)		412
Erlenmeyer flasks, 250-mL	259	
ethanol	794	
filter paper, cut in strips	35	
flashlight	508, 858	
flour		28
flower petals, different colors	669	
food coloring		50
freezer		326
fruits and vegetables		576
glasses, drinking		266, 772
graduated cylinders, 10-mL and 100-mL	259, 308	50, 266, 772
grape juice	669	
graph paper	346, 399, 448, 858	532, 840
grass stain on piece of white fabric	669	
grocery bag, plastic	448	
hairpins, metal		704

MINI LAB AND DISCOVER IT! MATERIALS LIST

Materials List continued

Material	Mini Lab	Discover It!
heat-proof gloves		704
heat-resistant surface	556	704
hot plate	697	
hydrochloric acid, 1.0*M* 20.5 mL of 12*M* HCl in 250 mL	259	
hydrochloric acid (HCl), 6*M* 125 mL of 12*M* HCl in 250 mL **Caution! Always add acid to water carefully and slowly!**	383	
hydrogen peroxide, 3% H_2O_2	669, 735	808
ice	286, 308	500, 532
index cards, 3 inch × 5 inch	62, 100	82, 266
ingredients labels from several products		132
iodine solution, 1% I_2 in 2%(m/v) KI 2.5 g I_2 plus 5.0 g KI in 250 mL	669	
iron(III) chloride, 0.1*M* 6.8 g $FeCl_3 \cdot 6H_2O$ in 250 mL	163	
jar, cylindrical with lid	508, 593	50, 840
knife	593	576
laundry detergent	593	
lead(II) nitrate, 0.1*M* 8.3 g $Pb(NO_3)_2$ in 250 mL	163	
lemon		676
lemon juice	593	
lithium chloride, 0.1*M* 1.1 g LiCl in 250 mL	383	
magnesium ribbon	259	
magnesium strip	224	
magnesium sulfate, 0.05*M* 1.5 g $MgSO_4$ in 250 mL	224	
magnifying glass		412
manganese dioxide	735	
marble	112	
masking tape	425, 508	28, 532
matches	735	
measuring cups	17, 593	28
measuring spoons		2, 28
meat, cooked		808
meat, red raw		808
medicine droppers	346, 593, 615, 669, 794	50, 808
methanal	794	
metric ruler, meter stick	35, 62, 448, 593, 826, 858	106, 412, 474
milk	593	
milk of magnesia	593	
mineral oil		742
mirror		412
mortar and pestle	615	
mouthwash	593	
nails, iron finishing		644
oxalic acid solution, 1%(m/v) 2.5 g in 250 mL	669	
paper, plain white	593	
paper clips, metal	448, 493	50, 170, 236
paper clips, vinyl coated, different colors		202, 236
paper towels	615	576, 644
pen	100	28, 532
pencil	35	82
pennies		840
pens, colored marking, green included	35, 826	326, 412
petroleum jelly	412	644, 742
plastic foam shapes	112	
plastic wrap	35, 593, 735	644
plate, large paper		576
plate, plastic		500, 808
pliers		360, 644, 704
poster board	858	
potassium permanganate, 0.05*M* 0.40 g $KMnO_4$ in 250 mL	669	
propanone	794	

Materials List continued

Material	Mini Lab	Discover It!
red cabbage leaves	593	
rubber band, No. 25 (about 2 inches long)	448, 493, 593	292
rubbing alcohol	35	50, 742
rusty water	669	
sandpaper, fine	224	644, 676
sandwich bag, clear plastic		236
saucepan	697	
scissors	858	644
shampoo	593	50
silver nitrate, 5% 12.5 g $AgNO_3$ in 250 mL	794	
silver nitrate, 0.05*M* 2.1 g $AgNO_3$ in 250 mL	224	
silver nitrate, 0.1*M* 4.2 g $AgNO_3$ in 250 mL	163	
sodium carbonate, 0.1*M* 2.65 g Na_2CO_3 in 250 mL	163	
sodium chloride (NaCl)	17, 383, 425, 593	412, 500, 644
sodium chloride, 0.05*M* 0.7 g NaCl in 250 mL	224	
sodium hydrogen carbonate ($NaHCO_3$)	508, 615, 697	
sodium hydroxide, 1*M* 10 g NaOH in 250 mL	794	
sodium hydroxide, 0.1*M* 1 g NaOH in 250 mL	163	
sodium hypochlorite (NaOCl), household liquid bleach	669	
sodium phosphate, 0.1*M* 9.5 g $Na_3PO_4 \cdot 12H_2O$ in 250 mL	163	
sodium sulfate ($Na_2SO_4 \cdot 10H_2O$)	195	
sodium thiosulfate solution, 0.2*M* 12.4 g $Na_2S_2O_3 \cdot 5H_2O$ in 250 mL	669	
spatula	195	
spot plate	669	
steel wool (0000 weight)	556	
straw, drinking	17	
string		436, 500
sucrose (table sugar)	17, 425	28, 412
sunlight		326
tablespoon	697	28, 50, 412, 772
tape measure	346	326
tape, transparent	35, 593	106, 360
tarnished silver item	697	
teaspoon	17, 508, 593, 615	474
test tube holder	195	
test tube rack	163, 195, 383, 794	
test tubes, medium	195, 735	
test tubes, small	163, 383, 794	
thermometer	308	532
thumbtack	826	
tissue paper	556	
tongs	383, 556, 697, 735	
toothpaste	593	
toothpicks, wooden	735, 758, 826	742
universal indicator	615	
vegetable oil	493	742
vinegar, white (CH_3COOH)	425, 593, 615	612
voltmeter		676
waxed paper		474
white glue		772
wintergreen mints, 3 different brands		360
wire for flame tests (10-cm length of nichrome wire)	383	
zinc strip	224	644, 676
zinc sulfate, 0.05*M* 2.0 g $ZnSO_4$ in 250 mL	224	

AUDIOVISUAL AND TECHNOLOGY SUPPLIERS

Audiovisual Distributors

AIMS Multimedia
9710 DeSoto Avenue
Chatsworth, CA 91311-4409
800-328-6700

Charles W. Clark Co., Inc.
4540 Preslyn Drive
Raleigh, NC 27616-3177
800-247-7009

Chem Study
Ward's Natural Sciences Est., Inc.
5100 West Henrietta Road
West Henrietta, NY 14586-9729
800-962-2660

Churchill Films/SVE
6677 North Northwest Highway
Chicago, IL 60631-1304
800-624-1678

Coronet/MTI Films & Video
108 Wilmot Road
Deerfield, IL 60015
800-777-8100

Encyclopedia Brittanica Edu. Corp.
310 S. Michigan Ave.
Chicago, IL 60604-9839
800-621-3900

Films for the Humanities
& Science, Inc.
Box 2053
Princeton, NJ 08543-2053
800-257-5126

The Media Guild
11722 Sorrento Valley Rd., Suite E
San Diego, CA 92121
858-755-9191

National Film Board of Canada
350 Fifth Avenue, Suite 4820
New York, NY 10118
212-629-8890

National Geographic Society
Educational Services
1145 17th St. NW
Washington, DC 20036
800-627-5162

Prentice Hall
One Lake Street
Upper Saddle River, NJ 07458
800-848-9500

University of California Extension
Media Center
2000 Center St., Suite 400
Berkeley, CA 94704
510-642-4124

Equipment Distributors

Flinn Scientific
P.O. Box 219
Batavia, IL 60510
800-452-1261
E-mail: flinn@flinnsci.com
www.flinnsci.com

PASCO® scientific
P.O. Box 619011
10101 Foothills Boulevard
Roseville, CA 95661-9011
800-772-8700
www.pasco.com

Science Kit & Boreal Laboratories
P.O. Box 5003
Tonawanda, NY 14151-5003
800-828-7777
www.sciencekit.com

Technology Distributors

Cambridge Development Lab, Inc.
214 Third Avenue
Waltham, MA 02154
800-637-0047

JCE Software
Department of Chemistry
University of Wisconsin, Madison
1101 University Avenue
Madison, WI 53706-1396
800-991-5534

Falcon Software, Inc.
One Hollis St.
Wellesley, MA 02482
781-235-1767

Project Seraphim
Department of Chemistry
University of Wisconsin, Madison
1101 University Avenue
Madison, WI 53706-1396
608-263-2837

Sunburst Technology
101 Castleton Street
Pleasantville, NY 10570
800-321-7511

Texas Instruments
800-TI-CARES (800-842-2737)

Vernier Software & Technology
13979 SW Millikan Way
Beaverton, OR 97005-2886
503-277-2299
E-mail: info@vernier.com
www.vernier.com

In ordering software, remember to check your system type and system requirements, memory, drives (number and type), and color or monochrome monitor. Check with the distributor about the availability of lab packs, network versions, site licensing, individual program disks versus multiple-disk packages, and the availability of preview disks.

ANTONY C. WILBRAHAM

DENNIS D. STALEY

MICHAEL S. MATTA

EDWARD L. WATERMAN

Needham, Massachusetts
Upper Saddle River, New Jersey
Glenview, Illinois

Program Components

Student Edition
Teacher's Edition
Teaching Resources
- Chapter Review Modules
- Solutions Manual for Chapter Reviews
- Laboratory Practicals
- Laboratory Recordsheets
- Graphing Calculator Problems
- Overhead Transparency Preview
- Spanish Supplement
- Block Scheduling Guide
- How to Assess Student Work

Overhead Transparency Package
Laboratory Manual, ATE
Laboratory Manual
Small-Scale Chemistry Laboratory Manual, ATE
Small-Scale Chemistry Laboratory Manual
Probeware Lab Manual for Computers and Calculators
Guided Reading and Study Workbook, ATE
Guided Reading and Study Workbook
Prentice Hall Assessment System
Color Transparencies
Chem ASAP! CD-ROM
ActivChemistry CD-ROM
Resource Pro CD-ROM
Assessment Resources CD-ROM
Web site at www.phschool.com
SCIENCE NEWS at www.phschool.com
The Chemistry Place Internet Site
Chemistry Alive! Videodisc
Small-Scale Lab Video and Videodisc
CHEMedia™ Videodiscs
Problem Pro CD-ROM

Cover photographs: Clockwise from top left: Test tube with zinc in acid, Richard Megna, Fundamental Photographs; Molecular structure of buckyball, Ken Eward, Photo Researchers; Bunsen burner flame, and flask containing precipitate of lead iodide, Richard Megna, Fundamental Photographs; Scanning tunneling microscope image, Fran Heyl Associates.

About the cover: The collage of images on the cover illustrates the beauty and excitement of chemistry: the old and the new; the macro view and the micro view; the applied and the theoretical. Cover designed by Amanda Kavanagh/Ark Design, New York.

STAFF CREDITS

The people who made up the **Addison-Wesley Chemistry** team—representing design services, editorial, editorial services, electronic publishing technology, manufacturing & inventory planning, market research, marketing services, online services & multimedia development, planning & budgeting, product planning, production services, project office, and publishing processes—are listed below. Bold type denotes the core team members.

Barbara A. Bertell, **Peggy Bliss**, **Kristen Cetrulo Braghi**, **Kathy Dempsey**, **Frederick Fellows**, **Thomas Ferreira**, **Joel Gendler**, **Jerry Hooten**, **Ellen Levinger**, Diahanne Lucas, Marie Mathis, Paul Murphy, Jane Orner, **Matthew Raycroft**, Wendy Chadbourne Simpson, **Dori Steinhauff**, Chris Wawack, Helen Young

ISBN 0-13-054384-5
3 4 5 6 7 8 9 10 05 04 03 02

Brief Contents

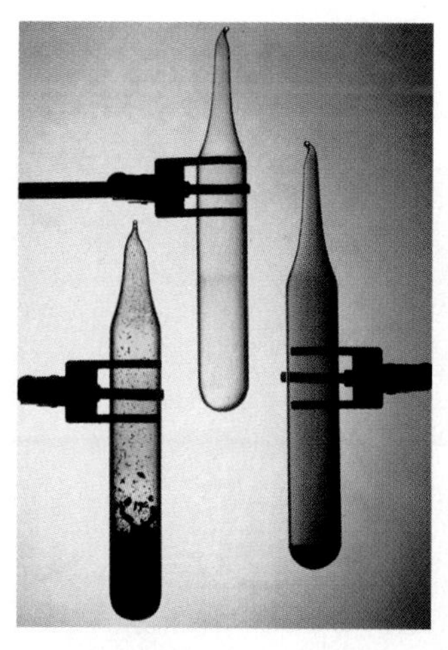

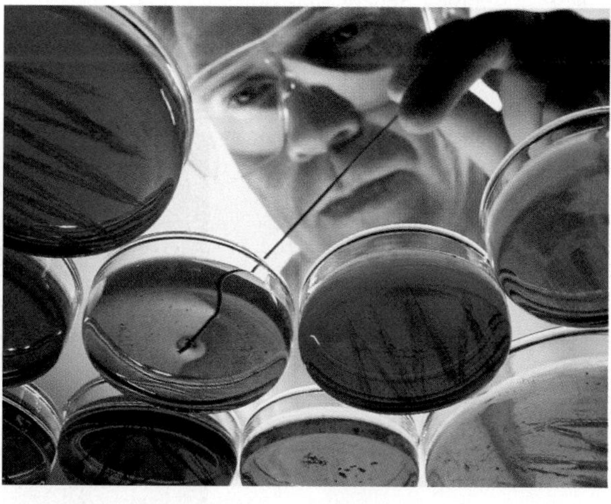

Contents

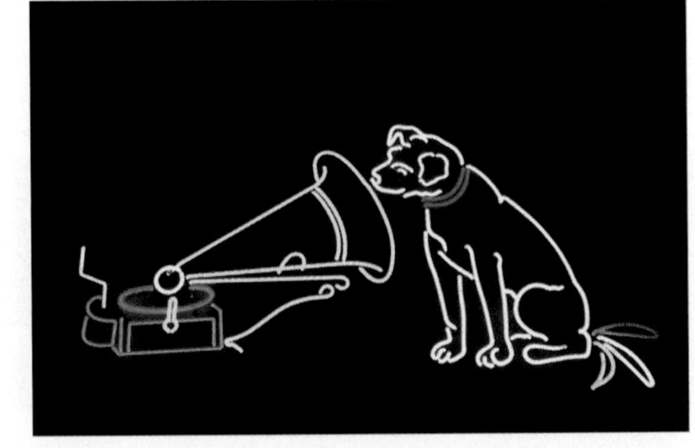

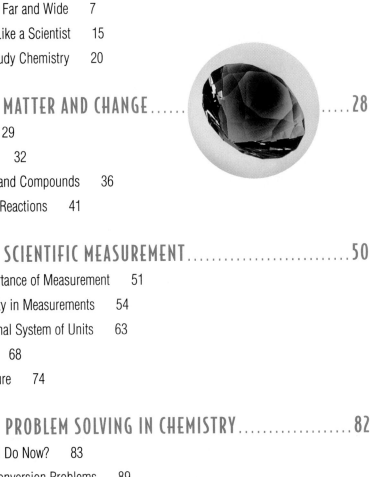

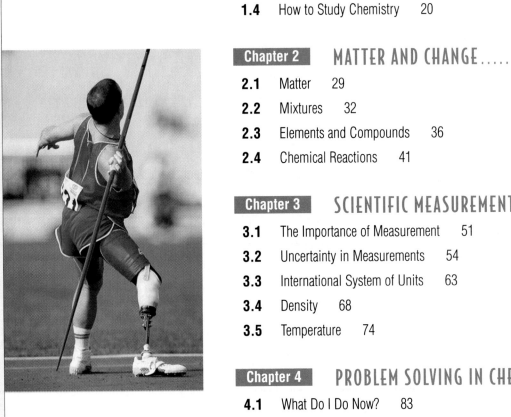

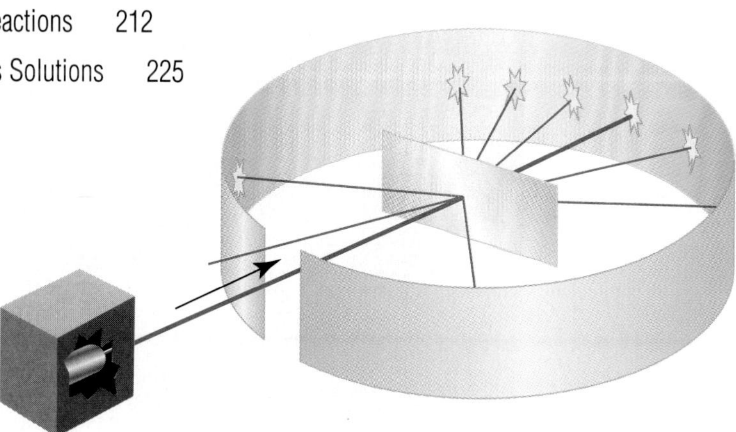

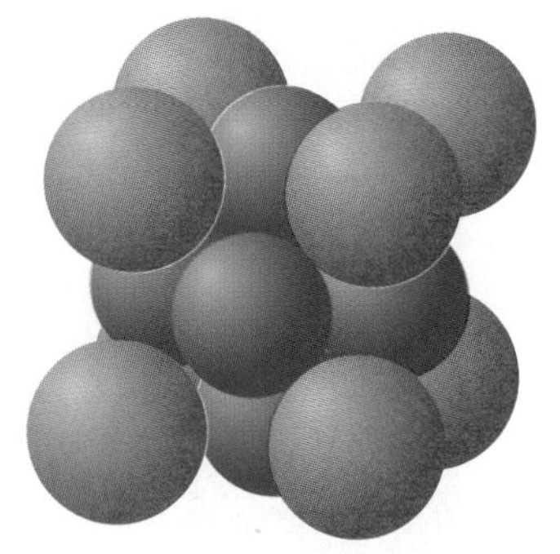

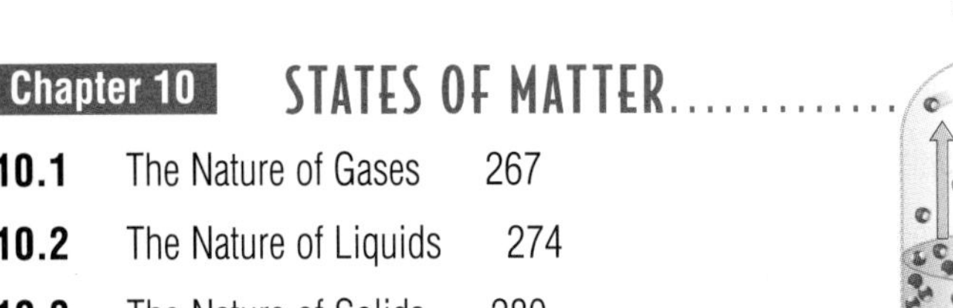

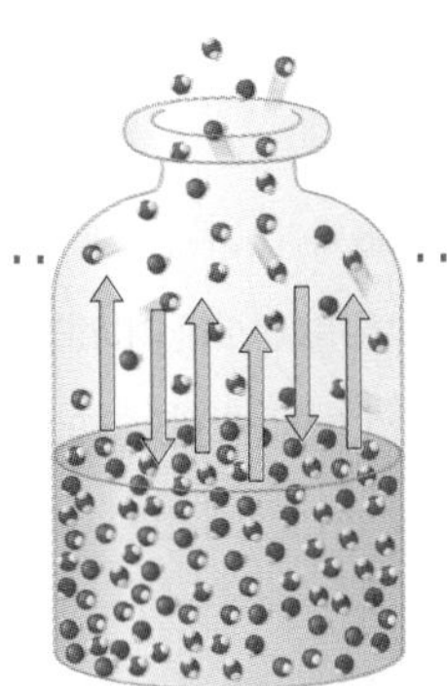

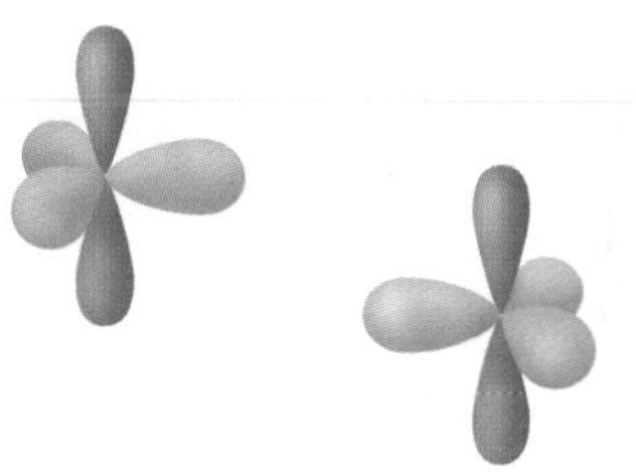

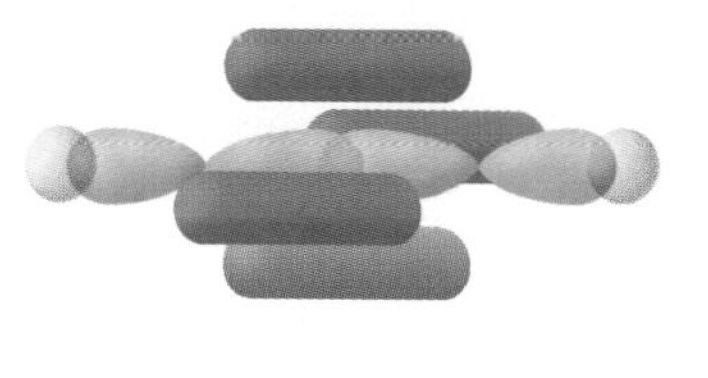

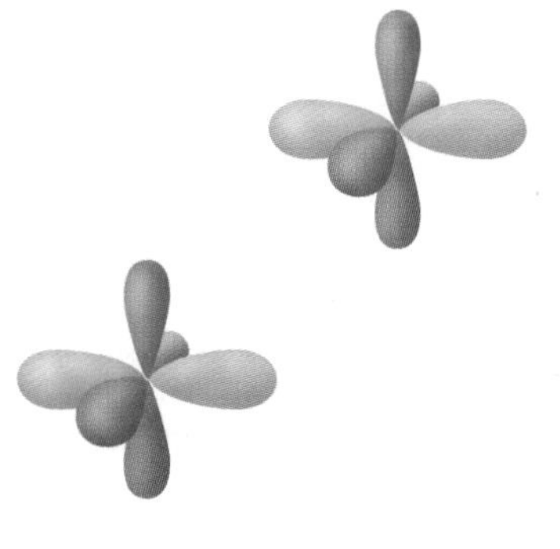

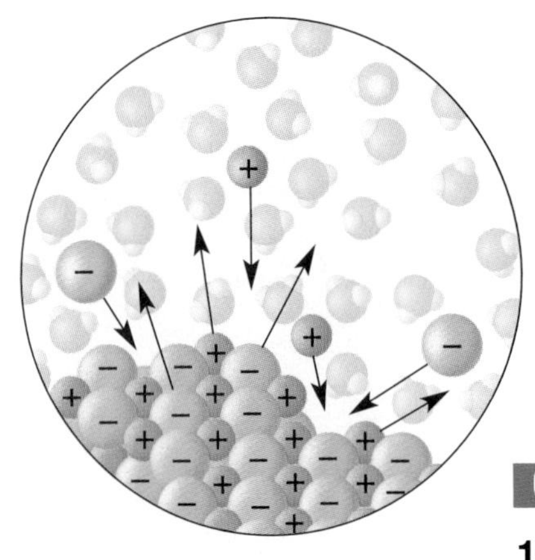

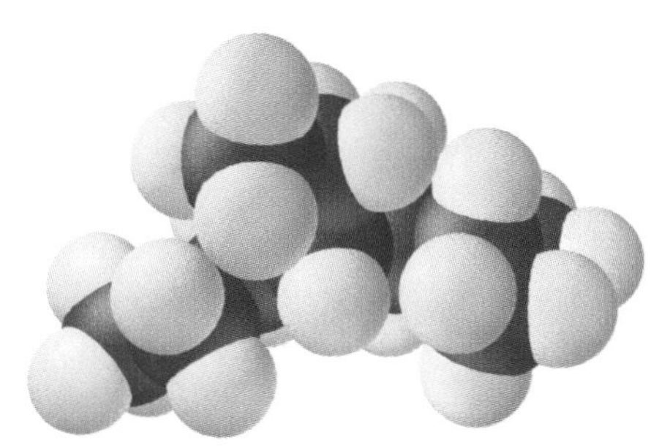

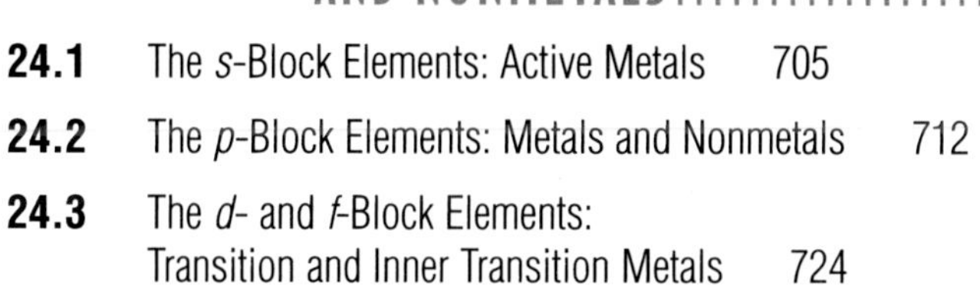

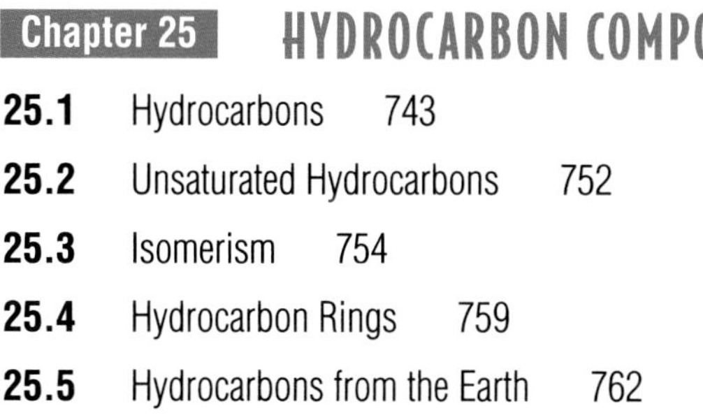

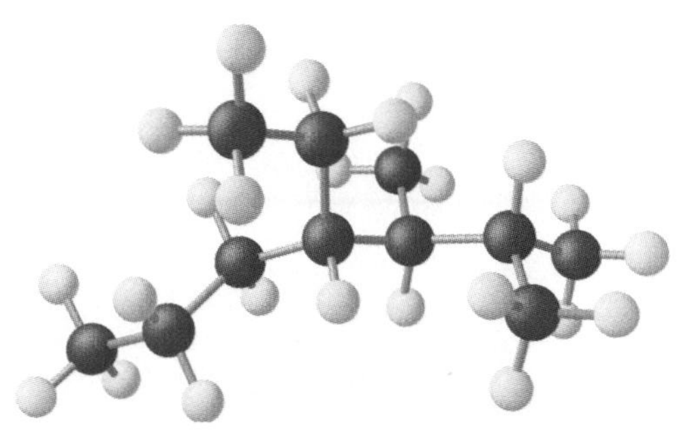

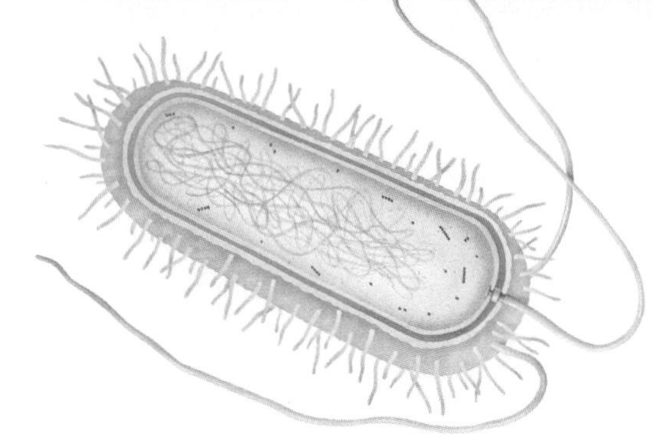

Labs & Activities

DISCOVER IT!

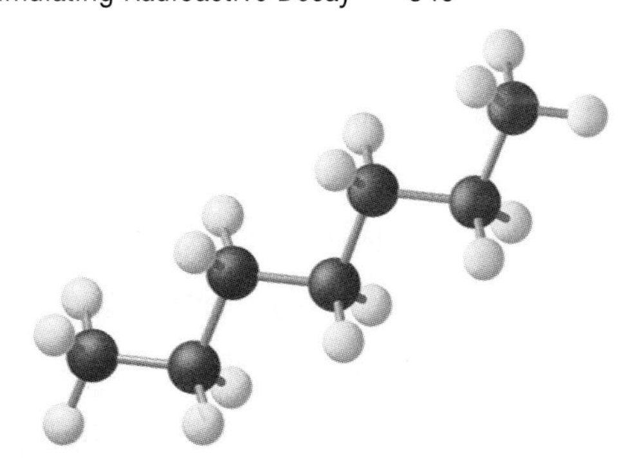

MINI LAB

SMALL-SCALE LAB

Probe or sensor versions available in the Probeware Lab Manual.

Features

Chemistry Serving...

Society

The Consumer

The Environment

Industry

LINK TO

CHEMISTRY IN CAREERS

CHEMath

CHEMISTRY FOR THE NEW CENTURY!

Addison-Wesley Chemistry is a comprehensive and motivating program that builds problem-solving skills and stresses the relevance and application of chemistry to your present and future life.

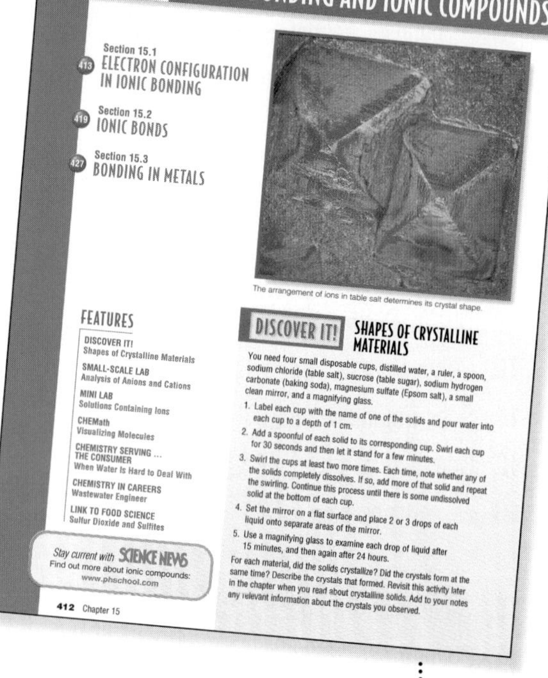

Chapter 15 IONIC BONDING AND IONIC COMPOUNDS

The arrangement of ions in table salt determines its crystal shape.

FEATURES

DISCOVER IT! Shapes of Crystalline Materials

SMALL-SCALE LAB Analysis of Anions and Cations

MINI LAB Solutions Containing Ions

CHEMath Visualizing Molecules

CHEMISTRY SERVING ... THE CONSUMER When Water Is Hard to Deal With

CHEMISTRY IN CAREERS Wastewater Engineer

LINK TO FOOD SCIENCE Sulfur Dioxide and Sulfites

Stay current with SCIENCE NEWS Find out more about ionic compounds: www.phschool.com

DISCOVER IT! SHAPES OF CRYSTALLINE MATERIALS

You need four small disposable cups, distilled water, a ruler, a spoon, sodium chloride (table salt), sucrose (table sugar), sodium hydrogen carbonate (baking soda), magnesium sulfate (Epsom salt), a small clean mirror, and a magnifying glass.

1. Label each cup with the name of one of the solids and pour water into each cup to a depth of 1 cm.
2. Add a spoonful of each solid to its corresponding cup. Swirl each cup for 30 seconds and then let it stand for a few minutes.
3. Swirl the cups at least two more times. Each time, note whether any of the solids completely dissolves. If so, add more of that solid and repeat the swirling. Continue this process until there is some undissolved solid at the bottom of each cup.
4. Set the mirror on a flat surface and place 2 or 3 drops of each liquid onto separate areas of the mirror.
5. Use a magnifying glass to examine each drop of liquid after 15 minutes, and then again after 24 hours.

For each material, did the solids crystallize? Did the crystals form at the same time? Describe the crystals that formed. Revisit this activity later in the chapter when you read about crystalline solids. Add to your notes any relevant information about the crystals you observed.

412 Chapter 15

SCIENCE NEWS

Keep current with *Science News!* Find weekly science updates on the web, brought to you by *Science News* magazine. **www.phschool.com**

Section Openers

Each section begins with an opportunity to learn about the applications and uses of chemistry in your everyday world. These short readings end with a question that you will be able to answer by reading the text in that section.

Discover It!

These activities provide opportunities to explore ideas and discover concepts of chemistry—before you read about them.

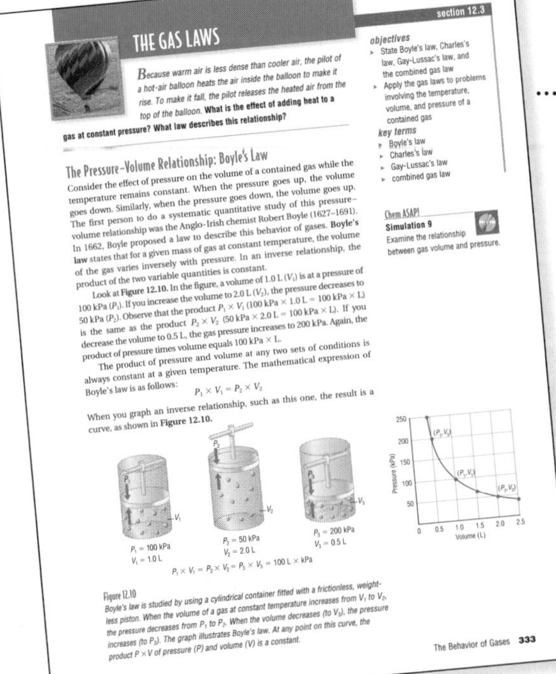

section 12.3

THE GAS LAWS

Because warm air is less dense than cooler air, the pilot of a hot-air balloon heats the air inside the balloon to make it rise. To make it fall, the pilot releases the heated air from the top of the balloon. **What is the effect of adding heat to a gas at constant pressure? What law describes this relationship?**

objectives
- State Boyle's law, Charles's law, Gay-Lussac's law, and the combined gas law
- Apply the gas laws to problems involving the temperature, volume, and pressure of a contained gas

key terms
- Boyle's law
- Charles's law
- Gay-Lussac's law
- combined gas law

The Pressure–Volume Relationship: Boyle's Law

Consider the effect of pressure on the volume of a contained gas while the temperature remains constant. When the pressure goes up, the volume goes down. Similarly, when the pressure goes down, the volume goes up. The first person to do a systematic quantitative study of this pressure–volume relationship was the Anglo-Irish chemist Robert Boyle (1627–1691). In 1662, Boyle proposed a law to describe this behavior of gases. **Boyle's law** states that for a given mass of gas at constant temperature, the volume of the gas varies inversely with pressure. In an inverse relationship, the product of the two variable quantities is constant.

Look at **Figure 12.10.** In the figure, a volume of 1.0 L (V_1) is at a pressure of 100 kPa (P_1). If you increase the volume to 2.0 L (V_2), the pressure decreases to 50 kPa (P_2). Observe that the product $P_1 \times V_1$ (100 kPa × 1.0 L = 100 kPa × L) is the same as the product $P_2 \times V_2$ (50 kPa × 2.0 L = 100 kPa × L). If you decrease the volume to 0.5 L, the gas pressure increases to 200 kPa. Again, the product of pressure times volume equals 100 kPa × L.

The product of pressure and volume at any two sets of conditions is always constant at a given temperature. The mathematical expression of Boyle's law is as follows:

$$P_1 \times V_1 = P_2 \times V_2$$

When you graph an inverse relationship, such as this one, the result is a curve, as shown in **Figure 12.10.**

Chem ASAP! **Simulation 9** Examine the relationship between gas volume and pressure.

Figure 12.10
Boyle's law is studied by using a cylindrical container fitted with a frictionless, weightless piston. When the volume of a gas at constant temperature increases from V_1 to V_2, the pressure decreases from P_1 to P_2. When the volume decreases (to V_3), the pressure increases (to P_3). The graph illustrates Boyle's law. At any point on this curve, the product P × V of pressure (P) and volume (V) is a constant.

The Behavior of Gases 333

Objectives and Key Terms

The objectives describe what you should focus on as you read the section. The key terms, printed in bold type, are the vocabulary words you should be able to define.

Recognize the Relevance

Chemistry Serving...the Environment

IT'S THE WATER

Earth is a watery planet. Oceans cover 72% of Earth's surface, and large amounts of water, in the form of water vapor, exist in Earth's atmosphere. Weather conditions around the planet are mostly the result of complex interactions between the water in Earth's oceans and in the atmosphere.

Large-scale weather patterns around the globe are somewhat stable. Some parts of the world usually receive large amounts of rainfall at certain times; other parts receive very little. These patterns are in part the result of how water in the oceans is heated and circulates around the planet.

Sometimes, for reasons scientists do not understand, water circulation in the oceans shifts away from the norm, causing major changes in the weather over large parts of Earth. One of these shifts, called El Niño, has now been observed and studied many times in the Pacific Ocean. It occurs, on average, every seven years.

El Niño begins in the waters just north of the equator, where the ocean surface water is heated rapidly by the nearly vertical rays of sunlight. Because of water's high heat capacity, this water tends to stay warm, and because warm water is less dense than cold water, it also tends to stay at the surface. Normally, trade winds blow from east to west across the Pacific and push the warm surface water to the west, toward Indonesia. If the trade winds die down, however, the warm water that usually pools in the western Pacific spreads toward the east. This shift is the El Niño effect.

Moving east, the warm water of El Niño reaches the coast of South America. Normally, cold, nutrient-rich waters rise from deep in the Pacific off the western coast of South America. The nutrients in the water fuel the growth of microscopic plants and animals, which are food for large fish populations. But as warmer waters move east during El Niño, the nutrient-rich colder waters are kept from rising. The microscopic ocean plants and animals have less food, and fish populations suffer.

When an El Niño occurs, among the first people to feel its effects are those who fish off the western coast of South America. They are also the ones who gave El Niño its name. They noted that the changes brought on by El Niño often begin around Christmas, the Christian celebration of the birth of Jesus, who is called "El Niño" (the Child) in South America.

At no other time is the ocean's influence on weather more obvious than during El Niño.

El Niño's effects spread well beyond South America. The warm water bumps up against the South American coast and spreads north to the western coast of North America, affecting the weather and fish populations there. In Indonesia, where the nearby ocean is cooler than normal, there is often a severe drought.

El Niño covers so much of the Pacific that it influences the movement of the jet stream, a high-altitude wind that steers storms through the atmosphere. The effect on the jet stream is unpredictable, but the usual result is wetter-than-normal weather in parts of South America, North America, and Asia, and drier-than-normal weather in other parts of the world. At no other time is the ocean's influence on global weather more obvious than during El Niño.

CHEMISTRY IN CAREERS

OCEANOGRAPHER Study ocean surface-water temperature, salt content, and nutrient levels. See page 876.

TAKE IT TO THE NET Find out more about career opportunities: www.phschool.com

CHEMIST NEEDED Enviro. Lab, GC Must know

CHEMICAL SPECIALIST Local food service distributor seeks responsible self-motivated individ-

494 Chapter 17

354 Chapter 12

862 Chapter 28

76 Chapter 3

Chemistry Serving...

This feature is designed to illustrate the important roles chemistry plays in serving the **Environment, Industry, Society,** and the **Consumer.**

CHEMISTRY IN CAREERS

OCEANOGRAPHER

Oceanography is the study of oceans. Because oceans are complex systems, oceanography is broken down into several interrelated subfields.

Physical oceanographers study the movement of seawater by waves, currents, and tides, and the interactions between the ocean and the atmosphere. Chemical oceanographers are concerned with the composition of seawater and the interactions between seawater and the organic and inorganic material in it. Geological oceanographers study the structure, features, and evolution of the ocean floor. Biological oceanographers study the diverse plants and animals of the sea and their relationship to their environment.

Oceanographers use many different tools to explore the oceans. Some use submersible vehicles to observe conditions in and take samples from the deepest parts of the ocean. Others wear scuba gear and visit the ocean floor. Oceanographers also use aircraft and satellites to study water temperatures and wave heights. When they are not in the field, oceanographers use computers to analyze data and create theoretical models.

Most oceanographers have advanced degrees. They begin their college education by obtaining a bachelor's degree in biology, marine biology, chemistry, or geology. They then obtain a master's degree or doctorate in a more specialized field, such as marine geology, meteorology, marine biology, or aquatic chemistry.

NEPHROLOGY NURSE

Because renal disease (disease of the kidneys) can be so debilitating to patients who have it, caring for these patients is an important task. Nephrology nurses are health-care professionals who specialize in the treatment and care of renal disease patients. The word nephrology comes from nephron, the name given to each of the more than one million tiny blood-processing units contained in each kidney. Once a patient has been diagnosed with renal disease, a nephrology nurse will assist the patient's doctor in assessing the disease's impact on the patient. Nephrology nurses are often the health-care professionals who help patients during hemodialysis. A major goal of nephrology nurses is to help patients regain the highest level of functioning while undergoing lengthy dialysis treatments. Duties include connecting patients to the dialysis machine, answering questions about treatment and progress, and providing moral support to patients who must endure hours of treatment. Nephrology nurses also assist in renal surgery such as kidney transplants.

Nephrology nurses are usually registered nurses (RNs) who specialize in nephrology. This level of certification requires an undergraduate degree in nursing plus special training in nephrology. Many vocational nurses are also involved in nephrology, and must undergo several years of training after high school.

876 Chemistry in Careers

Chemistry in Careers

Highlighted on each **Chemistry Serving...** page, and featured on pages 868–881, Chemistry in Careers discusses career opportunities that require an understanding of chemistry.

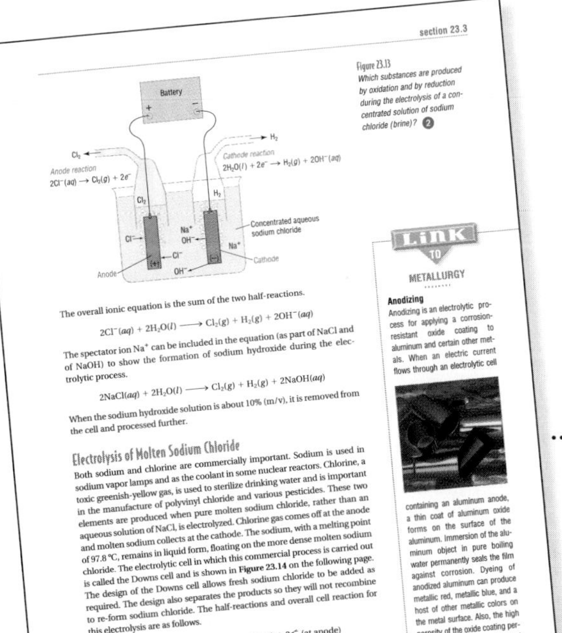

section 23.3

Figure 23.13 Which substances are produced by oxidation and by reduction during the electrolysis of a concentrated solution of sodium chloride (brine)?

The overall ionic equation is the sum of the two half-reactions.

$$2Cl^-(aq) + 2H_2O(l) \longrightarrow Cl_2(g) + H_2(g) + 2OH^-(aq)$$

The spectator ion Na^+ can be included in the equation (as part of NaCl and of NaOH) to show the formation of sodium hydroxide during the electrolytic process.

$$2NaCl(aq) + 2H_2O(l) \longrightarrow Cl_2(g) + H_2(g) + 2NaOH(aq)$$

When the sodium hydroxide solution is about 10% (m/v), it is removed from the cell and processed further.

Electrolysis of Molten Sodium Chloride

Both sodium and chlorine are commercially important. Sodium is used in sodium vapor lamps and as the coolant in some nuclear reactors. Chlorine, a toxic greenish-yellow gas, is used to sterilize drinking water and is important in the manufacture of polyvinyl chloride and various pesticides. These two elements are produced when pure molten sodium chloride, rather than an aqueous solution of NaCl, is electrolyzed. Chlorine gas comes off at the anode and molten sodium collects at the cathode. The sodium, with a melting point of 97.8 °C, remains in liquid form, floating on the more dense molten sodium chloride. The electrolytic cell in which this commercial process is carried out is called the Downs cell and is shown in **Figure 23.14** on the following page. The design of the Downs cell allows fresh sodium chloride to be added as required. The design also separates the products so they will not recombine to re-form sodium chloride. The half-reactions and overall cell reaction for this electrolysis are as follows.

Oxidation: $2Cl^-(l) \longrightarrow Cl_2(g) + 2e^-$ (at anode)

Reduction: $2Na^+(l) + 2e^- \longrightarrow 2Na(l)$ (at cathode)

Overall cell reaction: $2NaCl(l) \longrightarrow 2Na(l) + Cl_2(g)$

LINK TO METALLURGY

Anodizing

Anodizing is an electrolytic process for applying a corrosion-resistant oxide coating to aluminum and certain other metals. When an electric current flows through an electrolytic cell containing an aluminum anode, a thin coat of aluminum oxide forms on the surface of the aluminum. Immersion of the aluminum object in pure boiling water permanently seals the film against corrosion. Dyeing of anodized aluminum can produce metallic red, metallic blue, and a host of other metallic colors on the metal surface. Also, the high porosity of the oxide coating permits it to absorb pigments and lubricants.

Electrochemistry 695

Link To...

This feature illustrates how concepts in chemistry relate to both science and non-science disciplines.

Appreciate the Process

Probeware Lab Manual

This supplementary lab manual provides exciting opportunities for high-tech lab work.

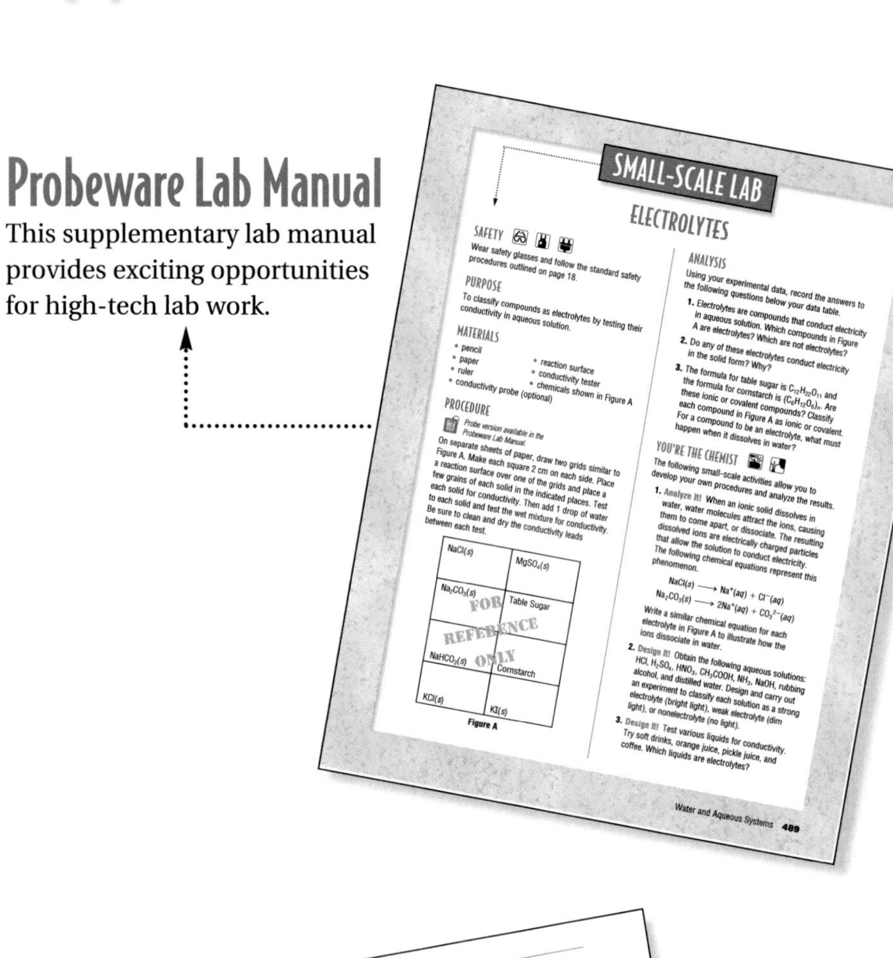

SMALL-SCALE LAB

ELECTROLYTES

SAFETY

Wear safety glasses and follow the standard safety procedures outlined on page 18.

PURPOSE

To classify compounds as electrolytes by testing their conductivity in aqueous solution.

MATERIALS

- pencil
- paper
- ruler
- conductivity probe (optional)
- reaction surface
- conductivity tester
- chemicals shown in Figure A

PROCEDURE

Probe version available in the Probeware Lab Manual.

On separate sheets of paper, draw two grids similar to Figure A. Make each square 2 cm on each side. Place a reaction surface over one of the grids and place a few grains of each solid in the indicated places. Test each solid for conductivity. Then add 1 drop of water to each solid and test the wet mixture for conductivity. Be sure to clean and dry the conductivity leads between each test.

$NaCl(s)$	$MgSO_4(s)$
$Na_2CO_3(s)$	Table Sugar
$NaHCO_3(s)$	Cornstarch
$KCl(s)$	$KI(s)$

Figure A

ANALYSIS

Using your experimental data, record the answers to the following questions below your data table.

1. Electrolytes are compounds that conduct electricity in aqueous solution. Which compounds in Figure A are electrolytes? Which are not electrolytes?
2. Do any of these electrolytes conduct electricity in the solid form? Why?
3. The formula for table sugar is $C_{12}H_{22}O_{11}$ and the formula for cornstarch is $(C_6H_{12}O_6)_n$. Are these ionic or covalent compounds? Classify each compound in Figure A as ionic or covalent. For a compound to be an electrolyte, what must happen when it dissolves in water?

YOU'RE THE CHEMIST

The following small-scale activities allow you to develop your own procedures and analyze the results.

1. **Analyze It!** When an ionic solid dissolves in water, water molecules attract the ions, causing them to come apart, or dissociate. The resulting dissolved ions are electrically charged particles that allow the solution to conduct electricity. The following chemical equations represent this phenomenon.

   $NaCl(s) \longrightarrow Na^+(aq) + Cl^-(aq)$

   $Na_2CO_3(s) \longrightarrow 2Na^+(aq) + CO_3^{2-}(aq)$

   Write a similar chemical equation for each electrolyte in Figure A to illustrate how the ions dissociate in water.
2. **Design It!** Obtain the following aqueous solutions: HCl, H_2SO_4, HNO_3, CH_3COOH, NH_3, NaOH, rubbing alcohol, and distilled water. Design and carry out an experiment to classify each solution as a strong electrolyte (bright light), weak electrolyte (dim light), or nonelectrolyte (no light).
3. **Design It!** Test various liquids for conductivity. Try soft drinks, orange juice, pickle juice, and coffee. Which liquids are electrolytes?

Water and Aqueous Systems 489

Small-Scale Labs

These labs require inexpensive, locally available equipment to carry out sophisticated and diverse chemical investigations on a small scale. You will use a variety of skills to hypothesize, experiment, observe, analyze, conclude, and apply your conclusions to other situations.

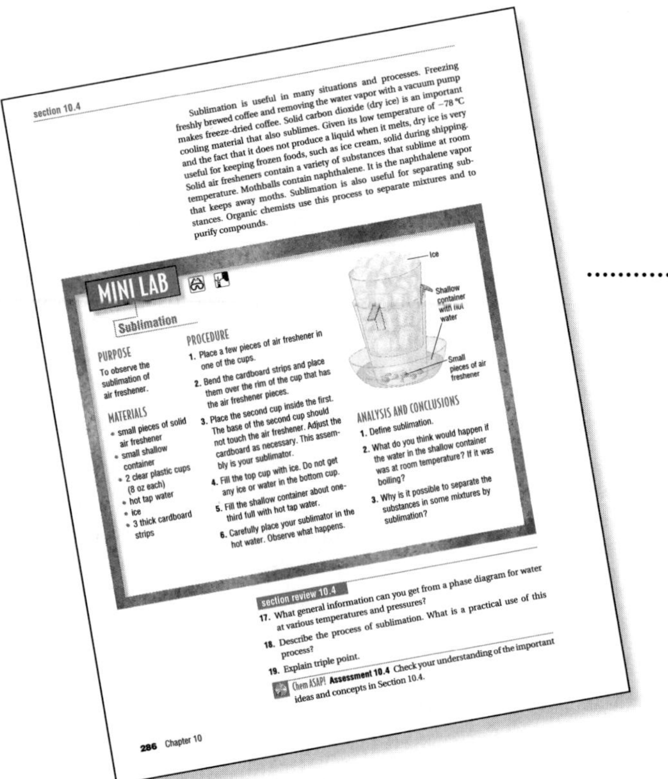

section 10.4

Sublimation is useful in many situations and processes. Freezing freshly brewed coffee and removing the water vapor with a vacuum pump makes freeze-dried coffee. Solid carbon dioxide (dry ice) is an important cooling material that also sublimes. Given its low temperature of −78 °C and the fact that it does not produce a liquid when it melts, dry ice is very useful for keeping frozen foods, such as ice cream, solid during shipping. Solid air fresheners contain a variety of substances that sublime at room temperature. Mothballs contain naphthalene. It is the naphthalene vapor that keeps away moths. Sublimation is also useful for separating substances. Organic chemists use this process to separate mixtures and to purify compounds.

MINI LAB

Sublimation

PURPOSE

To observe the sublimation of air freshener.

MATERIALS

- small pieces of solid air freshener
- small shallow container
- 2 clear plastic cups (8 oz each)
- hot tap water
- ice
- 3 thick cardboard strips

PROCEDURE

1. Place a few pieces of air freshener in one of the cups.
2. Bend the cardboard strips and place them over the rim of the cup that has the air freshener pieces.
3. Place the second cup inside the first. The base of the second cup should not touch the air freshener. Adjust the cardboard as necessary. This assembly is your sublimator.
4. Fill the top cup with ice. Do not get any ice or water in the bottom cup.
5. Fill the shallow container about one-third full with hot tap water.
6. Carefully place your sublimator in the hot water. Observe what happens.

ANALYSIS AND CONCLUSIONS

1. Define sublimation.
2. What do you think would happen if the water in the shallow container was at room temperature? If it was boiling?
3. Why is it possible to separate the substances in some mixtures by sublimation?

section review 10.4

17. What general information can you get from a phase diagram for water at various temperatures and pressures?
18. Describe the process of sublimation. What is a practical use of this process?
19. Explain triple point.

Chem ASAP! **Assessment 10.4** Check your understanding of the important ideas and concepts in Section 10.4.

286 Chapter 10

Mini Lab

Mini Lab activities are quick and effective hands-on opportunities to apply chemistry concepts and skills. Requiring minimal equipment, they enhance the concepts they are located near.

Reinforce Skills

Sample Problems and Practice Problems

Problem solving is an essential skill in chemistry. Sample problems provide an opportunity to sharpen this skill as you **analyze** the problem, **calculate** the answer, **evaluate** the results, and **practice** similar problems.

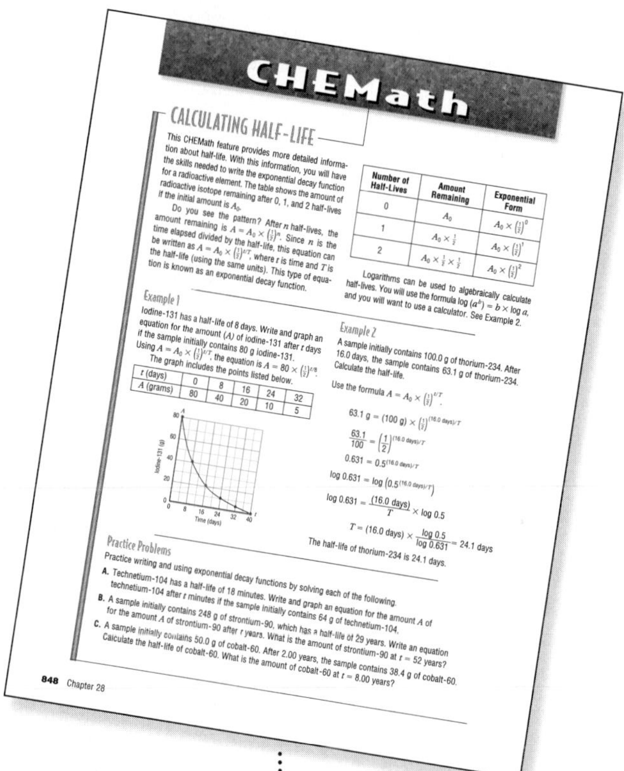

CHEMath

These features review specific math skills and explain why each skill is essential to success in chemistry. CHEMath features are distributed throughout the textbook to provide a point-of-use math refresher. CHEMath provides step-by-step explanations and practice problems.

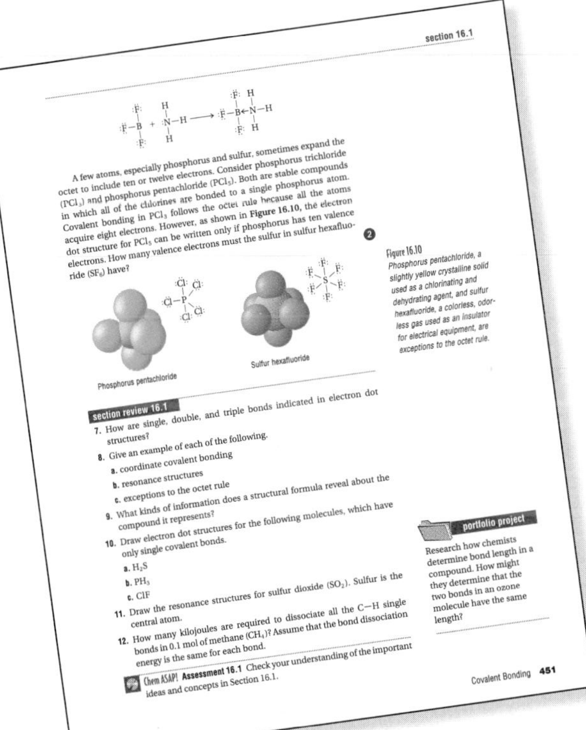

Section Review

Evaluate what you have learned in each section by answering the section review questions.

Understand the Concepts

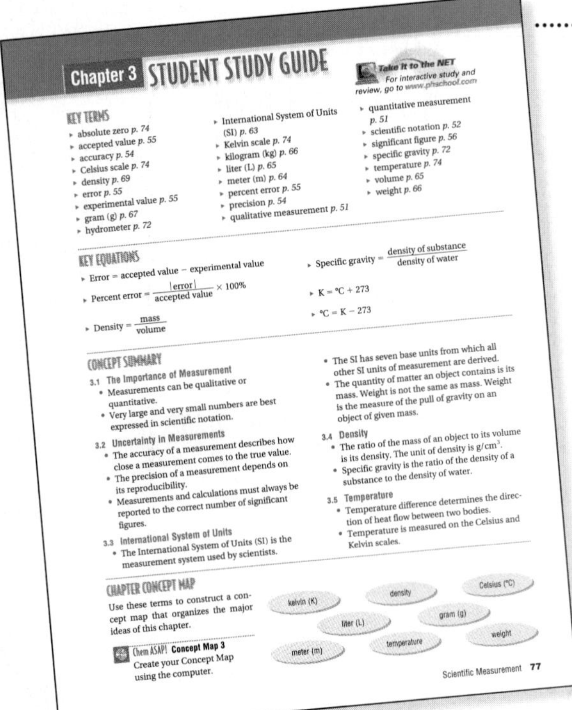

Chapter 3 STUDENT STUDY GUIDE

Take It to the NET
For interactive study and review, go to www.phschool.com

KEY TERMS

- absolute zero p. 74
- accepted value p. 55
- accuracy p. 54
- Celsius scale p. 74
- density p. 69
- error p. 55
- experimental value p. 55
- gram (g) p. 67
- hydrometer p. 72
- International System of Units (SI) p. 63
- Kelvin scale p. 74
- kilogram (kg) p. 66
- liter (L) p. 65
- meter (m) p. 64
- percent error p. 55
- precision p. 54
- qualitative measurement p. 51
- quantitative measurement p. 51
- scientific notation p. 52
- significant figure p. 56
- specific gravity p. 72
- temperature p. 74
- volume p. 65
- weight p. 66

KEY EQUATIONS

- Error = accepted value − experimental value
- $\text{Percent error} = \frac{|\text{error}|}{\text{accepted value}} \times 100\%$
- $\text{Density} = \frac{\text{mass}}{\text{volume}}$
- $\text{Specific gravity} = \frac{\text{density of substance}}{\text{density of water}}$
- K = °C + 273
- °C = K − 273

CONCEPT SUMMARY

3.1 The Importance of Measurement
- Measurements can be qualitative or quantitative.
- Very large and very small numbers are best expressed in scientific notation.

3.2 Uncertainty in Measurements
- The accuracy of a measurement describes how close a measurement comes to the true value.
- The precision of a measurement depends on its reproducibility.
- Measurements and calculations must always be reported to the correct number of significant figures.

3.3 International System of Units
- The International System of Units (SI) is the measurement system used by scientists.
- The SI has seven base units from which all other SI units of measurement are derived.
- The quantity of matter an object contains is its mass. Weight is not the same as mass. Weight is the measure of the pull of gravity on an object of given mass.

3.4 Density
- The ratio of the mass of an object to its volume is its density. The unit of density is g/cm^3.
- Specific gravity is the ratio of the density of a substance to the density of water.

3.5 Temperature
- Temperature difference determines the direction of heat flow between two bodies.
- Temperature is measured on the Celsius and Kelvin scales.

CHAPTER CONCEPT MAP

Use these terms to construct a concept map that organizes the major ideas of this chapter.

Chem ASAP! Concept Map 3
Create your Concept Map using the computer.

Student Study Guide

The Student Study Guide checks your understanding of **Key Terms** and **Key Equations and Relationships.** You will also find a **Summary** of important concepts and **Concept Maps** to further test your understanding. The Guided Study Workbook provides active reading and guided practice—a great way to prepare for tests!

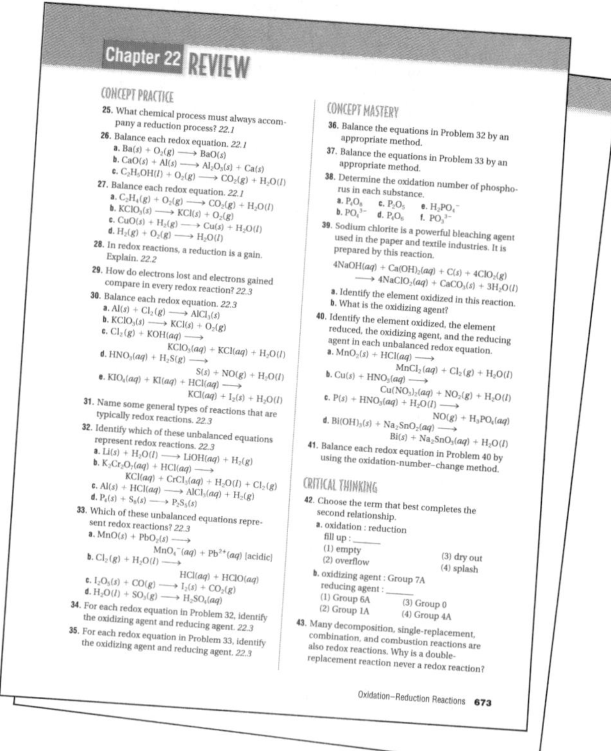

Chapter 22 REVIEW

CONCEPT PRACTICE

25. What chemical process must always accompany a reduction process? 22.1
26. Balance each redox equation. 22.1
    a. $Ba(s) + O_2(g) \longrightarrow BaO(s)$
    b. $CaO(s) + Al(s) \longrightarrow Al_2O_3(s) + Ca(s)$
    c. $C_2H_5OH(l) + O_2(g) \longrightarrow CO_2(g) + H_2O(l)$
27. Balance each redox equation. 22.1
    a. $C_2H_4(g) + O_2(g) \longrightarrow CO_2(g) + H_2O(l)$
    b. $KClO_3(s) \longrightarrow KCl(s) + O_2(g)$
    c. $CuO(s) + H_2(g) \longrightarrow Cu(s) + H_2O(l)$
    d. $H_2(g) + O_2(g) \longrightarrow H_2O(l)$
28. In redox reactions, a reduction is a gain. Explain. 22.2
29. How do electrons lost and electrons gained compare in every redox reaction? 22.3
30. Balance each redox equation. 22.3
    a. $Al(s) + Cl_2(g) \longrightarrow AlCl_3(s)$
    b. $KClO_3(s) \longrightarrow KCl(s) + O_2(g)$
    c. $Cl_2(g) + KOH(aq) \longrightarrow KClO_3(aq) + KCl(aq) + H_2O(l)$
    d. $HNO_3(aq) + H_2S(g) \longrightarrow S(s) + NO(g) + H_2O(l)$
    e. $KIO_4(aq) + KI(aq) + HCl(aq) \longrightarrow KCl(aq) + I_2(s) + H_2O(l)$
31. Name some general types of reactions that are typically redox reactions. 22.3
32. Identify which of these unbalanced equations represent redox reactions. 22.3
    a. $Li(s) + H_2O(l) \longrightarrow LiOH(aq) + H_2(g)$
    b. $K_2Cr_2O_7(aq) + HCl(aq) \longrightarrow KCl(aq) + CrCl_3(aq) + H_2O(l) + Cl_2(g)$
    c. $Al(s) + HCl(aq) \longrightarrow AlCl_3(aq) + H_2(g)$
    d. $P_4(s) + S_8(s) \longrightarrow P_2S_5(s)$
33. Which of these unbalanced equations represent redox reactions? 22.3
    a. $MnO(s) + PbO_2(s) \longrightarrow MnO_4^-(aq) + Pb^{2+}(aq)$ [acidic]
    b. $Cl_2(g) + H_2O(l) \longrightarrow HCl(aq) + HClO(aq)$
    c. $I_2O_5(s) + CO(g) \longrightarrow I_2(s) + CO_2(g)$
    d. $H_2O(l) + SO_3(g) \longrightarrow H_2SO_4(aq)$
34. For each redox equation in Problem 32, identify the oxidizing agent and reducing agent. 22.3
35. For each redox equation in Problem 33, identify the oxidizing agent and reducing agent. 22.3

CONCEPT MASTERY

36. Balance the equations in Problem 32 by an appropriate method.
37. Balance the equations in Problem 33 by an appropriate method.
38. Determine the oxidation number of phosphorus in each substance.
    a. P_4O_8 c. P_2O_5 e. $H_2PO_4^-$
    b. PO_3^{3-} d. P_4O_6 f. PO_4^{3-}
39. Sodium chlorite is a powerful bleaching agent used in the paper and textile industries. It is prepared by this reaction.
    $4NaOH(aq) + Ca(OH)_2(aq) + C(s) + 4ClO_2(g) \longrightarrow 4NaClO_2(aq) + CaCO_3(s) + 3H_2O(l)$
    a. Identify the element oxidized in this reaction.
    b. What is the oxidizing agent?
40. Identify the element oxidized, the element reduced, the oxidizing agent, and the reducing agent in each unbalanced redox equation.
    a. $MnO_2(s) + HCl(aq) \longrightarrow MnCl_2(aq) + Cl_2(g) + H_2O(l)$
    b. $Cu(s) + HNO_3(aq) \longrightarrow Cu(NO_3)_2(aq) + NO_2(g) + H_2O(l)$
    c. $P(s) + HNO_3(aq) + H_2O(l) \longrightarrow NO(g) + H_3PO_4(aq)$
    d. $Bi(OH)_3(s) + Na_2SnO_2(aq) \longrightarrow Bi(s) + Na_2SnO_3(aq) + H_2O(l)$
41. Balance each redox equation in Problem 40 by using the oxidation-number-change method.

CRITICAL THINKING

42. Choose the term that best completes the second relationship.
    a. oxidation : reduction
    fill up : ______
    (1) empty (3) dry out
    (2) overflow (4) splash
    b. oxidizing agent : Group 7A
    reducing agent : ______
    (1) Group 6A (3) Group 0
    (2) Group 1A (4) Group 4A
43. Many decomposition, single-replacement, combination, and combustion reactions are also redox reactions. Why is a double-replacement reaction never a redox reaction?

Chapter Review

The questions in **Concept Practice, Concept Mastery, Critical Thinking, Cumulative Review,** and **Concept Challenge** assess your comprehension of the chapter content. Selected questions identified by the calculator icon enhance proficiency in the use of this technology. Further instructions and practice problems are provided in the **Graphing Calculator** ancillary.

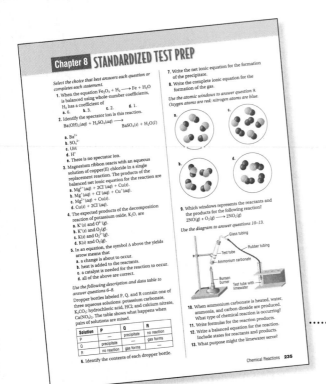

Chapter 8 STANDARDIZED TEST PREP

Select the choice that best answers each question or completes each statement.

1. When the equation $Fe_2O_3 + H_2 \longrightarrow Fe + H_2O$ is balanced using whole-number coefficients, H_2 has a coefficient of
   a. 6. b. 3. c. 2. d. 1.
2. Identify the spectator ion in this reaction.
   $Ba(OH)_2(aq) + H_2SO_4(aq) \longrightarrow BaSO_4(s) + H_2O(l)$
   a. Ba^{2+}
   b. SO_4^{2-}
   c. OH^-
   d. H^+
   e. There is no spectator ion.
3. Magnesium ribbon reacts with an aqueous solution of copper(II) chloride in a single replacement reaction. The products of the balanced net ionic equation for the reaction are
   a. $Mg^{2+}(aq) + 2Cl^-(aq) + Cu(s)$.
   b. $Mg^+(aq) + Cl^-(aq) + Cu^+(aq)$.
   c. $Mg^{2+}(aq) + Cu(s)$.
   d. $Cu(s) + 2Cl^-(aq)$.
4. The expected products of the decomposition reaction of potassium oxide, K_2O, are
   a. $K^+(s)$ and $O^{2-}(g)$.
   b. $K^+(s)$ and $O_2(g)$.
   c. $K(s)$ and $O_2^{2-}(g)$.
   d. $K(s)$ and $O_2(g)$.
5. In an equation, the symbol Δ above the yields arrow means that
   a. a change is about to occur.
   b. heat is added to the reactants.
   c. a catalyst is needed for the reaction to occur.
   d. all of the above are correct.

Use the following description and data table to answer questions 6–8.

Dropper bottles labeled P, Q, and R contain one of three aqueous solutions: potassium carbonate, K_2CO_3; hydrochloric acid, HCl; and calcium nitrate, $Ca(NO_3)_2$. The table shows what happens when pairs of solutions are mixed.

Solution	P	Q	R
P	—	precipitate	no reaction
Q	precipitate	—	gas forms
R	no reaction	gas forms	—

6. Identify the contents of each dropper bottle.
7. Write the net ionic equation for the formation of the precipitate.
8. Write the complete ionic equation for the formation of the gas.

Use the atomic windows to answer question 9. Oxygen atoms are red; nitrogen atoms are blue.

a. b. c. d.

9. Which windows represents the reactants and the products for the following reaction?
   $2NO(g) + O_2(g) \longrightarrow 2NO_2(g)$

Use the diagram to answer questions 10–13.

Glass tubing
Test tube
Rubber tubing
Ammonium carbonate
Bunsen burner
Test tube with limewater

10. When ammonium carbonate is heated, water, ammonia, and carbon dioxide are produced. What type of chemical reaction is occurring?
11. Write formulas for the reaction products.
12. Write a balanced equation for the reaction. Include states for reactants and products.
13. What purpose might the limewater serve?

Chemical Reactions 235

Standardized Test Prep

In each chapter, a practice test using a variety of formats prepares you for state and national exams. The Prentice Hall Assessment System has instruction in test-taking strategies and additional practice tests.

Incorporate Technology

Chem ASAP!

The Chem ASAP! CD-ROM is an easy-to-use learning and reference tool directly tied into the textbook. It is designed to help you understand chemistry concepts and master problem-solving skills through the use of **Animations, Simulations, Assessment** questions, and guided **Problem-Solving** tutorials.

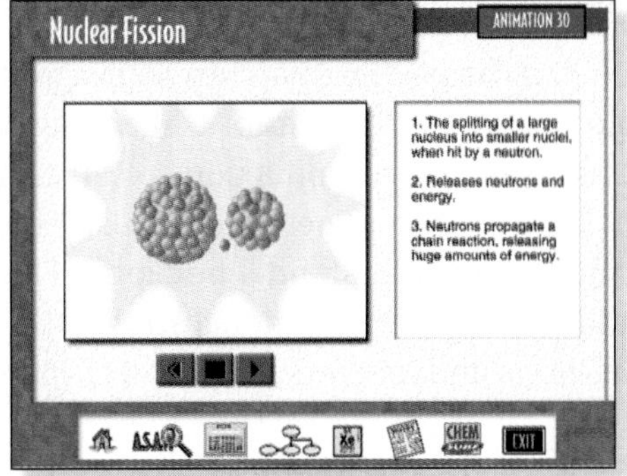

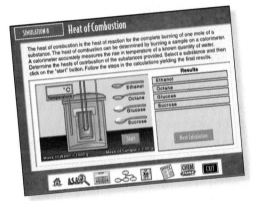

Animations

Each narrated animation brings the textbook to life as it illustrates important chemistry concepts, chemical processes, and applications of technology.

Simulations

Each simulation reinforces your understanding of important concepts and allows you to manipulate variables, observe results, and draw conclusions.

Assessment

More than 500 interactive assessment questions allow you to check your understanding of key concepts and ideas from each section of the textbook.

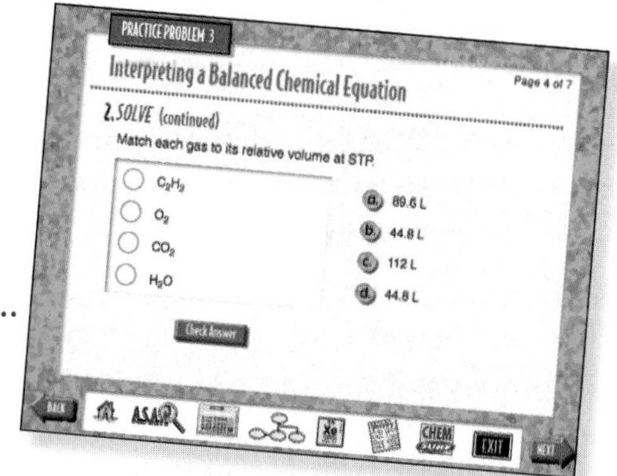

Problem-Solving

Interactive tutorials guide you through detailed solutions of nearly 100 Practice Problems found in the textbook. Each tutorial follows the 3-step problem-solving approach and concludes with an additional related Practice Problem.

Additional Features

- **Periodic Table**—contains a wealth of easily accessed information
- **Glossary**—includes the complete textbook glossary along with audio pronunciations
- **Concept Mapping Tool**—easily creates custom concept maps relating key ideas
- **Take It to The Net**—directly connect to www.phschool.com from Chem ASAP!

. . . and More Technology

Overhead Transparencies

The Overhead Transparency Package contains 80 color transparencies and a Teaching Guide. The transparencies, reproductions of art from the textbook, have been selected to help students visualize difficult-to-understand concepts. The Teaching Guide provides instructional strategies that include questions and answers for each transparency.

Videodiscs and Videotapes

Chemistry Alive! Videodisc This informative series of 27 fascinating chemistry demonstrations performed by master teachers Lee Marek and Robert Lewis brings fast-paced fun and solid chemistry instruction into the classroom.

Small-Scale Lab Video and Videodisc Author Ed Waterman demonstrates small-scale techniques and discusses the advantages of small-scale labs.

CHEMedia™ Videodiscs Five videodiscs add real-world connections to chemistry topics through demos, field trips, animations, and explorations.

Internet

SCIENCE NEWS Keep current with *Science News!* Weekly science updates keep you informed with the latest chemistry news and discoveries.
www.phschool.com

www.phschool.com
Enhance every chapter with Internet activities, self-tests, dynamic links, and career information.

The Chemistry Place This premier Internet chemistry site for students and teachers includes the latest chemistry research presented by prestigious resident faculty, extensive learning activities, members' forum, and more.

CD-ROMs

Chem ASAP! CD-ROM An easy-to-use learning and reference tool provided on CD-ROM to every student, Chem ASAP! is designed to help students master chemistry concepts and problem-solving skills. Nearly 300 in-text references to Animations, Simulations, Assessment questions, and Problem-Solving tutorials contained in Chem ASAP! ensure that the CD-ROM is easily and effectively integrated into daily instruction.

Resource Pro CD-ROM The complete Teacher's Resource Package is available in a customizable CD-ROM format that also includes an integrated comprehensive lesson planner. Supplementary materials can be accessed by chapter or by title. Planning can be done for a day, week, term, or year for regular or block schedules.

ActivChemistry™ CD-ROM A virtual chemistry laboratory on CD-ROM, ActivChemistry lets students simulate laboratory experiments on the computer. The CD-ROM includes 17 lessons correlated to Addison-Wesley Chemistry and can be used as an authoring tool to create custom experiments and lessons.

Assessment Resources Traditional and alternative testing for every section in the textbook is provided by this easy-to-use-and-edit CD-ROM for Windows and Macintosh.

Problem Pro CD-ROM Includes options for customizing and generating a virtual endless number of practice–problem worksheets.

Probeware Lab Manual for Computers and Calculators Whether you are using Vernier, TI, or PASCO probes and sensors, you will find easy-to-use labs with step-by-step instructions. CD-ROM included for quick start-up and technical support.

The Authors...

Antony C. Wilbraham

Antony Wilbraham is an Emeritus Professor of Chemistry. He taught college-level chemistry for more than twenty-five years at Southern Illinois University at Edwardsville, Illinois, where he was also Director of Hazardous Waste Management. In 1978, he received the University Teaching Excellence Award. Dr. Wilbraham has been writing high school and college-level chemistry textbooks and related ancillaries for more than twenty years and has published extensively in science journals for over thirty years. He is a member of the American Chemical Society, a member of the National Science Teachers Association, and a Fellow of the Royal Society of Chemistry.

Dennis D. Staley

Dennis Staley is an Instructor in the Department of Chemistry and the Office of Science and Math Education at Southern Illinois University at Edwardsville, Illinois. He has been teaching high school and college-level chemistry for more than twenty years. In 1981 he received the University Teaching Excellence Award. Mr. Staley has been writing high school and college-level chemistry textbooks and related ancillaries for more than twenty years. He is a member of the American Chemical Society and a member of the National Science Teachers Association.

Michael S. Matta

Michael Matta is an Emeritus Professor of Chemistry. He was a professor of chemistry at Southern Illinois University at Edwardsville, Illinois, from 1969 to 1996, and served as Department Chair from 1980 to 1983. In 1973, Dr. Matta received the University Teaching Excellence Award. He has been developing and writing high school and college-level chemistry textbooks and related ancillaries for more than twenty years and has published extensively in scientific journals. He is a member of the American Chemical Society, the National Science Teachers Association, and the American Association for the Advancement of Science.

Edward L. Waterman

Ed Waterman has taught chemistry, advanced placement chemistry, and organic and biochemistry since 1976 at Rocky Mountain High School in Fort Collins, Colorado. Mr. Waterman conducts workshops for teachers on such topics as small-scale chemistry laboratory, advanced placement chemistry, block scheduling, Internet for the chemistry classroom, and virtual chemistry on CD-ROM. Mr. Waterman holds a Bachelor of Science degree in chemistry from Montana State University and a Master of Science degree in chemistry from Colorado State University.

Content Reviewers

Dr. John I. Gelder
Oklahoma State University
Stillwater, OK

Dr. Jameica Hill
Wofford College
Spartanburg, SC

Dr. Samuel Kounaves
Tufts University
Medford, MA

Dr. Doris Lewis
Suffolk University
Boston, MA

Dr. Lorraine Rellick
Capital University
Columbus, OH

Dr. David Taylor
Slippery Rock University
Slippery Rock, PA

Anita Dunn Thurwachter
Austin Community College
Austin, TX

Teacher Reviewers

Sidney Jay Abramowitz
Stamford High School
Stamford, CT

Christy Alexander
Permian High School
Odessa, TX

Max Ceballos
Edinburg High School
Edinburg, TX

Charles DiSapio
Stamford High School
Stamford, CT

Ron Dunaway
Monterey High School
Lubbock, TX

Lewis Hitzrot
Phillips Exeter Academy
Exeter, NH

Judy Mosher
Indian River High School
Chesapeake, VA

Terrie Reed
Permian High School
Odessa, TX

Mark Sandomir
Mountain View High School
Mesa, AZ

Victoria Sanftner
Montwood High School
El Paso, TX

Pamela Shlachtman
South Dade High School
Miami, FL

Sharron Story
Monterey High School
Lubbock, TX

Dorothy Thomas
Forth Worth ISD
Fort Worth, TX

David Tuskey
Clover Hill High School
Midlothian, VA

Kevin Williams
Edgewater High School
Orlando, FL

Senior Texas Consultant
Susan M. Cory
Cypress Fairbanks ISD
Houston, TX

Safety Reviewer

Sally Busboom
Austin ISD
Austin, TX

Planning Guide

SECTION OBJECTIVES	ACTIVITIES/FEATURES	MEDIA & TECHNOLOGY
1.1 Chemistry ● ■ ◇ ▸ Define chemistry and differentiate among its traditional divisions ▸ List several reasons to study chemistry	**SE** **Discover It!** *Solid, Liquid, or What?*, p. 2 **SE** **Link to Food Science** *Chemistry in the Kitchen*, p. 6 **TE** **DEMO**, p. 4	**ASAP** Assessment 1.1
1.2 Chemistry Far and Wide ● ■ ◇ ▸ Summarize ways in which chemistry affects your daily life ▸ Describe the impact of chemistry on various fields of science	**SE** **Link to Atmospheric Science** *Predicting the Formation of the Ozone Hole*, p. 13 **TE** **DEMO**, p. 9	**ASAP** Assessment 1.2 **CHM** Side 1, 2: *Connections to Our World* **CHM** Side 1, 10: *Restoring the Sistine Chapel*
1.3 Thinking Like a Scientist ● ■ ◇ ▸ Describe the steps involved in the scientific method ▸ Distinguish between a theory and scientific law	**SE** **Mini Lab** *Bubbles!*, p. 17) (LRS 1-1) **SE** **Small-Scale Lab** *Introduction to Small-Scale Chemistry*, p. 18 (LRS 1-2) **LM** 1: *Observing and Inferring* **TE** **DEMO**, p. 16	**ASAP** Assessment 1.3 **CA** *Safety* **CA** *Ira Remsen Story* **SSV** 1: *What is Small-Scale Chemistry?* **SSV** 2: *Setting Up a Small-Scale Laboratory* **OT** 1: Scientific Method and Law
1.4 How to Study Chemistry ● ■ ◇ ▸ Explain why learning chemistry requires daily effort ▸ Describe the importance of writing in the study of chemistry	**SE** **Chemistry Serving . . . Society** *Chemistry in the Information Age*, p. 23 **SE** **Chemistry in Careers** *Chemist*, p. 23	**ASAP** Assessment 1.4 **ASAP** Concept Map 1 **RP** Lesson Plans, Resource Library **AR** Computer Test 1 **www** Activities, Self-Tests, SCIENCE NEWS updates **TCP** The Chemistry Place Web Site

KEY

● Conceptual (concrete concepts)	**AR** Assessment Resources	**GRS** Guided Reading and Study Workbook
○ Conceptual (more abstract/math)	**ASAP** Chem ASAP! CD-ROM	**LM** Laboratory Manual
■ Standard (core content)	**ACT** ActivChemistry CD-ROM	**LP** Laboratory Practicals
□ Standard (extension topics)	**CHM** CHEMedia Videodiscs	**LRS** Laboratory Recordsheets
◆ Honors (core content)	**CA** Chemistry Alive! Videodiscs	**SSLM** Small-Scale Lab Manual
◇ Honors (options to accelerate)	**GCP** Graphing Calculator Problems	

PRACTICE		ASSESSMENT	
GRS	Section 1.1	**SE**	Section Review
RM	Practice Problems 1.1	**RM**	Section Review 1.1
RM	Interpreting Graphics	**RM**	Chapter 1 Quiz
GRS	Section 1.2	**SE**	Section Review
RM	Practice Problems 1.2	**RM**	Section Review 1.2
		RM	Chapter 1 Quiz
GRS	Section 1.3	**SE**	Section Review
RM	Practice Problems 1.3	**RM**	Section Review 1.3
		RM	Chapter 1 Quiz
GRS	Section 1.4	**SE**	Section Review
RM	Practice Problems 1.4	**RM**	Section Review 1.4
		RM	Vocabulary Review 1
		SE	Chapter Review
		SM	Chapter 1 Solutions
		SE	Standardized Test Prep
		PHAS	Chapter 1 Test Prep
		RM	Chapter 1 Quiz
		RM	Chapter 1 A & B Test

PLANNING FOR ACTIVITIES

STUDENT EDITION

Discover It! p. 2
- measuring spoons
- cornstarch
- small bowls (cereal bowls)
- water

Mini Lab, p. 17
- plastic cups
- liquid dish detergent
- measuring cups
- water
- sugar
- salt
- drinking straws

TEACHER'S EDITION

Teacher Demo, p. 4
- slide projector
- slides that show highlights of natural and human-created phenomena

Teacher Demo, p. 9
- hydrogen gas
- small balloon
- extender
- flame source

Activity, p. 11
- local and national newspapers

Activity, p. 12
- labels from several aerosol and refrigerant containers
- literature published by the Environmental Protection Agency regarding what chemicals are used in place of Freon™ today

Teacher Demo, p. 16
- 250 mg phenolphthalein in 250 mL of 70% 2-propanol
- small artist's paintbrush
- photocopy paper or white butcher paper
- spray bottle of household glass cleaner containing ammonia

Activity, p. 20
- arrange for students to speak to an art conservator

OT	Overhead Transparency	**SE**	Student Edition
PHAS	PH Assessment System	**SM**	Solutions Manual
PLM	Probeware Lab Manual	**SSV**	Small-Scale Video/Videodisc
PP	Problem Pro CD-ROM	**TCP**	www.chemplace.com
RM	Review Module	**TE**	Teacher's Edition
RP	Resource Pro CD-ROM	**www**	www.phschool.com

Chapter 1

Key Terms

1.1 chemistry, organic chemistry, inorganic chemistry, analytical chemistry, physical chemistry, biochemistry
1.3 scientific method, observation, hypothesis, experiment, theory, scientific law

DISCOVER IT!

When a student pushes slowly into the mixture with a finger or a fist, either one is likely to reach the bottom of the bowl, but the fist meets more resistance.
When a student jabs the mixture, it feels solid. Neither the finger nor the fist is likely to reach the bottom of the bowl.
When a ball of the mixture is squeezed, it behaves like a solid. When the pressure is released, the ball behaves more like a liquid.
Pressure is the factor that appears to determine whether the mixture behaves more like a solid or more like a liquid.

Chapter 1 INTRODUCTION TO CHEMISTRY

The Galileo spacecraft collects data about Jupiter and its moons.

FEATURES

DISCOVER IT!
Solid, Liquid, or What?

SMALL-SCALE LAB
Introduction to Small-Scale Chemistry

MINI LAB
Bubbles!

CHEMISTRY SERVING ... SOCIETY
Chemistry in the Information Age

CHEMISTRY IN CAREERS
Chemist

LINK TO FOOD SCIENCE
Chemistry in the Kitchen

LINK TO ATMOSPHERIC SCIENCE
Predicting the Formation of the Ozone Hole

Stay current with SCIENCE NEWS
Find out more about chemistry:
www.phschool.com

DISCOVER IT! SOLID, LIQUID OR WHAT?

You need a measuring spoon, cornstarch, a small bowl (such as a cereal bowl), and water.

1. Add 3 heaping tablespoons of cornstarch to the bowl, and then add 3 tablespoons of water.
2. Stir the contents of the bowl thoroughly and let stand for 5 minutes.
3. Slowly push your finger into the mixture. Repeat with your fist.
4. Quickly jab your finger into the mixture. Repeat with your fist.
5. Take a handful of the mixture and form a ball. Squeeze and release the ball several times.

What happens when you slowly push into the mixture with your finger or fist? When you jab it? When you squeeze the ball? When you release the ball? What condition seems to determine whether the mixture behaves like a solid or a liquid? When you have completed the chapter, return to this activity. What hypotheses can you form? How can you test them?

1.1 ● ■ ◇
1.2 ● ■ ◇
1.3 ● ■ ◇
1.4 ● ■ ◇

More time spent initially focusing on safe lab practices will make lab time more efficient throughout the year.

Conceptual The photos in this chapter help students visualize the role of chemistry in daily life. Section 1.2 is a survey; students do not need to understand each topic in detail. Use Figure 1.19 to help students visualize the steps used in scientific methods. The Guided Reading and Study Workbook will help students make efficient use of their study time.

Standard Determine the pace of this chapter based on students' prior knowledge. Emphasize that scientific investigation is a systematic, logical approach, not a single method. Take the time to help students set up a realistic and effective study plan, which will save time later in the course.

Honors Much of this chapter may be assigned for homework or self-study because most concepts will be familiar. Use the Chemistry Serving feature to emphasize that all information sources are not equally reliable.

CHEMISTRY

The Galileo spacecraft was placed in orbit around Jupiter to collect data about the planet and its moons. Instruments aboard Galileo analyzed the atmosphere of the moon Io and found large amounts of sulfur and sulfur dioxide. The presence of these chemicals verified that the volcanos on Io's surface are active. Chemistry allowed scientists to study the geology of a distant object in the solar system. **What is chemistry, and what are some of its different branches?**

objectives

- Define chemistry and differentiate among its traditional divisions
- List several reasons to study chemistry

key terms

- chemistry
- organic chemistry
- inorganic chemistry
- analytical chemistry
- physical chemistry
- biochemistry

What Is Chemistry?

Anyone who has ever watched a toddler explore a new environment knows that people are born with a natural curiosity about the world: a world of vivid colors and different textures, of simple and complex objects, of things that move and things that are permanently fixed. As the toddler grows, he or she learns to name observable things and, as much as possible, to figure out how they work. The desire to know more and more about the world grows as well. This desire, a lifelong attribute, is what compels humans to discover, build, and invent.

Chemistry is one branch of knowledge that grew from human curiosity about the world. **Chemistry** is the study of the composition of matter—the stuff things are made of—and the changes that matter undergoes. In fact, as you can see in **Figure 1.2** on page 4, the Japanese symbols that make up the word for chemistry mean "change study."

Much of chemistry is very practical and has obvious applications to everyday life. Perhaps you are wearing clothes made of synthetic fibers or natural fibers that have been dyed. The pan your dinner was cooked in may have a nonstick surface. Or perhaps you recently used nail-polish remover or hair spray. Other aspects of chemistry are theoretical, or without everyday application. But what is theoretical one day may become practical the next. For example, fifty years ago it would have been difficult to conceive of the impact of plastics or computers on our everyday lives.

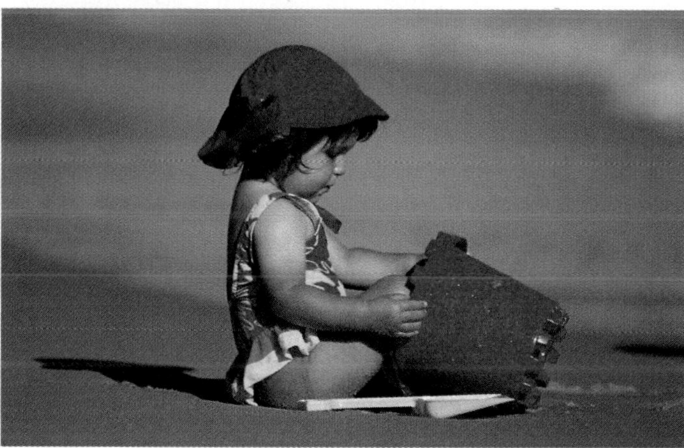

Figure 1.1
Children learn by watching and doing. This toddler is learning about sand and ocean water, as well as exploring the uses of a pail, as she experiences a day at the beach.

1 Engage

Use the Visual

Have students study the photograph and read the text that opens the section. Ask:

- **Why would a spacecraft sent to study Jupiter be named for Galileo?** (Students may know that Galileo was the first to use a telescope to observe Jupiter. He discovered that the planet had four moons.) **Why would large amounts of sulfur and sulfur dioxide be a sign that the volcanoes on Io's surface are active?** (Students may know that molten sulfur and sulfur dioxide gas can be found in molten magma.)
- **What chemical do scientists look for when considering whether a planet or moon could support organisms similar to those on Earth?** (Scientists look for water.)
- **What is chemistry, and what are some of its different branches?** (Student are likely to know that chemists study the composition of matter and the changes that matter undergoes. After Section 1.1, they will know that there are five major branches of chemistry: organic chemistry, inorganic chemistry, analytical chemistry, physical chemistry, and biochemistry.)

Check Prior Knowledge

To find out if students know the scope of the category *matter*, ask students to classify items as matter on non-matter.

Student Resources

From the Teacher's Resource Package, use:

- Section Review 1.1, Ch. 1 Practice Problems, Interpreting Graphics, and Quizzes from the Review Module (Ch. 1–4)

Technology Resources

Relevant technology resources include:

- Chem ASAP! CD-ROM
- Resource Pro CD-ROM

2 Teach

TEACHER DEMO

Students are often more enthusiastic about studying chemistry once they realize its significance to their own interests and goals. Organize a short slide show that highlights a variety of natural and human-created phenomena. Natural examples might include lakes, forests, mountains, farms, volcanos, lightning storms, icebergs, glaciers, limestone formations in caves, and supernovae.

Human-created phenomena include fireworks displays, space shuttle launches, pieces of art such as paintings, statues, and pottery, consumer products, plastics, sports equipment, electric lights (neon lights), medicines, nutritional supplements, and electron-microscope images of viruses and bacteria. Public libraries and visitor-information bureaus have a large selection of photos that can be borrowed. Discuss some of the ways in which the study of chemistry has benefited and harmed society. Emphasize that chemistry is a dynamic science that is continually evolving.

化学

Figure 1.2
The Japanese characters for chemistry literally mean "change study." Chemistry concerns the changes that matter undergoes, such as the rusting of metal, the transformation of liquid water to steam, and the burning of a match.

Traditionally, chemistry has been divided into five major areas of study. Organic chemistry was originally the study of substances from living organisms. Today, with a few exceptions, **organic chemistry** is the study of essentially all substances containing carbon. **Inorganic chemistry** specializes primarily in substances that do not contain carbon. These are mainly substances from nonliving things. **Analytical chemistry** is concerned with the composition of substances. Finding minute quantities of a particular medication in blood requires the practice of analytical chemistry. **Physical chemistry** is concerned with theories and experiments that describe the behavior of chemicals. For example, the stretching of nylon can be explained using the concepts of physical chemistry. Finally, **biochemistry** is the study of the chemistry of living organisms. The processes of digestion, blood clotting, and respiration, to name just a few, are explained by biochemistry. These five subdivisions of chemistry often overlap; for example, one cannot learn the composition of an organic or inorganic substance without being skilled in analytical chemistry.

Chemistry is central to modern science and to almost all human endeavors, as you can see from **Figure 1.4.** Today, many chemists work in teams with other scientists—biologists, geologists, physicists, physiologists, physicians, and environmental scientists—as well as with engineers.

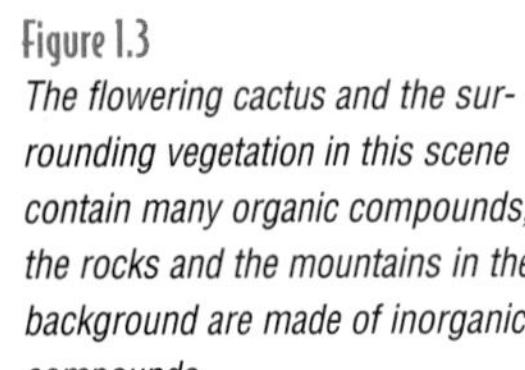

Figure 1.3
The flowering cactus and the surrounding vegetation in this scene contain many organic compounds; the rocks and the mountains in the background are made of inorganic compounds.

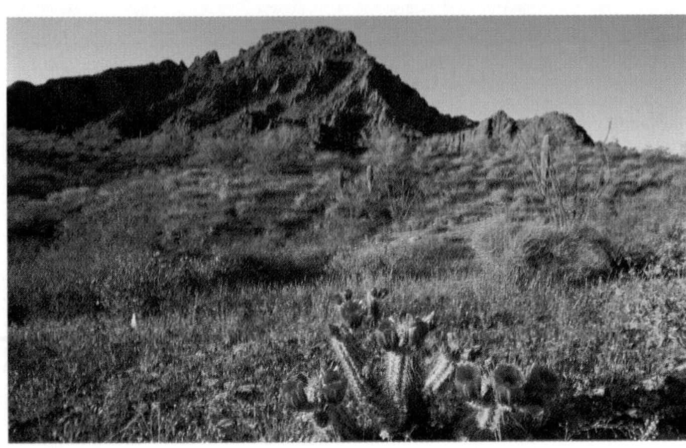

MEETING DIVERSE NEEDS

Gifted Challenge students to conduct research on the specialized areas of study in chemistry. Encourage students to set up an interview with a chemist at a nearby college, laboratory, or online. Ask students to share their research with the class.

Why Study Chemistry?

All the wonderful people and things that fill the world around you involve chemistry in one way or another! You are made from chemicals, and you use chemicals every day—when you breathe, drink a glass of water, wash your hair, or eat a snack. These common activities—and many, many

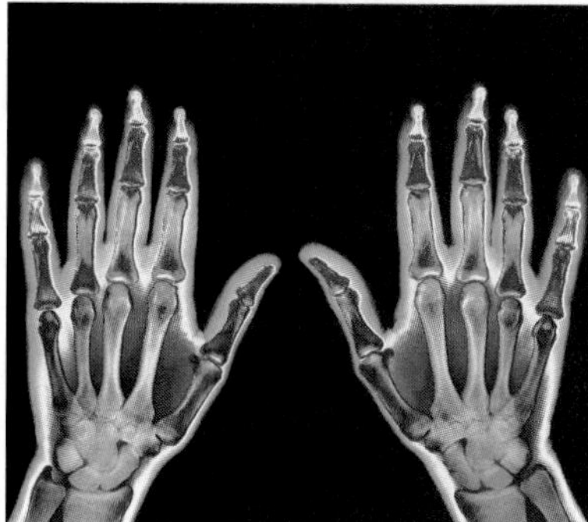

Figure 1.4
Chemistry is everywhere in your everyday life. In what ways do these photographs "say" chemistry to you? ❶

Use The Visual

Have students study the photographs in Figure 1.4. Ask students to consider some of the ways in which chemical technology has improved the quality of our lives. Ask students to point out scene(s) in which there is a clear indication that a chemical reaction is occurring. Ask:

- **What are the signs of a chemical reaction?** (Answers will vary but may include: heat, light, and generation of gas(es).)

Point out that, in many cases, chemical reactions release energy.

Discuss

Ask students to consider their activities during a single day. Ask them to give examples of things they do that involve chemical processes or contact with chemical products. Then ask if they can think of four items or activities that do not in some way involve chemistry. (It will be hard to find any that do not! Students should consider what they eat, their clothing, the products they use for personal hygiene, their normal mode of transportation, their residence, the purity of their tap water, and the types of medications they use when ill.)

Answer

❶ Answers will vary, but students should relate the examples to the different branches of chemistry.

3 Assess

Evaluate Understanding

To determine students' understanding about the field of chemistry, ask:

- **In your own words, how would you define chemistry?** (Chemistry is the study of the composition of matter and the changes it undergoes.)
- **What are chemicals?** (The substances that comprise matter.)
- **What do chemists do?** (Chemists study the composition and behavior of matter. Their skills and knowledge allow them to modify materials for new purposes.)

The study of chemistry developed out of our curiosity about natural phenomena. Ask students to describe two or three natural processes or materials encountered in everyday life that they find intriguing.

Reteach

Emphasize that chemistry is not just a collection of facts. It is an ongoing human activity. Chemists not only acquire knowledge, they play an important role in society based on that knowledge. For example, they teach students, conduct research, and develop new products. Point out that learning about the principles of chemistry will enable students to better understand the modern world. Chemistry is also involved in many environmental issues such as toxic waste, acid rain, and air and water pollution.

FOOD SCIENCE

Chemistry in the Kitchen

Chemical reactions underlie your most wonderful successes (or dismal disasters) in the kitchen. The cut surfaces of apples, pears, and bananas turn brown

upon exposure to oxygen in the air. Light, fluffy cakes result when baking powder reacts with acid in the batter to produce carbon dioxide. When sugar is heated, it forms dark-brown caramel. Heating the protein structure in eggs changes their texture from slimy to firm when they are boiled, scrambled, or fried. The tasty brown coating on the surface of roasted and grilled meats is formed by a complex series of chemical changes. Although tradition governs most of our behavior in the kitchen, food scientists are constantly searching for new and improved methods to prepare, present, and preserve food.

others—all involve chemicals and chemistry. Your natural desire to understand how these things work is perhaps the best reason to study chemistry. Chemistry, with its focus on the workings of the natural world, helps satisfy your need to know and understand.

Every day, you make choices: what to eat, what to wear, when and how much to study. As a citizen of this planet you will also be asked to act on questions of much greater importance. Is nuclear power acceptable or are there better alternative energy sources? What are appropriate responses to the problems of global warming and ozone depletion? Given limited resources, what deserves more support: the space program or finding a cure for cancer? Should experiments that involve manipulating heredity be regulated or forbidden? Knowledge of the basics of chemistry and other sciences can help you arrive at informed opinions and take appropriate actions on these questions.

Although relatively few people become professional chemists, a career in chemistry can be a satisfying one, indeed. This introductory course can help you decide whether chemistry is a career for you. As a chemist, you might develop new products such as textiles, paints, medicines, or cosmetics. Perhaps you would find methods to reduce pollution, clean up the environment, or prevent the destruction of the ozone layer. You could share your knowledge through teaching, analyzing pharmaceuticals, or checking the quality of manufactured goods.

Very often, chemistry is used to attain a specific goal. The goal might be to formulate a new paint or adhesive or to set up procedures for a new medical test. Such goals are examples of applied chemistry, or chemical technology. In applied chemistry, scientific knowledge is used in ways that can either benefit or harm people and/or the environment. Political and social debates about the uses of scientific knowledge are often really debates about the risks and benefits of technology. In addition to applied chemistry, there is pure chemistry. Like other pure sciences, pure chemistry accumulates knowledge for its own sake. As you read on in this textbook, you will recognize many ways in which chemistry affects your life. And perhaps you will consider a career in chemistry!

section review 1.1

1. Match the numbered terms in the right column with the lettered terms in the left column.

a. technology	(1) life
b. organic chemistry	(2) matter and its changes
c. biochemistry	(3) carbon
d. chemistry	(4) applications
e. analytical chemistry	(5) composition

2. List three reasons to study chemistry.
3. Explain the difference between pure and applied chemistry.

 Chem ASAP! **Assessment 1.1** Check your understanding of the important ideas and concepts in Section 1.1.

Answers

SECTION REVIEW 1.1

1. **a.** 4 **b.** 3 **c.** 1 **d.** 2 **e.** 5
2. Reasons to study chemistry are to better understand how the world works, to become a better-informed citizen, and to understand what chemistry might entail as a life's work.
3. Pure chemistry examines chemical aspects of the world for the sake of knowledge alone; applied chemistry uses this knowledge to solve real-world problems.

CHEMISTRY FAR AND WIDE

section 1.2

objectives

- Summarize ways in which chemistry affects your daily life
- Describe the impact of chemistry on various fields of science

Every autumn thousands of visitors travel to New England to view the breathtakingly beautiful colors of the foliage. These colors—vivid oranges, yellows, reds, and purples—appear as the trees approach the winter months when growth no longer takes place. The bright pigments are produced by a complex chemical process, which depends on changes in temperature and hours of sunlight. This process is only one of many that occur all around you—and even inside you!
What aspects of life involve chemistry?

Materials

People have used chemistry to their advantage throughout history. Almost 3000 years ago, early chemists produced iron from iron ore by heating the ore with carbon. The invention of steel followed, in about 500 B.C. Steel and other mixtures of metals, such as brass and bronze, are important structural materials, as are the many recently developed ceramics. But today is primarily the age of plastics. Plastics, also called polymers, are gigantic molecules with many important properties. Chemists have learned to tailor these properties to meet specific needs. The most important property of structural plastics is their high strength-to-weight ratio. A piece of plastic with the same weight as a piece of steel may be as much as five or six times stronger than the steel! A high strength-to-weight ratio makes polymers perfect for the protective weaving in bullet-resistant vests and for structural automobile body panels. The resulting reduction in the weight of the vehicle increases fuel efficiency. An automobile contains more than 500 pounds of polymers, if one counts the rubber, paint, sealants, lubricants, and upholstery.

Figure 1.5
People have always used the materials of chemistry. Bronze is a mixture that has been used since ancient times. Ceramic tiles on a space shuttle help to protect it from overheating. Light passing through optical fibers is used to transmit telephone messages. Optical fibers are made mainly from silicon dioxide (SiO_2), which is found in sand.

STUDENT RESOURCES

From the Teacher's Resource Package, use:

- Section Review 1.2, Ch. 1 Practice Problems and Quizzes from the Review Module (Ch. 1–4)

TECHNOLOGY RESOURCES

Relevant technology resources include:

- Chem ASAP! CD-ROM
- ResourcePro CD-ROM

1.2

1 Engage

Use the Visual

Discuss with students how the study of chemistry leads to a greater understanding about living things. Explain that all living things are made of chemicals. Have students study the photograph and read the text that opens the section. Ask:

- **What form of energy do plants use?** (light energy)

Explain that photosynthetic organisms are able to convert solar energy into chemical energy because they contain light-absorbing chemicals, or pigments, such as chlorophylls and carotenoids. Tell students they will learn more about photosynthesis in Chapter 27. Ask:

- **What aspects of life can you name that involve chemistry?** (Answers may include: The breakdown of chemicals for energy, the production of new chemicals to grow, heal, and reproduce.)

Motivate and Relate

Point out that science is not an activity restricted to professional scientists in laboratories. Any person with a curiosity about natural materials or phenomena can participate in scientific inquiry. Scientists make careful observations of processes and events to gain a greater understanding of a phenomenon. Have students consider how they use past and present observations of common events to analyze and solve daily problems. Point out that chemists use the same logical reasoning to solve chemistry problems that the students use to solve everyday problems, such as why a car fails to start or what provokes an allergic response.

2 Teach

CHEMISTRY AND SCIENCE HISTORY

Carotenoids are responsible for the red, orange, and yellow colors sometimes visible in leaves during autumn. In the fall, factors such as temperature change and the gradual reduction in daylight hours trigger both the breakdown of chlorophyll and the production of pigments such as anthocyanin. Different soil conditions and combinations of pigments lead to variations in leaf color from one plant species to another.

Discuss

To illustrate the difference between renewable and non-renewable energy sources, compare hydrogen and fossil fuels. When hydrogen burns in air, water is produced. With an expenditure of energy, this process can be reversed to produce hydrogen. In contrast, the combustion of fossil fuels is not a reversible process. When fossil fuels burn, their supply is permanently reduced. Refer to Section 25.5 for a description of the materials, conditions, and time required for the formation of fossil fuels.

Your home contains a dazzling array of plastic items. Do you refrigerate leftovers in plastic food wrap, use nonstick cooking utensils, or water the lawn with a plastic hose? Perhaps you walk on a polyester or nylon carpet. You may be wearing wrinkle-resistant clothing woven from polyester fibers mixed with natural fibers of cotton or wool. The plumbing in your home could be made from polyvinyl chloride (PVC) pipe, and your walls may contain polystyrene insulation covered with a polyethylene vapor barrier. You have probably seen lawn and patio furniture made from plastic, as well as kitchen counters covered with heat- and moisture-resistant polymers. To repair a piece of furniture, you might use an epoxy resin polymer to produce a bond stronger than the wood itself. Fabrics such as nylon and Dacron™ make up many of the clothes you wear everyday. Bowling balls, billiard balls, tennis rackets, and other sports equipment are made from polymers. None of these materials would exist if it were not for the efforts of chemists.

But plastics are not the only materials produced by chemists. Black-and-white and color photographs are produced by the interaction of light with chemicals. Personal computers rely on silicon memory chips. Telephone communications are transmitted over lines made from optical fibers. These are only a few examples, but surely you get the idea.

Energy

Population growth and increased standards of living around the globe create ever-greater demands for energy to power homes and factories. Where will this energy come from? There are only two ways to provide it: conserve energy and produce more energy. Chemistry plays an essential role in both of these options. Some methods of conserving energy are obvious, others less so. But from energy-efficient fuels to safe and effective insulation materials to roadways lighted with sodium lamps, chemistry is creating new and exciting energy conservation methods.

Figure 1.6
In a continuing effort to conserve energy, efficient outdoor lighting methods, such as high intensity discharge, have been a focus of chemists. Chemists are also helping to develop renewable energy resources, such as solar power and wind power, to replace the ever-shrinking supply of fossil fuels.

Figure 1.7
The leaf of a green plant is an excellent solar collector; the energy of the sun is used to build the chemicals for life. What is this process called? 1

Most of our energy needs are filled by burning fossil fuels. But the supply of fossil fuels is limited. Because they are a nonrenewable resource, someday they will run out. Much of the world's petroleum is difficult or impossible to extract from Earth. However, an enormous reserve of petroleum could become available if tar sands and oil shales were coaxed into giving up the oil they contain. In addition, coal, lignite, and peat are potential sources of petroleum products. Even garbage can be converted into natural gas. Chemists are currently working on these and other exciting possibilities.

Scientists are also working to provide new and perhaps inexhaustible sources of energy. Sunlight is the greatest source of energy available on Earth. Plants capture some of the sun's energy in the food-making process called photosynthesis. But only about 0.1% of the energy that strikes Earth is used by all the plant life in the world. Scientists are working to develop devices that efficiently convert the energy of sunlight into electrical energy. If they succeed, the electrical energy might be used to decompose ordinary water into hydrogen and oxygen. Hydrogen could then be burned as a renewable and nonpolluting fuel.

The development of batteries for the storage of energy is another goal of chemists. Early flashlight batteries, known as dry cells, did not last long in use. Dry cells were subsequently replaced by alkaline batteries, an improved design offering significantly longer life. The lithium–iodine battery is a recent advancement in the ongoing evolution of the battery. The lithium–iodine battery is an important improvement. Used in cardiac pacemakers, lithium–iodine batteries are small and efficient and can last as long as ten years. Electric automobiles will need powerful storage batteries to make them a practical alternative to vehicles with internal-combustion engines. Fuel cells under development are a promising source of pollution-free energy.

Figure 1.8
From entertainment devices to watches to medical appliances such as hearing aids, batteries have numerous applications. The development of practical electric automobiles is on the horizon. Just imagine what life would be like without batteries!

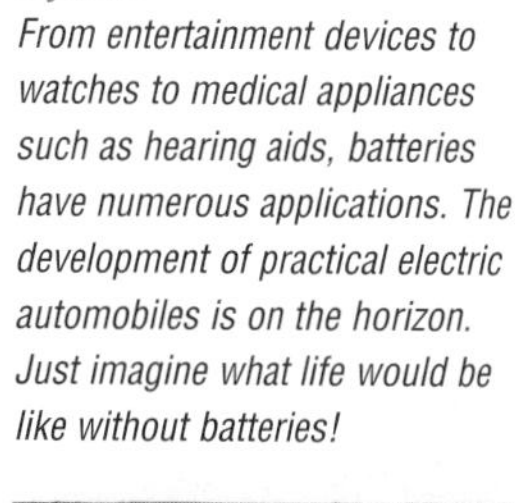

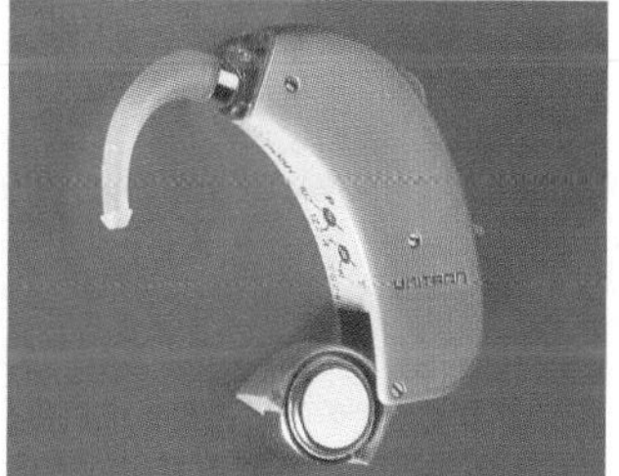

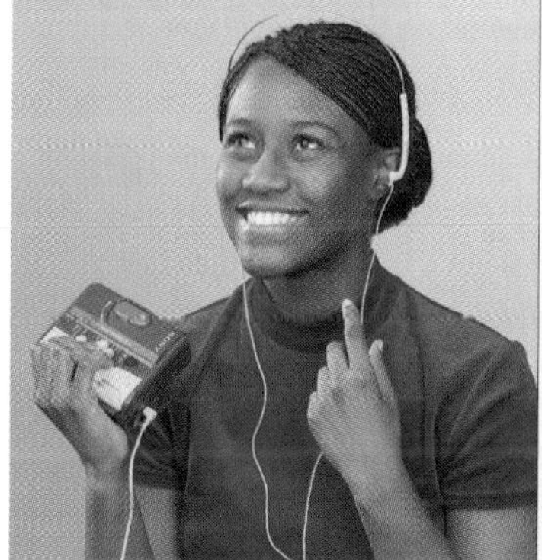

1.2

TEACHER DEMO

If hydrogen gas is available, demonstrate the energy that is made available through the burning of hydrogen. Fill a *small* balloon with hydrogen gas and use an extender to hold a flame against the balloon and ignite the hydrogen. **CAUTION!** Flammable substances, including clothing and solvents, should be kept at a safe distance away from the balloon during the experiment. Have students record and describe their observations.

- What evidence do you have that a chemical reaction occurred? (Answers will vary.)

Think Critically

Solar energy can be harvested by a number of different methods. One interesting method is the use of so-called "solar ponds" to gather heat from the sun and convert the heat to electricity using specially designed turbines. Have students do research on the use of solar ponds in different parts of the world. Students should be prepared to describe the materials and conditions necessary to build a working solar pond and the amount of electricity a typical pond could deliver.

Answer

1 photosynthesis

Discuss

The study of chemistry has revolutionized the treatment of diseases by the medical establishment. Discuss the introduction and use of vaccines, antibiotics, and recombinant proteins (insulin for example) to treat and prevent bacterial and viral infection, and diabetes. You may wish to have students debate some of the ethical issues involved with the latest revolution in medical technology, the use of gene therapy to treat and prevent genetic diseases. Gene therapy refers to those methods used to alter human genetic information. Have students research the pros and cons of gene therapy.

Discuss

Discuss some of the ways that chemical technology is applied in the kitchen. Challenge students to name some commercial food products that are used during preparation and cooking to improve or alter the flavors of foods. Examples include products such as meat tenderizers, additives to help break down lactose in dairy products, flavor enhancers, and many others. You may wish to bring in some of these products to show the class. Discuss the use and properties of these products.

Figure 1.9
The sun's energy is produced by nuclear fusion. Nuclear power plants produce energy by nuclear fission.

Nuclear fission and nuclear fusion may also provide energy for the future. Nuclear fission, the process used in nuclear power plants, has been responsible for some serious nuclear accidents. However, if scientists learn enough about containing and controlling nuclear reactions and disposing of the waste, nuclear power may again achieve prominence as an energy source. Nuclear fusion, the process that produces the sun's energy, could be pollution-free. Research into producing energy from nuclear fusion on Earth is in its infancy. Scientists are currently concerned with methods for producing and containing the energy of nuclear fusion.

Medicine and Biotechnology

Every year, the chemical industry produces millions of pounds of such medically important substances as vitamin C, penicillin, and aspirin. However, the synthesis of vitamins and medicines is far from the only role that chemistry plays in the health and biological sciences. No endeavors have benefited more from advances in chemistry than medicine and biotechnology. Much of this benefit derives from the ability of scientists to determine the spatial arrangement of atoms in complex biological molecules such as proteins. Many medicines are effective because they interact in a very specific way with biological molecules. Knowledge of the molecular structure of the target biological molecule greatly assists the design of safe and effective drug molecules. Medicines to reduce high blood pressure, high blood cholesterol, and some cancers have been developed in this way.

Chemistry is also providing new materials for medical applications. Diseased or weakened arteries can be replaced surgically with tubes made of Dacron™ polymers. Hipbones are replaced by substitutes made from metal alloys, or mixtures, of tungsten and cobalt. Much progress has been made in the development of synthetic blood and skin.

Genes, the units of heredity in living organisms, are composed of deoxyribonucleic acid (DNA). The discovery of DNA's molecular structure and of many of the details of its role in heredity has resulted in major advances in biology and biotechnology. With the information gathered thus far and the continuing efforts of the Human Genome Project, scientists worldwide expect to be able to identify and determine the molecular structures of all human genes. Such knowledge will have significant applications in medicine, genetics, agriculture, and even economics!

Figure 1.10

You are who you are because of genes, which are segments of DNA. Molecular models, such as this intricate segment of DNA, help scientists unravel the mysteries of the world around us.

Among other things, biotechnology involves the transfer of genes from one organism to another. Using this technology, human genes have been inserted into bacteria that then become factories for the production of human substances such as insulin, which is used to treat diabetes. In a few cases, a human genetic disease has been cured or its symptoms alleviated by replacing a faulty gene with a normal gene. Through biotechnology, completely normal sheep and cattle have been produced from a single adult cell of these animals by a process called cloning. The offspring of cloning, called a clone, is an exact genetic copy of its parent. Cloned cattle can produce human proteins in their milk. If in the future, cloned animals can provide a continuous supply of proteins that can be used to treat human diseases, one of the goals of biotechnology will have been realized.

Figure 1.11

More than an identical twin? Dolly was cloned from a single cell from her mother.

Agriculture

Chemistry plays an important role in efforts to increase the world's food supply and to protect crops. Developing hardier and more productive plants is a key to these efforts. Studies of photosynthesis, the food-making process in plants, and nitrogen fixation, the process by which atmospheric nitrogen gas is converted into usable nitrogen compounds, may lead to the development of plants that use these processes more efficiently. Why do you think such plants would be valuable? Advances in the understanding ❶ of plant hormones, which regulate plant growth, represent another opportunity to increase the strength and viability of plants and, thus, increase the world's food supply.

1.2

Discuss

Explain that genetic engineering methods allow scientists to alter the genomes of plants and animals by stitching together DNA from genes of different organisms. Plants and animals that have been altered in this way are called transgenic organisms. Ask students if they have ever seen any transgenic fruits or vegetables in the supermarket. (Example: tomatoes) **Point out that cattle are sometimes fed growth hormones produced by genetically engineered bacteria. If possible, bring a transgenic tomato or tomato plant to class. Students may wish to compare the taste of transgenic tomatoes to normal tomatoes. Discuss any concerns that students may have towards growing or eating transgenic foods. You may wish to have students collect and post articles about genetically engineered foods on a bulletin board in the classroom.**

ACTIVITY

Have students scan local and national newspapers for current news stories related to chemistry. Ask students to cut out one article and prepare a short presentation on the content of the article and how it relates to chemistry. Students may wish to specify which field(s) of chemistry (organic, inorganic, analytical, physical, or biochemistry) the news story is most closely connected to.

Answer

❶ Such plants could produce more food with fewer resources.

ACTIVITY

Have students do research to find out what chemicals are used in place of Freon™ today. Students might start by looking at labels on several aerosol and refrigerant containers, by contacting the companies that manufacture these products, and by looking up literature published by the Environmental Protection Agency. Additional questions students should answer include:

- Why are these alternative chemicals safer?
- What are some of the disadvantages, if any, of using these alternative chemicals?

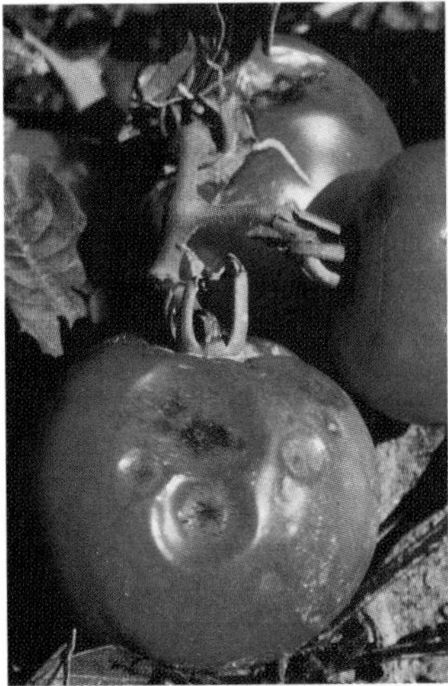

Figure 1.12
Insecticides, herbicides, and fungicides are used to protect crops from pests and disease. The effects of insect infestation, corn smut, and wheat rust are apparent in these ruined crops.

In addition to developing hardier and more productive plants, scientists are also focusing on safer and more effective ways to protect crops. In the past, insecticides, herbicides, and fungicides were quite nonspecific. Today, the trend is toward chemicals that are more specific for the condition they are designed to treat. For example, many newer insecticides are similar in molecular structure to natural protectants produced by the plants themselves. The discovery of more than 100 pheromones, which include insect sex attractants, is also being used in the battle against pests. These molecules have proved effective in attracting and trapping such pests as gypsy moths and boll weevils. Chemists are also working to create more pest-resistant and blight-resistant plants. Recombinant DNA technology is helping to breed plants with genetic resistance to their natural enemies.

Figure 1.13
Do not take clean air for granted! As our roads grow more crowded, smog pollution becomes a problem chemists and others work hard to combat.

The Environment

Progress has its price: Every new development has both risks and benefits. Society, often through the political process, must decide whether the benefits of a new development outweigh its risks. This is especially true for environmental protection.

Chemists work with environmental scientists to identify pollutants, prevent or minimize environmental pollution, and clean up toxic wastes. One example of this teamwork involves the battle against the irritating smog that envelops many of the world's cities. Nitrogen oxides emitted by the internal-combustion engines of automobiles are a major contributor to smog. To combat these pollutants, scientists designed catalytic converters that remove nitrogen oxides from automobile exhaust. Pollution caused by burning fossil fuels such as coal is another example of a problem chemists and environmental scientists are tackling together. Burning coal produces sulfur compounds. Gaseous sulfur compounds are components of acid rain, which as you probably know is harmful to forests and lakes. Electrostatic precipitators and scrubbers help remove sulfur compounds from the gas emitted by power plants and factories.

MEETING DIVERSE NEEDS

Gifted Have students research the elemental composition of the universe, Earth (its crust, oceans, and atmosphere), and the human body. These data are readily available in library reference books or in textbooks. Have students construct pie graphs that show the relative percent by mass of elements in the universe, in Earth's regions, and in the human body. By comparing the graphs, students should recognize that the chemical compositions of the universe, Earth, and human body are dramatically different. Have students propose other questions to consider throughout the year.

(Examples: Why isn't helium abundant on Earth? Why is carbon more abundant in humans than in Earth's crust. What determines whether an element will be found in seawater?)

Figure 1.14
Acid rain changed the appearance of this marble statue of George Washington from what it was 60 years ago (left) to what it is today (right). What causes acid rain? ❷

The most abundant pollutant generated by humans, carbon dioxide, is produced by the combustion of fossil fuels. Earth's surface is warmed naturally by the greenhouse effect, which is similar to the heating of an automobile's interior on a sunny day. The increasing amounts of carbon dioxide released into the atmosphere each year trap more of the sun's heat on Earth's surface than is natural, or even normal. The result of this increased amount of heat is global warming. Global warming could drastically alter weather patterns, sea levels, and agriculture around the world. In a worldwide effort, nations are working independently and collectively to reduce carbon dioxide emissions.

Scientists are also concerned about the thinning of the ozone layer. The ozone layer protects Earth from the sun's harmful ultraviolet rays. Why is that important? ❶ One reason for the depletion of the ozone layer is the presence of chlorofluorocarbons, most notably Freon, which are a family of fluorine-chlorine-containing hydrocarbons that were originally used in refrigeration and air-conditioning equipment. Methyl bromide, used in agriculture, may also contribute to the thinning of the ozone layer. When depletion of the ozone layer was discovered, chemists provided an explanation of the processes causing the deterioration. With this knowledge, they then began to design new compounds to replace the chlorofluorocarbons.

ATMOSPHERIC SCIENCE

Predicting the Formation of the Ozone Hole

In the 1970s, F. Sherwood Rowland and Mario Molina predicted that certain chlorine-containing compounds, such as chlorofluorocarbons (CFCs), might damage the ozone layer. The use of these compounds in aerosol sprays was banned in 1978. A decade later, British researchers showed that the ozone layer over the South Pole disappeared each year in the Antarctic spring. Since then, thinning of the ozone layer has been observed in other places around the globe. Work by atmospheric chemists has confirmed the idea of Rowland and Molina, and scientists now believe that a single chlorine atom can destroy millions of ozone molecules. In 1995, Rowland, Molina, and Paul Crutzen, a German atmospheric chemist who had also studied ozone depletion, received the Nobel Prize in chemistry for their work.

Global mean temperature has risen fairly steadily since 1880.

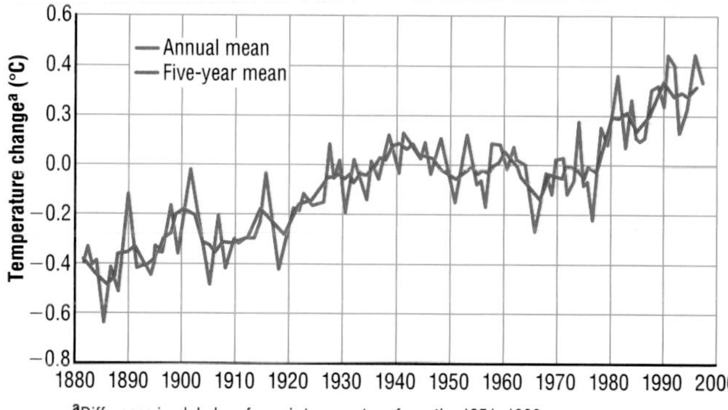

[a]Difference in global surface-air temperature from the 1951–1980 average.
Source: National Aeronautics and Space Administration's Goddard Institute for Space Studies
C&E News Nov. 17, 1997

Figure 1.15
There has been a steady rise in the global mean temperature since 1880. How would you describe the overall trend? ❸

1.2

Use the Visual

Have students study the photograph and read the caption to Figure 1.14. Explain that acid rain is a result of automobile and industrial emissions. When fossil fuels are burned, sulfur oxide and nitrogen oxide gases are released into the air where they can combine with water to form corrosive chemicals called acids. Ask students to name other types of damage caused by acid rain. (Examples include ecological damage, damage to urban structures, and effects on the health of individuals.) **Point out that the gaseous products released from the combustion of fossil fuels are carried by wind currents and can travel large distances across state and national borders.**

Answers

❶ Ultraviolet rays are known to cause skin cancers.
❷ Sulfur compounds produced during the burning of coal are the main components of acid rain.
❸ The trend is described as global warming.

3 Assess

Evaluate Understanding

Ask students to name some of the most recent technological advances that were made possible through the study of chemistry. Discuss the impact of chemistry on various fields of science. Explain that very often chemistry is applied towards achieving a specific goal. Ask:

- What are some of the goals of modern day engineers and scientists?

Have students compare the technology available during their parents' childhoods and today. Ask:

- What trends, if any, have you observed in technological advances?
- What are some of the problems in your community that have been solved by chemical technology?

Reteach

Explain that most of the phenomena that occur in the world around us involve chemical changes. Substances can sometimes combine to give new substances which have different properties. Illustrate chemical change by performing one or two precipitation reactions. Choose combinations of ionic compounds that produce colored precipitates. Examples include mixing $Pb(NO_3)_2(aq)$ with $KI(aq)$ to produce $PbI_2(s)$, and mixing $AgNO_3(aq)$ with $KCl(aq)$ to produce $AgCl(s)$. As additional examples, you may wish to show the reaction of $NaHCO_3$ with HCl to produce CO_2 gas, and the combustion of magnesium ribbon to produce heat, light, and MgO.

Figure 1.16
This astronaut is walking on the moon to collect samples of the lunar surface, an example of which is shown in the inset photograph.

Figure 1.17
Samples of the surface of Mars were analyzed by remote control. This photograph, taken by the Mars Pathfinder, shows the Sojourner making observations of the rock named Yogi. Why is such information valuable? 

Astronomy and Space Exploration

In the early nineteenth century, scientists first learned that by analyzing light transmitted to Earth by stars and other celestial objects, they could gain a "window" into the chemical composition of these objects. Obtaining similar information about the composition of the moon and planets is more difficult because these objects do not generate their own light. However, space exploration of Earth's moon and Mars is now providing some of this vital information. More than 850 pounds of moon rocks brought to Earth by various missions have been chemically analyzed. The moon rocks collected by Apollo astronauts were formed from volcanic material, suggesting that vast oceans of molten lava once covered the moon's surface. Chemical analysis of the atmosphere of Mars reveals the presence of primarily carbon dioxide, with smaller amounts of nitrogen, argon, oxygen, carbon monoxide, water vapor, and other gases. The Sojourner robotic vehicle, delivered to the surface of Mars by the Mars Pathfinder spacecraft, has analyzed and determined the chemical composition of Mars rocks. The composition of these rocks, particularly those known as conglomerates, indicates that a large amount of water once existed on the surface of Mars. Geologists are using this information to learn more about the formation of Earth and other planets, as well as to determine whether life as we know it could exist elsewhere in the solar system.

section review 1.2

4. Explain how chemistry affects your daily life.
5. Name at least three areas of science in which chemistry plays an important role.
6. How might plant crops improve as a result of agricultural research? Why would such improvement be important?
7. Name at least three items that are made completely or partially of plastic.

Chem ASAP! **Assessment 1.2** Check your understanding of the important ideas and concepts in Section 1.2.

Answers

1. Information about the chemical composition of rocks on Mars enables scientists to learn about Mars and its history, as well as to learn more about other planets such as Earth.
2. Fleming discovered penicillin.

SECTION REVIEW 1.2

4. Answers will vary
5. agriculture, medicine and biotechnology, and materials science
6. Crops might produce higher yields and be more blight-resistant. This would allow farmers to produce more food with fewer resources.
7. Answers will vary, but might include toothbrush, comb, compact disc, and many others.

THINKING LIKE A SCIENTIST

section 1.3

In March 1989, two chemists reported the discovery of an ordinary laboratory procedure for producing a nuclear fusion reaction. Their "cold fusion" reaction was relatively simple, using apparatus similar to that used for the electrolysis of water, yet it caused much excitement in the scientific community. Nuclear fusion, you see, is a thermonuclear reaction; that is, it involves the combination of two nuclei to produce one nucleus of heavier mass and the release of extraordinary amounts of energy. In fact, the energy released from the sun is the result of nuclear fusion. The prospect of producing huge quantities of energy to meet society's increasing demands by rather ordinary means ignited much interest and excitement. Many scientists rushed to duplicate the experiment. But after many failures, the dream of using cold fusion to obtain inexpensive, unlimited energy died. According to the scientific method, the results of an experiment must be reproducible if they are to be accepted. **What are the steps of the scientific method?**

objectives

- Describe the steps involved in the scientific method
- Distinguish between a theory and a scientific law

key terms

- scientific method
- observation
- hypothesis
- experiment
- theory
- scientific law

The Scientific Method

As you read the previous section, perhaps you found yourself thinking, "Who does these exciting and important things?" Scientists may seem to be special people because they obtain valuable—and sometimes spectacular—results. Nevertheless, as a Nobel Prize winner in science has said, science is about "ordinary people doing ordinary things."

An important discovery may involve some luck, but one must be prepared to recognize the lucky event. Alexander Fleming noticed that infectious bacteria failed to grow on portions of a bacterial culture where mold was growing. Fleming was very observant and scientifically prepared to realize the importance of this chance observation, which led to one of the greatest scientific achievements of all time. Do you know what Fleming's discovery was? Most scientific advances, however, involve little or no luck. A logical, systematic approach to the solution of a difficult problem is often the best method, as well as the most powerful tool that ordinary people—you and scientists included—have at their disposal. 2

The **scientific method** is one logical approach to the solution of scientific problems. The scientific method is useful for solving many kinds of problems because it is closely related to ordinary common sense. Suppose you want to use a flashlight, but when you turn it on it does not light. You have made an **observation**—that is, you have used your senses to obtain information directly. In this case, your observation is that the flashlight does not light. This raises a question: What is wrong with the flashlight? You guess that the flashlight batteries are probably dead. By guessing, you have proposed a reason for your observation. You have made a **hypothesis,** or a proposed explanation or reason for what is observed. In the scientific method, scientists typically first observe something of scientific interest and then propose a hypothesis.

Returning to your flashlight problem, you will probably want to test your hypothesis with an experiment. An **experiment** is a means to test a hypothesis. Most likely, you will put new batteries in the flashlight. If the

Figure 1.18
When you make observations, keep both eyes open.

STUDENT RESOURCES

From the Teacher's Resource Package, use:

- Section Review 1.3, Ch. 1 Practice Problems and Quizzes from the Review Module (Ch 1–4)
- Laboratory Recordsheets 1-1 and 1-2
- Laboratory Manual: Experiment 1

TECHNOLOGY RESOURCES

Relevant technology resources include:

- Chem ASAP! CD-ROM
- ResourcePro CD-ROM
- Chemistry Alive! Videodisc: *Safety, Ira Remsen Story*
- Small-Scale Chemistry Lab Video or Videodisc: *What is Small-Scale Chemistry?, Setting Up a Small-Scale Laboratory*

1.3

1 Engage

Use the Visual

Have students study the photograph and read the text that opens the section. Because fusion ordinarily requires very high temperatures, people were skeptical when the "cold fusion" results were announced. Peer review assures that unexpected results are not accepted without confirmation. Ask:

- **What are the steps of the scientific method?** (Make an observation, propose a hypothesis, design an experiment to test the hypothesis)

Motivate and Relate

Students often associate the methods of science with professional scientists working in laboratories. Remind students that the scientific method is merely a technique for approaching and solving problems. Point out, by way of example, that students unconsciously use the scientific method routinely to solve everyday problems.

An excellent example of "ordinary people" applying the scientific method is the story of how Augusto and Michaela Odone, on their own, developed an oil that relieved the symptoms of adrenoleukodystrophy (ALD), an inherited, neurological disease afflicting their son.

This story is told in the movie *Lorenzo's Oil*, starring Nick Nolte and Susan Sarandon. You might suggest that students rent or borrow this film from the local library and examine it as a case study in the application of the scientific method. Students should take notes as events unfold, and report on how Augusto and Michaela Odone use observation, hypothesis, and experiment to help their son.

2 Teach

TEACHER DEMO

Prepare a solution of phenolphthalein in alcohol (250 mg of phenolphthalein in 250 mL of 70% 2-propanol). Use a small artist's paintbrush to letter messages on sheets of standard copy paper or sheets of white butcher paper. Try messages such as "CHEM-IS-TRY," and "This is a LABOR-atory, not a lab-ORATORY."

Allow the papers to dry until the messages are invisible. Then post them around the class in well-ventilated areas. At intervals, spray each sheet with household glass cleaner containing ammonia.

Ask the students to hypothesize about why the pink messages appear, disappear after a few minutes, and then reappear when the paper is sprayed again. Students are likely to infer that something in the cleaner caused a reversible change to something on the paper. They may infer that the material in the cleaner is volatile.

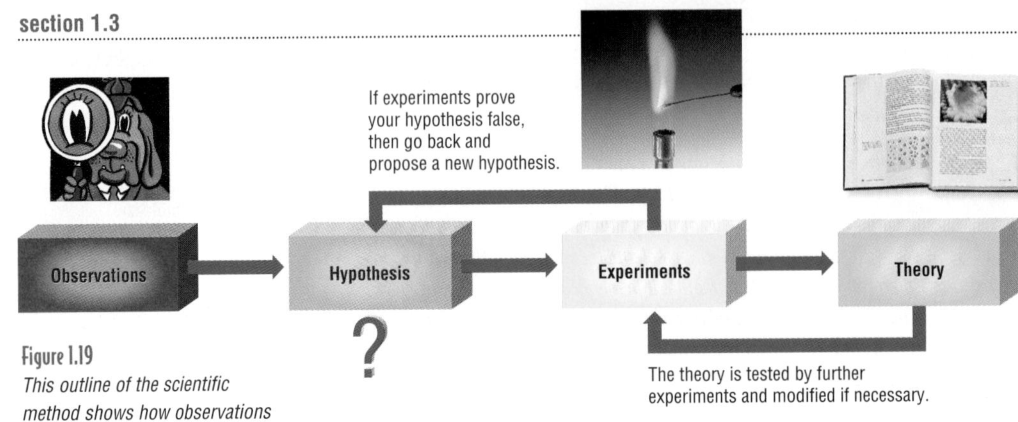

Figure 1.19
This outline of the scientific method shows how observations lead to the development of hypotheses and theories. Note that if experiments prove a hypothesis false, a new hypothesis must be proposed.

flashlight lights, you can be fairly certain from the one experiment that your hypothesis is true. Scientists also perform experiments to test their hypotheses. For the results of an experiment to be accepted, the experiment must produce the same result no matter how many times it is repeated, or by whom. The repeatability of scientific experiments distinguishes science from nonscientific fields.

Many different kinds of experiments may be needed to learn whether a hypothesis is valid. A scientific hypothesis is useful only if it accounts for what scientists observe in many situations. Suppose the flashlight does not work after you replace the batteries. What does this indicate about your hypothesis? When experimental data do not fit a hypothesis, the hypothesis must be rejected or changed. The new or refined hypothesis is then subjected to further experimental testing. Again, returning to the flashlight problem, you might now decide to replace the light bulb because replacing the batteries was not helpful. The original, false hypothesis (dead batteries) has led to a new hypothesis (burnt-out bulb) and a new experiment to test it. The scientific method of observing, hypothesizing, and experimenting is repeated until the hypothesis fits all the observed experimental facts. **Figure 1.19** outlines the major features of the scientific method.

Once a scientific hypothesis meets the test of repeated experimentation, it may be elevated to a higher level of ideas. It may become a theory. A **theory** is a broad and extensively tested explanation of why experiments give certain results. A theory can never be proved because it is always possible that a new experiment will disprove it. But theories are very useful because they help you form mental pictures of objects or processes that cannot be seen. Moreover, theories give you the power to predict the behavior of natural systems.

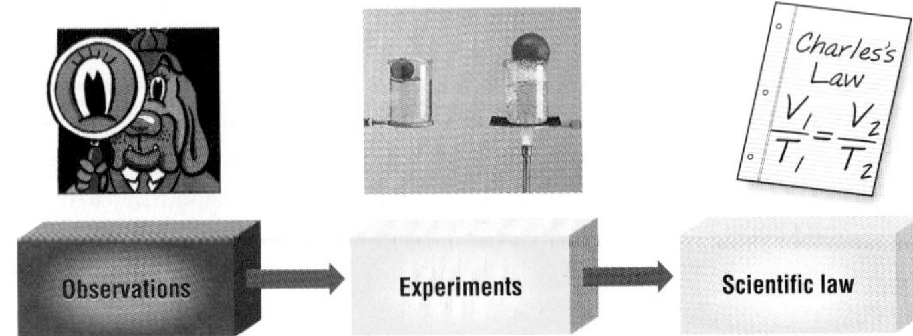

Figure 1.20
A scientific law summarizes the results of many observations and experiments.

MEETING DIVERSE NEEDS

LEP Have LEP students work with a partner to describe, act out, or illustrate in a drawing the different steps that make up the scientific method. Challenge students to identify how they used each step to solve problems in the past. Suggest that the students record the steps in their notebooks in both English and their native language.

Chemistry Alive!

Ira Remsen Story

Play

MINI LAB

Bubbles!

PURPOSE

To test the hypothesis that bubble making can be affected by adding chemicals to a bubble-blowing mixture.

MATERIALS

- 3 plastic cups
- 3 teaspoons liquid dish detergent
- measuring cup
- 150 mL water
- $\frac{1}{2}$ teaspoon sugar
- $\frac{1}{2}$ teaspoon salt
- drinking straw

PROCEDURE

1. Label three cups 1, 2, and 3. Measure 1 teaspoon liquid dish detergent into each cup. Use the measuring cup to add 50 mL water to each cup, then swirl the cups to make a clear mixture.
2. Add $\frac{1}{2}$ teaspoon table sugar to cup 2 and add $\frac{1}{2}$ teaspoon table salt to cup 3. Swirl each cup for 1 minute.
3. Dip the drinking straw into cup 1, withdraw it, and blow gently into the straw to make the largest bubble you can. Practice making bubbles until you feel you have reasonable control over your bubble production.
4. Repeat Step 3 with the mixtures in cups 2 and 3.

ANALYSIS AND CONCLUSIONS

1. Did you observe any differences in your bubble making using the mixtures in cup 1 and cup 2?
2. Did you observe any differences in your bubble making using the mixtures in cup 1 and cup 3?
3. What can you conclude about the effects of table sugar and table salt on your ability to produce bubbles?
4. Propose another hypothesis based on bubble making and design an experiment to test your hypothesis.

Scientific Laws

A **scientific law** is a concise statement that summarizes the results of many observations and experiments. A scientific law describes a natural phenomenon without attempting to explain it. Scientific laws can often be expressed by simple mathematical relationships. Gay-Lussac's law summarizes what happens when a sealed, gas-filled container is heated: The pressure of a gas is directly proportional to the Kelvin temperature at constant volume. Pressure can thus rise dangerously and lead to an explosion. Gay-Lussac's law states what happens; a theory explains why.

section review 1.3

8. Describe the steps involved in the scientific method. Give an example.
9. Distinguish among a theory, a hypothesis, and a scientific law.
10. Why should a hypothesis be developed before experiments take place?
11. In Chapter 2, you will learn that matter is neither created nor destroyed in any chemical change. Is this statement a theory or a law?

Chem ASAP! **Assessment 1.3** Check your understanding of the important ideas and concepts in Section 1.3.

portfolio project

Antoine Lavoisier used the scientific method to disprove a popular explanation for burning called the phlogiston theory. Research the history of this theory. Describe the experiments Lavoisier used to demonstrate that phlogiston did not exist.

Answers

SECTION REVIEW 1.3

8. Answers will vary. Sample:
   Observation: My stomach is growling.
   Hypothesis: Maybe I am hungry.
   Experiment: Eat something and see what happens.
9. A law is a concise statement that summarizes the results of a broad variety of observations and experiments. A hypothesis is a proposed explanation or reason for what is observed. A theory is an extensively tested explanation for experimental results.
10. The design of the experiments will be guided by the proposed hypothesis.
11. a law; The statement summarizes facts; it does not give an explanation.

1.3

Discuss

Students often think that an experiment is a failure if they do not get the "right" (expected) results. Help students analyze results that do not fit a hypothesis or vary widely from those of other students. Often, you can identify experimental errors that explain deviations. Also point out that scientists can gain important insights from "failed" experiments.

MINI LAB

Bubbles!

ANALYSIS AND CONCLUSIONS

1. no
2. Yes; no bubbles formed from the liquid in cup 3.
3. Sugar has no effect on bubble production, but salt stops it completely.
4. Answers will vary but could include examining the effect of temperature or dilution of the bubble-making mixture. For example, diluting the mixture can reverse the salt effect.

Portfolio Project

The ancient Greeks thought that flammable objects contained the element fire, which Georg Stahl (1660–1734) called phlogiston. During burning, phlogiston was transferred to air. Phlogiston-rich air (now called nitrogen) did not support burning; objects burned brightly in phlogiston-poor air. Lavoisier measured the mass of metals before and after heating in a closed container. He showed that the mass gained by the metal was lost by the air. Thus, the process of burning involved a gain of matter, not a loss of phlogiston. Lavoisier named the portion of air that supported combustion *oxygen*.

SMALL-SCALE LAB

INTRODUCTION TO SMALL-SCALE CHEMISTRY

OVERVIEW

Procedure Time: 40 minutes

This lab is designed to introduce students to those procedures that will provide a safe laboratory experience.

SAFETY ISSUES

Though small-scale quantities and techniques are far safer than those used in conventional chemistry labs, and the experiments in this book were carefully constructed with safety in mind, some issues remain. The following safety issues are particularly important.

- Never allow students to use small-scale pipets as squirt guns.
- Stress cleaning the small-scale reaction surface with paper towels and avoiding direct contact with solutions. Insist that students wash their hands thoroughly when they finish cleaning up.
- Do not let students handle chemical solutions with concentrations greater than 1.0*M*. Always dilute concentrated acids yourself by slowly adding them to water.
- Follow an approved method for storing chemicals.
- Always model good safety practices, such as wearing safety glasses.

After you discuss the safety rules, have students sign a safety contract. For an example, see the contract on p. 15 of the *Small-Scale Chemistry Laboratory Manual*. For additional instructions on safety procedures, see pages T21–22 in this textbook and the *Small-Scale Chemistry Laboratory Manual*.

Chemistry Alive!

Safety

Play

SMALL-SCALE LAB

INTRODUCTION TO SMALL-SCALE CHEMISTRY

PURPOSE

The small-scale chemistry experiments in this textbook are designed to help you learn chemistry by doing chemistry. Each experiment provides the opportunity to interact with matter, make observations, and interpret what you see. As you solve problems, you will become more inventive. You will be encouraged to ask questions and find creative ways to answer them. You may even be able to contribute some original ideas and discoveries to chemistry. Most of all, you will have fun while learning chemistry.

The small-scale experiments in each chapter have been carefully designed to minimize the risk of injury. However, safety is also your responsibility. You are expected to behave in a way that is consistent with safe laboratory practices.

The following rules are essential for keeping you safe in the laboratory. You are responsible for knowing these rules and following them at all times. Some rules address pre-lab preparation; other rules address proper laboratory procedures.

1. Recognize that all laboratory procedures involve some degree of risk. Read the entire procedure before you begin. Listen to all your teacher's instructions. When in doubt about a procedure, ask your teacher.
2. Do only the assigned experiments. Do any experiment only when your teacher is present and has given you permission to work.
3. Know the location and operation of the following safety equipment: fire extinguisher, fire blanket, emergency shower, and eye wash station.
4. Know the location of emergency exits and escape routes. To make it easy to exit quickly, do not block walkways with furniture. Keep your work area orderly and free of personal belongings such as coats and backpacks.
5. Protect your clothing and hair from chemicals and sources of heat. Tie back long hair and roll up loose sleeves when working in the laboratory. Avoid wearing bulky or loose-fitting clothing. Remove dangling jewelry. Wear closed-toe shoes at all times in the laboratory.
6. Report any accident, no matter how minor, to your teacher.
7. Wear safety glasses at all times when working in the laboratory. Safety glasses are designed to protect your eyes from injury. While working in the laboratory, do not rub your eyes, because chemicals are easily transferred from your hands to your eyes.

   If, despite these precautions, you get a chemical in your eye, remove any contact lenses and immediately wash your eye with a continuous stream of running water for at least 15 minutes.
8. To reduce danger, waste, and cleanup, always use minimal amounts of the chemicals specified for an experiment. This small-scale approach also saves time.
9. Use small-scale pipets for the controlled delivery of liquids, one drop at a time.
10. Never taste any chemical used in the laboratory, including food products that are the subject of an investigation. Treat all items as though they are contaminated with unknown chemicals that may be toxic. Keep all food and drink that is not part of an experiment out of the laboratory. Do not eat, drink, or chew gum in the laboratory.

    If you accidentally ingest a substance, notify your teacher immediately.
11. Do not use chipped or cracked glassware. Do not handle broken glass. If glassware does break, notify your teacher and nearby classmates. Discard broken glassware according to your teacher's instructions.

    If, despite these precautions, you receive a minor cut, allow it to bleed for a short time. Wash the injured area under cold running water and notify your teacher. More serious cuts or puncture wounds require immediate medical attention.

12. Do not handle hot glassware or equipment. You can prevent burns by being aware that hot and cold equipment can look exactly the same.

If you are burned, immediately run cold water over the burned area for several minutes until the pain is reduced. Cooling also helps the burn to heal. Ask a classmate to notify your teacher.

13. Recognize that the danger of an electrical shock is greater in the presence of water. Keep electrical appliances away from sinks and faucets to minimize the risk of electrical shock. Be careful not to spill water or other liquids in the vicinity of an electrical appliance.

If, despite these precautions, you spill water near an appliance, stand back, notify your teacher, and warn other students in the area.

14. Safety begins and continues with a clean laboratory. Report any chemical spills immediately to your teacher. Follow your teacher's instructions for cleaning up spills. Warn other students about the identity and location of spilled chemicals.

If, despite these precautions, a corrosive chemical gets on your skin or clothing, wash the affected area with cold running water for several minutes.

15. Dispose of chemicals in a way that protects you, your classmates, and the environment. Always follow your teacher's directions for cleanup and disposal. Clean your small-scale reaction surface by draining the contents onto a paper towel. Then wipe the surface with a damp paper towel and dry the surface completely. Dispose of the paper towels in the waste bin. Wash your hands thoroughly with soap and water before leaving the laboratory.

16. Take appropriate precautions whenever any one of the following safety symbols appears in an experiment.

Eye Hazard
Wear safety goggles.

Clothing Hazard
Wear a lab coat or apron when using corrosive chemicals or chemicals that can stain clothing.

Fire Hazard
Tie back hair and loose clothing.
Do not use a burner near flammable materials.

Inhalation Hazard
Avoid inhaling this substance.

Corrosive Substance Hazard
Wear safety goggles and a laboratory apron.
Do not touch chemicals.

Poison Hazard
Do not chew gum, drink, or eat in the laboratory. Keep your hands away from your face.

Electrical Hazard
Use care when using electrical equipment.

Thermal Burn Hazard
Do not touch hot equipment.

Breakage Hazard
Do not use chipped or cracked glassware.
Do not heat the bottom of a test tube.

Disposal Hazard
Dispose of this chemical only as directed.

Background

The small-scale chemistry experiments in this book are designed to help you teach students important chemical principles, not just process. For most experiments, the procedure is short and simple. In many cases students are asked to construct a grid, place a small-scale reaction surface over the grid, do the experiment, and record their results in a similar grid. Often the grid is marked with black Xs where students are to place drops of solutions. These Xs provide both black and white backgrounds against which students can observe reaction mixtures.

The ANALYSIS questions allow students to think and write about the main points of each experiment. You may want to have students work on these questions as they do the lab.

The extended experiments in the YOU'RE THE CHEMIST section ask students to apply what they learned in the basic experiment. Some of the YOU'RE THE CHEMIST activities could be used for performance-based assessment.

3 Assess

Evaluate Understanding

Ask students to choose an everyday type of problem and explain how they would use the scientific method to solve the problem.

Reteach

Students may mistakenly infer that a theory "grows" into a law by constant testing and refinement. Or, they may confuse a theory with a hypothesis. To clarify these concepts, ask students to describe *theory* and *law* in their own words.

1 Engage

Use the Visual

Have students study the photograph that opens the section. Ask:

- **What materials do you see in the photo?**
- **How could you find out more about the nature of a particular material?**
- **What are some resources in the library that are available for persons interested in studying chemistry?**

Help students identify journals, reference books, and Internet sites that are sources of reliable information.

- **How can you make your study of chemistry more effective and enjoyable?** (Answers will vary.)

ACTIVITY

Challenge students to name one type of professional career that requires a knowledge of chemistry. Examples include forensic science, conservation science, photographers, farmers, and physicians. Have students visit professionals and do library research to gain a better insight into the application of chemistry in a particular line of work. For those interested in art, it may be possible to arrange for an art conservator or conservation scientist to speak to a group of students at an art museum. Students should write reports on their findings.

section 1.4

HOW TO STUDY CHEMISTRY

objectives

- Explain why learning chemistry requires daily effort
- Describe the importance of writing in the study of chemistry

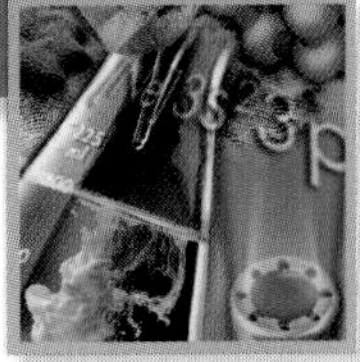

Do you recognize any of the images in this photograph of your textbook cover? The images were chosen to capture the beauty and excitement of chemistry: the old and the new; the macro view and the micro view; the applied and the theoretical. We wrote this textbook to help you understand and appreciate chemistry—a subject that explains your daily observations and experiences. **How can you make your study of chemistry more effective and enjoyable?**

Understanding and Applying Concepts

Like anything worth doing, learning chemistry requires effort on your part. But the reward is worth the effort. You have to read carefully and take notes. You need to study often rather than waiting until you are facing a test. For your study to be effective, find a quiet, well-lit place away from distractions such as the television or loud music. Then pick up a pencil and begin.

Chemistry deals with scientific facts—facts that can be discovered by making observations and doing experiments. Your chemistry experience will include a great deal of observing and experimenting. Because discovering scientific facts usually takes a lot of time and effort, you often will need to rely on information that others have discovered.

You will need to understand the chemical concepts behind some new vocabulary and many familiar words. For example, the word *diamond* contains many layers of chemical meaning. When most people hear the word, they think of a glittering gem. Did you know that diamond is one of the hardest known substances? Or that it is a form of the element carbon? Or that it has a highly ordered molecular structure? Chemistry will help you make connections between the visible, macroscopic world and the microscopic world—a world that is too small to be seen with the unaided eye.

To reinforce what you are learning in your classroom and laboratory, connect the chemical concepts with your experiences outside the classroom. Try to visualize reactions you observe at the level of atoms and molecules. Use resources such as newspapers, television, and the Internet to keep up with developments in chemistry.

Figure 1.21
The images shown here reveal a lot about the subject of chemistry. Can you explain the concepts they illustrate? When you have completed your study of this exciting subject, return to these images and find out exactly what you have learned! ❶

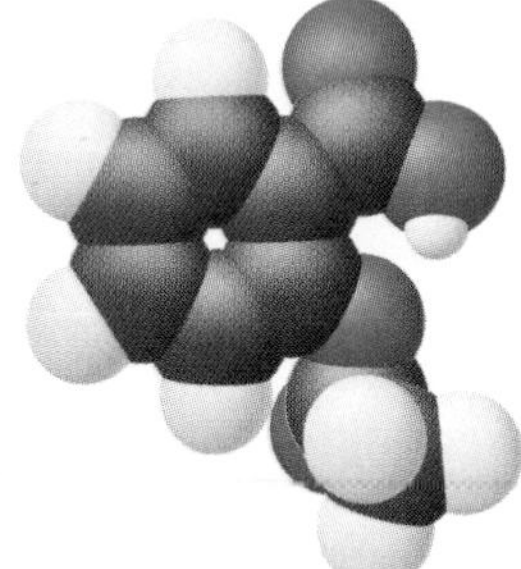

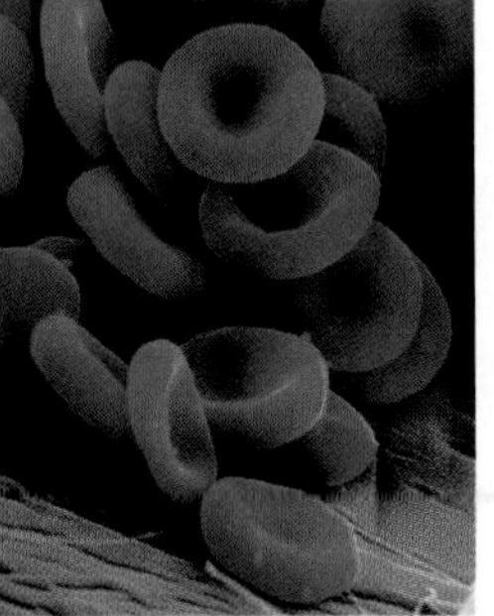

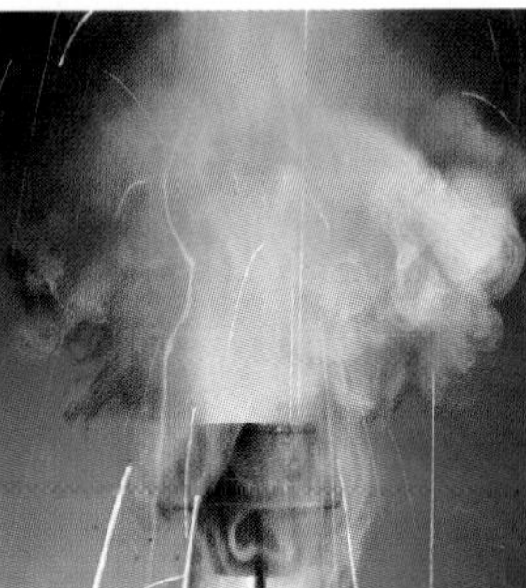

STUDENT RESOURCES

From the Teacher's Resource Package, use:

- Section Review 1.4, Ch. 1 Practice Problems, Vocabulary Review, Quizzes, and Tests from the Review Module (Ch. 1–4)

TECHNOLOGY RESOURCES

Relevant technology resources include:

- Chem ASAP! CD-ROM
- Resource Pro CD-ROM
- Assessment Resources CD-ROM: Chapter 1 Tests

❶ Answers will vary. Examples may include crystal structure, molecular bonding, cell structure and composition, and energy.

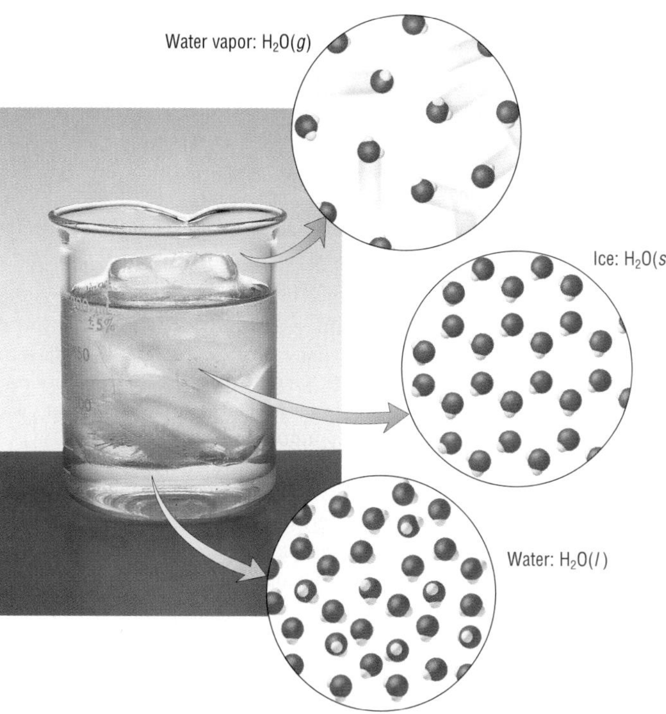

Figure 1.22
Drawings can help you visualize those parts of matter that are too small to be seen, even with a powerful microscope. Here you see the submicroscopic world of water, ice, and water vapor.

Using Your Textbook

This textbook will be your companion throughout your chemistry course. As its authors, we think the textbook is a "good read." But you should not read a textbook as you do a novel. Before reading a section in detail, read once through quickly to get a content overview. Take notes during the second, detailed reading. Combining reading with writing helps you retain information. Use the objectives and key terms to focus your reading. The tables, illustrations, and figure captions are an important part of the content. Visual representations like the one in **Figure 1.22** help to make concepts clearer. Solve any Practice Problems as they appear and compare your answers with those given in Appendix C.

Throughout your textbook, you will see references to a Chem ASAP! CD, which is designed to help you learn chemistry *A*s *S*oon *A*s *P*ossible. If you have access to this CD, use its animations and simulations to visualize the concepts presented in your textbook. CHEM ASAP! also provides a format that leads you through selected Practice Problems. After completing a chapter, use the section summaries in the Student Study Guide to review what you have learned. Test your knowledge by writing definitions for the key terms and answering chapter review problems assigned by your teacher. The answers to odd-numbered problems are in Appendix C. If you are unable to solve a problem or understand a concept after a reasonable amount of effort, move on to the next topic or problem. Ask your teacher or study group for help with the difficult problem or concept.

Figure 1.23
The Chem ASAP! CD-ROM has animations, simulations, assessment exercises, and problems that are designed to be used with your textbook.

Chem ASAP!

Animations
Simulations
Assessment
Problem-Solving

1.4

2 Teach

Discuss

Explain that the memorization of facts is a relatively small part of learning chemistry. A person who succeeds in chemistry has become a good problem solver. Trained chemists often excel in areas such as finance and marketing. Encourage the students to "think out loud" while working with classmates on practice problems. To enrich the learning process, suggest that students share problem-solving methods and techniques.

CHEMISTRY AND LITERATURE

The public's perception and expectations of science have changed dramatically over the last two hundred years. Society has become increasingly technologically advanced, raising new hopes and concerns among the public about the role that science can play in shaping our future. Many of these evolving social themes are reflected in the work of 19th and 20th century novelists such as H.G. Wells, Edward Bellamy, Jules Verne, George Orwell, and Mary Shelley. Have students select a novel by one of these novelists and critically examine it to determine how the author addressed the role of science, technology, and society in his or her era.

3 Assess

Evaluate Understanding

Ask students to describe useful ways to prepare for a test or a quiz. Guide students toward the realization that daily study can be the most effective approach.

Have students comment on the statement "Chemistry is mostly memorization."

Reteach

Discuss using classification schemes as tools for organizing information. Ask students to suggest classification schemes they encounter in their daily lives (the ways that goods are organized in stores or books in libraries or items in a kitchen cabinet).

Encourage students to draw diagrams whenever possible to help visualize concepts.

Portfolio Project

Encourage students to consider their own learning styles while developing their plans. The key is to develop plans that students can follow consistently. Have students compare strategies and offer constructive suggestions.

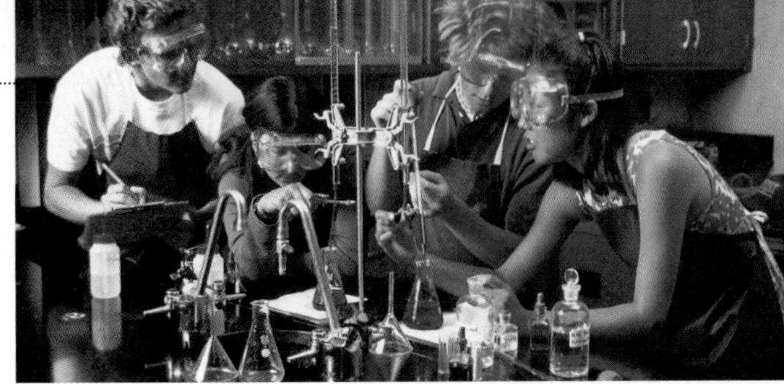

Figure 1.24
Working in a group often leads to the biggest discoveries!

On Your Own

You can be even more successful in your study of chemistry if you go beyond the textbook. For example, draw your own diagrams to illustrate important concepts and definitions. Or make flashcards for important formulas, equations, terms, and definitions. Quiz yourself and your classmates often to evaluate your understanding.

As you learned in Section 1.1, chemistry is often practiced as a group activity. Adopting this method might well enhance your study of chemistry. Many students form study groups with their classmates to work on difficult problems, talk through complex concepts, or gain a new perspective on what they are learning. Like writing, talking helps. If you cannot belong to a study group, explain chemistry to your friends, classmates, and family.

Tests and Quizzes

Learning chemistry can be a lot of fun! However, it is very likely that as part of your study, you will be required to take tests and quizzes. Cramming for a test at the last second is never a good idea. It is much more beneficial to set aside a certain amount of time to study every day. How much time you set aside depends on how long it takes you to do the work. When test time comes, reviewing the work that you have already completed should be part of your preparation. Remember, exhaustion is your mind's enemy. Get enough sleep the night before the test. When you get the test, read it over quickly. Determine which parts of the test you can answer immediately or work out rapidly. Complete those parts first. Reserve any difficult questions for last. Show all of your work when answering problem-solving questions. After you have solved a numerical problem, evaluate your answer to be certain that it makes sense. If you have time, go over the entire test to check for errors. Remember, the best way to avoid panic and unwelcome surprises is to be well prepared.

portfolio project

Design a study strategy or plan that you can use to learn chemistry. Discuss your plan with your teacher and share it with your classmates. Then use your plan throughout this course.

section review 1.4

12. Explain why it is important to study chemistry every day.
13. Why is chemistry best studied with a pencil in hand?
14. Explain the value of discussing chemistry with others.
15. Briefly describe what your strategy for studying chemistry will be throughout the upcoming year.

 Chem ASAP! **Assessment 1.4** Check your understanding of the important ideas and concepts in Section 1.4.

Answers

SECTION REVIEW 1.4

12. It is a good idea to study every day rather than cramming for tests at the last minute.
13. What you write is retained longer than what you only see.
14. If you can explain a concept clearly to someone else, you probably understand the concept. You can also learn from others.
15. study, write, explain

Chemistry Serving...Society

CHEMISTRY IN THE INFORMATION AGE

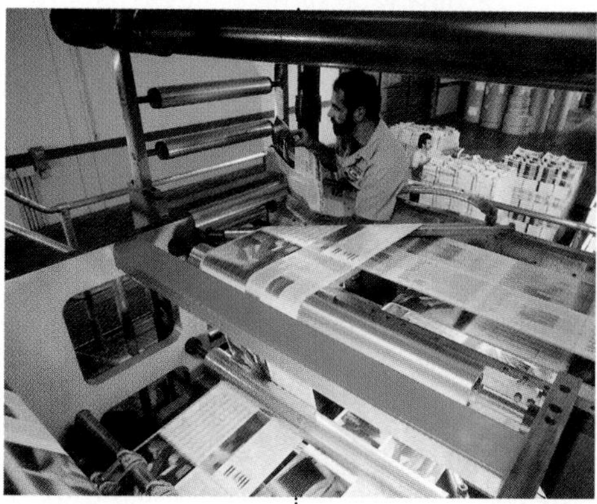

The general public needs to understand science. Citizens are often asked to make decisions about topics such as watershed protection, water quality, or waste disposal. These issues can affect the well-being of individuals and communities.

The way that scientists communicate with each other and with the general public has changed over the centuries. In earlier centuries, scientists exchanged ideas through letters. They also formed societies that met now and then to discuss the latest findings of their members. When societies began to publish journals, scientists could use the journals to keep up with experimental data. If results were published in an unfamiliar language or in an obscure journal, they might be lost for decades, as were Gregor Mendel's studies of heredity.

Nowadays, such a communication gap would be unlikely. Scientists exchange ideas with other scientists by phone, by e-mail, and at international conferences. Scientists still publish their results in scientific journals. These journals are the most reliable sources of information about new discoveries because articles are not published until they are reviewed by two or three experts in the author's field. The reviewers point out any errors in the author's hypothesis or experimental design. Reviewers may suggest that the author do more work to strengthen the argument before publishing. The review process can be painful for a scientist, but it is good for science as a whole because work that is not well founded is usually not published.

Because the review process can take from six months to a year, or even longer, some journals publish preliminary results on the Internet. Preliminary results should be treated with caution because such results are subject to change.

> To judge the reliability of information you find on the Internet, you have to consider the source.

One advantage of the Internet is that anyone can get access to its information. One disadvantage is that anyone can post information on the Internet without first having that information reviewed. To judge the reliability of information you find on the Internet, you have to consider the source.

You also need to treat the scientific articles you read in newspapers and magazines or the news you listen to on television with similar caution. Sometimes scientific reports in the popular media are a poor reflection of the actual work done by scientists. If a media outlet has a reporter who specializes in science, the chances are better that the report will not have major distortions, gaps, or errors. But even the best reporter may have limited time to prepare an article and limited space in which to describe the topic. Also, the reporter may need to simplify a complex issue to reach the widest possible audience. Should you avoid reading about science in the popular press? Absolutely not! There are too many important advances worthy of your attention. But you should look for reliable sources and approach the news with a certain amount of healthy skepticism. Links provided through the Prentice Hall web site are good sources of current data.

CHEMISTRY IN CAREERS

TAKE IT TO THE NET

Find out more about career opportunities:
www.phschool.com

CHEMIST

Solve mysteries, identify important chemicals, discover new medicines, and explain interesting phenomena.
See page 868.

CHEMICAL SPECIALIST

ocal food service distributor seeks esponsible self-motivated individ-

CHEMIST NEEDED

Chemistry Serving... Society

DISCUSS

Have students read the article and then compare the reliability of information published in newspapers, in journals, and on the Internet. Then ask

- Are there issues currently being discussed in your community where knowing chemistry would be useful?
- Are ordinary citizens today more likely to be literate in science than ordinary citizens were in earlier centuries?

One could reasonably support either possibility by comparing factors such as the existing means of communication, levels of education, and size of the body of scientific knowledge.

CHEMISTRY IN CAREERS

Have students connect to the Internet address shown to learn more about careers in chemistry. Students may also wish to read about the duties and responsibilities of a chemist on page 868 of this text.

Chapter 1 STUDENT STUDY GUIDE

4 Close

Summary

Ask the following questions that require students to summarize the information contained in the chapter.

- **What are some of the ways that chemistry has altered the environment? Include at least one positive and one negative effect.** (Answers will vary.)
- **Give an example of a problem a chemist might study and describe the approach he or she would take.** (Answers should include a discussion of the scientific method.)
- **What are some reasons for learning chemistry?** (Answers will vary.)

Extension

Have students do research to find examples of successful collaborative projects that involved a multidisciplinary approach to solving a scientific problem. One example is the successful series of NASA moon trips in the 1970s. What types of scientists contributed to this huge effort?

Looking Back . . . Looking Ahead . . .

This chapter defines chemistry and discusses in general terms technological advances made possible through chemistry. The chapter provides a formal definition of the scientific method. In the next chapter, students learn to characterize matter in terms of physical properties and to distinguish between physical and chemical changes.

Chapter 1 STUDENT STUDY GUIDE

Take It to the NET For interactive study and review, go to www.phschool.com

KEY TERMS

- analytical chemistry *p. 4*
- biochemistry *p. 4*
- chemistry *p. 3*
- experiment *p. 15*
- hypothesis *p. 15*
- inorganic chemistry *p. 4*
- observation *p. 15*
- organic chemistry *p. 4*
- physical chemistry *p. 4*
- scientific law *p. 17*
- scientific method *p. 15*
- theory *p. 16*

CONCEPT SUMMARY

1.1 Chemistry

- Chemistry is the study of the composition of matter and the changes that matter undergoes.
- Analytical chemistry, biochemistry, inorganic chemistry, organic chemistry, and physical chemistry are the traditional divisions of chemistry.
- Chemistry helps you to understand your world and to make informed decisions about scientific issues. Chemistry is an excellent career for some people.
- Applied chemistry, or chemical technology, is chemistry used to attain a specific goal. Pure chemistry accumulates knowledge for its own sake.

1.2 Chemistry Far and Wide

- Chemistry reaches into such diverse areas as materials science, energy, medicine and biotechnology, agriculture, environmental studies, and astronomy and space exploration.
- Progress has its price. Every new development has its risks and benefits. Society must decide in which direction the balance shifts.

1.3 Thinking Like a Scientist

- The scientific method provides a logical approach to the solution of scientific problems.
- The steps of the scientific method include observations, hypotheses, experiments, theories, and scientific laws.
- An observation is information gathered directly by using the senses. A hypothesis is a proposed explanation or reason for what is observed. An experiment is a means to test a hypothesis. A scientific experiment must be repeatable.
- A theory is a broad and extensively tested explanation of why experiments give certain results. A scientific law describes a natural phenomenon but does not explain it.

1.4 How to Study Chemistry

- You can study chemistry alone or in groups.
- It is important to learn the language and vocabulary of chemistry.
- Your textbook is an important aid in your studies.

CHAPTER CONCEPT MAP

Use these terms to construct a concept map that organizes the major ideas of this chapter.

 Chem ASAP! **Concept Map 1**
Create your Concept Map using the computer.

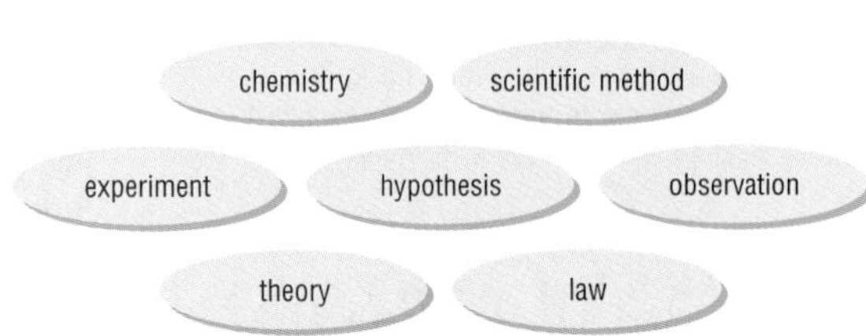

Take It to the Net

At **www.phschool.com** students will find for this chapter

- an Internet activity
- links to related chemistry sites
- an interactive quiz
- career links

Chapter 1 REVIEW

CONCEPT PRACTICE

16. Define chemistry. What is matter? *1.1*

17. Describe the difference between chemistry and chemical technology. *1.1*

18. Match each numbered term with a lettered term. *1.1*

a. physical chemistry	(1) composition
b. organic chemistry	(2) life
c. analytical chemistry	(3) carbon
d. inorganic chemistry	(4) behavior
e. biochemistry	(5) without carbon

19. Match each numbered photograph with the lettered field of chemistry it best represents. *1.2*

a. materials science
b. energy
c. medicine and biotechnology
d. agriculture
e. environment
f. astronomy and space

(1)

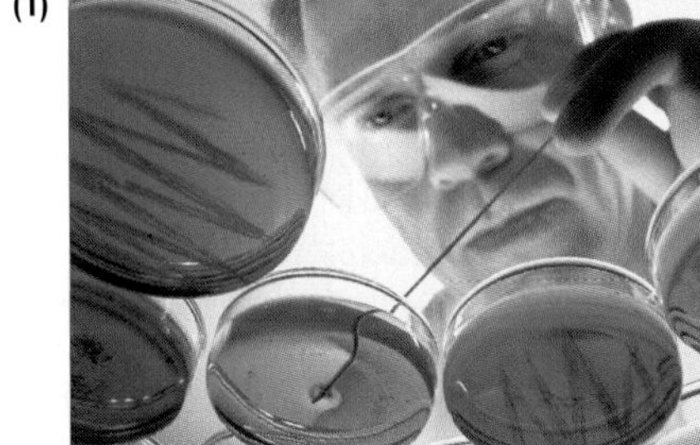

(2)

(3)

(4)

(5)

(6)

Chapter 1 REVIEW

Answers

16. Chemistry is the study of matter and the changes that matter undergoes. Matter is the stuff material things are made of.

17. Chemistry is the study of matter and its changes for the sake of understanding them; chemical technology is the application of this knowledge to attain specific goals.

18. **a.** 4 **b.** 3 **c.** 1 **d.** 5 **e.** 2

19. **a.** 2 **b.** 5 **c.** 1 **d.** 6 **e.** 3 **f.** 4

20. Energy research could help produce cleaner energy more efficiently and reduce the amount of nonrenewable fossil fuels expended in producing it.

21. The ozone layer protects Earth from harmful ultraviolet rays.

22. Some plastics are stronger than steel and do not rust. Automobiles with many plastic components are lighter than automobiles made completely from steel and are, therefore, more fuel-efficient.

23. Experiments are used to test hypotheses.

Answers

24. Answers will vary but might include diagnosing a car's problem, a computer, or an electric appliance.
25. If you are certain that your experiment is not flawed, you must revise your hypothesis.
26. c.
27. A language uses its vocabulary to express thoughts about anything—food, politics, religion, etc. The vocabulary of the sciences is used to express concepts and facts about the workings of the material world.
28. b.
29. Students' diagrams should show one string that is threaded through both holes at A and C. The string at hole B is a separate thread from the string passing through holes A and C.
30. The doctor's hypothesis is that you have a strep throat. She tests the hypothesis with an experiment to learn whether your throat is infected with strep bacteria.
31. The experiment may be flawed; if not, the results are evidence that the theory may need to be revised.
32. In general, the more you know, the better off you are. The study of chemistry helps you think logically and analytically. This skill is useful in any career.
33. **a.** 1 **b.** 3 **c.** 1
34. A hypothesis is retained if it is supported by the results of experiments; otherwise, it is rejected.
35. No; for any discovery to take place, the discoverer must have the knowledge to recognize the significance of the observation or data.
36. A theory can never be proven. Although a theory may be strongly supported by existing experiments, there is always the possibility that a new kind of experiment will prove the theory to be false.

20. How might advances in energy research affect your life? *1.2*
21. Why is the ozone layer surrounding Earth important? *1.2*
22. Give at least two benefits of replacing some of the steel in automobiles with plastics. *1.2*
23. What is the purpose of an experiment as part of the scientific method? *1.3*
24. Describe two situations in which you used at least part of the scientific method. *1.3*
25. You perform an experiment and get unexpected results. According to the scientific method, what should you do next? *1.3*
26. Which of the following is not a part of the scientific method? *1.3*
    a. hypothesis **c.** guess
    b. experiment **d.** theory
27. Compare and contrast the study of chemistry with the study of a language. *1.4*
28. Which statement about the study of chemistry is false? *1.4*
    a. The chemistry textbook is an important study aid.
    b. Chemistry is mostly memorization.
    c. It is better to study on a regular basis than to cram before a test.
    d. A study group is beneficial.
    e. Discussing chemistry with family and friends is helpful.

CONCEPT MASTERY

29. You find a sealed box with strings protruding from three holes, as shown in the diagram. When you tug string **a**, it becomes longer and string **c** becomes shorter. When you tug string **b**, it becomes longer, but strings **a** and **c** are unaffected. Make a diagram showing the arrangement of the strings inside the box.

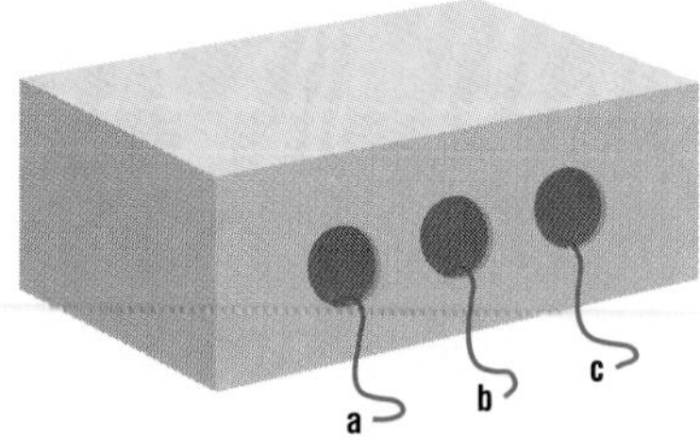

30. You have a sore throat, so you go to the doctor. The doctor examines your throat and says she thinks you have strep throat. She takes a sample to test for strep bacteria. What parts of the scientific method is the doctor applying?
31. You perform an experiment and find that the results do not agree with theory. Is something wrong with your experiment?
32. Explain why chemistry might be useful in any career.

CRITICAL THINKING

33. Choose the term that best completes the second relationship.
    a. seed:plant
    data:______
    (1) theory (3) experiment
    (2) techniques (4) scientific method
    b. hub:wheel
    chemistry:______
    (1) tire (3) science
    (2) hypothesis (4) theory
    c. part:whole
    theory:______
    (1) scientific method (3) observation
    (2) experiment (4) law
34. Comment on the idea that science accepts what works and rejects what does not work.
35. Occasionally you may read about important discoveries being made accidentally. Louis Pasteur said, "Chance favors the prepared mind." Are these two statements contradictory? Explain.
36. Criticize the statement, "Theories are proven by experiments."
37. Refer back to question 28 and provide an explanation for each statement that is true. Correct the statement that is false.

37. **a.** The textbook contains all of the major content and concepts covered in the course. The visuals, reading/studying guides, and practice/review features are essential elements in learning and understanding chemistry.
b. (false statement) Chemistry concepts are interwoven and are best understood as parts of a big picture. Seeing connections and building a framework imply understanding, not memorization.
c. Learning in small chunks is more effective. Studying on a regular basis helps build a big picture of chemistry into which individual facts can be placed. It assures understanding rather than memorization.
d. An exchange of ideas and perspectives is valuable.
e. Talking about chemistry aids understanding and retention.

Chapter 1 STANDARDIZED TEST PREP

Select the choice that best answers each question or completes each statement.

1. The branch of chemistry that studies most carbon-containing substances is ______ chemistry.
   a. physical c. analytical
   b. inorganic d. organic
2. An analytical chemist is most likely to
   a. explain why paint is stirred before it is used.
   b. explain what forces keep paint attached to the steel frame of an automobile.
   c. identify the type of paint chips found at the scene of a hit-and-run accident.
   d. investigate the effect of leaded paint on the development of a young child.
3. The greatest source of energy on Earth is
   a. petroleum. c. electricity.
   b. sunlight. d. nuclear power.
4. Chemists who work in the biotechnology field are most likely to work with
   a. x-ray technicians. c. physicians.
   b. geologists. d. physicists.
5. Increased levels of carbon dioxide in the atmosphere can cause
   a. smog. c. global warming.
   b. acid rain. d. ozone depletion.

Respond to each statement in questions 6–8.

6. Someone who wears contact lenses does not have to wear safety glasses.
7. Eating food that is left over from an experiment is an alternative to discarding the food.
8. For a student who has read the procedure, the teacher's pre-lab instructions are unnecessary.

Use the atomic windows to answer question 9.

9. The air you breathe is composed of about 20% oxygen and 80% nitrogen. Which atomic window best represents a sample of air?

a.
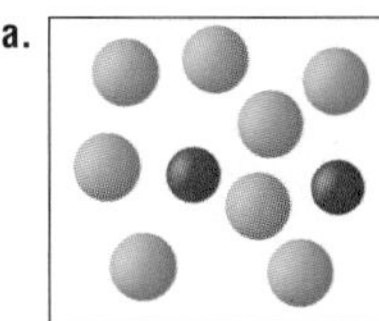

c.
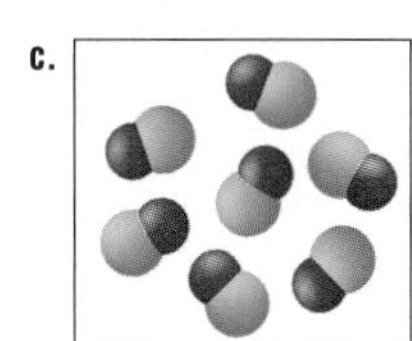

b.
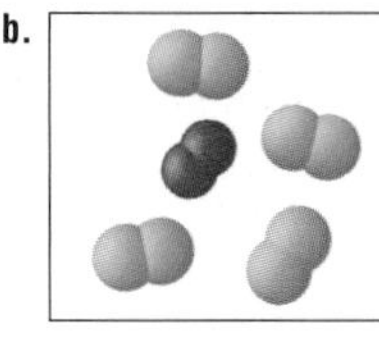

d.
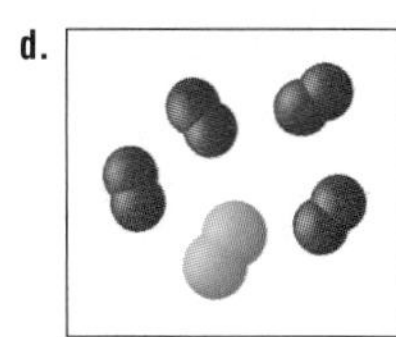

Oxygen Nitrogen

Use this paragraph to answer questions 10–12.

(1) On a cold morning, your car does not start. (2) You say, "Oh no! The battery is dead!". (3) Your friend who works on cars uses a battery tester and finds that the battery has a full charge. (4) Your friend notices a lot of corrosion on the battery terminals. (5) Your friend says "Maybe corrosion is causing a bad connection in the electrical circuit, preventing the car from starting." (6) Your friend cleans the terminals and the car starts.

10. Which statements are observations?
11. Which statements are hypotheses?
12. Which statement describes an experiment?

For each question there are two statements. Decide whether each statement is true or false. Then decide whether Statement II is a correct explanation for Statement I.

	Statement I		Statement II
13.	A hypothesis may be rejected after an experiment.	BECAUSE	Experiments are used to test hypotheses.
14.	Fossil fuels are nonrenewable resources.	BECAUSE	Only about 1% of the energy in sunlight is captured by plant life on Earth.
15.	Theories help you model objects or processes that cannot be seen.	BECAUSE	Theories summarize the results of many observations and experiments.
16.	Ideally, insecticides, herbicides, and fungicides should be nonspecific.	BECAUSE	Scientists are looking for safer, more effective ways to protect crops.
17.	All Internet sites that provide scientific information are equally reliable.	BECAUSE	All information on these sites is reviewed by qualified scientists.

Chapter 1 STANDARDIZED TEST PREP

1. d
2. c
3. b
4. c
5. c
6. Contact lenses increase the possibility of eye damage because chemicals can become trapped behind a lens. Safety glasses also protect the eyes from flying debris.
7. No food should be eaten in the laboratory because food may become contaminated with a toxic substance.
8. The teacher may modify the instructions to suit available equipment or chemicals. The teacher will stress any safety precautions. Accidents are more likely to happen when instructions or warnings are ignored.
9. b
10. 1 and 4
11. 2 and 5
12. 3
13. True, True, correct explanation
14. True, True
15. True, False
16. False, True
17. False, False

The modified true-false format used in problems 13–17 may be unfamiliar to students. We have designed this first set with one example of each possible outcome from True, True, correct explanation to False, False.

Planning Guide

SECTION OBJECTIVES	ACTIVITIES/FEATURES	MEDIA & TECHNOLOGY
2.1 Matter ● ■ ◇ ▸ Identify the characteristics of matter and substances ▸ Differentiate among the three states of matter ▸ Define physical property and list several common physical properties of substances	**SE** **Discover It!** *Classifying Matter,* p. 28 **SE** **Link to Physics** *Changing States of Matter,* p. 31 **LM** 2: *Physical and Chemical Change* **TE** **DEMOS**, pp. 29, 30	**ASAP** Animation 1 **ASAP** Assessment 2.1 **OT** 2: *Solid, Liquid, Gas* **OT** 3: *Matter*
2.2 Mixtures ● ■ ◆ ▸ Categorize a sample of matter as a substance or a mixture ▸ Distinguish between homogeneous and heterogeneous samples of matter	**SE** **Mini Lab** *Mixtures,* p. 35 (**LRS** 2-1) **TE** **DEMOS**, pp. 32, 33	**ASAP** Problem Solving 6 **ASAP** Assessment 2.2 **CA** *Making Sodium Chloride* **OT** 8: *Molecular Structures*
2.3 Elements and Compounds ● ■ ◆ ▸ Explain the difference between an element and a compound ▸ Identify the chemical symbols of common elements, and name common elements, given their symbols	**SE** **Link to Linguistics** *Origins of Element Names,* p. 40 **TE** **DEMOS**, pp. 36, 37, 39	**ASAP** Problem Solving 13 **ASAP** Assessment 2.3 **CHM** Side 1, 38: *Elements and Compounds*
2.4 Chemical Reactions ● ■ ◆ ▸ Differentiate between physical and chemical changes in matter ▸ Apply the law of conservation of mass	**SE** **Link to Engineering** *Chemical Engineers,* p. 42 **SE** **Small-Scale Lab** *1 + 2 + 3 = Black!,* p. 44 (**LRS** 2-2) **SE** **Chemistry Serving . . . Industry** *Barriers to Heat Flow,* p. 45 **SE** **Chemistry in Careers** *Materials Scientist,* p. 45 **LM** 2: *Physical and Chemical Properties* **LM** 3: *Observing a Chemical Reaction* **SSLM** 1: *A Study of Chemical Changes* **TE** **DEMOS**, pp. 41, 42	**ASAP** Assessment 2.4 **ASAP** Concept Map 2 **ACT** 2: *Matter and Change* **CHM** Side 1, 13: *Plastics Recycling* **RP** Lesson Plans, Resource Library **AR** Computer Test 2 **www** Activities, Self-Tests, *SCIENCE NEWS* updates **TCP** The Chemistry Place Web Site

KEY

● Conceptual (concrete concepts)	**AR** Assessment Resources	**GRS** Guided Reading and Study Workbook
○ Conceptual (more abstract/math)	**ASAP** Chem ASAP! CD-ROM	**LM** Laboratory Manual
■ Standard (core content)	**ACT** ActivChemistry CD-ROM	**LP** Laboratory Practicals
□ Standard (extension topics)	**CHM** CHEMedia Videodiscs	**LRS** Laboratory Recordsheets
◆ Honors (core content)	**CA** Chemistry Alive! Videodiscs	**SSLM** Small-Scale Lab Manual
◇ Honors (options to accelerate)	**GCP** Graphing Calculator Problems	

PRACTICE		ASSESSMENT	
GRS	Section 2.1	SE	Section Review
RM	Practice Problems 2.1	RM	Section Review 2.1
		RM	Chapter 2 Quiz
		LP	Lab Practical 2-1
SE	**Sample Problem** 2-1	SE	Section Review
GRS	Section 2.2	RM	Section Review 2.2
RM	Practice Problems 2.2	RM	Chapter 2 Quiz
SE	**Sample Problem** 2-2	SE	Section Review
GRS	Section 2.3	RM	Section Review 2.3
RM	Practice Problems 2.3	RM	Chapter 2 Quiz
RM	Interpreting Graphics		
GRS	Section 2.4	SE	Section Review
RM	Practice Problems 2.4	RM	Section Review 2.4
RM	Vocabulary Review 2	RM	Chapter Quiz
		LP	Lab Practical 2-1
		RM	Vocabulary Review 2
		SE	Chapter Review
		SM	Chapter 2 Solutions
		SE	Standardized Test Prep
		PHAS	Chapter 2 Test Prep
		RM	Chapter 2 Quiz
		RM	Chapter 2 A & B Test

OT	Overhead Transparency	SE	Student Edition
PHAS	PH Assessment System	SM	Solutions Manual
PLM	Probeware Lab Manual	SSV	Small-Scale Video/Videodisc
PP	Problem Pro CD-ROM	TCP	www.chemplace.com
RM	Review Module	TE	Teacher's Edition
RP	Resource Pro CD-ROM	www	www.phschool.com

PLANNING FOR ACTIVITIES

STUDENT EDITION

Discover It! p. 28
- small clear containers
- sugar
- baking powder
- flour, baking soda
- half-cup measures
- half-teaspoon measures
- spoons
- tape or self-adhesive labels

Mini Lab, p. 35
- green marking pens
- strips of filter paper
- metric rulers
- clear plastic tape
- rubbing alcohol
- clear plastic drinking cups
- clear plastic wrap
- pencil
- water

Small-Scale Lab, p. 44
- pencils, rulers, paper
- $NaOCl$, H_2O_2, $CuSO_4$, KIO_3, KI, $FeCl_3$, $KMnO_4$, $NaNO_2$
- starch
- cereal
- reaction surfaces
- pipets

TEACHER'S EDITION

Teacher Demo, p. 29
- pieces of solid chocolate
- small beaker
- boiling water

Teacher Demo, p. 30
- empty syringe (50 mL or bigger)
- balance that reads to 0.01 g
- stopper
- funnel
- Erlenmeyer flask
- water

Teacher Demo, p. 32
- iron-fortified breakfast cereal
- stirring bar
- 400-mL beaker
- distilled water
- magnetic stirrer

Teacher Demo, p. 33
- sand
- salt
- iron filings
- water

Teacher Demo, p. 36
- safety goggles
- fume hood or well-ventilated area
- 25 g powdered sugar
- 100-mL beaker
- 10-15 mL of 18*M* sulfuric acid
- glass rod

Teacher Demo, p. 37
- Hoffman apparatus
- glowing splint
- match

Activity, p. 38
- poster board
- masking tape

Teacher Demo, p. 39
- samples of elements from a chemical supply company

Teacher Demo, p. 41
- 5 g of iron filings
- 5 g of sulfur
- glass plate
- spatula
- one magnet
- test tube
- beaker
- gas burner
- matches
- safety goggles
- fume hood
- cold water

Teacher Demo, p. 42
- Bunsen burner
- tongs
- magnesium ribbon
- 10 mL of 0.1*M* $AgNO_3$
- two test tubes
- small amounts of 0.1*M* $NaCl$ and 0.1*M* K_2CrO_4
- 10 mL of 3*M* H_2SO_4
- mossy zinc
- marble chip
- cobalt blue glass

Chapter 2

Key Terms

2.1 matter, mass, substance, physical property, solid, liquid, gas, vapor, physical change
2.2 mixture, heterogeneous mixture, homogeneous mixture, solutions, phase, distillation
2.3 elements, compounds, chemical symbol
2.4 chemical reaction, reactants, products, chemical property, law of conservation of mass

DISCOVER IT!

After adding the solids to water without stirring, students can distinguish baking powder because it produces bubbles and flour because it floats instead of sinking. When students stir the contents, the sugar and baking soda should begin to dissolve. Students may not use appropriate scientific terms to describe the distinguishing characteristics they have observed. Accept any reasonable substitutes.

Chapter 2 MATTER AND CHANGE

Hot lava from a volcano changes ocean water to steam.

FEATURES

Stay current with SCIENCE NEWS
Find out more about matter and change:
www.phschool.com

DISCOVER IT! CLASSIFYING MATTER

You need four small clear containers, sugar, baking powder, flour, baking soda, water, a half-cup measure (about 120 mL), a half-teaspoon measure (about 2.5 mL), a spoon, tape or self-adhesive labels, and a pen.

1. Label the containers with the names of the four white solids (sugar, baking powder, flour, baking soda).
2. Pour a half-cup of water into each of the four containers.
3. Without stirring, add a half-teaspoon of each white solid to its corresponding container. Take note of any reaction between the solid and the water. Wipe the spoon between each addition.
4. Stir the contents of each container for at least 45 seconds, observing what happens to each of the white solids. Rinse the spoon with water between stirrings.

Which solid could you distinguish from the other three after the initial mixing with water? How? Can you distinguish completely among the remaining three solids based on the results of Step 4? Why or why not? Make a list of the distinguishing characteristics you have just explored, and add to it as you read this chapter.

2.1 ● ■ ◇
2.2 ● ■ ◆
2.3 ● ■ ◆
2.4 ● ■ ◆

Conceptual The brief discussion of phases will be expanded in Chapter 10 after kinetic theory has been introduced. The Mini Lab gives students a concrete example of separating mixtures. Use Figure 2.8 to help students visualize the classification of matter. Students must understand conservation of mass to successfully master Chapters 6–9.

Standard Introduce the three-step problem-solving method. Use the guided problem-solving tutorials on the CD-ROM to save class instruction time, especially Chem ASAP! for students with weaker problem solving skills. Solution chemistry will be covered in more detail in Chapter 18. If students are familiar with the classification of substances, use Figure 2.8 as review.

Honors Section 2.1 may be accelerated. Use Table 2.1 to review physical properties. Briefly review classification of substances. Use Table 2.4 and the Link to Linguistics to help students master nomenclature.

MATTER

section 2.1

As incredibly hot molten lava from a volcano reaches the ocean, the thick liquid lava cools to form solid rock. At the same time, some of the liquid water is heated to form a gas called steam. **What are the characteristics of solids, liquids, and gases, and how do their shapes and volumes differ?**

objectives
- Identify the characteristics of matter and substances
- Differentiate among the three states of matter
- Define physical property and list several common physical properties of substances

key terms
- matter
- mass
- substance
- physical property
- solid
- liquid
- gas
- vapor
- physical change

Properties of Matter

Look around you. All the things you see are examples of matter. But what exactly is matter? What forms does matter take? What can cause matter to change its form? **Matter** is defined as anything that has mass and takes up space. The **mass** of an object is the amount of matter the object contains. A golf ball has a greater mass than a table-tennis ball. The golf ball, therefore, contains more matter.

Everything is made up of matter. However, materials may differ in terms of the kind of matter they contain. For example, table sugar is one particular kind of matter, with the chemical name sucrose. Table sugar is always 100% sucrose. It always has the same makeup, or chemical composition. Matter that has a uniform and definite composition is called a **substance.** Substances, also referred to as pure substances, contain only one kind of matter. Lemonade is not a substance because not all samples of lemonade are identical. Lemonade contains more than one kind of matter and the relative amounts of each kind may differ. For example, different pitchers of lemonade may have varying amounts of sugar, lemon juice, or water, and may taste different. Thus they have different compositions.

All samples of a substance have identical physical properties. For example, all crystals of sucrose taste sweet and dissolve completely in water. A **physical property** is a quality or condition of a substance that can be observed or measured without changing the substance's composition. Physical properties include color, solubility, odor, hardness, density, melting point, and boiling point.

Scan the physical properties of the common substances listed in **Table 2.1.** Such physical properties help chemists identify the substances.

Table 2.1

Physical Properties of Some Common Substances						
Substance	**Formula**	**State**	**Color**	**Melting point (°C)**	**Boiling point (°C)**	**Density (g/cm³)**
Neon	Ne	gas	colorless	−249	−246	0.0009
Oxygen	O_2	gas	colorless	−218	−183	0.0014
Chlorine	Cl_2	gas	greenish-yellow	−101	−34	0.0032
Ethanol	C_2H_5OH	liquid	colorless	−117	78	0.789
Mercury	Hg	liquid	silvery-white	−39	357	13.5
Bromine	Br_2	liquid	red-brown	−7	59	3.12
Water	H_2O	liquid	colorless	0	100	1.00
Sulfur	S	solid	yellow	113	445	2.07
Sucrose	$C_{12}H_{22}O_{11}$	solid	white	185	d*	1.59
Sodium chloride	$NaCl$	solid	white	801	1413	2.17

d*, decomposes on heating

STUDENT RESOURCES

From the Teacher's Resource Package, use:
- Section Review 2.1, Ch. 2 Practice Problems and Quizzes from the Review Module (Ch 1–4)
- Laboratory Manual: Experiment 2
- Laboratory Practical 2–1

TECHNOLOGY RESOURCES

Relevant technology resources include:
- Chem ASAP! CD-ROM
- Resource Pro CD-ROM
- Chemistry Alive! Videodisc: *Phase Change*

2.1

1 Engage

Use the Visual

Have students examine the photograph that opens the section. Ask:

- **What states of matter are represented in the photograph?** (*Liquid* lava releases heat to form *solid rock; liquid* water absorbs heat to form a *gas*.)
- **What must you do to a substance to change its physical state?** (add or remove heat)

Help students to recognize that solids have a fixed, rigid shape and volume. Liquids have a defined volume, but no fixed shape. Gases have no defined shape or volume.

Check Prior Knowledge

Explain that chemistry is the study of matter and the changes it undergoes. To assess students' prior knowledge about matter, ask:

- **What is matter?** (anything that takes up space and has mass)
- **What is mass and how does it compare to weight?** (Weight is a measure of the pull of gravity on an object; mass is the amount of matter the object contains.)
- **What word is used to describe the amount of space an object takes up?** (volume)

TEACHER DEMO

Show students pieces of solid chocolate. Place them in a small beaker surrounded by boiling water in a larger beaker. After a few minutes, ask:

- How many physical states can you discern in the chocolate mixture? (Three: solid, liquid, and vapor. Remind students that their ability to smell the chocolate means that some components of the mixture must have vaporized.)

2 Teach

TEACHER DEMO

Ask students to consider whether or not air takes up space and has mass. Show them that it does in two different ways. Determine the mass of an empty syringe (50 mL or bigger) on a balance that reads to 0.01 g. Determine the mass when the syringe is full of air. Subtract to find the mass of air. (0.050 L × 1.18 g/L = 0.060 g) Fit a stopper and funnel into an Erlenmeyer flask. The equipment must be airtight. Try to pour water through the funnel. Some water will get into the flask (air is compressible), but not very much, because air takes up the space.

Table 2.2

Important Properties of the States of Matter			
Property	**Solid**	**Liquid**	**Gas or vapor**
Shape	definite	indefinite	indefinite
Volume	definite	definite	indefinite
Expansion on heating	very slight	moderate	great
Compressibility	almost incompressible	almost incompressible	readily compressible

For example, a colorless liquid that was found to boil at 100 °C and melt at 0 °C would likely be water. A colorless liquid that boiled at 78 °C and melted at −117 °C would most certainly not be water. Which substance in **Table 2.1** has these physical properties? ❶

Chem ASAP!
Animation 1
Relate the states of matter to kinetic energy and temperature.

States of Matter

You are very familiar with the substance called water. Under certain circumstances, however, you use the words ice and steam to refer to what is really the same substance. You use these three names because water—like most other substances—can exist in three different physical states: solid, liquid, and gas. The physical state of a substance is a physical property of that substance. Certain characteristics, summarized in **Table 2.2**, distinguish the three states of matter.

Coal, sugar, ice, and iron are examples of materials that are solids. A **solid** is matter that has a definite shape and volume. The shape of a solid does not depend on the shape of its container. The particles in a solid are packed tightly together, as shown in **Figure 2.1a.** As a result, solids are almost incompressible—that is, they cannot be squashed into a smaller volume. In addition, they expand only slightly when heated.

Water, milk, and blood are examples of liquids. The particles in a liquid are in close contact with one another, but unlike solid particles, they are not rigidly packed. A liquid can flow; that is, it can take the shape of the container in which it is placed. The amount of space, or volume, occupied by a sample of a liquid is the same no matter what shape it takes. This

Figure 2.1
*Compare the structured arrangement of the particles in a solid, a liquid, and a gas. **(a)** The particles in a solid are packed closely together in a rigid arrangement. A solid thus has a definite shape and volume. **(b)** The particles in a liquid contact one another yet have more freedom to move than the particles in a solid. A liquid can flow and takes on the shape of its container. **(c)** The particles in a gas are relatively far apart and are free to move anywhere inside their container. A gas thus has an indefinite shape and indefinite volume.*

Answers

❶ ethanol
❷ Because gallium is shown in its liquid state, the hand temperature must be greater than 30 °C.

MEETING DIVERSE NEEDS

Gifted Have students research the fourth state of matter (plasma) to determine why substances on Earth are not normally found in that state. Have them prepare an oral or written report.

Chemistry Alive!

Phase Change

Play

unchanging volume is said to be fixed or constant. A **liquid** is thus a form of matter that flows, has a fixed volume, and takes the shape of its container. Liquids are almost incompressible, but they tend to expand when heated.

Like liquids, gases flow to take the shape of the container that holds them. The particles in a gas are spaced far apart. Unlike liquids, gases expand without limit to fill any space. Thus a **gas** is a form of matter that takes both the shape and volume of its container. Gases are also easily compressed.

The words gas and vapor should not be used interchangeably; there is a difference. The term gas is limited to those substances that exist in the gaseous state at ordinary room temperature. For example, air is a mixture of gases including oxygen and nitrogen. The word **vapor** describes the gaseous state of a substance that is generally a liquid or solid at room temperature. Steam, the gaseous form of water, is referred to as a vapor because water is a liquid at room temperature. Moist air contains water vapor.

Physical Changes

Matter can be changed in many ways without altering the chemical composition of the material. Such a change, which alters a given material without changing its composition, is called a **physical change.** Cutting, grinding, and bending a material are examples of physical changes. A change in temperature may also bring about a physical change, such as the melting of the metal gallium, shown in **Figure 2.2.** The melting of ice, the freezing of water, the conversion of water to steam, and the condensation of steam to water are all examples of physical changes—changes that do not alter the chemical composition of the water.

Words such as boil, freeze, dissolve, melt, condense, break, split, crack, grind, cut, crush, and bend usually signify a physical change. Table salt (sodium chloride) is a white solid at room temperature. It can be made to undergo physical changes. For example, it can melt to form a liquid or boil to form a vapor. The temperatures at which these changes of state occur in sodium chloride are very different from the corresponding changes for water. Sodium chloride melts (becomes a liquid) at 801 °C and boils (becomes a vapor) at 1413 °C—high temperatures compared with water's melting and boiling points of 0 °C and 100 °C, respectively.

section review 2.1

1. Is every sample of matter a substance? Explain.
2. Contrast the characteristics of the three states of matter.
3. Which of the following are physical changes?
   - **a.** making caramel from sugar
   - **b.** carving a wooden figurine
   - **c.** freezing mercury
   - **d.** dissolving salt in water
4. Use **Table 2.1** to answer the following questions.
   - **a.** Which of the liquids listed has the highest boiling point?
   - **b.** What two properties of sucrose distinguish it from sodium chloride?
   - **c.** What single property do neon, oxygen, and ethanol have in common?

 Chem ASAP! **Assessment 2.1** Check your understanding of the important ideas and concepts in Section 2.1.

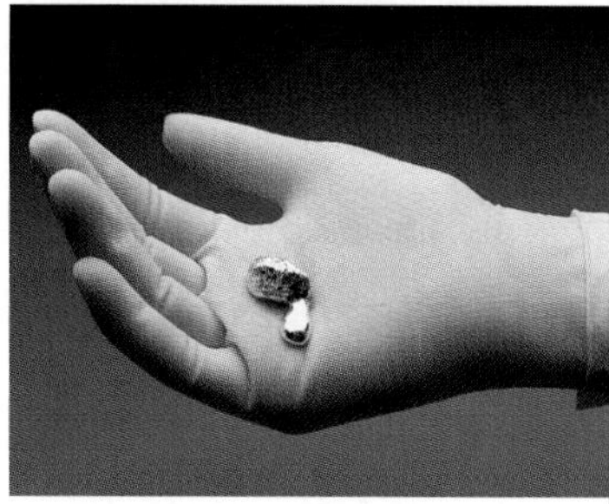

Figure 2.2
The melting point of gallium metal is 30 °C. What is the approximate Celsius temperature of the hand shown holding the gallium? How can you tell? 2

PHYSICS

Changing States of Matter
Changes of state are typically associated with changes in temperature. However, the state of matter can also be affected by other variables. Thixotropic materials are solidlike materials that liquefy when subjected to shearing forces. For example, many paints are thixotropic; they thin out when brushed on a surface and thicken when the brush strokes stop, thus keeping the paint from sliding off the wall! A shearing force has an opposite effect on quicksand. Quick movements "thicken" the quicksand and make it much more difficult for a person or animal trapped in it to move.

3 Assess

Evaluate Understanding

To assess students' knowledge of states of matter, ask:

- **How does burning a candle involve the three states of matter? Tell students to include the definitions for each state.** (Solid wax has a definite shape and volume, the melted liquid wax is shapeless and runs down the side of the candle, the vapor which forms above the wick, where it is the hottest, has no shape.)

Reteach

Display Table 2.1 on an overhead projector. Explain that chemists distinguish different substances from one another on the basis of their physical properties. Emphasize that every substance has a unique set of physical properties. No two substances have exactly the same melting point, boiling point, or density.

If possible, bring samples of some of the substances listed in Table 2.1 to class. Describe the physical properties of selected samples and have students practice identifying them by referring to Table 2.1. Have students take turns describing other substances to each other. If you have a Thiele tube available, show how the device is used to measure melting points.

Answers

SECTION REVIEW 2.1

1. No; a substance is a particular kind of matter that has a uniform and definite composition. Pure substances contain only one kind of matter. Matter is anything that has mass and occupies space.
2. Solids have definite shape and volume and are nearly incompressible. Liquids have definite volume but no definite shape, are nearly incompressible, and can flow. Gases have neither definite shape nor definite volume and are easily compressed.
3. **b.**, **c.**, and **d.**
4. **a.** mercury
   b. melting point and density
   c. All are colorless.

1 Engage

Use the Visual

Have students study the photograph and read the text that opens the section. Point out that the gold with the sand is a mixture. Ask:

- **What is a mixture, and how can it be separated?** (A mixture is a physical blend of two or more substances that can usually be separated without changing the chemical identities of the substances.)

Motivate and Relate

Ask students how they would distinguish between a mixture and a substance. Bring orange juice or coffee to class and compare the properties of these mixtures to those of pure water or pure NaCl. Explain that mixtures are variable in composition. For example, orange juice can be strong or weak depending on the amount of orange pulp mixed with water. On the other hand, pure substances have a uniform and constant composition.

TEACHER DEMO

Show students the following mixture. Obtain an iron-fortified breakfast cereal. Place a stirring bar in a 400-mL beaker. To the beaker, add approximately 30 g of cereal and distilled water until the beaker is about half full. Using a magnetic stirrer, mix gently for about 20 minutes. Carefully retrieve the stirring bar and observe the black iron filings attached to it. You might want to explain that stomach acid will change the iron into a form usable by the body.

section 2.2

MIXTURES

objectives

- Categorize a sample of matter as a substance or a mixture
- Distinguish between homogeneous and heterogeneous samples of matter

key terms

- mixture
- heterogeneous mixture
- homogeneous mixture
- solutions
- phase
- distillation

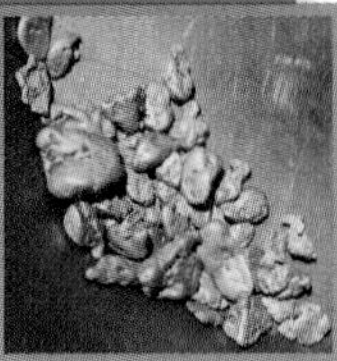

In 1848, gold was discovered in the foothills near Placerville, California. This discovery led to a massive gold rush in the following year. Many people in the California foothills still pan for gold as a hobby. Panning separates gold out of a mixture of gold and sand. **What is a mixture, and how can it be separated?**

Classifying Mixtures

You might prepare a salad by tossing lettuce, tomatoes, cucumbers, and celery with some vinegar and oil. The result is not only nutritious; it is also a mixture. In this section, you will learn how to identify and classify mixtures.

Most samples of matter are obviously mixtures. For example, you can easily recognize chicken noodle soup as a mixture of chicken pieces, noodles, and broth. Recognizing other materials as mixtures may be much harder. Air is a mixture of gases, but its components cannot be distinguished by eye, even through a microscope.

A **mixture** is a physical blend of two or more substances. One important characteristic of mixtures is that their compositions may vary. A dinner salad can have varying amounts of tomatoes or celery in it. The composition of air in a forest may differ from that in an industrial city, particularly in the amounts of pollutants. Blood, a mixture of water, various chemicals, and cells, varies somewhat in composition from one individual to another and, from time to time, in a given individual.

Mixtures can be of two basic kinds: heterogeneous or homogeneous. **Figure 2.3** gives examples of each kind. A **heterogeneous mixture** is one that is not uniform in composition. If you were to sample one portion of such a mixture, its composition would be different from that of another portion.
1 Why is the salad described above heterogeneous? A **homogeneous mixture** in contrast, is one that has a completely uniform composition. Its components are evenly distributed throughout the sample. A sample of salt water is the same throughout. Thus salt water is an example of a homogeneous mixture.

Figure 2.3
All of these items are mixtures. The bar of soap and the beverage are homogeneous mixtures; they have uniform compositions. The salad is a heterogeneous mixture; it consists of several phases containing components that are not evenly distributed. What other everyday items can you identify as either homogeneous or heterogeneous mixtures? 2

STUDENT RESOURCES

From the Teacher's Resource Package, use:

- Section Review 2.2, Ch. 2 Practice Problems and Quizzes from the Review Module (Ch 1–4)
- Laboratory Recordsheet 2–1

TECHNOLOGY RESOURCES

Relevant technology resources include:

- Chem ASAP! CD-ROM
- Resource Pro CD-ROM

Table 2.3

Some Common Types of Solutions	
System	**Examples**
Gas–gas	Carbon dioxide and oxygen in nitrogen (air)
Liquid–gas	Water vapor in air (moist air)
Gas–liquid	Carbon dioxide in water (soda water)
Liquid–liquid	Acetic acid in water (vinegar)
Solid–liquid	Sodium chloride in water (brine)
Solid–solid	Copper in silver (sterling silver, an alloy)

Homogeneous mixtures are so important in chemistry that chemists give them the special name of **solutions.** As **Table 2.3** shows, solutions may be gases, liquids, or solids. If you were to take a sample from any portion of a solution of sugar in water, you would find that it has the same composition as any other portion. Any part of a system with uniform composition and properties is called a **phase.** Thus a homogeneous mixture consists of a single phase. A heterogeneous mixture consists of two or more phases. A mixture of vinegar and oil is an example of a heterogeneous mixture with two phases. When the mixture is left unshaken, the separate phases are visible; the oil phase floats on the water phase.

Separating Mixtures

Some mixtures can be separated into their components by simple physical methods. You might use a fork to separate taco filling into meat, lettuce, cheese, and tomatoes. But separating the grayish mixture of powdered yellow sulfur and black iron filings, shown in **Figure 2.4,** is not so simple. The individual particles of sulfur and iron can be readily distinguished from one another under a microscope, so the mixture is heterogeneous. What property of iron makes using a magnet an effective way to separate the mixture? 3

Tap water is a homogeneous mixture of water plus other substances that are dissolved in it. How would you separate the components in tap water? One method is called distillation. In **distillation,** a liquid is boiled to produce a vapor that is then condensed again to a liquid. **Figure 2.5** on page 34 shows an apparatus that can be used to perform a distillation. When water containing dissolved solids is distilled, it is first heated in a flask to form steam that enters a glass tube. The solid substances that originally dissolved in the water remain in the distillation flask because they do not change into a vapor. The steam cools and forms droplets of water inside the tube. The water drips into a receiver, where it is collected. The resulting distilled water is pure except for the dissolved gases it contains. Water from which even the dissolved gases are removed is a pure substance.

Figure 2.4
The mixture of iron filings and sulfur can be separated using a magnet.

2.2

2 Teach

Discuss

Discuss the criteria used to distinguish between heterogeneous and homogeneous mixtures. Write the names *homogenous* and *heterogeneous* on the board and the names of several types of mixtures on large index cards. Hold the cards up to the class one at a time, and ask the students to classify the mixtures as heterogeneous or homogeneous. Use masking tape to arrange the cards into the separate categories. Examples of mixtures include air, salt water, tea, brass, vinegar, hydrogen peroxide, steel, salad dressing, an apple, sand, paint, granite, laundry detergent, and cereal. Point out that most forms of matter that we encounter in our everyday lives are mixtures.

Discuss

Review key terms such as substance, homogeneous mixtures, and heterogeneous mixtures. Introduce the definition of a solution. Remind students that mixtures can usually be separated by simple physical methods.

TEACHER DEMO

Mix a small amount of sand together with salt, iron filings, and water. Ask:

- What are some of the differences in the physical properties of these substances that make it possible to separate them? (solubility in water, magnetic properties) Have students design a method to separate the mixture.

Answers

1 Each mouthful of salad will not necessarily contain the same amount of celery and tomatoes.
2 Answers will vary.
3 Iron is magnetic.

Meeting Diverse Needs

LEP Encourage LEP students to preview the chapter looking for words in bold print and other key terms. Have students compile a glossary in which they define each term in English and in their native language. Suggest that students also include illustrations when appropriate.

Use the Visual

Have students study Figure 2.5. Review the distillation process, and describe the components of the distillation apparatus. If possible, do a distillation using a colored solution such as aqueous potassium permanganate. Students can observe the separation of a clear liquid (H_2O) from a colored mixture.

Practice Problems Plus

Related Chapter Review Problem
Chapter Review problem 30 relates to Sample Problem 2-1.

Additional Practice Problem
The Chem ASAP! CD-ROM contains the following problem: Classify the following mixtures as homogeneous or heterogeneous.
a. granite rock **c.** paint
b. salt water **d.** a silver ring
(Homogeneous—**b** and **d**; Heterogeneous—**a** and **c**)

CHEMISTRY AND SCIENCE HISTORY

A working distillation apparatus was described in the writings of Maria of Alexandria, an alchemist who lived and worked nearly two thousand years ago. The city of Alexandria, located on the Nile River in North Africa, was a world center of science and culture at that time. Maria of Alexandria is also credited with inventing other chemical apparatus, such as the water bath, which to this day bears her name: the *bain marie.*

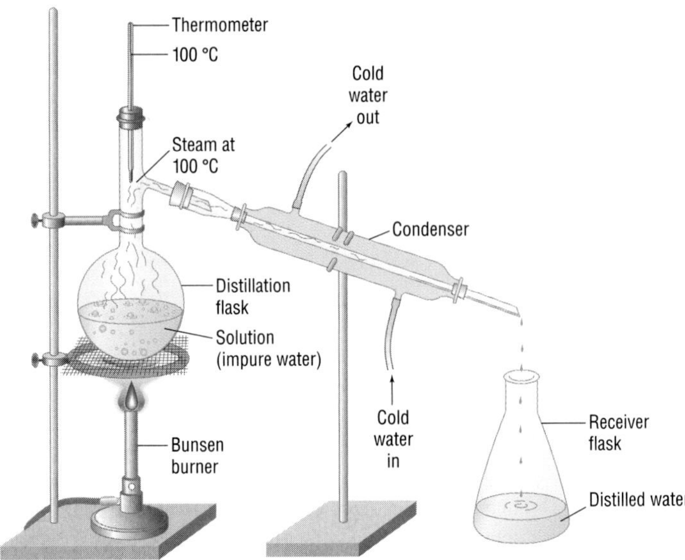

Figure 2.5
A solution of impure water is being distilled. As the solution boils, the water turns into steam, leaving the impurities behind in the distillation flask. As the steam passes through the water-cooled condenser, it turns to liquid. The distilled water is collected in the receiver flask.

Sample Problem 2-1

How can a mixture of iron filings and aluminum filings be separated?

1. ***ANALYZE*** ***Plan a problem-solving strategy.***
List the properties of iron and aluminum and look for something that would be useful in separating the mixture.

2. ***SOLVE*** ***Apply the problem-solving strategy.***

Iron:	*Aluminum:*
• metal	• metal
• grayish	• grayish
• not soluble in water	• not soluble in water
• attracted to a magnet	• not attracted to a magnet

Make use of a property that differentiates the metals; use a magnet to attract the iron filings.

3. ***EVALUATE*** ***Does the result make sense?***
Because the magnet attracts iron but not aluminum filings, the iron would be removed while the aluminum would be left behind.

Practice Problems

5. What physical properties could be used to separate iron filings from salt?

6. Which of the following are homogeneous? Heterogeneous?
a. spaghetti sauce
b. glass
c. muddy water
d. cough syrup
e. mixture of nitrogen gas and helium gas

Chem ASAP!
Problem-Solving 6

Solve Problem 6 with the help of an interactive guided tutorial.

Meeting Diverse Needs

Gifted Panning is one method for separating gold from sand. Have students do research on the method that uses mercury to separate gold from sand. Write a report explaining the physical properties used in the separation. Be sure to have them include the hazards of the method.

Answers

SECTION REVIEW 2.2

7. Heterogeneous mixtures have a non-uniform composition consisting of two or more phases. Homogeneous mixtures have a uniform composition throughout the sample.

8. Add water to dissolve the salt. Pour the resulting mixture onto a piece of closely-woven cloth. The sand will remain on the cloth, and

MINI LAB

MIXTURES

PURPOSE

To separate a mixture using paper chromatography.

MATERIALS

- green marking pens
- strips of filter paper
- metric ruler
- clear plastic tape
- pencil
- rubbing alcohol
- clear plastic drinking cups
- clear plastic wrap
- water

PROCEDURE

1. Use the marking pen to draw a horizontal line across the width of a strip of filter paper, 2 cm from one end of the strip.
2. Tape the unmarked end of the filter paper to the center of a pencil so that the strip hangs down when the pencil is held horizontally.
3. Working in a well-ventilated room, pour rubbing alcohol into a plastic cup to a depth of 1 cm.
4. Rest the pencil on the rim of the cup so that the end of the paper strip with the ink mark is just barely in contact with the rubbing alcohol. Carefully cover the top of the cup with plastic wrap.
5. Observe for 15 minutes.
6. If time permits, repeat this lab using different brands and different colors of pens. Also, try using water in place of rubbing alcohol.

ANALYSIS AND CONCLUSIONS

1. What appeared on the filter paper?
2. What did the results indicate about the nature of the green ink? Is the ink a mixture?

section review 2.2

7. What is the difference between a heterogeneous and a homogeneous mixture?
8. Describe a procedure that could be used to separate a mixture consisting of sand and salt.
9. Classify each of the following as a substance or a mixture.
   a. silver
   b. alphabet soup
   c. textbook
   d. table salt (sodium chloride)
10. Describe in your own words the difference between a pure substance and a mixture.
11. Describe ways in which the various components of a mixture can be separated.
12. Explain the term phase as it relates to homogeneous and heterogeneous mixtures.

 Chem ASAP! **Assessment 2.2** Check your understanding of the important ideas and concepts in Section 2.2.

MINI LAB

Mixtures

SAFETY

Rubbing alcohol is poisonous and flammable. It is also an irritant when inhaled. Keep containers covered and away from heat.

ANALYSIS AND CONCLUSIONS

1. Bands of other colors.
2. Green ink is a mixture of colors.

3 Assess

Evaluate Understanding

Have students identify five items that fit each of the following categories:
a. pure substance
b. homogeneous mixture
c. heterogeneous mixture
d. solution
Have students select three items (one each from categories b, c, and d). For each item selected, students should outline a method for separating the components.

Reteach

Set up a lab practical and assign lab partners. At numbered lab stations, place common examples of elements, mixtures, and compounds. Have students classify each sample as a substance, homogeneous mixture, or heterogeneous mixture.

the salt solution will pass through. Use evaporation to remove the water from the solution, leaving solid salt behind.

9. a. substance
   b. mixture
   c. mixture
   d. substance
10. A pure substance contains only one kind of matter. A mixture contains two or more kinds of matter that may or may not be uniform in composition.
11. The components of a mixture can be separated by physical means such as filtration or distillation.
12. A phase is any part of a system with uniform composition. A homogenous mixture consists of one phase; a heterogeneous mixture consists of two or more phases.

2.3

1 Engage

Use the Visual

Have students study the photograph and read the text that opens the section. Ask:

- **Why does a dessert taste sweet if the ingredients that make up sugar don't taste sweet?** (Elements combine to form compounds that have different properties. C, H, and O do not taste sweet, but the compound $C_{12}H_{22}O_{11}$ does.)
- **Distinguish between an element and a compound.** (Elements are the simplest forms of matter that can exist under normal laboratory conditions. Compounds are substances that can be separated into simpler substances by chemical means.)

TEACHER DEMO

Perform the following demonstration in a fume hood or in a well-ventilated area and be sure to wear safety goggles. Place about 25 g of powdered sugar in a 100 mL beaker. Carefully add about 10–15 mL of 18*M* sulfuric acid and stir rapidly with a glass rod. The mixture darkens and a column of carbon rises out of the beaker. Point out that the compound sugar has been separated into an element (carbon) and a compound (water). Explain that this is a chemical change. **CAUTION!** This is an extremely exothermic reaction. Place the cool beaker and its contents in a large beaker and add water. Neutralize with sodium hydroxide and pour the liquid down the drain, flushing with excess water. Dispose of the solid in the trash.

section 2.3

objectives

- Explain the difference between an element and a compound
- Identify the chemical symbols of common elements, and name common elements, given their symbols

key terms

- elements
- compounds
- chemical symbol

ELEMENTS AND COMPOUNDS

Most people like sugar, a compound made of the elements carbon, hydrogen, and oxygen. Pure carbon is a black solid substance, pure hydrogen is a flammable gas, and pure oxygen is a gas that supports combustion. Yet these three elements combine in a particular way to form the white, sweet-tasting compound called sugar. **What is the difference between an element and a compound?**

Distinguishing Elements and Compounds

As you learned in the preceding section, a mixture can be physically separated into its components. Such a separation may yield pure substances, which have uniform and definite compositions. Substances themselves can be classified into two groups: elements and compounds. **Elements** are the simplest forms of matter that can exist under normal laboratory conditions. Elements cannot be separated into simpler substances by chemical means. They are the building blocks for all other substances. Oxygen, hydrogen, and carbon are examples of elements.

Two or more elements can combine chemically with one another to form compounds. For example, oxygen, hydrogen, and carbon can combine to produce the compound sucrose, or common table sugar. **Compounds** are substances that can be separated into simpler substances only by chemical means. There are a variety of chemical processes that can be used to separate compounds into simpler substances. Heating, as the photographs in **Figure 2.6** show, is one of these processes.

Heating a thin layer of sugar in a skillet demonstrates the difference between elements and compounds. With gentle heating, the sugar turns light brown. With continued heating, it turns black, breaking down completely into carbon and water vapor. This experiment shows that sugar is a compound, not an element. The chemical changes caused by strong heating break down the sugar into two different substances: carbon and water. But the question remains: Can the water and the carbon that are produced also be broken down, or are they elements? It turns out that water can be broken down into hydrogen and oxygen by another chemical change. Thus water, like sugar, is a compound. However, the carbon that is

(a)

(b)

(c)

Figure 2.6
*To be separated into their component elements, compounds must undergo a chemical change. Here sugar is heated in a process known as caramelization. The intermediate and final products in photos **(b)** and **(c)** look very different from the sugar in photo **(a)**. As water is completely removed, all that remains is charred, hardened carbon.*

STUDENT RESOURCES

From the Teacher's Resource Package, use:

- Section Review 2.3, Ch. 2 Practice Problems, Interpreting Graphics, and Quizzes from the Review Module (Ch 1–4)
- Laboratory Recordsheet 2–2
- Small-Scale Chemistry Lab Manual: Experiment 1

TECHNOLOGY RESOURCES

Relevant technology resources include:

- Chem ASAP! CD-ROM
- Resource Pro CD-ROM
- Chemistry Alive! Videodisc: *Making Sodium Chloride*

produced by the heating process cannot be broken down into simpler substances. This shows that carbon is an element, not a compound. The following diagram illustrates the overall process just described.

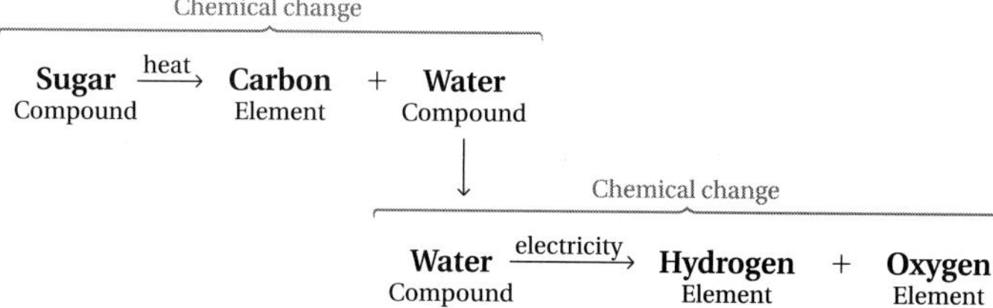

In general, the properties of compounds are quite different from those of their component elements. For example, the sugar placed in the skillet was a sweet, white solid, but the carbon that remained was a black, tasteless solid. Water is a colorless liquid, but oxygen and hydrogen are colorless gases. Table salt (sodium chloride) is a compound of the elements sodium and chlorine. Sodium is a soft metal that reacts explosively with water. Chlorine is a pale yellow-green poisonous gas. **Figure 2.7** shows how the physical appearances of sodium, chlorine, and sodium chloride differ.

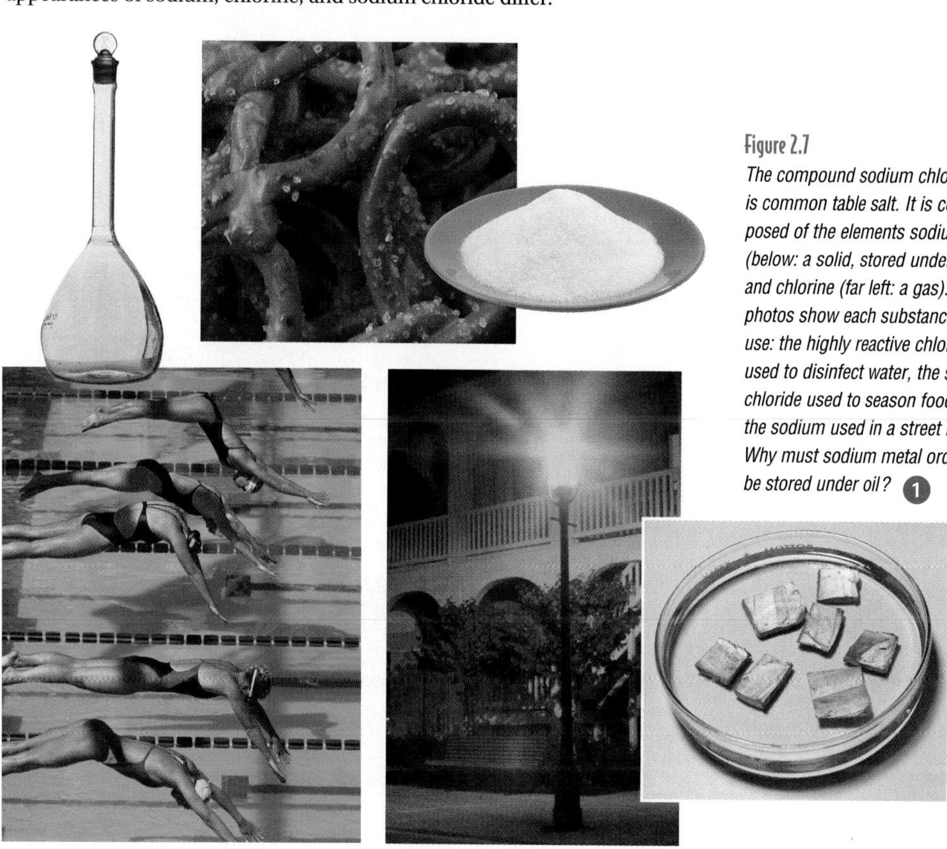

Figure 2.7

The compound sodium chloride is common table salt. It is composed of the elements sodium (below: a solid, stored under oil) and chlorine (far left: a gas). The photos show each substance in use: the highly reactive chlorine used to disinfect water, the sodium chloride used to season food, and the sodium used in a street lamp. Why must sodium metal ordinarily be stored under oil? 1

2 Teach

TEACHER DEMO

Use a Hoffman apparatus to separate water into hydrogen and oxygen. Ask:

- Why is there nearly twice as much of one gas compared to the other? (2 parts H for each O, H_2O)
- What are some properties of each gas that enable you to determine which gas is hydrogen and which is oxygen? (Oxygen supports combustion and hydrogen explodes.) Be sure to collect some of each gas and test for oxygen with a glowing splint and for hydrogen with a match.

Discuss

In the two previous demonstrations, a chemical compound was broken down into simpler substances: sugar into carbon and water, and water into hydrogen and oxygen. Explain to students that they have witnessed chemical changes. The resulting materials have different chemical and physical properties from the original substances.

Answer

1 Sodium reacts vigorously with oxygen and water.

Meeting Diverse Needs

Gifted Other forms of energy besides electricity can be used to bring about the decomposition of water into hydrogen and oxygen. Ask students to consider whether heat could be used for this purpose. Have students do research on the conditions necessary to decompose water using heat. What hazards, if any, might preclude the routine use of this method?

Chemistry Alive!

Making Sodium Chloride

Play

ACTIVITY

To establish the definition of "substance", use an inductive approach. Make signs with the names of each of the 20 items listed below. 1 oxygen, 2 neon, 3 apple, 4 sand, 5 iron, 6 water, 7 air, 8 paint, 9 sodium chloride, 10 sucrose, 11 carbon dioxide, 12 granite, 13 laundry detergent, 14 citric acid, 15 cereal, 16 salad, 17 salad dressing, 18 copper, 19 salt water, 20 gold.
Using masking tape and the chalkboard or wall, create two columns. Fix items 1 and 2 under column A and items 3 and 4 under column B. Ask the students to think about the criteria for each column and then ask:

- Where would you place iron? (Put item 5 in column A.) Students may have to change their criteria. Ask:
- Where does water go? (Put item 6 in column A.) Ask the students what they think the criteria is now. Continue, in this fashion, putting one item up at a time until students understand the criteria. Everything in column A has a uniform and definite composition. Column A is for substances. Column A : 1, 2, 5, 6, 9, 10, 11, 14, 18, 20. Column B is for mixtures. Column B: 3, 4, 7, 8, 12, 13, 15, 16, 17, 19.

Deciding whether a sample of matter is a substance or a mixture can sometimes be difficult. After all, a homogeneous mixture looks like a substance. In some cases, you can decide by considering whether the material in question is always a single kind of material. For example, how would you classify gasoline? Based on its physical appearance, you might conclude that gasoline is a pure substance. However, it must be a mixture, because it exists in many different grades. Gasoline can have many different octane ratings and it may or may not contain alcohol. **Figure 2.8** summarizes some information about elements, compounds, and mixtures.

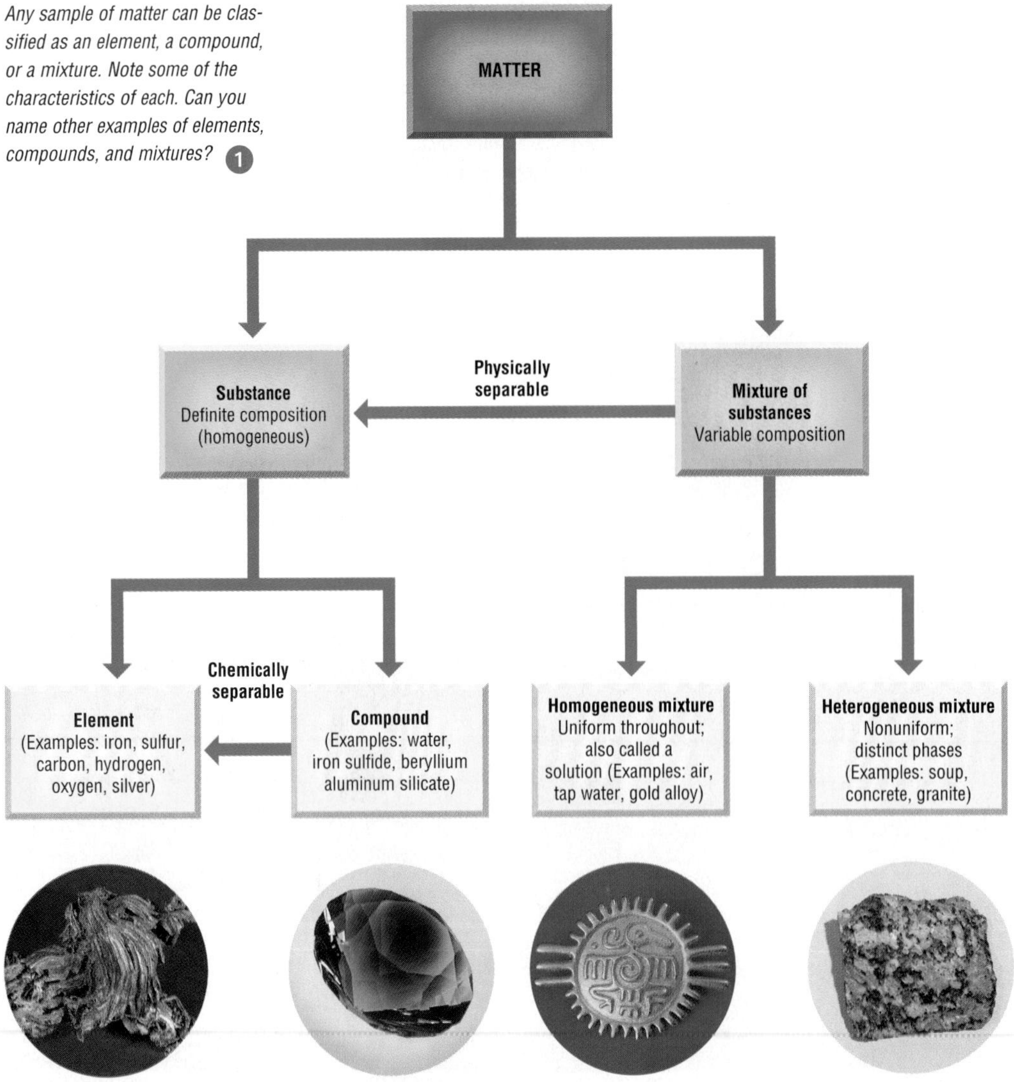

Figure 2.8
Any sample of matter can be classified as an element, a compound, or a mixture. Note some of the characteristics of each. Can you name other examples of elements, compounds, and mixtures? (1)

Meeting Diverse Needs

LEP and At Risk Have students make flash cards to help learn the names and symbols of at least 30 commonly used elements in chemistry.

Answers

(1) Answers will vary. Elements include helium, neon, and potassium. Compounds include sodium chloride and hydrogen peroxide. Mixtures include seawater, cereal, and granite.
(2) Yb

Sample Problem 2-2

When a blue solid is heated in the absence of air, two other substances—a colorless gas and a white solid—are formed. Which of these substances are elements and which are compounds? Is it possible to tell? Explain.

1. ***ANALYZE*** ***Plan a problem-solving strategy.***
   List the known facts and determine if there is enough information to identify the substances by type (element or compound).
   - A blue solid is changed into two substances (a colorless gas and a white solid) when heated in the absence of air.
   - Compounds can be chemically broken down to simpler substances, but elements cannot be.
   - Heating can cause a chemical change.
2. ***SOLVE*** ***Apply the problem-solving strategy.***
   The blue solid was separated into two different substances by heating. Therefore, it must be a compound. The two resulting substances may be either elements or compounds—it is impossible to tell based on the information given.
3. ***EVALUATE*** ***Does the result make sense?***
   Given the limited amount of known information, the conclusions reached are reasonable.

Practice Problem

13. A clear liquid in an open container is allowed to evaporate. After three days, a solid residue is left. Was the original liquid an element, a compound, or a mixture? How do you know?

Chem ASAP!

Problem-Solving 13
Solve Problem 13 with the help of an interactive guided tutorial.

Symbols and Formulas

Carbon, hydrogen, oxygen, sodium, and chlorine are only a few of the more than 100 known elements. All matter in the universe is composed of these elements. Each element is represented by a one- or two-letter **chemical symbol.** The symbols for most elements consist of the first one or two letters of the element's name. **Table A.1** in the appendix gives the names and symbols for all of the elements. What is the symbol for the element ytterbium? ❷

Note that the first letter of a chemical symbol is always capitalized. If a second letter is used, it is lowercase. Some chemical symbols are derived from Latin or Greek names for the element. In those cases, the symbol does not resemble the common name. **Table 2.4** lists such cases.

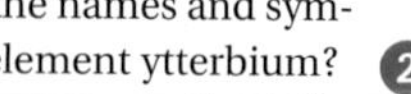

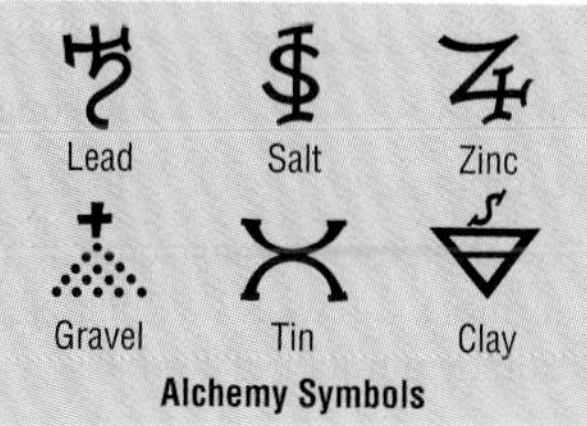

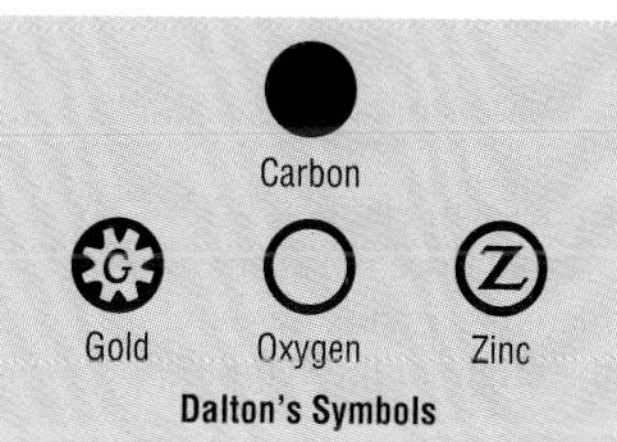

Figure 2.9
Many different symbols have been developed over time to represent chemicals, processes, and phenomena. This figure shows symbols used by the important eighteenth-century English chemist John Dalton, as well as some Chinese and alchemist symbols.

Practice Problems Plus

Related Chapter Review Problem
Chapter Review problem 32 relates to Sample Problem 2-2.

Additional Practice Problem
The Chem ASAP! CD-ROM contains the following problem: Classify the following as an element, compound, or mixture. Give reasons for your answers.
a. salt (NaCl)
b. saltwater
c. sodium (Na)
(**a.** compound, **b.** mixture, **c.** element)

Discuss

Point out that the use of chemical symbols is an example of the "international language of chemistry." Discuss why people all over the world use the same set of chemical symbols. Show how symbols are used in chemical formulas. Emphasize that a compound is always made up of the same elements in the same proportions. Thus, every compound has its own unique formula.

TEACHER DEMO

Set up a display of elements in the classroom to help students relate abstract chemical symbols to real substances. Students can refer to the display when they present oral reports about the physical properties of elements.

Meeting Diverse Needs

LEP Help students to create a list of the most commonly used elements in the chemistry class. Ask them to compare these names and symbols with the names and symbols for the same elements in their native countries.

3 Assess

Evaluate Understanding

Ask students to explain the difference between an element and a compound.

Reteach

Explain to students that most elements are found combined in compounds in nature. The rare exceptions include gold, silver, copper, and sulfur. Bring samples of these elements to show the class. Have students identify at least five items found at home that are either elements or compounds. Challenge the students to give the chemical symbol or chemical formula for each item. Have the students work in groups of four. Each group should compile a single list of the "best" items and prepare a report for the class.

Portfolio Project

This project allows students to explore the history of science. You could extend the project and have students work as a group to make a timeline for the discovery of elements. Students could use the timeline to detect shifts in naming conventions.

LINGUISTICS

Origins of Element Names
Many elements are named for the people who discovered them or the places where they were discovered. Some elements were given descriptive names taken from classical Latin or Greek. Others were named for figures in mythology. Polonium is named for Poland, the native land of Marie Curie, the discoverer of radium. Californium was discovered at the University of California. The word chlorine comes from the Greek *chloros,* meaning greenish-yellow; chlorine is a greenish-yellow gas. The name calcium is derived from the Latin *calx,* meaning lime. Calcium is a major component of limestone. You can find the origins of elements' names by consulting a dictionary or encyclopedia.

Table 2.4

Symbols and Name Origins for Some Elements

Name	Symbol	Latin or other name
Sodium	Na	*natrium*
Potassium	K	*kalium*
Antimony	Sb	*stibium*
Copper	Cu	*cuprum*
Gold	Au	*aurum*
Silver	Ag	*argentum*
Iron	Fe	*ferrum*
Lead	Pb	*plumbum*
Mercury	Hg	*hydrargyrum* (from Greek)
Tin	Sn	*stannum*
Tungsten	W	*wolfram* (from German)

Chemical symbols provide a shorthand way to write the chemical formulas of compounds. The compound water is composed of the elements hydrogen (H) and oxygen (O). The formula for water is H_2O. The formula for sucrose, or table sugar, is $C_{12}H_{22}O_{11}$. Sucrose is composed of the elements carbon (C), hydrogen (H), and oxygen (O). The subscript numbers in chemical formulas represent the proportions of the various elements in the compound. The elements that make up a compound are always present in the same proportions. Thus, in the case of water, there are always two parts of hydrogen for each part of oxygen. A specific compound is always made up of the same elements in the same proportions. Thus, the formula for a specific chemical compound is always the same.

portfolio project

Choose an element whose symbol does not seem to match its name. Research the origin of the symbol.

section review 2.3

14. How can you distinguish between an element and a compound?

15. Write the chemical symbols for each of the following elements.

a. copper **b.** oxygen **c.** phosphorus **d.** silver **e.** sodium **f.** helium

16. Name the chemical elements represented by the following symbols.

a. Sn **b.** Ca **c.** S **d.** Cd **e.** P **f.** Cl

17. Classify each of these samples of matter as an element, a compound, or a mixture.

a. spaghetti sauce **b.** glass **c.** table sugar **d.** river water **e.** cough syrup **f.** nitrogen

18. What elements make up the pain reliever acetaminophen, chemical formula $C_8H_9O_2N$? Which element is present in the greatest proportion by number of atoms?

Chem ASAP! **Assessment 2.3** Check your understanding of the important ideas and concepts in Section 2.3.

Answers

SECTION REVIEW 2.3

14. Compounds can be separated by chemical means into elements. Elements cannot be separated into simpler substances by chemical techniques.

15. a. Cu **b.** O **c.** P **d.** Ag **e.** Na **f.** He

16. a. tin **b.** calcium **c.** sulfur **d.** cadmium **e.** phosphorus **f.** chlorine

17. a. mixture **b.** mixture **c.** compound **d.** mixture **e.** mixture **f.** element

18. carbon, hydrogen, oxygen, and nitrogen; Hydrogen is present in the greatest proportion by number of atoms.

CHEMICAL REACTIONS

section 2.4

Over a period of time, objects made out of iron will rust if they are left exposed to air. Rust is the product of a chemical reaction involving the reactants iron and oxygen. **What is a chemical reaction, and what are reactants and products?**

objectives

- Differentiate between physical and chemical changes in matter
- Apply the law of conservation of mass

key terms

- chemical reaction
- reactants
- products
- chemical property
- law of conservation of mass

Changing Reactants to Products

Just as every substance has physical properties, each also has properties that relate to the kinds of chemical changes it can undergo. For example, iron has the property of being able to combine with oxygen to form rust. Such a change is an example of a chemical reaction. In a **chemical reaction,** one or more substances change into new substances. The original substances iron and oxygen combine to form a new substance, iron oxide, or rust. In chemical reactions, the starting substances are called **reactants,** and the substances formed are called **products.** What are the reactants in the reaction just described? What is the product? 1

The ability of a substance to undergo a chemical reaction and to form new substances is called a **chemical property.** Rusting is a chemical property of iron. Chemical properties are observed only when a substance undergoes a chemical change. A chemical change always results in a change in chemical composition of the substances involved. Words such as burn, rot, rust, decompose, ferment, explode, and corrode usually signify a chemical change.

To help distinguish physical changes from chemical changes, recall the mixture of sulfur and iron filings discussed earlier. The separation of these substances by means of a magnet was an example of a physical change. If the same mixture is heated, however, a chemical reaction takes place. The sulfur and iron change into a new substance, iron sulfide. This change can be written in shorthand form as follows:

$$\text{iron} + \text{sulfur} \xrightarrow{\text{heat}} \text{iron sulfide}$$

Figure 2.10

The fizz of an antacid tablet dropped into a glass of water, the brilliance of fireworks exploding in the night sky, and the magnificent colors of autumn leaves are all examples of the chemical reactions you can see all around you.

STUDENT RESOURCES

- Section Review 2.4, Ch. 2 Practice Problems, Vocabulary Review, Quizzes, and Tests from the Review Module (Ch 1–4)
- Laboratory Manual: Experiments 2 and 3
- Laboratory Practical 2–1
- Laboratory Recordsheet 2–2
- Small-Scale Lab Manual: Experiment 1

TECHNOLOGY RESOURCES

Relevant technology resources include:

- Chem ASAP! CD-ROM
- Resource Pro CD-ROM
- ActivChemistry CD-ROM: *Matter and Change*
- Chemistry Alive! Videodisc: *Thermite Reaction*
- Assessment Resources CD-ROM: Chapter 2 Tests

1 iron and oxygen; iron oxide, rust

2.4

1 Engage

Use the Visual

Have students study the photograph and read the text that opens the section. Explain that some chemical reactions occur rapidly while others occur slowly. Ask:

- **What evidence do you see in this picture that a chemical reaction has occurred?** (The metal has changed color.)
- **What is a chemical reaction, and what are reactants and products?** (In a chemical reaction, one or more substances change into new substances. Reactants are the starting substances and products are the new substances that are formed.)

TEACHER DEMO

Demonstrate that mixtures can be separated into simpler components by physical means, but pure substances cannot. Explain that chemical reactions are needed to separate a compound into its elements or simpler compounds. You will need 5 g of iron filings, 5 g of sulfur, a glass plate, a spatula, a magnet, a test tube, a beaker, water, a gas burner, matches, and safety goggles. Mix the iron and sulfur on the plate. Use the magnet to separate out the iron. Place iron filings from the magnet and sulfur from the pile into the test tube. Heat until the reaction is complete. **CAUTION!** Use a fume hood. Put the hot test tube in a beaker of cold water to shatter the tube. Remove the reaction product from the water. Using the magnet, try to separate the iron and test for magnetism. Dispose of the broken glass and iron sulfide in the school's broken glass container.

2 Teach

TEACHER DEMO

Have students practice identifying chemical changes. Perform each of the following.

a. Light a Bunsen burner. (heat and light produced)
b. Using tongs, hold a piece of magnesium ribbon in a burner flame until it ignites. Remove from heat and observe. (This product is a white powder; heat and light are produced.) **CAUTION!** Students should not look directly at burning magnesium without cobalt blue glass filters!
c. Put 5 mL 0.1 *M* $AgNO_3$ (silver nitrate) in each of two test tubes. Add a small amount of 0.1 *M* NaCl (sodium chloride) to one tube and 0.1 *M* K_2CrO_4 (potassium chromate) to the other. (color change, formation of a precipitate) **CAUTION!** Silver nitrate stains skin.
d. Put 5 mL of 3 *M* H_2SO_4 (sulfuric acid) in each of two test tubes. Add a piece of mossy zinc to one tube and a marble chip ($CaCO_3$) in the other. (formation of a gas)

Use the Visual

Have students study Figure 2.12. Ask them to explain what occurs in the flash bulb. (Magnesium burns.) **Ask:**

- **What evidence is there that a chemical reaction occurred?** (light)

Have them write the shorthand equation for the reaction. (magnesium and oxygen → magnesium oxide + energy)

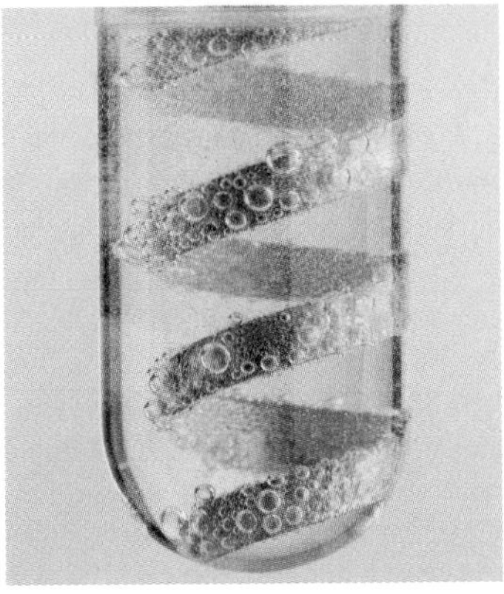

(a) (b)

Figure 2.11
*The formation of a gas or a solid from a liquid or the production of a color change are common indications that a chemical change may be taking place. **(a)** Zinc metal reacts with sulfuric acid solution to release hydrogen gas. **(b)** A red solid of silver chromate forms when a yellow solution of sodium chromate is added to a colorless solution of silver nitrate.*

LINK TO ENGINEERING

Chemical Engineers
Chemical engineers are employed by many industries, including companies that produce fuels, metals, plastics, chemicals, cosmetics, drugs, rubber, paper, paints, and foods. Chemical engineers must determine whether a reaction can be done in large enough amounts for mass production. They plan the layout of an industrial plant, design or select the equipment, and supervise the plant's construction and operation. They may add safety or pollution control features to comply with federal and state regulations. Chemical engineers are responsible for ensuring that the plant operates efficiently. They must also be aware of production costs and stay within a budget. Many engineering and technical schools offer degrees in chemical engineering. The course of study emphasizes chemistry, physics, mathematics, economics, writing, and computers. Specialized engineering courses are also required.

The arrow in the formula stands for the words "change into" or "produce." The reactants are written to the left of the arrow, the products to the right. How would you represent, in shorthand form, the reaction in ❶ which hydrogen and oxygen produce water? Several common chemical reactions are illustrated in **Figure 2.10** on page 41. You are probably familiar with many other chemical reactions.

How can you tell whether a chemical reaction has taken place? In general, there are several clues that may serve as a guide. The first is that energy is always absorbed or given off in chemical reactions. When you cook food, chemical changes that involve an absorption of heat take place. When you burn natural gas on a stove (another chemical reaction), heat is given off. However, energy is also absorbed or given off when the physical state of matter changes. Therefore, an energy change in itself is not proof of a chemical change. Other clues include a change in color or odor or the production of a gas or a solid from a liquid. See **Figure 2.11** for examples of chemical reactions that produce such changes. Once again, however, such clues are not always proof of a chemical change. For example, gas or vapor formation can be the result of a change of physical state and not a chemical change. When water boils, it changes from a liquid to a vapor, but the change is physical only; its chemical composition remains the same. A final indicator of chemical change is irreversibility. Physical changes, especially those involving a change of state, are usually reversible. Water can be melted and then refrozen. In contrast, most chemical changes are not easily reversed. For example, once iron has reacted with oxygen to form rust (iron oxide), as happens on a car, you cannot easily reverse the process and turn the rust back into iron.

Answers

❶ hydrogen + oxygen → water
❷ Mass is the same before and after the reaction. Mass is conserved during chemical reactions.

Answers

SECTION REVIEW 2.4

19. a. In a chemical change, the chemical composition of the reactants changes as one or more different products is formed. In a physical change, the chemical composition of the substance remains the same even if its physical appearance changes. Indicators are (1) a change in color or odor, or production of a gas; (2) energy released or absorbed;

Conservation of Mass

Combustion, or burning, is an example of one of the most familiar chemical changes. When you burn a lump of coal, atmospheric oxygen combines with the coal. The reaction produces carbon dioxide gas, water vapor, and a small ash residue. This change seems to involve a reduction in the amount of matter. A sizable piece of matter seems to have produced only a trace of ash. However, careful measurements show that the total mass of the reactants (the coal and the oxygen consumed) equals the total mass of the products (the carbon dioxide, water vapor, and ash) when the gases involved are taken into account.

During any chemical reaction, the quantity of matter is unchanged. The mass of products is always equal to the mass of reactants. Constancy of mass also holds for physical changes. For example, when 10 grams of ice melt, 10 grams of liquid water are obtained. Again, in this physical process, mass remains the same. Similar observations have been recorded for all chemical and physical changes studied. The **law of conservation of mass** is reflected in these observations and states that in any physical change or chemical reaction, mass is neither created nor destroyed; it is conserved. In every case, the mass of the products equals the mass of the reactants. One example of the conservation of mass is shown in **Figure 2.12.**

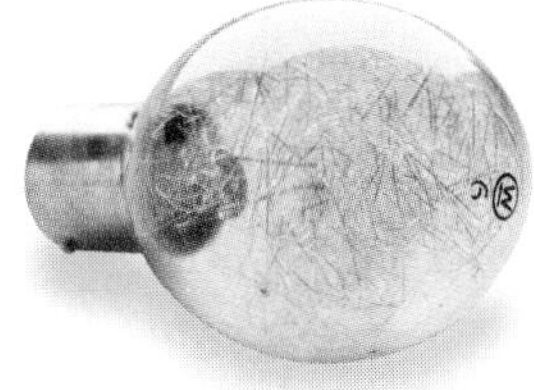

(a) 

(b) 

Figure 2.12
*Magnesium wire inside an old-fashioned flashbulb burns in oxygen gas to produce magnesium oxide. How do the mass readings for the unused bulb **(a)** and the used bulb **(b)** compare? What does this show about mass during the chemical reaction?* 2

section review 2.4

19. **a.** State the difference between a physical change and a chemical change, and list three likely indications that a chemical change has occurred. Which indication is most suggestive of a chemical reaction?
    b. State the law of conservation of mass. How does the mass of reactants compare with the mass of products in a given reaction?
20. Classify the following changes as physical or chemical.
    a. Cookies are baked. **d.** A firefly emits light.
    b. Water boils. **e.** Milk spoils.
    c. Salt dissolves in water. **f.** A metal chair rusts.
21. Consider the law of conservation of mass as you answer this problem. When ammonium nitrate (NH_4NO_3) breaks down explosively, it forms nitrogen gas (N_2), oxygen gas (O_2), and water (H_2O). When 40 grams of ammonium nitrate explode, 14 grams of nitrogen and 8 grams of oxygen are formed. How many grams of water are formed?
22. State several physical or chemical properties that could be used to distinguish between each of the following pairs of substances and mixtures.
    a. gasoline and water **c.** water and a saltwater solution
    b. copper and silver **d.** aluminum and steel
23. Hydrogen and oxygen react chemically to form water. How much water would be formed if 4.8 grams of hydrogen reacted with 38.4 grams of oxygen?

 Chem ASAP! **Assessment 2.4** Check your understanding of the important ideas and concepts in Section 2.4.

2.4

3 Assess

Evaluate Understanding

To determine students' knowledge about chemical reactions, ask the students to explain the terms "reactants" and "products." After completing the four demonstrations in the preceding Teacher Demo, ask students to write the shorthand equations for the reactions performed.

a. methane + oxygen → carbon dioxide + water + energy
b. magnesium + oxygen → magnesium oxide + energy
c. sodium chloride + silver nitrate → sodium nitrate + silver chloride
potassium chromate + silver nitrate → potassium nitrate + silver chromate
d. sulfuric acid + zinc → hydrogen gas + zinc sulfate
sulfuric acid + calcium carbonate → calcium sulfate + carbon dioxide + water

Reteach

Use the example of magnesium reacting with oxygen to produce magnesium oxide to remind students that mass is conserved in a physical or chemical change. The difference between the mass of the magnesium and the mass of the magnesium oxide is the mass of the oxygen with which magnesium combines.

SECTION REVIEW 2.4 CONTINUED

(3) irreversibility. (3) is the most reliable indicator of chemical change.
b. In any physical change or chemical reaction, mass is neither created nor destroyed; it is conserved. The mass of the products equals the mass of the reactants in a chemical reaction.

20. **a.** chemical **d.** chemical
    b. physical **e.** chemical
    c. physical **f.** chemical
21. 18 g
22. **a.** color, odor, reaction upon heating, boiling point
    b. color, melting point, reactions with other substances, hardness, brittleness, strength
    c. boiling point, freezing point, density
    d. density, melting point, magnetism
23. 43.2 g

 Chemistry Alive!

Thermite Reaction

Play

SMALL-SCALE LAB

1 + 2 + 3 = BLACK

OVERVIEW

Procedure Time: 30 minutes
This lab allows students with little knowledge of chemistry to familiarize themselves with the small-scale equipment and methodology.

MATERIALS

Solution	Preparation
0.1*M* KI	4.2 g in 250 mL
0.2*M* $CuSO_4$	12.5 g $CuSO_4 \cdot 5H_2O$ in 250 mL
1% NaOCl	50 mL household bleach in 200 mL
3% H_2O_2	Use undiluted household supply.
20% starch	50 mL liquid starch in 200 mL
0.1*M* $FeCl_3$	6.8 g $FeCl_3 \cdot 6H_2O$ in 25 mL of 1.0 *M* NaCl; dilute to 250 mL
0.1*M* KIO_3	5.35 g in 250 mL
0.02*M* $KMnO_4$	0.8 g in 250 mL
0.4*M* $NaNO_2$	6.9 g in 250 mL

Household products: Samples of cereal, laundry detergents, iodized and non-iodized salt, paper, dish-washing liquid, and cleansers.

TEACHING SUGGESTIONS

Explain that using a mixture of KI and NaOCl to test for starch is an example of qualitative analysis. Ask students to bring in small samples of cereal and laundry detergents.

ANALYSIS

1. yellow
2. The mixture turns a blue-black color.
3. Yes; It turns KI black in the presence of starch.
4. The paper turns blue-black, which suggests the presence of starch.
5. The black color indicates that cereal contains starch.

SMALL-SCALE LAB

1 + 2 + 3 = BLACK!

SAFETY

Wear your safety glasses and follow standard safety procedures as outlined on page 18.

PURPOSE

To make macroscopic observations of chemical reactions and use them to solve problems.

MATERIALS

- pencil
- ruler
- chemicals shown in Figure A
- paper
- reaction surface
- pipette

PROCEDURE

Draw two grids similar to Figure A on separate sheets of paper. Make each square 2 cm on each side. Place a reaction surface over one of the grids and add one drop, one piece, or a few grains of each chemical, as shown in Figure A. Stir by blowing air through an empty pipette. Use the second grid as a data table to record your observations for each solution.

	NaOCl	H_2O_2	$CuSO_4$
KI	**yellow**	**yellow**	**brown ppt**
KI + Starch	**black**	**black**	**black**
KI + Paper	**black**	**black**	**black**
KI + Cereal	**black**	**black**	**black**

Figure A

ANALYSIS

Using your experimental data, record the answers to the following questions below your data table.

1. What color is a mixture of NaOCl and KI?
2. What happens when you mix NaOCl, KI, and starch?
3. NaOCl is a powerful bleaching agent as indicated by its reaction with KI and starch. Is hydrogen peroxide (H_2O_2) a bleaching agent? Explain.
4. What happens when you add NaOCl and KI to paper? What ingredient does this suggest the paper contains?
5. What ingredient is contained in cereal? How do you know?

YOU'RE THE CHEMIST!

The following small-scale activities allow you to develop your own procedures and analyze the results.

1. **Design It!** Design and carry out an experiment to see which foods contain starch.
2. **Design It!** Read the label on a package of iodized salt. How much KI does iodized salt contain? Design an experiment to prove the presence of KI in iodized salt and the absence of KI in regular salt.
3. **Design It!** Laundry detergents, automatic dish washing liquids, and cleansers sometimes contain a bleaching agent similar to NaOCl. The purpose of the bleach is to whiten clothes and remove stains. Design an experiment to decide which laundry detergents and cleansers contain bleach.
4. **Design It!** Antacid tablets and other pharmaceuticals often contain starch as a binder to hold the ingredients in the tablet together. Design and carry out an experiment to explore various antacid tablets to see if they contain starch.
5. **Analyze It!** Mix one drop of NaOCl on a piece of colored construction paper. What happens? Try inventing a technique that will create some original "bleach art."
6. **Analyze It!** Other bleaching agents include $FeCl_3$, KIO_3, $KMnO_4$, and $NaNO_2$. Try mixing these with KI and starch to see what happens. What is the best agent to detect the presence of KI in table salt?

YOU'RE THE CHEMIST

1. Add KI + NaOCl to various foods. A black color indicates the presence of starch.
2. Most table salt contains 0.01% KI. Wet only a portion of a small pile of salt with starch. Add $CuSO_4$ or H_2O. A black color indicates the presence of KI.
3. Add KI and starch to the sample being tested. A black color indicates the presence of a bleaching agent.
4. Add KI and starch to antacid tablets. A black color indicates the presence of starch. Neither baking soda nor baking powder contains starch.
5. NaOCl decolorizes the construction paper. Try adding HCl for a faster reaction.
6. $KMnO_4$ and $FeCl_3$ are more powerful bleaching (oxidizing) agents than KIO_3 and $NaNO_2$, which require the addition of dilute HCl (1*M*) before they turn starch and KI black.

Chemistry Serving...Industry

BARRIERS TO HEAT FLOW

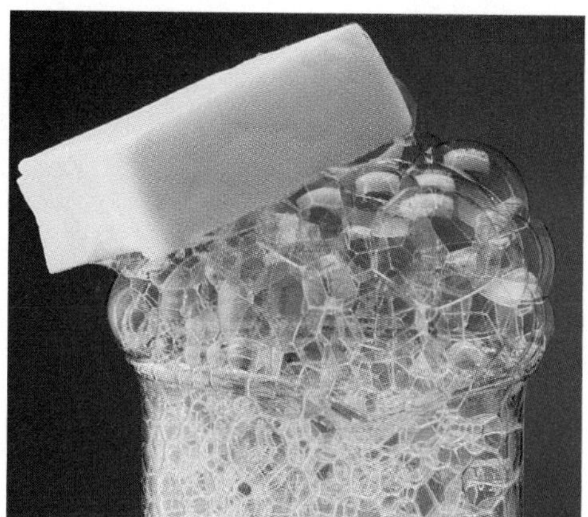

Will a hot drink stay hotter in a paper cup or in a foam cup? You probably know from experience that the foam cup will work better. In fact, the same foam cup will also do a good job of keeping a cold drink cold. The foam cup is effective at these jobs because polystyrene foam is a good thermal insulator.

A thermal insulator is a material that works as a barrier to the movement of thermal energy through matter. Thermal energy spreads from areas of higher temperature to areas of lower temperature, often very rapidly. This means that something hot, such as tea, will tend to lose its thermal energy to its surroundings. It also means that something cold, such as iced tea, will tend to be warmed by its surroundings. An insulator can slow the movement of thermal energy that causes the undesirable change in temperature.

Thermal insulators are everywhere in human society. They are used to keep houses warm and freezers cold. They keep you warm on a cold day and allow you to pick up a hot pan without getting burned. Because of the importance of insulators, chemists are always at work manipulating matter to make materials with good insulating properties.

> Chemists are always at work manipulating matter to make materials with good insulating properties.

Many of the best insulators are materials in which air is trapped. Air is a good barrier to the movement of thermal energy as long as the air is not moving, and trapped pockets of air satisfy this requirement well.

Using insulation that has a network of thin fibers is an effective way to trap air. Consider fiberglass insulation, which is often used to insulate the walls and ceilings of houses. Fiberglass is just what its name implies—thin fibers of glass (silicon dioxide) that have been woven together. A wool sweater or fleece jacket insulates your body in the same way as fiberglass.

Foams are another type of insulating material. Foams differ from fiber networks in that their air pockets are completely enveloped by the solid material that makes up the framework of the foam.

Traditional polystyrene foam and other similar foams are excellent insulators, but researchers knew they could take these foams a step further. Their goal was to make foams with as little mass as possible for a given volume. The researchers succeeded in developing several new high-tech foams, which are very effective insulators.

One of these new foams, SEAgel, is made of agar, a carbohydrate material that comes from seaweed. SEAgel begins as a gelatin-like mixture of agar and water. Then it is freeze-dried to remove the water. What is left is a honeycomb of dried agar filled with air. The sample in the photograph is light enough to float on soap bubbles. In fact, SEAgel is among the least dense solids known with a density approximately equal to that of air! SEAgel is a very lightweight and efficient insulator. Because it is made entirely of biological material, it is completely biodegradable.

CHEMISTRY IN CAREERS

MATERIALS SCIENTIST
Want to design your own solid substances?
See page 868.

TAKE IT TO THE NET
Find out more about career opportunities:
www.phschool.com

CHEMIST NEEDED
Enviro. Lab. GC Must know EPA methods, troubleshooting;

CHEMICAL SPECIALIST
Local food service distributor seeks responsible self-motivated individ

Chemistry Serving... Industry

DISCUSS

Have students name the kinds of properties that are important for a material to be an effective thermal insulator. Ask:

- Why is air an effective barrier to the transfer of heat as compared to a liquid or solid?

Challenge students to name other situations in which insulation is an important factor. For example:

- How do marine and land mammals protect themselves against warm or cold climates?
- What types of materials are used to insulate space vehicles against the harsh conditions in space and provide protection during reentry into the atmosphere?

Ask students if heat can be transferred through a vacuum. (Insulated containers are constructed with a vacuum tight layer between the inner and outer shells.)

CHEMISTRY IN CAREERS

Have students read about materials scientists on page 868. If you have access to the Internet in your classroom or school, you may wish to suggest that students connect to the address shown and look up key terms related to the field.

4 Close

Summary

Ask students to summarize, in their own words, the main concepts of the chapter. This can be done as a class discussion. The students could take notes and synthesize one summary for the whole class. Afterwards, have students compare the summary they developed with the chapter summary on this page.

Extension

Have students research the purpose of baking soda or baking powder in cooking.

- Why is it part of a cake recipe?
- What chemical does the baking soda or baking powder react with?

Looking Back . . . Looking Ahead . . .

The previous chapter contained an introduction to the study of chemistry. It discussed the scientific method and presented some strategies for learning chemistry. The current chapter describes the ways in which matter can be classified and characterized. Students are taught how to describe matter in terms of its physical and chemical properties. The next chapter introduces students to the skills necessary to take accurate measurements and demonstrates the value of having reliable measurements and data in chemical studies.

Chapter 2 STUDENT STUDY GUIDE

Take It to the NET
For interactive study and review, go to www.phschool.com

KEY TERMS

- chemical property *p. 41*
- chemical reaction *p. 41*
- chemical symbol *p. 39*
- compound *p. 36*
- distillation *p. 33*
- element *p. 36*
- gas *p. 31*
- heterogeneous mixture *p. 32*
- homogeneous mixture *p. 32*
- law of conservation of mass *p. 43*
- liquid *p. 31*
- mass *p. 29*
- matter *p. 29*
- mixture *p. 32*
- phase *p. 33*
- physical change *p. 31*
- physical property *p. 29*
- product *p. 41*
- reactant *p. 41*
- solid *p. 30*
- solution *p. 33*
- substance *p. 29*
- vapor *p. 31*

KEY RELATIONSHIP

- Law of conservation of mass:

  mass of reactants = mass of products

CONCEPT SUMMARY

2.1 Matter

- Matter has mass and occupies space.
- The three common states of matter are solid, liquid, and gas.
- A pure substance contains one kind of matter.

2.2 Mixtures

- A mixture is a physical combination of two or more substances that can be separated by physical means.
- Heterogeneous mixtures are not uniform in composition.
- Homogeneous mixtures, also called solutions, have uniform properties throughout and may be gases, liquids, or solids.

2.3 Elements and Compounds

- Elements are the simplest forms of matter that exist under normal conditions.
- Elements are always present in the same ratio in a given compound.
- Properties of a compound are usually different from those of the elements composing it.
- Chemical methods are required to separate compounds into their constituent elements.
- Each element is represented by a one- or two-letter chemical symbol. Chemical symbols of the elements are used as a shorthand method of writing chemical formulas of compounds.

2.4 Chemical Reactions

- A physical change is a change in the physical properties of a substance without a change in chemical composition.
- A chemical change is a change in the chemical composition of a substance.
- In a chemical change (chemical reaction), reactants are converted to products. Mass is conserved in any physical or chemical change.

CHAPTER CONCEPT MAP

Use these terms to construct a concept map that organizes the major ideas of this chapter.

chemical reaction · matter · substance · element · law of conservation of mass · heterogeneous mixture · homogeneous mixture · physical property · compound · physical change · chemical property

Chem ASAP! Concept Map 2
Create your Concept Map using the computer.

Take It to the Net

At **www.phschool.com** students will find for this chapter

- an Internet activity
- links to related chemistry sites
- an interactive quiz
- career links

Chapter 2 REVIEW

CONCEPT PRACTICE

24. List three physical properties of an iron nail. *2.1*
25. What is the physical state of each of the following items at room temperature? *2.1*
    a. gold d. paraffin wax
    b. gasoline e. rubbing alcohol
    c. helium f. mercury
26. In which state of matter do the following exist at room temperature? *2.1*
    a. diamond d. mercury
    b. oxygen e. clay
    c. cooking oil f. neon
27. Fingernail-polish remover (mostly acetone) is a liquid at room temperature. Would you describe acetone in the gaseous state as a vapor or a gas? Explain. *2.1*
28. List three substances that you have experienced in at least two physical states. *2.1*
29. Use **Table 2.1** to identify four substances that undergo a physical change if the temperature is decreased from 50 °C to −50 °C. Describe the nature of the physical change. *2.1*
30. Classify each of the following as homogeneous or heterogeneous mixtures. *2.2*
    a. blood
    b. chocolate-chip ice cream
    c. brass (a blend of copper and zinc)
    d. motor oil
    e. black coffee
31. How many phases does every solution have? Explain. *2.2*
32. Classify each of the following as an element or a mixture. *2.3*
    a. silver d. oxygen
    b. pine tree e. iced tea
    c. orange juice f. air
33. Name the elements found in each of the following compounds. *2.3*
    a. ammonium chloride (NH_4Cl)
    b. potassium permanganate ($KMnO_4$)
    c. isopropyl alcohol (C_3H_7OH)
    d. calcium iodide (CaI_2)
34. List four indications that a chemical change has probably taken place. *2.4*
35. Classify each of the following as a physical or chemical change. *2.4*
    a. bending a piece of wire
    b. burning coal
    c. cooking a steak
    d. cutting grass
36. When powdered iron is left exposed to the air, it rusts. Explain why the rust weighs more than the original powdered iron. *2.4*
37. A friend observes a burning candle and comments that the wax is lost as the candle burns. Having recently studied the law of conservation of mass, how would you correct your friend? *2.4*

CONCEPT MASTERY

38. Devise a way to separate sand from a mixture of charcoal, sand, sugar, and water.
39. Imagine first standing in the kitchen of your home and then in the middle of a park. When you view the surroundings in each location do you see mostly elements, compounds, or mixtures?
40. Use **Table 2.1** to answer each question.
    a. Which property most easily distinguishes sulfur from the other solid substances?
    b. How many of these substances are elements?
    c. Which compound has the highest boiling point?
    d. The solids are gradually heated. Which one will melt first?
41. Identify each of the following as a mixture or a compound. For the mixtures, classify each as homogeneous or heterogeneous.
    a. soda e. egg
    b. candle wax f. ice
    c. fog g. gasoline
    d. ink h. blood
42. Classify the following properties of the element silicon as chemical or physical properties.
    a. blue-gray color
    b. brittle
    c. insoluble in water
    d. melts at 1410 °C
    e. reacts vigorously with fluorine

Answers

24. solid, metallic luster, gray color, high melting point, malleable
25. a. solid c. gas e. liquid
    b. liquid d. solid f. liquid
26. a. solid c. liquid e. solid
    b. gas d. liquid f. gas
27. a vapor; The term "vapor" is used to refer to the gaseous state of a substance which normally exists as a liquid or solid at room temperature.
28. water, gasoline, acetone (fingernail polish remover), aromatic salves such as those used in vaporizers, butter
29. chlorine, mercury, bromine, and water; Chlorine condenses, and mercury, bromine and water all freeze when the temperature drops within the stated range.

30. a. heterogeneous
    b. heterogeneous
    c. homogeneous
    d. homogeneous
    e. homogeneous
31. one; A solution is a system with uniform composition and properties. Solutions are homogeneous mixtures, consisting of a single phase.
32. a. element d. element
    b. mixture e. mixture
    c. mixture f. mixture
33. a. nitrogen, hydrogen, chlorine
    b. potassium, manganese, oxygen
    c. carbon, hydrogen, oxygen
    d. calcium, iodine
34. color change; energy absorbed or released; gas produced; odor change
35. a. physical c. chemical
    b. chemical d. physical
36. The iron combines with oxygen in the air, and oxygen has mass.
37. As the wax burns, the chemical composition of the wax changes, producing the products water and carbon dioxide, which are released into the surrounding air.
38. Add sufficient water to dissolve all of the sugar. Separate the charcoal and sand from the sugar water by filtration. Large pieces of charcoal could be separated on the basis of color. Small pieces of charcoal could be burned.
39. a. mixtures b. mixtures
40. a. color c. sodium chloride
    b. six d. sulfur
41. a. homogeneous mixture
    b. homogeneous mixture
    c. heterogeneous mixture
    d. homogeneous mixture
    e. heterogeneous mixture
    f. compound
    g. homogeneous mixture
    h. heterogeneous mixture
42. a. physical
    b. physical
    c. physical
    d. physical
    e. chemical

Answers

43. a. color and odor change
b. gas is produced
c. formation of a precipitate (solid)
d. color and texture change
e. energy change, odor, irreversible

44. a. (1) product
b. (3) compound

45. In gases, particles are far apart; in liquids, particles are in contact; in solids, particles are tightly packed.

46. The appearance of a substance will change during a change of state, which is a physical change.

47. a. Yes; because the graph is a straight line, the proportion of iron to oxygen is a constant, which is true for a compound.
b. No; plotting these values on the graph would not give a point on the line indicating that the mass ratio of iron to oxygen is different from the other four samples.

48. a. oxygen and calcium
b. silicon, aluminum, and iron
c. Different; the second most abundant element in Earth's crust, silicon, is not present in the human body, and the second most abundant element in the human body, carbon, is not among the most abundant elements of Earth's crust. If the elements are different, then the compounds must also be different.

49. a. two, mercury and sulfur
b. Sulfur melts at 113 °C and boils at 445 °C. Between 113 °C and 445 °C, it exists as a liquid. Mercury melts at –39 °C, and boils at 357 °C. In between these temperatures, it exists as a liquid.
c. Possibilites include: by color, alphabetically, by boiling point, or by density.

50. Many answers are possible.

43. How do you know that each of these is a chemical change?
a. Food spoils.
b. A foaming antacid tablet fizzes in water.
c. A ring of scum forms around your bathtub.
d. Iron rusts.
e. A firecracker explodes.

CRITICAL THINKING

44. Choose the numbered term that best completes the second relationship.

a. initial : final — reactant: _______
(1) product (3) matter
(2) mixture (4) compound

b. words : sentence — elements: _______
(1) reactant (3) compound
(2) theory (4) substance

45. Compare the relationships among individual particles in the three states of matter.

46. Explain why this statement is false. "Because there is no change in composition during a physical change, the appearance of the substance will not change."

CONCEPT CHALLENGE

47. The mass of the elements iron and oxygen were measured in four samples of a rust-colored substance believed to be a compound. The amount of iron and oxygen found in each sample is shown on the graph.

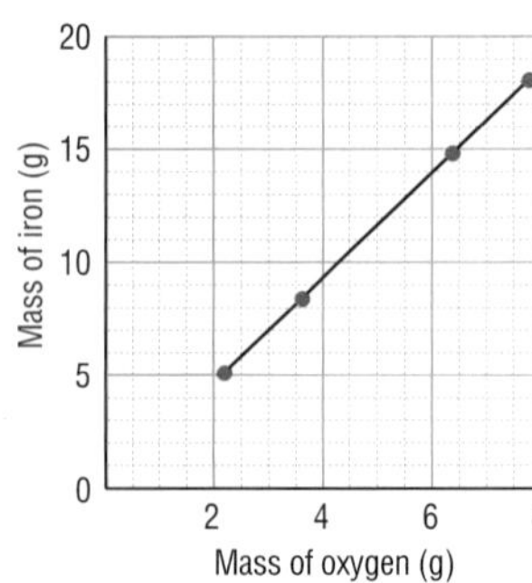

a. Do you think each sample is of the same compound? Explain.
b. Another sample of similar material was found to contain 9.9 g of iron and 3.4 g of oxygen. Is this the same substance as the other four? Explain.

48. Five elements make up 97.9% of the mass of the human body. These elements are oxygen (64.8%), carbon (18.1%), hydrogen (10.0%), nitrogen (3.1%), and calcium (1.9%). Compare these data with those in the pie graph below, which shows the five most abundant elements by mass in Earth's crust, oceans, and atmosphere.

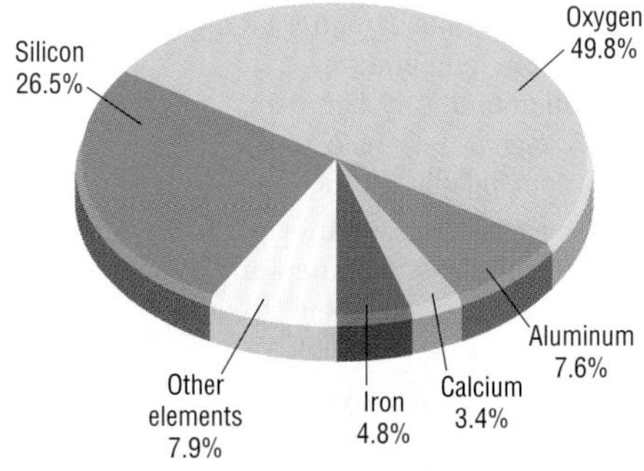

a. Which elements are abundant in both Earth's surface and the human body?
b. Which elements are abundant in Earth's surface but not in the human body?
c. Would you expect the compounds making up the human body to be the same as or different from those found in rocks, water, and air? Explain your answer based on the evidence in the pie graph and the data above.

49. These questions refer to the substances in **Table 2.1**.
a. How many of these substances are in the liquid state at 125 °C?
b. Describe the physical properties of one of these substances that led you to the answer.
c. The substances in the table are listed in order of increasing melting point. Propose another way that these data could be arranged.

50. Each day of your life you encounter some chemical changes that are helpful and some that are harmful. Cite three examples of each. For each example, list the indications that identified the change as chemical.

STANDARDIZED TEST PREP

Select the choice that best answers each question or completes each statement.

1. Which of the following is *not* a chemical change?
   a. paper being shredded
   b. steel rusting
   c. charcoal burning
   d. a newspaper yellowing in the sun
2. Which phrase best describes an apple?
   a. heterogeneous mixture
   b. homogeneous compound
   c. heterogeneous substance
   d. homogeneous mixture
3. Which element is paired with the wrong symbol?
   a. sulfur, S   c. nitrogen, N
   b. potassium, P   d. silicon, Si
4. Which of these properties could *not* be used to distinguish between table salt and table sugar?
   a. boiling point   c. density
   b. melting point   d. color
5. The state of matter characterized by a definite volume and an indefinite shape is a
   a. solid.   c. liquid.
   b. mixture.   d. gas.

Use the atomic windows to answer question 6.

6. The species in window (a) react. Use the law of conservation of mass to determine which window best represents the reaction products.

a.
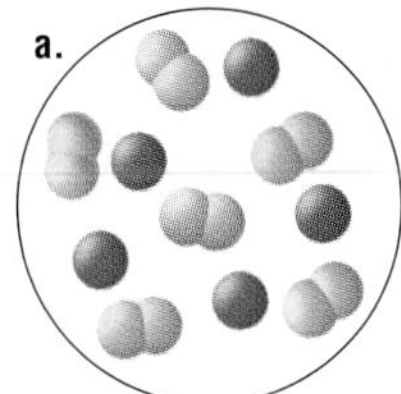

c.
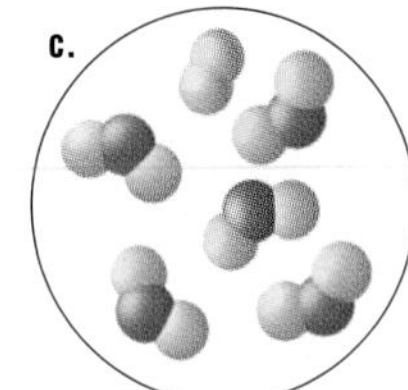

b.
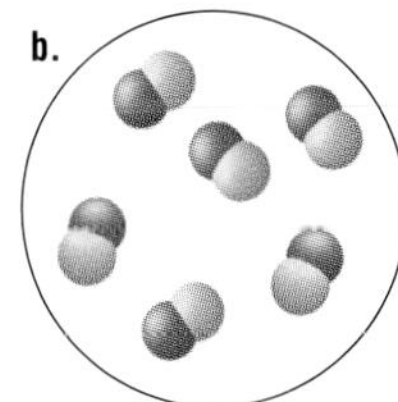

d.
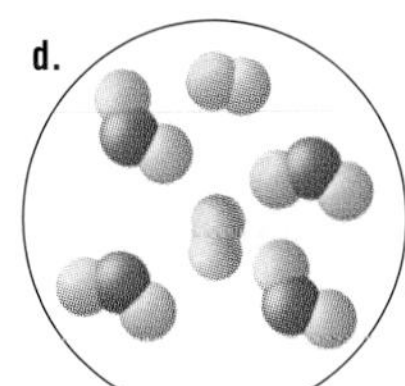

The lettered choices below refer to questions 7–10. A lettered choice may be used once, more than once, or not at all.

(A) compound
(B) heterogeneous mixture
(C) element
(D) homogeneous mixture

Which description correctly identifies each of the following materials?

7. air
8. carbon monoxide
9. zinc
10. mushroom pizza

Use the data table to answer questions 11–14.

Mass of magnesium (g)	Mass of oxygen (g)	Mass of magnesium oxide (g)
5.0	3.3	8.3
6.5	(a)	10.8
13.6	9.0	(b)
(c)	12.5	31.5

11. Magnesium metal burns vigorously in oxygen to produce the compound magnesium oxide. Use the law of conservation of mass to identify the masses labeled (a), (b), and (c) in the table.
12. Use the data in the completed table to construct a graph with mass of magnesium on the x-axis and mass of magnesium oxide on the y-axis.
13. How many grams of magnesium oxide form when 8.0 g of magnesium are burned?
14. How many grams of magnesium and oxygen react to form 20.0 g of magnesium oxide?

Use the data table to answer questions 15–17.

Substance	Color	Melting point (°C)	Boiling point (°C)
bromine	red-brown	−7	59
chlorine	green-yellow	−101	−34
ethanol	colorless	−117	78
mercury	silver-white	−39	357
neon	colorless	−249	−246
sulfur	yellow	113	445
water	colorless	0	100

15. Which colorless substance is a liquid at −30 °C?
16. Which colorless substance is a gas at 60 °C?
17. Which substance is a solid at 7 °C?

Chapter 2
STANDARDIZED TEST PREP

1. a
2. a
3. b
4. d
5. c
6. c
7. D
8. A
9. C
10. B
11. (a) is 4.3 g; (b) is 22.6 g; (c) is 19.0 g
12.

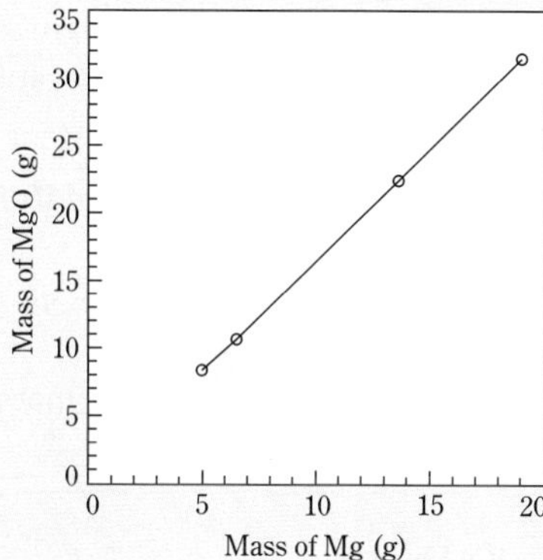

13. 13.3 g magnesium oxide
14. 12.1 g magnesium and 7.9 g oxygen
15. ethanol
16. neon
17. sulfur

Planning Guide

SECTION OBJECTIVES	ACTIVITIES/FEATURES	MEDIA & TECHNOLOGY
3.1 The Importance of Measurement ○ ■ ◆ ▸ Distinguish between quantitative and qualitative measurements ▸ Convert measurements to scientific notation	SE **Discover It!** *Exploring Density,* p. 50 SE **Link to Art** *Measurement and Art,* p. 53 SSLM 2: *Design and Construction of a Small-Scale Balance*	ASAP Assessment 3.1 SSV 9: *Small-Scale Balances*
3.2 Uncertainty in Measurements ○ ■ ◆ ▸ Distinguish among the accuracy, precision, and error of a measurement ▸ Identify the number of significant figures in a measurement and in the result of a calculation	SE **Link to Engineering** *Computer-Aided Design,* p. 56 SE **Mini-Lab** *Accuracy and Precision,* p. 62 (**LRS** 3-1)	ASAP Animation 2 ASAP Problem Solving 7, 10, 12 ASAP Assessment 3.2
3.3 International System of Units ● ■ ◇ ▸ List SI units of measurement and common SI prefixes ▸ Distinguish between the mass and weight of an object	LM 4: *Mass, Volume, and Density* SSLM 3: *Design and Construction of a Set of Standardized Weights* TE **DEMOS**, pp. 64, 65	ASAP Assessment 3.3 CHM Side 1, 20: *International System of Units*
3.4 Density ● ■ ◆ ▸ Calculate the density of an object from experimental data ▸ List some useful applications of the measurement of specific gravity	SE **CHEMath** *What Is an Equation?,* p. 70 SE **Small-Scale Lab** *Measurement,* p. 73 (**LRS** 3-2) LM 4: *Mass, Volume, and Density* TE **DEMO**, p. 69	ASAP Simulation 1 ASAP Problem Solving 24 ASAP Assessment 3.4 ACT 3: *Exploring Density* CA *Giant Density Column* CA *Methane Bubbles* CHM Side 1, 30: *Density Column* CHM Side 1, 23: *Why Does Ice Float?* CHM Side 1, 36: *Matter with No Mass?* OT 4: *Volume and Density*
3.5 Temperature ● ■ ◆ ▸ Convert between the Celsius and Kelvin temperature scales	SE **Chemistry Serving . . . the Consumer** *These Standards Are Worth the Weight,* p. 76 SE **Chemistry in Careers** *Analytical Chemist,* p. 76 SE **Sample Problem** 3-6	ASAP Assessment 3.5 ASAP Problem Solving 31 ASAP Concept Map 3 CA *Superconductivity* CHM Side 1, 32: *A Weird Thermometer* CHM Side 1, 34: *Boiling Point Reduction* PP Chapter 3 Problems RP Lesson Plans, Resource Library AR Computer Test 3 www Activities, Self-Tests, *SCIENCE NEWS* updates TCP The Chemistry Place Web Site

KEY

- ● Conceptual (concrete concepts)
- ○ Conceptual (more abstract/math)
- ■ Standard (core content)
- □ Standard (extension topics)
- ◆ Honors (core content)
- ◇ Honors (options to accelerate)

AR	Assessment Resources
ASAP	Chem ASAP! CD-ROM
ACT	ActivChemistry CD-ROM
CHM	CHEMedia Videodiscs
CA	Chemistry Alive! Videodiscs
GCP	Graphing Calculator Problems
GRS	Guided Reading and Study Workbook
LM	Laboratory Manual
LP	Laboratory Practicals
LRS	Laboratory Recordsheets
SSLM	Small-Scale Lab Manual

PRACTICE		ASSESSMENT	
GRS	Section 3.1	SE	Section Review
RM	Practice Problems 3.1	RM	Section Review 3.1
		RM	Chapter 3 Quiz
SE	**Sample Problems** 3-1 to 3-4	SE	Section Review
GRS	Section 3.2	RM	Section Review 3.2
RM	Practice Problems 3.2	RM	Chapter 3 Quiz
RM	Interpreting Graphics		
RM	Practice Problems 3.3	SE	Section Review
GRS	Section 3.3	RM	Section Review 3.3
		RM	Chapter 3 Quiz
		RM	Lab Practical 3-1
SE	**Sample Problem** 3-5	SE	Section Review
GRS	Section 3.4	RM	Section Review 3.4
RM	Practice Problems 3.4	RM	Chapter 3 Quiz
		LP	Lab Practical 3-2
SE	**Sample Problem** 3-6	SE	Section Review
GRS	Section 3.5	RM	Section Review 3.5
RM	Practice Problems 3.5	RM	Vocabulary Review 3
GCP	Chapter 3	SE	Chapter Review
		SM	Chapter 3 Solutions
		SE	Standardized Test Prep
		PHAS	Chapter 3 Test Prep
		RM	Chapter 3 Quiz
		RM	Chapter 3 A & B Test

OT	Overhead Transparency	SE	Student Edition
PHAS	PH Assessment System	SM	Solutions Manual
PLM	Probeware Lab Manual	SSV	Small-Scale Video/Videodisc
PP	Problem Pro CD-ROM	TCP	www.chemplace.com
RM	Review Module	TE	Teacher's Edition
RP	Resource Pro CD-ROM	www	www.phschool.com

PLANNING FOR ACTIVITIES

STUDENT EDITION

Discover It! p. 50
- beakers
- graduated cylinders
- colorless cylindrical glasses or plastic containers
- shampoo
- water
- rubbing alcohol
- red food coloring
- medicine droppers
- spoons or stirring rods
- small solid objects

Mini Lab, p. 62
- index cards
- metric rulers

Small-Scale Lab, p. 73
- pencils
- paper
- rulers
- balances
- rectangular blocks
- calculators

Activity, p. 64
- gram balance
- U.S. pennies minted between 1970 and today

Teacher Demo, p. 65
- meter stick with English units

Activity, p. 65
- objects with irregular volumes, e.g., small pebbles
- water
- graduated cylinder
- stopwatch
- object for dropping

Activity, p. 66
- water, tablespoon
- 12 oz. can of soda
- half gallon of juice or milk (or empty food containers filled with water)
- graduated cylinders

TEACHER'S EDITION

Activity, p. 52
- kitchen scale
- hard cheese

Activity, p. 55
- triple-beam balance
- high-quality balance
- beaker
- buret
- water
- small objects

Activity, p. 56
- winning Olympic times for the men's and women's 100-meter dashes

Teacher Demo, p. 64
- packages from consumer products with metric system measurements

Teacher Demo, p. 69
- graduated cylinder
- water
- large fishing weight with known mass
- balsa wood with known mass

Teacher Demo, p. 69
- 3 or 4 wooden, metal or marble cubes
- chalkboard or overhead projector

Activity, p. 72
- hydrometer
- salt or sugar
- water
- diet and regular colas
- balance
- graduated cylinder

Chapter 3

Key Terms

3.1 qualitative measurements, quantitative measurements, scientific notation
3.2 accuracy, precision, accepted value, experimental value, error, percent error, significant figures
3.3 International System of Units (SI), meter (m), volume, liter (L), weight, kilogram (kg), gram (g)
3.4 density, specific gravity, hydrometer
3.5 temperature, Celsius scale, Kelvin scale, absolute zero

DISCOVER IT!

Students should explain that the shampoo, which makes up the bottom layer, is denser than the water, which is denser than the rubbing alcohol. They should also make clear that the order in which the liquids are poured does not make a difference as long as the pouring is done carefully, so that mixing does not occur.

Chapter 3 SCIENTIFIC MEASUREMENT

Cranberries float on water in a cranberry bog.

FEATURES

Stay current with SCIENCE NEWS
Find out more about measurement:
www.phschool.com

DISCOVER IT! EXPLORING DENSITY

You need three beakers, a graduated cylinder, a colorless cylindrical glass or plastic container, shampoo, water, rubbing alcohol, red food coloring, a medicine dropper, a spoon or stirring rod, and small solid objects (such as a paper clip, plastic soda-bottle cap, chalk, ball of aluminum foil, aluminum nail, piece of uncooked pasta, dried bean, or piece of cork).

1. Pour about 250 mL each of shampoo, water, and rubbing alcohol into separate beakers.
2. Stir a drop of food coloring into the rubbing alcohol and the shampoo.
3. Carefully pour each liquid (in the following order: shampoo, water, rubbing alcohol) down the inside of the colorless cylindrical container to a depth of 2.5 cm (about 1 in.).
4. Record the positions of the three layers of liquids.
5. Gently add each small solid object to the layered liquids and record its resting location.

Propose an explanation for the order of layering of the liquids. Do you think it would make any difference if you added the liquids in a different order? Why or why not? After completing this chapter, reevaluate your explanation and revise it if necessary.

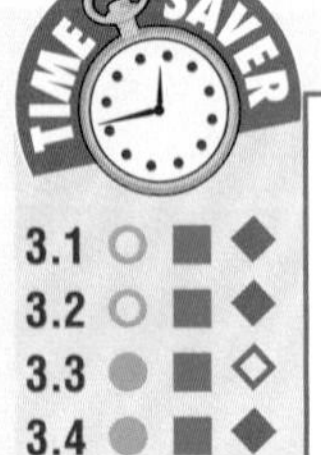

3.1 ○ ■ ◆
3.2 ○ ■ ◆
3.3 ● ■ ◇
3.4 ● ■ ◆
3.5 ● ■ ◆

Conceptual Students need to understand the concepts of accuracy, precision, significant figures, and error. However, students do not need to focus on significant figures in calculations or calculating percent error. Depending on the level of problem solving expected in the course, specific gravity may be omitted. SI units can be taught without emphasizing scientific notation.

Standard Use Figures 3.4 and 3.5 to save time in explaining accuracy and precision. Density may be used to assess student proficiency with ratios. No concepts in later chapters require a knowledge of specific gravity. Sample Problem 3-6 is important if students will be doing gas law problems in Chapter 12, which require temperature conversions.

Honors Section 3.3 can be accelerated if students are familiar with SI units.

THE IMPORTANCE OF MEASUREMENT

In horse racing, the finish is sometimes so close that the winner can only be determined by a photograph taken at the instant the horses cross the finish line. In the 130th Belmont Stakes, for example, the horse named Real Quiet, the winner of both the Kentucky Derby and the Preakness, was beaten by only the length of a nose in the final race for the Triple Crown. Chemistry also requires making accurate and often very small measurements. **What types of measurements are made in chemistry?**

objectives
- Distinguish between quantitative and qualitative measurements
- Convert measurements to scientific notation

key terms
- qualitative measurements
- quantitative measurements
- scientific notation

Qualitative and Quantitative Measurements

Everyone makes and uses measurements. For example, you decide how to dress in the morning based on the temperature outside. You measure the ingredients for your favorite cookie recipe. If you were building a cabinet for your stereo system, you would carefully measure the length of each piece of wood.

Measurements are fundamental to the experimental sciences as well. For that reason, it is important to be able to make measurements and to decide whether a measurement is correct. In chemistry, you will use the International System of Measurements (SI).

Not all measurements give the same amount of information. For example, how might you determine whether someone who is sick has a fever? You might simply touch the person's forehead and think, "Yes! This person feels feverish." This is an example of a qualitative measurement. **Qualitative measurements** give results in a descriptive, nonnumerical form. If several people touch the sick person's forehead, their qualitative measurements may not agree with yours. One reason for this is that a person's own temperature influences his or her perception of how warm another person feels. By using a thermometer, however, each person can eliminate this personal bias. The temperature takers will each report a numerical value, or quantitative measurement. **Quantitative measurements** give results in a definite form, usually as numbers and units. For example, the thermometer might reveal the person's temperature to be 39.2 °C (102.5 °F). This measurement has a definite value that can be compared with the person's temperature at a later time to check for changes. Yet this measurement, however definite, can be no more reliable than the instrument used to make the measurement and the care with which it is used and read.

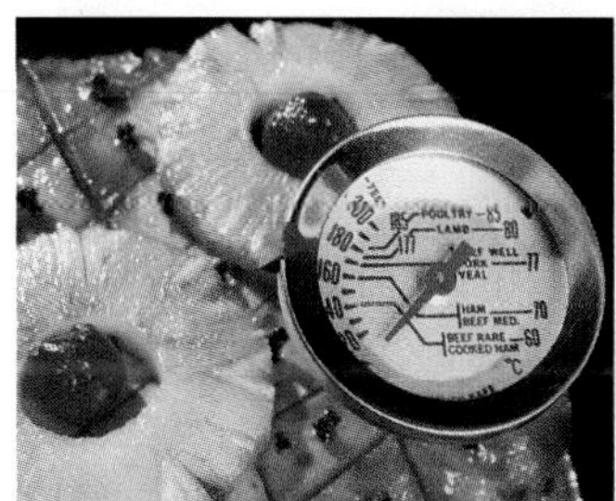

Figure 3.1
Instruments are needed to make quantitative measurements. A meat thermometer, a radar gun, and a grocery scale are common instruments used to measure temperature, speed, and weight, respectively.

3.1

1 Engage

Use the Visual

Have students study the figure and read the text that opens the section. Explain that making measurements is an important part of the scientific process. Discuss the distinguishing features of qualitative and quantitative measurements. Point out that for a quantitative measurement to be meaningful, it must consist of a number and a unit such as meters, degrees Celsius, or seconds. Ask:

- **Is the method used to determine the Kentucky Derby winner a qualitative measurement or a quantitative measurement?** (Qualitative; the winner is determined by visual inspection.)
- **What types of measurements are made in chemistry?** (length, temperature, mass, volume)

Check Prior Knowledge

To assess students' knowledge about measurements, ask students to think of everyday activities that involve measuring. Examples include buying consumer products, doing sports activities, and cooking. Ask students to recall which units of measure are related to each of the examples they give. Have some students estimate their height in inches. Have other students measure the same students with a yard stick or tape measure. Compare the estimates with measured values. Have students convert the height measurements to centimeters using the relation: 1 inch = 2.54 cm. Ask:

- **How many of you know your weight in pounds? In kilograms?**
- **How many pounds equal one kilogram?** (2.2 lbs)

STUDENT RESOURCES

From the Teacher's Resource Package, use:
- Section Review 3.1, Ch. 3 Practice Problems and Quizzes from the Review Module (Ch. 1–4)
- Small-Scale Chemistry Lab Manual: Experiment 2

TECHNOLOGY RESOURCES

Relevant technology resources include:
- Chem ASAP! CD-ROM
- ResourcePro CD-ROM

2 Teach

ACTIVITY

To demonstrate a practical use of measurement, show students different-sized portions of hard cheese. Have them estimate the weight of each portion. Use a kitchen scale to weigh the items. Note that people on portion-controlled diets use the scale to choose the correct portions.

Ask students to compare a kitchen scale and a triple-beam balance in terms of accuracy. Would the kitchen scale be a useful scientific tool? (no, limited accuracy) Ask students if they think the standard portions used in Calorie lists (e.g., 1/2 cup cereal) or recipes (serves 6) are realistic given the average person's eating habits.

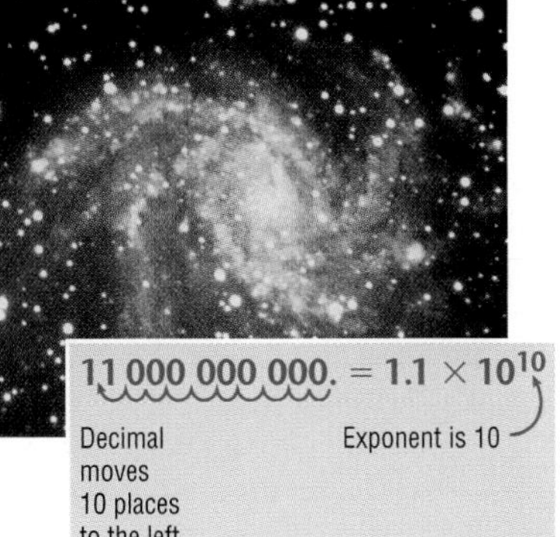

Figure 3.2
Expressing very large numbers, such as the estimated number of stars in a galaxy, is easier if scientific notation is used. The number of decimal places moved is equal to the exponent.

Scientific Notation

In chemistry, you will often encounter very small and very large numbers. The mass of an atom of gold, for example, is 0.000 000 000 000 000 000 000 327 gram. A gram of hydrogen contains 602 000 000 000 000 000 000 000 hydrogen atoms. Writing and using such small or large numbers is very cumbersome. You can work more easily with these numbers by writing them in scientific, or exponential, notation.

In **scientific notation,** a number is written as the product of two numbers: a coefficient and 10 raised to a power. For example, the number 36 000 is written in scientific notation as 3.6×10^4. The coefficient in this number is 3.6. In scientific notation, the coefficient is always a number greater than or equal to one and less than ten. The power of 10, or exponent, in this example is 4. The exponent indicates how many times the coefficient 3.6 must be multiplied by 10 to equal the number 36 000.

$$3.6 \times 10^4 = 3.6 \times 10 \times 10 \times 10 \times 10 = 36\,000$$

For numbers greater than ten, the exponent is positive and equals the number of places the original decimal point has been moved to the left to write the number in scientific notation. Numbers less than ten have a negative exponent. For example, the number 0.0081 written in scientific notation is 8.1×10^{-3}. The exponent -3 indicates that the coefficient 8.1 must be divided by 10 three times to equal 0.0081, as shown in **Figure 3.3.**

$$8.1 \times 10^{-3} = \frac{8.1}{10 \times 10 \times 10} = 0.0081$$

For numbers less than ten, the value of the exponent equals the number of places the original decimal point has been moved to the right to write the number in scientific notation. Can you express the mass of a single atom of gold, which is 0.000 000 000 000 000 000 000 327 gram, in scientific notation? Try it! ①

Multiplication and Division Using scientific notation makes calculating more straightforward. To multiply numbers written in scientific notation, multiply the coefficients and add the exponents. For example,

$$(3.0 \times 10^4) \times (2.0 \times 10^2) = (3.0 \times 2.0) \times 10^{4+2} = 6.0 \times 10^6$$

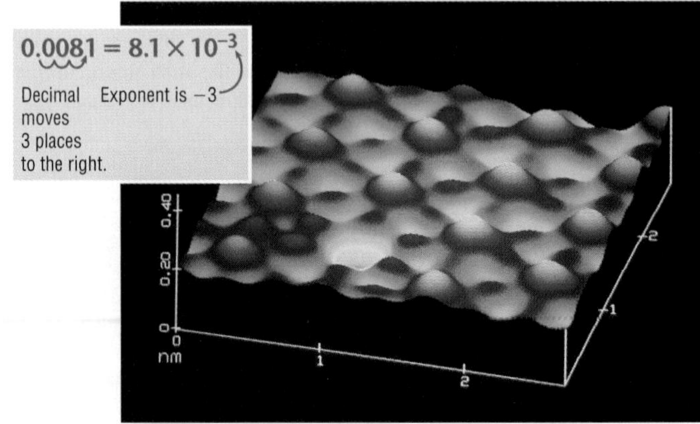

Figure 3.3
A tiny distance, such as that between the atoms in this electron micrograph, can be expressed conveniently in scientific notation. The direction in which the decimal point is moved determines the sign of the exponent. What is the sign of the exponent in this case? ②

Answers

① Yes: 3.27×10^{-22} gram
② negative

To divide numbers written in scientific notation, first divide the coefficients. Then subtract the exponent in the denominator (bottom) from the exponent in the numerator (top). For example,

$$\frac{3.0 \times 10^4}{2.0 \times 10^2} = \frac{3.0}{2.0} \times 10^{4-2} = 1.5 \times 10^2$$

Addition and Subtraction Before you add or subtract numbers written in scientific notation, you must make the exponents the same because the exponents determine the locations of the decimal points in the original numbers. The decimal points must be aligned before you add two numbers. For numbers in scientific notation, making the exponents the same aligns the decimal points. For example, when adding 5.40×10^3 to 6.0×10^2, you must adjust the exponents to make them the same. You can choose to adjust 6.0×10^2 to 0.60×10^3.

$$\begin{array}{r} 5.40 \times 10^3 \\ +\,0.60 \times 10^3 \\ \hline 6.00 \times 10^3 \end{array}$$

section review 3.1

1. a. What is the difference between a qualitative measurement and a quantitative measurement?
   b. How is a number converted to scientific notation?
2. Classify each statement as either qualitative or quantitative.
   a. The basketball is brown.
   b. The diameter of the basketball is 31 centimeters.
   c. The air pressure in the basketball is 12 pounds per square inch.
   d. The surface of the basketball has indented seams.
3. Write each measurement in scientific notation.
   a. the length of a football field, 91.4 meters
   b. the diameter of a carbon atom, 0.000 000 000 154 meter
   c. the radius of Earth, 6 378 000 meters
   d. the diameter of a human hair, 0.000 008 meter
   e. the average distance between the center of the sun and the center of Earth, 149 600 000 000 meters
4. Solve each problem, and express each answer in correct scientific notation.
   a. $(4 \times 10^7) \times (2 \times 10^{-3})$
   b. $\dfrac{(6.3 \times 10^{-2})}{(2.1 \times 10^4)}$
   c. $(4.6 \times 10^3) - (1.8 \times 10^3)$
   d. $(7.1 \times 10^{-2}) + (5 \times 10^{-3})$

 Chem ASAP! **Assessment 3.1** Check your understanding of the important ideas and concepts in Section 3.1.

LINK TO ART

Measurement and Art

Although some people believe that art and science are very different, artists and scientists share many of the same methods in their work. For example, artists and scientists are both concerned with measurements. A sculptor designing a realistic statue of a human figure must be certain that the head and its features, as well as the arms, legs, and body, are in proportion to each other. An artist may use a pantograph in making scale drawings. A pantograph is a device in which a pencil traced over an original drawing creates a new drawing that is larger or smaller in scale than the original drawing.

Answers

SECTION REVIEW 3.1

1. a. Qualitative measurements are expressed in descriptive, nonnumerical form. Quantitative measurements are expressed in a definite form, usually numerical.
   b. It is written as the product of a coefficient greater than or equal to one and less than ten and 10 raised to a power.
2. a. qualitative
   b. quantitative
   c. quantitative
   d. qualitative
3. a. 9.14×10^1 meters
   b. 1.54×10^{-10} meter
   c. 6.378×10^6 meters
   d. 8×10^{-6} meter
   e. 1.496×10^{11} meters
4. a. 8×10^4
   b. 3.0×10^{-6}
   c. 2.8×10^3
   d. 7.6×10^{-2}

Think Critically

Why is it important to use scientific notation? A number that chemists use is called Avogadro's Number, named after the Italian chemist Amedeo Avogadro. This number is 6 followed by 23 zeros.

- **Have students write this number with all the zeros.**
- **Now have students multiply 6 with 23 zeros by 23 000. Compare results to check consistency.**
- **Finally, have students do the above operation using scientific notation.**

3 Assess

Evaluate Understanding

To determine the students' knowledge of scientific notation, have them:

- **convert the number 0.000137 to scientific notation** (1.37×10^{-4})
- **multiply 3.9×10^7 by 1.2×10^2** (4.7×10^9)
- **divide 8.4×10^{-3} by 4.7×10^4** (1.8×10^{-7})
- **add 2.3×10^2 and 1.9×10^1 to 1.7×10^3** (1.9×10^3)
- **subtract 3.1×10^{-3} from 8.5×10^{-2}** (8.2×10^{-2})

Reteach

Have students find several very large or very small numbers, such as U.S. budget figures or atomic dimensions, and help them convert these numbers to scientific notation. Point out to students the importance of quantitative measurement. Discuss with students some quantitative measurements that are important in their lives. (Examples include correct weights and volumes when buying foods, and correct medical laboratory reports for diagnoses by physicians.)

1 Engage

Use the Visual

Have students study the photograph and read the text that opens the section. Ask:

- **How do scientists ensure the accuracy and precision of their measurements?** (Using the most precise equipment available, they make multiple measurements. Scientists use samples with known values to check the reliability of the equipment. If possible, they devise more than one method for measuring a given value.

Check Prior Knowledge

Show students a regular meter stick and a shortened meter stick. (Saw off the first 5–10 cm and relabel the intervals so that the stick appears to be 1 m long.) With each stick, make three measurements of an object with a known length. Write the values on the board. (Values should be precise.) Ask:

- **Compare the accuracy of the first meter stick to that of the second meter stick.** (The shortened meter stick is less accurate.)
- **Does using a defective meter stick affect the precision?** (No, as long as the measurements are read in a consistent manner.)

section 3.2

objectives

- Distinguish among the accuracy, precision, and error of a measurement
- Identify the number of significant figures in a measurement and in the result of a calculation

key terms

- accuracy
- precision
- accepted value
- experimental value
- error
- percent error
- significant figures

UNCERTAINTY IN MEASUREMENTS

On July 4, 1997, the Mars Pathfinder spacecraft landed on Mars. Shortly after landing, the spacecraft released a small robotic rover called Sojourner to explore the Martian surface around the landing site in Ares Vallis. NASA scientists had to make thousands of precise calculations to ensure that Pathfinder reached its destination safely. All measurements have some uncertainty. **How do scientists ensure the accuracy and precision of their measurements?**

Accuracy, Precision, and Error

Your success in the chemistry lab and in many of your daily activities depends on your ability to make reliable measurements. Ideally, measurements are both correct and reproducible.

Correctness and reproducibility relate to the concepts of accuracy and precision, two words that mean the same thing to many people. In chemistry, however, their meanings are quite different. **Accuracy** is a measure of how close a measurement comes to the actual or true value of whatever is measured. **Precision** is a measure of how close a series of measurements are to one another. Note that the precision of a measurement depends on more than one measurement.

By contrast, an individual measurement may be accurate or inaccurate. Darts on a dartboard illustrate accuracy and precision in measurement. Let the bull's-eye of the dartboard represent the true, or correct, value of what you are measuring. The closeness of a dart to the bull's-eye corresponds to the degree of accuracy. The closer it comes to the bull's-eye, the more accurately the dart was thrown. The closeness of several darts to one another corresponds to the degree of precision. The closer together the darts are, the greater the precision and the reproducibility. Look at **Figure 3.4** as you consider the following outcomes.

(a) All of the darts land close to the bull's-eye and to one another. Closeness to the bull's-eye means that the degree of accuracy is great. Each dart in the bull's-eye corresponds to an accurate measurement of a value. Closeness of the darts to one another indicates high precision.

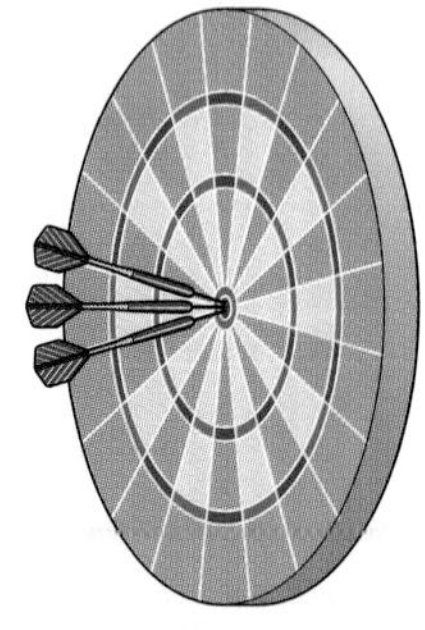

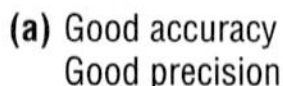

(a) Good accuracy Good precision

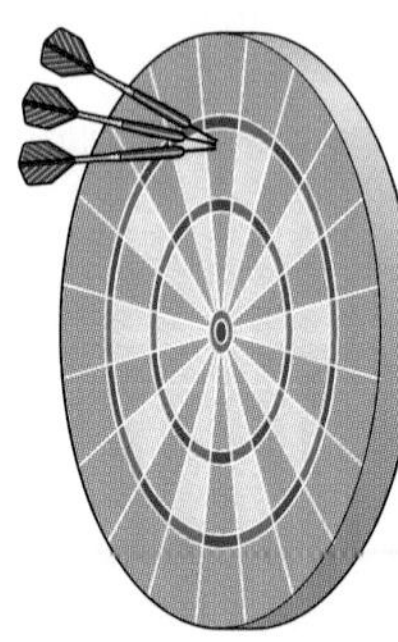

(b) Poor accuracy Good precision

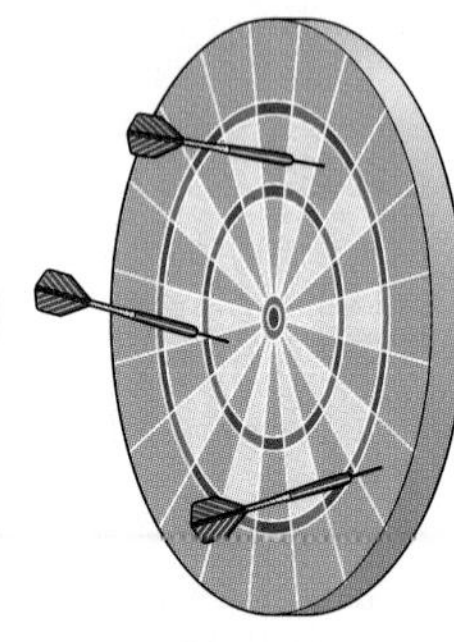

(c) Poor accuracy Poor precision

Figure 3.4
The distribution of darts illustrates the difference between accuracy and precision. ***(a)*** *Good accuracy and good precision: The darts are close to the bull's-eye and to one another.* ***(b)*** *Poor accuracy and good precision: The darts are far from the bull's-eye but close to one another.* ***(c)*** *Poor accuracy and poor precision: The darts are far from the bull's-eye and from one another.*

STUDENT RESOURCES

From the Teacher's Resource Package, use:

- Section Review 3.2, Ch. 1 Practice Problems, Interpreting Graphics, and Quizzes from the Review Module (Ch. 1–4)
- Laboratory Recordsheet 3-1

TECHNOLOGY RESOURCES

Relevant technology resources include:

- Chem ASAP! CD-ROM
- Resource Pro CD-ROM

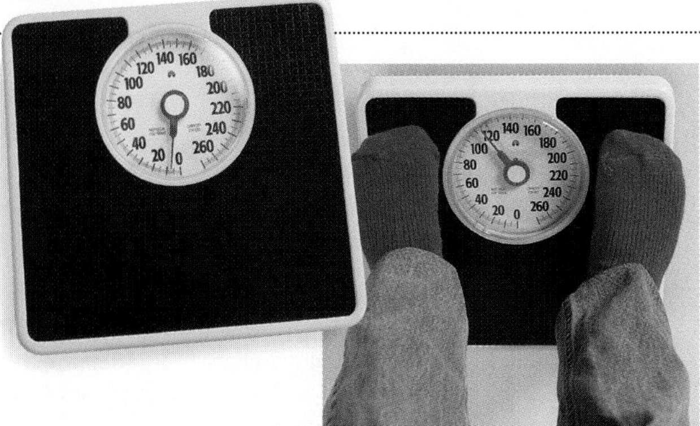

Figure 3.5

This scale has not been properly zeroed. So the reading obtained for the person's weight is inaccurate. There is a difference between the person's correct weight and the measured value.

(b) All of the darts land close to one another but far from the bull's-eye. The precision is high because of the closeness of grouping and thus the high level of reproducibility. The results are inaccurate, however, because of the distance of the darts from the bull's-eye.

(c) The darts land far from one another and from the bull's-eye. The results are both inaccurate and imprecise.

To evaluate the accuracy of a measurement, it must be compared with the correct value. Suppose you use a thermometer to measure the boiling point of pure water at standard atmospheric pressure. The thermometer reads 99.1 °C. You probably know that the true or accepted value of the boiling point of pure water under these conditions is actually 100.0 °C. There is a difference between the **accepted value,** which is the correct value based on reliable references, and the **experimental value,** the value measured in the lab. The difference between the accepted value and the experimental value is called the **error.**

$$\text{Error} = \text{experimental value} - \text{accepted value}$$

Error can be positive or negative depending on whether the experimental value is greater than or less than the accepted value.

For the boiling-point measurement, the error is 99.1 °C − 100.0 °C, or −0.9 °C. The magnitude of the error shows the amount by which the experimental value is too high or too low, compared with the accepted value. Often, it is useful to calculate the relative error, or percent error. The **percent error** is the absolute value of the error divided by the accepted value, multiplied by 100%.

$$\text{Percent error} = \frac{|\text{error}|}{\text{accepted value}} \times 100\%$$

Using the absolute value of the error means that the percent error will always be a positive value. For the boiling-point measurement, the percent error is calculated as follows.

$$\begin{aligned}\text{Percent error} &= \frac{|99.1\ °\text{C} - 100.0\ °\text{C}|}{100.0\ °\text{C}} \times 100\% \\ &= \frac{0.9\ \cancel{°\text{C}}}{100.0\ \cancel{°\text{C}}} \times 100\% \\ &= 0.009 \times 100\% \\ &= 0.9\%\end{aligned}$$

2 Teach

ACTIVITY

Place a set of small objects near a triple-beam balance. Set a deadline by which each student will have determined the mass of each object. After everyone has had an opportunity, have students compile a summary of all the measurements for each object. Use the summary to illustrate the concept of precision.

To illustrate accuracy, determine the mass of each object on a high quality balance; compare these values to the values found by the students. To further illustrate the effect of equipment on accuracy, pour about 20 mL of water into a graduated beaker and about 20 mL into a buret; then have students take volume readings.

Discuss

Explain the importance of using absolute values for calculating percent error. Students often have trouble with this calculation. Use the measurements made in the preceding activity to do additional percent error calculations.
Be sure to point out that significant figures are the numbers you can measure plus one more estimated number. For example, on a metric ruler that is divided into millimeters, measurements can be estimated to the nearest 0.1 millimeter. Be sure to review the role of zeros in determining the number of significant figures. When adding or subtracting numbers expressed in scientific notation, remind students that the numbers must all have the same exponent.

ACTIVITY

Have students look up the winning Olympic times for the men's and women's 100-meter dashes. Ask:

- How many digits to the right of the decimal point were used to record the winning time in each of the Olympics? (Only tenths of seconds were recorded up until 1968. Beginning in 1972, hundredths of seconds were recorded.)
- Why were more digits eventually needed to record the times of the runners in the 100-meter dash?

This question should provoke discussion concerning the precision that is sometimes necessary to discern certain features of natural phenomena.

Significant Figures in Measurements

If you use a liquid-filled thermometer that is calibrated in 1 °C intervals, you can easily read the temperature to the nearest degree. With such a thermometer, however, you can also estimate the temperature to about the nearest tenth of a degree by noting the closeness of the liquid inside to the calibrations. Suppose you estimate a temperature that lies between 23 °C and 24 °C to be 24.3 °C. This estimated number has three digits. The first two digits (2 and 4) are known with certainty. But the rightmost digit has been estimated and involves some uncertainty. These reported digits all convey useful information, however, and are called significant figures. The **significant figures** in a measurement include all of the digits that are known, plus a last digit that is estimated. Measurements must always be reported to the correct number of significant figures because, as you will soon learn, calculated answers depend upon the number of significant figures in the values used in the calculation.

Suppose you take someone's temperature with a thermometer that is calibrated in 0.1 °C intervals. You can read the temperature with virtual certainty to the nearest 0.1 °C, and you can estimate it to the nearest 0.01 °C. You might report the temperature as 35.82 °C. This measurement has four significant figures, the rightmost of which is uncertain. Instruments differ in the number of significant figures that can be obtained from them and thus in the precision of measurements. The three meter sticks in **Figure 3.6** can be used to make successively more precise measurements of the board.

To determine whether a digit in a measured value is significant, you need to apply the following rules.

1. Every nonzero digit in a reported measurement is assumed to be significant. The measurements 24.7 meters, 0.743 meter, and 714 meters each express a measure of length to three significant figures.

LINK TO ENGINEERING

Computer-Aided Design
Measurement is important, but making or specifying a large number of measurements can be very time consuming. An engineer designing a complex machine must specify hundreds of dimensions. Today, computer-aided design (CAD) programs have replaced many tedious aspects of technical design. A CAD program usually permits a designer to begin with a few important elements and measurements of the object being designed. As the design progresses, the program computes the dimensions of added elements and produces drawings of the object from any perspective the designer desires. When one dimension in a design is changed, CAD programs automatically adjust all of the other dimensions in proportion. The introduction of CAD programs has reduced the time needed to design complex items as well as the number of errors that plague big projects.

Chem ASAP!
Animation 2
See how the accuracy of a calculated result depends on the sensitivity of the measuring instruments.

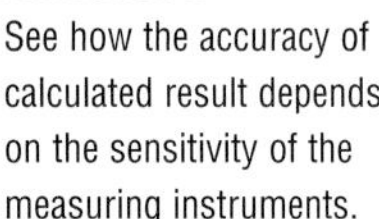

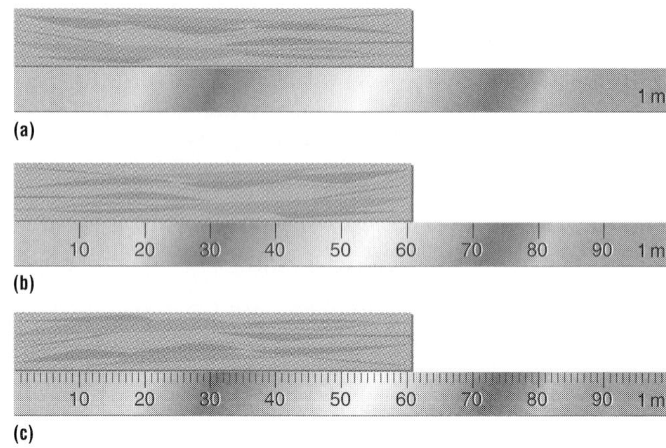

Figure 3.6
Three differently calibrated meter sticks can be used to measure the length of a board. What measurement is obtained in each case? Are there differences in the number of significant figures in the three measurements? Explain. ❶

Answers

❶ The readings are: **a.** 0.6 m, **b.** 0.61 m, **c.** 0.607 m. Yes; the greater the number of divisions on the meter stick, the more precision with which it can be read and the greater the number of significant figures in the measurement.
❷ The cylinder on the right. The volume can be read to a greater number of significant figures.

MEETING DIVERSE NEEDS

Gifted Have students prepare a display on current applications of CAD. Local colleges or technical schools that teach CAD are good sources of information. Other possibilities include library research or articles in magazines or newspapers.

2. Zeros appearing between nonzero digits are significant. The measurements 7003 meters, 40.79 meters, and 1.503 meters each have four significant figures.

3. Leftmost zeros appearing in front of nonzero digits are not significant. They act as placeholders. The measurements 0.0071 meter, 0.42 meter, and 0.000 099 meter each have only two significant figures. The zeros to the left are not significant. By writing the measurements in scientific notation, you can get rid of such placeholding zeros: in this case, 7.1×10^{-3} meter, 4.2×10^{-1} meter, and 9.9×10^{-5} meter.

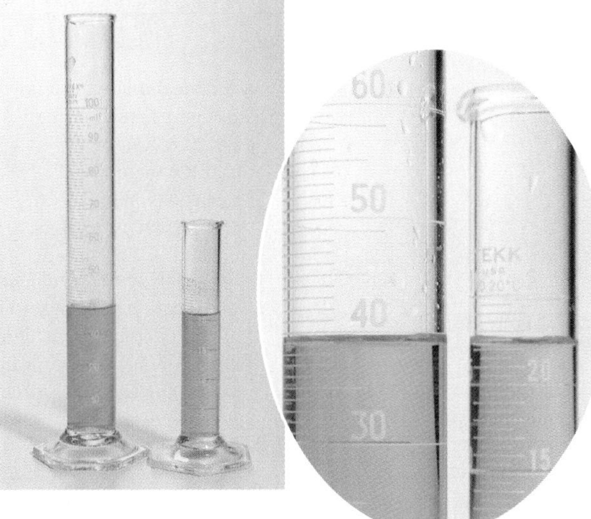

Figure 3.7
Two differently calibrated graduated cylinders are used to measure the volume of a liquid. Which cylinder would give more precise measurements? 2

4. Zeros at the end of a number and to the right of a decimal point are always significant. The measurements 43.00 meters, 1.010 meters, and 9.000 meters each have four significant figures.

5. Zeros at the rightmost end of a measurement that lie to the left of an understood decimal point are not significant if they serve as placeholders to show the magnitude of the number. The zeros in the measurements 300 meters, 7000 meters, and 27 210 meters are not significant. The numbers of significant figures in these values are one, one, and four, respectively. If such zeros were known measured values, however, then they would be significant. For example, if the value of 300 meters resulted from a careful measurement rather than a rough, rounded measurement, the zeros would be significant. Ambiguity is avoided if measurements are written in scientific notation. For example, if all of the zeros in the measurement 300 meters were significant, writing the value as 3.00×10^2 meters makes it clear that these zeros are significant.

6. There are two situations in which measurements have an unlimited number of significant figures. The first involves counting. If you carefully count that there are 23 people in your classroom, then there are exactly 23 people, not 22.9 or 23.1. This measurement can only be a whole number and has an unlimited number of significant figures, in the form of zeros understood to be to the right of the decimal point; thus, 23.000 000 ... is understood. The second situation of unlimited significant figures involves exactly defined quantities, such as those usually used within a system of measurement. When, for example, you write 60 minutes = 1 hour, these quantities have an unlimited number of significant figures; there are exactly 60 minutes in an hour, by definition. It is important to recognize when quantities are exact and to round calculated answers correctly in problems involving such values.

3.2

Discuss

Point out that the concept of significant figures applies only to measured quantities. If students ask why an estimated digit is considered significant, tell them a significant figure is one that is known to be reasonably reliable. A careful estimate fits this definition.

Use the Visual

Have students study Figure 3.6. Explain that because of the limitations of a measuring device and the limited ability of individuals to distinguish quantitative differences, every measurement carries a degree of uncertainty or error. Even the most precise measuring devices available have some degree of uncertainty. Significant figures are the digits in a measurement that are known with certainty plus one digit that is uncertain, the estimated figure.

Point out that, using Ruler (a), one can be certain that the length of the stick is between 0 and 1 m, and one can say that the actual length is closer to 1 m. Thus, one can estimate the length as 0.6 m. Ruler (b) has more subdivisions. Using Ruler (b), one can say with certainty that the length is between 60 and 70 cm. Because the length is very close to 60 cm, one should estimate the length as 61 cm or 0.61 m. Have students study Ruler (c) and use similar reasoning to describe the measurement and estimation process.

Meeting Diverse Needs

At Risk Have students write in their own words the rules for determining the number of significant digits. Help them if necessary. Have them measure several lengths, masses, and volumes; then have them use their rules to determine the correct number of significant figures for each measurement.

ACTIVITY

To give students additional practice in applying rules for zeros, have them search their textbooks and other sources for length, mass, volume, or temperature measurements that contain zeros. Include some examples written in scientific notation. Ask students to identify which of the rules for zeros applies to each measurement. Remind them that more than one rule can apply if a measurement contains more than one zero. Then have students determine the number of significant figures in each example.

Discuss

The rules for rounding off calculated numbers can be compared with the old adage, "A chain is only as strong as its weakest link." Explain that an answer cannot be more precise than the least precise value used to calculate the answer. In addition and subtraction, the least precise value is the measurement with the fewest digits to the right of the decimal point. In multiplication and division, the least precise value is the measurement with the fewest significant figures.

Practice Problems Plus

Related Chapter Review Problem
Chapter Review problem 41 is related to Sample Problem 3-1.

Sample Problem 3-1

An engineer made the following measurements. How many significant figures are in each measurement?

a. 123 meters
b. 0.123 meter
c. 40 506 meters
d. 9.8000×10^4 meters
e. 30.0 meters
f. 22 meter sticks
g. 0.070 80 meter
h. 98 000 meters

1. ***ANALYZE Plan a problem-solving strategy.***
The location of each zero in the measurement and the location of the decimal point determine which of the rules apply for determining significant figures. These locations are known by examining each measurement value.

2. ***SOLVE Apply the problem-solving strategy.***
Examine each measurement and apply the rules for determining significant figures. All nonzero digits are significant (rule 1). Use rules 2–5 to determine if the zeros are significant.
a. 3 (rule 1)
b. 3 (rule 3)
c. 5 (rule 2)
d. 5 (rule 4)
e. 3 (rule 4)
f. unlimited (rule 6)
g. 4 (rules 2, 3, 4)
h. 2 (rule 5)

3. ***EVALUATE Do the results make sense?***
The rules for determining significant digits have been correctly applied in each case.

Practice Problems

5. Determine the number of significant figures in each measurement.
   a. 0.057 30 meter
   b. 8765 meters
   c. 0.000 73 meter
   d. 40.007 meters
6. How many significant figures are in each measurement?
   a. 143 grams
   b. 0.074 meter
   c. 8.750×10^{-2} gram
   d. 1.072 meters

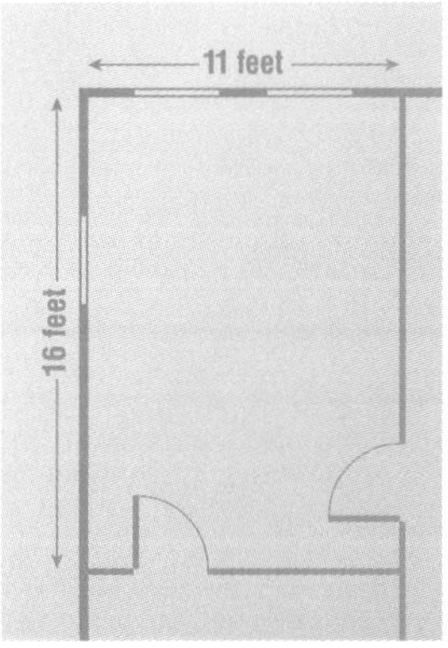

Figure 3.8
The room shown in the blueprint measures 11 feet by 16 feet. What is the calculated area of the room to the correct number of significant figures? ❶

Significant Figures in Calculations

Rounding Suppose you use a calculator to find the area of a floor that measures 7.7 meters by 5.4 meters. The calculator would give an answer of 41.58 square meters. The calculated area is expressed to four significant figures. However, each of the measurements used in the calculation is expressed to only two significant figures. It is important to know that the calculated area cannot be more precise than the measured values used to obtain it.

The calculated area must be rounded to make it consistent with the measurements from which it was calculated. In general, an answer cannot be more precise than the least precise measurement from which it was calculated.

To round a number, you must first decide how many significant figures the answer should have. This decision depends on the given measurements and on the mathematical process used to arrive at the answer. Once you know the number of significant figures your answer should have, round to that many digits, counting from the left. If the digit immediately to the right of the last significant digit is less than 5, it is simply dropped and the value of the last significant digit stays the same. If the digit in question is 5 or greater, the value of the digit in the last significant

Answer

❶ 1.8×10^2 ft^2

place is increased by 1. For example, rounding 56.312 meters to four significant figures produces the result 56.31 meters because 2, the digit to the right of the last significant digit, is less than 5. Rounding 56.316 meters gives the result 56.32 meters because 6, the digit to the right of the last significant figure, is greater than 5.

Sample Problem 3-2

Round each measurement to the number of significant figures shown in parentheses. Write the answers in scientific notation.

a. 314.721 meters (4)
b. 0.001 775 meter (2)
c. 64.32×10^{-1} meters (1)
d. 8792 meters (2)

1. *ANALYZE Plan a problem-solving strategy.*
Using the rules for determining significant figures, round each number. Then, apply the rules for expressing numbers in scientific notation.

2. *SOLVE Apply the problem-solving strategy.*
Count from the left and apply the rule to the digit immediately to the right of the digit to which you are rounding. The arrows point to the digit immediately following the last significant digit. (The number of significant figures for each is shown in parentheses.)

a. 314.721 meters (arrow under 2)
2 is less than 5, so do not round up.
314.7 meters (4) = 3.147×10^2 meters

b. 0.001 775 meter (arrow under the second 7)
7 is greater than 5, so round up.
0.0018 meter (2) = 1.8×10^{-3} meter

c. 64.32×10^{-1} meters (arrow under 4)
4 is less than 5, so do not round up.
60×10^{-1} meters (1) = 6 meters

(Note that 6 meters can be expressed in scientific notation as 6×10^0 meters.)

d. 8792 meters (arrow under 9)
9 is greater than 5, so round up.
8800 meters (2) = 8.8×10^3 meters

3. *EVALUATE Do the results make sense?*
The rules for rounding and for writing numbers in scientific notation have been correctly applied.

Practice Problems

7. Round each measurement to three significant figures. Write your answers in scientific notation.
a. 87.073 meters
b. 4.3621×10^8 meters
c. 0.015 52 meter
d. 9009 meters
e. 1.7777×10^{-3} meter
f. 629.55 meters

8. Round each measurement in Practice Problem 7 to one significant figure. Write your answers in scientific notation.

Chem ASAP!
Problem-Solving 7
Solve Problem 7 with the help of an interactive guided tutorial.

CHEMISTRY AND SCIENCE HISTORY

Antoine Lavoisier worked hard to establish the importance of accurate measurement in scientific inquiry. Lavoisier devised an experiment to test the Greek scientists' idea that when water was heated, it could turn into earth. For 100 days, Lavoisier boiled water in a glass flask constructed to allow steam to condense without escaping. He weighed the water and the flask separately before and after boiling. He found that the mass of the water had not changed. The flask, however, lost a small mass equal to the sediment he found in the bottom of it. Lavoisier proved that the sediment was not earth, but part of the flask etched away by the boiling water.

Practice Problems Plus

Related Chapter Review Problem
Chapter Review problem 42 is related to Sample Problem 3-2.

Additional Practice Problem
The Chem ASAP! CD-ROM contains the following problem: Round each measurement to two significant figures. Write your answers in scientific notation.
a. 94.592 grams
b. 2.4232×10^3 grams
c. 0.007 438 grams
d. 54 752 grams
e. 6.0289×10^{-3} grams
f. 405.11 grams
(**a.** 9.5×10^1 grams **b.** 2.4×10^3 grams **c.** 7.4×10^{-3} grams **d.** 5.5×10^4 grams **e.** 6.0×10^{-3} grams **f.** 4.1×10^2 grams)

3.2

Discuss

Many students may experience difficulty deciding how many significant figures to write when converting decimal notation to scientific notation. Point out that the decimal and scientific notation forms of a number or measurement always contain the same number of significant figures. Have students convert 0.00058, 0.000580, and 0.0005800 to scientific notation. (5.8×10^{-4}, 5.80×10^{-4}, 5.800×10^{-4}) **Point out that the decimal point is moved to a position just after the first nonzero digit.**

Discuss

Ask students to use a calculator to find the volume of one gram of aluminum if 15 grams have a volume of 41 cm^3. (2.7333333 cm^3) **Ask students if they think they know this volume to the nearest ten-millionth of a gram, the precision of the calculator answer. This exercise shows how important it is to have rules that specify which digits should be kept in an answer. Because calculators are not programmed to display the correct number of significant figures, they usually display more than the acceptable number of figures.**

Practice Problems Plus

Related Chapter Review Problems
Chapter Review problems 44a, 44c, and 44f are related to Sample Problem 3-3.

Additional Practice Problem
The Chem ASAP! CD-ROM contains the following problem: Find the total mass of four stones that weigh 10.32 grams, 11.81 grams, 124.678 grams, and 0.9129 gram. (147.72 grams)

Addition and Subtraction The answer to an addition or subtraction calculation should be rounded to the same number of decimal places (not digits) as the measurement with the least number of decimal places. Work through Sample Problem 3-3 below which provides examples of rounding in addition and subtraction calculations.

Sample Problem 3-3

Perform the following addition and subtraction operations. Give each answer to the correct number of significant figures.

a. 12.52 meters + 349.0 meters + 8.24 meters
b. 74.626 meters − 28.34 meters

1. *ANALYZE* ***Plan a problem-solving strategy.***
Perform the required math operation and then analyze each measurement to determine the number of decimal places required in the answer.

2. *SOLVE* ***Apply the problem-solving strategy.***
Round the answers to match the measurement with the least number of decimal places.

a. Align the decimal points and add the numbers.

	12.52	meters
	349.0	meters
+	8.24	meters
	369.76	meters

The second measurement (349.0 meters) has the least number of digits (one) to the right of the decimal point. Thus the answer must be rounded to one digit after the decimal point. The answer is rounded to 369.8 meters, or 3.698×10^2 meters.

b. Align the decimal points and subtract the numbers.

	74.626	meters
−	28.34	meters
	46.286	meters

The answer must be rounded to two digits after the decimal point to match the second measurement. The answer is 46.29 meters, or 4.629×10^1 meters.

3. *EVALUATE* ***Do the results make sense?***
The mathematical operations have been correctly carried out and the resulting answers are reported to the correct number of decimal places.

Practice Problems

9. Perform each operation. Give your answers to the correct number of significant figures.
a. 61.2 meters + 9.35 meters + 8.6 meters
b. 9.44 meters − 2.11 meters
c. 1.36 meters + 10.17 meters
d. 34.61 meters − 17.3 meters

10. Find the total mass of three diamonds that have masses of 14.2 grams, 8.73 grams, and 0.912 gram.

Chem ASAP!
Problem-Solving 10
Solve Problem 10 with the help of an interactive guided tutorial.

Multiplication and Division In calculations involving multiplication and division, you need to round the answer to the same number of significant figures as the measurement with the least number of significant figures.

You can see in **Figure 3.9** that the calculator answer (5.7672) must be rounded to three significant figures because each measurement used in the calculation has only three significant figures.

The position of the decimal point has nothing to do with the rounding process when multiplying and dividing measurements. The position of the decimal point is important only in rounding the answers of addition or subtraction problems.

Figure 3.9

This calculator was used to multiply the length and width measurements of a bolt of fabric, 3.24 meters by 1.78 meters, each of which has three significant figures. The area of the fabric is really not known with the precision suggested by the calculator. What is the product when correctly rounded? ❶

Sample Problem 3-4

Perform the following operations. Give the answers to the correct number of significant figures.

a. 7.55 meters × 0.34 meter
b. 2.10 meters × 0.70 meter
c. 2.4526 meters ÷ 8.4
d. 0.365 meter ÷ 0.0200

1. *ANALYZE* ***Plan a problem-solving strategy.***
Perform the required math operation and then analyze each of the original numbers to determine the correct number of significant figures required in the answer.

2. *SOLVE* ***Apply the problem-solving strategy.***

Round the answers to match the measurement with the least number of significant figures.

a. 7.55 meters × 0.34 meter = 2.567 square meters = 2.6 square meters
(0.34 meter has two significant figures.)

b. 2.10 meters × 0.70 meter = 1.47 square meters = 1.5 square meters
(0.70 meter has two significant figures.)

c. 2.4526 meters ÷ 8.4 = 0.291 976 meter = 0.29 meter
(8.4 has two significant figures.)

d. 0.365 meter ÷ 0.0200 = 18.25 meters = 18.3 meters
(Both numbers have three significant figures.)

3. *EVALUATE* ***Do the results make sense?***
The mathematical operations have been performed correctly, and the resulting answers are reported to the correct number of places.

Practice Problems

11. Solve each problem. Give your answers to the correct number of significant figures and in scientific notation.
a. 8.3 meters × 2.22 meters
b. 8432 meters ÷ 12.5
c. 35.2 seconds × 1 minute/60 seconds

12. Calculate the volume of a warehouse that has inside dimensions of 22.4 meters by 11.3 meters by 5.2 meters. (Volume = $l \times w \times h$)

Chem ASAP!

Problem-Solving 12

Solve Problem 12 with the help of an interactive guided tutorial.

Answer

❶ 5.77 square meters

3.2

Discuss

Many calculations in chemistry involve more than one operation. Students may be unsure about when to round off. Write a number of practice problems on the board that involve more than one operation. Examples include $(3.56 \times 10^{-3}) \times (7.325) \div 5.1$ and $16 - 9.27 + 150.13$. Remind students that when you perform a calculation on your calculator, the number of digits displayed is usually greater than the correct number of significant figures in the answer. When you use a calculator to do a problem with multiple steps, carry all the digits until the final answer and round off just once at the end.

Practice Problems Plus

Related Chapter Review Problems
Chapter Review problems 44b, 44d, 44e, and 45 are related to Sample Problem 3-4.

Additional Practice Problem
The Chem ASAP! CD-ROM contains the following problem: Calculate the volume of a house that has dimensions of 12.52 m by 36.86 m by 2.46 m. ($1.14 \times 10\ m^3$)

MINI LAB

Accuracy and Precision

ANALYSIS AND CONCLUSIONS

1. 4 for length; 3 for width
2. Measured values should be similar, but not necessarily identical for all students.
3. Significant digits for rounded-off answers are: area, 3; perimeter, 4. Some students may not round to the proper number of digits.
4. Answers will vary.

3 Assess

Evaluate Understanding

Ask students to explain the difference between precision and accuracy. Write the accepted value and one experimental value for the boiling point of a liquid on the board. Ask:

- **What is the percent error for this measurement?**

Ask students to state the rounding rules in their own words.

Reteach

On the board, demonstrate several different calculations for students. Remind them how to report the correct number of significant figures when using a calculator to carry out the operations. Review the definition of significant figures. Review with students the relationship between the number of places they move a decimal point in a number and the value of the exponent when they write the number in scientific notation.

Portfolio Project

If video equipment is available, students could record their interviews to share with the class. This approach would be useful if workers have measuring devices to display and describe.

MINI LAB

Accuracy and Precision

PURPOSE

To measure the dimensions of an object as accurately and precisely as possible and to apply rules for rounding answers calculated from the measurements.

MATERIALS

- index card (3″ × 5″)
- metric ruler

PROCEDURE

1. Use a metric ruler to measure in centimeters the length and width of an index card as accurately and precisely as you can. The hundredths place in your measurement should be estimated.
2. Calculate the perimeter [2 × (length + width)] and the area (length × width) of the index card. Write both your unrounded answers and your correctly rounded answers on the chalkboard.

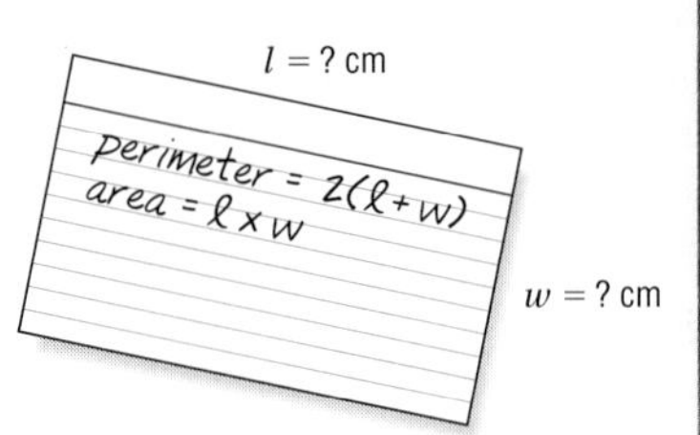

ANALYSIS AND CONCLUSIONS

1. How many significant figures are in your measurements of length and of width?
2. How do your measurements compare with those of your classmates?
3. How many significant figures are in your calculated value for the area? In your calculated value for the perimeter? Do your rounded answers have as many significant figures as your classmates' measurements?
4. Assume that the correct (accurate) length and width of the card are 12.70 cm and 7.62 cm, respectively. Calculate the percent error for each of your two measurements.

portfolio project

Interview workers who construct buildings or highways about the importance of accurate measurements.

section review 3.2

13. Explain the differences between accuracy, precision, and error of a measurement.
14. Determine the number of significant figures in each of the following measurements and calculation results.
    a. 12 basketball players
    b. 0.010 square meter
    c. 507 thumbtacks
    d. 0.070 020 meter
    e. 10 800 meters
    f. 5.00 cubic meters
15. Solve the following and express each answer in scientific notation.
    a. $(5.3 \times 10^4) + (1.3 \times 10^4)$
    b. $(7.2 \times 10^{-4}) \div (1.8 \times 10^3)$
    c. $10^4 \times 10^{-3} \times 10^6$
    d. $(9.12 \times 10^{-1}) - (4.7 \times 10^{-2})$
    e. $(5.4 \times 10^4) \times (3.5 \times 10^9)$
    f. $(1.2 \times 10^2) \times (8.9 \times 10^2)$
16. A technician experimentally determined the boiling point of octane to be 124.1 °C. The actual boiling point of octane is 125.7 °C. Calculate the error and the percent error.

 **Chem ASAP! Assessment 3.2** Check your understanding of the important ideas and concepts in Section 3.2.

Answers

SECTION REVIEW 3.2

13. Accuracy is a measure of how close a measurement is to the true value. Precision is a measure of how close a series of measurements are to one another. Error is the difference between the accepted value and the experimental value.
14. a. unlimited
    b. 2
    c. unlimited
    d. 5
    e. 3
    f. 3
15. a. 6.6×10^4
    b. 4.0×10^{-7}
    c. 10^7
    d. 8.65×10^{-1}
    e. 1.9×10^{14}
    f. 1.1×10^5
16. error: 1.6 °C; percent error: 1.3%

INTERNATIONAL SYSTEM OF UNITS

section 3.3

Weights and measures were among the earliest human inventions. Length units were used for construction; units of weight were needed for trading goods. Body parts made convenient units of length. An inch was measured with a thumb, a foot was the length of a foot, and a yard was the distance from one's nose to the tip of an extended arm. Archeologists think that polished stones found in the ruins of ancient Babylonian cities were standards of weight. The carat measures the mass of gold, silver, and gems. Arabs based the carat on the seed of the carob plant. **What system do modern scientists use for measurement?**

objectives

- List SI units of measurement and common SI prefixes
- Distinguish between the mass and weight of an object

key terms

- International System of Units (SI)
- meter (m)
- volume
- liter (L)
- weight
- kilogram (kg)
- gram (g)

Units of Length

When you make a measurement, you must assign the correct units to the numerical value. Without the units, it is impossible to communicate the measurement clearly to others. Imagine the confusion that would follow if someone told you to "walk five in that direction." Your immediate response would be "Five what? Feet, meters, yards, miles?"

All measurements depend on units that serve as reference standards. The standards of measurement used in science are those of the metric system. The metric system is important because of its simplicity and ease of use. All metric units are based on 10 or multiples of 10. As a result, you can convert between units easily. The metric system was originally established in France in 1790. The **International System of Units** (abbreviated **SI,** after the French name, *Le Systéme International d'Unités*) is a revised version of the metric system. The SI was adopted by international agreement in 1960. There are seven SI base units, which are listed in **Table 3.1.** From these base units, all other SI units of measurement can be derived. Derived units are used for measurements such as volume, density, and pressure.

Table 3.1

Units of Measurement

Quantity	SI base unit or SI derived unit	Symbol	Non-SI unit	Symbol
Length	meter*	m		
Volume	cubic meter	m^3	liter	L
Mass	kilogram*	kg		
Density	grams per cubic centimeter	g/cm^3		
	grams per milliliter	g/mL		
Temperature	kelvin*	K	degree Celsius	°C
Time	second*	s		
Pressure	pascal	Pa	atmosphere	atm
			millimeter of mercury	mm Hg
Energy	joule	J	calorie	cal
Amount of substance	mole*	mol		
Luminous intensity	candela*	cd		
Electric current	ampere*	A		

*denotes SI base unit

3.3

1 Engage

Use the Visual

Have students study the photograph and read the text that opens the section. Ask:

- **Compare the ancient Roman balance to the balances you use in lab.** (Answers will vary, but should include the need for standard masses to balance the object being measured.)
- **What system do modern scientists use for measurement?** (SI)

Explain that the SI units are preferred for scientific work. With the exception of those working in the sciences, the United States is the only country not using the metric system as a routine method of measurement. Have students give a few examples of applications in which the metric system is used in the United States. Examples may include 35-mm film; meter distances in track and field; the dimensions of tools (wrenches) and hardware (bolts); and road signs.

Check Prior Knowledge

To assess students' prior knowledge about the metric system, ask:

- **Where have you used the metric system in everyday life?**
- **Where have you seen English units replaced by metric units in everyday life?**
- **Where have you seen English and metric units displayed side by side?**

STUDENT RESOURCES

- Section Review 3.3, Ch. 3 Practice Problems and Quizzes from the Review Module (Ch. 1–4)
- Laboratory Manual: Experiment 4
- Laboratory Practical 3-1
- Small-Scale Chemistry Lab Manual: Experiment 3

TECHNOLOGY RESOURCES

Relevant technology resources include:

- Chem ASAP! CD-ROM
- Resource Pro CD-ROM

2 Teach

TEACHER DEMO

Bring in, or ask students to bring in, packages from consumer products with metric system measurements. Discuss which units are used on each package, and why each unit is appropriate.

Discuss

Discuss with students the importance of the SI units for scientists and how it relates to the use of scientific notation. (The units are always in increments of ten and can be expressed as powers of ten.)

Use the feature on p. 76 to explain the origin of the metric system and its development. Technological advancements have allowed scientists to produce more accurate standards for base units, e.g., a meter based on the speed of light.

ACTIVITY

On a gram balance that reads at least 0.01 g, have students measure the mass of several U.S. pennies minted between 1970 and today. Have students plot the mass of the pennies versus the year the penny was minted. Students should observe variations from year to year due to chance error, and a dramatic decrease in mass between 1982 and 1983. Ask students to propose a hypothesis for their findings and to design additional experiments to test their hypotheses.

Table 3.2

Commonly Used Prefixes in the Metric System				
Prefix	**Symbol**	**Meaning**	**Factor**	**Scientific notation**
mega	M	1 million times larger than the unit it precedes	1 000 000	10^6
kilo	k	1000 times larger than the unit it precedes	1000	10^3
deci	d	10 times smaller than the unit it precedes	1/10	10^{-1}
centi	c	100 times smaller than the unit it precedes	1/100	10^{-2}
milli	m	1000 times smaller than the unit it precedes	1/1000	10^{-3}
micro	μ	1 million times smaller than the unit it precedes	1/1 000 000	10^{-6}
nano	n	1000 million times smaller than the unit it precedes	1/1 000 000 000	10^{-9}
pico	p	1 trillion times smaller than the unit it precedes	1/1 000 000 000 000	10^{-12}

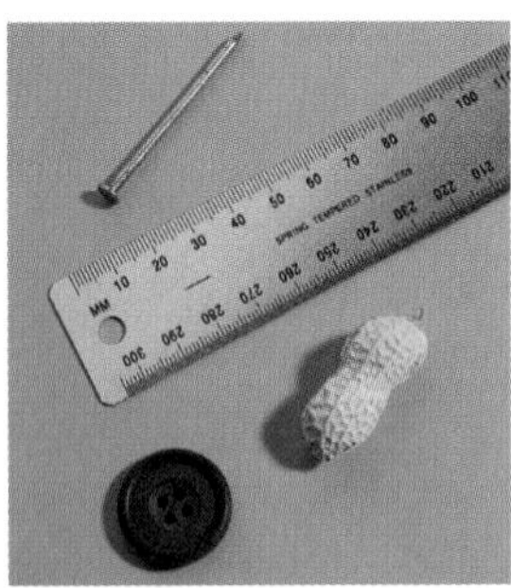

Figure 3.10
A meter stick is divided into 100 centimeters (cm). Each centimeter is further divided into 10 millimeters (mm). How many millimeters are there in a meter? ❶

All measured quantities can be reported in SI units. Sometimes, however, non-SI units are preferred for convenience or for practical reasons. **Table 3.1** also lists some derived SI units and non-SI units of measurement that are used in this textbook.

Size is an important property of matter. In SI, the basic unit of length, or linear measure, is the **meter (m)**. All measurements of length can be expressed in meters. (The length of a page in this book is about one-fourth of a meter.) For very large and very small lengths, however, it may be more convenient to use a unit of length that has a prefix. **Table 3.2** lists the prefixes in common use. For example, the prefix *milli-* means 1/1000, so a millimeter (mm) is 1/1000 of a meter, or 0.001 m. A hyphen (-) measures about 1 mm.

For large distances, it is usually most appropriate to express measurements in kilometers (km). The prefix *kilo-* means 1000, so 1 km equals 1000 m. A standard marathon distance race of about 42 000 m is more conveniently expressed as 42 km (42 × 1000 m). **Table 3.3** summarizes the relationships among units of length.

Table 3.3

Metric Units of Length			
Unit	**Symbol**	**Relationship**	**Example**
Kilometer	km	1 km = 10^3 m	length of about five city blocks ≈ 1 km
Meter	m	base unit	height of doorknob from the floor ≈ 1 m
Decimeter	dm	10^1 dm = 1 m	diameter of large orange ≈ 1 dm
Centimeter	cm	10^2 cm = 1 m	width of shirt button ≈ 1 cm
Millimeter	mm	10^3 mm = 1 m	thickness of dime ≈ 1 mm
Micrometer	μm	10^6 μm = 1 m	diameter of bacterial cell ≈ 1 μm
Nanometer	nm	10^9 nm = 1 m	thickness of RNA molecule ≈ 1 nm

MEETING DIVERSE NEEDS

LEP Have students search magazines and newspapers for examples of SI units. Have them report seasonal day/night temperature ranges for their countries of origin.

Gifted Have students research the efforts to make the metric system standard in the United States. Who has promoted the change? What progress has been made? What obstacles remain?

Answer

❶ 1000

Figure 3.11

These photographs should give you some idea of the relative sizes of some different units of volume. The volume of 20 drops of liquid from a medicine dropper is approximately 1 milliliter (mL). A sugar cube is 1 centimeter (cm) on each edge and has a volume of 1 cubic centimeter (cm^3). These two volumes are equivalent; that is, a volume of 1 mL is the same as a volume of 1 cm^3. A gallon of milk has about twice the volume of a two-liter bottle of soda. Note that a gallon is a non-SI unit and it is used here only for comparison purposes.

Units of Volume

The space occupied by any sample of matter is called its **volume.** You calculate the volume of any cubic or rectangular solid by multiplying its length by its width by its height. The unit for volume is thus derived from units of length. The SI unit of volume is the amount of space occupied by a cube that is 1 m along each edge. This volume is a cubic meter (m^3). An automatic dishwasher has a volume of about 1 m^3.

A more convenient unit of volume for everyday use is the liter, a non-SI unit. A **liter** (L) is the volume of a cube that is 10 centimeters (cm) along each edge (10 cm $\times$ 10 cm $\times$ 10 cm $=$ 1000 cm^3 $=$ 1 L). A decimeter (dm) is equal to 10 cm, so 1 L is also equal to 1 cubic decimeter (dm^3). A smaller non-SI unit of volume is the milliliter (mL); 1 mL is 1/1000 of a liter. Thus there are 1000 mL in 1 L. Because 1 L is defined as 1000 cm^3, 1 mL and 1 cm^3 are the same volume. The units milliliter and cubic centimeter are thus used interchangeably. **Table 3.4** summarizes the most commonly used relationships among units of volume.

Table 3.4

Metric Units of Volume								
Unit	**Symbol**	**Relationship**			**Example**		
Liter	L	base unit			quart of milk	≈	1 L
Milliliter	mL	10^3 mL	=	1 L	20 drops of water	≈	1 mL
Cubic centimeter	cm^3	1 cm^3	=	1 mL	cube of sugar	≈	1 cm^3
Microliter	μL	10^6 μL	=	1 L	crystal of table salt	≈	1 μL

3.3

TEACHER DEMO

Show students how to convert some common English units into SI units. Examples:
1 inch = 2.54 cm
1 kilogram = 2.2 pounds
1 kilometer = 0.60 miles
1 foot = 30.5 cm
1 ounce = 28 grams
For length measurements, use a meter stick with English units on it.

ACTIVITY

Have students measure objects with irregular volumes, by measuring the amount of water displaced in a graduated cylinder. Small pebbles can be used. In another activity, have students use a stopwatch to measure the time required for an object to drop a few meters. Compare the precision of several measurements.

Think Critically

Ask students how they might infer that one cubic centimeter and one milliliter represent the same volume. Have them write out a numbered list of statements of fact or observations they would need in order to make that inference. You may wish to suggest that students work in groups outside class to design and construct a decimeter cube. The cube should be lined with a plastic bag and be strong enough to hold water. On an assigned day, have each group bring its cube to class and measure the volume of water it holds.

Discuss

Stress the need to choose an appropriate measurement unit. Data can be reported accurately in various related units (mm, cm, m), but it is best to choose the unit closest in scale to the object being measured.

- **The unit of mass most often used by chemists is the gram; not the SI base unit of mass, the kilogram. Kilograms of chemicals are too large for laboratory experiments.**
- **The density of most gases is measured in grams per liter; the densities of most solids and liquids are measured in grams per cubic centimeter.**
- **The milliliter is the most common unit of volume used by chemists. The cubic meter is much too large a volume for most laboratory experiments.**

ACTIVITY

Students need to practice estimating with unfamiliar SI units to develop an intuitive sense of their size. Have students use mL and L to estimate the volumes of a tablespoon of water, a 12 oz. can of soda, and a half-gallon of juice or milk. (Note: you could use empty food containers and fill them with water.) Then use graduated cylinders to measure the volume of each sample. This also provides an opportunity for students to practice converting milliliters to liters.

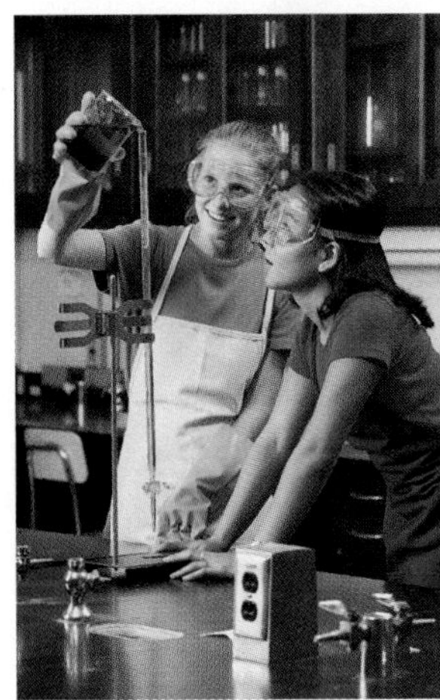

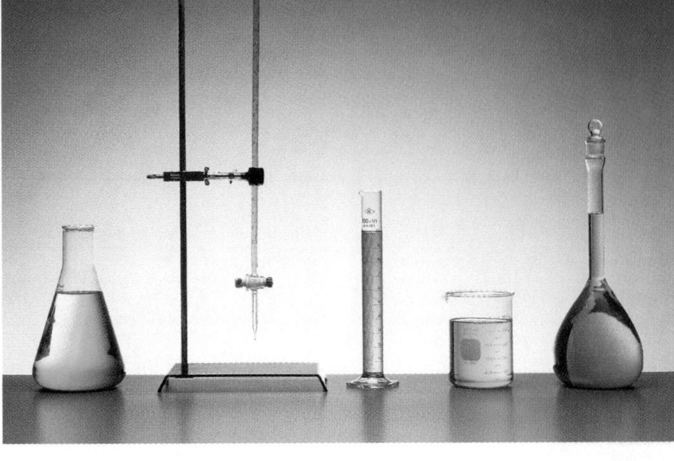

Figure 3.12

Shown left to right are five types of glassware used to measure volume: Erlenmeyer flask, buret, graduated cylinder, beaker, and volumetric flask. Beakers, Erlenmeyer flasks, and graduated cylinders are used to measure approximate volumes. Volumetric flasks and burets allow more precise measurements. What would you use to measure a large volume, such as 1 L, accurately? 1

Figure 3.13

An astronaut's weight on the moon is one-sixth as much as it is on Earth. Earth exerts six times the force of gravity as the moon. The astronaut's lowered weight accounts for his ability to jump high on the moon. How does his mass on the moon compare with his mass on Earth? 2

There are many devices for measuring the volume of a liquid. Examples of volumetric glassware are shown in **Figure 3.12.** A graduated cylinder is useful for dispensing approximate volumes. A pipet or buret must be used, however, when accuracy is important. A volumetric flask contains a specified volume of liquid when it is filled to the calibration mark. Volumetric flasks are available in many sizes. A syringe is used to measure small volumes of liquids for injection.

The volume of any solid, liquid, or gas will change with temperature (although the change is much more dramatic for gases). Accurate volume-measuring devices are calibrated at a given temperature—usually 20 degrees Celsius (20 °C), which is about normal room temperature.

Units of Mass

The astronaut shown saluting on the surface of the moon in **Figure 3.13** weighs one-sixth of what he weighs on Earth. The reason for this difference is that the force of gravity on Earth is about six times what it is on the moon. **Weight** is a force that measures the pull on a given mass by gravity. Weight is different from mass, which is a measure of the quantity of matter. Although the weight of an object can change with its location, its mass remains constant regardless of its location—whether it is on Earth or on the moon. Objects can thus become weightless, but they can never become massless.

The mass of an object is measured in comparison to a standard mass of 1 **kilogram (kg)**, which is the basic SI unit of mass. A kilogram was originally defined as the mass of 1 L of liquid water at 4 °C. A cube of water

Table 3.5

Metric Units of Mass			
Unit	**Symbol**	**Relationship**	**Example**
Kilogram (base unit)	kg	1 kg = 10^3 g	small textbook ≈ 1 kg
Gram	g	1 g = 10^{-3} kg	dollar bill ≈ 1 g
Milligram	mg	10^3 mg = 1 g	ten grains of salt ≈ 1 mg
Microgram	μg	10^6 μL = 1 g	particle of baking powder ≈ 1 μg

at 4 °C measuring 10 cm on each edge would have a volume of 1 L and a mass of 1000 grams (g), or 1 kg. A **gram (g)** is 1/1000 of a kilogram and is a more commonly used unit of mass in chemistry. The mass of 1 cm^3 of water at 4 °C is 1 g. The relationships among units of mass are shown in **Table 3.5.**

You can use a platform balance to measure the mass of an object. The object is placed on one side of the balance, and standard masses are added to the other side until the balance beam is level, as shown in **Figure 3.14.** The unknown mass is equal to the sum of the standard masses. Laboratory balances range from very sensitive instruments with a maximum capacity of only a few milligrams to devices for measuring quantities in kilograms. An analytical balance is used to measure objects of less than 100 g and can determine mass to the nearest 0.0001 g (0.1 mg).

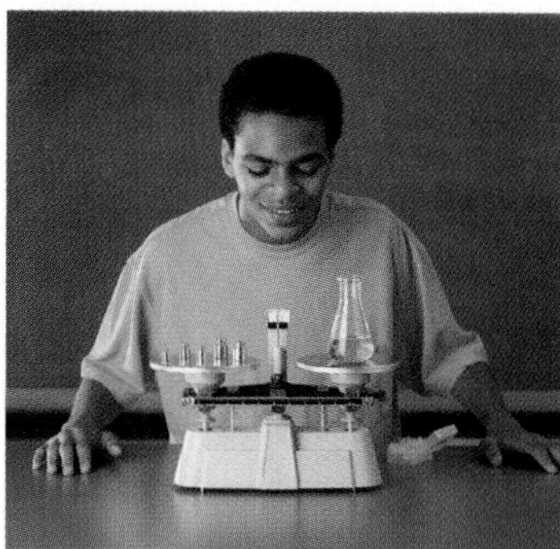

Figure 3.14
A platform balance compares an unknown mass with a known mass. The mass of the water and the flask is the same as the sum of the standard masses.

section review 3.3

17. Name the quantity measured by each of the following SI units and give the SI symbol of the unit.
    - a. mole
    - b. kilogram/cubic meter
    - c. second
    - d. pascal
    - e. meter
    - f. kilogram
18. State the difference between mass and weight.
19. What is the symbol and meaning of each prefix?
    - a. *milli-*
    - b. *nano-*
    - c. *deci-*
    - d. *centi-*
20. As you climbed a mountain and the force of gravity decreased, would your weight increase, decrease, or remain constant? How would your mass change?
21. What is the volume of a paperback book 21 cm tall, 12 cm wide, and 3.5 cm thick?
22. List these units in order, from largest to smallest.
    - a. 1 dm^3
    - b. 1 μL
    - c. 1 mL
    - d. 1 L
    - e. 1 cL
    - f. 1 dL

Chem ASAP! **Assessment 3.3** Check your understanding of the important ideas and concepts in Section 3.3.

3.3

3 Assess

Evaluate Understanding

To determine students intuitive understanding of the size of metric units, have students compare some English and metric units. For example, have students relate liters, grams, and kilometers to quarts, pounds, and miles, respectively. Display Table 3.2 on an overhead projector. Ask:

- **What is an SI unit for mass?** (gram)

To see if students understand the relative magnitude of a one-gram mass, have them estimate the mass in grams of objects such as a book of stamps, a paperclip, a contact lens, a packet of ketchup, a rubber ball, a lemon seed, a quarter, a plastic cup, or a roll of paper towels. Determine the actual mass of the selected items. Then have students convert the results to kilograms or milligrams to test their knowledge of prefixes.

Reteach

Review Tables 3.3, 3.4, and 3.5 with students. Display each on an overhead projector and go over the pronunciations of the prefixes as well as their numerical relationships.

Hold up and describe as many different pieces and sizes of glassware used for volume measurement as possible, including pipets, burets, graduated cylinders, and volumetric flasks. To help students develop a feel for metric volumes, fill the glassware with water and transfer the liquid to plastic pint, quart, or gallon containers from home.

Answers

1 a volumetric flask
2 His mass is the same, only his weight differs.

SECTION REVIEW 3.3

17. a. amount of substance, mol
    b. density, kg/m^3
    c. time, s
    d. pressure, Pa
    e. length, m
    f. mass, kg
18. Mass is a measure of the amount of matter in an object. Weight is a measure of the force of gravity on an object.
19. a. m, 10^{-3}
    b. n, 10^{-9}
    c. d, 10^{-1}
    d. c, 10^{-2}
20. Your weight would decrease; your mass would remain constant.
21. 8.8×10^2 cm^3
22. **a.** and **d., f., e., c., b.**

1 Engage

Use the Visual

Have students study the photograph and read the text that opens the section. Ask:

- **Why do cranberries float in water?** (They are less dense than water—have less mass per unit volume. Note that a person can float on water because his/her mass is distributed over a large area of water.)
- **What measurements would you need to make to determine whether an object would float in water?** (Measure the object's volume and mass; then compute its density and compare the result to the density of water.)

Check Prior Knowledge

To assess students' prior knowledge about density, hold up a piece of lead in one hand and a similar size piece of aluminum in the other hand and ask:

- **Why does this piece of lead seem heavier than this piece of aluminum?** (Lead is more dense than aluminum.)

Next, hold up a bottle of oil and vinegar salad dressing and ask:

- **Why does oil float on water? Why doesn't water float on oil?** (Oil is less dense than water.)

Finally, hold up a glass of water with ice cubes floating on top and ask:

- **Why does ice float on liquid water?** (Ice is less dense than liquid water.)

section 3.4

DENSITY

objectives

- Calculate the density of an object from experimental data
- List some useful applications of the measurement of specific gravity

key terms

- density
- specific gravity
- hydrometer

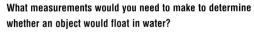

Have you ever wondered why some objects float in water, while others sink? If you think that these cranberries float because they are lightweight, you are only partially correct. **What measurements would you need to make to determine whether an object would float in water?**

Determining Density

Has anyone ever tried to trick you with this question: "Which is heavier, a pound of lead or a pound of feathers?" Most people would not give the question much thought and would incorrectly answer "lead." Of course, a pound of lead has the same mass as a pound of feathers. What concept, instead of mass, are people really thinking of when they answer this question?

Most people are incorrectly applying a perfectly correct idea: namely, that if a piece of lead and a feather of the same volume are weighed, the lead would have a greater mass than the feather. It would take a much larger volume of feathers to equal the mass of a given volume of lead.

Table 3.6

Relationship Between Volume and Density for Identical Masses of Common Substances

Substance	Cube of substance (face shown actual size)	Mass (g)	Volume (cm^3)	Density (g/cm^3)
Lithium		10	19	0.53
Water		10	10	1.0
Aluminum		10	3.7	2.7
Lead		10	0.88	11.4

Student Resources

From the Teacher's Resource Package, use:

- Section Review 3.4, Ch. 3 Practice Problems and Quizzes from the Review Module (Ch. 1–4)
- Laboratory Recordsheet 3-2
- Laboratory Manual: Experiment 4
- Laboratory Practical 3-2

Technology Resources

Relevant technology resources include:

- Chem ASAP! CD-ROM
- Resource Pro CD-ROM
- ActivChemistry CD-ROM: *Exploring Density*
- Chemistry Alive! Videodisc: *Giant Density Column, Methane Bubbles*

Table 3.7

Densities of Some Common Materials			
Solids and Liquids		Gases	
Material	Density at 20 °C (g/cm^3)	Material	Density at 20 °C (g/L)
Gold	19.3	Chlorine	2.95
Mercury	13.6	Carbon dioxide	1.83
Lead	11.4	Oxygen	1.33
Aluminum	2.70	Air	1.20
Table sugar	1.59	Nitrogen	1.17
Water (4 °C)	1.000	Neon	0.84
Corn oil	0.922	Ammonia	0.718
Ice (0 °C)	0.917	Methane	0.665
Ethanol	0.789	Helium	0.166
Gasoline	0.66–0.69	Hydrogen	0.084

The important relationship in this case is between the object's mass and its volume. This relationship is called density. **Density** is the ratio of the mass of an object to its volume.

$$\text{Density} = \frac{\text{mass}}{\text{volume}}$$

A 10.0-cm^3 piece of lead, for example, has a mass of 114 g. What, then, is the density of lead? If you substitute the mass and volume into the equation above, you can see that the density of lead is as follows.

$$\frac{114\ \text{g}}{10.0\ \text{cm}^3} = 11.4\ \text{g/cm}^3$$

Note that when mass is measured in grams, and volume in cubic centimeters, density has units of grams per cubic centimeter (g/cm^3).

Table 3.6 compares the density of four substances. Why does each 10-g sample have a different volume? The volumes vary because the substances have different densities. Density is a characteristic property that depends only on the composition of a substance, not on the size of the sample. With a mixture, density can vary because the composition of a mixture can vary.

What do you think will happen if glycerine (glycerol) and corn oil are poured into a beaker of water? Using **Table 3.7,** you can see that the density of corn oil is less than the density of water. For that reason, the corn oil floats on top of the water. As you can see in **Figure 3.15,** the glycerol sinks below the water because its density is greater than the density of water. Other liquids with different densities are also shown.

You have probably seen a helium-filled balloon rapidly rise to the ceiling when it is released. Whether a gas-filled balloon will sink or rise when released depends on how the density of the gas compares with the density of air. The densities of various gases are given in **Table 3.7.** Would a carbon dioxide-filled balloon sink or rise in air? 

What happens to the density of a substance as its temperature increases? Experiments show that the volume of most substances increases as the temperature increases. Meanwhile, the mass remains the same

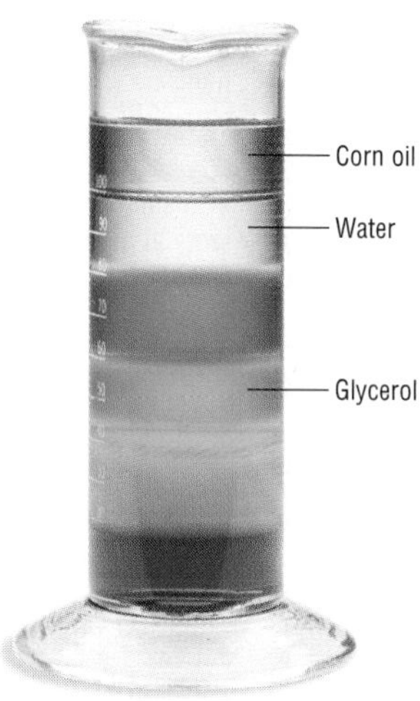

Figure 3.15
Because of differences in density, the blue-colored glycerol sinks below the surface of the water, and the amber-colored corn oil floats on top of the water. Is the red-colored liquid more or less dense than glycerol? How do you know? 2

Chem ASAP!
Simulation 1
Rank materials according to their densities.

3.4

TEACHER DEMO

Pour water into a graduated cylinder until it is half full. Read the volume. Immerse a large fishing weight with known mass in the water. Measure the volume of water displaced. Use 1 mL = 1 g to determine the mass of the water. Ask:

- Is this mass greater than or less than the mass of the sinker?

Drop a piece of balsa wood with known mass in the container and again measure the mass of the water that is displaced. Ask:

- Is this mass greater than or less than the mass of the balsa wood?

2 Teach

TEACHER DEMO

Measure the volume and mass of three or four cubes made from a material such as wood, metal, or marble. The cubes should have different volumes so you can show the constant ratio of mass to volume. On the board or an overhead projector, make a plot of mass versus volume. Show students how they can divide mass by volume to find density and how they can use the slope of the graph to find density. If you get noticeable variation in density with wood or marble cubes, explain that density is a physical constant only for pure substances.

Answers

1 It will sink.
2 Less dense; it is floating above the glycerol.

Meeting Diverse Needs

At Risk and **LEP** Have students write definitions of density in English and, if appropriate, in a native language. Make sure they include the expression density = mass/volume. Have them find photos of objects containing substances from Table 3.7. They could label each photo with the substance name in multiple languages and the density, whose units are part of a universal language.

CHEMath

Teaching Tips

Write the example equations on the board. Have students complete a similar problem which asks them to solve for the height of a triangle given its area. Some students may have trouble with the rearrangement and isolation of variables.

Use a platform balance as a model for an equation. Start with equal sets of masses on both sides. Show how adding or subtracting mass from one side requires an identical addition or subtraction from the other side to maintain the balance or equality.

Answers to Practice Problems

A. Money earned equals rate per hour times number of hours worked.
B. °C = K − 273
C. K = 473
D. mass = density × volume
E. 2.24 g/cm^3
F. volume = $\frac{\text{mass}}{\text{density}}$
G. $y = 2x^2z - \frac{3}{2}x^2$

CHEMath

WHAT IS AN EQUATION?

You will encounter numerous equations as you read through this textbook. You should consider every equation as a powerful and easy-to-use tool for simplifying and applying complex relationships. Furthermore, once you understand the basics of what equations are and how they are used, you will find they are usually solved using simple math.

The Equation An equation is a mathematical sentence that uses an equal sign (=) to represent two different expressions that have the same value. In addition to the equal sign, an equation may contain a combination of words, numbers, variables, or mathematical functions. An example of an equation that contains numbers is:

$$5 + 4 = 9$$

Each side of the equation is separated by the equal sign and is equivalent; 5 + 4 is simply another way to write 9. An example of a word equation is:

The perimeter of a square is equal to the sum of the lengths of its four equal sides.

Although the above statement is correct, it can be stated much more concisely by the following mathematical equation:

$$\text{perimeter} = 4 \times \text{length of a side}$$

This can be further simplified by replacing the words with variables that represent the words. Letting p represent the perimeter, and L represent the length of each side yields,

$$p = 4 \times L$$
$$p = 4L$$

Units in Equations 5 + 4 = 9 is an example of an equation that contains no units. Many equations do involve units, however, and you can use the units as a way to check that the equation is correct. Because both sides of an equation are equal, the units on each side of an equation must also be equal. Consider the following equation:

$$\text{mass}_{\text{total}} = \text{mass}_{\text{beaker}} + \text{mass}_{\text{chemical}}$$

Both sides of this equation must contain units of mass, such as grams or kilograms. If the units on both sides of the equation are not the same, the answer must be incorrect.

Solving Equations It is very important to realize that both sides of an equation are always equal to each other. This is important because you can use this fact to help solve for an unknown value.

You only need to remember a single rule when solving an equation for a specific quantity: whatever you do on one side of the equation must be done on the other side of the equation. For example, reconsider the perimeter equation.

$$p = 4L$$

If you need to calculate the length of a single side of the square (L), you must solve the equation for L. In other words the variable L must be isolated from everything else. To isolate L you can divide the right side of the equation by 4. However, remember you must also divide the left side by 4 as shown below.

$$p = 4L \quad \frac{p}{4} = \frac{4L}{4} \quad \frac{p}{4} = \frac{\cancel{4}L}{\cancel{4}} \quad \frac{p}{4} = L$$
$$L = \frac{p}{4}$$

Practice Problems

A. Write a word equation for money earned each week from a part-time job.

B. Solve the equation K = °C + 273 for °C.

C. Using the equation K = °C + 273, calculate K if °C = 200.

D. Solve the equation density = $\frac{\text{mass}}{\text{volume}}$ for mass.

E. Using the equation density = $\frac{\text{mass}}{\text{volume}}$, calculate density if mass = 228 g and volume = 102 cm^3.

F. Repeat problem D, but solve for volume.

G. Solve the following equation for y.

$$\frac{2y}{x^2} + 3 = 4z$$

Figure 3.16
Solid paraffin sinks in melted paraffin (left), but ice floats in water (right). What does this tell you about the densities of each of these two substances in the solid and liquid states? ❶

despite the temperature and volume changes. Because density is mass divided by volume, the density of a substance generally decreases as its temperature increases. As you will learn in Chapter 17, water is an important exception. Over a certain range of temperatures, the volume of water increases as its temperature decreases. Ice, or solid water, floats because it is less dense than liquid water. Compare the densities of the substances shown in **Figure 3.16.**

Sample Problem 3-5

A copper penny has a mass of 3.1 g and a volume of 0.35 cm^3. What is the density of copper?

1. *ANALYZE* ***List the knowns and the unknown.***

Knowns:
- mass = 3.1 g
- volume = 0.35 cm^3

Unknown:
- density = ? g/cm^3

Use the known values and the definition of density,

$$\text{density} = \frac{\text{mass}}{\text{volume}},$$

to calculate the density of copper.

2. *CALCULATE* ***Solve for the unknown.***

The equation is already set up to solve for the unknown. Substitute the known values for mass and volume, and calculate the density.

$$\text{density} = \frac{\text{mass}}{\text{volume}} = \frac{3.1 \text{ g}}{0.35 \text{ cm}^3} = 8.8571 \text{ g/cm}^3$$

$$= 8.9 \text{ g/cm}^3 \text{ (rounded to two significant figures)}$$

3. *EVALUATE* ***Does the result make sense?***

A piece of copper with a volume of about 0.3 cm^3 has a mass of about 3 g. Thus, about three times that volume of copper, 1 cm^3, should have a mass three times larger, about 9 g. This estimate agrees with the calculated result.

Practice Problems

23. A student finds a shiny piece of metal that she thinks is aluminum. In the lab, she determines that the metal has a volume of 245 cm^3 and a mass of 612 g. Calculate the density. Is the metal aluminum?
24. The density of silver at 20 °C is 10.5 g/cm^3. What is the volume of a 68-g bar of silver?

Chem ASAP!

Problem-Solving 24

Solve Problem 24 with the help of an interactive guided tutorial.

3.4

Use the Visual

Have students study the photograph and read the caption in Figure 3.15. The caption states that glycerol is more dense than water, but its density is not given in Table 3.7. Have students check a handbook to verify the statement made in the caption. Explain why density values have units but specific gravity values do not.

Discuss

Show how the density equation can be used to determine the volume or mass of an object. For example,
Density = Mass/Volume,
Mass = Density × Volume,
or Volume = Mass/Density.

Practice Problems Plus

Related Chapter Review Problem

Chapter Review problem 61 is related to Sample Problem 3-5.

Additional Practice Problem

The Chem ASAP! CD-ROM contains the following problem: The density of gold is 19.3 g/cm^3 at 20 °C. What is the mass of a 0.18-cm^3 gold ring? (3.5 g)

Chemistry Alive!

Giant Density Column — Play

Methane Bubbles — Play

Answer

❶ Solid paraffin is denser than its liquid. Ice, however, is less dense than liquid water.

ACTIVITY

Show students how to use a hydrometer to measure the specific gravity of a liquid. Ask:

- How does the density of a solution of salt (or sugar) water compare to a solution of pure water? (The density of salt water is greater than the density of pure water.)

Have students measure the specific gravity of diet and regular colas using the hydrometer. Ask students to hypothesize about their findings and to propose another measurement that could be made to test their hypothesis. (Measure the densities of both liquids using a balance and a graduated cylinder.)

3 Assess

Evaluate Understanding

Ask:

- **What are the most common units for expressing the density of a substance?** (g/cm^3)
- **How can density be used to determine whether a metal sample is aluminum or zinc?** (Measure and compare the densities of the samples to the accepted reference values.)

Reteach

Remind students that the density equation can be rewritten as Mass = Density × Volume, or Volume = Mass/Density. Have students examine the application of these alternate equations by asking them to find the thickness of a square piece of aluminum. Provide students with aluminum squares and ask them to measure the area and mass of each. Have students use the density of aluminum in Table 3.7 to calculate the thickness of the foil.

Specific Gravity

Specific gravity is a comparison of the density of a substance with the density of a reference substance, usually at the same temperature. Water at 4 °C, which has a density of 1 g/cm^3, is commonly used as the reference substance. In that case,

$$\text{Specific gravity} = \frac{\text{density of substance } (\text{g}/\cancel{\text{cm}^3})}{\text{density of water } (\text{g}/\cancel{\text{cm}^3})}$$

Note that the units in the equation cancel — specific gravity has no units.

The specific gravity of a liquid can be measured with a device called a **hydrometer.** As is shown in **Figure 3.17a,** the depth to which the hydrometer sinks depends on the specific gravity of the liquid tested. The calibration mark on the hydrometer stem at the surface of the liquid indicates the specific gravity of the liquid.

Specific-gravity measurements are commonly used for various practical purposes. A physician uses the measured specific gravity of a patient's urine to help diagnose certain diseases, such as diabetes. You can check the condition of the antifreeze in your car by measuring the specific gravity of the solution in the radiator. The hydrometer in **Figure 3.17** is being used to measure the specific gravity of the acid in an automobile battery.

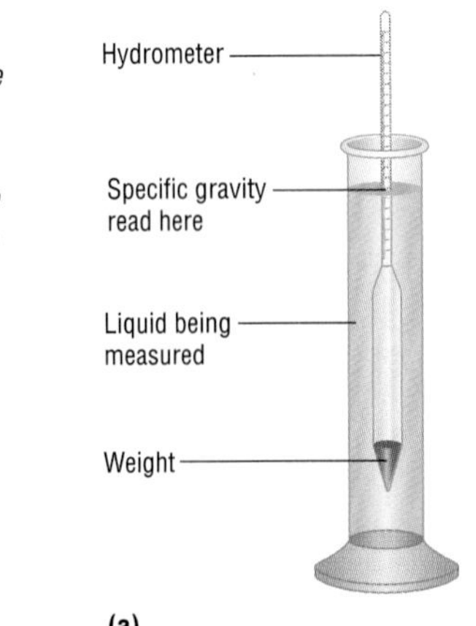

Figure 3.17
(a) *A hydrometer is a sealed tube with a weight in the bottom. It is used to measure the specific gravity of a liquid. The higher the hydrometer floats, the higher the specific gravity of the liquid tested.* ***(b)*** *In the photograph, the concentration of acid in an automobile battery is checked with a hydrometer.*

section review 3.4

25. How is density calculated from measured data?

26. A weather balloon is inflated to a volume of 2.2×10^3 L with 37.4 g of helium. What is the density of helium, in grams per liter?

27. List some applications of the measurement of specific gravity.

28. A plastic ball with a volume of 19.7 cm^3 has a mass of 15.8 g. What is its density? Would this ball sink or float in a container of gasoline?

29. Given samples of gold, gasoline, ice, mercury, lead, and aluminum, which substances have the highest and lowest specific gravities?

Chem ASAP! **Assessment 3.4** Check your understanding of the important ideas and concepts in Section 3.4.

Answers

SECTION REVIEW 3.4

25. Mass is divided by volume.

26. **a.** 1.7×10^{-2} g/L

27. checking urine to diagnose patients, measuring acid concentration in a car battery, checking antifreeze solution, etc.

28. 0.802 g/cm^3; It would sink.

29. highest: **a.** gold; lowest: **b.** gasoline

SMALL-SCALE LAB

MEASUREMENT

PURPOSE

To make precise and accurate measurements and to use fundamental data to calculate derived quantities.

MATERIALS

- pencil
- paper
- ruler
- balance
- rectangular block
- calculator

PROCEDURE

1. Use your ruler to measure the length, width, and height of a rectangular block in centimeters. Be sure to estimate between the lines of the ruler's smallest markings and report each measurement so that it includes the estimated number. Draw a diagram of the block and label its dimensions.
2. Measure and record the mass of the block. Estimate between the lines of the smallest markings of the balance and report your measurement in grams. (*Note:* If you are using a digital balance, the last number often blinks, or oscillates, between two or more numbers. Recording this last digit is similar to estimating between the markings on a balance.)

ANALYSIS

Using your experimental data, record the answers to the following questions.

1. How many significant figures do each of your measurements have?
2. Calculate the perimeter of each face of the rectangular block. Report the answer to the correct number of significant figures. (The perimeter of a face is the sum of the four edges of that face.)
3. Calculate the area of each face. Include the proper unit with your calculated value for area. Round your value to the correct number of significant figures. (Area = length × width)
4. Calculate the volume of the block in units of cubic centimeters. Round your value to the correct number of significant figures. (Volume = length × width × height)
5. Express the volume of your block in liters. ($1\ \text{L} = 1000\ \text{cm}^3$)
6. Express the mass of your block in milligrams. (1 g = 1000 mg)

YOU'RE THE CHEMIST

The following small-scale activities allow you to develop your own procedures and analyze the results.

1. **Analyze It!** Density is a measure of how tightly matter is packed together.

$$\text{Density} = \frac{\text{mass}}{\text{volume}}$$

Calculate the density of the block in units of grams per cubic centimeter (g/cm^3). Convert your calculated density into units of grams per liter (g/L), milligrams per cubic centimeter (mg/cm^3), and kilograms per liter (kg/L). (1 kg = 1000 g)

2. **Design It!** Make the necessary measurements to calculate the volume and density of a cylindrical block. The volume is determined from the following equation.

$$V = \pi r^2 h$$

$\pi \approx 3.1416$
r = radius
h = height

3. **Design It!** Make the necessary measurements to calculate the volume and density of a block that has had one or more holes drilled in it. Assume each hole is cylindrical and account for them in your calculations. What approximations must you make?

SMALL-SCALE LAB

MEASUREMENT

OVERVIEW

Procedure Time: 20 minutes
This lab is designed to give students practice in measurement and practice in using significant figures in calculations involving measured data.

TEACHING SUGGESTIONS

Use painted wooden blocks of various sizes available at toy stores.

PROCEDURE

1.

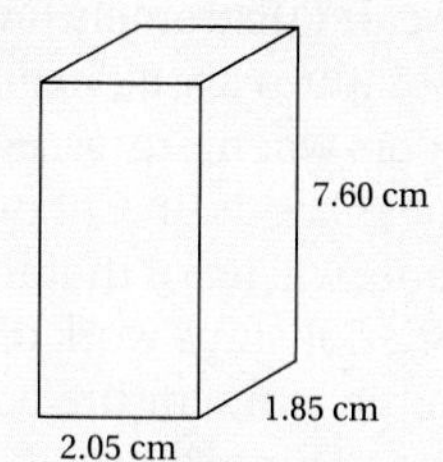

2. Mass = 27.092 g.

ANALYSIS

1. Each length has three significant figures. The mass has five.
2. $P = 7.60\ \text{cm} + 7.60\ \text{cm} + 2.05\ \text{cm} + 2.05\ \text{cm} = 19.30\ \text{cm}$
   $P = 7.60\ \text{cm} + 7.60\ \text{cm} + 1.85\ \text{cm} + 1.85\ \text{cm} = 18.90\ \text{cm}$
   $P = 2.05\ \text{cm} + 2.05\ \text{cm} + 1.85\ \text{cm} + 1.85\ \text{cm} = 7.80\ \text{cm}$
3. $A = 7.60\ \text{cm} \times 2.05\ \text{cm} = 15.6\ \text{cm}^2$
   $A = 7.60\ \text{cm} \times 1.85\ \text{cm} = 14.1\ \text{cm}^2$
   $A = 1.85\ \text{cm} \times 2.05\ \text{cm} = 3.79\ \text{cm}^2$
4. $V = 7.60\ \text{cm} \times 2.05\ \text{cm} \times 1.85\ \text{cm} = 28.8\ \text{cm}^3$
5. $x\ \text{L} = 28.8\ \cancel{\text{cm}^3} \times 1\ \text{L} / 1000\ \cancel{\text{cm}^3} = 0.0288\ \text{L}$
6. $x\ \text{mg} = 27.092\ \cancel{\text{g}} \times 1000\ \text{mg}/\cancel{\text{g}} = 27\ 092\ \text{mg}$

YOU'RE THE CHEMIST

1. $d = 27.092\ \text{g}/28.8\ \text{cm}^3 = 0.941\ \text{g/cm}^3$
   $x\ \text{g/L} = 0.941\ \text{g}/\cancel{\text{cm}^3} \times 1000\ \cancel{\text{cm}^3}/\text{L} = 941\ \text{g/L}$
   $x\ \text{mg/cm}^3 = 0.941\ \cancel{\text{g}}/\text{cm}^3 \times 1000\ \text{mg}/\cancel{\text{g}} = 941\ \text{mg/cm}^3$
   $x\ \text{kg/L} = 0.941\ \cancel{\text{g}}/\cancel{\text{cm}^3} \times 1000\ \cancel{\text{cm}^3}/\text{L} \times 1\ \text{kg}/1000\ \cancel{\text{g}} = 0.941\ \text{kg/L}$
2. $h = 8.85\ \text{cm};\ r = 1.10\ \text{cm};\ m = 15.153\ \text{g}$
   $V = 3.1416(1.10\ \text{cm})^2(8.85\ \text{cm}) = 33.6\ \text{cm}^3$
   $d = 15.153\text{g}/33.6\ \text{cm}^3 = 0.451\ \text{g/cm}^3$
3.

diameter = 1.10 cm
2.50 cm
2.50 cm
2.50 cm

$V = 15.6\ \text{cm}^3$; r of hole $= 0.550\ \text{cm}$
V of hole $=$
$3.1416(0.550\ \text{cm})^2(2.50\ \text{cm}) = 2.38\ \text{cm}^3$; V of block excluding hole $=$
$15.6\ \text{cm}^3 - 2.38\ \text{cm}^3 = 13.2\ \text{cm}^3$

3.5

1 Engage

Use the Visual

Have students study the photograph and read the text that opens the section. Explain that temperature is an important parameter in many physical and chemical processes. Ask:

- **What temperature scale has its zero point at absolute zero?** (the Kelvin scale)
- **Why do you think scientists use more than one temperature scale?** (Extremely low temperatures are all negative numbers when expressed in degrees Celsius. In Chapter 12, students will learn that the gas law relationships work only when the temperature is in kelvins.) This text uses 273, not 273.15, to convert between temperature scales.

Check Prior Knowledge

To assess students' prior knowledge about temperature scales, ask:

- **What temperature scale do you use to measure the temperature in your home?** (most probably degrees Fahrenheit)
- **How is temperature reported on weather forecasts?** (in the United States, always degrees Fahrenheit, and sometimes degrees Celsius)
- **What is your normal body temperature on the Fahrenheit scale? The Celsius scale?** (98 °F and 37 °C, respectively)
- **What is the approximate temperature in this classroom?** (After students have estimated the temperature, measure the actual temperature in degrees Fahrenheit and Celsius. Ask students if they know what the temperature is in kelvins.)

section 3.5

objective

- Convert between the Celsius and Kelvin temperature scales

key terms

- temperature
- Celsius scale
- Kelvin scale
- absolute zero

TEMPERATURE

In 1911, a Dutch physicist named Heike Kamerlingh Onnes discovered that when mercury is cooled to near absolute zero (−460 °F), it loses all resistance to electrical current. This state is known as superconductivity. If a material could be found that would superconduct at room temperature (around 64 °F), it would revolutionize the computer industry and other electronics industries. **Which temperature scale has its zero point at absolute zero?**

Measuring Temperature

When you hold a glass of hot water, the glass feels hot because heat transfers from the glass to your hand. When you hold an ice cube, it feels cold because heat transfers from your hand to the ice cube. The **temperature** of an object determines the direction of heat transfer. When two objects at different temperatures are in contact, heat moves from the object at the higher temperature to the object at the lower temperature. In Chapter 10, you will learn how the temperature of an object is related to the energy and motion of particles.

Almost all substances expand with an increase in temperature and contract as the temperature decreases (a very important exception is water). These properties are the basis for the common mercury-in-glass thermometer, shown in **Figure 3.19.** The liquid mercury in the thermometer expands and contracts more than the volume of the glass bulb that holds it, producing changes in the column height of the mercury.

Figure 3.18

Temperatures on Earth range from the scorching heat of a desert, to the frigid cold of the Antarctic, to the normal human body temperature of 37 °C. What is the hottest temperature ever recorded? The coldest? 1

Temperature Scales

Several temperature scales have been devised. The Celsius scale of the metric system is named after the Swedish astronomer Anders Celsius (1701–1744). It uses two readily determined temperatures as reference temperature values: the freezing point and the boiling point of water. The **Celsius scale** sets the freezing point of water at 0 °C and the boiling point of water at 100 °C. The distance between these two fixed points is divided into 100 equal intervals, or degrees Celsius (°C).

Another temperature scale used in the physical sciences is the Kelvin, or absolute, scale. This scale is named for Lord Kelvin (1824–1907), a Scottish physicist and mathematician. On the Kelvin scale, the freezing point of water is 273.15 kelvins (K), and the boiling point is 373.15 K. Notice that with the Kelvin scale, the degree sign is not used. You can use **Figure 3.19** to compare the Celsius and Kelvin scales. The zero point on the Kelvin scale, 0 K, or **absolute zero,** is equal to −273.15 °C. For problems in this text, you can round −273.15 °C to −273 °C. Because one degree on the Celsius scale is equivalent to one kelvin on the Kelvin scale, converting from one temperature scale to the other is easy. You simply add or subtract 273, as shown in the following equations.

$$K = °C + 273$$

$$°C = K - 273$$

STUDENT RESOURCES

From the Teacher's Resource Package, use:

- Section Review 3.5, Ch. 3 Practice Problems, Vocabulary Review, Quizzes, and Tests from the Review Module (Ch. 1–4)

TECHNOLOGY RESOURCES

Relevant technology resources include:

- Chem ASAP! CD-ROM
- Resource Pro CD-ROM
- Chemistry Alive! Videodisc: *Superconductivity*
- Assessment Resources CD-ROM: Chapter 3 Tests

Figure 3.19
These thermometers show a comparison of the Celsius and Kelvin temperature scales. Note that a 1 °C change on the Celsius scale is equal to a 1 K change on the Kelvin scale. What is a change of 10 K equivalent to on the Celsius scale? ❷

Sample Problem 3-6

Normal human body temperature is 37 °C. What is that temperature in kelvins?

1. **ANALYZE** ***List the knowns and the unknown.***

Known:
- Temperature in °C = 37 °C

Unknown:
- Temperature in K = ? K

Use the known value and the equation K = °C + 273 to calculate the temperature in kelvins.

2. **CALCULATE** ***Solve for the unknown.***

Substitute the known value for the Celsius temperature into the equation and solve.

$$K = °C + 273$$
$$= 37 + 273 = 310\ K$$

3. **EVALUATE** ***Does the result make sense?***

You should expect the Kelvin temperature to be in this range because the freezing point of water is 273.15 K and the boiling point of water is 373.15 K; normal body temperature is between these two values.

Practice Problems

30. Liquid nitrogen boils at 77.2 K. What is this temperature in degrees Celsius?

31. The element silver melts at 960.8 °C and boils at 2212 °C. Express these temperatures in kelvins.

Chem ASAP!
Problem-Solving 31
Solve Problem 31 with the help of an interactive guided tutorial.

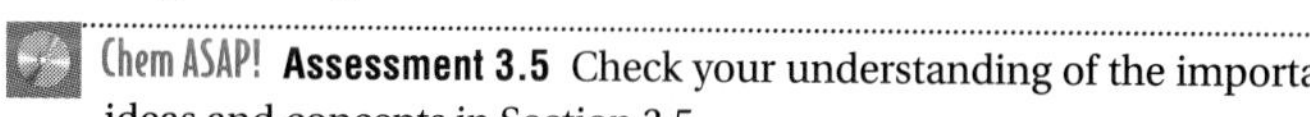

section review 3.5

32. State the relationship between degrees Celsius and kelvins.

33. Chocolate cookies are baked at 190 °C. Express this temperature in kelvins.

34. Surgical instruments may be sterilized by heating at 170 °C for 1.5 hours. Convert 170 °C to kelvins.

35. The boiling point of the element argon is 87 K. What is the boiling point of argon in degrees Celsius?

Chem ASAP! **Assessment 3.5** Check your understanding of the important ideas and concepts in Section 3.5.

3.5

2 Teach

Practice Problems Plus

Related Chapter Review Problem
Chapter Review problem 65 is related to Sample Problem 3-6.

Additional Practice Problem
The Chem ASAP! CD-ROM contains this problem:
The boiling point of water on top of Mount Everest is 343 K, while at the bottom of Death Valley, California, water boils at 373.3 K. Express these temperatures in°C. (Mount Everest: 70.0 °C; Death Valley: 100.0 °C)

3 Assess

Evaluate Understanding

Test students on their abilities to convert between degrees Celsius and kelvins. Ask:

- **What are the freezing and boiling point temperatures of water in degrees Celsius?** (0 °C and 100 °C, respectively)
- **What are these temperatures on the Kelvin scale?** (273.15 K and 373.15 K, respectively)

Reteach

Point out that zero degrees on the Celsius scale does not represent the lowest possible value of temperature. Therefore, some Celsius temperatures are negative numbers. However, on the Kelvin scale the lowest possible temperature is 0 K.

Meeting Diverse Needs

Gifted Have students study normal human body temperature. Questions they might ask include:

- Does normal body temperature vary between individuals?
- Does normal body temperature vary during the course of a day, for example between morning, afternoon, and night?

Answers

❶ hottest: 58°C in El Azizia, Libya; coldest: −89°C in Antarctica
❷ 10°C

SECTION REVIEW 3.5

32. °C = K − 273
33. 463 K
34. 443 K
35. −186°C

Chemistry Alive!

Superconductivity Play

Chemistry Serving... the Consumer

DISCUSS

This section discusses the development of modern standards for the fundamental SI units of length, time, and mass. Point out that the U.S. still uses the old English units of length and mass such as the foot and pound. Also ask students how metric units have been used in the U.S. since its founding. (money)

CHEMISTRY IN CAREERS

Analytical chemists ask the important question "How much?" They must make very careful measurements because they deal with small quantities of material. Analytical chemists are employed in a wide range of industrial and research occupations. Encourage students to connect to the Internet address shown on this page to find out more about careers in analytical chemistry. See also the information on page 869.

Chemistry Serving...the Consumer

THESE STANDARDS ARE WORTH THE WEIGHT

Imagine the confusion if every soft drink manufacturer defined a liter differently, or if the mass of a kilogram depended on where you live. As a consumer, you rely on units of mass, time, and length being unchanging and consistent, whether you buy a kilogram of ice cream, a half-hour of Internet access, or 10 000 square meters of land.

Luckily, you usually do not have to question the consistency of units of measurement because standards for each unit have been defined and agreed upon. But humans have not always had as good a system of standards as we have today. Throughout history, standards of measurement have become more precise and more consistent, and they continue in this direction today.

Using units that were based on body measurements was not a very reliable system of measurement. There could be great variation in the size of people's thumbs or the length of people's feet. In other words, thumbs and feet were not consistent units. People needed standards that did not change.

> As a consumer, you rely on units of mass, time, and length being unchanging and consistent.

In 1790, with the establishment of the metric system, the French became the first to adopt measurement standards that were close to being precise. The meter was defined as one ten-millionth of the distance from the equator to the North Pole along the meridian that passes through Paris. The second was defined as $\frac{1}{86\,400}$ of the average day.

As scientific measurement techniques became more precise, the definitions of these and other base units also became more precise. For example, the meter is now defined as the distance traveled by light in a vacuum in $\frac{1}{299\,792\,458}$ of a second. The second itself is defined in terms of the number of cycles of radiation given off by a specific isotope of the element cesium.

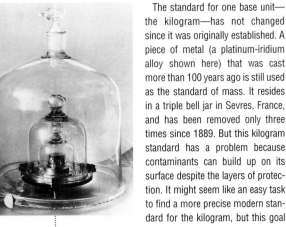

The standard for one base unit—the kilogram—has not changed since it was originally established. A piece of metal (a platinum-iridium alloy shown here) that was cast more than 100 years ago is still used as the standard of mass. It resides in a triple bell jar in Sevres, France, and has been removed only three times since 1889. But this kilogram standard has a problem because contaminants can build up on its surface despite the layers of protection. It might seem like an easy task to find a more precise modern standard for the kilogram, but this goal has so far been a challenge to scientists.

How about defining a kilogram as a given number of atoms of a particular element? Scientists thought of that idea a few years ago, using ultra-pure spheres of silicon as the standard. However, the presence of lattice holes (empty spaces in the crystalline structure of silicon) led to different estimates of the number of atoms in a given mass. This problem has almost been solved, and the "atom counters" are getting close to the goal of defining a new kilogram standard. Another group of researchers, the "force measurers," have a competing idea. Their method uses a Watt balance (a type of balance that uses electromagnetic force) to express the mass of a kilogram in terms of the base units of voltage and current. Either way, the old kilogram may soon have a more modern definition.

CHEMISTRY IN CAREERS

ANALYTICAL CHEMIST

Interested in a career with a focus on measurement and problem solving?

See page 869.

CHEMIST NEEDED

Enviro. Lab. GC Must know

TAKE IT TO THE NET

Find out more about career opportunities: www.phschool.com

CHEMICAL SPECIALIST

Local food service distributor seeks responsible self-motivated individ-

Chapter 3 STUDENT STUDY GUIDE

Take It to the NET
For interactive study and review, go to www.phschool.com

KEY TERMS

- absolute zero *p. 74*
- accepted value *p. 55*
- accuracy *p. 54*
- Celsius scale *p. 74*
- density *p. 69*
- error *p. 55*
- experimental value *p. 55*
- gram (g) *p. 67*
- hydrometer *p. 72*
- International System of Units (SI) *p. 63*
- Kelvin scale *p. 74*
- kilogram (kg) *p. 66*
- liter (L) *p. 65*
- meter (m) *p. 64*
- percent error *p. 55*
- precision *p. 54*
- qualitative measurement *p. 51*
- quantitative measurement *p. 51*
- scientific notation *p. 52*
- significant figure *p. 56*
- specific gravity *p. 72*
- temperature *p. 74*
- volume *p. 65*
- weight *p. 66*

KEY EQUATIONS

- Error = experimental value − accepted value
- $\text{Percent error} = \dfrac{|\text{error}|}{\text{accepted value}} \times 100\%$
- $\text{Density} = \dfrac{\text{mass}}{\text{volume}}$
- $\text{Specific gravity} = \dfrac{\text{density of substance}}{\text{density of water}}$
- K = °C + 273
- °C = K − 273

CONCEPT SUMMARY

3.1 The Importance of Measurement
- Measurements can be qualitative or quantitative.
- Very large and very small numbers are best expressed in scientific notation.

3.2 Uncertainty in Measurements
- The accuracy of a measurement describes how close a measurement comes to the true value.
- The precision of a measurement depends on its reproducibility.
- Measurements and calculations must always be reported to the correct number of significant figures.

3.3 International System of Units
- The International System of Units (SI) is the measurement system used by scientists.
- The SI has seven base units from which all other SI units of measurement are derived.
- The quantity of matter an object contains is its mass. Weight is not the same as mass. Weight is the measure of the pull of gravity on an object of given mass.

3.4 Density
- The ratio of the mass of an object to its volume is its density. The unit of density is g/cm^3.
- Specific gravity is the ratio of the density of a substance to the density of water.

3.5 Temperature
- Temperature difference determines the direction of heat flow between two bodies.
- Temperature is measured on the Celsius and Kelvin scales.

CHAPTER CONCEPT MAP

Use these terms to construct a concept map that organizes the major ideas of this chapter.

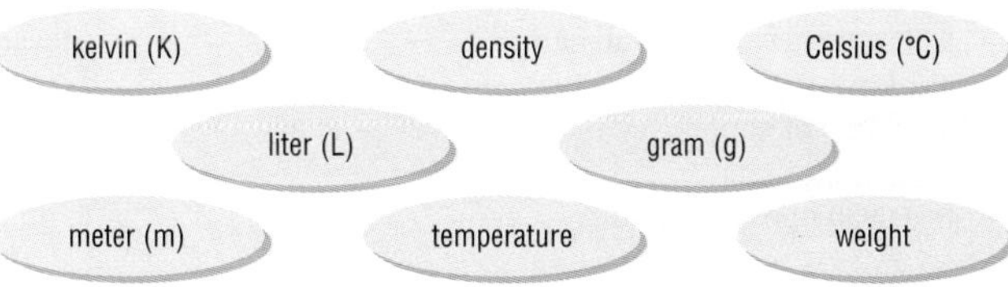

Chem ASAP! **Concept Map 3**
Create your Concept Map using the computer.

Take It to the Net

At **www.phschool.com** students will find for this chapter

- an Internet activity
- links to related chemistry sites
- an interactive quiz
- career links

Chapter 3
STUDENT STUDY GUIDE

4 Close

Summary

Hand out a worksheet that lists qualitative and quantitative data for an "unknown" element. Pick an element from Table 3.7 or Table 2.1. Include a volume in milliliters, a mass in grams, and melting and boiling points in kelvins. Pick values that differ slightly from those in the table. Have students identify the element by answering the following questions.

- **What is the SI base unit for each measurement?**
- **What are the melting and boiling points of the unknown substance in degrees Celsius?**
- **What is the density of the substance in g/cm^3?**
- **What is the percent error for each measurement?**

Finally, have students identify the substance on the basis of its physical properties.

Extension

Have students create a list of measurement standards used in everyday life. Examples include labels on grocery items, gasoline pumps, and global positioning satellites.

Looking Back . . . Looking Ahead . . .

The preceding chapter discussed the classification and properties of matter and the changes that matter undergoes. The current chapter describes the importance of accurate measurements and calculations in the study of chemistry. The next chapter explains methods and techniques used to solve problems in chemistry.

Chapter 3 REVIEW

CONCEPT PRACTICE

36. Identify the following as quantitative or qualitative measurements. *3.1*
a. A flame is hot.
b. A candle has a mass of 90 g.
c. Wax is soft.
d. A candle's height decreases 4.2 cm/hr.

37. Which of these synonyms or characteristics apply to the concept of accuracy? Which apply to the concept of precision? *3.2*
a. multiple measurements
b. correct
c. repeatable
d. reproducible
e. single measurement
f. true value

38. Under what circumstances could a series of measurements of the same quantity be precise but inaccurate? *3.2*

39. Three students made multiple weighings of a copper cylinder, each using a different balance. The correct mass of the cylinder had been previously determined to be 47.32 g. Describe the accuracy and precision of each student's measurements. *3.2*

Mass of Cylinder (g)			
	Lissa	Lamont	Leigh Anne
Weighing 1	47.13	47.45	47.95
Weighing 2	47.94	47.39	47.91
Weighing 3	46.83	47.42	47.89
Weighing 4	47.47	47.41	47.93

40. Comment on the accuracy and precision of these basketball free-throw shooters. *3.2*
a. 99 of 100 shots are made.
b. 99 of 100 shots hit the front of the rim and bounce off.
c. 33 of 100 shots are made; the rest miss.

41. How many significant figures are in each underlined measurement? *3.2*
a. <u>60 s</u> = 1 min
b. <u>9 innings</u> in a baseball game
c. 1 km = <u>1000 m</u>
d. <u>47.70 g</u> of copper
e. <u>25 computers</u>
f. <u>0.0950 m</u> of gold chain

42. Round each of these measurements to three significant figures. *3.2*
a. 98.473 L
b. 0.000 763 21 cg
c. 57.048 m
d. 12.17 °C
e. 0.007 498 3 $\times$ 10^4 mm
f. 1764.9 mL

43. Write each of the rounded measurements in Problem 42 in scientific notation. *3.2*

44. Round each of the answers correctly. *3.2*
a. 8.7 g + 15.43 g + 19 g = 43.13 g
b. 4.32 cm $\times$ 1.7 cm = 7.344 cm^2
c. 853.2 L − 627.443 L = 225.757 L
d. 38.742 kg ÷ 0.421 = 92.023 75 kg
e. 5.40 m $\times$ 3.21 m $\times$ 1.871 m = 32.431 914 m^3
f. 5.47 m^3 + 11 m^3 + 87.300 m^3 = 103.770 m^3

45. Express each of the rounded answers in Problem 44 in scientific notation. *3.2*

46. How are the error and the percent error of a measurement calculated? *3.2*

47. Why is the percent error of a measurement always positive? *3.2*

48. A student estimated the volume of a liquid in a beaker as 200 mL. When she poured the liquid into a graduated cylinder, she measured the volume as 208 mL. What is the percent error of the estimated volume from the beaker, taking the measurement in the graduated cylinder as the accepted value? *3.2*

49. Water with a mass of 35.4 g is added to an empty flask with a mass of 87.432 g. The mass of the flask and the water is 146.72 g after a rubber stopper is added. Express the mass of the stopper to the correct number of significant figures. *3.2*

50. List at least two advantages of using SI units for measuring. *3.3*

51. List the SI base unit of measurement for each of these quantities. *3.3*
a. time
b. length
c. temperature
d. mass

52. Use the tables in this chapter to order these lengths from smallest to largest. Give each measurement in terms of meters. *3.3*
a. centimeter
b. micrometer
c. kilometer
d. millimeter
e. meter
f. nanometer
g. decimeter
h. picometer

Answers

36. a. qualitative
b. quantitative
c. qualitative
d. quantitative

37. a. precision
b. accuracy
c. precision
d. precision
e. accuracy
f. accuracy

38. when using an improperly calibrated measuring device

39. Lissa: inaccurate and imprecise
Lamont: accurate and precise
Leigh Anne: inaccurate but precise

40. a. accurate and precise
b. inaccurate but precise
c. inaccurate and imprecise

41. a. infinite
b. infinite
c. infinite
d. 4
e. infinite
f. 3

42. a. 98.5 L
b. 0.000 763 cg
c. 57.0 m
d. 12.2°C
e. 0.007 50 $\times$ 10^4 mm
f. 1760 mL

43. a. 9.85 $\times$ 10^1 L
b. 7.63 $\times$ 10^{-4} cg
c. 5.70 $\times$ 10^1 m
d. 1.22 $\times$ 10^1 °C
e. 7.50 $\times$ 10^1 mm
f. 1.76 $\times$ 10^3 mL

44. a. 43 g
b. 7.3 cm^2
c. 225.8 L
d. 92.0 kg
e. 32.4 m^3
f. 104 m^3

45. a. 4.3 $\times$ 10^1 g
b. 7.3 $\times$ 10^0 cm^2
c. 2.258 $\times$ 10^2 L
d. 9.20 $\times$ 10^1 kg
e. 3.24 $\times$ 10^1 m^3
f. 1.04 $\times$ 10^2 m^3

46. The error is the difference between the accepted and the experimental values. The percent error is the absolute value of the error divided by the accepted value multiplied by 100%.

47. The absolute value of the error is used.

48. 4%

49. 23.9 g

50. Possible answers are: Units are based on multiples of ten; prefixes have the same meaning when attached to different units of measure.

51. a. second, s
b. meter, m
c. kelvin, K
d. kilogram, kg

52. picometer (10^{-12} m), nanometer (10^{-9} m), micrometer (10^{-6} m), millimeter (10^{-3} m), centimeter (10^{-2} m), decimeter (10^{-1} m), meter, kilometer (10^3 m)

53. Measure each dimension using a unit with the appropriate prefix. *3.3*
a. the height of this letter: I
b. the width of **Table 3.3**
c. the height of this page

54. From what unit is a measure of volume derived? *3.3*

55. What is the volume of a glass cylinder with an inside diameter of 6.0 cm and a height of 28 cm? (The volume of a cylinder equals pi × radius squared × height, or $V = 3.14\, r^2h$.) *3.3*

56. Match the approximate volume with each item. *3.3*

a. orange	(1) 30 m^3
b. basketball	(2) 200 cm^3
c. van	(3) 20 L
d. aspirin tablet	(4) 200 mm^3

57. How many grams are in each of these quantities? *3.3*
a. 1 cg **b.** 1 μg **c.** 1 kg **d.** 1 mg

58. Astronauts in space are said to have apparent weightlessness. Explain why it is incorrect to say that they are massless. *3.3*

59. Match the approximate mass with each item. *3.3*

a. peanut	(1) 400 cg
b. pear	(2) 50 mg
c. stamp	(3) 60 kg
d. person	(4) 150 g

60. Would the density of a person be the same on the surface of Earth as it is on the surface of the moon? Explain. *3.4*

61. A shiny, gold-colored bar of metal weighing 57.3 g has a volume of 4.7 cm^3. Is the metal bar pure gold? *3.4*

62. Why doesn't a measure of specific gravity have a unit? *3.4*

63. Use the values in **Table 3.7** to calculate the specific gravity of the following substances at 20 °C. *3.4*
a. aluminum
b. mercury
c. ice

64. Three balloons filled with neon, carbon dioxide, and hydrogen are released into the atmosphere. Using the data in **Table 3.7**, describe the movement of each balloon. *3.4*

65. Which would melt first, germanium with a melting point of 1210 K or gold with a melting point of 1064 °C? *3.5*

CONCEPT MASTERY

66. List two possible reasons for precise, but inaccurate, measurements.

67. Rank these numbers from smallest to largest.
a. 5.3×10^4 **d.** 0.0057
b. 57×10^3 **e.** 5.1×10^{-3}
c. 4.9×10^{-2} **f.** 0.0072×10^2

68. Criticize this statement: "When two measurements are added together, the answer can have no more significant figures than the measurement with the least number of significant figures."

69. Fahrenheit is a third temperature scale. Use the data in the table or the graph to derive an equation for the relationship between the Fahrenheit and Celsius temperature scales.

	°C	°F
Melting point of selenium	221	430
Boiling point of water	100	212
Normal body temperature	37	98.6
Freezing point of water	0	32
Boiling point of chlorine	−34.6	−30.2

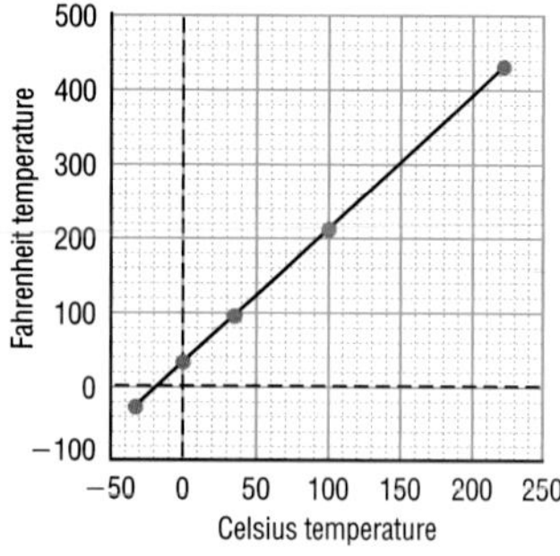

70. Which is larger?
a. 1 centigram or 1 milligram
b. 1 liter or 1 centiliter
c. 1 calorie or 1 kilocalorie
d. 1 millisecond or 1 centisecond
e. 1 microliter or 1 milliliter
f. 1 cubic millimeter or 1 cubic decimeter

53. a. 2.4 mm
b. 17.6 cm
c. 27.6 cm
54. Volume is a length unit cubed.
55. a. 7.9×10^2 cm^3
56. a. 2
b. 3
c. 1
d. 4
57. a. 0.01 g
b. 0.000 001 g
c. 1000 g
d. 0.001 g
58. The mass of an object is constant. The weight of an object varies with location.
59. a. 1
b. 4
c. 2
d. 3
60. Yes, neither mass nor volume changes with location.
61. No; the density of the metal bar is 12 g/cm^3, but the density of gold is 19 g/cm^3.
62. Specific gravity is a ratio of two density measurements, so the density units cancel.
63. a. 2.70
b. 13.6
c. 0.917
64. The carbon dioxide-filled balloon would sink. The neon- and hydrogen-filled balloons would rise, the hydrogen at a much faster rate.
65. germanium
66. improper calibration or improper use of the measuring device
67. e., d., c., f., a., b.
68. Significant figures in the answer of an addition problem depend on the measurement with the least number of decimal places.
69. °F = 1.8 °C + 32
70. a. cg
b. L
c. kcal
d. cs
e. mL
f. 1 dm^3

Answers

71. The digit to the right of the last significant figure is dropped if it is less than five.

72. **a.** 0.69 – 0.789 g/cm^3

73. Yes; an object can change position such that the force of gravity acting on it changes, causing its weight to change.

74. The egg is floating at the juncture of two liquids of different densities.

75. **a.** 2
b. 1

76. It is not possible when measurements are being compared to only one true value. Accurate measurements would all come close to the true value; therefore, they would also be precise.

77. You do not change your estimate. The extra 15 ducks are negligible compared to 850 000 ducks.

78. Gasoline is a mixture and has a variable composition.

79. to understand the world around us

80. **a.** Na **d.** Cu
b. Al **e.** S
c. Cl **f.** Sr

81. **a.** chemical **d.** chemical
b. physical **e.** chemical
c. physical **f.** physical

82. Add water to dissolve the salt. Then decant or filter.

83. The components of a homogeneous mixture can be separated by physical means; a compound must be chemically broken apart into its components. A homogeneous mixture has a variable composition; a compound has a definite composition.

84. Answers will vary. Lakes would freeze solid from the bottom up; aquatic life would be killed; possible climate changes.

85. volume of iron = 45.1 cm^3
mass of lead = 514 g Pb

86. **a.** corn oil on top of water on top of mercury
b. The density of sugar is greater than the density of water and less than the density of mercury; it floats between the layers of mercury and water.
c. The sugar cube will dissolve in the water over time.

87. density of sulfur = 2.1 g/cm^3; student graphs should be plotted as shown at right.

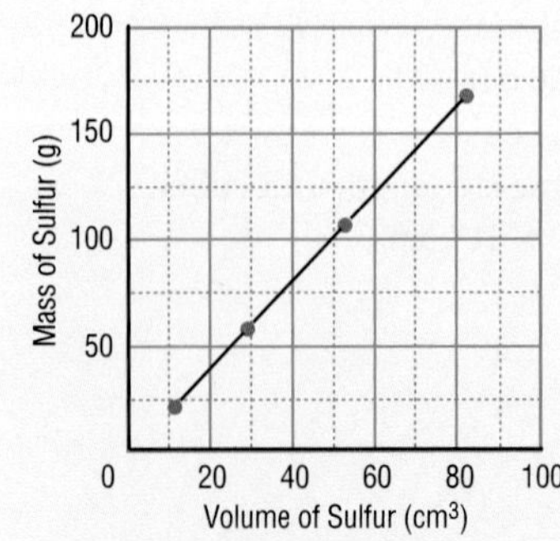

71. Criticize this statement: "When a number is rounded, the last significant figure is dropped if it is less than 5."

72. A piece of wood sinks in gasoline but floats in ethanol. Give a range of possible densities for the wood.

73. Is it possible for an object to lose weight but at the same time not lose mass? Explain.

74. How has the egg been suspended in the clear liquid?

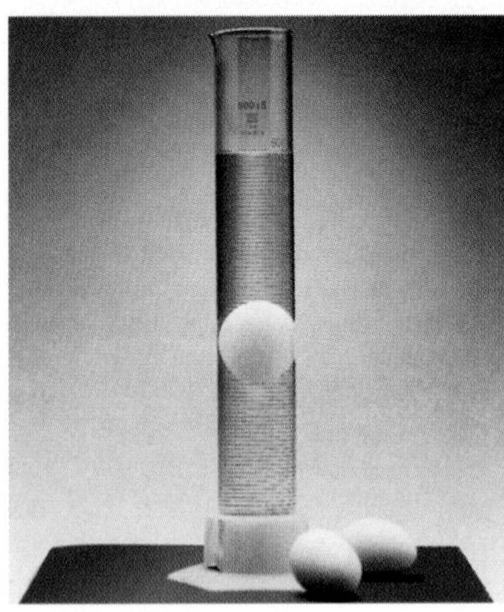

CRITICAL THINKING

75. Choose the term that best completes the second relationship.
a. mass:kilograms money: ______
(1) hot (3) coins
(2) dollars (4) spending
b. foot:inch meter: ______
(1) millimeter (3) kilometer
(2) liter (4) kilogram

76. Is it possible for experimental data to be accurate but imprecise?

77. You are hired to count the number of ducks on three northern lakes during the summer. In the first lake, you estimate 500 000 ducks; in the second, 250 000 ducks; and in the third, 100 000 ducks. You record that you have counted 850 000 ducks. As you drive away, you see 15 ducks fly in from the south and land on the third lake. Do you change the number of ducks that you report? Justify your answer.

78. Why is there a range of values given for the density of gasoline on **Table 3.8**?

CUMULATIVE REVIEW

79. What is a general goal of the scientific method?

80. What is the correct symbol for each element?
a. sodium **d.** copper
b. aluminum **e.** sulfur
c. chlorine **f.** strontium

81. Classify each of the following as a chemical or physical change.
a. grass growing
b. sugar dissolving in water
c. crushing a rock
d. cooking potatoes
e. bleaching clothes
f. boiling water

82. How would you separate a mixture of ground glass and salt?

83. How can you distinguish between a homogeneous mixture and a compound?

CONCEPT CHALLENGE

84. If ice were denser than water, it would certainly be easier to pour water from a pitcher of ice cubes and water. Can you imagine situations of more consequence?

85. The mass of a cube of iron is 355 g. Iron has a density of 7.87 g/cm^3. What is the mass of a cube of lead that has the same dimensions?

86. Equal amounts of mercury, water, and corn oil are added to a beaker.
a. Describe the arrangement of the layers of liquids in the beaker.
b. A small sugar cube is added to the beaker. Describe its location.
c. What change will occur to the sugar cube over time?

87. Plot these data that show how the mass of sulfur increases with an increase in volume. Determine the density of sulfur from the slope of the line.

Mass of sulfur (g)	Volume of sulfur (cm^3)
23.5	11.4
60.8	29.2
115	55.5
168	81.1

Chapter 3 STANDARDIZED TEST PREP

Select the choice that best answers each question or completes each statement.

1. Which of these measurements have four significant figures?

I. 3.045 g
II. 0.0450 g
III. 3.0×10^4 g
IV. 4.050 g

a. I and III only
b. II and IV only
c. I and IV only
d. I, II, and IV only

2. A graduated cylinder contains 44.2 mL of water. A 48.6-g piece of metal is dropped into the cylinder. When the metal is completely covered with water, water rises to the 51.3 mL mark. What is the density of the metal?

a. 0.95 g/mL　　**c.** 7.1 g/mL
b. 6.8 g/mL　　**d.** 0.15 g/mL

3. Which of these series of units is ordered from smallest to largest?

a. μg, cg, mg, kg　　**c.** μs, ns, cs, s
b. mm, dm, m, km　　**d.** nL, mL, dL, cL

4. Which answer represents the measurement 0.00428 g rounded off to two significant figures?

a. 4.28×10^3 g　　**c.** 4.3×10^3 g
b. 4.3×10^{-3} g　　**d.** 4.0×10^{-3} g

Use these data to answer questions 5 and 6.

At 20 °C, the density of air is 1.20 g/L. Nitrogen's density is 1.17 g/L. Oxygen's density is 1.33 g/L.

5. Will balloons filled with oxygen and balloons filled with nitrogen rise or sink in air?

6. Air is mainly a mixture of nitrogen and oxygen. Which gas is the main component? Explain.

Use the data table to answer questions 7–10.

Body density (g/cm³)	Percent body fat (%)
1.01	38.3
1.03	29.5
1.05	21.0
1.07	12.9
1.09	5.07

7. What is the relationship between the density of a person's body and the percent of body fat?

8. Plot a graph with body density on the x-axis and percent body fat on the y-axis.

9. Are the variables directly or inversely related?

10. Use the graph to estimate the percent body fat of a person with a body density of 1.06 g/cm³.

Use the atomic windows below to answer questions 11 and 12.

a.
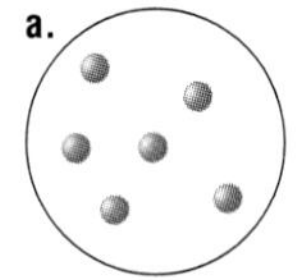
b.
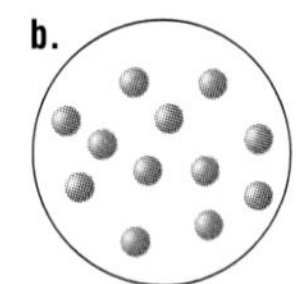
c.
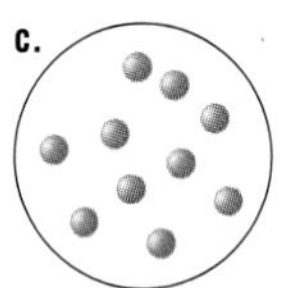

The atomic windows represent particles of the same gas occupying the same volume at the same temperature. The systems differ only in the number of gas particles per unit volume.

11. List the windows in order of decreasing density.

12. Compare the density of the gas in window (a) to the density of the gas in window (b).

For each question there are two statements. Decide whether each statement is true or false. Then decide whether Statement II is a correct explanation for Statement I.

Statement I		Statement II
13. When a marshmallow is squashed, its mass and density do not change.	BECAUSE	Squashing an object does not change the amount of matter present.
14. There are five significant figures in the measurement 0.00450 m.	BECAUSE	All zeros to the right of a decimal point in a measurement are significant.
15. Precise measurements will always be accurate measurements.	BECAUSE	A value that is measured 10 times in a row must be accurate.
16. A temperature in kelvins is always numerically larger than the same temperature in degrees Celsius.	BECAUSE	A temperature in kelvins equals a temperature in degrees Celsius plus 273.

Chapter 3
STANDARDIZED TEST PREP

1. c
2. b
3. b
4. b
5. Nitrogen-filled balloons will rise; oxygen-filled balloons will sink.
6. Nitrogen because its density is much closer to that of air than is oxygen's density.
7. The higher the body density, the lower the percent body fat.
8.

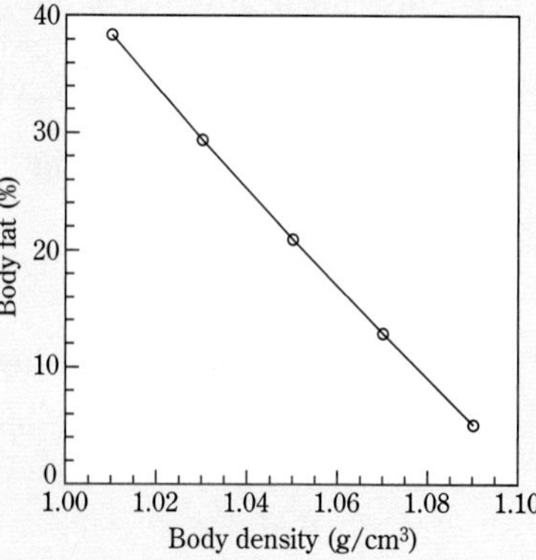

9. The variables are inversely related.
10. about 17%
11. b, c, a
12. The density in a is one half the density in b.
13. False, True
14. False, False
15. False, False
16. True, True, correct explanation

Planning Guide

SECTION OBJECTIVES	ACTIVITIES/FEATURES	MEDIA & TECHNOLOGY
4.1 What Do I Do Now? ● ■ ◇ ▸ List several useful problem-solving skills ▸ Describe the three-step problem-solving approach	**SE** Discover It! *Testing Problem-Solving Skills,* p. 82 **SE** **CHEMath** *Calculator Skills,* p. 85 **TE** DEMO, p. 83	**ASAP** Assessment 4.1
4.2 Simple Conversion Problems ○ ■ ◆ ▸ Construct conversion factors from equivalent measurements ▸ Apply the techniques of dimensional analysis to a variety of conversion problems	**SE** **Link to Business** *Monetary Exchange Rates,* p. 90 **SE** Small-Scale Lab *Now What Do I Do?,* p. 96 (LRS 4-1)	**ASAP** Animation 3 **ASAP** Problem Solving 12, 13 **ASAP** Assessment 4.2
4.3 More-Complex Problems □ ◆ ▸ Solve problems by breaking the solution into steps ▸ Convert complex units, using dimensional analysis	**SE** Mini Lab *Dimensional Analysis,* p. 100 (LRS 4-2) **SE** **Chemistry Serving . . . the Consumer** *Nature's Medicine Cabinet,* p. 101 **SE** **Chemistry in Careers** *Medical Laboratory Technician,* p. 101 SSLM 4: *Weighing Activities for a Small-Scale Balance* **TE** DEMO, p. 98	**ASAP** Problem Solving 23, 25 **ASAP** Assessment 4.3 **ASAP** Concept Map 4 **PP** Chapter 4 Problems **RP** Lesson Plans, Resource Library **AR** Computer Test 4 **www** Activities, Self-Tests, *SCIENCE NEWS* updates **TCP** The Chemistry Place Web Site

KEY

- ● Conceptual (concrete concepts)
- ○ Conceptual (more abstract/math)
- ■ Standard (core content)
- □ Standard (extension topics)
- ◆ Honors (core content)
- ◇ Honors (options to accelerate)

AR Assessment Resources
ASAP Chem ASAP! CD-ROM
ACT ActivChemistry CD-ROM
CHM CHEMedia Videodiscs
CA Chemistry Alive! Videodiscs
GCP Graphing Calculator Problems

GRS Guided Reading and Study Workbook
LM Laboratory Manual
LP Laboratory Practicals
LRS Laboratory Recordsheets
SSLM Small-Scale Lab Manual

PRACTICE		ASSESSMENT	
SE	**Sample Problems** 4.1 to 4.2	SE	Section Review
GRS	Section 4.1	RM	Section Review 4.1
RM	Practice Problems 4.1	RM	Chapter 4 Quiz
SE	**Sample Problems** 4-3 to 4-7	SE	Section Review
GRS	Section 4.2	RM	Section Review 4.2
RM	Practice Problems 4.2	RM	Chapter 4 Quiz
SE	**Sample Problems** 4-8 to 4-10	SE	Section Review
GRS	Section 4.3	RM	Section Review 4.3
RM	Practice Problems 4.3	RM	Vocabulary Review 4
RM	Interpreting Graphics	SE	Chapter Review
GCP	Chapter 4	SM	Chapter 4 Solutions
		SE	Standardized Test Prep
		PHAS	Chapter 4 Test Prep
		RM	Chapter 4 Quiz
		RM	Chapter 4 A & B Test

PLANNING FOR ACTIVITIES

STUDENT EDITION

Discover It! p. 82
- index cards
- paper
- pencils

Small-Scale Lab, p. 96
- pencil
- paper
- meter stick
- balance
- dice
- aluminum can
- calculator
- small-scale pipet
- water
- pre- and post-1982 pennies
- 8-well strip
- plastic cup

Mini Lab, p. 100
- index cards
- pen

TEACHER'S EDITION

Teacher Demo, p. 83
- ice cubes
- empty soda bottle
- bowl

Activity, p. 91
- recipe for simple cake
- teaspoon
- tablespoon
- 1/4 and 1/2 and 1 cup
- lists of equivalents and conversions found in many cookbooks

Activity, p. 92
- media guides for sports franchises
- conversion table for metric and U.S. units

Teacher Demo, p. 91
- teaspoon

OT	Overhead Transparency	SE	Student Edition
PHAS	PH Assessment System	SM	Solutions Manual
PLM	Probeware Lab Manual	SSV	Small-Scale Video/Videodisc
PP	Problem Pro CD-ROM	TCP	www.chemplace.com
RM	Review Module	TE	Teacher's Edition
RP	Resource Pro CD-ROM	www	www.phschool.com

Chapter 4

Key Terms

4.2 conversion factor, dimensional analysis

DISCOVER IT!

To solve the problem, students need to determine what the problem is asking for; that is, they need to identify the unknown. Sketching the cylinder and labeling its height as a variable, say *h*, is a logical first step. Students can then determine what data are known that they can use to solve for *h*. Algebraically, the problem should be set up as follows:

$$\text{height} = 24\text{ cm} + \frac{1}{2}(\text{height}),\text{ or}$$

$$h = 24\text{ cm} + \frac{1}{2}h$$

$$h - \frac{1}{2}h = 24\text{ cm}$$

$$\frac{1}{2}h = 24\text{ cm}$$

$$h = 48\text{ cm}$$

Chapter 4 PROBLEM SOLVING IN CHEMISTRY

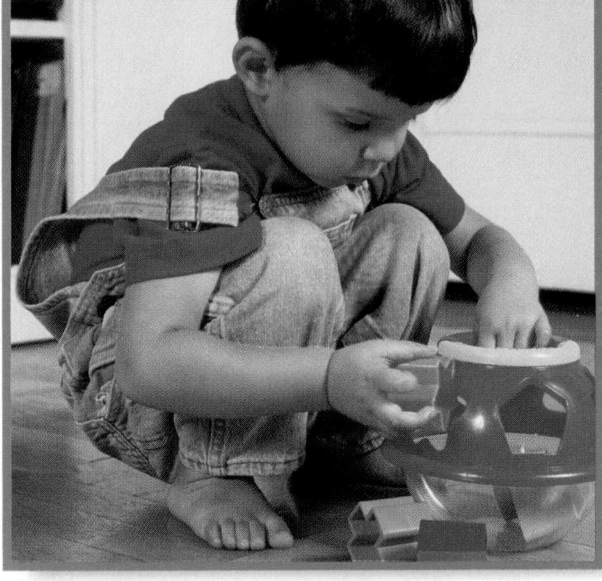

A child practices problem solving with shape-sorter toys.

FEATURES

DISCOVER IT! TESTING PROBLEM-SOLVING SKILLS

You need five index cards, or five sheets of paper, and five pencils.

1. Copy the following problem onto each of the index cards: "The height of a graduated cylinder is 24 cm plus one-half of its height. How tall is the graduated cylinder?"
2. Work the problem yourself and then compare your answer with the solution provided by your teacher. How did you do?
3. Ask four family members or friends (not in your chemistry class) to work the problem.
4. How did they do?

Congratulate those who got the correct answer on their first try. Ask them how they got the answer. Point out to those who could not solve the problem that if they had checked their work they might have caught their initial error in reasoning. Show them that sketching the problem statement would have been useful in solving this problem. As you solve the problems in this chapter, make note of the problem-solving skills you use.

Stay current with **SCIENCE NEWS**
Find out more about problem solving:
www.phschool.com

4.1 ● ■ ◇
4.2 ○ ■ ◆
4.3 □ ◆

Conceptual Students need to know what a conversion factor is to be able to follow sample problems throughout the text. Depending on course expectations, students may not need extensive practice applying dimensional analysis and, thus, can omit Sample Problems 4-6 and 4-7. To save time, omit the complex problems in Section 4.3.

Standard In Section 4.1, emphasize the Evaluate step to help students become more self-reliant problem solvers. Use the practice problems to ensure that students become proficient with conversion factors; this effort early in the course will save time throughout the year. Emphasis on Section 4.3 will depend on the degree to which students will be expected to perform multi-step problems.

Honors Because honors students tend to be strong problem solvers, Section 4.1 may be assigned as homework and then discussed briefly in class. In Section 4.3, the Mini Lab may be assigned as homework.

4.1

1 Engage

Use the Visual

Have students study the photograph and read the text that opens the section. Ask students how they would approach the situation pictured. (Identify the goal as matching an object with a hole; compare the shape of an object with the available locations; place an object in a matching hole.) **Ask students to generalize the approach.** (Identify the objective; gather data; analyze data; solve the problem.) **Ask:**

- **What is the most effective way to solve problems?** (Approaches vary depending on the given problem; effective approaches are generally organized and well-planned.)

Check Prior Knowledge

Have students share their experiences with problem solving in other subject areas, such as math. Have them compare problems that were easy to solve and problems that were difficult to solve. (solving equations vs. solving word problems, perhaps) **Ask them to identify factors that made some problems easier than others.**

2 Teach

TEACHER DEMO

Place a bowl of ice cubes and an empty soda bottle in front of the classroom. Explain that the problem is to transfer the ice to the inside of the bottle. Have the students analyze the problem and suggest different approaches. (Crush the ice so that pieces are small enough to fit through the mouth of the bottle, melt the ice, pour the water into the bottle, and refreeze.)

WHAT DO I DO NOW?

section 4.1

objectives

- List several useful problem-solving skills
- Describe the three-step problem-solving approach

Shape-sorter toys fascinate young children. Typically, the children take one shape and try it in every hole over and over again until they find the right one. As adults, we can use information we have gathered about the distinction between shapes to evaluate the toy and place the star in the star-shaped hole. So although trial and error is one method of problem solving, it is usually not the best one. **What is the most effective way to solve problems?**

Skills Used in Solving Problems

Do you like word problems? You know the kind: "Sally is twice as old as Sara, who is three years younger than Suzy. What will Sally's age be one year from now if Suzy is one-third as old as twelve-year-old Sonja will be nine years from now?" (The answer is nine years old.) If problems such as this seem confusing, rest assured: None of the problems you will encounter in this textbook are that difficult. However, problem solving may be an area in which you feel less than proficient. This chapter will teach you how to become a good problem solver—and perhaps even to enjoy the process.

One way to become a better problem solver is to learn and use various problem-solving techniques. These techniques will turn out to apply to many situations, not just to your chemistry studies. You probably already use many problem-solving skills in everyday situations as well as in classes like algebra. Can you think of an instance in which you have used such skills today? ❶

Problem solving involves developing a plan. Suppose one of your good friends moved to a distant state and then invited your family for a visit. You would face the problem of getting to the new location. You might fly to the airport closest to your friend's house and then rent a car to drive the rest of the way. But before you drove there, you would need to identify your destination on a road map like that in **Figure 4.1.** On the map, you see several alternate routes between the airport and your friend's town. You would have to select and plan a specific route to get from your starting point to your destination. Without the proper planning, your trip might be difficult.

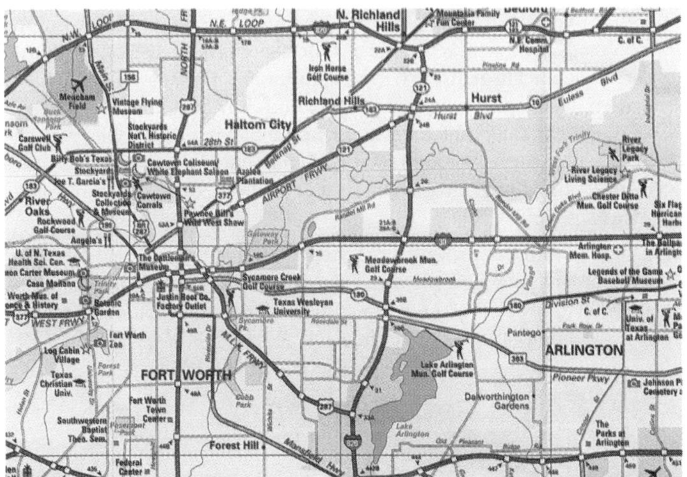

Figure 4.1
Carefully planning your route helps ensure a successful trip. What are some possible routes for traveling from Meacham Field to Arlington? ❷

STUDENT RESOURCES

From the Teacher's Resource Package, use:

- Section Review 4.1, Ch. 4 Practice Problems and Quizzes from the Review Module (Ch. 1–4)

❶ An example might be deciding which clothes to put on in the morning. You could check the weather forecast to help make the decision.

TECHNOLOGY RESOURCES

Relevant technology resources include:

- Chem ASAP! CD-ROM
- ResourcePro CD-ROM

❷ Answers will vary. One route is the NE Loop 820 to 303 E.

4.1

Discuss

Emphasize that a trial-and-error method of problem solving is an approach that is likely to be unsuccessful and frustrating. An orderly approach takes time and practice to develop, but is really the easiest way to get consistently correct answers. The chapter explains methods used to solve both qualitative and quantitative problems in chemistry.
The three-step technique (analyze, calculate, and evaluate) provides a common-sense approach to solving word problems successfully. Instruct students to break up problems into steps and to write down the name of each step:

- **Analyze: List knowns and unknowns.**
- **Calculate: Solve for unknowns.**
- **Evaluate: Does the result make sense?**

Students should continue with this discipline until the process becomes ingrained.

Figure 4.2
The three-step approach to solving problems in chemistry is a useful and effective strategy that you should employ throughout this course—and in other problem-solving situations.

A Three-Step Problem-Solving Approach

As you may have noted in Chapter 3, the techniques used in this book to solve problems are conveniently organized into a three-step problem-solving approach. While this approach is not the only strategy for solving problems, we believe it is the most helpful and effective. Thus we recommend that you follow it when working on the problems in this textbook. You will now examine the steps in detail.

Step 1. *ANALYZE* Solving a word problem is not too different from taking a trip to a new place. You must determine where you are starting from (identify the known), where you are going (identify the unknown), and how you are going to get there (plan a solution). What is known in a word problem may include a measurement and one or more relationships or equations that link measurements. You identify the unknown by reading the problem carefully to be sure that you understand what the problem is asking you to find. If the problem is long, you might need to read it several times. If the problem is to have a numerical answer, write down the units that the answer should have before you begin your solution.

Planning a solution—figuring out how to get from the known to the unknown—is obviously at the heart of problem solving. Sketching a picture that represents the problem may help you visualize a relationship between the known and the unknown. It might also suggest a way to break the problem down into two or more simpler problems. The solutions to the simpler problems can then be combined to solve the more complex problem. At this point, you might need to use resources such as tables or figures to find other facts or relationships. For example, you might need to look up a constant or an equation that relates a known measurement to an unknown measurement.

Step 2. *CALCULATE* If you have done a good job planning your solution, doing the calculation is usually straightforward. The calculation may involve substituting known quantities and doing the arithmetic needed to solve an equation for the unknown. Sometimes more than one equation may be needed in the solution. In some problems, you might need to convert a measurement to a different form.

Step 3. *EVALUATE* Once you have done the calculation, you must evaluate your answer. Does the answer make sense? Is it reasonable? Is it written in the correct unit? Solving the problem should give an answer with the correct unit. If it does not, you may have set up the equation incorrectly. The answer should also be expressed to the correct number of significant figures. Where appropriate, the answer should be written in scientific notation. Most importantly, you must check your work. Reread the problem to make sure that what the problem asked for is what you have found. Did you copy down the given facts correctly? Check your math. Often, you can estimate an appropriate answer as a quick check.

With slight modifications, this three-step approach can be used for both numeric and nonnumeric problems. With some practice, you will soon find yourself applying it without difficulty to all sorts of situations. Sample Problem 4-1 on page 86 illustrates how the approach is used.

Meeting Diverse Needs

At Risk Some students are convinced they cannot do word problems because they are too "hard." The three-step, problem-solving technique can provide a framework. Ask students to describe the process verbally as they attempt to solve a problem. The verbal descriptions will help uncover stumbling blocks that are hindering progress.

CHEMath

CALCULATOR SKILLS

Chemistry problems often involve calculations that are too complicated to do by hand. Thus the correct use of a calculator is essential to your success in this course. You will also find your calculator is very useful for solving a variety of problems in your everyday life.

The following instructions will help you use a typical scientific calculator. Be aware, however, that your calculator may vary slightly from the instructions that follow.

Addition, subtraction, multiplication, and division are entered much as they appear on paper. Use the change-sign key, [+/-], to enter negative numbers, and use parentheses as appropriate. For example:

Enter 4×1.25 as 4 [×] 1 [.] 25 [=].

Enter $10 \div (-0.4)$ as 10 [÷] 0 [.] 4 [+/-] [=].

Enter $3 \times (2 + 3)$ as 3 [×] [(] 2 [+] 3 [)] [=].

In chemistry, you will encounter many numbers written in scientific notation (see page 52). You can enter these numbers using the [EXP] key (labeled [EE] or [E] on some calculators). For example, the mass of an electron is 9.11×10^{-28} g, which you would enter as 9 [.] 11 [EXP] [+/-] 28. How would you calculate the mass of three electrons?

Different calculators may have different square root keys. Some calculators have a [$\sqrt{x}$] or [SQRT] key, but others require a two-key process using [INV] [x^2]. To find the square root of 16, enter 16 [$\sqrt{x}$], 16 [SQRT], or 16 [INV] [x^2], depending on your calculator.

To calculate powers, use the [y^x] key. For example, find 5^3 by pressing 5 [y^x] 3 [=]. When the exponent is 2, you can use the [x^2] key. To calculate 8^2, enter 8 [x^2] [=].

Example 1

Use a calculator to evaluate $4^2 \times (10 + \sqrt{10\,609})$.

Enter 4 [x^2] [×] [(] 10 [+] 10 609 [$\sqrt{x}$] [)] [=].

The expression is equal to 1808.

Example 2

One atom of gold has a mass of 3.271×10^{-22} g. What is the mass of 158 atoms of gold?

The mass of 158 atoms of gold is given by $(158 \text{ atoms gold}) \times \frac{3.271 \times 10^{-22}\text{ g}}{1 \text{ atom gold}}$.

158 [×] 3.271 [EXP] [+/-] 22 [=] 5.16818×10^{-20}

Rounding to four significant figures, the mass is 5.168×10^{-20} g.

Practice Problems

Prepare for upcoming problems in this chapter by using a calculator to solve each of the following problems.

A. Evaluate 7.823×15.76.

B. Evaluate $89^2 + 17^3 + 5^6$.

C. Evaluate $\frac{35 + \sqrt{529}}{2.9 \times 10^{17}}$.

D. The volume of a sphere with radius r is given by $\frac{4}{3}\pi r^3$. Find the volume of a sphere with a radius of 3.00 cm.

E. Find the number of atoms in 7.000 g of gold by evaluating $(7.000 \text{ g}) \times \frac{1 \text{ atom gold}}{3.271 \times 10^{-22}\text{ g}}$.

MEETING DIVERSE NEEDS

At Risk Proficiency with a calculator is an essential tool for successfully solving quantitative problems. If students are having difficulty working with a calculator, show them how to enter numbers and functions in the correct order. Demonstrate how scientific notation allows students to solve problems with very large or small numbers.

CHEMath

Teaching Tips

Write the example problems on the board and carry out the three-step problem-solving approach. Stress the discipline of writing down the steps that make the solution process possible. You may wish to use chalk or markers of different colors for the knowns and unknowns to make the process clearer. Stess the importance of assessing whether the result makes sense. Students should make an initial mental math estimate to compare with the final result displayed by the calculator.

Because the order in which numbers and functions are entered will vary for different brands of calculators, check to see what types of calculators are being used by students. You may wish to have students with similar types of calculators work together to complete the practice problems. At first, some students may need individual help to find the correct keys and to learn the proper sequence of key strokes needed to complete the operations.

Answers to Practice Problems

A. 123.3

B. 28 459

C. 2.0×10^{-16}

D. 113 cm^3

E. 2.140×10^{22} atoms

Discuss

Point out to students that in the laboratory, as well as during their normal everyday routines, they are often presented with more data than is needed to solve a given problem. Explain that they will need to sort out essential data from extraneous data. In cases with more than the required data, sorting the data into knowns and unknowns helps students to determine the answer to "What data do I need to solve this problem?"

ACTIVITY

Have students consider the following problem as an example of the need to separate relevant data. Suppose the class is to do an experiment that requires each lab pair to use a 1.0 g piece of copper wire. Your teacher informs you that you will need to cut the wire from a spool of copper wire that has a total mass of 84.5 g. One cm of the copper wire has a mass of 0.034 g and the spool contains 933 cm of wire that is uniform in diameter. The density of copper is 8.96 g/cm^3. Ask students to describe the procedure they should use to cut a 1.0 g piece of wire from the spool. (Use the fact that 1 cm Cu wire = 0.034 g Cu wire; thus, 1.0 g Cu wire = 29 cm Cu wire. Students should cut a 29 cm piece of wire from the spool. All other data supplied by the teacher are irrelevant.)

Figure 4.3
The three-step problem-solving approach is much like the approach used in many non-science situations. Here, the golfer analyzes the situation, assessing what is known about variables such as the ball's position and distance from the hole. Then she acts on her analysis by attempting her putt. Her evaluation reveals she has been successful! Careful analysis and execution have produced the desired outcome.

Sample Problem 4-1

What is the mass, in grams, of a piece of lead that has a volume of 19.84 cm^3?

1. ***ANALYZE*** ***List the knowns and the unknown.***

   Knowns:
   - volume of lead = 19.84 cm^3
   - density = relationship between the mass and volume of a substance = $\frac{\text{mass}}{\text{volume}}$

     The density of lead is not given in the problem, so you must look it up. According to **Table 3.7** on page 69, the density of lead is 11.4 g/cm^3.
   - density of lead = 11.4 g/cm^3

   Unknown:
   - mass = ? g

   Using the known values for the volume and density of lead and the equation for density, density = $\frac{\text{mass}}{\text{volume}}$, the unknown mass can be determined.

2. ***CALCULATE*** ***Solve for the unknown.***

   The definition for density is an algebraic relationship that includes the unknown variable (mass) and two variables whose values are known in this problem (volume and density). Solve the equation for the unknown variable (mass) by rearranging the equation to isolate mass on one side of the equation. To do so, multiply both sides of the equation by the volume and cancel like terms.

$$\text{density} = \frac{\text{mass}}{\text{volume}}$$

$$\frac{\text{volume}}{1} \times \text{density} = \frac{\text{mass}}{\text{volume}} \times \frac{\text{volume}}{1}$$

Sample Problem 4-1 (cont.)

Canceling like terms yields:

$$\frac{\text{volume}}{1} \times \text{density} = \frac{\text{mass}}{\cancel{\text{volume}}} \times \frac{\cancel{\text{volume}}}{1}$$

$$\text{volume} \times \text{density} = \text{mass}$$

$$(\text{or, mass} = \text{volume} \times \text{density})$$

Substitute the known values for volume and density, and carry out the calculation.

$$\text{mass} = 19.84\ \cancel{\text{cm}^3} \times 11.4\ \text{g}/\cancel{\text{cm}^3} = 226.176\ \text{g}$$

Finally, round the answer to the correct number of significant figures (three) to match the number of figures in 11.4 g/cm^3.

$$\text{mass} = 226\ \text{g}$$

3. ***EVALUATE*** ***Does the result make sense?***
Evaluating the answer involves a number of checks. Has the unknown been found? Yes, the problem asked for the mass of lead. Do the units in the solution cancel so that the calculated answer has the correct units? Yes, the answer is given in grams, a unit of mass. Does the numerical value of the answer make sense? Yes, estimating an answer based on the approximate volume and the approximate density gives an estimated mass close to the calculated result. Is the number of significant figures correct? Yes, the answer to a multiplication calculation can have no more significant figures than the measurement with the smallest number of significant figures.

Here is another Sample Problem. Although it is somewhat different from Sample Problem 4-1, the problem-solving process is the same. Notice that this problem includes Practice Problems for you to solve on your own.

Sample Problem 4-2

What is the volume, in cubic centimeters, of a sample of cough syrup that has a mass of 50.0 g? The density of the cough syrup is 0.950 g/cm^3.

1. ***ANALYZE*** ***List the knowns and the unknown.***

Knowns:
- mass of cough syrup = 50.0 g
- density of cough syrup = 0.950 g/cm^3
- density = $\frac{\text{mass}}{\text{volume}}$

Unknown:
- volume of cough syrup = ? cm^3

Using the known values for the mass and density and the equation for density, the unknown volume can be determined.

Practice Problems Plus

Related Chapter Review Problem
Chapter Review problem 33 is related to Sample Problem 4-1. Chapter Review problem 62 is related to Sample Problem 4-2.

Discuss

Remind students that spending a few minutes looking over the results of a calculation allows them to catch inadvertent errors. Tell students to round the numbers in the problem to simple whole numbers, and then mentally estimate the answer. If a calculated answer differs considerably from an estimated answer, students should suspect an error.

3 Assess

Evaluate Understanding

To determine student understanding of the three-way process for problem solving, ask students to suggest several ways to evaluate the accuracy of an answer. (Reread the problem to be sure the answer supplies the requested unknown; make sure the known data was copied correctly; to assess whether the answer makes sense, round off the numbers and do a quick mental estimate.)

Reteach

Focus on calculations involved in problem solving. Reinforce required math skills, especially the use of scientific notation with calculators. Remind students that estimation is an excellent technique for revealing inadvertent errors. If the calculated answer differs widely from the estimate, students should suspect a math error. Another clue to a wrong answer is units that do not cancel as expected.

Sample Problem 4-2 (cont.)

2. CALCULATE ***Solve for the unknown.***
The equation for density must be solved for volume. Multiplying both sides of the equation by volume yields:

$$\text{volume} \times \text{density} = \frac{\text{mass}}{\cancel{\text{volume}}} \times \cancel{\text{volume}}$$

$$\text{volume} \times \text{density} = \text{mass}$$

Dividing both sides of the equation by density yields the desired result—the equation solved for volume.

$$\frac{\text{volume} \times \cancel{\text{density}}}{\cancel{\text{density}}} = \frac{\text{mass}}{\text{density}}$$

$$\text{volume} = \frac{\text{mass}}{\text{density}}$$

Substituting the known values and solving yields the volume.

$$\text{volume} = \frac{50.0\ \cancel{g}}{0.950\ \cancel{g}/\text{cm}^3} = 52.6316\ \text{cm}^3$$

Rounding to three significant figures gives 52.6 cm^3.

3. EVALUATE ***Does the result make sense?***
Substituting the known mass and the calculated volume into the equation for density yields a result of 0.950 g/cm^3. This agrees with the known density. The unit obtained is the desired unit of volume. The answer has three significant figures as required.

Practice Problems

1. The density of silicon is 2.33 g/cm^3. What is the volume of a piece of silicon that has a mass of 62.9 g?
2. Helium has a boiling point of 4 K. This is the lowest boiling point of any liquid. Express this temperature in degrees Celsius.

section review 4.1

3. List three useful problem-solving skills.
4. State in your own words the three suggested steps for solving word problems.
5. Identify the statements that correctly complete the sentence: Good problem solvers
   a. read a problem only once.
   b. check their work.
   c. break complex problems down into one or more simpler problems.
   d. look for relationships among pieces of information.
6. A small piece of gold has a volume of 1.35 cm^3.
   a. What is the mass, given that the density of gold is 19.3 g/cm^3?
   b. What is the value of this piece if the market value of gold is \$11/g?
7. What is normal body temperature (37 °C) on the Kelvin scale?
8. Match the steps taken in the trip described on page 83 with the three problem-solving steps used in chemistry problems.

Chem ASAP! **Assessment 4.1** Check your understanding of the important ideas and concepts in Section 4.1.

Answers

SECTION REVIEW 3.1

3. Answers will vary.
4. Answers should include (1) Analyze: list the knowns and the unknown; (2) Calculate: solve for the unknown; (3) Evaluate: check that the answer makes sense.
5. **b., c.,** and **d.** are correct.
6. **a.** 26.1 g
   b. \$287
7. 310 K
8. Step 1: Identify starting location and destination. Then choose the best route considering relative distances and quality of roads. Step 2: Make the trip over the planned route. Step 3: At the destination, check the street address.

SIMPLE CONVERSION PROBLEMS

Perhaps you have traveled abroad or are planning to do so. If so, you know—or will soon discover—that different countries have different currencies. As a tourist, exchanging money is essential to the enjoyment of your trip. After all, you must pay for your meals, hotel, transportation, gift purchases, and tickets to exhibits and events. Because each country's currency compares differently with the U.S. dollar, knowing how to convert currency units correctly is very important.

Is there a simple and effective method for making conversions?

objectives

- Construct conversion factors from equivalent measurements
- Apply the techniques of dimensional analysis to a variety of conversion problems

key terms

- conversion factor
- dimensional analysis

Conversion Factors

If you think about any number of everyday situations, you will realize that a quantity can usually be expressed in several different ways. For example, 1 dollar = 4 quarters = 10 dimes = 20 nickels = 100 pennies. These are all expressions, or measurements, of the same amount of money. The same thing is true of scientific quantities. For example, 1 meter = 10 decimeters = 100 centimeters = 1000 millimeters. These are different ways to express the same length.

Whenever two measurements are equivalent, a ratio of the two measurements will equal 1, or unity. For example, you can divide both sides of the equation 1 m = 100 cm by 1 m or by 100 cm.

$$\frac{1\text{ m}}{1\text{ m}} = \frac{100\text{ cm}}{1\text{ m}} = 1 \quad \text{or} \quad \frac{1\text{ m}}{100\text{ cm}} = \frac{100\text{ cm}}{100\text{ cm}} = 1$$

conversion factors

A ratio of equivalent measurements, such as $\frac{100\text{ cm}}{1\text{ m}}$ or $\frac{1\text{ m}}{100\text{ cm}}$, is called a **conversion factor.** In a conversion factor, the measurement in the numerator (on the top) is equivalent to the measurement in the denominator (on the bottom). The conversion factors above are read "one hundred centimeters per meter" and "one meter per hundred centimeters." **Figure 4.4** illustrates another way to look at the relationship in a conversion factor.

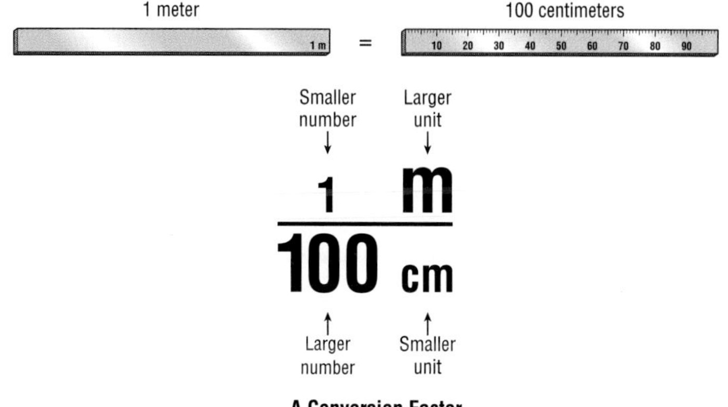

Figure 4.4

The two parts of a conversion factor, the numerator and the denominator, are equal. The smaller number is part of the quantity with the larger unit; for example, a meter is physically larger than a centimeter. The larger number is part of the quantity with the smaller unit.

Chem ASAP!

Animation 3

Learn how to select the proper conversion factor and how to use it.

STUDENT RESOURCES

From the Teacher's Resource Package, use:

- Section Review 4.2, Ch. 4 Practice Problems and Quizzes from the Review Module (Ch. 1–4)
- Laboratory Recordsheet 4-1

TECHNOLOGY RESOURCES

Relevant technology resources include:

- Chem ASAP! CD-ROM
- ResourcePro CD-ROM

1 Engage

Use the Visual

Have students examine the photograph of the currency that opens the section. Then ask if any of them has ever noticed a chart or table in a bank or in the financial section of the newspaper relating the values of foreign currency to the U.S. dollar. Explain that these are conversion tables that allow people to relate one currency to another. Ask:

- **How would you decide which amount of money would be worth more—75 Dutch guilders or 75 German marks?** (Convert these values to a familiar currency—U.S. dollars.)

Note that converting between chemical units is similar to exchanging currency.

- **Is there a simple and effective method for making conversions?** (Students may suggest setting up ratios based on their experiences with unit conversions in math.)

Check Prior Knowledge

To assess student understanding of the conversion between units, ask them to describe situations in daily life that call for such conversions. (currency: among different denominations of bills or between bills and coins; cooking: when adjusting recipes for different numbers of servings, between teaspoons and tablespoons or cups and pints; measurements: from inches to feet or yards; travel: from kilometers to miles)

2 Teach

Discuss

Have students read Link to Business. Explain that currency exchange rates vary from day to day. The daily exchange rates affect all international monetary transactions. Each time one type of money is exchanged for another, the current exchange rate serves as a *conversion factor.*

International currency traders keep track of exchange rates 24 hours a day through a linked computer network. Ask students why it is important for travelers to know the exchange rates of the countries they plan to visit. Ask why it is also important to investors and businesses. Have students follow the exchange rate of one country of their choice for one week using the financial section of the newspaper. Ask students to convert $100 into the foreign currency each day; have them prepare a table or a graph summarizing what happens to their purchasing power as the value of the dollar rises or falls from day to day.

BUSINESS

Monetary Exchange Rates

The conversion of chemical units is similar to the exchange of currency. Americans who travel outside the United States must exchange U.S. dollars for foreign currency at a given rate of exchange. The exchange rate for one U.S. dollar might be 8.57 Mexican pesos, 0.855 Euro, or 129.32 Japanese yen. Exchange rates change daily. Each time an exchange is made, the current exchange rate serves as a conversion factor. For example, travelers to Britain might receive £0.6067 (British pound) for every U.S. dollar they exchange. Suppose, for example, that a traveler has saved $3500.00 to spend during a vacation in Britain. How many pounds does that represent? The conversion factors that relate dollars and pounds are

$$\frac{£0.6067}{\$1.000} \text{ and } \frac{\$1.000}{£0.6067}.$$

The conversion factor that allows dollars to cancel and gives an answer in pounds is the correct form.

$$\frac{\$3500.00}{1} \times \frac{£0.6067}{\$1.0000} = £2123.45$$

Conversion factors have many important applications!

Conversion factors are useful in solving problems in which a given measurement must be expressed in some other unit of measure. When a measurement is multiplied by a conversion factor, the numerical value is generally changed, but the actual size of the quantity measured remains the same. For example, even though the numbers in the measurements 1 g and 10 dg (decigrams) differ, both measurements represent the same mass. In addition, conversion factors within a system of measurement are defined or exact quantities. Therefore, they have an unlimited number of significant figures. How would a conversion factor affect the rounding of a calculated answer? ❶

Here are some additional examples of pairs of conversion factors written from equivalent measurements. The relationship between grams and kilograms is 1000 g = 1 kg. Possible conversion factors are:

$$\frac{1000\text{ g}}{1\text{ kg}} \quad \text{and} \quad \frac{1\text{ kg}}{1000\text{ g}}$$

The relationship between nanometers and meters is 10^9 nm = 1 m. The possible conversion factors are:

$$\frac{10^9\text{ nm}}{1\text{ m}} \quad \text{and} \quad \frac{1\text{ m}}{10^9\text{ nm}}$$

What conversion factors can you write based on the relationship 1 L = 1000 mL? ❷

Dimensional Analysis

No one method is best for solving every type of problem. Several good approaches are available, and generally one of the best is dimensional analysis. **Dimensional analysis** is a way to analyze and solve problems using the units, or dimensions, of the measurements. The best way to explain this problem-solving technique is to use it to solve an everyday situation and then to apply the technique to a chemistry problem.

Sample Problem 4-3 uses dimensional analysis and the problem-solving techniques you learned in Section 4.1.

Sample Problem 4-3

Your school club has sold 600 tickets to a chili-supper fundraising event, and you have volunteered to make the chili. You have a chili recipe that serves ten. The recipe calls for two teaspoons of chili powder. How much chili powder do you need for 600 servings?

1. *ANALYZE* ***List the knowns and the unknown.***

Knowns:	*Unknown:*
• servings = 600	• amount of chili powder = ? tsp
• 10 servings = 2 tsp chili powder	

The correct conversion factor has the known unit in the denominator and the unknown unit in the numerator.

Answers

❶ Conversion factors based on defined quantities have an unlimited number of significant figures; they do not affect rounding of a calculated answer.
❷ possible conversion factors: 1L/1000 mL or 1000 mL/1L

Sample Problem 4-3 (cont.)

2. ***CALCULATE*** ***Solve for the unknown.***

The known measurement (600 servings) must be converted to teaspoons. Use the relationship between servings of chili and teaspoons of chili powder from the recipe to write a conversion factor that has the unit of the known in the denominator. The known unit will cancel, resulting in an answer that has the unit of the unknown.

The correct conversion factor is $\frac{2 \text{ tsp chili powder}}{10 \text{ servings}}$.

$$600 \cancel{\text{servings}} \times \frac{2 \text{ tsp chili powder}}{10 \cancel{\text{servings}}} = 120 \text{ tsp chili powder}$$

3. ***EVALUATE*** ***Does the result make sense?***

The known measurement has been multiplied by a conversion factor (which equals unity). This gives another measurement: 120 teaspoons of chili powder. Because the size of the recipe was increased by a factor of 60, the amount of chili powder must also increase by a factor of 60. The unit of the known (servings) has canceled leaving the answer with the correct unit (tsp).

The answer to this problem has now generated a problem of a different sort. Do you really take the time to measure out 120 teaspoons of chili powder or do you just estimate and pour in several cans? Obviously, the first option would be tedious and the second could be dangerous! Why not measure out the chili powder by the cup? The question then becomes: How many cups are there in 120 teaspoons of chili powder? To solve this problem, you need to know the relationship between teaspoons and cups. This information and other volume relationships can be found in a cookbook. The given relationships do not include the one between teaspoons and cups, but they do include the following:

3 teaspoons = 1 tablespoon

16 tablespoons = 1 cup

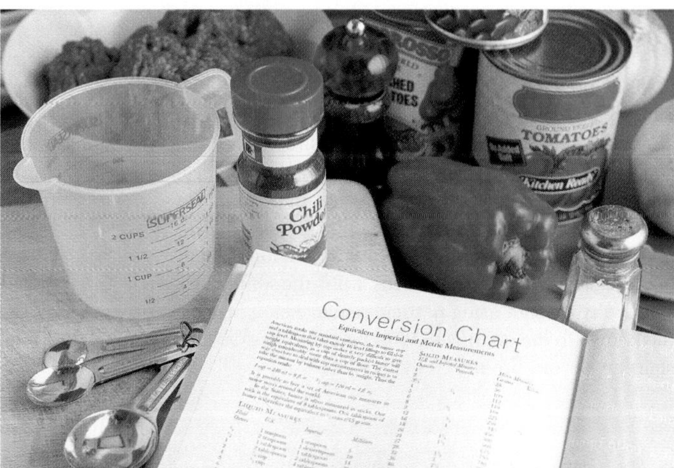

Figure 4.5

When you prepare chili, you need to know how to measure and convert units. Relationships among various measurements of volume used in cooking can be found in a conversion table in a cookbook.

4.2

ACTIVITY

Write the recipe for a simple cake on the board. Explain that you have a problem because the recipe only makes a cake large enough for 5 people and that you would like it to feed 30. Use the three-step problem-solving process to adjust the recipe for 30 people. Have the students look at the quantities of ingredients necessary. Point out that it would be tedious to have to measure out a particular ingredient (pick out one) in teaspoons or tablespoons. It would be easier to measure in larger units. At this point show the different measuring devices available to the cook: teaspoon, tablespoon, 1/4 cup, 1/2 cup, and 1 cup, and explain that many cookbooks have lists of equivalents and conversions. Pass out such a list and have the students suggest alternatives to make the measuring easier. Finish by writing this new recipe on the board. Emphasize that it is equivalent to the original recipe in that the proportions of the ingredients have not changed.

Practice Problems Plus

Related Chapter Review Problem

Chapter Review problem 36 is related to Sample Problem 4-3.

ACTIVITY

Ask local sports franchises for media guides containing vital statistics such as heights and weights for their players. Have students work in small groups. Give each group a metric versus U.S. conversion chart and a list of vital statistics.

Assign each group a set of players; ask the group to convert heights and weights into meters and kilograms. Have them document their approach, including dimensional analysis expressions, conversion factors, and calculations.

Use the Visual

Ask students to examine the photograph in Figure 4.6. Illustrate on the board how you can use dimensional analysis to solve this problem. (Divide the $4.00 by the cost per pound, $1.39/lb. Be sure to include the units and manipulate them along with the numbers.)

Practice Problems Plus

Related Chapter Review Problem
Chapter Review problem 38 is related to Sample Problem 4-4.

Discuss

Explain that measurements are often made using one unit and then converted into a related unit before being used in calculations. For example, students might measure volume in cubic centimeters in the laboratory, but calculate volume in liters or milliliters. Explain to the students that conversions are done using conversion factors (ratios of mathematical equivalents).

Sample Problem 4-4

How many cups are in 120 teaspoons of chili powder?

1. ANALYZE ***List the knowns and the unknown.***

Knowns:

- amount of chili powder = 120 teaspoons
- 3 teaspoons = 1 tablespoon
- 16 tablespoons = 1 cup

Unknown:

- amount of chili powder = ? cups

The first conversion factor must be written with the unit teaspoons in the denominator. The second conversion factor must be written with the unit tablespoons in the denominator. This will provide the desired unit in the answer.

2. CALCULATE ***Solve for the unknown.***

Start with the known, 120 tsp chili powder. Use the first relationship to write a conversion factor that expresses 120 tsp as tablespoons. The unit teaspoons must be in the denominator so that the known unit will cancel. Then use the second conversion factor to change the unit tablespoons into the unit cups. This conversion factor must have the unit tablespoons in the denominator. The two conversion factors can be used together in a simple overall calculation.

$$120\ \cancel{\text{teaspoons}} \times \frac{1\ \cancel{\text{tablespoon}}}{3\ \cancel{\text{teaspoons}}} \times \frac{1\ \text{cup}}{16\ \cancel{\text{tablespoons}}} = 2.5\ \text{cups}$$

3. EVALUATE ***Does the result make sense?***

The numerical result makes sense because there are approximately 50 tsp in a cup, and 120 tsp should be between 2 and 3 cups. The unit in the solution is the desired unit (cups). Before you do the actual arithmetic, it is a good idea to make sure that the units cancel and that the numerator and denominator of each conversion factor are equal to each other.

Figure 4.6
How many pounds of apples can you buy with $4.00 if the price of apples is $1.39/lb? Solve this problem using dimensional analysis. 2

There is usually more than one way to solve a problem. When you first read the previous Sample Problems, you may have thought about different and equally correct ways to approach and solve some of the problems. Some problems are easily worked with simple algebra. Dimensional analysis provides you with a different approach. In each case, you should choose the problem-solving method that works best. What are some of the advantages of mastering more than one approach to problem solving? 1

In solving a problem, you should try to focus on the reasoning you are using at each step. In other words, think the solution through. Try to understand why you are using particular conversion factors in progressing from the known to the unknown. With perseverance, you can learn to apply and sharpen all your problem-solving skills.

Answers

1 For a particular problem one approach may be more effective than another. Also, one approach may be used to solve the problem and the other approach to check the solution.

2 $\cancel{\$}4.00 \times \frac{1\ \text{lb apples}}{\cancel{\$}1.39} = 2.88\ \text{lb apples}$

3 $100\ \cancel{\text{students}} \times \frac{7.5\ \cancel{\text{dg}}}{1\ \cancel{\text{student}}} \times \frac{1\ \text{g}}{10\ \cancel{\text{dg}}} = 75\ \text{g Mg}$

4 $41\ \cancel{\text{mm}} \times \frac{2\ \text{m}}{10\ \cancel{\text{mm}}} = 8.2\ \text{m}$

Sample Problem 4-5

The directions for an experiment ask each student to measure out 1.84 g of copper (Cu) wire. The only copper wire available is a spool with a mass of 50.0 g. How many students can do the experiment before the copper runs out?

1. *ANALYZE* ***List the knowns and the unknown.***

   Knowns:
   - mass of copper available = 50.0 g Cu
   - Each student needs 1.84 grams of copper, or $\frac{1.84 \text{ g Cu}}{\text{student}}$.

   Unknown:
   - number of students = ? students

   From the known mass of copper, calculate the number of students that can do the experiment. The desired conversion is mass of copper ⟶ number of students.

2. *CALCULATE* ***Solve for the unknown.***

   Because students is the desired unit for the answer, the conversion factor should be written with students in the numerator.

   $$50.0 \text{ g Cu} \times \frac{1 \text{ student}}{1.84 \text{ g Cu}} = 27.174 \text{ students} = 27 \text{ students}$$

   Note that because students cannot be fractional, the result is shown rounded to a whole number.

3. *EVALUATE* ***Does the result make sense?***

   The number of students (27) seems to be a reasonable answer. The approximate calculation using the ratio $\frac{1 \text{ student}}{2 \text{ g Cu}}$ gives an approximate answer of 25 students. The unit of the answer (students) is the one desired.

Practice Problems

9. An experiment requires that each student use an 8.5-cm length of magnesium ribbon. How many students can do the experiment if there is a 570-cm length of magnesium ribbon available?
10. A 1.00-degree increase on the Celsius scale is equivalent to a 1.80-degree increase on the Fahrenheit scale. If a temperature increases by 48.0 °C, what is the corresponding temperature increase on the Fahrenheit scale?

Converting Between Units

In chemistry, as in many other subjects, you often need to express a measurement in a unit different from the one given or measured initially. For example, suppose a laboratory experiment requires 7.5 dg of magnesium metal, and 100 students will do the experiment. How many grams of magnesium should your teacher have on hand? This is a typical conversion problem because you need to express a given measurement in a different unit. As you have seen, conversion problems are easily solved using dimensional analysis. ❸

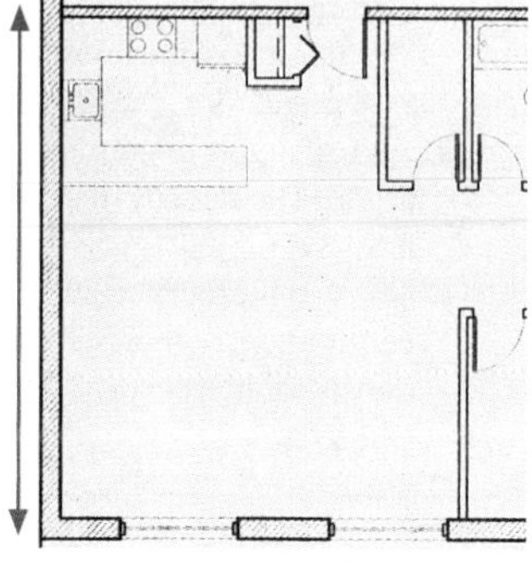

Figure 4.7
Architects use many conversion factors to draw buildings to scale accurately. If 10 mm on a scaled drawing equals 2 actual meters, what is the length of this room, front to back, as indicated by the red arrow? ❹

4.2

Cooperative Learning

Have the students try working in groups of two as a "problem-solving pair." One student thinks aloud while trying to solve a problem. The partner keeps a careful record of the process. Then, they reverse roles and work on another problem. This procedure helps students see the mental process another person goes through while solving a problem. It can provide insight about the approach.

Practice Problems Plus

Related Chapter Review Problem

Chapter Review problem 40 is related to Sample Problem 4-5.

Discuss

Have students work one of the practice problems involving unit conversions while you work the same problem on the board. Ask them to suggest the appropriate conversion factors.

Students should use dimensional analysis as a tool for solving all of the problems they encounter in chemistry. Their first question about any quantity should be "What are the units of this quantity?" By comparing the units of various quantities in a problem, students can discover whether they need to perform any unit conversions before proceeding.

Practice Problems Plus

Additional Practice Problem
The Chem ASAP! CD-ROM contains the following problem related to Sample Problem 4-6: Make the following conversions.
a. 0.045 L to cubic centimeters (4.5×10^1 cm^3)
b. 14.3 mg to grams (1.43×10^{-2} g)
c. 0.0056 m to micrometers (5.6×10^3 μm)
d. 0.035 cm to millimeters (3.5×10^{-1} mm)

Discuss

Explain that dimensional analysis is an extremely powerful problem-solving tool. Learning this method requires extra effort on the part of students. They must often use multiple conversion factors. The extra effort can be justified because the proper manipulation of the units assures accurate manipulation of the numbers.

Practice Problems Plus

Additional Practice Problem
The Chem ASAP! CD-ROM contains the following problem related to Sample Problem 4-7: Make the following conversions.
a. 2.53 cm^3 of gold to grams (density of gold = 19.3 g/cm^3) (48.8 g)
b. 1.6 g of oxygen to liters (density of oxygen = 1.33 g/L) (1.2 L)
c. 25.0 g of ice to cubic centimeters (density of ice = 0.917 g/cm^3) (27.3 cm^3)

As you work through the examples that follow, notice that the three-step problem-solving approach is applied to each.

Sample Problem 4-6

Express 750 dg in grams.

1. ***ANALYZE*** ***List the knowns and the unknown.***

Knowns:
- mass = 750 dg
- 1 g = 10 dg

Unknown:
- mass = ? g

The desired conversion is decigrams ⟶ grams. Using the expression relating the units, 10 dg = 1 g, multiply the given mass by the proper conversion factor.

2. ***CALCULATE*** ***Solve for the unknown.***

The correct conversion factor is $\frac{1\text{ g}}{10\text{ dg}}$ because the known unit is in the denominator and the unknown unit is in the numerator.

$$750\ \cancel{\text{dg}} \times \frac{1\text{ g}}{10\ \cancel{\text{dg}}} = 75\text{ g}$$

3. ***EVALUATE*** ***Does the result make sense?***

Because the unit gram represents a larger mass than the unit decigram, it makes sense that the number of grams is less than the given number of decigrams. The unit of the known (dg) cancels, and the answer has the correct unit (g). The answer also has the correct number of significant figures.

Practice Problems

11. Using tables from Chapter 3, convert the following.
- **a.** 0.044 km to meters
- **b.** 4.6 mg to grams
- **c.** 8.9 m to decimeters
- **d.** 0.107 g to centigrams

12. Convert the following.
- **a.** 15 cm^3 to liters
- **b.** 7.38 g to kilograms
- **c.** 0.67 s to milliseconds
- **d.** 94.5 g to micrograms

Chem ASAP!
Problem-Solving 12
Solve Problem 12 with the help of an interactive guided tutorial.

Sample Problem 4-7

What is the volume of a pure silver coin that has a mass of 14 g? The density of silver (Ag) is 10.5 g/cm^3.

1. ***ANALYZE*** ***List the knowns and the unknown.***

Knowns:
- mass of coin = 14 g
- density of silver = 10.5 g/cm^3

Unknown:
- volume of coin = ? cm^3

This problem can be solved by using algebra, or it can be solved by using a conversion factor. Convert the given mass of the coin into a corresponding volume. The density of silver gives the needed relationship between mass and volume. Multiply the mass of the coin by the proper conversion factor to yield an answer in cm^3.

Practice Problems

13. Use dimensional analysis and the given densities to make the following conversions.
- **a.** 14.8 g of boron to cubic centimeters of boron. The density of boron is 2.34 g/cm^3.
- **b.** 2.8 L of argon to grams of argon. The density of argon is 1.78 g/L.
- **c.** 4.62 g of mercury to cubic centimeters of mercury. The density of mercury is 13.5 g/cm^3.

Meeting Diverse Needs

LEP Students may benefit from a reminder that certain key words and phrases in each word problem indicate the unknown quantity and its units. Some of these phrases are:
- How much
- What is
- How long
- Determine
- Find

Sample Problem 4-7 (cont.)

2. ***CALCULATE*** ***Solve for the unknown.***

The correct conversion factor is $\frac{1\ \text{cm}^3\ \text{Ag}}{10.5\ \text{g Ag}}$ because the known unit is in the denominator and the unknown unit is in the numerator.

$$14\ \text{g Ag} \times \frac{1\ \text{cm}^3\ \text{Ag}}{10.5\ \text{g Ag}} = 1.3\ \text{cm}^3\ \text{Ag}$$

3. ***EVALUATE*** ***Does the result make sense?***

Because a mass of 10.5 g of silver has a volume of 1 cm^3, it makes sense that 14.0 g of silver should have a volume slightly larger than 1 cm^3. The answer has two significant figures because the given mass has two significant figures.

Practice Problems (cont.)

14. Rework the preceding problems by applying the equation:

$$\text{density} = \frac{\text{mass}}{\text{volume}}.$$

Chem ASAP!

Problem-Solving 13

Solve Problem 13 with the help of an interactive guided tutorial.

section review 4.2

15. What conversion factor would you use to convert between these pairs of units?

a. minutes to hours

b. grams of water to cubic centimeters of water

c. grams to milligrams

d. cubic decimeters to milliliters

16. Make the following conversions. Express your answers in scientific notation.

a. 36 cm to meters

b. 14.8 g to micrograms

c. 1.44 kL to liters

d. 68.9 m to decimeters

e. 3.72×10^{-3} kg to grams

f. 66.3 L to cubic centimeters

g. 0.0371 m to kilometers

17. A 2.00-kg sample of bituminous coal is composed of 1.30 kg of carbon, 0.20 kg of ash, 0.15 kg of water, and 0.35 kg of volatile (gas-forming) material. Using this information, determine how many kilograms of carbon are in 125 kg of this coal.

18. Which of the following linear measures is the longest?

a. 6×10^4 cm

b. 6×10^6 mm

c. 0.06 km

d. 6×10^9 nm

19. An atom of gold has a mass of 3.271×10^{-22} g. How many atoms of gold are in 5.00 g of gold?

Chem ASAP! **Assessment 4.2** Check your understanding of the important ideas and concepts in Section 4.2.

portfolio project

Develop four charts of conversion factors for units of length, volume, area, and mass. Include common non-SI units such as miles, gallons, acres, and tons. Provide both the names and the abbreviations of the units.

Answers

SECTION REVIEW 4.2

15. a. $\frac{1\ \text{hour}}{60\ \text{min}}$

b. $\frac{1\ \text{cm}^3\ H_2O}{1\ \text{g}\ H_2O}$

c. $\frac{10^3\ \text{mg}}{1\ \text{g}}$

d. $\frac{10^3\ \text{mL}}{1\ \text{dm}^3}$

16. a. 3.6×10^{-1} m

b. 1.48×10^7 µg

c. 1.44×10^3 L

d. 6.89×10^2 dm

e. 3.72 g

f. 6.63×10^4 cm^3

g. 3.71×10^{-5} km

17. 81.3 kg carbon

18. b.

19. 1.53×10^{22} atoms of gold

3 Assess

Evaluate Understanding

To determine students' grasp of conversion factors, ask:

- **What is the relationship between the numerator and the denominator of any conversion factor?** (They are equivalent so that the ratio of numerator to denominator equals 1.)

Have students summarize the dimensional analysis process with a series of sentence statements.

Reteach

Remind students that a number without a unit is not an acceptable measure and that units are the key to dimensional analysis. Conversion factors are chosen so that the units cancel out until only the desired unit remains. If students are having trouble doing unit conversions, suggest the following method for evaluating conversion factors. Suppose they need to convert from liters to milliliters and cannot decide whether to multiply or divide by 1000. Explain that in a conversion factor such as 1L = 1000 mL, the larger unit (L) should have the smaller number (1) and the smaller unit (mL) should have the larger number (1000).

Portfolio Project

Once students have made their tables of conversion factors, you can extend this project by having students collect examples from the news of situations where these factors could be applied.

SMALL-SCALE LAB

NOW WHAT DO I DO?

OVERVIEW

Procedure Time: 20 minutes
Students practice solving problems in divergent ways.

PROCEDURE

1. Empty cup = 2.79 g
   Cup + 50 water drops = 3.74 g
   Mass of 50 drops of water =
   3.74 g − 2.79 g = 0.95 g
   Mass of one drop of water =
   0.95 g/50 drops = 0.019 g/drop
2. Mass of pre-1982 penny = 3.11 g
   Mass of post-1982 penny = 2.50 g

ANALYSIS

1. 19 mg
2. 0.019 cm^3; 0.019 mL; 19 μL
3. 1000 mg/cm^3; 1000 mg/mL
4. 2.95 g Cu; 0.16 g Zn
5. 0.060 g Cu; 2.44 g Zn
6. The new penny is mostly zinc, which has a lower density than copper.

YOU'RE THE CHEMIST

1. at 90°, 1 drop = 0.019 g
   at 45°, 1 drop = 0.0218 g
   at 0°, 1 drop = 0.0242 g
   Pipets give different results.
2. The best angle is 90° because the pipet is easiest to control. Expel the air bubble so that the first drop will be the same size as the rest.
3. Find mass of can and divide by density of aluminum. Sample answer: one can = 14.77 g; d_{Al} − 2.70 g/cm^3; V = 5.47 cm^3
4. (1) Measure the mass before and after you fill the can with water. Use the mass and density of water to find the volume.
   (2) Measure the height and radius and calculate volume.
   $V = \pi r^2 h$
   (Can is not a perfect cylinder.)
   (3) Read label: 12 oz = 355 mL
5. Sample answer:
   V = 16.5 m × 3.0 m × 12.8 m
   = 634 m^3 × 1000 L/m^3
   = 634 000 L
   Assume 30 people with an average weight of 130 lb (1 kg = 2.2 lb) and a density of about 1.0 kg/L.
   Volume of 30 people = 30 × 130 lb ×
   1 kg/ 2.2 lb × 1 L/1.0 kg = 1800 L
   The volume of 30 chairs, 15 tables, and 2 desks is about that of 30 people or 1800 L. The volume of people and furniture is 3600 L.
   % error = (3600/634 000)(100) = 0.57%.
6. If die measures 1.55 cm on a side:
   V = (1.55 cm)^3 = 3.72 cm^3
   A die has 21 holes that are hemispheres with a radius of 0.20 cm.
   V of hemisphere = $2/3\pi r^3$ = 0.017 cm^3
   V of 21 hemispheres = 0.36 cm^3
   V of die is 3.72 cm^3 − 0.36 cm^3 = 3.36 cm^3
   Error = 0.36 cm^3
   % error = (0.36 cm^3/3.36 cm^3)(100) = 11%
   Note: The holes in some dice are cones. The volume of a cone is $1/3\pi r^2 h$.
7. Find weight in pounds and convert to kg. Assume the density is about 1.00 kg/L.
   V = weight × 1 kg/2.2 lb × 1L/1.00 kg

SMALL-SCALE LAB

NOW WHAT DO I DO?

PURPOSE

To solve problems by making accurate measurements and applying mathematics.

MATERIALS

- pencil
- paper
- meter stick
- balance
- pair of dice
- aluminum can
- calculator
- small-scale pipet
- water
- a pre- and a post-1982 penny
- 8-well strip
- plastic cup

PROCEDURE

1. Determine the mass, in grams, of one drop of water. To do this, measure the mass of an empty cup. Add 50 drops of water from a small-scale pipet to the cup and measure its mass again. Subtract the mass of the empty cup from the mass of the cup with water in it. To determine the average mass in grams of a single drop, divide the mass of the water by the number of drops (50). Repeat this experiment until your results are consistent.
2. Determine the mass of a pre-1982 penny and a post-1982 penny.

ANALYSIS

Using your experimental data, record the answers to the following questions.

1. What is the average mass of a single drop of water in milligrams? (1 g = 1000 mg)
2. The density of water is 1.00 g/cm^3. Calculate the volume of a single drop in cm^3 and mL. (1 mL = 1 cm^3) What is the volume of a drop in microliters (μL)? (1000 μL = 1 mL)
3. What is the density of water in units of mg/cm^3 and mg/mL? (1 g = 1000 mg)
4. Pennies made before 1982 consist of 95.0% copper and 5.0% zinc. Calculate the mass of copper and the mass of zinc in the pre-1982 penny.
5. Pennies made after 1982 are made of zinc with a thin copper coating. They are 97.6% zinc and 2.4% copper. Calculate the mass of copper and the mass of zinc in the newer penny.
6. Why does one penny have less mass than the other?

YOU'RE THE CHEMIST

The following small-scale activities allow you to develop your own procedures and analyze the results.

1. **Design It!** Design an experiment to determine if the size of drops varies with the angle at which they are delivered from the pipet. Try vertical (90°), horizontal (0°), and halfway between (45°). Repeat until your results are consistent.
2. **Analyze It!** What is the best angle to hold a pipet for ease of use and consistency of measurement? Explain. Why is it important to expel the air bubble before you begin the experiment?
3. **Design It!** Make the necessary measurements to determine the volume of aluminum used to make an aluminum soda can. *Hint:* Look up the density of aluminum in your textbook.
4. **Design It!** Design and carry out some experiments to determine the volume of liquid that an aluminum soda can will hold.
5. **Design It!** Measure a room and calculate the volume of air it contains. Estimate the percent error associated with not taking into account the furniture and people in the room.
6. **Design It!** Make the necessary measurements and do the necessary calculations to determine the volume of a pair of dice. First ignore the volume of the dots on each face, and then account for the volume of the dots. What is your error and percent error when you ignore the holes?
7. **Design It!** Design an experiment to determine the volume of your body. Write down what measurements you would need to make and what calculations you would do. What additional information might be helpful?

MORE-COMPLEX PROBLEMS

section 4.3

Architectural wonders like the Guggenheim Museum in Bilbao, Spain, are a testament to the vision, knowledge, and skills of all those associated with the construction. Solving the myriad problems that accompany the building of such structures is an essential first step that often involves converting complex units.

How can complex conversions be calculated?

objectives

- Solve problems by breaking the solution into steps
- Convert complex units, using dimensional analysis

Multistep Problems

Many complex tasks in your everyday life are best handled by breaking them down into manageable parts. For example, if you were cleaning a car, you might first vacuum the inside, then wash the exterior, then dry the exterior, and finally put on a fresh coat of wax. Similarly, many complex word problems are more easily solved by breaking the solution down into steps.

When converting between units, it is often necessary to use more than one conversion factor. Sample Problem 4-8 illustrates the use of multiple conversion factors.

Sample Problem 4-8

What is 0.073 cm in micrometers?

1. *ANALYZE* ***List the knowns and the unknown.***

Knowns:

- length = 0.073 cm = 7.3×10^{-2} cm
- 10^2 cm = 1 m
- 1 m = 10^6 μm

Unknown:

- length = ? μm

The desired conversion is from centimeters to micrometers. The problem can be solved in a two-step conversion.

2. *CALCULATE* ***Solve for the unknown.***

First, change centimeters to meters; then change meters to micrometers: centimeters ⟶ meters ⟶ micrometers. Each conversion factor must be written so that the unit in the denominator cancels the unit in the numerator of the previous factor.

$$7.3 \times 10^{-2}\ \cancel{\text{cm}} \times \frac{1\ \cancel{\text{m}}}{10^2\ \cancel{\text{cm}}} \times \frac{10^6\ \mu\text{m}}{1\ \cancel{\text{m}}} = 7.3 \times 10^2\ \mu\text{m}$$

3. *EVALUATE* ***Does the result make sense?***

Because a micrometer is a much smaller unit than a centimeter, the answer should be numerically larger than the given measurement. The units have canceled correctly, and the answer has the correct number of significant figures.

Practice Problems

20. The radius of a potassium atom is 0.227 nm. Express this radius in centimeters.
21. Earth's diameter is 1.3×10^4 km. What is the diameter expressed in decimeters?

Student Resources

- Section Review 4.3, Ch. 4 Practice Problems, Interpreting Graphics, Vocabulary Review, Quizzes, and Tests from the Review Module (Ch. 1–4)
- Laboratory Recordsheet 4-2
- Small-Scale Chemistry Lab Manual: Experiment 4

Technology Resources

Relevant technology resources include:

- Chem ASAP! CD-ROM
- Resource Pro CD-ROM
- Assessment Resources CD-ROM

4.3

1 Engage

Use the Visual

Have students study the photograph and read the text that opens the section. Point out that constructing a building requires solving numerous problems involving complex mathematical conversions. Ask:

- **How can complex conversions be calculated?** (by breaking problems into simpler steps)
- **Describe the steps involved in calculating the total cost of titanium for a structure if you know the square yards needed, the number of centimeters per foot, and the cost of titanium in dollars per square meter.** (Multiply square yards of titanium twice by number of feet per yard; multiply the result twice by number of centimeters per foot; divide the result twice by the number of centimeters per meter; finally, multiply the result by the cost per meter squared.)

Check Prior Knowledge

To assess knowledge about significant figures and rounding off results of computations, have students evaluate each of the following and write the answer to the appropriate number of significant figures.

- **210.4 + 22.9 + 506.07 =** (739.4)
- **42.998 + 24.04 − 0.8852 =** (66.15)
- **Round 1.245 to three significant figures.** (1.25)
- **How many significant figures are in 0.702 350?** (6)

Practice Problems Plus

Related Chapter Review Problems

Chapter Review problems 43c and 43d are related to Sample Problem 4-8.

2 Teach

Use the Visual

Have the students examine the series of photographs in Figure 4.8. Ask the students to hypothesize about the steps that occur in the building of a house. List some steps on the board. (find a site, get permits, design house, purchase materials, start construction) **Explain that as complex as this sounds, each of the steps actually represents a series of simpler steps.** (construction broken down into framing, plumbing, electrical, and so on)

TEACHER DEMO

Tell students you want to make a scale model of the classroom. The model should be 1/20th the size of the classroom. Ask students to consider how they might determine the dimensions of the model. Have the students start by measuring the lengths of the four walls in feet; then have them determine the conversion factor to use for designing the model. On the board, show the conversion of these measurements to the scale model dimensions. Repeat this process for the height of the walls. Explain that this process would have to be followed for all of the objects in the classroom.

Practice Problems Plus

Related Chapter Review Problems
Chapter Review problems 43a, 44a, and 49 are related to Sample Problem 4-9.

Additional Practice Problem
The Chem ASAP! CD-ROM contains the following problem: At Earth's farthest point from the sun, sunlight takes 8.5 minutes to reach Earth. How many weeks is this? (8.4×10^{-4} weeks)

Figure 4.8
A complex process, such as building a house, must be broken down into several simpler steps. Each of those steps is generally executed in logical order so that the product of the process is the desired one. Obviously one of the steps in this example involved the removal of a tree!

Here is another example of a multistep problem. You probably do not know how many seconds are in one day. However, you can easily calculate this quantity by using dimensional analysis and equivalent expressions of time that you do know.

Sample Problem 4-9

How many seconds are there in exactly one day?

1. **ANALYZE** ***List the knowns and the unknown.***

Knowns:
- length of time = 1 day
- 1 day = 24 hr
- 1 hr = 60 min
- 1 min = 60 s

Unknown:
- length of time = ? s

The desired conversion is days ⟶ seconds.

2. **CALCULATE** ***Solve for the unknown.***

As identified in the first step, the conversion required is from days to seconds. This conversion can be carried out by the following sequence of conversions: day ⟶ hours ⟶ minutes ⟶ seconds. Each conversion factor in the series must be written so that the unit in the denominator cancels the unit in the numerator of the previous factor.

$$1\,\cancel{\text{day}} \times \frac{24\,\cancel{\text{h}}}{\cancel{\text{day}}} \times \frac{60\,\cancel{\text{min}}}{\cancel{\text{h}}} \times \frac{60\ \text{s}}{\cancel{\text{min}}} = 8.6400 \times 10^{4}\ \text{s}$$

3. **EVALUATE** ***Does the result make sense?***

A large number is expected because a second is a much smaller unit than a day. The numerator and denominator of the conversion factors have canceled correctly to yield the desired unit.

Practice Problems

22. How many minutes are there in exactly one week?
23. How many seconds are in exactly a 40-hr work week?

Chem ASAP!
Problem-Solving 23
Solve Problem 23 with the help of an interactive guided tutorial.

Meeting Diverse Needs

At Risk Kinesthetic learners may find it useful to maintain a set of personal conversion-factor cards. These students can continue to use the cards as they work through problems in later chapters, until the conversion process becomes familiar.

Answer

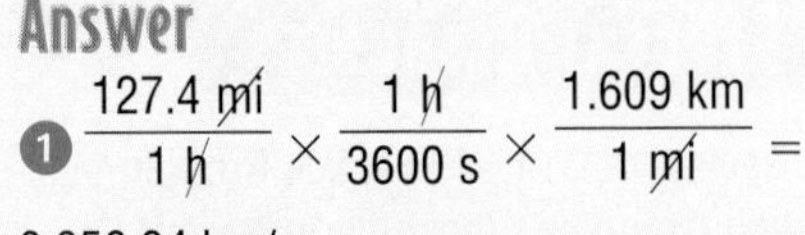

1. $\frac{127.4\ \cancel{\text{mi}}}{1\ \cancel{\text{h}}} \times \frac{1\ \cancel{\text{h}}}{3600\ \text{s}} \times \frac{1.609\ \text{km}}{1\ \cancel{\text{mi}}} =$

0.056 94 km/s

Converting Complex Units

Many common measurements are expressed as a ratio of two units. For example, the results of international car races often give average lap speeds in kilometers per hour. You measure the densities of solids and liquids in grams per cubic centimeter. You measure the gas mileage in a car in miles per gallon of gasoline. If you use dimensional analysis, converting these complex units is just as easy as converting single units. It will just take multiple steps to arrive at an answer.

Figure 4.9
Different units are used in different settings. The speed of Venus Williams's record-setting serve was displayed in mi/h. What would this service speed be in km/s? 1

Sample Problem 4-10

The density of manganese is 7.21 g/cm^3. What is the density of manganese expressed in units of kg/m^3?

1. ***ANALYZE*** ***List the knowns and the unknown.***

Knowns:
- density of manganese = 7.21 g/cm^3
- 10^3 g = 1 kg
- 10^6 cm^3 = 1 m^3

Unknown:
- density of manganese = ? kg/m^3

The desired conversion is $g/cm^3 \longrightarrow kg/m^3$. The mass unit in the numerator must be changed from grams to kilograms: $g \longrightarrow kg$. In the denominator, the volume unit must be changed from cubic centimeters to cubic meters: $cm^3 \longrightarrow m^3$. Note that the relationship between cm^3 and m^3 was determined from the relationship between cm and m. Cubing the relationship 10^2 cm = 1 m yields $(10^2\ cm)^3 = (1\ m)^3$, or $10^6\ cm^3 = 1\ m^3$.

2. ***CALCULATE*** ***Solve for the unknown.***

$$\frac{7.21\ \cancel{g}}{1\ \cancel{cm^3}} \times \frac{1\ kg}{10^3\ \cancel{g}} \times \frac{10^6\ \cancel{cm^3}}{1\ m^3} = 7.21 \times 10^3\ kg/m^3$$

3. ***EVALUATE*** ***Does the result make sense?***

Because the physical size of the volume unit m^3 is so much larger than cm^3 (10^6 times), the calculated value of the density should be larger than the given value even though the mass unit is also larger (10^3 times). The units cancel, the conversion factors are correct, and the answer has the correct ratio of units.

Practice Problems

24. Gold has a density of 19.3 g/cm^3. What is the density in kilograms per cubic meter?
25. There are 7.0×10^6 red blood cells (RBC) in 1.0 mm^3 of blood. How many red blood cells are in 1.0 L of blood?

Chem ASAP!

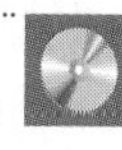

Problem-Solving 25
Solve Problem 25 with the help of an interactive guided tutorial.

4.3

Discuss

Give students a few minutes to try and solve a complex problem such as "how many seconds are there in a school day?" without writing anything down. Solve the problem on the board using dimensional analysis. (Write the length of the school day in hours. Multiply by the conversion factors for minutes per hour and seconds per minute.)

Use the Visual

Have students examine Figure 4.9 and note the units used to measure speed. Ask:

- **What units of speed are shown on the speedometers of passenger cars?** (Most newer cars show speed in mi/h and km/h.)

Converting a complex unit can require one conversion factor or more based on the conversion (mi/h to km/h vs. mi/h to km/s).

Discuss

Ask the students to think of common measurements that have complex units. (Prices such as cost per pound, kilogram, or liter are familiar examples.) **Point out that students must use multiple conversion factors when working with complex units.**

Practice Problems Plus

Related Chapter Review Problems
Chapter Review problems 44b, 44c, and 47 are related to Sample Problem 4-10.

Additional Practice Problem
The Chem ASAP! CD-ROM contains the following problem: 1.00 L of neon gas contains 2.69×10^{22} neon atoms. How many neon atoms are in 1.00 mm^3 of neon gas under the same conditions? (2.69×10^{16} atoms)

Meeting Diverse Needs

At Risk Students vary in their ability to think logically. Those with poor skills in this area need individual help. Provide as much class time as possible for students to work on problem assignments in cooperative learning groups. Have students explore their own problem-solving styles within the context of the three-step method. Encourage students to draw a diagram or picture of the problem to be solved whenever possible. Some students may want to read the problem out loud, or have a partner read it to them. Some may want to work with symbols and equations.

MINI LAB

Dimensional Analysis

ANALYSIS AND CONCLUSIONS

1. The measurement is converted to a different unit.
2. a. 0.785 m
   b. 56 cm^3
   c. 7.7×10^7 mg
   d. 4.54×10^3 mg H_2O
   e. 8.7×10^{-10} dm
   f. 7.85×10^{-2} L H_2O
   g. 9.6×10^3 μm
   h. 6.7×10^3 nm

3 Assess

Evaluate Understanding

Have students set up the conversion path for the following three-conversion problem using complex units: Gold has a density of 19.3 g/cm^3. What is the mass, in kilograms, of a cubic meter of gold? (1.93×10^4 kg)

Reteach

Remind students to develop a written plan, or "path," for the solution of complex problems. This plan should include the known, all intermediate units, and the unknown. Once the steps are in place, students can select the correct conversion factors and put them in the proper order. To help students identify the correct form of a conversion factor, point out that units in the path always appear as the units in the numerators of the fractions. Remind students that working with complex units involves developing two conversion paths, one for the numerator of the measurement and one for the denominator.

MINI LAB

DIMENSIONAL ANALYSIS

PURPOSE

To apply the problem-solving technique of dimensional analysis to conversion problems.

MATERIALS

- 3 inch × 5 inch index cards or paper cut to approximately the same size
- pen

PROCEDURE

A conversion factor is a ratio of equivalent measurements. For any relationship, you can write two ratios. On a conversion factor card you can write one ratio on each side of the card.

1. Make a conversion factor card for each metric relationship shown in **Tables 3.3, 3.4,** and **3.5.** Show the inverse of the conversion factor on the back of each card.
2. Make a conversion factor card showing the mass–volume relationship for water (1.00 g H_2O = 1.00 mL H_2O). Show the inverse of the conversion factor on the back of the card.
3. Use the appropriate conversion factor cards to set up solutions to Sample Problems 4-6 and 4-8. Notice that in each solution, the unit in the denominator of the conversion factor cancels the unit in the numerator of the previous conversion factor.
4. Use conversion factor cards to set up solutions to Problem 28, below.

ANALYSIS AND CONCLUSIONS

1. What is the effect of multiplying a given measurement by one or more conversion factors?
2. Use your conversion factor cards to set up solutions to these problems.
   a. 78.5 cm = ? m
   b. 0.056 L = ? cm^3
   c. 77 kg =? mg
   d. 4.54 mL H_2O = ? mg H_2O
   e. 0.087 nm = ? dm
   f. 78.5 g H_2O = ? L H_2O
   g. 0.96 cm = ? μm
   h. 0.0067 mm = ? nm

section review 4.3

26. How can you solve a complicated problem more easily?
27. How are complex units dealt with in calculations?
28. Convert the following. Express your answers in scientific notation.
   a. 7.5×10^4 nm to kilometers
   b. 3.9×10^5 mg to decigrams
   c. 0.764 km to centimeters
   d. 2.21×10^{-4} dL to microliters
29. Light travels at a speed of 3.00×10^{10} cm/s. What is the speed of light in kilometers per hour?
30. A bar of gold measures 4.5 cm by 6.5 cm by 1.6 dm. Calculate the mass of the gold bar in kilograms. The density of gold is 19.3 g/cm^3.
31. What is the mass, in kilograms, of 14.0 L of gasoline? (Assume that the density of gasoline is 0.680 g/cm^3.)

Chem ASAP! **Assessment 4.3** Check your understanding of the important ideas and concepts in Section 4.3.

Answers

SECTION REVIEW 4.3

26. Answers should include a description of a step-by-step dimensional analysis.
27. Answers should include a description of multiple conversion factors.
28. a. 7.5×10^{-8} km
   b. 3.9×10^3 dg
   c. 7.64×10^4 cm
   d. 2.21×10^1 μL
29. 1.08×10^9 km/h
30. 9.0 kg
31. 9.52 kg

Chemistry Serving...the Consumer

NATURE'S MEDICINE CABINET

You have probably taken drugs such as aspirin and decongestants several times in your life. Have you ever wondered whom you should thank for the welcome relief these drugs provide? In some cases, the "inventor" of the drug was not a white-coated chemist but a green-leafed plant!

In many ways, plants are the master chemists of the world. For millions of years, they have been busily evolving the ability to manufacture an array of chemical compounds. Many of these compounds have obvious uses for the plant — defense against plant-eating insects, for example. Others are by-products of growth processes of plants. A sizable number of plant compounds have some kind of property or chemical activity that makes them useful to humans as medicine.

Humans have known about the medicinal properties of plants for a long time. Before modern science, humans used plant extracts to treat illnesses and injuries, without understanding the chemical basis of the treatment. Today, chemists and medical researchers continue to look to plants as either the source or inspiration for new drugs.

The story of aspirin is a good example of the ongoing link between plant compounds and drugs for humans. The first aspirin-like drug was an extract from the bark of willow trees, which was used by native North Americans to alleviate pain and fever. In the 1800s, chemists begin to search for the active ingredient in willow bark. They isolated a substance that had the fever-reducing effect and called it salicin, after *Salix*, the Latin name for the willow. Then chemists worked on synthesizing a compound in the laboratory with the same effects. They succeeded in making salicylic acid, which worked well to reduce fever but was very irritating to the stomach. Finally, in 1899, a German company began marketing a derivative of salicylic acid that did not upset the stomach but that still reduced fever and pain. They called it aspirin.

In many ways, plants are the master chemists of the world.

Many other common drugs have their roots in plant chemistry. The active ingredient in many decongestants, for example, is closely related to a compound in the ephedra plant called ephedrine. Plants are also a continuing source of new drugs. For example, a drug called taxol, extracted from the bark of the Pacific yew tree, has been shown to be effective in fighting both breast and ovarian cancer.

There are also some plant compounds that have been known for a long time to have medicinal properties but that are only now beginning to be recognized as effective drugs by scientists. One of these is contained in the plant called St. John's wort. Extracts of this plant have been shown to counteract depression, one of the most common mental disorders. The active ingredient in St. John's wort is related to a new class of drugs called selective serotonin re-uptake inhibitors (SSRIs), which are also used to treat depression. St. John's wort and the SSRI drugs both work by changing how the body uses serotonin, a hormone that plays a key part in mood regulation.

The next time you take a stroll in the woods, think about the wealth of chemical compounds contained in the flora around you. One of those compounds might someday offer a cure for what ails you.

CHEMISTRY IN CAREERS

TAKE IT TO THE NET

Find out more about career opportunities: **www.phschool.com**

MEDICAL LABORATORY TECHNICIAN

Process sera for blood tests; determine cholesterol and enzyme levels.

See page 869.

CHEMICAL SPECIALIST

Local food service distributor seeks responsible self-motivated individ-

CHEMIST NEEDED

Chemistry Serving... the Consumer

DISCUSS

Discuss the growing popularity of herbal remedies. These products are widely available at local markets and drug stores. Echinacea, for example, is believed to stimulate the immune system for treatment and prevention of colds and flu; gingko biloba is believed to improve circulation and help with memory. The "smoothie," a popular fruit drink available in juice bars, often contains natural so-called "boosters," such as ginseng and ginko biloba.

Have students browse their local market for these products. Encourage students to read the labels and record any plant sources listed as well as their reputed therapeutic benefits. Remind students that these herbal remedies have not gone through the rigorous requirements for FDA approval, and do not always have clinical data to back up their claims. People often forget that herbal remedies may have significant drug interactions with prescription medications.

CHEMISTRY IN CAREERS

Discuss with students the role that analytical chemists play in identifying the biologically active components of naturally occurring products. Have students read about careers in the field of medical technology on page 869 of this text.

4 Close

Summary

Ask students to review the chapter sections and list the main ideas. Discuss their lists and compile a class summary. Ask:

- **Name in order the three steps used to solve problems in this chapter.** (analyze, calculate, and evaluate)
- **Explain what conversion factors are and how they are used in dimensional analysis.** (Conversion factors provide a comparison for any two units. In dimensional analysis, conversion factors are used to help solve problems.)

Extension

Explain that systems of measurement rely on the existence of a set of standards for a few fundamental quantities. The definition of metric standards has evolved as scientists have needed to make more accurate and precise measurements. Have students prepare a written report about the definition of a fundamental SI standard.

Looking Back . . . Looking Ahead . . .

The previous chapter discussed the importance of measurement and the use of significant figures in measurements and calculations. It introduced the International System of Units, SI, and the various temperature scales. The current chapter discusses how to use a three-step, problem-solving process. It also discusses conversion factors and the technique of dimensional analysis. The next chapter discusses atomic structure, and introduces the periodic table of elements.

Chapter 4 STUDENT STUDY GUIDE

Take It to the NET
For interactive study and review, go to www.phschool.com

KEY TERMS

- conversion factor *p. 89*
- dimensional analysis *p. 90*

CONCEPT SUMMARY

4.1 What Do I Do Now?

- Problem solving is a skill learned through practice. The more you practice, the more proficient you become.
- Problem solving involves developing a plan. In this textbook, a three-step approach is the plan used. The three steps are: analyze, calculate, and evaluate.
- To analyze a word problem, identify both the knowns and the unknown. Then plan a solution and do the calculations.
- In the final step, evaluate whether the answer to a word problem seems reasonable and whether the answer has the correct number of significant figures.
- No problem-solving method can replace the need for you to read carefully and to think through the steps needed to solve a given problem.
- A good problem-solving strategy can be applied to all sorts of situations—not just chemistry problems.

4.2 Simple Conversion Problems

- Any two measurements that are equal to one another but expressed in different units can be written as a ratio.
- A ratio of equivalent measurements is called a conversion factor and is equal to unity.
- Conversion factors are used in the problem-solving technique of dimensional analysis.
- A ratio of equivalent measurements has two forms. The correct conversion factor for solving a particular problem will have the known unit in the denominator and the unknown unit in the numerator.
- Conversion problems in which you are asked to express a measurement in some other unit are easily solved using dimensional analysis.
- In dimensional analysis, units are used to help write the solution to a problem.

4.3 More-Complex Problems

- Many complex problems, whether they be in chemistry or in your daily life, can be successfully solved by breaking the solution down into steps.
- More than one conversion factor may be required in some more-complex conversion problems.
- The given measurement in a rate problem has a ratio of units.
- Rate problems are solved by converting the unit in the numerator followed by converting the unit in the denominator.

CHAPTER CONCEPT MAP

Use these terms to construct a concept map that organizes the major ideas of this chapter.

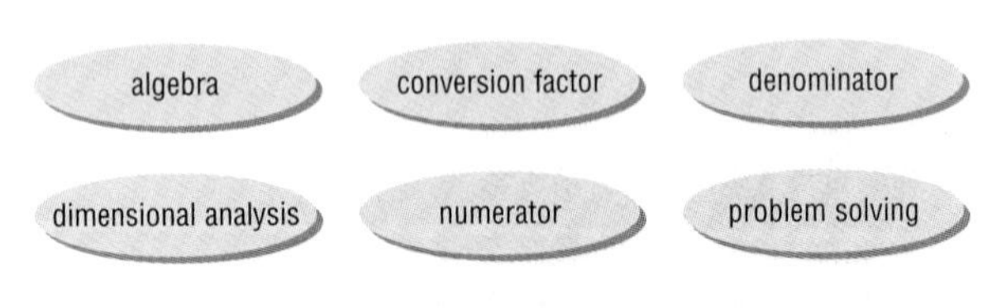

 Chem ASAP! **Concept Map 4**
Create your Concept Map using the computer.

Take It to the Net

At **www.phschool.com** students will find for this chapter

- an Internet activity
- links to related chemistry sites
- an interactive quiz
- career links

Chapter 4 REVIEW

CONCEPT PRACTICE

32. In which step of the three-step problem-solving approach is a problem-solving strategy developed? *4.1*

33. A volume of 5.00 mL of mercury is added to a beaker that has a mass of 87.3 g. What is the mass of the beaker with the added mercury? *4.1*

34. What is the name given to a ratio of two equivalent measurements? *4.2*

35. One measure of area is the hectare, which is equal to 10 000 m^2. What does the ratio $\frac{10^4 \text{ m}^2}{\text{hectare}}$ equal? *4.2*

36. Write six conversion factors involving these units of measure: 1 g = 100 cg = 10^3 mg. *4.2*

37. What must be true for a ratio of two measurements to be a conversion factor? *4.2*

38. One of the first mixtures of metals used by dentists for tooth fillings consisted of 26.0 g of silver, 10.8 g of tin, 2.4 g of copper, and 0.8 g of zinc. How much silver is in a 25.0 g sample of this amalgam? *4.2*

39. How do you know which unit of a conversion factor must be in the denominator (on the bottom)? *4.2*

40. The density of dry air measured at 25 °C is 1.19×10^{-3} g/cm^3. What is the volume of 50.0 g of air? *4.2*

41. List at least two things you should do after you have calculated the answer to a problem on your calculator. *4.2*

42. Have you ever found yourself sitting through a terrible movie, counting the minutes that remain? If a movie has 0.20 hour remaining, how many seconds of movie remain? *4.3*

43. Make the following conversions. *4.3*
- **a.** 157 cs to seconds
- **b.** 42.7 L to milliliters
- **c.** 261 nm to millimeters
- **d.** 0.065 km to decimeters
- **e.** 642 cg to kilograms
- **f.** 8.25×10^2 cg to nanograms

44. Make the following conversions. *4.3*
- **a.** 0.44 mL/min to microliters per second
- **b.** 7.86 g/cm^2 to milligrams per square millimeter
- **c.** 1.54 kg/L to grams per cubic centimeter

45. How many milliliters are contained in 1 m^3? *4.3*

46. Complete this table so that the measurements in each row have the same value. *4.3*

mg	g	cg	kg
		28.3	
6.6×10^3			
	2.8×10^{-4}		

47. A cheetah can run 112 km/h over a 100-m distance. What is this speed in meters per second? *4.3*

CONCEPT MASTERY

48. A flask that can hold 158 g of water at 4 °C can hold only 127 g of ethanol. What is the density of ethanol?

49. A watch loses 0.15 s every minute. How many minutes will the watch lose in 1 day?

50. A tank measuring 28.6 cm by 73.0 mm by 0.72 m is filled with olive oil that has a mass of 1.38×10^4 g. What is the density of olive oil in kilograms per liter?

51. Alkanes are a class of molecules that have the general formula C_nH_{2n+2}, where *n* is an integer. The table below gives the boiling points for the first five alkanes with an odd number of carbon atoms. Using the table, construct a graph with number of carbon atoms on the *x*-axis.

Boiling point (°C)	Number of carbon atoms
−162.0	1
−42.0	3
36.0	5
98.0	7
151.0	9

- **a.** What are the approximate boiling points for the C_2, C_4, C_6, and C_8 alkanes?
- **b.** Which of these nine alkanes are gases at room temperature (20 °C)?
- **c.** How many of these nine alkanes are liquids at 350 K?
- **d.** What is the approximate increase in boiling point per additional carbon atom in this series of alkanes?

Answers

32. step 1. ANALYZE.

33. 155.3 g

34. conversion factor

35. one

36. $\frac{1\text{ g}}{100\text{ cg}}$; $\frac{100\text{ cg}}{1\text{ g}}$; $\frac{1\text{ g}}{1000\text{ mg}}$; $\frac{1000\text{ mg}}{1\text{ g}}$; $\frac{1000\text{ mg}}{100\text{ cg}}$; $\frac{100\text{ cg}}{1000\text{ mg}}$

37. They must equal one another.

38. 16.3 g silver

39. The unit of the conversion factor in the denominator must be identical to the unit in the given measurement.

40. 4.20×10^4 cm^3

41. Estimate to see if your calculator answer makes sense; round off to the correct number of significant figures.

42. 720 s

43. **a.** 1.57 s
b. 4.27×10^4 mL
c. 2.61×10^{-4} mm
d. 6.5×10^2 dm
e. 6.42×10^{-3} kg
f. 8.25×10^9 ng

44. **a.** 7.3 μL/s
b. 78.6 mg/mm^2
c. 1.54 g/cm^3

45. 10^6 mL

46. 2.83×10^2 mg, 2.83×10^{-1} g, 2.83×10^{-4} kg; 6.6 g, 6.6×10^2 cg, 6.6×10^{-3} kg; 2.8×10^{-1} mg, 2.8×10^{-2} cg, 2.8×10^{-7} kg

47. 31.1 m/s

48. 0.804 g/cm^3

49. 3.6 min lost

50. 0.92 kg/L

51. **a.** C_2 = −90 °C, C_4 = 0 °C, C_6 = 70 °C, C_8 = 125 °C
b. C_1 through C_4
c. three
d. From C_1 through C_9, the increase is approximately 38 °C per additional carbon. Over the range C_3 through C_9, the increase is approximately 32 °C per additional carbon.

Answers

52. 8.3 min
53. 73 g
54. 5.52 kg/dm^3
55. 24.0 kg of water
56. a. 2
    b. 3
57. Answers will vary.
58. Units can help set the correct path to conversion problems.
59. The mass of fuel plus mass of oxygen must equal the mass of ash plus gaseous products.
60. Possible answers include:
    a. solid, transparent, hard, colorless, insoluble in water
    b. liquid, mixture, colorless
    c. solid, less dense than soda, colorless
    d. gas, clear, tasteless, odorless
61. a. mixture
    b. mixture
    c. compound
    d. element
62. 1235 K
63. a. centigram
    b. kiloliter
    c. millisecond
    d. cubic decimeter
    e. micrometer
64. Answers will vary.
65. a. 3
    b. 4
    c. 2
    d. 4
    e. 5
    f. 4
66. a. 5.1 g
    b. 3.5×10^6 kg
    c. 7.8×10^{-5} dm^3
    d. 4.5×10^{-2} mm
    e. 9.9×10^2 K
    f. 65 s
67. 8.0 g Sr
68. 1.19
69. 32 cm
70. 1.8×10^3 kg
71. a. 85 g
    b. 1.3 g/mL

52. Earth is approximately 1.5×10^8 km from the sun. How many minutes does it take light to travel from the sun to Earth? The speed of light is 3.0×10^8 m/s.

53. What is the mass of a cube of aluminum that is 3.0 cm on each edge? The density of aluminum is 2.7 g/cm^3.

54. The average density of Earth is 5.52 g/cm^3. Express this density in units of kg/dm^3.

55. How many kilograms of water (at 4 °C) are needed to fill an aquarium that measures 40.0 cm by 20.0 cm by 30.0 cm?

CRITICAL THINKING

56. Choose the term that best completes the second relationship.
    a. journey:route problem:______
       (1) unknown (3) known
       (2) plan (4) calculate
    b. meter:100 cm gram:______
       (1) 0.001 kL (3) 1000 mg
       (2) 100 cm (4) 100 kg

57. You have solved many word problems up to this point. Review the techniques for solving word problems. Which step is most difficult for you? What kind of problems do you find most difficult to solve?

58. Why are units so important in working word problems?

CUMULATIVE REVIEW

59. Describe how the law of conservation of mass applies to a burning campfire.

60. List three physical properties of each of the following objects in the accompanying photo.
    a. the glass
    b. the soda
    c. the ice cube
    d. the bubbles

61. Classify each of the following as an element, compound, or mixture.
    a. an egg c. dry ice (CO_2)
    b. a cake d. iron powder

62. The melting point of silver is 962 °C. Express this temperature in kelvins.

63. Identify the larger quantity in each of these pairs of measurements.
    a. centigram, milligram
    b. deciliter, kiloliter
    c. millisecond, microsecond
    d. cubic decimeter, milliliter
    e. micrometer, nanometer

64. Name three physical and three chemical changes that you have seen today.

65. How many significant figures are in each of these measurements?
    a. 5.12 g d. 0.045 04 mm
    b. 3.456×10^6 kg e. 985.20 K
    c. 0.000 078 dm^3 f. 65.02 s

66. Round each of the measurements in Problem 65 to two significant figures.

CONCEPT CHALLENGE

67. Sea water contains 8.0×10^{-1} cg of the element strontium per kilogram of sea water. Assuming that all the strontium could be recovered, how many grams of strontium could be obtained from one cubic meter of sea water? Assume the density of sea water is 1.0 g/mL.

68. When 121 g of sulfuric acid is added to 400 mL of water, the resulting solution's volume is 437 mL. What is the specific gravity of the resulting solution?

69. How tall is a rectangular block of balsa wood measuring 4.4 cm wide and 3.5 cm deep that has a mass of 98.0 g? The specific gravity of balsa wood is 0.20.

70. The density of dry air at 20 °C is 1.20 g/L. What is the mass of air, in kilograms, of a room that measures 25.0 m by 15.0 m by 4.0 m?

71. Different volumes of the same liquid were added to a flask on a balance. After each addition of liquid, the mass of the flask with the liquid was measured. Graph the data using mass as the dependent variable. Use the graph to answer these questions.

Volume (mL)	14	27	41	55	82
Mass (g)	103.0	120.4	139.1	157.9	194.1

a. What is the mass of the flask?
b. What is the density of the liquid?

Chapter 4 STANDARDIZED TEST PREP

Select the choice that best answers each question or completes each statement.

1. Which of these conversion factors is *not* correct?
   a. 1 m/10^2 mm
   b. 10^9 ns/1 s
   c. 1 dm^3/1 L
   d. 1 g/10^6 μg
2. An over-the-counter medicine has 325 mg of its active ingredient per tablet. How many grams does this mass represent?
   a. 325 000 g
   b. 32.5 g
   c. 3.25 g
   d. 0.325 g
3. The density of zinc is 9.394 g/cm^3 at 20 °C. What is the volume of a sphere of zinc that has a mass of 15.6 g? ($V = 4/3\pi r^3$; $\pi = 3.14$)
   a. 1.66 cm^3
   b. 6.21 cm^3
   c. 0.602 cm^3
   d. 147 cm^3
4. If 10^4 μm = 1 cm, how many μm^3 = 1 cm^3?
   a. 10^4
   b. 10^6
   c. 10^8
   d. 10^{12}
5. How many meters does a car moving at 95 km/h travel in 1.0 s?
   a. 1.6 m
   b. 340 m
   c. 1600 m
   d. 26 m
6. If a substance contracts when it freezes, its
   a. density will remain the same.
   b. density will increase.
   c. density will decrease.
   d. change in density cannot be predicted.

For questions 7–10, identify the known and the unknown. Include units in your answers.

7. The density of water is 1.0 g/mL. How many deciliters of water will fill a 0.5-L bottle.
8. A 34.5-g gold nugget is dropped into a graduated cylinder containing water. By how many milliliters does the measured volume increase? The density of water is 1.0 g/mL. The density of gold is 19.3 g/cm^3.
9. A watch loses 0.2 s every minute. How many minutes will the watch lose in a day?
10. Eggs shipped to market are packed 12 eggs to a carton and 20 cartons to a box. A crate holds 4 boxes. Crates are stacked on a truck 5 crates wide, 6 crates deep, and 5 crates high. How many eggs are there in 5 truckloads?
11. Based on the prices displayed on the gas pumps, which station offers a lower price for gasoline? (1 gal = 3.79 L)

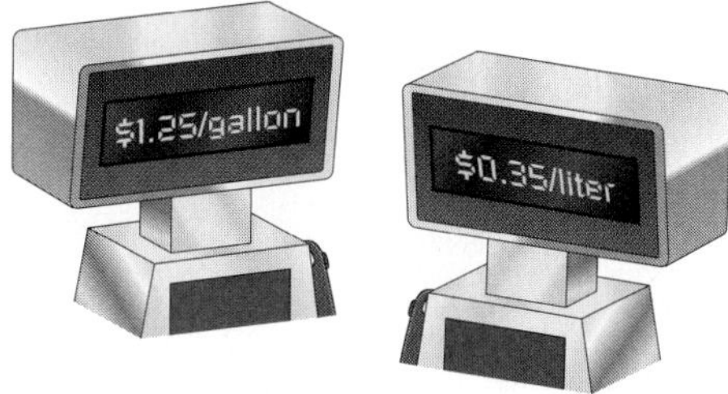

The lettered choices below refer to questions 12–15. A lettered choice may be used once, more than once, or not at all.

(A) 60 min/h
(B) 1 min/60 s
(C) 1 h/60 min
(D) 60 s/1 min

Which conversion factors are needed to do each of the following conversions?

12. number of seconds to cook a 3-minute egg
13. number of hours in 1000 s
14. number of minutes in "Wait just a second."
15. number of minutes in an 8-hour workday

Use the drawing to answer questions 16–18. The scale models show the relative sizes of a helium atom (left) and a xenon atom (right).

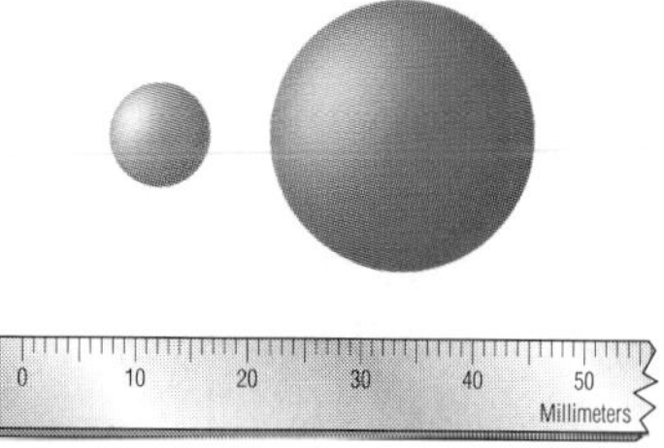

16. Use the metric ruler provided to determine the diameter of each model atom in millimeters.
17. Calculate the ratio of the diameter of the helium atom to the diameter of the xenon atom.
18. Calculate the ratio of the volume of the helium atom to the volume of the xenon atom. Assume that the atoms are perfect spheres. ($V = 4/3\pi r^3$; $\pi = 3.14$)

Chapter 4 STANDARDIZED TEST PREP

1. a
2. d
3. a
4. d
5. d
6. b
7. Known: volume in liters; unknown: volume in deciliters
8. Known: density of gold in g/cm^3 and mass of gold in grams; unknown: volume of gold in milliliters
9. Known: lost time in seconds; unknown: lost time in minutes per day
10. Known: eggs per carton, cartons per box, boxes per crate, and how crates are stacked on a truck; unknown: crates per truck and number of eggs per 5 truck-loads.
11. $1.25/gal
12. (D)
13. (B) and (C)
14. (B)
15. (A)
16. diameter of "helium atom" is 9.0 mm; diameter of "xenon atom" is 23.5 mm
17. He : Xe = 9.0 mm : 23.5 mm = 1 : 2.6
18. volume of He is 382 mm^3; volume of Xe is 6792 mm^3; He : Xe = 382 mm^3 : 6796 mm^3 = 1 : 17.8

Planning Guide

SECTION OBJECTIVES	ACTIVITIES/FEATURES	MEDIA & TECHNOLOGY
5.1 Atoms ● ■ ◇ ▸ Summarize Dalton's atomic theory ▸ Describe the size of an atom	**SE** Discover It! *Electric Charge,* p. 106	**ASAP** Assessment 5.1 **CHM** Side 2, 2: *Designing Fireworks*
5.2 Structure of the Nuclear Atom ● ■ ◆ ▸ Distinguish among protons, electrons, and neutrons in terms of relative mass and charge ▸ Describe the structure of an atom, including the location of the protons, electrons, and neutrons with respect to the nucleus	**SE** Mini Lab *Using Inference: The Black Box,* p. 112 (LRS 5-1) LM 5: *Atomic Structure: Rutherford's Experiment* **TE** DEMO, p. 110	**ASAP** Animation 4 **ASAP** Assessment 5.2 **OT** 5: Cathode Ray Tubes **OT** 6: Rutherford's Experiment **CHM** Side 2, 20: *Rutherford's Experiment* **CHM** Side 2, 30: *Electric Charges*
5.3 Distinguishing Between Atoms ○ ■ ◆ ▸ Explain how the atomic number identifies an element ▸ Use the atomic number and mass number of an element to find the numbers of protons, electrons, and neutrons ▸ Explain how isotopes differ and why the atomic masses of elements are not whole numbers ▸ Calculate the average atomic mass of an element from isotope data	**SE** **CHEMath** *Using Positive and Negative Numbers,* p. 114 **SE** **Link to Humanities** *Philosophy of Science,* p. 119 **SE** Small-Scale Lab *The Atomic Mass of Candium,* p. 122 (LRS 5-2) SSLM 5: *Isotopes and Atomic Mass*	**ASAP** Problem Solving 9, 13, 17 **ASAP** Assessment 5.3
5.4 The Periodic Table: Organizing the Elements ● ■ ◇ ▸ Describe the origin of the periodic table ▸ Identify the position of groups, periods, and the transition metals in the periodic table	**SE** **Chemistry Serving . . . Society** *Ask an Artifact for a Date!,* p. 127 **SE** **Chemistry in Careers** *Archaeologist,* p. 127 **TE** DEMO, p. 124	**ASAP** Assessment 5.4 **ASAP** Concept Map 5 **RP** Lesson Plans, Resource Library **AR** Computer Test 5 **www** Activities, Self-Tests, *SCIENCE NEWS* updates **TCP** The Chemistry Place Web Site

KEY

- ● Conceptual (concrete concepts)
- ○ Conceptual (more abstract/math)
- ■ Standard (core content)
- □ Standard (extension topics)
- ◆ Honors (core content)
- ◇ Honors (options to accelerate)

AR Assessment Resources
ASAP Chem ASAP! CD-ROM
ACT ActivChemistry CD-ROM
CHM CHEMedia Videodiscs
CA Chemistry Alive! Videodiscs
GCP Graphing Calculator Problems
GRS Guided Reading and Study Workbook
LM Laboratory Manual
LP Laboratory Practicals
LRS Laboratory Recordsheets
SSLM Small-Scale Lab Manual

PRACTICE		ASSESSMENT	
GRS	Section 5.1	**SE**	Section Review
RM	Practice Problems 5.1	**RM**	Section Review 5.1
		RM	Chapter 5 Quiz
GRS	Section 5.2	**SE**	Section Review
RM	Practice Problems 5.2	**RM**	Section Review 5.2
		RM	Chapter 5 Quiz
		LP	Lab Practical 5-1
SE	**Sample Problems** 5-1 to 5-5	**SE**	Section Review
GRS	Section 5.3	**RM**	Section Review 5.3
RM	Section Review 5.3	**RM**	Chapter 5 Quiz
RM	Practice Problems 5.3		
GRS	Section 5.4	**SE**	Section Review
RM	Practice Problems 5.4	**RM**	Section Review 5.4
RM	Interpreting Graphics	**RM**	Vocabulary Review 5
		SE	Chapter Review
		SM	Chapter 5 Solutions
		SE	Standardized Test Prep
		PHAS	Chapter 5 Test Prep
		RM	Chapter 5 Quiz
		RM	Chapter 5 A & B Test

PLANNING FOR ACTIVITIES

STUDENT EDITION

Discover It! p. 106
- metric rulers
- 25-cm lengths of clear plastic tape

Small-Scale Lab, p. 122
- pencils and paper
- three kinds of candy
- balance
- plastic cups

Mini Lab, p. 112
- boxes
- marbles
- geometric foam shapes

TEACHER'S EDITION

Teacher Demo, p. 110
- cathode ray tube

Teacher Demo, p. 124
- variety of chemical elements that were known in ancient times: copper, sulfur, silver, carbon, gold, and tin

OT	Overhead Transparency	**SE**	Student Edition
PHAS	PH Assessment System	**SM**	Solutions Manual
PLM	Probeware Lab Manual	**SSV**	Small-Scale Video/Videodisc
PP	Problem Pro CD-ROM	**TCP**	www.chemplace.com
RM	Review Module	**TE**	Teacher's Edition
RP	Resource Pro CD-ROM	**www**	www.phschool.com

Chapter 5

Key Terms

5.1 Dalton's atomic theory, atom
5.2 electrons, cathode ray, protons, neutrons, nucleus
5.3 atomic number, mass number, isotopes, atomic mass unit (amu), atomic mass
5.4 periodic table, periods, periodic law, group, representative elements, metals, alkali metals, alkaline earth metals, transition metals, inner transition metals, nonmetals, halogens, noble gases, metalloids

DISCOVER IT!

When tape is pulled from the desk, it receives a negative charge. When tape is pulled through the fingers, it receives a positive charge. Steps 1 and 2: The pieces of tape with like charges are pushed apart. Step 3: Predictions may vary. The pieces of tape are drawn together. Students should conclude that the pieces of tape from Steps 1 and 2 do not have the same charge because objects of like charge would have been pushed apart in Step 3.

Chapter 5 ATOMIC STRUCTURE AND THE PERIODIC TABLE

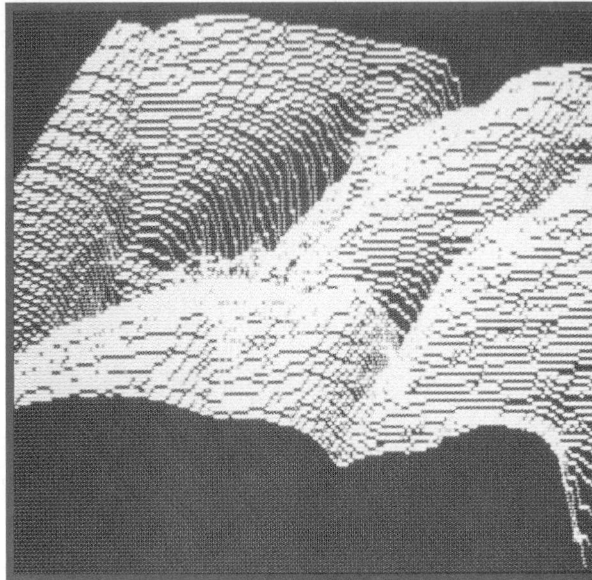

Each bright spot on this image represents one gold atom.

FEATURES

DISCOVER IT!
Electric Charge

SMALL-SCALE LAB
The Atomic Mass of Candium

MINI LAB
Using Inference: The Black Box

CHEMath
Using Positive and Negative Numbers

CHEMISTRY SERVING ... SOCIETY
Ask an Artifact for a Date!

CHEMISTRY IN CAREERS
Archaeologist

LINK TO HUMANITIES
Philosophy of Science

Stay current with SCIENCE NEWS
Find out more about atomic structure:
www.phschool.com

DISCOVER IT! ELECTRIC CHARGE

You need four 25-cm lengths of clear plastic tape and a metric ruler.

1. Firmly stick two of the 25-cm pieces of tape side-by-side, about 10 cm apart, on your desk top. Leave 2 to 3 cm of tape sticking over the edge of the desk. Grasp the free ends of the tapes and pull sharply upward to peel the tape pieces off of the desk. Slowly bring the pieces, which have similar charges, toward one another. What do you observe?
2. Pull the third and fourth pieces of tape between your thumb and forefinger several times, as if you were trying to clean each one. Slowly bring these two pieces of tape, which now have similar charges, toward one another. What do you observe?
3. Predict what might happen if you brought a piece of tape pulled from your desk top close to a piece of tape pulled between your fingers. Try it! What happens?

Do you think the pieces of tape used in Step 1 have the same charge as the pieces used in Step 2? Explain. After reading about charged particles in this chapter, return to this activity and re-evaluate your answer.

5.1 ● ■ ◇
5.2 ● ■ ◆
5.3 ○ ■ ◆
5.4 ● ■ ◇

Conceptual Although students need to understand atomic mass and why atomic mass numbers are not whole numbers, the concept of relative abundance is theoretically and mathematically challenging. Sample Problems 5-3, 5-4, and 5-5 may be omitted. The CHEMath feature will help students who are having difficulty working with negative numbers.

Standard Section 5-4 is a brief introduction; the periodic table will be explored in much greater depth in Chapter 14 after electron configurations are introduced. Focusing on the historical development and key experiments with subatomic particles will help students to remember properties of those particles.

Honors Accelerate section 5.1 if students have encountered Dalton's atomic theory in prior courses. Section 5.4 is a descriptive section, which may be assigned as homework reading. Periodic properties and trends will be covered in more depth in Chapter 14.

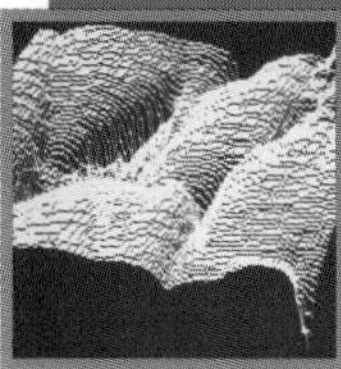

ATOMS

section 5.1

***In** 1981, Swiss scientists Gerd Binnig and Heinrich Rohrer completed the construction of the scanning tunneling microscope. Their device, which they first tested using a specimen of gold, produces an image of individual atoms, often seen as rows of bright spots on a monitor as seen here. The scanning tunneling microscope is used today to study how atoms are arranged on the surface of many different materials. Binnig and Rohrer won the Nobel Prize for Physics in 1986 for their invention.* **Why was an image of individual atoms considered such an important breakthrough?**

objectives

- Summarize Dalton's atomic theory
- Describe the size of an atom

key terms

- Dalton's atomic theory
- atom

Early Models of the Atom

Have you ever been asked to believe in something you could not see? Using your unaided eyes, you cannot see the tiny fundamental particles that make up matter. Yet all matter is composed of such particles. Democritus of Abdera, a teacher who lived in Greece during the fourth century B.C., first suggested the existence of these particles, which he called atoms. He believed that these atoms were indivisible and indestructible. Although Democritus's ideas agreed with later scientific theory, they were not useful in explaining chemical behavior. They also lacked experimental support because scientific testing was unknown at that time.

The real nature of atoms and the connection between observable changes and events at the atomic level were not established for more than 2000 years after Democritus. The modern process of discovery regarding atoms began with John Dalton (1766–1844), an English schoolteacher. Unlike Democritus, Dalton performed experiments to test and correct his atomic theory. Dalton studied the ratios in which elements combine in chemical reactions. Based on the results of his experiments, Dalton formulated hypotheses and theories to explain his observations. The result was **Dalton's atomic theory,** which includes the ideas illustrated in **Figure 5.1** and listed below.

1. All elements are composed of tiny indivisible particles called atoms.
2. Atoms of the same element are identical. The atoms of any one element are different from those of any other element.

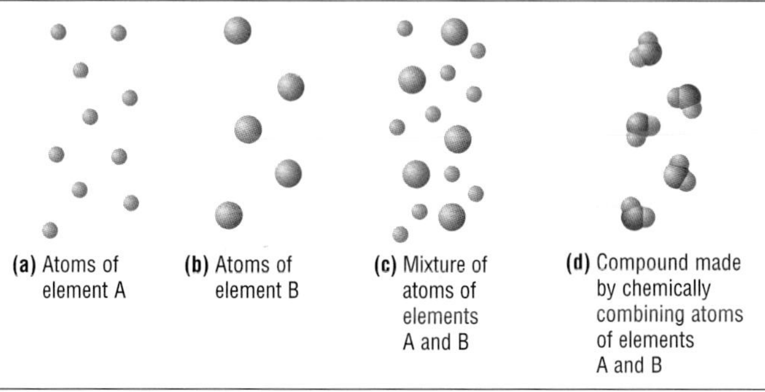

Figure 5.1
According to Dalton's atomic theory, an element is composed of only one kind of atom, and a compound is composed of particles that are chemical combinations of different kinds of atoms.

5.1

1 Engage

Use the Visual

Have students study the photograph and read the text that opens the section. Ask:

- **How does a person see an object?** (Sensors in the eye detect light reflected off an object and convert the data to electrical impulses that are transmitted to the brain.)
- **Why can't an optical microscope be used to see atoms?** (Atoms are so small relative to the wavelengths of visible light that light flows by the atoms as a water wave flows by pebbles on a beach.)

Scanning microscopes use properties such as electrical conductivity, magnetic force, and friction to map the atoms on the surface of a material.

- **Why was an image of individual atoms considered such an important breakthrough?** (It provided direct evidence for scientific theories about the structure of matter.)

Check Prior Knowledge

To assess students' prior knowledge about atoms, ask students to:

- **Describe what atoms are made of.**
- **Estimate the size of atoms.**
- **Draw a picture of a simple atom on a sheet of paper.**

Post the outcomes.

Student Resources

From the Teacher's Resource Package, use:

- Section Review 5.1, Ch. 5 Practice Problems and Quizzes from the Review Module (Ch. 5–8)

Technology Resources

Relevant technology resources include:

- Chem ASAP! CD-ROM
- Resource Pro CD-ROM

2 Teach

Discuss

Discuss the development of atomic theory as an example of the scientific method. Present the material in a historical context to show how the perception of the atom has changed from the early Greek model to the present model. Explain that John Dalton's work, published in 1808, became the basis for modern atomic theory. This model represented the atom as a simple sphere with no internal structure. The way that atoms are held together in compounds was poorly understood. Point out how experimental data have been used to test and refine atomic theory over time.

3 Assess

Evaluate Understanding

Have students evaluate and criticize the following statements according to Dalton's theory:

- **"All atoms are identical."** (Dalton said: "All atoms *of a given element* are identical.")
- **"Chemical reactions occur when atoms of one element change into atoms of another element."** (False. Chemical reactions occur when atoms are separated, joined, or rearranged. The elemental identity of atoms does not change during chemical reactions.)

Reteach

Review the Greek and Dalton models of the atom and discuss the differences between them.

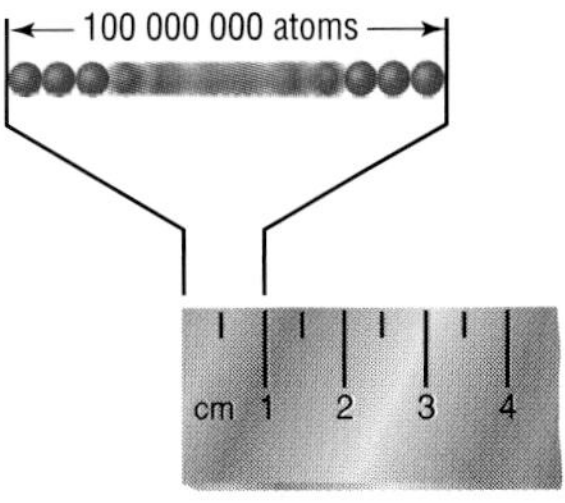

Figure 5.2
If 100 000 000 copper atoms were placed side by side, they would form a line 1 cm long. What is this number of atoms written in scientific notation? ❶

3. Atoms of different elements can physically mix together or can chemically combine with one another in simple whole-number ratios to form compounds.
4. Chemical reactions occur when atoms are separated, joined, or rearranged. Atoms of one element, however, are never changed into atoms of another element as a result of a chemical reaction.

Just How Small Is an Atom?

A coin the size of a penny and composed of pure copper (Cu) illustrates Dalton's concept of the atom. Imagine grinding the copper coin into a fine dust. Each speck in the small pile of shiny red dust would still have the properties of copper. If by some means you could continue to make the copper dust particles smaller, you would eventually come upon a particle of copper that could no longer be divided and still have the properties of copper. This final particle is called an **atom,** defined in its modern sense as the smallest particle of an element that retains the properties of that element.

Copper atoms are very small. A pure copper coin the size of a penny contains about 2.4×10^{22} atoms. By comparison, Earth's population is only about 6×10^{9} people. There are about 4×10^{12} as many atoms in the coin as there are people on Earth. If you could line up 100 000 000 copper atoms side by side, they would produce a line only 1 cm long, as shown in **Figure 5.2.**

Does seeing individual atoms seem impossible? Despite their small size, individual atoms are observable with the proper instrument. As you read in the introduction to this section, and as **Figure 5.3** shows, a scanning tunneling microscope provides a visual image of individual atoms. Individual atoms can even be moved around and arranged in patterns. The ability to move individual atoms holds future promise for the creation of atomic-sized electronic devices, such as circuits and computer chips. This atomic-scale technology could someday be applied to communications and space exploration.

Figure 5.3
A scanning tunneling microscope can be used to view the surface of individual atoms, such as the gold atoms shown here.

section review 5.1

1. In your own words, state the main ideas of Dalton's atomic theory.
2. Characterize the size of an atom.
3. Democritus and Dalton both proposed that matter consists of atoms. How did their approaches to reaching that conclusion differ?

 Chem ASAP! **Assessment 5.1** Check your understanding of the important ideas and concepts in Section 5.1.

Answers

❶ 1×10^{8}

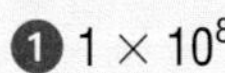

SECTION REVIEW 5.1

1. Answers will vary but should include the ideas that all matter is composed of atoms, atoms of different elements differ, and chemical change involves a rearrangement of atoms.
2. Answers will vary but should emphasize the smallness of atoms.
3. Dalton used reasoning based on the results of scientific experiments; Democritus used mental reasoning only.

STRUCTURE OF THE NUCLEAR ATOM

section 5.2

Atoms are so small they can only be visualized with a scanning tunneling microscope. How do scientists study the even smaller particles that make up atoms? Scientists study the makeup of atoms by breaking them apart! The atoms are accelerated to tremendous speeds—nearly the speed of light—in giant devices called particle accelerators. Then the atoms are smashed into one another, causing the particles within them to be released. Although scientists cannot actually see the particles, they can use the device shown here, called a bubble chamber, to see the tracks these particles make. **What are the component particles that scientists have discovered within atoms?**

objectives

- Distinguish among protons, electrons, and neutrons in terms of relative mass and charge
- Describe the structure of an atom, including the location of the protons, electrons, and neutrons with respect to the nucleus

key terms

- electrons
- cathode ray
- protons
- neutrons
- nucleus

Electrons

Much of Dalton's atomic theory is accepted today. One important change, however, is that atoms are now known to be divisible. They can be broken down into even smaller, more fundamental particles. Dozens of kinds of subatomic particles are unleashed when powerful devices known as atom smashers are used to fracture atoms. You will now learn about three kinds of subatomic particles.

Electrons are negatively charged subatomic particles. The English physicist J. J. Thomson (1856–1940) discovered electrons in 1897. Thomson performed experiments that involved passing electric current through gases at low pressure. He sealed the gases in glass tubes fitted at both ends with metal disks called electrodes. **Figure 5.4** shows the kind of apparatus he used. The electrodes were connected to a source of high-voltage electricity. One electrode, the anode, became positively charged. The other electrode, the cathode, became negatively charged. A glowing beam formed between the electrodes. This beam, which traveled from the cathode to the anode, is called a **cathode ray.**

Thomson found that cathode rays are attracted to metal plates that have a positive electrical charge. Plates that carry a negative electrical charge repel the rays. **Figure 5.5** on page 110 shows the deflection of cathode rays. Thomson knew that opposite charges attract and like charges repel, so he proposed that a cathode ray is a stream of tiny negatively charged

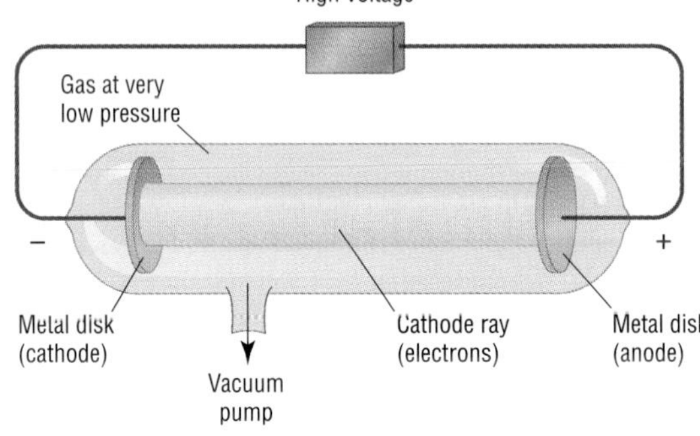

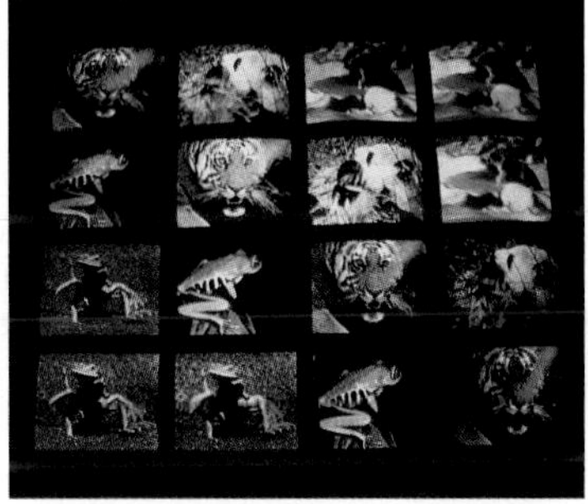

Figure 5.4
In a cathode-ray tube, electrons travel as a ray from the cathode (–) to the anode (+). A television tube is a specialized type of cathode-ray tube.

5.2

1 Engage

Use the Visual

Have students study the photograph and read the text that opens the section. Explain that particle accelerators are instruments for investigating the subatomic particles that make up atoms. Ask:

- **What are the component particles that scientists have discovered within atoms?** (electrons, protons, neutrons)

Explain that it took another eighty years after the Dalton model for scientists to begin gathering clues about the structure of atoms. By 1887, the British scientist Crookes knew that metal atoms contained negatively charged particles (by using the device shown in Figures 5.4 and 5.5). In a similar device containing hydrogen gas at low pressure, hydrogen was found to contain positive charges. Scientists knew that the mass of the atom was greater than its proton content. They inferred that this extra mass was due to neutral particles, which were called neutrons. The evidence for neutrons was not confirmed until 1932 by Chadwick.

STUDENT RESOURCES

From the Teacher's Resource Package, use:

- Section Review 5.2, Ch. 5 Practice Problems and Quizzes from the Review Module (Ch. 5–8)
- Laboratory Manual: Experiment 5
- Laboratory Practical 5-1
- Laboratory Recordsheet 5-1

TECHNOLOGY RESOURCES

Relevant technology resources include:

- Chem ASAP! CD-ROM
- ResourcePro CD-ROM

2 Teach

Discuss

Be sure that students understand the importance of the experiments done by Thomson, Millikan, Rutherford, Moseley, and Chadwick.

- **Thomson measured the charge-to-mass ratio (e/m) of the electron.**
- **Millikan measured the charge on the electron.**
- **Moseley used an X-ray technique to measure the number of protons in an atom.**
- **Rutherford employed alpha-particle scattering to discover the massive nature of metal nuclei. Explain Figure 5.6.**
- **Chadwick used his own experiments and those of Irene Joliot-Curie to demonstrate the existence of the neutron.**

TEACHER DEMO

If possible, demonstrate a cathode ray tube in class. Show students how the beam of particles is deflected by a magnet. Point out that J.J. Thomson discovered that atoms contain electrons by using a similar device. Review the components of a cathode ray tube and discuss the connection to television picture tubes and computer monitors.

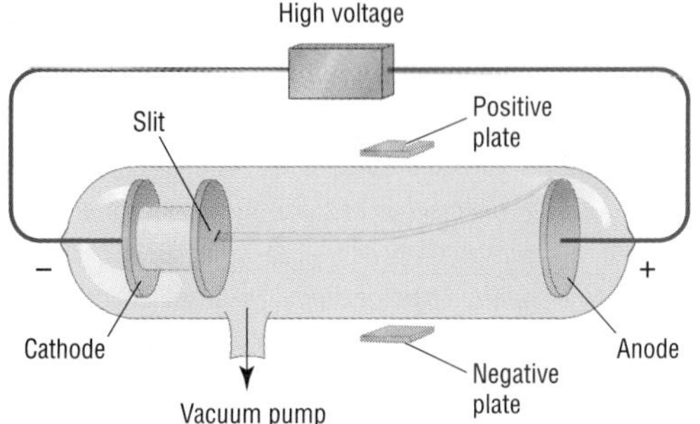

Figure 5.5
Cathode rays are deflected by a magnet and attracted by a positively charged plate. This shows that the particles that make up the rays are negatively charged.

particles moving at high speed. Thomson named these particles electrons. Thomson also showed that the production of cathode rays did not depend on the kind of gas in the cathode-ray tube or the type of metal used for the electrodes. He concluded that electrons must be parts of the atoms of all elements. By 1900, Thomson and others had determined that an electron's mass is about 1/2000 the mass of a hydrogen atom.

The American scientist Robert A. Millikan (1868–1953) carried out experiments that allowed him to find the quantity of charge carried by an electron. He also determined the ratio of the charge to the mass of an electron. Millikan used these two values to calculate an accurate value for the mass of the electron. Millikan's values for electron charge and mass, reported in 1916, are very similar to those accepted today. An electron carries exactly one unit of negative charge, and its mass is 1/1840 the mass of a hydrogen atom.

Protons and Neutrons

If cathode rays are electrons given off by atoms, what remains of the atoms that have lost the electrons? For example, after a hydrogen atom (the lightest kind of atom) loses an electron, what is left? You can think through this problem using four simple ideas about matter and electric charges. First, atoms have no net electric charge; they are electrically neutral. (One important piece of evidence for electrical neutrality is that you do not receive an electric shock every time you touch something!) Second, electric charges are carried by particles of matter. Third, electric charges always exist in whole-number multiples of a single basic unit; that is, there are no fractions of charges. Fourth, when a given number of negatively charged particles combines with an equal number of positively charged particles, an electrically neutral particle is formed.

Considering all of this information, it follows that a particle with one unit of positive charge should remain when a typical hydrogen atom loses an electron. Evidence for such a positively charged particle was found in 1886, when E. Goldstein observed a cathode-ray tube and found rays traveling in the direction opposite to that of the cathode rays. He called these rays canal rays and concluded that they were composed of positive particles. Such positively charged subatomic particles are called **protons.** Each proton has a mass about 1840 times that of an electron.

In 1932, the English physicist James Chadwick (1891–1974) confirmed the existence of yet another subatomic particle: the neutron. **Neutrons** are

Meeting Diverse Needs

Gifted Have students use the Internet or library to find the original papers for the discoveries described in this chapter and write a report on what they have learned. (Example: Chadwick's paper on the discovery of the neutron is available on the Internet.)

Table 5.1

Properties of Subatomic Particles				
Particle	**Symbol**	**Relative electrical charge**	**Relative mass (mass of proton = 1)**	**Actual mass (g)**
Electron	e^-	1−	1/1840	9.11×10^{-28}
Proton	p^+	1+	1	1.67×10^{-24}
Neutron	n^0	0	1	1.67×10^{-24}

subatomic particles with no charge but with a mass nearly equal to that of a proton. Thus the fundamental building blocks of atoms are the electron, the proton, and the neutron. **Table 5.1** summarizes the properties of these subatomic particles.

The Atomic Nucleus

When subatomic particles were discovered, scientists wondered how these particles were put together in an atom. This was a difficult question to answer, given how tiny atoms are. Most scientists thought it likely that the electrons were evenly distributed throughout an atom filled uniformly with positively charged material. In 1911, Ernest Rutherford (1871–1937) and his coworkers at the University of Manchester, England, decided to test this theory of atomic structure. Their test used relatively massive alpha particles, which are helium atoms that have lost their two electrons and have a double positive charge because of the two remaining protons. In the experiment, illustrated in **Figure 5.6,** Rutherford directed a narrow beam of alpha particles at a very thin sheet of gold foil. According to the prevailing theory, the alpha particles should have passed easily through the gold, with only a slight deflection due to the positive charge thought to be spread out in the gold atoms.

Chem ASAP!

Animation 4

Take a look at Rutherford's gold-foil experiment, its results, and its conclusions.

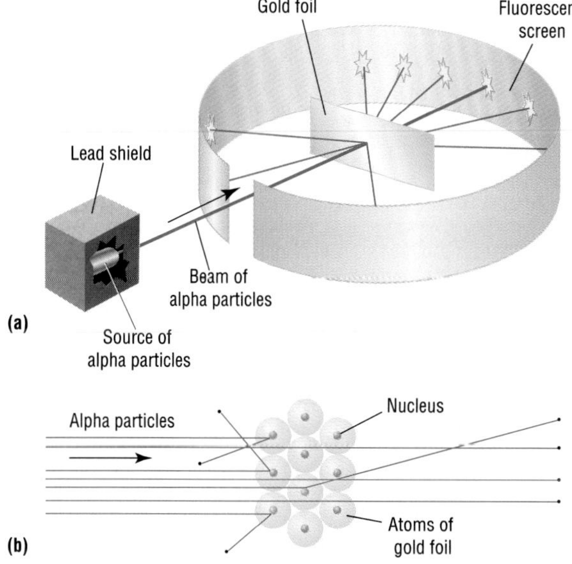

Figure 5.6

(a) *To learn more about the nature of the atom, Rutherford and his coworkers aimed a beam of alpha particles at a sheet of gold foil surrounded by a fluorescent screen. They found that most of the particles passed through the foil with no deflection at all. A few particles were greatly deflected.* ***(b)*** *Rutherford concluded that most of the alpha particles pass through the gold foil because the atom is mostly empty space. The mass and positive charge are concentrated in a small region of the atom. Rutherford called this region the nucleus. Particles that approach the nucleus closely are greatly deflected.*

5.2

Discuss

Use Thomson's experiment to introduce the concept that the neutral atom is composed of positively and negatively charged particles. Emphasize that cathode rays are negatively charged electrons stripped from atoms of a sample gas by a high voltage. Left behind are positively charged gas ions. The negative electrons move toward the anode (positive electrode), and the positive gas ions move toward the cathode.

Use the Visual

Have students compare the masses and charges of the three elementary particles shown in Table 5.1. Explain that the mass of an atom is due mainly to the number of protons and neutrons. Point out that the assigned charges 1+ and 1− for the proton and the electron are relative charges. The absolute charge can also be measured (1.6×10^{-19} coulomb).

Discuss

To help students visualize how small the nucleus is compared to the entire atom, explain that if a housefly sitting on second base in baseball stadium represented the nucleus of an atom, the rest of the atom would be the size of the stadium. Also, be sure students are aware that in electrically neutral atoms, the number of electrons and protons must be equal.

Discuss

Point out that protons and neutrons are made of still smaller particles called quarks. If possible, arrange a tour of a laboratory that studies subatomic particles.

MINI LAB

Using Inference: The Black Box

LAB PREP AND PLANNING

Cut geometric shapes-such as a triangle, circle, or L, from a sheet of 1-inch plastic foam. Students will attempt to determine the shape of the hidden object by analyzing the rebound paths of a marble rolled at the object.

ANALYSIS AND CONCLUSIONS

1. Answers will vary.
2. The activity simulates the strategy that Ernest Rutherford used to probe the structure of metal atoms. Like the students, Rutherford and his coworkers were faced with the problem of identifying properties of an object not visible to the eye.

3 Assess

Evaluate Understanding

Have students describe the discoveries of Thomson, Millikan, and Rutherford, and relate the importance of these discoveries to the current understanding of the structure of atoms. Ask students to compare and contrast electrons, protons, and neutrons.

Reteach

Review with students the following items:
(1) the distribution of the electric charge in neutral atoms,
(2) the concentration of positive charge and mass in the nucleus,
(3) the relative size of the atom and the nucleus, and
(4) the names of the subatomic particles that make up an atom.

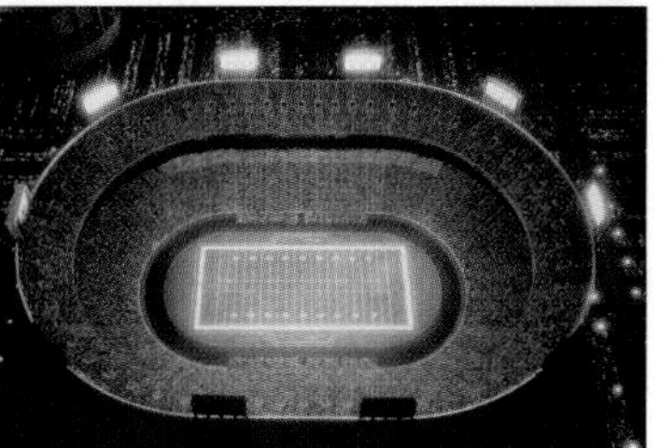

Figure 5.7
If an atom were the size of this stadium, then its nucleus would be about the size of a marble!

To everyone's surprise, the great majority of alpha particles passed straight through the gold atoms, without deflection. Even more surprisingly, a small fraction of the alpha particles bounced off the gold foil at very large angles. Some even bounced straight back toward the source. Rutherford later recollected, "It was about as credible as if you had fired a 15-inch shell at a piece of tissue paper, and it came back and hit you."

Based on the experimental results, Rutherford suggested a new theory of the atom. He proposed that the atom is mostly empty space, thus explaining the lack of deflection of most of the alpha particles. He concluded that all the positive charge and almost all the mass are concentrated in a small region that has enough positive charge to account for the great deflection of some of the alpha particles. He called this region the nucleus. The **nucleus** is the central core of an atom and is composed of protons and neutrons. It is tiny compared with the atom as a whole. See **Figure 5.7.**

MINI LAB

Using Inference: The Black Box

PURPOSE

To determine the shape of a fixed object inside a sealed box without opening the box.

MATERIALS

- box containing a regularly shaped object fixed in place and a loose marble

PROCEDURE

1. Do not open the box.
2. Carefully manipulate the box so that the marble moves around the fixed object.
3. Gather data (clues) that describe the movement of the marble.
4. Sketch a picture of the object in the box, showing its shape, size, and location within the box.
5. Repeat this activity with a different box containing a different object.

ANALYSIS AND CONCLUSIONS

1. Find a classmate who had the same lettered box that you had. Compare your findings and try to come to agreement about the shape and location of the fixed object.
2. What experiment that contributed to a better understanding of the atom does this activity remind you of?

section review 5.2

4. What are the charges and relative masses of the three main subatomic particles?
5. Describe the basic structure of an atom.
6. Describe Thomson's, Millikan's, and Rutherford's contributions to atomic theory. Include their experiments if appropriate.

 Chem ASAP! **Assessment 5.2** Check your understanding of the important ideas and concepts in Section 5.2.

Answers

SECTION REVIEW 5.2

4. proton, positive charge, relative mass = 1; electron, negative charge, relative mass = 1/1840; neutron, no charge, relative mass = 1
5. An atom has a central core composed of protons and neutrons, called the nucleus. Electrons surround the nucleus and occupy most of the volume of the atom.
6. Thomson passed electric current through sealed glass tubes filled with gases. The resulting glowing beam was described as a stream of tiny negatively charged particles moving at high speed. He concluded that electrons must be parts of the atoms of all elements. Millikan determined the charge and mass of the electron. Rutherford's gold-foil experiments indicated that the atom had a positively charged, dense nucleus which is tiny compared to the atom as a whole.

DISTINGUISHING BETWEEN ATOMS

section 5.3

"Rose is a rose is a rose is a rose," wrote the famous author Gertrude Stein. Of course, this is not true when it comes to the color of roses, which can range from the familiar red or white to yellow or even lavender. In all, there are more than 13 000 colorful varieties of roses. Just as roses come in different varieties,a given chemical element can come in different "varieties" called isotopes.
What is an isotope, and how does one isotope of an element differ from another?

objectives

- Explain how the atomic number identifies an element
- Use the atomic number and mass number of an element to find the numbers of protons, electrons, and neutrons
- Explain how isotopes differ and why the atomic masses of elements are not whole numbers
- Calculate the average atomic mass of an element from isotope data

key terms

- atomic number
- mass number
- isotopes
- atomic mass unit (amu)
- atomic mass

Atomic Number

Atoms are composed of electrons, protons, and neutrons. Protons and neutrons make up the small, dense nucleus. Electrons surround the nucleus and occupy most of the volume of the atom. How, then, are atoms of hydrogen, for example, different from atoms of oxygen? Examine **Table 5.2.** Compare the entries for hydrogen and oxygen. You should notice that a hydrogen atom has one proton in its nucleus, but an oxygen atom has eight protons in its nucleus. Elements are different because they contain different numbers of protons.

The **atomic number** of an element is the number of protons in the nucleus of an atom of that element. Because all hydrogen atoms have one proton, the atomic number of hydrogen is 1. Similarly, because all oxygen atoms have eight protons, the atomic number of oxygen is 8. The atomic number identifies an element.

Look again at **Table 5.2.** For each element listed, the number of protons equals the number of electrons. Remember that atoms are electrically neutral. Thus the number of electrons (negatively charged particles) in an atom must equal the number of protons (positively charged particles) in the nucleus. A hydrogen atom has one electron, and an oxygen atom has eight electrons. How is the number of electrons for a neutral atom of a given element related to the atomic number of that element? ❶

Table 5.2

Atoms of the First Ten Elements

Name	Symbol	Atomic number	Composition of the nucleus		Mass number	Number of electrons
			Protons	Neutrons*		
Hydrogen	H	1	1	0	1	1
Helium	He	2	2	2	4	2
Lithium	Li	3	3	4	7	3
Beryllium	Be	4	4	5	9	4
Boron	B	5	5	6	11	5
Carbon	C	6	6	6	12	6
Nitrogen	N	7	7	7	14	7
Oxygen	O	8	8	8	16	8
Fluorine	F	9	9	10	19	9
Neon	Ne	10	10	10	20	10

* Number of neutrons in the most abundant isotope. Isotopes are introduced later in Section 5.3.

5.3

1 Engage

Use the Visual

Have students study the photograph and read the text that opens the section. Explain that roses are a family of plants with similar genetic characteristics. However, roses have a variety of colors and petal patterns. Although all atoms share a set of characteristics, i.e., they are made from the same fundamental particles, atoms do vary. Ask:

- **How are the atoms of one element distinguished from the atoms of another element?** (The atoms of different elements contain different numbers of protons.)
- **What is an isotope?** (Isotopes are atoms of an element that have the same number of protons but different numbers of neutrons.)

There are about ninety different naturally occurring elements. In 1913, Mosley used X-rays to count the number of protons in the atomic nuclei of different atoms. Point out that there are many elements with more than two isotopes. Ask:

- **How does one isotope of an element differ from another?** (Isotopes differ by the number of neutrons in the nucleus.)

2 Teach

Use the Visual

Have students examine Table 5.2. Point out that the atomic number is equal to the number of protons for each element. Ask:

- **Why must the number of electrons equal the number of protons for each element?** (Atoms are electrically neutral.)

STUDENT RESOURCES

From the Teacher's Resource Package, use:

- Section Review 5.3, Ch. 5 Practice Problems and Quizzes from the Review Module (Ch. 5–8)
- Laboratory Recordsheet 5-2
- Small-Scale Chemistry Lab Manual: Experiment 5

TECHNOLOGY RESOURCES

Relevant technology resources include:

- Chem ASAP! CD-ROM
- Resource Pro CD-ROM

❶ The atomic number of a neutral atom is equal to the number of electrons.

CHEMath

Teaching Tips

Some students may have difficulties determining the sign of the sum of two integers. This section gives students practice adding and subtracting positive and negative numbers. Use a number line to show students how to find the sum and the difference of two numbers. This technique may be especially useful for explaining *origin* and *absolute value*.

Answers to Practice Problems

A. 8
B. 5
C. −40
D. 3
E. −19
F. −2.6
G. −0.62
H. −6.25
I. −20.1
J. 9.9
K. 3−
L. 1+
M. 1−
N. 2−

CHEMath

USING POSITIVE AND NEGATIVE NUMBERS

In this chapter you are introduced to protons and electrons, particles with charges of 1+ and 1− respectively. In addition to these charges, you will encounter a variety of positive and negative numbers throughout this course. The table below offers a review of basic mathematical operations related to positive and negative numbers.

BASIC CONCEPTS	EXAMPLE
Positive and negative numbers are used in many applications, including temperature scales. On a number line, the positive numbers are to the right of the *origin* (zero), and the negative numbers are to the left of the origin.	Negative Numbers · Origin · Positive Numbers −4 −3 −2 −1 0 +1 +2 +3 +4
The *absolute value* of any number is its distance from the origin. Absolute values are always positive (or zero).	$\|3\|$ = (absolute value of 3) = 3 $\|-2\|$ = (absolute value of −2) = 2
The *opposite* of any number is a number having the same absolute value but the opposite sign.	The opposite of 8 is −8. The opposite of −5 is −(−5), or 5.
ADDITION	
If the two numbers are both positive, add them as usual.	5 + 16 = 21
If the two numbers are both negative, add their absolute values and take the opposite.	To find −6 + (−5), note that 6 + 5 = 11. Therefore, −6 + (−5) = −11.
It the two numbers have opposite signs, subtract the smaller absolute value from the larger absolute value. Use the sign of the number with the larger absolute value.	To find 6 + (−3): 6 − 3 = 3, so 6 + (−3) = 3. To find 6 + (−8): 8 − 6 = 2, so 6 + (−8) = −2. To find (−6) + 3: 6 − 3 = 3, so (−6) + 3 = −3. To find (−6) + 8: 8 − 6 = 2, so (−6) + 8 = 2.
SUBTRACTION	
Subtracting any number is the same as adding its opposite.	12 − 23 = 12 + (−23) = −11 6 − (−4) = 6 + 4 = 10
MULTIPLICATION AND DIVISION	
If the two numbers have the same sign (both positive or both negative), the product or quotient is positive.	3 × 5 = 15 −24 ÷ (−8) = 3
If the two numbers have opposite signs, the product or quotient is negative.	4 × (−3) = −12 −20 ÷ 5 = −4
CHEMISTRY CONNECTION	
An *ion* is a charged atom. The net charge of an ion is (number of protons) − (number of electrons).	If an ion has 3 protons and 5 electrons, its net charge is 3 − 5 = −2.

Practice Problems

Practice using positive and negative numbers by performing each calculation.

A. 12 + (−4)
B. −5 − (−10)
C. 4 × (−10)
D. −21 ÷ (−7)
E. −36 + 17
F. 8.7 − 11.3
G. −3.1 × 0.20
H. 22.5 ÷ (−3.60)
I. −24.8 − (−4.7)
J. −5.5 × (−1.8)
K. Find the net charge of an ion with 15 protons and 18 electrons.
L. Find the net charge of an ion with 9 protons and 8 electrons.
M. An ion with a charge of 3− loses 2 electrons. What is its charge?
N. An ion with a charge of 1+ gains 3 electrons. What is its charge?

Sample Problem 5-1

The element nitrogen (N) has an atomic number of 7. How many protons and how many electrons are in a neutral nitrogen atom?

1. ***ANALYZE*** ***List the known and the unknowns.***

   Known:
   - atomic number = 7

   Unknowns:
   - number of protons = ?
   - number of electrons = ?

   The atomic number gives the number of protons, which in a neutral atom equals the number of electrons.

   atomic number = number of protons = number of electrons

2. ***CALCULATE*** ***Solve for the unknowns.***

   atomic number = 7 = number of protons
   = number of electrons

3. ***EVALUATE*** ***Do the results make sense?***

   The relationships have been applied correctly. The seven electrons are needed to balance the seven protons.

Practice Problems

7. How many protons and electrons are in each atom?
   a. fluorine
   b. aluminum
   c. calcium
8. Complete the table.

Element	Atomic number	Protons	Electrons
K	19	____	19
____	____	____	5
____	16	____	____
____	____	23	____

Mass Number

You know that most of the mass of an atom is concentrated in its nucleus and depends on the number of protons and neutrons. Look again at **Table 5.2** and note the number of protons and neutrons in helium and in carbon. The total number of protons and neutrons in an atom is called the **mass number.** A helium atom has two protons and two neutrons, so its mass number is 4. A carbon atom, which has six protons and six neutrons, has a mass number of 12.

If you know the atomic number and mass number of an atom of any element, you can determine the atom's composition. **Table 5.2** shows that an oxygen atom has an atomic number of 8 and a mass number of 16. Because the atomic number equals the number of protons, which equals the number of electrons, an oxygen atom has eight protons and eight electrons. The mass number of oxygen is 16 and is equal to the number of protons plus the number of neutrons. The oxygen atom, then, has eight neutrons, which is the difference between the mass number and the atomic number ($16 - 8 = 8$). For any atom, the number of neutrons can be determined from the following equation.

$$\text{Number of neutrons} = \text{mass number} - \text{atomic number}$$

The composition of any atom can be represented in shorthand notation using atomic number and mass number. **Figure 5.8 on page 116** shows how an atom of gold is represented using this notation. The chemical symbol Au appears with two numbers written to its left. The atomic number is the subscript (a number positioned lower). The mass number is the superscript (a number positioned higher). How many neutrons does this gold (Au) atom have? 

Discuss

Remind students that virtually the entire mass of an atom is determined by the protons and neutrons. Therefore, the mass number of an element is defined as the number of protons and neutrons. Explain that chemists have arbitrarily assigned a value of 1 atomic mass unit to represent the mass of a proton or neutron. Because these particles occur in the nucleus, they are called nucleons.
mass number = atomic number + number of neutrons

Note that it is always possible to find the number of neutrons in an atom by subtracting the atomic number from the mass number. For example, carbon has a mass number of 12 and an atomic number of 6. Therefore, carbon has six neutrons.

Practice Problems Plus

Related Chapter Review Problem
Chapter Review problem 39 is related to Sample Problem 5-1.

Answer

1 This gold atom has 118 neutrons

Cooperative Learning

Have students work in pairs to answer Practice Problems 9, 10, and 11. In addition, have students rewrite the chemical symbols in Practice Problem 9 as descriptions such as oxygen-16 and silver-108.

Practice Problems Plus

Related Chapter Review Problem
Chapter Review problem 42 is related to Sample Problem 5-2.

Additional Practice Problem
The Chem ASAP! CD-ROM contains the following problem: Calculate the number of neutrons in each of the following radioactive isotopes.

a. ${}^{14}_{6}C$ (8)
b. ${}^{40}_{19}K$ (21)
c. ${}^{238}_{92}U$ (146)
d. ${}^{99}_{42}Mo$ (57)

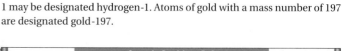

Figure 5.8
Au is the chemical symbol for gold. How many electrons does a gold atom have? ❶

You can also use the mass number and the name of the element to designate atoms. For example, atoms of hydrogen with a mass number of 1 may be designated hydrogen-1. Atoms of gold with a mass number of 197 are designated gold-197.

Sample Problem 5-2

How many protons, electrons, and neutrons are in the following atoms?

	Atomic number	Mass number
a. Beryllium (Be)	4	9
b. Neon (Ne)	10	20
c. Sodium (Na)	11	23

1. ***ANALYZE** List the knowns and the unknowns.*

Knowns:
- For each atom, the atomic number and mass number are known.

Unknowns:
- number of protons = ?
- number of electrons = ?
- number of neutrons = ?

Use the definitions of atomic number and mass number to calculate the numbers of protons, electrons, and neutrons.

2. ***CALCULATE** Solve for the unknowns.*

number of electrons = atomic number
a. 4 **b.** 10 **c.** 11
number of protons = atomic number
a. 4 **b.** 10 **c.** 11
number of neutrons = mass number − atomic number
a. 9 − 4 = 5 **b.** 20 − 10 = 10 **c.** 23 − 11 = 12

3. ***EVALUATE** Do the results make sense?*

The relationships among atomic number, number of protons, number of neutrons, number of electrons, and mass number have been applied correctly.

Practice Problems

9. How many neutrons are in each atom?
a. ${}^{16}_{8}O$ **c.** ${}^{108}_{47}Ag$ **e.** ${}^{207}_{82}Pb$
b. ${}^{32}_{16}S$ **d.** ${}^{80}_{35}Br$

10. Use **Table 5.2** and **Figure 5.8** to express the composition of each atom in shorthand form.
a. carbon-12
b. fluorine-19
c. beryllium-9

11. For each atom in Problem 9, identify the number of electrons.

Chem ASAP!
Problem-Solving 9
Solve Problem 9 with the help of an interactive guided tutorial.

Isotopes

Figure 5.9 shows that there are three different kinds of neon atoms. How do these atoms differ? All have the same number of protons (10) and electrons (10), but they each have different numbers of neutrons. Atoms that have the same number of protons but different numbers of neutrons are called **isotopes.** Because isotopes of an element have different numbers of neutrons, they also have different mass numbers. Despite these differences, isotopes are chemically alike because they have identical numbers of protons and electrons, which are the subatomic particles responsible for

Answers

❶ 79 electrons
❷ Dalton's atomic theory states that all the atoms of an element are identical. Although isotopes are chemically identical, they have dissimilar masses.
❸ Numbers of protons and electrons are equal, but number of neutrons is not equal.

Neon-20
10 protons
10 neutrons
10 electrons

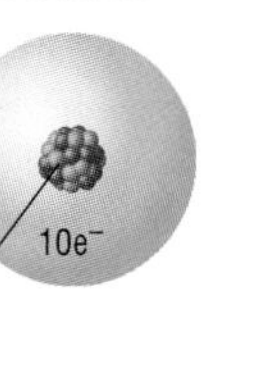

Neon-21
10 protons
11 neutrons
10 electrons

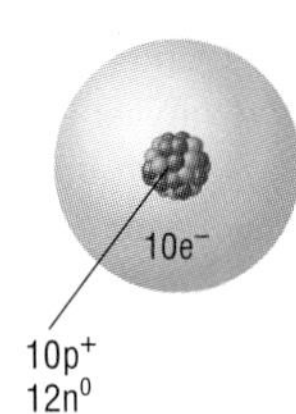

Neon-22
10 protons
12 neutrons
10 electrons

10e⁻
$10p^+$
$12n^0$

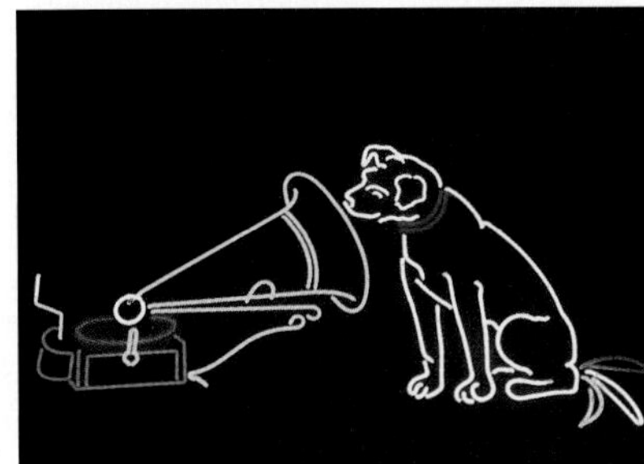

chemical behavior. How does the discovery of isotopes contradict Dalton's atomic theory? ❷

There are three known isotopes of hydrogen. Each isotope of hydrogen has one proton in its nucleus. The most common hydrogen isotope has no neutrons. It has a mass number of 1 and is called hydrogen-1 ($^{1}_{1}H$) or simply hydrogen. The second isotope has one neutron and a mass number of 2. It is called either hydrogen-2 ($^{2}_{1}H$) or deuterium. The third isotope has two neutrons and a mass number of 3. This isotope is called hydrogen-3 ($^{3}_{1}H$) or tritium.

Figure 5.9
Neon-20, neon-21, and neon-22 are three isotopes of neon, a gaseous element used in lighted signs. How are these isotopes different? How are they the same? ❸

Sample Problem 5-3

Two isotopes of carbon are carbon-12 and carbon-13. Write the symbol for each isotope using superscripts and subscripts to represent the mass number and the atomic number.

1. ***ANALYZE** Plan a problem-solving strategy.*

   Knowns:
   - two isotopes of carbon: carbon-12 and carbon-13

   Unknowns:
   - each isotope's symbol

   Write the symbol for carbon and place the mass number to the left of the symbol as a superscript. Place the atomic number to the left of the symbol as a subscript.

2. ***SOLVE** Apply the problem-solving strategy.*

   Based on **Table 5.2,** the symbol for carbon is C and the atomic number is 6. The mass number for each isotope is given by its name. For carbon-12, the symbol is $^{12}_{6}C$. For carbon-13, the symbol is $^{13}_{6}C$.

3. ***EVALUATE** Do the results make sense?*

   The concepts of atomic number and mass number have been applied correctly, and the information is positioned properly next to the chemical symbols.

Practice Problems

12. Three isotopes of oxygen are oxygen-16, oxygen-17, and oxygen-18. Write the complete symbol for each, including the atomic number and mass number.

13. The three isotopes of chromium are chromium-50, chromium-52, and chromium-53. How many neutrons are in each isotope, given that chromium always has an atomic number of 24?

Chem ASAP!
Problem-Solving 13
Solve Problem 13 with the help of an interactive guided tutorial.

5.3

Discuss

Explain that isotopes are atoms that differ by the number of neutrons in the nucleus. The existence of isotopes was not predicted by Dalton's work. More sophisticated experimental techniques were needed to detect them. Isotopes of an element are chemically the same, because the electrons, not the neutrons, determine the atom's chemical properties.

Practice Problems Plus

Additional Practice Problem
The Chem ASAP! CD-ROM contains the following problem related to Sample Problem 5-3: The element strontium has three isotopes, strontium-86, strontium-88, and strontium-90. Given that strontium has an atomic number of 38, how many neutrons are in each of these isotopes? (48, 50, 52)

ACTIVITY

Have students do research in the library and on the Internet to find an isotope of an element that has a practical use.
Examples are:
(1) carbon-14 (archaeological carbon dating)
(2) americium-241 (smoke alarm)
(3) iodine-131 (treatment of thyroid disorders)
(4) cobalt-60 (treatment of cancer)

Use the Visual

Students should be aware that the relative atomic masses listed in Table 5.3 are average atomic masses based on the masses of *stable* isotopes and their percent abundance in Earth's crust. Point out to students that some elements in the periodic table are radioactive and their nuclei decay so rapidly that an atomic mass cannot be accurately measured. In fact, two elements, technetium and promethium, are no longer present in Earth's crust. Have students study the atomic masses of various elements in the periodic table. Ask them to infer which elements exist predominantly as one natural isotope. (those with atomic masses closest to a whole number)

Atomic Mass

A glance back at **Table 5.1** on page 111 shows that the actual mass of a proton or a neutron is very small (1.67×10^{-24} g). The mass of an electron is 9.11×10^{-28} g, which is negligible in comparison. Given these values, the mass of even the largest atom is incredibly small. Since the 1920s, it has been possible to determine these tiny masses by using a mass spectrometer. With this instrument, the mass of a fluorine atom was found to be 3.155×10^{-23} g, and the mass of an arsenic atom was found to be 1.244×10^{-22} g. Such data about the actual masses of individual atoms can provide useful information, but, in general, these values are inconveniently small and impractical to work with. Instead, it is more useful to compare the relative masses of atoms using a reference isotope as a standard. The isotope chosen is carbon-12. This isotope of carbon was assigned a mass of exactly 12 atomic mass units. An **atomic mass unit (amu)** is defined as one-twelfth the mass of a carbon-12 atom. Using these units, a helium-4 atom, with a mass of 4.0026 amu, has about one-third the mass of a carbon-12 atom. How many carbon-12 atoms would have about the same mass as a
1 nickel-60 atom?

A carbon-12 atom has six protons and six neutrons in its nucleus, and its mass is set as 12 amu. The twelve protons and neutrons account for nearly all of this mass. Therefore the mass of a single proton or a single neutron is about one-twelfth of 12 amu, or about 1 amu. Because the mass of any single atom depends mainly on the number of protons and neutrons in the nucleus of the atom, you might predict that the atomic mass of an element should be a whole number. However, that is not usually the case. For example, the atomic mass of chlorine (Cl) is 35.453 amu. How can such an atomic mass be explained? The explanation involves the relative abundance of the naturally occurring isotopes of the element.

In nature, most elements occur as a mixture of two or more isotopes. Each isotope of an element has a fixed mass and a natural percent abundance. Consider the three isotopes of hydrogen discussed earlier in this section. According to **Table 5.3,** almost all naturally occurring hydrogen (99.985%) is hydrogen-1. The other two isotopes are present in trace amounts. Notice that

Figure 5.10

Chlorine is a reactive element used to disinfect swimming pools. It is made up of two isotopes: chlorine-35 and chlorine-37. Because there is more chlorine-35 than chlorine-37, the atomic mass of chlorine, 35.453 amu, is closer to 35 than to 37; it is a weighted average.

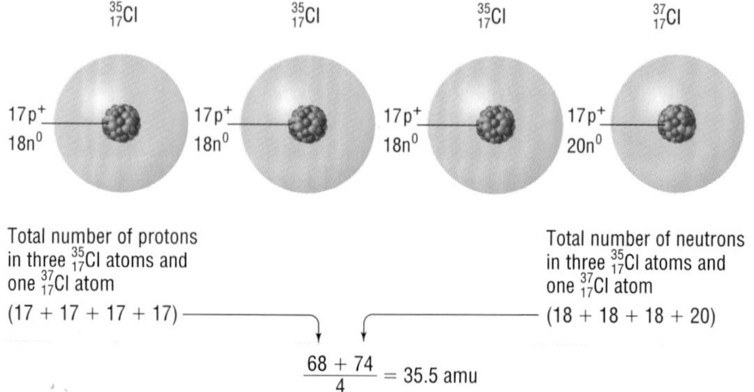

Weighted Average Mass of a Chlorine Atom

Answer

1 About 5 carbon-12 atoms equal the mass of 1 nickel-60 atom.

Table 5.3

Natural Percent Abundance of Stable Isotopes of Some Elements				
Name	Symbol	Natural percent abundance	Mass (amu)	"Average" atomic mass
Hydrogen	${}^{1}_{1}H$	99.985	1.0078	
	${}^{2}_{1}H$	0.015	2.0141	1.0079
	${}^{3}_{1}H$	negligible	3.0160	
Helium	${}^{3}_{2}He$	0.0001	3.0160	4.0026
	${}^{4}_{2}He$	99.9999	4.0026	
Carbon	${}^{12}_{6}C$	98.89	12.000	12.011
	${}^{13}_{6}C$	1.11	13.003	
Nitrogen	${}^{14}_{7}N$	99.63	14.003	14.007
	${}^{15}_{7}N$	0.37	15.000	
Oxygen	${}^{16}_{8}O$	99.759	15.995	
	${}^{17}_{8}O$	0.037	16.995	15.999
	${}^{18}_{8}O$	0.204	17.999	
Sulfur	${}^{32}_{16}S$	95.002	31.972	
	${}^{33}_{16}S$	0.76	32.971	32.06
	${}^{34}_{16}S$	4.22	33.967	
	${}^{36}_{16}S$	0.014	35.967	
Chlorine	${}^{35}_{17}Cl$	75.77	34.969	35.453
	${}^{37}_{17}Cl$	24.23	36.966	
Zinc	${}^{64}_{30}Zn$	48.89	63.929	
	${}^{66}_{30}Zn$	27.81	65.926	
	${}^{67}_{30}Zn$	4.11	66.927	65.38
	${}^{68}_{30}Zn$	18.57	67.925	
	${}^{70}_{30}Zn$	0.62	69.925	

the atomic mass of hydrogen in **Table 5.3** (1.0079 amu) is very close to the mass of hydrogen-1 (1.0078 amu). The slight difference takes into account the larger masses, but smaller amounts, of the other two isotopes of hydrogen.

Now consider the two stable isotopes of chlorine listed in **Table 5.3:** chlorine-35 and chlorine-37. If you calculate the arithmetic mean of these two masses, you would get an average atomic mass of 35.968 amu ((34.969 amu + 36.966 amu)/2). A quick comparison with the value in **Table 5.3** indicates that 35.968 amu is higher than the actual value. To explain this, you need to know the natural percent abundance of the isotopes of chlorine. Chlorine-35 accounts for 75% of the naturally occurring chlorine atoms; chlorine-37 accounts for only 25%. See **Figure 5.10.**

HUMANITIES

Philosophy of Science

Modern philosophers search for wisdom in many areas of human life, such as medicine, the arts, and the sciences. Today's philosophers of science are primarily concerned with the critical analysis of scientific concepts and the ways in which these concepts are expressed. These philosophers analyze such concepts as number, space, force, and organism. The search for wisdom leads philosophers of science to debate questions that textbooks may lead you to believe are settled. For example, is the modern scientific method the only correct way to examine and explain the natural world? Or is there a better way, as yet unknown? In spite of the successes of the scientific method in explaining nature, philosophers of science are still studying such questions. These studies could greatly affect science if an improved scientific method were suggested. For this reason, philosophical questions and answers about the methods and concepts of science could be an important aid to scientific progress.

Discuss

Discuss the *Link to Humanities* segment on the philosophy of science. Have students search for first-hand accounts of scientists' discoveries. (Example: *The Double Helix* by James Watson) **These accounts show that scientists are not bound by a rigid application of "the scientific method". Intuition and luck can also play a part.**

The scientific method as described in textbooks is a formal method that enables other scientists to confirm new theories and discoveries. If another better method can be developed to test new discoveries, it will probably replace the one scientists are now using. Discuss with students the difference between "common sense" and scientific method.

Practice Problems Plus

Additional Practice Problem

This problem can be used with Sample Problem 5-4: Argon has three isotopes with mass numbers 36, 38, and 40, respectively. Which of these isotopes is the most abundant? (argon-40)

ACTIVITY

To reinforce the concept of weighted averages, have students consider an analogous situation where teachers evaluate a student's performance based on the weighted average of three separate assessments: a term paper worth 20%, a midterm worth 30%, and a final exam worth 50%. If an A = 4.0, a B = 3.0, and a C = 2.0, have students calculate a grade for a student who receives a B on the term paper, a C on the midterm, and a B on the final exam. (The student would receive a score of 2.7, a C+.)

Practice Problems Plus

Related Chapter Review Problems

Chapter Review problems 45 and 53 are related to Sample Problem 5-5.

Additional Practice Problem

The Chem ASAP! CD-ROM contains the following problem: Chlorine has two isotopes, chlorine-35 (atomic mass = 34.97 amu, relative abundance = 75.77%) and chlorine-37 (atomic mass = 36.97 amu, relative abundance = 24.23%). Calculate the atomic mass of chlorine. (35.45 amu)

The **atomic mass** of an element is a weighted average mass of the atoms in a naturally occurring sample of the element. A weighted average mass reflects both the mass and the relative abundance of the isotopes as they occur in nature.

Sample Problem 5-4

Which isotope of copper is more abundant: copper-63 or copper-65? (The atomic mass of copper is 63.546 amu.)

1. *ANALYZE* ***Plan a problem-solving strategy.***

Knowns:

- isotopes of copper: copper-63 and copper-65
- atomic mass of copper = 63.546 amu

Unknown:

- isotope that is more abundant

This problem can be solved without any calculations by analyzing the atomic mass of copper relative to the masses of the two isotopes.

2. *SOLVE* ***Apply the problem-solving strategy.***

The atomic mass of 63.546 amu is closer to 63 than to 65. Thus, because the atomic mass is a weighted average of the isotopes, copper-63 must be more abundant than copper-65.

3. *EVALUATE* ***Does the result make sense?***

The relative abundance is reflected in the atomic mass, so it is reasonable that copper-63 must be more abundant.

Practice Problems

14. Boron has two isotopes: boron-10 and boron-11. Which is more abundant, given that the atomic mass of boron is 10.81?
15. There are three isotopes of silicon; they have mass numbers of 28, 29, and 30. The atomic mass of silicon is 28.086 amu. Comment on the relative abundance of these three isotopes.

Now that you know that the atomic mass of an element is a weighted average of the masses of its isotopes, you can calculate atomic mass based on relative abundance. To do this, you must know three values:

- the number of stable isotopes of the element,
- the mass of each isotope, and
- the natural percent abundance of each isotope.

Table 5.3 on the previous page shows these values for a few elements. For other elements, you can use standard chemistry reference books. Once you have these values for an element, multiply the atomic mass of each isotope by its abundance, expressed as a decimal, then add the results. Sample Problem 5-5 illustrates this method.

Sample Problem 5-5

Element X has two natural isotopes. The isotope with a mass of 10.012 amu (^{10}X) has a relative abundance of 19.91%. The isotope with a mass of 11.009 amu (^{11}X) has a relative abundance of 80.09%. Calculate the atomic mass of this element.

Answers

SECTION REVIEW 5.3

18. All atoms of an element have the same number of protons even though the number of neutrons may vary. The protons and electrons are responsible for the chemical behavior of atoms. In a neutral atom, both the number of protons and the number of electrons equal the atomic number.
19. In a neutral atom, the atomic number is the number of protons and the number of electrons. The mass number is the total number of neutrons plus protons. To find the number of neutrons, subtract the atomic number from the mass number.
20. **a.** mass number

    b. $^{195}_{78}Pt$
21. Isotopes of the same element have identical numbers of protons; they have different

Sample Problem 5-5 (cont.)

1. *ANALYZE* ***List the knowns and the unknown.***

Knowns:
- isotope ^{10}X:
  mass = 10.012 amu
  relative abundance = 19.91% = 0.1991
- isotope ^{11}X:
  mass = 11.009 amu
  relative abundance = 80.09% = 0.8009

Unknown:
- atomic mass of element X = ?

The mass each isotope contributes to the element's atomic mass can be calculated by multiplying the isotope's mass by its relative abundance. The atomic mass of the element is the sum of these contributions.

2. *CALCULATE* ***Solve for the unknown.***

for ^{10}X:	10.012 amu × 0.1991 =	1.993 amu
for ^{11}X:	11.009 amu × 0.8009 =	8.817 amu
for element X:	atomic mass =	10.810 amu

3. *EVALUATE* ***Does the result make sense?***
The calculated value is closer to the mass of the more abundant isotope, as would be expected.

Practice Problems

16. The element copper has naturally occurring isotopes with mass numbers of 63 and 65. The relative abundance and atomic masses are 69.2% for mass = 62.93 amu, and 30.8% for mass = 64.93 amu. Calculate the average atomic mass of copper.
17. Calculate the atomic mass of bromine. The two isotopes of bromine have atomic masses and relative abundance of 78.92 amu (50.69%) and 80.92 amu (49.31%).

Chem ASAP!

Problem-Solving 17
Solve Problem 17 with the help of an interactive guided tutorial.

section 5.3 review

18. Explain how the atomic number of an element identifies the element.
19. How can atomic number and mass number be used to find the numbers of protons, electrons, and neutrons?
20. An atom is identified as platinum-195.
    a. What does the number represent?
    b. Symbolize this atom using superscripts and subscripts.
21. How are isotopes of the same element alike? How are they different?
22. Determine the number of protons, electrons, and neutrons in each of the five isotopes of zinc.
23. List the number of protons, neutrons, and electrons in each pair of isotopes.
    a. $^{6}_{3}Li$, $^{7}_{3}Li$ b. $^{42}_{20}Ca$, $^{44}_{20}Ca$ c. $^{78}_{34}Se$, $^{80}_{34}Se$
24. The atomic masses of elements are generally not whole numbers. Explain why.
25. How is the atomic mass of an element calculated from isotope data?
26. Using the data for nitrogen listed in **Table 5.3,** calculate the average atomic mass of nitrogen. Show your work.

Chem ASAP! **Assessment 5.3** Check your understanding of the important ideas and concepts in Section 5.3.

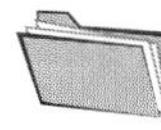
portfolio project

An instrument called a mass spectrometer can be used to determine the masses and relative abundances of isotopes. Find out how this instrument works and make a poster to summarize your findings.

masses, mass numbers, and numbers of neutrons.

22. zinc-64: 30 p^+, 30 e^-, 34 n^0
zinc-66: 30 p^+, 30 e^-, 36 n^0
zinc-67: 30 p^+, 30 e^-, 37 n^0
zinc-68: 30 p^+, 30 e^-, 38 n^0
zinc-70: 30 p^+, 30 e^-, 40 n^0
23. a. lithium-6: 3 p^+, 3 e^-, 3 n^0
lithium-7: 3 p^+, 3 e^-, 4 n^0
b. calcium-42: 20 p^+, 20 e^-, 22 n^0
calcium-44: 20 p^+, 20 e^-, 24 n^0
c. selenium-78: 34 p^+, 34 e^-, 44 n^0
selenium-80: 34 p^+, 34 e^-, 46 n^0
24. The atomic mass is the weighted average of the masses of its isotopes.
25. The mass of the isotopes in a sample of the element are averaged based on relative abundance. The result is the element's atomic mass.
26. $^{14}_{7}N$: 14.003 amu; 99.63%
$^{15}_{7}N$: 15.000 amu; 0.37%
average atomic mass = 14.01 amu

5.3

3 Assess

Evaluate Understanding

Write the symbols for isotopes of an element not previously described in the chapter. Ask students what the superscripts and subscripts refer to and the differences between the atoms shown. Ask students what information they would need to calculate the average atomic mass for the element. Ask:

- **How would changing the value in the subscript change the chemical properties of the atom?** (The subscript designates the number of protons in the atoms of that isotope. Changing the number of protons would change the chemical identity of the isotope to that of another element.)

Reteach

Review with students the methods for adding and subtracting integers and how they apply to calculating atomic mass number and the number of neutrons in an atom. Review the concept of weighted averages.

Work through the following calculations. If 75% of chlorine atoms are ^{35}Cl species and 25% are ^{37}Cl species, this implies that for a sample of 100 atoms, 75 atoms are ^{35}Cl and 25 atoms are ^{37}Cl species. The combined masses of these atoms would be 75 × 35 (amu) + 25 × 37 (amu) = 3550 amu for 100 atoms, or 35.5 amu for one atom.

Portfolio Project

Mass spectrometers operate like cathode ray tubes. A vaporized sample is bombarded by high-energy electrons. A magnetic field separates the positively charged products based on their mass/charge ratios.

SMALL-SCALE LAB

THE ATOMIC MASS OF CANDIUM

OVERVIEW

Procedure Time: 30 minutes
Students make measurements to calculate the relative abundances of three types of candy in a mixture. They use their data to calculate the average mass of a candium particle. This exercise is analogous to determining atomic mass.

SAFETY

Discourage students from eating the candies after the experiment. Contamination can easily occur in a lab even if you have taken every precaution to keep the candy free of contamination.

TEACHING SUGGESTIONS

Prepare in advance a large mixture of the three candies and half fill a clean 3 1/2-ounce plastic cup for each student. Each sample will contain about 50 total pieces. This lab is similar to the longer Small-Scale Lab "Isotopes and Atomic Mass" found in the *Small-Scale Chemistry Laboratory Manual.*

YOU'RE THE CHEMIST

1. Any differences are probably due to small variations in the numbers of each kind of candy in the samples, which affects the relative abundances.
2. The larger the samples, the better the results with any of the methods. Mass is likely to provide better results than volume.

SMALL-SCALE LAB

THE ATOMIC MASS OF CANDIUM

PURPOSE

To analyze the isotopes of candium and to calculate its atomic mass.

MATERIALS

- sample of candium
- balance
- pencil
- paper

PROCEDURE

Obtain a sample of candium. Separate the three isotopes (m&m's®, Skittles®, and Reese's Pieces®) and measure the mass of each isotope. Count the numbers of m&m's®, Skittles®, and Reese's Pieces®. Make a table similar to Figure A to record your measured and calculated data.

	m&m's®	Skittles®	Reese's Pieces®	Totals
Total mass (grams)	**13.16 g**	**13.83 g**	**15.40 g**	**42.39 g**
Number	**15**	**13**	**20**	**48**
Average mass (grams)	**0.8773 g**	**1.064 g**	**0.7700 g**	**0.8831 g**
Percent abundance	**31.25**	**27.08**	**41.67**	**100.0**
Relative abundance	**0.3125**	**0.2708**	**0.4167**	**1.000**
Relative mass	**0.2742 g**	**0.2883 g**	**0.3208 g**	**0.8833 g**

Figure A

ANALYSIS

Using the experimental data, record the answers to the following questions below your data table.

1. Calculate the average mass of each isotope by dividing its total mass by the number of particles of that isotope.
2. Calculate the percent abundance of each isotope by dividing its number of particles by the total number of particles and multiplying by 100.
3. Calculate the relative abundance of each isotope by dividing the percent abundance from Step 2 by 100.
4. Calculate the relative mass of each isotope by multiplying its relative abundance from Step 3 by its average mass.
5. Calculate the average mass of all candium particles by adding the relative masses. This average mass is the atomic mass of candium.
6. Explain the difference between percent abundance and relative abundance. What is the result when you total the individual percent abundances? The individual relative abundances?
7. The percent abundance of each kind of candy tells you how many of each kind of candy there are in every 100 particles. What does relative abundance tell you?
8. Compare the total values for rows 3 and 6 in the table. Explain why the totals differ and why the value in row 6 best represents atomic mass.
9. Explain any differences between the atomic mass of your candium sample and that of your neighbor. Explain why the difference would be smaller if larger samples were used.

YOU'RE THE CHEMIST

The following small-scale activities allow you to develop your own procedures and analyze the results.

1. **Analyze It!** Determine the atomic mass of a second sample of candium. How does it compare with the first? Suggest reasons for any differences between the samples.
2. **Design It!** Design and test methods to produce identical samples of candium. Try measuring mass or volume as a means of counting. Test these methods by counting each kind of candy in each sample you produce. Which method of sampling gives the most consistent results?

ANALYSIS

Sample data for answers to questions 1–5 are listed above.

6. Percent abundance is parts per hundred. Relative abundance is parts per one, or the decimal form of percent. The individual percent abundances add up to 100. The individual relative abundances add up to 1.
7. Relative abundance tells you the decimal fraction of particles.
8. The total in row 3 is an average that ignores the relative abundances of particles. The total in row 6 is a weighted average that best represents atomic mass because it considers differences in mass and abundance among the particles.
9. Another student might not have had the same relative abundance of each candy.

THE PERIODIC TABLE: ORGANIZING THE ELEMENTS

How do you know where to find products in the supermarket? From your experience, you probably know that different types of products are arranged according to similar characteristics in aisles or sections of aisles. Such a classification structure makes finding and comparing products easy. **Is there a way of arranging more than 100 known elements?**

objectives

- Describe the origin of the periodic table
- Identify the position of groups, periods, and the transition metals in the periodic table

key terms

- periodic table
- periods
- periodic law
- group
- representative elements
- metals
- alkali metals
- alkaline earth metals
- transition metals
- inner transition metals
- nonmetals
- halogens
- noble gases
- metalloids

Development of the Periodic Table

About 70 elements had been discovered by the mid-1800s, but until the work of the Russian chemist Dmitri Mendeleev (1834–1907), no one had found a way to relate the elements in a systematic, logical way. Mendeleev listed the elements in columns in order of increasing atomic mass. He then arranged the columns so that the elements with the most similar properties were side by side. He thus constructed the first **periodic table,** an arrangement of the elements according to similarities in their properties. As you can see in **Figure 5.11,** Mendeleev left blank spaces in the table because there were no known elements with the appropriate properties and masses.

Mendeleev and others were able to predict the physical and chemical properties of the missing elements. Eventually these elements were discovered and were found to have properties similar to those predicted.

In 1913, Henry Moseley (1887–1915), a British physicist, determined the atomic number of the atoms of the elements. Moseley arranged the elements in a table by order of atomic number instead of atomic mass. That is the way the periodic table is arranged today.

			Ti = 50	Zr = 90	? = 180.
			V = 51	Nb = 94	Ta = 182.
			Cr = 52	Mo = 96	W = 186.
			Mn = 55	Rh = 104,4	Pt = 197,4
			Fe = 56	Ru = 104,4	Ir = 198.
			Ni = Co = 59	Pl = 106,6	Os = 199
H = 1			Cu = 63,4	Ag = 108	Hg = 200
	Be = 9,4	Mg = 24	Zn = 65,2	Cd = 112	
	B = 11	Al = 27,4	? = 68	Ur = 116	Au = 197?
	C = 12	Si = 28	? = 70	Sn = 118	
	N = 14	P = 31	As = 75	Sb = 122	Bi = 210
	O = 16	S = 32	Se = 79,4	Te = 128?	
	F = 19	Cl = 35,5	Br = 80	I = 127	
Li = 7	Na = 23	K = 39	Rb = 85,4	Cs = 133	Tl = 204
		Ca = 40	Sr = 87,6	Ba = 137	Pb = 207.
		? = 45	Ce = 92		
		?Er = 56	La = 94		
		?Yt = 60	Di = 95		
		?In = 75,6	Th = 118?		

Figure 5.11
A version of Dmitri Mendeleev's periodic table is shown here.

5.4

1 Engage

Use the Visual

Have students study the photograph and read the text that opens the section. Explain that just as items in a supermarket are arranged by food type, such as breads, dairy products, and produce, the chemical elements in the periodic table are arranged in groups that have similar chemical properties. Explain that the periodic table is a chemical road map that can be used to predict chemical behaviors of elements.

2 Teach

Use the Visual

Have students study Mendeleev's version of the periodic table in Figure 5.11. Show students how the elements with similar chemical properties were arranged in horizontal rows. Note that the table also included gaps where elements that had not yet been discovered would be placed.

STUDENT RESOURCES

From the Teacher's Resource Package, use:

- Section Review 5.4, Ch. 5 Practice Problems, Interpreting Graphics, Vocabulary Review, Quizzes, and Tests from the Review Module (Ch. 5–8)

TECHNOLOGY RESOURCES

Relevant technology resources include:

- Chem ASAP! CD-ROM
- Resource Pro CD-ROM
- Assessment Resources CD-ROM: Chapter 5 Tests

TEACHER DEMO

Show students a variety of chemical elements that were known in ancient times, such as copper, sulfur, silver, carbon, gold, and tin. Ask students to infer why these elements were familiar long before gases, such as oxygen and nitrogen, or metals such as iron or aluminum.

Discuss

To emphasize the concept that elements in the same group of the periodic table have similar properties, point out one group and describe the properties of the elements in that group. You can point to the group on the far right, and explain that all of these elements are gases that do not normally react with other elements. Using a large poster-size display of the periodic table, trace one period across the table. Describe some physical and chemical properties of each element in the period. Then have students compare these elements to those below them in the next period.

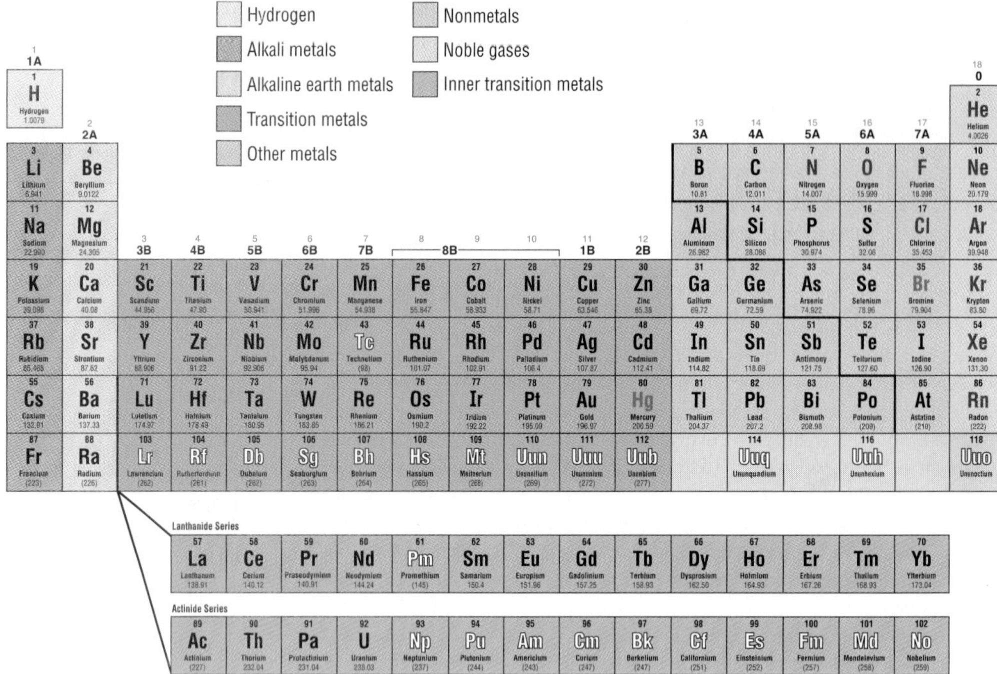

Figure 5.12
Elements are arranged in the modern periodic table in order of atomic number. The symbols are color coded according to the natural state of the elements: red for gases, black for solids, green for liquids (mercury and bromine), and white for elements that do not occur naturally.

The Modern Periodic Table

The most commonly used form of the modern periodic table, sometimes called the long form, is shown in **Figure 5.12.** Each element is identified by its symbol placed in a square. The atomic number of the element is shown centered above the symbol. The atomic mass and the name of the element are shown below the symbol. Notice that the elements are listed in order of increasing atomic number, from left to right and from top to bottom. Hydrogen (H), the lightest element, is in the top left corner. Helium (He), atomic number 2, is at the top right. Lithium (Li), atomic number 3, is at the left end of the second row.

The horizontal rows of the periodic table are called **periods.** There are seven periods. The number of elements per period ranges from 2 (hydrogen and helium) in period 1, to 32 in period 6. The properties of the elements within a period change as you move across it from element to element. The pattern of properties within a period repeats, however, when you move from one period to the next. This gives rise to the **periodic law:** When the elements are arranged in order of increasing atomic number, there is a periodic repetition of their physical and chemical properties. The arrangement of the elements into periods has an important consequence. Elements that have similar chemical and physical properties end up in the same column in the periodic table.

Each vertical column of elements in the periodic table is called a **group,** or family. The elements in any group of the periodic table have similar physical and chemical properties. Each group is identified by a number and

MEETING DIVERSE NEEDS

At Risk Have students make a list of familiar elements and describe at least one use for each element. (Examples are nitrogen, oxygen, argon, iron, aluminum, carbon, and chromium.) Then have students combine their lists into a master list on the chalkboard.

LEP Have students select a few of the more common elements from the periodic table such as gold. Write the symbols for these elements on the chalkboard. Have students list as many names as possible for these elements in different languages. Ask why some elements are less likely to have multiple names.

the letter A or B. Look at the first column on the left. It includes the elements H, Li, Na, K, Rb, Cs, and Fr. This first column is designated Group 1A. Except for hydrogen, all of the Group 1A elements react vigorously, even explosively, with water. The next column to the right, Group 2A, starts with Be. Next comes Group 3A, toward the right of the table. The Group A elements are made up of Group 1A through Group 7A and Group 0 (the group at the far right). Group A elements are called the **representative elements** because they exhibit a wide range of both physical and chemical properties.

The representative elements can be divided into three broad classes. The first are **metals,** which have a high electrical conductivity and a high luster when clean. They are ductile (able to be drawn into wires) and malleable (able to be beaten into thin sheets). Except for hydrogen, the representative elements on the left side of the periodic table are metals. The Group 1A elements are called the **alkali metals,** and the Group 2A elements are called the **alkaline earth metals.** Most of the remaining elements that are not Group A elements are also metals. These include the **transition metals** and the **inner transition metals,** which together make up the Group B elements. Copper, silver, gold, and iron are familiar transition metals. The inner transition metals, which appear below the main body of the periodic table, are also called the rare-earth elements. Approximately 80% of all of the elements are metals. With one exception, all metals are solids at room temperature. **Figure 5.14** on page 126 shows the exception to this rule. What is the name, symbol, and physical state of this element? ❶

Figure 5.13

The elements in the periodic table vary greatly in their properties.

Sodium emits bright yellow light during a flame test.

Potassium reacts violently with water.

Chromium, the principal element in chrome plating, resists corrosion.

Iodine exists as a solid and a vapor at 25°C.

Because of its malleability, silver is easily stamped into coins.

5.4

ACTIVITY

Have each student choose an element without revealing the choice to other students. Each student should write a short description of the chosen element to read to the class. Ask the other students to identify the element from its description. Encourage students to be as specific as possible in their descriptions. Example: bromine is a reddish-brown liquid.

Discuss

Point out to students the following features of the modern periodic table.

- **Metals**
- **Non-metals**
- **Metalloids**
- **Transition metals**
- **Inner-transition metals**
- **Representative elements**
- **Alkali metals**
- **Alkaline earth metals**
- **Halogens**
- **Noble (inert) gases**

Answers

❶ mercury, Hg, liquid

3 Assess

Evaluate Understanding

Have students draw a concept map relating the following terms: groups, horizontal rows, periods, periodic law, periodic table, sequence of change, similar properties, vertical columns. Write the names of selected elements on the board and have students write the chemical symbols for the elements. Ask them to point out the elements on the periodic table and to classify them as metals, nonmetals, or metalloids.

Reteach

Remind students why chemists need a classification system for elements. Emphasize the "road map" character of the periodic table and how to locate elements with similar chemical and physical properties. If possible, show a video of the chemical properties of a family, such as the alkali metals. Recognizing patterns of physical and chemical properties makes it possible for chemists to predict with some reliability the outcomes of events.

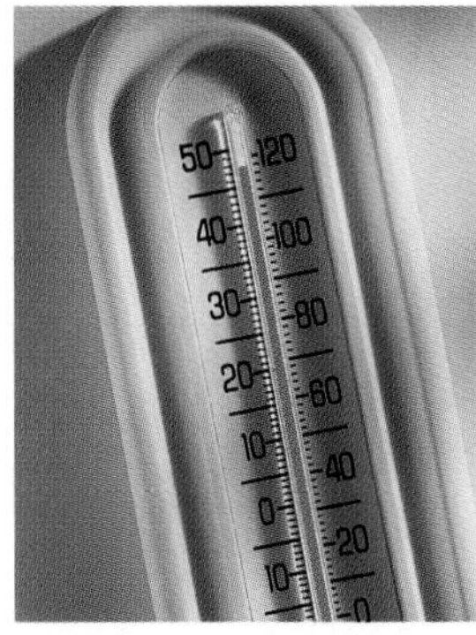

Figure 5.14
Mercury, a transition metal, is the only metallic element that is a liquid at room temperature. It is used in thermometers and barometers and as the electrical contact in a thermostat.

The nonmetals occupy the upper-right corner of the periodic table. **Nonmetals** are elements that are generally nonlustrous and that are generally poor conductors of electricity. Some of these elements, such as oxygen and chlorine, are gases at room temperature. Others, such as sulfur, shown in **Figure 5.15,** are brittle solids. One element, bromine, is a fuming dark-red liquid at room temperature. Two groups of nonmetals are given special names. The nonmetals of Group 7A are called the **halogens,** which include chlorine and bromine. The nonmetals of Group 0 are known as the **noble gases,** which are sometimes called the inert gases because they undergo few chemical reactions. The noble gas neon is used to fill the glass tubes of neon lights.

Notice the heavy stair-step line in **Figure 5.12.** This line divides the metals from the nonmetals. Most of the elements that border this line are **metalloids,** elements with properties that are intermediate between those of metals and nonmetals. Silicon and germanium are two important metalloids that are used in the manufacture of computer chips and solar cells.

Without the help of the periodic table, it would be quite difficult to learn and remember the chemical and physical properties of the more than 100 elements. Instead of memorizing their properties separately, you need only learn the general behavior and trends within the major groups. This gives you a useful working knowledge of the properties of most elements.

Figure 5.15
Sulfur is a low-melting point nonmetallic element that occurs as a crystalline solid or in the amorphous (formless) state. It is often mined through a process involving the pumping of hot water, which melts the sulfur. Sulfur is used primarily in the manufacture of sulfuric acid.

section 5.4 review

27. Describe how the periodic table was developed.

28. What criteria did Mendeleev use to construct his periodic table of the elements?

29. Relate group, period, and transition metals to the periodic table.

30. Identify each element as a metal, metalloid, or nonmetal.

a. gold **b.** silicon **c.** manganese **d.** sulfur **e.** barium

31. Which of the elements listed in the preceding question are representative elements?

32. Name two elements that have properties similar to those of the element calcium.

 Chem ASAP! **Assessment 5.4** Check your understanding of the important ideas and concepts in Section 5.4.

Answers

SECTION REVIEW 5.4

27. Mendeleev observed trends in properties and grouped similar elements together. Then he arranged the groups so that the elements were in order of increasing mass. There were blank spaces in the arrangement that were filled in as elements were discovered. Moseley rearranged the elements according to increasing atomic number.

28. increasing atomic mass and similarities of properties

29. A group is a vertical column. A period is a horizontal row. A transition metal is a Group B element.

30. **a.** metal **b.** metalloid **c.** metal **d.** nonmetal **e.** metal

31. silicon, sulfur, barium

32. beryllium, magnesium, strontium, barium

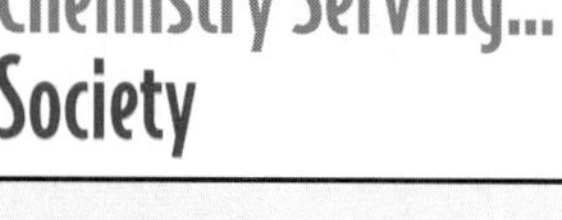

Chemistry Serving...Society

ASK AN ARTIFACT FOR A DATE!

A human skeleton is discovered when construction begins on a new school. Is it the remains of someone who died within the past 200 years? Perhaps it is the remains of an early Native American who lived in the area 9000 years ago. An archaeologist is called to the site. She examines the skeleton and finds evidence that it may be very old. To determine its age, she uses a technique called radiometric dating.

Radiometric dating is based on two important facts: each element exists as more than one isotope, and some of these isotopes undergo radioactive decay. As you know, isotopes are atoms with the same number of protons and electrons but with different numbers of neutrons. Isotopes with an unstable ratio of protons and neutrons are radioactive. An atom of a radioactive isotope will emit radiation, changing into an atom of a different element. Although it is impossible to predict when a single atom of a radioactive isotope will decay, a large group of such atoms has a regular and predictable rate of decay. The amount of time it takes for half the atoms in a sample of a certain isotope to decay is called the half-life of that isotope.

When scientists want to date an artifact that was once part of a living organism, they often use a kind of radiometric dating called carbon-14 ($^{14}_{6}C$) dating. This method involves measuring the amount of the isotope carbon-14 in the artifact.

How is the amount of carbon-14 in an artifact related to its age? Carbon-14 is a radioactive isotope of carbon, with a half-life of 5730 years. It is present in small amounts in the environment, along with the two stable and more common isotopes of carbon, $^{12}_{6}C$ and $^{13}_{6}C$. The ratio of $^{14}_{6}C$ to the other carbon isotopes is relatively constant throughout the environment because $^{14}_{6}C$ is produced at a constant rate in the upper atmosphere by high-energy cosmic rays and is spread evenly throughout the biosphere. Living organisms all have the same ratio of $^{14}_{6}C$ to stable carbon in their bodies because they are constantly exchanging carbon with the environment.

Carbon-14 dating can give accurate ages of artifacts that are up to about 40 000 years old.

When an organism dies, however, it stops exchanging carbon with the environment. The radioactive $^{14}_{6}C$ atoms in the remains of the organism decay at the rate characteristic of $^{14}_{6}C$ without being replaced by new ones. Therefore, the ratio of $^{14}_{6}C$ to stable carbon in an organism begins to change in a regular, predictable way at the moment the organism dies.

To determine the age of once-living remains, a scientist can find the ratio of $^{14}_{6}C$ to stable carbon in the remains and compare it with the ratio that should have been in the organism when it was alive. If a once-living artifact contains half the $^{14}_{6}C$ of a living organism, for example, it must be about 5730 years old—the amount of time equal to one half-life of carbon-14.

Carbon-14 dating can give accurate ages of artifacts that are up to about 40 000 years old. When objects are much older than that, the amount of $^{14}_{6}C$ they contain is so small that the use of carbon-14 dating is impractical.

CHEMISTRY IN CAREERS

TAKE IT TO THE NET
ind out more about
areer opportunities:
www.phschool.com

CHEMICAL SPECIALIST
ocal food service distributor seeks
esponsible self-motivated individ-

ARCHAEOLOGIST
Use carbon-14 dating and other chemical techniques to estimate the age and composition of artifacts.
See page 870.

CHEMIST NEEDED

Chemistry Serving... Society

DISCUSS

Point out that all living organisms contain carbon-12 and carbon-14 in a fixed ratio. However, after an organism dies, this ratio changes as the carbon-14 decays. Archaeologists use this fact to establish the age of fossils and ancient artifacts. Scientists compare the $^{14}C/^{12}C$ ratio of dead organisms to the $^{14}C/^{12}C$ ratio of living organisms to determine the age.

CHEMISTRY IN CAREERS

Have students read about archaeology on page 870. Encourage students to look up key terms from the article on the Internet.

4 Close

Summary

Ask the following questions that require students to summarize information contained in the chapter.

- **How was Dalton's model of the atom revised to fit with Rutherford's experiments?** (Rutherford showed that atoms consist of extremely small, dense nuclei and that the majority of the volume of an atom is empty space.)
- **How are chemical reactions explained by Dalton's atomic theory?** (Chemical reactions occur when atoms are separated, joined, or rearranged.)
- **How was Mendeleev able to predict the properties of elements before they were discovered?** (Mendeleev grouped elements according to similar properties. He left spaces for what he felt were undiscovered elements.)

Extension

Have students use a timeline to trace the development of the atomic model. Have them note the data that led to each change in the model. Students may wish to extend the model to include particles such as quarks.

Looking Back . . . Looking Ahead . . .

The previous chapter presented a three-step approach to solving problems in chemistry. This chapter describes experiments that led to the discovery of protons, electrons, and neutrons. The next chapter discusses the role of electrons in the chemical bonds that hold atoms together in compounds.

Chapter 5 STUDENT STUDY GUIDE

Take It to the NET
For interactive study and review, go to www.phschool.com

KEY TERMS

- alkali metal *p. 125*
- alkaline earth metal *p. 125*
- atom *p. 108*
- atomic mass *p. 120*
- atomic mass unit (amu) *p. 118*
- atomic number *p. 113*
- cathode ray *p. 109*
- Dalton's atomic theory *p. 107*
- electron *p. 109*
- group *p. 124*
- halogen *p. 126*
- inner transition metal *p. 125*
- isotope *p. 116*
- mass number *p. 115*
- metal *p. 125*
- metalloid *p. 126*
- neutron *p. 110*
- noble gas *p. 126*
- nonmetal *p. 126*
- nucleus *p. 112*
- period *p. 124*
- periodic law *p. 124*
- periodic table *p. 123*
- proton *p. 110*
- representative element *p. 125*
- transition metal *p. 125*

KEY EQUATIONS AND RELATIONSHIPS

- atomic number = number of protons = number of electrons
- number of neutrons = mass number − atomic number

CONCEPT SUMMARY

5.1 Atoms
- Elements are composed of atoms, which are the basic building blocks of matter.
- The atoms of a given element are different from the atoms of all other elements.

5.2 Structure of the Nuclear Atom
- Atoms contain positively charged protons, negatively charged electrons, and electrically neutral neutrons.
- The nucleus of an atom is composed of protons and neutrons. The electrons surround the nucleus and occupy most of the volume of the atom.

5.3 Distinguishing Between Atoms
- The number of protons in an atom's nucleus is the atomic number of that element.
- Because atoms are electrically neutral, an atom has the same number of protons and electrons.
- The sum of the number of protons and number of neutrons is the mass number of an atom.
- Atoms with the same number of protons but different numbers of neutrons are called isotopes.
- The atomic mass of an element is expressed in atomic mass units (amu).
- The atomic mass of an element is a weighted average of all the naturally occurring isotopes of that element.

5.4 The Periodic Table: Organizing the Elements
- In the periodic table, the elements are organized into groups (vertical columns) and periods (horizontal rows) in order of increasing atomic number.
- Elements that have similar chemical properties are in the same group.
- Elements in the periodic table are classified as metals, nonmetals, or metalloids.

CHAPTER CONCEPT MAP

Use these terms to construct a concept map that organizes the main ideas of this chapter.

atom | electron | nucleus
isotope | atomic mass | mass number | periodic table
proton | neutron | atomic number

Chem ASAP! Concept Map 5
Create your Concept Map using the computer.

Take It to the Net

At **www.phschool.com** students will find for this chapter
- an Internet activity
- links to related chemistry sites
- an interactive quiz
- career links

CONCEPT PRACTICE

33. With which of these statements would John Dalton have agreed in the early 1800s? For each, explain why or why not? *5.1*
- **a.** Atoms are the smallest particles of matter.
- **b.** The mass of an iron atom is different from the mass of a copper atom.
- **c.** Every atom of silver is identical to every other atom of silver.
- **d.** A compound is composed of atoms of two or more different elements.

34. What experimental evidence did Thomson have for each statement? *5.2*
- **a.** Electrons have a negative charge.
- **b.** Atoms of all elements contain electrons.

35. Would you expect two electrons to attract or repel each other? *5.2*

36. How did the results of Rutherford's gold foil experiment differ from his expectations? *5.2*

37. What is the charge, positive or negative, of the nucleus of every atom? *5.2*

38. Why is an atom electrically neutral? *5.3*

39. What does the atomic number of each atom represent? *5.3*

40. How many protons are in the nuclei of the following atoms? *5.3*
- **a.** phosphorus
- **b.** molybdenum
- **c.** aluminum
- **d.** cadmium
- **e.** chromium
- **f.** lead

41. What is the difference between the mass number and the atomic number of an atom? *5.3*

42. Complete this table. *5.3*

Atomic number	Mass number	Number of protons	Number of neutrons	Number of electrons	Symbol of element
9			10		
		14	15		
	47		25		
	55	25			

43. Name two ways that isotopes of an element differ. *5.3*

44. How can there be more than 1000 different atoms when there are only about 100 different elements? *5.3*

45. What data must you know about the isotopes of an element to calculate the atomic mass of the element? *5.3*

46. What is the atomic mass of an element? *5.3*

47. Look up the word *periodic* in the dictionary. Propose a reason for the naming of the periodic table. *5.4*

48. How did Moseley's arrangement of the elements differ from that of Mendeleev? *5.4*

49. Give the symbol of each element. *5.4*
- **a.** the nonmetal in Group 4A
- **b.** the inner transition metal with the lowest atomic number
- **c.** all of the nonmetals for which the atomic number is a multiple of five
- **d.** the two elements that are liquid at room temperature
- **e.** the metal in Group 5A

CONCEPT MASTERY

50. Compare the relative size and relative density of an atom with its nucleus.

51. Imagine you are standing on the top of a boron-11 nucleus. Describe the numbers and kinds of subatomic particles you would see looking down into the nucleus, and those you would see looking out from the nucleus.

52. What parts of Dalton's atomic theory no longer agree with the current picture of the atom?

53. The four isotopes of lead are shown below, each with its percent by mass abundance and the composition of its nucleus. Using these data, calculate the approximate atomic mass of lead.

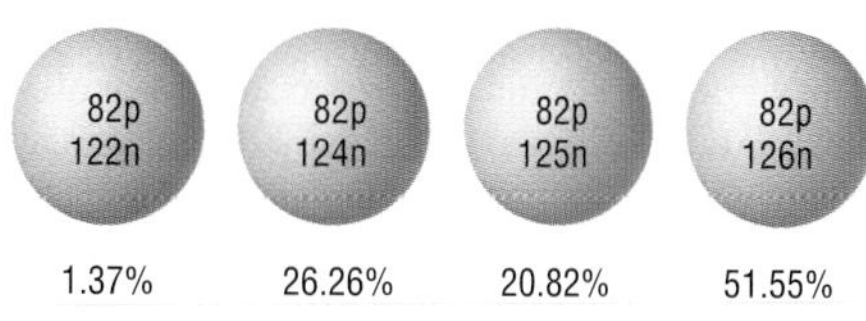

54. Dalton's atomic theory was not correct in every detail. Should this be taken as a criticism of Dalton as a scientist? Explain.

55. Why are atoms considered the basic building blocks of matter even though smaller particles, such as protons and electrons, exist?

Answers

33. Dalton would agree with all four statements because they all fit his atomic theory.

34. **a.** A beam of electrons (cathode rays) is deflected by an electric field toward the positively charged plate.
b. The cathode rays were always composed of electrons regardless of the metal used in the electrodes or the gas used in the cathode-ray tube.

35. repel

36. He did not expect any alpha particles to be deflected over a large angle.

37. Every atomic nucleus is positively charged.

38. It has equal numbers of protons and electrons.

39. number of protons in the nucleus

40. **a.** 15 **b.** 42 **c.** 13 **d.** 48 **e.** 24 **f.** 82

41. The atomic number is the number of protons. The mass number is the sum of the number of protons and number of neutrons.

42.

9	19	9	10	9	F
14	29	14	15	14	Si
22	47	22	25	22	Ti
25	55	25	30	25	Mn

43. mass numbers, atomic masses, number of neutrons, relative abundance

44. because of the existence of isotopes

45. which isotopes exist, their masses, and their natural percent abundance

46. The atomic mass is the weighted average of the masses of all of its isotopes.

47. Answers will vary.

48. Moseley arranged the elements in order of increasing atomic number, not atomic mass.

49. **a.** C **b.** La **c.** B, Ne, P, Br **d.** Hg, Br **e.** Bi

50. The nucleus is very small and very dense compared with the atom.

51. five protons and six neutrons in the nucleus; five electrons outside the nucleus

52. All atoms of the same element are not identical (isotopes). The atom is not the smallest particle of matter.

53. 207 amu

54. No; in general, he proposed a valid theory in line with the experimental evidence he had available to him.

55. An atom is the smallest particle of an element that retains the properties of that element.

Answers

56. **a.** 92.90%
b. 99.89%
c. 0.00993%

57. **a.** 3
b. 1
c. 4

58. Change the metal used as a target and account for differences in deflection patterns.

59. The following are reasonable hypotheses: The space in an individual atom is large relative to the volume of the atom, but very small relative to an object the size of a hand. There are many layers of atoms in a wall or a desk. The space that exists is distributed evenly throughout the solid, similar to the distribution of air pockets in foam insulation.

60. The theory must be modified and then retested.

61. Answers will vary.

62. 48 g

63. **a.** 4
b. 3
c. 3
d. 4

64. In a chemical change, atoms are not created or destroyed, they are rearranged.

65. 4.84×10^5 cm^3

66. 122 g

67. **a.** element
b. mixture
c. mixture
d. mixture
e. mixture
f. mixture

68. Because diamond is more dense than graphite, pressure could be used to squeeze the carbon atoms closer together.

69. 92.5%

70. 4×10^{-25} g

56. The following table shows some of the data collected by Rutherford and his colleagues during their gold foil experiment.

Angle of deflection (degrees)	Number of deflections
5	8 289 000
10	502 570
15	120 570
30	7800
45	1435
60	477
75	211
>105	198

a. What percentage of the alpha particle deflections were 5° or less?
b. What percentage of the deflections were 15° or less?
c. What percentage of the deflections were 60° or greater?

CRITICAL THINKING

57. Choose the term that best completes the second relationship.

a. female:male — proton:_______
(1) atom (3) electron
(2) neutron (4) quark

b. cow:horse — neutron:_______
(1) proton (3) atom
(2) nucleus (4) quark

c. atom:proton — house:_______
(1) school (3) planet
(2) nucleus (4) brick

58. How could you modify Rutherford's experimental procedure to determine the relative sizes of different nuclei?

59. Rutherford's atomic theory proposed a dense nucleus surrounded by very small electrons. This implies that atoms are composed mainly of empty space. If all matter is mainly empty space, why is it impossible to walk through walls or pass your hand through your desk?

60. This chapter illustrates the scientific method in action. What happens when new experimental results cannot be explained by the existing theory?

61. Do you think there are more elements left to be discovered? Explain your answer.

CUMULATIVE REVIEW

62. Oxygen and hydrogen react explosively to form water. In one reaction, 6 g of hydrogen combines with oxygen to form 54 g of water. How much oxygen was used?

63. How many significant figures are in each measurement?
a. 4.607 mg
b. 4.35×10^4 km
c. 0.001 50 mL
d. 60.09 kg

64. The law of conservation of mass was introduced in Chapter 2. Use Dalton's atomic theory to explain this law.

65. An aquarium measures 55.0 cm × 1.10 m × 80.0 cm. How many cm^3 of water will this aquarium hold?

66. What is the mass of 5.42 cm^3 of platinum? The density of platinum is 22.5 g/cm^3.

67. Classify each as an element, a compound, or a mixture.
a. sulfur
b. salad oil
c. newspaper
d. orange
e. cardboard
f. apple juice

CONCEPT CHALLENGE

68. Diamond and graphite are both composed of carbon atoms. The density of diamond is 3.52 g/cm^3. The density of graphite is 2.25 g/cm^3. In 1955, scientists successfully made diamond from graphite. Using the relative densities, imagine what happens at the atomic level when this change occurs. Then suggest how this synthesis may have been accomplished.

69. Lithium has two naturally occurring isotopes. Lithium-6 has an atomic mass of 6.015 amu; lithium-7 has an atomic mass of 7.016 amu. The atomic mass of lithium is 6.941 amu. What is the percentage of naturally occurring lithium-7?

70. When the masses of the particles that make up an atom are added together, the sum is always larger than the actual mass of the atom. The missing mass, called the mass defect, represents the matter converted into energy when the nucleus was formed from its component protons and neutrons. Calculate the mass defect of a chlorine-35 atom by using the data in **Table 5.1.** The actual mass of a chlorine-35 atom is 5.81×10^{-23} g.

Chapter 5 STANDARDIZED TEST PREP

Select the choice that best answers each question or completes each statement.

1. An atom composed of 16 protons, 16 electrons, and 16 neutrons is
   a. $^{48}_{16}S$ b. $^{16}_{32}Ge$ c. $^{32}_{16}S$ d. $^{16}_{32}S$.
2. Which element is *not* a transition metal?
   a. aluminum c. iron
   b. silver d. zirconium
3. Which of these subatomic particle descriptions is *incorrect?*
   a. proton: positive charge, in nucleus, mass of ≈ 1 amu
   b. electron: negative charge, mass of ≈ 0 amu, in nucleus
   c. neutron: mass of ≈ 1 amu, no charge
4. Which of these statements about the periodic table are correct?
   I. Elements are arranged in order of increasing atomic mass.
   II. A period is a horizontal row.
   III. Nonmetals are located on the right side of the table.
   a. I only
   b. I and II only
   c. I, II, and III
   d. I and III only
   e. II and III only
5. Thallium has two isotopes, thallium-203 and thallium-205. Thallium's atomic number is 81 and its atomic mass is 204.38 amu. Which statement about the thallium isotopes is true?
   a. There is more thallium-203 in nature.
   b. Atoms of both isotopes have 81 protons.
   c. Thallium-205 atoms have fewer neutrons.
   d. The most common atom of thallium has a mass of 204.38 amu.

Use the art to answer question 6.

6. How many nitrogen-14 atoms (^{14}N) would you need to place on the right pan to balance the three calcium-42 atoms (^{42}Ca) on the left pan of the "atomic balance" below? Describe the method you used to determine your answer, including any calculations.

Use the diagram below to answer questions 7–9. The diagram shows gold atoms being bombarded with fast-moving alpha particles.

7. The large yellow spheres represent gold atoms. What do the small gray spheres represent?
8. List at least two characteristics of the small gray spheres.
9. Which subatomic particle cannot be found in the area represented by the gray spheres?

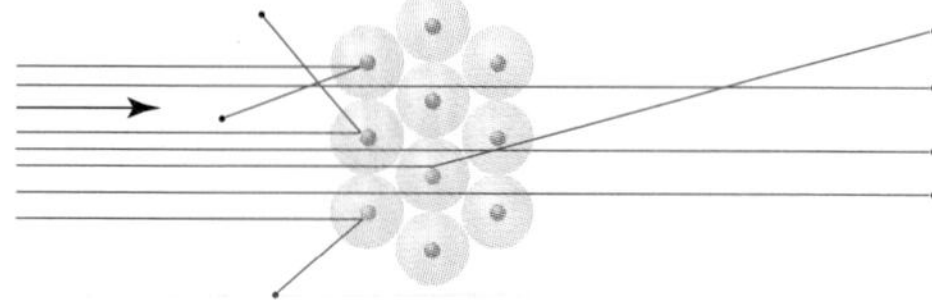

For each question there are two statements. Decide whether each statement is true or false. Then decide whether Statement II is a correct explanation for Statement I.

	Statement I		Statement II
10.	Every aluminum-27 atom has 27 protons and 27 electrons.	BECAUSE	The mass number of aluminum-27 is 27.
11.	Isotopes of an element have different atomic masses.	BECAUSE	The nuclei of an element's isotopes contain different numbers of protons.
12.	An electron is repelled by a negatively charged particle.	BECAUSE	An electron has a negative charge.
13.	The element hydrogen is a metal.	BECAUSE	Hydrogen is on the left in the periodic table.

Chapter 5 STANDARDIZED TEST PREP

1. c
2. a
3. b
4. e
5. b
6. 9; three ^{42}Ca atoms have an approximate mass of $3 \times 42 =$ 126 amu; one ^{14}N atom has an approximate mass of 14 amu; $126/14 = 9$ ^{14}N atoms with an approximate mass of 126 amu.
7. the nucleus of an atom
8. very small volume; almost all the mass of the atom; high density; positive charge
9. electron
10. False, True
11. True, False
12. True, True, correct explanation
13. False, True

Planning Guide

SECTION OBJECTIVES	ACTIVITIES/FEATURES	MEDIA & TECHNOLOGY
6.1 Introduction to Chemical Bonding ● ■ ◆ ▸ Distinguish between ionic and molecular compounds ▸ Define cation and anion and relate them to metal and nonmetal	**SE** **Discover It!** *Element Name Search,* p. 132 **TE** **DEMO**, p. 134	**ASAP** Animation 5 **ASAP** Problem Solving 1 **ASAP** Assessment 6.1 **OT** 7: *Atoms and Ions*
6.2 Representing Chemical Compounds ● ■ ◆ ▸ Distinguish among chemical formulas, molecular formulas, and formula units ▸ Use experimental data to show that a compound obeys the law of definite proportions	**SSLM** 6: *Chemical Names and Formulas* **TE** **DEMO**, p. 139	**ASAP** Problem Solving 11 **ASAP** Assessment 6.2 **CA** *Making Sodium Chloride* **OT** 8: *Molecular Structures*
6.3 Ionic Charges ● ■ ◆ ▸ Use the periodic table to determine the charge on an ion ▸ Define a polyatomic ion and give the names and formulas of the most common polyatomic ions	**LM** 6: *Identification of Anions and Cations in Solution* **LM** 7: *Precipitation Reactions* **TE** **DEMO**, p. 144	**ASAP** Problem Solving 16 **ASAP** Assessment 6.3
6.4 Ionic Compounds ● ■ ◆ ▸ Apply the rules for naming and writing formulas for binary ionic compounds ▸ Apply the rules for naming and writing formulas for ternary ionic compounds	**SE** **CHEMath** *Prefixes and Suffixes,* p. 152 **SE** **Small-Scale Lab** *Names and Formulas of Ionic Compounds,* p. 157 (**LRS** 6-1) **TE** **DEMOS**, pp. 150, 154	**ASAP** Simulation 2 **ASAP** Problem Solving 25, 28 **ASAP** Assessment 6.4
6.5 Molecular Compounds and Acids ● ■ ◆ ▸ Apply the rules for naming and writing formulas for binary molecular compounds ▸ Name and write formulas for common acids	**SE** **Link to Environmental Awareness** *A Toxic Environmental Gas,* p. 160	**ASAP** Problem Solving 37 **ASAP** Assessment 6.5
6.6 Summary of Naming and Formula Writing ● ■ ◆ ▸ Use the flowchart in **Figure 6.21** to write the name of a compound when given its chemical formula ▸ Use the flowchart in **Figure 6.23** to write a chemical formula when given the name of a compound	**SE** **Mini Lab** *Making Ionic Compounds,* p. 163 (**LRS** 6-2) **SE** **Chemistry Serving . . . the Consumer** *What's in a Name?,* p. 164 **SE** **Chemistry in Careers** *Pharmacist,* p. 164	**ASAP** Assessment 6.6 **ASAP** Concept Map 6 **OT** 9: *Naming Chemical Compounds* **PP** Chapter 6 Problems **RP** Lesson Plans, Resource Library **AR** Computer Test 6 **www** Activities, Self-Tests **TCP** The Chemistry Place Web Site

KEY

● Conceptual (concrete concepts)	**AR** Assessment Resources	**GRS** Guided Reading and Study Workbook
○ Conceptual (more abstract/math)	**ASAP** Chem ASAP! CD-ROM	**LM** Laboratory Manual
■ Standard (core content)	**ACT** ActivChemistry CD-ROM	**LP** Laboratory Practicals
□ Standard (extension topics)	**CHM** CHEMedia Videodiscs	**LRS** Laboratory Recordsheets
◆ Honors (core content)	**CA** Chemistry Alive! Videodiscs	**SSLM** Small-Scale Lab Manual
◇ Honors (options to accelerate)	**GCP** Graphing Calculator Problems	

PRACTICE		ASSESSMENT	
SE	**Sample Problem** 6-1	SE	Section Review
GRS	Section 6.1	RM	Section Review 6.1
RM	Practice Problems 6.1	RM	Chapter 6 Quiz
SE	**Sample Problem** 6-2	SE	Section Review
GRS	Section 6.2	RM	Section Review 6.2
RM	Practice Problems 6.2	RM	Chapter 6 Quiz
SE	**Sample Problems** 6-3 to 6-4	SE	Section Review
GRS	Section 6.3	RM	Section Review 6.3
RM	Practice Problems 6.3	RM	Chapter 6 Quiz
RM	Interpreting Graphics	LP	Lab Practical 6-2, 6-3
SE	**Sample Problems** 6-5 to 6-8	SE	Section Review
GRS	Section 6.4	RM	Section Review 6.4
RM	Practice Problems 6.4	RM	Chapter 6 Quiz
SE	**Sample Problem** 6-9	SE	Section Review
GRS	Section 6.5	RM	Section Review 6.5
RM	Practice Problems 6.5	RM	Chapter 6 Quiz
GRS	Section 6.6	SE	Section Review
RM	Practice Problems 6.6	RM	Section Review 6.6
		RM	Vocabulary Review 6
		SE	Chapter Review
		SM	Chapter 6 Solutions
		SE	Standardized Test Prep
		PHAS	Chapter 6 Test Prep
		RM	Chapter 6 Quiz
		RM	Chapter 6 A & B Test

OT	Overhead Transparency	SE	Student Edition
PHAS	PH Assessment System	SM	Solutions Manual
PLM	Probeware Lab Manual	SSV	Small-Scale Video/Videodisc
PP	Problem Pro CD-ROM	TCP	www.chemplace.com
RM	Review Module	TE	Teacher's Edition
RP	Resource Pro CD-ROM	www	www.phschool.com

PLANNING FOR ACTIVITIES

STUDENT EDITION

Discover It! p. 132
- variety of ingredient labels

Small-Scale Lab, p. 157
- pencils and paper
- rulers
- reaction surfaces
- solutions of $AgNO_3$, $Pb(NO_3)_2$, $CaCl_2$, Na_2CO_3, Na_3PO_4, NaOH, Na_2SO_4, $FeCl_3$, $MgSO_4$, $CuSO_4$

Mini Lab, p. 163
- small test tubes
- test tube rack
- plastic dropper bottles
- $0.1M FeCl_3$, $AgNO_3$, NaOH, $Pb(NO_3)_2$, Na_3PO_4, Na_2CO_3

Activity, p. 144
- copy of periodic table on a wall chart or overhead

Activity, p. 145
- photocopy of a blank periodic table for each student

Activity, p. 147
- styrofoam balls
- wooden rods
- string

TEACHER'S EDITION

Teacher Demo, p. 134
- molecular model kit

  or
- styrofoam balls and toothpicks

Activity, p. 135
- 24 pieces of paper, half with "+" and half with "−"

Teacher Demo, p. 139
- large models of ammonia
- large cards with molecular and structural formulas for ammonia

Teacher Demo, p. 144
- large beakers
- prepared solutions containing cations from soluble metal salts (such as $MnCl_2$, $FeCl_2$, $CoCl_2$, $NiCl_2$, $CuCl_2$ and $ZnCl_2$)

Teacher Demo, p. 150
- fume hood
- safety goggles
- 1 g of powdered zinc (Zn)
- 4 g of iodine (I_2)
- watch glass
- eyedropper
- 8 mL of water
- plastic or cardboard container

Activity, p. 153
- formulas for various binary ionic compounds written on slips of paper
- large box or basket

Teacher Demo, p. 154
- goggles
- gloves
- 50 mL each of $0.1M$ solutions of lead(II) nitrate and sodium carbonate
- beaker
- 10 g solid NaCl
- filter
- plastic or cardboard container

Chapter 6

Key Terms

6.1 molecule, molecular compounds, ions, cation, anions, ionic compounds
6.2 chemical formula, molecular formula, formula unit, law of definite proportions, law of multiple proportions
6.3 monatomic ions, polyatomic ions
6.4 binary compounds, ternary compound

DISCOVER IT!

Students should be able to find compounds that contain sodium, whose salts are in processed foods and in toiletries. Students may be able to infer that chlorides, sulfates, oxides, and phosphates contain the nonmetals chlorine, sulfur, oxygen, and phosphorus, respectively. Students will be unlikely to discern which elements are in compounds with names such as hexylene glycol or benzoic acid.

Chapter 6 CHEMICAL NAMES AND FORMULAS

FEATURES

Stay current with SCIENCE NEWS
Find out more about chemical formulas:
www.phschool.com

The first cola formula was made and sold in Waco, Texas.

DISCOVER IT! ELEMENT NAME SEARCH

You need a variety of ingredient labels from items such as breads, cake mixes, baking soda, cereals, soap, shampoo, conditioner, toothpaste, and antacids.

1. Select a variety of ingredient labels from the suggested items or from similar ones.
2. List all the ingredients that you recognize. Do not be discouraged by the many unfamiliar names—there are thousands upon thousands of substances used in everyday products.
3. With the help of a periodic table, make another list of all the ingredient names that contain the name of an element.

Did you find the same element name in more than one product? What element names are on both lists? Did all the ingredients have complex chemical-sounding names? How do the names of element-containing compounds compare with the other names on your list? Return to your list after completing this chapter and identify which are common names and which are systemic names.

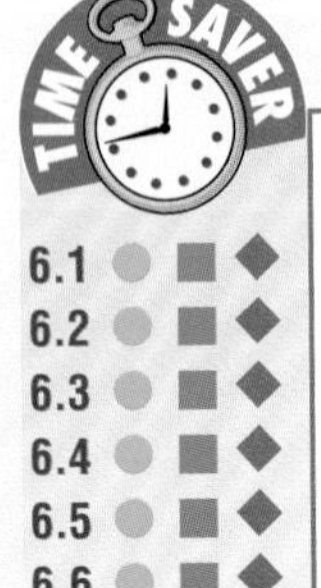

6.1 ● ■ ◆
6.2 ● ■ ◆
6.3 ● ■ ◆
6.4 ● ■ ◆
6.5 ● ■ ◆
6.6 ● ■ ◆

Conceptual The representations of molecules in Figure 6.7 will be used throughout the text. Figures 6.8 and 6.9 will help students visualize the difference between molecular and ionic compounds. In Section 6.2, the concepts of definite and multiple proportions are important, but Sample Problem 6-2 may be omitted.

Standard Use Figures 6.8 and 6.9 to help students understand the difference between molecular and ionic compounds. Students do not need to memorize Tables 6.3 and 6.4. Allowing students to refer to the tables as they solve problems is more efficient than memorization. In Section 6.5, naming common acids may be omitted. The flow charts in Figures 6.21 and 6.22 will help students maintain their focus as they practice naming compounds.

Honors Section 6.6 summarizes the chapter and may be assigned as homework if students have demonstrated proficiency in naming compounds and writing formulas.

INTRODUCTION TO CHEMICAL BONDING

objectives
- Distinguish between ionic and molecular compounds
- Define cation and anion and relate them to metal and nonmetal

key terms
- molecule
- molecular compounds
- ions
- cation
- anions
- ionic compounds

Globally, there are about 5 million lightning strikes each day during thunderstorms, snowstorms, and dust storms. Lightning is often observed during volcanic eruptions. Lightning strikes have been responsible for loss of life, forest fires, house fires, power outages, and aircraft accidents. A lightning strike is an electrical discharge that usually occurs between the negatively charged bottom of a cloud and the oppositely charged Earth below. Positive and negative charges are important in chemistry. **What role do positive and negative charges play in the formation of ions?**

Molecules and Molecular Compounds

Elements are the building materials of the substances that make up all living and nonliving things. Although there are only about 100 different elements, there are millions of different compounds made from their atoms. When scientists communicate, they must know which compounds they are discussing. Thus, naming compounds is an essential skill in chemistry.

In nature, only the noble gas elements, such as helium and neon, tend to exist as isolated atoms. They are monatomic; that is, they consist of single atoms, as shown in **Figure 6.1.** Many elements found in nature are in the

Figure 6.1

The noble gases, including helium and neon, are monatomic. That means they exist as single atoms. Helium, which is less dense than air, is often used to inflate balloons. The colors produced in what we commonly call neon lights are a result of passing an electric current through one or more noble gases. Neon actually produces an orange-red color. Helium gives a yellowish light, argon shines lavender, krypton produces a whitish light, and xenon gives blue.

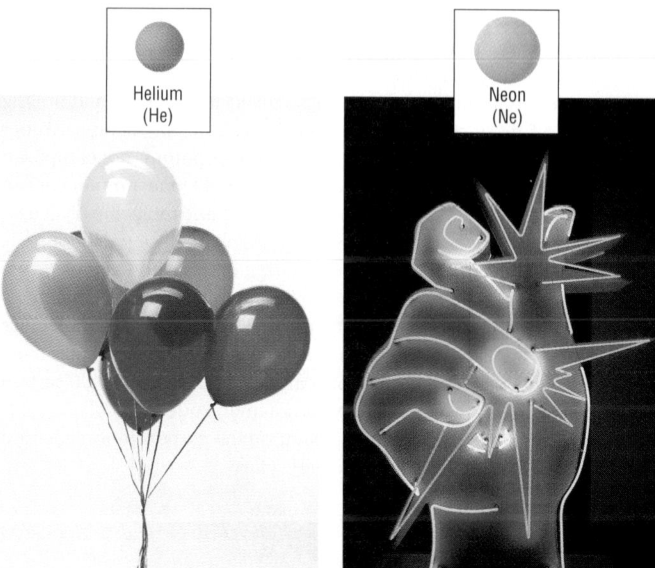

6.1

1 Engage

Use the Visual

Have students read the section-opener paragraph and examine the photograph. Compare the discharge of electricity in lightning to the spark that jumps between a metal doorknob and a hand in winter. Note that the difference is one of scale. Remind students that electrons in a lightning bolt are identical to the electrons in an electron cloud. Ask students to hypothesize about why an electron might be more likely than a proton or neutron to escape from an atom. (its location outside the nucleus and its kinetic energy)

Check Prior Knowledge

Ask:

- **What does it mean to say that an atom is neutral?** (It contains an equal number of protons and electrons.)
- **What would happen to the charge on an atom if an electron were removed? if an electron were added?** (The atom would have a positive charge; the atom would have a negative charge.)
- **What is the name for an atom that has a charge?** (an ion)
- **What is a molecule?** (a particle that contains more than one atom)
- **How do the molecules of elements and compounds differ?** (The molecules of a compound contain atoms from more than one element.)

STUDENT RESOURCES

From the Teacher's Resource Package, use:

- Section Review 6.1, Ch. 6 Practice Problems and Quizzes from the Review Module (Ch. 5–8)

TECHNOLOGY RESOURCES

Relevant Technology Resources include:

- Chem ASAP! CD-ROM
- ResourcePro CD-ROM
- Chemistry Alive! Videodisc: *Making Sodium Chloride*

2 Teach

TEACHER DEMO

Prepare the following pairs of molecular models using a model kit or styrofoam balls and toothpicks: O_2 and O_3; CO and CO_2; H_2O and H_2O_2. Show the class each pair and ask them to state how many of each type of atom is in each molecule. Write the corresponding chemical formulas on the chalkboard. Explain that the addition of one more oxygen atom to O_2, CO, and H_2O results in new molecules that have very different physical and chemical characteristics.

Think Critically

Have students determine the accuracy of the following statements:

1) All molecular compounds are composed of atoms of different elements. (True)

2) All compounds contain molecules. (False)

3) No elements are composed of molecules. (False)

4) Most molecular compounds are composed of two or more nonmetallic elements. (True)

Figure 6.2
The diatomic molecule oxygen and the triatomic molecule ozone are both composed of atoms of oxygen. Ozone is produced during electrical discharges, such as lightning, and is found in a layer of Earth's stratosphere.

form of molecules. A **molecule** is the smallest electrically neutral unit of a substance that still has the properties of the substance. Molecules are made up of two or more atoms that act as a unit, as shown in **Figure 6.2.** For example, the oxygen gas in the air you breathe consists of oxygen molecules that contain two oxygen atoms each. Oxygen is an example of a diatomic molecule. Ozone is another molecular form of oxygen. In Earth's atmosphere, ozone absorbs harmful ultraviolet radiation from the sun. Each molecule of ozone is composed of three oxygen atoms. Therefore, ozone is a triatomic molecule.

Atoms of different elements may combine chemically to form compounds. In many compounds, the atoms combine to form molecules. Compounds composed of molecules are called **molecular compounds.**

Molecular compounds tend to have relatively low melting and boiling points. Many of these compounds thus exist as gases or liquids at room temperature. The molecules in most molecular compounds are composed of the atoms of two or more nonmetals. For example, one atom of carbon can combine with one atom of oxygen to produce one molecule of a compound known as carbon monoxide. Carbon monoxide is a poisonous gas produced by the burning of gasoline in internal-combustion engines. Two hydrogen atoms can combine with one oxygen atom to produce one molecule of the compound water. The molecules of a given molecular compound are all the same. However, they differ from the molecules of all other molecular compounds. As you can see in **Figure 6.3,** all carbon monoxide molecules are identical, but carbon monoxide molecules are different from water molecules. Do the compounds carbon monoxide and
1 water consist of diatomic or triatomic molecules?

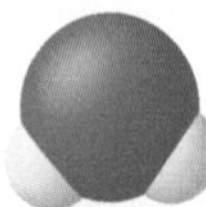

Water (H_2O)

Carbon monoxide (CO)

Figure 6.3
Water is a molecular compound with molecules composed of two hydrogen atoms and one oxygen atom. Carbon monoxide is a molecular compound with molecules composed of one carbon atom and one oxygen atom. Most molecular compounds are gases, liquids, or low-melting solids.

Answers

1. Carbon monoxide is diatomic; water is triatomic.
2. cations and anions

Ions and Ionic Compounds

Not all compounds are molecular. Many compounds are composed of particles called ions. **Ions** are atoms or groups of atoms that have a positive or negative charge. An ion forms when an atom or group of atoms loses or gains electrons. Recall that an atom is electrically neutral because it has equal numbers of protons and electrons. For example, an atom of sodium (Na) has 11 positively charged protons and 11 negatively charged electrons. The net charge on a sodium atom is zero $[11 + (-11) = 0]$. In forming a chemical compound, however, an atom of sodium tends to lose one of its electrons. The number of electrons in the atom is then no longer equal to the number of protons. The atom of sodium becomes an ion. Because there are more positive charges (protons) than negative charges (electrons), the sodium ion has a positive charge.

Atoms of the metallic elements, such as sodium, tend to form ions by losing one or more electrons. Such ions are called cations. A **cation** is any atom or group of atoms that has a positive charge. A cation has fewer electrons than the electrically neutral atom from which it formed. An ionic charge is written as a number followed by a sign. Look at **Figure 6.4.** Because a sodium cation has 11 protons but only 10 electrons, it must have a charge of 1+. The number 1 is usually omitted when writing the complete symbol for the ion; thus Na^{1+} and Na^{+} are equivalent. Magnesium (Mg) is another example of an atom that tends to form cations. Magnesium does so by losing two electrons. Therefore, a magnesium cation has a charge of 2+ because it has 12 protons but only 10 electrons. Its symbol is Mg^{2+}.

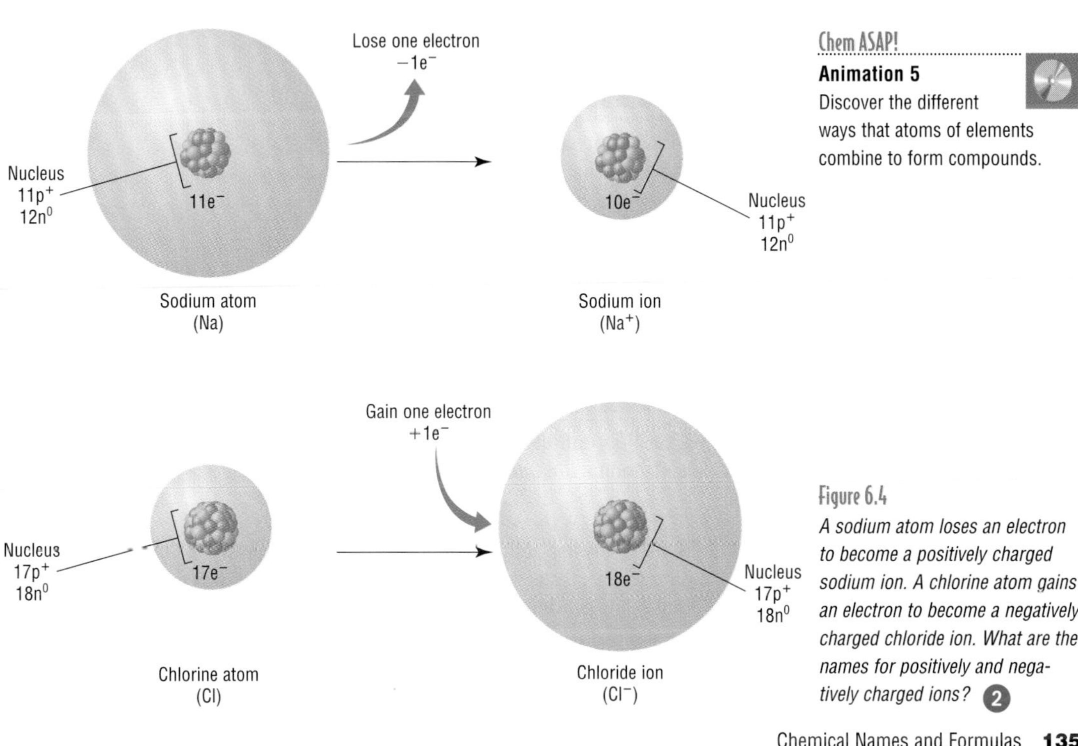

Chem ASAP!

Animation 5

Discover the different ways that atoms of elements combine to form compounds.

Figure 6.4

A sodium atom loses an electron to become a positively charged sodium ion. A chlorine atom gains an electron to become a negatively charged chloride ion. What are the names for positively and negatively charged ions? 2

6.1

ACTIVITY

To reinforce the role of electron transfer in ion formation, have students act out the formation of LiF. Give 12 students + signs to identify them as "protons" and give 12− signs to identify them as "electrons." (Explain that you are ignoring neutrons for this exercise.) Have the remaining students arrange the "protons" and "electrons" into a lithium atom on one side of the room and a fluorine atom on the other. (Note if they remember to place the protons in the middle of the group.) Have them move the lithium atom group slowly toward the fluorine atom. Just short of collision, have the organizers transfer one electron from the lithium atom to the fluorine atom.

Use the Visual

Have the students examine Figure 6.4 and discuss how sodium and chlorine atoms can be converted to ions by the loss or gain of electrons. Emphasize that ion formation always involves electrons and that the number of protons is always fixed. Ask why the resulting sodium and chloride ions would favorably interact to form the ionic compound sodium chloride. (They are oppositely charged and therefore attracted to each other.)

Chemistry Alive!

Making Sodium Chloride

Play

Building Writing Skills

Have the students construct definitions of the following terms in their own words: molecule, molecular compound, ionic compound, anion, and cation. Have them provide a chemical example for each term.

Think Critically

To focus on the fundamental difference between ionic and molecular compounds, ask students to hypothesize about what happens to the electrons when a molecular compound forms. (Students may be able to infer that the electrons are somehow shared between atoms rather than transferred.) **Tell students that they will learn much more about ionic and molecular bonds in Chapters 15 and 16.**

Practice Problems Plus

Related Chapter Review Problems
Chapter Review problems 46 and 47 are related to Sample Problem 6-1.

Additional Practice Problem
The Chem ASAP! CD-ROM contains the following problem: Give the name and symbol of the ion formed when:
a. a nitrogen atom gains three electrons
b. a lead atom loses four electrons
c. a fluorine atom gains one electron
(**a.** nitride ion, N^{3-}, **b.** lead ion, Pb^{4+}, **c.** fluoride ion, F^{-})

For metallic elements, the name of a cation is the same as the name of the element. As you have already seen, a sodium atom forms a sodium cation. Likewise, a lithium atom (Li) forms a lithium cation (Li^{+}), and an aluminum atom (Al) forms an aluminum cation (Al^{3+}). How many electrons does an aluminum atom lose to form an aluminum cation? ❶

Although their names are the same, there are many important chemical differences between metals and their cations. Sodium metal, for example, reacts explosively with water. By contrast, sodium cations are quite unreactive. As you may know, they are a component of table salt, a compound that is very stable in water.

Atoms of nonmetallic elements tend to form ions by gaining one or more electrons. In this way they form **anions,** which are atoms or groups of atoms that have a negative charge. An anion has more electrons than the electrically neutral atom from which it formed. Look again at **Figure 6.4** on the preceding page to see what happens when a chlorine atom (Cl) gains one electron to form an anion. The chloride anion has 17 protons and 18 electrons. Therefore, it has an ionic charge of 1−. The chloride anion is written as Cl^{-}. Here again, the number 1 is usually omitted when writing the complete symbol for the ion. Another common anion is the oxide anion. An oxygen atom gains two electrons in forming this ion, so the oxide anion has an ionic charge of 2−. It is written O^{2-}. Notice that the name of an anion of a nonmetallic element is not the same as the element name. The name of the anion typically ends in *-ide*. Thus a sulfur atom (S) forms a sulfide anion (S^{2-}), and a bromine atom (Br) forms a bromide anion (Br^{-}).

Sample Problem 6-1

Write the symbol and name of the ion formed when
a. a strontium atom loses two electrons.
b. an iodine atom gains one electron.

1. **ANALYZE** ***Plan a problem-solving strategy.***
   a. An atom that loses electrons forms a positively charged ion (cation). The name of a cation of a metallic element is the same as the name of the element.
   b. An atom that gains electrons forms a negatively charged ion (anion). The name of an anion of a nonmetallic element ends in *-ide.*
2. **SOLVE** ***Apply the problem-solving strategy.***
   a. Sr^{2+}, strontium ion (a cation)
   b. I^{-}, iodide ion (an anion)
3. **EVALUATE** ***Do the results make sense?***
   It makes sense that a metal atom that loses two negatively charged electrons should form a cation with a 2+ charge. It also makes sense that a nonmetal atom that gains one electron should form an anion with a 1− charge. The names conform to the naming convention.

Practice Problems

1. Give the name and symbol of the ion formed when
   a. a sulfur atom gains two electrons.
   b. an aluminum atom loses three electrons.
   c. a calcium atom loses two electrons.
2. How many electrons are lost or gained in forming each ion?
   a. Ba^{2+} **b.** As^{3-} **c.** Cu^{2+}

Chem ASAP!
Problem-Solving 1

Solve Problem 1 with the help of an interactive guided tutorial.

Meeting Diverse Needs

Gifted Have students discover the arbitrary nature of the assignment of charges to electrons and protons by researching the work of William Gilbert, Charles François de Cisternay du Fay, and Benjamin Franklin (who coined the terms *positive* and *negative charge*).

Answer

❶ An aluminum atom loses 3 electrons.

Compounds composed of cations and anions are called **ionic compounds.** Ionic compounds are usually composed of metal cations and nonmetal anions. One example of an ionic compound is sodium chloride, or table salt, which is composed of sodium cations and chloride anions. Another example is calcium oxide, commonly known as lime, which is composed of calcium cations and oxide anions. Although they are composed of ions, ionic compounds are electrically neutral. The total positive charge of the cations equals the total negative charge of the anions. Ionic compounds are usually solid crystals at room temperature, and they melt at high temperatures. **Table 6.1** summarizes some of the important differences between the properties of molecular compounds and ionic compounds. Knowing the differences in these properties can often be helpful in distinguishing between molecular and ionic compounds.

Table 6.1

Characteristics of Molecular and Ionic Compounds

Characteristic	Molecular compound	Ionic compound
Representative unit	Molecule	Formula unit (balance of oppositely charged ions)
Type of elements	Nonmetallic	Metallic combined with nonmetallic
Physical state	Solid, liquid, or gas	Solid
Melting point	Low (usually below 300 °C)	High (usually above 300 °C)

section review 6.1

3. List three characteristics that distinguish ionic compounds from molecular compounds.
4. What is a cation? What is an anion? Relate the two to metals and to nonmetals.
5. What does the presence of an *-ide* ending on the name of an ion tell you about that ion?
6. What are the only elements that exist in nature as isolated atoms? What term is used to describe such elements?
7. What is a molecule? What is the difference between a diatomic molecule and a triatomic molecule? Give an example of each.
8. Write the symbol and name for the cation formed when
   a. a potassium atom loses one electron.
   b. a zinc atom loses two electrons.
9. Write the symbol and name for the anion formed when
   a. a fluorine atom gains one electron.
   b. a sulfur atom gains two electrons.

Chem ASAP! **Assessment 6.1** Check your understanding of the important ideas and concepts in Section 6.1.

6.1

3 Assess

Evaluate Understanding

Ask students to explain how a sodium ion differs from a sodium atom and how a chloride ion differs from a chlorine atom. Ask students to describe the origin and nature of a nitride. Present the names of compounds such as calcium bromide and sulfur dioxide; ask students to predict which compounds are ionic and which are molecular.

Reteach

Emphasize the difference in properties between molecular and ionic compounds by noting that it is relatively easy to melt a spoonful of table sugar, a molecular compound, over an open flame, but that it is virtually impossible to do so with table salt, an ionic compound. Solid molecular compounds generally have much lower melting points than do ionic ones.

Answers

SECTION REVIEW 6.1

3. Ionic compounds are usually solids with high melting points formed from a metal and a nonmetal.
4. Metals tend to form positively charged cations; nonmetals tend to form negatively charged anions.
5. It is an anion.
6. They are monatomic noble gases.
7. A molecule is the smallest neutral particle of a substance that retains all the substance's properties; diatomic O_2 has two atoms; triatomic O_3 has three atoms.
8. a. potassium ion, K^+
   b. zinc ion, Zn^{2+}
9. a. fluoride ion, F^-
   b. sulfide ion, S^{2-}

1 Engage

Use the Visual

Have students study the photograph and read the text that opens the section. Ask:

- **How can two elements combine to form more than one chemical compound?** (Just as the letters in the alphabet can be combined in many different ways to form different words, the atoms of two or more elements can also be combined in different ways to form more than one type of chemical compound. Consider elements A and B. One could have AB, or A_2B_2, or AB_2, and many other possible combinations.)

Check Prior Knowledge

To assess students' prior knowledge about information contained in chemical formulas, ask:

- **What does the subscript 2 represent in the chemical formula H_2O?** (A molecule of water contains two atoms of hydrogen.)
- **What information is provided by a physical model of H_2O that the chemical formula alone does not provide?** (Both the chemical formula and model indicate that there are 2 hydrogen and 1 oxygen atoms in a water molecule; only the model provides information concerning the spatial arrangement of these atoms.)
- **If there is 16 g of oxygen and 2 g of hydrogen in 18 g of pure water, how much oxygen is there in 9 g of water?** (8 g)

section 6.2

objectives

- Distinguish among chemical formulas, molecular formulas, and formula units
- Use experimental data to show that a compound obeys the law of definite proportions

key terms

- chemical formula
- molecular formula
- formula unit
- law of definite proportions
- law of multiple proportions

REPRESENTING CHEMICAL COMPOUNDS

On the surface, identical twins may appear to be exactly the same. After all, they were formed from the same fertilized egg. On closer inspection, however, twins can be very different. Twins share some physical characteristics but are never completely alike. The same is true for chemical compounds made from the same elements. There are pairs of compounds that contain the same elements but have very different physical and chemical properties. **How can two elements combine to form more than one chemical compound?**

Chemical Formulas

Chemists have identified more than ten million chemical compounds. Some of these are molecular compounds, such as the proteins and hormones in your body. Others, such as the salts in body fluids, are ionic compounds. No two of these compounds have identical properties. In this section, you will learn how a chemical formula represents the composition of each of these chemical substances.

A **chemical formula** shows the kinds and numbers of atoms in the smallest representative unit of the substance. You can represent the chemical formulas of monatomic elements by means of their atomic symbols. For example, helium and neon have the chemical formulas He and Ne, respectively. If the molecules of the element each have more than one atom, a number is used as a subscript. For example, the diatomic form of oxygen has the formula O_2 (read as "oh two"). The names and chemical formulas of the seven diatomic elements pictured in **Figure 6.5** are hydrogen (H_2), fluorine (F_2), oxygen (O_2), nitrogen (N_2), chlorine (Cl_2), bromine (Br_2), and iodine (I_2). Because of the importance of these elements, you should remember their names and formulas. The ozone molecule, which is composed of three atoms of oxygen, has the formula O_3.

Hydrogen (H_2)

Fluorine (F_2)

Oxygen (O_2)

Nitrogen (N_2)

Chlorine (Cl_2)

Bromine (Br_2)

Iodine (I_2)

Figure 6.5
These models represent seven nonmetallic elements that exist as diatomic molecules. What does the subscript 2 indicate in each case? ❶

Molecular Formulas

The chemical formula of a molecular compound is called a molecular formula. A **molecular formula** shows the kinds and numbers of atoms present in a molecule of a compound. A water molecule is a tightly bound unit of two hydrogen atoms and one oxygen atom. The molecular formula of water is H_2O. Notice that a subscript written after the symbol indicates the number of atoms of each element. If there is only one atom, the subscript 1 is omitted. The molecular formula of carbon dioxide is CO_2. This formula represents a molecule containing one carbon atom and two oxygen atoms. As shown in **Figure 6.6,** ethane, a component of natural gas, is also a molecular compound. The molecular formula for ethane is C_2H_6. According to this formula, one molecule of ethane contains two carbon atoms and six hydrogen atoms.

Although a molecular formula shows the composition of a molecule, it tells you nothing about the molecule's structure. In other words, it does not show the arrangement of the various atoms. As shown in **Figure 6.7,** a variety of diagrams and molecular models can be used to

STUDENT RESOURCES

From the Teacher's Resource Package, use:

- Section Review 6.2, Ch. 6 Practice Problems and Quizzes from the Review Module (Ch. 5–8)
- Small-Scale Chemistry Lab Manual: Experiment 6

TECHNOLOGY RESOURCES

Relevant Technology Resources include:

- Chem ASAP! CD-ROM
- Resource Pro CD-ROM
- Chemistry Alive! Videodisc: *Making Sodium Chloride*

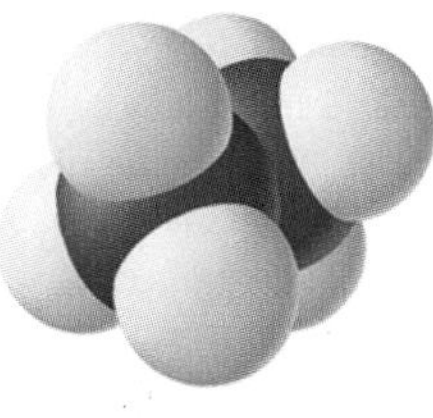

Figure 6.6
Ethane is a component of natural gas. What information is given in its molecular formula, C_2H_6? ❸

show the arrangement of the atoms in a molecule of ammonia gas (NH_3). Diagrams and molecular models such as these will be used throughout the textbook to illustrate the arrangement of atoms in a molecule.

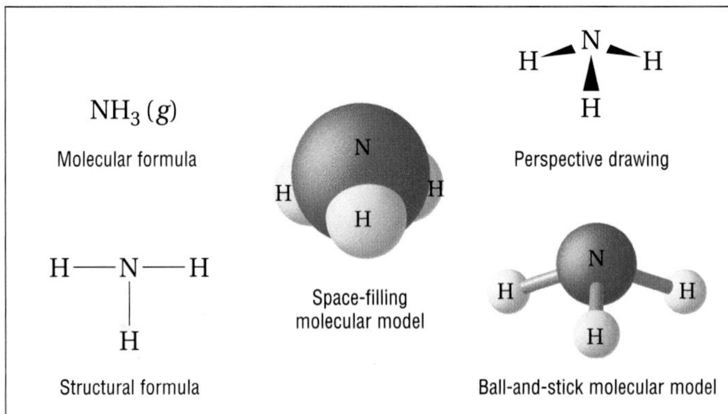

Figure 6.7
Ammonia (NH_3) is used in solution as a cleaning agent. Here are some ways the ammonia molecule can be represented and visualized.

Look at **Figure 6.8** to see the chemical formulas and structures of some other molecular compounds. On the basis of its molecular structure, what do you know about a molecule of ethanol? ❷

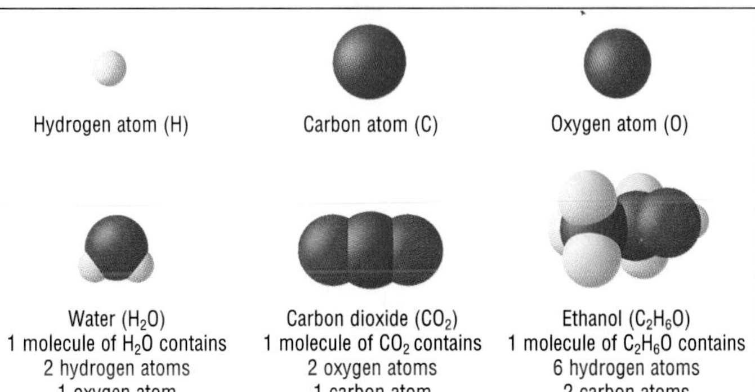

Figure 6.8
The formula of a molecular compound indicates the numbers and kinds of atoms in a molecule of the compound. The arrangement of the atoms within a molecule is called the molecular structure. Which of these molecules has the greatest number of oxygen atoms? ❹

6.2

Motivate and Relate

Point out that the recipe for a cake provides vital information about what type and how much of each ingredient is needed. Varying the types and amounts of ingredients leads to cakes that have very different tastes and textures. Ask students how a chemical formula provides similar information about the "ingredients" of a compound.

2 Teach

TEACHER DEMO

Prepare large models of ammonia similar to those in Figure 6.7. Also, write the molecular and structural formulas for ammonia on large cards. Hold the models and cards up one by one and point out the information that each model or formula provides. Make a summary list on the chalkboard.

ACTIVITY

Have students copy a chemical formula, such as $MgCl_2$, and annotate the formula with word balloons to point out the kind of information they can get from a chemical formula.

Answers

❶ There are two atoms in each molecule.
❷ An ethanol molecule is composed of 2 carbon atoms, 6 hydrogen atoms, and 1 oxygen atom.
❸ It has two carbon and six hydrogen atoms in each of its molecules.
❹ carbon dioxide

CHEMISTRY AND SCIENCE HISTORY

Before modern chemical symbols were adopted, many different types of chemical symbols were used. For example, in Dalton's time, the symbol for nitrogen was a circle bisected by one vertical line. The symbol for sulfur was a circle with a cross inside. Symbols for other elements known at that time were also based on the circle.

Have students do library research to discover the many different chemical symbols that have been used throughout history. Have students prepare a visual display to summarize their findings.

Think Critically

Write the formulas NaCl and $MgCl_2$ on the chalkboard. Ask students to predict the formulas for compounds resulting from the following combinations of elements: Ca and Cl; K and Cl; Na and Br; Mg and I. Have students explain how they arrived at their predictions.

Formula Units

Chemical formulas can also be written for ionic compounds. In this case, however, the formula does not represent a molecule. Recall that sodium chloride (table salt) is an ionic compound. Sodium chloride is composed of equal numbers of sodium cations (Na^+) and chloride anions (Cl^-). As you can see in **Figure 6.9,** the ions in solid NaCl are arranged in an orderly pattern. There are no separate molecular units, only a continuous array of ions. To represent an ionic compound, chemists use a **formula unit,** which is the lowest whole-number ratio of ions in the compound. For sodium chloride, the lowest whole-number ratio of the ions is 1:1 (one Na^+ to each Cl^-). Thus the formula unit for sodium chloride is NaCl. Notice that although ionic charges are used to derive the correct formula, they are not shown when you write the formula unit of the compound.

The ionic compound magnesium chloride contains magnesium cations (Mg^{2+}) and chloride anions (Cl^-). In magnesium chloride, the ratio of magnesium cations to chloride anions is 1:2 (one Mg^{2+} to two Cl^-). So its formula unit is $MgCl_2$. Because there are twice as many chloride anions (each with a 1− charge) as magnesium cations (each with a 2+ charge), the compound is electrically neutral. What is the ratio of aluminum to chloride
1 ions in $AlCl_3$?

Remember, there is no such thing as a molecule of sodium chloride or magnesium chloride. Instead, these compounds exist as collections of positively and negatively charged ions arranged in repeating three-dimensional patterns.

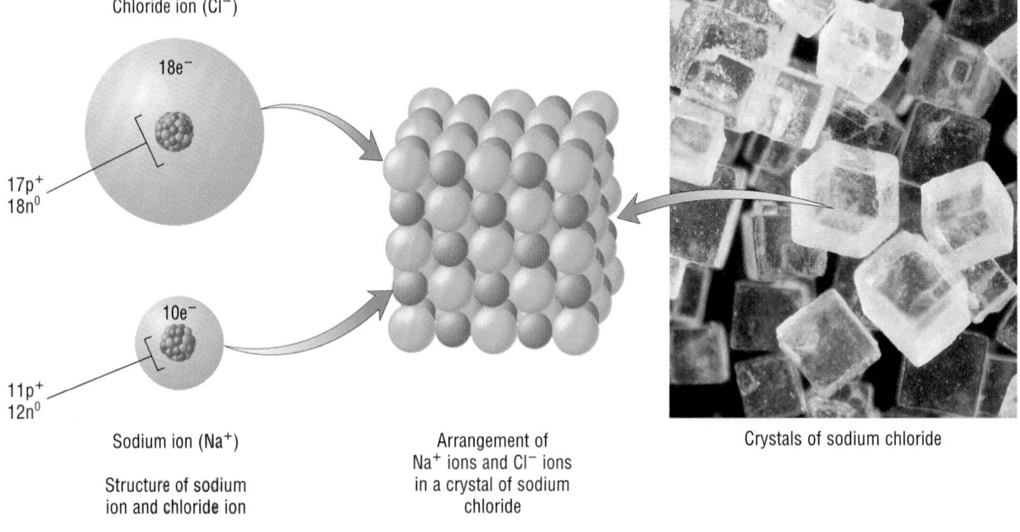

Figure 6.9
Sodium cations and chloride anions form a repeating three-dimensional array in sodium chloride (NaCl). Chemists use a formula unit to represent ionic compounds such as sodium chloride. How many ions make up a single formula unit of this compound? 2

Answers

1 1:3

2 two: one sodium ion and one chloride ion

3 H_2O_2

(a)

(b)

Figure 6.10
*Water and hydrogen peroxide both contain only atoms of hydrogen and oxygen. Nevertheless, they have different chemical and physical properties. Water **(a)** does not bleach dyes, whereas hydrogen peroxide **(b)** bleaches most dyes. What is the formula of hydrogen peroxide?* 3

The Laws of Definite and Multiple Proportions

Consider the compound magnesium sulfide (MgS). It is produced by the combination of two kinds of ions, magnesium cations and sulfide anions. If you could take 100.00 g of magnesium sulfide and break it down into its elements, you would always obtain 43.13 g of magnesium and 56.87 g of sulfur. The ratio of these masses is 43.13/56.87 or 0.7584:1. The way the magnesium sulfide formed or the size of the sample does not change this mass ratio. Like all compounds, MgS obeys the **law of definite proportions,** which states that in samples of any chemical compound, the masses of the elements are always in the same proportions. This law is consistent with Dalton's atomic theory. If atoms combine in simple whole-number ratios, as Dalton postulated, it follows that their proportions by mass must always be the same.

In the early 1800s, Dalton and others studied pairs of compounds that contain the same elements but have different physical and chemical properties. For example, **Figure 6.10** shows two compounds—water (H_2O) and hydrogen peroxide (H_2O_2)—formed by the elements hydrogen and oxygen. The formulas of these compounds were not known in Dalton's time. Each compound obeys the law of definite proportions. In every sample of hydrogen peroxide, 16.0 g of oxygen are present for each 1.0 g of hydrogen. The mass ratio of oxygen to hydrogen is always 16 to 1. In every sample of water, the mass ratio of oxygen to hydrogen is always 8 to 1. If a sample of hydrogen peroxide has the same mass of hydrogen as a sample of water, the ratio of the masses of oxygen in the two compounds is exactly 2:1.

$$\frac{16\text{ g O (in } H_2O_2 \text{ sample that has 1 g H)}}{8\text{ g O (in } H_2O \text{ sample that has 1 g H)}} = \frac{16}{8} = \frac{2}{1} = 2:1$$

Using the results from these kinds of studies, Dalton stated the **law of multiple proportions:** Whenever two elements form more than one compound, the different masses of one element that combine with the same mass of the other element are in the ratio of small whole numbers. **Figure 6.11** illustrates the law of multiple proportions.

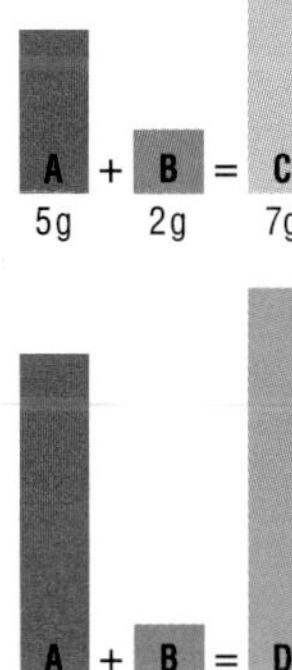

Figure 6.11
This is an example of Dalton's law of multiple proportions. Two compounds, C and D, contain equal masses of B. The ratio of the masses of A in these compounds is 5:10 or 1:2 (a small whole-number ratio).

6.2

Discuss

Ask:

- **Given the formulas CO and CO_2, what do you know about these compounds based on the law of definite proportions?** (The ratio by mass of carbon to oxygen within each compound will be constant.)
- **What do you know about these compounds based on the law of multiple proportions?** (The ratio of oxygen in CO_2 to oxygen in CO is 2:1.)

Meeting Diverse Needs

At Risk Students who have difficulty with mathematics may need extra help with this section. Assign such students to work with students who have strong math skills. Encourage students with strong math skills to develop ways of sharing their problem-solving techniques with other students.

Chemistry Alive!

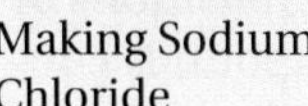

Making Sodium Chloride

Play

Practice Problems Plus

Related Chapter Review Problem
Chapter Review problem 50 is related to Sample Problem 6-2.

Additional Practice Problem
The Chem ASAP! CD-ROM contains the following problem: In the compound iron oxide, also known as rust, the mass ratio of iron to oxygen is 7:3. A 33-g sample of a compound composed of iron and oxygen contains 10 g of oxygen. Is the sample iron oxide? (Yes, because the ratio in the compound is 2.3:1 and the ratio in iron oxide is also 2.3:1.)

3 Assess

Evaluate Understanding

In order to evaluate the students' understanding of this section, ask:

- **What information is provided by the chemical formulas for molecular or ionic compounds?** (the kinds of atoms in the compound and the relative numbers)
- **Why is the molecular formula of hydrogen peroxide H_2O_2 and not HO? Conversely, why is the formula unit of sodium chloride NaCl and not Na_2Cl_2?** (A molecular formula shows the actual number of atoms in the molecule; the formula unit shows the lowest possible whole number ratio.)

Reteach

Make sure that students do not confuse the laws of definite and multiple proportions. Stress that the law of definite proportions describes the composition of one compound, while the law of multiple proportions compares the compositions of two different compounds containing the same elements.

Sample Problem 6-2

Carbon reacts with oxygen to form two compounds. Compound A contains 2.41 g of carbon for each 3.22 g of oxygen. Compound B contains 6.71 g of carbon for each 17.9 g of oxygen. What is the lowest whole-number mass ratio of carbon that combines with a given mass of oxygen?

1. ANALYZE ***Plan a problem-solving strategy.***
The law of multiple proportions applies to two different compounds composed of the same two elements. For each compound, find the grams of carbon that combine with each 1.00 g of oxygen by dividing the mass of carbon by the mass of oxygen. Then find the ratio of the masses of carbon in the two compounds by dividing one into the other. Confirm that the ratio is a low whole-number ratio.

2. SOLVE ***Apply the problem-solving strategy.***

Compound A $\frac{2.41 \text{ g C}}{3.22 \text{ g O}} = \frac{0.748 \text{ g C}}{1.00 \text{ g O}}$

Compound B $\frac{6.71 \text{ g C}}{17.9 \text{ g O}} = \frac{0.375 \text{ g C}}{1.00 \text{ g O}}$

Compare the masses of carbon per gram of oxygen.

$$\frac{0.748 \text{ g C (in compound A)}}{0.375 \text{ g C (in compound B)}} = \frac{1.99}{1} = \text{roughly } \frac{2}{1} = 2{:}1$$

The mass ratio of carbon per gram of oxygen in the compounds is 2:1.

3. EVALUATE ***Does the result make sense?***
The ratio is a low whole-number ratio, as expected. For a given mass of oxygen, there is twice the mass of carbon in compound A as in compound B.

Practice Problems

10. Lead forms two compounds with oxygen. One compound contains 2.98 g of lead combined with 0.461 g of oxygen. The other compound contains 9.89 g of lead combined with 0.763 g of oxygen. What is the lowest whole-number mass ratio of lead in the two compounds that combines with a given mass of oxygen?

11. In the compound nitrous oxide, also known as laughing gas and used as an anesthetic in dentistry, the mass ratio of nitrogen to oxygen is 7:4. A 68-g sample of a compound composed of nitrogen and oxygen contains 42 g of nitrogen. Is the sample nitrous oxide? Explain.

Chem ASAP!

Problem-Solving 11
Solve Problem 11 with the help of an interactive guided tutorial.

section review 6.2

12. Differentiate among a chemical formula, a molecular formula, and a formula unit.

13. How are experimental data used to show that a compound obeys the law of definite proportions?

14. Which law is illustrated by this statement: "In every sample of carbon monoxide, the mass ratio of carbon to oxygen is 3:4"?

15. Which law is illustrated by this statement: "When carbon and oxygen form the compounds carbon monoxide and carbon dioxide, the different masses of carbon that combine with the same mass of oxygen are in the ratio of 2:1"?

Chem ASAP! **Assessment 6.2** Check your understanding of the important ideas and concepts in Section 6.2.

Answers

SECTION REVIEW 6.2

12. A chemical formula shows the kinds and numbers of atoms in the smallest representative unit of the substance. A molecular formula shows the number of each kind of atom in a molecule of the compound. The formula unit shows the lowest whole-number ratio of ions in a compound.

13. In every sample of the compound, the elements are shown to combine in the same ratio by mass.

14. law of definite proportions

15. law of multiple proportions

IONIC CHARGES

section 6.3

In the play Romeo and Juliet, William Shakespeare wrote, "What's in a name? That which we call a rose/By any other name would smell as sweet." What we call a rose in English is rosa in Spanish, warda in Arabic, and julab in Hindi. To truly understand someone from another culture, you must first learn their language. Similarly, to understand chemistry, you must learn its language. Learning the language of chemistry involves learning the names of ionic compounds. **What are the rules for naming monatomic and polyatomic ions?**

objectives

- Use the periodic table to determine the charge on an ion
- Define a polyatomic ion and give the names and formulas of the most common polyatomic ions

key terms

- monatomic ions
- polyatomic ions

Monatomic Ions

To write chemical formulas for ionic compounds, you must know the types of ions that atoms tend to form. That means you must know the ionic charges typically obtained by the elements. For **monatomic ions,** or ions consisting of only one atom, the ionic charges can often be determined by using the periodic table.

Recall that metallic elements tend to lose electrons. Lithium, sodium, and potassium (in Group 1A) form cations. These ions have a 1+ charge (Li^+, Na^+, and K^+), as do all the other Group 1A ions. Magnesium and calcium are Group 2A metals. They form cations with a 2+ charge (Mg^{2+} and Ca^{2+}), as do all the other Group 2A metals. Aluminum is the only common Group 3A metal. As you might expect, it forms a 3+ cation (Al^{3+}). Thus the metals in Groups 1A, 2A, and 3A lose electrons when they form cations. The ionic charge is positive and is numerically equal to the group number.

The numerical charge of an ion of a Group A nonmetal is determined by subtracting the group number from 8. Because nonmetals tend to gain electrons and form anions, the sign of the charge is negative. For example, the elements in Group 7A form anions with a 1− charge ($8 - 7 = 1$). The anions for this group include fluoride (F^-) and chloride (Cl^-). Anions of nonmetals in Group 6A have a 2− charge ($8 - 6 = 2$); the oxide anion (O^{2-}) is an example. The three nonmetals in Group 5A can form anions with a 3− charge ($8 - 5 = 3$); one example is the nitride ion (N^{3-}). **Table 6.2** summarizes the ionic charges of representative elements that can be obtained from the periodic table. What is the charge of an iodide ion? 1

Table 6.2

Ionic Charges of Representative Elements							
1A	**2A**	**3A**	**4A**	**5A**	**6A**	**7A**	**0**
Li^+	Be^{2+}			N^{3-}	O^{2-}	F^-	
Na^+	Mg^{2+}	Al^{3+}		P^{3-}	S^{2-}	Cl^-	
K^+	Ca^{2+}			As^{3-}	Se^{2-}	Br^-	
Rb^+	Sr^{2+}					I^-	
Cs^+	Ba^{2+}						

1 Engage

Use the Visual

Have students read the text that opens the section. Explain that some elements can form ions with more than one charge; chemists need to have a naming system that distinguishes between two or more common ions of the same element. Ask:

- **What are the rules for naming monatomic and polyatomic ions?** (Ions are named according to their elemental composition and charge.)

Check Prior Knowledge

To assess students' prior knowledge about ions, ask:

- **What kind of charge do ions formed from metals have?** (positive) **Nonmetals?** (negative)
- **What is the difference between Cu and Cu^{2+}?** (Cu is the neutral atom of copper; Cu^{2+} is the cation of Cu, which has lost 2 electrons.)
- **What is the difference between Cu^{2+} and Cu^+?** (Cu^{2+} indicates the loss of 2 electrons; Cu^+ indicates the loss of 1 electron.)
- **What does the ionic formula for hypochlorite, ClO^-, indicate?** (that the atoms in a hypochlorite ion act as a unit with an overall charge of 1−)

STUDENT RESOURCES

From the Teacher's Resource Package, use:

- Section Review 6.3, Ch. 6 Practice Problems, Interpreting Graphics, and Quizzes from the Review Module (Ch. 5–8)
- Laboratory Manual: Experiments 6 and 7
- Laboratory Practicals 6-2 and 6-3

TECHNOLOGY RESOURCES

Relevant Technology Resources include:

- Chem ASAP! CD-ROM
- Resource Pro CD-ROM

 1−

2 Teach

TEACHER DEMO

In large beakers, prepare solutions containing cations from soluble metal salts (a good series is: $MnCl_2$, $FeCl_2$, $CoCl_2$, $NiCl_2$, $CuCl_2$, and $ZnCl_2$). Write the symbols for the metal ions on the chalkboard and have students describe the colors of the solutions. Have students use Table 6.3 to write both the Stock and classical names of the cations. Ask students to suggest why chemists currently use only the Stock system for naming ions.

ACTIVITY

Display a copy of the periodic table on a wall chart or overhead projector. Point to elements on the table and see how rapidly students can identify the charge for the ion that each element forms.

Figure 6.12
Many compounds containing the Fe^{2+} ion are blue-green, whereas those compounds containing the Fe^{3+} ion tend to be yellow-brown. Note that the ions can be named by means of the Stock system as iron(II) and iron(III) or by means of the classical system as ferrous (the lower charge) and ferric (the higher charge).

The majority of the elements in the two remaining representative groups, 4A and 0, usually do not form ions. The elements in Group 0 rarely form compounds. Ordinarily, the two nonmetals in Group 4A, carbon and silicon, are found in molecular compounds.

Unlike the cations of the Group 1A, 2A, and 3A metals, many of the cations of transition metals have more than one common ionic charge. (This is also a characteristic of cations of tin and lead, the two metals in Group 4A.) For example, the transition metal iron forms two common cations, Fe^{2+} and Fe^{3+}. There are two methods of naming such cations. The preferred method is called the Stock system. As part of the name of the element, a Roman numeral in parentheses indicates the numerical value of the charge. As you will discover if you look at **Table 6.3,** when using the Stock system, the cation Fe^{2+} is named the iron(II) ion. This is read as the "iron two ion." Note that there is no space between the element name and the first parenthesis.

An older, less preferred method of naming these cations uses a root word with different suffixes at the end of the word. The older, or classical, name of the element is used to form the root word—for example *ferr-*, from *ferrum,* which is Latin for iron. The suffix *-ous* is used to name the cation with the lower of the two ionic charges. The suffix *-ic* is used with the higher of the two ionic charges. Using this system, Fe^{2+} is the ferr*ous* cation, and Fe^{3+} is the ferr*ic* cation, as shown in **Table 6.3.** Notice that you can usually identify an element from what may be an unfamiliar classical name by looking for the symbol of the element in the name. Thus *fe*rrous (Fe) is iron; *cu*prous (Cu) is copper; and *stan*nous (Sn) is tin. A major disadvantage of using the classical name for ions is that they do not tell you the actual charge of the ion. The name tells you only that the cation has either the smaller (*-ous*) or the larger (*-ic*) charge of the pair.

Table 6.3

Formulas and Names of Common Metal Ions with More than One Ionic Charge

Formula	Stock name	Classical name
Cu^{+}	Copper(I) ion	Cuprous ion
Cu^{2+}	Copper(II) ion	Cupric ion
Fe^{2+}	Iron(II) ion	Ferrous ion
Fe^{3+}	Iron(III) ion	Ferric ion
$^{*}Hg_2^{2+}$	Mercury(I) ion	Mercurous ion
Hg^{2+}	Mercury(II) ion	Mercuric ion
Pb^{2+}	Lead(II) ion	Plumbous ion
Pb^{4+}	Lead(IV) ion	Plumbic ion
Sn^{2+}	Tin(II) ion	Stannous ion
Sn^{4+}	Tin(IV) ion	Stannic ion
Cr^{2+}	Chromium(II) ion	Chromous ion
Cr^{3+}	Chromium(III) ion	Chromic ion
Mn^{2+}	Manganese(II) ion	Manganous ion
Mn^{3+}	Manganese(III) ion	Manganic ion
Co^{2+}	Cobalt(II) ion	Cobaltous ion
Co^{3+}	Cobalt(III) ion	Cobaltic ion

*A diatomic elemental ion.

MEETING DIVERSE NEEDS

At Risk For some students, the charges that result when electrons are transferred is counterintuitive—additions result in a (−) charge, subtractions in a (+) charge. Give students different colored disks to represent protons and electrons in neutral atoms. Remove or add electrons and ask students to find the charge on the resulting "ion."

A few transition metals have only one ionic charge. The names of these cations do not have a Roman numeral. These exceptions include silver, with cations that nearly always have a 1+ charge (Ag^+), and cadmium and zinc, with cations that nearly always have a 2+ charge (Cd^{2+} and Zn^{2+}). As **Figure 6.13** shows, many transition metal compounds are used as pigments.

Figure 6.13
Many transition metals form brightly colored compounds that are used in making artists' paints.

Sample Problem 6-3

What is the charge of the ion typically formed by each element? (For transition metals with more than one common ionic charge, the number of electrons lost is indicated.)

a. sulfur
b. lead, 4 electrons lost
c. strontium
d. argon
e. bromine
f. copper, 1 electron lost

1. ***ANALYZE*** ***Plan a problem-solving strategy.***
a.–f. The charge of the ion of a representative (Group A) element is determined by that element's location in the periodic table. Group A metals have a positive charge equal to the group number. The charge of a Group A nonmetal is negative and numerically equal to 8 minus the group number. Ionic charges of transition metals cannot usually be read from the periodic table but can be determined from the number of electrons lost in forming the ion.

2. ***SOLVE*** ***Apply the problem-solving strategy.***
a. 2− (nonmetal in Group 6A; $8 - 6 = 2$)
b. 4+ (4 electrons lost)
c. 2+ (metal in Group 2A)
d. no ion formed (in Group 0)
e. 1− (nonmetal in Group 7A; $8 - 7 = 1$)
f. 1+ (1 electron lost)

3. ***EVALUATE*** ***Do the results make sense?***
As expected, the metals formed positively charged ions (cations) by losing one or more electrons. The nonmetals formed negatively charged ions (anions) by gaining one or more electrons.

Practice Problems

16. What is the charge of the typical ion of each element?
a. selenium **c.** cesium
b. barium **d.** phosphorus

17. How many electrons does the neutral atom gain or lose when each ion forms?
a. Fe^{3+} **c.** Cu^+
b. O^{2-} **d.** Cd^{2+}

Chem ASAP!
Problem-Solving 16
Solve Problem 16 with the help of an interactive guided tutorial.

ACTIVITY

Give students a photocopy of a blank periodic table. Have them work in pairs, referring to their textbooks or other resource materials. Have students write the symbols for the cations and anions of as many elements as possible in the boxes. Have them write the name of each ion in the boxes as well. Have them use their tables to identify any trends in ionic charges.

Cooperative Learning

Divide students into research teams to gather data about paint pigments from different sources. Possible assignments: (1) Examine the pigments available at an art-supply store and try to tell from the names what they contain. (2) Interview an art teacher about the pigments that he or she prefers. (3) Use the library or Internet to research which pigments were available and popular during earlier eras; look for information about natural sources of pigments. (4) Invite a curator or art restorer from a local museum to talk about using pigments to repair damaged art, authenticate art, or unmask a forgery.

Practice Problems Plus

Related Chapter Review Problem
Chapter Review problem 53 is related to Sample Problem 6-3.

Additional Practice Problem
The Chem ASAP! CD-ROM contains the following problem: Write the formula (including charge) for each ion.
a. arsenic
b. beryllium
c. astatine
d. gallium
(**a.** As^{3-}, **b.** Be^{2+}, **c.** At^-, **d.** Ga^{3+})

Discuss

Write the following formulas on the chalkboard: NaOH, H_2SO_4, NH_4OH, Na_2CO_3. Ask students what these compounds have in common. After students have offered their ideas, explain that all of these compounds contain ions that are made up of more than one type of atom.

Practice Problems Plus

Related Chapter Review Problem
Chapter Review Problem 54 is related to Sample Problem 6-4.

Sample Problem 6-4

Name the ions in Sample Problem 6-3. Classify each as a cation or an anion.

1. ***ANALYZE Plan a problem-solving strategy.***
   a.–f. Apply the different rules for naming cations and anions. The names of nonmetallic anions end in *-ide*. Metallic cations take the name of the metal. If the metal has more than one common ionic charge, a Roman numeral or classical name with a suffix is used.

2. ***SOLVE Apply the problem-solving strategy.***
   a. sulfide ion, anion
   b. lead(IV) or plumbic ion, cation
   c. strontium ion, cation
   d. a noble gas, no ion formed
   e. bromide ion, anion
   f. copper(I) or cuprous ion, cation

3. ***EVALUATE Do the results make sense?***
   The rules have been applied correctly. Each name uniquely identifies an ion and provides sufficient information for the formula of the ion to be written.

Practice Problems

18. Name each ion in Practice Problem 16. Identify each as an anion or a cation.
19. Name each ion in Practice Problem 17.

Polyatomic Ions

All the ions mentioned to this point are monatomic ions; that is, they are composed of only one atom. Unlike monatomic ions, the sulfate anion is composed of one sulfur atom and four oxygen atoms. These five atoms together form a sulfate anion. It has a 2− charge and is written SO_4^{2-}. **Polyatomic ions,** such as sulfate, are tightly bound groups of atoms that behave as a unit and carry a charge. You can see the structures of some polyatomic ions in **Figure 6.14.**

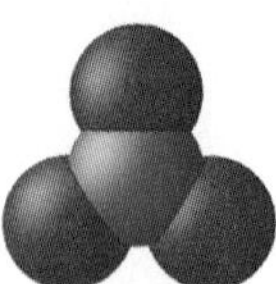
Nitrate ion (NO_3^-)

Phosphate ion (PO_4^{3-})

Sulfate ion (SO_4^{2-})

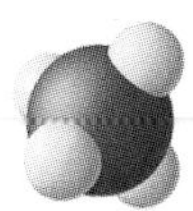
Ammonium ion (NH_4^+)

Figure 6.14
These models show the arrangement of atoms in four common polyatomic ions.

Table 6.4 lists the names and formulas of some common polyatomic ions. The ions are grouped by charge. Observe that the names of most of the polyatomic anions end in *-ite* or *-ate*. However, there are three important exceptions: the positively charged ammon*ium* cation (NH_4^+) and two polyatomic anions that end in *-ide,* the cyan*ide* ion (CN^-) and the hydrox*ide* ion (OH^-).

Examine the relationships among the polyatomic ions below for which there is an *-ite*/*-ate* pair. Look at the charge on each ion in the pair, the number of oxygen atoms, and the endings on each name. You should be able to discern a pattern in the naming convention.

-ite	*-ate*
SO_3^{2-}, sulfite	SO_4^{2-}, sulfate
NO_2^-, nitrite	NO_3^-, nitrate
ClO_2^-, chlorite	ClO_3^-, chlorate

The charge on each polyatomic ion in a given pair is the same. The *-ite* ending indicates one less oxygen atom than the *-ate* ending. However, the ending does not tell you the actual number of oxygen atoms in the ion. For example, the nitr*ite* ion has two oxygen atoms and the sulf*ite* ion has three oxygen atoms.

When the formula for a polyatomic ion begins with H (hydrogen), you can imagine the H to represent a hydrogen ion (H^+) combined with another polyatomic ion. For example, HCO_3^- is a combination of H^+ and CO_3^{2-}.

Table 6.4

Common Polyatomic Ions

1− charge		2− charge		3− charge	
Formula	**Name**	**Formula**	**Name**	**Formula**	**Name**
$H_2PO_4^-$	Dihydrogen phosphate	HPO_4^{2-}	Hydrogen phosphate	PO_3^{3-}	Phosphite
$C_2H_3O_2^-$	Acetate	$C_2O_4^{2-}$	Oxalate	PO_4^{3-}	Phosphate
HSO_3^-	Hydrogen sulfite	SO_3^{2-}	Sulfite		
HSO_4^-	Hydrogen sulfate	SO_4^{2-}	Sulfate		
HCO_3^-	Hydrogen carbonate	CO_3^{2-}	Carbonate	**1+ charge**	
NO_2^-	Nitrite	CrO_4^{2-}	Chromate	**Formula**	**Name**
NO_3^-	Nitrate	$Cr_2O_7^{2-}$	Dichromate	NH_4^+	Ammonium
CN^-	Cyanide	SiO_3^{2-}	Silicate		
OH^-	Hydroxide				
MnO_4^-	Permanganate				
ClO^-	Hypochlorite				
ClO_2^-	Chlorite				
ClO_3^-	Chlorate				
ClO_4^-	Perchlorate				

Cooperative Learning

Have students prepare flash cards with the name of a polyatomic ion on one side and its formula on the other side. Pairs of students can use the flash cards to quiz each other on the names and formulas of polyatomic ions. You can set aside a few minutes at the beginning or end of several class periods for this activity.

Think Critically

Write on the chalkboard the names and formulas of at least ten different polyatomic ions. Purposely mismatch the names so that they do not appear with the correct formulas. Challenge students to see how quickly they can rearrange the names and formulas so that they are paired correctly.

ACTIVITY

Have students work in pairs to research the three-dimensional structure of one of the polyatomic ions in Table 6.4. Have the students prepare a ball-and-stick model of the ion using styrofoam balls and wooden rods. The students may wish to differentiate atoms by size and/or color. Make sure that they pay attention to correct geometric placement of atoms within the ion. Attach strings to the models and hang them from the ceiling.

3 Assess

Evaluate Understanding

Write the chemical symbols of the following atoms or groups of atoms on the chalkboard: K, NH_3, CN, Mg, O, OH, Co, NH_4, Pb, PO_4, Ar, ClO, H_2O, and Zn. Ask students to identify which species can normally exist as ions by simply attaching a superscript. (They are: K, CN, Mg, O, OH, Co, NH_4, Pb, PO_4, ClO, and Zn.) **For these ions, ask the students to identify the charges and to name the ion formed** (note: some of the species can have multiple charges). **For the remaining species, ask why they cannot be transformed into ions.** (NH_3 and H_2O are molecules; Ar is an atom that does not gain or lose electrons easily.)

Reteach

Students may think that polyatomic ions can exist on their own, like compounds. Remind students that one of the most important properties of a compound is that it is electrically neutral. Because polyatomic ions carry a charge, they are not found free in nature, but are found combined with other ions in compounds.

Portfolio Project

Sodium is the major cation in extra-cellular fluids, such as plasma. Potassium is the major cation within cells. During the transmission of nerve impulses, potassium ions flow out of the cell and sodium ions flow in. Compounds attached to the cell membrane transport the ions across the membrane to restore the initial concentrations.

(a)

(b)

(c)

Figure 6.15
(a) This cloth shows the effect of sodium hypochlorite (NaClO) when it is used as a laundry bleach. (b) Ammonium phosphate ($(NH_4)_3PO_4$) is used as a fertilizer. (c) Some antacids contain the hydrogen carbonate anion.

Note that the charge on the new ion is the algebraic sum of the ionic charges.

$H^+ + CO_3^{2-}$ (carbonate)	$\longrightarrow$	HCO_3^- (hydrogen carbonate)
$H^+ + PO_4^{3-}$ (phosphate)	$\longrightarrow$	HPO_4^{2-} (hydrogen phosphate)
$H^+ + HPO_4^{2-}$ (hydrogen phosphate)	$\longrightarrow$	$H_2PO_4^-$ (dihydrogen phosphate)

The hydrogen carbonate anion (HCO_3^-), the hydrogen phosphate anion (HPO_4^{2-}), and the dihydrogen phosphate ($H_2PO_4^-$) anion are all essential components of living systems. In contrast, the cyanide ion (CN^-) is extremely poisonous to living systems because it blocks a cell's means of producing energy. Most laundry and household bleaches contain the hypochlorite anion (ClO^-). According to **Table 6.4,** what is the formula for
1 the hydrogen sulfite polyatomic ion?

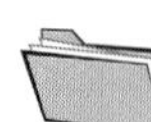

portfolio project

Sodium ions (Na^+) and potassium ions (K^+) are needed for the human body to function. Research where these ions are most likely to be found in the body and the roles they play.

section review 6.3

20. How can the periodic table be used to determine the charge of an ion? Use a specific example to explain.
21. Explain what is meant by a polyatomic ion.
22. Using only the periodic table, name and write the formula for the typical ion of each representative element.
    - a. potassium
    - b. sulfur
    - c. argon
    - d. bromine
    - e. beryllium
    - f. sodium
23. Write the formula (including charge) for each ion.
    - a. ammonium ion
    - b. tin(II) ion
    - c. chromate
    - d. nitrate ion
    - e. cyanide ion
    - f. iron(III) ion
    - g. permanganate ion
    - h. manganese(II) ion

Chem ASAP! **Assessment 6.3** Check your understanding of the important ideas and concepts in Section 6.3.

Answers

1 HSO_3^-

SECTION REVIEW 6.3

20. by noting the group in which the element is found
21. Polyatomic ions contain two or more tightly bound atoms that behave as a unit.
22. a. potassium, K^+ b. sulfide, S^{2-} c. no ion d. bromide, Br^- e. beryllium, Be^{2+} f. sodium, Na^+
23. a. NH_4^+ b. Sn^{2+} c. CrO_4^{2-} d. NO_3^- e. CN^- f. Fe^{3+} g. MnO_4^- h. Mn^{2+}

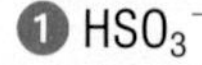

IONIC COMPOUNDS

Dr Pepper®, the oldest of the major brand soft drinks in the United States, was first made and sold in Waco, Texas, in 1885. How does Dr Pepper continue to maintain the unique flavor associated with it? A specific formula, listing exact amounts of each ingredient, must be followed. Understandably, the formula for Dr Pepper is a closely guarded secret. In chemistry, however, there are no secret ingredients! Once you know the rules, you can write the formula for any chemical compound. **How do you write the formulas for ionic compounds?**

objectives

- Apply the rules for naming and writing formulas for binary ionic compounds
- Apply the rules for naming and writing formulas for ternary ionic compounds

key terms

- binary compounds
- ternary compound

Writing Formulas for Binary Ionic Compounds

In the days before the science of chemistry developed, the person who discovered a new compound often got to name it anything he or she wished. It was not uncommon for the name to describe some property of the substance or its source. For example, a common name for potassium carbonate (K_2CO_3) is potash. The name evolved because the compound was obtained by boiling wood ashes in iron pots. Laughing gas is the common name for the gaseous compound dinitrogen monoxide (N_2O) because people may laugh when they inhale it. Baking soda is added to batter to make cakes rise as they bake. Plaster of paris, gypsum, and lye are other common names of chemical compounds. Unfortunately, however, such names do not tell you anything about the chemical composition of the compound.

As the science of chemistry developed, it became obvious that some systematic method of naming chemical compounds was needed for consistency and clarity. A chemical compound's name should indicate its composition, behavior, and how it is related to other compounds. The rest of this chapter is devoted to examining the system of naming inorganic compounds, which are generally compounds that do not contain the element carbon.

The ionic compound potassium chloride (KCl) is composed of potassium cations (K^+) and chloride anions (Cl^-). Compounds composed of two elements are called **binary compounds.** Potassium chloride is a binary compound. In writing the formula for potassium chloride (or any other ionic compound), the positive charge of the cation must balance the negative charge of the anion. In other words, the net ionic charge of the formula must be zero. In potassium chloride, the charge of each K^+ cation is balanced by the charge of each Cl^- anion. The potassium and chloride ions combine in a 1:1 ratio. For each K^+ there is one Cl^-. Thus the formula for potassium chloride is KCl. The net ionic charge of the formula unit is zero. Notice that in this formula, and in the formulas for essentially all ionic compounds, the cation is written first: K before Cl.

The binary ionic compound calcium bromide is composed of calcium cations (Ca^{2+}) and bromide anions (Br^-). Each calcium ion with its 2+ charge must combine with (or be balanced by) two bromide ions, each with a 1− charge. The ions must combine in a 1:2 ratio. For each Ca^{2+} there are two Br^-. The formula for calcium bromide is $CaBr_2$, with a net ionic charge of zero.

Chem ASAP!

Simulation 2

Combine ions and derive the chemical formulas for several ionic compounds.

6.4

1 Engage

Use the Visual

Have students study the photograph and read the text that opens the section. Ask:

- **How do you write the formulas for ionic compounds?** (Use the compound name to identify the cation and anion; then determine the ionic charges. Finally, balance the formula so that the total positive charge equals the total negative charge.)

Check Prior Knowledge

To assess students' prior knowledge of formula units and the naming of ionic compounds, ask

- **Why isn't the formula unit for sodium chloride $NaCl_2$ or Na_2Cl_2?** ($NaCl_2$ would have a net negative charge; Na_2Cl_2 is not the lowest possible whole-number ratio of cations to anions.)
- **If a metallic element can form multiple cations, what is the preferred naming system?** (element name with Roman numeral in parentheses to indicate the charge)
- **Why isn't K^+ referred to as potassium(I)?** (Potassium has only one type of cation.)
- **Name the ionic compound with the formula unit $NaNO_3$. From what ions is it formed?** (sodium nitrate; Na^+ and NO_3^-)

STUDENT RESOURCES

From the Teacher's Resource Package, use:

- Section Review 6.4, Ch. 6 Practice Problems and Quizzes from the Review Module (Ch. 5–8)
- Laboratory Recordsheet 6-1

TECHNOLOGY RESOURCES

Relevant Technology Resources include:

- Chem ASAP! CD-ROM
- Resource Pro CD-ROM

Motivate and Relate

Ask students: Does someone you know have a nickname? Does the nickname tell you something about the person? Guide students to recognize that nicknames are somewhat like the common names used to describe compounds.
In terms of compounds, use the example of potassium carbonate, K_2CO_3. The common name, potash, evolved because the compound was separated by boiling wood *ashes* in iron *pots*. Note that the systematic name, potassium carbonate, provides information about the compound that "potash" does not.

2 Teach

TEACHER DEMO

Do the following demonstration in a fume hood. Wearing safety goggles, mix 1 g of powdered zinc (Zn) and 4 g of iodine (I_2) in a watch glass. With an eyedropper, carefully add 8 mL of water, one drop at a time. After the reaction is complete, show students the zinc iodide that was prepared. Write the formula unit for zinc iodide, ZnI_2, along with its name on the chalkboard. Ask:

- What happened to elemental Zn and I_2 during the reaction? (Zinc and iodine reacted to form an ionic compound composed of Zn^{2+} and I^- ions.)
- Use the criss-cross method to show how the formula unit for zinc iodide was derived.
- Why is ZnI_2 named zinc iodide and not zinc(II) iodide?

Place the zinc iodide in a plastic or cardboard container and bury it in an approved landfill site.

Figure 6.16
The rusting of a tractor produces iron(III) oxide, or rust, which is a binary ionic compound.

Figure 6.16 shows an example of the formation of rust, or iron(III) oxide. What is the formula for rust? Recall that a Roman numeral shows the charge of the metal ion. Thus this compound consists of Fe^{3+} cations combined with oxide anions (O^{2-}). Writing a balanced formula for this compound involves finding the least common multiple of the charges. Because $3 \times 2 = 6$, two Fe^{3+} cations (a 6+ charge) will balance three O^{2-} anions (a 6− charge). The balanced formula, then, is Fe_2O_3.

Another approach to writing a balanced formula for a compound is to use the crisscross method. In this method, the numerical charge of each ion is crossed over and used as a subscript for the other ion. The signs of the numbers are dropped.

$$Fe^{3+} \quad O^{2-}$$
$$Fe_2O_3$$
$$2(3+) + 3(2-) = 0$$

If you use the crisscross method to write the formula for a compound such as calcium sulfide (which contains Ca^{2+} and S^{2-}), however, you would obtain the result Ca_2S_2. Remember that formulas for ionic compounds show the lowest whole-number ratio of ions. The formula for calcium sulfide, written correctly, is thus reduced to CaS.

$$Ca^{2+} \quad S^{2-}$$
$$Ca_2S_2 \text{ reduces to } CaS$$
$$1(2+) + 1(2-) = 0$$

Of course, if the magnitudes of the charges of the cation and anion are the same, as they are in this case, the charges will balance. There is no reason even to use the crisscross method. The ions will simply combine in a 1:1 ratio.

Meeting Diverse Needs

Gifted Have students research the possible mechanisms for the ability of lithium ions to alleviate symptoms of bi-polar disorder. Challenge students to explain why other Group IA ions such as sodium and potassium, which are present in high amounts in the body naturally, do not have the same effects as lithium ions even though they have the same charge.

Sample Problem 6-5

Write formulas for these binary ionic compounds.

a. copper(II) sulfide

b. potassium nitride

1. *ANALYZE* ***Plan a problem-solving strategy.***

a.–b. Binary ionic compounds are composed of a monatomic cation and a monatomic anion. The ionic charges in an ionic compound must balance (add up to zero), and there must be a lowest whole-number ratio of ions. The formula for the cation appears first in the formula for the compound.

2. *SOLVE* ***Apply the problem-solving strategy.***

Write the formula (symbol and charge) for each ion in each compound.

a. Cu^{2+} and S^{2-} b. K^{+} and N^{3-}

Balance the formula using appropriate subscripts.

CuS

K_3N

3. *EVALUATE* ***Do the results make sense?***

The ions are in the lowest whole-number ratio and the net ionic charge is zero: 1(2+) + 1(2−) = 0 and 3(1+) + 1(3−) = 0.

Practice Problems

24. Write formulas for compounds formed from these pairs of ions.

a. Ba^{2+}, S^{2-} c. Ca^{2+}, N^{3-}

b. Li^{+}, O^{2-} d. Cu^{2+}, I^{-}

25. Write formulas for these compounds.

a. sodium iodide

b. stannous chloride

c. potassium sulfide

d. calcium iodide

Chem ASAP!

Problem-Solving 25

Solve Problem 25 with the help of an interactive guided tutorial.

Naming Binary Ionic Compounds

So far, you have written the formulas for ionic compounds when given their names. What if you wanted to do the reverse? Suppose, for example, you wanted to write the name of the binary ionic compound that has the formula of CuO. Your first guess might be to write the name of the metallic cation (copper) followed by the name of the nonmetallic anion (oxide). However, the name copper oxide is incomplete. Recall that copper commonly forms two cations: Cu^{+} and Cu^{2+}, called copper(I) and copper(II), respectively. How can you tell which of these cations forms the compound CuO? Working backward will help. The formula indicates that the copper cation and the oxide anion combine in a 1:1 ratio. You know that the oxide anion always has a 2− charge. The charge of the copper cation must therefore be 2+ (to balance the 2− charge). The compound CuO must be copper(II) oxide. Recall that iron is another element that forms two cations: Fe^{2+} and Fe^{3+}.

It is helpful to know the common ionic charges of transition metals: for example, tin forms cations with 2+ and 4+ charges. However, you must still determine which of the cations is in a particular compound. Thus the compound SnO_2 must be tin(IV) oxide, because only the tin(IV) cation (Sn^{4+}) can balance the two oxide anions, each of which has a 2− charge. **Figure 16.17,** on page 153, shows examples of compounds consisting of these ions. What is the name of the compound that has the formula SnO? 

Answer

1 tin(II) oxide, or stannous oxide

6.4

Build Writing Skills

Ask students to imagine that they are writing instructions for another student to use. Ask them to write a short explanation of how to write formulas for binary ionic compounds, without using their textbook.

Practice Problems Plus

Related Chapter Review Problem

Chapter Review problem 58 is related to Sample Problem 6-5.

Additional Practice Problem

The Chem ASAP! CD-ROM contains the following problem: Write formulas for these compounds.

a. lithium fluoride

b. aluminum chloride

c. sodium nitride

d. ferric oxide

(a. LiF, b. $AlCl_3$, c. Na_3N, d. Fe_2O_3)

CHEMath

Teaching Tips

You may wish to begin this feature by writing several everyday terms on the chalkboard. Help students to analyze the words in terms of prefixes and suffixes. For example, speedo*meter*, *anti*histamine, *pre*natal, and bio*graphy*. Then add several simple science terms to the list. For example, *kilo*meter, *pre*hensile, *anti*oxidant, and photo*graphy*. This may help students to look at related words in terms of their prefixes and sufffixes.

If you have time, expand the content of the feature by discussing the names of acids. Tell students that the name of an acid usually comes from the name of its anion. If the anion ends in the suffix *ide* (such as chloride), the name of the acid begins with the prefix *hydro*. In addition, the suffix of the anion changes from *ide* to *ic*. So the acid with the formula HCl would be hydrochloric acid. An acid that does not contain oxygen is named by adding the prefix *hydro-* and the suffix *-ic* to the root of the anion. For example, HBr is named *hydro*brom*ic* acid.

Acids that contain oxygen are named a little differently. The suffixes *-ic* and *-ous* are used in naming acids that contain oxygen. An acid derived from an *-ite* ion is given the suffix *-ous*. An acid derived from an *-ate* ion is given the suffix *-ic*. For example, the acid derived from nitr*ite* is named nitr*ous* acid and the acid derived from nitr*ate* is named nitr*ic* acid.

Answers to Practice Problems

A. milli
B. four chlorine atoms
C. dinitrogen monoxide
D. nitrogen dioxide
E. chlorite is ClO_2^- and chlorate is ClO_3^-.

CHEMath

PREFIXES AND SUFFIXES

Throughout this textbook you will encounter many new words. Some of them may seem intimidating at first, but don't be discouraged. You can get an idea of what a word means by analyzing its prefix or suffix. Prefixes and suffixes are one or more letters added to the beginning or end of a root word to modify the root word's meaning. Prefixes are added to the beginning of a root word and suffixes are added to the end. You can use what you learn about prefixes and suffixes to interpret and write the names of chemical compounds.

Measurements You already use prefixes to modify the magnitude of SI units. By adding prefixes to such root words as meter, liter, and gram, you can increase or decrease the magnitude of the unit. For example, if you write the prefix *kilo-* before the root word meter, you get kilometer. There are one thousand meters in a kilometer. If instead you write the prefix *centi-* before the root word meter, you get the word centimeter. There are one hundred centimeters in a meter. **Table 3.2** lists common prefixes used with SI units.

Chemical Compounds Prefixes and suffixes are also found in the names of chemical compounds. All binary compounds, for example, end in the suffix *-ide*. So whenever you see the name of a compound with the suffix *-ide*, such as carbon monoxide, you know that the compound is generally a binary compound. What is a binary compound? The prefix *bi-* means two, so a binary compound is one that contains only two elements.

In one type of binary compound, known as a molecular compound, prefixes are used to indicate the number of atoms in the chemical formula. In the compound carbon monoxide (CO), the prefix *mono-* indicates that there is one oxygen atom. In the compound carbon dioxide (CO_2), the prefix *di-* indicates that there are two oxygen atoms. **Table 6.5** lists the prefixes used to count up to ten atoms.

As you read this chapter, you will also encounter compounds whose names have the suffixes *-ite* or *-ate*. Compounds such as these contain polyatomic ions. The prefix *poly-* means many. Most polyatomic ions consist of three or more atoms.

The suffixes *-ite* or *-ate* relate to the number of oxygen atoms in the ion. An ion whose name ends in *-ite* has one less oxygen atom than the ion that ends in *-ate*. For example, the nitrite ion, NO_2^-, has two oxygen atoms and the nitrate ion, NO_3^-, has three. Neither suffix, however, indicates the specific number of oxygen atoms in the ion. **Table 6.4** lists several examples of common polyatomic ions.

Example 1

Give the name for BCl_3.

This is a binary compound so its name ends with the suffix *-ide*. Attach the suffix to the second element, giving the name chloride. Then use prefixes to indicate the relative number of atoms in the formula. There are three atoms of chlorine (*tri-*) and one atom of boron (*mono-*). The *mono-* prefix is omitted at the beginning of a name, so the name of the compound is boron trichloride.

Example 2

Write the formula for dinitrogen pentoxide.

First, use the prefixes to determine the subscript of each element in the formula. The prefix *di-* indicates two nitrogen atoms and the prefix *penta-* indicates five oxygen atoms. Next, write the correct symbols for the elements with the appropriate subscripts. The symbol for nitrogen is N and the symbol for oxygen is O. So the formula is N_2O_5.

Practice Problems

Prepare for upcoming problems in this chapter by solving each of the following.

A. According to **Table 3.2,** what prefix would you use to write 0.001 gram?
B. Use **Table 6.5** to determine the number of chlorine atoms that are in a molecule of carbon tetrachloride.
C. Which of the following is a binary compound: dinitrogen monoxide or calcium carbonate?
D. Give the name for NO_2.
E. Which ion is chlorite and which is chlorate: ClO_2^-, ClO_3^-.

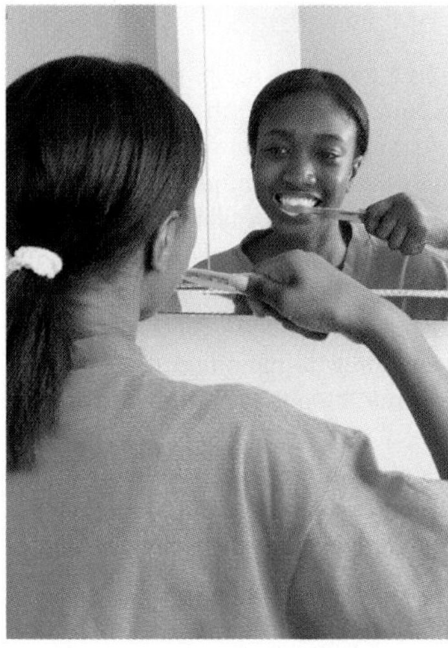

Figure 6.17
Tin forms two ions: Sn^{2+} (tin(II) or stannous) and Sn^{4+} (tin(IV) or stannic). Stannous fluoride is used in some toothpastes to toughen enamel. Stannic sulfide is used to mimic the appearance of gold. What are the formulas of these two tin compounds? 1

Sample Problem 6-6

Name these binary ionic compounds.

a. CoI_2
b. Cs_2O

1. *ANALYZE Plan a problem-solving strategy.*
a.–b. Confirm that the compound is a binary ionic compound; that is, it is composed of a one-element metallic cation and a one-element nonmetallic anion. Name the compound by naming the ions in the order written in the formula: cation name followed by anion name. The name of a transition metal ion that has more than one common ionic charge must include a Roman numeral indicating the charge.

2. *SOLVE Apply the problem-solving strategy.*
Name the cation, followed by the name of the anion, using the crisscross method as needed. Use Roman numerals where appropriate.
a. cobalt(II) iodide or cobaltous iodide

b. cesium oxide

3. *EVALUATE Do the results make sense?*
As a check, try to write the formula for the original compound using the ions from the name you have written. The check leads to the correct formulas: Co^{2+} and I^- make CoI_2, and Cs^+ and O^{2-} make Cs_2O.

Practice Problems

26. Write names for these binary ionic compounds.
a. ZnS c. BaO
b. KCl d. $CuBr_2$

27. Write names for these binary ionic compounds.
a. CaO c. FeS
b. Cu_2Se d. AlF_3

Answer

1 stannous fluoride (SnF_2); stannic sulfide (SnS_2)

6.4

Think Critically

Challenge students to identify charges on ions in a binary compound from the formula unit of that compound. For example, ask students to identify the charge on lead in PbI_4. By remembering that the anion of iodine is always 1^- and that the overall formula unit must show electric neutrality, they can deduce that the lead cation must be Pb^{4+} and not Pb^{2+}.

Discuss

Write the formula Fe_2O_3 on the chalkboard. Ask students to name the compound. Record all responses. If the correct name is given, ask students to explain why that name is appropriate. If the correct name is not given, use the incorrect responses to review the naming of binary compounds.

ACTIVITY

Write formulas for various binary ionic compounds on slips of paper; place the slips in a large box or basket. Have each student draw a slip of paper at random and then write the name of the compound whose formula appears on the paper.

Practice Problems Plus

Additional Practice Problem

This problem can be used with Sample Problem 6-6: Name the following binary compounds:

a. $SrCl_2$
b. SnS_2
c. CrI_2
d. Li_2Se

(**a.** strontium chloride; **b.** tin(IV) sulfide; **c.** chromium(II) iodide; **d.** lithium selenide)

TEACHER DEMO

Because lead(II) nitrate is toxic, wear goggles and gloves for the following demonstration. Mix the following two solutions: 50 mL each of 0.1 *M* solutions of lead(II) nitrate and sodium carbonate. Make sure students see the precipitate of lead carbonate that forms.

Write the following four ions on the chalkboard: Pb^{2+}, NO_3^-, Na^+, CO_3^{2-}. State that the reaction mixture contained these four ions originally. State that the precipitate is a ternary ionic compound formed by association of two of these ions. Ask students to list all the possible formula units of the solid and ask them to name each possibility. Identify the solid as lead(II) carbonate, $PbCO_3$. Disposal: In a beaker, combine the reaction mixture with 10 g solid NaCl. Stir and allow to settle. Filter or decant to isolate the precipitate. Place the dry precipitate in a plastic or cardboard container, and bury it in an approved landfill. Flush the filtrate down the drain with excess water.

ACTIVITY

Have students annotate the formula for $Al_2(SO_4)_3$ with word balloons to show the types and numbers of atoms present, and the types and numbers of cations and anions present. For example, the word balloons could point out that there are 2 aluminum cations and 3 sulfate anions, and that, within the polyatomic anions, there are a total of 3 sulfur atoms and 12 oxygen atoms.

Ternary Ionic Compounds

The pearl and the oyster shell shown in **Figure 6.18** are both made of calcium carbonate ($CaCO_3$), an ionic compound. Calcium carbonate is an example of a **ternary compound,** or a compound that contains atoms of three different elements. Usually, ternary ionic compounds contain a polyatomic ion. How would you write the formula for calcium nitrate? The procedure is the same as that for binary ionic compounds. First, write the formula (symbol and charge) for each ion. Then balance the charges. Remember that an *-ate* or *-ite* ending on the name of a compound indicates that the compound contains a polyatomic anion that includes oxygen. The only two common polyatomic ions that end in *-ide* are the hydroxide and cyanide ions.

Calcium nitrate is composed of calcium cations (Ca^{2+}) and polyatomic nitrate anions (NO_3^-). In calcium nitrate, two nitrate anions, each with a 1− charge, are needed to balance the 2+ charge of each calcium cation, as shown below.

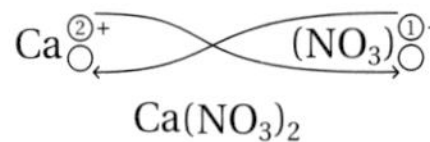

$Ca(NO_3)_2$

Notice that parentheses are used around the nitrate ion in the formula because more than one nitrate anion is needed. The 2 that follows the parentheses shows that two nitrate anions are needed.

When more than a single polyatomic ion are needed to balance a formula, parentheses in the formula set off the polyatomic ion. This is the only time parentheses are used. For example, the formula for strontium sulfate is simply $SrSO_4$ because only a single polyatomic sulfate anion is needed to balance the strontium cation.

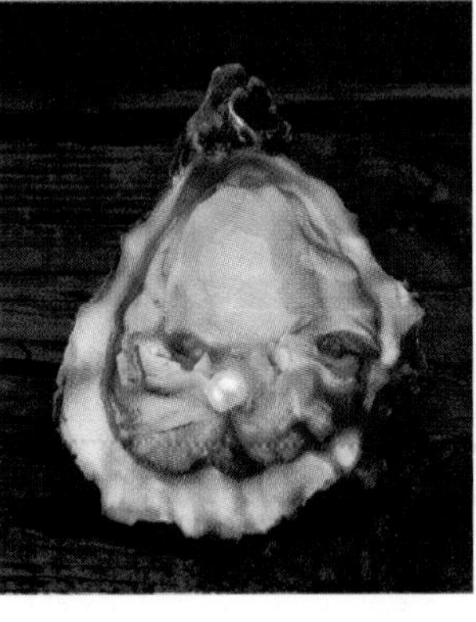

Figure 6.18
Some examples of ternary ionic compounds are shown here. Oysters produce calcium carbonate as they form their shells and produce pearls, which develop around irritant particles, such as grains of sand. The mineral celestite is composed of strontium sulfate, which usually occurs as white crystals but is sometimes blue. Most strontium compounds are produced from celestite. Ammonium dichromate is often used to sensitize silkscreens to light, creating a negative image that can be used to make the positive print. Here you can see the varied results of the silkscreening process.

Lithium carbonate is a ternary compound composed of lithium cations (Li^+) and polyatomic carbonate anions (CO_3^{2-}).

Li_2CO_3

In lithium carbonate, two lithium cations, each with a 1+ charge, are needed to balance the 2− charge of one carbonate anion. Parentheses are not needed to set off the polyatomic carbonate anion. Lithium carbonate can be prescribed to patients who have mood disorders, such as manic-depressive, or bipolar, disorder. A person with bipolar disorder experiences distressing mood swings, from elation to depression and back again. Lithium ions probably exert their mood-stabilizing effects on neurotransmission, which is how messages are sent between nerve cells, including those in the brain.

Sample Problem 6-7

Write formulas for these ternary ionic compounds.

a. potassium sulfate **b.** magnesium hydroxide

1. ANALYZE ***Plan a problem-solving strategy.***

a.–b. The general rules for writing a formula for an ionic compound apply. Write the formula for each ion in the order listed in the name (cation followed by anion), then use subscripts to balance the charges (net charge = 0). If more than one polyatomic ion is needed to balance a formula, place the polyatomic ion formula in parentheses, followed by a subscript showing the number needed.

2. SOLVE ***Apply the problem-solving strategy.***

a. 

K_2SO_4

Two potassium cations with 1+ charges are needed to balance the 2− charge on one sulfate anion. The formula for potassium sulfate is thus K_2SO_4.

b. 

$Mg(OH)_2$

Two hydroxide anions with 1− charges are needed to balance the 2+ charge on one magnesium cation. The formula for magnesium hydroxide must make use of parentheses.

3. EVALUATE ***Do the results make sense?***

The ionic charges correctly add up to zero: 2(1+) + 1(2−) = 0 and 1(2+) + 2(1−) = 0.

Practice Problems

28. Write formulas for compounds formed from these pairs of ions.

a. NH_4^+, SO_3^{2-}

b. calcium ion, phosphate ion

c. Al^{3+}, NO_3^-

d. potassium ion, chromate ion

29. Write formulas for these compounds.

a. lithium hydrogen sulfate

b. chromium(III) nitrite

c. mercury(II) bromide

d. ammonium dichromate

Chem ASAP!

Problem-Solving 28

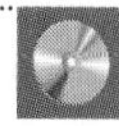

Solve Problem 28 with the help of an interactive guided tutorial.

Cooperative Learning

Divide the class into groups of two or three students. Have each group practice writing formula units and names for ternary ionic compounds using randomly chosen complex cations and anions. The students in a group should take turns selecting the cation and anion for the others to combine. The students should discuss their answers.

Practice Problems Plus

Related Chapter Review Problems

Chapter Review problems 61 and 62 are related to Sample Problem 6-7.

Additional Practice Problem

The Chem ASAP! CD-ROM contains the following problem: Write formulas of compounds formed from these pairs of ions:

a. Pb^{2+}, NO_3^-

b. iron(III) ion and sulfate ion

c. Cr^{3+}, OH^-

d. sodium ion and hydrogen phosphate ion

(**a.** $Pb(NO_3)_2$, **b.** $Fe_2(SO_4)_3$, **c.** $Cr(OH)_3$ **d.** Na_2HPO_4)

3 Assess

Evaluate Understanding

Write the following cations on one side of the chalkboard: NH_4^+, Mg^{2+}, Pb^{4+}. Write the following anions on the other side of the chalkboard: Br^-, OH^-, PO_4^{3-}. Ask students to write formula units for and name all the possible ionic compounds that these ions could form. Ask them to explain how they arrived at their answers.

Reteach

Point out that there are three things to consider when naming an ionic compound: (1) identification of the ions, (2) the order of the names, and (3) the possibility of multiple charges for some cations.

(1) For binary compounds, ions can be easily identified from their symbols. Students must include the suffix *-ide* when naming the nonmetal element. For ternary compounds, students must locate the polyatomic ion from Table 6.4.

(2) Order is the easiest issue. The name of the cation always precedes the name of the anion.

(3) For many cations, students must include the correct charge by adding a Roman numeral in parentheses directly after the cation name.

Use FeN and $Mg(NO_3)_2$ as examples of binary and ternary compounds. Have students use the three-step naming procedure to name these compounds—iron(III) nitride and magnesium nitrate.

Practice Problems Plus

Related Chapter Review Problems

Chapter Review problems 67 and 69 are related to Sample Problem 6-8.

When naming ternary ionic compounds from their formulas, you must first recognize the polyatomic ions. Like binary ionic compounds, ternary ionic compounds are named by naming the ions, cation first. For example, the compound $K_2Cr_2O_7$ consists of two K^+ ions combined with one $Cr_2O_7^{2-}$ polyatomic ion. Its name is potassium dichromate.

Sample Problem 6-8

Name these compounds.

a. LiCN **c.** $(NH_4)_2C_2O_4$
b. $Sr(H_2PO_4)_2$ **d.** $Fe(ClO_3)_3$

1. ANALYZE ***Plan a problem-solving strategy.***

a.–d. If a formula has three or more different elements, the first thing to do is see whether you can recognize a polyatomic ion in the formula for the compound. Name a compound that has one or more polyatomic ions by naming the ions in the order listed in the formula, cation followed by anion.

2. SOLVE ***Apply the problem-solving strategy.***

a. lithium cyanide
b. strontium dihydrogen phosphate
c. ammonium oxalate
d. iron(III) chlorate (Iron is a transition metal that can form more than one ion; its name must include a Roman numeral.)

3. EVALUATE ***Do the results make sense?***

Reversing the process as a check and using the names written to write formulas lead correctly back to the given formulas.

Practice Problems

30. Name each ternary ionic compound.
a. CaC_2O_4 **c.** $KMnO_4$
b. KClO **d.** Li_2SO_3

31. Write names for these compounds.
a. $Al(OH)_3$ **c.** $Sn_3(PO_4)_2$
b. $NaClO_3$ **d.** Na_2CrO_4

section review 6.4

32. How are formulas written for binary ionic compounds, given their names? How is the reverse done?

33. How are formulas written for ternary ionic compounds, given their names? How is the reverse done?

34. Write the name or formula, as appropriate.
a. chromium(III) nitrite **d.** sodium perchlorate
b. $Mg_3(PO_4)_2$ **e.** $Pb(C_2H_3O_2)_2$
c. LiF **f.** magnesium hydrogen carbonate

35. When are parentheses used in writing a chemical formula?

36. What condition must be met in writing a balanced formula for an ionic compound?

Chem ASAP! **Assessment 6.4** Check your understanding of the important ideas and concepts in Section 6.4.

Answers

SECTION REVIEW 6.4

32. The formula must be written so the net ionic charge is zero. The cation is written first, the anion second. To write the name from the formula, name the cation followed by the anion.

33. Write the formula (symbol and charge) for each ion. Then use the crisscross method. To do the reverse, name the cation followed by the anion.

34. a. $Cr(NO_2)_3$
b. magnesium phosphate
c. lithium fluoride
d. $NaClO_4$
e. lead(II) acetate, plumbous acetate
f. $Mg(HCO_3)_2$

35. when more than one polyatomic ion is needed to balance a formula

36. The net ionic charge must be zero.

SMALL-SCALE LAB

NAMES AND FORMULAS OF IONIC COMPOUNDS

SAFETY    

Wear safety glasses and follow the standard safety procedures, as outlined on page 18.

PURPOSE

To observe the formation of compounds, and to write their names and formulas.

MATERIALS

- pencil
- chemicals shown in Figure A
- paper
- ruler
- reaction surface

PROCEDURE

On separate sheets of paper, draw two grids similar to Figure A. Make each square 2 cm on each side. Draw black Xs on only one of the grids. Place a reaction surface over the grid with black Xs and add the chemicals as shown in Figure A. Use the other grid as a data table to record your observations for each solution.

	$AgNO_3$ (Ag^+)	$Pb(NO_3)_2$ (Pb^{2+})	$CaCl_2$ (Ca^{2+})
Na_2CO_3 (CO_3^{2-})	a milky white ppt	e cloudy tan ppt	i grainy white ppt
Na_3PO_4 (PO_4^{3-})	b cloudy white ppt	f milky white ppt	j milky white ppt
NaOH (OH^-)	c muddy brown ppt	g milky white ppt	k cloudy white ppt
Na_2SO_4 (SO_4^{2-})	d no visible reaction	h milky white ppt	l grainy white ppt

Figure A

ANALYSIS

Using the experimental data, record the answers to the following questions below your data table.

1. Describe each precipitate that forms as milky, grainy, cloudy, or gelatinous. Which mixture(s) did not form a precipitate?
2. Write the formulas and names of the chemical compounds produced in the mixings.

YOU'RE THE CHEMIST

The following small-scale activities allow you to develop your own procedures and analyze the results.

1. **Analyze It!** Repeat the above experiment, using the chemicals in Figure B. Identify the precipitates, write their formulas, and name them.

	$FeCl_3$ (Fe^{3+})	$MgSO_4$ (Mg^{2+})	$CuSO_4$ (Cu^{2+})
Na_2CO_3 (CO_3^{2-})	a orange ppt	e white ppt	i blue ppt
Na_3PO_4 (PO_4^{3-})	b orange ppt	f white ppt	j blue ppt
NaOH (OH^-)	c orange ppt	g white ppt	k blue ppt
Na_2SO_4 (SO_4^{2-})	d no visible reaction	h no visible reaction	l no visible reaction

Figure B

2. **Analyze It!** In ionic equations, the precipitate is written to the right of the arrow, and the ions that produced it are to the left.

   Write ionic equations for the precipitates formed from the reactions related to Figure B.

SMALL-SCALE LAB

NAMES AND FORMULAS OF IONIC COMPOUNDS

OVERVIEW

Procedure Time: 30 minutes
Students mix solutions of aqueous ions and observe precipitates. They write the formulas for and name the precipitates.

MATERIALS

Solution	Preparation
0.05*M* $AgNO_3$	2.1 g in 250 mL
0.2*M* $Pb(NO_3)_2$	16.6 g in 250 mL
0.5*M* $CaCl_2$	13.9 g in 250 mL
1.0*M* Na_2CO_3	26.5 g in 250 mL
0.1*M* Na_3PO_4	9.5 g $Na_3PO_4 \cdot 12H_2O$ in 250 mL
0.5*M* NaOH	20.0 g in 1.0 L
0.2*M* Na_2SO_4	7.1 g in 250 mL
0.2*M* $CuSO_4$	12.5 g $CuSO_4 \cdot H_2O$ in 250 mL
0.2*M* $MgSO_4$	6.0 g in 250 mL
0.1*M* $FeCl_3$	6.8 g $FeCl_3 \cdot 6H_2O$ in 25 mL of 1.0*M* NaCl (for stability); dilute to 250 mL

TEACHING SUGGESTIONS

Emphasize that Figures A and B give the ions that can form precipitates in parentheses.

DISPOSAL

Flush drops of dilute silver and lead solutions down the drain.

ANALYSIS

1. $Na_2SO_4 + AgNO_3$ did not form a precipitate.
2. **a.** Ag_2CO_3, silver carbonate
   b. Ag_3PO_4, silver phosphate
   c. AgOH, silver hydroxide (Note: This is Ag_2O, silver oxide.)
   d. no visible reaction
   e. $PbCO_3$, lead(II) carbonate
   f. $Pb_3(PO_4)_2$, lead(II) phosphate
   g. $Pb(OH)_2$, lead(II) hydroxide
   h. $PbSO_4$, lead(II) sulfate
   i. $CaCO_3$, calcium carbonate
   j. $Ca_3(PO_4)_2$, calcium phosphate
   k. $Ca(OH)_2$, calcium hydroxide
   l. $CaSO_4$, calcium sulfate

YOU'RE THE CHEMIST

1. **a.** $Fe_2(CO_3)_3$, iron(III) carbonate
   b. $FePO_4$, iron(III) phosphate
   c. $Fe(OH)_3$, iron(III) hydroxide
   e. $MgCO_3$, magnesium carbonate
   f. $Mg_3(PO_4)_2$, magnesium phosphate
   g. $Mg(OH)_2$, magnesium hydroxide
   i. $CuCO_3$, copper(II) carbonate
   j. $Cu_3(PO_4)_2$, copper(II) phosphate
   k. $Cu(OH)_2$, copper(II) hydroxide
   d, h, l. no visible reaction
2. **a.** $2Fe^{3+} + 3CO_3^{2-} \rightarrow Fe_2(CO_3)_3(s)$
   b. $Fe^{3+} + PO_4^{3-} \rightarrow FePO_4(s)$
   c. $Fe^{3+} + 2OH^- \rightarrow Fe(OH)_2(s)$
   e. $Mg^{2+} + CO_3^{2-} \rightarrow MgCO_3(s)$
   f. $3Mg^{2+} + 2PO_4^{3-} \rightarrow Mg_3(PO_4)_2(s)$
   g. $Mg^{2+} + 2OH^- \rightarrow Mg(OH)_2(s)$
   i. $Cu^{2+} + CO_3^{2-} \rightarrow CuCO_3(s)$
   j. $3Cu^{2+} + 2PO_4^{3-} \rightarrow Cu_3(PO_4)_2(s)$
   k. $Cu^{2+} + 2OH^- \rightarrow Cu(OH)_2(s)$

6.5

1 Engage

Use the Visual

Have students read the section-opener paragraph and examine the photograph. Ask:

- **What do prefixes tell you about the composition of binary molecular compounds?** (Based on experience, students should infer that prefixes refer to the quantities of atoms.)
- **Name some common words that include prefixes such as bi-, tri- and mono-.** (Possibilities include bicycle, bifocal, triangle, triathlete, monorail, and monotone.)

Explain that most systematic names of molecular compounds formed from nonmetals use prefixes to describe the ratio of atoms in a molecule.

Check Prior Knowledge

To assess students' prior knowledge about naming molecular compounds, ask:

- **How is a molecular compound generally defined?** (It is a substance composed of molecules in which two or more nonmetallic elements are joined.)
- **Why is "carbon oxide" an insufficient name for a compound composed of carbon and oxygen atoms?** (Carbon and oxygen, like most nonmetals, can combine in different proportions.)

section 6.5

MOLECULAR COMPOUNDS AND ACIDS

objectives

- Apply the rules for naming and writing formulas for binary molecular compounds
- Name and write formulas for common acids

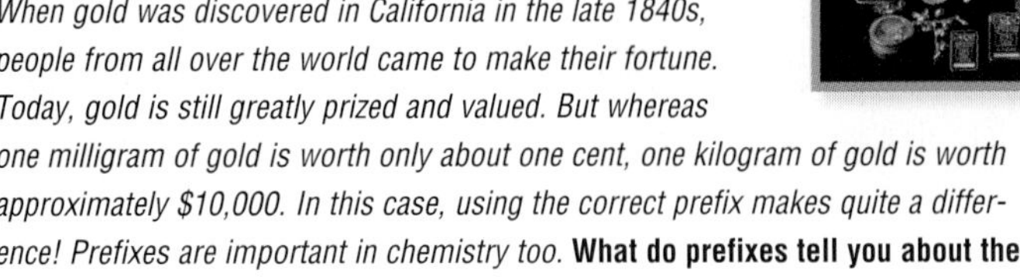

Gold was one of the first metals to attract human attention. When gold was discovered in California in the late 1840s, people from all over the world came to make their fortune. Today, gold is still greatly prized and valued. But whereas one milligram of gold is worth only about one cent, one kilogram of gold is worth approximately $10,000. In this case, using the correct prefix makes quite a difference! Prefixes are important in chemistry too. **What do prefixes tell you about the composition of binary molecular compounds?**

Binary Molecular Compounds

Unlike binary ionic compounds, binary molecular compounds are composed of two nonmetallic elements. This difference affects the naming of these compounds, as well as the writing of their formulas. Because binary molecular compounds are composed of molecules, ionic charges are not used to assign formulas or names to them. In addition, when two nonmetallic elements combine, they often can do so in more than one way. For example, the elements carbon and oxygen combine to form two different gaseous compounds, CO and CO_2. (They also form two different polyatomic ions, CO_3^{2-} and $C_2O_4^{2-}$.) How would you name a binary compound formed by the combination of carbon and oxygen atoms? It might seem satisfactory to call it carbon oxide. However, the two carbon oxides, CO and CO_2, are very different compounds. Sitting in a room with small amounts of the carbon oxide CO_2 in the air would not present any problems. You exhale CO_2 as a product of your body chemistry. Thus it is normally present in the air you breathe. On the other hand, if the same amount of the other carbon oxide, CO, were in the room, you could very well die of asphyxiation as a result. The binary compound CO is a poisonous gas that interferes with the blood's

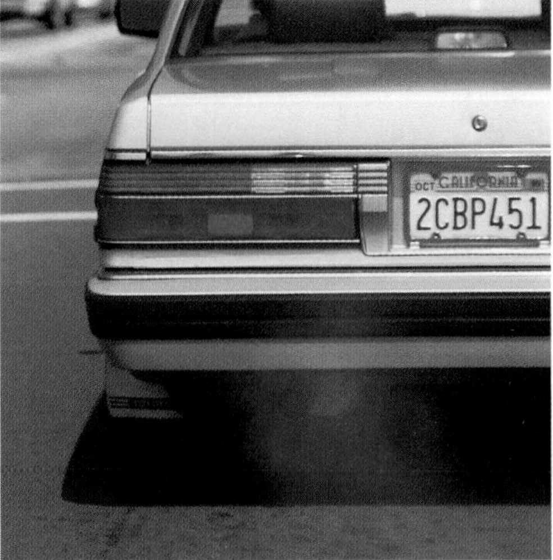

Figure 6.19
Carbon dioxide (CO_2) and carbon monoxide (CO) are both binary molecular compounds of carbon and oxygen, but they differ considerably in properties. Carbon dioxide is a nonpoisonous product of respiration, but carbon monoxide is an extremely toxic product of incomplete burning, such as occurs in automobile gasoline engines.

Student Resources

From the Teacher's Resource Package, use:

- Section Review 6.5, Ch. 6 Practice Problems and Quizzes from the Review Module (Ch. 5–8)

Technology Resources

Relevant Technology Resources include:

- Chem ASAP! CD-ROM
- ResourcePro CD-ROM

ability to carry oxygen to body cells. Obviously, a naming system that distinguishes between these two compounds is needed.

Prefixes help distinguish between different compounds. **Table 6.5** lists the prefixes used to name binary molecular compounds. These prefixes tell how many atoms of each element are present in each molecule (and in the formula) of the compound. According to the table, the prefix *mono-* indicates the presence of the single oxygen atom in CO. The prefix *di-* indicates the presence of the two oxygen atoms in CO_2. The two compounds of carbon and oxygen, CO and CO_2, are thus named carbon monoxide and carbon dioxide, respectively. Notice that the second element in the name ends with *-ide*. The names of all binary molecular compounds end in *-ide,* just as was the case for binary ionic compounds. Also note that the vowel at the end of a prefix often is dropped when the name of the element begins with a vowel. For CO, you would write monoxide, not monooxide. If there is just a single atom of the first element in the name, omit the prefix *mono-* for that element. How many chlorine atoms are there in carbon tetrachloride, a toxic substance once used in the dry cleaning of clothes? ❶

Writing formulas for binary molecular compounds, given their names, is quite easy. Use the prefixes to tell you the subscript of each element in the formula. Then write the correct symbols for the two elements, with the appropriate subscripts. For example, the formula for tetraiodine nonoxide is I_4O_9. The prefix *tetra-* indicates four iodine atoms, and the prefix *nona-* indicates nine oxygen atoms in the molecule. How would you write formulas for sulfur trioxide and phosphorus pentafluoride? ❷

Table 6.5

Prefixes Used in Naming Binary Molecular Compounds

Prefix	Number
mono-	1
di-	2
tri-	3
tetra-	4
penta-	5
hexa-	6
hepta-	7
octa-	8
nona-	9
deca-	10

Sample Problem 6-9

Name these binary molecular compounds.

a. N_2O **b.** PCl_3 **c.** SF_6

1. ***ANALYZE** Plan a problem-solving strategy.*

a.–c. Confirm that the compound is a binary molecular compound—that is, the compound is composed of two nonmetals. The name must reflect the composition of the compound's molecule. Name the elements in the order listed in the formulas. Prefixes indicate the number of each kind of atom. The suffix of the name of the second element is *-ide*.

2. ***SOLVE** Apply the problem-solving strategy.*

a. The formula N_2O contains two nitrogen atoms (prefix *di-*) and one oxygen atom (prefix *mono-*). The name of the compound is dinitrogen monoxide.

b. phosphorus trichloride

c. sulfur hexafluoride

3. ***EVALUATE** Do the results make sense?*

Each name provides sufficient information to regenerate the correct formula.

Practice Problems

37. Name these binary molecular compounds.
a. OF_2 **b.** Cl_2O_8 **c.** SO_3

38. Write formulas for the following binary molecular compounds.
a. nitrogen trifluoride
b. disulfur dichloride
c. dinitrogen tetroxide

Chem ASAP!

Problem-Solving 37

Solve Problem 37 with the help of an interactive guided tutorial.

6.5

2 Teach

Discuss

Lead a class discussion on the procedure for naming binary molecular compounds. Have students make a chart with three columns: element name, number of atoms, and prefix. Use N_2O as an example.

Nitrogen	2	di-
Oxygen	1	mono-

The name is thus dinitrogen monoxide. Write the molecular formulas for several more compounds on the chalkboard, and have students add them to their charts. Some possibilities are CCl_4, PBr_5, and P_4S_3.

Think Critically

Challenge the students to name the following pairs of compounds and to identify what each pair has in common.

PBr_3 and $CrBr_3$
N_2O and Na_2O
CI_4 and PbI_4
P_2O_3 and Fe_2O_3

(Each pair is a binary molecular compound and binary ionic compound with one element in common.)

Practice Problems Plus

Related Chapter Review Problems
Chapter Review problems 64 and 66 are related to Sample Problem 6-9.

Additional Practice Problem
The Chem ASAP! CD-ROM contains the following problem: Name these binary molecular compounds.
a. SBr_2
b. N_2O_5
c. XeF_4
d. O_2F_2
(**a.** sulfur dibromide,
b. dinitrogen pentoxide,
c. xenon tetrafluoride,
d. dioxygen difluoride)

Meeting Diverse Needs

Gifted The binary molecular compound CO_2 has received a lot of attention in recent years with regard to its importance in global warming. Large quantities of CO_2 are released by burning fossil fuels. Once released into the atmosphere, CO_2 acts as a "greenhouse gas," trapping solar energy. Ask students to learn more about this topic and to prepare a report or poster.

Answers

❶ 4
❷ SO_3, PF_5

3 Assess

Evaluate Understanding

Ask students to write the formulas and names for several molecular compounds using the following combinations of elements: nitrogen and oxygen; sulfur and oxygen; carbon and oxygen. (NO, NO_2, N_2O, N_2O_4, CO, CO_2, SO_2, SO_3)

Reteach

Point out that because common acids all contain hydrogen ions, they must be distinguished from one another by their anions. For example, consider the following pairs of acids:
hydro*chloric* acid and hydro*bromic* acid
***nitric* acid and *sulfuric* acid**

Figure 6.20
Acids are used for many purposes. The nitric acid shown here is being used to etch a metal plate that will be used to create an artist's print.

ENVIRONMENTAL AWARENESS

A Toxic Environmental Gas
The binary molecular compound carbon monoxide is a colorless, odorless, flammable gas used to manufacture numerous organic and inorganic chemicals. This gas is also highly toxic to humans. Upon inhalation, carbon monoxide binds to the hemoglobin molecules in red blood cells, where it replaces oxygen. Carbon monoxide binds to hemoglobin about 200 times more effectively than oxygen does. This keeps the red blood cells from carrying oxygen to body tissues. Therefore, at very low levels carbon monoxide is a fast-acting poison that may cause serious illness or death. At peak traffic times, the level of carbon monoxide around streets and highways may reach as high as 100 parts per million (ppm) of air. In the United States, automobiles must have catalytic converters, which change toxic carbon monoxide to less toxic carbon dioxide. Carbon monoxide is also present in cigarette smoke. It takes several hours to replace the carbon monoxide in a smoker's blood after only one cigarette.

Naming Common Acids

Acids are a special group of compounds. As you will see in Chapter 20, there are several definitions of acids. For now, it is sufficient to know that under the more common definition, acids are compounds that produce hydrogen ions when dissolved in water. When naming acids and writing their formulas, you can consider them to be combinations of anions connected to as many hydrogen ions (H^+) as are needed to make the molecule electrically neutral.

Many industrial processes, including steel and fertilizer manufacturing, use acids. There are a few acids that you will regularly use in the laboratory. You should know the names and formulas of these acids:

Hydrochloric acid	HCl
Sulfuric acid	H_2SO_4
Nitric acid	HNO_3
Acetic acid	$HC_2H_3O_2$
Phosphoric acid	H_3PO_4
Carbonic acid	H_2CO_3

section review 6.5

39. How are formulas written for binary molecular compounds, given their names? How is the reverse done?

40. Give the name or formula for these common acids.
a. H_2SO_4 **b.** H_2CO_3 **c.** nitric acid **d.** phosphoric acid

41. Write the formula or name for these compounds.
a. CS_2 **c.** carbon tetrabromide
b. Cl_2O_7 **d.** diphosphorus trioxide

42. What element typically appears in the formula of a common acid?

 Chem ASAP! **Assessment 6.5** Check your understanding of the important ideas and concepts in Section 6.5.

Answers

SECTION REVIEW 6.5

39. Use the prefixes to determine the subscript for each element in the formula. Write the correct symbols for the two elements. When there are multiple atoms of an element, place a prefix before the element name. The name of the second element will end in *-ide*.

40. a. sulfuric acid **c.** HNO_3
b. carbonic acid **d.** H_3PO_4

41. a. carbon disulfide
b. dichlorine heptoxide
c. CBr_4
d. P_2O_3

42. hydrogen

SUMMARY OF NAMING AND FORMULA WRITING

objectives

- Use the flowchart in **Figure 6.21** to write the name of a compound when given its chemical formula
- Use the flowchart in **Figure 6.23** to write a chemical formula when given the name of a compound

In the average home, there are probably hundreds of chemicals, including cleaning products, drugs, and pesticides. Most people would not know what to do if some of these chemicals accidentally mixed together and began to react, or, even worse, if a small child ingested one of these chemicals. A phone call to a poison-control center can provide life-saving information to victims of such poisonings. On the inside front cover of most phone books is a list of phone numbers for regional centers and other emergency numbers. **Why would knowing the correct name of a chemical be important to a poison-control center worker?**

Practicing Skills: Follow the Arrows

In this chapter, you have learned two basic skills: writing chemical formulas and naming chemical compounds. If this is the first time you have tried to master these skills, you may feel a little overwhelmed at this point. For example, you may find it difficult to know when to use prefixes and Roman numerals in a name, and when not to do so. Or you may have trouble determining if a compound's name should end in *-ate*, *-ide*, or *-ite*. The flowchart in **Figure 6.21** is designed to help you name compounds correctly. Follow the arrows on the flowchart to find directions for naming a particular compound. By using the flowchart while working exercises, you will increase your skill at naming compounds.

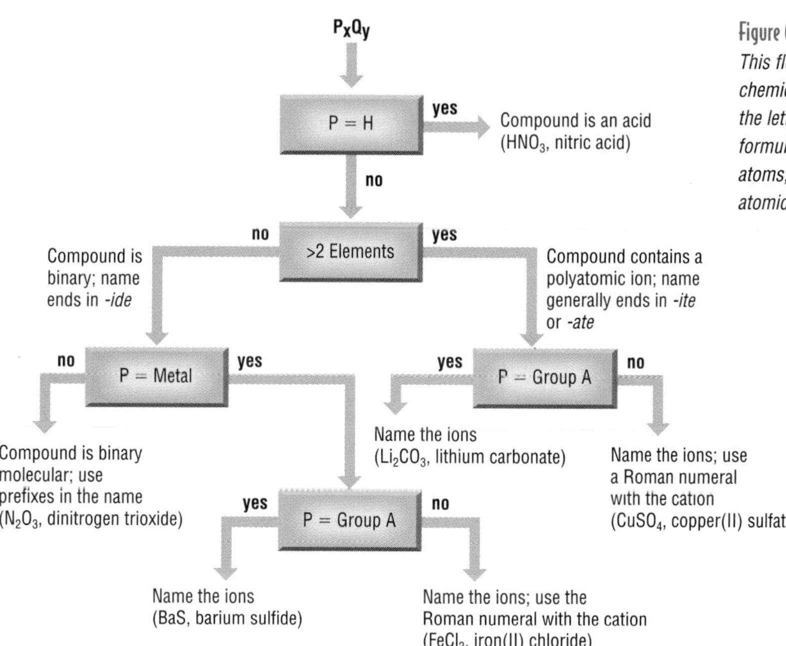

Figure 6.21

This flowchart will help you name chemical compounds. Begin with the letters P and Q in the general formula P_xQ_y. P and Q can be atoms, monatomic ions, or polyatomic ions.

6.6

1 Engage

Use the Visual

Have students read the section-opener paragraph and examine the photograph. Discuss the importance of naming conventions for compounds. Explain that chemists need to be able to understand papers written by colleagues and to reproduce experimental results. Focus on the problems that could occur if common names in multiple languages were the norm. Ask:

- **Why would knowing the correct name for a chemical be important to a poison-control center worker?** (The worker needs to know the specific chemical before offering a solution. Administering an antidote for the wrong chemical could cause further harm.)

Motivate and Relate

Ask students to think of other situations where specific names are important for accurate or fast results. Ask students why many bureaucratic agencies prefer to use social security or driver's license numbers instead of people's names.

STUDENT RESOURCES

From the Teacher's Resource Package, use:

- Section Review 6.6, Ch. 6 Practice Problems, Vocabulary Review, Quizzes and Tests from the Review Module (Ch. 5–8)
- Laboratory Recordsheet 6-2

TECHNOLOGY RESOURCES

Relevant Technology Resources include:

- Chem ASAP! CD-ROM
- ResourcePro CD-ROM
- Assessment Resources CD-ROM

2 Teach

Discuss

Have students summarize the rules for writing formulas and naming compounds. Point out that naming compounds and writing formulas becomes easier with practice because chemical names and formulas are systematized. Ask students to consider how difficult the task would be if the names were assigned randomly. After students have read the *Link to Information Science,* point out that to quickly locate information on physical constants such as melting point, boiling point, and density, many scientists use chemical handbooks such *Lange's Handbook of Chemistry, The Merck Index,* and the *CRC Handbook of Chemistry and Physics. The Merck Index* is particularly helpful to organic chemists interested in the synthesis and identification of a particular compound. Encourage students to look for these books in their local public library.

3 Assess

Evaluate Understanding

Give students formulas for compounds and have them use the flow chart to figure out their names. Then give them names for different compounds and have them use the flow chart to figure out their formulas.

Figure 6.22
Artists Palette at Death Valley, in California, boasts many colorful rock formations. The colors indicate the presence of salts of iron, potassium, aluminum, magnesium, and manganese. These salts, some of which are complex compounds, are named and their formulas are written according to the rules you have learned here.

In writing a chemical formula from a chemical name, it is helpful to remember the following rules.

1. In an ionic compound, the net ionic charge is zero.
2. An *-ide* ending generally indicates a binary compound.
3. An *-ite* or *-ate* ending means there is a polyatomic ion that includes oxygen in the formula.
4. Prefixes in a name generally indicate that the compound is molecular. They show the number of each kind of atom in the molecule.
5. A Roman numeral after the name of a cation shows the ionic charge of the cation.

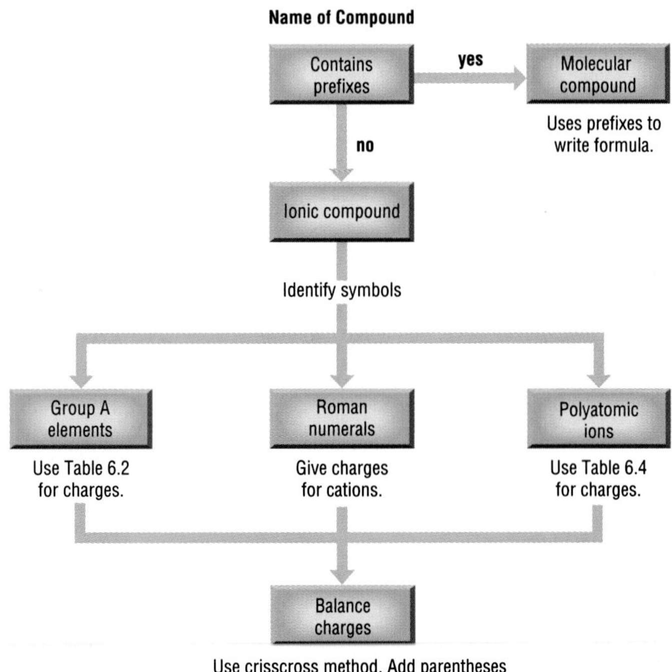

Figure 6.23 ▶
This flowchart will help you write a chemical formula when given a chemical name.

Meeting Diverse Needs

Gifted Challenge advanced students to write a computer program for naming chemical compounds, using the flow chart as a guide.

Making Ionic Compounds

PURPOSE

Mix cations and anions to make ionic compounds.

MATERIALS

- 9 small test tubes
- test tube rack
- 6 solutions in plastic dropper bottles containing the following ions:
  Solution A (Fe^{3+} ion)
  Solution B (Ag^{+} ion)
  Solution C (Pb^{2+} ion)
  Solution X (CO_3^{2-} ion)
  Solution Y (OH^{-} ion)
  Solution Z (PO_4^{3-} ion)

PROCEDURE

1. Label three test tubes A, three test tubes B, and three test tubes C.
2. Add 10 drops (0.5 mL) of solutions A, B, and C to each of their respective labeled test tubes.
3. Add 10 drops of solution X to one test tube of A, 10 drops to one test tube of B, and 10 drops to one test tube of C. Observe each for the formation of a solid material.
4. Make a 3-by-3 grid in which to record your observations. Label the rows A, B, and C. Label the columns X, Y, and Z. Write a short description of the solid material you observed forming in each test tube.
5. Repeat Steps 3 and 4, adding 10 drops of solution Y to test tubes A, B, and C.
6. Repeat Steps 3 and 4, adding 10 drops of solution Z to test tubes A, B, and C.

ANALYSIS AND CONCLUSIONS

1. Can mixing a cation and an anion lead to the formation of an insoluble (does not dissolve in water) ionic compound?
2. Write the correct formula for each ionic compound formed.
3. Name each ionic compound formed.
4. Do you think mixing together a solution of any cation with a solution of any anion will always lead to the formation of an insoluble ionic compound?

section review 6.6

43. Name these compounds, using **Figure 6.21** as an aid if necessary.
    a. $CaCO_3$ c. $PbCrO_4$ e. $SnCr_2O_7$
    b. $KMnO_4$ d. $CaHPO_4$ f. Mg_3P_2
44. Write formulas for these compounds, using **Figure 6.23** as an aid if necessary.
    a. tin(II) hydroxide
    b. barium fluoride
    c. tetraiodine nonoxide
    d. iron(III) oxalate
    e. calcium sulfide
    f. aluminum hydrogen carbonate

 Chem ASAP! **Assessment 6.6** Check your understanding of the important ideas and concepts in Section 6.6.

6.6

Reteach

Compare the two flow charts. Focus on steps that are exact opposites of each other, e.g., using the Roman numeral in a name to determine the charge to be used in the crisscross method versus using the charge determined from a reverse crisscross procedure to establish which Roman numeral to use in a name.

MINI LAB

Making Ionic Compounds

SAFETY

Students should wear safety goggles and avoid skin contact with these chemicals.

LAB PREP AND PLANNING

Prepare 0.1 *M* solutions of $FeCl_3$, $AgNO_3$, $Pb(NO_3)_2$, Na_2CO_3, NaOH, and Na_3PO_4.

LAB TIPS

Place the tubes on a white sheet of paper to help students detect precipitates.

ANALYSIS AND CONCLUSIONS

1. Yes; an insoluble solid formed in every tube.
2. $Fe(OH)_3$, $Fe_2(CO_3)_3$, $FePO_4$, Ag_2CO_3, AgOH, Ag_3PO_4, $Pb(OH)_2$, $PbCO_3$, $Pb_3(PO_4)_2$
3. iron(III) hydroxide, iron(III) carbonate, iron(III) phosphate, silver carbonate, silver hydroxide, silver phosphate, lead(II) hydroxide, lead(II) carbonate, lead(II) phosphate
4. No; consider sodium chloride, for example.

Answers

SECTION REVIEW 6.6

43. a. calcium carbonate
    b. potassium permanganate
    c. lead(II) chromate
    d. calcium hydrogen phosphate
    e. tin(II) dichromate
    f. magnesium phosphide
44. a. $Sn(OH)_2$ d. $Fe_2(C_2O_4)_3$
    b. BaF_2 e. CaS
    c. I_4O_9 f. $Al(HCO_3)_3$

Chemistry Serving... the Consumer

DISCUSS

After students read the text, explain that the naming of pharmaceuticals by manufacturers is a process driven more by marketing than by any concern for consistency. Emphasize that, no matter how many common names are assigned to a given drug, there is only one acceptable systematic name for the generic compound.

Ask a local druggist for a list of the top ten prescription drugs sold by volume. Have students do research to find the systematic names for those ten drugs and to determine their main applications.

CHEMISTRY IN CAREERS

Have students read more about the responsibilities and duties of a pharmacist on page 870 of this text. If you have access to the Internet in your classroom or school, you may wish to have students connect to the address shown.

Chemistry Serving...the Consumer

WHAT'S IN A NAME?

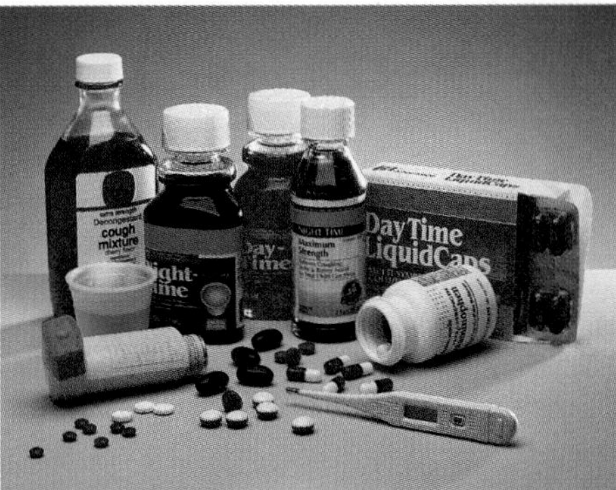

When it comes to drugs and medicines, chemical naming can get very complex and even confusing. Consider the example of a drug commonly prescribed for lowering blood pressure. Its chemical name is 4-[2'-hydroxy-3'-[(1-methylethyl) amino] propoxy]-benzeneacetamide, but a pharmacist is not likely to know the drug by this name. So a doctor would write Tenormin® on a prescription. Alternatively, the doctor could write Atenolol, Alinor, Altol, Antipressan, Atcardil, AteHexal, Atenil, Betablok, Betacard, Felo-Bits, or Myocord. Or if the doctor did not like any of those names, he or she could put down Normiten, Oraday, Premorine, Prenolol, Seles, Tenolol, or Vericordin. Any of these names would get the patient the same chemical compound—4-[2'-hydroxy-3'-[(1-methylethyl) amino] propoxy]-benzeneacetamide.

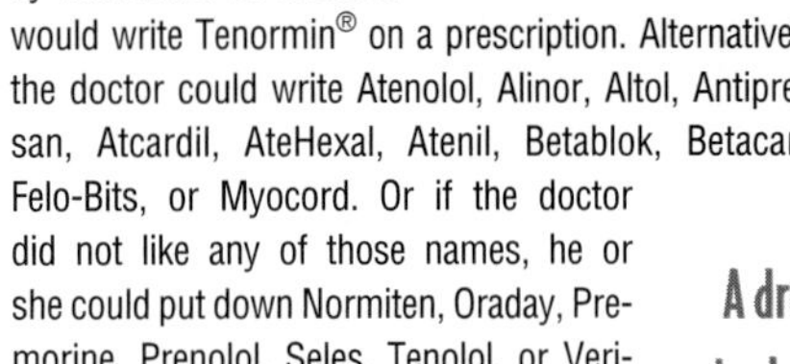

A drug is assigned a single generic name by an international council that includes the World Health Organization.

Why are there so many names for the same drug? A drug is assigned a single generic name, such as Atenolol, by an international council that includes the World Health Organization. There are many brand names, such as Ternormin®, which are assigned by the companies that develop and produce the drug.

When a manufacturer develops a drug, it puts to work teams of chemists and pharmacists who spend many years doing expensive research. Then the company must test the drug and prove it safe and effective, another lengthy and costly process. Eventually, the company submits the drug to the Food and Drug Administration (FDA) for review. If the FDA approves the drug, the company can finally market it. At this point, the marketing department of the manufacturer names the drug. The name, which then becomes a brand name, is often descriptive of what the drug does but usually has nothing to do with the actual chemical name of the drug.

Because manufacturers invest so much time and money in developing a new drug, they must protect their investment. They do this by patenting the drug. The patent gives the company the exclusive right to produce and market the drug for 17 years.

After the patent expires, other companies can manufacture the drug. Because these companies do not have to do all the development and testing, they can sell the drug at a lower cost than the original manufacturer can. Although other companies can manufacture the drug after the patent expires, they cannot market their versions under the same brand name. Each company gives its version of the drug a different brand name. If the market for the drug is large, many different companies may be motivated to produce and sell the drug. Thus, there can be many brand names for a single drug. Doctors often write prescriptions using only the generic name. The pharmacist fills the prescription with whatever brand is in stock. The FDA assures that all available drugs are safe and effective.

To fill prescriptions correctly, pharmacists must keep track of all the different brand names and generic names. To do this, pharmacists use a variety of resources, such as the *Physicians Desk Reference.* Most larger pharmacies have computer programs that suggest substitutions automatically.

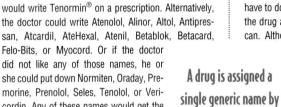

CHEMISTRY IN CAREERS

PHARMACIST

Fill prescriptions, determining generic-drug substitutions and advising customers on recommended dosage and usage.

See page 870.

CHEMIST NEEDED

TAKE IT TO THE NET

Find out more about career opportunities: **www.phschool.com**

CHEMICAL SPECIALIST

Local food service distributor see responsible self-motivated indivi

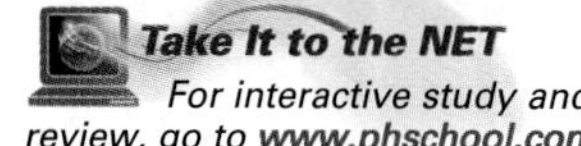

For interactive study and review, go to www.phschool.com

KEY TERMS

- anion *p. 136*
- binary compound *p. 149*
- cation *p. 135*
- chemical formula *p. 138*
- formula unit *p. 140*
- ion *p. 135*
- ionic compound *p. 137*
- law of definite proportions *p. 141*
- law of multiple proportions *p. 141*
- molecular compound *p. 134*
- molecular formula *p. 138*
- molecule *p. 134*
- monatomic ion *p. 143*
- polyatomic ion *p. 146*
- ternary compound *p. 154*

KEY RELATIONSHIP

- For ionic compounds, net ionic charge = 0.

CONCEPT SUMMARY

6.1 Introduction to Chemical Bonding

- Every substance is either an element or a compound.
- A compound consists of more than one kind of atom.
- A compound is either molecular or ionic in nature.

6.2 Representing Chemical Compounds

- Molecular compounds are composed of two or more nonmetals.
- A molecular formula shows the number and kinds of atoms present in a molecule of a compound.
- Ionic compounds are composed of oppositely charged ions (cations and anions) combined in electrically neutral groupings.
- A formula unit gives the lowest whole-number ratio of ions in the compound.

6.3 Ionic Charges

- The charges of the ions of the representative elements can be determined by the position of these elements in the periodic table.
- Most transition metals have more than one common ionic charge.
- A polyatomic ion is a group of atoms that behaves as a unit and has a charge.

6.4 Ionic Compounds

- Binary (two-element) ionic compounds are named by writing the name of the cation followed by the name of the anion. Binary compounds end in *-ide*.
- When a cation can have more than one ionic charge, a Roman numeral is used in the name.
- Ternary ionic compounds contain at least one polyatomic ion. The names of these compounds generally end in *-ite* or *-ate*.

6.5 Molecular Compounds and Acids

- Binary molecular compounds are composed of two nonmetallic elements.
- The name of a binary molecular compound always ends in *-ide*.
- Prefixes are used to show how many atoms of each element are present in a molecule of the compound.

6.6 Summary of Naming and Formula Writing

- The use of a flowchart is helpful in naming compounds.

CHAPTER CONCEPT MAP

Use these terms to construct a concept map that organizes the major ideas of this chapter.

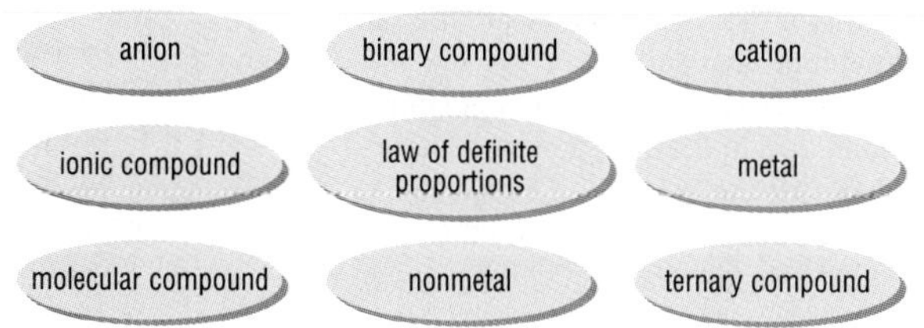

 **Chem ASAP! Concept Map 6**
Create your Concept Map using the computer.

Take It to the Net

At **www.phschool.com** students will find for this chapter

- an Internet activity
- links to related chemistry sites
- an interactive quiz
- career links

Chapter 6 STUDENT STUDY GUIDE

4 Close

Summary

Ask:

- **What are the charges and names of ions formed from these elements: Sr, Cu, Zn, Se, F, Ar?** (Sr^{2+}: strontium; Cu^{+}: copper(I); Cu^{2+}: copper(II); Zn^{2+}: zinc; Se^{2-}: selenide; F^{-}: fluoride; Ar does not form ions readily.)
- **Identify these compounds as molecular or ionic and name them: Li_2S, $SeCl_6$, $Co(NO_3)_3$, PbO_2, N_2O_5, Ag_3PO_4** (ionic: lithium sulfide, cobalt(III) nitrate, lead(IV) oxide, silver phosphate; molecular: selenium hexachloride, dinitrogen pentoxide)
- **Write the formulas for these compounds:**
  beryllium sulfide (BeS)
  beryllium sulfate ($BeSO_4$)
  carbon tetrafluoride (CF_4)
  iron(II) cyanide ($Fe(CN)_2$)

Extension

Have students research rules for naming simple alkanes. Have them draw structures for the compounds and describe how this way of naming compounds differs from what they have learned thus far. Explain that the chemistry of carbon compounds will be studied in depth in Chapter 26.

Looking Back . . . Looking Ahead . . .

The previous chapter discussed atomic structure and introduced the periodic table. The current chapter explains how ions are formed, formulas are written, and compounds are named. The next chapter introduces the concept of the mole, and relates it to mass and volume.

Answers

45. Ions are formed when atoms lose or gain electrons.

46. **a.** 1 gained **b.** 1 lost **c.** 3 gained **d.** 2 lost **e.** 1 lost **f.** 1 gained

47. **a.** bromide ion, anion **b.** sodium ion, cation **c.** arsenide ion, anion **d.** calcium ion, cation **e.** copper(I) ion, cation **f.** hydride ion, anion

48. The net positive charge on the cations is exactly balanced by the net negative charge on the anions.

49. **a.** ionic **b.** molecular **c.** ionic **d.** molecular **e.** ionic **f.** molecular

50. **a.** Yes; the ratio of the mass of the colorless gas to the mass of the white powder is a constant. **b.** law of definite proportions **c.** 3.5 g colorless gas

51. **a.** carbon 6, hydrogen 8, oxygen 6 **b.** carbon 5, hydrogen 8, oxygen 4, sodium 1 **c.** carbon 12, hydrogen 22, oxygen 11 **d.** carbon 7, hydrogen 5, nitrogen 3, oxygen 6 **e.** nitrogen 2, hydrogen 4, oxygen 3

52. ionic; high melting point

53. **a.** O^{2-} **b.** Pb^{2+} **c.** Li^{+} **d.** N^{3-} **e.** Cu^{2+} **f.** F^{-}

54. **a.** barium ion **b.** iodide ion **c.** silver ion **d.** mercury(II) ion **e.** phosphide ion **f.** tin(IV) ion

55. cyanide (CN^-) and hydroxide (OH^-)

56. **a.** hydroxide **b.** lead(IV) **c.** sulfate **d.** oxide **e.** hydrogen phosphate **f.** dichromate **g.** aluminum **h.** chlorite

57. The net ionic charge is zero. Ionic compounds are electrically neutral.

58. **a.** Na_2O **b.** SnS_2 **c.** KCl **d.** Mg_3N_2

Chapter 6 REVIEW

CONCEPT PRACTICE

45. Describe two ways that an ion forms from an atom. *6.1*

46. State the number of electrons either lost or gained in forming each ion. *6.1*
a. Br^- **b.** Na^+ **c.** As^{3-} **d.** Ca^{2+} **e.** Cu^+ **f.** H^-

47. Name each ion in Problem 46. Identify each as an anion or a cation. *6.1*

48. If ionic compounds are composed of charged particles (ions), why isn't every ionic compound either positively or negatively charged? *6.2*

49. Would you expect the following pairs of atoms to combine chemically to give an ionic or a molecular compound? *6.2*
a. Li and S **b.** O and S **c.** Al and O **d.** F and Cl **e.** I and K **f.** H and N

50. Four students heat different masses of a blue substance. The blue substance decomposes into a white powder and a colorless gas. The mass of the white powder and the mass of the colorless gas for each sample is plotted on the graph below. *6.2*

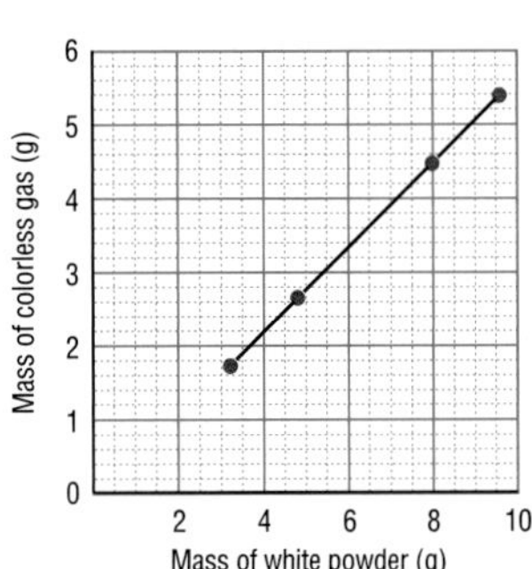

a. Does the graph support the idea that all four samples of blue substance are the same material? Explain.
b. Assuming the blue powder is a compound, what law do the data presented here demonstrate?
c. On heating, a fifth sample of the same blue substance gives 6.2 g of white powder. How many grams of colorless gas are also produced?

51. Identify the number and kinds of atoms present in a molecule of each compound. *6.2*
a. ascorbic acid (vitamin C), $C_6H_8O_6$
b. monosodium glutamate (MSG), $C_5H_8O_4Na$
c. sucrose (table sugar), $C_{12}H_{22}O_{11}$
d. trinitrotoluene (TNT), $C_7H_5N_3O_6$
e. ammonium nitrate (fertilizer), NH_4NO_3

52. The melting point of a compound is 1240 °C. Is this compound an ionic or a molecular compound? *6.2*

53. Write the symbol for each ion. Be sure to include the charge. *6.3*
a. oxide ion **b.** lead(II) ion **c.** lithium ion **d.** nitride ion **e.** cupric ion **f.** fluoride ion

54. Name the following ions. Use Table 6.3 if necessary. *6.3*
a. Ba^{2+} **b.** I^- **c.** Ag^+ **d.** Hg^{2+} **e.** P^{3-} **f.** Sn^{4+}

55. Write the names and formulas of the two polyatomic anions in **Table 6.4** with names that do not end in *-ite* or *-ate*. *6.3*

56. Without consulting **Table 6.4,** name the following ions. *6.3*
a. OH^- **b.** Pb^{4+} **c.** SO_4^{2-} **d.** O^{2-} **e.** HPO_4^{2-} **f.** $Cr_2O_7^{2-}$ **g.** Al^{3+} **h.** ClO_2^-

57. What is the net ionic charge of every ionic compound? Explain. *6.4*

58. Write formulas for compounds composed of these pairs of ions. *6.4*
a. Na^+, O^{2-} **b.** Sn^{4+}, S^{2-} **c.** K^+, Cl^- **d.** Mg^{2+}, N^{3-}

59. How do you determine the charge of a transition metal cation from the formula of an ionic compound? *6.4*

60. When must parentheses be used in a formula? *6.4*

61. Write the formulas for these compounds. *6.4*
a. calcium carbonate
b. barium hydrogen carbonate
c. lithium hypochlorite
d. tin(IV) dichromate

62. Complete the table on the following page by writing correct formulas for the compounds formed by combining positive and negative ions. Then name each compound. *6.4*

59. By knowing the number of each ion in the formula and the ionic charge of the anion; net ionic charge is zero.

60. Parentheses are used to indicate more than one polyatomic ion.

61. **a.** $CaCO_3$ **b.** $Ba(HCO_3)_2$ **c.** LiClO **d.** $Sn(Cr_2O_7)_2$

62. NH_4NO_3, ammonium nitrate
$(NH_4)_2CO_3$, ammonium carbonate
NH_4CN, ammonium cyanide
$(NH_4)_3PO_4$, ammonium phosphate
$Sn(NO_3)_4$, tin(IV) nitrate
$Sn(CO_3)_2$, tin(IV) carbonate
$Sn(CN)_4$, tin(IV) cyanide
$Sn_3(PO_4)_4$, tin(IV) phosphate
$Fe(NO_3)_3$, iron(III) nitrate
$Fe_2(CO_3)_3$, iron(III) carbonate
$Fe(CN)_3$, iron(III) cyanide
$FePO_4$, iron(III) phosphate
$Mg(NO_3)_2$, magnesium nitrate
$MgCO_3$, magnesium carbonate
$Mg(CN)_2$, magnesium cyanide
$Mg_3(PO_4)_2$, magnesium phosphate

	NO_3^-	CO_3^{2-}	CN^-	PO_4^{3-}
NH_4^+				
Sn^{4+}				
Fe^{3+}				
Mg^{2+}				

63. What are the components of a binary molecular compound? *6.5*

64. Write the formula or name for these compounds. *6.5*
a. boron trichloride **c.** N_2O_5
b. dinitrogen tetrahydride **d.** CCl_4

65. Give the name or the formula for these acids. *6.5*
a. HCl **c.** sulfuric acid
b. HNO_3 **d.** acetic acid

66. Which prefix indicates each of the following numbers of atoms in a molecular compound formula? *6.5*
a. 3 **b.** 1 **c.** 2 **d.** 6 **e.** 5 **f.** 4

CONCEPT MASTERY

67. Name these compounds.
a. $NaClO_3$ **d.** AlI_3 **g.** $KHSO_4$
b. Hg_2Br_2 **e.** SnO_2 **h.** CaH_2
c. K_2CrO_4 **f.** $Fe(C_2H_3O_2)_3$

68. Write formulas for these compounds.
a. potassium permanganate
b. calcium hydrogen carbonate
c. dichlorine heptoxide
d. trisilicon tetranitride
e. sodium dihydrogen phosphate
f. phosphorus pentabromide
g. carbon tetrachloride

69. Name each substance. Use **Figure 6.21** if necessary.
a. $LiClO_4$ **d.** CaO **g.** $SrSO_4$
b. Cl_2O **e.** $Ba_3(PO_4)_2$ **h.** $CuC_2H_3O_2$
c. HgF_2 **f.** I_2

70. Write formulas for these compounds.
a. magnesium sulfide **e.** sulfite ion
b. nitrogen gas **f.** calcium carbonate
c. barium hydroxide **g.** sodium bromide
d. copper(II) nitrite **h.** ferric sulfate

71. Name each compound.
a. $Mg(MnO_4)_2$ **d.** N_2H_4 **g.** PI_3
b. $Be(NO_3)_2$ **e.** LiOH **h.** ZnO
c. K_2CO_3 **f.** BaF_2

72. The United States produces thousands of different kinds of inorganic chemicals. The following data table shows the amounts (in billions of kg) of the top ten inorganic chemicals produced in a recent year.

Chemical	Amount produced (billions of kg)
Sulfuric acid	39.4
Nitrogen	26.9
Oxygen	17.7
Ammonia	16.5
Lime	16.3
Phosphoric acid	11.2
Sodium hydroxide	11.0
Chlorine	10.3
Sodium carbonate	9.3
Nitric acid	6.8

a. For what percentage of the total production of the top ten did lime (calcium oxide) account?
b. Three diatomic gases are on the list. What are their names and what was their total production?
c. For what percentage of the total production of the top ten did the three acids account?
d. Write formulas for the top ten inorganic chemicals.

73. Write formulas for these compounds.
a. calcium bromide **e.** tin(IV) cyanide
b. silver chloride **f.** lithium hydride
c. aluminum carbide **g.** strontium acetate
d. nitrogen dioxide **h.** sodium silicate

CRITICAL THINKING

74. Choose the term that best completes the second relationship.
a. black : white
cation : ______
(1) anion (3) sodium ion
(2) metal (4) iron ion
b. plant : oak tree
anion : ______
(1) sodium ion (3) cation
(2) magnesium ion (4) chloride ion
c. molecular compound : molecule
ionic compound : ______
(1) cation (3) anion
(2) formula unit (4) metalloid

63. two nonmetals
64. a. BCl_3 **c.** dinitrogen pentoxide
b. N_2H_4 **d.** carbon tetrachloride
65. a. hydrochloric acid **c.** H_2SO_4
b. nitric acid **d.** $HC_2H_3O_2$
66. a. tri- **d.** hexa-
b. mono- **e.** penta-
c. di- **f.** tetra-
67. a. sodium chlorate
b. mercury(I) bromide
c. potassium chromate
d. aluminum iodide
e. tin(IV) oxide
f. iron(III) acetate
g. potassium hydrogen sulfate
h. calcium hydride
68. a. $KMnO_4$ **e.** NaH_2PO_4
b. $Ca(HCO_3)_2$ **f.** PBr_5
c. Cl_2O_7 **g.** CCl_4
d. Si_3N_4
69. a. lithium perchlorate
b. dichlorine monoxide
c. mercury(II) fluoride
d. calcium oxide
e. barium phosphate
f. iodine
g. strontium sulfate
h. copper(I) acetate
70. a. MgS **e.** SO_3^{2-}
b. N_2 **f.** $CaCO_3$
c. $Ba(OH)_2$ **g.** NaBr
d. $Cu(NO_2)_2$ **h.** $Fe_2(SO_4)_3$
71. a. magnesium permanganate
b. beryllium nitrate
c. potassium carbonate
d. dinitrogen tetrahydride
e. lithium hydroxide
f. barium fluoride
g. phosphorus triiodide
h. zinc oxide
72. a. 9.85%
b. nitrogen, oxygen, chlorine; 54.9 kg
c. 34.7%
d. H_2SO_4, N_2, O_2, NH_3, CaO, H_3PO_4, NaOH, Cl_2, Na_2CO_3, HNO_3
73. a. $CaBr_2$ **e.** $Sn(CN)_4$
b. AgCl **f.** LiH
c. Al_4C_3 **g.** $Sr(C_2H_3O_2)_2$
d. NO_2 **h.** Na_2SiO_3
74. a. 1 **b.** 4 **c.** 2

Answers

75. A molecular formula shows the number of each kind of atom in a molecule of the compound. The formula unit shows the lowest whole-number ratio of ions in a compound.
76. on the right side
77. Common names vary in different languages and might vary in one language.
78. The statement is true for the representative metals, but not for transition metals, which often have multiple charges.
79. Possible answers include: cations always come before anions; when a cation has more than one ionic charge, the charge is indicated by a Roman numeral; monatomic anions use an *-ide* ending.
80. **a.** 7.75×10^5 μL
b. 208 K
c. 0.832 cg
81. **a.** 12 protons, 10 electrons
b. 35 protons, 36 electrons
c. 38 protons, 36 electrons
d. 16 protons, 18 electrons
82. Answers may include color and state (physical), ability to burn or rust (chemical).
83. 0.538 g/cm^3
84. See *Solutions Manual* for answers.
85. **a.** Potassium carbonate has a much greater water solubility than $CaCO_3$.
b. The copper compound is blue; the iron compound is white.
c. Add water to dissolve the NH_4Cl; then filter out the insoluble $BaSO_4$.
d. chlorine (nonmetal), sulfur (nonmetal), bromine (nonmetal), barium (metal), iodine (nonmetal), mercury (metal)
e. barium sulfate, calcium carbonate, potassium carbonate, copper(II) sulfate pentahydrate, iron(II) sulfate pentahydrate, ammonium chloride
f. 639 g
g. 7.54 cm^3
h. color, density, melting point, or boiling point

75. Compare and contrast the information conveyed by a molecular formula with that given by a formula unit of a compound.

76. Where on the periodic table will you find the two elements in a binary molecular compound?

77. Why is it important for chemists to have a scientific system of writing chemical names and formulas?

78. Criticize this statement: "The ionic charge of any metal can be determined from the position of the element in the periodic table."

79. Summarize the rules that chemists use for naming ionic compounds. What is the purpose for each rule?

CUMULATIVE REVIEW

80. Make the following conversions.
a. 775 mL to microliters (μL)
b. −65 °C to K
c. 8.32 mg Ag to centigrams of silver (cg Ag)

81. How many protons and electrons are in each ion?
a. magnesium ion
b. bromide ion
c. strontium ion
d. sulfide ion

82. List five properties of the chair you are sitting on. Classify each as physical or chemical.

83. A student finds that 6.62 g of a substance occupies a volume of 12.3 cm^3. What is the density of the substance?

CONCEPT CHALLENGE

84. The *Handbook of Chemistry and Physics* is a reference work that contains a wealth of information about elements and compounds. Two sections of this book you might use are "Physical Constants of Inorganic Compounds" and "Physical Constants of Organic Compounds." To familiarize yourself with this work, make a table with these headings: Name, Formula, Crystalline form or color, Density, Melting point (°C), Boiling point (°C), and Solubility in water. Enter these substances in the body of the table: ammonium chloride, barium, barium sulfate, bromine, calcium carbonate, chlorine, copper(II) sulfate pentahydrate, iodine, iron(II) sulfate pentahydrate, mercury, potassium carbonate, and sulfur. Use the *Handbook* to complete the table.

85. Use the table you prepared for Problem 84 to answer the following questions.
a. You have two unlabeled bottles, each containing a white powder. One of the substances is calcium carbonate, and the other is potassium carbonate. Describe a simple physical test you could carry out to distinguish between these two compounds.
b. How would you distinguish between samples of copper(II) sulfate pentahydrate and iron(II) sulfate pentahydrate?
c. A bottle contains a mixture of ammonium chloride and barium sulfate. What method could you use to separate these two compounds?
d. List the elements in the table in order of increasing density. Identify the elements as metals or nonmetals.
e. List the compounds in the table in order of decreasing density.
f. Calculate the mass of 47.0 cm^3 of mercury.
g. Calculate the volume of 16.6 g of sulfur.
h. How would you distinguish among the Group 7A elements (halogens) listed in the table?

Chapter 6 STANDARDIZED TEST PREP

Select the choice that best answers each question or completes each statement.

1. Identify the pair in which the formula does *not* match the name.
   a. sulfite, SO_3^{2-}
   b. dichromate, $Cr_2O_7^{2-}$
   c. hydroxide, OH^-
   d. nitrite, NO_3^-
   e. perchlorate, ClO_4^-
2. Which of these compounds are ionic?
   I. $CaSO_4$
   II. N_2O_4
   III. NH_4NO_3
   IV. CaS
   a. I and II only
   b. II and III only
   c. III and IV only
   d. I, III, and IV only
   e. I, II, III, and IV
3. A sample of aspirin contains 6.00 g of carbon and 3.56 g of oxygen. A second sample of aspirin contains 4.20 g of carbon. How many grams of oxygen are in the second sample?
   a. 2.49 g
   b. 5.09 g
   c. 3.08 g
   d. 2.44 g
4. The Roman numeral in manganese(IV) sulfide indicates the
   a. group number on the periodic table.
   b. positive charge on the manganese ion.
   c. number of manganese ions in the formula.
   d. the number of sulfide ions needed in the formula.
5. Which of these statements does *not* describe every binary molecular compound?
   a. Molecules of binary molecular compounds are composed of two atoms.
   b. The names of binary molecular compounds contain prefixes.
   c. The names of binary molecular compounds end in the suffix *–ide.*
   d. Binary molecular compounds are composed of two nonmetals.
6. What is the formula of ammonium carbonate?
   a. NH_4CO_3
   b. NH_4C
   c. $(NH_4)_2CO_3$
   d. NH_3CO_4
   e. $(NH_3)_2CO_4$

The lettered choices below refer to questions 7–11. A lettered choice may be used once, more than once, or not at all.

(A) PQ
(B) PQ_2
(C) P_2Q
(D) P_2Q_3

Which formula shows the correct ratio of ions in the compound formed by each pair of elements?

	Element P	*Element Q*
7.	aluminum	sulfur
8.	potassium	oxygen
9.	lithium	chlorine
10.	strontium	bromine
11.	sodium	sulfur

Use the data table to answer questions 12–13.

Cation	Anion			
	A	B	C	D
M	MA_2	(1)	(2)	MD
N	(3)	N_2B	(4)	(5)
P	PA_3	(6)	PC	$P_2(D)_3$

The table gives formulas for some of the ionic compounds formed when cations (M, N, P) combine with anions (A, B, C, D).

12. Use the given formulas to determine the ionic charge of each cation and anion.
13. Write formulas for compounds (1) through (6).

Use the atomic windows to answer questions 14–15.

a.
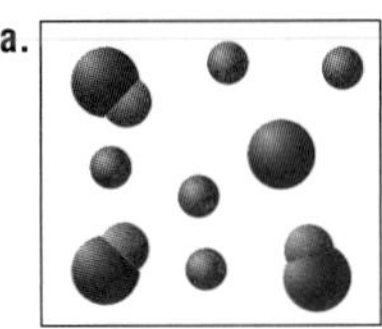

c.
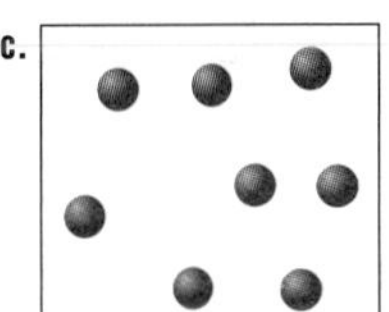

b.
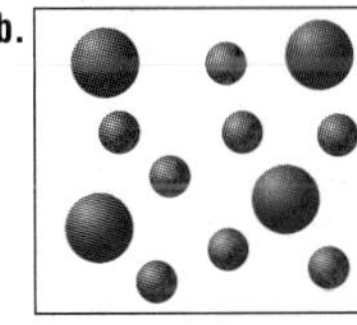

d.
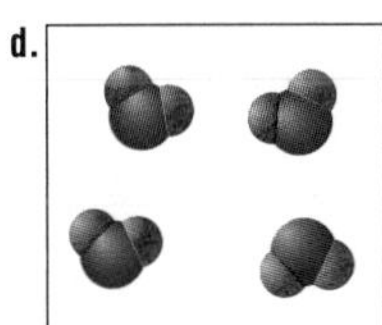

14. Identify the contents of each atomic window as a substance or a mixture.
15. Classify the contents as elements only, compounds only, or elements and compounds.

Chapter 6 STANDARDIZED TEST PREP

1. d
2. d
3. a
4. b
5. a
6. c
7. D
8. C
9. A
10. B
11. C
12. cations: M^{2+}, N^{+}, P^{3+}; anions: A^{-}, B^{2-}, C^{3-}, D^{2-}
13. (1) MB, (2) M_3C_2, (3) NA, (4) N_3C, (5) N_2D, (6) P_2B_3
14. C and D are substances; A and B are mixtures.
15. A contains elements and compounds; B and C contain only elements; D contains only a compound.

Planning Guide

SECTION OBJECTIVES	ACTIVITIES/FEATURES	MEDIA & TECHNOLOGY
7.1 The Mole: A Measurement of Matter ○ ■ ◆ ▸ Describe how Avogadro's number is related to a mole of any substance ▸ Calculate the mass of a mole of any substance	**SE** Discover It! *Counting By Measuring Mass,* p. 170 **TE** DEMOS, pp. 173, 178	**ASAP** Animation 6 **ASAP** Problem Solving 1, 4 **ASAP** Assessment 7.1 **OT** 10: *Gram Formula Mass* **OT** 11: *Gram Molecular Mass* **CHM** Side 4, 20: *Visualizing Moles*
7.2 Mole-Mass and Mole-Volume Relationships ○ ■ ◆ ▸ Use the molar mass to convert between mass and moles of a substance ▸ Use the mole to convert among measurements of mass, volume, and number of particles	**SE** Small-Scale Lab *Measuring Mass As a Means of Counting,* p. 187 (LRS 7-1) LM 8: *The Masses of Equal Volumes of Gases* SSLM 7: *Weighing: A Means of Counting*	**ASAP** Simulation 3 **ASAP** Problem Solving 16, 20, 22 **ASAP** Assessment 7.2 **OT** 12: *Mole Conversion*
7.3 Percent Composition and Chemical Formulas ■ ◆ ▸ Calculate the percent composition of a substance from its chemical formula or experimental data ▸ Derive the empirical formula and the molecular formula of a compound from experimental data	**SE** **CHEMath** *Fractions, Ratios, and Percent,* p. 190 **SE** Mini Lab *Percent Composition,* p. 195 (LRS 7-2) **SE** **Chemistry Serving . . . the Environment** *Water Worth Drinking,* p. 196 **SE** **Chemistry in Careers** *Ecologist,* p. 196 LM 9: *Empirical Formula Determination*	**ASAP** Problem Solving 30, 34, 36, 37 **ASAP** Assessment 7.3 **ASAP** Concept Map 7 **CHM** Side 4, 30: *Percent Composition* **PP** Chapter 7 Problems **RP** Lesson Plans, Resource Library **AR** Computer Test 7 **www** Activities, Self-Tests, *SCIENCE NEWS* updates **TCP** The Chemistry Place Web Site

KEY

- ● Conceptual (concrete concepts)
- ○ Conceptual (more abstract/math)
- ■ Standard (core content)
- □ Standard (extension topics)
- ◆ Honors (core content)
- ◇ Honors (options to accelerate)

AR	Assessment Resources
ASAP	Chem ASAP! CD-ROM
ACT	ActivChemistry CD-ROM
CHM	CHEMedia Videodiscs
CA	Chemistry Alive! Videodiscs
GCP	Graphing Calculator Problems
GRS	Guided Reading and Study Workbook
LM	Laboratory Manual
LP	Laboratory Practicals
LRS	Laboratory Recordsheets
SSLM	Small-Scale Lab Manual

PRACTICE		ASSESSMENT	
SE	**Sample Problems** 7-1 to 7-5	SE	Section Review
GRS	Section 7.1	RM	Section Review 7.1
RM	Practice Problems 7.1	RM	Chapter 7 Quiz
RM	Chapter Quiz		
SE	**Sample Problems** 7-6 to 7-9	SE	Section Review
GRS	Section 7.2	RM	Section Review 7.2
RM	Practice Problems 7.2	RM	Chapter 7 Quiz
SE	**Sample Problems** 7-10 to 7-14	SE	Section Review
GRS	Section 7.3	RM	Section Review 7.3
RM	Practice Problems 7.3	LP	Lab Practical 7-1
RM	Interpreting Graphics	RM	Vocabulary Review 7
GCP	Chapter 7	SE	Chapter Review
		SM	Chapter 7 Solutions
		SE	Standardized Test Prep
		PHAS	Chapter 7 Test Prep
		RM	Chapter 7 Quiz
		RM	Chapter 7 A & B Test

PLANNING FOR ACTIVITIES

STUDENT EDITION

Discover It! p. 170
- paper clips
- centigram balance

Small-Scale Lab, p. 187
- pencils and paper
- plastic spoons
- balances
- rulers
- water
- sodium chloride
- calcium carbonate

Mini Lab, p. 195
- centigram balances
- burners
- medium-sized test tubes
- test tube holders
- test tube racks
- spatulas
- hydrated salts $CuSO_4$, $CaCl_2$, Na_2SO_4

TEACHER'S EDITION

Activity, p. 171
- bag of beans, each containing a multiple of 12 beans

Teacher Demo, p. 173
- 1 mole each of various molecular and ionic chemicals displayed in sealed containers

Teacher Demo, p. 178
- display prepared for Teacher Demo on p. 173

Activity, p. 189
- labels from packaged foods

OT	Overhead Transparency	SE	Student Edition
PHAS	PH Assessment System	SM	Solutions Manual
PLM	Probeware Lab Manual	SSV	Small-Scale Video/Videodisc
PP	Problem Pro CD-ROM	TCP	www.chemplace.com
RM	Review Module	TE	Teacher's Edition
RP	Resource Pro CD-ROM	www	www.phschool.com

Chapter 7

Key Terms

7.1 mole (mol), Avogadro's number, representative particle, gram atomic mass (gam), gram molecular mass (gmm), gram formula mass (gfm)
7.2 molar mass, standard temperature and pressure (STP), molar volume
7.3 percent composition, empirical formula

DISCOVER IT!

The number of paper clips determined by mass in Step 2 should equal the number of paper clips counted in Step 3. Any difference could be traced to a measurement error or some variation in the masses of individual paper clips. To measure out 185 paper clips without counting, students should multiply the average mass of a paper clip (from Step 1) by 185. Then they can use the balance to find the corresponding mass of paper clips.
By using a larger sample in Step 1, students reduce the effect of any variation in the masses of individual paper clips on the average mass. However, more time is needed to count out the paper clips.

Chapter 7 CHEMICAL QUANTITIES

This sand sculpture contains millions of grains of sand.

FEATURES

DISCOVER IT! COUNTING BY MEASURING MASS

You need 100 paper clips of the same size and a centigram balance.

1. Find the mass of 25 paper clips. Divide the total mass by 25 to find the average mass of a paper clip. Repeat this step using 25 different paper clips until your average masses agree.
2. Select about 75% of your paper clips and find their mass. Without counting, calculate the number of paper clips in your sample.
3. Count the number of paper clips in your sample.
4. Repeat Steps 2 and 3 with a different sample size.

Did the number of paper clips you counted in the sample (Step 3) equal the number you calculated by mass (Step 2)? Explain how you would use the balance to count out 185 paper clips. What is the advantage of using a larger sample size in Step 1? What is a disadvantage? After completing this chapter, revisit this activity and suggest other ways you might count your paper clips.

Stay current with SCIENCE NEWS
Find out more about chemical quantities:
www.phschool.com

7.1 ○ ■ ◆
7.2 ○ ■ ◆
7.3 ■ ◆

Conceptual Visualize the mole with Table 7.1 and Figures 7.8, 7.9, and 7.10. Note that all problems in the chapter depend on using conversion factors. Because percent composition and empirical formulas are mathematically challenging topics, omit Section 7.3.

Standard It may save time to present the mole road map (Figure 7.13) before having students solve problems. All of the problems need to be covered in a standard curriculum. The CHEMath feature will help students review calculations involving fractions, ratios, and percents. In section 7.3, percent composition and empirical formulas are important concepts but the emphasis on sample problems may be adjusted based on course requirements.

Honors In Section 7.2, three pairs of sample problems are similar except for the unknown; assign the second of each pair as homework: Sample Problems 7-3, 7-5, and 7-7. In Section 7.3, use the Mini Lab as an introduction if students are proficient with conversion factors.

THE MOLE: A MEASUREMENT OF MATTER

Every year, contestants from all over the world travel to Harrison Hot Springs in British Columbia, Canada to compete in the world championship sand sculpture contest. Each contestant creates a beautiful work of art out of millions of tiny grains of sand. **If you assume that sand is pure silicon dioxide (SiO_2), what chemical unit could you use to measure the amount of sand in a sand sculpture?**

section 7.1

objectives

- Describe how Avogadro's number is related to a mole of any substance
- Calculate the mass of a mole of any substance

key terms

- mole (mol)
- Avogadro's number
- representative particle
- gram atomic mass (gam)
- gram molecular mass (gmm)
- gram formula mass (gfm)

What Is a Mole?

You live in a quantitative world. The grade you got on your last exam, the number of times you heard your favorite song on the radio yesterday, and the cost of a bicycle you would like to own are all important quantities to you. These are quantities that answer such questions as "how much?" or "how many?" Scientists spend time answering similar questions. How many kilograms of iron can be obtained from one kilogram of iron ore? How many grams of the elements hydrogen and nitrogen must be combined to make 200 grams of the fertilizer ammonia (NH_3)? These two questions illustrate that chemistry is a quantitative science. In your study of chemistry, you will analyze the composition of samples of matter. You will also perform chemical calculations relating quantities of reactants and products to chemical equations. To solve these and other problems, you will have to be able to measure the amount of matter you have.

How do you measure matter? One way is to count how many of something you have. For example, you can count the CDs in your collection or the number of pins you knock down when bowling. Another way to measure matter is to determine its mass or weight. You can buy potatoes by the kilogram or pound and gold by the gram or ounce. You can also measure matter by volume. For instance, people buy gasoline by the liter or gallon and take cough medicine by the milliliter or teaspoon. Often, more than one method of measurement—a count, a mass, a volume—can be used. For example, you can buy soda by the six-pack or by the liter. **Figure 7.2** on the following page shows how some everyday items are measured.

Figure 7.1
Jellybeans are often sold by either weight or mass.

STUDENT RESOURCES

From the Teacher's Resource Package, use:

- Section Review 7.1, Ch. 7 Practice Problems and Quizzes from the Review Module (Ch. 5–8)

TECHNOLOGY RESOURCES

Relevant technology resources include:

- Chem ASAP! CD-ROM
- ResourcePro CD-ROM

7.1

1 Engage

Use The Visual

If possible, bring some sand to class. As you pour it from your hand, ask students how you could measure the sand. Ask:

- **Is it practical to count each grain of sand?** (Students should realize that it is not practical to measure sand by counting individual grains.)
- **How else might you measure or quantify, the sand?** (find its mass or volume)

Lead students to see that just as a small amount of sand contains millions of smaller particles, so too do chemical substances.

Check Prior Knowledge

Have students review the definitions of terms they will be working with in this chapter. Be sure they know the meaning of mass and volume and the units used to describe them. Ask them what they know about the particles that make up substances. Have students define key terms such as element, compound, atom, molecule, and ion.

ACTIVITY

Students usually need to review scientific notation and multiplication and division of exponents. Write several numbers on the board using standard notation. Have students rewrite the numbers using scientific notation. Then pass around bags of beans. In each bag have a multiple of 12 beans. Ask the students how they would go about describing the quantity of beans in the bags. They could count, weigh, or find the volume of the beans. With multiples of 12 in each bag, students may be guided to how the word "dozen" can be used to describe the quantity of beans.

2 Teach

Discuss

The major concepts presented in this section include the use of conversion factors in the factor-label method of problem solving. Ask students to write an equality that is also a definition, such as 3 feet = 1 yard. Write their equalities on the board. (60 min = 1 h; 12 in. = 1 ft; 60 s = 1 min; 2.54 cm = 1 in.; 100 cm = 1 m) **Point out that equalities can be written in either direction, so 60 min = 1 h can be written as 1 h = 60 min.**
Ask:

- **How many seconds are in 2.5 hours?** (9000 s)

Show students how to set up the factor-label method to do this problem.

Some of the units used when measuring always indicate a specific number of items. For example, a pair always means two. A pair of shoes is two shoes, and a pair of aces is two aces. Similarly, a dozen always means 12 (except for a baker's dozen which is 13). A dozen eggs is 12 eggs, a dozen pens is 12 pens, and a dozen donuts is 12 donuts.

Apples are commonly measured in three different ways. At a fruit stand, apples are often sold by the *count* (5 for \$2.00). In a supermarket, you usually buy apples by weight (\$0.89/pound) or *mass* (\$1.95/kg). At an orchard, you can buy apples by *volume* (\$9.00/bushel). Each of these different ways to measure apples—by count, by mass, and by volume—can be equated to a dozen apples.

By count:
1 dozen apples = 12 apples

For average-sized apples the following approximations can be used.

By mass:
1 dozen apples = 2.0 kg apples

By volume:
1 dozen apples = 0.20 bushel apples

Knowing how the count, mass, and volume of apples relate to a dozen apples allows you to convert between these units. For example, you could calculate the mass of a bushel of apples or the mass of 90 average-sized apples using conversion factors based on the unit relationships given above.

In chemistry, you will do calculations using a measuring unit called a mole. The mole, the SI unit that measures the amount of substance, is a unit just like the dozen. The mole can be related to the number of particles (a count), the mass, and the volume of an element or a compound just as a dozen was related to these three units for apples.

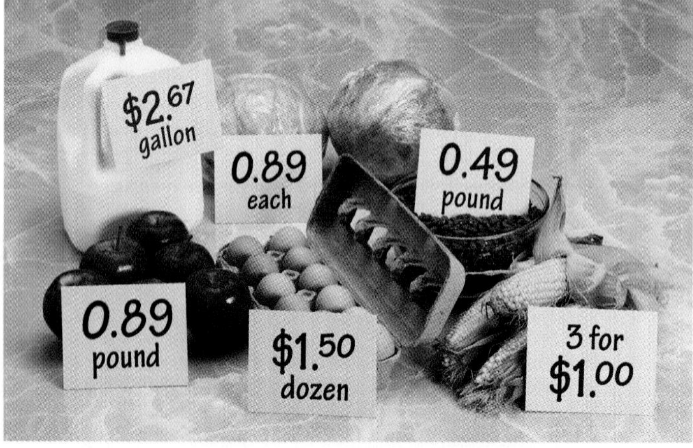

Figure 7.2
Items are often sold by different types of measurements, such as a count, a weight or mass, or a volume. Which of these common supermarket items are being sold by weight? By volume? By count? ❶

Meeting Diverse Needs

At Risk Students will experience difficulty with the factor-label method and should be given additional simple conversion problems for practice before they start using the mole. Review the use of parentheses and the fraction bar as grouping symbols that indicate the order of operations. Stress the importance of labels.

Answer

❶ apples and beans; milk; eggs, lettuce, corn

Sample Problem 7-1

What is the mass of 90 average-sized apples?

1. ***ANALYZE*** ***List the knowns and the unknown.***

Knowns:
- number of apples = 90 apples
- 12 apples = 1 dozen apples
- 1 dozen apples = 2.0 kg apples

Unknown:
- mass of 90 apples = ? kg

The desired conversion is:

number of apples ⟶ mass of apples

This conversion can be carried out by performing the following sequence of conversions:

number ⟶ dozens ⟶ mass of apples

2. ***CALCULATE*** ***Solve for the unknown.***

The first conversion factor is $\frac{\text{1 dozen apples}}{\text{12 apples}}$.

The second conversion factor is $\frac{\text{2.0 kg apples}}{\text{1 dozen apples}}$.

Multiplying the original number of apples by these two conversion factors yields the answer in kilograms,

$$90 \cancel{\text{apples}} \times \frac{1 \cancel{\text{dozen apples}}}{12 \cancel{\text{apples}}} \times \frac{\text{2.0 kg apples}}{1 \cancel{\text{dozen apples}}}$$

$= 15$ kg apples

3. ***EVALUATE*** ***Does the result make sense?***

Because a dozen apples has a mass of 2.0 kg and 90 apples is less than 10 dozen apples, the mass should be less than 20 kg of apples (10 dozen × 2.0 kg/dozen).

Practice Problems

1. What is the mass of 0.50 bushel of apples?
2. Assume that a variety of apples has 8 seeds in each apple. How many apple seeds are in 14 kg of apples?

Chem ASAP!

Problem-Solving 1

Solve Problem 1 with the help of an interactive guided tutorial.

The Number of Particles in a Mole

In Chapter 2, you learned that matter is composed of different kinds of particles. One way to measure the amount of a substance is to count the number of particles in that substance. Because atoms, molecules, and ions are exceedingly small, the number of individual particles in a sample (even a very small sample) of any substance is very large. Counting the particles is not practical. However, you can count particles if you introduce a term that represents a specified number of particles. Just as a dozen eggs represents 12 eggs, a **mole (mol)** of a substance represents 6.02×10^{23} representative particles of that substance. The experimentally determined number 6.02×10^{23} is called **Avogadro's number,** in honor of Amedeo Avogadro di Quaregna (1776–1856).

The term **representative particle** refers to the species present in a substance: usually atoms, molecules, or formula units (ions). The representative particle of most elements is the atom. Iron is composed of iron atoms. Helium is composed of helium atoms. Seven elements, however, normally exist as diatomic molecules (H_2, N_2, O_2, F_2, Cl_2, Br_2, and I_2). The

Figure 7.3

Although Amedeo Avogadro clarified the difference between atoms and molecules, he did not calculate the value that is named after him. Avogadro's number was given his name to honor his contributions to science.

TEACHER DEMO

Create a display for students to examine and study. Measure out 1 mole each of a variety of common chemicals. Place them in sealed containers so that students can handle them and see the difference in the mass and volume of one mole of different substances.

Your collection should have at least two substances from each of the following categories: molecules = sucrose, water, and paradichlorobenzene; formula units = cobalt(II) chloride, potassium hydroxide, potassium dichromate; atoms = sulfur, iron, carbon, mercury.

On each container, mark the mass and number of moles of each element. For example: Sucrose ($C_{12}H_{22}O_{11}$), 342.0 g, 1 mol sucrose molecules, 12 mol carbon atoms, 22 mol hydrogen atoms, and 11 mol oxygen atoms. Cobalt(II) chloride ($CoCl_2$), 129.9 g, 1 mol cobalt ions, 2 mol chloride ions.

Point out that each container holds a mole of a chemical, no matter whether the representative particles are molecules, atoms, or formula units. Note that a mole of each substance has a different mass, but the same number of representative particles—6.02×10^{23}.

Practice Problems Plus

Additional Practice Problem

The Chem ASAP! CD-ROM contains the following problem: How many bushels of apples are in 1.0 kg of apples? (0.10 bushels)

Meeting Diverse Needs

Gifted Students may wish to investigate different methods for determining Avogadro's number. They could present their findings to the class.

Sight Impaired For Teacher Demo, mark the levels of the filled containers so that sight-impaired students can tell by handling the displays that moles of different substances occupy different volumes and have different masses.

LEP and **At Risk** Gram atomic mass, gram molecular mass, and gram formula mass all sound alike. Have students compile a glossary in which they define each term in English and in their native language. Encourage students to include an example beside each definition. Students should be able to use the periodic table and calculate the mass of a mole of any substance.

7.1

Use the Visual

Have students study Figure 7.4 and read the text that discusses the number of particles in a mole on page 175. Point out that the mole represents a number of items just as dozen, gross, and ream all represent a number of items. Ask:

- **How many items in a dozen? A gross? A ream?** (12, 144, 500) **The smaller the item you are counting, the larger the number should be. A mole = 6.02×10^{23} items and is used to measure extremely small objects.**

Practice Problems Plus

Related Chapter Review Problems
Chapter Review problems 47, 48, and 49 are related to Sample Problem 7-2.

Additional Practice Problems

1. How many moles are equal to 2.41×10^{24} formula units of sodium chloride (NaCl)? (4.00 moles)

2. How many moles are equal to 9.03×10^{24} atoms of mercury (Hg)? (15.0 moles)

3. How many atoms are equal to 4.50 moles of copper (Cu)? (2.71×10^{24})

4. How many molecules are equal to 10.0 moles of carbon dioxide (CO_2)? (6.02×10^{24})

The Chem ASAP! CD-ROM contains the following problem: How many formula units are 0.670 mol of sodium chloride? (4.03×10^{23})

Figure 7.4
There are words other than mole used to describe a number of something—for example, a dozen (12) eggs, a gross (144) of pencils, and a ream (500 sheets) of paper.

representative particle of these elements and of all molecular compounds is the molecule. The molecular compounds water (H_2O) and sulfur dioxide (SO_2) are composed of H_2O and SO_2 molecules, respectively. For ionic compounds, the formula unit is the representative particle. The ionic compound calcium chloride is composed of $CaCl_2$ formula units. Calcium ions and chloride ions are present in a one to two ratio in this formula unit. **Table 7.1** summarizes the relationship between representative particles and moles of substances. Remember that a mole of any substance always contains 6.02×10^{23} representative particles.

Table 7.1

Representative Particles and Moles

Substance	Representative particle	Chemical formula	Representative particles in 1.00 mol
Atomic nitrogen	Atom	N	6.02×10^{23}
Nitrogen gas	Molecule	N_2	6.02×10^{23}
Water	Molecule	H_2O	6.02×10^{23}
Calcium ion	Ion	Ca^{2+}	6.02×10^{23}
Calcium fluoride	Formula unit	CaF_2	6.02×10^{23}
Sucrose	Molecule	$C_{12}H_{22}O_{11}$	6.02×10^{23}

Sample Problem 7-2

How many moles of magnesium is 1.25×10^{23} atoms of magnesium?

1. *ANALYZE* ***List the knowns and the unknown.***

Knowns:
- number of atoms = 1.25×10^{23} atoms Mg
- 1 mol Mg = 6.02×10^{23} atoms Mg

Unknown:
- moles = ? mol Mg

The desired conversion is:

$$\text{atoms} \longrightarrow \text{moles}$$

2. *CALCULATE* ***Solve for the unknown.***

The conversion factor is $\dfrac{1 \text{ mol Mg}}{6.02 \times 10^{23} \text{ atoms Mg}}$.

Multiplying atoms of Mg by the conversion factor yields the answer.

$$1.25 \times 10^{23} \cancel{\text{atoms Mg}} \times \frac{1 \text{ mol Mg}}{6.02 \times 10^{23} \cancel{\text{atoms Mg}}}$$

$$= 2.08 \times 10^{-1} \text{ mol Mg}$$

3. *EVALUATE* ***Does the result make sense?***

Because the given number of atoms is less than one-fourth of Avogadro's number, the answer should be less than one-fourth mole of atoms. The answer should have three significant figures.

Practice Problems

3. How many moles is 2.80×10^{24} atoms of silicon?

4. How many molecules is 0.360 mol of water?

Chem ASAP!
Problem-Solving 4
Solve Problem 4 with the help of an interactive guided tutorial.

Now suppose you want to determine how many atoms are in a mole of a compound. To do this, you must know how many atoms are in a representative particle of the compound. This number is determined from the chemical formula. For example, each molecule of carbon dioxide (CO_2) is composed of three atoms: one carbon atom and two oxygen atoms. A mole of carbon dioxide contains Avogadro's number of CO_2 molecules. Thus a mole of carbon dioxide contains three times Avogadro's number of atoms. A molecule of carbon monoxide (CO) consists of two atoms, so a mole of carbon monoxide contains two times Avogadro's number of atoms. To find the number of atoms in a mole of a compound, you must first determine the number of atoms in a representative particle of that compound and then multiply that number by Avogadro's number. **Figure 7.5** illustrates this idea with marbles (atoms) in cups (molecules).

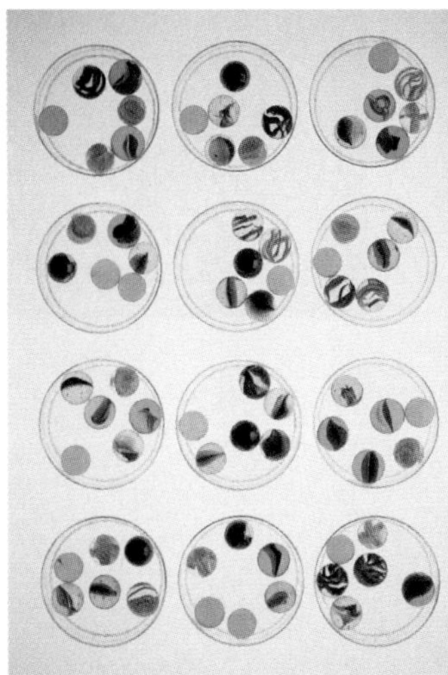

Figure 7.5
A dozen cups of marbles contain more than a dozen marbles. Similarly, a mole of molecules contains more than a mole of atoms. How many atoms are in one mole of molecules if each molecule consists of six atoms? 1

Sample Problem 7-3

How many atoms are in 2.12 mol of propane (C_3H_8)?

1. *ANALYZE List the knowns and the unknown.*

Knowns:
- number of moles = 2.12 mol C_3H_8
- 1 mol C_3H_8 = 6.02×10^{23} molecules C_3H_8
- 1 molecule C_3H_8 = 11 atoms (3 carbon atoms and 8 hydrogen atoms)

Unknown:
- number of atoms = ? atoms

The desired conversion is:

moles ⟶ molecules ⟶ atoms

Using the relationships among units given above, the desired conversion factors can be written.

2. *CALCULATE Solve for the unknown.*

The first conversion factor is $\dfrac{6.02 \times 10^{23} \text{ molecules } C_3H_8}{1 \text{ mol } C_3H_8}$.

The second conversion factor is $\dfrac{11 \text{ atoms}}{1 \text{ molecule } C_3H_8}$.

Multiplying the moles of C_3H_8 by the proper conversion factors yields the answer.

$$2.12 \cancel{\text{mol } C_3H_8} \times \frac{6.02 \times 10^{23} \cancel{\text{molecules } C_3H_8}}{1 \cancel{\text{mol } C_3H_8}} \times \frac{11 \text{ atoms}}{1 \cancel{\text{molecule } C_3H_8}}$$

$= 1.4039 \times 10^{25}$ atoms $= 1.40 \times 10^{25}$ atoms

3. *EVALUATE Does the result make sense?*

Because there are 11 atoms in each molecule of propane and more than 2 mol of propane, the answer should be more than 20 times Avogadro's number of propane molecules. The answer has three significant figures based on the three significant figures in the given measurement.

Practice Problems

5. How many atoms are there in 1.14 mol SO_3?
6. How many moles are there in 4.65×10^{24} molecules of NO_2?

7.1

Use the Visual

Have students examine the photograph in Figure 7.5. Note that each dish contains exactly 6 marbles. A dozen dishes provide you with 72 marbles. Ask the students to think of everyday items that come in a "package" of more than one. (gum = 5 sticks, shoes = 2, batteries = 2, 4, 6) **A dozen packages of gum would be 5 dozen sticks of gum or 60 sticks of gum. Point out that this idea can be applied to molecules. For example, in a mole of water (H_2O) there are 2 moles of H and 1 mole of O. There are 1.2×10^{24} atoms of hydrogen and 6.02×10^{23} atoms of oxygen.**

Practice Problems Plus

These problems are related to Sample Problem 7-3.

1. How many atoms are in 1.00 mole of sucrose, $C_{12}H_{22}O_{11}$? (2.71×10^{25} atoms)

2. How many atoms of C are in 2.0 moles of $C_{12}H_{22}O_{11}$? (1.4×10^{25}atoms)

3. How many atoms of H are in 2.00 moles of $C_{12}H_{22}O_{11}$? (2.65×10^{25} atoms)

4. How many atoms of O are in 3.65 moles of $C_{12}H_{22}O_{11}$? (2.42×10^{25} atoms)

Answer

1 3.61×10^{24} atoms

7.1

Discuss

Explain that the gram atomic masses of any two elements contain the same number of atoms, because the atomic masses of the elements are relative values. Present this idea by saying to the class, "Suppose that the mass of an atom of element X is twice as great as the mass of an atom of element Y." Now suppose that you have 10 grams of element X and 10 grams of element Y. Would you expect both samples to contain the same number of atoms? (no) **Why not?** (Because atoms of element X are twice as massive as atoms of element Y; therefore, the sample of X would contain only half as many atoms as the sample of Y.) **What would you have to do to get the same number of atoms in both samples?** (Double the mass of element X to twice the mass of element Y.)

Discuss

Point out to students that the mass of a single atom can be expressed in atomic mass units, but it is not realistic to work with single atoms. Chemists work with large numbers of atoms for which the mass can be determined in grams. In this text, the gram atomic masses are rounded to one place after the decimal point. Have students solve a given problem multiple times using a different rounding rule each time so they can see how rounding affects the answer.

Figure 7.6
How big is a mole?

Perhaps you are wondering just how large Avogadro's number is. The SI unit, the mole, is not related to the small burrowing animal of the same name shown in **Figure 7.6.** However, you can use this animal to help develop an appreciation for the size of the number 6.02×10^{23}. Assume that an average animal-mole is 15 cm long, 5 cm tall, and has a mass of 150 g. Based on this information, what is the mass of 6.02×10^{23} animal-moles?

$$6.02 \times 10^{23}\ \cancel{\text{animal-mole}} \times \frac{150\text{ g}}{1\ \cancel{\text{animal mole}}} = 9.03 \times 10^{25}\text{g}$$
$$= 9.03 \times 10^{22}\text{ kg}$$

The mass of animal-moles is equivalent to

- more than 1% of Earth's mass.
- more than 1.3 times the mass of the moon.
- more than 60 times the combined mass of Earth's oceans.

If spread over the entire surface of Earth, Avogadro's number of animal-moles would form a layer more than 8 million animal-moles thick.

What about the length of 6.02×10^{23} animal-moles?

$$6.02 \times 10^{23}\ \cancel{\text{animal-mole}} \times \frac{15\text{ cm}}{1\ \cancel{\text{animal mole}}} = 9.03 \times 10^{24}\text{cm}$$
$$= 9.03 \times 10^{19}\text{ km}$$

If lined up end-to-end, 6.02×10^{23} animal-moles would stretch from Earth to the nearest star, Alpha Centauri, more than two million times.

Suppose you could convince Avogadro's number of animal-moles to line up in 6 billion equal columns. Further suppose that each of the approximately 6 billion people on Earth counted the animal-moles in one column at the rate of 1000 animal-moles per second. Even with that many people counting that fast, it would still take more than 3000 years to count 6.02×10^{23} animal-moles! Are you beginning to understand how enormous Avogadro's number is?

The Mass of a Mole of an Element

Chem ASAP!
Animation 6
Find out how Avogadro's number is based on the relationship between the amu and the gram.

You are always working with large numbers of atoms even if you are using microgram quantities. Even one billion atoms would be a very small amount of a substance. Working with grams of atoms is much easier. The **gram atomic mass (gam)** is the atomic mass of an element expressed in grams. For carbon, the gram atomic mass is 12.0 g. For atomic hydrogen, the gram atomic mass is 1.0 g. **Figure 7.7** shows one gram atomic mass of carbon, sulfur, mercury, and iron. Compare the gram atomic masses in the figure to the atomic masses in your periodic table. Notice that the gram atomic masses were rounded off to one place after the decimal point. All the examples and problems in this text use gram atomic masses that are rounded off in the same way. If your teacher uses a different rounding rule for gram atomic masses, your answers to problems may differ slightly from the answers given in the text.

You learned previously that the atomic mass of an element (the mass of a single atom) is expressed in atomic mass units (amu). Remember that atomic masses of atoms are relative values. The atomic masses of elements found in the periodic table are weighted average masses of the isotopes of

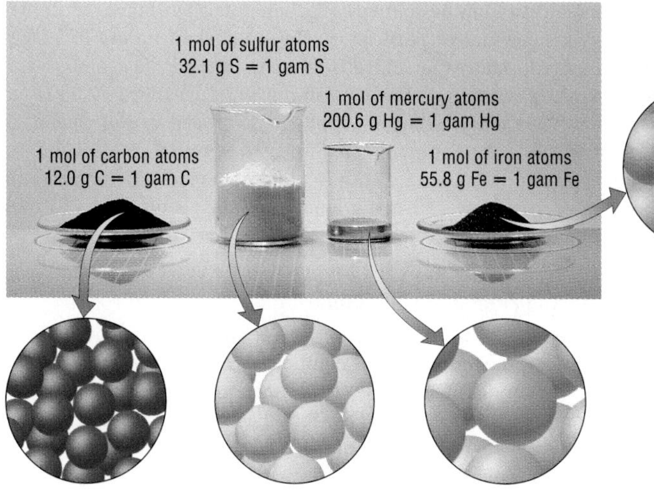

Figure 7.7
One gram atomic mass (gam) is shown for carbon, sulfur, mercury, and iron. Each of these quantities contains one mole, or 6.02×10^{23} atoms, of that substance.

that element. As you can see in **Figure 7.8,** an average carbon atom (C) with an atomic mass of 12.0 amu is 12 times heavier than an average hydrogen atom (H) with an atomic mass of 1.0 amu. Therefore 100 carbon atoms are 12 times heavier than 100 hydrogen atoms. In fact, any number of carbon atoms is 12 times heavier than the same number of hydrogen atoms, as **Figure 7.8** demonstrates. Therefore 12.0 g of carbon atoms and 1.0 g of hydrogen atoms contain the same number of atoms.

The gram atomic masses of any two elements must contain the same number of atoms. If you were to compare 12.0 g of carbon atoms with 16.0 g of oxygen atoms, you would find they contain the same number of atoms. How many atoms are contained in the gram atomic mass of an element? You are now familiar with this quantity—the gram atomic mass of any element contains 1 mol of atoms (6.02×10^{23} atoms) of that element.

Figure 7.8
The mass ratio of equal numbers of carbon atoms to hydrogen atoms is always 12 to 1.

CARBON ATOMS		HYDROGEN ATOMS		MASS RATIO
Number	Mass (amu)	Number	Mass (amu)	$\frac{\text{Mass carbon}}{\text{Mass hydrogen}}$
	12		1	$\frac{12\text{ amu}}{1\text{ amu}} = \frac{12}{1}$
	24 [2 × 12]		2 [2 × 1]	$\frac{24\text{ amu}}{2\text{ amu}} = \frac{12}{1}$
	120 [10 × 12]		10 [10 × 1]	$\frac{120\text{ amu}}{10\text{ amu}} = \frac{12}{1}$
	600 [50 × 12]		50 [50 × 1]	$\frac{600\text{ amu}}{50\text{ amu}} = \frac{12}{1}$
Avogadro's number	$(6.02 \times 10^{23}) \times (12)$	Avogadro's number	$(6.02 \times 10^{23}) \times (1)$	$\frac{(6.02 \times 10^{23}) \times (12)}{(6.02 \times 10^{23}) \times (1)} = \frac{12}{1}$

Use the Visual

Have students study the photograph in Figure 7.7. Have students verify the stated values for gram atomic mass by looking up the values for atomic mass in the periodic table. Point out that the gram atomic mass of an element is numerically equal to the atomic mass. Prepare a similar display of other elements in labeled containers for students to view. Ask students to write equations, like those in Figure 7.7, for each of the elements displayed.

Use the Visual

Have students study Figure 7.8 which will reinforce the concept that since the atomic masses are relative, no matter how many atoms you compare, the relationship will stay the same. Be sure students understand the definition of a mole as "the amount of substance that contains as many representative particles as the number of atoms in 12.0 g of carbon-12."

TEACHER DEMO

Use the display you prepared for the Teacher Demo on p. 173. Select a molecular compound such as water and show students how to calculate the mass of 1 mole using the gam of the elements contained in water. When students have finished the calculation, point out that they have determined the gram molecular mass (gmm) of water. Repeat this exercise for other molecular and ionic compounds in the display. Ask:

- What is the representative particle of an ionic compound? (a formula unit)
- What term is used to express the mass of one mole of an ionic compound? (gram formula mass)

The mole can now be defined as the amount of substance that contains as many representative particles as the number of atoms in 12.0 g of carbon-12. You know that 12.0 g is the gram atomic mass of carbon-12. Because 12.0 g of carbon is the gram atomic mass of carbon, 12.0 g is 1 mol of carbon. The same relationship applies to hydrogen as well; that is, 1.0 g of hydrogen is 1 mol of hydrogen. Similarly, because 24.3 g is the gram atomic mass of magnesium, 24.3 g is 1 mol of magnesium (or 6.02×10^{23} atoms of magnesium). Thus the gram atomic mass is the mass of 1 mol of atoms of any element.

The Mass of a Mole of a Compound

What is the mass of a mole of a compound? To answer this question, you must first know the formula of the compound. The formula of a compound tells you the number of atoms of each element in a representative particle of that compound. For example, the formula of the molecular compound sulfur trioxide is SO_3. A molecule of SO_3 is composed of one atom of sulfur and three atoms of oxygen.

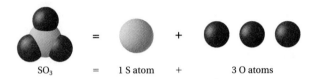

SO_3 = 1 S atom + 3 O atoms

You can calculate the mass of a molecule of SO_3 by adding the atomic masses of the atoms making up the molecule. From the periodic table, the atomic mass of sulfur (S) is 32.1 amu. The mass of three atoms of oxygen is three times the atomic mass of a single oxygen atom (O): 3×16.0 amu = 48.0 amu. Thus the molecular mass of SO_3 is 32.1 amu + 48.0 amu = 80.1 amu.

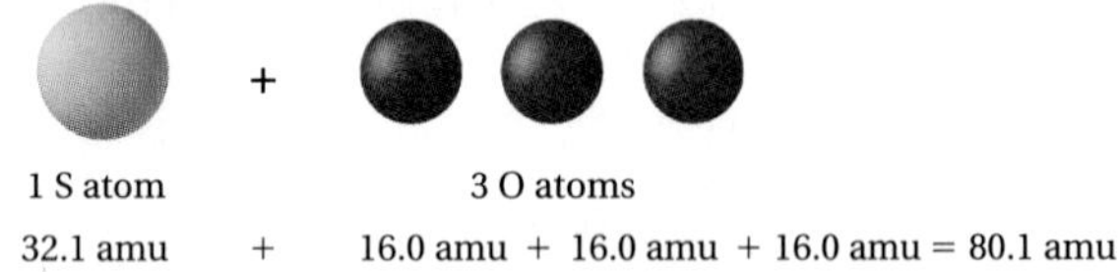

If you now substitute the unit grams for atomic mass units, you will have the gram molecular mass of SO_3. The **gram molecular mass (gmm)** of any molecular compound is the mass of 1 mol of that compound. The gmm equals the molecular mass expressed in grams. Thus 1 mol of SO_3 has a mass of 80.1 g.

Gram molecular masses may be calculated directly from gram atomic masses. For each element in a compound, find the number of grams of that element per mole of the compound. Then sum the masses of the elements in the compound. The gram molecular masses of the molecular compounds in **Figure 7.9** were obtained in this way. Try calculating the gram molecular mass values shown in **Figure 7.9** yourself. Do you get the same values? If not, review the calculation for sulfur trioxide shown above, then try again. Calculating gram molecular mass values is an important skill that you will use often in your study of chemistry.

Figure 7.9
One gram molecular mass (gmm) is shown for each of three molecular compounds. Each of these quantities contains 6.02×10^{23} molecules. Do they each contain the same number of atoms? ❶

Sample Problem 7-4

The molecular formula of hydrogen peroxide is H_2O_2. What is its gram molecular mass?

1. ***ANALYZE*** ***List the knowns and the unknown.***

Knowns:
- molecular formula = H_2O_2
- 1 gam H = 1 mol H = 1.0 g H
- 1 gam O = 1 mol O = 16.0 g O

Unknown:
- gmm = ? g

The molecular formula gives the number of moles of each element in 1 mol of hydrogen peroxide: 2 mol of hydrogen atoms and 2 mol of oxygen atoms. Moles of atoms are converted to grams by using conversion factors (g/mol) based on the gram atomic mass of each element. The sum of the masses of the elements gives the gram molecular mass.

2. ***CALCULATE*** ***Solve for the unknown.***

Use the proper conversion factors to convert moles of hydrogen and oxygen to grams of hydrogen and oxygen. Adding the results gives the answer.

$$2\,\cancel{\text{mol H}} \times \frac{1.0\text{ g H}}{1\,\cancel{\text{mol H}}} = 2.0\text{ g H}$$

$$2\,\cancel{\text{mol O}} \times \frac{16.0\text{ g O}}{1\,\cancel{\text{mol O}}} = 32.0\text{ g O}$$

$$\text{gram molecular mass of } H_2O_2 = 34.0\text{ g}$$

3. ***EVALUATE*** ***Does the result make sense?***

The answer reflects the number of moles of atoms of each element and the gram atomic mass of each element. The answer is expressed to the tenth's place because the numbers being added are expressed to the tenth's place.

Practice Problems

7. Find the gram molecular mass of each compound.
   a. C_2H_6
   b. PCl_3
   c. C_3H_7OH
   d. N_2O_5
8. What is the mass of 1.00 mol of each substance?
   a. chlorine
   b. nitrogen dioxide
   c. carbon tetrabromide
   d. silicon dioxide

Answer

❶ No. Each molecule contains a different number of atoms.

7.1

Practice Problems Plus

Related Chapter Review Problems
Chapter Review problems 50, 51, 52, and 53 are related to Sample Problems 7-4 and 7-5. For additional practice, have students determine the mass of one mole of the following compounds: CO_2 (44.0 g), SO_3 (80.1 g), Br_2 (159.8 g), H_2 (2.0 g), N_2 (28.0 g), NaOH (40.0 g), $Al_2(SO_4)_3$ (342.3 g), and $Ba(NO_3)_2$ (261.3 g).

Discuss

Remind students that the three terms they have learned—*gram atomic mass* (*gam*), *gram molecular mass* (*gmm*), and *gram formula mass* (*gfm*)—describe the same number of representative particles, but differ in which representative particle is being described.

Figure 7.10
One gram formula mass (gfm) is shown for each of three ionic compounds. Each of these quantities contains 6.02×10^{23} formula units. Which of these compounds contains the greatest number of atoms? ❷

It is inappropriate to calculate the gram molecular mass of calcium iodide (CaI_2) because it is an ionic compound. The representative particle of an ionic compound is a formula unit, not a molecule. The mass of one mole of an ionic compound is the **gram formula mass (gfm).** The gfm equals the formula mass expressed in grams. A gram formula mass is calculated the same way as a gram molecular mass: Simply sum the atomic masses of the ions in the formula of the compound. For example, the gram formula mass of calcium iodide is the gram atomic mass of calcium plus two times the gram atomic mass of iodine.

$$40.1 \text{ g Ca} + (2 \times 126.9 \text{ g I}) = 293.9 \text{ g } CaI_2$$

There are 293.9 g CaI_2 in 1 gfm or 1 mol CaI_2.

Figure 7.10 shows one gram formula mass of three ionic compounds. ❶ How many formula units are in each sample in the figure?

Sample Problem 7-5

What is the gram formula mass of ammonium carbonate ($(NH_4)_2CO_3$)?

1. ***ANALYZE* List the knowns and the unknown.**

Knowns:
- formula unit = $(NH_4)_2CO_3$
- 1 gam N = 1 mol N = 14.0 g N
- 1 gam H = 1 mol H = 1.0 g H
- 1 gam C = 1 mol C = 12.0 g C
- 1 gam O = 1 mol O = 16.0 g O

Unknown:
- gfm = ? g

The formula shows that a mole of this ionic compound is composed of 2 mol of nitrogen atoms, 8 mol of hydrogen atoms, 1 mol of carbon atoms, and 3 mol of oxygen atoms. Moles of atoms are converted to grams by using conversion factors based on the gram atomic masses. The sum of the masses of the elements gives the gram formula mass.

Answers

❶ 6.02×10^{23} formula units
❷ potassium dichromate

Sample Problem 7-5 (cont.)

2. *CALCULATE Solve for the unknown.*
Using the proper conversion factors and adding the results gives the answer.

$$2\,\cancel{mol\,N} \times \frac{14.0\text{ g N}}{1\,\cancel{mol\,N}} = 28.0\text{ g N}$$

$$8\,\cancel{mol\,H} \times \frac{1.0\text{ g H}}{1\,\cancel{mol\,H}} = 8.0\text{ g H}$$

$$1\,\cancel{mol\,C} \times \frac{12.0\text{ g C}}{1\,\cancel{mol\,C}} = 12.0\text{ g C}$$

$$3\,\cancel{mol\,O} \times \frac{16.0\text{ g O}}{1\,\cancel{mol\,O}} = 48.0\text{ g O}$$

gram formula mass of $(NH_4)_2CO_3$ = 96.0 g

3. *EVALUATE Does the result make sense?*
The answer reflects the number of moles of atoms of each element and the gram atomic mass of each element. The answer is expressed to the tenth's place because the numbers being added are expressed to the tenth's place.

Practice Problems

9. Calculate the gram formula mass of each ionic compound.
   a. K_2O
   b. $CaSO_4$
   c. CuI_2
10. Find the gram formula mass of each compound.
   a. barium fluoride
   b. strontium cyanide
   c. sodium hydrogen carbonate
   d. aluminum sulfite

section review 7.1

11. Describe the relationship between Avogadro's number and one mole of any substance.
12. Find the gram formula mass of each compound.
   a. Li_2S
   b. $FeCl_3$
   c. $Ca(OH)_2$
13. How many oxygen atoms are in a representative particle of each substance?
   a. ammonium nitrate (NH_4NO_3), a fertilizer
   b. acetylsalicylic acid ($C_8H_8O_4$), the fever-reducing compound aspirin
   c. ozone (O_3), a disinfectant
   d. nitroglycerine ($C_3H_5(NO_3)_3$), an explosive
14. How many moles is each of the following?
   a. 1.50×10^{23} molecules NH_3
   b. 1 billion (1×10^9) molecules O_2
   c. 6.02×10^{22} molecules Br_2
   d. 4.81×10^{24} atoms Li
15. Distinguish among gram atomic mass, gram molecular mass, and gram formula mass.

Chem ASAP! Assessment 7.1 Check your understanding of the important ideas and concepts in Section 7.1.

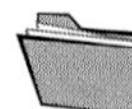

portfolio project

Research the history of Avogadro's number. What elements other than carbon have been used to define a mole? Write a report that summarizes your findings.

Answers

SECTION REVIEW 7.1

11. One mole of any substance contains Avogadro's number (6.02×10^{23}) of representative particles.
12. a. 45.9 g
   b. 162.3 g
   c. 74.1 g
13. a. 3 c. 3
   b. 4 d. 9
14. a. 2.49×10^{-1} mol NH_3
   b. 2×10^{-15} mol O_2
   c. 0.100 mol Br_2
   d. 7.99 mol Li
15. Gram atomic mass is the atomic mass of an element expressed in grams. Gram molecular mass is the mass of a mole of a molecular element or compound. Gram formula mass is the mass of a mole of an ionic compound.

7.1

3 Assess

Evaluate Understanding

To determine the students' knowledge of chemical quantities, ask them to:

- **explain what a mole of a substance represents** (6.02×10^{23} representative particles of that substance)
- **demonstrate how to convert the number of atoms or molecules of a substance to moles** (Divide the number of atoms or molecules by 6.02×10^{23}.)
- **define gram atomic mass, gam** (atomic mass of an element expressed in grams); **gram molecular mass, gmm** (mass of 1 mole of a molecular compound); **and gram formula mass, gfm** (the mass of one mole of an ionic compound)
- **determine the molar mass of $Al_2(CO_3)_3$** (234.0 g).

Reteach

Set up cooperative learning groups of 3 or 4 students with one student in each group who can assume leadership and help with the calculations. Provide the groups with problems related to Section 7-1.

Portfolio Project

Dalton set up a table of atomic masses based on the incorrect assumption that atoms always paired up one to one. He set the atomic mass of hydrogen as 1. As better tables were devised, it became inconvenient to have oxygen's value at 15.9 amu. Oxygen was made the arbitrary standard at 16.0000 amu. At an international meeting in 1961, physicists and chemists agreed to the current standard based on a single isotope, carbon-12.

1 Engage

Use the Visual

Have the students study the photograph that opens the section. Ask:

- **What factors can affect the relationship between average weight and height?** (ratio of muscle tissue to fat; bone structure)
- **What molar relationship do chemists use to solve problems?** (molar mass is a substance's mass in grams)

Check Prior Knowledge

To assess students' knowledge about the properties of solids, liquids, and gases, ask students to characterize the three states of matter. (Solids have a definite volume and definite shape, liquids have a definite volume but no definite shape, and gases have no definite volume and no definite shape.)

Practice Problems Plus

Related Chapter Review Problem
Chapter Review problem 56 is related to Sample Problem 7-6.

Additional Practice Problem
The Chem ASAP! CD-ROM contains the following problem: What is the mass, in g, of 1.72 mol $CaCl_2$?
(191 g)

section 7.2

objectives
- Use the molar mass to convert between mass and moles of a substance
- Use the mole to convert among measurements of mass, volume, and number of particles

key terms
- molar mass
- standard temperature and pressure (STP)
- molar volume

MOLE-MASS AND MOLE-VOLUME RELATIONSHIPS

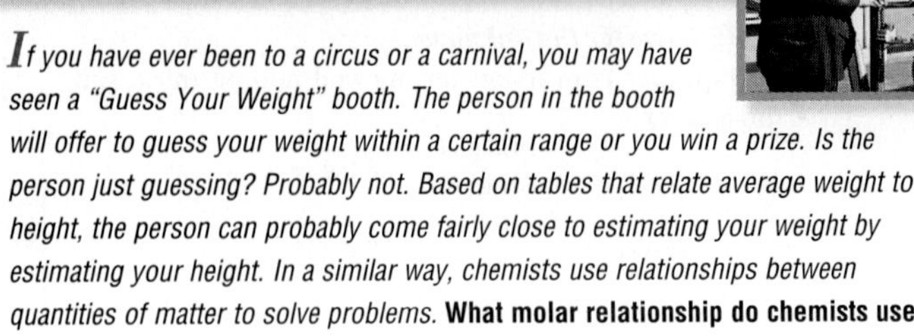

If you have ever been to a circus or a carnival, you may have seen a "Guess Your Weight" booth. The person in the booth will offer to guess your weight within a certain range or you win a prize. Is the person just guessing? Probably not. Based on tables that relate average weight to height, the person can probably come fairly close to estimating your weight by estimating your height. In a similar way, chemists use relationships between quantities of matter to solve problems. **What molar relationship do chemists use to solve problems?**

The Molar Mass of a Substance

In the previous section, you learned three new terms: gram atomic mass (gam), gram molecular mass (gmm), and gram formula mass (gfm). Each is used to represent a mole of a particular kind of substance. The gram atomic mass of an element contains a mole of atoms. The gram molecular mass of a molecular compound contains a mole of molecules. The gram formula mass of an ionic compound contains a mole of formula units. Although these three terms have different specific meanings, we can use the broader term molar mass to refer to a mole of an element, a molecular compound, or an ionic compound. The **molar mass** of any substance is the mass (in grams) of one mole of the substance.

There are situations in which the term molar mass is unclear. Consider this example: What is the molar mass of oxygen? How you answer this question depends on your interpretation of it. If you assume the oxygen in the question is molecular oxygen (O_2), then the molar mass is 32.0 g (2×16.0 g)—its gram molecular mass. If you assume that the question is asking for the mass of a mole of oxygen atoms (O), then the answer is 16.0 g—its gram atomic mass. Throughout this textbook, the term molar mass is used unless there is the potential for confusion. In that case, a more specific term will be used or the formula of the substance will be given.

In the following Sample Problems, the molar mass of an element or compound is used to convert between grams and moles of a substance.

Sample Problem 7-6

How many grams are in 9.45 mol of dinitrogen trioxide (N_2O_3)?

1. *ANALYZE* ***List the knowns and the unknown.***

Knowns:
- number of moles = 9.45 mol N_2O_3
- 1 mol N_2O_3 = 76.0 g N_2O_3

Unknown:
- mass = ? g N_2O_3

The number of grams of the compound must be calculated from the known number of moles of the compound. The desired conversion is moles $\longrightarrow$ grams.

STUDENT RESOURCES

From the Teacher's Resource Package, use:
- Section Review 7.2, Ch. 7 Practice Problems and Quizzes from the Review Module (Ch. 5–8)
- Laboratory Recordsheet 7-1
- Laboratory Manual: Experiment 8
- Small-Scale Lab Manual: Experiment 7

TECHNOLOGY RESOURCES

Relevant technology resources include:
- Chem ASAP! CD-ROM
- Resource Pro CD-ROM

Sample Problem 7-6 (cont.)

2. ***CALCULATE*** ***Solve for the unknown.***

Multiply the given number of moles by the proper conversion factor relating moles of N_2O_3 to grams of N_2O_3.

$$9.45\ \cancel{mol\ N_2O_3} \times \frac{76.0\ g\ N_2O_3}{1.00\ \cancel{mol\ N_2O_3}} = 718.2\ g\ N_2O_3$$

$$= 718\ g\ N_2O_3$$

3. ***EVALUATE*** ***Does the result make sense?***

Because 1 mol N_2O_3 has a mass of 76.0 g, and there are almost ten moles of the compound, the answer should be about 700. The answer has been rounded to the correct number of significant figures.

Practice Problems

16. Find the mass, in grams, of each.
a. 3.32 mol K
b. 4.52×10^{-3} mol $C_{20}H_{42}$
c. 0.0112 mol K_2CO_3

17. Calculate the mass, in grams, of 2.50 mol of each substance.
a. sodium sulfate
b. iron(II) hydroxide

Chem ASAP!

Problem-Solving 16
Solve Problem 16 with the help of an interactive guided tutorial.

Whereas Sample Problem 7-6 used a conversion factor based on the molar mass to convert moles to grams, the following Sample Problem does the reverse, using a conversion factor based upon the molar mass to convert grams to moles.

Sample Problem 7-7

Find the number of moles in 92.2 g of iron(III) oxide (Fe_2O_3).

1. ***ANALYZE*** ***List the knowns and the unknown.***

Knowns:	*Unknown:*
• mass = 92.2 g Fe_2O_3	• number of moles =
• 1 mol Fe_2O_3 = 159.6 g Fe_2O_3	? mol Fe_2O_3

From a known number of grams of a compound, the unknown number of moles of the compound must be calculated. The desired conversion is grams ⟶ moles.

2. ***CALCULATE*** ***Solve for the unknown.***

Multiply the given mass by the proper conversion factor relating mass of Fe_2O_3 to moles of Fe_2O_3.

$$92.2\ \cancel{g\ Fe_2O_3} \times \frac{1.00\ mol\ Fe_2O_3}{159.6\ \cancel{g\ Fe_2O_3}} = 0.5776\ mol\ Fe_2O_3$$

$$= 0.578\ mol\ Fe_2O_3$$

3. ***EVALUATE*** ***Does the result make sense?***

Because the given mass (about 90 g) is slightly larger than the mass of one-half mole of Fe_2O_3 (about 160 g), the answer should be slightly larger than one-half (0.5) of a mole.

Practice Problems

18. Find the number of moles in each quantity.
a. 3.70×10^{-1} g B
b. 27.4 g TiO_2
c. 847 g $(NH_4)_2CO_3$

19. Calculate the number of moles in 75.0 g of each substance.
a. dinitrogen trioxide
b. nitrogen gas
c. sodium oxide

2 Teach

Discuss

After students have read the text, point out that the term "molar mass" can be substituted for gam, gmm, and gfm most of the time. Review the mathematical conversions of grams to moles and moles to grams. Stress that using the factor-label method of problem solving allows them to calculate these problems without having to memorize the process. Be sure students always use units when solving problems.

Practice Problems Plus

Related Chapter Review Problem
Chapter Review problem 55 is related to Sample Problem 7-7.

Additional Practice Problems

1. Calculate the mass in grams for 0.250 moles of:
a. sucrose
b. sodium chloride
c. potassium permanganate

2. Calculate the number of moles in 1.00×10^2 g of
a. sucrose
b. sodium chloride
c. potassium permanganate

(**1a.** 85.5 g; **1b.** 14.6 g; **1c.** 39.5 g; **2a.** 0.292 mol; **2b.** 1.71 mol; **2c.** 0.633 mol)

7.2

Discuss

Ask students to name the units that determine the mass of a mole. (grams per mole) **Ask students which units should be used to describe the volume of a mole.** (mL per mole or L per mole) **Students should understand that unlike solids and liquids, the molar volume of gases is more predictable and is affected by temperature and pressure. Ask:**

- **How does temperature affect the volume of a gas?** (increase temperature: particles move faster and take up more space; decrease temperature: particles move slower and take up less space)
- **How does pressure affect the volume of a gas?** (Increased pressure pushes the particles closer together and decreases the volume, while decreased pressure allows the particles to spread out and increases the volume.)

Explain that when comparing the molar volumes of gases, it is necessary to have the gases at the same conditions of temperature and pressure. If the gases are at 0 °C and 101.3 kPa, they are at standard conditions of temperature and pressure (STP). **At STP, 1 mole of any gas occupies 22.4 L. The molar volume of any gas at STP is 22.4 L/mole.**

Practice Problems Plus

Related Chapter Review Problems
Chapter Review problems 57 and 59 are related to Sample Problem 7-8.

Additional Practice Problem
The Chem ASAP! CD-ROM contains the following problem: Determine the volume, in liters, of 2.74 mol He at standard temperature and pressure. (61.4 L)

Figure 7.11
The volume of eleven 2-liter soda bottles is 22 L. The volume of 1 mole of any gas at STP is a little more, 22.4 L.

The Volume of a Mole of Gas

If you look back at **Figure 7.9** on page 179, you will see that the volumes of one mole of different solid and liquid substances are not the same. For example, the volume of a mole of glucose (blood sugar) and one mole of paradichlorobenzene (moth crystals) are much larger than that of a mole of water. Unlike liquids and solids, the volumes of moles of gases are much more predictable under the same physical conditions.

The volume of a gas varies with a change in temperature or a change in pressure. Because of this variation, the volume of a gas is usually measured at a **standard temperature and pressure (STP).** Standard temperature is 0 °C. Standard pressure is 101.3 kPa, or 1 atmosphere (atm). At STP, 1 mol of any gas occupies a volume of 22.4 L. **Figure 7.11** should give you an idea of the volume occupied by 22.4 L. This quantity, 22.4 L, is known as the **molar volume** of a gas and is measured at STP. Because 1 mol of any substance contains Avogadro's number of particles, 22.4 L of any gas at STP contains 6.02×10^{23} representative particles of that gas. Would these values differ for gaseous elements compared with gaseous compounds? **(1)**

Would 22.4 L of one gas also have the same mass as 22.4 L of another gas at STP? Probably not. A mole of a gas (22.4 L at STP) has a mass equal to its molar mass. Only gases with the same molar masses would have equal masses for equal volumes at STP.

Practice Problems

20. What is the volume at STP of these gases?
a. 3.20×10^{-3} mol CO_2
b. 0.960 mol CH_4
c. 3.70 mol N_2

21. Assuming STP, how many moles are in these volumes?
a. 67.2 L SO_2
b. 0.880 L He
c. 1.00×10^3 L C_2H_6

Chem ASAP!
Problem-Solving 20
Solve Problem 20 with the help of an interactive guided tutorial.

Sample Problem 7-8

Determine the volume, in liters, of 0.60 mol SO_2 gas at STP.

1. *ANALYZE* ***List the knowns and the unknown.***

Knowns:	*Unknown:*
• moles = 0.60 mol SO_2	• volume = ? L SO_2
• 1 mol SO_2 = 22.4 L SO_2	

The known is the number of moles and the unknown is the number of liters of SO_2. Use the relationship 1.00 mol SO_2 = 22.4 L SO_2 (at STP) to write the conversion factor needed to perform the conversion of moles ⟶ liters.

2. *CALCULATE* ***Solve for the unknown.***

$$0.60\ \cancel{\text{mol } SO_2} \times \frac{22.4\ \text{L } SO_2}{1\ \cancel{\text{mol } SO_2}} = 13.44 = 13\ \text{L } SO_2$$

3. *EVALUATE* ***Does the result make sense?***
Because 1 mol of any gas at STP has a volume of 22.4 L, 0.60 mol should have a volume slightly larger than 22.4/2 = 11.2 L. The answer should have two significant figures.

Answers

(1) No; one mole of a gaseous compound and one mole of a gaseous element both occupy a volume of 22.4 L (at STP) and have Avogadro's number of representative particles.
(2) two conversion factors
(3) The density of the gas in these balloons is greater than the density of helium.

The density of a gas is usually measured in the units grams per liter (g/L). The experimentally determined density of a gas at STP is used to calculate the molar mass of that gas. The gas can be an element or a compound. As you can see in **Figure 7.12,** whether a gas-filled balloon sinks or floats is determined by the density of the gas in the balloon compared with the density of the surrounding air.

Sample Problem 7-9

The density of a gaseous compound containing carbon and oxygen is 1.964 g/L at STP. Determine the molar mass of the compound.

1. **ANALYZE** ***List the knowns and the unknown.***

Knowns:
- density = 1.964 g/L
- 1 mol (gas at STP) = 22.4 L

Unknown:
- molar mass = ? g/mol

Use the relationship 1 mol (gas at STP) = 22.4 L to write the conversion factor needed to perform the required conversion.

$$\frac{\text{g}}{\text{L}} \longrightarrow \frac{\text{g}}{\text{mol}}$$

2. **CALCULATE** ***Solve for the unknown.***

$$\frac{1.964\text{ g}}{1\text{ L}} \times \frac{22.4\text{ L}}{1\text{ mol}} = 43.9936 = 44.0\text{ g/mol}$$

3. **EVALUATE** ***Does the result make sense?***

The ratio of the calculated mass (44.0 g) to the volume (22.4 L) is about two, which is close to the known density. The answer should have three significant figures.

Practice Problems

22. A gaseous compound composed of sulfur and oxygen that is linked to the formation of acid rain has a density of 3.58 g/L at STP. What is the molar mass of this gas?

23. What is the density of krypton gas at STP?

Chem ASAP!

Problem-Solving 22

Solve Problem 22 with the help of an interactive guided tutorial.

The Mole Road Map

You have now examined a mole in terms of particles, mass, and volume of gases at STP. **Figure 7.13** on page 186 summarizes these relationships and illustrates the importance of the mole. To convert from one unit to another, you use the mole as an intermediate step. The form of the conversion factor depends on whether you are going from moles or to moles. You use the mole conversion factor in the same way you used the unit dozen to convert among mass, volume, and number of apples in Section 7.1. According to **Figure 7.13** on the following page, how many conversion factors are needed to convert from the mass of a gas to the volume of a gas (at STP)? 2

Figure 7.12

Helium is less dense than air. Balloons filled with helium must be tied to a heavy object to prevent them from floating away. The balloons sitting on the table are not tied down. How does the density of the gas in these balloons compare with the density of helium? 3

7.2

Discuss

Review the concept of density as a ratio of mass to volume. Discuss the units that go with density. (g/mL, g/cm^3, or g/L) **Ask:**

- **If you had a mole of gas at STP, how could you calculate the density?** (molar mass/22.4 L = density)
- **What information do you need to calculate the molar mass of a gas?** (density and molar volume)

Use the Visual

Have students examine Figure 7.12. Ask:

- **How does the density of the gas in the balloons sitting on the table compare to the density of air?** (more dense than air)
- **What can you assume about the gas in balloons that float in air?** (less dense than air)

Practice Problems Plus

Related Chapter Review Problem

Chapter Review problem 58 is related to Sample Problem 7-9.

Additional Practice Problem

The Chem ASAP! CD-ROM contains the following problem: Determine the molar mass of a gas that has a density of 0.179 g/L at standard temperature and pressure. (4.01 g/mol)

Meeting Diverse Needs

LEP and **At Risk** To strengthen both math and language skills, have students write the conversions found in Figure 7.13 on 3 × 5 cards. For example, on opposite sides of one card, write these related conversion ratios:

$$\frac{\text{molar mass (in grams)}}{1.00\text{ mole}} \qquad \frac{1.00\text{ mole}}{\text{molar mass (in grams)}}$$

Also make cards with the words "grams," "moles," "liters," and "particles." When solving problems, these students should set the problem up using these cards and using only conversion factors at first. Then when the units cancel out correctly, fill in the actual numbers and do the problem.

7.2

Use the Visual

Have students study Figure 7.13. Guide students through the various mole conversions, showing them how simple the process is.
For example: start with 50.0 g of a compound or element and convert it to moles, and then to particles or volume. Alternately, start with a certain volume and convert it to mass or particles.

3 Assess

Evaluate Understanding

Have students work problems in which they use molar mass and molar volume to calculate the densities of gases. Have volunteers come to the board to show their calculations. Have the class evaluate if the calculations are correct.

Reteach

Review the concept of density as a ratio of mass to volume. Ask students, "If you know the molar volume of a gas, how could density help you determine the molar mass?" Through the discussion, students should realize that if you know the density of the gas, you can solve the equation:

$$\text{density} = \frac{\text{molar mass}}{\text{molar volume}}$$

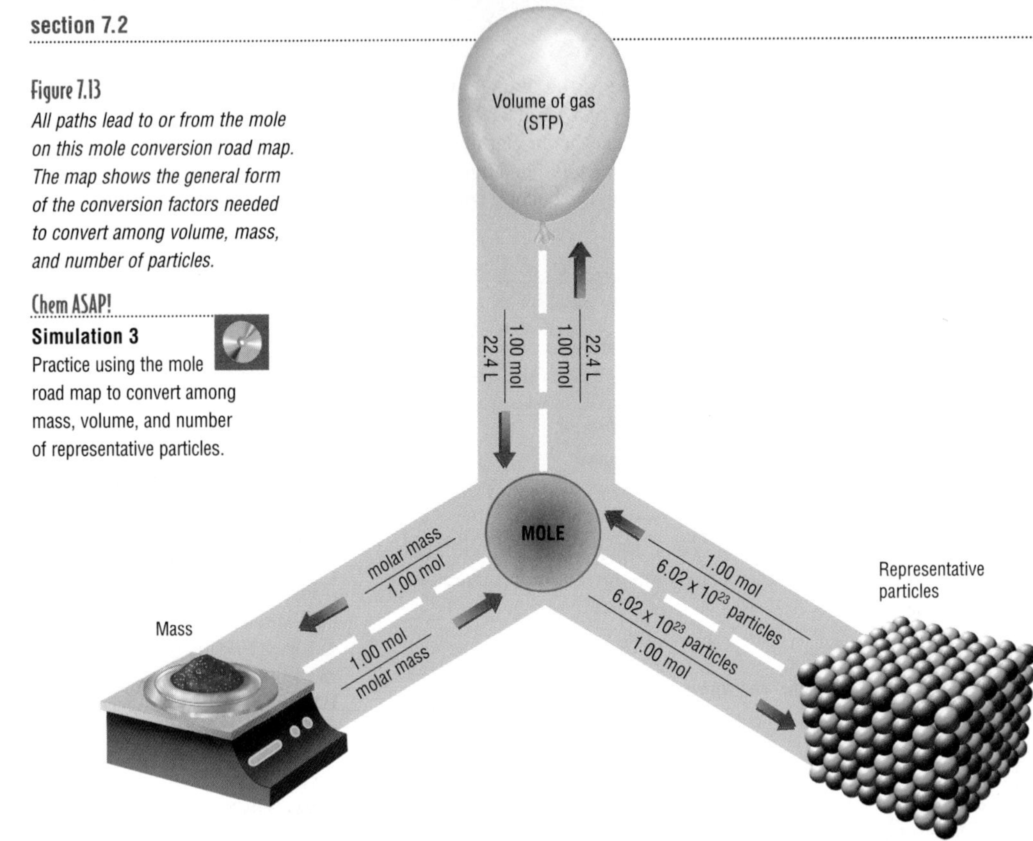

Figure 7.13
All paths lead to or from the mole on this mole conversion road map. The map shows the general form of the conversion factors needed to convert among volume, mass, and number of particles.

Chem ASAP!
Simulation 3
Practice using the mole road map to convert among mass, volume, and number of representative particles.

section review 7.2

24. Find the mass in grams of each quantity.
    - a. 0.720 mol Be
    - b. 2.40 mol N_2
    - c. 0.160 mol H_2O_2
    - d. 5.08 mol $Ca(NO_3)_2$
25. a. Calculate the number of molecules in 60.0 g NO_2.
    b. Calculate the volume, in liters, of 3.24×10^{22} molecules Cl_2 (STP).
    c. Calculate the mass, in grams, of 18.0 L CH_4 (STP).
26. Would three balloons, each containing the same number of molecules of a different gas at STP, have the same mass or the same volume? Explain.
27. Find the number of moles in each quantity.
    - a. 5.00 g hydrogen molecules
    - b. 0.000264 g Li_2HPO_4
    - c. 187 g Al
    - d. 333 g SnF_2
28. The densities of gases A, B, and C are 1.25 g/L, 2.86 g/L, and 0.714 g/L, respectively. Calculate the molar mass of each substance. Identify each substance as ammonia (NH_3), sulfur dioxide (SO_2), chlorine (Cl_2), nitrogen (N_2), or methane (CH_4).

Chem ASAP! Assessment 7.2 Check your understanding of the important ideas and concepts in Section 7.2.

Answers

SECTION REVIEW 7.2

24. a. 6.5 g Be
    b. 67.2 g N_2
    c. 5.44 g H_2O_2
    d. 8.34×10^2 g $Ca(NO_3)_2$
25. a. 7.85×10^{23} molecules NO_2
    b. 1.21 L Cl_2
    c. 12.9 g CH_4
26. They would have the same volume but different masses; equal volumes of gases have the same number of molecules at the same temperature and pressure.
27. a. 2.5 mol H_2
    b. 2.40×10^{-6} mol Li_2HPO_4
    c. 6.93 mol Al
    d. 2.13 mol SnF_2
28. gas A: 28.0 g, nitrogen
    gas B: 64.1 g, sulfur dioxide
    gas C: 16.0 g, methane

SMALL-SCALE LAB

MEASURING MASS AS A MEANS OF COUNTING

SAFETY 

Wear your safety glasses and follow standard safety procedures as outlined on page 18.

PURPOSE

To determine the mass of several chemical compound samples and use the data to count atoms.

MATERIALS

- pencil
- paper
- plastic spoon
- chemicals shown in Figure A
- balance
- ruler

PROCEDURE

Measure the mass of one level teaspoon of sodium chloride (NaCl), water (H_2O), and calcium carbonate ($CaCO_3$). Make a table similar to Figure A to record your measured and calculated data.

	$H_2O(l)$	$NaCl(s)$	$CaCO_3(s)$
Mass (grams)	**4.30**	**5.09**	**9.68**
Molar Mass (g/mol)	**18.0**	**58.5**	**100.1**
Moles of each compound	**0.239**	**0.0870**	**0.0967**
Moles of each element	**0.239 O** **0.478 H**	**0.0870 Na** **0.0870 Cl**	**0.0967 Ca** **0.0967 C** **0.290 O**
Atoms of each element	**1.44×10^{23} O** **2.88×10^{23} H**	**5.24×10^{22} Na** **5.24×10^{22} Cl**	**5.82×10^{22} Ca** **5.82×10^{22} C** **1.75×10^{23} O**

Figure A

ANALYSIS

Using your data, record the answers to the following questions in or below your data table.

1. Calculate the moles of NaCl contained in one level teaspoon and record the result in your table.

$$\text{moles of NaCl} = ?\text{ g NaCl} \times \frac{1\text{ mol NaCl}}{58.5\text{ g}}$$

2. Repeat Step 1 for the other compounds in Figure A. Use the periodic table if necessary to calculate the molar mass of water and calcium carbonate.
3. Calculate the moles of each element present in the teaspoon-sized sample of H_2O.

$$\text{moles of H} = ?\text{ mol } H_2O \times \frac{2\text{ mol H}}{1\text{ mol } H_2O}$$

Repeat for all the other compounds in your table.

4. Calculate the number of atoms of each element present in the teaspoon-sized sample of H_2O.

$$\text{atoms of H} = ?\text{ mol H} \times \frac{6.02 \times 10^{23}\text{ atoms H}}{1\text{ mol } H_2O}$$

Repeat for all the other compounds in your table.

5. Which of the three teaspoon-sized samples contains the greatest number of moles?
6. Which of the three compounds contains the most atoms?

YOU'RE THE CHEMIST!

The following small-scale activities allow you to develop your own procedures and analyze the results.

1. **Design It!** Can you use the technique of measuring volume as a means of counting? Design and carry out an experiment to do it!
2. **Design It!** Design an experiment that will determine the number of atoms of calcium, carbon, and oxygen it takes to write your name on the chalkboard with a piece of chalk. Assume chalk is 100 percent calcium carbonate, $CaCO_3$.

SMALL-SCALE LAB

MEASURING MASS AS A MEANS OF COUNTING

OVERVIEW

Procedure Time: 30 minutes

Students measure masses of common chemicals and convert their data to moles and atoms. They explore the quantitative chemical compositions of common objects.

MATERIALS

solid NaCl, solid $CaCO_3$, water, chalk

TEACHING SUGGESTIONS

Remind students that determining mass is the best way to count atoms and molecules if the chemical composition of the sample is known. A similar, more complete version of this lab appears in the Small-Scale Laboratory Manual.

YOU'RE THE CHEMIST

1. Try finding the mass of 100 drops of water and then calculate the mass in grams per drop.
2. Find the mass of a piece of chalk. Write your name and find the mass of the chalk again. Convert the mass difference to moles and atoms.

ANALYSIS

Student data may vary slightly from the data in the table.

1. $5.09\text{ g NaCl} \times \frac{1\text{ mol}}{58.5\text{ g}} = 0.0870\text{ mol NaCl}$
2. See data table for answers.
3. See data table for answers.
4. $0.478\text{ mol H} \times \frac{6.02 \times 10^{23}\text{ atoms}}{1\text{ mol H}} = 2.88 \times 10^{23}\text{ atoms H}$

   See data table for additional answers.
5. Water has the greatest number of moles in one teaspoon.
6. Water has the greatest total number of atoms.

1 Engage

Use the Visual

Have students study the photograph and read the text that opens the section. Point out that in order for chemists to write the formula of a compound, they must identify the kinds and measure the quantities of elements present in a fixed amount of the compound. In other words, they must determine the composition of the compound. Ask:

- **How do chemists calculate the percent composition of a substance?** (For a given mass of a substance, this information can be found by comparing the mass of each element in the substance to the mass of the substance. Each of these quotients is then converted to percent by multiplying by 100%.)

Check Prior Knowledge

To assess students' knowledge about mathematical calculations involving percents, have students solve a number of sample problems. For example, ask:

- **What is 73% of 150?** (110)
- **What percent of 6.5 is 3.1?** (48%)

Use the Visual

Have students study Figure 7.15 and read the text on percent composition. Notice that the three numbers in the pie graphs add up to a total of 100%. Ask:

- **Which compound is the better source of potassium?** (K_2CrO_4)

section 7.3

objectives

- Calculate the percent composition of a substance from its chemical formula or experimental data
- Derive the empirical formula and the molecular formula of a compound from experimental data

key terms

- percent composition
- empirical formula

PERCENT COMPOSITION AND CHEMICAL FORMULAS

Growing healthy plants requires more than just a green thumb. Lawn-care professionals, as well as most home gardeners, know the importance of fertilizers in maintaining healthy lawns and gardens. Lawn and garden fertilizers are labeled with three numbers, such as 15-10-15, which represent the relative amounts of nitrogen, phosphorus, and potassium, respectively. Relative quantities are similar to percentages and can be used to calculate percent composition. **What is the percent composition of a substance, and how is it calculated?**

Figure 7.14
Percent compositions, such as those found on this label, are frequently used in everyday life.

Calculating the Percent Composition of a Compound

If you have had any experience with lawn care, then you know that the amount of fertilizer you put on a lawn is important. The relative amount, or the percent, of each nutrient in the fertilizer is also important. In the spring, you may use a fertilizer that has a relatively high percent of nitrogen to "green" the grass. In the fall, you may want to use a fertilizer with a higher percent of potassium to strengthen the root system of the grass. Knowing the relative amounts of the components of a mixture or compound is often useful or important.

The relative amounts of each element in a compound are expressed as the **percent composition,** or the percent by mass of each element in a compound. The percent composition of a compound has as many percent values as there are different elements in the compound. As you can see in **Figure 7.15,** the percent composition of K_2CrO_4 is K = 40.3%, Cr = 26.8%, and O = 32.9%. These percents must total 100% (40.3% + 26.8% + 32.9% = 100%).

The percent by mass of an element in a compound is the number of grams of the element divided by the number of grams of the compound, multiplied by 100%.

$$\%\text{ mass of element E} = \frac{\text{grams of element E}}{\text{grams of compound}} \times 100\%$$

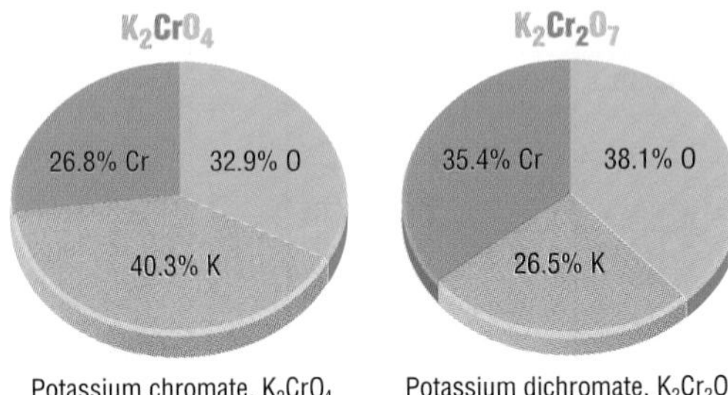

Figure 7.15
Potassium chromate (K_2CrO_4) is composed of 40.3% potassium, 26.8% chromium, and 32.9% oxygen. How does this differ from the percent composition of potassium dichromate ($K_2Cr_2O_7$), a compound composed of the same three elements? ❶

STUDENT RESOURCES

- Section Review 1.4, Ch. 1 Practice Problems, Interpreting Graphics, Vocabulary Review, Quizzes, and Tests from the Review Module (Ch. 5–8)
- Laboratory Manual: Experiment 9
- Laboratory Practical 7-1
- Laboratory Recordsheet 7-2

TECHNOLOGY RESOURCES

Relevant technology resources include:

- Chem ASAP! CD-ROM
- Resource Pro CD-ROM
- Assessment Resources CD-ROM: Chapter 7 Tests

Sample Problem 7-10

An 8.20-g piece of magnesium combines completely with 5.40 g of oxygen to form a compound. What is the percent composition of this compound?

1. *ANALYZE* ***List the knowns and the unknown.***

Knowns:	*Unknowns:*
• mass of magnesium = 8.20 g Mg	• percent Mg = ? % Mg
• mass of oxygen = 5.40 g O	• percent O = ? % O
• mass of compound = 8.20 g + 5.40 g = 13.60 g	

The percent by mass of an element in a compound is the mass of that element divided by the mass of the compound multiplied by 100%.

2. *CALCULATE* ***Solve for the unknown.***

$$\% \text{ Mg} = \frac{\text{mass of Mg}}{\text{grams of compound}} \times 100\%$$

$$= \frac{8.2 \text{ g}}{13.6 \text{ g}} \times 100\% = 60.29412 = 60.3\%$$

$$\% \text{ O} = \frac{\text{mass of O}}{\text{grams of compound}} \times 100\%$$

$$= \frac{5.4 \text{ g}}{13.6 \text{ g}} \times 100\% = 39.70588 = 39.7\%$$

3. *EVALUATE* ***Does the result make sense?***

The percents of the elements add up to 100%: 60.3% + 39.7% = 100%.

Practice Problems

29. a. 9.03 g Mg combine completely with 3.48 g N to form a compound. What is the percent composition of this compound?

b. 29.0 g Ag combine completely with 4.30 g S to form a compound. What is the percent composition of this compound?

30. When a 14.2-g sample of mercury(II) oxide is decomposed into its elements by heating, 13.2 g Hg is obtained. What is the percent composition of this compound?

Chem ASAP!

Problem-Solving 30

Solve Problem 30 with the help of an interactive guided tutorial.

To calculate the percent composition of a known compound, use the chemical formula to calculate the molar mass. This gives the mass of one mole of the compound. Then for each element, calculate the percent by mass in one mole of the compound. This is done by dividing the mass of each element in one mole of the compound by the molar mass and multiplying the result by 100%. Remember that the subscripts in the formula of the compound are used to calculate the grams of each particular element in a mole of that compound.

$$\% \text{ mass} = \frac{\text{grams of element in 1 mol compound}}{\text{molar mass of compound}} \times 100\%$$

Figure 7.16

The percent composition of water is always the same, regardless of the volume of the water sample. A sample of water is always 11.1% H and 88.9% O by mass.

2 Teach

ACTIVITY

Have students bring in labels from packaged foods and work in groups to answer the following questions.

- Using the Daily Value on the label, how many grams of fat should you eat in a day? (Based on a 2000-Calorie diet you should have fewer than 65 g of fat per day. Some people are restricted to as few as 5 g.)
- What is the recommended daily intake of sodium? (about 2400 mg sodium)
- Calculate the grams of carbohydrates allowed per day. (about 310 g of carbohydrates)
- What other information can you learn from reading the label? (Answers should include information about necessary vitamins and minerals.)
- Which food label listed the most grams of fat?
- Which food was highest in sodium? Which food provided the most protein per serving?

Practice Problems Plus

To give students additional practice doing problems like Sample Problems 7-10 and 7-11, ask: What is the percent composition of the compound formed when 2.64 g of aluminum combine with oxygen to form 5.10 g of aluminum oxide? (51.8% Al, 48.2% O)

Additional Practice Problem

The Chem ASAP! CD-ROM contains the following problem: Calculate the percent composition when 13.3 g Fe combine completely with 5.7 g O. (70% Fe, 30% O)

Meeting Diverse Needs

At Risk These students may experience more difficulty with mathematical concepts. To help the students learn the many calculations in this chapter, they should write out a step-by-step procedure in their own words for doing each of the sample problems. Then they should ask a study partner to read/correct their procedures.

Answer

❶ The percent composition of potassium dichromate is 26.5% K, 35.4% Cr, 38.1% O. Potassium chromate contains more potassium, but less chromium and oxygen.

CHEMath

Teaching Tips

Provide students with a number of different scenarios in which a collection of different objects such as colored marbles can be represented by a sum of fractions. For example, suppose you have 3 red marbles, 6 green marbles, 3 black marbles, and 12 blue marbles. Have students express the number of different colored marbles as fractions and percentages of the whole collection. For example, ask:

- What percentage of the population do the red marbles represent? (12.5%)

Show them that the sums of the fractions and percentages are equal to 1 and 100% respectively. Finally, have students express the ratio red:green:black:blue marbles in lowest terms. (1:2:1:4) This activity can be extended if the different colored marbles are assigned "molar masses." Have students assume that each color represents an atom of a different element. Ask:

- What is the empirical formula of a hypothetical "compound" that consists of 25% red marbles and 75% green marbles?

Answers to Practice Problems

A. $\frac{5}{6}, \frac{8}{39}, \frac{9}{25}$

B. 12.9 g

C. 3.81 g

D. 2 to 5

CHEMath

FRACTIONS, RATIOS, AND PERCENT

In this chapter, you will use percents and ratios when calculating the composition of a substance or determining empirical and molecular formulas. Percents are also used in routine activities such as shopping—stores often have "percent-off" sales. To understand percents and ratios, you must first understand fractions.

A fraction is a special quotient that defines a quantity in terms of equal parts of a whole. The quotient $3 \div 4$ can be written as the fraction $\frac{3}{4}$, or three-fourths, which means three of the four equal parts that make up a whole. The top number of a fraction is the *numerator* and the bottom number is the *denominator*.

Fractions that represent the same quantity are *equivalent*. A fraction is in *lowest terms* if the numerator and denominator have no common factors. So, $\frac{4}{5}$ and $\frac{16}{20}$ are equivalent, but only $\frac{4}{5}$ is in lowest terms. To express a fraction in lowest terms, divide the numerator and denominator by their greatest common factor (GCF).

A *ratio* is a comparison of two quantities. Ratios are often written as fractions. If there are 10 dogs and 15 cats, the ratio of dogs to cats is 10 to 15, or $\frac{10}{15}$, or $\frac{2}{3}$. You can multiply or divide both quantities in a ratio by the same nonzero quantity without changing the value of the ratio. For example, multiplying quantities in the ratio $\frac{2}{3}$ by 5 yields $\frac{10}{15}$ (the original ratio).

A *percent* compares a number to 100. Thus, a ratio of $\frac{73}{100}$ can be written as 73%. 100% represents a whole, so $100\% = 1$.

You can think of 100% as a *conversion factor* that converts to percent. To convert a fraction or decimal to a percent, multiply by 100%. So, $\frac{3}{5}$ as a percentage is $\frac{3}{5} \times 100\% = 60\%$. This is the idea behind the formula for % mass on page 189.

Example 1

Sulfur makes up 26.7% of the mass of $NaHSO_4$. Find the mass of the sulfur in 16.8 g of $NaHSO_4$.

Since $NaHSO_4$ is 26.7% sulfur, 100 g of the compound would contain 26.7 g sulfur. So, you can use

$$\frac{26.7 \text{ g S}}{100 \text{ g NaHSO}_4}$$

as a conversion factor.

Conversion: grams $NaHSO_4 \longrightarrow$ grams S

$$16.8 \text{ g NaHSO}_4 \times \frac{26.7 \text{ g S}}{100 \text{ g NaHSO}_4} = 4.49 \text{ g S}$$

The mass of the sulfur is 4.49 g.

Example 2

100 g of a compound contains 1.88 mol O and 1.25 mol Fe. Write the mole ratio of oxygen to iron. Use whole numbers in lowest terms. (*Hint:* Each whole number is less than 10.)

Notice that the quantities are approximate, with three significant figures. The ratio is:

$$\frac{1.88}{1.25} = \frac{1.88 \div 1.25}{1.25 \div 1.25} = \frac{1.50}{1.00} = \frac{1.50 \times 2}{1.00 \times 2} = \frac{3.00}{2.00}$$

The mole ratio of oxygen to iron is 3 to 2.

Practice Problems

Prepare for upcoming problems in this chapter by solving each of the following.

A. Express the fractions $\frac{35}{42}$, $\frac{16}{78}$, and $\frac{36}{100}$ in lowest terms.

B. Aluminum makes up 52.9% of the mass of Al_2O_3. Find the mass of the aluminum in 24.3 g of Al_2O_3.

C. Hydrogen makes up 2.06% of the mass of H_2SO_4. Find the mass of the hydrogen in 185 g H_2SO_4.

D. 100 g of a compound contains 1.85 mol N and 4.63 mol O. Write the mole ratio of nitrogen to oxygen. Use whole numbers in lowest terms. (*Hint:* Each whole number is less than 10.)

Sample Problem 7-11

Calculate the percent composition of propane (C_3H_8).

1. *ANALYZE* ***List the knowns and the unknowns.***

Knowns:
- molar mass C_3H_8 = 44.0 g/mol
- mass of C in 1 mol C_3H_8 = 36.0 g
- mass of H in 1 mol C_3H_8 = 8.0 g

Unknowns:
- percent C = ? % C
- percent H = ? % H

The percent mass of each element can be calculated by dividing the mass of that element in one mole of the compound by the molar mass of the compound and multiplying by 100%.

2. *CALCULATE* ***Solve for the unknowns.***

$$\% \text{ C} = \frac{\text{mass of C}}{\text{grams of compound}} \times 100\%$$

$$= \frac{36.0 \text{ g}}{44.0 \text{ g}} \times 100\% = 0.818181 = 81.8\%$$

$$\% \text{ H} = \frac{\text{mass of H}}{\text{grams of compound}} \times 100\%$$

$$= \frac{8.0 \text{ g}}{44.0 \text{ g}} \times 100\% = 0.181818 = 18\%$$

3. *EVALUATE* ***Does the result make sense?***

The percents of the elements add up to 100% after the answers are expressed to two significant figures.

Practice Problems

31. Calculate the percent composition of these compounds.
a. ethane (C_2H_6)
b. sodium bisulfate ($NaHSO_4$)
c. ammonium chloride (NH_4Cl)

32. Calculate the percent nitrogen in these common fertilizers.
a. $CO(NH_2)_2$
b. NH_3
c. NH_4NO_3

Using Percent as a Conversion Factor

You can use percent composition to calculate the number of grams of an element contained in a specific amount of a compound. To do this, you multiply the mass of the compound by a conversion factor that is based on the percent composition. In Sample Problem 7-11, you found that propane is 81.8% C. Thus in a 100-g sample, you would have 81.8 g C. Sample Problem 7-12 shows how the ratio 81.8 g C/100 g C_3H_8 can be used to solve for the mass of carbon contained in a specific amount of propane (C_3H_8).

Sample Problem 7-12

Calculate the mass of carbon in 82.0 g of propane (C_3H_8).

1. *ANALYZE* ***List the knowns and the unknown.***

Knowns:
- mass C_3H_8 = 82.0 g
- percent by mass of C in C_3H_8 = 81.8% (from Sample Problem 7-11)

Unknown:
- mass of carbon = ? g C

The conversion is grams $C_3H_8 \longrightarrow$ grams C.

Practice Problems

33. Using Problem 31, calculate the mass of hydrogen in each of the following.
a. 350 g C_2H_6
b. 20.2 g $NaHSO_4$
c. 2.14 g NH_4Cl

Practice Problems Plus

Related Chapter Review Problem
Chapter Review problem 61 is related to Sample Problem 7-12. For additional practice, have students determine the percent composition of the following oxides:
a. Fe_2O_3 (69.9% Fe, 30.1%O)
b. HgO (92.6% Hg, 7.39%O)
c. Ag_2O (93.1% Ag, 6.90 %O)
d. Na_2O (74.2% Na, 25.8%O)

Additional Practice Problem
The Chem ASAP! CD-ROM contains the following problem: Calculate the grams of oxygen in 90.0 g of Cl_2O. (16.6 g)

MEETING DIVERSE NEEDS

LEP For this chapter, team up a limited English proficient student with a student who has strong mathematical skills. Encourage them to ask their partner for help with any difficulty they are having understanding the terms and how to solve the problems.

Discuss

Using examples such as H_2O and H_2O_2, discuss the differences between empirical formulas, molecular formulas, and structural formulas. Ask students, "Suppose that you are given a sample of a substance that contains hydrogen and oxygen. You are told that the ratio of moles of hydrogen to moles of oxygen is 1 : 1. What is the formula of the substance?" (Many students will probably say the formula would be HO.) **Challenge this response by asking, "Why not H_2O_2 or H_3O_3?" Use these examples to introduce the concept of empirical formula. Lead the class through Sample Problems 7-13 and 7-14.**

Practice Problems (cont.)

34. Calculate the grams of nitrogen in 125 g of each fertilizer.

a. $CO(NH_2)_2$

b. NH_3

c. NH_4NO_3

Chem ASAP!

Problem-Solving 34

Solve Problem 34 with the help of an interactive guided tutorial.

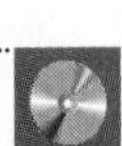

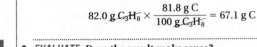

Sample Problem 7-12 (cont.)

2. *CALCULATE* ***Solve for the unknown.***

A conversion factor based on the percent mass of carbon in propane is used. Sample Problem 7-11 showed that the percent composition of propane is 81.8% C and 18.2% H. Remember that 81.8% C means 81.8 g C per 100 g C_3H_8. This is the desired conversion factor.

$$82.0\ \text{g}\ \cancel{C_3H_8} \times \frac{81.8\ \text{g C}}{100\ \text{g}\ \cancel{C_3H_8}} = 67.1\ \text{g C}$$

3. *EVALUATE* ***Does the result make sense?***

Because carbon constitutes about 82% by mass of propane, it is logical that in this sample, carbon has a mass of about 82% of 82.0 g, or 67.1 g.

Calculating Empirical Formulas

Determining the percent composition of a compound has an important application. Percent composition data can be used to calculate the empirical formula of a compound. The **empirical formula** gives the lowest whole-number ratio of the atoms of the elements in a compound. For example, a compound may have the empirical formula CO. The empirical formula provides valuable information about the kinds and relative count of atoms or moles of atoms in molecules or formula units of a compound. As you can see in **Figure 7.17**, empirical formulas may be interpreted at the microscopic (atoms) or macroscopic (moles of atoms) level.

If you use the mole interpretation, an empirical formula is the lowest whole-number ratio of moles of atoms that combine to form a compound. An empirical formula may or may not be the same as a molecular formula. If the formulas are different, the molecular formula is a simple multiple of the empirical formula. The empirical formula of hydrogen peroxide (H_2O_2) for example, is HO.

For carbon dioxide, the empirical and molecular formulas are the same—CO_2. Dinitrogen tetrahydride, whose molecular formula is N_2H_4, has an empirical formula of NH_2 because this is the simplest ratio of nitrogen to hydrogen in the compound. **Figure 7.18** shows two compounds of carbon that have the same empirical formula but different molecular formulas.

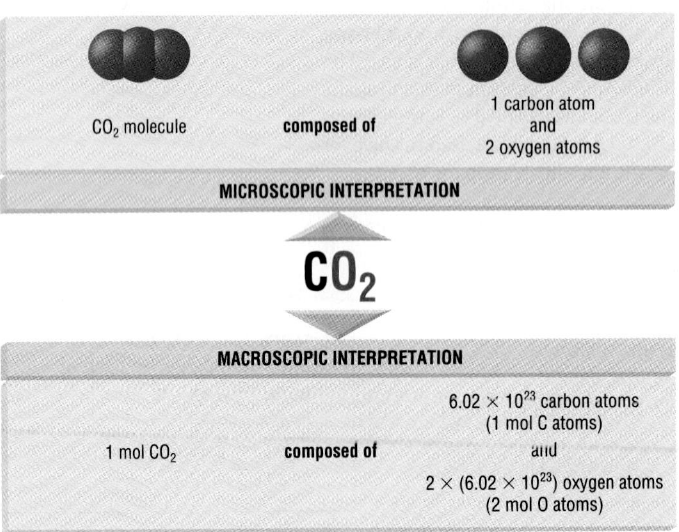

Figure 7.17
A formula can be interpreted on a microscopic level in terms of atoms or on a macroscopic level in terms of moles of atoms.

Sample Problem 7-13

What is the empirical formula of a compound that is 25.9% nitrogen and 74.1% oxygen?

1. *ANALYZE* *List the knowns and the unknown.*

Knowns:
- percent of nitrogen = 25.9% N
- percent of oxygen = 74.1% O

Unknown:
- empirical formula = $N_?O_?$

The lowest whole-number ratio of moles of nitrogen atoms to oxygen atoms, or the empirical formula, is to be calculated. The percent composition tells the ratio of masses of nitrogen atoms to oxygen atoms in the compound. The ratio of masses is changed to a ratio of moles by using conversion factors based on the molar mass of each element. This mole ratio is then reduced to the lowest whole-number ratio.

2. *CALCULATE* *Solve for the unknown.*

In 100.0 g of the compound, there are 25.9 g N and 74.1 g O. These values are used to convert to moles.

$$25.9 \text{ g N} \times \frac{1 \text{ mol N}}{14.0 \text{ g N}} = 1.85 \text{ mol N}$$

$$74.1 \text{ g O} \times \frac{1 \text{ mol O}}{16.0 \text{ g O}} = 4.63 \text{ mol O}$$

The mole ratio of nitrogen to oxygen is $N_{1.85}O_{4.63}$. This is not the correct empirical formula because it is not the lowest whole-number ratio. The correct values can be obtained by dividing both molar quantities by the smaller number of moles. This will give a 1 for the element with the smaller number of moles. However, this step may not give whole numbers for the other elements.

$$\frac{1.85 \text{ mol N}}{1.85} = 1 \text{ mol N}$$

$$\frac{4.63 \text{ mol O}}{1.85} = 2.50 \text{ mol O}$$

Is the final answer $N_1O_{2.5}$? Obviously not. This formula still does not represent the lowest whole-number ratio. To obtain such a ratio, each part of the ratio is multiplied by a number (in this case 2) that converts the fraction to a whole number.

$$1 \text{ mol N} \times 2 = 2 \text{ mol N}$$
$$2.5 \text{ mol O} \times 2 = 5 \text{ mol O}$$

The empirical formula is N_2O_5.

3. *EVALUATE* *Does the result make sense?*

This is an empirical formula. The subscripts are whole numbers. As a further check, the percent composition of this empirical formula can be calculated. The percents of each element should equal the percents given in the original problem.

Figure 7.18
Ethyne (C_2H_2), also called acetylene, is a gas used in welder's torches. Styrene (C_8H_8) is used in making polystyrene. These two compounds have the same empirical formula. What is the empirical formula? ❶

Practice Problems

35. Calculate the empirical formula of each compound.
- **a.** 94.1% O, 5.9% H
- **b.** 79.8% C, 20.2% H
- **c.** 67.6% Hg, 10.8% S, 21.6% O
- **d.** 27.59% C, 1.15% H, 16.09% N, 55.17% O

36. 1,6-diaminohexane is used to make nylon. What is the empirical formula of this compound if it is 62.1% C, 13.8% H, and 24.1% N?

Chem ASAP!

Problem-Solving 36
Solve Problem 36 with the help of an interactive guided tutorial.

Discuss

Emphasize that the first step in calculating an empirical formula is to find the mole-to-mole ratio of the elements in the compound. Point out that if the numbers in this ratio are both whole numbers, these will be the subscripts of the elements in the formula. If either or both numbers are not whole numbers, numbers in the ratio must be multiplied by the same number to yield whole number subscripts.

Think Critically

Have students apply their problem-solving skills to this question: Students heated a mixture of potassium chlorate and manganese dioxide, producing 0.377 L of oxygen gas at STP. What was the mass of the gas collected? (0.539 g)

Practice Problems Plus

Related Chapter Review Problem
Chapter Review problem 64 is related to Sample Problem 7-13.

Additional Practice Problem
The Chem ASAP! CD-ROM contains the following problem: What is the empirical formula of a compound that is 3.7% H, 44.4% C, and 51.9% N? (HCN)

Answer

❶ empirical formula = CH

MINI LAB

Percent Composition

SAFETY

Students should wear safety goggles and tie back loose hair. Caution students that while heating test tubes, they should not aim the opening of the tube toward anyone. Tell them to move the test tube in the flame and not to heat one spot excessively. **CAUTION!** Be sure that students allow the tubes to cool completely before they touch them. Hot glass looks exactly like cold glass!

LAB PREP AND PLANNING

Place the chemicals near the balances in plainly labeled jars. Label the spatulas so that the chemicals will not become contaminated.

LAB TIPS

For best results, students should do a second heating and cooling to determine whether or not all of the water has been driven off.

ANALYSIS AND CONCLUSIONS

4. The hydrated salt of sodium sulfate lost the greatest percent. The hydrated salt of calcium chloride lost the least percent.

Table 7.2

Comparison of Empirical and Molecular Formulas		
Formula (name)	**Classification of formula**	**Molar mass**
CH	Empirical	13
C_2H_2 (ethyne)	Molecular	26 (2 × 13)
C_6H_6 (benzene)	Molecular	78 (6 × 13)
CH_2O (methanal)	Empirical and Molecular	30
$C_2H_4O_2$ (ethanoic acid)	Molecular	60 (2 × 30)
$C_6H_{12}O_6$ (glucose)	Molecular	180 (6 × 30)

Calculating Molecular Formulas

Look at the two series of compounds in **Table 7.2.** Ethyne and benzene have the same empirical formula—CH. Glucose, ethanoic acid, and methanal have the same empirical formula—CH_2O. But as you can see, the compounds in each series have different molar masses. These molar masses are simple whole-number multiples of the molar mass of the empirical formulas, CH and CH_2O. The molecular formula of a compound is either the same as its experimentally determined empirical formula, or it is a simple whole-number multiple of it.

You can determine the molecular formula of a compound if you know its empirical formula and its molar mass. From the empirical formula, you can calculate the empirical formula mass (efm). This is simply the molar mass of the empirical formula. The known molar mass is then divided by the empirical formula mass. This gives the number of empirical formula units in a molecule of the compound and is the multiplier to convert the empirical formula to the molecular formula.

Practice Problems

37. Find the molecular formula of each compound given its empirical formula and molar mass.
    a. ethylene glycol (CH_3O), used in antifreeze, molar mass = 62 g/mol
    b. *p*-dichlorobenzene (C_3H_2Cl), mothballs, molar mass = 147 g/mol
38. Which pair of molecules has the same empirical formula?
    a. $C_2H_4O_2$, $C_6H_{12}O_6$
    b. $NaCrO_4$, $Na_2Cr_2O_7$

Chem ASAP!

Problem-Solving 37

Solve Problem 37 with the help of an interactive guided tutorial.

Sample Problem 7-14

Calculate the molecular formula of the compound whose molar mass is 60.0 g and empirical formula is CH_4N.

1. *ANALYZE* ***List the knowns and the unknown.***

   Knowns:
   - empirical formula = CH_4N
   - molar mass = 60.0 g

   Unknown:
   - molecular formula = ?

2. *CALCULATE* ***Solve for the unknown.***

   First calculate the empirical formula mass. Then divide the empirical formula mass into the molar mass. Multiply formula subscripts by this value to get the molecular formula.

Empirical formula	efm	Molar mass / efm	Molecular formula
CH_4N	30.0	60.0/30.0 = 2	$C_2H_8N_2$

3. *EVALUATE* ***Does the result make sense?***

   The molecular formula has the molar mass of the compound and can be reduced to the empirical formula.

ANSWERS TO MINI-LAB QUESTIONS 1–3

	$CuSO_4 \cdot 5H_2O$ (g)	$CaCl_2 \cdot 2H_2O$(g)	$Na_2SO_4 \cdot 10H_2O$ (g)
Test tube + hydrate (before heating)	23.88	23.60	23.92
Empty test tube	21.19	21.25	21.17
Mass of hydrate	2.69	2.35	2.75
Test tube + salt (after heating)	22.88	23.07	22.71
Empty test tube	21.19	21.25	21.17
Mass of anhydrous salt	1.69	1.82	1.54
Mass of water lost	1.00	0.53	1.21
Percent water (experimental)	37.2%	22.6%	44.0%
Percent water (theoretical)	36.1%	24.5%	55.9%

MINI LAB

Percent Composition

PURPOSE

To measure the percent of water in a series of crystalline compounds called hydrates.

MATERIALS

- centigram balance
- burner
- 3 medium-sized test tubes
- test tube holder
- test tube rack
- spatula
- hydrated salts of copper(II) sulfate, calcium chloride, and sodium sulfate

PROCEDURE

1. Label each test tube with the name of a salt. Weigh and record its mass.
2. Add 2–3 g of salt (a good-sized spatula full) to the appropriately labeled test tube. Weigh and record the mass of each test tube plus the salt.
3. Hold one of the tubes at a 45° angle and gently heat its contents over the burner, slowly passing it in and out of the flame. Note any change in the appearance of the solid salt.
4. As moisture begins to condense in the upper part of the test tube, gently heat the entire length of the tube. Continue heating until all of the moisture is driven from the tube. This may take 2–3 minutes. Repeat Steps 3 and 4 for the other two tubes.
5. Allow each tube to cool. Then weigh and record the mass of each test tube and the heated salt.

ANALYSIS AND CONCLUSIONS

1. Set up a data table so you can subtract the mass of the empty tube from the mass of the salt and the test tube, both before and after heating.
2. Calculate the difference between the mass of each salt before and after heating. This difference represents the amount of water lost by the hydrate on heating.
3. Calculate the percent by mass of water lost by each compound.
4. Which compound lost the greatest percent by mass of water? The least?

section review 7.3

39. Calculate the percent composition of the compounds that are formed from these reactions.
    a. 222.6 g N combines completely with 77.4 g O.
    b. 2.62 g Na and 4.04 g Cl are formed by the decomposition of table salt.
40. The compound methyl butanoate smells like apples. Its percent composition is 58.8% C, 9.8% H, and 31.4% O. Methyl butanoate's molar mass is 102 g/mol. What is its molecular formula?
41. Calculate the percent composition of each compound.
    a. calcium acetate ($Ca(C_2H_3O_2)_2$) b. hydrogen cyanide (HCN)
42. Using the results of Problem 41, calculate the amount of hydrogen in the following amounts of these compounds.
    a. 124 g $Ca(C_2H_3O_2)_2$ b. 378 g HCN
43. Which of the following molecular formulas are also empirical formulas?
    a. ribose ($C_5H_{10}O_5$)
    b. ethyl butyrate ($C_6H_{12}O_2$)
    c. chlorophyll ($C_{55}H_{72}MgN_4O_5$)
    d. DEET ($C_{12}H_{17}ON$)

Chem ASAP! Assessment 7.3 Check your understanding of the important ideas and concepts in Section 7.3.

portfolio project

Find out whether your state sets standards for the composition of bottled water. If so, is all the bottled water sold tested or just water bottled in your state?

SECTION REVIEW 7.3

39. a. 74.2% N, 25.8% O
    b. 39.3% Na, 60.7% Cl
40. $C_5H_{10}O_2$
41. a. 25.4% Ca, 30.4% C, 3.8% H, 40.5% O
    b. 3.7% H, 44.4% C, 51.9% N
42. a. 4.7 g H
    b. 14 g H
43. a. molecular c. molecular and empirical
    b. molecular d. molecular and empirical

7.3

Practice Problems Plus

Related Chapter Review Problems
Chapter Review problems 63, 65, and 66 are related to Sample Problem 7-14.

Additional Practice Problem
The Chem ASAP! CD-ROM contains the following problem: What is the molecular formula of a compound that is 56.6% K, 8.7% C, and 34.7% O? (K_2CO_3)

3 Assess

Evaluate Understanding

Have students list the steps they would take to calculate the molecular formula in each of the following situations:

- **The empirical formula and molar mass are known.**
- **The percent composition and molar mass are known.**

Reteach

Point out to students that when they know the percent composition and molar mass of a compound, they must first use the percent composition to calculate the empirical formula. They can then calculate the empirical formula mass and compare it to the molar mass to determine the molecular formula.

Portfolio Project

Students may be surprised to find that the standards set for tap water do not necessarily apply to bottled water.

Chemistry Serving... the Environment

DISCUSS

Point out that scientists express concentrations of extremely dilute solutions in units of parts per million (ppm) or parts per billion (ppb).

$$\text{ppm} = \frac{\text{mg of substance}}{\text{L of water}}$$

$$\text{ppb} = \frac{\mu\text{g of substance}}{\text{L of water}}$$

Show students that 1 part of substance per million parts of water can be expressed as 1 gram of substance per 10^6 grams of water or as 1 milligram of substance per liter of water (10^6 milligrams water = 1 L). Ask students to express the concentration of 2 ppm mercury in units of mg/L. (2 mg/L) Ask:

- How many moles of chlorine (Cl_2) are in 1 liter of a solution that is 0.5 ppm chlorine? (0.000007 moles or 7 micromoles)
- How many molecules of chlorine is this? (4×10^{18})

Have students complete similar problems using units of parts per billion. Note that 1 ppb is to a billion as a matchhead is to the volume of water in a swimming pool.

CHEMISTRY IN CAREERS

Encourage students to connect to the Internet address shown on this page to find out more about careers in ecology. There is also information on page 871 of this text.

Chemistry Serving...the Environment

WATER WORTH DRINKING

When you turn on your faucet to get a glass of water, you might assume that the water coming out is safe to drink. You probably do not think about the many substances other than water that may be in your glass or how much of each of these substances the water contains.

Worrying about the quality of drinking water is the job of your local water supplier and the federal government's Environmental Protection Agency (EPA). The EPA has established drinking water standards that state and local governments must follow to ensure that water is safe. A drinking water standard sets an upper limit on a potentially harmful substance if that substance is present in the water. In addition to setting these standards, the EPA also requires that local water providers test their public water supplies regularly. Levels of minerals, chlorine and microorganisms such as bacteria are monitored.

> You probably do not think about the many substances other than water that may be in your glass.

Water standards are established by scientific study. Any naturally occurring or synthetic compounds that may be harmful when present in drinking water are examined by the EPA and by state and local water labs. Compounds found to have negative effects on human health and the environment are studied to determine at what level they begin to produce their harmful effects. Based on this research, standards are set for each contaminant. These regulations help ensure that good quality drinking water reaches your faucet.

The regulation and control of water quality requires measuring the concentration of each contaminant that may be contained in the water. Some contaminants are measured in exceedingly small units called parts per billion (ppb). Trace metals, such as mercury, and organic compounds, such as trihalomethane, are both measured in ppb. Such small units are used because very small amounts of these compounds can make humans ill or can have an adverse effect on the environment. Another unit commonly used is parts per million (ppm). Dissolved minerals such as sodium and calcium, chlorine, and levels of microorganisms, such as coliform bacteria, are measured in ppm.

The number of microorganisms in water is one of the most important aspects of water quality. If left unchecked, microorganisms could cause outbreaks of bacterial infections and other water-borne diseases. Local and state officials use an indicator species, coliform bacteria, to monitor microbiological water quality. If periodic water samples indicate too high a level of coliform bacteria, disinfectants such as chlorine must be added to the water distribution system to destroy potentially dangerous microorganisms.

Where water is treated with chlorine, tests are done frequently to ensure that levels of this disinfectant do not fall below 0.5 ppm. By contrast, the amount of sodium in drinking water may vary depending on the source of the water.

The mineral content of drinking water can vary widely by location. To find out the percent composition of your drinking water, check with your local government office or agriculture bureau. You may be surprised to find out what you are drinking!

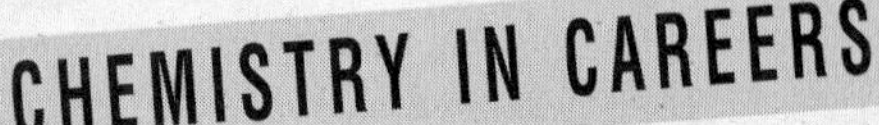

CHEMISTRY IN CAREERS

ECOLOGIST

Knowledge of water chemistry and aquatic life. Training in ecology and chemistry required.

See page 871.

CHEMIST NEEDED

TAKE IT TO THE NET

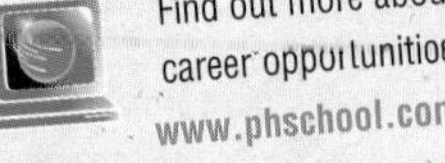

Find out more about career opportunities: www.phschool.com

CHEMICAL SPECIALIST

Local food service distributor seeks responsible self-motivated indivi...

Chapter 7 STUDENT STUDY GUIDE

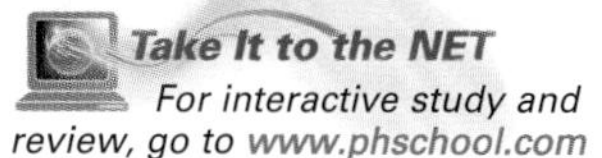
Take It to the NET
For interactive study and review, go to www.phschool.com

KEY TERMS

- Avogadro's number *p. 173*
- empirical formula *p. 192*
- gram atomic mass (gam) *p. 176*
- gram formula mass (gfm) *p. 180*
- gram molecular mass (gmm) *p. 178*
- molar mass *p. 182*
- molar volume *p. 184*
- mole (mol) *p. 173*
- percent composition *p. 188*
- representative particle *p. 173*
- standard temperature and pressure (STP) *p. 184*

CONCEPT SUMMARY

7.1 The Mole: A Measurement of Matter

- The SI unit that measures the amount of substance is the mole.
- A mole of any substance is composed of Avogadro's number (6.02×10^{23}) of representative particles.
- The representative particle of most elements is the atom.
- A molecule is the representative particle of diatomic elements and molecular compounds.
- The representative particle of ionic compounds is a formula unit.
- The gram atomic mass (gam), gram molecular mass (gmm), and gram formula mass (gfm) are the mass of one mole of an element, a molecular compound, and an ionic compound, respectively.
- The molar mass of a substance is the mass in grams of one mole of that substance.
- One mole of any substance contains the same number of representative particles as one mole of any other substance.

7.2 Mole–Mass and Mole–Volume Relationships

- Multiplying the number of moles of a substance by its molar mass gives the mass of the substance.
- Dividing the mass of a substance by its molar mass gives the number of moles of that substance.
- One mole of any gas at STP (1 atm pressure and 0 °C) occupies a volume of 22.4 L.
- The density of any gas at STP is its molar mass divided by 22.4 L.

7.3 Percent Composition and Chemical Formulas

- The percent composition is the percent by mass of each element in a compound.
- The percent by mass of any element in a given compound is calculated by dividing the element's mass by the mass of the compound and multiplying by 100%.
- An empirical formula is the simplest whole-number ratio of atoms of the elements in a compound.

CHAPTER CONCEPT MAP

Use these terms to construct a concept map that organizes the major ideas of this chapter.

Chem ASAP! **Concept Map 7** Create your Concept Map using the computer.

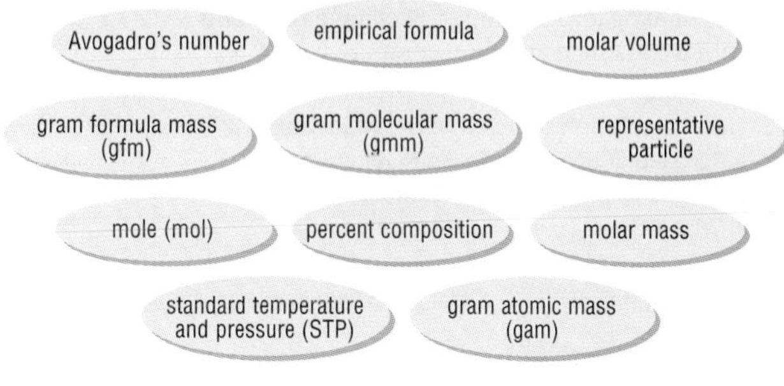

Take It to the Net

At **www.phschool.com** students will find for this chapter

- an Internet activity
- links to related chemistry sites
- an interactive quiz
- career links

4 Close

Summary

Ask students to summarize the main concepts of the chapter by writing one or two sentences about each section. They should include samples of the calculations that were introduced. You may wish to do this as a class activity, in which groups of students each contribute a section to the overall chapter review. Read through the class's chapter review to be sure it meets the chapter objectives.

Extension

Some common metal ores are chromite, chalcopyrite, cuprite, galena, pentlandite, cassiterite, rutile, sphalerite, and cobalite. Have students research these ores to learn about the metals they contain and the customary methods used to separate out the metals. They should include the percent composition for the ore they investigate.

Looking Back . . . Looking Ahead . . .

The previous chapter discussed covalent and ionic bonding and the formation of molecules. This chapter provided instruction on different means for measuring the quantity of molecules, atoms, or ions. In the next chapter, students learn how molecules react with each other. Students are shown how to write balanced chemical equations from descriptions of chemical reactions.

Answers

44. number, mass, or volume; Examples will vary.

45. **a.** molecule **c.** molecule
b. formula unit **d.** atom

46. **a.** 3 **c.** 9
b. 2 **d.** 10

47. All contain 6.02×10^{23} molecules.

48. 1.00 mol C_2H_6

49. **a.** 1.81×10^{24} atoms Sn
b. 2.41×10^{23} formula units KCl
c. 4.52×10^{24} molecules SO_2
d. 2.89×10^{21} formula units NaI

50. **a.** 98.0 g **d.** 132.1 g
b. 76.0 g **e.** 89.0 g
c. 100.1 g **f.** 159.8 g

51. **a.** 60.1 g **c.** 106.8 g
b. 28.0 g **d.** 63.5 g

52. Answers will vary but should include:
(1) Determine the number of moles of each atom from the formula.
(2) Look up the atomic mass of each element.
(3) Multiply the number of moles of each atom by its molar mass.
(4) Sum these products.

53. 71.0 g Cl_2

54. Answers will vary.

55. **a.** 0.258 mol SiO_2
b. 4.80×10^{-4} mol AgCl
c. 1.12 mol Cl_2
d. 0.106 mol KOH
e. 5.93 mol $Ca(C_2H_3O_2)_2$
f. 2.00×10^{-2} mol Ca

56. **a.** 108 g C_5H_{12}
b. 547 g F_2
c. 71.8 g $Ca(CN)_2$
d. 238 g H_2O_2
e. 224 g NaOH
f. 1.88 g Ni

57. **a.** 1.7×10^2 L Ar **c.** 26.9 L O_2
b. 9.9 L C_2H_6

58. **a.** 1.96 g/L **c.** 2.05 g/L
b. 0.902 g/L

59. **a.** 234 L SO_3
b. 2.99×10^{-22} g $C_9H_8O_4$
c. 3.13×10^{25} atoms

60. **a.** 5.9% H, 94.1% S
b. 22.6% N, 6.5% H, 19.4% C, 51.6% O
c. 41.7% Mg, 54.9% O, 3.4% H
d. 42.1% Na, 18.9% P, 39.0% O

61. **a.** 3.33 g S **c.** 40.6 g Mg
b. 5.65 g N **d.** 152 g P

62. **d.** 77.7% Fe in FeO

63. H_2O_2

64. **a.** molecular **d.** molecular
b. molecular **e.** empirical
c. empirical **f.** empirical

65. **a.** $C_3H_6O_3$
b. Hg_2Cl_2
c. $C_6H_{10}O_4$

66. **a.** H_2O_2
b. $C_4H_8O_4$

Chapter 7 REVIEW

CONCEPT PRACTICE

44. List three common ways that matter is measured. Give examples of each. *7.1*

45. Name the representative particle (atom, molecule, or formula unit) of each substance. *7.1*
a. oxygen **c.** sulfur dioxide
b. sodium sulfide **d.** potassium

46. How many hydrogen atoms are in a representative particle of each substance? *7.1*
a. $Al(OH)_3$ **c.** $(NH_4)_2HPO_4$
b. $H_2C_2O_4$ **d.** $C_4H_{10}O$

47. Which contains more molecules: 1.00 mol H_2O_2, 1.00 mol C_2H_6, or 1.00 mol CO? *7.1*

48. Which contains more atoms: 1.00 mol H_2O_2, 1.00 mol C_2H_6, or 1.00 mol CO? *7.1*

49. Find the number of representative particles in each substance. *7.1*
a. 3.00 mol Sn **c.** 7.50 mol SO_2
b. 0.400 mol KCl **d.** 4.80×10^{-3} mol NaI

50. Calculate the molar mass of each substance. *7.1*
a. H_3PO_4 **c.** $CaCO_3$ **e.** $C_4H_9O_2$
b. N_2O_3 **d.** $(NH_4)_2SO_4$ **f.** Br_2

51. Calculate the mass of 1.00 mol of each of these substances. *7.1*
a. silicon dioxide (SiO_2)
b. diatomic nitrogen (N_2)
c. iron(III) hydroxide ($Fe(OH)_3$)
d. copper (Cu)

52. List the steps you would take to calculate the molar mass of any compound. *7.1*

53. What is the gram molecular mass (gmm) of chlorine? *7.1*

54. Construct a numerical problem to illustrate the size of Avogadro's number. Exchange problems with a classmate and then compare your answers. *7.1*

55. How many moles is each of the following? *7.2*
a. 15.5 g SiO_2 **d.** 5.96 g KOH
b. 0.0688 g AgCl **e.** 937 g $Ca(C_2H_3O_2)_2$
c. 79.3 g Cl_2 **f.** 0.800 g Ca

56. Find the mass of each substance. *7.2*
a. 1.50 mol C_5H_{12}
b. 14.4 mol F_2
c. 0.780 mol $Ca(CN)_2$
d. 7.00 mol H_2O_2
e. 5.60 mol NaOH
f. 3.21×10^{-2} mol Ni

57. Calculate the volume of each of the following gases at STP. *7.2*
a. 7.6 mol Ar
b. 0.44 mol C_2H_6
c. 1.20 mol O_2

58. What is the density of each of the following gases at STP? *7.2*
a. C_3H_8 **b.** Ne **c.** NO_2

59. Find each of the following quantities. *7.2*
a. the volume, in liters, of 835 g SO_3 at STP
b. the mass, in grams, of a molecule of aspirin ($C_9H_8O_4$)
c. the number of atoms in 5.78 mol NH_4NO_3

60. Calculate the percent composition of each compound. *7.3*
a. H_2S **c.** $Mg(OH)_2$
b. $(NH_4)_2C_2O_4$ **d.** Na_3PO_4

61. Using your answers from Problem 60, calculate the number of grams of these elements. *7.3*
a. sulfur in 3.54 g H_2S
b. nitrogen in 25.0 g $(NH_4)_2C_2O_4$
c. magnesium in 97.4 g $Mg(OH)_2$
d. phosphorus in 804 g Na_3PO_4

62. Which of the following compounds has the highest iron content? *7.3*
a. $FeCl_2$ **c.** $Fe(OH)_2$
b. $Fe(C_2H_3O_2)_3$ **d.** FeO

63. You find that 7.36 g of a compound has decomposed to give 6.93 g of oxygen. The only other element in the compound is hydrogen. If the molar mass of the compound is 34.0 g/mol, what is its molecular formula? *7.3*

64. Classify each formula as an empirical or a molecular formula. *7.3*
a. S_2Cl_2 **c.** Na_2SO_3 **e.** $C_{17}H_{19}NO_3$
b. $C_6H_{10}O_4$ **d.** $C_5H_{10}O_5$ **f.** $(NH_4)_2CO_3$

65. What is the molecular formula for each compound? Each compound's empirical formula and molar mass is given. *7.3*
a. CH_2O, 90 g/mol
b. HgCl, 472.2 g/mol
c. $C_3H_5O_2$, 146 g/mol

66. Determine the molecular formula for each compound. *7.3*
a. 94.1% O and 5.9% H; molar mass = 34 g
b. 40.0% C, 6.6% H, and 53.4% O; molar mass = 120 g

CONCEPT MASTERY

67. How can you determine the molar mass of a gaseous compound if you do not know its molecular formula?

68. A series of compounds has the empirical formula CH_2O. The graph shows the relationship between the molar mass of the compounds and the mass of carbon in each compound.

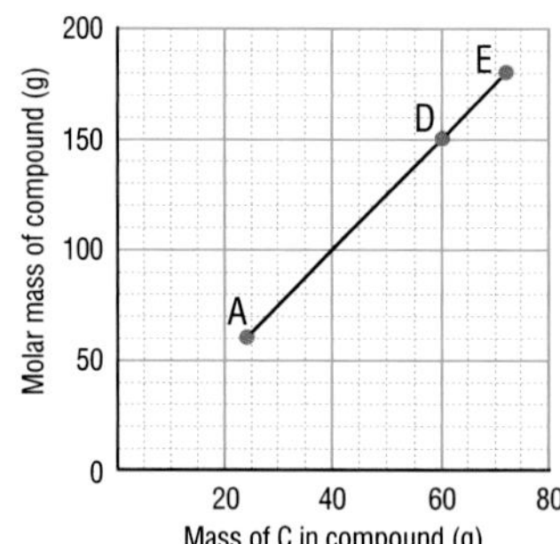

a. What are the molecular formulas for the compounds represented by data points A, D, and E?
b. Find the slope of the line. Is this value consistent with the empirical formula?
c. There are two other valid data points that fall on the line between points A and D. What are the *x*, *y* values for these data points?

69. Explain what is wrong with each statement.
a. One mole of any substance contains the same number of atoms.
b. The gram atomic mass of a compound is the atomic mass expressed in grams.
c. One gram molecular mass of CO_2 contains Avogadro's number of atoms.

70. Which of the following contains the largest number of atoms?
a. 82.0 g Kr
b. 0.842 mol C_2H_4
c. 36.0 g N_2

71. What is the total mass of a mixture of 3.50×10^{22} formula units Na_2SO_4, 0.500 mol H_2O, and 7.23 g AgCl?

72. Determine the empirical formula of each.
a. 42.9% C and 57.1% O
b. 32.00% C, 42.66% O, 18.67% N, and 6.67% H
c. 71.72% Cl, 16.16% O, and 12.12% C

73. An imaginary "atomic balance" is shown below. Fifteen atoms of boron on the left side of the balance are balanced by six atoms of an unknown element E on the right side.

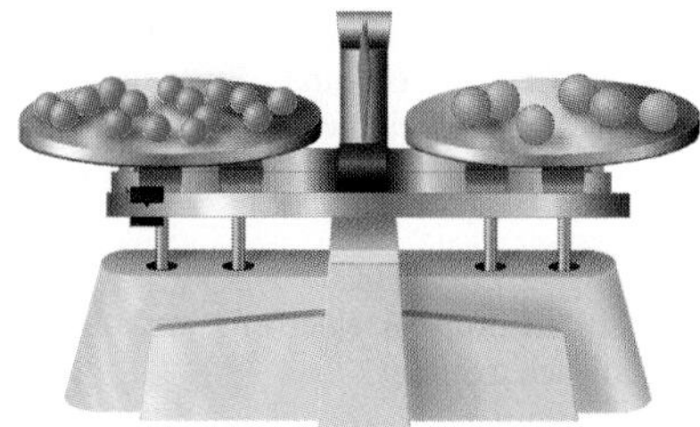

a. What is the atomic mass of element E?
b. What is the identity of element E?

74. A typical virus is 5×10^{-6} cm in diameter. If Avogadro's number of these virus particles were laid in a row, how many kilometers long would the line be?

75. Calculate the empirical formula for each compound.
a. compound consisting of 0.40 mol Cu per 0.80 mol Br
b. compound with 4 atoms of carbon for every 12 atoms of hydrogen

CRITICAL THINKING

76. Choose the term that best completes the second relationship.
a. dozen : eggs
mole : ______
(1) atoms
(2) 6.02×10^{23}
(3) size
(4) grams
b. gam : element
gmm : ______
(1) formula unit
(2) molecule
(3) ionic compound
(4) molecular compound
c. mole : Avogadro's number
molar volume : ______
(1) mole
(2) water
(3) STP
(4) 22.4 L

Chapter 7 REVIEW

72. a. CO
b. $C_2O_2NH_5$
c. Cl_2OC
73. a. 27 amu
b. aluminum
74. 3.01×10^{13} km
75. a. $CuBr_2$
b. CH_3
76. a. 1
b. 4
c. 4

67. You can measure the mass of 22.4 L of the compound at STP; this is the molar volume of the gas. The mass of the molar volume is the molar mass.
68. a. A: $C_2H_4O_2$; D: $C_5H_{10}O_5$; E: $C_6H_{12}O_6$
b. The slope = 2.5/1, which is the ratio of the molar mass of the empirical formula to the mass of carbon in the empirical formula: $= \frac{30}{12} = \frac{2.5}{1}$.
c. mass of carbon = 36; molar mass = 90; mass of carbon = 48; molar mass = 120
69. a. A molecule is composed of two or more atoms.
b. A compound has a gram molecular mass; not a gram atomic mass.
c. A mole of CO_2 has 3 times Avogadro's number of atoms.
70. b. 0.842 mol C_2H_4
71. 24.5 g

Answers

77. A molecular formula is a whole number multiple of its empirical formula.
78. Sulfur atoms have a greater atomic mass. Most sulfur atoms have 16 protons, 16 electrons, and 16 neutrons; carbon is composed of 6 protons, 6 electrons, and 6 neutrons. Therefore, 6.02×10^{23} sulfur atoms will have a greater mass than the same number of carbon atoms.
79. Gas molecules are separated by so much empty space, their own volumes are insignificant when considering how much space a certain quantity of gas occupies.
80. **a.** 40, 40, 50
b. 46, 46, 62
c. 35, 35, 46
d. 51, 51, 72
81. 1.59 mol Pt
82. A molecule is composed of two or more atoms.
83. **a.** 4.72×10^3 mg
b. 97 km/hr
c. 4.4×10^{-2} dm
84. **a.** iron(III) hydroxide
b. ammonium iodide
c. sodium carbonate
d. carbon tetrachloride
85. **a.** KNO_3
b. CuO
c. Mg_3N_2
d. AgF
86. $C_3H_5O_9N_3$
87. 21.9 cm^3
88. 3.54×10^{23} O_2 molecules
89. **a.**

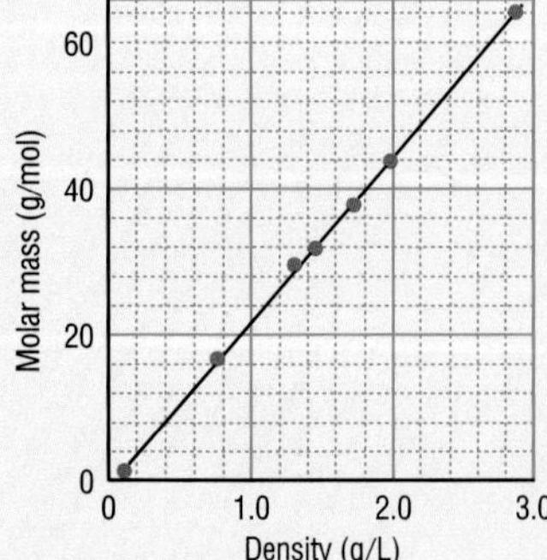

b. 22.4 L/mol
c. 24.6 g/mol
d. 2.5 g/L

77. How are the empirical and molecular formulas of a compound related?
78. Why does one mole of carbon have a smaller mass than one mole of sulfur? How are the atomic structures of these elements different?
79. One mole of any gas at STP equals 22.4 L of that gas. It is also true that different elements have different atomic volumes, or diameters. How can you reconcile these two statements?

CUMULATIVE REVIEW

80. How many protons, electrons, and neutrons are in each isotope?
a. zirconium-90
b. palladium-108
c. bromine-81
d. antimony-123
81. How many moles is 14.5 cm^3 of platinum? The density of platinum is 21.45 g/cm^3.
82. How does a molecule differ from an atom?
83. Convert each of the following.
a. 4.72 g to mg
b. 2.7×10^3 cm/s to km/h
c. 4.4 mm to dm
84. Name these compounds
a. $Fe(OH)_3$ **c.** Na_2CO_3
b. NH_4I **d.** CCl_4
85. Write formulas for these compounds.
a. potassium nitrate
b. copper(II) oxide
c. magnesium nitride
d. silver fluoride

CONCEPT CHALLENGE

86. Nitroglycerine contains 60% as many carbon atoms as hydrogen atoms; three times as many oxygen atoms as nitrogen atoms; and the same number of carbon and nitrogen atoms. The number of moles of nitroglycerine in 1 g is 0.00441. What is the molecular formula of nitroglycerine?
87. The density of nickel is 8.91 g/cm^3. How large a cube, in cm^3, would contain 2.00×10^{24} atoms of nickel?
88. Dry air is about 20.95% oxygen by volume. Assuming STP, how many oxygen molecules are in a 75.0-g sample of air? The density of air is 1.19 g/L.
89. The table below gives the molar mass and density of seven gases at STP.

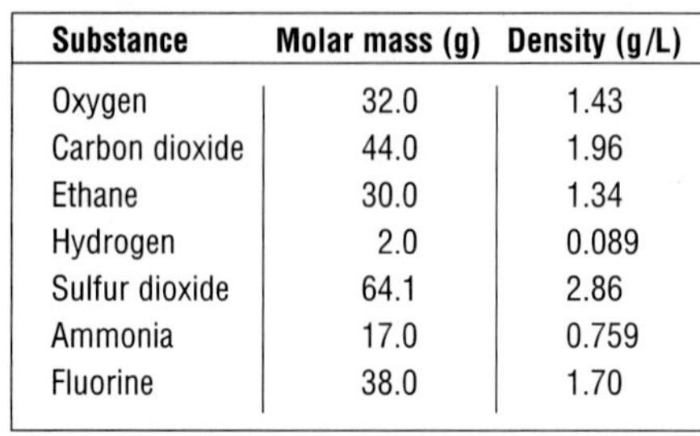

Substance	Molar mass (g)	Density (g/L)
Oxygen	32.0	1.43
Carbon dioxide	44.0	1.96
Ethane	30.0	1.34
Hydrogen	2.0	0.089
Sulfur dioxide	64.1	2.86
Ammonia	17.0	0.759
Fluorine	38.0	1.70

a. Plot these data, with density on the *x*-axis.
b. What is the slope of the straight-line plot?
c. What is the molar mass of a gas at STP that has a density of 1.10 g/L?
d. A mole of a gas at STP has a mass of 56.0 g. Use the graph to determine its density.
90. The element gold has properties that have made it much sought after through the ages. A cubic meter of ocean water has 6×10^{-6} g gold. If the total mass of the water in Earth's oceans is 4×10^{20} kg, how many kilograms of gold are distributed throughout the oceans? (Assume that the density of seawater is 1 g/cm^3.) How many liters of seawater would have to be processed to recover 1 kg of gold (which has a value of about \$8850 at 2001 prices)? Do you think this recovery operation is feasible?
91. Have you ever wondered how Avogadro's number was determined? Actually, Avogadro's number has been independently determined by about 20 different methods. In one approach, Avogadro's number is calculated from the volume of a film of a fatty acid floating on water. In another method, it is determined by comparing the electric charge of one electron with the electric charge of a mole of electrons. In a third method, the spacing between ions in an ionic substance can be determined by using a technique called x-ray diffraction. In the x-ray diffraction of sodium chloride, it has been determined that the distance between adjacent Na^+ and Cl^- ions is 2.819×10^{-8} cm. The density of solid sodium chloride is 2.165 g/cm^3. By calculating the molar mass to four significant figures, you can determine Avogadro's number. What value do you obtain?

90. 2.4×10^9 kg Au; 2×10^{11} L H_2O; not feasible
91. 6.025×10^{23} formula units/mol

Chapter 7 STANDARDIZED TEST PREP

Select the choice that best answers each question or completes each statement.

1. Calculate the molar mass of ammonium phosphate, $(NH_4)_3PO_4$.
   a. 149.0 c. 242.0
   b. 113.0 d. 121.0
2. Based on the structural formula below, what is the empirical formula for tartaric acid, a compound found in grape juice?

   HO—CH—COOH
   |
   HO—CH—COOH

   a. $C_2H_3O_3$
   b. $C_4H_6O_6$
   c. CHO
   d. $C_1H_{1.5}O_{1.5}$
3. How many hydrogen atoms are in six molecules of ethylene glycol, $HOCH_2CH_2OH$?
   a. 6
   b. 36
   c. $6 \times 6.02 \times 10^{23}$
   d. $36 \times 6.02 \times 10^{23}$
4. Which of these statements is true of a balloon filled with 1.00 mol $N_2(g)$ at STP?
   I. The balloon has a volume of 22.4 L.
   II. The contents of the balloon have a mass of 14.0 g.
   III. The balloon contains 6.02×10^{23} molecules.
   a. I only
   b. I and II only
   c. I and III only
   d. II and III only
   e. I, II, and III
5. Which of these compounds has the largest percent by mass of nitrogen?
   a. N_2O
   b. NO
   c. NO_2
   d. N_2O_3
   e. N_2O_4
6. Allicin, $C_6H_{10}S_2O$, is the compound that gives garlic its odor. A sample of allicin contains 3.0×10^{21} atoms of carbon. How many atoms of hydrogen does this sample contain?
   a. 1.8×10^{21} atoms
   b. 10 atoms
   c. 5.0×10^{21} atoms
   d. 1.0×10^{21} atoms

The lettered choices below refer to questions 7–11. A lettered choice may be used once, more than once, or not at all.

(A) CH
(B) CH_2
(C) C_2H_5
(D) CH_3
(E) C_2H_3

Which of the formulas is the empirical formula for each of the following compounds?

7. C_8H_{12}
8. C_6H_6
9. $C_{12}H_{24}$
10. C_2H_6
11. C_4H_{10}

Use the ball-and-stick models to answer questions 12–14. In the models, carbon is black, hydrogen is light blue, oxygen is red, and nitrogen is dark blue. Write the molecular formula for each compound. Then calculate its molar mass.

12.

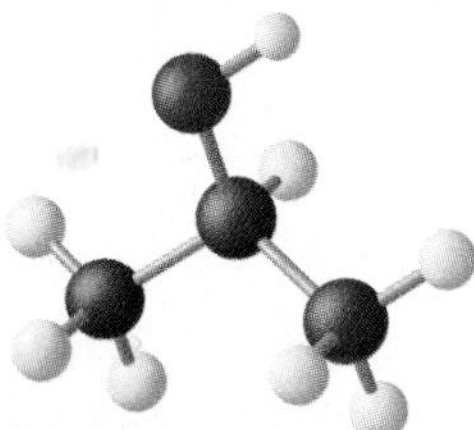

Isopropanol

13.

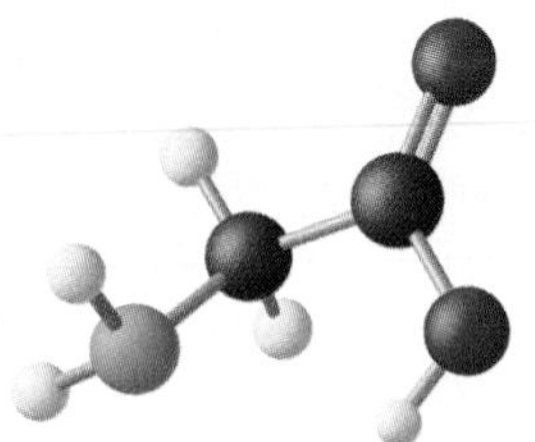

Glycine

14.

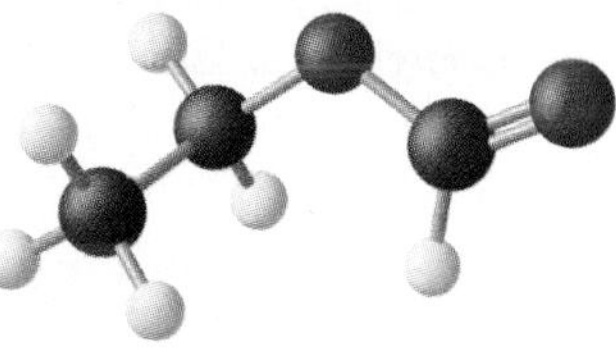

Ethyl formate

Chapter 7
STANDARDIZED TEST PREP

1. a
2. a
3. b
4. c
5. a
6. c
7. E
8. A
9. B
10. D
11. C
12. C_3H_8O, 60.0 g/mol
13. $C_2H_5NO_2$, 75.0 g/mol
14. $C_3H_6O_2$, 74.0 g/mol

Planning Guide

SECTION OBJECTIVES	ACTIVITIES/FEATURES	MEDIA & TECHNOLOGY
8.1 Describing Chemical Change ● ■ ◆ ▸ Write equations describing chemical reactions using appropriate symbols ▸ Write balanced chemical equations when given the names or formulas of the reactants and products in a chemical reaction	**SE** Discover It! *Modeling Chemical Reactions,* p. 202 **SE** **Link to Physiology** *Hydrogen Peroxide,* p. 207 SSLM 8: *Chemical Equations* **TE** DEMOS, pp. 205, 208	ASAP Animation 7, Simulation 4 ASAP Problem Solving 4, 8 ASAP Assessment 8.1 SSV 3: *Chemical Equations* CHM Side 3, 2: The *Writing on the Wall* OT 13: *Balancing Chemical Equation*
8.2 Types of Chemical Reactions ● ■ ◆ ▸ Identify a reaction as combination, decomposition, single-replacement, double-replacement, or combustion ▸ Predict the products of combination, decomposition, single-replacement, double-replacement, and combustion reactions	**SE** Mini Lab *Activity Series of Metals,* p. 224 (LRS 8-1) LM 10: *Qualitative Analysis* LM 11: *Types of Chemical Reactions* SSLM 9: *Balancing Chemical Equations* **TE** DEMOS, pp. 213, 216, 217, 218, 219	ASAP Simulation 5 ASAP Problem Solving 14, 17, 18, 21 ASAP Assessment 8.2 CA *Making Sodium Chloride* CA *Elephant Toothpaste* CA *Watermelon Surprise* CA *Superheated Steam* CA *Thermite Reaction* CA *Carbide Cannon* CA *Burning Magnesium* CA *Big Burner* CA *Methane Bubbles* CA *Ira Remsen Story* CHM Side 2, 38: *Reactivity of Magnesium* OT 14: *Single-Replacement Reactions* OT 15: *Double-Replacement Reactions*
8.3 Reactions in Aqueous Solution ○ ■ ◆ ▸ Write and balance net ionic equations ▸ Use solubility rules to predict the precipitates formed in double-replacement reactions	**SE** Small-Scale Lab *Precipitation Reactions: Formation of Solids,* p. 229 (LRS 8-2) **SE** **Chemistry Serving . . . Society** *Combating Combustion,* p. 230 **SE** **Chemistry in Careers** *Firefighter,* p. 230 **TE** DEMOS, pp. 226, 227	ASAP Problem Solving 25, 26 ASAP Assessment 8.3 ASAP Concept Map 8 RP Lesson Plans, Resource Library AR Computer Test 8 www Activities, Self-Tests, *SCIENCE NEWS* updates TCP The Chemistry Place Web Site

KEY

● Conceptual (concrete concepts)	AR Assessment Resources	GRS Guided Reading and Study Workbook
○ Conceptual (more abstract/math)	ASAP Chem ASAP! CD-ROM	LM Laboratory Manual
■ Standard (core content)	ACT ActivChemistry CD-ROM	LP Laboratory Practicals
□ Standard (extension topics)	CHM CHEMedia Videodiscs	LRS Laboratory Recordsheets
◆ Honors (core content)	CA Chemistry Alive! Videodiscs	SSLM Small-Scale Lab Manual
◇ Honors (options to accelerate)	GCP Graphing Calculator Problems	

PRACTICE		ASSESSMENT	
SE	**Sample Problems** 8-1 to 8-4	SE	Section Review
GRS	Section 8.1	RM	Section Review 8.1
RM	Practice Problems 8.1	RM	Chapter 8 Quiz
SE	**Sample Problems** 8-5 to 8-9	SE	Section Review
GRS	Section 8.2	RM	Section Review 8.2
RM	Practice Problems 8.2	RM	Chapter 8 Quiz
		LP	Lab Practicals 8-1, 8-2
SE	**Sample Problems** 8-10 to 8-11	SE	Section Review
GRS	Section 8.3	RM	Section Review 8.3
RM	Practice Problems 8.3	RM	Vocabulary Review 8
RM	Interpreting Graphics	SE	Chapter Review
		SM	Chapter 8 Solutions
		SE	Standardized Test Prep
		PHAS	Chapter 8 Test Prep
		RM	Chapter 8 Quiz
		RM	Chapter 8 A & B Test

OT	Overhead Transparency	SE	Student Edition
PHAS	PH Assessment System	SM	Solutions Manual
PLM	Probeware Lab Manual	SSV	Small-Scale Video/Videodisc
PP	Problem Pro CD-ROM	TCP	www.chemplace.com
RM	Review Module	TE	Teacher's Edition
RP	Resource Pro CD-ROM	www	www.phschool.com

PLANNING FOR ACTIVITIES

STUDENT EDITION

Discover It! p. 202
- paper clips (3 different colors)

Mini Lab p. 224
- 100-mL beakers
- strips of Cu, Mg, Zn
- steel wool or fine sand-paper
- 0.05*M* solutions of $CuSO_4$, $MgSO_4$, NaCl, $AgNO_3$, $ZnSO_4$

Small-Scale Lab, p. 229
- pencils and paper
- reaction surfaces
- solutions of $AgNO_3$, $Pb(NO_3)_2$, $CaCl_2$, Na_2CO_3, Na_3PO_4, NaOH, Na_2SO_4, KI
- rulers

TEACHER'S EDITION

Teacher Demo, p. 205
- 150-mL Pyrex beaker
- goggles
- sugar
- 18*M* H_2SO_4
- fume hood
- lab apron
- face shield
- protective gloves
- plexiglas shield
- 500-mL beaker
- base with molarity $< 1M$

Teacher Demo, p. 208
- 200 mL of 0.1*M* $CuCl_2(aq)$
- 400-mL beaker
- 5-cm square piece of aluminum foil

Teacher Demo, p. 213
- 5- to 7-cm strip of magne-sium ribbon
- large crucible
- crucible tongs
- Bunsen burner
- cobalt blue glass

Teacher Demo, p. 216
- small piece (2 cm) of magnesium metal
- 10 mL of 6*M* $HCl(aq)$
- small piece of sodium
- cold water
- iron nail
- 50 mL of 1*M* $CuCl_2$

Teacher Demo, p. 217
- 50 mL of 1.0*M* potassium chromate (K_2CrO_4)
- 250-mL Erlenmeyer flask and stopper
- 5 mL of dilute silver nitrate solution ($AgNO_3$)
- small test tube (one that fits inside flask)
- balance

Teacher Demo, p. 218
- 50 mL each of 0.1*M* $AgNO_3(aq)$ and $NaCl(aq)$
- beakers
- 10 g of solid NaCl to aid in disposal

Activity, p. 218
- resources to research fuels used to propel the space shuttle orbiter

Teacher Demo, p. 219
- superfine steel wool pad
- plastic sandwich bag
- balance
- ring stand
- utility clamp
- matches

Teacher Demo, p. 226
- 5–10 mL of $Pb(NO_3)_2$
- beaker
- 3–5 mL of KI solution

Teacher Demo, p. 227
- 5 mL of 0.1*M* $BaCl_2(aq)$
- 5 mL of 0.1*M* $Na_2SO_4(aq)$
- beaker
- 5 mL of 0.1*M* $K_2CrO_4(aq)$
- 5 mL of 0.1*M* $Ba(NO_3)_2(aq)$

Key Terms

8.1 chemical equation, skeleton equation, catalyst, coefficients, balanced equation
8.2 combination reaction, decomposition reaction, single-replacement reaction, activity series of metals, double-replacement reactions, combustion reaction
8.3 complete ionic equation, spectator ions, net ionic equation

DISCOVER IT!

Step 3: $2H_2 + O_2 \rightarrow 2H_2O$
Step 5: $CH_4 + 2O_2 \rightarrow CO_2 + 2H_2O$
Dalton proposed that elements contain indivisible atoms that combine into compounds according to fixed ratios. For the reactions modeled in Steps 2 and 4, the number of reactant atoms per element equals the number of product atoms per element. This conservation of atoms supports Dalton's theory that atoms are neither created nor destroyed during a reaction.

Chapter 8 CHEMICAL REACTIONS

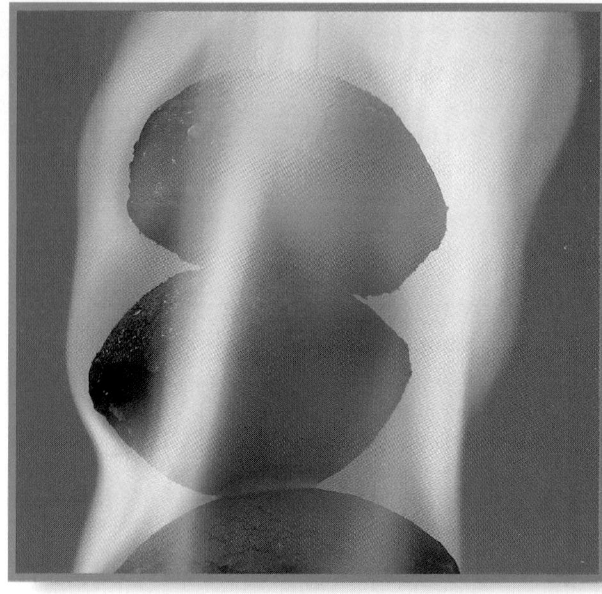
Combustion of carbon in charcoal briquettes produces heat and light.

FEATURES

DISCOVER IT!
Modeling Chemical Reactions

SMALL-SCALE LAB
Precipitation Reactions: Formation of Solids

MINI LAB
Activity Series of Metals

CHEMISTRY SERVING ... SOCIETY
Combating Combustion

CHEMISTRY IN CAREERS
Firefighter

LINK TO PHYSIOLOGY
Hydrogen Peroxide

Stay current with SCIENCE NEWS
Find out more about chemical reactions:
www.phschool.com

DISCOVER IT! MODELING CHEMICAL REACTIONS

You need 36 paper clips (12 each of 3 different colors).

1. Each paper clip represents a single atom. Designate a different color of paper clip to represent atoms of oxygen (O), hydrogen (H), or carbon (C). Make two molecules each of hydrogen (H_2) and methane (CH_4) and six molecules of oxygen (O_2).
2. "React" one H_2 with one O_2 by splitting the molecules and joining one oxygen atom to two hydrogen atoms. Because there is an unreacted oxygen atom, you must react it with another hydrogen molecule to form a second water molecule.
3. Summarize what happened in this reaction by placing the number of each molecule reacted or formed in front of its formula.
$$\_\_H_2 + \_\_O_2 \longrightarrow \_\_H_2O$$
4. Now "react" methane (CH_4) with oxygen (O_2) to produce carbon dioxide (CO_2) and water (H_2O). Start with one molecule of methane and one molecule of oxygen. Continue reacting molecules of CH_4 and O_2 until all the reactant atoms have been used to form products.
5. Summarize what happened in this reaction.
$$\_\_CH_4 + \_\_O_2 \longrightarrow \_\_CO_2 + \_\_H_2O$$

Summarize Dalton's atomic theory. How might the numbers and kinds of atoms to the left and right of the arrow ($\longrightarrow$) in the model reactions you performed support this theory?

8.1 ● ■ ◆
8.2 ● ■ ◆
8.3 ○ ■ ◆

Conceptual Use the paper-clip reactions in Discover It! as a concrete model for balancing equations. Emphasize that formulas cannot be changed as students balance equations; they must focus on coefficients. Section 8.2 is a long but important section. The prediction of reaction products at the end of the section can serve as a brief review if the general equations are omitted. Sample Problems 8-10 and 8-11 may be omitted in Section 8.3.

Standard Encourage students to use models and draw particle representations when balancing simple equations. In Section 8.2, use the figures to help students connect the macroscopic and particle views of reactions, which will help them remember the different types of reactions.

Honors Use the figures and activities in this chapter to help students approach reactions from multiple perspectives: mathematical and symbolic representations, macroscopic observations, and the particle view. You may want to use the Mini Lab to introduce Section 8.2 and the Small-Scale Lab to introduce Section 8.3.

DESCRIBING CHEMICAL CHANGE

On May 6, 1937, the huge airship Hindenburg was heading for its landing site in Lakehurst, New Jersey, after an uneventful trans-Atlantic crossing. Suddenly, to the horror of observers on the ground, the airship erupted into a fireball as 210 000 cubic meters of hydrogen burst into flame. The chemical reaction that caused the disaster can be described simply as "hydrogen combines explosively with oxygen to produce water." **How do chemists describe chemical reactions?**

objectives

- Write equations describing chemical reactions using appropriate symbols
- Write balanced chemical equations when given the names or formulas of the reactants and products in a chemical reaction

key terms

- chemical equation
- skeleton equation
- catalyst
- coefficients
- balanced equation

Word Equations

Every minute of the day chemical reactions take place—both inside you and around you. After a meal, a series of complex chemical reactions take place as your body digests food. Likewise, plants use sunlight to drive the photosynthetic processes needed to produce plant growth. Although the chemical reactions involved in photosynthesis and digestion are quite different, both of these chemical reactions are necessary to sustain life. What are some other reactions necessary for life? 1

Not every chemical reaction involves living things directly. The useful chemical reaction that produces electrical energy in a car battery is one example. The energy produced is used to start the car. At the same time, an undesirable chemical reaction might be occurring between the iron in the car fender and the oxygen in the air. This reaction produces rust. Many common reactions occur when food is grown or cooked. All chemical reactions, whether simple or complex, desirable or undesirable, involve changing substances.

It will be important for you to be able to describe a chemical reaction in writing. In a reaction, one or more substances (the reactants) change into one or more new substances (the products). Recall from Chapter 2 the shorthand method for writing a brief description of what happens in a chemical reaction. In writing reactions, separate the reactants from the products by writing an arrow ($\rightarrow$), which means yields, gives, or reacts to produce.

$$\text{Reactants} \longrightarrow \text{products}$$

You may remember that John Dalton proposed an explanation for the way in which substances change in a chemical reaction. His atomic theory holds true today. As reactants are converted to products, the bonds holding the atoms together are broken and new bonds are formed. It is

Figure 8.1
Flour, sugar, salt, yeast, and water are the main ingredients for making leavened bread. The reactants (ingredients) undergo physical and chemical changes to form the product (baked bread). What are some indications in the pictures below that chemical reactions are taking place? 2

STUDENT RESOURCES

From the Teacher's Resource Package, use:

- Section Review 8.1, Ch. 8 Practice Problems and Quizzes from the Review Module (Ch. 5–8)
- Small-Scale Lab Manual: Experiment 8

1 Answers will vary.
2 formation of gas as dough rises, increase in volume, change in color, irreversibility

TECHNOLOGY RESOURCES

Relevant technology resources include:

- Chem ASAP! CD-ROM
- Resource Pro CD-ROM
- Chemistry Alive! Videodisc: *Making Sodium Chloride*
- Small-Scale Chemistry Laboratory Video or Videodisc: *Chemical Equations*

1 Engage

Use The Visual

Have students study the photograph and read the text that opens the section. Explain that airships during the 1930s often contained hydrogen because it is the lightest element. However, when hydrogen is ignited in air, a vigorous reaction occurs. Have students describe the results as shown in the photograph. Write the word equation for the reaction of hydrogen with oxygen to produce water. Ask:

- **Do you think there is a simpler way to write chemical equations?** (Yes, describe the reaction using chemical symbols of the elements.)

Check Prior Knowledge

To assess students' prior knowledge about chemical compounds, ask:

- **What are the differences between molecular compounds and ionic compounds?** (Molecular compounds are electrically neutral. Ionic compounds are composed of positive cations and negative anions.)
- **How are chemical compounds represented?** (By chemical formulas that show the kinds and numbers of atoms in each molecule or formula unit of the substance.)
- **What is the difference between a monatomic ion and a polyatomic ion?** (A monatomic ion, such as Na^+, is formed when a single atom gains or loses one or more electrons. A polyatomic ion, such as CO^{2-}, is a tightly bound group of atoms that behaves as a unit and carries a charge.)

2 Teach

Use the Visual

Have students study the photographs in Figure 8.1 on page 203. Ask:

- **What are some indications in the pictures shown that a chemical reaction is taking place?** (increase in volume and change in color)

Explain that bakers use yeast to make bread dough rise. Yeasts are unicellular organisms that are able to extract energy from sugar in the absence of oxygen, a process known as fermentation. (Refer to Figure 8.2.) Point out that yeasts produce proteins that *catalyze* or facilitate the breakdown of carbohydrates in the dough, producing CO_2 and ethanol. The bubbles of trapped CO_2 cause the dough to rise. As the dough bakes, the ethanol evaporates. Have students name other commercial uses of yeast. (wine, beer) **Most students are familiar with the taste or smell of sour milk. Point out that the sour taste is due to a fermentation product *catalyzed* by bacterial proteins. Yogurts and sauerkraut are also produced with the help of bacterial fermentation.**
Note: The compound basic copper carbonate has the formula $CuCO_3{\cdot}Cu(OH)_2$.

important to note that the atoms themselves are neither created nor destroyed; they are merely rearranged. This part of Dalton's theory explains the law of conservation of mass. Recall that this law states that in any physical or chemical change, mass is neither created nor destroyed. The atoms in the products are the same atoms that were in the reactants—they are just arranged differently.

Chemical reactions can be described in many different ways. For the example of rusting iron, you could say: "Iron reacts with oxygen to produce iron(III) oxide (rust)." Alternatively, you could identify the reactants and product in this reaction by writing a word equation.

$$\text{Iron} + \text{oxygen} \longrightarrow \text{iron(III) oxide}$$

In a word equation, the reactants are written to the left of the arrow, and the products are written to the right. Notice that the reactants are separated by a plus sign. Had there been two or more products, they also would have been separated by a plus sign.

Consider another example of a simple chemical reaction, one involving hydrogen peroxide, a common antiseptic. When you pour hydrogen peroxide on an open cut, bubbles of oxygen gas rapidly form. The production of a new substance, a gas, is visible evidence of a chemical change. Actually, two new substances are produced in this reaction, oxygen gas and liquid water. The statement "hydrogen peroxide reacts to form water and oxygen gas" describes this reaction. You can write this chemical reaction as a word equation.

$$\text{Hydrogen peroxide} \longrightarrow \text{water} + \text{oxygen}$$

A third common reaction is the burning of methane. Methane is the major component of natural gas, a common fuel for residential heating. When methane is burned to produce energy, a chemical reaction occurs. The word equation for this reaction must include all of the reactants and products. Burning a substance typically requires oxygen. The products of this chemical reaction are water and carbon dioxide. Thus the word equation is as follows.

$$\text{Methane} + \text{oxygen} \longrightarrow \text{carbon dioxide} + \text{water}$$

Figure 8.2
Some common chemical reactions. ***(a)*** *Copper exposed to moist air forms a pale green coating of basic copper(II) carbonate. This type of coating is called a patina.* ***(b)*** *Glucose is fermented by yeast (shown) to form ethanol and carbon dioxide.* ***(c)*** *Plants carry out photosynthesis—the creation of glucose and oxygen from carbon dioxide, water, and sunlight.*

(a) Copper + carbon dioxide + water $\longrightarrow$ basic copper(II) carbonate

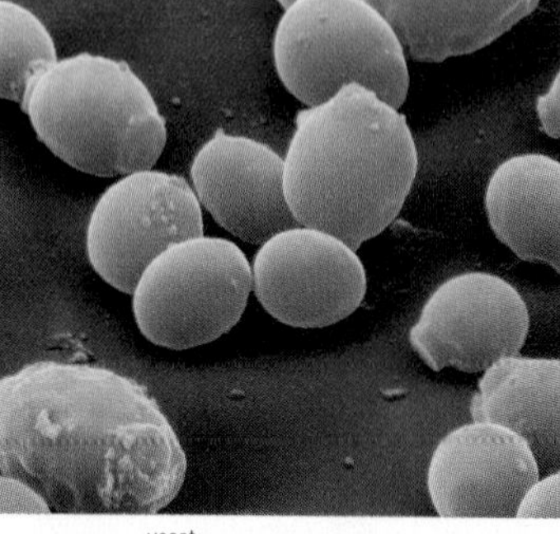

(b) Glucose $\xrightarrow{\text{yeast}}$ ethanol + carbon dioxide

(c) Carbon dioxide + water $\xrightarrow{\text{light}}$ glucose + oxygen

Chemistry Alive!

Making Sodium Chloride

Play

Meeting Diverse Needs

LEP Spanish-speaking students with limited English proficiency may find it helpful to consult the chapter summary and key terms in the Spanish supplement in the Teacher's Resource Package.
At Risk and Sight Impaired Polystyrene models can help students understand the meaning of subscripts and coefficients when balancing chemical reactions. They can also help illustrate how atoms are rearranged during a chemical reaction. Use a knife or razor blade to cut polystyrene balls of various sizes in half. Paint the hemispheres different colors to stand for various elements. Glue a small, flat magnet to the flat side of each of the hemispheres. Using the spheres, show on a metal board how formulas of reactants become formulas of products in various types of reactions.

Chemical Equations

Although word equations adequately describe chemical reactions, they are cumbersome. To communicate more effectively, you can use chemical formulas to write equations. In a **chemical equation,** the arrow separates the formulas of the reactants (on the left) from the formulas of the products (on the right). For instance, here is the chemical equation for rusting.

$$Fe + O_2 \longrightarrow Fe_2O_3$$

Such equations, which show just the formulas of the reactants and products, are skeleton equations. A **skeleton equation** is a chemical equation that does not indicate the relative amounts of the reactants and products involved in the reaction. Writing a skeleton equation with the correct formulas for the reactants and products is an important first step in obtaining a correct chemical equation.

You can indicate the physical state of a substance in the equation by putting a symbol after each formula. Use (*s*) for a solid, (*l*) for a liquid, (*g*) for a gas, and (*aq*) for an aqueous solution (a substance dissolved in water). Indicating the states in the chemical equation for rusting yields

$$Fe(s) + O_2(g) \longrightarrow Fe_2O_3(s)$$

In many chemical reactions, a catalyst is employed. A **catalyst** is a substance that speeds up the rate of a reaction but that is not used up in the reaction. Because a catalyst is neither a reactant nor a product, its formula is written above the arrow in a chemical equation. For example, **Figure 8.3** shows that the compound manganese(IV) oxide (MnO_2)(*s*) catalyzes the decomposition of an aqueous solution of hydrogen peroxide (H_2O_2)(*aq*) into water and oxygen. The skeleton equation is

$$H_2O_2(aq) \xrightarrow{MnO_2} H_2O(l) + O_2(g)$$

Many of the symbols commonly used in writing chemical equations are listed in **Table 8.1** on the following page.

Figure 8.3

Hydrogen peroxide slowly decomposes to form water and oxygen gas (left). The addition of manganese(IV) oxide (MnO_2) greatly speeds up the reaction, causing the rapid evolution of oxygen gas (right). Because it is not used up in the reaction, MnO_2 is a catalyst. The reaction produces enough heat to cause the water to boil; the white "smoke" in the photo is condensed water vapor.

Teacher Demo

To present a dramatic example of chemical change, fill a 150-mL Pyrex beaker to the 30-mL mark with table sugar. Place the beaker in a hood and add 30 mL of concentrated sulfuric acid, which will dehydrate the sugar. The carbon left behind will expand to form a porous, foam cylinder as the water vapor tries to escape.
The hood will remove any sulfur dioxide or carbon monoxide, which are possible by-products of the reaction.
CAUTION! Wear goggles, a face shield, a lab apron, and protective gloves. Use a plexiglas shield to separate the students from the reaction. As a further precaution, have them wear goggles. While wearing protective gear, place the beaker and contents into a 500-mL beaker half full of water. Swirl the small beaker in the water and then neutralize using a base with a molarity of less than 1M. Decant the liquid and fill the large beaker halfway with water once again. Let stand overnight in a protected location, test for pH, and neutralize. Repeat if necessary. Wrap the carbon in newspaper before placing in a wastebasket.

Discuss

Note that writing an equation to represent a chemical reaction is somewhat like writing a sentence. Each activity is governed by a set of rules. In a sentence, you must have a subject and a verb. You must begin the sentence with a capital letter and end with a punctuation mark. In an equation, you must have reactants and products. The equation must follow the law of conservation of mass.

Use the Visual

Use an overhead projector to display Table 8.1. Make sure students understand the meaning and role of each symbol. Point out that the phase symbols provide important clues about reactions. For example, a reaction that takes place when solids dissolve in water will probably not occur if the dry solids are mixed together. Note that the items placed above the yield arrow represent conditions that must be met before the reaction can take place at a reasonable pace.

Practice Problems Plus

Related Chapter Review Problem
Chapter Review problem 35 is related to Sample Problem 8-1.

Chem ASAP!

Animation 7

Relate chemical symbols and formulas to the information they communicate.

Table 8.1

Symbols Used in Chemical Equations	
Symbol	**Explanation**
$+$	Used to separate two reactants or two products
$\longrightarrow$	"Yields," separates reactants from products
$\rightleftharpoons$	Used in place of $\longrightarrow$ for reversible reactions
(s)	Designates a reactant or product in the solid state; placed after the formula
(l)	Designates a reactant or product in the liquid state; placed after the formula
(g)	Designates a reactant or product in the gaseous state; placed after the formula
(aq)	Designates an aqueous solution; the substance is dissolved in water; placed after the formula
$\xrightarrow{\Delta}$ $\xrightarrow{\text{heat}}$	Indicates that heat is supplied to the reaction
$\xrightarrow{\text{Pt}}$	A formula written above or below the yield sign indicates its use as a catalyst (in this example, platinum)

Sample Problem 8-1

Write a skeleton equation for this chemical reaction: Solid sodium hydrogen carbonate reacts with hydrochloric acid to produce aqueous sodium chloride, water, and carbon dioxide gas. Include appropriate symbols.

1. *ANALYZE* ***Plan a problem-solving strategy.***
Write the correct formula for each substance in the reaction, separate the reactants from the products, and indicate the state of each substance.

2. *SOLVE* ***Apply the problem-solving strategy.***
First, write the formulas and states of the reactants.
- solid sodium hydrogen carbonate: $NaHCO_3(s)$
- hydrochloric acid: $HCl(aq)$

Then do the same for the products.
- aqueous sodium chloride: $NaCl(aq)$
- water: $H_2O(l)$
- carbon dioxide gas: $CO_2(g)$

Finally, write the equation.
$$NaHCO_3(s) + HCl(aq) \longrightarrow NaCl(aq) + H_2O(l) + CO_2(g)$$

3. *EVALUATE* ***Does the result make sense?***
The rules for writing a skeleton equation have been properly applied. Formulas for the reactants are written first, followed by an arrow, followed by the formulas of the products.

Practice Problems

1. Write a skeleton equation for each chemical reaction. Include appropriate symbols from **Table 8.1.**
   a. Sulfur burns in oxygen to form sulfur dioxide.
   b. Heating potassium chlorate in the presence of the catalyst manganese dioxide produces oxygen gas. Potassium chloride is left as a solid.
2. Write a sentence that describes each chemical reaction.
   a. $KOH(aq) + H_2SO_4(aq) \longrightarrow H_2O(l) + K_2SO_4(aq)$
   b. $Na(s) + H_2O(l) \longrightarrow NaOH(aq) + H_2(g)$

Balancing Chemical Equations

You can write word equations similar to chemical equations for many everyday processes. For instance, you could write a word equation for the manufacture of bicycles. What do you need to make a bicycle? To simplify your task, limit yourself to four major components: frames, wheels, handlebars, and pedals. Your word equation for making a bicycle probably reads like this.

$$\underset{\text{(reactants)}}{\text{Frame + wheel + handlebar + pedal}} \longrightarrow \underset{\text{(product)}}{\text{bicycle}}$$

Your word equation is qualitatively correct. It shows the reactants (the kinds of parts) you need to make a product (a bicycle). If you were responsible for ordering parts to make a bicycle, however, this word equation would be inadequate. Why? It does not show a quantitative relationship; that is, it does not indicate the quantity of each part needed to make one bicycle.

A standard bicycle is composed of one frame (F), two wheels (W), one handlebar (H), and two pedals (P). Using these symbols, the formula for a bicycle is FW_2HP_2. The unbalanced skeleton equation for bicycle assembly is written as follows.

$$F + W + H + P \longrightarrow FW_2HP_2$$

An equation that does not indicate the quantity of the reactants needed to make the product is called an unbalanced equation. A complete description of the reaction would include both the kinds and the quantities of parts required. An equation that gives the correct quantity of each reactant and product is called a balanced equation. Numbers called **coefficients** are placed in front of the symbols for the respective parts. When no coefficient is written, it is assumed to be 1. The balanced equation for making a bicycle is shown below. If you had ten of each of the parts, how many bicycles could you make? Would you have any parts left over? If so, list them. ❶

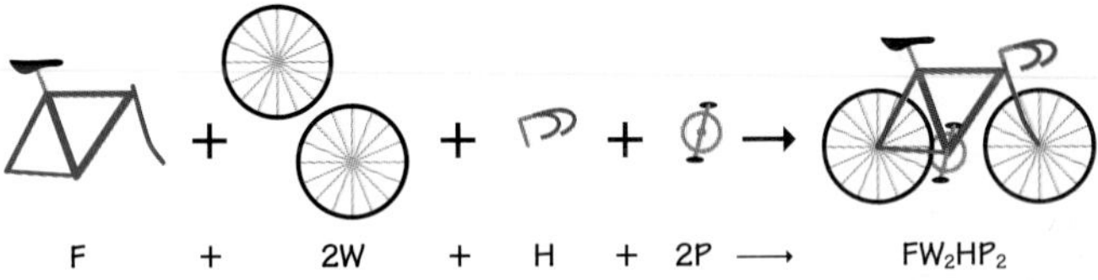

For chemical equations to represent a chemical reaction correctly, you must first write the correct formulas for the reactants and the products using what you know about formulas for elements and compounds. This is always the first step in writing a chemical equation. The next step is to balance the chemical equation so that it is quantitatively correct, indicating the amounts of the reactants and products.

Balancing the chemical equation is necessary so that it obeys the law of conservation of mass. Remember, in a chemical reaction, atoms are not created or destroyed; they are simply rearranged. In every **balanced equation**, each side of the equation has the same number of atoms of each element.

PHYSIOLOGY

Hydrogen Peroxide

Why does hydrogen peroxide decompose to water and oxygen when it comes into contact with blood? This rapid, foaming reaction is an example of chemical catalysis by a biological molecule. The enzyme catalase, found in living cells, catalyzes the breakdown of hydrogen peroxide. Catalase is a protein that contains an iron(II) ion at its center. The iron(II) ion is the actual site on the catalase molecule where the reaction takes place. When hydrogen peroxide is poured on a cut, it reacts with the iron(II) ions of catalase. This triggers the rapid release of atomic oxygen from the hydrogen peroxide. These energetic oxygen atoms produce the antiseptic effect of hydrogen peroxide.

Answer

❶ five bicycles; Yes; five frames and five handlebars would be left over.

8.1

Discuss

Have students read the *Link to Physiology.* Point out that although hydrogen peroxide contains the same kinds of elements as water, it has very different properties. Hydrogen peroxide is much less stable than water. Have students write down the balanced chemical equation for the reaction described in *Link to Physiology.* Students should indicate the Fe^{2+} catalyst above the reaction arrow. Write the correct equation on the board, and explain that this is an example of a decomposition reaction—single compound is broken down into two or more products. Explain that although hydrogen peroxide is an unstable compound that readily decomposes to produce water and oxygen, it does so slowly at room temperature. Therefore, hydrogen peroxide must come into contact with a catalyst in order for the decomposition reaction to be observable. Challenge students to propose a hypothesis about why our bodies produce an enzyme that helps catalyze the decomposition of hydrogen peroxide. (to prevent the accumulation of hydrogen peroxide, a toxic byproduct of reactions involving oxygen, in the body) Students may be interested to learn that hydrogen peroxide is sometimes used to restore the clarity of old paintings. Lead-based paints darken with time (PbS). Hydrogen peroxide converts PbS to $PbSO_4$.

TEACHER DEMO

Add 200 mL of 0.1 *M* copper(II) chloride solution to a 400-mL beaker. **CAUTION!** $CuCl_2$ is an irritant. Cut a 5-cm square piece of aluminum foil. This is sufficient for one demonstration. Crumple the piece of aluminum foil into a loose ball and place it in the copper solution. Ask students to observe and note any evidence of chemical change. The solution will begin to produce gas and a red-brown precipitate of copper will fall to the bottom of the beaker. Write a sentence on the board that describes the chemical reaction. Have students write a skeleton equation for the reaction in their notebooks. $CuCl_2(aq) + Al(s) \rightarrow AlCl_3(aq) + Cu(s)$. Be sure students include the appropriate symbols from Table 8.1. Then have students balance the equation. $3CuCl_2(aq) + 2Al(s) \rightarrow 2AlCl_3(aq) + 3Cu(s)$. Decant the liquid and flush down the drain with excess water. You may wish to dry the copper residue and use it in later experiments.

Discuss

Review with students the law of conservation of mass. Emphasize that mass, or matter, cannot be created or destroyed. Explain that it is because of this law that a chemical equation is not complete until it is balanced.

Sometimes when you write the formulas for the reactants and products in an equation, the equation may already be balanced. One example of this is the equation for the burning of carbon in the presence of oxygen to produce carbon dioxide.

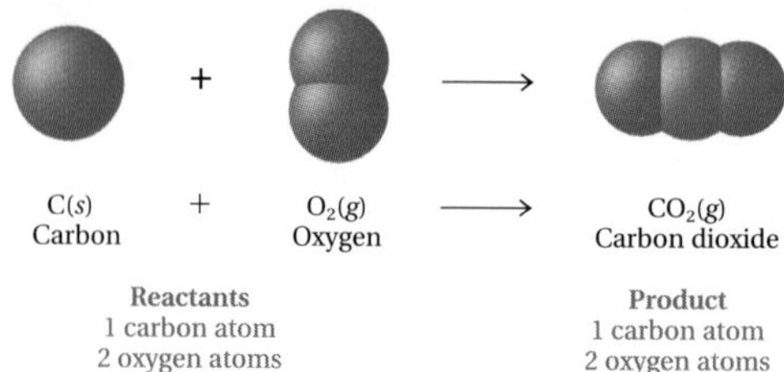

This equation is balanced. One carbon atom is on each side of the equation, and two oxygen atoms are on each side. You do not need to change the coefficients; they are all understood to be 1.

Another equation you have seen is the reaction of hydrogen with oxygen to form water.

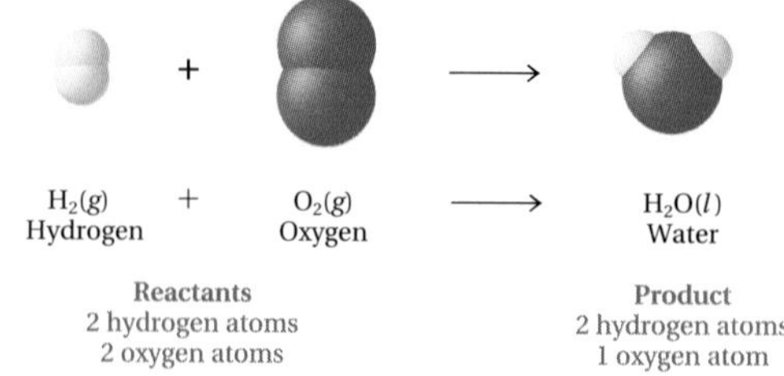

This equation is not balanced even though the formulas for all of the reactants and products are correct. Count the atoms on both sides of the equation. There are two oxygen atoms on the reactant (left) side of the equation and only one oxygen atom on the product (right) side. As written, the equation does not obey the law of conservation of mass. What can you do to balance it? Many chemical equations can be balanced by trial and error. A few guidelines, however, will speed up the process.

Rules for Balancing Equations

1. Determine the correct formulas for all the reactants and products in the reaction, using what you know about elements and compounds. In some cases, you may also indicate in parentheses the state in which the reactants and products exist.
2. Write the formulas for the reactants on the left and the formulas for the products on the right with a yields sign (→) in between. If two or more reactants or products are involved, separate their formulas with plus signs. When finished, you will have a skeleton equation.
3. Count the number of atoms of each element in the reactants and products. For simplicity, a polyatomic ion appearing unchanged on both sides of the equation is counted as a single unit.

4. Balance the elements one at a time by using coefficients. When no coefficient is written, it is assumed to be 1. It is best to begin the balancing operation with elements that appear only once on each side of the equation. You must not attempt to balance an equation by changing the subscripts in the chemical formula of a substance.
5. Check each atom or polyatomic ion to be sure that the equation is balanced.
6. Finally, make sure all the coefficients are in the lowest possible ratio that balances.

Now use these rules to balance the equation in Sample Problem 8-2 for the formation of water from hydrogen and oxygen.

Chem ASAP!
Simulation 4
Sharpen your skills by balancing chemical equations.

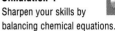

Sample Problem 8-2

Hydrogen and oxygen react to form water. Write a balanced equation for this reaction.

1. *ANALYZE* ***Plan a problem-solving strategy.***
Apply the rules for balancing equations to the word equation describing the reaction.

2. *SOLVE* ***Apply the problem-solving strategy.***
Write correct formulas to give the skeleton equation.

$$H_2(g) + O_2(g) \longrightarrow H_2O(l)$$

Counting the number of each atom shows that hydrogen is balanced but oxygen is not. If you put a coefficient of 2 in front of H_2O, the oxygen becomes balanced.

$$H_2(g) + O_2(g) \longrightarrow 2H_2O(l)$$

Now there are twice as many hydrogen atoms in the product as there are in the reactants. To correct this, put a coefficient of 2 in front of H_2. The equation is now balanced.

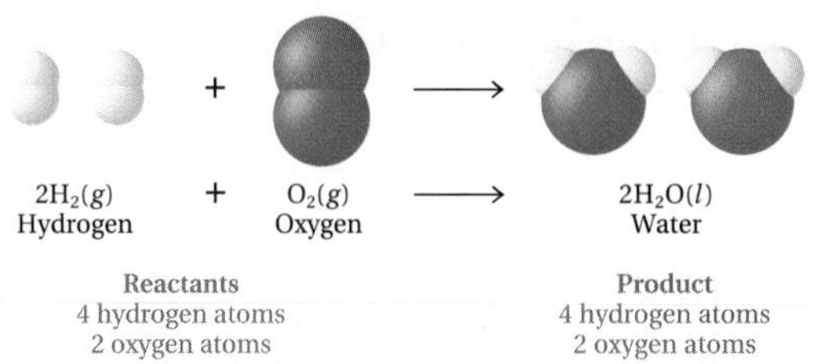

$2H_2(g)$ Hydrogen	+	$O_2(g)$ Oxygen	$\longrightarrow$	$2H_2O(l)$ Water
Reactants 4 hydrogen atoms 2 oxygen atoms				Product 4 hydrogen atoms 2 oxygen atoms

3. *EVALUATE* ***Does the result make sense?***
There are four hydrogen atoms and two oxygen atoms on each side of the chemical equation. The chemical formulas for the elements and compounds involved are correct, and the coefficients are in their lowest possible ratio: $2(H_2)$, $1(O_2)$, and $2(H_2O)$.

Practice Problems

3. Balance each equation.
   a. $AgNO_3 + H_2S \longrightarrow Ag_2S + HNO_3$
   b. $MnO_2 + HCl \longrightarrow MnCl_2 + H_2O + Cl_2$
   c. $Zn(OH)_2 + H_3PO_4 \longrightarrow Zn_3(PO_4)_2 + H_2O$
4. Rewrite these word equations as balanced chemical equations.
   a. hydrogen + sulfur ⟶ hydrogen sulfide
   b. iron(III) chloride + calcium hydroxide ⟶ iron(III) hydroxide + calcium chloride

Chem ASAP!
Problem-Solving 4
Solve Problem 4 with the help of an interactive guided tutorial.

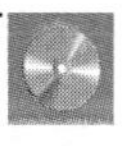

8.1

Discuss

Initiate a discussion with students about the kinds of chemical reactions that take place in nature. Mention that quite frequently the products of one reaction become the reactants of a subsequent reaction.

An example is the combustion of octane (C_8H_{18}), the primary ingredient of gasoline, which produces carbon dioxide and water vapor. The carbon dioxide can combine with atmospheric water vapor to produce carbonic acid (H_2CO_3), which is a component of acid rain.

Similarly, the burning of coal that contains sulfur, will produce CO_2, SO_2, and SO_3. The oxides of sulfur can combine with atmospheric water vapor to produce sulfurous acid (H_2SO_3) and sulfuric acid (H_2SO_4). Ask students if they can think of other such processes. (One is the production of glucose by plants and its subsequent use during respiration to produce carbon dioxide and water.)

Practice Problems Plus

Related Chapter Review Problem
Chapter Review problem 39 is related to Sample Problem 8-2.

Additional Practice Problem
The Chem ASAP! CD-ROM contains the following problem: Rewrite the following word equation as a balanced chemical equation.
aluminum sulfate + calcium hydroxide → aluminum hydroxide + calcium sulfate
($Al_2(SO_4)_3 + 3Ca(OH)_2 \rightarrow 2Al(OH)_3 + 3CaSO_4$)

Cooperative Learning

Have students, working in pairs, take turns writing equations for each other to balance. Two pairs of students may enjoy getting together for a "doubles match" in which they work as a team to balance equations that the opposing team creates.

Discuss

Students often think that they can balance an equation by changing the subscripts in one or more of the formulas. Use examples to show why this approach is incorrect. ($H_2(g) + O_2(g) \rightarrow H_2O(l)$ could be balanced by changing the formula of the product to H_2O_2. But hydrogen peroxide is not the same substance as water. Therefore, the equation would describe a different reaction.)

To help students overcome this misconception, have them draw a box around each symbol or formula for the reactants and products before they start to balance the equation. Tell them that the boxes are "off-limits" or "out of bounds." They cannot change any number that appears inside a box.

Practice Problems Plus

Related Chapter Review Problem
Chapter Review problem 54 is related to Sample Problem 8-3 and 8-4.

Additional Practice Problem
The Chem ASAP! CD-ROM contains the following problem: Rewrite the following word equation as a balanced chemical equation.
phosphoric acid + sodium hydroxide → sodium phosphate + water
($H_3PO_4 + 3NaOH \rightarrow Na_3PO_4 + 3H_2O$)

Figure 8.4
Notice the deposit of silver crystals on the copper wire when copper reacts with silver nitrate in aqueous solution.

Sample Problem 8-3

Figure 8.4 shows the reaction of copper metal and an aqueous solution of silver nitrate. Balance the equation for this reaction.

$$AgNO_3(aq) + Cu(s) \longrightarrow Cu(NO_3)_2(aq) + Ag(s)$$

1. **ANALYZE** ***Plan a problem-solving strategy.***
Apply the rules for balancing equations. Because the nitrate polyatomic ion appears as a reactant and a product, this ion can be balanced as a unit.

2. **SOLVE** ***Apply the problem-solving strategy.***
Put a coefficient of 2 in front of $AgNO_3(aq)$ to balance the nitrate ion.
$$2AgNO_3(aq) + Cu(s) \longrightarrow Cu(NO_3)_2(aq) + Ag(s)$$
By inspection, the silver is not balanced. Put a coefficient of 2 in front of $Ag(s)$.
$$2AgNO_3(aq) + Cu(s) \longrightarrow Cu(NO_3)_2(aq) + 2Ag(s)$$

3. **EVALUATE** ***Does the result make sense?***
The equation is balanced, and the coefficients are in the lowest whole-number ratio.

Practice Problems

5. Balance the equation.
$CO + Fe_2O_3 \longrightarrow Fe + CO_2$
6. Write the balanced chemical equation for the reaction of carbon with oxygen to form carbon monoxide.

Sample Problem 8-4

Aluminum reacts with oxygen in the air to form a thin protective coat of aluminum oxide. Balance the equation for this reaction.

$$Al(s) + O_2(g) \longrightarrow Al_2O_3(s)$$

1. **ANALYZE** ***Plan a problem-solving strategy.***
Apply the rules for balancing equations.

2. **SOLVE** ***Apply the problem-solving strategy.***
First balance the aluminum by placing a coefficient of 2 in front of $Al(s)$.
$$2Al(s) + O_2(g) \longrightarrow Al_2O_3(s)$$
There is now a situation that occurs quite frequently in balancing equations. It can be called the even-odd problem. Any whole-number coefficient placed in front of the O_2 will give an even number of oxygen atoms on the left. This is because the coefficient is always being multiplied by the subscript 2. How can the odd number of oxygen atoms in the product (right side) be made even to balance with the even number on the left? The simplest way to do this is to begin by multiplying the formula with the odd number of oxygen atoms by 2.
$$2Al(s) + O_2(g) \longrightarrow 2Al_2O_3(s)$$

Practice Problems

7. Balance each equation.
   a. $FeCl_3 + NaOH \longrightarrow Fe(OH)_3 + NaCl$
   b. $CS_2 + Cl_2 \longrightarrow CCl_4 + S_2Cl_2$
   c. $CH_4 + Br_2 \longrightarrow CH_3Br + HBr$

Sample Problem 8-4 (cont.)

Now there are six oxygen atoms on the right. Balance the oxygens on the left with a 3 and rebalance the aluminum on the left with a 4.

$$4Al(s) + 3O_2(g) \longrightarrow 2Al_2O_3(s)$$

3. **EVALUATE** ***Does the result make sense?***
There are equal numbers of each kind of atom in the reactants and product (4 aluminum and 6 oxygen). Also, the coefficients are in the lowest possible ratio. Suppose the equation for the formation of aluminum oxide was written

$$8Al(s) + 6O_2(g) \longrightarrow 4Al_2O_3(s)$$

Because this equation obeys the law of conservation of mass, it appears to be correct. It is incorrect, however, because the coefficients are not in their lowest possible ratio. Each of the coefficients can be divided by 2 to give the previous equation, which has the lowest whole-number ratio of coefficients.

Practice Problems (cont.)

8. Write the balanced chemical equations.
   a. sodium + water $\longrightarrow$ sodium hydroxide + hydrogen
   b. calcium hydroxide + sulfuric acid $\longrightarrow$ calcium sulfate + water

Chem ASAP!

Problem-Solving 8
Solve Problem 8 with the help of an interactive guided tutorial.

section review 8.1

9. Write word equations for the following chemical reactions.
   a. Pure copper can be produced by heating copper(II) sulfide in the presence of diatomic oxygen from the air. Sulfur dioxide gas is also produced in this reaction.
   b. Water is formed by the explosive reaction between hydrogen gas and oxygen gas.
   c. When baking soda (sodium hydrogen carbonate) is heated, it decomposes, forming the products sodium carbonate, carbon dioxide, and water.
10. Balance the following equations.
   a. $SO_2 + O_2 \longrightarrow SO_3$
   b. $Fe_2O_3 + H_2 \longrightarrow Fe + H_2O$
   c. $P + O_2 \longrightarrow P_4O_{10}$
   d. $Al + N_2 \longrightarrow AlN$
11. Write formulas and other symbols for these substances.
   a. sulfur trioxide gas
   b. potassium nitrate dissolved in water
   c. heat supplied to a chemical reaction
   d. metallic copper
   e. liquid mercury
   f. zinc chloride as a catalyst
12. How is the law of conservation of mass related to the balancing of a chemical equation?

Chem ASAP! **Assessment 8.1** Check your understanding of the important ideas and concepts in Section 8.1.

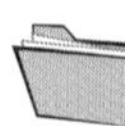

portfolio project

Some products are marketed as biodegradable. What does *biodegradable* mean? Identify some biodegradable products. How do these products benefit the environment?

8.1

3 Assess

Evaluate Understanding

To evaluate students' understanding of how to write and interpret chemical equations, put some word equations on the board and have students produce skeleton equations, using symbols from Table 8.1. Write some unbalanced chemical equations on the board; have students balance them and describe them in words. Write some equations on the board and begin to balance them by changing subscripts. When students object, ask why this approach is incorrect.

Reteach

Review with students the key steps in writing balanced equations. Have them make a flow chart to describe the best way to subdivide the task. Steps should include identifying reactants and products; using the periodic table to write accurate molecular or ionic formulas; indicating physical states and any conditions necessary for reaction; if needed, adding coefficients to conserve mass.

Portfolio Project

Student answers will vary but should reflect an understanding of the term *biodegradable* as applied to materials that can and will decompose through natural organic processes.

Answers

SECTION REVIEW 8.1

9. a. copper(II) sulfide + oxygen → copper + sulfur dioxide
   b. hydrogen + oxygen → water
   c. sodium hydrogen carbonate → sodium carbonate + carbon dioxide + water
10. a. $2SO_2 + O_2 \rightarrow 2SO_3$
   b. $Fe_2O_3 + 3H_2 \rightarrow 2Fe + 3H_2O$
   c. $4P + 5O_2 \rightarrow P_4O_{10}$
   d. $2Al + N_2 \rightarrow 2AlN$
11. a. $SO_3(g)$
   b. $KNO_3(aq)$
   c. $\xrightarrow{\Delta}$
   d. $Cu(s)$
   e. $Hg(l)$
   f. $\xrightarrow{ZnCl_2}$
12. When balancing an equation, the number and kinds of atoms in the reactants equals those in the products. Thus atoms (and mass) are conserved.

8.2

1 Engage

Use the Visual

Have students study the photograph and read the text that opens the section. Ask:

- **What type of reaction is shown in the photograph?** (combustion) **Explain that an important product of combustion reactions is energy, which can be in the form of heat and/or light.**
- **Where else can you observe combustion reactions on a daily basis?** (natural gas stoves and heaters, butane lighters, acetylene torches, automobile engines)
- **How can you identify the different types of chemical reactions?** (For synthesis and decomposition, compare the number of reactants and products; for combustion, check for oxygen; for single- and double-replacements, look for a cation swap or the formation of a precipitate.)

Check Prior Knowledge

To assess students' prior knowledge of names and formulas for common inorganic compounds, write the names of ionic and molecular compounds on the chalkboard. Examples include aluminum oxide, barium chloride, calcium phosphate, carbon dioxide, sulfur trioxide, dinitrogen tetroxide, ammonia, iron(II) oxide, iron(III) oxide, sodium bromide, potassium hydroxide, boron trichloride, carbon dioxide, nitrogen dioxide, calcium phosphate, magnesium nitrate, and sulfur hexafluoride. Have students provide formulas for the names shown. Remind students about the Stock System for naming ionic compounds.

section 8.2

objectives

- Identify a reaction as combination, decomposition, single-replacement, double-replacement, or combustion
- Predict the products of combination, decomposition, single-replacement, double-replacement, and combustion reactions

key terms

- combination reaction
- decomposition reaction
- single-replacement reaction
- activity series of metals
- double-replacement reactions
- combustion reaction

TYPES OF CHEMICAL REACTIONS

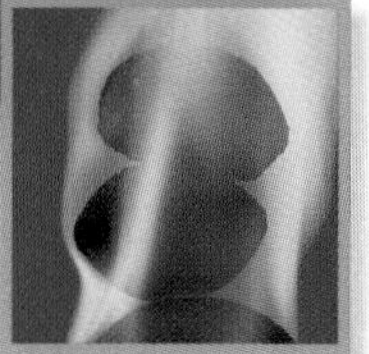

The largest barbeque on record served more than 50 000 people! Charcoal briquettes fuel barbeque grills through the combustion of carbon. Have you ever felt the heat and smelled the smoke coming from a burning barbeque grill? The heat and smoke are actually the products of a combustion reaction. Combustion is one of many types of chemical reactions. **How can you identify the different types of chemical reactions?**

Classifying Reactions

Consider for a moment the number of possible chemical reactions. Because there are millions of known chemical compounds, it is logical to expect that millions of different chemical reactions can occur in nature or be carried out in the laboratory. Just as you learned to name many compounds in Chapter 6 by following a system of nomenclature rather than memorizing common names for thousands of different compounds, you can learn to recognize patterns of chemical behavior that allow you to predict the products in many chemical reactions.

There are several ways to categorize chemical reactions. One classification scheme identifies five general types of reactions: combination, decomposition, single-replacement, double-replacement, and combustion. Not all chemical reactions, however, fit uniquely into only one of these classes. Occasionally, a reaction may fit equally well into two of these categories. In spite of these exceptions, recognizing a reaction as a particular type is useful. Patterns of behavior in reactions will become apparent and allow you to predict the products of these reactions.

Combination Reactions

The first reaction type you will look at is the combination, or synthesis, reaction. As shown in **Figure 8.5,** magnesium metal and oxygen gas combine to form a new compound, magnesium oxide. In a **combination reaction,** two or more substances combine to form a single substance. The reactants of most common combination reactions are either two elements or two compounds. The product of a combination reaction is always a compound. 1 What determines whether the compound formed is ionic or molecular? Several examples of the general types of combination reactions are explained in the following paragraphs.

When a Group A metal and a nonmetal react, the product is a compound that is a combination of the metal cation and the nonmetal anion. You should be able to write the correct formula of the product by considering the charges of the anions and cations. For example,

$$2K(s) + Cl_2(g) \longrightarrow 2KCl(s)$$

As shown in the two examples below, when two nonmetals react by combination, more than one product is often possible.

$$S(s) + O_2(g) \longrightarrow SO_2(g) \text{ sulfur dioxide}$$

$$2S(s) + 3O_2(g) \longrightarrow 2SO_3(g) \text{ sulfur trioxide}$$

Student Resources

From the Teacher's Resource Package, use:

- Section Review 8.2, Ch. 8 Practice Problems and Quizzes from the Review Module (Ch. 5–8)
- Laboratory Recordsheet 8-1
- Laboratory Manual: Experiments 10 and 11
- Laboratory Practicals 8-1 and 8-2
- Small-Scale Lab Manual: Experiment 9

Technology Resources

Relevant technology resources include:

- Chem ASAP! CD-ROM
- Resource Pro CD-ROM
- Chemistry Alive! Videodisc: *Big Burner, Methane Bubbles, Burning Magnesium, Superheated Steam, Thermite Reaction, Ira Remsen Story, Carbide Cannon, Elephant Toothpaste, Watermelon Surprise*

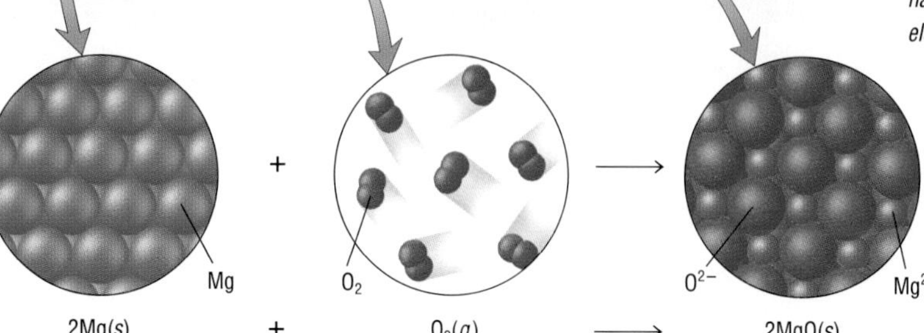

$2Mg(s)$ Magnesium + $O_2(g)$ Oxygen ⟶ $2MgO(s)$ Magnesium oxide

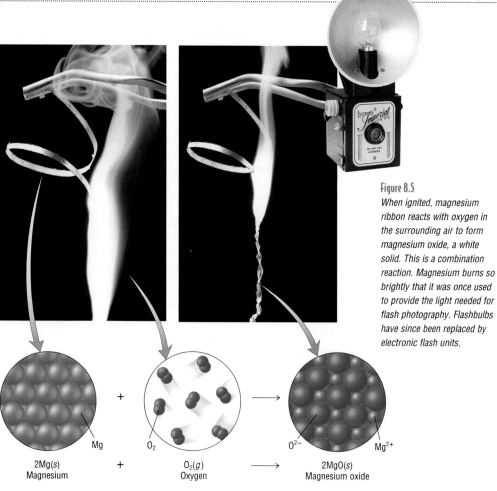

Figure 8.5
When ignited, magnesium ribbon reacts with oxygen in the surrounding air to form magnesium oxide, a white solid. This is a combination reaction. Magnesium burns so brightly that it was once used to provide the light needed for flash photography. Flashbulbs have since been replaced by electronic flash units.

More than one product may also result from the combination reaction of a transition metal and a nonmetal. (Remember Dalton's law of multiple proportions from Chapter 6.)

$$Fe(s) + S(s) \longrightarrow FeS(s) \text{ iron(II) sulfide}$$
$$2Fe(s) + 3S(s) \longrightarrow Fe_2S_3(s) \text{ iron(III) sulfide}$$

Some nonmetal oxides react with water to produce an acid, a compound that produces hydrogen ions in aqueous solution. You will learn about acids in chapter 20.

$$SO_2(g) + H_2O(l) \longrightarrow H_2SO_3(aq) \text{ sulfurous acid}$$

Some metallic oxides react with water to give a base, or a compound containing hydroxide ions. Again in this case, you can use the ionic charges to derive the formula for the product.

$$CaO(s) + H_2O(l) \longrightarrow Ca(OH)_2(aq) \text{ calcium hydroxide}$$

Answer

❶ The types of reactants determine whether the compound formed is ionic or molecular. A metal and a nonmetal react to form an ionic compound. Compounds composed of two or more nonmetals are typically molecular.

2 Teach

TEACHER DEMO

Explain to students that the word "synthesis" means "a combination of parts into a whole." Relate the word's definition to the fact that in a combination, or synthesis, reaction, the reaction "parts" combine to make one new substance, which is the product. Demonstrate a combination reaction. Measure out a 5 to 7 cm strip of magnesium ribbon. Obtain a large crucible and Bunsen burner. **CAUTION!** Warn students that they should not look directly at burning magnesium. It is extremely bright during the reaction. It is advisable that you have pieces of cobalt blue glass available so that students can observe the reaction safely. Light the burner and hold one end of the magnesium ribbon with a pair of crucible tongs. Ignite the magnesium and hold it above the crucible. Ask students to note any evidence of chemical change. When the crucible is cool, allow students to pass it around and observe the residue. Have them note the condition of the residue compared to the original magnesium. The product may be disposed of in the trash.
Ask:

- What are the reactants in this combination reaction? (Mg and O_2)
- What is the product of this reaction? (MgO)
- Does this reaction obey the law of conservation of mass? Explain. (Yes, the sum of the masses of magnesium and oxygen is equal to the mass of the magnesium oxide formed.) Have students write the balanced equation for the reaction.

ACTIVITY

To extend the demonstration on p. 213, challenge students to write the balanced equation for the synthesis of magnesium fluoride from its elements.

Practice Problems Plus

Related Chapter Review Problem
Chapter Review problem 43 is related to Sample Problem 8-5.

Additional Practice Problem
The Chem ASAP! CD-ROM contains the following problem: Write and balance an equation for the formation of aluminum chloride ($AlCl_3$) from its elements.
($2Al + 3Cl_2 \rightarrow 2AlCl_3$)

Discuss

Write a number of combination reactions on the board and have students practice balancing the reactions. Point out that many combination reactions release large amounts of energy. Rewrite some of the equations with the product on the reactant side and the reactants on the product side. Ask:

- **Is it possible to reverse the process?** (yes)
- **What is the name of the reverse process?** (a decomposition reaction)
- **Why is energy usually needed for a decomposition reaction to occur?** (Decomposition involves the breaking of bonds, and this requires energy.)

Chemistry Alive!

Making Sodium Chloride

Play

Sample Problem 8-5

Using the previous examples as guidelines, complete the following combination reactions.

a. $Al(s) + O_2(g) \longrightarrow$

b. $Cu(s) + S(s) \longrightarrow$ (two reactions possible)

c. $SO_3(g) + H_2O(l) \longrightarrow$

1. *ANALYZE Plan a problem-solving strategy.*

a. Because aluminum is a Group A metal combining with a nonmetal, write the correct formula for the compound aluminum oxide by using ionic charges. Then balance the equation.

b. Two reactions are possible because copper is a transition metal and has more than one common ionic charge (Cu^+ and Cu^{2+}). Balance the two possible equations.

c. Reactions between water and nonmetal oxides usually yield an acid in solution when the two molecules combine.

2. *SOLVE Apply the problem-solving strategy.*
For each reaction, write the skeleton equation first, then apply the rules for balancing equations.

a. $Al(s) + O_2(g) \longrightarrow Al_2O_3(s)$
$4Al(s) + 3O_2(g) \longrightarrow 2Al_2O_3(s)$ (balanced)

b. For copper(II):
$Cu(s) + S(s) \longrightarrow CuS(s)$ (balanced)
For copper(I):
$Cu(s) + S(s) \longrightarrow Cu_2S(s)$
$2Cu(s) + S(s) \longrightarrow Cu_2S(s)$ (balanced)

c. $SO_3(g) + H_2O(l) \longrightarrow H_2SO_4(aq)$ (balanced)

3. *EVALUATE Do the results make sense?*
In **a.** and **b.** a metal reacts with a nonmetal and the formula of the product is written using ionic charges. In **c.** a nonmetal oxide combines with water to form an acid by "adding" the molecules together.

Practice Problems

13. Complete and balance these combination reactions.
a. $Be + O_2 \longrightarrow$
b. $SO_2 + H_2O \longrightarrow$

14. Write and balance an equation for the formation of each compound from its elements.
a. strontium iodide (SrI_2)
b. magnesium nitride (Mg_3N_2)

Chem ASAP!
Problem-Solving 14
Solve Problem 14 with the help of an interactive guided tutorial.

Decomposition Reactions

When calcium carbonate is heated, it decomposes or breaks down into two simpler compounds. This reaction is used to make lime (calcium oxide) from limestone (primarily calcium carbonate).

$$CaCO_3(s) \xrightarrow{\text{heat}} CaO(s) + CO_2(g)$$

Many other compounds also decompose with the addition of energy. The decomposition of mercury(II) oxide to yield mercury and oxygen is shown in **Figure 8.6**.

MEETING DIVERSE NEEDS

Gifted Have students choose a particular chemical compound and research how it is produced. For example, NaOH, H_2SO_4, or CH_3OH. A report and project should be presented describing in as much detail as possible the different reactions that are necessary to produce the compound.

LEP and At Risk Make sure that students understand clearly the five terms that describe general types of chemical reactions: combination, decomposition, single-replacement, double-replacement, and combustion. Help them to spell and pronounce each term as well as to define it. Encourage them to ask questions about any term they do not fully understand.

Hg^{2+} O^{2-} $\xrightarrow{heat}$ Hg + O_2

$2HgO(s)$ $\xrightarrow{heat}$ $2Hg(l)$ + $O_2(g)$

Mercury(II) oxide — Mercury — Oxygen

Figure 8.6
When it is heated, orange-colored mercury(II) oxide decomposes into its constituent elements: liquid mercury and gaseous oxygen.

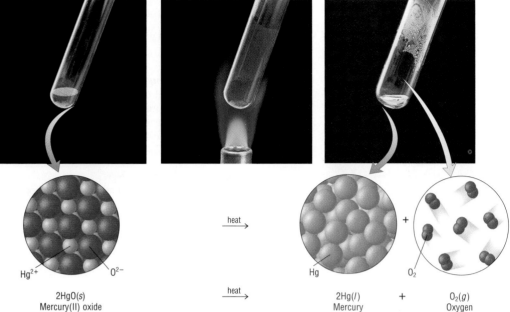

In a **decomposition reaction,** a single compound is broken down into two or more products. These products can be any combination of elements and compounds. It is usually very difficult to predict the products of decomposition reactions. However, as shown in **Figure 8.6,** when a simple binary compound breaks down you know the products will be the constituent elements. Extremely rapid decomposition reactions that produce gaseous products and heat are often the cause of explosions. As seen in the two reactions shown in **Figure 8.6** and **Figure 8.7,** most decomposition reactions require energy in the form of heat, light, or electricity.

Figure 8.7
Decomposition in action! The powerful explosive force of dynamite comes from a decomposition reaction. What is usually added to the reactants of a decomposition reaction to start the reaction? ❶

Answer

❶ heat, electricity, or light

8.2

Use the Visual

Have students study the photograph in Figure 8.7. Explain that the explosive properties of dynamite are due to the rapid production of large amounts of gases. A related reaction is the decomposition of trinitrotoluene (TNT). Write the equation for the decomposition of TNT on the board.

$2C_7H_5N_3O_6(s) \rightarrow 3N_2(g) + 7CO(g) + 5H_2O(g) + 7C(s)$. **Point out to students that for every 2 moles of TNT that decompose, 15 moles of hot expanding gases are produced.**

Discussion

Help students create a list of criteria for identifying decomposition and combination reactions. For example, decomposition reactions are characterized by a large molecule on the reactant side and smaller molecules or elements on the product side. Students should supplement their lists later, when they study single-replacement, double-replacement, and combustion reactions.

Chemistry Alive!

Elephant Toothpaste Play

Watermelon Surprise

Superheated Steam Play

TEACHER DEMO

Show three single-replacement reactions. Place a 2-cm piece of magnesium in 10 mL of 6*M* HCl. **CAUTION!** HCl(*aq*) is corrosive and can cause severe burns. Place a small piece of sodium in some cold water. **CAUTION!** The piece of sodium should be no larger than a match head. Wear plastic gloves to avoid contact between the sodium and your skin. Place an iron nail in about 50 mL of 1*M* $CuCl_2$. (Note: This reaction may not show results for 24 hours.) Show double-replacement reactions by mixing equimolar solutions of ionic compounds such as $BaCl_2$ and Na_2SO_4, Na_3PO_4 and $CaCl_2$, or $Pb(NO_3)_2$ and KI.

Practice Problems Plus

Related Chapter Review Problem
Chapter Review problem 44 is related to Sample Problem 8-6.

Use the Visual

With Table 8.2 displayed on an overhead projector, review the definition of a single-replacement reaction on the board. Explain that single-replacement reactions can be compared to partners cutting in on each other at a dance. A person who is alone approaches a couple and cuts in. The person replaces one member of the couple, who is now left alone.

In a chemical reaction, however, only certain substances can replace other substances in a given compound. The activity series of metals tells us which metals can replace other metals in a given compound. Review the reactions shown at the bottom of page 216, showing students how to predict the products of the reaction using Table 8.2.

Sample Problem 8-6

Write a balanced equation for each decomposition reaction.

a. $H_2O(l) \xrightarrow{\text{electricity}}$ **b.** lead(IV) oxide $\xrightarrow{\text{heat}}$

1. *ANALYZE* *Plan a problem-solving strategy.*

a. Water, a binary compound, breaks down into its elements. Balance the equation, remembering that hydrogen and oxygen are both diatomic molecules.

b. Lead(IV) oxide decomposes into its constituent elements. Balance the equation.

2. *SOLVE* *Apply the problem-solving strategy.*

For each reaction, write the skeleton equation first, then apply the rules for balancing equations.

a. $H_2O(l) \xrightarrow{\text{electricity}} H_2(g) + O_2(g)$

$2H_2O(l) \xrightarrow{\text{electricity}} 2H_2(g) + O_2(g)$ (balanced)

b. $PbO_2(s) \xrightarrow{\text{heat}} Pb(l) + O_2(g)$ (balanced)

3. *EVALUATE* *Do the results make sense?*

The formulas for the reactants and products are correct. The balanced equations have the same number and kinds of atoms on each side of the yields sign.

Practice Problems

15. Complete and balance these decomposition reactions.

a. $HI \longrightarrow$

b. $Mg(ClO_3)_2 \longrightarrow MgCl_2 +$

16. Write the formula for the binary compound that decomposes to each set of products.

a. $H_2 + Br_2$

b. $Na + Cl_2$

Single-Replacement Reactions

Dropping a small piece of potassium into a beaker of water creates a violent reaction, as shown in **Figure 8.8.** The reaction produces hydrogen gas and a large quantity of heat. The hydrogen gas released can ignite explosively.

Figure 8.8
The alkali metal potassium displaces hydrogen from water and forms a solution of potassium hydroxide in a single-replacement reaction. The heat of the reaction is often sufficient to ignite the hydrogen. Why are alkali metals stored under mineral oil or kerosene? ❶

Answer

❶ to keep them from reacting with the water vapor or oxygen in the air

The reaction of potassium with water is a single-replacement reaction. Similar but much less spectacular reactions can occur when certain elements are put into an aqueous solution of a compound. In a **single-replacement reaction,** one element replaces a second element in a compound. Single-replacement reactions are also called single-displacement reactions. In the reaction shown in **Figure 8.8,** potassium replaces hydrogen, which is released as a flammable gas.

Whether one metal will displace another metal from a compound can be determined by the relative reactivities of the two metals. The **activity series of metals,** given in **Table 8.2,** lists metals in order of decreasing reactivity. A reactive metal will replace any metal listed below it in the activity series. Thus, as shown in **Figure 8.9,** iron will displace copper from a copper compound in solution. Magnesium displaces zinc from a zinc compound in solution, as well as silver from a silver compound in solution. By contrast, magnesium does not displace lithium or calcium from aqueous solutions of their compounds.

$$Mg(s) + Zn(NO_3)_2(aq) \longrightarrow Mg(NO_3)_2(aq) + Zn(s)$$

$$Mg(s) + 2AgNO_3(aq) \longrightarrow Mg(NO_3)_2(aq) + 2Ag(s)$$

$$Mg(s) + LiNO_3(aq) \longrightarrow \text{no reaction}$$

A nonmetal can also replace another nonmetal from a compound. This replacement is usually limited to the halogens (F_2, Cl_2, Br_2, and I_2). The activity of the halogens decreases as you go down Group 7A of the periodic table.

Table 8.2

Activity Series of Metals

	Name	Symbol
Decreasing reactivity ↓	Lithium	Li
	Potassium	K
	Calcium	Ca
	Sodium	Na
	Magnesium	Mg
	Aluminum	Al
	Zinc	Zn
	Iron	Fe
	Lead	Pb
	(Hydrogen)	(H)*
	Copper	Cu
	Mercury	Hg
	Silver	Ag

*Metals from Li to Na will replace H from acids and water; from Mg to Pb they will replace H from acids only.

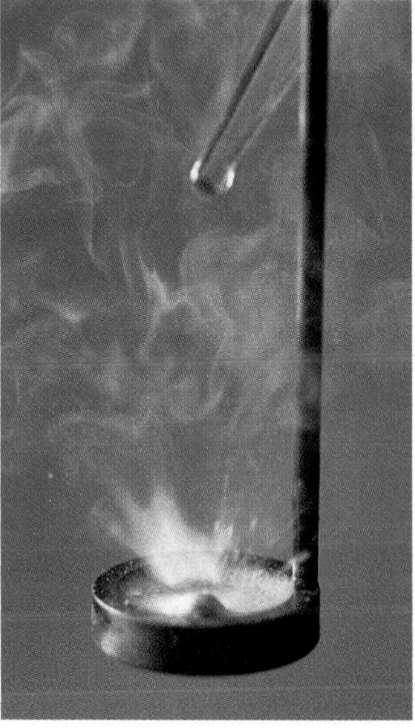

Figure 8.9
Some single-replacement reactions take place quietly, as when iron displaces copper from a compound in solution (left); the copper is deposited on the iron bolt. Other single-replacement reactions are violent, as when water is dropped onto sodium (right).

TEACHER DEMO

It is possible to demonstrate the law of conservation of mass using a double-replacement reaction. Place 50 mL of 1.0*M* potassium chromate (K_2CrO_4) in a 250-mL Erlenmeyer flask. Place about 5 mL of dilute silver nitrate solution ($AgNO_3$) in a small test tube that will fit inside the flask. **CAUTION!** Avoid skin contact with silver nitrate.
Stopper the flask and determine its mass. Invert the flask sufficiently so that the two solutions mix. Ask students to note any evidence of chemical changes.
A bright red precipitate should form. Measure the mass of the flask after the reaction. Point out that the mass has not changed. Flush the products down the drain with excess water.

Discuss

Explain to students that in order for a double-replacement reaction to occur, the reactants must be two ionic compounds in aqueous solution.

Use the Visual

Have students study the photos in Figure 8.9. Ask students to describe the reactions. Write the equations for each reaction on the board. Remind students why the reactions are classified as single-replacement reactions. Use representations of displacement reactions in which the atoms are drawn as spheres to help students visualize the process.

Chemistry Alive!

Thermite Reaction

Play

ACTIVITY

Decomposition and displacement reactions supply the energy needed for launching the space shuttle into space. Have students research what fuels are used to propel the space shuttle orbiter. Have volunteers describe the reactions that occur and how the products provide the thrust necessary to launch vehicles.

Practice Problems Plus

Related Practice Review Problem

Chapter Review problem 47 is related to Sample Problem 8-7.

Additional Practice Problem

The Chem ASAP! CD-ROM contains the following problem: Complete the equation for this single-replacement reaction that takes place in aqueous solution. Balance the equation. If a reaction does not occur (use activity series), write "no reaction."
$K(s) + H_2SO_4(aq) \rightarrow$
$[2K(s) + H_2SO_4(aq) \rightarrow K_2SO_4(aq) + H_2(g)]$

TEACHER DEMO

Demonstrate a double-replacement reaction. Prepare 0.1*M* solutions of silver nitrate ($AgNO_3$) and sodium chloride (NaCl). Mix equal volumes (50 mL each) of the two solutions. Observe the formation of a precipitate, silver chloride (AgCl). Have students write the balanced equation for the reaction. **CAUTION!** Avoid skin contact with silver nitrate. For disposal, add 10 g of solid NaCl to the reactant mixture. Stir and filter. Place the dry precipitate in a plastic or cardboard container. Flush the filtrate down the drain with excess water.

Sample Problem 8-7

Write a balanced chemical equation for each single-replacement reaction.

a. $Zn(s) + H_2SO_4(aq) \longrightarrow$
b. $Na(s) + H_2O(l) \longrightarrow$
c. $Sn(s) + NaNO_3(aq) \longrightarrow$
d. $Cl_2(g) + NaBr(aq) \longrightarrow$

1. *ANALYZE* *Plan a problem-solving strategy.*

a. According to the activity series of metals, zinc displaces hydrogen from an acid and takes its place. Balance the equation, remembering that elemental hydrogen is diatomic.

b. Sodium displaces hydrogen from water. As you balance the equation, you might find it helpful to write water as HOH and visualize it as being made of H^+ and OH^- ions.

c. No displacement reaction occurs because tin is less reactive than sodium.

d. Chlorine is more reactive than bromine and displaces bromine from its compounds. Balance the equation, remembering that bromine is diatomic and a liquid.

2. *SOLVE* *Apply the problem-solving strategy.*

For each reaction, write the skeleton equation first, then apply the rules for balancing equations.

a. $Zn(s) + H_2SO_4(aq) \longrightarrow ZnSO_4(aq) + H_2(g)$ (balanced)

b. $Na(s) + HOH(l) \longrightarrow NaOH(aq) + H_2(g)$
$2Na(s) + 2HOH(l) \longrightarrow 2NaOH(aq) + H_2(g)$ (balanced)

c. no reaction

d. $Cl_2(g) + NaBr(aq) \longrightarrow NaCl(aq) + Br_2(l)$
$Cl_2(g) + 2NaBr(aq) \longrightarrow 2NaCl(aq) + Br_2(l)$ (balanced)

3. *EVALUATE* *Do the results make sense?*

A single-replacement reaction occurs only when a more active metal (or nonmetal) displaces a less active metal (or nonmetal) from a compound. After ionic charges were used to write correct formulas for ionic products, the equations were balanced.

Practice Problem

17. Complete the equations for these single-replacement reactions that take place in aqueous solution. Balance each equation. If a reaction does not occur (use the activity series), write "no reaction."
a. $Fe(s) + Pb(NO_3)_2(aq) \longrightarrow$
b. $Cl_2(g) + NaI(aq) \longrightarrow$
c. $Ca(s) + H_2O(l) \longrightarrow$

Chem ASAP!

Problem-Solving 17

Solve Problem 17 with the help of an interactive guided tutorial.

Double-Replacement Reactions

Many ionic compounds can dissolve in water to form solutions. When two solutions of ionic compounds are mixed, the result still may be a solution. This is what happens when solutions of sodium chloride and calcium nitrate are mixed. A second possible outcome is that ions in the solutions

MEETING DIVERSE NEEDS

Gifted Encourage gifted students to devise general statements to represent each type of chemical reaction discussed in this section. For example: X + Y = XY, AB + CD = AD + CB, X + CD = XD + C.

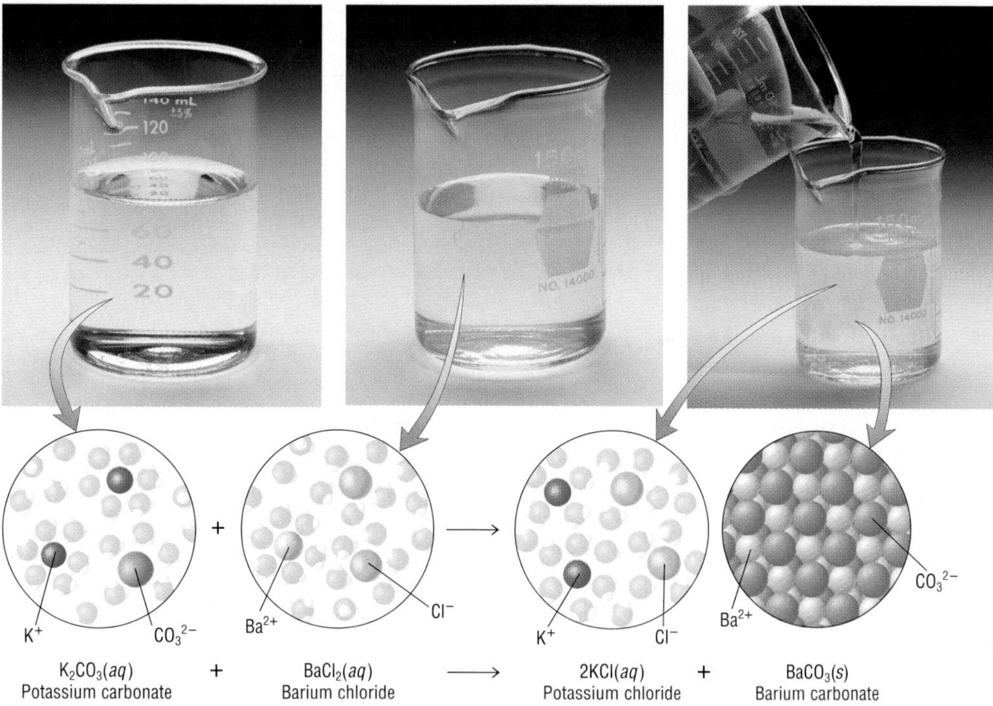

Figure 8.10
Aqueous solutions of barium chloride and potassium carbonate react in a double-replacement reaction to form barium carbonate, the white precipitate. Potassium chloride, the other product of the reaction, remains in solution. (In this figure, the polyatomic ion CO_3^{2-} is represented as a single sphere.)

react. **Figure 8.10** shows that mixing aqueous solutions of barium chloride and potassium carbonate results in a chemical reaction. Barium carbonate is the white precipitate formed. Potassium chloride, the other product of the reaction, remains in solution. **Double-replacement reactions** involve an exchange of positive ions between two reacting compounds. Such reactions generally take place between two ionic compounds in aqueous solution and are often characterized by the production of a precipitate. For a double-replacement reaction to occur, one of the following is usually true.

1. One product is only slightly soluble and precipitates from solution. For example, the reaction of aqueous solutions of sodium sulfide and cadmium nitrate produces a yellow precipitate of cadmium sulfide.

$$Na_2S(aq) + Cd(NO_3)_2(aq) \longrightarrow CdS(s) + 2NaNO_3(aq)$$

2. One product is a gas that bubbles out of the mixture. For example, hydrogen cyanide gas is produced when aqueous sodium cyanide is mixed with sulfuric acid.

$$2NaCN(aq) + H_2SO_4(aq) \longrightarrow 2HCN(g) + Na_2SO_4(aq)$$

3. One product is a molecular compound such as water. Combining solutions of calcium hydroxide and hydrochloric acid produces water as one of the products.

$$Ca(OH)_2(aq) + 2HCl(aq) \longrightarrow CaCl_2(aq) + 2H_2O(l)$$

8.2

ACTIVITY

On the chalkboard, write the products only of several double-replacement reactions; challenge students to apply their understanding of the concepts presented in this section by filling in the reactants. Examples include:
$\rightarrow AgBr(s) + KNO_3(aq)$
$\rightarrow BaSO_4(s) + 2NaCl(aq)$
$\rightarrow KCl(aq) + CaSO_4(s)$.
Remind students that in a double-replacement reaction, the positive ions (cations) of one compound trade places with the positive ions of another compound.

TEACHER DEMO

Demonstrate the combustion of iron. Place a superfine steel wool pad in a plastic sandwich bag and measure the mass and write the value on the board. Take the steel wool out of the bag and unfold it to full length. Clamp it to a ring stand with a utility clamp. Use a match to set the steel wool on fire. **CAUTION!** Steel wool burns readily. Ask students to predict what effect the burning will have on the mass of the sample. As soon as the wool stops burning and is cool, place the remains in the same plastic sandwich bag and measure the mass again. Write the new value on the board next to the original mass. Discuss why the sample gained mass. (Two elements, $Fe(s)$ and $O_2(g)$, combined in a combustion reaction to form a binary compound, Fe_2O_3.)

Chemistry Alive!

Carbide Cannon

Play

Practice Problems Plus

Related Chapter Review Problem
Chapter Review problem 48 is related to Sample Problem 8-8.

Additional Practice Problem
The Chem ASAP! CD-ROM contains the following problem: Write the products for this double-replacement reaction. Then balance the equation.
$Pb(NO_3)_2(aq) + NaI(aq) \rightarrow$ (lead iodide is a precipitate)
[$Pb(NO_3)_2(aq) + 2NaI(aq) \rightarrow PbI_2(s) + 2NaNO_3(aq)$]

Discuss

Point out to students that an important product of most combustion reactions is energy, which is usually in the form of heat or light. Emphasize that one of the reactants must be oxygen. Relate this fact to everyday experience by recalling how removing oxygen from a combustion reaction—for example, snuffing out a candle—causes the reaction to stop.

Think Critically

Ask students to infer why it is important that combustion reactions, such as those used to heat a home or run an automobile, take place in properly ventilated areas. (Without proper ventilation and enough available oxygen, the combustion may be incomplete, and poisonous carbon monoxide may be produced.)

Chemistry Alive!

Burning Magnesium Play

Sample Problem 8-8

Write a balanced chemical equation for each double-replacement reaction.

a. $BaCl_2(aq) + K_2CO_3(aq) \longrightarrow$ (A precipitate of barium carbonate is formed.)

b. $FeS(s) + HCl(aq) \longrightarrow$ (Hydrogen sulfide gas (H_2S) is formed.)

1. ***ANALYZE Plan a problem-solving strategy.***

a. The driving force behind the reaction is the formation of a precipitate. Write correct formulas of the products using ionic charges. Then balance the equation.

b. A gas is formed in this double-replacement reaction. Use ionic charges to write the correct formula of the other product. Then balance the equation.

2. ***SOLVE Apply the problem-solving strategy.***

For each reaction, write the skeleton equation first, then apply the rules for balancing equations.

a. $BaCl_2(aq) + K_2CO_3(aq) \longrightarrow BaCO_3(s) + KCl(aq)$
$BaCl_2(aq) + K_2CO_3(aq) \longrightarrow BaCO_3(s) + 2KCl(aq)$ (balanced)

b. $FeS(s) + HCl(aq) \longrightarrow H_2S(g) + FeCl_2(aq)$
$FeS(s) + 2HCl(aq) \longrightarrow H_2S(g) + FeCl_2(aq)$ (balanced)

3. ***EVALUATE Do the results make sense?***

After correct formulas of the ionic products were written using ionic charges, the equations were balanced. Also, each reaction follows one of the rules for double-replacement reactions.

Practice Problems

18. Write the products for these double-replacement reactions. Then balance each equation.

a. $NaOH + Fe(NO_3)_3 \longrightarrow$ (Iron hydroxide is a precipitate.)

b. $Ba(NO_3)_2 + H_3PO_4 \longrightarrow$ (Barium phosphate is a precipitate.)

19. Write a balanced equation for each reaction.

a. $KOH(aq) + H_3PO_4(aq) \longrightarrow$

b. $H_2SO_4 + Al(OH)_3 \longrightarrow$

Chem ASAP!
Problem-Solving 18
Solve Problem 18 with the help of an interactive guided tutorial.

Combustion Reactions

In a **combustion reaction,** an element or a compound reacts with oxygen, often producing energy as heat and light. Combustion reactions commonly involve hydrocarbons, compounds of hydrogen and carbon. The complete combustion of a hydrocarbon produces the compounds carbon dioxide and water. If the supply of oxygen is insufficient during a reaction, combustion will be incomplete. Elemental carbon and toxic carbon monoxide may be additional products. The complete combustion of a hydrocarbon releases a large amount of energy as heat. Thus hydrocarbons such as methane (CH_4), propane (C_3H_8), and butane (C_4H_{10}) are important fuels. The combustion reaction for methane is shown in **Figure 8.11.**

The reaction between some elements and oxygen is also an example of a combustion reaction. For example, both magnesium and sulfur will burn by reaction with oxygen. (1) As what other type can these two reactions be classified?

$$2Mg(s) + O_2(g) \longrightarrow 2MgO(s)$$

$$S(s) + O_2(g) \longrightarrow SO_2(g)$$

Answers

(1) combination reactions
(2) energy in the form of heat and light

Figure 8.11
Methane gas reacts with oxygen from the surrounding air in a combustion reaction to produce carbon dioxide and water. What else is produced in this reaction? 2

$CH_4(g)$ Methane + $2O_2(g)$ Oxygen ⟶ $CO_2(g)$ Carbon dioxide + $2H_2O(g)$ Water

Sample Problem 8-9

Write balanced equations for the complete combustion of these compounds.

a. benzene $(C_6H_6)(l)$
b. methanol $(CH_3OH)(l)$

1. *ANALYZE* ***Plan a problem-solving strategy.***
Remember that oxygen is the other reactant in these combustion reactions. Also, because the reactions are complete, the only products are CO_2 and H_2O. Write the skeleton equation for each reaction, then balance the equation.

2. *SOLVE* ***Apply the problem-solving strategy.***
For each reaction, write the skeleton equation first, then apply the rules for balancing equations.

a. $C_6H_6(l) + O_2(g) \longrightarrow CO_2(g) + H_2O(g)$
$2C_6H_6(l) + 15O_2(g) \longrightarrow 12CO_2(g) + 6H_2O(g)$ (balanced)

b. $CH_3OH(l) + O_2(g) \longrightarrow CO_2(g) + H_2O(g)$
$2CH_3OH(l) + 3O_2(g) \longrightarrow 2CO_2(g) + 4H_2O(g)$ (balanced)

3. *EVALUATE* ***Do the results make sense?***
Each balanced equation has the lowest whole-number set of coefficients. Each side of the balanced equation has the same number and kinds of atoms.

Practice Problems

20. Write a balanced equation for the complete combustion of each compound.
a. formic acid (HCOOH)
b. heptane (C_7H_{16})

21. Write a balanced equation for the complete combustion of glucose $(C_6H_{12}O_6)$.

Chem ASAP!
Problem-Solving 21

Solve Problem 21 with the help of an interactive guided tutorial.

8.2

Use the Visual

Have students study Figure 8.11. Remind students that the complete combustion of a hydrocarbon, such as methane, always produces water, carbon dioxide, heat, and light. Use a Bic lighter to show students the combustion of butane. Ask students to write the balanced equation for the combustion of butane, C_4H_8. Have students use the diagram and chemical equation at the top of page 221 as an aid.

Practice Problem Plus

Related Chapter Review Problem
Chapter Review problem 50 is related to Sample Problem 8-9.

Additional Practice Problem
The Chem ASAP! CD-ROM contains this problem: Write a balanced equation for the complete combustion of ethanol, C_2H_5OH. $(C_2H_5OH + 3O_2 \rightarrow 2CO_2 + 3H_2O)$

Discuss

Explain that many combustion reactions are a special type of combination reaction in which an element or a compound combines with oxygen to form a single product plus energy. For example: $2Mg(s) + O_2(g) \rightarrow 2MgO(s)$, and $4Fe(s) + 3O_2(g) \rightarrow 2Fe_2O_3(s)$.

Chemistry Alive!

Big Burner

Play

Methane Bubbles

Play

Discuss

Review with students the five types of general reactions that they have studied. These reaction types are decomposition, combustion, single-replacement, double-replacement, and combination. Have students identify the important characteristics of each reaction type, and give examples of specific reactions that illustrate these characteristics. Help students organize the important characteristics of each of the five reaction types into a chart, which they can then use as a guide when predicting the products of a reaction.

Chemistry Alive!

Ira Remsen Story

Play

Chem ASAP!

Simulation 5

Practice classifying reactions according to reaction type.

Predicting Products of a Chemical Reaction

Now that you have an understanding of some of the basic reaction types, you can predict the products of many reactions. To predict the products of a chemical reaction successfully, you must recognize the possible type of reaction that the reactants can undergo. Keys to recognizing reaction types and predicting products are shown in **Figure 8.12.**

Some reactions do not fit any one of the five general types discussed so far. Another very important type of chemical reaction is an oxidation–reduction, or redox, reaction. This type of reaction is discussed in Chapters 22 and 23. Recognizing types of reactions is important, but the only way to determine the products of any reaction with certainty is to carry out the reaction in the laboratory.

Figure 8.12

The five types of chemical reactions discussed in this chapter are summarized here.

Combination Reaction

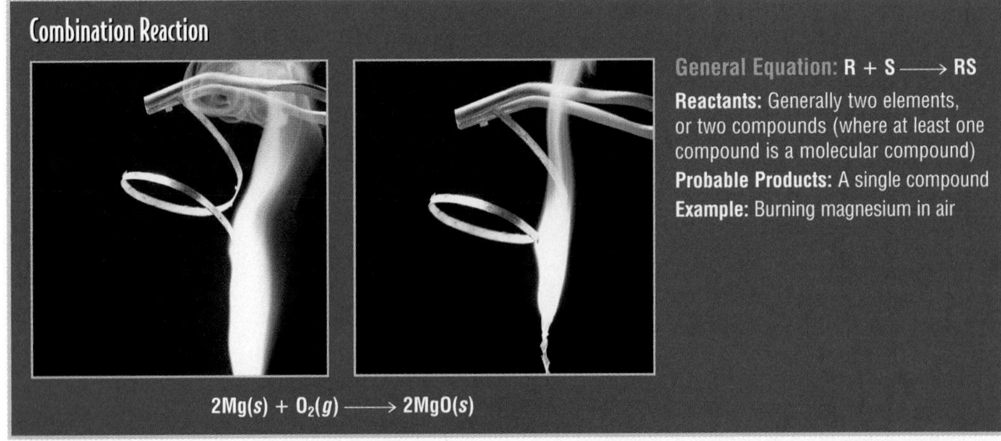

General Equation: $R + S \longrightarrow RS$

Reactants: Generally two elements, or two compounds (where at least one compound is a molecular compound)

Probable Products: A single compound

Example: Burning magnesium in air

$$2Mg(s) + O_2(g) \longrightarrow 2MgO(s)$$

Decomposition Reaction

General Equation: $RS \longrightarrow R + S$

Reactants: Generally a single binary or ternary compound

Probable Products: Two elements (for a binary compound), or two or more elements and/or compounds (for a ternary compound)

Example: Heating mercury(II) oxide

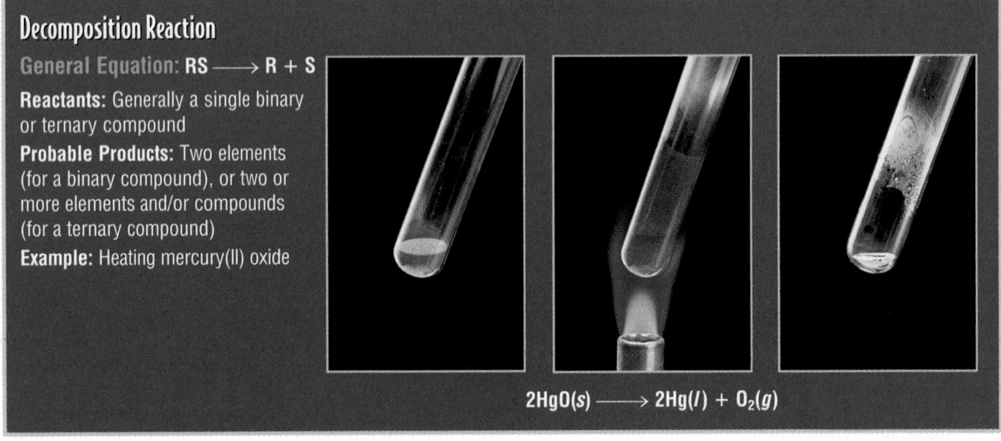

$$2HgO(s) \longrightarrow 2Hg(l) + O_2(g)$$

Single-Replacement Reaction

General Equation:
T + RS ⟶ TS + R

Reactants: An element and a compound. In a single-replacement reaction, an element replaces another element from a compound in aqueous solution. For a single-replacement reaction to occur, the element that is displaced must be less active than the element that is doing the displacing.

Probable Products: A different element and a new compound

Example: Potassium in water

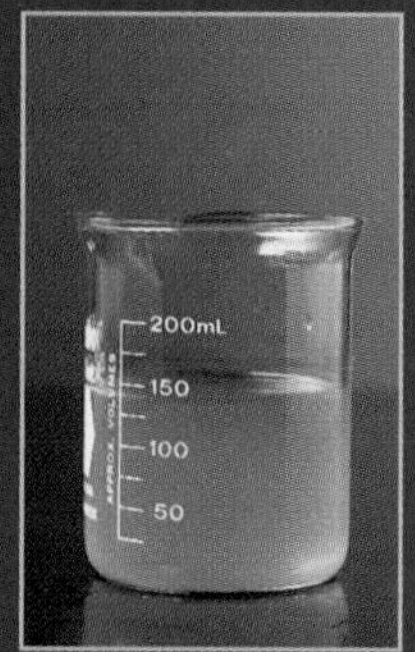

$2K(s) + 2H_2O(l) \longrightarrow 2KOH(aq) + H_2(g)$

Double-Replacement Reaction

General Equation:
$R^+S^- + T^+U^- \longrightarrow R^+U^- + T^+S^-$

Reactants: Two ionic compounds. In a double-replacement reaction, two ionic compounds react by exchanging cations to form two different compounds.

Probable Products: Two new compounds. Double-replacement reactions are driven by the formation of a precipitate, a gaseous product, or water.

Example: Reaction of aqueous solutions of barium chloride and potassium carbonate.

$K_2CO_3(aq) + BaCl_2(aq) \longrightarrow 2KCl(aq) + BaCO_3(s)$

Combustion Reaction

General Equation:
$C_xH_y + (x + y/4)\,O_2 \longrightarrow xCO_2 + (y/2)H_2O$

Reactants: Oxygen and a compound of C, H, (O). When oxygen reacts with an element or compound, combustion may occur.

Probable Products: CO_2 and H_2O With incomplete combustion, C and CO may also be products.

Example: The combustion of methane gas in air

$CH_4(g) + 2O_2(g) \longrightarrow CO_2(g) + 2H_2O(g)$

3 Assess

Evaluate Understanding

To evaluate students' understanding of chemical reactions, ask students to give an example of each type of reaction discussed in this section. Write several chemical equations on the board and ask students to classify them. In addition, write only the products of combination, decomposition, and displacement reactions, and challenge students to fill in the reactants. Have students refer to Table 8.2 to answer the following questions.

- **What would happen if a piece of iron were placed in a solution of lead(II) nitrate?** (Iron will displace lead.)
- **What would happen if a piece of aluminum were placed in a solution of calcium chloride?** (nothing since Al will not displace Ca)

MINI LAB

Activity Series of Metals

SAFETY

Silver nitrate solutions will stain skin and clothing. Avoid skin contact with this chemical.

ANALYSIS AND CONCLUSIONS

1. Copper metal reacts with $AgNO_3(aq)$. Zinc metal reacts with $AgNO_3(aq)$ and $CuSO_4(aq)$. Magnesium metal reacts with $ZnSO_4(aq)$, $CuSO_4(aq)$, and $AgNO_3(aq)$

2. $Cu(s) + 2AgNO_3(aq) \rightarrow 2Ag(s) + Cu(NO_3)_2(aq)$

$Zn(s) + 2AgNO_3(aq) \rightarrow 2Ag(s) + Zn(NO_3)_2(aq)$

$Zn(s) + CuSO_4(aq) \rightarrow Cu(s) + ZnSO_4(aq)$

$Mg(s) + 2AgNO_3(aq) \rightarrow 2Ag(s) + Mg(NO_3)_2(aq)$

$Mg(s) + CuSO_4(aq) \rightarrow Cu(s) + MgSO_4(aq)$

$Mg(s) + ZnSO_4(aq) \rightarrow Zn(s) + MgSO_4(aq)$

3. sodium, magnesium, zinc, copper, silver

Reteach

Help students develop a branched flow chart similar to those used in qualitative analysis. Start by asking if there is a single reactant. If so, the reaction is a decomposition. If not, proceed to the next step. Then ask if oxygen is one of the reactants. If so, the reaction is combustion. If not, proceed to the next step.

MINI LAB

Activity Series of Metals

PURPOSE

To develop an activity series of metals.

MATERIALS

- 15 100-mL beakers
- 5 small strips each of copper, magnesium, and zinc metal
- steel wool or fine sandpaper
- aqueous solutions of the following:
  $CuSO_4$ (0.05*M*)
  $MgSO_4$ (0.05*M*)
  NaCl (0.05*M*)
  $AgNO_3$ (0.05*M*)
  $ZnSO_4$ (0.05*M*)

PROCEDURE

1. Clean each strip of metal with the steel wool or sandpaper.
2. Label each beaker with the name of one of the solutions. There should be three beakers for each solution.
3. Pour 20.0 mL of each solution into each of its labeled beakers.
4. Divide the beakers into three groups, each with one beaker of each solution. Label one group of beakers copper, one magnesium, and one zinc. Place a strip of the corresponding metal into each beaker.
5. After 1, 5, and 15 minutes make observations of the metal strips and the solutions.

ANALYSIS AND CONCLUSIONS

1. In which solutions did the appearance of the metal change? (Ignore the formation of bubbles.)
2. Write a balanced equation for each reaction between a metal and a solution.
3. Based on the results of your experiment, construct an activity series for these five metals. Put the most active metal first.

section review 8.2

22. Balance each equation. Identify each as to type.

- **a.** $Pb(NO_3)_2 + K_2CrO_4 \longrightarrow PbCrO_4 + KNO_3$
- **b.** $Cl_2 + KI \longrightarrow KCl + I_2$
- **c.** $C_3H_6 + O_2 \longrightarrow CO_2 + H_2O$
- **d.** $Al(OH)_3 \longrightarrow Al_2O_3 + H_2O$
- **e.** $Li + O_2 \longrightarrow Li_2O$
- **f.** $HCl + Fe_2O_3 \longrightarrow FeCl_3 + H_2O$
- **g.** $MgCO_3 \longrightarrow MgO + CO_2$
- **h.** $Ba(CN)_2 + H_2SO_4 \longrightarrow BaSO_4 + HCN$

23. Complete and balance an equation for each reaction.

- **a.** $CaI_2 + Hg(NO_3)_2 \longrightarrow$
- **b.** $Al + Cl_2 \longrightarrow$
- **c.** $Ag + HCl \longrightarrow$
- **d.** $C_2H_2 + O_2 \longrightarrow$
- **e.** $MgCl_2 \longrightarrow$
- **f.** $H_2O_2 \longrightarrow$

24. What three types of products drive double-replacement reactions?

Chem ASAP! **Assessment 8.2** Check your understanding of the important ideas and concepts in Section 8.2.

Answers

SECTION REVIEW 8.2

22. a. $Pb(NO_3)_2 + K_2CrO_4 \rightarrow PbCrO_4 + 2KNO_3$, Double replacement

b. $Cl_2 + 2KI \rightarrow 2KCl + I_2$, Single replacement

c. $2C_3H_6 + 9O_2 \rightarrow 6CO_2 + 6H_2O$, Combustion

d. $2Al(OH)_3 \rightarrow Al_2O_3 + 3H_2O$, Decomposition

e. $4Li + O_2 \rightarrow 2Li_2O$, Combination

f. $6HCl + Fe_2O_3 \rightarrow 2FeCl_3 + 3H_2O$, Double replacement

g. $MgCO_3 \rightarrow MgO + CO_2$, Decomposition

h. $Ba(CN)_2 + H_2SO_4 \rightarrow BaSO_4 + 2HCN$, Double replacement

23. a. $CaI_2 + Hg(NO_3)_2 \rightarrow HgI_2 + Ca(NO_3)_2$

b. $2Al + 3Cl_2 \rightarrow 2AlCl_3$

c. no reaction

d. $2C_2H_2 + 5O_2 \rightarrow 4CO_2 + 2H_2O$

e. $MgCl_2 \xrightarrow{\text{electricity}} Mg + Cl_2$

f. $2H_2O_2 \rightarrow 2H_2O + O_2$

24. formation of a gas, a precipitate, or a molecular compound

REACTIONS IN AQUEOUS SOLUTION

section 8.3

The beauty of a limestone cavern is the direct result of chemical reactions involving water. Limestone caverns form as calcium carbonate reacts with carbon dioxide dissolved in water to form soluble calcium hydrogen carbonate. Additional carbon dioxide then converts the calcium hydrogen carbonate back into calcium carbonate, which precipitates out, leaving behind dramatic stalactites and stalagmites. **What causes the formation of a precipitate?**

objectives

- Write and balance net ionic equations
- Use solubility rules to predict the precipitate formed in double-replacement reactions

key terms

- complete ionic equation
- spectator ions
- net ionic equation

Net Ionic Equations

Our world is water-based. More than 70% of Earth's surface is covered by water, and about 66% of the adult human body is water. It is not surprising, therefore, that many important chemical reactions take place in water—that is, in aqueous solution.

The reaction of aqueous solutions of silver nitrate with sodium chloride to form solid silver chloride and aqueous sodium nitrate is a double-replacement reaction.

$$AgNO_3(aq) + NaCl(aq) \longrightarrow AgCl(s) + NaNO_3(aq)$$

As written, this equation continues the practice of representing ionic compounds by their formula units. This equation is more realistically written, however, if you recognize that most ionic compounds dissociate, or separate, into cations and anions when they dissolve in water. For example, when sodium chloride dissolves in water it separates into sodium ions $(Na^+)(aq)$ and chloride ions $(Cl^-)(aq)$. Similarly, silver nitrate dissociates into silver ions $(Ag^+)(aq)$ and nitrate ions $(NO_3^-)(aq)$.

You can now write a **complete ionic equation,** an equation that shows dissolved ionic compounds as their free ions.

$$Ag^+(aq) + NO_3^-(aq) + Na^+(aq) + Cl^-(aq) \longrightarrow AgCl(s) + Na^+(aq) + NO_3^-(aq)$$

The equation can be simplified and made more useful by eliminating the ions that do not participate in the reaction. You do this by canceling the ions that appear on both sides of the equation.

$$Ag^+(aq) + \cancel{NO_3^-(aq)} + \cancel{Na^+(aq)} + Cl^-(aq) \longrightarrow AgCl(s) + \cancel{Na^+(aq)} + \cancel{NO_3^-(aq)}$$

Ions that are not directly involved in a reaction are called **spectator ions.** Notice that the spectator ions appear on both sides of the equation. You can rewrite this equation leaving out the spectator ions. This gives you the **net ionic equation,** the equation that indicates only those particles that actually take part in the reaction.

$$Ag^+(aq) + Cl^-(aq) \longrightarrow AgCl(s)$$

In writing balanced net ionic equations, you must also balance the ionic charge. For the previous reaction, the net ionic charge on each side of the equation is zero and is therefore balanced. In other reactions, this may not be the case. Consider the following reaction.

$$Pb(s) + AgNO_3(aq) \longrightarrow Ag(s) + Pb(NO_3)_2(aq)$$

STUDENT RESOURCES

From the Teacher's Resource Package, use:

- Section Review 8.3, Ch. 8 Practice Problems, Interpreting Graphics, Vocabulary Review, Quizzes, and Tests from the Review Module (Ch. 5–8)
- Laboratory Recordsheet 8-2

TECHNOLOGY RESOURCES

Relevant technology resources include:

- Chem ASAP! CD-ROM
- Resource Pro CD-ROM
- Assessment Resources CD-ROM: Chapter 8 Tests

8.3

1 Engage

Use the Visual

Have students study the photograph and read the text that opens the section. Point out that the formation of stalactites and stalagmites is a precipitation reaction. Ask:

- **Have you ever seen evidence of a precipitation reaction in everyday life?** (The deposits in water pipes are the result of calcium ions combining with other ions in water.)
- **What causes the formation of a precipitate?** (Some ions combine in solution to form insoluble solids.)

Check Prior Knowledge

To assess students' prior knowledge about precipitation reactions, ask:

- **What type of reaction is a precipitation reaction?** (double replacement)
- **What evidence is there that a double-replacement reaction has occurred?** (formation of a precipitate, a gas, or a molecular compound like water)

Motivate and Relate

If any students have visited underground caves such as Carlsbad Caverns or Luray Caverns, have them describe what they saw. You may wish to locate Internet sites to show students or use photos from earth science textbooks. Explain that the caverns form when carbonic acid dissolves the calcium carbonate (calcite) in limestone. The icicle-like deposits on the roof of the cavern are called *stalactites.* The lumpy deposits on the floor are called *stalagmites.* Both form when dissolved calcium carbonate precipitates from groundwater.

2 Teach

TEACHER DEMO

Demonstrate a precipitation reaction. Place 10 mL of $Pb(NO_3)_2$ in a beaker. Add 5 mL of KI solution to the $Pb(NO_3)_2$ solution. A bright yellow precipitate will form. Explain that not all ionic compounds are soluble in water. The solid formed in this reaction is PbI_2. Point out that no one can predict for sure whether a precipitate will form until they mix the solutions and make observations, but scientists have developed solubility tables that make informed predictions possible. Demonstrate that not all combinations of ionic compounds lead to the formation of an insoluble precipitate. Mix NaCl with K_2SO_4.

Practice Problems Plus

Related Chapter Review Problem
Chapter Review problem 52 is related to Sample Problem 8-10.

Additional Practice Problem
The Chem ASAP! CD-ROM contains the following problem: Write a balanced net ionic equation for the reaction between sulfuric acid (H_2SO_4) and sodium hydroxide (NaOH).
[$H^+(aq) + OH^-(aq) \rightarrow H_2O(l)$]

Discuss

Explain to students that a net ionic equation differentiates between ions that react to form a solid precipitate, a gas, or water, and ions that simply remain in aqueous solution. It is important to note, however, that spectator ions are not completely unaffected by the reaction. Although they remain in solution, they end up paired with different anions or cations than they were paired with when the reaction began.

The nitrate ion is the spectator ion in this reaction. The net ionic equation is

$$Pb(s) + Ag^+(aq) \longrightarrow Ag(s) + Pb^{2+}(aq) \text{ (unbalanced)}$$

Although this equation is balanced with respect to atoms, the ionic charges do not balance. On the reactant side of the equation there is a single unit of positive charge. On the product side there are two units of positive charge. Placing the coefficient 2 in front of $Ag^+(aq)$ balances the charge. A coefficient of 2 in front of Ag(s) rebalances the atoms.

$$Pb(s) + 2Ag^+(aq) \longrightarrow 2Ag(s) + Pb^{2+}(aq) \text{ (balanced)}$$

Of the five types of reactions identified in this chapter, both single- and double-replacement reactions generally can be written as net ionic equations.

Sample Problem 8-10

Identify the spectator ions and write balanced net ionic equations for these reactions.

a. $HCl(aq) + ZnS(aq) \longrightarrow H_2S(g) + ZnCl_2(aq)$
b. $Cl_2(g) + NaBr(aq) \longrightarrow Br_2(l) + NaCl(aq)$
c. Aqueous solutions of iron(III) chloride and potassium hydroxide are mixed, forming a precipitate of iron(III) hydroxide.

1. *ANALYZE* ***Plan a problem-solving strategy.***
Write the complete ionic equation for each reaction, showing any soluble ionic compounds as individual ions. Any aqueous ions that appear as both reactants and products are spectator ions and are eliminated to give the net ionic equation. Finally, balance the net ionic equation.

2. *SOLVE* ***Apply the problem-solving strategy.***
a. Write the complete ionic equation.
$$H^+(aq) + Cl^-(aq) + Zn^{2+}(aq) + S^{2-}(aq) \longrightarrow H_2S(g) + Zn^{2+}(aq) + Cl^-(aq)$$
The spectator ions are Zn^{2+} and Cl^-. The balanced net ionic equation is
$$2H^+(aq) + S^{2-}(aq) \longrightarrow H_2S(g)$$
b. Write the complete ionic equation.
$$Cl_2(g) + Na^+(aq) + Br^-(aq) \longrightarrow Br_2(l) + Na^+(aq) + Cl^-(aq)$$
The spectator ion is Na^+. The balanced net ionic equation is
$$Cl_2(g) + 2Br^-(aq) \longrightarrow Br_2(l) + 2Cl^-(aq)$$
c. Write the complete ionic equation.
$$Fe^{3+}(aq) + 3Cl^-(aq) + 3K^+(aq) + 3OH^-(aq) \longrightarrow Fe(OH)_3(s) + 3K^+(aq) + 3Cl^-(aq)$$
The spectator ions are K^+ and Cl^-. The balanced net ionic equation is
$$Fe^{3+}(aq) + 3OH^-(aq) \longrightarrow Fe(OH)_3(s)$$

Practice Problems

25. Write balanced net ionic equations for each reaction.
a. $Pb(ClO_4)_2(aq) + NaI(aq) \longrightarrow PbI_2(s) + NaClO_4(aq)$
b. $Zn(s) + HCl(aq) \longrightarrow ZnCl_2(aq) + H_2(g)$
c. $Ca(OH)_2(aq) + H_3PO_4(aq) \longrightarrow Ca_3(PO_4)_2(aq) + H_2O(l)$

Chem ASAP!
Problem-Solving 25
Solve Problem 25 with the help of an interactive guided tutorial.

Meeting Diverse Needs

Gifted Magnesium metal is an important component of alloys used to make consumer materials. The main commercial source of Mg*(s)* is seawater; Mg^{2+} ion is the third most abundant dissolved ion in the ocean. A process for isolating magnesium from seawater depends on the fact that Mg^{2+} will precipitate when OH^- is added. Have students research this method.

Sample Problem 8-10 (cont.)

3. *EVALUATE* ***Do the results make sense?***
In each balanced net ionic equation, the number and kinds of atoms in the reactants equal the number and kinds of atoms in the product. The net charge of all the reactants equals the net charge of all the products.

Predicting the Formation of a Precipitate

You have seen that mixing solutions of two ionic compounds can sometimes result in the formation of an insoluble salt called a precipitate. The formation of a precipitate is predicted by using the general rules for solubility of ionic compounds, shown in **Table 8.3.** Should a precipitate form when aqueous solutions of $Na_2SO_4(aq)$ and $Ba(NO_3)_2(aq)$ are mixed? When these four ions are mixed, the two new compounds that might form are $NaNO_3$ and $BaSO_4$. These are the only two new combinations of cation and anion possible.

$$2Na^+(aq) + SO_4^{2-}(aq) + Ba^{2+}(aq) + 2NO_3^-(aq) \longrightarrow ?$$

Sodium nitrate will not form a precipitate because alkali metal salts and nitrate salts are soluble (Rules 1 and 2). Rule 4 indicates that barium sulfate is insoluble and therefore will precipitate. In this reaction Na^+ and NO_3^- are spectator ions. The net ionic equation for this reaction is as follows.

$$Ba^{2+}(aq) + SO_4^{2-}(aq) \longrightarrow BaSO_4(s)$$

The mixing of solutions of sodium sulfate and barium nitrate is shown in **Figure 8.13.** In this example, why is the formation of a precipitate from Na^+ and SO_4^{2-} or Ba^{2+} and NO_3^- not a concern? ❶

Table 8.3

Solubility Rules for Ionic Compounds

Compounds	Solubility	Exceptions
1. Salts of alkali metals and ammonia	Soluble	Some lithium compounds
2. Nitrate salts and chlorate salts	Soluble	Few exceptions
3. Sulfate salts	Soluble	Compounds of Pb, Ag, Hg, Ba, Sr, and Ca
4. Chloride salts	Soluble	Compounds of Ag and some compounds of Hg and Pb
5. Carbonates, phosphates, chromates, sulfides, and hydroxides	Most are insoluble	Compounds of the alkali metals and of ammonia

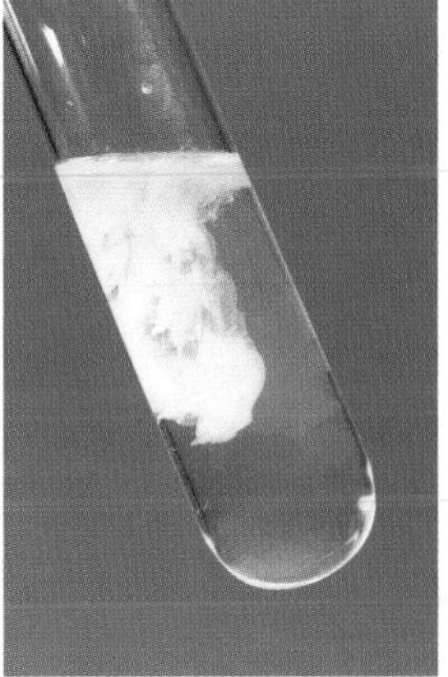

Figure 8.13
A precipitate forms when aqueous solutions of sodium sulfate and barium nitrate are mixed.

8.3

TEACHER DEMO

Demonstrate a series of precipitation reactions. Mix about 5 mL of 0.1 *M* $BaCl_2$ solution with about 5 mL of 0.1 *M* Na_2SO_4 solution in a beaker. A white precipitate should appear. Mix about 5 mL of 0.1 *M* K_2CrO_4 solution with about 5 mL of 0.1 *M* $Ba(NO_3)_2$ solution in a beaker. A yellow, dense precipitate should appear. The reaction products can be flushed down the drain with excess water.

ACTIVITY

As an extension of the preceding demonstration, have students form groups and write the complete ionic and net ionic equations for each of the reactions shown in the demonstration. When writing complete ionic equations, students may find it helpful to identify spectator ions by writing them in a different color, or by circling them.

Discuss

Initiate a discussion with students about precipitation reactions that occur in everyday life. (Examples include kidney stones, soap scum in bathtubs and showers, and limestone deposits.) Explain the value of being able to write net ionic equations as a very direct and simple way to describe chemical change. Relate the concept of net ionic equations to net income. Spectator ions can be compared to two people who go to a dance club on a double date and then watch while their partners get together and dance.

Answer

❶ You know that these combinations of ions will not form a precipitate because they were the initial aqueous reactants.

Practice Problems Plus

Additional Practice Problem

The Chem ASAP! CD-ROM contains this problem related to Sample Problem 8-11: Identify the precipitate formed when solutions of these ionic compounds are mixed. Write a balanced net ionic equation for the reaction.

$AgNO_3(aq) + K_2SO_4(aq) \rightarrow$
$[2Ag^+(aq) + SO_4^{2-}(aq) \rightarrow Ag_2SO_4(s)]$

3 Assess

Evaluate Understanding

To evaluate students' understanding of complete ionic equations, net ionic equations, and the formation of precipitates, write some sentences for precipitation reactions on the board. Include at least one example for which no reaction occurs. Have students write complete and net ionic equations.

Reteach

Review with students the writing and balancing of complete and net ionic equations and the use of the rules of solubility to predict the outcome of double-replacement reactions. Stress that it is important to note the physical states of reactants and products in precipitation reactions.

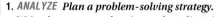

Sample Problem 8-11

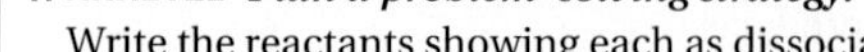

Use the solubility rules in **Table 8.3** to identify the precipitate formed and write the net ionic equation for the reaction of aqueous potassium carbonate with aqueous strontium chloride.

1. *ANALYZE* ***Plan a problem-solving strategy.***
Write the reactants showing each as dissociated free ions. Then look at possible new pairings of cation and anion that give an insoluble substance. Finally, eliminate the spectator ions and write the net ionic equation.

2. *SOLVE* ***Apply the problem-solving strategy.***

$$2K^+(aq) + CO_3^{2-}(aq) + Sr^{2+}(aq) + 2NO_3^-(aq) \longrightarrow ?$$

Of the two possible combinations, KNO_3 is soluble (Rules 1 and 2) and $SrCO_3$ is insoluble (Rule 5). Thus the net ionic equation is

$$Sr^{2+}(aq) + CO_3^{2-}(aq) \longrightarrow SrCO_3(s)$$

3. *EVALUATE* ***Does the result make sense?***
$SrCO_3$ is the only combination of cation and anion that gives a precipitate. The equation is balanced in terms of atoms and charge.

Practice Problems

26. Identify the precipitate formed when solutions of these ionic compounds are mixed. Write a net ionic equation.
$NH_4Cl(aq) + Pb(NO_3)_2(aq) \longrightarrow$

27. Write a complete ionic equation and a net ionic equation for the reaction of aqueous solutions of iron(III) nitrate and sodium hydroxide.

Chem ASAP!
Problem-Solving 26
Solve Problem 26 with the help of an interactive guided tutorial.

section review 8.3

28. Write a balanced net ionic equation for each reaction.
- **a.** $Pb(NO_3)_2(aq) + H_2SO_4(aq) \longrightarrow PbSO_4(s) + HNO_3(aq)$
- **b.** $Pb(C_2H_3O_2)_2(aq) + HCl(aq) \longrightarrow PbCl_2(s) + HC_2H_3O_2(aq)$
- **c.** $Na_3PO_4(aq) + FeCl_3(aq) \longrightarrow NaCl(aq) + FePO_4(s)$
- **d.** $(NH_4)_2S(aq) + Co(NO_3)_2(aq) \longrightarrow CoS(s) + NH_4NO_3(aq)$

29. Write a net ionic equation for each reaction.
- **a.** $HCl(aq) + AgNO_3(aq) \longrightarrow$
- **b.** $Pb(C_2H_3O_2)_2(aq) + LiCl(aq) \longrightarrow$
- **c.** $Na_3PO_4(aq) + CrCl_3(aq) \longrightarrow$

30. Identify the spectator ions in each reaction in Problem 29.

31. Identify the precipitate formed when solutions of these ionic compounds are mixed.
- **a.** $H_2SO_4 + BaCl_2 \longrightarrow$
- **b.** $Al_2(SO_4)_3 + NH_4OH \longrightarrow$
- **c.** $AgNO_3 + H_2S \longrightarrow$
- **d.** $CaCl_2 + Pb(NO_3)_2 \longrightarrow$
- **e.** $Ca(NO_3)_2 + Na_2CO_3 \longrightarrow$

Chem ASAP! **Assessment 8.3** Check your understanding of the important ideas and concepts in Section 8.3.

Answers

SECTION REVIEW 8.3

28. a. $Pb^{2+}(aq) + SO_4^{2-}(aq) \rightarrow PbSO_4(s)$
b. $Pb^{2+}(aq) + 2Cl^-(aq) \rightarrow PbCl_2(s)$
Because acetic acid is a weak acid, the net ionic equation could be expanded to include this soluble product: $Pb^{2+}(aq) + 2C_2H_3O_2^-(aq) + H^+(aq) + 2Cl^-(aq) \rightarrow PbCl_2(s) + HC_2H_3O_2(aq)$.
c. $Fe^{3+}(aq) + PO_4^{3-}(aq) \rightarrow FePO_4(s)$
d. $Co^{2+}(aq) + S^{2-}(aq) \rightarrow CoS(s)$

29. a. $Ag^+(aq) + Cl^-(aq) \rightarrow AgCl(s)$
b. $Pb^{2+}(aq) + 2Cl^-(aq) \rightarrow PbCl_2(s)$
c. $Cr^{3+}(aq) + PO_4^{3-}(aq) \rightarrow CrPO_4(s)$

30. a. H^+ and NO_3^-
b. Li^+ and $C_2H_3O_2^-$
c. Na^+ and Cl^-

31. a. $BaSO_4(s)$ **d.** $PbCl_2(s)$
b. $Al(OH)_3(s)$ **e.** $CaCO_3(s)$
c. $Ag_2S(s)$

SMALL-SCALE LAB

PRECIPITATION REACTIONS: FORMATION OF SOLIDS

SAFETY

Wear safety glasses and follow standard safety procedures outlined on page 18.

PURPOSE

To observe, identify, and write balanced equations for precipitation reactions.

MATERIALS

- pencil
- paper
- ruler
- reaction surface
- chemicals shown in Figure A

	$AgNO_3$ (Ag^+)	$Pb(NO_3)_2$ (Pb^{2+})	$CaCl_2$ (Ca^{2+})
Na_2CO_3 (CO_3^{2-})	a white ppt	f white ppt	k white ppt
Na_3PO_4 (PO_4^{3-})	b tan ppt	g white ppt	l white ppt
NaOH (OH^-)	c brown ppt	h white ppt	m white ppt
Na_2SO_4 (SO_4^{2-})	d no visible reaction	i white ppt	n no visible reaction
NaCl (Cl^-)	e white ppt	j white ppt	o no visible reaction

Figure A

PROCEDURE

Draw two grids similar to Figure A on separate sheets of paper. Make each square 2 cm on each side. Draw large black Xs on only one of the grids. Place a reaction surface over the grid with black Xs and add the chemicals as shown in Figure A. Use the other grid as a data table to record your observations for each solution.

ANALYSIS

Using the experimental data, record the answers to the following questions below your data table.

1. Translate the following word equations into balanced chemical equations and explain how the equations represent what happens in grid spaces a and g.
   a. In grid space a, sodium carbonate reacts with silver nitrate to produce sodium nitrate and solid silver carbonate.
   b. In grid space g, sodium phosphate reacts with lead(II) nitrate to produce sodium nitrate and solid lead(II) phosphate.
2. Write a word equation to represent what happens in grid space m.
3. What happens in grid space d? Which other mixings gave similar results? Is it necessary to write an equation when no reaction occurs? Explain.
4. Write balanced equations for the other precipitation reactions you observed.
5. Write balanced net ionic equations for the other precipitation reactions you observed.

YOU'RE THE CHEMIST

The following small-scale activities allow you to develop your own procedures and analyze the results.

1. **Analyze It!** Mix a solution of potassium iodide (KI) with silver nitrate and then separately with lead(II) nitrate to discover two distinctively colored precipitates. Write balanced equations and net ionic equations for each reaction.
2. **Design It!** Table salt is mostly sodium chloride. Design and carry out an experiment to find out if table salt will form a precipitate with either lead(II) or silver nitrate.
3. **Design It!** Design and carry out an experiment to show that iodized table salt contains potassium iodide (KI).

SMALL-SCALE LAB

PRECIPITATION REACTIONS: FORMATION OF SOLIDS

OVERVIEW

Procedure Time: 30 minutes
Students gain practice in writing balanced chemical equations.

MATERIALS

Solution	Preparation
0.05*M* $AgNO_3$	2.1 g in 250 mL
0.2*M* $Pb(NO_3)_2$	16.6 g in 250 mL
0.5*M* $CaCl_2$	13.9 g in 250 mL
1.0*M* Na_2CO_3	26.5 g in 250 mL
0.1*M* Na_3PO_4	9.5 g $Na_3PO_4 \cdot 12H_2O$ in 250 mL
0.5*M* NaOH	20.0 g in 1.0 mL
0.2*M* Na_2SO_4	7.1 g in 250 mL
1.0*M* NaCl	14.6 g in 250 mL

DISPOSAL

Flush all chemicals down the drain with excess water.

YOU'RE THE CHEMIST

1. $KI + AgNO_3 \rightarrow KNO_3 + AgI(s)$
   $Ag^+ + I^- \rightarrow AgI(s)$
   $2KI + Pb(NO_3)_2 \rightarrow 2KNO_3 + PbI_2(s)$
   $Pb^{2+} + 2I^- \rightarrow PbI_2(s)$
2. Add a drop of $Pb(NO_3)_2$ or $AgNO_3$ to a few grains of salt. Look for white crystals.
3. Place a drop of $Pb(NO_3)_2$ on a small pile of dry table salt. Keep part of the pile dry. Look for yellow lead iodide.

ANALYSIS

1. $Na_2CO_3 + 2AgNO_3 \rightarrow 2NaNO_3 + Ag_2CO_3(s)$
   $2Na_3PO_4 + 3Pb(NO_3)_2 \rightarrow 6NaNO_3 + Pb_3(PO_4)_2(s)$
2. Sodium hydroxide and calcium chloride form sodium chloride and solid calcium hydroxide.
3. Mixings d, n, and o do not react.
4. $Na_3PO_4 + 3AgNO_3 \rightarrow 3NaNO_3 + Ag_3PO_4(s)$
   $NaOH + AgNO_3 \rightarrow NaNO_3 + AgOH(s)$
   (actual products: $NaNO_3 + Ag_2O(s) + H_2O$)
   $NaCl + AgNO_3 \rightarrow NaNO_3 + AgCl(s)$
   $Na_2CO_3 + Pb(NO_3)_2 \rightarrow 2NaNO_3 + PbCO_3(s)$
   $2NaOH + Pb(NO_3)_2 \rightarrow 2NaNO_3 + Pb(OH)_2(s)$
   $Na_2SO_4 + Pb(NO_3)_2 \rightarrow 2NaNO_3 + PbSO_4(s)$
   $2NaCl + Pb(NO_3)_2 \rightarrow 2NaNO_3 + PbCl_2(s)$
   $Na_2CO_3 + CaCl_2 \rightarrow 2NaCl + CaCO_3(s)$
   $2Na_3PO_4 + 3CaCl_2 \rightarrow 6NaCl + Ca_3(PO_4)_2(s)$
   $2NaOH + CaCl_2 \rightarrow 2NaCl + Ca(OH)_2(s)$
5. $2Ag^+ + CO_3^{2-} \rightarrow Ag_2CO_3(s)$
   $3Ag^+ + PO_4^{3-} \rightarrow Ag_3PO_4(s)$
   $Ag^+ + OH^- \rightarrow AgOH(s)$
   $Ag^+ + Cl^- \rightarrow AgCl(s)$
   $Pb^{2+} + CO_3^{2-} \rightarrow PbCO_3(s)$
   $3Pb^{2+} + 2PO_4^{3-} \rightarrow Pb_3(PO_4)_2(s)$
   $Pb^{2+} + 2OH^- \rightarrow Pb(OH)_2(s)$
   $Pb^{2+} + SO_4^{2-} \rightarrow PbSO_4(s)$
   $Pb^{2+} + 2Cl^- \rightarrow PbCl_2(s)$
   $Ca^{2+} + CO_3^{2-} \rightarrow CaCO_3(s)$
   $3Ca^{2+} + 2PO_4^{3-} \rightarrow Ca_3(PO_4)_2(s)$
   $Ca^{2+} + 2OH^- \rightarrow Ca(OH)_2(s)$

Chemistry Serving... Society

DISCUSS

After students have read the article, ask:

- What three things are necessary for a fire to burn? (fuel, oxygen, and energy to initiate combustion)
- Why is it not safe to use a single kind of fire extinguisher on all fires? (A fire extinguisher that controls one type of fire may actually enhance other types of combustion reactions. For example, water is not sprayed on burning magnesium, because the intense heat can decompose the water, producing flammable hydrogen and oxygen gases.)

Have students contact their local fire department to obtain statistics on how many class A, B, C, and D fires have occurred in their area during the past year or six months. As a class activity, have students graph the data and discuss any conclusions that can be drawn from the data.

CHEMISTRY IN CAREERS

Have students connect to the Internet address shown to learn more about careers in fire science. Students may also wish to read about the duties and responsibilities of a firefighter on page 871 of this text.

Chemistry Serving...Society

COMBATING COMBUSTION

Many of the materials in the world around you—such as wooden structures, paper, trees, gasoline—are combustible. All it takes is a source of ignition to start a serious fire. In North America alone, fires kill more than 20 000 people and cause millions of dollars in property damage each year. Fighting fires is an important application of chemistry.

Fire, or combustion, is the rapid reaction of a fuel with oxygen. The fuel is heated to start the fire. Once started, a fire can be put out in three ways: (1) by cooling the fuel, (2) by cutting off the supply of oxygen, or (3) by trapping the combustion intermediates so they cannot react with oxygen.

Firefighters are called upon most often to fight fires involving solid fuel, such as the wooden frame of a house. This kind of fire is known as a Class A fire. In a Class A fire the burning materials produce flammable gases that sustain and spread the fire. Water can be used to extinguish Class A fires because (1) water cools the fuel to below its combustion temperature and (2) when liquid water vaporizes, the vapor dilutes the oxygen content of the surrounding air.

Wood, of course, is not the only material that can catch fire and burn. Cooking oil, greases, and flammable liquids are all extremely combustible, and when they burn the fire cannot easily be extinguished with water. A stream of water, in fact, might actually spread the burning liquid. These kinds of fires, called Class B, are often fought with carbon dioxide fire extinguishers. Because it is more dense than air, the carbon dioxide blankets the burning fire and extinguishes the fire by cutting off the oxygen supply. Class B fires can also be fought with chemical powders. Portable dry powder extinguishers found in homes contain monoammonium phosphate (MAP). MAP is useful for Class A, B, and C fires. MAP coats hot surfaces with a sticky residue that excludes oxygen and stops combustion.

In North America alone, fires kill more than 20 000 people and cause millions of dollars in property damage each year.

Another type of fire is one involving an energized electrical circuit. Known as Class C fires, these fires cannot be fought with water because water almost always contains enough dissolved salts to conduct electricity. Instead, Class C fires are fought with the same types of extinguishing agents as Class B fires.

As you may remember, certain metals, such as magnesium, can also burn. Fires involving metals are called Class D fires and must be extinguished with special dry chemical powder fire extinguishers. Applying water or carbon dioxide to a Class D fire is not wise because both compounds can actually react with the burning material.

When a whole forest of trees is burning, techniques somewhat different from those used for fires involving buildings are called for. Water can be used to extinguish a forest fire, but getting the water to the fire can be a problem. Airplanes or helicopters are used to dump fire retardant chemicals on brush or forest fires. The dry chemicals are usually a mixture of MAP, ammonium sulfate, and iron oxide. These chemicals form a sticky residue on hot surfaces. The residue acts as a barrier to oxygen. The chemicals also act as nutrients that encourage new growth of trees after the fire.

CHEMISTRY IN CAREERS

FIREFIGHTER

Are you ready for action? Do you want to control combustion reactions? Firefighting may be just the career for you. See page 871.

CHEMIST NEEDED

TAKE IT TO THE NET

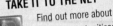

Find out more about career opportunities: www.phschool.com

CHEMICAL SPECIALIST

Local food service distributor see responsible, self-motivated indiv

Chapter 8 STUDENT STUDY GUIDE

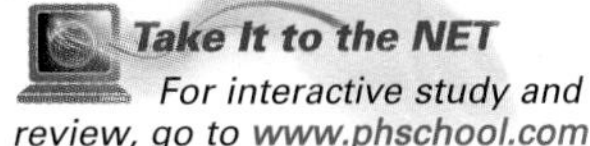

Take It to the NET *For interactive study and review, go to www.phschool.com*

KEY TERMS

- activity series of metals *p. 217*
- balanced equation *p. 207*
- catalyst *p. 205*
- chemical equation *p. 205*
- coefficient *p. 207*
- combination reaction *p. 212*
- combustion reaction *p. 220*
- complete ionic equation *p. 225*
- decomposition reaction *p. 215*
- double-replacement reaction *p. 219*
- net ionic equation *p. 225*
- single-replacement reaction *p. 217*
- skeleton equation *p. 205*
- spectator ion *p. 225*

CONCEPT SUMMARY

8.1 Describing Chemical Change

- A chemical reaction can be concisely represented by a chemical equation.
- The substances that undergo a chemical change are the reactants. The new substances formed are the products.
- Special symbols are written after formulas in equations to show a substance's state. The designations for a solid, liquid, or gas are (*s*), (*l*), and (*g*) respectively. A substance dissolved in water is designated (*aq*).
- A catalyst is a substance that increases reaction rate without being used up by the reaction. If a catalyst is used, its formula is written above the arrow.
- In accordance with the law of conservation of mass, a chemical equation must be balanced. In balancing an equation, coefficients are used so the same number of atoms of each element are on each side of the equation.

8.2 Types of Chemical Reactions

- In a combination reaction, there is always a single product. The reactants are two or more elements and/or compounds that combine.
- A decomposition reaction involves the breakdown of a single compound into two or more simpler substances.
- In a single-replacement reaction, the reactants and products are an element and a compound. The activity series of metals can be used to predict whether single-replacement reactions will take place.
- A double-replacement reaction involves the exchange of cations (or anions) between two compounds. This reaction generally takes place between two ionic compounds in aqueous solution.
- A combustion reaction always involves oxygen. The products of the complete combustion of a hydrocarbon are carbon dioxide and water.

8.3 Reactions in Aqueous Solution

- A complete ionic equation shows all dissolved ionic compounds as their free ions. It includes spectator ions as well as ions involved in the reaction.
- Both single- and double-replacement reactions can be written as net ionic equations in which spectator ions are deleted from both sides of the equation.
- The precipitate formed in a double-replacement reaction can be identified using a table of solubilities.

CHAPTER CONCEPT MAP

Use these terms to construct a concept map that organizes the major ideas of this chapter.

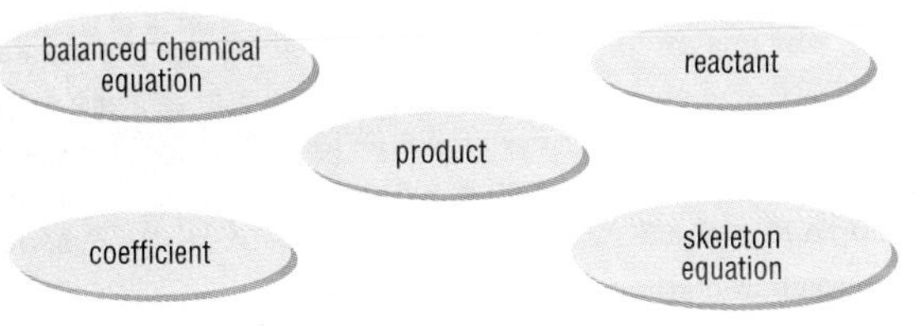

Chem ASAP! **Concept Map 8**
Create your Concept Map using the computer.

Take It to the Net

At **www.phschool.com** students will find for this chapter

- an Internet activity
- links to related chemistry sites
- an interactive quiz
- career links

Chapter 8 STUDENT STUDY GUIDE

4 Close

Summary

Ask the following questions which require students to summarize the information contained in the chapter.

- **Why must chemical equations be balanced?** (In a chemical reaction, atoms are neither created nor destroyed. All atoms present in the reactants must be accounted for in the products.)
- **What are the five main types of chemical reactions?** (combination, combustion, single-replacement, double-replacement, and decomposition reactions)
- **What is the activity series and how is it used?** (The activity series of metals is a list of metals in order of decreasing reactivity. The activity series of metals can be used to predict whether a single-replacement reaction will take place.)

Looking Back . . . Looking Ahead . . .

The preceding chapter introduced the students to the mole as the basic SI unit for measuring the amount of a substance. In this chapter students learn how to write chemical equations from descriptions of chemical reactions. In the next chapter, balanced chemical equations are used to solve mass-mass, mole-mole, and various types of stoichiometric problems. The concepts of limiting reagent and percent yield are developed.

Answers

32. **a.** reactants: sodium and water; products: hydrogen and sodium hydroxide
b. reactants: carbon dioxide and water; products: oxygen, glucose

33. Dalton said that the atoms of reactants are rearranged to form new substances as products.

34. The arrow separates the reactants from the products and indicates a reaction that progresses in a forward direction. A plus sign separates individual reactants and individual products from one another.

35. **a.** Gaseous ammonia and oxygen react in the presence of platinum to produce nitrogen monoxide gas and water vapor
b. Aqueous solutions of sulfuric acid and barium chloride are mixed to produce a precipitate of barium sulfate and aqueous hydrochloric acid.
c. The gas dinitrogen trioxide reacts with water to produce an aqueous solution of nitrous acid.

36. A catalyst speeds up a chemical reaction.

37. **a.** $C + 2F + 2G \rightarrow CF_2G_2$
b. $F + 3W + S + 2P \rightarrow FW_3SP_2$

38. A formula is a unique identifier of a substance. A different formula would indicate a different substance, not the one that is taking part in the reaction you are trying to balance.

39. **a.** $2PbO_2 \rightarrow 2PbO + O_2$
b. $2Fe(OH)_3 \rightarrow Fe_2O_3 + 3H_2O$
c. $(NH_4)_2CO_3 \rightarrow 2NH_3 + H_2O + CO_2$
d. $2NaCl + H_2SO_4 \rightarrow Na_2SO_4 + 2HCl$
e. $4H_2 + Fe_3O_4 \rightarrow 3Fe + 4H_2O$
f. $2Al + 3CuSO_4 \rightarrow Al_2(SO_4)_3 + 3Cu$

40. It helps in predicting products of reactions.

41. Use ionic charges to write an electrically neutral formula.

42. a single product

43. **a.** $2Mg + O_2 \rightarrow 2MgO$
b. $4P + 5O_2 \rightarrow 2P_2O_5$
c. $Ca + S \rightarrow CaS$
d. $2Fe + O_2 \rightarrow 2FeO$
e. $N_2O_5 + H_2O \rightarrow 2HNO_3$

44. **a.** $2Ag_2O \xrightarrow{\Delta} 4Ag + O_2$
b. $NiCO_3 \xrightarrow{\Delta} NiO + CO_2$
c. $NH_4NO_3 \xrightarrow{\Delta} N_2O + 2H_2O$

45. a single reactant

46. **a.** sodium displaces iron **b.** copper displaces silver **c.** zinc displaces hydrogen

47. **a.** no reaction
b. $Zn(s) + 2AgNO_3(aq) \rightarrow Zn(NO_3)_2(aq) + 2Ag(s)$
c. $2Al(s) + 3H_2SO_4(aq) \rightarrow Al_2(SO_4)_3(aq) + 3H_2(g)$
d. no reaction
e. $2Al(s) + 3CuSO_4(aq) \rightarrow Al_2(SO_4)_3(aq) + 3Cu(s)$

48. **a.** $2HCl(aq) + Ca(OH)_2(aq) \rightarrow CaCl_2(aq) + 2H_2O(l)$
b. $3Ag_2SO_4(aq) + 2AlCl_3(aq) \rightarrow 6AgCl(s) + Al_2(SO_4)_3(aq)$
c. $H_2C_2O_4(aq) + 2KOH(aq) \rightarrow K_2C_2O_4(aq) + 2H_2O(l)$
d. $CdBr_2(aq) + Na_2S(aq) \rightarrow CdS(s) + 2NaBr(aq)$

49. oxygen

Chapter 8 REVIEW

CONCEPT PRACTICE

32. Identify the reactants and products in each chemical reaction. *8.1*
a. Hydrogen gas and sodium hydroxide are formed when sodium is dropped into water.
b. In photosynthesis, carbon dioxide and water react to form oxygen gas and glucose.

33. How did John Dalton explain a chemical reaction using his atomic theory? *8.1*

34. What is the function of an arrow (⟶) and a plus sign (+) in a chemical equation? *8.1*

35. Write sentences that completely describe each of the chemical reactions shown in these skeleton equations. *8.1*
a. $NH_3(g) + O_2(g) \xrightarrow{Pt} NO(g) + H_2O(g)$
b. $H_2SO_4(aq) + BaCl_2(aq) \longrightarrow BaSO_4(s) + HCl(aq)$
c. $N_2O_3(g) + H_2O(l) \longrightarrow HNO_2(aq)$

36. What is the purpose of a catalyst? *8.1*

37. Balance equations for each item. The formula for each product (object) is given. *8.1*
a. a basketball team
center + forward + guard ⟶ team
$C + F + G \longrightarrow CF_2G_2$
b. a tricycle
frame + wheel + seat + pedal ⟶ tricycle
$F + W + S + P \longrightarrow FW_3SP_2$

38. The equation for the formation of water from its elements, $H_2(g) + O_2(g) \longrightarrow H_2O(l)$, can easily be "balanced" by changing the formula of the product to H_2O_2. Explain why this is incorrect. *8.1*

39. Balance the following equations. *8.1*
a. $PbO_2 \longrightarrow PbO + O_2$
b. $Fe(OH)_3 \longrightarrow Fe_2O_3 + H_2O$
c. $(NH_4)_2CO_3 \longrightarrow NH_3 + H_2O + CO_2$
d. $NaCl + H_2SO_4 \longrightarrow Na_2SO_4 + HCl$
e. $H_2 + Fe_3O_4 \longrightarrow Fe + H_2O$
f. $Al + CuSO_4 \longrightarrow Al_2(SO_4)_3 + Cu$

40. Explain why it is useful to classify reactions by their type. *8.2*

41. How do you predict the correct formula for the combination reaction between a nonmetal and a Group A metal? *8.2*

42. What is a characteristic of every combination reaction? *8.2*

43. Write balanced chemical equations for the following combination reactions. *8.2*
a. $Mg + O_2 \longrightarrow$
b. $P + O_2 \longrightarrow$ diphosphorus pentoxide
c. $Ca + S \longrightarrow$
d. $Fe + O_2 \longrightarrow$ iron(II) oxide
e. $N_2O_5 + H_2O \longrightarrow$

44. Write a balanced chemical equation for each decomposition reaction. *8.2*
a. $Ag_2O(s) \xrightarrow{\Delta}$?
b. nickel(II) carbonate $\xrightarrow{\Delta}$ nickel(II) oxide + ?
c. ammonium nitrate $\xrightarrow{\Delta}$ dinitrogen monoxide + water

45. What is a distinguishing feature of every decomposition reaction? *8.2*

46. For each of the following pairs, predict which element as an atom would displace the other element as an ion from a compound in aqueous solution. *8.2*
a. iron and sodium
b. silver and copper
c. zinc and hydrogen (in HCl)

47. Use the activity series of metals to write a balanced chemical equation for each single-replacement reaction. *8.2*
a. $Au(s) + KNO_3(aq) \longrightarrow$
b. $Zn(s) + AgNO_3(aq) \longrightarrow$
c. $Al(s) + H_2SO_4(aq) \longrightarrow$
d. $Cu(s) + H_2O(l) \longrightarrow$
e. $Al(s) + CuSO_4(aq) \longrightarrow$

48. Write a balanced equation for each of the following reactions. *8.2*
a. $HCl(aq) + Ca(OH)_2(aq) \longrightarrow$
b. $Ag_2SO_4(aq) + AlCl_3(aq) \longrightarrow$
(Silver chloride is a precipitate.)
c. $H_2C_2O_4(aq) + KOH(aq) \longrightarrow$
d. $CdBr_2(aq) + Na_2S(aq) \longrightarrow$
(Cadmium sulfide is a precipitate.)

49. What substance is common to all combustion reactions? *8.2*

50. Write a balanced equation for the complete combustion of each compound. *8.2*
a. butene (C_4H_8)
b. octane (C_8H_{18})
c. glycerol ($C_3H_8O_3$)
d. acetone (C_3H_6O)

51. Balance each equation and identify its type. *8.2*
a. $Hf + N_2 \longrightarrow Hf_3N_4$
b. $Mg + H_2SO_4 \longrightarrow MgSO_4 + H_2$
c. $C_2H_6 + O_2 \longrightarrow CO_2 + H_2O$
d. $Pb(NO_3)_2 + NaI \longrightarrow PbI_2 + NaNO_3$
e. $Fe + O_2 \longrightarrow Fe_3O_4$
f. $Pb(NO_3)_2 \longrightarrow PbO + NO_2 + O_2$
g. $Hg(NO_3)_2 + NH_4SCN \longrightarrow Hg(SCN)_2 + NH_4NO_3$
h. $(NH_4)_2SO_4 + NaOH \longrightarrow NH_3 + H_2O + Na_2SO_4$
(*Hint:* There are two stepwise reactions in this equation.)

52. Complete each equation and then write a net ionic equation. *8.3*
a. $Al(s) + H_2SO_4(aq) \longrightarrow$
b. $HCl(aq) + Ba(OH)_2(aq) \longrightarrow$
c. $Au(s) + HCl(aq) \longrightarrow$

53. What is a spectator ion? *8.3*

CONCEPT MASTERY

54. Each of these equations is "balanced" but incorrect. Find the errors and correctly balance each equation.
a. $Cl_2 + NaI \longrightarrow NaCl_2 + I$
b. $NH_3 \longrightarrow N + H_3$
c. $Na + O_2 \longrightarrow NaO_2$
d. $2Mg + H_2SO_4 \longrightarrow Mg_2SO_4 + H_2$
e. $MgCl + CaOH \longrightarrow MgOH + CaCl$
f. $H_2 + Cl_2 \longrightarrow H_2Cl_2$

55. Write a balanced chemical equation for each reaction. Use the necessary symbols from **Table 8.1** to describe the reaction completely.
a. Bubbling chlorine gas through a solution of potassium iodide gives elemental iodine and a solution of potassium chloride.
b. Bubbles of hydrogen gas and aqueous iron(III) chloride are produced when metallic iron is dropped into hydrochloric acid.
c. Solid tetraphosphorus decoxide reacts with water to produce phosphoric acid.
d. Solid silver oxide can be heated to give silver and oxygen gas.
e. Iodine crystals react with chlorine gas to form solid iodine trichloride.
f. Mercury metal is produced by heating a mixture of mercury(II) sulfide and calcium oxide. Additional products are calcium sulfide and calcium sulfate.

56. Write balanced chemical equations for these double-replacement reactions that occur in aqueous solution.
a. Zinc sulfide is added to sulfuric acid.
b. Sodium hydroxide reacts with nitric acid.
c. Solutions of potassium fluoride and calcium nitrate are mixed.

57. Write a balanced chemical equation for each combination reaction.
a. sodium oxide + water $\longrightarrow$
b. hydrogen + bromine $\longrightarrow$
c. dichlorine heptoxide + water $\longrightarrow$

58. Write a balanced chemical equation for each single-replacement reaction that takes place in aqueous solution. Write "no reaction" if a reaction does not occur.
a. A piece of steel wool (iron) is placed in sulfuric acid.
b. Mercury is poured into an aqueous solution of zinc nitrate.
c. Bromine reacts with aqueous barium iodide.

59. Pieces of sodium and magnesium are dropped into separate water-filled test tubes (A and B). Tube A bubbles vigorously; Tube B does not.
a. Which tube contains the sodium metal?
b. Write an equation for the reaction in the tube containing the sodium metal. What type of reaction is occurring in this tube?

60. Write a balanced equation for the complete combustion of each compound. Assume that the products are carbon dioxide and water.
a. octane (C_8H_{18})
b. glucose ($C_6H_{12}O_6$)
c. acetic acid ($HC_2H_3O_2$)

61. Write balanced chemical equations for these decomposition reactions.
a. Aluminum can be obtained from aluminum oxide with the addition of a large amount of electrical energy.
b. Heating tin(IV) hydroxide gives tin(IV) oxide and water.
c. Silver carbonate decomposes into silver oxide and carbon dioxide when it is heated.

62. Write a balanced net ionic equation for each reaction. The product that is not in solution is given.
a. $H_2C_2O_4 + KOH \longrightarrow$ $[H_2O]$
b. $CdBr_2 + Na_2S \longrightarrow$ $[CdS]$
c. $NaOH + Fe(NO_3)_3 \longrightarrow$ $[Fe(OH)_3]$

50. a. $C_4H_8 + 6O_2 \rightarrow 4CO_2 + 4H_2O$
b. $2C_8H_{18} + 25O_2 \rightarrow 16CO_2 + 18H_2O$
c. $2C_3H_8O_3 + 7O_2 \rightarrow 6CO_2 + 8H_2O$
d. $C_3H_6O + 4O_2 \rightarrow 3CO_2 + 3H_2O$

51. a. $3Hf + 2N_2 \rightarrow Hf_3N_4$; combination
b. $Mg + H_2SO_4 \rightarrow MgSO_4 + H_2$; single-replacement
c. $2C_2H_6 + 7O_2 \rightarrow 4CO_2 + 6H_2O$; combustion
d. $Pb(NO_3)_2 + 2NaI \rightarrow PbI_2 + 2NaNO_3$; double-replacement
e. $3Fe + 2O_2 \rightarrow Fe_3O_4$; combination
f. $2Pb(NO_3)_2 \rightarrow 2PbO + 4NO_2 + O_2$; decomposition
g. $Hg(NO_3)_2 + 2NH_4SCN \rightarrow Hg(SCN)_2 + 2NH_4NO_3$; double-replacement
h. $(NH_4)_2SO_4 + 2NaOH \rightarrow 2NH_3 + 2H_2O + Na_2SO_4$; double-replacement then decomposition

52. a. $2Al(s) + 6H^+(aq) \rightarrow 2Al^{3+}(aq) + 3H_2(g)$
b. $H^+(aq) + OH^-(aq) \rightarrow H_2O(l)$
c. no reaction

53. an ion that does not participate in the reaction

54. a. $Cl_2 + 2NaI \rightarrow 2NaCl + I_2$
b. $2NH_3 \rightarrow N_2 + 3H_2$
c. $4Na + O_2 \rightarrow 2Na_2O$
d. $Mg + H_2SO_4 \rightarrow MgSO_4 + H_2$
e. $MgCl_2 + Ca(OH)_2 \rightarrow Mg(OH)_2 + CaCl_2$
f. $H_2 + Cl_2 \rightarrow 2HCl$

55. a. $Cl_2(g) + 2KI(aq) \rightarrow I_2(s) + 2KCl(aq)$
b. $2Fe(s) + 6HCl(aq) \rightarrow 2FeCl_3(aq) + 3H_2(g)$
c. $P_4O_{10}(s) + 6H_2O(l) \rightarrow 4H_3PO_4(aq)$
d. $2Ag_2O(s) \xrightarrow{\Delta} 4Ag(s) + O_2(g)$
e. $I_2(s) + 3Cl_2(g) \rightarrow 2ICl_3(s)$
f. $4HgS(s) + 4CaO(s) \xrightarrow{\Delta} 4Hg(l) + 3CaS(s) + CaSO_4(s)$

56. a. $ZnS(aq) + H_2SO_4(aq) \rightarrow H_2S(g) + ZnSO_4(aq)$
b. $NaOH(aq) + HNO_3(aq) \rightarrow H_2O(l) + NaNO_3(aq)$
c. $2KF(aq) + Ca(NO_3)_2(aq) \rightarrow CaF_2(s) + 2KNO_3(aq)$

57. a. $Na_2O(s) + H_2O(l) \rightarrow 2NaOH(aq)$
b. $H_2(g) + Br_2(g) \rightarrow 2HBr(g)$
c. $Cl_2O_7(l) + H_2O(l) \rightarrow 2HClO_4(aq)$

58. a. $Fe(s) + H_2SO_4(aq) \rightarrow FeSO_4(aq) + H_2(g)$
b. no reaction
c. $Br_2(l) + BaI_2(aq) \rightarrow BaBr_2(aq) + I_2(s)$

59. a. A
b. $2Na(s) + 2H_2O(l) \rightarrow 2NaOH(aq) + H_2(g)$ single-replacement

60. a. $2C_8H_{18} + 25O_2 \rightarrow 16CO_2 + 18H_2O$
b. $C_6H_{12}O_6 + 6O_2 \rightarrow 6CO_2 + 6H_2O$
c. $HC_2H_3O_2 + 2O_2 \rightarrow 2CO_2 + 2H_2O$

61. a. $2Al_2O_3 \xrightarrow{\text{energy}} 4Al + 3O_2$
b. $Sn(OH)_4 \xrightarrow{\Delta} SnO_2 + 2H_2O$
c. $Ag_2CO_3 \xrightarrow{\Delta} Ag_2O + CO_2$

62. a. $H^+(aq) + OH^-(aq) \rightarrow H_2O(l)$
b. $Cd^{2+}(aq) + S^{2-}(aq) \rightarrow CdS(s)$
c. $3OH^-(aq) + Fe^{3+}(aq) \rightarrow Fe(OH)_3(s)$

Answers

63. **a.** 4
b. 2
c. 1

64. Smoking is not permitted near an oxygen source because a fire will burn faster in an area of high oxygen concentration. However, if a match were struck in a room full of oxygen and isolated from combustible material, it would only burn more vigorously.

65. **a.** $C_5H_{12} + 8O_2 \rightarrow 5CO_2 + 6H_2O$
$C_9H_{20} + 14O_2 \rightarrow 9CO_2 + 10H_2O$
b. $2C_{12}H_{26} + 37O_2 \rightarrow 24CO_2 + 26H_2O$
$C_{17}H_{36} + 26O_2 \rightarrow 17CO_2 + 18H_2O$
c. $n = CO_2$; $(n + 1) = H_2O$

66. **a.** 2.41 mol
b. 6.91×10^{-2} mol
c. 0.934 mol
d. 7.09 mol

67. 425 g Au

68. $C_8H_{10}O_2N_4$

69. **a.**

mol $CaCl_2$	mol H_2O
0.156	0.312
0.439	0.878
1.12	2.24
3.03	6.06

b.

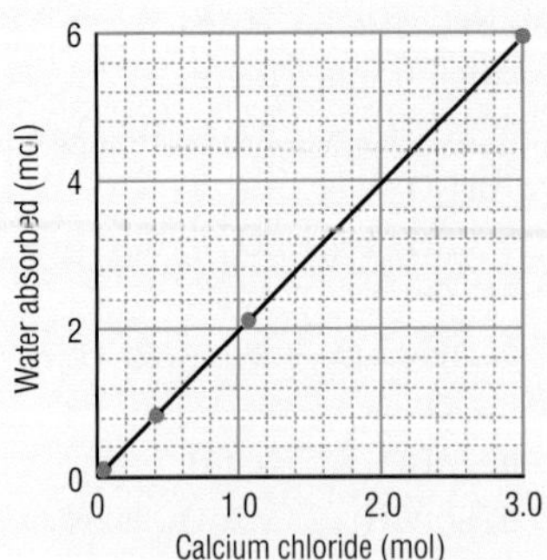

c. Two molecules of water are absorbed by each formula unit of $CaCl_2$.

70. **a.** $3NaI + H_3PO_4 \rightarrow 3HI + Na_3PO_4$; double-replacement
b. $K_2O + H_2O \rightarrow 2KOH$; combination
c. $2H_2SO_4 \xrightarrow{\Delta} 2H_2O + O_2 + 2SO_2$; decomposition
d. $2Al + 3H_2SO_4 \rightarrow 3H_2 + Al_2(SO_4)_3$; single-replacement
e. $C_5H_{12} + 8O_2 \rightarrow 5CO_2 + 6H_2O$; combustion

71. **a.** 1.01 g/mol protons, 5.48×10^{-4} g/mol electrons
b. 1.84×10^3 electrons/proton

CRITICAL THINKING

63. Choose the term that best completes the second relationship.

a. charcoal:ashes — reactant: _______
(1) equation (3) heat
(2) reaction (4) product

b. tool:hammer — catalyst: _______
(1) reaction (3) sulfur
(2) MnO_2 (4) combustion

c. east:west — combination: _______
(1) decomposition (3) coefficient
(2) reaction (4) replacement

64. Why is smoking not permitted near an oxygen source? What would happen if a match were struck in a room filled with oxygen?

65. Alkanes are hydrocarbon molecules that have the general formula C_nH_{2n+2}. The graph shows the number of oxygen, carbon dioxide, and water molecules needed to balance the equations for the complete combustion of every alkane having from one to ten carbon atoms.

$$C_nH_{2n+2} + \_\_\_ O_2 \longrightarrow \_\_\_ CO_2 + \_\_\_ H_2O$$

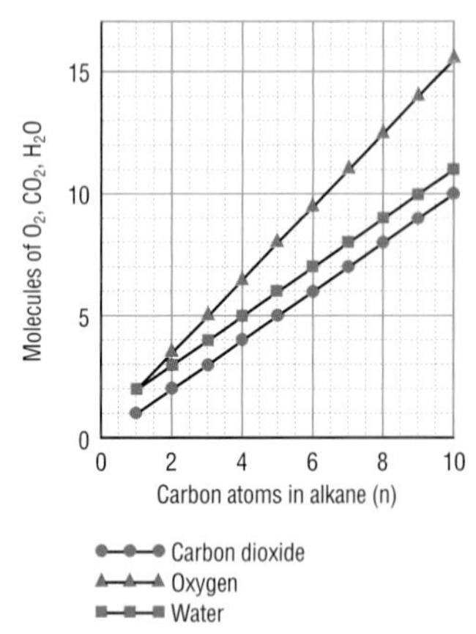

a. Use the graph to write balanced equations for the combustion of C_5H_{12} and C_9H_{20}.

b. Extrapolate the graph and write balanced equations for the combustion of $C_{12}H_{26}$ and $C_{17}H_{36}$.

c. The coefficient for O_2 in the general equation is $n + \sqrt{\frac{n+1}{2}}$. What are the coefficients for CO_2 and H_2O?

CUMULATIVE REVIEW

66. Calculate the number of moles in each substance.
a. 54.0 L of nitrogen dioxide (at STP)
b. 1.68 g of magnesium ions
c. 69.6 g of sodium hypochlorite
d. 4.27×10^{24} molecules of carbon monoxide

67. What mass of gold (density = 19.3 g/cm^3) occupies the same volume as 2.20 mol of aluminum (density = 2.70 g/cm^3)?

68. Many coffees and colas contain the stimulant caffeine. The percent composition of caffeine is 49.5% C, 5.20% H, 16.5% O, and 28.9% N. What is the molecular formula of caffeine if its molar mass is 194.1 g/mol?

69. The white solid calcium chloride ($CaCl_2$) is used as a drying agent. The maximum amount of water absorbed by different quantities of $CaCl_2$ is given in the table below.

$CaCl_2$ (g)	$CaCl_2$ (mol)	H_2O (g)	H_2O (mol)
17.3		5.62	
48.8		15.8	
124		40.3	
337		109	

a. Complete the table.
b. Plot the moles of water absorbed (y-axis) versus the moles of $CaCl_2$.
c. Based on your graph, how many molecules of water does each formula unit of $CaCl_2$ absorb?

CONCEPT CHALLENGE

70. Write a balanced chemical equation for each reaction. Classify each as to type.
a. Sodium iodide reacts with phosphoric acid.
b. Potassium oxide reacts with water.
c. Heating sulfuric acid produces water, oxygen, and sulfur dioxide.
d. Aluminum reacts with sulfuric acid.
e. Pentane (C_5H_{12}) reacts with oxygen.

71. The mass of a proton and an electron are 1.67×10^{-24} g and 9.11×10^{-28} g, respectively.
a. What is the mass of a mole of protons and a mole of electrons?
b. How many electrons are equal in mass to one proton?

Chapter 8 STANDARDIZED TEST PREP

Select the choice that best answers each question or completes each statement.

1. When the equation $Fe_2O_3 + H_2 \longrightarrow Fe + H_2O$ is balanced using whole-number coefficients, H_2 has a coefficient of
   a. 6. b. 3. c. 2. d. 1.
2. Identify the spectator ion is this reaction.
   $Ba(OH)_2(aq) + H_2SO_4(aq) \longrightarrow BaSO_4(s) + H_2O(l)$
   a. Ba^{2+}
   b. SO_4^{2-}
   c. OH^-
   d. H^+
   e. There is no spectator ion.
3. Magnesium ribbon reacts with an aqueous solution of copper(II) chloride in a single replacement reaction. The products of the balanced net ionic equation for the reaction are
   a. $Mg^{2+}(aq) + 2Cl^-(aq) + Cu(s)$.
   b. $Mg^+(aq) + Cl^-(aq) + Cu^+(aq)$.
   c. $Mg^{2+}(aq) + Cu(s)$.
   d. $Cu(s) + 2Cl^-(aq)$.
4. The expected products of the decomposition reaction of potassium oxide, K_2O, are
   a. $K^+(s)$ and $O^{2-}(g)$.
   b. $K^+(s)$ and $O_2(g)$.
   c. $K(s)$ and $O_2^{2-}(g)$.
   d. $K(s)$ and $O_2(g)$.
5. In an equation, the symbol Δ above the yields arrow means that
   a. a change is about to occur.
   b. heat is added to the reactants.
   c. a catalyst is needed for the reaction to occur.
   d. all of the above are correct.

Use the following description and data table to answer questions 6–8.

Dropper bottles labeled P, Q, and R contain one of three aqueous solutions: potassium carbonate, K_2CO_3; hydrochloric acid, HCl; and calcium nitrate, $Ca(NO_3)_2$. The table shows what happens when pairs of solutions are mixed.

Solution	P	Q	R
P	—	precipitate	no reaction
Q	precipitate	—	gas forms
R	no reaction	gas forms	—

6. Identify the contents of each dropper bottle.
7. Write the net ionic equation for the formation of the precipitate.
8. Write the complete ionic equation for the formation of the gas.

Use the atomic windows to answer question 9. Oxygen atoms are red; nitrogen atoms are blue.

a.
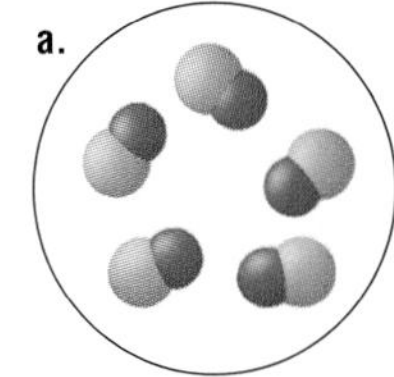

c.
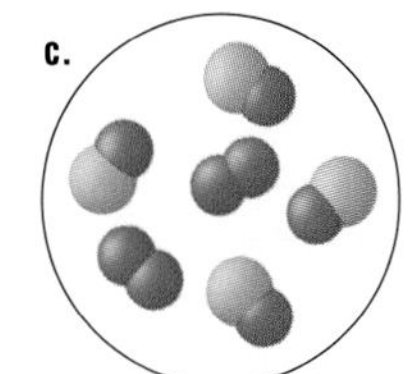

b.
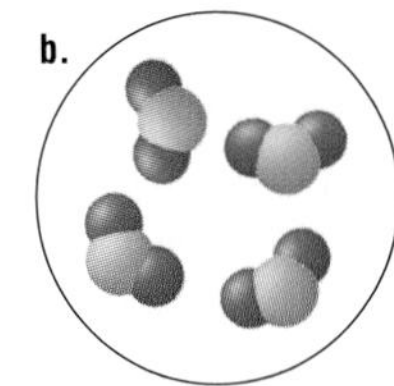

d.
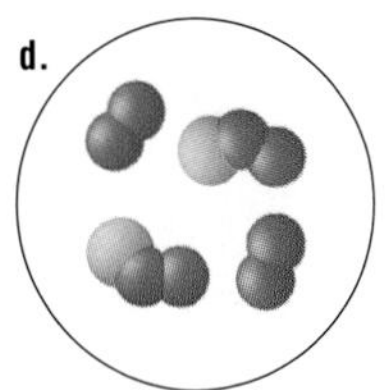

9. Which windows represents the reactants and the products for the following reaction?
   $2NO(g) + O_2(g) \longrightarrow 2NO_2(g)$

Use the diagram to answer questions 10–13.

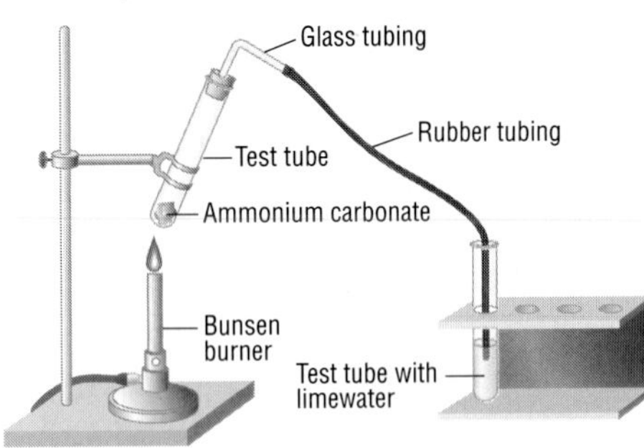

10. When ammonium carbonate is heated, water, ammonia, and carbon dioxide are produced. What type of chemical reaction is occurring?
11. Write formulas for the reaction products.
12. Write a balanced equation for the reaction. Include states for reactants and products.
13. What purpose might the limewater serve?

Chapter 8
STANDARDIZED TEST PREP

1. b
2. e
3. c
4. d
5. b
6. P is calcium nitrate, Q is potassium carbonate, and R is hydrochloric acid.
7. $Ca^{2+}(aq) + CO_3^{2-}(aq) \rightarrow CaCO_3(s)$
8. $2K^+(aq) + CO_3^{2-}(aq) + 2H^+(aq) + 2Cl^-(aq) \rightarrow 2K^+(aq) + 2Cl^-(aq) + H_2O(l) + CO_2(g)$
   Question 8 is a challenging question. You may need to tell students that a carbonate and an acid produce water and carbon dioxide.
9. (c) represents the reactants; (b) represents the products.
10. decomposition reaction
11. NH_3, CO_2, and H_2O
12. $(NH_4)_2CO_3(s) \rightarrow 2NH_3(g) + CO_2(g) + H_2O(g)$
13. to react with and identify the carbon dioxide

Planning Guide

SECTION OBJECTIVES	ACTIVITIES/FEATURES	MEDIA & TECHNOLOGY
9.1 The Arithmetic of Equations ● ■ ◇ ▸ Calculate the amounts of reactants required or product formed in a nonchemical process ▸ Interpret balanced chemical equations in terms of interacting moles, representative particles, masses, and gas volume at STP	**SE** Discover It! *How Much Can You Make?*, p. 236 LM 12: *Quantitative Analysis* **TE** DEMOS, pp. 240, 241	ASAP Simulation 6 ASAP Problem Solving 4 ASAP Assessment 9.1 ACT 4: *Stoichiometry* CA *Carbide Cannon* OT 16: *Formation of Ammonia*
9.2 Chemical Calculations ■ ◆ ▸ Construct mole ratios from balanced chemical equations and apply these ratios in mole-mole stoichiometric calculations ▸ Calculate stoichiometric quantities from balanced chemical equations using units of moles, mass, representative particles, and volumes of gases at STP	**SE** **CHEMath** *Dimensional Analysis*, p. 243 **SE** **Link to Agriculture** *Ammonia in the Nitrogen Cycle*, p. 246 **SE** Small-Scale Lab *Analysis of Baking Soda*, p. 251 (LRS 9-1) PLM *Analysis of Baking Soda* LM 12: *Quantitative Analysis* SSLM 10: *Titration: Determining How Much Acid is in a Solution* SSLM 11: *Weight Titrations: Measuring Molar Concentrations* **TE** DEMO, p. 245	ASAP Problem Solving 11, 16 ASAP Assessment 9.2 ACT 4: *Stoichiometry* CHM Side 4, 13: *Space Shuttle Air Systems* OT 17: *Stoichiometric Calculations*
9.3 Limiting Reagent and Percent Yield □ ◆ ▸ Identify and use the limiting reagent in a reaction to calculate the maximum amount of product(s) produced and the amount of excess reagent ▸ Calculate theoretical yield, actual yield, or percent yield given appropriate information	**SE** Mini Lab *Limiting Reagents*, p. 259 (LRS 9-2) **SE** **Chemistry Serving . . . Society** *Just the Right Volume of Gas*, p. 260 **SE** **Chemistry in Careers** *Quality Control Chemist*, p. 260 LM 13: *Balanced Chemical Equations* **TE** DEMO, p. 253	ASAP Animation 8 ASAP Problem Solving 24, 28 ASAP Assessment 9.3 ASAP Concept Map 9 CHM Side 4, 23: *Limiting Reactants Animation* CHM Side 4, 32: *Limit those Half-Baked Ideas* OT 18: *Limiting Reagents* PP Chapter 9 Problems RP Lesson Plans, Resource Library AR Computer Test 9 www Activities, Self-Tests, *SCIENCE NEWS* updates TCP The Chemistry Place Web Site

KEY

● Conceptual (concrete concepts)	AR Assessment Resources	GRS Guided Reading and Study Workbook
○ Conceptual (more abstract/math)	ASAP Chem ASAP! CD-ROM	LM Laboratory Manual
■ Standard (core content)	ACT ActivChemistry CD-ROM	LP Laboratory Practicals
□ Standard (extension topics)	CHM CHEMedia Videodiscs	LRS Laboratory Recordsheets
◆ Honors (core content)	CA Chemistry Alive! Videodiscs	SSLM Small-Scale Lab Manual
◇ Honors (options to accelerate)	GCP Graphing Calculator Problems	

PRACTICE / ASSESSMENT

Practice		Assessment	
SE	**Sample Problems** 9-1 to 9-2	SE	Section Review
GRS	Section 9.1	RM	Section Review 9.1
RM	Practice Problems 9.1	RM	Chapter 9 Quiz
SE	**Sample Problems** 9-3 to 9-7	SE	Section Review
GRS	Section 9.2	RM	Section Review 9.2
RM	Practice Problems 9.2	RM	Chapter 9 Quiz
SE	**Sample Problems** 9-8 to 9-10	SE	Section Review
GRS	Section 9.3	RM	Section Review 9.3
RM	Practice Problems 9.3	LP	Lab Practical 9-1, 9-2
RM	Interpreting Graphics	RM	Vocabulary Review 9
		SE	Chapter Review
		SM	Chapter 9 Solutions
		SE	Standardized Test Prep
		PHAS	Chapter 9 Test Prep
		RM	Chapter 9 Quiz
		RM	Chapter 9 A & B Test

PLANNING FOR ACTIVITIES

STUDENT EDITION

Discover It! p. 236
- metal paper clips
- vinyl-coated paper clips
- plastic sandwich bags

Small-Scale Lab, p. 251
- baking soda
- baking powder
- soda straws
- plastic cups
- pipets of HCl, NaOH, and thymol blue
- balances

Mini Lab, p. 259
- graduated cylinders
- balances
- 250-mL Erlenmeyer flasks
- rubber balloons
- magnesium ribbon
- 1.0*M* hydrochloric acid

TEACHER'S EDITION

Teacher Demo, p. 240
- 2 tsp lemon juice
- sweetener
- one small bottle of carbonated water
- small cups
- clean, non-lab container
- tray

Teacher Demo, p. 241
- strip of Mg approx. 2.5 cm to 3.5 cm long
- 50 mL of 1*M* HCl(*aq*)
- 100-mL beaker

Teacher Demo, p. 245
- 0.1*M* KI(*aq*) and $Pb(NO_3)_2(aq)$
- 50.0 mL of $Pb(NO_3)_2$
- 150 mL of KI
- two 250-mL beakers

Teacher Demo, p. 253
- 15 plastic bottles
- 30 plastic caps to fit bottles
- 6 containers to hold 5 caps each

Activity, p. 257
- 3 Styrofoam™ cups
- thermometer
- 100 mL of 1.0*M* HCl
- 200 mL of 1.0*M* NaOH
- safety goggles

OT	Overhead Transparency	**SE**	Student Edition
PHAS	PH Assessment System	**SM**	Solutions Manual
PLM	Probeware Lab Manual	**SSV**	Small-Scale Video/Videodisc
PP	Problem Pro CD-ROM	**TCP**	www.chemplace.com
RM	Review Module	**TE**	Teacher's Edition
RP	Resource Pro CD-ROM	**www**	www.phschool.com

Chapter 9

Key Terms

9.1 stoichiometry
9.3 limiting reagent, excess reagent, theoretical yield, actual yield, percent yield

DISCOVER IT!

Answers will vary, but students should recognize that the number of product molecules formed depends on the ratio of M_2 to C_2 molecules drawn from the bag. After students have done the experiment twice, ask them what combination of M_2 and C_2 molecules would produce the maximum number of product molecules. Given that one molecule of M_2 reacts with three molecules of C_2, and given that there are only ten molecules of each reactant available, three M_2 molecules can react with nine C_2 molecules to form a total of six MC_3 molecules.

Chapter 9 STOICHIOMETRY

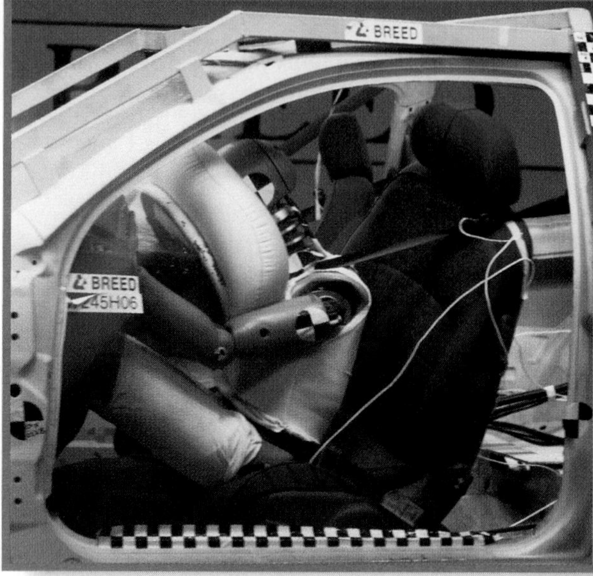

A chemical reaction produces just enough gas to inflate an air bag.

FEATURES

DISCOVER IT! HOW MUCH CAN YOU MAKE?

You need twenty metal paper clips (symbol M), twenty identically colored vinyl-coated paper clips (symbol C), and one plastic sandwich bag.

1. Join together pairs of paper clips of the same color to form models representing 10 diatomic molecules of each reactant. Place these molecules in the plastic bag.
2. Without looking, choose 15 molecules from the plastic bag.
3. Line up the M_2 and C_2 molecules in two adjacent vertical rows.
4. Pair up reactant molecules in the 1 : 3 M_2-to-C_2 ratio as shown in the equation $M_2 + 3C_2 \longrightarrow 2MC_3$.
5. Make the molecules "react" by taking them apart and forming two molecules of the product.
6. Continue making M_2 and C_2 react in a 1:3 ratio until you run out of one of the reactants.

List the number of each type of reactant molecule that was drawn from the bag. How many molecules of the product could you form? Which reactant molecule did you run out of first? How many molecules of each reactant remained at the completion of the reaction? Repeat the experiment and compare the results. Use what you learn in this chapter to provide an explanation for your observations.

9.1 ● ■ ◇
9.2 ■ ◆
9.3 □ ◆

Conceptual Figure 9.2 and Sample Problem 9-1 can be used to give students a sense of what stoichiometry can accomplish. Figure 9.4 will help students to visualize different interpretations of equations, a skill that will help them throughout the course. Sections 9.2 and 9.3 are mathematically challenging and may be omitted.

Standard It may help to review the mole road map (Figure 7.13) before starting Section 9.2. The CHEMath feature will help students who have struggled with conversion factors. Figures 9.9 and 9.10 could be used to give students an overview of stoichiometry before they attempt the sample and practice problems. Use Figure 9.14 to help students visualize limiting reactants. Percent yield can be omitted if not included in course requirements.

Honors Assign Section 9.1 as homework and use the sample and practice problems to ensure that students have mastered this material.

THE ARITHMETIC OF EQUATIONS

section 9.1

objectives

- Calculate the amount of reactants required or product formed in a nonchemical process
- Interpret balanced chemical equations in terms of interacting moles, representative particles, masses, and gas volume at STP

key term

- stoichiometry

Silk, one of the most beautiful and luxurious of all fabrics, is spun from the cocoons of tiny silkworms. Silkworms have the unique ability to transform the leaves of mulberry trees into silk thread, which they use to weave their cocoons. More than 3000 cocoons are needed to produce enough silk to make just one elegant Japanese kimono. **How do chemists calculate the amount of reactants and products in chemical reactions?**

Using Everyday Equations

Nearly everything you use is manufactured from chemicals—soaps, shampoos and conditioners, cassette tapes, cosmetics, medicines, and clothes. Obviously, for the manufacturer to make a profit, the cost of making any of these items cannot exceed the money paid for them. Therefore, the chemical processes used in manufacturing must be carried out economically. This is where balanced chemical equations help.

Equations are the recipes that tell chemists what amounts of reactants to mix and what amounts of products to expect. You can determine the quantities of reactants and products in a reaction from the balanced equation. When you know the quantity of one substance in a reaction, you can calculate the quantity of any other substance consumed or created in the reaction. Quantity usually means the amount of a substance expressed in grams or moles. But quantity could just as well be in liters, tons, or molecules. Can you name some other units you might use to measure the amount of matter? ❶

The calculation of quantities in chemical reactions is a subject of chemistry called **stoichiometry.** Calculations using balanced equations are called stoichiometric calculations. For chemists, stoichiometry is a form of bookkeeping.

When you bake cookies, you probably use a recipe. A cookie recipe tells you the amounts of ingredients to mix together to make a certain number of cookies. If you need a larger number of cookies than the recipe provides, the amounts of ingredients can be doubled or tripled. In a way, a cookie recipe provides the same kind of information that a balanced chemical equation does. The ingredients are the reactants; the cookies are the products.

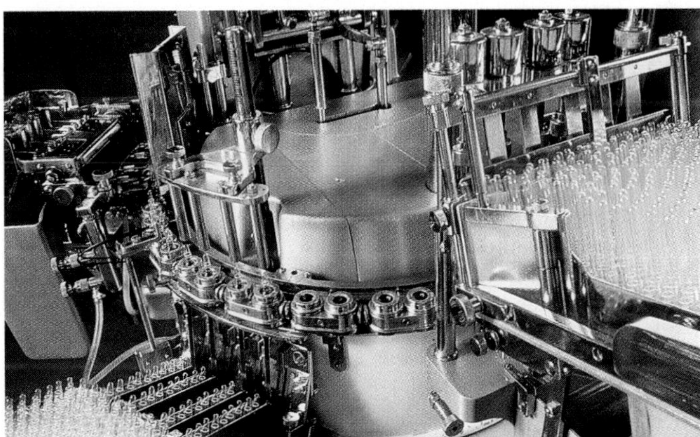

Figure 9.1
Just like cooking, manufacturing requires specific amounts of ingredients to get a certain number of products.

9.1

1 Engage

Use the Visual

Have students study the photograph and read the text that opens the section. Ask:

- **How many cocoons would be required to produce enough silk for two Japanese kimonos?** (twice as many, or 6000 cocoons)
- **How did you calculate the number of cocoons?** (by multiplying the number needed for one kimono by two)

Remind students that during a chemical reaction, atoms are rearranged into new combinations and groupings. It is somewhat similar to changing silk cocoons into a kimono. Just as a chemical reaction has reactants and products, the cocoon is the reactant and the kimono is the product. Much like having the proper supply of silk cocoons, chemists must have an adequate supply of reactants for a chemical reaction. Ask:

- **How do chemists calculate the amount of reactants and products in chemical reactions?** (Chemists use a quantity known as the mole to calculate amounts of reactants and products. Molar ratios can then be used to describe quantities in terms of mass or volume.)

Student Resources

From the Teacher's Resource Package, use:

- Section Review 9.1, Ch. 9 Practice Problems and Quizzes from the Review Module (Ch. 9–12)
- Laboratory Manual: Experiment 12

Technology Resources

Relevant technology resources include:

- Chem ASAP! CD-ROM
- Resource Pro CD-ROM
- Chemistry Alive! Videodisc: *Carbide Cannon*
- ActivChemistry CD-ROM: *Stoichiometry*

❶ Other units are grams, kilograms, moles, ounces, pounds.

Check Prior Knowledge

To assess students' prior knowledge about writing and balancing chemical reactions, write several unbalanced equations on the board and ask students to balance them. Some examples are:
$3CuO(s) + 2NH_3(aq) \rightarrow 3Cu(s) + 3H_2O(l) + N_2(g)$;
$4NH_3(g) + 5O_2(g) \rightarrow 4NO(g) + 6H_2O(g)$;
$2KClO_3(s) \rightarrow 2KCl(s) + 3O_2(g)$;

- **Why is it not correct to balance an equation by changing the subscripts in one or more of the formulas?** (Changing the subscripts in a formula changes the chemical identity of the substance.)
- **What does STP stand for?** (Standard temperature and pressure –0 °C and 101.3 kPa; because the volume of a gas varies with temperature and pressure, the volumes of gases are usually measured at STP.)
- **What is the molar volume of any gas at STP? How many particles does it contain?** (22.4 L/mol; 22.4 L of any ideal gas at STP contains 6.02×10^{23} particles of that gas.)

Practice Problems Plus

Related Chapter Review Problem
Chapter Review problem 60 is related to Sample Problem 9-1.

Here is another example, this time from the business world rather than from the world of cooking. Imagine you are in charge of manufacturing for the Tiny Tyke Tricycle Company. The business plan for Tiny Tyke requires the production of 128 custom-made tricycles each day. One of your responsibilities is to be sure that there are enough parts available at the start of each day to make these tricycles. To simplify this discussion, assume that the major components of the tricycle are the frame (F), the seat (S), the wheels (W), the handlebars (H), and the pedals (P). The finished tricycle has a "formula" of FSW_3HP_2. The balanced equation for the production of a tricycle is

$$F + S + 3W + H + 2P \longrightarrow FSW_3HP_2$$

This equation gives you the "recipe" to make a single tricycle: Making a tricycle requires one frame, one seat, three wheels, one handlebar, and two pedals.

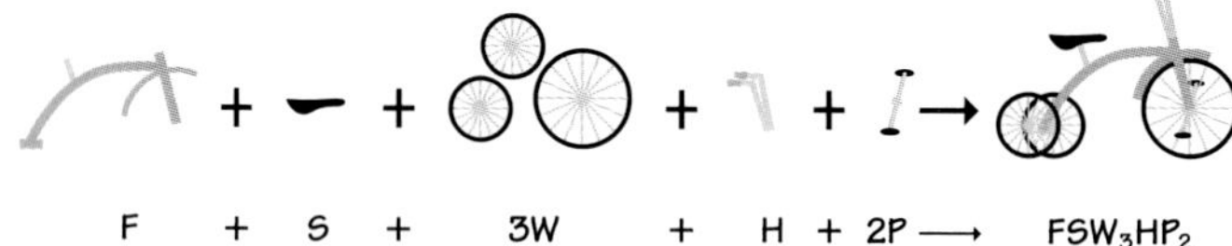

Figure 9.2
A balanced equation can be thought of as a recipe. In the equation shown here, the tricycle parts are the reactants and the assembled tricycle is the product. How many pedals are needed to make four tricycles? ❶

Sample Problem 9-1

In a five-day workweek, Tiny Tyke is scheduled to make 640 tricycles. How many wheels should be in the plant on Monday morning to make these tricycles?

1. *ANALYZE List the knowns and the unknown.*

Knowns:
- number of tricycles = 640 tricycles
- $1\ FSW_3HP_2 = 3\ W$ (from balanced equation)

Unknown:
- number of wheels = ? wheels

Use the conversion factor $\frac{3\ W}{1\ FSW_3HP_2}$ to calculate the unknown.

2. *CALCULATE Solve for the unknown.*

$$640\ \cancel{FSW_3HP_2} \times \frac{3\ W}{1\ \cancel{FSW_3HP_2}} = 1920\ W$$

3. *EVALUATE Does the result make sense?*

If 3 wheels are required for each tricycle, and a total of more than 600 tricycles are being made, then a number of wheels in excess of 1800 is a logical answer. The unit of the known cancels with the unit in the denominator of the conversion factor, and the answer is in the unit of the unknown. The conversion factor is exact and does not affect the rounding of the answer.

Practice Problems

1. How many tricycle seats, wheels, and pedals are needed to make 288 tricycles?
2. Write an equation that gives your own "recipe" for making a puppet or a piece of furniture.

Answers

❶ 8 pedals
❷ No, there are 4 moles of reactants, but only 2 moles of product.
❸ 10 molecules of NH_3

Interpreting Chemical Equations

As you may recall from Chapter 7, ammonia is widely used as a fertilizer. Ammonia is produced industrially by the reaction of nitrogen with hydrogen.

$$N_2(g) + 3H_2(g) \longrightarrow 2NH_3(g)$$

What kinds of information can be derived from this equation?

1. **Particles** One molecule of nitrogen reacts with three molecules of hydrogen to produce two molecules of ammonia. Nitrogen and hydrogen will always react to form ammonia in this 1:3:2 ratio of molecules. So if you could make 10 molecules of nitrogen react with 30 molecules of hydrogen, you would expect to get 20 molecules of ammonia. Of course, it is not possible to count such small numbers of molecules and allow them to react. You could, however, take Avogadro's number of nitrogen molecules and make them react with three times Avogadro's number of hydrogen molecules. This would be the same 1:3 ratio of molecules of reactants. The reaction would form two times Avogadro's number of ammonia molecules.
2. **Moles** You know that Avogadro's number of representative particles is one mole of a substance. On the basis of the particle interpretation you just read, the equation tells you the number of moles of reactants and products. One mole of nitrogen molecules reacts with three moles of hydrogen molecules to form two moles of ammonia molecules. The coefficients of a balanced chemical equation indicate the relative numbers of moles of reactants and products in a chemical reaction. This is the most important information that a balanced chemical equation provides. Using this information, you can calculate the amounts of reactants and products. Does the number of moles of reactants equal the number of moles of product in this reaction? (2)

Figure 9.3
Gardeners use ammonium salts as fertilizer. The nitrogen in these salts is essential to plant growth.

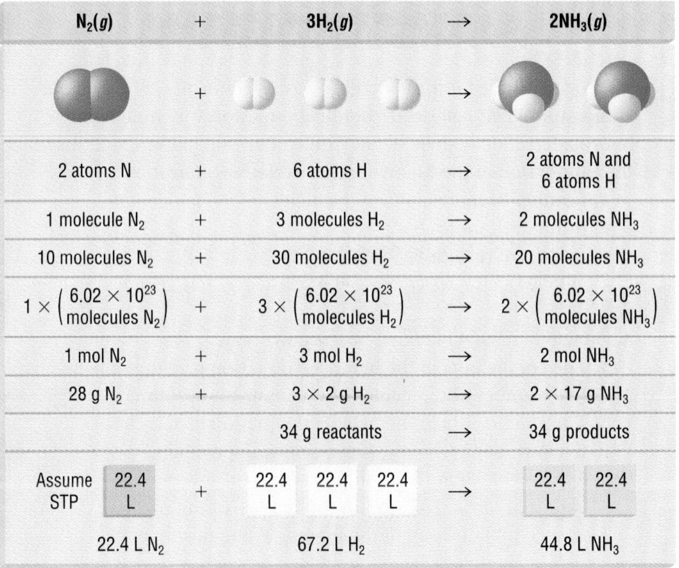

$N_2(g)$	+	$3H_2(g)$	→	$2NH_3(g)$
	+		→	
2 atoms N	+	6 atoms H	→	2 atoms N and 6 atoms H
1 molecule N_2	+	3 molecules H_2	→	2 molecules NH_3
10 molecules N_2	+	30 molecules H_2	→	20 molecules NH_3
$1 \times \left(\begin{array}{c}6.02 \times 10^{23}\\ \text{molecules } N_2\end{array}\right)$	+	$3 \times \left(\begin{array}{c}6.02 \times 10^{23}\\ \text{molecules } H_2\end{array}\right)$	→	$2 \times \left(\begin{array}{c}6.02 \times 10^{23}\\ \text{molecules } NH_3\end{array}\right)$
1 mol N_2	+	3 mol H_2	→	2 mol NH_3
28 g N_2	+	3×2 g H_2	→	2×17 g NH_3
		34 g reactants	→	34 g products
Assume STP 22.4 L	+	22.4 L 22.4 L 22.4 L	→	22.4 L 22.4 L
22.4 L N_2		67.2 L H_2		44.8 L NH_3

Figure 9.4
The balanced chemical equation for the formation of ammonia can be interpreted in several ways. How many molecules of NH_3 could be made from 5 molecules N_2 and 15 molecules H_2? (3)

2 Teach

Use the Visual

Have students study the diagram in Figure 9.2. Explain that the parts of the tricycle on the left-hand side of the arrow represent reactants, the arrow represents the assembly process (the reaction), and the tricycle represents product. Ask:

- **How many wheels are needed to make four tricycles?** (12)

Show students how to set up proportions to solve these and related problems.

Discuss

Write the following sentence and equation on the board: "Nitrogen gas and hydrogen gas react to form ammonia gas. $N_2(g) + H_2(g) \rightarrow NH_3(g)$ Initiate a discussion as to whether this equation correctly represents the reaction. Have students balance the equation correctly and explain why they had to balance it. Have students interpret the equation on each of the different possible levels.

Meeting Diverse Needs

At Risk Have students construct a table like the one shown in Figure 9.4 for the reaction of hydrogen gas with oxygen gas to form water. Students should begin by writing the balanced equation. Encourage students to draw pictures as shown in Figure 9.4 to represent reactants and products. Have students use the completed table to answer questions such as, "How many moles of water are produced by reacting 4 moles of hydrogen gas with excess oxygen?" Other reactions for students to work with include $CO(g) + 2H_2(g) \rightarrow CH_3OH(l)$ and $CH_4(g) + 2O_2(g) \rightarrow CO_2(g) + 2H_2O(g)$.

Chemistry Alive!

Carbide Cannon Play

TEACHER DEMO

In a clean plastic or glass container, combine 2 tsp of lemon juice, sweetener, and one small bottle of carbonated water. Taste can be adjusted as long as the amount of each component in the mixture is recorded. Place the container and several dozen small water cups on a tray and keep it out of sight. Tell students that you found a great recipe for lemonade fizz. Write the recipe that you have developed on the board. Then pour some fizz into cups, and pass them out to the students. After passing out the fizz, ask students how they each could have had a larger drink. Remind students how the relative amounts of the ingredients in a recipe can be adjusted proportionally to increase or decrease the number of servings without altering the taste of the final product. Relate the information in food recipes to the stoichiometry implied by a balanced chemical equation.

Practice Problems Plus

Related Chapter Review Problems
Chapter Review Problems 33, 34 and 35 are related to Sample Problem 9-2.

Additional Practice Problem
The Chem ASAP! CD-ROM contains the following problem: Balance the equation for the combustion of pentane: $C_5H_{12}(g) + O_2(g) \rightarrow CO_2(g) + H_2O(g)$ Interpret the balanced equation in terms of relative number of moles, volumes of gas at STP, and masses of reactants and products.
(1 mol $C_5H_{12}(g)$ + 8 mol $O_2(g) \rightarrow$ 5 mol $CO_2(g)$ + 6 mol $H_2O(g)$; 22.4 L $C_5H_{12}(g)$ + 179 L $O_2(g) \rightarrow$ 112 L $CO_2(g)$ + 134 L $H_2O(g)$; 328 g reactants $\rightarrow$ 328 g products)

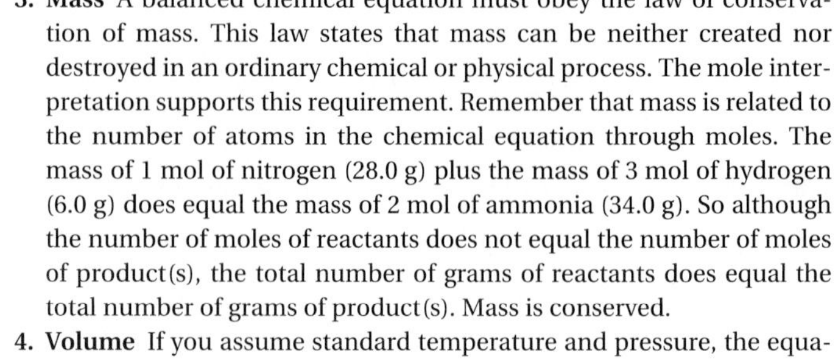

3. **Mass** A balanced chemical equation must obey the law of conservation of mass. This law states that mass can be neither created nor destroyed in an ordinary chemical or physical process. The mole interpretation supports this requirement. Remember that mass is related to the number of atoms in the chemical equation through moles. The mass of 1 mol of nitrogen (28.0 g) plus the mass of 3 mol of hydrogen (6.0 g) does equal the mass of 2 mol of ammonia (34.0 g). So although the number of moles of reactants does not equal the number of moles of product(s), the total number of grams of reactants does equal the total number of grams of product(s). Mass is conserved.
4. **Volume** If you assume standard temperature and pressure, the equation also tells you about the volumes of gases. Recall that 1 mol of any gas at STP occupies a volume of 22.4 L. It follows that 22.4 L of nitrogen reacts with 67.2 L (3×22.4 L) of hydrogen to form 44.8 L (2×22.4 L) of ammonia.

Look at **Figure 9.4** on the previous page. Do you see that mass and atoms are conserved in this chemical reaction? Mass and atoms are conserved in every chemical reaction. The mass of the reactants equals the mass of the products. The number of atoms of each reactant equals the number of atoms of that reactant in the product(s). Unlike mass and atoms, however, molecules, formula units, moles, and volumes of gases will not necessarily be conserved—although they may. Consider, for example, the formation of hydrogen iodide.

$$H_2(g) + I_2(g) \longrightarrow 2HI(g)$$

In this reaction, molecules, moles, and volume are all conserved. But in the majority of chemical reactions (including the reaction for the formation of ammonia), they are not. Only mass and atoms are conserved in every chemical reaction.

Figure 9.5
Hydrogen sulfide (H_2S) smells like rotten eggs. It escapes from the ground in volcanic areas.

Sample Problem 9-2

Hydrogen sulfide, a foul-smelling gas, is found in nature in volcanic areas. The balanced chemical equation for the burning of hydrogen sulfide is given below. Interpret this equation in terms of the interaction of the following three relative quantities.

a. number of representative particles
b. number of moles
c. masses of reactants and products

$$2H_2S(g) + 3O_2(g) \longrightarrow 2SO_2(g) + 2H_2O(g)$$

1. *ANALYZE* ***Plan a problem-solving strategy.***
   a. The coefficients in the balanced equation give the relative number of molecules of reactants and products.
   b. The coefficients in the balanced equation give the relative number of moles of reactants and products.
   c. A balanced chemical equation obeys the law of conservation of mass. The sum of the masses of the reactants must equal the sum of the masses of the products.

Practice Problems

3. Interpret the equation for the formation of water from its elements in terms of numbers of molecules and moles and volumes of gases at STP.
$$2H_2(g) + O_2(g) \longrightarrow 2H_2O(g)$$

Meeting Diverse Needs

At Risk Encourage students to find a method of problem solving that capitalizes on their strengths. For example, an artistically inclined student might do better drawing pictures of the reactants and products, instead of just writing the symbols. A concrete thinker might use a similar method for different reasons. Make molecular model kits available.

Sample Problem 9-2 (cont.)

2. ***SOLVE*** *Apply the problem-solving strategy.*
   a. 2 molecules H_2S react with 3 molecules O_2 to form 2 molecules SO_2 and 2 molecules H_2O.
   b. 2 mol H_2S react with 3 mol O_2 to produce 2 mol SO_2 and 2 mol H_2O.
   c. Multiply the number of moles of each reactant and product by its molar mass: 2 mol H_2S + 3 mol $O_2 \longrightarrow$ 2 mol SO_2 + 2 mol H_2O.

$$\left(2\ \cancel{\text{mol}} \times 34.1\ \frac{\text{g}}{\cancel{\text{mol}}}\right) + \left(3\ \cancel{\text{mol}} \times 32.0\ \frac{\text{g}}{\cancel{\text{mol}}}\right) \longrightarrow$$

$$\left(2\ \cancel{\text{mol}} \times 64.1\ \frac{\text{g}}{\cancel{\text{mol}}}\right) + \left(2\ \cancel{\text{mol}} \times 18.0\ \frac{\text{g}}{\cancel{\text{mol}}}\right)$$

$$68.2\text{ g } H_2S + 96.0\text{ g } O_2 \longrightarrow 128.2\text{ g } SO_2 + 36.0\text{ g } H_2O$$

$$164.2\text{ g} = 164.2\text{ g}$$

3. ***EVALUATE*** *Do the results make sense?*
   Because all the substances in this reaction are molecular, the mole ratio of reactants and products equals the molecular ratio of reactants and products (2:3:2:2). The sum of the masses of the reactants equals the sum of the masses of the products.

Practice Problems (cont.)

4. Balance the equation for the combustion of acetylene:

$$C_2H_2(g) + O_2(g) \longrightarrow CO_2(g) + H_2O(g)$$

Interpret the equation in terms of relative numbers of moles, volumes of gas at STP, and masses of reactants and products.

Chem ASAP!

Problem-Solving 4

Solve Problem 4 with the help of an interactive guided tutorial.

section review 9.1

5. Your school club has "adopted" a local nursing home and provides welcoming packages to new residents. Each welcoming package contains a toothbrush (B), three washcloths (W), a hand mirror (M), two decks of cards (C), and four small bottles of skin lotion (L).
   a. Write a balanced equation for preparing a welcoming package ($BW_3MC_2L_4$).
   b. Calculate the number of each item needed for 45 packages.
6. Balance this equation: $C_2H_5OH(l) + O_2(g) \longrightarrow CO_2(g) + H_2O(g)$.
   a. Interpret the equation in terms of numbers of molecules and moles.
   b. Show that the balanced equation obeys the law of conservation of mass.
7. Explain this statement: "Mass and atoms are conserved in every chemical reaction, but moles will not necessarily be conserved."
8. Interpret the following equation in terms of relative numbers of representative particles, numbers of moles, and masses of reactants and products.

$$2K(s) + 2H_2O(l) \longrightarrow 2KOH(aq) + H_2(g)$$

Chem ASAP! **Assessment 9.1** Check your understanding of the important ideas and concepts in Section 9.1.

9.1

TEACHER DEMO

Prior to class, determine the mass of a strip of Mg (approximately 2.5 cm to 3.5 cm long). Place 50 mL of 1*M* HCl(*aq*) in a 100-mL beaker. Place the Mg strip in the 1*M* HCl(*aq*). Have students write a balanced chemical equation for the observed reaction. Ask them to interpret the equation in terms of particles, moles, and mass.

3 Assess

Evaluate Understanding

To evaluate students' understanding of how to interpret equations, write several equations on the board. Examples include: $CO(g) + H_2(g) \rightarrow CH_3OH(l)$; $H_2(g) + I_2(g) \rightarrow 2HI(g)$; $2\text{ g } H_2O(l) \rightarrow 2\text{ g } H_2(g) + 1\text{ g } O_2(g)$. Have students explain whether each equation is correct or incorrect. (Only the second equation is correct as written.) **Ask them to describe the quantitative relationships between reactants and products in terms of mass, moles, particles, and volumes. Ask why it is important to be able to interpret equations in several different ways.**

Reteach

Remind students that the coefficients in a balanced chemical equation state the relationships among substances involved in the reaction. The molar relationships among reactants and products are especially useful. Emphasize that the amount of a product formed during a chemical reaction depends, in large part, on the relative amounts of reactants.

Answers

SECTION REVIEW 9.1

5. a. $B + 3W + M + 2C + 4L \rightarrow BW_3MC_2L_4$
   b. 45 toothbrushes, 135 wash cloths, 45 mirrors, 90 decks of cards, and 180 bottles of skin lotion
6. $C_2H_5OH(l) + 3O_2(g) \rightarrow 2CO_2(g) + 3H_2O(l)$
   a. 1 molecule C_2H_5OH + 3 molecules $O_2 \rightarrow$ 2 molecules CO_2 + 3 molecules H_2O; 1 mol C_2H_5OH + 3 mol $O_2 \rightarrow$ 2 mol CO_2 + 3 mol H_2O
   b. 46.0 g C_2H_5OH + 96.0 g $O_2 \rightarrow$ 88.0 g CO_2 + 54.0 g H_2O
   142.0 g reactants = 142.0 g products
7. The number of moles of reactants and products depends on the chemical reaction. For some reactions, moles of reactants and products are equal, but this is not generally the case.
8. 2 atoms K + 2 molecules $H_2O \rightarrow$ 2 formula units KOH + 1 molecule H_2;
   2 mol K + 2 mol $H_2O \rightarrow$ 2 mol KOH + 1 mol H_2;
   114.2 g reactants $\rightarrow$ 114.2 g products

1 Engage

Use the Visual

Have students study the photograph and read the text that opens the section. Write the equation for the decomposition of sodium azide on the board with heat as one of the products ($2NaN_3(s) \rightarrow 2Na(s) + 3N_2(g) +$ heat). **Ask:**

- **How can stoichiometry be used to calculate the volume of a gas produced in this reaction?** (The number of moles of nitrogen gas formed by this reaction depends on the number of moles of sodium azide that decompose.)

Point out that the heat produced by a reaction can also be measured and related to the amount of reactant(s) consumed. In this case, the amount of heat produced depends on the mass of reactant (sodium azide) that decomposes. It is possible to convert grams of reactant to moles of reactant to heat produced.

Check Prior Knowledge

To assess students' prior knowledge about chemical quantities, ask:

- **What is the gram molecular mass of hydrogen peroxide (H_2O_2)?** (34.0 g)
- **What is the gram formula mass of ammonium carbonate, $(NH_4)_2CO_3$?** (96.0 g)
- **What is the difference between the gram molecular mass (gmm) and the gram formula mass (gfm)?** (The gmm is used to express the mass of one mole of a molecular compound. The gfm is used to express the mass of one ionic compound.)

section 9.2

CHEMICAL CALCULATIONS

objectives

- Construct mole ratios from balanced chemical equations and apply these ratios in mole-mole stoichiometric calculations
- Calculate stoichiometric quantities from balanced chemical equations using units of moles, mass, representative particles, and volumes of gases at STP

Chem ASAP!

Simulation 6

Strengthen your analytical skills by solving stoichiometric problems.

Air bags inflate almost instantaneously upon impact. The effectiveness of air bags is based on the rapid conversion of a small mass of sodium azide into a large volume of gas. **How can stoichiometry be used to calculate the volume of gas produced in this reaction?**

Mole-Mole Calculations

As you just learned, a balanced chemical equation provides a wealth of quantitative information relating representative particles (atoms, molecules, formula units), moles of substances, and masses. Most important, a balanced chemical equation is essential for all calculations involving amounts of reactants and products: If you know the number of moles of one substance, the balanced chemical equation allows you to determine the number of moles of all other substances in the reaction.

Figure 9.6
Manufacturing plants produce ammonia by combining nitrogen with hydrogen. Ammonia is used in cleaning products, fertilizers, and in the manufacture of other chemicals.

Look again at the production of ammonia from nitrogen and hydrogen. The balanced equation for the reaction is

$$N_2(g) + 3H_2(g) \longrightarrow 2NH_3(g)$$

The most important interpretation of this equation is that 1 mol of nitrogen reacts with 3 mol of hydrogen to form 2 mol of ammonia. With this interpretation, you can relate moles of reactants to moles of product. The coefficients from the balanced equation are used to write conversion factors called mole ratios. The mole ratios are used to calculate the number of moles of product from a given number of moles of reactant or to calculate the number of moles of reactant from a given number of moles of product. Three of the mole ratios for this equation are

$$\frac{1 \text{ mol } N_2}{3 \text{ mol } H_2} \qquad \frac{2 \text{ mol } NH_3}{1 \text{ mol } N_2} \qquad \frac{3 \text{ mol } H_2}{2 \text{ mol } NH_3}$$

1 What are the other three mole ratios?

Student Resources

From the Teacher's Resource Package, use:

- Section Review 9.2, Ch. 9 Practice Problems and Quizzes from the Review Module
- Laboratory Recordsheet 9-1
- Laboratory Manual: Experiment 12
- Small-Scale Chemistry Lab Manual: Experiments 10 and 11

Technology Resources

Relevant technology resources include:

- Chem ASAP! CD-ROM
- Resource Pro CD-ROM
- ActivChemistry CD-ROM: *Stoichiometry*
- Chemistry Alive! Videodisc: *Carbide Cannon*

CHEMath

DIMENSIONAL ANALYSIS

Converting units is an essential skill for solving problems in chemistry and other sciences. In Section 4.2, you learned how to convert units using *dimensional analysis.* This page provides a chance for you to recall and brush up your skills prior to plunging into Chapter 9—a chapter requiring extensive use of dimensional analysis.

A *conversion factor* is a ratio of equivalent measurements. (Mole ratios, which are used frequently in Chapter 9, are a special type of conversion factor.) By definition, any conversion factor is equal to 1, so you can always multiply by a conversion factor without changing the value of an expression. By carefully choosing conversion factors and canceling units, you can convert any measurement into one expressed in the desired units.

For example, to convert 17 minutes into seconds, you would multiply by a conversion factor that relates the two units. From the relationship 1 minute = 60 seconds, the correct conversion factor can be written and applied.

$$17 \text{ min} \times \frac{60 \text{ s}}{1 \text{ min}} = 1020 \text{ s}$$

Dimensional analysis is also a powerful tool for preventing mistakes. The variables and constants in scientific formulas usually include units, and dimensional analysis can be used to ensure that the units on each side of an equation are the same. For example, consider the formula

$$\text{distance} = \text{velocity} \times \text{time}$$

$$\text{or } d = vt.$$

Notice the typical units in this equation:

$$\text{meters} = \frac{\text{meter}}{\text{second}} \times \text{seconds}$$

After canceling, both sides of the equation have the same units, meters. But if you were to write an incorrect equation using d, v, and t, the units would *not* balance and you would know, immediately, that your formula was incorrect!

Everyone makes mistakes. But if you pay attention to the units in all of your calculations, you will be able to identify and correct many of your mistakes before they cost you a few points and a lower grade.

Example 1

Ingrid drove 18 minutes at 40 km/h. Use $d = vt$ to find out how far she traveled.

Apply the formula: $d = vt = \frac{40 \text{ km}}{1 \text{ h}} \times 18 \text{ min}$

Notice that the time units do not cancel. (Minutes must be converted to hours.) Inserting the proper conversion factor and using dimensional analysis yields,

$$d = vt = \frac{40 \text{ km}}{1 \text{ h}} \times \frac{1 \text{ h}}{60 \text{ min}} \times 18 \text{ min} = 12 \text{ km}$$

She traveled a distance of 12 km.

Example 2

One mole of copper has a mass of 63.5 g. Find the number of moles in 6.00 kg of copper.

This problem requires two conversion factors. First, kilograms are converted to grams, and then grams are converted to moles.

$$6.00 \text{ kg Cu} \times \frac{1000 \text{ g Cu}}{1 \text{ kg Cu}} \times \frac{1 \text{ mol Cu}}{63.5 \text{ g Cu}} = 94.5 \text{ mol Cu}$$

There are 94.5 moles in 6.00 kg of copper.

Practice Problems

A. Convert a volume of 38.5 L to milliliters.

B. Boris rode a bicycle for 35 minutes at a 24 km/h. How far did he travel?

C. Convert a speed of 90 km/h to meters per second.

D. Convert a speed of 66 ft/s to mi/h. (1 mi = 5280 ft)

E. Convert a density of 158.7 kg/m^3 to g/cm^3.

F. One mole of sulfur has a mass of 32.1 g. Find the mass, in kilograms, of 64.8 moles of sulfur.

G. One mole of 2,2-dichlorohexane ($C_6H_{12}Cl_2$) has a mass of 155.1 g. Find the number of moles in 18.74 kg of 2,2-dichlorohexane.

Answer

1 $\frac{3 \text{ mol } H_2}{1 \text{ mol } N_2}, \frac{1 \text{ mol } N_2}{2 \text{ mol } NH_3}, \frac{2 \text{ mol } NH_3}{3 \text{ mol } H_2}$

CHEMath

Teaching Tips

Remind students that they often convert from one unit to another, both inside and outside of chemistry class. They convert money from cents to dollars and time from minutes to hours. Start out by giving them practice with everyday examples. Ask:

- A chicken needs to be cooked 20 minutes for each pound it weighs. How long should the chicken be cooked if it weighs 4.5 pounds? (4.5 lb × 20 min/lb = 90 min; 90 min × 1 h/60 min = 1.5 h. Most students will automatically relate 90 minutes to 1.5 hours. This may help them become comfortable with the process.)

If students are having difficulty with conversion factors, you may wish to have them list several conversion factors on the chalkboard. Divide the class in half and have each group challenge the other to write the conversion factor given two related units. Remind them that each conversion factor can appear in two forms depending on which value they put in the numerator. However, for dimensional analysis they must choose the form that has the unit they are converting from in the denominator.

Once they are comfortable with conversion factors and the process of dimensional analysis, relate it to examples they see in this chapter. Work through Example 2 and the Practice Problems.

Answers to Practice Problems

A. 3.85×10^4 mL

B. 14 km

C. 25 m/s

D. 45 mi/h

E. 0.1587 g/cm^3

F. 2.08 kg

G. 120.8 mol

2 Teach

Use the Visual

Direct students back to Figure 9.6 on page 242. Ask them to consider the sizes of the containers shown relative to the sizes of containers used in a classroom laboratory. Ask them to imagine they are in charge of the manufacturing facility pictured.

- **What factors would they need to consider to meet demands for ammonia?** (the number of customers, the number of cylinders per customer, the amount of ammonia per cylinder, and the amount of H_2 and N_2 needed to produce that quantity of NH_3)

Point out that sometimes it is not easy to reduce an entire manufacturing process to a simple equation.

Practice Problems Plus

Related Chapter Review Problem
Chapter Review problem 37 is related to Sample Problem 9-3.

Chemistry Alive!

Carbide Cannon Play

Sample Problem 9-3

How many moles of ammonia are produced when 0.60 mol of nitrogen reacts with hydrogen?

1. ***ANALYZE*** ***List the known and the unknown.***

Known:
- moles of nitrogen = 0.60 mol N_2

Unknown:
- moles of ammonia = ? mol NH_3

The conversion is mol $N_2 \longrightarrow$ mol NH_3. According to the balanced equation, 1 mol N_2 combines with 3 mol H_2 to produce 2 mol NH_3. To determine the number of moles of NH_3, the given quantity of N_2 is multiplied by the form of the mole ratio from the balanced equation that allows the given unit to cancel.

2. ***CALCULATE*** ***Solve for the unknown.***

$$0.60\ \cancel{\text{mol N}_2} \times \frac{2\ \text{mol NH}_3}{1\ \cancel{\text{mol N}_2}} = 1.2\ \text{mol NH}_3$$

3. ***EVALUATE*** ***Does the result make sense?***

The balanced chemical equation shows that two moles of ammonia are produced for each mole of nitrogen reacted. Note that mole ratios from balanced equations are considered to be exact (defined numbers). They do not enter into the determination of significant figures in the answer.

Practice Problems

9. This equation shows the formation of aluminum oxide.
$4Al(s) + 3O_2(g) \longrightarrow 2Al_2O_3(s)$

a. Write the six mole ratios that can be derived from this equation.

b. How many moles of aluminum are needed to form 3.7 mol Al_2O_3?

10. According to the equation in Problem 9:

a. How many moles of oxygen are required to react completely with 14.8 mol Al?

b. How many moles of Al_2O_3 are formed when 0.78 mol O_2 reacts with aluminum?

In the mole ratio below, W is the unknown quantity. The values of a and b are the coefficients from the balanced equation. Thus a general solution for a mole-mole problem, such as Sample Problem 9-3, is given by

$$\underset{\text{Given}}{x\ \cancel{\text{mol } G}} \times \underset{\text{Mole ratio}}{\frac{b\ \text{mol } W}{a\ \cancel{\text{mol } G}}} = \underset{\text{Calculated}}{\frac{xb}{a}\ \text{mol } W}$$

Figure 9.7
To determine the number of moles in a sample of a compound, first measure the mass of the sample. Then use the molar mass to calculate the number of moles in that mass.

Mass-Mass Calculations

No laboratory balance can measure substances directly in moles. Instead, as is shown in **Figure 9.7,** the amount of a substance is usually determined by measuring its mass in grams. From the mass of a reactant or product, the mass of any other reactant or product in a given chemical equation can be calculated. The mole interpretation of a balanced equation is the basis for this conversion. If the given sample is measured in grams, the mass can be converted to moles by using the molar mass. Then the mole ratio from the balanced equation can be used to calculate the number of moles of the unknown. If it is the mass of the unknown that needs to be determined, the number of moles of the unknown can be multiplied by the molar mass. As in mole-mole calculations, the unknown can be either a reactant or a product.

Sample Problem 9-4

Calculate the number of grams of NH_3 produced by the reaction of 5.40 g of hydrogen with an excess of nitrogen. The balanced equation is

$$N_2(g) + 3H_2(g) \longrightarrow 2NH_3(g)$$

1. ***ANALYZE*** ***List the knowns and the unknown.***

Knowns:
- mass of hydrogen = 5.40 g H_2
- 3 mol H_2 = 2 mol NH_3 (from balanced equation)
- 1 mol H_2 = 2.0 g H_2 (molar mass)
- 1 mol NH_3 = 17.0 g NH_3 (molar mass)

Unknown:
- mass of ammonia = ? g NH_3

The mass in grams of hydrogen will be used to find the mass in grams of ammonia: g $H_2 \longrightarrow$ g NH_3.
The coefficients in the balanced equation show that 3 mol H_2 reacts with 1 mol N_2 to produce 2 mol NH_3. The following calculations need to be done:

$$\text{g } H_2 \longrightarrow \text{mol } H_2 \longrightarrow \text{mol } NH_3 \longrightarrow \text{g } NH_3$$

2. ***CALCULATE*** ***Solve for the unknown.***

Convert the given (5.40 g H_2) to moles by using the molar mass of hydrogen.

$$5.40 \,\cancel{\text{g } H_2} \times \frac{1 \text{ mol } H_2}{2.0 \,\cancel{\text{g } H_2}} = 2.7 \text{ mol } H_2$$

Use the mole ratio from the balanced equation to calculate the number of moles of NH_3.

$$2.7 \,\cancel{\text{mol } H_2} \times \frac{2 \text{ mol } NH_3}{3 \,\cancel{\text{mol } H_2}} = 1.8 \text{ mol } NH_3$$

Convert the moles of NH_3 to grams of NH_3 by using the molar mass of ammonia.

$$1.8 \,\cancel{\text{mol } NH_3} \times \frac{17.0 \text{ g } NH_3}{1 \,\cancel{\text{mol } NH_3}} = 31 \text{ g } NH_3$$

This series of calculations can be combined:

$$\text{g } H_2 \longrightarrow \text{mol } H_2 \longrightarrow \text{mol } NH_3 \longrightarrow \text{g } NH_3$$

$$\underset{\text{Given quantity}}{5.40 \,\cancel{\text{g } H_2}} \times \underset{\text{Change given unit to moles}}{\frac{1 \,\cancel{\text{mol } H_2}}{2.0 \,\cancel{\text{g } H_2}}} \times \underset{\text{Mole ratio}}{\frac{2 \,\cancel{\text{mol } NH_3}}{3 \,\cancel{\text{mol } H_2}}} \times \underset{\text{Change moles to grams}}{\frac{17.0 \text{ g } NH_3}{1 \,\cancel{\text{mol } NH_3}}} = 31 \text{ g } NH_3$$

3. ***EVALUATE*** ***Does the result make sense?***

Because there are three conversion factors involved in this solution, it is more difficult to estimate an answer. However, because the molar mass of NH_3 is substantially greater than the molar mass of H_2, the answer should have a larger mass than the given mass. The answer should have two significant figures.

Practice Problems

11. Acetylene gas (C_2H_2) is produced by adding water to calcium carbide (CaC_2).

$$CaC_2(s) + 2H_2O(l) \longrightarrow C_2H_2(g) + Ca(OH)_2(aq)$$

How many grams of acetylene are produced by adding water to 5.00 g CaC_2?

12. Using the same equation, determine how many moles of CaC_2 are needed to react completely with 49.0 g H_2O.

Chem ASAP!

Problem-Solving 11

Solve Problem 11 with the help of an interactive guided tutorial.

Meeting Diverse Needs

Gifted In living organisms, nitrogen is necessary for the production of nucleic acids, such as DNA and RNA, and amino acids, which form proteins. However, animals are unable to store excess reserves of nitrogen in their bodies. It must be obtained from foods on a steady basis. Have students work in groups to develop a schematic diagram illustrating how nitrogen is cycled among the atmosphere, the soil, and living systems. Make an overhead transparency of the diagram and have students explain the diagram to the class.

9.2

TEACHER DEMO

Prepare 0.1 *M* solutions of potassium iodide and lead(II) nitrate. Measure out 50.0 mL of $Pb(NO_3)_2$ and 150 mL of KI in separate 250-mL beakers. Tell students that you are going to mix 0.005 moles of lead(II) nitrate with excess potassium iodide and observe the result. In a 250-mL beaker, combine both solutions. A bright yellow precipitate will form. Have students write a balanced chemical equation for the observed reaction. [$2KI(aq) + Pb(NO_3)_2(aq) \rightarrow 2KNO_3(aq) + PbI_2(s)$] Have students predict the number of moles of product produced. (0.005 moles PbI_2 assuming the reaction was complete)

Note that, in an actual reaction, the amounts of reactants often are not present in the mole ratios predicted by the coefficients in a balanced equation. Explain the importance of the mole ratios in an equation for calculating the relative quantities of reactant consumed and product produced. Have students calculate the mass of lead(II) nitrate reacted (1.66 g) and the mass of lead(II) iodide produced (2.30 g).

Practice Problems Plus

Related Chapter Review Problems

Chapter Review problems 38, 39, and 40 are related to Sample Problem 9-4.

Additional Practice Problem

The Chem ASAP! CD-ROM contains the following problem: Rust (Fe_2O_3) is produced when iron (Fe) reacts with oxygen (O_2).

$4Fe(s) + 3O_2(g) \rightarrow 2Fe_2O_3(s)$

How many grams of Fe_2O_3 are produced when 12.0 g of iron rusts? (17.2 g)

Cooperative Learning

Divide the class into groups of 2 to 4 students and have them work on the practice problems. Try to ensure that there is variety in the level of ability within each group. Tell the class that you will assign each group at random a different problem to present to the class. This will help to ensure that all of the groups do all of the problems.

Discuss

Students sometimes try to do mass-mass conversions in the same way they convert units of mass. That is, they try to convert the mass of one substance in a balanced equation directly to the mass of another substance, bypassing the mole ratio. Stress that because the number of grams in one mole of a substance varies with its molar mass, a mole conversion is a necessary intermediate step in stoichiometric calculations.

Use examples of balanced equations in which the mole ratio is not 1:1 to emphasize why the mole ratio is necessary. Also, remind students that the units in a problem will not cancel as they should unless the mole ratio expression is present. Have students check their work as they go along, keeping in mind that each mass-mass stoichiometric problem must have three conversion steps.

AGRICULTURE

Ammonia in the Nitrogen Cycle

Ammonia is part of the nitrogen cycle in nature. Earth's atmosphere contains 0.01 parts per million of ammonia, and small amounts of ammonia occur in volcanic gases. Most ammonia cycles through the living world without returning to the atmosphere. Ammonia plays a role in several stages of the nitrogen cycle. Nitrogen-fixing bacteria form nodules, or swellings, on the roots of plants in the legume family, such as beans and clover plants. Here the bacteria change atmospheric nitrogen into ammonia molecules or ammonium ions. Other bacteria break down the nitrogenous material in dead plants and animals into ammonia molecules. Certain soil bacteria oxidize these molecules into nitrate ions, the form readily absorbed by plant roots. When a plant dies, this cycle begins again.

Figure 9.8
In this Hubble Space Telescope image, clouds of condensed ammonia are visible covering the surface of Saturn.

If the law of conservation of mass is true, how is it possible to make 31 g NH_3 from only 5.40 g H_2? Looking back at the equation for the reaction, you will see that hydrogen is not the only reactant. Another reactant, nitrogen, is also involved. If you were to calculate the number of grams of nitrogen needed to produce 31 g NH_3 and then compare the total masses of reactants and products, you would have an answer to this question. Go ahead and try it!

You can see from Sample Problem 9-4 that mass-mass problems can be solved in basically the same way as mole-mole problems. **Figure 9.9** reviews the steps for the mass-mass conversion of any given mass *(G)* and any wanted mass *(W)*.

1. The mass G is changed to moles of G (mass G $\longrightarrow$ mol G) by using the molar mass of G.

$$\text{mass } G \times \frac{1 \text{ mol } G}{\text{molar mass } G} = \text{mol } G$$

2. The moles of *G* are changed to moles of *W* (mol *G* $\longrightarrow$ mol *W*) by using the mole ratio from the balanced equation.

$$\text{mol } G \times \frac{b \text{ mol } W}{a \text{ mol } G} = \text{mol } W$$

3. The moles of W are changed to grams of *W* (mol *W* $\longrightarrow$ mass *W*) by using the molar mass of *W*.

$$\text{mol } W \times \frac{\text{molar mass } W}{1 \text{ mol } W} = \text{mass } W$$

Figure 9.9 also shows the steps for doing mole-mass and mass-mole stoichiometric calculations. For a mole-mass problem, the first conversion (from mass to moles) is skipped. For a mass-mole problem, the last conversion (from moles to mass) is skipped. You can use parts of the three-step process shown in **Figure 9.9** as they are appropriate to the problem you are solving.

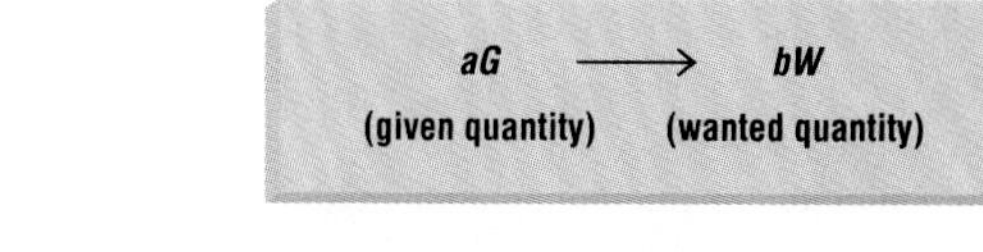

mass of G × $\frac{1 \text{ mol } G}{\text{mass } G}$ → mol G × $\frac{b \text{ mol } W}{a \text{ mol } G}$ → mol W × $\frac{\text{mass } W}{1 \text{ mol } W}$ → mass of W

Mass-mole conversion | Mole ratio from balanced equation | Mole-mass conversion

Figure 9.9

This general solution diagram indicates the steps necessary to solve a mass-mass stoichiometry problem: convert mass to moles, use the mole ratio, and then convert moles to mass. Is the given always a reactant? ❶

Other Stoichiometric Calculations

As you already know, a balanced chemical equation indicates the relative number of moles of reactants and products. From this foundation, stoichiometric calculations can be expanded to include any unit of measurement that is related to the mole. The given quantity can be expressed in numbers of representative particles, units of mass, or volumes of gases at STP. The problems can include mass-volume, volume-volume, and particle-mass calculations. In any of these problems, the given quantity is first converted to moles. Then the mole ratio from the balanced equation is used to calculate the number of moles of the wanted substance. Once this has been determined, the moles are converted to any other unit of measurement related to the unit mole, as the problem requires. **Figure 9.10** summarizes these steps for a typical stoichiometric problem.

Figure 9.10

With your knowledge of conversion factors and this problem-solving approach you can solve a variety of stoichiometric problems. What conversion factor is used to convert moles to representative particles? ❷

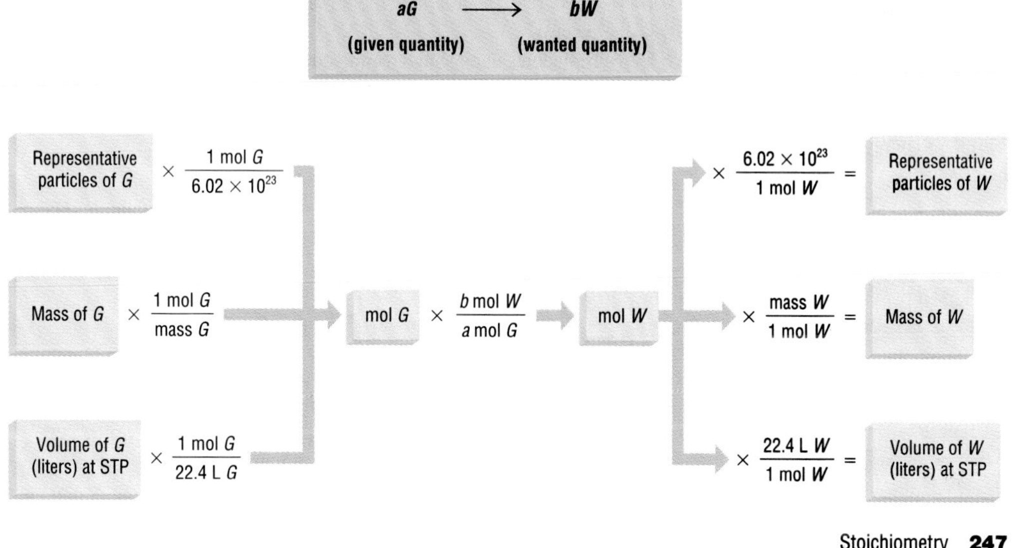

Answers

❶ No; it may be a product in which case you may be asked to find the amount of reactant necessary to produce that amount of product.

❷ Avogadro's number/mol

9.2

Build Writing Skills

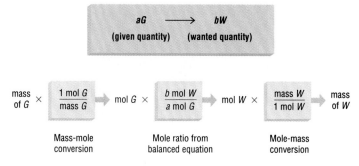

Have students study the Link to Agriculture. Ask students to research how the reaction utilized by bacteria to produce ammonia differs from the industrial process carried out in fertilizer factories. Ask students to research the energy costs for producing ammonia by traditional methods, and how alternative methods such as bacterial production could serve as cheaper sources of ammonia. What efforts are being made in that direction? Students should present their findings in a written report.

Discuss

Initiate a discussion with students by asking whether the law of conservation of mass is always true. If not, ask them to give an example. (Nuclear reactions are the only cases where this law does not hold true.) Ask:

- **Why isn't there a "law of conservation of moles"?** (Because reactions involve rearrangements of atoms, reactants can combine or decompose to produce fewer or greater numbers of moles of product. Although the total mass of reactants and products is constant, the number of moles of particles can increase or decrease depending on the final grouping of atoms.)
- **Give an example of a reaction in which the number of moles of products is greater than the number of moles of reactants.** ($2H_2O(l) \rightarrow 2H_2(g) + O_2(g)$)
- **Give an example of a reaction in which the number of moles of products is less than the number of moles of reactants.** ($2Mg(s) + O_2(g) \rightarrow 2MgO(s)$)

Discuss

On the chalkboard, write equations for reactions in which the reactants are both gases or are a gas and a solid. Ask students how the reactants and products in each reation would most likely be measured. Have students relate these measurements to the concept of a mole.

Think Critically

Have computer literate students use the sample calculations in the sample problems as general algorithms to write computer programs that solve stoichiometric problems. Have students demonstrate and explain their programs to interested students.

Discuss

Point out that mass-mass, mass-volume, volume-volume, and particle-mass calculations follow the same basic pattern. These calculations involve

- converting a given quantity to moles;
- using the mole ratio to find moles of the wanted substance;
- and changing moles of the wanted substance to a measured quantity.

Practice Problems Plus

Additional Practice Problem

This problem is related to Sample Problem 9-5: Hydrogen gas can be made by reacting methane (CH_4) with high-temperature steam.

$CH_4(g) + H_2O(g) \rightarrow CO(g) + 3H_2(g)$

How many hydrogen molecules are produced when 158 g of methane reacts with steam? (1.78×10^{25} hydrogen molecules)

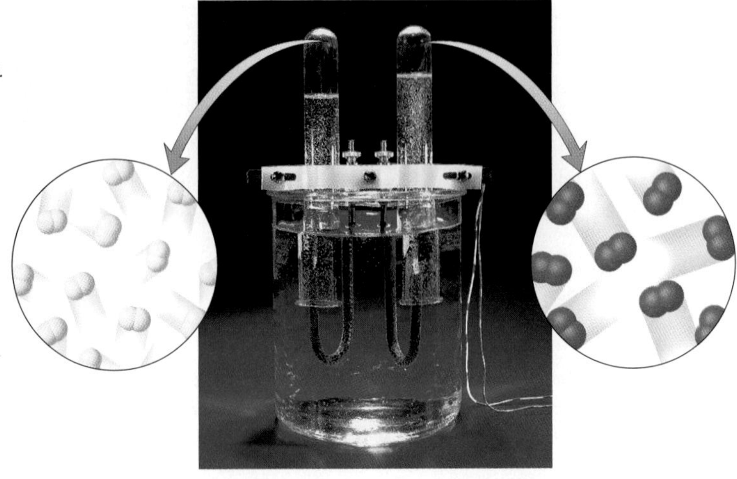

Figure 9.11
The electrolysis of water causes it to decompose into hydrogen and oxygen.

Sample Problem 9-5

How many molecules of oxygen are produced when a sample of 29.2 g of water is decomposed by electrolysis according to this balanced equation?

$$2H_2O(l) \xrightarrow{\text{electricity}} 2H_2(g) + O_2(g)$$

1. ANALYZE ***List the knowns and the unknown.***

Knowns:

- mass of water = 29.2 g H_2O
- 2 mol H_2O = 1 mol O_2 (from balanced equation)
- 1 mol H_2O = 18.0 g H_2O (molar mass)
- 1 mol O_2 = 6.02×10^{23} molecules O_2

Unknown:

- molecules of oxygen = ? molecules O_2

Use appropriate conversion factors to convert the given quantity in grams to number of molecules.

2. CALCULATE ***Solve for the unknown.***

$$29.2\ \text{g H}_2\text{O} \times \frac{1\ \text{mol H}_2\text{O}}{18.0\ \text{g H}_2\text{O}} \times \frac{1\ \text{mol O}_2}{2\ \text{mol H}_2\text{O}} \times \frac{6.02 \times 10^{23}\ \text{molecules O}_2}{1\ \text{mol O}_2}$$

Given quantity | Change to moles | Mole ratio | Change to molecules

$$= 4.88 \times 10^{23}\ \text{molecules O}_2$$

3. EVALUATE ***Does the result make sense?***

The given mass of water should produce a little less than 1 mol of oxygen, or a little less than Avogadro's number of molecules. The answer should have three significant figures.

Practice Problems

13. How many molecules of oxygen are produced by the decomposition of 6.54 g of potassium chlorate ($KClO_3$)?

$$2KClO_3(s) \longrightarrow 2KCl(s) + 3O_2(g)$$

14. The last step in the production of nitric acid is the reaction of nitrogen dioxide with water.

$$3NO_2(g) + H_2O(l) \longrightarrow 2HNO_3(aq) + NO(g)$$

How many grams of nitrogen dioxide must react with water to produce 5.00×10^{22} molecules of nitrogen monoxide?

Sample Problem 9-6

Assuming STP, how many liters of oxygen are needed to produce 19.8 L SO_3 according to this balanced equation?

$$2SO_2(g) + O_2(g) \longrightarrow 2SO_3(g)$$

1. ***ANALYZE*** *List the knowns and the unknown.*

   Knowns:
   - volume of sulfur trioxide = 19.8 L
   - 2 mol SO_3 = 1 mol O_2 (from balanced equation)
   - 1 mol SO_3 = 22.4 L SO_3 (at STP)
   - 1 mol O_2 = 22.4 L O_2 (at STP)

   Unknown:
   - volume of oxygen = ? L O_2

2. ***CALCULATE*** *Solve for the unknown.*

$$19.8\,\cancel{L\,SO_3} \times \frac{1\,\cancel{mol\,SO_3}}{22.4\,\cancel{L\,SO_3}} \times \frac{1\,\cancel{mol\,O_2}}{2\,\cancel{mol\,SO_3}} \times \frac{22.4\,L\,O_2}{1\,\cancel{mol\,O_2}} = 9.90\,L\,O_2$$

Given quantity — Change to moles — Mole ratio — Change to liters

3. ***EVALUATE*** *Does the result make sense?*

   Because 2 mol SO_3 is produced for each 1 mol O_2 that reacts, the volume of O_2 should be half the volume of SO_3. The answer should have three significant figures.

Practice Problems

15. The equation for the combustion of carbon monoxide is
    $2CO(g) + O_2(g) \longrightarrow 2CO_2(g)$
    How many liters of oxygen are required to burn 3.86 L of carbon monoxide?
16. Phosphorus and hydrogen can be combined to form phosphine (PH_3).
    $P_4(s) + 6H_2(g) \longrightarrow 4PH_3(g)$
    How many liters of phosphine are formed when 0.42 L of hydrogen reacts with phosphorus?

Chem ASAP!

Problem-Solving 16
Solve Problem 16 with the help of an interactive guided tutorial.

In Sample Problem 9-6, did you notice that the 22.4 L/mol factors canceled out? This will always be true in a volume-volume problem. The coefficients in a balanced chemical equation indicate the relative numbers of moles. The coefficients also indicate the relative volumes of interacting gases. The volume can be expressed in any unit. What are some other units of volume? 

Sample Problem 9-7

Nitrogen monoxide and oxygen gas combine to form the brown gas nitrogen dioxide. How many milliliters of nitrogen dioxide are produced when 3.4 mL of oxygen reacts with an excess of nitrogen monoxide? Assume conditions of STP.

$$2NO(g) + O_2(g) \longrightarrow 2NO_2(g)$$

1. ***ANALYZE*** *List the knowns and the unknown.*

   Knowns:
   - volume of oxygen = 3.4 mL O_2
   - 1 mL O_2 = 2 mL NO_2 (from balanced equation)

   Unknown:
   - volume of nitrogen dioxide = ? mL NO_2

Figure 9.12
The brown gas nitrogen dioxide is a component of photochemical smog, which builds up when still air hangs over a city or other pollution source. Persistent smog can pose a health danger.

Discuss

Construct a diagram on the chalkboard or overhead projector showing the relationships that are useful for solving stoichiometry problems. One simple model reaction is A → B. Use double-headed arrows to connect the terms: *Particles of A, Moles of A, Grams of A, Moles of B, Particles of B,* and *Grams of B.* Above the appropriate arrows, write: *Avogadro's number, equation coefficients,* and *Molar mass.* Explain that the only "transitions" allowed are between quantities connected by arrows. Point out that the required conversion factor to make a "transition" is written above each arrow. Have students refer to the diagram when working practice problems.

Practice Problems Plus

Related Chapter Review Problems
Chapter Review problems 41 and 56 are related to Sample Problems 9-6 and 9-7.

Additional Practice Problem
The Chem ASAP! CD-ROM contains the following problem: Ammonia (NH_3) reacts with oxygen (O_2) to produce nitrogen monoxide (NO) and water.

$4NH_3(g) + 5O_2(g) \rightarrow 4NO(g) + 6H_2O(l)$

How many liters of NO are produced when 1.40 L of oxygen reacts with ammonia? (1.12 L)

Answer

1 milliliters (mL), cubic centimeters (cm^3), deciliters (dL), and cubic meters (m^3)

3 Assess

Evaluate Understanding

To evaluate students' understanding of chemical calculations, write several balanced equations on the board and have students write all the different mole ratios for each reaction. Have students choose the correct mole ratio for each problem you present. Include gases in at least one equation so that the students can calculate the volume of a gas produced or consumed at STP.

Reteach

Review the importance of mole ratios for solving stoichiometric problems. Use visual aids such as molecular models to illustrate the stoichiometric relationships between reactants and products.

Portfolio Project

Answers will vary, but the cost and precision of an analytical balance should be related. Have students hypothesize about this relationship and then contact a supplier by phone or e-mail to test their explanations.

Sample Problem 9-7 (cont.)

2. CALCULATE ***Solve for the unknown.***

$$3.4\ \text{mL O}_2 \times \frac{2\ \text{mL NO}_2}{1\ \text{mL O}_2} = 6.8\ \text{mL NO}_2$$

Given quantity — Volume ratio

3. EVALUATE ***Does the result make sense?***

The conversion is mL $O_2 \longrightarrow$ mL NO_2. The given quantity of O_2 is multiplied by the volume ratio from the balanced equation that allows the given unit to cancel. Because the volume ratio is 2 volumes NO_2 to 1 volume O_2, the calculated volume of NO_2 should be twice the given volume of O_2. The answer should have two significant figures.

Practice Problems

Consider this equation.

$$CS_2(l) + 3O_2(g) \longrightarrow CO_2(g) + 2SO_2(g)$$

17. Calculate the volume of sulfur dioxide produced when 27.9 mL O_2 reacts with carbon disulfide.

18. How many deciliters of carbon dioxide are produced when 0.38 L SO_2 is formed?

section review 9.2

19. Isopropyl alcohol (C_3H_7OH) burns in air according to this equation:

$$2C_3H_7OH(l) + 9O_2(g) \longrightarrow 6CO_2(g) + 8H_2O(g)$$

a. Calculate the moles of oxygen needed to react with 3.40 mol C_3H_7OH.

b. Find the moles of each product formed when 3.40 mol C_3H_7OH reacts with oxygen.

20. What ratio is used to carry out each conversion?

a. mol CH_4 to g CH_4

b. L $CH_4(g)$ to mol $CH_4(g)$ (at STP)

c. molecules CH_4 to mol CH_4

21. The combustion of acetylene gas is represented by this equation:

$$2C_2H_2(g) + 5O_2(g) \longrightarrow 4CO_2(g) + 2H_2O(g)$$

a. How many grams of CO_2 and grams of H_2O are produced when 52.0 g C_2H_2 burns?

b. How many grams of oxygen are required to burn 52.0 g C_2H_2?

c. Use the answers from **a** and **b** to show that this equation obeys the law of conservation of mass.

22. Tin(II) fluoride, formerly found in many kinds of toothpaste, is formed in this reaction:

$$Sn(s) + 2HF(g) \longrightarrow SnF_2(s) + H_2(g)$$

a. How many liters of HF are needed to produce 9.40 L H_2 at STP?

b. How many molecules of H_2 are produced by the reaction of tin with 20.0 L HF at STP?

c. How many grams of SnF_2 can be made by reacting 7.42×10^{24} molecules of HF with tin?

Chem ASAP! **Assessment 9.2** Check your understanding of the important ideas and concepts in Section 9.2.

portfolio project

Using the Internet or the library, research analytical balances. How small a mass or large a mass can these types of balances measure? Compare different balances to find out whether there is a relationship between precision and cost.

Answers

SECTION REVIEW 9.2

19. a. 15.3 mol O_2

b. 10.2 mol CO_2, 13.6 mol H_2O

20. a. $\frac{16.0\ \text{g CH}_4}{1\ \text{mol CH}_4}$

b. $\frac{1\ \text{mol CH}_4}{22.4\ \text{L CH}_4}$

c. $\frac{1\ \text{mol CH}_4}{6.02 \times 10^{23}\ \text{molecules CH}_4}$

21. a. 176 g CO_2, 36.0 g H_2O

b. 1.60×10^2 g O_2

c. 212 g = 212 g

22. a. 18.8 L HF

b. 2.69×10^{23} molecules H_2

c. 966 g SnF_2

SMALL-SCALE LAB

ANALYSIS OF BAKING SODA

SAFETY    

Wear safety glasses and follow the standard safety procedure as outlined on page 18.

PURPOSE

To determine the mass of sodium hydrogen carbonate in a sample of baking soda using stoichiometry.

MATERIALS

- baking soda
- 3 plastic cups
- pipets of HCl, NaOH, and thymol blue
- soda straw
- balance
- pH sensor (optional)

PROCEDURE

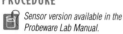

Sensor version available in the Probeware Lab Manual.

Prepare and analyze a sample of baking soda. Take care to write down the results of each step.

A. Measure the mass of a clean, dry plastic cup.

B. Using the straw as a scoop, fill one end with baking soda to a depth of about 1 cm. Add the sample to the cup and measure its mass again.

C. Place two HCl pipets that are about $\frac{3}{4}$ full into a clean cup and measure the mass of the system.

D. Transfer the contents of both HCl pipets to the cup containing baking soda. Swirl until the fizzing stops. Wait 5–10 minutes to be sure the reaction is complete. Measure the mass of the two empty HCl pipets in their cup again.

E. Add 5 drops of thymol blue to the plastic cup.

F. Place two full NaOH pipets in a clean cup and measure the mass of the system.

G. Add NaOH slowly to the baking soda/HCl mixture until the pink color just disappears. Measure the mass of the NaOH pipets in their cup again.

ANALYSIS

Using your experimental data, record the answers to the following questions below your data table.

1. Write a balanced equation for the reaction between baking soda ($NaHCO_3$) and HCl.
2. Calculate the mass in grams of the baking soda.

   (Step B − Step A)
3. Calculate the total mmol of 1*M* HCl.
   Note: Every gram of HCl contains 1 mmol.

   (Step C − Step D) × 1.00 mmol/g
4. Calculate the total mmol of 0.5*M* NaOH.
   Note: Every gram of NaOH contains 0.5 mmol.

   (Step F − Step G) × 0.500 mmol/g
5. Calculate the mmol of HCl that reacted with the baking soda. *Note:* The NaOH measures the amount of HCl that did not react.

   (Step 3 − Step 4)
6. Calculate the mass of the baking soda from the reaction data.

   (0.084 g/mmol × Step 5)
7. Calculate the percent error of the experiment.

   $$\frac{(\text{Step 2} - \text{Step 6})}{\text{Step 2}} \times 100\%$$

YOU'RE THE CHEMIST

The following small-scale activities allow you to develop your own procedures and analyze the results.

1. **Analyze It!** For each calculation you did, substitute each quantity (number and unit) into the equation and cancel the units to explain why each step gives the quantity desired.
2. **Design It!** Baking powder consists of a mixture of baking soda, sodium hydrogen carbonate, and a solid acid, usually calcium dihydrogen phosphate ($Ca(H_2PO_4)_2$). Design and carry out an experiment to determine the percentage of baking soda in baking powder.

SAMPLE DATA

A. 2.83 g **B.** 3.28 g **C.** 10.70 g
D. 4.29 g **F.** 10.53 g **G.** 8.78 g

ANALYSIS

1. $HCl + NaHCO_3(s) \rightarrow CO_2(g) + H_2O + NaCl$
2. 3.28 g − 2.83 g = 0.45 g
3. (10.70 − 4.29)~~g~~ × 1.00 mmol/~~g~~ = 6.41 mmol
4. (10.53 − 8.78)~~g~~ × 0.500 mmol/~~g~~ = 0.875 mmol (0.875 mmol HCl unreacted)
5. 6.41 mmol total − 0.875 mmol unreacted = 5.53 mmol (5.53 mmol $NaHCO_3$)
6. (0.0840 g/~~mmol~~) × 5.53 ~~mmol~~ = 0.46 g
7. $100\% \times \frac{(0.46\ \text{g} - 0.45\ \text{g})}{0.45\ \text{g}} = 2.2$ percent error

   (assuming baking soda is 100% $NaHCO_3$).

SMALL-SCALE LAB

ANALYSIS OF BAKING SODA

OVERVIEW

Procedure Time: 30 minutes

See Probeware Lab Manual for Vernier, TI, and PASCO versions.

Students react a known mass of baking soda with a measured amount of HCl and use stoichiometry to determine the mass of $NaHCO_3$ in baking soda.

MATERIALS

Solution	Preparation
0.5*M* NaOH	20.0 g in 1.0 L
1.0*M* HCl	82 mL of 12*M* in 1.0 L **Caution!** Always add acid to water carefully and slowly!
0.04% TB	100 mg in 21.5 mL of 0.01*M* NaOH; dilute to 250 mL

TEACHING SUGGESTIONS

Stress that the procedure measures the amount of excess HCl that is not reacted with the baking soda (Step 4). Because this excess HCl reacts with the NaOH in a 1 : 1 mole ratio, the moles of NaOH equal the moles of HCl in excess. Subtracting the excess moles of HCl from the total moles used in the experiment (Step 5) yields the moles reacted with the baking soda, which is 100% $NaHCO_3$. If the mixture does not turn red when thymol blue is added, the student should find the mass of a third pipet and add just enough HCl to turn the mixture cherry red. Then the student should find the mass of the half-empty pipet so the mass of HCl added can be calculated and added to the total mass used.

YOU'RE THE CHEMIST

1. (See Steps 2–7.)
2. Repeat Steps A–G and 1–7 using baking powder instead of baking soda. The percent error is the percent of baking soda in baking powder, assuming no other errors.

9.3

1 Engage

Use the Visual

Have students study the photograph and read the text that opens the section. Explain that a chemical process is like an assembly line in a factory—the amount of product that can be produced is limited by the quantity of materials or reactant(s) available. Ask:

- **How does a limiting reagent affect a chemical reaction?** (The limiting reagent determines the maximum amount of a product or products that can be produced.)

Check Prior Knowledge

To assess students' prior knowledge about chemical quantities and both mole-mole and mass-mass conversions, write two or three balanced equations on the board. Have students solve the problems for reactant amounts or product amounts. Point out that the equations are assumed to be stoichiometrically perfect. That is, all reactants and products are present in exactly the right amounts.

2 Teach

Use the Visual

Have students examine Figure 9.13. Ask:

- **Where else do you see examples of processes or activities being "limited" on an everyday basis?** (Answers will vary. Examples include the number of people you can fit into a car, the number of cookies that can be made from a given amount of dough, the number of miles a car can drive on one tank of gasoline.)

section 9.3

objectives

- Identify and use the limiting reagent in a reaction to calculate the maximum amount of product(s) produced and the amount of excess reagent
- Calculate theoretical yield, actual yield, or percent yield given appropriate information

key terms

- limiting reagent
- excess reagent
- theoretical yield
- actual yield
- percent yield

LIMITING REAGENT AND PERCENT YIELD

If a carpenter had two table tops and seven table legs, he would have difficulty building more than one functional four-legged table. The first table would require four of the legs, leaving just three legs for the second table. In this case, the number of table legs is the limiting factor in the construction of four-legged tables. A similar concept applies in chemistry when knowing the exact amounts of reactants and products in a chemical reaction is crucial. **How does a limiting reagent affect a chemical reaction?**

What Is a Limiting Reagent?

Perhaps you know from your own experience that many cooks follow a recipe when making a new dish. They know that sufficient quantities of all the ingredients must be available. Suppose, for example, that you are preparing to make lasagna and you have more than enough meat, tomato sauce, ricotta cheese, eggs, mozzarella cheese, spinach, and seasoning on hand. However, you have only half a box of lasagna noodles. The amount of lasagna you can make will be limited by the amount of noodles you have. Thus the noodles are the limiting reagent in this baking venture. **Figure 9.13** illustrates another example of a limiting reagent in the kitchen. A chemist often faces a similar situation. It is impossible for a chemist to make a certain amount of a desired compound if there is an insufficient quantity of any of the required reactants.

As you know, a balanced chemical equation is a chemist's recipe—a recipe that can be interpreted on a microscopic scale (interacting particles) or on a macroscopic scale (interacting moles). The coefficients used to write

Figure 9.13
The amount of product is determined by the quantity of the limiting reagent. In this example, the rolls are the limiting reagent. No matter how much of the other ingredients you have, with two rolls you can make only two sandwiches.

STUDENT RESOURCES

- Section Review 9.3, Ch. 9 Practice Problems, Interpreting Graphics, Vocabulary Review, Quizzes, and Tests from the Review Module (Ch. 9–12)
- Laboratory Recordsheet 9-2
- Laboratory Manual: Experiment 13
- Laboratory Practicals 9-1 and 9-2

TECHNOLOGY RESOURCES

Relevant technology resources include:

- Chem ASAP! CD-ROM
- Resource Pro CD-ROM
- Assessment Resources CD-ROM: Chapter 9 Tests

the balanced equation give both the ratio of representative particles and the mole ratio. Recall the equation for the preparation of ammonia.

$$N_2(g) + 3H_2(g) \longrightarrow 2NH_3(g)$$

When one molecule (mole) of N_2 reacts with three molecules (moles) of H_2, two molecules (moles) of NH_3 are produced. What would happen if two molecules (moles) of N_2 reacted with three molecules (moles) of H_2? Would more than two molecules (moles) of NH_3 be formed? **Figure 9.14** shows both the particle and the mole interpretations of this problem.

Before the reaction takes place, nitrogen and hydrogen are present in a 2:3 molecule (mole) ratio. The reaction takes place according to the balanced equation. One molecule (mole) of N_2 reacts with three molecules (moles) of H_2 to produce two molecules (moles) of NH_3. At this point, all the hydrogen has been used up, and the reaction stops. One molecule (mole) of unreacted nitrogen is left in addition to the two molecules (moles) of NH_3 that have been produced by the reaction.

In this reaction, only the hydrogen is completely used up. It is called the limiting reagent. As the name implies, the **limiting reagent** limits or determines the amount of product that can be formed in a reaction. The reaction occurs only until the limiting reagent is used up. By contrast, the reactant that is not completely used up in a reaction is called the **excess reagent.** In this example, nitrogen is the excess reagent because some nitrogen will remain unreacted. You probably know that if you put a glass over a burning candle, the candle goes out. In this example of the combustion of candle wax, what is the limiting reagent? ❶

Sometimes in problems, the given quantities of reactants are expressed in units other than moles. In such cases, the first step in the solution is to convert each reactant to moles. Then the limiting reagent can be identified, as in Sample Problem 9-8. Finally, the amount of product can be determined from the given amount of limiting reagent.

Chem ASAP!

Animation 8
Apply the limiting reagent concept to the production of iron from iron ore.

Chemical Equations

	$N_2(g)$	+	$3H_2(g)$	→	$2NH_3(g)$
"Microscopic recipe"	1 molecule N_2	+	3 molecules H_2	→	2 molecules NH_3
"Macroscopic recipe"	1 mol N_2	+	3 mol H_2	→	2 mol NH_3

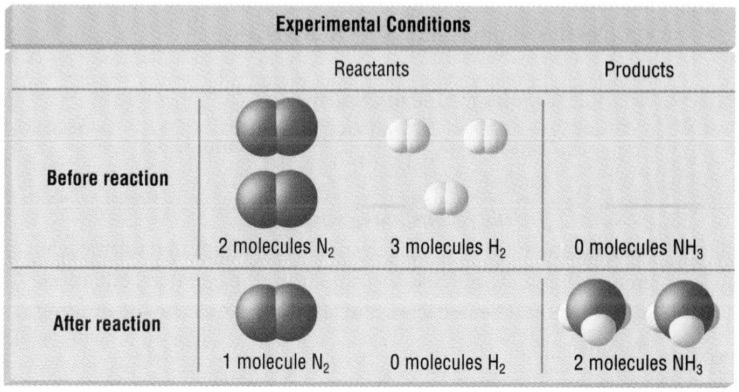

Experimental Conditions

	Reactants		Products
Before reaction	2 molecules N_2	3 molecules H_2	0 molecules NH_3
After reaction	1 molecule N_2	0 molecules H_2	2 molecules NH_3

Figure 9.14
The "recipe" calls for 3 molecules of H_2 for every 1 molecule of N_2. In this particular experiment, H_2 is the limiting reagent and N_2 is in excess. Would the amount of products formed change if you started with 4 molecules of N_2 and 3 molecules of H_2? ❷

TEACHER DEMO

Illustrate the concept of a limiting reagent using plastic bottles and caps. You will need 15 plastic bottles, 30 plastic caps to fit the bottles, and 6 containers to hold 5 caps each. Set out all 15 plastic bottles. Bring out one container of caps. Ask students:

- How many closed plastic bottles can we have?
- What limits the number?

Bring out another container of caps. Again, ask the students what will happen. Continue bringing out caps and matching them to the bottles. With three containers of caps, there should be a one-to-one correspondence. With the remaining containers, the caps should be in excess and the bottles should be the limiting factor.

MEETING DIVERSE NEEDS

At Risk and LEP Help students understand the concept of a limiting reagent by comparing it to everyday situations. For example, suppose a person is planning a dinner party. They have 6 glasses, 8 plates, and 4 chairs. How many people can they invite? Or, how many letters can you send if you have 24 pieces of stationery, 12 envelopes, and 10 stamps?

Answers

❶ oxygen, $O_2(g)$
❷ No, the amount would remain the same; 2 molecules of NH_3 would form.

Practice Problems Plus

Related Chapter Review Problems
Chapter Review problems 44, 45 and 46 are related to Sample Problem 9-8.

Additional Practice Problem
The Chem ASAP! CD-ROM contains the following problem: Phosphoric acid reacts with sodium hydroxide according to the equation:

$H_3PO_4(aq) + 3NaOH(aq) \rightarrow Na_3PO_4(aq) + 3H_2O(l)$

If 1.75 mol H_3PO_4 is made to react with 5.00 mol NaOH:

a. Identify the limiting reagent. (5.25 mol NaOH required; NaOH is the limiting reagent)

b. Calculate the moles of Na_3PO_4 produced. (1.67 mol)

c. Calculate the moles of excess reagent remaining. (0.08 mol H_3PO_4)

Use the Visual

Have students examine Figure 9.15. Write the balanced chemical equation for the reaction on the board:

$2Na(s) + Cl_2(g) \rightarrow 2NaCl(s)$

Ask a series of questions that require students to predict the amount of product formed based on the reactant quantities given. As students solve the problems, create a list of numbers under each reactant and product showing how the molar quantities are related for each set of conditions. In each case, ask students to state which reactant is the limiting reagent and circle the value in the list to designate it as the limiting quantity. Use small whole numbers that students can easily manipulate without calculators. Solve the first problem on the board, and then challenge students to do the mole-mole conversions using mental math.

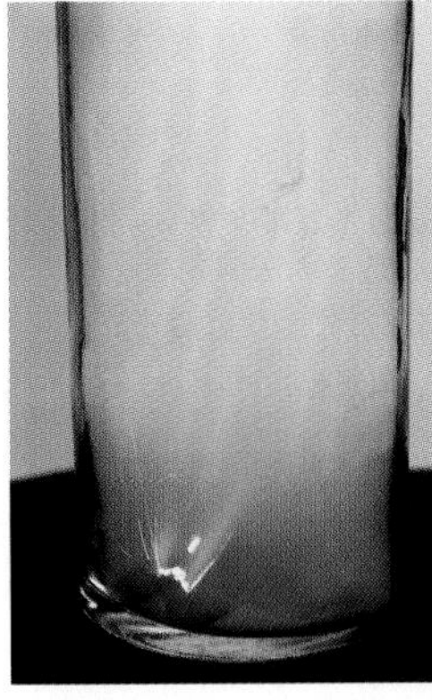

Figure 9.15
Igniting sodium metal in chlorine gas produces a white smoke of NaCl and bright yellow light.

Sample Problem 9-8

Sodium chloride can be prepared by the reaction of sodium metal with chlorine gas.

$$2Na(s) + Cl_2(g) \longrightarrow 2NaCl(s)$$

Suppose that 6.70 mol Na reacts with 3.20 mol Cl_2.

a. What is the limiting reagent?
b. How many moles of NaCl are produced?

1. ANALYZE ***List the knowns and the unknown for* a.**

Knowns:
- moles sodium = 6.70 mol Na
- moles chlorine = 3.20 mol Cl_2
- 2 mol Na = 1 mol Cl_2 (from balanced equation)

Unknown:
- limiting reagent = ?

The known amount of one of the reactants is multiplied by the mole ratio from the balanced equation to calculate the required amount of the other reactant. Sodium is chosen arbitrarily here: mol Na $\longrightarrow$ mol Cl_2.

2. CALCULATE ***Solve for the unknown.***

$$\underset{\text{Given amount}}{6.70\ \cancel{\text{mol Na}}} \times \underset{\text{Mole ratio}}{\frac{1\ \text{mol } Cl_2}{2\ \cancel{\text{mol Na}}}} = \underset{\text{Required amount}}{3.35\ \text{mol } Cl_2}$$

This calculation indicates that 3.35 mol Cl_2 is needed to react with 6.70 mol Na. Because only 3.20 mol Cl_2 is available, however, chlorine becomes the limiting reagent. Sodium, then, must be in excess.

1. ANALYZE ***List the knowns and the unknown for* b.**

Knowns:
- amount of limiting reagent = 3.20 mol Cl_2
- 1 mol Cl_2 = 2 mol NaCl (from balanced equation)

Unknown:
- moles of sodium chloride = ? mol NaCl

2. CALCULATE ***Solve for the unknown.***

$$3.20\ \cancel{\text{mol } Cl_2} \times \frac{2\ \text{mol NaCl}}{1\ \cancel{\text{mol } Cl_2}} = 6.40\ \text{mol NaCl}$$

3. EVALUATE ***Do the results make sense?***
Because the ratio of the given moles of sodium to chlorine was greater than 2:1, which is the ratio from the balanced equation, sodium should be in excess and chlorine should be the limiting reagent.

Practice Problems

23. The equation for the complete combustion of ethene (C_2H_4) is

$$C_2H_4(g) + 3O_2(g) \longrightarrow 2CO_2(g) + 2H_2O(g)$$

If 2.70 mol C_2H_4 is reacted with 6.30 mol O_2,
a. identify the limiting reagent.
b. calculate the moles of water produced.

24. The equation for the incomplete combustion of ethene (C_2H_4) is

$$C_2H_4(g) + 2O_2(g) \longrightarrow 2CO(g) + 2H_2O(g)$$

If 2.70 mol C_2H_4 is reacted with 6.30 mol O_2,
a. identify the limiting reagent.
b. calculate the moles of water produced.

Chem ASAP!

Problem-Solving 24
Solve Problem 24 with the help of an interactive guided tutorial.

Sample Problem 9-9

As illustrated in **Figure 9.16,** the properties of copper(I) sulfide are very different from the properties of the elements copper and sulfur.

$$2Cu(s) + S(s) \longrightarrow Cu_2S(s)$$

a. What is the limiting reagent when 80.0 g Cu reacts with 25.0 g S?

b. What is the maximum number of grams of Cu_2S that can be formed?

1. ANALYZE ***List the knowns and the unknown for* a.**

Knowns:
- mass of copper = 80.0 g Cu
- mass of sulfur = 25.0 g S
- 2 mol Cu = 1 mol S (from balanced equation)

Unknown:
- limiting reagent = ?

The number of moles of each reactant must first be found:

$$\text{g Cu} \longrightarrow \text{mol Cu}$$
$$\text{g S} \longrightarrow \text{mol S}$$

The balanced equation is used to calculate the number of moles of one reactant needed to react with the given amount of the other reactant:

$$\text{mol Cu} \longrightarrow \text{mol S}$$

2. CALCULATE ***Solve for the unknown.***

$$80.0\ \cancel{\text{g Cu}} \times \frac{1\ \text{mol Cu}}{63.5\ \cancel{\text{g Cu}}} = 1.26\ \text{mol Cu}$$

$$25.0\ \cancel{\text{g S}} \times \frac{1\ \text{mol S}}{32.1\ \cancel{\text{g S}}} = 0.779\ \text{mol S}$$

$$\underset{\text{Given quantity}}{1.26\ \cancel{\text{mol Cu}}} \times \underset{\text{Mole ratio}}{\frac{1\ \text{mol S}}{2\ \cancel{\text{mol Cu}}}} = \underset{\text{Needed amount}}{0.630\ \text{mol S}}$$

Comparing the amount of sulfur needed (0.630 mol S) with the given amount (0.779 mol S) indicates that sulfur is in excess. Thus copper is the limiting reagent.

1. ANALYZE ***List the knowns and the unknown for* b.**

Knowns:
- limiting reagent = 1.26 mol Cu
- 2 mol Cu = 1 mol Cu_2S (from balanced equation)
- 1 mol Cu_2S = 159.1 g Cu_2S (molar mass)

Unknown:
- mass copper(I) sulfide = ? g Cu_2S

The limiting reagent, which was determined in the last step, is used to calculate the maximum amount of Cu_2S formed: mol Cu $\longrightarrow$ mol Cu_2S $\longrightarrow$ g Cu_2S

Figure 9.16
Black crystalline copper(I) sulfide (bottom) is formed as a product when the reactants, sulfur (top left) and copper (top right), are heated together.

Practice Problems

25. Hydrogen gas can be produced in the laboratory by the reaction of magnesium metal with hydrochloric acid.

$$Mg(s) + 2HCl(aq) \longrightarrow MgCl_2(aq) + H_2(g)$$

a. Identify the limiting reagent when 6.00 g HCl reacts with 5.00 g Mg.

b. How many grams of hydrogen can be produced when 6.00 g HCl is added to 5.00 g Mg?

Cooperative Learning

Have students work in pairs to solve some limiting reactant practice problems. Problems should be restricted to molar quantities until students fully understand the concept of limiting reactant. Later, extend these types of problems to include grams.

Explain to students that, in a chemical reaction, the reactant which is consumed first is the *limiting reactant.* Remind students that it is not possible to tell which of the reactants will be depleted first by comparing the number of grams of each reactant. Some students find it helpful to depict atoms and molecules as spheres and connected spheres instead of as chemical symbols to help determine the limiting reactant. Have the students develop some of their own limiting reactant problems that can be shared with the rest of the class.

Practice Problems Plus

Related Chapter Review Problem

Chapter Review problem 51 is related to Sample Problem 9-9.

9.3

Discuss

Discuss with students practical applications of limiting and excess reagents. Point out the importance of the yield in a chemical reaction. Note that actual yields are almost always less than theoretical yields. For industrial chemists or chemical engineers, the goal is to find cost-effective methods for converting reactants into products. Industrial chemists and chemical engineers want to achieve the maximum product yield at the lowest cost. One way to control costs is to minimize the amount of excess reagent by calculating stoichiometric quantities of the reactants.

Sample Problem 9-9 (cont.)

2. **CALCULATE** ***Solve for the unknown.***

$$1.26 \text{ mol Cu} \times \frac{1 \text{ mol Cu}_2\text{S}}{2 \text{ mol Cu}} \times \frac{159.1 \text{ g Cu}_2\text{S}}{1 \text{ mol Cu}_2\text{S}} = 1.00 \times 10^2 \text{ g Cu}_2\text{S}$$

The given quantity of copper, 80.0 g, could have been used for this step instead of the moles of copper, which were calculated in the very first step of the solution.

3. **EVALUATE** ***Do the results make sense?***

Copper is the limiting reagent in this reaction. The maximum number of grams of Cu_2S produced should be more than the amount of copper that initially reacted because copper is combining with sulfur. The amount of Cu_2S produced should be less than the sum of copper and sulfur (80.0 g Cu + 25.0 g S = 105.0 g) because sulfur was in excess.

Practice Problems (cont.)

26. Acetylene (C_2H_2) will burn in the presence of oxygen.

$$2C_2H_2(g) + 5O_2(g) \longrightarrow 4CO_2(g) + 2H_2O(g)$$

How many grams of water can be produced by the reaction of 2.40 mol C_2H_2 with 7.4 mol O_2?

Calculating the Percent Yield

In theory, when a teacher gives an exam to the class, every student should get a grade of 100%. For a variety of reasons, this generally does not occur. Instead, the performance of the class is usually spread over a range of grades. Your exam grade, expressed as a percentage, is a quantity that shows how well you did on the exam (questions answered correctly) compared with how well you could have done if you had answered all the questions correctly (100%). This calculation is analogous to the percent yield calculation that you do in the laboratory when the product from a chemical reaction is less than you expected based on the balanced chemical equation.

So far in this chapter, you have probably assumed that when doing stoichiometric problems things do not go wrong in chemical reactions. This assumption is as faulty as assuming that all students will score 100% on an exam. When an equation is used to calculate the amount of product that will form during a reaction, a value representing the theoretical yield is obtained. The **theoretical yield** is the maximum amount of product that could be formed from given amounts of reactants. In contrast, the amount of product that actually forms when the reaction is carried out in the laboratory is called the **actual yield.** The actual yield is often less than the theoretical yield. The **percent yield** is the ratio of the actual yield to the theoretical yield expressed as a percent. The percent yield measures the efficiency of the reaction.

$$\text{Percent yield} = \frac{\text{actual yield}}{\text{theoretical yield}} \times 100\%$$

Figure 9.17

The productivity of a farm is measured in yield. Because growing conditions may vary from year to year, the actual yield often differs from the theoretical yield.

A percent yield should not normally be larger than 100%. Many factors cause percent yields to be less than 100%. Reactions do not always go to completion; when this occurs, less than the calculated amount of product is formed. Impure reactants and competing side reactions may cause unwanted products to form. Actual yield can also be lower than the theoretical yield due to a loss of product during filtration or in transferring

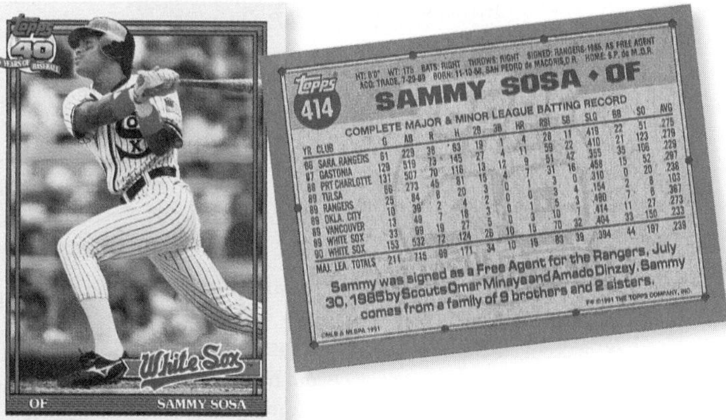

Figure 9.18
A batting average is actually a percent yield.

between containers. Moreover, if reactants or products have not been carefully measured, a percent yield of 100% is unlikely.

An actual yield is an experimental value. **Figure 9.19** shows a typical laboratory procedure for determining the actual yield of a product of a decomposition reaction. If you do not do an experiment, you cannot calculate a percent yield unless you are given the value of an actual yield. For reactions in which percent yields have been determined, you can calculate and therefore predict an actual yield if the reaction conditions remain the same. A farmer's crop yield could also be expressed as a percent yield. What factors would a farmer use to predict a theoretical yield? ❶

Figure 9.19
(a) *The mass of sodium hydrogen carbonate ($NaHCO_3$), the reactant, is measured.* ***(b)*** *The reactant is heated.* ***(c)*** *The mass of one of the products, sodium carbonate (Na_2CO_3), the actual yield, is measured after the reaction is completed. The percent yield can be calculated once the actual yield has been determined. What are the other products of this reaction?* ❷

(a)

(b)

(c)

Answers

❶ amount of rainfall, number of acres under cultivation, average number of sunny days per year, amount of fertilizer

❷ $H_2O(g) + CO_2(g)$

9.3

ACTIVITY

Many chemical reactions absorb or release heat as they occur. Thus, the change in temperature of a reaction mixture can be used to follow the progress of a reaction and serve as an indirect measure of the amount of product formed (the actual yield of a reaction). Have students measure the reaction between HCl and NaOH. Write the balanced chemical equation on the board for students to use as a reference. [$HCl(aq) + NaOH(aq) \rightarrow H_2O(l) + NaCl(aq)$ + heat] Point out that 1 mole of HCl reacts with 1 mole of NaOH to form 1 mole of NaCl and 1 mole of water.

Provide students with 3 styrofoam cups, a thermometer, 100 mL of 1.0*M* HCl, and approximately 200 mL of 1.0*M* NaOH. To begin, have students transfer 30 mL of the HCl to the styrofoam cup and measure the temperature. Next, add 10 mL of NaOH and gently stir the contents with the thermometer. Students should report the highest temperature reached as the reaction proceeds. Using new cups each time, repeat the measurement with 30 mL of HCl and 30 mL of NaOH, and with 30 mL of HCl and 60 mL of NaOH. Ask:

- Which reaction produced the greatest temperature change? (The one in which the reactants are present in stoichiometric amounts; 1:1 mole ratio.)
- Which reaction produces the greatest amount of product? (Again, the reaction for which the temperature change was the greatest, when both reactants are completely used up.)

CAUTION! Students should wear safety goggles. HCl and NaOH are corrosive! Caution students to avoid skin contact with these chemicals. Neutralized solutions can be flushed down the drain with excess water.

Practice Problems Plus

Related Chapter Review Problems
Chapter Review problems 47, 50, 54, 55, and 58 are related to Sample Problem 9-10.

Additional Practice Problem
The Chem ASAP! CD-ROM contains the following problem: If 75.0 g of siderite ore ($FeCO_3$) is heated with an excess of oxygen, 45.0 g of ferric oxide (Fe_2O_3) is produced.

$$4FeCO_3(s) + O_2(g) \rightarrow 2Fe_2O_3(s) + 4CO_2(g)$$

What is the percent yield of this reaction? (87.0%)

MINI LAB

Limiting Reagents

SAFETY

The balloons contain hydrogen gas and should be kept away from heat and open flames during and after the experiment. Caution students about the corrosive nature of HCl. Students should wear safety goggles. Once the reaction in a flask is complete, carefully remove the balloon. Disperse the gas in a fume hood or outside. Neutralize remaining HCl(*aq*) with baking soda before flushing down the drain.

LAB PREP AND PLANNING

Prepare an adequate amount of 1*M* HCl prior to the experiment.

ANALYSIS AND CONCLUSIONS

1. The approximate volumes of the balloons are 0.60 g Mg, 0.5 L H_2; 1.2 g Mg, 1 L H_2; 2.4 g Mg, 1 L H_2. Most students predict a doubling of the volume with each doubling of the mass of magnesium.
2. $Mg + 2HCl \rightarrow MgCl_2 + H_2$

Sample Problem 9-10

Calcium carbonate is decomposed by heating, as shown in the following equation.

$$CaCO_3(s) \xrightarrow{\Delta} CaO(s) + CO_2(g)$$

a. What is the theoretical yield of CaO if 24.8 g $CaCO_3$ is heated?

b. What is the percent yield if 13.1 g CaO is produced?

1. *ANALYZE List the knowns and the unknown for* a.

Knowns:
- mass of calcium carbonate = 24.8 g $CaCO_3$
- 1 mol $CaCO_3$ = 1 mol CaO (from balanced equation)
- 1 mol $CaCO_3$ = 100.1 g $CaCO_3$ (molar mass)
- 1 mol CaO = 56.1 g CaO (molar mass)

Unknown:
- theoretical yield of calcium oxide = ? g CaO

The theoretical yield can be calculated using the mass of the reactant:

$$\text{g } CaCO_3 \longrightarrow \text{mol } CaCO_3 \longrightarrow \text{mol CaO} \longrightarrow \text{g CaO}$$

2. *CALCULATE Solve for the unknown.*

First, the theoretical yield of the reaction is calculated.

$$24.8 \text{ g } CaCO_3 \times \frac{1 \text{ mol } CaCO_3}{100.1 \text{ g } CaCO_3} \times \frac{1 \text{ mol CaO}}{1 \text{ mol } CaCO_3} \times \frac{56.1 \text{ g CaO}}{1 \text{ mol CaO}}$$

$$= 13.9 \text{ g CaO}$$

1. *ANALYZE List the knowns and the unknown for* b.

Knowns:
- actual yield = 13.1 g CaO
- theoretical yield = 13.9 g CaO (from **a**)
- percent yield = $\frac{\text{actual yield}}{\text{theoretical yield}} \times 100\%$

Unknown:
- percent yield = ? %

2. *CALCULATE Solve for the unknown.*

$$\text{percent yield} = \frac{\text{actual yield}}{\text{theoretical yield}} \times 100\%$$

$$\text{percent yield} = \frac{13.1 \text{ g CaO}}{13.9 \text{ g CaO}} \times 100\% = 94.2\%$$

3. *EVALUATE Does the result make sense?*

In this example, the actual yield is slightly less than the theoretical yield. Therefore, the percent yield should be slightly less than 100%. The answer should have three significant figures.

Practice Problems

27. When 84.8 g of iron(III) oxide reacts with an excess of carbon monoxide, 54.3 g of iron is produced.

$$Fe_2O_3(s) + 3CO(g) \longrightarrow 2Fe(s) + 3CO_2(g)$$

What is the percent yield of this reaction?

28. If 50.0 g of silicon dioxide is heated with an excess of carbon, 27.9 g of silicon carbide is produced.

$$SiO_2(s) + 3C(s) \longrightarrow SiC(s) + 2CO(g)$$

What is the percent yield of this reaction?

Chem ASAP!

Problem-Solving 28
Solve Problem 28 with the help of an interactive guided tutorial.

ANALYSIS AND CONCLUSIONS CONTINUED

3. A mass of 0.60 g Mg is 0.025 mol Mg. Because Mg and HCl react in a 1:2 mol ratio, HCl is in excess and Mg is limiting. According to the balanced equation, 0.025 mol Mg should produce 0.025 mol H_2. A mass of 1.2 g Mg is 0.050 mol Mg. According to the balanced equation, 0.050 mol Mg will react with 0.1 mol HCl to produce 0.05 mol H_2.
4. A mass of 2.4 g Mg is 0.10 mol Mg. Because there is only 0.1 mol HCl in the flask and, according to the equation, 0.20 mol HCl is needed to react with 0.10 mol Mg, HCl is limiting. Therefore, only 0.05 mol H_2 is produced.

MINI LAB

Limiting Reagents

PURPOSE

To illustrate the concept of a limiting reagent in a chemical reaction.

MATERIALS

- graduated cylinder
- balance
- 3 250-mL Erlenmeyer flasks
- 3 rubber balloons
- 4.2 g magnesium ribbon
- 300 mL 1.0*M* hydrochloric acid

PROCEDURE

1. Add 100 mL of the hydrochloric acid solution to each flask.
2. Weigh out 0.6 g, 1.2 g, and 2.4 g of magnesium ribbon, and place each sample into its own balloon.
3. Stretch the end of each balloon over the mouth of each flask. Do not allow the magnesium ribbon in the balloon to fall into the flask.
4. Magnesium reacts with hydrochloric acid to form hydrogen gas. When you mix the magnesium with the hydrochloric acid in the next step, you will generate a certain volume of hydrogen gas. How do you think the volume of hydrogen produced in each flask will compare?
5. Lift up on each balloon and shake the magnesium metal down into each flask. Observe the volume of gas produced until the reaction in each flask is completed.

0.6 g Mg 1.2 g Mg 2.4 g Mg

ANALYSIS AND CONCLUSIONS

1. How did the volumes of hydrogen gas produced, as measured by the size of the balloons, compare? Did the results agree with your prediction?
2. Write a balanced equation for the reaction between magnesium metal and hydrochloric acid.
3. The 100 mL of hydrochloric acid contained 0.10 mol HCl. Show by calculation why the balloon with 1.2 g Mg inflated to about twice the size of the balloon with 0.60 g Mg.
4. Show by calculation why the balloons with 1.2 g and 2.4 g Mg inflated to approximately the same volume. What was the limiting reagent when 2.4 g Mg was added to the acid?

section review 9.3

29. What is a limiting reagent? An excess reagent?

30. What is the percent yield if 4.65 g of copper is produced when 1.87 g of aluminum reacts with an excess of copper(II) sulfate?

$$2Al(s) + 3CuSO_4(aq) \longrightarrow Al_2(SO_4)_3(aq) + 3Cu(s)$$

31. What is the difference between an actual yield and a theoretical yield? Which yield is larger for a given reaction? How are these values used to determine percent yield?

32. How many grams of SO_3 are produced when 20.0 g FeS_2 reacts with 16.0 g O_2 according to this balanced equation?

$$4FeS_2(s) + 15O_2(g) \longrightarrow 2Fe_2O_3(s) + 8SO_3(g)$$

 Chem ASAP! **Assessment 9.3** Check your understanding of the important ideas and concepts in Section 9.3.

3 Assess

Evaluate Understanding

To evaluate students' understanding of limiting and excess reagents, write several balanced equations on the board. Give students different amounts of reactants and ask them to calculate theoretical yields. Give the students "hypothetical" actual yields of products and ask them to calculate percent yields. Place a small piece of zinc in a large beaker of dilute hydrochloric acid. Explain the reaction to the students.
[$Zn(s) + 2HCl(aq) \rightarrow ZnCl_2(aq) + H_2(g)$]

Ask:

- **Which of the two reactants is the limiting reagent?** (Answers will vary. If the zinc completely disappears after some time, students may conclude that the zinc is used up first, and is, therefore, the limiting reagent.)

After students respond, ask how they could test their hypothesis. (Measure the volume of H_2 produced for various amounts of zinc added to a fixed volume of HCl. Determine the stoichiometric quantity of zinc, the amount which reacts to give the greatest volume of H_2.)

Reteach

Emphasize the importance of having a correctly balanced equation in order to properly calculate the maximum theoretical yield for a reaction. Explain that the way to decide which substance, if any, is the limiting reagent, is to compare the mole ratios of the given substances to the required mole ratio, as shown in the balanced chemical equation.

Answers

SECTION REVIEW 9.3

29. Limiting reagent determines the amount of product possible from a reaction; excess reagent is not used up by the reaction.

30. 70.5%

31. The actual yield is an experimentally determined value; a theoretical yield is a value calculated using the balanced chemical equation. The theoretical yield is normally larger than the actual yield. The percent yield is a ratio of the actual yield over the theoretical yield expressed as a percentage.

32. 21.4 g SO_3

Chemistry Serving... Society

DISCUSS

After students have read the article, have them write down the equations

$2NaN_3(s) \rightarrow 2Na(s) + 3N_2(g)$

$10Na(s) + 2KNO_3(s) \rightarrow K_2O(s) + 5Na_2O(s) + N_2(g)$

in their notebooks. Ask them how many moles of potassium nitrate must be included in the reaction mixture to consume the sodium produced by the decomposition of one mole of sodium azide. (0.2 mol KNO_3) Ask:

- How many liters of N_2 are produced at STP if 1.0 mole of sodium azide and 0.20 mole of potassium nitrate react? (36 L)

Ask students to speculate about how the pressure of the gas inside the airbag depends on the number of moles of nitrogen produced and the temperature inside the car.

CHEMISTRY IN CAREERS

Have students read about careers in quality control on page 872 of this text. If you have access to the Internet in your classroom or school, you may wish to have students connect to the address shown to learn more about recycling and reclamation of Earth's resources.

Chemistry Serving...Society

JUST THE RIGHT VOLUME OF GAS

The steering wheel shown here displays the letters SRS, which stand for Supplemental Restraint System and indicate that the steering wheel is equipped with an air bag. If you are lucky, you will never have to experience an air bag. But if you do have a front-end collision while driving, an air bag may save your life. A sensor in the front of the car detects the sudden deceleration and sends a signal to a cylinder containing a mixture of chemicals. In the cylinder, an igniter goes off, starting a series of chemical reactions that release a large volume of nitrogen gas. The gas fills the air bag, and you hit the soft bag instead of the steering wheel or dashboard. All this happens in less than a second.

The design of the air bag may sound fairly simple, but it was actually quite a challenge. Engineers had to come up with a system that was safe and effective and that would still work after many years of not being used. Perhaps most importantly, the bag had to inflate in less than a tenth of a second, and it had to inflate with exactly the right amount of gas. If it underinflated it would not provide enough protection; if it overinflated it would cause injury or it might even rupture. To design an air bag inflation system that met these requirements, engineers had to choose the right chemical reactants and pay close attention to the stoichiometry of the reactions.

> The engineers had to determine the exact quantity of each reactant to include in the reaction mixture and to allow for the change in gas density caused by the heat released in the reaction.

The inflation system that engineers settled on uses a series of chemical reactions that take place very quickly. The first reaction—set off by the igniter—is the decomposition of sodium azide into sodium metal and nitrogen gas

$$2NaN_3(s) \longrightarrow 2Na(s) + 3N_2(g).$$

By itself, this reaction cannot fill the air bag fast enough, and the sodium metal that is produced is dangerously reactive. To solve these problems, the engineers included potassium nitrate in the mixture of reactants. The potassium nitrate reacts with the sodium produced in the first reaction, releasing even more nitrogen gas

$$10Na + 2KNO_3 \longrightarrow K_2O + 5Na_2O + N_2$$

The heat released by this reaction raises the temperature of the gaseous product, helping the bag inflate even faster. The heat causes all the solid reaction products to fuse together with SiO_2, powdered sand, which is also part of the reaction mixture. This forms a safe, unreactive glass.

The volume of gas produced by these reactions is the key factor in getting the proper bag inflation. The volume depends on both the quantity of reactants and the temperature of the gas. Thus, the engineers had to determine the exact quantity of each reactant to include in the reaction mixture and to allow for the change in gas pressure caused by the heat released in the reaction. Even the quantity of potassium nitrate included in the mixture had to be precisely determined to make sure no sodium metal was left over from the first reaction. So if an air bag ever saves you from injury in a front-end collision, be thankful that the chemical engineers who designed the air bag's inflation system knew their stoichiometry!

CHEMISTRY IN CAREERS
QUALITY CONTROL CHEMIST
Interested in a career maintaining high-quality standards?
See page 872.
CHEMIST NEEDED
TAKE IT TO THE NET
Find out more about career opportunities:
www.phschool.com
CHEMICAL SPECIALIST
Local food service distributor seek
responsible self-motivated indivic

Chapter 9 STUDENT STUDY GUIDE

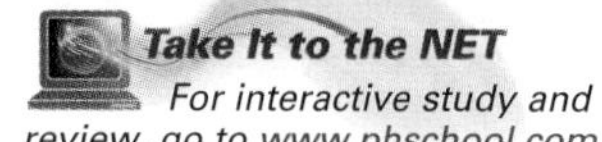

For interactive study and review, go to www.phschool.com

KEY TERMS

- actual yield *p. 256*
- excess reagent *p. 253*
- limiting reagent *p. 253*
- percent yield *p. 256*
- stoichiometry *p. 237*
- theoretical yield *p. 256*

KEY EQUATIONS AND RELATIONSHIPS

- mole-mole relationship for $aG \longrightarrow bW$:

$$\underset{\text{Given}}{x \text{ mol } G} \times \underset{\text{Mole ratio}}{\frac{b \text{ mol } W}{a \text{ mol } G}} = \underset{\text{Calculated}}{\frac{xb}{a} \text{ mol } W}$$

- $\text{percent yield} = \dfrac{\text{actual yield}}{\text{theoretical yield}} \times 100\%$

CONCEPT SUMMARY

9.1 The Arithmetic of Equations

- The coefficients in a balanced chemical equation tell the relative number of moles of reactants and products.
- Chemists use moles to do chemical arithmetic, or stoichiometry.
- All stoichiometric calculations involving chemical reactions begin with a balanced equation because mass is conserved in every chemical reaction.
- The number and kinds of atoms in the reactants equal the number and kinds of atoms in the products.

9.2 Chemical Calculations

- Stoichiometric problems are solved using conversion factors derived from a balanced chemical equation.
- A conversion factor called a mole ratio relates the moles of a given substance to the moles of the desired substance.
- Units such as mass, volume of gases (at STP), and particles are converted to moles when working stoichiometry problems.

9.3 Limiting Reagent and Percent Yield

- Whenever quantities of two or more reactants are given in a stoichiometry problem, the limiting reagent must be identified.
- A limiting reagent is completely used up in a chemical reaction.
- The amount of limiting reagent determines the amount of product formed in a chemical reaction.
- If there is a single limiting reagent in a reaction, all the other reactants are in excess.
- A theoretical yield is the maximum amount of product that can be obtained from a given amount of reactants in a chemical reaction.
- An actual yield is the amount of product obtained when the reaction is carried out in the laboratory.
- A ratio of the actual yield to the theoretical yield, expressed as a percentage, is the percent yield of a reaction.

CHAPTER CONCEPT MAP

Use these terms to construct a concept map that organizes the major ideas of this chapter.

Chem ASAP! **Concept Map 9**
Create your Concept Map using the computer.

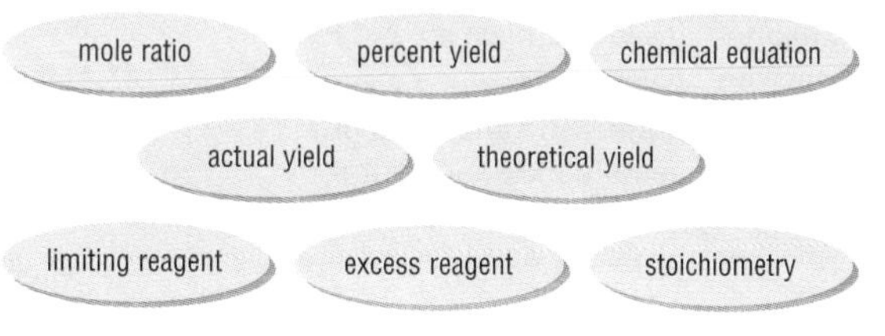

Take It to the Net

At **www.phschool.com** students will find for this chapter
- an Internet activity
- links to related chemistry sites
- an interactive quiz
- career links

4 Close

Summary

Ask the following questions, which require students to summarize information contained in the chapter:

- **What information do the coefficients in a balanced equation provide?** (the quantitative or stoichiometric relationships among reactants and products)
- **What is meant by the term *mole ratio?*** (It refers to the relative quantities of two components in a chemical equation. It is used as a conversion factor to relate the moles of one substance to the moles of another.)
- **Why are limiting reagents important?** (The actual yield of a reaction depends on the initial quantity of the limiting reagent. Ideally, to minimize waste, no reagent should be limiting.)
- **What is the percent yield of a reaction?** (the ratio of actual yield to theoretical yield expressed as a percentage)

Extension

Have students research chemical manufacturing processes to determine the importance of percent yields and waste.

Looking Back . . . Looking Ahead . . .

The previous chapter discussed reactions in general. This chapter focuses on the quantitative aspects of reactions. The next chapter presents kinetic theory to explain physical properties of gases, liquids, and solids.

Answers

33. a. Two formula units $KClO_3$ decompose to form 2 formula units KCl and 3 molecules O_2.
b. Four molecules NH_3 react with 6 molecules NO to form 5 molecules N_2 and 6 molecules H_2O.
c. Four atoms K react with 1 molecule O_2 to form 2 formula units K_2O.

34. a. Two mol $KClO_3$ decompose to form 2 mol KCl and 3 mol O_2.
b. Four mol NH_3 react with 6 mol NO to form 5 mol N_2 and 6 mol H_2O.
c. Four mol K react with 1 mol O_2 to form 2 mol K_2O.

35. a. 245.2 g **b.** 248.0 g **c.** 188.4 g All obey the law of conservation of mass.

36. Answers will vary but should include the idea of writing a ratio using the coefficients of two substances from a balanced equation as the number of moles of each substance reacting or being formed.

37. a. 0.54 mol
b. 13.6 mol
c. 0.984 mol
d. 236 mol

38. a. 11.3 mol CO, 22.5 mol H_2
b. 112 g CO, 16.0 g H_2
c. 11.4 g H_2

39. a. 372 g F_2
b. 1.32 g NH_3
c. 123 g N_2F_4

40. The coefficients indicate the relative number of moles (or particles) of reactants and products.

41. a. 51.2 g H_2O
b. 5.71×10^{23} molecules NH_3
c. 23.2 g Li_3N

42. The amount of the limiting reagent determines the maximum amount of product that can be formed. The excess reagent is only partially consumed in the reaction.

43. To identify the limiting reagent, express quantities of reactants as moles; compare to the mole ratios from the balanced equation.

Chapter 9 REVIEW

CONCEPT PRACTICE

33. Interpret each chemical equation in terms of interacting particles. *9.1*

a. $2KClO_3(s) \longrightarrow 2KCl(s) + 3O_2(g)$

b. $4NH_3(g) + 6NO(g) \longrightarrow 5N_2(g) + 6H_2O(g)$

c. $4K(s) + O_2(g) \longrightarrow 2K_2O(s)$

34. Interpret each equation in Problem 33 in terms of interacting numbers of moles of reactants and products. *9.1*

35. Calculate and compare the mass of the reactants with the mass of the products for each equation in Problem 33. Show that each balanced equation obeys the law of conservation of mass. *9.1*

36. Explain the term mole ratio in your own words. When would you use this term? *9.2*

37. Carbon disulfide is an important industrial solvent. It is prepared by the reaction of coke with sulfur dioxide. *9.2*

$$5C(s) + 2SO_2(g) \longrightarrow CS_2(l) + 4CO(g)$$

a. How many moles of CS_2 form when 2.7 mol C reacts?

b. How many moles of carbon are needed to react with 5.44 mol SO_2?

c. How many moles of carbon monoxide form at the same time that 0.246 mol CS_2 forms?

d. How many mol SO_2 are required to make 118 mol CS_2?

38. Methanol (CH_3OH) is used in the production of many chemicals. Methanol is made by reacting carbon monoxide and hydrogen at high temperature and pressure. *9.2*

$$CO(g) + 2H_2(g) \longrightarrow CH_3OH(g)$$

a. How many moles of each reactant are needed to produce 3.60×10^2 g CH_3OH?

b. Calculate the number of grams of each reactant needed to produce 4.00 mol CH_3OH.

c. How many grams of hydrogen are necessary to react with 2.85 mol CO?

39. The reaction of fluorine with ammonia produces dinitrogen tetrafluoride and hydrogen fluoride. *9.2*

$$5F_2(g) + 2NH_3(g) \longrightarrow N_2F_4(g) + 6HF(g)$$

a. If you have 66.6 g NH_3, how many grams of F_2 are required for complete reaction?

b. How many grams of NH_3 are required to produce 4.65 g HF?

c. How many grams of N_2F_4 can be produced from 225 g F_2?

40. What information about a chemical reaction is derived from the coefficients in a balanced equation? *9.2*

41. Lithium nitride reacts with water to form ammonia and aqueous lithium hydroxide. *9.2*

$$Li_3N(s) + 3H_2O(l) \longrightarrow NH_3(g) + 3LiOH(aq)$$

a. What mass of water is needed to react with 32.9 g Li_3N?

b. When the above reaction takes place, how many molecules of NH_3 are produced?

c. Calculate the number of grams of Li_3N that must be added to an excess of water to produce 15.0 L NH_3 (at STP).

42. What is the significance of the limiting reagent in a chemical process? What happens to the amount of any reagent that is present in an excess? *9.3*

43. How would you identify a limiting reagent in a chemical reaction? *9.3*

44. For each balanced equation, identify the limiting reagent for the given combination of reactants. *9.3*

a. $2Al$ (3.6 mol) $+ 3Cl_2$ (5.3 mol) $\longrightarrow 2AlCl_3$

b. $2H_2$ (6.4 mol) $+ O_2$ (3.4 mol) $\longrightarrow 2H_2O$

c. $2P_2O_5$ (0.48 mol) $+ 6H_2O$ (1.52 mol) $\longrightarrow 4H_3PO_4$

d. $4P$ (14.5 mol) $+ 5O_2$ (18.0 mol) $\longrightarrow 2P_2O_5$

45. For each reaction in Problem 44, calculate the number of moles of product formed. *9.3*

46. For each reaction in Problem 44, calculate the number of moles of excess reagent remaining after the reaction. *9.3*

47. Heating an ore of antimony (Sb_2S_3) in the presence of iron gives the element antimony and iron(II) sulfide.

$$Sb_2S_3(s) + 3Fe(s) \longrightarrow 2Sb(s) + 3FeS(s)$$

When 15.0 g Sb_2S_3 reacts with an excess of Fe, 9.84 g Sb is produced. What is the percent yield of this reaction? *9.3*

48. In an experiment, varying masses of sodium metal are reacted with a fixed initial mass of chlorine gas. The amounts of sodium used and the amounts of sodium chloride formed are shown on the following graph. *9.3*

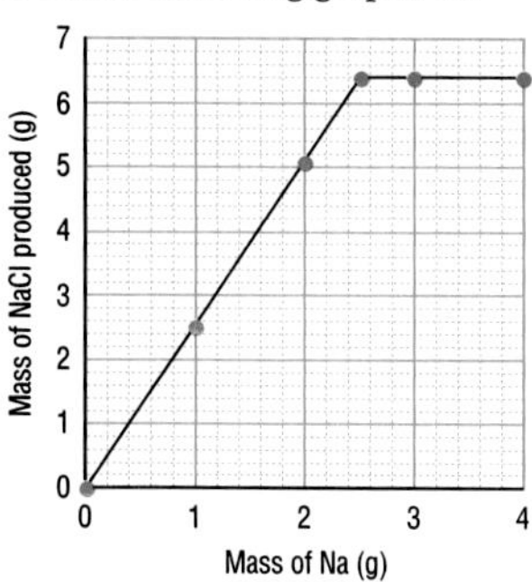

a. Explain the general shape of the graph.
b. Estimate the amount of chlorine gas used in this experiment at the point where the curve becomes horizontal.

49. What does the percent yield of a chemical reaction measure? *9.3*

50. The manufacture of compound F requires five separate chemical reactions. The initial reactant, compound A, is converted to compound B, compound B is converted to compound C, and so forth. The diagram below summarizes the stepwise manufacture of compound F, including the percent yield for each step. Provide the missing quantities or missing percent yields. Assume that the reactant and product in each step react in a one-to-one mole ratio. *9.3*

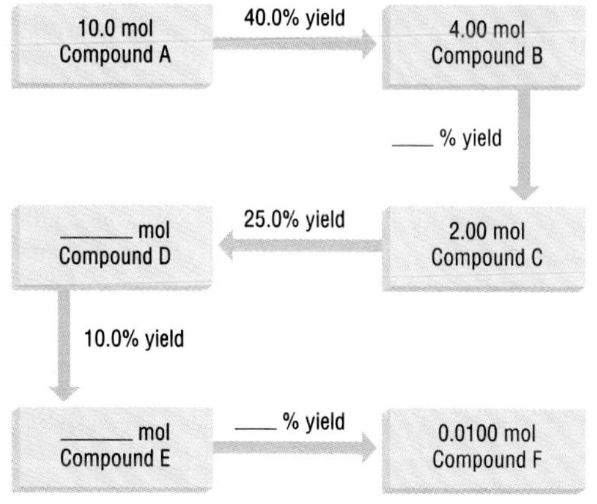

CONCEPT MASTERY

51. Calcium carbonate reacts with phosphoric acid to produce calcium phosphate, carbon dioxide, and water.

$$3CaCO_3(s) + 2H_3PO_4(aq) \longrightarrow Ca_3(PO_4)_2(aq) + 3CO_2(g) + 3H_2O(l)$$

a. How many grams of phosphoric acid react with excess calcium carbonate to produce 3.74 g $Ca_3(PO_4)_2$?
b. Calculate the number of grams of CO_2 formed when 0.773 g H_2O is produced.

52. Nitric acid and zinc react to form zinc nitrate, ammonium nitrate, and water.

$$4Zn(s) + 10HNO_3(aq) \longrightarrow 4Zn(NO_3)_2(aq) + NH_4NO_3(aq) + 3H_2O(l)$$

a. How many atoms of zinc react with 1.49 g HNO_3?
b. Calculate the number of grams of zinc that must react with an excess of HNO_3 to form 29.1 g NH_4NO_3.

53. Hydrazine (N_2H_4) is used as rocket fuel. It reacts with oxygen to form nitrogen and water.

$$N_2H_4(l) + O_2(g) \longrightarrow N_2(g) + 2H_2O(g)$$

a. How many liters of N_2 (at STP) form when 1.0 kg N_2H_4 reacts with 1.0 kg O_2?
b. How many grams of the excess reagent remain after the reaction?

54. When 50.0 g of silicon dioxide is heated with an excess of carbon, 32.2 g of silicon carbide is produced.

$$SiO_2(s) + 3C(s) \longrightarrow SiC(s) + 2CO(g)$$

a. What is the percent yield of this reaction?
b. How many grams of CO gas are made?

55. If the reaction below proceeds with a 96.8% yield, how many kilograms of $CaSO_4$ are formed when 5.24 kg SO_2 reacts with an excess of $CaCO_3$ and O_2?

$$2CaCO_3(s) + 2SO_2(g) + O_2(g) \longrightarrow 2CaSO_4(s) + 2CO_2(g)$$

56. Ammonium nitrate will decompose explosively at high temperatures to form nitrogen, oxygen, and water vapor.

$$2NH_4NO_3(s) \longrightarrow 2N_2(g) + 4H_2O(g) + O_2(g)$$

What is the total number of liters of gas formed when 228 g NH_4NO_3 is decomposed? (Assume STP.)

44. a. Cl_2 c. P_2O_5
b. H_2 d. O_2

45. a. 3.5 mol $AlCl_3$
b. 6.4 mol H_2O
c. 0.96 mol H_3PO_4
d. 7.20 mol P_2O_5

46. a. 0.1 mol Al
b. 0.2 mol O_2
c. 0.08 mol H_2O
d. 0.1 mol P

47. 91.5%

48. a. Initially, the amount of NaCl formed increases as the amount of Na used increases. For this part of the curve sodium is the limiting reagent. Beyond a mass of about 2.5 g of Na, the amount of product formed remains constant because chlorine is now the limiting reagent.
b. Chlorine becomes the limiting reagent when the mass of sodium exceeds 2.5 g. This corresponds to a mass of about 3.9 g chlorine.

49. The efficiency of a reaction; the actual yield/theoretical yield expressed as a percent.

50. 50.0% yield; 0.500 mol; 0.0500 mol 20.0% yield

51. a. 2.36 g H_3PO_4
b. 1.89 g CO_2

52. a. 5.70×10^{21} atoms Zn
b. 95. 2 g Zn

53. a. 7.0×10^2 L N_2
b. no reagent in excess

54. a. 96.4%
b. 45.0 g

55. 10.7 kg $CaSO_4$

56. 224 L gas

Chapter 9

REVIEW

Answers

57. a. 2
b. 4

58. The percent yield is 115%; such a yield could be attributed to experimental error, to unreacted starting material, or to outside materials contaminating the product.

59. Yes; a net ionic equation is balanced and thus obeys the law of conservation of mass.

60. a. 29 frames b. 58 wheels
c. 174 pedals d. 87 seats

61. a. $2Pb(NO_3)_2 \rightarrow 2PbO + 4NO_2 + O_2$
b. $2C_3H_7OH + 9O_2 \rightarrow 6CO_2 + 8H_2O$
c. $2Al + 3FeO \rightarrow 3Fe + Al_2O_3$

62. 7.38 g Be

63. a. 1, 1, 1, 2
b. 1, 3, 3, 1
c. 1, 1, 1, 2

64. a. $Ba^{2+}(aq) + SO_4^{2-}(aq) \rightarrow BaSO_4(s)$
b. $Ag^+(aq) + Cl^-(aq) \rightarrow AgCl(s)$
c. $H^+(aq) + OH^-(aq) \rightarrow H_2O(l)$

65. a. sodium ion and nitrate ion
b. aluminum ion and nitrate ion
c. magnesium ion and sulfate ion

66. a. 22,22,25
b. 50,50,70
c. 8,8,10
d. 12,12,14

67. 1.30×10^{-22} g C_6H_6

68. a. $Al_2(CO_3)_3$
b. SiO_2
c. K_2S
d. $MnCrO_4$
e. NaBr

69. $C_2H_2O_4$

70. 1.1×10^6 L air

71. 13 days

72. 1.96 g/L

73. 87.4% $CaCO_3$

CRITICAL THINKING

57. Choose the term that completes the second relationship.

a. equation:coefficients — balance:_______
(1) moles (3) weight
(2) standard masses (4) atoms

b. actual:theoretical — experimental:_____
(1) excess (3) real
(2) limiting (4) calculated

58. Given a certain quantity of reactant, you calculate that a particular reaction should produce 55 g of a product. When you perform the reaction, you find that you have produced 63 g of product. What is your percent yield? What could have caused a percent yield greater than 100%?

59. Would the law of conservation of mass hold in a net ionic equation? Explain.

60. A bicycle-built-for-three has a frame, two wheels, six pedals, and three seats. The balanced equation for this bicycle is

$$F + 2W + 6P + 3S \longrightarrow FW_2P_6S_3$$

How many of each part are needed to make 29 bicycles-built-for-three?
a. frames b. wheels c. pedals d. seats

CUMULATIVE REVIEW

61. Write a balanced chemical equation for each reaction.
a. When heated, lead(II) nitrate decomposes to form lead(II) oxide, nitrogen dioxide, and molecular oxygen.
b. The complete combustion of isopropyl alcohol (C_3H_7OH) produces carbon dioxide and water vapor.
c. When a mixture of aluminum and iron(II) oxide is heated, metallic iron and aluminum oxide are produced.

62. How many grams of beryllium are in 147 g of the mineral beryl ($Be_3Al_2Si_6O_{18}$)?

63. Balance each equation.
a. $Ba(NO_3)_2(aq) + Na_2SO_4(aq) \longrightarrow BaSO_4(s) + NaNO_3(aq)$
b. $AlCl_3(aq) + AgNO_3(aq) \longrightarrow AgCl(s) + Al(NO_3)_3(aq)$
c. $H_2SO_4(aq) + Mg(OH)_2(aq) \longrightarrow MgSO_4(aq) + H_2O(l)$

64. Write a net ionic equation for each reaction in Problem 63.

65. Identify the spectator ions in each reaction in Problem 63.

66. How many electrons, protons, and neutrons are in an atom of each isotope?
a. titanium-47 c. oxygen-18
b. tin-120 d. magnesium-26

67. What is the mass, in grams, of a molecule of benzene (C_6H_6)?

68. Write the formula for each compound.
a. aluminum carbonate
b. silicon dioxide
c. potassium sulfide
d. manganese(II) chromate
e. sodium bromide

69. What is the molecular formula of oxalic acid, molar mass 90 g/mol? Its percent composition is 26.7% C, 2.2% H, and 71.1% O.

CONCEPT CHALLENGE

70. A car gets 9.2 kilometers to a liter of gasoline. Assuming that gasoline is 100% octane (C_8H_{18}), which has a specific gravity of 0.69, how many liters of air (21% oxygen by volume at STP) will be required to burn the gasoline for a 1250-km trip? Assume complete combustion.

71. Ethyl alcohol (C_2H_5OH) can be produced by the fermentation of glucose ($C_6H_{12}O_6$). If it takes 5.0 h to produce 8.0 kg of alcohol, how many days will it take to consume 1.0×10^3 kg of glucose? (An enzyme is used.)

$$C_6H_{12}O_6 \xrightarrow{\text{enzyme}} 2C_2H_5OH + 2CO_2$$

72. A 1004.0-g sample of $CaCO_3$ that is 95.0% pure gives 225 L CO_2 at STP when reacted with an excess of hydrochloric acid.

$$CaCO_3 + 2HCl \longrightarrow CaCl_2 + CO_2 + H_2O$$

What is the density (in g/L) of the CO_2?

73. The white limestone cliffs of Dover, England, contain a large percentage of calcium carbonate ($CaCO_3$). A sample of limestone weighing 84.4 g reacts with an excess of hydrochloric acid to form calcium chloride.

$$CaCO_3 + 2HCl \longrightarrow CaCl_2 + H_2O + CO_2$$

The mass of calcium chloride formed is 81.8 g. What is the percentage of calcium carbonate in the limestone?

Chapter 9 STANDARDIZED TEST PREP

Select the choice that best answers each question or completes each statement.

1. Nitric acid is formed by the reaction of nitrogen dioxide with water.
   $3NO_2(g) + H_2O(l) \longrightarrow NO(g) + 2HNO_3(aq)$
   How many mol of water are needed to react with 8.4 mol NO_2?
   a. 2.8 mol c. 8.4 mol
   b. 3.0 mol d. 25 mol
2. Phosphorus trifluoride is formed from its elements.
   $P_4(s) + 6F_2(g) \longrightarrow 4PF_3(g)$
   How many grams of fluorine are needed to react with 6.20 g of phosphorus?
   a. 2.85 g c. 11.4 g
   b. 5.70 g d. 37.2 g
3. Which of the following are needed to calculate the percent yield of a reaction?
   I. theoretical yield
   II. excess yield
   III. actual yield
   a. I only
   b. I and II only
   c. I and III only
   d. II and III only
   e. I, II, and III
4. According to the balanced equation for the combustion of ethane, C_2H_6,
   $2C_2H_6(g) + 7O_2(g) \longrightarrow 4CO_2(g) + 6H_2O(l)$,
   for every molecule of ethane that reacts,
   a. 2 molecules of carbon dioxide are produced.
   b. 3 molecules of water are produced.
   c. Both (a) and (b) are correct.
   d. Neither (a) nor (b) is correct.
5. Magnesium nitride is formed in the reaction of magnesium metal with nitrogen gas.
   $3Mg(s) + N_2(g) \longrightarrow Mg_3N_2(s)$
   The reaction of 4.0 mol of nitrogen with 6.0 mol of magnesium produces
   a. 2.0 mol of Mg_3N_2 and 2.0 mol of excess N_2.
   b. 4.0 mol of Mg_3N_2 and 1.0 mol of excess Mg.
   c. 6.0 mol of Mg_3N_2 and 3.0 mol of excess N_2.
   d. no product because the reactants are not in the correct mole ratio.

Questions 6 and 7 involve the reaction between diatomic element P and diatomic element Q to form the compound P_3Q.

P_2

Q_2

P_3Q

6. Write a balanced equation for the reaction between element P and element Q.
7. Based on the atomic windows below, identify the limiting reagent.

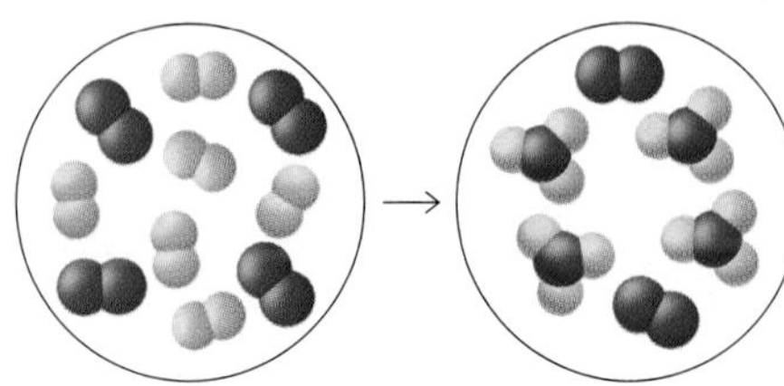

For each question there are two statements. Decide whether each statement is true or false. Then decide whether Statement II is a correct explanation for Statement I.

Statement I		Statement II
8. Every stoichiometry calculation uses a balanced equation.	BECAUSE	Every chemical reaction obeys the law of conservation of mass.
9. A percent yield is always greater than 0% and less than 100%.	BECAUSE	The actual yield in a reaction is never more than the theoretical yield.
10. The amount of the limiting reagent left after a reaction is zero.	BECAUSE	The limiting reagent is completely used up in a reaction.
11. The coefficients in a balanced equation represent the relative masses of the reactants and products.	BECAUSE	The mass of the reactants must equal the mass of the products in a chemical reaction.
12. A mole ratio is always written with the larger number in the numerator.	BECAUSE	A mole ratio will always be greater than one.

Chapter 9 STANDARDIZED TEST PREP

1. a
2. c
3. c
4. c
5. a
6. $3P_2 + Q_2 \rightarrow 2P_3Q$
7. P_2 is the limiting reagent
8. True, True, correct explanation
9. False, True
10. True, True, correct explanation
11. False, True
12. False, False

Planning Guide

SECTION OBJECTIVES	ACTIVITIES/FEATURES	MEDIA & TECHNOLOGY
10.1 The Nature of Gases ○ ■ ◆ ▸ Describe the motion of gas particles according to the kinetic theory ▸ Interpret gas pressure in terms of kinetic theory	**SE** Discover It! *Observing Gas Pressure*, p. 266 **SE** **CHEMath** *Making and Interpreting Graphs*, p. 270 **SE** **Link to Medicine** *Cryogenics*, p. 272 **SE** Small-Scale Lab *Kinetic Theory in Action*, p. 273 (LRS 10-1) **TE** DEMOS, pp. 268, 271	**ASAP** Animation 9 **ASAP** Problem Solving 1 **ASAP** Assessment 10.1 **ACT** 5: *States of Matter* **CA** *Superheated Steam* **CHM** Side 5, 5: *Imploding Can* **SSV** 7: *Chemistry of Gases* **OT** 19: *Kinetic Theory* **OT** 20: *Barometers* **OT** 21: *Kinetic Energy and Temperature*
10.2 The Nature of Liquids ● ■ ◆ ▸ Describe the nature of a liquid in terms of the attractive forces between the particles ▸ Differentiate between evaporation and boiling of a liquid, using kinetic theory	SSLM 12: *Absorption of Water by Paper Towels: A Consumer Lab* **TE** DEMO, p. 276	**ASAP** Animation 10, 11 **ASAP** Assessment 10.2 **ACT** 5: *States of Matter* **CHM** Side 5, 22: *Why a Solid Is a Solid* **CHM** Side 5, 25: *Viscosity Derby* **OT** 22: *Evaporation* **OT** 23: *The Manometer and Vapor Pressure* **OT** 24: *Boiling Point and Pressure*
10.3 The Nature of Solids ○ ■ ◆ ▸ Describe how the degree of organization of particles distinguishes solids from gases and liquids ▸ Distinguish between a crystal lattice and a unit cell ▸ Explain how allotropes of an element differ	**TE** DEMOS, p. 281, 283	**ASAP** Assessment 10.3 **ACT** 5: *States of Matter* **OT** 25: *Crystal Lattices* **OT** 26: *Cubic Chapter Cells*
10.4 Changes of State □ ◆ ▸ Interpret the phase diagram of water at any given temperature and pressure ▸ Describe the behavior of solids that change directly to the vapor state and recondense to solids without passing through the liquid state	**SE** **Link to Food Science** *Freeze-Drying*, p. 285 **SE** Mini Lab *Sublimation*, p. 286 (LRS 10-2) **SE** **Chemistry Serving . . . the Environment** *Gases Can Alter Global Temperature*, p. 287 **SE** **Chemistry in Careers** *Climatologist*, p. 287 LM 14: *Changes of Physical State*	**ASAP** Simulation 7 **ASAP** Assessment 10.4 **ASAP** Concept Map 10 **ACT** 5: *States of Matter* **CA** *Phase Change* **CHM** Side 5, 20: *Liquefaction of CO_2* **OT** 27: *Phase Diagram of Water* **RP** Lesson Plans, Resource Library **AR** Computer Test 10 **www** Activities, Self-Tests, *SCIENCE NEWS* updates **TCP** The Chemistry Place Web Site

KEY

● Conceptual (concrete concepts)
○ Conceptual (more abstract/math)
■ Standard (core content)
□ Standard (extension topics)
◆ Honors (core content)
◇ Honors (options to accelerate)

AR	Assessment Resources
ASAP	Chem ASAP! CD-ROM
ACT	ActivChemistry CD-ROM
CHM	CHEMedia Videodiscs
CA	Chemistry Alive! Videodiscs
GCP	Graphing Calculator Problems
GRS	Guided Reading and Study Workbook
LM	Laboratory Manual
LP	Laboratory Practicals
LRS	Laboratory Recordsheets
SSLM	Small-Scale Lab Manual

PRACTICE / ASSESSMENT

PRACTICE	ASSESSMENT
SE **Sample Problem** 10-1 GRS Section 10.1 RM Practice Problems 10.1	SE Section Review RM Section Review 10.1 RM Chapter 10 Quiz
GRS Section 10.2 RM Practice Problems 10.2	SE Section Review RM Section Review 10.2 RM Chapter 10 Quiz
GRS Section 10.3 RM Practice Problems 10.3	SE Section Review RM Section Review 10.3 LP Lab Practical 10-1 RM Chapter 10 Quiz
GRS Section 10.4 RM Practice Problems 10.4 RM Interpreting Graphics	SE Section Review RM Section Review 10.4 RM Vocabulary Review 10 SE Chapter Review SM Chapter 10 Solutions SE Standardized Test Prep PHAS Chapter 10 Test Prep RM Chapter 10 Quiz RM Chapter 10 A & B Test

PLANNING FOR ACTIVITIES

STUDENT EDITION

Discover It! p. 266
- index cards
- graduated cylinders
- small glasses
- water

Small-Scale Lab, p. 273
- plastic cups or Petri dishes
- pencils, paper
- rulers
- cotton swabs
- medicine droppers
- reaction surfaces
- $NaHSO_3$, NH_4Cl, KI, $NaNO_2$, NaOH, HCl
- bromthymol blue

Mini Lab, p. 286
- solid air fresheners
- small shallow containers
- clear plastic cups (8 oz)
- hot tap water
- ice
- thick cardboard strips

TEACHER'S EDITION

Teacher Demo, p. 268
- "Newtonian Cradle"

Activity, p. 269
- small cardboard containers
- ball bearings

Teacher Demo, p. 271
- empty aluminum can
- hot plate
- pan of cool water

Teacher Demo, p. 276
- manometer
- two-hole stopper
- 1-L flask
- separatory funnel
- 50 mL of acetone
- glass tubing bent at 90° angle
- rubber hose
- ice water

Activity, p. 278
- thermometers
- beaker of water
- heat source

Teacher Demo, p. 281
- aqueous solution of detergent
- watch glass
- straw
- overhead projector

Activity, p. 282
- books about rocks and minerals

Teacher Demo, p. 283
- hammer
- lead, salt, ice

OT	Overhead Transparency	SE	Student Edition
PHAS	PH Assessment System	SM	Solutions Manual
PLM	Probeware Lab Manual	SSV	Small-Scale Video/Videodisc
PP	Problem Pro CD-ROM	TCP	www.chemplace.com
RM	Review Module	TE	Teacher's Edition
RP	Resource Pro CD-ROM	www	www.phschool.com

Chapter 10

Key Terms

10.1 kinetic energy, kinetic theory, gas pressure, vacuum, atmospheric pressure, barometers, pascal (Pa), standard atmosphere (atm)
10.2 vaporization, evaporation, vapor pressure, boiling point, normal boiling point
10.3 melting point, crystal, unit cell, allotropes, amorphous solids, glasses
10.4 phase diagram, triple point, sublimation

DISCOVER IT!

Students should observe that the index card sticks to the glass and the water remains in the inverted glass. The number of grams of water equals the number of milliliters. After students have studied Section 10.1, they should infer that the external atmospheric pressure is greater than the pressure of the water in the inverted glass. Have students repeat the activity, substituting a saucer or pan of water for the index card. Compare the assembly to a barometer.

As an extension, ask students to calculate the total mass of water that could be held in the inverted glass. Explain that atmospheric pressure at sea level is about 14.7 psi (pounds per square inch). Have students express their results in both pounds and kilograms.

Chapter 10 STATES OF MATTER

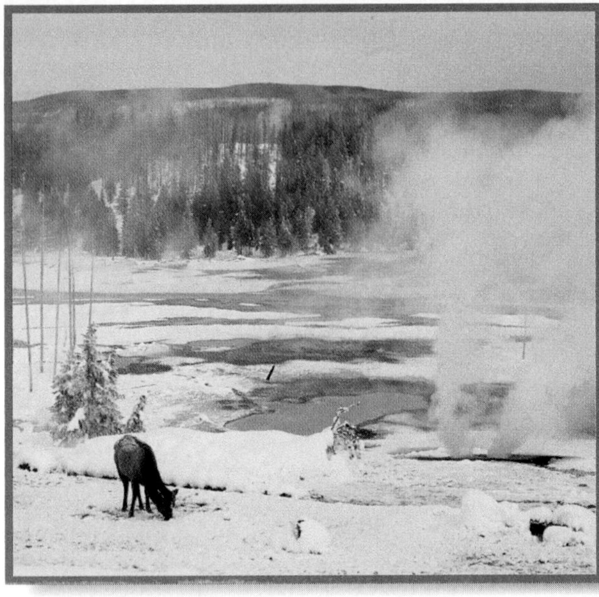

Elk experience three states of water at Yellowstone National Park.

FEATURES

Stay current with SCIENCE NEWS
Find out more about states of matter:
www.phschool.com

DISCOVER IT! OBSERVING GAS PRESSURE

You need an index card, a graduated cylinder, water, and a small glass with a smooth, even rim.

1. Fill the glass to its rim with water.
2. Place the index card on top of the glass.
3. Working over a sink, use one hand to press the index card firmly to the top of the glass. Then quickly invert the glass, keeping your hand in place.
4. Remove your hand from the index card.

What did you observe when you removed your hand? Measure the volume of water in the glass. How many grams of water are in the glass? (Remember that 1 mL H_2O = 1 g H_2O.) What keeps the water inside the glass? As you read about the properties of gases in this chapter, re-examine this activity and see if you can improve your explanation of what occurred.

10.1 ○ ■ ◆
10.2 ● ■ ◆
10.3 ○ ■ ◆
10.4 □ ◆

Conceptual Kinetic theory is important for students to visualize, but the more abstract illustrations are optional: energy distribution diagram (Figure 10.4), U-tube manometers (Figure 10.9), vapor pressure versus temperature (Figure 10.11), and crystal systems (Figure 10.14). Sample Problem 10-1 is optional if gas law calculations will not be done in Chapter 12. Section 10.4 may be omitted.

Standard Kinetic theory is an important prerequisite for Chapters 12 and 19. In Section 10.3, crystal structures do not need thorough coverage because students will revisit this topic in more detail in Chapter 15. In Section 10.4, sublimation is an important concept but phase diagrams may be omitted.

Honors In Section 10.1, accelerate the discussion of gas pressure. In Sections 10.2 and 10.3, A Model for Liquids and A Model for Solids are suitable for self-study. Use the phase diagram for water (Figure 10-18) to synthesize chapter concepts.

THE NATURE OF GASES

You are walking your dog in the woods, enjoying the great outdoors. Suddenly your dog begins to bark at what you believe is a black cat. But before you realize what it actually is, the damage is done. It's too late; the skunk has released its spray! Within seconds you smell that all-too recognizable smell. **How do the gaseous odor molecules travel from one place to another?**

section 10.1

objectives

- Describe the motion of gas particles according to the kinetic theory
- Interpret gas pressure in terms of kinetic theory

key terms

- kinetic energy
- kinetic theory
- gas pressure
- vacuum
- atmospheric pressure
- barometers
- pascal (Pa)
- standard atmosphere (atm)

Kinetic Theory

An ice cube is solid water; tap water is liquid water; and steam is gaseous water. Most substances commonly exist in only one of the three states of matter: solid, liquid, or gas. However, a substance in one state may change to another state with a change in temperature. You know that when cooled, water freezes to become ice, and steam condenses to form water. You also know that soon after you put on perfume, anyone nearby will be able to smell it. Obviously, the molecules of the perfume have moved through the air. How do you account for these behaviors? A model called the kinetic theory will help you find the answer.

The word kinetic refers to motion. The energy an object has because of its motion is called **kinetic energy.** The **kinetic theory** states that the tiny particles in all forms of matter are in constant motion. The following are the basic assumptions of the kinetic theory as it applies to gases.

1. A gas is composed of particles, usually molecules or atoms. These particles are considered to be small, hard spheres that have insignificant volume and are relatively far apart from one another. Between the particles there is empty space. No attractive or repulsive forces exist between the particles.
2. The particles in a gas move rapidly in constant random motion. They travel in straight paths and move independently of each other. As a result, gases fill their containers regardless of the shape and volume of the containers; uncontained gases diffuse into space without limit. The gas particles change direction only when they rebound from collisions with one another or with other objects. Measurements indicate that the average speed of oxygen molecules in air at 20 °C is an amazing 1700 km/h! At these high speeds, the odor molecules from a hot cheese pizza in Washington, D.C., should reach Mexico City in about 106 minutes. That does not happen, however, because the odor molecules are constantly striking molecules of air and rebounding in other directions. Their path of uninterrupted travel in a straight line is very short. The aimless path the gas molecules take is called a random walk. **Figure 10.1** on the following page illustrates a typical random walk.
3. All collisions are perfectly elastic. This means that during collisions kinetic energy is transferred without loss from one particle to another, and the total kinetic energy remains constant.

As you will learn next, the kinetic theory of gases is very helpful in explaining gas pressure.

STUDENT RESOURCES

From the Teacher's Resource Package, use:

- Section Review 10.1, Ch. 10 Practice Problems and Quizzes from the Review Module (Ch. 9–12)
- Laboratory Recordsheet 10-1

TECHNOLOGY RESOURCES

Relevant technology resources include:

- Chem ASAP! CD-ROM
- Resource Pro CD-ROM
- ActivChemistry CD-ROM: *States of Matter*
- Chemistry Alive! Videodisc: *Superheated Steam*

10.1

1 Engage

Use the Visual

Have students study the photograph and read the text that opens the section. Point out that Earth's atmosphere is a mixture of gases. A sample of this mixture passes through the nose and lungs with each breath. Ask:

- **How do gaseous odor molecules travel from one place to another?** (Once emitted by a skunk, each scent molecule has an equal probability of traveling in any direction. The odor is detected once one or more of the particles collide with and bind to a chemoreceptor in a person's nose.)

Check Prior Knowledge

To assess students' prior knowledge about the properties of gases, ask:

- **What physical property is used to describe the amount of gas in a closed container such as a car tire?** (pressure)
- **What units are normally used to express the pressure of a gas?** (Pounds per square inch, or psi; point out that pressure is defined as force per unit area.)

Motivate and Relate

Because many problems in this section involve unit conversions, you may wish to review this skill using practice problems related to gases. Explain that scientists sometimes use a unit of pressure called an atmosphere: one atm equals 14.69 psi. Ask:

- **The pressure in a bicycle tire is 32 psi. What is this pressure in atmospheres?** (2.2 atm)

2 Teach

Discuss

Create a condensed version of the postulates presented on page 267 and display it on an overhead projector. Review each assumption in turn. Remind students that all forms of matter are composed of particles, which are in constant random motion. To facilitate comprehension of subsequent topics, encourage students to memorize the postulates.

TEACHER DEMO

To explain the difference between "elastic" and "perfectly elastic" collisions, use a "Newtonian Cradle"—a device in which small steel balls are suspended by thin nylon tethers to horizontal wooden sticks. Pull back one or more of the balls and let it (them) fall into the other balls. After the balls come to rest, ask:

- Were the collisions between the balls elastic? (Yes, because kinetic energy was transferred with each collision.) Use collisions between croquet balls or pool balls as examples of "elastic" collisions.
- Why did the balls eventually stop swinging? (The collisions were not perfectly elastic; some kinetic energy was lost as heat during each collision.)

Explain that, at constant temperatures and low to moderate pressures, collisions between gas particles are perfectly elastic.

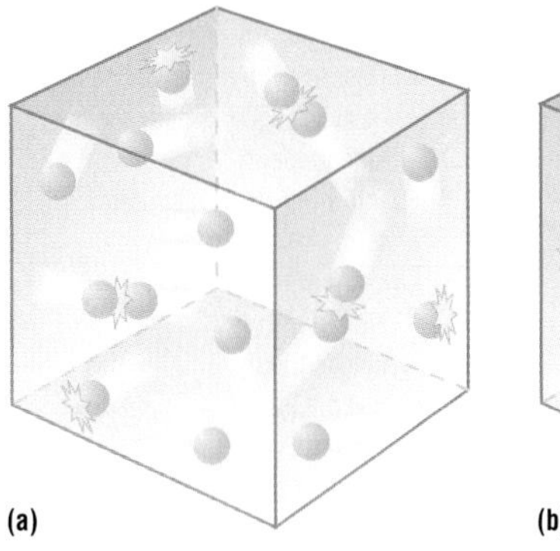

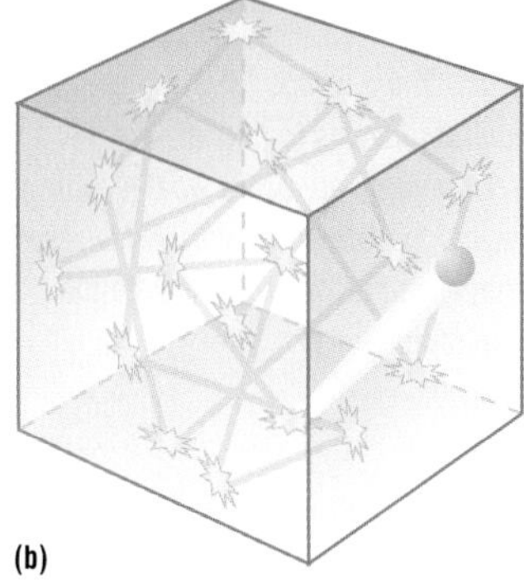

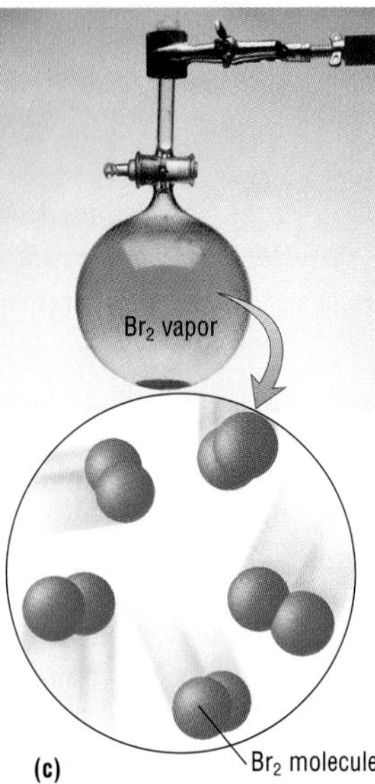

Figure 10.1
(a) Gas particles move randomly. They are constantly colliding with one another and with the walls of their containers. The collisions cause the particles to change direction frequently. (b) Every gas particle travels in a straight-line path until it collides with another particle, object, or wall. (c) A molecular view of a gas shows that the particles are relatively far apart and fill all available space. In this example, the gas is bromine vapor (Br_2).

Gas Pressure

Moving bodies exert forces when they collide with other bodies. Although a single gas particle is a moving body, the force it exerts is extremely small. Yet it is not hard to imagine that simultaneous collisions involving many such particles would produce a measurable force on an object. Gas pressure is the result of simultaneous collisions of billions of gas particles with an object. **Gas pressure** is defined as the force exerted by a gas per unit surface area of an object. If there are no gas particles present, there cannot be collisions, and there is consequently no pressure. Such an empty space, with no particles and no pressure, is called a **vacuum.**

Air exerts pressure on Earth because gravity holds air molecules in Earth's atmosphere. **Atmospheric pressure** results from the collisions of air molecules with objects. Atmospheric pressure decreases as you climb a mountain because the air layer around Earth thins out as elevation increases. **Barometers** are devices commonly used to measure atmospheric pressure. This pressure is dependent on weather.

The SI unit of pressure is the **pascal (Pa).** Atmospheric pressure at sea level is about 101.3 kilopascals (kPa). Two older units of pressure are millimeters of mercury (mm Hg) and atmospheres (atm). The origin of these units is the early use of mercury barometers similar to the one shown in **Figure 10.2.** Such a barometer was a straight glass tube filled with mercury and closed at one end. The tube was placed in a dish of mercury so that the open end was below the surface of mercury in the dish. The height of mercury in the tube depended on the pressure created by collisions of air molecules with the surface of the mercury in the dish. At sea level, this pressure is sufficient to support a mercury column about 760 mm high. One **standard atmosphere (atm)** is the pressure required to support 760 mm of mercury in a mercury barometer at 25 °C. That is, 1 atm = 760 mm Hg = 101.3 kPa.

MEETING DIVERSE NEEDS

Gifted Encourage students to examine critically how barometric pressure is reported in local newspapers and on television news weather forecasts. Ask:

- What units are commonly used by the news media to express barometric pressure? (Unfortunately, units are rarely stated by the news media; pressure is usually given in inches of mercury. Students may infer this based on the numerical values of the stated pressures.)

Bring in the weather sections of local and national newspapers and have students determine today's pressure in mm Hg, in kPa, and in atm. Students may wish to track the daily pressure and display the changes on a graph.

Vacuum
760 mm Hg (barometeric pressure)
Atmospheric pressure
Mercury
Sea level

Vacuum
253 mm Hg
Atmospheric pressure
Mercury
On top of Mount Everest

Figure 10.2
Typical atmospheric pressure at sea level in fair weather is 1 atm, which equals 760 mm Hg, or 101.3 kPa. Such atmospheric pressure supports a 760-mm column of mercury (101.3 kPa). On top of Mount Everest (at an altitude of 9000 m), the air exerts only enough pressure to support a 253-mm column of mercury. Express this pressure in kilopascals. 2

Many modern barometers do not contain mercury and are called aneroid barometers. In these devices, atmospheric pressure is related to the number of collisions of air molecules with a sensitive metal diaphragm. The diaphragm controls the movement of a pointer, which in turn indicates the pressure reading.

In the case of gases, it is important to be able to relate measured values to standards. Standard conditions are defined as a temperature of 0 °C and a pressure of 101.3 kPa, or 1 atm. This set of conditions is called standard temperature and pressure, or STP. What is standard temperature expressed on the Kelvin temperature scale? 1

Kinetic Energy and Kelvin Temperature

What happens when a substance is heated? The particles of the substance absorb energy, some of which is stored within the particles. This stored portion of the energy, or potential energy, does not raise the temperature of the

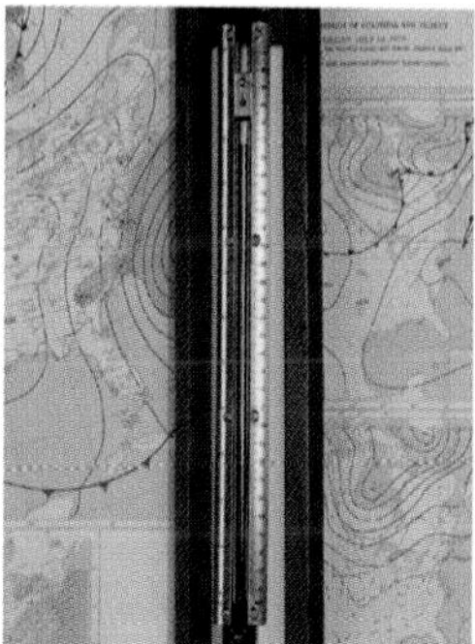

Figure 10.3
The standard lab barometer is still commonly used (left). An aneroid barometer (right) does not contain mercury.

Answers

1 273 kelvin
2 33.7 kPa

10.1

Think Critically

Hold up an inflated balloon and ask students to describe how the pressure inside the balloon would change if collisions between gas molecules were not perfectly elastic. (The total kinetic energy of the gas would decrease over time. Therefore, the average speed of the molecules would decrease. The gas would lose pressure and collapse. Such spontaneous collapses have never been observed.)

ACTIVITY

To help students develop an intuitive sense of the kinetic molecular theory as it applies to gases, set up small cardboard containers (e.g., milk cartons) filled with different numbers of ball bearings on a bench top in order of increasing concentration. Have students shake the containers in order and report their observations. Students should hear and feel the differences in "pressure" due to the frequency of collisions. Ask students to use what they learn from this activity to explain the relationship between the number of gas particles in a container and the pressure in the container. The effect of temperature on the kinetic energy, and hence on pressure, can be developed by supposing that the force with which a box is shaken represents the amount of heat transferred to the "gas particles." The "particles" have a greater average kinetic energy at higher shaking speeds. Ask students how the pressure changes with increasing speed of shaking.

CHEMath

Teaching Tips

Ask students to consider some of the advantages to presenting data in a graphical format. Point out that a graph is a way to summarize data in a format that is easy to read and understand. Graphs are tools that investigators use to discover trends and/or relationships that may exist among two sets of data and to quickly convey their findings to others.

Have students examine Figure 10.11. Explain that to obtain graphs, it is necessary to measure the vapor pressures of each liquid at many different temperatures. However, once the graphs are constructed, it is possible to predict the vapor pressure at other temperatures not directly examined in the laboratory.

Thus, in many instances, graphs are used to *interpolate* between data and *extrapolate* beyond data. Have students practice constructing graphs by hand and by using graphing calculators if available. Refer students to Appendix B.2 for more about making graphs.

Answers to Practice Problems

A. Range 0 °C to 120 °C; Interval 20 °C

B. 75 °C

C. 30 kPa

D.

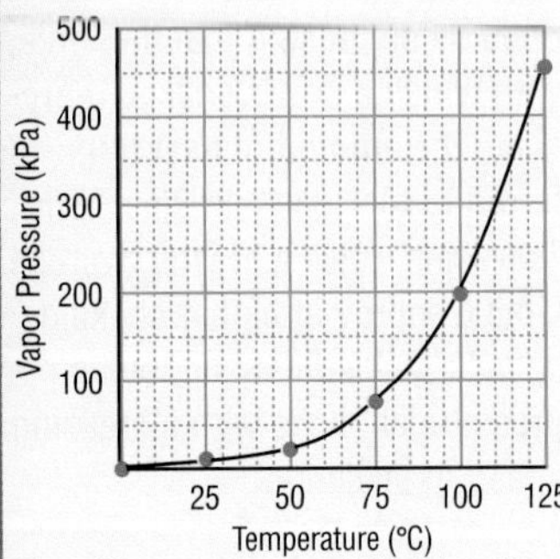

Vapor Pressure of Isopropyl Alcohol vs. Temperature

CHEMath

MAKING AND INTERPRETING GRAPHS

A graph is used to display relationships or data visually. You will encounter many graphs in your study of chemistry. Graphs are also used in economics, physics, social studies, and other fields. Your local newspaper probably includes several graphs each week.

In a graph, the horizontal axis (or x-axis) represents the *independent variable* and the vertical axis (or y-axis) represents the *dependent variable.* For example, in the graph shown below, the independent variable is temperature, and the dependent variable is vapor pressure. So, this graph shows vapor pressure vs. temperature. When stating a relationship, you should always give the dependent variable first.

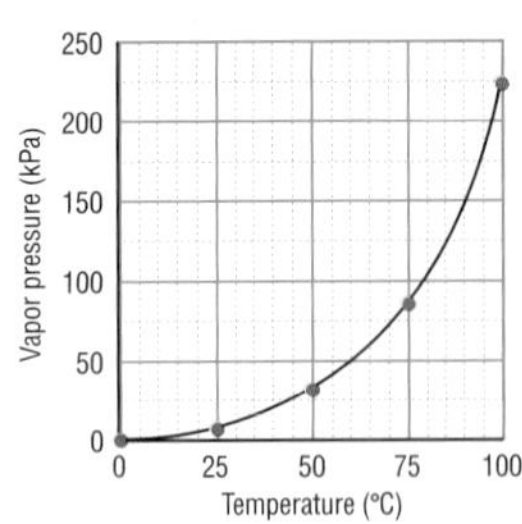

Vapor Pressure of Ethanol vs. Temperature

The *range* and *interval* for each axis determine the appearance of the graph. The range is given by the minimum and maximum values represented in the graph, and the interval is the distance between grid lines or tick marks. In the graph shown, the temperature range is 0 °C to 100 °C, and the temperature interval is 5 °C.

To make a graph, first choose appropriate ranges and scales. Choose the range for each axis so that all important data is included. The intervals should be convenient numbers (such as 1, 5, or 20) that result in an easy-to-read graph. After you draw and label the axes, plot the data. Each data point should be aligned with the correct values on the axes. When appropriate, connect the data with a smooth curve.

In Section 10.4, you will learn about the *phase diagram* for a substance (see Figure 10.18). A phase diagram is a pressure vs. temperature graph that has been divided into regions for each of the solid, liquid, and vapor phases. For a given pressure and temperature, you can determine from the graph if the substance is a solid, a liquid, or a gas.

Example 1

Refer to the graph above. Give the range and interval for vapor pressure. Then find the boiling point of ethanol when the atmospheric pressure is 60 kPa.

The vertical axis includes values from 0 kPa to 225 kPa with the grid lines 25 kPa apart. So, the vapor pressure range is 0 kPa to 225 kPa and the vapor pressure interval is 25 kPa.

To find the boiling point of ethanol when the atmospheric pressure is 60 kPa, notice that the curve appears to include the point (65 °C, 60 kPa). Therefore, the boiling point is about 65 °C at 60 kPa.

Practice Problems

Prepare for upcoming problems in this chapter by solving the following problems.

A. For the graph in Figure 10.11, what are the range and interval for temperature?

B. Refer to the graph above. Find the boiling point of ethanol when the atmospheric pressure is 90 kPa.

C. Refer to the graph above. What is the vapor pressure of ethanol when its temperature is 50 °C?

D. The table gives the vapor pressure of isopropyl alcohol (rubbing alcohol) at various temperatures. Graph the data. Connect the data points using a smooth curve.

Temperature (°C)	0	25	50	75	100	125
Vapor pressure (kPa)	1.11	6.02	23.9	75.3	198	452

Sample Problem 10-1

A gas is at a pressure of 1.50 atm. Convert this pressure to
a. kilopascals. **b.** millimeters of mercury.

1. ANALYZE ***List the knowns and the unknowns.***

Knowns:	*Unknowns:*
• pressure = 1.50 atm	**a.** pressure = ? kPa
• 1 atm = 101.3 kPa	**b.** pressure = ? mm Hg
• 1 atm = 760 mm Hg	

The given pressure is converted into the desired unit by multiplying by the proper conversion factor.

2. CALCULATE ***Solve for the unknowns.***

a. For the conversion atm ⟶ kPa, the conversion factor is $\frac{101.3 \text{ kPa}}{1 \text{ atm}}$.

$$1.50 \cancel{\text{atm}} \times \frac{101.3 \text{ kPa}}{1 \cancel{\text{atm}}} = 151.95 \text{ kPa} = 1.52 \times 10^2 \text{ kPa}$$

b. For the conversion atm ⟶ mm Hg, the conversion factor is $\frac{760 \text{ mm Hg}}{1 \text{ atm}}$.

$$1.50 \cancel{\text{atm}} \times \frac{760 \text{ mm Hg}}{1 \cancel{\text{atm}}} = 1140 \text{ mm Hg} = 1.14 \times 10^3 \text{ mm Hg}$$

3. EVALUATE ***Do the results make sense?***
Because the conversion factor for calculating mm Hg is larger than the one for kPa, it makes sense that the value expressed in mm Hg is larger than the value in kPa. In each case the given unit cancels, and the answer has the desired unit. Each answer has the three significant figures required.

Practice Problems

1. What pressure, in kilopascals and in atmospheres, does a gas exert at 385 mm Hg?
2. The pressure at the top of Mount Everest is 33.7 kPa. Is that pressure greater or less than 0.25 atm?

Chem ASAP!

Problem-Solving 1
Solve Problem 1 with the help of an interactive guided tutorial.

substance. The remaining absorbed energy speeds up the particles—that is, increases their average kinetic energy—which results in an increase in temperature. The particles in any collection of atoms or molecules at a given temperature have a wide range of kinetic energies, from very low to very high. Most of the particles have kinetic energies somewhere in the middle of this range. Therefore average kinetic energy is used when discussing the kinetic energy of a collection of particles in a substance. **Figure 10.4** on the following page shows the distribution of kinetic energies of gas particles at two different temperatures. Notice that at the higher temperature there is a wider range of kinetic energies.

An increase in the average kinetic energy of particles causes the temperature of a substance to rise. As a substance cools, the particles tend to move more slowly, and their average kinetic energy declines. You could reasonably expect the particles of all substances to stop moving at some very low temperature. The particles would have no kinetic energy at that temperature because they would have no motion. Absolute zero (0 K, or

10.1

TEACHER DEMO

Most students are unaware of the tremendous pressure exerted by Earth's atmosphere. Relate this pressure to the pressure a scuba diver feels in deep water. Explain to students that they live at the bottom of a very deep "ocean" of air. Fill an empty aluminum beverage container with water to a depth of about 1 to 2 cm. With the top open to the atmosphere, set the can on a hot plate and bring the water to a boil. Allow the water to boil vigorously for several minutes. Point out that the majority of the volume in the can is occupied by water vapor. Remove the can from the heat source, seal the opening, and allow the can to cool at room temperature. Alternatively, immediately invert the can in a pan of cool water. Once the water vapor inside the can condenses, the can is crushed by atmospheric pressure. Ask students to interpret their observations in terms of the particle theory of gases.

Practice Problems Plus

Related Chapter Review Problems
Chapter Review problems 22, 23, and 24 are related to Sample Problem 10-1.

Additional Practice Problem
The Chem ASAP! CD-ROM contains the following problem: What pressure, in mm Hg and atmospheres, does a sample of neon gas exert at 75.0 kPa? (563 mm Hg, 0.740 atm)

MEETING DIVERSE NEEDS

Gifted Unlike traditional barometers, aneroid barometers do not balance atmospheric pressure against a liquid of known density. In fact, they contain no liquid and are independent of gravity. If possible, bring one to class and explain its components. Have students research the advantages of aneroid barometers. (They are compact, self enclosed, and easily transported. They can operate under conditions not suited to mercury barometers.) Interested students may wish to construct an aneroid barometer and demonstrate its use.

Chemistry Alive!

Superheated Steam

Play

10.1

3 Assess

Evaluate Understanding

To assess students' comprehension of kinetic theory as it applies to gases, ask:

- **What is kinetic energy?** (energy due to the motion of an object)
- **How does the average kinetic energy of a collection of particles change with temperature?** (Average kinetic energy is directly proportional to the Kelvin temperature. Higher temperatures reflect a greater average kinetic energy and, thus, a greater average speed of particles.)

Reteach

Use Figure 10.4 to help summarize the relationship between temperature and the kinetic energy of a gas. Point out that the peak of each distribution curve corresponds to the most probable kinetic energy in that particular sample. The peak of the curve shifts to a higher value as the temperature of the gas is increased.

Use an air-filled balloon and a small pan of liquid nitrogen to illustrate the effect of temperature on the pressure of a gas. Relate the relative pressures in the balloon at room temperature and at the temperature of liquid nitrogen (−196 °C) to the energy distributions shown in Figure 10.4.

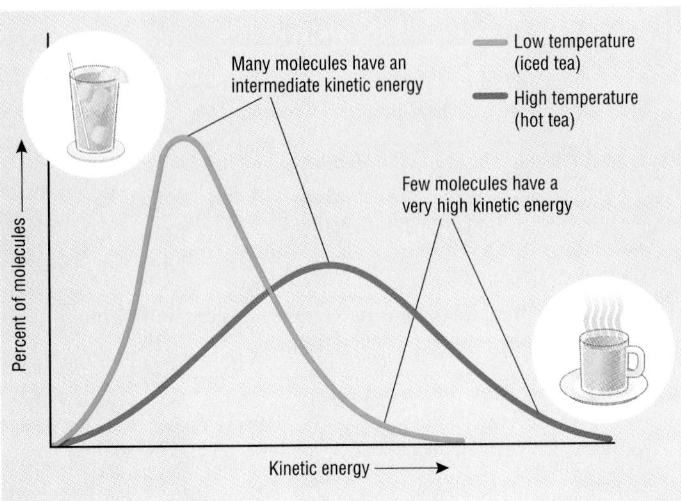

Figure 10.4
The blue curve shows the kinetic energy distribution of a typical collection of molecules at a fairly low temperature, such as the water molecules in iced tea. Notice that most molecules have intermediate energies. The red curve shows the energy distribution of molecules at a higher temperature, such as the water molecules in hot tea. How do the average kinetic energies of the two liquids compare? 2

Chem ASAP!
Animation 9
Observe particles in motion and discover the connection between temperature and kinetic energy.

MEDICINE

Cryogenics
Cryogenics is the science of producing very low temperatures and studying the behavior of matter at such temperatures. Biological reactions slow down or even stop at very low temperatures. Biologists use this fact to study reactions more easily. In medicine, some tissues can be removed from the body and preserved by rapid freezing. They can later be used when thawed. Blood and cartilage are examples of such materials. Doctors also perform cryosurgery by freezing tissue instead of cutting it. Cryosurgery minimizes bleeding during operations, and healing is rapid with minimal scarring.

−273.15 °C) is the temperature at which the motion of particles theoretically ceases. Absolute zero has never been produced in the laboratory, although temperatures of about 0.000 001 K have been achieved. Would 1 you expect to find negative temperatures on the Kelvin temperature scale?

The Kelvin temperature scale reflects the relationship between temperature and average kinetic energy. The Kelvin temperature of a substance is directly proportional to the average kinetic energy of the particles of the substance. For example, the particles in helium gas at 200 K have twice the average kinetic energy as the particles in helium gas at 100 K. As you will see later in this chapter, the effects of temperature on particle motion in liquids and solids are more complex than in gases. Nevertheless, at any given temperature the particles of all substances, regardless of physical state, have the same average kinetic energy.

section review 10.1

3. According to the assumptions of kinetic theory, how do the particles in a gas move?
4. Use kinetic theory to explain what causes gas pressure.
5. Express the pressure 545 mm Hg in kilopascals.
6. How can you raise the average kinetic energy of the water molecules in a glass of water?
7. A cylinder of oxygen gas is cooled from 300 K (27 °C) to 150 K (−123 °C). By what factor does the average kinetic energy of the oxygen molecules in the cylinder decrease?

Chem ASAP! **Assessment 10.1** Check your understanding of the important ideas and concepts in Section 10.1.

Answers

1 No; absolute zero is the lowest possible temperature on the Kelvin scale.
2 The molecules in hot tea have higher average kinetic energies.

SECTION REVIEW 10.1

3. Answers will vary but should include the idea that particles exhibit rapid, random motion with elastic collisions.
4. The collision of gas particles with an object causes gas pressure.
5. 72.6 kPa
6. Heat the water.
7. by one-half

SMALL-SCALE LAB

KINETIC THEORY IN ACTION

SAFETY   

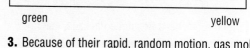

Wear safety glasses and follow the standard safety procedures as outlined on page 18.

PURPOSE

To observe color changes in the chemical reactions of gases and to interpret these changes in terms of kinetic theory.

MATERIALS

- clear plastic cup or Petri dish
- pencil
- paper
- ruler
- cotton swab
- medicine dropper
- reaction surface
- chemicals shown in Figure A and KI, $NaNO_2$, NH_4Cl, and NaOH

PROCEDURE

Use a clear plastic cup or Petri dish as a template to draw a large circle, as shown in Figure A. Place a reaction surface over the grid and add small drops of BTB (bromthymol blue) in the pattern shown by the small circles. Be sure the drops do not touch one another. Mix one drop each of HCl and $NaHSO_3$ in the center of the pattern. Place a clear plastic cup or Petri dish over the grid and observe what happens. Do not clean up until after you complete any assigned YOU'RE THE CHEMIST activities.

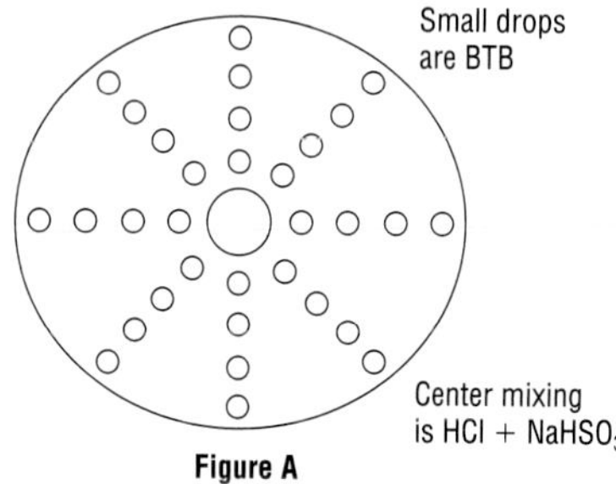

Figure A

ANALYSIS

Using your experimental data, record the answers to the following questions below your data table.

1. Describe in detail the changes you observed in the drops of BTB over time. Draw pictures to illustrate the changes.
2. Draw a series of pictures showing how one of the BTB drops might look over time if you could view the drop from the side.
3. The BTB changed even though you added nothing to it. If the mixture in the center circle produced a gas, would this explain the change in the drops of BTB? Use kinetic theory to explain your answer.
4. Translate the following word equation into a balanced chemical equation: Sodium hydrogen sulfite reacts with hydrochloric acid to produce sulfur dioxide gas, water, and sodium chloride.

YOU'RE THE CHEMIST

The following small-scale activities allow you to develop your own procedures and analyze the results.

1. **Analyze It!** Carefully absorb the center mixture of the original experiment onto a cotton swab and replace it with one drop of NaOH and one drop of NH_4Cl. What happens? Explain in terms of kinetic theory. Ammonium chloride reacts with sodium hydroxide to produce ammonia gas, water, and sodium chloride. Write and balance a chemical equation to describe this reaction.
2. **Design It!** Design an experiment to observe the effect of the size of the BTB drops on the rate at which they change. Explain your results in terms of kinetic theory.
3. **Analyze It!** Repeat the original experiment, except use KI in place of BTB and mix sodium nitrite ($NaNO_2$) with hydrochloric acid (HCl) at the center. Record your results. Write and balance an equation: Sodium nitrite reacts with hydrochloric acid to produce nitrogen monoxide gas, water, sodium nitrate, and sodium chloride.

ANALYSIS

1. The drops near the center change immediately. As the gas diffuses, all the drops change. The color change begins at the edge of each drop.

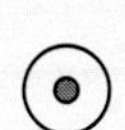

yellow ... green

2. green ... yellow

3. Because of their rapid, random motion, gas molecules produced in the center diffuse through the container and react with BTB.
4. $NaHSO_3 + HCl \rightarrow SO_2(g) + H_2O + NaCl$

SMALL-SCALE LAB

KINETIC THEORY IN ACTION

OVERVIEW

Procedure Time: 30 minutes
Students generate different gases and use indicators to visualize gas diffusion. They interpret their results in terms of kinetic molecular theory.

MATERIALS

Solution	Preparation
1.0*M* HCl	82 mL of 12*M* in 1.0 L
CAUTION! Always add acid to water slowly and carefully!	
0.04% BTB	100 mg in 16.0 mL 0.01*M* NaOH, dilute to 250 mL
0.5*M* NaOH	20.0 g in 1.0 L
1.0*M* NH_4Cl	13.4 g in 250 mL
0.4*M* $NaNO_2$	6.9 g in 250 mL
0.1*M* KI	4.2 g in 250 mL
1.0*M* $NaHSO_3$	26 g in 250 mL

TEACHING SUGGESTIONS

Remind students to set up the drops of BTB before they mix the HCl and $NaHSO_3$. To keep the gas contained, students must quickly cover the array. Encourage students to watch carefully as changes occur. If you want students to do You're The Chemist, remind them not to clean up until they do activities 1 and 2. If your stock BTB solution is not green, add a drop of HCl and then slowly add NaOH with stirring until the solution is green.

YOU'RE THE CHEMIST

1. As ammonia diffuses, BTB changes from yellow to blue. $NH_4Cl + NaOH \rightarrow NH_3(g) + H_2O + NaCl$
2. Vary the size of the BTB drops from "pin-heads" to "puddles." Tiny drops are better able to detect small quantities of gas.
3. KI drops change from colorless to yellow as iodine forms. $3NaNO_2 + 2HCl \rightarrow 2NO(g) + H_2O + NaNO_3 + 2NaCl$

1 Engage

Use the Visual

Have students study the photograph and read the text that opens the section. Ask:

- **What makes a liquid different from a solid?** (The particles in a liquid are relatively free to move and slide past one another; the particles in a solid are held together in a more rigid and ordered arrangement. Liquids flow; they can be poured and will assume the shape of their containers. In solids, the intermolecular forces are strong enough to hold particles in fixed positions; thus, solids retain their shape and volume.)

2 Teach

Use the Visual

Have students examine Figure 10.5. Point out that the term *vapor* is used to refer to the gaseous state of a substance that is normally a liquid or solid at room temperature. Explain that liquids and gases can be viewed as fluids because they both flow and change shape in response to slight outside perturbations. One property used to describe a fluid is *viscosity.* The viscosity of a fluid is a measure of its resistance to flow and depends on the ease with which individual particles of a fluid can move past each other. For example, motor oils are rated according to their relative viscosities. The viscosity of a fluid decreases with increasing temperature because the greater kinetic energy of the particles overcomes the intermolecular forces that oppose the motion of the particles.

section 10.2

THE NATURE OF LIQUIDS

objectives

- Describe the nature of a liquid in terms of the attractive forces between the particles
- Differentiate between evaporation and boiling of a liquid, using kinetic theory

key terms

- vaporization
- evaporation
- vapor pressure
- boiling point
- normal boiling point

The Kilauea volcano in Hawaii is the most active volcano in the world—it has been erupting for centuries. The hot lava oozes and flows, scorching everything in its path, occasionally including nearby houses. When the lava cools, it solidifies into rock. **What makes a liquid different from a solid?**

A Model for Liquids

You have just learned from the kinetic theory that gas pressure can be explained by assuming that gas particles have motion and that there is no attraction between the particles. The particles that make up liquids are also in motion, as **Figure 10.5** shows. Liquid particles are free to slide past one another. For that reason, both liquids and gases can flow, as you can see in **Figure 10.6.** However, the particles in a liquid are attracted to each other, while according to kinetic theory, those in a gas are not. The attractive forces between the molecules are called intermolecular forces.

The particles that make up liquids vibrate and spin while they move from place to place. All of these motions contribute to the average kinetic energy of the particles. Even so, most of the particles do not have enough kinetic energy to escape into the gaseous state. To do so, a particle must have sufficient kinetic energy to overcome the intermolecular forces that hold it together with the other particles. The intermolecular forces also reduce the amount of space between the particles in a liquid. Thus liquids are much more dense than gases. Increasing the pressure on a liquid has hardly any effect on its volume. The same is true of solids. For that reason, liquids and solids are known as condensed states of matter.

The interplay between the disruptive motions of particles and the attractive forces between them determines many of the physical properties of liquids. You will now explore two of these properties: vapor pressure and boiling point.

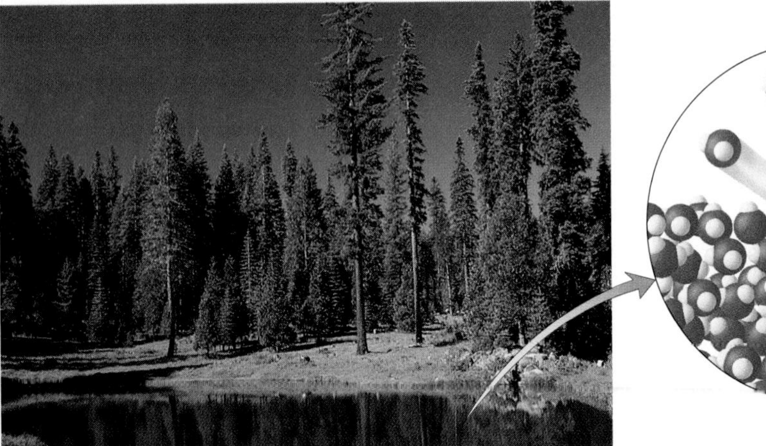

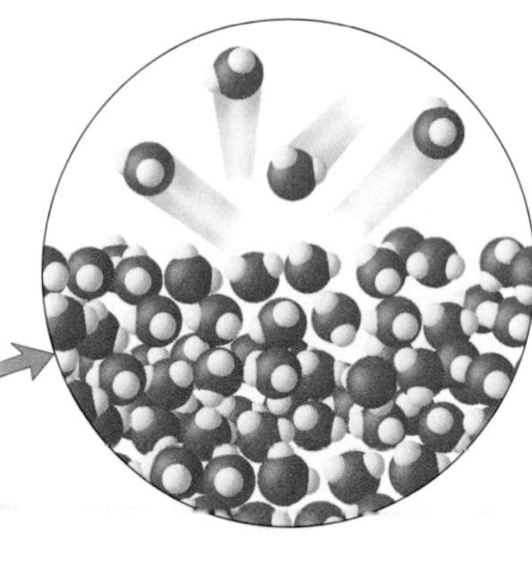

Figure 10.5
The particles in a liquid, such as the water in this lake, are close together but can move and slide around one another. The attractive forces between the particles generally prevent most of the particles from escaping the liquid and entering into the vapor state.

Student Resources

From the Teacher's Resource Package, use:

- Section Review 10.2, Ch. 10 Practice Problems and Quizzes from the Review Module (Ch. 9–12)
- Small-Scale Chemistry Lab Manual: Experiment 12

Technology Resources

Relevant technology resources include:

- Chem ASAP! CD-ROM
- Resource Pro CD-ROM
- ActivChemistry CD-ROM: *States of Matter*

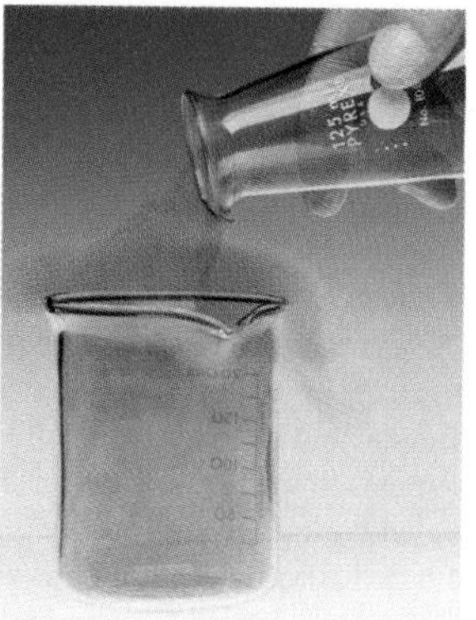

Figure 10.6
Both liquids and gases can flow. The liquid on the left is colored water. The gas on the right is bromine vapor. If a gas is denser than air, it can be poured from one container into another. These pictures were taken in a fume hood because bromine is toxic and corrosive. What will eventually happen to the gas in the uncovered beaker? 1

Evaporation

As you probably know from experience, water in an open vessel or a puddle eventually goes into the air. **See Figure 10.7a.** The conversion of a liquid to a gas or vapor is called **vaporization.** When such a conversion occurs at the surface of a liquid that is not boiling, the process is called **evaporation.** In evaporation, some molecules in the liquid break away and enter the gas or vapor state. Only those molecules of the liquid with a certain minimum kinetic energy can break away from the surface. Even some of the particles that do escape collide with air molecules and rebound into the liquid.

You may have noticed that a liquid evaporates faster when heated. This occurs because the added heat increases the average kinetic energy of the liquid's particles, enabling more of the particles to overcome the attractive forces keeping them in the liquid state. However, evaporation itself is a cooling process. The cooling occurs because the particles with the highest kinetic energy tend to escape first. The particles left in the liquid have a lower average kinetic energy than the particles that have escaped. Thus the liquid's temperature decreases. The process is similar to removing the fastest runner from a race. The remaining runners have a lower average speed.

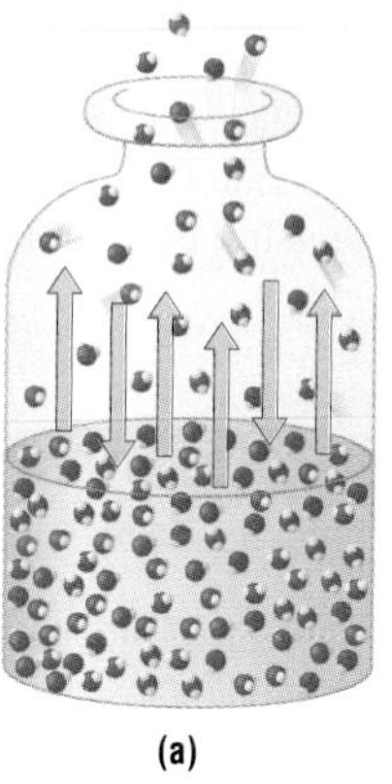

(a)

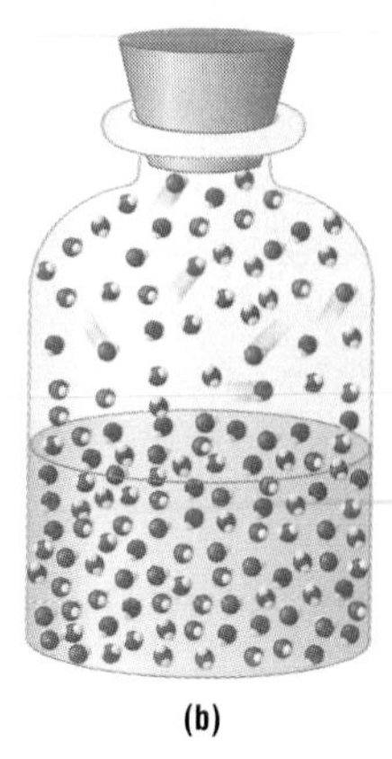

(b)

Figure 10.7
(a) *In an open container, water molecules evaporate from the liquid state and escape from the container. Equilibrium cannot be established in this system.* ***(b)*** *In a closed container, water molecules go into the vapor state, but they cannot escape from the system. They collect as a vapor above the liquid. A dynamic equilibrium between the water molecules in the vapor state and in the liquid state can be established.*

Chem ASAP!

Animation 10
Observe the phenomenon of evaporation from a molecular perspective.

10.2

Discuss

Draw simple diagrams of a collection of particles that show the relative amounts of space between particles in the gas, liquid, and solid states. The first diagram should show a disordered gaseous state in which particles are relatively far apart. The second diagram should show the liquid state in which particles are closer together but still somewhat disordered; gravity draws particles to the bottom of the container. The third diagram should show the ordered, fixed arrangement of particles in a solid. The spaces depicted between particles should be smallest in the solid. The diagrams should be shown side-by-side so that students can directly compare the three states of matter.

Use the diagrams to discuss physical properties of solids, liquids, and gases, including the relative densities of substances in the three physical states. Explain how the properties of a liquid lie between those of a solid and a gas. Ask students whether they think a liquid more closely resembles a gas or a solid. The diagrams may also be useful later when comparing the influence of intermolecular forces on the physical properties of a substance in each of the physical states.

Meeting Diverse Needs

LEP To help students relate vocabulary to concepts discussed in this section, help them make a concept map with these terms: *vaporization, contained liquid, uncontained liquid, evaporation, condensation, dynamic equilibrium,* and *vapor pressure.* Have them study Figure 10.7 and then add connecting words to the concept map explaining the relationships between the terms.

Answer

1 It will dissipate into the atmosphere.

TEACHER DEMO

Demonstrate how the vapor pressure of a liquid changes with temperature. You will need to obtain or construct a manometer and assemble the following apparatus. Insert a two-hole stopper into a 1-L flask. Into one hole of the stopper insert a separatory funnel capable of containing 50 mL of acetone. Into the other hole insert a short piece of glass tubing bent at a 90° angle. Connect the bent piece of glass tubing to one end of the manometer using a rubber hose. Be sure the hose fits both connections tightly. Measure the initial difference in pressure indicated by the levels of mercury on either side of the U-shaped manometer. Open the valve on the separatory funnel, let the acetone drain into the funnel and have students observe that the pressure in the funnel rises immediately. The maximum pressure observed is the vapor pressure of acetone at room temperature. Next, place the flask in ice water and have students observe the change in pressure. Help students interpret their observations in terms of the kinetic theory of matter.

Figure 10.8
This sealed terrarium is in liquid–vapor equilibrium. The rate of vaporization of the liquid water equals the rate of condensation of the water vapor. Condensation appears on the inner surface of the glass.

You have observed the effects of evaporative cooling on hot days. When you perspire, water molecules in your perspiration absorb heat from your body and evaporate from your skin's surface. This evaporation leaves the remaining perspiration cooler. The perspiration that remains cools you further by absorbing more body heat. Can you relate this feeling to the chill
1 you may possibly feel when getting out of a swimming pool on a warm, windy day?

The evaporation of a liquid in a closed container is somewhat different. **Figure 10.7b** on the previous page shows that no particles can escape into the outside air. When a partially filled container of liquid is sealed, some of the particles in the liquid vaporize. These particles collide with the walls of the sealed container and produce a **vapor pressure,** or a force due to the gas above the liquid. As time passes, the number of particles entering the vapor increases. Eventually, some of the particles will return to the liquid, or condense. The following equation illustrates the two opposing processes.

$$\text{Liquid} \underset{\text{condensation}}{\overset{\text{evaporation}}{\rightleftharpoons}} \text{Vapor (gas)}$$

After a time, the number of vapor particles condensing will equal the number of liquid particles vaporizing. The space in the container above the liquid is now saturated with vapor, and a dynamic equilibrium exists between the gas and the liquid. At the point of equilibrium,

$$\text{Rate of evaporation of liquid} = \text{Rate of condensation of vapor}$$

At equilibrium, the particles in the system continue to evaporate and condense, but there is no net change in the number of either liquid or vapor particles. The sealed terrarium in **Figure 10.8** is an example of a closed container at equilibrium. The moisture on the inner walls of the terrarium is a sign that equilibrium has been established. Particles that once evaporated are condensing, but other particles are evaporating to take their place. What
2 would happen if the cover of the terrarium were punctured?

An increase in the temperature of a contained liquid increases the vapor pressure. This happens because the particles in the warmed liquid have increased kinetic energy. As a result, more of the particles will have the minimum kinetic energy necessary to escape the surface of the liquid. The particles escape the liquid and collide with the walls of the container at a greater frequency. **Table 10.1** gives the vapor pressures of some common liquids at various temperatures. The vapor pressure of a liquid can be determined by means of a device called a manometer.

Table 10.1

Vapor Pressures (kPa) of Several Substances at Various Temperatures

	0 °C	20 °C	40 °C	60 °C	80 °C	100 °C
Water	0.61	2.33	7.37	19.92	47.34	101.33
Ethanol	1.63	5.85	18.04	47.02	108.34	225.75
Diethyl ether	24.70	58.96	122.80	230.65	399.11	647.87

MEETING DIVERSE NEEDS

At Risk Have students write a narrative about the molecular events that take place during the evaporation of a liquid and condensation of a gas. The story should include as many of the key terms from this section as possible and describe the cyclical nature of the evaporation-condensation equilibrium for a liquid in a closed container.

Answers

1 Wind causes water on the skin to evaporate, a cooling process.

2 There would be a net loss of water from the terrarium.

3 Vapor pressure increases.

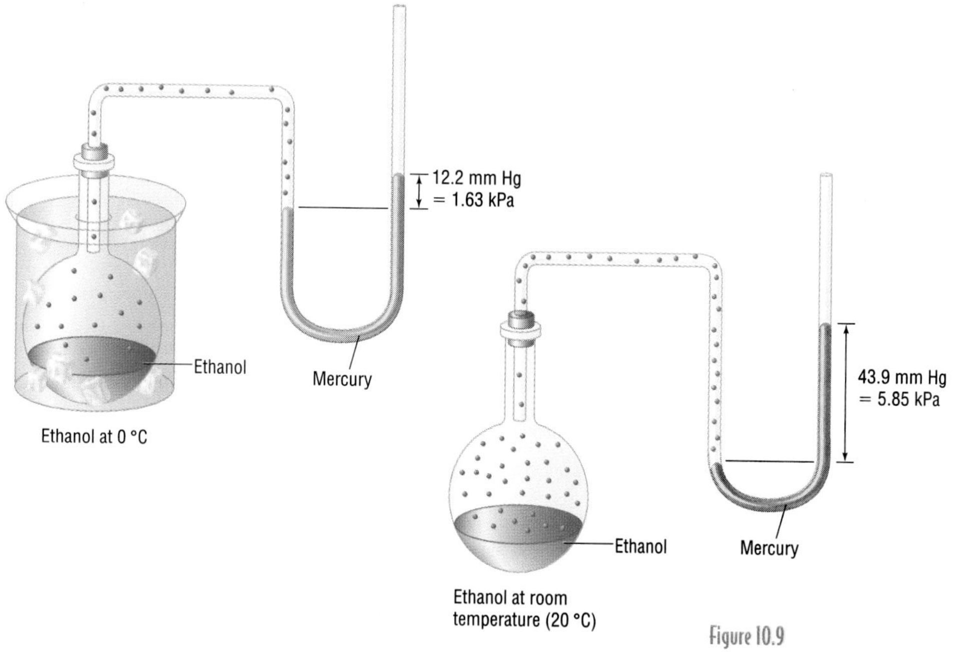

Figure 10.9

The height difference of the mercury in the two arms of the U-tube equals the vapor pressure of the contained liquid at that temperature. What happens to the vapor pressure as temperature increases? 3

In a simple manometer, one end of a U-shaped glass tube containing mercury is attached to a container. The other end of the tube is open to the surrounding atmosphere. When the container is empty, the mercury levels on both sides of the U-tube are the same because both sides of the tube are at the same pressure. When a liquid is added to the container, the pressure in the container increases. The increase in pressure is due to the vapor pressure of the liquid. The vapor pressure of the liquid pushes the mercury on the container side of the U-tube; the levels of mercury in the U-tube are no longer the same. The vapor pressure of the liquid is thus equal to the difference in the levels of mercury. **Figure 10.9** illustrates the use of a manometer to determine the effect of temperature on the vapor pressure of a liquid.

Boiling Point

You know that the rate of evaporation of a liquid from an open container increases as heat is added. Heating allows larger numbers of particles at the liquid's surface to overcome the attractive forces keeping them in the liquid state. The remaining particles in the liquid move faster and faster as they absorb the heat that is being added. Consequently, their average kinetic energy increases. Therefore the temperature of the liquid increases. When the liquid is heated to a high enough temperature, many of the particles throughout the liquid have enough kinetic energy to vaporize. At that point, boiling (or vaporization throughout the liquid) occurs. The **boiling point** (bp) is the temperature at which the vapor pressure of the liquid is just equal to the external pressure. Bubbles of vapor form throughout the liquid, rise to the surface, and escape into the air as the liquid boils.

10.2

Think Critically

Use cotton balls to simultaneously dab spots of water and rubbing alcohol onto the chalkboard. Have the class observe what happens to the spots. Ask students to infer which liquid has the greater vapor pressure at room temperature and explain how they made their inference.

Discuss

Explain that boiling is different from evaporation because evaporation is entirely a surface effect that can occur at any temperature. Boiling is a phenomenon in which bubbles of gas form anywhere within the interior of the liquid and then rise to the surface. Boiling takes place at specific temperatures and pressures. Atmospheric pressure opposes the formation of bubbles in a liquid. The atmosphere pushes on the liquid, reducing the likelihood of bubbles being formed. Bubbles form only when the vapor pressure inside a bubble is equal to atmospheric pressure. Under such conditions, boiling occurs.

10.2

ACTIVITY

Many students think that the temperature of a liquid increases as it boils. To correct this misconception, have students measure the temperature every 30 seconds for several minutes as water is heated and then boiled. Have students use their data to construct a graph of temperature versus time. Ask students to write a short paragraph explaining their results.

Use the Visual

Display Figure 10.11 on an overhead projector. Ask:

- **What happens to the vapor pressure of a liquid as the temperature increases? Explain.** (The vapor pressure increases with higher temperature because more of the particles in the warmed liquid will have the minimum kinetic energy needed to escape the surface of the liquid.)
- **What is the normal boiling of ethanoic acid?** (approximately 119 °C)

Ask students to construct a similar graph for diethyl ether using the data in Table 10.1. Have them use the graph to determine the normal boiling point of diethyl ether (34.5 °C).

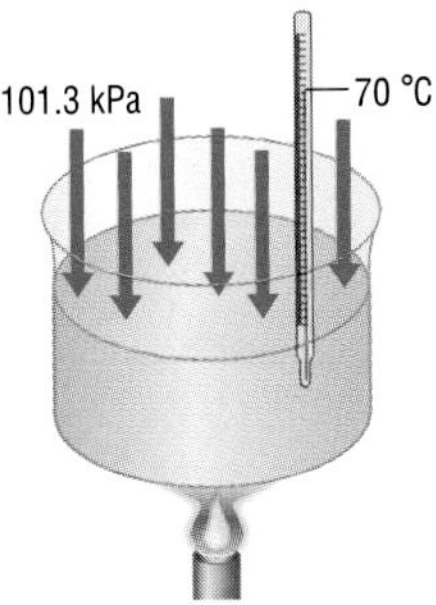

Sea Level

At 70 °C and 101.3 kPa, the atmospheric pressure at the surface of the liquid water is greater than the vapor pressure of the liquid. Bubbles of vapor cannot form in the liquid, and the liquid does not boil.

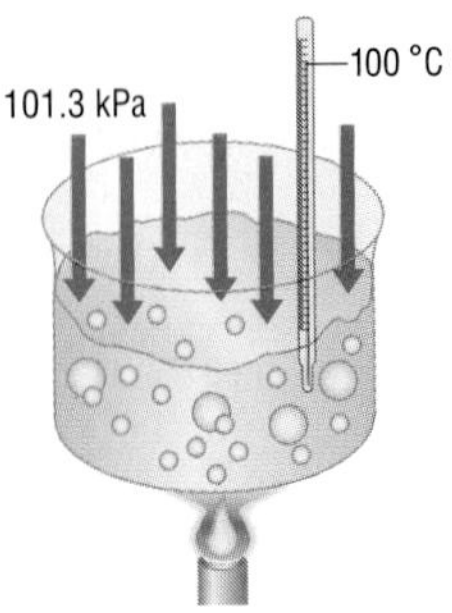

Sea Level

At the boiling point, the vapor pressure is equal to atmospheric pressure. Bubbles of vapor form in the liquid, and the liquid boils.

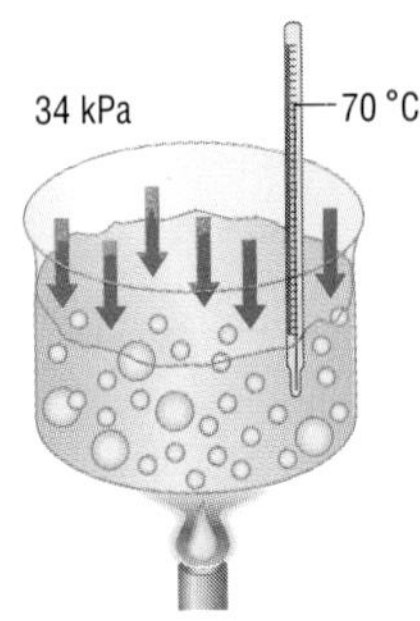

Atop Mount Everest

At high altitudes, the atmospheric pressure is lower than it is at sea level. Thus the liquid boils at a lower temperature.

Figure 10.10

A liquid boils when the vapor pressure of particles within the liquid equals the atmospheric pressure. What would happen to the boiling point of water at pressures above 101.3 kPa? ❶

As you just learned, the boiling point is the temperature at which the vapor pressure of the liquid is just equal to the external pressure. Thus the boiling point of a liquid changes if the external pressure changes. **Figure 10.10** illustrates this point, using water as the example. Therefore the **normal boiling point** is defined as the boiling point of a liquid at a pressure of 101.3 kPa. **Table 10.2** shows the normal boiling points of various substances. It follows then that the normal boiling point decreases at lower pressures and increases at higher pressures. The vapor-pressure diagram of **Figure 10.11** illustrates this fact. At lower pressures, the boiling point decreases because the particles need less kinetic energy to escape the liquids. Similarly, at higher external pressures, a liquid's boiling point increases because the particles in the liquid need more kinetic energy to escape.

The normal boiling point of water is 100 °C. In Denver, however, water boils at about 95 °C. This is because Denver is 1600 m above sea level and has an average atmospheric pressure of only 85.3 kPa (or 640 mm Hg, or 0.84 atm). You may know about pressure cookers, which reduce cooking time. At the high pressure in the cooker, water boils well above 100 °C, so food cooks faster. Autoclaves, devices often used to sterilize medical instruments, operate in a similar way.

Table 10.2

The Normal Boiling Points of Several Common Substances

Name and formula	Boiling point (°C)
Carbon disulfide (CS_2)	46.0
Chloroform ($CHCl_3$)	61.7
Ethanol (C_2H_6O)	78.5
Water (H_2O)	100.0

Answers

❶ It would increase above 100 °C.
❷ Steam; vapor molecules contain more potential energy than the liquid molecules.
❸ about 75 °C

MEETING DIVERSE NEEDS

Gifted A linear plot of vapor pressure versus temperature can be constructed using the algebraic relationship between vapor pressure and temperature given by the *Clausius-Clapeyron* equation. Ask students to use the data in Table 10.1 to plot log P versus $1/T$, where P is the vapor pressure in mm Hg and T is the absolute temperature.

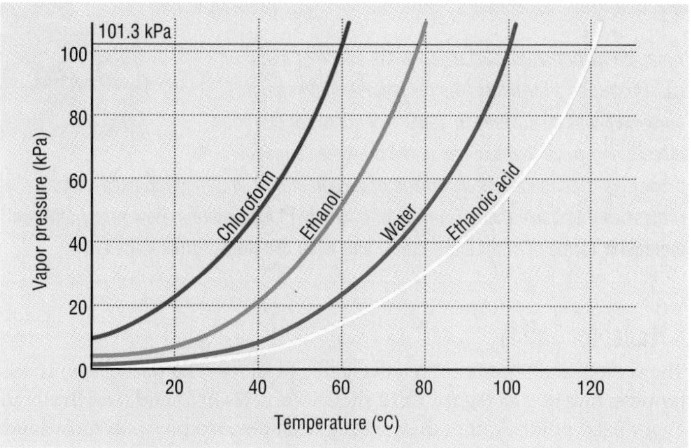

Figure 10.11
The intersection of a curve with the 101.3-kPa line indicates the normal boiling point of that substance. At about what temperature would ethanol boil in Denver, where the atmospheric pressure is 85.3 kPa? 3

Chem ASAP!

Animation 11
Relate vapor pressure and boiling point to intermolecular attractive forces.

Boiling is a cooling process similar to evaporation. In boiling, as in evaporation, the particles with the highest kinetic energy escape first when the liquid is at the boiling point. Turning off the source of external heat drops the liquid's temperature below its boiling point. Supplying more heat allows more particles to acquire enough kinetic energy to escape. However, the temperature of the boiling liquid never rises above its boiling point. If heat is supplied at a greater rate, the liquid only boils faster. The vapor produced is at the same temperature as that of the boiling liquid. It is important to note that although the vapor has the same kinetic energy, its potential (or stored energy) is much higher. Using this fact, which type of burn do you think is more severe: one from boiling water or one from an equal mass of steam at the same temperature? 2

section review 10.2

8. Describe the nature of liquids. Refer to the role of attractive forces in your answer.

9. Use kinetic theory to explain the differences between the particles in a gas and those in a liquid.

10. Use kinetic theory to explain the difference between evaporation and boiling of a liquid.

11. Use **Figure 10.11** to determine the boiling point of each liquid.
  a. ethanoic acid at 200 mm Hg
  b. chloroform at 600 mm Hg
  c. ethanol at 400 mm Hg

12. Explain why the boiling point of a liquid varies with atmospheric pressure.

13. Why does evaporation lower the temperature of a liquid?

 **Chem ASAP! Assessment 10.2** Check your understanding of the important ideas and concepts in Section 10.2.

portfolio project

Look for high-altitude cooking directions on packages of cake, brownie, or muffin mixes. Check the Internet for additional tips on high-altitude cooking.

10.2

3 Assess

Evaluate Understanding

To summarize section content ask:

- **What are some of the physical properties that distinguish a liquid from a gas?** (Gases can expand to fill their containers. Gases are easily compressed when pressure is applied. Liquids conform to the shape of their containers, but do not expand to fill them. Liquids are not as compressible as gases because the spaces between particles in a liquid are much smaller than those in a gas.)
- **What is meant by the term *vapor pressure?*** (Vapor pressure is the pressure exerted by gas molecules above a liquid when the liquid and its vapor are in dynamic equilibrium.)

Reteach

Remind students that the temperature at which the vapor pressure of a liquid is equal to the atmospheric pressure is the boiling point of the liquid. The exact temperature at which boiling occurs depends on the strength of the intermolecular forces between molecules in the liquid and the magnitude of the atmospheric pressure. Explain that when the atmospheric pressure is reduced below the vapor pressure of water, water can boil at room temperature.

Portfolio Project

Answers will vary. Examples of recommended adjustments include adding more flour or greasing non-stick muffin tins.

Answers

SECTION REVIEW 10.2

8. Liquid molecules slide past one another despite intermolecular forces.

9. Liquid particles are more attracted to one another and have little space between them.

10. Evaporation, which happens only at the surface, can occur below the boiling point.

11. a. about 76 °C **b.** about 50 °C **c.** about 62 °C

12. At boiling, vapor pressure equals external pressure. As external pressure increases, the temperature needed to produce an equivalent vapor pressure—the boiling point—must increase.

13. When the molecules with the highest kinetic energy escape, average kinetic energy of the liquid is lowered.

10.3

1 Engage

Use the Visual

Have students study the photograph and read the text that opens the section. Point out the resemblance between the shape of buckminsterfullerene and a geodesic dome. Explain that for some elements, the atoms can be arranged in more than one way to produce different molecular forms of the element. Ask:

- **How many different molecular forms of carbon are there, and what are these forms called?** (Three that have been discovered thus far: graphite, diamond, and buckminsterfullerene.)

Check Prior Knowledge

Check students' knowledge about molecular and ionic compounds. Ask:

- **Name some molecular and ionic compounds.** (molecular: H_2O, CH_4, NH_3, CO_2; ionic: NaCl, KBr, $MgCl_2$, CaF_2)
- **Compare molecular and ionic compounds.** (Ionic compounds are composed of negative and positive ions and are usually formed from a metallic and nonmetallic element. Most ionic compounds are crystalline solids at room temperature with relatively high melting points. Molecular compounds are composed of molecules formed from two or more nonmetallic elements; they tend to exist as gases or liquids at room temperature. Molecular solids have much lower melting points than ionic solids.)

section 10.3

THE NATURE OF SOLIDS

objectives

- Describe how the degree of organization of particles distinguishes solids from gases and liquids
- Distinguish between a crystal lattice and a unit cell
- Explain how allotropes of an element differ

key terms

- melting point
- crystal
- unit cell
- allotropes
- amorphous solids
- glasses

Imagine carbon atoms arranged in the shape of a soccer ball. Just such a structure, the buckminsterfullerene or buckyball, was discovered in 1985. Scientists are currently researching ways to make use of the buckyball's unique properties. Diamonds are also made of carbon atoms, and they too have unique properties that make them valuable for a variety of applications. **How many different molecular forms of carbon are there, and what are these forms called?**

A Model for Solids

The particles in liquids are relatively free to move. The particles in solids, however, are not, as **Figure 10.12** shows. Particles in a solid tend to vibrate about fixed points, rather than sliding from place to place. In most solids the particles are packed against one another in a highly organized pattern. Solids tend to be dense and incompressible. Because of the fixed positions about which their particles vibrate, solids do not flow to take the shape of their containers.

When you heat a solid, its particles vibrate more rapidly as their kinetic energy increases. The organization of particles within the solid breaks down and eventually the solid melts. The **melting point** (mp) is the temperature at which a solid turns into a liquid. At this temperature, the disruptive vibrations of the particles are strong enough to overcome the interactions that hold them in fixed positions. The melting process can be reversed by cooling the liquid so it freezes. The freezing of a liquid is thus the reverse of melting the corresponding solid. The melting and freezing points of a substance are at the same temperature. At that temperature, the liquid and solid substance are in equilibrium with each other.

$$\text{Solid} \underset{\text{freezing}}{\overset{\text{melting}}{\rightleftharpoons}} \text{Liquid}$$

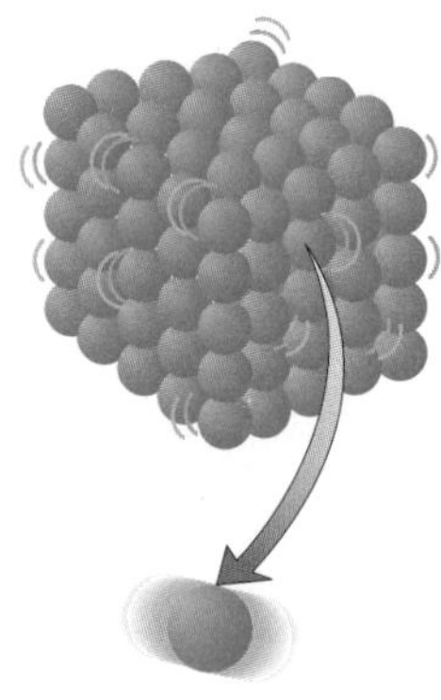

Figure 10.12
The particles in a solid vibrate about fixed points. A solid melts when its temperature is raised to a level at which the vibrations of the particles become so intense that they disrupt the ordered structure.

In general, ionic solids have high melting points. This is because relatively strong forces hold them together. Sodium chloride, an ionic compound, has a rather high melting point of 801 °C. By contrast, molecular solids have relatively low melting points. For example, hydrogen chloride, a molecular compound, melts at −112 °C. What is the freezing point of
1 hydrogen chloride? Not all solids melt, however. Wood and cane sugar, for example, decompose when heated.

Crystal Structure and Unit Cells

Most solid substances are crystalline. In a **crystal,** such as the one shown in **Figure 10.13,** the atoms, ions, or molecules that make up the solid substance are arranged in an orderly, repeating, three-dimensional pattern called the crystal lattice. All crystals have a regular shape. The shape of a crystal reflects the arrangement of the particles within the solid. The type of bonding that exists between the atoms determines the melting points of crystals.

Student Resources

From the Teacher's Resource Package, use:

- Section Review 10.3, Ch. 10 Practice Problems and Quizzes from the Review Module (Ch. 9–12)

TECHNOLOGY RESOURCES

Relevant technology resources include:

- Chem ASAP! CD-ROM
- Resource Pro CD-ROM
- ActivChemistry CD-ROM: *States of Matter*

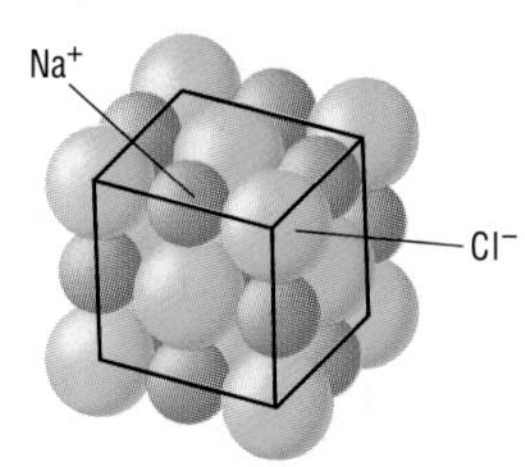

Figure 10.13

The regular shape of a sodium chloride crystal (left) can be explained by the arrangement of sodium ions and chloride ions within the crystal (right). The ions are close together, packed in a regular array, and move very little relative to each other. What happens to the arrangement of the particles if the solid is heated sufficiently? ②

A crystal has sides, or faces. The angles at which the faces of a crystal intersect are always the same for a given substance and are characteristic of that substance. Crystals are classified into seven groups, or crystal systems that have the characteristic shapes shown in **Figure 10.14.** The seven crystal systems differ in terms of the angles between the faces and in the number of edges of equal length on each face.

Figure 10.14

Crystals are classified into the seven crystal systems shown here.

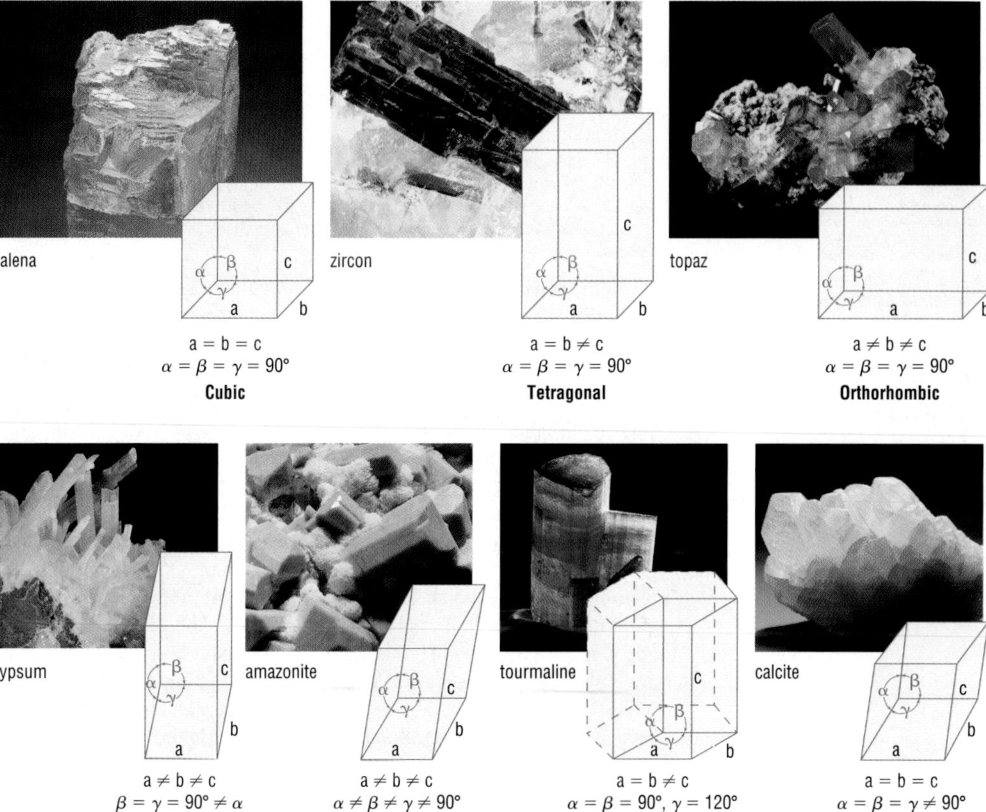

Answers

① −112°C

② The motion of the particles increases and eventually disrupts the rigid structure.

2 Teach

Discuss

Explain that the three-dimensional arrangement of particles in crystals is similar to the repetitive two-dimensional patterns found in wallpaper, tiles, and other decorative materials. These patterns consist of one or more basic motifs (the equivalent of unit cells) that are repeated to create the overall effect. The same principles that limit the arrangement of repetitive patterns in wallpaper can be used to classify crystal forms. If possible, bring in samples of different wallpaper patterns (preferably geometric designs) and have students examine them to determine the "unit cell" in each case. Students should be allowed to write on the paper. Show students how to obtain a set of "lattice points" by choosing the same point in each basic unit of the repeating pattern. The collection of lattice points shows the fundamental arrangement of the units in the pattern. Ask students if the same unit cell can be arranged in any other way to create a different design.

TEACHER DEMO

A model for a crystalline solid can be created using an aqueous solution of detergent. Place the solution in a watch glass on an overhead projector and use a straw to blow bubbles into the solution. The bubbles form a regular, repeating pattern that is analogous to the arrangement of unit cells in a crystalline solid.

Discuss

Emphasize that the arrangement of particles in solids is very orderly in comparison to the arrangement of particles in liquids and gases. Although the particles in solids are restricted to fixed positions, they are still in motion; instead of moving from place to place, they vibrate in position. Just as with liquids and gases, the kinetic energy of the vibrating particles varies directly with temprature.

ACTIVITY

Have students research the different types of crystal structures that occur in nature. A good place to find this kind of information is in a book about rocks and minerals. Ask students to make drawings or three-dimensional models of each crystal type and give examples of compounds that display each type.

Discuss

Ask students if they know of any other elements that exist in different allotropic forms. (Elemental oxygen occurs as O_2 and O_3, both of which are gases; sulfur can exist as monoclinic or orthorhombic crystalline solids; phosphorous can exist as red or white phosphorous, both solids at room temperature.)

The shape of a crystal depends on the arrangement of the particles within it. The smallest group of particles within a crystal that retains the geometric shape of the crystal is known as a **unit cell.** A crystal lattice is a repeating array of any one of fourteen kinds of unit cells. Each crystal system has from one to four types of unit cells that can be associated with that crystal system. **Figure 10.15** shows the three kinds of unit cells that can make up a cubic crystal system.

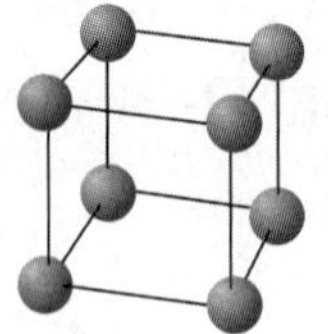

In a **simple cubic** unit cell, the atoms or ions are arranged at the corners of an imaginary cube.

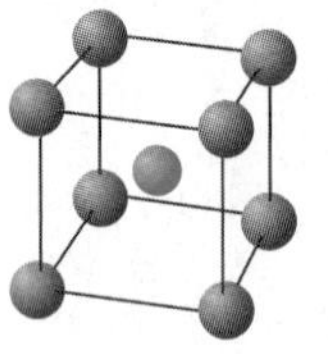

In a **body-centered cubic** unit cell, the atoms or ions are at the corners and in the center of an imaginary cube.

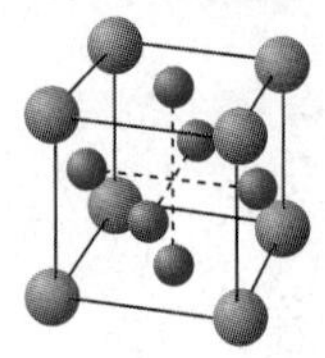

In a **face-centered cubic** unit cell, the atoms or ions are in the center of each face of the imaginary cube (but there is no atom or ion at the center).

Figure 10.15

The top row shows spheres representing the atoms or ions of three different cubic lattices. The bottom row illustrates the cubic unit cells for these lattices.

Some solid substances can exist in more than one form. A good example is the element carbon. Diamond is one crystalline form of carbon. It forms when carbon crystallizes under tremendous pressure (thousands of atmospheres). A different crystalline form of carbon is graphite. The lead in a pencil is not really lead; it is graphite. In graphite, the carbon atoms are packed in sheets rather than in the extended three-dimensional array characteristic of diamond.

In 1985, a third form of carbon—buckminsterfullerene, or the buckyball—was discovered. The buckyball is a molecule with 60 or more carbon atoms arranged in a pattern that looks like a soccer ball. It is found in ordinary soot. **Figure 10.16** shows the arrangement of carbon atoms in diamond, graphite, and a buckyball. The physical properties of diamond, graphite, and buckyballs are quite different. Diamond has a high density and is very hard; graphite has a relatively low density and is soft. Buckyballs are hollow cages with great strength and rigidity. Diamond, graphite, and buckyballs are called allotropes of carbon because all are made of carbon atoms and all are solids. **Allotropes** are two or more different molecular forms of the same element in the same physical state.

Not all solids are crystalline in form; some solids are amorphous. **Amorphous solids** lack an ordered internal structure. Rubber, plastic, and asphalt are amorphous solids. Their atoms are randomly arranged.

Meeting Diverse Needs

Gifted Have students do a literature search on the crystalline structure and formation of natural, synthetic, and simulated diamonds. Simulated diamonds are known as YAGs (for *yttrium aluminum garnet*). Have students write a short paper comparing the three kinds of crystals.

Gifted Liquid-crystal displays (LCDs) are commonly used in calculators and wristwatches. Liquid crystals display properties between a liquid and a solid. Have students do research to find out how liquid crystal displays work, with emphasis on their physical properties.

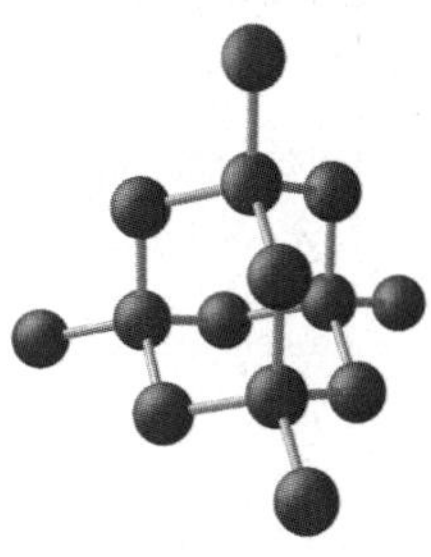

In **diamond,** each carbon atom in the interior of the diamond is strongly bonded to four others, creating a rigid compact array. Diamond is the hardest known material.

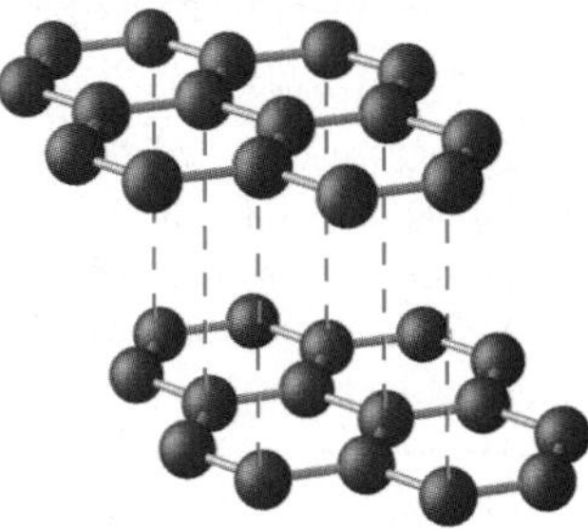

In **graphite,** the carbon atoms are linked in widely spaced layers of hexagonal (six-sided) arrays. Weak bonds between these layers allow them to slide over one another, making graphite very soft.

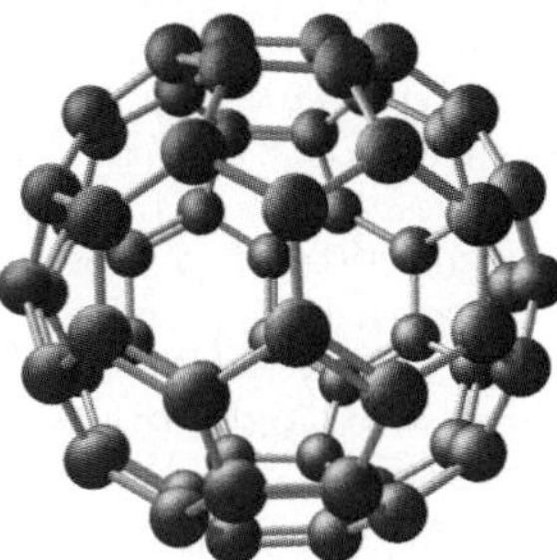

In **buckminsterfullerene,** the so-called buckyball, 60 carbon atoms form a hollow sphere; the carbons are arranged in pentagons and hexagons as on a soccer ball.

Figure 10.16
The differing arrangements of carbon atoms dramatically affect the properties of diamond, graphite, and buckminsterfullerene—the three forms of carbon. What are such different forms of an element called? ❶

Other examples of amorphous solids are **glasses**—transparent fusion products of inorganic substances that have cooled to a rigid state without crystallizing. Glasses are sometimes called supercooled liquids. The irregular internal structures of glasses are intermediate between those of a crystalline solid and a free-flowing liquid. Glasses do not melt at a definite temperature but gradually soften when heated. This softening with temperature is critical to the glassblower's art, as shown in **Figure 10.17.** When a crystalline solid is shattered, the fragments tend to have the same surface angles as the original solid. By contrast, when an amorphous solid, such as glass, is shattered, the fragments have irregular angles and jagged edges.

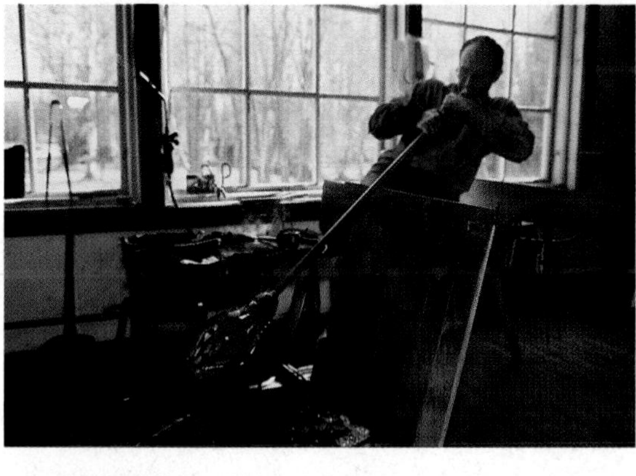

Figure 10.17
Because it is an amorphous solid, glass does not have a definite melting point. This glassblower is heating glass to make a beautiful vase.

section review 10.3

14. Explain the nature of solids and tell why they differ from liquids. Refer to the organization of particles in your answer.

15. How does the crystal lattice of a solid differ from its unit cell?

16. How do allotropes of an element differ?

 Chem ASAP! **Assessment 10.3** Check your understanding of the important ideas and concepts in Section 10.3.

10.3

TEACHER DEMO

Demonstrate the response of different crystalline solids when struck with a hammer. Compare lead (e.g., a fishing weight), salt (a large piece of rock salt or block from a pet supply store), and ice. Molecular and ionic crystals are brittle and fracture along crystal planes. Metallic crystals are soft and do not fracture when struck. The atoms in a metal slide past each other like ball bearings immersed in an oil. **CAUTION!** Students should wear safety goggles.

3 Assess

Evaluate Understanding

Ask students to describe the distinguishing characteristics of crystalline solids and amorphous solids. Have students name three types of crystal systems observed for crystalline solids. (Crystalline solids are characterized by an orderly, repeating three-dimensional arrangement of atoms, ions, or molecules. All crystals have a regular shape that reflects the arrangement of particles in the solid. Types of crystal systems include monoclinic, hexagonal, and cubic. Amorphous solids lack a well-defined arrangement of basic units.)

Reteach

Describe how a higher degree of organization and stronger intermolecular forces between particles distinguish solids from gases and liquids.

Answers

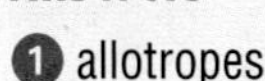

❶ allotropes

SECTION REVIEW 10.3

14. Particles in solids are packed tightly together and vibrate about fixed points; in liquids, particles are freer to move relative to one another.

15. A crystal lattice is a solid, regular array of positive and negative ions. A unit cell is the smallest group of particles that still displays the crystal's shape.

16. Allotropes of an element have different crystalline structures.

1 Engage

Use the Visual

Have students study the photograph and read the text that opens the section. Explain that rain represents the condensation product of water vapor in the atmosphere. To elicit the relationship among the three states of water, ask:

- **How are the solid, liquid, and vapor states related?** (Liquid and solid forms of water on Earth's surface are in dynamic equilibrium with water vapor in the atmosphere. The amount of water vapor in the air, and hence the position of this equilibrium, varies with the temperature. Higher temperatures favor greater amounts of vapor in the air. If air that is saturated with water vapor is cooled, water vapor condenses, or precipitates, as liquid water or rain.)

Motivate and Relate

The flux of water between the solid, liquid, and vapor phases has a large influence on Earth's climate. One climatological phenomenon of importance for communities located near large bodies of water is the dew point. Heavy fog results when a mass of water-saturated air encounters a cooler air mass. Fog is usually heaviest when the air temperature approaches the dew point. Have students do research to find out more about the dew point and how it is related to the changes of state discussed in this section.

section 10.4

objectives

- Interpret the phase diagram of water at any given temperature and pressure
- Describe the behavior of solids that change directly to the vapor state and recondense to solids without passing through the liquid state

key terms

- phase diagram
- triple point
- sublimation

CHANGES OF STATE

All life on Earth requires water, which cycles through our ecosystem in three forms—solid ice, liquid water, and water vapor. You are familiar with these states in a variety of ways: raindrops in a spring shower, snowflakes from a winter blizzard, and water vapor making a summer day humid. **How are the solid, liquid, and vapor states related?**

Phase Diagrams

The relationships among the solid, liquid, and vapor states (or phases) of a substance in a sealed container are best represented in a single graph called a phase diagram. A **phase diagram** gives the conditions of temperature and pressure at which a substance exists as solid, liquid, and gas (vapor). **Figure 10.18** shows the phase diagram for water. Each of the three regions represents a pure phase of water. A line that separates any two regions gives the conditions at which those two phases exist in equilibrium. The curving line that separates water's vapor phase from its liquid phase reveals the equilibrium conditions for liquid and vapor; it also illustrates how the vapor pressure of water varies with temperature. Similarly, the other two curving lines give the conditions for equilibrium between liquid water and ice and between water vapor and ice. A unique feature of the diagram is the point at which all three curves meet. This meeting point, called the **triple point,** describes the only set of conditions at which all three phases can exist in equilibrium with one another. For water, the triple

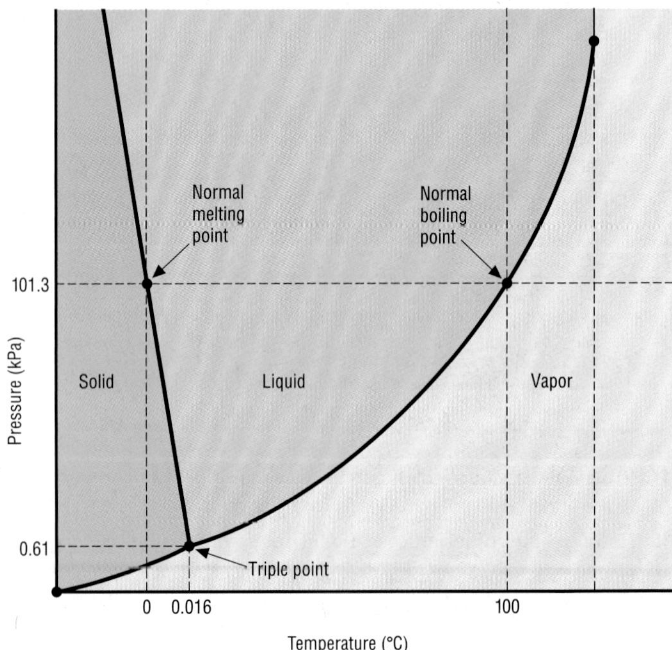

Figure 10.18
The phase diagram for water shows the relationship among pressure, temperature, and physical state. At the triple point, ice, liquid water, and water vapor can exist at equilibrium. Freezing, melting, boiling, and condensation can all occur at the same time, as shown in the photograph.

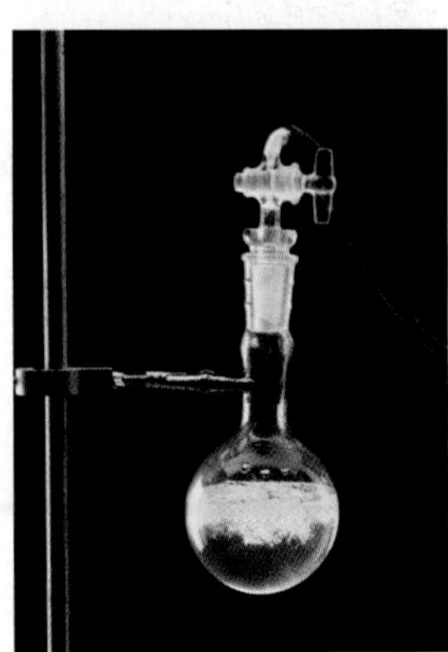

STUDENT RESOURCES

- Section Review 10.4, Ch. 10 Practice Problems, Interpreting Graphics, Vocabulary Review, Quizzes and Tests from the Review Module (Ch. 9–13)
- Laboratory Recordsheet 10-2
- Laboratory Manual: Experiment 14
- Laboratory Practical 10-1

TECHNOLOGY RESOURCES

Relevant technology resources include:

- Chem ASAP! CD-ROM
- Resource Pro CD-ROM
- ActivChemistry CD-ROM: *States of Matter*
- Chemistry Alive! Videodisc: *Phase Change*
- Assessment Resources CD-ROM: Chapter 10 Tests

point is a temperature of 0.016 °C and a pressure of 0.61 kPa (0.0060 atm). The photograph in **Figure 10.18** shows that the triple point for water can be achieved in the laboratory.

Using a phase diagram, you can determine the changes in the melting point and boiling point of a substance with changes in external pressure. As shown in **Figure 10.18,** at a pressure of 101.3 kPa (1 atm), the normal boiling point and melting point of water are 100 °C and 0 °C, respectively. What happens if boiling and melting are carried out at pressures less than 101.3 kPa? A decrease in pressure lowers the boiling point and raises the melting point. What happens to the boiling point of water if the pressure is raised above atmospheric pressure? ❶

When ice skaters move along the ice, the blades of their skates exert pressure which lowers the melting point of the ice. Consequently, the ice melts and a film of water forms under the blades of the skates. This film acts as a lubricant, enabling the skaters to glide gracefully over the ice.

Sublimation

If you hang wet laundry on a clothesline on a very cold day, the water in the clothes quickly freezes to ice. Eventually, however, the clothes become dry although the ice never thaws. The ice changes directly to water vapor without melting and passing through the liquid state. This is because solids, like liquids, have a vapor pressure. The vapor pressure of some solids can be high enough that they pass to a gas or vapor without becoming liquid. The change of a substance from a solid to a vapor without passing through the liquid state is called **sublimation.** Have you ever noticed that ice cubes left in the freezer for a long time get smaller? Can you propose an explanation? ❷

Iodine is another example of a substance that undergoes sublimation. This violet-black solid ordinarily changes into a purple vapor without passing through a liquid state. Notice in **Figure 10.19** how dark crystals of iodine deposit on the underside of a watch glass placed on top of a beaker containing solid iodine that is being heated. The iodine vapor sublimes from iodine crystals in the bottom of the beaker and condenses to form crystals on the watch glass.

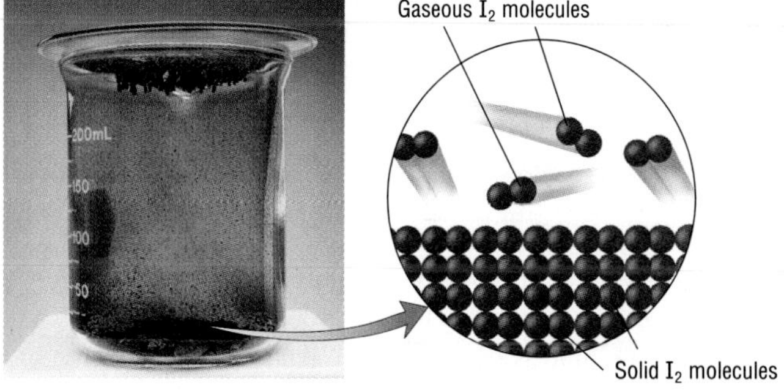

Figure 10.19
When solid iodine is heated, the crystals sublime, going directly from the solid to the vapor state. When the purple vapor cools, solid crystals re-form without passing through the liquid state.

Chem ASAP!

Simulation 7
Predict the physical states present at different points on a phase diagram.

FOOD SCIENCE

Freeze-Drying
Astronauts and backpackers alike enjoy the benefits of foods preserved by freeze-drying. Freeze-dried foods are lightweight and conveniently reconstituted by adding water. They need no refrigeration because bacteria cannot multiply in the absence of moisture. The ready sublimation of ice at low pressure makes the process of freeze-drying possible. Food is frozen and then placed in a chamber attached to a vacuum pump. By lowering the pressure in the chamber below the vapor pressure of ice, the ice crystals sublime and the food is dried without the loss of its flavor.

10.4

2 Teach

Use the Visual

Display Figure 10.18 on an overhead projector. Explain that a phase diagram is a convenient way to summarize the conditions of temperature and pressure under which the solid, liquid, and gas phases of a substance are most stable. Point out that each line in the diagram represents the set of all possible temperature-pressure values at which different phases are in dynamic equilibrium with one another. For example, in a closed system, at 100 °C and 101.3 kPa, liquid water is in dynamic equilibrium with water vapor. (Note that the *x*- and *y*-axes are not drawn to scale so that the triple point and normal boiling point can fit on the graph.) Ask:

- **At what temperature and pressure are the liquid and solid phases of water in dynamic equilibrium? The solid, liquid and gas phases?** (There are many possible combinations, however, the points labeled on the graph are the most obvious: solid/liquid = 0 °C, 101.3 kPa; solid/liquid/gas = 0.016 °C, 0.61 kPa)

Show students that the liquid/gas line in Figure 10.18 is the same as the graph of the vapor pressure of water at different temperatures, which is shown in Figure 10.10. The solid/liquid line in Figure 10.18 represents the melting point of water at different pressures.

MEETING DIVERSE NEEDS

At Risk The ability to interpret data in graphical format is fundamental to a course in chemistry. To reinforce these skills as applied in this section, have students compare the pressure-temperature graph in Figure 10.11 to the graph in Figure 10.18. Ask students how many and which phases of each substance are being illustrated in Figure 10.11. (two: liquid and gas)

Answers

❶ The boiling point increases.
❷ Some of the ice sublimes.

Chemistry Alive!

Phase Change

Play

MINI LAB

Sublimation

SAFETY

Perform this activity in a well-ventilated room.

ANSWERS

1. Sublimation is the vaporization of a solid.
2. At room temperature, there would be sublimation; at the boiling point, there would be more sublimation.
3. A substance that sublimes leaves the mixture; other components remain as residue.

3 Assess

Evaluate Understanding

List a series of temperature–pressure values on the chalkboard. Have students state which phase of water is most stable under each set of conditions. Next, assign students one pair of the temperature–pressure values. Then ask how they would adjust one of the variables (temperature or pressure) to carry out a phase conversion if the other variable is held constant.

Reteach

Stress that every substance has a unique set of properties that includes melting point, boiling point, and triple point. The conditions of pressure and temperature under which the phases of a substance are most stable depends on its chemical composition. Convey this message by comparing the optimum conditions for sublimation of iodine and ice.

Sublimation is useful in many situations and processes. Freezing freshly brewed coffee and removing the water vapor with a vacuum pump makes freeze-dried coffee. Solid carbon dioxide (dry ice) is an important cooling material that also sublimes. Given its low temperature of −78 °C and the fact that it does not produce a liquid when it melts, dry ice is very useful for keeping frozen foods, such as ice cream, solid during shipping. Solid air fresheners contain a variety of substances that sublime at room temperature. Mothballs contain naphthalene. It is the naphthalene vapor that keeps away moths. Sublimation is also useful for separating substances. Organic chemists use this process to separate mixtures and to purify compounds.

MINI LAB

Sublimation

PURPOSE

To observe the sublimation of air freshener.

MATERIALS

- small pieces of solid air freshener
- small shallow container
- 2 clear plastic cups (8 oz each)
- hot tap water
- ice
- 3 thick cardboard strips

PROCEDURE

1. Place a few pieces of air freshener in one of the cups.
2. Bend the cardboard strips and place them over the rim of the cup that has the air freshener pieces.
3. Place the second cup inside the first. The base of the second cup should not touch the air freshener. Adjust the cardboard as necessary. This assembly is your sublimator.
4. Fill the top cup with ice. Do not get any ice or water in the bottom cup.
5. Fill the shallow container about one-third full with hot tap water.
6. Carefully place your sublimator in the hot water. Observe what happens.

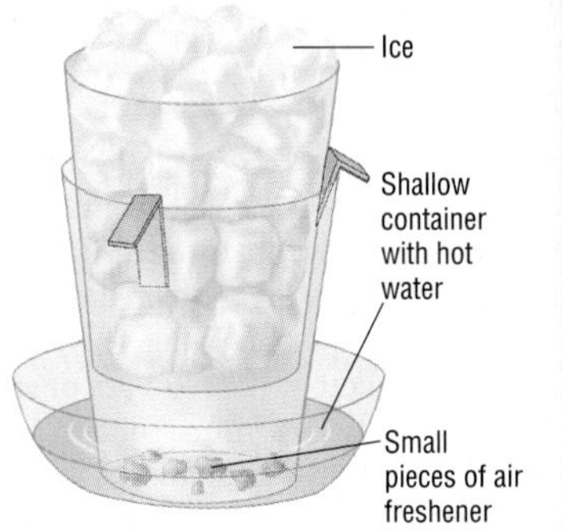

ANALYSIS AND CONCLUSIONS

1. Define sublimation.
2. What do you think would happen if the water in the shallow container was at room temperature? If it was boiling?
3. Why is it possible to separate the substances in some mixtures by sublimation?

section review 10.4

17. What general information can you get from a phase diagram for water at various temperatures and pressures?
18. Describe the process of sublimation. What is a practical use of this process?
19. Explain triple point.

 Chem ASAP! **Assessment 10.4** Check your understanding of the important ideas and concepts in Section 10.4.

Meeting Diverse Needs

At Risk Use Figure 10.18 to help students understand how the freeze-drying process works. Point out the conditions of temperature and pressure on the diagram that promote sublimation, or "freeze-drying" of moisture-laden foods. Ask students why, in most cases, this process does not remove the flavors in the food.

SECTION REVIEW 10.4

17. the state of water at given temperatures and pressures and the conditions at which phase changes occur
18. During sublimation, a substance changes directly from solid to vapor; drying without extensive heating.
19. The triple point is the temperature and pressure at which an equilibrium exists among all three phases of a substance.

Chemistry Serving...the Environment

GASES CAN ALTER GLOBAL TEMPERATURE

Gases in Earth's atmosphere called greenhouse gases act much like the glass ceiling of a greenhouse. Earth's heat radiates skyward where it is absorbed and re-emitted by the greenhouse gases in the atmosphere back toward the surface of Earth. This process is very similar to how glass traps heat inside a greenhouse. If greenhouse gases were not present in the atmosphere, Earth's heat would escape into space—Earth would be a lifeless planet! Without greenhouse gases, Earth would have an average surface temperature of only −18 °C. With these gases in place, however, Earth's average surface temperature is 15 °C.

The main greenhouse gases are water vapor, carbon dioxide, ozone, methane, nitrous oxide, and halogenated substances including chlorofluorocarbons (CFCs). Many activities release greenhouse gases. Some of these are natural processes while others are directly related to human actions. Human activities are adding larger and larger amounts of greenhouse gases into the atmosphere, resulting in an excess of trapped heat. This unnatural trapping of heat in the atmosphere is called global warming.

Without greenhouse gases, Earth would have an average surface temperature of only −18°C.

Carbon dioxide is the largest contributor to global warming. Processes such as cellular respiration produce this gas naturally. In addition, the burning of fossil fuels and forests contributes heavily to the concentration of this gas.

CFCs are also contributors to global warming. These compounds are produced almost entirely from human actions. The manufacture of plastic foams and the evaporation of industrial solvents release CFCs. Some aerosol propellants give off CFCs, and CFC-based air conditioners and refrigerators can leak CFCs into the atmosphere. CFC molecules have a much greater effect on global warming than do carbon dioxide molecules.

Methane is another greenhouse gas capable of trapping more heat than carbon dioxide. The natural process of decomposition releases methane into the atmosphere. Landfills further contribute to methane's atmospheric concentration. In addition, methane is released as underground deposits of fossil fuels and natural gas are brought to the surface.

If the release of greenhouse gases by human activities continues unchecked, the potential for overall global climate changes increases, as does the threat to the survival of many species. The predicted changes in climate could cause droughts in some regions, floods in others, loss of biodiversity, and rising sea levels. These gases have the power to affect Earth's climate and its animal and plant life.

CHEMISTRY IN CAREERS

TAKE IT TO THE NET

Find out more about career opportunities:

www.phschool.com

CHEMICAL SPECIALIST

Local food service distributor seeks responsible self-motivated individ-

CLIMATOLOGIST

Study the weather of the future and the past using supercomputers and ice core sampling.

See page 872.

CHEMIST NEEDED

Enviro. Lab. GC Must know

Chemistry Serving... the Environment

DISCUSS

Initiate a class discussion on the pros and cons of restricting the production of greenhouse gases worldwide. Ask:

- What steps are societies taking to curb the release of greenhouse gases?
- How can more technologically advanced nations help developing countries develop without contributing to the greenhouse effect?

Explain that much of the carbon dioxide generated dissolves in the ocean, is incorporated into carbonates, or is used by plants for photosynthesis. However, society is producing CO_2 faster than it can be absorbed. Scientists can determine how CO_2 levels in the atmosphere have changed over several centuries by analyzing samples of air trapped in cores of ice taken from Antarctica.

Have students do research on the Internet or at the library to find data for the concentration of atmospheric CO_2 over the past 100–200 years. Ask students to plot CO_2 levels versus time and discuss any trends observed. In addition, ask students to collect data, for the same time frame, for average global temperatures and to plot these data on the same graph. What, if any, correlations do students observe in the collected data?

CHEMISTRY IN CAREERS

Encourage students to connect to the Internet address shown on this page to find out more about careers in climatology. Also see the career feature on page 872 of this text.

4 Close

Summary

On the chalkboard, draw a stoppered, sidearm flask containing water and a chunk of ice. Note that the sidearm is connected to a vacuum pump. Tell students that the diagram illustrates ice, liquid water, and water vapor in equilibrium. Have volunteers add temperature labels for each phase to the diagram and write an equation with double arrows showing the equilibrium between solid, liquid, and vapor. Students should write the words *condensation, evaporation, freezing,* and *melting* above and below the appropriate arrows to describe the equilibrium processes.

Have other students sketch schematic representations of the molecular motions for each phase.

Extension

A phase diagram describes the set of conditions under which all forms of an element are stable. Have students research the phase diagram of carbon.

Looking Back . . . Looking Ahead . . .

The previous chapter discussed how chemical equations are used to solve stoichiometric problems. The current chapter describes the particle nature of gases, liquids, and solids and the influence of temperature and pressure on changes of state. The next chapter introduces heat changes that occur during physical and chemical processes.

Chapter 10 STUDENT STUDY GUIDE

Take It to the NET For interactive study and review, go to www.phschool.com

KEY TERMS

- allotrope *p. 282*
- amorphous solid *p. 282*
- atmospheric pressure *p. 268*
- barometer *p. 268*
- boiling point *p. 277*
- crystal *p. 280*
- evaporation *p. 275*
- gas pressure *p. 268*
- glass *p. 283*
- kinetic energy *p. 267*
- kinetic theory *p. 267*
- melting point *p. 280*
- normal boiling point *p. 278*
- pascal (Pa) *p. 268*
- phase diagram *p. 284*
- standard atmosphere (atm) *p. 268*
- sublimation *p. 285*
- triple point *p. 284*
- unit cell *p. 282*
- vacuum *p. 268*
- vaporization *p. 275*
- vapor pressure *p. 276*

CONCEPT SUMMARY

10.1 The Nature of Gases

- The kinetic theory describes the motion of particles (atoms, ions, or molecules) in matter and the forces of attraction between them.
- The kinetic theory assumes that the volume occupied by a gas is mostly empty space and that the particles of a gas are far apart, move rapidly, and have random motion.
- The pressure of a gas results from the collisions of the gas particles with an object.
- Standard conditions are 0 °C and 1 atm.
- The temperature of a gas is directly proportional to the average kinetic energy of the particles.

10.2 The Nature of Liquids

- When the temperature of a gas is lowered enough, the gas condenses to a liquid.
- A liquid boils when its vapor pressure equals the external pressure.
- When a liquid that is not boiling changes to a vapor or gas, the process is called evaporation.
- The normal boiling point of a liquid is the temperature at which the vapor pressure is equal to 1 atm.

10.3 The Nature of Solids

- The movement of particles in a solid is restricted to vibrations about fixed points.
- Solids melt when the vibrations become greater than the forces holding the particles together.
- Most solids are crystalline. The particles are arranged in a repeating three-dimensional pattern known as a crystal lattice.
- The smallest subunit of a crystal lattice is the unit cell.
- An element may have two or more different molecular forms, called allotropes, in the same physical state.

10.4 Changes of State

- Most substances change their physical state and melt or vaporize as the temperature increases. Substances condense or freeze as the temperature decreases.
- While the physical state of a substance changes during a phase change, the temperature of the system remains constant.
- In sublimation, a solid can change directly to a gas or vapor without first becoming a liquid.

CHAPTER CONCEPT MAP

Use these terms to construct a concept map that organizes the major ideas of this chapter.

Chem ASAP! Concept Map 10
Create your Concept Map using the computer.

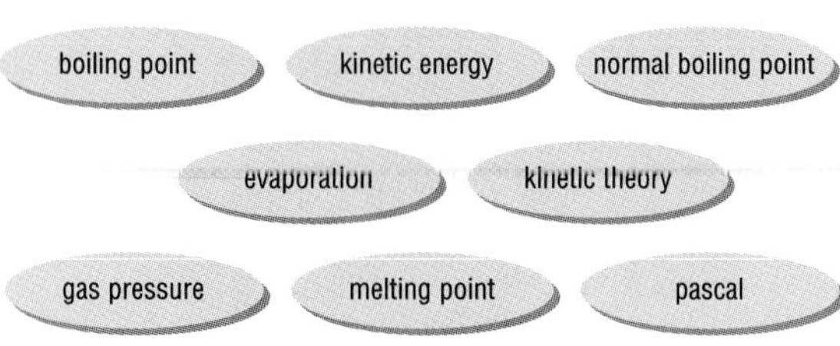

Take It to the Net

At **www.phschool.com** students will find for this chapter

- an Internet activity
- links to related chemistry sites
- an interactive quiz
- career links

Chapter 10 REVIEW

CONCEPT PRACTICE

20. What is meant by elastic collision? *10.1*

21. List the various units used to measure pressure, and identify the SI unit. *10.1*

22. Change 1656 kPa to atm. *10.1*

23. Convert 190 mm Hg to the following. *10.1*
a. kilopascals
b. atmospheres of pressure

24. How much pressure (in mm Hg) does a gas exert at 3.1 atm? *10.1*

25. Explain the relationship between the absolute temperature of a substance and the kinetic energy of its particles. *10.1*

26. How is the average kinetic energy of water molecules affected when you pour hot water from a kettle into cups at the same temperature as the water? *10.1*

27. What does the abbreviation STP represent? *10.1*

28. Express standard temperature in kelvins and standard pressure in kilopascals and in millimeters of mercury. *10.1*

29. What is significant about the temperature absolute zero? *10.1*

30. By what factor does the average kinetic energy of gas molecules in an aerosol container increase when the temperature is raised from 27 °C (300 K) to 627 °C (900 K)? *10.1*

31. A liquid is a condensed state of matter. Explain. *10.2*

32. Explain why liquids and gases differ *10.2*
a. in physical state.
b. in compressibility.

33. Compare the evaporation of a contained liquid with that of an uncontained liquid. *10.2*

34. Explain vapor pressure and dynamic equilibrium. *10.2*

35. Explain why increasing the temperature of a liquid increases its rate of evaporation. *10.2*

36. Would you expect an equilibrium vapor pressure to be reached above a liquid in an open container? Why? *10.2*

37. Describe the effect that increasing temperature has on the vapor pressure of a liquid. *10.2*

38. Distinguish between the boiling point and the normal boiling point of a liquid. *10.2*

39. Use the graph to answer each question. *10.2*

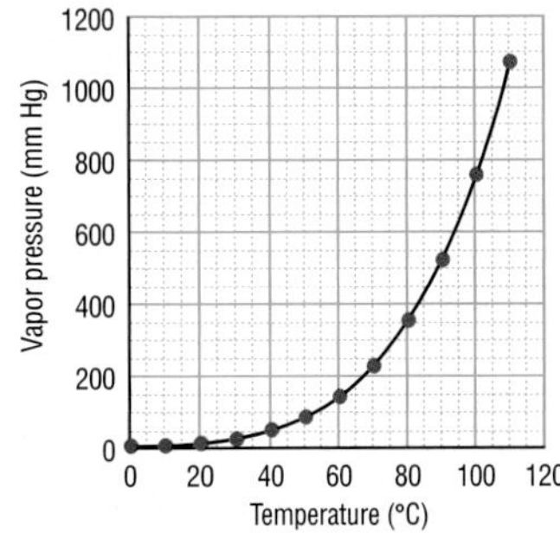

a. What is the vapor pressure of water at 40 °C?
b. At what temperature is the vapor pressure of water 600 mm Hg?
c. What is the significance of the vapor pressure of water at 100 °C?

40. Use **Figure 10.11** to determine the temperature at which water will boil in an open vessel when the atmospheric pressure is 400 mm Hg. *10.2*

41. At the top of Mount Everest, water boils at only 69 °C. Use **Figure 10.11** to estimate the atmospheric pressure at the top of this mountain. *10.2*

42. Explain how boiling is a cooling process. *10.2*

43. Name at least one physical property that would permit you to distinguish a molecular solid from an ionic solid. *10.3*

44. Describe what happens when a solid is heated to its melting point. *10.3*

45. Molecular solids usually have lower melting points than ionic solids. Why? *10.3*

46. When you remove the lid from a food container that has been left in a freezer for several months, you discover a large collection of ice crystals on the underside of the lid. Explain what has happened. *10.4*

47. Any liquid stays at a constant temperature while it is boiling. Why? *10.4*

CONCEPT MASTERY

48. Describe evaporation, vapor pressure, and boiling point.

49. Mount McKinley (6194 m) in Alaska is the tallest peak in North America. The atmospheric pressure at its peak is 330 mm Hg. Find the boiling point of water there. Use **Figure 10.11**.

Answers

20. An elastic collision transfers energy from one particle to another. There is no change in the total kinetic energy.

21. pascal (Pa), millimeter of mercury (mm Hg), atmosphere (atm); Pa is the SI unit.

22. 16.35 atm

23. **a.** 25 kPa
b. 0.25 atm

24. 2.4×10^3 mm Hg

25. Kinetic energy is directly proportional to the absolute temperature.

26. The average kinetic energy is unaffected.

27. standard temperature and pressure, 0 °C and 1 atm

28. 273 K; 101.3 kPa, 760 mm Hg

29. Average kinetic energy of particles is zero.

30. The average kinetic energy triples.

31. Its volume is minimally affected by an increase in pressure.

32. **a.** Because of attractive forces between particles, liquids are denser than gases.
b. The molecules of liquids are in close contact and cannot be squeezed together. The molecules of gases are far apart and, thus, compressible.

33. In both cases, particles with sufficient kinetic energy move from the liquid to the vapor phase. In a container, a dynamic equilibrium is set up between the liquid and its vapor.

34. Vapor pressure results from collisions of vapor particles with the container's walls. A dynamic equilibrium exists when the rate of evaporation of the liquid equals the rate of condensation of the vapor.

35. More molecules have enough energy to escape attractions within the liquid.

36. No; vapor continuously leaves the surface of the liquid preventing dynamic equilibrium from being established.

37. It increases the kinetic energy, which increases the vapor pressure.

38. The boiling point is the temperature at which the vapor pressure equals the external pressure. At the normal boiling point, external pressure is 1 atm.

39. **a.** about 50 mm Hg
b. about 93 °C
c. 760 mm Hg is standard pressure.

40. about 82 °C

41. about 35 kPa

42. Escaping molecules have more kinetic energy than remaining molecules, which lowers the average kinetic energy.

43. Ionic compounds have crystalline structures with relatively high melting points.

44. Its molecules gain sufficient kinetic energy to overcome attractive forces.

45. The forces between the molecules of molecular solids are weaker.

Answers

46. Moisture in the food has sublimed and then resolidified on the container lid.

47. Molecules use the added heat to escape the liquid; average kinetic energy remains the same.

48. Evaporation is the conversion of a liquid to a vapor at temperatures below the boiling point. Vapor pressure is the force per unit area exerted by the vapor particles on the container walls. Boiling point is the temperature at which vapor pressure equals external pressure.

49. about 77 °C

50. The mass of air pressing on molecules below it; the mass of air above a mountain is less than the mass at sea level.

51. Although there is an ongoing exchange of particles, net amounts of vapor molecules and liquid molecules remain constant.

52. At −196 °C, the kinetic energy of the air particles decreases drastically as does the pressure the particles exert, thus reducing the balloon's flexible volume. The kinetic energy of the air particles increases as the balloon warms back to room temperature.

53. It increases.

54. 819 K

55. There is sufficient average kinetic energy to disrupt the forces holding the solid crystal together.

56. **a.** 2 **b.** 1

57. Possible answers include: odors travel through a room; ink will move throughout a beaker of water.

58. They are the same because the temperature is the same.

59. No; collisions between nonatomic objects involve the conversion of kinetic energy into heat.

50. What causes atmospheric pressure, and why is it much lower on the top of a mountain than it is at sea level?

51. Why is the equilibrium that exists between a liquid and its vapor in a closed container called a dynamic equilibrium?

52. Pouring liquid nitrogen onto a balloon decreases the volume of the balloon dramatically, as shown in the photograph. Afterward, the balloon reinflates. Use kinetic theory to explain this sequence of events. The temperature of liquid nitrogen is −196 °C.

53. What happens to the average kinetic energy of the water molecules in your body when you get a fever?

54. The temperature of the gas in an aerosol container is 0 °C (273 K). To what temperature must this gas be raised to increase the average kinetic energy of the gas molecules by a factor of three?

55. Explain what happens at the melting point of a substance.

CRITICAL THINKING

56. Choose the term that best completes the second relationship.

a. temperature : thermometer
atmospheric pressure : _______
(1) valve (3) volume
(2) barometer (4) gauge

b. evolution : organisms
kinetic theory : _______
(1) particles (3) heat
(2) temperature (4) pressure

57. What everyday evidence suggests that all matter is in constant motion?

58. Is the average kinetic energy of the particles in a block of ice at 0 °C the same as or different from the average kinetic energy of the particles in a gas-filled weather balloon at 0 °C? Explain.

59. Are perfectly elastic collisions possible between objects that you can see?

60. How does perspiration help cool your body on a hot day?

61. Why do different liquids have different boiling points?

CUMULATIVE REVIEW

62. Balance these equations.

a. $V_2O_5 + H_2 \longrightarrow V_2O_3 + H_2O$
b. $(NH_4)_2Cr_2O_7 \longrightarrow Cr_2O_3 + N_2 + H_2O$
c. $NH_3 + O_2 \longrightarrow NO + H_2O$
d. $C_6H_{14} + O_2 \longrightarrow CO + H_2O$

63. What is the density of krypton gas at STP?

64. Hydrogen reacts with ethene (C_2H_4) to form ethane (C_2H_6).

$$C_2H_4 + H_2 \longrightarrow C_2H_6$$

What is the limiting reagent when 40.0 g C_2H_4 reacts with 3.0 g H_2?

65. How many moles is each substance?

a. 888 g of sulfur dioxide
b. 2.84×10^{22} molecules of ammonia
c. 0.47 L of carbon dioxide (at STP)

66. Perchloric acid forms by the reaction of water with dichlorine heptoxide.

$$Cl_2O_7 + H_2O \longrightarrow 2HClO_4$$

a. How many grams of Cl_2O_7 must react with an excess of H_2O to form 56.2 g $HClO_4$?
b. How many mL of water are needed to form 3.40 mol $HClO_4$?

CONCEPT CHALLENGE

67. If the volume of the container in which there is a liquid–vapor equilibrium is changed, the vapor pressure is not affected. Why?

68. The ions in sodium chloride are arranged in a face-centered cubic pattern. Sketch a layer of the ions in a crystal of sodium chloride.

69. Using **Figure 10.14**, identify the crystal system described by these characteristics.

a. three unequal axes mutually perpendicular
b. three equal axes making equal angles with each other
c. two equal axes and one unequal axis mutually perpendicular
d. three unequal axes intersecting obliquely
e. three axes equal and mutually perpendicular

60. The energy needed to evaporate perspiration comes from skin cells, which helps lower body temperature.

61. Some compounds have stronger intermolecular forces than others.

62. **a.** 1, 2, 1, 2 **c.** 4, 5, 4, 6
b. 1, 1, 1, 4 **d.** 2, 13, 12, 14

63. $3.74\ g/L = 0.003\ 74\ g/cm^3$

64. C_2H_4

65. **a.** 13.9 mol SO_2 **c.** 0.021 mol CO_2
b. 0.0472 mol NH_3

66. **a.** 51.2 g Cl_2O_7 **b.** 30.6 mL H_2O

67. Vapor pressure depends only on the kinetic energy of the escaping molecules.

68. $Na^+Cl^-\ Na^+Cl^-$
$Cl^-\ Na^+Cl^-\ Na^+$
$Na^+Cl^-\ Na^+Cl^-$
$Cl^-\ Na^+Cl^-\ Na^+$

69. **a.** orthorhombic **d.** triclinic
b. rhombohedral **e.** cubic
c. tetragonal

Chapter 10 STANDARDIZED TEST PREP

Select the choice that best answers each question or completes each statement.

1. To double the average kinetic energy of helium atoms in a balloon at 27 °C, the temperature of the gas must be
a. decreased to 150 K.
b. increased to 600 K.
c. increased to 450 K.
d. increased to 54 °C.

2. Which sequence has the states of CH_3OH correctly ordered in terms of increasing average kinetic energy?
a. $CH_3OH(s)$, $CH_3OH(g)$, $CH_3OH(l)$
b. $CH_3OH(g)$, $CH_3OH(l)$, $CH_3OH(s)$
c. $CH_3OH(l)$, $CH_3OH(g)$, $CH_3OH(s)$
d. $CH_3OH(s)$, $CH_3OH(l)$, $CH_3OH(g)$

3. Which statement is *not* part of kinetic theory?
a. Collisions between gas particles are elastic.
b. All gas particles are in constant motion.
c. All particles have identical kinetic energies.
d. The motion of gas particles is random.

Use the drawing to answer questions 4–6. The same liquid is in each flask.

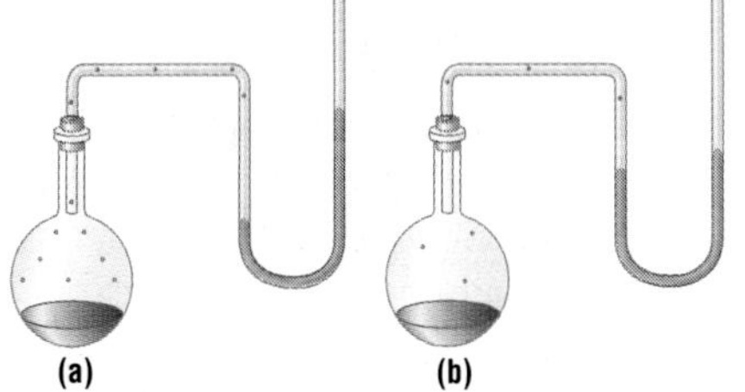

4. In which flask is the vapor pressure lower? Give a reason for your answer.

5. In which flask is the liquid at the higher temperature? Explain your answer.

6. How can the vapor pressure in each flask be determined?

Use the graph to answer questions 7–9.

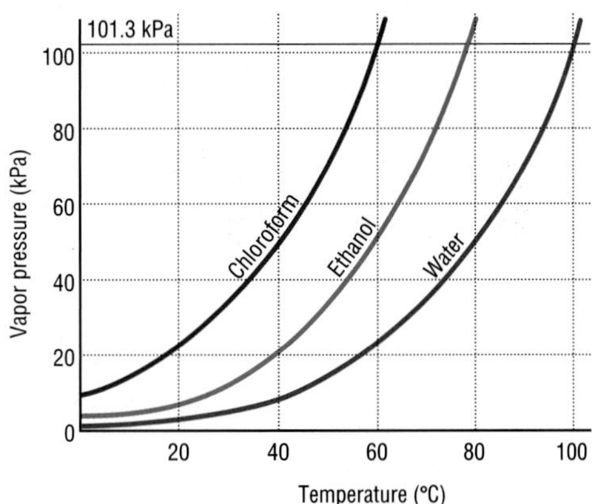

7. What is the normal boiling point of ethanol?

8. Can chloroform be heated to 90 °C in an open container?

9. To what temperature must water be heated so that its vapor pressure equals the vapor pressure of ethanol at 60 °C?

For each question there are two statements. Decide whether each statement is true or false. Then decide whether Statement II is a correct explanation for Statement I.

Statement I		Statement II
10. In an open container, the rate of evaporation of a liquid always equals the rate of condensation.	BECAUSE	A dynamic equilibrium exists between a liquid and its vapor in an open container.
11. Water boils at a temperature below 100 °C on top of a mountain.	BECAUSE	Atmospheric pressure decreases with an increase in altitude.
12. The temperature of a substance always increases as heat is added to the substance.	BECAUSE	The average kinetic energy of the particles in a substance increase with an increase in temperature.
13. Solids have a fixed volume.	BECAUSE	Particles in a solid cannot move.
14. Gases are more compressible than liquids.	BECAUSE	There is more space between particles in a gas than between particles in a liquid.

Chapter 10 STANDARDIZED TEST PREP

1. b
2. d
3. c
4. Vapor pressure is lower in flask (b) because the level of mercury in the left arm of (b) is higher, which indicates less pressure on the mercury.
5. Temperature is higher in flask (a) because vapor pressure increases with temperature.
6. by measuring the difference in heights between the two arms
7. ≈78 °C
8. no
9. ≈82 °C
10. False, False
11. True,True, correct explanation
12. False, True
13. True, False
14. True, True, correct explanation

Planning Guide

SECTION OBJECTIVES	ACTIVITIES/FEATURES	MEDIA & TECHNOLOGY
11.1 The Flow of Energy—Heat ○ ■ ◆ ▸ Explain the relationship between energy and heat ▸ Distinguish between heat capacity and specific heat	SE Discover It! *Observing Heat Flow*, p. 292 SE **Link to Physiology** *Dietary Calories*, p. 296 SE **CHEMath** *Review of Significant Figures*, p. 298 TE DEMO, p. 294	ASAP Problem Solving 3 ASAP Assessment 11.1 ACT 6: *Thermochemistry* CA *Burning Magnesium* CA *Exploding Balloons* CA *Superheated Steam* CA *Thermite Reaction* CA *Fountain of Light* CA *Crushing Cans* CHM Side 4, 36: *Heat Capacity* CHM Side 4, 2: *Balancing Diet and Exercise* CHM Side 4, 34: *Enthalpy Changes*
11.2 Measuring and Expressing Heat Changes ○ ■ ◆ ▸ Construct equations that show the heat changes for chemical and physical processes ▸ Calculate heat changes in chemical and physical processes	SE **Link to Biology** *Warmth from Fat*, p. 306 LM 15: *The Specific Heat of a Metal* TE DEMO, p. 305	ASAP Simulation 8 ASAP Problem Solving 12, 14 ASAP Assessment 11.2 ACT 6: *Thermochemistry* CHM Side 4, 10: *Measuring Energy Expenditure*
11.3 Heat in Changes of State ○ □ ◆ ▸ Classify, by type, the heat changes that occur during melting, freezing, boiling, and condensing ▸ Calculate heat changes that occur during melting, freezing, boiling, and condensing	SE Mini Lab *Heat of Fusion of Ice*, p. 308 (LRS 11-1) PLM *Heat of Fusion of Ice* SSLM 13: *Heat of Fusion of Ice* TE DEMO, p. 310	ASAP Animation 12 ASAP Problem Solving 21, 23 , 25 ASAP Assessment 11.3 ACT 6: *Thermochemistry* CHM Side 5, 27: *Evaporative Cooling* CA *Phase Change* CA *Superheated Steam*
11.4 Calculating Heat Changes ◆ ▸ Apply Hess's law of heat summation to find heat changes for chemical and physical processes ▸ Calculate heat changes using standard heats of formation	SE Small-Scale Lab *Heat of Combustion of a Candle*, p. 319 (LRS 11-2) PLM *Heat of Combustion of a Candle* SE **Chemistry Serving . . . the Environment** *Harnessing Solar Energy*, p. 320 SE **Chemistry in Careers** *Solar Engineer*, p. 320 LM 16: *Heats of Reaction* TE DEMO, p. 316	ASAP Problem Solving 30 ASAP Assessment 11.4 ASAP Concept Map 11 ACT 6: *Thermochemistry* OT *28: Hess's Law* PP Chapter 11 Problems RP Lesson Plans, Resource Library AR Computer Test 11 www Activities, Self-Tests, *SCIENCE NEWS* updates TCP The Chemistry Place Web Site

KEY

● Conceptual (concrete concepts)	AR Assessment Resources	GRS Guided Reading and Study Workbook
○ Conceptual (more abstract/math)	ASAP Chem ASAP! CD-ROM	LM Laboratory Manual
■ Standard (core content)	ACT ActivChemistry CD-ROM	LP Laboratory Practicals
□ Standard (extension topics)	CHM CHEMedia Videodiscs	LRS Laboratory Recordsheets
◆ Honors (core content)	CA Chemistry Alive! Videodiscs	SSLM Small-Scale Lab Manual
◇ Honors (options to accelerate)	GCP Graphing Calculator Problems	

PRACTICE		ASSESSMENT	
SE	**Sample Problem** 11-1	SE	Section Review
GRS	Section Review 11.1	RM	Section Review 11.1
RM	Practice Problems 11.1	RM	Chapter 11 Quiz
SE	**Sample Problems** 11-2 to 11-3	SE	Section Review
GRS	Section Review 11.2	RM	Section Review 11.2
RM	Practice Problems 11.2	RM	Chapter 11 Quiz
		LP	Lab Practical 11-1
SE	**Sample Problems** 11-4 to 11-6	SE	Section Review
GRS	Section 11.3	RM	Section Review 11.3
RM	Practice Problems 11.3	RM	Chapter 11 Quiz
RM	Interpreting Graphics		
SE	**Sample Problem** 11-7	SE	Section Review
GRS	Section 11.4	RM	Section Review 11.4
RM	Practice Problems 11.4	LP	Lab Practical 11-2
GCP	Chapter 11	RM	Vocabulary Review 19
		SE	Chapter Review
		SM	Chapter 11 Solutions
		SE	Standardized Test Prep
		PHAS	Chapter 11 Test Prep
		RM	Chapter 11 Quiz
		RM	Chapter 11 A & B Test

PLANNING FOR ACTIVITIES

STUDENT EDITION

Discover It! p. 292
- rubber bands

Mini Lab, p. 308
- ice
- foam cups
- graduated cylinders
- thermometers
- hot water

Small-Scale Lab, p. 319
- candle
- aluminum foil
- ruler
- balance

TEACHER'S EDITION

Teacher Demo, p. 294
- 32 g of barium hydroxide octahydrate
- 11 g of ammonium chloride
- 250-mL Erlenmeyer flask fitted with a stopper
- wet piece of wood
- fume hood

Activity, p. 294
- glue
- 13 cm × 18 cm piece of sheet metal
- 13 cm × 18 cm piece of Styrofoam™
- 40 cm × 20 cm piece of wood

Activity, p. 301
- 100 mL of 1*M* HCl
- Styrofoam™ cup
- thermometer
- 0.5 g of mossy zinc *or* 0.5 g of magnesium turnings
- protective goggles and aprons for students
- fume hood

Teacher Demo, p. 305
- 100 g calcium oxide
- water
- metal tray
- thin aluminum pan
- fresh egg

Activity, p. 307
- thermometer
- water
- boiling chips
- ice
- Styrofoam™ cup
- heat source

Teacher Demo, p. 310
- 3 wide-mouth test tubes each containing 20 grams of sodium thiosulfate pentahydrate
- hot water bath
- cool water bath
- thermometer
- seed crystals

Activity, p. 312
- hot and cold packs used by athletes

Teacher Demo, p. 316
- 2 clay flower pots (as small as possible)
- commercial thermite
- commercial thermite starter
- teaspoon
- glycerine in dropper bottle
- potassium permanganate $KMnO_4$
- a few 2-inch or smaller pieces of magnesium ribbon
- ring stand and ring
- large bucket of sand
- paper towel

OT	Overhead Transparency	SE	Student Edition
PHAS	PH Assessment System	SM	Solutions Manual
PLM	Probeware Lab Manual	SSV	Small-Scale Video/Videodisc
PP	Problem Pro CD-ROM	TCP	www.chemplace.com
RM	Review Module	TE	Teacher's Edition
RP	Resource Pro CD-ROM	www	www.phschool.com

Chapter 11

Key Terms

11.1 thermochemistry, energy, chemical potential energy, heat, system, surroundings, universe, law of conservation of energy, endothermic process, exothermic process, calorie, joule, heat capacity, specific heat capacity, specific heat

11.2 calorimetry, calorimeter, enthalpy *(H)*, thermochemical equation, heat of reaction, heat of combustion

11.3 molar heat of fusion, molar heat of solidification, molar heat of vaporization, molar heat of condensation, molar heat of solution

11.4 Hess's law of heat summation, standard heat of formation

DISCOVER IT!

The rubber band feels warmer after it is stretched and cooler after it is allowed to relax. Students may infer that heat is related to the observed temperature changes. From their experiences, students may realize that heat flows from a warm object to a cool object. The stretching of rubber is exothermic. The reverse process (relaxation of the stretched rubber) is endothermic.

Chapter 11 THERMOCHEMISTRY—HEAT AND CHEMICAL CHANGE

FEATURES

DISCOVER IT!
Observing Heat Flow

SMALL-SCALE LAB
Heat of Combustion of a Candle

MINI LAB
Heat of Fusion of Ice

CHEMath
Review of Significant Figures

CHEMISTRY SERVING ... THE ENVIRONMENT
Harnessing Solar Energy

CHEMISTRY IN CAREERS
Solar Engineer

LINK TO PHYSIOLOGY
Dietary Calories

LINK TO BIOLOGY
Warmth from Fat

Stay current with SCIENCE NEWS
Find out more about thermochemistry:
www.phschool.com

Heat released by a burning match is absorbed by air.

DISCOVER IT! OBSERVING HEAT FLOW

You need a clean, medium-sized rubber band.

1. Hook your index fingers through each end of the rubber band. Without stretching the rubber band, place it against your upper lip or forehead. Note the temperature of the rubber band.
2. Move the rubber band away from your skin, quickly stretch and hold it, and then place it back against your skin. Note any temperature change.
3. Fully stretch the rubber band, and then allow it to return to its original shape. Place it against your skin and note any temperature change.
4. Repeat Steps 2 and 3 until you are certain of the temperature change in each step.

Did the rubber band feel cool or warm after it was stretched in Step 2? Did the rubber band feel cool or warm after it returned to its original shape in Step 3? Think about the temperature changes you observed and form some initial answers to the following questions. What is heat? In what direction does heat flow? As you work through this chapter, return to these questions.

11.1 ○ ■ ◆
11.2 ○ ■ ◆
11.3 ○ □ ◆
11.4 ◆

Conceptual This chapter is mathematically challenging and includes many abstract concepts. Use Figure 11.3 to explain exothermic and endothermic processes and Figure 11.5 to help explain the concept of heat capacity. Figure 11.15 may help students visualize energy conversions during phase changes. Section 11.4 may be omitted.

Standard The sample problems in Section 11.3 may be omitted, but students should be exposed to the heats of fusion, solidification, vaporization, and condensation. Figure 11.13 can be used to concisely summarize the enthalpy changes associated with phase changes. The heating curve for water (Figure 11.15) should also be covered. Section 11.4 may be omitted.

Honors Figures 11.11, 11.13, and 11.17 through 11.20 will help students visualize energy changes as they work through the problems in this chapter. The Mini Lab can be used to introduce Section 11.3. Standard heats of formation will be required in Section 19.4.

THE FLOW OF ENERGY—HEAT

section 11.1

Lava flowing out of an erupting volcano is very hot. Its temperature ranges from 550 °C to 1400 °C. As lava flows down the side of a volcano, it loses heat and begins to cool slowly. In some instances, the lava may flow into the ocean where it cools rapidly. **Why does lava cool more quickly in water than on land?**

objectives

- Explain the relationship between energy and heat
- Distinguish between heat capacity and specific heat

key terms

- thermochemistry
- energy
- chemical potential energy
- heat
- system
- surroundings
- universe
- law of conservation of energy
- endothermic process
- exothermic process
- calorie
- joule
- heat capacity
- specific heat capacity
- specific heat

Energy Transformations

Glowing campfires, the sun's rays, and rubbing your hands together all produce heat. However, other activities, such as melting ice and boiling water, absorb heat. **Thermochemistry** is concerned with the heat changes that occur during chemical reactions. In this chapter, you will examine heat and its effects on a number of chemical and physical processes. First, however, it is important to understand energy transformations.

When you buy gasoline, you are buying the stored potential energy it contains. This energy is used to do work, most often to propel a car. The controlled explosions of the gasoline in the car's engine transform the potential energy into useful work. Work is done when a force is used to move an object. **Energy** is the capacity for doing work or supplying heat. Unlike matter, energy is weightless, odorless, and tasteless. Energy is detected only because of its effects. Energy stored within the structural units of chemical substances is called **chemical potential energy.** Gasoline contains a significant amount of chemical potential energy. Different substances store different amounts of energy. The kinds of atoms and their arrangement in the substance determine the amount of energy stored in the substance.

Heat, represented by q, is energy that transfers from one object to another because of a temperature difference between them. Heat, itself, cannot be detected by the senses or by instruments. Only changes caused by heat can be detected. One of the effects of adding heat is a rise in the temperature of objects. It is the radiant heat of the sun's rays that makes a summer day hot. In this example, air is the object that absorbs heat and

(a)

(b)

Figure 11.1
*Chemical potential energy is stored within the bonds of gasoline molecules **(a)**. As the gasoline burns, the energy is released and is used to do work. In this example, the work is to propel race cars around the track **(b)**.*

11.1

1 Engage

Use the Visual

Have students study the photograph that opens the section. Ask:

- **What phase change occurs when molten magma is exposed to air or to Earth's surface?** (liquid to solid)
- **Does the air experience a phase change? What happens to air molecules that come in contact with magma?** (No; their average kinetic energy increases because they absorb heat and thus, the temperature of the air rises.)
- **Why does lava cool more quickly in water than on land?** (Water has a greater capacity to absorb heat than does air; the temperature difference between the lava and water may be greater than between the lava and the air.)

Check Prior Knowledge

To assess whether students know the difference between temperature and energy, ask:

- **What does a thermometer measure?** (average kinetic energy)
- **Does a thermometer measure heat?** (no)

Suppose two identical candles are used to heat two samples of water. One sample is a cup of water; the other is 10 gallons of water in a large drum. Ask:

- **How will the change in temperature of the samples compare?** (practically no change in the drum; a large increase in the cup)
- **How will the amount of heat received by each container compare?** (Both containers receive the same amount.)

STUDENT RESOURCES

From the Teacher's Resource Package, use:

- Section Review 11.1, Ch. 11 Practice Problems and Quizzes from the Review Module (Ch. 9–12)

TECHNOLOGY RESOURCES

Relevant technology resources include:

- Chem ASAP! CD-ROM
- Resource Pro CD-ROM
- ActivChemistry CD-ROM: *Thermochemistry*
- Chemistry Alive! Videodisc: *Burning Magnesium, Exploding Balloons, Superheated Steam, Thermite Reaction, Fountain of Light*

2 Teach

ACTIVITY

Glue a 13 cm × 18 cm piece of sheet metal and a 13 cm × 18 cm piece of Styrofoam™ to a piece of wood that is about 40 cm × 20 cm. Pass the wood around the classroom and ask students to put their hands on each of the three surfaces and describe the temperature of each. Usually students indicate that the metal feels the coldest and the Styrofoam™ feels the same temperature as their hands. Explain that all of the surfaces are colder than their hand, and that heat is being transferred from their hand to those surfaces. Metals conduct heat away from the hand more efficiently than Styrofoam™, so they feel colder than Styrofoam™.

TEACHER DEMO

Demonstrate an endothermic reaction between two solids. Mix 32 g of barium hydroxide octahydrate ($Ba(OH)_2 \cdot 8H_2O$) with 11 g of ammonium chloride (NH_4Cl) in a 250-mL Erlenmeyer flask fitted with a stopper. Swirl to mix. **CAUTION!** The flask becomes extremely cold. Place the flask on a wet piece of wood and the flask will freeze to the wood. **CAUTION!** Barium salts are poisonous if ingested. Inhalation of concentrated ammonia vapors can be dangerous. Perform this demo in a fume hood or a well-ventilated room. Disposal: Flush the solution down the drain with excess water. (Reference: *Chemical Demonstrations* by Bassam Z. Shakhashiri, The University Press, 1983)

increases in temperature. Heat always flows from a warmer object to a cooler object. If two objects remain in contact, heat will flow from the warmer object to the cooler object until the temperature of both objects is the same.

Figure 11.2
The part of the universe being studied is the system. What constitutes the surroundings? 1

Exothermic and Endothermic Processes

Essentially all chemical reactions and changes in physical state involve either the release or the absorption of heat. In studying heat changes, it is useful to define a **system** as the part of the universe on which you focus your attention. The **surroundings** include everything else in the universe. In thermochemical experiments, it is a good approximation to consider the region in the immediate vicinity of the system as the surroundings. In **Figure 11.2,** for example, the mixture of chemicals undergoing a reaction is the system, and everything else is the surroundings. Together, the system and its surroundings constitute the **universe.** A major goal of studying thermochemistry is to examine the flow of heat from the system to its surroundings, or the flow of heat from the surroundings to the system. The **law of conservation of energy** states that in any chemical or physical process, energy is neither created nor destroyed. All of the energy involved in a process can be accounted for as work, stored energy, or heat.

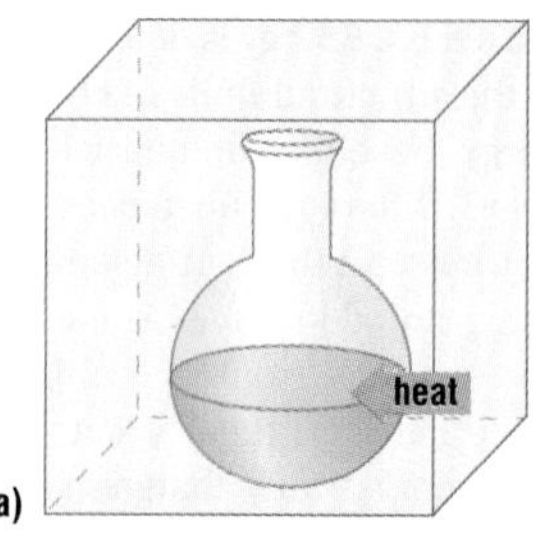

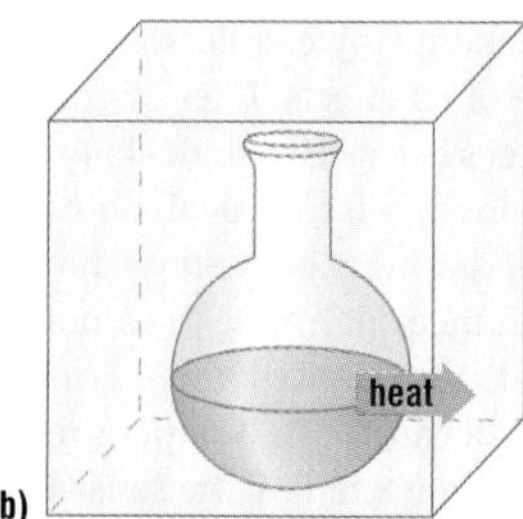

Figure 11.3
There are two directions in which the heat of a system can flow. ***(a)*** *In an endothermic process, heat flows into the system from the surroundings. The system absorbs heat.* ***(b)*** *In an exothermic process, heat flows from the system to the surroundings. The system loses heat.*

In thermochemical calculations the direction of the heat flow is given from the point of view of the system. Look at **Figure 11.3a.** Heat flowing into a system from its surroundings is defined as positive; q has a positive value. A process that absorbs heat from the surroundings is called an **endothermic process.** In an endothermic process, the system gains heat as the surroundings cool down. In **Figure 11.3b,** heat flows out of the system into its surroundings. This type of heat flow is given a negative value; q is negative because the system is losing heat. A process that releases heat to its surroundings is called an **exothermic process.** In an exothermic process, the system loses heat as the surroundings heat up. **Table 11.1** explains the sign convention for heat changes.

Table 11.1

Heat Change Sign Convention		
Direction of heat flow	**Sign**	**Reaction type**
Heat flows out of the system	Heat change < 0 (negative)	Exothermic
Heat flows into the system	Heat change > 0 (positive)	Endothermic

MEETING DIVERSE NEEDS

LEP and **At Risk** Have students write definitions for each of the key terms in this chapter. Special attention should be given to those terms that are defined by mathematical relationships. Example calculations showing how to apply the formulas should be included in their notes.

Figure 11.4

A fire helps keep you warm when you are out in the cold. If your body is the system, what is the fire? The human body cools itself by giving off heat when perspiration evaporates. What is the system here? Which of these processes is exothermic and which is endothermic? 

In the photograph on the left in **Figure 11.4,** the system (the people) gains heat from its surroundings (the fire). As shown in the inset illustration, heat flows into the system from its surroundings. What kind of process is this? In the photograph on the right, the system (the body) cools as perspiration evaporates from the skin and heat flows to the surroundings. What kind of process is shown in this illustration? 2

Heat Capacity and Specific Heat

You have probably heard of someone exercising to burn off fat and calories. What does it mean to "burn calories"? During exercise your body generates heat, and this heat is measured in units called calories. The heat is generated as your body breaks down sugars and fats into carbon dioxide and water. Although there is not an actual fire burning the sugars and fats within your body, chemical reactions accomplish the same result. In breaking down 10 g of sugar, for example, your body generates a certain amount of heat. The same amount of heat would be produced if 10 g of sugar were completely burned in a fire, producing carbon dioxide and water.

A **calorie** is defined as the quantity of heat needed to raise the temperature of 1 g of pure water 1 °C. There is an important difference, however, between a calorie and a Calorie. The calorie, written with a small c, is defined above and is used except when referring to the energy contained in food. The dietary Calorie, written with a capital C, always refers to the energy in food. One dietary Calorie is actually equal to one kilocalorie, or 1000 calories.

$$1 \text{ Calorie} = 1 \text{ kilocalorie} = 1000 \text{ calories}$$

The statement "10 g of sugar has 41 Calories" means that 10 g of sugar releases 41 kilocalories of heat when completely burned to produce carbon dioxide and water.

11.1

Discuss

Discuss with students the distinction between kinetic and potential energy. Kinetic energy is the energy associated with an object because of its motion. Potential energy is the energy associated with an object because of its position in a field of force or due to its particular chemical composition. Explain that the potential energy of a reactant or a product in a chemical reaction is determined by the strengths of the attractive and repulsive forces between atoms. In a chemical reaction, atoms are rearranged into new groupings that have different relative potential energies. The change in potential energy is either the result of absorption of energy from the surroundings (endothermic reaction) **or the release of energy to the surroundings** (exothermic reaction).

Discuss

Make sure students understand the terms introduced in Section 11.1 and how to convert from one to another. Students may have difficulty with the distinction between calorie and Calorie because everyday usage of the terms is often not accurate. The distinction between heat capacity and specific heat can be hard to remember because the terms are so closely related and sound so similar. Stress that specific heats of different materials can be compared because the quantity of matter involved (1 g) is *specified.*

Answers

1 Everything else constitutes the surroundings.
2 endothermic; exothermic
3 surroundings; the body (not including the perspiration); the first is endothermic (body absorbs heat from the fire); the second is exothermic (body gives off heat, causing perspiration to evaporate).

Chemistry Alive!

Burning Magnesium Play

11.1

Discuss

Have students read Link to Physiology. Explain that caloric ratings of activities, such as running and jumping, are calculated from metabolic functions such as oxygen consumption. You may wish to have students do research to compare the energy expended (in Calories per minute) during several types of sports- and non-sports-related activities.

Another useful measurement of energy usage is *power.* Power is the rate at which energy is converted into useful work. The unit of power is the watt (W), which is defined as 1 joule per second. Using the values given in the feature and in the text, have students compare their own power ratings for sitting at a desk, for running, or any other exercise. Point out that a joule is a small amount of energy. About 200 000 joules are required to operate a small color TV for an hour.

Chemistry Alive!

Exploding Balloons Play

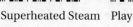

Superheated Steam Play

Figure 11.5
Which food will warm you up more when you are cold: a bowl of hot soup, or two slices of hot buttered toast? Why? ❶

The calorie is also related to the **joule,** the SI unit of heat and energy named after the English physicist James Prescott Joule (1818–1889). A joule is slightly less than one-fourth of a calorie. One joule of heat raises the temperature of 1 g of pure water 0.2390 °C. You can convert between calories and joules using the following relationships.

$$1\ \text{J} = 0.2390\ \text{cal} \qquad 4.184\ \text{J} = 1\ \text{cal}$$

The amount of heat needed to increase the temperature of an object exactly 1 °C is the **heat capacity** of that object. The heat capacity of an object depends on its mass as well as its chemical composition. The greater the mass of the object, the greater its heat capacity. A massive steel girder, for example, requires much more heat to raise its temperature 1 °C than a small steel nail does. Similarly, a cup of water has a much greater heat capacity than a drop of water. Besides varying with mass, the heat capacity of an object also depends on its chemical composition. It follows that different substances with the same mass may have different heat capacities.

On a sunny day, a 20 kg puddle of water may be cool, while a nearby 20 kg iron sewer cover may be too hot to touch. This situation illustrates how different heat capacities affect the temperature of objects. Assuming that both the water and the iron absorb the same amount of radiant energy from the sun, the temperature of the water changes less than the temperature of the iron because the specific heat capacity of water is larger.

Link to Physiology

Dietary Calories

Your proper caloric intake depends on your level of physical activity. In an eight-hour day at a desk, you burn about 800 Calories. This is about the number of Calories in two helpings of spaghetti. When exercising, however, you become a relative biochemical blast furnace. In vigorous activities such as running and jumping, you expend 7–10 Calories per minute, or 420–600 Calories per hour. At these rates, a runner who covers a 26-mile marathon course in 3 hours might expend 1800 Calories, or the equivalent of 4.5 helpings of spaghetti.

Table 11.2

Specific Heat Capacities of Some Common Substances

Substance	Specific heat capacity J/(g × °C)	Specific heat capacity cal/(g × °C)
Water	4.18	1.00
Grain alcohol	2.4	0.58
Ice	2.1	0.50
Steam	1.7	0.40
Chloroform	0.96	0.23
Aluminum	0.90	0.21
Glass	0.50	0.12
Iron	0.46	0.11
Silver	0.24	0.057
Mercury	0.14	0.033

Answers

❶ If students interpret "warm you" to refer to tissues in the mouth and upper digestive tract, they should choose the soup because it is served at a hotter temperature and because a water-based substance has a greater heat capacity, and thus, more heat content per mass to release.
❷ The heat capacity of an object is equal to its specific heat times its mass in grams.

The **specific heat capacity,** or simply the **specific heat,** of a substance is the amount of heat it takes to raise the temperature of 1 g of the substance 1 °C. What is the relationship between specific heat and heat capacity? **Table 11.2** gives specific heats for some common substances.  Water has a very high specific heat compared with the other substances in the table. You can see from the table that one calorie of heat raises the temperature of 1 g of water 1 °C. Metals, however, have low specific heats. One calorie of heat raises the temperature of 1 g of iron 9 °C. Thus water has a specific heat nine times that of iron. Heat affects the temperature of objects with a high specific heat much less than the temperature of those with a low specific heat. Just as it takes a lot of heat to raise the temperature of water, water also releases a lot of heat as it cools. Water in lakes and oceans absorbs heat from the air on hot days and releases it back into the air on cool days. As illustrated in **Figure 11.6,** this property of water is responsible for moderate climates in coastal areas. The specific heat of water is often used by farmers to protect their crops. In freezing weather, citrus crops are often sprayed with water to protect the fruit from damage. As the water freezes, it releases heat, which helps to prevent the fruit from freezing. The results of this procedure are shown in **Figure 11.7.**

To calculate the specific heat of a substance, you divide the heat input by the temperature change times the mass of the substance. The equation for specific heat (C) follows, where q is heat and m is mass. The symbol ΔT (read "delta T") in the equation represents the change in temperature. ΔT is calculated from the equation $\Delta T = T_f - T_i$, where T_f is the final temperature and T_i is the initial temperature.

$$C = \frac{q}{m \times \Delta T} = \frac{\text{heat (joules or calories)}}{\text{mass (g)} \times \text{change in temperature (°C)}}$$

As you can see from this equation, specific heat may be expressed in terms of joules or calories. Therefore, the units of specific heat are either J/(g × °C) or cal/(g × °C).

Figure 11.6
San Francisco is located on the Pacific coast. The high specific heat of the ocean helps keep the temperature in San Francisco much more moderate than that of towns and cities farther inland.

Figure 11.7
Water must give off a lot of heat in order to freeze. Sometimes farmers use water's high specific heat capacity to their advantage. In freezing weather, orange groves are often sprayed with water to protect the fruit from frost damage.

Thinking Critically

Refer to Figure 11.6 and ask students to explain a remark once made by Mark Twain: "The coldest winter I ever spent was the summer I spent in San Francisco." (The relatively high heat capacity of the large bodies of water near San Francisco help moderate local temperatures. During summer, the ocean water warms slowly as it absorbs large amounts of heat from warm air masses. Onshore breezes off the cool water keep coastal land temperatures from rising to match those of inland areas. Condensation of water vapor in the air produces a cool fog, which probably enhanced Mark Twain's impression of the unseasonable climate conditions.)

Use the Visual

Draw students' attention to Figure 11.7. Discuss how spraying fruit and flooding orchards with a few inches of water can keep the fruit from freezing. (As water cools and freezes, it releases energy that can warm the air and protect the fruit.)

 **Chemistry Alive!**

Thermite Reaction Play

Fountain of Light Play

CHEMath

Teaching Tips

Point out that the concept of significant figures applies only to measured quantities. Students may wonder why an estimated digit can be considered significant. Tell them a significant figure is one that is known to be reasonably reliable. A careful estimate fits this criterion.

Take some time to be sure students understand what can be estimated and what cannot. In the example given, students can only estimate to the hundredths place. The estimated digit depends on the number of subdivisions on the measuring device. On the chalkboard, illustrate different examples—such as a liquid in a graduated cylinder or beaker—and discuss where the estimated digit can occur.

Go over the 4 rules for determining which digits are significant and give examples of each. Stress that zeros that serve only as placeholders are not significant. For example, the numbers 3200 and 0.0032 both have 2 significant figures.

Answers to Practice Problems

1. A. 3
B. 1
C. 4
D. 3
E. 3
F. 33.9
G. 8.4
H. 63.6
I. 17.0
J. 6.046×10^7 or 60 460 000
K. 1.4×10^{11}
L. $0.4838 \dfrac{\text{J}}{\text{g} \times °\text{C}}$

CHEMath

REVIEW OF SIGNIFICANT FIGURES

This page offers you a chance to refresh your knowledge of significant figures before beginning this chapter—a chapter which involves solving problems containing numerous numerical values. As you learned in Section 3.2, a value in science must be reported using the correct number of *significant figures* (or digits). Answers cannot be more precise than the given data, as they would be misleading.

The rules for determining significant figures are summarized below. (Note that this list is a condensed version of the rules listed in Chapter 3.)

1. Nonzero digits are always significant.
2. A zero is significant only if it is
   a. at the right end of a number and after a decimal point, or
   b. between digits that are significant according to rule **1** or **2a.**

   Zeros to the left of nonzero digits or at the end of a quantity written as a whole number are "placeholders" and are *not* significant.
3. If a quantity is known to be exact, it has an unlimited number of significant figures.
4. If a quantity is written in scientific notation, all digits of the coefficient are significant.

For example, the significant digits in each number below are shown in blue.

30 400	3 significant figures
150.0	4 significant figures
2401	4 significant figures
168.030	6 significant figures
0.0058	2 significant figures
3.010×10^8	4 significant figures

When working with significant digits, *round up* the final significant digit *only if* the next digit is 5 or greater. When you do a calculation, round your final answer (not the intermediate steps) according to these rules:

Addition and Subtraction Round the result to the same number of decimal places as the measurement with the fewest decimal places.

Multiplication and Division Round the result to the same number of significant figures as the measurement with the fewest significant figures.

Example 1

Evaluate 2.34 + 1.2 and express the answer with the correct number of significant figures.

2.34 has 2 decimal places and 1.2 has 1 decimal place. So, the answer should be rounded to 1 decimal place.

Without rounding, 2.34 + 1.2 = 3.54.

After rounding 3.54 to 1 decimal place, the final answer is 3.5.

Example 2

Use the formula $C = \dfrac{q}{m \times \Delta T}$ to calculate C, the specific heat capacity, if $q = 516$ J, $m = 6.4$ g, and $\Delta T = 25.80$ °C.

Because 6.4 g has only 2 significant figures, the answer should be rounded to 2 significant figures.

$$C = \frac{516 \text{ J}}{6.4 \text{ g} \times 25.80 \text{ °C}} = 3.125 \frac{\text{J}}{\text{g} \times °\text{C}} = 3.1 \frac{\text{J}}{\text{g} \times °\text{C}}$$

Practice Problems

Find the number of significant figures in each quantity.

A. 1340 **B.** 0.06 **C.** 3.400×10^4 **D.** 0.00350 **E.** 16.0

Round each answer to the appropriate number of significant figures.

F. 16.382 + 17.5
G. 1.4×6.03
H. 128.0 − 64.37
I. 317.04 ÷ 18.7
J. $(6.030 \times 10^7) + (1.64 \times 10^5)$
K. $(3.0 \times 10^{15}) \div (2.19 \times 10^4)$
L. Use the formula $C = \dfrac{q}{m \times \Delta T}$ to calculate C if $q = 14.80$ J, $m = 3.056$ g, and $\Delta T = 10.01$ °C.

Sample Problem 11-1

The temperature of a piece of copper with a mass of 95.4 g increases from 25.0 °C to 48.0 °C when the metal absorbs 849 J of heat. What is the specific heat of copper?

1. ***ANALYZE*** ***List the knowns and the unknown.***

   Knowns:
   - $m_{Cu} = 95.4$ g
   - $\Delta T = (48.0\ °C - 25.0\ °C) = 23.0\ °C$
   - $q = 849$ J

   Unknown:
   - $C_{Cu} = ?\ \frac{J}{g \times °C}$

   Use the known values and the definition of specific heat, $C = \frac{q}{m \times \Delta T}$, to calculate the unknown value C_{Cu}.

2. ***CALCULATE*** ***Solve for the unknown.***

   Substitute the known values into the equation for the specific heat and solve.

   $$C_{Cu} = \frac{q}{m \times \Delta T}$$

   $$C_{Cu} = \frac{849\ J}{95.4\ g \times 23.0\ °C} = 0.387\ \frac{J}{g \times °C}$$

3. ***EVALUATE*** ***Does the result make sense?***

   Remember that water has a very high specific heat (4.18 J/(g × °C)). Metals, however, have low specific heats—values less than 4.18 J/(g × °C). Thus the calculated value of 0.387 J/(g × °C) seems reasonable.

Practice Problems

1. When 435 J of heat is added to 3.4 g of olive oil at 21 °C, the temperature increases to 85 °C. What is the specific heat of olive oil?
2. A 1.55-g piece of stainless steel absorbs 141 J of heat when its temperature increases by 178 °C. What is the specific heat of the stainless steel?
3. How much heat is required to raise the temperature of 250.0 g of mercury 52 °C?

Chem ASAP!

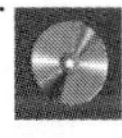

Problem-Solving 3
Solve Problem 3 with the help of an interactive guided tutorial.

section review 11.1

4. Define energy and explain how energy and heat are related.
5. Explain the difference between heat capacity and specific heat.
6. Will the specific heat of 50 g of a substance be the same as, or greater than, the specific heat of 10 g of the same substance?
7. On a sunny day, why does the concrete deck around an outdoor swimming pool become hot, while the water stays cool?
8. Using calories, calculate how much heat 32.0 g of water absorbs when it is heated from 25.0 °C to 80.0 °C. How many joules is this?
9. A chunk of silver has a heat capacity of 42.8 J/°C. If the silver has a mass of 181 g, calculate the specific heat of silver.
10. How many kilojoules of heat are absorbed when 1.00 L of water is heated from 18 °C to 85 °C?

Chem ASAP! **Assessment 11.1** Check your understanding of the important ideas and concepts in Section 11.1.

Practice Problems Plus

Additional Practice Problem

The Chem ASAP! CD-ROM contains the following problem related to Sample Problem 11-1: How much heat is required to raise the temperature of 400.0 g of silver 45 °C? (4.3×10^3 J)

3 Assess

Evaluate Understanding

▸ Place an ice cube in a beaker of cool water. Have students discuss the flow of heat when the ice and water are the system and surroundings, respectively, and when the ice/water mixture and a 37 °C room are the system and surroundings, respectively.

Reteach

Emphasize that substances vary in their response to an input of heat. A given amount of heat raises the temperature of some substances (such as metals) far more than others (such as nonmetals). Point out that specific heat is a property of a substance. It is a measure of the ability of a substance to store heat.

Answers

SECTION REVIEW 11.1

4. Energy is the capacity for doing work or supplying heat. Heat is energy that transfers between objects across a temperature gradient.
5. The specific heat of a substance is independent of its mass. The heat capacity of an object is proportional to its mass.
6. It will be the same.
7. Water has a higher heat capacity than concrete. The sun's heat raises the temperature of the concrete more than that of the water.
8. 1.76×10^3 cal (1.76 kcal); 7.36×10^3 J (7.36 kJ)
9. 2.36×10^{-1} J/(g × °C)
10. 2.8×10^2 kJ

11.2

1 Engage

Use the Visual

Have students study the photograph that opens the section. Ask:

- **A match won't ignite unless you strike it and add the heat produced from friction. Is the burning of a match an endothermic reaction?** (No; the reaction releases more energy, in the form of heat and light, than the amount of energy it absorbs to start.)
- **Is there a way to measure how much heat is released from a burning match?** (Yes, but only indirectly. If the reaction were confined, then any temperature changes in the surroundings could be attributed to heat transfer from the reaction.)

Motivate and Relate

Show students typical utility bills for electricity, gas, or oil. Stress that utility bills mainly reflect energy used to control the temperature of a home, to heat water, and for cooking. Ask why the size of the bill would vary from one household to another. (floor space, number of occupants, thermostat setting, insulation, type of fuel used) **Ask why the bill might vary from month to month.** (changes in the temperature gradient between internal and external temperatures; fluctuations in the price of fossil fuels)

section 11.2

objectives

- Construct equations that show the heat changes for chemical and physical processes
- Calculate heat changes in chemical and physical processes

key terms

- calorimetry
- calorimeter
- enthalpy (*H*)
- thermochemical equation
- heat of reaction
- heat of combustion

MEASURING AND EXPRESSING HEAT CHANGES

As you know, a burning match gives off heat. When you strike a match, heat is released to the surroundings in all directions. As you have learned, heat cannot be detected by the senses or by instruments. **Is there a way to measure exactly how much heat is released from a burning match?**

Calorimetry

Energy changes occur in many systems, from the inner workings of a clock, to the eruption of volcanoes, to the formation of the solar system. Most chemical and physical changes you will encounter occur at constant atmospheric pressure. For example, a reaction in an open beaker, the formation of ice in a lake, and the reactions in many living organisms all occur at constant atmospheric pressure. By defining a thermodynamic variable called enthalpy—a variable that takes constant pressure into account—you can measure the energy changes that accompany chemical and physical processes.

Heat that is released or absorbed during many chemical reactions can be measured by calorimetry. **Calorimetry** is the accurate and precise measurement of heat change for chemical and physical processes. In calorimetry, the heat released by the system is equal to the heat absorbed by
1 its surroundings. What law describes this relationship? To measure heat changes accurately and precisely, the processes must be carried out in an insulated container. The insulated device used to measure the absorption or release of heat in chemical or physical processes is called a **calorimeter.**

Foam cups, which keep hot drinks hot and cold drinks cold, are excellent heat insulators. Because they do not let much heat in or out, they can be used as simple calorimeters. In fact, the heat change for many chemical reactions can be measured in a constant-pressure calorimeter similar to the one shown in **Figure 11.8.** Because most chemical reactions and physical

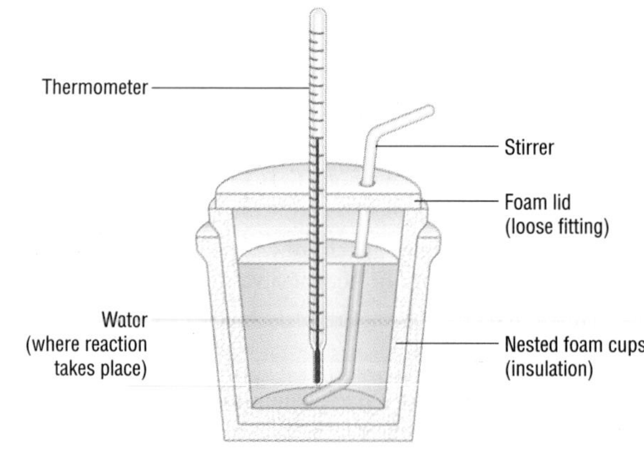

Figure 11.8
A simple constant-pressure calorimeter is shown here. In a calorimeter, the thermometer measures the temperature change of the chemicals as they react in water. The stirrer is used to keep the solution at a uniform temperature. The chemical substances that react in solution constitute the system. Is the water, in which the chemicals dissolved, part of the system or part of the surroundings? 

Student Resources

From the Teacher's Resource Package, use:

- Section Review 11.2, Ch. 11 Practice Problems and Quizzes from the Review Module (Ch. 9–12)
- Laboratory Manual: Experiment 15
- Laboratory Practical 11-1

TECHNOLOGY RESOURCES

Relevant technology resources include:

- Chem ASAP! CD-ROM
- Resource Pro CD-ROM
- ActivChemistry CD-ROM: *Thermochemistry*

Table 11.3

Enthalpy Sign Convention	
Exothermic reaction	ΔH is negative ($\Delta H < 0$)
Endothermic reaction	ΔH is positive ($\Delta H > 0$)

changes carried out in the laboratory are open to the atmosphere, these changes occur at constant pressure. For systems at constant pressure, the heat content is the same as a property called the **enthalpy** *(H)* of the system. Heat changes for reactions carried out at constant pressure are the same as changes in enthalpy, symbolized as ΔH (read "delta H"). Because the reactions presented in this textbook occur at constant pressure, the terms heat and enthalpy are used interchangeably. In other words, $q = \Delta H$. Recalling the equation for specific heat, we can write the following relationship for the heat change in a chemical reaction carried out in aqueous solution.

$$q = \Delta H = m \times C \times \Delta T$$

ΔH is the heat change; m is the mass of the water; C is the specific heat capacity of water; and $\Delta T = T_f - T_i$. The sign of ΔH is negative for an exothermic reaction and positive for an endothermic reaction. **Table 11.3** summarizes the sign convention for enthalpy.

To measure the heat change for a reaction in aqueous solution in a foam cup calorimeter, you dissolve the reacting chemicals (the system) in known volumes of water (the surroundings). Then measure the initial temperature of each solution and mix the solutions in the foam cup. After the reaction is complete, measure the final temperature of the mixed solutions. Because you know the initial and final temperatures and the heat capacity of water, you can calculate the heat released or absorbed in the reaction using the equation for specific heat.

Calorimetry experiments can also be performed at constant volume using a device called a bomb calorimeter. A bomb calorimeter, similar to the one shown in **Figure 11.9,** measures the heat released from burning a compound. The calorimeter is a closed system; that is, the mass of the system is constant.

Chem ASAP!

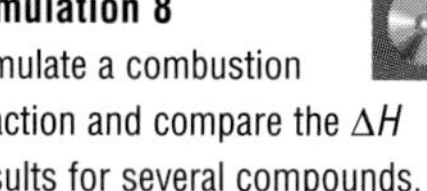

Simulation 8

Simulate a combustion reaction and compare the ΔH results for several compounds.

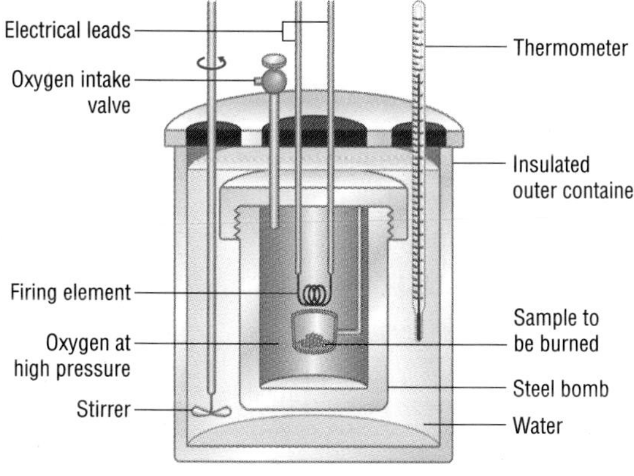

Figure 11.9
In a bomb calorimeter, a sample is burned in a constant-volume chamber in the presence of oxygen at high pressure. The heat that is released warms the water surrounding the chamber. By measuring the temperature increase of the water, it is possible to calculate the quantity of heat released during the combustion reaction.

11.2

2 Teach

ACTIVITY

Have students make calorimetry measurements and calculations. Two separate but related oxidation-reduction reactions can be studied by different groups of students. The results can then be pooled and compared to provide insight into how calorimetry can be used as a tool to investigate the physical and chemical properties of substances.

Pour 100 mL of 1*M* HCl into a Styrofoam™ cup and determine the temperature to the nearest tenth of a degree Celsius. Add 0.5 g of mossy zinc or 0.5 g of magnesium turnings. Students must wear protective goggles and aprons. Remind them of the safety precautions needed for handling HCl. Use a fume hood if possible. Students should not be allowed to dispense concentrated HCl themselves.

For calculations, ignore the mass of the solids added. Assume that HCl is primarily water and has the same specific heat and density of water. The calculations then become: $\Delta H = 100$ g $\times$ 4.18 J/(g $\times$ °C) $\times$ ΔT. Discuss sources of error such as heat lost to surroundings.

ΔH_{rxn} Zn = −150 kJ/mole
ΔH_{rxn} Mg = −460 kJ/mole

References:

1. Shakhashiri, B.Z. *Chemical Demonstrations,* Vol. 1, University Press, 1983, 1.10
2. Alyea, H.N. *J.Chem. Educ.* 1970, 47, A-387

Answers

1 The Law of Conservation of Energy
2 surroundings

Discuss

If possible, show photographs or actual examples of the types of calorimeter(s) used in laboratories today. Discuss the use of calorimetry as an analytical tool; for example, in investigating the caloric content of foods. Make sure students understand all the terms in the equation: $q = \Delta H = m \times C \times \Delta T$. Emphasize that the delta notation (Δ) represents an algebraic operation involving two separate measurements. When determining the algebraic value of ΔT, the algebraic signs of the final and initial temperatures must be included in the operation.

Practice Problems Plus

Additional Practice Problem

The Chem ASAP! CD-ROM contains the following problem related to Sample Problem 11-2: A lead mass is heated and placed in a foam cup calorimeter containing 40.0 mL of water at 17.0 °C. The water reaches a temperature of 20.0 °C. How many joules of heat were released by the lead? (502 J)

Sample Problem 11-2

To study the amount of heat released during a neutralization reaction (you will learn about neutralization in Chapter 21), 25.0 mL of water containing 0.025 mol HCl is added to 25.0 mL of water containing 0.025 mol NaOH in a foam cup calorimeter. At the start, the solutions and the calorimeter are all at 25.0 °C. During the reaction, the highest temperature observed is 32.0 °C. Calculate the heat (in kJ) released during this reaction. Assume the densities of the solutions are 1.00 g/mL.

1. *ANALYZE* ***List the knowns and the unknown.***

Knowns:

HCl solution:
- $V_{HCl} = 25.0$ mL
- solution contains 0.025 mol HCl

NaOH solution:
- $V_{NaOH} = 25.0$ mL
- solution contains 0.025 mol NaOH

- $V_{final} = V_{HCl} + V_{NaOH}$
  $= 25.0 \text{ mL} + 25.0 \text{ mL} = 50.0 \text{ mL}$
- $T_i = 25.0$ °C
- $T_f = 32.0$ °C
- $C_{water} = 4.18$ J/(g × °C)
- $\text{Density}_{solution} = 1.00$ g/mL

Unknown:
- $\Delta H = ?$ kJ

The equation requires the mass of the water used in the experiment, but the mass is not known. Use dimensional analysis to determine the mass of the water. ΔT must also be calculated. Once m, C, and ΔT are known, use $\Delta H = m \times C \times \Delta T$ to solve for ΔH of the water.

2. *CALCULATE* ***Solve for the unknown.***

First, calculate the total mass of the water. Only the final volume of the solution (V_f) is needed to make the calculation.

$$m = (50.0 \cancel{\text{mL}}) \times \left(\frac{1.00 \text{ g}}{\cancel{\text{mL}}}\right) = 50.0 \text{ g}$$

Now calculate ΔT.

$$\Delta T = T_f - T_i \qquad \Delta T = 32.0 \text{ °C} - 25.0 \text{ °C} = 7.0 \text{ °C}$$

Substitute the values for m, C_{water}, and ΔT into the equation and solve for the unknown (ΔH).

$$\Delta H = m \times C \times \Delta T$$
$$= (50.0 \text{ g})(4.18 \text{ J/(g} \times \text{°}\cancel{\text{C}}))(7.0 \text{ °}\cancel{\text{C}})$$
$$= 1463 \text{ J} = 1.5 \times 10^3 \text{ J}$$

Convert joules to kilojoules.

$$\Delta H = (1.5 \times 10^3 \cancel{\text{J}})\left(\frac{1 \text{ kJ}}{1000 \cancel{\text{J}}}\right) = 1.5 \text{ kJ}$$

Practice Problems

11. A student mixed 50.0 mL of water containing 0.50 mol HCl at 22.5 °C with 50.0 mL of water containing 0.50 mol NaOH at 22.5 °C in a foam cup calorimeter. The temperature of the resulting solution increased to 26.0 °C. How much heat in kilojoules (kJ) was released by this reaction?
12. A small pebble is heated and placed in a foam cup calorimeter containing 25.0 mL of water at 25.0 °C. The water reaches a maximum temperature of 26.4 °C. How many joules of heat were released by the pebble?

Chem ASAP!

Problem-Solving 12

Solve Problem 12 with the help of an interactive guided tutorial.

Meeting Diverse Needs

Gifted A large percentage of the electricity used in the United States is produced by the burning of fossil fuels. Power plants generate electricity by transferring heat to water during the combustion of fossil fuels. The heated water turns to steam, which drives turbines that generate electricity. In order to condense and recycle the steam, a source of coolant water is also needed. Consequently, power plants (including nuclear power plants) are often built near a source of water, such as oceans, lakes, or rivers. However, when the coolant water is returned to its source, it is usually warmer. This increase in temperature is called *thermal pollution.* Have students do research to find out if this label is justified.

Sample Problem 11-2 (cont.)

3. ***EVALUATE* Does the result make sense?**
The sign of ΔH for the water is positive; the water absorbs 1.5 kJ of heat. Therefore, this neutralization reaction releases 1.5 kJ of heat into the water in the calorimeter, so the sign of ΔH for the reaction is negative. About 4 J of heat is required to raise the temperature of 1 g of water 1 °C. Thus it would take about 200 J to raise the temperature of 50 g of water 1 °C. Further, about 1400 J, or 1.4 kJ, is needed to raise the temperature of 50 g of water 7 °C. This estimated answer is very close to the calculated value of ΔH for the neutralization reaction.

Figure 11.10
The reaction between iron(III) oxide and aluminum, called the thermite reaction, releases so much heat that the iron produced is in the molten state.
$Fe_2O_3(s) + 2Al(s) \longrightarrow Al_2O_3(s) + 2Fe(l)$

Thermochemical Equations

If you mix calcium oxide with water, an exothermic reaction takes place. The water in the mixture becomes warm. This reaction occurs when cement, which contains calcium oxide, is mixed to make concrete. When 1 mol of calcium oxide reacts with 1 mol of water, 1 mol of calcium hydroxide forms and 65.2 kJ of heat is released. You can show this in the chemical equation by including heat change as a product of the reaction. The diagram in **Figure 11.11a** shows the heat change that occurs in this exothermic reaction.

$$CaO(s) + H_2O(l) \longrightarrow Ca(OH)_2(s) + 65.2\text{ kJ}$$

You can treat heat change in a chemical reaction like any other reactant or product in a chemical equation. An equation that includes the heat change is called a **thermochemical equation.** A **heat of reaction** is the heat change for the equation exactly as it is written. You will usually see heats of reaction reported as ΔH, which is the heat change at constant pressure. The physical state of the reactants and products must also be given. The standard conditions are that the reaction is carried out at 101.3 kPa (1 atmosphere) and that the reactants and products are in their usual physical states at 25 °C.

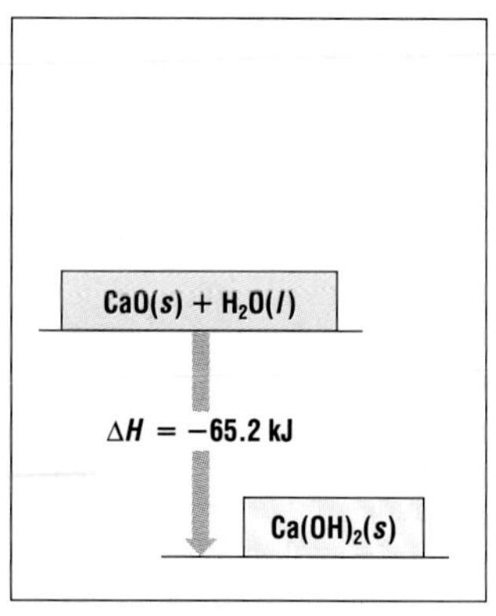

(a) Exothermic Reaction

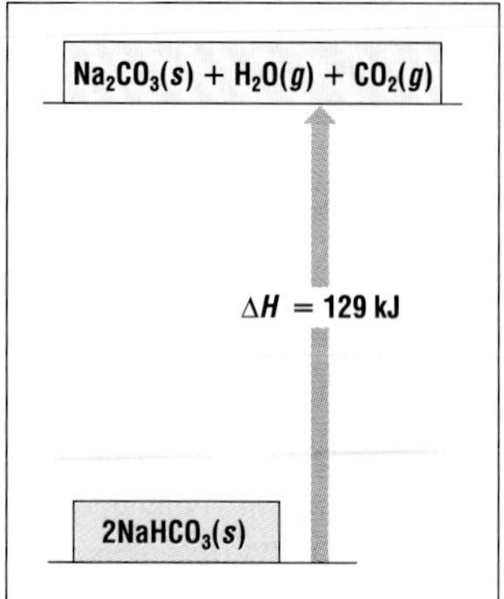

(b) Endothermic Reaction

Figure 11.11
*These enthalpy diagrams show exothermic and endothermic processes: **(a)** the reaction of calcium oxide and water and **(b)** the decomposition of sodium hydrogen carbonate. In which case is the enthalpy of the reactant(s) higher than that of the product(s)?* ❶

Answer

❶ for the exothermic reaction: calcium oxide + water

11.2

Discuss

To illustrate writing a thermochemical equation that includes energy, refer to two familiar reactions. The combustion of methane in a Bunsen burner adds energy to the surroundings. Thus, the burning of methane has a negative ΔH.
$\Delta H = -890$ kJ/mol
$CH_4(g) + 2O_2(g) \rightarrow CO_2(g) + 2H_2O(g) + 890\text{ kJ}$
During the electrolysis of water, electrical energy is used to decompose the water into hydrogen gas and oxygen gas. Energy is added to the system from the surroundings. Thus, the electrolysis of water has a positive ΔH.
$\Delta H = +241.8$ kJ/mol
$241.8\text{ kJ} + 2H_2O(l) \rightarrow 2H_2(g) + O_2(g)$
Remind students that the sign of ΔH indicates whether the reaction is endothermic (+) or exothermic (−). Although by convention, an exothermic reaction has a $-\Delta H$, when heat is shown as a "product" in an equation, it is expressed as an absolute value.

Use the Visual

Ask students to study Figure 11.11 and draw similar diagrams for the burning of CH_4 and the electrolysis of H_2O. For each reaction, ask:

- **Is the potential energy of the reactants higher or lower than the products?** (higher for CH_4, lower for H_2O)

Practice Problems Plus

Additional Practice Problem
The Chem ASAP! CD-ROM contains the following problem related to Sample Problem 11-3: The burning of magnesium is a highly exothermic reaction. How many kilojoules of heat are released when 0.75 mol of Mg burn in an excess of O_2? (450 kJ)
$2Mg(s) + O_2(g) \rightarrow 2MgO(s) + 1204\ kJ$

The heat of reaction, or ΔH, in the above example is -65.2 kJ. Each mole of calcium oxide and water that react to form calcium hydroxide produces 65.2 kJ of heat.

$$CaO(s) + H_2O(l) \longrightarrow Ca(OH)_2(s) \qquad \Delta H = -65.2\ kJ$$

Other reactions absorb heat from the surroundings. For example, baking soda (sodium hydrogen carbonate) decomposes when it is heated, making it useful in baking. The carbon dioxide released in the reaction causes a cake to rise while baking. This process is endothermic, and the heat of reaction is 129 kJ.

$$2NaHCO_3(s) + 129\ kJ \longrightarrow Na_2CO_3(s) + H_2O(g) + CO_2(g)$$

Remember that ΔH is positive for endothermic reactions. Therefore, you can write the reaction as follows.

$$2NaHCO_3(s) \longrightarrow Na_2CO_3(s) + H_2O(g) + CO_2(g) \qquad \Delta H = 129\ kJ$$

Sample Problem 11-3

Using the equation for the reaction above, calculate the kilojoules of heat required to decompose 2.24 mol $NaHCO_3(s)$.

1. ***ANALYZE*** ***List the knowns and the unknown.***

Knowns:
- 2.24 mol $NaHCO_3(s)$ decomposes
- $\Delta H = 129kJ$

Unknown:
- $\Delta H = ?$ kJ

Use the thermochemical equation,

$$2NaHCO_3(s) + 129\ kJ \longrightarrow Na_2CO_3(s) + H_2O(g) + CO_2(g),$$

to write a conversion factor relating kilojoules of heat and moles of $NaHCO_3$. Then use the conversion factor to determine ΔH for 2.24 mol $NaHCO_3$.

2. ***CALCULATE*** ***Solve for the unknown.***

The thermochemical equation indicates that 129 kJ are needed to decompose 2 mol $NaHCO_3(s)$. Using this relationship, the conversion factor is

$$\frac{129\ kJ}{2\ mol\ NaHCO_3(s)}$$

Using dimensional analysis, solve for ΔH.

$$\Delta H = 2.24\ \cancel{mol\ NaHCO_3(s)} \times \frac{129\ kJ}{2\ \cancel{mol\ NaHCO_3(s)}}$$
$$= 144\ kJ$$

3. ***EVALUATE*** ***Does the result make sense?***

Because the ΔH of 129 kJ refers to the decomposition of 2 mol $NaHCO_3(s)$, the decomposition of 2.24 mol should absorb about 10% more heat than 129 kJ, or slightly more than 142 kJ. The answer of 144 kJ is consistent with this estimate.

Practice Problems

13. When carbon disulfide is formed from its elements, heat is absorbed. Calculate the amount of heat (in kJ) absorbed when 5.66 g of carbon disulfide is formed.

$$C(s) + 2S(s) \longrightarrow CS_2(l)$$
$$\Delta H = 89.3\ kJ$$

14. The production of iron and carbon dioxide from iron(III) oxide and carbon monoxide is an exothermic reaction. How many kilojoules of heat are produced when 3.40 mol Fe_2O_3 reacts with an excess of CO?

$$Fe_2O_3(s) + 3CO(g) \longrightarrow 2Fe(s) + 3CO_2(g) + 26.3\ kJ$$

Chem ASAP!

Problem-Solving 14

Solve Problem 14 with the help of an interactive guided tutorial.

Meeting Diverse Needs

Gifted Iron wool will burn rapidly in high concentrations of oxygen. It will oxidize slowly at the concentrations normally available in air. Rusting of iron is essentially a slow combustion reaction. Have students design an experiment to determine the heat of reaction for the rusting of iron wool. (Wrapping moist iron wool around the bulb of a thermometer demonstrates the heat released by this reaction. A foam cup calorimeter can be used to quantify the thermodynamic parameters of the reaction.)

Chemistry problems involving enthalpy changes are similar to stoichiometry problems. The amount of heat released or absorbed during a reaction depends on the number of moles of the reactants involved. The decomposition of 2 mol of sodium hydrogen carbonate, for example, requires 129 kJ of heat. Therefore, the decomposition of 4 mol of the same substance would require twice as much heat, or 258 kJ. **Figure 11.11b** on page 303 shows the heat changes for this reaction. In this and other endothermic processes, the potential energy of the product(s) is higher than the potential energy of the reactant(s).

The physical state of the reactants and products in a thermochemical reaction must also be stated. To see why, compare the following two equations for the decomposition of 1 mol of water.

$$H_2O(l) \longrightarrow H_2(g) + \tfrac{1}{2}O_2(g) \qquad \Delta H = 285.8 \text{ kJ}$$
$$H_2O(g) \longrightarrow H_2(g) + \tfrac{1}{2}O_2(g) \qquad \Delta H = 241.8 \text{ kJ}$$
$$\text{difference} = 44.0 \text{ kJ}$$

Although the two equations are very similar, the different physical states of the H_2O result in different ΔH values. In one case, the reactant is a liquid; in the other case, the reactant is a gas. The vaporization of 1 mole of liquid water to water vapor at 25 °C requires an extra 44.0 kJ of heat. Notice also that fractional coefficients are used here for O_2 because 1 mol H_2O is being decomposed.

$$H_2O(l) \longrightarrow H_2O(g) \qquad \Delta H = 44.0 \text{ kJ}$$

Table 11.4 lists heats of combustion for some common substances. The **heat of combustion** is the heat of reaction for the complete burning of one mole of a substance.

Table 11.4

Heats of Combustion at 25 °C		
Substance	**Formula**	**ΔH (kJ/mol)**
Hydrogen	$H_2(g)$	−286
Carbon	$C(s)$, graphite	−394
Carbon monoxide	$CO(g)$	−283
Methane	$CH_4(g)$	−890
Methanol	$CH_3OH(l)$	−726
Acetylene	$C_2H_2(g)$	−1300
Ethanol	$C_2H_5OH(l)$	−1368
Propane	$C_3H_8(g)$	−2220
Benzene	$C_6H_6(l)$	−3268
Glucose	$C_6H_{12}O_6(s)$	−2808
Octane	$C_8H_{18}(l)$	−5471
Sucrose	$C_{12}H_{22}O_{11}(s)$	−5645

Meeting Diverse Needs

Gifted Many metals react with oxygen to produce metal oxides. Have students research the combustion reactions of a series of selected metals (for example, Mg, Ca, Fe). Students should write the complete thermochemical equations for each reaction and compare the relative amounts of heat released in each case. Ask students to describe any trend(s) they observe.

Teacher Demo

Demonstrate an exothermic reaction. Tell students that you are going to mix calcium oxide, $CaO(s)$, with water and observe what happens. Place 100 g of CaO in a metal tray. **CAUTION!** CaO is corrosive. Place a thin aluminum pan or pie tin in the tray so that its bottom is completely in contact with the calcium oxide. Pour just enough water into the tray to cover the calcium oxide. Then break a fresh egg into the aluminum pan. Students should observe that enough heat is liberated by the reaction to cook the egg. Ask:

- Is this an exothermic or endothermic reaction? (Exothermic; point out that 65.2 kJ of heat is released per mole of CaO.)

Have students write a thermochemical equation to describe the reaction they have observed. Help students by writing a word equation on the board: Calcium oxide reacts with water to produce calcium hydroxide and 65.2 kJ of heat [$CaO(s) + H_2O(l) \rightarrow Ca(OH)_2(s)$; $\Delta H = -65.2$ kJ]

Check to see that students express the enthalpy of reaction as a negative value. Remind students about the importance of including phase labels in a thermochemical equation because ΔH depends on the phase of the substances. Have students use the thermochemical equation to answer stoichiometric questions such as:

- How much heat is released when 100.0 g of calcium oxide reacts with excess water? (116 kJ)

3 Assess

Evaluate Understanding

To determine students' knowledge about measuring and expressing heat changes:

- **Ask students to explain the difference between H and ΔH.** (H is enthalpy or heat content, ΔH represents a change in heat content.)
- **Have students use the data in Table 11.4 to write the themo-chemical equation for the combustion of propane.** [$C_3H_8(g) + 5O_2(g) \rightarrow 3CO_2(g) + 4H_2O(g); \Delta H = -2220$ kJ]
- **Have them draw an enthalpy diagram for this reaction like those in Figure 11.11.**
- **Have students calculate the amount of heat produced when 64.0 grams of propane burn completely.** (3.23×10^3 kJ)

Reteach

Select a substance from Table 11.4 and write the themochemical equation on the chalkboard. Discuss the ΔH for the reaction, why it is negative or positive, and draw the enthalpy diagram for the reaction. Remind students that heat is the energy that flows into or out of a thermodynamic system because of differences in temperature between the system and its suroundings. Remind students that the total energy (potential and kinetic) of the system and the surroundings must remain the same in any chemical or physical process.

BIOLOGY

Warmth from Fat

How do animals such as polar bears and seals survive the cold land and water temperatures where they live? A good coat of fur helps, but it is not enough to keep them warm. These animals also have special fat cells that help generate heat. These special cells are in tissue called brown fat. The cells of brown fat are unlike other fat cells in the animal's body. Most other cells store chemical energy from the breakdown of carbohydrates and fatty acids in adenosine triphosphate (ATP). ATP acts as the central source of energy for the activities and growth of all animals. Heat is mostly a waste product in ATP-producing cells. The heat generated by the brown fat tissue, however, helps the animal keep relatively comfortable even at subzero temperatures.

Figure 11.12
The combustion of natural gas is an exothermic reaction. As bonds in methane, the main component of natural gas, and oxygen are broken and bonds in carbon dioxide and water are formed, large amounts of energy are released. This energy powers the engine of automobiles. Natural gas is an alternative to gasoline because it substantially reduces air pollution while increasing engine life.

The combustion of natural gas, which is mostly methane, is an exothermic reaction used to heat many homes around the country.

$$CH_4(g) + 2O_2(g) \longrightarrow CO_2(g) + 2H_2O(l) + 890 \text{ kJ}$$

This can also be written as follows.

$$CH_4(g) + 2O_2(g) \longrightarrow CO_2(g) + 2H_2O(l) \qquad \Delta H = -890 \text{ kJ}$$

Burning 1 mol of methane releases 890 kJ of heat. The heat of combustion (ΔH) for this reaction is -890 kJ per mole of carbon burned.

Like other heats of reaction, heats of combustion are reported as the enthalpy changes when the reactions are carried out at 101.3 kPa of pressure and the reactants and products are in their physical states at 25 °C.

section review 11.2

15. When 2 mol of solid magnesium (Mg) combines with 1 mole of oxygen gas (O_2), 2 mol of solid magnesium oxide (MgO) is formed and 1204 kJ of heat is released. Write the thermochemical equation for this combustion reaction.
16. Gasohol contains ethanol ($C_2H_5OH)(l)$, which when burned reacts with oxygen to produce $CO_2(g)$ and $H_2O(g)$. How much heat is released when 12.5 g of ethanol burns?
    $$C_2H_5OH(l) + 3O_2(g) \longrightarrow 2CO_2(g) + 3H_2O(g) \qquad \Delta H = -1235 \text{ kJ}$$
17. Explain the term heat of reaction.
18. Hydrogen gas and fluorine gas react to produce hydrogen fluoride. Calculate the heat change (in kJ) for the conversion of 15.0 g of hydrogen gas to hydrogen fluoride gas at constant pressure.
    $$H_2(g) + F_2(g) \longrightarrow 2HF(g) \qquad \Delta H = -536 \text{ kJ}$$
19. Why is it important to give the physical state of a substance in a thermochemical reaction?

Chem ASAP! Assessment 11.2 Check your understanding of the important ideas and concepts in Section 11.2.

Answers

SECTION REVIEW 11.2

15. $2Mg(s) + O_2(g) \rightarrow 2MgO(s) + 1204$ kJ
16. 3.36×10^2 kJ
17. the heat released or absorbed in a chemical change
18. -4.02×10^3 kJ
19. because a phase change involves an energy change

HEAT IN CHANGES OF STATE

When your body heats up, you start to sweat. The evaporation of sweat is your body's way of cooling itself to a normal temperature. **Why does the evaporation of sweat from your skin help to rid your body of excess heat?**

objectives

- Classify, by type, the heat changes that occur during melting, freezing, boiling, and condensing
- Calculate heat changes that occur during melting, freezing, boiling, and condensing

key terms

- molar heat of fusion
- molar heat of solidification
- molar heat of vaporization
- molar heat of condensation
- molar heat of solution

Heats of Fusion and Solidification

What happens if you place an ice cube on a table in a warm room? The ice cube is the system, and the table and air around it are the surroundings. The ice absorbs heat from its surroundings and begins to melt. The temperature of the ice and the water produced remains at 0 °C until all of the ice has melted. The temperature of the water begins to increase only after all of the ice has melted. In this section you will learn about heat changes that occur during changes of state.

Like ice cubes, all solids absorb heat as they melt to become liquids. The heat absorbed by one mole of a substance in melting from a solid to a liquid at a constant temperature is the **molar heat of fusion** (ΔH_{fus}). The heat lost when one mole of a liquid solidifies at a constant temperature is the **molar heat of solidification** (ΔH_{solid}). The quantity of heat absorbed by a melting solid is exactly the same as the quantity of heat lost when the liquid solidifies; that is, $\Delta H_{fus} = -\Delta H_{solid}$, as shown in **Figure 11.13.** Why is this true? **Table 11.5** on the following page gives heats of fusion of some substances. ❶

The melting of 1 mol of ice at 0 °C to 1 mol of water at 0 °C requires the absorption of 6.01 kJ of heat. This quantity of heat is the molar heat of fusion. Likewise, the conversion of 1 mol of water at 0 °C to 1 mol of ice at 0 °C releases 6.01 kJ. This quantity of heat is the molar heat of solidification.

$$H_2O(s) \longrightarrow H_2O(l) \qquad \Delta H_{fus} = 6.01\ \text{kJ/mol}$$

$$H_2O(l) \longrightarrow H_2O(s) \qquad \Delta H_{solid} = -6.01\ \text{kJ/mol}$$

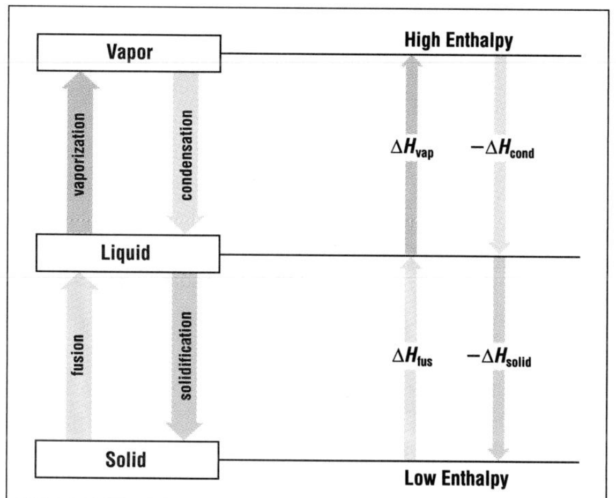

Chem ASAP!

Animation 12

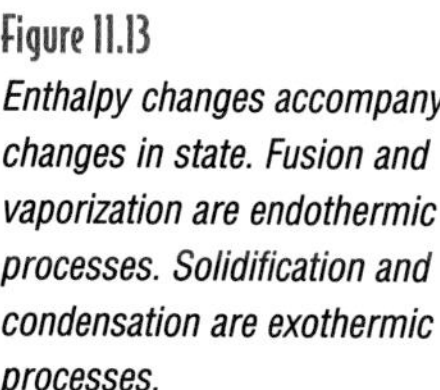

Observe the phase changes as ice is converted to steam when heat is added.

Figure 11.13

Enthalpy changes accompany changes in state. Fusion and vaporization are endothermic processes. Solidification and condensation are exothermic processes.

1 Engage

Use the Visual

Have students study the photograph and read the text that opens the section. Remind them that evaporation is a physical change from a liquid state to a gas (vapor). **Ask:**

- **On a molecular level, what is the difference between liquid water and water vapor?** (In the vapor phase, the water molecules are farther apart and not as strongly attracted to each other.)
- **Because of the attraction between molecules in a liquid, what must happen for a liquid to change into a gas?** (Energy must be added to overcome the attraction.)
- **Is evaporation endothermic or exothermic?** (endothermic)

ACTIVITY

Have students record the temperature of gently boiling water (use boiling chips) and of melting ice in a Styrofoam™ cup. Students should record the temperatures every 30 seconds. Even if the thermometers do not record exactly 100 °C or 0 °C, students should note that the temperature does not change during boiling and melting.

STUDENT RESOURCES

From the Teacher's Resource Package, use:

- Section Review 11.3, Ch. 11 Practice Problems, Interpreting Graphics, and Quizzes from the Review Module (Ch. 9–12)
- Laboratory Recordsheet 11-1
- Small-Scale Chemistry Lab Manual: Experiment 13

TECHNOLOGY RESOURCES

Relevant technology resources include:

- Chem ASAP! CD-ROM
- Resource Pro CD-ROM
- ActivChemistry CD-ROM: *Thermochemistry*
- Chemistry Alive! Videodisc: *Phase Change, Superheated Steam*

❶ because energy is conserved in all chemical and physical changes

MINI LAB

Heat of Fusion of Ice

SAFETY

Students should wear safety goggles. Warn students not to use the thermometer as a stirring rod and discuss what students should do if the thermometer breaks.

LAB PREP AND PLANNING

Probeware Lab Manual has Vernier, TI, and PASCO labs.

Provide students with a source of water between 50 °C and 60 °C.

LAB TIPS

Remind students to make all temperature measurements to the nearest tenth of a degree and all volume measurements to the nearest milliliter. Be sure students realize that the volume of cooled water is equal to the initial volume of the warm water plus the volume of the melted ice.

ANALYSIS AND CONCLUSIONS

1. The values should not differ significantly. However, if the experiment is not done quickly, loss of heat to a cooler classroom may cause ΔT to be too large. Consequently, students would get a larger ΔH_{fus}. Careful reading of the thermometer and graduated cylinder is crucial. Some water may adhere to the thermometer and cup; thus, the actual volumes may be different from the recorded volumes.
2. Answers should include the following: to determine the mass of warm water cooled and the mass of ice melted, measure the mass of (1) the empty cup, (2) the cup plus the hot water, and (3) the cup after the ice has melted. Use a covered calorimeter to decrease the influence of the room temperature.

Table 11.5

Substance	Formula	Freezing point (K)	ΔH_{fus} (kJ/mol)	Boiling point (K)	ΔH_{vap} (kJ/mol)
Acetone	CH_3COCH_3	177.8	5.72	329.4	29.1
Ammonia	NH_3	195.3	5.65	239.7	23.4
Argon	Ar	83.8	1.2	87.3	6.5
Benzene	C_6H_6	278.7	9.87	353.3	30.8
Ethanol	C_2H_5OH	158.7	4.60	351.5	43.5
Helium	He	3.5	0.02	4.22	0.08
Hydrogen	H_2	14.0	0.12	20.3	0.90
Methane	CH_4	90.7	0.94	111.7	8.2
Methanol	CH_3OH	175.5	3.16	337.2	35.3
Neon	Ne	24.5	0.33	27.1	1.76
Nitrogen	N_2	63.3	0.72	77.4	5.58
Oxygen	O_2	54.8	0.44	90.2	6.82
Water	H_2O	273.2	6.01	373.2	40.7

Heats of Physical Change

MINI LAB

Heat of Fusion of Ice

PURPOSE

To estimate the heat of fusion of ice.

MATERIALS

- ice
- foam cup
- graduated cylinder
- thermometer
- hot water
- temperature probe (optional)

PROCEDURE

Probe version available in the Probeware Lab Manual.

1. Fill a 100-mL graduated cylinder with hot tap water. Allow the filled cylinder to stand for 1 minute. Pour the water into the sink.
2. Use the graduated cylinder to measure 70 mL of hot water. Pour the water into the foam cup. Measure the temperature of the water.
3. Add a small ice cube to the cup of water and gently swirl the cup. Measure the temperature of the water immediately after the ice cube has completely melted.
4. Pour the water into the graduated cylinder and measure the volume.
5. Calculate the heat of fusion of ice (kJ/mol) by dividing the heat given up from the water by the moles of ice melted. *Hint:* The mass of ice melted is the same as the increase in the volume of the water: 1 g H_2O = 1 mL H_2O.

ANALYSIS AND CONCLUSIONS

1. Compare your experimental value for the heat of fusion of ice with the accepted value of 6.01 kJ/mol. Account for any error in your value.
2. Suggest some changes in this procedure that would improve the accuracy of the results.

Figure 11.14
Ice is commonly used to refrigerate perishable foods. What happens to the temperature of the ice as it begins to melt? ❶

Sample Problem 11-4

How many grams of ice at 0 °C and 101.3 kPa could be melted by the addition of 2.25 kJ of heat?

1. ***ANALYZE*** ***List the knowns and the unknown.***

Knowns:

- Initial conditions are 0 °C and 101.3 kPa.
- $\Delta H_{fus} = 6.01$ kJ/mol
- $\Delta H = 2.25$ kJ

Unknown:

- $m_{ice} = ?$ g

The conditions 0 °C and 101.3 kPa indicate that the standard conditions for the fusion of ice have been met. Use the chemical equation $H_2O(s) + 6.01\text{ kJ} \longrightarrow H_2O(l)$ to find the number of moles of ice that can be melted by the addition of 2.25 kJ of heat. Convert moles of ice to grams of ice.

2. ***CALCULATE*** ***Solve for the unknown.***

The required conversion factors come from ΔH_{fus} and the molar mass of ice. The conversion factors are

$$\frac{1\text{ mol ice}}{6.01\text{ kJ}} \quad \text{and} \quad \frac{18.0\text{ g ice}}{1\text{ mol ice}}$$

Multiply the known heat change (2.25 kJ) by the conversion factors

$$m_{ice} = 2.25\text{ kJ} \times \frac{1\text{ mol ice}}{6.01\text{ kJ}} \times \frac{18.0\text{ g ice}}{1\text{ mol ice}}$$

$$= 6.74\text{ g ice}$$

3. ***EVALUATE*** ***Does the result make sense?***

6.01 kJ is required to melt 1 mol of ice. Because only about one-third of this amount of heat (roughly 2 kJ) is available, only about one-third mol of ice, or $\frac{18.0\text{ g}}{3} = 6$ g, should melt. This estimate and the calculated answer are similar.

Practice Problems

20. How many grams of ice at 0 °C and 101.3 kPa could be melted by the addition of 0.400 kJ of heat?
21. How many kilojoules of heat are required to melt a 10.0 g popsicle at 0 °C and 101.3 kPa? Assume the popsicle has the same molar mass and heat capacity as water.

Chem ASAP!

Problem-Solving 21

Solve Problem 21 with the help of an interactive guided tutorial.

Answer

❶ The ice and water are both at 0 °C. The temperature will not rise above 0 °C until all of the ice has melted.

2 Teach

Discuss

Ask:

- **At what temperature does water freeze? At what temperature does ice melt?** (Many students do not understand that melting and freezing temperatures are the same, 0 °C.)

Propose a hypothetical experiment: Place a beaker of water at precisely 0 °C and a block of ice at precisely 0 °C in a refrigerator that is regulated to maintain exactly 0 °C. After several hours, reexamine these two substances. What will you see? (The water is still water and the ice is still ice.) **Why?** (In order for ice to melt, there must be a source of energy to provide 6.01 kJ/mole of ice. Because no temperature gradient exists between the ice and the air in the refrigerator, no heat (energy) flows into the ice. In order for water to freeze, 6.01 kJ heat/mole water must be released. Again, since no temperature gradient exists between the liquid water and its surroundings, no heat flows from the water to the refrigerator.) **Point out that the amount of energy that it takes to melt ice is exactly the same as the amount of energy it takes to freeze the same quantity of water. Use Figure 11.13 to reinforce this concept.**

Practice Problems Plus

Related Chapter Review Problems

Chapter Review problems 53a and 84 are related to Sample Problem 11-4.

Additional Practice Problem

The Chem ASAP! CD-ROM contains the following problem: How much heat must be removed to freeze a tray of ice cubes if the water has a mass of 50.0 g? (16.7 kJ)

TEACHER DEMO

Set three wide-mouth test tubes each containing about 20 grams of sodium thiosulfate pentahydrate ($Na_2S_2O_3 \cdot 5H_2O$) in a hot water bath until the chemical is completely melted. Transfer the tubes to a cool water bath at about 30°C. Let the test tubes sit undisturbed until they are close to 30°C and feel lukewarm to touch. Give the tubes to three student volunteers at different locations in the classroom so that many students can see and feel what happens. Give each volunteer a thermometer and a seed crystal. Ask them to quickly take the temperature of the liquid, drop in the seed crystal and stir gently. Have them describe any changes they can observe by touching the test tube and reading the thermometer. This is an exceptionally good demo to illustrate the release of heat when a liquid solidifies.

Heats of Vaporization and Condensation

When liquids absorb heat at their boiling points, they become vapors. Vaporization of a liquid, through boiling or evaporation, cools the environment around the liquid as heat flows from the surroundings to the liquid. The amount of heat necessary to vaporize one mole of a given liquid is called its **molar heat of vaporization.** **Table 11.5** on page 308 gives some values of molar heats of vaporization for various compounds. The values are derived at standard conditions: the reactions are carried out at one atmosphere pressure and the reactants and products are in their usual physical states at the same temperature.

The molar heat of vaporization of water is 40.7 kJ/mol. This means that in order to vaporize 1 mol of water, 40.7 kJ of energy must be supplied. This energy converts 1 mol of water molecules in the liquid state to 1 mol of water molecules in the vapor state, given the same temperature and 1 atm pressure. This process is described in the thermochemical equation below.

$$H_2O(l) \longrightarrow H_2O(g) \qquad \Delta H_{vap} = 40.7 \text{ kJ/mol}$$

Diethyl ether ($C_4H_{10}O$) is a low-boiling-point liquid (bp = 34.6 °C) that is a good solvent and was formerly used as an anesthetic. If diethyl ether is poured into a beaker on a warm, humid day, the ether will absorb heat from the beaker walls and evaporate very rapidly. If the beaker loses enough heat, the water vapor in the air may condense and freeze on the beaker walls. If so, a coating of frost will form on the outside of the beaker. Diethyl ether has a molar heat of vaporization (ΔH_{vap}) of 15.7 kJ/mol. Is this
1 an endothermic or an exothermic process?

$$C_4H_{10}O(l) \longrightarrow C_4H_{10}O(g) \qquad \Delta H_{vap} = 15.7 \text{ kJ/mol}$$

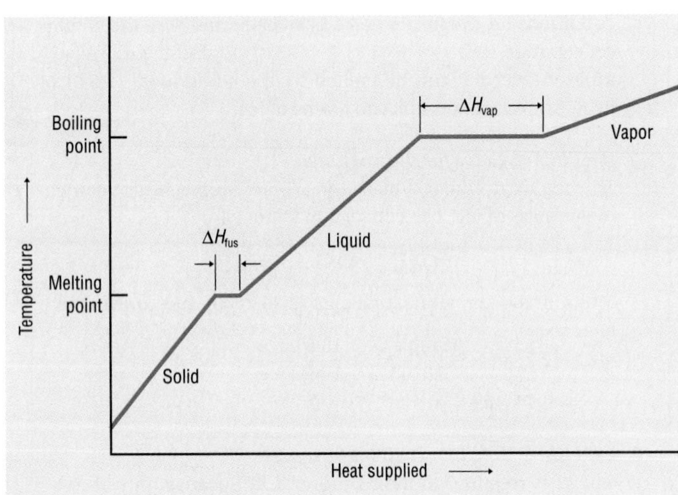

Figure 11.15
This graph shows the heating curve for water. Notice that the temperature remains constant during melting and vaporization. Notice also that it requires much more energy to vaporize liquid water than it does to melt the same mass of ice.

Answer

1 endothermic

Condensation is the exact opposite of vaporization. Therefore, the amount of heat released when 1 mol of vapor condenses is called its **molar heat of condensation** (ΔH_{cond}). This value is numerically the same as the corresponding molar heat of vaporization, however the value has the opposite sign. Because energy is conserved in a physical change, $\Delta H_{vap} = -\Delta H_{cond}$. **Figure 11.15** summarizes the heat changes that occur as a solid is heated to a liquid and then to a gas. You should be able to identify certain trends regarding the temperature during changes of state and the energy requirements that accompany these changes from the graph. The large values for ΔH_{vap} and ΔH_{cond} are the reason hot vapors such as steam can be very dangerous. You can receive a scalding burn from steam when the heat of condensation is released as it touches your skin.

$$H_2O(g) \longrightarrow H_2O(l) \qquad \Delta H_{cond} = -40.7 \text{ kJ/mol}$$

Sample Problem 11-5

How much heat (in kJ) is absorbed when 24.8 g $H_2O(l)$ at 100 °C is converted to steam at 100 °C?

1. *ANALYZE* *List the knowns and the unknown.*

Knowns:
- mass of water converted to steam = 24.8 g
- $\Delta H_{vap} = 40.7$ kJ/mol

Unknown:
- $\Delta H = ?$ kJ

The ΔH_{vap} in the following equation is given in kJ/mol, but the quantity of water is given in grams. Thus the first steps in the problem solution are to convert grams of water to moles of water and then multiply by ΔH_{vap}.

$$H_2O(l) + 40.7 \text{ kJ/mol} \longrightarrow H_2O(g)$$

2. *CALCULATE* *Solve for the unknown.*

The required conversion factors come from ΔH_{vap} and the molar mass of water.

$$\frac{1 \text{ mol } H_2O(l)}{18.0 \text{ g } H_2O(l)} \quad \text{and} \quad \frac{40.7 \text{ kJ}}{1 \text{ mol } H_2O(l)}$$

Multiply the known mass of water in grams by the conversion factors.

$$\Delta H = 24.8 \cancel{\text{g } H_2O(l)} \times \frac{1 \cancel{\text{mol } H_2O(l)}}{18.0 \cancel{\text{g } H_2O(l)}} \times \frac{40.7 \text{ kJ}}{1 \cancel{\text{mol } H_2O(l)}}$$
$$= 56.1 \text{ kJ}$$

3. *EVALUATE* *Does the result make sense?*

Knowing the molar mass of water is 18.0 g/mol, 24.8 g $H_2O(l)$ can be estimated to be somewhat less than 1.5 mol H_2O. Thus the calculated heat change should be somewhat less than 1.5 $\cancel{\text{mol}} \times 40$ kJ/$\cancel{\text{mol}}$ = 60 kJ, and it is.

Practice Problems

22. How much heat (in kJ) is absorbed when 63.7 g $H_2O(l)$ at 100 °C is converted to steam at 100 °C?

23. How many kilojoules of heat are absorbed when 0.46 g of chloroethane (C_2H_5Cl, bp 12.3 °C) vaporizes at its boiling point? The molar heat of vaporization of chloroethane is 26.4 kJ/mol.

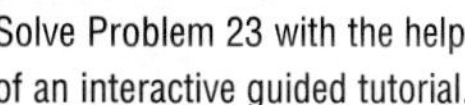

Problem-Solving 23
Solve Problem 23 with the help of an interactive guided tutorial.

ACTIVITY

Have students look up the molar heat of vaporization, the molar heat of fusion, the boiling point, the melting point, and the specific heat capacity for ethanol in Tables 11.2 and 11.5. Have students summarize the heat changes that occur as solid ethanol is heated to a liquid and then to a gas by constructing a heating curve like the one shown in Figure 11.15 for water. Ask students to find the amount of heat necessary to convert 1 mole of solid ethanol to gaseous ethanol.

Practice Problems Plus

Related Chapter Review Problems
Chapter Review problems 53b and 53d are related to Sample Problem 11-5.

Additional Practice Problem
The Chem ASAP! CD-ROM contains the following problem: How many kilojoules of heat are required to vaporize 50.0 g of ethanol, C_2H_5OH? The boiling point of ethanol is 78.5 °C. Its molar heat of vaporization is 43.5 kJ/mol. (47.3 kJ)

ACTIVITY

Purchase a number of hot and cold packs used by athletes and show the class how they work. Most instant cold and hot packs operate by utilizing the heat evolved or absorbed when certain substances dissolve in water. Hot packs usually contain calcium chloride, $CaCl_2(s)$, while cold packs usually contain ammonium nitrate, $NH_4NO_3(s)$. If the labels specify the amount of salt contained in each package, have students use the molar heats of solution provided in the text to calculate the amount of heat absorbed and released in each case.

Heat of Solution

Heat changes can also occur when a solute dissolves in a solvent. The heat change caused by dissolution of one mole of substance is the **molar heat of solution** (ΔH_{soln}). Sodium hydroxide provides a good example of an exothermic molar heat of solution. When 1 mol of sodium hydroxide $(NaOH)(s)$ is dissolved in water, the solution can become so hot that it steams. The heat from this process is released as the sodium ions and the hydroxide ions separate and interact with the water. The temperature of the solution increases, releasing 445.1 kJ of heat as the molar heat of solution.

$$NaOH(s) \xrightarrow{H_2O(l)} Na^+(aq) + OH^-(aq)$$

$$\Delta H_{soln} = -445.1 \text{ kJ/mol}$$

A practical application of an exothermic reaction is how a hot pack works. A hot pack mixes calcium chloride ($CaCl_2$) and water, which produces the heat characteristic of an exothermic reaction.

$$CaCl_2(s) \xrightarrow{H_2O(l)} Ca^{2+}(aq) + 2Cl^-(aq)$$

$$\Delta H_{soln} = -82.8 \text{ kJ/mol}$$

The dissolution of ammonium nitrate $(NH_4NO_3)(s)$ is an example of an endothermic process. When ammonium nitrate dissolves in water, the solution becomes so cold that frost may form on the outside of the container. Is heat absorbed or released as the ammonium and nitrate ions separate and
1 interact with the water? Heat is released from the water and the temperature of the solution decreases. The cold pack allows water and ammonium nitrate (NH_4NO_3) to mix, producing an endothermic reaction.

$$NH_4NO_3(s) \xrightarrow{H_2O(l)} NH_4^+(aq) + NO_3^-(aq)$$

$$\Delta H_{soln} = 25.7 \text{ kJ/mol}$$

Figure 11.16 illustrates a practical application of heats of solution.

Figure 11.16
Cold packs and hot packs are available for a variety of medical uses.

Chemistry Alive!

Phase Change Play

Superheated Steam Play

MEETING DIVERSE NEEDS

At Risk Provide at least two brands of cold or hot packs. Have students activate the packs at exactly the same time and then measure the maximum or minimum temperature achieved. Have students determine how long the packs remain effective. Ask students to decide, based on the data and the cost of each pack, which brand provided the best value.

Answer

1 absorbed

Sample Problem 11-6

How much heat (in kJ) is released when 2.500 mol NaOH(*s*) is dissolved in water?

1. ***ANALYZE*** ***List the knowns and the unknown.***

Knowns:
- $\Delta H_{soln} = -445.1$ kJ/mol
- amount of NaOH(*s*) dissolved: 2.500 mol

Unknown:
- ΔH = ? kJ

Use the heat of solution from the following chemical equation to solve for the amount of heat released (ΔH).

$$NaOH(s) + H_2O(l) \longrightarrow Na^+(aq) + OH^-(aq) + 445.1 \text{ kJ/mol}$$

2. ***CALCULATE*** ***Solve for the unknown.***

Multiplying the number of mol NaOH and ΔH_{soln} will yield the value of the unknown (ΔH).

$$\Delta H = 2.500 \cancel{\text{mol NaOH}(s)} \times \frac{-445.1 \text{ kJ}}{1 \cancel{\text{mol NaOH}(aq)}} = -1113 \text{ kJ}$$

3. ***EVALUATE*** ***Does the result make sense?***

By inspection, ΔH is 2.5 times greater than ΔH_{soln}, as it should be. Also, the heat of solution in the answer is negative, indicating an exothermic reaction. This is consistent with the heat of solution of sodium hydroxide discussed earlier in the text.

Practice Problems

24. How much heat (in kJ) is released when 0.677 mol NaOH(*s*) is dissolved in water?
25. How many moles of $NH_4NO_3(s)$ must be dissolved in water so that 88.0 kJ of heat is released from the water?

Chem ASAP!

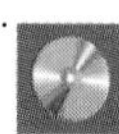

Problem-Solving 25

Solve Problem 25 with the help of an interactive guided tutorial.

section review 11.3

26. Identify each heat change by name and classify each change as exothermic or endothermic.
    - **a.** 1 mol $C_3H_8(l) \longrightarrow$ 1 mol $C_3H_8(g)$
    - **b.** 1 mol NaCl(*s*) + 3.88 kJ/mol $\longrightarrow$ 1 mol NaCl(*aq*)
    - **c.** 1 mol NaCl(*s*) $\longrightarrow$ 1 mol NaCl(*l*)
    - **d.** 1 mol $NH_3(g) \longrightarrow$ 1 mol $NH_3(l)$
    - **e.** 1 mol Hg(*l*) $\longrightarrow$ 1 mol Hg(*s*)
27. Heavy water, in which the hydrogens are hydrogen-2 instead of the more common hydrogen-1, is called deuterium oxide (D_2O). Solid D_2O melts at 3.78 °C. The molar heat of fusion of $D_2O(s)$ is 6.34 kJ/mol. How much heat is released when 8.46 g $D_2O(l)$ solidifies at its melting point?
28. Why is a burn from steam potentially far more serious than a burn from very hot water?
29. Why does an ice cube melt at room temperature?

Chem ASAP! **Assessment 11.3** Check your understanding of the important ideas and concepts in Section 11.3.

11.3

Practice Problem Plus

Related Chapter Review Problem

Chapter Review problem 53c is related to Sample Problem 11-6.

Additional Practice Problem

The Chem ASAP! CD-ROM contains the following problem: How many kilojoules of heat are released when 25.0 g of NaOH are dissolved in water? (−278 kJ)

3 Assess

Evaluate Understanding

Ask students to sketch a heating curve for 1 mol of ice being heated from −10 °C to 110 °C. Have students write brief explanations of their graphs. Check to see that students' graphs contain plateaus at the correct melting/freezing and vaporization/condensation temperatures. The relative lengths of the plateaus should accurately depict the relative amounts of heat required to accomplish each phase change. Ask students to calculate how much energy would be required to change 54.0 g of ice at −10.0 °C to water vapor at 110. °C. (165 kJ)

Reteach

Display Figure 11.15 on an overhead projector. Remind students that during endothermic phase changes (melting or evaporation), energy absorbed does not increase the temperature because the energy is being used to overcome attractions between particles. The energy can be released back to the surroundings during the opposing exothermic phase change (condensation or freezing).

Answers

SECTION REVIEW 11.3

26. **a.** molar heat of vaporization, endothermic
    b. molar heat of solution, endothermic
    c. molar heat of fusion, endothermic
    d. molar heat of condensation, exothermic
    e. molar heat of solidification, exothermic
27. 2.68 kJ
28. When a mole of steam condenses it releases a substantial amount of heat, its molar heat of condensation.
29. The ice absorbs sufficient heat from the surroundings to change from the solid to the liquid state.

11.4

1 Engage

Use the Visual

Have students examine the photograph that introduces this section. Ask:

- **Why do you think gemstones are so expensive?** (Students may comment on the relative rarity of the stones in addition to their beauty.)
- **Given that an emerald is too valuable to destroy, is there a way to determine the heat of reaction without actually performing the reaction?** (Yes. The heat of reaction might be determined indirectly by studying the enthalpy changes for related reactions. These reactions could involve less valuable forms of the structures found in an emerald crystal.)

Have students recall how the overall enthalpy change for converting one mole of ice to water vapor can be calculated by summing the enthalpy changes for each step in the process. A similar strategy can be used for determining the enthalpy change for a chemical reaction.

section 11.4

objectives

- Apply Hess's law of heat summation to find heat changes for chemical and physical processes
- Calculate heat changes using standard heats of formation

key terms

- Hess's law of heat summation
- standard heat of formation

CALCULATING HEAT CHANGES

Emeralds are beautiful gemstones composed of the elements chromium, aluminum, silicon, oxygen, and beryllium. If you were interested in the heat changes that occur when an emerald is converted to its elements, it is more than likely that you would not want to destroy the emerald by measuring the heat changes directly. **Is there a way to determine the heat of reaction without actually performing the reaction?**

Hess's Law

It is possible to talk in general terms about the heat changes that take place in chemical reactions. However, most reactions occur in a series of steps. Suppose, for example, you need to know the heat of reaction for an intermediate step, but it is impossible for you to obtain the value directly. Fortunately, Hess's law makes it possible to measure a heat of reaction indirectly. Thus even when a direct measurement cannot be made, the heat of reaction can still be determined.

Elemental carbon exists as both graphite and diamond at 25 °C. Because graphite is more stable than diamond, you would expect the following process to take place.

$$\text{C(diamond)} \longrightarrow \text{C(graphite)}$$

Fortunately for people who own diamonds, the conversion of diamond to graphite takes millions and millions of years. This enthalpy change cannot be measured directly because the reaction is far too slow. Hess's law, however, provides a way to calculate the heat of reaction. Hess's law is expressed by a simple rule: If you add two or more thermochemical equations to give a final equation, then you can also add the heats of reaction to give the final heat of reaction. This rule is **Hess's law of heat summation.**

You can use Hess's law to find the enthalpy changes for the conversion of diamond to graphite by using the following combustion reactions and **Figure 11.17.**

$$\textbf{a.}\ \text{C}(s,\text{ graphite}) + O_2(g) \longrightarrow CO_2(g) \qquad \Delta H = -393.5\text{ kJ}$$

$$\textbf{b.}\ \text{C}(s,\text{ diamond}) + O_2(g) \longrightarrow CO_2(g) \qquad \Delta H = -395.4\text{ kJ}$$

Write equation **a** in reverse to give:

$$\textbf{c.}\ CO_2(g) \longrightarrow \text{C}(s,\text{ graphite}) + O_2(g) \qquad \Delta H = 393.5\text{ kJ}$$

When you write a reverse reaction, you must also change the sign of ΔH. If you now add equations **b** and **c,** you get the equation for the conversion of diamond to graphite. The $CO_2(g)$ and $O_2(g)$ terms on both sides of the summed equations cancel, just as they do in algebra. Now if you also add the values of ΔH for equations **b** and **c,** you get the heat of reaction for this conversion.

$$\text{C}(s,\text{ diamond}) + \cancel{O_2(g)} \longrightarrow \cancel{CO_2(g)} \qquad \Delta H = -395.4\text{ kJ}$$

$$\cancel{CO_2(g)} \longrightarrow \text{C}(s,\text{ graphite}) + \cancel{O_2(g)} \qquad \Delta H = 393.5\text{ kJ}$$

$$\text{C}(s,\text{ diamond}) \longrightarrow \text{C}(s,\text{ graphite}) \qquad \Delta H = -1.9\text{ kJ}$$

STUDENT RESOURCES

From the Teacher's Resource Package, use:

- Section Review 11.4, Ch. 11 Practice Problems, Vocabulary Review, Quizzes and Tests from the Review Module (Ch. 9–12)
- Laboratory Recordsheet 11-2
- Laboratory Manual: Experiment 16
- Laboratory Practical 11-2

TECHNOLOGY RESOURCES

Relevant technology resources include:

- Chem ASAP! CD-ROM
- Resource Pro CD-ROM
- ActivChemistry CD-ROM: *Thermochemistry*
- Assessment Resources CD-ROM: Chapter 11 Tests

C(*s*, diamond) + $O_2(g)$

? $\Delta H = -1.9$ kJ

C(*s*, graphite) + $O_2(g)$

$\Delta H = -395.4$ kJ

$\Delta H = -393.5$ kJ

$CO_2(g)$

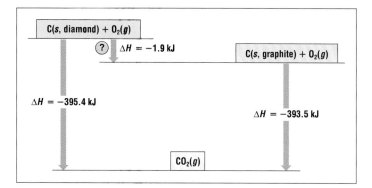

Figure 11.17
Hess's law is used to determine the enthalpy change of a very slow chemical process.
C(s, diamond) ⟶ C(s, graphite)

Thus the conversion of diamond to graphite is an exothermic process; its heat of reaction has a negative sign. Conversely, the change of graphite to diamond is an endothermic process. Some reactions give other products in addition to the product of interest. Suppose you want to determine the enthalpy change for the formation of carbon monoxide from its elements. The reaction is:

$$C(s, \text{graphite}) + \tfrac{1}{2}O_2(g) \longrightarrow CO(g) \qquad \Delta H = ?$$

Although it is easy to write the equation, carrying out the reaction in the laboratory as written is virtually impossible. Carbon dioxide (a "side product") is produced along with carbon monoxide (the "desired product"). Therefore, any measured heat of reaction is related to the formation of both $CO(g)$ and $CO_2(g)$, and not $CO(g)$ alone. You can solve the problem, however, using Hess's law and two reactions that can be carried out in the laboratory. The reactions are:

a. $C(s, \text{graphite}) + O_2(g) \longrightarrow CO_2(g) \qquad \Delta H = -393.5 \text{ kJ}$

b. $CO(g) + \tfrac{1}{2}O_2(g) \longrightarrow CO_2(g) \qquad \Delta H = -283.0 \text{ kJ}$

Writing the reverse of equation **b** and changing the sign of ΔH yields equation **c.**

c. $CO_2(g) \longrightarrow CO(g) + \tfrac{1}{2}O_2(g) \qquad \Delta H = 283.0 \text{ kJ}$

As shown on the following page, adding equations **a** and **c** gives the expression for the formation of $CO(g)$ from its elements. Notice that only $\tfrac{1}{2}O_2(g)$ cancels in the final equation. See **Figure 11.18.**

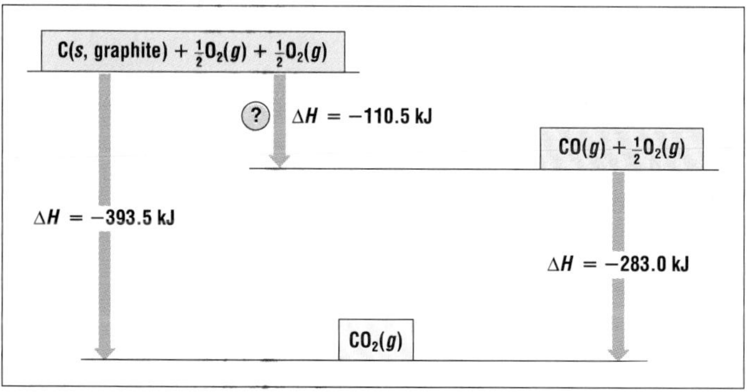

Figure 11.18
Hess's law is used to determine the enthalpy change for the reaction:
C(s, graphite) + ½O₂(g) ⟶ CO(g)

11.4

Motivate and Relate

In this section, students learn how to relate the quantity of heat involved in a given reaction to the quantities of heat involved in other reactions. Note that the enthalpy change for a chemical reaction is independent of the pathway by which the products are formed. This fact was discovered by the Russian chemist Germain Henri Hess in 1840. Provide the following analogy. A group of hikers is attempting to reach the summit of a mountain. Two routes are possible: one is a short, steep path that goes directly to the top; the other is a long path that is less steep but involves a number of changes in direction. Although the distances covered by hikers using the two routes would be different, the final altitude would be the same. The change in altitude is analogous to the enthalpy change for a reaction. The distances traveled by the hikers on each path is analogous to the number of steps required to form a given set of products. No matter how one goes from a given set of reactants to a given set of products, in one step or in several, the overall enthalpy change is the same.

2 Teach

TEACHER DEMO

Thermite is a mixture of aluminum powder and iron(III) oxide that reacts to produce molten iron, which solidifies at 25 °C.
$Fe_2O_3(s) + 2Al(s) \rightarrow 2Fe(s) + Al_2O_3(s)$; $\Delta H = -853.9$ kJ
After the demonstration, have students calculate the amount of heat evolved, using Hess's law of heat summation and these equations:
(1) $2Al(s) + 3/2O_2(g) \rightarrow Al_2O_3(s)$ $\Delta H = -1676.0$ kJ
(2) $2Fe(s) + 3/2O_2(g) \rightarrow Fe_2O_3(s)$ $\Delta H = -822.1$ kJ
Materials: 2 very small clay flower pots nested together, commercial thermite, commercial thermite starter, glycerin in a dropper bottle, potassium permanganate, $KMnO_4$, a few 2-inch or smaller pieces of magnesium ribbon, ring stand and ring that will hold the flower pots, a large bucket of sand positioned 10–12 inches under the flower pots.
Procedure: If possible, perform the demonstration outside of the classroom. Cover the hole in the uppermost flowerpot with a piece of paper towel. Spoon in about 20–25 mL of thermite mixture. Make a well in the top of the thermite; place about one teaspoonful of commercial starter in the depression. Place a few ribbons of magnesium into the thermite pile to ensure even heating. Make a depression in the starter and fill with potassium permanganate. **Caution students to look away when the light becomes very bright or use cobalt blue glass.** With students standing at a safe distance and with nothing flammable in the vicinity, add a few drops of glycerin to potassium permanganate. Move away and observe. Hot molten iron falls through the hole in the flower pot onto the bucket of sand.

$$\begin{array}{lr}
C(s,\text{ graphite}) + O_2(g) \longrightarrow \cancel{CO_2(g)} & \Delta H = -393.5\text{ kJ} \\
\cancel{CO_2(g)} \longrightarrow CO(g) + \tfrac{1}{2}O_2(g) & \Delta H = 283.0\text{ kJ} \\
\hline
C(s,\text{ graphite}) + \tfrac{1}{2}O_2(g) \longrightarrow CO(g) & \Delta H = -110.5\text{ kJ}
\end{array}$$

The formation of $CO(g)$ is exothermic; 110.5 kJ of heat is given off when 1 mol $CO(g)$ is formed from its elements.

Standard Heats of Formation

Sometimes it is hard to measure the heat change for a reaction. What are (1) some examples where this might be the case? In such cases you can calculate the heat of reaction from standard heats of formation. The **standard heat of formation** (ΔH_f^0) of a compound is the change in enthalpy that accompanies the formation of one mole of a compound from its elements with all substances in their standard states at 25 °C.

The ΔH_f^0 of a free element in its standard state is arbitrarily set at zero. For example, $\Delta H_f^0 = 0$ for the diatomic molecules $H_2(g)$, $N_2(g)$, $O_2(g)$, $F_2(g)$, $Cl_2(g)$, $Br_2(l)$, and $I_2(s)$. Similarly, $\Delta H_f^0 = 0$ for the graphite form of carbon, $C(s,$ graphite$)$.

Many values of ΔH_f^0 have been measured. **Table 11.6** lists ΔH_f^0 values for some common substances. Standard heats of formation of compounds are handy for calculating heats of reaction at standard conditions. The standard heat of reaction (ΔH^0) is the difference between the standard heats of formation of all the reactants and products. This relationship can be expressed by the following equation.

$$\Delta H^0 = \Delta H_f^0(\text{products}) - \Delta H_f^0(\text{reactants})$$

Table 11.6

Standard Heats of Formation (ΔH_f^0) at 25 °C and 101.3 kPa

Substance	ΔH_f^0 (kJ/mol)	Substance	ΔH_f^0 (kJ/mol)	Substance	ΔH_f^0 (kJ/mol)
$Al_2O_3(s)$	−1676.0	$Fe(s)$	0.0	$NO(g)$	90.37
$Br_2(g)$	30.91	$Fe_2O_3(s)$	−822.1	$NO_2(g)$	33.85
$Br_2(l)$	0.0	$H_2(g)$	0.0	$Na_2CO_3(s)$	−1131.1
$C(s,$ diamond$)$	1.9	$H_2O(g)$	−241.8	$NaCl(s)$	−411.2
$C(s,$ graphite$)$	0.0	$H_2O(l)$	−285.8	$O_2(g)$	0.0
$CH_4(g)$	−74.86	$H_2O_2(l)$	−187.8	$O_3(g)$	142.0
$CO(g)$	−110.5	$HCl(g)$	−92.31	$P(s,$ white$)$	0.0
$CO_2(g)$	−393.5	$H_2S(g)$	−20.1	$P(s,$ red$)$	−18.4
$CaCO_3(s)$	−1207.0	$I_2(g)$	62.4	$S(s,$ rhombic$)$	0.0
$CaO(s)$	−635.1	$I_2(s)$	0.0	$S(s,$ monoclinic$)$	0.30
$Cl_2(g)$	0.0	$N_2(g)$	0.0	$SO_2(g)$	−296.8
$F_2(g)$	0.0	$NH_3(g)$	−46.19	$SO_3(g)$	−395.7

Answers

(1) Materials may not be available; the reaction may be too slow for an enthalpy change to be measured; the chemicals may be explosive, toxic, or expensive.
(2) lower; because it takes energy to decompose water

Figure 11.19 is a diagram similar to others you have seen, except that it displays the standard heats of formation of the reactants hydrogen and oxygen and the product water. The heat difference between the reactants and products, −285.8 kJ/mol, is the standard heat of formation of liquid water from the gases hydrogen and oxygen. Does water have a lower or higher enthalpy than the elements from which it is formed? On what other basis can you account for your answer? 2

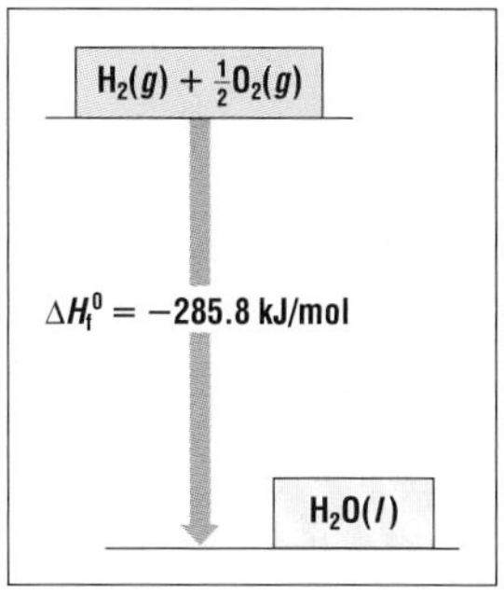

Figure 11.19
This enthalpy diagram shows the standard heat of formation of water.

Sample Problem 11-7

What is the standard heat of reaction (ΔH^0) for the reaction of gaseous carbon monoxide with oxygen to form gaseous carbon dioxide?

1. ***ANALYZE*** ***List the knowns and the unknown.***

Knowns (from **Table 11.6**):

- $\Delta H_f^0 O_2(g) = 0$ kJ/mol (free element)
- $\Delta H_f^0 CO(g) = -110.5$ kJ/mol
- $\Delta H_f^0 CO_2(g) = -393.5$ kJ/mol

Unknown:

- $\Delta H^0 = ?$ kJ

Balance the equation of the reaction of $CO(g)$ with $O_2(g)$ to form $CO_2(g)$. Then determine ΔH^0 using the standard heats of formation of the reactants and products.

2. ***CALCULATE*** ***Solve for the unknown.***

First, write the balanced equation.

$$2CO(g) + O_2(g) \longrightarrow 2CO_2(g)$$

Next, find and sum the ΔH_f^0 of all of the reactants, taking into account the number of moles of each.

$$\Delta H_f^0(\text{reactants}) = 2 \text{ mol } \cancel{CO(g)} \times \frac{-110.5 \text{ kJ}}{1 \text{ mol } \cancel{CO(g)}} + 1 \text{ mol } \cancel{O_2(g)} \times \frac{0 \text{ kJ}}{1 \text{ mol } \cancel{O_2(g)}}$$

$$= -221.0 \text{ kJ}$$

Then, find the ΔH_f^0 of the product in a similar way.

$$\Delta H_f^0(\text{product}) = 2 \text{ mol } \cancel{CO_2(g)} \times \frac{-393.5 \text{ kJ}}{1 \text{ mol } \cancel{CO_2(g)}}$$

$$= -787.0 \text{ kJ}$$

Finally, find the difference between ΔH_f^0(products) and ΔH_f^0(reactants).

$$\Delta H^0 = \Delta H_f^0(\text{products}) - \Delta H_f^0(\text{reactants})$$

$$\Delta H^0 = (-787.0 \text{ kJ}) - (-221.0 \text{ kJ})$$

$$\Delta H^0 = -566.0 \text{ kJ}$$

Practice Problems

30. Use the standard heats of formation to calculate the standard heats of reaction (ΔH^0) for these reactions.
    a. $Br_2(g) \longrightarrow Br_2(l)$
    b. $CaCO_3(s) \longrightarrow CaO(s) + CO_2(g)$
    c. $2NO(g) + O_2(g) \longrightarrow 2NO_2(g)$
31. With one exception, the standard heats of formation of $Na(s)$, $O_2(g)$, $Br_2(l)$, $CO(g)$, $Fe(s)$, and $He(g)$ are identical. What is the exception? Explain.

Chem ASAP!

Problem-Solving 30
Solve Problem 30 with the help of an interactive guided tutorial.

11.4

Use the Visual

With Table 11.6 displayed on an overhead projector, define *standard heat of formation* (the change in enthalpy that accompanies the formation of 1 mole of a compound from its elements with all substances in their standard states at 25 °C and 101.3 kPa) **Point out that the standard heat of formation of a free element in its standard state is defined as 0. Demonstrate how to use standard heats of formation to calculate heats of reaction at standard conditions.**

Practice Problems Plus

Related Chapter Review Problems
Chapter Review problems 56–59, 68, 69, 74, and 83 are related to Sample Problem 11-7.

Additional Practice Problem
The Chem ASAP! CD-ROM contains the following problem: Calculate the standard heat of reaction for the following reaction:
$2SO_2(g) + O_2(g) \rightarrow 2SO_3(g)$
(−197.8 kJ)

3 Assess

Evaluate Understanding

Have students explain two ways that scientists can determine the heat of a reaction indirectly.
1. Add two or more thermochemical equations to give the desired equation.
2. Find the difference between the standard heats of formation for all reactants and the standard heats of formation for all products in the reaction.

Reteach

Work a heat-of-reaction problem using both of the methods described above.

Portfolio Project

In comparison to a stainless steel skillet, a cast iron skillet has a greater heat capacity and takes longer to reach a desired cooking temperature. Once a cast iron skillet is heated, its temperature remains fairly constant as food is added or removed. A stainless steel skillet can reach the desired cooking temperature quickly, but it cools down just as quickly when cold food is added to the skillet.

Sample Problem 11-7 (cont.)

3. ***EVALUATE*** ***Does the result make sense?***
The ΔH^0 is negative. Therefore, the reaction is exothermic, as illustrated in **Figure 11.20.** This makes sense because the oxidation of carbon monoxide is a combustion reaction. Combustion reactions always release heat.

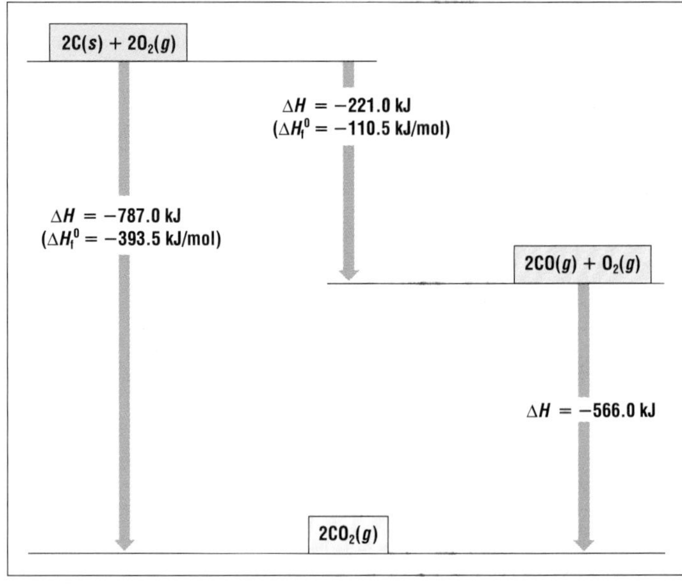

Figure 11.20
Hess's law is used to determine the enthalpy change for the reaction of carbon monoxide and oxygen.
$2CO(g) + O_2(g) \longrightarrow 2CO_2(g)$

portfolio project

Some chefs use heavy cast iron skillets; others use lightweight stainless steel. Use what you have learned about heat capacity to list the advantages and drawbacks of these skillets. Write an advertisement for each skillet stressing only its positive features.

section review 11.4

32. Calculate the enthalpy change (ΔH) in kJ for the following reaction.

$$2Al(s) + Fe_2O_3(s) \longrightarrow 2Fe(s) + Al_2O_3(s)$$

Use the enthalpy changes for the combustion of aluminum and iron:

$$2Al(s) + 1\tfrac{1}{2}O_2(g) \longrightarrow Al_2O_3(s) \qquad \Delta H = -1669.8 \text{ kJ}$$

$$2Fe(s) + 1\tfrac{1}{2}O_2(g) \longrightarrow Fe_2O_3(s) \qquad \Delta H = -824.2 \text{ kJ}$$

33. What is the standard heat of reaction (ΔH^0) for the decomposition of hydrogen peroxide?

$$2H_2O_2(l) \longrightarrow 2H_2O(l) + O_2(g)$$

34. State Hess's law of heat summation in your own words. Explain its usefulness.
35. What happens to the sign of ΔH when the reverse of a chemical reaction is written? Why?

 Chem ASAP! **Assessment 11.4** Check your understanding of the important ideas and concepts in Section 11.4.

Answers

SECTION REVIEW 11.4

32. -8.456×10^2 kJ
33. -1.960×10^2 kJ
34. Answers will vary, but should include the idea that chemical equations can be added algebraically along with their enthalpies to obtain the enthalpy of a different chemical reaction.
35. The sign of ΔH must be changed.

SMALL-SCALE LAB

HEAT OF COMBUSTION OF A CANDLE

SAFETY  

Wear safety glasses and follow the standard safety procedures outlined on page 18. Keep the burning candle away from combustible materials.

PURPOSE

To observe a burning candle and calculate the heat associated with the combustion reaction.

MATERIALS

- candle
- aluminum foil
- safety matches
- ruler
- balance
- temperature probe (optional)

PROCEDURE

Probe version available in the Probeware Lab Manual.

Measure and record the length of a candle in centimeters. Place the candle on a small piece of aluminum foil and measure the mass of the foil–candle system. Note the time as you light the candle. Let the candle burn for about five minutes. While you wait, begin answering the ANALYSIS questions. After about 5 minutes, extinguish the candle and record the time. Measure the mass of the foil–candle system again. Do not try to measure the mass while the candle is burning.

ANALYSIS

Using your experimental data, answer the following questions.

1. Observe the candle burn and draw a picture of what you see.
2. Examine the flame closely. Is it the wax or the wick that burns?
3. If you said the wax, how does the wax burn without touching the flame? If you said the wick, what is the function of the wax?
4. If you could measure the temperature near the flame, you would find that the air is much hotter above the flame than it is beside it. Why? Explain.
5. Scientists have often wondered if a candle would burn well in zero gravity. How would zero gravity change the shape of the flame?
6. How much length and mass did the candle lose? Are these data more consistent with the wax or the wick burning?
7. Keeping in mind that 'wick' is also a verb, explain how a candle works.
8. The formula for candle wax can be approximated as $C_{20}H_{42}$. Write and balance an equation for the complete combustion of the candle wax.
9. Calculate the moles of candle wax burned in the experiment.
10. Calculate the heat of combustion of candle wax in kJ/mol. The standard heat of formation of candle wax ($C_{20}H_{42}$) is -2230 kJ/mol. The standard heats of formation of carbon dioxide and water are -394 kJ/mol and -242 kJ/mol, respectively. The heat of combustion of candle wax equals the sum of the heats of formation of the products minus the sum of the heats of formation of the reactants.
11. Calculate the kilojoules of heat released in your reaction. *Hint:* Multiply the number of moles of candle wax burned in the experiment by the heat of combustion of candle wax.

YOU'RE THE CHEMIST!

The following small-scale activities allow you to develop your own procedures and analyze the results.

1. **Design It!** Design an experiment to show that the candle wax does not burn with complete combustion.
2. **Design It!** Design an experiment to show that water is a product of the combustion of a candle.

SMALL-SCALE LAB

HEAT OF COMBUSTION OF A CANDLE

OVERVIEW

Procedure Time: 20 minutes

See Probeware Lab Manual for Vernier, TI, and PASCO versions.

Students make simple observations of a candle, explore how a candle works, and apply basic principles of thermodynamics.

SAFETY

For best results and greater safety, use short, stout votive candles. Remind students to wear safety glasses and to keep the flame away from any combustible material.

TEACHING SUGGESTIONS

You can use this lab to clear up common misconceptions about how candles work. Questions 1 and 2 prompt students to wrongly conclude that the wick burns and the wax just slows the rate of burning. Questions 5 and 6 focus students on the key experiment, the determination of the mass loss of the candle, and allow them to revise their opinions.

YOU'RE THE CHEMIST

1. Soot will appear on a glass Petri dish held over the flame.
2. Water will condense on the bottom of an ice-filled, glass Petri dish held over the flame.

ANALYSIS

1. Drawings should include an indication of the melted wax.
2. Many students choose the wick.
3. The wick draws melted wax to the flame. Those who think the wick burns may suggest that the wax slows the rate of burning.
4. The hot gases expand and rise.
5. The flame might be round in zero gravity.
6. Data will vary, but a few millimeters and a few tenths of a gram are reasonable answers. The mass loss is consistent with burning wax.
7. Heat from the combustion melts wax, which is drawn up by the wick, evaporated, and burned.
8. $C_{20}H_{42} + \frac{61}{2}O_2 \rightarrow 20CO_2 + 21H_2O$
9. 0.0018 mol
10. −10 700 kJ/mol
11. 19 kJ

Chemistry Serving... the Environment

DISCUSS

Ask students to make a list of the different forms of energy discussed in the article and to construct a diagram or concept map depicting the conversions of solar energy into other forms of energy. For example, solar energy can be used to heat water. Heated water can then be used to generate electricity. Ask students if they know of other methods that are used to harness and store solar energy in addition to those mentioned in the article.

Ask students to write the balanced thermochemical equation for the reaction of methane with carbon dioxide to produce hydrogen gas and carbon monoxide. Have students look up the enthalpy change for this reaction and state whether it is endothermic or exothermic. Ask:

- What are the major technical challenges that must be met before solar energy can be a viable and efficient energy source?

CHEMISTRY IN CAREERS

Have students read more about the responsibilities and activities of a Solar Engineer on page 873.

Chemistry Serving...the Environment

HARNESSING SOLAR ENERGY

When you consider energy sources other than oil or coal, you may think of solar power. Solar power is energy from the sun, and it is free, clean, and readily available.

In theory, solar power is surely the most plentiful energy source available. The enormous power of the sun's rays can be harnessed by solar, or photovoltaic, cells. In reality, this process has been faced with problems that have prevented solar power from becoming a more prevalent part of your life. First, the sun does not always shine on a desired location with maximum power. The tilt of Earth on its axis, as well as weather conditions, can alter the amount of sunlight falling upon different locations. Second, solar cells are not very efficient at converting a "packet" of light energy, called a photon, into electricity. Finally, storing the electricity is not very practical because small batteries that can hold a lot of charge for extended periods of time have not yet been developed.

In theory, solar power is the most plentiful energy source available. The enormous power of the sun's rays can be harnessed by solar, or photovoltaic, cells.

Instead of using the sun's energy to generate electricity, what about using it to heat an object directly? The sun's rays can be concentrated onto very small areas. Anyone who has used a magnifying glass to burn a hole in a sheet of paper knows this! The heat collected can be transferred to water, or oil, or even molten NaCl to be stored, or piped, to a remote area. This stored heat energy can then be converted to electricity. One of the drawbacks of this scheme is that some heat will always be lost because the heat-storing system is not at thermal equilibrium with its surroundings.

Scientists have also begun to take advantage of storing solar energy in the form of chemical energy. Think of it this way: It "costs" a certain amount of energy to break a chemical bond in a molecule; when the process is reversed, the energy "cost" is refunded in the form of heat energy. The first process is an example of an endothermic reaction, which requires heat to be supplied to the system for the reaction to occur. The second (reverse) reaction is exothermic, during which heat is released as a reaction product.

One system that uses endothermic and exothermic reactions has been successfully tested. In this system, the sun's rays are concentrated to heat a "solar reactor" to about 800 °C, where methane (CH_4) and carbon dioxide (CO_2) absorb the heat to form carbon monoxide (CO) and hydrogen gas (H_2). The product gases (CO and H_2) are cooled to room temperature and piped to a destination where they are recombined to release the energy that is stored in chemical bonds. This approach is significant because the product gases are not hot, so they do not lose heat to the surroundings. After the CO and H_2 react, heat, CH_4, and CO_2 are produced. The gases are used to start the process over again, and the heat is used to generate electricity.

Solar energy is a very dynamic field. Many researchers are actively seeking new and better ways to harness the world's most plentiful energy resource.

CHEMISTRY IN CAREERS

SOLAR ENGINEER
Must have knowledge of photovoltaics. Use of experimental techniques to test cells and modules required.
See page 873.

CHEMIST NEEDED

TAKE IT TO THE NET
Find out more about career opportunities:
www.phschool.com

CHEMICAL SPECIALIST
Local food service distributor seek responsible self-motivated individ

Chapter 11 STUDENT STUDY GUIDE

Take It to the NET
For interactive study and review, go to www.phschool.com

KEY TERMS

- calorie *p. 295*
- calorimeter *p. 300*
- calorimetry *p. 300*
- chemical potential energy *p. 293*
- endothermic process *p. 294*
- energy *p. 293*
- enthalpy (*H*) *p. 301*
- exothermic process *p. 294*
- heat *p. 293*
- heat capacity *p. 296*
- heat of combustion *p. 305*
- heat of reaction *p. 303*
- Hess's law of heat summation *p. 314*
- joule *p. 296*
- law of conservation of energy *p. 294*
- molar heat of condensation *p. 311*
- molar heat of fusion *p. 307*
- molar heat of solidification *p. 307*
- molar heat of solution *p. 312*
- molar heat of vaporization *p. 310*
- specific heat *p. 297*
- specific heat capacity *p. 297*
- standard heat of formation *p. 316*
- surroundings *p. 294*
- system *p. 294*
- thermochemical equation *p. 303*
- thermochemistry *p. 293*
- universe *p. 294*

KEY EQUATIONS AND RELATIONSHIPS

- 1 Calorie = 1 kilocalorie = 1000 calories
- 1 J = 0.239 cal and 4.184 J = 1 cal
- $C = \frac{q}{m \times \Delta T} = \frac{\text{heat (joules or calories)}}{\text{mass (g)} \times \text{change in temperature (°C)}}$
- $\Delta H = m \times C \times \Delta T$, where $\Delta T = T_f - T_i$
- $\Delta H^0 = \Delta H_f^0(\text{products}) - \Delta H_f^0(\text{reactants})$

CONCEPT SUMMARY

11.1 The Flow of Energy—Heat

- Energy is the capacity to do work or to supply heat. The law of conservation of energy states that energy cannot be created or destroyed.
- A process is exothermic if heat flows from the system to the surroundings and endothermic if heat flows from the surroundings to the system.
- An object's heat capacity is the amount of heat it takes to change the object's temperature by exactly 1 °C. The specific heat capacity of a substance is the amount of heat it takes to raise the temperature of 1 g of the substance 1 °C.

11.2 Measuring and Expressing Heat Changes

- Thermochemical equations show the accompanying heat of reaction at constant pressure.
- Calorimetry measures heat changes associated with chemical reactions and phase changes.

11.3 Heat in Changes of State

- The molar heats of fusion, solidification, vaporization, and condensation describe the quantity of heat transferred to or from a system when one mole of substance undergoes a change of state at constant temperature.

11.4 Calculating Heat Changes

- Hess's law states that in a reaction that is the sum of two or more other reactions, ΔH for the overall process is the sum of the ΔH values for all of the constituent reactions.
- The enthalpy for the formation of one mole of a compound from its elements in standard states is the enthalpy of formation (H_f^0).
- The standard enthalpy change for a reaction (ΔH^0) can be calculated from the H_f^0 of the reactants and products.

CHAPTER CONCEPT MAP

Use these terms to construct a concept map that organizes the major ideas of this chapter.

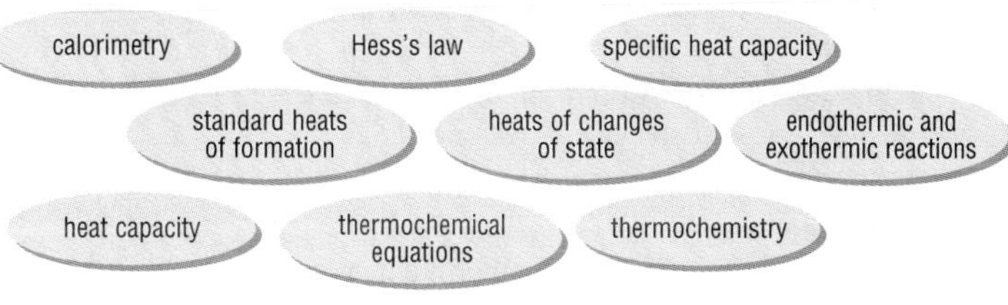

Chem ASAP! **Concept Map 11**
Create your Concept Map using the computer.

Take It to the Net

At **www.phschool.com** students will find for this chapter

- an Internet activity
- links to related chemistry sites
- an interactive quiz
- career links

Chapter 11
STUDENT STUDY GUIDE

4 Close

Summary

Ask the following questions to help students summarize chapter content.

- **How does the law of conservation of energy apply to endo- and exothermic reactions.** (Endothermic: energy that flows into the system is stored as potential energy in the products; exothermic: potential energy stored in reactants is released into the surroundings.)
- **How do you determine the heat change associated with a chemical reaction or change of state in aqueous solutions?** (In a calorimeter, the system under study absorbs energy from the water or releases energy into the water. $q(\text{energy}) = \Delta H = m \times C \times \Delta T$)

Extension

The law of conservation of energy applies to the uptake and expenditure of energy by the human body. Explain that *metabolism* refers to all of the endothermic and exothermic reactions taking place in the body. Have students research *catabolism* and *anabolism* and construct a simple diagram showing how these interrelated, but opposite processes control the flow of energy within living systems.

Looking Back . . . Looking Ahead . . .

The previous chapter discussed the properties of solids, liquids, and gases. This chapter explains how to quantify the energy flow during exothermic and endothermic processes. The next chapter examines the physical properties of gases.

Answers

36. Answers will vary, but should include the idea that energy is conserved in every physical and chemical process.
37. Heat flows from the object at the higher temperature to the object at the lower temperature.
38. Potential energy is energy stored in a substance because of its chemical composition.
39. the chemical composition of the substance and its mass
40. 1 Cal = 1000 cal = 1 kcal
41. a. 8.50×10^{-1} Calorie
    b. 1.86×10^{3} J
    c. 1.8×10^{3} J
    d. 1.1×10^{2} cal
42. Answers will vary, but should mention that thermochemistry measures heat flow across the boundary between the system and the surroundings.
43. A negative sign is given to heat flow from the system to the surroundings. A positive sign is given to heat flow to the system from the surroundings.
44. a. exothermic
    b. The immediate surroundings are the glass beaker and the air. If one or more of the substances is an aqueous solution, the water is also considered part of the surroundings.
45. a. exothermic
    b. endothermic
    c. exothermic
    d. endothermic
46. enthalpy
47. A calorimeter is an instrument used to measure heat changes in physical or chemical processes.
48. The foam cup will absorb heat. Some heat will be lost to the air. If the reactants are not completely mixed, temperature measurements will be inaccurate.
49. bomb calorimeter
50. one atmosphere pressure (101.3 kPa); all reactants and products in their normal physical state
51. amount of heat released or absorbed in the chemical change at constant pressure
52. Heat is being used to melt the ice.
53. a. -2.10×10^{1} kJ
    b. -1.8×10^{1} kJ
    c. -5.56×10^{2} kJ
    d. 6.5 kJ
54. increase
55. It allows the calculation of the enthalpy of a reaction from the known enthalpies of two or more other reactions.
56. a. -1.676×10^{3} kJ
    b. exothermic
57. 3.02×10^{1} kJ
58. -5.68×10^{1} kJ

Chapter 11 REVIEW

CONCEPT PRACTICE

36. Explain in your own words the law of conservation of energy. *11.1*
37. What always happens when two objects of different temperatures come in contact? Give an example from your own experience. *11.1*
38. Define potential energy in terms of chemistry. *11.1*
39. What factors determine the heat capacity of an object? *11.1*
40. What is the relationship between a calorie and a Calorie? *11.1*
41. Make the following conversions. *11.1*
    a. 8.50×10^{2} cal to Calories
    b. 444 cal to joules
    c. 1.8 kJ to joules
    d. 4.5×10^{-1} kJ to calories
42. Why do you think it is important to define the system and the surroundings? *11.1*
43. Describe the sign convention that is used in thermochemical calculations. *11.1*
44. Two substances in a glass beaker chemically react, and the glass beaker becomes too hot to touch. *11.1*
    a. Is this an exothermic or endothermic reaction?
    b. If the two substances are defined as the system, what constitutes the surroundings?
45. Classify these processes as exothermic or endothermic. *11.1*
    a. condensing steam
    b. evaporating alcohol
    c. burning alcohol
    d. baking a potato
46. What special name is given to a heat change at constant pressure? *11.2*
47. What is the function of a calorimeter? *11.2*
48. There are some obvious sources of error in experiments that use foam cups as calorimeters. Name at least three. *11.2*
49. What special device would you use to measure the heat released at constant volume? *11.2*
50. Give the standard conditions for heat of combustion. *11.2*
51. What information is given in a thermochemical equation? *11.2*
52. Explain why ice melts at 0 °C without an increase of temperature, even though heat is flowing from the surroundings into the system (the ice). *11.3*
53. Calculate the quantity of heat gained or lost in the following changes. *11.3*
    a. 3.50 mol of water freezes at 0 °C.
    b. 0.44 mol of steam condenses at 100 °C.
    c. 1.25 mol NaOH(*s*) dissolves in water.
    d. 0.15 mol $C_2H_5OH(l)$ vaporizes at 78.3 °C.
54. Sodium acetate dissolves readily in water according to the following equation. *11.3*

$$NaC_2H_3O_2(s) \longrightarrow NaC_2H_3O_2(aq) \quad \Delta H = -17.3 \text{ kJ/mol}$$

Would this process increase or decrease the temperature of the water?

55. Explain the usefulness of Hess's law of heat summation in thermochemistry. *11.4*
56. A considerable amount of heat is required for the decomposition of aluminum oxide. *11.4*

$$2Al_2O_3(s) \longrightarrow 4Al(s) + 3O_2(g) \quad \Delta H = 3352 \text{ kJ}$$

    a. What is the heat change for the formation of 1 mol of aluminum oxide from its elements?
    b. Is the reaction exothermic or endothermic?

57. Calculate the heat change for the formation of lead(IV) chloride by the reaction of lead(II) chloride with chlorine. *11.4*

$$PbCl_2(s) + Cl_2(g) \longrightarrow PbCl_4(l) \quad \Delta H = ?$$

Use the following thermochemical equations.

$$Pb(s) + 2Cl_2(g) \longrightarrow PbCl_4(l) \quad \Delta H = -329.2 \text{ kJ}$$

$$Pb(s) + Cl_2(g) \longrightarrow PbCl_2(s) \quad \Delta H = -359.4 \text{ kJ}$$

58. From the following reactions: *11.4*

$$\tfrac{1}{2}N_2(g) + \tfrac{1}{2}O_2(g) \longrightarrow NO(g) \quad \Delta H = 90.4 \text{ kJ/mol}$$

$$\tfrac{1}{2}N_2(g) + O_2(g) \longrightarrow NO_2(g) \quad \Delta H = 33.6 \text{ kJ/mol}$$

determine the heat of reaction for.

$$NO(g) + \tfrac{1}{2}O_2(g) \longrightarrow NO_2(g) \quad \Delta H = ?$$

59. Calculate the heat change for the formation of copper(I) oxide from its elements. *11.4*

$$Cu(s) + \frac{1}{2}O_2(g) \longrightarrow CuO(s)$$

Use the following thermochemical equations to make the calculation.

$$CuO(s) + Cu(s) \longrightarrow Cu_2O(s)$$
$$\Delta H = -11.3 \text{ kJ}$$
$$Cu_2O(s) + \frac{1}{2}O_2(g) \longrightarrow 2CuO(s)$$
$$\Delta H = -114.6 \text{ kJ}$$

60. What is the standard heat of formation of a free element in its standard state? *11.4*

61. Consider the statement, "the more negative the value of ΔH_f^0, the more stable the compound." Is this statement true or false? Explain. *11.4*

62. Calculate the change in enthalpy (in kJ) for the following reactions. *11.4*
   a. $CH_4(g) + 2O_2(g) \longrightarrow CO_2(g) + 2H_2O(l)$
   b. $2CO(g) + O_2(g) \longrightarrow 2CO_2(g)$

63. What is the standard heat of formation of a compound? *11.4*

CONCEPT MASTERY

64. Equal masses of two substances absorb the same amount of heat. The temperature of substance A increases twice as much as the temperature of substance B. Which substance has the higher specific heat? Explain.

65. If 3.20 kcal of heat is added to 1.00 kg of ice at 0 °C, how much water at 0 °C is produced, and how much ice remains?

66. The amounts of heat required to change different quantities of carbon tetrachloride $(CCl_4)(l)$ into vapor are given in the table.

Mass of CCl_4 (g)	Heat (J)	Heat (cal)
2.90	652	156
7.50	1689	404
17.0	3825	915
26.2	5894	1410
39.8	8945	2140
51.0	11453	2740

   a. Graph the data, using heat as the dependent variable.
   b. What is the slope of the line?
   c. The heat of vaporization of $CCl_4(l)$ is 53.8 cal/g. How does this value compare with the slope of the line?

67. Calculate the heat change in calories when 45.2 g of steam at 100 °C condenses to water at the same temperature. What is the heat change in joules?

68. Find the enthalpy change for the formation of phosphorus pentachloride from its elements.

$$2P(s) + 5Cl_2(g) \longrightarrow 2PCl_5(s)$$

Use the following thermochemical equations.

$$PCl_5(s) \longrightarrow PCl_3(g) + Cl_2(g)$$
$$\Delta H = 87.9 \text{ kJ}$$
$$2P(s) + 3Cl_2(g) \longrightarrow 2PCl_3(g)$$
$$\Delta H = -574 \text{ kJ}$$

69. Use standard heats of formation (ΔH_f^0) to calculate the change in enthalpy for these reactions.
   a. $2C(s, \text{graphite}) + O_2(g) \longrightarrow 2CO(g)$
   b. $2H_2O_2(l) \longrightarrow 2H_2O(l) + O_2(g)$
   c. $4NH_3(g) + 5O_2(g) \longrightarrow 4NO(g) + 6H_2O(g)$

70. An ice cube with a mass of 40.0 g melts in water originally at 25.0 °C.
   a. How much heat does the ice cube absorb from the water when it melts? Report your answer in calories, kilocalories, and joules.
   b. Calculate the number of grams of water that can be cooled to 0 °C by the melting ice cube.

71. The molar heat of vaporization of ethanol $(C_2H_5OH(l))$ is 43.5 kJ/mol. Calculate the heat required to vaporize 25.0 g of ethanol at its boiling point.

72. An orange contains 445 kJ of energy. What mass of water could this same amount of energy raise from 25.0 °C to the boiling point?

73. The combustion of ethane (C_2H_4) is an exothermic reaction.

$$C_2H_4(g) + 3O_2(g) \longrightarrow 2CO_2(g) + 2H_2O(l)$$
$$\Delta H = -1.39 \times 10^3 \text{ kJ}$$

Calculate the amount of heat liberated when 4.79 g C_2H_4 reacts with excess oxygen.

74. Calculate the heat change (ΔH) for the formation of nitrogen monoxide from its elements.

$$N_2(g) + O_2(g) \longrightarrow 2NO(g)$$

Use these thermochemical equations.

$$4NH_3(g) + 3O_2(g) \longrightarrow 2N_2(g) + 6H_2O(l)$$
$$\Delta H = -1.53 \times 10^3 \text{ kJ}$$
$$4NH_3(g) + 5O_2(g) \longrightarrow 4NO(g) + 6H_2O(l)$$
$$\Delta H = -1.17 \times 10^3 \text{ kJ}$$

CHAPTER 11 REVIEW

59. -1.259×10^2 kJ
60. zero
61. This statement is true, because stability implies lower energy. The greater the release of heat, the more stable is the compound relative to its elements (all of which have $\Delta H_f^0 = 0$).
62. **a.** -8.902×10^2 kJ
   b. -5.660×10^2 kJ
63. ΔH_f^0 for the formation of one mole of a compound from its elements
64. substance B; For equal masses, the substance with the greater heat capacity undergoes the smaller temperature change.
65. 4.00×10^1 g water; 9.60×10^2 g ice
66. **a.** See students' graphs.
   b. about 54 cal/g
   c. The two values are essentially the same.
67. 2.44×10^4 cal; 1.02×10^5 J
68. -7.50×10^2 kJ
69. **a.** -2.21×10^2 kJ
   b. -1.96×10^2 kJ
   c. -9.046×10^2 kJ
70. **a.** 3.19×10^3 cal, 3.19 kcal, 1.34×10^4 J
   b. 1.28×10^2 g H_2O
71. 2.36×10^1 kJ
72. 1.42×10^3 g
73. 2.38×10^2 kJ
74. 1.8×10^2 kJ

Answers

75. 6.72×10^1 kJ
76. **a.** (2) volume
b. (3) endothermic
c. (4) heat
77. The region denoted by ΔH_{fus} represents the coexistence of solid and liquid. The region denoted by Δh_{vap} represents the coexistence of liquid and vapor.
78. 0.40
79. 1.20×10^{24} H_2 molecules
80. $Ag^+(aq) + Cl^-(aq) \rightarrow AgCl(s)$
81. 1.18×10^1 g O_2
82. **a.** 3.24×10^1 kcal, 1.36×10^2 kJ
b. 8.13 kg
83. -1.37×10^2 kJ
84. 9.6 g
85. 45.4°C
86. ΔH_{vap} for water at 70°C is approximately 42 kJ/mol. 1 L of water (.1000 mL) has a mass of 1000 g and contains 55.6 mol water. Therefore, the amount of heat required is 42 kJ/mole × 55.6 mol = 2.34×10^3 kJ.

75. How much heat must be removed from a 45.0-g sample of naphthalene ($C_{10}H_8$) at its freezing point to bring about solidification? The heat of fusion of naphthalene is 191.2 kJ/mol.

CRITICAL THINKING

76. Choose the term that best completes the second relationship.

a. kilojoules:heat
cm³:_______
(1) mass (3) energy
(2) volume (4) weight

b. right:left
exothermic:_______
(1) combustion (3) endothermic
(2) heat (4) joule

c. thermometer:temperature
calorimeter:_______
(1) constant pressure (3) endothermic
(2) reactants (4) heat

77. Refer to **Figure 11.15.** Which region of the graph represents the coexistence of solid and liquid? Liquid and vapor?

CUMULATIVE REVIEW

78. What fraction of the average kinetic energy of hydrogen gas at 100 K does hydrogen gas have at 40 K?

79. How many hydrogen molecules are in 44.8 L $H_2(g)$ at STP?

80. Write the net ionic equation for the reaction of aqueous solutions of sodium chloride and silver acetate.

81. How many grams of oxygen are formed by the decomposition of 25.0 g of hydrogen peroxide?

$$2H_2O_2(l) \longrightarrow 2H_2O(l) + O_2(g)$$

CONCEPT CHALLENGE

82. The temperature of a person with an extremely high fever can be lowered with a sponge bath of isopropyl alcohol (C_3H_7OH). The heat of vaporization of this alcohol is 11.1 kcal/mol.

a. How many kilocalories of heat are removed from a person's skin when 175 g of isopropyl alcohol evaporates? How many kilojoules?

b. How many kilograms of water would this energy loss cool in lowering the temperature from 40.0 °C to 36.0 °C?

83. Ethane ($C_2H_6(g)$) can be formed by the reaction of ethene ($C_2H_4(g)$) with hydrogen gas.

$$C_2H_4(g) + H_2(g) \longrightarrow C_2H_6(g)$$

Use the heats of combustion for the following reactions to calculate the heat change for the formation of ethane from ethene and hydrogen.

$$2H_2(g) + O_2(g) \longrightarrow 2H_2O(l)$$
$$\Delta H = -5.72 \times 10^2 \text{ kJ}$$
$$C_2H_4(g) + 3O_2(g) \longrightarrow 2H_2O(l) + 2CO_2(g)$$
$$\Delta H = -1.401 \times 10^3 \text{ kJ}$$
$$2C_2H_6(g) + 7O_2(g) \longrightarrow 6H_2O(l) + 4CO_2(g)$$
$$\Delta H = -3.100 \times 10^3 \text{ kJ}$$

84. An ice cube at 0 °C was dropped into 30.0 g of water in a cup at 45.0 °C. At the instant that all of the ice was melted, the temperature of the water in the cup was 19.5 °C. What was the mass of the ice cube?

85. 41.0 g of glass at 95 °C is placed in 175 g of water at 21 °C in an insulated container. They are allowed to come to the same temperature. What is the final temperature of the glass-water mixture? The specific heat of glass is 2.1 cal/(g × °C).

86. The enthalpy of vaporization of water at various temperatures is given in the graph. From this graph, estimate the amount of heat required to convert 1 L of water to steam on the summit of Mount Everest (29 002 ft), where the boiling temperature of water is 70 °C.

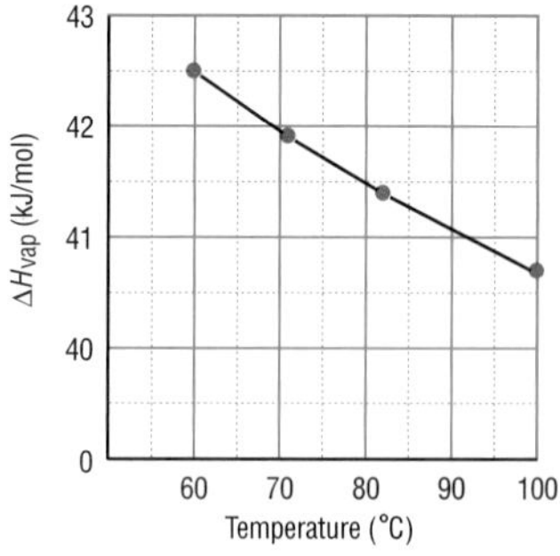

STANDARDIZED TEST PREP

Select the choice that best answers each question or completes each statement.

1. The ΔH_{fus} of ethanol (C_2H_5OH) is 4.60 kJ/mol. How many kilojoules are required to melt 24.5 g of ethanol at its freezing point?
   a. 2.45 kJ **c.** 245 kJ
   b. 5.33 kJ **d.** 18.8 kJ
2. How much heat, in kilojoules, must be added to 178 g of water to increase the temperature of the water by 5.0 °C ?
   a. 890 kJ **c.** 3.7 kJ
   b. 36 kJ **d.** 0.093 kJ
3. The standard heat of formation of a free element in its standard state is always
   a. zero.
   b. positive.
   c. negative.
   d. higher for solids than gases.
4. If ΔH for the reaction $2HgO(s) \longrightarrow 2Hg(l) + O_2(g)$ is +181.66 kJ, then ΔH for the reaction $Hg(l) + 1/2O_2(g) \longrightarrow HgO(s)$ is
   a. 90.83 kJ. **c.** 181.66 kJ.
   b. –90.83 kJ. **d.** –181.66 kJ.
5. The specific heat capacity of grain alcohol is ten times larger than the specific heat capacity of silver. A hot bar of silver with a mass of 55 g is dropped into an equal mass of cool alcohol. If the temperature of the silver bar drops 45 °C, the temperature of the alcohol
   a. increases 45 °C.
   b. decreases 4.5 °C.
   c. increases 4.5 °C.
   d. decreases 45 °C.

The lettered choices below refer to questions 6–9. A lettered choice may be used once, more than once, or not at all.

(A) kJ/mol
(B) (J × °C)/g
(C) J/(g × °C)
(D) kJ
(E) kJ/°C

Which unit is appropriate for each of the following measurements?

6. heat of reaction
7. heat capacity
8. molar heat of fusion
9. specific heat capacity

Use the data table to answer questions 10–12. Alkanes are a class of compounds containing carbon and hydrogen that are often used as fuels.

Alkane	Heat of combustion (kJ/mol)
Methane, CH_4	–890
Ethane, C_2H_6	–1560
Propane, C_3H_8	–2220
Butane, C_4H_{10}	
Pentane, C_5H_{12}	–3540
Hexane, C_6H_{14}	–4160
Heptane, C_7H_{16}	–4810

10. Graph the data. Make the number of carbon atoms per molecule the independent variable.
11. Estimate the heat of combustion for butane.
12. Estimate the heat of combustion for the next alkane in the series, octane, C_8H_{18}.

Use the graph and table to answer questions 13–16. Assume 1.00 mol of substance in each container.

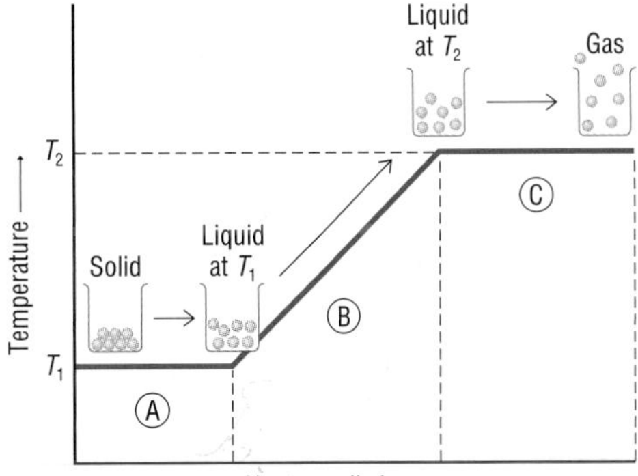

Substance	Freezing point (K)	ΔH_{fus} (kJ/mol)	Boiling point (K)	ΔH_{vap} (kJ/mol)
Ammonia	195.3	5.65	239.7	23.4
Benzene	278.7	9.87	353.3	30.8
Methanol	175.5	3.16	337.2	35.3
Neon	24.5	0.33	27.1	1.76

13. Calculate heat absorbed in region A for neon.
14. Calculate heat absorbed in region C for benzene.
15. Calculate heat absorbed in regions B and C for methanol. [specific heat = 81.6 J/(g × °C)]
16. Calculate heat absorbed in regions A, B, and C for ammonia. [specific heat = 35.1 J/(g × °C)]

STANDARDIZED TEST PREP

1. a
2. c
3. a
4. b
5. c
6. (D)
7. (E)
8. (A)
9. (C)
10.

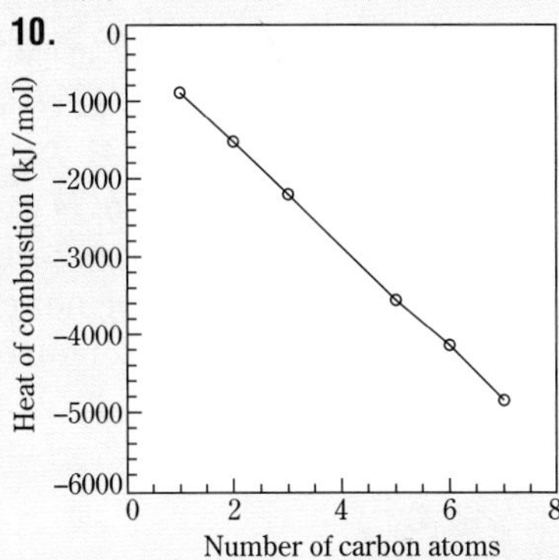

11. Heat of combustion of butane is about –3000 kJ/mol from the graph. The accepted value is –2877 kJ/mol.
12. Extrapolation gives a heat of combustion for octane of about –5400 kJ/mol. The accepted value is –5471 kJ/mol.
13. 0.33 kJ
14. 30.8 kJ
15. region B, 422 kJ; region C, 35.3 kJ
16. region A 5.65 kJ; region B, 26.5 kJ; region C, 23.4 kJ

Planning Guide

SECTION OBJECTIVES	ACTIVITIES/FEATURES	MEDIA & TECHNOLOGY
12.1 The Properties of Gases ● ■ ◇ ▸ Describe the properties of gas particles ▸ Explain how the kinetic energy of gas particles relates to Kelvin temperature	**SE** Discover It! *Observing Volume Changes,* p. 326 **SE** Small-Scale Lab *Reactions of Acids with Carbonates,* p. 329 (LRS 12-1) **SSLM** 14: *Synthesis and Qualitative Analysis of Gases*	**ASAP** Assessment 12.1 **CA** *Methane Bubbles* **SSV** 7: *Chemistry of Gases* **CHM** Side 5, 2: *Racing Hot Air Balloons*
12.2 Factors Affecting Gas Pressure ● ■ ◇ ▸ Explain how the amount of gas and the volume of the container affect gas pressure ▸ Infer the effect of temperature changes on the pressure exerted by a contained gas	**TE** DEMO, p. 330	**ASAP** Assessment 12.2 **CA** *Crushing Cans, Big Bottle Shake* **CHM** Side 5, 5: *Imploding Can* **OT** 29: *Adding and Removing Gases*
12.3 The Gas Laws ○ ■ ◆ ▸ State Boyle's Law, Charles's Law, Gay-Lussac's Law, and the combined gas law ▸ Apply the gas laws to problems involving the temperature, volume, and pressure of a contained gas	**SE** **CHEMath** *Solving Equations,* p. 334 **LM** 17: *Pressure-Volume Relationship for Gases* **LM** 18: *Temperature-Volume Relationship for Gases* **TE** DEMOS, pp. 333, 338	**ASAP** Simulation 9, 10, 11 **ASAP** Problem Solving 11, 13, 15, 17 **ASAP** Assessment 12.3 **ACT** 7, 8: *Behavior of Gases* **CA** *Crushing Cans, Easter Bunnies, Wok and Balloons* **CHM** Side 5, 9: *Scuba Diving* **CHM** Side 5, 12: *Lighter-than-Air Craft* **OT** 30: *Pressure-Volume Relationship for a Gas* **OT** 31: *Charles's Law*
12.4 Ideal Gases ■ ◆ ▸ Calculate the amount of gas at any specified conditions of pressure, volume, and temperature ▸ Distinguish between ideal and real gases	**SE** **Link to Physics** *Cryostats,* p. 345 **SE** Mini Lab *Carbon Dioxide from Antacid Tablets,* p. 346 (LRS 12-2) **PLM** *Carbon Dioxide from Antacid Tablets* **TE** DEMOS, pp. 341, 343	**ASAP** Problem Solving 24 **ASAP** Assessment 12.4 **ACT** 7, 8: *Behavior of Gases* **OT** 32: *Ideal vs Real Gases*
12.5 Gas Molecules: Mixtures and Movements ○ ■ ◆ ▸ State Avogadro's hypothesis, Dalton's Law, and Graham's Law ▸ Calculate moles, masses, and volumes of gases at STP ▸ Calculate partial pressures and rates of effusion	**SE** **Chemistry Serving . . . Industry** *Diving Can be a Gas,* p. 354 **SE** **Chemistry in Careers** *Commercial Diver,* p. 354 **LM** 19: *Diffusion of Gases* **TE** DEMOS, pp. 347, 352	**ASAP** Animation 13, 14 **ASAP** Problem Solving 34, 38 **ASAP** Assessment 12.5 **ASAP** Concept Map 12 **OT** 33: *Partial Pressures* **PP** Chapter 12 Problems **RP** Lesson Plans, Resource Library **AR** Computer Test 12 **www** Activities, Self-Tests **TCP** The Chemistry Place Web Site

KEY

● Conceptual (concrete concepts)	**AR** Assessment Resources	**GRS** Guided Reading and Study Workbook
○ Conceptual (more abstract/math)	**ASAP** Chem ASAP! CD-ROM	**LM** Laboratory Manual
■ Standard (core content)	**ACT** ActivChemistry CD-ROM	**LP** Laboratory Practicals
□ Standard (extension topics)	**CHM** CHEMedia Videodiscs	**LRS** Laboratory Recordsheets
◆ Honors (core content)	**CA** Chemistry Alive! Videodiscs	**SSLM** Small-Scale Lab Manual
◇ Honors (options to accelerate)	**GCP** Graphing Calculator Problems	

PRACTICE / ASSESSMENT

PRACTICE	ASSESSMENT
GRS Section 12.1 **RM** Practice Problems 12.1	**SE** Section Review **RM** Section Review 12.1 **RM** Chapter 12 Quiz
GRS Section 12.2 **RM** Practice Problems 12.2	**SE** Section Review **RM** Section Review 12.2 **RM** Chapter 12 Quiz
SE **Sample Problems** 12-1 to 12-4 **GRS** Section 12.3 **RM** Practice Problems 12.3	**SE** Section Review **RM** Section Review 12.3 **RM** Chapter 12 Quiz **LP** Lab Practicals 12-1, 12-2
SE **Sample Problems** 12-5 to 12-6 **GRS** Section 12.4 **RM** Practice Problems 12.4	**SE** Section Review **RM** Section Review 12.4 **RM** Chapter 12 Quiz
SE **Sample Problems** 12-7 to 12-10 **GRS** Section 12.5 **RM** Practice Problems 12.5 **RM** Interpreting Graphics **GCP** Chapter 12	**SE** Section Review **RM** Section Review 12.5 **RM** Vocabulary Review 12 **SE** Chapter Review **SM** Chapter 12 Solutions **SE** Standardized Test Prep **PHAS** Chapter 12 Test Prep **RM** Chapter 12 Quiz **RM** Chapter 12 A & B Test

OT	Overhead Transparency	**SE**	Student Edition
PHAS	PH Assessment System	**SM**	Solutions Manual
PLM	Probeware Lab Manual	**SSV**	Small-Scale Video/Videodisc
PP	Problem Pro CD-ROM	**TCP**	www.chemplace.com
RM	Review Module	**TE**	Teacher's Edition
RP	Resource Pro CD-ROM	**www**	www.phschool.com

PLANNING FOR ACTIVITIES

STUDENT EDITION

Discover It! p. 326
- spherical balloons
- marking pens
- tape measures
- freezer
- sunny window

Small-Scale Lab, p. 329
- pencils, paper, rulers
- reaction surfaces
- Na_2CO_3
- $NaHCO_3$
- baking soda
- HCl
- vinegar (CH_3COOH)
- citric acid ($H_3C_6H_5O_7$)
- $FeCl_3$
- household products

Mini Lab, p. 346
- pencils
- graph paper
- tape measures
- clocks or watches
- rubber balloons
- plastic medicine droppers
- antacid tablets, effervescent

TEACHER'S EDITION

Teacher Demo, p. 330
- aerosol spray can of room freshener

Teacher Demo, p. 333
- vacuum pump
- bell jar
- marshmallows

Teacher Demo, p. 338
- 2 large vats
- ice water
- hot water
- bicycle tire

Teacher Demo, p. 341
- piece of dry ice

Teacher Demo, p. 343
- large glass beaker
- small glass beaker
- marbles

Activity, p. 344
- plastic soda bottle fitted with an automobile tire valve
- hand bicycle pump
- tire gauge

Teacher Demo, p. 347
- two identical glass containers
- small beads
- large beads

Activity, p. 348
- pairs of balloons—one filled with helium and one with an equal volume of nitrogen

Teacher Demo, p. 352
- two identical balloons—one filled with helium and the other with air

Chapter 12

Key Terms

12.1 compressibility
12.3 Boyle's law, Charles's law, Gay-Lussac's law, combined gas law
12.4 ideal gas constant, ideal gas law
12.5 Avogadro's hypothesis, partial pressure, Dalton's law of partial pressures, diffusion, effusion, Graham's law of effusion

DISCOVER IT!

Students should observe that the balloon shrinks when it is cooled, and expands when it is warmed. This demonstration shows that the volume of a gas increases as the temperature is increased (assuming the pressure is constant).

Chapter 12 THE BEHAVIOR OF GASES

A hot-air balloon rises when air inside the balloon is heated.

FEATURES

Stay current with SCIENCE NEWS
Find out more about gas behavior:
www.phschool.com

DISCOVER IT! OBSERVING VOLUME CHANGES

You need a spherical balloon, a marking pen, a tape measure, a freezer, and a sunny window.

1. Inflate the balloon and tie it closed. Use the marking pen to draw a line around the middle of the balloon. This is the circumference of the balloon.
2. Place the balloon in the freezer for half an hour. Remove it and quickly measure and record the circumference of the balloon in centimeters.
3. Place the balloon in a sunny window for half an hour. Measure and record the circumference of the balloon in centimeters.
4. Assume the balloon is a perfect sphere and calculate the volume for each circumference recorded. Use the following formula:

$$\text{Volume} = \frac{4\pi r^3}{3}$$

$$\text{Note that } r = \frac{\text{circumference}}{2\pi}$$

What do you observe about the volume of the balloon at the two temperatures? Use your results to suggest a relationship between temperature and volume when pressure remains constant. Refer back to your suggested relationship when you read about Charles's law later in this chapter.

TIME SAVER

12.1 ● ■ ◇
12.2 ● ■ ◇
12.3 ○ ■ ◆
12.4 ■ ◆
12.5 ○ ■ ◆

Conceptual In Section 12.3, you may want to start with Charles's Law, a direct relationship, and then teach Boyle's Law, an inverse relationship. In Section 12.5, Avogadro's hypothesis and Dalton's law of partial pressures may be visualized with Figures 12.21 and 12.22; the sample problems do not need to be covered. Graham's Law may be omitted.

Standard Use the illustrations in the chapter to help students visualize a particle view for each gas law. The CHEMath feature will help students manipulate equations to solve for an unknown. Emphasize the need to list knowns and unknowns when solving gas-law problems.

Honors Assign Section 12.1, which reviews kinetic theory, as homework. Section 12.2 can be accelerated if students have encountered the effects of pressure and temperature in a previous course.

THE PROPERTIES OF GASES

You have probably seen huge, colorful, helium-filled balloons like this one in holiday parades. But what's wrong with Bullwinkle? As you can see, Bullwinkle isn't looking too good—he is deflating. The balloon is leaking helium and is beginning to sag and list. Soon Bullwinkle may collapse altogether. **How does kinetic theory explain why losing helium causes the balloon to sag and collapse?**

objectives

- Describe the properties of gas particles
- Explain how the kinetic energy of gas particles relates to Kelvin temperature

key terms

- compressibility

Kinetic Theory Revisited

This chapter discusses the ways in which simple kinetic molecular theory is used to explain gas behavior. You may have already observed some everyday examples of gas behavior. For example, you may have noticed that a sealed bag of potato chips bulges at its seams when placed in a sunny window. The air inside the bag exerts greater pressure as its temperature increases. Why does this happen? Kinetic theory can explain this and other gas behavior.

Recall from Chapter 10 that the kinetic theory of gases makes several basic assumptions. It assumes that gases consist of hard, spherical particles that have the following properties. First, the gas particles are so small in relation to the distances between them that their individual volumes can be assumed to be insignificant. The large relative distances between the gas particles means that there is considerable empty space between the particles. This assumption that gas particles are far apart explains the important property of gas compressibility. **Compressibility** is a measure of how much the volume of matter decreases under pressure. Gases, unlike solids or liquids, are easily compressed because of the space between the particles. Because gases are compressible, they are used in automobile airbags and other safety devices designed to absorb the energy of an impact. The energy of a collision is absorbed when the gas particles are forced closer together. As seen in **Figure 12.1**, gases can be used to absorb a considerable amount of energy.

Figure 12.1
An air bag cushions the impact of this crash dummy. How will the compressibility of gases protect the dummy from being broken? ❶

12.1

1 Engage

Use the Visual

Have students examine the photograph and read the text that accompanies it. Ask:

- **How does kinetic theory explain why losing helium causes the balloon to sag and collapse?** (The balloon deflates due to loss of pressure. The pressure exerted by the gas inside the balloon is due to the collisions of gas particles with the inner walls of the balloon. The magnitude of the pressure is determined by the frequency and force of these collisions. As the balloon leaks gas, the number of particles inside the balloon decreases and, consequently, the frequency of collisions against the inner walls of the balloon decreases, resulting in loss of pressure.)

Check Prior Knowledge

Point out that in this chapter students will build upon their knowledge of kinetic theory to describe the behavior of gases. Ask students to define kinetic energy. (energy of motion) **Ask how temperature and kinetic energy are related.** (Temperature is a measure of average kinetic energy.)

STUDENT RESOURCES

From the Teacher's Resource Package, use:

- Section Review 12.1, Ch. 12 Practice Problems and Quizzes from the Review Module (Ch. 9–12)
- Laboratory Recordsheet 12-1
- Small-Scale Chemistry Lab Manual: Experiment 14

TECHNOLOGY RESOURCES

Relevant technology resources include:

- Chem ASAP! CD-ROM
- Resource Pro CD-ROM
- Chemistry Alive! Videodisc: *Methane Bubbles*

❶ The compressible nature of gases allows the energy of the collision to be absorbed by the air bag instead of the dummy.

Chemistry Alive!

Methane Bubbles

Play

2 Teach

Discuss

Lead the class in a discussion of the four variables, i.e., volume, temperature, pressure, and number of moles, used to describe a gas. Discuss how the properties of gases lead to some unique uses of them. For example, since gas particles are far apart, gases are poor thermal conductors and are used for insulation between the glass panes in double-paned windows.

3 Assess

Evaluate Understanding

To determine students' understanding of the kinetic theory of gases, ask:

- **What assumption does the kinetic theory make regarding individual gas particles?** (Gas particles are considered to have negligible volume compared to the distances between the particles.)
- **What assumption explains why a gas expands?** (No attractive or repulsive forces exist between the particles of a gas.)
- **What does the kinetic theory say about the collisions between gas particles?** (Collisions are completely elastic. Kinetic energy is transferred, but the total amount of kinetic energy remains constant at a constant temperature.)

Reteach

Help students construct a list of the properties of gases assumed by the kinetic theory of gases. Review the list and ask for a behavior of gases relating to each assumption.

Figure 12.2
According to kinetic theory, the particles of air in this balloon are not affected by attractive or repulsive forces. As a result, the air expands to fill the interior of the balloon evenly.

The second property of gas particles assumed by the kinetic theory is that no attractive or repulsive forces exist between the particles. As a result, gases are free to move inside their containers. In fact, a gas expands until it takes the shape and volume of its container.

The third assumption is that gas particles move rapidly in constant random motion. The particles travel in straight paths and move independently of each other. As you can see in **Figure 12.3,** only when a particle collides with another particle or object does it deviate from its straight-line path. Kinetic theory assumes further that these collisions between gas particles are perfectly elastic, which means that during a collision the total amount of kinetic energy remains constant and that the kinetic energy is transferred without loss from one particle to another. You should also recall that the average kinetic energy of a collection of gas particles is directly proportional to the Kelvin temperature of the gas.

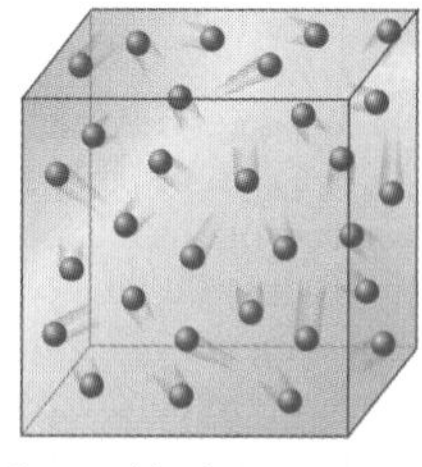

Gas particles in constant random motion

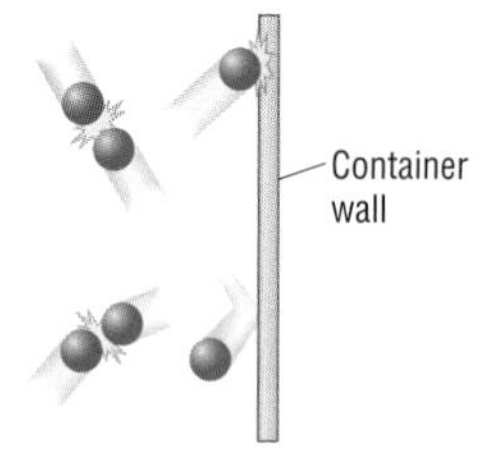

Detailed look at gas particles colliding

Figure 12.3
Notice the random motion of the gas particles. Changes in direction occur only when particles collide with each other or with another object.

Variables That Describe a Gas

Four variables are generally used to describe a gas. The variables and their common units are pressure (P) in kilopascals, volume (V) in liters, temperature (T) in kelvins, and number of moles (n). The gas laws that you will learn about in this chapter will enable you to predict gas behavior at specific conditions. Understanding the gas laws will help you understand everyday applications of gases such as in automobile airbags, scuba-diving equipment, and hot-air balloons, among many others.

section review 12.1

1. State the main assumptions of kinetic theory regarding gas particles.
2. Describe what happens to kinetic energy during gas particle collisions and as Kelvin temperature increases.
3. How does kinetic theory explain the compressibility of gases?
4. What variables and units are used to describe a gas?

 **Chem ASAP! Assessment 12.1** Check your understanding of the important ideas and concepts in Section 12.1.

Answers

SECTION REVIEW 12.1

1. The volume of gas particles is negligible; gas particles do not interact; gas particles move rapidly; the average kinetic energy of the gas particles is directly proportional to Kelvin temperature.
2. Kinetic energy is transferred from one gas particle to another without loss of energy when gas particles collide. As Kelvin temperature increases, the kinetic energy increases in direct proportion.
3. According to kinetic theory, there is considerable space between gas particles, which allows gases to be compressed.
4. pressure in kilopascals (kPa), Kelvin temperature (K), volume in liters (L), and amount (moles)

SMALL-SCALE LAB

REACTIONS OF ACIDS WITH CARBONATES

SAFETY   

Wear safety glasses and follow the standard safety procedures outlined on page 18.

PURPOSE

To observe and identify the reactions of carbonates that produce carbon dioxide gas.

MATERIALS

- pencil
- paper
- ruler
- reaction surface
- chemicals shown in Figure A

PROCEDURE

On separate sheets of paper, draw two grids similar to Figure A. Make each square 2 cm on each side. Place a reaction surface over one of the grids and add the chemicals as shown in Figure A. Use the second grid as a data table to record your observations for each reaction.

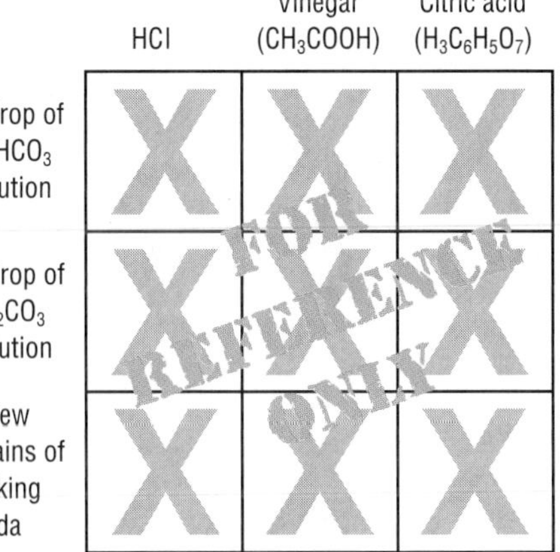

Figure A

ANALYSIS

Using your experimental data, record the answers to the following questions below your data table.

1. What common observation can you make for each of the mixings?
2. Write the name and formula of the gas produced in each reaction. Identify one other product formed in all of the reactions.
3. Hydrochloric acid reacts with sodium hydrogen carbonate to yield carbon dioxide, water, and sodium chloride. Convert this word equation into a chemical equation.
4. Write complete chemical equations for the reactions you observed in this experiment.

YOU'RE THE CHEMIST 

The following small-scale activities allow you to develop your own procedures and analyze the results.

1. **Analyze It!** Add some HCl to $NaHCO_3$ and then add some HCl to Na_2CO_3. What do you observe? Is it possible to use HCl to distinguish between $NaHCO_3$ and Na_2CO_3? Explain.
2. **Analyze It!** Add some $FeCl_3$ to $NaHCO_3$ and then add some $FeCl_3$ to Na_2CO_3. Look carefully. What do you observe? Is it possible to use $FeCl_3$ to distinguish between $NaHCO_3$ and Na_2CO_3? Explain.
3. **Design It!** Most laundry detergents contain carbonates such as $NaHCO_3$ and Na_2CO_3. Design and carry out an experiment to determine which laundry detergents contain carbonate in some form. Using your method, is it possible to tell exactly which carbonate a detergent contains? Explain.
4. **Design It!** Experiment with chalk, antacid tablets, seashells, limestone, and marble to determine if they contain carbonates. How can you tell?
5. **Design It!** Experiment with powdered soft drink mixes to see if they contain citric acid. Record your results.

ANALYSIS

1. All acids make all the carbonates bubble.
2. carbon dioxide, CO_2, and water, H_2O
3. $HCl + NaHCO_3 \rightarrow CO_2 + H_2O + NaCl$
4. $HCl + NaHCO_3 \rightarrow CO_2 + H_2O + NaCl$
   $2HCl + Na_2CO_3 \rightarrow CO_2 + H_2O + 2NaCl$
   $CH_3COOH + NaHCO_3 \rightarrow CO_2 + H_2O + NaCH_3COO$
   $2CH_3COOH + Na_2CO_3 \rightarrow CO_2 + H_2O + 2NaCH_3COO$
   $H_3C_6H_5O_7 + 3NaHCO_3 \rightarrow 3CO_2 + 3H_2O + Na_3C_6H_5O_7$
   $2H_3C_6H_5O_7 + 3Na_2CO_3 \rightarrow 3CO_2 + 3H_2O + 2Na_3C_6H_5O_7$

SMALL-SCALE LAB

REACTIONS OF ACIDS WITH CARBONATES

OVERVIEW

Procedure Time: 20 minutes
Students explore the reactions of acids with carbonates and apply what they learn to the analysis of household chemicals.

MATERIALS

Solution	**Preparation**
1.0*M* HCl	82 mL of 12*M* in 1.0 L
CAUTION! Always add acid to water carefully and slowly!	
vinegar	Use white vinegar.
0.2*M* citric acid	9.6 g in 250 mL
1.0*M* $NaHCO_3$	21 g in 250 mL
1.0*M* Na_2CO_3	26.5 g in 250 mL
0.1*M* $FeCl_3$	6.8 g $FeCl_3 \cdot 6H_2O$ in 25 mL of 1.0*M* NaCl; dilute to 250 mL

Samples of detergents, chalk, antacid tablets, sea shells, baking powder, limestone, and marble.

TEACHING SUGGESTIONS

Emphasize that acids and carbonates always produce carbon dioxide and water. The presence of bubbles is a good analytical clue to the presence of acid when you add carbonate and the presence of carbonate when you add acid.

YOU'RE THE CHEMIST

1. HCl cannot distinguish between $NaHCO_3$ and Na_2CO_3 because they both produce CO_2.
2. $NaHCO_3$ and Na_2CO_3 both form orange precipitates. $NaHCO_3$ produces bubbles; Na_2CO_3 does not.
3. Acid added to a detergent will detect a carbonate but will not identify a specific compound.
4. Samples that bubble when acid is added contain carbonates.
5. Add $NaHCO_3$ to soft drinks and look for bubbles.

12.2

1 Engage

Use the Visual

Ask students to examine the photograph that opens the section. Ask:

- **What factors affect the gas pressure inside the raft and its resulting rigidity?** (the average kinetic energy, or temperature, of the gas, the amount of gas contained in the raft, and the volume enclosed by the raft)

Check Prior Knowledge

To assess students' knowledge of the kinetic theory, ask them to describe pressure. (Pressure is defined as the force per unit area of surface.) Ask:

- **How does increasing the number of gas particles affect the pressure?** (the more particles colliding with the walls of a container, the higher the pressure)

2 Teach

TEACHER DEMO

Take an aerosol spray can of room freshener and begin spraying. Ask students to hypothesize about the pressure inside the can relative to the pressure outside the can. (Pressure is greater inside the can.) Ask:

- How is the pressure inside the can changing as I keep spraying? (As you keep spraying, you are removing gas particles, which results in fewer collisions inside the container and a lower pressure inside the can.)

section 12.2

FACTORS AFFECTING GAS PRESSURE

objectives

- Explain how the amount of gas and the volume of the container affect gas pressure
- Infer the effect of temperature changes on the pressure exerted by a contained gas

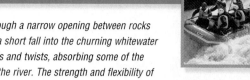

Your raft blasts through a narrow opening between rocks and plummets over a short fall into the churning whitewater below. The raft bends and twists, absorbing some of the pounding energy of the river. The strength and flexibility of the raft are impressive. **What factors affect the gas pressure inside the raft and its resulting rigidity?**

Amount of Gas

Using the kinetic theory, you can predict and explain how gases will respond to a change of conditions. When you pump up a tire, for example, you should expect the pressure inside it to increase. See **Figure 12.4.** Collisions of gas particles with the inside walls of the tire result in the pressure that is exerted by the enclosed gas. By adding gas, you increase the number of gas particles, thus increasing the number of collisions, which explains why the gas pressure increases.

Low pressure: Fewer gas particles inside tire

High pressure: More gas particles inside the tire

Figure 12.4
Using a pump to force more air particles into a partially deflated bicycle tire increases the pressure inside the tire.

Now look at **Figure 12.5.** As long as gas temperature does not change, doubling the number of gas particles doubles the pressure. Tripling the number of gas particles triples the pressure, and so forth. With a powerful pump and a strong container, you can generate very high pressures by adding more and more gas. Once the pressure exceeds the strength of the container, however, the container will rupture.

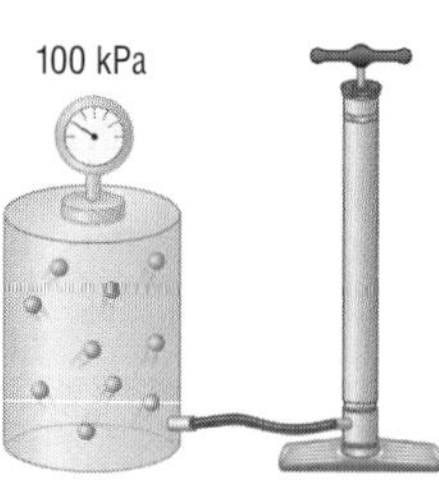

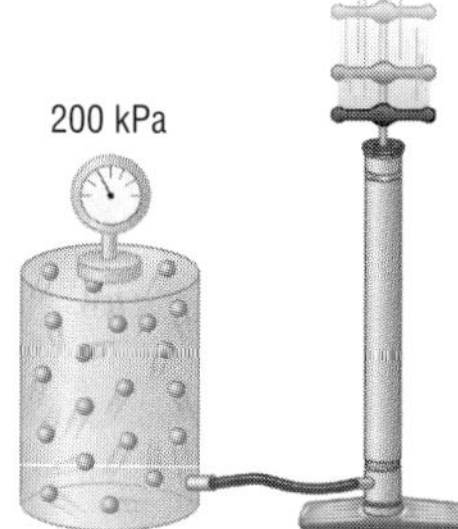

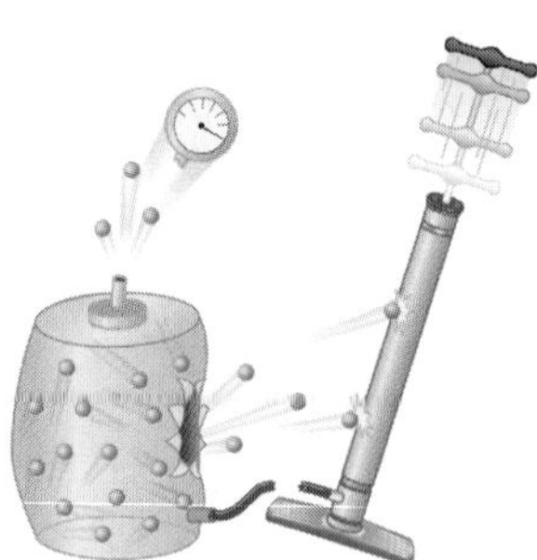

Figure 12.5
When a gas is pumped into a closed rigid container, the pressure increases in proportion to the number of gas particles added. If the number of gas particles doubles, the pressure doubles.

STUDENT RESOURCES

From the Teacher's Resource Package, use:

- Section Review 12.2, Ch. 12 Practice Problems and Quizzes from the Review Module (Ch. 9–12)

TECHNOLOGY RESOURCES

Relevant technology resources include:

- Chem ASAP! CD-ROM
- Resource Pro CD-ROM
- Chemistry Alive! Videodisc: *Crushing Cans, Big Bottle Shake*

200 kPa Decreasing Pressure 100 kPa

Figure 12.6
Gas pressure inside this constant-volume container decreases as gas particles escape. The gas at 100 kPa has exactly half the number of particles as the gas at 200 kPa.

In a similar way, letting the air out of a tire decreases the pressure inside the tire. The fewer particles inside exert less pressure. As you can see by looking at **Figure 12.6,** halving the number of gas particles in a given volume decreases the pressure by half.

When a sealed container of gas under pressure is opened, gas inside moves from the region of higher pressure to the region of lower pressure outside. This is the principle used in aerosol cans. You have probably used many aerosol products, from whipped cream, to hair spray and mousse, to the spray paint shown in **Figure 12.7.** The can of spray paint contains a gas at high pressure that acts as the propellant. The air outside the can is at a lower pressure. Pushing the spray button creates an opening between the inside of the can and the air outside. The high-pressure propellant gas flows to the lower pressure region outside, forcing paint out with it. As the propellant gas is depleted, the pressure inside the aerosol can decreases.

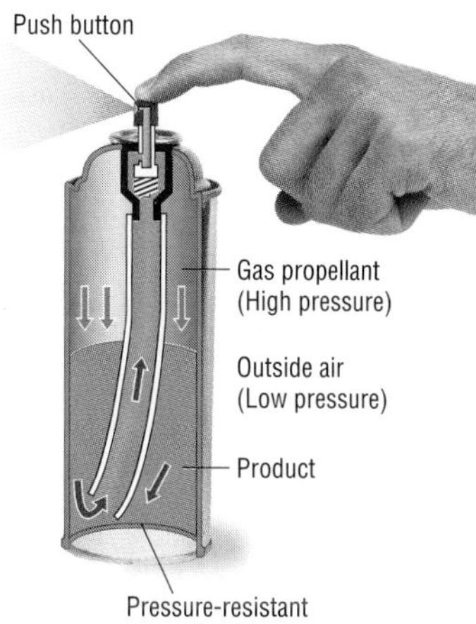

Figure 12.7
The difference in pressure between the inside of the spray can and the air outside allows aerosol sprays to work. What is the pressure inside the can when the aerosol will no longer spray? ❶

Volume

There are other ways to increase gas pressure. For example, you can raise the pressure exerted by a contained gas by reducing its volume. The more the gas is compressed, the greater is the pressure it exerts inside the container. Reducing the volume of a contained gas by half doubles the pressure, as **Figure 12.8** demonstrates. Increasing the volume of the contained gas has just the opposite effect. By doubling the volume, you halve the gas pressure because the same number of gas particles occupy a volume twice the original size.

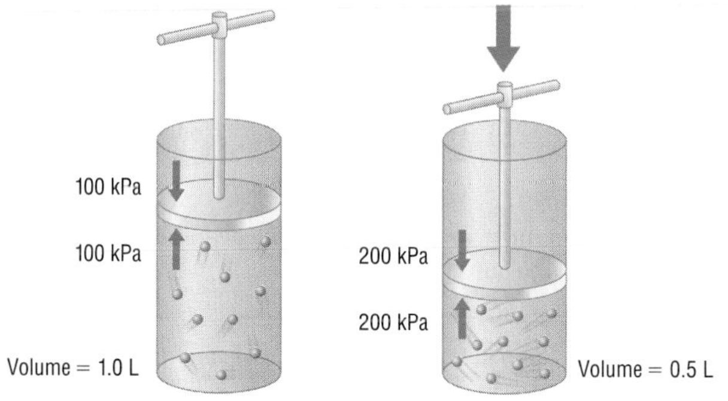

Figure 12.8
A piston, such as those used in engines, forces a gas in a cylinder to reduce in volume. A decrease in volume by half doubles the pressure the gas exerts.

12.2

Discuss

Explain that gas pressure is the result of the simultaneous collisions of billions and billions of gas particles with the walls of a container. By increasing the number of particles in a given volume, the number of simultaneous collisions with the walls of the container will increase. Therefore, the pressure in the container will increase. (Refer to Figures 12.4 and 12.5.) The reverse is also true as you reduce the number of particles in a given volume. (Refer to Figure 12.7.)

Use the Visual

Discuss Figure 12.8 and ask the students to speculate about what will happen to the pressure in the piston as the volume is decreased. (Pressure will increase.) **Point out that the number of gas particles in both diagrams is the same. Then ask:**

- **If this is true, why does the pressure of a contained gas double when the volume is reduced by one-half?** (Students should invoke the kinetic theory of gases. Reducing the volume by one-half doubles the number of simultaneous collisions of gas particles with the walls of the piston, doubling the pressure.)

Chemistry Alive!

Crushing Cans

Play

Big Bottle Shake

Play

Answer

❶ The pressure inside the container is equal to the pressure outside the container.

Meeting Diverse Needs

At Risk Have students construct a concept map to illustrate how the pressure of a contained gas changes when the amount of gas and the size of the container are changed. It should have the words: *Gas Pressure* in the center. It should use the phrases: *Adding Gas, Removing Gas, Increasing Volume, Decreasing Volume, Increases Gas Pressure,* and *Decreases Gas Pressure.*

12.2

Discuss

Explain that gas molecules have momentum, the product of their mass and velocity (mv). The pressure exerted by a gas in a given volume is proportional to the average kinetic energy of its molecules ($1/2\ mv^2$). A rise in temperature reflects a rise in average kinetic energy, corresponding to an increase in momentum and pressure.

3 Assess

Evaluate Understanding

To determine students' understanding of gas pressure, ask:

- **Describe what effect tripling the number of gas particles in a closed container would have on the pressure exerted.** (The pressure would triple.)
- **Describe what effect doubling the volume of an enclosed gas would have on the pressure.** (It would decrease by half.)
- **Describe how the pressure of an enclosed gas changes with increasing temperature.** (The number and force of collisions increase with temperature. The pressure would increase.)

Reteach

Help students make a table to summarize how gas pressure changes as variables increase or decrease. Label the rows Volume, Temperature, and Number of Particles. Label the columns Increase and Decrease. Have students use up or down arrows to complete the table.

Portfolio Project

Tire pressure is measured in pounds per square inch. It should be read when tires are cool. Underinflation can cause increased tire wear, lower gas mileage, and poor handling.

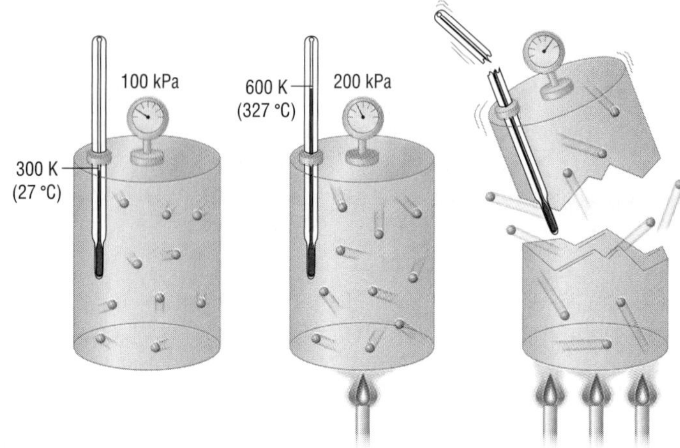

Figure 12.9
When a gas in a container is heated from 300 K (27 °C) to 600 K (327 °C), the kinetic energy of the gas particles doubles. The doubling of the kinetic energy results in a doubling of the pressure exerted by the gas. Pressure buildup due to high temperatures may cause the container to explode.

Temperature

Raising the temperature of an enclosed gas provides yet another way to increase gas pressure. (Remember the potato chip bag mentioned in Section 12.1.) The speed and kinetic energy of gas particles increase as the particles absorb thermal energy. The faster-moving particles impact the walls of their container with more energy, exerting greater pressure, as shown in **Figure 12.9.** If the average kinetic energy of a gas doubles, the Kelvin temperature doubles and the pressure of the enclosed gas also doubles. A gas in a sealed container may thus generate enormous pressure when heated. For that reason, an aerosol can, even an empty one, carelessly thrown onto a fire is an explosion hazard.

By contrast, as the temperature of an enclosed gas decreases, the particles move more slowly and have less kinetic energy. They strike the container walls with less force. Halving the Kelvin temperature of a gas in a rigid container decreases the gas pressure by half.

portfolio project

Contact local tire stores or car dealers to find out what factors determine recommended tire pressure. Find out how and under what conditions to use a tire gauge. What are the typical units of pressure for a tire gauge? What can happen if tires are not properly inflated?

section review 12.2

5. What effects do changes in the amount of gas and in the volume of the container have on gas pressure?
6. What is the effect of temperature change on the pressure of a contained gas?
7. What would you have to do to the volume of a gas to reduce its pressure to one-quarter of the original value, assuming that the gas is at a constant temperature?
8. Keeping temperature constant, how could you increase the pressure in a container by one hundredfold?
9. The manufacturer of an aerosol deodorant packaged in a 150-mL container wishes to produce a container of the same size that will hold twice as much gas. How will the pressure of the gas in the new product compare with that of the gas in the original container?

 Chem ASAP! **Assessment 12.2** Check your understanding of the important ideas and concepts in Section 12.2.

Answers

SECTION REVIEW 12.2

5. As the number of gas molecules increases in a constant-volume container, the pressure increases. As the volume of the container decreases, pressure increases, if other variables are constant.
6. As the temperature of a contained gas increases, the pressure increases, and vice versa.
7. Increase the volume by a factor of four.
8. Add 100 times as much gas or decrease the volume by a factor of 100.
9. The pressure will double.

THE GAS LAWS

Because warm air is less dense than cooler air, the pilot of a hot-air balloon heats the air inside the balloon to make it rise. To make it fall, the pilot releases the heated air from the top of the balloon. **What is the effect of adding heat to a gas at constant pressure? What law describes this relationship?**

objectives

- State Boyle's law, Charles's law, Gay-Lussac's law, and the combined gas law
- Apply the gas laws to problems involving the temperature, volume, and pressure of a contained gas

key terms

- Boyle's law
- Charles's law
- Gay-Lussac's law
- combined gas law

The Pressure-Volume Relationship: Boyle's Law

Consider the effect of pressure on the volume of a contained gas while the temperature remains constant. When the pressure goes up, the volume goes down. Similarly, when the pressure goes down, the volume goes up. The first person to do a systematic quantitative study of this pressure–volume relationship was the Anglo-Irish chemist Robert Boyle (1627–1691). In 1662, Boyle proposed a law to describe this behavior of gases. **Boyle's law** states that for a given mass of gas at constant temperature, the volume of the gas varies inversely with pressure. In an inverse relationship, the product of the two variable quantities is constant.

Look at **Figure 12.10.** In the figure, a volume of 1.0 L (V_1) is at a pressure of 100 kPa (P_1). If you increase the volume to 2.0 L (V_2), the pressure decreases to 50 kPa (P_2). Observe that the product $P_1 \times V_1$ (100 kPa $\times$ 1.0 L = 100 kPa $\times$ L) is the same as the product $P_2 \times V_2$ (50 kPa $\times$ 2.0 L = 100 kPa $\times$ L). If you decrease the volume to 0.5 L, the gas pressure increases to 200 kPa. Again, the product of pressure times volume equals 100 kPa $\times$ L.

The product of pressure and volume at any two sets of conditions is always constant at a given temperature. The mathematical expression of Boyle's law is as follows:

$$P_1 \times V_1 = P_2 \times V_2$$

When you graph an inverse relationship, such as this one, the result is a curve, as shown in **Figure 12.10.**

Chem ASAP!

Simulation 9

Examine the relationship between gas volume and pressure.

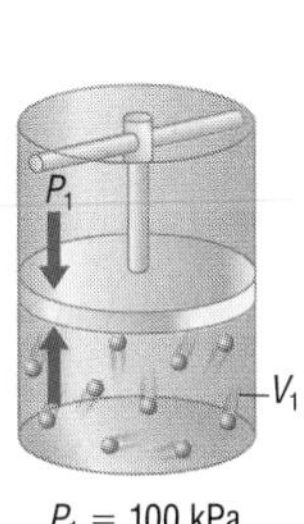

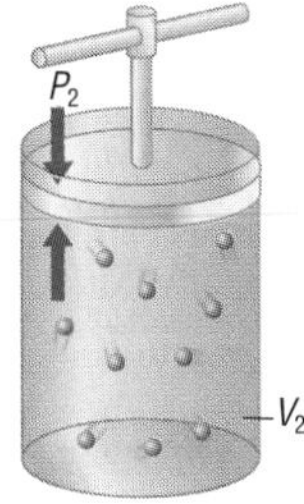

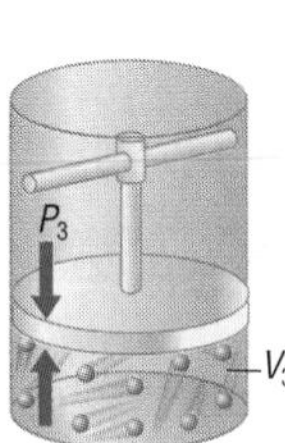

$P_1 \times V_1 = P_2 \times V_2 = P_3 \times V_3 = 100\text{ L} \times \text{kPa}$

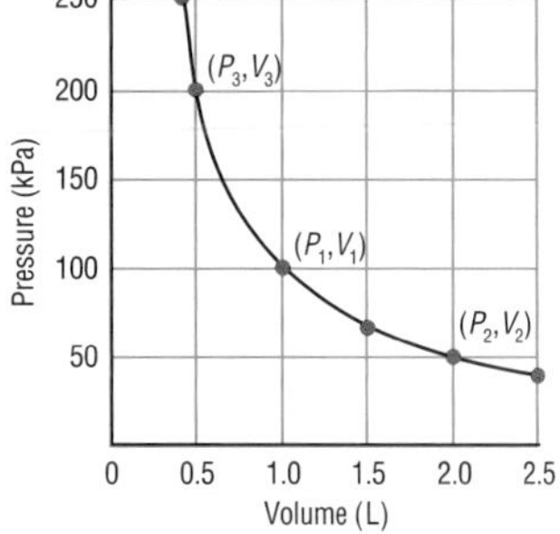

Figure 12.10

Boyle's law is studied by using a cylindrical container fitted with a frictionless, weightless piston. When the volume of a gas at constant temperature increases from V_1 to V_2, the pressure decreases from P_1 to P_2. When the volume decreases (to V_3), the pressure increases (to P_3). The graph illustrates Boyle's law. At any point on this curve, the product $P \times V$ of pressure (P) and volume (V) is a constant.

12.3

1 Engage

Use the Visual

Have students study the photograph and read the text that opens the section. Ask:

- **What is the effect of adding heat to a gas at constant pressure?** (The kinetic energy of the enclosed gas increases with temperature. As the kinetic energy increases, the volume also increases. As the balloon expands, the density of the gas inside the balloon is lowered relative to the gas outside the balloon and the balloon rises. The opposite effect occurs when the gas inside the balloon is allowed to cool.)
- **What law describes this relationship?** (Charles's law: The volume of a gas, at constant pressure, is directly proportional to the temperature.)

TEACHER DEMO

Using a vacuum pump, bell jar, and marshmallows, demonstrate the effect that changing pressure has on the volume of a gas. First, explain that the marshmallows contain trapped air. After placing several marshmallows in the bell jar, pull a vacuum in the jar. The removal of the air outside the marshmallows reduces the pressure on them. Air trapped inside the marshmallows can expand to a greater volume. The pressure inside the marshmallows is reduced to correspond to the lowered pressure in the container.

STUDENT RESOURCES

From the Teacher's Resource Package, use:

- Section Review 12.3, Ch. 12 Practice Problems and Quizzes from the Review Module (Ch. 9–12)
- Laboratory Manual: Experiments 17 and 18
- Laboratory Practicals 12-1 and 12-2

TECHNOLOGY RESOURCES

Relevant technology resources include:

- Chem ASAP! CD-ROM
- Resource Pro CD-ROM
- ActivChemistry CD-ROM: *Behavior of Gases*
- Chemistry Alive! Videodisc: *Crushing Cans, Wok and Balloons, Easter Bunnies*

CHEMath

Teaching Tips

Write the example equations on the board and carry out the cancellations and calculations step by step. Stress the rearrangement and isolation of variables that make the solution process possible. You may wish to use chalk or markers of different colors for the different variables to make the process clearer. Substitution of values for the variables can also make the solution methods more concrete for the students.

Answers to Practice Problems

A. $V_2 = \dfrac{P_1 \times V_1}{P_2}$

B. $T_2 = \dfrac{V_2 \times T_1}{V_1}$

C. $P_2 = P_{total} - P_1 - P_3$

D. $T_1 = \dfrac{P_1 \times V_1 \times T_2}{P_2 \times V_2}$

E. $T = \dfrac{P \times V}{nR}$

F. molar $mass_B = \left(\dfrac{Rate_A \times \sqrt{\text{molar } mass_A}}{Rate_B}\right)^2$

CHEMath

SOLVING EQUATIONS

Knowing which equation to use is the key first step in solving many problems. However, you must also master the skill of solving the equation for the unknown variable.

To solve an equation for one of its variables, isolate that variable on one side of the equal sign. In other words, the other variables and numbers in the equation must be on the other side of the equal sign.

For example, you are probably familiar with the equation for the area of a rectangle:

Area = base × height, or $A = bh$

As written, the equation solves for area (A). Notice that the variable A is on one side of the equal sign, while all of the other variables are on the other side of the equal sign. Variable A is therefore isolated.

If the variable you need to solve for is not already isolated, you will use addition, subtraction, multiplication, and division to isolate it. The key to the process is remembering that whatever you do on one side of the equation to isolate the variable, you must also do on the other side.

Example 1

Solve the equation A = bh for the variable b.

You must isolate the variable b. Because b is multiplied by h, divide both sides of the equation by h.

$$\frac{A}{h} = \frac{bh}{h}$$

Canceling terms, $\dfrac{A}{h} = \dfrac{b\cancel{h}}{\cancel{h}}$ yields $\dfrac{A}{h} = b$. Thus $b = \dfrac{A}{h}$.

Example 2

This example involves one of the equations you will learn and use in this chapter.

Solve the equation $P_1 \times V_1 = P_2 \times V_2$ for the variable P_1.

$$P_1 \times V_1 = P_2 \times V_2$$

You must isolate the variable P_1. Because P_1 is multiplied by V_1, divide both sides by V_1.

$$\frac{P_1 \times V_1}{V_1} = \frac{P_2 \times V_2}{V_1}$$

Canceling terms, $\dfrac{P_1 \times \cancel{V_1}}{\cancel{V_1}} = \dfrac{P_2 \times V_2}{V_1}$ yields $P_1 = \dfrac{P_2 \times V_2}{V_1}$.

Practice Problems

Prepare for upcoming problems in this chapter by solving the following equations for the variable indicated.

A. Solve $P_1 \times V_1 = P_2 \times V_2$ for V_2.

B. Solve $\dfrac{V_1}{T_1} = \dfrac{V_2}{T_2}$ for T_2.

C. Solve $P_{total} = P_1 + P_2 + P_3$ for P_2.

D. Solve $\dfrac{P_1 \times V_1}{T_1} = \dfrac{P_2 \times V_2}{T_2}$ for T_1.

E. Solve $P \times V = n \times R \times T$ for T.

F. Solve $\dfrac{Rate_A}{Rate_B} = \dfrac{\sqrt{\text{molar } mass_B}}{\sqrt{\text{molar } mass_A}}$ for molar $mass_B$.

Sample Problem 12-1

A high-altitude balloon contains 30.0 L of helium gas at 103 kPa. What is the volume when the balloon rises to an altitude where the pressure is only 25.0 kPa? (Assume that the temperature remains constant.)

1. ***ANALYZE*** ***List the knowns and the unknown.***

Knowns:	Unknown:
• $P_1 = 103$ kPa	• $V_2 = ?$ L
• $V_1 = 30.0$ L	
• $P_2 = 25.0$ kPa	

Use the known values and Boyle's law ($P_1 \times V_1 = P_2 \times V_2$) to calculate the unknown value (V_2).

2. ***CALCULATE*** ***Solve for the unknown.***

Rearrange the expression for Boyle's law to isolate V_2.

$$V_2 = \frac{V_1 \times P_1}{P_2}$$

Substitute the known values for P_1, V_1, and P_2 into the equation and solve.

$$V_2 = \frac{30.0\ \text{L} \times 103\ \cancel{\text{kPa}}}{25.0\ \cancel{\text{kPa}}}$$

$$= 1.24 \times 10^2\ \text{L}$$

3. ***EVALUATE*** ***Does the result make sense?***

Using kinetic theory, a decrease in pressure at constant temperature must correspond to a proportional increase in volume. The calculated result agrees with both the kinetic theory and the pressure–temperature relationship. Also, the units have canceled correctly and the answer is expressed to the proper number of significant figures.

Practice Problems

10. The pressure on 2.50 L of anesthetic gas changes from 105 kPa to 40.5 kPa. What will be the new volume if the temperature remains constant?

11. A gas with a volume of 4.00 L at a pressure of 205 kPa is allowed to expand to a volume of 12.0 L. What is the pressure in the container if the temperature remains constant?

Chem ASAP!

Problem-Solving 11

Solve Problem 11 with the help of an interactive guided tutorial.

The Temperature–Volume Relationship: Charles's Law

In 1787, the French physicist and balloonist Jacques Charles (1746–1823) investigated the quantitative effect of temperature on the volume of a gas at constant pressure. In every experiment, he observed an increase in the volume of a gas with an increase in temperature, and a decrease in volume with a decrease in temperature. In practice, the temperature–volume relationship for any gas can be measured only over a limited range because at low temperatures gases condense to form liquids.

From his quantitative studies, Charles observed that at constant pressure the graph of gas volume versus temperature yields a straight line. **Figure 12.11** shows such a graph for three different gas samples in balloons. In addition to the straight lines, another important feature emerged. The lines extended (extrapolated) to zero volume ($V = 0$) all intersect the temperature axis at the same point, −273.15 °C.

Figure 12.11

This graph shows the direct relationship between volume and temperature for three different gas samples in balloons at constant pressure. What is the significance of the temperature −273.15 °C? ❶

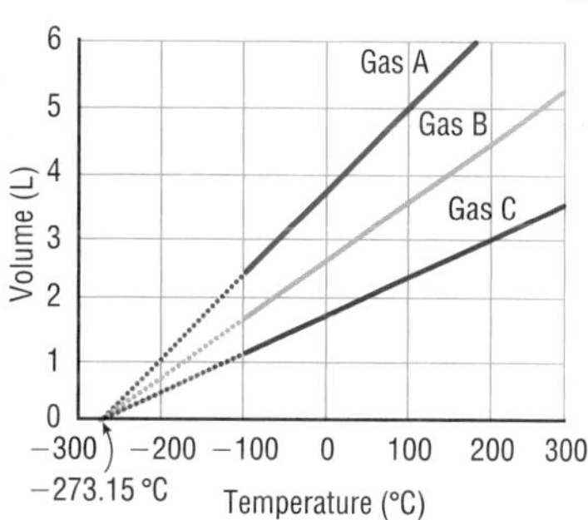

Answer

❶ This is the lowest temperature that gases can reach.

Meeting Diverse Needs

Gifted Explain to gifted students that Charles summarized his observations of the relationship between the volume and temperature of a gas in the following equation: $V = V_0(1 + aT)$ where V_0 is the volume of the gas at 0 °C, T is its temperature expressed in °C, and a is a constant for all gases. Have students show that the numerical value of a is approximately 1/273.

2 Teach

Practice Problems Plus

Related Chapter Review Problem

Chapter Review problem 53 is related to Sample Problem 12-1.

Additional Practice Problem

The Chem ASAP! CD-ROM contains the following problem: The volume of a gas is 4.23 L at 99.6 kPa and 24 °C. What volume will it occupy at 93.3 kPa and 24 °C? (4.52 L)

Cooperative Learning

Consider having students work in pairs to solve the practice problems in this chapter. Match up students who have mastery of algebraic equations with students who have a better grasp of abstract concepts such as the kinetic theory.

Discuss

Have the students consider what will happen when a helium-filled balloon is released into the sky. (Assume temperature remains constant.) Remind the students that as the elevation increases, the atmospheric pressure decreases. If the balloon contains 30 L of gas at 100 kPa, what would be the volume at 25 kPa? (120 L)

Discuss

Have students keep in mind that in kinetic theory, the new volume of a gas is equal to its original volume times a quotient that indicates whether the gas is compressed or allowed to expand. If the gas is compressed ($P_2 > P_1$), the new volume is smaller, so the quotient (P_1/P_2) has to be less than one. If the gas expands ($P_2 < P_1$), the new volume will be greater, so the quotient (P_1/P_2) must be greater than one.

Use the Visual

Have students study the graph in Figure 12.11 and note the direct relationship between volume and temperature for a gas at constant pressure. Have students use the graph to find the temperature at which a gas will, theoretically, have zero volume. (Students should extrapolate the straight lines to the x-axis and observe that they all intersect at −273.15 °C.) **Explain that −273.15 °C is known as absolute zero. Ask:**

- **What is this temperature on the Kelvin scale?** (0 K)

Point out that when the pressure and amount of a gas are unchanged, the ratio of the volume of the gas to the absolute temperature of the gas is a constant. Mathematically, this constant can be expressed as $V_1/T_1 = V_2/T_2$ and is known as Charles's law.

CHEMISTRY AND SCIENCE HISTORY

Jaques Charles was one of three passengers in the world's second balloon ascension that carried humans. The hydrogen-filled balloon was launched on December 1, 1783 in Paris, France. More than 400 000 curious onlookers attended.

Chemistry Alive!

Crushing Cans

Play

Wok and Balloons

Play

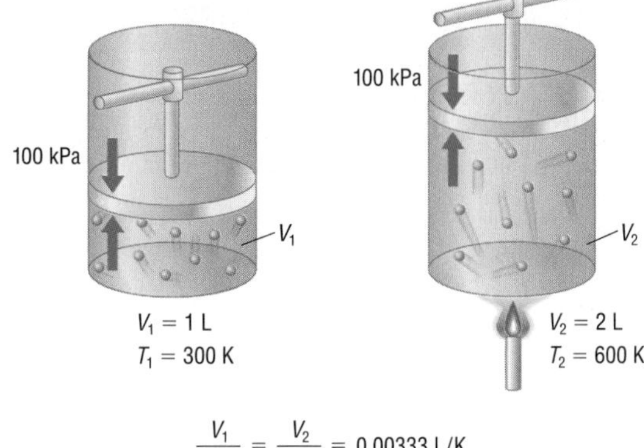

Figure 12.12
This apparatus illustrates Charles's law. When a gas is heated at constant pressure, the volume increases. When a gas is cooled at constant pressure, the volume decreases.

$$\frac{V_1}{T_1} = \frac{V_2}{T_2} = 0.00333 \text{ L/K}$$

William Thomson (Lord Kelvin) realized the significance of this temperature value. He identified this value as absolute zero, the lowest possible temperature. It is the temperature at which the average kinetic energy of gas particles would theoretically be zero. This was the basis for the absolute temperature scale established by Kelvin in 1848. This scale is now called the Kelvin temperature scale. On the Kelvin temperature scale, 0 K corresponds to −273.15 °C. Throughout this textbook, absolute zero expressed in degrees Celsius is rounded to −273. ❶ What is 0 °C in kelvins?

Charles's law summarizes Charles's observations and the findings of Kelvin. **Charles's law** states that the volume of a fixed mass of gas is directly proportional to its Kelvin temperature if the pressure is kept constant. In a direct relationship, such as this one, the ratio of the two quantities that change is a constant. In **Figure 12.12,** for example, a 1-L sample of gas (V_1) is at a temperature of 300 K (T_1). (Note that when solving gas-law problems, the temperature must always be expressed in kelvins.) When the temperature is increased to 600 K (T_2), the volume increases to 2 L (V_2). The ratio V_1/T_1 is equal to the ratio V_2/T_2. Moreover, at constant pressure, the ratio of volume to Kelvin temperature for a gas sample at any two sets of conditions is constant. Thus you can write Charles's law as follows:

$$\frac{V_1}{T_1} = \frac{V_2}{T_2}$$

Chem ASAP!
Simulation 10
Examine the relationship between gas volume and temperature.

The graph of a relationship such as Charles's law that is a direct proportion is a straight line. **Figure 12.13** provides a detailed look at the Charles's law relationship.

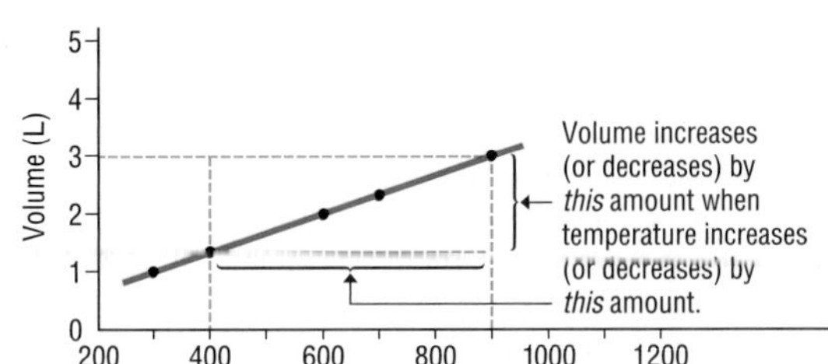

Figure 12.13
This graph illustrates Charles's law. At any point on this line, the ratio of volume (V) to temperature (T) is constant; in this case, V/T = 0.00333 L/K.

Answers

❶ 273 K
❷ The volume of the gas inside the balloon decreases at lower temperatures.

Figure 12.14
The volume of a balloon in a beaker of ice water is less than it is in a beaker of hot water. Why? 2

Sample Problem 12-2

A balloon inflated in a room at 24 °C has a volume of 4.00 L. The balloon is then heated to a temperature of 58 °C. What is the new volume if the pressure remains constant?

1. *ANALYZE* ***List the knowns and the unknown.***

Knowns:
- $V_1 = 4.00$ L
- $T_1 = 24$ °C
- $T_2 = 58$ °C

Unknown:
- $V_2 = ?$ L

Use the known values and Charles's law ($V_1/T_1 = V_2/T_2$) to calculate the unknown value (V_2).

2. *CALCULATE* ***Solve for the unknown.***

Because the gas laws will be applied, express the temperatures in kelvins.

$$T_1 = 24\text{ °C} + 273 = 297\text{ K}$$
$$T_2 = 58\text{ °C} + 273 = 331\text{ K}$$

Rearrange the expression for Charles's law to isolate V_2.

$$V_2 = \frac{V_1 \times T_2}{T_1}$$

Substitute the known values for T_1, V_1, and T_2 into the equation and solve.

$$V_2 = \frac{4.00\text{ L} \times 331\ \cancel{\text{K}}}{297\ \cancel{\text{K}}} = 4.46\text{ L}$$

3. *EVALUATE* ***Does the result make sense?***

From kinetic theory, the volume should increase with an increase in temperature (at constant pressure). This result agrees with the kinetic theory and Charles's law. The volume does increase with increasing temperature.

Practice Problems

12. If a sample of gas occupies 6.80 L at 325 °C, what will be its volume at 25 °C if the pressure does not change?
13. Exactly 5.00 L of air at −50.0 °C is warmed to 100.0 °C. What is the new volume if the pressure remains constant?

Chem ASAP!

Problem-Solving 13
Solve Problem 13 with the help of an interactive guided tutorial.

Meeting Diverse Needs

Gifted Have students do research on the use of refrigerants in cooling devices such as household refrigerators. Have students write a brief explanation of the refrigeration cycle. Students should also draw schematic diagrams representing the refrigeration cycle. Have them label each change of state, expansion, and compression of the refrigerant.

Discuss

Have students keep in mind that in using kinetic theory to solve Charles's law problems, the new volume of a gas is equal to its original volume times a quotient. The value of the quotient indicates whether the gas is heated or cooled. If the gas is heated ($T_2 > T_1$), the new volume is greater because the gas expands. So, the quotient has to be greater than one (T_2/T_1). If the gas is cooled ($T_2 < T_1$), the new volume is smaller because the gas contracts. So, the quotient must be less than one (T_1/T_2).

Practice Problems Plus

Related Chapter Review Problem
Chapter Review problem 54 is related to Sample Problem 12-2.

Additional Practice Problem
The Chem ASAP! CD-ROM contains the following problem: The volume of a gas is 0.80 L at 101.3 kPa and 0 °C. What volume will it occupy at 101.3 kPa and 24 °C? (0.87 L)

Chemistry and Technology

Refrigerators keep food cold by moving heat from inside the refrigerator to the outside. In a refrigerator, a fluid called a refrigerant goes through a cycle of changes. Heat inside the refrigerator causes the refrigerant, a liquid at low pressure, to boil into a vapor, absorbing heat in the process. The low-pressure vapor is then compressed. The high-pressure vapor condenses into a liquid at high pressure outside the refrigerator, giving off heat in the process. The refrigerant is then expanded into a low-pressure liquid, and the cycle begins again.

TEACHER DEMO

Fill a large vat with ice water and a second with hot water. Have students squeeze a bicycle tire to assess its firmness. Next, immerse the tire in ice water and have them feel its firmness. Finally, immerse the tire in hot water and assess the firmness. Have the students hypothesize about the relationship between pressure and temperature at constant volume. (There is a direct relationship between pressure and temperature at a constant volume.)

Discuss

Have students discuss the relationships among gas pressure, average kinetic energy of gas particles, and Kelvin temperature. Ask students to explain why auto tire manufacturers recommend checking for proper inflation before driving the car more than a mile. (The tires get warm as the car moves, increasing the pressure inside the tires.) **There is a direct relationship between pressure and temperature at constant volume. Mathematically it can be expressed as $P_1/T_1 = P_2/T_2$, and is known as Gay-Lussac's law.**

Practice Problems Plus

Related Chapter Review Problems

Chapter Review problems 51 and 55 are related to Sample Problem 12-3.

Additional Practice Problem

The Chem ASAP! CD-ROM contains the following problem: The pressure in a sealed plastic container is 108 kPa at 41 °C. What is the pressure when its temperature reaches 22 °C? (Assume that the volume has not changed.) (101 kPa)

The Temperature–Pressure Relationship: Gay-Lussac's Law

On a hot summer day, the pressure in a car tire increases. This increase illustrates a relationship that was discovered in 1802 by Joseph Gay-Lussac (1778–1850), a French chemist. **Gay-Lussac's law** states that the pressure of a gas is directly proportional to the Kelvin temperature if the volume remains constant. See **Figure 12.15.** Because Gay-Lussac's law involves direct proportions, the ratios P_1/T_1 and P_2/T_2 are equal at constant volume. Therefore, assuming that the volume remains constant, you can write Gay-Lussac's law as follows:

$$\frac{P_1}{T_1} = \frac{P_2}{T_2}$$

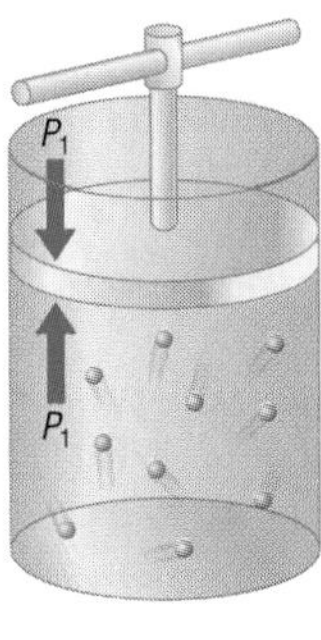

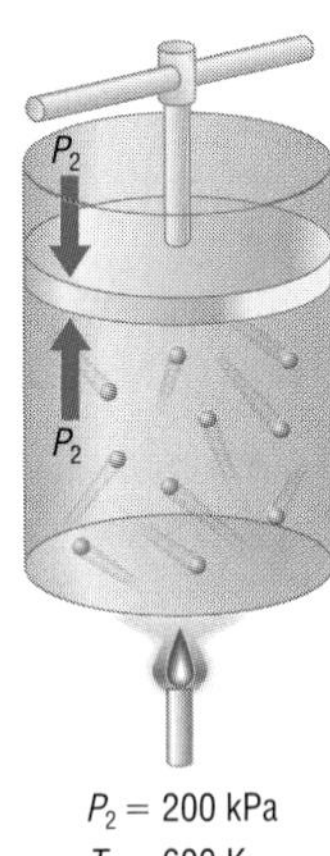

Figure 12.15
When a gas is heated at constant volume, the pressure increases. When a gas is cooled at constant volume, the pressure decreases.

Chem ASAP!
Simulation 11
Explore the relationship between gas temperature and pressure.

Sample Problem 12-3

The gas left in a used aerosol can is at a pressure of 103 kPa at 25 °C. If this can is thrown onto a fire, what is the pressure of the gas when its temperature reaches 928 °C? (Calculating the answer to this problem will show you why it is dangerous to dispose of aerosol cans in a fire. Most aerosol cans carry warnings on their labels that clearly say not to incinerate (burn) or to store above a certain temperature.)

1. *ANALYZE* ***List the knowns and the unknown.***

Knowns:
- P_1 = 103 kPa
- T_1 = 25 °C
- T_2 = 928 °C

Unknown:
- P_2 = ? kPa

Use the known values and Gay-Lussac's law ($P_1/T_1 = P_2/T_2$) to calculate the unknown (P_2). Remember, because this problem involves temperatures and a gas law, the temperatures must be expressed in kelvins.

Practice Problems

14. A gas has a pressure of 6.58 kPa at 539 K. What will be the pressure at 211 K if the volume does not change?

Meeting Diverse Needs

Gifted Gases such as oxygen, nitrogen, helium, and hydrogen can be liquefied by lowering the temperature and increasing the pressure. But different gases must be cooled to different temperatures in order to condense, no matter how high the pressure is raised. This temperature is called the "critical temperature." Have students look up the critical temperatures for the four gases listed above. Ask:

- What do these temperatures tell us about the relative strengths of the intermolecular attractive forces for these gases? (the lower the critical temperature, the lower the intermolecular attractive forces)

Sample Problem 12-3 (cont.)

2. ***CALCULATE*** ***Solve for the unknown.***
First convert degrees Celsius to kelvins.

$$T_1 = 25\ °C + 273 = 298\ K$$
$$T_2 = 928\ °C + 273 = 1201\ K$$

Rearrange Gay-Lussac's law to isolate P_2.

$$P_2 = \frac{P_1 \times T_2}{T_1}$$

Substitute the known values for P_1, T_2, and T_1 into the equation and solve.

$$P_2 = \frac{103\ kPa \times 1201\ \cancel{K}}{298\ \cancel{K}} = 415\ kPa$$
$$= 4.15 \times 10^2\ kPa$$

3. ***EVALUATE*** ***Does the result make sense?***
From the kinetic theory, one would expect the increase in temperature of a gas to produce an increase in pressure if the volume remains constant. The calculated value does show such an increase.

Practice Problems (cont.)

15. The pressure in an automobile tire is 198 kPa at 27 °C. At the end of a trip on a hot sunny day, the pressure has risen to 225 kPa. What is the temperature of the air in the tire? (Assume that the volume has not changed.)

Chem ASAP!

Problem-Solving 15
Solve Problem 15 with the help of an interactive guided tutorial.

The Combined Gas Law

If you have been wondering how to remember the individual expressions for the gas laws, there is really no need. A single expression, called the **combined gas law,** combines the three gas laws, as follows:

$$\frac{P_1 \times V_1}{T_1} = \frac{P_2 \times V_2}{T_2}$$

The other laws can be obtained from this law by holding one quantity (pressure, volume, or temperature) constant.

To illustrate, suppose you hold temperature constant ($T_1 = T_2$). Rearrange the combined gas law to get the two temperature terms on the same side of the equation and then cancel.

$$P_1 \times V_1 = \frac{P_2 \times V_2 \times \cancel{T_1}}{\cancel{T_2}}$$
$$P_1 \times V_1 = P_2 \times V_2$$

As you can see, you are left with Boyle's law. The same kind of process yields Charles's law when pressure remains constant and Gay-Lussac's law when volume remains constant.

In addition to providing a useful way of recalling the three previous gas laws, the combined gas law also enables you to do calculations for situations in which none of the variables are constant. In Sample Problem 12-4, on the following page, you will see how to use the combined gas law to calculate the new volume of a gas that results from changing both the pressure and the temperature.

Discuss

Have students keep in mind that in using kinetic theory to solve Gay-Lussac's law problems, the new pressure of the gas is equal to its original pressure times a quotient. The value of the quotient indicates whether the gas is heated or cooled. If the gas is heated ($T_2 > T_1$), the new pressure is greater. So, the quotient has to be greater than one (T_2/T_1). If the gas is cooled ($T_2 < T_1$), the new pressure is less. So, the quotient must be less than one (T_1/T_2).

Chemistry Alive!

Easter Bunnies

Play

12.3

Practice Problems Plus

Related Chapter Review Problem
Chapter Review problem 53 is related to Sample Problem 12-4.

Additional Practice Problem
The Chem ASAP! CD-ROM contains the following problem: The volume of a gas at 26 °C and 75 kPa is 10.5 L. What final temperature would be required to reduce the volume to 9.5 L if the pressure were increased to 116 kPa? (418 K or 145 °C)

3 Assess

Evaluate Understanding

Sketch two balloons (one twice as large as the other) on the chalkboard. Label the smaller balloon V_1 and the other V_2, and indicate that $V_2 = 2V_1$. Copy the following temperature table onto the chalkboard listing only those values in boldface type. Have students complete the table.

T_1	T_2
256 K	**512 K**
40 °C	353 °C
T_1	$2T_1$
−136 °C	**1 °C**

Reteach

In many experiments, a sample of gas is simultaneously subjected to pressure and temperature changes that have opposite effects on volume. Consider a weather balloon rising through the atmosphere. The higher it gets, the colder the temperature and the lower the volume. At the same time, atmospheric pressure decreases, allowing the gas to expand. The combined gas law can handle two variables at once. It determines which variable has the greater effect on the original volume.

Sample Problem 12-4

The volume of a gas-filled balloon is 30.0 L at 40 °C and 153 kPa pressure. What volume will the balloon have at standard temperature and pressure (STP)?

1. *ANALYZE* ***List the knowns and the unknown.***

Knowns:
- $V_1 = 30.0$ L
- $T_1 = 40$ °C
- $T_2 = 273$ K (standard temperature)
- $P_1 = 153$ kPa
- $P_2 = 101.3$ kPa (standard pressure)

Unknown:
- $V_2 = ?$ L

Use the known values and the combined gas law to calculate the unknown (V_2).

2. *CALCULATE* ***Solve for the unknown.***

Convert degrees Celsius to kelvins.

$$T_1 = 40\text{ °C} + 273 = 313\text{ K}$$

Rearrange the combined gas law to isolate V_2.

$$V_2 = \frac{V_1 \times P_1 \times T_2}{P_2 \times T_1}$$

Substitute the known quantities into the equation and solve.

$$V_2 = \frac{30.0\text{ L} \times 153\cancel{\text{ kPa}} \times 273\cancel{\text{ K}}}{101.3\cancel{\text{ kPa}} \times 313\cancel{\text{ K}}} = 39.5\text{ L}$$

3. *EVALUATE* ***Does the result make sense?***

The temperature decreases; therefore, the temperature ratio is less than 1 (273 K/313 K). The pressure decreases, so the pressure ratio is greater than 1 (153 kPa/101.3 kPa). Recalculate by multiplying the initial volume of gas by these two ratios.

$$V_2 = 30.0\text{ L} \times \frac{153\cancel{\text{ kPa}}}{101.3\cancel{\text{ kPa}}} \times \frac{273\cancel{\text{ K}}}{313\cancel{\text{ K}}} = 39.5\text{ L}$$

The result is the same.

Practice Problems

16. A gas at 155 kPa and 25 °C occupies a container with an initial volume of 1.00 L. By changing the volume, the pressure of the gas increases to 605 kPa as the temperature is raised to 125 °C. What is the new volume?
17. A 5.00-L air sample at a temperature of −50 °C has a pressure of 107 kPa. What will be the new pressure if the temperature is raised to 102 °C and the volume expands to 7.00 L?

Chem ASAP!
Problem-Solving 17
Solve Problem 17 with the help of an interactive guided tutorial.

section review 12.3

18. State Boyle's law, Charles's law, and Gay-Lussac's law.
19. Briefly explain how the combined gas law can be reduced to the other three gas laws.
20. Write the mathematical equation for Boyle's law and explain the symbols. What must be true about the temperature?
21. A given mass of air has a volume of 6.00 L at 101 kPa. What volume will it occupy at 25.0 kPa if the temperature does not change?

Chem ASAP! **Assessment 12.3** Check your understanding of the important ideas and concepts in Section 12.3.

Answers

SECTION REVIEW 12.3

18. Boyle's law: $P_1 \times V_1 = P_2 \times V_2$

Charles's law: $\frac{V_1}{T_1} = \frac{V_2}{T_2}$

Gay-Lussac's law: $\frac{P_1}{T_1} = \frac{P_2}{T_2}$

19. When one condition (P, V, or T) is held constant, its variable cancels out, and one of the other three laws is left.
20. $P_1 \times V_1 = P_2 \times V_2$
P = pressure; V = volume; The subscript 1 represents the starting conditions; the subscript 2, the final conditions.
Temperature is constant, so it is not in the expression.
21. 24.2 L

Ideal or Real?

IDEAL GASES

What if someone told you that all of the gas laws you just learned were wrong! In some ways, the person would be correct. The gas laws assume that gases behave in an ideal way, obeying the assumptions of kinetic theory. As it turns out, gases do not behave exactly this way. **What then is an ideal gas, and under what conditions do the gas laws apply?**

objectives

- Calculate the amount of gas at any specified conditions of pressure, volume, and temperature
- Distinguish between ideal and real gases

key terms

- ideal gas constant (R)
- ideal gas law

Ideal Gas Law

Up to this point in this textbook, you have worked with three variables regarding gas behavior: pressure, volume, and temperature. There is a fourth variable still to be considered: the amount of gas in the system, expressed in terms of the number of moles. Suppose you want to calculate the number of moles (n) of a gas in a fixed volume at a known temperature and pressure, as illustrated in **Figure 12.16.** The calculation of moles is possible by modifying the combined gas law. You can understand the modification by recognizing that the volume occupied by a gas at a specified temperature and pressure must depend on the number of gas particles. The number of moles of gas is directly proportional to the number of particles. Hence, moles must be directly proportional to volume as well. Therefore you can introduce moles into the combined gas law by dividing each side of the equation by n.

$$\frac{P_1 \times V_1}{T_1 \times n_1} = \frac{P_2 \times V_2}{T_2 \times n_2}$$

This equation shows that $(P \times V)/(T \times n)$ is a constant. This constancy holds for what are called ideal gases. A gas behaves ideally if it conforms to the gas laws. Ideal behavior depends upon certain conditions, as you will see later in this section.

If you could evaluate the constant $(P \times V)/(T \times n)$, you could then calculate the number of moles of gas at any specified values of P, V, and T. This constant is symbolized as R.

You can find the actual value of R, given an important fact about gases: 1 mol of every gas occupies 22.4 L at STP (101.3 kPa and 273 K). Inserting the values of P, V, T, and n into the equation:

$$R = \frac{P \times V}{T \times n} = \frac{101.3 \text{ kPa} \times 22.4 \text{ L}}{273 \text{ K} \times 1 \text{ mol}} = 8.31 \text{ (L} \times \text{kPa)/(K} \times \text{mol)}$$

The **ideal gas constant** ***(R)*** has the value 8.31 (L × kPa) / (K × mol). Rearranging the equation for R, you obtain the usual form of the **ideal gas law:**

$$R = \frac{P \times V}{T \times n}$$

$$\text{or } P \times V = n \times R \times T$$

An advantage of the ideal gas law over the combined gas law is that it permits you to solve for the number of moles of a contained gas when P, V, and T are known.

Figure 12.16
When the temperature, pressure, and volume of a gas are known, you can use the ideal gas law to calculate the number of moles of the gas.

STUDENT RESOURCES

From the Teacher's Resource Package, use:

- Section Review 12.4, Ch. 12 Practice Problems and Quizzes from the Review Module (Ch. 9–12)
- Laboratory Recordsheet 12-2

TECHNOLOGY RESOURCES

Relevant technology resources include:

- Chem ASAP! CD-ROM
- Resource Pro CD-ROM
- ActivChemistry CD-ROM: *Behavior of Gases*

1 Engage

Use the Visual

Have students study the photograph and read the text that opens the section. Ask:

- **What is an ideal gas, and under what conditions do the gas laws apply?** (An ideal gas is one whose particles exhibit no attractive or repulsive forces. In addition, the volume of the individual particles of an ideal gas is assumed to be negligible compared to the total volume occupied by the gas. These assumptions are valid so long as gas behavior is studied at relatively low pressure and/or moderate to high temperature. The exact conditions will depend on the gas being investigated.)

Motivate and Relate

Write the equation for the combined gas law on the board. Ask:

- **What variable is missing from this equation that is used to describe a gas?** (n, number of particles or moles of gas)

TEACHER DEMO

Bring in a piece of dry ice and place it on the desk. Explain that dry ice is solid carbon dioxide. Ask students how the existence of dry ice violates the assumptions about ideal gases on which the gas laws are based. (The gas laws assume that the volume occupied by gas particles is negligible and that there are no attractive forces between the particles. Dry ice has volume and its particles appear to have an attraction for one another, given that it is a solid.)

2 Teach

ACTIVITY

Expose students to as many types of ideal gas law problems as possible. Work out a number of different problems on the board. Show students how the ideal gas law can be used to find the molar mass and density of a gas—two physical properties that students have already studied. Combine an ideal gas law problem with a stoichiometry problem. Once students have seen some variations, have the students work out one or two problems on their own.

Practice Problems Plus

Additional Practice Problem

The following problem is related to Sample Problem 12-5: At 34.0 °C, the pressure inside a nitrogen-filled tennis ball with a volume of 148 cm^3 is 212 kPa. How many moles of N_2 are in the tennis ball? (1.23×10^{-2} mol N_2)

Critical Thinking

Using the Ideal gas law,

$$\frac{P_1 \times V_1}{T_1 \times n_1} = \frac{P_2 \times V_2}{T_2 \times n_2}$$

Have students demonstrate how the other gas law equations can be derived from the ideal gas law. (When you hold one variable constant, it is identical on both sides of the equation and cancels to yield the other gas laws.) **Have students derive the value for the ideal gas constant. Explain that one mole of any gas at STP occupies 22.4 L. Standard temperature is 273 K and standard pressure is 101.3 kPa.** $\left(R = 8.31 \frac{L \times kPa}{K \times mol}\right)$

Sample Problem 12-5

You fill a rigid steel cylinder that has a volume of 20.0 L with nitrogen gas (N_2)(*g*) to a final pressure of 2.00×10^4 kPa at 28 °C. How many moles of $N_2(g)$ does the cylinder contain?

1. *ANALYZE* ***List the knowns and the unknown.***

   Knowns:
   - $P = 2.00 \times 10^4$ kPa
   - $V = 20.0$ L
   - $T = 28$ °C

   Unknown:
   - $n = ?$ mol $N_2(g)$

   Use the known values and the ideal gas law to calculate the unknown (*n*).

2. *CALCULATE* ***Solve for the unknown.***

   Convert degrees Celsius to kelvins.

   $$28\ °C + 273 = 301\ K$$

   Rearrange the ideal gas law to isolate *n*.

   $$n = \frac{P \times V}{R \times T}$$

   Substitute the known quantities *P*, *V*, *R*, and *T* into the equation and solve.

   $$n = \frac{2.00 \times 10^4\ \cancel{kPa} \times 20.0\ \cancel{L}}{8.31 \frac{\cancel{L} \times \cancel{kPa}}{\cancel{K} \times mol} \times 301\ \cancel{K}} = 160\ mol\ N_2(g)$$

   $$= 1.60 \times 10^2\ mol\ N_2(g)$$

3. *EVALUATE* ***Does the result make sense?***

   The gas is at a high pressure, but the volume is not large. This means that a large number of moles of gas must be compressed into the volume. The large answer is thus reasonable, and the units have canceled correctly.

Practice Problems

22. When the temperature of a rigid hollow sphere containing 685 L of helium gas is held at 621 K, the pressure of the gas is 1.89×10^3 kPa. How many moles of helium does the sphere contain?
23. What pressure will be exerted by 0.450 mol of a gas at 25 °C if it is contained in a 0.650-L vessel?

Figure 12.17
Natural gas (mainly methane) is often found in underground pockets near petroleum reserves. In the past, excess gas was simply burned off, as shown here, involving a great waste of this energy resource.

Sample Problem 12-6

A deep underground cavern contains 2.24×10^6 L of methane gas $(CH_4)(g)$ at a pressure of 1.50×10^3 kPa and a temperature of 42 °C. How many kilograms of CH_4 does this natural-gas deposit contain?

1. ***ANALYZE*** ***List the knowns and the unknown.***

Knowns:
- $P = 1.50 \times 10^3$ kPa
- $V = 2.24 \times 10^6$ L
- $T = 42$ °C

Unknown:
- $m = ?$ kg CH_4

Calculate the number of moles (n) using the ideal gas law. Convert moles to grams, using the molar mass of methane, and then convert grams to kilograms.

2. ***CALCULATE*** ***Solve for the unknown.***

Convert degrees Celsius to kelvins.

$$42\ °C + 273 = 315\ K$$

Rearrange the equation for the ideal gas law to isolate n, the number of moles of methane.

$$n = \frac{P \times V}{R \times T}$$

Substitute the known quantities into the equation to find the number of moles of methane.

$$n = \frac{1.50 \times 10^3\ \cancel{kPa} \times (2.24 \times 10^6\ \cancel{L})}{8.31\ \frac{\cancel{L} \times \cancel{kPa}}{\cancel{K} \times mol} \times 315\ \cancel{K}} = 1.28 \times 10^6\ mol\ CH_4$$

Convert moles of methane to grams.

$$\text{molar mass } CH_4 = \left(4\ \cancel{mol\ H} \times \frac{1.0\ g\ H}{1\ \cancel{mol\ H}}\right) + \frac{12.0\ g\ C}{1\ mol\ C}$$
$$= 16.0\ g\ CH_4/mol\ CH_4$$

A mole–mass conversion gives the number of grams of methane.

$$1.28 \times 10^6\ \cancel{mol\ CH_4} \times \frac{16.0\ g\ CH_4}{1\ \cancel{mol\ CH_4}} = 20.5 \times 10^6\ g\ CH_4$$
$$= 2.05 \times 10^7\ g\ CH_4$$

Convert this answer to kilograms.

$$2.05 \times 10^7\ \cancel{g}\ CH_4 \times \frac{1\ kg}{1000\ \cancel{g}} = 2.05 \times 10^4\ kg\ CH_4$$

3. ***EVALUATE*** ***Does the result make sense?***

The volume and pressure of the methane are very large. It is reasonable that the cavern contains the large mass of methane gas found as the solution to the problem. Also, the units canceled correctly, and the answer is expressed to the proper number of significant figures.

Practice Problems

24. A child has a lung capacity of 2.20 L. How many grams of air do her lungs hold at a pressure of 102 kPa and a normal body temperature of 37 °C? Air is a mixture, but you may assume an average molar mass of 29 g/mol for air because air is about 20% O_2 (molar mass 32) and 80% N_2 (molar mass 28).

25. What volume will 12.0 g of oxygen gas $(O_2)(g)$ occupy at 25 °C and a pressure of 52.7 kPa?

Chem ASAP!

Problem-Solving 24
Solve Problem 24 with the help of an interactive guided tutorial.

12.4

TEACHER DEMO

Fill both a large and a small glass beaker with an identical number of marbles. Ask students to imagine that both beakers represent sealed containers and that the marbles represent gas particles that are floating freely in the containers. Ask:

- Given that the volume occupied by the individual particles of an ideal gas is negligible, which model more closely represents an ideal gas? (the large beaker because the gas particles occupy a smaller volume fraction)
- Which gas is under higher pressure? (The gas in the smaller container has more simultaneous collisions with the walls of the container.)

Practice Problems Plus

Related Chapter Review Problems
Chapter Review problems 62 and 63 are related to Sample Problem 12-6.

Additional Practice Problem
The Chem ASAP! CD-ROM contains the following problem: A helium-filled balloon contains 0.16 mol He at 101 kPa and a temperature of 23 °C. What is the volume of the gas in the balloon? (3.9 L)

Discuss

Discuss how the ideal gas law works well for gases with large volumes and at low pressures. In an ideal gas there are no molecular attractions and the particles have no volume. Real gases can be liquefied (liquid propane) and sometimes solidified (dry ice) due to attractive forces between the particles. These attractive forces only become significant when the particles are closer together (high pressure).

ACTIVITY

Have students use the ideal gas law to calculate the molar mass of air. You will need a device with a known volume that can safely hold air under pressure. You could use a plastic soda bottle fitted with an automobile tire valve like the one described by Kavanah and Zipp in *J. Chem. Educ.* 1998, 75, 1405–1406. Use a hand bicycle pump to fill the bottle with air to a given pressure—say 40 psi. Measure the pressure with a tire gauge and find the mass of the bottle. (The pressure measured by the gauge should be corrected by adding 14.7 psi.) Release a small amount of air by depressing the valve, and measure the pressure and mass once again. Repeat this process to give several mass–pressure values. Plot the data and then draw a best-fit line. Students should extrapolate to zero pressure (on the *x*-axis) to find the mass of the "evacuated" bottle. Subtract this mass from the mass of the bottle at any other pressure to give the mass of air in the bottle at any other pressure. Finally, use the equation

$$\text{Molar Mass}_{air} = \frac{gRT}{PV},$$ where

g is the mass of the air in the bottle at pressure *P* and temperature *T*, and *V* is the volume of the bottle, to find the molar mass of air. Show students how to derive this form of the ideal gas law. (The molar mass of air should be between 27 g and 28 g.)

Figure 12.18
Carbon dioxide freezes at –78.5 °C. The solid that forms (dry ice) can damage skin if touched. The dry ice in the photo is under water. As it sublimes, carbon dioxide gas bubbles up through the water and into the air. The cold gas causes water vapor in air to condense and form a white fog. Why is dry ice not as messy to use as ice? ❶

The Ideal Gas Law and Kinetic Theory

The previous discussions of kinetic theory and the gas laws assumed that the gases were ideal gases. A truly ideal gas is one that follows the gas laws at all conditions of pressure and temperature. Such a gas would have to conform precisely to the assumptions of kinetic theory. Its particles could thus have no volume and could not be attracted to each other at all. As you probably suspect, there is no gas for which this is true. An ideal gas does not exist. Nevertheless, at many conditions of temperature and pressure, real gases behave very much like an ideal gas.

An important behavior of real gases that differs from that of a hypothetical ideal gas is that real gases can be liquefied and sometimes solidified by cooling and by applying pressure. Ideal gases cannot be. For example, when water vapor is cooled below 100 °C at standard atmospheric pressure, it condenses to a liquid. The behavior of other real gases is similar, although lower temperatures and greater pressures may be required. In some cases, such as carbon dioxide (as you learned in Chapter 10), the gas may be converted directly to a solid. Solid carbon dioxide sublimes readily back to the gas state, as shown in **Figure 12.18.**

Departures from the Ideal Gas Law

A gas that adheres very closely to the gas laws at some conditions of temperature and pressure is said to exhibit ideal behavior under those conditions. No gas behaves ideally at all temperatures and pressures.

You can analyze how much a gas departs from ideal behavior by considering the ratio $(P \times V)/(n \times R \times T)$. According to the ideal gas law, this ratio for an ideal gas equals 1 (to see that, just divide both sides of the ideal gas law by $n \times R \times T$). This ratio plotted against pressure gives a horizontal line for an ideal gas because the ratio is constant. See **Figure 12.19.** For real gases at high pressures, the ratio $(P \times V)/(n \times R \times T)$ may depart widely from the ideal constant value of 1. The ratio may be greater or less than 1; the deviation may be positive (above the line) or negative (below the line). The explanation for these departures, or deviations, from the ideal is based on two factors: the attractions between molecules and the volume of gas molecules.

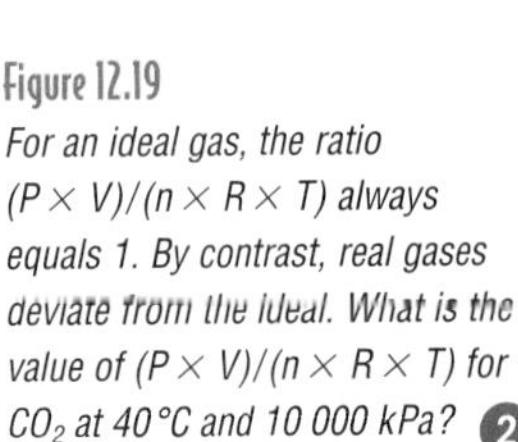

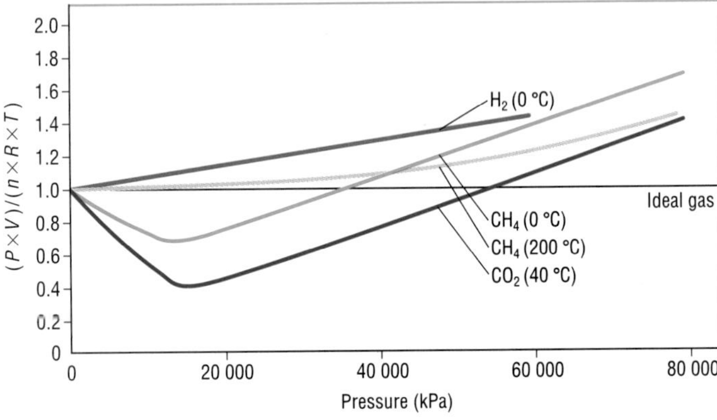

Figure 12.19
For an ideal gas, the ratio $(P \times V)/(n \times R \times T)$ always equals 1. By contrast, real gases deviate from the ideal. What is the value of $(P \times V)/(n \times R \times T)$ for CO_2 at 40 °C and 10 000 kPa? ❷

Answers

❶ Under normal conditions, dry ice sublimes; it goes directly from the solid to gaseous state without passing through the liquid state.
❷ about 0.6

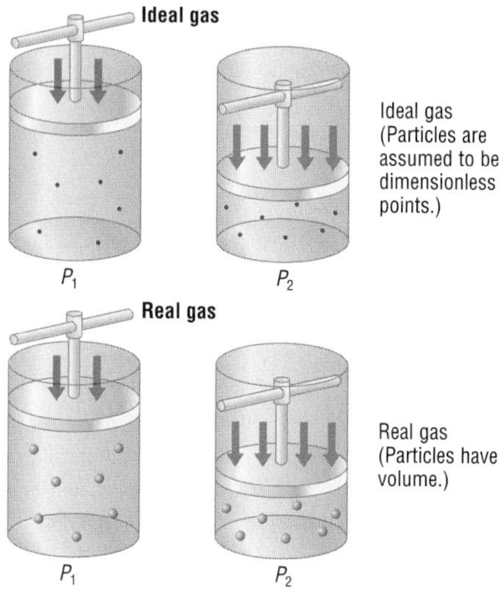

Figure 12.20
As pressure increases, the actual volumes of individual gas molecules in a real gas are no longer insignificant. It becomes increasingly difficult to compress the gas beyond a certain point, no matter what pressure is applied.

As you have read, simple kinetic theory assumes that gas particles are not attracted to each other and that the particles have no volume. These assumptions are incorrect. Gases and vapors could not be liquefied if there were no attractions between molecules. Also real gases are made up of actual physical particles, which do have a volume, as **Figure 12.20** illustrates.

The intermolecular forces that tend to hold the particles in a gas together effectively reduce the distance between particles. The gas therefore occupies less volume than is expected by the no-attractions assumption of the kinetic theory. This fact, considered alone, causes the $(P \times V)/(n \times R \times T)$ ratio to tend to be less than 1. At the same time, the molecules themselves occupy some volume, thus contradicting the zero-volume assumption of kinetic theory. This fact, on its own, causes the $(P \times V)/(n \times R \times T)$ ratio to tend to be greater than 1. One or the other of these two effects will usually dominate. In portions of the curves below the line, the intermolecular attractions dominate, causing the total volume to be less than ideal. In portions of the curves above the line, the effect of volume of the molecules dominates, causing the total volume to be greater than ideal. The temperature of the gas determines which of these two effects is the dominant one.

Compare the curves for $CH_4(g)$ at 0 °C and at 200 °C. At 0 °C, the methane molecules are moving relatively slowly. The attractions between the molecules are sufficiently strong so that, at low pressures, the curve is below the $(P \times V)/(n \times R \times T) = 1$ line. At higher pressures, the space between the molecules is reduced. The actual physical volume of the methane molecules now becomes important, and the curve is above the $(P \times V)/(n \times R \times T) = 1$ line. Raising the temperature to 200 °C increases the average kinetic energy of the methane molecules sufficiently to overcome the weak intermolecular attractive effects. Thus $(P \times V)/(n \times R \times T)$ is nearly equal to 1 at lower pressure at the elevated temperature. The ratio increases to greater than 1 only when the volume of the individual gas particles becomes important, as it does at high pressure.

PHYSICS

Cryostats

The containers that store and transport liquefied gases are called cryostats. Their design prevents heat from being transferred from the environment to

the very cold liquid inside. The most widely used cryostats are called Dewar flasks, named after the Scottish scientist Sir James Dewar, who invented the containers in 1892. Dewar flasks are double-walled vessels with a vacuum between the walls, similar to the familiar Thermos™ bottle used to carry hot or cold drinks. Cryostats are very lightweight in comparison with compressed-gas cylinders. A given amount of a substance has a much smaller volume as a liquid than as a gas, even if the gas is under pressure. For these reasons, many gases are stored and transported in their liquefied form rather than as gases.

Use the Visual

Have students study Figure 12.19. Point out that for constants *n*, *R*, and *T*, the denominator of the fraction *PV/nRT* is constant. Any deviation from unity must occur in the numerator, which is the product of *P* and *V*. Because the pressure is a measured value, the manipulated variable must be the volume. If $PV/nRT < 1$, the value of *V* is less than would be expected from the ideal gas law. The real gas is not occupying as much space as the ideal gas. The real gas is being confined internally. If $PV/nRT > 1$, *V* is greater than expected. The real gas is occupying more volume than the ideal gas. The difference takes into account the volumes of the real gas particles.

3 Assess

Evaluate Understanding

Have students apply the ideal gas law to a sample of gas in a balloon. Ask them to explain why in this system *n* and *R* are constants and *P*, *V*, and *T* are variables.

Ask students to explain why a real gas can become a liquid, but an ideal gas cannot. (Particles of an ideal gas have no volume and are not attracted to each other.)

MINI LAB

Carbon Dioxide from Antacid Tablets

LAB TIPS

Probeware Lab Manual has Vernier, TI, and PASCO labs.

Students should make sure that the balloons and their hands are dry before beginning the activity. They might use pestles and mortars to crush the effervescent tablets, and powder funnels to get the crushed tablets into the balloons.

ANALYSIS AND CONCLUSIONS

1. Answers should indicate that the volume of the balloon is directly proportional to the number of tablets reacted.
2. Answers will vary, but the masses and numbers of moles should be in ratios of 1 : 2 : 3 for the three balloons.
3. Answers will vary. For example, 2.0 g of $NaHCO_3$ (molar mass = 84.01 g) should yield about 1.2×10^{-2} mol of CO_2.

Reteach

Point out to students that one advantage of the ideal gas law is that it enables them to find the number of moles of a gas by measuring its temperature, pressure, and volume.

MINI LAB 

Carbon Dioxide from Antacid Tablets

PURPOSE

To measure the amount of carbon dioxide gas given off when antacid tablets dissolve in water.

MATERIALS

- pencil
- graph paper
- tape measure
- clock or watch
- 3 rubber balloons (round)
- plastic medicine dropper
- water
- 6 antacid tablets
- pressure sensor (optional)

PROCEDURE

Sensor version available in the Probeware Lab Manual.

1. Break six antacid tablets into small pieces. Keep the pieces from each tablet in a separate pile. Put the pieces from one tablet into the first balloon. Put the pieces from two tablets into a second balloon. Put the pieces from three tablets into a third balloon.
2. After you use the medicine dropper to squirt about 5 mL of cold water into each balloon, immediately tie off each balloon.
3. Shake the balloons to mix the contents. Allow the contents to warm to room temperature.
4. Carefully measure and record the circumference of each balloon several times during the next 20 minutes.
5. Use the maximum circumference of each balloon to calculate the volume of each balloon. Assume that the balloons are spherical. (*Hint:* Volume of a sphere $= \frac{4\pi r^3}{3}$)

ANALYSIS AND CONCLUSIONS

1. Make a graph of balloon volume versus number of tablets. According to your graph, describe the relationship between the number of tablets used and the volume of the balloon.
2. Assume that the balloon is filled with carbon dioxide gas at 20 °C and standard pressure. Calculate the mass and the number of moles of CO_2 in each balloon at maximum inflation.
3. If a typical antacid tablet contains 2.0 g of sodium hydrogen carbonate, show how your carbon dioxide results compare with the theoretical values.

section review 12.4

26. How is it possible to calculate the amount of gas in a sample at given conditions of temperature, pressure, and volume?
27. What is the difference between an ideal gas and a real gas?
28. Explain the meaning of this statement: "No gas exhibits ideal behavior at all temperatures and pressures." At what conditions do real gases behave like ideal gases? Why?
29. Determine the volume occupied by 0.582 mol of a gas at 15 °C if the pressure is 81.8 kPa.
30. If 28.0 g of methane gas (CH_4) are introduced into an evacuated 2.00-L gas cylinder at a temperature of 35 °C, what is the pressure inside the cylinder? Note that the volume of the gas cylinder is constant.

 Chem ASAP! **Assessment 12.4** Check your understanding of the important ideas and concepts in Section 12.4.

Answers

SECTION REVIEW 12.4

26. by using the ideal gas law, $PV = nRT$
27. An ideal gas obeys the assumptions of the kinetic molecular theory of gases. A real gas will deviate from ideal behavior except within a small range of $P - V - T$ conditions.
28. Real gases have attractions between molecules, and their molecules have volume. At low temperatures, the attractions between molecules pull them together and reduce the gas volume. At high pressures, the volume occupied by the molecules is a significant part of the total volume because the molecules are close together.
29. 17.0 L
30. 2.24×10^3 kPa

GAS MOLECULES: MIXTURES AND MOVEMENTS

section 12.5

objectives

- State Avogadro's hypothesis, Dalton's law, and Graham's law
- Calculate moles, masses, and volumes of gases at STP
- Calculate partial pressures and rates of effusion

key terms

- Avogadro's hypothesis
- partial pressure
- Dalton's law of partial pressures
- diffusion
- effusion
- Graham's law of effusion

The top of Mount Everest is more than 29 000 feet above sea level. In addition to tents, food, warm clothes, and ropes, an expedition to climb Mount Everest requires cylinders of oxygen. Humans need a partial pressure of oxygen of at least 10.67 kPa to survive—continual exposure to less oxygen will result in death! **What is meant by the partial pressure of oxygen, and why does it decrease as altitude increases?**

Avogadro's Hypothesis

The particles that make up different gases are not the same size. For example, chlorine molecules have large numbers of electrons, protons, and neutrons. They are bigger and occupy more volume than hydrogen molecules, which have only two protons and two electrons. Early scientists recognized that there must be such size differences and assumed that collections of larger molecules must have larger volumes than collections of an equal number of small molecules. Thus many scientists reacted in disbelief in 1811 when they heard of **Avogadro's hypothesis:** Equal volumes of gases at the same temperature and pressure contain equal numbers of particles. It was as if Avogadro were suggesting that two rooms of the same size could be filled by the same number of objects, regardless of whether the objects were marbles or basketballs.

What Avogadro had in mind is not really so mysterious if you consider that the particles in a gas are very far apart, with nothing but space in between. Thus a collection of relatively large particles does not require much more space than the same number of relatively small particles. On average, there would be large expanses of space between the particles in either case, as shown in **Figure 12.21.** This was Avogadro's great insight, today easily demonstrated by experiment. At STP, 1 mol (6.02×10^{23}) of particles of any gas, regardless of the size of the particles, occupies a volume of 22.4 L.

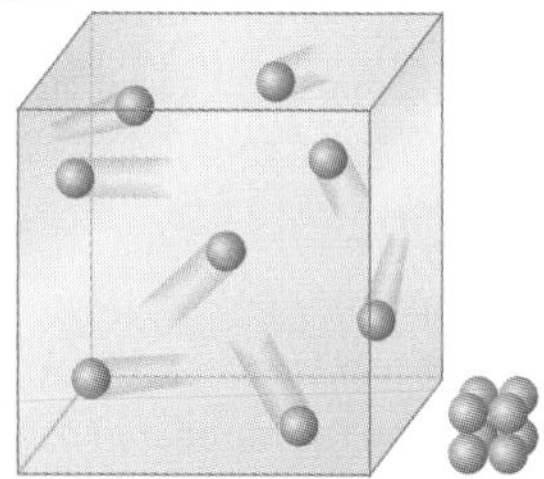

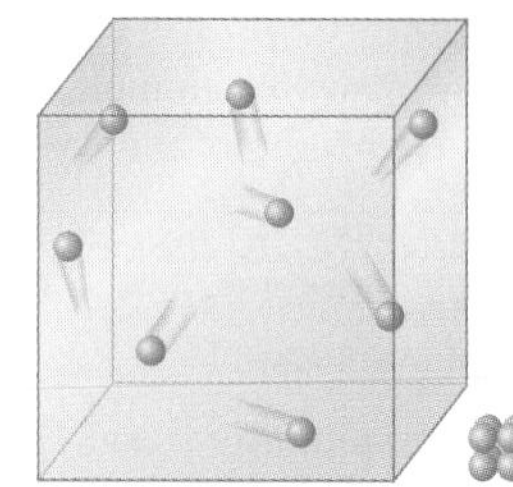

Figure 12.21
The volume of a container easily accommodates the same number of relatively large or small particles, as long as the particles are not tightly packed. There is a great deal of empty space compared with the volume occupied by the particles. When the particles are tightly packed, large particles take up more space than small particles.

STUDENT RESOURCES

From the Teacher's Resource Package, use:

- Section Review 12.5, Ch. 12 Practice Problems, Interpreting Graphics, Vocabulary Review, Quizzes and Tests from the Review Module (Ch. 9–12)
- Laboratory Manual: Experiment 19

TECHNOLOGY RESOURCES

Relevant technology resources include:

- Chem ASAP! CD-ROM
- Resource Pro CD-ROM
- Assessment Resources CD-ROM: Chapter 12 Tests

12.5

1 Engage

Use the Visual

Have students study the photograph and read the text that opens the section. Ask students to explain why high-altitude climbers carry a supply of oxygen. Ask:

- **What is meant by the partial pressure of oxygen, and why does it decrease at high altitudes?** (Explain that as elevation increases, atmospheric pressure is reduced so that the partial pressure of oxygen is also reduced to the point where it is insufficient for respiration.)

Check Prior Knowledge

Assess students' knowledge about gas molecules by asking them to compare the number of gas molecules in two identical, sealed containers of propane and helium gas at the same temperature and pressure. (Equal volumes of gases at the same temperature and pressure contain equal numbers of particles.)

2 Teach

TEACHER DEMO

Take two identical glass containers and fill approximately 1/8 the volume of one with beads. Fill the other container with an identical number of larger beads. Have the students assume that the beads represent gas particles in sealed containers at the same temperature. Ask the students to hypothesize on the relative pressures in the two containers assuming the gases are ideal. (Both will have the same pressure, because both have the same number of gas particles with the same average kinetic energy.)

Discuss

Have students discuss how they would determine the mass of a balloon full of helium gas at STP without making any mass measurements. (According to Avogadro's hypothesis, one mole of a gas has a volume of 22.4 liters at standard temperature and pressure. Calculate the volume of the balloon in liters at STP. Then, convert from liters to moles. To find the mass, multiply the number of moles by the atomic mass of helium.)

ACTIVITY

To demonstrate Avogadro's hypothesis, pass around a few pairs of balloons. Fill one balloon in each pair with about 2.0–2.5 L of helium; fill the second balloon with an approximately equal volume of nitrogen. Tell the students to assume that each balloon contains 0.10 mol of gas. Have students calculate the volume of the balloons (from the volume of a sphere, $V = 4\pi r^3/3$). Based upon this activity, what can be said about equal molar quantities of gases at the same temperature and pressure? (Equal molar quantities of gases at the same temperature and pressure will occupy the same volume.)

Practice Problems Plus

Additional Practice Problem

The following problem is related to Sample Problem 12-7: Find the volume of a weather balloon filled with 86 mol of helium gas at STP. (1.9×10^3 L)

You can also understand Avogadro's hypothesis by thinking about the modern explanation for gas pressure. Equal numbers of particles of different gases in equal volumes at the same temperature should exert the same pressure because the particles have the same average kinetic energy and are contained within equal volumes. Thus whenever you have equal volumes of gases at the same temperature and pressure, the volumes should contain equal numbers of particles.

Sample Problem 12-7

Determine the volume (in L) occupied by 0.202 mol of a gas at standard temperature and pressure (STP).

1. ***ANALYZE*** ***List the known and the unknown.***

   Known:
   - $n = 0.202$ mol

   Unknown:
   - $V = ?$ L

   The number of moles of gas at STP is given. Convert from moles to the volume (V).

   moles $\longrightarrow$ volume

   The conversion factor for moles $\longrightarrow$ volume at STP is 22.4 L/1 mol.

2. ***CALCULATE*** ***Solve for the unknown.***

   Multiplying the known value by the conversion factor yields:

   $$V = 0.202\ \cancel{\text{mol}} \times \frac{22.4\ \text{L}}{1\ \cancel{\text{mol}}} = 4.52\ \text{L}$$

3. ***EVALUATE*** ***Does the result make sense?***

   Because 1 mol of the gas occupies 22.4 L at STP, 0.202 mol of the gas should occupy about one-fifth of that volume, or about 4.5 L.

Practice Problems

31. What is the volume occupied by 0.250 mol of a gas at STP?
32. What volume does 0.742 mol of argon gas occupy at STP?

Sample Problem 12-8

How many oxygen molecules are in 3.36 L of oxygen gas at standard temperature and pressure (STP)?

1. ***ANALYZE*** ***List the known and the unknown.***

   Known:
   - $V = 3.36$ L

   Unknown:
   - ? molecules O_2

   The volume of O_2 at STP is known. To find the number of oxygen molecules, make the following conversion.

   volume $\longrightarrow$ moles $\longrightarrow$ molecules

   The conversion factors are 1 mol O_2/22.4 L O_2 and 6.02×10^{23} molecules O_2/1 mol O_2, respectively.

Practice Problems

33. How many nitrogen molecules are in 5.12 L of the gas at STP?

Sample Problem 12-8 (cont.)

2. *CALCULATE* ***Solve for the unknown.***
Multiplying the known value by the conversion factors yields:

$$3.36\ \cancel{L\ O_2} \times \frac{1\ \cancel{mol\ O_2}}{22.4\ \cancel{L\ O_2}} \times \frac{6.02 \times 10^{23}\ \text{molecules}\ O_2}{1\ \cancel{mol\ O_2}}$$

$$= 9.03 \times 10^{22}\ \text{molecules}\ O_2$$

3. *EVALUATE* ***Does the result make sense?***
The answer is reasonable: 3.36 L O_2 should contain about one-seventh mole of O_2 molecules. One-seventh of approximately 6×10^{23} molecules is about 9×10^{22} molecules.

Practice Problems (cont.)

34. What volume is occupied by 4.02×10^{22} molecules of helium gas at STP?

Chem ASAP!

Problem-Solving 34

Solve Problem 34 with the help of an interactive guided tutorial.

Sample Problem 12-9

Determine the volume (in L) occupied by 14.0 g of nitrogen gas at STP.

1. *ANALYZE* ***List the known and the unknown.***

Known:	*Unknown:*
• mass = 14.0 g $N_2(g)$	• $V = ?$ L

The mass of nitrogen is known. Convert the known mass to volume.

mass ⟶ moles ⟶ volume

The conversion factors are 1 mol N_2/molar mass N_2 and 22.4 L N_2/1 mol N_2, respectively. The conversion factors require the value of the molar mass of N_2.

2. *CALCULATE* ***Solve for the unknown.***
First find the molar mass of N_2.

$$\text{molar mass } N_2 = 2\ \cancel{mol\ N} \times \frac{14.0\ g\ N}{1\ \cancel{mol\ N}}$$

$$= 28.0\ g\ N_2$$

Substitute the molar mass into the conversion factor, then solve for the volume of nitrogen gas.

$$14.0\ \cancel{g\ N_2} \times \frac{1\ \cancel{mol\ N_2}}{28.0\ \cancel{g\ N_2}} \times \frac{22.4\ L\ N_2}{1\ \cancel{mol\ N_2}}$$

$$= 11.2\ L\ N_2$$

3. *EVALUATE* ***Does the result make sense?***
The mass of nitrogen corresponds to one-half mole of N_2. One-half mole of gas at STP should occupy half of 22.4 L, or 11.2 L.

Practice Problems

35. What is the volume of a container that holds 8.80 g of carbon dioxide at STP?

36. A container holds 6.92 g of hydrogen gas at STP. What is the volume of the container?

Practice Problems Plus

Additional Practice Problem
The Chem ASAP! CD-ROM contains this problem related to Sample Problem 12-8: How many molecules are in 4.5 L CO_2 at STP? (1.21×10^{23} molecules CO_2)

Discuss

Explain that Avogadro's hypothesis makes it possible to relate the molar quantity of a gas to its temperature, volume, and pressure. The hypothesis assumes that as long as gas particles are not tightly packed, equal volumes of gases at the same temperature and pressure contain equal numbers of particles. Remind students that it is the average kinetic energy of its particles that determines the pressure exerted by a confined gas. Emphasize that Avogadro's hypothesis works only for gases. Similar statements do not apply to solids and liquids. The factor that makes gases so different from solids and liquids is the very large portion of their volume that is empty space. At low pressure, the volumes of individual gas molecules are negligible compared to the volume of the container holding the gas. The volume of a gas depends on the number of gas particles present, not their size.

Practice Problems Plus

Additional Practice Problem
The following problem is related to Sample Problem 12-9: At STP, a balloon is inflated with 5.00 g of helium gas. How many grams of oxygen gas are needed to inflate an identical balloon to the same volume at STP? (40.0 g)

Discuss

Point out that the particles of each kind of gas in a mixture exert their own pressure as if they were by themselves. The presence of one gas does not affect the partial pressure of another gas. By Dalton's law the total pressure, at constant volume and temperature, exerted by a mixture of gases is equal to the sum of the partial pressures of the component gases. $P_{total} = P_1 + P_2 + P_3$

Think Critically

Ask students to form a generalization explaining why the relative partial pressure exerted by a gas in a mixture of gases does not vary with temperature, pressure, or volume of the mixture. Students should realize that the ratio of n_1/n (the relative amount of the gas in the mixture) remains the same. Because the value of n_1/n remains constant, the relative partial pressure (P_1/P) exerted by the gas remains the same.

Dalton's Law

The gases listed in **Table 12.1** make up a mixture called air. The particles in a gas mixture at the same temperature have the same average kinetic energy. Gas pressure depends only on the number of gas particles in a given volume and on their average kinetic energy—the kind of particle is unimportant. Each particle makes the same contribution to the pressure. Thus if you know the pressure exerted by each gas in a mixture, you can add the individual pressures to get the total gas pressure.

Table 12.1

Composition of Dry Air		
Component	**Volume (%)**	**Partial pressure (kPa)**
Nitrogen	78.08	79.11
Oxygen	20.95	21.22
Carbon dioxide	0.04	0.04
Argon and others	0.93	0.95
	100.00	101.32

Chem ASAP!
Animation 13
Observe the behavior of a mixture of nonreacting gases.

The contribution each gas in a mixture makes to the total pressure is called the **partial pressure** exerted by that gas. In a mixture of gases, the total pressure is the sum of the partial pressures of the gases.

$$P_{total} = P_1 + P_2 + P_3 + \ldots$$

This equation is one mathematical form of **Dalton's law of partial pressures:** At constant volume and temperature, the total pressure exerted by a mixture of gases is equal to the sum of the partial pressures of the component gases. The individual gases in containers a, b, and c of **Figure 12.22** are combined in container T. What partial pressure does each individual
1 gas contribute to the total pressure of the mixture?

The fractional contribution to pressure exerted by each gas in a mixture does not change as the temperature, pressure, or volume changes. This fact has important implications for aviators and mountain climbers. For example, on top of Mount Everest, the total atmospheric pressure is reduced to 33.73 kPa (about one-third of its value at sea level). The partial pressure of oxygen is reduced by the same factor, to only 7.06 kPa (one-third the partial pressure of oxygen at sea level). This reduced oxygen

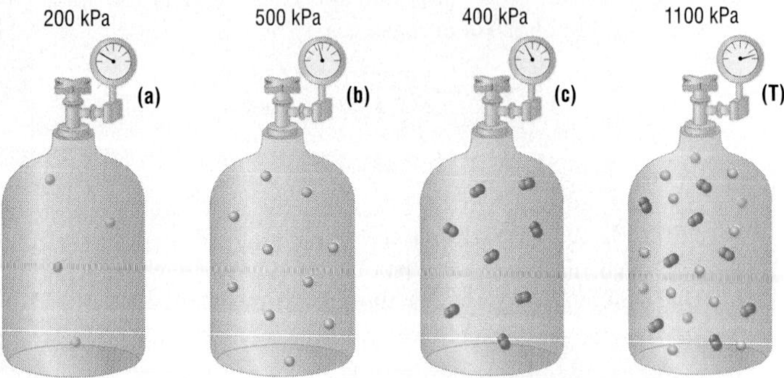

Figure 12.22
The sum of the pressures exerted by the gas in each of the three containers on the left is the same as the total pressure exerted by a mixture of the gases in the same volume, assuming the temperature stays the same. Dalton's law of partial pressures holds true because each gas exerts its own pressure independent of the pressure exerted by the other gases.

Answer

1 200 kPa; 500 kPa; 400 kPa; $P_T = P_a + P_b + P_c = 200\ kPa + 500\ kPa + 400\ kPa = 1100\ kPa$

pressure is insufficient for respiration because humans need an oxygen partial pressure of at least 10.67 kPa; some individuals need a higher partial pressure. **Figure 12.23** shows some of the steps jet pilots and mountaineers take to counteract high-altitude conditions.

Figure 12.23
High-altitude pilots and mountaineers must have supplemental oxygen supplies available when at high altitudes.

Sample Problem 12-10

Air contains oxygen, nitrogen, carbon dioxide, and trace amounts of other gases. What is the partial pressure of oxygen (P_{O_2}) at 101.30 kPa of total pressure if the partial pressures of nitrogen, carbon dioxide, and other gases are 79.10 kPa, 0.040 kPa, and 0.94 kPa, respectively.

1. **ANALYZE** ***List the knowns and the unknown.***

Knowns:
- $P_{N_2} = 79.10$ kPa
- $P_{CO_2} = 0.040$ kPa
- $P_{others} = 0.94$ kPa
- $P_{total} = 101.30$ kPa

Unknown:
- $P_{O_2} = ?$ kPa

Use the known values and Dalton's law of partial pressures ($P_{total} = P_{O_2} + P_{N_2} + P_{CO_2} + P_{others}$) to calculate the unknown value (P_{O_2}).

2. **CALCULATE** ***Solve for the unknown.***

Rearrange the expression for Dalton's law to isolate P_{O_2}. Substitute the values for the partial pressures and solve the equation.

$$P_{O_2} = P_{total} - (P_{N_2} + P_{CO_2} + P_{others})$$
$$= 101.30 \text{ kPa} - (79.10 \text{ kPa} + 0.040 \text{ kPa} + 0.94 \text{ kPa})$$
$$= 21.22 \text{ kPa}$$

3. **EVALUATE** ***Does the result make sense?***

The partial pressure of oxygen must be smaller than that of nitrogen because P_{total} is only 101.30 kPa. The other partial pressures are small, so an answer of 21.22 kPa seems reasonable.

Practice Problems

37. Determine the total pressure of a gas mixture that contains oxygen, nitrogen, and helium if the partial pressures of the gases are as follows: $P_{O_2} =$ 20.0 kPa, $P_{N_2} =$ 46.7 kPa, and $P_{He} = 26.7$ kPa.
38. A gas mixture containing oxygen, nitrogen, and carbon dioxide has a total pressure of 32.9 kPa. If $P_{O_2} = 6.6$ kPa and $P_{N_2} = 23.0$ kPa, what is P_{CO_2}?

Chem ASAP!

Problem-Solving 38
Solve Problem 38 with the help of an interactive guided tutorial.

12.5

Discuss

Sketch two identical-volume containers on the board. Tell students one container is filled with neon gas and the other with helium at the same temperature and pressure. Ask the students to use the ideal gas law to find the equation for the number of atoms in each container.

$$n_{He} = \frac{PV}{RT} \qquad n_{Ne} = \frac{PV}{RT}$$

Students will see that $n_{He} = n_{Ne}$. Point out that this equality indicates that the pressure of a gas sample depends only on the number of particles present, not the type. For the mixture of He and Ne:

$$P_{total} = n_{total}\frac{RT}{V}$$

$$n_{total} = n_{He} + n_{Ne}$$

The term on the right yields the pressure caused by the atoms of helium and neon present in the sample.

$$P_{total} = P_{He} + P_{Ne}$$

Therefore, the total pressure of a mixture of helium and neon is the sum of the partial pressures of each gas.

Practice Problems Plus

Additional Practice Problem

The Chem ASAP! CD-ROM contains the following problem related to Sample Problem 12-10: The pressure in an automobile tire filled with air is 245.0 kPa. The $P_{O_2} =$ 51.3 kPa, $P_{CO_2} = 0.10$ kPa, and $P_{others} = 2.3$ kPa. What is the P_{N_2}? (191.3 kPa)

Meeting Diverse Needs

At Risk Have students write summaries of the chapter by relating key terms in the Student Study Guide on page 355. Have students work in groups of 3 or 4. Ask them to share their summaries with each other. Alternatively, have students hand in their summaries. Review them for clarity and check for any misconceptions.

Use the Visual

Have students study the photograph in Figure 12.25 illustrating the diffusion of bromine gas through the air in a cylinder. Explain that the diffusion of a gas is a relatively slow process whose rate is inversely proportional to the square root of the gas' molar mass.

TEACHER DEMO

One day prior to this lesson, fill two identical balloons with equal volumes of gases: one with helium and the other with air. Calculate the volume of the balloons (from the volume of a sphere, $V = 4\pi r^3/3$). On the day of the lesson recalculate the volumes of the two balloons and have the students record their observations. Which balloon lost gas at a faster rate? (the helium balloon)

Discuss

Explain that diffusion is the tendency of molecules to move towards areas of lower concentration until the concentration is uniform. Graham's law states that the rate of diffusion of a gas is inversely proportional to the square root of its molar mass.

$$\frac{rate_1}{rate_2} = \frac{\sqrt{molar\ mass_2}}{\sqrt{molar\ mass_1}}$$

According to Graham's law, for two gases at the same temperature, the gas with the lighter molecular mass moves faster.

Figure 12.24
White vapors of ammonium chloride form as the vapors from aqueous ammonia (NH_3)(aq) and hydrochloric acid (HCl)(aq) diffuse into the air. Aqueous ammonia used to be called ammonium hydroxide, a name seldom used today. What are the names and formulas of the molecules in the vapors? Which molecule diffuses faster? Why? 1

Chem ASAP!
Animation 14
Examine the processes of gas effusion and diffusion.

Graham's Law

When you open a perfume bottle inside a room, the perfume molecules eventually spread throughout the room, and you can smell them everywhere. **Diffusion** is the tendency of molecules to move toward areas of lower concentration until the concentration is uniform throughout. **Figure 12.25** illustrates the diffusion process for bromine vapor. The concentrated bromine vapor tends to move outside the graduated cylinder toward an area where the bromine vapor concentration is less.

Much of the early work on diffusion was done in the 1840s by the Scottish chemist Thomas Graham (1805–1869). Graham measured the rates of **effusion,** the process in which a gas escapes through a tiny hole in its container. Graham noticed that gases of lower molar mass effuse faster than gases of higher molar mass. From his observations, he proposed **Graham's law of effusion:** The rate of effusion of a gas is inversely proportional to the square root of the gas's molar mass. Subsequently, this relationship was also shown to be true for the diffusion of gases. Thus the rate of diffusion of a gas is also inversely proportional to the square root of its molar mass.

To understand Graham's law, examine the relationship of the mass and speed of a moving body to the kinetic energy the body transfers when it strikes a stationary object (assuming perfectly elastic collisions). The mathematical expression that relates the mass (m) and the speed or velocity (v) of a body to its kinetic energy (KE) is KE $= \frac{1}{2}mv^2$. Suppose a small ball with a mass of 2 g traveling at 5 m/s has just enough kinetic energy to shatter a pane of glass. A ball bearing with a mass of only 1 g would need to travel faster (slightly more than 7 m/s) to have the same kinetic energy as the ball and to be able to shatter the same pane of glass.

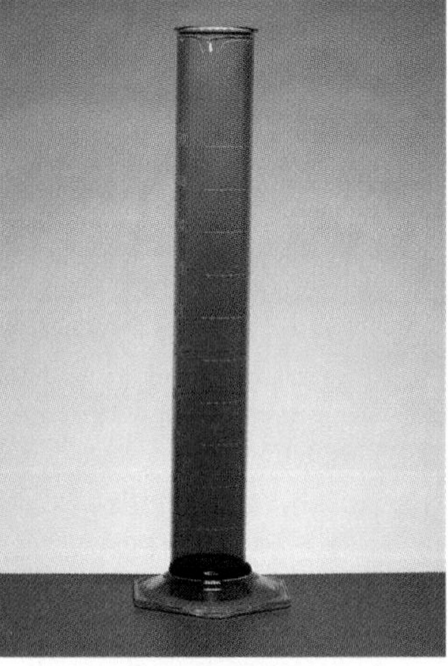

Figure 12.25
The diffusion of one substance through another is a relatively slow process. Here, bromine vapor is diffusing upward through the air in a graduated cylinder. After several hours, bromine vapors will mingle with air outside the cylinder.

Answer

1 hydrogen chloride (HCl)(*g*) and ammonia (NH_3)(*g*); NH_3; NH_3 is lighter than HCl.

Answers

SECTION REVIEW 12.5

39. Avogadro's hypothesis: Equal volumes of gases at the same temperature and pressure contain equal numbers of particles. Dalton's law: At constant volume and temperature, the total pressure exerted by a mixture of gases equals the sum of the partial pressures of the compo-

There is an important principle here. If two bodies of different masses have the same kinetic energy, the lighter body must move faster. You already know that the particles of two different gases at the same temperature have the same average kinetic energy. Thus a gas particle of low mass should move faster than a gas particle of high mass if the gases are at the same temperature. The gas of lower molar mass should therefore diffuse and effuse faster.

It is easy to show that the above phenomenon is true by comparing two balloons: one filled with helium and the other filled with air. There are pores in a balloon that, although tiny, are still large enough for both helium atoms and molecules in air to pass through freely. Balloons filled with air stay inflated longer, because the main components of air—oxygen molecules and nitrogen molecules—are more massive than helium atoms. The air particles move more slowly; therefore, they diffuse and effuse more slowly than helium atoms. The fast-moving helium atoms, with molar masses of only 4 g, rapidly effuse through the pores in the balloon. The rate of effusion or diffusion is related only to the particle's speed. In a mathematical form, Graham's law can be written as follows for two gases, A and B.

$$\frac{\text{Rate}_A}{\text{Rate}_B} = \frac{\sqrt{\text{molar mass}_B}}{\sqrt{\text{molar mass}_A}}$$

In other words, the rates of effusion of two gases are inversely proportional to the square roots of their molar masses. Now compare the rates of effusion of the air component nitrogen (molar mass = 28.0 g) and helium (molar mass = 4.0 g):

$$\frac{\text{Rate}_{He}}{\text{Rate}_{N_2}} = \frac{\sqrt{28.0\text{ g}}}{\sqrt{4.0\text{ g}}} = \frac{5.3\,\cancel{g}}{2.0\,\cancel{g}} = 2.7$$

You can see that helium effuses and diffuses nearly three times faster than nitrogen at the same temperature.

section review 12.5

39. In your own words, briefly state Avogadro's hypothesis, Dalton's law, and Graham's law.

40. How are moles, masses, or volumes of gases calculated from one another for a gas at STP?

41. Calculate the number of liters occupied at STP.
- **a.** 1.7 mol $H_2(g)$
- **b.** 1.8×10^{-2} mol $N_2(g)$
- **c.** 2.5×10^{2} mol $O_2(g)$

42. How is the partial pressure of a gas in a mixture calculated? How is the rate of effusion of a gas calculated?

43. What is the significance of the volume 22.4 L?

44. At the same temperature, the rates of diffusion of carbon monoxide and nitrogen gas are virtually identical. Explain.

Chem ASAP! **Assessment 12.5** Check your understanding of the important ideas and concepts in Section 12.5.

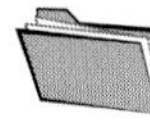

portfolio project

To estimate the amount of CO_2 in a soft drink, obtain an unopened bottle and a large spherical balloon. Open the bottle and quickly place the balloon over its neck. Fasten the balloon tightly and shake the bottle gently for at least 5 minutes. Use the volume of a sphere to estimate the volume of CO_2. Use the gas laws to find the number of moles of CO_2. Compare different brands. Identify possible sources of error in this experiment.

3 Assess

Evaluate Understanding

Tell students that the partial pressures of oxygen and hydrogen gases in a container are both 100 kPa. Ask which gas has more molecules present. Which gas has molecules with greater average kinetic energy? (Both gases have the same number of molecules and kinetic energy.) **Ask the students if Graham's law applies when one of the gases is heated.** (No, heating increases the average kinetic energy of molecules of the gas.) **Review the demo with the balloons filled with helium and air. Have the students assume that the air consists entirely of oxygen. What is the calculated relative rate of diffusion?**

$$\frac{rate_{\text{helium}}}{rate_{\text{oxygen}}} = \frac{32g}{4g} = 2.8$$

(The helium diffuses almost three times faster.)

Reteach

Remind students that diffusion is a general term that applies to molecules moving away from a region of high concentration. Effusion is a specific example of diffusion in which molecules pass through a narrow opening. Graham's law applies to both.

Portfolio Project

Different drinks have different amounts of carbon dioxide. Sources of error include the gas lost before the balloon can be fastened, elasticity of the balloon, estimated volume, and temperature differences. Pressure in the balloon will keep some gas in solution.

SECTION REVIEW 12.5 CONTINUED

nent gases. Graham's law: The rate of effusion of a gas is inversely proportional to the square root of the molar mass of the gas.

40. by using Avogadro's hypothesis and the molar mass and molar volume of the gas

41. **a.** 38 L
b. 0.40 L
c. 5600 L

42. Rearrange the equation $P_{\text{total}} = P_1 + P_2 + P_3 \ldots P_x$ to isolate the desired pressure term. The rate of effusion of a gas can be calculated by using the equation:

$$\frac{\text{Rate}_A}{\text{Rate}_B} = \frac{\sqrt{\text{molar mass}_B}}{\sqrt{\text{molar mass}_A}}$$

43. This is the volume of 1 mol of any gas at STP.

44. Carbon monoxide and nitrogen have identical molar masses of 28.0 g.

Chemistry Serving... Industry

DISCUSS

Discuss the content of the article in the context of Dalton's law of partial pressures. With Table 12.1 displayed on an overhead projector, point out that the air we normally breathe at sea level is about 21% oxygen and 78% nitrogen. The values for the partial pressures given in Table 12.1 reflect conditions at sea level, or 1 atm = 101.3 kPa. Have students verify the values for the partial pressures of each gas given in the table. Have students calculate the partial pressures of each gas that a scuba diver, breathing the same air mixture, experiences at depths of 100 feet (approximately 4 atm) and 300 feet (approximately 10 atm). Write the algebraic form of Dalton's law of partial pressures on the chalkboard, and remind students that, according to Dalton's law of partial pressures, the total pressure exerted by a mixture of gases is equal to the sum of the partial pressures exerted by each of the different gases in the mixture. The fractional contribution to pressure exerted by each gas does not change as the temperature, pressure, or volume changes as long as the composition of the mixture is constant. Supplement Table 12.1 with new columns showing the partial pressures of each gas at each depth. Tabulate the students' calculations. Ask students to use Dalton's law of partial pressures to explain how changing the air mixture in the tanks used by divers can help prevent problems.

CHEMISTRY IN CAREERS

Have students read about commercial divers on page 873 of this text. Encourage students to find out more about issues and topics related to the article by connecting to the web site listed on this page.

Chemistry Serving...Industry

DIVING CAN BE A GAS

A lot of important work goes on under the ocean's surface, and much of it is done by commercial scuba divers. Divers repair and inspect ships, help build offshore oil platforms, salvage sunken ships, and repair undersea cables and pipelines.

Professional divers often have to do their work at great depths, where the water exerts extreme pressure. How can divers journey safely into this foreign world? Only by bringing with them special breathing equipment and mixtures of gases can they compensate for the effects of high pressure.

If you have ever dived below the surface, you are aware of the pressure exerted by water. The pressure is high enough, even a few meters under water, that if you tried to breathe through a hose connected to the surface you would not be able to. Your lungs would be unable to expand. So how are scuba divers able to breathe at even greater depths?

Divers can breathe deep under water because the gases they take with them are under pressure.

Divers can breathe deep under water because the gases they take with them are under pressure. A device attached to a diver's air tanks, called a regulator, automatically adjusts the pressure of the air coming out of the tanks and into the diver's lungs. The regulator works to equalize the pressure inside and outside the lungs.

Regulating the pressure of the air they breathe, however, is not divers' main concern. At the high pressures that exist even at moderate depths, more gas can be dissolved in divers' blood than can be dissolved at normal pressures. Any gas present in divers' air tanks, therefore, dissolves in their blood in greater-than-normal concentrations.

The main component of air is molecular nitrogen (N_2). When dissolved nitrogen reaches high concentrations in the blood, it can cause two different problems. The first is nitrogen narcosis, a condition with effects similar to those produced by ingesting too much alcohol. The second problem arises only when the diver begins to ascend toward the surface. As the pressure is reduced, the nitrogen dissolved in the blood tends to come out of solution and to form tiny bubbles. If this happens, it causes a condition called the bends, which is extremely painful and potentially fatal. Divers prevent the bends by gradually letting the nitrogen come out of the bloodstream at stops during the ascent.

For deep dives, it is especially important to take measures to prevent nitrogen narcosis and the bends. For these dives, commercial divers often use air tanks containing a special blend of gases instead of pressurized air.

One of these blends is a mixture of nitrogen and oxygen, called Nitrox, which has less nitrogen than pressurized air. By breathing Nitrox, a diver reduces the potential for nitrogen-related problems. For very deep dives, a mixture of helium and oxygen, called Heliox, is often used. This mixture eliminates nitrogen's negative effects, but it can lead to high-pressure nervous syndrome, in which the diver shakes uncontrollably. A third combination of gases, called Trimix, is used for extremely deep dives. Trimix is Heliox with a small amount of nitrogen added to prevent the shaking problem.

COMMERCIAL DIVER

Are you good at working under pressure? Consider being a commercial diver.

See page 873.

CHEMIST NEEDED

Enviro. Lab. GC Must know

TAKE IT TO THE NET

Find out more about career opportunities: www.phschool.com

CHEMICAL SPECIALIST

Local food service distributor seek responsible self-motivated indivic

Chapter 12 STUDENT STUDY GUIDE

Take It to the NET
For interactive study and review, go to www.phschool.com

KEY TERMS

- Avogadro's hypothesis *p. 347*
- Boyle's law *p. 333*
- Charles's law *p. 336*
- combined gas law *p. 339*
- compressibility *p. 327*
- Dalton's law of partial pressures *p. 350*
- diffusion *p. 352*
- effusion *p. 352*
- Gay-Lussac's law *p. 338*
- Graham's law of effusion *p. 352*
- ideal gas constant (*R*) *p. 341*
- ideal gas law *p. 341*
- partial pressure *p. 350*

KEY EQUATIONS

- Boyle's law: $P_1 \times V_1 = P_2 \times V_2$
- Charles's law: $\frac{V_1}{T_1} = \frac{V_2}{T_2}$
- Gay-Lussac's law: $\frac{P_1}{T_1} = \frac{P_2}{T_2}$
- Combined gas law: $\frac{P_1 \times V_1}{T_1} = \frac{P_2 \times V_2}{T_2}$
- Ideal gas law: $P \times V = n \times R \times T$
- Dalton's law: $P_{total} = P_1 + P_2 + P_3 + \ldots$
- Graham's law: $\frac{\text{Rate}_A}{\text{Rate}_B} = \frac{\sqrt{\text{molar mass}_B}}{\sqrt{\text{molar mass}_A}}$

CONCEPT SUMMARY

12.1 The Properties of Gases

- Kinetic molecular theory can be used to explain gas pressure, volume, and temperature.
- The average kinetic energy of a collection of gas particles is directly proportional to the Kelvin temperature of the gas.

12.2 Factors Affecting Gas Pressure

- The collision of gas particles with the walls of the container constitutes gas pressure.
- In general, increasing the volume of a container decreases gas pressure; decreasing the volume of a container increases gas pressure.
- In general, increasing the temperature of a contained gas increases its pressure; decreasing the temperature decreases its pressure.

12.3 The Gas Laws

- The pressure and volume of a fixed mass of gas are inversely related (Boyle's law).
- The volume of a gas at constant pressure is directly related to its Kelvin temperature (Charles's law).
- The pressure of a fixed volume of gas is directly related to its Kelvin temperature (Gay-Lussac's law).

12.4 Ideal Gases

- The ideal gas law relates the moles of a gas to its pressure, temperature, and volume.
- Real gases differ from ideal gases because intermolecular forces tend to reduce the distance between real gas particles and because real gas particles have volume.

12.5 Gas Molecules: Mixtures and Movements

- Avogadro stated that equal volumes of gases, at the same temperature and pressure, contain an equal number of particles.
- The total pressure in a mixture of gases is equal to the sum of the partial pressures of each gas present (Dalton's law).
- Gases diffuse from a region of high gas concentration to one of lower concentration. The smaller the molar mass of a gas, the greater its rate of diffusion (Graham's law).

CHAPTER CONCEPT MAP

Use these terms to construct a concept map that organizes the major ideas of this chapter.

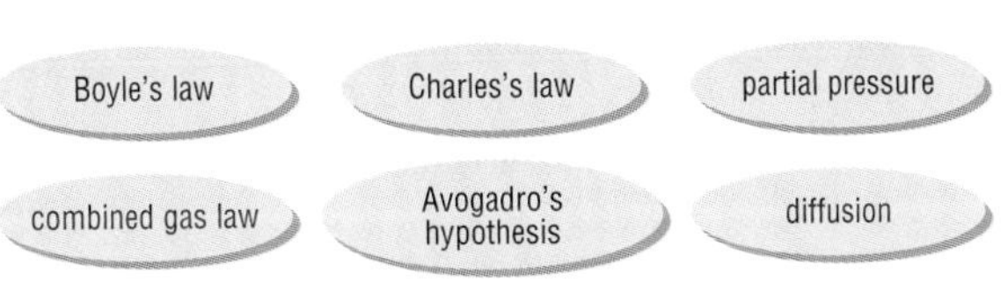

Chem ASAP! Concept Map 12
Create your Concept Map using the computer.

Take It to the Net

At **www.phschool.com** students will find for this chapter

- an Internet activity
- links to related chemistry sites
- an interactive quiz
- career links

4 Close

Summary

Ask the following questions that require students to summarize information contained in the chapter.

- **What assumptions does the kinetic theory make about gas particles?** (They have negligible volume, do not attract or repel each other, and are in constant, random, straight-line motion.)
- **State Boyle's law and Charles's law.** (Boyle's: $P_1V_1 = P_2V_2$; Charles's: $\frac{V_1}{T_1} = \frac{V_2}{T_2}$)
- **Rank the relative rates of diffusion at a constant temperature for carbon dioxide, helium, and nitrogen.** (Rates of diffusion are: helium > nitrogen > carbon dioxide.)

Extension

Have students research how the gas laws and kinetic theory figure into space shuttle flights. They should prepare a report on their findings.

Looking Back . . . Looking Ahead . . .

The previous chapter discussed exothermic and endothermic reactions and the use of a calorimeter to measure entropy changes in chemical reactions. The current chapter deals with energy and temperature as they relate to kinetic theory and the behavior of gases. The next chapter discusses how quantum theory is used to describe the structure of the atom.

Chapter 12
REVIEW

Answers

45. The increased kinetic energy of the gas particles causes collisions to occur with more force.
46. The gas particles become closer together.
47. The pressure doubles.
48. The pressure quadruples.
49. Temperatures measured on the Kelvin scale are always positive and directly proportional to the average kinetic energy of the gaseous particles.
50. The volume decreases. The molecules have less kinetic energy, which causes less pressure on the inside of the balloon.
51. 1.00×10^2 kPa
52. 1.80 L
53. 1.8×10^1 L
54. 846 K (573 °C)
55. High temperatures can sufficiently increase the pressure of the gas remaining in the container to cause it to explode.
56. $\frac{P_1 \times V_1}{T_1} = \frac{P_2 \times V_2}{T_2}$
57. 1.10×10^3 kPa
58. $\frac{P_1 \times V_1}{T_1} = \frac{P_2 \times V_2}{T_2}$
    When the pressure is constant, $P_1 = P_2$, so the pressure terms cancel, leaving the equation for Charles's law.
59. Gas particles have a finite volume and are attracted to one another, especially at low temperatures.
60. Its particles have no volume, no forces between them, and elastic collisions. It follows the gas laws at all temperatures and pressures.
61. At low temperatures, gas particles are attracted to one another; the finite volume of gas particles is significant at high pressures.
62. 3.60×10^2 kPa
63. **a.** 5.6×10^1 L **b.** 6.7 L **c.** 7.84 L
64. equal numbers of particles

Chapter 12 REVIEW

CONCEPT PRACTICE

45. Heating a contained gas that is held at a constant volume increases its pressure. Why? *12.1*
46. What happens to gas particles when a gas is compressed? *12.1*
47. A metal cylinder contains 1 mol of nitrogen gas at STP. What will happen to the pressure if another mole of gas is added to the cylinder, but the temperature and volume do not change? *12.2*
48. If a gas is compressed from 4 L to 1 L and the temperature remains constant, what happens to the pressure? *12.2*
49. Why is Kelvin temperature specified in calculations that involve gases? *12.2*
50. Describe what happens to the volume of a balloon when it is taken outside on a cold winter day. Explain why this happens. *12.2*
51. The gas in a closed container has a pressure of 3.00×10^2 kPa at 30 °C (303 K). What will the pressure be if the temperature is lowered to −172 °C (101 K)? *12.3*
52. Calculate the volume of a gas (in L) at a pressure of 1.00×10^2 kPa if its volume at 1.20×10^2 kPa is 1.50×10^3 mL. *12.3*
53. A gas with a volume of 4.0 L at 90.0 kPa expands until the pressure drops to 20.0 kPa. What is the new volume if the temperature remains constant? *12.3*
54. A gas with a volume of 3.00×10^2 mL at 150.0 °C is heated until its volume is 6.00×10^2 mL. What is the new temperature of the gas if the pressure remains constant during the heating process? *12.3*
55. Why do aerosol containers display the warning, "Do not incinerate"? *12.3*
56. State the combined gas law. 12.3
57. A sealed cylinder of gas contains nitrogen gas at 1.00×10^3 kPa pressure and a temperature of 20 °C. The cylinder is left in the sun, and the temperature of the gas increases to 50 °C. What is the new pressure in the cylinder? *12.3*
58. Show how Charles's law can be derived from the combined gas law. *12.3*
59. Explain why it is impossible for an ideal gas to exist. *12.4*
60. Describe an ideal gas. *12.4*
61. Explain the reasons why real gases deviate from ideal behavior. *12.4*
62. If 4.50 g of methane gas (CH_4) is introduced into an evacuated 2.00-L container at 35 °C, what is the pressure in the container? *12.4*
63. Calculate the number of liters occupied at STP. *12.5*
    a. 2.5 mol $N_2(g)$
    b. 0.600 g $H_2(g)$
    c. 0.350 mol $O_2(g)$
64. How would the number of particles of two gases compare if their partial pressures in a container were identical? *12.5*
65. Which gas effuses faster: hydrogen or chlorine? How much faster? *12.5*
66. Which gas effuses faster at the same temperature: molecular oxygen or atomic argon? *12.5*
67. Calculate the ratiò of the velocity of helium atoms to the velocity of neon atoms at the same temperature. *12.5*
68. Calculate the ratio of the velocity of helium atoms to fluorine molecules at the same temperature. *12.5*

CONCEPT MASTERY

69. What can you conclude about the nature of the relationship between two variables with a quotient that is a constant?
70. A certain gas effuses four times as fast as oxygen (O_2). What is the molar mass of the gas?
71. A 3.50-L gas sample at 20 °C and a pressure of 86.7 kPa expands to a volume of 8.00 L. The final pressure of the gas is 56.7 kPa. What is the final temperature of the gas, in degrees Celsius?
72. During an effusion experiment, it took 75 seconds for a certain number of moles of an unknown gas to pass through a tiny hole. Under the same conditions, the same number of moles of oxygen gas passed through the hole in 30 seconds. What is the molar mass of the unknown gas?

CRITICAL THINKING

73. Choose the term that best completes the second relationship.

a. ideal gas : real gas
fiction : _______
(1) biography
(2) novel
(3) movie
(4) nonfiction

b. Charles's law : temperature
Boyle's law : _______
(1) pressure
(2) volume
(3) ideal mass
(4) mass

c. volume : Charles's law
pressure : _______
(1) Boyle's law
(2) combined gas law
(3) Gay-Lussac's law
(4) temperature

d. inverse relationship : Boyle's law
direct relationship : _______
(1) absolute zero
(2) Avagadro's hypothesis
(3) Charles's law
(4) ideal gas law

e. kelvins : degrees Celsius
kilopascals : _______
(1) atmospheric pressure
(2) atmospheres
(3) pressure
(4) absolute zero

74. Gases will expand to fill a vacuum. Why do Earth's atmospheric gases not escape into the near-vacuum of space?

75. How does the vacuum used in Thermos™ bottles prevent heat transfer?

76. What real gas comes closest to having the characteristics of an ideal gas? Why?

CUMULATIVE REVIEW

77. Calculate the molar mass of each substance.
a. $Ca(CH_3CO_2)_2$
b. H_3PO_4
c. $C_{12}H_{22}O_{11}$
d. $Pb(NO_3)_2$

78. Name each compound.
a. $SnBr_2$
b. $BaSO_4$
c. $Mg(OH)_2$
d. IF_5

79. An atom of lead-206 weighs 17.16 times as much as an atom of carbon.
a. What is the molar mass of this isotope of lead?
b. How many protons, electrons, and neutrons are in this atom of lead?

80. How many kilojoules and kilocalories of heat are required to raise 40.0 g of water from −12 °C to 130 °C?

81. Write a balanced equation for each chemical reaction.
a. Calcium reacts with water to form calcium hydroxide and hydrogen gas.
b. Tetraphosphorus decoxide reacts with water to form phosphoric acid.
c. Mercury and oxygen are prepared by heating mercury(II) oxide.
d. Aluminum hydroxide and hydrogen sulfide form when aluminum sulfide reacts with water.

82. Classify each of the reactions in Problem 81 based on type.

83. Calculate the molecular formula of each of the following compounds.
a. The empirical formula is C_2H_4O and the molar mass = 88 g/mol.
b. The empirical formula is CH and the molar mass = 104 g/mol.
c. The molar mass = 90 g/mol. The percent composition is 26.7% C, 71.1% O, and 2.2% H.

84. A piece of metal has a mass of 9.92 g and measures 4.5 cm × 1.3 cm × 1.6 mm. What is the density of the metal?

85. Calculate the percent composition of 2-propanol (C_3H_7OH).

Chapter 12 REVIEW

65. At any temperature, hydrogen gas diffuses faster than chlorine gas by an approximate factor of six.
66. oxygen
67. 2.25:1
68. 3.08
69. The variables are directly proportional.
70. 2.0 g
71. 165 °C
72. 2.0×10^2 g
73. **a.** (4)
b. (1)
c. (1)
d. (3)
e. (2)
74. The gases that make up the atmosphere, just like any other form of matter, are held near Earth by the force of gravity.
75. A vacuum contains no matter to allow the transfer of kinetic energy between molecules.
76. Helium gas; it is composed of small, monatomic atoms with little attraction for each other.
77. **a.** 1.58×10^2 g
b. 9.80×10^1 g
c. 3.42×10^2 g
d. 3.31×10^2 g
78. **a.** tin(II) bromide
b. barium sulfate
c. magnesium hydroxide
d. iodine pentafluoride
79. **a.** 2.06×10^2 g
b. 82 protons, 82 electrons, 124 neutrons
80. 1.24×10^2 kJ; 2.96×10^1 kcal
81. **a.** $Ca + 2H_2O \rightarrow Ca(OH)_2 + H_2$
b. $P_4O_{10} + 6H_2O \rightarrow 4H_3PO_4$
c. $2HgO \rightarrow 2Hg + O_2$
d. $Al_2S_3 + 6H_2O \rightarrow 2Al(OH)_3 + 3H_2S$
82. **a.** single-replacement
b. combination
c. decomposition
d. double-replacement
83. **a.** $C_4H_8O_2$
b. C_8H_8
c. $C_2H_2O_4$
84. 11 g/cm^3
85. 60.0% C, 13.3% H, 26.7% O

86. Aluminum oxide is formed from its elements.

$$Al(s) + O_2(g) \longrightarrow Al_2O_3(s)$$

a. Balance the equation.
b. How many grams of each reactant are needed to form 583 g $Al_2O_3(s)$?

CONCEPT CHALLENGE

87. Oxygen is produced in the laboratory by heating potassium nitrate (KNO_3). Use the data table below, which gives the volume of oxygen produced at STP from varying quantities of KNO_3, to determine the mole ratio by which KNO_3 and O_2 react.

Mass of KNO_3 (g)	Volume of O_2 (cL)
0.84	9.3
1.36	15.1
2.77	30.7
4.82	53.5
6.96	77.3

88. The following reaction takes place in a sealed 40.0-L container at a temperature of 120 °C.

$$4NH_3(g) + 5O_2(g) \longrightarrow 4NO(g) + 6H_2O(g)$$

a. When 34.0 g $NH_3(g)$ reacts with 96.0 g $O_2(g)$, what is the partial pressure of $NO(g)$ in the container?
b. What is the total pressure in the container?

89. A mixture of ethyne (C_2H_2)(g) and methane (CH_4)(g) occupied a certain volume at a total pressure of 16.8 kPa. Upon burning the sample to form $CO_2(g)$ and $H_2O(g)$, the $CO_2(g)$ was collected and its pressure found to be 25.6 kPa in the same volume and at the same temperature as the original mixture. What percentage of the original mixture was methane?

90. A 0.10-L container holds 3.0×10^{20} molecules H_2 at 100 kPa and 0 °C.
a. If the volume of a hydrogen molecule is 6.7×10^{-24} mL, what percentage of the volume of the gas is occupied by its molecules?
b. If the pressure is increased to 100 000 kPa, the volume of the gas is 1×10^{-4} L. What fraction of the total volume do the hydrogen molecules now occupy?

91. Many gases that have small molecules, such as $N_2(g)$ and $O_2(g)$, have the expected molar volume of 22.41 L at STP. However, other gases behave in a very nonideal manner, even if extreme pressures and temperatures are not involved. The molar volumes of $CH_4(g)$, $CO_2(g)$, and $NH_3(g)$ at STP are 22.37 L, 22.26 L, and 22.06 L, respectively. Explain the reasons for these large departures from the ideal.

Answers

86. a. $4Al + 3O_2 \rightarrow 2Al_2O_3$
b. 3.09×10^2 g Al; 2.74×10^2 g O_2

87. 2 mol KNO_3 for each 1 mol O_2

88. a. 1.63×10^2 kPa
b. 4.48×10^2 kPa

89. 46% CH_4

90. a. 2.0×10^{-3}%
b. 2.0%

91. Because attractions between molecules in gases such as nitrogen and oxygen are insignificant, these gases have the molar volume of an ideal gas—22.41 L at STP. Based on their molar volumes at STP, attractions between molecules increase in strength and effect from methane to carbon dioxide to ammonia.

Chapter 12 STANDARDIZED TEST PREP

Select the choice that best answers each question or completes each statement.

1. A gas in a container has a volume of 120.0 mL at −123 °C. What volume does this gas occupy at 27.0 °C?
a. 60.0 mL **c.** 26.5 mL
b. 240.0 mL **d.** 546 mL

2. If the Kelvin temperature of a gas is tripled and the volume is doubled, the new pressure will be
a. 1/6 the original pressure.
b. 2/3 the original pressure.
c. 3/2 the original pressure.
b. 5 times the original pressure.

3. Which of these gases effuses fastest?
a. Cl_2 **c.** NH_3
b. NO_2 **d.** N_2

4. All the oxygen gas from a 10.0-L container at a pressure of 202 kPa is added to a 20.0-L container of hydrogen at a pressure of 505 kPa. After the transfer, what are the partial pressures of oxygen and hydrogen?
a. oxygen is 101 kPa; hydrogen is 505 kPa
b. oxygen is 202 kPa; hydrogen is 505 kPa
c. oxygen is 101 kPa; hydrogen is 253 kPa
d. oxygen is 202 kPa; hydrogen is 253 kPa

5. Which of the following changes would increase the pressure of a gas in a closed container?
I. Part of the gas is removed.
II. The container size is decreased.
III. Temperature is increased.
a. I and II only
b. II and III only
c. I and III only
d. I, II, and III

6. A real gas behaves most nearly like an ideal gas
a. at high pressure and low temperature.
b. at low pressure and high temperature.
c. at low pressure and low temperature.
d. at high pressure and high temperature.

Use this description to answer questions 7 and 8.

A teacher adds a 1 mL of water to an empty metal soda can. The teacher heats the can over a burner and then quickly plunges it upside down in an ice-water bath. The can immediately collapses inward as though crushed in a trash compactor.

7. Use kinetic theory to explain why the can collapsed inward.

8. If the experiment were done with a dry can, would the results be similar? Explain.

Use the graphs to answer questions 9-12. A graph may be used once, more than once, or not at all.

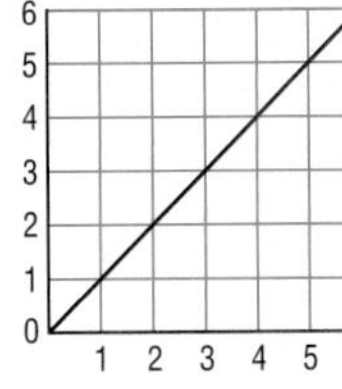

a.

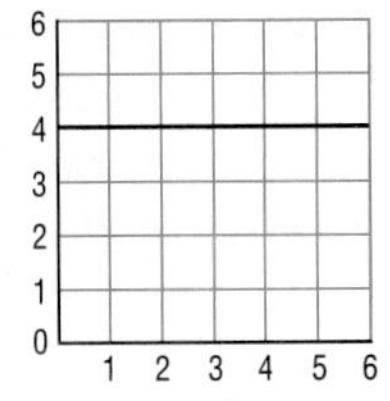

b.

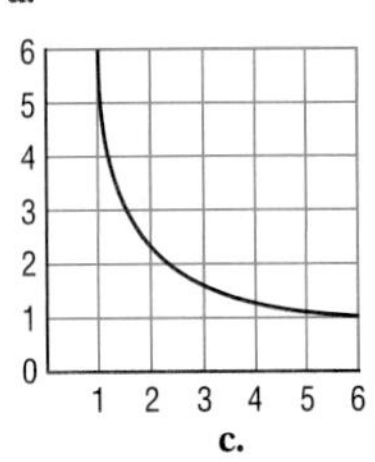

c.

Which graph shows each of the following?

9. directly proportional relationship
10. graph with slope = 0
11. inversely proportional relationship
12. graph with a constant slope

Use the drawing to answer questions 13 and 14.

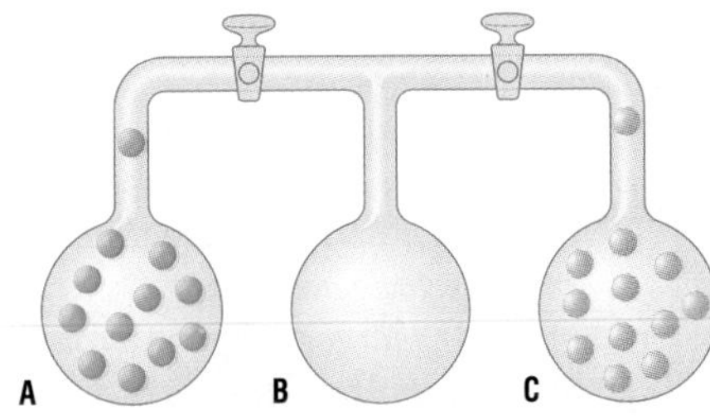

13. Bulbs A and C contain different gases. Bulb B contains no gas. If the valves between the bulbs are opened, how will the particles of gas be distributed when the system reaches equilibrium? For this exercise, assume none of the particles are in the tubes that connect the bulbs.

14. *You Draw it!* Make a similar three-bulb drawing. Place 6 blue spheres in bulb A, 9 green spheres in bulb B, and 12 red spheres in bulb C. Then draw the setup to represent the distribution of gases after the valves are opened and the system reaches equilibrium.

STANDARDIZED TEST PREP

1. b
2. c
3. c
4. a
5. b
6. b
7. Boiling the water fills the can with steam. When the can is plunged upside down into the cold water the steam is trapped and quickly condenses, reducing gas pressure inside the can. Because the sides of the can are not very strong, the comparatively high atmospheric pressure crushes the can.
8. The results would be much less dramatic. The change in volume of heated air (and internal pressure) is much less than when steam condenses to a liquid.
9. a
10. b
11. c
12. a and b
13. There will be four green and four red spheres in each bulb.
14. Each bulb will contain 2 blue, 3 green, and 4 red spheres.

Planning Guide

SECTION OBJECTIVES	ACTIVITIES/FEATURES	MEDIA & TECHNOLOGY
13.1 Models of the Atom ● ■ ◆ ▸ Summarize the development of atomic theory ▸ Explain the significance of quantized energies of electrons as they relate to the quantum mechanical model of the atom	**SE** **Discover It!** *Observing Light Emission from Wintergreen Mints*, p. 360 **TE** **DEMO**, p. 362	**ASAP** Animation 15 **ASAP** Assessment 13.1 **ACT** 9: *Electron Configurations* **OT** 34: *Development of Atomic Models* **OT** 35: *Atomic Orbitals* **OT** 36: *Atomic Orbitals*
13.2 Electron Arrangement in Atoms ○ ■ ◆ ▸ Apply the aufbau principle, the Pauli exclusion principle, and Hund's rule in writing the electron configurations of elements ▸ Explain why the electron configurations for some elements differ from those assigned using the aufbau principle	**SE** **Small-Scale Lab** *Electron Configurations of Atoms and Ions*, p. 371 (**LRS** 13-1)	**ASAP** Simulation 12 **ASAP** Problem Solving 6 **ASAP** Assessment 13.2 **ACT** 9: *Electron Configurations* **OT** 37: *Electron Configurations* **CHM** Side 2, 23: *Electron Orbital Shape*
13.3 Physics and the Quantum Mechanical Model ○ □ ◆ ▸ Calculate the wavelength, frequency, or energy of light, given two of these values ▸ Explain the origin of the atomic emission spectrum of an element	**SE** **Link to Astronomy** *The Discovery of Helium*, p. 376 **SE** **CHEMath** *Constants*, p. 378 **SE** **Mini Lab** *Flame Tests*, p. 383 (**LRS** 13-2) **SE** **Chemistry Serving . . . Society** *Lasers at Work*, p. 384 **SE** **Chemistry in Careers** *Laser Technician*, p. 384 **LM** 20: *Flame Tests for Metals* **LM** 21: *Introduction to the Spectrophotometer* **LM** 22: *Energies of Electrons* **SSLM** 15: *Design and Construction of a Quantitative Spectroscope* **SSLM** 16: *Visible Spectra and the Nature of Light and Color*	**ASAP** Simulation 13, 14 **ASAP** Animation 16 **ASAP** Problem Solving 12, 14 **ASAP** Assessment 13.3 **ASAP** Concept Map 13 **ACT** 9: *Electron Configurations* **SSV** 10: *Small-Scale Spectroscope* **CHM** Side 2, 10: *Neon Lights* **CHM** Side 2, 34: *Spectra of Various Salts* **OT** 38: *Waves* **OT** 39: *Emission Spectrum of an Element* **OT** 40: *The Photoelectric Effect* **OT** 41: *Hydrogen Spectral Lines* **PP** Chapter 13 Problems **RP** Lesson Plans, Resource Library **AR** Computer Test 13 **www** Activities, Self-Tests, *SCIENCE NEWS* updates **TCP** The Chemistry Place Web Site

KEY

● Conceptual (concrete concepts)	**AR** Assessment Resources	**GRS** Guided Reading and Study Workbook
○ Conceptual (more abstract/math)	**ASAP** Chem ASAP! CD-ROM	**LM** Laboratory Manual
■ Standard (core content)	**ACT** ActivChemistry CD-ROM	**LP** Laboratory Practicals
□ Standard (extension topics)	**CHM** CHEMedia Videodiscs	**LRS** Laboratory Recordsheets
◆ Honors (core content)	**CA** Chemistry Alive! Videodiscs	**SSLM** Small-Scale Lab Manual
◇ Honors (options to accelerate)	**GCP** Graphing Calculator Problems	

PRACTICE / ASSESSMENT

PRACTICE	ASSESSMENT
GRS Section 13.1 **RM** Practice Problems 13.1	**SE** Section Review **RM** Section Review 13.1 **RM** Chapter 13 Quiz
SE **Sample Problem** 13-1 **GRS** Section 13.2 **RM** Practice Problems 13.2	**SE** Section Review **RM** Section Review 13.2 **RM** Chapter 13 Quiz
SE **Sample Problems** 13-2 to 13-3 **GRS** Section 13.3 **RM** Practice Problems 13.3 **RM** Interpreting Graphics	**SE** Section Review **RM** Section Review 13.3 **LP** Lab Practicals 13-1, 13-2 **RM** Vocabulary Review 13 **SE** Chapter Review **SM** Chapter 13 Solutions **SE** Standardized Test Prep **PHAS** Chapter 13 Test Prep **RM** Chapter 13 Quiz **RM** Chapter 13 A & B Test

OT	Overhead Transparency	**SE**	Student Edition
PHAS	PH Assessment System	**SM**	Solutions Manual
PLM	Probeware Lab Manual	**SSV**	Small-Scale Video/Videodisc
PP	Problem Pro CD-ROM	**TCP**	www.chemplace.com
RM	Review Module	**TE**	Teacher's Edition
RP	Resource Pro CD-ROM	**www**	www.phschool.com

PLANNING FOR ACTIVITIES

STUDENT EDITION

Discover It! p. 360
- wintergreen mints, 3 brands
- pliers
- transparent tape
- dark room

Small-Scale Lab, p. 371
- solutions of NaCl, $MgSO_4$, $AlCl_3$, $FeCl_3$, $CaCl_2$, $NiSO_4$, $CuSO_4$, $ZnCl_2$, $AgNO_3$, NaOH, Na_2CO_3
- pencils, paper, rulers
- reaction surfaces

Mini Lab, p. 383
- Bunsen burners
- tongs
- small beakers
- small test tubes
- test tube racks
- flame test wires
- solutions of HC1, NaCl, $CaCl_2$, LiCl, $CuCl_2$, $BaCl_2$

TEACHER'S EDITION

Teacher Demo, p. 362
- several musical instruments brought in by students

Activity, p. 372
- several sealed boxes containing different common items

Activity, p. 377
- list of physicists who contributed to modern theories concerning light and atoms

Chapter 13

Key Terms

13.1 energy level, quantum, quantum mechanical model, atomic orbitals
13.2 electron configurations, aufbau principle, Pauli exclusion principle, Hund's rule
13.3 electromagnetic radiation, amplitude, wavelength (λ), frequency (ν), hertz (Hz), spectrum, atomic emission spectrum, Planck's constant *(h)*, photons, photoelectric effect, ground state, de Broglie's equation, Heisenberg uncertainty principle

DISCOVER IT!

Crushing the mint creates an unequal division of electrons: positive charge accumulates on one piece, negative charge on the other. When an electric spark jumps between them, nitrogen atoms in the air absorb the energy, emitting it as ultraviolet light. The wintergreen flavor molecules absorb part of this energy and release it as visible blue-green light. All wintergreen mints should emit light. To see the light, students must allow enough time for their eyes to adapt to the dark room.

Chapter 13 ELECTRONS IN ATOMS

A "neon" sign's colors depend on the gases in its light tubes.

FEATURES

DISCOVER IT! OBSERVING LIGHT EMISSION FROM WINTERGREEN MINTS

You need 3 different brands of wintergreen mints, a pair of pliers, transparent tape, and a dark room.

1. Break each mint in half.
2. Wrap the jaws of the pliers with transparent tape.
3. Turn off the lights and allow your eyes to adjust to the darkness.
4. Watch the exposed edge of the mint as you carefully crush it between the jaws of the pliers. Note the color and brightness of any emitted light.
5. Repeat Step 4 for the other two mints.

What did you observe when you crushed the mints? Did all the mints emit light? Propose an explanation for your observations. After completing this chapter, return to this activity to revise and expand your explanation.

Stay current with **SCIENCE NEWS**
Find out more about electrons in atoms:
www.phschool.com

13.1 ● ■ ◆
13.2 ○ ■ ◆
13.3 ○ □ ◆

Conceptual In section 13.1, the discussion of atomic orbitals may be limited; Figures 13.4 and 13.5 will help students to visualize orbitals. The rules for determining electron configurations are challenging, but Section 13.2 is important for understanding the periodic table in Chapter 14. Exceptional electron configurations may be omitted. Other than the electromagnetic spectrum, topics in Section 13.3 may be omitted.

Standard Have students reread Section 5.2, the Structure of the Nuclear Atom, before starting this chapter. Use Figure 13.2 to review the historical development of the atomic model. Quantum mechanics can be omitted if not included in course requirements.

Honors Section 13.3 includes many abstract concepts and will be challenging even for honors students. Use the flame-test Mini Lab to illustrate and synthesize chapter concepts.

MODELS OF THE ATOM

Aeronautical engineers use wind tunnels and scale models to simulate and test the aerodynamic drag forces that will act on a proposed design. The scale model shown is a physical model. However, not all models are physical. In fact, several theoretical models of the atom have been developed over the past several hundred years. **What is the current model of the atom, and how was it developed?**

objectives
- Summarize the development of atomic theory
- Explain the significance of quantized energies of electrons as they relate to the quantum mechanical model of the atom

key terms
- energy level
- quantum
- quantum mechanical model
- atomic orbitals

The Evolution of Atomic Models

Thus far, the atomic model presented in this textbook has considered atoms as combinations of protons and neutrons making up a nucleus that is surrounded by electrons. Although this model has worked very well, it has outlived its usefulness because it explains only a few simple properties of atoms. It does not explain, for example, why metals or compounds of metals give off characteristic colors when heated in a flame. A more sophisticated atomic model is needed. As it turns out, the chemical properties of atoms, ions, and molecules are related to the arrangement of the electrons within them. Therefore, in this chapter, models of atomic structure will be expanded, with an emphasis on the electrons in atoms.

For about 50 years past the time of John Dalton (1766–1844), the atom was considered a solid indivisible mass. Dalton's atomic theory was a great advance in explaining the nature of chemical reactions. However, the discovery of subatomic particles shattered every theory scientists had about indivisible atoms.

The discoverer of the electron, J. J. Thomson (1856–1940), realized that the accepted model of an indivisible atom did not take electrons into account. Thomson, therefore, proposed a revised model, referred to as the plum-pudding atom. **Figure 13.2** on the following page shows this and other proposed models of the atom. The plum-pudding atom had negatively charged electrons stuck into a lump of positively charged material, similar to raisins stuck in dough. The plum-pudding model explained some electrical properties of atoms. It said nothing, however, about the number of protons and electrons, their arrangements in the atom, or the ease with which atoms are stripped of electrons to form ions.

Figure 13.1
Each of these scientists contributed to the development of atomic models.

John Dalton
1766–1844

J. J. Thomson
1856–1940

Ernest Rutherford
1871–1937

Niels Bohr
1885–1962

13.1

1 Engage

Use the Visual

Have students study the photograph and read the text that opens the section. Discuss the value of computer models for designing cars, crash test dummies, and scale models for making movies such as *Titanic*. Ask:

- **What key property of atoms has limited the types of atomic models that scientists have proposed?** (Because atoms are not visible to the unaided eye, scientists have inferred the structure of atoms from observed behaviors.)
- **What is the current model of the atom, and how was it developed?** (The current model describes a dense, positively charged nucleus surrounded by electrons in constant motion. The model was developed to explain experimental data.)

Motivate and Relate

Pick a manufactured item (automobile, computer, or television) and discuss its evolution over time. Compare this evolution to the development of the atomic model, in particular the role of scientific discoveries (semiconductors or subatomic particles) and new technology (assembly lines/robots or particle accelerators/electron microscopes).

STUDENT RESOURCES

From the Teacher's Resource Package, use:

- Section Review 13.1, Ch. 13 Practice Problems and Quizzes from the Review Module (Ch. 13–16)

TECHNOLOGY RESOURCES

Relevant technology resources include:

- Chem ASAP! CD-ROM
- ResourcePro CD-ROM
- ActivChemistry CD-ROM
  Electron Configurations

Check Prior Knowledge

To assess students' knowledge of atomic structure, ask:

- **What are the three major subatomic particles that comprise atoms?** (the electron, proton, and neutron)
- **What are the relative electrical charges associated with each of these particles?** (electron: −1; proton: +1; neutron: 0)
- **Describe the structure of the nuclear atom in terms of the locations of each of the subatomic particles.** (Every atom consists of a small, extremely dense nucleus, the central, positively charged core of an atom, composed of protons and neutrons (except for hydrogen−1 which contains only a single proton in its nucleus) which accounts for most of the mass of an atom. The negatively charged electrons surround the nucleus and occupy most of the volume of an atom, but contribute very little to the mass of an atom.)

2 Teach

TEACHER DEMO

A musical instrument, such as a trumpet or trombone, can be used to demonstrate the concept of quantized energy. Have a student blow into a trumpet. Challenge the student to elicit as many different notes as he or she can without depressing any valves. It is not possible to play the entire scale without changing the valve positions. In the open position, no matter how much energy is put in, the instrument will only accept certain specific amounts of energy and produce certain notes. The instrument accepts quantized packages of energy.

Figure 13.2
These illustrations show how the atomic model has changed as scientists have learned more about the atom's structure. What would a picture of Dalton's model of the atom look like? ❷

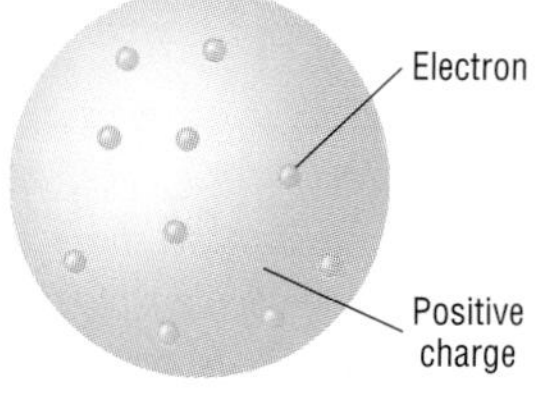

Thomson model
In the nineteenth century, Thomson described the atom as a ball of positive charge containing a number of electrons.

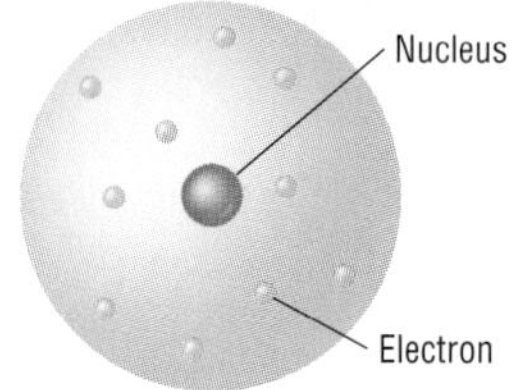

Rutherford model
In the early twentieth century, Rutherford showed that most of an atom's mass is concentrated in a small, positively charged region called the nucleus.

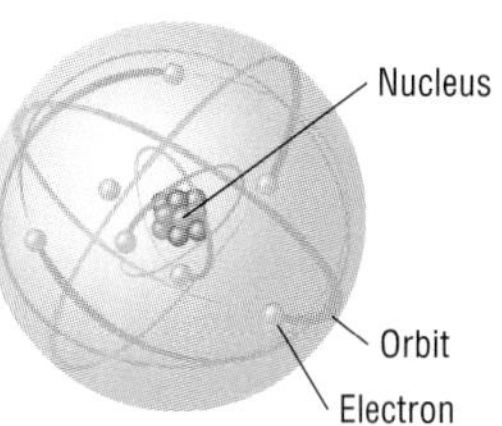

Bohr model
After Rutherford's discovery, Bohr proposed that electrons travel in definite orbits around the nucleus.

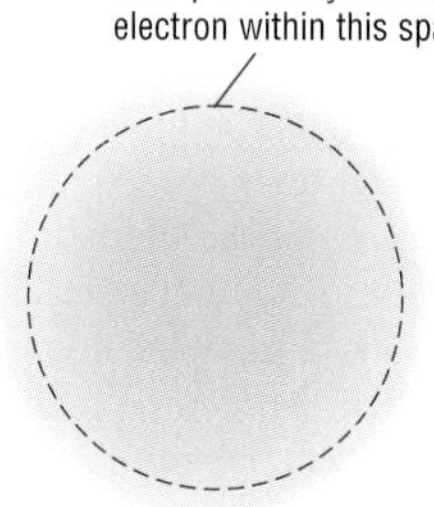

Quantum mechanical model
Modern atomic theory describes the electronic structure of the atom as the probability of finding electrons within certain regions of space.

Based on his discovery of the nucleus, Ernest Rutherford (1871–1937) proposed the nuclear atom, in which electrons surround a dense nucleus. He thought of the rest of the atom as empty space. Later experiments showed that the nuclei of atoms are composed of protons and neutrons. ❶ What kind of charge does the nucleus of every atom carry?

As you know, oppositely charged particles attract each other. Thus one could argue that the negative electrons should be drawn into a positive nucleus, causing the atom to collapse. However, this does not occur.

In 1913, Niels Bohr (1885–1962), a young Danish physicist and a student of Rutherford's, came up with a new atomic model. He proposed that electrons are arranged in concentric circular paths, or orbits, around the nucleus. This model, patterned after the motions of the planets around the sun, is often referred to as the planetary model. Bohr answered in a novel way the question of what prevents electrons from falling into the nucleus. He proposed that the electrons in a particular path have a fixed energy; the electrons do not lose energy and cannot fall into the nucleus. The **energy level** of an electron is the region around the nucleus where the electron is likely to be moving.

The fixed energy levels of electrons are analogous to the rungs of a ladder. The lowest rung of the ladder corresponds to the lowest energy level. A person can climb up or down a ladder by going from rung to rung. Similarly, an electron can jump from one energy level to another. A person on a ladder cannot stand between the rungs. Similarly, the electrons in an atom

Answers

❶ positive
❷ solid sphere
❸ Answers will vary, but may include the idea that the electrons are farther away from the positively charged nucleus.
❹ 2 in 3, or 66%

cannot exist between energy levels. To move from one rung to another, a person climbing a ladder must move just the right distance. To move from one energy level to another, an electron must gain or lose just the right amount of energy.

A **quantum** of energy is the amount of energy required to move an electron from its present energy level to the next higher one. The energies of electrons are said to be quantized. The term quantum leap, used to describe an abrupt change, comes from this concept. In general, the higher an electron is on the energy ladder, the farther it is from the nucleus.

The amount of energy gained or lost by an electron is not always the same. Unlike the rungs of a ladder, the energy levels in an atom are not equally spaced. In fact, the energy levels become more closely spaced the farther they are from the nucleus. Instead of a ladder, you might prefer to think of the spacing of energy levels in terms of garden steps like those illustrated in **Figure 13.3.** The steps become closer together as you climb higher. This makes it easier to step off at the top. Similarly, the higher the energy level occupied by an electron, the easier the electron escapes from the atom. Why do you think this is true? 3

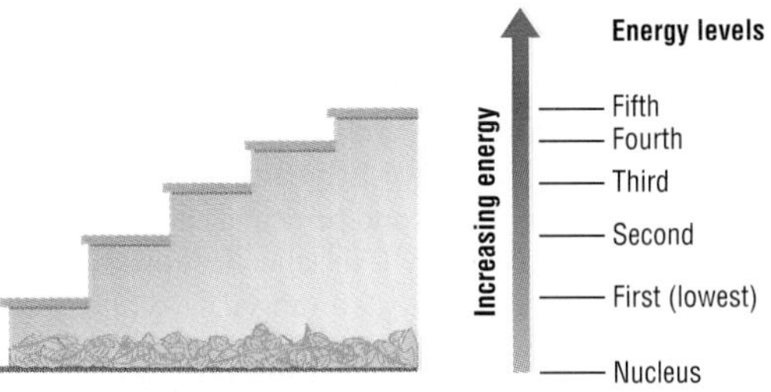

Figure 13.3
These garden steps are analogous to the energy levels in an atom. The higher the energy level occupied by an electron, the more energetic it is, and the farther it usually is from the nucleus.

The Quantum Mechanical Model

In 1926, the Austrian physicist Erwin Schrödinger (1887–1961) took atomic models one step further. He used the new quantum theory to write and solve a mathematical equation describing the location and energy of an electron in a hydrogen atom. The modern description of the electrons in atoms, the **quantum mechanical model,** comes from the mathematical solutions to the Schrödinger equation. Previous models were essentially physical models based on the motion of large objects. In contrast, the quantum mechanical model is primarily mathematical. It has few analogies in the visible world.

Like the Bohr model, the quantum mechanical model of the atom restricts the energy of electrons to certain values. In a radical departure from the Bohr model, however, the quantum mechanical model does not define an exact path an electron takes around the nucleus. Rather, it estimates the probability of finding an electron in a certain position. Probability is a concept that you may know of as chance. If you place three red marbles and one green marble into a box and then pick a marble without looking, the probability, or chance, of picking the green marble is one in four, or 25%. The chance of picking a red marble is three in four, or 75%. What is the chance of picking a red marble after one red marble has already been taken from the box? 4

13.1

Discuss

Have students imagine a used dart board that has concentric circles of decreasing value. (The analogy will not work for professional dart boards.) Ask:

- **Describe the distribution of holes.** (Usually there are more holes close to the bull's-eye than near the edge, which is analogous to the locations of electrons in an atom. Probability diminishes with distance from the nucleus; theoretically, some small probability remains until it reaches zero at infinity.)
- **If the dart board is a model for an atom, what are its major failures?** (2-dimensional model for a 3-dimensional atom; distribution of holes may not decrease uniformly)
- **Assume the holes do decrease uniformly. Compare the probability for two holes on a circle at a specified distance from the center.** (equal) **This rule holds true for electrons that are equidistant from the nucleus.**

Chemistry and Science History

Schrödinger's equation accounted mathematically for de Broglie's discovery of the wave-like properties of the electron described on page 381. These developments led to a major break from the Bohr model of the atom. Schrödinger extended de Broglie's work by considering the movement of a particle in an electromagnetic field, which established an area of physics known as wave mechanics.

Meeting Diverse Needs

At Risk Have students make a poster showing the evolution of the model of the atom. Start with Dalton's model and end with the current quantum mechanical model. Include on the poster some information about each model and the experiments that contributed to the development of the model.

Discuss

Present the orbital concept as an outgrowth of the quantum mechanical model and describe the shapes of the *s* and *p* orbitals. Stress that the quantum mechanical model predicts the shapes of the various orbitals. The fact that experiments affirm the models predictions is validation that the model is correct. Remind students that an orbital is only a region of mathematical probability, not a concrete item that can be seen or felt. An orbital does not have the sharp boundaries models and diagrams would seem to indicate.

In the quantum mechanical model of the atom, the probability of finding an electron within a certain volume of space surrounding the nucleus can be represented as a fuzzy cloud. The cloud is more dense where the probability of finding the electron is high. The cloud is less dense where the probability of finding the electron is low. Although it is unclear where the cloud ends, there is at least a slight chance of finding the electron a considerable distance from the nucleus. Therefore, attempts to show probabilities as a fuzzy cloud are usually limited to the volume in which the electron is found 90% of the time. To visualize an electron probability cloud, imagine that you could mold a sack around the cloud so that the electron was inside the sack 90% of the time. The shape of the sack would then give you a useful picture of the shape of the cloud. Illustrations of electron clouds typically show the shape of the space in which the electron is found 90% of the time.

Atomic Orbitals

As in the Bohr atom, the quantum mechanical model designates energy levels of electrons by means of principal quantum numbers (n). Each principal quantum number refers to a major, or principal, energy level in an atom. These principal energy levels are assigned values in order of increasing energy: $n = 1, 2, 3, 4$, and so forth. The average distance of the electron from the nucleus increases with increasing values of n. Electrons in the third principal energy level have a greater average distance from the nucleus than electrons in the second principal energy level.

Within each principal energy level, the electrons occupy energy sublevels, much as people in theater seats arranged in sections (principal energy levels) occupy rows within those sections (energy sublevels). **Table 13.1** gives the number of sublevels within each principal energy level. Notice that the number of energy sublevels is the same as the principal quantum number. ❶ How many energy sublevels would be in the fifth principal energy level?

Where are the electrons in the various sublevels located in relation to the nucleus? You may recall that the quantum mechanical model limits the description of an electron's position to an area within an electron probability cloud. Because the electron is not confined to a fixed circular path, as it is in the Bohr atom, these regions in which electrons are likely to be found

Table 13.1

Summary of Principal Energy Levels, Sublevels, and Orbitals

Principal energy level	Number of sublevels	Type of sublevel
$n = 1$	1	1*s* (1 orbital)
$n = 2$	2	2*s* (1 orbital), 2*p* (3 orbitals)
$n = 3$	3	3*s* (1 orbital), 3*p* (3 orbitals), 3*d* (5 orbitals)
$n = 4$	4	4*s* (1 orbital), 4*p* (3 orbitals), 4*d* (5 orbitals), 4*f* (7 orbitals)

Answers

❶ five

❷ They both lie in the x–y plane. The lobes in the d_{xy} orbital lie between the axes. The lobes in the $d_{x^2-y^2}$ orbital lie on the axes.

s orbital

p orbitals

Composite of orbitals

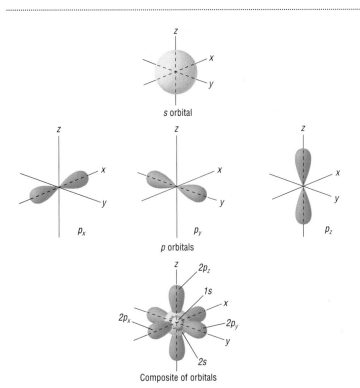

Figure 13.4
The electron probability clouds for the s orbital, the p orbitals, and a composite of one 1s, one 2s, and three 2p orbitals are shown here.

Chem ASAP!
Animation 15 
Observe the characteristics of atomic orbitals.

cannot be called orbits. In the quantum mechanical model, these regions are called **atomic orbitals.** Letters denote the atomic orbitals. As **Figure 13.4** shows, *s* orbitals are spherical, and *p* orbitals are dumbbell-shaped. The three kinds of *p* orbitals have different orientations in space. **Figure 13.5** shows the shapes of *d* orbitals. Four of the five kinds of *d* orbitals have clover-leaf shapes. The shapes of *f* orbitals are very complex and hard to visualize. Notice that in *p* orbitals and *d* orbitals there are regions close to the nucleus where the probability of finding the electron is very low. These regions are called nodes.

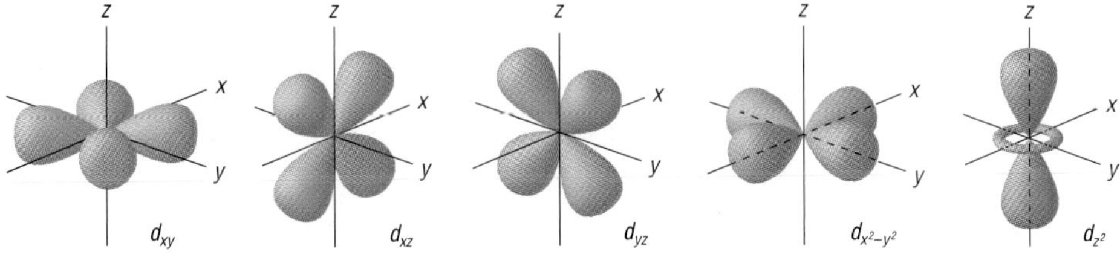

Figure 13.5
The d orbitals are illustrated here. Four of the five d orbitals have the same shape but different orientations in space. How are the orientations of the d_{xy} and $d_{x^2-y^2}$ orbitals similar? How are they different? 2

13.1

Discuss

Direct students' attention to Table 13.1 on page 364. Explain that every atom has an infinite number of energy levels. However, in the ground state there are no elements in the periodic table that require more than 7 energy levels to describe the probability regions (orbitals) where their electrons could be found. Ask:

- **How many sublevels would be in level 5?** (5) **in level 6?** (6) **in level 7?** (7).

Review the type of sublevels that would be in level 5. [5 *s* (1 orbital), **5*p*** (3 orbitals), **5*d*** (5 orbitals), **5*f*** (7 orbitals) **and 5*g*** (9 orbitals)] **It is important that students see the mathematical relationship between type and number of orbitals** (1, 3, 5, 7, 9 and so on). **Ask:**

- **How many orbitals are in the 1st level?** (1) **in the 2nd level?** (4) **in the 3rd level?** (9) **in the 4th level?** (16)
- **Do you see a pattern emerging?** (The number of orbitals is the square of the energy level, n^2.)
- **How many orbitals could one find in the 7th level?** (49)

Stress that orbitals are mathematical probability areas where electrons may be located. The equation for determining how many electrons can be in each energy level is $2n^2$ where *n* represents the energy level.

Meeting Diverse Needs

LEP and At Risk Have students expand Table 13.1 to include two more columns: number of orbitals (n^2) and number of electrons ($2n^2$). Make sure students do not confuse the orbitals shown with predictable physical locations. Use *s* and *p* orbitals, which are key to understanding the behavior of representative elements, to review the concept of probability clouds.

3 Assess

Evaluate Understanding

Ask students to discuss the parts of the Bohr model that are valid and the parts that are no longer considered accurate. Ask students to explain what a quantum of energy is and how an electron moves from one energy level to another in an atom. Ask students to explain the relationship between energy levels, sublevels, orbitals, and electrons.

Reteach

If students are having trouble remembering the mathematical relationships relative to electron location, review the following information. Energy levels are designated with the letter n and are numbered from 1 to 7 starting with the level closest to the nucleus. Each energy level contains as many sublevels as the number of the energy level. When $n = 2$, there are 2 sublevels. Each energy level contains n^2 orbitals. When $n = 3$, there are 3 sublevels and 9 orbitals ($1s + 3p + 5d = 9$). Because each orbital contains 2 electrons, energy level 3 contains 18 electrons; $2n^2$. When $n = 4$, there are 4 sublevels, $n^2 = 16$ orbitals ($1s + 3p + 5d + 7f = 16$) and $2n^2 = 32$ electrons.

The numbers and kinds of atomic orbitals depend on the energy sublevel. The lowest principal energy level ($n = 1$) has only one sublevel, called 1s. Because an s orbital is spherical, at any given distance from the nucleus, the probability of finding an electron does not vary with direction. This is not true for p, d, or f orbitals.

The second principal energy level ($n = 2$) has two sublevels, $2s$ and $2p$. The $2s$ orbital is spherical, and the $2p$ orbitals are dumbbell-shaped. The $2p$ sublevel is of higher energy than the $2s$ and consists of three p orbitals of equal energy. The long axis of each dumbbell-shaped p orbital is perpendicular to the other two. It is convenient to label the axes $2p_x$, $2p_y$, and $2p_z$. Thus the second principal energy level has four orbitals: $2s$, $2p_x$, $2p_y$, and $2p_z$.

The third principal energy level ($n = 3$) has three sublevels, called $3s$, $3p$, and $3d$. The $3d$ sublevel consists of five d orbitals of equal energy. Thus the third principal energy level has nine orbitals (one $3s$, three $3p$, and five $3d$ orbitals).

The fourth principal energy level ($n = 4$) has four sublevels, called $4s$, $4p$, $4d$, and $4f$. The $4f$ sublevel consists of seven f orbitals of equal energy. The fourth principal energy level, then, has 16 orbitals (one $4s$, three $4p$, five $4d$, and seven $4f$ orbitals).

As mentioned, the principal quantum number always equals the number of sublevels within that principal energy level. The maximum number of electrons that can occupy a principal energy level is given by the formula $2n^2$, where n is the principal quantum number. The number of electrons allowed in each of the first four energy levels is as follows.

	Increasing energy (increasing distance from nucleus) →			
Energy level n	1	2	3	4
Maximum number of electrons allowed	2	8	18	32

section review 13.1

1. List, in chronological order, a major contribution of each of these scientists to the understanding of the atom: Dalton, Thomson, Bohr, Schrödinger, Rutherford.
2. In general terms, explain how the quantum mechanical model of the atom describes the electron structure of an atom.
3. The energies of electrons are said to be quantized. Explain what this means.
4. How many orbitals are in the following sublevels?
   a. $3p$ sublevel
   b. $2s$ sublevel
   c. $4f$ sublevel
   d. $4p$ sublevel
   e. $3d$ sublevel

 Chem ASAP! **Assessment 13.1** Check your understanding of the important ideas and concepts in Section 13.1.

Answers

SECTION REVIEW 13.1

1. Dalton: elements are composed of atoms; Thomson: discovered electron; Rutherford: discovered nucleus; Bohr: quantized energies of electrons; Schrödinger: quantum mechanical model
2. The quantum mechanical states that electrons can have only fixed energy levels. Electrons are located in orbitals that may be visualized as clouds of various shapes at different distances from the nucleus.
3. In an atom, electrons can only exist in certain fixed energy levels. To move from one energy level to another requires the emission or absorption of an exact amount of energy, or quantum.
4. **a.** 3 **b.** 1 **c.** 7 **d.** 3 **e.** 5

ELECTRON ARRANGEMENT IN ATOMS

section 13.2

Does this scene look natural to you? Surprisingly, it is. This structure was formed by natural processes. Arrangements like this are rare in nature because they are unstable. Unstable arrangements, whether the grains of sand in a sandcastle or the rock formation shown here, lose energy in an attempt to become more stable. The tendency to become stable also greatly affects how electrons are arranged in atoms. **What role do energy and stability play in the way in which electrons are configured in an atom?**

objectives

- Apply the aufbau principle, the Pauli exclusion principle, and Hund's rule in writing the electron configurations of elements
- Explain why the electron configurations for some elements differ from those assigned using the aufbau principle

key terms

- electron configurations
- aufbau principle
- Pauli exclusion principle
- Hund's rule

Electron Configurations

In most natural phenomena, change proceeds toward the lowest possible energy. High-energy systems are unstable. Unstable systems lose energy to become more stable. In the atom, electrons and the nucleus interact to make the most stable arrangement possible. The ways in which electrons are arranged around the nuclei of atoms are called **electron configurations.**

Three rules—the aufbau principle, the Pauli exclusion principle, and Hund's rule—tell you how to find the electron configurations of atoms. The three rules are as follows.

1. **Aufbau principle:** *Electrons enter orbitals of lowest energy first.* The various orbitals within a sublevel of a principal energy level are always of equal energy. Further, within a principal energy level the *s* sublevel is always the lowest-energy sublevel. Yet the range of energy levels within a principal energy level can overlap the energy levels of an adjacent principal level. Electrons enter the orbitals of lowest energy first. Note

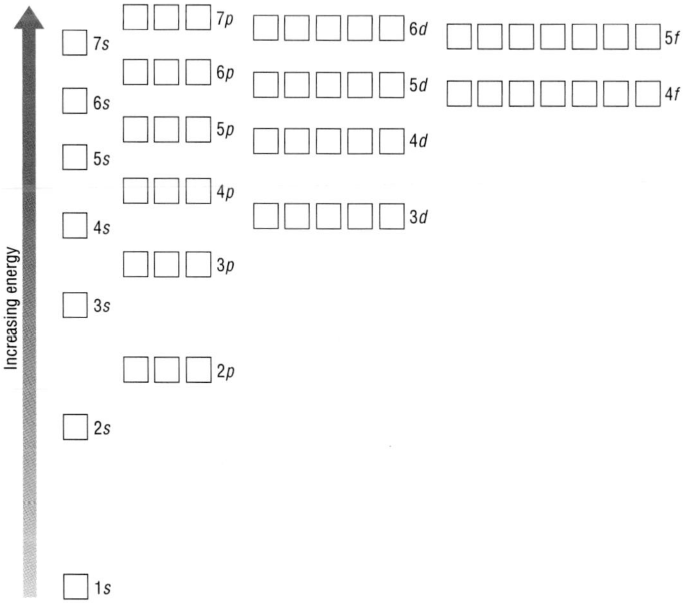

Increasing energy

7s 7p 6d 5f
6s 6p 5d 4f
5s 5p 4d
4s 4p 3d
3s 3p
2s 2p
1s

Figure 13.6

This aufbau diagram shows the energy levels of the various atomic orbitals. Orbitals of greater energy are higher on the diagram. Which is of higher energy, a 4d or a 5s orbital? ❶

Chem ASAP!

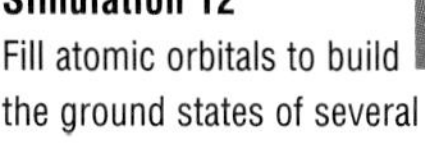

Simulation 12

Fill atomic orbitals to build the ground states of several atoms.

13.2

1 Engage

Use the Visual

Have students study the photograph and read the text that opens the section. Because of gravity, rocks such as the one shown are less stable than rocks on level ground. Electrons have a comparable ground level—a position as close to the nucleus as possible. Ask:

- **What role do energy and stability play in the way that electrons are configured in an atom?** (In the most stable atom, electrons occupy the lowest possible energy orbitals, those nearest the nucleus.)

2 Teach

Discuss

Develop the electron configurations for several of the simpler elements; introduce each rule governing the process as needed. Begin with hydrogen. Use the aufbau diagram to explain that electrons enter orbitals of lowest energy first. Show how the orbital notation ($1s^1$) describes the energy level, the orbital, and the number of electrons. Repeat the process for helium. Then, continue with lithium, beryllium, and boron. Apply the Pauli exclusion principle to explain why additional orbitals must be used. When you reach carbon, explain and apply Hund's rule. Complete the configurations for the second period elements.

STUDENT RESOURCES

From the Teacher's Resource Package, use:

- Section Review 13.2, Ch. 13 Practice Problems and Quizzes from the Review Module (Ch. 13–16)
- Laboratory Recordsheet 13-1

TECHNOLOGY RESOURCES

Relevant technology resources include:

- Chem ASAP! CD-ROM
- Resource Pro CD-ROM
- ActivChemistry CD-ROM *Electron Configurations*

❶ 4d

ACTIVITY

Students usually need a great deal of practice before they are familiar with the correct order for filling orbitals. Have students work with a partner to develop the electron configurations for the third period elements. When the exercise is complete, ask students to compare the configurations of second and third period elements in preparation for upcoming discussions on periodic trends.

that the filling of atomic orbitals does not follow a simple pattern beyond the second energy level. For example, the 4*s* orbital is lower in energy than a 3*d* orbital. This is shown on the aufbau diagram of **Figure 13.6** on the previous page. Each box (□) represents an atomic orbital. Is a 4*f* orbital higher or lower in energy than a 5*d* orbital? ❶

2. **Pauli exclusion principle:** *An atomic orbital may describe at most two electrons.* For example, either one or two electrons can occupy an *s* orbital or a *p* orbital. To occupy the same orbital, two electrons must have opposite spins; that is, the electron spins must be paired. Spin is a quantum property of electrons and may be clockwise or counterclockwise. A vertical arrow indicates an electron and its direction of spin (↑ or ↓). An orbital containing paired electrons is written as [↑↓].
3. **Hund's rule:** *When electrons occupy orbitals of equal energy, one electron enters each orbital until all the orbitals contain one electron with parallel spins.* For example, three electrons would occupy three orbitals of equal energy as follows: [↑][↑][↑]. Second electrons then add to each orbital so their spins are paired with the first electrons. Thus each orbital can eventually have two electrons with paired spins.

Consider the electron configurations of atoms of the nine elements in **Table 13.2.** An oxygen atom contains eight electrons. The orbital of lowest energy, 1*s*, gets one electron, then a second electron of opposite spin. The next orbital to fill is 2*s*. It also gets one electron, then a second electron of opposite spin. One electron then goes into each of the three 2*p* orbitals of equal energy. The remaining electron now pairs with an electron occupying one of the 2*p* orbitals. The other two 2*p* orbitals remain only half filled with one electron each.

Table 13.2

Electron Configurations for Some Selected Elements

Element	Orbital filling						Electron configuration
	1*s*	2*s*	$2p_x$	$2p_y$	$2p_z$	3*s*	
H	[↑]	[ ]	[ ]	[ ]	[ ]	[ ]	$1s^1$
He	[↑↓]	[ ]	[ ]	[ ]	[ ]	[ ]	$1s^2$
Li	[↑↓]	[↑]	[ ]	[ ]	[ ]	[ ]	$1s^22s^1$
C	[↑↓]	[↑↓]	[↑]	[↑]	[ ]	[ ]	$1s^22s^22p^2$
N	[↑↓]	[↑↓]	[↑]	[↑]	[↑]	[ ]	$1s^22s^22p^3$
O	[↑↓]	[↑↓]	[↑↓]	[↑]	[↑]	[ ]	$1s^22s^22p^4$
F	[↑↓]	[↑↓]	[↑↓]	[↑↓]	[↑]	[ ]	$1s^22s^22p^5$
Ne	[↑↓]	[↑↓]	[↑↓]	[↑↓]	[↑↓]	[ ]	$1s^22s^22p^6$
Na	[↑↓]	[↑↓]	[↑↓]	[↑↓]	[↑↓]	[↑]	$1s^22s^22p^63s^1$

Answer

❶ lower

There is a convenient shorthand method for showing the electron configuration of an atom. This method involves writing the energy level and the symbol for every sublevel occupied by an electron. A superscript indicates the number of electrons occupying that sublevel. For hydrogen, with one electron in a 1*s* orbital, the electron configuration is written $1s^1$. For helium, with two electrons in a 1*s* orbital, the configuration is $1s^2$. For oxygen, with two electrons in a 1*s* orbital, two electrons in a 2*s* orbital, and four electrons in 2*p* orbitals, it is $1s^2 2s^2 2p^4$. Note that the sum of the superscripts equals the number of electrons in the atom.

Sample Problem 13-1

Use **Figure 13.6** to write the electron configurations of these atoms.

a. phosphorus **b.** nickel

1. ***ANALYZE*** ***Plan a problem-solving strategy.***
Phosphorus has 15 electrons; nickel has 28 electrons. Using **Figure 13.6,** start placing electrons in the orbitals with the lowest energy (1*s*). Remember, there is a maximum of two electrons per orbital. Electrons do not pair up within an energy sublevel (orbitals of equal energy) until each orbital already has one electron.

2. ***SOLVE*** ***Apply the problem-solving strategy.***
a. phosphorus **b.** nickel

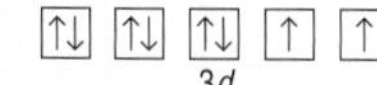

4s 4s

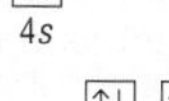

↑↓ 3s

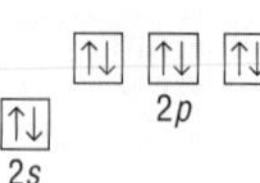

↑↓ ↑↓ ↑↓ 2p

↑↓ 2s

↑↓ 1s

P $1s^2 2s^2 2p^6 3s^2 3p^3$ Ni $1s^2 2s^2 2p^6 3s^2 3p^6 3d^8 4s^2$

3. ***ANALYZE*** ***Do the results make sense?***
The superscripts add up to the number of electrons. When the configurations are written, the sublevels within the same principal energy level are written together. This is not always the same order as given on the aufbau diagram. In this case, the 3*d* sublevel is written before the lower-energy 4*s* sublevel.

Practice Problems

5. Write the complete electron configuration for each atom.
a. carbon
b. argon
6. Write the electron configuration for each atom. How many unpaired electrons does each atom have?
a. boron
b. silicon

Chem ASAP!

Problem-Solving 6

Solve Problem 6 with the help of an interactive guided tutorial.

Meeting Diverse Needs

LEP and At Risk Encourage students to draw the aufbau diagram (Figure 13.6) on a 5 x 7 card and keep it handy when doing electron configurations. They could also make flash cards with the name and atomic number of the element on one side and the correct electron configuration on the other side.

13.2

Discuss

Students may ask how two negative electrons can occupy the same orbital, or why an orbital never contains more than two electrons. The answer lies in the quantum property known as spin. A spinning electron acts as a tiny magnet, with a north pole at one end and a south pole at the other. If electrons with opposite spins enter one orbital, the attraction of their opposite magnetic poles counteracts some of the electrical repulsive force. If a third electron tries to enter the orbital, its spin will always be the same as one of the existing electrons, and it will be repelled.

Practice Problems Plus

Related Chapter Review Problems
Chapter Review problems 28 and 33 are related to Sample Problem 13-1.

Additional Practice Problem
The Chem ASAP! CD-ROM contains the following problem: Write electron configurations for atoms of the following elements. How many unpaired electrons do these atoms have?
a. argon; **b.** sulfur
(**a.** $1s^2 2s^2 2p^6 3s^2 3p^6$; no unpaired electrons,
b. $1s^2 2s^2 2p^6 3s^2 3p^4$; two unpaired electrons)

Discuss

Students may find it helpful to use a diagonal diagram to write electron configurations until they become familiar with the order in which sublevels fill.

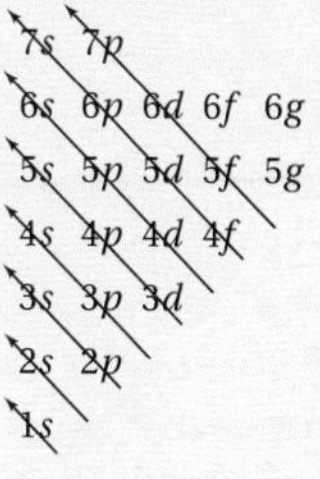

Discuss

Among the transition elements there are some exceptions to the filling rules. Exceptions can be explained by the atom's need to keep the energy of its electrons as low as possible. These exceptions help explain the unexpected chemical behavior of transition elements.

3 Assess

Evaluate Understanding

Make the class a set of small cards, each with a different element on it, starting with neodymium, atomic number 60. Have students choose a card and then write out the electron configuration for the element shown on the card.

Reteach

Have students work in small groups, using the set of element cards to practice writing more electron configurations. Have them refer to Figure 13.6.

Portfolio Project

For the magnets to approach, their north poles must be facing in opposite directions. The placement of the magnets is analagous to two electrons with opposing spins in the same orbital. Electrons with opposing spins can overcome repulsion.

Figure 13.7
The hood ornament has a high shine and resists corrosion because it is plated with chromium (left). Copper (right) is a good conductor of electricity and is commonly used in electrical wiring.

Exceptional Electron Configurations

You can obtain correct electron configurations for the elements up to vanadium (atomic number 23) by following the aufbau diagram for orbital filling. If you were to continue in that fashion, however, you would assign chromium and copper the following incorrect configurations.

Cr $1s^2 2s^2 2p^6 3s^2 3p^6 3d^4 4s^2$
Cu $1s^2 2s^2 2p^6 3s^2 3p^6 3d^9 4s^2$

The correct electron configurations are

Cr $1s^2 2s^2 2p^6 3s^2 3p^6 3d^5 4s^1$
Cu $1s^2 2s^2 2p^6 3s^2 3p^6 3d^{10} 4s^1$

These arrangements give chromium a half-filled *d* sublevel and copper a filled *d* sublevel. Filled energy sublevels are more stable than partially filled sublevels. Half-filled levels are not as stable as filled levels, but they are more stable than other configurations. Therefore, chromium atoms and copper atoms are more stable with only one electron in the 4*s* sublevel.

portfolio project

Use two bar magnets to model and explain the Pauli exclusion principle. Compare the property that determines whether two magnets attract or repel with the property that determines whether two electrons can occupy the same orbital.

section review 13.2

7. Write the complete electron configuration for each atom.
   a. lithium
   b. fluorine
   c. rubidium
8. Explain why the actual electron configurations for chromium and copper differ from those assigned using the aufbau diagram.
9. Arrange the following sublevels in order of decreasing energy: 2*p*, 4*s*, 3*s*, 3*d*, and 3*p*.
10. Why does one electron in a potassium atom go into the fourth energy level instead of squeezing into the third energy level along with the eight already there?

Chem ASAP! **Assessment 13.2** Check your understanding of the important ideas and concepts in Section 13.2.

Answers

SECTION REVIEW 13.2

7. a. $1s^2 2s^1$
   b. $1s^2 2s^2 2p^5$
   c. $1s^2 2s^2 2p^6 3s^2 3p^6 3d^{10} 4s^2 4p^6 5s^1$
8. Half-filled energy sublevels are more stable than partially filled sublevels.
9. 3*d*, 4*s*, 3*p*, 3*s*, 2*p*
10. The 3*s* and 3*p* orbitals are already filled; therefore the last electron must go to the next energy sublevel, which is 4*s*.

SMALL-SCALE LAB

ELECTRON CONFIGURATIONS OF ATOMS AND IONS

SAFETY

Wear safety glasses and follow the standard safety procedures outlined on page 18.

PURPOSE

To make observations of metal ion solutions and relate them to electron configurations.

MATERIALS

- pencil
- paper
- ruler
- reaction surface
- solutions of chemicals shown in Figure A

PROCEDURE

On separate sheets of paper, draw two grids similar to Figure A. Make each square 2 cm on each side. Draw large black **X**s on only one of the grids. Place a reaction surface over the grid with black **X**s and add one drop of each chemical as shown in Figure A. Use the other grid as a data table to record your observations.

$NaCl$ colorless giving no change	$MgSO_4$ colorless changing to white ppt	$AlCl_3$ colorless changing to white ppt
$FeCl_3$ yellow changing to orange ppt	$CaCl_2$ colorless changing to white ppt	$NiSO_4$ green changing to blue ppt
$CuSO_4$ blue changing to blue ppt	$ZnCl_2$ colorless changing to white ppt	$AgNO_3$ colorless changing to white ppt

Figure A

ANALYSIS

Using your experimental data, record the answers to the following questions below your data table.

1. Write the electron configurations of Na, Mg, and Al.
2. Metal ions form when metal atoms lose valence electrons. The number of electrons lost equals the ion's charge. Write the electron configurations of Na^+, Mg^{2+}, and Al^{3+}. What do they have in common?
3. How many electrons does Cl^- have? Write its electron configuration.
4. Transition-metal ions having partially filled *d* orbitals usually have a color. Which solutions contain transition-metal ions with partially filled *d* orbitals? Transition metals usually lose *s* orbital electrons first. Write electron configurations for Fe and Fe^{3+} and Ni and Ni^{2+}.
5. Write the exceptional electron configurations of Cu and Ag.
6. The solutions of both Ag^+ and Zn^{2+} ions have no color. What does this suggest about their electron configurations? Write them.
7. Predict which of the following transition metal ions has a color: Cr^{3+}, Cd^{2+}, Hg^{2+}, V^{2+}.

YOU'RE THE CHEMIST

The following small-scale activities allow you to develop your own procedures and analyze the results.

1. **Analyze It!** Predict which of the metal cations in this experiment will form colored precipitates upon the addition of NaOH. Carry out an experiment to find out. What color are the precipitates?
2. **Analyze It!** Predict which of the metal cations in this experiment will form colored precipitates upon the addition of Na_2CO_3.
3. **Design It!** Design and carry out an experiment to find out which metal ions form precipitates with sodium carbonate. What color are the precipitates?
4. **Analyze It!** Write chemical equations for all the reactions in Steps 1 and 3.

ANALYSIS

1. Na: $1s^22s^22p^63s^1$; Mg: $1s^22s^22p^63s^2$
   Al: $1s^22s^22p^63s^23p^1$
2. Na^+, Mg^{2+}, and Al^{3+} ions have the same electron configuration: $1s^22s^22p^6$
3. Cl^- has 18 electrons: $1s^22s^22p^63s^23p^6$
4. $CuSO_4$, $NiSO_4$ and $FeCl_3$
   Fe: $1s^22s^22p^63s^23p^64s^23d^6$
   Fe^{3+}: $1s^22s^22p^63s^23p^63d^5$
   Ni: $1s^22s^22p^63s^23p^64s^23d^8$
   Ni^{2+}: $1s^22s^22p^63s^23p^63d^8$
5. Cu: $1s^22s^22p^63s^23p^64s^13d^{10}$
   Ag: $1s^22s^22p^63s^23p^64s^23d^{10}4p^65s^14d^{10}$
6. They have filled *d* orbitals.
   Ag^+: $1s^22s^22p^63s^23p^64s^23d^{10}4p^64d^{10}$
   Zn^{2+}: $1s^22s^22p^63s^23p^63d^{10}$
7. Cr^{3+} and V^{2+} are colored because they contain partially filled *d* orbitals.

SMALL-SCALE LAB

ELECTRON CONFIGURATIONS OF ATOMS AND IONS

OVERVIEW

Procedure Time: 20 minutes
Students write electron configurations for metal atoms and ions; they learn that colored ions are associated with partially filled *d* orbitals.

MATERIALS

Solution	Preparation
1.0*M* NaCl	14.6 g in 250 mL
0.2*M* $MgSO_4$	6.0 g in 250 mL
0.2*M* $AlCl_3$	12.1 g $AlCl_3 \cdot 6H_2O$ in 250 mL
0.1*M* $FeCl_3$	6.8 g $FeCl_3 \cdot 6H_2O$ in 25 mL of 1.0*M* NaCl; dilute to 250 mL
0.5*M* $CaCl_2$	13.9 g in 250 mL
0.2*M* $NiSO_4$	2.94 g in 250 mL
0.2*M* $CuSO_4$	12.5 g $CuSO_4 \cdot 5H_2O$ in 250 mL
0.2*M* $ZnCl_2$	6.8 g in 250 mL
0.05*M* $AgNO_3$	2.1 g in 250 mL
0.5*M* NaOH	20.0 g in 1.0 L
1.0*M* Na_2CO_3	26.5 g in 250 mL

YOU'RE THE CHEMIST

1. $CuSO_4$, $NiSO_4$, and $FeCl_3$ form precipitates with NaOH. $Cu(OH)_2$ and $Ni(OH)_2$ are blue. $Fe(OH)_3$ is orange.
2. $CuSO_4$, $NiSO_4$, and $FeCl_3$ form precipitates with Na_2CO_3.
3. $CuCO_3$ and $NiCO_3$ are blue. $Fe_2(CO_3)_3$ is orange.
4. $CuSO_4 + 2NaOH \rightarrow Cu(OH)_2(s) + Na_2SO_4$
   $NiSO_4 + 2NaOH \rightarrow Ni(OH)_2(s) + Na_2SO_4$
   $FeCl_3 + 3NaOH \rightarrow Fe(OH)_3(s) + 3NaCl$
   $CuSO_4 + Na_2CO_3 \rightarrow CuCO_3(s) + Na_2SO_4$
   $NiSO_4 + Na_2CO_3 \rightarrow NiCO_3(s) + Na_2SO_4$
   $2FeCl_3 + 3Na_2CO_3 \rightarrow Fe_2(CO_3)_3(s) + 6NaCl$

1 Engage

Use the Visual

Have students study the photograph and read the text that opens the section. Ask:

- **What causes an element such as neon to emit light when heated by an electric current?** (Energy absorbed by electrons is emitted as light.)

ACTIVITY

To show students how information about unseen objects can be determined, set up several sealed boxes in which you have placed different common items. Pass the boxes around the room and record student observations and guesses about what is inside each "black box." Allow students to move the boxes and listen to what happens. Point out that they are adding kinetic energy and recording how the hidden object responds. This method is analagous to the way scientists add energy to atoms and observe their resulting behavior. If the items are carefully selected, students will be amazed at how much they can determine without opening the boxes. (Examples include rubber bands, golf, tennis, or ping-pong balls, cans of different sizes, or different shapes cut from blocks of wood or foam.) This activity could also be done as a group activity with the groups reporting what they find.

section 13.3

objectives

- Calculate the wavelength, frequency, or energy of light, given two of these values
- Explain the origin of the atomic emission spectrum of an element

key terms

- electromagnetic radiation
- amplitude
- wavelength (λ)
- frequency (ν)
- hertz (Hz)
- spectrum
- atomic emission spectrum
- Planck's constant (h)
- photons
- photoelectric effect
- ground state
- de Broglie's equation
- Heisenberg uncertainty principle

PHYSICS AND THE QUANTUM MECHANICAL MODEL

Neon signs are exciting in more than one sense. In addition to being exciting to see, neon signs contain gas that emits light when its atoms become excited by electricity. **What causes an element such as neon to emit light when excited by an electric current?**

Light and Atomic Spectra

The previous sections in this chapter introduced you to some ideas about atomic structure. You also learned how to write electron configurations for atoms. In the remainder of this chapter, you will backtrack a bit to delve further into the work that led to the development of Schrödinger's equation and the quantum mechanical model of the atom. Rather curiously, this model grew out of the study of light. Isaac Newton (1642–1727) thought of light as consisting of particles. By the year 1900, however, most scientists accepted the idea that light was a wave phenomenon.

According to the wave model, light consists of electromagnetic waves. **Electromagnetic radiation** includes radio waves, microwaves, infrared waves, visible light, ultraviolet waves, x-rays, and gamma rays. **Figure 13.8** shows a typical wave. All electromagnetic waves travel in a vacuum at a speed of 3.0×10^{10} cm/s (3.0×10^{8} m/s). Each complete wave cycle begins at the origin, then returns to the origin. The **amplitude** of a wave is the wave's height from the origin to the crest. The **wavelength** (**λ**, the Greek letter lambda) is the distance between the crests. **Frequency** (**ν**, the Greek letter nu) is the number of wave cycles to pass a given point per unit of time.

The frequency and wavelength of all waves, including light, are inversely related. As the wavelength of light increases, for example, the frequency decreases. **Figure 13.9** shows this relationship. The product of frequency and wavelength always equals a constant (c), the speed of light; that is,

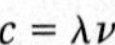

$$c = \lambda\nu$$

The units of frequency are usually cycles per second. The SI unit of cycles per second is called a **hertz (Hz).** A hertz can also be expressed as a reciprocal second (s^{-1}) where the term "cycles" is assumed to be understood. Most problems are easiest to solve when frequency is expressed as s^{-1}.

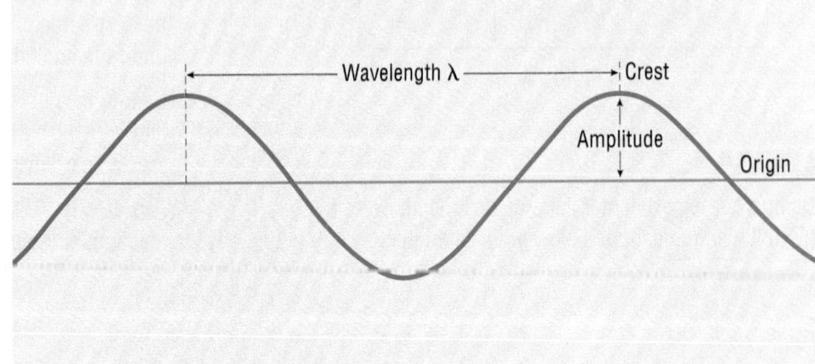

Figure 13.8
The amplitude of a wave is the height of the wave from the origin to the crest. The wavelength is the distance between crests. How does a change in wavelength affect the amplitude of a wave? 1

Chem ASAP!

Simulation 13
Explore the properties of electromagnetic radiation.

STUDENT RESOURCES

- Section Review 13.3, Ch. 13 Practice Problems, Interpreting Graphics, Vocabulary Review, Quizzes and Tests from the Review Module (Ch. 13–17)
- Laboratory Recordsheet 13-2
- Laboratory Manual: Experiments 20, 21, 22
- Laboratory Practicals 13-1 and 13-2
- Small-Scale Lab Manual: Experiments 15, 16

TECHNOLOGY RESOURCES

Relevant technology resources include:

- Chem ASAP! CD-ROM
- Resource Pro CD-ROM
- ActivChemistry CD-ROM: *Electron Configurations*
- Assessment Resources CD-ROM: Chapter 13 Tests

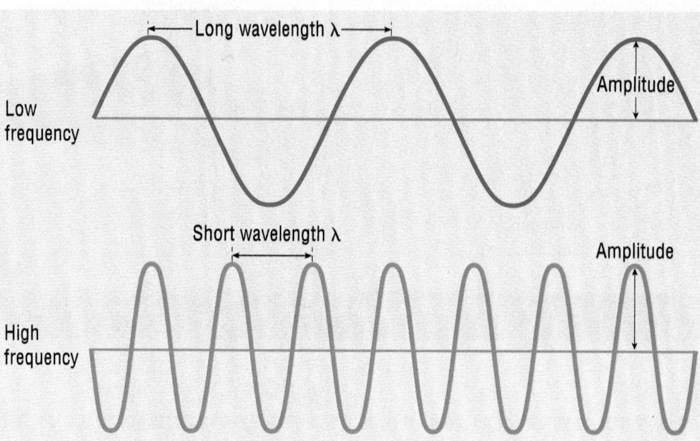

Figure 13.9
The frequency and wavelength of light waves are inversely related. As the wavelength increases, the frequency decreases. The wavelength and frequency do not affect the amplitude.

Sunlight consists of light with a continuous range of wavelengths and frequencies. The wavelength and frequency of each color of light are characteristic of that color. When sunlight passes through a prism, the different wavelengths separate into a **spectrum** of colors. A rainbow is an example of this phenomenon. Each tiny droplet of water acts as a prism to produce a spectrum. Each color blends into the next in the order red, orange, yellow, green, blue, indigo, and violet. In the visible spectrum, as shown in **Figure 13.10,** red light has the longest wavelength and the lowest frequency. Which part of the visible spectrum has the shortest wavelength and highest frequency? ②

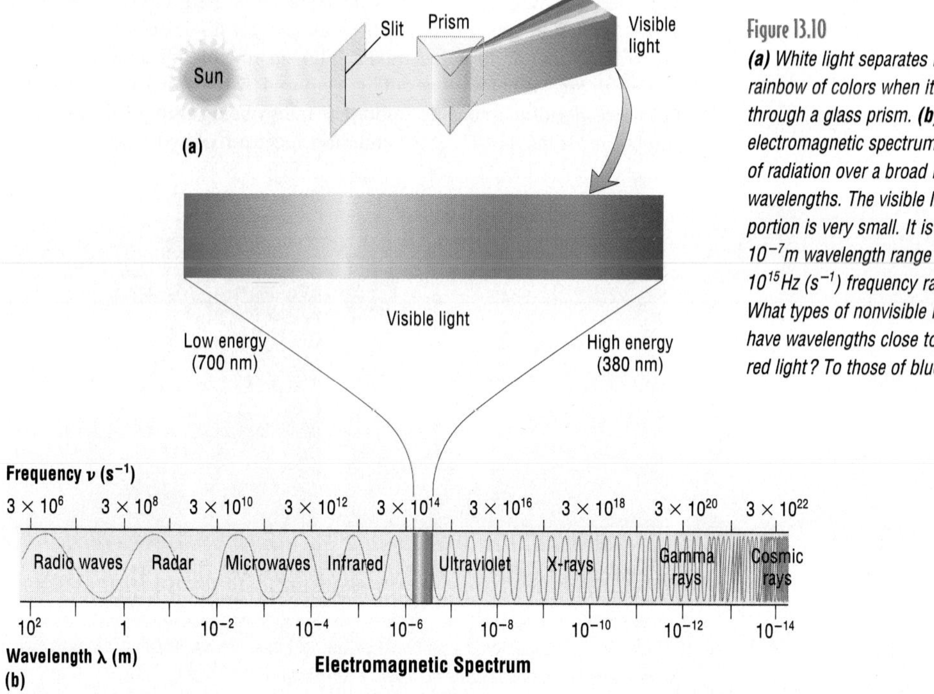

Figure 13.10
(a) *White light separates into a rainbow of colors when it passes through a glass prism.* ***(b)*** *The electromagnetic spectrum consists of radiation over a broad band of wavelengths. The visible light portion is very small. It is in the 10^{-7}m wavelength range and 10^{15}Hz (s^{-1}) frequency range. What types of nonvisible radiation have wavelengths close to those of red light? To those of blue light?* ③

13.3

2 Teach

CHEMISTRY AND SCIENCE HISTORY

Isaac Newton carried out many important experiments with light in 1666. He proposed the idea that light consists of tiny particles called corpuscles. Twelve years later, a Dutch scientist, Christian Huygens, suggested a wave theory to explain the properties of light. For more than 200 years, scientists argued about these seemingly contradictory theories. By the year 1900, most scientists accepted the idea that light was a wave.

Discuss

Discuss the continuous spectrum. Expand the discussion from the visible spectrum to include the whole electromagnetic spectrum. Ask:

- **Name some types of radiation that are not visible.** (infrared, ultraviolet, x-ray, gamma rays)
- **Which of these invisible rays are more energetic than visible light? Offer evidence.** (UV causes suntan or sunburns, x-rays penetrate flesh, gamma rays can even penetrate bones.)
- **What do all the rays in the electromagnetic spectrum have in common?** (All travel at the speed of light.)

Answers

① It has no effect.
② violet
③ red, infrared; blue, ultraviolet

Use the Visuals

Refer students to Figures 13.11 and 13.12. Explain that the high voltage required to measure the emission spectrum of a diatomic element, such as hydrogen, causes some of the molecules to break apart into atoms. The excited atoms release the excess energy in discrete packets of energy that are observed as a line spectrum. Scientists that were trying to understand the inner architecture of the atom realized that there was a relationship between the emission spectrum and atomic structure. Ask:

- **Why is hydrogen a good model for studying the correlation between emission spectra and atomic spectra?** (Hydrogen has only one electron. It has the simplest electron configuration of all elements. Larger more complex atoms produce many more emission lines.)

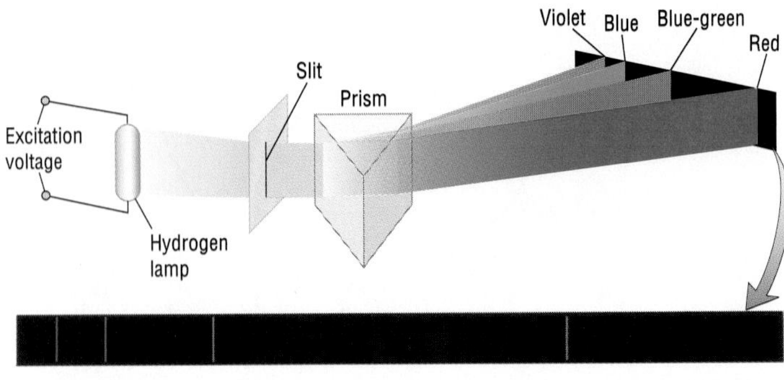

Figure 13.11

The emission spectrum of an element consists of a series of discrete spectral lines that are characteristic of the element. The emission spectrum of atomic hydrogen shown here consists of a series of widely spaced lines, the most prominent of which are violet, blue, blue-green, and red. Which of the bands has the highest frequency? ①

Chem ASAP!

Animation 16

Learn about atomic emission spectra and about how neon lights work.

Every element emits light when it is excited by the passage of an electric discharge through its gas or vapor. The atoms first absorb energy, then lose the energy as they emit light. Passing the light emitted by an element through a prism gives the **atomic emission spectrum** of the element. The instrument used to obtain the emission spectra is called an emission spectrograph. The emission spectra of elements are quite different from the spectrum of white light. While white light gives a continuous spectrum, atomic emission spectra consist of relatively few lines and are called line spectra or discontinuous spectra. Each line in an emission spectrum corresponds to one exact frequency of light emitted by the atom. Therefore each line corresponds to a specific amount of energy being emitted. **Figure 13.11** shows the visible portion of the emission spectrum of hydrogen.

Sodium light

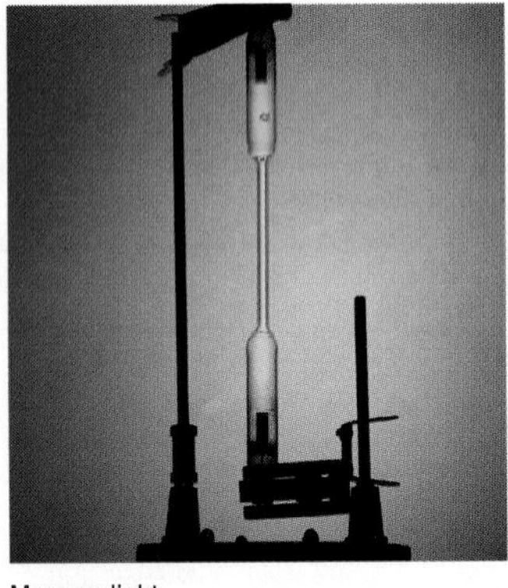

Mercury light

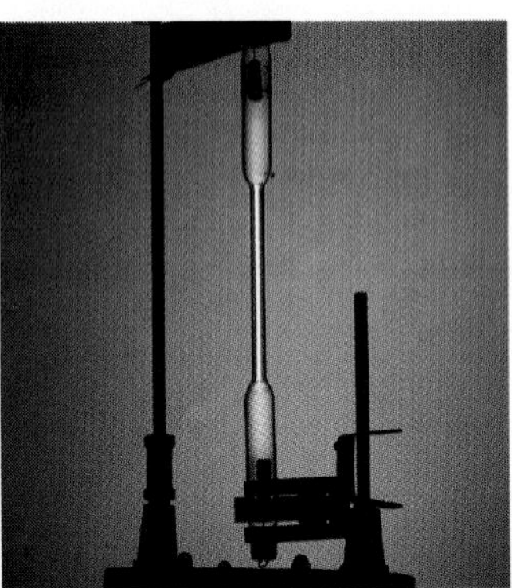

Nitrogen light

Figure 13.12

Sodium vapor lamps produce a bright yellow light; mercury vapor lamps produce a blue glow; and nitrogen gas gives off a yellowish-orange light.

Answer

① violet

Figure 13.13
When a noble gas is energized, it gives off light of a characteristic color. Helium produces a pink light, and argon produces lavender. Krypton gives off a whitish light, xenon produces blue, and neon shines orange-red.

The emission spectrum of each element is unique to that element. See **Figure 13.12** and **Figure 13.13.** This makes such spectra extremely useful for the identification of unknown or inaccessible substances. Much of the knowledge of the composition of the universe comes from studying the atomic spectra of the stars, which are hot glowing bodies of gases. A telescope gathers light from the star of interest. This light is then transmitted to an emission spectrograph to record the atomic emission spectrum.

Sample Problem 13-2

Calculate the wavelength of the yellow light emitted by a sodium lamp if the frequency of the radiation is 5.10×10^{14} Hz ($5.10 \times 10^{14}\ s^{-1}$).

1. ***ANALYZE*** ***List the knowns and the unknown.***

Knowns:
- frequency (ν) = $5.10 \times 10^{14}\ s^{-1}$
- $c = 3.00 \times 10^{10}$ cm/s
- $c = \lambda\nu$

Unknown:
- wavelength (λ) = ? cm

Solve the equation $c = \lambda\nu$ for λ, then substitute the known values and solve.

2. ***CALCULATE*** ***Solve for the unknown.***

Solving for λ yields

$$\lambda = \frac{c}{\nu}$$

Substituting values for c and ν yields

$$\lambda = \frac{c}{\nu} = \frac{3.00 \times 10^{10}\ \text{cm/s}}{5.10 \times 10^{14}\ \text{s}^{-1}} = 5.88 \times 10^{-5}\ \text{cm}$$

3. ***EVALUATE*** ***Does the result make sense?***

The magnitude of the frequency is much larger than the numerical value of the speed of light, so the answer should be much less than 1. The answer has three significant figures, as it should, because the original known value had three significant figures.

Practice Problems

11. What is the wavelength of radiation with a frequency of $1.50 \times 10^{13}\ s^{-1}$? Does this radiation have a longer or shorter wavelength than red light?

12. What frequency is radiation with a wavelength of 5.00×10^{-6} cm? In what region of the electromagnetic spectrum is this radiation?

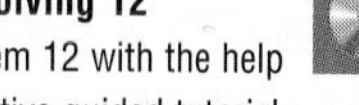

Chem ASAP!

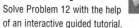

Problem-Solving 12
Solve Problem 12 with the help of an interactive guided tutorial.

Practice Problems Plus

Related Chapter Review Problems
Chapter Review problems 49 and 50 are related to Sample Problem 13-2.

Additional Practice Problem
The Chem ASAP! CD-ROM contains the following problem: What frequency is radiation whose wavelength is 7.00×10^{-5}cm? In what region of the electromagnetic spectrum is this radiation? ($4.29 \times 10^{14}s^{-1}$; infrared)

Meeting Diverse Needs

LEP and At Risk When introducing the units and formulas in this section, be sure to review mathematical operations for multiplication and division using scientific notation. Also be sure to show that $1/10 = 0.10 = 10^{-1}$. This will help students understand the units used for frequency: cycles per second = cycles/sec = cycles $\times\ s^{-1}$ or hertz (Hz).

The Quantum Concept and the Photoelectric Effect

The laws of classical physics state that there is no limit to how small the energy gained or lost by an object may be. According to these laws, when a gaseous or vaporized element is excited by electricity, the spectrum of the emitted light should be continuous. Thus classical physics does not explain the emission spectra of atoms, which consist of lines. The seed of an idea that explained atomic spectra came in 1900 from the German physicist Max Planck (1858–1947).

Planck was trying to describe quantitatively why a body such as a chunk of iron appears to change color as it is heated. First it appears black, then red, yellow, white, and blue as its temperature increases, as shown in **Figure 13.14.** Planck found he could explain the color changes if he assumed that the energy of a body changes only in small discrete units. You can think of these discrete units as bricks used to build a wall—a brick wall can only increase or decrease in size in increments of one or more bricks.

Planck showed mathematically that the amount of radiant energy (E) absorbed or emitted by a body is proportional to the frequency of the radiation (ν).

$$E \propto \nu \quad \text{or} \quad E = h \times \nu$$

The constant *(h)* in the above equation is called **Planck's constant,** which has a value of 6.6262×10^{-34} J $\times$ s (J is the joule, the SI unit of energy). The energy of a quantum equals $h \times \nu$. Any attempt to increase or decrease the energy of a system by a fraction of $h \times \nu$ must fail. The size of an emitted or absorbed quantum depends on the size of the energy change. A small energy change involves the emission or absorption of low-frequency radiation. A large energy change involves the emission or absorption of high-frequency radiation.

Planck's proposal that quanta of energy are absorbed or emitted was revolutionary. Everyday experience had led people to believe that there was no limitation to the smallness of permissible energy changes in a system. It appears, for example, that thermal energy may be continuously supplied to heat liquid water to any temperature between 0 °C and 100 °C. Actually, the water temperature increases by infinitesimally small steps, which occurs as individual molecules absorb quanta of energy. An ordinary thermometer is unable, however, to detect such small changes in temperature. Thus your everyday experience gives you no clue to the fact that energy is quantized.

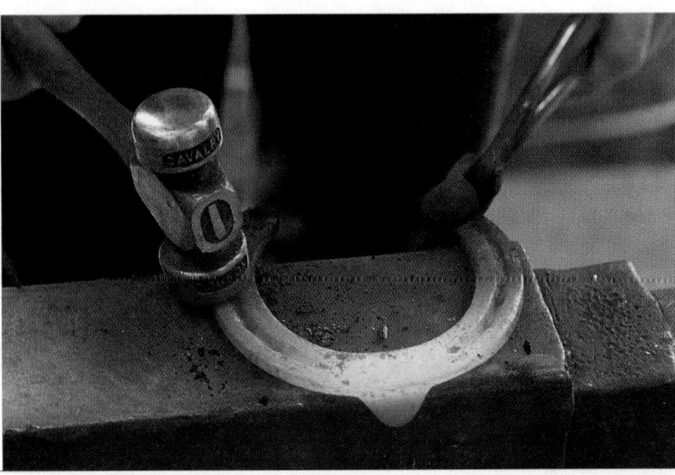

Figure 13.14
All objects change color when heated. Although the color change of this horseshoe looks like a continuous process, the atoms in the iron are giving off energy in small discrete units.

Chemistry and Science History

Max Planck's quantum theory was a major contribution to the understanding of atomic structure and behavior. His proposal that energy comes in discrete packages and his ability to relate the energy of those quanta to their frequency was revolutionary. Many prominent physicists did not accept the idea. However, when Albert Einstein used Planck's theory to explain the photoelectric effect, a situation that was not explainable by either the wave or particle theory, he validated the new theory and opened the door for its acceptance. The present day concept of light describes light as packages of energy (quanta or photons) that travel with wave behavior. Albert Einstein, Max Planck, Louis de Broglie, and Werner Heisenberg all contributed to today's understanding of light.

Astronomy

The Discovery of Helium
Sometimes discoveries in one area of science, such as chemistry, become important in solving problems in another, such as astronomy. In 1868,

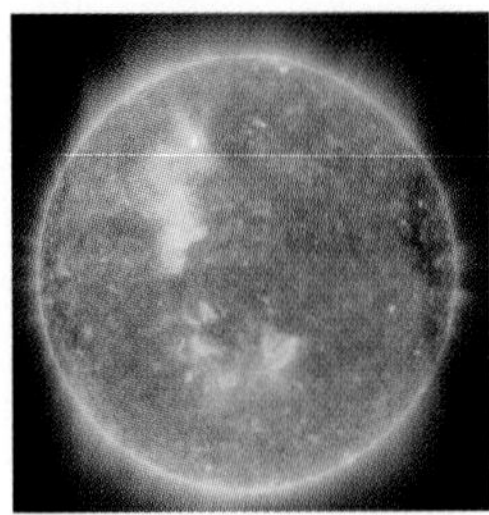

Pierre Janssen and Joseph Norman Lockyer discovered an emission spectrum from gases on the surface of the sun that was unmatched by any known element on Earth. In 1895, William Ramsay discovered the existence of helium on Earth. The emission spectrum of helium was found to be identical to that of the unknown gas observed by Janssen and Lockyer almost thirty years earlier. Thus by combining two discoveries from different fields of science, a new discovery was made: Earth and the stars have some elements in common.

Light
Electron
Nucleus
Metal

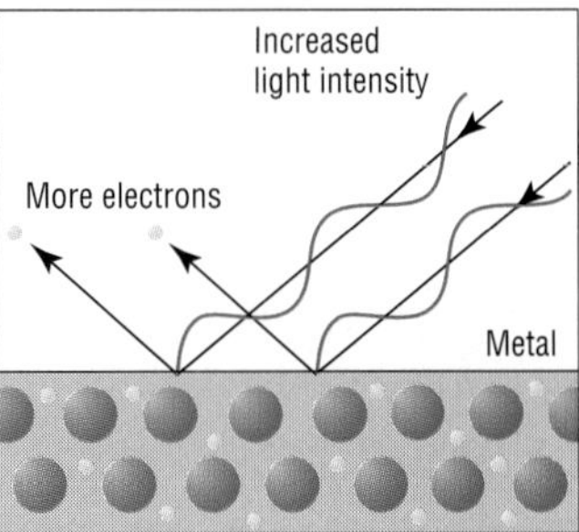

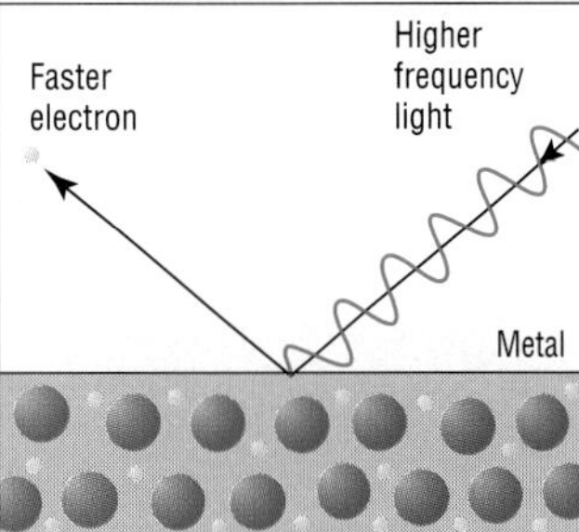

(a) When light strikes a metal surface, electrons are ejected.

(b) If the threshold frequency has been reached, increasing the intensity only increases the number of electrons ejected.

(c) If the frequency is increased, the ejected electrons will travel faster.

Figure 13.15
The photoelectric effect, illustrated here, was explained by Einstein's dual wave-particle theory of light.

In 1905 Albert Einstein, then a patent examiner in Zurich, Switzerland, returned to Newton's idea of particles of light. Einstein proposed that light could be described as quanta of energy that behave as if they were particles. Light quanta are called **photons.** The energy of photons is quantized according to the equation $E = h \times \nu$.

The dual wave–particle behavior of light was difficult for scientists trained in classical physics to accept. However, it was even more difficult to dispute because it provided an explanation for the previously mysterious photoelectric effect. In the **photoelectric effect,** metals eject electrons called photoelectrons when light shines on them. The alkali metals (Li, Na, K, Rb, and Cs) are particularly subject to the effect. Not just any frequency of light will cause the photoelectric effect. Red light ($\nu = 4.3 \times 10^{14}\ s^{-1}$ to $4.6 \times 10^{14}\ s^{-1}$), for example, will not cause the ejection of photoelectrons from potassium, no matter how intense the light. Yet even a very weak yellow light ($\nu = 5.1 \times 10^{14}\ s^{-1}$ to $5.2 \times 10^{14}\ s^{-1}$) shining on potassium begins the effect.

Chem ASAP!

Simulation 14

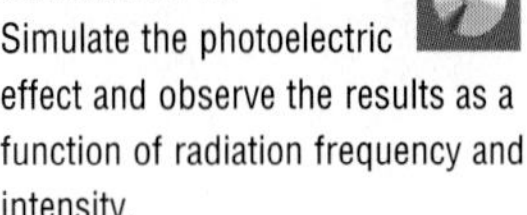

Simulate the photoelectric effect and observe the results as a function of radiation frequency and intensity.

The photoelectric effect could not be explained by classical physics, which had no quantum concept. Classical physics correctly viewed light as a form of energy. It assumed, however, that under weak light of any wavelength, an electron in a metal should eventually collect enough energy to be ejected. Obviously, the photoelectric effect presented a serious problem for the classical wave theory of light.

Einstein used his particle theory of light to explain the photoelectric effect, which is demonstrated in **Figure 13.15.** He recognized that there is a threshold value of energy below which the photoelectric effect does not occur. Because $E = h \times \nu$, all the photons in a beam of monochromatic light (light of only one frequency) have the same energy. If the frequency, and therefore the energy, of the photons is too low, then no photoelectrons will be ejected. It does not matter whether a single photon or a steady stream of low-energy photons strikes an electron in the metal. Only if the frequency of light is above the threshold frequency will the photoelectric effect occur. The photoelectric effect has practical applications in photoelectric cells such as in the car shown in **Figure 13.16,** which uses the energy of light to generate electricity. The operation of numerous electronic devices, such as solar-powered calculators, is also based on the photoelectric effect.

Figure 13.16
Photoelectric cells convert light energy into the electrical energy that will be used to power this car.

13.3

ACTIVITY

There were numerous physicists that contributed to modern theories concerning light and atoms. Dalton, Becquerel, Roentgen, Rutherford, Thomson, Millikan, Planck, Einstein, de Broglie, Heisenberg, Schrödinger, Bohr, Chadwick, Balmer, Huygens, Newton, and Foucault all contributed something. Have students work in teams of two or three. Assign scientists and have students prepare a brief oral discussion highlighting how their work contributed to the understanding of either light or the atom. Encourage students to consider other scientists whose work they encounter while doing research on the main contributors. Contributions at the periphery can help students understand the interactive and collaborative nature of scientific research.

CHEMath

Teaching Tips

Explain that many physical properties of matter can be described using mathematical relationships. To describe how two or more properties vary with respect to one another it is often necessary to use constants of proportionality. For example, one can say that the pressure of a gas in a closed container is proportional to the temperature of the gas, but to relate these two variables in a mathematical equation, a constant of proportionality, *k*, is needed: $P = kT$. Constants make it possible to equate two or more variables. Some variables, such as the pressure and volume of a gas, are *directly* proportional to one another, others are *inversely* proportional. Ask students if they can think of any inverse relationships that they have studied. ($P = \frac{k}{V}$)

- What constant makes it possible to relate the pressure, volume, temperature, and quantity of an ideal gas? (*R*, the ideal gas constant)

Answers to Practice Problems

A. 90.3 cm^2
B. 1.4×10^8 s or 4.3 yr
C. 5.92×10^{-18} J
D. 0.00302 kg

CHEMath

USING CONSTANTS

You have seen many constants in your study of mathematics and chemistry. In this chapter you are introduced to Planck's constant, h. But what exactly are constants and how are they used?

In mathematics, a *constant* is a quantity that never changes. For example, the numbers 2, 10, and π are constants you have used in mathematics and in your everyday life. A constant is distinguished from a *variable,* which is a quantity that can vary. Treat a constant the same as any other numerical value used in a problem.

Some examples of constants are shown in the table below. Notice that some constants include units, but some do not. Constants with units are often expressed in different but equivalent forms. For example, the constant representing the speed of light can be written as 2.998×10^{10} cm/s or 2.998×10^8 m/s—different in form but equivalent in value. When solving a problem involving a constant, always pay attention to the units and use dimensional analysis to ensure that you use the correct form of the constant.

Also pay attention to whether the constant used in a formula is exact or approximate. For example, if you use 3.14 for π or 6.02×10^{23} for Avogadro's number, your approximation is accurate to 3 significant figures. However, a conversion factor such as $\frac{60\text{ s}}{1\text{ min}}$ is exact and has an unlimited number of significant figures.

Constant	Description	Approximate value
π	Pi, the ratio of circumference to diameter of a circle	3.1416
N_A	Avogadro's number, the number of representative particles in 1 mole	6.02×10^{23}
m_p	The mass of a proton	1.673×10^{-27} kg
R	Ideal gas constant, used in the ideal gas law: $PV = nRT$	$8.31 \frac{L \times kPa}{K \times mol}$
c	The speed of light in a vacuum	2.998×10^8 m/s
h	Planck's constant, used to calculate the radiant energy absorbed or emitted by a body: $E = h \times v$.	6.626×10^{-34} J × s

Example

Use the formula $A = \pi r^2$ to find the area of a circle with radius 8.43 cm.

Write the formula. $A = \pi r^2$

Substitute the known values for the constant π and the variable r. $A = 3.1416 \times (8.43\text{ cm})^2$

Evaluate and round to 3 significant figures. $A = 223\text{ cm}^2$

The area of the circle is 223 cm^2.

Practice Problems

Prepare for upcoming problems in this chapter by using the constants in the table above to solve each of the following problems.

A. Use the formula $A = \pi r^2$ to calculate the area of a circle with radius $r = 5.36$ cm.

B. Use the formula $t = d/c$ to calculate the time it takes for light to travel from Alpha Centauri to Earth, a distance of $d = 4.1 \times 10^{16}$ meters.

C. Use the formula $E = h \times v$ to calculate the energy E (in J) of a photon with a frequency of $v = 8.93 \times 10^{15}\text{ s}^{-1}$.

D. Find the mass, in kilograms, of 3.00 moles of protons.

Sample Problem 13-3

Calculate the energy (in J) of a quantum of radiant energy (the energy of a photon) with a frequency of $5.00 \times 10^{15}\ s^{-1}$.

1. *ANALYZE* ***List the knowns and the unknown.***

   Knowns:
   - frequency (ν) = $5.00 \times 10^{15}\ s^{-1}$
   - $h = 6.626 \times 10^{-34}\ J \times s$

   Unknown:
   - energy (E) = ? J

   Substitute the known values for h and ν into the equation $E = h \times \nu$ and solve.

2. *CALCULATE* ***Solve for the unknown.***

   Substitute values for h and ν and calculate E.

   $$E = h \times \nu$$
   $$E = (6.626 \times 10^{-34}\ J \times \cancel{s}) \times (5.00 \times 10^{15}\ \cancel{s^{-1}})$$
   $$= 3.31 \times 10^{-18}\ J$$

3. *EVALUATE* ***Does the result make sense?***

   Individual photons have very small energies, so the answer seems reasonable. The answer has the proper number of significant figures.

Practice Problems

13. What is the energy of a photon of microwave radiation with a frequency of $3.20 \times 10^{11}\ s^{-1}$?

14. The threshold photoelectric effect in tungsten is produced by light of wavelength 260 nm. Give the energy of a photon of this light in joules.

Chem ASAP!

Problem-Solving 14

Solve Problem 14 with the help of an interactive guided tutorial.

An analogous situation to the photoelectric effect occurs when a table-tennis ball strikes a billiard ball. The table-tennis ball simply does not have enough energy to budge the stationary billiard ball. No matter how many times the table-tennis ball collides with the billiard ball, the billiard ball will not move. By contrast, one golf ball moving at the same speed as the table-tennis ball sets the target billiard ball in motion. The golf ball is above the energy threshold. With the photoelectric effect, any excess energy of a photon beyond that needed to eject a photoelectron causes the ejected electron to travel faster. Increasing the intensity of light will increase the number of photons striking the metal and the number of electrons ejected; however, it will not effect the speed of the ejected electrons.

An Explanation of Atomic Spectra

Bohr's application of quantum theory to electron energy levels in atoms resulted in an explanation of the hydrogen spectrum. The lines observed in the spectrum are consistent with the idea that quantization limits the possible energies that an electron in a hydrogen atom can attain. Consider the lone electron of a hydrogen atom in its lowest energy level, or **ground state.** In the ground state, the quantum number (n) is 1. Excitation of the electron raises it to an excited state so that n = 2, 3, 4, 5, or 6, and so forth. If the energy levels are quantized, it takes a quantum of energy ($h \times \nu$) to raise the electron from the ground state to an excited state. The same amount of energy is emitted as a photon when the electron drops from the excited state to the ground state. Only electrons in transition from higher to lower energy levels lose energy and emit light.

Practice Problems Plus

Related Chapter Review Problems

Chapter Review problems 38 and 51 are related to Sample Problem 13-3.

Additional Practice Problem

The Chem ASAP! CD-ROM contains the following problem: The wavelength of an x-ray is typically about 0.10 nm. Give the energy of an x-ray photon in joules. (2.0×10^{-15} J)

Think Critically

Ask students to compare the transition of an electron with the positioning of an Earth satellite. Have them note differences as well as similarities in the two processes.

Discuss

Refer students to Figure 13.17. Ask:

- **Suppose an electron is excited enough to jump to energy level two. When it returns to the ground state, what type of radiation will it emit?** (UV radiation)
- **If you observed a hydrogen gas discharge tube through a diffraction grating, would you be able to see the line corresponding to this emission?** (No; the human eye cannot detect radiation in the UV range.)
- **Which series of lines would you be able to detect?** (the Balmer series, which correspond to visible light)
- **Compare the energy of the Paschen and Balmer series.** (The Paschen series has lower energy.)
- **What do you notice about the spacing of the energy levels from $n = 1$ to $n = 7$?** (The levels are not evenly spaced. The lines appear to get closer together as the distance from the nucleus increases.) **Note that each set of lines is a converging series that approaches, but never reaches, the ionization energy of the hydrogen ion.**

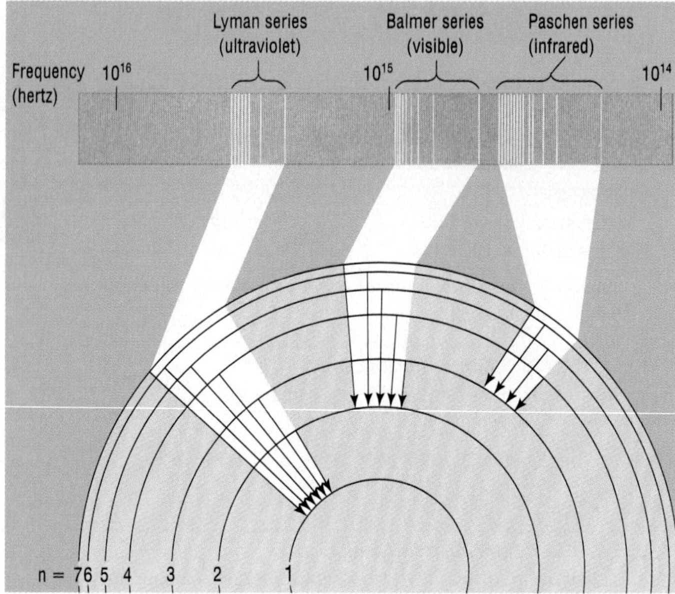

Figure 13.17

The three groups of lines in the hydrogen spectrum correspond to the transition of electrons from higher energy levels to lower energy levels. The Lyman series corresponds to the transition to the $n = 1$ energy level. The Balmer series corresponds to the transition to the $n = 2$ energy level. The Paschen series corresponds to the transition to the $n = 3$ energy level.

Figure 13.17 shows the explanation for the three groups of lines observed in the emission spectrum of hydrogen atoms. The lines at the ultraviolet end of the hydrogen spectrum are the Lyman series. These match expected values for the emission due to the transition of electrons from higher energy levels to $n = 1$. The lines in the visible spectrum are the Balmer series. The lines in the Balmer series result from transitions from higher energy levels to $n = 2$. The lines in the infrared spectrum are the Paschen series. They correspond to transitions from higher energy levels to $n = 3$. Spectral lines for the transitions from higher energy levels to $n = 4$ and $n = 5$ also exist. Note that the spectral lines in each group become more closely spaced at increased values of n. This means that the energy difference between any two higher energy levels is smaller than that between two lower levels. There is an upper limit to the frequency of emitted light for each set of lines. The upper limit exists because a very excited electron completely escapes the atom.

Bohr's theory of the atom was only partially satisfactory. It explained only the emission spectra of atoms and ions containing one electron. Moreover, it was of no help in understanding how atoms bond to form molecules. Eventually a new and better model displaced the Bohr model of the atom. The new model is based on the description of the motion of material objects as waves.

Quantum Mechanics

Such strange goings on! Energy absorbed or emitted in packages. Light behaving as waves *and* particles. Stranger things were yet to come. In 1924 Louis de Broglie (1892–1987), a French graduate student, asked an important question: Given that light behaves as waves and particles, can particles of matter behave as waves? De Broglie derived an equation that described the wavelength (λ) of a moving particle:

$$\lambda = \frac{h}{mv}$$

Here h is Planck's constant, m is the mass of the particle, and v is the velocity of the particle. From this equation it is easy to calculate the wavelength of a moving electron. With a mass of 9.11×10^{-28} g and moving at nearly the speed of light, an electron has a wavelength of about 2×10^{-10} cm. How does this value compare with the diameter of a typical atom? ❶

Indeed, **de Broglie's equation** predicts that all matter exhibits wavelike motions. You have certainly seen wave motion in water, as in **Figure 13.18.** Why then are people generally unaware of the wave motion of all matter? As with quanta, the answer is that the motion depends on the size of the object. Wavelengths of objects visible to the unaided eye are too small to measure. Objects with measurable wavelengths cannot be seen by the unaided eye. A 200-g baseball moving at 30 m/s has a wavelength of approximately 10^{-32} cm, which is too small to detect by any experiment that you could perform. By contrast, an electron moving at the same speed has a wavelength of about 2×10^{-3} cm. This distance is large enough to be measured with appropriate scientific instruments.

Figure 13.18
De Broglie stated that all matter in motion exhibits wavelike properties. This phenomenon is more apparent in some forms of matter than in others. In what other forms of moving matter do you notice wavelike properties? ❷

Think Critically

Have students use de Broglie's equation to show why the wavelength of a photon decreases with increasing mass.

Discuss

Tell students that, according to quantum mechanics, scientists cannot observe something without changing the object in the process. The change produced by the method of observation is the basis of the uncertainty principle. Present the following analogy: Suppose you want to find out if there is a car in a long tunnel. In quantum mechanics, the only sort of experiment you could do would be to send another car into the tunnel and listen for a crash. Although it is possible to detect the presence of a car with this method, it is obvious that the car will be changed by the crash.

Answers

❶ It is about the same size as the diameter of an atom.
❷ Answers will vary.

CHEMISTRY AND SCIENCE HISTORY

Albert Einstein and Niels Bohr had different views of quantum mechanics. Although Einstein accepted quantum mechanical theory as the best explanation available at the time, he was convinced the theory was incomplete. He believed in the existence of "hidden variables" which, if known, would eliminate problems such as duality and uncertainty, Bohr, on the other hand, urged physicists to accept quantum mechanics without qualification.

De Broglie's prediction that matter would exhibit both wave and particle properties set the stage for an entirely new method of describing the motions of subatomic particles, atoms, and molecules. Because mechanics is the study of the motion of bodies, the new method is called quantum mechanics. Here is a summary of the most important differences between classical mechanics and quantum mechanics.

1. Classical mechanics adequately describes the motions of bodies much larger than the atoms that they comprise. It appears that such a body gains or loses energy in any amount.
2. Quantum mechanics describes the motions of subatomic particles and atoms as waves. These particles gain or lose energy in packages called quanta.

Another feature of quantum mechanics not found in classical mechanics is the uncertainty principle, derived by the German physicist Werner Heisenberg (1901–1976) in 1927. The **Heisenberg uncertainty principle** states that it is impossible to know exactly both the velocity and the position of a particle at the same time. The more precisely the velocity is measured, the less precise the measurement of the position must become. Conversely, if the position of a moving particle is known precisely, the velocity is less precisely known. **Figure 13.19** illustrates and explains the Heisenberg uncertainty principle.

The uncertainty principle is much more obvious with small bodies like electrons than with large objects like baseballs. The uncertainty in the position of a baseball traveling at 30 m/s is only about 10^{-21} cm, which is not measurable. The uncertainty in the position of an electron with a mass of 9.11×10^{-28} g is nearly a billion centimeters! Schrödinger's quantum mechanical description of the electrons in atoms shaped the concept of electron orbitals and configurations and incorporated matter's wavelike motion and the uncertainty principle.

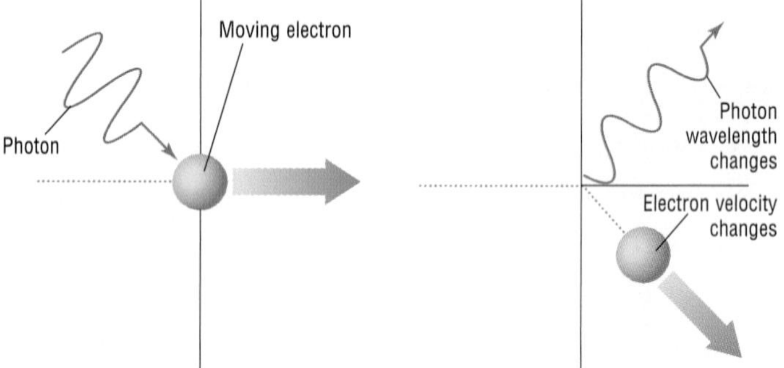

Before collision
The position of an electron can be determined only if it is struck by another particle, such as a photon.

After collision
After the impact, the electron's velocity changes. Thus it is impossible to know both the position and velocity of the electron at the same time.

Figure 13.19

The Heisenberg uncertainty principle, illustrated here, states that it is impossible to know exactly both the velocity and the position of a particle at the same time.

MINI LAB

Flame Tests

PURPOSE

Use the flame test to determine the identity of the cation in an unknown solution based on its characteristic color.

MATERIALS

- Bunsen burner
- small beaker
- 6 small test tubes
- test tube rack
- tongs
- flame test wire (10-cm length of nichrome wire)
- 6*M* HCl
- 0.1*M* NaCl
- 0.1*M* $CaCl_2$
- 0.1*M* LiCl
- 0.1*M* $CuCl_2$
- 0.1*M* $BaCl_2$
- unknown solution

PROCEDURE

1. Make a two-column data table. Label column one Cation. Label column two Flame Color. Enter the name of the cation of each salt solution in column one.
2. Label each of 5 test tubes with the name of a salt solution; label the sixth tube Unknown. Add 1 mL of each salt solution to the appropriately labeled test tube. Add 5 mL HCl to the beaker.
3. Grasp one end of the nichrome wire with the tongs. Clean the wire by alternately heating it in the hot burner flame and dipping it into the hydrochloric acid. Repeat several times until the flame remains almost colorless when the wire is heated.
4. Dip the clean wire into the sodium chloride solution and then hold it in the hot burner flame. Record the color of the flame. Repeat Step 3 to clean the wire.
5. Repeat Step 4 for each of the remaining salt solutions.
6. Perform a flame test with the unknown solution. Note the color of the flame.

ANALYSIS AND CONCLUSIONS

1. What is the identity of the cation in the unknown?
2. Each known salt solution has a unique color. Would you expect this based on the modern view of the atom? Explain.
3. Some commercially available fireplace logs burn with a red and/or green flame. On the basis of your data, what elements could be responsible for these colored flames?
4. Aerial fireworks contain gunpowder and chemicals that produce colors. What elements would you include to produce the following colors?
   a. crimson red **b.** yellow

section review 13.3

15. A hydrogen lamp emits several lines in the visible region of the spectrum. One of these lines has a wavelength of 6.56×10^{-5} cm. What are the color and frequency of this radiation?
16. Explain the origin of the atomic emission spectrum of an element.
17. Can classical physics explain the photoelectric effect? Explain your answer.
18. Compare the ground state and the excited state of an electron.
19. Arrange the following in order of decreasing wavelength.
    a. infrared radiation from a heat lamp
    b. dental x-rays
    c. signal from a shortwave radio station

Chem ASAP! **Assessment 13.3** Check your understanding of the important ideas and concepts in Section 13.3.

Answers

SECTION REVIEW 13.3

15. orange-red; $4.57 \times 10^{14} s^{-1}$
16. Electrons in atoms absorb energy, then lose the energy and emit it as light.
17. Metals eject electrons when certain wavelengths of light shine on them. Classical physics assumed any wavelength of light could cause the photoelectric effect. However, only light with some minimum frequency and threshold energy can cause an electron to be ejected.
18. The ground state is the lowest energy level of an electron. The excited state is an energy state higher than the ground state.
19. **c., a., b.**

13.3

MINI LAB

Flame Tests

SAFETY

Students should wear safety goggles. Warn students that hydrochloric acid is corrosive and can cause severe burns.

ANALYSIS AND CONCLUSIONS

Data table:
Na^+–bright yellow
Ca^{2+}–orange-red
Li^+–bright crimson red
Cu^{2+}–blue-green
Ba^{2+}–pale yellow-green

1. Answers will vary.
2. Yes; because the composition of every atom is different, the amount of energy needed for transitions between energy sublevels is different. These energy differences are reflected in the different line spectra.
3. copper (green); lithium (red) or calcium (orange-red)
4. **a.** lithium
   b. sodium

3 Assess

Evaluate Understanding

Have students draw and label wave diagrams that illustrate the relationship between wavelength and frequency. Ask students to explain the origin of the lines in the Lyman, Balmer, and Paschen series in the atomic emission spectrum of hydrogen.

Reteach

Have students go back over the section and write each scientist's name. Then have them write a brief description of each scientist's contribution to the quantum mechanical theory of the atom.

Chemistry Serving... Society

DISCUSS

Point out that, in addition to the uses mentioned in the article, lasers are also good for measuring distances. Because the beam does not spread out or lose energy, it can be reflected off a distant object. The time it takes for the beam to make the round trip can be measured to calculate the distance to the object. Scientists have actually made a laser beam travel from Earth to the moon and back by bouncing it off a mirror placed on the moon by astronauts! Ask students to find out how different kinds of lasers differ in their materials and construction.

CHEMISTRY IN CAREERS

Encourage students to connect to the Internet address shown on this page to find out more about a career as a laser technician. Information may also be found on page 874 of this textbook.

Chemistry Serving...Society

LASERS AT WORK

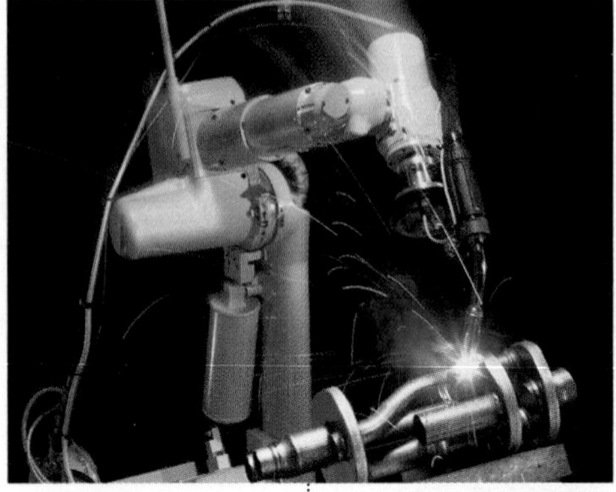

Once chemists and physicists understood the relationship between electrons and light, they could begin applying this knowledge in practical ways. One of these applications is the laser. When they were first conceived, lasers were described as "death rays" in science fiction stories, but today lasers can be found in a variety of everyday technologies.

What exactly is a laser? It is a device that emits a beam of light very different from ordinary light. Ordinary white light, you will recall, is composed of waves (or photons) of many different wavelengths and frequencies; moreover, all these waves are traveling in different directions, making ordinary light chaotic. In contrast, the waves of light emitted by a laser are all of the same frequency, and they are all traveling parallel to one another. This type of light is called coherent light.

> The waves of light emitted by a laser are all of the same frequency, and they are all traveling parallel to one another.

How is a laser able to create a beam of coherent light? The main part of a gas laser is a tube filled with an inert gas, such as neon, argon, or krypton. An electrical current is passed through the tube. The electrical energy excites the electrons of the gas atoms to higher energy states. As the electrons return to their normal energy states, they emit photons of light. The same process is at work in neon signs.

But unlike a neon sign, the gas tube of a laser is not made of glass and it has mirrors at both ends, one of which is semitransparent. Photons that happen to be traveling parallel to the walls of the tube will begin reflecting back and forth between the mirrors. As they do so, they strike already excited gas atoms, causing each of those atoms to emit a photon with the same energy and direction as the original photon. As a result, the number of photons traveling parallel to the walls of the tube increases. At the same time, some of these photons escape through the semitransparent mirror, forming a laser beam. Because all the photons/waves emerging from the laser are parallel, the laser beam does not spread out and lose intensity with increasing distance from the source, unlike the beam of a flashlight.

The coherent light of a laser has enormous power and possibilities, which is why lasers have many applications in society. One place you will find lasers is the supermarket. Many cash registers now have low-intensity lasers to scan the bar codes on packages and cans. The coherent light of the laser reflects off the bar code clearly enough to be interpreted by an optical reader, which translates the code into a number unique to the product. Lasers are used in a similar way to read the digital code on compact discs.

Lasers of higher intensity can cut through matter with great precision. High-intensity lasers directed by computers are used in the automobile and aircraft industries, for example, to cut metals for parts. Lasers have also proven invaluable in the medical field. Because everything about a laser beam can be finely controlled—its energy, width, and movement—it can be used for surgery. Laser surgery is sometimes called bloodless surgery because the laser burns away small areas, sealing the blood vessels as it goes.

CHEMISTRY IN CAREERS

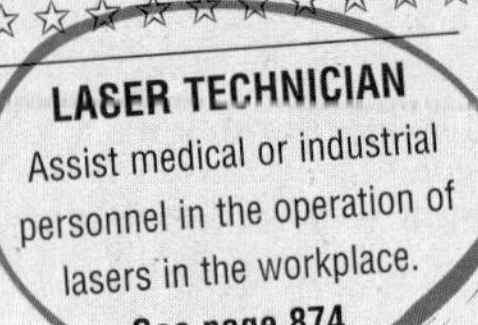

LASER TECHNICIAN
Assist medical or industrial personnel in the operation of lasers in the workplace.
See page 874.

CHEMIST NEEDED
Enviro. Lab. GC Must know

TAKE IT TO THE NET
Find out more about career opportunities:
www.phschool.com

CHEMICAL SPECIALIST
Local food service distributor see responsible self-motivated indivi

Chapter 13 STUDENT STUDY GUIDE

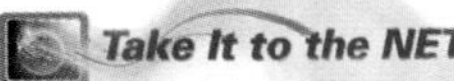
Take It to the NET
For interactive study and review, go to www.phschool.com

KEY TERMS

- amplitude *p. 372*
- atomic emission spectrum *p. 374*
- atomic orbital *p. 365*
- aufbau principle *p. 367*
- de Broglie's equation *p. 381*
- electromagnetic radiation *p. 372*
- electron configuration *p. 367*
- energy level *p. 362*
- frequency (ν) *p. 372*
- ground state *p. 379*
- Heisenberg uncertainty principle *p. 382*
- hertz (Hz) *p. 372*
- Hund's rule *p. 368*
- Pauli exclusion principle *p. 368*
- photoelectric effect *p. 377*
- photon *p. 377*
- Planck's constant (h) *p. 376*
- quantum *p. 363*
- quantum mechanical model *p. 363*
- spectrum *p. 373*
- wavelength (λ) *p. 372*

KEY EQUATIONS

- $c = \lambda\nu$
- $E = h \times \nu$
- $\lambda = \dfrac{h}{m\nu}$

CONCEPT SUMMARY

13.1 Models of the Atom

- Rutherford pictured the atom as a dense nucleus surrounded by electrons.
- In the Bohr model of the atom, the electrons move in fixed circular paths around a dense, positively charged nucleus.
- The energies of electrons in an atom are quantized in the modern quantum mechanical model of the atom.
- Current theory predicts the probability of finding an electron in terms of a cloud of negative charge. The atomic orbital, or regions in which electrons are likely to be found, can be calculated from a mathematical expression.

13.2 Electron Arrangement in Atoms

- The ways in which electrons are arranged around the nuclei of atoms are called electron configurations.
- Correct electron configurations for atoms may be written using the aufbau principle, the Pauli exclusion principle, and Hund's rule.

13.3 Physics and the Quantum Mechanical Model

- For all electromagnetic waves, the product of frequency and wavelength always equals the speed of light.
- The concept of quantized electron energy levels in atoms grew out of the study of the interaction of light and matter.
- The line emission spectra of atoms are best explained by quantized energy levels.
- The quantum concept developed in part from Planck's studies of light radiation from heated objects and from Einstein's explanation of the photoelectric effect.
- De Broglie proposed that all matter in motion has wavelike properties.
- Schrödinger devised the most successful of these early quantum mechanical models.
- In de Broglie's equation, the wavelength of an object equals Planck's constant divided by the product of the mass and velocity of the object.

CHAPTER CONCEPT MAP

Use these terms to construct a concept map that organizes the major ideas of this chapter.

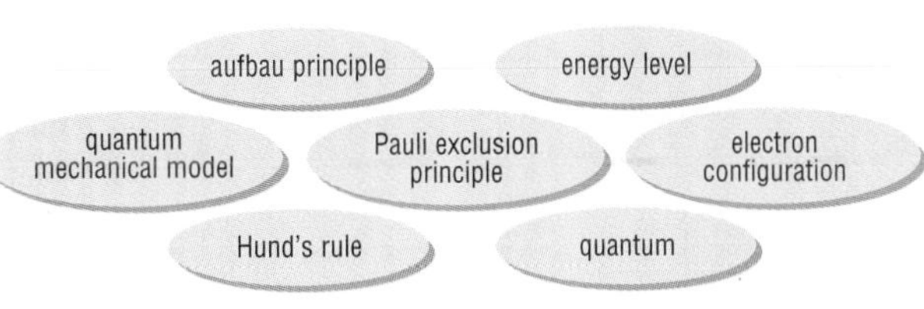

Chem ASAP! Concept Map 13
Create your Concept Map using the computer.

Take It to the Net

At **www.phschool.com** students will find for this chapter

- an Internet activity
- links to related chemistry sites
- an interactive quiz
- career links

4 Close

Summary

Have students work in small groups to develop an outline of the main chapter concepts. Then develop a composite outline, discussing each main concept as it is added.

- **Have students compare Schrödinger's and Bohr's models of the atom.** (Both models restrict the energy of electrons to certain levels. However the exact path of an orbiting electron cannot be determined in Schrödinger's model. Rather, it estimates the probability of finding an electron in a certain location.)

Extension

Suggest that students make a presentation of the visible spectrum using a prism, a diffraction grating, or a compact disk. Students may wish to extend the presentation by including the parts of the spectrum beyond the visible range. If so, have students relate various points on the spectrum to familiar experiences. For example, they could mark the point that corresponds to their favorite radio station.

Looking Back . . . Looking Ahead . . .

The previous chapter discussed the properties of gases and the laws that govern gas behavior. This chapter summarized the development of atomic theory and explained how electrons are arranged in atoms. The next chapter discusses the connection between electron configuration and periodicity.

Answers

20. electrons and positively charged particles

21. Electrons have fixed energies. To move between energy levels, they must emit or absorb specific amounts of energy (quanta).

22. In Rutherford's model, negatively charged electrons surround a dense, positively charged nucleus. In Bohr's model, electrons are assigned to circular orbits of fixed energy.

23. 90% of the time an electron is inside this boundary.

24. a region beyond the nucleus where there is a high probability of finding an electron

25. The 1*s* orbital is spherical. The 2*s* orbital is spherical with a diameter larger than that of the 1*s* orbital. The dumb-bell shaped 2*p* orbitals reach beyond the diameter of the 2*s* orbital.

26. **a.** 2 **b.** 1 **c.** 3 **d.** 6

27. Aufbau principle: electrons occupy the lowest possible energy levels; Pauli exclusion principle: an orbital can hold at most two electrons; Hund's rule: before pairing of electrons occurs, one electron occupies each of a set of orbitals with equal energies.

28. **a.** $1s^22s^22p^63s^23p^3$
b. $1s^22s^22p^63s^2$
c. $1s^22s^22p^5$
d. $1s^22s^22p^63s^23p^6$

29. The *p* orbitals in the third quantum level have three electrons.

30. **b** and **c** are invalid.

31. **a.** 2 **b.** 6 **c.** 2 **d.** 10 **e.** 6 **f.** 2 **g.** 14 **h.** 6

32. **a.** 8 **b.** 8 **c.** 8

33. **a.** $1s^22s^22p^63s^23p^63d^{10}4s^24p^4$
b. $1s^22s^22p^63s^23p^63d^34s^2$
c. $1s^22s^22p^63s^23p^63d^84s^2$
d. $1s^22s^22p^63s^23p^64s^2$

34. violet, indigo, blue, green, yellow, orange, red

35. Frequency is the number of wave cycles that pass a given point per unit time. Frequency units are cycles, reciprocal seconds (s^{-1}), or hertz. Frequency and wavelength are inversely related.

Chapter 13 REVIEW

CONCEPT PRACTICE

20. Which subatomic particles did Thomson include in the plum-pudding model of the atom? *13.1*

21. How did Bohr answer the objection that an electron traveling in a circular orbit would radiate energy and fall into the nucleus? *13.1*

22. Describe Rutherford's model of the atom and compare it with the model proposed by his student Niels Bohr. *13.1*

23. What is the significance of the boundary of an electron cloud? *13.1*

24. What is an atomic orbital? *13.1*

25. Sketch 1*s*, 2*s*, and 2*p* orbitals using the same scale for each. *13.1*

26. How many electrons are in the highest occupied energy level of these atoms? *13.2*
a. barium
b. sodium
c. aluminum
d. oxygen

27. What are the three rules that govern the filling of atomic orbitals by electrons? *13.2*

28. Write electron configurations for the elements that are identified only by these atomic numbers. *13.2*
a. 15
b. 12
c. 9
d. 18

29. What is meant by $3p^3$? *13.2*

30. Which of these orbital designations are invalid? *13.2*
a. 4*s*
b. 3*f*
c. 2*d*
d. 3*d*

31. What is the maximum number of electrons that can go into each of the following sublevels? *13.2*
a. 2*s*
b. 3*p*
c. 4*s*
d. 3*d*
e. 4*p*
f. 5*s*
g. 4*f*
h. 5*p*

32. How many electrons are in the second energy level of an atom of each element? *13.2*
a. chlorine
b. phosphorus
c. potassium

33. Write electron configurations for atoms of these elements. *13.2*
a. selenium
b. vanadium
c. nickel
d. calcium

34. List the colors of the visible spectrum in order of increasing wavelength. *13.3*

35. What is meant by the frequency of a wave? What are the units of frequency? Describe the relationship between frequency and wavelength. *13.3*

36. Use a diagram to illustrate each term. *13.3*
a. wavelength
b. amplitude
c. wave cycle

37. Explain the difference between the laws of classical physics and the quantum concept when describing the energy lost or gained by an object. *13.3*

38. What is the energy of a photon of green light with a frequency of $5.80 \times 10^{14}\ s^{-1}$? *13.3*

39. How did Planck influence the development of modern atomic theory? *13.3*

40. What will happen if the following occur? *13.3*
a. Monochromatic light shining on the alkali metal cesium is just above the threshold frequency.
b. The intensity of light increases, but the frequency remains the same.
c. Monochromatic light of a shorter wavelength is used.

41. Explain the difference between a photon and a quantum. *13.3*

42. What happens when a hydrogen atom absorbs a quantum of energy? *13.3*

43. When white light is viewed through sodium vapor in a spectroscope, the spectrum is continuous except for a dark line at 589 nm. How can you explain this observation? *13.3*

44. What is the wavelength of a 2500-kg truck traveling at a rate of 75 km/h? *13.3*

36.

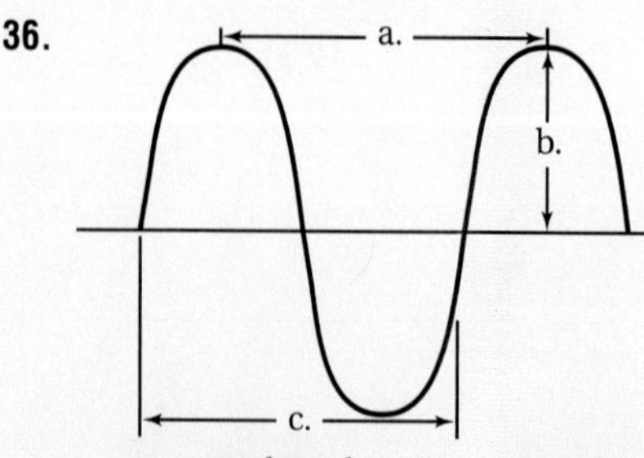

a. wavelength **b.** amplitude **c.** wave cycle

37. Classical physics viewed energy changes as continuous, occurring in any quantity. In the quantum concept, energy changes occur in discrete units called quanta.

38. 3.84×10^{-19} J

39. He showed that the quanta of radiation emitted depends on the frequency of radiation.

40. **a.** Emitted electrons have a low velocity.
b. More electrons are emitted; velocity stays low.
c. Emitted electrons have higher velocity.

CONCEPT MASTERY

45. Give the symbol for the atom that corresponds to each electron configuration.
    a. $1s^2 2s^2 2p^6 3s^2 3p^6$
    b. $1s^2 2s^2 2p^6 3s^2 3p^6 3d^{10} 4s^2 4p^6 4d^7 5s^1$
    c. $1s^2 2s^2 2p^6 3s^2 3p^6 3d^{10} 4s^2 4p^6 4d^{10} 4f^7 5s^2 5p^6 5d^1 6s^2$
46. Write the electron configuration for an arsenic atom. Calculate the total number of electrons in each energy level and state which energy levels are not full.
47. How many paired electrons are there in an atom of each element?
    a. helium
    b. boron
    c. sodium
    d. oxygen
48. An atom of an element has two electrons in the first energy level and five electrons in the second energy level. Write the electron configuration for this atom and name the element. How many unpaired electrons does an atom of this element have?
49. Suppose your favorite AM radio station broadcasts at a frequency of 1150 kHz. What is the wavelength, in centimeters, of the radiation from the station?
50. A mercury lamp, such as the one below, emits radiation with a wavelength of 4.36×10^{-7} m.

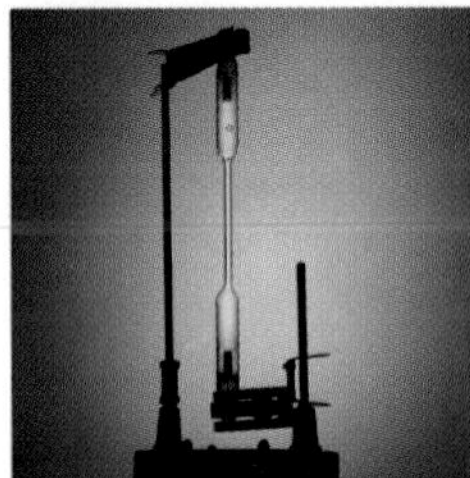

    a. What is the wavelength of this radiation in centimeters?
    b. In what region of the electromagnetic spectrum is this radiation?
    c. Calculate the frequency of this radiation.
51. Calculate the energy of a photon of red light with a wavelength of 6.45×10^{-5} cm. Compare your answer with the answer to Problem 38. Is red light of higher or lower energy than green light?
52. Give the symbol and name of the elements that correspond to these configurations.
    a. $1s^2 2s^2 2p^6 3s^1$
    b. $1s^2 2s^2 2p^3$
    c. $1s^2 2s^2 2p^6 3s^2 3p^2$
    d. $1s^2 2s^2 2p^4$
    e. $1s^2 2s^2 2p^6 3s^2 3p^6 4s^1$
    f. $1s^2 2s^2 2p^6 3s^2 3p^6 3d^2 4s^2$

CRITICAL THINKING

53. Choose the term that best completes the second relationship.
    a. orbital : energy level
       apartment : ______
       (1) floor (3) building
       (2) room (4) stairway
    b. water : container
       electrons : ______
       (1) frequency (3) nuclei
       (2) sublevel (4) orbitals
    c. electromagnetic radiation : visible light
       student body : ______
       (1) eleventh graders (3) teachers
       (2) principal (4) textbooks
54. Traditional cooking methods make use of infrared radiation (heat). Microwave radiation cooks food faster. Could radio waves be used for cooking? Explain.
55. Think about the currently accepted models of the atom and of light. In what ways do these models seem strange to you? Why are these models not exact or definite?

CUMULATIVE REVIEW

56. A potassium atom has a diameter of about 0.406 nm. Express this in meters and micrometers. If the measurement is always the same, explain why the two numbers are different.
57. Balance the following chemical equations.
    a. $KNO_3 + H_2SO_4 \longrightarrow K_2SO_4 + HNO_3$
    b. $Cu_2O + H_2 \longrightarrow Cu + H_2O$
    c. $NO + Br_2 \longrightarrow NOBr$
    d. $SnO_2 + CO \longrightarrow Sn + CO_2$
58. Calculate the volume of O_2 at STP required for the complete combustion of 5.00 L of acetylene (C_2H_2) at STP.

$$2C_2H_2(g) + 5O_2 \longrightarrow 4CO_2 + 2H_2O(l)$$

53. a. 1 b. 4 c. 1
54. Answers will vary. Students may note that radio waves have the lowest energy in the electromagnetic spectrum, and thus would not be energetic enough to cook food. Others may reason that if microwaves cook food faster than infrared radiation, then radio waves would cook food even faster.
55. The model of the atom uses the abstract idea of probability; light is considered a particle and a wave at the same time. Atoms and light cannot be compared with familiar objects because humans cannot experience them directly. Because matter and energy behave differently at the atomic level than at the level humans can observe directly.
56. 4.06×10^{-10} m; 4.06×10^{-4} μm; The units are different, but the values are all equal.
57. a. 2, 1, 1, 2
    b. 1, 1, 2, 1
    c. 2, 1, 2
    d. 1, 2, 1, 2
58. 12.5 L

41. A quantum is a discrete amount of energy. Photons are light quantas.
42. Its electron is raised to a higher energy level.
43. The outermost electron of sodium absorbs photons of wavelength 589 nm.
44. 1.3×10^{-38} m
45. a. Ar b. Ru c. Gd
46. $1s^2 2s^2 2p^6 3s^2 3p^6 4s^2 3d^{10} 4p^3$; Total = 33; the first three levels are full; the fourth level is partially filled.
47. a. 2 b. 4 c. 10 d. 6
48. $1s^2 2s^2 2p^3$; nitrogen; 3 unpaired electrons
49. 2.61×10^4 cm
50. a. 4.36×10^{-5}cm
    b. visible
    c. $6.88 \times 10^{14} s^{-1}$
51. 3.08×10^{-19} J; lower energy
52. a. Na, sodium d. O, oxygen
    b. N, nitrogen e. K, potassium
    c. Si, silicon f. Ti, titanium

Answers

59. **a.** 46.8% Si, 53.2% O
b. 34.4% Fe, 65.6% Cl
c. 11.1% H, 88.9% O
d. 2.1% H, 32.7% S, 65.2% O
e. 40% Ca, 12% C, 48% O

60. **a.** Fe^{3+}
b. Hg^{2+}
c. N^{3-}
d. HCO_3^-
e. O^{2-}
f. MnO_4^-

61. **a.** 55 protons, 55 electrons
b. 47 protons, 46 electrons
c. 48 protons, 46 electrons
d. 34 protons, 36 electrons

62. 1.46 atm

63. 154 g; 0.154 kg

64. **a.** 5.20×10^{12}, 4.40×10^{13}
9.50×10^{13}, 1.70×10^{14}
2.20×10^{14}, 4.70×10^{14}
b.

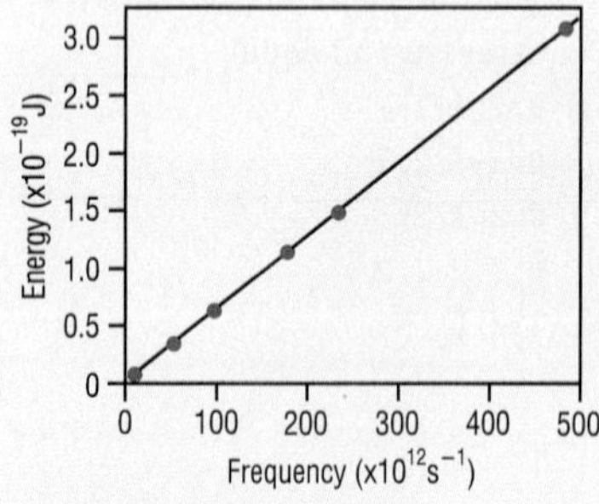

c. 6.3×10^{-34} J × s
d. The slope is Planck's constant.

65. 6.93×10^2 s

66. Outermost electrons are ejected first. Then electrons at lower energy levels, which require greater escape energy, are ejected.

67. H: 1312 kJ/mol ($n = 1$); 328 kJ ($n = 2$)
Li^{2+}: 1.18×10^4 kJ ($n = 1$)

59. Calculate the percent composition of each compound.
a. SiO_2
b. $FeCl_3$
c. H_2O
d. H_2SO_4
e. $CaCO_3$

60. Write symbols for the following ions.
a. iron(III)
b. mercury(II)
c. nitride
d. hydrogen carbonate
e. oxide
f. permanganate

61. Give the number of protons and electrons in each of the following.
a. Cs
b. Ag^+
c. Cd^{2+}
d. Se^{2-}

62. The temperature of a gas at STP is changed to 125 °C at constant volume. Calculate the final pressure of the gas in atmospheres.

63. The density of gold is 19.3 g/cm³. What is the mass, in grams, of a cube of gold 2.00 cm on each edge? In kilograms?

CONCEPT CHALLENGE

64. The energy of a photon is related to its wavelength and its frequency.
a. Complete the following table.
b. Plot the energy of the photon (y-axis) versus the frequency (x-axis).
c. Determine the slope of the line.
d. What is the significance of this slope?

Energy of photon (J)	Frequency (s^{-1})	Wavelength (cm)
3.45×10^{-21}		5.77×10^{-3}
2.92×10^{-20}		6.82×10^{-4}
6.29×10^{-20}		3.16×10^{-4}
1.13×10^{-19}		1.76×10^{-4}
1.46×10^{-19}		1.36×10^{-4}
3.11×10^{-19}		6.38×10^{-5}

65. The average distance between Earth and Mars is about 2.08×10^8 km. How long does it take to transmit television pictures from the *Mariner* spacecraft to Earth from Mars?

66. In a photoelectric experiment, a student shines light of greater than the threshold frequency on the surface of a metal. The student observes that after a long time the maximum energy of the ejected electrons begins to decrease. Can you explain why?

67. Bohr's atomic theory can be used to calculate the energy required to remove an electron from an orbit of a hydrogen atom or an ion containing only one electron. This is the ionization energy for that atom or ion. The formula for determining the ionization energy *(E)* is

$$E = Z^2 \times \frac{k}{n^2}$$

where Z is the atomic number, k is 1312 kJ mol, and n is the energy level. What is the energy required to eject an electron from a hydrogen atom when the electron is in the ground state ($n = 1$)? In the second energy level? How much energy is required to eject a ground state electron from the species Li^{2+}?

Chapter 13 STANDARDIZED TEST PREP

Select the choice that best answers each question or completes each statement.

1. Select the correct electron configuration for silicon, atomic number 14.
- **a.** $1s^22s^22p^23s^23p^24s^23d^2$
- **b.** $1s^22s^22p^43s^23p^4$
- **c.** $1s^22s^62p^6$
- **d.** $1s^22s^22p^63s^23p^2$

2. Which pair of orbitals has the same shape?
- **a.** $2s$ and $2p$
- **b.** $2s$ and $3s$
- **c.** $3p$ and $3d$
- **d.** More than one is correct.

3. Which of these statements characterize the nucleus of every atom?
- I. It has a positive charge.
- II. It is very dense.
- III. It is composed of protons, electrons, and neutrons.
- **a.** I and II only
- **b.** II and III only
- **c.** I and III only
- **d.** I, II, and III

4. As the wavelength of light increases
- **a.** the frequency increases.
- **b.** the speed of light increases.
- **c.** the energy decreases.
- **d.** the intensity increases.

5. In the third energy level,
- **a.** there are two energy sublevels.
- **b.** the f sublevel has 7 orbitals.
- **c.** there are three s orbitals.
- **d.** a maximum of 18 electrons are allowed.

The lettered choices below refer to questions 6–10. A lettered choice may be used once, more than once, or not at all.

(A) s^2p^6
(B) s^2p^2
(C) s^2
(D) s^4p^1
(E) s^2p^4

Which configuration is the outer shell electron configuration for each of these elements?

6. sulfur
7. germanium
8. beryllium
9. krypton
10. strontium

Use the atomic models to answer questions 11–13.

a.
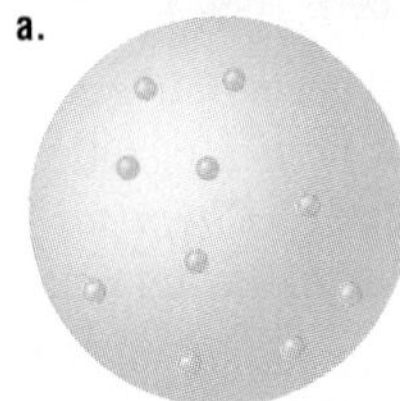

c.
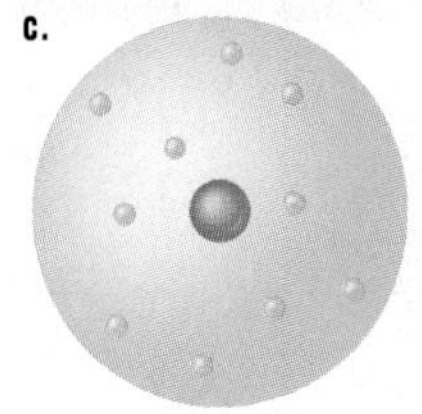

b.
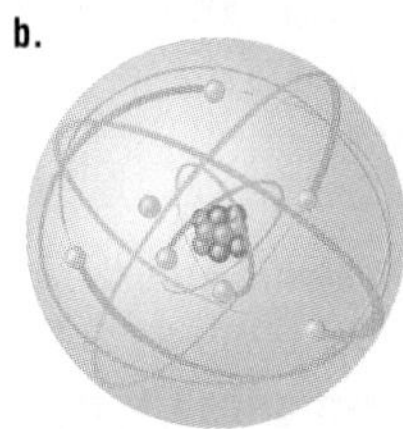

d.
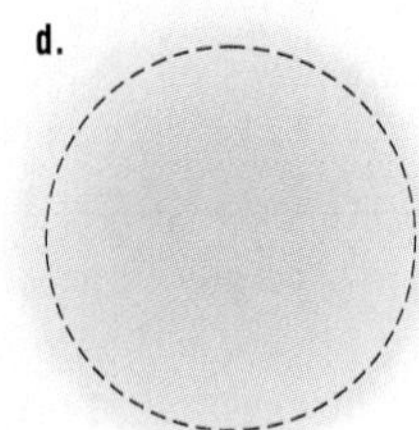

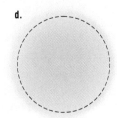

11. Which model of the atom most closely resembles a planetary model?

12. Which model was proposed based on data from Rutherford's gold foil experiment?

13. In which model are electrons in orbitals?

Use the drawings to answer questions 14–17. Each drawing represents an electromagnetic wave.

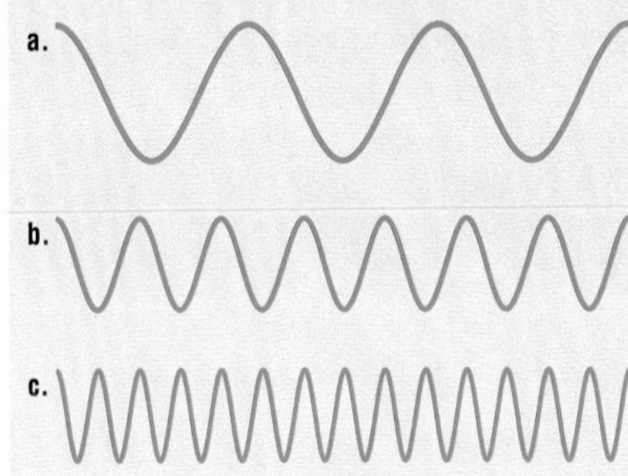

14. The wave has the longest wavelength.
15. The wave has the highest energy,
16. The wave has the lowest frequency.
17. The wave has the highest amplitude.

Write a short essay to answer question 18.

18. Explain the rules that determine how electrons are arranged around the nuclei of atoms.

Chapter 13 STANDARDIZED TEST PREP

1. d
2. b
3. a
4. c
5. d
6. (E)
7. (B)
8. (C)
9. (A)
10. (C)
11. b
12. c
13. d
14. a
15. c
16. a
17. a
18. According to the aufbau principle, electrons enter orbitals of lowest energy first. According to the Pauli exclusion principle, an orbital may contain at most two electrons. According to Hund's rule, one electron will enter each orbital of equal energy before electrons begin to pair up.

Planning Guide

SECTION OBJECTIVES	ACTIVITIES/FEATURES	MEDIA & TECHNOLOGY
14.1 Classification of the Elements ● ■ ◇ ▸ Explain why you can infer the properties of an element based on those of other elements in the periodic table ▸ Use electron configurations to classify elements as noble gases, representative elements, transition metals, or inner transition metals	**SE** Discover It! *Periodic Trends in Physical Properties,* p. 390 **SE** Small-Scale Lab *Chemical Properties of the Halides,* p. 397 (LRS 14-1) LM 23: *Periodic Properties* **TE** DEMOS, pp. 392, 394	**ASAP** Assessment 14.1 **ACT** 10: *Periodic Trends* **OT** 42, 43: *Periodic Table of the Elements* **OT** 44: *Element Sublevels*
14.2 Periodic Trends ● ■ ◆ ▸ Interpret group trends in atomic radii, ionic radii, ionization energies, and electronegativities ▸ Interpret periodic trends in atomic radii, ionic radii, ionization energies, and electronegativities	**SE** Mini Lab *Periodic Trends in Atomic Radii,* p. 399 (LRS 14-2) **SE** **Link to Astronomy** *The Big Bang,* p. 402 **SE** **Link to Music** *Newland's Octaves,* p. 405 **SE** **Chemistry Serving . . . Industry** *Big Jobs for Little Devices,* p. 407 **SE** **Chemistry in Careers:** *Solid State Chemist,* p. 407 SSLM 17: *Halogen Ions in Solution* **TE** DEMOS, pp. 400, 402, 403	**ASAP** Assessment 14.2 **ASAP** Concept Map 14 **ACT** 10: *Periodic Trends* **CHM** Side 2, 36: *Periodic Trends: Reactivity* **OT** 45: *Diatomic Molecules* **OT** 46: *Periodic Trends* **OT** 47: *Atomic Radius vs Atomic Number* **OT** 48: *Radii of Atoms and Ions* **OT** 49: *Summary of Periodic Trends* **PP** Chapter 14 Problems **RP** Lesson Plans, Resource Library **AR** Computer Test 14 **www** Activities, Self-Tests, *SCIENCE NEWS* updates **TCP** The Chemistry Place Web Site

KEY

● Conceptual (concrete concepts)	**AR** Assessment Resources	**GRS** Guided Reading and Study Workbook
○ Conceptual (more abstract/math)	**ASAP** Chem ASAP! CD-ROM	**LM** Laboratory Manual
■ Standard (core content)	**ACT** ActivChemistry CD-ROM	**LP** Laboratory Practicals
□ Standard (extension topics)	**CHM** CHEMedia Videodiscs	**LRS** Laboratory Recordsheets
◆ Honors (core content)	**CA** Chemistry Alive! Videodiscs	**SSLM** Small-Scale Lab Manual
◇ Honors (options to accelerate)	**GCP** Graphing Calculator Problems	

PRACTICE		ASSESSMENT	
SE	**Sample Problem** 14-1	**SE**	Section Review
GRS	Section 14.1	**RM**	Section Review 14.1
RM	Practice Problems 14.1	**RM**	Chapter 14 Quiz
RM	Interpreting Graphics	**LP**	Lab Practicals, 14-1, 14-2
GRS	Section 14.2	**SE**	Section Review
RM	Practice Problems 14.2	**RM**	Section Review 14.2
		RM	Vocabulary Review 14
		SE	Chapter Review
		SM	Chapter 14 Solutions
		SE	Standardized Test Prep
		PHAS	Chapter 14 Test Prep
		RM	Chapter 14 Quiz
		RM	Chapter 14 A & B Test

PLANNING FOR ACTIVITIES

STUDENT EDITION

Discover It! p. 390
- pencils and paper
- textbooks

Small-Scale Lab, p. 397
- pencils, paper, rulers
- reaction surfaces
- solutions of KF, KCl, KBr, KI, $AgNO_3$, $Pb(NO_3)_2$, NaOCl, HNO_3

Mini Lab, p. 399
- graph paper
- pencils

TEACHER'S EDITION

Teacher Demo, p. 392
- 0.2*M* solution of HCl
- 3 large test tubes
- test tube rack
- small pieces of cleaned magnesium, tin, copper
- safety goggles

Teacher Demo, p. 394
- as many samples of pure elements as possible
- consumer products made from alloys

Teacher Demo, p. 402
- two 50-mL beakers
- 40 mL of 6*M* hydrochloric acid
- overhead projector
- 20 cm of magnesium ribbon
- 1 g of calcium
- Table 14.1

Teacher Demo, p. 403
- overhead projector
- washers or poker chips

Activity, p. 404
- lists of elements

Activity, p. 405
- toothpicks cut to different lengths
- modeling clay

or

- 96-well reaction plate
- soda straws cut to different lengths
- model cement
- paper punch
- paper

OT	Overhead Transparency	**SE**	Student Edition
PHAS	PH Assessment System	**SM**	Solutions Manual
PLM	Probeware Lab Manual	**SSV**	Small-Scale Video/Videodisc
PP	Problem Pro CD-ROM	**TCP**	www.chemplace.com
RM	Review Module	**TE**	Teacher's Edition
RP	Resource Pro CD-ROM	**www**	www.phschool.com

Chapter 14

Key Terms

14.1 noble gases, representative elements, transition metals, inner transition metals
14.2 atomic radius, ionization energy, electronegativity

DISCOVER IT!

The melting points and boiling points of the alkali metals decrease with increasing atomic number. The melting points and boiling points of the halogens increase with increasing atomic number. The densities of the alkali metals generally increase with increasing atomic number. The densities of the halogens also increase with increasing atomic number. Bromine and iodine are a liquid and solid, respectively, and their densities are much higher than those of gaseous fluorine and chlorine.

Chapter 14 CHEMICAL PERIODICITY

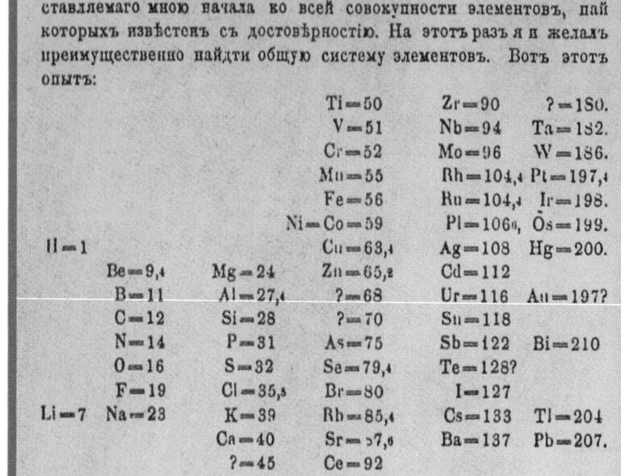
но въ ней, мнѣ кажется, уже ясно выражается примѣнимость выставляемаго мною начала ко всей совокупности элементовъ, пай которыхъ извѣстенъ съ достовѣрностію. На этотъ разъ я и желалъ преимущественно найдти общую систему элементовъ. Вотъ этотъ опытъ:

			Ti=50	Zr=90	?=180.
			V=51	Nb=94	Ta=182.
			Cr=52	Mo=96	W=186.
			Mn=55	Rh=104,4	Pt=197,4
			Fe=56	Ru=104,4	Ir=198.
			Ni=Co=59	Pl=106,6,	Os=199.
H=1			Cu=63,4	Ag=108	Hg=200.
	Be=9,4	Mg=24	Zn=65,2	Cd=112	
	B=11	Al=27,4	?=68	Ur=116	Au=197?
	C=12	Si=28	?=70	Sn=118	
	N=14	P=31	As=75	Sb=122	Bi=210
	O=16	S=32	Se=79,4	Te=128?	
	F=19	Cl=35,5	Br=80	I=127	
Li=7	Na=23	K=39	Rb=85,4	Cs=133	Tl=204
		Ca=40	Sr=87,6	Ba=137	Pb=207.
		?=45	Ce=92		
		?Er=56	La=94		
		?Yt=60	Di=95		
		?In=75,6	Th=118?		

This early periodic table was designed by Dmitri Mendeleev.

FEATURES

Stay current with **SCIENCE NEWS**
Find out more about the periodic table:
www.phschool.com

DISCOVER IT! PERIODIC TRENDS IN PHYSICAL PROPERTIES

You need paper, a pencil, and this textbook.

1. Make a table with five columns. In the first column, list the alkali metals Li, Na, K, Rb, and Cs in order. Title the other four columns Atomic number, Melting point, Boiling point, and Density.
2. Make a similar table, but list the halogens F, Cl, Br, and I in the first column.
3. Complete each table using Table A.1 in Appendix A of this textbook. Include the appropriate units.
4. Make a single histogram graph showing melting point and boiling point versus atomic number of the alkali metals.
5. Repeat Step 4 for the halogens.

For the alkali metals, what are the trends in melting points and boiling points as the atomic number increases? Are the trends the same or different for the halogens? What is the general trend in densities of the alkali metals with increasing atomic number? Although there is a trend in the densities of the halogens, some of the values differ dramatically. Explain. After completing the chapter, return to this activity and provide detailed explanations for the trends you observed.

14.1 ● ■ ◇
14.2 ● ■ ◆

Conceptual Have students work in small groups to become more familiar with the periodic table by creating flash-card quizzes or labeling blank copies of the periodic table. In Section 14.2, students will need time to interpret the graphs of periodic trends. Illustrations of trends, such as Figures 14.7 and 14.14, may be less challenging.

Standard In Section 14.1, emphasize electron configurations to help students understand the organization of the periodic table. Encourage students to use numerical data, graphs, and particle representations to visualize trends. Electronegativity is an important concept, which will be used to classify bonds in Chapter 16.

Honors Section 14.1 can be accelerated because much of it is a review of Section 5.4. Show students how to use the noble gases as a shorthand for electron configurations. Use Figure 14.16 to summarize and discuss patterns in periodic trends.

CLASSIFICATION OF THE ELEMENTS

section 14.1

Chlorine and bromine are used in swimming pools to disinfect the water. Copper and silver, two relatively soft metals, are excellent conductors of heat and electricity. Each of these pairs of elements have similar chemical properties and are listed in the same column on the modern periodic table. This order is not coincidental, but by design. **How is the early periodic table, shown here, similar to the modern periodic table, and what does an element's position on the table indicate about its properties?**

objectives

- Explain why you can infer the properties of an element based on those of other elements in the periodic table
- Use electron configurations to classify elements as noble gases, representative elements, transition metals, or inner transition metals

key terms

- noble gases
- representative elements
- transition metals
- inner transition metals

The Periodic Table Revisited

In the second half of the 1800s, before physicists began to uncover the structure of the atom, chemists were busy exploring the nature of solutions and establishing the basis of kinetic molecular theory. The science of organic chemistry was born, as were the fertilizer, bleaching, glassmaking, dyestuff, and soap industries, to name just a few. Of crucial importance to this growth in chemistry was a tool for organizing the numerous facts being discovered about the behavior of elements. Among the many attempts that were made, the most successful organizing scheme was the one devised by the Russian chemist Dmitri Mendeleev in 1871. As you learned in Chapter 5, Mendeleev arranged all of the then-known elements into a periodic table—the forerunner of the modern periodic table displayed in current chemistry textbooks. Do you recall on what basis Mendeleev ordered the elements in his table? **Figure 14.2,** on the following pages, shows a detailed periodic table of the elements. In addition to the chemical symbol, atomic number, average atomic mass, and physical state of each element, the table includes group numbers, electron configurations, and the recently accepted names and symbols for elements 104 through 109. (1)

The periodic table is probably the most important tool in chemistry. Among other things, it is very useful for understanding and predicting the properties of the elements. For example, if you know the physical and chemical properties of one element in a group or family (vertical column) of the periodic table, you can make a good guess about the physical and chemical properties of the other elements in the same group—and perhaps even of the elements in neighboring groups.

You already know that elements are arranged in the periodic table by order of increasing atomic number. In this chapter, you will learn how the periodic table also relates to atomic structure and trends among the elements.

Classifying Elements by Electron Configuration

Of the three major subatomic particles, the electron plays the most significant role in determining the physical and chemical properties of an element. The arrangement of elements in the periodic table depends on these properties. Thus there should be some relationship between the electron configurations of the elements and their placement in the table.

Figure 14.1
Dmitri Mendeleev proposed the basis for the periodic table that is used today.

14.1

1 Engage

Use the Visual

Have students study the photograph and read the text that opens the section. Ask:

- **How is the early periodic table shown here similar to the modern periodic table, and what does an element's position on the table indicate about its properties?** (In both the early and modern periodic tables, elements are arranged in rows and columns. Early periodic tables ordered the elements according to atomic mass. Modern periodic tables order elements according to atomic number.

  An element's position on the periodic table indicates whether it has the general properties of a metal, nonmetal, or noble gas. Furthermore, the properties of an element are similar to those of other elements in the same column.

Check Prior Knowledge

To assess students' prior knowledge about the organization of the periodic table, ask:

- **Going from left to right along a row in the periodic table, how are the elements arranged?** (by increasing atomic number)
- **Elements belong to what three general sections of the periodic table?** (metals, nonmetals, and noble gases)
- **What is the defining chemical property of the noble gases?** (They are relatively unreactive elements.)

STUDENT RESOURCES

From the Teacher's Resource Package, use:

- Section Review 14.1, Ch. 14 Practice Problems, Interpreting Graphics, and Quizzes from the Review Module (Ch. 13–16)
- Laboratory Recordsheet 14-1
- Laboratory Manual: Experiment 23
- Laboratory Practicals 14-1 and 14-2

TECHNOLOGY RESOURCES

Relevant technology resources include:

- Chem ASAP! CD-ROM
- Resource Pro CD-ROM
- ActivChemistry CD-ROM: *Periodic Trends*

(1) By increasing atomic mass in vertical rows; by similarities in chemical and physical properties in horizontal rows.

14.1

Motivate and Relate

Have students examine the modern periodic table shown here. Then challenge them to find similarities between the organization of the periodic table and a monthly calendar.

2 Teach

TEACHER DEMO

Prepare a 0.2*M* HCl solution. Place three large test tubes in a test tube rack. To each test tube, add a small piece of a different cleaned metal: magnesium, tin, and copper. While wearing safety goggles, carefully add some of the 0.2*M* HCl solution to each test tube. Point out the appearance of bubbles and the disappearance of metal in the tubes containing magnesium and tin. Nothing will happen in the test tube containing copper.

For a more dramatic demo, trap the hydrogen gas produced in an inverted test tube and carefully ignite the hydrogen, which reacts with an explosive pop.

Ask students to predict how silver and gold, the other elements in the same group with copper, would react with acid. Because the elements in this group are relatively unreactive, they often have been used to make long-lasting, valuable items such as jewelry and coins.

Figure 14.2

Periodic Table of the Elements

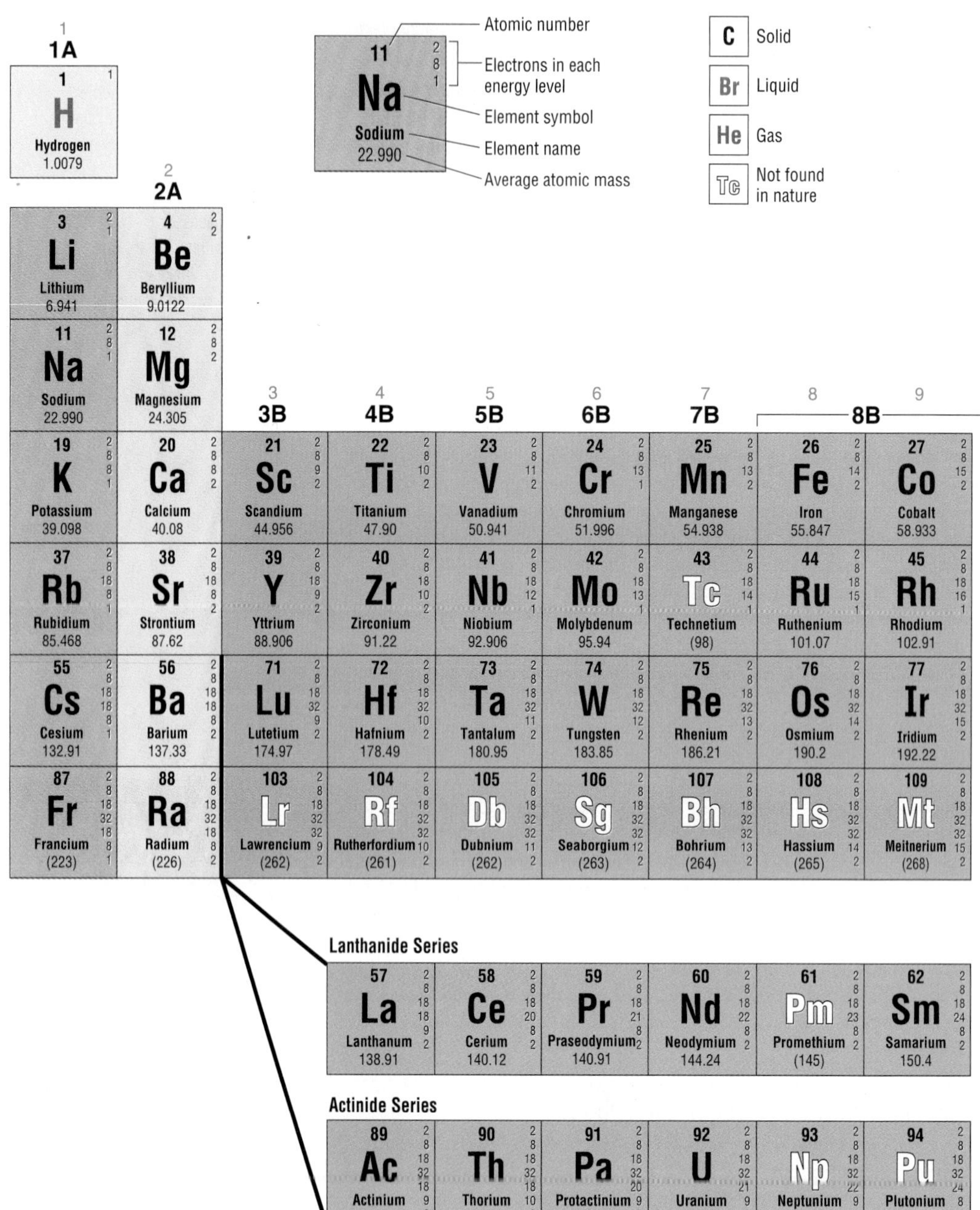

- Hydrogen
- Alkali metals
- Alkaline earth metals
- Transition metals
- Other metals
- Nonmetals
- Noble gases
- Inner transition metals

10	11 1B	12 2B	13 3A	14 4A	15 5A	16 6A	17 7A	18 0
								2 **He** Helium 4.0026 (2)
			5 **B** Boron 10.81 (2, 3)	6 **C** Carbon 12.011 (2, 4)	7 **N** Nitrogen 14.007 (2, 5)	8 **O** Oxygen 15.999 (2, 6)	9 **F** Fluorine 18.998 (2, 7)	10 **Ne** Neon 20.179 (2, 8)
			13 **Al** Aluminum 26.982 (2, 8, 3)	14 **Si** Silicon 28.086 (2, 8, 4)	15 **P** Phosphorus 30.974 (2, 8, 5)	16 **S** Sulfur 32.06 (2, 8, 6)	17 **Cl** Chlorine 35.453 (2, 8, 7)	18 **Ar** Argon 39.948 (2, 8, 8)
28 **Ni** Nickel 58.71 (2, 8, 16, 2)	29 **Cu** Copper 63.546 (2, 8, 18, 1)	30 **Zn** Zinc 65.38 (2, 8, 18, 2)	31 **Ga** Gallium 69.72 (2, 8, 18, 3)	32 **Ge** Germanium 72.59 (2, 8, 18, 4)	33 **As** Arsenic 74.922 (2, 8, 18, 5)	34 **Se** Selenium 78.96 (2, 8, 18, 6)	35 **Br** Bromine 79.904 (2, 8, 18, 7)	36 **Kr** Krypton 83.80 (2, 8, 18, 8)
46 **Pd** Palladium 106.4 (2, 8, 18, 18)	47 **Ag** Silver 107.87 (2, 8, 18, 18, 1)	48 **Cd** Cadmium 112.41 (2, 8, 18, 18, 2)	49 **In** Indium 114.82 (2, 8, 18, 18, 3)	50 **Sn** Tin 118.69 (2, 8, 18, 18, 4)	51 **Sb** Antimony 121.75 (2, 8, 18, 18, 5)	52 **Te** Tellurium 127.60 (2, 8, 18, 18, 6)	53 **I** Iodine 126.90 (2, 8, 18, 18, 7)	54 **Xe** Xenon 131.30 (2, 8, 18, 18, 8)
78 **Pt** Platinum 195.09 (2, 8, 18, 32, 17, 1)	79 **Au** Gold 196.97 (2, 8, 18, 32, 18, 1)	80 **Hg** Mercury 200.59 (2, 8, 18, 32, 18, 2)	81 **Tl** Thallium 204.37 (2, 8, 18, 32, 18, 3)	82 **Pb** Lead 207.2 (2, 8, 18, 32, 18, 4)	83 **Bi** Bismuth 208.98 (2, 8, 18, 32, 18, 5)	84 **Po** Polonium (209) (2, 8, 18, 32, 18, 6)	85 **At** Astatine (210) (2, 8, 18, 32, 18, 7)	86 **Rn** Radon (222) (2, 8, 18, 32, 18, 8)
110 ***Uun** Ununnilium (269) (2, 8, 18, 32, 32, 17, 1)	111 ***Uuu** Unununium (272) (2, 8, 18, 32, 32, 18, 1)	112 ***Uub** Ununbium (277) (2, 8, 18, 32, 32, 18, 2)		114 ***Uuq** Ununquadium		116 ***Uuh** Ununhexium		118 ***Uuo** Ununoctium

*Name not officially assigned.

63 **Eu** Europium 151.96 (2, 8, 18, 25, 8, 2)	64 **Gd** Gadolinium 157.25 (2, 8, 18, 25, 9, 2)	65 **Tb** Terbium 158.93 (2, 8, 18, 27, 8, 2)	66 **Dy** Dysprosium 162.50 (2, 8, 18, 28, 8, 2)	67 **Ho** Holmium 164.93 (2, 8, 18, 29, 8, 2)	68 **Er** Erbium 167.26 (2, 8, 18, 30, 8, 2)	69 **Tm** Thulium 168.93 (2, 8, 18, 31, 8, 2)	70 **Yb** Ytterbium 173.04 (2, 8, 18, 32, 8, 2)
95 **Am** Americium (243) (2, 8, 18, 32, 25, 8, 2)	96 **Cm** Curium (247) (2, 8, 18, 32, 25, 9, 2)	97 **Bk** Berkelium (247) (2, 8, 18, 32, 27, 8, 2)	98 **Cf** Californium (251) (2, 8, 18, 32, 28, 8, 2)	99 **Es** Einsteinium (252) (2, 8, 18, 32, 29, 8, 2)	100 **Fm** Fermium (257) (2, 8, 18, 32, 30, 8, 2)	101 **Md** Mendelevium (258) (2, 8, 18, 32, 31, 8, 2)	102 **No** Nobelium (259) (2, 8, 18, 32, 32, 8, 2)

Meeting Diverse Needs

Gifted Show students how to use the noble gases to produce a shorthand notation for electron configurations. The symbol of a noble gas in brackets can stand for its electron configuration. Thus, the configuration of sodium can be written [Ne] $3s^1$; fluorine becomes [He] $2s^2 2p^5$; and chromium becomes [Ar] $3d^5 4s^1$.

14.1

Discuss

Names suggested by those who create new elements must be approved by the International Union of Pure and Applied Chemistry (IUPAC). The process can be lengthy. For example, the names chosen for elements 104–108 by the nomenclature committee of IUPAC in 1994 were not those endorsed by the American Chemical Society (ACS). After years of negotiation, a compromise resulted in the names that appear in Figure 14.2. Elements beyond 109 have been discovered but are not yet named.

Use the Visual

Have students examine the periodic table and find the types of information that the table provides. Summerize suggestions on the chalkboard. Examples include element name, atomic number, average atomic mass, and physical state at atmospheric pressure and room temperature. Point out the sections of the table that correspond to metals, nonmetals, and noble gases.

14.1

Think Critically

Have students examine the periodic table and determine the accuracy of this statement: Atomic mass always increases as atomic number increases. (The trend is generally true, but there are exceptions. For example, the atomic numbers of Co, Ni, and Cu increase by one unit (27, 28, 29), but Ni has the lowest average atomic mass of these three elements.)

TEACHER DEMO

Display as many samples of pure elements as possible. If elements are not available, use pictures, slides, or color transparencies. Supplement with consumer items made from alloys (silverware, coins, metal cookware). Organize the elements by groups or periods. Make lists of their properties and encourage students to find similarities among elements in the same group.

Discuss

Have students create a list of all elements found in the body that are essential for metabolism. Explain that many elements are present in *trace* amounts, but are, nevertheless, essential for survival. Many trace elements play important roles in the catalytic activity of enzymes (proteins). For example, cobalt is an essential component of vitamin B_{12}, which is used to help metabolize foods in the body. Have students do research on such elements as copper, chromium, iodine, and manganese to find out more about their biological roles. Encourage students to correlate the biological roles of the elements with their chemical properties.

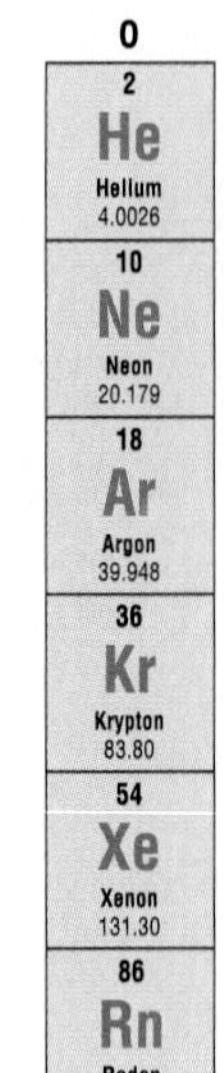

(a)

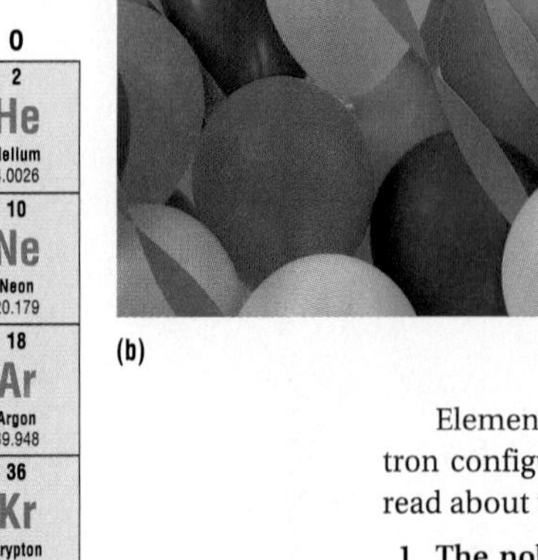

(b)

(c)

Figure 14.3
*(a) Group 0 of the periodic table contains the noble gases. **(b)** Why do helium-filled balloons rise in air?* 2
(c) *Passing an electric current through a glass tube filled with any noble gas creates the bright glow of "neon" lights.*

Elements can be classified into four categories according to their electron configurations. You will find it useful to refer to **Figure 14.2** as you read about these classifications.

1. **The noble gases.** *These are elements in which the outermost s and p sublevels are filled.* The noble gases belong to Group 0. The elements in this group are sometimes called the inert gases because they do not participate in many chemical reactions. The electron configurations for the first four noble-gas elements are listed below. Notice that these elements have filled outermost *s* and *p* sublevels.

Helium	$1s^2$
Neon	$1s^22s^22p^6$
Argon	$1s^22s^22p^63s^23p^6$
Krypton	$1s^22s^22p^63s^23p^63d^{10}4s^24p^6$

2. **The representative elements.** *In these elements, the outermost s or p sublevel is only partially filled.* The representative elements are usually called the Group A elements. (Some definitions of the representative elements may also include the noble gases.) Three groups of representative elements that have been named are the Group 1A elements, called the alkali metals; the Group 2A elements, called the alkaline earth metals; and the nonmetallic elements of Group 7A, called the halogens. For any representative element, the group number equals the number of electrons in the outermost energy level. For example, the elements in Group 1A (lithium, sodium, potassium, rubidium, and cesium) have one electron in the outermost energy level.

Lithium	$1s^22s^1$
Sodium	$1s^22s^22p^63s^1$
Potassium	$1s^22s^22p^63s^23p^64s^1$

Carbon, silicon, and germanium, in Group 4A, have four electrons in the outermost energy level.

Carbon	$1s^22s^22p^2$
Silicon	$1s^22s^22p^63s^23p^2$
Germanium	$1s^22s^22p^63s^23p^63d^{10}4s^24p^2$

How many electrons are in the outermost energy level of the Group 2A elements magnesium and calcium? In the outermost energy level of
1 the Group 5A elements phosphorus and arsenic?

3. **The transition metals.** *These are metallic elements in which the outermost s sublevel and nearby d sublevel contain electrons.* The transition elements, called the Group B elements, are characterized by addition of electrons to the *d* orbitals.

Answers

1 Group 2A elements have two electrons in the outermost energy level. Group 5A elements have five electrons in the outermost energy level.

2 Helium balloons rise in air because helium is less dense than air.

3 There are five electrons in the *p* sublevel of the outermost energy level of every halogen.

1A	2A	3A	4A	5A	6A	7A
1 H Hydrogen 1.0079						
3 Li Lithium 6.941	4 Be Beryllium 9.0122	5 B Boron 10.81	6 C Carbon 12.011	7 N Nitrogen 14.007	8 O Oxygen 15.999	9 F Fluorine 18.998
11 Na Sodium 22.990	12 Mg Magnesium 24.305	13 Al Aluminum 26.982	14 Si Silicon 28.086	15 P Phosphorus 30.974	16 S Sulfur 32.06	17 Cl Chlorine 35.453
19 K Potassium 39.098	20 Ca Calcium 40.08	31 Ga Gallium 69.72	32 Ge Germanium 72.59	33 As Arsenic 74.922	34 Se Selenium 78.96	35 Br Bromine 79.904
37 Rb Rubidium 85.468	38 Sr Strontium 87.62	49 In Indium 114.82	50 Sn Tin 118.69	51 Sb Antimony 121.75	52 Te Tellurium 127.60	53 I Iodine 126.90
55 Cs Cesium 132.91	56 Ba Barium 137.33	81 Tl Thallium 204.37	82 Pb Lead 207.2	83 Bi Bismuth 208.98	84 Po Polonium (209)	85 At Astatine (210)
87 Fr Francium (223)	88 Ra Radium 226.03					

(a) (b) (c) (d)

Figure 14.4
Elements in Groups 1A–7A are called representative elements. ***(a)*** *Recyclable cans are often made of aluminum.* ***(b)*** *Pure sodium, a very reactive metal, is so soft that it can be cut with a knife.* ***(c)*** *The element carbon accounts for 0.08% of the mass of Earth's crust. Graphite, diamond, and buckminsterfullerene are allotropes of carbon.* ***(d)*** *Sulfur, a Group 6A element, is present in petroleum and coal, causing environmental pollution during combustion of these fossil fuels.*

4. **The inner transition metals.** *These are metallic elements in which the outermost s sublevel and nearby f sublevel generally contain electrons.* The inner transition metals are characterized by the filling of *f* orbitals.

If you consider both the electron configurations and the positions of the elements in the periodic table, another pattern emerges. As you can see in **Figure 14.5,** the periodic table can be divided into sections, or blocks, that correspond to the sublevels that are filled with electrons.

The *s* block is the part of the periodic table that contains the elements with s^1 and s^2 outer electron configurations. It is composed of the elements in Groups 1A and 2A and the noble gas helium. The *p* block is composed of the elements in Groups 3A, 4A, 5A, 6A, 7A, and 0 with the exception of helium. The transition metals belong to the *d* block, and the inner transition metals belong to the *f* block.

Electron configurations of elements can be determined by using the periodic table in **Figure 14.5.** Simply read the periodic table as you would read a book, from left to right and top to bottom, until you reach the element of interest. Each period number on the periodic table corresponds to

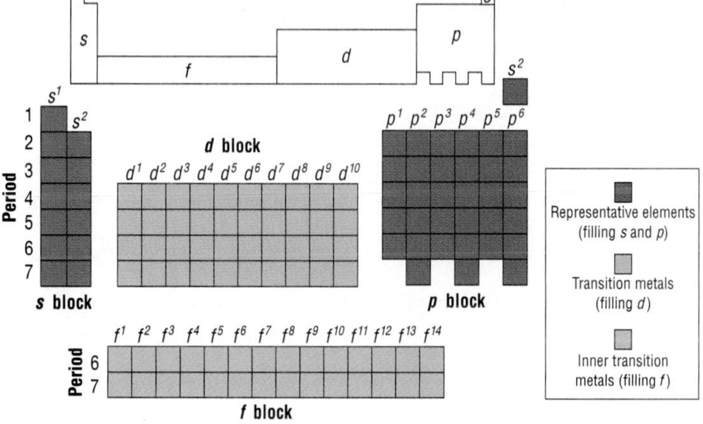

Figure 14.5
This block diagram identifies groups of elements according to the sublevels filled with electrons. How many electrons are in the p sublevel of every halogen? 3

14.1

Discuss

Lead a class discussion on the four categories of elements in the periodic table according to electron configuration. For each category, select some elements and have students write out the electron configurations for those elements. Have students compare the electron configurations for all the selected elements in a single category. Ask students to identify similarities.

Cooperative Learning

Divide the class into groups of two or three students. Have each group write electron configurations for a portion of the periodic table. (Include representative elements, noble gases, and selected examples of the transition elements, lanthanide series, and actinide series.) Have students arrange the electron configurations into periods and groups. Do this on the chalkboard or an overhead transparency so that everyone in the class can see.

Discuss

Students may be surprised to learn that the lanthanides are widely used in industry. Color television pictures are brighter, for example, because phosphor dots on the screen are made of europium-activated yttrium compounds. Have students do library research on the uses of these elements, and compile a list of consumer products that contain lanthanides or require lanthanides for processing.

Meeting Diverse Needs

At Risk Hand out photocopies of the periodic table. Have students color sections of the table corresponding to the following four categories: noble gases, representative elements, transition metals, and inner transition metals. Have them label rows with periodic numbers and columns with energy sublevels being filled with electrons. Ask them to write a paragraph describing how they can use their table to write the electron configuration for any element. Ask them to include a couple of specific examples.

Practice Problems Plus

Related Chapter Review Problems
Chapter Review problems 11, 13, and 15 are related to Sample Problem 14-1.

3 Assess

Evaluate Understanding

Ask the students to examine the periodic table. Call out pairs of elements in the same group and have the students write the electron configurations for each element. This activity can be made into a game where groups of students work together and compete with other groups to come up with the answer first. The students should eventually be able to write the electron configurations quickly if they understand how to use the position in the group and period of the periodic table to assign the electron configurations. Ask what is the similarity in the outer shell electron configurations of each pair of elements.

Reteach

To reinforce the relationship between configurations and position on the periodic table, provide configurations and ask students to identify and locate the corresponding elements on the periodic table. Ask students to explain which parts of a configuration proved most useful for determining the identity and location of an element.

the principal energy level. The number of electrons in each sublevel is determined by counting over to the element, again from left to right. For the transition elements, electrons are added to a *d* sublevel with a principal energy level that is one less than the period number. For the inner transition metals, the principal energy level of the *f* sublevel is two less than the period number. Following this procedure gives the correct electron configurations for most atoms.

Sample Problem 14-1

Use the periodic table in **Figure 14.5** to write the electron configurations of these elements.

a. nitrogen **b.** nickel

1. *ANALYZE* ***Plan a problem-solving strategy.***
Apply the procedure of using position in the periodic table to determine electron configurations of elements. The atomic number equals the number of electrons. The period in which the element is located corresponds to the highest occupied principal energy level. The number of electrons in the highest occupied sublevel is related to the group location.

2. *SOLVE* ***Apply the problem-solving strategy.***
a. Nitrogen has 7 electrons. The periodic table in **Figure 14.5** shows that the first period is $1s^2$ and the second period is $2s^22p^3$. There are 3 electrons in the $2p$ sublevel because nitrogen is the third element in the $2p$ block.
b. Nickel has 28 electrons. From **Figure 14.5,** the first three periods are $1s^22s^22p^63s^23p^6$. Next is $4s^2$ and finally $3d^8$. Remember that the principal energy level number for the *d* block is always one less than the period number. The complete configuration is $1s^22s^22p^63s^23p^63d^84s^2$.

3. *EVALUATE* ***Do the results make sense?***
In each case, the sum of the superscripts equals the total number of electrons in, or the atomic number of, the atom.

Practice Problems

1. Use **Figure 14.5** to write the electron configurations of these elements.
   a. carbon
   b. vanadium
   c. strontium
2. What are the symbols for all the elements that have the following outer configurations?
   a. s^2
   b. s^2p^5
   c. s^2d^2

section review 14.1

3. Why do the elements potassium and sodium have similar chemical and physical properties?
4. Categorize each element as a representative element, a transition metal, or a noble gas.
   a. $1s^22s^22p^63s^23p^63d^{10}4s^24p^64d^{10}5s^1$
   b. $1s^22s^22p^63s^23p^63d^{10}4s^24p^6$
   c. $1s^22s^22p^63s^23p^63d^54s^1$
   d. $1s^22s^22p^63s^23p^2$
5. Which of the following are transition metals: Cu, Sr, Cd, Au, Al, Ge, Co?

 Chem ASAP! **Assessment 14.1** Check your understanding of the important ideas and concepts in Section 14.1.

Answers

SECTION REVIEW 14.1

3. Sodium and potassium have similar chemical and physical properties because they have similar electron configurations with a single electron in the outermost *s* sublevel.
4. **a.** transition metal (Ag)
   b. noble gas (Kr)
   c. transition metal (Cr)
   d. representative element (Si)
5. Cu, Cd, Au, Co; These elements have partially filled *d* sublevels.

SMALL-SCALE LAB

CHEMICAL PROPERTIES OF THE HALIDES

SAFETY   

Wear safety glasses and follow the standard safety procedures outlined on page 18.

PURPOSE

To observe some properties of halide ions and to use these data to predict trends.

MATERIALS

- pencil
- reaction surface
- paper
- ruler
- chemicals shown in Figure A along with HNO_3 and NaOCl

PROCEDURE

On separate sheets of paper, draw two grids similar to Figure A, one with large Xs and one without. Make each square 2 cm on each side. Place a reaction surface over the grid with the Xs and add the chemicals as shown in Figure A. Use the second grid as a data table to record your observations.

	$AgNO_3$ (Ag^+)	$Pb(NO_3)_2$ (Pb^{2+})
KF (F^-)	no visible reaction	white ppt
KCl (Cl^-)	white ppt	white ppt
KBr (Br^-)	white ppt	white ppt
KI (I^-)	lime-green ppt	bright yellow ppt

Figure A

ANALYSIS

Using your experimental data, record the answers to the following questions below your data table.

1. List the formulas and charges of each halide ion.
2. Write the electron configuration of each halide ion. What do the electron configurations of all the halide ions have in common?
3. Which halide ions formed precipitates? What color was each precipitate? Write and balance a net ionic equation for each precipitation reaction you observed.
4. Which silver halide did not form a precipitate? This halide is the most soluble. Assuming the solubility of the silver halides follows a group trend, list the silver halides in order of decreasing solubility.
5. Can you make the same sort of prediction about the relative solubilities of the lead halides? Why or why not?

YOU'RE THE CHEMIST

The following small-scale activities allow you to develop your own procedures and analyze the results.

1. **Design It!** Design and carry out an experiment to determine how halide ions react with sodium hypochlorite in the presence of nitric acid. Which halide ions react with NaOCl? Which do not?
2. **Analyze It!** Use your data to predict a trend in reactivity of the halide ions.
3. **Analyze It!** OCl^- reacts with I^- and H^+ to form I_2 according to the following chemical equation.

   $$OCl^- + 2H^+ + 2I^- \longrightarrow H_2O + Cl^- + I_2$$

   Write a similar equation for the other halide ion that reacts with OCl^-.
4. **Analyze It!** The value of the reduction potential (E^0) is a measure of the relative ease with which halide ions react to form halogens. In general, the more positive the reduction potential, the less likely a halide will form a halogen. Look up the reduction potential values of I_2, Br_2, Cl_2, and F_2 in Section 23.2. Are these reduction potential values consistent with your prediction? Explain.

SMALL-SCALE LAB

CHEMICAL PROPERTIES OF THE HALIDES

OVERVIEW

Procedure Time: 30 minutes
Students observe some chemical properties of halide ions and use their results to predict family trends.

MATERIALS

Solution	**Preparation**
0.05*M* $AgNO_3$	2.1 g in 250 mL
0.2*M* $Pb(NO_3)_2$	16.6 g in 250 mL
0.1*M* KI	4.2 g in 250 mL
0.2*M* KBr	6.0 g in 250 mL
0.4*M* KCl	7.5 g in 250 mL
0.4*M* KF	5.8 g in 250 mL
1% NaOCl	50 mL household bleach in 200 mL
1.0*M* HNO_3	63 mL of 15.8*M* in 1.0 L

CAUTION! Always add acid to water carefully and slowly!

TEACHING SUGGESTIONS

Make sure students understand why they can predict a solubility trend for silver halides but not for lead halides. YOU'RE THE CHEMIST activity 4 is most appropriate for honors students because reduction potentials are not discussed until Chapter 23.

YOU'RE THE CHEMIST

1. Add a drop of each halide to one drop of NaOCl and one drop of HNO_3. KBr and KI react with NaOCl. KCl and KF do not.
2. The reactivity of the halides with NaOCl is KI > KBr > KCl > KF.
3. $OCl^- + 2H^+ + 2Br^- \rightarrow H_2O + Cl^- + Br_2$
4. $I_2 + 2e^- \rightarrow 2I^-$ $E° = +0.54V$
   $Br_2 + 2e^- \rightarrow 2Br^-$ $E° = 1.07V$
   $Cl_2 + 2e^- \rightarrow 2Cl^-$ $E° = 1.36V$
   $F_2 + 2e^- \rightarrow 2F^-$ $E° = +2.87V$
   These reduction potential values are consistent with the prediction made in activity 2.

ANALYSIS

1. F^-, Cl^-, Br^-, I^-
2. F^-: $1s^22s^22p^6$
   Cl^-: $1s^22s^22p^63s^23p^6$
   Br^-: $1s^22s^22p^63s^23p^64s^23d^{10}4p^6$
   I^-: $1s^22s^22p^63s^23p^64s^23d^{10}4p^65s^24d^{10}5p^6$
   All halides have 6 *p* valence electrons.
3. $Ag^+ + Cl^- \rightarrow AgCl(s)$
   $Ag^+ + Br^- \rightarrow AgBr(s)$
   $Ag^+ + I^- \rightarrow AgI(s)$
   $Pb^{2+} + F^- \rightarrow PbF_2(s)$
   $Pb^{2+} + Cl^- \rightarrow PbCl_2(s)$
   $Pb^{2+} + Br^- \rightarrow PbBr_2(s)$
   $Pb^{2+} + I^- \rightarrow PbI_2(s)$
4. AgF did not form a precipitate. The trend for solubility is AgF > AgCl > AgBr > AgI.
5. Because all of the lead halides formed precipitates, students can't derive a trend for solubility from these data.

1 Engage

Use the Visual

Have students look at the opening photograph and describe the types of characteristics that exist in members of the same family that might be used to characterize these families. Then ask them to describe characteristics that they think can be different in different members of the same family. (Be sensitive to students who are not biologically related to their immediate family members.) Discuss the analogy as it pertains to groupings in the periodic table—elements can share traits but also display individual traits as well. Ask:

- **What trends in physical and chemical properties exist among the families and periods of the periodic table?** (Elements within the different sections of the periodic table show shared characteristics of metals, nonmetals, or noble gases. Elements in the same group show similar chemical and physical properties. This section discusses chemical and physical properties such as atomic radii, ionization energy, and electronegativity that show group and period trends.)

Motivate and Relate

As an analogy to positions and properties on the periodic table, use seating charts and pricing data from local theaters or sports venues to discover trends. Ask students to determine patterns that relate the position of a seat to its price. Students should discover that variables such as distance from the stage or field, location relative to the center of the action, and whether the view will be obstructed, all affect price.

section 14.2

PERIODIC TRENDS

objectives
- Interpret group trends in atomic radii, ionic radii, ionization energies, and electronegativities
- Interpret period trends in atomic radii, ionic radii, ionization energies, and electronegativities

key terms
- atomic radius
- ionization energy
- electronegativity

Have you ever noticed physical similarities between relatives at a family reunion? The relatives may have similar-shaped noses or faces, or have lots of freckles. These characteristics generally indicate a relationship among the family members. As you know, the elements also belong to families—chemical families. **What trends in physical and chemical properties exist among the families and periods of the periodic table?**

Trends in Atomic Size

You know from the quantum mechanical model, which you read about in Chapter 13, that an atom does not have a sharply defined boundary that sets the limit of its size. Therefore, the radius of an atom cannot be measured directly. There are, however, several ways to estimate the relative sizes of atoms. If the atoms are in a solid crystalline structure, a technique called x-ray diffraction can provide an estimate of the distance between the nuclei. For elements that exist as diatomic molecules, the distance between the nuclei of the atoms bonded in the molecule can be estimated. The **atomic radius** is one-half of the distance between the nuclei of two like atoms in a diatomic molecule. Look at **Figure 14.7,** which shows the distance between the nuclei in the diatomic molecules of seven elements. The separation between the nuclei in a diatomic bromine molecule (Br_2) is 228 pm (1 pm = 1 picometer = 1×10^{-12} m). As you can see, because the atomic radius is one-half the distance between the nuclei, a value of 114 pm (228/2) is assigned as the radius of the bromine atom. **Figure 14.8** shows atomic radii for most of the representative elements. Remember, the atomic radius of an element indicates its relative size.

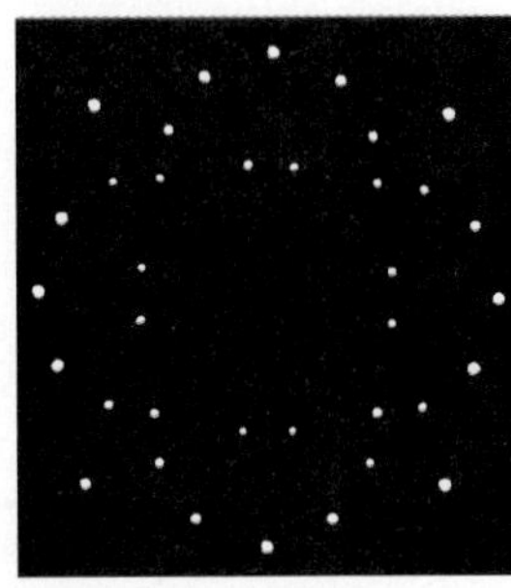

Figure 14.6
Analysis of this x-ray diffraction pattern of NaCl will reveal the distance between the two nuclei in the crystalline structure.

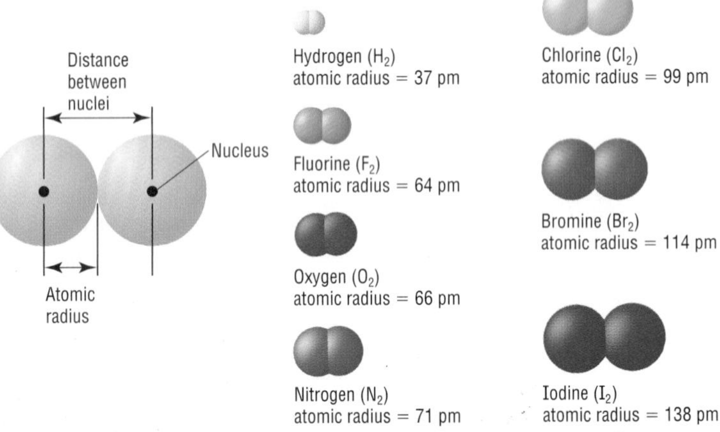

Figure 14.7
Seven elements exist as diatomic molecules. In the bromine molecule (Br_2), the distance between the nuclei is 228 pm and the atomic radius is 114 pm. What is the atomic radius, in meters, of a bromine atom? What is the diameter, in nanometers?
1

STUDENT RESOURCES

From the Teacher's Resource Package, use:
- Section Review 14.2, Ch. 14 Practice Problems, Vocabulary Review, Quizzes and Tests from the Review Module (Ch.13–16)
- Laboratory Recordsheet 14-2
- Small-Scale Chemistry Lab Manual: Experiment 17

TECHNOLOGY RESOURCES

Relevant technology resources include:
- Chem ASAP! CD-ROM
- Resource Pro CD-ROM
- ActivChemistry CD-ROM: *Periodic Trends*
- Assessment Resources CD-ROM: Chapter 14 Tests

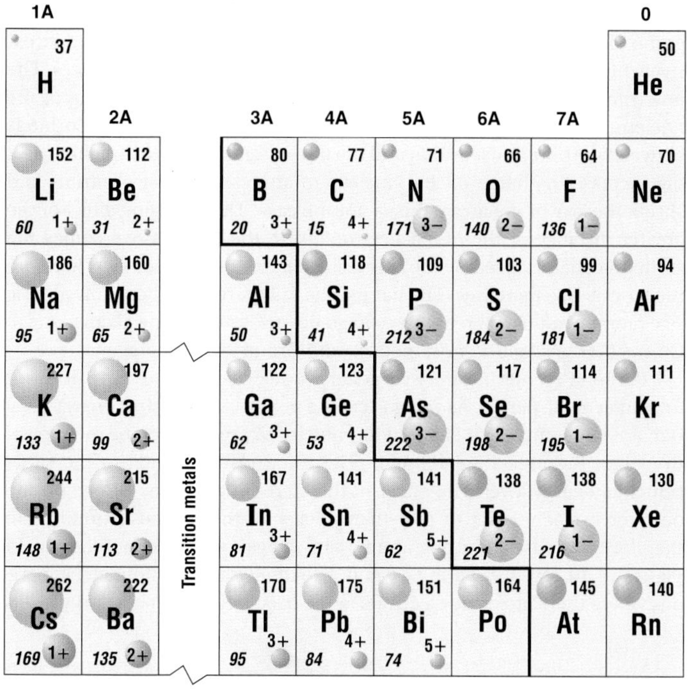

1A	2A		3A	4A	5A	6A	7A	0
H 37								He 50
Li 152; 60 1+	Be 112; 31 2+		B 80; 20 3+	C 77; 15 4+	N 71; 171 3−	O 66; 140 2−	F 64; 136 1−	Ne 70
Na 186; 95 1+	Mg 160; 65 2+	Transition metals	Al 143; 50 3+	Si 118; 41 4+	P 109; 212 3−	S 103; 184 2−	Cl 99; 181 1−	Ar 94
K 227; 133 1+	Ca 197; 99 2+		Ga 122; 62 3+	Ge 123; 53 4+	As 121; 222 3−	Se 117; 198 2−	Br 114; 195 1−	Kr 111
Rb 244; 148 1+	Sr 215; 113 2+		In 167; 81 3+	Sn 141; 71 4+	Sb 141; 62 5+	Te 138; 221 2−	I 138; 216 1−	Xe 130
Cs 262; 169 1+	Ba 222; 135 2+		Tl 170; 95 3+	Pb 175; 84 4+	Bi 151; 74 5+	Po 164	At 145	Rn 140

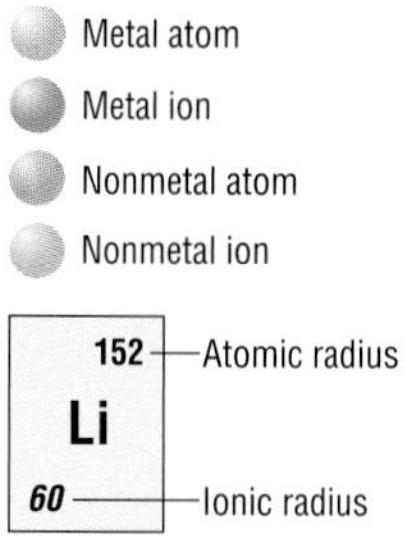

Metal atom
Metal ion
Nonmetal atom
Nonmetal ion

152 — Atomic radius
Li
60 — Ionic radius

Figure 14.8
Atomic and ionic radii of the representative elements are given here in picometers. Transition metals are omitted from the figure because they show many exceptions to the general trend.

MINI LAB

Periodic Trends in Atomic Radii

PURPOSE

To graph ionic radius versus atomic number for the representative elements in periods 2–5 and to examine the graph for periodic and group trends.

MATERIALS

- graph paper
- pencil

PROCEDURE

Use the information presented in **Figure 14.8** to plot ionic radius versus atomic number.

ANALYSIS AND CONCLUSIONS

Ionic Radii vs. Atomic Number

Ionic radii (pm): 0, 50, 100, 150, 200, 250
Atomic number: 0, 10, 20, 30, 40, 50, 60

1. Comment on the sizes of cations compared with the sizes of anions. How do these sizes compare with those of the atoms?
2. Are the general trends shown for periods 2, 3, 4, and 5 similar or different? 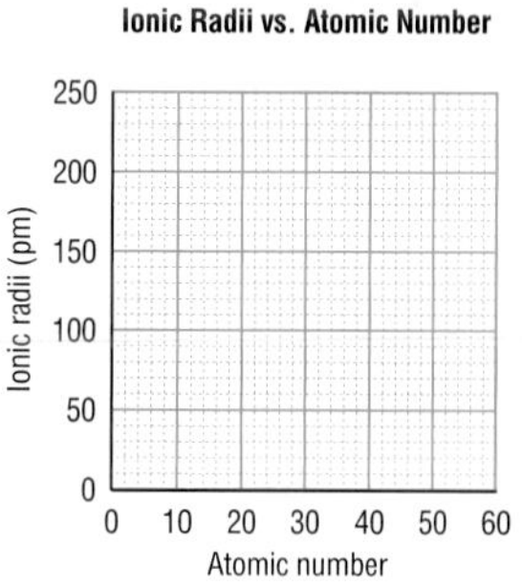
3. Describe and explain the shape of each period's portion of the graph.
4. How do the radii for anions and cations change as you go down a group? Explain.

Answer

❶ 1.14×10^{-10} m; 0.228 nm

Discuss

Emphasize the key roles electrical attraction and repulsion play within atoms and ions. Review the effects of increasing nuclear charge and changes in the shielding effect of electrons on the size of an atom. Have students use these effects to describe the size changes for atoms within a period, as well as for groups of atoms.

MINI LAB

Periodic Trends in Atomic Radii

LAB PREP AND PLANNING

Review basic graphing procedure before students begin working.

ANALYSIS AND CONCLUSIONS

1. Cations are smaller than anions. Cations are smaller than the neutral atoms; anions are larger.
2. Similar
3. The lines slope downward, then rise sharply and resume a downward trend. The abrupt changes in slope occur each time a new period begins.
4. Radii for both increase, because electrons begin to fill the next higher energy level with each new period.

TEACHER DEMO

Use the following demo to clarify the concepts of effective nuclear charge and electron shielding. Choose four students to be "protons" and four to be "electrons." Construct a lithium "nucleus" by having three "protons" stand together at the front of the room. Note that for purposes of this demo you are ignoring the neutrons. Place two "electrons" together at a short distance from the nucleus to represent the 1*s* electrons. Place the third "electron" a bit farther away to represent the 2*s* electron. You should be able to draw a line from the nucleus through the 1*s* electrons to the 2*s* electron. Point out that there are no other electrons between the inner electrons and the nucleus; thus, the 1*s* electrons experience the full impact of the 3+ charge. Because the third electron's "view" of the nucleus is partially blocked, it is shielded somewhat from the full force of the 3+ charge.

Group Trends Atomic size generally increases as you move down a group of the periodic table. As you descend, electrons are added to successively higher principal energy levels and the nuclear charge increases. The outermost orbital is larger as you move downward. The shielding of the nucleus by electrons also increases with the additional occupied orbitals between the outermost orbital and the nucleus. Although you might expect the increase in charge on the nucleus to attract the outer electrons and shrink the size of the atom, this is not the case. The enlarging effect of the greater distance of the outer electrons from the nucleus overcomes the shrinking effect caused by the increasing charge of the nucleus. Therefore the atomic size increases. The bar graphs in **Figure 14.9** show how atomic size (atomic radius) increases as you go down Group 1A (the alkali metals), Group 2A (the alkaline earth metals), and Group 7A (the halogens).

Periodic Trends Atomic size generally decreases as you move from left to right across a period. As you go across a period, the principal energy level remains the same. Each element has one more proton and one more electron than the preceding element. The electrons are added to the same principal energy level. The effect of the increasing nuclear charge on the outermost electrons is to pull them closer to the nucleus. Atomic size therefore decreases. Plotting atomic radius against atomic number, as in **Figure 14.10,** reveals a periodic trend. The trend is less pronounced in periods where there are more electrons in the occupied principal energy levels

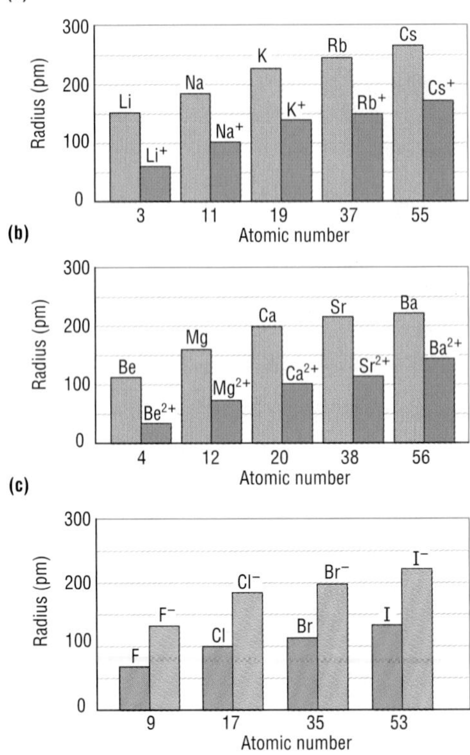

Figure 14.9
*The atomic radii of **(a)** Group 1A, **(b)** Group 2A, and **(c)** Group 7A elements increase as you go down the group, or as the atomic number increases. The ions (cations) in **(a)** and **(b)** are smaller than the neutral atoms. In contrast, the ions (anions) in **(c)** are larger than the neutral atoms. Why is a potassium atom larger than a potassium ion?* ❶

Answers

❶ A potassium atom is larger than the potassium cation since removal of an electron results in increased attraction by the nucleus for fewer electrons.

❷ The atomic radius of a period-2 alkaline earth metal is smaller than that of a period-4 alkaline earth metal.

❸ An alkali metal has a larger atomic radius than a halogen within the same period.

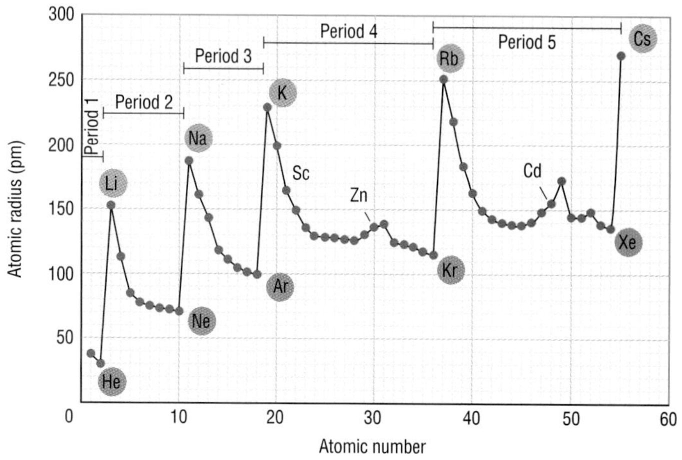

Figure 14.10

This graph of atomic radius versus atomic number shows a periodic trend.

between the nucleus and the outermost electrons. This is because these inner electrons help shield the outermost electrons and the nucleus from each other. In any given period, however, the number of electrons between the nucleus and the outermost electrons is the same for all elements. Consequently, the shielding effect of these electrons on the nucleus is constant within a period.

Figure 14.11 summarizes the group and period trends in atomic size. How would you describe the atomic radius of a period-2 alkaline earth metal compared with that of a period-4 alkaline earth metal? 2

Trends in Ionization Energy

When an atom gains or loses an electron, it becomes an ion. The energy required to overcome the attraction of the nuclear charge and remove an electron from a gaseous atom is called the **ionization energy.** Removing one electron results in the formation of a positive ion with a 1+ charge.

$$Na(g) \longrightarrow Na^+(g) + e^-$$

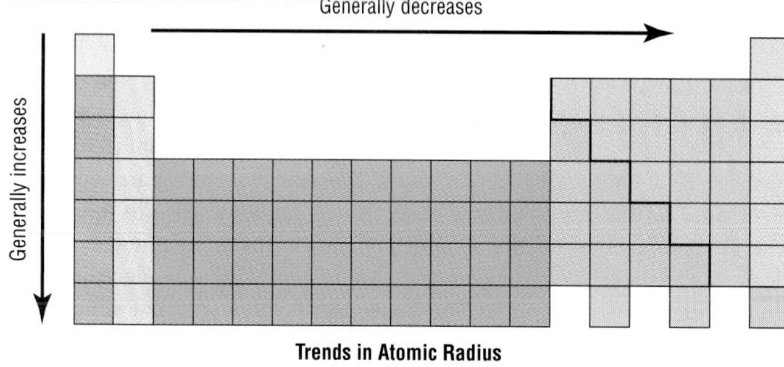

Figure 14.11

Atomic radii generally decrease across periods and increase down groups. Which has the larger atomic radius within the same period: a halogen or an alkali metal? 3

14.2

Build Writing Skills

Have students write paragraphs comparing periodic and group trends in atomic and ionic size.

Cooperative Learning

Divide the class into eight groups, and have them make three-dimensional models of atoms in the representative groups of the first five periods of the periodic table. Before work begins, have a committee made up of one member from each group agree on an appropriate scale to show atomic size. (Most of the information they need is in Figure 14.9.) Each group of students should select its own format for the models. In addition to atomic size, each model should show the composition of the nucleus and a generalized arrangement of electrons. Arrange the completed models as in the periodic table, and hang them by strings from the ceiling of the classroom.

Discuss

Explain that ionization energy is a measure of the difficulty in removing an electron from the outermost energy level of an atom. Two factors affect ionization energy: nuclear charge and distance from the nucleus.

Meeting Diverse Needs

Gifted Have students research and describe the phenomenon of the *lanthanide contraction.* Ask them to discuss how the lanthanide contraction accounts for the fact that zirconium and hafnium have virtually the same atomic radius even though hafnium is below zirconium in Group 4B of the periodic table.

TEACHER DEMO

Demonstrate the relative reactivities of magnesium and calcium. Pour 20 mL of 6*M* HCl into each of two 50-mL beakers. Then, place the beakers on an overhead projector. Coil about 20 cm of magnesium ribbon and drop it into one beaker. Drop 1 g of calcium into the other beaker. Compare the reaction rates in the two beakers. Have students record their observations. Point out the positions of the two elements in the periodic table, and relate the difference in chemical reactivity to their first and second ionization energies. Students know that Group 2A metals form ions with a 2+ charge. Using Table 14.1, show students how this fact is reflected by a large increase in ionization energy between the second and third ionization energy for both Mg and Ca. Ask students to predict the relative reactivities of other combinations of elements in Groups 1A and 2A. For example: Na vs. Li, and Ba vs. Ca.

Discuss

Ask students to describe group and periodic trends in ionization energy. Ask why the first ionization energy of a nonmetal such as chlorine, is much higher than that of an active metal such as potassium. Ask students if they think the same trends would apply to the tendency for atoms to attract electrons.

ASTRONOMY

The Big Bang

Astronomers have evidence that the material universe began with an event of indescribable energy. At the moment of this event, called the Big Bang, the

temperature was many billions of degrees. As a result of the Big Bang, the elements formed. Neutrons, protons, and electrons may have formed within 10^{-4} second after the Big Bang, and the lightest nuclei formed within 3 minutes. At this time, the temperature was still probably 70 times the temperature of Earth's sun. Matter was in the form of plasma, a sea of positive nuclei and negative electrons. It took an estimated 500 000 years for electrons and nuclei to cool enough to form atoms. According to the Big Bang theory, planet Earth, with its wealth of chemical elements, is the debris of supernova explosions. It is this literal star dust that contains all the elements essential for life.

Table 14.1

Ionization Energies of the First 20 Elements

Symbol of element	Ionization energy (kJ/mol) First	Second	Third
H	1 312		
He (noble gas)	2 371	5 247	
Li	520	7 297	11 810
Be	900	1 757	14 840
B	800	2 430	3 659
C	1 086	2 352	4 619
N	1 402	2 857	4 577
O	1 314	3 391	5 301
F	1 681	3 375	6 045
Ne (noble gas)	2 080	3 963	6 276
Na	495.8	4 565	6 912
Mg	737.6	1 450	7 732
Al	577.4	1 816	2 744
Si	786.2	1 577	3 229
P	1 012	1 896	2 910
S	999.6	2 260	3 380
Cl	1 255	2 297	3 850
Ar (noble gas)	1 520	2 665	3 947
K	418.8	3 069	4 600
Ca	589.5	1 146	4 941

The energy required to remove this first outermost electron is called the first ionization energy. To remove the outermost electron from the gaseous 1+ ion requires an amount of energy called the second ionization energy, and so forth. **Table 14.1** gives the first three ionization energies of the first 20 elements.

You can use the concept of ionization energy to predict ionic charges. Look at the three Group 1A metals in **Table 14.1.** Do you see a large increase in energy between the first and second ionization energies? It is relatively easy to remove one electron from a Group 1A metal to form an ion with a 1+ charge. It is very difficult, however, to remove an additional electron. For the three Group 2A metals, the large increase in ionization energy occurs between the second and third ionization energies. What does this tell you about the relative ease of removing one electron from
1 these metals? Two electrons? Three electrons? You know that aluminum, in Group 3A, forms a 3+ ion. The large increase in ionization energy for aluminum occurs after the third electron is removed.

Group Trends As you can see from **Table 14.1,** the first ionization energy generally decreases as you move down a group of the periodic table. This is because the size of the atoms increases as you descend, so the

Answers

1 Removing one or two electrons is relatively easy in Group 2A metals; removing three electrons is much more difficult.
2 Oxygen has the highest first ionization energy in Group 6A. Neon has the highest first ionization energy in period 2.
3 Ionization energy decreases going down the group.

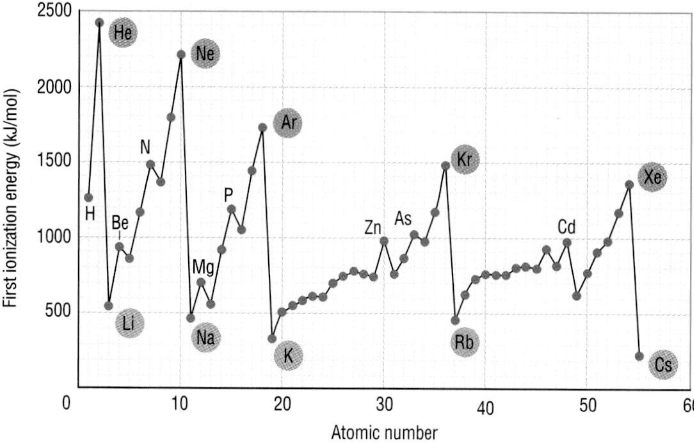

Figure 14.12
This graph of first ionization energy versus atomic number shows a periodic trend. Notice the ease with which Group 1A elements are ionized and the difficulty of ionizing noble gases. What is the group trend for the noble gases? 3

TEACHER DEMO

On an overhead projector, make a circle of washers or poker chips to represent an electron cloud in a neutral atom. Leave no space between the items in the circle. Place items in the center to represent the nucleus. Add or subtract "electrons" to mimic the formation of ions. With each change, adjust the circle so that the items still touch. Explain that the change in diameter of the circle is analogous to the change in the effective attraction of the nuclear charge.

outermost electron is farther from the nucleus. The outermost electron should be more easily removed, and the element should have a lower ionization energy.

Periodic Trends For the representative elements, the first ionization energy generally increases as you move from left to right across a period. See **Figure 14.12**. The nuclear charge increases and the shielding effect is constant as you move across. A greater attraction of the nucleus for the electron leads to the increase in ionization energy.

Figure 14.13 summarizes the group and periodic trends in first ionization energies. Which element in Group 6A has the highest first ionization energy? In period 2? 2

Trends in Ionic Size

The atoms of metallic elements have low ionization energies. They form positive ions easily. By contrast, the atoms of nonmetallic elements readily form negative ions. Let's look at how the loss or gain of electrons affects the size of the ion formed.

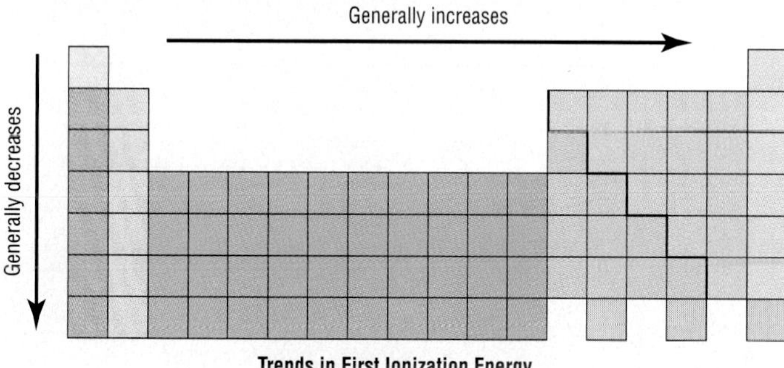

Figure 14.13
First ionization energies generally increase across periods and decrease down groups.

ACTIVITY

Give students a list of elements. Ask them to locate each element in the periodic table, and decide whether its atoms will form positive or negative ions. Have students make a list of elements that form positive ions and another list of elements that form negative ions. Have them generalize about how the size of a neutral atom will compare with that of the ion for each group of elements.

Discuss

Relate the periodic trends in ionic size to those discussed earlier for atomic size. Explain that the *effective nuclear charge* experienced by an electron in the outermost orbital of an atom or ion is equal to the total nuclear charge (the number of protons) **minus the shielding or screening effect due to any intervening electrons. The effective nuclear charge determines the atomic and ionic radii. As you proceed from left to right in any given period, the principal quantum number, *n*, of the outer orbital remains constant but the effective nuclear charge increases. Therefore, atomic and ionic radii decrease as you move to the right in a period. In contrast, within any group, as you proceed from top to bottom, the effective nuclear charge remains nearly constant, but the principal quantum number, *n*, gets larger; consequently, atomic and ionic radii increase as you move down in a group.**

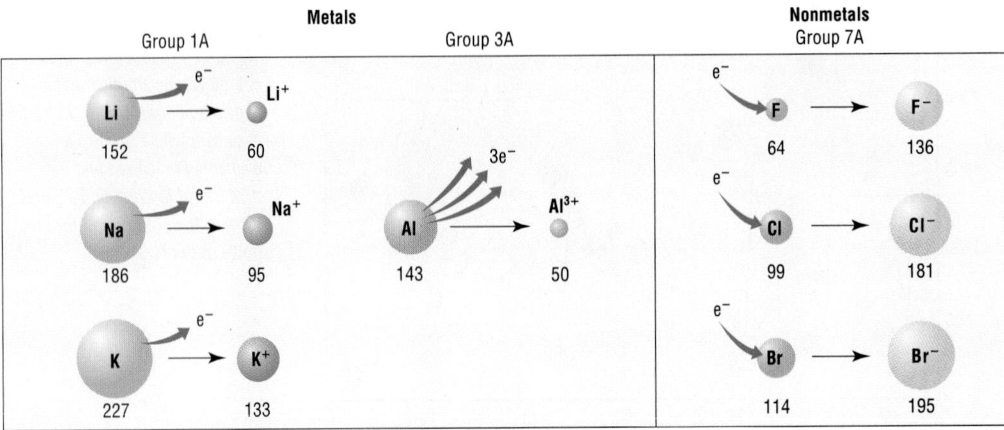

Figure 14.14
The relative sizes (radii) of atoms and ions for selected metals and nonmetals are given in picometers.

Group Trends Positive ions (cations) are always smaller than the neutral atoms from which they form. This is because the loss of outer-shell electrons results in increased attraction by the nucleus for the fewer remaining electrons. As you can see in **Figure 14.14,** the radius of the Na^+ ion (95 pm) is only about one-half that of the Na atom (186 pm). The radius of the Al^{3+} ion is only about one-third that of the Al atom. In contrast, negative ions (anions) are always larger than the neutral atoms from which they form. This is because the effective nuclear attraction is less for an increased number of electrons. The radius of the Cl^- ion (181 pm) is about twice that of the Cl atom (99 pm). The bar graphs in **Figure 14.9** on page 400, which show trends in atomic radii, also give the group trends in ionic size for Group 1A, 2A, and 7A elements.

Periodic Trends A periodic relationship for the ionic radii of the elements is shown in **Figure 14.8** on page 399. Going from left to right across a row, there is a gradual decrease in the size of positive ions. Then beginning with Group 5A, the negative ions, which are much larger, gradually decrease in size as you continue to move right. The ionic radii of both anions and cations increase as you go down each group.

The group and periodic trends in ionic size are summarized in **Figure 14.15.** How does the ionic radius of sodium compare with that of cesium? Boron with that of fluoride? 1

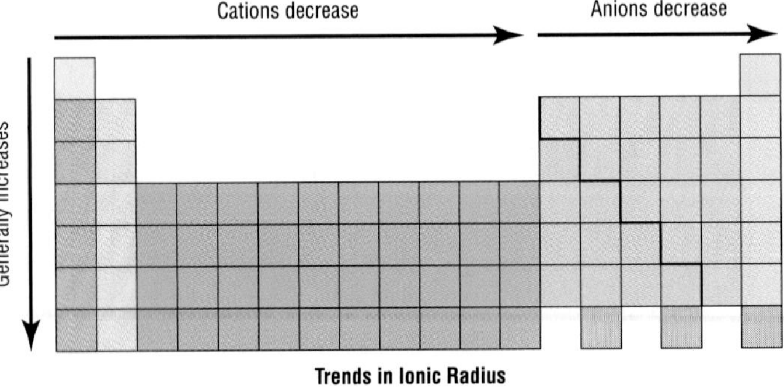

Figure 14.15
Cationic and anionic radii decrease across periods and increase down groups.

Answer

1 The ionic radius of sodium is smaller than that of cesium. The ionic radius of boron is smaller than that of fluorine.

Trends in Electronegativity

The **electronegativity** of an element is the tendency for the atoms of the element to attract electrons when they are chemically combined with atoms of another element. Electronegativities have been calculated for the elements and are expressed in arbitrary units on the Pauling electronegativity scale. This numerical scale is based on a number of factors, including the ionization energies of the elements.

The electronegativities of selected elements, arranged in the form of the periodic table, are presented in **Table 14.2.** Note that the noble gases are omitted because they do not form many compounds. With their exception, each element is assigned an electronegativity number in units of Paulings. As you can see, electronegativity generally decreases as you move down a group. As you go across a period from left to right, the electronegativity of the representative elements increases. The metallic elements at the far left of the periodic table have low electronegativities. By contrast, the nonmetallic elements at the far right (excluding the noble gases) have high electronegativities. The trends in electronegativities among the transition metals are not so regular; these numbers are not included in the table.

The electronegativity of cesium, the least electronegative element, is 0.7; the electronegativity of fluorine, the most electronegative element, is 4.0. Because fluorine has such a strong tendency to attract electrons, when it is chemically bonded to any other element it either attracts the shared electrons or forms a negative ion. In contrast, cesium has the least tendency to attract electrons. It loses the electron "tug-of-war" and forms a positive ion.

As you will learn in the following chapters on ionic bonding and covalent bonding, electronegativity values help predict the type of bonding that can exist between atoms in compounds.

Table 14.2

Electronegativity Values for Atoms of Selected Elements						
H 2.1						
Li 1.0	Be 1.5	B 2.0	C 2.5	N 3.0	O 3.5	F 4.0
Na 0.9	Mg 1.2	Al 1.5	Si 1.8	P 2.1	S 2.5	Cl 3.0
K 0.8	Ca 1.0	Ga 1.6	Ge 1.8	As 2.0	Se 2.4	Br 2.8
Rb 0.8	Sr 1.0	In 1.7	Sn 1.8	Sb 1.9	Te 2.1	I 2.5
Cs 0.7	Ba 0.9	Tl 1.8	Pb 1.9	Bi 1.9		

MUSIC

Newlands's Octaves

Mendeleev's publication of his first periodic table in 1869 was far from the first attempt to organize the elements according to their properties. In 1863, John Newlands (1838–1898), an English chemist, arranged the elements in order of increasing atomic masses. He noted that the properties of the elements repeated when the elements were arranged by increasing atomic mass in groups of eight. For example, the chemical properties of lithium and sodium are very similar. Newlands referred to his arrangement as the law of octaves. This is similar to the musical scale, which repeats every eighth tone. Although the law of octaves fails for elements beyond calcium, Newlands's work was a step in the right direction in the classification of the elements.

Discuss

Lead a class discussion on periodic and group trends in electronegativities. Ask why the noble gases are not included in a discussion on electronegativity. Ask which stable element in the entire periodic table is the most electronegative and which is the least electronegative. Point out that electronegativity values help chemists predict the type of bonding that exists between atoms in compounds.

ACTIVITY

Have students model trends in electronegativity with arrangements of toothpicks cut to different lengths. Stick the toothpicks into a flattened piece of modeling clay, and inscribe the symbol of each element below the appropriate toothpick. For a more durable model, use a 96-well reaction plate (also called a cell-well tissue culture plate) and soda straws cut to appropriate lengths. Write symbols on paper, and then cut them into circles with a standard paper punch. Put the symbols into the wells, and position the straws. For a permanent display, put a bit of model cement on each straw before placing it in the well.

Use the Visual

Display Table 14.2 on an overhead projector and discuss the relationship between electronegativity and the tendency of an atom to gain or lose electrons. Show how the electronegativity values correlate with the classification of elements as metals or nonmetals. Have students compare the ionization energies and electronegativities of the elements listed in Table 14.2.

14.2

3 Assess

Evaluate Understanding

Have students compare two elements in the same group in terms of atomic radius, ionic radius, ionization energy, and electronegativity. Repeat the exercise with a metal and nonmetal from the same period. Have students write general statements to summarize the trends revealed by these comparisons.

Reteach

Emphasize the key roles electrical attraction and repulsion play within atoms and ions. Review the effects of increasing nuclear charge and changes in the shielding effect of electrons on the size of an atom. Review with students how these effects can be used to describe the size changes for atoms within a period, as well as for groups of atoms. Remind students about the relationship between electronegativity and the tendency of atoms to gain or lose electrons. The interaction between atoms that results in chemical bonding can be understood in terms of the relative electronegativities of the atoms.

Portfolio Project

Students should discover that grades and prices are assigned to chemicals on the basis of their purity. Price differences between grades will vary from one chemical to another. When deciding which grade to purchase, a buyer must consider the intended application.

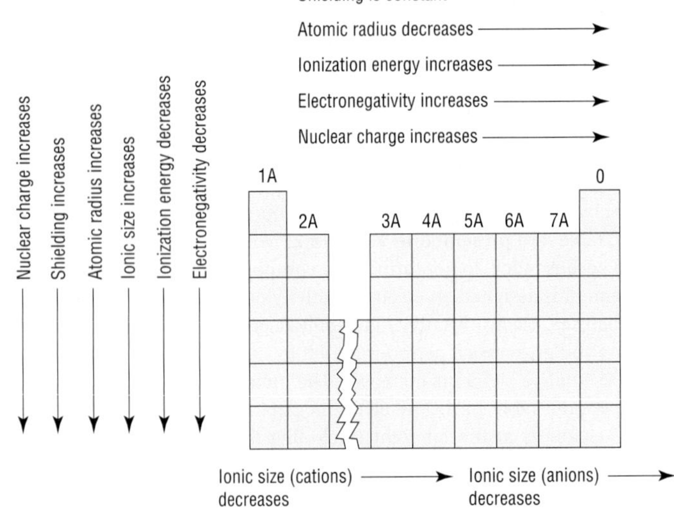

Figure 14.16
Periodic trends vary as you move across and down the periodic table. Properties that show periodic trends include atomic radius, ionic size, ionization energy, nuclear charge, shielding effect, and electronegativity of the elements.

Summary of Periodic Trends

You have now seen that a number of periodic trends exist among the elements and that these trends can be explained by looking at variations in atomic structure. Remember, trends occur within groups and within periods. **Figure 14.16** summarizes the trends in atomic radius, ionization energy, ionic size, and electronegativity that you have just learned about. Which is the only property that shows a decreasing trend as you move from left to
1 right across the periodic table?

section review 14.2

6. For which of these properties does lithium have a larger value than potassium?
   a. first ionization energy
   b. atomic radius
   c. electronegativity
   d. ionic radius
7. Arrange these elements in order of decreasing atomic size: sulfur, chlorine, aluminum, and sodium. Does your arrangement demonstrate a periodic trend or a group trend?
8. How does the ionic radius of a typical anion compare with the radius for the corresponding neutral atom?
9. Which element in each pair has the larger ionization energy?
   a. sodium, potassium
   b. magnesium, phosphorus

 **Chem ASAP! Assessment 14.2** Check your understanding of the important ideas and concepts in Section 14.2.

portfolio project

Research what grades of chemicals are available from suppliers. What is the difference between reagent grade and laboratory grade in terms of purity and cost? How pure are the chemicals that you use for your lab experiments?

Answers

1 The only property that shows a decreasing trend as you move from left to right across the periodic table is atomic radius.

SECTION REVIEW 14.2

6. a., c.
7. sodium, aluminum, sulfur, chlorine; periodic trend
8. The anion is larger than its atom.
9. a. sodium b. phosphorus

Chemistry Serving...Industry

BIG JOBS FOR LITTLE DEVICES

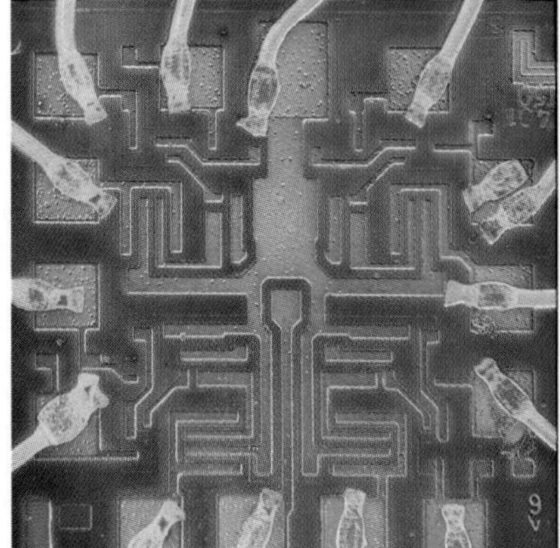

In 1946 at the University of Pennsylvania, a team of engineers built the Electronic Numerical Integrator and Calculator (ENIAC). A complex mass of wires and vacuum tubes, ENIAC occupied 2000 square feet of floor space, almost half the size of a basketball court. It was the first electronic computer, designed to perform high-speed calculations for the government.

Although ENIAC was an amazing achievement for its time, it would be outperformed today by a calculator the size of a credit card! Within the small space of the calculator's circuitry is the equivalent of ENIAC's thousands of vacuum tubes and miles of wires. This incredible miniaturization was made possible by semiconductor technology.

Chemically speaking, semiconductors are elements that conduct electricity better than insulators but less well than conductors. The elements silicon (Si), germanium (Ge), and gallium (Ga) are semiconductors.

In their pure forms, semiconductors are not very useful. But when contaminated with atoms of other elements, their properties of conductivity change dramatically. Because of the different ways they can affect the movement of electrons within them, "contaminated" semiconductors can be used to make tiny electronic devices. One such device is the transistor, which amplifies an electrical signal.

A transistor smaller than the size of the period at the end of this sentence can do the work of one of the many four-inch-tall vacuum tubes contained in ENIAC.

The basis of semiconductor technology is the process called doping. In doping, some of the atoms that make up a crystal of pure silicon are replaced with atoms of either arsenic (As) or boron (B). Doping with arsenic creates a donor, or n-type semiconductor. Each atom of arsenic contains five electrons in its outer shell, compared with the four electrons of the neighboring silicon atoms. The extra electron from the arsenic atom is free to move around.

An integrated circuit containing millions of components can be placed on a semiconductor wafer smaller than your fingernail!

Doping with boron has the opposite effect, creating an acceptor, or p-type semiconductor. Because boron has only three electrons in its outer shell, a positive "hole" is created for every boron atom. These holes can move and thus conduct.

Miniature electronic components are built from different combinations of n-type and p-type semiconductors placed together. Transistors are just one example. Integrated circuits, such as the one shown in the photograph, are another. An integrated circuit containing millions of components can be placed on a semiconductor wafer smaller than your fingernail! The resulting "chip" can control a computer, portable CD player, television, calculator, or automobile fuel system.

Miniature electronic circuits have revolutionized the electronics and computer industries. Electronic equipment is now smaller, more complex, and less expensive than ever. Current research is focusing on finding ways to make even smaller and faster chips.

CHEMISTRY IN CAREERS

TAKE IT TO THE NET

Find out more about career opportunities:

www.phschool.com

CHEMICAL SPECIALIST

ocal food service distributor seeks esponsible self-motivated individ- al to service established San Jose

SOLID STATE CHEMIST

Do you have a "solid" understanding of chemistry? If so, consider this career.

See page 874.

CHEMIST NEEDED

Enviro. Lab. GC Must know EPA methods

Chemistry Serving... Industry

DISCUSS

Explain that a semiconducting element is one that has low electrical conductivity at low temperatures but whose electrical conductivity rises with temperature or with the addition of other selected elements. Point out silicon's position in the periodic table and explain that silicon is a metalloid, an element having both metallic and nonmetallic properties. Silicon is the second most abundant element in Earth's crust, existing as silicon dioxide (SiO_2). Remind students about some of the properties that distinguish metals from nonmetals. Ask students to name other elements that are considered metalloids. (Ge, Sb, and As to name a few)

CHEMISTRY IN CAREERS

Have students read about the role of professional chemists in the design and manufacture of solid state electronic devices on page 874. Students may also wish to connect to the Internet address shown on this page to find out about careers in the electronics industry.

4 Close

Summary

Ask the following questions which require students to summarize information contained in the chapter.

- **What is true about the electron configurations of elements in the same group?** (The outermost energy levels of all the elements within a group have similar electron configurations.)
- **Based on periodic and group trends, which has the larger atomic radius, Ca or Se?** (Ca) **C or Sn?** (Sn)
- **Based on periodic and group trends, which is more electronegative, Cl or Br?** (Cl) **N or F?** (F)

Extension

Have students prepare a report on one of the representative groups of elements. Have them include information about shared properties and distinctive properties for elements in the chosen group.

Looking Back . . . Looking Ahead . . .

Chapter 13 introduced the development of the periodic table with an emphasis on Mendeleev's contribution. The current chapter shows that shared properties of elements in groups are based on electron configurations. The utility of the periodic table is demonstrated by the discussion of periodic and group trends. Knowledge of elemental trends such as ionization energy and electronegativity lay the groundwork for the discussion of bonding in Chapters 15 and 16.

Chapter 14 STUDENT STUDY GUIDE

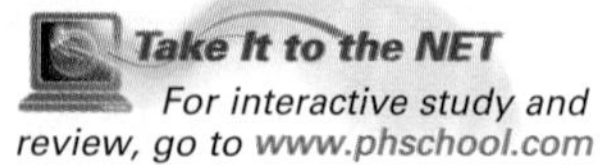

For interactive study and review, go to www.phschool.com

KEY TERMS

- atomic radius *p. 398*
- electronegativity *p. 405*
- inner transition metal *p. 395*
- ionization energy *p. 401*
- noble gas *p. 394*
- representative element *p. 394*
- transition metal *p. 394*

CONCEPT SUMMARY

14.1 Classification of the Elements

- Elements that have similar properties also have similar electron configurations and are members of the same group.
- The atoms of the noble gas elements have filled outermost *s* and *p* sublevels.
- The outermost *s* and *p* sublevels of the representative elements are only partially filled.
- The outermost *s* and nearby *d* sublevels of transition metals contain electrons.
- The outermost *s* and nearby *f* sublevels of inner transition metals contain electrons.

14.2 Periodic Trends

- Regular changes in the electron configuration of the elements cause gradual changes in both the physical and chemical properties of the elements within a group and within a period.
- Atomic radii generally decrease as you move from left to right in a given period because there is an increase in the nuclear charge while the number of inner electrons, and hence the shielding effect, remains constant.
- Ionization energy, the energy required to remove an electron from an atom, generally increases as you move from left to right across a period. Ionization energy decreases as you move down a group.
- Atomic radii generally increase within a given group because the outer electrons are farther from the nucleus as you go down the group. The attractive force of the increased nuclear charge is unable to overcome the effect of the greater distance, which acts in opposition.
- Ionic radii decrease for cations and anions as you move from left to right across a period and increase as you move down a group.
- Electronegativity measures the ability of a bonded atom to attract electrons to itself. It generally increases as you move from left to right across a period. It decreases as you move down a group.

CHAPTER CONCEPT MAP

Use these terms to construct a concept map that organizes the major ideas of this chapter.

Chem ASAP! Concept Map 14
Create your Concept Map using the computer.

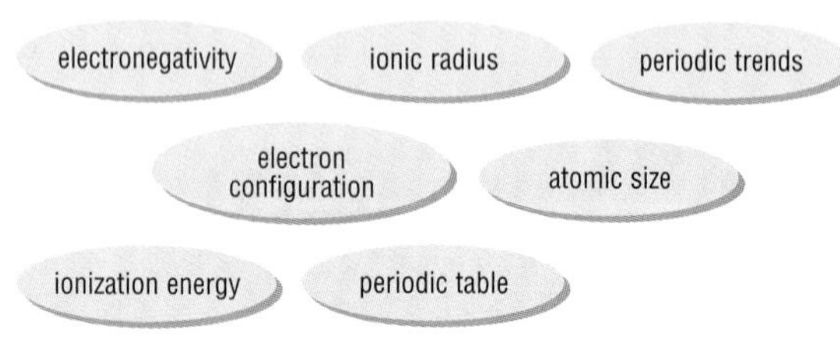

Take It to the Net

At **www.phschool.com** students will find for this chapter

- an Internet activity
- links to related chemistry sites
- an interactive quiz
- career links

CONCEPT PRACTICE

10. What are the noble gases, the representative elements, the transition elements, and the inner transition elements? *14.1*

11. Use **Figure 14.2** to write the electron configuration of these elements. *14.1*
a. boron **b.** magnesium **c.** arsenic

12. Which of the following are representative elements: Na, Mg, Fe, Ni, Cl? *14.1*

13. Write the electron configuration of these elements. *14.1*
a. the inert gas in period 3
b. the element in Group 4A, period 4
c. the element in Group 2A, period 6

14. Explain how an element's outer electron configuration is related to its position in the periodic table. *14.1*

15. Use **Figure 14.2** to write the electron configuration of these atoms. *14.1*
a. fluorine **b.** zinc **c.** aluminum **d.** tin

16. What are the symbols for all the elements with the following outer configurations? *14.1*
a. s^1 **b.** s^2p^4 **c.** s^2d^{10}

17. Explain why fluorine has a smaller atomic radius than both oxygen and chlorine. *14.2*

18. Indicate which element in each pair has the greater atomic radius. *14.2*
a. sodium, lithium
b. strontium, magnesium
c. carbon, germanium
d. selenium, oxygen

19. Distinguish between the first and second ionization energy of an atom. *14.2*

20. Indicate which element in each pair has the greater first ionization energy. *14.2*
a. lithium, boron
b. magnesium, strontium
c. cesium, aluminum

21. Would you expect metals or nonmetals to have higher ionization energies? Why? *14.2*

22. Arrange the following elements in order of increasing ionization energy. *14.2*
a. Be, Mg, Sr **b.** Bi, Cs, Ba **c.** Na, Al, S

23. Why is there a large increase between the first and second ionization energies of the alkali metals? *14.2*

24. Which particle has the larger radius in each atom/ion pair? *14.2*
a. Na, Na^+ **b.** S, S^{2-} **c.** I, I^- **d.** Al, Al^{3+}

25. How does the ionic radius of a typical metallic atom compare with its atomic radius? *14.2*

26. Explain why the noble gases do not appear in **Table 14.2**. *14.2*

27. Which element is more electronegative? *14.2*
a. Cl, F **b.** C, N **c.** Mg, Ne **d.** As, Ca

CONCEPT MASTERY

28. The Mg^{2+} and Na^+ ions each have ten electrons surrounding the nucleus. Which ion would you expect to have the smaller radius? Why?

29. Explain why it takes more energy to remove a 4*s* electron from zinc than from calcium.

30. The graphs show the relationship between the electronegativities and first ionization energies for period 2 and period 3 elements.
a. State the general trend between these two values in each period.
b. Propose an explanation for this trend.

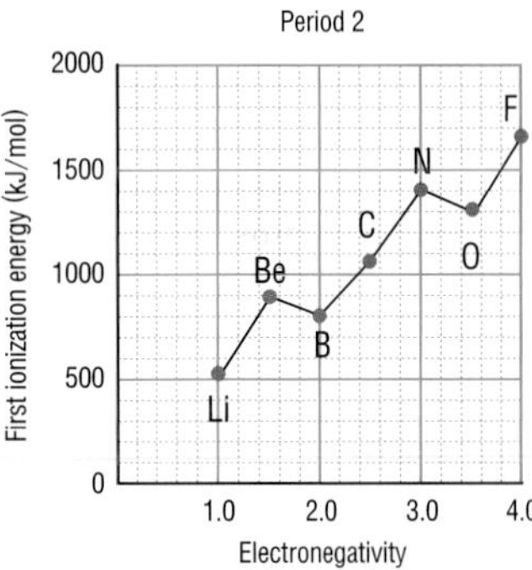

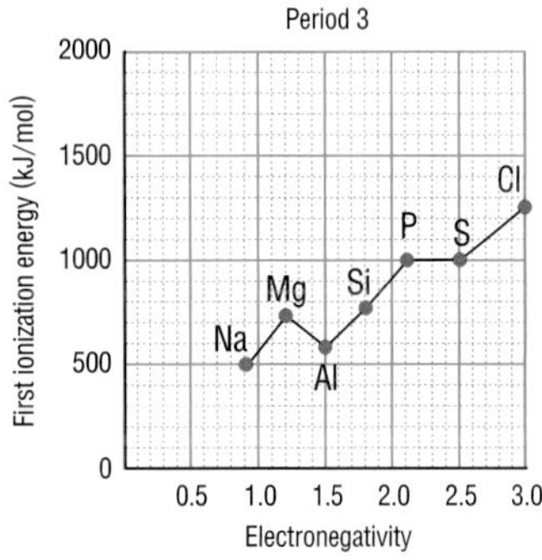

Answers

10. noble gases: Group 0; representative elements: Groups 1A–7A; transition elements: Groups 1B–8B; inner transition elements: a separate section between Groups 2A and 3B

11. a. $1s^22s^22p^1$
b. $1s^22s^22p^63s^2$
c. $1s^22s^22p^63s^23p^63d^{10}4s^24p^3$

12. Na, Mg, Cl

13. a. Ar: $1s^22s^22p^63s^23p^6$
b. Ge: $1s^22s^22p^63s^23p^63d^{10}4s^24p^2$
c. Ba: $1s^22s^22p^63s^23p^63d^{10}4s^24p^64d^{10}5s^25p^6 6s^2$

14. An element's outer electron configuration places it in a particular column (group) of the periodic table.

15. a. $1s^22s^22p^5$
b. $1s^22s^22p^63s^23p^63d^{10}4s^2$
c. $1s^22s^22p^63s^23p^1$
d. $1s^22s^22p^63s^23p^63d^{10}4s^24p^64d^{10}5s^25p^2$

16. a. H, Li, Na, K, Rb, Cs, Fr
b. O, S, Se, Te, Po
c. Zn, Cd, Hg, Uub

17. Fluorine has a smaller atomic radius than oxygen because fluorine has one more nuclear charge. Fluorine has a smaller radius than chlorine because fluorine has eight fewer electrons.

18. a. sodium **b.** strontium **c.** germanium **d.** selenium

19. The first ionization energy is the energy needed to remove the outermost electron. The second ionization energy is the energy needed to remove the second-outermost electron.

20. a. boron **b.** magnesium **c.** aluminum

21. nonmetals; The nuclear charge increases, but the shielding effect is the same, thus creating greater electron attraction.

22. a. Sr, Mg, Be **b.** Cs, Ba, Bi **c.** Na, Al, S

23. An atom of an alkali metal becomes stable by losing one electron. Removing the second electron involves removing an electron from an ion that has a stable noble-gas configuration. This requires much more energy.

24. a. Na **b.** S^{2-} **c.** I^- **d.** Al

25. The radius of a cation is smaller than the atom from which it forms.

26. Noble gases generally do not form compounds.

27. a. F **b.** N **c.** Mg **d.** As

28. Mg^{2+} has more protons in its nucleus; its electron attraction is therefore greater.

29. Zinc has more protons than calcium, thus attracting the 4*s* electrons more.

30. a. The general trend is that the first ionization energy increases as electronegativity increases. This is true for both period 2 and period 3.
b. A positive correlation is expected because both properties measure the interaction between the nucleus and the surrounding electrons.

Answers

31. **a.** potassium, K, [Ar]$4s^1$
b. aluminum, Al, [Ne]$3s^2 3p^1$
c. sulfur, S, [Ne]$3s^2 3p^4$
d. barium, Ba, [Xe]$6s^2$

32. **a.** Ca^{2+}
b. P^{3-}
c. Cu^+

33. scandium, Sc, [Ar]$3d^1 4s^2$
titanium, Ti, [Ar]$3d^2 4s^2$
vanadium, V, [Ar]$3d^3 4s^2$
chromium, Cr, [Ar]$3d^5 4s^1$
manganese, Mn, [Ar]$3d^5 4s^2$
iron, Fe, [Ar]$3d^6 4s^2$
cobalt, Co, [Ar]$3d^7 4s^2$
nickel, Ni, [Ar]$3d^8 4s^2$
copper, Cu, [Ar]$3d^{10} 4s^1$
zinc, Zn, [Ar]$3d^{10} 4s^2$

34. **a.** 2 **b.** 1 **c.** 3

35. Magnesium achieves a stable electron configuration by losing two electrons; aluminum achieves a stable electron configuration by losing three electrons.

36. **a.** 1851–1900: 25 elements
b. Mendeleev's periodic table helped scientists predict the existence of undiscovered elements.
c. None of these elements are found in nature.

37. **a.** Possible cations are Rb^+ and Sr^{2+}; possible anions are Br^-, Se^{2-}, and As^{3-}.
b. No; a cation is isoelectronic with the noble gas in the preceding period; an anion is isoelectronic with the noble gas in the same period.

38. $P_{CO_2} = 1.23$ atm; $P_{N_2} = 0.879$ atm

39. **a.** $2Ag + S \rightarrow Ag_2S$
b. $Na_2SO_4 + Ba(OH)_2 \rightarrow BaSO_4 + 2NaOH$
c. $Zn + 2HNO_3 \rightarrow Zn(NO_3)_2 + H_2$
d. $2H_2O + 2SO_2 + O_2 \rightarrow 2H_2SO_4$

40. 69.9 g Fe

41. **a.** Li_2SO_4
b. $Zn_3(PO_4)_2$
c. $KMnO_4$
d. $SrCO_3$

42. 7.60 L

43. The ionic radii decrease from S^{2-}, Cl^-, Ar, K^+, Ca^{2+}, to Sc^{3+} as the number of protons increases. The radii decrease from O^{2-}, F^-, Ne, Na^+, Mg^{2+}, to Al^{3+} for the same reason.

44. The table shows gradually increasing atomic masses and decreasing ionization energies for these elements. All need two atoms of chlorine or one atom of oxygen for compounds. These trends justify placing these elements in one group.

31. Give the electron configuration of the element found at each location in the periodic table.
a. Group 1A, period 4
b. Group 3A, period 3
c. Group 6A, period 3
d. Group 2A, period 6

32. In each pair, which ion is larger?
a. Ca^{2+}, Mg^{2+} **b.** Cl^-, P^{3-} **c.** Cu^+, Cu^{2+}

33. Give the names, symbols, and electron configurations for the ten period-4 transition metals.

CRITICAL THINKING

34. Choose the term that best completes the second relationship.

a. sister : brother — oxygen : _______
(1) hydrogen
(2) sulfur
(3) silicon
(4) Group 6A

b. potassium : cation — sulfur : _______
(1) anion
(2) nonmetal
(3) yellow
(4) solid

c. magnesium : *s* orbital — zinc : _______
(1) *s* orbital
(2) *p* orbital
(3) *d* orbital
(4) *f* orbital

35. There is a large jump between the second and third ionization energies of magnesium. The corresponding large jump is between the third and fourth ionization energies of aluminum. Explain.

36. The following graph shows how many elements were discovered before 1750 and in each 50-year period since then.
a. In which 50-year period were the most elements discovered?
b. How did Mendeleev's work contribute to the discovery of so many elements?
c. What characteristic do all the elements discovered since 1950 have in common?

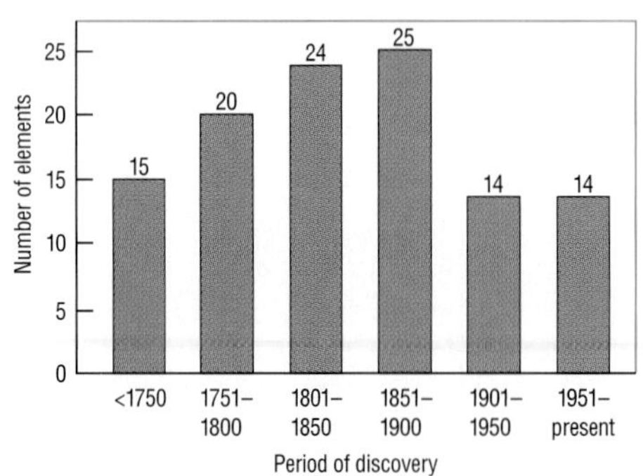

37. Atoms and ions with the same number of electrons are called isoelectronic.
a. Name a cation and an anion that are isoelectronic with krypton.
b. Is it possible for a cation to be isoelectronic with an anion from the same period? Explain.

CUMULATIVE REVIEW

38. A 2.00-L flask at 27 °C contains 4.40 g of carbon dioxide and 2.00 g of nitrogen gas. What is the pressure (in atm) of each of the two components?

39. Balance the following chemical equations.
a. $Ag + S \longrightarrow Ag_2S$
b. $Na_2SO_4 + Ba(OH)_2 \longrightarrow BaSO_4 + NaOH$
c. $Zn + HNO_3 \longrightarrow Zn(NO_3)_2 + H_2$
d. $H_2O + SO_2 + O_2 \longrightarrow H_2SO_4$

40. The smelting of iron ore consists of heating the ore with carbon.

$$2Fe_2O_3 + 3C \longrightarrow 4Fe + 3CO_2$$

What mass of iron can be obtained from 100 g of the ore?

41. Write chemical formulas for the following compounds.
a. lithium sulfate
b. zinc phosphate
c. potassium permanganate
d. strontium carbonate

42. If a gas sample at 25 °C occupies a volume of 2.93 L, what will be the volume at 500 °C if the pressure is unchanged?

CONCEPT CHALLENGE

43. The ions S^{2-}, Cl^-, K^+, Ca^{2+}, and Sc^{3+} have the same total number of electrons as the noble gas argon. How would you expect the radii of these ions to vary? Would you expect to see the same variation in the series O^{2-}, F^-, Na^+, Mg^{2+}, and Al^{3+}, in which each ion has the same total number of electrons as the noble gas neon? Why or why not?

44. Using a chemistry reference book, make a table for the Group 2A elements. Include densities, atomic masses, formulas of the chlorides and oxides, and first ionization energies. Can you justify placing these elements in one group on the basis of these data?

Chapter 14 STANDARDIZED TEST PREP

Select the choice that best answers each question or completes each statement.

1. Which of the following properties increases as you move across a period from left to right?
   I. electronegativity
   II. ionization energy
   III. atomic radius
   a. I and II only
   b. I and III only
   c. II and III only
   d. I, II, and III
2. List the symbols for sodium, sulfur, and cesium in order of increasing atomic radii.
   a. Na, S, Cs
   b. Cs, Na, S
   c. S, Na, Cs
   d. Cs, S, Na
3. The outer electron configuration for an element in the halogen group is
   a. s^2p^6.
   b. s^2p^5.
   c. s^2p^4.
   d. s^2p^2.
4. Which of these does *not* form an ion isoelectronic with krypton?
   a. argon
   b. bromine
   c. strontium
   d. selenium

Use the following data to answer questions 5 and 6. The ionization energies for the removal of the first six electrons in carbon are, starting with the first electron, 1086 kJ/mol, 2352 kJ/mol, 4619 kJ/mol, 6220 kJ/mol, 37 820 kJ/mol, and 47 260 kJ/mol.

5. Make a graph of ionization energy as a function of ionization number. The ionization number indicates which electron is lost.
6. Between which two ionization numbers does the ionization energy have the largest increase? Why is this behavior predictable?

Use the spheres to answer questions 7–9.

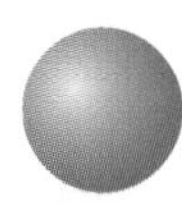

7. Which sphere would most likely represent a potassium atom, K?
8. Which sphere would most likely represent a potassium ion, K^+?
9. If the spheres represent an atom and an anion of the same element, which sphere represents the atom and which represents the anion?

Use the data table to answer questions 10–12.

Alkali metal	Atomic radius (pm)	First ionization energy (kJ/mol)	Electronegativity value
Li	152	520	1.0
Na	186	495.8	0.9
K	227	418.8	0.8
Rb	244	250	0.8
Cs	262	210	0.7

10. If you plotted atomic radius versus first ionization energy, would the graph reveal a direct or inverse relationship?
11. If you plotted atomic radius versus electronegativity, would the graph reveal a direct or inverse relationship?
12. If you plotted first ionization energy versus electronegativity, would the graph reveal a direct or inverse relationship?

For each question there are two statements. Decide whether each statement is true or false. Then decide whether Statement II is a correct explanation for Statement I.

Statement I		Statement II
13. A large increase between the first and second ionization energies is characteristic of the alkali metals.	BECAUSE	It takes considerable energy to remove an electron from a completely filled energy level.
14. Nonmetallic elements have higher electronegativity values than do metallic elements.	BECAUSE	Atoms of nonmetals are among the largest atoms.
15. A calcium atom is larger than a calcium ion.	BECAUSE	Ions are always larger than the atoms from which they are formed.
16. The ionization energy of the noble gas is always the smallest of all the elements in a period.	BECAUSE	Within any period, atomic radii tend to decrease moving from right to left.

Chapter 14 STANDARDIZED TEST PREP

1. a
2. c
3. b
4. a
5.

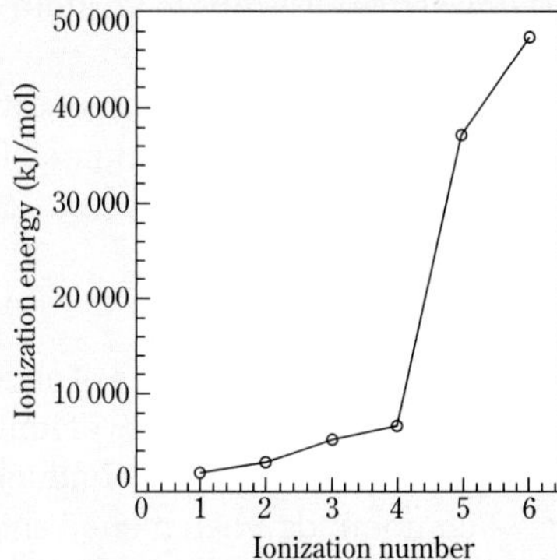

6. The largest increase is between ionization numbers 4 and 5 because carbon easily loses the first four electrons from the outermost (second) energy level. The fifth electron is removed from the first energy level.
7. the larger sphere
8. the smaller sphere
9. In this case, the smaller sphere represents the atom and the larger sphere represents the anion.
10. inverse
11. inverse
12. direct
13. True, True, correct explanation
14. True, False
15. True, False
16. False, True

Planning Guide

SECTION OBJECTIVES	ACTIVITIES/FEATURES	MEDIA & TECHNOLOGY
15.1 Electron Configuration in Ionic Bonding ● ■ ◆ ▸ Use the periodic table to infer the number of valence electrons in an atom and draw its electron dot structure ▸ Describe the formation of cations from metals and of anions from nonmetals	**SE** **Discover It!** *Shapes of Crystalline Materials,* p. 412 **SE** **Link to Food Science** *Sulfur Dioxide and Sulfites,* p. 418 **TE** **DEMO**, p. 414	**ASAP** Assessment 15.1 **OT** 50: *Cation Formation* **OT** 51: *Anion Formation* **ACT** 11: *Ions and Ionic Bonds*
15.2 Ionic Bonds ● ■ ◆ ▸ List the characteristics of an ionic bond ▸ Use the characteristics of ionic compounds to explain the electrical conductivity of ionic compounds when melted and when in aqueous solutions	**SE** **CHEMath** *Visualizing Molecules,* p. 420 **SE** **Mini Lab** *Solutions Containing Ions,* p. 425 (**LRS** 15-1) **PLM** *Solutions Containing Ions* **SE** **Small-Scale Lab** *Analysis of Anions and Cations,* p. 426 (**LRS** 15-2) **SSLM** 18: *Hard and Soft Water* **SSLM** 19: *Reactions of Aqueous Ionic Compounds* **SSLM** 20: *Identification of Eight Unknown Solutions* **LM** 24: *Crystal Structures* **TE** **DEMO**, p. 425	**ASAP** Animation 17 **ASAP** Simulation 15 **ASAP** Problem Solving 7 **ASAP** Assessment 15.2 **ACT** 11: *Ions and Ionic Bonds* **CHM** Side 3, 20: *Bonding* **CHM** Side 3, 30: *Ionic Bonding* **CHM** Side 3, 34: *Valence Electrons*
15.3 Bonding in Metals ○ □ ◆ ▸ Use the theory of metallic bonds to explain the physical properties of metals ▸ Describe the arrangements of atoms in some common metallic crystal structures	**SE** **Chemistry Serving . . . the Consumer** *When Water is Hard to Deal With,* p. 430 **SE** **Chemistry in Careers** *Wastewater Engineer,* p. 430 **TE** **DEMOS**, pp. 428, 429	**ASAP** Animation 18 **ASAP** Assessment 15.3 **ASAP** Concept Map 15 **CA** *Making Sodium Chloride* **CHM** Side 3, 32: *Metallic Bonding* **CHM** Side 6, 13: *The Mint* **OT** 52: *Metallic Crystals* **PP** Chapter 15 Problems **RP** Lesson Plans, Resource Library **AR** Computer Test 15 **www** Activities, Self-Tests, *SCIENCE NEWS* updates **TCP** The Chemistry Place Web Site

KEY

● Conceptual (concrete concepts)	**AR** Assessment Resources	**GRS** Guided Reading and Study Workbook
○ Conceptual (more abstract/math)	**ASAP** Chem ASAP! CD-ROM	**LM** Laboratory Manual
■ Standard (core content)	**ACT** ActivChemistry CD-ROM	**LP** Laboratory Practicals
□ Standard (extension topics)	**CHM** CHEMedia Videodiscs	**LRS** Laboratory Recordsheets
◆ Honors (core content)	**CA** Chemistry Alive! Videodiscs	**SSLM** Small-Scale Lab Manual
◇ Honors (options to accelerate)	**GCP** Graphing Calculator Problems	

PRACTICE		ASSESSMENT	
GRS	Section 15.1	**SE**	Section Review
RM	Practice Problems 15.1	**RM**	Section Review 15.1
		RM	Chapter 15 Quiz
SE	**Sample Problem** 15-1	**SE**	Section Review
GRS	Section 15.2	**RM**	Section Review 15.2
RM	Practice Problems 15.2	**RM**	Chapter 15 Quiz
RM	Interpreting Graphics	**LP**	Lab Practical 15-1
GRS	Section 15.3	**SE**	Section Review
RM	Practice Problems 15.3	**RM**	Section Review 15.3
GCP	Chapter 15	**RM**	Vocabulary Review 15
		SE	Chapter Review
		SM	Chapter 15 Solutions
		SE	Standardized Test Prep
		PHAS	Chapter 15 Test Prep
		RM	Chapter 15 Quiz
		RM	Chapter 15 A & B Test

PLANNING FOR ACTIVITIES

STUDENT EDITION

Discover It! p. 412
- small disposable cups
- distilled water
- rulers
- spoons
- table salt
- table sugar
- baking soda
- Epsom salts
- small clean mirrors
- magnifying glasses
- markers

Mini Lab, p. 425
- D batteries
- masking tape
- bell wire
- clear plastic cups
- distilled water
- tap water
- vinegar
- sucrose
- sodium chloride
- baking soda

Small-Scale Lab, p. 426
- pencils, paper, rulers
- medicine droppers
- cotton swabs
- reaction surfaces
- pipets
- solutions of Na_2SO_4, KI, HNO_3, Na_3PO_4, $AgNO_3$, HCl, $Pb(NO_3)_2$, NH_4Cl, $FeCl_3$, NaOH, KSCN
- staples
- solid fertilizer samples

TEACHER'S EDITION

Teacher Demo, p. 414
- plastic egg
- marbles

Activity, p. 422
- crystals of ionic compounds
- watch glasses
- magnifying glasses

Activity, p. 423
- NaCl, KCl, $MgCl_2$, or CsCl for preparing saturated solutions
- crystallizing dishes

Teacher Demo, p. 425
- foam balls and toothpicks

  or
- clay and soda straws

Teacher Demo, p. 428
- small sample of elemental copper or a copper alloy
- sample of a copper-containing crystalline ionic mineral such as chalcocite (Cu_2S)
- safety glasses
- hammer

Teacher Demo, p. 429
- Styrofoam™ balls of various sizes and colors
- toothpicks

OT	Overhead Transparency	**SE**	Student Edition
PHAS	PH Assessment System	**SM**	Solutions Manual
PLM	Probeware Lab Manual	**SSV**	Small-Scale Video/Videodisc
PP	Problem Pro CD-ROM	**TCP**	www.chemplace.com
RM	Review Module	**TE**	Teacher's Edition
RP	Resource Pro CD-ROM	**www**	www.phschool.com

Chapter 15

Key Terms

15.1 valence electrons, electron dot structures, octet rule, halide ions
15.2 ionic bonds, coordination number
15.3 metallic bonds

DISCOVER IT!

Students should observe that all of the solids crystallized. They should understand that crystals form at different rates and will have distinctive appearances.

Chapter 15 IONIC BONDING AND IONIC COMPOUNDS

The arrangement of ions in table salt determines its crystal shape.

FEATURES

Stay current with SCIENCE NEWS
Find out more about ionic compounds:
www.phschool.com

DISCOVER IT! SHAPES OF CRYSTALLINE MATERIALS

You need four small disposable cups, distilled water, a ruler, a spoon, sodium chloride (table salt), sucrose (table sugar), sodium hydrogen carbonate (baking soda), magnesium sulfate (Epsom salt), a small clean mirror, and a magnifying glass.

1. Label each cup with the name of one of the solids and pour water into each cup to a depth of 1 cm.
2. Add a spoonful of each solid to its corresponding cup. Swirl each cup for 30 seconds and then let it stand for a few minutes.
3. Swirl the cups at least two more times. Each time, note whether any of the solids completely dissolves. If so, add more of that solid and repeat the swirling. Continue this process until there is some undissolved solid at the bottom of each cup.
4. Set the mirror on a flat surface and place 2 or 3 drops of each liquid onto separate areas of the mirror.
5. Use a magnifying glass to examine each drop of liquid after 15 minutes, and then again after 24 hours.

For each material, did the solids crystallize? Did the crystals form at the same time? Describe the crystals that formed. Revisit this activity later in the chapter when you read about crystalline solids. Add to your notes any relevant information about the crystals you observed.

15.1 ● ■ ◆
15.2 ● ■ ◆
15.3 ○ □ ◆

Conceptual In Section 15.1, electron configuration diagrams may be omitted. Focus on the key concepts of valence electrons, the octet rule, and electron dot structures, which are essential to an understanding of bonding. Omit coordination numbers and unit cells in Section 15.2. The CHEMath feature will help students struggling with different representations of molecules. In Section 15.3, the crystalline structure of metals is optional.

Standard Encourage students to refer back to Tables 15.1 and 15.2 as they work through the Section Review and Chapter Review problems. Memorization of these details will follow after repeated practice. In Section 15.3, crystalline structure of metals is optional, but metallic bonding is an important concept, which should be covered.

Honors In Section 15.1, use electron configuration diagrams to predict bonding. In Sections 15.2 and 15.3, compare and contrast the crystalline structure, bonding, and physical properties of ionic compounds and metals.

ELECTRON CONFIGURATION IN IONIC BONDING

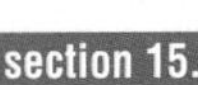

section 15.1

objectives

- Use the periodic table to infer the number of valence electrons in an atom and draw its electron dot structure
- Describe the formation of cations from metals and of anions from nonmetals

key terms

- valence electrons
- electron dot structures
- octet rule
- halide ions

Over time, underground salt deposits can rise through layers of sediment to form hollow, spherical structures called salt domes. More than 300 salt domes are found along a 2800-km stretch of the Gulf of Mexico. Some of these domes are more than 18 km deep and can be up to 10 km wide. As seen here, the salt found in salt domes exists as crystals of sodium chloride. **What characteristics of sodium and chlorine atoms allow them to form the stable compound sodium chloride, also known as table salt?**

Valence Electrons

In Chapters 5, 13, and 14, you learned about the electron structure of atoms and the organization of the periodic table. That knowledge will now help you understand the chemical bonding that occurs between atoms. For example, knowing the electron configurations of sodium and chlorine enables you to determine why these atoms combine to make sodium chloride. The configurations also explain why the formula unit for sodium chloride is NaCl and not Na_2Cl or $NaCl_2$. Questions about the properties of many compounds can be answered on the basis of their electron structure. Questions such as "Why does melted sodium chloride conduct electricity?" and "Why is sodium chloride a solid with a high melting point?" will be answered in this chapter.

You may recall that Mendeleev used similarities in the properties of elements to organize his periodic table. Scientists later learned that all of the elements within each group of the periodic table behave similarly because they have the same number of valence electrons. **Valence electrons** are the electrons in the highest occupied energy level of an element's atoms. The number of valence electrons largely determines the chemical properties of an element. One way to determine the number of valence electrons is to look at an element's electron configuration.

The number of valence electrons is also related to the group numbers in the periodic table. To find the number of valence electrons for a representative element, simply look at its group number. For example, the elements of Group 1A (hydrogen, lithium, sodium, potassium, and so forth) all have

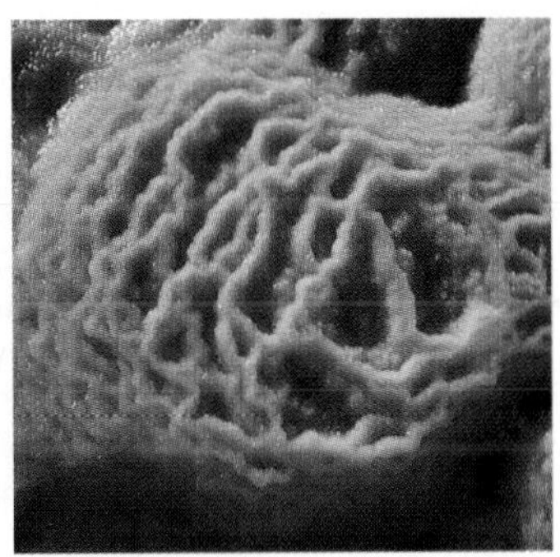

Figure 15.1
The sodium chloride in the sea water (left) has crystallized out of solution to form deposits on the rocks (right).

1 Engage

Use the Visual

Have students examine the section-opening photograph and Figure 15.1. Discuss how the reactive (and poisonous) elements sodium metal and chlorine gas can combine to form harmless table salt. Ask:

- **What characteristics of sodium and chlorine atoms allow them to form the stable compound sodium chloride, also known as table salt?** (Sodium atoms can lose an electron easily and chlorine atoms can accept an electron easily. The resulting ions can combine with the other oppositely charged ions.)

Remind students that NaCl is an example of an ionic compound. A crystal of NaCl contains equal numbers of Na^+ cations and Cl^- anions.

Check Prior Knowledge

To assess students' prior knowledge about the electronic configurations of atoms, ask:

- **How are elements arranged in the periodic table as described by Mendeleyev?** (They are organized into columns or 'groups' according to similarities in chemical properties.)
- **What are some examples of familiar ionic compounds?** (Have a volunteer make a list on the chalkboard.) **Which groups of the periodic table contain the majority of the elements in these compounds?** (Groups 1A and 2A; Groups 4A–7A) **What pattern of these groupings is evident in most of these ionic compounds?** (Ionic compounds are generally composed of elements on the left (metals) and the far right (nonmetals) of the periodic table.)

STUDENT RESOURCES

From the Teacher's Resource Package, use:

- Section Review 15.1, Ch. 15 Practice Problems and Quizzes from the Review Module (Ch. 13–16)

TECHNOLOGY RESOURCES

Relevant technology resources include:

- Chem ASAP! CD-ROM
- ResourcePro CD-ROM

2 Teach

TEACHER DEMO

Hold up a plastic egg containing marbles and shake it. State that the egg represents a sodium atom and that the marbles inside represent the ten stable, inner electrons making up the n=1 and n=2 energy levels. Emphasize the fact that the enclosed ten electrons cannot be "removed" without "breaking" the egg. Now hold up one additional marble and place it next to the egg. State that this marble represents the eleventh sodium electron, which is the only one in the n=3 energy level. Name this electron "sodium's valence electron" and emphasize that this electron is not like the others and can be pulled away or lost much more easily than the 'protected' inner electrons. Make sure to state that if this electron is lost, the resulting sodium atom has an overall 1+ charge.

Discuss

Lead a class discussion on writing electron dot structures for elements. Ask:

- **What is the significance of the electrons that are represented by dots?** (They are valence electrons.)
- **How are they placed?** (symmetrically around the atom to show placement in orbitals according to pairing rules)
- **Why are the non-valence electrons not shown.** (Generally, they are not available for the formation of chemical bonds.)

Table 15.1

Electron Dot Structures of Some Group A Elements								
	Group							
Period	1A	2A	3A	4A	5A	6A	7A	0
1	H·							He:
2	Li·	·Be·	·Ḃ·	·Ċ̣·	·Ṇ̈·	:Ö̤·	:F̤̈·	:N̤̈e:
3	Na·	·Mg·	·Ȧl·	·Ṩi·	·P̣̈·	:S̤̈·	:C̤̈l·	:Ä̤r:
4	K·	·Ca·	·Ġa·	·Ġ̣e·	·Ä̤s·	:S̤̈e·	:B̤̈r·	:K̤̈r:

one valence electron, corresponding to the 1 in 1A. Carbon and silicon, in Group 4A, have four valence electrons. Nitrogen and phosphorus, in Group 5A, have five valence electrons; and oxygen and sulfur, in Group 6A, have
1 six. How many valence electrons do the elements of Group 2A and Group 3A have? The noble gases (Group 0) are the only exceptions to the group-number rule: Helium has two valence electrons, and all of the other noble
2 gases have eight. Why does helium not have eight valence electrons? **Figure 15.2** shows several applications of Group 4A elements.

Valence electrons are usually the only electrons used in chemical bonds. Therefore, as a general rule only the valence electrons are shown in **electron dot structures.** Electron dot structures are diagrams that show valence electrons as dots. The inner electrons and the atomic nuclei are included in the letter symbol for the element being represented. **Table 15.1** shows electron dot structures for atoms of some Group A elements. Notice that all of the elements within a given group (with the exception of helium) have the same number of electron dots in their structures. How many elec-
3 tron dots would the Group 7A element iodine have?

Carbon (in the form of diamond)

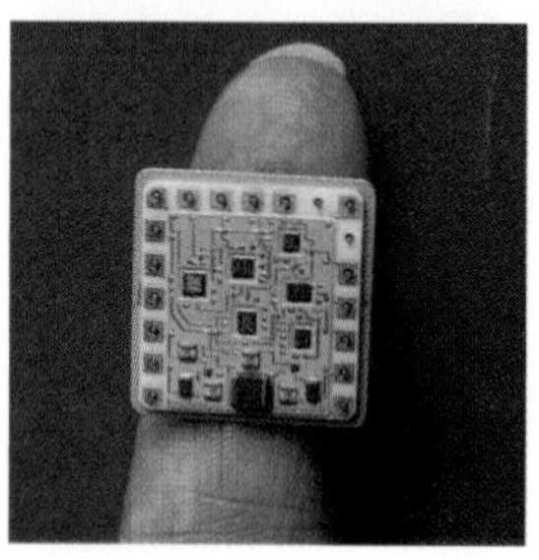

Silicon and germanium

Figure 15.2
This polishing bit (top) consists of carbon in the form of diamond. Silicon and germanium are used in computer chips (bottom). All of these elements are from Group 4A. How many valence electrons do they have? 4

Electron Configurations for Cations

You learned in Chapter 14 that noble gases, such as neon and argon, are unreactive in chemical reactions. In 1916, chemist Gilbert Lewis used this fact to explain why atoms form certain kinds of ions and molecules. He called his explanation the **octet rule:** In forming compounds, atoms tend to achieve the electron configuration of a noble gas. An octet is a set of eight. Recall that each noble gas (except helium) has eight electrons in its highest energy level and a general electron configuration of ns^2np^6. Thus the octet rule takes its name from this fact about noble gases. Atoms of the metallic elements tend to lose their valence electrons, leaving a complete octet in the next-lowest energy level. Atoms of some nonmetallic elements tend to gain electrons or to share electrons with another nonmetallic element to achieve a complete octet. Although there are exceptions, the octet rule applies to most atoms in compounds.

An atom's loss of valence electrons produces a cation, or a positively charged ion. The most common cations are those produced by the loss of valence electrons from metal atoms. Most of these atoms have one to three valence electrons, which are easily removed. Sodium, in Group 1A of the periodic table, is typical. Sodium atoms have a total of eleven electrons,

MEETING DIVERSE NEEDS

At Risk Ask students to work with a partner to practice drawing electron dot structures for Group A elements. Have one student randomly choose an element and write the symbol for that element on a piece of paper. Have another student fill in the electron dots for this element. Students can check their work by referring to Table 15.1. Make sure students understand that the electron dots represent only valence electrons, not the total electrons, and that valence electrons are the only electrons involved in chemical reactivity. Also, reinforce the stable octet dot structures for the noble gases and explain how ions of other elements will react to obtain this configuration.

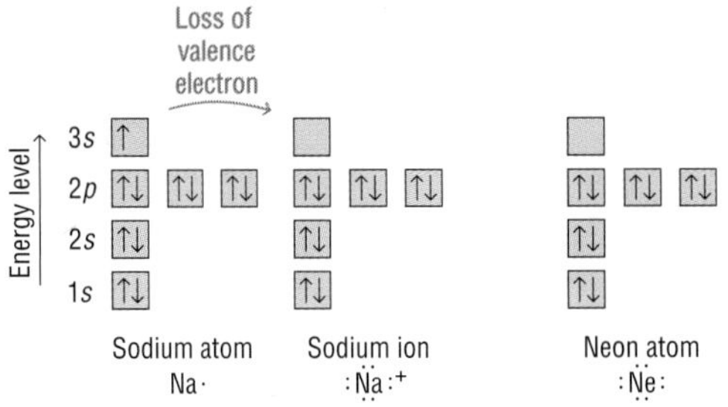

Figure 15.3

A sodium atom can lose an electron to become a positively charged sodium ion. The sodium ion has an electron configuration that is similar to the noble gas neon.

including one valence electron. When forming a compound, a sodium atom loses its one valence electron and is left with an octet (eight electrons) in what is now its highest energy level. Because the number of protons in the sodium nucleus is still eleven, the loss of one unit of negative charge (an electron) produces a sodium ion with a charge of 1+. You can represent the electron loss, or ionization, of the sodium atom by drawing the complete electron configuration of the atom and of the ion formed.

$$\text{Na } 1s^2 2s^2 2p^6 3s^1 \xrightarrow{-e^-} \text{Na}^+ \ 1s^2 \underbrace{2s^2 2p^6}_{\text{octet}}$$

As you can see in **Figure 15.3,** the electron configuration of the sodium ion is the same as that of the neon atom. How many electrons are in the highest energy levels of Na^+ and of Ne? 5

$$\text{Ne } 1s^2 \underbrace{2s^2 2p^6}_{\text{octet}}$$

Both the sodium ion and neon have eight electrons in their valance shells. Using an electron dot structure for the atom can show the ionization more simply.

$$\text{Na}\cdot \xrightarrow[\text{ionization}]{\text{loss of valence electron}} \text{Na}^+ + e^-$$

Sodium atom (electrically neutral, charge = 0) — Sodium ion (plus sign indicates one unit of positive charge) — electron (minus sign indicates one unit of negative charge)

Magnesium (atomic number 12) belongs to Group 2A of the periodic table. Therefore it has two valence electrons. A magnesium atom attains the electron configuration of neon by losing both valence electrons. The loss of the valence electrons produces a magnesium cation with twice the positive charge of a sodium ion.

$$\cdot\text{Mg}\cdot \longrightarrow \text{Mg}^{2+} + 2e^-$$

Magnesium atom (electrically neutral, charge = 0) — Magnesium ion (2+ indicates two units of positive charge) — (2 in front of e^- indicates two units of negative charge)

15.1

ACTIVITY

Have students identify the number of electrons and the number of valence electrons in randomly chosen elements from Table 15.1. Remind students that the number of valence electrons in an atom of a representative element is the same as the element's group number.

Think Critically

Have students determine the accuracy of this statement: All stable ions of elements result in electronic configurations that are isoelectronic with noble gases. (Most of the time this statement is true, but there are exceptions. Use Cu(I) as an example. Explain that a noble gas configuration is not generally possible with elements that would have to gain or lose too many electrons.)

Use the Visual

Display Figure 15.3 on an overhead projector. Point out that the interaction between atoms that produces bonding involves only the outermost electrons of the atoms. The inner electrons are locked tightly in filled energy levels and do not participate in bonding. Use a colored pen to circle the outermost electron in the sodium atom in Figure 15.3. Remind students that the outermost electrons are called valence electrons. Use a different colored pen to circle the octet of electrons in the sodium ion's highest energy level. Circle the corresponding octet of electrons in neon to show the similarity in electron configurations. Have students draw a similar diagram for calcium.

Answers

1 Group 2A and 3A elements have 2 and 3 valence electrons, respectively.
2 Helium has only 2 electrons.
3 7
4 4
5 Both the sodium ion and the neon atom have an octet, or eight electrons.

15.1

Discuss

One way to determine the number of valence electrons in an atom is to look at the electron configuration of the atom. Explain that any electron in an atom outside the noble-gas core or pseudo-noble-gas core is called a valence electron. Using diagrams such as those in Figures 15.3 and 15.5, show students several examples of how various atoms of the representative elements form ions and gain a noble-gas electron configuration. Indicate the noble-gas core and valence electrons in your diagrams.

Then show how the atoms of transition elements would have to gain or lose too many electrons to achieve a noble-gas electron configuration. For example, show how Ag, Zn, and Ga lose 1, 2, and 3 electrons, respectively to form pseudo noble-gas configurations. Remind students that no atom can lose an electron unless another atom is available to receive the electron.

Lay pieces of magnesium, zinc, copper, and, if possible, sodium metal on a dry surface in the lab to show that metals do not spontaneously decompose to form metal cations.

ACTIVITY

Have students write equations similar to those on page 417, showing the formation of metal cations from metal atoms. Students should show the electron dot structures for the metal atom and metal cation that is formed. In addition, you may want students to write out the electron configurations for the metal atom and cation.

Figure 15.4
Elemental copper's distinctive color is seen in this thirteenth century artifact (left). Basic copper(II) carbonate, containing Cu^{2+} ions, forms the greenish patina on this statue (right).

Copper

Copper(II) carbonate

You have seen that the cations of Group 1A elements always have a charge of 1+. Similarly, the cations of Group 2A elements always have a charge of 2+. This consistency can be explained in terms of the loss of valence electrons by metal atoms: The atoms lose enough electrons to attain the electron configuration of a noble gas. For example, all Group 2A elements have two valence electrons. In losing these two electrons, they form 2+ cations.

For transition metals, the charges of cations may vary. An atom of iron, for example, may lose two or three electrons. In the first case, it forms the iron(II), or ferrous, ion (Fe^{2+}). In the second case, it forms the iron(III), or ferric, ion (Fe^{3+}).

Some ions formed by transition metals do not have noble-gas electron configurations (ns^2np^6) and are therefore exceptions to the octet rule. Silver, with the electron configuration of $1s^22s^22p^63s^23p^63d^{10}4s^24p^64d^{10}5s^1$, is an example. To achieve the structure of krypton, which is the preceding noble gas, a silver atom would have to lose eleven electrons. To acquire the electron configuration of xenon, which is the following noble gas, silver would have to gain seven electrons. Ions with charges of three or greater are uncommon, and these possibilities are extremely unlikely. Thus silver does not achieve a noble-gas configuration. But if it loses its $5s^1$ electron, the configuration that results ($4s^24p^64d^{10}$), with 18 electrons in the outer energy level and all of the orbitals filled, is relatively favorable in compounds. Such a configuration is known as a pseudo noble-gas electron configuration. Silver forms a positive ion (Ag^+) in this way. Other elements that behave similarly to silver are found at the right of the transition metal block. Another example is shown in **Figure 15.5.** Copper(I) (Cu^+), gold(I) (Au^+), cadmium (Cd^{2+}), and mercury(II) (Hg^{2+}) ions all have pseudo noble-gas electron configurations.

Figure 15.5
By losing its lone 4s electron, copper attains a pseudo noble-gas electron configuration. Thus a copper atom can become a copper(I) ion (Cu^+).

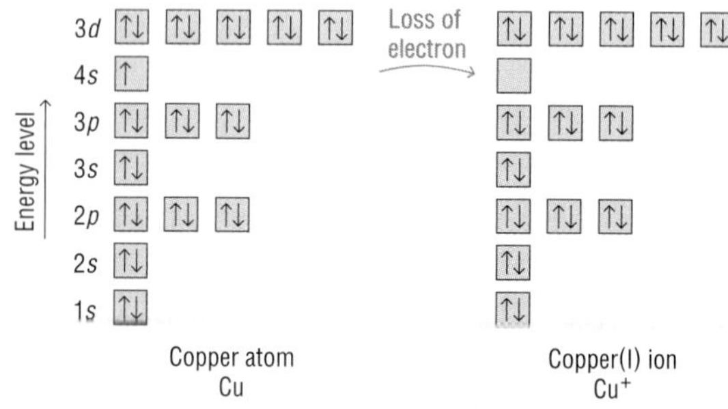

Electron Configurations for Anions

An anion is an atom or a group of atoms with a negative charge. The gain of negatively charged electrons by a neutral atom produces an anion. Because they have relatively full valance shells, atoms of nonmetallic elements attain noble-gas electron configurations more easily by gaining electrons than by losing them. For example, chlorine belongs to Group 7A (the halogen family) and has seven valence electrons. A gain of one electron gives chlorine an octet and converts a chlorine atom into a chloride ion. This ion is an anion with a single negative charge. Chlorine atoms, therefore, need one more valence electron to achieve the electron configuration of the nearest noble gas, argon.

$$\text{Cl} \quad 1s^22s^22p^63s^23p^5 \xrightarrow{+e^-} \text{Cl}^- \quad 1s^22s^22p^6\underbrace{3s^23p^6}_{\text{octet}}$$

The chloride ion has eight electrons, an octet, in its highest energy level, as shown in **Figure 15.6.** The ion has the same electron configuration as the noble gas argon.

$$\text{Ar} \quad 1s^22s^22p^6\underbrace{3s^23p^6}_{\text{octet}}$$

Electron configuration diagrams can be used to write an equation showing the formation of a chloride ion from a chlorine atom. Compare the electron configuration diagrams in **Figure 15.6** with the corresponding electron dot structures. What is the relationship between the number of electrons in the valance shells in the electron configuration diagrams and the number of dots in the electron dot structures? 1

$$:\!\ddot{\underset{..}{\text{Cl}}}\cdot \; + \; e^- \xrightarrow{\text{gain of one valence electron}} \; :\!\ddot{\underset{..}{\text{Cl}}}\!:^-$$

Chlorine atom — Chloride ion (Cl^-)

The ions that are produced when atoms of chlorine and other halogens gain electrons are called **halide ions.** All halogen ions have seven valence electrons and need to gain only one electron to achieve the electron configuration of a noble gas. Thus all halide ions (F^-, Cl^-, Br^-, and I^-) have a charge of 1–.

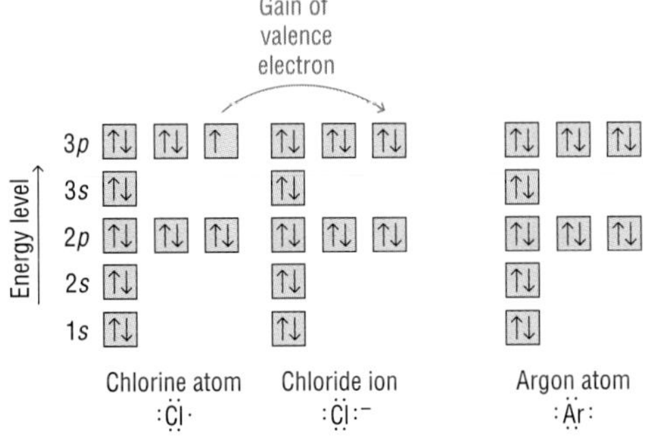

Figure 15.6
An atom of chlorine can gain an electron to become a negatively charged chloride ion. The chloride ion has an electron configuration similar to the noble gas argon. Both the chloride ion and the argon atom have an octet of electrons.

15.1

CHEMISTRY AND TECHNOLOGY

Chlorine, a highly reactive greenish gas, is one of the most useful elements for manufacturing chemical products such as paper, plastics, refrigerating gases, and medicines. In addition, chlorine is used to purify drinking water. Across the United States, chlorine is added to drinking water to kill a variety of microorganisms including those that carry diseases such as cholera and typhoid fever. But this extremely useful gas does have drawbacks, and many chemists are looking for alternatives to chlorine.

Researchers have found that chlorine reacts with organic substances in the water that it has purified to produce chlorinated compounds such as chloroacetonitrile. Chloroacetonitrile has been shown to cause inflammation of the digestive tract in laboratory animals. As an alternative to chlorine, some countries have begun purifying water with ozone. Ozone kills microorganisms even more effectively than chlorine. About one percent of the water supply in the United States is now purified with ozone. It is estimated that it would cost $6 billion to switch completely to ozone for treating all the drinking water supplies.

Answer

1 Each dot in the electron dot structure represents an electron in the valence shell in the electron configuration diagram.

3 Assess

Evaluate Understanding

Have students refer to the periodic table on page 124. To determine the students' knowledge about the formation of elemental anions and cations, ask the students to determine whether the following ions are likely to exist and why:

H^- (yes, isoelectronic with He); H^+ (yes, but without electrons, there is no comparable noble gas configuration); **Sr^{2+}** (yes, isoelectronic with Kr); **Al^{3+}** (yes, isoelectronic with Ne); **Xe^-** (no, cannot form ions easily due to stable electron configuration); **Zn^{6-}** (no, isoelectronic with Kr but formation would require a gain of too many electrons); **Zn^{2+}** (yes, not isoelectronic with a noble gas but has pseudo-noble-gas electron configuration with 18 electrons filling up the outer energy level: $3s^23p^63d^{10}$)

Reteach

Select groups from the periodic table in random order and ask students to predict the common ions that could be formed from elements of each group. Note that predicting is fairly easy for groups at the far left or far right of the table, but more difficult for groups in the center of the table, which have partially filled *d* and *f* orbitals.

To show one effect of the partially filled *d* orbitals compare ions of period-4 elements: colorless Ca^{2+}, red-violet Cr^{3+}, deep pink Co^{2+}, blue Cu^{2+}, green Ni^{2+}, pink Mn^{2+}, pale green Fe^{2+}, yellow Fe^{3+}, and colorless Zn^{2+}. Note that the two elements with colorless ions are calcium, which has no *d* electrons, and zinc, which has completely filled *d* orbitals.

Table 15.2

Some Common Anions									
1−				2−		3−	
F^-	fluoride	OH^-	hydroxide	O^{2-}	oxide	N^{3-}	nitride
Cl^-	chloride	ClO^-	hypochlorite	S^{2-}	sulfide	P^{3-}	phosphide
Br^-	bromide	NO_3^-	nitrate	SO_4^{2-}	sulfate	PO_4^{3-}	phosphate
I^-	iodide	HCO_3^-	hydrogen carbonate	CO_3^{2-}	carbonate		
		$C_2H_3O_2^-$	acetate				

Look at another example. Oxygen is in Group 6A, and oxygen atoms each have six valence electrons. Oxygen atoms attain the electron configuration of neon by gaining two electrons. The resulting oxide ions have charges of 2− and are written as O^{2-}. **Table 15.2** lists some common anions.

❶ How many electrons must a sulfur atom gain to form a sulfide (S^{2-}) ion?

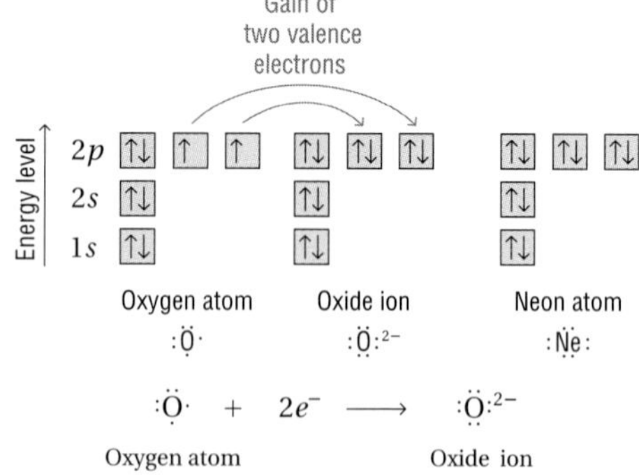

$$:\ddot{O}\cdot + 2e^- \longrightarrow :\ddot{O}:^{2-}$$

Oxygen atom — Oxide ion

LINK TO FOOD SCIENCE

Sulfur Dioxide and Sulfites
Sulfur dioxide (SO_2) has been used as a food additive since the time of the ancient Romans and Egyptians. Sulfur dioxide reacts with water to make hydrogen sulfite ions (HSO_3^-) and sulfite ions (SO_3^{2-}). The resulting mixture of these ions and sulfur dioxide is usually referred to simply as a sulfite. Sulfites are especially effective in preventing the deterioration of dried fruits and fruit juices. Spraying fresh lettuce and other produce with water containing low concentrations of sulfites keeps them from turning brown. The use of sulfites was once so common that they were often not labeled as an ingredient in foods. However, the use of sulfur dioxide is now being monitored more closely. Sulfur dioxide is a choking gas that plays a role in air pollution. Because of evidence that sulfites can trigger episodes of asthma in people who suffer from the illness, the use of sulfites to preserve foods is now being discouraged.

section review 15.1

1. How can the periodic table be used to infer the number of valence electrons in an atom?
2. Why do metals tend to form cations, and nonmetals tend to form anions?
3. How many valence electrons does each atom have?
   a. potassium **c.** magnesium
   b. carbon **d.** oxygen
4. Write the electron dot structure for each element in question 3.
5. Write electron configurations for the 1+ ion of copper and the 2+ ion of cadmium.
6. How many electrons will each element gain or lose in forming an ion?
   a. calcium (Ca) **c.** aluminum (Al)
   b. fluorine (F) **d.** oxygen (O)

Chem ASAP! **Assessment 15.1** Check your understanding of the important ideas and concepts in Section 15.1.

Answers

❶ 2

❷ Each aluminum atom has three valence electrons it can lose and each of the three bromine atoms can gain one valence electron.

SECTION REVIEW 15.1

1. The group number equals the number of valence electrons for representative elements.
2. It is easier for a metal to lose electrons and for a nonmetal to gain electrons to achieve the electron configurations of a noble gas.
3. **a.** 1 **b.** 4 **c.** 2 **d.** 6
4. **a.** K· **b.** ·Ċ· **c.** ·Mg· **d.** :Ö·
5. Cu^+: $1s^22s^22p^63s^23p^63d^{10}$
   Cd^{2+}: $1s^22s^22p^63s^23p^63d^{10}4s^24p^64d^{10}$
6. **a.** lose 2 **b.** gain 1 **c.** lose 3 **d.** gain 2

IONIC BONDS

section 15.2

Found naturally in Earth's crust, fluorite is a brittle, glasslike mineral. Although it is not durable enough to be used in fine jewelry, gem collectors often search for pure, colorless forms of fluorite to display as part of their collections. Like other crystalline solids, however, this beautiful crystal is very stable and melts at a very high temperature. **Why are crystalline solids such stable structures?**

objectives

- List the characteristics of an ionic bond
- Use the characteristics of ionic compounds to explain the electrical conductivity of ionic compounds when melted and when in aqueous solutions

key terms

- ionic bonds
- coordination number

Formation of Ionic Compounds

Anions and cations have opposite charges. They attract one another by electrostatic forces. The forces of attraction that bind these oppositely charged ions are called **ionic bonds.** Compounds that consist of electrically neutral groups of ions joined by electrostatic forces are called ionic compounds. In any sample of an ionic compound, the total positive charges of the cations must equal the total negative charges of the anions. This same principle of electrical neutrality was used (in Chapter 5) to write formulas of ionic compounds.

Sodium chloride provides a simple example of how ionic bonds are formed. Consider the reaction between a sodium atom and a chlorine atom. Sodium has a single valence electron that it can easily lose. (If the sodium atom loses its valence electron, it achieves the stable electron configuration of neon.) Chlorine has seven valence electrons and can easily gain one. (If the chlorine atom gains a valence electron, it achieves the stable electron configuration of argon.) When sodium and chlorine react to form a compound, the sodium atom gives its one valence electron to a chlorine atom. Thus sodium and chlorine atoms combine in a one-to-one ratio and both ions have stable octets.

$$\begin{array}{llcll} \text{Na}\cdot & :\ddot{\text{Cl}}: & \longrightarrow & \text{Na}^+ & :\ddot{\text{Cl}}:^- \\ 1s^22s^22p^6\,3s^1 & 1s^22s^22p^63s^23p^5 & & \underbrace{1s^22s^22p^6}_{\text{octet}} & \underbrace{1s^22s^22p^63s^23p^6}_{\text{octet}} \\ & & & \text{Ne} & \text{Ar} \\ & & & \underbrace{1s^22s^22p^6}_{\text{octet}} & \underbrace{1s^22s^22p^63s^23p^6}_{\text{octet}} \end{array}$$

The chemical formula for sodium chloride (NaCl) represents a formula unit. It indicates the lowest whole-number ratio of cations to anions in any sample of an ionic compound. The formula NaCl, for example, shows that one formula unit of sodium chloride contains one sodium ion and one chloride ion.

Figure 15.7 shows aluminum and bromine reacting to form the compound aluminum bromide. Each aluminum atom has three valence electrons to lose. Each bromine atom has seven valence electrons and readily gains one additional electron. Therefore, when aluminum and bromine react, three bromine atoms combine with each aluminum atom. The formula for aluminum bromide is $AlBr_3$.

Chem ASAP!

Animation 17

Take an atomic level look at the formation of KCl.

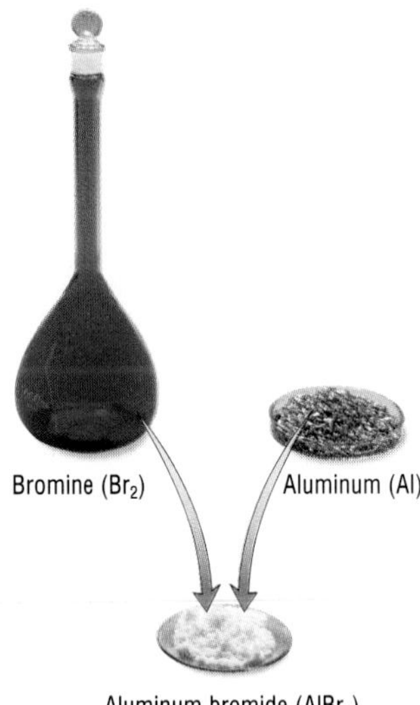

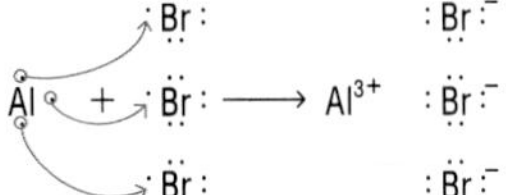

Figure 15.7

Aluminum metal and the nonmetal bromine combine to form the ionic solid, aluminum bromide. Why do three Br atoms combine with each Al atom? 2

STUDENT RESOURCES

- Section Review 15.2, Ch. 15 Practice Problems, Interpreting Graphics, and Quizzes from the Review Module
- Laboratory Recordsheets 15-1 and 15-2
- Laboratory Manual: Experiment 24
- Laboratory Practical 15-1
- Small-Scale Chemistry Lab Manual: Experiments 18, 19, 20

TECHNOLOGY RESOURCES

Relevant technology resources include:

- Chem ASAP! CD-ROM
- Resource Pro CD-ROM
- ActivChemistry CD-ROM: *Ions and Ionic Bonds*
- Chemistry Alive! Videodisc: *Making Sodium Chloride*

15.2

1 Engage

Use the Visual

Have students examine the mineral fluorite in the opening photograph. Discuss the ways in which Ca^{2+} and F^- can interact to form the regular fluorite crystal structure. Ask:

- **What property of the bonds between crystal atoms makes them so rigid and brittle?** (The crystal is rigid because it is held together by a specific three-dimensional array of relatively strong attractive forces between anions and cations, which is accompanied by minimal charge repulsion of like ions. The crystal is brittle because the attractive interactions are specifically arranged within the crystal structure. If this arrangement is perturbed as the result of a blow with a hammer, charge repulsion between ions of the same charge can force the crystal to fragment.)

Check Prior Knowledge

To assess students' prior knowledge about ionic bonds and crystals, ask:

- **What is an ionic bond?** (an interaction between ions of a different charge)
- **How do a polyatomic ion and a monatomic ion differ?** (A monatomic ion is an ion formed from a single atom; a polyatomic ion is a stable unit of two or more tightly bound atoms that carries a charge.)
- **Why are crystals of different ionic compounds different shapes?** (The shapes reflect different geometric arrangements of anions and cations with different sizes and charges.)

CHEMath

VISUALIZING MOLECULES

As you have seen by now, molecules can be described in a number of different ways. Each description provides specific information about the molecule.

Chemical Formulas A chemical formula indicates the kinds and numbers of atoms in a molecular compound, but does not give any information about the shape of the molecule. For example, the chemical formula for methane is

$$CH_4$$

Structural Formulas A structural formula indicates which atoms are bonded to each other in a molecule, but does not show the three-dimensional structure of the molecule. One type of structural formula, a Lewis electron-dot structure, use dots to represent each atom's valence electrons. In another type of structural formula, known as bond-line notation, lines are used to represent the covalent bonds in a molecule. Sometimes the bond lines are replaced with wedge-like shapes to provide a simple three-dimensional perspective. Various structural formulas for methane are shown here.

```
   H             H              H
 H:C:H       H — C — H          |
   H             H         H ⁄  C  ⁀ H
                                |
                                H
electron-dot   bond-line      wedge
 structure     structure     structure
```

Ball-and-Stick Models A ball-and-stick model uses spheres to represent an atom's nucleus and inner-level electrons, and sticks to represent bonds. A ball-and-stick model shows the three-dimensional shape of a molecule. The ball-and-stick model for methane is shown.

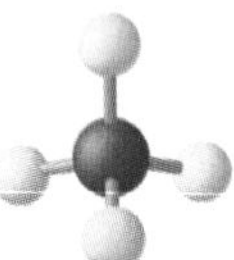

Space-Filling Models A space-filling model of a molecule uses spheres to show both the relative sizes of atoms and the three-dimensional shape of the molecule. The space-filling model does not use sticks to show the bonds. Space-filling models are the most realistic representations of molecules because they are large-scale versions of the actual molecules. The space-filling model for methane is shown.

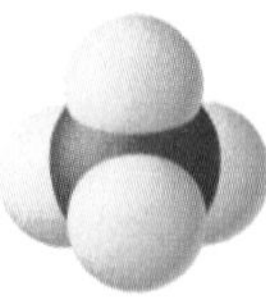

Examples

Several examples of the various molecular representations used throughout this text book are shown below.

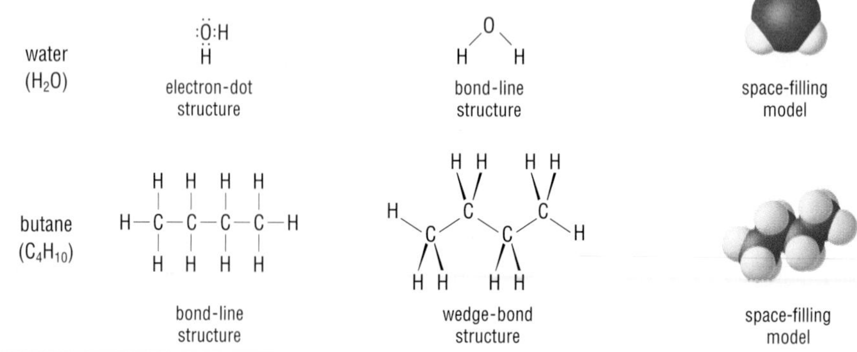

CHEMath

Teaching Tips

In this feature, students find a summary of the descriptions of molecules that they will see at different points throughout the textbook. Make sure they realize that no description is any more important than any of the others. Each description has a specific purpose. As an example of a situation in which different descriptions serve varying purposes, consider your house keys. While it is the composition of your house keys that makes them sturdy and resistant to rust, it is the shape of the key that enables you to open the door. In chemistry, both the composition and shape of a molecule provide important information.

Review the structures shown for methane, water, and butane. Show similar structures for other molecules, such as Cl_2 or CO_2. Point out that the Lewis electron-dot structure, bond-line structure, and wedge-bond structure are grouped together because they are very similar. The Lewis electron-dot structure and the bond-line structure differ only by whether they use dots or dashes for bonds. The wedge-bond structure differs from the bond-line structure only by thickened lines which give some shape to the molecule. Although the space-filling model is an actual model of the atom, the ball-and-stick model with help students visualize molecules when they learn VSEPR theory.

Challenge students to choose a molecule and prepare a chart similar to the one shown in the feature. Have them build the ball-and-stick model and space-filling models using foam spheres or other equipment.

Sample Problem 15-1

Use electron dot structures to predict the formulas of the ionic compounds formed from these elements.

a. potassium and oxygen
b. magnesium and nitrogen

1. ***ANALYZE*** ***Plan a problem-solving strategy.***
The correct electron dot structure of each atom in the compound must be written. Atoms of metals lose their valence electrons when forming an ionic compound. Atoms of nonmetals gain electrons. Enough atoms of each element must be used in the formula so that electrons lost equals electrons gained.

2. ***SOLVE*** ***Apply the problem-solving strategy.***
a. Start with the atoms.

$$K\cdot \quad \text{and} \quad \cdot\ddot{\underset{\cdot}{O}}:$$

In order to have a completely filled valence shell, oxygen requires the addition of two electrons. These electrons come from two potassium atoms, each of which loses one electron.

$$\begin{matrix} K\cdot \\ K\cdot \end{matrix} + \cdot\ddot{\underset{\cdot}{O}}: \longrightarrow \begin{matrix} K^+ \\ K^+ \end{matrix} :\ddot{\underset{\cdot\cdot}{O}}:^{2-}$$

Electrons lost now equals electrons gained. The formula of the compound formed (potassium oxide) is K_2O.

b. Start with the atoms.

$$\dot{Mg} \quad \text{and} \quad \cdot\dot{\underset{\cdot}{N}}:$$

Each nitrogen needs three electrons to have an octet, but each magnesium atom can lose only two electrons. Thus three magnesium atoms are needed for every two nitrogen atoms.

$$\begin{matrix} \dot{Mg} \\ \dot{Mg} \\ \dot{Mg} \end{matrix} + \begin{matrix} \cdot\dot{\underset{\cdot}{N}}: \\ \cdot\dot{\underset{\cdot}{N}}: \end{matrix} \longrightarrow \begin{matrix} Mg^{2+} \\ Mg^{2+} \\ Mg^{2+} \end{matrix} \begin{matrix} :\ddot{\underset{\cdot\cdot}{N}}:^{3-} \\ :\ddot{\underset{\cdot\cdot}{N}}:^{3-} \end{matrix}$$

The formula of the compound formed (magnesium nitride) is Mg_3N_2.

3. ***EVALUATE*** ***Do the results make sense?***
In each example, the number of electrons gained by the nonmetal balances the number of electrons lost by the metal. In addition, division by an integer cannot reduce the formulas of the compounds further. The formulas K_2O and Mg_3N_2 are correct.

Practice Problems

7. Use electron dot structures to determine chemical formulas of the ionic compounds formed when the following elements combine.
   a. potassium and iodine
   b. aluminum and oxygen
8. Name the compounds formed in Practice Problem 7.

Chem ASAP!

Problem-Solving 7

Solve Problem 7 with the help of an interactive guided tutorial.

15.2

2 Teach

Practice Problems Plus

Related Chapter Review Problem
Chapter Review problem 35 is related to Sample Problem 15-1.

Additional Practice Problem
The Chem ASAP! CD-ROM contains the following problem: Use electron dot structures to determine chemical formulas of the ionic compounds formed when the following elements combine.
a. magnesium and chlorine
b. aluminum and sulfur
(**a.** $MgCl_2$, **b.** Al_2S_3)

Discuss

Explain that the formation of positive ions and of negative ions are simultaneous and interdependent processes. An ionic compound is the result of the transfer of electrons from one set of atoms to another set of atoms. An ionic compound consists entirely of ions.

Meeting Diverse Needs

Gifted Have students research how minerals are categorized according to their ionic nature. Suggest that their written report include information concerning the physical properties of minerals and how these are used in mineral identification. Encourage students to include drawings, photos, or examples of minerals from each category.

Chemistry Alive!

Making Sodium Chloride

Play

ACTIVITY

Pass around crystals of ionic compounds of various types in watch glasses. Have the students examine the crystals with magnifying glasses and write down their observations. Make a list of these observations and then discuss them in terms of the underlying ionic lattice structures.

ACTIVITY

Have students write the formulas for ionic compounds formed from randomly chosen pairs of cations and anions. Include various complex cations and anions as well. Example:
1) K^+ and S^{2-}. (Answer: K_2S)
2) Ca^{2+} and HCO_3^-.
(Answer: $Ca(HCO_3)_2$)

CHEMISTRY AND THE ENVIRONMENT

Point out that the rusting of iron is the production of the ionic compound iron oxide from iron metal and oxygen gas. Discuss the formula of rust and point out that Fe^{3+} is a stable cation of Fe. (The students should know that O^{2-} is the stable anion of O; therefore, iron oxide is Fe_2O_3.) Almost all metals can react with oxygen to form metal oxides. Ask which ionic compound forms from the "rusting," of aluminum metal. (aluminum oxide, Al_2O_3) Point out that aluminum cans acquire a thin layer of Al_2O_3 which protects the aluminum metal underneath from undergoing further corrosion, since oxygen and water cannot penetrate this coating. This is why aluminum cans take a very long time to break down.

Chem ASAP!
Simulation 15
Simulate the formation of ionic compounds at the atomic level.

Properties of Ionic Compounds

At room temperature, most ionic compounds are crystalline solids. **Figure 15.8** shows the striking beauty of the crystals of some ionic compounds. The component ions in such crystals are arranged in repeating three-dimensional patterns. The composition of a crystal of sodium chloride is typical. In solid NaCl, each sodium ion is surrounded by six chloride ions, and each chloride ion is surrounded by six sodium ions. In this arrangement, each ion is attracted strongly to each of its neighbors and repulsions are minimized. The large attractive forces result in a very stable structure. This is reflected in the fact that NaCl and ionic compounds in general have high melting temperatures.

Figure 15.8
The beauty of crystalline solids, such as these, comes from the orderly arrangement of their component ions.

Aragonite ($CaCO_3$)

Barite ($BaSO_4$) and calcite ($CaCO_3$)

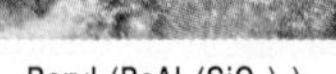

Beryl ($BeAl_2(SiO_3)_6$)

Franklinite ($(Zn,Mn^{2+},Fe^{2+})(Fe^{3+},Mn^{3+})_2O_4$)

Pyrite (FeS_2)

Hematite (Fe_2O_3)

Rutile (TiO_2)

Cinnabar (HgS)

MEETING DIVERSE NEEDS

Gifted Have students develop formulas for calculating the density of a metal given the atomic radius of the metal and the cubic unit cell packing arrangement of the metal atoms. Use sodium, potassium, iron, chromium, and tungsten for body-centered cubic metals. Use copper, silver, gold, aluminum, and lead for face-centered cubic metals. The students may need to make models to see geometric relationships more clearly.

LEP Encourage LEP students to look up and define terms used to describe metals. Students should define terms such as *ductile* and *malleable* in English and their native language. Emphasize how their strong but flexible bonding arrangement accounts for metals having the unique properties of ductility and malleability.

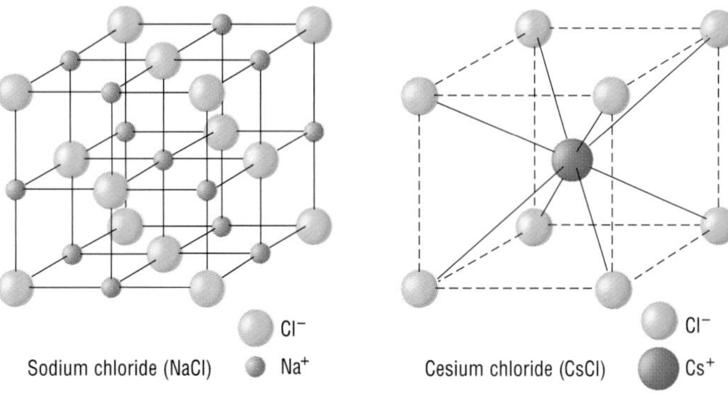

Figure 15.9
Sodium chloride and cesium chloride form clear, colorless cubic crystals. The unit cells of these cubic compounds differ. The unit cell of sodium chloride is face-centered cubic, but that of cesium chloride is simple cubic. The arrangement of the ions in the crystal structure are shown for sodium chloride (left) and cesium chloride (right). How many chloride ions surround each sodium ion? How many chloride ions surround each cesium ion? ❶

The **coordination number** of an ion is the number of ions of opposite charge that surround the ion in a crystal. **Figure 15.9** shows the three-dimensional arrangement of ions in NaCl. Because each Na^+ ion is surrounded by six Cl^- ions, Na^+ has a coordination number of 6. Each Cl^- ion is surrounded by six Na^+ ions and also has a coordination number of 6.

Cesium chloride (CsCl) has a formula unit that is similar to that of NaCl. Both compounds have cubic crystals, but their internal crystal structures are different. Each Cs^+ ion is surrounded by eight Cl^- ions, and each Cl^- ion is surrounded by eight Cs^+ ions. The anion and cation in cesium chloride each have a coordination number of 8.

The crystalline form of titanium dioxide (TiO_2) is called rutile. In this compound, the coordination number for the cation (Ti^{4+}) is 6. Each Ti^{4+} ion is surrounded by six O^{2-} ions. The coordination number of the anion (O^{2-}) is 3. Each O^{2-} ion is surrounded by three Ti^{4+} ions. **Figure 15.10** shows a specimen of a rutile crystal along with an illustration showing how the ions are arranged in the crystal.

Rutile (TiO_2)

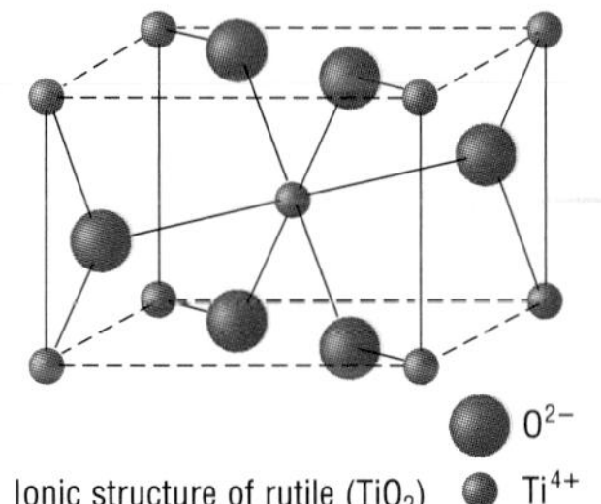

Ionic structure of rutile (TiO_2)

Figure 15.10
The shape of a crystal depends on its unit cell structure. Crystals of the mineral rutile (titanium dioxide) are tetragonal. The ionic structure of rutile is also shown.

15.2

ACTIVITY

Grow crystals in the lab. Students are used to seeing ionic compounds as granular powders, but under the right conditions, aqueous solutions of ionic salts can be induced to form large crystals that can be handled and examined with the unaided eye. For example, under the right conditions, NaCl(*aq*) will form beautiful rectangular prisms.

Have students prepare saturated solutions of innocuous salts such as NaCl, KCl, $MgCl_2$, or CsCl. Use analytical grade quality preparations if possible. Provide students with crystallizing dishes in which to incubate a portion of the saturated solution. The solution should then be stored for a number of days for the crystals to grow to full size.

Students can monitor growth of the crystals on a daily basis. The rate of crystallization will vary for different compounds. Once the crystals are formed, students should appreciate how the orderly three-dimensional arrangement of ions at the microscopic level gives rise to an impressive degree of symmetry at the macroscopic level.

Answer

❶ 6 Cl^- ions surround each sodium ion.
8 Cl^- ions surround each cesium ion.

Cooperative Learning

Divide the class into groups. Have each group choose a different class of ionic compounds to research and write about. For example, one group could work with oxides while another group worked with sulfides. Initially, each student should work alone to discover information such as where the compounds occur in nature, how they are produced, their physical and chemical properties, and any important uses. Finally, students in each group can pool their information to prepare a class display and/or report.

MINI LAB

Solutions Containing Ions

Safety Students should handle wires with caution. The wire may become hot during the activity.
Time 45 minutes
Group Size 2–4 students

LAB PREP AND PLANNING

Probeware Lab Manual has Vernier, TI, and PASCO labs.

You may want students to collect the gases in test tubes and test for hydrogen and oxygen.

ANALYSIS AND CONCLUSIONS

1. The acetic acid, sodium chloride, and sodium hydrogen carbonate solutions produced gas bubbles. Solutions of these compounds contain ions and therefore conduct electric current.
2. Tap water, distilled water, and sugar solution did not produce bubbles. Water and sugar solutions do not contain ions and therefore do no conduct an electric current.
3. Answers will vary but should indicate that a larger number of batteries will increase the current which will, in turn, cause the rate at which the bubbles appear to increase.

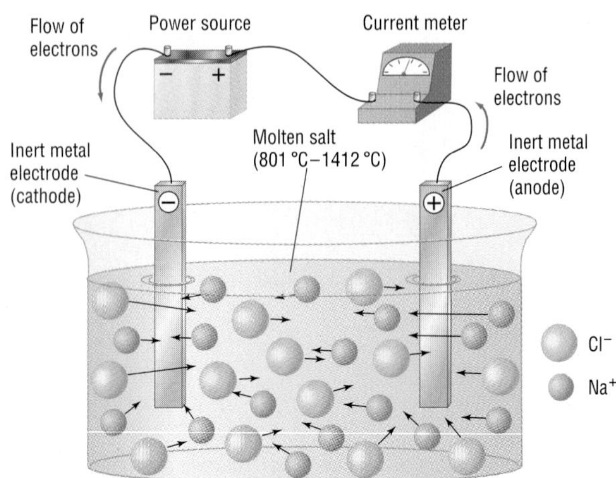

Figure 15.11
When sodium chloride melts, the sodium and chloride ions are free to move throughout the molten salt. If a voltage is applied, positive sodium ions move to the negative electrode (the cathode). At the same time, negative chloride ions move to the positive electrode (the anode).

The internal structures of crystals are determined by a technique called x-ray diffraction crystallography. X-rays that pass through a crystal are recorded on film. The pattern on the exposed film shows how ions in the crystal deflect the x-rays. The pattern is used to calculate the positions of ions in the crystal and to define the structure of the crystal.

When melted, ionic compounds can conduct an electric current. When sodium chloride is melted (melting point about 800 °C), the orderly crystal structure breaks down, as shown in **Figure 15.11.** If a voltage is applied across this molten mass, cations migrate freely to one electrode and anions migrate to the other. This ion movement produces a flow of electricity between the electrodes through an external wire. For a similar reason, ionic compounds also conduct electricity if they are dissolved in water. When dissolved, the ions are free to move about in the aqueous solution.

Figure 15.12
This solar facility uses molten NaCl to store thermal energy. Rather than using the compound for its ability to conduct electricity, the facility uses the molten NaCl for its ability to absorb and hold a large quantity of heat.

Meeting Diverse Needs

At Risk Have students break up into small groups and have them prepare models of the unit cells of two different ionic compounds. Use styrofoam balls of different sizes to represent the ions, and pipe cleaners to represent electrostatic interactions. Have the students determine coordination numbers based on the identification of "nearest neighbors."

MINI LAB

Solutions Containing Ions

PURPOSE

To show that ions in solution conduct an electric current.

MATERIALS

- 3 D-cell batteries
- masking tape
- 2 30-cm lengths of bell wire with ends scraped bare
- clear plastic cup
- distilled water
- tap water
- vinegar
- sucrose
- sodium chloride
- baking soda
- conductivity probe (optional)

PROCEDURE

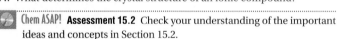

Probe version available in the Probeware Lab Manual.

1. Tape the batteries together so the positive end of one touches the negative end of another. Tape the bare end of one wire to the positive terminal of the battery assembly and the bare end of the other wire to the negative terminal.
2. Half fill the cup with distilled water. Hold the bare ends of the wires close together in the water. Look for the production of bubbles. They are a sign that the solution conducts electricity.
3. Repeat Step 2 with tap water, vinegar, and concentrated solutions of sucrose, sodium chloride, and baking soda (sodium hydrogen carbonate).

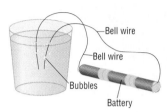

ANALYSIS AND CONCLUSIONS

1. Which solutions produced bubbles of gas? Explain.
2. Which samples did not produce bubbles of gas? Explain.
3. Would you expect the same results if you used only one battery? If you used six batteries? Explain your answer.

section review 15.2

9. What are the characteristics of ionic bonds?
10. Explain why ionic compounds can conduct electricity when melted and when in aqueous solutions.
11. Write the correct chemical formula (the formula unit) for the compounds formed from each pair of ions.
    a. K^+, S^{2-} **c.** Na^+, SO_4^{2-}
    b. Ca^{2+}, O^{2-} **d.** Al^{3+}, PO_4^{3-}
12. Write formulas for each compound.
    a. potassium nitrate **d.** lithium oxide
    b. barium chloride **e.** ammonium carbonate
    c. magnesium sulfate **f.** calcium phosphate
13. Which pairs of elements are likely to form ionic compounds?
    a. chlorine and bromine **c.** lithium and chlorine
    b. potassium and helium **d.** iodine and sodium
14. What determines the crystal structure of an ionic compound?

Chem ASAP! **Assessment 15.2** Check your understanding of the important ideas and concepts in Section 15.2.

portfolio project

Research x-ray diffraction crystallography. How are the samples prepared? How are the x-rays generated and detected? If possible, visit a site where x-ray crystal diffraction studies are done.

Answers

SECTION REVIEW 15.2

9. characterized by attraction between oppositely charged ions formed through electron transfer
10. Ionic compounds conduct electricity when melted and in aqueous solution because their ions are free to move about.
11. **a.** K_2S **b.** CaO **c.** Na_2SO_4 **d.** $AlPO_4$
12. **a.** KNO_3
    b. $BaCl_2$
    c. $MgSO_4$
    d. Li_2O
    e. $(NH_4)_2CO_3$
    f. $Ca_3(PO_4)_2$
13. **c.** and **d.**
14. the charges and relative sizes of the ions

15.2

TEACHER DEMO

Use foam balls and toothpicks or clay and straws to make models of the unit cells for several ionic compounds. Point out the coordination numbers of the ions in each cell. Stress that different packing arrangements depend on ionic size and charge.

Use the Visual

Have students examine the unit cell diagrams for NaCl and CsCl in Figure 15.9. Discuss how the overall crystal shape as seen at the macroscopic level is related to the arrangement of ions in the unit cells. Discuss why the packing arrangements are different for NaCl and CsCl. (The Cs^+ ion is larger than the Na^+ ion.)

3 Assess

Evaluate Understanding

Name ionic compounds and ask students to identify the cation, anion, and ratio of cations to anions in each compound.

Reteach

Emphasize that an ionic solid is a collection of independent ions. There is no joining of individual particles to form molecules. Each ion "belongs" as much to one of its nearest neighbors as it belongs to any other. The arrangement in an ionic crystal is such that each ion is surrounded by ions of opposite charge, which produces a strong bonding force.

Portfolio Project

Labs that do x-ray diffraction studies often will provide tours for student groups if arrangements are made in advance.

SMALL-SCALE LAB

QUALITATIVE ANALYSIS OF CHEMICAL FERTILIZERS

OVERVIEW

Procedure Time: 40 minutes
Students perform qualitative analysis on aqueous mixtures of ionic compounds and apply what they learn to the analysis of fertilizers.

MATERIALS

Solution	Preparation
$0.05M\ AgNO_3$	2.1 g in 250 mL
$0.2M\ Pb(NO_3)_2$	16.6 g in 250 mL
$0.1M\ Na_3PO_4$	9.5 g $Na_3PO_4 \cdot 12H_2O$ in 250 mL
$0.5M$ NaOH	20.0 g in 1.0 L
$0.2M\ Na_2SO_4$	7.1 g in 250 mL
$1.0M\ NH_4Cl$	13.4 g in 250 mL
$0.1M$ KSCN	2.4 g in 250 mL
$0.1M$ KI	4.2 g in 250 mL
$0.1M\ FeCl_3$	6.8 g $FeCl_3 \cdot 6H_2O$ in 25 mL of $1.0M$ NaCl; dilute to 250 mL
$1.0M$ HCl	82 mL of $12M$ in 1.0 L.
$1.0M\ HNO_3$	63 mL of $15.8M$ in 1.0 L.

CAUTION! Always add acid to water carefully and slowly!
Household products: samples of solid lawn and garden fertilizers; staples for solid Fe

TEACHING SUGGESTIONS

Labels on fertilizers have a sequence that indicates the percentages of the three major ingredients. For example, a 20-10-5 fertilizer is 20% nitrogen, 10% phosphorus, and 5% potassium. The nitrogen is usually in the form of nitrate ion, ammonium ion, or urea. Phosphorus is usually present as phosphate ions. Potassium ions are combined with either nitrate or phosphate ions. For the best results, use inexpensive lawn fertilizers that list iron on the package. Make unknown solutions by mixing chemicals from Figures A and B.

SMALL-SCALE LAB

ANALYSIS OF ANIONS AND CATIONS

SAFETY

Wear safety glasses and follow the standard safety procedures outlined on page 18.

PURPOSE

To develop tests for various ions and use the tests to analyze unknown substances.

MATERIALS

- pencil
- paper
- ruler
- reaction surface
- cotton swab
- pipet
- medicine droppers
- chemicals shown in Figures A and B

PROCEDURE

On one sheet of paper, draw grids similar to Figure A and Figure B. Draw similar grids on a second sheet of paper. Make each square 2 cm on each side. Place a reaction surface over the grids on one of the sheets of paper and add one drop of each solution or one piece of each solid as shown in Figures A and B. Stir each solution by blowing air through an empty pipet. Use the grids on the second sheet of paper as a data table to record your observations for each solution. Absorb the NH_4Cl and NaOH mixture onto a cotton swab and carefully determine its odor.

	Na_2SO_4 (SO_4^{2-})	HNO_3 (NO_3^-)	Na_3PO_4 (PO_4^{3-})
$AgNO_3$	NVR	NVR	light yellow ppt
HCl plus 1 piece of Fe(*s*)	bubbles	bubbles w/yellow solution	bubbles
$Pb(NO_3)_2$	white ppt	NVR	white ppt

Figure A
Anion Analysis

	KI (K^+)	NH_4Cl (NH_4^+)	$FeCl_3$ (Fe^{3+})
NaOH	NVR	smells like ammonia	rust ppt
KSCN	NVR	NVR	blood-red soln

Figure B
Cation Analysis

ANALYSIS

Using your experimental data, record the answers to the following questions below your data table.

1. Carefully examine the reaction of Fe(*s*) and HCl in the presence of HNO_3. What is unique about this reaction? How can you use it to identify nitrate ion?
2. Which solutions from Figure A are the best for identifying each anion? Explain.
3. Which solutions from Figure B are the best for identifying each cation? Explain.
4. Can your experiments conclusively identify potassium ions? Explain.

YOU'RE THE CHEMIST

The following small-scale activities allow you to develop your own procedures and analyze the results.

1. **Design It!** Obtain a set of unknown anion solutions from your teacher and design and carry out a series of tests that will identify each anion.
2. **Design It!** Obtain a set of unknown cation solutions from your teacher and design and carry out a series of tests that will identify each cation.
3. **Design It!** Obtain a set of unknown solid ionic compounds from your teacher and design and carry out a series of tests that will identify each cation and anion present.
4. **Analyze It!** Obtain an unknown fertilizer sample from your teacher and analyze it for cations and anions.

ANALYSIS

1. Nitrate ion is the only ion that produces a yellow color with iron in the presence of acid.
2. $AgNO_3$ identifies PO_4^{3-}; Fe and HCl identifies NO_3^-; $Pb(NO_3)_2$ identifies SO_4^{2-}. Each test produces a distinctive result.
3. NaOH identifies NH_4^+ by the odor of ammonia; KSCN identifies Fe^{3+} by a distinct red solution.
4. No; K^+ does not visibly react to any of the tests.

YOU'RE THE CHEMIST

1. Mix a drop of each unknown with one drop of each solution from Figure A. Compare the results to those with known solutions.
2. Mix a drop of each unknown with one drop of both solutions from Figure B. Compare the results to those with known solutions.
3. Mix an unknown with one drop of each solution from Figures A and B and compare results.
4. Use the approach from activity 3.

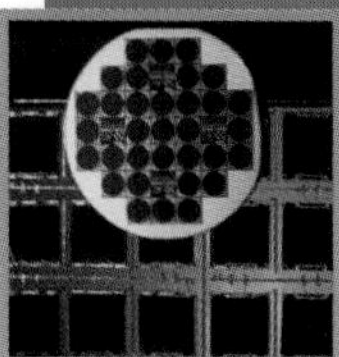

BONDING IN METALS

The melting point of gallium is only 30°C, which is very low for a metal. Because normal human body temperature is 37°C, gallium melts readily if you hold it in your hand. In both its liquid and solid forms, this metal is a strong conductor of electricity and is used in both transistors and solar cells. **What property makes metals good electrical conductors?**

objectives

- Use the theory of metallic bonds to explain the physical properties of metals
- Describe the arrangements of atoms in some common metallic crystal structures

key term

- metallic bonds

Metallic Bonds and Metallic Properties

Metals are made up of closely packed cations rather than neutral atoms. The cations are surrounded by mobile valence electrons, as shown in **Figure 15.13,** which can drift freely from one part of the metal to another. **Metallic bonds** consist of the attraction of the free-floating valence electrons for the positively charged metal ions. These bonds are the forces of attraction that hold metals together.

This model of metallic bonding explains many physical properties of metals. For example, metals are good conductors of electrical current because electrons can flow freely in them. As electrons enter one end of a bar of metal, an equal number leave the other end. Metals are ductile—that is, they can be drawn into wires, as shown in **Figure 15.14.** Is copper ductile? ❶ How do you know? Metals are also malleable, which means that they can be hammered or forced into shapes. Both the ductility and malleability of metals can be explained in terms of the mobility of valence electrons. A sea of drifting valence electrons insulates the metal cations from one another. When a metal is subjected to pressure, the metal cations easily slide past one another like ball bearings immersed in oil. In contrast, if an ionic crystal is struck with a hammer, the blow tends to push ions of like charge into contact. They repel, and the crystal shatters.

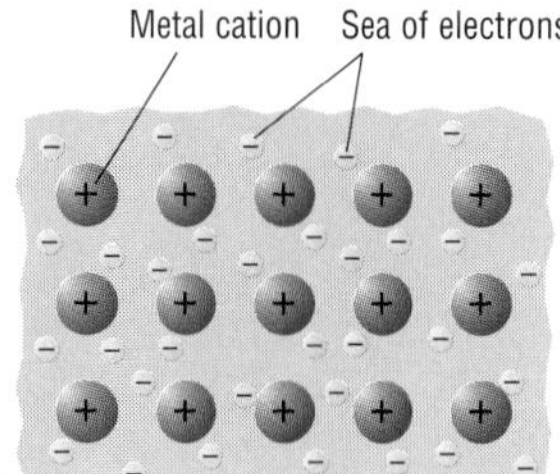

Figure 15.13
The valence electrons of metal atoms pool to form a sea of electrons. The metal is held together by the attractions between these electrons and the metal cations.

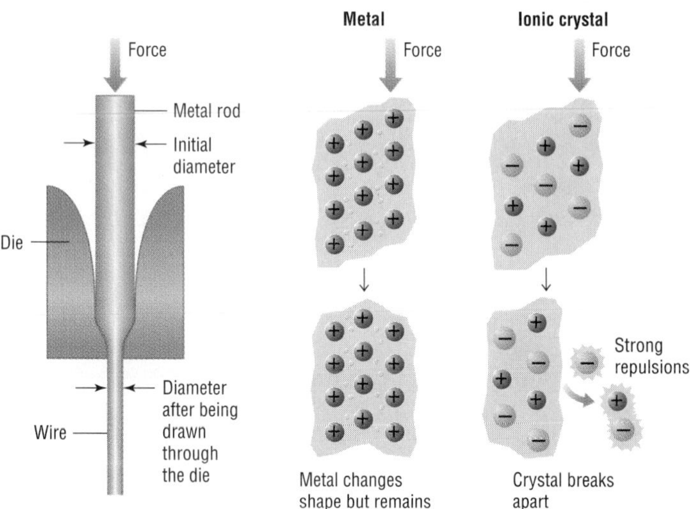

Chem ASAP!

Animation 18
See how metallic bonding explains some physical properties of metals.

Figure 15.14
A metal rod can be forced through a narrow opening in a die to produce wire (left). As this occurs, the metal changes shape but remains in one piece (center). If an ionic crystal were forced through the die it would shatter (right). Why is this so? ❷

15.3

1 Engage

Use the Visual

Have students study the photograph and read the text that opens the section. Ask:

- **What property makes metals good electrical conductors?** (Metal bonding involves highly mobile electrons, which are shared by all of the nuclei in a metallic solid. Electron mobility accounts for the high electrical conductivity of molten and solid metals.)

Check Prior Knowledge

To assess students' prior knowledge about metals, ask:

- **What are the properties of malleable and ductile metals?** (They can be hammered into different shapes and drawn into wires.)
- **Why is it unwise to stand in a lightning storm with a metal rod in your hand?** (A metal rod can conduct the flow of electrons to your body, which is also a conductor.)
- **What is an alloy? Give a common example.** (An alloy is a mixture of two or more elements, at least one of which is a metal. Examples include steel, bronze, and brass.)

STUDENT RESOURCES

From the Teacher's Resource Package, use:

- Section Review 15.3, Ch. 15 Practice Problems, Vocabulary Review, Quizzes, and Tests from the Review Module (Ch. 13–16)

❶ Yes; it is the most widely used metal for electrical wiring.

TECHNOLOGY RESOURCES

Relevant technology resources include:

- Chem ASAP! CD-ROM
- Resource Pro CD-ROM
- Assessment Resources CD-ROM

❷ Forcing an ionic crystal through the die pushes ions of like charge closer together. As the ions repel one another, the crystal breaks apart.

2 Teach

TEACHER DEMO

Show the class a small sample of elemental copper, or a copper alloy, and a sample of a copper-containing crystalline ionic mineral such as chalcocite (Cu_2S). Wearing safety glasses and standing far from the students, smash both samples with a hammer. The elemental copper will stay together and bend; the crystal will shatter. Discuss why the two substances respond differently to the stress of the hammer blow even though they both contain copper as a major component.

Use the Visual

Lead a class discussion on the concept of "closest packing" of metal cations in pure metals. Use the three different closest packing arrangements shown in Figure 15.16 as a reference. Describe how substitutional and interstitial alloys change these packing arrangements. Discuss how these alloys might change the observed properties of the pure metals using metal bonding theory. Emphasize that the concept of closest packing also relates to more than just metal atoms. Have the students describe other examples of closest packing, such as oranges stacked at a supermarket as shown in Figure 15.15.

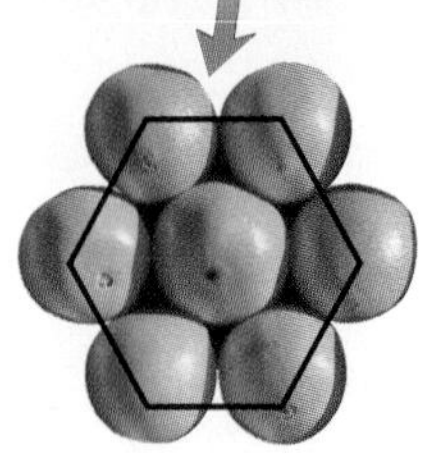

Figure 15.15
These oranges illustrate a hexagonal close-packed arrangement.

Crystalline Structure of Metals

Often brightly colored and transparent, the shiny crystals of many ionic compounds are among the most beautiful substances known. You may be surprised to learn, however, that metals are also crystalline. In fact, metals that contain just one kind of atom are among the simplest forms of all crystalline solids.

Metal atoms are arranged in very compact and orderly patterns, like tennis balls packed in a box. For spheres of identical size, such as metal atoms, there are several arrangements that give the closest packing possible. **Figure 15.16** shows three such arrangements: body-centered cubic, face-centered cubic, and hexagonal close-packed arrangements. Which of
1 these three arrangements are the least closely packed?

In a body-centered cubic structure, every atom (except those on the surface) has eight neighbors. The metallic elements sodium, potassium, iron, chromium, and tungsten crystallize in a body-centered cubic pattern. In a face-centered cubic arrangement, every atom has twelve neighbors. Among the metals that form a face-centered cubic lattice are copper, silver, gold, aluminum, and lead. In a hexagonal close-packed arrangement, every atom also has twelve neighbors. Because of its hexagonal shape, however, the pattern is different from the face-centered cubic arrangement. Metals that have the hexagonal close-packed crystal structure include magnesium, zinc, and cadmium. These arrangements are not limited strictly to metal atoms however. The next time you visit a grocery store, take a look at how the apples or oranges are stacked. More than likely, they will have a close-packed arrangement, as shown in **Figure 15.15.** What is
2 the purpose of stacking fruit in a hexagonal close-packed arrangement?

Chromium

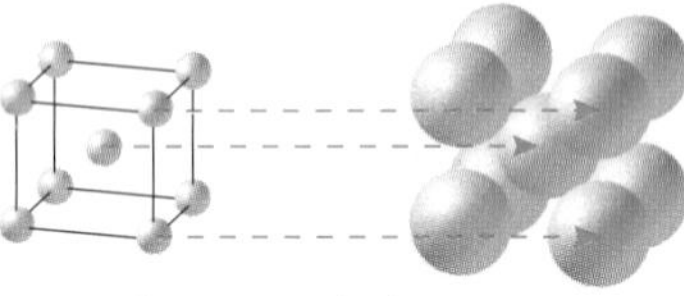

Body-centered cubic

Gold

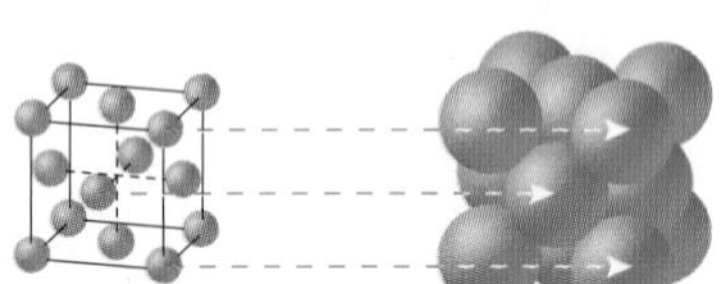

Face-centered cubic

Zinc

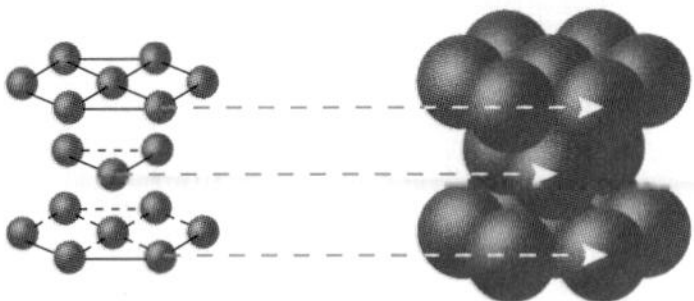

Hexagonal close-packed

Figure 15.16
The atoms of chromium (top) have a body-centered cubic arrangement. The atoms of gold (center) have a face-centered cubic arrangement. The atoms of zinc (bottom) have a hexagonal close-packed arrangement.

Answers

1 face-centered and body-centered cubic arrangements
2 to save space

Alloys

Although every day you use metallic items, such as spoons, very few of these objects are pure metals. Instead, most metals you encounter are alloys. Alloys are mixtures composed of two or more elements, at least one of which is a metal. Alloys are generally prepared by melting a mixture of the ingredients and then cooling the mixture. There are many different alloys. Brass, for example, is an alloy of copper and zinc.

Alloys are important because their properties are often superior to those of their component elements. Sterling silver (92.5% silver and 7.5% copper) is harder and more durable than pure silver but still soft enough to be made into jewelry and tableware. Bronze is an alloy generally containing seven parts of copper to one part of tin. Bronze is harder than copper and more easily cast. Nonferrous (noniron) alloys, such as bronze, copper-nickel, and aluminum alloys, are commonly used in coinage. **Figure 15.17** shows examples of alloys.

The most important alloys today are steels. The principal elements in most steel, in addition to iron and carbon, are boron, chromium, manganese, molybdenum, nickel, tungsten, and vanadium. Steels have a wide range of useful properties, such as corrosion resistance, ductility, hardness, and toughness. **Table 15.3** lists the composition of some common alloys.

Alloys can form from their component atoms in different ways. If the atoms of the components in an alloy are about the same size, they can replace each other in the crystal. This type of alloy is called a substitutional alloy. If the atomic sizes are quite different, the smaller atoms can fit into the interstices (spaces) between the larger atoms. Such an alloy is called an interstitial alloy. In the various types of steel, for example, carbon atoms occupy the spaces between the iron atoms. Thus, steels are interstitial alloys.

Table 15.3

Composition of Some Common Alloys

Name	Composition (by mass)
Sterling silver	Ag 92.5% Cu 7.5%
Coinage silver	Ag 90% Cu 10%
Cast iron	Fe 96% C 4%
Stainless steel	Fe 80.6% Cr 18.0% C 0.4% Ni 1.0%
Spring steel	Fe 98.6% Cr 1.0% C 0.4%
Surgical steel	Fe 67% Cr 18% Ni 12% Mo 3%

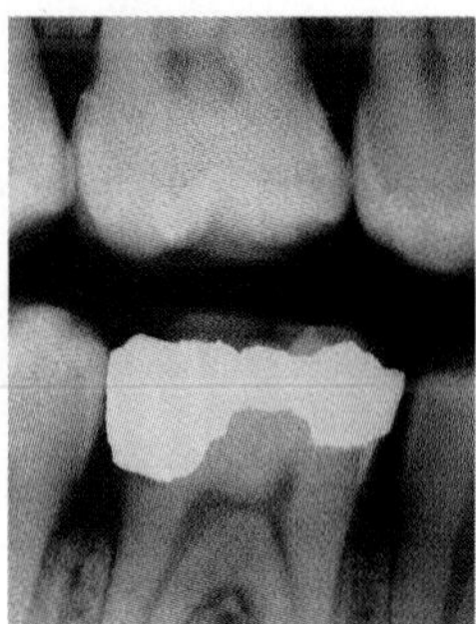

Figure 15.17

Alloys have many different uses. Chrome-moly, an alloy of steel, is commonly used for bicycle frames. Amalgams are alloys of mercury. Dental amalgam, an alloy of mercury, silver, and zinc, expands as it hardens to fill the cavity in a tooth.

section review 15.3

15. Use metallic bonding theory to explain the physical properties of metals.
16. Describe the arrangement of atoms in metallic crystal structures.
17. In your own words define metallic bond.
18. Describe what is meant by ductile and malleable.
19. Why is it possible to bend metals but not ionic crystals?

 Chem ASAP! **Assessment 15.3** Check your understanding of the important ideas and concepts in Section 15.3

TEACHER DEMO

Use styrofoam balls of various sizes and colors to illustrate the crystal structures of interstitial and substitutional alloys. Use toothpicks to hold the "atoms" together. Point out that brass is an example of a substitutional alloy in which copper atoms are replaced by similarly sized zinc atoms. Steel is an example of an interstitial alloy in which relatively small carbon atoms occupy "holes" or interstices between the closely packed iron atoms.

3 Assess

Evaluate Understanding

To determine the students' knowledge about metal bonding theory, ask the following:

- **What is the basic model?** (metal cations held together by a sea of mobile valence electrons)
- **How does this model explain the electrical conductivity of metals?** (Electrons can flow in and out of a metal sample since the valence electrons are not fixed.)
- **How are metal atoms arranged in pure metals?** (in closest packing geometries)

Reteach

Compare and contrast chemical bonding in ionic compounds and pure metals. Review the concepts of malleability or brittleness, and electrical conductivity for both models. Stress that the properties of metals can be explained only by a structure in which electrons are free to move around.

Answers

SECTION REVIEW 15.3

15. The sea of free-flowing electrons conduct electricity and heat; they also shield the cations from repulsive contact during physical stress.
16. Answers will vary but should include the idea of a solid containing tightly packed atoms (or ions in a sea of electrons) in a regular, repeating three-dimensional pattern.
17. Stationary positive metal ions attract mobile valence electrons.
18. Ductile: can be drawn into wires; malleable: can be hammered into different shapes
19. In an ionic compound, ions of like charge do not have mobile electrons as insulation. When forced into contact by physical stress, the ions of like charge repel, causing the crystal to shatter.

Chemistry Serving... the Consumer

DISCUSS

Explain that water hardness varies with location and source. Generally, water from groundwater sources is harder than water from surface sources. In the United States, most northeastern, southern, and northwestern states have predominantly soft water. Generally, hard water of varying degrees is found in the southwestern and midwestern states. Have students bring water samples from home to test for hardness. Test 2-mL samples as follows. Add three drops of potassium thiocyanate (KSCN) to the first sample. A red color from the iron(III) thiocyanate ion, $Fe(SCN)^{2+}$, indicates the presence of Fe^{3+} ions. Add three drops of dilute ethanoic acid, CH_3COOH, and three drops of sodium oxalate, $Na_2C_2O_4$, to the second sample. Mix well. A white precipitate of calcium oxalate, CaC_2O_4, indicates the presence of Ca^{2+} ions.

CHEMISTRY IN CAREERS

Encourage students to connect to the Internet address shown on this page to find out more about a career as a wastewater engineer. There is also career information on page 875 of this text.

Chemistry Serving...the Consumer

WHEN WATER IS HARD TO DEAL WITH

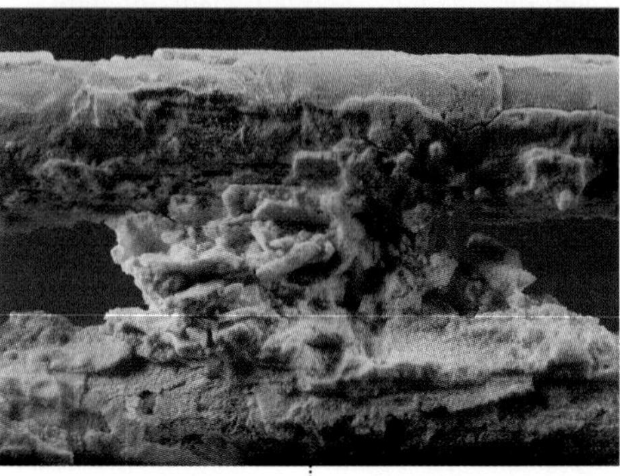

Have you ever found it difficult to get soap to lather? If so, you may have been trying to wash in hard water. Hard water is water that contains a significant concentration of calcium ions (Ca^{2+}) and often ions of magnesium (Mg^{2+}) and iron (Fe^{2+}) as well. These ions usually get into the water while it is underground, when dissolved CO_2 reacts with minerals such as limestone ($CaCO_3$).

The presence of these ions in water gives the water a number of annoying properties. In addition to deactivating soap, these ions can be deposited as a rock-hard "scale" in industrial boilers, hot-water heaters, pipes, and tea kettles. The photo shows an electrode from a hot-water heater that has been corroded by hard water. As a result, families and businesses spend large sums of money every year to replace clogged plumbing and water heaters and to buy extra soap.

To prevent these problems, it is possible to remove the guilty cations from the water. This is done with a home water softener. It works by replacing the calcium, magnesium, and iron ions with sodium ions, which do not give water the hardness property. Water entering a home passes through the water softener before it reaches faucets, the washing machine, or the water heater.

Inside the water softener is a matrix made of a material with negatively charged sites that attract positive cations. The matrix material may be the naturally occurring mineral zeolite (aluminum silicate) or beads of synthetic polystyrene resin. Before a water softener is put into use, the matrix is bathed with a brine solution (sodium chloride). Sodium ions from the brine are attracted and bond to the negatively charged sites on the matrix material.

When hard water enters the water softener, magnesium, calcium, and iron ions in the hard water exchange places with sodium ions on the matrix material.

When hard water enters the water softener, calcium, magnesium, and iron ions in the hard water exchange places with sodium ions on the matrix material. They can displace sodium ions because they are more strongly attracted than the sodium ions to the negative charges of the matrix material. As hard-water cations leave the water moving through the water softener, sodium cations enter the water. By the time the water leaves the water softener, it has lost nearly all its hard-water cations.

The sodium ions in softened water do not cause hardness, but they do have problems of their own. Drinking water high in sodium is linked to certain forms of high blood pressure. For this reason, health experts recommend that people with home water softeners use some source of unsoftened water for drinking.

The process of ion exchange in the water softener can continue for weeks or months because the matrix material has a huge surface area and many negatively charged sites. Eventually, however, all the sites become filled with hard-water cations and the water softener must be recharged. This is done by pouring a strong sodium chloride solution through the matrix to displace the calcium, magnesium, and iron cations.

CHEMISTRY IN CAREERS
WASTEWATER ENGINEER
Interested in monitoring water purity and providing safe drinking water?
See page 875.
CHEMIST NEEDED
Enviro. Lab. GC Must know
TAKE IT TO THE NET
Find out more about career opportunities:
www.phschool.com
CHEMICAL SPECIALIST
Local food service distributor see
responsible self-motivated indivi

Chapter 15 STUDENT STUDY GUIDE

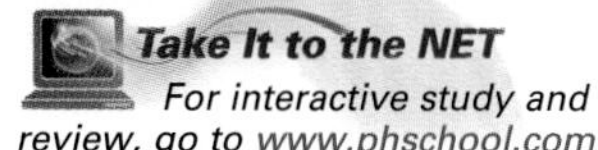

For interactive study and review, go to www.phschool.com

KEY TERMS

- coordination number *p. 423*
- electron dot structure *p. 414*
- halide ion *p. 417*
- ionic bond *p. 419*
- metallic bond *p. 427*
- octet rule *p. 414*
- valence electron *p. 413*

CONCEPT SUMMARY

15.1 Electron Configuration in Ionic Bonding

- Atoms in compounds are held together by chemical bonds. Chemical bonds result from the sharing or transfer of valence electrons between pairs of atoms.
- Bonded atoms attain the stable electron configuration of a noble gas. The noble gases themselves exist as isolated atoms because that is their most stable condition.
- For the representative elements, the number of valence electrons is equal to the element's group number in the periodic table.
- The transfer of one or more valence electrons between atoms produces positively and negatively charged ions, or cations and anions, respectively.

15.2 Ionic Bonds

- The attraction between an anion and a cation is an ionic bond. A substance with ionic bonds is an ionic compound.
- Nearly all ionic compounds are crystalline solids at room temperature. They generally have high melting points. The total positive charge in an ionic compound is balanced by the total negative charge; thus ionic compounds are electrically neutral.
- Ionic solids consist of positive and negative ions packed in an orderly arrangement. The coordination number of an ion indicates the number of ions of opposite charge that surround the ion in a crystal.
- When melted or in aqueous solution, ionic compounds can conduct electricity because the ions can move freely when a voltage is applied.

15.3 Bonding in Metals

- Metals are like ionic compounds in some ways. They consist of positive metal ions packed together and surrounded by a sea of valence electrons. This arrangement constitutes the metallic bond.
- The valence electrons in metals are mobile and can travel from one end of a piece of metal to the other. This electron mobility accounts for the excellent electrical conductivity of metals and helps explain why metals are malleable and ductile.
- Metals are among the simplest crystalline solids. The metal atoms are commonly packed in a body-centered cubic, a face-centered cubic, or a hexagonal close-packed arrangement.

CHAPTER CONCEPT MAP

Use these terms to construct a concept map that organizes the major ideas of this chapter.

Chem ASAP! Concept Map 15
Create your Concept Map using the computer.

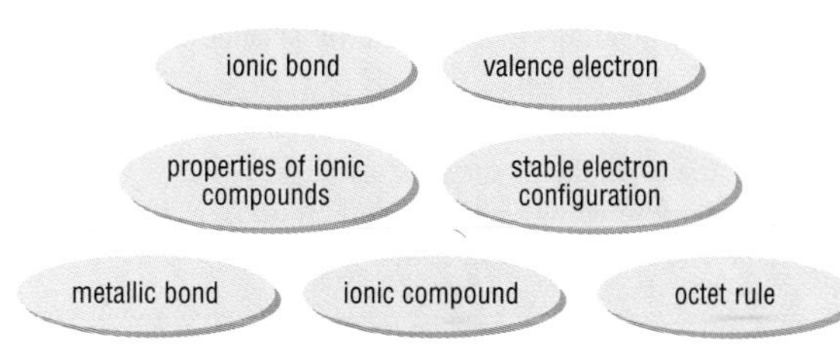

Take It to the Net

At **www.phschool.com** students will find for this chapter

- an Internet activity
- links to related chemistry sites
- an interactive quiz
- career links

4 Close

Summary

Ask the following questions, which require students to summarize the information contained in the chapter.

- **What information does the electron dot structure of an element provide?** (the number of valence electrons)
- **In most cases, what is the stable electron configuration of an elemental ion?** (In most cases, a stable configuration is characterized by an octet of valence electrons; the configuration is isoelectronic with that of a noble gas.)
- **Why does NaCl dissolved in water conduct electricity?** (The Na^+ and Cl^- ions dissociate in water, and are free to move in the solution.)
- **What is the theory of bonding in pure metals?** (cations held together by mobile electrons)

Extension

Have students research a specific ionic compound, pure metal, or alloy. They should draw or make a model of the packing arrangement in the chosen substance.

Looking Back . . . Looking Ahead . . .

The last chapter discussed the relationship between electron configurations and periodic trends. This chapter focused on the role of electron transfer in the formation of ionic bonds. The properties of metals were explained by the nature of metallic bonds. The next chapter will discuss the covalent bonds that form when atoms share electrons.

Chapter 15 REVIEW

Answers

20. electrons in the highest occupied energy level
21. fluorine (F), chlorine (Cl), bromine (Br), iodine (I); Group 7A; seven valence electrons
22. a. 7; 5A
    b. 3; 1A
    c. 15; 5A
    d. 56; 2A
23. a. :C̤l·
    b. :S̤·
    c. ·Ȧl·
    d. Li·
24. Their outermost occupied energy levels are filled.
25. a. 2 b. 3 c. 1 d. 2
26. a. Al^{3+}
    b. Li^{+}
    c. Ba^{2+}
    d. K^{+}
    e. Ca^{2+}
    f. Sr^{2+}
27. a. $1s^22s^22p^63s^23p^63d^3$
    b. $1s^22s^22p^63s^23p^63d^4$
    c. $1s^22s^22p^63s^23p^63d^5$
28. Most nonmetals gain 1, 2, or 3 electrons to achieve a noble-gas electron configuration.
29. a. S^{2-} b. Na^{+} c. F^{-} d. P^{3-}
30. a. 3 b. 2 c. 1 d. 3
31. a. Br^{-} b. H^{-} c. As^{3-} d. Se^{2-}
32. All are $1s^22s^22p^6$. All have the same configuration as neon.
33. The positive charges of the cations equal the negative charges of the anions.
34. a., b., and d.
35. a. $K^{+}Cl^{-}$
    b. $Ba^{2+}SO_4^{2-}$
    c. $Mg^{2+}Br^{-}$
    d. $Li^{+}CO_3^{2-}$
36. No; the packing of ions in a crystalline structure depends on a number of factors, including the relative sizes of the ions. The coordination number of an element can vary from compound to compound.
37. Their network of electrostatic attractions and repulsions forms a rigid structure.

Chapter 15 REVIEW

CONCEPT PRACTICE

20. Define valence electrons. *15.1*
21. Name the first four halogens. What group are they in, and how many valence electrons does each have? *15.1*
22. How many electrons does each atom have? What group is each in? *15.1*
    a. nitrogen c. phosphorus
    b. lithium d. barium
23. Write electron dot structures for each of the following elements. *15.1*
    a. Cl b. S c. Al d. Li
24. The atoms of the noble gas elements are stable. Explain. *15.1*
25. How many electrons must each atom lose to attain a noble-gas electron configuration? *15.1*
    a. Ca b. Al c. Li d. Ba
26. Write the formula for the ion formed when each of the following elements loses its valence electrons. *15.1*
    a. aluminum c. barium e. calcium
    b. lithium d. potassium f. strontium
27. Write electron configurations for the tripositive ions (3+) of these elements. *15.1*
    a. chromium b. manganese c. iron
28. Why do nonmetals tend to form anions when they react to form compounds? *15.1*
29. What is the formula of the ion formed when the following elements gain or lose valence electrons and attain noble-gas configurations? *15.1*
    a. sulfur c. fluorine
    b. sodium d. phosphorus
30. How many electrons must be gained by each of the following atoms to achieve a stable electron configuration? *15.1*
    a. N b. S c. Cl d. P
31. Write the formula for the ion formed when each element gains electrons and attains a noble-gas configuration. *15.1*
    a. Br b. H c. As d. Se
32. Write electron configurations for the following and comment on the result. *15.1*
    a. N^{3-} b. O^{2-} c. F^{-} d. Ne
33. Explain why ionic compounds are electrically neutral. *15.2*
34. Which of the following pairs of elements will not form ionic compounds? *15.2*
    a. sulfur and oxygen
    b. sodium and calcium
    c. sodium and sulfur
    d. oxygen and chlorine
35. Write the formula for the ions in the following compounds. *15.2*
    a. KCl b. $BaSO_4$ c. $MgBr_2$ d. Li_2CO_3
36. Can you predict the coordination number of an ion from the formula of an ionic compound? Explain. *15.2*
37. Most ionic substances are brittle. Why? *15.2*
38. Explain why molten $MgCl_2$ does conduct an electric current although crystalline $MgCl_2$ does not. *15.2*
39. Explain briefly why metals are good conductors of electricity. *15.3*
40. Name the three crystal arrangements of closely packed metal atoms. Give an example of a metal that crystallizes in each arrangement. *15.3*
41. Name some alloys that you have used or seen today. *15.3*
42. Why aren't the properties of all steels identical? Explain. *15.3*
43. The properties of all samples of brass are not identical. Explain. *15.3*

CONCEPT MASTERY

44. Which of the following substances are most likely not ionic?
    a. H_2O c. CO_2 e. NH_3
    b. Na_2O d. CaS f. SO_2
45. Construct a table that shows the relationship among the group number, valence electrons lost or gained, and the formula of the cation or anion produced for the following metallic and nonmetallic elements: Na, Ca, Al, N, S, Br.
46. Write electron dot formulas for the following atoms.
    a. C b. Be c. O d. F e. Na f. P
47. Show the relationship between the electron dot structure of an element and the location of the element in the periodic table.
48. In terms of electrons, why does a cation have a positive charge?

38. Ions are free to move in molten $MgCl_2$.
39. They have many mobile valence electrons. Electrons in the current replace the electrons leaving the metal.
40. body-centered cubic: Na, K, Fe, Cr, or W
    face-centered cubic: Cu, Ag, Au, Al, or Pb
    hexagonal closest-packed: Mg, Zn, or Cd
41. Answers will vary and could include tableware, steel in cars and buses, high-speed dental drill bits, solder in stereos and televisions, and structural steel in buildings.
42. The properties of steel vary according to its composition. In addition to iron, steel can contain varying amounts of carbon and such metals as chromium, nickel, and molybdenum.
43. Brass is a mixture of copper and zinc. The properties of a particular sample of brass will vary with the relative proportions of the two metals.
44. a., c., e., and f.

49. Why does an anion have a negative charge?

50. Metallic cobalt crystallizes in a hexagonal close-packed structure. How many neighbors will a cobalt atom have?

51. Write electron configurations for the dipositive ions (2+) of these elements.
a. Fe **b.** Co **c.** Ni

52. Write electron configurations for these atoms and ions, and comment on the result.
a. Ar **b.** Cl^- **c.** S^{2-} **d.** P^{3-}

53. Explain how hexagonal close-packed, face-centered cubic, and body-centered cubic unit cells are different from one another.

54. The spheres below represent the relative diameters of atoms or ions. Rearrange the sequences in **a.** and **b.** so the relative sizes of the particles correspond to the increasing size of the particles in the illustration.

a. oxygen atom, oxide ion, sulfur atom, sulfide ion
b. sodium atom, sodium ion, potassium atom, potassium ion

55. Write complete electron configurations for the following atoms and ions. For each group, comment on the results.
a. Ar, K^+, Ca^{2+} **b.** Ne, Na^+, Mg^{2+}, Al^{3+}

CRITICAL THINKING

56. Choose the term that best completes the second relationship.

a. cow : horse — ionic bond : _______
(1) anion (3) chemical bond
(2) cation (4) covalent bond

b. Cl : Cl^- — Mg : _______
(1) Mg^{2+} (3) Al^{3+}
(2) Mg^{2-} (4) Mn

c. pipe : water — metal : _______
(1) ions (3) electricity
(2) ionic bond (4) conductor

57. Compare the physical and chemical characteristics of metals and ionic compounds.

CUMULATIVE REVIEW

58. Hydrogen and oxygen react to produce water according to this equation.

$$2H_2 + O_2 \longrightarrow 2H_2O$$

How many liters of hydrogen at STP are needed to produce 0.50 mol H_2O?

59. Distinguish among gases, liquids, and solids with respect to shape, volume, relative density, and motion of particles.

60. What is the volume, in liters, occupied by 8.0 g of oxygen gas at STP?

61. If you raise the temperature of a gas, what actually happens to the gas particles?

62. A gas occupies 750 cm^3 at 27 °C and 1.6 kPa. Find its volume at STP.

63. Explain each of these observations on the basis of the kinetic theory and the forces of attraction that exist between the particles in matter.
a. Water evaporates faster at 40 °C than at 20 °C.
b. A burn from steam at 100 °C is worse than a burn from water at 100 °C.
c. A cap pops off a bottle of root beer that has been kept in the trunk of a car on a hot day.
d. Diethyl ether ($C_4H_{10}O$) has a vapor pressure of 58.9 kPa at 20 °C, whereas the vapor pressure of water at 20 °C is only 2.3 kPa.
e. A melting ice cube cools a glass of tea.
f. Foods are dehydrated (water is removed) by using low pressures rather than high heat.

64. Use the gas laws and kinetic theory to complete these statements. Unless otherwise stated, assume a constant amount of gas.
a. As the volume of a gas increases at constant temperature, its pressure _______.
b. As the temperature of a gas increases and its pressure decreases, its volume _______.
c. At constant pressure, a decrease in the volume of a gas is caused by a(n) _______ in its temperature.
d. An increase in the volume and pressure of a gas is caused by a(n) _______ in its temperature.
e. At constant volume, a decrease in the temperature of a gas causes the pressure to_______.
f. If the volume of a gas is increased while its temperature is increased, the pressure will _______.

Chapter 15 REVIEW

45.

Group number	1A	2A	3A	5A	6A	7A
Valence electrons lost or gained	1	2	3	3	2	1
Ion formula	Na^+	Ca^{2+}	Al^{3+}	N^{3-}	S^{2-}	Br^-

46. **a.** ·Ċ· **d.** :F̤·
b. ·Be· **e.** Na·
c. :Ö· **f.** ·P̤·

47. For the representative elements, the number of electrons in the electron dot structure is the group number.

48. It has lost valence electrons.

49. It has gained valance electrons.

50. 12

51. **a.** $1s^22s^22p^63s^23p^63d^6$
b. $1s^22s^22p^63s^23p^63d^7$
c. $1s^22s^22p^63s^23p^63d^8$

52. All have the noble gas configuration of $1s^22s^22p^63s^23p^6$

53. Hexagonal closest-packed unit cells have 12 neighbors for every atom or ion. Face-centered cubic unit cells also have 12 neighbors for every atom or ion, with an atom or ion in the center of each face. Body-centered cubic units cells have 8 neighbors for every atom or ion, with an atom or ion at the center of each cube.

54. **a.** oxygen atom, sulfur atom, oxide ion, sulfide ion
b. sodium ion, potassium ion, sodium atom, potassium atom

55. **a.** $1s^22s^22p^63s^23p^6$
b. $1s^22s^22p^6$
Each has a noble-gas electron configuration.

56. **a.** 4
b. 1
c. 3

57. Both are composed of ions and are held together by electrostatic bonds. Metals always conduct electricity, but ionic compounds conduct only when melted or in aqueous solution. Ionic compounds are composed of cations and anions, but metals are composed of cations and free-floating valence electrons. Metals are malleable, but ionic compounds are brittle.

58. 11 L

59. Gases and liquids assume the shapes of their containers. Solids have a definite shape. Liquids and solids have a definite volume; gases do not. Gases have a low density; liquids have an intermediate density; solids have a high density. The molecules in a gas move freely and randomly. The molecules of a liquid flow. The molecules of a solid vibrate and rotate around a fixed position.

60. 5.6 L

61. The average kinetic energy and the speed of the gas molecules increase.

62. 11 cm^3

Answers

63. a. At 40 °C more water molecules have energies that allow them to overcome intermolecular attractions than at 20 °C.
b. Steam at 100 °C has more potential energy.
c. The molecules' kinetic energy has increased, creating a greater vapor pressure.
d. Because water molecules require more energy to evaporate than do ether molecules, the attractions between water molecules must be greater than between ether molecules.
e. Ice absorbs energy from the tea. The kinetic energy of the tea decreases.
f. The goal is to change water into water vapor without destroying chemical bonds by heating; lowering the external pressure lowers the temperature at which liquid water molecules have enough energy to vaporize.

64. a. decreases b. increases
c. decrease d. increase
e. decrease
f. increase or decrease depending on the relative sizes of the changes of T and P.

65. 1.36×10^2 kPa

66. 596 K, or 323 °C

67. a. 0.0147
0.0373
0.0639
0.0879
b.

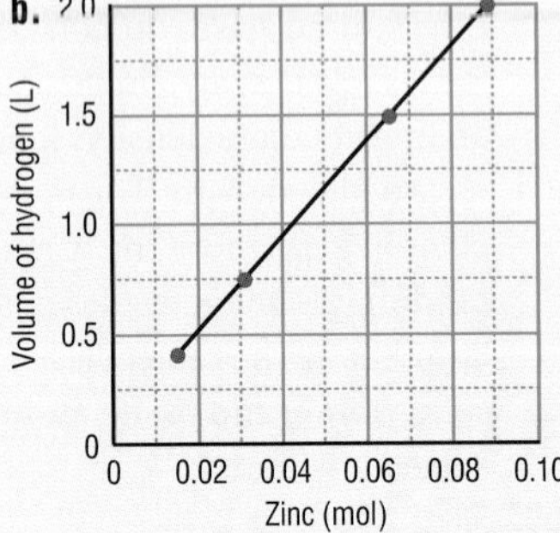

c. 22.4 L/mol d. 1:1 ratio

68. Na^+ and Cs^+ differ greatly in size. Na^+ and Cl^- are similar in size to Mn^{2+} and S^{2-}.

69. 0.1445 nm

65. A gas mixture contains 1.50 g O_2 and 1.50 g N_2. If the partial pressure of O_2 is 63.3 kPa, what is the total pressure of the mixture?

66. At what temperature will the average kinetic energy of a nitrogen molecule be twice the average kinetic energy of a nitrogen molecule at 25 °C?

67. The reaction of zinc with an acid produces hydrogen gas. The table below lists the amount of hydrogen gas produced at STP when various amounts of zinc are reacted with an excess of hydrochloric acid.

Mass of Zn (g)	Moles of Zn	Volume of H_2 (L)
0.960		0.329
2.44		0.835
4.18		1.43
5.75		1.96

a. Calculate the moles of zinc used.
b. Make a graph of the volume of hydrogen produced (y-axis) versus the moles of zinc reacting (x-axis).
c. Determine the slope of the line. Include the units of the slope.
d. Based on the value of the slope, what can you determine about the moles of hydrogen produced per mole of zinc that reacts?

CONCEPT CHALLENGE

68. The chemically similar alkali metal chlorides NaCl and CsCl have different crystal structures, whereas the chemically different NaCl and MnS have the same crystal structures. Why?

69. Silver crystallizes in a face-centered cubic unit cell. A silver atom is at the edge of each lattice point. The length of the edge of the unit cell is 0.4086 nm. What is the atomic radius of silver?

70. Classify each atom in the following list. Will each atom form a cation or an anion? Or is it chemically unreactive? For the atoms that do form ions during a chemical reaction, write the number of electrons the atom will gain or lose.
a. lithium
b. sodium
c. neon
d. magnesium
e. chlorine

71. Write the IUPAC name of the cation in each transition metal compound.
a. $FeCl_2$
b. Fe_2O_3
c. CuO
d. Cu_2O
e. $CoCl_3$

72. List the elements that are used to make each alloy.
a. brass
b. sterling silver
c. bronze
d. stainless steel
e. surgical steel
f. spring steel

70. a. cation: lose $1e^-$
b. cation: lose $1e^-$
c. unreactive
d. cation: lose $2e^-$
e. anion: gain $1e^-$

71. a. iron(II); Fe^{2+}
b. iron(III); Fe^{3+}
c. copper(II); Cu^{2+}
d. copper(I); Cu^+
e. cobalt(III); Co^{3+}

72. a. copper and zinc
b. silver and copper
c. copper and tin
d. iron, chromium, carbon, nickel
e. iron, chromium, nickel, molybdenum
f. iron, chromium, carbon

Select the choice that best answers each question or completes each statement.

1. Which of these is not an ionic compound?
   a. KF
   b. Na_2SO_4
   c. SiO_2
   d. Na_2O
2. Which statements are correct when barium and oxygen react to form an ionic compound?
   I. Barium atoms lose 2 electrons and form a cation.
   II. Oxygen atoms form oxide anions (O^{2-}).
   III. In the compound the ions are present in a one-to-one ratio.
   a. I and II only
   b. II and III only
   c. I and III only
   d. I, II, and III
3. How many valence electrons does arsenic have?
   a. 5 b. 4 c. 3 d. 2
4. For which compound name is the *incorrect* formula given?
   a. magnesium iodide, MgI_2
   b. potassium selenide, K_2Se
   c. calcium oxide, Ca_2O_2
   d. aluminum sulfide, Al_2S_3
5. Which electron configuration represents a nitride ion?
   a. $1s^2\,2s^2\,3s^2\,4s^2$
   b. $1s^2\,2s^2\,2p^3$
   c. $1s^2\,2s^2\,2p^6$
   d. $1s^2$
6. When a bromine atom gains an electron
   a. a bromide ion is formed.
   b. the ion formed has a 1− charge.
   c. the ion formed is an anion.
   d. all the above are correct.

The lettered choices below refer to questions 7–10. A lettered choice may be used once, more than once, or not at all.

(A) gains two electron
(B) loses two electrons
(C) gains three electrons
(D) loses one electron
(E) gains one electron

Which choice describes what happens as each of the following elements forms its ion?

7. iodine
8. magnesium
9. cesium
10. phosphorus

Use the description and the graph to answer questions 11–13.

Lattice energy is the energy required to change one mole of a crystalline, ionic solid to gaseous ions. The graph below shows the lattice energy for ionic compounds formed between selected alkali metals and halogens.

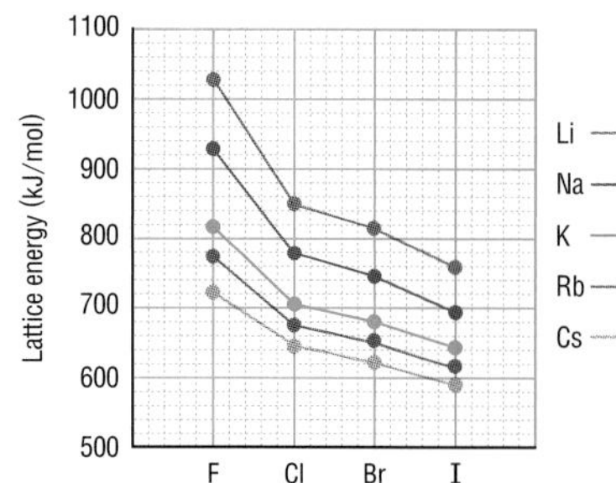

11. For a given alkali metal, what is the trend in lattice energy as the atomic radius of the halogens increases?
12. For a given halogen, what is the trend in lattice energy as the atomic radius of the alkali metals increases?
13. Complete this sentence. "As the atomic radius of either the halogen or the alkali metal increases, the lattice energy ______________."

Use the atomic windows to answer question 14.

14. How atoms and ions are arranged in crystals is not just dependent on size. The spheres in each window are identical in size. The windows have exactly the same area. In which window are the spheres more closely packed? Explain your reasoning.

a.
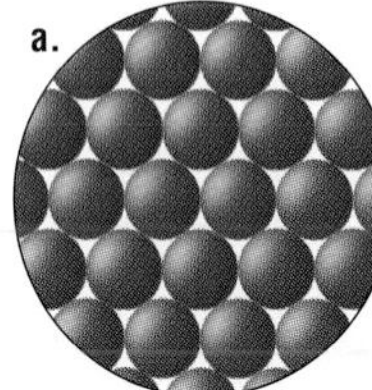
b.
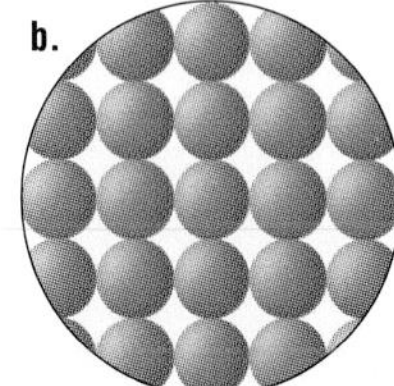

Chapter 15 STANDARDIZED TEST PREP

1. c
2. d
3. a
4. c
5. c
6. d
7. (E)
8. (B)
9. (D)
10. (C)
11. Because the atomic radius increases moving down the halogen group, the lattice energy decreases.
12. Lattice energy decreases as the atomic radius of the alkali metal increases.
13. decreases
14. The spheres are more closely packed in circle (a) because it has about 26 spheres compared to about 22 spheres in circle (b).

Planning Guide

SECTION OBJECTIVES	ACTIVITIES/FEATURES	MEDIA & TECHNOLOGY
16.1 The Nature of Covalent Bonding ● ■ ◆ ▸ Use electron dot structures to show the formation of single, double, and triple covalent bonds ▸ Describe and give examples of coordinate covalent bonding, resonance structures, and exceptions to the octet rule	**SE** Discover It! *Shapes of Molecules,* p. 436 **SE** **CHEMath** *Lewis Electron Dot Structures,* p. 441 **SE** Mini Lab *Strengths of Covalent Bonds,* p. 448 (LRS 16-1) PLM *Strengths of Covalent Bonds* **TE** DEMOS, pp. 447, 449	**ASAP** Simulation 16 **ASAP** Problem Solving 2, 4 **ASAP** Assessment 16.1 **ACT** 12: *Covalent Bonds* **CHM** Side 3, 36: *Paramagnetism* **CHM** Side 3, 38: *Decomposition of Nitrogen Triiodide* **OT** 53: *Covalent Bonds*
16.2 Bonding Theories □ ◆ ▸ Describe the molecular orbital theory of covalent bonding, including orbital hybridization ▸ Use VSEPR theory to predict the shapes of simple covalently bonded molecules	**SE** **Link to Geography** *Topographic Maps,* p. 453 **TE** DEMOS, p. 453	**ASAP** Simulation 17 **ASAP** Assessment 16.2 **CHM** Side 3, 23: *VSEPR Theory* **OT** 54: *Sigma and Pi Bonds* **OT** 55: *Molecular Shapes*
16.3 Polar Bonds and Molecules ● ■ ◆ ▸ Use electronegativity values to classify a bond as nonpolar covalent, polar covalent, or ionic ▸ Name and describe the weak attractive forces that hold groups of molecules together	**SE** **Link to Computer Science** *Molecular Modeling,* p. 464 **SE** Small-Scale Lab *Paper Chromatography of Food Dyes,* p. 467 (LRS 16-2) **SE** **Chemistry Serving . . . Society** *Blocks and Bonds,* p. 468 **SE** **Chemistry in Careers** *Oncologist,* p. 468 LM 25: *Molecular Models* SSLM 21: *Paper Chromatography* **TE** DEMOS, pp. 461, 462, 465	**ASAP** Animation 19 **ASAP** Simulation 18 **ASAP** Assessment 16.3 **ASAP** Concept Map 16 **SSV** 12: *Paper Chromatography* **CHM** Side 5, 22: *Why a Solid Is a Solid* **CHM** Side 3, 20: *Bonding* **CHM** Side 3, 10: *Synthetic Gems* **OT** 56: *Hydrogen Bonding* **PP** Chapter 16 Problems **RP** Lesson Plans, Resource Library **AR** Computer Test 16 **www** Activities, Self-Tests, *SCIENCE NEWS* updates **TCP** The Chemistry Place Web Site

KEY

- ● Conceptual (concrete concepts)
- ○ Conceptual (more abstract/math)
- ■ Standard (core content)
- □ Standard (extension topics)
- ◆ Honors (core content)
- ◇ Honors (options to accelerate)

AR	Assessment Resources
ASAP	Chem ASAP! CD-ROM
ACT	ActivChemistry CD-ROM
CHM	CHEMedia Videodiscs
CA	Chemistry Alive! Videodiscs
GCP	Graphing Calculator Problems
GRS	Guided Reading and Study Workbook
LM	Laboratory Manual
LP	Laboratory Practicals
LRS	Laboratory Recordsheets
SSLM	Small-Scale Lab Manual

PRACTICE		ASSESSMENT	
SE	**Sample Problems** 16-1 to 16-3	SE	Section Review
GRS	Section 16.1	RM	Section Review 16.1
RM	Practice Problems 16.1	RM	Chapter 16 Quiz
GRS	Section 16.2	SE	Section Review
RM	Practice Problems 16.2	RM	Section Review 16.2
RM	Interpreting Graphics	RM	Chapter 16 Quiz
SE	**Sample Problem** 16-4	SE	Section Review
GRS	Section 16.3	RM	Section Review 16.3
RM	Practice Problems 16.3	LP	Lab Practical 16-1
		RM	Vocabulary Review 16
		SE	Chapter Review
		SM	Chapter 16 Solutions
		SE	Standardized Test Prep
		PHAS	Chapter 16 Test Prep
		RM	Chapter 16 Quiz
		RM	Chapter 16 A & B Test

PLANNING FOR ACTIVITIES

STUDENT EDITION

Discover It! p. 436
- spherical balloons
- string

Mini Lab, p. 448
- 6-oz and 16-oz food cans
- No. 25 rubber bands
- metric rulers
- coat hangers
- plastic grocery bags
- paper clips
- pencils
- graph paper
- safety goggles

Small-Scale Lab, p. 467
- pencils, paper, rulers
- toothpicks
- scissors
- food coloring (4 colors)
- plastic cups
- 0.1% NaCl solution
- chromatography paper
- colored candy, markers, and powdered drinks
- 2-propanol, vinegar, ammonia
- coffee filters

TEACHER'S EDITION

Teacher Demo, p. 447
- 10- to 15-cm piece of magnesium ribbon
- tongs
- Bunsen burner
- cobalt blue glass filters
- charcoal

Teacher Demo, p. 449
- fume hood
- copper metal
- concentrated nitric acid
- evaporating dish

Teacher Demo, p. 453
- balloon
- tissue paper

Teacher Demo, p. 453
- clear plastic
- fine point marker
- metric ruler
- overhead projector

Teacher Demo, p. 461
- magnet toy (with iron)

Teacher Demo, p. 462
- 2 burets
- ring stand
- water
- turpentine
- 2 beakers
- safety goggles
- rubber and glass rods
- fur cloth

Teacher Demo, p. 465
- five 300-mL stoppered bottles, each containing 100 mL of a different liquid: glycerin, ethylene, glycol, water, ethanol

OT	Overhead Transparency	SE	Student Edition
PHAS	PH Assessment System	SM	Solutions Manual
PLM	Probeware Lab Manual	SSV	Small-Scale Video/Videodisc
PP	Problem Pro CD-ROM	TCP	www.chemplace.com
RM	Review Module	TE	Teacher's Edition
RP	Resource Pro CD-ROM	www	www.phschool.com

Chapter 16

Key Terms

16.1 single covalent bond, structural formulas, unshared pairs, double covalent bonds, triple covalent bonds, coordinate covalent bond, bond dissociation energy, resonance structures, diamagnetic, paramagnetic

16.2 molecular orbitals, bonding orbital, antibonding orbital, sigma bond, pi bond, tetrahedral angles, VSEPR theory, hybridization

16.3 nonpolar covalent bond, polar covalent bond, polar bond, polar molecule, dipole, van der Waals forces, dispersion forces, dipole interactions, hydrogen bonds, network solids

DISCOVER IT!

Students should observe that the balloons naturally tend to arrange themselves so that they are as far apart as possible. After students have completed this chapter, they should understand that the three clusters of balloons are similar to the arrangement of electron pairs around atoms in molecules. The electron pairs orient themselves so as to minimize the repulsion between like charges. The terms and clusters can be matched as follows: two balloons with *linear*, three with *trigonal planar*, and four with *tetrahedral*.

Chapter 16 COVALENT BONDING

Individual snowflakes can contain 100 snow crystals.

FEATURES

DISCOVER IT! SHAPES OF MOLECULES

You need nine spherical balloons and several short pieces of string.

1. Inflate the nine balloons to approximately the same size and tie them off.
2. Use short pieces of string to tie the ends of two balloons together to form a cluster. Tie the ends of three other balloons together to form a second cluster, and then tie the remaining four balloons to form a third cluster.
3. Arrange the balloons in each cluster so they are as far away from one another as possible. (Note: The balloons will probably be positioned this way already.)
4. Sketch the three-dimensional shape of each balloon cluster.

What are some geometric terms that you can use to describe the shapes? Some of the terms used by chemists to describe these shapes are trigonal planar, tetrahedral, and linear. Can you match each term with one of the shapes you made? After reading about molecular shapes later in this chapter, return to this activity and review your answers.

16.1 ● ■ ◆
16.2 □ ◆
16.3 ● ■ ◆

Conceptual The CHEMath feature can be used to review Lewis dot structures. In Section 16.1, omit bond dissociation energies, exceptions to the octet rule, coordinate covalent bonds, and resonance. Section 16.2 may be omitted. Section 16.3 lays the foundation for understanding aqueous systems in chapter 17.

Standard In Section 16.2, the discussion should focus on VSEPR theory and molecular shapes. Molecular orbital theory is not essential to understanding the geometry of molecules. Use Table 16.5 to summarize differences between ionic and covalent compounds.

Honors In Section 16.1, the discussion of single, double, and triple bonds may be accelerated. In Section 16.2, use the theme of attractive and repulsive forces to help students understand different approaches to molecular bonding. In Section 16.3, focus on the relationships among bond polarity, intermolecular forces, and the properties of molecular compounds.

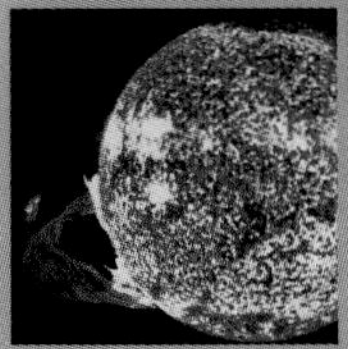

THE NATURE OF COVALENT BONDING

section 16.1

You know that without oxygen to breathe, you could not live. But did you know that oxygen plays another important role in your life? High in the atmosphere, a different kind of oxygen molecule, called ozone, forms a layer that filters out harmful radiation from the sun. **How are the bonds in molecules of ordinary oxygen similar to the bonds in molecules of ozone?**

objectives

- Use electron dot structures to show the formation of single, double, and triple covalent bonds
- Describe and give examples of coordinate covalent bonding, resonance structures, and exceptions to the octet rule

key terms

- single covalent bond
- structural formulas
- unshared pairs
- double covalent bonds
- triple covalent bonds
- coordinate covalent bond
- bond dissociation energy
- resonance structures
- diamagnetic
- paramagnetic

Single Covalent Bonds

Salts, such as sodium chloride (NaCl), are crystalline solids with high melting points. Other compounds, however, have very different properties. Hydrogen chloride (HCl), for example, is a gas at room temperature. Water (H_2O) is a liquid. These two compounds are so different from salts that you might suspect that electrostatic attraction between ions fails to explain their bonding. Your suspicions would be correct. Such compounds are not ionic. Their atoms do not give up electrons or accept electrons as readily as sodium does in combining with chlorine. Instead, a "tug of war" for electrons takes place between the atoms, ending more or less in a standoff. In this chapter, you will learn about atoms that share electrons to form a different kind of bond, called a covalent bond.

To begin your study of covalent bonding, consider hydrogen (H_2), a very simple molecule. A hydrogen atom has a single valence electron. A pair of hydrogen atoms shares electrons to form a diatomic hydrogen molecule.

$$\underset{\text{Hydrogen atom}}{H\cdot} + \underset{\text{Hydrogen atom}}{\cdot H} \longrightarrow \underset{\text{Hydrogen molecule}}{H{:}H}$$

shared pair of electrons

In this molecule, each hydrogen achieves the electron configuration of the noble gas helium, which has two valence electrons. The hydrogen atoms form a **single covalent bond,** a bond in which two atoms share a pair of electrons. **Figure 16.1** shows this process in terms of atomic orbitals.

When writing a formula for a covalent bond, the pair of electrons is represented as a dash, as in H—H for hydrogen. Notations of this type, called **structural formulas,** are chemical formulas that show the arrangement of atoms in molecules and polyatomic ions. Each dash between atoms in structural formulas indicates a pair of shared electrons. Such dashes are never used to show ionic bonds.

By looking at hydrogen (H_2), you can see that there is a difference between the formulas of ionic and covalent compounds. The chemical formulas of ionic compounds describe formula units, while the chemical formulas of covalent compounds describe molecules. Ionic compounds do not have molecular formulas because they are not composed of molecules. Ionic copper(II) oxide, for example, is composed of equal numbers of Cu^{2+} and O^{2-} ions in a crystal lattice. The formula unit, which is written as CuO, shows the lowest whole-number ratio of Cu^{2+} to O^{2-}, which is 1:1. The chemical formula CuO represents the smallest electrically neutral unit that is representative of copper(II) oxide.

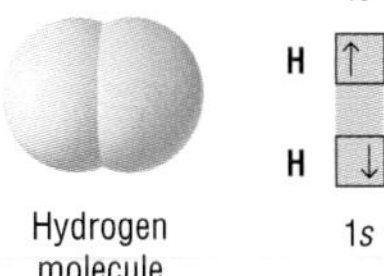

Figure 16.1
The bonding electrons in the hydrogen molecule come from the 1s atomic orbitals of the hydrogen atoms. How many electrons do the hydrogen atoms share? 1

16.1

1 Engage

Use The Visual

Have students read the text that opens the section. Have them recall the formation of ionic bonds. Write the equation for the reaction between Na and Cl, showing the transfer of electrons and the formation of ions. Help students recognize that large differences in electronegativity between atoms promote the formation of ionic bonds. Point out, however, that for many molecules, such as hydrogen (H_2) and chlorine (Cl_2), electronegativity differences do not exist. Write the electron dot structures for two chlorine atoms and ask:

- **Is there any other way for two atoms to form a chemical bond without giving up an electron?** (The atoms can share electrons between their nuclei. Show the formation of the Cl : Cl bond.)
- **How are the bonds in molecules of ordinary oxygen similar to molecules of ozone?** (The atoms in O_2 and O_3 are held together by the sharing of electrons between atoms.)

Check Prior Knowledge

Ask students to recall the definition of electronegativity (tendency for an atom to attract electrons from another atom) **and the periodic trends in electronegativity** (increases from left to right and decreases as you move down a group).

STUDENT RESOURCES

From the Teacher's Resource Package, use:

- Section Review 16.1, Ch. 16 Practice Problems and Quizzes from the Review Module (Ch. 13–16)
- Laboratory Recordsheet 16-1

TECHNOLOGY RESOURCES

Relevant technology resources include:

- Chem ASAP! CD-ROM
- ResourcePro CD-ROM
- ActivChemistry CD-ROM: *Covalent Bonds*

1 two

2 Teach

Discuss

Use the hydrogen molecule to introduce the shared nature of covalent bonding. Remind students that atoms bond to reach a more stable state, one in which the outermost electron levels of an atom are filled, as in the highly unreactive noble gases. A transfer of electrons between hydrogen atoms would not work; what factor could determine which hydrogen atom donated and which received an electron? However, if the hydrogen atoms share their electrons, they can each achieve the stable arrangement of a helium atom. On the chalkboard or overhead projector, show the various ways that the bonding in a hydrogen molecule can be represented: molecular formula, structural formula, electron-dot formula, and orbital diagram showing the overlap of the 1*s* orbitals. Then introduce the octet rule and repeat the exercise for a chlorine molecule.

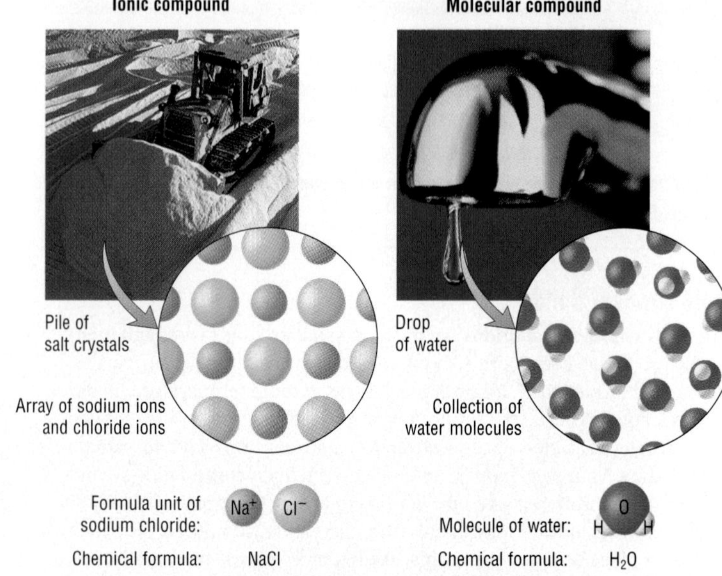

Figure 16.2
Sodium chloride—an ionic compound—and water—a molecular compound—are compared here. How do molecular compounds differ from ionic compounds? ❷

In contrast, individual hydrogen molecules actually do exist. Each contains two hydrogen atoms bonded together. The molecular formula of hydrogen is thus H_2. The correct formulas of molecular compounds reflect the actual number of atoms in each molecule, and their subscripts are not necessarily lowest whole-number ratios. **Figure 16.2** illustrates some essential differences between ionic and covalent compounds, using sodium chloride and water as examples.

Combinations of atoms of the nonmetallic elements in Groups 4A, 5A, 6A, and 7A of the periodic table are likely to form covalent bonds. Chemist Gilbert Lewis summarized this tendency in his formulation of the octet rule for covalent bonding: Sharing of electrons occurs if the atoms involved acquire the electron configurations of noble gases. Often the configurations contain eight valence electrons (an octet). Which covalent compound that ❶ has already been discussed in this chapter is a notable exception to this rule?

The halogens form single covalent bonds in their diatomic molecules. Fluorine is one example. Because a fluorine atom has seven valence electrons, it needs one more to attain the electron configuration of a noble gas. By sharing electrons and forming a single covalent bond, two fluorine atoms each achieve the electron configuration of neon. **Figure 16.3** illustrates this process in terms of atomic orbitals.

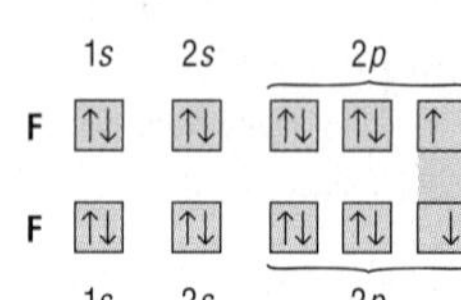

	1*s*	2*s*	2*p*		
F	↑↓	↑↓	↑↓	↑↓	↑
F	↑↓	↑↓	↑↓	↑↓	↓
	1*s*	2*s*	2*p*		

Figure 16.3
The bonding electrons in the fluorine molecule come from the 2p atomic orbitals of the fluorine atoms. How many electrons are needed to form the single covalent bond in the fluorine molecule? ❸

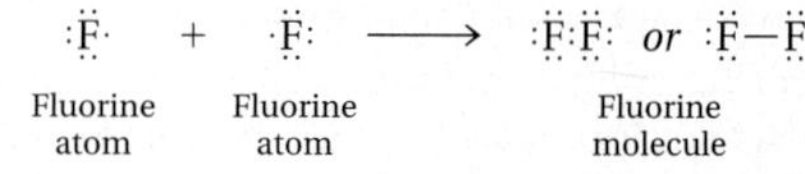

In the F_2 molecule, each fluorine atom contributes one electron to complete the octet. Notice that the two fluorine atoms share only one pair of valence electrons. Pairs of valence electrons that are not shared between atoms are called **unshared pairs,** also known as lone pairs or nonbonding pairs.

Answers

❶ H_2
❷ Molecular compounds share electron pairs.
❸ two
❹ One of carbon's 2*s* electrons is promoted to the 2*p* orbital.

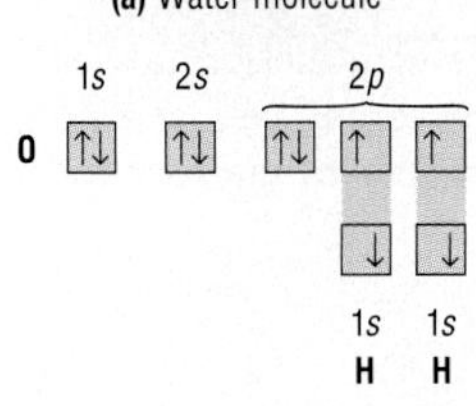

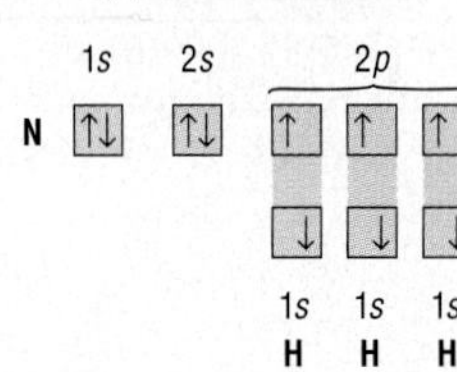
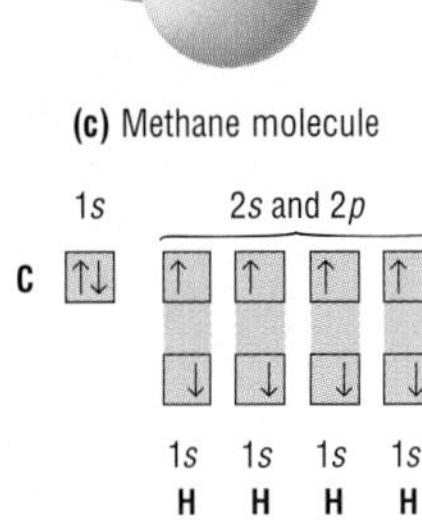

You can write electron dot formulas for molecules of compounds in much the same way that you write them for molecules of diatomic elements. Consider the examples for water, ammonia, and methane, illustrated in **Figure 16.4.**

Water (H_2O) is a triatomic molecule with two single covalent bonds. Two hydrogen atoms share electrons with one oxygen atom. The hydrogen and oxygen atoms attain noble-gas configurations by sharing electrons. As you can see in the electron dot structure below, the oxygen atom in water has two unshared pairs of valence electrons.

$$2H\cdot + :\dot{\ddot{O}}\cdot \longrightarrow :\ddot{O}:H \text{ or } :\ddot{O}-H$$

Hydrogen atoms — Oxygen atom — Water molecule

You can write the electron dot formula for ammonia (NH_3), a suffocating gas, in a similar way. The ammonia molecule has one unshared pair of electrons.

$$3H\cdot + :\dot{N}\cdot \longrightarrow H:\ddot{N}:H \text{ or } H-\ddot{N}-H$$

Hydrogen atoms — Nitrogen atom — Ammonia molecule

Methane (CH_4) contains four single covalent bonds. The carbon atom has four valence electrons and needs four more valence electrons to attain a noble-gas configuration. Each of the four hydrogen atoms contributes one electron to share with the carbon atom, forming four identical carbon-hydrogen bonds. As you can see in the electron dot structure below, methane has no unshared pairs of electrons.

$$4H\cdot + \cdot\dot{C}\cdot \longrightarrow H:\ddot{C}:H \text{ or } H-C-H$$

Hydrogen atoms — Carbon atom — Methane molecule

Figure 16.4

(a) In a water molecule, two hydrogen atoms form single covalent bonds with one oxygen atom.
(b) In an ammonia molecule, three hydrogen atoms form single covalent bonds with one nitrogen atom.
(c) A methane molecule has four carbon–hydrogen bonds. In each bond, carbon and hydrogen share a 1s electron from the hydrogen and an electron from the carbon. Why does carbon form four covalent bonds, and not two, with hydrogen? 4

Discuss

To further student understanding of covalent bonding, first review the molecular, structural, electron-dot, and orbital formulas for the fluorine, water, and ammonia molecules. If possible, display physical models.

Then, use the formulas for methane to present the concept of electron promotion, which allows carbon to form four single covalent bonds. Point out that elements in groups 13 and 14 promote electrons to increase their bonding capacity. For example, boron's electron configuration is $1s^2 2s^2 2p^1$. Based on this configuration, students might infer that boron forms only one covalent bond. However, the promotion of one 2*s* electron to the 2*p* orbital allows for the formation of three bonds. This promotion is supported by the formulas of compounds that boron forms with the halogens—BCl_3 rather than BCl.

ACTIVITY

Divide students into groups of three or four. Have them practice drawing molecular, structural, electron-dot, and orbital diagrams for molecules such as OF_2, SCl_2, N_2H_4, CCl_4, $CHCl_3$, and C_2H_6.

Meeting Diverse Needs

LEP and **At Risk** Students should get together with a study partner to quiz each other on the Lewis electron dot configurations for single atoms as well as for bonded atoms. Be sure they understand that the Group Number 1A–7A indicates the number of valence electrons.

Practice Problems Plus

Related Chapter Review Problem
Chapter Review problem 32 is related to Sample Problem 16-1.

Additional Practice Problem
The Chem ASAP! CD-ROM contains the following problem: The following covalent molecules have only single covalent bonds. Draw an electron dot structure for each.

a. NF_3
b. SBr_2
(**a.** :F̤:N̤:F̤: with :F̤: below N   **b.** :B̤r:S̤:B̤r:)

ACTIVITY

Ask students to draw Lewis electron–dot structures for each element in the second row of the periodic table: Li, Be, B, C, N, O, and F. Ask:

- Can you predict how many bonds each of these atoms will form in order to attain a noble gas configuration? (1, 2, 3, 4, 3, 2, 1)
- Can lithium form a covalent bond and reach stability? (no)
- Which elements can form covalent bonds and reach stability? (C, N, O, F)
- Can fluorine form an ionic bond? (yes)
- Would diatomic nitrogen molecules, N_2, contain ionic or covalent bonds? (covalent)

When carbon forms bonds with other atoms, it usually forms four bonds. You might not have predicted this based on carbon's electron configuration.

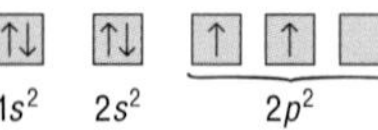

If you tried to form covalent C—H bonds for methane by combining the two $2p$ electrons of the carbon with two $1s$ electrons of hydrogen atoms, you would incorrectly predict a molecule with the formula CH_2 (instead of CH_4). The formation of four bonds by carbon can be simply explained. One of carbon's $2s$ electrons is promoted to the vacant $2p$ orbital to form the following.

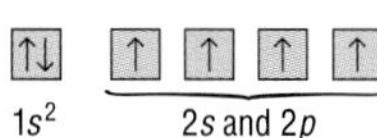

This electron promotion requires only a small amount of energy. As shown in **Figure 16.4c** on the previous page, the promotion provides four electrons of carbon that are capable of forming covalent bonds with four hydrogen atoms. Methane, the carbon compound formed by electron sharing of carbon with four hydrogen atoms, is much more stable than CH_2. The stability of the resulting methane more than compensates for the small energy cost of the electron promotion. Therefore, formation of methane (CH_4) is much preferred to formation of CH_2.

Sample Problem 16-1

Hydrogen chloride (HCl) is a diatomic molecule with a single covalent bond. Draw the electron dot structure for HCl.

1. ***ANALYZE*** ***Plan a problem-solving strategy.***
In a single covalent bond, a hydrogen and a chlorine atom must share a pair of electrons. Each must contribute one electron to the bond. First, write the electron dot structures for the two atoms. Then show the electron sharing in the compound they produce.

2. ***SOLVE*** ***Apply the problem-solving strategy.***

H·	+	·C̤̈l:	⟶	H:C̤̈l:
Hydrogen atom		Chlorine atom		Hydrogen chloride molecule

3. ***EVALUATE*** ***Does the result make sense?***
In the electron dot structures, the hydrogen atom and the chlorine atom are each correctly shown to have an unpaired electron. Through electron sharing, the hydrogen and chlorine atoms are shown to attain the electron configurations of the noble gases helium and argon, respectively.

Practice Problems

1. Draw electron dot structures for each molecule.
   a. chlorine
   b. bromine
   c. iodine
2. The following molecules have single covalent bonds. Draw an electron dot structure for each.
   a. H_2O_2
   b. PCl_3

Chem ASAP!
Problem-Solving 2
Solve Problem 2 with the help of an interactive guided tutorial.

LEWIS ELECTRON DOT STRUCTURES

As you have learned, it is the valence electrons of an atom that are involved in chemical bonding. Because valence electrons are significant, it is important to be able to study the valence electrons of atoms and compounds. Chemists use a notation in which valence electrons are identified as dots, and their bonds as dashes. This notation is known as a Lewis electron-dot structure in honor of its founder, Gilbert Lewis.

You were introduced to Lewis electron-dot structures in Chapter 15. Now is a good time to review these structures because they will be used to explain concepts in this and upcoming chapters. Understanding how to read and write electron-dot structures will help you understand how electrons are arranged in molecules, how they rearrange during chemical reactions, and how they contribute to determining the shape of a molecule.

Electron-Dot Structures for Atoms In an electron-dot structure, the symbol for the element represents the nucleus of the atom along with its inner electrons. Dots are placed around the symbol to represent the valence electrons.

To write an electron-dot structure for an atom, use the periodic table to find the number of valence electrons. Iodine, for example, has seven valence electrons. Write the symbol for the atom and then place the dots around the symbol. No more than two dots can be placed on any one side of a chemical symbol. The electron-dot structure for iodine is

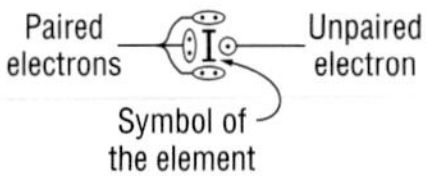

Paired electrons
I
Unpaired electron
Symbol of the element

Electron-Dot Structures for Molecules Follow these basic rules to write the electron-dot structure for molecules.

1. Add up the valence electrons for each atom in the molecule. Use the periodic table to determine the number of valence electrons for each atom. For example, in the case of a water molecule, H_2O,

   (Group 1) H H (Group 6) O

   $1 + 1 \quad + \quad 6 \quad = \quad 8$ valence electrons

2. Write the symbols for the atoms in the molecule. In simple molecules, one atom will be the central atom surrounded by the other atoms. The central atom is often the first atom in the formula. Hydrogen can form only one covalent bond, so it cannot be the central atom.

   H O H

3. Draw a dash between each pair of atoms covalently bonded together.

   H—O—H

4. For each dash you drew, subtract 2 from your total number of valence electrons. Then draw the remaining electrons as dots around the atoms. Arrange the dots so that most atoms have eight valence electrons, and hydrogen has two. In the case of water, there are two covalent bonds. So subtract 4 (2×2) electrons from the 8 original valence electrons. That leaves 4 valence electrons to add to the structure.

   $H—\ddot{\underset{..}{O}}—H$

5. If there are not enough electrons to give the atoms eight electrons, shift unbonded electrons as necessary or change single bonds to double or triple bonds. Verify that each atom has a noble gas structure (two electrons for hydrogen and eight for the others).

Practice Problems

Prepare for upcoming problems in this chapter by determining the total number of valence electrons for these compounds. Then draw an electron dot structure for each.

A. HCl **C.** PCl_3

B. CH_4 **D.** CO_2

Teaching Tips

It is important that students realize why they need to know how to determine and interpret information regarding the shapes of molecules. Have students think of everyday examples in which the shape of a substance is just as important as its composition. For example, several keys might be made of the same metals, but only one will fit into a particular lock. Tires are useful not only because they are made of rubber, but because they are round in shape.

You may wish to extend this feature by citing some biological examples where molecular shape affects behavior. For example, characteristics of DNA are related to its double helix. Antigens and antibodies join together because of a lock-and-key relationship between sites on each molecule. Some toxins can be deactivated when other chemicals are used to alter their shapes.

Answers to Practice Problems

A. 8,

H:Cl:

B. 8,

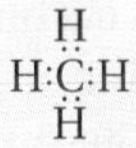

H
H:C:H
H

C. 26,

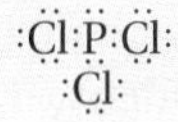

:Cl:P:Cl:
:Cl:

D. 16,

O::C::O

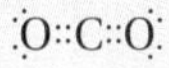

Discuss

Use the Lewis structure for the nitrogen molecule to introduce the discussion of multiple covalent bonds. Ask students to show how the structure of diatomic nitrogen satisfies the octet rule. Have them compare the bonding in ammonia and in nitrogen gas.

Then introduce the oxygen molecule. Ask students to draw a structure that would follow the octet rule. Ask:

- **Why doesn't oxygen form a triple bond?** (It needs only two electrons to achieve a stable electron configuration.)

Explain that their predicted structure for the oxygen molecule is not the one supported by experimental evidence. Actual oxygen molecules are one of the rare exceptions to the octet rule because they contain unpaired electrons. Note that an explanation for their existence will be given later in Section 16.1.

Have students try to draw the Lewis structure and orbital diagram for carbon dioxide. Ask:

- **What type of bonds does carbon form with oxygen in this case?** (two double covalent bonds) **Note that carbon can form single, double, and triple bonds. However, geometric restrictions prevent any atom from forming a quadruple bond. Have students draw diagrams for hydrogen cyanide (HCN) and formaldehyde (H_2CO). Ask:**
- **What kind of bonds does carbon form in each of these molecules?** (one single covalent and one triple covalent bond in HCN; two single and one double bond in H_2CO)

If possible, provide physical models for each compound discussed.

Double and Triple Covalent Bonds

Atoms sometimes share more than one pair of electrons to attain stable noble-gas electron configurations. **Double covalent bonds** are bonds that involve two shared pairs of electrons. **Triple covalent bonds** are bonds that involve three shared pairs of electrons.

Oxygen (O_2) is an example of a molecule that may seem as if it should have a double covalent bond according to the octet rule. You might assume that an oxygen atom, which has six valence electrons, would form a double bond by sharing two of its electrons with another oxygen atom.

:Ö: + :Ö: ⟶ :Ö::Ö: *or* :Ö=Ö:

Oxygen atom   Oxygen atom   Oxygen molecule

Chem ASAP!

Simulation 16

Simulate the covalent bonding between atoms of molecules.

In such an arrangement, all the electrons within the molecule would be paired. Experimental evidence, however, indicates that two of the electrons in O_2 are unpaired. Study **Figure 16.5** and keep oxygen in mind when you read about exceptions to the octet rule later in this section.

In contrast to the exceptional bonding in the oxygen molecule, the nitrogen molecule (N_2) does bond according to the octet rule. As shown in **Figure 16.5,** N_2 contains a triple covalent bond. In the nitrogen molecule, each nitrogen has one unshared pair of electrons. A single nitrogen atom has five valence electrons. How many electrons must each nitrogen atom in the nitrogen molecule acquire to attain the electron configuration of neon? (1)

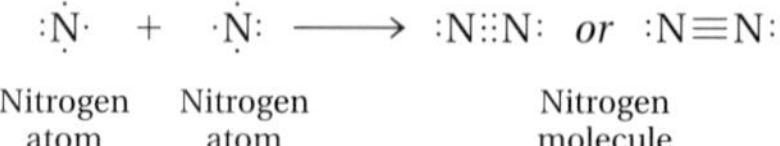

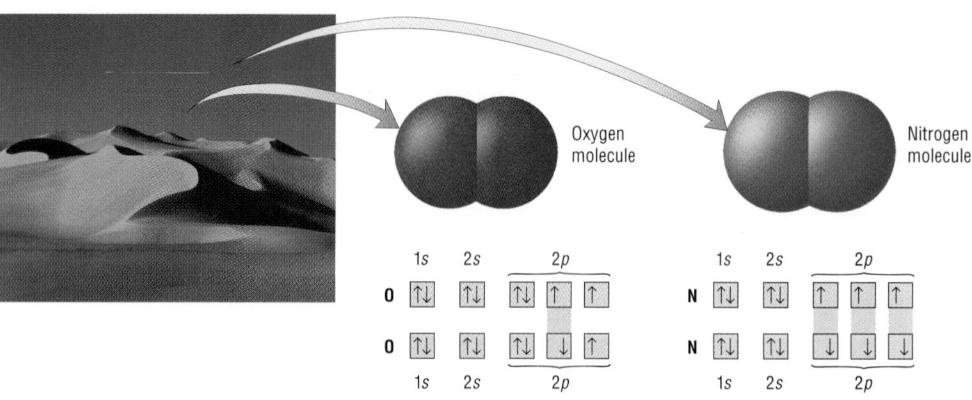

Figure 16.5

Oxygen and nitrogen are the main components of Earth's atmosphere. The oxygen molecule is an exception to the octet rule. It has two unpaired electrons. Three pairs of electrons are shared in a nitrogen molecule.

Answer

(1) three

Table 16.1

The Diatomic Elements			
Name	Chemical formula	Structure	Properties and uses
Fluorine	F_2	:F̤̈—F̤̈:	Greenish-yellow reactive toxic gas. Compounds of fluorine, a halogen, are added to drinking water and toothpaste to promote healthy teeth.
Chlorine	Cl_2	:C̤̈l—C̤̈l:	Greenish-yellow reactive toxic gas. Chlorine is a halogen used in household bleaching agents.
Bromine	Br_2	:B̤̈r—B̤̈r:	Dense red-brown liquid with pungent odor. Compounds of bromine, a halogen, are used in the preparation of photographic emulsions.
Iodine	I_2	:Ï̤—Ï̤:	Dense gray-black solid that produces purple vapors; a halogen. A solution of iodine in alcohol (tincture of iodine) is used as an antiseptic.
Hydrogen	H_2	H—H	Colorless, odorless, tasteless gas. Hydrogen is the lightest known element.
Nitrogen	N_2	:N≡N:	Colorless, odorless, tasteless gas. Air is almost 80% nitrogen by volume.
Oxygen	O_2	:Ö=Ö:	Colorless, odorless, tasteless gas that is vital for life. Air is about 20% oxygen by volume.

Up to this point in your textbook, the examples of multiple covalent bonds have all involved diatomic molecules. **Table 16.1** lists the properties and uses of the elements that exist as diatomic molecules. Multiple covalent bonds can also exist between unlike atoms. For example, consider carbon dioxide (CO_2), shown in **Figure 16.6.** The carbon dioxide molecule contains two oxygens, both of which share two electrons with carbon to form two carbon–oxygen double bonds.

$$:\ddot{O}\cdot + \cdot\dot{C}\cdot + \cdot\ddot{O}: \longrightarrow :\ddot{O}::C::\ddot{O}: \; or \; :\ddot{O}=C=\ddot{O}:$$

Oxygen atom   Carbon atom   Oxygen atom   Carbon dioxide molecule

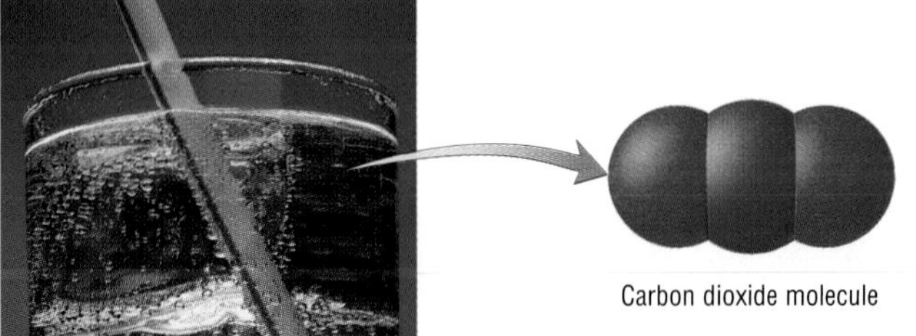

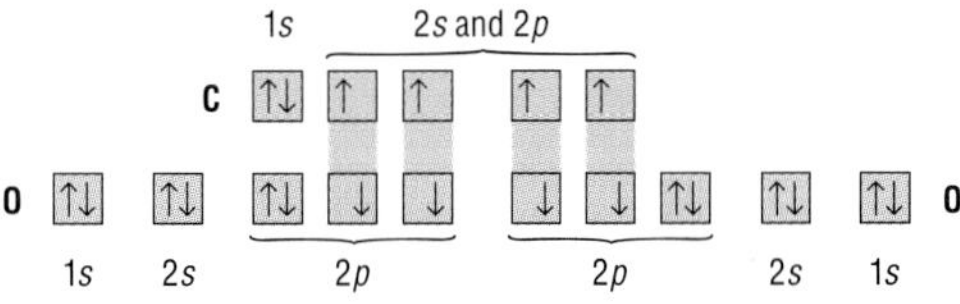

Figure 16.6
Carbon dioxide gas is soluble in water and is used to carbonate many beverages. A carbon dioxide molecule has two carbon–oxygen double bonds.

16.1 Think Critically

Although the atmosphere is about 80% nitrogen gas, there are surprisingly few nitrogen compounds. What explanation could you propose to explain why nitrogen is so unreactive? (The triple bond must be very strong, preventing nitrogen from reacting under normal conditions.)

Discuss

Coordinate covalent bonding is the exception to the rule that covalent bonding occurs between half-empty orbitals. In this case, one atom has an empty orbital and the other a filled orbital not yet involved in a chemical bond. Note that the bonding still involves only one pair of electrons and one pair of orbitals.

Discuss

Ask students to compare the Lewis dot formula for ammonia, NH_3 and for a hydrogen ion, H^+. With ammonia, there is a pair of electrons that are not involved in bonding; with the hydrogen ion, there are no electrons available for bonding. Ask:

- **How could you explain the existence of the ammonium ion, NH_4^+?** (The empty orbital of the H^+ must overlap the filled orbital in ammonia. The electrons from ammonia are attracted to both the nitrogen nucleus and the hydrogen nucleus. A bond forms when electrons are simultaneously attracted to two nuclei. This type of bond is called a coordinate covalent bond.)

Ask students to first explain the bonding in CO_2 and then explain the bonding in CO using coordinate covalent bonding.

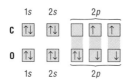

Carbon monoxide molecule

	1s	2s	2p		
C	↑↓	↑↓		↑	↑
O	↑↓	↑↓	↑↓	↓	↓
	1s	2s	2p		

Figure 16.7
In a coordinate covalent compound, one atom contributes both electrons of a bonding pair. In carbon monoxide, which atom contributes two electrons in one of the carbon–oxygen bonds? 1

Coordinate Covalent Bonds

Carbon monoxide (CO) is an example of a type of covalent-bonding different from that seen in water, ammonia, methane, and carbon dioxide. A carbon atom needs to gain four electrons to attain the electron configuration of neon. An oxygen atom needs two electrons. Yet it is possible for both atoms to achieve noble-gas electron configurations by a type of bonding called coordinate covalent bonding. To see how, begin by looking at the double covalent bond between carbon and oxygen.

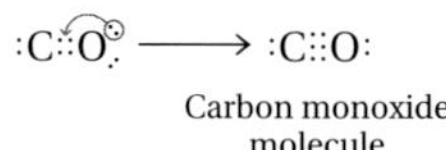

With the double bond in place, the oxygen has a stable configuration but the carbon does not. As shown in **Figure 16.7,** the dilemma is solved if the oxygen also donates one of its unshared pairs of electrons for bonding.

:C::O: ⟶ :C:::O:

Carbon monoxide molecule

A covalent bond in which one atom contributes both bonding electrons is called a **coordinate covalent bond.** In structural formulas, you can show coordinate covalent bonds as arrows that point from the atom donating the pair of electrons to the atom receiving them. The structural formula of carbon monoxide, with two covalent bonds and one coordinate covalent bond, is C ≡ O. Once formed, a coordinate covalent bond is like any other covalent bond. The only difference is the source of the bonding electrons.

The polyatomic ammonium ion (NH_4^+) has a coordinate covalent bond. This ion forms when a positively charged hydrogen ion (H^+) is attracted to the unshared electron pair of an ammonia molecule (NH_3) and attaches to it. As **Figure 16.8** shows, the ammonium ion is an important component of some nitrogen fertilizers.

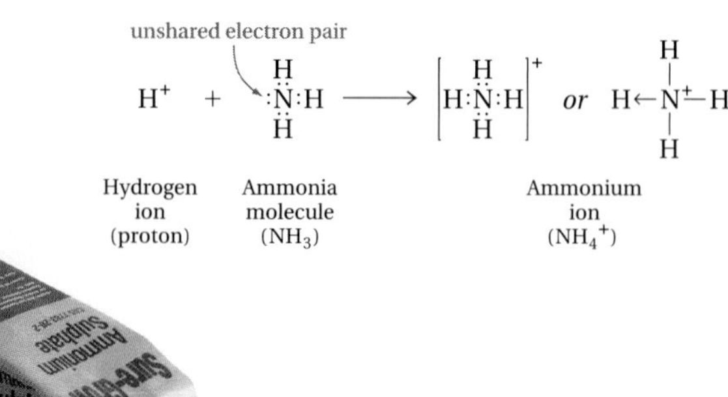

Figure 16.8
The polyatomic ammonium ion (NH_4^+), present in ammonium sulfate, is an important component of fertilizer for field crops, home gardens, and potted plants.

Most polyatomic cations and anions contain covalent and coordinate covalent bonds. **Table 16.2** lists some common examples. Which of these compounds are exceptions to the octet rule? 2

Answer

1 oxygen
2 Exceptions are nitric oxide (N has only 7) and nitrogen dioxide (O has only 7).

Table 16.2

Some Common Covalent Compounds

Name	Chemical formula	Structure	Properties and uses
Carbon monoxide	CO	:C≡O:	Colorless, highly toxic gas. It is a major air pollutant present in cigarette smoke and exhaust.
Carbon dioxide	CO_2	:Ö=C=Ö:	Colorless unreactive gas. This normal component of the atmosphere is exhaled in the breath of animals and is essential for plant growth.
Water	H_2O	H—Ö: —H	Colorless, odorless, tasteless liquid with melting point of 0 °C and boiling point of 100 °C. The human body is approximately 60% water.
Hydrogen peroxide	H_2O_2	H—Ö—Ö:—H	Colorless, unstable liquid when pure. It is used as rocket fuel. A 3% solution is used as a bleach and antiseptic.
Sulfur dioxide	SO_2	:Ö=S̈→Ö:	Oxides of sulfur are produced in combustion of petroleum products and coal. They are major air pollutants in industrial areas. Oxides of sulfur can lead to respiratory problems.
Sulfur trioxide	SO_3	:Ö=S(→Ö:)→Ö:	
Ammonia	NH_3	H—N̈—H, with H below N	Colorless gas with pungent odor; extremely soluble in water. Household ammonia is a solution of ammonia in water.
Nitric oxide	NO	:Ö=N̈·	Oxides of nitrogen are major air pollutants produced by the combustion of fossil fuels in automobile engines. They irritate the eyes, throat, and lungs. Nitrogen dioxide, a dark-brown gas, readily converts to colorless dinitrogen tetroxide.
Nitrogen dioxide	NO_2	:Ö=N·→Ö:	
Dinitrogen tetroxide	N_2O_4	(:Ö=)(:Ö←)N—N(→Ö:)(=Ö:)	
Nitrous oxide	N_2O	:Ö←N≡N:	Colorless sweet-smelling gas. It is used as an anesthetic commonly called laughing gas.
Hydrogen cyanide	HCN	H—C≡N:	Colorless toxic gas with the smell of almonds.
Hydrogen fluoride	HF	H—F̈:	Four hydrogen halides, all extremely soluble in water. Hydrogen chloride, a colorless gas with pungent odor, readily dissolves in water to give a solution called hydrochloric acid.
Hydrogen chloride	HCl	H—C̈l:	
Hydrogen bromide	HBr	H—B̈r:	
Hydrogen iodide	HI	H—Ï:	

16.1

Discuss

Have students use the three-step problem-solving approach to write the electron dot structure of SO_2. Emphasize that the structure should satisfy the bonding requirements of all three atoms. Students should find that, to satisfy the octet rule for all the atoms, they must write a structure in which one oxygen atom is double bonded to sulfur and the other is single bonded by a coordinate covalent bond in which the electrons are donated by sulfur. Point out that experimental evidence indicates that both sulfur-oxygen bonds are identical. Explain that this evidence indicates that the bonding in SO_2 must be some intermediate between a single and double bond. Ask:

- **How does the formation of a coordinate covalent bond differ from that of a covalent bond?** (In a covalent bond, each atom gives up one electron, in a coordinate covalent bond, both electrons come from the same atom.)

Discuss

Students may mistakenly infer that the atoms that make up polyatomic ions are bonded ionically because polyatomic ions are charged species. However, most polyatomic cations and anions contain covalent and coordinate covalent bonds.

Practice Problems Plus

Additional Practice Problem

The Chem ASAP! CD-ROM contains the following problem related to Sample Problem 16-2: Draw the electron dot structure of the polyatomic chlorate anion, ClO_3^-.

$$\left[:\ddot{O}:\ddot{Cl}:\ddot{O}:\right]^- \text{ with } :\ddot{O}: \text{ below Cl}$$

The atoms in polyatomic ions are covalently bonded. You can write electron dot structures for these ions. The negative charge of a polyatomic ion shows the number of electrons in addition to the valence electrons of the atoms present. Because a polyatomic ion is part of an ionic compound, the positive charge of the cation of the compound balances these additional electrons.

Sample Problem 16-2

The polyatomic hydronium ion (H_3O^+) contains a coordinate covalent bond. It forms when a hydrogen ion is attracted to an unshared electron pair in a water molecule. Write the electron dot structure of the hydronium ion.

1. ***ANALYZE*** ***Plan a problem-solving strategy.***
H_3O^+ forms by the addition of a hydrogen ion to a water molecule. Write the electron dot structure of the water molecule. Then, add the hydrogen ion. Oxygen must share a pair of electrons with the added hydrogen ion to form a coordinate covalent bond.

2. ***SOLVE*** ***Apply the problem-solving strategy.***

$$H^+ + :\ddot{O}:H \text{ (H above O)} \longrightarrow \left[H:\ddot{O}:H\right]^+ \text{ (H above O)} \quad \textit{or} \quad \left[H \leftarrow \ddot{O}-H\right]^+ \text{ (H above O)}$$

Hydrogen ion (proton) | Water molecule (H_2O) | Hydronium ion (H_3O^+)

3. ***EVALUATE*** ***Does the result make sense?***
The oxygen in the hydronium ion has eight valence electrons, and each hydrogen shares two valence electrons. This satisfies the needs of both hydrogen and oxygen for valence electrons. The water molecule is electrically neutral, and the hydrogen ion has a positive charge. The combination of these two species must have a charge of 1+, as is found in the hydronium ion.

Practice Problems

3. Draw the electron dot structure of the hydroxide ion (OH^-).
4. Draw the electron dot structure of the polyatomic boron tetrafluoride anion (BF_4^-).

Chem ASAP!

Problem-Solving 4

Solve Problem 4 with the help of an interactive guided tutorial.

Sample Problem 16-3

Draw the electron dot structure of the sulfite ion (SO_3^{2-}), in which sulfur is the central atom.

1. ***ANALYZE*** ***Plan a problem-solving strategy.***
The formula of the sulfite ion (SO_3^{2-}) is given, as is the information that the oxygen atoms are attached to the sulfur; that is, sulfur is the central atom. Start with the oxygen and sulfur atoms, their valence electrons, and the two extra electrons indicated by the charge.

Sample Problem 16-3 (cont.)

2. ***SOLVE*** ***Apply the problem-solving strategy.***
Draw the atoms, their valence electrons, and the two extra electrons.

$$:\ddot{O}\cdot\ \cdot\dot{S}:\ \cdot\ddot{O}:\ +\ \cdot\cdot$$
$$\cdot\ddot{O}:$$

Join two of the oxygens to sulfur by single covalent bonds.

$$:\ddot{O}:\ddot{S}:\ \cdot\ddot{O}:$$
$$\cdot\ddot{O}:$$

Join the remaining oxygen by a coordinate covalent bond and add the two extra electrons. Put brackets around the structure and indicate the 2− charge.

$$:\ddot{O}:\ddot{S}:\ddot{O}:\ \ + \cdot\cdot \longrightarrow \left[:\ddot{O}:\ddot{S}:\ddot{O}:\right]^{2-}$$

3. ***EVALUATE*** ***Does the result make sense?***
Each of the atoms of the completed structure has eight valence electrons, satisfying the octet rule. The two extra electrons were shown to be needed. Without them, the SO_3 grouping would be electrically neutral, and two of the oxygens would be electron-deficient.

Practice Problems

5. Draw the electron dot structures for sulfate (SO_4^{2-}) and carbonate (CO_3^{2-}). Sulfur and carbon are the central atoms, respectively.
6. Draw the electron dot structure for the hydrogen carbonate ion (HCO_3^-). Carbon is the central atom, and hydrogen is attached to oxygen in this polyatomic anion.

Bond Dissociation Energies

A large quantity of heat is liberated when hydrogen atoms combine to form hydrogen molecules. This release of heat suggests that the product is more stable than the reactants. The covalent bond in the hydrogen molecule is so strong that it requires 435 kJ of energy to dissociate 1 mol of hydrogen molecules to hydrogen atoms. The total energy required to break the bond between two covalently bonded atoms is known as the **bond dissociation energy.** Hydrogen molecules have a bond dissociation energy of 435 kJ/mol.

$$\text{H—H} + 435\ \text{kJ} \longrightarrow \text{H}\cdot + \cdot\text{H}$$

A typical carbon–carbon single covalent bond has a bond dissociation energy of about 347 kJ. The ability of carbon to form strong carbon–carbon bonds helps explain the stability of carbon compounds. Compounds with only C—C and C—H single covalent bonds, such as methane, are quite unreactive chemically. They are unreactive partly because the dissociation energy for each of these bonds is high. **Table 16.3** on the following page gives bond dissociation energies and bond lengths of several representative covalent bonds. What is the strongest bond listed? What is the weakest bond listed? ❶

TEACHER DEMO

It takes energy to break bonds, and energy is released when bonds form. The amount of energy necessary to break a bond, bond dissociation energy, is equal to the energy released when the bond forms.

Procedure:

1. Clean a 10- to 15-cm piece of magnesium ribbon. Hold with tongs and light in a Bunsen burner. **CAUTION!** Tell students not to look directly at burning Mg ribbon. Supply filters for viewing. Discuss with students the large amount of heat and light being given off. Write the balanced equation on the board.
$2Mg + O_2 \rightarrow 2MgO + \text{energy}$
2. Using tongs, place a small piece of charcoal in the Bunsen burner flame and try to ignite it. Write the balanced chemical equation.
$C + O_2 \rightarrow CO_2 + \text{energy}$

Note that energy is also given off, but not anywhere near as much as in the burning of magnesium. Ask:

- What kind of bonding occurs in reaction 1? (ionic)
- What kind of bonding occurs in reaction 2? (covalent) The bond energy for ionic bonds is, in general, greater than for covalent bonds.

Practice Problems Plus

Additional Practice Problem

This problem is related to Sample Problem 16-3: Draw the electron dot structure for the phosphite ion, PO_3^{3-}.

$$\left[:\ddot{O}:\ddot{P}:\ddot{O}:\right]^{3-}$$
$$:\ddot{O}:$$

Answer

❶ strongest: C ≡ O
weakest: O — O

MINI LAB

Strengths of Covalent Bonds

SAFETY

Have students wear safety goggles in case any rubber bands break or shoot off from the setup.

LAB PREP AND PLANNING

Probeware Lab Manual has Vernier, TI, and PASCO labs.

Obtain the necessary materials in advance. You may wish to ask students to help with this task by bringing cans of food, coat hangers, and plastic grocery bags to class.

LAB TIPS

Use tape to secure the hanger in place if necessary. You may wish to have groups of students record the data and then create a graph as a class on an overhead projector.

ANALYSIS AND CONCLUSIONS

1. Triple covalent bonds are stronger than double covalent bonds, which are stronger than single covalent bonds.
2. The change in bond dissociation energies in going from a carbon–carbon single bond to a carbon–carbon double bond to a carbon–carbon triple bond is nearly constant. The change in length of one, two, and three rubber bands, as given by the slopes of the lines, is not constant—it is large going from one to two rubber bands and small going from two to three rubber bands.

Table 16.3

Bond	Bond energy (kJ/mol)	Bond length (pm)
H—H	435	74
C—H	393	109
C—O	356	143
C=O	736	121
C≡O	1074	113
C—C	347	154
C=C	657	133
C≡C	908	121
C—N	305	147
Cl—Cl	243	199
N—N	209	140
O—H	464	96
O—O	142	132

Bond Dissociation Energies and Bond Lengths for Covalent Bonds

MINI LAB

Strengths of Covalent Bonds

PURPOSE

To compare and contrast the stretching of rubber bands and the dissociation energy of covalent bonds.

MATERIALS

- 1 170-g (6-oz) can of food
- 2 454-g (16-oz) cans of food
- 3 identical No. 25 rubber bands
- metric ruler
- coat hanger
- plastic grocery bag
- paper clip
- graph paper
- pencil
- motion detector (optional)

PROCEDURE

Sensor version available in the Probeware Lab Manual.

1. Bend the coat hanger to fit over the top of a door. The hook should hang down on one side of the door. Measure the length of the rubber bands (in cm). Hang a rubber band on the hook created by the coat hanger.
2. Place the 170-g can in the plastic bag. Use a paper clip to fasten the bag to the end of the rubber band. Lower the bag gently until it is suspended from the end of the rubber band. Measure and record the length of the stretched rubber band. Using different food cans, repeat this process three times with the following masses: 454 g, 624 g, and 908 g.
3. Repeat Step 2, first using two rubber bands to connect the hanger and the paper clip, and then using three.
4. Graph the length difference: (stretched rubber band) − (unstretched rubber band) on the *y*-axis versus mass (kg) on the *x*-axis for one, two, and three rubber bands. Draw an estimated best-fit straight line through the points for each set of data. (Your graph should have three separate lines.) The *x*-axis and *y*-axis intercepts of the lines should pass through zero, and the lines should extend past 1 kg on the *x*-axis. Determine the slope of each line in cm/kg.

ANALYSIS AND CONCLUSIONS

1. Assuming the rubber bands are models for covalent bonds, what can you conclude about the relative strengths of single, double, and triple bonds?
2. How does the behavior of the rubber bands differ from that of covalent bonds?

Resonance

Consider the two electron dot structures for ozone.

$$:\ddot{O}:\ddot{O}::\ddot{O}: \longleftrightarrow :\ddot{O}::\ddot{O}:\ddot{O}:$$

Notice that the structure on the left can be converted to the one on the right by shifting electron pairs without changing the positions of the oxygen atoms.

As drawn, these electron dot structures suggest that the bonding in ozone consists of one single coordinate covalent bond and one double covalent bond. Because earlier chemists imagined that the electron pairs rapidly flip back and forth, or resonate, between the various electron dot structures, they used double-headed arrows to indicate that two or more structures are in resonance.

Double covalent bonds are usually shorter than single bonds, so it was believed that the bond lengths in ozone were unequal. Experimental measurements show, however, that this is not the case. The two bonds in ozone are the same length. This result can be explained if you assume that the actual bonding in the ozone molecule is the average of the two electron dot structures. The electron pairs do not actually resonate back and forth. The actual bonding is a hybrid, or mixture, of the extremes represented by the resonance forms.

Thus the two electron dot formulas for ozone are examples of what are still referred to as resonance structures. **Resonance structures** are structures that occur when it is possible to write two or more valid electron dot formulas that have the same number of electron pairs for a molecule or ion. Resonance structures are simply a way to envision the bonding in certain molecules. Although no back-and-forth changes occur, double-headed arrows are used to connect resonance structures.

Exceptions to the Octet Rule

As you have seen, the octet rule provides guidance for writing electron dot structures. For some molecules or ions, however, it is impossible to write structures that satisfy the octet rule. This occurs whenever the total number of valence electrons is an odd number. The NO_2 molecule, for example, contains a total of seventeen valence electrons. Each oxygen contributes six electrons and the nitrogen contributes five. **Figure 16.9** illustrates bonding in the nitrogen dioxide molecule.

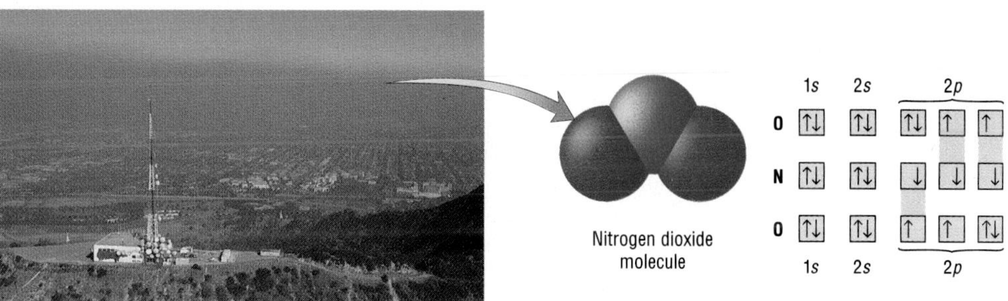

Figure 16.9
The nitrogen dioxide molecule, a major component of air pollution in some cities, is unusual in that it contains an unpaired electron.

TEACHER DEMO

In an efficient hood, where the poisonous nitrogen dioxide fumes will be drawn safely away, demonstrate the formation of NO_2 by reacting copper metal with concentrated nitric acid in an evaporating dish. Write the balanced equation for the reaction on the board.

$$Cu + 4HNO_3 \rightarrow Cu(NO_3)_2 + 2H_2O + 2NO_2 + \text{energy}$$

NO_2 is one of the pollutants in automobile exhaust. It gives smog its reddish-brown color. It is very reactive and poisonous.

Have students try to write the Lewis electron dot formula for NO_2. Note that N is the central atom to which each oxygen is attached. There are a total of 17 valence electrons. There is no way to follow the octet rule for all three atoms. Yet this chemical does exist and is reactive. NO_2 is considered a resonance hybrid. Look at the two structures at the top of p. 450. The unpaired electron may account for the high reactivity of NO_2.

Discuss

Have students look at the two resonance structures for diatomic oxygen molecules. Ask:

- **Would you consider oxygen a reactive gas?** (yes) **Note that the best explanation for the bonding in oxygen gas is a resonance hybrid.**
- **How could resonance account for the reactivity of oxygen?** (One of the resonance structures has an unpaired electron, which would make it very reactive.)

After a brief discussion of the other exceptions to the octet rule that are presented in this section, it is important to remind students that the large majority of bonds do follow the octet rule. At this level of chemistry, it is hoped that students will retain the rules rather than the exceptions.

Two plausible resonance structures can be written for the NO_2 molecule.

$$:\ddot{O}=\ddot{N}-\ddot{O}\cdot$$
$$\cdot\ddot{O}-\ddot{N}=\ddot{O}:$$

An unpaired electron is present in each of these structures, both of which fail to follow the octet rule. It is impossible to write a Lewis dot structure for NO_2 that satisfies the octet rule for all atoms. Yet, NO_2 does exist as a stable molecule.

The concept of electron pairing is important in understanding the bonding and properties of molecules. You can consider electrons as small, spinning, electric charges. These moving electric charges create magnetic fields, much as the current in an electric motor creates a magnetic field. You can think of paired electrons as spinning in opposite directions. The magnetic effects of paired electrons essentially cancel. Substances in which all of the electrons are paired are **diamagnetic.** Diamagnetic substances are weakly repelled by an external magnetic field. In contrast, **paramagnetic** substances are those that contain one or more unpaired electrons. Such substances show a strong attraction to an external magnetic field. Paramagnetism can be detected by measuring the mass of a compound in the absence and then in the presence of a magnetic field. If the compound is paramagnetic, its mass will appear to be greater in the magnetic field.

Paramagnetism should not be confused with ferromagnetism, the form of magnetism with which you are most familiar. Ferromagnetism is the strong attraction of iron, cobalt, and nickel to magnetic fields. The ions Fe^{2+}, Co^{2+}, and Ni^{2+} all have unpaired electrons. Normally, these ions are randomly dispersed in a sea of electrons throughout their metals. In a magnetic field, the ions line up in an orderly fashion with the field, which creates a strong magnetic attraction. Because the new arrangement of ions remains when the magnetic field is removed, the magnetism is permanent.

As you have learned in this section, it is possible to write a structure for the oxygen molecule in which both oxygen atoms are surrounded by eight electrons and all the electrons are paired. You have also learned, however, that such a structure is incorrect. The experimental evidence for this statement is that oxygen is paramagnetic. Given what you know about the
1 correct structure for O_2, can you explain the paramagnetism of oxygen?

The measured distance between the oxygen atoms in O_2 indicates that the molecule does have some multiple bond character. The idea that the oxygen molecule has some double-bond character is plausible. All this information taken together suggests that oxygen is a hybrid of the following two structures.

$$:\ddot{O}-\ddot{O}:$$
$$:\ddot{O}=\ddot{O}:$$

Several other molecules, such as some compounds of boron, also do not follow the octet rule. The boron atom in boron trifluoride (BF_3), for example, is deficient by two electrons. Boron trifluoride readily reacts with ammonia to make the compound $BF_3{\cdot}NH_3$. In doing so, the boron atom accepts the unshared electron pair from ammonia and completes the octet.

Meeting Diverse Needs

Gifted Have students research Linus Pauling and find out what he did to earn two Nobel Prizes. (Chemistry prize for work on the nature of chemical bonding and Peace prize for his efforts towards disarmament.)

Answers

1 One of its resonance structures has at least one unpaired electron.
2 In sulfur hexafluoride, the sulfur atom must have twelve valence electrons.

$$\mathrm{F_3B} + \mathrm{:NH_3} \longrightarrow \mathrm{F_3B{\leftarrow}NH_3}$$

A few atoms, especially phosphorus and sulfur, sometimes expand the octet to include ten or twelve electrons. Consider phosphorus trichloride (PCl_3) and phosphorus pentachloride (PCl_5). Both are stable compounds in which all of the chlorines are bonded to a single phosphorus atom. Covalent bonding in PCl_3 follows the octet rule because all the atoms acquire eight electrons. However, as shown in **Figure 16.10,** the electron dot structure for PCl_5 can be written so that phosphorus has ten valence electrons. How many valence electrons does the sulfur in sulfur hexafluoride (SF_6) have for the structure shown in **Figure 16.10?** 2

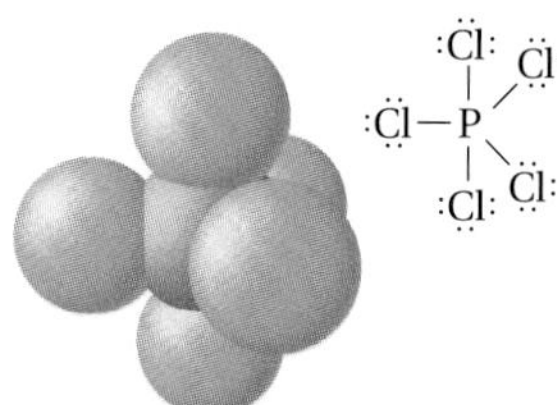

Phosphorus pentachloride

Sulfur hexafluoride

Figure 16.10

Phosphorus pentachloride, a slightly yellow crystalline solid used as a chlorinating and dehydrating agent, and sulfur hexafluoride, a colorless, odorless gas used as an insulator for electrical equipment, are exceptions to the octet rule.

section review 16.1

7. How are single, double, and triple bonds indicated in electron dot structures?
8. Give an example of each of the following.
   a. coordinate covalent bonding
   b. resonance structures
   c. exceptions to the octet rule
9. What kinds of information does a structural formula reveal about the compound it represents?
10. Draw electron dot structures for the following molecules, which have only single covalent bonds.
    a. H_2S
    b. PH_3
    c. ClF
11. Draw the resonance structures for sulfur dioxide (SO_2). Sulfur is the central atom.
12. How many kilojoules are required to dissociate all the C—H single bonds in 0.1 mol of methane (CH_4)? Assume that the bond dissociation energy is the same for each bond.

Chem ASAP! **Assessment 16.1** Check your understanding of the important ideas and concepts in Section 16.1.

portfolio project

Research how chemists determine bond length in a compound. How might they determine that the two bonds in an ozone molecule have the same length?

16.1

3 Assess

Check for Understanding

Ask student volunteers to write Lewis electron dot structural formulas for two molecules, such as SBr_2 and C_3H_8, on the chalkboard. Ask students how the formation of coordinate covalent bonds differs from that of covalent bonds. (A coordinate covalent bond is formed when one atom contributes both bonding electrons. In a normal covalent bond, each bonding atom contributes electrons.) **Ask students to explain what resonance formulas are.** (Resonance structures occur when two or more valid electron dot formulas can be written for a molecule.)

Reteach

Use a concept map to emphasize the relationship between the number of paired electrons and the number of covalent bonds in a molecule. Have students suggest words that will link these terms.

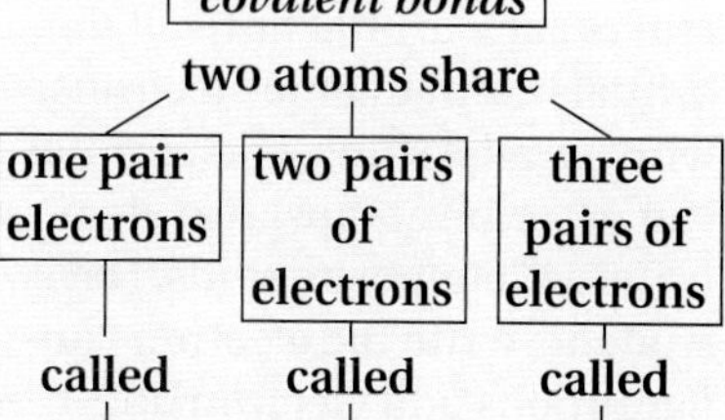

Portfolio Project

Research should reveal that two methods used to find bond lengths are electron diffraction and microwave spectroscopy.

Answers

SECTION REVIEW 16.1

7. A single bond is indicated by two dots or one line between two atoms. A double bond is indicated by four dots or two lines between atoms. A triple bond is indicated by six dots or three lines between atoms.
8. a. carbon monoxide (CO) (or similar)
   b. ozone (O_3) (or similar)
   c. nitrogen dioxide (NO_2) (or similar)
9. The structural formula identifies the atoms in the compound and their respective number and arrangement in the molecule.
10. a. H:S:H
    b. H:P:H with H below P
    c. :Cl:F:
11. :O::S:O: ↔ :O:S::O:
12. 157 kJ

16.2

1 Engage

Use the Visual

Have students study the photograph and read the text that opens the section. Ask:

- **How do attractive and repulsive forces influence molecular bonding?** (The bonding in molecules depends on the interactions between negatively charged electrons and positively charged nuclei. Attractive forces between electrons and nuclei that draw atoms together are balanced by repulsive forces between nuclei that force atoms apart.)

Motivate and Relate

To prepare students for a discussion of molecular orbitals, review atomic orbital theory. Display Figure 13.6 on an overhead projector and ask students to recall how atomic orbitals are filled in order of increasing energy. Have students refer to Table 13.2 to review the rules for writing electron configurations. When electrons occupy orbitals of equal energy, one electron enters each orbital until all the orbitals contain lone electrons with parallel spins. Each orbital can describe at most two electrons with opposite spins. Have students write the electron configurations and electron dot structures for carbon, oxygen, sulfur, and phosphorus. Have them arrange the atomic orbitals according to energy level as in Figure 13.6.

section 16.2

objectives

- Describe the molecular orbital theory of covalent bonding, including orbital hybridization
- Use VSEPR theory to predict the shapes of simple covalently bonded molecules

key terms

- molecular orbitals
- bonding orbital
- antibonding orbital
- sigma bond
- pi bond
- tetrahedral angle
- VSEPR theory
- hybridization

BONDING THEORIES

This car is being painted by a process called electrostatic spray painting. The paint droplets become negatively charged as they exit the spray gun. The negatively charged paint particles are attracted to the positively charged auto body. Painting with attractive and repulsive forces is very efficient, because almost all the paint is applied to the car body and very little is wasted. **How do attractive and repulsive forces influence molecular bonding?**

Molecular Orbitals

The model for covalent bonding that you have been working with so far is based on the assumption that the atomic orbitals of individual atoms are unchanged in bonded atoms. There is a quantum mechanical model of bonding, however, that describes the electrons in molecules by means of orbitals that exist only for groupings of atoms. When two atoms combine, this model assumes that their atomic orbitals overlap to produce **molecular orbitals,** or orbitals that apply to the entire molecule.

Some important parallels exist between atomic orbitals and molecular orbitals. Just as an atomic orbital belongs to a particular atom, a molecular orbital belongs to a molecule as a whole. Each atomic orbital is filled if it contains two electrons. Similarly, two electrons are required to fill a molecular orbital.

The molecular orbital model of bonding requires that the number of molecular orbitals equal the number of overlapping atomic orbitals. When two atomic orbitals overlap, for example, two molecular orbitals are created. One is called a **bonding orbital,** which is a molecular orbital with an energy that is lower than that of the atomic orbitals from which it formed. The other molecular orbital is called an **antibonding orbital,** and its energy is higher than that of the atomic orbitals from which it formed. The following illustration shows the relationship between atomic orbitals and molecular orbitals.

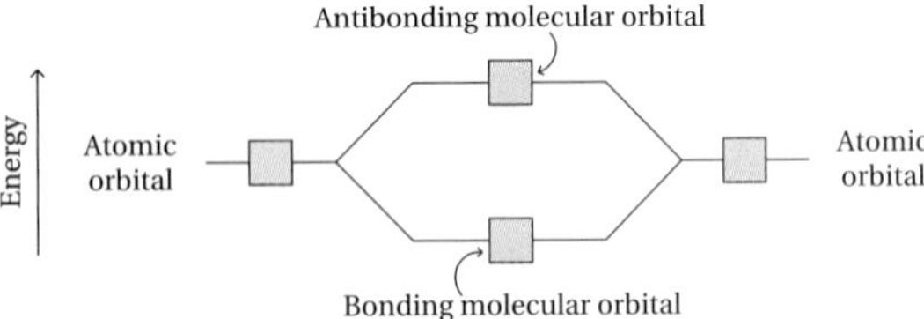

You can use the molecular orbital model of bonding to explain the bonding in the hydrogen molecule (H_2). In the formation of a hydrogen molecule, the 1*s* atomic orbitals of each of the two hydrogen atoms overlap. Two electrons, one from each hydrogen atom, are available for bonding. The energy of the electrons in the bonding molecular orbital is lower than the energy of the electrons in the atomic orbitals of separate hydrogen atoms. Because electrons seek the lowest energy, they fill the bonding molecular orbital. This makes a stable covalent bond between the hydrogens. Note that the

STUDENT RESOURCES

From the Teacher's Resource Package, use:

- Section Review 16.2, Ch. 16 Practice Problems, Investigating Graphics, and Quizzes from the Review Module (Ch. 13–16)

TECHNOLOGY RESOURCES

Relevant technology resources include:

- Chem ASAP! CD-ROM
- Resource Pro CD-ROM

⊕ represents the nucleus.

Bond axis

Sigma-antibonding molecular orbital

s atomic orbital + *s* atomic orbital

Bond axis

Sigma-bonding molecular orbital

Figure 16.11

In a bonding molecular orbital, the electron density between the nuclei is high. In an antibonding molecular orbital, the electron density between the nuclei is extremely low. Which orbital do the two electrons occupy in a hydrogen molecule? ❶

antibonding orbital is empty. **Figure 16.11** shows the formation of the bonding and antibonding molecular orbitals in a hydrogen molecule.

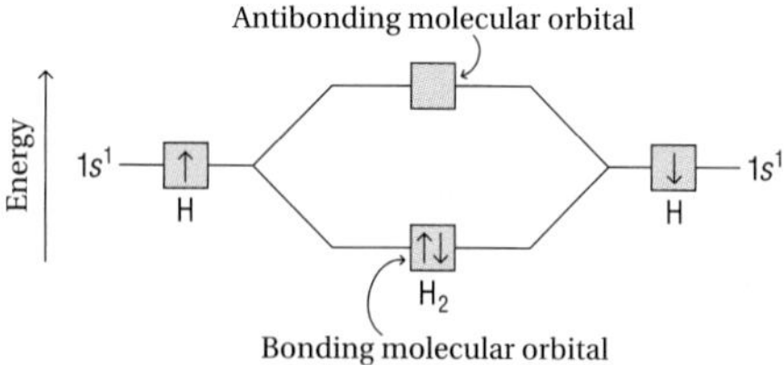

In the bonding molecular orbital of H_2, there is a high probability of finding the electrons between the nuclei of the bonded atoms. The orbital is symmetrical along the axis between the hydrogen atoms. When two atomic orbitals combine to form a molecular orbital that is symmetrical along the axis connecting two atomic nuclei, a **sigma bond** is formed. The symbol for this bond is the Greek letter sigma (σ).

In general, covalent bonding results from an imbalance between the attractions and repulsions of the nuclei and electrons involved. Being oppositely charged, the nuclei and electrons attract each other. Being similarly charged, nuclei repel other nuclei and electrons repel other electrons. In a hydrogen molecule, the nuclei repel each other, as do the electrons. In a bonding molecular orbital of hydrogen, however, the attractions between the hydrogen nuclei and the electrons are stronger than the repulsions. The balance of all the interactions between the hydrogen atoms is thus tipped in favor of the attractions. The result is a stable diatomic molecule of H_2.

In other cases, attempts to combine two atoms could lead to electron pairs in higher energy, antibonding molecular orbitals. Electrons in antibonding orbitals are not found between the nuclei, and the balance between attractions and repulsions would favor the repulsions. Such destabilizing electron repulsions would occur, for example, if two helium

GEOGRAPHY

Topographic Maps

Topographic maps are usually drawn as a series of lines tracing the elevations of a terrain. For example, a line might be drawn for every increase of 10 meters in

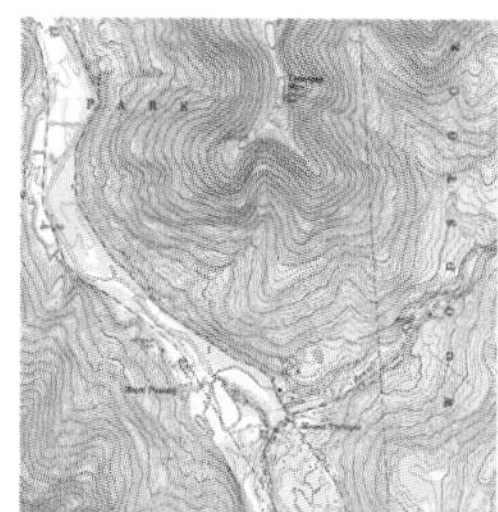

elevation. Thus the more closely spaced the lines are, the steeper is the terrain. Scientists who wish to know the positions of atoms in crystals of molecules may use electron densities that are plotted in much the same way that a terrain is shown on a topographic map. From x-ray diffraction data, scientists can map the electron density at any position in a crystal of interest. Where the lines are closest in the map, the electron density of the molecule is greatest. This method allows scientists to pinpoint positions of the atoms in the molecule with great accuracy.

16.2

2 Teach

TEACHER DEMO

Rub a balloon on your hair and bring it close to small pieces of tissue paper. Notice that the tissue paper sticks to the balloon. Ask:

- Why does the paper stick to the balloon? (Electrons are transferred from hair to the balloon, giving the balloon a net negative charge. The balloon *induces* a positive charge on the tissue paper. Opposite charges attract and the paper sticks to the balloon.)

TEACHER DEMO

Have students read the Link to Geography. Construct two cones from clear plastic. They should have equal diameters (about 10 cm), but different heights of 10 and 20 cm, respectively. Use a fine-point marker and metric ruler to draw circumference lines representing 2–cm rises in elevation on each cone. Place the cones side by side on the stage of an overhead projector. Have students note the difference in heights of the two cones and then compare densities of the elevation lines projected on the screen.

Answer

❶ the bonding molecular orbital

Discuss

Molecular orbital theory attempts to treat covalent bonds in terms of orbitals that involve an entire molecule. Molecular orbital theory assumes that the atomic orbitals of atoms are combined to give a new set of molecular orbitals, which have their own set of characteristics. These orbitals are arranged in order of increasing energy and the molecule's valence electrons are distributed among the available molecular orbitals. Electrons are placed in these orbitals following the same rules as for atomic orbitals. (two electrons per orbital, electrons in the lowest energy level, and Hund's rule) Show sigma and pi bonds using overhead transparencies. Then work out bonding diagrams for several examples on the chalkboard or overhead projector, showing the formation of sigma and pi bonds.

atoms were to combine into an He_2 molecule. Each He atom has two 1*s* electrons, so there would be a total of four electrons in an He_2 molecule. Two of these electrons can go into a low-energy bonding orbital, but the other two must go into a high-energy antibonding orbital.

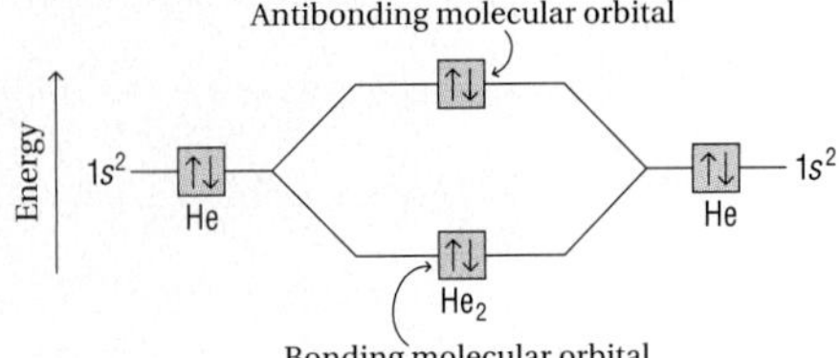

In this case, the overall repulsions are stronger than the overall attractions. Hence, a theoretical He_2 molecule is unstable compared with two separate helium atoms. For this reason, helium exists as individual atoms in nature.

Not only *s* orbitals overlap to form molecular orbitals. Atomic *p* orbitals can overlap as well. A fluorine atom, for example, has a half-filled 2*p* orbital. When two fluorine atoms combine, as shown in **Figure 16.12,** the *p* orbitals overlap to produce a bonding molecular orbital. There is a high probability of finding a pair of electrons in the bonding molecular orbital between the positively charged nuclei of the two fluorines. The fluorine nuclei are attracted to this region of high electron density. This attraction holds the atoms together in the fluorine molecule (F_2). The overlap of the 2*p* orbitals produces a bonding molecular orbital that is symmetrical when viewed along the F—F bond axis connecting the nuclei. Therefore, the F—F bond is a sigma bond.

In the sigma bond of the fluorine molecule, the *p* atomic orbitals overlap end-to-end. In some molecules, however, orbitals can overlap side-by-side. As shown in **Figure 16.13,** the side-by-side overlap of atomic *p* orbitals produces what are called pi molecular orbitals. When a pi molecular orbital is filled with two electrons, a pi bond results. In a **pi bond** (symbolized by the Greek letter π), the bonding electrons are most likely to be found in sausage-shaped regions above and below the bond axis of the bonded atoms. Orbital overlap in pi bonding is not as extensive as it is in sigma bonding. Therefore, pi bonds tend to be weaker than sigma bonds.

⊕ represents the nucleus.

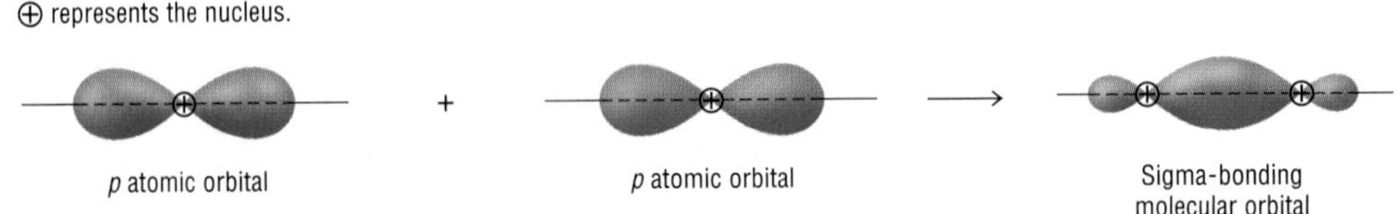

Figure 16.12
Two p atomic orbitals can combine to form a sigma-bonding molecular orbital, as in the case of fluorine (F_2). Notice that the sigma bond is symmetrical along the bond axis connecting the nuclei.

⊕ represents the nucleus.

p atomic orbital + p atomic orbital ⟶ Pi-bonding molecular orbital

Figure 16.13
The side-by-side overlap of two p atomic orbitals produces a pi-bonding molecular orbital. Together, the two sausage-shaped regions in which a bonding electron pair is most likely to be found constitute one pi-bonding molecular orbital.

VSEPR Theory

A photograph or sketch may fail to do justice to your appearance. Similarly, electron dot structures fail to reflect the three-dimensional shapes of molecules. The electron dot structure and structural formula of methane (CH_4), for example, show the molecule as if it were flat and merely two-dimensional.

```
   H              H
 H:C:H            |
   H          H — C — H
                  |
                  H
```

Methane (electron dot structure) Methane (structural formula)

In reality, methane molecules are three-dimensional. As **Figure 16.14** shows, the hydrogens in a methane molecule are at the four corners of a geometric solid called a regular tetrahedron. In this arrangement, all of the H—C—H angles are 109.5°, the **tetrahedral angle.**

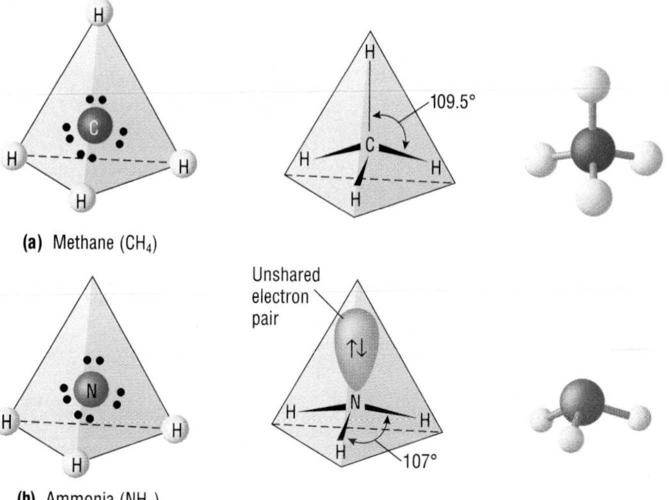

(a) Methane (CH_4)

(b) Ammonia (NH_3)

Figure 16.14
(a) *Methane is a good example of a tetrahedral molecule. The hydrogens in methane are at the four corners of a regular tetrahedron, and the bond angles are all 109.5°.* ***(b)*** *An ammonia molecule is pyramidal. The unshared pair of electrons repels the bonding pairs. Are the resulting H—N—H bond angles larger or smaller than the tetrahedral angle?* ❶

Answer

❶ smaller

16.2

Discuss

Ask students to write out the electron configuration for carbon when it is forming bonds. ($1s^2, 2s^1, 2p^3$) **Ask:**

- **How many valence electrons does carbon have?** (4)
- **How many bonds does carbon form?** (4)
- **What is the description of the three *p* orbitals?** (dumbbell–shaped orbitals containing two lobes; the three orbitals are perpendicular to each other, oriented on the *x*–, *y*–, and *z*–axes)

Point out that if carbon forms 3 of its 4 bonds using *p* orbitals, it must form molecules that are three–dimensional.

Discuss

To help students develop a sense of the variety of molecular shapes, allow students to handle and inspect three-dimensional ball-and-stick models of NH_3, H_2O, CH_4, and CO_2. Use these models to help students visualize the two-dimensional representations presented in Figures 16.14, 16.15, and 16.16. Point out that when using VSEPR theory to predict molecular shape, double and triple bonds are viewed as single bonds. Emphasize this point by drawing and showing a model for the structure of formaldehyde. Have students predict the shapes of H_2S and BF_3. (bent and trigonal planar respectively)

The valence-shell electron-pair repulsion theory, or VSEPR theory, explains the three-dimensional shape of methane. **VSEPR theory** states that because electron pairs repel, molecular shape adjusts so the valence-electron pairs are as far apart as possible. The methane molecule has four bonding electron pairs and no unshared pairs. The bonding pairs are farthest apart when the angle between the central carbon and its attached hydrogens is 109.5°. This is the H—C—H bond angle found experimentally. Any other arrangement tends to bring two bonding pairs of electrons closer together.

Unshared pairs of electrons are also important when you are trying to predict the shapes of molecules. The nitrogen in ammonia (NH_3) is surrounded by four pairs of valence electrons, so you might predict the tetrahedral angle of 109.5° for the H—N—H bond angle. However, one of the valence-electron pairs is an unshared pair. See **Figure 16.14** on the previous page. No bonding atom is vying for these unshared electrons. Thus they are held closer to the nitrogen than are the bonding pairs. The unshared pair strongly repels the bonding pairs, pushing them closer together than might be expected. The experimentally measured H—N—H
1 bond angle is only 107°. What is the shape of the ammonia molecule?

In a water molecule, oxygen forms single covalent bonds with two hydrogen atoms. The two bonding pairs and the two unshared pairs of electrons form a tetrahedral arrangement around the central oxygen. Thus the water molecule is planar (flat) but bent. With two unshared pairs repelling the bonding pairs, the H—O—H bond angle is compressed in comparison with the H—C—H bond angle in methane. The experimentally measured bond angle in water is about 105°, as shown in **Figure 16.15a.**

In contrast, the carbon in a carbon dioxide molecule has no unshared electron pairs. The double bonds joining the oxygens to the carbon are farthest apart when the O=C=O bond angle is 180°. See **Figure 16.15b.** Thus CO_2 is a linear molecule. A selection of various molecular shapes is shown in **Figure 16.16.**

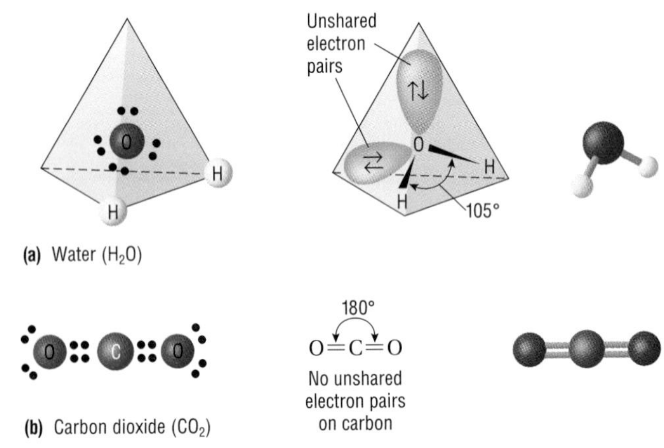

Figure 16.15
(a) The water molecule is bent because the two unshared pairs of electrons on oxygen repel the bonding electrons. (b) In contrast, the carbon dioxide molecule is linear. The carbon atom has no unshared electron pairs.

Meeting Diverse Needs

LEP and **At Risk** Be sure that these students have the opportunity to see as many three-dimensional models of molecules as possible. Remind them that, in general, the periodic table can help predict shapes once you have worked through one sample from each chemical family.

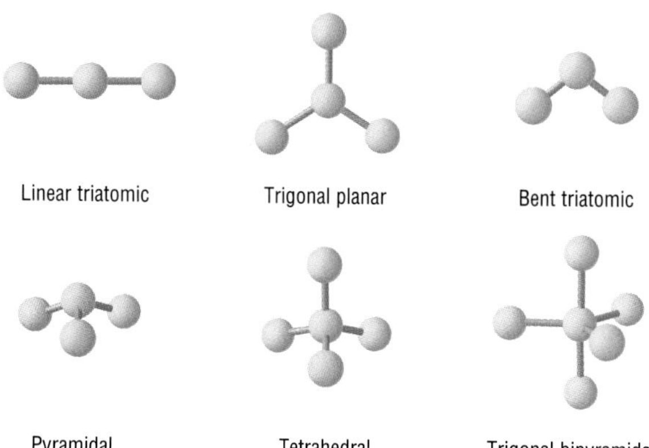

Figure 16.16
Shown here are several common molecular shapes.

Hybrid Orbitals

The VSEPR theory works well when accounting for molecular shapes, but it does not help much in describing the types of bonds formed. Another way to describe molecules, which is informative of both bonding and shape, is orbital hybridization. In **hybridization,** several atomic orbitals mix to form the same total number of equivalent hybrid orbitals.

Orbital hybridization can be used to describe the single covalent bonds of the methane molecule.

```
    H
    |
H — C — H
    |
    H
```

Methane

One 2*s* orbital and three 2*p* orbitals of a carbon atom mix to form four sp^3 hybrid orbitals. These are at the tetrahedral angle of 109.5°. As you can see in **Figure 16.17,** the four sp^3 orbitals of carbon overlap with the 1*s* orbitals of the four hydrogen atoms. The sp^3 orbitals extend farther into space than either *s* or *p* orbitals, allowing a great deal of overlap with the hydrogen 1*s* orbitals. The eight available valence electrons fill the molecular orbitals to form four C—H sigma bonds. The extent of overlap results in unusually strong covalent bonds.

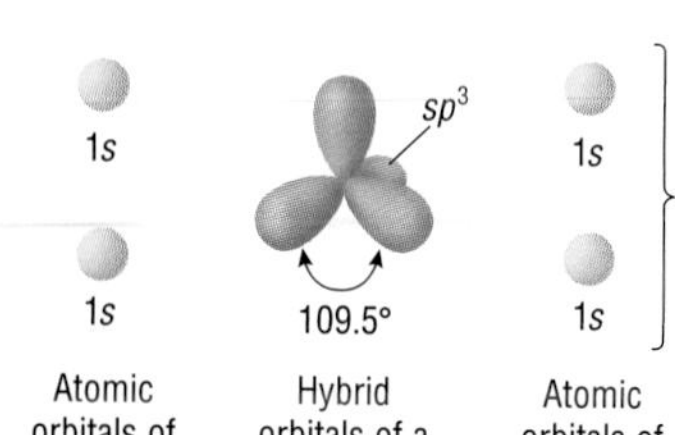

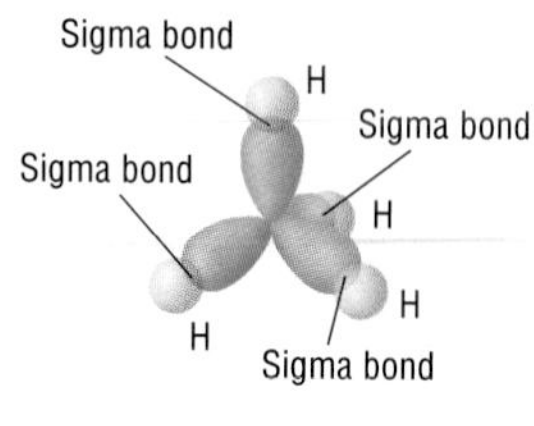

Figure 16.17
In methane, each of the four sp^3 hybrid orbitals of carbon overlaps with a 1s orbital of hydrogen. What type of bond results: sigma or pi? ②

16.2

Discuss

Have students compare the shapes of the carbon hybrid orbitals in methane, ethene, and ethyne molecules. Point out that the atomic orbitals that combine to form the hybrid orbital lose their identities. Emphasize that the shape and orientation of the hybrid orbitals cannot be predicted from the shapes of the combining atomic orbitals. The key concept in hybridization is that atomic orbitals blend together to form entirely new orbitals that have new characteristics. The hybrid orbitals are identical to one another in terms of shape and bonding properties.

Answers

① pyramidal
② sigma

Think Critically

Have students describe and compare the hybrid orbitals used by oxygen in H_2O and in CO_2. (sp^3 and sp^2 respectively)

CHEMISTRY AND AGRICULTURE

Ethene, $H_2C{=}CH_2$,is a hormone found in most plants. Tomatoes release ethene as they ripen. In turn, ethene causes other tomatoes to ripen more quickly. Because it is often convenient to pick and ship fruit before it is ripe, the produce industry uses ethene (also called ethylene) as a ripening agent.

Hybridization is also useful in describing double covalent bonds. Ethene is a relatively simple molecule that has one carbon–carbon double bond and four carbon–hydrogen single bonds.

$$H_2C{=}CH_2$$

Ethene

Experimental evidence indicates that the H—C—H bond angles in ethene are about 120°. In ethene, sp^2 hybrid orbitals form from the combination of one 2s and two 2p atomic orbitals of carbon. As you can see in **Figure 16.18,** each hybrid orbital is separated from the other two by 120°. Two sp^2 hybrid orbitals of each carbon form sigma-bonding molecular orbitals with the four available hydrogen 1s orbitals. The third sp^2 orbitals of each of the two carbons overlap to form a carbon–carbon sigma-bonding orbital. The nonhybridized 2p carbon orbitals overlap side-by-side to form a pi-bonding orbital. A total of twelve electrons fills six bonding molecular orbitals. Thus five sigma bonds and one pi bond hold the ethene molecule together. The sigma bonds and the pi bond are two-electron covalent bonds. Although they are drawn alike in structural formulas, pi bonds are weaker than sigma bonds. In chemical reactions that involve the cleavage of one bond of a carbon–carbon double bond, would the pi bond or the sigma bond be more likely to break? ❶

A third type of covalent bond is a triple bond, such as is found in ethyne (C_2H_2), also called acetylene.

$$H{-}C{\equiv}C{-}H$$

As with other molecules, the hybrid orbital description of ethyne is guided by an understanding of the properties of the molecule. Ethyne is a linear molecule. The best hybrid orbital description is obtained if a 2s atomic orbital of carbon mixes with only one of the three 2p atomic orbitals.

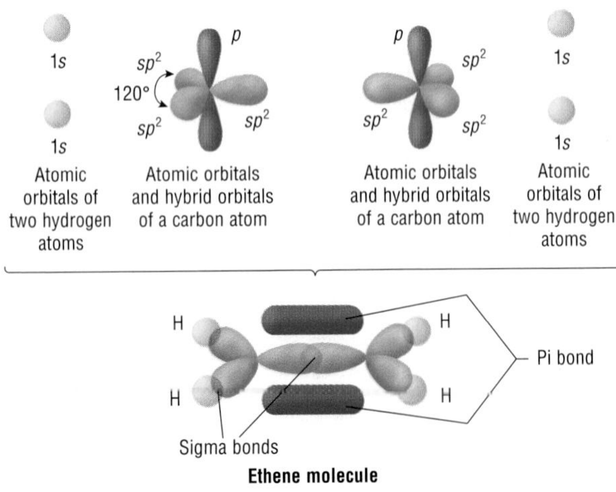

Figure 16.18 *In an ethene molecule, two sp^2 hybrid orbitals from each carbon overlap with a 1s orbital of hydrogen to form a sigma bond. The other sp^2 orbitals overlap to form a carbon–carbon sigma bond. The p atomic orbitals overlap to form a pi bond. What region of space does the pi bond occupy relative to the carbon atoms?* ❷

Answers

❶ pi bond is weaker
❷ the regions above and below the carbons
❸ They form two pi–bonding orbitals.

SECTION REVIEW 16.2

13. When two atoms combine, their atomic orbitals overlap to produce molecular orbitals. In hybridization, several atomic orbitals mix to form the same total number of equivalent hybrid orbitals.

14. a. In a methane molecule, the four valence electron pairs repel each other, forming the corners of a tetrahedron in which the pairs are equidistant from each other. The angle between the bonds is 109.5°.

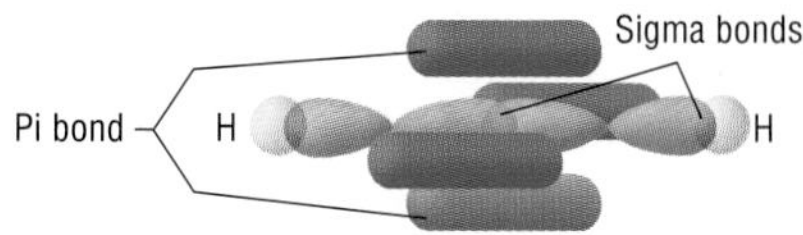

Ethyne molecule

Figure 16.19
In an ethyne molecule, one sp hybrid orbital from each carbon overlaps with a 1s orbital of hydrogen to form a sigma bond. The other sp hybrid orbital of each carbon overlaps to form a carbon–carbon sigma bond. The two p atomic orbitals from each carbon also overlap. What kind of bonds do they form? How many of these bonds do they form? 3

Chem ASAP!

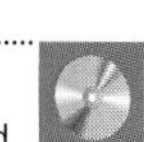

Simulation 17
Compare the *sp*, sp^2, and sp^3 hybrid orbitals.

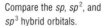

The result is two *sp* hybrid orbitals for each carbon. See **Figure 16.19.** A carbon–carbon sigma-bonding molecular orbital forms from the overlap of one *sp* orbital from each carbon. The other *sp* orbital of each carbon overlaps with the 1*s* orbital of each hydrogen, also forming sigma-bonding molecular orbitals. The remaining pair of *p* atomic orbitals on each carbon overlap side-by-side. They form two pi-bonding molecular orbitals that surround the central carbons. The ten available electrons completely fill five bonding molecular orbitals. The bonding of ethyne consists of three sigma bonds and two pi bonds.

section review 16.2

13. Use the molecular orbital theory to describe covalent bonding. What occurs during hybridization?
14. Explain how the VSEPR theory can be used to predict bond angles in the following covalently bonded molecules.
    a. methane
    b. ammonia
    c. water
15. What shape would you expect a simple carbon-containing compound to have if the carbon atom has the following hybridizations.
    a. sp^2 b. sp^3 c. sp
16. What is a sigma bond? Describe, with the aid of a diagram, how the overlap of two half-filled 1*s* orbitals produces a sigma bond.
17. How many sigma and how many pi bonds are in an ethyne molecule (C_2H_2)?
18. The BF_3 molecule is planar. The attachment of a fluoride ion to the boron in BF_3, through a coordinate covalent bond, creates the BF_4^- ion. What is the geometric shape of this ion?

Chem ASAP! **Assessment 16.2** Check your understanding of the important ideas and concepts in Section 16.2.

16.2

3 Assess

Evaluate Understanding

Have students compare and contrast the characteristics of sigma and pi bonds. (A sigma bond is formed when two atomic orbitals combine to form a hybrid molecular orbital that is symmetrical along the axis connecting two atomic nuclei. In a pi bond, the electrons are most likely forced into sausage shaped regions above and below the bond axis.)

Ask:

- **How do unshared pairs of electrons influence the shape of a molecule?** (The unshared pairs strongly repel the bonding pairs, pushing them closer together and decreasing the bond angle.)

Reteach

VSEPR theory explains the shapes of molecules. Because electron pairs repel, valence-electron pairs arrange themselves as far apart as possible. Sometimes the arrangements can be determined by using the electron–dot structures of the molecules or ions.

- **four bonding pairs + no nonbonding pairs = tetrahedral**
- **three bonding pairs + one nonbonding pair = pyramidal**
- **two bonding pairs + two nonbonding pairs = bent**
- **two bonding pairs + no nonbonding pair = linear**
- **one bonding pair + any number of nonbonding pairs = linear**

SECTION REVIEW 16.2 CONTINUED

b. In an ammonia molecule, the four valence-electron pairs repel each other, but the unshared pair repels the bonded pairs more strongly. The bond angle is 107°.
c. In a water molecule, the four valence-electron pairs repel each other, but the two unshared pairs repel the bonded pairs more strongly. The bond angle is 105°.

15. **a.** trigonal planar
b. tetrahedral
c. linear

16. A sigma bond is formed by the overlap of two *s* orbitals, the overlap of an *s* orbital with a *p* orbital, or the end-to-end overlap of two *p* orbitals. See **Figure 16.11.**

17. 3 sigma bonds and 2 pi bonds

18. tetrahedral

16.3

1 Engage

Use the Visual

Have students study the photograph and read the text that opens the section. Point out that in 1951, the Commission of Snow and Ice of the International Association of Hydrology divided solid precipitation into 10 separate classes comprising seven types of snow crystals (plates, stellars, columns, needles, spatial dendrites, capped columns, and irregular crystals). **Ask:**

- **How do the polar bonds of atoms in water molecules influence the distinctive geometry of snowflakes?** (Polar bonds in molecules influence the *intermolecular* forces between molecules. Water molecules are organized into a network that allows maximum interaction between polar bonds in different molecules. This interaction results in the distinctive crystalline lattices observed in snowflakes.)

Check Prior Knowlege

Ask:

- **What does the term *electronegativity* mean?** (It describes the attraction an atom has for electrons. The relative electronegativities of elements is based on a unitless scale from 0.0 to 4.0.)
- **Which element is the most electronegative? Which is the least?** (Fluorine, the element that has the greatest affinity for electrons, has a value of 4.0; cesium, the element that has the lowest affinity for electrons, has a value of 0.7.)

section 16.3

POLAR BONDS AND MOLECULES

objectives

- Use electronegativity values to classify a bond as nonpolar covalent, polar covalent, or ionic
- Name and describe the weak attractive forces that hold groups of molecules together

key terms

- nonpolar covalent bond
- polar covalent bond
- polar bond
- polar molecule
- dipole
- van der Waals forces
- dispersion forces
- dipole interactions
- hydrogen bonds
- network solids

Snow covers approximately 23 percent of Earth's surface. Each individual snowflake is formed from as many as 100 snow crystals. The size and shape of each crystal depends mainly on the air temperature and amount of water vapor in the air at the time the snow crystal forms. **How do the polar bonds in water molecules influence the distinctive geometry of snowflakes?**

Bond Polarity

Covalent bonds involve electron sharing between atoms. However, covalent bonds differ in terms of how the bonded atoms share the electrons. The character of the bonds in a given molecule depends on the kind and number of atoms joined together. These features, in turn, determine the molecular properties.

The bonding pairs of electrons in covalent bonds are pulled, as in a tug-of-war, between the nuclei of the atoms sharing the electrons. See **Figure 16.20.** When the atoms in the bond pull equally (as occurs when like atoms are bonded), the bonding electrons are shared equally, and the bond is a **nonpolar covalent bond.** Molecules of hydrogen (H_2), oxygen (O_2), and nitrogen (N_2) have nonpolar covalent bonds. Are diatomic halo-
1 gen molecules, such as Cl_2, also nonpolar?

When a covalent bond joins two atoms of different elements and the bonding electrons are shared unequally, the bond is a **polar covalent bond,** or simply a **polar bond.** The more electronegative atom will have the stronger electron attraction and will acquire a slightly negative charge. The less electronegative atom will acquire a slightly positive charge. Refer back to **Table 14.2** on page 405 to see the electronegativities of some common elements. The higher the electronegativity value, the greater the ability of an atom to attract electrons to itself.

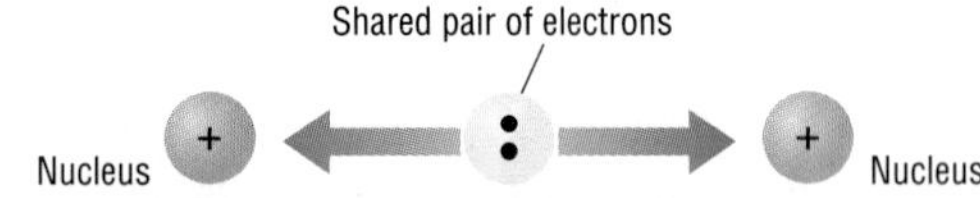

Figure 16.20
The nuclei of atoms pull on the bonding electrons between them, much as a rope is pulled toward opposing sides in a tug-of-war between two groups.

STUDENT RESOURCES

From the Teacher's Resource Package, use:

- Section Review 16.3, Ch. 16 Practice Problems, Vocabulary Review, Quizzes and Tests from the Review Module (Ch. 13–16)
- Laboratory Recordsheet 16-2
- Laboratory Manual: Experiment 25
- Laboratory Practical 16-1
- Small-Scale Lab Practical: Experiment 21

TECHNOLOGY RESOURCES

Relevant technology resources include:

- Chem ASAP! CD-ROM
- Resource Pro CD-ROM
- Small-Scale Chemistry Lab Video or Videodisc: *Paper Chromatography*
- Assessment Resources CD-ROM: Chapter 16 Tests

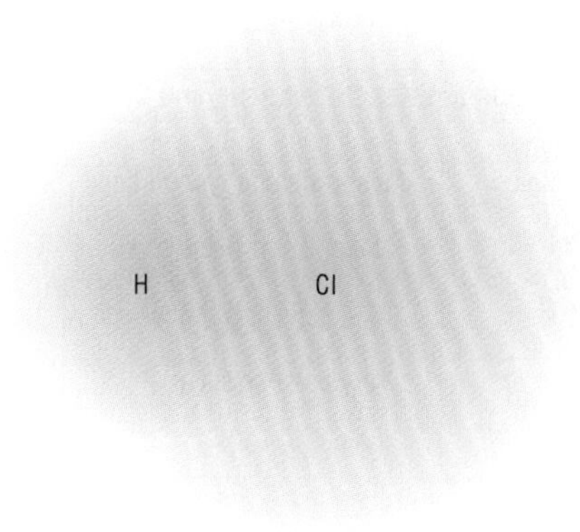

Figure 16.21

This electron-cloud picture of hydrogen chloride shows that the chlorine atom attracts the electron cloud more than the hydrogen atom does. Which atom is more electronegative, a chlorine atom or a hydrogen atom? ❸

Consider the hydrogen chloride molecule (HCl) shown in **Figure 16.21.** Hydrogen has an electronegativity of 2.1 and chlorine has an electronegativity of 3.0. These values are significantly different, so the covalent bond in hydrogen chloride is polar. The chlorine atom acquires a slightly negative charge. The hydrogen atom acquires a slightly positive charge. The lowercase Greek letter delta (δ) shows that atoms involved in the covalent bond acquire only partial charges, much less than 1+ or 1−.

$$\overset{\delta+}{\mathrm{H}}-\overset{\delta-}{\mathrm{Cl}}$$

The minus sign in this notation shows that chlorine has acquired a slightly negative charge. The plus sign shows that hydrogen has acquired a slightly positive charge. These partial charges are shown as clouds of electron density. The polarity of the bond may also be represented with an arrow pointing to the more electronegative atom, as shown here.

$$\overset{+\!\!\longrightarrow}{\mathrm{H}-\mathrm{Cl}}$$

The O—H bonds in the water molecule are also polar. The highly electronegative oxygen partially pulls the bonding electrons away from hydrogen. The oxygen acquires a slightly negative charge. What type of charge do the hydrogens acquire? ❷

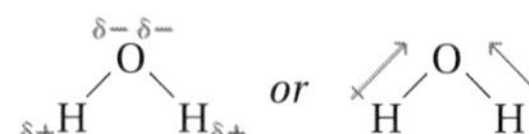

As shown in **Table 16.4** on the following page, the size of the electronegativity difference between two atoms is one indicator of the type of bond that the two atoms may form. For each range of electronegativity difference, the table shows the most probable type of bond. When you use the ranges in the table, it is important to remember that the boundary between ionic and covalent bonds is not distinct. As the electronegativity difference between two atoms increases, the polarity of the bond increases. If the electronegativity difference is greater than 2.0, there is a high probability that electrons will be pulled away completely by one of the atoms. In that case, an ionic bond will form.

Answers

❶ yes

❷ a slightly positive charge

❸ chlorine

16.3

2 Teach

Discuss

When students apply Table 16.4 to Sample Problem 16-4, they may be surprised to discover that bonds previously labeled ionic are suddenly categorized as polar covalent. The reality is that most bonds are neither totally ionic nor totally covalent. Most bonds fall somewhere on the polar–covalent continuum with electrons displaced more toward one atom than the other. Make sure that students understand the two uses of the Greek letter delta. A lowercase delta (δ) is used to denote partial charges; an uppercase delta (Δ) is used to denote a change in a variable.

TEACHER DEMO

If possible, show students a magnet toy with which you can arrange pieces of iron into shapes. Discuss polarity and magnetism and relate the observations made to the properties of polar molecules.

Discuss

Draw diagrams of molecules on the board or overhead projector, using electronegativity values to determine which element has the partial positive charge and which has the partial negative charge. Explain how the attraction of electrons can be described with a vector. The direction of the vector is determined by the shape of the molecule. If the sum of the vectors acting on the central atom in one direction is cancelled by the sum of the vectors acting in the other direction, the molecule has no dipole. If they don't cancel, then the molecule is polar.

TEACHER DEMO

Polar molecules show electrostatic attractions to charged molecules. Attach two burets to a ring stand. Fill one with water and the other with turpentine. Place a beaker beneath each. **CAUTION!** Turpentine is an irritant. Wear safety goggles. Rub a rubber rod with a fur cloth to charge it negatively. Bring the rod near the open end of the water-containing buret as you slowly release the water from the buret. Have students carefully observe the behavior of the stream. Repeat this procedure using turpentine. Then repeat using a glass rod for both buret systems. Point out that the polar water molecules are attracted to both negatively and positively charged materials, but the nonpolar turpentine molecules are not attracted to either. All materials can be saved and reused.

Practice Problems Plus

Related Chapter Review Problems
Chapter Review problems 51–53 are related to Sample Problem 16-4.

Table 16.4

Electronegativity Differences and Bond Types		
Electronegativity difference range	**Most probable type of bond**	**Example**
0.0–0.4	Nonpolar covalent	H—H (0.0)
0.4–1.0	Moderately polar covalent	$\overset{\delta+}{H}—\overset{\delta-}{Cl}$ (0.9)
1.0–2.0	Very polar covalent	$\overset{\delta+}{H}—\overset{\delta-}{F}$ (1.9)
≥ 2.0	Ionic	Na^+Cl^- (2.1)

Sample Problem 16-4

Which type of bond (nonpolar covalent, moderately polar covalent, very polar covalent, or ionic) will form between each of the following pairs of atoms?

a. N and H **b.** F and F **c.** Ca and O **d.** Al and Cl

1. *ANALYZE Plan a problem-solving strategy.*
In each case, the pairs of atoms involved in the bonding pair are given. The types of bonds formed depend on the electronegativity differences between the bonding elements. Use **Table 14.2** on page 405 to find the electronegativity difference, then use **Table 16.4** to determine the bond type.

2. *SOLVE Apply the problem-solving strategy.*
From **Tables 14.2** and **16.4,** the electronegativities, their differences, and the corresponding bond types are
a. N (3.0), H (2.1); 0.9; moderately polar covalent
b. F (4.0), F (4.0); 0.0; nonpolar covalent
c. Ca (1.0), O (3.5); 2.5; ionic
d. Al (1.5), Cl (3.0); 1.5; very polar covalent

3. *EVALUATE Do the results make sense?*
The answers are consistent with the magnitudes of the electronegativity differences.

Practice Problems

19. Identify the bonds between atoms of each pair of elements as nonpolar covalent, moderately polar covalent, very polar covalent, or ionic.
a. H and Br **d.** Cl and F
b. K and Cl **e.** Li and O
c. C and O **f.** Br and Br

20. Order the following covalent bonds from least to most polar.
a. H—Cl **d.** H—C
b. H—Br **e.** F—F
c. H—S

Polar Molecules

The presence of a polar bond in a molecule often makes the entire molecule polar. In a **polar molecule,** one end of the molecule is slightly negative and the other end is slightly positive. In the hydrogen chloride molecule, for example, the partial charges on the hydrogen and chlorine atoms are electrically charged regions or poles. A molecule that has two poles is called a dipolar molecule, or **dipole.** The hydrogen chloride molecule is a dipole. As shown in **Figure 16.22,** when such molecules are placed in an electric field, they become oriented with respect to the positive and negative plates.

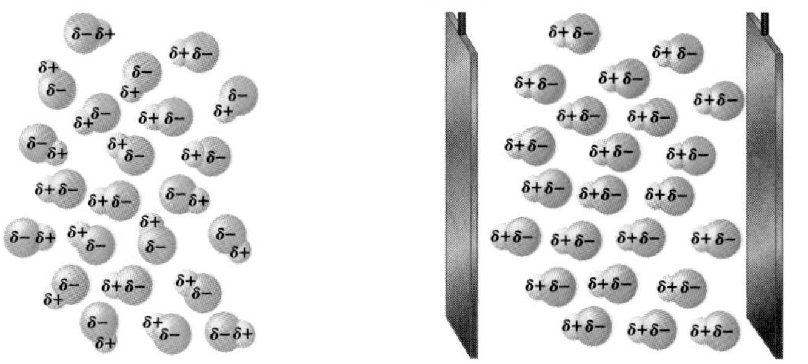

Negative plate

Positive plate

Electric field absent.
Polar molecules oriented randomly.

Electric field on.
Polar molecules line up.

Figure 16.22
When polar molecules, such as HCl, are placed in an electric field, the slightly negative ends of the molecules become oriented toward the positively charged plate and the slightly positive ends of the molecules become oriented toward the negatively charged plate. What would happen if carbon dioxide molecules were placed in such a field? Why? 2

The effect of polar bonds on the polarity of an entire molecule depends on the shape of the molecule and the orientation of the polar bonds. A carbon dioxide molecule, for example, has two polar bonds and is linear.

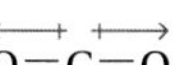

Note that the carbon and oxygens lie along the same axis. Therefore, the bond polarities cancel because they are in opposite directions. Carbon dioxide is thus a nonpolar molecule, despite the presence of two polar bonds.

The water molecule also has two polar bonds. However, the water molecule is bent rather than linear. Therefore, the bond polarities do not cancel. Is a water molecule polar or nonpolar? 1

Chem ASAP!

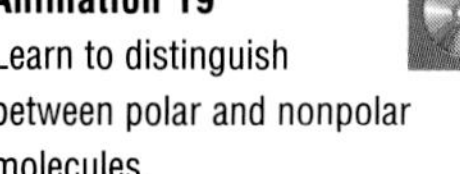

Animation 19
Learn to distinguish between polar and nonpolar molecules.

Attractions Between Molecules

Molecules are often attracted to each other by a variety of forces. These intermolecular attractions are weaker than either an ionic or a covalent bond. Nevertheless, you should not underestimate the importance of these forces. Among other things, these attractions are responsible for determining whether a molecular compound is a gas, a liquid, or a solid at a given temperature.

There are several kinds of molecular attractions. The weakest attractions are collectively called **van der Waals forces,** named after the Dutch chemist Johannes van der Waals (1837–1923). Van der Waals forces consist of two possible types: dispersion forces and dipole interactions. **Dispersion forces,** the weakest of all molecular interactions, are caused by the motion of electrons. The strength of dispersion forces generally increases as the number of electrons in a molecule increases. The halogen diatomic molecules are examples of molecules where the major attraction for one another is caused by dispersion forces. Fluorine and chlorine have relatively few electrons and are gases at STP because of their especially weak dispersion forces. The larger number of electrons in bromine generates larger dispersion forces. Bromine molecules are therefore sufficiently attracted to each other to make bromine a liquid at STP. Iodine, with a still larger number of electrons, is a solid at STP.

16.3

Discuss

The sharing of electrons ties atoms together in the form of molecules and isolates the electrons from further interactions. If there were no other forces present, then all substances made of covalently bonded molecules would be gases at any temperature. However, forces exist between covalently bonded molecules that cause them to form solids and liquids. These bonding forces include Van der Waals forces, dipole interactions, and hydrogen bonding.

CHEMISTRY AND SCIENCE HISTORY

Johannes van der Waals investigated the behavior of real gases. He surmised that their behavior differed from that of ideal gases because of weak attractive forces between their molecules. For his work in developing an understanding of intermolecular attractions, he was awarded the Nobel Prize in Physics in 1910.

Answers

1 polar

2 Carbon dioxide molecules would not change orientation because the CO_2 molecule is nonpolar.

Discuss

Construct a concept map using the following terms: intermolecular attractions, van der Waals forces, dispersion forces, dipole forces, and hydrogen bonding. Use the concept map to introduce and discuss each force in detail. Use the halogen family to demonstrate dispersion forces, carbon dioxide to discuss dipole interactions, and water to discuss hydrogen bonding. Ask:

- **Why does He have a lower boiling point than Rn?** (The strength of the intermolecular forces between helium atoms is less than those between radon atoms.)

Point out that hydrogen bonds only occur between molecules in which hydrogen is bonded to a strongly electronegative atom such as nitrogen, oxygen, or fluorine. Since living systems contain molecules rich in oxygen and nitrogen, hydrogen bonding plays an important role in the chemistry of living systems. Hydrogen bonding explains many of the physical properties of water. Tell students that they will study the properties of water in Chapter 17.

COMPUTER SCIENCE

Molecular Modeling

The widespread availability of fast computers has greatly benefited chemists interested in the shapes of molecules and in the interactions between molecules.

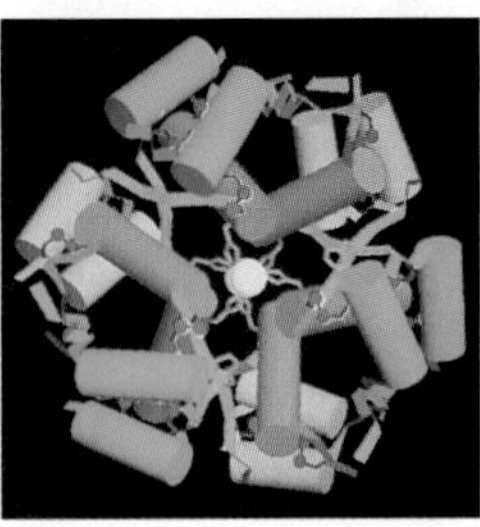

Molecular modeling computer programs, for example, are becoming an important tool in the pharmaceutical industry for designing drugs. These programs allow the creation of models of molecules composed of thousands of atoms. Using a special viewing apparatus, chemists can see three-dimensional representations of complex molecules on the computer monitor. On the monitor, it is possible to turn the molecule over to see the other side, to shrink it, or to cut it in pieces. Through a computer technology called virtual reality, it is likely that a chemist will soon be able to get virtually inside a molecular structure to examine it from every possible angle.

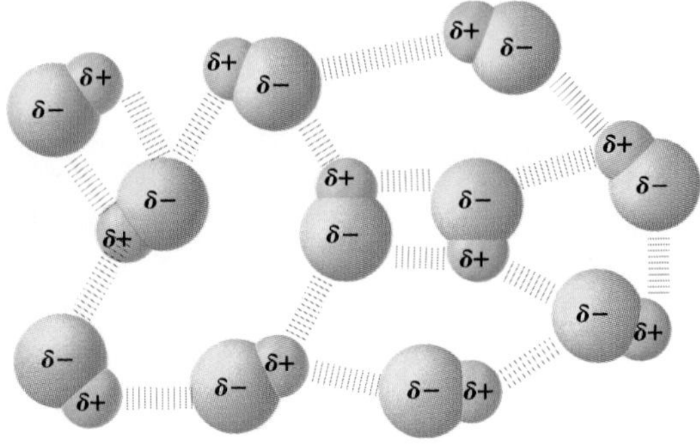

Figure 16.23
Polar molecules are attracted to one another by dipole interactions, a type of van der Waals force.

The second type of van der Waals force, **dipole interactions,** occurs when polar molecules are attracted to one another. The electrostatic attractions involved occur between the oppositely charged regions of dipolar molecules, as shown in **Figure 16.23.** The slightly negative region of a polar molecule is weakly attracted to the slightly positive region of another polar molecule. Dipole interactions are similar to but much weaker than ionic bonds.

The dipole interactions in water, for example, result in a weak attraction of water molecules for one another. Each O—H bond in the water molecule is highly polar, and the oxygen acquires a slightly negative charge because of its greater electronegativity. The hydrogens in water acquire a slightly positive charge. The positive region of one water molecule attracts the negative region of another water molecule. The dipolar attraction between the hydrogen of one water molecule and the oxygen of another water molecule is strong relative to other dipolar attractions. This relatively strong interaction, which is also found in hydrogen-containing molecules other than water, is given the special name hydrogen bond.

Hydrogen bonds are attractive forces in which a hydrogen covalently bonded to a very electronegative atom is also weakly bonded to an unshared electron pair of another electronegative atom. This other atom may be in the same molecule or in a nearby molecule. Hydrogen bonding always involves hydrogen—it is the only chemically reactive element with valence electrons that are not shielded from the nucleus by a layer of underlying electrons.

A very polar covalent bond forms when hydrogen bonds to an electronegative atom such as oxygen, nitrogen, or fluorine. This leaves the hydrogen nucleus very electron-deficient. The sharing of a nonbonding electron pair on a nearby electronegative atom compensates for the deficiency. The resulting hydrogen bond has about 5% of the strength of an average covalent bond. Hydrogen bonds are the strongest of the intermolecular forces. They are extremely important in determining the properties of water and biological molecules such as proteins. **Figure 16.24** shows hydrogen bonding between water molecules.

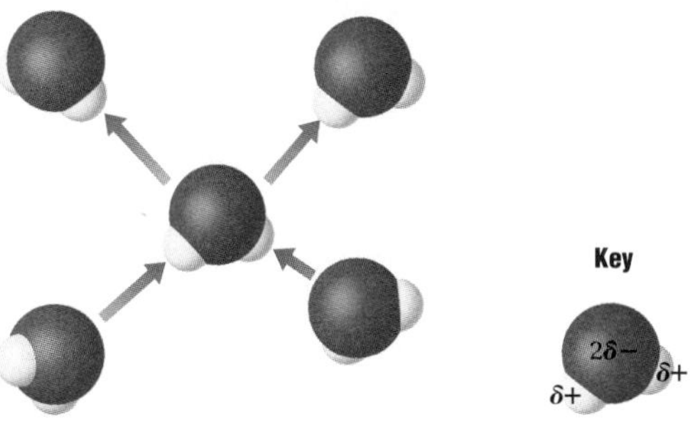

Figure 16.24
The strong hydrogen bonding between water molecules accounts for many properties of water, such as the fact that water is a liquid rather than a gas at ordinary temperatures. Although a hydrogen bond between water molecules has only about 5% of the strength of an average covalent bond, it is much stronger than the dipole interactions characteristic of other polar molecules.

Intermolecular Attractions and Molecular Properties

At room temperature, some compounds are gases, some are liquids, and some are solids. The physical properties of a compound depend on the type of bonding it displays—in particular, on whether it is ionic or covalent. **Table 16.5** summarizes some of the characteristic differences between ionic and covalent (molecular) substances. A great range of physical properties occurs among covalent compounds. This is mainly because of widely varying intermolecular attractions.

The melting and boiling points of most molecular compounds are low compared with those of ionic compounds. In most molecular solids, only the weak attractions between molecules need to be broken. However, a few

Table 16.5

Characteristics of Ionic and Covalent Compounds		
Characteristic	**Ionic compound**	**Covalent compound**
Representative unit	Formula unit	Molecule
Bond formation	Transfer of one or more electrons between atoms	Sharing of electron pairs between atoms
Type of elements	Metallic and nonmetallic	Nonmetallic
Physical state	Solid	Solid, liquid, or gas
Melting point	High (usually above 300 °C)	Low (usually below 300 °C)
Solubility in water	Usually high	High to low
Electrical conductivity of aqueous solution	Good conductor	Poor to nonconducting

Figure 16.25
Liquids tend to form drops because of intermolecular attractions. Molecular substances with weak intermolecular attractions are gases at room temperature. Molecular substances with very strong intermolecular attractions are generally found as solids.

TEACHER DEMO

Demonstrate how hydrogen bonding affects the surface tension and viscosity of fluids. Prepare five 300-mL stoppered bottles each containing 100 mL of a different liquid: glycerin, ethylene, glycol, water, and ethanol. Gently rotate one of the bottles so that its contents begin to swirl. Have students note the time interval between your ceasing to rotate the bottle and the disappearance of the swirling vortex that is formed. Repeat this procedure for each fluid. Be sure to rotate each bottle the same number of times. Point out that hydrogen bonding increases surface tension and viscosity. Thus, liquids with a relatively high degree of hydrogen bonding will stop swirling earlier. Have students rank the fluids according to increasing hydrogen bonding. Show the structural formulas of the compounds to demonstrate the origins of the hydrogen bonding.

MEETING DIVERSE NEEDS

Gifted Have students look up the terms hail, ice pellets, and graupel and report to the class how these three forms of solid precipitation differ. Other students may draw examples of snowflakes or investigate differences between regular ice and glacial ice.

3 Assess

Evaluate Understanding

Have students account for the generally low melting points of covalent compounds in terms of their structure. (Intermolecular attractions between covalent compounds are generally not as strong as ionic bonds.)

Reteach

Help students list the types of intermolecular attractions that operate between molecules. Give examples of at least five compounds that would fit into each category. Ask students to name other examples.

Portfolio Project

Most of the available programs for constructing molecules focus on organic molecules. One source for RasMol is www.rasmol.com

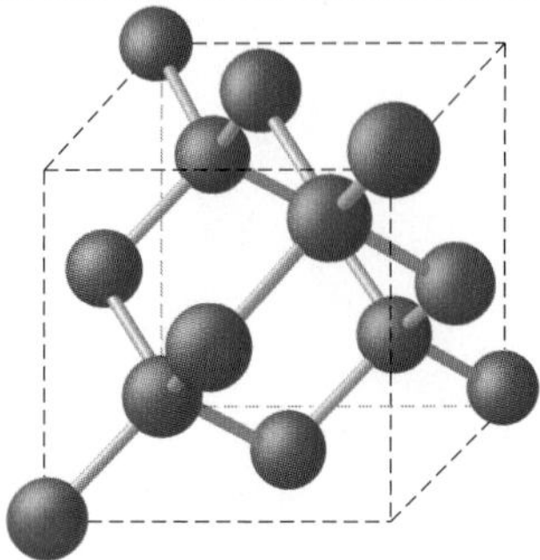

Figure 16.26

Diamond is a network-solid form of carbon. Diamond has a three-dimensional structure, with each carbon at the center of a tetrahedron.

Chem ASAP!

Simulation 18

Relate melting and boiling points to the strength of intermolecular forces.

molecular solids do not melt until the temperature reaches 1000 °C or higher, or they decompose without melting at all. Most of these very stable substances are **network solids** (or network crystals), solids in which all of the atoms are covalently bonded to each other. Diamond is an example of a network solid. As shown in **Figure 16.26,** each carbon atom in a diamond is covalently bonded to four other carbons. Cutting a diamond requires breaking a multitude of these bonds. Diamond does not melt; rather, it vaporizes to a gas at 3500 °C and above. Silicon carbide, with the formula SiC and a melting point of about 2700 °C, is also a network solid. Silicon carbide is so hard that it is used in grindstones and as an abrasive. How do you think the molecular structures of silicon carbide and diamond compare? ❶ You can think of samples of diamond, silicon carbide, and other network solids as single molecules. Their melting or decomposition requires that actual covalent bonds be broken.

portfolio project

RasMol is a free program for viewing the structures of molecules. Search for the RasMol site on the Internet. Download a copy of the RasMol package, which contains the program, documentation, and models of protein molecules. Prepare a demonstration of the software for your class.

section review 16.3

21. Explain how you can use electronegativity values to classify a bond as nonpolar covalent, polar covalent, or ionic.
22. Describe the three kinds of weak attractive forces that hold groups of molecules together. Rank these forces from weakest to strongest.
23. Not every molecule with polar bonds is polar. Explain this statement. Use CCl_4 as an example.
24. Draw the electron dot structure for each molecule. Identify polar covalent bonds by assigning slightly positive ($\delta+$) and slightly negative ($\delta-$) symbols to the appropriate atoms.
    a. HOOH b. BrCl c. HBr d. H_2O
25. How does a network solid differ from most other covalent compounds?
26. Which of the following are characteristic of most covalent compounds?
    a. high melting point
    b. shared bonding electrons
    c. low water solubility
    d. existence as molecules
    e. composed of a metal and a nonmetal

Chem ASAP! **Assessment 16.3** Check your understanding of the important ideas and concepts in Section 16.3.

Answer

❶ They are most likely quite similar based on their properties.

SECTION REVIEW 16.3

21. Find the difference in electronegativity values for the two atoms. Then use Table 16.4 to determine the most likely type.
22. Dispersion forces (the weakest of the three) are caused by the motion of electrons. Dipole interactions are the attractions between the oppositely charged ends of polar molecules. Hydrogen bonding (the strongest of the three) occurs when a hydrogen atom bonded to a more electronegative atom is attracted to another highly electronegative atom.
23. The atoms in CCl_4 are oriented so that the bond polarities cancel.
24. a. $\overset{\delta+}{H}:\overset{\delta-}{\underset{\cdot\cdot}{\overset{\cdot\cdot}{O}}}:\overset{\delta-}{\underset{\cdot\cdot}{\overset{\cdot\cdot}{O}}}:\overset{\delta+}{H}$ (δ+ δ− δ− δ+) c. $\overset{\delta+}{H}:\overset{\delta-}{Br}:$
    b. $:\overset{\delta+}{Br}:\overset{\delta-}{Cl}:$ d. $\overset{\delta+}{H}:\overset{\delta-}{O}:\overset{\delta+}{H}$
25. The atoms in a network solid are covalently bonded in a large array (or crystal), which can be thought of as a single molecule.
26. b., c., d.

SMALL-SCALE LAB

PAPER CHROMATOGRAPHY OF FOOD DYES

SAFETY

Use safe and proper laboratory procedures.

PURPOSE

To use paper chromatography to separate and identify food dyes in various samples.

MATERIALS

- pencil
- paper
- ruler
- scissors
- toothpicks
- 4 different colors of food coloring
- plastic cup
- 0.1% NaCl solution
- chromatography paper

PROCEDURE

Cut a 5 cm × 10 cm strip of chromatography paper and label it with a pencil as shown in Figure A. Use a different toothpick to place a spot of each of the four food colors on the Xs on your chromatography paper. Allow the spots to dry for a few minutes. Fill the plastic cup so its bottom is just covered with the solvent (0.1% NaCl solution). Wrap the chromatography paper around a pencil. Remove the pencil and place the chromatography paper, color-spot side down, in the solvent. When the solvent reaches the top of the chromatography paper, remove the paper and allow it to dry.

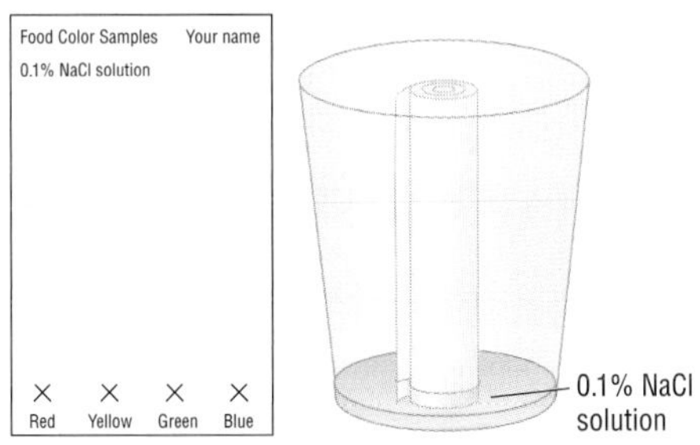

Figure A

ANALYSIS

Using your experimental data, record the answers to the following questions below your data table.

1. If a food color sample yields a single streak or spot, it is usually a pure compound. Which food colors consist of pure compounds?
2. Which food colors are mixtures of compounds?
3. Food colors often consist of a mixture of three colored dyes: Red No. 40, Yellow No. 5, and Blue No. 1. Read the label on the food color package. Which dyes do your food color samples contain?
4. Identify each spot or streak on your chromatogram as Red No. 40, Yellow No. 5, or Blue No. 1.
5. Paper chromatography separates polar covalent compounds on the basis of their relative polarities. The most polar dyes migrate the fastest and appear at the top of the paper. Which dye is the most polar? The least polar?

YOU'RE THE CHEMIST

The following small-scale activities allow you to develop your own procedures and analyze the results.

1. **Design It!** Design and carry out an experiment to identify the dyes in various colored candies.
2. **Design It!** Design and carry out an experiment to identify the dyes in various colored markers using the paper chromatography method.
3. **Design It!** Design and carry out an experiment to identify the dyes in various colored powdered drinks using the paper chromatography method.
4. **Analyze It!** Use different solvents, such as 2-propanol (rubbing alcohol), vinegar, and ammonia, to separate food colors. Does the solvent affect the results?
5. **Analyze It!** Explore the effect of different papers on your results. Try paper towels, notebook paper, and coffee filters. Report your results. Examine the relative positions of Blue No. 1 and Yellow No. 5. What do you observe?

SMALL-SCALE LAB

PAPER CHROMATOGRAPHY OF FOOD DYES

OVERVIEW

Procedure Time: 30 minutes
Students use paper chromatography to separate and identify common covalently bonded dyes.

MATERIALS

Solution	Preparation
0.1% NaCl	1.0 g in 1.0 L

TEACHING SUGGESTIONS

Jelly beans, small candy coated chocolates, and unsweetened powdered soft drinks are excellent sources of food dyes. Dissolve a package of unsweetened powdered soft drink in a small amount of water and have students use toothpicks to spot this solution on chromatography paper. You can substitute coffee filters if students are careful not to overload their samples. A similar, more complete lab on chromatography is found in the *Small-Scale Laboratory Manual.*

YOU'RE THE CHEMIST

1. Wet the candy. Blot to remove excess water. Press the wet candy onto chromatography paper to produce a spot. Develop in 0.1% NaCl.
2. Make a small spot with each pen on chromatography paper. Develop in solvent.
3. Use a toothpick to spot a solution of powdered drink on chromatography paper.
4. Rubbing alcohol runs more slowly and gives slightly better separation than 0.1% NaCl.
5. With some paper, the positions of the blue and yellow dyes are reversed because the water content of the paper can affect the variations in polarity.

ANALYSIS

1. Red, yellow, and blue are pure.
2. Green is Yellow No. 5 plus Blue No. 1.
3. Blue No. 1; Red No. 40 (No. 3 if supplies haven't been used up); Yellow No. 5 or Yellow No. 6 if the dye appears orange.
4. See the drawing.
5. Blue is the most polar; red, the least polar.

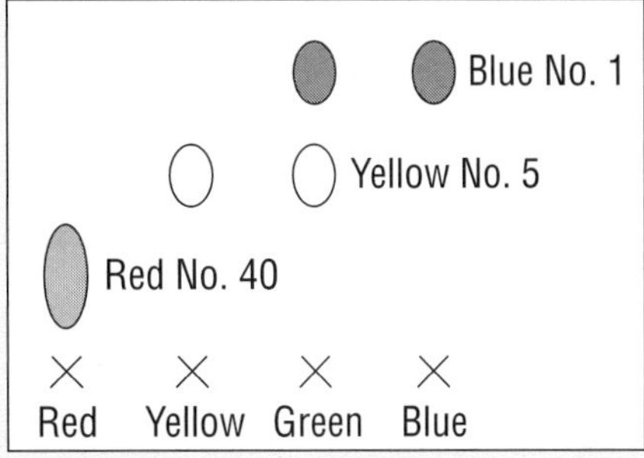

Chemistry Serving... Society

DISCUSS

Point out that the sun produces UV rays that are classified as UV-C (wavelengths of 200–290 nm), UV-B (290–320 nm), and UV-A (320–400 nm). To be effective, a sunscreen must block all types of UV radiation. Ask:

- How do sunscreens work? (Sunscreens can work physically or chemically. A physical sunscreen scatters and reflects UV radiation. Ask students which compounds they might have heard of that would work this way? (titanium dioxide and zinc oxide)

A chemical sunscreen works by *absorbing* UV radiation rather than by reflecting or scattering it. Ask:

- Which chemicals might be responsible for this? (Bring in a couple of tubes of sunscreens and help students identify those chemicals that absorb ultraviolet radiation. Point out that these chemicals all contain multiple bonds (π bonds) that absorb ultraviolet light before it can reach the skin.

Ask students what other precautions an individual might take to guard against UV radiation. (If possible, avoid exposure between 10 AM and 2 PM, when the amount of radiation is the greatest. Remind students that clouds can filter infrared radiation but do not block UV radiation. Other issues are the reflection of radiation off surfaces such as snow, water, and sand; and the increase of exposure with elevation and as one nears the equator.)

CHEMISTRY IN CAREERS

Have students connect to the Internet address shown to find out more about careers in oncology. Students may also wish to read about oncology careers on page 875.

Chemistry Serving...Society

BLOCKS AND BONDS

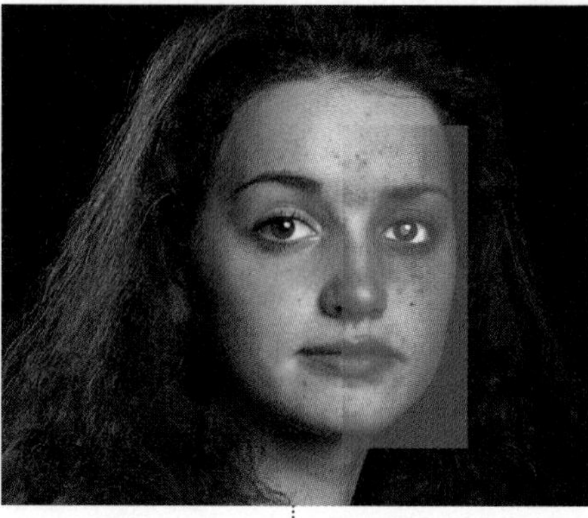

Is sunlight good for you or not? At one time, people thought you should get as much sun as possible. A dark tan was considered a sign of health. Today, however, it is well-known that sunlight can be very harmful, especially for people with light skin. The reason is that the ultraviolet (UV) radiation in sunlight can break covalent bonds in the molecules of your skin cells.

Sunlight, of course, is not all bad. Your skin cells require some sunlight to make vitamin D, a substance necessary for the formation of healthy bones and teeth. But unless they are protected in some way, skin cells are damaged by excessive exposure to UV radiation. In the short term, the damage shows itself as sunburn. Over the long term, skin that has received too much UV radiation becomes wrinkled and may develop cancer.

Over the long term, skin that has received too much UV radiation becomes wrinkled and may develop cancer.

UV radiation is capable of penetrating your skin to a depth of about 1 mm. When UV radiation penetrates your skin, it can damage the molecules of DNA in your skin cells by breaking the covalent bonds that hold the molecules together. Even slight damage to DNA can be significant, as DNA controls the functions of a cell. When a skin cell's DNA is damaged, it may no longer be able to function normally. One possible result is that it may begin to divide uncontrollably, making more and more cells with abnormal DNA. This uncontrolled cell division is the beginning stage of cancer.

In people with light skin, a suntan is visible evidence that the body is trying to protect itself from harm. The skin responds to UV exposure by making more of the dark pigment melanin. Melanin absorbs some of the damaging radiation and thus protects the molecular structure of the skin. Even with this protective mechanism, however, damage to the skin can occur. People with dark skin can withstand more UV exposure without cell damage because they have larger amounts of melanin, but even they are not immune to the effects of UV exposure.

The ozone layer in the upper atmosphere also helps to block the sun's harmful UV rays. Unfortunately, however, the ozone layer is being diminished by chemical reactions between ozone and pollutants such as chlorofluorocarbons (CFCs). As a result, the worldwide incidence of skin cancer is gradually increasing. Today, UV radiation from the sun must be considered an environmental hazard.

The best way to protect yourself from the effects of UV radiation is to remain out of the sun as much as possible. The next-best way is to use a sunscreen or sunblock on your skin. The covalent bonds of certain compounds in the sunscreen absorb many of the harmful wavelengths of UV radiation, thus protecting the covalent bonds in your skin.

All sunscreens are given a sun protection factor (SPF) value as a measure of their effectiveness. The SPF value indicates the amount of increase in UV protection you receive by using the product. If you use a sunscreen with an SPF of 6, for example, and stay in the sun for six hours, you will receive an amount of UV equivalent to one hour of exposure with no sunscreen. Most doctors recommend that light-skinned people use sunscreens with an SPF of 15 or higher.

CHEMISTRY IN CAREERS

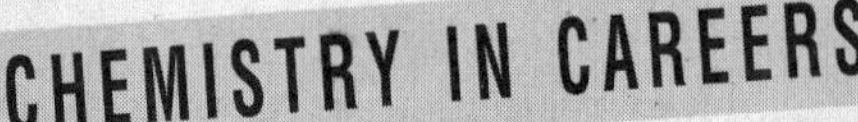

ONCOLOGIST

Use the latest medical treatments to save lives.

See page 875.

TAKE IT TO THE NET

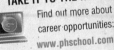

Find out more about career opportunities: **www.phschool.com**

CHEMIST NEEDED

Environment GC. Must know EPA methods, troubleshooting;

CHEMICAL SPECIALIST

Local food service distributor see responsible self-motivated indivi

Chapter 16 STUDENT STUDY GUIDE

Take It to the NET
For interactive study and review, go to www.phschool.com

KEY TERMS

- antibonding orbital *p. 452*
- bond dissociation energy *p. 447*
- bonding orbital *p. 452*
- coordinate covalent bond *p. 444*
- diamagnetic *p. 450*
- dipole *p. 462*
- dipole interaction *p. 464*
- dispersion force *p. 463*
- double covalent bond *p. 442*
- hybridization *p. 457*
- hydrogen bond *p. 464*
- molecular orbital *p. 452*
- network solid *p. 466*
- nonpolar covalent bond *p. 460*
- paramagnetic *p. 450*
- pi bond *p. 454*
- polar bond *p. 460*
- polar covalent bond *p. 460*
- polar molecule *p. 462*
- resonance structure *p. 449*
- sigma bond *p. 453*
- single covalent bond *p. 437*
- structural formula *p. 437*
- tetrahedral angle *p. 455*
- triple covalent bond *p. 442*
- unshared pair *p. 438*
- van der Waals force *p. 463*
- VSEPR theory *p. 456*

CONCEPT SUMMARY

16.1 The Nature of Covalent Bonding

- Atoms form covalent bonds when they share electrons to form an octet.
- A shared pair of valence electrons constitutes a single covalent bond. Sometimes two or three pairs of electrons may be shared to give double or triple covalent bonds.
- Sometimes one atom may contribute both bonding electrons in a covalent bond. This type of bond is called a coordinate covalent bond.
- Resonance structures help to visualize the bonding in molecules when more than one electron dot formula can be written.

16.2 Bonding Theories

- Molecular orbital theory is a logical extension of the quantum mechanical description of the electron structure of atoms. Covalent bonding is described in terms of sigma and pi bonds.
- The valence-shell electron-pair repulsion, or VSEPR, theory of molecular geometry states that, as a general rule, molecules adjust their three-dimensional shapes so the valence-electron pairs are as far apart as possible.
- In some instances, molecular geometry is adequately described by simple overlap of atomic orbitals. In others, a description of molecular shape that better fits experimental results is obtained from hybridized atomic orbitals.

16.3 Polar Bonds and Molecules

- When covalent bonds join like atoms, the bonding electrons are shared equally and the bond is nonpolar. When the atoms in a bond have different electronegativities, the bonding electrons are shared unequally and the bond is polar.
- Hydrogen bonds are attractive forces in which a hydrogen covalently bonded to a very electronegative atom is also weakly bonded to an unshared electron pair of another electronegative atom. Hydrogen bonds are strong relative to other dipole interactions.
- Weak intermolecular forces determine whether a covalent compound will be a solid, liquid, or gas. Weak attractions between molecules are called van der Waals forces and include dispersion forces, dipole interactions, and hydrogen bonds.

CHAPTER CONCEPT MAP

Use these terms to construct a concept map that organizes the major ideas of this chapter.

Chem ASAP! **Concept Map 16**
Create your Concept Map using the computer.

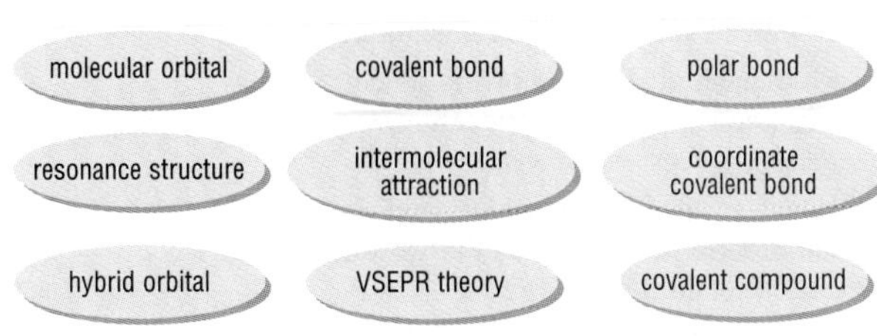

Take It to the Net

At **www.phschool.com** students will find for this chapter

- an Internet activity
- links to related chemistry sites
- an interactive quiz
- career links

Chapter 16
STUDENT STUDY GUIDE

4 Close

Summary

Write *molecular orbitals, VSEPR, hybrid orbitals, polar bonds, coordinate covalent bonds, and resonance structures* in six columns across the chalkboard. For each category, have students give examples of compounds with molecular structures explained by the concept.

Extension

Have students research how hydrogen bonding is involved in the structure of proteins and DNA.

Looking Back . . . Looking Ahead . . .

The previous chapter discussed ionic bonding and the properties of ionic substances. The current chapter dealt with covalent compounds and bonding. The next chapter focuses on how the chemical and physical properties of water are related to its molecular structure.

Answers

27. Neon has an octet of valence electrons. A chlorine atom achieves an octet by sharing an electron with another chlorine atom.

28. a. ionic c. covalent
b. ionic d. covalent

29. Ionic bonds depend on electrostatic attraction between ions. Covalent bonds depend on electrostatic attraction between shared electrons and nuclei of combining atoms.

30. A double covalent bond has four shared electrons; a triple covalent bond has six shared electrons.

31. A single atom of an element is usually the central atom.

32. a. :Ï:Ï: c. H:S̈:H
b. :F̈:Ö:F̈: d. :Ï:N̈:Ï: with :Ï: below N

33. a. :F̈:F̈: c. H:C:::C:H
b. H:C̈l: d. H:C:::N:

34. One atom contributes both electrons to a coordinate covalent bond as in CO.

35. An unshared pair of electrons is needed for a coordinate covalent bond. There are no unshared pairs in C—H or C—C bonds.

36. Molecules of each compound can be described by more than one electron dot structure.

37. $[:\ddot{O}::C:\ddot{O}:]^{2-}$ (with :Ö: below C) ⟷ $[:\ddot{O}:C::\ddot{O}:]^{2-}$ (with :Ö: below C) ⟷ $[:\ddot{O}:C:\ddot{O}:]^{2-}$ (with :O: double-bonded below C)

38. $[:\ddot{O}:\ddot{N}::\ddot{O}:]^{-} \leftrightarrow [:\ddot{O}::\ddot{N}:\ddot{O}:]^{-}$

39. The measured mass of a paramagnetic substance appears greater when measured in the presence of a magnetic field than when measured in the absence of a magnetic field.

40. a. diamagnetic
b. paramagnetic
c. paramagnetic
d. diamagnetic

41. (b) and (c); assuming only single bonds, the P and S atoms each have 10 valence electrons.

Chapter 16 REVIEW

CONCEPT PRACTICE

27. Explain why neon is monatomic but chlorine is diatomic. *16.1*

28. Classify the following compounds as ionic or covalent. *16.1*
a. $MgCl_2$ b. Na_2S c. H_2O d. H_2S

29. Describe the difference between an ionic and a covalent bond. *16.1*

30. How many electrons do two atoms in a double covalent bond share? How many in a triple covalent bond? *16.1*

31. Based on the examples given in Section 16.1, state a general rule for determining which atom is the central one in a binary molecular compound. *16.1*

32. Write plausible electron dot structures for the following substances. Each substance contains only single covalent bonds. *16.1*
a. I_2 b. OF_2 c. H_2S d. NI_3

33. Draw the electron dot structure of each of the following molecules. *16.1*
a. F_2 b. HCl c. HCCH d. HCN

34. Characterize a coordinate covalent bond and give an example. *16.1*

35. Explain why compounds containing C—N and C—O single bonds can form coordinate covalent bonds with H^+ but compounds containing only C—H and C—C single bonds cannot. *16.1*

36. What is true for the electron dot structures of all compounds that exhibit resonance? *16.1*

37. Draw resonance structures for the carbonate ion (CO_3^{2-}). Each oxygen is attached to the carbon. *16.1*

38. Using electron dot structures, draw at least two resonance structures for the nitrite ion (NO_2^-). The oxygens in NO_2^- are attached to the nitrogen. *16.1*

39. How can you experimentally determine whether a substance is paramagnetic? *16.1*

40. Predict whether the following species are diamagnetic or paramagnetic. *16.1*
a. BF_3 b. O_2^- c. NO_2 d. F_2

41. Which of these compounds would you expect to contain elements that do not follow the octet rule? Explain. *16.1*
a. NF_3 b. PCl_2F_3 c. SF_4 d. SCl_2

42. Which of the following species would you predict to be diamagnetic? Paramagnetic? *16.1*
a. NO_3^- b. OH^- c. H_2O d. SO_3

43. What is the relationship between the magnitude of a molecule's bond dissociation energy and its expected chemical reactivity? *16.1*

44. Explain what is meant by bond dissociation energy. *16.1*

45. Assume the total bond energy in a molecule is the sum of the individual bond energies. Calculate the total bond energy in a mole of ethyne (C_2H_2). (*Hint:* Write the electron dot structure to determine the kinds of bonds. Then refer to **Table 16.3.**) *16.1*

46. Draw molecular orbital diagrams for the possible diatomic molecule Li_2. Would you expect Li_2 to exist as a stable molecule? *16.2*

47. What is the total number of sigma bonds and pi bonds in each molecule? *16.2*
a. H—C≡N b. H—C(H)(H)—N=C=O

48. Use VSEPR theory to predict the shapes of the following species. *16.2*
a. CO_2 c. SO_3 e. CO
b. $SiCl_4$ d. SCl_2 f. H_2Se

49. The molecule CO_2 has two carbon–oxygen double bonds. Describe the bonding in the CO_2 molecule, which involves hybridized orbitals for carbon and oxygen. *16.2*

50. What types of hybrid orbitals are involved in the bonding of the carbon atoms in the following molecules? *16.2*
a. CH_4 c. HC≡CH
b. $H_2C{=}CH_2$ d. N≡C—C≡N

51. What must always be true if a covalent bond is to be polar? *16.3*

52. The bonds between the following pairs of elements are covalent. Arrange them according to polarity, naming the most polar bond first. *16.3*
a. H—Cl c. H—F e. H—H
b. H—C d. H—O f. S—Cl

53. Arrange the following bonds in order of increasing ionic character. *16.3*
a. Cl—F c. K—O e. S—O
b. N—N d. C—H f. Li—F

42. a. diamagnetic
b. diamagnetic
c. diamagnetic
d. diamagnetic

43. Increasing bond dissociation energy is linked to lower chemical reactivity.

44. Bond dissociation energy is defined as the energy needed to break one covalent bond.

45. H:C:::C:H 1694 kJ/mol
2H:C 2 × 393 kJ/mol = 786 kJ/mol
C:::C 908 kJ/mol

46.

$2s^1$ [↑] Li — [ ] / [↓↑] Li_2 — [↓] Li $2s^1$

The lithium molecule has two electrons in a bonding molecular orbital. Lithium theoretically should exist as a diatomic molecule. The Li_2 molecule is moderately stable in the gaseous state.

47. a. 2 sigma, 2 pi b. 6 sigma, 2 pi

54. Based on the information about molecular shapes in Section 16.2, which of these molecules would you expect to be polar? *16.3*

a. SO_2 **c.** CO_2
b. H_2S **d.** BF_3

55. Depict the hydrogen bonding between two ammonia molecules and between one ammonia molecule and one water molecule. *16.3*

56. Which compound in each pair exhibits the stronger intermolecular hydrogen bonding? *16.3*

a. H_2S, H_2O **c.** HBr, HCl
b. HCl, HF **d.** NH_3, H_2O

57. What is a hydrogen bond? *16.3*

58. Why do compounds with strong intermolecular attractive forces have higher boiling points than compounds with weak intermolecular attractive forces? *16.3*

CONCEPT MASTERY

59. Devise a hybridization scheme for PCl_3 and predict the molecular shape based on this scheme.

60. Write two electron dot structures for the molecule N_2O. Predict the magnetic properties and shape of this molecule.

61. The chlorine and oxygen atoms in thionyl chloride ($SOCl_2$) are bonded directly to the sulfur. Write an acceptable electron dot structure for thionyl chloride.

62. Explain why each electron dot structure is incorrect. Replace each structure with one that is more acceptable.

a. $[:C::\ddot{N}:]^-$ **c.** H:C::O (with :Cl: above C)
b. :F:P::F: (with :F: below P) **d.** :F:B:F: (with :F: above B)

63. Use VSEPR theory to predict the geometry of each of the following.

a. $SiCl_4$ **c.** CCl_4
b. CO_3^{2-} **d.** SCl_2

64. The following graph shows how the percent ionic character of a single bond varies according to the difference in electronegativity between the two elements forming the bond. Answer the following questions, using this graph and **Table 14.2.**

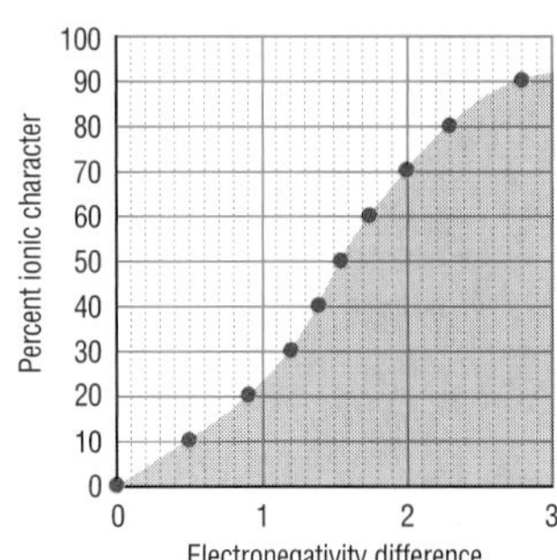

a. What is the relationship between the percent ionic character of single bonds and the electronegativity difference of their elements?
b. What electronegativity difference will result in a bond with a 50% ionic character?
c. Estimate the percent ionic character of the bonds formed between (1) lithium and oxygen, (2) nitrogen and oxygen, (3) magnesium and chlorine, and (4) nitrogen and fluorine.

65. Using bond dissociation energies, estimate ΔH for the following reaction.

$$CO(g) + 2H_2(g) \longrightarrow CH_3OH(g)$$

66. Give the angles between the orbitals of each hybrid.

a. sp^3 hybrids **b.** sp^2 hybrids **c.** sp hybrids

67. Describe the difference between a bonding molecular orbital and an antibonding molecular orbital. How do the energies of these orbitals compare?

CRITICAL THINKING

68. Choose the term that best completes the second relationship.

a. house : lumber
bond : _______
(1) atoms (3) valence electrons
(2) molecules (4) octet rule

b. globe : Earth
electron dots : _______
(1) atoms (3) valence electrons
(2) molecules (4) octet rule

69. Make a list of the elements found in **Table 16.2.** What do the elements that form covalent bonds have in common?

70. Is there a clear difference between a very polar covalent bond and an ionic bond? Explain.

48. a. linear **d.** bent
b. tetrahedral **e.** linear
c. trigonal planar **f.** bent

49. The 2*s* and the 2*p* orbitals form two sp^2 hybrid orbitals in the carbon atom. One sp^2 hybrid orbital from each oxygen atom forms a sigma bond with the carbon atom. Pi bonds between each oxygen atom and the carbon are formed by the unhybridized 2*p* orbitals.

50. a. sp^3 **c.** sp
b. sp^2 **d.** sp

51. The two atoms involved must have different electronegativities.

52. c., d., a., f., b., e.

53. b., d., e., a., c., f.

54. a. polar **c.** nonpolar
b. polar **d.** nonpolar

55. H—N(H)(H): ····H—N(H)(H):

:O(H)(H): ····H—N(H)(H):

56. a. H_2O **c.** HCl
b. HF **d.** H_2O

57. A hydrogen bond is formed by an electrostatic attraction between a hydrogen atom covalently bonded to a more electronegative atom and an unshared electron pair of a nearby atom.

58. They require more energy to separate the molecules.

59. The 3*s* and three 3*p* orbitals of phosphorus hybridize to form four sp^3 hybrid orbitals. The resulting shape is pyramidal with a bond angle of 107° between the sigma bonds.

60. :N::N::O: :N:::N:O:
The molecule is diamagnetic and linear.

61. :Cl:S:Cl: (with :O: below S)

62. a. Carbon does not have an ocet.
$[:C:::N:]^-$
b. A fluorine atom has 10 electrons.
:F:P:F: (with :F: below P)
c. Carbon has too many e^- (10) and oxygen too few (4).
H:C::O: (with :Cl: below C)
d. The atoms in boron trifluoride have 24 electrons to contribute, not 26.
:F:B:F: (with :F: below B)

63. a. tetrahedral, 109.5°
b. trigonal planar, 120°
c. tetrahedral, 109.5°
d. bent, 105°

64. a. The percent ionic character increases as the difference in electronegativities increases.
b. 1.6
c. (1) 85% (2) 10% (3) 62% (4) 23%

65. $\Delta H = -55$ kJ/mol

66. a. 109.5° **b.** 120° **c.** 180°

Answers

67. An antibonding orbital has a higher energy because the atomic orbitals overlap on opposite sides of the nuclei, not between them.

68. a. 3
b. 3

69. C, O, H, S, N, F, Cl, I, Br; These elements are all nonmetals.

70. Answers will vary. **Table 16.4** suggests there is no clear difference. The student's argument could be based on chemical properties, such as conductivity of the compound in the liquid state.

71. a., b., c., d.

72. 0.10 mol $BaSO_4$, 23 g $BaSO_4$

73. a. $1s^22s^22p^63s^1$
b. $1s^22s^22p^63s^23p^4$
c. $1s^22s^22p^63s^23p^3$
d. $1s^22s^22p^3$

74. a. 136.1 g/mol **c.** 394.7 g/mol
b. 98.1 g/mol **d.** 162.2 g/mol

75. a. 1.20×10^{22} **c.** 5.26×10^{15}
b. 9.27×10^{20} **d.** 1.81×10^{18}

76. a. manganese (Mn)
b. indium (In)
c. francium (Fr)
d. polonium (Po)

77.

H
H:C:H
H

90° 90°
H C H
H

109.5° 109.5°
H C H
109.5° 109.5°
H

The first sketch is tetrahedral. The second sketch is a tetrahedron. The bond angles in the first sketch are not all the same, with some being 90°. The bond angles in the second sketch are all 109.5°. The second sketch is correct. (Note: The wedge-shaped lines come out of the page; the dotted lines recede into the page.)

CUMULATIVE REVIEW

71. Which of the following ions have the same number of electrons as a noble gas?
a. Al^{3+} **b.** O^{2-} **c.** Br^- **d.** N^{3-}

72. A solution containing 0.10 mol $BaCl_2$ is mixed with a solution containing 0.20 mol Na_2SO_4 to give a precipitate of $BaSO_4$. What is the maximum yield in moles of $BaSO_4$? In grams?

73. Write correct electron configurations for atoms of the following elements.
a. sodium **c.** phosphorus
b. sulfur **d.** nitrogen

74. Calculate the gram formula mass of each substance.
a. $CaSO_4$ **c.** NI_3
b. H_2SO_4 **d.** $FeCl_3$

75. Give the number of representative particles in the following molar quantities.
a. 2.00 centimoles
b. 1.54 millimoles
c. 8.73 nanomoles
d. 3.00 micromoles

76. Name and give the symbol for the element in the following position in the periodic table.
a. Group 7B, period 4
b. Group 3A, period 5
c. Group 1A, period 7
d. Group 6A, period 6

CONCEPT CHALLENGE

77. The electron structure and geometry of the methane molecule (CH_4) can be described by a variety of models, including electron dot structure, simple overlap of atomic orbitals, and orbital hybridization of carbon. Write the electron dot structure of CH_4. Sketch two molecular orbital pictures of the CH_4 molecule. For your first sketch, assume that one of the paired $2s^2$ electrons of carbon has been promoted to the empty $2p$ orbital. Overlap each half-filled atomic orbital of carbon to a half-filled $2s$ orbital of hydrogen. What is the predicted geometry of the CH_4 molecule, using this simple overlap method? In your second sketch, assume hybridization of the $2s$ and $2p$ orbitals of carbon. Now what geometry would you predict for CH_4? Which picture is preferable based on the facts that all H—C—H bond angles in CH_4 are 109.5° and all C—H bond distances are identical?

78. There are some compounds in which one atom has more electrons than the corresponding noble gas. Examples are PCl_5, SF_6, and IF_7. Write the electron dot structures of P, S, and I atoms, and of these compounds. Considering the outer shell configuration of P, S, and I, develop an orbital hybridization scheme to explain the existence of these compounds.

78.

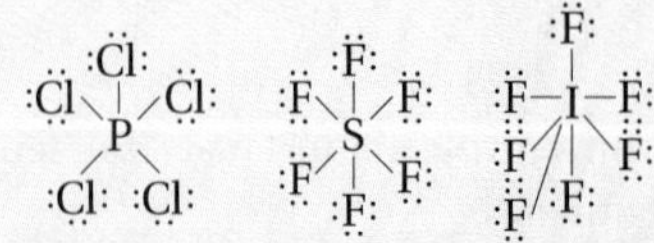

P forms 5 hybrid orbitals (dsp^3), S forms 6 hybrid orbitals (d^2sp^3), and I forms 7 hybrid orbitals (d^3sp^3).

Chapter 16 STANDARDIZED TEST PREP

Select the choice that best answers each question or completes each statement.

1. A bond in which two atoms share a pair of electrons is *not*
   a. a coordinate covalent bond.
   b. a polar covalent bond.
   c. an ionic bond.
   d. a nonpolar covalent bond.
2. How many valence electrons are in a molecule of phosphoric acid, H_3PO_4?
   a. 7 b. 16 c. 24 d. 32
3. Which of these molecules can form a hydrogen bond with a water molecule?
   a. N_2 b. NH_3 c. O_2 d. CH_4
4. Which substance contains both covalent and ionic bonds?
   a. NH_4NO_3 c. LiF
   b. CH_3OCH_3 d. $CaCl_2$
5. Which of these bonds is most polar?
   a. H—Cl c. H—F
   b. H—Br d. H—I

Use the description and data table below to answer questions 6–9.

Nonbonding pairs of electrons must be considered when predicting shapes of molecules. The table relates molecular shape to the number of bonding and nonbonding electron pairs in molecules.

Bonding pairs	Non-bonding pairs	Arrangement of electron pairs	Molecular shape	Example
4	0	tetrahedral	tetrahedral	CH_4
3	1	tetrahedral	pyramidal	NCl_3
2	2	tetrahedral	bent	H_2S
1	3	tetrahedral	linear	HF

6. Draw the electron dot structure for each example molecule.
7. Explain why the arrangement of electron pairs is tetrahedral in each molecule.
8. H_2S has two hydrogen atoms bonded to a sulfur atom. Why isn't the molecule linear?
9. What is the arrangement of electron pairs in PBr_3? Predict the molecular shape of a PBr_3 molecule.

For questions 10–12, identify the type of intermolecular bonding represented by the dotted lines in the drawings.

10. H_2O

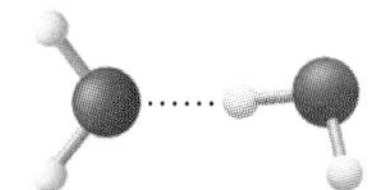

11. BrCl (bromine chloride)

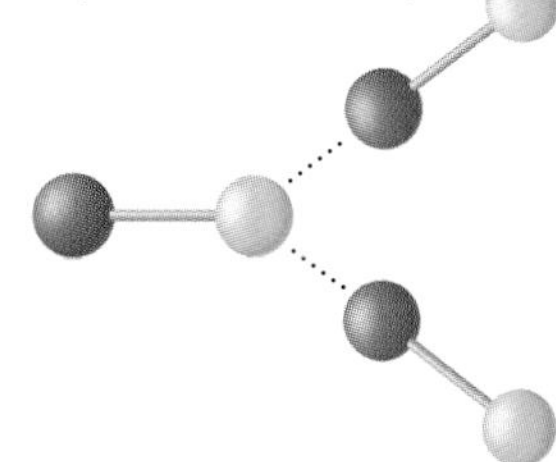

12. CH_3OH (methanol)

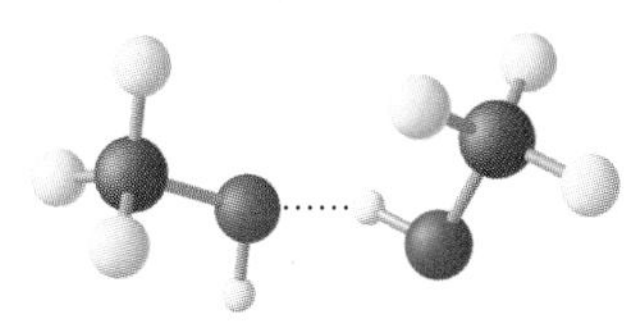

13. *You Draw It!* Draw two different pictures that show hydrogen bonding between a water molecule and a methanol molecule.

For each question there are two statements. Decide whether each statement is true or false. Then decide whether Statement II is a correct explanation for Statement I.

	Statement I		Statement II
14.	Water is a polar molecule.	BECAUSE	The covalent bonds in water are polar.
15.	A carbon monoxide molecule has a triple covalent bond.	BECAUSE	Carbon and oxygen atoms have an unequal number of valence electrons.
16.	Xenon has a lower boiling point than neon.	BECAUSE	Dispersion forces between xenon atoms are stronger than those between neon atoms.
17.	The nitrate ion has three resonance structures.	BECAUSE	The nitrate ion has three single bonds.

Chapter 16 STANDARDIZED TEST PREP

1. c
2. d
3. b
4. a
5. c
6. H:C:H (with H above and below C), Cl:N:Cl (with Cl above N), H:S:H, H:F:
7. Each central atom has four pairs of electrons that, according to VSEPR theory, assume a tetrahedral shape.
8. The two nonbonding pairs repel the bonding pairs; there are still four pairs of electrons around the sulfur atom.
9. The arrangement of electron pairs is tetrahedral. The electron dot structure shows three bonding electron pairs and one nonbonding electron pair; thus, the predicted molecular shape is pyramidal.
10. hydrogen bonding
11. primarily dispersion forces
12. hydrogen bonding
13. In these sample answers, black spheres represent carbon, gray spheres represent oxygen, and white spheres represent hydrogen.

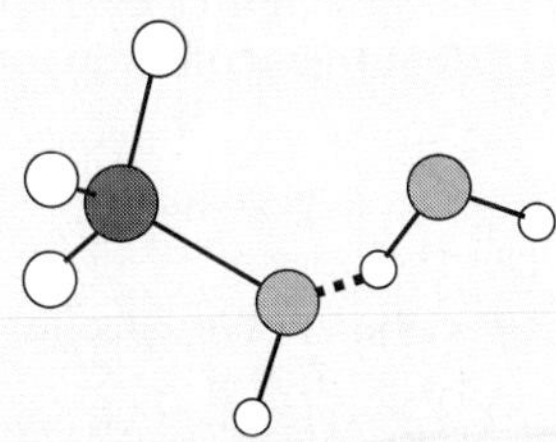

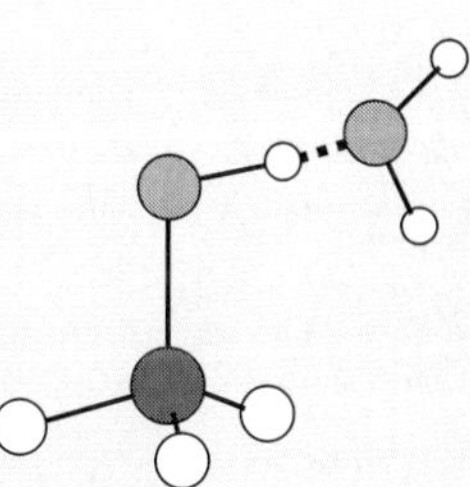

14. True, True
15. True, True
16. False, True
17. True, False

Planning Guide

SECTION OBJECTIVES	ACTIVITIES/FEATURES	MEDIA & TECHNOLOGY
17.1 Liquid Water and Its Properties ● ■ ◆ ▸ Describe the hydrogen bonding that occurs in water ▸ Explain the high surface tension and low vapor pressure of water in terms of hydrogen bonding	**SE** **Discover It!** *Observing Surface Tension,* p. 474 **SE** **Link to Health** *Water and Exercise,* p. 478	**ASAP** Animation 20, 21 **ASAP** Assessment 17.1 **OT** 57: *Bonding in Water*
17.2 Water Vapor and Ice ● ■ ◆ ▸ Account for the high heat of vaporization and the high boiling point of water in terms of hydrogen bonding ▸ Explain why ice floats in water		**ASAP** Assessment 17.2 **CHM** Side 1, 23: *Why Does Ice Float?*
17.3 Aqueous Solutions ● ■ ◆ ▸ Explain the significance of the statement "like dissolves like" ▸ Distinguish among strong electrolytes, weak electrolytes, and nonelectrolytes, giving examples of each	**SE** **Link to Sanitation** *Wastewater Treatment,* p. 485 **SE** **Small-Scale Lab** *Electrolytes,* p. 489 (**LRS** 17-1) **PLM** *Electrolytes* **LM** 26: *The Solvent Properties of Water* **LM** 27: *Distillation* **LM** 28: *Water of Hydration* **LM** 29: *Electrolytes and Nonelectrolytes* **TE** **DEMO**, p. 485	**ASAP** Simulation 19, 20 **ASAP** Problem Solving 9 **ASAP** Assessment 17.3 **ACT** 13: *Electrolytes* **OT** 58: *Solvation* **OT** 59: *Electrolytes and Nonelectrolytes* **CHM** Side 6, 30: *Polar and Nonpolar Solvents* **CHM** Side 6, 20: *Solutions*
17.4 Heterogeneous Aqueous Systems □ ◇ ▸ Explain how colloids and suspensions differ from solutions ▸ Describe the Tyndall effect	**SE** **Mini Lab** *Surfactants,* p. 493 (**LRS** 17-2) **SE** **Chemistry Serving . . . the Environment** *It's the Water,* p. 494 **SE** **Chemistry in Careers** *Oceanographer,* p. 494 **SSLM** 22: *Electrolytes*	**ASAP** Assessment 17.4 **ASAP** Concept Map 17 **RP** Lesson Plans, Resource Library **AR** Computer Test 17 **www** Activities, Self-Tests, *SCIENCE NEWS* updates **TCP** The Chemistry Place Web Site

KEY

● Conceptual (concrete concepts)	**AR** Assessment Resources	**GRS** Guided Reading and Study Workbook
○ Conceptual (more abstract/math)	**ASAP** Chem ASAP! CD-ROM	
■ Standard (core content)	**ACT** ActivChemistry CD-ROM	**LM** Laboratory Manual
□ Standard (extension topics)	**CHM** CHEMedia Videodiscs	**LP** Laboratory Practicals
◆ Honors (core content)	**CA** Chemistry Alive! Videodiscs	**LRS** Laboratory Recordsheets
◇ Honors (options to accelerate)	**GCP** Graphing Calculator Problems	**SSLM** Small-Scale Lab Manual

PRACTICE		ASSESSMENT	
GRS	Section 17.1	**SE**	Section Review
RM	Practice Problems 17.1	**RM**	Section Review 17.1
		RM	Chapter 17 Quiz
GRS	Section 17.2	**SE**	Section Review
RM	Practice Problems 17.2	**RM**	Section Review 17.2
RM	Interpreting Graphics	**RM**	Chapter 17 Quiz
SE	**Sample Problem** 17-1	**SE**	Section Review
GRS	Section 17.3	**RM**	Section Review 17.3
RM	Practice Problems 17.3	**RM**	Chapter 17 Quiz
		LP	Lab Practicals 17-1, 17-2
GRS	Section 17.4	**SE**	Section Review
RM	Practice Problems 17.4	**RM**	Section Review 17.4
GCP	Chapter 17	**RM**	Vocabulary Review 17
		SE	Chapter Review
		SM	Chapter 17 Solutions
		SE	Standardized Test Prep
		PHAS	Chapter 17 Test Prep
		RM	Chapter 17 Quiz
		RM	Chapter 17 A & B Test

PLANNING FOR ACTIVITIES

STUDENT EDITION

Discover It! p. 474
- waxed paper
- rulers
- teaspoons
- cups
- tap water
- liquid dish detergent

Small-Scale Lab, p. 489
- pencils, paper, rulers
- reaction surfaces
- conductivity testers
- solid NaCl, $MgSO_4$, Na_2CO_3, $NaHCO_3$, KCl KI, sucrose, cornstarch
- 1.0*M* HCl, H_2SO_4, NH_3, HNO_3, and 0.5*M* NaOH
- vinegar
- rubbing alcohol
- distilled water
- assorted household liquids

Mini Lab, p. 493
- shallow dishes or Petri dishes
- water
- paper clips
- rubber bands (2 in. long)
- vegetable oil
- liquid dish detergent

TEACHER'S EDITION

Activity, p. 480
- ice
- beaker
- room temperature water
- thermometer
- graph paper

Activity, p. 484
- zinc and copper strips
- lemon
- voltmeter

  or
- conductivity tester used in Teacher Demo, p. 485

Teacher Demo, p. 485
- light bulb (in a porcelain socket)
- 9V or lantern battery
- two copper metal strips
- lamp cord
- alligator clips
- 0.1*M* solutions of glucose ($C_6H_{12}O_6$) alanine ($HC_3H_6O_2N$) glycine ($HC_2H_4O_2N$) ascorbic acid ($HC_6H_7O_6$) malonic acid ($H_2C_3H_2O_4$) citric acid ($H_3C_6H_5O_7$) acetic acid ($HC_2H_3O_2$) hydrochloric acid (HCl)
- beakers for each solution
- marking pen

OT	Overhead Transparency	**SE**	Student Edition
PHAS	PH Assessment System	**SM**	Solutions Manual
PLM	Probeware Lab Manual	**SSV**	Small-Scale Video/Videodisc
PP	Problem Pro CD-ROM	**TCP**	www.chemplace.com
RM	Review Module	**TE**	Teacher's Edition
RP	Resource Pro CD-ROM	**www**	www.phschool.com

Chapter 17

Key Terms

17.1 surface tension, surfactant
17.3 aqueous solutions, solvent, solute, solvation, electrolytes, nonelectrolytes, weak electrolytes, strong electrolytes, water of hydration, effloresce, hygroscopic, desiccants, deliquescent
17.4 suspensions, colloids, Tyndall effect, Brownian motion, emulsions

DISCOVER IT!

Students should observe that the surface properties of water are different from those of a detergent–water mixture. On nonpolar surfaces, such as wax paper, water forms nearly spherical-shaped drops due to the surface tension resulting from intermolecular hydrogen bonding between water molecules. Because surfactants interfere with hydrogen bonding between water molecules, the beads collapse and spread out more evenly on the surface.

Chapter 17 WATER AND AQUEOUS SYSTEMS

Oceans cover about three quarters of Earth's surface.

FEATURES

DISCOVER IT!
Observing Surface Tension

SMALL-SCALE LAB
Electrolytes

MINI LAB
Surfactants

CHEMISTRY SERVING … THE ENVIRONMENT
It's the Water

CHEMISTRY IN CAREERS
Oceanographer

LINK TO HEALTH
Water and Exercise

LINK TO SANITATION
Wastewater Treatment

DISCOVER IT! OBSERVING SURFACE TENSION

You need waxed paper, a ruler, a teaspoon, a cup, tap water, and liquid dish detergent.

1. Place a 15-cm square of waxed paper on a flat surface.
2. Drop 1/3 teaspoon of water from a height of about 20 cm onto the center of the paper.
3. Add 1 drop of detergent to 1/2 cup of water and stir.
4. Repeat Steps 1 and 2 using the diluted detergent in place of water.
5. You may wish to repeat this activity with squares of aluminum foil, clear wrap, and writing paper.

What do you observe when you drop water onto the waxed paper surface? What do you observe when you use diluted detergent instead of water? Use your results to suggest how the detergent may have changed the physical properties of water. After completing this chapter, review your suggestion and modify it if necessary.

Stay current with SCIENCE NEWS
Find out more about aqueous systems:
www.phschool.com

TIME SAVER

17.1 ● ■ ◆
17.2 ● ■ ◆
17.3 ● ■ ◆
17.4 □ ◇

Conceptual In Section 17.1, specific heat capacity may be omitted. Sample Problem 17-1 may be omitted because students at this level would not be expected to solve percent composition problems. Section 17.4 may be omitted because colloids and suspensions are not prerequisites for the chapters that follow.

Standard Use the particle-view illustrations in Sections 17.1 and 17.2 to explain water's unique properties. You may want to use the Small-Scale Lab to introduce electrolytes with a concrete experience. Suspensions and colloids in Section 17.4 are appropriate extension topics if time permits.

Honors In Section 17.2, the discussion of evaporation and condensation may be accelerated depending on students' prior knowledge. Section 17-4 may be assigned as homework because colloids and suspensions are descriptive topics that are not prerequisites for later chapters.

LIQUID WATER AND ITS PROPERTIES

In December of 1968, when the Apollo 8 astronauts first saw their home planet from a distance of thousands of kilometers, they affectionately called it the big blue marble. As you can see in this photograph, water covers about three-quarters of Earth's surface. In addition to making up Earth's oceans, water forms the polar ice caps and cycles through the atmosphere. All life forms with which we are familiar are made mostly of water. **What properties of water make this unique substance essential to life on Earth?**

section 17.1

objectives

- Describe the hydrogen bonding that occurs in water
- Explain the high surface tension and low vapor pressure of water in terms of hydrogen bonding

key terms

- surface tension
- surfactant

The Water Molecule

Water is a unique compound. It is the foundation of all life on Earth. Without water neither plant nor animal life as we know it could exist. Liquid water in the form of oceans, lakes, rivers, ponds, and streams covers about three-quarters of Earth's surface. Immense aquifers store water deep underground. Ice, or solid water, dominates the vast polar regions of the globe. Water appears as icebergs in the oceans, and it blankets the temperate zones as snow in winter. Water vapor from the evaporation of surface water and from steam spouted out of geysers and volcanoes is ever-present in Earth's atmosphere.

Water is a simple triatomic molecule. Each O—H covalent bond in the water molecule is highly polar. Because of its greater electronegativity, oxygen attracts the electron pair of the covalent O—H bond and acquires a slightly negative charge. The less electronegative hydrogen atoms acquire a slightly positive charge. The water molecule has an H—O—H bond angle of 105°. Because of this bent shape, the two O—H bond polarities do not cancel, and the water molecule as a whole is polar. The net polarity of the water molecule is illustrated in **Figure 17.2.** The region around the oxygen has a slightly negative charge, and the region around the hydrogens has a slightly positive charge.

Figure 17.1
Water is vital to life. Animals that live on the grasslands depend on watering holes, such as the one shown here.

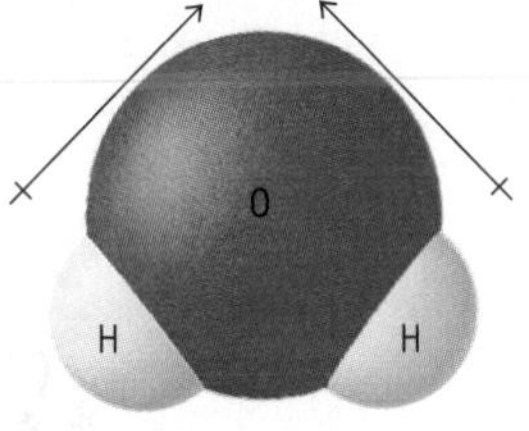

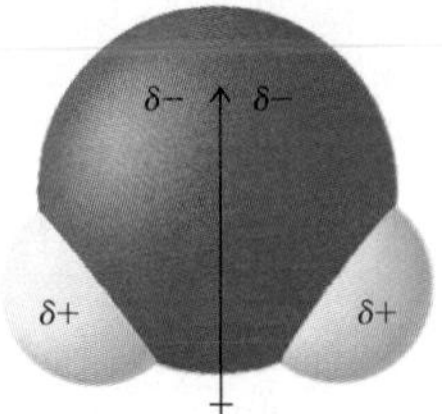

Polar bonds

Molecule has net polarity.

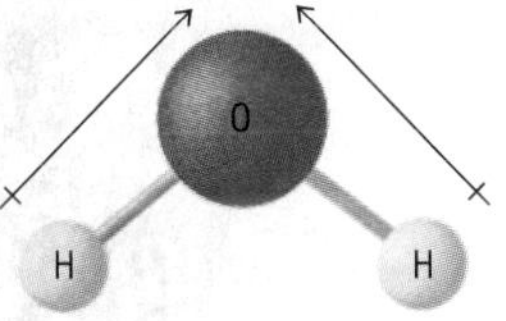

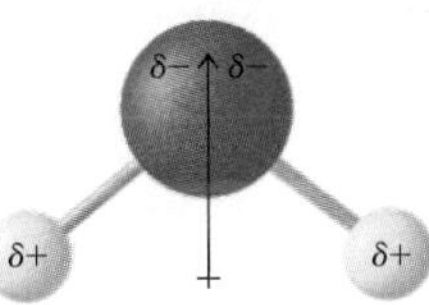

Figure 17.2
In a water molecule, the bond polarities are equal, but the two dipoles do not cancel each other because a water molecule is bent. There is a net polarity, and the molecule as a whole is polar.

Student Resources

From the Teacher's Resource Package, use:

- Section Review 17.1, Ch. 17 Practice Problems and Quizzes from the Review Module (Ch. 17–20)

TECHNOLOGY RESOURCES

Relevant technology resources include:

- Chem ASAP! CD-ROM
- Resource Pro CD-ROM

1 Engage

Use the Visual

Have students study the photograph and read the text that opens the section. Ask:

- **What properties of water make this ever-present substance so remarkable?** (Water has a set of properties that distinguish it from other substances. Because it is a polar molecule that forms hydrogen bonds, it is a powerful solvent. Reactions involving polar and ionic substances take place in aqueous solution. Water's solid phase is less dense than its liquid phase. Water has a large heat capacity.)

Water is vital to life. Because water has a melting/freezing point of 0 °C, and a boiling point of 100 °C, most living organisms have adapted to live somewhere within this temperature range; however, there are some thermophiles that exist in much hotter environments. Point out that water is not just a passive solvent. Because it can act as either a base or an acid and form highly ordered arrays, water is an active participant in many chemical reactions, especially biochemical reactions involving enzymes.

Check Prior Knowledge

To assess students' prior knowledge about the physical properties of water, ask:

- **What is meant by the term *polarity*?** (It refers to the net molecular dipole arising out of electronegativity differences between covalently bonded atoms)
- **What kind of substances dissolve in water?** (polar substances; ionic compounds)

17.1

2 Teach

Discuss

To help students understand why liquid water assumes a spherical shape on many surfaces, explain that nature tends to find the path of least resistance. Moving molecules takes work. A spherical shape provides the least surface area for a given volume; molecules expend the least energy possible to move into a spherical arrangement while maximizing their interactions with one another in the bulk liquid. This energy efficiency creates the surface tension. In order to adopt any other shape, more work would have to be done. The work "saved" is the surface tension.

Figure 17.3
(a) Due to their polarity, water molecules attract each other.
(b) They participate in hydrogen bonding. To form a hydrogen bond, what must be true about the hydrogen and the element to which it is hydrogen bonded? 

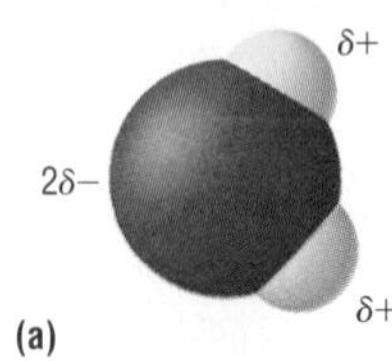

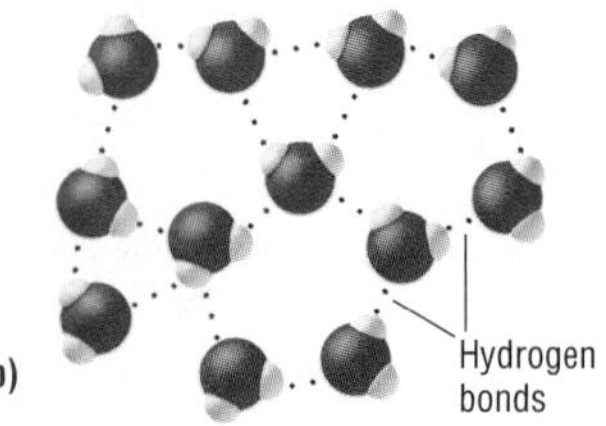

Chem ASAP!
Animation 20
See how hydrogen bonding results in the unique properties of water.

In general, polar molecules are attracted to one another by dipole interactions. In water, however, the attraction is the result of intermolecular hydrogen bonds, as **Figure 17.3** illustrates. Water molecules change position by sliding over one another. But because of hydrogen bonding, most water molecules do not have enough kinetic energy at room temperature to escape the surface of the liquid. Many unique and important properties of water—including its high surface tension, low vapor pressure, high specific heat capacity, high heat of vaporization, and high boiling point—result from hydrogen bonding.

Surface Properties

Have you ever seen a glass so filled with water that the water surface bulges above the rim? Or have you noticed that water forms nearly spherical droplets at the end of a medicine dropper or when sprayed on a greasy surface? Perhaps you have observed that a needle floats if placed on the surface of water, but the needle sinks immediately if it breaks through the surface.

The surface of water acts like a skin. The water strider shown in **Figure 17.4** takes advantage of this skinlike property of water's surface. This phenomenon, called surface tension, is explained by water's ability to form hydrogen bonds. The molecules within the liquid are surrounded by and hydrogen-bonded to adjacent water molecules. At the surface of the

Figure 17.4
Surface tension makes it possible for some insects, such as this water strider, to "walk" on water. The force of gravity pulling the insect downward is less than the forces of attraction between the water molecules.

Chem ASAP!
Animation 21
Discover how some insects can walk on water. 

Answers

❶ Only when hydrogen atoms are covalently bonded to either nitrogen, oxygen, or fluorine atoms is there a sufficiently large electronegativity difference to create intermolecular hydrogen bonds. Have students look up the electronegativity values for H, F, O, and N (H = 2.1, F = 4.0, O = 3.5, and N = 3.0) and calculate the differences.
❷ Each arrow represents the inward pull on water molecules at the periphery, or surface, of the liquid. The molecules are drawn toward the body of the liquid where they can form hydrogen bonds.
❸ water with detergent

liquid, however, water molecules experience an uneven attraction. As you can see in **Figure 17.5,** the water molecules are hydrogen-bonded on only one side of the drop. The water molecules are not attracted to the air because they cannot form hydrogen bonds with air molecules. As a result, water molecules at the surface tend to be drawn into the body of the liquid. The inward force, or pull, that tends to minimize the surface area of a liquid is called **surface tension.**

The surface tension of a liquid tends to hold a drop of liquid in a spherical shape, as shown in **Figure 17.6.** A sphere has the smallest surface area for a given volume. At the same time, the force of gravity tends to flatten the drop. A liquid that has strong intermolecular attractions has a high surface tension. The higher the surface tension, the more nearly spherical is the drop of that particular liquid.

All liquids have a surface tension, but water's surface tension is higher than most. This is why water tends to bead up on some surfaces rather than to spread out. It is possible to decrease the surface tension of water by adding a **surfactant,** which is a wetting agent such as soap or detergent that gets its name from *surf*ace *act*ive *agent.* When a detergent is added to beads of water on a greasy surface, the detergent molecules interfere with the hydrogen bonding between water molecules. As a result, surface tension is reduced, the beads collapse, and the water spreads out.

Hydrogen bonding between water molecules also explains water's unusually low vapor pressure. As you learned in Section 10.2, the vapor pressure of a liquid is caused by molecules that escape the surface of the liquid and enter the gas phase. Because hydrogen bonds hold water molecules to one another, the tendency of these molecules to escape is low. Imagine what would happen if it were not. All the lakes and oceans, with their large surface areas, would rapidly evaporate!

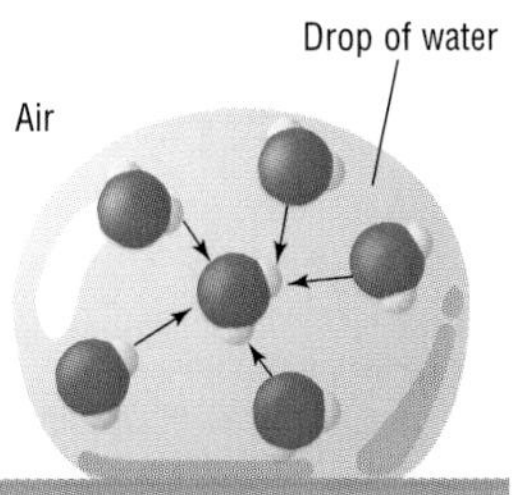

Figure 17.5
Water molecules at the surface of the water drop experience an uneven attraction. Because they cannot form hydrogen bonds with air molecules, they are drawn into the body of the liquid, producing surface tension. What does each arrow in the figure represent? 2

Figure 17.6
***(a)** Surface tension tends to hold drops of liquids in a spherical shape. At the same time, the force of gravity tends to flatten the drops. **(b)** All the drops shown are of equal volume. From left to right, they are mercury, water, and water with detergent. Which of these three liquids has the least surface tension?* 3

(a)

(b)

17.1

Discuss

Point out that the skin-like qualities of water are due to an exceptionally high surface tension that is created by an extensive network of hydrogen bonds. As an analogy, describe the following scene. A crowd of autograph seekers surrounds a celebrity. As each person approaches as closely as possible, an impenetrable circular barrier forms. This barrier will remain intact as long as there is a net attraction toward the circle's center. The process is dynamic. As one person wiggles closer to the celebrity, another is forced to retreat, but the overall shape doesn't change. Similarly within a drop of water, attractions between individual molecules may shift, but the overall shape remains constant as molecules continue to be drawn toward a central focal point.

Meeting Diverse Needs

Gifted Have students compare areas of population density and the distribution of water. Give them two different maps: one map should show the locations of cities and towns. The other should be a regional contour map showing water courses, rivers, lakes, and reservoirs. Why is it that most large cities or even historic sites are located near water sources? What happens when they're not? Consider the vast irrigation networks and water channels that crisscross the nation's agricultural areas. What happens to the concentration of salts in water in the Colorado River as it nears its end?

3 Assess

Evaluate Understanding

To assess students' understanding about how the molecular structure and chemical composition of water are related to its physical properties, ask:

- **Why is the surface tension of water so high compared to other liquids?** (extensive number of hydrogen bonds, in addition to dipole-dipole forces between molecules)
- **Why does water form a meniscus in a narrow tube?** (because water molecules have a greater attraction to each other than they do to the molecules on the surface of the glass)
- **Ask students to predict the relative late evening temperatures of the water puddle and sewer cover mentioned in the text. Ask them to explain any differences.** (Due to the high specific heat of water, the water will retain heat longer than the sewer cover. On a cool evening, the sewer cover will be cooler than the water.)

Reteach

Write the complete structural formula of water on the board. Using VSEPR theory, show students how the lone oxygen's non bonding electron pairs close the bond angle to 105° in a water molecule. Point out how the net molecular dipole and hydrogen bonding properties of water are due to water's bent shape and the large electronegativity differences between H and O. Point out that hydrogen bonding contributes to water's high heat capacity. Some of the energy water absorbs is used to break hydrogen bonds; some increases the average kinetic energy of water molecules.

LINK TO HEALTH

Water and Exercise

The normal human body contains about 66% water by mass. You lose water through perspiration, as water vapor from breathing, and in your body's waste products—mainly urine.

The loss of water during periods of inactivity is probably 25–100 mL/h. Obviously, you must replace this water to avoid dehydration. Medical experts recommend that you drink at least eight glasses of water every day. When you exercise vigorously for prolonged periods, your water loss through perspiration may increase to as much as 3000 mL/h, which heightens your risk of dehydration. Most sports physiologists agree that athletes should drink water when exercising to maintain good health and maximize performance.

Figure 17.7
Many people flock to the beach in hot weather because the shore is cooler than inland areas. How would you explain this phenomenon in terms of water's heat capacity? 2

Specific Heat Capacity

As you may recall, it takes 4.18 J (1 cal) of heat energy to raise the temperature of 1 g of water 1 °C. This is the specific heat capacity of water. The specific heat capacity of water is nearly constant at 4.18 J/(g × °C) at temperatures between 0 °C and 100 °C. Because of hydrogen bonding, the specific heat capacity of water is higher than that of most other substances. Iron has a specific heat capacity of 0.447 J/(g × °C). For the same increase in temperature, iron requires only about one-tenth as much heat as does an equal mass of water. On a sunny day, a puddle of water remains relatively cool, but a nearby sewer cover gets hot enough, it is said, to fry an egg. If the same amount of heat energy were added to identical masses of
1 water and iron, how would the increases in temperature compare?

Water's high specific heat capacity helps moderate daily air temperatures around large bodies of water. On a warm day, water absorbs heat from its warmer environment, lowering the air temperature. On a cool night, heat is transferred from the water to its cooler environment, raising the air temperature. Thus water is a good storage medium for solar energy.

section review 17.1

1. Describe the hydrogen bonding between molecules in water.
2. How is hydrogen bonding responsible for the high surface tension and low vapor pressure of water?
3. What is a surfactant? How does it affect the surface tension of water?
4. What is the specific heat capacity of water?

Chem ASAP! **Assessment 17.1** Check your understanding of the important ideas and concepts in Section 17.1.

Answers

1 Because iron has a lower specific heat capacity than water, the temperature of a mass of iron will rise higher than the temperature of an identical mass of water.

2 Because of high heat capacity, water is able to absorb a large amount of heat from warm air masses. This has a moderating effect on the air temperatures in areas surrounding large bodies of water.

SECTION REVIEW 17.1

1. A slightly positive hydrogen atom on one water molecule is attracted to a pair of unshared electrons on the oxygen atom of another.
2. Water molecules are hydrogen-bonded to each other, but not to air molecules. Net attraction is inward, minimizing water surface. Hydrogen bonding prevents escape of water molecules.
3. A wetting agent that reduces surface tension; it interferes with hydrogen bonding of water.
4. 4.18 J/(g × °C)

WATER VAPOR AND ICE

section 17.2

On October 27, 1915, the British Imperial Trans-Antarctic Expedition ship Endurance was crushed by pack ice and sank, leaving the members of the expedition to float on ice floes for five months. Today, icebreakers are used to open a safe path for navigation through floating pack ice. **What property of water allows its solid phase, ice, to float in its liquid phase?**

objectives

- Account for the high heat of vaporization and the high boiling point of water in terms of hydrogen bonding
- Explain why ice floats in water

Evaporation and Condensation

You are probably familiar with several examples of both the evaporation and condensation of water. Have you ever left a glass of water out on the kitchen counter overnight? What happens? What have you noticed about the air you exhale on a very cold winter morning? 

Because of hydrogen bonding, water absorbs a large amount of heat as it evaporates or vaporizes. The heat of vaporization is the amount of energy needed to convert 1 g of a substance from a liquid to a gas at the boiling point. It takes 2.26 kJ to convert 1 g of liquid water at 100 °C to 1 g of steam at 100 °C. An extensive network of hydrogen bonds tightly holds the molecules in liquid water together. These bonds must be broken before water changes from the liquid to the vapor state. The heat of vaporization of liquid ammonia, which is less hydrogen-bonded than water, is 1.37 kJ/g. Liquid methane, which has no hydrogen bonding, has a heat of vaporization of only 0.510 kJ/g.

The reverse of vaporization is condensation. When 1 g of steam at 100 °C condenses to 1 g of liquid water at 100 °C, 2.26 kJ of heat is given off. Thus the heat of condensation of water is 2.26 kJ/g. As expected, the heat of condensation of water is equal to the heat of vaporization of water. You can get a severe burn if steam condenses on your skin. Not only is the temperature of steam 100 °C, but the additional heat of condensation is also absorbed by your skin.

Figure 17.8
Mist rising from the rain forest, dew on morning grass, and clouds in the sky are all evidence of evaporation and condensation.

17.2

1 Engage

Use the Visual

Have students study the photograph and read the text that opens the section. Explain that at lower temperatures, when the kinetic energy and, therefore, the motion of water molecules decreases, hydrogen bonds act to organize and fix water molecules into an ordered hexagonal array known as ice. Ask:

- **What property of water allows water's solid phase, ice, to float in its liquid phase?** (Ice has a lower density than liquid water. In its crystalline or solid form, water exhibits a rigid, but relatively open framework of water molecules. Since there is more space between water molecules in the solid state than in the liquid state, there are fewer molecules in any given volume. Less mass per volume results in a lower relative density of ice as compared to water.)

Motivate and Relate

Ask how global warming might affect the distribution of water worldwide. (Water currently trapped in ice caps could melt and significantly increase the volume of the ocean, which would increase sea level and flood many acres of shoreline.)

Check Prior Knowledge

To assess students' knowledge about changes of state, ask:

- **What is the melting point of a substance? The boiling point?** (The melting point is the temperature at which a substance changes from solid to liquid; the boiling point is the temperature at which a substance changes from liquid to gas.)

STUDENT RESOURCES

From the Teacher's Resource Package, use:

- Section Review 17.2, Ch. 17 Practice Problems, Interpreting Graphics, and Quizzes from the Review Module (Ch. 17–20)

TECHNOLOGY RESOURCES

Relevant technology resources include:

- Chem ASAP! CD-ROM
- Resource Pro CD-ROM

3 The volume of water in the glass decreases. Air exhaled on a cold morning forms a visible "cloud."

2 Teach

Discuss

When water molecules separate, energy is required to break hydrogen bonds. During melting or evaporation, the temperature of water does not rise because the energy is used to accomplish the phase change. The amount of energy used during these phase changes is returned to the environment when the molecules reform hydrogen bonds as they condense or freeze.

ACTIVITY

Have students place ice in a beaker of room temperature water. Have them take the initial temperature of the water before the ice is added, during the melting of the ice, and at the lowest temperature reached. Encourage students to construct a graph of temperature versus time. Let students determine an appropriate time interval, and ask them why the frequency of measurements made on a system undergoing change is important to the design of an experiment. Ask:

- What happens to the temperature as the ice is melting? (The temperature drops as heat is removed from the liquid water to melt the ice.)
- What happens when all the ice has disappeared? (The temperature begins to rise, and returns to the original temperature. Ice regulates the temperature of the liquid water by absorbing heat from the warmer liquid. This heat is equal to the heat of fusion of ice.)

Table 17.1

Melting Points and Boiling Points of Some Substances with Low Molar Mass

Name of substance	Formula	Molar mass (g/mol)	Melting point (°C)	Boiling point (°C)
Methane	CH_4	16	−183	−164
Ammonia	NH_3	17	−77.7	−33.3
Water	H_2O	18	0	100
Neon	Ne	20	−249	−246
Methanol	CH_3OH	32	−93.9	64.9
Hydrogen sulfide	H_2S	34	−85.5	−60.7

The evaporation and condensation of water are important to regional temperatures on Earth. Temperatures in the tropics would be much higher if water did not absorb heat while evaporating from the surfaces of the surrounding oceans. Temperatures in the polar regions would be much lower if water vapor did not release its heat while condensing out of the air.

Molecular compounds of low molar mass are usually gases or liquids with low boiling points at normal atmospheric pressure. As you can see in **Table 17.1,** ammonia is a typical example of a molecular compound. Ammonia has a molar mass of 17 g/mol and boils at about −33 °C. Water, however, with a molar mass of 18 g/mol, is an important exception. Its boiling point is 100 °C. What do you think accounts for water's high boiling point? You are correct if your answer is hydrogen bonding. Hydrogen bonding is more extensive in water than in ammonia. Therefore it takes much more heat to disrupt the attractions between water molecules than those between ammonia molecules. If the hydrogen bonding in water were as weak as it is in ammonia, water would be a gas at the usual temperatures
1 found on Earth. What do you think this would mean for life as we know it? Compare the molar masses and boiling points of methanol and hydrogen sulfide provided in **Table 17.1.** Which of these two compounds has more
2 extensive hydrogen bonding?

Ice

As a typical liquid cools, it contracts slightly. Its density increases because its volume decreases while its mass stays constant. If the cooling continues, the liquid eventually solidifies. Because the density of a typical solid is greater than that of the corresponding liquid, the solid sinks in its own liquid. For example, lead shot sinks in molten lead.

As water cools, it first behaves like a typical liquid. It contracts slightly and its density gradually increases until the temperature reaches 4 °C. Below 4 °C, however, the density of water starts to decrease. Water no longer behaves like a typical liquid. Ice, which forms at 0 °C, has about 10% greater volume and therefore a lower density than liquid water at 0 °C. See **Table 17.2.** As a result of its lower density, ice floats in liquid water. This is certainly unusual behavior for a solid. Ice is one of only a few solids that float in their own liquid.

Table 17.2

Density of Liquid Water and Ice

Temperature (°C)	Density (g/cm^3)
100 (liquid water)	0.9584
50	0.9881
25	0.9971
10	0.9997
4	1.000*
0 (liquid water)	0.9998
0 (ice)	0.9168

*Most dense

Meeting Diverse Needs

Gifted The structure of ice was determined using X-ray diffraction. In a similar way, it was X-ray diffraction studies of DNA by Rosalind Franklin that contributed to the development of the helical structure by James Watson and Francis Crick. Have students research the role of Franklin and X-ray diffraction in this discovery.

Answers

1 Answers will vary. Students should predict that life on Earth would be very different if water could not exist as a liquid.
2 methanol
3 Answers will vary, but could include the breaking of a soft-drink container when the liquid inside it freezes or the weathering of rock.

Hydrogen bonds

Liquid water

Ice

Side view End view

Why does ice behave so differently? As you can see in **Figure 17.9**, the structure of ice is a very regular open framework of water molecules arranged like a honeycomb. Hydrogen bonding holds the water molecules in place. At low temperatures, the molecules do not have sufficient kinetic energy to break out of this rigid framework. When ice melts, the framework collapses. Consequently, the water molecules pack closer together, making liquid water more dense than ice.

The fact that ice floats has important consequences for organisms. A layer of ice on the top of a pond acts as an insulator for the water beneath, preventing it from freezing solid except under extreme conditions. Because the liquid water at the bottom of an otherwise frozen pond is warmer than 0 °C, aquatic life is able to survive. If ice were more dense than liquid water, bodies of water would tend to freeze solid during the winter months, destroying many types of aquatic life. What are some other phenomena that result from the expansion of water as it freezes to form ice? 3

Water molecules require a considerable amount of kinetic energy to return to the liquid state from the solid state. Ice melts at 0 °C, a high temperature for a molecule with such a low molar mass. The heat absorbed when 1 g of water changes from a solid to a liquid is 334 J/g. This heat-absorbing property can be used to great advantage, as you can see in **Figure 17.10.** This same amount of energy could also raise the temperature of 1 g of liquid water from 0 °C to 80 °C!

section review 17.2

5. Explain why water has a relatively high boiling point and heat of vaporization.
6. What is the difference between the structure of liquid water and the structure of ice? How does this explain why ice floats in water?
7. How much energy in kilojoules is required to change 47.6 g of ice at 0 °C to liquid water at the same temperature? In kilocalories?

Chem ASAP! Assessment 17.2 Check your understanding of the important ideas and concepts in Section 17.2.

Figure 17.9
The molecules in liquid water are hydrogen bonded but free to move about. Extensive hydrogen bonding in ice holds the water molecules farther apart in a more ordered arrangement than in liquid water. The hexagonal symmetry of snowflakes results from the hydrogen bonding of water molecules in ice.

Figure 17.10
Ice packs decrease swelling and pain from an injury. Ice absorbs a large amount of heat as it changes from a solid to a liquid at 0 °C.

3 Assess

Evaluate Understanding

Display Table 17.1 on an overhead projector. Ask students to predict which of the substances would have the highest vapor pressure in a sealed container at room temperature. Which would have the lowest vapor pressure? (Roughly speaking, the substance with the highest boiling point will have the lowest vapor pressure and the one with the lowest boiling point will have the highest vapor pressure.) Next, display Table 17.2 on the overhead projector. Based on these data, the density of liquid water increases with decreasing temperature.
Ask:

- **Why doesn't this trend continue below 4 °C?** (The maximum density of water is reached at 4 °C. Below 4 °C, the kinetic energy of the molecules cannot overcome the hydrogen bonding, which holds the water molecules in fixed positions. A more regular lattice forms with open spaces between molecules. Consequently, water occupies a greater volume per mass below 4 °C.)

Reteach

Use the production of sweat to review phase changes. Explain that sweating helps maintain body temperature; as the water in sweat evaporates, it absorbs heat from the body. Explain that sweating is an effective cooling mechanism because the breaking of hydrogen bonds during evaporation requires a relatively large amount of energy.

Answers

SECTION REVIEW 17.2

5. Hydrogen bonds hold liquid water molecules together, making it harder for them to escape to the gas state.
6. Ice has a honeycomb-like structure of hydrogen-bonded water molecules. Liquid water has a less rigid structure, in which water molecules are closer together. The density of ice is less than that of water because of ice's open crystal structure. As a result ice floats on water.
7. 15.9 kJ; 3.81 kcal

17.3

1 Engage

Use the Visual

Have students study the photograph and read the text that opens the section. Explain that pure water does not conduct electricity but can be made conductive when certain substances are dissolved in it.
Ask:

- **Which kind of solution conducts electricity?** (The solutions that conduct electricity are those that contain ions. The word "ion" comes from the Greek word for wanderer. Because ions are able to carry a charge between the electrodes of battery terminals, a current will flow through a solution in which they are dissolved.)

Check Prior Knowledge

To assess students' prior knowledge of ionic and molecular compounds, ask:

- **How do ionic and molecular compounds differ?** (The molecules in molecular compounds are electrically neutral, covalently bonded groups of atoms that behave as a unit. Ionic compounds are composed of positive and negative ions.)
- **What is a cation? An anion?** (A cation is any positively charged atom or group of atoms. An anion is any negatively charged atom or group of atoms.)
- **Why are ionic compounds generally soluble in water?** (The cations and anions are attracted to the charged ends of polar water molecules.)

section 17.3

objectives

- Explain the significance of the statement "like dissolves like"
- Distinguish among strong electrolytes, weak electrolytes, and nonelectrolytes, giving examples of each

key terms

- aqueous solutions
- solvent
- solute
- solvation
- electrolytes
- nonelectrolytes
- weak electrolyte
- strong electrolyte
- water of hydration
- effloresce
- hygroscopic
- desiccants
- deliquescent

AQUEOUS SOLUTIONS

Is it possible to read by the light of a glowing pickle? Although it sounds absurd, it is possible! The aqueous solution in a pickle will glow as an electric current is passed through it. **What kind of solution conducts electricity?**

Solvents and Solutes

Chemically pure water never exists in nature because water dissolves so many substances. Tap water contains varying amounts of dissolved minerals and gases, as does water from ponds, streams, rivers, lakes, and oceans. Water samples containing dissolved substances are called **aqueous solutions.** In a solution, the dissolving medium is called the **solvent,** and the dissolved particles are called the **solute.** When sodium chloride dissolves in water, water is the solvent and sodium chloride is the solute. If
1 you were making lemonade, what would the solvent be? The solutes?

As you may recall, solutions are homogeneous mixtures. They are also stable mixtures. For example, sodium chloride does not settle out when its solutions are allowed to stand, provided other conditions, such as temperature, remain constant. Solute particles can be either ionic or molecular, and their average diameters are usually less than 1.0 nm (10^{-9} m). Therefore, if you filter a solution through filter paper, both the solute and the solvent pass through the filter. See **Figure 17.11.** Solvents and solutes may be gases, liquids, or solids. It might be helpful for you to refer to **Table 2.3** on page 33 to review some common types of solutions.

Substances that dissolve most readily in water include ionic compounds and polar covalent molecules. Nonpolar covalent molecules, such as methane and those in oil, grease, and gasoline, do not dissolve in water. However, oil and grease will dissolve in gasoline. To understand this difference, you must know more about the structural features of the solvent and the solute and the attractions between them.

Figure 17.11
A solution cannot be separated by filtration. The small size of solute particles allows them to pass through filter paper.

Student Resources

From the Teacher's Resource Package, use:

- Section Review 17.3, Ch. 17 Practice Problems and Quizzes from the Review Module
- Laboratory Recordsheet 17-1
- Laboratory Manual: Experiments 26-29
- Laboratory Practicals 17-1 and 17-2
- Small-Scale Lab Manual: Experiment 22

Technology Resources

Relevant technology resources include:

- Chem ASAP! CD-ROM
- Resource Pro CD-ROM
- ActivChemistry CD-ROM: *Electrolytes*

The Solution Process

Water molecules are in continuous motion because of their kinetic energy. When a crystal of sodium chloride is placed in water, the water molecules collide with it. The solvent molecules (H_2O) attract the solute ions (Na^+, Cl^-). As individual sodium ions and chloride ions break away from the crystal, the sodium chloride dissolves. **Solvation** is the process that occurs when a solute dissolves. The negatively and positively charged ions become solvated; that is, they become surrounded by solvent molecules. **Figure 17.12** shows a model of the solvation of an ionic solid.

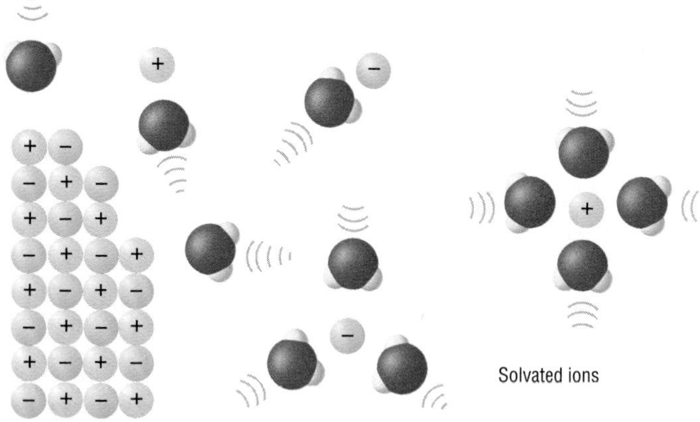

Figure 17.12
When an ionic solid dissolves, the ions become solvated and are surrounded by solvent molecules. Why do the water molecules orient themselves differently around the anions and the cations? 3

Chem ASAP!
Simulation 19
Explore the nature of solute–solvent interactions.

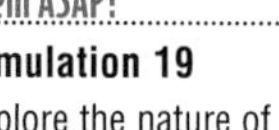

In some ionic compounds, the attractions between the ions in the crystals are stronger than the attractions exerted by water. These compounds cannot be solvated to any significant extent and are therefore nearly insoluble. Barium sulfate ($BaSO_4$) and calcium carbonate ($CaCO_3$) are examples of nearly insoluble ionic compounds.

What about dissolving oil in gasoline? Both oil and gasoline are composed of nonpolar molecules. They mix to form a solution, not because the solute and solvent are attracted, but because there are no repulsive forces between them. As a rule, polar solvents dissolve ionic compounds and polar molecules; nonpolar solvents dissolve nonpolar compounds. This relationship can be summed up in the expression "like dissolves like." What is the chemistry that explains the saying "oil and water don't mix"? 2

Figure 17.13
Oil and water do not mix.

Electrolytes and Nonelectrolytes

Compounds that conduct an electric current in aqueous solution or the molten state are called **electrolytes.** All ionic compounds are electrolytes. Sodium chloride, copper(II) sulfate, and sodium hydroxide are typical water-soluble electrolytes. Electrolytes conduct electricity in solution and in the molten state. Barium sulfate is an example of an ionic compound that conducts electricity in the molten state but not in aqueous solution because it is insoluble.

Answers

1 solvent = H_2O; solute = materials that make up lemon juice, sugar
2 Water is polar; oil is nonpolar.
3 The hydrogen end of the water molecule is slightly positive and orients itself toward the negatively charged anion. The oxygen end of the water molecule is slightly negative and orients itself toward the positively charged cation.

Meeting Diverse Needs

At Risk and LEP Have students preview the section by looking for vocabulary and other unfamiliar terms. Encourage students to write the terms, their phonetic respellings, and their definitions in their notebooks.

17.3

2 Teach

Discuss

Remind students that solutions are homogeneous mixtures containing a solvent and one or more solutes. Usually the solvent is defined as the component in the system that is present in the greatest amount. Point out that a water-soluble solute can be a solid, liquid, or gas. (Acetone is used to remove nail polish. Turpentine is used to thin or dissolve oil-based paints. Dry cleaners use organic solvents instead of water.) **Ask students to name some of the solutes in blood?** (Solutes in blood are typically ions such as sodium, potassium, calcium, chloride, hydrogen carbonate, and phosphate. Dissolved gases such as oxygen and carbon dioxide are also present in the blood, as is glucose, a sugar and nonelectrolyte.)

Use the Visual

Have students study the photograph in Figure 17.13. Ask:

- **Why doesn't oil mix with water?** (Oil is a mixture of nonpolar compounds. For it to mix with oil, water's hydrogen bonds must be broken and replaced by much weaker attractive forces. A greater number of strong attractive forces between molecules result if oil and water remain unmixed.)

Stress the idea that when a solid dissolves in water, it breaks into small pieces. For a solid to dissolve, the molecules of the solvent must be able to overcome the attractive forces that are holding the solid together.

Discuss

Some students may think that all polar molecules are nonelectrolytes. Point out that hydrogen chloride, ammonia, and acetic acid are polar molecules that dissociate in water to form ions.

ACTIVITY

Have students construct a 'lemon battery' by inserting a zinc and copper strip into a lemon and measuring the voltage with a voltmeter. A more elaborate voltaic cell can be constructed using copper and zinc solutions connected via a salt bridge. When connected to a light bulb in an external circuit, the oxidation-reduction reactions occurring at the electrodes will cause current to flow and illuminate the light bulb. Ask students how current is able to flow through the lemon. (The lemon contains natural electrolytes, in solution, which carry current between the metals.)

CHEMISTRY AND BIOLOGY

Electrolytes are essential to all metabolic processes. Sodium and potassium ions generate the action potential for nerve function. If renal function is impaired or malabsorption from the gut disturbs optimum sodium and potassium levels, then serious nervous-system problems arise. Loss of consciousness or ataxia (difficulty in maintaining muscle coordination) could result. Any condition that causes prolonged bouts of diarrhea can be life threatening because of the dramatic loss of electrolytes. During dehydration, excess electrolytes are lost through the skin via perspiration. This will lead to cramps and heat stroke unless electrolytes are replenished.

Compounds that do not conduct an electric current in either aqueous solution or the molten state are called **nonelectrolytes.** Because they are not composed of ions, many molecular compounds are nonelectrolytes. Most compounds of carbon, such as cane sugar and rubbing alcohol, are nonelectrolytes.

Some very polar molecular compounds are nonelectrolytes in the pure state, but become electrolytes when they dissolve in water. This occurs because such compounds ionize in solution. For example, neither ammonia (NH_3)(g) nor hydrogen chloride (HCl)(g) is an electrolyte in the pure state. Yet an aqueous solution of ammonia conducts electricity because ammonium ions (NH_4^+) and hydroxide ions (OH^-) form when ammonia dissolves in water.

$$NH_3(g) + H_2O(l) \longrightarrow NH_4^+(aq) + OH^-(aq)$$

Similarly, in aqueous solution, hydrogen chloride produces hydronium ions (H_3O^+) and chloride ions (Cl^-). An aqueous solution of hydrogen chloride conducts electricity and is therefore an electrolyte.

$$HCl(g) + H_2O(l) \longrightarrow H_3O^+(aq) + Cl^-(aq)$$

Not all electrolytes conduct an electric current to the same degree. In a simple conductivity test, shown in **Figure 17.14,** a bulb glows brightly when electrodes attached to it are immersed in a sodium chloride solution. By

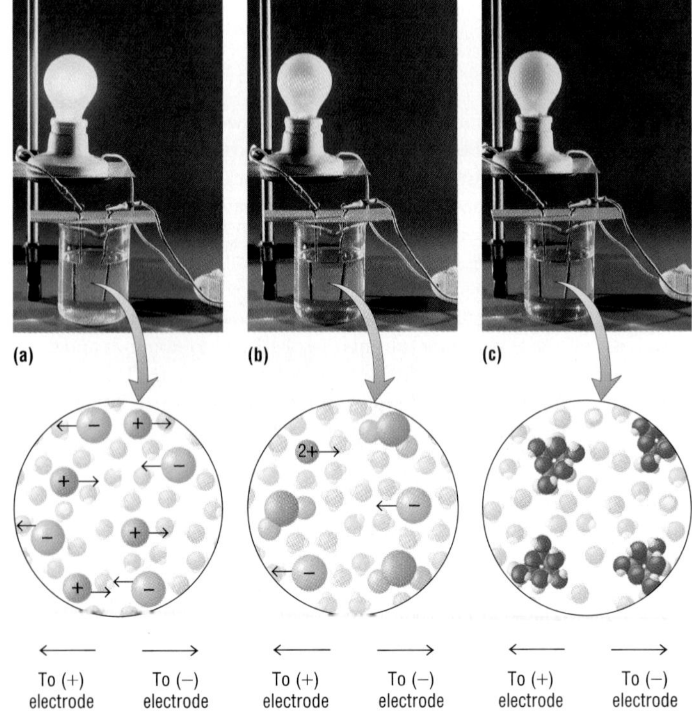

Figure 17.14
The presence of an electrolyte in solution can be determined by a conductivity test. ***(a)*** *Sodium chloride is a strong electrolyte; it dissociates nearly 100% in water. Its ions move in solution and readily conduct a current.* ***(b)*** *Mercury(II) chloride is a weak electrolyte; it only partially dissociates in water and conducts a small current.* ***(c)*** *Glucose is a nonelectrolyte; it does not dissociate in water. Would the molecular compound ethanol (C_2H_5OH) be an electrolyte?* 1

Chem ASAP!
Simulation 20
Explore the behavior of electrolytes and nonelectrolytes in a circuit and at the atomic level.

Answers

1 No; ethanol is a nonelectrolyte.
2 A molecule of sucrose does not ionize in solution.

contrast, the bulb glows dimly when the electrodes are immersed in a mercury(II) chloride solution. Mercury(II) chloride is a weak electrolyte. When a **weak electrolyte** is in solution, only a fraction of the solute exists as ions. In a solution of mercury(II) chloride, most of the solute exists as un-ionized $HgCl_2$. When a **strong electrolyte** is dissolved, almost all the solute exists as separate ions. Sodium chloride is a strong electrolyte. Nearly all the dissolved sodium chloride in a solution exists as separate Na^+ and Cl^- ions. The ions move in solution and conduct an electric current. In a solution of glucose, a nonelectrolyte, the bulb fails to glow. What is the representative particle of glucose? How does this explain why glucose is a nonelectrolyte? **Table 17.3** 2 lists some common electrolytes and nonelectrolytes.

Table 17.3

Some Examples of Strong Electrolytes, Weak Electrolytes, and Nonelectrolytes		
Strong electrolyte	**Weak electrolyte**	**Nonelectrolyte**
Acids (inorganic) HCl HBr HI HNO_3 H_2SO_4 $HClO_4$ Bases (inorganic) NaOH KOH Soluble salts KCl $MgSO_4$ $KClO_3$ $CaCl_2$	Heavy metal halides $HgCl_2$ $PbCl_2$ Bases (inorganic) NH_3 Acids (organic) acetic acid Bases (organic) aniline H_2O (very weak)	Most organic compounds Glucose Glycerol

Water of Hydration

Water molecules are an integral part of the crystal structure of many substances. The water in a crystal is called the **water of hydration** or water of crystallization. A compound that contains water of hydration is called a hydrate. When an aqueous solution of copper(II) sulfate is allowed to evaporate, deep-blue crystals of copper(II) sulfate pentahydrate are deposited. The chemical formula for this compound is $CuSO_4 \cdot 5H_2O$. In writing the formula of a hydrate, a dot is used to connect the formula of the compound and the number of water molecules per formula unit. Crystals of copper sulfate pentahydrate always contain five molecules of water for each copper and sulfate ion pair. The deep-blue crystals are dry to the touch. They are unchanged in composition or appearance in normally moist air. When heated above 100 °C, however, the crystals lose their water of hydration. As

SANITATION

Wastewater Treatment

Cleaning wastewater protects the environment from the harmful substances it carries. Wastewater treatment also helps recycle water, a resource that is in increasingly short supply. Wastewater treatment begins with the used water from your home, your school, or industry. The water collects in underground pipes called sewers. Gravity moves the wastewater through the sewers to the wastewater treatment plant. Once at the plant, the wastewater undergoes a number of purification stages. The primary treatment stage is mostly a physical process involving filtration and sedimentation. In secondary treatment, large populations of microorganisms convert the remaining dissolved organic matter into solids that separate as a sludge. Sometimes tertiary treatment is needed as well. At this level of treatment, nutrients such as nitrogen compounds and phosphorus compounds that produce unwanted growth of algae are removed.

TEACHER DEMO

Compare strong and weak electrolytes using a conductivity tester. The complete circuit consists of a light bulb (in a porcelain socket), a 9V or lantern battery, and two copper metal strips immersed in the aqueous solution to be tested. Connect the leads of the battery and lightbulb sockets using lamp cord and alligator clips. If possible, prepare 8 such testers so that students can directly compare the brightness of the lightbulbs for the following solutions. Prepare 0.1*M* solutions of 1) glucose ($C_6H_{12}O_6$), 2) alanine ($HC_3H_6O_2N$), 3) glycine ($HC_2H_4O_2N$), 4) ascorbic acid ($HC_6H_7O_6$), 5) malonic acid ($H_2C_3H_2O_4$), 6) citric acid ($H_3C_6H_5O_7$), 7) acetic acid ($HC_2H_3O_2$), and 8) hydrochloric acid (HCl).

Write the names and numbers of the solutions on the board. Pour an equal volume of each solution into a beaker; use the assigned numbers to label the beakers. Remind students that the solutions have the same concentration (0.1*M*).

Immerse a set of electrodes in each solution and have students compare the brightness of the light bulbs. Explain that the conductivity, and hence the brightness of the lightbulb, depends on the concentration of ions in solution. Ask students to list the substances in order from strongest to weakest electrolyte. Remind students that a strong electrolyte is a substance that exists in solution almost entirely as ions. A weak electrolyte is a substance that only partly dissociates into ions in solution. Students should recognize that the varying conductivities of each solution must be due to the fact that the different substances are dissociated, or ionized, to different extents.

MEETING DIVERSE NEEDS

Gifted The solubility of gases in water depends on the temperature of the water and partial pressure of the gas. Have students research the aqueous solubilities of various gases in the atmosphere, especially oxygen, nitrogen, and argon. What percentage of air is comprised of each of these gases? Are other gases present? Have students consider how the solubility of these gases in blood changes at high pressure. For example, what precautions must scuba divers take when ascending and why? How are the gases used by scuba divers changed to make diving safer? Students may consider how the solubility of gases in lakes and oceans is important for marine life and how a factor such as thermal pollution effects the amounts of dissolved gases.

Use the Visual

Have students study the photographs in Figure 17.15. Explain that a hydrate is a compound that contains weakly bound water molecules as part of the crystal structure. The dot in the compound formula of a hydrate between the salt and water indicates that water is an intimate part of the representative unit of the compound. The effect of the water on the structure of the crystal lattice can be observed by the change in the appearance of the hydrated substance.

Use the Visual

Display Table 17.4 on an overhead projector. Ask students if they recognize any of the common names listed. Review the nomenclature used to name salts and the use of prefixes (*mono-*, *di-*, *tri-*, *tetra-*, and so on) to indicate the number of water molecules in the crystalline hydrate. Write the formula $CoCl_2 \bullet 6H_2O$ on the board and ask students to name it. (cobalt(II) chloride hexahydrate).

Cooperative Learning

After students have studied the Practice Problem on page 488, have them work in groups to find the percent of mass of water for compounds in Table 17.4.

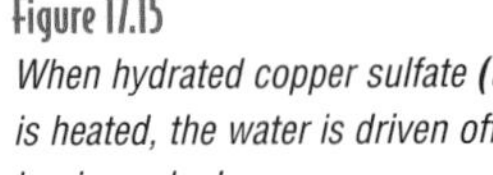

(a)

(b)

Figure 17.15
*When hydrated copper sulfate **(a)** is heated, the water is driven off, leaving anhydrous copper sulfate **(b)**.*

you can see in **Figure 17.15,** the blue crystals crumble to a white anhydrous powder with a formula of $CuSO_4$. If anhydrous copper sulfate is treated with water, the blue pentahydrate is regenerated.

$$CuSO_4 \cdot 5H_2O(s) \underset{-\text{heat}}{\overset{+\text{heat}}{\rightleftharpoons}} CuSO_4(s) + 5H_2O(g)$$

Cobalt(II) chloride is another compound that changes color in the presence of moisture. A piece of filter paper dipped in an aqueous solution of cobalt(II) chloride and dried is blue in color (anhydrous $CoCl_2$). If exposed to moist air, the paper turns pink due to the formation of the hexahydrate ($CoCl_2 \cdot 6H_2O$). See **Figure 17.16.** This paper is used to test for the presence of water.

Table 17.4 lists some familiar hydrates. They all contain a fixed quantity of water and have a definite composition. Do you know uses for any of
1 these compounds?

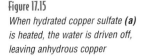

Figure 17.16
Paper treated with anhydrous cobalt(II) chloride is blue. In the presence of moisture the paper turns pink. How could you change the pink paper back to blue? 2

Table 17.4

Some Common Hydrates

Formula	Chemical name	Common name
$MgSO_4 \cdot 7H_2O$	magnesium sulfate heptahydrate	Epsom salt
$Ba(OH)_2 \cdot 8H_2O$	barium hydroxide octahydrate	
$CaCl_2 \cdot 2H_2O$	calcium chloride dihydrate	
$CuSO_4 \cdot 5H_2O$	copper(II) sulfate pentahydrate	blue vitriol
$Na_2SO_4 \cdot 10H_2O$	sodium sulfate decahydrate	Glauber's salt
$KAl(SO_4)_2 \cdot 12H_2O$	potassium aluminum sulfate dodecahydrate	alum
$Na_2B_4O_7 \cdot 10H_2O$	sodium tetraborate decahydrate	borax
$FeSO_4 \cdot 7H_2O$	iron(II) sulfate heptahydrate	green vitriol
$H_2SO_4 \cdot H_2O$	sulfuric acid hydrate (mp 8.6 °C)	

Answers

1 Answers will vary. Epsom salts are used as a cathartic; Glauber's salt is used as a cathartic or diuretic; alum is used as an astringent, as an emetic, and in the manufacture of baking powder, dyes, and paper; borax is used in the manufacture of glass, enamel, artificial gems, soaps, and antiseptics.
2 Heat it gently.
3 deliquescence

The forces holding the water molecules in hydrates are not very strong, which is apparent from the ease with which water is lost and regained. Because the water molecules are held by weak forces, hydrates often have an appreciable vapor pressure. If a hydrate has a vapor pressure higher than that of the water vapor in air, the hydrate will **effloresce** by losing the water of hydration. For example, $CuSO_4 \cdot 5H_2O$ has a vapor pressure of about 1.0 kPa at room temperature. The average pressure of water vapor at room temperature is usually about 1.3 kPa. This hydrate is stable until the humidity decreases. When the vapor pressure drops below 1.0 kPa, the hydrate effloresces. Washing soda, or sodium carbonate decahydrate ($Na_2CO_3 \cdot 10H_2O$), is efflorescent. As the crystals lose water of hydration, they effloresce and become coated with a white powder of anhydrous sodium carbonate.

Some hydrated salts that have a low vapor pressure remove water from moist air to form higher hydrates. Salts and other compounds that remove moisture from air are **hygroscopic.** Calcium chloride monohydrate is hygroscopic, as are sodium hydroxide and concentrated sulfuric acid.

$$CaCl_2 \cdot H_2O(s) \xrightarrow{\text{moist air}} CaCl_2 \cdot 2H_2O(s)$$

Hygroscopic substances are used as drying agents, or **desiccants.** One familiar example is silica gel, which is used to maintain a dry atmosphere in food packaging containers. Small packets of silica gel are also packaged with electronic equipment to absorb moisture from the air and to prevent damage to the equipment. Some compounds are so hygroscopic that they become wet when exposed to normally moist air. These compounds are deliquescent. **Deliquescent** compounds remove sufficient water from the air to dissolve completely and form solutions. As you can see in **Figure 17.17,** pellets of sodium hydroxide are deliquescent. Deliquescence occurs when the solution formed has a lower vapor pressure than that of the water in the air.

(a)

(b)

Figure 17.17
(a) *Dry sodium hydroxide pellets exposed to air will, over time, remove water from the air.* ***(b)*** *Eventually a solution will be formed. What is this process called?* 

17.3

Discuss

Many of the terms used to describe the properties of hydrates may be unfamiliar to students. Review the definition and pronunciation of the terms *effloresce, hygroscopic, desiccant,* and *deliquescent.* Have students write the definition for each term in their notebooks. Demonstrate some of these phenomena by setting out watch glasses containing fresh NaOH pellets, anhydrous calcium chloride, and anhydrous copper sulfate, respectively. Warn students not to touch the compounds. Point out which compounds are hygroscopic and which are deliquescent. Many students may have seen rice in salt shakers at restaurants. Ask them what purpose the rice serves. (Rice acts as a desiccant.) **Explain why hygroscopic agents are preferred over deliquescent agents for use as desiccants or drying agents. Emphasize the reversible properties of hydration. Desiccants can be recycled. If possible show them a desiccator and explain why desiccators are used in the laboratory to store hygroscopic chemicals.**

Practice Problems Plus

Related Chapter Review Problem
Chapter Review problem 49 is related to Sample Problem 17-1.

Additional Practice Problem
The Chem ASAP! CD-ROM contains the following problem: Magnesium sulfate forms a heptahydrate.

a. Write the equation for the formation of this hydrate from the anhydrous salt. ($MgSO_4 + 7H_2O \rightarrow MgSO_4 \bullet 7H_2O$)

b. Calculate the percent by mass of water. (51.2%)

3 Assess

Evaluate Understanding

To assess students' understanding of the properties of aqueous solutions, ask:

- **What types of substances will dissolve in water to form aqueous solutions?** (ionic compounds and polar covalent molecules)

Dissolve a small amount of nickel(II) chloride in 500 mL of distilled water and ask:

- **In this solution, what is the solute and what is the solvent?** (Water is the solvent and nickel(II) chloride is the solute.)
- **What type of electrolyte is nickel(II) chloride? Explain.** (Nickel(II) chloride is an ionic compound and is completely soluble in water; therefore, it is a strong electrolyte.)

Reteach

Project models of water molecules, cations, and anions on an overhead projector to show how water molecules orient their dipoles to solvate ions. Point out that cations and anions are attracted to different ends of the water molecule. Stress that the classification of water soluble substances as strong, weak, or nonelectrolytes is determined by the relative number of ions in solution. Nonelectrolytes do not produce any ions.

Portfolio Project

Sports drinks include Na^+, K^+, and Ca^{2+}. Electrolytes control nerve-impulse transmission and muscle contraction. They are excreted in sweat and, therefore, they need to be replenished after exercise.

Sample Problem 17-1

Calculate the percent by mass of water in washing soda, or sodium carbonate decahydrate ($Na_2CO_3 \cdot 10H_2O$).

1. ***ANALYZE*** ***List the known and the unknown.***

Known:
- formula of hydrate = $Na_2CO_3 \cdot 10H_2O$

Unknown:
- percent water = ? %

To determine the percent by mass, determine the mass of the hydrated compound and the mass of the water in the hydrate, then substitute these values into the following equation and solve.

$$\text{percent } H_2O = \frac{\text{mass of water}}{\text{mass of hydrate}} \times 100\%$$

2. ***CALCULATE*** ***Solve for the unknown.***

$$\text{mass of } 10H_2O = 180 \text{ g}$$

$$\text{mass of } Na_2CO_3 \cdot 10H_2O = 286.0 \text{ g}$$

$$\text{percent } H_2O = \frac{1.80 \times 10^2 \,\cancel{\text{g}}}{286.0 \,\cancel{\text{g}}} \times 100\% = 62.9\%$$

3. ***EVALUATE*** ***Does the result make sense?***

Because the mass of the water accounts for more than half the molar mass of the compound, a percentage greater than 50% should be expected. The answer should have three significant figures.

Practice Problems

8. What is the percent by mass of water in $CuSO_4 \cdot 5H_2O$?
9. Calcium chloride forms a hexahydrate.
   a. Write the equation for the formation of this hydrate from the anhydrous salt.
   b. Calculate the percent by mass of water in the hexahydrate.

Chem ASAP!
Problem-Solving 9
Solve Problem 9 with the help of an interactive guided tutorial.

portfolio project

Research and report on which electrolytes are found in sports drinks. What is the biological function of these electrolytes? How does exercise affect the body's electrolyte balance?

section review 17.3

10. What is the significance of the statement "like dissolves like"? What does "like" refer to?
11. Distinguish between an electrolyte and a nonelectrolyte. Distinguish between a weak electrolyte and a strong electrolyte. Give an example of each.
12. Identify the solvent and the solute in vinegar, a dilute aqueous solution of acetic acid.
13. Write an equation showing how calcium chloride dissociates in water.
14. Which of the following substances dissolve to a significant extent in water? Explain your answer in terms of polarity.

a. CH_4	d. $MgSO_4$
b. KCl	e. sucrose
c. He	f. $NaHCO_3$

Chem ASAP! **Assessment 17.3** Check your understanding of the important ideas and concepts in Section 17.3.

Answers

SECTION REVIEW 17.3

10. Polar solvents dissolve polar compounds, and nonpolar solvents dissolve nonpolar compounds. *Like* refers to the similar polarity of the substances.
11. Electrolytes (e.g., NaOH) can conduct an electric current in aqueous solution or in the molten state; nonelectrolytes (e.g., sucrose) cannot. Weak electrolytes (e.g., $HgCl_2$) do not exist mainly as separate ions in solution, but strong electrolytes (e.g., NaCl) do.
12. Acetic acid is the solute; water is the solvent.
13. $CaCl_2(s) \rightarrow Ca^{2+}(aq) + 2Cl^-(aq)$
14. a. no; nonpolar
    b. yes; ionic
    c. no; nonpolar
    d. yes; ionic
    e. yes; polar covalent
    f. yes; ionic

SMALL-SCALE LAB

ELECTROLYTES

SAFETY   

Wear safety glasses and follow the standard safety procedures outlined on page 18.

PURPOSE

To classify compounds as electrolytes by testing their conductivity in aqueous solution.

MATERIALS

- pencil
- paper
- ruler
- conductivity probe (optional)
- reaction surface
- conductivity tester
- chemicals shown in Figure A

PROCEDURE

 *Probe version available in the Probeware Lab Manual.*

On separate sheets of paper, draw two grids similar to Figure A. Make each square 2 cm on each side. Place a reaction surface over one of the grids and place a few grains of each solid in the indicated places. Test each solid for conductivity. Then add 1 drop of water to each solid and test the wet mixture for conductivity. Be sure to clean and dry the conductivity leads between each test.

$NaCl(s)$ **aqueous conducts**	$MgSO_4(s)$ **aqueous conducts**
$Na_2CO_3(s)$ **aqueous conducts**	Table Sugar **aqueous does not conduct**
aqueous conducts $NaHCO_3(s)$	**aqueous does not conduct** Cornstarch
aqueous conducts $KCl(s)$	**aqueous conducts** $KI(s)$

Figure A

ANALYSIS

Using your experimental data, record the answers to the following questions below your data table.

1. Electrolytes are compounds that conduct electricity in aqueous solution. Which compounds in Figure A are electrolytes? Which are not electrolytes?
2. Do any of these electrolytes conduct electricity in the solid form? Why?
3. The formula for table sugar is $C_{12}H_{22}O_{11}$ and the formula for cornstarch is $(C_6H_{12}O_6)_n$. Are these ionic or covalent compounds? Classify each compound in Figure A as ionic or covalent. For a compound to be an electrolyte, what must happen when it dissolves in water?

YOU'RE THE CHEMIST  

The following small-scale activities allow you to develop your own procedures and analyze the results.

1. **Analyze It!** When an ionic solid dissolves in water, water molecules attract the ions, causing them to come apart, or dissociate. The resulting dissolved ions are electrically charged particles that allow the solution to conduct electricity. The following chemical equations represent this phenomenon.

$$NaCl(s) \longrightarrow Na^+(aq) + Cl^-(aq)$$
$$Na_2CO_3(s) \longrightarrow 2Na^+(aq) + CO_3^{2-}(aq)$$

Write a similar chemical equation for each electrolyte in Figure A to illustrate how the ions dissociate in water.

2. **Design It!** Obtain the following aqueous solutions: HCl, H_2SO_4, HNO_3, CH_3COOH, NH_3, $NaOH$, rubbing alcohol, and distilled water. Design and carry out an experiment to classify each solution as a strong electrolyte (bright light), weak electrolyte (dim light), or nonelectrolyte (no light).
3. **Design It!** Test various liquids for conductivity. Try soft drinks, orange juice, pickle juice, and coffee. Which liquids are electrolytes?

SMALL-SCALE LAB

ELECTROLYTES

OVERVIEW

Procedure Time: 20 minutes

 *See Probeware Lab Manual for Vernier, TI, and PASCO versions.*

Students test solids and liquids for conductivity and find that water-soluble ionic compounds are electrolytes.

MATERIALS

Solution	Preparation
$1.0M$ HCl	82 mL of $12M$ in 1.0 L
$1.0M$ HNO_3	63 mL of $15.8M$ in 1.0 L
$1.0M$ H_2SO_4	56 mL of $18M$ in 1.0 L

CAUTION! Always add acid to water carefully and slowly!

$1.0M$ NH_3	67 mL of $15M$ in 1.0 L

CAUTION! Using proper ventilation, add ammonia to water carefully and slowly!

CH_3COOH	Use white vinegar.
$0.5M$ NaOH	20.0 g in 1.0 L

TEACHING SUGGESTIONS

There is a more complete lab on electrolytes in the *Small-Scale Chemistry Laboratory Manual* where you can find instructions for making a classroom set of conductivity devices.

ANALYSIS

1. electrolytes: NaCl, $MgSO_4$, Na_2CO_3, $NaHCO_3$, KCl, KI; nonelectrolytes: sugar, cornstarch
2. None of the electrolytes conduct electricity in the solid form because the ions are locked in a crystal lattice and cannot move.
3. Table sugar and cornstarch are covalent compounds. $NaHCO_3$, NaCl, $MgSO_4$, Na_2CO_3, KCl, and KI are ionic compounds. In general an electrolyte is a compound that dissociates into ions in solution.

YOU'RE THE CHEMIST

1. $MgSO_4(s) \rightarrow Mg^{2+}(aq) + SO_4^{2-}(aq)$
   $Na_2CO_3(s) \rightarrow 2Na^+(aq) + CO_3^{2-}(aq)$
   $NaHCO_3(s) \rightarrow Na^+(aq) + HCO_3^-(aq)$
   $KCl(s) \rightarrow K^+(aq) + Cl^-(aq)$
   $KI(s) \rightarrow K^+(aq) + I^-(aq)$
2. Test each solution with a conductivity device.
   strong electrolytes: HCl, H_2SO_4, HNO_3, NaOH
   weak electrolytes: CH_3COOH, NH_3
   nonelectrolytes: rubbing alcohol, distilled water
3. strong electrolytes: soft drinks, pickle juice
   weak electrolytes: orange juice, coffee

1 Engage

Use the Visual

Have students study the photograph and read the text that opens the section. Ask students to imagine how the size of a solute would affect the formation of a solution. (Larger particles might be less likely to dissolve. With undissolved particles, the mixture would not be homogeneous—by definition, not a solution.) **Size of particles is one way to classify heterogeneous mixtures. Ask:**

- **What is a suspension and what are its characteristics?** (Suspensions are mixtures in which the particles are large enough to settle out upon standing. Suspensions appear cloudy due to the scattering of light. Suspended particles can be removed by filtration.)
- **What is a colloid and what are its characteristics?** (Colloids are heterogeneous mixtures with particle sizes between that of true solutions and suspensions. The particles of a colloid cannot be filtered out and do not settle out over time.)

section 17.4

objectives
- Explain how colloids and suspensions differ from solutions
- Describe the Tyndall effect

key terms
- suspensions
- colloids
- Tyndall effect
- Brownian motion
- emulsions

HETEROGENEOUS AQUEOUS SYSTEMS

It wiggles and jiggles. It comes in many colors and flavors. When you pop it in your mouth, it dissolves. It is gelatin, one of the most popular desserts in the United States. In fact, more than a million packages of gelatin are purchased or eaten every day. Gelatin has even traveled into space. In 1996, American astronaut Shannon Lucid shared a gelatin dessert with her Russian crewmates. Gelatin is a type of mixture called a colloid. **What is a colloid, and what are its characteristics?**

Suspensions

So far, the emphasis of this chapter has been on homogeneous mixtures involving water. We will now focus on heterogeneous mixtures, as opposed to mixtures that are true solutions. Two such heterogeneous mixtures are called suspensions and colloids.

Figure 17.18
A suspension is a heterogeneous mixture. Suspended particles can be removed by filtration.

Suspensions are mixtures from which particles settle out upon standing. A piece of clay shaken with water forms a suspension. The clay breaks into fine particles that become suspended in the water, but they start to settle out when shaking stops. A suspension differs from a solution because the component particles of a suspension are much larger. The particles in a typical suspension have an average diameter greater than 100 nm. By contrast, the particle size in a solution is usually about 1 nm. Suspensions are heterogeneous because at least two substances can be clearly identified. In the example just given, the two substances are clay and water. As you can see in **Figure 17.18,** if muddy water is filtered, the suspended clay particles are trapped by the filter, and clear water passes through.

Colloids

Colloids are heterogeneous mixtures containing particles that are intermediate in size between those of suspensions and true solutions. Thus the particles range in size from 1 nm to 100 nm. The particles are the dispersed phase. They are spread throughout the dispersion medium, which can be a solid, liquid, or gas. The first colloids that were identified as such were glues. Other colloids include such mixtures as gelatin desserts, paint, aerosol sprays, and smoke. **Table 17.5** lists some common colloidal systems and gives examples of familiar colloids.

The properties of colloids differ from those of solutions and suspensions. Many colloids are cloudy or milky in appearance when concentrated,

STUDENT RESOURCES

From the Teacher's Resource Package, use:

- Section Review 17.4, Ch. 17 Practice Problems, Vocabulary Review, Quizzes and Tests from the Review Module (Ch. 17–20)
- Laboratory Recordsheet 17-2

TECHNOLOGY RESOURCES

Relevant technology resources include:

- Chem ASAP! CD-ROM
- Resource Pro CD-ROM
- Assessment Resources CD-ROM: Chapter 17 Tests

1 No; it is an aqueous solution.

Table 17.5

Some Colloidal Systems			
System			
Dispersed phase	Dispersion medium	Type	Example
gas	liquid	foam	whipped cream
gas	solid	foam	marshmallow
liquid	liquid	emulsion	milk, mayonnaise
liquid	gas	aerosol	fog, aerosols
solid	gas	smoke	dust in air
solid	liquid	sols and gels	egg white, jellies, paint, blood, colloidal gold, starch in water, gelatin

but they look clear or almost clear when they are very dilute. Intermediate-sized particles in a colloid cannot be retained by filter paper and do not settle out with time.

Colloidal particles exhibit the **Tyndall effect,** which is the scattering of visible light in all directions. You can see a beam of light passed through a colloid just as you see a sunbeam in a dusty room. Colloidal particles scatter light in the same way that dust particles scatter sunlight. Suspensions also exhibit the Tyndall effect, but solutions never do. Particles of solute in a true solution are too small to produce this effect. **Figure 17.19** on the following page summarizes the Tyndall effect with respect to solutions, colloids, and suspensions. Would carbonated mineral water exhibit the Tyndall effect? 

17.4

Motivate and Relate

Many foods and other familiar products are colloids or suspensions. Bring to class examples of these products. For example, compare orange juice to orange soda. Ask students to apply the criteria they have learned to determine the proper classification for these items. (The soda is a solution; the presence or absence of pulp affects the classification of the juice.)

Milk is a colloid of protein and fat. The curdling of milk occurs when bacteria produce enough lactic acid to cause dispersed protein and fat particles to coagulate into larger particles, which separate from the rest of the mixture. This process is comparable to the clotting of blood.

Ask students why paint needs to be stirred before it is used. Explain that latex paint is a complex mixture of binders, pigments, and drying agents. Some ingredients are water-soluble; others are present as dispersed particles.

2 Teach

Use the Visual

Display Table 17.5 on an overhead projector. Explain that colloids are characterized by the physical state of the *dispersed phase* and of the *continuous phase*—also called *dispersion medium.* Help students understand this two-part system by comparing it to that of a true solution which contains the *solute* and the *solvent,* respectively. Table 17.5 lists various types of colloids.

Meeting Diverse Needs

Gifted Ask students to compare the water purification processes used in wastewater treatment facilities to those used on drinking water before it is piped to homes to purification that takes place naturally during the water cycle. Natural purification primarily occurs in three ways: the water cycle of evaporation and precipitation is like a huge distillation unit removing any dissolved salts. Soil bacteria breakdown many organic compounds and render them harmless. Filtration through sandy loams (diatomaceous earth) removes suspended matter and may absorb colloidal material.

Discuss

Point out that the main difference between solutions, suspensions, and colloids is particle size. Solution particles are typically less than 1 nm in diameter. Colloid particles are between 1 nm and 100 nm in diameter. Suspension particles are typically larger than 100 nm. Smaller particles are less susceptible to the effects of gravity and are influenced more by the effects of Brownian movement—the chaotic motions of microscopic particles. The collisions of molecules with extremely small colloidal particles are sufficiently energetic to move colloidal particles in a random fashion that prevents their settling to the bottom.

3 Assess

Evaluate Understanding

Ask students to use Table 17.6 to answer the following questions:

- **In what way are colloids similar to solutions?** (In both cases dispersed particles are small enough to pass through standard filter paper and to withstand the gravitational pull toward the bottom.)
- **In what way are colloids similar to suspensions?** (Both types of mixtures produce the Tyndall effect.)

Ask students why it is usually recommended that drivers use low beams, when driving under foggy conditions at night? (Fog is a colloid that produces the Tyndall effect. Bright lights produce a higher degree of light scattering in all directions, including straight back into the driver's eyes.)

(a)

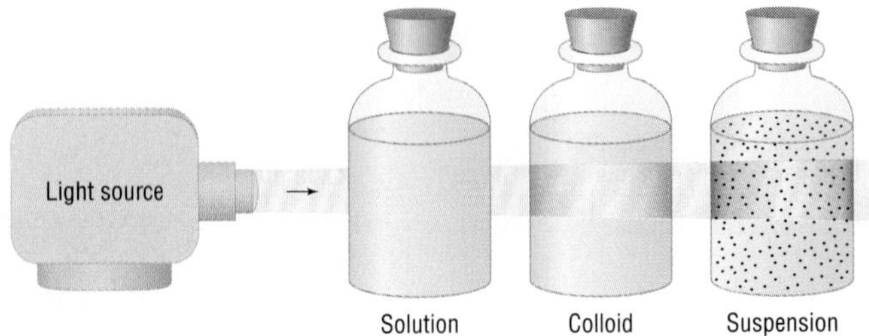

(b)

Figure 17.19
(a) *Fog is a colloid and thus exhibits the Tyndall effect.*
(b) *Particles in colloids and suspensions reflect light in all directions. This effect is not observed with true solutions.*

Flashes of light, or scintillations, are seen when colloids are studied under a microscope. Colloids scintillate because the particles reflecting and scattering the light move erratically. The chaotic movement of colloidal particles is called **Brownian motion**, which was first observed by the Scottish botanist Robert Brown (1773–1858). While he was studying water samples with a microscope, Brown noticed the random movement of pollen grains suspended in the water. Brownian motion is caused by collisions of the water molecules of the medium with the small, dispersed colloidal particles. These collisions prevent the colloidal particles from settling.

Colloidal particles may also adsorb (collect) ions from the surrounding medium onto their surface. Some colloidal particles can adsorb positively charged ions and become positively charged. Other colloidal particles can adsorb negatively charged ions and become negatively charged. All the particles in a particular colloidal system will have the same charge. The repulsion of the like-charged particles prevents the particles from forming aggregates, which keeps them dispersed throughout the medium. When ions of opposite charge are added to the colloidal system, charged colloidal particles form aggregates and precipitate from the dispersion. **Table 17.6** summarizes the properties of solutions, colloids, and suspensions.

Table 17.6

Properties of Solutions, Colloids, and Suspensions

Property	System		
	Solution	**Colloid**	**Suspension**
Particle type	ions, atoms, small molecules	large molecules or particles	large particles or aggregates
Particle size (approximate)	0.1–1 nm	1–100 nm	100 nm and larger
Effect of light	no scattering	exhibits Tyndall effect	exhibits Tyndall effect
Effect of gravity	stable, does not separate	stable, does not separate	unstable, sediment forms
Filtration	particles not retained on filter	particles not retained on filter	particles retained on filter
Uniformity	homogeneous	borderline	heterogeneous

Meeting Diverse Needs

At Risk and Sight Impaired Textures of foods can dramatically affect the "feel" and thus the pleasure of eating. Ice cream is a colloidal suspension of fats, which are whipped with milk, various stabilizers, and thickeners. Have students make ice cream the old fashioned way, by hand.

MINI LAB

Surfactants

PURPOSE

To observe the unusual surface properties of water due to hydrogen bonding.

MATERIALS

- shallow dish or Petri dish
- water
- paper clip
- rubber band, approximately 2 inches in diameter
- vegetable oil
- liquid dish detergent

PROCEDURE

1. Thoroughly clean and dry the dish.
2. Fill the dish most of the way with water. Dry your hands.
3. Being careful not to break the surface, gently place the paper clip on the water. Observe what happens.
4. Repeat Steps 1 and 2.
5. Gently place the open rubber band on the water.
6. Slowly add oil, drop by drop, onto the water encircled by the rubber band until that water is covered with a layer of oil. Observe for 15 seconds.
7. Allow one drop of dish detergent to fall onto the center of the oil layer. Observe the system for 15 seconds.

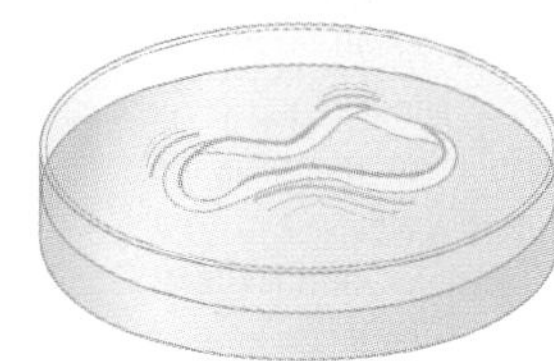

ANALYSIS AND CONCLUSIONS

1. What happened to the paper clip? Explain.
2. If a paper clip becomes wet, does it float? Explain your answer.
3. What shape did the rubber band take when the water inside it was covered with oil? Why did it take the observed shape?
4. Describe what happened when dish detergent was dropped onto the layer of oil.

Emulsions are colloidal dispersions of liquids in liquids. An emulsifying agent is essential for the formation of an emulsion and for maintaining the emulsion's stability. For example, oils and greases are not soluble in water. However, they readily form a colloidal dispersion if soap or detergent is added to the water. Soap and detergents are emulsifying agents. One end of the large soap or detergent molecule is polar and is attracted to water molecules. The other end of the soap or detergent molecule is nonpolar and is soluble in oil or grease. Soaps and other emulsifying agents thus allow the formation of colloidal dispersions between liquids that do not ordinarily mix.

Figure 17.20
Many salad dressings are liquid-in-liquid colloids, or emulsions. When vinegar and oil are shaken, a temporary emulsion forms. If an emulsifying agent, such as egg yolk, is used and the oil is added slowly to the egg–vinegar mixture, the result is mayonnaise, a stable emulsion of vinegar and oil.

section review 17.4

15. What is the basis for distinguishing among solutions, colloids, and suspensions?
16. What is the Tyndall effect?
17. What is Brownian motion?
18. What are emulsions? How are emulsions stabilized?

 Chem ASAP! **Assessment 17.4** Check your understanding of the important ideas and concepts in Section 17.4.

17.4

Reteach

On the board, draw the relative sizes of spherical solute, colloid, and suspension particles. Make an analogy to golf balls, baseballs, and basketballs. Ask students which "particles" would get caught in a sieve a bit smaller than a basketball hoop and which would pass through? Remind students that solution particles are typically less than 1 nm in diameter while colloid particles are ten to one hundred times larger (between 1 and 100 nm) and those of suspensions a magnitude larger again at more than 100 nm.

MINI LAB

Surfactants

LAB TIPS

For best results, be sure to have students dry their hands before touching the paper clips.

ANALYSIS AND CONCLUSIONS

1. The paper clip floats. The surface tension of the water supports the paper clip.
2. No. The surface tension is broken and the clip sinks, because it is made of metal that is more dense than water.
3. The rubber band takes on a circular shape. The surface tension of the surrounding water pulling the rubber band radially outward is not balanced by the surface tension of the oil.
4. The oil layer is broken, and the rubber band returns to its original shape.

Answers

SECTION REVIEW 17.4

15. particle size (smallest in solutions, largest in suspensions)
16. The Tyndall effect is the scattering of visible light by the particles in a colloid or suspension.
17. Brownian motion is chaotic movement of colloidal particles.
18. Emulsions are colloidal dispersions of liquids in liquids. An emulsifying agent interacts with both liquids and keeps the emulsion stable.

Chemistry Serving... the Environment

DISCUSS

Discuss the ways in which the surface temperature of the ocean affects climate. Ask students to explain how large bodies of water act to "buffer" large swings in temperature. You may wish to have students do research to find out more about ocean currents. Students could construct a diagram of ocean currents that shows, for example, the drift of warm waters between the tropics and the North Atlantic. The large heat capacity of water means that warm waters from the equator can travel long distances before they cool down again. These waters mix with the colder waters surrounding Britain and Scandinavia.

The boundary between large masses of cold and warm water in the ocean is quite sharp. This results in distinct domains of low and high density water, which has important consequences for animals and underwater vessels (submarines) that use sonar for navigation, and, in the case of marine mammals, to locate food. The speed of sound through a body of water depends on the density of the water. Scientists have used strategically placed sonar devices to make incredibly accurate measurements of the temperature of the world's oceans by measuring the velocity of sound in the water. Submarines exploit the transition zones between cold and warm masses of surface water to hide from enemy sonar detection devices.

CHEMISTRY IN CAREERS

Have students read more about careers in oceanography on page 876.

Chemistry Serving...the Environment

IT'S THE WATER

Earth is a watery planet. Oceans cover 72% of Earth's surface, and large amounts of water, in the form of water vapor, exist in Earth's atmosphere. Weather conditions around the planet are mostly the result of complex interactions between the water in Earth's oceans and in the atmosphere.

Large-scale weather patterns around the globe are somewhat stable. Some parts of the world usually receive large amounts of rainfall at certain times; other parts receive very little. These patterns are in part the result of how water in the oceans is heated and circulates around the planet.

Sometimes, for reasons scientists do not understand, water circulation in the oceans shifts away from the norm, causing major changes in the weather over large parts of Earth. One of these shifts, called El Niño, has now been observed and studied many times in the Pacific Ocean. It occurs, on average, every seven years.

El Niño begins in the waters just north of the equator, where the ocean surface water is heated rapidly by the nearly vertical rays of sunlight. Because of water's high heat capacity, this water tends to stay warm, and because warm water is less dense than cold water, it also tends to stay at the surface. Normally, trade winds blow from east to west across the Pacific and push the warm surface water to the west, toward Indonesia. If the trade winds die down, however, the warm water that usually pools in the western Pacific spreads toward the east. This shift is the El Niño effect.

Moving east, the warm water of El Niño reaches the coast of South America. Normally, cold, nutrient-rich waters rise from deep in the Pacific off the western coast of South America. The nutrients in the water fuel the growth of microscopic plants and animals, which are food for large fish populations. But as warmer waters move east during El Niño, the nutrient-rich colder waters are kept from rising. The microscopic ocean plants and animals have less food, and fish populations suffer.

When an El Niño occurs, among the first people to feel its effects are those who fish off the western coast of South America. They are also the ones who gave El Niño its name. They noted that the changes brought on by El Niño often begin around Christmas, the Christian celebration of the birth of Jesus, who is called "El Niño" (the Child) in South America.

> At no other time is the ocean's influence on weather more obvious than during El Niño.

El Niño's effects spread well beyond South America. The warm water bumps up against the South American coast and spreads north to the western coast of North America, affecting the weather and fish populations there. In Indonesia, where the nearby ocean is cooler than normal, there is often a severe drought.

El Niño covers so much of the Pacific that it influences the movement of the jet stream, a high-altitude wind that steers storms through the atmosphere. The effect on the jet stream is unpredictable, but the usual result is wetter-than-normal weather in parts of South America, North America, and Asia, and drier-than-normal weather in other parts of the world. At no other time is the ocean's influence on global weather more obvious than during El Niño.

CHEMISTRY IN CAREERS

OCEANOGRAPHER
Study ocean surface-water temperature, salt content, and nutrient levels.
See page 876.

CHEMIST NEEDED
Enviro. Lab. GC Must know

TAKE IT TO THE NET
Find out more ab
career opportunit
www.phschool.c

CHEMICAL SPECIALIS
Local food service distributor
responsible self-motivated ir

Chapter 17 STUDENT STUDY GUIDE

Take It to the NET
For interactive study and review, go to www.phschool.com

KEY TERMS

- aqueous solution *p. 482*
- Brownian motion *p. 492*
- colloid *p. 490*
- deliquescent *p. 487*
- desiccant *p. 487*
- effloresce *p. 487*
- electrolyte *p. 483*
- emulsion *p. 493*
- hygroscopic *p. 487*
- nonelectrolyte *p. 484*
- solute *p. 482*
- solvation *p. 483*
- solvent *p. 482*
- strong electrolyte *p. 485*
- surface tension *p. 477*
- surfactant *p. 477*
- suspension *p. 490*
- Tyndall effect *p. 491*
- water of hydration *p. 485*
- weak electrolyte *p. 485*

KEY EQUATION

- Percent $H_2O = \frac{\text{mass of water}}{\text{mass of hydrate}} \times 100\%$

CONCEPT SUMMARY

17.1 Liquid Water and Its Properties

- Water molecules are polar and they extensively hydrogen-bond with one another.
- Intermolecular hydrogen bonding explains the high freezing point (0 °C) and high boiling point (100 °C) of water.
- Hydrogen bonding is responsible for the high surface tension of water.

17.2 Water Vapor and Ice

- Extensive hydrogen bonding between water molecules explains water's high heat of vaporization.
- Ice floats in liquid water because it is less dense than liquid water.

17.3 Aqueous Solutions

- Aqueous solutions are homogeneous mixtures of ions or molecules in water. The solubility of a solute depends on solute–solvent interactions. A good rule to remember is "like dissolves like."
- Substances that are in solution as ions are electrolytes. A solute that is completely ionized in solution is a strong electrolyte. A solute that is only partially ionized is a weak electrolyte.
- Many crystals are hydrates; they contain water of hydration, which can be lost through heating. Loss of water of hydration from a hydrate when exposed to air is called efflorescence.
- Hygroscopic substances take up water from moist air. A deliquescent substance takes up so much moisture it forms an aqueous solution.

17.4 Heterogeneous Aqueous Systems

- Particles in a suspension have an average diameter greater than 100 nm and can be kept in suspension if the fluid (water) is kept agitated. Gravity or filtration, however, will separate the suspended particles from the liquid.
- The particles in a colloidal dispersion range in size from 1 nm to 100 nm. In general, they do not settle under gravity, and they pass through ordinary filter paper unchanged.
- Colloids and suspensions scatter light, as evidenced by the Tyndall effect. Colloidal dispersions also exhibit Brownian motion.

CHAPTER CONCEPT MAP

Use these terms to construct a concept map that organizes the major ideas of this chapter.

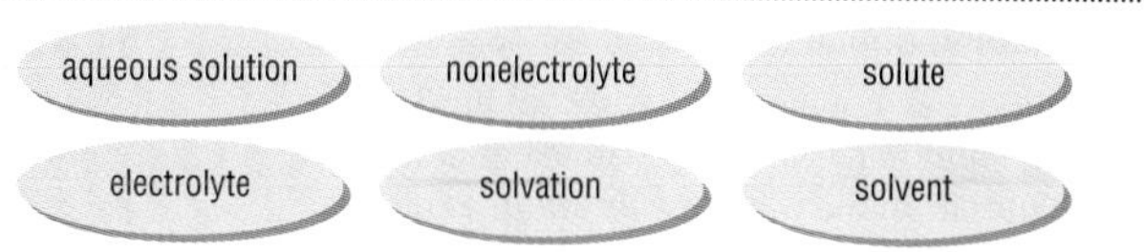

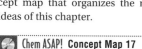

Chem ASAP! Concept Map 17
Create your Concept Map using the computer.

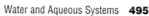

Take It to the Net

At **www.phschool.com** students will find for this chapter

- an Internet activity
- links to related chemistry sites
- an interactive quiz
- career links

Chapter 17
STUDENT STUDY GUIDE

4 Close

Summary

Prepare a series of aqueous systems (200 mL to 500 mL total volume). Use large glass jars with caps to hold ice, water, copper(II) sulfate(aq), acetic acid(aq), sucrose(aq), milk, or other colloids such as sodium thiosulfate in aqueous acid (2.5 g $Na_2S_2O_3 \cdot 5H_2O$ added to 500 mL of water containing 2.5 mL of concentrated HCl) and potting soil suspended in water by shaking. Label each jar and have students note and observe the contents. Have students use the Key Terms from this chapter to organize their observations of the jars.

Extension

Have students research the use of light scattering as a method for measuring the density of particles in colloids and suspensions. In addition, students may wish to study which wavelengths of light are scattered most effectively by a colloid.

Looking Back . . . Looking Ahead . . .

The preceding chapter discussed how covalent bonds are formed and how the three-dimensional shapes of molecules can be predicted using VSEPR theory. The current chapter relates the chemical and physical properties of water to its molecular structure. It also defines and classifies aqueous systems. The next chapter describes the factors that determine the solubility of a solute and the rate at which a water-soluble solute dissolves.

Answers

19. Oxygen is more electronegative than hydrogen. Because the water molecules are bent, the bond polarities do not cancel.
20. Surface molecules are attracted to the liquid molecules below but not to the air. Molecules inside the liquid are attracted in all directions.
21. Some water molecules have sufficient energy to leave the surface of the liquid. Vapor pressure is the pressure of the water vapor above the liquid.
22. Drops are spherical; objects denser than a liquid can float on the surface.
23. It physically interferes with hydrogen bonding and reduces surface tension.
24. No; specific heat is expressed per gram.
25. water, 2.0×10^4 cal; iron, 2.2×10^3 cal
26. 11 J
27. Ammonia molecules form hydrogen bonds, whereas methane molecules do not. Substances whose molecules form hydrogen bonds tend to have higher boiling points.
28. 1.3×10^4 cal
29. Ice floats in liquid water. Solids generally sink in their liquids because a substance is generally more dense in the solid state.
30. 8.0 kJ
31. Bodies of water would freeze from the bottom up. This would kill many forms of aquatic life.
32. Solutions are homogeneous mixtures in which a solute is dissolved in a solvent. Aqueous solutions are solutions that have water as the solvent.
33. The polar water molecules electrostatically attract ions and polar covalent molecules. Polar compounds will dissolve but nonpolar compounds are unaffected as they have no charges.
34. No; the molecules and ions are smaller than the pores of the filter.

Chapter 17 REVIEW

CONCEPT PRACTICE

19. Explain why water molecules are polar. *17.1*
20. Why do the particles at the surface of a liquid behave differently from those in the bulk of the liquid? *17.1*
21. Describe the origin of the vapor pressure of water. *17.1*
22. Describe some observable effects that can be produced by the surface tension of a liquid. *17.1*
23. What is a surfactant? Explain how it works. *17.1*
24. Does the specific heat capacity of water vary depending on the quantity of water? Explain. *17.1*
25. How many calories are required to heat 256 g of water from 20 °C to 99 °C? How many calories are required to heat the same mass of iron through the same range of temperature? *17.1*
26. How many joules are required to vaporize 5.0 mg of water at its boiling point? *17.2*
27. Why does ammonia have a much higher boiling point (−33 °C) than methane (−164 °C) even though their molar masses are almost the same? *17.2*
28. How many calories are liberated when 24 g of steam at 100 °C condenses to liquid water at 100 °C? *17.2*
29. What characteristic of ice distinguishes it from other solid substances? *17.2*
30. How many kilojoules are required to melt 24 g of ice at 0 °C? *17.2*
31. What would be some of the consequences if ice were denser than water? *17.2*
32. Distinguish between a solution and an aqueous solution. *17.3*
33. Why is water an excellent solvent for most ionic compounds and polar covalent molecules but not for nonpolar compounds? *17.3*
34. Suppose an aqueous solution contains both sugar and salt. Can you separate either of these solutes from the water by filtration? Explain your reasoning. *17.3*
35. Describe how an ionic compound dissolves in water. *17.3*
36. Which of the following substances dissolve appreciably in water? Give reasons for your choice. *17.3*
    a. HCl d. $MgSO_4$ f. $CaCO_3$
    b. NaI e. CH_4 g. gasoline
    c. NH_3
37. Why does molten sodium chloride conduct electricity? *17.3*
38. What is the main distinction between an aqueous solution of a strong electrolyte and an aqueous solution of a weak electrolyte? *17.3*
39. Describe the water of hydration of a substance. *17.3*
40. Write formulas for these hydrates. *17.3*
    a. sodium sulfate decahydrate
    b. magnesium sulfate heptahydrate
    c. barium hydroxide octahydrate
41. Name each hydrate. *17.3*
    a. $SnCl_4 \cdot 5H_2O$ c. $BaBr_2 \cdot 4H_2O$
    b. $FeSO_4 \cdot 7H_2O$ d. $FePO_4 \cdot 4H_2O$
42. Epsom salt ($MgSO_4 \cdot 7H_2O$) changes to the monohydrate form at 150 °C. Write an equation for this change. *17.3*
43. Name two ways to distinguish a suspension from a colloid. *17.4*
44. Solutions do not demonstrate the Tyndall effect. Why? *17.4*
45. What makes a colloidal dispersion stable? *17.4*
46. Define Brownian motion. *17.4*

CONCEPT MASTERY

47. Water has its maximum density at 4 °C. Explain why this is so, and discuss the consequences of this fact.
48. From your knowledge of intermolecular forces, arrange these liquids in order of increasing surface tension: water (H_2O), hexane (C_6H_{14}), ethanol (C_2H_5OH).
49. Name these hydrates and determine the percent by mass of water in each.
    a. $Na_2CO_3 \cdot H_2O$ b. $MgSO_4 \cdot 7H_2O$
50. If 5.00 g of steam (H_2O)(*g*) at 100.0 °C condenses to liquid water (H_2O)(*l*) that is cooled to 50.0 °C, how much heat is liberated in
    a. calories? b. kilocalories?

35. The positive ions are attracted by the negatively charged end of the polar water molecule; the negative ions are attracted by the positively charged end. As the ions are pulled away from the crystal, they are surrounded by the water molecules.
36. a. HCl (polar) will dissolve.
    b. NaI (ionic) will dissolve.
    c. NH_3 (polar) will dissolve.
    d. $MgSO_4$ (ionic) will dissolve.
    e. CH_4 (nonpolar) will not dissolve.
    f. $CaCO_3$ (strong ionic forces) will not dissolve.
    g. Gasoline (nonpolar) will not dissolve.
37. Its ions are free to move toward an electrode.
38. An aqueous solution of a strong electrolyte is almost totally ionized.
39. water in the crystal structure of a substance
40. a. $Na_2SO_4 \cdot 10H_2O$
    b. $MgSO_4 \cdot 7H_2O$
    c. $Ba(OH)_2 \cdot 8H_2O$
41. a. tin(IV) chloride pentahydrate
    b. iron(II) sulfate heptahydrate

51. Explain why ions become solvated in aqueous solution.

52. A block of ice at 0 °C has a mass of 176.0 g. How much heat must be added to change 25% of this mass of ice to liquid water at 0 °C? Express your answer in calories, kilocalories, joules, and kilojoules. What is the mass of ice remaining?

53. You have a solution containing either sugar or salt dissolved in water.
  a. Can you tell which it is by visual inspection?
  b. Give two ways by which you could easily tell which it is.

54. Water is a polar solvent; gasoline is a nonpolar solvent. Decide which compounds are more likely to dissolve in water and which are more likely to dissolve in gasoline.
  a. sucrose ($C_{12}H_{22}O_{11}$)
  b. Na_2SO_4
  c. methane (CH_4)
  d. KCl

55. Match each term with the following descriptions. A description may apply to more than one term.
  a. true solution
  b. colloidal
  c. suspension
  (1) does not settle out on standing
  (2) heterogeneous mixture
  (3) particle size less than 1.0 nm
  (4) particles can be filtered out
  (5) demonstrates Tyndall effect
  (6) particles are invisible to the unaided eye
  (7) homogenized milk
  (8) salt water
  (9) jelly

56. Explain which properties of water are responsible for these occurrences.
  a. Water in tiny cracks in rocks helps break up the rocks when it freezes.
  b. Water beads up on a newly waxed car.
  c. As you exercise and your body temperature increases, your body cools itself by producing sweat.
  d. Temperatures below 28 °F damage grapevines. When severe frost is predicted, grape growers spray a mist of water on their vines.
  e. An efficient way of heating a large building is to generate steam in a boiler and circulate it through pipes to radiators throughout the building.

57. Explain why ethanol (C_2H_5OH) will dissolve in both gasoline and water.

58. A 25.0 m × 10.0 m swimming pool is filled with fresh water to a depth of 1.7 m. The water temperature is initially 25 °C. How much heat must be removed from the water to change it all to ice at 0 °C? Express your answer in kilocalories and kilojoules.

59. Are all liquids soluble in each other? Explain.

60. Write equations to show how these substances ionize or dissociate in water.
  a. NH_4Cl
  b. $Cu(NO_3)_2$
  c. $HC_2H_3O_2$
  d. $HgCl_2$

61. The following graph shows the density of water over the temperature range 0 °C to 20 °C.

  a. What is the maximum density of water?
  b. At what temperature does the maximum density of water occur?
  c. Would it be correct to extend the smooth curve of the graph below 0 °C?

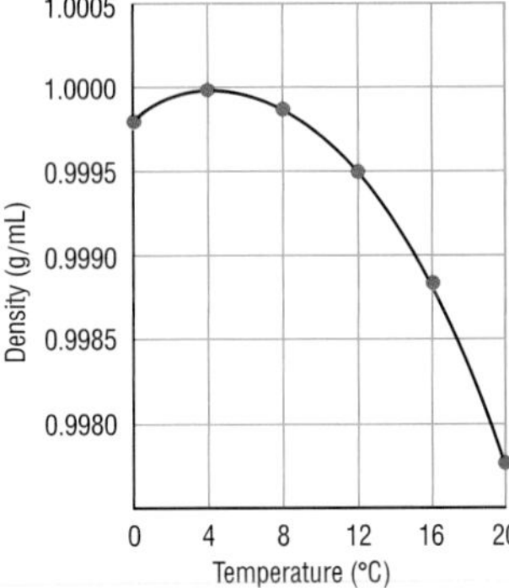

CRITICAL THINKING

62. Choose the term that best completes the second relationship.
  a. plant : green — water molecule : ______
  (1) polar
  (2) ionic
  (3) frozen
  (4) vapor
  b. east : west — condensation : ______
  (1) boiling
  (2) vapor pressure
  (3) vaporization
  (4) freezing
  c. colloid : emulsion — cat : ______
  (1) dog
  (2) Siamese
  (3) poodle
  (4) fox

c. barium bromide tetrahydrate
d. iron(III) phosphate tetrahydrate

42. $MgSO_4{\cdot}7H_2O \rightarrow MgSO_4{\cdot}H_2O + 6H_2O$

43. A suspension is a mixture with large particles that settle on standing. A colloid is a mixture with particles of intermediate size that do not settle.

44. The molecules or ions are too small to have reflective surfaces.

45. the random motion of the dispersion medium molecules

46. the random movement of colloidal particles

47. Water molecules at 4°C are tightly packed and have maximum density. Below 4°C the water molecules arrange in a regular network because of the attractions between them. As a result, ice has a lower density than water and floats.

48. hexane, ethanol, water

49. a. sodium carbonate monohydrate, 14.5% H_2O
  b. magnesium sulfate heptahydrate, 51.1% H_2O

Chapter 17 REVIEW

50. a. 2.95×10^3 cal b. 2.95 kcal

51. Ions in solution are surrounded by water molecules. Negative ions are attracted to the hydrogen atoms and positive ions are attracted to the oxygen atoms.

52. 3.52×10^3 cal, 3.52 kcal, 1.47×10^4 J, 14.7 kJ; 132.0 g H_2O

53. a. no
  b. drying to examine the crystals, testing for electrical conductivity, doing a flame test

54. a. water b. water c. gasoline d. water

55. 1. a, b 2. b, c 3. a 4. c 5. b, c 6. a, b 7. b 8. a 9. b

56. a. Water expands when it freezes to ice.
  b. Water is polar and wax is nonpolar, and water has a high surface tension.
  c. Sweat evaporates to carry away the heat of vaporization.
  d. Freezing damages plants because water inside them expands as it freezes. Water on the surface gives off the heat of fusion to the plants when it freezes.
  e. Steam carries the heat of vaporization.

57. Ethanol has a polar hydroxyl end (—OH) that dissolves in water, and a nonpolar hydrocarbon end (C_2H_5—) that dissolves in gasoline.

58. 4.5×10^7 kcal, 1.9×10^8 kJ

59. No; nonpolar molecules do not dissolve in polar molecules.

60. a. $NH_4Cl \rightarrow NH_4^+ + Cl^-$
  b. $Cu(NO_3)_2 \rightarrow Cu^{2+} + 2NO_3^-$
  c. $HC_2H_3O_2 \rightarrow H^+ + C_2H_3O_2^-$
  d. $HgCl_2 \rightarrow Hg^{2+} + 2Cl^-$

61. a. 1.00 g/mL b. 4°C
  c. No; the density of ice is 0.917 g/mL at 0°C. There would be a break in the curve at 0°C as liquid water at 0°C changes to solid water (ice) at 0°C.

62. a. 1
  b. 3
  c. 2

Chapter 17
REVIEW

Answers

63. The blood and all of the body's cells contain large amounts of water. Most of the important chemical reactions of life take place in aqueous solution inside cells.

64. The nonpolar liquid would form a layer in the water. The layer would be on top of the water if its density were less than that of water and on the bottom if its density were greater. A temporary emulsion would form if the mixture were shaken.

65. **a.** 195 g
b. 1.95×10^5 mg
c. 0.195 kg

66. 1.27 atm

67. **a.** raises the boiling point
b. lowers the boiling point

68. 3.60×10^{-2} g H_2O, 2.24×10^{-2} L O_2

69. 9.00 g H_2, 72.0 g O_2

70. 25.7 g H_2O

71. 636 g C_2H_4O

72. **a.** pink
b. pink
c. blue
d. 45.4%
e. water or water vapor

73. **a.** As the molar mass increases, the boiling point increases.
b. The boiling point more than doubles.

74. **a.** The volume of 1 g ice at 0°C is greater.
b. The volume of 1 g liquid water at 100°C is less.

75. **a.** hydrogen
b. 4.9×10^{-2} g H_2O
c. oxygen
d. 10 cm^3 O_2

63. When the humidity is low and the temperature high, humans must take in large quantities of water or face serious dehydration. Why do you think water is so important for the proper functioning of your body?

64. Describe as specifically as possible what would happen if a nonpolar molecular liquid were added to water. What would form if you shook this mixture vigorously?

CUMULATIVE REVIEW

65. A cylindrical vessel, 28.0 cm in height and 3.00 cm in diameter, is filled with water at 50 °C. The density of water is 0.988 g/cm^3 at this temperature. Express the mass of water in the vessel in the following units.
a. grams **b.** milligrams **c.** kilograms

66. The temperature of 1 L of steam at constant volume and 1.00 atm pressure is increased from 100 °C to 200 °C. Calculate the final pressure of the steam in atmospheres, assuming the volume does not change.

67. How do the following changes in the pressure on the surface of water affect the water's boiling point?
a. an increase in pressure
b. a decrease in pressure

68. The decomposition of hydrogen peroxide is given by this equation.

$$2H_2O_2(l) \longrightarrow 2H_2O(l) + O_2(g)$$

Calculate the mass of water (in grams) and the volume of oxygen at STP formed when 2.00×10^{-3} mol of hydrogen peroxide is decomposed.

69. How many grams each of hydrogen and oxygen are required to produce 4.50 mol of water?

70. Calculate the mass of water produced in the complete combustion of 8.00 L of propane (C_3H_8) at STP, given the unbalanced equation

$$C_3H_8 + O_2 \longrightarrow CO_2 + H_2O$$

71. Acetaldehyde (C_2H_4O) is produced commercially by the reaction of acetylene (C_2H_2) with water, as shown by the equation

$$C_2H_2 + H_2O \longrightarrow C_2H_4O$$

How many grams of C_2H_4O can be produced from 2.60×10^2 g H_2O, assuming sufficient C_2H_2 is present?

CONCEPT CHALLENGE

72. Cobalt chloride test paper is blue. This paper is made by soaking strips of paper in an aqueous solution of $CoCl_2 \cdot 6H_2O$. These paper strips are then dried in an oven.

$$CoCl_2 \cdot 6H_2O \text{ (pink)} + \text{heat} \longrightarrow CoCl_2 \text{ (blue)} + 6H_2O$$

a. When cobalt(II) chloride hexahydrate is dissolved in water, what is the color of the solution?
b. What is the color of wet cobalt chloride paper?
c. What is the color of dry cobalt chloride paper?
d. What is the percent by mass of water in the hexahydrate?
e. What does cobalt chloride test paper test for?

73. A graph of boiling point versus molar mass for the noble gases is given below.
a. Make a statement summarizing the observed trend relating these two physical properties.
b. How does a doubling of the molar mass (from Ne to Ar) affect the boiling point?

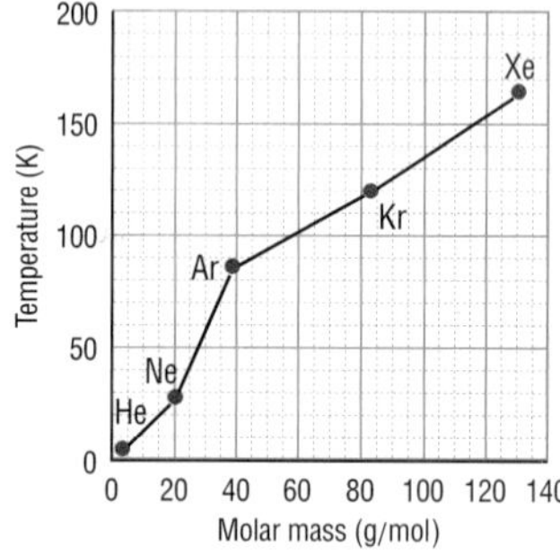

74. What relationships exist between the following volumes?
a. 1 g of ice at 0 °C and 1 g of liquid water at 0 °C
b. 1 g of liquid water at 100 °C and 1 g of steam at 100 °C

75. A mixture of 40 cm^3 of oxygen gas and 60 cm^3 of hydrogen gas at STP is ignited.
a. Which gas is the limiting reagent, oxygen or hydrogen?
b. What is the mass of water produced?
c. Which gas remains after reaction?
d. What is the volume, at STP, of the remaining gas?

Chapter 17 STANDARDIZED TEST PREP

Select the choice that best answers each question or completes each statement.

1. When a sugar cube completely dissolves in a glass of water, it forms
- **a.** a colloid.
- **b.** an emulsion.
- **c.** a suspension.
- **d.** a solution.

2. How many water molecules are tied up per formula unit of a compound that is an octahydrate?
- **a.** nine
- **b.** eight
- **c.** seven
- **d.** six

3. Which property is characteristic of water?
- **a.** relatively low boiling point
- **b.** relatively low vapor pressure
- **c.** relatively low surface tension
- **d.** relatively low heat of vaporization

4. Which statement is *false*?
- **a.** If less than 1% of an acid is ionized in water, the acid is a weak electrolyte.
- **b.** A weak electrolyte can become a strong electrolyte with the addition of more solute.
- **c.** Because melted wax does not conduct an electrical current, it is a nonelectrolyte.
- **d.** $(NH_4)_3PO_4(aq)$ is a strong electrolyte.

Use this description and the data table in the next column to answer questions 5–7.

A student used a conductivity meter to measure the ability of several aqueous solutions to conduct electric current. The magnitude of the conductivity value is proportional to the number of ions in the solution. The SI conductivity unit is the microsiemens/cm (μS/cm).The table gives the results reported by the student who tested six solutions plus distilled water. Each solution had a concentration of 0.02*M*.

Solution	Conductivity (μS/cm)
potassium chloride, KCl	2050
aluminum chloride, $AlCl_3$	4500
calcium chloride, $CaCl_2$	3540
sodium hydroxide, NaOH	2180
ethanol, C_2H_5OH	0
magnesium bromide, $MgBr_2$	3490
distilled water	0

5. Why do distilled water and the ethanol solution have zero conductivity?

6. Explain why two pairs of conducting solutions have similar conductivities.

7. The $AlCl_3$ solution has a conductivity that is about twice that of the KCl solution. Offer an explanation for this ratio.

Use the atomic windows to answer question 8.

8. Atomic window (a) represents solute particles in a given volume of solution. Which window represents the solute particles in the same volume of solution when the amount of solvent is doubled?

a.
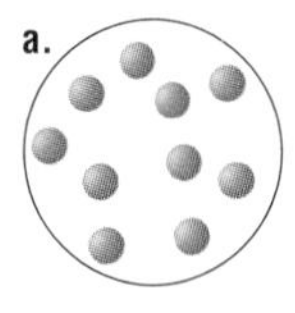

b.
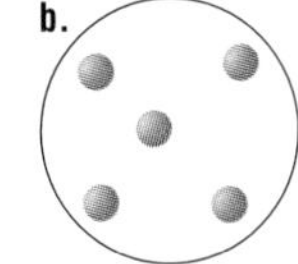

c.
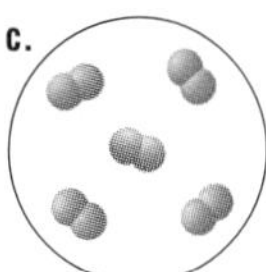

d.
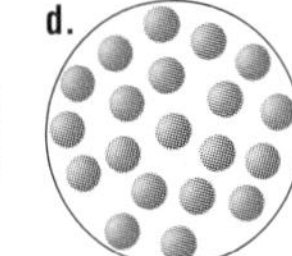

For each question there are two statements. Decide whether each statement is true or false. Then decide whether Statement II is a correct explanation for Statement I.

Statement I		Statement II
9. Water has a relatively high surface tension.	BECAUSE	Water molecules form strong hydrogen bonds with other water molecules.
10. Methanol, CH_3OH, is a strong electrolyte.	BECAUSE	Methanol molecules can form hydrogen bonds with water.
11. Particles in a colloid settle out faster than particles in a solution.	BECAUSE	Particles in a colloid are larger than particles in a solution.
12. Water molecules are polar.	BECAUSE	The bond between hydrogen and oxygen atoms in a water molecule is polar.

Chapter 17 STANDARDIZED TEST PREP

1. d
2. b
3. b
4. b
5. When water is purified by distillation, dissolved ions are removed. Ethanol is a molecular compound that does not ionize in water.
6. KCl and NaOH have similar conductivities because they provide two moles of ions per mole of compound. $CaCl_2$ and $MgBr_2$ have similar conductivities because they provide three moles of ions per mole of compound.
7. Aluminum chloride provides twice as many ions per mole as potassium chloride.
8. b
9. True, True, correct explanation
10. False, True
11. False, True
12. True, True

Planning Guide

SECTION OBJECTIVES	ACTIVITIES/FEATURES	MEDIA & TECHNOLOGY
18.1 Properties of Solutions ● ■ ◆ ▸ Identify the factors that determine the rate at which a solute dissolves ▸ Calculate the solubility of a gas in a liquid under various pressure conditions	**SE** **Discover It!** *Salt and the Freezing Point of Water,* p. 500 **SE** **Link to Agriculture** *Fertilizer Runoff,* p. 502 **SE** **Mini Lab** *Solutions and Colloids,* p. 508 (**LRS** 18-1) **LM** 30: *Factors Affecting Solution Formation* **LM** 31: *Supersaturation* **SSLM** 23: *Solubility Rules* **TE** **DEMO**, p. 505	**ASAP** Simulation 21 **ASAP** Problem Solving 2 **ASAP** Assessment 18.1 **CA** *Big Bottle Shake* **CHM** Side 6, 10: *Waste Water Treatment* **CHM** Side 6, 34: *Supersaturated Solutions* **CHM** Side 2, 13: *Fertilizers* **OT** 60: *Dynamic Equilibrium* **OT** 61: *Solubility and Temperature*
18.2 Concentrations of Solutions ○ ■ ◆ ▸ Solve problems involving the molarity of a solution ▸ Describe how to prepare dilute solutions from more-concentrated solutions of known molarity ▸ Explain what is meant by percent by volume (% (v/v)) and percent by mass (% (m/v)) solutions	**SE** **Link to Nursing** *Intravenous Solutions,* p. 510 **SE** **Small-Scale Lab** *Making a Solution,* p. 516 (**LRS** 18-2) **LM** 32: *Introduction to Chromatography* **TE** **DEMO**, p. 514	**ASAP** Problem Solving 11, 12, 15 **ASAP** Assessment 18.2 **ACT** 14: *Properties of Solutions* **CA** *Big Bottle Shake* **CHM** Side 6, 2: *What's in the Water?* **OT** 62: *Concentrated and Dilute Solutions*
18.3 Colligative Properties of Solutions ■ ◆ ▸ Explain on a particle basis why a solution has a lower vapor pressure than the pure solvent of that solution ▸ Explain on a particle basis why a solution has an elevated boiling point and a depressed freezing point compared to the pure solvent	**LM** 33: *Freezing Point*	**ASAP** Assessment 18.3 **ACT** 14: *Properties of Solutions* **CHM** Side 6, 32: *Colligative Properties*
18.4 Calculations Involving Colligative Properties □ ◆ ▸ Calculate the molality and mole fraction of a solution ▸ Calculate the molar mass of a molecular compound from the freezing-point depression or boiling-point elevation of a solution of the compound	**SE** **Chemistry Serving . . . Society** *A Solution for Kidney Failure,* p. 526 **SE** **Chemistry in Careers** *Nephrology Nurse,* p. 526 **LM** 33: *Freezing Point*	**ASAP** Simulation 22 **ASAP** Problem Solving 31, 33 **ASAP** Assessment 18.4 **ACT** 14: *Properties of Solutions* **ASAP** Concept Map 18 **PP** Chapter 18 Problems **RP** Lesson Plans, Resource Library **AR** Computer Test 18 **www** Activities, Self-Tests, *SCIENCE NEWS* updates **TCP** The Chemistry Place Web Site

KEY

● Conceptual (concrete concepts)	**AR** Assessment Resources	**GRS** Guided Reading and Study Workbook
○ Conceptual (more abstract/math)	**ASAP** Chem ASAP! CD-ROM	**LM** Laboratory Manual
■ Standard (core content)	**ACT** ActivChemistry CD-ROM	**LP** Laboratory Practicals
□ Standard (extension topics)	**CHM** CHEMedia Videodiscs	**LRS** Laboratory Recordsheets
◆ Honors (core content)	**CA** Chemistry Alive! Videodiscs	**SSLM** Small-Scale Lab Manual
◇ Honors (options to accelerate)	**GCP** Graphing Calculator Problems	

PRACTICE	ASSESSMENT
SE **Sample Problem** 18-1 **GRS** Section 18.1 **RM** Practice Problems 18.1 **RM** Interpreting Graphics	**SE** Section Review **RM** Section Review 18.1 **RM** Chapter 18 Quiz **LP** Lab Practical 18-1
SE **Sample Problems** 18-2 to 18-6 **GRS** Section 18.2 **RM** Practice Problems 18.2	**SE** Section Review **RM** Section Review 18.2 **RM** Chapter 18 Quiz
GRS Section 18.3 **RM** Practice Problems 18.3	**SE** Section Review **RM** Section Review 18.3 **RM** Chapter 18 Quiz
SE **Sample Problems** 18-7 to 18-10 **GRS** Section 18.4 **RM** Practice Problems 18.4 **GCP** Chapter 18	**SE** Section Review **RM** Section Review 18.4 **RM** Vocabulary Review 18 **SE** Chapter Review **SM** Chapter 18 Solutions **SE** Standardized Test Prep **PHAS** Chapter 18 Test Prep **RM** Chapter 18 Quiz **RM** Chapter 18 A & B Test

OT	Overhead Transparency	**SE**	Student Edition
PHAS	PH Assessment System	**SM**	Solutions Manual
PLM	Probeware Lab Manual	**SSV**	Small-Scale Video/Videodisc
PP	Problem Pro CD-ROM	**TCP**	www.chemplace.com
RM	Review Module	**TE**	Teacher's Edition
RP	Resource Pro CD-ROM	**www**	www.phschool.com

PLANNING FOR ACTIVITIES

STUDENT EDITION

Discover It! p. 500
- plastic plates
- string or narrow ribbon
- water
- ice cubes
- table salt

Mini Lab, p. 508
- sodium hydrogen carbonate
- cornstarch
- water
- flashlights
- masking tape
- jars with parallel sides
- teaspoons
- cups

Small-Scale Lab, p. 516
- NaCl (lab grade)
- water
- sucrose
- pipets
- 50-mL plastic volumetric bottles
- balances

Activity, p. 511
- water
- NaCl
- sucrose
- 50-mL, 100-ml, and 1-L volumetric flasks
- balances
- graph paper

Teacher Demo, p. 514
- four 20-mL test tubes
- 10 mL of 0.5*M* sucrose
- 15 mL water
- 10-mL pipet

Activity, p. 519
- ice water
- rock salt
- thermometer
- Styrofoam cup

Activity, p. 524
- different concentrations of sodium chloride or ethylene glycol solution
- thermometers
- heat sources
- beakers

TEACHER'S EDITION

Teacher Demo, p. 505
- bottle of warm soda
- bottle of cold soda
- 2 other bottles of soda

Chapter 18

Key Terms

18.1 saturated solution, solubility, unsaturated, miscible, immiscible, Henry's law, supersaturated solution
18.2 concentration, dilute solution, concentrated solution, molarity (*M*)
18.3 colligative properties, boiling-point elevation, freezing-point depression
18.4 molality (*m*), mole fraction, molal boiling-point elevation constant (K_b), molal freezing-point depression constant (K_f)

DISCOVER IT!

Students should observe that sprinkling salt along the string causes the ice beneath the string to melt. (The salt lowers the freezing point of the ice. Because the ice absorbs heat from the wet string, the string freezes to the ice.) Students may have seen rock salt used to freeze homemade ice cream. In areas that experience heavy snows, salt and gravel mixtures are spread on roads to prevent the formation of "black ice." Both sugar and baking soda will lower the freezing point of water, but to different degrees. In Section 18.3, students will learn that depression of the freezing point is a colligative property.

Chapter 18 SOLUTIONS

A sinkhole can form when groundwater dissolves limestone.

FEATURES

DISCOVER IT!
Salt and the Freezing Point of Water

SMALL-SCALE LAB
Making a Solution

MINI LAB
Solutions and Colloids

CHEMISTRY SERVING ... SOCIETY
A Solution for Kidney Failure

CHEMISTRY IN CAREERS
Nephrology Nurse

LINK TO AGRICULTURE
Fertilizer Runoff

LINK TO NURSING
Intravenous Solutions

Stay current with SCIENCE NEWS
Find out more about solutions:
www.phschool.com

DISCOVER IT! SALT AND THE FREEZING POINT OF WATER

You need a plastic plate, a 20-cm piece of string or narrow ribbon, water, an ice cube, and some table salt (sodium chloride).

1. Remove an ice cube from the freezer and place it on the plate.
2. Moisten the string or ribbon with water and place it in a straight line across the ice cube.
3. Sprinkle some salt along the length of the string on the ice.
4. Wait 1–2 minutes and then slowly lift the ends of the string, which should raise the ice cube off the plate.
5. If you are not successful in lifting the cube with the string, repeat Steps 3 and 4.

What effect does the salt have on the water/ice? Can you name some practical instances of salt being used to affect the freezing point of water? Do you think sugar or baking soda would work equally well? Try it! When you read about solution properties later in this chapter, look back at your findings and try to explain them.

18.1 ● ■ ◆
18.2 ○ ■ ◆
18.3 ■ ◆
18.4 □ ◆

Conceptual Henry's law is challenging because it relates gas solubility, which cannot be directly observed, to pressure, which can be measured directly. Sample Problem 18-1 may be omitted. Use Figure 18.12 to visualize the difference between dilute and concentrated solutions. Students need to be familiar with molarity, but the Sample Problems in Section 18.2 can be omitted. Percent solutions may also be omitted. Sections 18.3 and 18.4 are beyond the scope of a conceptual course.

Standard The illustrations in Section 18.3 will help students use particle interactions to explain colligative properties. Calculations with colligative properties in Section 18.4 is an appropriate extension topic, time permitting.

Honors After presenting concepts related to boiling point, have students extend these concepts to freezing point on their own. If students are successful with problems in class, assign some of the Sample Problems as homework.

PROPERTIES OF SOLUTIONS

It was there one minute and gone the next! An entire house—front yard and back yard included—swallowed up by the earth. A victim of a natural phenomenon of moving groundwater, the house had disappeared into a sinkhole! Groundwater, or the water that soaks into the ground from rain and melted snow, can have a significant effect on the rocks beneath the soil surface. As groundwater moves through layers of limestone, it can dissolve huge amounts over time. This process creates beautiful limestone caves, as well as destructive sinkholes. A sinkhole forms when an overlying layer of limestone, often the roof of a cave, weakens from being dissolved and suddenly collapses. One recorded sinkhole swallowed a house, several other buildings, five cars, and a swimming pool! **How does the solution process occur and what influences it?**

objectives

- Identify the factors that determine the rate at which a solute dissolves
- Calculate the solubility of a gas in a liquid under various pressure conditions

key terms

- saturated solution
- solubility
- unsaturated
- miscible
- immiscible
- Henry's law
- supersaturated solution

Solution Formation

If you have ever made ice cream, you know that salt is added to the ice–water mixture surrounding the ice-cream container. But do you know why the salt is needed? You may also have noticed that granulated sugar dissolves faster than sugar cubes in iced tea, and that both will dissolve faster in hot tea or with stirring. See **Figure 18.1.** You will be able to explain each of these observations once you have gained an understanding of the properties of solutions. 1

Recall that solutions are homogeneous mixtures that may be solid, liquid, or gaseous. In this chapter, you will study the formation of solutions and ways to express their composition. You will also examine how the properties of solutions change when the number of particles dissolved in them is changed.

Figure 18.1

The process by which a solute dissolves takes place at the surface of the solid. The photograph showing a sugar cube suspended in water illustrates this. The sugar particles near the cube's surface are dissolving. Agitation and heat increase the rate at which a solute dissolves. ***(a)*** *A cube of sugar in cold tea dissolves slowly.* ***(b)*** *Granulated sugar dissolves in cold water more quickly than a sugar cube, especially with stirring.* ***(c)*** *Granulated sugar dissolves very quickly in hot tea.*

(a)

(b)

(c)

18.1

1 Engage

Use the Visual

Have students examine the photograph and read the text that opens the section. Explain that the principal component of limestone is calcium carbonate, $CaCO_3$. The solution of $CaCO_3$ in ground water is one way that Ca^{2+} enters fresh water supplies, producing "hard water." Ask:

- **How does the solution process occur and what influences it?** (Solutions form when one substance disperses uniformly throughout another. Substances that dissolve in water do so because the attractions between the particles of solute and the water molecules are great enough to overcome any mutual attractions between solute particles. The extent and rate at which one substance dissolves in another depends on the nature of the solute and the solvent. Solution formation is also affected by agitation, temperature, and surface area of the solute. When the solute is a gas, pressure is an important consideration as well.)

Check Prior Knowledge

To assess students' knowledge about mixtures, ask them how solutions and mixtures are related. Students should recall from Chapter 2 that solutions are homogeneous mixtures. Ask students to give examples of solutions. If they do not include solid or gaseous solutions, ask if all solutions contain liquids.

STUDENT RESOURCES

- Section Review 18.1, Ch. 18 Practice Problems, Interpreting Graphics, and Quizzes from the Review Module (Ch. 17–20)
- Laboratory Recordsheet 18-1
- Laboratory Manual: Experiments 30 and 31
- Laboratory Practical 18-1
- Small-Scale Chemistry Lab Manual: Experiment 23

TECHNOLOGY RESOURCES

Relevant technology resources include:

- Chem ASAP! CD-ROM
- Resource Pro CD-ROM
- Chemistry Alive! Videodisc: *Big Bottle Shake*

1 The salt lowers the freezing point of the ice-water mixture.

Motivate and Relate

Prepare a bulletin board display using large pictures that illustrate the importance of solution processes in nature. Examples include the interior of a cave such as the Carlsbad Caverns in New Mexico, a farmer applying fertilizer to a field, a close-up of a plant, a person eating food, fish swimming, a volcanic eruption, natural crystals, and hard water deposits in a water pipe. Encourage students to contribute to the display by posting pictures they find in magazines or photos they may have taken themselves.

2 Teach

Discuss

Explain that solubility is somewhat like population density. Both terms express the concentration of objects. With solubility, the objects are molecules or ions dissolved in a given quantity of solvent. With population density, the objects are organisms per unit area. However, point out that the precise amount of solute that a particular solvent holds under given conditions is fixed. Population density is not limited in the same way.

Discuss

Have students read the Link to Agriculture and compare the contents of different solid and liquid fertilizers used for house plants, shrubs, or lawns. Have them try to discern any relationship between the contents and the intended use.

AGRICULTURE

Fertilizer Runoff

Most farmers use fertilizers that contain salts of one or more of three elements essential to plant growth: potassium, nitrogen, and phosphorus. When more fertilizer is applied than the soil can absorb, rain washes off the excess salts. Not only is this economically wasteful, it is also hazardous to the environment. The water containing these dissolved salts flows into streams and rivers. The dissolved nutrients contribute to the eutrophication of these waters. Eutrophied waters are rich in nutrients but deficient in dissolved oxygen. Signs of eutrophication include overgrowth of water plants, algal blooms, and bad odors resulting from the growth of bacteria that do not need oxygen. In recent years, the problem of fertilizer runoff has lessened with the introduction of slow-release nitrogenous fertilizers. Many farmers have adopted new fertilizer application cycles and cultivation methods to reduce runoff.

One factor that affects whether a substance will dissolve is the nature of the solvent and of the solute. Do you remember the definition of solvent
1 and solute? Several other factors determine how fast the substance dissolves. These factors are stirring (agitation), temperature, and surface area. Each factor involves the contact of the solute with the solvent. If a few sodium chloride crystals are placed in a flask of water, for example, the crystals will dissolve slowly. If the flask and its contents are shaken, however, the crystals dissolve quickly. The agitation has this effect because the dissolving process is a surface phenomenon. It occurs faster if fresh solvent is brought into contact with the surface of the solute, which is what occurs during shaking. This is why people stir coffee after they have added sugar. It is important to realize, however, that agitation affects only the rate at which a solute dissolves. It cannot influence the amount of solute that dissolves. An insoluble substance remains undissolved regardless of how vigorously or for how long the system is agitated.

Temperature is another factor that influences the rate at which solutes dissolve. Sugar dissolves much more rapidly in hot tea than in iced tea because the kinetic energy of the water molecules is greater at the higher temperature. This increase in kinetic energy leads to an increased frequency and force of collisions between water molecules and the surfaces of the sugar crystals.

Another factor that determines the rate at which a solute dissolves is its particle size. A fine powder dissolves more rapidly than do larger crystals, because finer particles expose a greater surface area to the colliding water molecules. Here again, the dissolving process is a surface phenomenon, and the more surface of the solute that is exposed, the faster is the rate of dissolving. Can you think of an example of how particle size affects the rate
2 of dissolving?

Solubility

If you add 36.0 g of sodium chloride to 100 g of water at 25 °C, all 36.0 g of salt dissolves. But if you add one more gram of salt and stir, no matter how vigorously or for how long, only 0.2 g of the last portion goes into solution. Why do crystals of salt remain out of solution? According to the kinetic theory, water molecules are in continuous motion. Therefore, they should continue to bombard the excess solid, removing and solvating the ions. As ions are solvated, they dissolve in the water. Based on this information, you might expect all of the sodium chloride to dissolve eventually. That does not happen, however, because an exchange process is occurring. New particles from the solid are solvated and enter into solution, but at the same time an equal number of already dissolved particles come out of solution. These particles become desolvated and are deposited as a solid. As ions in solution are desolvated, they crystallize. The mass of undissolved crystals remains constant.

Thus the particles move from the solid to the solvated state and back to the solid again. Yet there is no net change in the overall system. As shown in **Figure 18.2,** a state of dynamic equilibrium exists between the solution and the undissolved solute, provided that the temperature remains constant. Such a solution is said to be saturated. A **saturated solution** contains the maximum amount of solute for a given amount of solvent at a constant

Answers

1 A solvent is the dissolving medium. A solute is the substance being dissolved.
2 Answers will vary; sugar cube vs. granulated sugar is a good example.
3 Vinegar is polar; oil is nonpolar.

temperature. For example, 36.2 g of sodium chloride in 100 g of water produces a saturated solution at 25 °C. The **solubility** of a substance is the amount that dissolves in a given quantity of a solvent at a given temperature to produce a saturated solution. Solubility is often expressed in grams of solute per 100 g of solvent. A solution that contains less solute than a saturated solution is said to be **unsaturated.**

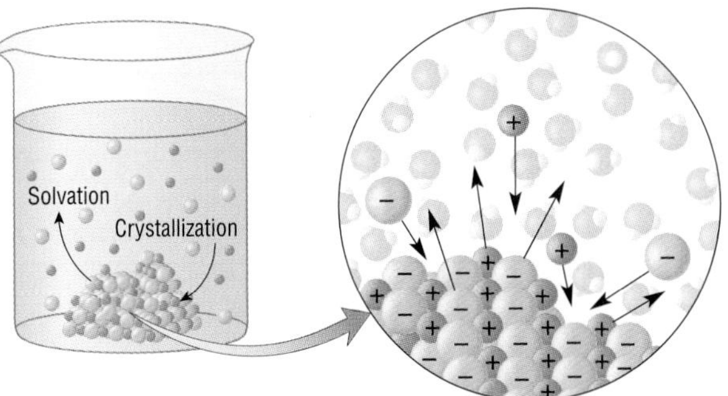

Figure 18.2

In a saturated solution, a state of dynamic equilibrium exists between the solution and the excess solute. At these conditions, the rate of solvation (dissolving) equals the rate of crystallization, so the total amount of dissolved solute remains constant.

Two liquids are said to be **miscible** if they dissolve in each other. Some liquids—for example, water and ethanol—are infinitely soluble in each other. Any amount of ethanol will dissolve in a given volume of water, and vice versa. Such a pair of liquids is said to be completely miscible. Liquids that are slightly soluble in each other—for example, water and diethyl ether—are partially miscible. Liquids that are insoluble in each other are **immiscible.** As you can see in **Figure 18.3**, oil and vinegar are immiscible, as are oil and water. Why? (*Hint:* Remember from Chapter 17 that like  dissolves like.)

Factors Affecting Solubility

As you have just read, the solubility of sodium chloride in water at 25 °C is 36.2 g per 100 g of water. At 100 °C, the solubility increases to 39.2 g NaCl per 100 g of water. As shown in **Figure 18.4** on the following page, the solubility of most solid substances increases as the temperature of the solvent increases. For a few substances, however, the reverse occurs. For example,

Figure 18.3

Vinegar, which is water-based, and oil are immiscible. The less-dense oil floats on the water phase. Light produces a rainbow of colors as it is reflected off an oil film on the surface of water.

18.1

Use the Visual

Display Figure 18.2 on an overhead projector. Use the solvation of an ionic compound to emphasize the dynamic equilibrium that exists between the hydrated and crystalline states. When an ionic compound dissolves in water, ions leave the surface of the crystal and enter solution (assuming the attractive forces of the water molecules exceed the forces within the crystal). As the hydrated ions move about in solution, they sometimes collide with the crystal and stick. The more ions in solution, the more frequent the collisions. As more and more salt dissolves, the rate at which ions enter solution equals the rate at which ions return to the crystalline state. The solution is saturated. Write the equation for the equilibrium in a saturated solution of NaCl.

$NaCl(s) \rightleftharpoons Na^{+}(aq) + Cl^{-}(aq)$

Tell students that the double arrow indicates that solvation is a reversible process.

Use the Visual

Have students study the photographs in Figure 18.3 and offer possible explanations as to why the two liquids do not mix. (For nonpolar liquids to mix with water, hydrogen bonds must be broken and replaced by much weaker forces between water and the nonpolar compound. In the cases shown, the maximum forces of attraction between molecules result if the liquids remain unmixed.)

Meeting Diverse Needs

At Risk Equilibrium may be a difficult concept for some students. Emphasize the *dynamic* nature of saturation. Initially, salt readily dissolves in water because the ions can be solvated by available water molecules. However the number of solvated ion-complexes that can inhabit a given volume of water is limited. If more ions are to join in the solvation, some ions must give up their places. Use substitutions in a basketball game as an analogy for the dynamic nature of saturation. Tell students that only five players from each team can be on the court at one time. Players on the bench must wait until a player comes off the court. In this analogy, the players on the sidelines are equivalent to ions in a crystal; those on the court, to dissolved ions.

ACTIVITY

Have students examine Figure 18.4. Explain that, in general, the solubility of a substance depends on the temperature of the solution. Ask students to identify the dependent and independent variables. (concentration and temperature) Have students find the aqueous solubilities of the solids in Figure 18.4 at 0 °C, 25 °C, 40 °C, and 80 °C. Encourage students to tabulate their data as in Table 18.1.
Have students construct a graph showing the solubility of $KClO_3$ at different temperatures. Check to see that students correctly label the *x*- and *y*-axes and that they indicate the units of measurement. Next, ask:

- Do all solids become more soluble as temperature increases? [$Yb_2(SO_4)_3$ is less soluble at higher temperatures. For NaCl, the solubility is relatively constant.]
- Which of the solids show the greatest change in solubility with temperature? (The solubility of KNO_3 increases significantly between 30 °C and 80 °C.)

Propose a general statement regarding any trends implied by the data in Figure 18.4 and Table 18.1. (Students should recognize that, for most of the ionic compounds, solubility increases with temperature.)
To relate this observation to endothermic and exothermic processes; ask:

- Which of the compounds in Figure 18.4 have negative molar heats of solution? Have students look up the ΔH_{soln} for each compound in Table 18.1. (CRC Handbook of Chemistry and Physics) Students should state how the solubility curve for a compound is related to its ΔH_{soln}.

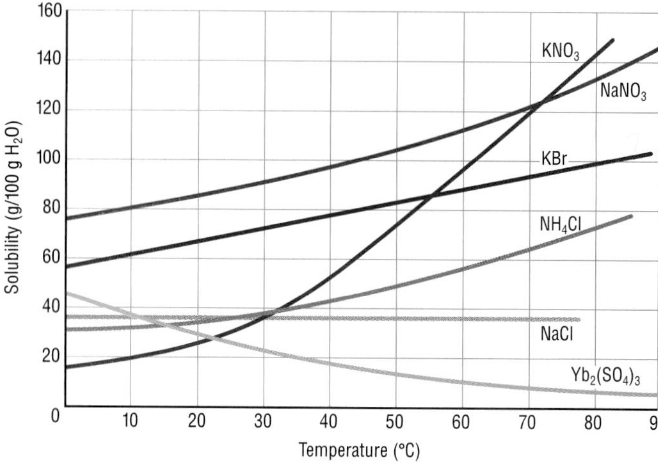

Figure 18.4
Changing the temperature may affect the solubility of a substance. Notice that increasing the temperature greatly increases the solubility of KNO_3 but decreases the solubility of $Yb_2(SO_4)_3$.

the solubility of ytterbium sulfate, $Yb_2(SO_4)_3$, in water drops from 44.2 g per 100 g of water at 0 °C to 5.8 g per 100 g of water at 90 °C. Which of the solids
1 has the lowest solubility at 40 °C? Which has the highest? **Table 18.1** lists the solubilities of some common substances at various temperatures.

Note that the solubilities of the gases listed in the table are greater in cold water than in hot water. You are probably familiar with this fact if you have ever heated water and observed bubbles forming before the water has reached its boiling point. These bubbles are dissolved atmospheric gases escaping from solution. As **Figure 18.5** shows, the two main components of

Table 18.1

Solubilities of Some Substances in Water at Various Temperatures

Substance	Formula	Solubility ($g/100\ g\ H_2O$) 0 °C	20 °C	50 °C	100 °C
Barium hydroxide	$Ba(OH)_2$	1.67	31.89	—	—
Barium sulfate	$BaSO_4$	0.00019	0.00025	0.00034	—
Calcium hydroxide	$Ca(OH)_2$	0.189	0.173	—	0.07
Lead(II) chloride	$PbCl_2$	0.60	0.99	1.70	—
Lithium carbonate	Li_2CO_3	1.5	1.3	1.1	0.70
Potassium chlorate	$KClO_3$	4.0	7.4	19.3	56.0
Potassium chloride	KCl	27.6	34.0	42.6	57.6
Sodium chloride	NaCl	35.7	36.0	37.0	39.2
Sodium nitrate	$NaNO_3$	74	88.0	114.0	182
Aluminum chloride	$AlCl_3$	30.84	31.03	31.60	33.32
Silver nitrate	$AgNO_3$	122	222.0	455.0	733
Lithium bromide	LiBr	143.0	166	203	266.0
Sucrose (cane sugar)	$C_{12}H_{22}O_{11}$	179	230.9	260.4	487
Hydrogen*	H_2	0.00019	0.00016	0.00013	0.0
Oxygen*	O_2	0.0070	0.0043	0.0026	0.0
Carbon dioxide*	CO_2	0.335	0.169	0.076	0.0

* Gas at 101 kPa total pressure

MEETING DIVERSE NEEDS

At Risk and **LEP** Have students draw and label a diagram illustrating the dynamic equilibrium of a saturated solution with arrows describing the motion of particles between the crystalline and solvated states.

air—oxygen and nitrogen—become less soluble in water as the temperature of the solution rises. When an industrial plant takes cool water from a lake and dumps the resulting heated water back into the lake, the temperature of the entire lake increases. Such a change in temperature is known as thermal pollution. Because the temperature increase lowers the concentration of dissolved oxygen in the lake water, aquatic animal and plant life can be severely affected.

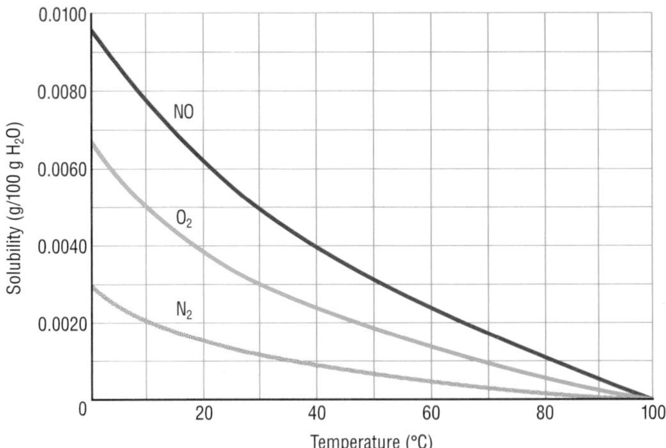

Figure 18.5

Gases have different solubilities in water at different temperatures. In general, as the temperature increases, the solubilities of gases decrease. Notice on the graph that nitrogen, oxygen, and nitric oxide become virtually insoluble as the temperature approaches 100 °C. Can you use the kinetic theory to explain this solubility behavior of gases? ❷

Chem ASAP!

Simulation 21

Observe the effect of temperature on the solubility of solids and gases in water.

The solubility of a gas is also affected by pressure. Gas solubility increases as the partial pressure of the gas above the solution increases. Carbonated beverages are a good example of this principle. These drinks contain large amounts of carbon dioxide (CO_2) dissolved in water. (It is dissolved CO_2 that makes the liquid fizz and your mouth tingle.) The drinks are bottled under a high pressure of CO_2 gas, which forces large amounts of CO_2 into solution. When a carbonated-beverage bottle is opened, the partial pressure of CO_2 above the liquid decreases, and the concentration of dissolved CO_2 decreases. Bubbles of CO_2 form in the liquid and escape from the open bottle, as shown in **Figure 18.6.** If the bottle is left open, the drink becomes flat as the solution loses its CO_2.

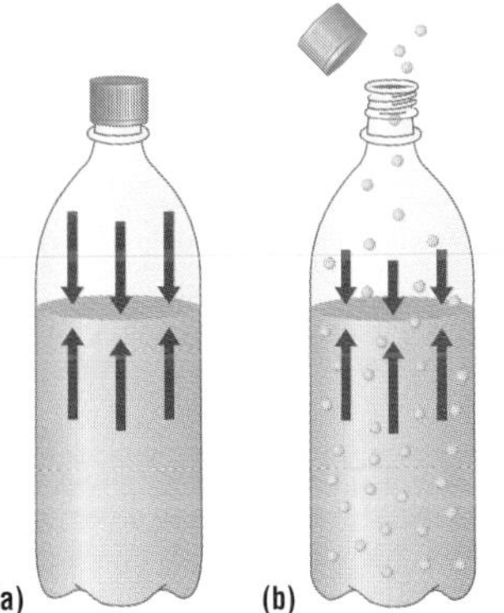

(c)

Figure 18.6

***(a)** When a carbonated-beverage bottle is sealed, the pressure of CO_2 above the liquid is high and the concentration of CO_2 in the liquid is also high. **(b)** When the cap is removed, the equilibrium is disturbed; the pressure of CO_2 gas above the liquid becomes low. **(c)** Carbon dioxide bubbles out of the carbonated beverage.*

Think Critically

Explain that, although the solvation of a substance can be observed, the quantitative data in Table 18.1 were obtained by other methods. Challenge students to design a method for determining the solubility of NaCl at a given temperature. (Measure the density of solution for various concentrations of NaCl. Plot density vs. concentration. The solubility limit could be estimated from the asymptote. Linear plots of the data could also be constructed for more accurate estimates.)

TEACHER DEMO

To illustrate the solubility of gases, open a bottle of warm soda and a bottle of cold soda and have students observe and compare the results. (Carbon dioxide gas comes out of solution much more quickly in the warm soda.) The rush of bubbles in a warm bottle of soda can be a model for the "bends," a malady that afflicts divers when dissolved nitrogen forms bubbles in the bloodstream. The model also shows how the escape of carbon dioxide from rising magma can trigger a volcanic eruption. In each case, high pressure keeps a gas in solution; reduced pressure allows it to escape. Repeat the demonstration, this time using a shaken and unshaken bottle of soda. Have students compare and contrast the effect of agitation on the solubility of gases and solids.

Answers

❶ lowest, $Yb_2(SO_4)_3$; highest, $NaNO_3$
❷ At higher temperatures, the gaseous molecules have more kinetic energy and escape the liquid more easily.

Discuss

Explain that the action of a gas pressing down on a liquid and increasing the solubility of the gas in the liquid is like the action of a hammer driving a nail into a piece of wood. When you hit the nail harder, more of it is driven into the wood. The hammer is analogous to pressure, the nail to solute gas, and the wood to solvent liquid.

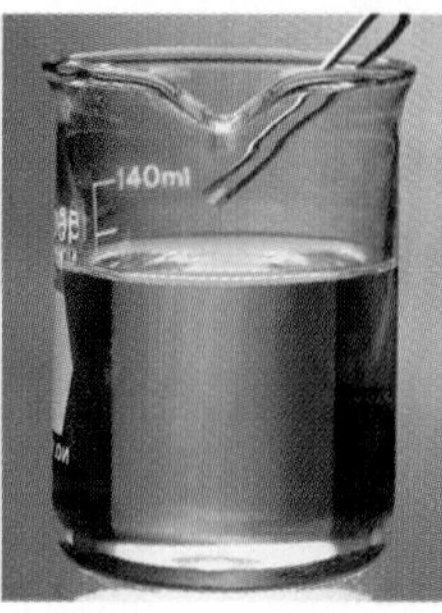

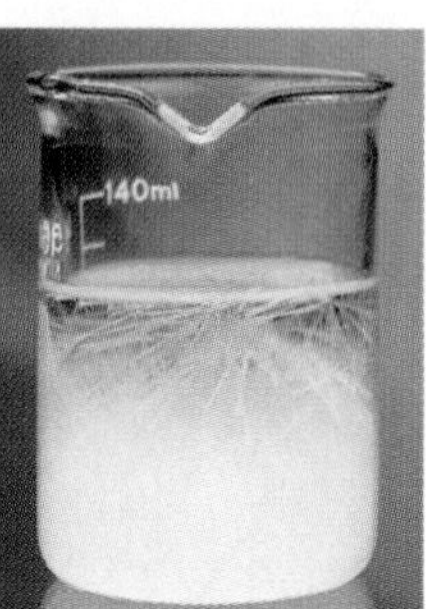

Figure 18.7
A supersaturated solution before a seed crystal is added (left). The same solution after the addition of the seed crystal (center). The seed crystal causes the excess solute to crystallize rapidly (right). How would you characterize the solution once crystallization has ceased? ❶

Henry's law states that at a given temperature the solubility (S) of a gas in a liquid is directly proportional to the pressure (P) of the gas above the liquid. In other words, as the pressure of the gas above the liquid increases, the solubility of the gas increases. Similarly, as the pressure of the gas decreases, the solubility of the gas decreases. You can write this relationship in mathematical form as follows.

$$\frac{S_1}{P_1} = \frac{S_2}{P_2}$$

When the temperature of a saturated solution in contact with a small excess of solid solute is raised, some or all of the excess solid generally dissolves. If the solution is then slowly cooled to its original temperature, the excess solute may not always immediately crystallize out of solution. A solution that contains more solute than it should theoretically continue to hold at a given temperature is called a **supersaturated solution.** Crystallization in a supersaturated solution can be initiated if a very small crystal, called a seed crystal, of the solute is added. The rate at which excess solute deposits upon the surface of a seed crystal can be very rapid, as shown in **Figure 18.7.** Crystallization can also occur if the inside of the container is scratched. The water of the hot spring in **Figure 18.8** is saturated with minerals. As the solution of minerals cools, it becomes supersaturated. The rocks over which the solution flows act as seed crystals that speed the formation of mineral deposits. Another example of crystallization in a supersaturated solution is the production of rock candy. The solution is supersaturated with sugar. Seed crystals cause the sugar to crystallize out of solution onto a string for you to enjoy!

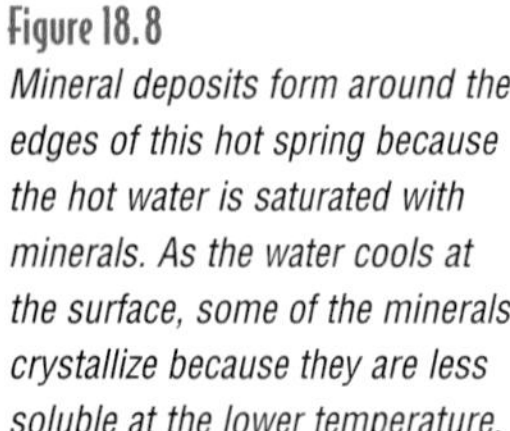

Figure 18.8
Mineral deposits form around the edges of this hot spring because the hot water is saturated with minerals. As the water cools at the surface, some of the minerals crystallize because they are less soluble at the lower temperature.

Answer

❶ The solution is saturated.

Chemistry Alive!

Big Bottle Shake Play

Figure 18.9
Clouds can be seeded with silver iodide to produce water droplets, causing rain to fall.

You may have heard of clouds being seeded to make them produce rain. Scientific rainmaking is done by seeding clouds, which contain masses of air supersaturated with water vapor. Tiny silver iodide (AgI) crystals are dusted onto a cloud to seed it, as shown in **Figure 18.9.** Water molecules are attracted to the ionic AgI particles and come together, forming droplets that act as seeds for other water molecules. The water droplets grow and eventually fall as rain when they are large enough.

Sample Problem 18-1

If the solubility of a gas in water is 0.77 g/L at 3.5 atm of pressure, what is its solubility (in g/L) at 1.0 atm of pressure? (The temperature is held constant at 25 °C.)

1. *ANALYZE* ***List the knowns and the unknown.***

Knowns:
- $P_1 = 3.5$ atm
- $S_1 = 0.77$ g/L
- $P_2 = 1.0$ atm
- Henry's law: $\frac{S_1}{P_1} = \frac{S_2}{P_2}$

Unknown:
- $S_2 = ?$ g/L

2. *CALCULATE* ***Solve for the unknown.***

Solving Henry's law for S_2 yields

$$S_2 = \frac{S_1 \times P_2}{P_1}$$

Substituting the known values and calculating yields

$$S_2 = \frac{0.77\ \text{g/L} \times 1.0\ \cancel{\text{atm}}}{3.5\ \cancel{\text{atm}}} = 0.22\ \text{g/L}$$

3. *EVALUATE* ***Does the result make sense?***

The pressure is reduced (from 3.5 atm to 1.0 atm), so the solubility of the gas should decrease, which it is shown to do. Because the new pressure is approximately one-third of the original pressure, the new solubility should be approximately one-third of the original, which it is. The answer is correctly expressed to two significant figures.

Practice Problems

1. The solubility of a gas in water is 0.16 g/L at 104 kPa of pressure. What is the solubility when the pressure of the gas is increased to 288 kPa? Assume the temperature remains constant.
2. A gas has a solubility in water at 0 °C of 3.6 g/L at a pressure of 1.0 atm. What pressure is needed to produce an aqueous solution containing 9.5 g/L of the same gas at 0 °C?

Chem ASAP!
Problem-Solving 2
Solve Problem 2 with the help of an interactive guided tutorial.

18.1

CHEMISTRY AND TECHNOLOGY

U.S. scientists Vincent Schaefer, Irving Langmuir, and Bernard Vonnegut carried out the first successful rainmaking experiments. They knew that water vapor in the atmosphere condenses around tiny nuclei, such as particles of soot. In 1946, Schaefer produced snow by dropping dry ice (solid carbon dioxide) pellets into clouds. A year later, Vonnegut discovered that silver iodide particles also serve as "precipitation nuclei." Since that time, seeding clouds with silver iodide has been the standard method.

Practice Problems Plus

Related Chapter Review Problem
Chapter Review problem 48 is related to Sample Problem 18-1.

Additional Practice Problem
The Chem ASAP! CD-ROM contains the following problem: The solubility of a gas is 1.04 g/L at 98.0 kPa and 25 °C. What is the solubility of the same gas at 125 kPa and 25 °C? (1.33 g/L)

3 Assess

Evaluate Understanding

Have students use the concept of solvation to explain how agitation, temperature, and particle size affect the rate of solution formation. Ask:

- **Which of these factors affect the solubility of a substance?** (temperature only)
- **How is the solubility of a gas affected by the pressure of the gas above the liquid?** (The solubility of the gas increases with increasing pressure.)

MEETING DIVERSE NEEDS

At Risk and **LEP** Have students observe crystal formation in two sealed jars containing glacial acetic acid. In one jar, have a few grains of sand or glass scrapings; in the other, just pure distilled glacial acetic acid. (The antiquated name "glacial" reflects the acid's icy appearance at very low temperatures.) Dunk the jars into an ice/salt bath. Crystals readily form at the site of the "seed" and take longer when no nucleation site is available. Students could time these processes and note the difference on several trial runs. Any variation in times could be used to reinforce the notion of statistical averages.

MINI LAB

Solutions and Colloids

LAB PREP AND PLANNING

You may want to have students test and classify various household mixtures using the Tyndall effect.

ANALYSIS AND CONCLUSIONS

1. undissolved cornstarch
2. Cornstarch particles reflect light.
3. Yes for a colloid; no for a suspension.
4. Because two solutions form, the beam of light would not be visible.
5. Because flour and milk do not dissolve, the beam of light will be visible.
6. Cannot. Must wait for the suspension to settle or use a filter.

Reteach

Review the factors that affect the aqueous solubilities of solid and gaseous substances. (temperature and pressure) **Make sure that students are able to distinguish between those factors that affect the solubility and those that affect the rate of solution, such as agitation and particle size. Tell students that solubility is the amount of a substance that will dissolve in a given amount of solvent, while rate of solution is the amount of solute entering a solution per unit time. Temperature is the only factor discussed that affects both the rate of solution and the solubility.**

MINI LAB  

Solutions and Colloids

The Tyndall effect

PURPOSE

To classify mixtures as solutions or colloids using the Tyndall effect.

MATERIALS

- sodium hydrogen carbonate
- cornstarch
- distilled water (or tap water)
- flashlight
- masking tape
- 3 jars with parallel sides
- teaspoon
- cup

PROCEDURE

1. In a cup, make a paste by mixing $\frac{1}{2}$ teaspoon cornstarch with 4 teaspoons water.
2. Fill one of the jars with water. Add $\frac{1}{2}$ teaspoon sodium hydrogen carbonate to a second jar and fill with water. Stir to mix. Add the cornstarch paste to the third jar and fill with water. Stir to mix.
3. Turn out the lights in the room. Shine the beam of light from the flashlight at each of the jars and record your observations.

ANALYSIS AND CONCLUSIONS

1. In which of the jars was it possible to see the path of the beam of light?
2. What made the light beam visible?
3. If a system that made the light beam visible were filtered, would the light beam be visible in the filtrate?
4. Predict what you would observe if you were to replace the sodium hydrogen carbonate with sucrose (cane sugar) or sodium chloride (table salt).
5. Predict what you would observe if you were to replace the cornstarch with flour or diluted milk.
6. Explain how you could use this method to distinguish a colloid from a suspension.

section review 18.1

3. Name three factors that influence the rate at which a solute dissolves in a solvent.
4. How can you calculate the solubility of a gas in a liquid under different pressure conditions?
5. What mass of NaCl can be dissolved in 7.50×10^2 g of water at 25 °C?
6. What could you do to change
   a. a saturated solution to an unsaturated solution?
   b. an unsaturated solution to a saturated solution?
7. Use the solid substances listed in **Table 18.1** on page 504 to make a general statement that relates a change in solubility of a solid to a change in temperature.

 Chem ASAP! **Assessment 18.1** Check your understanding of the important ideas and concepts in Section 18.1.

Answers

SECTION REVIEW 18.1

3. agitation, temperature, particle size
4. by using Henry's law
5. 272 g NaCl
6. a. Add solvent.
   b. Add solute until no more will dissolve.
7. Solubility usually increases with temperature.

CONCENTRATIONS OF SOLUTIONS

A supply of clean drinking water is important for all communities. What constitutes clean water? The federal government, along with state governments, has set standards limiting the amount of contaminants allowed in drinking water. These contaminants include metals, pesticides, bacteria, and even the byproducts of water treatment. Water must be tested continuously to ensure that the concentrations of these contaminants do not exceed established limits. **How is the concentration of a solute in a solvent calculated?**

objectives

- Solve problems involving the molarity of a solution
- Describe how to prepare dilute solutions from more concentrated solutions of known molarity
- Explain what is meant by percent by volume (% (v/v)) and percent by mass (% (m/v)) solutions

key terms

- concentration
- dilute solution
- concentrated solution
- molarity (*M*)

Molarity

So far, you have learned that at a certain set of conditions a particular substance dissolves to some extent in a particular solvent to form a solution. This section focuses on ways to express the actual extent of dissolving. The **concentration** of a solution is a measure of the amount of solute that is dissolved in a given quantity of solvent. A **dilute solution** is one that contains only a low concentration of solute. By contrast, a **concentrated solution** contains a high concentration of solute. An aqueous solution of sodium chloride containing 1 g NaCl per 100 g H_2O might be described as dilute compared with an NaCl solution containing 30 g NaCl per 100 g H_2O, but concentrated compared with another solution containing only 0.01 g NaCl per 100 g H_2O. The terms concentrated and dilute are thus only qualitative descriptions of the amount of a solute in solution.

There are ways to express concentration quantitatively. The most important unit of concentration in chemistry is molarity. **Molarity (*M*)** is the number of moles of a solute dissolved per liter of solution. Molarity is also known as molar concentration and when accompanying a numerical value is read as "molar." Note that the volume involved is the total volume of the resulting solution, not the volume of the solvent alone. **Figure 18.10** demonstrates how a solution with a molarity of 0.5 (or a 0.5-molar solution) can be made. **Figure 18.11** on the following page shows 1-mol samples of some common compounds.

(a)

(b)

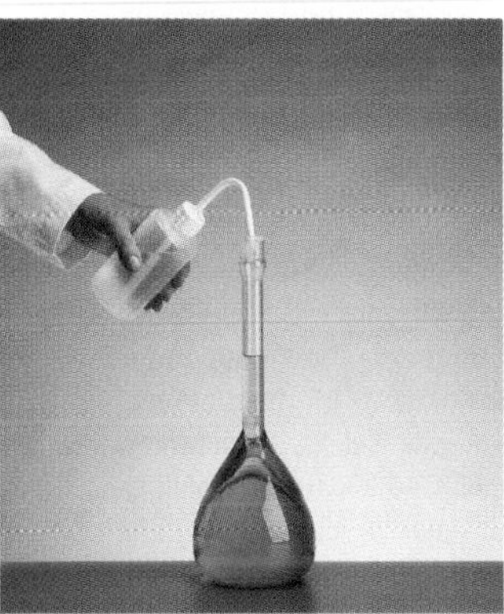

(c)

Figure 18.10

*To make a 0.5-molar (0.5M) solution, **(a)** add 0.5 mol of solute to a 1-L volumetric flask half filled with distilled water. **(b)** Swirl the flask carefully to dissolve the solute. **(c)** Then fill the flask with water exactly to the 1-L mark.*

STUDENT RESOURCES

From the Teacher's Resource Package, use:

- Section Review 18.2, Ch. 18 Practice Problems and Quizzes from the Review Module (Ch. 17–20)
- Laboratory Recordsheet 18-2
- Laboratory Manual: Experiment 32

TECHNOLOGY RESOURCES

Relevant technology resources include:

- Chem ASAP! CD-ROM
- Resource Pro CD-ROM
- ActivChemistry CD-ROM: *Properties of Solutions*
- Chemistry Alive! Videodisc: *Molarity*

1 Engage

Use the Visual

Have students study the photograph and read the text that opens the section. Explain that the use of electronic instrumentation with ever-increasing sensitivity has allowed detection of many substances that were once considered undetectable. Environmental chemists typically use parts per million (*ppm*) to measure small concentrations of air and water pollutants. For example, 20 *ppm* means 20 parts of a solute for every 1 000 000 parts of solvent. Ask:

- **How is the concentration of a solute in a solvent calculated?** (There are several ways to express the concentration of a solute in a solvent. Each expression is a relation that describes the quantity of solute dissolved in a given quantity of solvent. The amount of solute is expressed in terms of moles or mass. The quantity of solvent is expressed in terms of volume or mass.)

Check Prior Knowledge

To assess students' knowledge about chemical quantities, ask:

- **How many moles are in each quantity of the following substances:**
  A. 12.0 g NaCl
  (0.205 mol NaCl)
  B. 53.8 g KNO_3
  (0.532 mol KNO_3)
- **Find the mass in grams of each of these amounts of substances:**
  C. 1.50 mol NaOH (60.0 g)
  D. 0.575 mol $NaHCO_3$ (48.3 g)
- **Which contains more molecules: 1.00 mol SO_2 or 1.00 mol SO_3?** (numbers are equal) **Which contains more mass?** (1.00 mol SO_3)

2 Teach

Discuss

Students may think that solutions of given molarity can be correctly prepared by addition of the solute to a premeasured volume of solvent. Students often forget that the solute will take up some of the available space in the volumetric flask. Explain that the sequence of steps is important when making up standard solutions because the solute plus the solvent will take up more volume than just the solvent alone. It is important that the solute is added to some of the solvent and dissolved. Then solvent is added up to the 1 L mark on the volumetric flask. If not, the total volume of the mixture is likely to exceed the desired volume.

Practice Problems Plus

Related Chapter Review Problem
Chapter Review problem 50 is related to Sample Problem 18-2.

Discuss

Have students read the Link to Nursing. Either invite a nurse to address the class, or have students do research on IV drips. Possible questions to address include: Why are IV drips often used after operations? When else are they used? Why are the drips hung on poles? How is the rate of the drip controlled?

Chemistry Alive!

Molarity

Play

Figure 18.11
If 1 mol of a compound is dissolved in 1 L of a solution, it has a 1-molar (1M) concentration. These are 1-mol samples of several compounds.

NURSING

Intravenous Solutions
A patient in a hospital is often administered an intravenous (IV) drip containing an aqueous solution. The patient may need the solution to prevent dehydration

or to administer nutrients or medicines. Great care must be taken when fluids are introduced into the bloodstream. The balance of dissolved electrolytes inside and outside of the tissue cells could be disturbed. The effects of an imbalance in electrolytes and water can be harmful or could even cause the death of the patient. Aqueous solutions containing 0.85%–0.9% (mass/volume) of sodium chloride or 5% (mass/volume) of glucose prevent undesired changes in the electrolyte balance.

To determine the molarity of any solution, calculate the number of moles in 1 L of the solution, using the following equation.

$$\text{Molarity } (M) = \frac{\text{moles of solute}}{\text{liters of solution}}$$

For example, to calculate the molarity when 2 mol of glucose is dissolved in 5 L of solution, divide the number of moles by the volume in liters.

$$\frac{2 \text{ mol glucose}}{5 \text{ L solution}} = 0.4 \text{ mol/L} = 0.4M$$

In some problems, the number of moles of solute dissolved in a given volume of solution needs to be determined. This can be done if the molarity of the solution is known. For example, how many moles are in 2.00 L of 2.5M lithium chloride? Rearranging the formula to solve for number of moles gives the following equation.

$$\begin{aligned}\text{Moles of solute} &= \text{liters of solution} \times \text{molarity } (M)\\ &= 2.00 \text{ L} \times 2.5M = 2.00 \not{\text{L}} \times 2.5 \text{ mol}/\not{\text{L}}\\ &= 5.0 \text{ mol}\end{aligned}$$

Sample Problem 18-2

A saline solution contains 0.90 g NaCl in exactly 100 mL of solution. What is the molarity of the solution?

1. *ANALYZE* ***List the knowns and the unknown.***

Knowns:
- solution concentration = 0.90 g NaCl/100 mL
- molar mass NaCl = 58.5 g/mol

Unknown:
- solution concentration = ?M

Convert the concentration from g/100 mL to mol/L. The sequence is g/100 mL ⟶ mol/100 mL ⟶ mol/L.

Meeting Diverse Needs

Gifted Have students imagine they are writing a recipe for someone else to follow who does not know much about chemistry. Have them write out the steps necessary to make a 0.5M aqueous solution of KOH and 0.2M $MgCl_2$. Have them include tips to ensure success.

Sample Problem 18-2 (cont.)

2. *CALCULATE* ***Solve for the unknown.***
   First use the molar mass to convert g NaCl/100 mL to mol NaCl/100 mL. Then use the conversion factor between milliliters and liters to convert to mol/L, which is molarity.

$$\frac{0.90\ \cancel{\text{g NaCl}}}{100\ \cancel{\text{mL}}} \times \frac{1\ \text{mol NaCl}}{58.5\ \cancel{\text{g NaCl}}} \times \frac{1000\ \cancel{\text{mL}}}{1\ \text{L}}$$
$$= 0.15\ \text{mol/L} = 0.15M$$

3. *EVALUATE* ***Does the result make sense?***
   An answer less than 1*M* should be expected because a concentration of 0.90 g/100 mL is the same as 9.0 g/1000 mL (9.0 g/1 L), and 9.0 g is less than 1 mol NaCl. The answer is correctly expressed to two significant figures.

Practice Problems

8. A solution has a volume of 2.0 L and contains 36.0 g of glucose. If the molar mass of glucose is 180 g/mol, what is the molarity of the solution?
9. A solution has a volume of 250 mL and contains 0.70 mol NaCl. What is its molarity?

Sample Problem 18-3

How many moles of solute are present in 1.5 L of 0.24*M* Na_2SO_4?

1. *ANALYZE* ***List the knowns and the unknown.***
   Knowns:
   - volume of solution = 1.5 L
   - solution concentration = 0.24*M* Na_2SO_4

   Unknown:
   - moles solute = ? mol

   The conversion is volume of solution ⟶ moles of solute. Molarity has the units mol/L and is a conversion factor between moles of solute and volume of solution. Multiply the given volume by the molarity, expressed in mol/L.

2. *CALCULATE* ***Solve for the unknown.***

$$1.5\ \cancel{\text{L}} \times \frac{0.24\ \text{mol NaCl}}{1\ \cancel{\text{L}}} = 0.36\ \text{mol NaCl}$$

3. *EVALUATE* ***Does the result make sense?***
   An answer less than 1 mol should be expected because the solution concentration is less than 0.25 mol/L and the volume is less than 2 L. The answer is correctly expressed to two significant figures.

Practice Problems

10. How many moles of ammonium nitrate are in 335 mL of 0.425*M* NH_4NO_3?
11. How many moles of solute are in 250 mL of 2.0*M* $CaCl_2$? How many grams of $CaCl_2$ is this?

Chem ASAP!

Problem-Solving 11

Solve Problem 11 with the help of an interactive guided tutorial.

Making Dilutions

Solutions of certain standard molarities are usually available in the laboratory. However, you may need a dilute solution of a different concentration. You can make a solution less concentrated by diluting it with solvent.

ACTIVITY

Demonstrate the correct method for preparing a one molar solution of a soluble compound. Review the definition of 1*M* and how to calculate the molar mass of NaCl (58.44 g) from its atomic masses. Add 500 mL of water to a volumetric flask. Stir until the NaCl dissolves. Add water up to the 1-L mark. Ask students if the flask contains a liter of water. (No, because the dissolved NaCl takes up some volume.)
Have students prepare 0.0625*M*, 0.125*M*, 0.250*M*, and 0.500*M* solutions of sucrose using 100-mL and 50-mL volumetric flasks. Remind students that 100 mL = 0.100 L and 50 mL = 0.050 L. Have students determine the density of the solutions by measuring the mass of the flask before and after the addition of solution, subtracting, and dividing the difference by 100 or 50 mL.
Have students make a graph of density versus concentration; include the density of pure water (1 g/mL). Ask how density could be used to determine the concentration of an unknown solution.

Practice Problems Plus

Related Chapter Review Problem
Chapter Review problem 52 is related to Sample Problem 18-3.

Additional Practice Problem
The Chem ASAP! CD-ROM contains the following problem: How many grams of solute are in 2.40 L of 0.650*M* $HClO_2$? (107 g)

Discuss

Remind students that when substances react, they do so as particles and that moles of different substances contain the same number of representative particles. Because many reactions occur in solution, a unit of concentration based on moles (such as molarity) is needed to provide information about the number of reactant particles in solution.

Cooperative Learning

Divide students into groups of three and have the groups practice the calculations needed to make up solutions of various concentrations. One student should name a solute; the other two should each specify one of three quantities: *solute mass, solution volume,* or *molarity.* The unspecified quantity should then be calculated with input from each member. For example, the first student may say "sodium chloride," the second may say "12 grams," and the third may say "2.5 liters." The group would then have to calculate the molarity of the resulting solution. Each group should carry out at least one calculation involving each of the possible unknown quantities.

Discuss

Point out that many chemical reagents are shipped and stored in concentrated form, then diluted for use. For example, sodium hydroxide is sold as a 6*M* stock solution, but it is often used in the laboratory at a concentration of 0.1*M*.

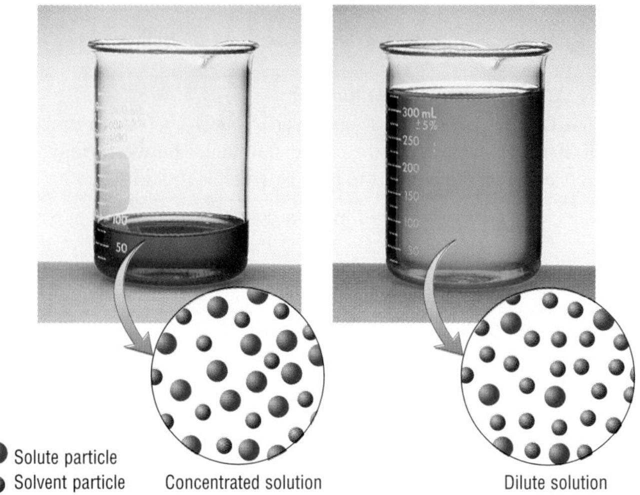

Figure 18.12
A concentrated solution contains a relatively large amount of solute for a given quantity of solution. A dilute solution contains a relatively small amount of solute. A concentrated solution has more solute particles per unit volume than does a dilute solution. Adding solvent to a concentrated solution lowers the concentration, but the total number of moles of solute present remains the same.

The dilution in **Figure 18.12** reduces the moles of solute per unit volume; however, the total moles of solute in solution does not change. Thus

Moles of solute before dilution = moles of solute after dilution

Rearranging the definition of molarity gives an expression for moles of solute in terms of molarity and volume.

$$\text{Molarity } (M) = \frac{\text{moles of solute}}{\text{liters of solution}}$$

$$\text{Moles of solute} = \text{molarity } (M) \times \text{liters of solution } (V)$$

The total number of moles of solute remains unchanged upon dilution, so you can write

$$\text{Moles of solute} = M_1 \times V_1 = M_2 \times V_2$$

where M_1 and V_1 are the initial solution's molarity and volume, and M_2 and V_2 are the final solution's molarity and volume. Volumes can be in liters or milliliters, as long as the same units are used for both V_1 and V_2. Sample Problem 18-4 shows how to calculate the volume needed in preparing a dilute solution. **Figure 18.13** shows a solution being diluted.

Figure 18.13
To prepare 100 mL of 0.40M $MgSO_4$ from a stock solution of 2.0M $MgSO_4$, transfer 20 mL of the stock solution to a 100-mL volumetric flask. Carefully add water to make 100 mL of solution.

Meeting Diverse Needs

At Risk Have students work in groups of three to explore dilution. Have each group bring in a shoe box and fill it with twenty red balls. These red balls will represent the solute molecules. Then add forty green balls. These balls will represent the solvent molecules. Students should see that there are twice as many green balls as red balls; thus, the concentration is 0.5 r/g. Now add another forty green balls. This is tantamount to a dilution from 0.5 to 0.25 r/g. Although the red concentration has been halved, the total number of red particles has not changed. If the 20 red balls represented moles of solute, there would still be *X* moles of solute in solution.

Sample Problem 18-4

How many milliliters of a stock solution of $2.00M$ $MgSO_4$ would you need to prepare 100.0 mL of $0.400M$ $MgSO_4$?

1. ***ANALYZE*** ***List the knowns and the unknown.***

Knowns:
- $M_1 = 2.00M$ $MgSO_4$
- $M_2 = 0.400M$ $MgSO_4$
- $V_2 = 100.0$ mL of $0.400M$ $MgSO_4$
- $M_1 \times V_1 = M_2 \times V_2$

Unknown:
- $V_1 = ?$ mL of $2.00M$ $MgSO_4$

2. ***CALCULATE*** ***Solve for the unknown.***

Solving for V_1 yields:

$$V_1 = \frac{M_2 \times V_2}{M_1}$$

$$= \frac{0.400M \times 100.0 \text{ mL}}{2.00M}$$

$$= 20.0 \text{ mL}$$

Thus 20.0 mL of the initial solution must be diluted by adding enough water to raise the volume to 100.0 mL.

3. ***EVALUATE*** ***Does the result make sense?***

The concentration of the initial (stock) solution is five times larger than the concentration of the solution to be prepared. Because the moles of solute in each solution are identical, it makes sense that the volume of the stock solution to be diluted should be one-fifth the final volume of the diluted solution. The answer is correctly expressed to three significant figures.

Practice Problems

12. How many milliliters of a stock solution of $4.00M$ KI would you need to prepare 250.0 mL of $0.760M$ KI?
13. Suppose you need 250 mL of $0.20M$ NaCl, but the only supply of sodium chloride you have is a solution of $1.0M$ NaCl. How do you prepare the required solution? Assume that you have the appropriate volume-measuring devices on hand.

Chem ASAP!

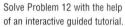

Problem-Solving 12

Solve Problem 12 with the help of an interactive guided tutorial.

You may have a variety of volume-measuring devices available to you in the laboratory. Your choice of which to use in a given dilution depends on how accurate the concentration of the solution you are making must be. Suppose the solution in Sample Problem 18-4 must be prepared very accurately. You might measure 20.0 mL of the $2.00M$ $MgSO_4$ with a 20-mL volumetric pipet and transfer the solution to a 100-mL volumetric flask. (You can also use a buret to measure volumes accurately.) You would then add distilled water to the flask up to the etched line and mix well. This dilutes the contents to exactly 100.0 mL of $0.400M$ $MgSO_4$.

Percent Solutions

If both the solute and the solvent are liquids, a convenient way to make a solution is to measure volumes. The concentration of the solute is then expressed as a percent of the solution by volume. For example, if 20 mL of pure alcohol is diluted with water to a total volume of 100 mL, the final solution is 20% alcohol by volume. The concentration can be expressed as

Discuss

If students are having difficulty with the concepts of molarity and dilution, have them consider a container of juice concentrate (frozen or liquid). No matter how much water is added to the concentrate, the initial quantity of concentrate does not change. What changes is the amount of concentrate in a given volume of the diluted solution.

Then have students consider using only a fraction of the original concentrate. For example, a gallon of juice prepared from half a can will contain half as much concentrate as a gallon prepared from the entire can. Thus, the molarity would be cut in half also.

Discuss

Write the dilution equation on the board. $M_1 \times V_1 = M_2 \times V_2$ Show students that when $\frac{\text{mol}}{V}$ is substituted for M, the volumes cancel out, leaving moles. Emphasize that, when diluting a solution, the number of moles you start with is the same as the number of moles you end with. Work through several examples using the dilution equation. For example, how many milliliters of a $6.00M$ stock solution of $NH_3(aq)$ would you need to prepare 100.0 mL of $0.30M$ $NH_3(aq)$? (5.0 mL)

Practice Problems Plus

Additional Practice Problem

The Chem ASAP! CD-ROM contains the following problem related to Sample Problem 18-4: What volume of $12.00M$ sulfuric acid is required to prepare 1.00 L of $0.400M$ sulfuric acid? (0.0333 L)

TEACHER DEMO

Remind students of the activity in which they prepared standard solutions of sucrose. Each solution was prepared independently. Ask if there is a more efficient way to prepare this same set of standard solutions. (Yes; make serial dilutions.) Demonstrate the process. Use four 20-mL test tubes. Place 10 mL of 0.5*M* sucrose in the first test tube and 5 mL water in each of the others. Using a 10-mL pipet, transfer 5 mL of the 0.5*M* solution to the second test tube; mix by pipeting up and down. Repeat this process for the third and fourth test tubes. Have students calculate the molarities of each of the resulting solutions. Serial dilutions are used to prepare extremely dilute solutions such as a 0.000001*M* solution of sucrose to eliminate the error that would arise from having to measure out such small volumes of stock solutions.

Discuss

Point out that percents are easy to calculate because there is no need to look up molar masses. Because a volume change can occur when two liquids mix, a percent composition can be misleading. When 20 mL of ethanol are mixed with 80 mL of water, the resulting volume is less than 100 mL; more water must be added.

Practice Problems Plus

Additional Practice Problem

The following problem is related to Sample Problem 18-5: Ethanol is mixed with gasoline to make gasohol. What is the percent by volume of ethanol in gasohol when 95 mL of ethanol is added to sufficient gasoline to make 1.0 L of gasohol? [9.5% (v/v)]

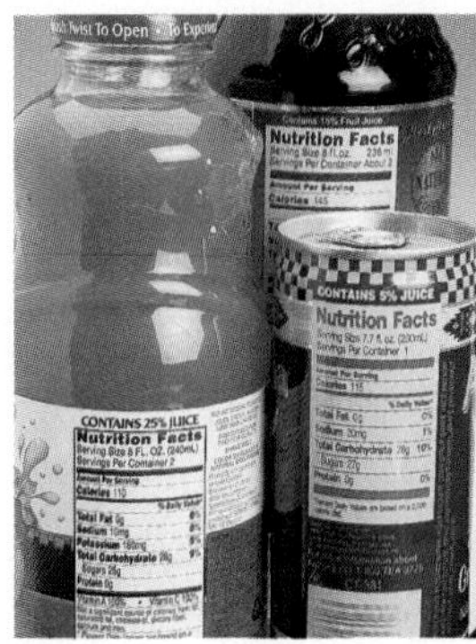

Figure 18.14

Many product labels give the amounts of their ingredients by percent. Without units, however, you cannot be certain how the percent composition was determined. What are the two kinds of percent solutions? ❶

20 percent (volume/volume), or 20% (v/v). The relationship between percent by volume and the volumes of solute and solution is

$$\text{Percent by volume (\% (v/v))} = \frac{\text{volume of solute}}{\text{solution volume}} \times 100\%$$

A commonly used relationship for solutions of solids dissolved in liquids is percent (mass/volume). It is usually convenient to weigh the solute in grams and to measure the volume of the resulting solution in milliliters. Percent (mass/volume) is the number of grams of solute per 100 mL of solution. For example, a solution containing 7 g of sodium chloride in 100 mL of solution is 7 percent (mass/volume), or 7% (m/v). The relationship among percent (mass/volume), solute mass, and solution volume is

$$\text{Percent (mass/volume) (\% (m/v))} = \frac{\text{mass of solute (g)}}{\text{solution volume (mL)}} \times 100\%$$

Because there are two kinds of percent solutions, information expressed as percent composition, such as on food labels, can often be misleading unless the units are given. See **Figure 18.14.** For example, a label stating that a candy bar contains 5% glucose is not conveying clear information. The number expressed is probably specifying percent (mass/mass), but this is not certain because the units are not given. When you use percentages to express concentration, be sure to state the units.

Sample Problem 18-5

What is the percent by volume of ethanol (C_2H_6O), or ethyl alcohol, in the final solution when 85 mL of ethanol is diluted to a volume of 250 mL with water?

1. *ANALYZE* ***List the knowns and the unknown.***

Knowns:
- volume of ethanol = 85 mL
- volume of solution = 250 mL
- $\% \text{ (v/v)} = \frac{\text{volume of solute}}{\text{volume of solution}} \times 100\%$

Unknown:
- % ethanol (v/v) = ? %

2. *CALCULATE* ***Solve for the unknown.***

Substitute the known values into the equation and solve for the unknown.

$$\% \text{ (v/v)} = \frac{\text{volume of solute}}{\text{volume of solution}} \times 100\%$$
$$= \frac{85 \text{ mL ethanol}}{250 \text{ mL solution}} \times 100\%$$
$$= 34\% \text{ ethanol (v/v)}$$

3. *EVALUATE* ***Does the result make sense?***

The volume of the solute is about one-third the volume of the solution, so the answer is reasonable. The answer is correctly expressed to two significant figures.

Practice Problems

14. If 10 mL of pure acetone is diluted with water to a total solution volume of 200 mL, what is the percent by volume of acetone in the solution?
15. A bottle of hydrogen peroxide antiseptic is labeled 3.0% (v/v). How many mL H_2O_2 are in a 400.0-mL bottle of this solution?

Chem ASAP!

Problem-Solving 15

Solve Problem 15 with the help of an interactive guided tutorial.

Answer

❶ percent by volume [%(v/v)] and percent (mass/volume)[%(m/v)]

Sample Problem 18-6

How many grams of glucose ($C_6H_{12}O_6$) would you need to prepare 2.0 L of 2.8% glucose (m/v) solution?

1. **ANALYZE** ***List the knowns and the unknown.***

Knowns:
- volume of solution = 2.0 L
- concentration of solution (% (m/v)) = 2.8% $C_6H_{12}O_6$ (m/v)

Unknown:
- mass of solute = ? g $C_6H_{12}O_6$

In a 2.8% glucose (m/v) solution, each 100 mL of the solution contains 2.8 g of glucose. Convert the given volume to milliliters, then use the concentration term to convert from milliliters of solution to grams of glucose:

liters of solution $\longrightarrow$ milliliters of solution $\longrightarrow$ grams of solute

2. **CALCULATE** ***Solve for the unknown.***

$$2.0\ \cancel{L} \times \frac{1000\ \cancel{mL}}{1\ \cancel{L}} \times \frac{2.8\ g}{100\ \cancel{mL}} = 56\ g\ C_6H_{12}O_6$$

3. **EVALUATE** ***Does the result make sense?***

Because one-tenth of a liter of solution contained 2.8 g, 1 L of solution should contain ten times as much solute (28 g) and 2 L should contain twice as much again (56 g of solute). The answer is correctly expressed to two significant figures.

Practice Problems

16. Calculate the grams of solute required to make 250 mL of 0.10% $MgSO_4$ (m/v).
17. A solution contains 2.7 g $CuSO_4$ in 75 mL of solution. What is the percent (mass/volume) of the solution?

section review 18.2

18. How are problems involving solution molarity solved?
19. Describe how dilute solutions are prepared from more concentrated solutions of known molarity.
20. Distinguish between percent (v/v) and percent (m/v) solutions.
21. Calculate the molarity of each solution.
    a. 400 g $CuSO_4$ in 4.00 L of solution
    b. 0.060 mol $NaHCO_3$ in 1500 mL of solution
22. You have the following stock solutions available: 2.00*M* NaCl, 4.0*M* KNO_3, and 0.50*M* $MgSO_4$. Calculate the volumes you must dilute to make the following solutions.
    a. 500.0 mL of 0.500*M* NaCl
    b. 2.0 L of 0.20*M* $MgSO_4$
    c. 50.0 mL of 0.20*M* KNO_3
23. What is the concentration, in percent (m/v), of a solution with 75 g K_2SO_4 in 1500 mL of solution?

Chem ASAP! **Assessment 18.2** Check your understanding of the important ideas and concepts in Section 18.2.

portfolio project

Write a report comparing the concentration of fruit juice in different types of juice drinks. Is the percent of fruit juice given by volume or by mass? Can the way these concentrations are reported be confusing or misleading?

Answers

SECTION REVIEW 18.2

18. Molarity is found by dividing the number of moles of solute by the number of liters of solution.
19. Solvent is added to the concentrated solution until the desired molarity is achieved.
20. Percent by volume equals the volume of solute per volume of solution. Percent (mass/volume) equals mass of solute (in g) per volume of solution (in mL).
21. a. 0.627*M* $CuSO_4$
    b. 0.040*M* $NaHCO_3$
22. a. 125 mL
    b. 0.80 L
    c. 2.5 mL
23. 5.0% K_2SO_4 (m/v)

18.2

Practice Problems Plus

Related Chapter Review Problems
Chapter Review problems 53 and 54 are related to Sample Problem 18-6.

Additional Practice Problem
The Chem ASAP! CD-ROM contains the following problem: Rubbing alcohol is 70.0% isopropyl alcohol by volume. How many milliliters of isopropyl alcohol are in a 250 mL bottle of this solution? (175 mL)

3 Assess

Evaluate Understanding

Have students compare the percent method and the molarity method for expressing the concentration of a solution. Have students write out a numbered list of steps they would follow to prepare a 1*M* solution of KCl in H_2O.

Reteach

Bring in products that list ingredients and use them to discuss whether the approach taken on the label matches any standard method for reporting concentration. Is data supplied for all ingredients or for just one such as "fat"?

Portfolio Project

Students may find it hard to tell which percent method is used on labels. Have students consider whether one or the other method might present the product in a more favorable light. Even the name "fruit drink" can be misleading.

SMALL-SCALE LAB

MAKING A SOLUTION

OVERVIEW

Procedure Time: 20 minutes
Students make a solution. Then they calculate its molarity, molality, percent by mass, and mole fraction.

TEACHING SUGGESTIONS

Household table salt contains a small amount of desiccant such as sodium silicate, which gives the salt solution an opaque appearance. Use lab grade NaCl instead. Remind students to use significant figures. Point out that the balance is probably more accurate than the graduated bottle. Calculations based only on the balance (molality, percent by mass, and mole fraction) will have a greater number of significant figures.

ANALYSIS

Sample data:
dry bottle = 15.98 g
bottle + NaCl = 22.88 g
bottle + NaCl + water = 69.09 g

1. a. 6.90 g
b. 46.21 g
c. 13.0%

2. a. 0.118 mol
b. 2.57 mol
c. 0.0439

3. 2.55m

4. a. 0.050 L
b. 2.4M

5. 1.1 g/mL

YOU'RE THE CHEMIST

1. Sample data:
dry bottle = 15.98 g
bottle + NaCl solution = 22.88 g
bottle + NaCl solution + water = 69.09 g
mass of NaCl solution = 2.09 g
mass of water = 47.77 g
mass of NaCl = 0.271 g
percent mass of NaCl = 0.544%
moles NaCl = 4.64×10^{-3} mol
mass of water = 49.59 g
moles of water = 2.76 mol
mole fraction = 1.68×10^{-3}
molality = 0.0936m
density = 1.0 g

SMALL-SCALE LAB

MAKING A SOLUTION

SAFETY  

Wear safety glasses and follow the standard safety procedures outlined on page 18.

PURPOSE

To make a solution and use carefully measured data to calculate the solution's concentration in various units.

MATERIALS

- solid NaCl
- water
- 50-mL plastic volumetric bottle
- balance

PROCEDURE

Weigh a clean, dry, plastic volumetric bottle. Add solid NaCl to approximately the 5-mL mark on the bottle. Weigh the bottle again. Half fill the bottle with water and shake it gently until all the NaCl dissolves. Fill the bottle with water to the 50-mL mark and weigh it again.

ANALYSIS

Using your experimental data, record the answers to the following questions below your data table.

1. Percent by mass tells how many grams of solute are present in 100 g of solution.

$$\% \text{ by mass} = \frac{\text{mass of solute}}{\text{mass of solute + solvent}} \times 100\%$$

a. Calculate the mass of the solute (NaCl).

b. Calculate the mass of the solvent (water).

c. Calculate the percent by mass of NaCl in the solution.

2. Mole fraction tells how many moles of solute are present for every 1 mol of total solution.

$$\text{Mole fraction} = \frac{\text{mol NaCl}}{\text{mol NaCl} + \text{mol } H_2O}$$

a. Calculate the moles of NaCl solute. Molar mass NaCl = 58.5 g/mol

b. Calculate the moles of water. Molar mass H_2O = 18 g/mol

c. Calculate the mole fraction of your solution.

3. Molality (m) tells how many moles of solute are present in 1 kg of solvent.

$$m = \frac{\text{mol NaCl}}{\text{kg } H_2O}$$

Calculate the molality of your solution.

4. Molarity (M) tells how many moles of solute are dissolved in 1 L of solution.

$$M = \frac{\text{mol NaCl}}{\text{L solution}}$$

a. Calculate the liters of solution. 1000 mL = 1 L

b. Calculate the molarity of the NaCl solution.

5. Density tells how many grams of solution are present in 1 mL of solution.

$$\text{Density} = \frac{\text{g solution}}{\text{mL solution}}$$

Calculate the density of the solution.

YOU'RE THE CHEMIST

The following small-scale activities allow you to develop your own procedures and analyze the results.

1. Analyze It! Use a small-scale pipet to extract a sample of your NaCl solution and deliver it to a pre-weighed empty plastic bottle. Weigh the bottle again and then fill the bottle with water to the 50-mL line. Weigh the bottle a third time and calculate the concentration of this dilute solution using the same units you used to calculate the concentration of the NaCl solution.

2. Design It! Design and carry out an experiment to make a solution of table sugar quantitatively. Calculate the concentration of the table sugar solution using the same units you used to calculate the concentration of the NaCl solution.

YOU'RE THE CHEMIST CONTINUED

2. Sample data:
mass of dry bottle = 16.72 g
mass of bottle + sugar = 20.85 g
mass of bottle + sugar + water = 69.53 g
mass of sugar = 4.13 g
mass of solvent = 48.68 g
percent mass of sugar = 7.82%
moles of sugar = 0.0121 mol
moles of water = 2.70 mol
mole fraction = 4.46×10^{-3}
molality =0.249m
molarity = 0.242M
density = 1.1 g

COLLIGATIVE PROPERTIES OF SOLUTIONS

section 18.3

objectives

- Explain on a particle basis why a solution has a lower vapor pressure than the pure solvent of that solution
- Explain on a particle basis why a solution has an elevated boiling point and a depressed freezing point compared with the pure solvent

key terms

- colligative properties
- boiling-point elevation
- freezing-point depression

The wood frog is a remarkable creature because it can survive being frozen. Scientists believe that a substance in the cells of this frog acts as a natural antifreeze, which prevents the cells from freezing. Although fluids surrounding the frog's cells may freeze, the cells themselves do not. **How can a solute change the freezing point of a solvent?**

Decrease in Vapor Pressure

The physical properties of a solution differ from those of the pure solvent used to make the solution. Some of these differences are due to the mere presence of solute particles in the solution, rather than to the specific identity of the solute. Such properties, which depend only on the number of particles dissolved in a given mass of solvent, are called **colligative properties.** Three important colligative properties of solutions are vapor-pressure lowering, boiling-point elevation, and freezing-point depression.

Recall that vapor pressure is the pressure exerted by a vapor that is in dynamic equilibrium with its liquid in a closed system. A solution that contains a solute that is nonvolatile (not easily vaporized) always has a lower vapor pressure than the pure solvent. Glucose, a molecular compound, and sodium chloride, an ionic compound, are examples of nonvolatile solutes. Both glucose and sodium chloride would lower the vapor pressure of a pure solvent. Consider an aqueous sodium chloride solution as an example. Sodium ions and chloride ions are dispersed throughout the liquid water. Both within the liquid and at the surface, the ions are surrounded by layers of shells of associated water, or shells of water of solvation. The formation of these shells of water of solvation reduces the number of solvent molecules that have enough kinetic energy to escape as vapor. See **Figure 18.15.** As a result, the solution has a lower vapor pressure than the pure solvent (water) would have on its own.

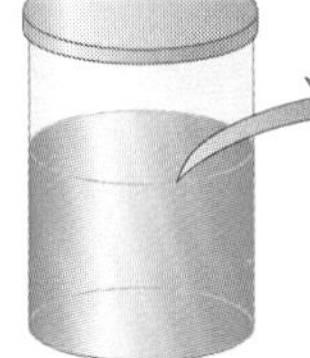

Figure 18.15

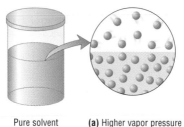

The vapor pressure of a solution of a nonvolatile solute is less than the vapor pressure of a pure solvent. ***(a)*** *Equilibrium is established between the liquid and the vapor in the pure solvent.* ***(b)*** *The equilibrium is disrupted when solute is added. Solvent particles form shells around the solute particles, thus reducing the number of free solvent particles able to escape the liquid. Equilibrium is eventually re-established at a lower vapor pressure.*

STUDENT RESOURCES

From the Teacher's Resource Package, use:

- Section Review 18.3 and Ch. 18 Practice Problems and Quizzes from the Review Module (Ch. 17–20)
- Laboratory Manual: Experiment 33

TECHNOLOGY RESOURCES

Relevant technology resources include:

- Chem ASAP! CD-ROM
- Resource Pro CD-ROM
- ActivChemistry CD-ROM: *Properties of Solutions*

18.3

1 Engage

Use the Visual

Have students study the photograph and read the text that opens the section. Mention that some fish and reindeer also benefit from so-called "natural antifreeze." Ask:

- **What happens to the molecules in water when water freezes?** (The molecules form a crystalline lattice.)
- **How can a solute change the freezing point of a solvent?** (It might slow down the formation of the crystal lattice.)

Explain that attractions between solvent and solute can interfere with the solvent–solvent attractions required to construct the lattice.

Check Prior Knowledge

To assess students' knowledge about the dissolution of electrolytes and nonelectrolytes in an aqueous solvent, have students write the chemical equations for the following examples.

- **hydrochloric acid in water** [$HCl(aq) \rightarrow H^+(aq) + Cl^-(aq)$]
- **magnesium chloride in water** [$MgCl_2(s) \rightarrow Mg^{2+}(aq) + 2Cl^-(aq)$]
- **glucose in water** [$C_6H_{12}O_6(s) \rightarrow C_6H_{12}O_6(aq)$]

Remind students that ionic compounds and certain molecular compounds, such as HCl, dissociate into two or more particles when they dissolve in water. Other molecular compounds, such as glucose, do not dissociate when they dissolve in water. For each formula unit of $MgCl_2$ that dissolves, 3 particles are formed in solution. Ask students how many particles are formed when $FeCl_3$ dissolves in water. (4)

2 Teach

Discuss

Emphasize that colligative properties do not depend on the kind of particles, but on their concentration. For colligative properties, a mole of one kind of particle has the same effect as a mole of any other kind of particle. Remind students that an ionic solid produces a greater change in colligative properties than a soluble molecular solid because it will produce two or more moles of ions for every mole of solid that dissolves.

CHEMISTRY AND MEDICINE

Another kind of colligative property, which is important in many biological processes, is *osmotic pressure:* the tendency for solvent to flow through a semipermeable membrane from a region of low solute concentration to a region of higher solute concentration. Osmotic pressure helps regulate the movement of fluids across cell membranes. Cells remain healthy so long as they are continually bathed in an *isotonic* media, which maintains the correct osmotic balance between the inside and outside of the cell membrane. Physicians make use of the osmotic effect to treat patients with kidney failure. Afflicted individuals undergo dialysis, a process in which the blood is circulated through a machine equipped with a semipermeable membrane. In the machine, osmotic pressure is used to separate waste materials from the blood. See Chemistry Serving...Society on page 526.

The decrease in the vapor pressure is proportional to the number of particles the solute makes in solution. A solute such as sodium chloride, which dissociates into several particles, has a greater effect on the vapor pressure than a nondissociating solute such as glucose. For example, the vapor-pressure lowering caused by 0.1 mol of sodium chloride in 1000 g of water is twice that caused by the same number of moles of glucose in the same quantity of water. This is because each formula unit of the ionic compound sodium chloride produces two particles in solution: a sodium ion and a chloride ion, as illustrated in **Figure 18.16.** In contrast, when glucose dissolves, the molecules do not dissociate. In the same way, 0.1 mol $CaCl_2$ in 1000 g of water produces three times the vapor-pressure lowering as 0.1 mol of glucose in the same quantity of water. This is because each formula unit of calcium chloride produces three particles in solution. How many particles in solution are produced by each formula unit of aluminum bromide? ❶

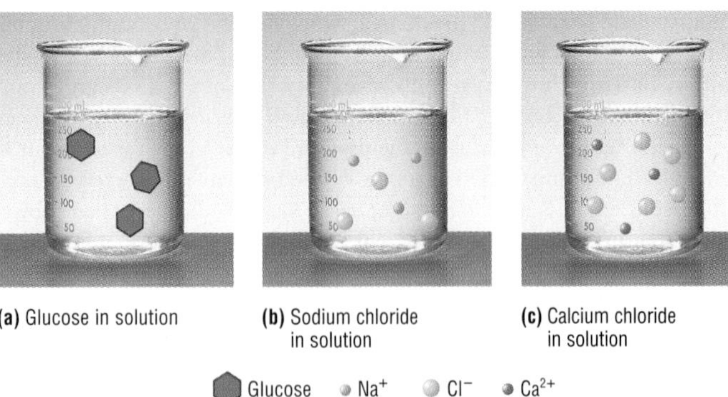

Figure 18.16
Particle concentrations for dissolved covalent and ionic compounds in water differ depending on the number of particles formed per formula unit. ***(a)*** *When 3 mol of glucose is dissolved in water, there are 3 mol of solute particles because glucose does not dissociate.* ***(b)*** *When 3 mol of sodium chloride is dissolved in water, there are 6 mol of solute particles because each formula unit of NaCl dissociates into two ions.* ***(c)*** *When 3 mol of calcium chloride dissolves in water, there are 9 mol of particles because each formula unit of $CaCl_2$ dissociates into three ions. How many moles of particles would 3 mol Na_3PO_4 give in solution?* ❷

Boiling-Point Elevation

The boiling point of a substance is the temperature at which the vapor pressure of the liquid phase equals the atmospheric pressure. As you just learned, adding a nonvolatile solute to a liquid solvent decreases the vapor pressure of the solvent. Because of the decrease in vapor pressure, additional kinetic energy must be added to raise the vapor pressure of the liquid phase of the solution to atmospheric pressure. Thus the boiling point of the solution is higher than the boiling point of the pure solvent. The **boiling-point elevation** is the difference in temperature between the boiling point of a solution and that of the pure solvent.

Answers

❶ 4
❷ 12 mol of particles
❸ −5.58 °C
❹ colligative property

You can also think about boiling-point elevation in terms of particles. Attractive forces exist between the solvent and solute particles. It takes additional kinetic energy for the solvent particles to overcome the attractive forces that keep them in the liquid. Thus the presence of a solute elevates the boiling point of the solvent. The magnitude of the boiling-point elevation is proportional to the number of solute particles dissolved in the solvent. For example, the boiling point of water increases by 0.512 °C for every mole of particles that the solute forms when dissolved in 1000 g of water. Thus boiling-point elevation is a colligative property: It depends on the concentration of particles, not on their identity.

Freezing-Point Depression

When a substance freezes, the particles of the solid take on an orderly pattern. The presence of a solute in water disrupts the formation of this pattern because of the shells of water of solvation. As a result, more kinetic energy must be withdrawn from a solution than from pure solvent for it to solidify. The **freezing-point depression** is the difference in temperature between the freezing point of a solution and that of the pure solvent. The magnitude of the freezing-point depression is proportional to the number of solute particles dissolved in the solvent and does not depend upon their identity. Freezing-point depression is a colligative property.

The addition of 1 mol of solute particles to 1000 g of water lowers the freezing point by 1.86 °C. For example, if you add 1 mol (180 g) of glucose to 1000 g of water, the solution freezes at −1.86 °C. However, if you add 1 mol (58.5 g) of sodium chloride to 1000 g of water, the solution freezes at −3.72 °C. This is because 1 mol NaCl produces 2 mol of particles, and thus doubles the freezing-point depression. What is the freezing point of a solution that contains 1 mol of calcium chloride in 1000 g of water? You take  advantage of the freezing-point depression of aqueous solutions when you sprinkle salt on icy sidewalks to make the ice melt. The ice melts and forms a solution with a lower freezing point than that of pure water. **Figure 18.17** shows another application of the freezing-point depression of an aqueous solution.

Figure 18.17

Temperatures below 0 °C are needed to make ice cream. Rock salt is added to ice in an ice-cream maker to produce a freezing-point depression. The temperature of the ice–water–rock salt mixture decreases to a few degrees below 0 °C. What kind of property is freezing-point depression? 4

section review 18.3

24. Why does a solution have a lower vapor pressure than the pure solvent of that solution?
25. Why does a solution have an elevated boiling point and a depressed freezing point compared with the pure solvent?
26. Would a dilute or a concentrated sodium fluoride solution have a higher boiling point? Explain.
27. An equal number of moles of KI and MgF_2 are dissolved in equal volumes of water. Which solution has the higher
    a. boiling point?
    b. vapor pressure?
    c. freezing point?

 Chem ASAP! **Assessment 18.3** Check your understanding of the important ideas and concepts in Section 18.3.

portfolio project

Salt is often spread on icy streets and highways. Research the environmental problems this causes. Find out what alternative approaches are available.

18.3

ACTIVITY

Have students observe the freezing point depression of ice by the addition of rock salt (NaCl) to a slurry of ice-water. Provide students with a thermometer, styrofoam cup, and ice. Ask students to measure the initial temperature of the ice-water slurry, and the lowest temperature reached after the addition of the rock salt. Explain that the ice-water/salt mixture melts at a much faster rate than ice-water alone. As a result, more heat is absorbed from the immediate surroundings per unit time.

3 Assess

Evaluate Understanding

Have students explain how the addition of solute particles produces a lower vapor pressure, a higher boiling point, and a lower freezing point than those of the pure solvent.

Reteach

Remind students that colligative properties are those physical properties of solutions that depend only on the number of particles of solute in solution and not on the chemical composition of the particles. The magnitudes of the physical changes observed are proportional to the quantity of solute particles in solution.

Portfolio Project

Environmental issues vary by region. To cut back on salt, some areas use more cinders, sawdust, and sand.

Answers

SECTION REVIEW 18.3

24. The introduction of solute molecules reduces the number of solvent molecules with enough kinetic energy to escape.
25. Because vapor pressure has been reduced, more kinetic energy is needed to reach the boiling point. For a solution to freeze, it must lose more kinetic energy than the pure solvent does.
26. Concentrated; the concentrated solution has more dissolved particles. Boiling-point elevation is proportional to the number of dissolved particles.
27. a. $MgF_2(aq)$
    b. $KI(aq)$
    c. $KI(aq)$

1 Engage

Use the Visual

Have students study the photograph and read the text that opens the section. Ask:

- **Would the resulting boiling point increase be enough to shorten the time required for cooking?** (Unless a cook uses an extremely large amount of salt, the effect of the small rise in boiling point on the time required for cooking would be negligible.) **Note that pressure cookers can have a noticeable effect on cooking time. These devices increase the external pressure, causing the boiling point to rise. Food does cook faster in a pressure cooker.**
- **How is the boiling-point elevation of an aqueous solution calculated?** (The magnitude of the boiling-point elevation of a solution relative to the pure solvent is directly proportional to the number of solute particles per fixed number of moles of solvent. Note that 1 kg of pure liquid water contains approximately 55.6 moles. Thus, the mass defines a fixed number of moles of solvent.)

section 18.4

objectives

- Calculate the molality and mole fraction of a solution
- Calculate the molar mass of a molecular compound from the freezing-point depression or boiling-point elevation of a solution of the compound

key terms

- molality (m)
- mole fraction
- molal boiling-point elevation constant (K_b)
- molal freezing-point depression constant (K_f)

CALCULATIONS INVOLVING COLLIGATIVE PROPERTIES

Cooking instructions for a wide variety of foods—from dried pasta to packaged beans to frozen fruits to fresh vegetables—often call for the addition of a small amount of salt to the cooking water. Most people like the flavor of food cooked with salt. What other effect can adding salt have on the cooking process? Recall that dissolved salt elevates the boiling point of water. Suppose you added a teaspoon of salt to 2 liters of water. A teaspoon of salt has a mass of about 20 g. Would the resulting boiling point increase be enough to shorten the time required for cooking? **How is the boiling-point elevation of an aqueous solution calculated?**

Molality and Mole Fraction

Colligative properties depend only on solute concentration, or the ratio of the number of solute particles to solvent particles. There are two convenient ways of expressing this ratio: in molality and in mole fractions. **Molality (m)** is the number of moles of solute dissolved per kilogram (1000 g) of solvent. Molality is also known as molal concentration.

$$\text{Molality} = \frac{\text{moles of solute}}{\text{kilogram of solvent}} = \frac{\text{moles of solute}}{\text{1000 g solvent}}$$

Note that molality is not the same as molarity. Molality refers to moles per kilogram of solvent rather than per liter of solution. In the case of water as the solvent, 1000 g equals a volume of 1000 mL, or 1 L.

You can prepare a solution that is 1.00 molal ($1m$) in glucose, for example, by adding 1.00 mol (180 g) of glucose to 1000 g of water. A solution prepared by dissolving 0.500 mol (29.25 g) of sodium chloride in 1.000 kg (1000 g) of water is 0.500 molal ($0.500m$) in NaCl. See **Figure 18.18.**

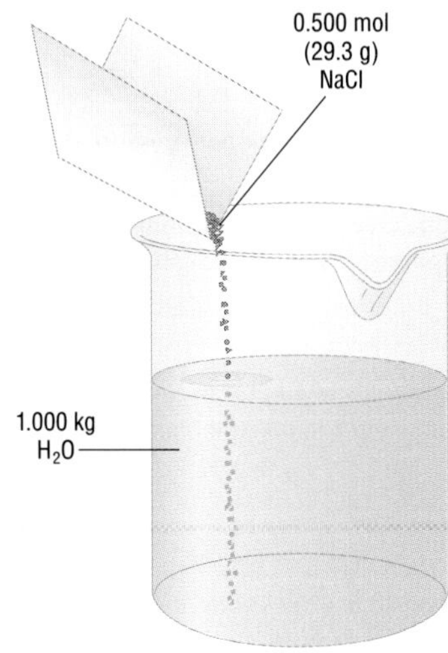

Figure 18.18
To make a 0.500m solution of NaCl, use a balance to measure 1.000 kg of water and add 0.500 mol (29.3 g) NaCl. What would be the molality if only 0.500 kg of water were used? 1

Student Resources

From the Teacher's Resource Package, use:

- Section Review 18.4 and Ch. 18 Practice Problems, Vocabulary Review, Quizzes and Tests from the Review Module (Ch. 17–20)
- Laboratory Manual: Experiment 33

Technology Resources

Relevant technology resources include:

- Chem ASAP! CD-ROM
- Resource Pro CD-ROM
- ActivChemistry CD-ROM: *Properties of Solutions*
- Assessment Resources CD-ROM: Chapter 18 Tests

Sample Problem 18-7

How many grams of potassium iodide must be dissolved in 500.0 g of water to produce a 0.060 molal KI solution?

1. ***ANALYZE*** ***List the knowns and the unknown.***

Knowns:
- mass of water = 500.0 g
- solution concentration = $0.060m$
- molar mass KI = 166.0 g/mol

Unknown:
- mass of solute = ? g KI

According to the definition of molal, the final solution must contain 0.060 mol KI per 1000 g H_2O. Use the molality as a conversion factor to convert from mass of water to moles of the solute (KI). Then use the molar mass of KI to convert from mol KI to g KI. The steps are mass of $H_2O \longrightarrow$ mol KI $\longrightarrow$ g KI.

2. ***CALCULATE*** ***Solve for the unknown.***

$$500.0\ \cancel{g\ H_2O} \times \frac{0.060\ \cancel{mol\ KI}}{1000\ \cancel{g\ H_2O}} \times \frac{166.0\ g\ KI}{1\ \cancel{mol\ KI}} = 5.0\ g\ KI$$

3. ***EVALUATE*** ***Does the result make sense?***

A 1 molal KI solution is made by dissolving a molar mass of KI (166.0 g) in 1000 g of water. The desired molal concentration ($0.060m$) is about $\frac{1}{20}$ of that value, so it should be expected that the mass of KI would be much less than the molar mass. The answer is correctly expressed to two significant figures.

Practice Problems

28. How many grams of sodium fluoride are needed to prepare a $0.400m$ NaF solution that contains 750.0 g of water?

29. Calculate the molality of a solution prepared by dissolving 10.0 g NaCl in 600 g of water.

Solution concentration can be expressed in yet another way. The ratio of the moles of solute in solution to the total number of moles of solvent and solute is the **mole fraction** of that solute. In a solution containing n_A mol of solute A and n_B mol of solvent B, the mole fraction of solute (X_A) and mole fraction of solvent (X_B) can be expressed as follows.

$$X_A = \frac{n_A}{n_A + n_B} \qquad X_B = \frac{n_B}{n_A + n_B}$$

Sample Problem 18-8

Compute the mole fraction of each component in a solution of 1.25 mol of ethylene glycol (EG) and 4.00 mol of water.

1. ***ANALYZE*** ***List the knowns and the unknowns.***

Knowns:
- moles of ethylene glycol (n_{EG}) = 1.25 mol
- moles of water (n_{H_2O}) = 4.00 mol H_2O

Unknowns:
- mole fraction EG (X_{EG}) = ?
- mole fraction H_2O (X_{H_2O}) = ?

2 Teach

Discuss

Write the expressions defining molarity and molality on the board. Compare the chemical quantities in each expression. Point out that molarity is denoted by *M* and molality by *m*. Explain that the molality of a solution does not vary with temperature because the mass of the solvent does not change. In contrast, the molarity of a solution does vary with temperature because the liquid can expand and contract. When studying colligative properties such as boiling point–elevation and freezing–point depression, it is preferable to use a concentration that does not depend on temperature.

Use the Visual

Display Figure 18.18 on an overhead projector. Ask students to write down the definition of molality in their notebooks. Show the step-by-step approach a chemist would use to prepare a $0.500m$ solution of NaCl. Have students confirm your calculations and the data given in the figure. Ask:

- **What would the molality be if only 0.500 kg water were used?** (1.00 m)

Practice Problems Plus

Additional Practice Problem

The following problem is related to Sample Problem 18-7: How many grams of lithium bromide must be dissolved in 444 g of water to prepare a $0.140m$ LiBr solution? (5.40 g LiBr)

Answer

❶ $1.00m$

Meeting Diverse Needs

LEP Have LEP students write sentences using the words *molarity* and *molality,* and have them circle the letter in each word that makes them distinct. The mnemonic, '*r*' for mola*r*ity and lite*r* may help.

Discuss

Explain that the mole fraction compares the number of moles of a solute to the total number of moles in the solution. Organic chemists, who frequently work with non-aqueous solvent systems, often use this method of expressing concentration. The mole fraction is also used when calculating the vapor pressure of a solution.

Practice Problems Plus

Related Chapter Review Problems
Chapter Review problems 65 and 69 are related to Sample Problem 18-8.

Additional Practice Problem
The Chem ASAP! CD-ROM contains the following problem: Calculate the mole fraction of each component in a solution of 42 g CH_3OH, 35 g C_2H_5OH, and 50 g C_3H_7OH. ($CH_3OH = 0.45$ $C_2H_5OH = 0.26$, $C_3H_7OH = 0.29$)

ACTIVITY

Have students design a diagram that describes and explains the various methods used to calculate the concentration of a solution. The diagram should include definitions of each type of concentration unit. More than one design is possible, but each should include the terms *concentration of a solution, percent by mass, percent by volume, mass-volume percent, molarity, molality,* and *mole fraction.* Students may wish to include additional detail by showing interconversions between *mass of solute* and *moles of solute.* Underneath each concentration unit, students should provide examples of when the unit would be used in the laboratory and in everyday life.

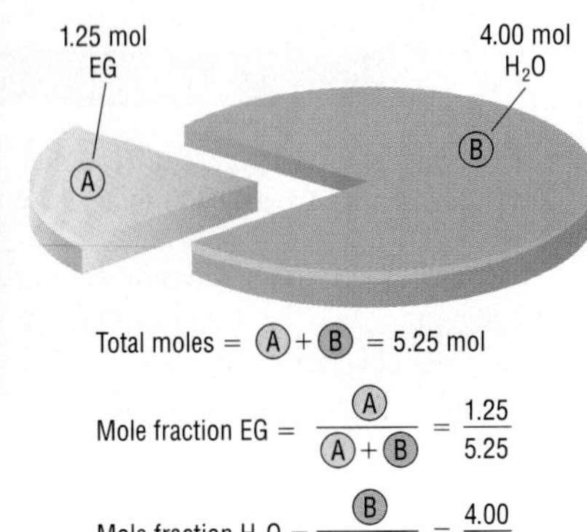

Figure 18.19
The mole fraction is the ratio of the number of moles of one substance to the total number of moles of all substances in the solution. Ethylene glycol (EG) is often added to water in a car radiator as an antifreeze. What is the sum of all mole fractions in a solution? ❶

Sample Problem 18-8 (cont.)

The mole fraction of ethylene glycol (X_{EG}) in the solution is the number of moles of ethylene glycol divided by the total number of moles in solution:

$$X_{EG} = \frac{n_{EG}}{n_{EG} + n_{H_2O}}$$

Similarly, the mole fraction of water (X_{H_2O}) is the number of moles of water divided by the total number of moles in solution:

$$X_{H_2O} = \frac{n_{H_2O}}{n_{EG} + n_{H_2O}}$$

2. ***CALCULATE*** ***Solve for the unknowns.***

$$X_{EG} = \frac{n_{EG}}{n_{EG} + n_{H_2O}} = \frac{1.25 \text{ mol}}{1.25 \text{ mol} + 4.00 \text{ mol}} = 0.238$$

$$X_{H_2O} = \frac{n_{H_2O}}{n_{EG} + n_{H_2O}} = \frac{4.00 \text{ mol}}{1.25 \text{ mol} + 4.00 \text{ mol}} = 0.762$$

3. ***EVALUATE*** ***Do the results make sense?***
Note that mole fraction is a dimensionless quantity. The sum of the mole fractions of all the components in a solution equals unity, or one, as should be expected. Each answer is correctly expressed to three significant figures.

Practice Problems

30. What is the mole fraction of each component in a solution made by mixing 300 g of ethanol (C_2H_5OH) and 500 g of water?

31. A solution contains 50.0 g of carbon tetrachloride (CCl_4) and 50.0 g of chloroform ($CHCl_3$). Calculate the mole fraction of each component in the solution.

Chem ASAP!
Problem-Solving 31
Solve Problem 31 with the help of an interactive guided tutorial.

Boiling-Point Elevation and Freezing-Point Depression

Elevations of boiling points and depressions of freezing points are usually quite small. Thus they can be measured accurately only with thermometers that measure temperature changes as small as 0.001 °C.

The boiling point of a solvent is raised by the addition of a nonvolatile solute, as shown in **Figure 18.20.** The magnitude of the boiling-point elevation (ΔT_b) is directly proportional to the molal concentration (m), assuming the solute is molecular, not ionic.

$$\Delta T_b \propto m$$

Answers

❶ 1
❷ 1 m $CaBr_2$

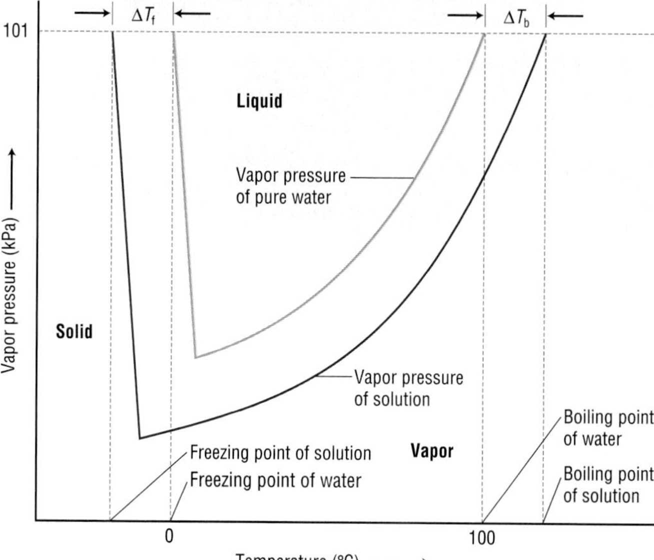

Figure 18.20
The addition of a nonvolatile solute to water lowers the water's vapor pressure. This results in an increase in the normal boiling point of the water and a decrease in the normal freezing point of the water.

Chem ASAP!
Simulation 22
Discover the principle underlying the colligative properties of solutions.

The change in the boiling temperature (ΔT_b) is the elevation of the boiling point of the solvent. It is the boiling point of the solution minus the boiling point of the pure solvent. The term m is the molal concentration of the solution. Adding a proportionality constant (K_b) to this relationship makes it an equality.

$$\Delta T_b = K_b \times m$$

This constant is the **molal boiling-point elevation constant (K_b)**, which is equal to the change in boiling point for a 1-molal solution of a nonvolatile molecular solute. K_b is dependent on the solvent, and its units are °C/m. **Table 18.2** lists the K_b values for water and some other solvents.

The boiling-point elevation for ionic solids depends on the number of ions produced by each formula unit. This number is used to calculate the effective molality. Which aqueous solution would have the larger boiling-point elevation: 1m KCl or 1m $CaBr_2$?

Table 18.2

K_b Values for Some Common Solvents

Solvent	K_b (°C/m)
Water	0.512
Ethanol	1.19
Benzene	2.53
Cyclohexane	2.79
Acetic acid	3.07
Phenol	3.56
Nitrobenzene	5.24
Camphor	5.95

Sample Problem 18-9

What is the boiling point of a 1.50m NaCl solution?

1. ***ANALYZE*** ***List the knowns and the unknown.***

Knowns:
- solution concentration (m) = 1.50m NaCl
- K_b for H_2O = 0.512 °C/m
- $\Delta T_b = K_b \times m$

Unknown:
- boiling point = ? °C

18.4

Use the Visual

Display Figure 18.20 on an overhead projector. Choose an arbitrary concentration for an aqueous solution of NaCl or ethylene glycol. Calculate the boiling-point elevation and freezing-point depression. Then write the temperatures on the horizontal axis. Show students how to read the data for the boiling points of the pure solvent and of the solution. Then have students read the corresponding values for the freezing points. Ask students to calculate the ΔT_b and ΔT_f for this solution. Point out that the solute affects both the freezing point and boiling point of a liquid. Ask:

- **Why would knowing the boiling-point elevation and freezing-point depression be important when choosing an antifreeze for car radiators?** (Despite its name, antifreeze protects against both freezing and overheating.)

ACTIVITY

Some students may think that freezing points and boiling points can be depressed or elevated without end. Explain that as the concentration of a solute, such as ethylene glycol, increases, there comes a point when the quantity of solute exceeds the quantity of solvent. Ethylene glycol becomes the solvent and water becomes the solute. The trends in colligative properties begin to reflect ethylene glycol instead of water. If the solute is a solid, such as NaCl, eventually the solution becomes saturated. But even before this saturation point, the magnitude of certain colligative properties may reach a maximum.

As an extension, assign students different concentrations of an NaCl or ethylene glycol solution. [(K_b and K_f for ethylene glycol are 2.26 (°C/m) and 3.11(°C/m), respectively.)] Have students calculate the boiling point and freezing point values for each concentration. Then have students prepare solutions with the assigned concentrations; use them to measure boiling-point elevation. Compare calculated and experimental values. Ask students to suggest reasons for any observed differences. (Choose concentrations that provide differences beyond those due to experimental error.)

Practice Problems Plus

Related Chapter Review Problems

Chapter Review problems 55, 62, and 68 are related to Sample Problem 18-9.

Additional Practice Problem

The Chem ASAP! CD-ROM contains the following problem: What is the boiling point of a solution containing 96.7 g of sucrose ($C_{12}H_{22}O_{11}$) in 250.0 g water at 1 atm? (100.579 °C)

Sample Problem 18-9 (cont.)

Each formula unit of NaCl is ionized into two particles.

$$NaCl(s) \longrightarrow Na^+(aq) + Cl^-(aq)$$

Therefore, the molality of total particles is $2 \times 1.50m = 3.00m$. Calculate the boiling-point elevation and then add it to 100 °C.

2. *CALCULATE* ***Solve for the unknown.***

$$\Delta T_b = K_b \times m$$
$$= 0.512\ °C/\cancel{m} \times 3.00\cancel{m}$$
$$= 1.54\ °C$$

The boiling point of this solution is

$$100\ °C + 1.54\ °C = 101.54\ °C$$

3. *EVALUATE* ***Does the result make sense?***

The boiling point increases about 0.5 °C for each mole of solute particles, so the total change is reasonable. The answer is correctly expressed to hundredths of a degree. Because the boiling point of water is defined as exactly 100 °C, this value does not limit the number of significant figures in the solution of the problem.

Practice Problems

32. What is the boiling point of a solution that contains 1.25 mol $CaCl_2$ in 1400 g of water?

33. What mass of NaCl would have to be dissolved in 1.000×10^3 g of water to raise the boiling point by 2.00 °C?

Chem ASAP!

Problem-Solving 33

Solve Problem 33 with the help of an interactive guided tutorial.

You can also calculate the freezing-point depression of a solution. To do so use the equation

$$\Delta T_f = K_f \times m$$

The change in the freezing temperature (ΔT_f) represents the freezing-point depression. The molality is m. The **molal freezing-point depression constant (K_f)** is equal to the change in the freezing point for a 1-molal solution of a nonvolatile molecular solute. The units of K_f are °C/m. **Table 18.3** gives the values of the constant K_f for some common solvents.

Table 18.3

K_f Values for Some Common Solvents

Solvent	K_f (°C/m)
Water	1.86
Acetic acid	3.90
Benzene	5.12
Nitrobenzene	7.00
Phenol	7.40
Cyclohexane	20.2
Camphor	37.7

Molar Mass

You can use the changes in boiling and freezing points to determine the molar mass of a substance. To do this, you would first add a known mass of solute to a known mass of solvent that has a known molal boiling-point elevation constant (K_b) or molal freezing-point depression constant (K_f). Next you would measure the change in the boiling or freezing point. Finally, you would calculate the molar mass for the solute. Sample Problem 18-10 illustrates this strategy.

Sample Problem 18-10

A solution of 7.50 g of a nonvolatile compound in 22.60 g of water boils at 100.78 °C at 760 mm Hg. What is the molar mass of the solute? Assume that the solute exists as molecules, not ions.

Sample Problem 18-10 (cont.)

1. ***ANALYZE*** ***List the knowns and the unknown.***

Knowns:
- mass of water = 22.60 g
- mass of solute = 7.50 g
- boiling-point elevation (ΔT_b) = (100.78 °C − 100 °C) = 0.78 °C
- K_b for H_2O = 0.512 °C/m
- $\Delta T_b = K_b \times m$

Unknown:
- molar mass = ? g/mol

2. ***CALCULATE*** ***Solve for the unknown.***

Solve $\Delta T_b = K_b \times m$ for m and substitute the known values.

$$m = \frac{\Delta T_b}{K_b} = \frac{0.78\ \cancel{°C}}{0.512\ \cancel{°C}/m} = 1.5m = 1.5\ \frac{\text{mol solute}}{\text{kg water}}$$

Calculate moles of solute.

$$22.60\ \cancel{\text{g water}} \times \frac{1\ \cancel{\text{kg water}}}{1000\ \cancel{\text{g water}}} \times \frac{1.5\ \text{mol solute}}{1\ \cancel{\text{kg water}}} = 0.034\ \text{mol solute}$$

$$\text{Molar mass of solute} = \frac{\text{mass of solute}}{\text{moles of solute}} = \frac{7.50\ \text{g}}{0.034\ \text{mol}} = 2.2 \times 10^2\ \text{g/mol}$$

3. ***EVALUATE*** ***Does the result make sense?***

A molality of 1.5 is reasonable because the boiling point was increased about 1.5 times K_b. The amount of water is about $\frac{1}{50}$ of a kilogram and the solution is 1.5m, so it makes sense that the moles of solute should be greater than $\frac{1}{50}$ of a mole. The actual number of moles was about $\frac{1}{30}$ of a mole, so the molar mass should be about 30 times the mass of the solute.

Practice Problems

34. The freezing point of the water is lowered to −0.390 °C when 3.90 g of a nonvolatile molecular solute is dissolved in 475 g of water. Calculate the molar mass of the solute.

35. A solution containing 16.9 g of a nonvolatile molecular compound in 250 g of water has a freezing point of −0.744 °C. What is the molar mass of the substance?

section review 18.4

36. How many kilograms of water must be added to 9.0 g of oxalic acid ($H_2C_2O_4$) to prepare a 0.025m solution?

37. One mole of a compound of iron and chlorine is dissolved in 1 kg of water. The boiling point of this aqueous solution is 102.05 °C. The freezing point of this aqueous solution is −7.44 °C. What is the formula of the solute compound?

38. How are boiling-point elevation and freezing-point depression related to molality?

39. Estimate the freezing point of a solution of 12.0 g of carbon tetrachloride dissolved in 750 g of benzene (which has a freezing point of 5.48 °C).

Chem ASAP! **Assessment 18.4** Check your understanding of the important ideas and concepts in Section 18.4.

3 Assess

Evaluate Understanding

Ask students:

- **Which solution has a higher boiling point, 1 mol of $Al(NO_3)_3$ in 1000 g of water or 1.5 mol of KCl in 1000 g of water? Have them explain their answer.** (The solution of $Al(NO_3)_3$ will have a higher boiling point because $Al(NO_3)_3$ dissociates into a greater number of particles.)
- **Why is it important to distinguish between nonvolatile and volatile compounds when discussing certain colligative properties?** (Volatile solutes would quickly evaporate at higher temperatures, which would change the *molal* concentration of the solution.)

Reteach

Remind students that when they do calculations involving *ionic* solids they must find the molarity of the solution in terms of the total number of particles. Work through several examples to be sure they understand this important point. Explain how colligative properties such as osmotic pressure and freezing-point depression can be used as a convenient and accurate method for estimating the molar mass of large molecular compounds such as proteins.

Practice Problems Plus

Related Chapter Review Problem
Chapter Review problem 79 is related to Sample Problem 18-10.

Answers

SECTION REVIEW 18.4

36. 4.0 kg H_2O

37. $FeCl_3$

38. The molality of a solution is directly proportional to its boiling-point elevation and freezing-point depression. $\Delta T_b = K_b m$ and $\Delta T_f = K_f m$

39. 4.95 °C

Chemistry Serving... Society

DISCUSS

After students have had a chance to read the article, ask them to name some of the molecular and ionic solutes found in the bloodstream. Create a table on the chalkboard that lists the different components in the bloodstream including: water, proteins, Na^+, K^+, Mg^{2+}, Ca^{2+}, Cl^-, HCO_3^-, SO_4^{2-}, HPO_4^{2-}, O_2, CO_2, red and white blood cells, and platelets. Create additional columns with the headings *concentration, possible causes of elevated levels, possible causes of depressed levels.* Explain that changes in the composition of blood, for example in the concentration of one or more of the ions can be an indication of a disease or disorder. Doctors frequently use blood tests to check on patients' health. Have students do research to find the concentrations of each of the components in "normal" blood and to find out what types of diseases are linked to changes in the concentrations of certain ions. For example, elevated levels of Na^+ are associated with overactive adrenal glands. Ask students which components of blood are suspended and which are dissolved.

CHEMISTRY IN CAREERS

Have students read about the duties and responsibilities of a nephrology nurse on page 876 of this text. Students may wish to connect to the Internet address shown to find out more about career opportunities in this field.

Chemistry Serving...Society

A SOLUTION FOR KIDNEY FAILURE

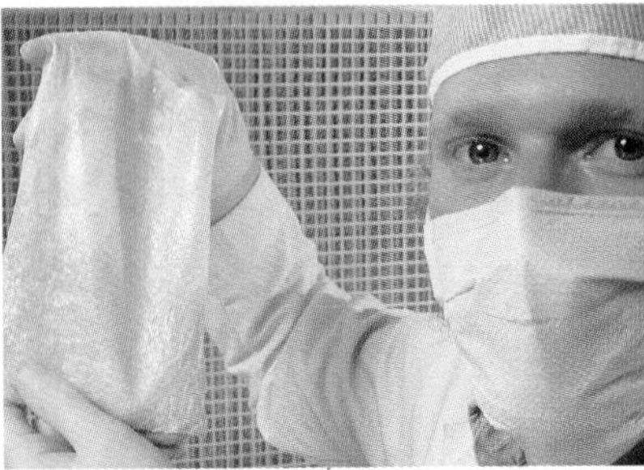

Your blood is a solution, a colloid, and a suspension all at the same time. Blood contains gases, ions, and nutrient molecules in solution. It also contains biologically important particles—protein molecules in colloidal dispersion and cells in suspension.

For your body to function properly, the blood's solutes must be kept at the right concentrations. Many of your organs participate in this process. Your liver and pancreas, for example, are responsible for regulating the concentration of sugar in your blood.

Similarly, your kidneys have the important but often under-appreciated job of keeping the concentration of dissolved waste products in your blood at a minimum. As part of their normal functioning, all of your body's cells continuously produce waste products and dump these wastes into the blood. If they were not removed from your blood, the dissolved waste products would build up and become toxic to your body.

For your body to function properly, the blood's solutes must be kept at the right concentrations. Many of your organs participate in this process.

The work of filtering waste products from your blood is done by thousands of small units in your kidneys called nephrons. Blood enters a nephron through a tiny blood vessel. Everything except blood cells and very large molecules can then pass out of the blood through a semipermeable membrane into a long collecting tube. As the fluid travels through this tube, all the necessary components of your blood—water, ions, and other solutes—pass through the walls of the tube and back into the blood. Waste products, however, remain in the tube. By the time the tube exits the nephron, it contains only urine. Blood that leaves the nephron has all its necessary components—minus the waste products in the urine.

What happens if the kidneys fail to function correctly? Many people suffer from kidney failure and must have their blood "cleaned" in some other way. Many of these people rely on the process called hemodialysis, a way of filtering the blood outside the body.

The hemodialysis machine is an artificial kidney. Blood leaves a patient's artery through a tube and enters the dialysis machine. Inside the machine, the blood passes through a semipermeable membrane similar to the one shown in the photograph. The membrane is submersed in a solution containing the same ions and other solutes as in human blood at the same concentrations. This solution, called dialysate, contains no waste products, so the waste products in the blood readily move out of the blood, across the dialyzing membrane, and into the dialysate. During the process of hemodialysis, the dialysate is continuously recirculated to prevent the return of wastes to the blood. After cleaning the blood, the machine returns it to the body through a vein.

The process of cleaning a person's entire blood supply takes several hours. The process is effective, but as soon as the patient is disconnected from the machine, waste products once again begin building up in the blood. For this reason, several visits to the dialysis clinic are needed every week. Today, a dialysis machine represents life-giving technology for those suffering from kidney failure.

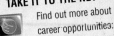

CHEMISTRY IN CAREERS

NEPHROLOGY NURSE

Work in a hospital or treatment facility to care for patients with various types of renal disease.

See page 876.

CHEMIST NEEDED

Enviro. Lab. GC Must know

TAKE IT TO THE NET

Find out more about career opportunities:

www.phschool.com

CHEMICAL SPECIALIST

Local food service distributor see responsible self-motivated indivi

Chapter 18 STUDENT STUDY GUIDE

Take It to the NET For interactive study and review, go to www.phschool.com

KEY TERMS

- boiling-point elevation *p. 518*
- colligative property *p. 517*
- concentrated solution *p. 509*
- concentration *p. 509*
- dilute solution *p. 509*
- freezing-point depression *p. 519*
- Henry's law *p. 506*
- immiscible *p. 503*
- miscible *p. 503*
- molal boiling-point elevation constant (K_b) *p. 523*
- molal freezing-point depression constant (K_f) *p. 524*
- molality (m) *p. 520*
- molarity (M) *p. 509*
- mole fraction *p. 521*
- saturated solution *p. 502*
- solubility *p. 503*
- supersaturated solution *p. 506*
- unsaturated *p. 503*

KEY EQUATIONS

- Henry's law: $\frac{S_1}{P_1} = \frac{S_2}{P_2}$
- Molarity $= \frac{\text{moles of solute}}{\text{liters of solution}}$
- $M_1 \times V_1 = M_2 \times V_2$
- Percent by volume $= \frac{\text{volume of solute}}{\text{solution volume}} \times 100\%$
- Percent (mass/volume) $= \frac{\text{mass of solute (g)}}{\text{solution volume (mL)}} \times 100\%$
- Molality $= \frac{\text{moles of solute}}{\text{kilogram of solvent}} = \frac{\text{moles of solute}}{\text{1000 g solvent}}$
- Mole fractions: $X_A = \frac{n_A}{n_A + n_B}$ $X_B = \frac{n_B}{n_A + n_B}$
- $\Delta T_b = K_b \times m$
- $\Delta T_f = K_f \times m$

CONCEPT SUMMARY

18.1 Properties of Solutions

- The rate at which a solute dissolves is influenced by a number of factors, including the temperature of the solvent, the particle size of the solute, and stirring of the solution.
- Two liquids are said to be miscible if they dissolve in each other; immiscible if they do not.
- Changes in temperature and pressure of a system affect the solubility of a solute.

18.2 Concentrations of Solutions

- The relative amounts of solute and solvent in a solution are described by molar concentration, percent composition, molal concentration, and mole fraction.

18.3 Colligative Properties of Solutions

- Colligative properties include vapor-pressure lowering, freezing-point depression, and boiling-point elevation.

18.4 Calculations Involving Colligative Properties

- The magnitude of every colligative property is directly proportional to the number of solute molecules or ions present.
- Molality is moles of solute per kilogram of solvent. The mole fraction of a substance in solution is equal to the number of moles of the substance divided by the total number of moles of all substances in the solution.

CHAPTER CONCEPT MAP

Use these terms to construct a concept map that organizes the major ideas of this chapter.

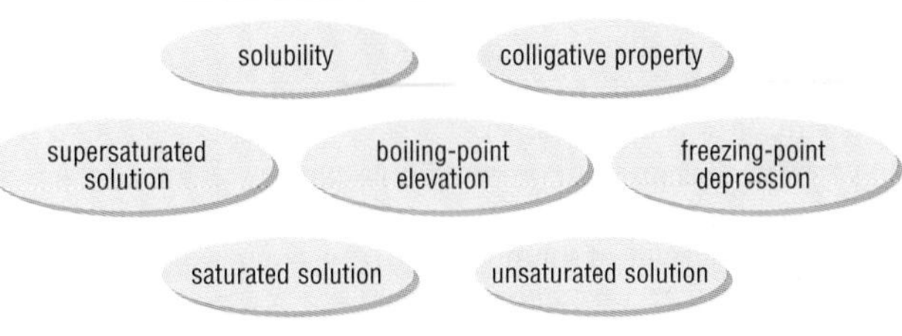

Chem ASAP! Concept Map 18
Create your Concept Map using the computer.

Take It to the Net

At **www.phschool.com** students will find for this chapter

- an Internet activity
- links to related chemistry sites
- an interactive quiz
- career links

Chapter 18
STUDENT STUDY GUIDE

4 Close

Summary

Ask the following questions to summarize chapter content.

- **What factors affect the solubility of a solute and the rate at which it dissolves?** (the chemical nature of the solute and the solvent, temperature, agitation, and particle size)
- **Compare unsaturated, saturated, and supersaturated solutions.** (Make sure students mention the dynamic equilibrium in saturated solutions.)
- **Name five ways to express solute concentration.** (%m/v, %v/v, molality, molarity, and mole fraction)
- **What are colligative properties?** (physical properties of a solution that depend solely on the number of solute particles)

Extension

Have students research how solubility and colligative properties play a role in human physiology, environmental science, plant growth and industry. (e.g., hemodialysis, desalination plants, plant growth)

Looking Back . . . Looking Ahead . . .

The previous chapter discussed the properties of water, and the role of hydrogen bonding in solvation. In this chapter, methods for expressing concentrations of solute are presented along with the effect of solute particles on physical properties. The next chapter describes the conditions that affect reaction rates.

Answers

40. The solvent is the substance in which the solute is dissolved.

41. Random collisions of the solvent molecules with the solute particles provide enough force to overcome gravity.

42. Miscible liquids dissolve in each other; immiscible liquids do not.

43. Solubility is the amount of solute dissolved in a given amount of solvent to form a saturated solution at a given temperature. A saturated solution contains the maximum possible amount of solute at that temperature. An unsaturated solution contains less dissolved solute than a saturated solution.

44. 555 g $AgNO_3$

45. Particles of solute crystallize.

46. No; if there were undissolved solute, the excess solute would come out of solution.

47. Solubility increases with pressure.

48. **a.** 0.016 g/L **b.** 0.047 g/L

49. Molarity provides the exact number of moles of solute per liter of solution. Dilute and concentrated are relative terms and are not quantitative.

50. **a.** 1.3*M* KCl
b. 0.33*M* $MgCl_2$

51. the number of moles of solute dissolved in one liter of solution

52. **a.** 0.50 mol NaCl, 29 g NaCl
b. 1.0 mol KNO_3, 1.0×10^2 g KNO_3
c. 0.025 mol $CaCl_2$, 2.8 g $CaCl_2$
d. 0.60 mol Na_2SO_4, 85 g Na_2SO_4

53. **a.** 23 g NaCl **b.** 2.0 g $MgCl_2$

54. **a.** 3.3% KCl
b. 1.6% $NaNO_3$

55. **a.** 100.26 °C **b.** 101.54 °C

56. Add 27.0 g H_2O to 32.0 g CH_3OH.

57. 1*M* solution: 1 mol of solute in 1 L of solution; 1*m* solution: 1 mol of solute in 1000 g of solvent

58. **a.** −4.46 °C **b.** −2.2 °C

59. **a.** −1.1 °C **b.** −0.74 °C **c.** −1.5 °C

60. mass of solute, mass of solvent, boiling point of solvent, boiling point of solution

61. **a.** The freezing-point depression is twice as great for solute B; solute B must provide twice as many particles in solution.
b. Solute A forms a saturated solution.

62. $\Delta T_f = -9.60$ °C; $\Delta T_b = +4.74$ °C

Chapter 18 REVIEW

CONCEPT PRACTICE

40. Name and distinguish between the two components of a solution. *18.1*

41. Explain why the dissolved component does not settle out of a solution. *18.1*

42. Explain miscible and immiscible. *18.1*

43. Define solubility, saturated solution, and unsaturated solution. *18.1*

44. What mass of $AgNO_3$ can be dissolved in 250 g of water at 20 °C? *18.1*

45. If a saturated solution of sodium nitrate is cooled, what change might you observe? *18.1*

46. Can a solution with undissolved solute be supersaturated? Explain. *18.1*

47. What is the effect of pressure on the solubility of gases in liquids? *18.1*

48. The solubility of methane, the major component of natural gas, in water at 20 °C and 1.00 atm pressure is 0.026 g/L. If the temperature remains constant, what will be the solubility of this gas at the following pressures? *18.1*
a. 0.60 atm
b. 1.80 atm

49. Having a measure of the molarity of a solution is more meaningful than knowing whether a solution is dilute or concentrated. Explain. *18.2*

50. Calculate the molarity of each solution. *18.2*
a. 1.0 mol KCl in 750 mL of solution
b. 0.50 mol $MgCl_2$ in 1.5 L of solution

51. Define molarity. *18.2*

52. Calculate the moles and grams of solute in each solution. *18.2*
a. 1.0 L of 0.50*M* NaCl
b. 5.0×10^2 mL of 2.0*M* KNO_3
c. 250 mL of 0.10*M* $CaCl_2$
d. 2.0 L of 0.30*M* Na_2SO_4

53. Calculate the grams of solute required to make the following solutions. *18.2*
a. 2.5 L of normal saline solution (0.90% NaCl (m/v))
b. 50.0 mL of 4.0% $MgCl_2$ (m/v)

54. What is the concentration (in % (m/v)) of the following solutions? *18.2*
a. 20.0 g KCl in 0.60 L of solution
b. 32 g $NaNO_3$ in 2.0 L of solution

55. What is the boiling point of each solution? *18.4*
a. 0.50 mol glucose in 1000 g H_2O
b. 1.50 mol NaCl in 1000 g H_2O

56. Describe how you would make an aqueous solution of methanol (CH_3OH) in which the mole fraction of methanol is 0.40. *18.4*

57. Distinguish between a 1*M* solution and a 1*m* solution. *18.4*

58. What is the freezing point of each solution? *18.4*
a. 1.40 mol Na_2SO_4 in 1750 g H_2O
b. 0.060 mol $MgSO_4$ in 100 g H_2O

59. Determine the freezing points of each 0.20*m* aqueous solution. *18.4*
a. K_2SO_4
b. $CsNO_3$
c. $Al(NO_3)_3$

60. What laboratory measurements must be made to find the molar mass of a solute in a boiling-point elevation experiment? *18.4*

CONCEPT MASTERY

61. Varying numbers of moles of two different solutes, A and B, were added to identical quantities of water. The graph shows the freezing point of each of the solutions formed with various amounts of solutes.
a. Explain the relative slopes of the two lines between 0 and 2 mol of solute added.
b. Why does the freezing point for solution A not continue to drop as amounts of solute A are added beyond 2.4 mol?

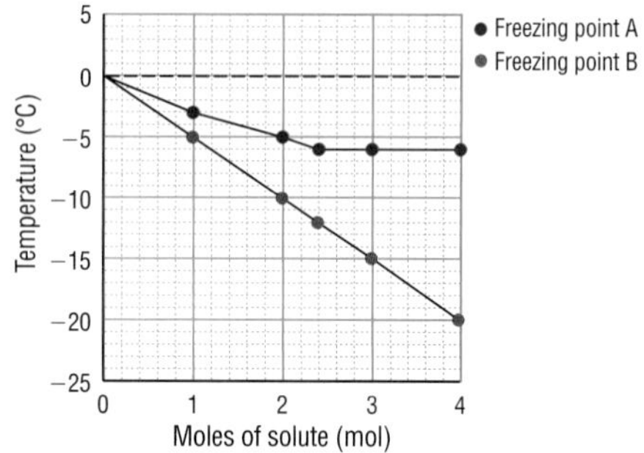

62. Calculate the freezing- and boiling-point changes for a solution containing 12.0 g of naphthalene ($C_{10}H_8$) in 50.0 g of benzene.

63. Describe how you would prepare an aqueous solution of acetone (CH_3COCH_3) in which the mole fraction of acetone is 0.25.

64. The solubility of sodium hydrogen carbonate ($NaHCO_3$) in water at 20 °C is 9.6 g/100 g H_2O. What is the mole fraction of $NaHCO_3$ in a saturated solution? What is the molality of the solution?

65. A solution is labeled 0.150*m* NaCl. What are the mole fractions of the solute and solvent in this solution?

66. You are given a clear aqueous solution containing KNO_3. How would you determine experimentally if the solution is unsaturated, saturated, or supersaturated?

67. A mixture of ethylene glycol (EG) and water is used as antifreeze in automobile engines. The freezing point and specific gravity of the mixture vary with the percent by mass of (EG) in the mixture. On the following graph, point A represents 20% (EG) by mass; point B, 40%; and point C, 60%.

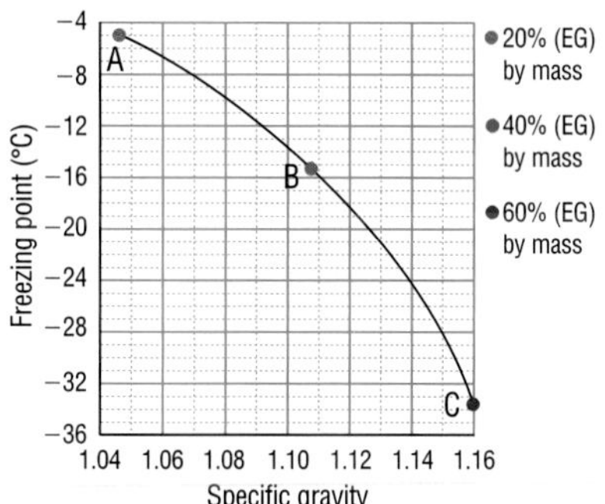

a. What is the specific gravity of the antifreeze mixture that freezes at −25 °C?
b. What is the freezing point of a mixture that has a specific gravity of 1.06?
c. Estimate the freezing point of a mixture that is 30% by mass (EG).

68. Calculate the freezing point and the boiling point of a solution that contains 15.0 g of urea (CH_4N_2O) in 250 g of water. Urea is a covalently bonded compound.

69. Calculate the mole fractions in a solution that is made of 25.0 g of ethanol (C_2H_5OH) and 40.0 g of water.

70. Estimate the freezing point of an aqueous solution of 20.0 g of glucose ($C_6H_{12}O_6$) dissolved in 500.0 g of water.

71. Plot a graph of solubility versus temperature for the three gases listed in **Table 18.1.**

72. The solubility of KCl in water is 34.0 g KCl/100 g H_2O at 20 °C. A warm solution containing 50.0 g KCl in 130 g H_2O is cooled to 20 °C.
a. How many grams of KCl remain dissolved?
b. How many grams came out of solution?

73. How many moles of ions are present when 0.10 mol of each compound is dissolved in water?
a. K_2SO_4
b. $Fe(NO_3)_3$
c. $Al_2(SO_4)_3$
d. $NiSO_4$

74. Complete the following table for aqueous solutions of glucose ($C_6H_{12}O_6$).

Mass solute	Moles solute	Volume of solution	Molarity
12.5 g		219 mL	
	1.08		0.519
		1.62 L	1.08

75. A solution contains 26.5 g NaCl in 75.0 g H_2O at 20 °C. Determine if the solution is unsaturated, saturated, or supersaturated. (The solubility of NaCl at 20 °C is 36.0 g/100 g H_2O.)

76. An aqueous solution freezes at −2.47 °C. What is its boiling point?

77. Hydrogen peroxide is often sold commercially as a 3.0% (m/v) aqueous solution.
a. If you buy a 250-mL bottle of 3.0% H_2O_2 (m/v), how many grams of hydrogen peroxide have you purchased?
b. What is the molarity of this solution?

78. How many grams of $NaNO_3$ will precipitate if a saturated solution of $NaNO_3$ in 200 g H_2O at 50 °C is cooled to 20 °C?

79. What is the molar mass of a nondissociating compound if 5.76 g of the compound in 750 g of benzene gives a freezing-point depression of 0.460 °C?

80. The molality of an aqueous solution of sugar ($C_{12}H_{22}O_{11}$) is 1.62*m*. Calculate the mole fractions of sugar and water.

71.

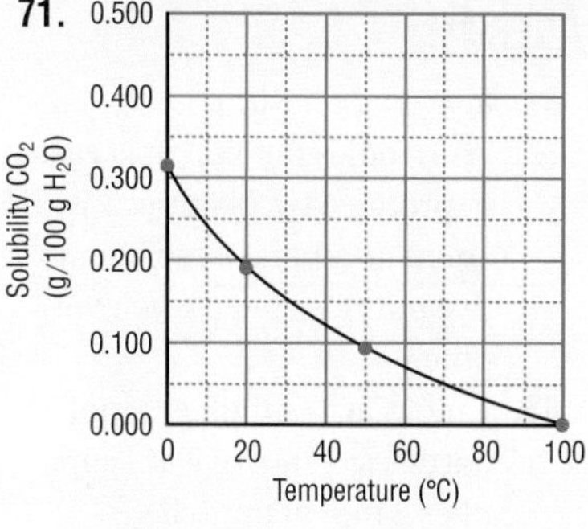

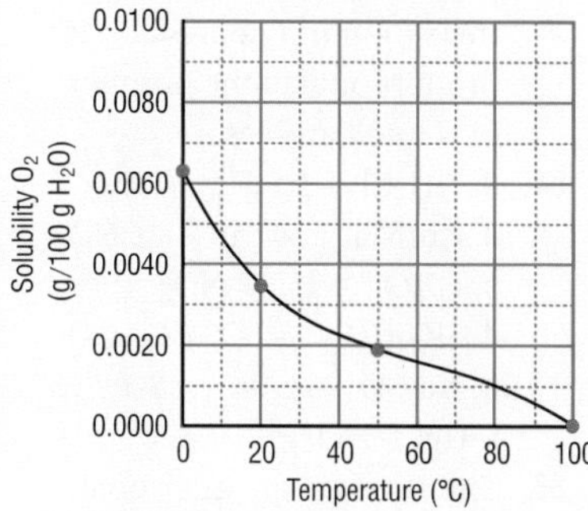

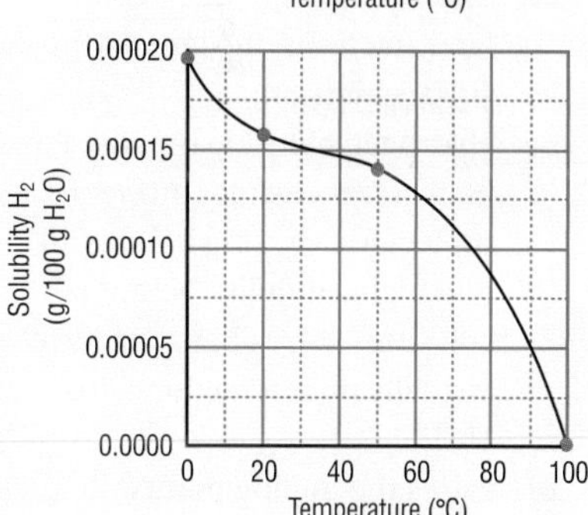

72. **a.** 44.2 g KCl **b.** 5.8 g KCl

73. **a.** 0.30 mol **c.** 0.50 mol
b. 0.40 mol **d.** 0.20 mol

74.

Mass solute	Moles solute	Volume of solution	Molarity
12.5 g	0.0694	219 mL	0.317
194 g	1.08	2.08 L	0.519
315 g	1.75	1.62 L	1.08

75. unsaturated

76. 100.680 °C

77. **a.** 7.5 g H_2O_2 **b.** 0.88*M*

78. 52 g $NaNO_3$

79. 85.5 g/mol

80. $X_{H_2O} = 0.972$; $X_{C_{12}H_{22}O_{11}} = 0.028$

63. Each gram of acetone requires 0.93 g of water.

64. The mole fraction of $NaHCO_3$ is 0.020; of water is 0.98. The solution is 1.1*m*.

65. The mole fraction of NaCl is 0.00269; the mole fraction of H_2O is 0.997.

66. If the solution is supersaturated, one crystal of KNO_3 causes crystallization. If it is saturated, the crystal does not dissolve: if unsaturated, the crystal dissolves.

67. **a.** about 1.14 **b.** about −7.2 °C
c. about −9.5 °C

68. fp = −1.86 °C; bp = 100.512 °C

69. $X_{C_2H_5OH} = 0.20$; $X_{H_2O} = 0.80$

70. −0.413 °C

Answers

81. a. 4 b. 1
82. With moles per liter, it is easier to produce two solutions with equal numbers of representative particles per a given volume.
83. 1 mol $CaCl_2$ supplies 3 mol of particles; 1 mol of NaCl supplies 2 mol of particles; freezing-point depression is proportional to the number of solute particles.
84. a. $NH_4Cl \rightarrow NH_4^+ + Cl^-$
b. $Cu(NO_3)_2 \rightarrow Cu^{2+} + 2NO_3^-$
c. $HNO_3 \rightarrow H^+ + NO_3^-$
d. $HC_2H_3O_2 \rightarrow H^+ + C_2H_3O_2^-$
e. $Na_2SO_4 \rightarrow 2Na^+ + SO_4^{2-}$
f. $HgCl_2 \rightarrow Hg^{2+} + 2Cl^-$
85. The stronger the intermolecular attractions, the greater the surface tension.
86. Hydrogen chloride is polar. Polar water molecules attract HCl molecules, causing ionization and high solubility. Nonpolar solvents such as benzene have virtually no interaction with HCl.
87. raises the boiling points
88. a solution because polar soap molecules attract polar water molecules
89. a. the substance that is dissolved
b. the substance used for dissolving
c. No additional solute can be dissolved.
d. More solute can be dissolved.
e. The solution contains more solute than a saturated solution.
f. a solute that cannot conduct electricity
g. a solute that conducts a small amount of electricity
h. a solute that conducts electricity well
90. 0.120*M* HCl
91. 1.10×10^2 mL HNO_3
92. a. 76 °C: 15 mol/kg; 33 °C: 5 mol/kg
b. 82 °C
c. 30 °C
93. 0.090*M* Na_2SO_4
94. To solve this problem:
(1) For each ion, multiply molar mass by molality to find mass per 1000 g of solvent.
(2) Sum the masses from (1) and add to 1000 g.
(3) Calculate the percent mass of each ion by dividing each answer in (1) by answer (2).
(4) Multiply the percent mass of each ion by the mass of 5.00 L of sea water (5120 g).

chloride: 103 g
sodium: 56.8 g
magnesium: 7.1 g
sulfate: 14 g
calcium: 2.3 g
potassium: 2.0 g
hydrogen carbonate: 0.6 g

CRITICAL THINKING

81. Choose the term that best completes the second relationship.

a. volume : liter — concentration : ____
(1) moles (3) solution
(2) volume (4) molarity

b. molarity : L of solution — molality : ______
(1) kg of solvent (3) concentration
(2) L of solvent (4) mol of solute

82. Why is molarity measured in moles per liter of solution instead of in grams per liter?

83. Why might calcium chloride spread on icy roads be more effective at melting ice than sodium chloride?

CUMULATIVE REVIEW

84. Indicate by simple equations how the following substances ionize or dissociate in water.
a. NH_4Cl c. HNO_3 e. Na_2SO_4
b. $Cu(NO_3)_2$ d. $HC_2H_3O_2$ f. $HgCl_2$

85. What relationship exists between surface tension and intermolecular attractions in a liquid?

86. The solubility of hydrogen chloride gas in the polar solvent water is much greater than its solubility in the nonpolar solvent benzene. Why?

87. Discuss the effect that hydrogen bonding has on the boiling points of compounds that have low formula mass.

88. When soap is shaken with water, do you get a solution, a suspension, or a colloid? Explain.

89. Explain each term as it applies to solutions.
a. solute e. supersaturated
b. solvent f. nonelectrolyte
c. saturated g. weak electrolyte
d. unsaturated h. strong electrolyte

CONCEPT CHALLENGE

90. When an excess of zinc is added to 800 mL of a hydrochloric acid solution, the solution evolves 1.21 L of hydrogen gas measured over water at 21 °C and 747.5 mm Hg. What was the molarity of the acid? The vapor pressure of water at 21 °C is 18.6 mm Hg.

91. How many milliliters of 1.50*M* HNO_3 contain enough nitric acid to dissolve an old copper penny with a mass of 3.94 g?

$$3Cu + 8HNO_3 \longrightarrow 3Cu(NO_3)_2 + 2NO + 4H_2O$$

92. One way to express the solubility of a compound is in terms of moles of compound that will dissolve in 1 kg of water. Solubility depends on temperature. Plot a graph of the solubility of potassium nitrate (KNO_3) from the following data.

Temperature (°C)	Solubility (mol/kg)
0	1.61
20	2.80
40	5.78
60	11.20
80	16.76
100	24.50

From your graph estimate
a. the solubility of KNO_3 at 76 °C and at 33 °C.
b. the temperature at which its solubility is 17.6 mol/kg of water.
c. the temperature at which the solubility is 4.24 mol/kg of water.

93. A 250-mL sample of Na_2SO_4 is reacted with an excess of $BaCl_2$. If 5.28 g $BaSO_4$ is precipitated, what is the molarity of the Na_2SO_4 solution?

94. The following table lists the most abundant ions in sea water and their molal concentrations.

Ion	Molality (*m*)
Chloride	0.568
Sodium	0.482
Magnesium	0.057
Sulfate	0.028
Calcium	0.011
Potassium	0.010
Hydrogen carbonate	0.002

Calculate the mass (in g) of each component ion contained in 5.00 L of sea water. The density of sea water is 1.024 g/mL.

Chapter 18 STANDARDIZED TEST PREP

Select the choice that best answers each question or completes each statement.

1. An aqueous solution is 65% (v/v) rubbing alcohol. How many milliliters of water are in a 95-mL sample of this solution?
   a. 62 mL c. 33 mL
   b. 1.5 mL d. 30 mL
2. Which of these actions will cause more sugar to dissolve in a saturated sugar water solution?
   I. Add more sugar while stirring.
   II. Add more sugar and heat the solution.
   III. Grind the sugar to a powder; then add while stirring.
   a. I only
   b. II only
   c. III only
   d. I and II only
   e. II and III only
3. How many moles of sodium nitrate are in 650 mL of a 0.28*M* $NaNO_3$ solution?
   a. 0.18 mol c. 0.43 mol
   b. 2.3 mol d. 18 mol
4. Compared to pure water, an aqueous solution of potassium bromide has a
   a. higher boiling point and freezing point.
   b. lower vapor pressure and higher boiling point.
   c. lower freezing point and higher vapor pressure.
   d. lower boiling point and vapor pressure.
5. When 2.0 mol of methanol is dissolved in 45 g of water, the mole fraction of methanol is
   a. 0.44. c. 2.25.
   b. 0.043. d. 0.55.

The lettered choices below refer to questions 6–9. A lettered choice may be used once, more than once, or not at all.

(A) moles/liter of solution
(B) grams/mole
(C) moles/kilogram of solvent
(D) °C/molal
(E) no units

Which of the above units is appropriate for each measurement?

6. molality
7. mole fraction
8. molar mass
9. molarity

Use the description and the data table to answer questions 10–13.

A student measured the freezing points of three different aqueous solutions at five different concentrations. The data table summarizes the data the student collected.

Molarity (*M*)	Freezing Point Depression (°C)		
	NaCl	$CaCl_2$	C_2H_5OH
0.5	1.7	2.6	0.95
1.0	3.5	5.6	2.0
1.5	5.3	8.3	3.0
2.0	7.2	11.2	4.1
2.5	9.4	14.0	5.3

10. Graph the data for all three solutes on the same graph, using molarity as the independent variable.
11. Summarize the relationship between molarity and freezing point depression.
12. Compare the slopes of the three lines and explain any difference.
13. If you collected similar data for KOH and added a fourth line to your graph, which existing line would the new line approximate?

Use the atomic windows to answer questions 14–16. The windows show water and two aqueous solutions with different concentrations. The purple spheres represent the solute particles; the blue spheres represent water.

a. 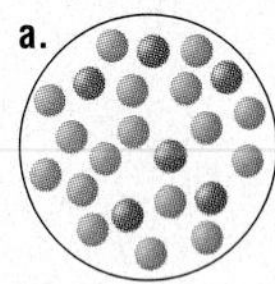
b. 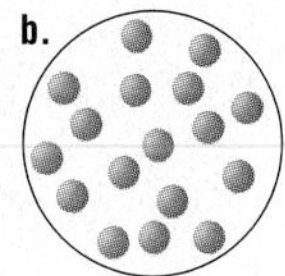
c. 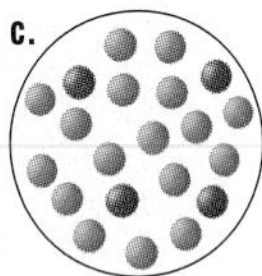

14. Which system has the highest vapor pressure?
15. Which system has the lowest vapor pressure?
16. Which solution has the lowest boiling point?

Write a brief essay to answer question 17.

17. Describe how you would prepare 100 mL of 0.50*M* KCl starting with a stock solution that is 2.0*M* KCl.

STANDARDIZED TEST PREP

1. c
2. b
3. a
4. b
5. a
6. (C)
7. (E)
8. (B)
9. (A)
10.

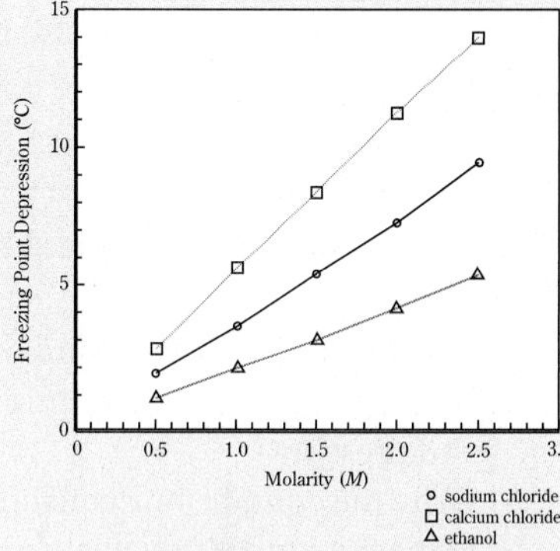

11. As the molarity increases the freezing point depression increases.
12. The slopes are in an approximate 1 : 2 : 3 ratio that reflects the relative number of particles per mole of each solute in solution.
13. NaCl
14. b
15. a
16. b
17. Transfer 25 mL of the stock solution to a 100-mL volumetric flask and add water to make 100 mL of solution.

Planning Guide

SECTION OBJECTIVES	ACTIVITIES/FEATURES	MEDIA & TECHNOLOGY
19.1 Rates of Reaction ● ■ ◆ ▸ Explain what is meant by the rate of a chemical reaction ▸ Using collision theory, explain how the rate of a chemical reaction is influenced by the reaction conditions	**SE** **Discover It!** *Temperature and Reaction Rates,* p. 532 **SE** **Link to Auto Shop** *Auto Body Repair,* p. 538 **LM** 34: *Factors Affecting Reaction Rates* **LM** 35: *The Clock Reaction* **SSLM** 24: *Factors Affecting the Rate of a Chemical Reaction* **TE** **DEMOS**, pp. 534, 536, 537	**ASAP** Simulation 23, Animation 22 **ASAP** Assessment 19.1 **CA** *Grain Elevator Explosion, Oscillating Clock* **CHM** Side 9, 2: *Fighting Fire* **CHM** Side 9, 20: *Collision Theory* **CHM** Side 9, 23: *Catalysts at Work* **CHM** Side 9, 30: *Effect of Temperature on Rate* **OT** 63: *Collision Theory* **OT** 64: *Activation Energy and Catalysts*
19.2 Reversible Reactions and Equilibrium ○ □ ◆ ▸ Predict changes in the equilibrium position due to changes in concentration, temperature, and pressure ▸ Write the equilibrium-constant expression for a reaction and calculate its value from experimental data	**LM** 36: *Disturbing Equilibrium* **SSLM** 25: *Le Châtelier's Principle and Chemical Equilibrium* **TE** **DEMOS**, pp. 541, 545	**ASAP** Animation 23, Simulation 24 **ASAP** Problem Solving 7, 8, 12 **ASAP** Assessment 19.2 **SSV** 19: *Equilibrium and Kinetics* **CHM** Side 6, 36: *Le Châtelier's Principle* **OT** 65: *Reversible Reactions and Equilibrium*
19.3 Determining Whether a Reaction Will Occur □ ◆ ▸ Define entropy and free energy, and characterize reactions as spontaneous or nonspontaneous ▸ Describe how heat change and entropy change determine the spontaneity of a reaction	**SE** **Mini Lab** *Does Steel Burn?* p. 556 (**LRS** 19-2) **SE** **Small-Scale Lab** *Enthalpy and Entropy,* p. 557 (**LRS** 19-2) **TE** **DEMO**, p.552	**ASAP** Simulation 25 **ASAP** Assessment 19.3 **CHM** Side 9, 10: *Fireworks* **CHM** Side 9, 13: *Refining Ores* **CHM** Side 9, 38: *Carbon Disulfide and Phosphorus* **OT** 66: *Entropy* **OT** 67: *Heat, Entropy, and Free Energy*
19.4 Calculating Entropy and Free Energy ◆ ▸ Calculate the standard entropy changes that accompany chemical and physical processes ▸ Calculate the free-energy changes that accompany chemical and physical processes	**SE** **CHEMath** *Balancing Equations,* p. 559 **SE** **Link to Geology** *Weathering of Rocks,* p. 562 **TE** **DEMO**, p.562	**ASAP** Problem Solving 25, 27, 29 **ASAP** Assessment 19.4
19.5 The Progress of Chemical Reactions ◆ ▸ Interpret experimental rate data to deduce the rate laws for simple chemical reactions ▸ Given an energy diagram for a reaction, analyze the mechanism for the reaction	**SE** **Chemistry Serving . . . the Consumer** *Don't Let Good Food Go Bad,* p. 570 **SE** **Chemistry in Careers** *FDA Inspector,* p. 570	**ASAP** Animation 24 **ASAP** Problem Solving 34 **ASAP** Assessment 19.5 **PP** Chapter 19 Problems **RP** Lesson Plans, Resource Library **AR** Computer Test 19 **www** Activities, Self-Tests, *SCIENCE NEWS* updates **TCP** The Chemistry Place Web Site

KEY

- ● Conceptual (concrete concepts)
- ○ Conceptual (more abstract/math)
- ■ Standard (core content)
- □ Standard (extension topics)
- ◆ Honors (core content)
- ◇ Honors (options to accelerate)

AR Assessment Resources
ASAP Chem ASAP! CD-ROM
ACT ActivChemistry CD-ROM
CHM CHEMedia Videodiscs
CA Chemistry Alive! Videodiscs
GCP Graphing Calculator Problems

GRS Guided Reading and Study Workbook
LM Laboratory Manual
LP Laboratory Practicals
LRS Laboratory Recordsheets
SSLM Small-Scale Lab Manual

PRACTICE		ASSESSMENT	
GRS	Section 19.1	SE	Section Review
RM	Practice Problems 19.1	RM	Section Review 19.1
		RM	Chapter 19 Quiz
		LP	Lab Practical 19-1
SE	**Sample Problems** 19-1 to 19-4	SE	Section Review
GRS	Section 19.2	RM	Section Review 19.2
RM	Practice Problems 19.2	RM	Chapter 19 Quiz
GRS	Section 19.3	SE	Section Review
RM	Practice Problems 19.3	RM	Section Review 19.3
RM	Interpreting Graphics	RM	Chapter 19 Quiz
SE	**Sample Problems** 19-5 to 19-8	SE	Section Review
GRS	Section 19.4	RM	Section Review 19.4
RM	Practice Problems 19.4	RM	Chapter 19 Quiz
RM	Interpreting Graphics 19-2		
SE	**Sample Problem** 19-9	SE	Section Review
GRS	Section 19.5	RM	Section Review 19.5
RM	Practice Problems 19.5	RM	Vocabulary Review 19
GCP	Chapter 19	SE	Chapter Review
		SM	Chapter19 Solutions
		SE	Standardized Test Prep
		PHAS	Chapter 19 Test Prep
		RM	Chapter 19 Quiz
		RM	Chapter 19 A & B Test

OT	Overhead Transparency	SE	Student Edition
PHAS	PH Assessment System	SM	Solutions Manual
PLM	Probeware Lab Manual	SSV	Small-Scale Video/Videodisc
PP	Problem Pro CD-ROM	TCP	www.chemplace.com
RM	Review Module	TE	Teacher's Edition
RP	Resource Pro CD-ROM	www	www.phschool.com

PLANNING FOR ACTIVITIES

STUDENT EDITION

Discover It! p. 532
- masking tape
- plastic cups
- hot and cold tap water
- ice
- thermometers
- effervescent antacid tablets
- clocks or watches with second hand
- graph paper
- pens or pencils

Mini Lab, p. 556
- steel wool pads (0000)
- tissue paper
- tongs
- Bunsen burners
- heat-resistant pads
- pencils
- paper

Small-Scale Lab, p. 557
- alcohol thermometers
- 1-oz plastic cups
- NaCl, NH_4Cl, $CaCl_2$, KCl, $NaHCO_3$, $NaCO_3$, $NaPO_4$
- plastic spoons
- crushed ice
- water

TEACHER'S EDITION

Teacher Demo, p. 536
- 5 mL of ethanoic acid
- 5 mL of isoamyl alcohol
- two test tubes
- granules of anhydrous calcium sulfate
- concentrated sulfuric acid
- boiling water bath

Teacher Demo, p. 537
- cornstarch
- water glass
- match
- Bunsen burner
- spatula
- goggles
- Plexiglas shield

Teacher Demo, p. 541
- materials from demo, p.536

Teacher Demo, p. 545
- fume hood
- small amount (3 g or less) of copper turnings
- test tube
- 10 mL of 6*M* nitric acid
- one-hole stopper fitted with a glass delivery tube and rubber hose
- two Pyrex test tubes with stoppers
- ice bath
- warm water bath

Teacher Demo, p. 552
- block of ice
- beaker
- water
- heat source

Teacher Demo, p. 562
- 0.1*M* solution of $KMnO_4$ in 0.1*M* HCl
- 1000-mL beaker
- thermometer
- graduated cylinder
- 20 mL of 30% H_2O_2
- watch glass
- glowing splint

Chapter 19

Key Terms

19.1 rates, collision theory, activation energy, activated complex, transition state, catalyst, inhibitor
19.2 reversible reactions, chemical equilibrium, equilibrium position, Le Châtelier's principle, equilibrium constant
19.3 free energy, spontaneous reactions, nonspontaneous reactions, entropy, law of disorder
19.4 standard entropy, Gibbs free-energy change (ΔG)
19.5 rate law, specific rate constant, first-order reaction, elementary reaction, reaction mechanism, intermediate

DISCOVER IT!

Antacid tablets contain sodium hydrogen carbonate ($NaHCO_3$) and citric acid ($H_3C_6H_5O_7$). The acid reacts with hydrogen carbonate anions to produce carbon dioxide gas. Students should observe that the reaction proceeds faster in warmer water. The increased temperatures provide increased energy for the reaction to occur. After completing the chapter, students should understand that increasing the surface area by crushing the tablets would cause each of the reactions to proceed faster, but the overall pattern would be the same.

Chapter 19 REACTION RATES AND EQUILIBRIUM

FEATURES

Stay current with SCIENCE NEWS
Find out more about reaction rates:
www.phschool.com

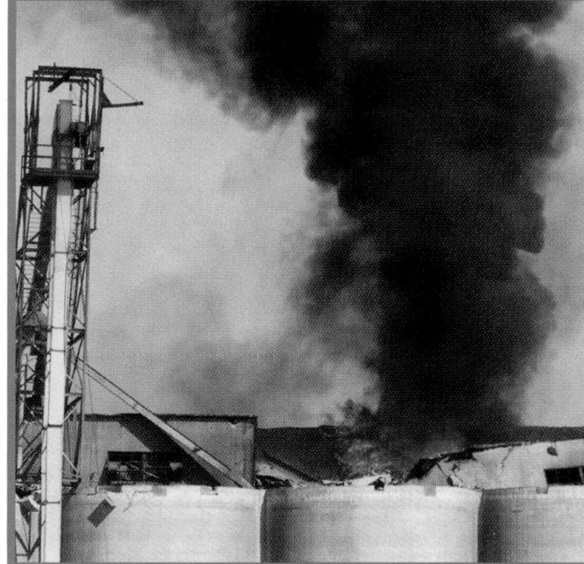

The energy stored in oil is wasted during uncontrolled combustion.

DISCOVER IT! TEMPERATURE AND REACTION RATES

You need masking tape, four plastic cups, hot and cold tap water, ice, a thermometer, four effervescent antacid tablets, a clock or watch with a second hand, graph paper, and a pen or pencil.

1. Mark an A, B, C, or D on four separate pieces of masking tape. Label each plastic cup with one of the lettered pieces of tape.
2. Fill each cup three-quarters full as follows: (A) cold tap water plus some ice, (B) cold tap water, (C) half-and-half mixture of cold tap water and hot tap water, (D) hot tap water.
3. Measure and record the temperature of the water in each cup.
4. Drop a tablet into cup A and measure the time it takes for the reaction to go to completion. Repeat this procedure using the remaining three cups.
5. Draw a line or bar graph of temperature versus reaction time.

Do the tablets react faster at higher temperatures? Explain why or why not. What results would you expect if you crushed the tablets into a powder before adding them to the water? Explain your prediction. If time permits, check this prediction experimentally. As you read this chapter, return to this activity and revise your explanations on the basis of the new information.

19.1 ● ■ ◆
19.2 ○ □ ◆
19.3 □ ◆
19.4 ◆
19.5 ◆

Conceptual Use Figure 19.1 together with Figure 19.3 to help students visualize reaction rates. In Section 19.2, use Figure 19.9 to illustrate reversible reactions and Figure 19.13 to give students a sense of Le Châtelier's principle. Equilibrium constants may be omitted. The last three sections of the chapter involve abstract concepts beyond the scope of most conceptual courses.

Standard Equilibrium is a challenging concept students must master before they tackle Le Châtelier's principle and K_{eq}. Use Figures 19.9 and 19.10 to help students visualize equilibrium from two different perspectives (particle view and graphs). If K_{sp} will be addressed in Chapter 21, then K_{eq} needs to be introduced in Section 19.2. Figure 19.20 can be used to summarize how enthalpy and entropy relate to free energy and spontaneity.

Honors If students are successful with Sample Problems 19-5 and 19-7, and the associated Practice Problems, assign Sample Problems 19-6 and 19-8 as homework.

RATES OF REACTION

Although corrosion can damage many different types of metal objects, the process can be useful to hungry soldiers, truck drivers, and others who want a hot meal but do not have a place to cook it. Products now on the market use the heat given off by the corrosion reaction of an iron-magnesium alloy with salt water to produce a hot meal. Normally the reaction takes place at a very slow rate, so that the heat generated has ample time to dissipate. But when the rate is increased, as it is in this case by the addition of salt water, heat is produced very rapidly. **What are some ways the rate of a reaction can be increased?**

objectives

- Explain what is meant by the rate of a chemical reaction
- Using collision theory, explain how the rate of a chemical reaction is influenced by the reaction conditions

key terms

- rates
- collision theory
- activation energy
- activated complex
- transition state
- catalyst
- inhibitor

Collision Theory

The amount of time required for a chemical reaction to come to completion can vary tremendously, depending on the reaction. When you strike a match, it seems to erupt into flame almost instantaneously. Many other reactions, however, occur much more slowly. Did you know that coal is produced from dead plants by a process involving heat and pressure that takes millions of years? In this chapter, you will examine the factors that make chemical reactions proceed at different speeds.

When exposed to damp air, iron turns to rust. However, you know the reverse process does not readily occur. Bringing a rusty chair in out of the damp weather cannot make it look like new again. In this chapter, you will learn why some substances react readily, whereas others seem inert.

The concept of speed is probably already familiar to you. A fast sprinter may cover 100 meters in about 11.5 seconds; a slower sprinter may take 15.0 seconds to cover the same distance. On average, the first sprinter covers 100 meters in 11.5 seconds, or runs at a speed of 8.70 m/s. The second sprinter covers 100 meters in 15.0 seconds, or runs at a speed of 6.67 m/s. Both 8.70 m/s and 6.67 m/s express rates of travel. **Rates** measure the speed of any

Start — Distance (m) — Finish

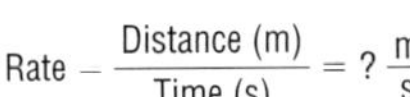

Figure 19.1
A rate tells you how much something changes in a specified amount of time. Speeds are measured as changes of distance in a given interval of time. A world-class sprinter might cover 100 meters in 11.5 seconds; the speed, or rate, is 100 m/11.5 s = 8.70 m/s.

1 Engage

Use the Visual

Have students read the section-opening paragraph. Explain that all chemical reactions occur at rates specific to the nature of the reactants. In the example shown in the photograph, salt water is being used to increase the reaction rate. Ask:

- **What are some ways the rate of a reaction can be increased?** (temperature, concentration of reactants, particle size, pressure, catalysts)

Check Prior Knowledge

Ask students if every chemical reaction for which an equation can be written will actually take place. First review how the reactants chosen affect the likelihood of a reaction. Write Na + $H_2O \rightarrow$ and Au + $H_2O \rightarrow$ on the board; ask students to complete and balance the equations. ($2Na + 2H_2O \rightarrow 2NaOH + H_2$; $2Au + 2H_2O \rightarrow 2AuOH + H_2$) **Students should be able to point out that gold doesn't react with water.**

Next show how the conditions chosen can affect the likelihood of a reaction. Write the equations for the synthesis and decomposition of water on the board. ($2H_2 + O_2 \rightarrow 2H_2O$; $2H_2O \rightarrow 2H_2 + O_2$) **Ask students what conditions are necessary for each reaction to take place.** (Synthesis is explosive after an initial spark; decomposition requires a steady application of electricity.)

STUDENT RESOURCES

From the Teacher's Resource Package, use:

- Section Review 19.1, Ch. 19 Practice Problems and Quizzes from the Review Module (Ch. 17–20)
- Laboratory Manual: Experiments 34 and 35
- Laboratory Practical 19-1
- Small-Scale Laboratory Manual: Experiment 24

TECHNOLOGY RESOURCES

Relevant technology resources include:

- Chem ASAP! CD-ROM
- Resource Pro CD-ROM
- Chemistry Alive! Videodisc: *Grain Elevator Explosion, Oscillating Clock*

2 Teach

TEACHER DEMO

Tell the students that the time required for a chemical change to take place may be $<10^{-6}$ seconds or $>10^{6}$ years. Draw three vertical columns on the board and label them "very fast," "moderate," and "very slow." Ask the students to suggest chemical changes that fit into each of the three categories. If they have trouble thinking of reactions that fit in the "moderate" category, suggest reactions that take place over several hours or days, such as food spoilage or the rusting of iron.

Use the Visual

Ask the students to examine Figure 19.3 and read the accompanying text. Explain that the rates of chemical reactions are often measured by the decrease in the concentration of one of the reactants or the increase in the concentration of one of the products.

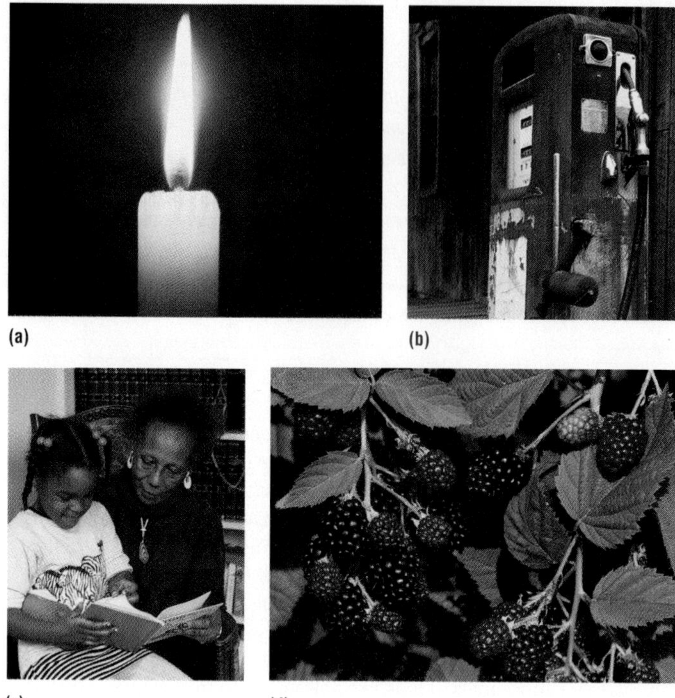

Figure 19.2
***(a)** Burning, **(b)** rusting, **(c)** aging, and **(d)** ripening are four processes. Which occurs fastest (that is, at the highest rate) and which occurs slowest?* ❶

change that occurs within an interval of time. The interval of time may range from fractions of a second to centuries. **Figure 19.2** shows some familiar examples of rates of change. In chemistry, rates of chemical change usually are expressed as the amount of reactant changing per unit time. The rate at which an iron plate containing 1 mol of iron rusts might be expressed as 0.5 mol/yr. **Figure 19.3** illustrates the progress of a typical reaction.

Visible changes caused by chemical reactions are related to changes in the properties of individual atoms, ions, or molecules. For example, atoms of shiny silver-colored sodium metal react with molecules of yellow-green chlorine gas to produce colorless crystals of sodium chloride. The properties of sodium atoms and chlorine molecules are different from the

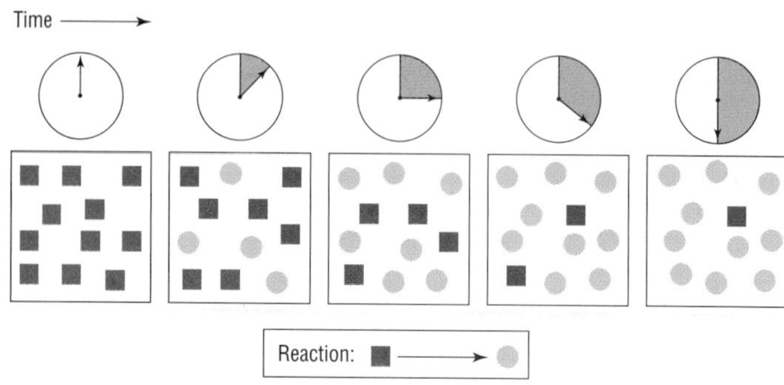

Figure 19.3
As time passes, the amount of reactant (red squares) decreases and the amount of product (blue spheres) increases. Rates of chemical reactions are often measured as a change in the number of moles during a given interval of time.

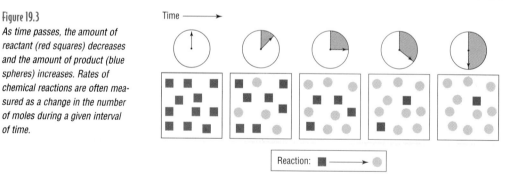

Answer

❶ Burning occurs fastest; the low rates of aging and rusting are relative (it depends on the organisms or objects involved).

(a)

(b)

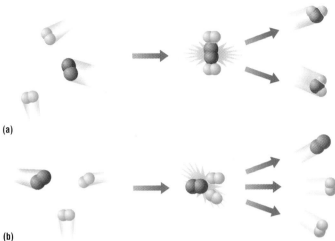

Figure 19.4
If colliding particles have enough kinetic energy and collide at the right orientation, they can react to form a new product. ***(a)*** *An effective collision of reactant molecules produces product molecules.*
(b) *An ineffective collision of reactant molecules produces no reaction, and the reactants bounce apart unchanged.*

properties of sodium ions and chloride ions in sodium chloride. Rates of chemical reactions are related to the properties of atoms, ions, and molecules through a model called collision theory. According to **collision theory,** atoms, ions, and molecules can react to form products when they collide, provided that the particles have enough kinetic energy. Particles lacking the necessary kinetic energy to react bounce apart when they collide. This process is illustrated in **Figure 19.4.**

You can use two balls of soft modeling clay to illustrate further the concept of collision theory. If you throw the balls of clay together gently, they do not stick to each other, which is analogous to a lack of reaction between colliding particles of low energy. The same balls of clay thrown together with great force, however, stick tightly to each other, which is analogous to the chemical combination of two particles that collide with sufficient energy.

You can also use modeling clay to illustrate another point about chemical reactions. If you roll the clay into a rope and vigorously shake one end, the clay rope will eventually break. Similarly, the bonds holding molecules together can break apart. Substances supplied with enough energy decompose to simpler substances or reorganize themselves into new substances. The minimum amount of energy that particles must have in order to react is called the **activation energy.** In a sense, activation energy is a barrier that reactants must cross to be converted to products, as shown in **Figure 19.5.** During a reaction, particles that are neither reactants nor products form momentarily if there is sufficient

Figure 19.5
The activation-energy barrier must be crossed before reactants are converted to products. The activated complex is a temporary arrangement of particles that has sufficient energy to become either reactants or products.

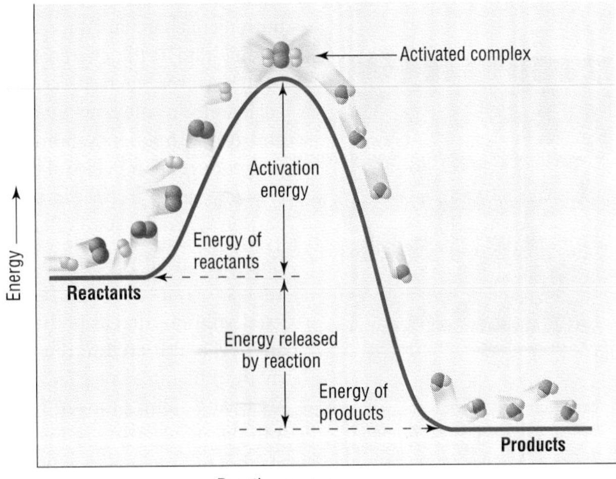

19.1

Discuss

Because the rest of the concepts in Chapter 19 are based on collision theory, mastery of this theory is crucial. Students may think that particles always react when they collide. Stress that particles form bonds when they collide *only if* the particles have sufficient kinetic energy. In any given reaction, only a certain fraction of the particles have enough energy to react upon collision.

Use the Visual

Have the students study the reaction-energy diagram in Figure 19.5. Explain that the diagram graphically depicts the energy changes that occur during a reaction. Point out that the activation-energy barrier must be crossed before reactants are converted to products.
Explain that the activation energy determines how rapidly a reaction will occur at a given temperature. A large activation energy results in a slow reaction because few molecules will collide with enough energy to form an activated complex. Ask:

- **Is the energy level of the products lower or higher than the energy level of the reactants?** (lower)
- **Is this reaction endothermic or exothermic?** (exothermic)

Meeting Diverse Needs

Gifted Tell students that the collision theory uses two-particle collisions to explain reaction mechanisms because the probability of three or more particles colliding simultaneously is relatively remote. Have the students research typical reactions that require the collision of three or more particles and summarize their findings in a report that they will present to the class.

Chemistry Alive!

Grain Elevator Explosion

Play

TEACHER DEMO

Mix 5 mL of ethanoic acid (acetic acid) and 5 mL of isoamyl alcohol in each of two test tubes. Add a few granules of anhydrous calcium sulfate to each test tube. To one test tube, add three drops of concentrated sulfuric acid and heat this test tube in a boiling water bath for a few minutes. Write the general chemical equation for the formation of an ester on the board. (organic acid + alcohol → ester + water) Inform students that esters are responsible for the characteristic odors of many fruits and may be prepared by reacting an organic acid with an alcohol. Have the students smell the odor emanating from both test tubes by carefully wafting the odor towards their noses.

- Ask the students for their observations? (The solution containing sulfuric acid that was heated smells like bananas while the other solution has a strong odor of ethanoic acid (vinegar); a change in odor is a sign of reaction.)
- What purpose does the sulfuric acid and heat serve in this reaction? (Both are used to speed up the reaction rate. The sulfuric acid is a catalyst for the reaction.)

energy and if the atoms are oriented properly. An **activated complex** is the arrangement of atoms at the peak of the activation-energy barrier. The lifetime of an activated complex is typically about 10^{-13} s. The very unstable activated complex is as likely to re-form reactants as it is to form products. Thus the activated complex is sometimes called the **transition state.** Can you explain why? ①

Collision theory explains why some naturally occurring reactions are immeasurably slow at room temperature. For example, carbon and oxygen react when charcoal burns, but this reaction has a high activation energy. At room temperature, the collisions of oxygen and carbon molecules are not energetic enough to break the O—O and C—C bonds. These bonds must be broken to form the activated complex. Thus the reaction rate of carbon with oxygen at room temperature is essentially zero.

Factors Affecting Reaction Rates

Every chemical reaction proceeds at its own rate. Some reactions are naturally fast and some are naturally slow under the same conditions. However, by varying the conditions of the reaction, the rate of almost any reaction can be modified. Collision theory helps explain why these rate changes occur and how they depend on conditions such as temperature, concentration, and particle size.

Temperature Usually, raising the temperature speeds up reactions, while lowering the temperature slows down reactions. At higher temperatures, the motions of the reactant particles are faster and more chaotic than they are at lower temperatures. The main effect of increasing the temperature is to increase the number of particles that have enough kinetic energy to react when they collide. At high temperatures, more colliding particles are energetic enough to slip over the activation-energy barrier to become products. The frequency of high-energy collisions between reactant particles also increases. An increase in temperature, therefore enables products to form faster.

A familiar example of the effect of temperature on reaction rate is the burning of charcoal. At room temperature, charcoal does not burn at a measurable rate. If the charcoal is supplied with enough energy in the form of heat, however, the result is dramatic. When a starter flame touches the charcoal, atoms of the reactants (carbon and oxygen) collide with higher energy and greater frequency. Some collisions are at a high enough energy to form the product (carbon dioxide). The heat released by the reaction then supplies enough energy to get more carbon and oxygen over the activation-energy barrier. When the starter flame is removed, the reaction continues.

Concentration The number of reacting particles in a given volume also affects the rate at which reactions occur. Cramming more particles into a fixed volume increases the concentration of reactants, the collision frequency, and therefore the reaction rate. The lighted splint in **Figure 19.6** vividly illustrates this effect. Such a splint glows in air, which is 20% oxygen. A glowing splint plunged into pure oxygen, however, immediately bursts into flame because the increased concentration of oxygen greatly speeds up the combustion reaction. This is why smoking is forbidden in areas where bottled oxygen is in use.

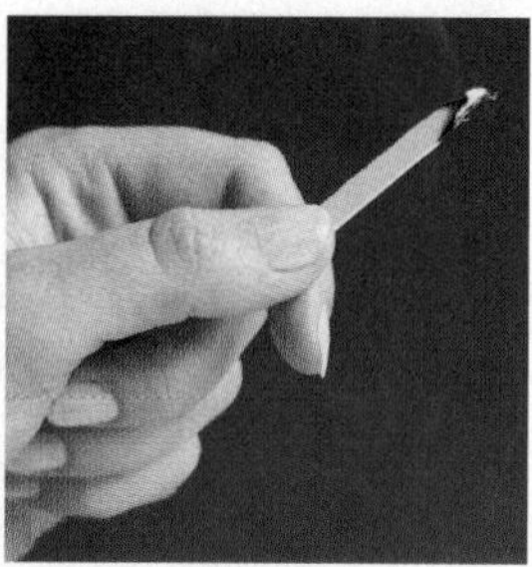

(a)

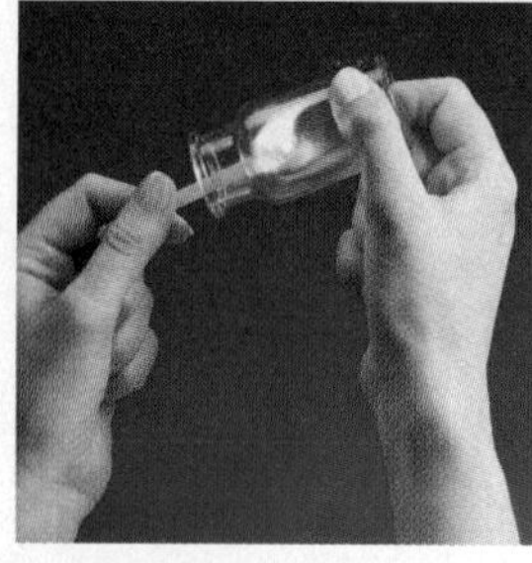

(b)

Figure 19.6
*Why will a glowing splint **(a)** burst into flame **(b)** if it is placed in a bottle of pure oxygen?* ②

Answers

① The activated complex is called the transition state because it is an unstable intermediate structure that connects reactants to products.
② Oxygen supports combustion. A smoldering splint bursts into flame in pure oxygen because of the high concentration of the gas, which increases the collision rate and the likelihood of reaction.
③ The sticks of kindling have a larger total surface area than the log.

Chemistry Alive!

Oscillating Clock

Play

Particle Size The smaller the particle size, the larger is the surface area for a given mass of particles. The total surface area of a solid or liquid reactant has an important effect on the reaction rate: An increase in surface area increases the amount of the reactant exposed for reaction, which further increases the collision frequency and the reaction rate. Why would a bundle of kindling burn faster than a single log of equal mass? 3

One way to increase the surface area of solid reactants is to dissolve them. This process separates the particles and makes them more accessible to other reactants. Another way to increase the surface area is to grind solids into a fine powder. Small dustlike particles, however, can be very dangerous. As coal miners know, large chunks of coal pose very little hazard, but coal dust suspended in the air is potentially very reactive and presents an explosion hazard. The same is true of dust in a flour mill, as the dramatic photograph in **Figure 19.7** shows.

Chem ASAP!

Animation 22

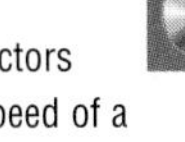

Explore several factors that control the speed of a chemical reaction.

Figure 19.7

In 1932, an explosion destroyed this Chicago grain elevator. The minute size of the reactant particles (grain dust), and the mixture of the grain dust with oxygen in the air caused the reaction to be explosive.

Catalysts An increase in temperature is not always the best way to increase the rate of a reaction. A catalyst, if one can be found, is often better. A **catalyst** is a substance that increases the rate of a reaction without being used up itself during the reaction. Catalysts permit reactions to proceed at a lower energy than is normally required. **Figure 19.8** on the following page shows how catalysts lower the activation-energy barrier. With a lower activation-energy barrier, more reactants can form products within a given time. For instance, the reaction rate of hydrogen and oxygen at room temperature is negligible. With a trace of catalyst added, however, the reaction is rapid. Finely divided platinum (Pt) catalyzes this reaction, as illustrated in the following equation.

$$2H_2(g) + O_2(g) \xrightarrow{Pt} 2H_2O(l)$$

Because catalysts are not consumed during a reaction, they do not appear as reactants or products in the chemical equation. To convey this information, without suggesting that the catalyst is chemically altered, catalysts are often written above the yield arrow, as in the preceding equation.

Chem ASAP!

Simulation 23

Explore the effects of concentration, temperature, and a catalyst on reaction rate.

19.1

Discuss

Point out that an increase in temperature causes an increase in average kinetic energy. There is an increase in the number of collisions because molecules are moving faster. More importantly, there is the increase in the number of effective collisions. Ask:

- **How do refrigerators prevent food from spoiling?** (The kinetic energy of reactants is reduced; fewer particles possess sufficient energy to react.)

Catalysts reduce the activation energy required for a reaction. With catalysts, reactions can be more efficient at lower temperatures; this is critical for biological reactions, which take place at body temperature.

TEACHER DEMO

Demonstrate the influence that particle size, and hence, surface area has on the rate of a chemical reaction. Pile some cornstarch on a watch glass and try to light it with a match. Next, dust the starch into a Bunsen burner flame with a spatula. The starch dust flares up. **CAUTION!** Stand back from the burner flame. Wear safety goggles and place a plexiglass shield between the reaction and the students.

MEETING DIVERSE NEEDS

At Risk Explain that an important part of understanding how and why substances undergo chemical change requires careful study of the rates of reactions. By investigating what factors affect the speed of a reaction, chemists are able to piece together the molecular events that transpire during the conversion of reactants to products. Every reaction proceeds at a unique rate. Some take years, others just seconds. As an analogy, discuss with students the reasons why food recipes specify the time and temperature needed to completely cook the ingredients of a particular dish. Someone had to know how long different changes take before writing the instructions.

3 Assess

Evaluate Understanding

Have students use the activated complex theory (Figure 19.5) to explain why hydrogen and oxygen do not react at room temperature but do combine explosively in the presence of a spark or flame. (The explosive reaction indicates that the product (water) has less energy than the reactants. The reaction does not occur at room temperature because the reaction has a very high activation energy.)

Have students use collision theory to explain the effects of temperature, concentration, and particle size on reaction rates. Ask:

- **How does a catalyst increase the rate of a reaction?** (A catalyst lowers the activation energy required for the reaction to occur.)

Reteach

Remind students that the signs of a reaction visible at the macroscopic level (release of energy, color change, and so on) **result from individual reactions between particles. Therefore, factors that affect collisions between particles are the factors that influence the rate of reaction.**

Note that increasing the concentration or surface area of reactants increases the frequency of collisions. So does an increase in temperature; however, the main effect of a temperature rise is an increase in the number of particles with enough energy to produce an activated complex.

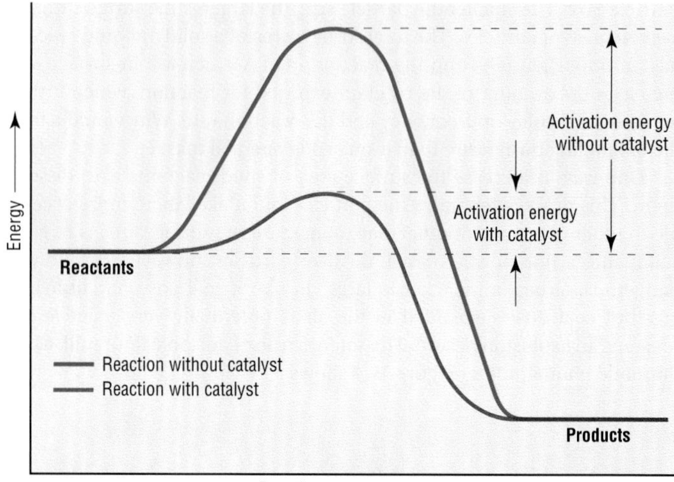

Figure 19.8
A catalyst increases the rate of a reaction by lowering the activation-energy barrier.

AUTO SHOP

Auto Body Repair

Auto body repair technicians often repair rust-damaged automobiles. Rust damage usually occurs where water is able to catalyze the reaction of oxygen

with iron to produce iron(III) oxide, or rust. Removing the rust and repainting the damaged area is effective for small repairs, but extensive damage may require the replacement of body panels. Due to the extensive use of galvanized steel and noncorroding plastics in newer automobiles, body rust is much less common today than it was in the past.

Catalysts are crucial for many life processes. Your body temperature is only 37 °C and cannot be raised significantly without danger. Without catalysts, few reactions in the body would proceed fast enough at that temperature. Enzymes, biological catalysts, increase the rates of biological reactions. You could not live without enzymes to catalyze these reactions. When you eat a meal containing protein, enzymes in your digestive tract break down the protein molecules in a few hours. Without enzymes, the digestion of protein at 37 °C would take many years!

An **inhibitor** is a substance that interferes with the action of a catalyst. Some inhibitor molecules work by reacting with, or "poisoning," the catalyst itself. Thus the inhibitor reduces the amount of functional catalyst available. Reactions slow or even stop when a catalyst is poisoned.

section review 19.1

1. What is meant by the rate of a chemical reaction?
2. How does each factor affect the rate of a chemical reaction?
   a. temperature
   b. concentration
   c. particle size
   d. an inhibitor
3. Does every collision between reacting particles lead to products? Explain.
4. Suppose a thin sheet of zinc containing 0.2 mol of the metal is completely converted in air to zinc oxide (ZnO) in one month. How would you express the rate of conversion of the zinc?
5. Refrigerated food stays fresh for long periods. The same food stored at room temperature quickly spoils. Why?

Chem ASAP! **Assessment 19.1** Check your understanding of the important ideas and concepts in Section 19.1.

Answers

SECTION REVIEW 19.1

1. number of atoms, ions, or molecules that react per given unit of time to form products
2. **a.** A temperature increase usually speeds up a reaction.
   b. An increase in reactant concentration speeds up a reaction.
   c. Smaller particle size speeds up a reaction.
   d. An inhibitor slows down a reaction.
3. No; the collision must have sufficient energy to break and form bonds.
4. Rate = 0.2 mol Zn/month
5. Chemical reactions involved in food spoilage occur faster at higher temperatures because more energy is available.

REVERSIBLE REACTIONS AND EQUILIBRIUM

For years, scientists tried to produce nitrogen compounds that would be useful as fertilizers. Unfortunately, none of these efforts produced compounds in quantities sufficient to be commercially successful. Finally, in the early 1900s, two German chemists, Fritz Haber and Karl Bosch refined the process of making ammonia for use as a fertilizer. Their success came from controlling the temperature and pressure under which the ammonia synthesis reaction takes place. **How can changing the reaction conditions influence the yield of a chemical reaction?**

objectives

- Predict changes in the equilibrium position due to changes in concentration, temperature, and pressure
- Write the equilibrium-constant expression for a reaction and calculate its value from experimental data

key terms

- reversible reactions
- chemical equilibrium
- equilibrium position
- Le Châtelier's principle
- equilibrium constant (K_{eq})

Reversible Reactions

From the previous discussions of chemical reactions, you may have inferred that chemical reactions go completely from reactants to products as written. This is not always the case, however. Some reactions are reversible. In **reversible reactions,** the reactions occur simultaneously in both directions. One example of a reversible reaction is the reaction of sulfur dioxide with oxygen to give sulfur trioxide. The double arrows in the equation below show that the reaction is reversible.

$$\underset{\text{Sulfur dioxide}}{2SO_2(g)} + \underset{\text{Oxygen}}{O_2(g)} \rightleftharpoons \underset{\text{Sulfur trioxide}}{2SO_3(g)}$$

Figure 19.9 shows that this equation represents two opposite reactions. In the first reaction, which is read from left to right, sulfur dioxide and oxygen produce sulfur trioxide. In the second reaction, which is read from right to left, sulfur trioxide decomposes into oxygen and sulfur dioxide. The first reaction is called the forward reaction. The second reaction is called the reverse reaction.

Consider what happens when sulfur dioxide and oxygen gases are mixed in a sealed chamber. Initially, there is no sulfur trioxide. The rate of the reverse process—formation of sulfur dioxide and oxygen from sulfur trioxide—is zero. Sulfur trioxide, the product of the forward reaction, then

Chem ASAP!

Animation 23

Take a close look at a generalized reversible reaction.

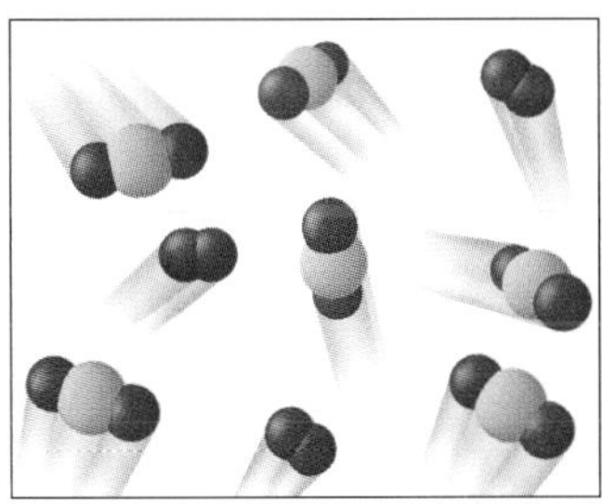

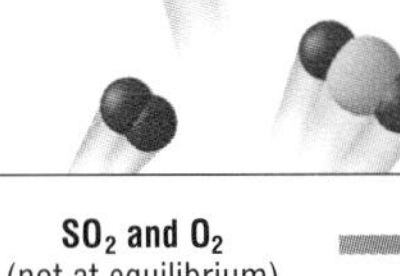

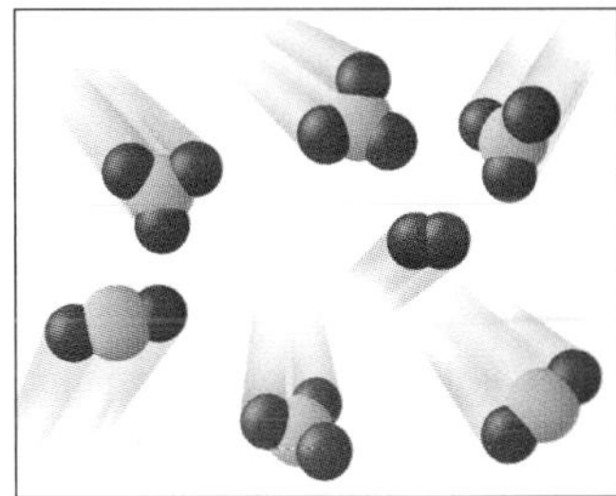

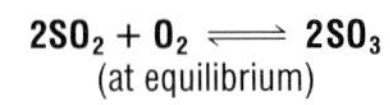

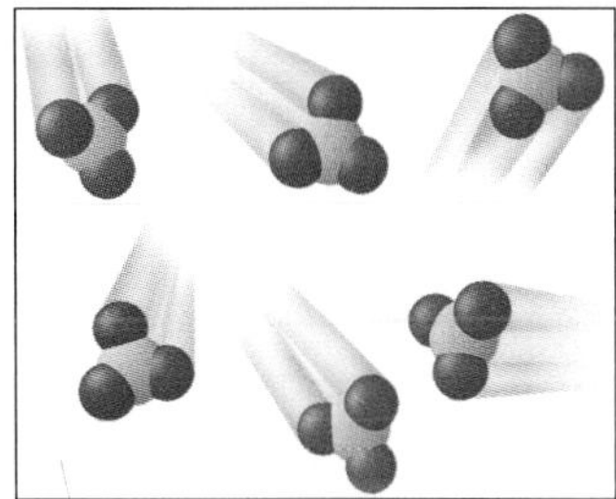

SO_2 and O_2 (not at equilibrium) → $2SO_2 + O_2 \rightleftharpoons 2SO_3$ (at equilibrium) ← SO_3 (not at equilibrium)

Figure 19.9

Molecules of SO_2 and O_2 react to give SO_3. Molecules of SO_3 decompose to give SO_2 and O_2. At equilibrium, all three types of molecules are present in the mixture.

19.2

1 Engage

Use the Visual

Ask the students to examine the photograph and to read the text that opens the section. Ask:

- **How can changing the reaction conditions influence the yield of a chemical reaction?** (The amount of product formed by a chemical reaction depends on the nature of the reactants involved and on the surrounding conditions. Remind students that every chemical and physical process absorbs or releases heat energy. Thus, the *yield* of a product in a particular reaction very often depends on the amount of heat that is applied or removed from a chemical system during the reaction. In some instances, where a gas is consumed or produced during a reaction, the yield of a product may be affected by the external pressure applied to the system.)

Student Resources

From the Teacher's Resource Package, use:

- Section Review 19.2, Ch. 19 Practice Problems and Quizzes from the Review Module (Ch. 17–20)
- Laboratory Manual: Experiment 36
- Small-Scale Chemistry Lab Manual: Experiment 25

Technology Resources

Relevant technology resources include:

- Chem ASAP! CD-ROM
- ResourcePro CD-ROM

Motivate and Relate

You can use a discussion of what happens at the line of scrimmage in a football game to introduce the concept of opposing reactions and equilibrium. The offensive and defensive lines are trying to move in opposite directions. If the offensive and defensive lines are fairly evenly matched, it is hard for either team to push the other off the line of scrimmage. If one team is stronger, its players may be able to penetrate into the other team's backfield. When you discuss factors that affect equilibrium, you could extend the analogy by comparing the removal of an injured player to the removal of a product; the introduction of a new blocking scheme to changes in pressure; or a blitz package to a catalyst.

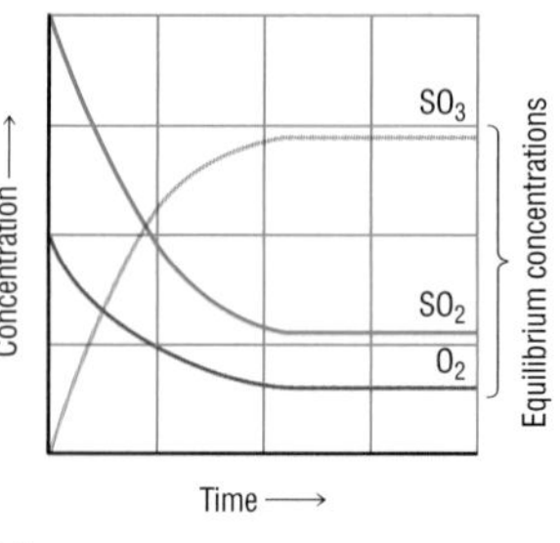

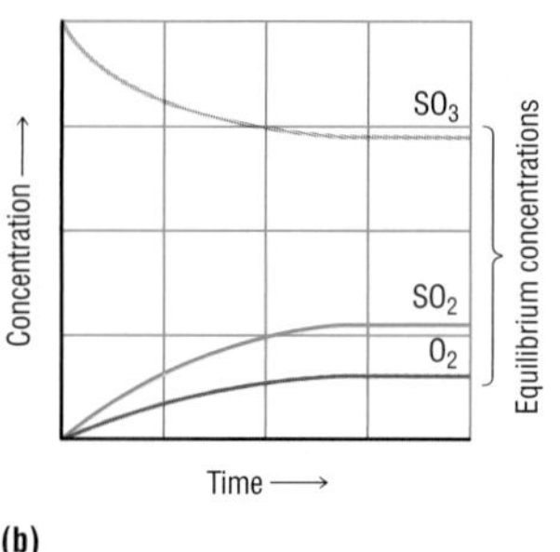

Figure 19.10
*These graphs show the variation with time of the concentration of O_2, SO_2, and SO_3. **(a)** Initially, SO_2 concentration is twice that of O_2; SO_3 is absent. At equilibrium, a mixture of all three gases is obtained. **(b)** Initially only SO_3 is present. At equilibrium, the amounts of O_2, SO_2, and SO_3 are the same as in the equilibrium shown in **(a)**.*

begins to form. As the sulfur trioxide concentration increases, a small amount slowly reverts to sulfur dioxide and oxygen by the reverse reaction. As the concentration of sulfur trioxide becomes higher and higher, the reverse reaction speeds up. The forward reaction is simultaneously slowing down because sulfur dioxide and oxygen are being used up. As you can see in **Figure 19.10,** eventually sulfur trioxide decomposes to sulfur dioxide and oxygen as fast as sulfur dioxide and oxygen form sulfur trioxide. At equilibrium, ❶ what substance has the highest concentration? The reaction has reached **chemical equilibrium,** a state in which the forward and reverse reactions take place at the same rate. At chemical equilibrium, there is no net change in the actual amounts of the components of the system.

Although the rates of the forward and reverse reactions are equal at chemical equilibrium, the concentrations of the components on both sides of the chemical equation are not necessarily the same. In fact, they can be dramatically different. The escalators in **Figure 19.11** are like the double arrows in a dynamic equilibrium equation. The number of people using the up escalator must equal the number of people using the down escalator for the number of people on each floor to remain constant.

The **equilibrium position** of a reaction is given by the relative concentrations of the system's components at equilibrium. The equilibrium position indicates whether the components on the left or right side of a reversible reaction are at a higher concentration. If A reacts to give B and the equilibrium mixture contains significantly more of B—say 1% A and 99% B—then the formation of B is said to be favored.

$$\underset{1\%}{A} \rightleftharpoons \underset{99\%}{B}$$

On the other hand, if the mixture contains 99% A and 1% B at equilibrium, then the formation of A is favored.

$$\underset{99\%}{A} \rightleftharpoons \underset{1\%}{B}$$

Note that the longer of the two double arrows indicates the favored direction of a reaction.

In principle, almost all reactions are reversible to some extent under the right conditions. In practice, one set of components is often so favored at equilibrium that the other set cannot be detected. If one set of components

Figure 19.11
If the rate at which shoppers move from the first floor to the second is equal to the rate at which shoppers move from the second floor to the first, then the number of shoppers on each floor will remain constant, although not necessarily equal to each other. The two populations of shoppers will be in a state of dynamic equilibrium.

Meeting Diverse Needs

Gifted Have gifted students investigate how a catalytic converter on an automobile works. Ask them what role the spatial arrangements in the converter play.

Answer

❶ SO_3

(reactants) is completely converted to new substances (products), you can say that the reaction has gone to completion or is irreversible. When you mix chemicals expecting to get a reaction but no products can be detected, you can say there is no reaction. Reversible reactions occupy a middle ground between the theoretical extremes of irreversibility and no reaction.

A catalyst speeds up forward and reverse reactions equally because the reverse reaction is exactly the opposite of the forward reaction. Therefore, the catalyst reduces the energy needed for the reaction by the same amount in both the forward and reverse directions. Catalysts do not affect the amounts of reactants and products present at equilibrium; they simply decrease the time it takes to establish equilibrium.

Factors Affecting Equilibrium: Le Châtelier's Principle

A delicate balance exists in a system at equilibrium. Changes of almost any kind can disrupt this balance. A system whose equilibrium has been disturbed makes adjustments to restore equilibrium. When equilibrium has been restored, however, the position of equilibrium is different from its original position; that is, the amount of products or reactants may have increased or decreased. Such a difference is called a shift in the position of equilibrium.

The French chemist Henri Le Châtelier (1850–1936) studied shifts in the position of equilibrium that result from changing conditions. He proposed **Le Châtelier's principle:** If a stress is applied to a system in dynamic equilibrium, the system changes to relieve the stress. Stresses that upset the equilibrium of a chemical system include changes in the concentration of reactants or products, changes in temperature, and changes in pressure.

The following examples of applications of Le Châtelier's principle all involve reversible reactions. Therefore, the product of the forward reaction is the reactant of the reverse reaction, and vice versa. For simplicity and clarity, however, the components to the left of the reaction arrow will be considered the reactants and the components to the right of the reaction arrow will be considered the products. Blue arrows indicate the shifts resulting from additions to or removals from the system. The arrows always point in the direction of the resulting shift in the equilibrium position—that is, toward the favored side. In each case, the equilibrium shift compensates for the disturbance that caused it.

Concentration Changing the amount, or concentration, of any reactant or product in a system at equilibrium disturbs the equilibrium. The system adjusts to minimize the effects of the change. Consider an equilibrium involving carbonic acid (H_2CO_3) in aqueous solution, which decomposes to form carbon dioxide and water. At equilibrium the amount of carbonic acid is less than 1%.

$$\underset{< 1\%}{H_2CO_3(aq)} \underset{\substack{\text{Remove } CO_2 \\ \text{Direction of shift} \rightarrow}}{\overset{\substack{\text{Add } CO_2 \\ \leftarrow \text{Direction of shift}}}{\rightleftharpoons}} \underset{> 99\%}{CO_2(aq) + H_2O(l)}$$

Adding more carbon dioxide disturbs the equilibrium. For example, at the instant of the addition, the added carbon dioxide changes the ratio of carbon

2 Teach

Use the Visuals

Ask the students to compare Figures 19.10 and 19.11. Explain that chemical reactions do not always go completely to products, but are often reversible. Once the product(s) form they may decompose/react to form the reactants.

Discuss

Refer back to the formation of banana oil, which was demonstrated in Section 19.1. Write the general word equation for the formation of an ester on the board (organic acid + alcohol → ester + water). Add a reverse arrow to show the possibility of the opposite reaction.
Remind students that the reaction was effective in the test tube to which anhydrous calcium sulfate had been added. Calcium sulfate reacted with water to form a hydrate, which limited the reverse reaction by removing a product.

TEACHER DEMO

You may want to repeat the banana oil demonstration. This time, add concentrated sulfuric acid to both test tubes before heating. However, only add the anhydrous calcium sulfate to one tube. Ask students to predict which reaction will yield the most product. Then compare the results qualitatively by strength of odor or quantitatively if you have access to gas phase chromatography (GC) equipment.

Discuss

Point out that, in many cases, the concept of turning reactants into products is an oversimplification. Many reactions reach an equilibrium state in which the reaction mixture contains both reactant and product particles. The percentage of reactants converted to products varies considerably. When hydrogen and oxygen combine to form water, for example, nearly all of the molecules react. In the reaction of hydrogen and nitrogen to form ammonia, the percentage of reactants converted to product is much smaller. Stress that equilibrium refers to the rates of the forward or reverse reactions, not to the quantities of reactants and products.

Discuss

Give examples of the changes (stresses) that can be applied to various systems in equilibrium. Discuss adding or subtracting products or reactants, heating or cooling the system, changing the concentration of the reactants or products, and raising or lowering the pressure on the system. Explain that, according to Le Châtelier's Principle, at equilibrium the system always goes in the direction that relieves the stress.

Figure 19.12
The rapid exhalation of CO_2 during and after vigorous exercise helps reestablish the body's correct $CO_2:H_2CO_3$ equilibrium. This keeps the acid concentration in the blood within a safe range.

dioxide to carbonic acid ($CO_2:H_2CO_3$) from 99:1 to 99.5:0.5. Although the change in the ratio may look small, the amount of carbon dioxide has doubled from the original 99:1 ratio to a ratio of 199:1 (99.5/0.5 = 199:1). As soon as the carbon dioxide is added, however, it reacts with water to form more carbonic acid. The system shifts to the left to use up some of the added CO_2 and thus minimize the stress. The ratio shifts toward the original 99:1. What occurs is typical of such shifts in equilibrium—adding a product always pushes a reversible reaction in the direction of reactants.

If, on the other hand, carbon dioxide is removed, the $CO_2:H_2CO_3$ ratio momentarily decreases to less than 99:1. Carbonic acid decomposes to minimize the imposed stress, and the system readjusts itself toward a $CO_2:H_2CO_3$ ratio of 99:1. Removing a product always pulls a reversible reaction in the direction of products.

Removal of products is often used to increase the yield of a desired product. As products are removed from a reaction mixture, the system continually changes to restore equilibrium by producing more products. Because the products are being removed as fast as they form, however, the reaction never builds up enough of them to reestablish an equilibrium. The system continues readjusting, and the reaction continues to give products until the reactants are completely used up. Farmers use a similar technique to increase the yield of eggs laid by hens. Hens lay eggs and then proceed to hatch them. When the eggs are removed after they are laid (removing the product), the hen will lay more eggs (increasing the yield). Another example of this concept is found in the body. Blood contains dissolved carbonic acid in equilibrium with carbon dioxide and water. The body uses the removal of products to keep carbonic acid at low, safe concentrations in the blood. When the athletes in **Figure 19.12** exhale carbon dioxide, the equilibrium shifts toward carbon dioxide and water, thus reducing the amount of carbonic acid and keeping the acid concentration of the blood within a safe range. The same principle applies to adding or removing reactants. When a reactant is added to a system at equilibrium, the reaction shifts in the direction of the formation of products. When a reactant is removed, the reaction shifts in the direction of formation of reactants.

Temperature Increasing the temperature causes the equilibrium position of a reaction to shift in the direction that absorbs heat. The heat absorption reduces the applied temperature stress. For example, consider the following exothermic reaction that occurs when SO_3 is produced from the reaction of SO_2 and O_2.

$$2SO_2(g) + O_2(g) \underset{\text{Remove heat (cool) Direction of shift}\rightarrow}{\overset{\text{Add heat}\ \leftarrow\text{Direction of shift}}{\rightleftharpoons}} 2SO_3(g) + \text{heat}$$

The heat can be considered to be a product, just like SO_3. Heating the reaction mixture at equilibrium pushes the equilibrium position to the left, which favors the reactants. As a result, the product yield decreases. Cooling, or removing heat, pulls the equilibrium to the right, and the product yield increases.

Pressure A change in the pressure on a system affects only an equilibrium that has an unequal number of moles of gaseous reactants and products. An example is the equilibrium established between ammonia gas and the gaseous elements from which it forms. Imagine those gases are at equilibrium in a cylinder that has a piston attached to a plunger, similar to a bicycle pump with the hose sealed. A catalyst has been included to speed up the reaction. What happens to the pressure when you push the plunger down? The pressure on the gases momentarily increases because the same number of molecules is contained in less than the original volume. The system can relieve some of the pressure increase by reducing the number of gas molecules. For every two molecules of ammonia made, four molecules of the reactants are used up (three molecules of hydrogen and one of nitrogen). Therefore, the equilibrium position shifts to make more ammonia. There are then fewer molecules in the system. The pressure decreases, although it will not decrease all the way to the original pressure. As you can see in **Figure 19.13,** increasing the pressure on the system results in a shift in the equilibrium position that favors the formation of product. What will happen to the equilibrium position when the pressure is restored to its initial value? 1

$$N_2(g) + 3H_2(g) \underset{\text{Reduce pressure}\ \leftarrow\text{Direction of shift}}{\overset{\text{Increase pressure Direction of shift}\rightarrow}{\rightleftharpoons}} 2NH_3(g)$$

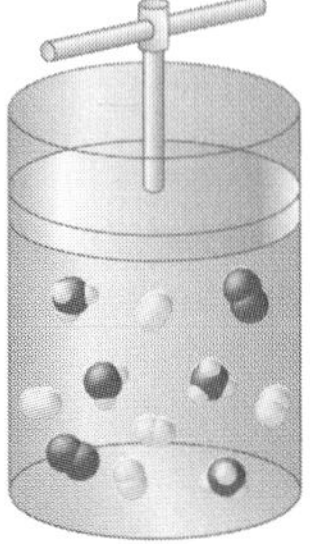

(a) Initial equilibrium condition (11 gas molecules)

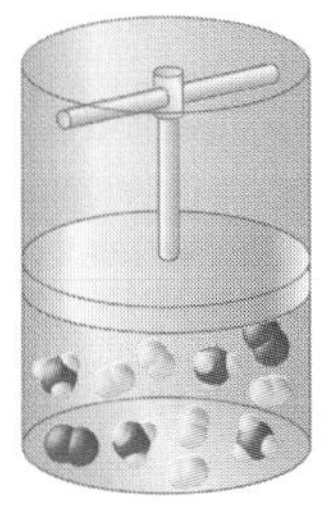

(b) Pressure increased, equilibrium disturbed

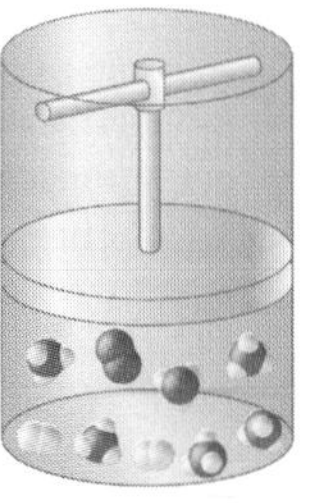

(c) New equilibrium condition at increased pressure (9 gas molecules)

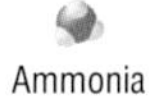

Ammonia molecule (NH_3)

Hydrogen molecule (H_2)

Nitrogen molecule (N_2)

Chem ASAP!

Simulation 24
Simulate Le Châtelier's principle for the synthesis of ammonia.

Figure 19.13
*Pressure affects a mixture of nitrogen, hydrogen, and ammonia in equilibrium. **(a)** The system is in equilibrium. **(b)** Equilibrium is disturbed by an increase in pressure. **(c)** A new equilibrium condition is created, more hydrogen and nitrogen have reacted and formed ammonia, reducing the total number of molecules and partially offsetting the pressure increase in the cylinder.*

CHEMISTRY AND BIOLOGY

Molecules such as carbon monoxide (CO) that are structurally similar to the oxygen molecule can bind to the hemoglobin molecule. Therefore, they compete with oxygen for the same binding locations. This competition is the cause of carbon monoxide poisoning. Because their affinity for the heme group is greater, carbon monoxide molecules can displace oxygen molecules from hemoglobin. $HbO_2 + CO \rightleftharpoons HbCO + O_2$. The equilibrium in this reaction lies far to the right because the affinity of carbon monoxide for heme is approximately 150 times stronger than that of oxygen.

Use the Visual

Have the students study Figure 19.13 and write the equation for the dissociation of H_2CO_3 on the board.
($H_2CO_3 \rightleftharpoons CO_2 + H_2O$)
Explain that their bodies expel CO_2 to maintain the correct equilibrium between CO_2 and H_2CO_3. During and after vigorous exercise, more CO_2 needs to be removed to maintain equilibrium. Ask:

- **What are some signs that the body is trying to reestablish equilibrium after exercising?** (an increase in heart and breathing rates)

Answer

1 If the pressure is restored to its initial condition, the equilibrium position will shift in favor of formation of reactants. Ammonia particles will decompose to form nitrogen and hydrogen particles, partially offsetting the decrease in pressure.

Critical Thinking

The industrial production of ammonia provides a practical application of Le Châtelier's principle. Place the equilibrium equation on the board. Ask:

- **Based on the equilibrium equation, select ideal temperature and pressure conditions for the highest possible yield of ammonia.** (high pressure and low temperature)

Ask students to research the process devised by Carl Bosch, which uses high temperatures and pressures plus an iron-based catalyst. After students report back, ask:

- **Why doesn't Bosch's process reflect the ideal conditions they predicted?** (A higher-than-ideal temperature (plus the catalyst) is needed to increase the reaction rate.)

Point out that an efficient industrial process must often strike a balance between multiple goals, in this case, yield and reaction time.

Practice Problems Plus

Related Chapter Review Problem

Chapter Review problem 70 is related to Sample Problem 19-1.

Additional Practice Problem

The Chem ASAP! CD-ROM contains the following problem: The gases dinitrogen tetroxide and nitrogen dioxide exist in equilibrium according to the equation $N_2O_4(g) + 58\ kJ \rightleftharpoons 2NO_2(g)$ What effect do the following changes have on the equilibrium position?

a. addition of heat
b. decrease in pressure
c. addition of NO_2
d. removal of N_2O_4

(**a.** favors products **b.** favors products **c.** favors reactants **d.** favors reactants)

The equilibrium position for this reaction can be made to favor the reactants instead of the product. Imagine pulling the plunger of the piston device back up so the volume containing the gases increases. This increase of volume decreases the pressure on the system. To restore the higher starting pressure, the system needs more gas molecules, which are produced by decomposition of some of the ammonia molecules. Decomposition of two molecules of gaseous ammonia produces a total of four molecules of reactants (three molecules of hydrogen and one molecule of nitrogen). Pressure at the new equilibrium is higher than when the pressure was first decreased, although it is not as high as it was at the starting equilibrium. Lowering the pressure on the system thus results in a shift of equilibrium to favor the reactants.

Sample Problem 19-1

What effect do each of the following changes have on the equilibrium position for this reversible reaction?

$$PCl_5(g) + heat \rightleftharpoons PCl_3(g) + Cl_2(g)$$

a. addition of Cl_2
b. increase in pressure
c. removal of heat
d. removal of PCl_3 as it forms

1. ***ANALYZE*** ***Plan a problem-solving strategy.***

a.–d. In each part of the problem, the stress placed on the system is known. The effect of the stress is unknown. By Le Châtelier's principle, the equilibrium system will shift in a direction that minimizes the imposed stress. Analyze the effect of each change on the reaction.

2. ***SOLVE*** ***Apply the problem-solving strategy.***

a. The addition of Cl_2, a product, shifts the equilibrium to the left, forming more PCl_5.
b. The equation shows 2 mol of gaseous product and 1 mol of gaseous reactant. The increase in pressure is relieved if the equilibrium shifts to the left because a decrease in the number of moles of gaseous substances gives a decrease in pressure.
c. The removal of heat causes the equilibrium to shift to the left because the reverse reaction is heat-producing.
d. The removal of PCl_3 causes the equilibrium to shift to the right to produce more PCl_3.

3. ***EVALUATE*** ***Do the results make sense?***

The answers follow Le Châtelier's principle: If a stress is applied to a system in a dynamic equilibrium, the system changes to relieve the stress.

Practice Problems

6. How is the equilibrium position of this reaction affected by the following changes?

$$C(s) + H_2O(g) + heat \rightleftharpoons CO(g) + H_2(g)$$

a. lowering the temperature
b. increasing the pressure
c. removing H_2
d. adding H_2

7. What effect does each change have on the equilibrium position of this reaction?

$$N_2(g) + 3H_2(g) \rightleftharpoons 2NH_3(g) + 92\ kJ$$

a. addition of heat
b. increase in pressure
c. addition of catalyst
d. removal of heat

Chem ASAP!

Problem-Solving 7

Solve Problem 7 with the help of an interactive guided tutorial.

Equilibrium Constants

Chemists generally express the position of equilibrium in terms of numerical values. These values relate the amounts of reactants to products at equilibrium. Consider the hypothetical reaction in which *a* mol of reactant A and *b* mol of reactant B react to give *c* mol of product C and *d* mol of product D at equilibrium.

$$aA + bB \rightleftharpoons cC + dD$$

The **equilibrium constant** (K_{eq}) is the ratio of product concentrations to reactant concentrations at equilibrium, with each concentration raised to a power equal to the number of moles of that substance in the balanced chemical equation. Thus the equilibrium-constant expression has the general form

$$K_{eq} = \frac{[C]^c \times [D]^d}{[A]^a \times [B]^b}$$

The exponents in the equilibrium-constant expression are the coefficients from the balanced chemical equation. The square brackets indicate the concentrations of substances in moles per liter (mol/L). The value of K_{eq} for a reaction depends on the temperature. If the temperature changes, the value of K_{eq} also changes.

Equilibrium constants provide valuable chemical information. Among other things, they show whether products or reactants are favored at equilibrium. Because an equilibrium constant is always written as a ratio of products to reactants, a value of K_{eq} greater than 1 means that products are favored over reactants. Conversely, a value of K_{eq} less than 1 means that reactants are favored over products. In short,

$K_{eq} > 1$, products favored at equilibrium

$K_{eq} < 1$, reactants favored at equilibrium

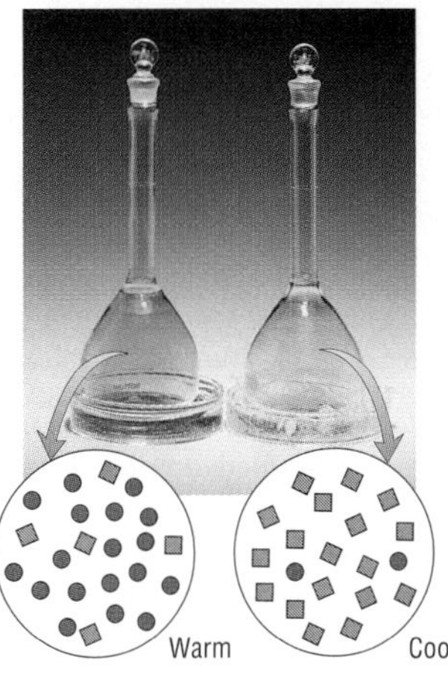

● Nitrogen dioxide (NO_2)

■ Dinitrogen tetroxide (N_2O_4)

Figure 19.14

Dinitrogen tetroxide is a colorless gas, whereas nitrogen dioxide is brown. The flask on the left is in a dish of hot water; the flask on the right is in ice. How does temperature affect the equilibrium of a mixture of these gases? ❶

Sample Problem 19-2

Dinitrogen tetroxide (N_2O_4), a colorless gas, and nitrogen dioxide (NO_2), a dark-brown gas, exist in equilibrium with each other, as shown in **Figure 19.14.**

$$N_2O_4(g) \rightleftharpoons 2NO_2(g)$$

A liter of the gas mixture at 10 °C at equilibrium contains 0.0045 mol N_2O_4 and 0.030 mol NO_2. Write the expression for the equilibrium constant and calculate the equilibrium constant (K_{eq}) for the reaction.

1. ***ANALYZE*** ***List the knowns and the unknowns.***

Knowns:
- $[N_2O_4] = 0.0045$ mol/L
- $[NO_2] = 0.030$ mol/L
- $K_{eq} = \frac{[C]^c \times [D]^d}{[A]^a \times [B]^b}$

Unknowns:
- K_{eq} (algebraic expression) = ?
- K_{eq} (numerical value) = ?

Practice Problems

8. Analysis of an equilibrium mixture of nitrogen, hydrogen, and ammonia contained in a 1-L flask at 300 °C gives the following results: hydrogen, 0.15 mol; nitrogen, 0.25 mol; ammonia, 0.10 mol. Calculate K_{eq} for the reaction
$N_2(g) + 3H_2(g) \rightleftharpoons 2NH_3(g)$

Answer

❶ A high temperature shifts the position of equilibrium to favor nitrogen dioxide.

TEACHER DEMO

Demonstrate the effect of temperature changes on the equilibrium position of an exothermic reaction. **CAUTION!** Use a fume hood. Nitric acid is corrosive and NO_2 is an extremely toxic gas. Place a small amount (3 g or less) of copper turnings in a test tube, and add 10 mL of 6*M* nitric acid (HNO_3). Insert a one-hole stopper, fitted with a glass delivery tube and rubber hose, into the test tube. Fill two sturdy, Pyrex test tubes with the brown gas that forms. When the color is the same as in the reacting test tube, stopper the tubes. Once sealed, the tubes can be saved indefinitely for repeated use or discharged in a fume hood.

Set up an ice bath and warm water bath (70–80 °C). Hold up the tubes for students to see and tell students that each of the tubes contains NO_2 gas. Place one tube in warm water and the other in cold water. After a few minutes, remove the tubes and compare their colors. The cooled tube should be light brown, and the heated tube should be dark brown. Then place the tube that was previously in the warm water bath into the ice water and the tube that was previously in the ice water into the warm water. After a few minutes, hold the tubes up for students to observe.

Write the reaction equation on the board: $2NO_2(g) \rightleftharpoons N_2O_4(g) +$ heat. Explain that nitrogen dioxide is a brown gas and that dinitrogen tetroxide is a colorless gas. Ask students to explain the effect of changing temperature. (The reaction is exothermic. Adding heat shifts the equilibrium in the direction that absorbs heat, to the left. Removing heat shifts the equilibrium in the direction that produces heat, to the right.)

Discuss

Point out that whenever a chemical system reaches equilibrium, the reactants and products have a fixed numerical relationship. This relationship is stated in the equilibrium constant expression. The value of the equilibrium constant can be used to predict the position of equilibrium. Because the equilibrium constant refers to an expression in which products are divided by reactants, a larger K_{eq} value favors the products. Conversely, a smaller K_{eq} value favors the reactants.

Practice Problems Plus

Related Chapter Review Problems

Chapter Review problems 48, 49, 50, 71, and 72 are related to Sample Problems 19-2 and 19-3.

Additional Practice Problem

The Chem ASAP! CD-ROM contains this problem related to Sample Problem 19-2: Analysis of an equilibrium mixture of hydrogen, iodine, and hydrogen iodide contained in a 10.0-L flask at a certain temperature gives the following results: hydrogen, 0.15 mol; iodine, 0.15 mol; hydrogen iodide, 0.87 mol. Calculate K_{eq} for the reaction: $H_2(g) + I_2(g) \rightleftharpoons 2HI(g)$ ($K_{eq} = 34$)

Sample Problem 19-2 (cont.)

2. CALCULATE ***Solve for the unknowns.***

At equilibrium, there is no net change in the amount of N_2O_4 or NO_2 at any given instant. The only product of the reaction is NO_2, which has a coefficient of 2 in the balanced equation. According to the rule for writing the equilibrium-constant expression, the concentration of NO_2, raised to the second power, is placed in the numerator. The only reactant is N_2O_4, with a coefficient of 1. The concentration of this substance, raised to the first power, is placed in the denominator. The equilibrium-constant expression and its calculated value are

$$K_{eq} = \frac{[NO_2]^2}{[N_2O_4]}$$

$$= \frac{(0.030\ mol/L)^2}{0.0045\ mol/L}$$

$$= \frac{(0.030\ mol/L \times 0.030\ \cancel{mol/L})}{0.0045\ \cancel{mol/L}}$$

$$= 0.20\ mol/L$$

3. EVALUATE ***Do the results make sense?***

Each concentration is raised to the correct power. The numerical value of the constant is correctly rounded to two significant figures.

Practice Problems (cont.)

9. Assume you have the mixture described in Problem 8 with the same volume, temperature, and equilibrium concentrations.

a. Calculate K_{eq} for the reaction $2NH_3(g) \rightleftharpoons 3H_2(g) + N_2(g)$

b. Based on your answers to Problem 8 and Problem 9a, how is the K_{eq} for a forward reaction related to the K_{eq} for a reverse reaction?

Chem ASAP!

Problem-Solving 8

Solve Problem 8 with the help of an interactive guided tutorial.

Sample Problem 19-3

One mol of colorless hydrogen gas and 1 mol of violet iodine vapor are sealed in a 1-L flask and allowed to react at 450 °C. At equilibrium, 1.56 mol of colorless hydrogen iodide is present, together with some of the reactant gases. Calculate K_{eq} for the reaction.

$$H_2(g) + I_2(g) \rightleftharpoons 2HI(g)$$

1. ANALYZE ***List the knowns and the unknowns.***

Knowns:

- $[H_2]$ (initial) = 1 mol/L
- $[I_2]$ (initial) = 1 mol/L
- [HI] (equilibrium) = 1.56 mol/L
- $K_{eq} = \frac{[C]^c \times [D]^d}{[A]^a \times [B]^b}$

Unknowns:

- K_{eq} (algebraic expression) = ?
- K_{eq} (numerical value) = ?

Practice Problems

10. Suppose the following system reaches equilibrium at a high temperature.

$$N_2(g) + O_2(g) \rightleftharpoons 2NO(g)$$

An analysis of the equilibrium mixture in a 1-L flask gives the following results: nitrogen, 0.50 mol; oxygen, 0.50 mol; nitrogen monoxide, 0.020 mol. Calculate K_{eq} for the reaction.

Meeting Diverse Needs

At Risk Show students how to set up and solve problems involving equilibrium constants. Stress the need to clearly establish what is given and what is being asked. Break down the process into easily followed steps. Point out that the only quantities that can go into the brackets used in the equilibrium law expression are concentrations, or quantities like partial pressures, which are related to concentrations. Warn students not to forget to subtract the number of moles that react from the number originally present in order to calculate the number remaining at equilibrium. Point out that it is this value for the concentration of a reactant that is substituted in the denominator of the equation for K_{eq}.

Sample Problem 19-3 (cont.)

2. ***CALCULATE*** *Solve for the unknowns.*
   The balanced equation indicates that 1.00 mol of hydrogen and 1.00 mol of iodine are needed to form 2.00 mol of hydrogen iodide. Making 1.56 mol of hydrogen iodide, therefore, consumes $\frac{1}{2} \times 1.56$ mol of each: 0.78 mol of hydrogen and 0.78 mol of iodine. First, calculate how much H_2 and I_2 remain in the flask at equilibrium.

   $$\text{mol } H_2 = \text{mol } I_2 = (1.00 \text{ mol} - 0.78 \text{ mol}) = 0.22 \text{ mol}$$

   Write the expression for K_{eq}.

   $$K_{eq} = \frac{[HI]^2}{[H_2] \times [I_2]}$$

   Now substitute the concentrations of the reactants and products into the equation for K_{eq}.

   $$K_{eq} = \frac{(1.56 \text{ mol/L})^2}{0.22 \text{ mol/L} \times 0.22 \text{ mol/L}} = \frac{1.56 \cancel{\text{mol/L}} \times 1.56 \cancel{\text{mol/L}}}{0.22 \cancel{\text{mol/L}} \times 0.22 \cancel{\text{mol/L}}}$$

   $$= 5.0 \times 10^1$$

3. ***EVALUATE*** *Do the results make sense?*
   Each concentration is raised to the correct power. The numerical value of the constant is correctly rounded to two significant figures. Note that because $K_{eq} > 1$, the product (HI) is favored at equilibrium.

Practice Problems (cont.)

11. At 750 °C the following reaction reaches equilibrium in a 1-L container.

$H_2(g) + CO_2(g) \rightleftharpoons H_2O(g) + CO(g)$

An analysis of the equilibrium mixture gives the following results: hydrogen, 0.053 mol; carbon dioxide, 0.053 mol; water vapor, 0.047 mol; carbon monoxide, 0.047 mol. Calculate K_{eq} for the reaction.

Sample Problem 19-4

Bromine chloride (BrCl) decomposes to form chlorine and bromine.

$$2BrCl(g) \rightleftharpoons Cl_2(g) + Br_2(g)$$

At a certain temperature, the equilibrium constant for the reaction is 11.1, and the equilibrium mixture contains 4.00 mol Cl_2. How many moles of Br_2 and BrCl are present in the equilibrium mixture? Assume that initially only pure BrCl existed and that the container has a volume of 1 L.

1. ***ANALYZE*** *List the knowns and the unknowns.*

   Knowns:
   - $K_{eq} = 11.1$
   - quantity of Cl_2 (at equilibrium) = 4.00 mol
   - $K_{eq} = \frac{[C]^c \times [D]^d}{[A]^a \times [B]^b}$

   Unknowns:
   - $[Br_2]$ = ? mol/L
   - [BrCl] = ? mol/L

Practice Problems

12. The decomposition of hydrogen iodide at 450 °C in a 1-L container produces an equilibrium mixture that contains 0.50 mol of hydrogen. The equilibrium constant is 0.020 for the reaction. How many moles of iodine and of hydrogen iodide are present in the equilibrium mixture? The equation is

$2HI(g) \rightleftharpoons H_2(g) + I_2(g)$

Discuss

If students have trouble writing the correct equilibrium expression from a balanced equation, write a list of steps on the chalkboard that describes how to construct an equilibrium expression. Use as an example the equilibrium process for the production and decomposition of ozone in the stratosphere. Explain that stratospheric ozone screens out about 95% of the UV radiation from the sun. Within the ozone layer, ozone is continually formed and consumed through an equilibrium process: $3O_2(g) \rightleftharpoons 2O_3(g)$. To construct the equilibrium expression, 1) write the balanced equation, 2) label the coefficients, reactants, and products, 3) write a ratio with the concentration of the products in the numerator and the concentration of the reactants in the denominator, 4) and attach the coefficients as powers.

$$K_{eq} = \frac{[O_3]^2}{[O_2]^3}$$

Practice Problems Plus

Additional Practice Problem

The Chem ASAP! CD-ROM contains this problem related to Sample Problem 19-4: Gaseous sulfur dioxide reacts with oxygen in a 0.50-L container at 600 °C to form gaseous sulfur trioxide. The equilibrium mixture contains 0.75 mol SO_2 and 0.62 mol O_2. If the equilibrium constant is 4.36 for the reaction, how many moles of SO_3 are present in the equilibrium mixture? The equation is: $2SO_2(g) + O_2(g) \rightleftharpoons 2SO_3(g)$ (1.7 mol SO_3)

3 Assess

Evaluate Understanding

Write several equilibrium chemical equations on the board. Have students use Le Châtelier's principle to improve the yield of a reaction product. Have the students write the equation for calculating an equilibrium constant using the generalized expression: aA + bB ↔ cC + dD.

$$K_{eq} = \frac{[C]^c \times [D]^d}{[A]^a \times [B]^b}$$

Reteach

Review the important concepts in this section:

- **Equilibrium is a dynamic process.**
- **The only things "equal" in an equilibrium are the rates of the opposing reactions.**
- **A catalyst does not affect the equilibrium position, only how fast it is reached. When a catalyst lowers the activation energy for one reaction, it also lowers the activation energy for the opposing reaction.**
- **Adding a reactant or removing a product can increase the yield of a reaction.**
- **Increasing temperature causes an equilibrium shift toward the direction that absorbs heat.**
- **For reactions in which the mole ratio of gaseous reactants and products is unequal, increasing pressure shifts the equilibrium toward the fewest molecules.**

Sample Problem 19-4 (cont.)

2. ***CALCULATE*** ***Solve for the unknowns.***
The equilibrium expression that corresponds to the chemical equation is

$$K_{eq} = \frac{[Cl_2] \times [Br_2]}{[BrCl]^2}$$

According to the equation, when BrCl breaks down, equal numbers of moles of Cl_2 and Br_2 form. The measured equilibrium quantity of Cl_2 was given as 4.00 mol. Thus there must also be 4.00 mol Br_2 in the equilibrium mixture. The volume of the container is 1 L, so $[Br_2]$ and $[Cl_2]$ are each 4.00 mol/L. Substitute the known values in the equilibrium expression for the reaction.

$$11.1 = \frac{(4.00\ mol/L) \times (4.00\ mol/L)}{[BrCl]^2}$$

Rearrange the equation to solve for $[BrCl]^2$, and then for [BrCl].

$$[BrCl]^2 = \frac{(4.00\ mol/L) \times (4.00\ mol/L)}{11.1}$$

$$= \frac{16.0\ mol^2/L^2}{11.1} = 1.44\ mol^2/L^2$$

$$[BrCl] = \sqrt{1.44\ mol^2/L^2} = 1.20\ mol/L$$

3. ***EVALUATE*** ***Do the results make sense?***
Each concentration is raised to the correct power. The result has the proper units and is correctly rounded to three significant figures. Note that $K_{eq} > 1$; the products Cl_2 and Br_2 are slightly favored at equilibrium.

Practice Problems (cont.)

13. Write the expression for the equilibrium constant for each.

a. $2HBr(g) \rightleftharpoons H_2(g) + Br_2(g)$
b. $2SO_3(g) \rightleftharpoons 2SO_2(g) + O_2(g)$
c. $CO_2(g) + H_2(g) \rightleftharpoons CO(g) + H_2O(g)$
d. $4NH_3(g) + 5O_2(g) \rightleftharpoons 6H_2O(g) + 4NO(g)$

Chem ASAP!

Problem-Solving 12
Solve Problem 12 with the help of an interactive guided tutorial.

section review 19.2

14. How can changes in the equilibrium position be predicted from changes in concentration, temperature, and pressure?

15. How can a balanced chemical equation and experimental data be used to write an equilibrium-constant expression and to calculate its value?

16. What is the significance of double arrows in an equation?

17. How do the amounts of reactants and products change once a reaction has achieved chemical equilibrium?

18. Can a pressure change shift the equilibrium position in every reversible reaction? Explain your answer.

19. Imagine you have determined the following equilibrium constants for several reactions. In which of these reactions are the products favored over the reactants? Why?

a. $K_{eq} = 1 \times 10^2$ **c.** $K_{eq} = 3.5$
b. $K_{eq} = 0.003$ **d.** $K_{eq} = 6 \times 10^{-4}$

Chem ASAP! **Assessment 19.2** Check your understanding of the important ideas and concepts in Section 19.2.

Answers

SECTION REVIEW 19.2

14. by applying Le Châtelier's principle

15. An equilibrium constant expression is a ratio. In the numerator, product concentrations are multiplied. In the denominator, reactant concentrations are multiplied. Each concentration is raised to a power equal to the coefficient for that species in the balanced equation. To calculate a numerical value for the equilibrium constant, you need to substitute concentration data from an experiment into the expression.

16. They show that the reaction is reversible.

17. At equilibrium, the concentrations of reactants and products do not change.

18. no; only in reversible reactions in which the mole ratios of gaseous reactants and products are unequal

19. Products are favored in reactions **a.** and **c.** because $K_{eq} > 1$.

DETERMINING WHETHER A REACTION WILL OCCUR

Fires in homes, barns, and other places sometimes appear to start on their own, without heat applied from an external source. These types of fires are caused by spontaneous combustion. Inside a pile of oily rags or a stack of hay that has not been thoroughly dried, oxidation causes heat to build up. If the existing conditions do not allow the heat to escape, the temperature can build up enough to cause a fire. **What are the conditions that will produce a spontaneous chemical reaction?**

section 19.3

objectives

- Define entropy and free energy, and characterize reactions as spontaneous or nonspontaneous
- Describe how heat change and entropy change determine the spontaneity of a reaction

key terms

- free energy
- spontaneous reactions
- nonspontaneous reactions
- entropy
- law of disorder

Free Energy and Spontaneous Reactions

Many chemical and physical processes release energy that can be used to bring about other changes. For example, some of the energy liberated in a reaction can be harnessed to do work, such as driving the pistons of an internal-combustion engine. **Free energy** is energy that is available to do work. Just because free energy is available to do work, however, does not mean it can be used efficiently. The internal-combustion engine in a car is only about 30% efficient; that is, only about 30% of the free energy released by burning gasoline is used to propel the car. The remaining 70% is lost as friction and waste heat. Efforts to conserve energy include attempts to increase the efficiency of engines and other mechanical devices. There are limits to such efficiency increases, however. No process can be made 100% efficient. Even in living things, which are among the most efficient users of free energy, processes are seldom more than 70% efficient. The temperature of the human body is maintained at 37 °C by the available free energy that is wasted as heat.

In addition to energy efficiency, there is another complication. Energy can be obtained from a reaction only if the reaction actually—rather than theoretically—occurs. In other words, although a balanced equation can

Figure 19.15
Fireworks displays are the result of highly favored spontaneous reactions. A large quantity of free energy is released.

STUDENT RESOURCES

From the Teacher's Resource Package, use:

- Section Review 19.3, Ch. 19 Practice Problems, Interpreting Graphics, and Quizzes from the Review Module (Ch. 17–20)
- Laboratory Recordsheets 19-1 and 19-2

TECHNOLOGY RESOURCES

Relevant technology resources include:

- Chem ASAP! CD-ROM
- Resource Pro CD-ROM

19.3

1 Engage

Use the Visual

Have the students study the photograph at the beginning of the section and read the accompanying text. Ask:

- **What conditions produce a spontaneous chemical reaction?** (Based on prior knowledge, students may suggest reactions that have a low activation energy, reactions that involve active elements such as alkali metals, combustion reactions above the kindling temperature, or reactions in solution that produce a precipitate.)

Motivate and Relate

Ask students to consider the general meaning of the term *spontaneous* as it relates to human behavior. Spontaneous action is often described as free flowing and "spur of the moment," not requiring external incitement. Spontaneous behavior can be unexpected or unpredictable. You may want to refer back to the question of spontaneous generation that students studied in biology (and Pasteur so cleverly refuted) and then address examples of spontaneous combustion.

Check Prior Knowledge

Place a list of various chemical reactions on the board and ask the students to categorize them as spontaneous or nonspontaneous. Example reactions might be: burning wood, rusting iron, reaction between diamond (carbon) and oxygen to produce carbon dioxide, and the decomposition of water to hydrogen and oxygen. Tell students that all of the reactions, with the exception of the decomposition of water, are spontaneous.

2 Teach

Use the Visual

Have the students examine Figure 19.15. Explain that the fireworks display is the result of highly spontaneous reactions that release a large quantity of free energy. Also explain that this reaction requires a very high initiation temperature to produce the fast reaction rate observed.

Discuss

Ask students to discuss what is meant by work. Point out that work is done on a gas when a force is applied to reduce its volume. Work is done by rapidly expanding gases such as those produced in an explosion. Ask:

- **Is work done on molecules when the temperature is raised or when chemical bonds are broken?** (Work is done in these situations.)
- **How is energy related to work?** (Point out that work and energy have the same units.)

Commercially, chemical reactions are usually carried out to manufacture a chemical substance or to produce energy. The commercial value of a chemical reaction depends on the conditions under which it occurs spontaneously and on the reaction rate. Emphasize that all spontaneous reactions release free energy.

be written for a chemical system, it does not mean the reaction will really take place. For example, you can write an equation for the decomposition of carbon dioxide to carbon and oxygen.

$$CO_2(g) \longrightarrow C(s) + O_2(g)$$

This equation, which represents the reverse of combustion, is balanced. However, experience tells you that the reaction represented by the equation does not tend to occur. Carbon and oxygen burn to make carbon dioxide, not the reverse. So the world of balanced chemical equations is really divided into two groups. One group contains equations representing reactions that actually occur; the other contains equations representing reactions that do not tend to occur, or at least not efficiently. The first group involves processes that are spontaneous. **Spontaneous reactions** are reactions that occur naturally and that favor the formation of products at the specified conditions. In other words, spontaneous reactions actually give substantial amounts of products at equilibrium. All spontaneous reactions release free energy. The colorful fireworks display shown in **Figure 19.15** on the previous page is due to spontaneous reactions. In contrast, **nonspontaneous reactions** are reactions that do not favor the formation of products at the specified conditions. Nonspontaneous reactions do not give substantial amounts of products at equilibrium.

Consider again the reversible decomposition of carbonic acid in water.

$$\underset{<1\%}{H_2CO_3(aq)} \rightleftharpoons \underset{>99\%}{CO_2(g) + H_2O(l)}$$

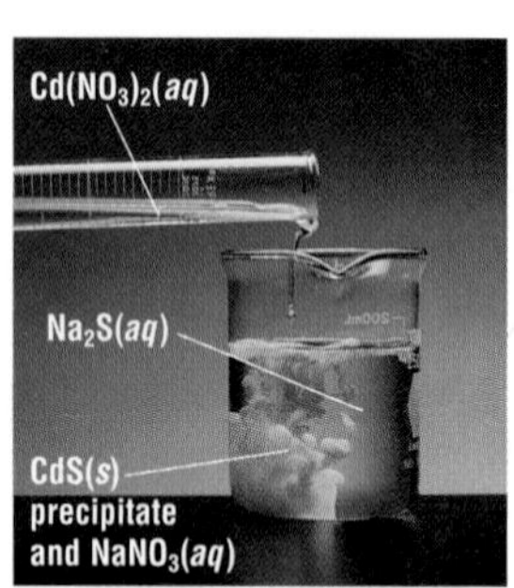

Figure 19.16
A yellow precipitate of cadmium sulfide (CdS) forms spontaneously when clear colorless solutions of sodium sulfide (Na_2S) and cadmium nitrate ($Cd(NO_3)_2$) are mixed.

Carbonic acid is the reactant of the forward reaction. If you could start with pure carbonic acid in water and let the system come to equilibrium, more than 99% of the reactant would be converted to the products carbon dioxide and water. These products are highly favored at equilibrium. There is a natural tendency for carbonic acid to decompose to carbon dioxide and water. Thus the forward reaction is spontaneous and releases free energy. In the reverse reaction, carbon dioxide and water are the reactants, and carbonic acid is the product. If you permit a solution of carbon dioxide in water to come to equilibrium, less than 1% of the reactants combines to form carbonic acid. There is little natural tendency for the reactants to go to products. Does this mean that the combination of carbon dioxide and
1 water to form carbonic acid is spontaneous or nonspontaneous? In nearly all reversible reactions, one reaction is favored over the other.

Another example, shown in **Figure 19.16**, is the spontaneous reaction of aqueous cadmium nitrate with aqueous sodium sulfide to produce aqueous sodium nitrate and solid yellow cadmium sulfide. Cadmium sulfide is highly favored in this equilibrium. The reverse reaction, or the production of cadmium nitrate and sodium sulfide from cadmium sulfide and sodium nitrate, is nonspontaneous.

$$Cd(NO_3)_2(aq) + Na_2S(aq) \rightleftharpoons CdS(s) + 2NaNO_3(aq)$$

It is important to note that the terms spontaneous and nonspontaneous do not refer to how fast reactants go to products. Some spontaneous reactions go so slowly that they appear to be nonspontaneous. The reaction

Answer

1 The reverse reaction is nonspontaneous.

of sugar and oxygen, for example, produces carbon dioxide and water. However, a bowl of sugar on a table does nothing, and you might assume that the equilibrium among sugar, oxygen, carbon dioxide, and water greatly favors the sugar and oxygen. The equilibrium does favor the products, but at room temperature this reaction is so slow that it takes thousands of years to reach equilibrium. When you supply energy in the form of heat, the reaction is fast. Only then is it obvious that, at equilibrium, the formation of carbon dioxide and water is highly favored. The combustion of the sugar goes to completion.

Some reactions that are nonspontaneous at one set of conditions may be spontaneous at other conditions. Changing the temperature or pressure, for example, may determine whether or not a reaction is spontaneous. The nonspontaneous photosynthesis reaction summarized in **Figure 19.17** could not be driven to completion without the energy supplied by sunlight.

Sometimes a nonspontaneous reaction can be made to occur if it is coupled to a spontaneous reaction—a reaction that releases free energy. Coupled reactions are a common feature of the complex biological processes that take place in living organisms. Within cells, a series of spontaneous reactions release the energy stored in glucose. As you will learn in Chapter 27, there are molecules in a cell that can capture and transfer free energy to nonspontaneous reactions, such as the formation of proteins.

$$6CO_2 + 6H_2O \xrightarrow{\text{light energy}} C_6H_{12}O_6 + 6O_2$$

Figure 19.17
Many nonspontaneous reactions are required for a plant to grow. Nonspontaneous reactions can occur when they are coupled with spontaneous reactions.

Entropy

You may recall that heat (enthalpy) changes accompany most chemical and physical processes. The combustion of carbon, for example, is exothermic and spontaneous. The heat released during this reaction is 393.5 kJ for each mole of carbon (graphite) burned.

$$C(s, \text{graphite}) + O_2(g) \longrightarrow CO_2(g) + 393.5 \text{ kJ/mol}$$

You might expect that only exothermic reactions would be spontaneous. Some processes, however, may be spontaneous even though they absorb heat. For example, consider the physical process of melting ice to water. As it turns from a solid to a liquid, 1 mol of ice at 25 °C absorbs 6.0 kJ of heat from its surroundings.

$$H_2O(s) + 6.0 \text{ kJ/mol} \longrightarrow H_2O(l)$$

Ice melts spontaneously at 25 °C, even though the water produced contains more heat than the ice from which it forms. Considering only heat changes, the energy of the water seems higher than the energy of the ice. That would seem to violate the rule that in spontaneous processes, the direction of the change in energy is from higher energy to lower energy; that is, free energy is released. Yet this process occurs anyway. Some factor other than the heat change must help determine whether a physical or chemical process is spontaneous.

Discuss

Students may think that a nonspontaneous reaction is one in which no products are formed. Remind them that, in a reversible reaction under a given set of conditions, one reaction is always spontaneous and the other reaction is nonspontaneous. Therefore, the difference between spontaneous and nonspontaneous reactions is the *amount* of products given at equilibrium.

Chemistry and Science History

The term *entropy* was coined by the German physicist Rudolf Clausius in 1865. He used the term to describe the portion of free energy that is unavoidably lost to the environment as waste heat. Clausius pointed out that in any process involving the flow of energy, there is always some loss, so that the entropy of the universe is continually increasing.

Meeting Diverse Needs

Gifted Pose the following problem: The reaction of iron, air, and water to form rust has a negative value for the change in entropy. Why is this reaction spontaneous? An ordered solid (Fe) and a less-ordered gas (O_2) release enough energy to balance the production of a more-ordered solid (Fe_2O_3) in the presence of a catalyst (H_2O).

Use the Visual

Have the students examine the two illustrations of bedrooms in Figure 19.18. Explain that entropy is measured as the disorder in a system. Ask:

- **Which bedroom has higher entropy?** (the messy room)

Ask students to describe other human-made or natural environments that can become disordered over time. (possible examples: weeds in a garden, books on library shelves, a table of sweaters on a sale day, sand castles or ice sculptures, the contents of a house after a flood or earthquake)

TEACHER DEMO

Place a block of ice, beaker of water, and a rapidly boiling beaker of water in front of the classroom. Ask:

- Which item has the least entropy? The most? Why? (Students should know that in solid ice the water molecules are very highly ordered into a lattice structure. In liquid water, despite the hydrogen bonding between the water molecules, they are relatively free to move around. In the water vapor produced by the boiling water, there is a great increase in randomness as the water molecules are able to move freely.)

Figure 19.18
The entropy of a recently cleaned and orderly room is low. However, over time a room tends to become disorderly. Entropy has a tendency to increase.

The other factor is related to order. You are probably familiar with everyday ideas about order and disorder. For example, a handful of marbles is relatively ordered in the sense that all the marbles are collected in one place. It is improbable that when permitted to fall the marbles will end up in the same neat arrangement. Instead, the marbles will scatter on the ground. The marbles will become disordered. The disorder of a system is measured as **entropy.** Scattered marbles have a higher entropy than gathered marbles.

The concept that physical and chemical systems attain the lowest possible energy has a companion idea called the law of disorder. The **law of disorder** states that processes move in the direction of maximum disorder or randomness. You already know something about this natural tendency toward disorder. More than likely, your bedroom is neat and clean at the beginning of the week. But unless you clean it regularly, your room probably becomes messy by the end of the week. Do the photographs in **Figure 19.18** look familiar to you? The law of disorder also operates at the level of atoms and molecules. On an atomic and molecular scale, the following information can be applied.

(a) For a given substance, the entropy of the gas is greater than the entropy of the liquid or the solid. Similarly, the entropy of the liquid is greater than that of the solid. Thus entropy increases in reactions in which solid reactants form liquid or gaseous products. Entropy also increases when liquid reactants form gaseous products.

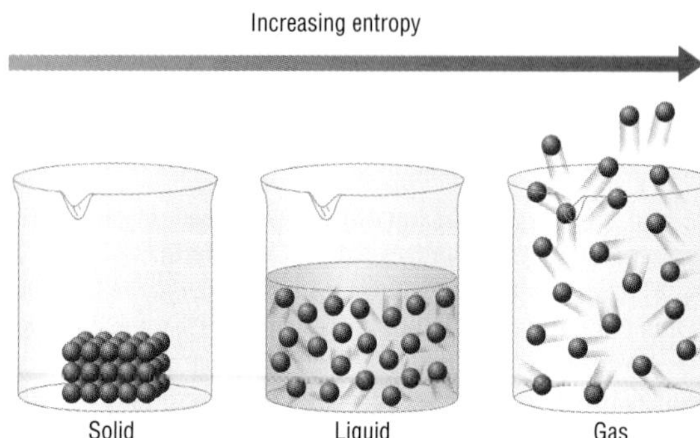

Figure 19.19
Entropy is a measure of the disorder of a system. ***(a)*** *The entropy of a solid is less than the entropy of the corresponding liquid, which is less than the corresponding gas.* ***(b)*** *Entropy increases when a substance is divided into parts.* ***(c)*** *Entropy tends to increase in reactions in which the number of molecules increases.* ***(d)*** *Entropy also increases with an increase in temperature.*

(b) Entropy increases when a substance is divided into parts. For instance, entropy increases when a crystalline ionic compound, such as sodium chloride, dissolves in water. This is because the solute particles—sodium ions and chloride ions—are more separated in solution than they are in the crystal form. What happens to the entropy of sodium and chloride ions when a solution of salt water evaporates? 1

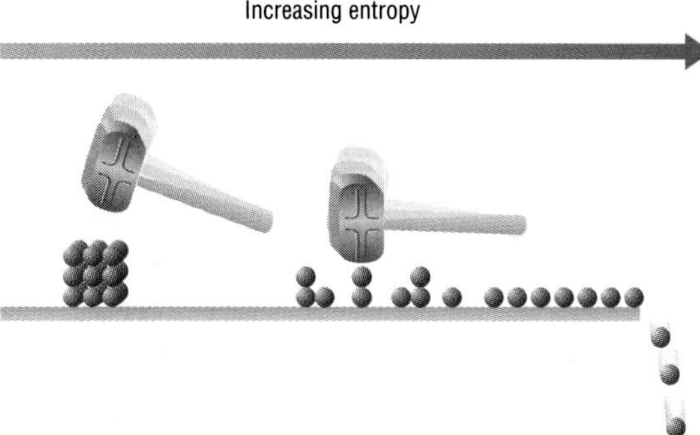

(c) Entropy tends to increase in chemical reactions in which the total number of product molecules is greater than the total number of reactant molecules.

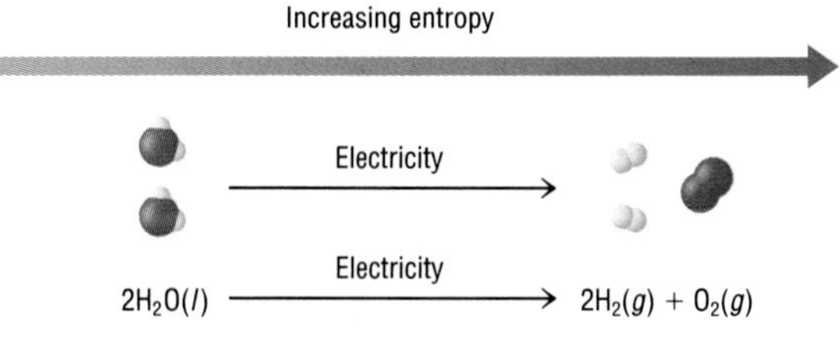

Electrolysis of water

(d) Entropy tends to increase when temperature increases. As the temperature increases, the molecules move faster and faster, which increases the disorder.

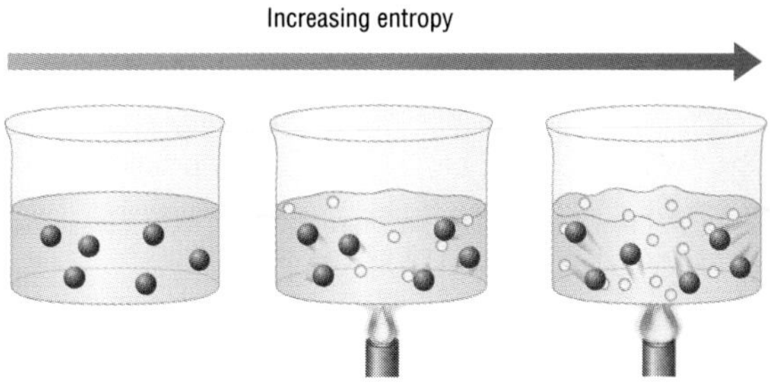

19.3

Think Critically

To develop the skill of making a mental model, have the students imagine an orderly arrangement such as an organized bookcase, cut vegetables arranged on a platter around a central bowl of dip, or a well-tended formal garden. Ask students to note what makes each arrangement orderly. (regular, geometrical, or symmetrical structure; each object in its designated place relative to the other objects)

Emphasize that they have made mental models of comparatively low-entropy systems. Next, have the students visualize a source of disorder—such as a mischievous person or a frisky animal—that enters their imaginary picture. The creature proceeds to destroy the order by knocking books off of the shelves, overturning the plate, or digging up and trampling the garden. Point out that their modified mental models represent an increase in entropy.

Answer

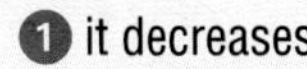

1 it decreases

Use the Visuals

Have students compare the charts in Table 19.1 and Figure 19.20. Note that these charts contain the general information students need to explain why some reactions are spontaneous and others are not. Ask:

- **What changes can occur during a reaction to favor spontaneity?** (changes that produce minimum energy and maximum entropy)
- **Under what conditions is a reaction least likely to be spontaneous?** (when products have both more energy than reactants and less disorder)
- **What happens when one change favors spontaneity and one change does not?** (The relative strengths of the opposing changes determine the outcome.)

For reactions that lead to equilibrium, the relative strengths of the enthalpy and entropy changes determine the point at which the opposing reactions are balanced.

Chem ASAP!

Simulation 25

See how enthalpy and entropy changes combine to determine the spontaneity of a reaction.

Heat, Entropy, and Free Energy

The size and direction of heat (enthalpy) changes and entropy changes together determine whether a reaction is spontaneous; that is, whether it favors products and releases free energy. An exothermic, or heat-releasing, reaction accompanied by an increase in entropy, or an increase in disorder, is definitely spontaneous because both factors are favorable. The combustion of carbon, for example, is exothermic. The entropy also increases as solid carbon is converted to gaseous carbon dioxide. Because both factors are favorable, the reaction must be spontaneous. The reverse reaction—production of carbon and oxygen from carbon dioxide—is nonspontaneous; neither the heat nor the entropy change is favorable.

A reaction will also be spontaneous if a decrease in entropy is offset by a large release of heat. Similarly, an endothermic, or heat-absorbing, reaction will be spontaneous if an entropy increase offsets the heat absorption. For example, heat change and entropy change work in opposition when ice melts. The melting of ice is endothermic but is still spontaneous above 0 °C. At such temperatures, the absorption of heat is more than offset by a favorable entropy change. Why is the freezing of water to ice a nonspontaneous process at 10 °C? (1)

Table 19.1 summarizes how entropy and heat changes affect the spontaneity of chemical reactions. Notice that either of the two variables, but not both, can be unfavorable for a spontaneous process. **Figure 19.20** summarizes the relationship among heat, entropy, and free-energy changes for spontaneous reactions.

A nonspontaneous reaction, one in which the products are not favored, has heat changes, entropy changes, or both working against it. The desired reaction might not be able to overcome a large decrease in entropy and might also be highly endothermic. In that case, both these changes work against the formation of products. On the other hand, a nonspontaneous reaction may be exothermic but involve a decrease in entropy large enough to offset the favorable heat change. Alternatively, an increase in entropy might be too small to overcome an endothermic heat change. **Figure 19.20** also summarizes the relationships for nonspontaneous reactions.

Table 19.1

How Changes in Heat and Entropy Affect Reaction Spontaneity

Heat change	Entropy	Spontaneous reaction?
Decreases (exothermic)	Increases (more disorder in products than in reactants)	Yes
Increases (endothermic)	Increases	Only if unfavorable heat change is offset by favorable entropy change
Decreases (exothermic)	Decreases (less disorder in products than in reactants)	Only if unfavorable entropy change is offset by favorable heat change
Increases (endothermic)	Decreases	No

Answer

(1) The favorable release of heat would be more than offset by the unfavorable decline in entropy.

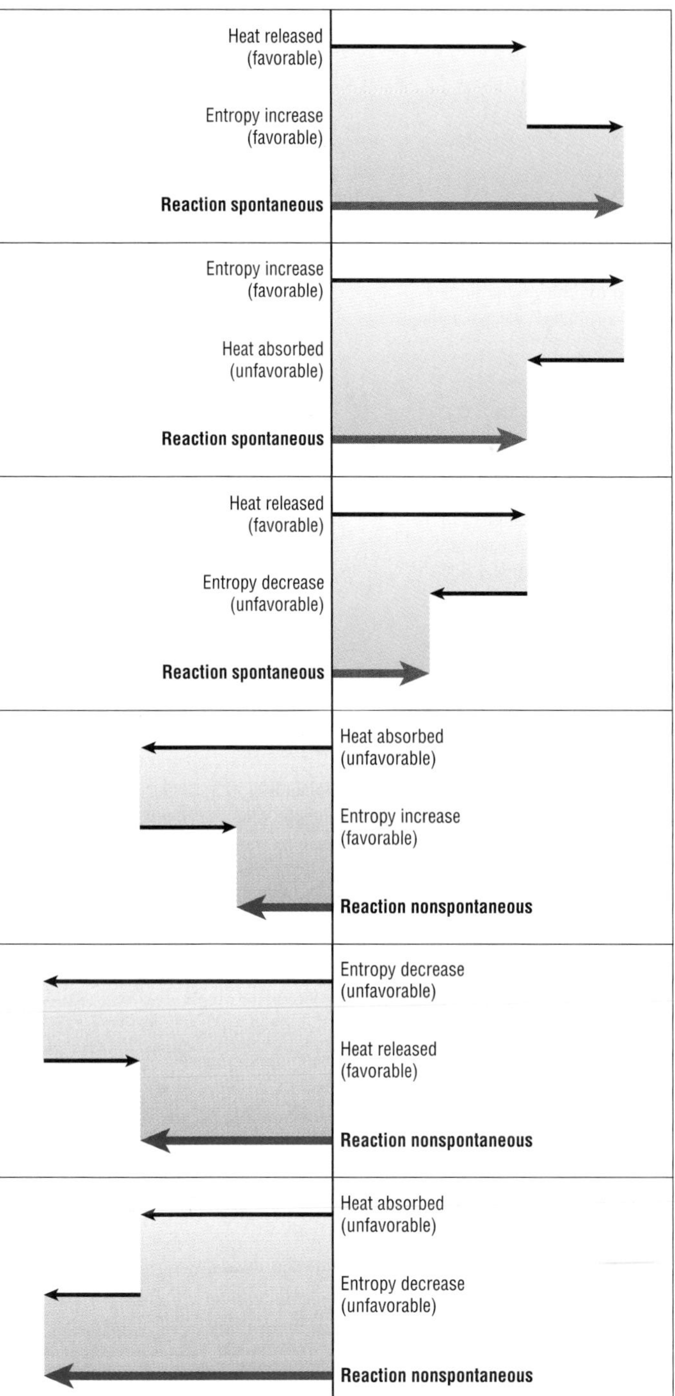

Figure 19.20
Free energy is released in spontaneous reactions. The reaction may be exothermic, the entropy of the system may increase, or both may occur. A reaction may be nonspontaneous because the reaction is endothermic or the entropy of the system decreases, or both may occur.

3 Assess

Evaluate Understanding

Have the students define free energy and explain the relationship between free energy and the efficiency of a machine. (Free energy is the energy available to do work. Free energy is not necessarily used efficiently and in machines some may be lost in friction, heat, etc.)

Ask the students to use an example to explain the relationship between reversible reactions and spontaneous and nonspontaneous reactions? (One reaction, forward or reverse, is always spontaneous and the other is always nonspontaneous).

Have the students summarize in their own words the four relationships between heat content and entropy that determine whether or not a reaction is spontaneous.

19.3

MINI LAB

Does Steel Burn?

SAFETY

Steel wool burns readily. Use proper precautions.

ANALYSIS AND CONCLUSIONS

1. Except for the glowing of a few loose steel fibers on the surface, there are no apparent changes when the tight ball is heated. The loose ball probably glowed. The loose fibers burst into flame almost immediately. Each ball acts as a single large particle of steel that is in contact with atmospheric oxygen only at its relatively small surface. The loose fibers act as many small particles, with larger total surface area, and each small particle is exposed to atmospheric oxygen.
2. $4Fe(s) + 3O_2(g) \rightarrow 2Fe_2O_3(s)$
3. Both reactions involve oxidations of iron. The combustion of the steel wool is faster because the high temperature overcomes the activation-energy barrier and the small particle size exposes more of the iron surface to atmospheric oxygen.
4. Steel wool is a hazard around heat, flames, and sparks because of the danger that the steel wool may ignite.

Reteach

Remind students that no mechanical or chemical process is ever 100% efficient. Discuss the comparative efficiencies of fluorescent and incandescent light bulbs. (With fluorescent bulbs, much less energy is lost as heat.) **To assure that students understand that the terms *spontaneous* and *nonspontaneous* do not refer to reaction rate, discuss what occurs in a compost heap. Although decomposition of organic matter is favored by both enthalpy and entropy changes, the process is slow.**

MINI LAB

Does Steel Burn?

PURPOSE

To determine whether steel will burn.

MATERIALS

- 0000 steel wool pad
- tissue paper
- tongs
- Bunsen burner
- heat-resistant pad
- pencil and paper

PROCEDURE

1. Roll a small piece of steel wool into a very tight, pea-sized ball. Use the tissue paper to protect your fingers from small steel fibers while handling the steel wool.
2. Holding the ball with tongs, heat the steel wool for no longer than 10 seconds in the blue-tip flame of the burner. In this and subsequent steps, place pieces of heated steel wool on the heat-resistant pad to cool. Observe all appropriate safety precautions when working with the burner and the heated materials. Record your observations.
3. Gently roll a second piece of steel wool into a loose, pea-sized ball. Holding the loose ball with the tongs, heat the wool for no longer than 10 seconds in the burner flame. Record your observations.
4. Pull a few individual fibers of steel wool from the pad. Hold one end of the loose fibers with the tongs and heat them for no longer than 10 seconds in the flame of the burner. Again record your observations.

ANALYSIS AND CONCLUSIONS

1. What differences did you observe when the tight ball, the loose ball, and the loose fibers were heated in the flame? Give a reason for any differences you observed.
2. Write the balanced equation for any chemical reaction you may have observed (assume that the steel wool is composed mainly of iron).
3. How do your results differ from those observed in the rusting of an automobile body?
4. Explain why steel wool is a hazard in shops where there are hot plates, open flames, or sparking motors.

section review 19.3

20. Explain what is meant by the following:
    a. entropy
    b. free energy
    c. spontaneous reaction
    d. nonspontaneous reaction
21. What two factors determine whether a reaction is spontaneous?
22. Where does lost free energy typically end up? Does free energy lost as heat ever serve a useful function? Explain.
23. What can change a reaction from nonspontaneous to spontaneous?
24. Suppose the products in a spontaneous process are more ordered than the reactants. Is the entropy change favorable or unfavorable?

 Chem ASAP! **Assessment 19.3** Check your understanding of the important ideas and concepts in Section 19.3.

Answers

SECTION REVIEW 19.3

20. a. Entropy is a measure of a system's disorder.
    b. Free energy is energy available to do work.
    c. A spontaneous reaction occurs naturally.
    d. A nonspontaneous reaction must be forced to occur; it does not happen naturally.
21. the change in heat content, or enthalpy, and the change in a system's entropy
22. As heat; waste heat from chemical reactions in the body maintains body temperature at 37°C. Similarly, waste heat from industrial processes and electrical generation can be used to heat water, buildings, and so on.
23. Depending on the reaction, a change in temperature or pressure.
24. unfavorable

SMALL-SCALE LAB

ENTHALPY AND ENTROPY

SAFETY    

Wear safety glasses and follow the standard safety procedures outlined on page 18.

PURPOSE

To observe and measure energy changes during the formation of a solution and to describe and explain these changes in terms of entropy and enthalpy.

MATERIALS

- alcohol thermometer
- four 1-oz plastic cups
- solid chemicals shown in Figure A
- plastic spoon
- crushed ice
- water

PROCEDURE

Place two level spoonfuls of water in a plastic cup and measure the temperature (T_1) of the water. Add one level spoonful of solid NaCl to the cup, stir gently, and measure the highest or lowest temperature (T_2). Record this information on a separate sheet of paper in a table similar to Figure A. Repeat the experiment twice more, substituting NaCl with NH_4Cl the first time and $CaCl_2$ the second time. Record the temperatures.

Mixture	T_1	T_2	ΔT
a. $NaCl(s) + H_2O(l)$			
b. $NH_4Cl(s) + H_2O(l)$			
c. $CaCl_2(s) + H_2O(l)$			

Figure A

ANALYSIS

Using your experimental data, record the answers to the following questions in or below your data table.

1. Calculate ΔT for each mixture. $\Delta T = T_2 - T_1$.
2. An exothermic process gives off heat (warms up). An endothermic process absorbs heat (cools off). Which solutions are endothermic and which are exothermic? What is the sign of ΔH in each case?
3. Which solution(s) had little or no change in temperature?
4. When sodium chloride dissolves in water, the ions dissociate.

$$NaCl(s) \longrightarrow Na^+(aq) + Cl^-(aq)$$

Write two ionic equations, similar to the one above, that describe how NH_4Cl and $CaCl_2$ each dissociate as they dissolve in water. Include heat as a reactant or product in each equation.

5. Which solids in this experiment rapidly dissolved in water? Does the dissolving process usually occur with an increase or decrease in entropy? What is the sign of ΔS in each case?
6. Consider the equation $\Delta G = \Delta H - T\Delta S$. For each dissolving process, substitute the sign of ΔS and ΔH into the equation and determine the sign of ΔG. For which process might ΔG be either positive or negative?

YOU'RE THE CHEMIST

The following small-scale activities allow you to develop your own procedures and analyze the results.

1. **Analyze It!** Mix a tablespoon of ice (solid water) with a tablespoon of solid NaCl. Stir gently, then measure and record the lowest temperature reached. Compare this mixture with the mixture of solid NaCl and liquid water. What is the difference in temperature?
2. **Analyze It!** Is the process of melting ice exothermic or endothermic? Does this explain the difference in temperature in Activity 1? Explain.
3. **Design It!** What effect does mixing other salts with crushed ice have on temperature? Try substituting different salts in Activity 1.
4. **Design It!** What effect does dissolving other salts in liquid water have on temperature? Try dissolving the following salts in liquid water: potassium chloride (KCl), sodium hydrogen carbonate ($NaHCO_3$), sodium carbonate (Na_2CO_3), and sodium phosphate (Na_3PO_4).

SMALL-SCALE LAB

ENTHALPY AND ENTROPY

OVERVIEW

Procedure Time: 40 minutes
Students investigate the energy changes associated with chemical reactions and relate them to exothermic and endothermic reactions.

MATERIALS

Solids: $CaCl_2$, NH_4Cl, NaCl, ice, KCl, $NaHCO_3$, Na_2CO_3, and Na_3PO_4 (optional)

SAFETY

Use small alcohol-based thermometers. If students are careful, they can use the thermometers to gently stir the contents of the plastic cup, Remind students of the hazard posed by broken glass. **CAUTION!** Never use mercury thermometers. They could break, releasing their poisonous contents. Look for a research lab that will trade alcohol thermometers for mercury thermometers.

YOU'RE THE CHEMIST

1. The temperature can drop as much as 15 °C.
2. The melting of ice is endothermic. Endothermic processes absorb heat and cool the environment.
3. Both $CaCl_2$ and NH_4Cl depress the freezing point and cause a drop in temperature.
4. Salts such as KCl, $NaHCO_3$, Na_2CO_3, and Na_3PO_4 dissolve endothermically, that is, with little or no change in temperature.

ANALYSIS

1. Sample data for ΔT:
   a. 0 °C; **b.** −16 °C; **c.** +32 °C
2. $NH_4Cl + H_2O$ is endothermic. ΔH is positive. $CaCl_2 + H_2O$ is exothermic. ΔH is negative.
3. $NaCl + H_2O(l)$ ΔH is close to 0.
4. $heat + NH_4Cl(s) \rightarrow NH_4^+(aq) + Cl^-(aq)$
   $CaCl_2(s) \rightarrow Ca^{2+}(aq) + 2Cl^-(aq) + heat$
5. All of the solids dissolved rapidly. Entropy usually increases during solvation. ΔS is positive in each case.
6. $\Delta G = \Delta H - T\Delta S$
   $\Delta G = (0) - T(+)$ for $NaCl(s)$; ΔG is −.
   $\Delta G = (+) - T(+)$ for $NH_4Cl(s)$; ΔG is + or −.
   $\Delta G = (-) - T(+)$ for $CaCl_2(s)$; ΔG is −.

19.4

1 Engage

Use the Visual

Have students study the photograph and read the text that opens the section. Remind students about the relationship between heat energy and temperature change. The heat required to raise the temperature of a substance is called the *heat capacity.* Write the thermochemical equation for the combustion of propane ($\Delta H = -2044$ kJ), and point out that in some reactions, such as the combustion of propane, heat is a product, while in others heat is a reactant. The amount of heat released by a reaction depends on the particular reactants involved. Ask:

- **How do chemists calculate the amount of energy that a chemical reaction will release?** (The heat released during a chemical or physical change can be measured using a calorimeter, a device which measures temperature changes under a set of controlled conditions. The heat released by the reaction is then calculated from the temperature change caused by the reaction.)

Check Prior Knowledge

Ask students to define the concept of entropy and enthalpy. (Entropy is a quantitative measure of the disorder of a system and enthalpy is a measure of the heat content of the system.)**Explain that in this section a new quantitative measure, *G*, will be introduced called the Gibbs free energy.**

section 19.4

objectives

- Calculate the standard entropy changes that accompany chemical and physical processes
- Calculate the free-energy changes that accompany chemical and physical processes

key terms

- standard entropy (S^0)
- Gibbs free-energy change (ΔG)

CALCULATING ENTROPY AND FREE ENERGY

Energy, in the form of heat, is required to cook a hot dog. Would you want to use a welding torch to cook your hot dog? A reaction that releases the same amount of energy over a longer period of time, such as the combustion of charcoal in a barbeque grill, might be more appropriate. **How do chemists calculate the amount of energy that a chemical reaction will release?**

Entropy Calculations

Entropy is a quantitative measure of the disorder of a system. The symbol for the entropy of a substance is *S*, with units of J/K. When entropy (*S*) is given per mole of substance, it has units of J/(K × mol), which is also written as J/K·mol. The **standard entropy** of a liquid or solid substance at 25 °C is designated S^0. The pressure at S^0 for gaseous substances is 101.3 kPa. The theoretical entropy of a perfect crystal at 0 K is zero. Other substances have positive entropies, even at absolute zero. **Table 19.2** lists standard entropies for some common substances.

Standard entropy change (ΔS^0) can be calculated from standard entropies using the following equation.

$$\Delta S^0 = S^0(\text{products}) - S^0(\text{reactants})$$

Table 19.2

Standard Entropies (S^0) at 25 °C and 101.3 kPa			
Substance	S^0 (J/K·mol)	Substance	S^0 (J/K·mol)
$Al_2O_3(s)$	50.99	$HCl(g)$	186.7
$Br_2(g)$	245.3	$H_2S(g)$	205.6
$Br_2(l)$	152.3	$I_2(g)$	260.6
C(*s*, diamond)	2.439	$I_2(s)$	117
C(*s*, graphite)	5.694	$N_2(g)$	191.5
$CH_4(g)$	186.2	$NH_3(g)$	192.5
$CO(g)$	197.9	$NO(g)$	210.6
$CO_2(g)$	213.6	$NO_2(g)$	240.5
$CaCO_3(s)$	88.7	$Na_2CO_3(s)$	136
$CaO(s)$	39.75	$NaCl(s)$	72.4
$Cl_2(g)$	223.0	$O_2(g)$	205.0
$F_2(g)$	203	$O_3(g)$	238
$Fe(s)$	27.2	P(*s*, white)	44.4
$Fe_2O_3(s)$	90.0	P(*s*, red)	29
$H_2(g)$	130.6	S(*s*, rhombic)	31.9
$H_2O(g)$	188.7	S(*s*, monoclinic)	32.6
$H_2O(l)$	69.94	$SO_2(g)$	248.5
$H_2O_2(l)$	92	$SO_3(g)$	256.2

STUDENT RESOURCES

From the Teacher's Resource Package, use:

- Section Review 19.4, Ch. 19 Practice Problems and Quizzes from the Review Module (Ch. 17–20)

TECHNOLOGY RESOURCES

Relevant technology resources include:

- Chem ASAP! CD-ROM
- ResourcePro CD-ROM

BALANCING EQUATIONS

You will find that many of the problems in this chapter rely on properly balanced chemical equations. This page offers a review of the skills needed to balance these equations. For a more thorough review, see Chapter 8. Balancing an equation is the process of inserting coefficients into a skeleton equation so that each side of the equation has the same number of atoms of each element. A correctly balanced equation is consistent with the principle that atoms are neither created nor destroyed in a chemical reaction, they are just rearranged into new groupings.

Balancing equations is largely a matter of logical thinking combined with trial and error.

1. Write a skeleton equation in the form
   reactants ⟶ products
   using chemical formulas separated by plus signs. Leave space for coefficients.
2. Choose one element or polyatomic ion to balance. It is best to start with elements that appear only once on each side of the equation. Adjust the coefficients as necessary so the number of atoms of the chosen element is the same on both sides of the equation. When no coefficient is written, it is assumed to be 1.
3. Repeat Step 2, choosing different elements until all the elements are balanced.
4. Check your work by making sure each element is balanced. This step is very important because the elements that you balance in the early steps can become "unbalanced" as you continue your work. Then make sure the coefficients are in the lowest possible ratio; that is, they should have a greatest common factor of 1.

Example 1

Balance the equation $Na(s) + Cl_2(g) \longrightarrow NaCl(s)$.

Sodium is balanced (1 atom on each side), but chlorine is not (2 on the left, 1 on the right). Balance the chlorine:

$$Na(s) + Cl_2(g) \longrightarrow \mathbf{2}NaCl(s)$$

Now chlorine is balanced (2 atoms on each side), but sodium is not (1 on the left, 2 on the right). Balance the sodium:

$$\mathbf{2}Na(s) + Cl_2(g) \longrightarrow 2NaCl(s)$$

Check:
Sodium is balanced (2 atoms on each side).
Chlorine is balanced (2 atoms on each side).

The equation is $2Na(s) + Cl_2(g) \longrightarrow 2NaCl(s)$.

Example 2

Balance the equation

$$NaOH(aq) + H_2SO_4(aq) \longrightarrow Na_2SO_4(aq) + H_2O(l)$$

Note that the polyatomic ion SO_4 appears on each side of the equation, so count it as a single unit. Balance the sodium:

$$\mathbf{2}NaOH(aq) + H_2SO_4(aq) \longrightarrow Na_2SO_4(aq) + H_2O(l)$$

Oxygen is not balanced (2 on the left, 1 on the right, not counting the SO_4). Balance the oxygen:

$$2NaOH(aq) + H_2SO_4(aq) \longrightarrow Na_2SO_4(aq) + \mathbf{2}H_2O(l)$$

The equation appears to be balanced. Check:
Na is balanced (2 atoms on each side).
O is balanced (2 atoms on each side).
H is balanced (4 atoms on each side).
SO_4 is balanced (1 ion on each side).

Practice Problems

Practice balancing chemical equations by writing a balanced equation for each reaction.

A. $Zn(s) + HCl(aq) \longrightarrow ZnCl_2(aq) + H_2(g)$

B. $CaCO_3(s) + HCl(aq) \longrightarrow H_2CO_3(aq) + CaCl_2(aq)$

C. $PbCl_2(aq) + K_2SO_4(aq) \longrightarrow PbSO_4(s) + KCl(aq)$

D. $C_4H_{10}(g) + O_2(g) \longrightarrow CO_2(g) + H_2O(g)$

E. Solid iron metal reacts with oxygen gas to produce solid iron(III) oxide (rust).

F. Solid tungsten(VI) oxide reacts with hydrogen gas to produce solid tungsten metal and water vapor.

CHEMath

Teaching Tips

The ability to write a balanced chemical equation is a necessary prerequisite to writing the correct equilibrium expression for a reaction. Some students may never have mastered this skill in earlier chapters. A quick review is beneficial at this stage. You can reinforce these fundamentals by beginning each of the problems you discuss in class with an unbalanced equation and having students balance it. Reassure students that, when using the trial-and-error method to balance equations, it is okay to make mistakes before finding the right coefficients to balance an equation. Encourage students to check their work by counting the number of atoms on both sides of the equation. Point out that the balanced equation shows the relative numbers of reactant and product atoms, molecules, and/or ions involved in a reaction.

Answers to Practice Problems

A. $Zn(s) + 2HCl(aq) \rightarrow ZnCl_2(aq) + H_2(g)$

B. $CaCO_3(s) + 2HCl(aq) \rightarrow H_2CO_3(aq) + CaCl_2(aq)$

C. $PbCl_2(aq) + K_2SO_4(aq) \rightarrow PbSO_4(s) + 2KCl(aq)$

D. $2C_4H_{10}(l) + 13O_2(g) \rightarrow 8CO_2(g) + 10H_2O(g)$

E. $4Fe(s) + 3O_2(g) \rightarrow 2Fe_2O_3(s)$

F. $WO_3(s) + 3H_2(g) \rightarrow W(s) + 3H_2O(g)$

2 Teach

Cooperative Learning

Have groups of three or four students carry out entropy calculations using Table 19.2. Each student should choose a reaction and write a balanced equation for it using the substances listed in the table as reactants and products. All of the students in the group can work together to calculate the entropy changes.

Discuss

Explain that in a game of chance, the probable event is one that can happen in many ways. The same is true of order and disorder. An ordered arrangement conforms to some prechosen requirements; all other arrangements are called disordered. Because far more arrangements are possible in the latter category, disorder is the natural condition of things. Explain to students that they will learn to calculate changes in entropy for chemical reactions as "the entropy of the products minus the entropy of the reactants."

Practice Problems Plus

Additional Practice Problem

The Chem ASAP! CD-ROM contains this problem related to Sample Problem 19-5: The standard entropies are given below for some substances at 25 °C.

$Al_2O_3(s)$ $S° = 50.99$ J/K·mol
$Al(s)$ $S° = 28.0$ J/K·mol
$O_2(g)$ $S° = 205.0$ J/K·mol

Calculate $\Delta S°$ for this reaction.

$Al_2O_3(s) \rightarrow 2Al(s) + 3/2O_2(g)$

(312.5 J/K·mol)

Sample Problem 19-5

Calculate the standard entropy change (ΔS^0) that occurs when 1 mol $H_2O(g)$ at 25 °C and 101.3 kPa condenses to 1 mol $H_2O(l)$ at the same temperature.

1. ***ANALYZE*** ***List the knowns and the unknown.***

Knowns (from ***Table 19.2****):*

- $H_2O(g)$ $S^0 = 188.7$ J/K·mol
- $H_2O(l)$ $S^0 = 69.94$ J/K·mol

Unknown:

- ΔS^0(condensation) = ?

$$\Delta S^0 = S^0(\text{products}) - S^0(\text{reactants})$$

The change in physical state is $H_2O(g) \longrightarrow H_2O(l)$. Substitute the known values into the equation and solve for the unknown.

2. ***CALCULATE*** ***Solve for the unknown.***

Only 1 mol of each reactant and product is involved; therefore,

$$\Delta S^0 = 69.94 \text{ J/K·mol} - 188.7 \text{ J/K·mol} = -118.8 \text{ J/K·mol}$$

The negative sign indicates that entropy decreases.

3. ***EVALUATE*** ***Does the result make sense?***

As would be expected, the condensation of 1 mol of water vapor to liquid water at 25 °C results in a large decrease in entropy. The result is expressed in the proper units.

Practice Problem

25. The standard entropies for some substances at 25 °C are

$KBrO_3(s)$ $S^0 = 149.2$ J/K·mol
$KBr(s)$ $S^0 = 96.4$ J/K·mol
$O_2(g)$ $S^0 = 205.0$ J/K·mol

Calculate ΔS^0 for the reaction

$KBrO_3(s) \longrightarrow KBr(s) + \frac{3}{2}O_2(g)$

Chem ASAP!

Problem-Solving 25

Solve Problem 25 with the help of an interactive guided tutorial.

Sample Problem 19-6

$NO(g)$ reacts with O_2 to form $NO_2(g)$. What is the standard change in entropy for this reaction when reactants and products are in the specified physical states at 101.3 kPa and 25 °C?

$$NO(g) + O_2(g) \longrightarrow NO_2(g)$$

1. ***ANALYZE*** ***List the knowns and the unknown.***

Knowns (from ***Table 19.2****):*

- reactants, $NO(g)$ and $O_2(g)$
- product, NO_2
- $NO(g)$ $S^0 = 210.6$ J/K·mol
- $O_2(g)$ $S^0 = 205.0$ J/K·mol
- $NO_2(g)$ $S^0 = 240.5$ J/K·mol
- $\Delta S^0 = S^0(\text{products}) - S^0(\text{reactants})$

Unknown:

- ΔS^0(reaction) = ?

First, balance the chemical equation. Obtain S^0 for the reactants, then substitute the known S^0 values into the ΔS^0 equation and solve.

Sample Problem 19-6 (cont.)

2. ***CALCULATE*** ***Solve for the unknown.***

Balancing the chemical equation yields:

$$2NO(g) + O_2(g) \longrightarrow 2NO_2(g)$$

Sum the S^0 of the reactants, taking into account the number of moles.

$$S^0 \text{ for } NO_2 = 2\,\cancel{\text{mol NO}} \times 210.6\ \text{J/K}\cdot\cancel{\text{mol NO}} = 421.2\ \text{J/K}$$
$$S^0 \text{ for } O_2 = 1\,\cancel{\text{mol } O_2} \times 205.0\ \text{J/K}\cdot\cancel{\text{mol } O_2} = 205.0\ \text{J/K}$$
$$S^0(\text{reactants}) = 626.2\ \text{J/K}$$

Calculate the S^0 of the products.

$$S^0 \text{ for } NO_2 = 2\,\cancel{\text{mol } NO_2} \times 240.5\ \text{J/K}\cdot\cancel{\text{mol } NO_2} = 481.0\ \text{J/K}$$

Calculate the change in entropy by subtracting.

$$\Delta S^0 = S^0(\text{products}) - S^0(\text{reactants})$$
$$= (481.0\ \text{J/K}) - (626.2\ \text{J/K}) = -145.2\ \text{J/K}$$

3. ***EVALUATE*** ***Does the result make sense?***

The entropy decreases as expected because three reactant molecules are reorganized to only two product molecules.

Practice Problem

26. Based on the data in **Table 19.2,** calculate ΔS^0 for converting the monoclinic form of sulfur to the rhombic form.

Figure 19.21
What characteristics of an explosion would lead you to think that an increase in entropy is occurring? ❶

Free-Energy Calculations

You may recall that in every spontaneous process, some energy becomes available to do work. This energy, called the **Gibbs free-energy change** (ΔG), is the maximum amount of energy that can be coupled to another process to do useful work. The change in Gibbs free energy is related to the change in entropy (ΔS) and the change in enthalpy (ΔH) of the system by the free-energy equation

$$\Delta G = \Delta H - T\Delta S,$$

where temperature (T) is in kelvins.

Answer

❶ The ordered arrangement of the exploding item becomes extremely disordered.

19.4

Discuss

Free energy refers to more than just the enthalpy change, ΔH, that occurs in a chemical reaction. Free energy combines ΔH with the change in entropy, ΔS. ΔH and ΔS are usually assumed to be constant for the same reactions at different temperatures. Therefore, the contribution to the free energy from the $T\Delta S$ term comes mostly from the temperature itself.

Chemistry and Physics

Generally, the spontaneity of a reaction is increased by heating the reactants to increase their temperature. In certain reactions however, other types of energy are more effective than heat. The aurora borealis, or northern lights, provide a good example of the importance of using the right kind of energy to control the outcome of a chemical reaction. During periods of intense solar activity, bombardment by charged particles causes a small fraction of nitrogen molecules in the upper atmosphere to dissociate, yielding excited nitrogen atoms. The green light of the aurora borealis results from the fluorescence of molecular nitrogen, which is produced when two excited nitrogen atoms recombine. Scientists who have heated samples of air in the laboratory have found that no amount of thermal energy is sufficient to reproduce the reaction that causes the green emission.

Practice Problems Plus

Related Chapter Review Problem
Chapter Review problem 77 is related to Sample Problem 19-6.

TEACHER DEMO

Demonstrate a spontaneous reaction with a negative ΔH and a positive ΔS. Prepare a 0.1 *M* solution of $KMnO_4$ in 0.1 *M* HCl. **CAUTION!** Potassium permanganate is a powerful oxidizing agent and a strong skin irritant. Transfer 100 mL of the $KMnO_4$ solution to a 1000-mL beaker. Record its temperature. With a graduated cylinder, measure 20 mL of 30% H_2O_2. (Use a more dilute solution for a less vigorous reaction.) Record its temperature. Add the peroxide to the permanganate and cover the beaker with a watch glass. Record the temperature once the most vigorous reaction has subsided. Test the gas produced with a glowing splint. Note that the reaction is spontaneous so ΔG must be negative. The entropy change must be positive because a gas is produced. From the Gibbs-Helmholtz equation, $\Delta G = \Delta H - T\Delta S$, and the temperature change, it is obvious that ΔH must be negative at room temperature. The reaction is $2MnO_4^-(aq) + 5H_2O_2(l) + 6H^+(aq) \rightarrow 2Mn^{2+}(aq) + 8H_2O(l) + 5O_2(g)$
The manganese ion is considered hazardous waste and must be disposed of in compliance with local regulations.

GEOLOGY

Weathering of Rocks

The weathering of rock usually occurs so slowly that the process is imperceptible over many human lifetimes. In spite of its low rate, the weathering

of rocks has shaped, and continues to shape, Earth's surface. Chemical and physical processes induced by water produce much of the weathering. Some mountains in the United States show the effects of weathering over long periods of time. The relatively new Rocky Mountains (100 million years) are high and craggy. By contrast, the old southern Appalachians (345 million years) are low and smooth because of an additional 245 million years of weathering!

All spontaneous processes release free energy. The numerical value of ΔG is negative in spontaneous processes because the system loses free energy. Nonspontaneous processes require that work be expended to make them go forward at the specified conditions. Therefore, the numerical value of ΔG is positive for a nonspontaneous process.

You can calculate the standard free-energy change (ΔG^0) to determine whether a chemical reaction is spontaneous when ΔH^0 and ΔS^0 are known. The following equation is used to calculate the standard free-energy change.

$$\Delta G^0 = \Delta H^0 - T\Delta S^0$$

Alternatively, you can determine ΔH^0 or ΔS^0 for a reaction if the other two quantities in the equation are known. **Table 19.3** lists values of ΔG^0 for some common reactions.

When ΔH^0 and ΔS^0 are unknown, you can use ΔG_f^0, the standard free-energy change for the formation of substances from their elements, to calculate ΔG^0 for a given reaction.

$$\Delta G^0 = \Delta G_f^0(\text{products}) - \Delta G_f^0(\text{reactants})$$

This is similar to how standard heats of formation (ΔH_f^0) are used to calculate heats of reaction. For what other kind of change have you calculated the
1 value in a similar way?

Table 19.4 lists values of ΔG_f^0 for some spontaneous processes. Note that $\Delta G_f^0 = 0$ for elemental substances. If you calculate ΔG^0 to determine whether a reaction is spontaneous, the result applies only to reactants and products in their standard states. Also, a reaction that is nonspontaneous under one set of conditions may be spontaneous under another set of conditions. Why does bromine gas (Br_2) have a greater free energy than
2 bromine liquid?

Table 19.3

Free Energy Values (ΔG^0) for Some Spontaneous Processes at 25 °C

Reaction	Free energy	
	Kilocalories	**Kilojoules**
$H_2(g) + Cl_2(g) \longrightarrow 2HCl(g)$ Hydrogen, Chlorine → Hydrogen chloride	−45.6	−191
$S(s) + O_2(g) \longrightarrow SO_2(g)$ Sulfur, Oxygen → Sulfur dioxide	−71.7	−300
$2N_2O_5(s) \longrightarrow 4NO_2(g) + O_2(g)$ Dinitrogen pentoxide → Nitrogen dioxide, Oxygen	−7.2	−30
$C_6H_{12}O_6(s) + 6O_2(g) \longrightarrow 6CO_2(g) + 6H_2O(l)$ Glucose, Oxygen → Carbon dioxide, Water	−686.0	−2868

MEETING DIVERSE NEEDS

LEP Terms such as *entropy, enthalpy, exothermic,* and *exergonic* will be confusing for students with limited English proficiency. Have them equate a descriptive phrase, such as molecular disorder or heat content, with a confusing term. Substitute the phrase when needed to help them understand the concepts. You may want to use both terms.

Answers

1 The standard entropy change (ΔS_f^0) is calculated in a similar way.

2 The standard free energies of formation of elements in their most stable states are arbitrarily assigned a value of zero. Bromine is most stable as a liquid.

Cooperative Learning

Divide students into groups of six, and subdivide each group into subgroups of three each. Assign each group the problem of deciding whether a reaction of your choosing is spontaneous (Base your choice of reactions on the data available in Tables 11.6, 19.2, and 19.4.) One subgroup should use the $\Delta G_f°$ method and the values in Table 19.4 in its calculation of ΔG. The other subgroup should use the values of $\Delta H_f°$ and $\Delta S°$ from Tables 11.6 and 19.2.

Table 19.4

Standard Gibbs Free Energies of Formation (ΔG_f^0) at 25 °C and 101.3 kPa

Substance	ΔG_f^0 (kJ/mol)	Substance	ΔG_f^0 (kJ/mol)
$Al_2O_3(s)$	−1576.4	$HCl(g)$	−95.27
$Br_2(g)$	3.14	$H_2S(g)$	−33.02
$Br_2(l)$	0.0	$I_2(g)$	19.4
C(s, diamond)	2.866	$I_2(s)$	0.0
C(s, graphite)	0.0	$N_2(g)$	0.0
$CH_4(g)$	−50.79	$NH_3(g)$	−16.64
$CO(g)$	−137.3	$NO(g)$	86.69
$CO_2(g)$	−394.4	$NO_2(g)$	51.84
$CaCO_3(s)$	−1127.7	$Na_2CO_3(s)$	−1048
$CaO(s)$	−604.2	$NaCl(s)$	−384.03
$Cl_2(g)$	0.0	$O_2(g)$	0.0
$F_2(g)$	0.0	$O_3(g)$	163.4
Fe(s)	0.0	P(s, white)	0.0
$Fe_2O_3(s)$	−741.0	P(s, red)	−14
$H_2(g)$	0.0	S(s, rhombic)	0.0
$H_2O(g)$	−228.6	S(s, monoclinic)	0.096
$H_2O(l)$	−237.2	$SO_2(g)$	−300.4
$H_2O_2(l)$	−114.0	$SO_3(g)$	−370.4

It may be helpful to consider a real chemical reaction to see the effect of temperature on the spontaneity of a reaction. Solid calcium carbonate decomposes to give calcium oxide and carbon dioxide.

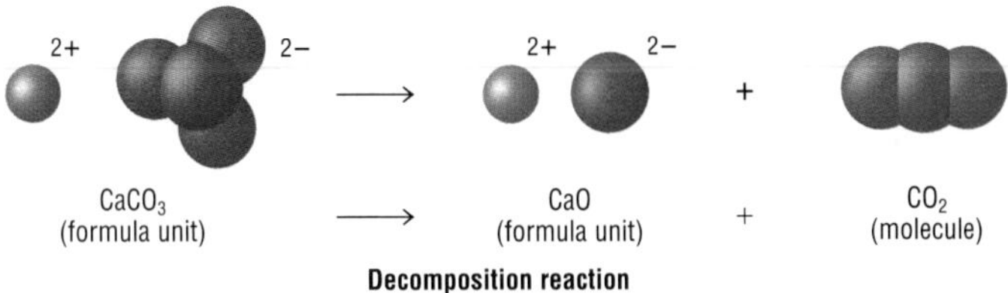

$CaCO_3$ (formula unit) ⟶ CaO (formula unit) + CO_2 (molecule)

Decomposition reaction

In this reaction, the entropy increases because one of the products formed from the solid reactant is a gas. The entropy increase is not great enough, however, for the reaction to be spontaneous at ordinary temperatures because the reaction is endothermic. The enthalpy of the reactants is lower than that of the products. The effect of an entropy increase, however, is magnified as the temperature increases. At temperatures above 850 °C, the favorable entropy–temperature term $T\Delta S^0$ outweighs the unfavorable enthalpy term ΔH^0. Under these conditions, the identical reaction then becomes spontaneous.

MEETING DIVERSE NEEDS

Gifted Have students consider the combustion of glucose to produce carbon dioxide and water. Point out that the reverse of this reaction is the overall reaction for photosynthesis, the process by which plants make glucose using carbon dioxide and water. Ask the students to consider the magnitude of the free energy value for the photosynthesis reaction, +2868 kJ/mol. They should realize that the reaction is highly nonspontaneous under ordinary conditions. Have the students research the process of photosynthesis and prepare a report explaining how it occurs. (Bright light provides the large amount of energy needed for the highly endothermic reaction, which also requires chlorophyll, a molecule that can capture light energy.)

CHEMISTRY AND THE ENVIRONMENT

There is evidence that Earth is being warmed because of the greenhouse effect, which is largely due to the release of carbon dioxide (CO_2) through combustion. One aspect of the greenhouse effect that relates to the topic of spontaneity is the increase in the quantity of atmospheric carbon dioxide gas that would occur if the oceans became warmer. Carbon dioxide is quite soluble in water. The oceans currently contain a great deal of dissolved CO_2. If the oceans become warmer, more carbon dioxide would be released into the atmosphere, because CO_2, like all other gases, is less soluble in warm ocean water than in cool ocean water. The process whereby CO_2 dissolves in ocean water would become nonspontaneous, and the reverse process, that of the release of CO_2 gas, would become spontaneous. The increased temperature would release more CO_2, which would further increase temperature, and so on.

Practice Problems Plus

Related Chapter Review Problem

Chapter Review problem 60 is related to Sample Problem 19-7.

Additional Practice Problem

The Chem ASAP! CD-ROM contains the following problem: When gaseous oxygen reacts with solid iron to form rust, Fe_2O_3, ΔG_f° equals -741.0 kJ/mol. The chemical equation is

$4Fe(s) + 3O_2(g) \rightarrow 2Fe_2O_3(s)$

Using the S° values of Table 19.2, calculate the ΔH_f° for the formation of rust.

(-822 kJ/mol)

Sample Problem 19-7

Using values for ΔH and S^0, determine whether this reaction is spontaneous at 25 °C.

$$C(s, \text{graphite}) + O_2(g) \longrightarrow CO_2(g)$$

1. ***ANALYZE*** ***List the knowns and the unknown.***

*Knowns at 25 °C (from **Tables 11.6** and **19.2**):*

Substance	ΔH_f^0 (kJ/mol)	S^0 (J/K·mol)
C(s, graphite)	0.0	5.69
$O_2(g)$	0.0	205
$CO_2(g)$	−393.5	214

Unknown:

- $\Delta G^0(\text{reaction}) = ?$

2. ***CALCULATE*** ***Solve for the unknown.***

The S^0 values listed in **Table 19.2** are in J/K·mol. Convert them to kJ/K·mol to match the units of ΔH_f^0.

$$C(s, \text{graphite})\ S^0 = 5.7\,\frac{\cancel{J}}{K\cdot mol} \times \frac{1\ kJ}{1000\,\cancel{J}} = 0.0057\ kJ/K\cdot mol$$

$$O_2(g)\ S^0 = 205\,\frac{\cancel{J}}{K\cdot mol} \times \frac{1\ kJ}{1000\,\cancel{J}} = 0.205\ kJ/K\cdot mol$$

$$CO_2(g)\ S^0 = 214\,\frac{\cancel{J}}{K\cdot mol} \times \frac{1\ kJ}{1000\,\cancel{J}} = 0.214\ kJ/K\cdot mol$$

Calculate the change in entropy (ΔS^0).

$$\Delta S^0 = S^0(\text{products}) - S^0(\text{reactants})$$

$$= 0.214\,\frac{kJ}{K\cdot mol} - (0.0057 + 0.205)\,\frac{kJ}{K\cdot mol} = 0.003\,\frac{kJ}{K\cdot mol}$$

Calculate the standard enthalpy change for the reaction in a similar way.

$$\Delta H^0 = \Delta H_f^0(\text{products}) - \Delta H_f^0(\text{reactants})$$

$$\Delta H^0 = -393.5\ kJ/mol - (0.0\ kJ/mol + 0.0\ kJ/mol)$$

$$= -393.5\ kJ/mol$$

Calculate the value of ΔG^0 from the changes in enthalpy and the changes in entropy.

$$\Delta G^0 = \Delta H^0 - T\Delta S^0$$

where $T = 273.15 + 25\ °C = 298.15\ K$

$$\Delta G^0 = -393.5\ kJ/mol - \left(298.15\,\cancel{K} \times 0.003\,\frac{kJ}{\cancel{K}\cdot mol}\right)$$

$$= -394.4\ kJ/mol$$

3. ***EVALUATE*** ***Does the result make sense?***

The reaction is spontaneous due to the decrease in ΔH^0 and the increase in entropy when solid graphite is converted to gaseous CO_2. In other words, both enthalpy and entropy changes work in favor of reaction spontaneity.

Practice Problem

27. When gaseous nitrogen and hydrogen are converted to gaseous ammonia, $\Delta G^0 = -16.64$ kJ/mol. The chemical equation is $N_2(g) + 3H_2(g) \longrightarrow 2NH_3(g)$ Using the S^0 values in **Table 19.2,** calculate the ΔH_f^0 for the formation of ammonia.

Chem ASAP!

Problem-Solving 27

Solve Problem 27 with the help of an interactive guided tutorial.

Sample Problem 19-8

Using the ΔG_f^0 for the reactants and products, determine whether the reaction in Sample Problem 19-7 is spontaneous.

1. *ANALYZE* *List the knowns and the unknown.*

*Knowns (from **Table 19.4**):*
- C(*s*, graphite) $\Delta G_f^0 = 0$ kJ/mol
- $O_2(g)$ $\Delta G_f^0 = 0$ kJ/mol
- $CO_2(g)$ $\Delta G_f^0 = -394.4$ kJ/mol
- $\Delta G^0 = \Delta G_f^0(\text{products}) - \Delta G_f^0(\text{reactants})$

Unknown:
- $\Delta G_f^0(\text{reaction}) = ?$

The chemical equation is C(*s*, graphite) + $O_2(g) \longrightarrow CO_2(g)$.

2. *CALCULATE* *Solve for the unknown.*

Sum the ΔG_f^0 of the reactants and sum the ΔG_f^0 of the products.

$$\Delta G_f^0(\text{reactants}) = 0.0 \text{ kJ/mol}$$

$$\Delta G_f^0(\text{products}) = -394.4 \text{ kJ/mol}$$

The ΔG_f^0 for the reaction is the difference between the ΔG_f^0 of the products and the ΔG_f^0 of the reactants.

$$\Delta G^0 = \Delta G_f^0(\text{products}) - \Delta G_f^0(\text{reactants})$$

$$\Delta G^0 = -394.4 \text{ kJ/mol} - 0.0 \text{ kJ/mol}$$

$$= -394.4 \text{ kJ/mol}$$

3. *EVALUATE* *Does the result make sense?*

The results obtained in this problem and Sample Problem 19-7 are identical, as they should be. The reaction is spontaneous, with a large release of free energy. This is consistent with experience—charcoal, which is primarily carbon, does burn readily.

Practice Problems

28. Use the data in **Table 19.4** to calculate the standard free-energy change for each reaction. Is the reaction spontaneous or nonspontaneous?

a. $Cl_2(g) + H_2O(g) \longrightarrow 2HCl(g) + \frac{1}{2}O_2(g)$

b. $4NH_3(g) + 7O_2(g) \longrightarrow 4NO_2(g) + 6H_2O(g)$

29. Use the data in **Table 19.4** to calculate the standard free-energy change for each reaction. Is the reaction spontaneous or nonspontaneous?

a. $2CO_2(g) \longrightarrow 2CO(g) + O_2(g)$

b. $H_2S(g) \longrightarrow H_2(g) + S(s, \text{rhombic})$

Chem ASAP!

Problem-Solving 29
Solve Problem 29 with the help of an interactive guided tutorial.

section review 19.4

30. Calculate the change in entropy for these reactions.

a. $CaCO_3(s) \longrightarrow CaO(s) + CO_2(g)$

b. $2H_2(g) + O_2(g) \longrightarrow 2H_2O(l)$

c. $H_2(g) + Cl_2(g) \longrightarrow 2HCl(g)$

d. $CH_4(g) + 2O_2(g) \longrightarrow CO_2(g) + 2H_2O(g)$

31. How does the magnitude of the standard entropy of an element in the gaseous state compare with the standard entropy of the same element in the liquid state?

32. Determine whether the following reactions are spontaneous. Assume all substances are at 25 °C and 101.3 kPa.

a. $2Na(s) + Cl_2(g) \longrightarrow 2NaCl(s)$

b. $4Al(s) + 3O_2(g) \longrightarrow 2Al_2O_3(s)$

Chem ASAP! **Assessment 19.4** Check your understanding of the important ideas and concepts in Section 19.4.

Practice Problems Plus

Related Chapter Review Problem

Chapter Review problem 66 is related to Sample Problem 19-8.

Additional Practice Problem

The Chem ASAP! CD-ROM contains the following problem: Use the data Table 19.4 to calculate the standard free energy change for the production of rust and tell whether the reaction is spontaneous or nonspontaneous.
$4Fe(s) + 3O_2(g) \rightarrow 2Fe_2O_3(s)$
(−1482 kJ, spontaneous)

3 Assess

Evaluate Understanding

Have students explain the significance of positive and negative changes in entropy. (A negative change in entropy corresponds to an increase in order. A positive change in entropy reflects a decrease in order.) **Have students explain how the values of ΔG, ΔH, and ΔS determine whether or not a reaction is spontaneous. (The quantities ΔG, ΔH, and ΔS are related by the equation: $\Delta G = \Delta H - T\Delta S$ A reaction is spontaneous if ΔG is negative for the reaction at a specific temperature.)**

Reteach

Review the meaning of positive and negative entropy changes. Be sure students interpret a negative change in entropy as an increase in order. To assure that students understand the importance of ΔG in determining whether or not a reaction is spontaneous, discuss the following statements:

- **When ΔG is negative, the reaction is spontaneous.**
- **When ΔH and ΔS are both negative, the reaction is spontaneous at low temperatures.**

Answers

SECTION REVIEW 19.4

30. a. 164.7 J/K
b. −326.3 J/K
c. 19.8 J/K
d. −5.2 J/K

31. The standard entropy in the gaseous state is greater than in the liquid state.

32. a. spontaneous $\Delta G_f^0 = -768.06$ kJ
b. spontaneous $\Delta G_f^0 = -3152.8$ kJ

1 Engage

Use the Visual

Have students study the photograph at the start of Section 19.5 and hypothesize how the Tour de France bicycle race is like the reaction-progress curve for a complex reaction.

- **What are the energy requirements for a chemical reaction to progress from reactants to products?** (It is necessary that a fraction of the particles have enough kinetic energy to form an activation complex at collision that is, to overcome the activation energy barrier for the reaction.)

Check Prior Knowledge

Have students recall the four factors that can influence the rate of a chemical reaction. (temperature, concentration, particle size, and the presence of a catalyst) **Ask:**

- **What effect does the concentration of reactants have on the rate of a chemical reaction?** (Increasing their concentration increases the reaction rate.)

section 19.5

objectives

- Interpret experimental rate data to deduce the rate laws for simple chemical reactions
- Given an energy diagram for a reaction, analyze the mechanism for the reaction

key terms

- rate law
- specific rate constant
- first-order reaction
- elementary reaction
- reaction mechanism
- intermediate

THE PROGRESS OF CHEMICAL REACTIONS

The Tour de France is one of the most famous bicycle races in the world. It is held from mid-July to early August. During the course of the race, cyclists travel almost 4000 kilometers and cross steep mountains at heights of 1600 meters or more. To cross these high mountains requires extra amounts of energy. A chemical reaction can be thought of as a process involving peaks and valleys also. **What are the energy requirements for a chemical reaction to progress from initial reactants to final products?**

Rate Laws

The rate of a reaction depends in part on the concentrations of the reactants. For a reaction in which reactant A reacts to form product B in one step, you can write a simple equation. A one-step reaction is a reaction with only one activated complex between the reactants and products.

$$\mathrm{A} \longrightarrow \mathrm{B}$$

The rate at which A transforms to B is the change in concentration of A with time. Mathematically, you can express the rate as the change in A (ΔA) with respect to the change in time (Δt).

$$\text{Rate} = \frac{\Delta \mathrm{A}}{\Delta t}$$

The rate of disappearance of A is proportional to the molar concentration of A.

$$\frac{\Delta \mathrm{A}}{\Delta t} \propto [\mathrm{A}]$$

The proportionality can be expressed as a constant (k) multiplied by [A].

$$\text{Rate} = \frac{\Delta \mathrm{A}}{\Delta t} = k \times [\mathrm{A}]$$

This equation is an example of a **rate law,** an expression relating the rate of a reaction to the concentration of reactants. The **specific rate constant** (k) for a reaction is a proportionality constant relating the concentrations of reactants to the rate of the reaction. The magnitude of the specific rate constant depends on the conditions at which the reaction is conducted. If reactant A reacts to form product B quickly, the value of k will be large. If reactant A reacts to form product B slowly, the value of k will be small.

The order of a reaction is the power to which the concentration of a reactant must be raised to give the experimentally observed relationship between concentration and rate. In a **first-order reaction,** the reaction rate is directly proportional to the concentration of only one reactant. The conversion of A to B in a one-step reaction is an example of a first-order reaction; the reaction rate is proportional to the concentration of A raised to the first power: $[\mathrm{A}]^1 = [\mathrm{A}]$. As a first-order reaction progresses, the rate of reaction decreases, as shown in **Figure 19.22.** This decrease occurs because the concentration of reactant decreases. On a graph, the rate ($\Delta \mathrm{A}/\Delta t$) at any

STUDENT RESOURCES

From the Teacher's Resource Package, use:

- Section Review 19.5, Ch. 19 Practice Problems, Vocabulary, Quizzes and Tests from the Review Module (Ch. 17–20)

TECHNOLOGY RESOURCES

Relevant technology resources include:

- Chem ASAP! CD-ROM
- ResourcePro CD-ROM
- Assessment Resources CD-ROM: Chapter 19 Tests

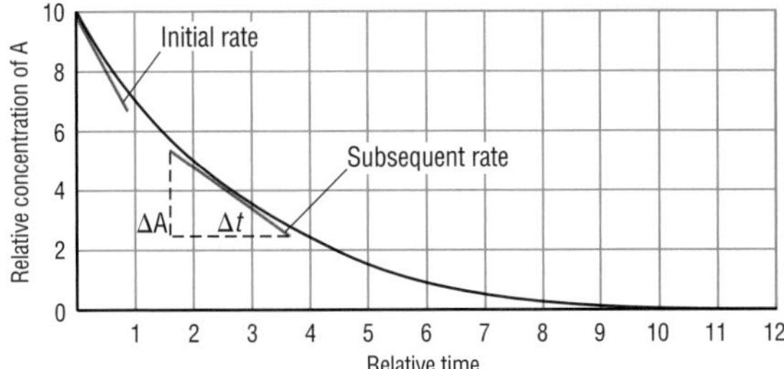

Figure 19.22
The rate of a first-order reaction decreases in direct proportion to the concentration of one reactant. The curved line shows the decrease in concentration of reactant A with time. The short colored lines illustrate the reaction rates at two distinct points of time.

point equals the slope of the tangent to the curve at that point. For a first-order reaction, a halving of [A] results in a halving of the reaction rate.

In some kinds of reactions, such as double-replacement, two substances react to give products. The coefficients in the equation for such a reaction are represented by lowercase letters.

$$aA + bB \longrightarrow cC + dD$$

For a one-step reaction of A with B, the rate of reaction is dependent on the concentrations of both A and B.

$$\text{Rate} = k[A]^a[B]^b$$

When each of the exponents a and b in the above rate law equals 1, the reaction is said to be first-order in A, first-order in B, and second-order overall. The overall order of a reaction is the sum of the exponents for the individual reactants.

For any one-step reaction, the experimentally determined exponents in the rate law will be the same as the coefficients a and b in the chemical equation. However, the exponents in the rate law and the coefficients in the equation do not correspond in most real reactions because most reactions are more complex than the one-step reactions used in the examples. Remember, the actual kinetic order of a reaction must be determined by experiment. There is not necessarily a relationship between the kinetic order and the coefficients in the overall chemical equations for complex reactions.

Chem ASAP!

Animation 24
Observe the characteristics of a first-order reaction.

Sample Problem 19-9

The rate law for the one-step reaction $aA \longrightarrow B$ is of the form Rate $= k[A]^a$. From the data in the following table, find the kinetic order of the reaction with respect to A and the overall order of the reaction.

Initial concentration of A (mol/L)	Initial rate (mol/L·s)
0.050	3.0×10^{-4}
0.10	12×10^{-4}
0.20	48×10^{-4}

19.5

2 Teach

Use the Visual

Have the students study Figure 19.22 and read the accompanying text. Explain that at the start of a reaction when the concentration of the reactants is at its highest, the rate is at its fastest. As the concentration of the reactants decreases so does the reaction rate. Inform the students that the shape of the curve tells you something about the dependence of the reaction rate on the concentration of the reactants.

Discuss

Discuss how the effect of concentration on the reaction rate can be expressed as a quantitative relationship. Each rate expression contains a rate constant, *k*, which is specific to the reaction. A knowledge of this dependence not only allows prediction of the actual rate, but also gives clues to the step-by-step mechanism by which the reaction occurs.
Demonstrate how to set up the rate equation for a chemical reaction such as the formation of water from hydrogen and oxygen.

ACTIVITY

Demonstrate how to calculate the rate of reaction by determining the slope of the tangent line for each of the two points shown along the graph in Figure 19.22.

- Have the students calculate the initial rate of reaction for the reaction graphed in Figure 19.22. (rate = (10 − 7.5)/(0.5 − 0) = 5 concentration/time units)
- Next have the students draw a tangent at time = 5 and calculate the rate at this point? (rate = (3 − 0)/(7− 3) = 0.75 concentration/time units) Point out that the rate is decreasing, as expected, as the concentration of reactant A decreases.

Practice Problems Plus

Related Chapter Review Problems
Chapter Review problems 62, 63, and 65 are related to Sample Problem 19-9.

Discuss

Explain to the students that even complex reactions occur as a series of simple steps. Each step usually involves the collision of no more than two different particles. Use the reaction between hydrogen and bromine as an example. The overall reaction $H_2 + Br_2 \rightarrow 2HBr$ occurs as a sequence of three elementary steps:
$Br_2 \rightarrow 2Br$
$Br + H_2 \rightarrow HBr + H$
$H + Br \rightarrow HBr$

Sample Problem 19-9 (cont.)

1. *ANALYZE* ***Plan a problem-solving strategy.***
The initial concentration, the initial rate data, and the form of the rate law are given. This is a one-step reaction. Use these data to find the desired kinetic orders by inspection.

2. *SOLVE* ***Apply the problem-solving strategy.***
If a reaction is first-order in A, then $a = 1$ in the rate equation. In that case, the initial rate would be directly proportional to the initial [A]. However, the reaction cannot be first-order in A because a doubling of A causes the rate to increase by four times. This suggests that $a = 2$, because $2^2 = 4$. Increasing [A] fourfold should, therefore, increase the rate sixteenfold ($4^2 = 16$). When the initial concentration of A is increased from 0.050 mol/L to 0.20 mol/L, the initial rate increases from 3.0×10^{-4} mol/L·s to 48×10^{-4} mol/L·s, which is a sixteenfold increase. The reaction is thus second-order in A; it must be second-order overall because A is the only reactant.

3. *EVALUATE* ***Does the result make sense?***
The hypothetical reaction given in this problem represents a special case, in that it is a one-step reaction. Remember, for a one-step reaction, the experimentally determined exponents in the rate law are the same as the coefficients a and b in the chemical equation. That is not generally true for most real reactions. For reactions in general, determine the kinetic order from the kinetic data and not from the coefficients in the chemical equation.

Practice Problems

33. Show that the unit of k for a first-order reaction is a reciprocal unit of time, such as a reciprocal second (s^{-1}).

34. Suppose a first-order reaction initially proceeds at a rate of 0.5 mol/L·s. What will be the rate when half the starting material remains? When one-fourth of the starting material remains?

Chem ASAP!
Problem-Solving 34
Solve Problem 34 with the help of an interactive guided tutorial.

Reaction Mechanisms

If enough information were available, you could graph all the energy changes that occur as reactants are converted to products in a chemical reaction. Such a graph would constitute a reaction progress curve. The simplest reaction progress curve would be obtained for an **elementary reaction,** one in which reactants are converted to products in a single step (a one-step reaction). Such a reaction has only one activated complex between reactants and products and thus only one activation-energy peak. Most chemical reactions, however, consist of a number of elementary reactions. A **reaction mechanism** includes all the elementary reactions of a complex reaction. For a complex reaction, the reaction progress curve resembles a series of hills and valleys. Study **Figure 19.23.** The hills correspond to the energies of the activated complexes. Each valley corresponds to the energies of an **intermediate** product, or a product of a reaction that becomes a reactant of another reaction. Intermediates have a significant lifetime compared with an activated complex. They have real ionic or molecular structures and some stability. They are reactive enough, however, to react further to give the final product of the reaction.

MEETING DIVERSE NEEDS

At Risk Some students may not understand the graph in Figure 19.22. Explain the concept of slope, using a straight line, as the ratio of the rise *y*, or in this case A, to the run *x*, or in this case *t*. Demonstrate how to calculate the slope for several straight lines and then for a tangent to the graph in Figure 19.22.

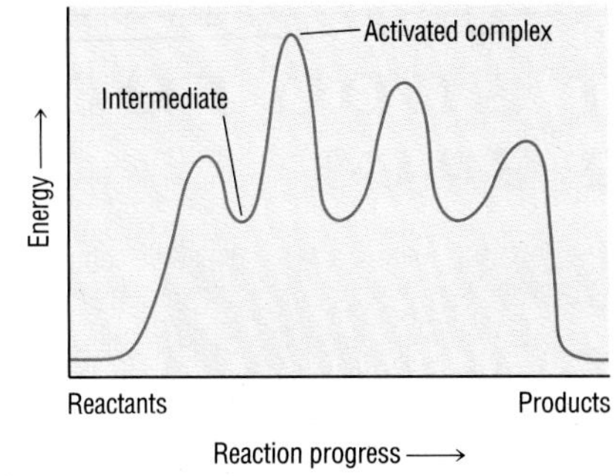

Figure 19.23
A reaction progress curve shows an activation-energy peak for each elementary reaction. Valleys indicate the formation of intermediates. How many elementary reactions are part of this reaction? How many intermediates are formed? 1

Intermediates do not appear in the overall chemical equation for a reaction. The decomposition of nitrous oxide (N_2O), for example, is believed to occur in two elementary steps:

$$N_2O(g) \longrightarrow N_2(g) + O(g)$$
$$N_2O(g) + O(g) \longrightarrow N_2(g) + O_2(g)$$
$$2N_2O(g) \longrightarrow 2N_2(g) + O_2(g)$$

Notice that oxygen atoms are intermediates in this reaction. They disappear when the individual reactants are summed to give the final chemical equation. This illustrates an important general point: The overall chemical equation for a complex reaction gives no information about the reaction mechanism.

section review 19.5

35. Ammonium ions and nitrite ions react in water to form nitrogen gas.

$$NH_4^+(aq) + NO_2^-(aq) \longrightarrow N_2(g) + 2H_2O(l)$$

From the following data, decide the kinetic order of the reaction with respect to NH_4^+, NO_2^-, and the overall order of the reaction.

Initial concentration of NO_2^- (mol/L)	Initial concentration of NH_4^+ (mol/L)	Initial rate (mol/L·s)
0.0100	0.200	5.4×10^{-7}
0.0200	0.200	10.8×10^{-7}
0.0400	0.200	21.5×10^{-7}
0.0600	0.200	32.3×10^{-7}
0.200	0.0202	10.8×10^{-7}
0.200	0.0404	21.6×10^{-7}
0.200	0.0606	32.4×10^{-7}
0.200	0.0808	43.3×10^{-7}

36. Define each term as it applies to chemical reactions.

a. elementary reaction
b. intermediate
c. reaction mechanism

37. How can an energy diagram be used to analyze a reaction mechanism?

Chem ASAP! **Assessment 19.5** Check your understanding of the important ideas and concepts in Section 19.5.

portfolio project

Research and report on the reaction mechanisms for the depletion of ozone in the stratosphere caused by the presence of CFCs.

19.5

3 Assess

Evaluate Understanding

**Ask the students to construct a reaction rate equation for the production of isoamyl acetate (banana oil):
isoamyl alcohol + ethanoic acid ↔ isoamyl acetate + H_2O**
(rate = $k \times$ [isoamyl alcohol] [ethanoic acid])
Have the students analyze the reaction mechanism for a given reaction from its potential energy diagram.

Reteach

Point out that most reactions are more complex than the one-step reactions used as examples in the text. In most cases, reaction order cannot be predicted on the basis of a balanced equation; it must be determined experimentally. This is because the rate depends primarily on the slowest step in the reaction mechanism. Have the students note the difference between rate determination and the calculation of an equilibrium constant, which can be determined from the overall reaction. Stress that there is no correlation between the equilibrium constant and the rate of a chemical reaction. Remind students that many familiar reactions have multiple steps. Ask them to suggest possible mechanisms for the following reactions:
$C + O_2 \rightarrow CO_2$
$2H_2 + O_2 \rightarrow 2H_2O$
$CH_4 + 2O_2 \rightarrow CO_2 + 2H_2O$

Portfolio Project

Many books and articles have been written on this complex topic. Students will discover that UV radiation plays a key role in the formation and the destruction of ozone.

Answers

1 There are four elementary reactions and three intermediates in the overall reaction.

SECTION REVIEW 19.5

35. The reaction is first-order in NO_2^-, first-order in NH_4^+, and second-order overall.

36. a. converts reactants to products in a single step
b. a product of one reaction that becomes a reactant in a second reaction
c. the sequence of elementary reactions of a complex reaction

37. The peaks and valleys in an energy diagram show the relative positions of activated complexes, intermediates, reactants, and products. Each of these energy positions is accounted for in a possible mechanism.

Chemistry Serving... the Consumer

DISCUSS

Because oxygen has six valence electrons in its highest occupied energy level, it has a strong tendency to combine with atoms that can donate or share electrons. Antioxidants interfere with this tendency because they are more easily oxidized than the substances they protect. In effect, they serve as *competitive inhibitors.* Carotenes, vitamin C, and vitamin E are naturally occurring antioxidants. Ask students to name foods that are rich in these substances. (broccoli, carrots, corn, squash, and tomatoes) Encourage students to find as many products as they can that contain the antioxidants described in this article. Students should make a table listing the products and the antioxidants they contain. Ask students to state any similarities they see in the products. Finally, have students research the pros and cons of using chemical preservatives in food.

CHEMISTRY IN CAREERS

Have students read the career feature about FDA inspectors on page 877 of this text. Students may wish to connect to the Internet address shown to discover more about career opportunities in this field.

Chemistry Serving...the Consumer

DON'T LET GOOD FOOD GO BAD!

The practice of preserving food is not new. For tens of thousands of years, people have salted or smoked meats and have dried fruits and vegetables to preserve them. Food preservation keeps food from spoiling and allows it to be stored for later use. Today, methods for preserving food also include the addition of preservatives, which are inhibitors of chemical and physical processes that cause food to spoil.

Food preservatives are classified into two basic groups—antimicrobials and antioxidants. Antimicrobials prevent the growth of bacteria, yeast, and molds. Antioxidants keep food from oxidizing. When food oxidizes, or reacts with oxygen, it becomes rancid and turns brown. Antioxidants also help reduce the decay of some essential amino acids and the loss of some vitamins.

Food preservatives are classified into two basic groups—antimicrobials and antioxidants.

Preservatives enable food to be produced, transported, purchased by the consumer, and used before it goes bad. Many foods are grown or produced in one location and then sent across the country or even overseas. Without preservatives, these foods would spoil long before they reached their destinations.

The Food and Drug Administration (FDA), a branch of the federal government, is responsible for approving the safety and use of preservatives. As part of the approval process, the manufacturer must demonstrate to the FDA that the chemical is safe. The FDA considers factors such as the quantity to be added to food, the long-term effects of the additive, and the potential toxicity. If the preservative is deemed safe, it is approved for use. If, at some later time, the preservative is found to be dangerous, the FDA can suspend its use.

The FDA requires that manufacturers include preservatives in the list of ingredients on food labels. Some common preservatives are BHT, BHA, sulfites, sulfur dioxide, and benzoic acid. BHT (butylated hydroxytoluene) and BHA (butylated hydroxyanisole) are often found in foods that are high in fats and oils. These two antioxidants slow the oxidation reactions that cause undesired flavors, odors, or colors in fats and oils. BHT and BHA are used in foods at a concentration of up to 0.02% of the fat or oil content. For example, if food has 1 g of fat, it can contain only 0.2 mg of either BHT or BHA.

Sulfites and sulfur dioxide have been used in foods for thousands of years. The ancient Egyptians and Romans used these compounds as preservatives. Sulfites and sulfur dioxide are antioxidants and antimicrobials, used mostly on fruit and in fruit juices to prevent discoloration from oxidation and to inhibit the growth of microbes. Sulfites can appear as many different compounds on food labels. Common sulfites include sulfur dioxide, sodium sulfite, sodium bisulfite, potassium bisulfite, sodium metabisulfite, and potassium metabisulfite.

Some people are allergic or sensitive to preservatives. It is especially important for these people to read food labels and avoid foods that have certain preservatives.

CHEMISTRY IN CAREERS

FDA INSPECTOR

Help prevent outbreaks of food-borne illnesses.

See page 877.

TAKE IT TO THE NET

Find out more about career opportunities: www.phschool.com

CHEMIST NEEDED

Enviro. Lab. GC Must know FDA methods, troubleshooting;

CHEMICAL SPECIALIST

Local food service distributor se responsible self-motivated indiv

Chapter 19 STUDENT STUDY GUIDE

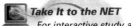

Take It to the NET
For interactive study and review, go to www.phschool.com

KEY TERMS

- activated complex *p. 536*
- activation energy *p. 535*
- catalyst *p. 537*
- chemical equilibrium *p. 540*
- collision theory *p. 535*
- elementary reaction *p. 568*
- entropy *p. 552*
- equilibrium constant (K_{eq}) *p. 545*
- equilibrium position *p. 540*
- first-order reaction *p. 566*
- free energy *p. 549*
- Gibbs free-energy change (ΔG) *p. 561*
- inhibitor *p. 538*
- intermediate *p. 568*
- law of disorder *p. 552*
- Le Châtelier's principle *p. 541*
- nonspontaneous reaction *p. 550*
- rate *p. 533*
- rate law *p. 566*
- reaction mechanism *p. 568*
- reversible reaction *p. 539*
- specific rate constant *p. 566*
- spontaneous reaction *p. 550*
- standard entropy (S^0) *p. 558*
- transition state *p. 536*

KEY EQUATIONS

- $K_{eq} = \dfrac{[C]^c \times [D]^d}{[A]^a \times [B]^b}$
- $\Delta G^0 = \Delta H^0 - T\Delta S^0$
- $\Delta S^0 = S^0(\text{products}) - S^0(\text{reactants})$
- $\Delta G^0 = \Delta G_f^0(\text{products}) - \Delta G_f^0(\text{reactants})$

CONCEPT SUMMARY

19.1 Rates of Reaction

- The rate at which a chemical reaction proceeds is determined by the number of collisions between reacting particles and the energy with which the particles collide.

19.2 Reversible Reactions and Equilibrium

- In principle, all reactions are reversible. Reactants go to products in the forward direction and products go to reactants in the reverse direction.
- The point at which the rates of conversion of reactants to products and of products to reactants are equal is the equilibrium position.
- Changes in the equilibrium position can be predicted by applying Le Châtelier's principle.

19.3 Determining Whether a Reaction Will Occur

- The natural tendency for all things to go to lower heat content (enthalpy) and greater randomness (entropy) determines whether a reaction will occur (is spontaneous).

19.4 Calculating Entropy and Free Energy

- Standard entropy and free-energy changes can be calculated from standard entropies (S^0), standard enthalpies of formation (H_f^0), and standard free energies of formation (G_f^0).

19.5 The Progress of Chemical Reactions

- The rate of a chemical reaction is dependent on the rate constant (k) and the kinetic order of the reaction.

CHAPTER CONCEPT MAP

Use these terms to construct a concept map that organizes the major ideas of this chapter.

Chem ASAP! Concept Map 19
Create your Concept Map using the computer.

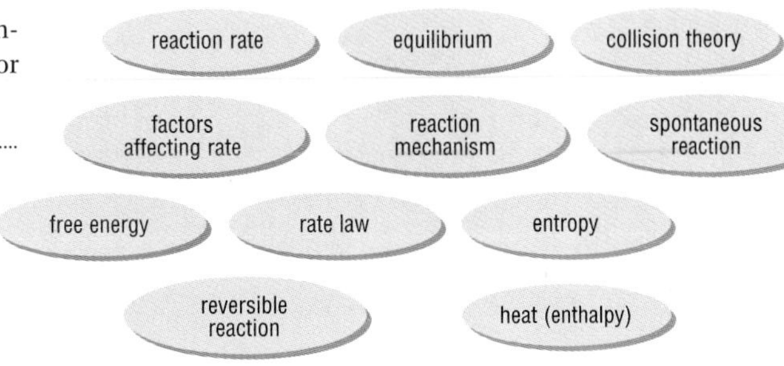

Take It to the Net

At **www.phschool.com** students will find for this chapter

- an Internet activity
- links to related chemistry sites
- an interactive quiz
- career links

4 Close

Summary

Ask students questions to help them summarize chapter content.

- **Given that reaction rates depend on effective collisions, which changes best help to speed up reactions?** (a temperature increase or adding a catalyst)
- **What is true of forward and reverse reaction rates at equilibrium?** (The rates are equal.)
- **According to Le Châtelier's principle, what factors affect the equilibrium point?** (changes in concentration, temperature, or pressure)
- **What two factors can combine to assure a spontaneous reaction at a given temperature?** (a decrease in enthalpy and an increase in entropy)
- **How is free energy calculated?** ($\Delta G = \Delta H - T\Delta S$) **What value must ΔG have for a reaction to be spontaneous?** (ΔG must be negative.)

Extension

Have the students list factors that affect whether a reaction will be spontaneous, the rate of the reaction, and the position of the equilibrium. Discuss which of these variables can be controlled for a specific chemical reaction.

Looking Back . . . Looking Ahead . . .

The previous chapter discussed the properties of solutions. The current chapter discussed factors affecting reaction rates, equilibrium, and spontaneity. The next chapter describes the properties of acids and bases.

Answers

38. Chemical reactions require collisions with sufficient energy to break and form bonds.

39. Reactant particles must have a certain minimum amount of energy to react to form product, just as it takes a certain minimum amount of energy to climb over a wall or barrier.

40. Above the reaction arrow; A catalyst is neither a reactant nor a product of a reaction.

41. A catalyst increases the rate of reactions by providing an alternative reaction mechanism with a lower activation energy.

42. **c.**

43. Gas molecules and oxygen molecules mix readily but do not have enough energy to react at room temperature. The flame raises the temperature and the energy of collisions, so the reaction rate is increased. The heat released by the reaction maintains the high temperature, and the reaction continues spontaneously.

44. In a reversible reaction, reactants are continuously forming products and products are continuously forming reactants.

45. The rate of formation of products from reactants and the rate of formation of reactants from products are equal.

46. The rates are equal.

47. A system in dynamic equilibrium changes to relieve the stress applied to it. Carbonated drinks in closed containers have achieved a state of dynamic equilibrium between the carbon dioxide in the liquid and gas states. When the containers are opened, carbon dioxide gas escapes. Carbon dioxide from the liquid goes into the gas state in an attempt to re-establish equilibrium.

48. $K_{eq} = \dfrac{[NH_3]^2}{[H_2]^3 \times [N_2]}$

Chapter 19 REVIEW

CONCEPT PRACTICE

38. Explain the collision theory of reactions. *19.1*

39. How is the activation energy of a reaction like a wall or barrier? *19.1*

40. Where is the formula of a catalyst written in a chemical equation? Why? *19.1*

41. How is the rate of a reaction influenced by a catalyst? How do catalysts make this possible? *19.1*

42. Which of these statements is true? *19.1*
- **a.** All chemical reactions can be sped up by increasing the temperature.
- **b.** Once a chemical reaction gets started, the reacting particles no longer have to collide for products to form.
- **c.** Enzymes are biological catalysts.

43. When the gas to a stove is turned on, the gas does not burn unless lit by a flame. Once lit, however, the gas burns until turned off. Explain these observations in terms of the effect of temperature on reaction rate. *19.1*

44. In your own words, define a reversible reaction. *19.2*

45. A reversible reaction has reached a state of dynamic chemical equilibrium. What does this information tell you? *19.2*

46. How do the rates of the forward and reverse reactions compare at a state of dynamic chemical equilibrium? *19.2*

47. What is Le Châtelier's principle? Use it to explain why carbonated drinks go flat when their containers are left open. *19.2*

48. Give the equilibrium-constant expression for the formation of ammonia from hydrogen and nitrogen. *19.2*

$$N_2(g) + 3H_2(g) \rightleftharpoons 2NH_3(g)$$

49. Write the expression for the equilibrium constant for each reaction. *19.2*
- **a.** $4H_2(g) + CS_2(g) \rightleftharpoons CH_4(g) + 2H_2S(g)$
- **b.** $PCl_5(g) \rightleftharpoons PCl_3(g) + Cl_2(g)$
- **c.** $2NO(g) + O_2(g) \rightleftharpoons 2NO_2(g)$
- **d.** $CO(g) + H_2O(g) \rightleftharpoons H_2(g) + CO_2(g)$

50. Comment on the favorability of product formation in each reaction. *19.2*
- **a.** $H_2(g) + F_2(g) \rightleftharpoons 2HF(g)$; $K_{eq} = 1 \times 10^{13}$
- **b.** $SO_2(g) + NO_2(g) \rightleftharpoons NO(g) + SO_3(g)$; $K_{eq} = 1 \times 10^2$

51. What is free energy? How can knowing the free energy of a reaction help predict whether the reaction will be spontaneous? *19.3*

52. The reaction of hydrogen and oxygen gas proceeds with a large decrease of free energy. The reaction is very slow at room temperature but occurs with explosive rapidity in the presence of a flame or ignition wire. Explain. *19.3*

53. The products in a spontaneous process are more ordered than the reactants. Is this entropy change favorable or unfavorable? *19.3*

54. What is the meaning of entropy? *19.3.*

55. Which system has the lower entropy? *19.3*
- **a.** completed jigsaw puzzle or separate jigsaw pieces
- **b.** 50 mL of liquid water or 50 mL of ice
- **c.** 10 g of sodium chloride crystals or a solution containing 10 g of sodium chloride

56. Predict the direction of the entropy change in each reaction. *19.3*
- **a.** $CaCO_3(s) \longrightarrow CaO(s) + CO_2(g)$
- **b.** $NH_3(g) + HCl(g) \longrightarrow NH_4Cl(s)$
- **c.** $2NaHCO_3(s) \longrightarrow Na_2CO_3(s) + H_2O(g) + CO_2(g)$
- **d.** $CaO(s) + CO_2(g) \longrightarrow CaCO_3(s)$

57. Is it true that all spontaneous processes are exothermic? Explain your answer. *19.3*

58. At normal atmospheric pressure, steam condenses to liquid water even though this is an unfavorable entropy change. Explain. *19.3*

59. What two factors together determine whether a reaction is spontaneous? *19.3*

60. For the decomposition of $CaCO_3(s)$ to $CaO(s)$ and $CO_2(g)$ at 298 K the ΔH_f^0 is 178.5 kJ/mol and the ΔS^0 is 161.6 J/K·mol. Is the reaction spontaneous or nonspontaneous at this temperature? *19.4*

61. What is meant by each term? *19.5*
- **a.** specific rate constant
- **b.** first-order reaction
- **c.** rate law

62. The rate law for the reaction

$$NO(g) + O_3(g) \longrightarrow NO_2(g) + O_2(g)$$

is first-order in NO and O_3, and second-order overall. Write the complete rate law for this reaction. *19.5*

49.
- **a.** $K_{eq} = \dfrac{[H_2S]^2 \times [CH_4]}{[H_2]^4 \times [CS_2]}$
- **b.** $K_{eq} = \dfrac{[PCl_3] \times [Cl_2]}{[PCl_5]}$
- **c.** $K_{eq} = \dfrac{[NO_2]^2}{[NO]^2 \times [O_2]}$
- **d.** $K_{eq} = \dfrac{[H_2] \times [CO_2]}{[CO] \times [H_2O]}$

50. **a.** highly favorable **b.** slightly favorable

51. the energy in a system available to do work; a spontaneous reaction has a negative free energy

52. A flame or ignition wire provides sufficient energy for activation. Once started, the spontaneous reaction can proceed rapidly.

53. unfavorable

54. a measure of the disorder of a system

55. **a.** completed jigsaw puzzle
b. 50 mL of ice
c. sodium chloride crystals

63. Half of the reactant in a first-order reaction has disappeared in 50 minutes. How many minutes are required for this particular reaction to be 75% complete? *19.5*

64. Sketch a potential-energy diagram for the overall reaction with the following mechanism.

$$2NO(g) \longrightarrow N_2O_2(g) \text{ (fast)}$$

$$N_2O_2(g) + O_2(g) \longrightarrow 2NO_2(g) \text{ (slow)}$$

Write the balanced equation for the overall reaction. *19.5*

CONCEPT MASTERY

65. Consider the decomposition of N_2O_5 in carbon tetrachloride (CCl_4) at 45 °C.

$$2N_2O_5(soln) \rightleftharpoons 4NO_2(g) + O_2(g)$$

The reaction is first-order in N_2O_5, with the specific rate constant 6.08×10^{-4}/s. Calculate the reaction rate at these conditions.
a. $[N_2O_5] = 0.200$ mol/L
b. $[N_2O_5] = 0.319$ mol/L

66. Which pieces of information are sufficient to determine whether a reaction will be spontaneous?
a. The reaction is exothermic.
b. Entropy is increased in the reaction.
c. Free energy is released in the reaction.

67. For the reaction $A + B \rightleftharpoons C$, the activation energy of the forward reaction is 5 kJ and the total energy change is −20 kJ. What is the activation energy of the reverse reaction?

68. A large box is divided into two compartments with a door between them. Equal quantities of two different monatomic gases are placed in the compartments, as shown in **(a)**. The door between the compartments is opened and the gas particles immediately start to mix. See **(b)**. Why would it be highly unlikely for the situation in **(b)** to progress to the situation shown in **(c)**?

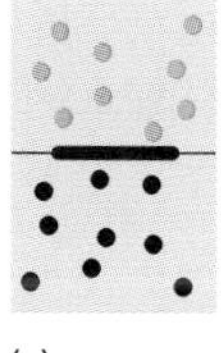
(a)

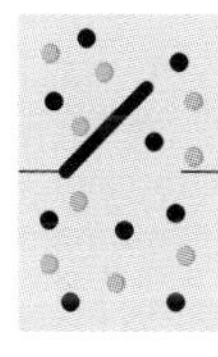
(b)

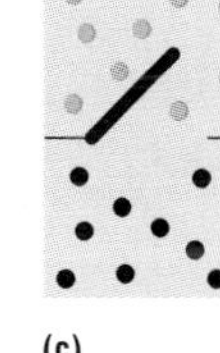
(c)

69. Would you expect the entropy to increase in each of the following reactions?
a. $C(s) + O_2(g) \longrightarrow CO_2(g)$
b. $Al_2O_3(s) \longrightarrow 2Al(s) + \frac{3}{2}O_2(g)$
c. $2N(g) \longrightarrow N_2(g)$
d. $N_2(g) \longrightarrow 2N(g)$

70. What would be the effect on the equilibrium position if the volume is decreased in each reaction?
a. $4HCl(g) + O_2(g) \rightleftharpoons 2Cl_2(g) + 2H_2O(g)$
b. $CO_2(s) \rightleftharpoons CO_2(g)$
c. $CaCO_3(s) \rightleftharpoons CaO(s) + CO_2(g)$

71. A mixture at equilibrium at 827 °C contains 0.552 mol CO_2, 0.552 mol H_2, 0.448 mol CO, and 0.448 mol H_2O. The balanced equation is $CO_2(g) + H_2(g) \rightleftharpoons CO(g) + H_2O(g)$. What is the value of K_{eq}?

72. Write the equilibrium-constant expression for each reaction.
a. $I_2(g) + Cl_2(g) \rightleftharpoons 2ICl(g)$
b. $2NO_2(g) \rightleftharpoons 2NO(g) + O_2(g)$
c. $2SO_2(g) + O_2(g) \rightleftharpoons 2SO_3(g)$
d. $Cl_2(g) + PCl_3(g) \rightleftharpoons PCl_5(g)$

73. The freezing of liquid water at 0 °C can be represented as follows.

$$H_2O(l, d = 1.00 \text{ g/cm}^3) \rightleftharpoons H_2O(s, d = 0.92 \text{ g/cm}^3)$$

Explain why the application of pressure causes ice to melt.

74. Sketch an energy profile curve for this gas-phase reaction.

$$F(g) + H_2(g) \rightleftharpoons HF(g) + H(g)$$

The reaction has an activation energy of 22 kJ, and the total energy change is −103 kJ.

75. The reaction between diamond (carbon) and oxygen is spontaneous. What can you say about the speed of this reaction?

$$C(s, \text{diamond}) + O_2(g) \longrightarrow CO_2(g)$$

76. Predict what will happen if a catalyst is added to a slow reversible reaction. What happens to the equilibrium position?

77. Use the standard entropy data of **Table 19.2** to calculate the standard entropy change (ΔS^0) for this reaction.

$$2NO(g) + 3H_2O(g) \longrightarrow 2NH_3(g) + \tfrac{5}{2}O_2(g)$$

56. **a.** Entropy increases.
b. Entropy decreases.
c. Entropy increases.
d. Entropy decreases.

57. No; some endothermic processes are spontaneous because of their favorable change in entropy.

58. The favorable exothermic change of the condensation process offsets the unfavorable entropy change.

59. In combination, the change in heat content (enthalpy) and the change in entropy determine whether a process is spontaneous.

60. $\Delta G_f^0 = 130.3$ kJ/mol; nonspontaneous

61. **a.** the proportionality constant relating the concentrations of the reactants to the reaction rate
b. a chemical reaction with a rate proportional to the concentration of one reactant
c. an equation relating the reaction rate to the concentrations of the reactants

62. Rate $= k\,[NO][O_3]$

63. 100 minutes

64. $2NO + O_2 \Leftrightarrow 2NO_2$

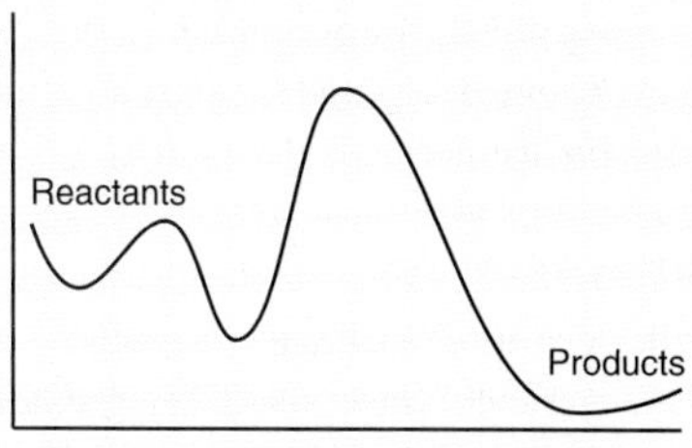

65. **a.** 1.22×10^{-4} mol/(L · s)
b. 1.94×10^{-4} mol/(L · s)

66. **c.**

67. 25 kJ

68. The change from Figure a to Figure b is spontaneous, favored by an increase in entropy. The change from Figure b to Figure c, however, will not occur because it would result in a decrease in entropy, causing the process to be nonspontaneous.

69. **a.** yes **b.** yes **c.** no **d.** yes

70. **a.** increase in products
b. decrease in products
c. decrease in products

71. $K_{eq} = 0.659$

72. **a.** $K_{eq} = \dfrac{[ICl]^2}{[I_2] \times [Cl_2]}$

b. $K_{eq} = \dfrac{[NO]^2 \times [O_2]}{[NO_2]^2}$

c. $K_{eq} = \dfrac{[SO_3]^2}{[SO_2]^2 \times [O_2]}$

d. $K_{eq} = \dfrac{[PCl_5]}{[Cl_2] \times [PCl_3]}$

73. Increasing pressure tends to reduce volume and increase density, so the system responds by favoring production of the liquid, which has the greater density.

74.

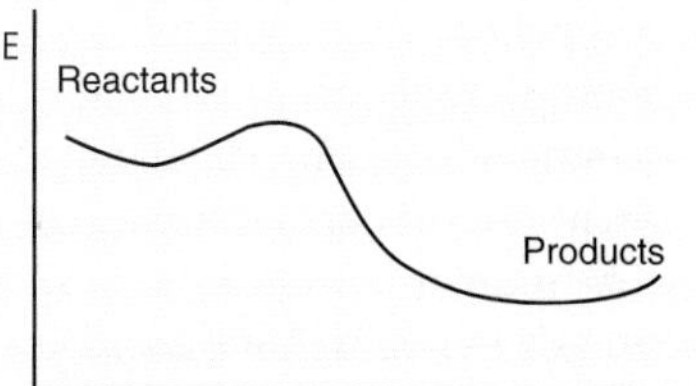

75. The reaction is slow (activation energy is high).

Answers

76. A catalyst will help establish equilibrium more quickly, but it does not change the equilibrium position.

77. −89.8 J/K

78. a. 3 **b.** 1

79. A catalyst increases the efficiency of the collisions; a greater number of collisions results in the formation of the product.

80. Possible answers: using a blow dryer, flushing the toilet, mowing the lawn, cooking breakfast, driving a car, and simply breathing.

81. a. fluoride ion (anion)
b. copper(II) ion (cation)
c. phosphide ion (anion)
d. hydrogen ion (cation)
e. sodium ion (cation)
f. iodide ion (anion)
g. oxide ion (anion)
h. magnesium ion (cation)

82. Crystalline substances have a regular, repeating pattern of atoms, ions, or molecules in three dimensions. Amorphous substances have an irregular arrangement of particles.

83. Solid potassium chloride is an ionic compound of K^+ and Cl^-, not of KCl molecules. Each ion is surrounded by six ions of opposite charge in a simple cubic unit cell crystal.

84. a. sodium perchlorate, ClO_4^-
b. potassium permanganate, MnO_4^-
c. calcium phosphate, PO_4^{3-}
d. magnesium carbonate, CO_3^{2-}
e. sodium sulfate, SO_4^{2-}
f. potassium dichromate, $Cr_2O_7^{2-}$

85. crystalline solids, high melting points; insulators when solid, conductors when molten; not malleable or ductile; often soluble in water

86. a. $1s^22s^22p^63s^23p^63d^{10}4s^24p^2$ $\cdot\dot{\text{Ge}}\cdot$
b. $1s^22s^22p^63s^23p^63d^{10}4s^2$ Ca:
c. $1s^22s^22p^4$ $\cdot\ddot{\text{O}}\cdot$
d. $1s^22s^22p^63s^23p^6$ $:\ddot{\text{Ar}}:$
e. $1s^22s^22p^63s^23p^63d^{10}4s^24p^5$ $:\ddot{\text{Br}}\cdot$
f. $1s^22s^22p^63s^23p^3$ $\cdot\dot{\text{P}}\cdot$

CRITICAL THINKING

78. Choose the term that best completes the second relationship.

a. reaction : intermediate — life span : ______
(1) birth (3) adolescence
(2) products (4) 75 years

b. catalyst : inhibitor — sunshine : ______
(1) clouds (3) wind
(2) photosynthesis (4) energy

79. An increase in temperature raises the energy of the collisions between reactant molecules. An increase in the concentration of reactants increases the number of collisions. What is the effect of a catalyst on the collisions between molecules?

80. Make a list of five things you did today that resulted in an increase in entropy.

CUMULATIVE REVIEW

81. Name each ion and identify it as an anion or a cation.
a. F^- **c.** P^{3-} **e.** Na^+ **g.** O^{2-}
b. Cu^{2+} **d.** H^+ **f.** I^- **h.** Mg^{2+}

82. Explain how crystalline and amorphous substances differ.

83. What is wrong with saying that solid potassium chloride is composed of KCl molecules?

84. Name the following compounds and give the charge on the anion for each.
a. $NaClO_4$ **c.** $Ca_3(PO_4)_2$ **e.** Na_2SO_4
b. $KMnO_4$ **d.** $MgCO_3$ **f.** $K_2Cr_2O_7$

85. List some properties that are typical of ionic compounds.

86. Write electron configurations and draw electron dot structures for the following elements.
a. Ge **b.** Ca **c.** O **d.** Ar **e.** Br **f.** P

87. Write the formula for each ionic compound using electron dot structures.
a. LiBr **b.** $AlCl_3$ **c.** MgF_2 **d.** Na_2S

88. Write the electron configuration and draw the electron dot structure for each.
a. Ca^{2+} **b.** Li^+ **c.** Br^- **d.** S^{2-}

89. Which atoms from the following list would you expect to form positive ions, and which would you expect to form negative ions?
a. Cl **c.** P **e.** Cu **g.** K **i.** N
b. Ca **d.** Se **f.** Sn **h.** Fe **j.** Ni

CONCEPT CHALLENGE

90. Is eating sugar candy really bad for your teeth? Tooth decay is the result of the dissolving of tooth enamel ($Ca_5(PO_4)_3OH$). In the mouth the following equilibrium is established.

$$Ca_5(PO_4)_3OH(s) \rightleftharpoons 5Ca^{2+}(aq) + 3PO_4^{3-}(aq) + OH^-(aq)$$

When sugar ferments on the teeth, H^+ is produced. What effect does this increased H^+ have on tooth enamel?

91. The following data were collected for the decomposition of compound AB into its elements. The reaction is first-order in AB.

[AB] (mol/L)	Time (s)
0.300	0
0.246	50
0.201	100
0.165	150
0.135	200
0.111	250
0.090	300
0.075	350

a. Make a graph of concentration (y-axis) versus time (x-axis).
b. Determine the rate of this reaction at t = 100 seconds and t = 250 seconds.

92. When table sugar, or sucrose, is dissolved in an acidic solution, the sucrose slowly decomposes into two simpler sugars: fructose and glucose. Use the graph to answer these questions.

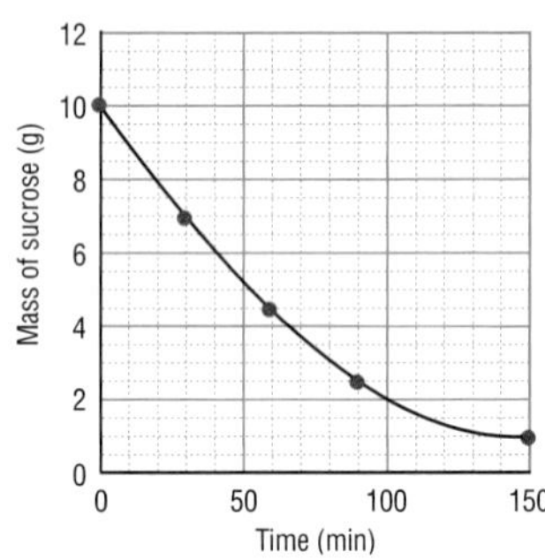

a. How many grams of sucrose decompose in the first 30 minutes?
b. How many grams of sucrose decompose in the 30-minute interval between 90 and 120 minutes?

87. a. $Li^+ :\ddot{\text{Br}}:^-$ **b.** $^-:\ddot{\text{Cl}}:Al^{3+}:\ddot{\text{Cl}}:^-$ (with $:\ddot{\text{Cl}}:^-$ above)
c. $^-:\ddot{\text{F}}:Mg^{2+}:\ddot{\text{F}}:^-$ **d.** $Na^+:\ddot{\text{S}}:^{2-}Na^+$

88. a. $1s^22s^22p^63s^23p^6$ $:\ddot{\text{Ca}}:^{2+}$
b. $1s^2$ $:Li^+$
c. $1s^22s^22p^63s^23p^63d^{10}4s^24p^6$ $:\ddot{\text{Br}}:^-$
d. $1s^22s^22p^63s^23p^6$ $:\ddot{\text{S}}:^{2-}$

89. positive ions: **b., e., f., g., h.,** and **j.**; negative ions: **a., c., d.,** and **i.**

90. The additional H^+ reacts with the OH^-, shifting the reaction toward more products. As a result, more of the reactant, tooth enamel, is broken down.

Select the choice that best answers each question or completes each statement.

1. Which reaction is represented by the following expression for an equilibrium constant?

$$K_{eq} = \frac{[CO]^2 \times [O_2]}{[CO_2]^2}$$

   a. $2CO_2 \rightleftharpoons O_2 + 2CO$
   b. $CO_2{}^2 \rightleftharpoons O_2 + 2CO^2$
   c. $O_2 + 2CO \rightleftharpoons 2CO_2$
   d. $O_2 + CO_2 \rightleftharpoons CO_2{}^2$
2. In which process is there an entropy increase?
   a. Dry ice sublimes.
   b. Liquid wax solidifies.
   c. Dew forms from water vapor.
   d. Liquid water forms from gaseous hydrogen and oxygen.
3. Based on the equilibrium equation, which change will *not* increase the production of $CH_3OH(g)$?

$$CO(g) + 2H_2(g) \rightleftharpoons CH_3OH(g) + \text{heat}$$

   a. adding heat
   b. removing CH_3OH
   c. adding CO
   d. increasing the pressure
4. Which of the following statements is true?
   a. A rate law relates the rate of a reaction to the concentration of the reactants.
   b. A reaction with a very large activation energy will occur very rapidly.
   c. When a reaction reaches equilibrium, the forward reaction stops.
   d. Any reactant particles that collide will react to form product.

Use the table to answer questions 5 and 6.

ΔS	ΔH	ΔG	Spontaneous?
+	–	(a)	Yes
+	(b)	+ or –	At high T
(c)	+	+	No
–	–	(d)	At low T

5. The value of ΔG depends on the enthalpy (ΔH) and entropy (ΔS) terms for a reaction. The value of ΔG also varies as a function of temperature. Use the data in the table to identify the missing entries (a), (b), (c), and (d).
6. Which of these reactions would you expect to be spontaneous at relatively low temperatures? At relatively high temperatures?
   (1) $H_2O(l) \longrightarrow H_2O(s)$
   (2) $H_2O(g) \longrightarrow H_2O(l)$
   (3) $H_2O(s) \longrightarrow H_2O(l)$

Use the atomic windows to answer question 7.

7. The atomic windows represent different degrees of entropy. Arrange the windows in order of increasing entropy.

a. 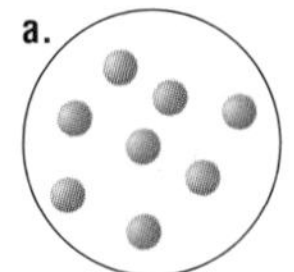
b. 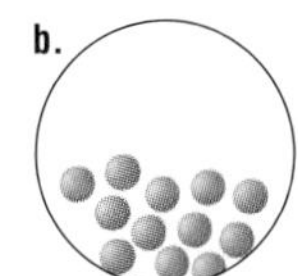
c. 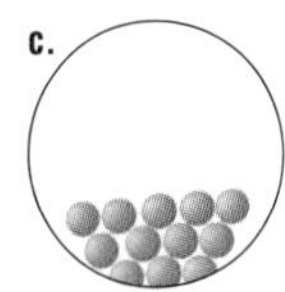

For each question there are two statements. Decide whether each statement is true or false. Then decide whether Statement II is a correct explanation for Statement I.

	Statement I		Statement II
8.	A catalyst lowers the activation energy for a reaction.	BECAUSE	A catalyst makes a reaction more exothermic.
9.	An exothermic reaction is always a spontaneous reaction.	BECAUSE	Exothermic reactions release heat to the surroundings.
10.	The entropy of ice is greater than the entropy of steam.	BECAUSE	The density of ice is greater than the density of steam.
11.	The rate of a chemical reaction is affected by a change in temperature.	BECAUSE	The kinetic energy of particles is related to the temperature.
12.	A very large value for an equilibrium constant indicates that products are favored at equilibrium.	BECAUSE	The ratio of products to reactants at equilibrium is always > 1.

Chapter 19

STANDARDIZED TEST PREP

1. a
2. a
3. a
4. a
5. (a): –; (b): +; (c): –; (d): + or –
6. (1) and (2) at low temperatures; (3) at high temperatures
7. c, b, a
8. True, False
9. False, True
10. False, True
11. True, True
12. True, False

91. a.

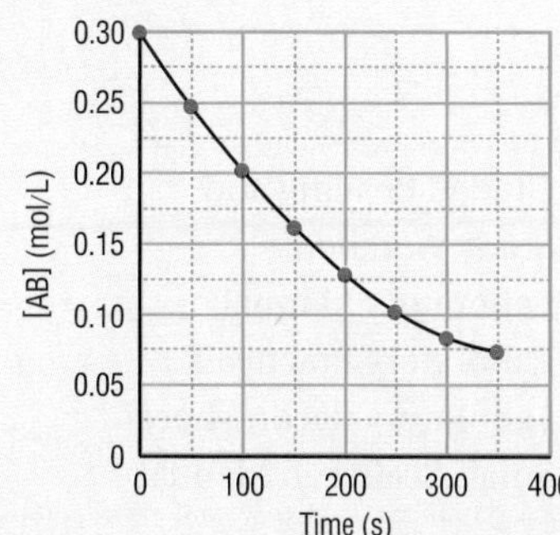

b. The rate when $t = 100$ s is 8×10^{-4} mol/L·s. The rate when $t = 250$ s is 4×10^{-4} mol/L·s.

92. a. 3 g

b. 1.3 g

Planning Guide

SECTION OBJECTIVES	ACTIVITIES/FEATURES	MEDIA & TECHNOLOGY
20.1 Describing Acids and Bases ● ■ ◇ ▸ List the properties of acids and bases ▸ Name an acid or base when given the formula	**SE** Discover It! *Effect of Foods on Baking Soda,* p. 576 **TE** DEMO, p. 577	**ASAP** Problem Solving 2 **ASAP** Assessment 20.1 **ACT** 15: *Acids and Bases* **CA** *Milk of Magnesia* **CA** *Titration* **CA** *Ira Remsen Story* **CHM** Side 7, 2: *Food from the 'Hood* **CHM** Side 7, 30: *Acids and Metals*
20.2 Hydrogen Ions and Acidity ○ ■ ◆ ▸ Given the hydrogen-ion or hydroxide-ion concentration, classify a solution as neutral, acidic, or basic ▸ Convert hydrogen-ion concentrations into values of pH and hydroxide-ion concentrations into values of pOH	**SE** **CHEMath** *Using Logarithms,* p. 585 **SE** Mini Lab *Indicators from Natural Sources,* p. 593 (**LRS** 20-1) **LM** 37: *Estimation of pH* **SSLM** 26: *A Small-Scale Colorimetric pH Meter* **TE** DEMOS, pp. 581, 588, 589	**ASAP** Problem Solving 9,13 **ASAP** Assessment 20.2 **ACT** 15: *Acids and Bases* **CA** *Milk of Magnesia* **CA** *Titration* **SSV** 11: *Small-Scale pH-Meter* **CHM** Side 7, 10: *pH in Aquaria* **OT** 68: *pH*
20.3 Acid-Base Theories ○ ■ ◆ ▸ Compare and contrast acids and bases as defined by the theories of Arrhenius, Brønsted-Lowry, and Lewis ▸ Identify conjugate acid–base pairs in acid–base reactions	**SE** **Link to Library Science** *Chemistry Rescues Crumbling Books,* p. 596 **SSLM** 27: *Lewis Acids: Electron Pair Acceptors*	**ASAP** Animation 25 **ASAP** Assessment 20.3 **ACT** 15: *Acids and Bases* **CHM** Side 7, 20: *Defining Acids and Bases*
20.4 Strengths of Acids and Bases □ ◆ ▸ Define strong acids and weak acids ▸ Calculate an acid dissociation constant (K_a) from concentration and pH measurements ▸ Arrange acids by strength according to their acid dissociation constants (K_a) ▸ Arrange bases by strength according to their base dissociation constants (K_b)	**SE** **Link to Health** *Tooth Decay,* p. 603 **SE** Small-Scale Lab *Ionization Constants of Weak Acids,* p. 606 **SE** **Chemistry Serving . . . the Environment** *Rain Like Vinegar,* p. 607 **SE** **Chemistry in Careers** *Stone Conservator,* p. 607 **LM** 38: *Reactions of Acids* **SSLM** 28: *Strong and Weak Acids and Bases* **TE** DEMO, p. 601	**ASAP** Problem Solving 27 **ASAP** Assessment 20.4 **ASAP** Concept Map 20 **CA** *Titration* **CHM** Side 7, 32: *Acid Strength* vs. *Concentration* **OT** 69: *Dissociation of Strong and Weak Acids* **PP** Chapter 20 Problems **RP** Lesson Plans, Resource Library **AR** Computer Test 20 **www** Activities, Self-Tests, *SCIENCE NEWS* updates **TCP** The Chemistry Place Web Site

KEY

● Conceptual (concrete concepts)	**AR** Assessment Resources	**GRS** Guided Reading and Study Workbook
○ Conceptual (more abstract/math)	**ASAP** Chem ASAP! CD-ROM	**LM** Laboratory Manual
■ Standard (core content)	**ACT** ActivChemistry CD-ROM	**LP** Laboratory Practicals
□ Standard (extension topics)	**CHM** CHEMedia Videodiscs	**LRS** Laboratory Recordsheets
◆ Honors (core content)	**CA** Chemistry Alive! Videodiscs	**SSLM** Small-Scale Lab Manual
◇ Honors (options to accelerate)	**GCP** Graphing Calculator Problems	

PRACTICE		ASSESSMENT	
SE	**Sample Problem** 20-1	**SE**	Section Review
GRS	Section 20.1	**RM**	Section Review 20.1
RM	Practice Problems 20.1	**RM**	Chapter 20 Quiz
SE	**Sample Problems** 20-2 to 20-6	**SE**	Section Review
GRS	Section 20.2	**RM**	Section Review 20.2
RM	Practice Problems 20.2	**RM**	Chapter 20 Quiz
RM	Interpreting Graphics	**LP**	Lab Practical 20-1
SE	**Sample Problem** 20-7	**SE**	Section Review
GRS	Section 20.3	**RM**	Section Review 20.3
RM	Practice Problems 20.3	**RM**	Chapter 20 Quiz
SE	**Sample Problem** 20-8	**SE**	Section Review
GRS	Section 20.4	**RM**	Section Review 20.4
RM	Practice Problems 20.4	**LP**	Lab Practical 20-2
GCP	Chapter 20	**RM**	Vocabulary Review 20
		SE	Chapter Review
		SM	Chapter 20 Solutions
		SE	Standardized Test Prep
		PHAS	Chapter 20 Test Prep
		RM	Chapter 20 Quiz
		RM	Chapter 20 A & B Test

OT	Overhead Transparency	**SE**	Student Edition
PHAS	PH Assessment System	**SM**	Solutions Manual
PLM	Probeware Lab Manual	**SSV**	Small-Scale Video/Videodisc
PP	Problem Pro CD-ROM	**TCP**	www.chemplace.com
RM	Review Module	**TE**	Teacher's Edition
RP	Resource Pro CD-ROM	**www**	www.phschool.com

PLANNING FOR ACTIVITIES

STUDENT EDITION

Discover It! p. 576
- baking soda
- large paper plates
- knives
- paper towels
- variety of fruits and vegetables

Mini Lab, p. 593
- red cabbage leaves
- white vinegar
- household ammonia
- baking soda
- 1-cup measures
- clear plastic cups
- jars
- tape
- knives
- pencils
- teaspoons
- rulers
- clean white cloths
- hot water
- plain white paper
- medicine droppers
- clear plastic wrap
- rubber bands
- salt, borax, milk, lemon juice, detergents, milk of magnesia, mouthwash, toothpaste, shampoo, carbonated beverages

Small-Scale Lab, p. 606
- pencils, rulers, paper
- reaction surfaces
- indicators such as thymol blue, bromcresol green, bromthymol blue, phenolphthalein
- boric acid, citric acid
- Na_3PO_4, H_3PO_4

TEACHER'S EDITION

Teacher Demo, p. 577
- 5–10 mL of 1*M* HCl
- test tube
- small pieces of zinc
- match

Teacher Demo, p. 581
- several test tubes
- HCl, NaOH
- household products such as lemon juice, vinegar, shampoo, liquid detergent
- pH meter

Teacher Demo, p. 588
- lemon juice
- tea
- Alka-Seltzer tablets
- 10% unsweetened grape juice
- household ammonia

Teacher Demo, p. 589
- 0.1% solutions of thymol blue, methyl red, bromothymol blue, phenolphthalein
- dilute aqueous buffers, e.g., acetic acid/acetate ion, citric acid/citrate ion, carbonate ion/hydrogen carbonate ion, mono-sodium phosphate ion/ disodium phosphate ion
- test tubes
- pH meter

Activity, p. 590
- chemical supply catalogs

Teacher Demo, p. 601
- two test tubes
- sucrose
- eye dropper
- concentrated sulfuric acid
- concentrated ethanoic acid

Chapter 20

Key Terms

20.1 acid, base
20.2 hydroxide ion (OH^-), hydronium ion (H_3O^+), self-ionization, neutral solution, ion-product constant for water (K_w), acidic solution, basic solution, alkaline solutions, pH
20.3 monoprotic acids, diprotic acids, triprotic acids, hydrogen-ion donor, hydrogen-ion acceptor, conjugate acid, conjugate base, conjugate acid–base pair, amphoteric, Lewis acid, Lewis base
20.4 strong acids, weak acids, acid dissociation constant (K_a), strong bases, weak bases, base dissociation constant (K_b)

DISCOVER IT!

Students should observe that fruits and vegetables with sour or tart tastes react with baking soda to produce bubbles of (carbon dioxide) gas. After completing the chapter, students should understand that the acids in the foods react with the base in the baking soda to produce water and a salt.

Chapter 20 ACIDS AND BASES

FEATURES

Stay current with **SCIENCE NEWS**
Find out more about acids and bases:
www.phschool.com

These tart-tasting fruits contain a weak acid called citric acid.

DISCOVER IT! EFFECT OF FOODS ON BAKING SODA

You need baking soda (sodium hydrogen carbonate, $NaHCO_3$), a large paper plate, a knife, paper towels, and a variety of fruits and vegetables (e.g., a celery stalk, a banana, a grape, a tomato, a lemon, an apple, an orange, a grapefruit).

1. Carefully cut the fruits and vegetables into small pieces and place them on the plate. Make sure the pieces are well separated from each other. Be sure to wipe any juice off the knife after cutting each fruit.
2. Sprinkle a pinch of baking soda on each sample.

What do you observe? Is there any relationship between what you observe and which foods you know from experience have a sour taste? After reading about acids and bases, provide an explanation for what you observed.

20.1 ● ■ ◇
20.2 ○ ■ ◆
20.3 ○ ■ ◆
20.4 □ ◆

Conceptual In Section 20.2, omit calculations of pH in Sample Problems 20-3 through 20-6. Use Table 20.2 to help students visualize pH. If normality will be covered in Chapter 21, then diprotic and triprotic acids need to be introduced in Section 20.3.

Standard The CHEMath feature will help students who are struggling with logarithms. In Section 20.4, focus on the distinction between strong and weak acids and bases versus concentrated and dilute acids and bases. Calculating dissociation constants may be omitted.

Honors Section 20.1 is a descriptive section that may be suitable for self-study. If students are successful with Sample Problems 20-3 and 20-4, assign Sample Problems 20-5 and 20-6 as homework.

DESCRIBING ACIDS AND BASES

section 20.1

Some ants can give painful stings when threatened or disturbed. Certain ant species called formicines have poison glands that produce venom containing formic acid. Formicines protect themselves by spraying this venom on their predators. Formic acid can stun or even kill the ants' most common enemies. A formicine attack on a human, however, is much less severe, usually resulting only in blistered skin. **What are some of the properties of acids?**

objectives

- List the properties of acids and bases
- Name an acid or base when given the formula

key terms

- acid
- base

Properties of Acids and Bases

Did you know that acids and bases play a central role in much of the chemistry that affects your daily life? Most manufacturing processes use acids or bases. Your body needs acids and bases to function properly. Vinegar, carbonated drinks, and foods such as citrus fruits contain acids. The electrolyte in a car battery is an acid. Bases are present in many commercial products, including antacid tablets and household cleaning agents. **Figure 20.1** shows some of the many products that contain acids and bases. As you read this chapter, you will learn about the qualitative and quantitative aspects of acids and bases. You will see how these two classes of compounds ionize or dissociate in water. And you will learn how pH is used to describe the concentrations of acidic and basic solutions.

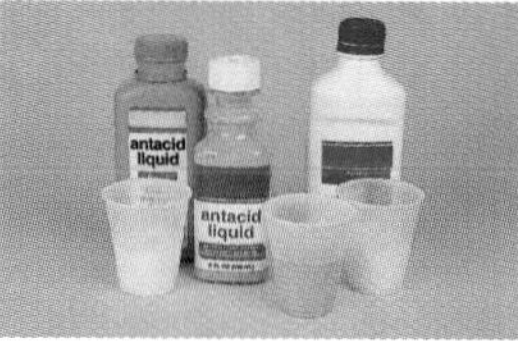

Figure 20.1
All of these items contain acids or bases, or produce acids or bases upon dissolving in water. Tomatoes contain ascorbic acid; tea contains tannic acid. The base calcium hydroxide is a component of mortar, which was used to build the Great Wall of China. Antacids use a variety of bases to neutralize excess stomach acid.

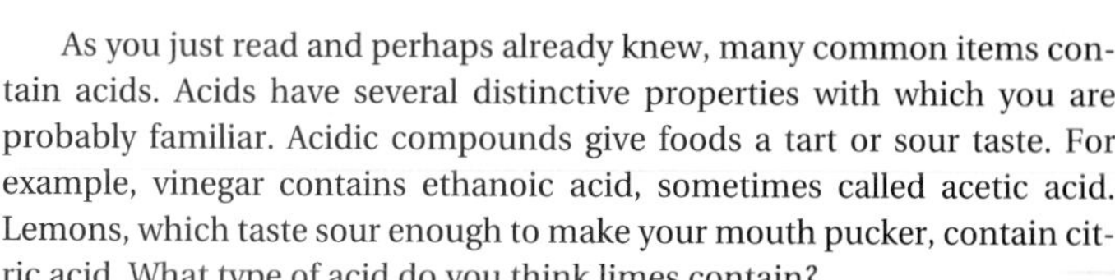

As you just read and perhaps already knew, many common items contain acids. Acids have several distinctive properties with which you are probably familiar. Acidic compounds give foods a tart or sour taste. For example, vinegar contains ethanoic acid, sometimes called acetic acid. Lemons, which taste sour enough to make your mouth pucker, contain citric acid. What type of acid do you think limes contain? 1

Aqueous solutions of acids are electrolytes. Recall from Chapter 17 that electrolytes conduct electricity. Some acid solutions are strong electrolytes and others are weak electrolytes. Acids cause certain chemical dyes, called indicators, to change color. Many metals, such as zinc and magnesium, react with aqueous solutions of acids to produce hydrogen gas. Acids react with compounds containing hydroxide ions to form water and a salt.

20.1

1 Engage

Use the Visual

Have the students read the section-opening paragraph and examine the photograph. Explain that there are many naturally occurring acids and that they serve a variety of functions. For example, these ants use formic acid for protection. Ask:

- **What are some of the properties of acids?** (Acids have a sour taste and can react with compounds such as baking soda. Based on the description of formic acid, acids can be corrosive.)

Check Prior Knowledge

On the board, make a two-column table. Label the columns *acid* and *base*. Have students list substances they think contain either acids or bases. Ask the students how they identified a substance as an acid or a base. Use their responses to compile a list of general properties for each of the two groups.

2 Teach

Teacher Demo

Add 5–10 mL of 1 *M* HCl to a test tube and add a few small pieces of zinc metal. Have the students note the gas bubbles being liberated from the solution. Explain to the students that many metals react with aqueous acids to produce hydrogen gas. Remind students that hydrogen is a highly flammable gas. Demonstrate the liberation of the gas by placing a lit match at the mouth of the test tube to ignite the hydrogen gas.

STUDENT RESOURCES

From the Teacher's Resource Package, use:

- Section Review 20.1, Ch. 20 Practice Problems and Quizzes from the Review Module (Ch. 17–20)

1 Limes and other citrus fruits contain citric acid.

TECHNOLOGY RESOURCES

Relevant technology resources include:

- Chem ASAP! CD-ROM
- ResourcePro CD-ROM
- Chemistry Alive! Videodisc: *Milk of Magnesia, Titration, Ira Remsen Story*
- ActivChemistry CD-ROM: *Acids and Bases*

Cooperative Learning

Divide the class into groups of four or five students. Group members should find products that contain acids or bases and prepare an informational display for the rest of the class. Each student in a group should contribute at least three items to the display. Evaluate the displays on the basis of accuracy, clarity, and creativity.

Discuss

Have students review Table 20.1. Make sure students understand the relationship between the suffix of an anion and the corresponding acid name. Focus on the critical connections between *-ide* and *hydro-* + *-ic*; *-ite* and *-ous*; *-ate* and *-ic*. Ask:

- **How are anions with the suffix *-ide* different from anions with the suffixes *-ite* or *-ate*?** (The latter are polyatomic ions containing oxygen.)

Explain that bases are named the same way as any other ionic compound; the name of the cation is followed by the name of the anion (hydroxide for bases).

Chemistry Alive!

Milk of Magnesia

Play

Titration

Play

Ira Remsen Story

Play

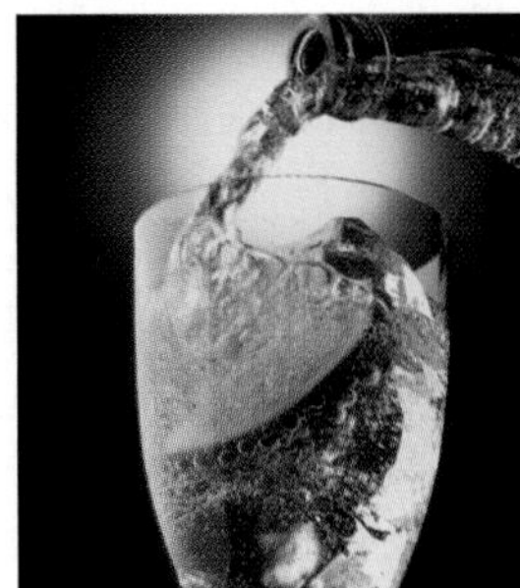
(a)

(b)

Figure 20.2
(a) Carbonated soft drinks contain carbonic acid (H_2CO_3), and many also contain phosphoric acid (H_3PO_4). These two acids give a drink its fizz and sharp, tangy taste. (b) Sodium hydroxide (NaOH) is used to prepare wood pulp for the manufacture of paper.

Bases are compounds that react with acids to form water and a salt. Milk of magnesia (a suspension of magnesium hydroxide in water) is a base used to treat the problem of excess stomach acid. Aqueous solutions of bases taste bitter and feel slippery. Like acids, bases will change the color of an acid–base indicator and can be strong or weak electrolytes.

Names and Formulas of Acids and Bases

An **acid** is a compound that produces hydrogen ions when dissolved in water. Therefore, the chemical formulas of acids are of the general form HX, where X is a monatomic or polyatomic anion. When the compound HCl(*g*) (hydrogen chloride) dissolves in water to form HCl(*aq*), it is named as an acid. How, then, is an acid named? To illustrate the naming of an acid, consider the following three rules as applied to the acid HX dissolved in water. Notice that the rules focus on the name of the anion, in particular the suffix of the anion name. **Table 20.1** summarizes these rules.

1. When the name of the anion (X) ends in *-ide*, the acid name begins with the prefix *hydro-*. The stem of the anion has the suffix *-ic* and is followed by the word *acid*. Therefore, HCl(*aq*) (X = chloride) is named *hydro*chlor*ic acid*. $H_2S(aq)$ (X = sulfide) is named *hydro*sulfur*ic acid*.
2. When the anion name ends in *-ite*, the acid name is the stem of the anion with the suffix *-ous*, followed by the word *acid*. Thus $H_2SO_3(aq)$ (X = sulfite) is named sulfur*ous acid*.
3. When the anion name ends in *-ate*, the acid name is the stem of the anion with the suffix *-ic*, followed by the word *acid*. Thus $HNO_3(aq)$ (X = nitrate) is named nitr*ic acid*.

These rules can be used in reverse fashion to write the formulas of acids when given their names. For example, what is the formula of chloric acid? According to Rule 3, chloric acid (*-ic* ending) must be a combination of hydrogen ion (H^+) and chlor*ate* ion (ClO_3^-). The formula of chloric acid is $HClO_3$. What is the formula of hydrobromic acid? Following Rule 1, hydrobromic acid (*hydro-* prefix and *-ic* suffix) must be a combination of hydrogen ion and bromide ion (Br^-). The formula of hydrobromic acid is HBr. What is the formula for phosphorous acid? Using Rule 2, hydrogen ion and phosphite ion (PO_3^{3-}) must be the components of phosphorous acid. The formula of phosphorous acid is H_3PO_3. (*Note:* Do not confuse *phosphorous* with *phosphorus*, the element name.)

Table 20.1

Naming Acids			
Anion ending	**Example**	**Acid name**	**Example**
-ide	Cl^- chlor*ide*	*hydro*-(stem)-*ic acid*	*hydro*chlor*ic acid*
-ite	SO_3^{2-} sulf*ite*	(stem)-*ous acid*	sulfur*ous acid*
-ate	NO_3^- nitr*ate*	(stem)-*ic acid*	nitr*ic acid*

A **base** is a compound that produces hydroxide ions when dissolved in water. Ionic compounds that are bases are named in the same way as any other ionic compound—the name of the cation followed by the name of the anion. For example, NaOH, a base used in making paper pulp, detergents, and soap, is called sodium hydroxide. What would you call $Ca(OH)_2$?  You write the formulas for bases by balancing the ionic charges, just as you do for any ionic compound.

Sample Problem 20-1

Name these compounds as acids.
a. HClO **b.** HCN **c.** H_3PO_4

1. *ANALYZE* ***Plan a problem-solving strategy.***
The rules for naming acids can be applied because the formulas of the acids are given.

2. *SOLVE* ***Apply the problem-solving strategy.***
a. Use Rule 2. The anion name (hypochlorite) ends in *-ite,* so add the suffix *-ous* to the anion stem, followed by the word *acid.* The correct name is hypochlorous acid.
b. Use Rule 1. The anion name (cyanide) ends in *-ide,* so this acid name begins with the prefix *hydro-* and ends with the suffix *-ic,* followed by the word *acid.* The correct name is hydrocyanic acid.
c. Use Rule 3. The anion name (phosphate) ends in *-ate.* So add the suffix *-ic* to the anion stem (in this case, modified slightly to *phosphor*), followed by the word *acid.* The correct name is phosphoric acid.

3. *EVALUATE* ***Do the results make sense?***
The names are consistent with the indicated rules.

Practice Problems

1. Name each acid or base.
a. HF **c.** KOH
b. HNO_3 **d.** H_2SO_4

2. Write formulas for each acid or base.
a. chromic acid
b. iron(II) hydroxide
c. hydriodic acid
d. lithium hydroxide

Chem ASAP!
Problem-Solving 2
Solve Problem 2 with the help of an interactive guided tutorial.

section review 20.1

3. Identify each property as applying to an acid, a base, or both.
a. bitter taste **c.** indicator color change
b. electrolyte **d.** sour taste

4. Write the formula for each acid or base.
a. barium hydroxide **c.** rubidium hydroxide
b. hydrobromic acid **d.** hydroselenic acid

5. Name each acid or base.
a. HF **c.** H_2CO_3
b. $HClO_3$ **d.** $Al(OH)_3$

 Chem ASAP! **Assessment 20.1** Check your understanding of the important ideas and concepts in Section 20.1.

Answers

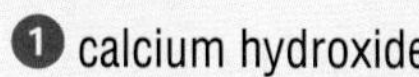

❶ calcium hydroxide

SECTION REVIEW 20.1

3. a. base
b. both
c. both
d. acid

4. a. $Ba(OH)_2$
b. HBr
c. RbOH
d. H_2Se

5. a. hydrofluoric acid
b. chloric acid
c. carbonic acid
d. aluminum hydroxide

20.1

Practice Problems Plus

Related Chapter Review Problem
Chapter Review problem 34 is related to Sample Problem 20-1.

Additional Practice Problem
The Chem ASAP! CD-ROM contains the following problem: Write formulas for the following acids and bases.
a. hydrobromic acid
b. boric acid
c. strontium hydroxide
d. arsenic acid
e. potassium hydroxide
f. sulfurous acid
(**a.** HBr **b.** H_3BO_3 **c.** $Sr(OH)_2$ **d.** H_3AsO_4 **e.** KOH **f.** H_2SO_3)

3 Assess

Evaluate Understanding

Have students name and write the formulas for: $HClO_4$ (perchloric acid), chloric acid ($HClO_3$), $HClO_2$ (chlorous acid), and hypochlorous acid (HClO).

Reteach

Review the Arrhenius definition of an acid and a base given in this section. Help students develop a flow chart for naming acids. Students should begin to recognize that most common acids can be grouped into two main categories: those that contain oxygen and those that do not. On the chalkboard, start the flow chart with the question: "Does the acid contain oxygen?" If "No", name as: *hydro-* + *element stem* + *-ic* + *acid.* If "Yes", then ask: "What does the name of the anion end in?" If "*-ite*", name as: *anion stem* + *-ous* + *acid.* If "*-ate*", name as: *anion stem* + *-ic* + *acid.* Using the flow chart, help students generate lists of formulas and names for common acids. Group the acids into oxygen-containing and oxygen-free acids.

1 Engage

Use the Visual

Ask students to examine the photograph and read the text that opens the section.

- **Ask the students how the pH scale is used to indicate the acidity of a solution and why this scale is used?** (The pH scale is related to the hydrogen-ion concentration in molarity. Because the $[H^+]$ in an aqueous solution is usually very small, chemists use the pH scale as a convenient way to express solution acidity. Values in chemistry that are much less than 1 often are represented using a "p" scale.)

Check Prior Knowledge

To assess students' knowledge of hydrogen ions and acidity, ask:

- **If a water molecule were to separate into ions, what ions might form?** (Show students that H^+ and OH^- ions represent a dissociated water molecule.)

Use the Visual

Ask students to examine Figure 20.3. Illustrate the reaction on the chalkboard: show how two water molecules can react to yield a hydronium ion and a hydroxide ion.

section 20.2

HYDROGEN IONS AND ACIDITY

objectives

- Given the hydrogen-ion or hydroxide-ion concentration, classify a solution as neutral, acidic, or basic
- Convert hydrogen-ion concentrations into values of pH and hydroxide-ion concentrations into values of pOH

key terms

- hydroxide ion (OH^-)
- hydronium ion (H_3O^+)
- self-ionization
- neutral solution
- ion-product constant for water (K_w)
- acidic solution
- basic solution
- alkaline solutions
- pH

A patient is brought to a hospital unconscious and with a fruity odor on his breath. The doctor suspects the patient has fallen into a diabetic coma. To confirm her diagnosis, she orders several tests, including one of the acidity of the patient's blood. The results from this test will be expressed in units of pH, not molar concentration. **How is the pH scale used to indicate the acidity of a solution, and why is this scale used?**

Hydrogen Ions from Water

As you already know, water molecules are highly polar and are in continuous motion, even at room temperature. Occasionally, the collisions between water molecules are energetic enough to transfer a hydrogen ion from one water molecule to another. See **Figure 20.3.** A water molecule that loses a hydrogen ion becomes a negatively charged **hydroxide ion (OH^-).** A water molecule that gains a hydrogen ion becomes a positively charged **hydronium ion (H_3O^+).**

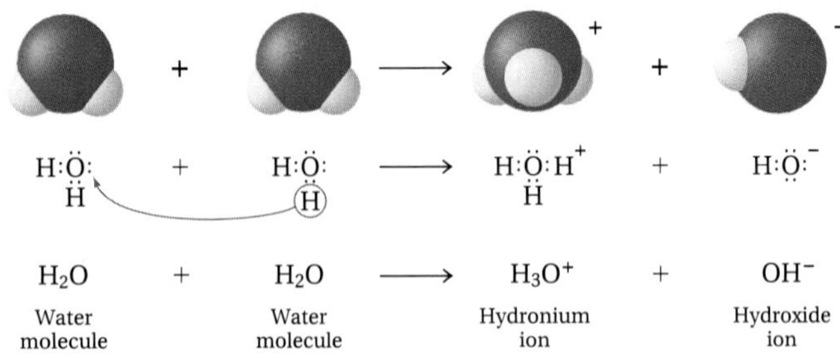

Figure 20.3
What is the name and formula of the particle that results when a water molecule gains a hydrogen ion? How is a hydroxide ion formed from a water molecule? 1

The reaction in which two water molecules produce ions is called the **self-ionization** of water. This reaction can also be written as a simple dissociation.

$$H_2O(l) \rightleftharpoons \underset{\text{Hydrogen ion}}{H^+(aq)} + \underset{\text{Hydroxide ion}}{OH^-(aq)}$$

In water or aqueous solution, hydrogen ions (H^+) are always joined to water molecules as hydronium ions (H_3O^+). The hydronium ions are themselves solvated to form species such as $H_9O_4^+$. Hydrogen ions in aqueous solution have several names. Some chemists call them protons. Others prefer to call them hydrogen ions, hydronium ions, or solvated protons. In this textbook, either H^+ or H_3O^+ is used to represent hydrogen ions in aqueous solution.

The self-ionization of water occurs to a very small extent. In pure water at 25 °C, the concentration of hydrogen ions ($[H^+]$) and the concentration of hydroxide ions ($[OH^-]$) are each only $1.0 \times 10^{-7}M$. This means that the concentrations of H^+ and OH^- are equal in pure water. Any aqueous solution in which $[H^+]$ and $[OH^-]$ are equal is described as a **neutral solution.**

Student Resources

- Section Review 20.2, Ch. 20 Practice Problems, Interpreting Graphics, and Quizzes from the Review Module
- Laboratory Recordsheet 20-1
- Laboratory Manual: Experiment 37
- Laboratory Practical 20-1
- Small-Scale Chemistry Lab Manual: Experiment 26

Technology Resources

Relevant technology resources include:

- Chem ASAP! CD-ROM
- Resource Pro CD-ROM
- Chemistry Alive! Videodisc: *Milk of Magnesia, Titration*
- ActivChemistry CD-ROM: *Acids and Bases*

In any aqueous solution, $[H^+]$ and $[OH^-]$ are interdependent. In other words, when $[H^+]$ increases, $[OH^-]$ decreases. When $[H^+]$ decreases, $[OH^-]$ increases. Le Châtelier's principle, which you learned about in Chapter 19, applies here. If additional ions (either hydrogen ions or hydroxide ions) are added to a solution, the equilibrium shifts. The concentration of the other type of ion decreases. More water molecules are formed in the process.

$$H^+(aq) + OH^-(aq) \rightleftharpoons H_2O(l)$$

For aqueous solutions, the product of the hydrogen-ion concentration and the hydroxide-ion concentration equals 1.0×10^{-14}.

$$[H^+] \times [OH^-] = 1.0 \times 10^{-14}$$

The product of the concentrations of the hydrogen ions and hydroxide ions in water is called the **ion-product constant for water (K_w).**

$$K_w = [H^+] \times [OH^-] = 1.0 \times 10^{-14}$$

Not all solutions are neutral. When some substances dissolve in water, they release hydrogen ions. For example, when hydrogen chloride dissolves in water, it forms hydrochloric acid. How does hydrochloric acid differ from hydrogen chloride? 2

$$HCl(g) \xrightarrow{H_2O} H^+(aq) + Cl^-(aq)$$

In such a solution, the hydrogen-ion concentration is greater than the hydroxide-ion concentration. The hydroxide ions are present from the self-ionization of water. An **acidic solution** is one in which $[H^+]$ is greater than $[OH^-]$. Therefore, the $[H^+]$ of an acidic solution is greater than $1.0 \times 10^{-7}M$.

When sodium hydroxide dissolves in water, it forms hydroxide ions in solution.

$$NaOH(s) \xrightarrow{H_2O} Na^+(aq) + OH^-(aq)$$

In such a solution, the hydrogen-ion concentration is less than the hydroxide-ion concentration. The hydrogen ions are present from the self-ionization of water. A **basic solution** is one in which $[H^+]$ is less than $[OH^-]$. Therefore, the $[H^+]$ of a basic solution is less than $1.0 \times 10^{-7}M$. Basic solutions are also known as **alkaline solutions.** Look at **Figure 20.4.** What are the names of the acids and bases shown? 3

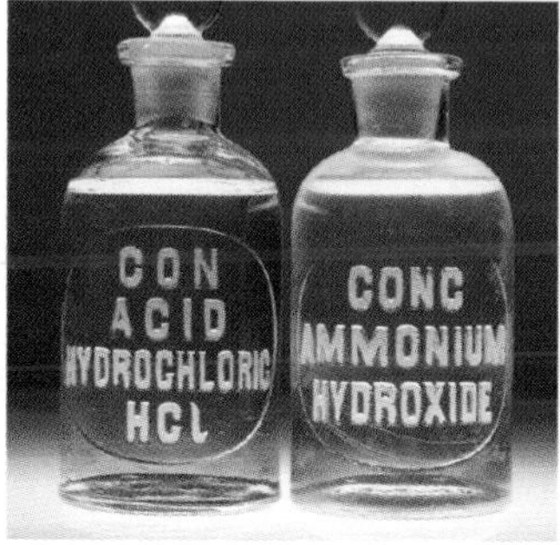

(a)

(b)

(c)

Figure 20.4
(a) *One of these two chemical reagents is an acid and the other is a base. Which of these reagents would increase the hydrogen-ion concentration when added to an aqueous solution? Which would increase the hydroxide-ion concentration?* ***(b)*** *Unrefined hydrochloric acid, commonly known as muriatic acid, is used to clean stone buildings and swimming pools.* ***(c)*** *Sodium hydroxide, or lye, is commonly used as a drain cleaner.* 4

20.2

2 Teach

TEACHER DEMO

Set up test tubes with acids and bases of varying strength. In addition to HCl and NaOH, include household products such as lemon juice, vinegar, shampoo, and liquid detergent. Use a pH meter to measure the pH of each solution. Explain that the pH meter measures the concentration of hydrogen ions.

Think Critically

Have students recall what they learned about the properties of acids and bases in Section 20.1. Ask students whether water meets the operational definitions of an acid or a base as presented in Section 20.1.

MEETING DIVERSE NEEDS

At Risk Students who do not have a good understanding of scientific notation may have trouble understanding the relationship between hydrogen- and hydroxide-ion concentrations. Draw diagrams to help students visualize the shift in equilibrium that occurs when one or the other kind of ion is added or removed.

Answers

1 hydronium ion (H_3O^+); A water molecule loses a hydrogen ion to form a hydroxide ion.
2 Hydrogen chloride is a gas [$HCl(g)$]; hydrochloric acid is aqueous hydrogen chloride [$HCl(aq)$], which ionizes to form $H^+(aq)$ and $Cl^-(aq)$
3 hydrochloric acid, ammonium hydroxide
4 HCl increases the hydrogen-ion concentration. Ammonium hydroxide (really aqueous ammonia) increases the hydroxide-ion concentration.

Discuss

Explain to the students that all aqueous systems contain both hydrogen and hydroxide ions due to self-ionization of water. About 1 molecule of water out of 550 000 000 will dissociate. This amounts to 1 g of hydrogen ions and 17 g of hydroxide ions in 10 000 000 L of water. No matter how small the concentration of ions, both ions are always present in a water solution.

The relationship between the concentrations of H^+ and OH^- in aqueous solutions at constant temperature is similar to the relationship between the pressure and volume of a gas at constant temperature. In both cases, the relationship is inverse. The product of the two quantities is a constant; as one quantity increases the other decreases.

Critical Thinking

Pure water self-ionizes to form hydrogen and hydroxide ions, yet it does not conduct electric current well. Ask students to infer a possible reason for this phenomenon? (Pure water is a poor conductor because the concentrations of the ions are very low.)

Practice Problems Plus

Additional Practice Problem

This problem is related to Sample Problem 20-2: The $[OH^-]$ of a carbonated soft drink is $1.0 \times 10^{-11}M$. What is the $[H^+]$ of this solution? Is the solution acidic or basic? ($1.0 \times 10^{-3}M$; acidic)

Sample Problem 20-2

If the $[H^+]$ in a solution is $1.0 \times 10^{-5}M$, is the solution acidic, basic, or neutral? What is the $[OH^-]$ of this solution?

1. *ANALYZE* ***List the knowns and the unknowns.***

   Knowns:
   - $[H^+] = 1.0 \times 10^{-5}M$
   - Ion-product constant for water: $K_w = [H^+] \times [OH^-] = 1 \times 10^{-14}$

   Unknowns:
   - solution = acidic, basic, or neutral?
   - $[OH^-] = ?M$

2. *CALCULATE* ***Solve for the unknowns.***

   $[H^+] = 1.0 \times 10^{-5}M$. Because this is greater than $1.0 \times 10^{-7}M$, the solution is acidic. By definition, $K_w = [H^+] \times [OH^-]$.

   Therefore, $[OH^-] = \frac{K_w}{[H^+]}$.

   Substituting the known numerical values, $[OH^-]$ is computed as follows.

   $$[OH^-] = \frac{1.0 \times 10^{-14}}{1.0 \times 10^{-5}M} = 1.0 \times 10^{-9}M$$

3. *EVALUATE* ***Do the results make sense?***

   If $[H^+]$ is greater than $1.0 \times 10^{-7}M$, then $[OH^-]$ must be less than $1.0 \times 10^{-7}M$. At $1 \times 10^{-9}M$, $[OH^-]$ is less than $1 \times 10^{-7}M$. Notice that when the concentration of the acid is 1.0×10^{-x} and the concentration of the base is 1.0×10^{-y}, $x + y = 14$. In this special case, you can easily find the value of $[H^+]$ when you know $[OH^-]$, or vice versa. To do so, subtract the exponent of the known, either $[H^+]$ or $[OH^-]$, from 14. For example, if

   $$[H^+] = 1.0 \times 10^{-9}M,$$
   $$\text{then } [OH^-] = 1.0 \times 10^{-(14-9)}M = 1.0 \times 10^{-5}M.$$

Practice Problems

6. If the hydroxide-ion concentration of an aqueous solution is $1.0 \times 10^{-3}M$, what is the $[H^+]$ in the solution? Is the solution acidic, basic, or neutral?
7. Classify each solution as acidic, basic, or neutral.
   a. $[H^+] = 6.0 \times 10^{-10}M$
   b. $[OH^-] = 3.0 \times 10^{-2}M$
   c. $[H^+] = 2.0 \times 10^{-7}M$
   d. $[OH^-] = 1.0 \times 10^{-7}M$

The pH Concept

Expressing hydrogen-ion concentration in molarity is cumbersome. A more widely used system for expressing $[H^+]$ is the pH scale, proposed in 1909 by the Danish scientist Søren Sørensen (1868–1939). On the pH scale, which ranges from 0 to 14, neutral solutions have a pH of 7. A pH of 0 is strongly acidic. (1) What is a solution with a pH of 14? Calculating the pH of a solution is straightforward. The **pH** of a solution is the negative logarithm of the hydrogen-ion concentration. The pH may be represented mathematically using the following equation.

$$pH = -\log[H^+]$$

Answers

(1) strongly basic
(2) pH 4.0

In a neutral solution, $[H^+] = 1 \times 10^{-7}M$. The pH of a neutral solution is 7.0.

$$\begin{aligned} pH &= -\log(1 \times 10^{-7}) \\ &= -(\log 1 + \log 10^{-7}) \\ &= -(0.0 + (-7.0)) \\ &= 7.0 \end{aligned}$$

To summarize, the pH of pure water or a neutral aqueous solution is 7.0. A solution in which $[H^+]$ is greater than $1 \times 10^{-7}M$ has a pH less than 7.0 and is acidic. A solution with a pH greater than 7 is basic and has a $[H^+]$ of less than $1 \times 10^{-7}M$. See **Figure 20.5** for a visual representation of this information.

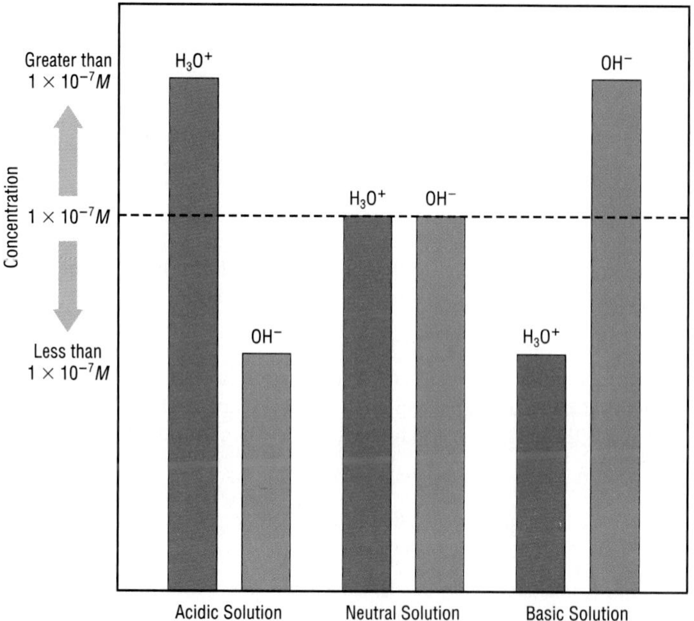

Figure 20.5
In acidic solutions, $[H^+]$ is greater than $[OH^-]$. In basic solutions, $[OH^-]$ is greater than $[H^+]$. Neutral solutions are those in which $[H^+]$ is equal to $[OH^-]$.

- Acidic solution: $pH < 7.0$ $[H^+]$ greater than $1 \times 10^{-7}M$
- Neutral solution: $pH = 7.0$ $[H^+]$ equals $1 \times 10^{-7}M$
- Basic solution: $pH > 7.0$ $[H^+]$ less than $1 \times 10^{-7}M$

The pH values of several common aqueous solutions are listed in **Table 20.2** on the following page. The table also summarizes the relationship among $[H^+]$, $[OH^-]$, and pH. You may notice that pH can sometimes be read from the value of $[H^+]$. If $[H^+]$ is written in scientific notation and has a coefficient of 1, then the pH of the solution equals the exponent, with the sign changed from minus to plus. For example, a solution with $[H^+] = 1 \times 10^{-2}M$ has a pH of 2.0. What is the pH of a solution with $[H^+] = 1 \times 10^{-4}M$?

20.2

Use the Visual

Have students study Figure 20.5 and the text that follows. Explain that the pH scale shows the relationship between pH and the hydrogen-ion concentration. Write the expression for the ion-product constant for water on the board, and remind students about the relationship between the concentrations of hydronium ion and hydroxide ion in an aqueous solution. The product of these concentrations in aqueous solutions is always $1 \times 10^{-14}M^2$. Arbitrarily assign concentration values to each of the hydronium-ion bars in Figure 20.5 for the acidic and basic solutions (for the neutral solution $[H^+]$ is $1 \times 10^{-7}M$); then ask students to calculate the pH of those solutions. Ask:

- **What is $[OH^-]$ for these solutions?**

Have students refer to Table 20.2 to find examples of aqueous systems with these $[H^+]$ and $[OH^-]$.

Meeting Diverse Needs

Gifted Have students identify occupations in which people need to measure pH. Interested students may select a local person in an occupation from the list and ask to interview that person about his or her use of pH measurement. Have students write up their interviews as if they were preparing them for a newspaper article.

At Risk Some students may need help working with logarithms. Pair these students with students who have strong math skills. Be sure less-experienced students are directly involved in solving the problems, and receive assistance only as needed.

Chemistry Alive!

Milk of Magnesia

Play

Titration

Play

Discuss

Introduce pH as a simpler way to express hydrogen-ion concentration. Point out that it is easier to say the pH of a solution is 3.00 than to say that the hydrogen ion concentration is equal to 1.0×10^{-3} moles per liter. For students who have no concept of logarithms, explain that pH is found by taking the negative of the power (exponent) of the hydrogen ion concentration and expressing it as a whole number. Most pH values are positive numbers, but negative values are also possible. The pH of 10*M* HCl solution is −1.00, for example.

Critical Thinking

Have the students compare the pH and pOH of a solution by posing a series of questions such as the following:

- **In an acidic solution with a pH of 3.25, are there any OH^- ions?** (Yes, there are always some OH^- ions in an aqueous solution.)
- **How do you know this?** (The product of the hydrogen ion concentration and the hydroxide ion concentration must always equal K_w.)
- **What is the pOH of this solution?** (pH + pOH = 14; pOH = 14 − 3.25 = 10.75)

Table 20.2

Relationship among $[H^+]$, $[OH^-]$, and pH

	$[H^+]$ (mol/L)	$[OH^-]$ (mol/L)	pH	Aqueous system
Increasing acidity ↑	1×10^{0}	1×10^{-14}	0.0	1*M* HCl
	1×10^{-1}	1×10^{-13}	1.0	0.1*M* HCl, Gastric juice
	1×10^{-2}	1×10^{-12}	2.0	Lemon juice
	1×10^{-3}	1×10^{-11}	3.0	
	1×10^{-4}	1×10^{-10}	4.0	Tomato juice
	1×10^{-5}	1×10^{-9}	5.0	Black coffee
	1×10^{-6}	1×10^{-8}	6.0	Milk
Neutral	1×10^{-7}	1×10^{-7}	7.0	Pure water
Increasing basicity ↓	1×10^{-8}	1×10^{-6}	8.0	Blood, Sodium hydrogen carbonate, sea water
	1×10^{-9}	1×10^{-5}	9.0	
	1×10^{-10}	1×10^{-4}	10.0	Milk of magnesia
	1×10^{-11}	1×10^{-3}	11.0	Household ammonia
	1×10^{-12}	1×10^{-2}	12.0	Washing soda
	1×10^{-13}	1×10^{-1}	13.0	0.1*M* NaOH
	1×10^{-14}	1×10^{0}	14.0	1*M* NaOH

In a definition similar to that of pH, the pOH of a solution equals the negative logarithm of the hydroxide-ion concentration.

$$pOH = -\log[OH^-]$$

A neutral solution has a pOH of 7. A solution with a pOH less than 7 is basic. A solution with a pOH greater than 7 is acidic. A simple relationship between pH and pOH makes it easy to find either one when the other is known.

$$pH + pOH = 14$$
$$pH = 14 - pOH$$
$$pOH = 14 - pH$$

For pH calculations, you should express the hydrogen-ion concentration in scientific notation. For example, a hydrogen-ion concentration of 0.0010*M*, rewritten as $1.0 \times 10^{-3}M$ in scientific notation, has two significant figures. The pH of this solution is 3.00, with the two numbers to the right of the decimal point representing the two significant figures in the concentration. A solution with a pH of 3.00 is acidic, as shown in
1 **Figure 20.6.** How many significant figures are indicated in a pH of 7.61?

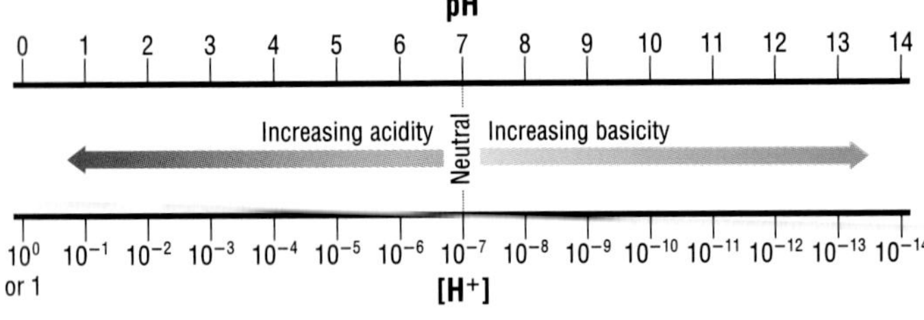

Figure 20.6
The pH scale shows the relationship between pH and the hydrogen-ion concentration. Notice that acids have lower pHs than bases.

Answer

1 pH 7.61 has two significant figures, indicated by the underlined digits.

USING LOGARITHMS

Logarithms are used throughout mathematics and science. Examples include the decibel scale for loudness, the Richter scale for earthquakes, and the pH scale for acidity. Because the pH scale is a logarithmic scale, it allows a tremendous range of values (from 1 to 10^{-14}) to be expressed as a number between 1 and 14.

The common logarithm of a number is the exponent to which ten must be raised to produce the given number: $\log x = y$ if $x = 10^y$. Thus $\log 0.01 = -2$ because $0.01 = 10^{-2}$, and $\log 10\,000 = 4$ because $10\,000 = 10^4$. The common logarithm is also known as the base-10 logarithm. Logarithms can have other bases, but this textbook only uses common logarithms.

Logarithms of numbers between 1 and 10 can be evaluated directly using **Table B.1** in Appendix B of this textbook. For example, $\log 7.21 = 0.8579$, as shown below.

x	0	1	2
7.1	.8513	.8519	.8525
7.2	.8573	.8579	.8585
7.3	.8633	.8639	.8645

To find the logarithm of a number that is less than 1 or greater than 10, write the number in scientific notation and use the formulas below. See Example 1.

$$\log (a \times b) = \log a + \log b, \text{ and } \log (10^x) = x$$

A logarithm is rounded so that its number of decimal places equals the number of significant figures in the original number.

The antilog of a number x is the number y whose logarithm is x. Because $\log y = x$ means $y = 10^x$, the antilog of x is the same as 10^x. Antilogs of numbers between 0 and 1 can be found directly from **Table B.1.** To find the antilog of any number, write it as a sum of its decimal part (between 0 and 1) and its integer part (less than or equal to the given number). Then use the formula $10^{a+b} = 10^a \times 10^b$. See Example 2.

Many calculators can be used to find logarithms and antilogs. In general, use [LOG] for a logarithm, or [10^x] for an antilog. The exact keystrokes depend on your calculator. Do not confuse the [LOG] key with the [LN] key. The [LN] key gives another kind of logarithm, called the natural logarithm, whose base is the constant $e = 2.718$.

Example 1

Evaluate log 0.0000721.

$$\begin{aligned}\log 0.0000721 &= \log (7.21 \times 10^{-5}) \\ &= \log 7.21 + \log (10^{-5}) \\ &= 0.8579 + (-5) \\ &= -4.1421\end{aligned}$$

The original number had 3 significant figures, so the result should be rounded to 3 decimal places.

$$\log 0.0000721 = -4.142$$

Example 2

Find the antilog of -8.375.

Write -8.375 as a sum. Note that the decimal part is 0.625, not -0.375 or 0.375.

$$-8.375 = (-8.375 + 9) + (-9) = 0.625 + (-9)$$

In **Table B.1,** the number nearest to 0.625 is 0.6253, and its antilog is 4.22. So,

$$\begin{aligned}\text{antilog } (-8.375) &= 10^{-8.375} = 10^{0.625+(-9)} \\ &= 10^{0.625} \times 10^{-9} \\ &= 4.22 \times 10^{-9}\end{aligned}$$

Practice Problems

Prepare for upcoming problems in this chapter by using **Table B.1** in Appendix B to evaluate the following expressions. Then check your work using a calculator.

A. log 1.68
B. log (3.57×10^4)
C. log (2.18×10^{-8})
D. antilog -8
E. antilog 0.969
F. antilog ($-5 + 0.782$)
G. log 17 800
H. log 0.0067
I. log 0.0000738
J. antilog 6.281
K. antilog (-3.192)
L. antilog (-5.936)

CHEMath

Teaching Tips

Explain to students that the logarithm of a number is a convenient, shorthand method for expressing exceptionally small or large numbers. Check to see that students have a firm grasp of the numerical relationship between numbers written in scientific notation before going on to calculate logarithms. Have students construct a number line, similar to Figure 20.6, which shows where various numbers, expressed as negative and positive powers of 10, lie on the number line. Provide students with a list of 5–10 numbers in scientific notation and have them graph these on a number line. Then have students use the number line to answer questions about these same numbers such as: "Is 2.0×10^{-4} greater than or less than 5.0×10^{-5}?" (greater than)

As an extension, show students how to use semilogarithmic graph paper to graph the data in Figure 20.6. The trend shown by the graph should help reinforce the concept that larger, more positive pH values reflect lower concentrations of hydronium ions. Students can refer to the graph when solving problems involving integer and non-integer pH values. Point out that each 1 unit change in pH is equal to a tenfold change in the hydronium-ion concentration.

Answers to Practice Problems

A. 0.225
B. 4.553
C. -7.662
D. 1×10^{-8}
E. 9.31
F. 6.05×10^{-5}
G. 4.250
H. -2.17
I. -4.132
J. 1.91×10^6
K. 6.43×10^{-4}
L. 1.16×10^{-6}

Practice Problems Plus

Related Chapter Review Problems
Chapter Review problems 39a, 39d, 41c, and 41d are related to Sample Problem 20-3.

Additional Practice Problem
The Chem ASAP! CD-ROM contains the following problem: What are the pH values of the following three solutions, based on their hydrogen ion concentrations?
a. $[H^+] = 1.0 \times 10^{-11}M$
b. $[H^+] = 1.0 \times 10^{-8}M$
c. $[H^+] = 0.00001M$
(**a.** 11.00 **b.** 8.00 **c.** 5.0)

CHEMISTRY AND SCIENCE HISTORY

Søren Sørensen, a Danish biochemist, conceived the concept of pH while working on problems connected with the brewing of beer, a process in which the control of acidity is very important. After concluding that there must be a simpler way to express hydrogen-ion concentration, the idea of using the "hydrogen ion exponent" struck him. Sørensen first wrote the symbol pH^+ but later simplified it to pH. The letter "p" stood for the French word *puissance,* the German word *potenz,* and the English word *power,* and the letter "H" represented the hydrogen ion.

Sample Problem 20-3

What is the pH of a solution with a hydrogen-ion concentration of $1.0 \times 10^{-10}M$?

1. ***ANALYZE*** ***List the knowns and the unknown.***

Knowns:	*Unknown:*
• $[H^+] = 1.0 \times 10^{-10}M$	• pH = ?
• $pH = -\log [H^+]$	

2. ***CALCULATE*** ***Solve for the unknown.***

$$\begin{aligned} pH &= -\log [H^+] \\ &= -\log(1.0 \times 10^{-10}) \\ &= -(\log 1.0 + \log 10^{-10}) \\ &= -(0.00 + (-10.00)) \\ &= -(-10.00) \\ &= 10.00 \end{aligned}$$

3. ***EVALUATE*** ***Does the result make sense?***
$[H^+]$ is three orders of magnitude (1000 times) less than $1 \times 10^{-7}M$ (pH 7). Each order-of-magnitude decrease in $[H^+]$ equals an increase of 1 pH unit, so the unknown solution should have a pH of $7 + 3 = 10$. This problem can also be solved by inspection. The coefficient of $[H^+]$ is 1.0; therefore, the pH is the value of the exponent (-10) with the sign changed from minus to plus. However, the answer based on experimental data must be reported as 10.00 to show the two significant figures of the original value of $[H^+]$.

Practice Problems

8. Find the pH of each solution.
a. $[H^+] = 1.0 \times 10^{-4}M$
b. $[H^+] = 0.0010M$
c. $[H^+] = 1.0 \times 10^{-9}M$

9. What are the pH values of the following three solutions, based on their hydrogen-ion concentrations?
a. $[H^+] = 1.0 \times 10^{-12}M$
b. $[H^+] = 0.010M$
c. $[H^+] = 1.0 \times 10^{-4}M$

Chem ASAP!

Problem-Solving 9
Solve Problem 9 with the help of an interactive guided tutorial.

Sample Problem 20-4

The pH of an unknown solution is 6.00. What is its hydrogen-ion concentration?

1. ***ANALYZE*** ***List the knowns and the unknown.***

Knowns:	*Unknown:*
• pH = 6.00	• $[H^+] = ?M$
• $pH = -\log [H^+]$	

2. ***CALCULATE*** ***Solve for the unknown.***
First, rearrange the equation for the definition of pH to solve for the unknown.

$$-\log [H^+] = pH$$

Next, substitute the value of pH.

$$-\log [H^+] = 6.00$$

Sample Problem 20-4 (cont.)

Change the signs on both sides of the equation.

$$\log[H^+] = -6.00$$

Finally, determine the number that has a log of −6.00. The antilog of −6.00 is 1.0×10^{-6}. Therefore,

$$[H^+] = 1.0 \times 10^{-6} M.$$

3. ***EVALUATE*** ***Does the result make sense?***
This problem can be solved by inspection. Any integral value of pH, such as 3.00, 5.00, and so forth, can easily be converted to $[H^+]$. The sign of the pH value is changed from plus to minus and is used as the exponent in the numerical expression of $[H^+]$. A decimal point and zeros are attached to the coefficient (1) to obtain the proper number of significant figures. For example, for pH 4.00, $[H^+] = 1.0 \times 10^{-4} M$; for pH 11.00, $[H^+] = 1.0 \times 10^{-11} M$.

Practice Problems

10. Calculate $[H^+]$ for each solution.
a. pH = 5.00 **c.** pH = 6.00
b. pH = 7.00 **d.** pH = 3.00

11. What are the hydrogen-ion concentrations for solutions with the following pH values?
a. 4.00 **b.** 11.00 **c.** 8.00

Calculating pH Values

Most pH values are not whole numbers. For example, **Table 20.2** on page 584 shows that milk of magnesia has a pH of 10.5. Using the definition of pH, this means that $[H^+]$ must equal $1 \times 10^{-10.5} M$. Thus $[H^+]$ must be less than $1 \times 10^{-10} M$ (pH 10.0), but greater than $1 \times 10^{-11} M$ (pH 11.0).

If $[H^+]$ is written in scientific notation but its coefficient is not 1, then you need a table of common logarithms or a hand calculator with a log function key to calculate pH. A four-place table of common logarithms is provided in Appendix B. Sample Problem 20-5 shows how to make such a pH calculation.

Sample Problem 20-5

What is the pH of a solution if $[OH^-] = 4.0 \times 10^{-11} M$?

1. ***ANALYZE*** ***List the knowns and the unknown.***

Knowns:
- $[OH^-] = 4.0 \times 10^{-11} M$
- $K_w = [OH^-] \times [H^+] = 1 \times 10^{-14}$
- $pH = -\log[H^+]$

Unknown:
- pH = ?

2. ***CALCULATE*** ***Solve for the unknown.***
To calculate pH, first calculate $[H^+]$ by using the definition of K_w.

$$K_w = [OH^-] \times [H^+]$$

$$[H^+] = \frac{K_w}{[OH^-]} = \frac{1.0 \times 10^{-14}}{4.0 \times 10^{-11} M}$$

$$= 0.25 \times 10^{-3} M$$

$$= 2.5 \times 10^{-4} M$$

Discuss

Point out to students that the pH scale can be compared to the Richter scale, which measures the strength of earthquakes. On each scale, a change of one unit represents a tenfold change in the value being measured. On the Richter scale, for example, a tremor measuring 4.0 is ten times stronger than one measuring 3.0. On the pH scale, the hydrogen-ion concentration of a solution with a pH of 3.0 is ten times greater than that of a solution with a pH of 4.0. Remind students that the lower the pH value, the more acidic the solution.

Practice Problems Plus

Related Chapter Review Problems
Chapter Review problems 40 and 42 are related to Sample Problem 20-4.

Discuss

Remind students that if they know the pH, the hydrogen-ion concentration, or the hydroxide-ion concentration, they can use K_w, the ion-product constant for water, to find the other two.

Meeting Diverse Needs

Gifted Remind students that strong bases such as sodium hydroxide are used in drain and oven cleaners. Have students find out how a particular brand of drain or oven cleaner works and prepare a written or class report.

Gifted Have students investigate toothpastes that contain baking soda. Ask them to describe and write equations for the chemical reactions that make such toothpastes effective. (Baking soda neutralizes the lactic acid produced by bacteria in the mouth. At low pH, the tooth enamel breaks down).

TEACHER DEMO

Without explanation, use familiar substances to demonstrate a color change resulting from a change in pH. You could add lemon juice to tea, for example, or add an Alka-Seltzer tablet to a glass of 10% unsweetened grape juice, or a few drops of household ammonia to another glass of juice. Have the students try to explain the changes they observe.

Use the Visual

Have students study Figures 20.7 and 20.8, which show the ranges for selected acid–base indicators. Compare the usefulness of acid–base indicators and pH meters. Beyond the obvious difference between qualitative and quantitative data, there are questions of cost and training required for implementation. The time of response is also an issue. A gardener may use an acid-base indicator to test soil; feedback is immediate. The gardener could choose to send a soil sample to a regional agricultural extension site for more accurate analysis; feedback is delayed.

Practice Problems Plus

Related Chapter Review Problems

Chapter Review problems 39b and 39c are related to Sample Problem 20-5. Chapter Review problems 41a, 41b, and 42 are related to Sample Problem 20-6.

Additional Practice Problem

The Chem ASAP! CD-ROM contains this problem related to Sample Problem 20-5: Calculate the pH for each solution.

a. $[OH^-] = 2.5 \times 10^{-11}M$
b. $[OH^-] = 6.4 \times 10^{-8}M$
c. $[OH^-] = 8.8 \times 10^{-5}M$
(**a.** 3.40 **b.** 6.81 **c.** 9.94)

Sample Problem 20-5 (cont.)

With the value of $[H^+]$ determined, use the definition of pH to solve for the pH.

$$\begin{aligned} pH &= -\log[H^+] \\ &= -\log(2.5 \times 10^{-4}) \\ &= -(\log 2.5 + \log 10^{-4}) \end{aligned}$$

A log table or calculator indicates that log 2.5 = 0.40, and $\log 10^{-4} = -4$. Insert these values to find the pH.

$$\begin{aligned} pH &= -(0.40) - (-4) \\ &= -0.40 + 4 \\ &= 3.60 \end{aligned}$$

3. ***EVALUATE*** ***Does the result make sense?***
A solution in which $[OH^-]$ is less than $1 \times 10^{-7}M$ would be acidic because $[H^+]$ would be greater than $1 \times 10^{-7}M$. $[OH^-]$ is less than $10^{-10}M$ but greater than $10^{-11}M$. Therefore, the solution is somewhere between pOH 10 and pOH 11. Because pOH + pH = 14, the pH should be less than 4 and greater than 3.

Practice Problems

12. Calculate the pH of each solution.
a. $[H^+] = 5.0 \times 10^{-6}M$
b. $[H^+] = 8.3 \times 10^{-10}M$
c. $[H^+] = 2.7 \times 10^{-7}M$

13. Calculate the pH of each solution.
a. $[OH^-] = 4.3 \times 10^{-5}M$
b. $[OH^-] = 2.0 \times 10^{-5}M$
c. $[OH^-] = 4.5 \times 10^{-11}M$

Chem ASAP!

Problem-Solving 13

Solve Problem 13 with the help of an interactive guided tutorial.

You can calculate the hydrogen-ion concentration of a solution if you know the pH. For example, if the solution has a pH of 3.00, then $[H^+] = 1.0 \times 10^{-3}M$. When the pH is not a whole number, you will need log tables or a calculator with a y^x function key to calculate the hydrogen-ion concentration. For example, if the pH is 3.70, the hydrogen-ion concentration must be greater than $1.0 \times 10^{-4}M$ (pH 4.0) and less than $1.0 \times 10^{-3}M$ (pH 3.0). To get an accurate value, use log tables or a calculator.

Sample Problem 20-6

What is $[H^+]$ of a solution if the pH = 3.70?

1. ***ANALYZE*** ***List the knowns and the unknown.***
Knowns:
- pH = 3.70
- $pH = -\log[H^+]$

Unknown:
- $[H^+] = ?M$

2. ***CALCULATE*** ***Solve for the unknown.***
First rearrange the equation $pH = -\log[H^+]$.

$$\log[H^+] = -pH = -3.70$$

A log table cannot be used directly to find a number that has a negative log. To avoid this problem, add and then subtract the whole number that is closest to and larger than the negative log. In this case, the negative log is 3.70, and the whole number is 4.

Sample Problem 20-6 (cont.)

$$\log[\text{H}^+] = (-3.70 + 4) - 4$$
$$= 0.30 - 4$$
$$[\text{H}^+] = 10^{(0.30-4)}$$
$$= 10^{0.30} \times 10^{-4}$$

From the log table, the number with a log of 0.30 is 2.0; the antilog of 0.30 is thus 2.0. The number with a log of -4 is 10^{-4}. Therefore, $[\text{H}^+] = 2.0 \times 10^{-4}M$. A calculator with a y^x function key can be used because $[\text{H}^+] = 10^{-\text{pH}}$; that is, $[\text{H}^+]$ is of the form y^x. Thus change the sign of the given pH, in this case to -3.70. Enter $y = 10$, $x = -3.70$, and press the y^x key in the order required by the calculator. The readout gives 1.995×10^{-4}. Rounded to two significant figures, $[\text{H}^+] = 2.0 \times 10^{-4}M$.

3. *EVALUATE* ***Does the result make sense?***
A solution of pH 3.70 should have a $[\text{H}^+]$ between $1 \times 10^{-4}M$ (pH 4) and $1 \times 10^{-3}M$ (pH 3), as found. Because $[\text{H}^+]$ is greater than $1 \times 10^{-7}M$, the solution is acidic.

Practice Problems

14. What is the molarity of $[\text{H}^+]$ in each solution?
a. pH = 7.30 **c.** pH = 7.05
b. pH = 1.80 **d.** pH = 6.70

15. Calculate the value of $[\text{OH}^-]$ in each solution in Practice Problem 14.

Measuring pH

People need to be able to measure the pH of the solutions they use. From maintaining the correct acid–base balance in a swimming pool, to creating soil conditions ideal for plant growth, to making medical diagnoses, pH measurement has valuable applications. For preliminary pH measurements and for small-volume samples, indicators such as the ones shown in **Figure 20.7** are often used. For precise and continuous measurements, a pH meter is preferred.

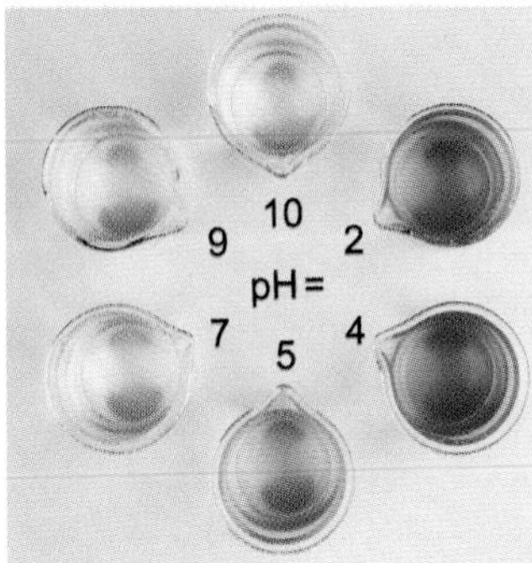

Methyl red

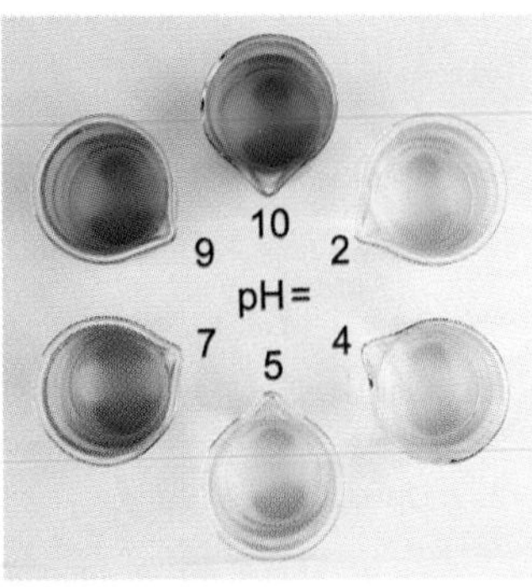

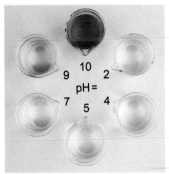

Phenolphthalein

Bromthymol blue

Figure 20.7
Acid–base indicators respond to pH changes over a specific range. Methyl red (left) changes from red to yellow at pH 5–7. Phenolphthalein (center) changes from colorless to pink at pH 7–9. Bromthymol blue (right) changes from yellow to blue at pH 5–7.

20.2

TEACHER DEMO

Demonstrate how pH indicators react to the acidity of their environment. Explain that many natural and synthetic pigments are weak acids that change color with varying pH. Prepare 0.1% solutions of the indicators thymol blue, methyl red, bromothymol blue, and phenolphthalein to test the pH of a series of dilute aqueous buffers. (acetic acid/acetate, citric acid/citrate, monosodium phosphate/disodium phosphate, and carbonate/hydrogen carbonate. Adjust the pH to desired values using HCl. Either create a range of buffers spanning pH 4 to pH 10 or higher, or purchase standard buffers with different pH values.)

Transfer a small volume of each buffer to a test tube. Label the tubes with their respective pH values. Add 5 drops of indicator to each test tube. Set up a separate rack of tubes for each indicator and ask students to infer which pH range each indicator is most suitable for. Students should note the colors and pH of the tubes. In addition, add 5 drops of the indicators to two or three tubes containing a solution with an "unknown" pH and have students estimate its pH value using the reference tubes. Ask:

- Why would an investigator prefer to use a pH meter to measure pH? (Individual indicators are only responsive to pH changes in narrow ranges. Many different indicators are needed to span the entire pH range. A pH meter allows an investigator to collect *quantitative* values on a continuous basis throughout the entire pH range. The precision and accuracy of pH meters are superior to standard indicators. pH meters can be calibrated and used reliably at different temperatures.)

ACTIVITY

Provide students with a variety of chemical supply catalogs and have them compare the costs and capabilities of pH meters vs. other qualitative acid–base indicators. Some pH meters are hand-held devices; others are larger, bench instruments, which can be connected to computer hardware and are intended for more detailed analyses. Students should compare the resolution and precision of each of the instruments and pH indicators. Ask:

- What are some advantages and uses of hand-held devices?
- Why would a chemist want to have a pH meter that can be calibrated at different temperatures?

Acid–Base Indicators An indicator (In) is an acid or a base that undergoes dissociation in a known pH range. An indicator is a valuable tool for measuring pH because its acid form and base form have different colors in solution. The following generalized equation represents the dissociation of an indicator (HIn)

$$\underset{\text{Acid form}}{HIn(aq)} \underset{H^+}{\overset{OH^-}{\rightleftharpoons}} \underset{\text{Base form}}{H^+(aq) + In^-(aq)}$$

The acid form dominates the dissociation equilibrium at low pH (high $[H^+]$), and the base form dominates the equilibrium at high pH (high $[OH^-]$). For each indicator, the change from dominating acid form to dominating base form occurs in a narrow range of approximately two pH units. Within this narrow range, the color of the solution is a mixture of the colors of the acid and the base forms. Knowing the pH range over which this color change occurs can give you a rough estimate of the pH of a solution. At all pH values below this range, you would see only the color of the acid form. At all pH values above this range, you would observe only the color of the base form. You could eventually zero in on a more precise estimate of the pH of the solution by repeating the experiment with indicators that have different pH ranges for their color changes. Many different indicators are needed to span the entire pH spectrum. **Figure 20.8** shows the pH ranges of some commonly used indicators.

Figure 20.8
Each indicator changes color at a different pH. Which indicator would you choose to show that a reaction solution has changed from pH 3 to pH 4? 1

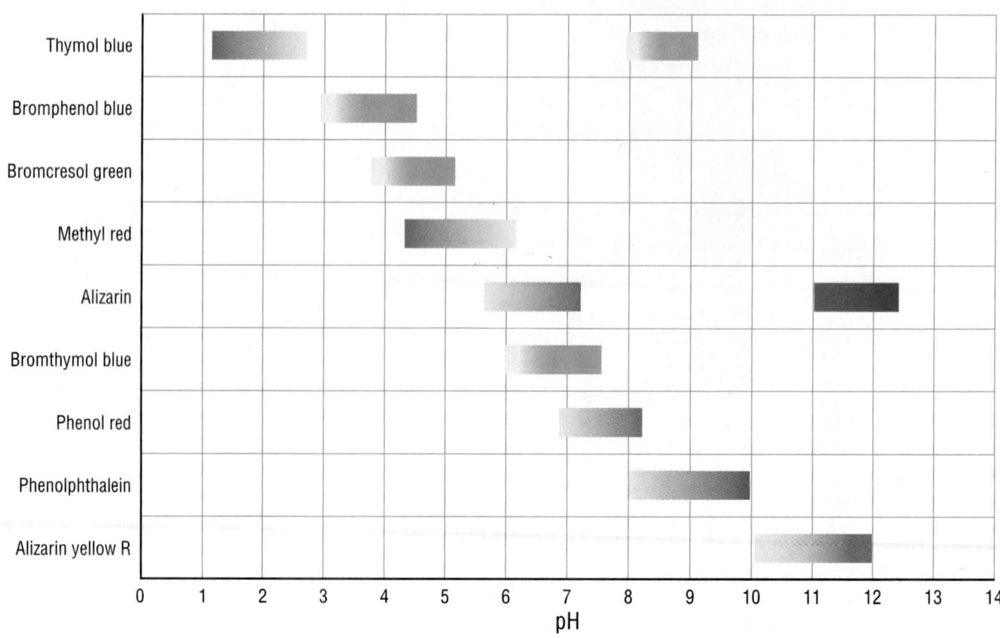

Answers

1 bromphenol blue
2 vinegar: pH about 3; soda water: pH about 4; ammonia: pH about 10
3 milk of magnesia

(a)

(b)

Figure 20.9
You can determine the approximate pH of a substance by testing it with pH indicator and comparing the color with standards. ***(a)*** *Universal indicator solution has been added to solutions of known pH in the range from 1 to 12 to produce a set of reference colors.* ***(b)*** *Universal indicator has been added to samples of some familiar household products: vinegar, soda water, and ammonia solution. Use the colors in* ***(a)*** *to assign pH values to the items in* ***(b)****.* 2

Although indicators are useful tools, they do have certain characteristics that limit their usefulness. The listed pH values of indicators are usually given for 25 °C. At other temperatures, an indicator may change color at a different pH. If the solution being tested is not colorless, the color of the indicator may be distorted. Dissolved salts in a solution may also affect the indicator's dissociation. Often, using indicator strips can help overcome these problems. An indicator strip is a piece of paper or plastic impregnated with an indicator. The paper is dipped into an unknown solution and compared with a color chart to measure the pH. Some indicator paper is impregnated with multiple indicators that give a palette of colors over a wide pH range. See **Figure 20.9.**

pH Meters A pH meter is used to make rapid, accurate pH measurements. Most chemistry laboratories have a pH meter. If a pH meter is connected to a computer or chart recorder, it can be used to make a continuous recording of pH changes.

Figure 20.10 shows a pH meter. The combination electrode—a glass electrode and a reference electrode—is connected to a millivoltmeter.

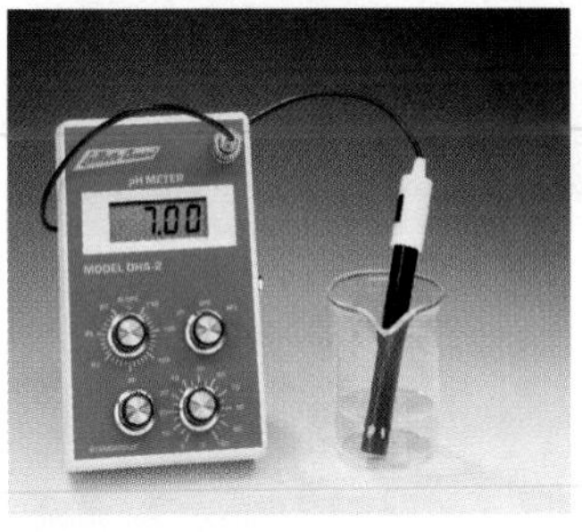

(a)

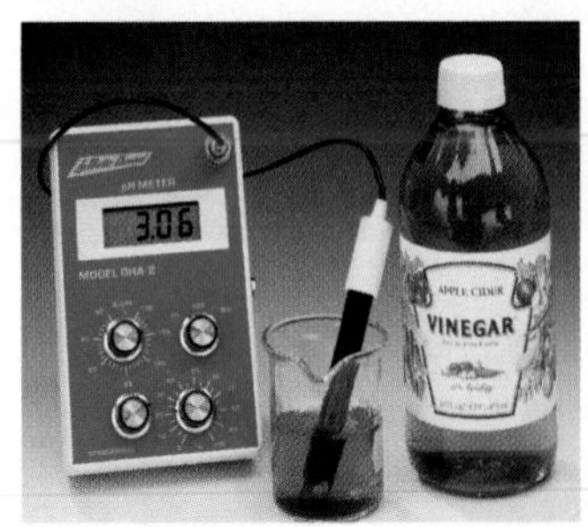

(b)

(c)

Figure 20.10
(a) *A pH meter is used to measure hydrogen ion concentrations. The instrument gives the user a direct reading of pH. For continuous monitoring, the pH meter may be connected to a strip-chart recorder.* ***(b)*** *The pH of vinegar, a dilute aqueous solution of ethanoic (acetic) acid, is about 3.* ***(c)*** *The pH of milk of magnesia, an aqueous suspension of magnesium hydroxide, is 10.5. In which of the test solutions is* $[H^+] < 1 \times 10^{-7} M$? 3

20.2

Cooperative Learning

Have students do research on natural substances that can serve as pH indicators, such as those from carrot stems, beet roots, hollyhocks, rhubarb, and cherries. Additional acid–base indicators include turmeric or curry powder dissolved in alcohol; or a solution containing crushed laxative tablets (a source of phenolphthalein). Divide the class into groups and ask each group to choose an indicator to investigate. Groups should meet briefly to assign responsibilities for securing the indicator material and gathering the necessary equipment. They should choose a time and place to prepare their indicator. On an appointed day, each group should bring its indicator to class and demonstrate its acid-base color change. You may wish to have students perform this activity after they have completed the MINI LAB on page 593.

3 Assess

Evaluate Understanding

Have students use equations to describe the relationship between the concentrations of hydrogen and hydroxide ions in pure water and to show what happens to the equilibrium when HCl is added to water. Repeat the procedure for NaOH. (When acid is added, the $[H^+]$ increases and the $[OH^-]$ decreases. When base is added, the $[OH^-]$ increases and $[H^+]$ decreases.)

Ask students to describe the pH scale and explain the relationship between $[H^-]$, $[OH^-]$, and pH. ($pH + pOH = 14$)

Ask students to explain how they would calculate the $[OH^-]$ when the pH of a solution is given. (From the pH they can get $[H^+]$. They can then solve the equation $K_w = [OH^-][H^+]$ for $[OH^-]$).

The reference electrode has a constant voltage. The voltage of the glass electrode changes with the $[H_3O^+]$ of the solution in which it is dipped. The pH meter makes an electrical measurement of pH by measuring the voltage between the two electrodes. As you can see in **Figure 20.10** on the previous page, the pH meter gives a direct readout of pH. Would you describe the two common household items shown in this figure as acidic
1 or basic?

Before you use a pH meter, you must first calibrate it by immersing the electrodes in a solution of known pH. With the electrodes in the solution, you adjust the readout of the millivoltmeter to this known pH. Then rinse the electrodes with distilled water and dip them into the solution of unknown pH. To make continuous pH readings, you can leave the electrodes in the solution.

A pH meter is a valuable instrument. In many situations it is easier to use than liquid indicators or indicator strips. Measurements of pH obtained with a pH meter are typically accurate to within 0.01 pH unit of the true pH. The color and cloudiness of the unknown solution do not affect the accuracy of the pH value obtained. Hospitals use pH meters to find small but meaningful changes of pH in blood and other body fluids. Sewage and industrial effluents and soil pH are also easily monitored with a pH meter.

(a) (b) (c)

Figure 20.11

(a) *Altering soil pH can affect the development of plants. In acidic soils, hydrangeas produce blue flowers. In basic soils, they produce pink flowers.* ***(b)*** *Spreading lime on lawns and gardens neutralizes acidic soils.* ***(c)*** *Blood pH is used to help diagnose illness.*

Answer

1 Vinegar is acidic; milk of magnesia is basic

MINI LAB

Indicators from Natural Sources

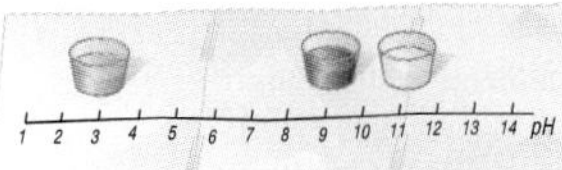

PURPOSE

To measure the pH of various household materials by using a natural indicator to make an indicator chart.

MATERIALS

- red cabbage leaves
- white vinegar (CH_3COOH)
- household ammonia
- baking soda ($NaHCO_3$)
- 1-cup measure
- 10 clear plastic cups
- 2 jars
- tape
- knife
- pencil
- teaspoon
- ruler
- clean white cloth
- hot water
- 3 sheets of plain white paper
- medicine dropper
- clear plastic wrap
- rubber band
- various household items listed in Step 5

PROCEDURE

1. Put $\frac{1}{2}$ cup of finely chopped red cabbage leaves in a jar and add $\frac{1}{2}$ cup of hot water. Stir and crush the leaves with a spoon. Continue the extraction until the water is distinctly colored.
2. Strain the extract through a piece of cloth into a clean jar. This liquid is your natural indicator.
3. Tape three sheets of paper end to end. Draw a line along the center and label it at 5 cm intervals with the numbers 1 to 14. This is your pH scale.
4. Pour your indicator to about 1 cm depth into each of three plastic cups. To one cup, add several drops of vinegar, to the second add a pinch of baking soda, and to the third add several drops of ammonia. The resulting colors indicate pH values of about 3, 9, and 11, respectively. Place these colored positions on your pH scale.
5. Repeat Step 4 for household items such as table salt, borax, milk, lemon juice, laundry detergent, dish detergent, milk of magnesia, mouthwash, toothpaste, shampoo, and carbonated beverages.

ANALYSIS AND CONCLUSIONS

1. What was the color of the indicator at acidic, neutral, and basic conditions?
2. What chemical changes were responsible for the color changes?
3. Label the materials you tested as acidic, basic, or neutral.
4. Which group contains items used for cleaning or for personal hygiene?

section review 20.2

16. What is true about the relative concentrations of hydrogen ions and hydroxide ions in each kind of solution?

    a. basic b. acidic c. neutral

17. Determine the pH of each solution.

    a. $[H^+] = 1.0 \times 10^{-6}M$ d. $[OH^-] = 1.0 \times 10^{-11}M$

    b. $[H^+] = 0.00010M$

    c. $[OH^-] = 1.0 \times 10^{-2}M$

18. What are the hydroxide-ion concentrations for solutions with the following pH values?

    a. 6.00 b. 9.00 c. 12.00

 Chem ASAP! **Assessment 20.2** Check your understanding of the important ideas and concepts in Section 20.2.

Answers

SECTION REVIEW 20.2

16. a. Hydroxide-ion concentration is greater.
    b. Hydrogen-ion concentration is greater.
    c. The concentrations are equal.
17. a. 6.00 b. 4.00
    c. 12.00 d. 3.00
18. a. $1.0 \times 10^{-8}M$
    b. $1.0 \times 10^{-5}M$
    c. $1.0 \times 10^{-2}M$

20.2

Reteach

Review the self-ionization of water. Show the equilibrium equation for this process, and emphasize that, in any aqueous solution, $[H^+]$ and $[OH^-]$ are interdependent species. If $[H^+]$ increases by a certain amount, then $[OH^-]$ decreases by the same amount. Remind students about Le Chatêlier's principle. If additional H^+ or OH^- ions are added to the solution, the equilibrium shifts. In an acidic solution, the $[H^+]$ is greater than the $[OH^-]$.

MINI LAB

Indicators from Natural Sources

SAFETY

Perform this lab in a well-ventilated room.

LAB PREP AND PLANNING

If the period allotted to lab work is short, spread the work over two days. Students can prepare the indicator the first day, and do the testing the next day. The indicator keeps reasonably well if stored, covered, in a refrigerator overnight. The water soluble pigments in red cabbage are called anthocyanins. These same pigments color fruits such as strawberries and blueberries, as well as many kinds of flowers and leaves. Anthocyanins are used commercially as colorants in acid solutions such as soft drinks.

ANALYSIS AND CONCLUSIONS

1. The initial color was purple. In acid the indicator was red; in a neutral solution it was blue-purple; in a basic solution it was green.
2. Changes in the relative number of H^+ and OH^- present.
3. Answers will vary.
4. Personal hygiene items mainly test neutral; cleaning materials such as soap mainly test basic.

1 Engage

Use the Visual

Have the students study the photograph and read the text that opens the section. Ask:

- **Why is ammonia considered a base?** (Aqueous solutions of ammonia have a pH > 7.)

Check Prior Knowledge

Have the students recall the operational definitions of acids and bases from Section 20.1.

- **Ask students about the chemical characteristics common to acids and bases in aqueous solution.** (Usually, acids contain ionizable hydrogen atoms and bases contain hydroxide ions.)

2 Teach

Use the Visual

Have students examine Table 20.3 and read the accompanying text. Note that the acids in Table 20.3 are listed in order of decreasing tendency to produce hydrogen ions. Ask:

- **Do diprotic acids produce more hydrogen ions in solution than monoprotic acids?** (No; there is no direct relationship between number of hydrogen atoms in a formula and strength of the acid. Hydrochloric acid, which is monoprotic, produces many more hydrogen ions than does carbonic acid, which is diprotic.)

ACID–BASE THEORIES

objectives

- Compare and contrast acids and bases as defined by the theories of Arrhenius, Brønsted-Lowry, and Lewis
- Identify conjugate acid–base pairs in acid–base reactions

key terms

- monoprotic acids
- diprotic acids
- triprotic acids
- hydrogen-ion donor
- hydrogen-ion acceptor
- conjugate acid
- conjugate base
- conjugate acid–base pair
- amphoteric
- Lewis acid
- Lewis base

Bracken Cave, near San Antonio, Texas, is home to twenty to forty million bats, which is probably the largest colony of mammals in the world. Visitors to the cave must wear protective goggles and respirators to protect themselves from the dangerous levels of ammonia in the cave. Ammonia is a by-product of the bats' urine. **Why is ammonia considered a base?**

Arrhenius Acids and Bases

In Section 20.1, you learned the definition of an acid and of a base. Although chemists had recognized the properties of these two groups of compounds for many years, they were not able to explain the chemical theory of this behavior. Then, in 1887, the Swedish chemist Svante Arrhenius (1859–1927) proposed a revolutionary way of defining and thinking about acids and bases. He said that acids are hydrogen-containing compounds that ionize to yield hydrogen ions (H^+) in aqueous solution. He also said that bases are compounds that ionize to yield hydroxide ions (OH^-) in aqueous solution.

Table 20.3 lists some common acids. Acids that contain one ionizable hydrogen, such as nitric acid (HNO_3), are called **monoprotic acids.** Acids that contain two ionizable hydrogens, such as sulfuric acid (H_2SO_4), are called **diprotic acids.** Acids that contain three ionizable hydrogens, such as phosphoric acid (H_3PO_4), are called **triprotic acids.** Not all compounds that contain hydrogen are acids, however. Also, not all the hydrogens in an acid may be released as hydrogen ions. Only the hydrogens in very polar bonds are ionizable. In such bonds, hydrogen is joined to a very electronegative element. When a compound that contains such bonds dissolves in water, it releases hydrogen ions because the hydrogen ions are stabilized by solvation. An example is the hydrogen chloride molecule, shown in **Figure 20.13.** Hydrogen chloride is a polar covalent molecule. It ionizes to form hydrochloric acid in aqueous solution. Is hydrochloric acid a monoprotic,
1 diprotic, or triprotic acid?

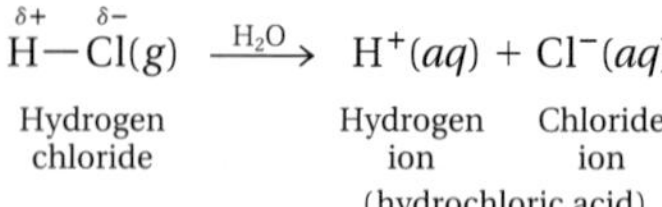

$$\overset{\delta+}{H}-\overset{\delta-}{Cl}(g) \xrightarrow{H_2O} H^+(aq) + Cl^-(aq)$$

Hydrogen chloride → Hydrogen ion + Chloride ion (hydrochloric acid)

Table 20.3

Some Common Acids	
Name	**Formula**
Hydrochloric acid	HCl
Nitric acid	HNO_3
Sulfuric acid	H_2SO_4
Phosphoric acid	H_3PO_4
Ethanoic acid	CH_3COOH
Carbonic acid	H_2CO_3

Figure 20.12
Sea slugs (left) contain diprotic H_2SO_4, which discourages predators. Some products that can keep rust spots on a car (center) from rusting further contain a triprotic acid, phosphoric acid. Monoprotic lactic acid builds up in muscles during prolonged exercise (right).

STUDENT RESOURCES

From the Teacher's Resource Package, use:

- Section Review 20.3, Ch. 20 Practice Problems and Quizzes from the Review Module (Ch. 17–20)
- Small-Scale Chemistry Lab Manual: Experiment 27

TECHNOLOGY RESOURCES

Relevant technology resources include:

- Chem ASAP! CD-ROM
- Resource Pro CD-ROM
- ActivChemistry CD-ROM: *Acids and Bases*

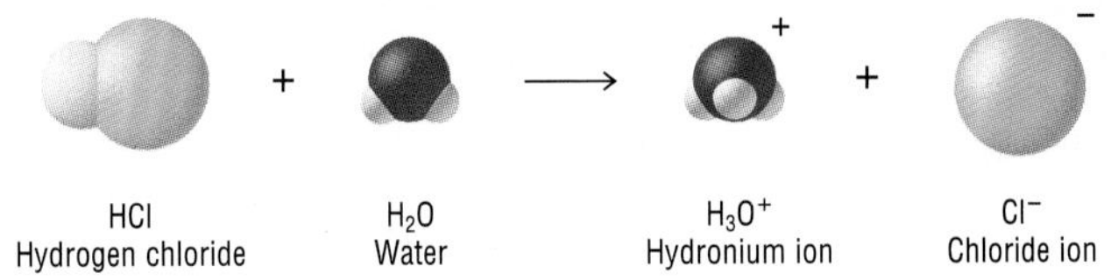

HCl Hydrogen chloride + H_2O Water ⟶ H_3O^+ Hydronium ion + Cl^- Chloride ion

Figure 20.13

Hydrochloric acid is actually an aqueous solution of hydrogen chloride. The hydrogen forms hydronium ions, making this compound an acid.

In contrast, the four hydrogens in methane (CH_4) are attached by weakly polar C—H bonds. Methane has no ionizable hydrogens and is not an acid. Ethanoic (acetic) acid (CH_3COOH), used in the manufacture of plastics, pharmaceuticals, and photographic chemicals, is different. Although this molecule contains four hydrogens, ethanoic acid is a monoprotic acid. The structural formula shows why.

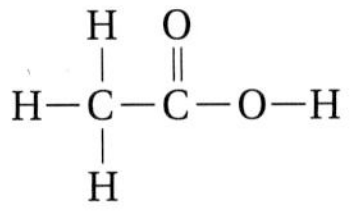

```
    H  O
    |  ||
 H—C—C—O—H
    |
    H
```

Ethanoic acid
(CH_3COOH)

The three hydrogens attached to the carbon are in weakly polar bonds. They do not ionize. Only the hydrogen attached to the highly electronegative oxygen is able to be ionized. As you gain more experience looking at written formulas for acids, you will be able to recognize which hydrogen atoms can be ionized.

Table 20.4 lists some common bases. The base with which you are perhaps most familiar is sodium hydroxide (NaOH). Sodium metal reacts violently with water to form sodium hydroxide. The following equation illustrates the reaction of sodium metal with water.

$$2Na(s) + 2H_2O(l) \longrightarrow 2NaOH(aq) + H_2(g)$$

Sodium metal — Water — Sodium hydroxide — Hydrogen

Because of its extremely caustic nature, sodium hydroxide, commonly known as lye, is a major component of consumer products used to clean clogged drains.

Table 20.4

Some Common Bases		
Name	**Formula**	**Solubility in water**
Potassium hydroxide	KOH	High
Sodium hydroxide	NaOH	High
Calcium hydroxide	$Ca(OH)_2$	Very low
Magnesium hydroxide	$Mg(OH)_2$	Very low

Discuss

Acid–base concepts have been a part of chemistry for more than 300 years. Over time, theories have been developed to explain the observed behavior of acids and bases. Arrhenius' s theory addressed compounds whose formulas contain obvious clues to their acidic or basic nature.

Answer

1 monoprotic acid

Discuss

Have students read the *Link to Library Science.* Explain that when aluminum sulfate $[Al_2(SO_4)_3]$ combines with water vapor in the atmosphere, small amounts of sulfuric acid form on the pages of books, causing the paper to turn yellow and brittle. Humidity has an important influence on how much acid forms and how quickly books deteriorate. Therefore, it is important for libraries to have adequate climate control systems. After students have read the feature, ask:

- **What are the advantages and disadvantages of having the reference materials in your library available as bound books? On microfilm? On audiotapes?**

Discuss

Brønsted arrived at his new definition of acids and bases through his work in kinetics and thermodynamics. Brønsted noticed that a great many compounds that would not be classified as acids and bases by the Arrhenius definition behaved like acids and bases in his experiments. He found that these compounds were capable of hydrogen-ion transfer reactions. This led him to define acids as hydrogen-ion donors and bases as hydrogen-ion acceptors in chemical reactions. Some students may think the conjugate base of an acid in a given reaction appears on the same side of the equation as the acid. Explain that an acid does not react with its conjugate base but instead produces it or is produced by it in a reaction.

LIBRARY SCIENCE

Chemistry Rescues Crumbling Books

Millions of books printed since the mid-nineteenth century slowly decay as they sit on library shelves. Their pages are so fragile, they crumble at a touch. The cause is the acidity of the paper, which originates from alum (aluminum sulfate). For the past 150 years, alum has been used to prevent inks from soaking into paper. Libraries throughout the world are working with chemists to find ways to stop the acid deterioration of books. One process, carried out in a vacuum chamber, is called mass deacidification. After the air is pumped out of the chamber and most of the moisture is removed from the books, the gas diethyl zinc is introduced. The gas penetrates the closed pages of the books and completely neutralizes any acid that is present. The gas also forms zinc oxide, which protects the paper against future acid attack. For an estimated cost of \$6 to \$10 per book, hundreds of volumes can be treated at once. If books had to be treated page by page, deacidification could cost \$1000 per book!

Potassium metal reacts vigorously with water to produce potassium hydroxide (KOH). Both sodium hydroxide and potassium hydroxide are ionic solids. What happens to the ions in these compounds when they dissolve in water? 1

$$\underset{\text{Sodium hydroxide}}{NaOH(s)} \xrightarrow{H_2O} \underset{\text{Sodium ion}}{Na^+(aq)} + \underset{\text{Hydroxide ion}}{OH^-(aq)}$$

Sodium and potassium are Group 1A elements. Elements in Group 1A, the alkali metals, react with water to produce alkaline solutions. Metal oxides also react with water to produce alkaline solutions. Both sodium hydroxide and potassium hydroxide are very soluble in water. Concentrated solutions of these compounds can be readily prepared. Such solutions, like other basic solutions, would have a bitter taste and slippery feel. Note, however, that they are extremely caustic to the skin and can cause deep, painful, slow-healing wounds if not immediately washed off.

Calcium hydroxide ($Ca(OH)_2$) and magnesium hydroxide ($Mg(OH)_2$) are hydroxides of Group 2A metals. These hydroxides are not very soluble in water. Consequently, their solutions are always very dilute, even when saturated. The concentration of hydroxide ions in such solutions is correspondingly low. A saturated solution of calcium hydroxide contains only 0.165 g $Ca(OH)_2$ per 100 g of water. Magnesium hydroxide is much less soluble than calcium hydroxide. A saturated solution contains only 0.0009 g $Mg(OH)_2$ per 100 g of water. Suspensions of magnesium hydroxide in water contain low concentrations of hydroxide ion. People take these suspensions internally as milk of magnesia, which is an antacid and a mild laxative.

Brønsted-Lowry Acids and Bases

The Arrhenius definition of acids and bases is not a very comprehensive one. It defines acids and bases rather narrowly and does not include certain substances that have acidic or basic properties. For example, aqueous solutions of ammonia ($NH_3(aq)$) and sodium carbonate ($Na_2CO_3(aq)$) are basic. Neither of these compounds is a hydroxide, however, and neither would be classified as a base under the Arrhenius definition. In 1923, the Danish chemist Johannes Brønsted (1879–1947) and the English chemist Thomas Lowry (1874–1936) independently proposed a new definition. The Brønsted-Lowry theory defines an acid as a **hydrogen-ion donor.** Similarly, a Brønsted-Lowry base is a **hydrogen-ion acceptor.** All the acids and bases included in the Arrhenius theory are also acids and bases according to the Brønsted-Lowry theory. Some compounds not included in the Arrhenius theory are classified as bases in the Brønsted-Lowry theory.

The behavior of ammonia as a base can be understood by using the Brønsted-Lowry theory. Ammonia gas is very soluble in water. When ammonia dissolves in water, it acts as a base because it accepts a hydrogen ion from water.

$$\underset{\substack{\text{Ammonia}\\\text{(hydrogen-ion}\\\text{acceptor, Brønsted-}\\\text{Lowry base)}}}{NH_3(aq)} + \underset{\substack{\text{Water}\\\text{(hydrogen-ion}\\\text{donor, Brønsted-}\\\text{Lowry acid)}}}{H_2O(l)} \rightleftharpoons \underset{\substack{\text{Ammonium}\\\text{ion}}}{NH_4^+(aq)} + \underset{\substack{\text{Hydroxide ion}\\\text{(makes the}\\\text{solution basic)}}}{OH^-(aq)}$$

Answers

1 They dissociate completely into solvated metal ions and hydroxide ions.

2 Ammonia is not classified as an Arrhenius base because it does not directly dissociate to give hydroxide ions in aqueous solution.

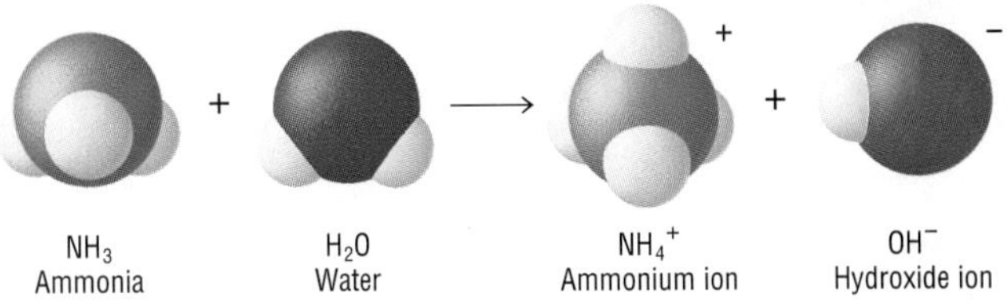

NH_3 Ammonia + H_2O Water ⟶ NH_4^+ Ammonium ion + OH^- Hydroxide ion

Figure 20.14
Ammonia dissolves in water to form ammonium ions and hydroxide ions. In this reaction, the water molecule donates a hydrogen ion to the ammonia molecule. Why is ammonia not classified as an Arrhenius base? 2

In this reaction, ammonia is the hydrogen-ion acceptor. Therefore, it is a Brønsted-Lowry base. Water, the hydrogen-ion donor, is a Brønsted-Lowry acid. Hydrogen ions are transferred from water to ammonia, as is shown in **Figure 20.14.** This causes the hydroxide-ion concentration to be greater than it is in pure water. As a result, solutions of ammonia are basic.

Heating an aqueous solution of ammonia releases ammonia gas. Do not try this at home, however; ammonia is a poisonous gas! As ammonia gas leaves the solution, the equilibrium in the equation shifts to the left. The ammonium ion (NH_4^+) reacts with OH^- to form NH_3 and H_2O. When the reaction goes from right to left, NH_4^+ gives up a hydrogen ion; it acts as a Brønsted-Lowry acid. The hydroxide ion accepts an H^+; it acts as a Brønsted-Lowry base. Overall, then, this equilibrium has two acids and two bases.

$$\underset{\text{Base}}{NH_3(aq)} + \underset{\text{Acid}}{H_2O(l)} \rightleftharpoons \underset{\text{Conjugate acid}}{NH_4^+(aq)} + \underset{\text{Conjugate base}}{OH^-(aq)}$$

When ammonia dissolves, NH_4^+ is the conjugate acid of the base NH_3. A **conjugate acid** is the particle formed when a base gains a hydrogen ion. Similarly, OH^- is the conjugate base of the acid water. A **conjugate base** is the particle that remains when an acid has donated a hydrogen ion. Conjugate acids and bases are always paired with a base or an acid, respectively. A **conjugate acid–base pair** consists of two substances related by the loss or gain of a single hydrogen ion. The ammonia molecule and ammonium ion are a conjugate acid–base pair. The water molecule and hydroxide ion are also a conjugate acid–base pair.

$$\underset{\text{Base}}{NH_3(aq)} + \underset{\text{Acid}}{H_2O(l)} \rightleftharpoons \underset{\text{Conjugate acid}}{NH_4^+(aq)} + \underset{\text{Conjugate base}}{OH^-(aq)}$$

The Brønsted-Lowry theory also applies to acids. Consider the ionization of hydrogen chloride in water.

$$\underset{\text{Acid}}{HCl(g)} + \underset{\text{Base}}{H_2O(l)} \rightleftharpoons \underset{\text{Conjugate acid}}{H_3O^+(aq)} + \underset{\text{Conjugate base}}{Cl^-(aq)}$$

Table 20.5

Several Conjugate Acid–Base Pairs

Acid (lose H^+ ⇄ gain H^+)	Base
HCl	Cl^-
H_2SO_4	HSO_4^-
H_3O^+	H_2O
HSO_4^-	SO_4^{2-}
CH_3COOH	H_3COO^-
H_2CO_3	HCO_3^-
HCO_3^-	CO_3^{2-}
NH_4^+	NH_3
H_2O	OH^-

Use the Visual

Have the students study the Brønsted-Løwry acid–base conjugate pairs in Table 20.5. Point out that the acids lose hydrogen ions and the bases gain hydrogen ions.

Discuss

Point out that Lewis was the first scientist to discuss the significance of electron pairs in bonding (Lewis electron-dot diagrams). His theory of acids and bases was an extension of his concept of electron pairs. A Lewis acid accepts a pair of electrons to form a covalent bond and a Lewis base donates a pair of electrons to form a covalent bond.

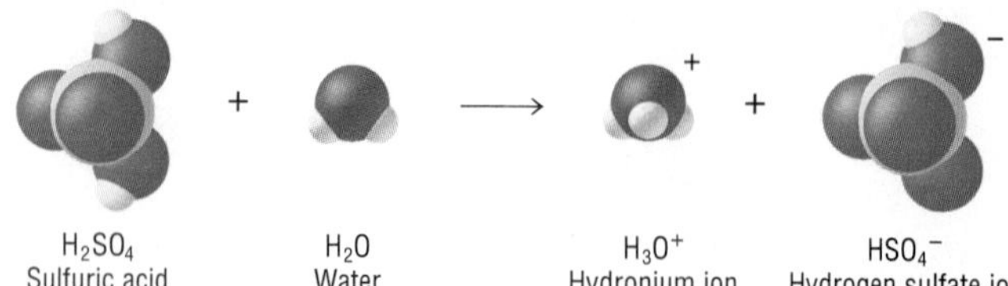

Figure 20.15
When sulfuric acid dissolves in water, it forms hydronium ions and hydrogen sulfate ions. Which ion is the conjugate acid and which is the conjugate base? 2

In this reaction, hydrogen chloride is the hydrogen-ion donor. Thus it is a Brønsted-Lowry acid. Water is the hydrogen-ion acceptor. It is a Brønsted-Lowry base. The chloride ion is the conjugate base of the acid HCl. The hydronium ion is the conjugate acid of the base water. Compare this reaction with what happens when sulfuric acid dissolves in water, as illustrated in **Figure 20.15.**

Sometimes water accepts a hydrogen ion. At other times, it donates a hydrogen ion. A substance that can act as both an acid and a base is said to be **amphoteric.** Water is amphoteric. In the reaction with HCl, water accepts a proton and is therefore a base. In the reaction with NH_3, is water
1 an acid or a base?

Lewis Acids and Bases

A third theory of acids and bases was proposed by Gilbert Lewis (1875–1946). In his definition, Lewis focused on the donation or acceptance of a pair of electrons during a reaction. This concept is more general than either the Arrhenius theory or the Brønsted-Lowry theory. A **Lewis acid** is a substance that can accept a pair of electrons to form a covalent bond. A **Lewis base** is a substance that can donate a pair of electrons to form a covalent bond. A hydrogen ion (Brønsted-Lowry acid) can accept a pair of electrons in forming a bond. A hydrogen ion, therefore, is also a Lewis acid. A Brønsted-Lowry base, or a substance that accepts a hydrogen ion, must have a pair of electrons available and, therefore, is also a Lewis base. Consider the reaction of H^+ and OH^-.

$$\underset{\text{Lewis acid}}{H^+} + \underset{\text{Lewis base}}{^-\!:\ddot{O}-H} \longrightarrow H-\ddot{O}-H$$

Chem ASAP!

Animation 25

Compare the three important definitions of acids and bases.

In this reaction, a hydroxide ion is a Lewis base. It is also a Brønsted-Lowry base. The hydrogen ion is both a Lewis acid and a Brønsted-Lowry acid. The Lewis definition also includes some compounds not classified as Brønsted-Lowry acids or bases.

Table 20.6

Acid–Base Definitions		
Type	**Acid**	**Base**
Arrhenius	H^+ producer	OH^- producer
Brønsted-Lowry	H^+ donor	H^+ acceptor
Lewis	electron-pair acceptor	electron-pair donor

Answers

1 Water donates a proton and is thus an acid.

2 H_3O^+ is the acid; HSO_4^- is the base.

SECTION REVIEW 20.3

21. **a.** In the Brønsted–Lowry theory, acids are hydrogen-ion donors and bases are hydrogen-ion acceptors.
b. It explains why such compounds, which lack hydroxide ions, can behave like bases.
c. The Lewis theory is the most general of the three. A Lewis acid is an electron-pair acceptor, and a Lewis base is an electron-pair donor.

22. Two substances that are related by the loss or gain of a single hydrogen ion

23. See answer on p. 599.

24. **a.** diprotic—two ionizable hydrogens
b. triprotic—three ionizable hydrogens
c. monoprotic—one ionizable hydrogen
d. diprotic—two ionizable hydrogens

25. **a.** $2K + 2H_2O \rightleftharpoons 2KOH + H_2$
b. $Ca + 2H_2O \rightleftharpoons Ca(OH)_2 + H_2$

Sample Problem 20-7

Identify the Lewis acid and the Lewis base in this reaction.

```
  H   F             H  F
  |   |             |  |
H—N: + B—F  ⟶  H—N—B—F
  |   |             |  |
  H   F             H  F
```

1. ***ANALYZE* Plan a problem-solving strategy.**
The Lewis acid–Lewis base definitions, which are to be used in solving the problem, are based on the acceptance and donation of a pair of electrons.

2. ***SOLVE* Apply the problem-solving strategy.**
Ammonia is donating a pair of electrons. Boron trifluoride is accepting a pair of electrons. Lewis bases donate electrons, so ammonia is acting as a Lewis base. Lewis acids accept electrons, so boron trifluoride is acting as a Lewis acid.

3. ***EVALUATE* Does the result make sense?**
In this reaction, ammonia and boron trifluoride fit the definitions of Lewis bases and Lewis acids, respectively.

Practice Problems

19. Would you predict PCl_3 to be a Lewis acid or a Lewis base in typical reactions? Explain your prediction.
20. Identify the Lewis acid and Lewis base in each reaction
a. $H^+ + H_2\ddot{O}: \longrightarrow H_3O^+$
b. $AlCl_3 + Cl^- \longrightarrow AlCl_4^-$

section review 20.3

21. a. How are acids and bases defined by the Brønsted-Lowry theory?
b. What advantage does this theory have in terms of accounting for the properties of compounds such as ammonia?
c. How do the Arrhenius and Brønsted-Lowry theories compare with the Lewis theory of acids and bases?
22. What is a conjugate acid–base pair?
23. Write equations for the ionization of HNO_3 in water and the reaction of CO_3^{2-} with water. For each equation, identify the hydrogen-ion donor and hydrogen-ion acceptor. Then label the conjugate acid–base pairs in each equation.
24. Identify the following acids as monoprotic, diprotic, or triprotic. Explain your reasoning.
a. H_2CO_3
b. H_3PO_4
c. HCl
d. H_2SO_4
25. Write a balanced equation for each reaction.
a. Potassium metal reacts with water.
b. Calcium metal reacts with water.

Chem ASAP! **Assessment 20.3** Check your understanding of the important ideas and concepts in Section 20.3.

portfolio project

Household drain cleaners contain pellets of sodium hydroxide (NaOH) and small metal particles. Use the library or Internet to find out how drain cleaners work. Include the identity of the metal particles in your written report.

3 Assess

Evaluate Understanding

Have students define Arrhenius acids and bases and give an example of each. Then pose this problem: When $NaHCO_3$ is heated, H_2CO_3 and Na_2CO_3 are formed by the following reaction:
$2NaHCO_3 \rightarrow H_2CO_3 + Na_2CO_3$
Show how this reaction can be treated as an acid–base reaction, with HCO_3^- acting as both acid and base. (HCO_3^- donates a hydrogen ion to form CO_3^{2-} and accepts a hydrogen ion to form H_2CO_3.) **Finally, have students explain the Lewis theory of acids and bases.**

Reteach

Emphasize that the observable properties of Arrhenius acids and bases are due to the release of H^+ or OH^- ions. Compounds such as CH_3OH and CH_4 do not qualify because they do not release H^+ or OH^- ions in solution. Every reaction between Brønsted-Lowry acids and bases produces a conjugate acid and conjugate base. Note that a Lewis acid and base are like a lock and a key. The electron pair of the base "fits into" the empty orbital of the acid.

Portfolio Project

Drain cleaners contain caustic NaOH. Reaction with aluminum generates heat, which softens greases and oils, and also generates hydrogen, which agitates the mixture.

SECTION REVIEW 20.3 CONTINUED

23.

HNO_3	+	H_2O	⇌	H_3O^+	+	NO_3^-
Acid (hydrogen-ion donor)		Base (hydrogen-ion acceptor)		Conjugate acid		Conjugate base
CO_3^{2-}	+	H_2O	⇌	HCO_3^-	+	OH^-
Base (hydrogen-ion acceptor)		Acid (hydrogen-ion donor)		Conjugate acid		Conjugate base

Practice Problems Plus

Related Chapter Review Problems
Chapter Review problems 51 and 65 are related to Sample Problem 20-7.

1 Engage

Use the Visual

Have students study and read the text that opens the section. Ask:

- **What makes some acids weak acids and other acids strong acids?** (Strong acids are essentially 100% ionized in water; weak acids are only partially ionized as defined by their dissociation constants, K_a.)

Check Prior Knowledge

Remind students of common acids that require no safety precautions.

- **Ask the student to name some of these "safe" acids?** (possible answers: ethanoic acid in vinegar and citric acid in lemon juice)
- **Ask what it feels like to have one of these "safe" acids in contact with an open cut?** (Contact with even a weak acid can be painful if the skin is punctured.)
- **Which common acids do they know that require the exercise of proper safety precautions?** (sulfuric acid in car batteries)

Have the students hypothesize on the difference between "safe" and "unsafe" acids.

section 20.4

objectives

- Define strong acids and weak acids
- Calculate an acid dissociation constant (K_a) from concentration and pH measurements
- Arrange acids by strength according to their acid dissociation constants (K_a)
- Arrange bases by strength according to their base dissociation constants (K_b)

key terms

- strong acids
- weak acids
- acid dissociation constant (K_a)
- strong bases
- weak bases
- base dissociation constant (K_b)

STRENGTHS OF ACIDS AND BASES

Lemons and grapefruits have a sour taste because they contain citric acid. When you make lemonade, or cut up a grapefruit, you probably do not wear safety goggles or chemical-resistant clothing even though you are working with an acid. However, some acids require just such precautions. For example, sulfuric acid is a widely used, industrial chemical that can quickly cause severe burns if it comes into contact with the skin. **What makes some acids weak acids and other acids strong acids?**

Strong and Weak Acids and Bases

Acids are classified as strong or weak depending on the degree to which they ionize in water. For practical purposes, **strong acids** are completely ionized in aqueous solution. Hydrochloric acid and sulfuric acid are strong acids.

$$HCl(g) + H_2O(l) \longrightarrow H_3O^+(aq) + Cl^-(aq) \text{ (100\% ionized)}$$

Weak acids ionize only slightly in aqueous solution. The ionization of ethanoic acid (acetic acid), a typical weak acid, is not complete.

$$\underset{\text{Ethanoic acid}}{CH_3COOH(aq)} + \underset{\text{Water}}{H_2O(l)} \rightleftharpoons \underset{\text{Hydronium ion}}{H_3O^+(aq)} + \underset{\text{Ethanoate ion}}{CH_3COO^-(aq)}$$

Fewer than 1% of ethanoic acid molecules are ionized at any instant. **Table 20.7** gives the relative strength of some common acids and bases.
1 Which is the weakest acid in the table? The weakest base?

Table 20.7

Relative Strengths of Common Acids and Bases		
Substance	**Formula**	**Relative Strength**
Hydrochloric acid	HCl	Strong acids
Nitric acid	HNO_3	
Sulfuric acid	H_2SO_4	
Phosphoric acid	H_3PO_4	↑ Increasing strength of acid
Ethanoic acid	CH_3COOH	
Carbonic acid	H_2CO_3	
Hydrosulfuric acid	H_2S	
Hypochlorous acid	$HClO$	
Boric acid	H_3BO_3	
		Neutral solution
Sodium cyanide	$NaCN$	↓ Increasing strength of base
Ammonia	NH_3	
Methylamine	CH_3NH_2	
Sodium silicate	Na_2SiO_3	
Calcium hydroxide	$Ca(OH)_2$	Strong bases
Sodium hydroxide	$NaOH$	
Potassium hydroxide	KOH	

Student Resources

- Section Review 20.3, Ch. 20 Practice Problems, Vocabulary Review, Quizzes, and Tests from the Review Module (Ch. 17–20)
- Laboratory Manual: Experiment 38
- Laboratory Practical 20-2
- Laboratory Recordsheet 20-2
- Small-Scale Chemistry Lab Manual: Experiment 28

Technology Resources

Relevant technology resources include:

- Chem ASAP! CD-ROM
- Resource Pro CD-ROM
- Chemistry Alive! Videodisc: *Titration*
- Assessment Resources CD-ROM

You can write the equilibrium-constant expression from a balanced chemical equation. What is the general equation for setting up an equilibrium constant? Using this equation, the equilibrium-constant expression for ethanoic acid is 

$$K_{eq} = \frac{[H_3O^+] \times [CH_3COO^-]}{[CH_3COOH] \times [H_2O]}$$

For dilute solutions, the concentration of water is a constant. It can be combined with K_{eq} to give an acid dissociation constant. An **acid dissociation constant** (K_a) is the ratio of the concentration of the dissociated (or ionized) form of an acid to the concentration of the undissociated (nonionized) form. The dissociated form includes both the H^+ and the anion.

$$K_{eq} \times [H_2O] = K_a = \frac{[H^+] \times [CH_3COO^-]}{[CH_3COOH]}$$

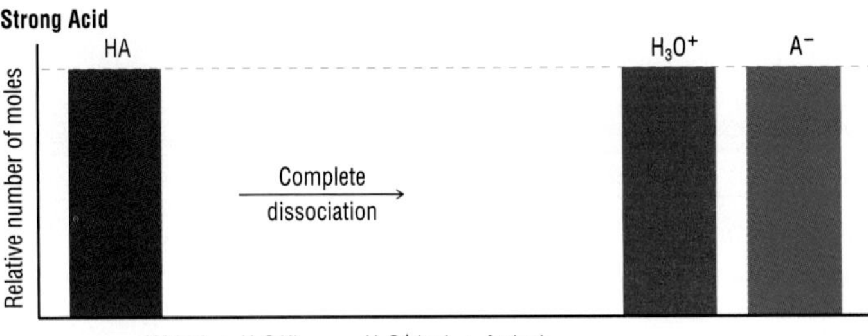

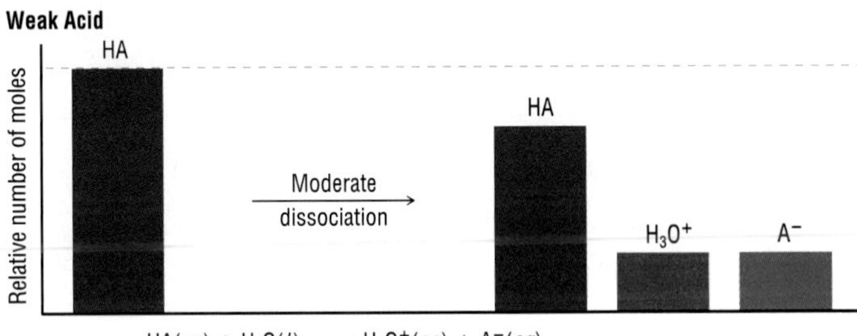

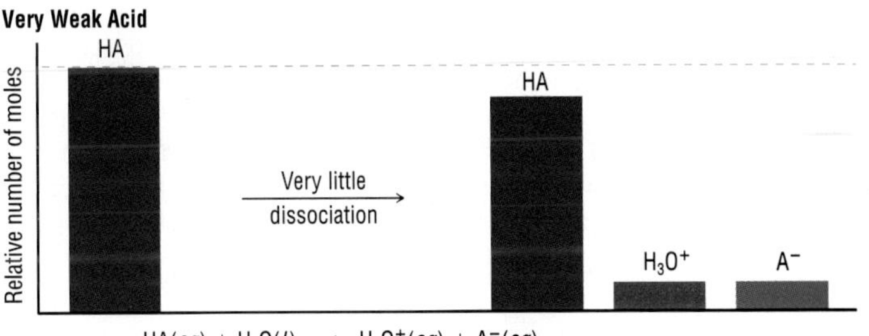

Figure 20.16

Dissociation of an acid (HA) in water yields H_3O^+ and A^-. The bar graphs compare the extent of dissociation of strong and weak acids. A strong acid completely dissociates in water; $[H_3O^+]$ is high. Weak acids remain largely undissociated; $[H_3O^+]$ is low. Give the name and formula for a strong acid, a weak acid, and a very weak acid. For each example, give the name and formula for the anion that results from the dissociation reaction. 3

2 Teach

TEACHER DEMO

Set up two test tubes at the front of the class, each containing several grams of sucrose. Explain that strong acids will react with the sugar. Using an eye dropper, place several drops of concentrated sulfuric acid into one test tube of sugar, and concentrated ethanoic acid (acetic acid) into the other. The sulfuric acid reacts violently with the sugar while the acetic acid doesn't. Have the students infer which is a strong acid and which is a weak acid.

Use the Visual

Have the students study Figure 20.16 and the accompanying text. Explain that the bar graphs compare the extent of ionization or dissociation of strong to weak acids. Explain that the graphs for the weak acid and the very weak acid exaggerate the concentration of products. Realistic concentrations—less than 1%—would not work visually.

- **Ask students what the primary difference is between strong and weak acids?** (Strong acids are nearly 100% dissociated, while weak acids are only partially dissociated.)

Answers

1 boric acid; sodium cyanide

2 For $aA + bB \rightleftharpoons cC + dD$,

$$K_{eq} = \frac{[C]^c[D]^d}{[A]^a[B]^b}.$$

3 Answers will vary; strong acid (and its anion): hydrochloric acid, HCl (Cl^-); weak acid (and its anion): ethanoic acid, CH_3COOH (CH_3COO^-); very weak acid (and its anion): hypochlorous acid, HClO (ClO^-)

Discuss

Remind students that, in solution, the negative ion of an acid is a base. Explain that strong acids, such as nitric acid, dissociate almost completely because the negative ion is a very weak base. That is, its tendency to combine with a hydrogen ion is slight. In contrast, the dissociation of a weak acid such as carbonic acid, is limited, because its negative ion is a relatively strong base. These ions combine with most of the available hydrogen ions to form the undissociated form of the acid again.

The terms *strong* and *weak*, as applied to acids, are often confused with the concept of concentration. Stress that the dissociation of an acid or base into ions involves the establishment of an equilibrium. The terms *strong* and *weak* refer to the position of the equilibrium. When a strong acid or base dissolves, the equilibrium favors the products. When a weak acid or base dissolves, the equilibrium favors the reactants.

Show students that the extent to which products or reactants are favored can be determined from the values of K_a or K_b. Also show them how to use values of K_a and K_b to compare the strengths of acids and bases.

The acid dissociation constant (K_a) reflects the fraction of an acid in the ionized form. For this reason, dissociation constants are sometimes called ionization constants. If the value of the dissociation constant is small, then the degree of dissociation or ionization of the acid in the solution is small. Weak acids have small K_a values. A larger value of K_a means the dissociation or ionization of the acid is more complete. The stronger an acid is, the larger is its K_a value. For example, nitrous acid has a K_a of 4.4×10^{-4}, whereas ethanoic acid (acetic acid) has a K_a of 1.8×10^{-5}. This means that nitrous acid is more ionized in solution than ethanoic acid. Nitrous acid is a stronger acid than ethanoic acid.

Diprotic and triprotic acids lose their hydrogens one at a time. Each ionization reaction has a separate dissociation constant. Thus phosphoric acid (H_3PO_4) has three dissociation constants to go with its three ionizable hydrogens. **Table 20.8** shows the ionization reactions and dissociation constants of some common weak acids, ranked by the value of the first dissociation constant of each acid. What is the second dissociation constant
❶ for the ionization of phosphoric acid?

Table 20.8

Dissociation Constants of Weak Acids		
Acid	**Ionization**	**K_a (25 °C)**
Oxalic acid	$HOOCCOOH(aq) \rightleftharpoons H^+(aq) + HOOCCOO^-(aq)$	5.6×10^{-2}
	$HOOCCOO^-(aq) \rightleftharpoons H^+(aq) + OOCCOO^{2-}(aq)$	5.1×10^{-5}
Phosphoric acid	$H_3PO_4(aq) \rightleftharpoons H^+(aq) + H_2PO_4^-(aq)$	7.5×10^{-3}
	$H_2PO_4^-(aq) \rightleftharpoons H^+(aq) + HPO_4^{2-}(aq)$	6.2×10^{-8}
	$HPO_4^{2-}(aq) \rightleftharpoons H^+(aq) + PO_4^{3-}(aq)$	4.8×10^{-13}
Methanoic acid	$HCOOH(aq) \rightleftharpoons H^+(aq) + HCOO^-(aq)$	1.8×10^{-4}
Benzoic acid	$C_6H_5COOH(aq) \rightleftharpoons H^+(aq) + C_6H_5COO^-(aq)$	6.3×10^{-5}
Ethanoic acid	$CH_3COOH(aq) \rightleftharpoons H^+(aq) + CH_3COO^-(aq)$	1.8×10^{-5}
Carbonic acid	$H_2CO_3(aq) \rightleftharpoons H^+(aq) + HCO_3^-(aq)$	4.3×10^{-7}
	$HCO_3^-(aq) \rightleftharpoons H^+(aq) + CO_3^{2-}(aq)$	4.8×10^{-11}

Just as there are strong acids and weak acids, there are also strong bases and weak bases. All the hydroxides in **Table 20.4** on page 595 are classified as strong bases. **Strong bases** dissociate completely into metal ions and hydroxide ions in aqueous solution. Some strong bases, such as calcium hydroxide and magnesium hydroxide, are not very soluble in water. The small amounts of these bases that do dissolve dissociate completely. **Table 20.7** on page 600 lists the relative strengths of some bases.

Weak bases react with water to form the hydroxide ion and the conjugate acid of the base. Ammonia is an example of a weak base.

$$\underset{\text{Ammonia}}{NH_3(aq)} + \underset{\text{Water}}{H_2O(l)} \rightleftharpoons \underset{\text{Ammonium ion}}{NH_4^+(aq)} + \underset{\text{Hydroxide ion}}{OH^-(aq)}$$

Answers

❶ 6.2×10^{-8}

❷ Le Châtelier's principle

The equilibrium of this equation greatly favors the reverse reaction. Only about 1% of the ammonia is present as NH_4^+, the conjugate acid of NH_3. The concentrations of NH_4^+ and OH^- are low and equal. Interestingly, the compound ammonium hydroxide (NH_4OH) cannot be isolated from solutions of aqueous ammonia. To isolate NH_4OH, you would need to remove the NH_3 and H_2O. If you began to remove the ammonia first, the equilibrium would shift even further in the reverse direction (to the left) to compensate for the loss of NH_3. What principle does this demonstrate? 2 This cycle of loss of NH_3 and reduction of the amount of NH_4^+ and OH^- would continue until all the ammonia had been removed from the water. An attempt to remove the water first or the ammonia and water together would also fail to isolate any NH_4OH. Bottles of aqueous ammonia used in laboratories are sometimes labeled ammonium hydroxide, but this compound has never been isolated. The equilibrium-constant expression for the reaction of ammonia with water is

$$K_{eq} = \frac{[NH_4^+] \times [OH^-]}{[NH_3] \times [H_2O]}$$

As you just learned in the discussion of K_a, the concentration of water is constant in dilute solutions. It can be combined with K_{eq} to give a base dissociation constant (K_b).

$$K_b = \frac{[NH_4^+] \times [OH^-]}{[NH_3]}$$

In general, the **base dissociation constant (K_b)** is the ratio of the concentration of the conjugate acid times the concentration of the hydroxide ion to the concentration of the conjugate base. The following is the general form of this equation.

$$K_b = \frac{[\text{conjugate acid}] \times [OH^-]}{[\text{conjugate base}]}$$

The magnitude of K_b indicates the ability of a weak base to compete with the very strong base OH^- for hydrogen ions. Because bases such as ammonia are weak relative to the hydroxide ion, K_b for such bases is usually small. The K_b for ammonia is 1.8×10^{-5}. The smaller the value of K_b, the weaker is the base.

The words concentrated and dilute indicate how much of an acid or base is dissolved in solution. These terms refer to the number of moles of the acid or base in a given volume. The words strong or weak refer to the extent of ionization or dissociation of an acid or base. They indicate how many of the particles ionize or dissociate into ions. Hydrochloric acid (HCl)(*aq*) is a strong acid; it is completely dissociated into ions. Gastric juice in the stomach is a dilute solution of hydrochloric acid. A relatively small number of HCl molecules are present in a given volume of gastric juice, but they are all dissociated into ions. A sample of hydrochloric acid added to a large volume of water becomes more dilute, but it is still a strong acid. Vinegar is a dilute solution of a weak acid, ethanoic acid. Pure ethanoic acid (glacial acetic acid) is still a weak acid, even though it is highly concentrated. Solutions of ammonia can be dilute or concentrated, depending on the amount of ammonia dissolved in a given volume of

HEALTH

Tooth Decay

The chief cause of tooth decay is the weak acid called lactic acid ($C_3H_6O_3$), K_a 1.4×10^{-4}. Lactic acid is formed in the mouth by the action of specific bacteria, such as *Streptococcus mutans,* on sugars present in sticky plaque on tooth surfaces. Under normal conditions, saliva keeps the plaque at pH 6.8. A pH of 6.0 or higher causes no problems. When lactic acid lowers the pH of plaque to 5.5 or less, however, the tooth enamel can begin to break down. Once the enamel is penetrated, dental caries, or cavities, result. The damage can progress to the underlying dentin and pulp, which contains blood vessels and nerves, and can cause a toothache.

20.4

Cooperative Learning

Shampoos labeled for use on normal, dry, or oily hair are formulated by controlling the strength and amount of the synthetic detergent. The quantity of the active ingredient controls the "defatting" action, which removes oil from the hair. Have groups of students conduct a survey of shampoo products. Each group should gather data on a different brand and type of shampoo. Decide as a class what information should be included in the survey and how the data will be reported. To assure uniform pH testing, have groups prepare a 1% shampoo solution and use universal indicator paper.

Chemistry Alive!

Titration

Play

Discuss

Explain how the dissociation constant of an acid or base can be determined experimentally from the concentration of the solution and its pH. The calculation is based on the assumption that the acid molecule forms an equal number of hydrogen ions and negative ions when it dissociates. This is true only if there is no additional source of either hydrogen ions or the negative ions.

Calculating K_a is also much more complicated in the case of a polyprotic substance. The dissociation of H_3PO_4, for example, involves three separate ionizations, each with its own K_a.

Practice Problems Plus

Related Chapter Review Problems
Chapter Review problems 55, 56, 59, and 62 are related to Sample Problem 20-8.

Additional Practice Problem
The Chem ASAP! CD-ROM contains the following problem: A solution of a weak acid, exactly 0.500*M*, has a $[H^+] = 5.77 \times 10^{-6}\,M$.
a. What is the pH of this solution?
b. What is the value of K_a for this acid?
(**a.** 5.239 **b.** 6.66×10^{-11})

Table 20.9

Concentrations of Some Common Laboratory Acids and Bases		
	Concentration	
Acid or base	Moles/liter (molarity)	Grams/liter
Concentrated hydrochloric acid	12	438
Dilute hydrochloric acid	6	219
Concentrated sulfuric acid	18	1764
Dilute sulfuric acid	6	588
Concentrated phosphoric acid	15	1470
Concentrated nitric acid	16	1008
Dilute nitric acid	6	378
Ethanoic acid, glacial	17	1020
Ethanoic acid, dilute	6	360
Dilute sodium hydroxide	6	240
Concentrated aqueous ammonia	15	255
Dilute aqueous ammonia	6	102

water. In any solution of ammonia, however, whether concentrated or dilute, ammonia will be a weak base because the amount of ionization will be small. **Table 20.9** lists the concentrations of acids and bases commonly found in the laboratory. Identify each acid and base in **Table 20.9** as either
1 weak or strong.

Calculating Dissociation Constants

You can calculate the acid dissociation constant (K_a) of a weak acid from experimental data. To do this, you need to measure the equilibrium concentrations of all the substances present at equilibrium. For a weak acid, you can determine these concentrations experimentally if you know two conditions. First, you must know the initial molar concentration of the acid. Second, you must know the pH (or $[H^+]$) of the solution at equilibrium.

Sample Problem 20-8

A 0.1000*M* solution of ethanoic acid is only partially ionized. From measurements of the pH of the solution, $[H^+]$ is determined to be $1.34 \times 10^{-3}M$. What is the acid dissociation constant (K_a) of ethanoic acid?

1. *ANALYZE* *List the knowns and the unknown.*

Knowns:
- [ethanoic acid] = 0.1000*M*
- $[H^+] = 1.34 \times 10^{-3}M$
- $CH_3COOH(aq) \rightleftharpoons H^+(aq) + CH_3COO^-(aq)$
- $K_a = \dfrac{[H^+] \times [CH_3COO^-]}{[CH_3COOH]}$

Unknown:
- $K_a = ?$

Answers

1 HCl: strong
H_2SO_4: strong
H_3PO_4: weak
HNO_3: strong
CH_3COOH: weak
NaOH: strong
NH_3: weak

SECTION REVIEW 20.4

28. A strong acid is completely ionized in aqueous solution. A weak acid is ionized only slightly in aqueous solution. The strongest acid in the table is oxalic acid; the weakest acid is carbonic acid.

29. $K_a = 1.5 \times 10^{-3}$; weak

30. boric acid

31. The [HX] is much greater than the $[H^+]$.

Sample Problem 20-8 (cont.)

2. *CALCULATE Solve for the unknown.*

Each molecule of CH_3COOH that ionizes gives an H^+ and a CH_3COO^- ion. Therefore, at equilibrium $[H^+] = [CH_3COO^-] = 1.34 \times 10^{-3}M$. The equilibrium concentration of CH_3COOH is the initial concentration minus the concentration of the ionized acid or $(0.1000 - 0.00134)M = 0.0987M$. The following table summarizes these data. (All concentrations are in *M*.)

Concentration	$[CH_3COOH]$	$[H^+]$	$[CH_3COO^-]$
Initial	0.1000	0	0
Change	-1.34×10^{-3}	1.34×10^{-3}	1.34×10^{-3}
Equilibrium	0.0987	1.34×10^{-3}	1.34×10^{-3}

The equilibrium values can now be substituted into the expression for K_a.

$$K_a = \frac{[H^+] \times [CH_3COO^-]}{[CH_3COOH]}$$

$$= \frac{(1.34 \times 10^{-3}) \times (1.34 \times 10^{-3})}{0.0987}$$

$$= 1.82 \times 10^{-5}$$

3. *EVALUATE Does the result make sense?*

The value of K_a is consistent with that of a weak acid.

Practice Problems

26. For a solution of methanoic acid (formic acid) exactly $0.1M$, $[H^+] = 4.2 \times 10^{-3}M$. Calculate the K_a of methanoic acid.

27. For a solution of a monoprotic weak acid exactly $0.2M$, $[H^+] = 9.86 \times 10^{-4}M$.

a. What is the pH of this solution?

b. What is the value of K_a for this acid?

Chem ASAP!

Problem-Solving 27

Solve Problem 27 with the help of an interactive guided tutorial.

section review 20.4

28. What is the definition of a strong acid? A weak acid? Based on the first ionization constant, which acid in **Table 20.8** on page 602 is the strongest? Which is the weakest?

29. A $0.18M$ solution of monoprotic chloroacetic acid ($CH_2ClCOOH$) has a pH of 1.80. Calculate the acid dissociation constant for this acid. Is this a strong or weak acid?

30. Which acid in **Table 20.7** on page 600 would you expect to have the lowest ionization constant?

31. Acid HX has a very small value of K_a. How do the relative amounts of H^+ and HX compare at equilibrium?

32. Write the equations for the ionization or dissociation of the following acids and bases in water.

a. nitric acid
b. ethanoic acid
c. ammonia
d. magnesium hydroxide

33. Compare strong/weak and concentrated/dilute as they pertain to acids and bases.

Chem ASAP! **Assessment 20.4** Check your understanding of the important ideas and concepts in Section 20.4.

20.4

3 Assess

Evaluate Understanding

Have students draw a concept map for this section, starting with the main heading, "Ways of Classifying Acids and Bases."

- **Ask the students to list the two conditions they need to know to calculate the dissociation constant of a weak acid?** (You must know the initial molar concentration of the acid and the pH or $[H^+]$ of the solution at equilibrium.)

Reteach

Table 20.8, Dissociation Constants of Weak Acids, may pose difficulties for some students. Use an example to explain the information in the table. Point out that the K_a for each acid is equal to the product of the concentrations of the ions on the right of the equation, divided by the concentration of the non-ionized acid. Also explain that the K_a value is a quantitative indication of acid strength. A lower K_a value means the reactants are favored in the equilibrium; thus, the acid is weaker. As a class activity, have students prepare a numbered list of steps for calculating dissociation constants. Give them a different example than those in Sample Problem 20-8, and have them use the numbered procedure to work through the problem.

SECTION REVIEW 20.4 CONTINUED

32. a. $HNO_3 + H_2O \rightarrow H_3O^+ + NO_3^-$

b. $CH_3COOH + H_2O \rightleftharpoons CH_3COO^- + H_3O^+$

c. $NH_3 + H_2O \rightleftharpoons NH_4^+ + OH^-$

d. $Mg(OH)_2 \rightarrow Mg^{2+} + 2OH^-$

33. Strong acids and bases ionize or dissociate completely in aqueous solution. Weak acids and bases ionize or dissociate only slightly in aqueous solution. Concentrated acid or base solutions contain large amounts (high concentrations) of acid or base. Dilute acid or base solutions contain small amounts (low concentration) of acid or base.

SMALL-SCALE LAB

IONIZATION CONSTANTS OF WEAK ACIDS.

OVERVIEW

Procedure Time: 30 minutes
Students measure the pH at which common indicators change colors and use their data to calculate the ionization constants of weak acids.

MATERIALS

Solution	Preparation
*0.04% BTB	100 mg in 16.0 mL 0.01*M* NaOH, dilute to 250 mL
*0.04% BCG	100 mg in 14.3 mL 0.01*M* NaOH, dilute to 250 mL
*0.04% BPB	100 mg in 14.9 mL 0.01*M* NaOH, dilute to 250 mL
*0.04% MCP	100 mg in 26.2 mL 0.01*M* NaOH, dilute to 250 mL
*0.04% TB	100 mg in 21.5 mL 0.01*M* NaOH, dilute to 250 mL
0.1%	phenolphthalein 250 mg in 250 mL of 70% 2-propanol
0.05% MO	125 mg in 250 mL
0.02% AYR	50 mg in 250 mL

*Can dissolve sodium salt directly in 250 mL of distilled water.
Solution A: 14.9 g boric acid and 12.6 g citric acid in 1200 mL water
Solution B: 45.6 g sodium phosphate in 1200 mL water
Prepare the buffer solutions, pH 2–12, by mixing the stock solutions according to the following table: (Use 1.0*M* phosphoric acid for pH 1.)

pH	mL of Solution A	mL of Solution B
2	195	5
3	176	24
4	155	45
5	134	66
6	118	82
7	99	101
8	85	115
9	69	131
10	54	146
11	44	156
12	17	183

SMALL-SCALE LAB

IONIZATION CONSTANTS OF WEAK ACIDS

SAFETY  

Wear safety glasses and follow the standard safety procedures outlined on page 18.

PURPOSE

To measure ionization constants of weak acids such as bromocresol green (BCG).

MATERIALS

- pencil
- paper
- ruler
- reaction surface
- chemicals shown in Figure A

PROCEDURE

On separate sheets of paper, draw two grids similar to Figure A. Make each square 2 cm on each side. Place a reaction surface over one of the grids and place one drop of BCG in each square. Place one drop of pH buffer in each square corresponding to its pH value. Use the second grid as a data table to record your observations for each solution.

1 drop pH buffer + 1 drop BCG

pH

1 yellow	2 yellow	3 yellow
4 green	5 blue	6 blue
7 blue	8 blue	9 blue
10 blue	11 blue	12 blue

Figure A

ANALYSIS

Using your experimental data, record the answers to the following questions below your data table.

1. What is the color of the lowest pH solutions?
2. What is the color of the highest pH solutions?
3. At which pH does the bromcresol green change from one color to the other? At which pH does an intermediate color exist?

An acid–base indicator is usually a weak acid with a characteristic color. Because bromcresol green is an acid, it is convenient to represent its rather complex formula as HBCG. HBCG ionizes in water according to the following equation.

$$\underset{\text{(yellow)}}{HBCG} + H_2O = \underset{\text{(blue)}}{BCG^-} + H_3O^+$$

The K_a expression is

$$K_a = \frac{[BCG^-] \times [H_3O^+]}{[HBCG]}$$

When $[BCG^-] = [HBCG]$, $K_a = [H_3O^+]$.

4. What color is the conjugate acid of BCG^-?
5. What color is the conjugate base of HBCG?
6. What color is an equal mixture of the conjugate acid and conjugate base of bromcresol green? At what pH does this equal mixture occur?

YOU'RE THE CHEMIST

The following small-scale activities allow you to develop your own procedures and analyze the results.

1. **Design It!** Design and carry out an experiment to measure the K_a of several more acid–base indicators. Record the colors of the conjugate acids (low pH) and the conjugate bases (high pH). Determine the K_a for each acid.
2. **Analyze It!** Explain in your own words how to measure the K_a of a colored acid–base indicator. Explain what you would do and how you would interpret your results.

606 Chapter 20

ANALYSIS

1. pH solutions 1–3 are yellow.
2. pH solutions 5–12 are blue.
3. pH solution 4 is green (an intermediate between yellow and blue).
4. The conjugate acid, HBCG, is yellow.
5. The conjugate base, BCG^-, is blue.
6. The equal mixture of HBCG and BCG^- is green at pH = 4.

YOU'RE THE CHEMIST

1. Results will vary depending on the indicator chosen.
2. To measure the K_a of a colored weak acid, mix one drop of the weak acid with one drop of each pH 1–12 buffer solution. Look for the pH of the color change. This pH is the K_a of the acid.

Chemistry Serving...the Environment

RAIN LIKE VINEGAR

In Canada and the eastern part of the United States, lakes once full of fish and frogs are now nearly lifeless. Acres and acres of pines and other forest trees are dying in the Appalachian Mountains. In Rome and Athens, ancient carved stonework has been weathered so badly that the details of the carvings have been erased. All this damage has a single cause—acid rain.

Acid rain is rain with a pH lower than that of "normal" precipitation. Normal precipitation is itself mildly acidic, with a pH of about 5.6. This mild acidity comes from small amounts of carbon dioxide in the atmosphere that dissolve in tiny water droplets to form carbonic acid (H_2CO_3). The acidity of normal rainfall poses no problems because it is so mild and because animals and plants have adapted to it. But when the pH of rainfall gets significantly lower than 5.6, negative effects begin to appear. The rain from a 1981 storm in Baltimore was measured at pH 2.7, about the same pH as vinegar! Where it occurs, acid rain causes enormous damage to stone structures, paint, metal, and the environment. It leaches nutrients from the soil and may even be a direct threat to human health.

The rain from a 1981 storm in Baltimore was measured at pH 2.7, about the same pH as vinegar!

What causes acid rain? Acid rain results when oxides of sulfur (SO_x) and oxides of nitrogen (NO_x) are emitted into the atmosphere. Although there are some natural sources of these pollutants (such as volcanoes), most of them come from human sources, such as coal- and oil-fired power plants. In the atmosphere, sulfur oxides form sulfuric acid (H_2SO_4), and nitrogen oxides form nitric acid (HNO_3). Eventually, these acids fall to Earth in raindrops or snowflakes.

Fortunately, acid rain can be prevented, and chemistry is playing a big part in the effort. Because power plants are one of the main sources of SO_x and NO_x, chemists have focused on developing devices called scrubbers that remove these pollutants from power plant smoke before it leaves the smokestack.

In one type of scrubber now in common use, the smoke passes through an aqueous suspension of lime (CaO). The lime reacts with the gaseous SO_2 to make calcium sulfite ($CaSO_3$), a solid that can be disposed of. To remove NO_x, a process called Selective Catalytic Reduction (SCR) is often used. In an SCR chamber, ammonia (NH_3) is sprayed onto a surface covered with a mixture of catalysts. As the smoke passes through the chamber, NO and NO_2 in the smoke react with the ammonia and oxygen to form harmless nitrogen gas (N_2) and water (H_2O).

To eliminate the need for separate scrubbers for SO_x and NO_x, researchers have also developed ways of removing the two types of oxides from power plant smoke at the same time. In one technique, the smoke is passed through an alkaline suspension of yellow phosphorus. A variety of chemical reactions occur that convert almost all the SO_x and NO_x into valuable by-products, including some of the important plant nutrients that make up fertilizers.

The use of scrubbers at fossil-fuel-burning power plants has begun to have an impact on the pH of rainfall in many parts of the world, but much more remains to be done to solve the acid rain problem.

CHEMISTRY IN CAREERS

TAKE IT TO THE NET

Find out more about career opportunities:

www.phschool.com

STONE CONSERVATOR

Would you like to help preserve historic buildings and sculpture?

See page 877.

CHEMICAL SPECIALIST

ocal food service distributor seeks esponsible self-motivated individ-

CHEMIST NEEDED

Enviro. Lab. GC Must know

Chemistry Serving... the Environment

DISCUSS

Have students do research about the sources and causes of acid rain. (Acid rain is thought to be due principally to the release of sulfur and nitrogen oxides into the atmosphere. For example, SO_2 is formed by the burning of fossil fuels, especially coal, that are rich in sulfur. Nitrogen oxides are produced as a result of industrial activities and the use of automobiles.)

Ask:

- How do weather patterns affect the distribution of environmental damage due to acid rain? (Prevailing winds generally carry pollutants from west to east.)
- What types of vegetation, if any, are most resistant to acid rain, or tend to thrive in acidic soils? (Students may wish to speak to staff at nurseries for help answering this question.)
- What types of conservation efforts are used to adjust the pH of lakes and rivers that become too acidic? (A special slurry of $CaCO_3$, or limestone, is sometimes spread over lakes to control pH.)

CHEMISTRY IN CAREERS

Have students learn about the duties and responsibilities of a stone conservator by reading the career feature on page 877 or connecting to the Internet address shown.

4 Close

Summary

Ask the following questions to help students summarize chapter content.

- **What are the concentrations of hydrogen and hydroxide ions in pure water at 25°C?** (1×10^{-7} mol/L)
- **What is the pH of acidic, neutral, and basic solutions?** (acidic: pH < 7, neutral: pH $= 7$, basic: pH > 7.)

Describe the three acid–base theories. (Arrhenius: Acids produce H^+ ions and bases produce OH^- ions in aqueous solution. Brønsted-Løwry: Acids donate protons; bases accept protons. Lewis: Acids donate electron pairs; bases accept electron pairs.)

- **How is the strength of an acid related to its K_a?** (A strong acid has a much larger K_a than does a weak acid.)

Extension

Have students list the defining characteristics of Arrhenius, Brønsted-Løwry, and Lewis acids and bases. Ask them to compare and contrast the characteristics they have listed.

Looking Back . . . Looking Ahead . . .

The previous chapter discussed the factors affecting reaction rates and equilibrium. Students learned how enthalpy and entropy changes determine whether a reaction is spontaneous. The current chapter described properties of acids and bases, including strength. The next chapter discusses neutralization reactions and the properties of the salts that are formed.

Chapter 20 STUDENT STUDY GUIDE

Take It to the NET
For interactive study and review, go to www.phschool.com

KEY TERMS

- acid *p. 578*
- acid dissociation constant (K_a) *p. 601*
- acidic solution *p. 581*
- alkaline solution *p. 581*
- amphoteric *p. 598*
- base *p. 579*
- base dissociation constant (K_b) *p. 603*
- basic solution *p. 581*
- conjugate acid *p. 597*
- conjugate acid–base pair *p. 597*
- conjugate base *p. 597*
- diprotic acid *p. 594*
- hydrogen-ion acceptor *p. 596*
- hydrogen-ion donor *p. 596*
- hydronium ion (H_3O^+) *p. 580*
- hydroxide ion (OH^-) *p. 580*
- ion-product constant for water (K_w) *p. 581*
- Lewis acid *p. 598*
- Lewis base *p. 598*
- monoprotic acid *p. 594*
- neutral solution *p. 580*
- pH *p. 582*
- self-ionization *p. 580*
- strong acid *p. 600*
- strong base *p. 602*
- triprotic acid p. *594*
- weak acid *p. 600*
- weak base *p. 602*

KEY EQUATIONS AND RELATIONSHIPS

- $K_w = [H^+] \times [OH^-] = 1.0 \times 10^{-14} M^2$
- $pH = -\log[H^+]$
- $pOH = -\log[OH^-]$
- $pH + pOH = 14$

CONCEPT SUMMARY

20.1 Describing Acids and Bases

- Acids taste sour, are electrolytes, react with active metals to produce hydrogen, react with bases to form water and salts, and cause indicators to change color.
- Bases taste bitter, are electrolytes, react with acids to form water and salts, and cause indicators to change color.

20.2 Hydrogen Ions and Acidity

- Water molecules dissociate into hydrogen ions (H^+) and hydroxide ions (OH^-).
- On the pH scale, 0 is strongly acidic, 14 is strongly basic, and 7 is neutral. Pure water at 25 °C has a pH of 7.

20.3 Acid–Base Theories

- A Brønsted-Lowry acid is a proton donor, and a Brønsted-Lowry base is a proton acceptor.
- An Arrhenius acid yields hydrogen ions in aqueous solution. An Arrhenius base yields hydroxide ions in aqueous solution.
- A Lewis acid is an electron-pair acceptor, and a Lewis base is an electron-pair donor.
- A conjugate acid–base pair consists of two substances related by the loss or gain of a single hydrogen ion.

20.4 Strengths of Acids and Bases

- The strength of an acid or a base is determined by its degree of ionization in solution. The acid dissociation constant (K_a) is a quantitative measure of acid strength.
- The base dissociation constant (K_b) is a quantitative measure of base strength.

CHAPTER CONCEPT MAP

Use these terms to construct a concept map that organizes the major ideas of this chapter.

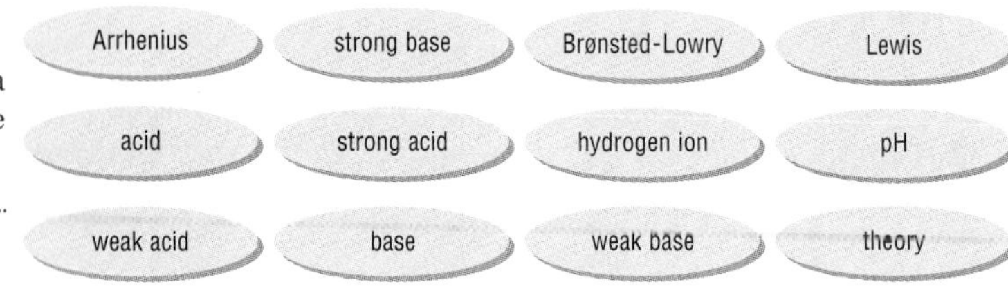

Chem ASAP! **Concept Map 20**
Create your Concept Map using the computer.

Take It to the Net

At **www.phschool.com** students will find for this chapter

- an Internet activity
- links to related chemistry sites
- an interactive quiz
- career links

CONCEPT PRACTICE

34. Write formulas for these compounds. *20.1*
    a. nitrous acid
    b. aluminum hydroxide
    c. hydroselenic acid
    d. strontium hydroxide
    e. phosphoric acid
    f. ethanoic acid
35. Write an equation showing the ionization of water. *20.2*
36. What are the concentrations of H^+ and OH^- in pure water at 25 °C? *20.2*
37. How is the pH of a solution calculated? *20.2*
38. Why is the pH of pure water at 25 °C equal to 7.00? *20.2*
39. Calculate the pH for the following solutions and indicate whether each solution is acidic or basic. *20.2*
    a. $[H^+] = 1.0 \times 10^{-2}M$
    b. $[OH^-] = 1.0 \times 10^{-2}M$
    c. $[OH^-] = 1.0 \times 10^{-8}M$
    d. $[H^+] = 1.0 \times 10^{-6}M$
40. What are the hydroxide-ion concentrations for solutions with the following pH values? *20.2*
    a. 4.00
    b. 8.00
    c. 12.00
41. Calculate the pH or $[H^+]$ for each solution. *20.2*
    a. $[H^+] = 2.4 \times 10^{-6}M$
    b. $[H^+] = 9.1 \times 10^{-9}M$
    c. pH = 13.20
    d. pH = 6.70
42. A soft drink has a pH of 3.80. What is the hydrogen-ion concentration in the drink? *20.2*
43. Write the reaction for the dissociation of each compound in water. *20.3*
    a. potassium hydroxide
    b. magnesium hydroxide
44. How did Arrhenius describe acids and bases? *20.3*
45. Classify each as an Arrhenius acid or an Arrhenius base. *20.3*
    a. $Ca(OH)_2$ d. C_2H_5COOH
    b. HNO_3 e. HBr
    c. KOH f. H_2SO_4
46. Identify each acid in Problem 45 as monoprotic, diprotic, or triprotic. *20.3*
47. Write balanced equations for the reaction of each metal with water. *20.3*
    a. lithium
    b. barium
48. Identify each reactant in the following equations as a hydrogen-ion donor (acid) or a hydrogen-ion acceptor (base). *20.3*
    a. $HNO_3 + H_2O \rightleftharpoons H_3O^+ + NO_3^-$
    b. $CH_3COOH + H_2O \rightleftharpoons H_3O^+ + CH_3COO^-$
    c. $NH_3 + H_2O \rightleftharpoons NH_4^+ + OH^-$
    d. $H_2O + CH_3COO^- \rightleftharpoons CH_3COOH + OH^-$
49. Label the conjugate acid–base pairs in each equation in Problem 48. *20.3*
50. What is an amphoteric substance? *20.3*
51. What is a Lewis acid? A Lewis base? In what sense is the Lewis theory more general than the Arrhenius and Brønsted-Lowry theories? *20.3*
52. Identify each compound as a strong or weak acid or base. *20.4*
    a. NaOH
    b. HCl
    c. NH_3
    d. H_2SO_4
53. Would a strong acid have a large or a small K_a? Explain. *20.4*
54. Why are $Mg(OH)_2$ and $Ca(OH)_2$ considered to be strong bases even though their saturated solutions are only mildly basic? *20.4*
55. Write the expression for K_a for each acid. Assume only one hydrogen is ionized. *20.4*
    a. HF
    b. H_2CO_3
56. A 0.0250*M* solution of $KHCrO_4$ has a pH of 3.50. Calculate the K_a for the equilibrium between $HCrO_4^-$ and CrO_4^{2-}. *20.4*

CONCEPT MASTERY

57. Is it possible to have a concentrated weak acid? Explain.
58. Write equations showing that the hydrogen phosphate ion (HPO_4^{2-}) is amphoteric.
59. The pH of a 0.50*M* HNO_2 solution is 1.83. What is the K_a of this acid?
60. Write the formula and name of the conjugate base of each Brønsted-Lowry acid.
    a. HCO_3^- c. NH_4^+
    b. HI d. H_2SO_3

Answers

34. a. $HNO_2(aq)$ d. $Sr(OH)_2(aq)$
    b. $Al(OH)_3(aq)$ e. $H_3PO_4(aq)$
    c. $H_2Se(aq)$ f. $CH_3COOH(aq)$
35. $H_2O \rightleftharpoons H^+ + OH^-$
36. 1.0×10^{-7} *M* for both H^+ and OH^- at 25°C
37. the negative logarithm of the $[H^+]$
38. The hydrogen-ion concentration of pure water at 25°C is 1.0×10^{-7} *M*. The negative logarithm, or pH, of this concentration is 7.00.
39. a. pH = 2.00, acidic c. pH = 6.00, acidic
    b. pH = 12.00, basic d. pH = 6.00, acidic
40. a. 1.0×10^{-10} *M*
    b. 1.0×10^{-6} *M*
    c. 1.0×10^{-2} *M*
41. a. 5.62 c. 6.3×10^{-14} *M*
    b. 8.04 d. 2.0×10^{-7} *M*
42. 1.6×10^{-4} *M*
43. a. $KOH \rightarrow K^+ + OH^-$
    b. $Mg(OH)_2 \rightarrow Mg^{2+} + 2OH^-$

44. In aqueous solution, acids produce hydrogen ions and bases produce hydroxide ions.
45. a. base d. acid
    b. acid e. acid
    c. base f. acid
46. monoprotic: **b.**, **d.**, and **e.**; diprotic: **f.**
47. a. $2Li + 2H_2O \rightarrow 2LiOH + H_2$
    b. $Ba + 2H_2O \rightarrow Ba(OH)_2 + H_2$
48. a. HNO_3, acid; H_2O, base
    b. CH_3COOH, acid; H_2O, base
    c. H_2O, acid; NH_3, base
    d. H_2O, acid; CH_3COO^-, base
49. a. HNO_3 with NO_3^-, H_2O with H_3O^+
    b. CH_3COOH with CH_3COO^-, H_2O with H_3O^+
    c. H_2O with OH^-, NH_3 with NH_4^+
    d. H_2O with OH^-, CH_3COO^- with CH_3COOH
50. An amphoteric substance can act as an acid or a base.
51. A Lewis acid accepts a pair of electrons to form a covalent bond. A Lewis base donates a pair of electrons to form a covalent bond. The Lewis theory includes as acids or bases many substances that do not donate or accept hydrogen ions.
52. a. strong base c. weak base
    b. strong acid d. strong acid
53. A strong acid is completely dissociated; K_a must be large.
54. They have high K_b ratios. Their concentration in saturated solution is low because of their low solubility.
55. a. $K_a = \frac{[H^+][F^-]}{[HF]}$
    b. $K_a = \frac{[H^+][HCO_3^-]}{[H_2CO_3]}$
56. 4.0×10^{-6}
57. Yes; acetic acid dissolves well but ionizes poorly.
58. $HPO_4^{2-} \rightleftharpoons H^+ + PO_4^{3-}$
    $HPO_4^{2-} + H^+ \rightleftharpoons H_2PO_4^-$
59. 4.4×10^{-4}
60. a. CO_3^{2-}, carbonate ion
    b. I^-, iodide ion
    c. NH_3, ammonia
    d. HSO_3^-, hydrogen sulfite ion

Chapter 20

REVIEW

Answers

61. a. $HClO_2$, chlorous acid
b. H_3PO_4, phosphoric acid
c. H_3O^+, hydronium ion
d. NH_4^+, ammonium ion

62. 1.6×10^{-5}

63. a. $[OH^-] = 4.0 \times 10^{-10}M$
b. $[OH^-] = 2.0 \times 10^{-5}M$
c. pH = 12.26
d. pH = 5.86

64. $H_3PO_4 \rightleftharpoons H^+ + H_2PO_4^-$
$H_2PO_4^- \rightleftharpoons H^+ + HPO_4^{2-}$
$HPO_4^{2-} \rightleftharpoons H^+ + PO_4^{3-}$

65. a. KOH is the base; HBr is the acid
b. HCl is the acid; H_2O is the base

66. a. 3 **b.** 2 **c.** 1

67. Answers will vary; students are likely to consider the Arrhenius theory the easiest to understand. All three theories provide definitions and describe accepted behavior of a certain group of compounds. Because it is more general, the Brønsted–Lowry theory includes a greater number of compounds than the Arrhenius theory. The Lewis theory includes the greatest number of compounds because it is the most general. Each theory has advantages in certain circumstances.

68. Because $\log(a \times b) = \log a + \log b$, and $K_w = [H^+] \times [OH^-] = 1.0 \times 10^{-14}$, taking the log of each concentration term gives:
$\log[H^+] + \log[OH^-] = -14$
$-\log[H^+] - \log[OH^-] = 14$
pH + pOH = 14

69. The *y*-axis might correspond to $[H^+]$ or pOH because HCl is a strong acid.

70. Add 15.3 g KOH to distilled water and dissolve. Then bring the volume of the solution up to 400.0 mL.

71. 0.47 L

72. a. increases products (shifts right)
b. no change
c. increases products (shifts right)
d. increases products (shifts right)
e. increases products (shifts right)

73. a. $K_{eq} = \dfrac{[CO]^2[O_2]}{[CO_2]^2}$
b. $K_{eq} = \dfrac{[NH_3]^2}{[H_2]^3[N_2]}$

74. pH = 10.66

75. $K_w = K_aK_b = \dfrac{[H^+][A^-]}{[HA]} \times \dfrac{[HA][OH^-]}{[A^-]} = [H^+][OH^-]$

76. a. 7.4721, 7.2675, 7.0835, 6.9165, 6.7675, 6.6310

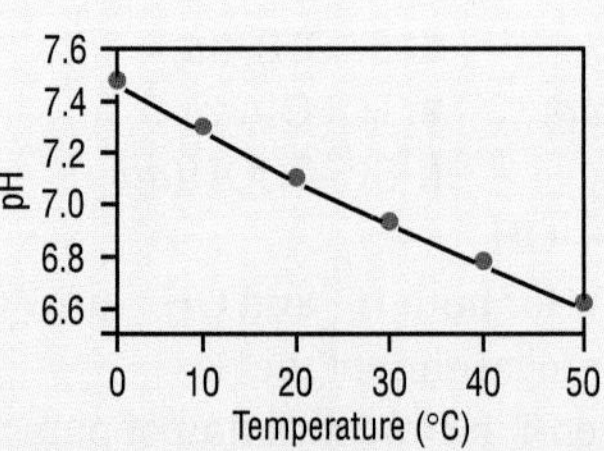

b. 7.37
c. 35°C

61. Write the formula and name of the conjugate acid of each Brønsted-Lowry base.
a. ClO_2^- **c.** H_2O
b. $H_2PO_4^-$ **d.** NH_3

62. You have determined that 1.40% of a 0.080*M* solution of a weak acid is ionized. Calculate the K_a for this acid.

63. Calculate the $[OH^-]$ or pH of each solution.
a. pH = 4.60 **c.** $[OH^-] = 1.8 \times 10^{-2}M$
b. pH = 9.30 **d.** $[OH^-] = 7.3 \times 10^{-9}M$

64. Write the three equations for the stepwise ionization of phosphoric acid.

65. Use the Brønsted-Lowry and Lewis definitions of acids and bases to identify each reactant as an acid or a base.
a. $KOH + HBr \rightleftharpoons KBr + H_2O$
b. $HCl + H_2O \rightleftharpoons Cl^- + H_3O^+$

CRITICAL THINKING

66. Choose the term that best completes the second relationship.

a. sky : blue — acid : ______
(1) metallic (3) sour
(2) basic (4) bitter

b. animals : carbon dioxide — base : ______
(1) electrons (3) bitter
(2) hydroxide ions (4) protons

c. thermometer : degrees — indicator : ______
(1) pH (3) H^+
(2) acid (4) base

67. Arrhenius, Brønsted-Lowry, and Lewis all developed theories to explain acids and bases.
a. Which theory is easiest for you to understand?
b. How can all three theories be accepted by chemists?

68. Prove the relationship pH + pOH = 14. (*Hint:* Use the expression for K_w.)

69. Which quantity might correspond to the *y*-axis: $[H^+]$, pH, $[OH^-]$, or pOH? Explain your answer.

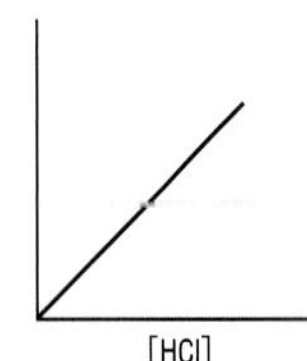

CUMULATIVE REVIEW

70. How would you prepare 400.0 mL of a 0.680*M* KOH solution?

71. How many liters of 8.0*M* HCl are needed to prepare 1.50 L of 2.5*M* HCl?

72. How would each change affect the position of equilibrium of this reaction?

$$2H_2(g) + O_2(g) \rightleftharpoons 2H_2O(g) + \text{heat}$$

a. increasing the pressure
b. adding a catalyst
c. increasing the concentration of $H_2(g)$
d. cooling the reaction mixture
e. removing water vapor from the container

73. Write an equilibrium-constant expression for each equation.
a. $2CO_2(g) \rightleftharpoons 2CO(g) + O_2(g)$
b. $N_2(g) + 3H_2(g) \rightleftharpoons 2NH_3(g)$

CONCEPT CHALLENGE

74. Calculate the pH of a 0.010*M* solution of sodium cyanide (NaCN). The K_b of CN^- is 2.1×10^{-5}.

75. Show that for any conjugate acid–base pair $K_a \times K_b = K_w$.

76. The K_w of water varies with temperature, as indicated in the table below.

Temperature (°C)	K_w	pH
0	1.137×10^{-15}	
10	2.917×10^{-15}	
20	6.807×10^{-15}	
30	1.469×10^{-14}	
40	2.917×10^{-14}	
50	5.470×10^{-14}	

a. Find the pH of water for each temperature in the table. Use these data to prepare a graph of pH versus temperature.
b. Using the graph, estimate the pH of water at 5 °C.
c. At what temperature is the pH of water approximately 6.85?

77. Suppose you slowly add 0.1*M* NaOH to 50.0 mL of 0.1*M* HCl. What volume of NaOH must be added for the resulting solution to become neutral (pH = 7)? Explain your reasoning.

STANDARDIZED TEST PREP

Select the choice that best answers each question or completes each statement.

1. An acid has a measured K_a of 3×10^{-6}.
   a. The acid is a strong acid.
   b. An aqueous solution of the acid would have a pH < 7.
   c. The acid is a strong electrolyte.
   d. All of the above are correct
2. The pH of a sample of orange juice is 3.5. A sample of tomato juice has a pH of 4.5. Compared to the $[H^+]$ of orange juice, the $[H^+]$ of tomato juice is
   a. 1.0 times higher.
   b. 10 times lower.
   c. 10 times higher.
   d. 1.0 times lower.
3. Which species is the conjugate base of an ammonium ion, NH_4^+?
   a. H_2O
   b. OH^-
   c. NH_3
   d. H_3O^+
4. A solution with a hydrogen ion concentration of $2.3 \times 10^{-8} M$ has a pH between
   a. 2 and 3.
   b. 3 and 4.
   c. 7 and 8.
   d. 8 and 9.

Use the description and data table to answer questions 5–8.

Ethanoic acid (acetic acid) is a weak acid with a K_a of 1.8×10^{-5} at room temperature. A student measured the pH of ethanoic acid solutions at different concentrations. She used the pH values to calculate the percent of ethanoic acid molecules that were ionized in each solution.

Concentration of ethanoic acid (*M*)	Percent of molecules ionized
0.050	1.9
0.040	2.1
0.030	2.5
0.020	3.1
0.010	4.0

5. Write the equilibrium equation for the ionization of ethanoic acid in water.
6. Use the data in the table to construct a graph. Make the molarity of ethanoic acid the independent variable.
7. Does the relationship between the variables make sense in terms of Le Châtelier's principle? Explain your answer.
8. Use the graph to estimate the percent of molecules ionized at a concentration of 0.015*M*.

Use the drawings to answer questions 9–11. They show what happens when acids form aqueous solutions. Water molecules have been omitted from the solution windows.

Undissociated acid Conjugate base  Hydronium ion

a.

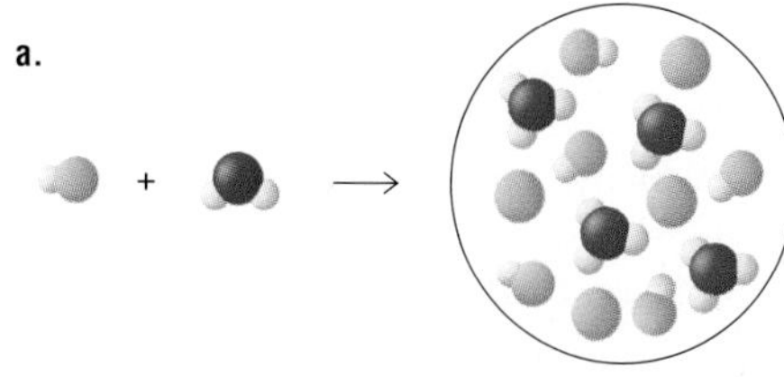

b.

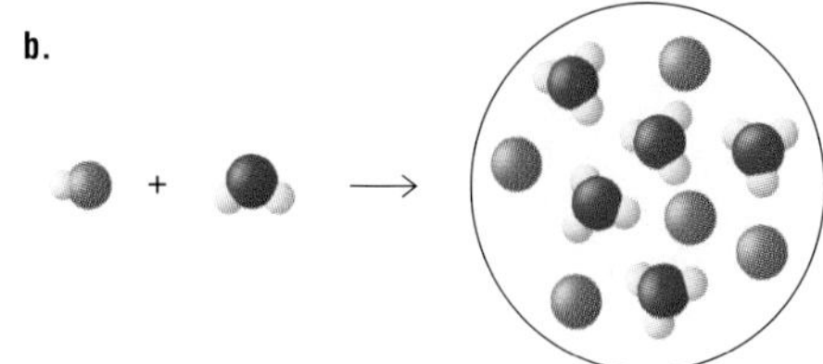

c.

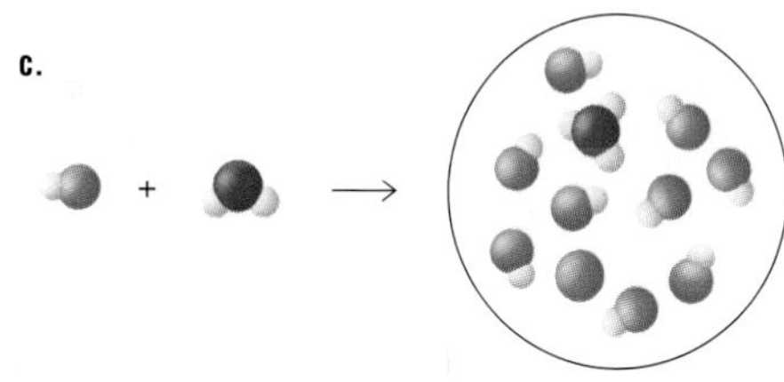

9. Put the acids in order of increasing strength.
10. How many of the acids are strong acids?
11. Which acid is dissociated about 10%?

Write a few sentences in response to each of the statements in questions 12–14.

12. Indicators such as methyl red provide accurate and precise measurements of pH.
13. According to the Arrhenius definition of acids and bases, ammonia qualifies as a base.
14. The strength of an acid or base changes as its concentration changes.

77. 50.0 mL; The pH = 7 when $[H^+] = [OH^-]$. Because HCl is a strong acid that supplies one hydrogen ion per formula unit and NaOH is a strong base that supplies one hydroxide ion per formula unit, $[H^+] = [OH^-]$ when equal volumes of solutions of the same molarity are combined.

STANDARDIZED TEST PREP

1. b
2. b
3. c
4. c
5. $CH_3COOH(aq) + H_2O(l) \rightleftharpoons CH_3COO^-(aq) + H_3O^+(aq)$
6.

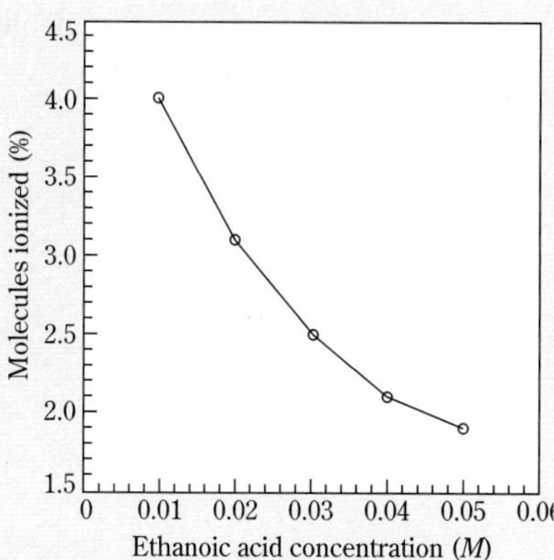

7. Increasing the water concentration puts a stress on the system, which is relieved by a shift to the right. As the concentration of ethanoic acid decreases, the percent of molecules ionized increases.
8. about 3.4%
9. c, a, b
10. one
11. c
12. Indicators give a quick approximation of the pH. For accurate and precise pH measurements, you need a pH meter.
13. Ammonia does not produce hydroxide ions in solution; thus, it does not qualify as a base according to Arrhenius.
14. *Weak* and *strong* refer to the degree of dissociation of an acid or base. *Dilute* and *concentrated* refer to the quantity of an acid or base dissolved in solution.

Planning Guide

SECTION OBJECTIVES	ACTIVITIES/FEATURES	MEDIA & TECHNOLOGY
21.1 Neutralization Reactions ● ■ ◆ ▸ Explain how acid–base titration is used to calculate the concentration of an acid or a base ▸ Explain the concept of equivalence in neutralization reactions	**SE** Discover It! *Reaction of an Acid with an Egg,* p. 612 **SE** Mini Lab *The Neutralizing Power of Antacids,* p. 615 (LRS 21-1) PLM *The Neutralizing Power of Antacids* **SE** **Link to Cosmetology** *Neutral Curls from Permanent Waves,* p. 619 **SE** Small-Scale Lab *Small-Scale Titrations,* p. 625 (LRS 21-2) PLM *Small-Scale Titrations* LM 39: *Neutralization Reactions* LM 40: *Acid-Base Titrations* SSLM 29: *Titration Curves* **TE** DEMOS, pp. 614, 617, 622	**ASAP** Simulation 26 **ASAP** Problem Solving 1, 4, 8, 11, 14 **ASAP** Assessment 21.1 ACT 16: *Acid-Base Titrations* CA *Milk of Magnesia* CA *Titration* CHM Side 7, 34: *Rainbow Indicators* CHM Side 7, 38: *Ammonia Fountain* SSV 8: *Titrations* SSV 11: *Small-Scale pH-Meter*
21.2 Salts in Solution □ ◆ ▸ Demonstrate with equations how buffers resist changes in pH ▸ Calculate the solubility product constant of a slightly soluble salt	**SE** **Link to Biology** *Blood Buffers,* p. 628 **SE** **CHEMath** *Algebra and K_{sp},* p. 633 **SE** **Chemistry Serving . . . Industry** *Extreme Organisms' Valuable Enzymes,* p. 638 **SE** **Chemistry in Careers** *Microbiologist,* p. 638 LM 41: *Salt Hydrolysis* LM 42: *Buffers* LM 43: *A Solubility Product Constant* SSLM 30: *Buffers* **TE** DEMOS, pp. 627, 628, 634	**ASAP** Animation 26 **ASAP** Problem Solving 26, 32 **ASAP** Assessment 21.2 **ASAP** Concept Map 21 CA *Milk of Magnesia* CHM Side 7, 36: *Does Aspirin Buffer?* PP Chapter 21 Problems RP Lesson Plans, Resource Library AR Computer Test 21 www Activities, Self-Tests, *SCIENCE NEWS* updates TCP The Chemistry Place Web Site

KEY

- ● Conceptual (concrete concepts)
- ○ Conceptual (more abstract/math)
- ■ Standard (core content)
- □ Standard (extension topics)
- ◆ Honors (core content)
- ◇ Honors (options to accelerate)

AR	Assessment Resources
ASAP	Chem ASAP! CD-ROM
ACT	ActivChemistry CD-ROM
CHM	CHEMedia Videodiscs
CA	Chemistry Alive! Videodiscs
GCP	Graphing Calculator Problems
GRS	Guided Reading and Study Workbook
LM	Laboratory Manual
LP	Laboratory Practicals
LRS	Laboratory Recordsheets
SSLM	Small-Scale Lab Manual

PRACTICE

SE	**Sample Problems** 21-1 to 21-8
GRS	Section 21.1
RM	Practice Problems 21.1

ASSESSMENT

SE	Section Review
RM	Section Review 21.1
RM	Chapter 21 Quiz

PRACTICE

SE	**Sample Problems** 21-9 to 21-13
GRS	Section 21.2
RM	Practice Problems 21.2
RM	Interpreting Graphics

ASSESSMENT

SE	Section Review
RM	Section Review 21.2
LP	Lab Practical 21-1
RM	Vocabulary Review 21
SE	Chapter Review
SM	Chapter 21 Solutions
SE	Standardized Test Prep
PHAS	Chapter 21 Test Prep
RM	Chapter 21 Quiz
RM	Chapter 21 A & B Test

PLANNING FOR ACTIVITIES

STUDENT EDITION

Discover It! p. 612
- eggs
- small containers
- white vinegar
- pencils and paper
- clocks

Mini Lab, p. 615
- antacids (several brands)
- mortar and pestle
- paper towels
- measuring teaspoons
- clear plastic cups
- medicine droppers
- universal indicator
- white vinegar
- $NaHCO_3$

Small-Scale Lab, p. 625
- plastic cups *or* well strips
- pipets
- NaOH, HNO_3, HCl, H_2SO_4, CH_3COOH
- phenolphthalein

TEACHER'S EDITION

Teacher Demo, p. 614
- equimolar solutions of HCl and NaOH
- phenolphthalein
- safety goggles and apron
- electric stirrer with a magnetic stir bar
- beaker
- table salt

Teacher Demo, p. 617
- equimolar solutions of HCl and NaOH
- phenolphthalein
- safety goggles and apron
- beaker
- buret
- pH meter
- graph paper

Teacher Demo, p. 622
- 1.0*M* HCl
- pH meter
- distilled water
- 10-mL pipet
- 25-mL, 50-mL, and 100-mL volumetric flasks

Teacher Demo, p. 627
- 1*M* solutions of NH_4NO_3, KCl, $NaHCO_3$, Na_2SO_3
- pH meter *or* universal pH paper

Teacher Demo, p. 628
- antacid tablets, buffered aspirin, unbuffered aspirin
- flasks and burets
- 0.5*M* HCl and NaOH
- distilled water
- methyl red, bromthymol blue

Teacher Demo, p. 634
- saturated solutions of silver chloride and silver nitrate
- 2 test tubes
- medicine dropper

OT	Overhead Transparency	SE	Student Edition
PHAS	PH Assessment System	SM	Solutions Manual
PLM	Probeware Lab Manual	SSV	Small-Scale Video/Videodisc
PP	Problem Pro CD-ROM	TCP	www.chemplace.com
RM	Review Module	TE	Teacher's Edition
RP	Resource Pro CD-ROM	www	www.phschool.com

Chapter 21

Key Terms

21.1 neutralization reactions, standard solution, titration, end point, equivalent (equiv), gram equivalent mass, normality (N), equivalence point

21.2 salt hydrolysis, buffers, buffer capacity, solubility product constant (K_{sp}), common ion, common-ion effect

DISCOVER IT!

Almost immediately after placing the egg in vinegar, students should observe the formation of gas bubbles indicating a reaction between the vinegar and the egg shell. Upon prolonged exposure (24 to 72 hours), the egg is completely decalcified, leaving only a thin membrane surrounding the egg white. After completing the chapter, students should understand that the reaction represents the neutralization of an acid (acetic acid) by a base (calcium carbonate).

Chapter 21 NEUTRALIZATION

Carbon dioxide forms when antacid tablets are dropped in water.

FEATURES

DISCOVER IT!
Reaction of an Acid with an Egg

SMALL-SCALE LAB
Small-Scale Titrations

MINI LAB
The Neutralizing Power of Antacids

CHEMath
Algebra and K_{sp}

CHEMISTRY SERVING ... INDUSTRY
Extreme Organisms' Valuable Enzymes

CHEMISTRY IN CAREERS
Microbiologist

LINK TO COSMETOLOGY
Neutral Curls from Permanent Waves

LINK TO BIOLOGY
Blood Buffers

Stay current with SCIENCE NEWS
Find out more about neutralization:
www.phschool.com

DISCOVER IT! REACTION OF AN ACID WITH AN EGG

You need an egg, a small container, white vinegar, paper, a pen or pencil, and a clock.

1. Carefully place the egg in the container.
2. Pour vinegar into the container so that the vinegar level is about 3 cm above the egg.
3. Note the time.
4. Observe the system periodically over 24 hours. Record the elapsed time and your observations.

When did you first notice something happening to the egg? What did you observe? How long did it take for the reaction to cease? Describe the condition of the egg after 24 hours. Hypothesize about what has occurred. After completing the chapter, review your hypothesis and suggest ways in which it might be modified.

21.1 ● ■ ◆
21.2 □ ◆

Conceptual While it is important for students to understand titrations, at this level they do not need to solve titration problems on their own. Sample Problem 21-2 may be omitted because it is a multi-step conversion problem. Calculations of equivalents based on normality may also be omitted (Sample Problems 21-6 through 21-8).

Standard In Section 21.2, help students understand what causes a precipitate to form rather than relying solely on a calculation of K_{sp} to determine whether a precipitate forms.

The CHEMath feature will help students who are struggling with algebraic manipulation of equilibrium expressions. The common ion effect may be omitted.

Honors You may use the Small-Scale Lab to introduce Section 21.1. If students successfully work through the first several sample problems of each section in class, the remaining sample problems may be assigned as homework.

NEUTRALIZATION REACTIONS

section 21.1

Nearly all of the adult population suffers from acid indigestion at some time. Although hydrochloric acid is always present in the stomach, an excess can cause heartburn and a feeling of nausea. A common way to relieve the pain of acid indigestion is to take antacids to neutralize the stomach acid. The active ingredient in many antacids is sodium hydrogen carbonate, aluminum hydroxide, or magnesium hydroxide. **What is a neutralization reaction?**

objectives

- Explain how acid–base titration is used to calculate the concentration of an acid or a base
- Explain the concept of equivalence in neutralization reactions

key terms

- neutralization reactions
- standard solution
- titration
- end point
- equivalent (equiv)
- gram equivalent mass
- normality (*N*)
- equivalence point

Acid-Base Reactions

In general, the reaction of an acid with a base produces water and one of a class of compounds called salts. When you hear the word salt, you may think of the substance that flavors your french fries or scrambled eggs. Indeed, table salt (sodium chloride) is an example of this class of compounds, but there are many more. Salts are compounds consisting of an anion from an acid and a cation from a base.

The properties of acids, bases, and salts help explain many diverse phenomena. The usefulness of antacids, for example, depends on the process of acid–base neutralization. Farmers use a similar process to control the pH of soil. The formation of caves and of stalactites in caves is caused by changes in the solubilities of salts. Within the human body, the formation of kidney stones from salts is also related to solubility. Neutralization reactions and the solubilities of salts are the topics of this chapter.

If you mix a solution of a strong acid containing hydronium (hydrogen) ions with a solution of a strong base that has an equal number of hydroxide ions, a neutral solution results. The final solution has properties that are characteristic of neither an acidic nor a basic solution. Consider these examples:

$$HCl(aq) + NaOH(aq) \longrightarrow NaCl(aq) + H_2O(l)$$

$$H_2SO_4(aq) + 2KOH(aq) \longrightarrow K_2SO_4(aq) + 2H_2O(l)$$

Figure 21.1
Geodes, such as this amethyst specimen (left), begin as cavities in limestone. Carbonic acid in ground water dissolves silicate minerals in rock. The silicate-rich water fills the cavities. Over time, the silicates in solution deposit as crystals of silicon dioxide, or quartz. Amethyst consists of quartz crystals containing traces of manganese. For fish and other aquatic animals to survive (right), the water in which they live must be maintained at the proper pH.

21.1

1 Engage

Use the Visual

Have students read the section-opener paragraph and examine the photograph. Ask:

- **What is a neutralization reaction?** (possible student answer: the reaction between an acid and a base, resulting in a solution with a pH of 7)
- **Why not use a stronger base to provide quicker relief from excess stomach acid?** (The stronger base would be too corrosive for the digestive tract.)

You may want to mention the discovery of bacteria that cause many stomach ulcers. These ulcers can often be treated with antibiotics.

Check Prior Knowledge

Ask students if they have witnessed or read a news report about people trying to clean up a chemical spill on a highway or at a factory. Ask:

- **How were the people treating the spill dressed? Why were they dressed in that manner?** (in protective body suits to avoid contact with poisons or corrosive chemicals)
- **If a spill involves an acid, what type of treatment is possible?** (A weak base in powder form, such as sodium carbonate, can be spread over the spill to absorb, react with, and neutralize the acid.)

Write an equation for a neutralization reaction on the board. Ask students:

- **Into what general category of reactions would you place this neutralization reaction?** (Double replacement; the positive ions are exchanged between the acid and base reactants; the products are a salt and water.)

STUDENT RESOURCES

From the Teacher's Resource Package, use:

- Section Review 21.1, Ch. 21 Practice Problems and Quizzes from the Review Module
- Laboratory Recordsheets 21-1 and 21-2
- Laboratory Manual: Experiments 39 and 40
- Laboratory Practical 21-1
- Small-Scale Chemistry Lab: Experiment 29

TECHNOLOGY RESOURCES

Relevant technology resources include:

- Chem ASAP! CD-ROM
- ResourcePro CD-ROM
- ActivChemistry CD-ROM: *Acid-Base Titrations*
- Chemistry Alive! Videodisc: *Milk of Magnesia, Titration*

21.1

2 Teach

TEACHER DEMO

Write the following equation on the board: $HCl(aq) + NaOH(aq) \rightarrow NaCl(aq) + H_2O(l)$
Prepare equimolar solutions of HCl and NaOH. Add a few drops of phenolphthalein indicator to the NaOH solution. Wearing protective goggles and a laboratory apron, slowly add the HCl to the NaOH while carefully stirring. If possible, use an electric stirrer with a magnetic stir bar. Ask students:

- What happens to the color of the solution as the acid is added? (The deep pink color slowly fades until the solution becomes colorless.) Remind students that phenolphthalein is an acid–base indicator that is colorless at pH lower than 8.

Have a few volunteers come up and gently touch the outside of the reaction beaker. Ask them to describe their observations to the class. (The beaker will be warm to the touch.) Ask:

- Is the neutralization reaction endothermic or exothermic? (exothermic)

Display a quantity of table salt equivalent to the amount of salt that could be retrieved if the neutral solution were evaporated.

Chemistry Alive!

Milk of Magnesia   Play

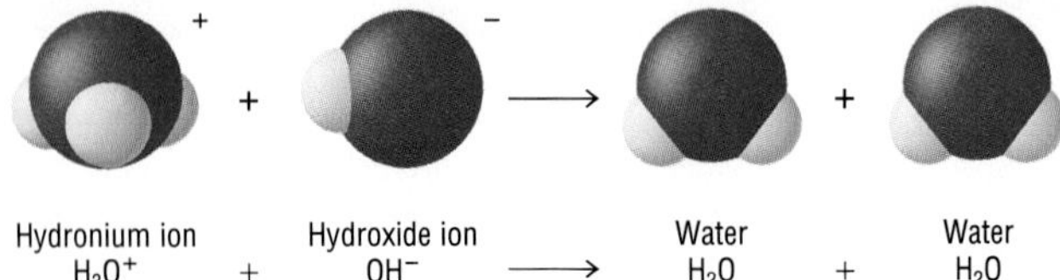

Figure 21.2
In a neutralization reaction, hydronium ions (H_3O^+) combine with hydroxide ions (OH^-) to form neutral water.

In each example, a strong acid reacts with a strong base. If solutions of these substances are mixed in the mole ratios specified by the balanced equation, neutral solutions will result. Similar reactions of weak acids and/or weak bases do not usually produce neutral solutions. In general, however, reactions in which an acid and a base react in an aqueous solution to produce a salt and water are called **neutralization reactions. Figure 21.2** illustrates the formation of water in a neutralization reaction. All neutralization reactions are double-replacement reactions.

Neutralization reactions are one way to prepare pure samples of salts. You could prepare potassium chloride, for example, by mixing equal molar quantities of hydrochloric acid and potassium hydroxide. An aqueous solution of potassium chloride would result. You could heat the solution to evaporate the water, leaving the salt potassium chloride. **Table 21.1** lists some common salts and their applications. What acid and base could you
1 mix to make sodium sulfate? To make lithium nitrate?

Table 21.1

Some Salts and Their Applications

Name	Formula	Applications
Ammonium sulfate	$(NH_4)_2SO_4$	Fertilizer
Barium sulfate	$BaSO_4$	Gastrointestinal studies; white pigment
Calcium chloride	$CaCl_2$	De-icing roadways and sidewalks
Calcium sulfate dihydrate (gypsum)	$CaSO_4 \cdot 2H_2O$	Plasterboard
Copper sulfate pentahydrate (blue vitriol)	$CuSO_4 \cdot 5H_2O$	Dyeing, fungicide
Calcium sulfate hemihydrate	$CaSO_4 \cdot \frac{1}{2} H_2O$	Plaster casts
Potassium chloride	KCl	Sodium-free salt substitute
Potassium permanganate	$KMnO_4$	Disinfectant and fungicide
Silver nitrate	$AgNO_3$	Cauterizing agent
Silver bromide	AgBr	Photographic emulsions
Sodium hydrogen carbonate (baking soda)	$NaHCO_3$	Antacid
Sodium carbonate decahydrate (washing soda)	$Na_2CO_3 \cdot 10H_2O$	Glass manufacture; water softener
Sodium chloride (table salt)	NaCl	Body electrolyte; chlorine manufacture
Sodium thiosulfate (hypo)	$Na_2S_2O_3$	Fixing agent in photographic process

Answer

1 Sodium sulfate: sodium hydroxide and sulfuric acid. Lithium nitrate: lithium hydroxide and nitric acid.

MINI LAB  

The Neutralizing Power of Antacids

PURPOSE

To measure the effectiveness of antacids at neutralizing excess stomach acid.

MATERIALS

- several brands of antacids
- mortar and pestle
- paper towels
- measuring teaspoon
- clear plastic cups
- dropper
- universal indicator
- white vinegar (5% ethanoic acid)
- sodium hydrogen carbonate
- pH sensor (optional)

PROCEDURE

 *Sensor version available in the Probeware Lab Manual.*

1. Using the mortar and pestle, grind a brand of antacid tablet into powder. Place the powder on a paper towel labeled with the brand name. Repeat this process for each brand of tablet.
2. Put 2 teaspoons of vinegar in a plastic cup to represent excess stomach acid.
3. Add 5 drops of indicator to the vinegar. Note the color.
4. Add small amounts of sodium hydrogen carbonate until the system is neutral; the indicator will be blue.
5. Estimate how much sodium hydrogen carbonate was needed to neutralize the acid. This is your reference standard.
6. Repeat Steps 2, 3, and 4, but substitute various antacid powders for the sodium hydrogen carbonate.
7. Compare the estimated amounts of each brand of antacid required to neutralize the vinegar with your reference standard.

ANALYSIS AND CONCLUSIONS

1. List the antacid brands tested and indicate their effectiveness with respect to the reference standard.
2. Which ingredients listed on antacid packages neutralize stomach acid?
3. What is the purpose of the non-neutralizing components?
4. Write a balanced equation for the neutralization of ethanoic acid, CH_3COOH, by $NaHCO_3$.
5. As described, this activity will give you only qualitative results. Explain what this means. Design your own activity to give quantitative results.

Titration

Acids and bases sometimes, but not always, react in a 1:1 mole ratio.

$$\underset{1\text{ mol}}{HCl(aq)} + \underset{1\text{ mol}}{NaOH(aq)} \longrightarrow \underset{1\text{ mol}}{NaCl(aq)} + \underset{1\text{ mol}}{H_2O(l)}$$

When sulfuric acid reacts with sodium hydroxide, however, the ratio is 1:2. Two moles of the base sodium hydroxide are required to neutralize one mole of H_2SO_4.

$$\underset{1\text{ mol}}{H_2SO_4(aq)} + \underset{2\text{ mol}}{2NaOH(aq)} \longrightarrow \underset{1\text{ mol}}{Na_2SO_4(aq)} + \underset{2\text{ mol}}{2H_2O(l)}$$

Similarly, hydrochloric acid and calcium hydroxide react in a 2:1 ratio.

$$\underset{2\text{ mol}}{2HCl(aq)} + \underset{1\text{ mol}}{Ca(OH)_2(aq)} \longrightarrow \underset{1\text{ mol}}{CaCl_2(aq)} + \underset{2\text{ mol}}{2H_2O(l)}$$

MINI LAB

The Neutralizing Power of Antacids

SAFETY

Caution the students never to taste any materials used in the laboratory. Students should wear safety goggles.

LAP PREP AND PLANNING

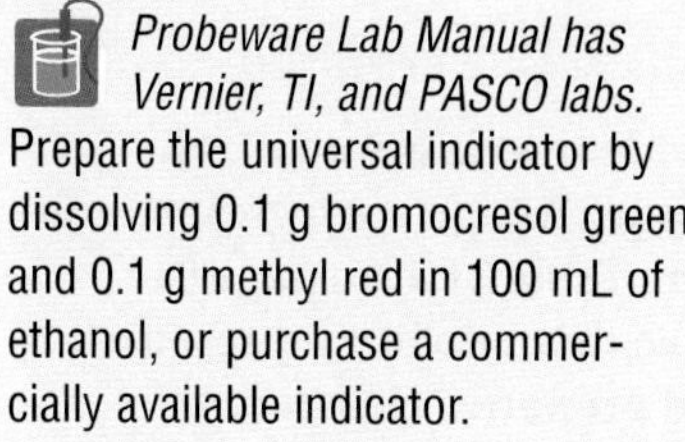

Probeware Lab Manual has Vernier, TI, and PASCO labs.

Prepare the universal indicator by dissolving 0.1 g bromocresol green and 0.1 g methyl red in 100 mL of ethanol, or purchase a commercially available indicator.

ANALYSIS AND CONCLUSIONS

1. Answers will vary.
2. Active ingredients include $NaHCO_3$, $CaCO_3$, $MgCO_3$, $Mg(OH)_2$, $Al(OH)_3$, and $AlNa(OH)_2CO_3$.
3. Simethicone breaks up gas bubbles; other ingredients act as preservatives, coating agents, colors, flavors, sweeteners, buffers, lubricants, and solvents.
4. $CH_3COOH + NaHCO_3 \rightarrow CH_3COONa + H_2O + CO_2$
5. Qualitative results are descriptive, not numeric. For quantitative results, students should suggest a procedure for measuring the amount of acid of known concentration that can be neutralized by an antacid.

Discuss

Point out that neutralization is a process that occurs whenever an acid reacts with a base in the mole ratios specified by the balanced equation. Explain that not all neutralization reactions produce neutral solutions. The strength of the reactants determines whether the solution will be acidic, basic, or neutral.

Cooperative Learning

Divide the class into small groups. Assign each group a salt from Table 21.1 on page 614. Have them do library research and prepare a classroom display. The display cards should include the name and formula of the salt, an equation showing how it can be prepared by neutralization, and a list of important uses. When the displays are ready, provide a small sample of each salt in a sealed glass container for students to add to their displays.

Practice Problems Plus

Additional Practice Problem

This problem is related to Sample Problem 21-1: How many moles of aluminum hydroxide are needed to neutralize 2.30 mol of sulfuric acid? (1.53 mol)

Sample Problem 21-1

How many moles of sulfuric acid are required to neutralize 0.50 mol of sodium hydroxide?

1. *ANALYZE* ***List the knowns and the unknown.***

Knowns:

- mol NaOH = 0.50 mol
- $H_2SO_4(aq) + 2NaOH(aq) \longrightarrow Na_2SO_4(aq) + 2H_2O(l)$
- $\frac{\text{mol } H_2SO_4}{\text{mol NaOH}} = \frac{1}{2}$ (from the balanced equation)

Unknown:

- mol H_2SO_4 = ? mol

2. *CALCULATE* ***Solve for the unknown.***

The mole ratio of H_2SO_4 to NaOH is 1:2. The necessary number of moles of H_2SO_4 is calculated using this ratio.

$$0.50 \text{ mol NaOH} \times \frac{1 \text{ mol } H_2SO_4}{2 \text{ mol NaOH}} = 0.25 \text{ mol } H_2SO_4$$

3. *EVALUATE* ***Does the result make sense?***

Because the mole ratio of H_2SO_4 to NaOH is 1:2, the expected number of moles of H_2SO_4 should be half the given number of moles of NaOH. The answer should have two significant figures.

Practice Problems

1. How many moles of potassium hydroxide are needed to completely neutralize 1.56 mol of phosphoric acid?
2. How many moles of sodium hydroxide are required to neutralize 0.20 mol of nitric acid?

Chem ASAP!

Problem-Solving 1

Solve Problem 1 with the help of an interactive guided tutorial.

The concentration of acid (or base) in a solution can be determined by performing a neutralization reaction. An appropriate acid–base indicator is used to show when neutralization has occurred. As you can see in **Figure 21.3**, the juice of the red cabbage is an acid–base indicator. In the laboratory, phenolphthalein is often the preferred indicator for acid–base neutralization reactions. Solutions that contain phenolphthalein turn from colorless to deep pink as the pH of the solution changes from acidic to basic. In slightly basic solutions, the indicator is very faintly pink. What is
1 the color of phenolphthalein at pH = 3?

Figure 21.3
The juice of a red cabbage can be used as an acid–base indicator. When a solution is highly acidic, the juice turns red. As the acidity of the solution decreases, or as the solution becomes more basic, the color changes from red to violet to green to yellow. Would the yellow solution have a high or low pH? 2

Answers

1 colorless
2 It would have a high pH.
3 The hydrogen-ion and hydroxide-ion concentrations are equal.

Chemistry Alive!

Titration

Play

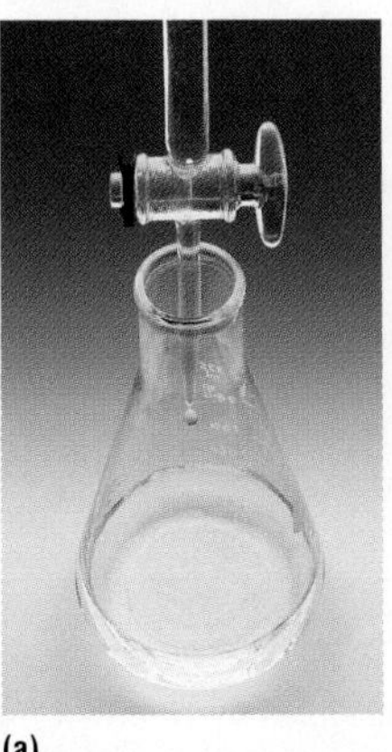

(a) (b) (c)

Figure 21.4
*The titration of an acid with a base is shown here. **(a)** A known volume of an acid (plus a few drops of phenolphthalein indicator) in a flask is placed beneath a buret filled with a base of known concentration. **(b)** Base is slowly added from the buret to the acid while the flask is gently swirled. **(c)** A change in the color of the indicator signals that neutralization has occurred.*

Chem ASAP!
Simulation 26
Titrate several acids and bases and observe patterns in the pH at equivalence.

The steps in a neutralization reaction are as follows.

1. A measured volume of an acid solution of unknown concentration is added to a flask.
2. Several drops of the indicator are added to the solution.
3. Measured volumes of a base of known concentration are mixed into the acid until the indicator just barely changes color.

The solution of known concentration is called the **standard solution.** The standard solution is added using a buret. This process of adding a known amount of solution of known concentration to determine the concentration of another solution is called **titration.** A titration is continued until the indicator shows that neutralization has just occurred. The point at which the indicator changes color is the **end point** of the titration. The titration of an acid of unknown concentration with a standard base is shown in **Figure 21.4.** A similar procedure could be used to find the concentration of a base using a standard acid.

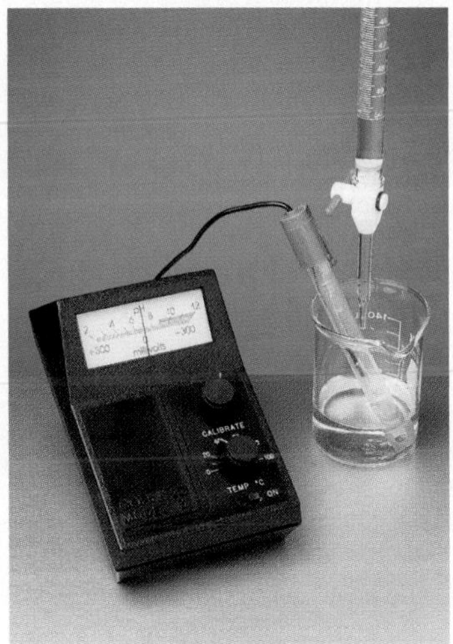

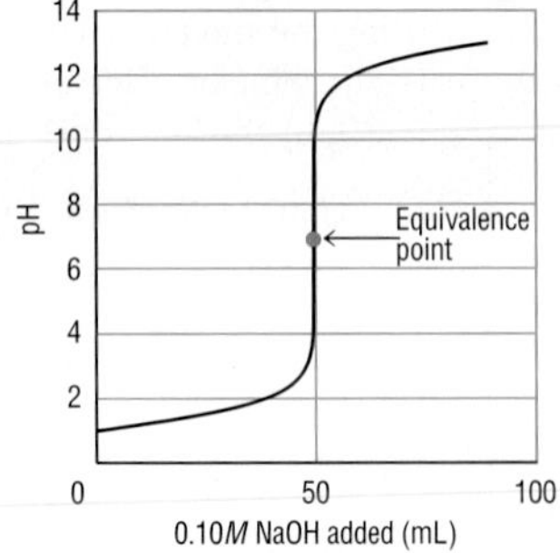

Figure 21.5
In this titration of a strong acid with a strong base, 0.10M NaOH is slowly added from a buret to 50.0 mL of 0.10M HCl in the beaker. The pH of the solution is measured with a pH meter and is recorded as a function of the volume of NaOH added. The equivalence point, the midpoint on the vertical portion of the pH titration curve, occurs at 50.0 mL of NaOH added. The pH at the equivalence point is 7, which is the pH of a neutral solution. What is true concerning $[H^+]$ and $[OH^-]$ at the equivalence point? 3

21.1

Discuss

Discuss the procedures that students suggested as answers to question 5 of the MINI LAB on page 615. Explain that a quantitative procedure called *titration* can be used to measure the concentration of an acid or base. For titration to work, the concentration of one of the solutions, the standard solution, must be known. An acid–base indicator or a pH meter can be used to detect the end point of a titration—the point at which the solution contains equal numbers of H^+ and OH^- ions.

TEACHER DEMO

Repeat the demo described on page 614. This time place the base in a beaker and the acid in a buret. Use a pH meter to monitor the titration. Construct a graph similar to the one in Figure 21.5. Ask:

- What happens to the graph as the titration nears its equivalence point? (The slope becomes very steep.)
- What would happen to the pH if a small amount of base were added to the beaker after the equivalence point is reached? (The pH would increase dramatically.) Add some base to the beaker and observe the pH. Then ask:
- What can be done to regain the equivalence point? (Add acid from the buret.)

Meeting Diverse Needs

At Risk If students are having trouble doing Sample Problem 21-1, point out that the acid–base mole ratio is a conversion factor for these problems.

Cooperative Learning

Have students do research using the library or Internet on the practical applications of neutralization reactions and titration. Decide who will research each topic and how the results will be shared or displayed. Possible applications: monitoring the chlorine content of swimming pools; adjusting the pH of soil; the crucial role pH plays in the growth and purification of biotechnology products; the use of titration in hospital laboratories; and tests done by regulatory agencies to monitor air or water quality.

Practice Problems Plus

Related Chapter Review Problems
Chapter Review problems 39 and 56 are related to Sample Problem 21-2.

Additional Practice Problem
The Chem ASAP! CD-ROM contains the following problem: What is the molarity of sulfuric acid if 25.0 mL of the solution is neutralized by 25.5 mL of 0.50*M* KOH? (0.26*M*)

Figure 21.5 on the previous page shows how the pH of a solution changes during the titration of a strong acid with a strong base. The pH of the initial acid solution is low. As the base is added, the pH increases because some of the acid is neutralized. As the titration approaches the point of neutralization, at a pH of 7, the pH increases dramatically as hydrogen ions are used up. Once past the point of neutralization, additional base produces a further increase of pH. The point of neutralization is the end point of the titration. At this point, the contents of the beaker consist of only H_2O and NaCl, which is the salt derived from the strong acid HCl and the strong base NaOH, plus a trace of indicator.

Sample Problem 21-2

A 25-mL solution of H_2SO_4 is completely neutralized by 18 mL of 1.0*M* NaOH using phenolphthalein as an indicator. What is the concentration of the H_2SO_4 solution?

1. *ANALYZE* *List the knowns and the unknown.*

Knowns:
- molarity base = 1.0*M* NaOH
- volume base = 18 mL = 0.018 L
- volume acid = 25 mL = 0.025 L
- $H_2SO_4(aq) + 2NaOH(aq) \longrightarrow Na_2SO_4(aq) + 2H_2O(l)$
- mole ratio = $\frac{1 \text{ mol } H_2SO_4}{2 \text{ mol NaOH}}$

Unknown:
- molarity acid = ?*M* H_2SO_4

The conversion steps are

$$\text{L NaOH} \longrightarrow \text{mol NaOH} \longrightarrow \text{mol } H_2SO_4 \longrightarrow M\ H_2SO_4$$

2. *CALCULATE* *Solve for the unknown.*

First convert the volume of base to the number of moles of base, using the molarity. Then use the mole ratio to find moles of H_2SO_4. Finally, calculate the molarity by dividing the moles of H_2SO_4 by the liters of H_2SO_4.

$$0.018 \text{ L NaOH} \times \frac{1.0 \text{ mol NaOH}}{1 \text{ L NaOH}} \times \frac{1 \text{ mol } H_2SO_4}{2 \text{ mol NaOH}}$$

$$= 0.0090 \text{ mol } H_2SO_4$$

$$\text{molarity} = \frac{\text{moles}}{\text{liters}} = \frac{0.0090 \text{ mol}}{0.025 \text{ L}} = 0.36M$$

The $[H_2SO_4]$ is 0.36*M*.

3. *EVALUATE* *Does the result make sense?*

Because H_2SO_4 has two ionizable hydrogens, the concentration of the acid should be half that of the base if equal volumes of the reactants were used in the neutralization. Because the volume of acid was greater than the volume of base, the concentration is less than 0.5*M*. The answer should have two significant figures.

Practice Problems

3. How many milliliters of 0.45*M* hydrochloric acid must be added to 25.0 mL of 1.00*M* potassium hydroxide to make a neutral solution?
4. What is the molarity of phosphoric acid if 15.0 mL of the solution is completely neutralized by 38.5 mL of 0.150*M* NaOH?

Chem ASAP!
Problem-Solving 4
Solve Problem 4 with the help of an interactive guided tutorial.

Meeting Diverse Needs

LEP Students with limited English proficiency may be confused by the terms *equivalents* and *equivalence,* which sound alike and have similar spellings. Have them write out both words and underline or circle the differences in the endings. Help them write out meaningful definitions for each word.

Equivalents

In any neutralization reaction, one mole of hydrogen ions reacts with one mole of hydroxide ions. This does not mean, however, that one mole of any acid will neutralize one mole of any base. Why? For some acids, one mole of the acid gives one mole of hydrogen ions. HCl, HNO_3, and CH_3COOH are such acids.

$$HCl \longrightarrow H^+ + Cl^-$$
$$HNO_3 \longrightarrow H^+ + NO_3^-$$
$$CH_3COOH \longrightarrow H^+ + CH_3COO^-$$

For other acids, one mole of the acid gives two or more moles of hydrogen ions. One mole of H_2SO_4, for example, gives two moles of hydrogen ions.

$$H_2SO_4 \longrightarrow 2H^+ + SO_4^{2-}$$

One mole of H_3PO_4 gives three moles of hydrogen ions.

$$H_3PO_4 \longrightarrow 3H^+ + PO_4^{3-}$$

The same is true for the number of hydroxide ions given by bases. The bases $Ca(OH)_2$ and $Al(OH)_3$, for example, give two and three moles of hydroxide ions, respectively, for each mole of base.

$$Ca(OH)_2 \longrightarrow Ca^{2+} + 2OH^-$$
$$Al(OH)_3 \longrightarrow Al^{3+} + 3OH^-$$

Calculating the amount of acid needed to neutralize a given amount of base is made simpler by the existence of a unit called an equivalent. One **equivalent (equiv)** is the amount of acid (or base) that will give one mole of hydrogen (or hydroxide) ions.

One mole of HCl is one equivalent of HCl. One mole of H_2SO_4 is two equivalents of H_2SO_4. One mole of NaOH is one equivalent of NaOH. How many equivalents of $Ca(OH)_2$ is one mole of $Ca(OH)_2$? One mole of HCl will **1** neutralize one mole of NaOH. This is because one mole of each compound is also one equivalent.

$$HCl(aq) + NaOH(aq) \longrightarrow NaCl(aq) + H_2O(l)$$

One mole of HCl will not neutralize one mole of $Ca(OH)_2$. Two moles of the acid are required to neutralize one mole of this base because two moles of HCl contain two equivalents of acid, and one mole of $Ca(OH)_2$ contains two equivalents of base.

$$2HCl(aq) + Ca(OH)_2(aq) \longrightarrow CaCl_2(aq) + 2H_2O(l)$$

In any neutralization reaction, the equivalents of acid must equal the equivalents of base. How many equivalents of base are in 2 mol $Ca(OH)_2$? **2**

The mass of one equivalent of a substance is its **gram equivalent mass.** One mole of HCl is one equivalent of HCl. Its gram equivalent mass is equal to its molar mass (36.5 g/mol). One mole of H_2SO_4 is two equivalents of H_2SO_4. Its gram equivalent mass is only half of its molar mass (98.1 g/mol), or roughly 49.0 g/equiv.

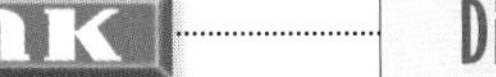

LINK TO COSMETOLOGY

Neutral Curls from Permanent Waves

Curly hair from a permanent wave involves an acid–base neutralization reaction. The waving solution is basic and has a pH of about 9.20. The active ingredient in the waving solution is often an ammonium compound that is applied once the hair has been rolled. After a specified time, the hair is rinsed to remove the excess base. Then a neutralizer is applied. The neutralizer is an acidic solution that has a pH range of 2 to 6. If the hair is exposed to the waving solution for too long a period of time, the hair's texture and appearance are damaged.

Discuss

Write formulas for three acids on the board: HNO_3, H_2SO_4, and H_3PO_4. Tell students to assume they have 0.6*M* solutions of each acid. (If necessary, review the definition of molarity.) Ask:

- **Suppose you have a 0.6*M* solution of NaOH. For each acid, what ratio of acid to base will produce a complete neutralization reaction?**
  (for HNO_3, 1–1; for H_2SO_4, 1–2; for H_3PO_4, 1–3)

Write formulas for three strong bases on the board: KOH, $Ca(OH)_2$, and $Al(OH)_3$. Tell students to assume they have 0.6*M* solutions of each base. Ask:

- **Suppose you have a 0.6 *M* solution of HCl. For each base, what ratio of base to acid will produce a complete neutralization reaction?**
  (for KOH, 1–1; for $Ca(OH)_2$, 1–2; for $Al(OH)_3$, 1–3)

Answers

1 2 equivalents
2 Each mole contains 2 equivalents. 2 moles = 4 equivalents.

Practice Problems Plus

Related Chapter Review Problem
Chapter Review problem 40 is related to Sample Problem 21-3.

Practice Problems Plus

Related Chapter Review Problem
Chapter Review problem 41 is related to Sample Problem 21-4.

Additional Practice Problem
The Chem ASAP! CD-ROM contains the following problem: Determine the number of equivalents in 50.0 g H_2Se. (1.24 equiv H_2Se)

Sample Problem 21-3

What is the mass of 1 equivalent of calcium hydroxide?

1. *ANALYZE* *List the known and the unknown.*

Known:

- molar mass $Ca(OH)_2$ = 74 g/mol

Unknown:

- gram equivalent mass $Ca(OH)_2$ = ? g/equiv

To find the gram equivalent mass, the molar mass must be divided by the number of equivalents/mole.

2. *CALCULATE* *Solve for the unknown.*

Because each formula unit of $Ca(OH)_2$ gives two hydroxide ions, the molar mass must be divided by 2 equiv/mol to find the gram equivalent mass.

$$\text{gram equivalent mass } Ca(OH)_2 = \frac{74 \text{ g}/\cancel{\text{mol}}}{2 \text{ equiv}/\cancel{\text{mol}}} = 37 \text{ g/equiv}$$

3. *EVALUATE* *Does the result make sense?*

Because 1 mol of calcium hydroxide is 2 equiv/mol, the equivalent mass is half the molar mass.

Practice Problems

5. What is the mass of 1 equivalent of each compound?
   a. H_3PO_4
   b. NaOH
6. Determine the gram equivalent mass of each compound.
   a. $Mg(OH)_2$
   b. CH_3COOH

Sample Problem 21-4

How many equivalents is 14.6 g of sulfuric acid?

1. *ANALYZE* *List the knowns and the unknown.*

Knowns:

- mass H_2SO_4 = 14.6 g H_2SO_4
- molar mass H_2SO_4 = 98.1 g/mol

Unknown:

- equivalents H_2SO_4 = ? equiv

Because H_2SO_4 has two ionizable hydrogens, its equivalent mass is half its molar mass. This relationship is the conversion factor used to convert mass to equivalents.

2. *CALCULATE* *Solve for the unknown.*

$$\text{gram equivalent mass } H_2SO_4 = \frac{98.1 \text{ g}/\cancel{\text{mol}}}{2 \text{ equiv}/\cancel{\text{mol}}} = 49.1 \text{ g/equiv}$$

$$14.6 \text{ g } \cancel{H_2SO_4} \times \frac{1 \text{ equiv } H_2SO_4}{49.1 \text{ g } \cancel{H_2SO_4}} = 0.297 \text{ equiv } H_2SO_4$$

3. *EVALUATE* *Does the result make sense?*

Because the equivalent mass of H_2SO_4 is approximately 50 g, and 14.6 is about one-third of 50, the given number of grams should be approximately one-third the mass of an equivalent.

Practice Problems

7. How many equivalents are in each sample of compound?
   a. 20 g NaOH
   b. 7.4 g $Ca(OH)_2$
8. Determine the number of equivalents in each sample.
   a. 9.8 g H_3PO_4
   b. 19.6 g H_2SO_4

Chem ASAP!

Problem-Solving 8

Solve Problem 8 with the help of an interactive guided tutorial.

Normality

As you have learned, the concentrations of acids and bases can be stated in terms of molarity. However, it is often more useful to know how many equivalents of acid or base a solution contains. An older, non-SI unit used to express the equivalents of acids and bases is normality. A solution containing 1.0 equivalent of an acid or base per liter has a normality of 1.0. That is, the solution is 1.0 normal (1.0N). The **normality (N)** of a solution is the concentration expressed as the number of equivalents of solute in one liter of solution.

$$\text{Normality } (N) = \text{equiv/L}$$

The numerical values of normality and molarity are equal for acids and bases that give 1 equivalent of H^+ or OH^- per mole. A solution containing 1 mole of NaOH per liter, for example, is 1M and also 1N. A solution containing 1 mole of H_2SO_4 per liter is 1M, but is 2N because H_2SO_4 contains 2 equivalents per mole. What is the normality of a 0.015M $Ca(OH)_2$ solution? What is the normality of a 1.0M solution of phosphoric acid (H_3PO_4)? 1

The number of equivalents of an acid or base in a known volume of a solution of known normality can be calculated as follows.

$$\text{Number of equivalents} = \text{volume (liters)} \times \text{normality of solution}$$

$$\text{equiv} = V\text{(L)} \times N$$

1M HCl contains 1 equiv HCl/L

1M H_2SO_4 contains 2 equiv H_2SO_4/L

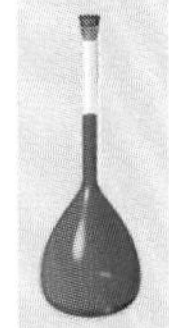

1M H_3PO_4 contains 3 equiv H_3PO_4/L

Figure 21.6
Color intensity is used to show the relationship between molarity and equivalents in monoprotic, diprotic, and triprotic acids.

Sample Problem 21-5

How many equivalents are in 2.5 L of 0.60N H_2SO_4?

1. *ANALYZE* ***List the knowns and the unknown.***

   Knowns:
   - normality H_2SO_4 = 0.60N H_2SO_4
   - volume H_2SO_4 = 2.5 L
   - equiv = V(L) × N

   Unknown:
   - equivalents H_2SO_4 = ? equiv

2. *CALCULATE* ***Solve for the unknown.***

   Substitute the given volume and normality (equiv/L) into the equation to calculate the number of equivalents.

   $$\text{equiv} = 2.5\text{ L} \times 0.60N$$
   $$= 2.5\,\cancel{\text{L}} \times \frac{0.60\text{ equiv}}{\cancel{\text{L}}}$$
   $$= 1.5\text{ equiv}$$

3. *EVALUATE* ***Does the result make sense?***

   Because 2 L of a 0.5N solution is 1 equiv, an answer of greater than 1 should be expected. The answer should have two significant figures.

Practice Problems

9. How many equivalents are in the following?
   a. 0.55 L of 1.8N NaOH
   b. 1.6 L of 0.50N H_3PO_4
   c. 250 mL of 0.28N H_2SO_4
10. What is the normality of each solution?
    a. 20.0 g NaOH in 1.0 L of solution
    b. 4.9 g H_2SO_4 in 500 mL of solution
    c. 15.0 g HCl in 0.400 L of solution
    d. 88.0 g H_3PO_4 in 1.50 L of solution

21.1

Thinking Critically

Have the students write a one-page paper comparing and contrasting the various methods of expressing concentration described in this text.

Discuss

Use the flasks in Figure 21.6 to remind students that some substances contain more than one mole of ionizable hydrogen or hydroxide ions in each mole of the substance. Explain that for the purposes of titration, it is more useful to prepare solutions that have equal numbers of reacting particles, rather than an equal number of moles of solute. Point out that this kind of equality is accomplished through the use of equivalents and normality. Thus, acid and base solutions often are labeled using normality instead of molarity. If students confuse molarity and normality, explain that the normality of a solution is the concentration expressed as the number of equivalents of solute in 1 L of solution. Show that normality must be used in titration calculations because different acids and bases produce different numbers of moles of hydrogen or hydroxide ions per mole of solute. Use acids such as phosphoric acid and bases such as barium hydroxide to show how a failure to convert molarity to normality will lead to error in titration calculations.

Practice Problems Plus

Related Chapter Review Problem
Chapter Review problem 58 is related to Sample Problem 21-5.

Answer

1 1 mole = 2 equivalents. 0.015M = 0.030 N;
1 mole = 3 equivalents. 1.0M = 3.0 N

TEACHER DEMO

Have students study Figure 21.7. Demonstrate the same dilution in the classroom using 1.0*M* HCl. Have students predict the pH of the resulting diluted solutions. Illustrate the dilution by using a pH meter to measure the pH of each diluted solution. Work through the calculations with the students to determine the new concentrations. Do several dilution demonstrations until the students are comfortable with the concept.

Cooperative Learning

Divide the class into groups of two or three students. Have each group work on the practice problems involving normality of acids and bases. Students in each group should take turns explaining the approach to the others in the group.

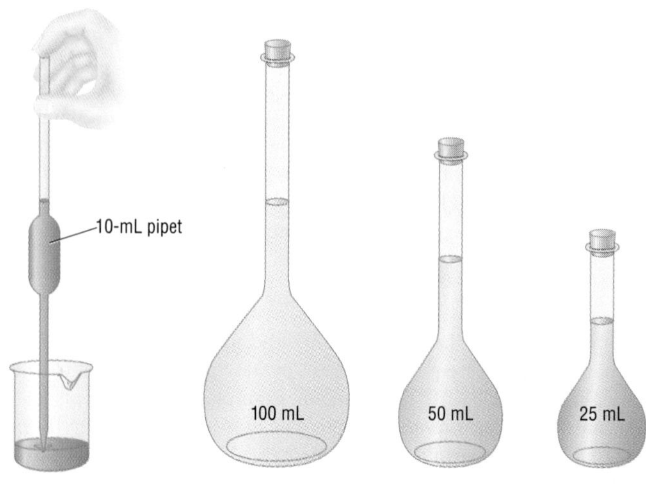

Figure 21.7

A solution with a given normality can be diluted to make a solution with a lower normality. The standard solution in the beaker contains 1.00 mole HCl per liter. The solution contains 1.00 equivalent of acid per liter. What is the normality of the acid? A volumetric pipet was used to transfer 10 mL of the standard solution to each of three volumetric flasks. How many equivalents of acid were added to each flask? The contents of each flask were then diluted with distilled water to their marks—100 mL, 50 mL, and 25 mL, respectively. How many equivalents of acid are in each flask? What is the normality of HCl in each flask? ❶

Solutions of known normality can be made less concentrated by diluting them with water. As you can see in **Figure 21.7,** a solution of known normality is used to make three solutions, each of known normality. The changes in concentration can be calculated by using the following relationship.

$$N_1 \times V_1 = N_2 \times V_2$$

Here N_1 and V_1 are the normality and volume of the initial (standard) solution, and N_2 and V_2 are the normality and volume of the final solution.

Titration calculations are often accomplished more easily using normality instead of molarity because normality takes into account the number of ionizable hydrogens in an acid, whereas molarity does not. In a titration, the point of neutralization is called the **equivalence point.** At the equivalence point, the number of equivalents of acid and base are equal. It is thus possible to calculate the number of equivalents of acid or base in an unknown sample. If N_A and N_B represent the normalities of the acid and base solutions, and V_A and V_B represent the volumes of the acid and base solutions required to give a neutral solution, then

$$\text{Equivalents of acid} = N_A \times V_A$$

$$\text{Equivalents of base} = N_B \times V_B$$

And because the number of equivalents of acid and base are equal at the equivalence point,

$$N_A \times V_A = N_B \times V_B$$

Answer

❶ 1 *N*; $\frac{1.00 \text{ equiv}}{1000 \text{ mL}} \times 10 \text{ mL} = 0.01$ equivalents;

0.01 equivalents; 100 mL: 0.10 *N*; 50 mL: 0.20 *N*; 25 mL: 0.40 *N*

Sample Problem 21-6

You need to make 250 mL of 0.10*N* sodium hydroxide from a stock solution that is 2.0*N* sodium hydroxide. How many milliliters of the stock solution must you dilute to 250 mL to get the required solution?

1. **ANALYZE** ***List the knowns and the unknown.***

Knowns:
- $N_1 = 2.0N$
- $N_2 = 0.10N$
- $V_2 = 250$ mL
- $N_1 \times V_1 = N_2 \times V_2$

Unknown:
- $V_1 = ?$ mL

2. **CALCULATE** ***Solve for the unknown.***

To find V_1, rearrange the equation and then substitute the values of N_1, N_2, and V_2.

$$V_1 = \frac{N_2 \times V_2}{N_1} = \frac{0.10\cancel{N} \times 250 \text{ mL}}{2.0\cancel{N}} = 13 \text{ mL}$$

Dilute 13 mL of 2.0*N* NaOH to 250 mL to make 0.10*N* NaOH.

3. **EVALUATE** ***Does the result make sense?***

Because the concentration decreases by a factor of 20, the volume of the stock solution (initial solution) to be diluted should be about one-twentieth of the volume of the final solution. The answer should have two significant figures.

Practice Problems

11. How many milliliters of 3.0*N* KOH are needed to prepare 870 mL of 0.20*N* KOH?
12. How would you prepare 500 mL of 0.20*N* sulfuric acid from a stock solution of 4.0*N* sulfuric acid?

Chem ASAP!

Problem-Solving 11

Solve Problem 11 with the help of an interactive guided tutorial.

Sample Problem 21-7

If 35.0 mL of 0.20*N* hydrochloric acid are needed to neutralize 25.0 mL of an unknown base, what is the normality of the base?

1. **ANALYZE** ***List the knowns and the unknown.***

Knowns:
- $V_A = 35.0$ mL
- $N_A = 0.20N$
- $V_B = 25.0$ mL
- $N_A \times V_A = N_B \times V_B$

Unknown:
- $N_B = ?N$

2. **CALCULATE** ***Solve for the unknown.***

To solve for N_B, rearrange the equation and then substitute the values of V_A, N_A, and V_B.

$$N_B = \frac{V_A \times N_A}{V_B} = \frac{35.0 \cancel{\text{mL}} \times 0.20N}{25.0 \cancel{\text{mL}}} = 0.28N$$

3. **EVALUATE** ***Does the result make sense?***

Because a smaller volume of the base is required, the normality of the base must be larger than the normality of the acid.

Practice Problems

13. If you need 50.0 mL of 0.152*N* hydrochloric acid to neutralize 29.2 mL $Ca(OH)_2$, what is the normality of the base?
14. What is the normality of a solution of a base if 25 mL is neutralized by 75 mL of 0.40*N* acid?

Chem ASAP!

Problem-Solving 14

Solve Problem 14 with the help of an interactive guided tutorial.

21.1

Practice Problems Plus

Related Chapter Review Problem

Chapter Review problem 63 is related to Sample Problem 21-6.

Additional Practice Problem

The Chem ASAP! CD-ROM contains the following problem: How many milliliters of 2.5*N* HNO_3 are needed to prepare 450 mL of 0.40*N* HNO_3? (72 mL 2.5*N* HNO_3)

Practice Problems Plus

Related Chapter Review Problem

Chapter Review problem 45 is related to Sample Problem 21-7.

Additional Practice Problem

The Chem ASAP! CD-ROM contains the following problem: What is the normality of a solution of an acid if 40.0 mL is neutralized by 10.0 mL of 0.75*N* base? (0.19*N* acid)

Practice Problems Plus

Additional Practice Problem
This problem is related to Sample Problem 21-8: How many milliliters of 2.50N KOH are needed to neutralize 35.0 mL of 1.10N HCl? (15.4 mL)

3 Assess

Evaluate Understanding

- **Ask students to summarize the titration process in their own words.** (Answers will vary, but should include the following: during a titration, equivalent volumes of an acid and of a base are determined. One is a standard solution with a known concentration; the concentration of the other solution is not known.)
- **Have students explain the difference between normality and molarity using H_2SO_4 as an example.** (Molarity is the number of moles in a 1 L solution. Normality is the number of equivalents in a 1 L solution. For H_2SO_4, a 1M solution is equivalent to a 2N solution.)

Reteach

Remind students that balanced equations are needed to determine correct mole ratios and that an equivalent represents the amount of an acid or base that gives one mole of H^+ or OH^- ions in solution.

Portfolio Project

If the pH is too high, dilute HCl(aq) or solid $NaHSO_4$ can be added. If the pH is too low, some acid can be neutralized with either $NaCO_3$ or $NaHCO_3$.

Sample Problem 21-8

How many milliliters of 0.500N sulfuric acid are needed to neutralize 50.0 mL of 0.200N potassium hydroxide?

1. ***ANALYZE** List the knowns and the unknown.*

Knowns:
- $V_B = 50.0$ mL
- $N_B = 0.200N$
- $N_A = 0.500N$
- $N_A \times V_A = N_B \times V_B$

Unknown:
- $V_A = ?$ mL

2. ***CALCULATE** Solve for the unknown.*

To solve for V_A, rearrange the equation and then substitute the values of N_B, V_B, and N_A.

$$V_A = \frac{V_B \times N_B}{N_A} = \frac{50.0 \text{ mL} \times 0.200\cancel{N}}{0.500\cancel{N}} = 20.0 \text{ mL}$$

3. ***EVALUATE** Does the result make sense?*

The normality of the acid is more than twice the normality of the base, so the volume of the acid needed for neutralization should be less than half the volume of base. The answer should have three significant figures.

Practice Problems

15. How many milliliters of 0.850N hydrochloric acid would you need to neutralize 25.0 mL of 0.480N sodium hydroxide?

16. How many milliliters of 1.05N NaOH are needed to neutralize 50.0 mL of 0.620N H_2SO_4?

portfolio project

Hypochlorite salts are used to disinfect swimming pools. Hypochlorite ions react with water.

$$OCl^-(aq) + H_2O(l) \rightleftharpoons HOCl(aq) + OH^-(aq)$$

Hypochlorous acid, HOCl, prevents the growth of algae and bacteria. Research and prepare a report describing how the pH of pool water is regulated to maintain the necessary concentration of HOCl.

section review 21.1

17. How is acid–base titration used to calculate the concentration of an acid or a base?

18. Write complete balanced equations for these acid–base reactions. Give the names of the salts produced.

a. $H_2SO_4(aq) + KOH(aq) \longrightarrow$

b. $HCl(aq) + LiOH(aq) \longrightarrow$

c. $H_3PO_4(aq) + Ca(OH)_2(aq) \longrightarrow$

d. $HNO_3(aq) + Mg(OH)_2(aq) \longrightarrow$

19. How many moles of hydrochloric acid are required to neutralize aqueous solutions of these bases?

a. 0.2 mol NaOH

b. 2 mol NH_3

c. 0.1 mol $Ca(OH)_2$

20. What is an equivalent? How is it related to normality?

21. Why are equivalents of acid and equivalents of base always equal in a neutralization reaction?

22. How many milliliters of 0.20N sodium hydroxide must be added to 75 mL of 0.050N hydrochloric acid to make a neutral solution?

 **Chem ASAP! Assessment 21.1** Check your understanding of the important ideas and concepts in Section 21.1.

Answers

SECTION REVIEW 21.1

17. To a measured quantity of an acid (or base) of unknown concentration, drops of a base (or acid) of known concentration are added until the equivalents of base equal the equivalents of acid. The unknown concentration is calculated from the following formula.
$V_1 \times N_1 = V_2 \times N_2$

18. **a** $H_2SO_4 + 2KOH \rightarrow 2H_2O + K_2SO_4$; potassium sulfate
b. $HCl + LiOH \rightarrow H_2O + LiCl$; lithium chloride
c. $2H_3PO_4 + 3Ca(OH)_2 \rightarrow 6H_2O + Ca_3(PO_4)_2$; calcium phosphate
d. $2HNO_3 + Mg(OH)_2 \rightarrow 2H_2O + Mg(NO_3)_2$; magnesium nitrate

19. **a.** 0.2 mol
b. 2 mol
c. 0.2 mol

SMALL-SCALE LAB

SMALL-SCALE TITRATIONS

SAFETY     

Wear safety glasses and follow the standard safety procedures outlined on page 18.

PURPOSE

To measure and compare the molar concentrations of acids using various titration techniques.

MATERIALS

- plastic cup
- well strip
- 2 pipets
- NaOH
- HCl
- phenolphthalein
- pH sensor (optional)

PROCEDURE

Sensor version available in the Probeware Lab Manual.

1. Perform a simple titration of HCl with NaOH.

a. Add 5 drops of HCl to a clean, dry plastic cup. Add 1 drop of phenolphthalein to the cup.

b. Slowly add NaOH to the cup, counting the number of drops that you add. As you titrate, swirl the cup gently. Continue to slowly add NaOH to the cup until the solution turns a very pale pink. Stop and record the total number of drops you added.

c. Carefully repeat the experiment to verify your answer.

2. Perform another titration. This time account for the differences in drop size.

a. Hold an HCl pipet vertically, expel the air bubble, and count the number of drops required to fill one well in a well strip. Repeat this procedure until you obtain consistent results.

b. Repeat the above step using a NaOH pipet.

c. Add 10 drops of HCl to a clean, dry plastic cup. Then add one drop of phenolphthalein to the cup.

d. Slowly add NaOH to the cup while counting the number of drops added. As you titrate, swirl the cup gently. Add NaOH slowly until the solution turns a very pale pink. Record the number of drops added. Repeat the procedure until you obtain consistent results.

3. Perform a mass titration.

a. Determine the mass of a clean, dry cup to which a drop of phenolphthalein has been added.

b. Add about $\frac{1}{3}$ of a pipet of HCl to the cup and determine its mass.

c. Slowly add NaOH from a pipet to the HCl until the solution turns a very pale pink. Determine the mass of the cup and its contents.

d. Clean and dry the same cup and repeat steps **a** through **d** until you obtain consistent results.

ANALYSIS

Using your experimental data, record the answers to the following questions below your data table.

1. Calculate the molar concentration (M) of HCl from the data in Step 1.

$$M\,\text{HCl} = 0.5M \times \frac{\text{b}}{\text{a}}$$

2. Calculate the molar concentration of HCl using the data from Step 2.

$$M\,\text{HCl} = 0.5M \times \frac{\text{d}}{\text{c}} \times \frac{\text{a}}{\text{b}}$$

3. Calculate the molar concentration of HCl from the data in Step 3.

$$M\,\text{HCl} = 0.5M \times \frac{\text{c} - \text{b}}{\text{b} - \text{a}}$$

YOU'RE THE CHEMIST!

The following small-scale activity allows you to develop your own procedures and analyze the results.

1. Analyze It! For each equation you used to calculate the molar concentration of HCl, what were the assumptions that allowed you to cancel all units except molar concentration?

SMALL-SCALE LAB

SMALL-SCALE TITRATIONS

OVERVIEW

See Probeware Lab Manual for Vernier, TI, and PASCO versions.

Procedure Time: 40 minutes

Students will use volumetric and mass titrations to measure the molar concentrations of acid solutions.

MATERIALS

Solution	Preparation
0.50M NaOH	20.0 g in 1.0 L
1.0M HCl	82 mL of 12M HCl in 1.0 L

CAUTION! Always add acid to water carefully and slowly!

0.1% phenolphthalein	250 mg in 250 mL of 70% 2-propanol

TEACHING SUGGESTIONS

Emphasize that the three different experimental designs give different results because of the assumptions made, the techniques involved, and the accuracy of the equipment used. None are "right" or "wrong"; they all have limitations.

YOU'RE THE CHEMIST

Step 1: Drops of HCl and NaOH cancel out in the calculation. It is assumed that the drop sizes for the two pipets are equal.

Step 2: HCl drops are different in size from NaOH drops, but each type of drop cancels out during the calculation based on this step.

Step 3: The calculation based on data from this step assumes dilute HCl and dilute NaOH solutions have the same density.

SECTION REVIEW 21.1 CONTINUED

20. An equivalent of an acid is the amount of acid that produces one mole of hydrogen ions. An equivalent of a base is the amount of base that produces one mole of hydroxide ions. An equivalent is related to normality by definition: Normality is equivalents per liter.

21. The number of H^+ equals the number of OH^-.

22. 19 mL NaOH

ANALYSIS

Sample answers:

1. $0.5M \times \frac{11 \text{ drops NaOH}}{5 \text{ drops HCl}} = 1M\,\text{HCl}$

2. $0.5M \times \frac{11 \text{ drops NaOH}}{5 \text{ drops HCl}} \times \frac{23 \text{ drops HCl}}{24 \text{ drops NaOH}} = 1M\,\text{HCl}$

3. $0.5M \times \frac{(3.474 \text{ g} - 2.258 \text{ g})}{(6.568 \text{ g} - 5.962 \text{ g})} = 1M\,\text{HCl}$

1 Engage

Use the Visual

Have students study the photograph and read the text that opens the section. Explain that the pH inside blood cells is very carefully regulated and that even relatively small changes can have drastic effects—even death. Human blood is regulated between a pH of 6.8 and 7.8. Ask:

- **What chemical process helps to ensure that blood is kept at the appropriate pH?** (The body contains buffer systems. The most important blood buffer consists of carbonic acid and hydrogen carbonate ions.
  $HCO_3^-(aq) + H^+(aq) \rightleftharpoons H_2CO_3(aq)$
  The buffer maintains the desired blood pH by removing or providing hydrogen ions.)

Check Prior Knowledge

Ask students:

- **Describe one method for forming salts.** (Salts may be formed by a reaction between an acid and a base.) **Remind students that a salt contains an anion from an acid and a cation from a base.**
- **What would you expect the pH of a salt solution to be?** (Students may think all salt solutions are neutral, but they may be neutral, basic, or acidic. In general, their pH would not be in the range of strong acids or bases.)

section 21.2

objectives

- Demonstrate with equations how buffers resist changes in pH
- Calculate the solubility product constant of a slightly soluble salt

key terms

- salt hydrolysis
- buffers
- buffer capacity
- solubility product constant (K_{sp})
- common ion
- common ion effect

SALTS IN SOLUTION

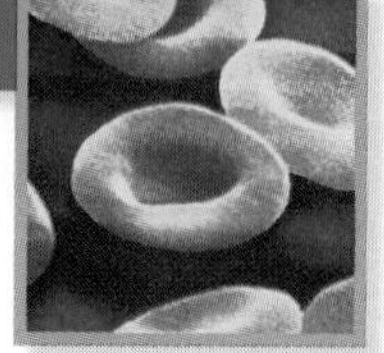

The internal pH of most living cells is close to 7. Because the chemical processes of the cell are very sensitive to pH levels, even a slight change in pH can be harmful. Human blood, for example, normally maintains a pH very close to 7.4. A person cannot survive for more than a few minutes if the blood pH drops to 6.8 or rises to 7.8. **What chemical process helps ensure that blood is kept at the appropriate pH?**

Salt Hydrolysis

As you learned in Section 21.1, a salt consists of an anion from an acid and a cation from a base, and it forms as a result of a neutralization reaction. Although solutions of many salts are neutral, there are some that are acidic and others that are basic. Solutions of sodium chloride and of potassium sulfate are neutral. A solution of ammonium chloride is acidic. A solution of sodium ethanoate (sodium acetate) is basic. Look at **Figure 21.8.** The titration curve **(a)** was obtained by adding a solution of sodium hydroxide, a strong base, to a solution of ethanoic (acetic) acid, a weak acid. An aqueous solution of sodium ethanoate exists at the equivalence point.

$$\underset{\text{Ethanoic acid}}{CH_3COOH(aq)} + \underset{\text{Sodium hydroxide}}{NaOH(aq)} \longrightarrow \underset{\text{Sodium ethanoate}}{CH_3COONa(aq)} + \underset{\text{Water}}{H_2O(l)}$$

The pH at the equivalence point is 8.7—basic. For a strong acid–strong base titration, the pH at the equivalence point was 7, or neutral. (Refer to **Figure 21.5** on page 617.) Why does this difference exist? The answer lies in the fact that some salts promote hydrolysis. In **salt hydrolysis,** the cations or anions of the dissociated salt remove hydrogen ions from or donate hydrogen ions to water. Depending on the direction of the hydrogen-ion transfer, solutions containing hydrolyzing salts may be either acidic or basic. Hydrolyzing salts are usually derived from a strong acid and a weak base, or from a weak acid and a strong base. What distinguishes a strong
1 acid from a weak acid?

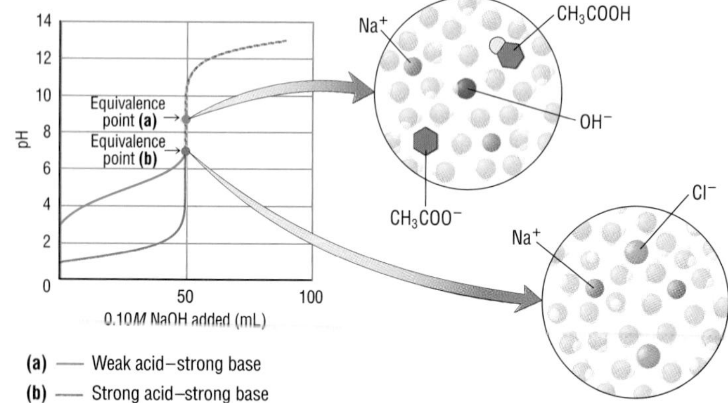

Figure 21.8
*The titration curve for a weak acid and a strong base **(a)** is compared with the titration curve for a strong acid and a strong base **(b)**. In the weak acid–strong base titration, 0.10M NaOH is added slowly from a buret to 50.0 mL of 0.10M CH_3COOH. The pH of the solution is measured with a pH meter and recorded as a function of the volume of NaOH added. The equivalence point occurs at 50.0 mL of NaOH added. The pH at the equivalence point is 8.7, which is the pH of a basic solution, which means that $[OH^-] > [H^+]$. Note that for the strong acid–strong base titration, the pH is 7 at the equivalence point, and $[H^+]$ is equal to $[OH^-]$.*

Student Resources

- Section Review 21.2, Ch. 21 Practice Problems, Interpreting Graphics, Vocabulary, Tests and Quizzes from the Review Module (Ch. 21–24)
- Laboratory Manual: Experiments 41, 42, 43
- Small-Scale Chemistry Lab Manual: Experiment 30

Technology Resources

Relevant technology resources include:

- Chem ASAP! CD-ROM
- Resource Pro CD-ROM
- Chemistry Alive! Videodisc: *Milk of Magnesia*
- Assessment Resources CD-ROM: Chapter 21 Tests

Sodium ethanoate (CH_3COONa) is the salt of a weak acid (ethanoic acid, CH_3COOH) and a strong base (sodium hydroxide, NaOH). In solution, the salt is completely ionized.

$$\underset{\text{Sodium ethanoate}}{CH_3COONa(aq)} \longrightarrow \underset{\text{Ethanoate ion}}{CH_3COO^-(aq)} + \underset{\text{Sodium ion}}{Na^+(aq)}$$

The ethanoate ion is a Brønsted-Lowry base, which means it is a hydrogen-ion acceptor. It establishes an equilibrium with water, forming electrically neutral ethanoic acid and negative hydroxide ions.

$$\underset{\text{(hydrogen-ion acceptor, Brønsted-Lowry base)}}{CH_3COO^-(aq)} + \underset{\text{(hydrogen-ion donor, Brønsted-Lowry acid)}}{H_2O(l)} \rightleftharpoons CH_3COOH(aq) + \underset{\text{(makes the solution basic)}}{OH^-(aq)}$$

This process is called hydrolysis because it splits a hydrogen ion off a water molecule. The resulting solution contains a hydroxide-ion concentration greater than the hydrogen-ion concentration. Thus the solution is basic.

Ammonium chloride (NH_4Cl) is the salt of a strong acid (hydrochloric acid, HCl) and a weak base (ammonia, NH_3). It is completely ionized in solution.

$$NH_4Cl(aq) \longrightarrow NH_4^+(aq) + Cl^-(aq)$$

The ammonium ion (NH_4^+) is a strong enough acid to donate a hydrogen ion to a water molecule, although the equilibrium is strongly to the left.

$$\underset{\text{(hydrogen-ion donor, Brønsted-Lowry acid)}}{NH_4^+(aq)} + \underset{\text{(hydrogen-ion acceptor, Brønsted-Lowry base)}}{H_2O(l)} \rightleftharpoons NH_3(aq) + \underset{\text{(makes the solution acidic)}}{H_3O^+(aq)}$$

This process is also called hydrolysis. It results in the formation of un-ionized ammonia and hydronium (hydrogen) ions. The $[H_3O^+]$ is greater than the $[OH^-]$. Thus a solution of ammonium chloride is acidic. Look at **Figure 21.10** and remember the following rules:

Strong acid + Strong base $\longrightarrow$ Neutral solution

Strong acid + Weak base $\longrightarrow$ Acidic solution

Weak acid + Strong base $\longrightarrow$ Basic solution

Figure 21.9
Vapors of the strong acid HCl(aq) and the weak base $NH_3(aq)$ combine to form the acidic white salt ammonium chloride (NH_4Cl).

Figure 21.10
A few drops of universal indicator solution have been added to each of these 0.10M aqueous salt solutions. The color of the indicator shows that ammonium chloride (NH_4Cl), at left, is acidic (pH about 5.3); sodium chloride (NaCl), center, is neutral (pH 7); and sodium ethanoate (CH_3COONa), at right, is basic (pH about 8.7).

Answer

❶ A strong acid is essentially 100% ionized in aqueous solution while a weak acid is only partially ionized in solution.

2 Teach

TEACHER DEMO

Prepare 1*M* solutions of the following salts: NH_4NO_3, KCl, $NaHCO_3$, Na_2SO_3. Have students speculate about the pH of these solutions and explain their reasoning. After some discussion, check the pH of the solutions with a pH meter or universal pH paper. After you obtain the experimental results, have students try to explain the data. (NH_4NO_3 is acidic because the salt results from a strong acid and a weak base; KCl is neutral because the salt results from a strong acid and a strong base; $NaHCO_3$ is basic because the salt results from a strong base and a weak acid; Na_2SO_3 is basic because the salt results from a weak acid and a strong base.)

Use the Visual

Have students compare Figures 21.8 and 21.5. Explain that the equivalence point is not always at pH = 7. The salts that form may hydrolyze, which results in a reversible reaction that produces a higher concentration of hydrogen or hydroxide ions. Ask the students:

- **What is different about neutralization reactions that produce a neutral solution at the equivalence point and those that produce acidic or basic solutions?** (The strengths of the reactants; when a strong acid and base react, the solution is neutral; when a strong base and a weak acid react, the solution is basic; when a strong acid and a weak base react, the solution is acidic.)

Discuss

Point out that the salts featured in the demo on page 627 and in the illustrations are classified as Brønsted-Lowry acids and bases because the cations and anions of these salts accept or donate hydrogen ions from water (hydrolysis). Have students learn the following rules:
Strong acid + Strong base → Neutral solution
Strong acid + Weak base → Acidic solution
Weak acid + Strong base → Basic solution

Use the Visual

Have students study Figure 21.11. Explain that buffers are solutions in which the pH stays relatively constant when small amounts of either acid or base are added. Point out that in the buffered solution on the left, there is little change in the solution pH.

TEACHER DEMO

To demonstrate that the neutralizing capacities of commercial buffers differ, compare antacid tablets, buffered aspirin, and aspirin. Set up pairs of flasks containing 100 mL of distilled water to test each product. Dissolve equal masses of the tablets in each flask. Add a few drops of methyl red to one of the paired flasks and titrate with 0.5*M* NaOH. To the other flask, add a few drops of bromothymol blue and titrate with 0.5*M* HCl. Titrate until the end point is reached. Explain that the buffering capacity is directly related to the amount of acid or base required to reach the end point. Have students compare the buffering capacities of the tablets tested.

(a)

(b)

Figure 21.11
***(a)** The universal indicator shows that the buffered solution on the left and the unbuffered solution on the right are basic—pH about 8.* ***(b)** After the addition of 1.0 mL of 0.01M HCl solution, the pH of the buffered solution shows no visible change. The pH of the unbuffered solution, however, is now about 3—the solution is acidic.*

Chem ASAP!
Animation 26
Discover the chemistry behind buffer action.

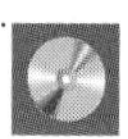

BIOLOGY

Blood Buffers
Your body functions properly only when the pH of your blood lies between 7.35 and 7.45. A blood pH outside these narrow limits can be life-threatening! To keep your blood at the proper pH, your blood contains buffers. The most important blood buffer consists of carbonic acid and hydrogen carbonate ions. Hydrogen ions produced by chemical reactions in your body are the main threat to your blood's pH. This buffer maintains the desired blood pH by removing excess hydrogen ions.

$$\underset{\text{Hydrogen carbonate ion}}{HCO_3^-(aq)} + \underset{\text{Hydrogen ion}}{H^+(aq)} \rightleftharpoons \underset{\text{Carbonic acid}}{H_2CO_3(aq)}$$

As long as there are hydrogen carbonate ions available, the excess hydrogen ions are removed, and the pH of the blood changes very little.

Buffers

The addition of 10 mL of 0.10*M* sodium hydroxide to 1 L of pure water increases the pH by 4.0 pH units (from 7.0 to 11.0). A solution containing 0.20 mol/L each of ethanoic acid and sodium ethanoate has a pH of 4.76. When moderate amounts of either acid or base are added to this solution, however, the pH changes little. The addition of 10 mL of 0.10*M* sodium hydroxide to 1 L of this solution, for example, increases the pH by only 0.01 pH unit, from 4.76 to 4.77. Another example of this behavior is illustrated in **Figure 21.11.**

The solution of ethanoic acid and sodium ethanoate is an example of a typical buffer. **Buffers** are solutions in which the pH remains relatively constant when small amounts of acid or base are added. A buffer is a solution of a weak acid and one of its salts, or a solution of a weak base and one of its salts.

A buffer solution is better able to resist drastic changes in pH than is pure water. **Figure 21.12** illustrates how a buffer works. Ethanoic acid (CH_3COOH) and its anion (CH_3COO^-) act as reservoirs of neutralizing power. They react with any hydroxide ions or hydrogen ions added to the solution. For example, consider the buffer solution in which the sodium ethanoate (CH_3COONa) is completely ionized.

$$\underset{\text{Sodium ethanoate}}{CH_3COONa(aq)} \longrightarrow \underset{\text{Ethanoate ion}}{CH_3COO^-(aq)} + \underset{\text{Sodium ion}}{Na^+(aq)}$$

When an acid is added to the solution, the ethanoate ions (CH_3COO^-) act as a hydrogen-ion "sponge." This creates ethanoic acid, which does not ionize extensively in water; thus the pH does not change appreciably.

$$\underset{\text{Ethanoate ion}}{CH_3COO^-(aq)} + \underset{\text{Hydrogen ion}}{H^+(aq)} \rightleftharpoons \underset{\text{Ethanoic acid}}{CH_3COOH(aq)}$$

A base is a source of hydroxide ions. When a base is added to the solution, the ethanoic acid and the hydroxide ions react to produce water and the ethanoate ion.

$$\underset{\text{Ethanoic acid}}{CH_3COOH(aq)} + \underset{\text{Hydroxide ion}}{OH^-(aq)} \rightleftharpoons \underset{\text{Ethanoate ion}}{CH_3COO^-(aq)} + \underset{\text{Water}}{H_2O(l)}$$

The ethanoate ion is not a strong enough base to accept hydrogen ions from water extensively. Again, the pH does not change very much.

Figure 21.12

The buffer described here is made of ethanoic acid (CH_3COOH) and sodium ethanoate, which is the source of ethanoate ions (CH_3COO^-). To begin, the concentrations of ethanoic acid and ethanoate ions are equal. When H^+ is added, it combines with CH_3COO^- to form CH_3COOH. The [CH_3COO^-] decreases and the [CH_3COOH] increases. When OH^- is added, it reacts with CH_3COOH to produce CH_3COO^- and H_2O. The [CH_3COOH] decreases and the [CH_3COO^-] increases. In both situations, the ratio of [CH_3COOH] to [CH_3COO^-] changes very little and, consequently, the pH changes very little.

An ethanoate buffer solution cannot control the pH when too much acid is added, because no more ethanoate ions are present to accept hydrogen ions. The ethanoate buffer also becomes ineffective when too much base is added. Do you know why this is so? When too much base is added, no more ethanoic acid molecules are present to donate hydrogen ions. When too much acid or base is added, the buffer capacity of a solution is exceeded. The **buffer capacity** is the amount of acid or base that can be added to a buffer solution before a significant change in pH occurs.

Two buffer systems are crucial in maintaining human blood pH within a very narrow range (pH 7.35–7.45). One is the carbonic acid–hydrogen carbonate buffer system. The other is the dihydrogen phosphate–monohydrogen phosphate buffer system. **Table 21.2** lists several important buffer systems.

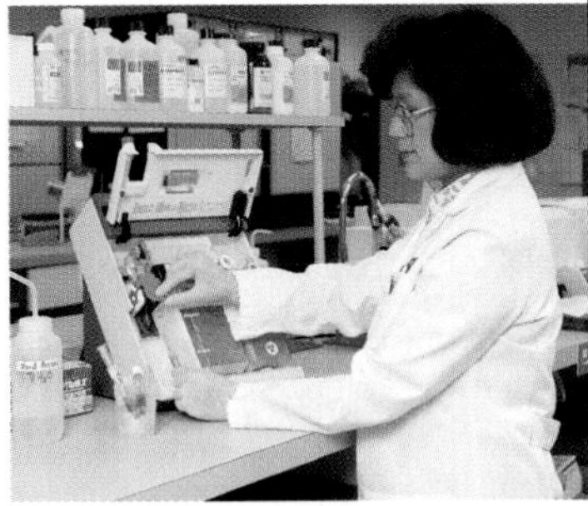

Figure 21.13

This device measures the pH of human blood. Changes in blood pH are minimized by very effective blood buffering systems.

Table 21.2

Important Buffer Systems

Buffer name	Buffer species	Buffer pH (components 0.1*M*)
Ethanoic acid–ethanoate ion	CH_3COOH/CH_3COO^-	4.76
Dihydrogen phosphate ion–hydrogen phosphate ion	$H_2PO_4^-/HPO_4^{2-}$	7.20
Carbonic acid–hydrogen carbonate (solution saturated with CO_2)	H_2CO_3/HCO_3^-	6.46
Ammonium ion–ammonia	NH_4^+/NH_3	9.25

21.2

Discuss

Explain that buffers present a practical application of concepts learned in Chapters 20 and 21. The limited dissociation of weak acids and bases gives these substances the ability to act as buffers. This ability is greatly expanded if the salt of the acid or base is added to the solution. Buffer action utilizes the equilibrium established when a weak acid or base is combined with its salt. When a base is added to a buffered solution, the acidic form removes hydroxide ions from the solution. When an acid is added to a buffered solution, the basic form removes hydrogen ions from the solution.

Discuss

Have students read the Link to Biology. Point out that inhaled air contains about 0.04% CO_2; exhaled air contains 4% CO_2. CO_2 is expelled to maintain the correct equilibrium between CO_2 and H_2CO_3.

$$H_2CO_3 \rightleftharpoons CO_2 + H_2O$$

Have students hypothesize how this equilibrium combined with the reaction of hydrogen carbonate ions and hydrogen ions work together to maintain the pH of blood.

Chemistry Alive!

Milk of Magnesia

Play

Practice Problems Plus

Related Chapter Review Problem
Chapter Review problem 59 is related to Sample Problem 21-9.

Think Critically

CO_2 is just one example of an aqueous soluble gas that can disturb the pH of an aqueous environment. Have students think of other gases that can influence the pH of an aqueous solution by promoting the hydrolysis of water. Ask students to write the molecular equation describing how the gas alters pH. (Sulfur and nitrogen oxides produce H_2SO_3, H_2SO_4, HNO_2, and HNO_3, the principal species in acid rain.)

Sample Problem 21-9

Show how the carbonic acid–hydrogen carbonate buffer can "mop up" added hydrogen ions and hydroxide ions.

1. *ANALYZE* ***Plan a problem solving strategy.***
The definition of a buffer, in terms of its ability to act as a reservoir of neutralizing power, should be illustrated. One component in a buffer can react with H^+; this component is a proton acceptor. Another component can react with OH^-; this component is a proton donor.

2. *CALCULATE* ***Apply the problem-solving strategy.***
The carbonic acid–hydrogen carbonate buffer is a solution of carbonic acid (H_2CO_3) and hydrogen carbonate ions (HCO_3^-). When a base is added to this buffer, it reacts with H_2CO_3 (a proton donor) to produce neutral water. The pH changes very little.

$$\underset{\text{Carbonic acid}}{H_2CO_3(aq)} + \underset{\text{Hydroxide ion}}{OH^-(aq)} \rightleftharpoons \underset{\text{Hydrogen carbonate ion}}{HCO_3^-(aq)} + \underset{\text{Water}}{H_2O(l)}$$

When an acid is added to the buffer, it reacts with HCO_3^- (a proton acceptor) to produce un-ionized carbonic acid. Again, the pH changes very little.

$$\underset{\text{Hydrogen carbonate ion}}{HCO_3^-(aq)} + \underset{\text{Hydrogen ion}}{H^+(aq)} \rightleftharpoons \underset{\text{Carbonic acid}}{H_2CO_3(aq)}$$

3. *EVALUATE* ***Does the result make sense?***
The carbonic acid–hydrogen carbonate system is a good buffer solution because it can neutralize small quantities of strong acid or strong base.

Practice Problems

23. Write reactions to show what happens when the following occur.
    a. Acid is added to a solution of HPO_4^{2-}.
    b. Base is added to a solution of $H_2PO_4^-$.
24. Write an equation that shows what happens when acid is added to the ethanoic acid–ethanoate buffer.

The Solubility Product Constant

Salts differ in their solubilities. In general, compounds of the alkali metals are soluble in water. More than 35 g of sodium chloride, for example, will dissolve in only 100 mL of water. Many classes of ionic compounds, however, are insoluble. For example, many compounds that contain phosphate, sulfide, sulfite, or carbonate ions are insoluble. Exceptions are compounds in which these ions are combined with ammonium ions or alkali metal ions. **Table 21.3** summarizes the solubilities of many ionic compounds in water. Would you expect copper(II) hydroxide to be very soluble? ❶

Most insoluble salts will actually dissolve to some extent in water. These salts are said to be slightly, or sparingly, soluble in water. For example, **Figure 21.14** shows what happens when the "insoluble" salt silver chloride is mixed with water.

Answers

❶ No, most hydroxides are insoluble.
❷ No, since the test tube already contains undissolved silver chloride (the solution is saturated).

Table 21.3

Solubilities of Ionic Compounds in Water		
Compounds	**Solubility**	**Exceptions**
Salts of Group 1A metals and ammonia	Soluble	Some lithium compounds
Ethanoates, nitrates, chlorates, and perchlorates	Soluble	Few exceptions
Sulfates	Soluble	Compounds of Pb, Ag, Hg, Ba, Sr, and Ca
Chlorides, bromides, and iodides	Soluble	Compounds of Ag and some compounds of Hg and Pb
Sulfides and hydroxides	Most are insoluble	Alkali metal sulfides and hydroxides are soluble. Compounds of Ba, Sr, and Ca are slightly soluble.
Carbonates, phosphates, and sulfites	Insoluble	Compounds of the alkali metals and of ammonium ions

When silver chloride is mixed with water a very small amount of silver chloride dissolves in the water.

$$\text{AgCl}(s) \rightleftharpoons \text{Ag}^+(aq) + \text{Cl}^-(aq)$$

You can write an equilibrium expression for this process.

$$K_{eq} = \frac{[\text{Ag}^+][\text{Cl}^-]}{[\text{AgCl}]}$$

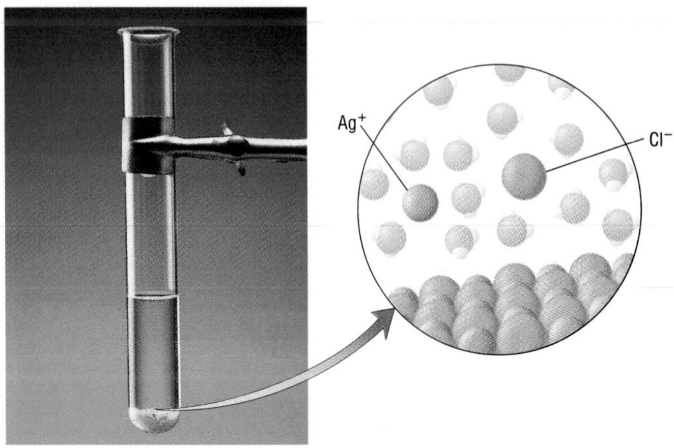

Figure 21.14
Silver chloride is slightly soluble in water. Would adding solid silver chloride to this test tube disturb the equilibrium? 2

21.2

Use the Visual

Have students examine Figure 21.14. Explain that the solubility of salts in water varies from very soluble to insoluble. Silver chloride is only slightly soluble and excess silver chloride remains in equilibrium with the ions in solution.

Discuss

Any ionic solid placed in water establishes an equilibrium between its dissociated ions and its undissociated (undissolved) form. The solubility product constant is an equilibrium constant defined specifically for equilibria between solids and their respective ions in solution. Be sure the students understand the difference between the terms *solubility* and *solubility product.*

Cooperative Learning

Divide the class into groups of four students. Ask the groups to create color-coded tables illustrating the solubilities of salts containing various anions. For example, students can design separate tables for nitrates, sulfates, chlorides, sulfides, hydroxides, and carbonates. Have them use different colors in each table to show which elements form soluble compounds containing the given anion and which form insoluble compounds. Students should divide the tasks of looking up solubilities, designing the tables, and drawing the final tables.

Think Critically

Have students recall what they have learned about the bonding of salts and the solution process. Ask them why some salts are soluble while others are not. (In insoluble salts, the electrostatic forces holding ions together are stronger than the forces of hydration.)

Table 21.4

Solubility Product Constants (K_{sp}) at 25 °C					
Salt	K_{sp}	Salt	K_{sp}	Salt	K_{sp}
Halides		**Sulfates**		**Hydroxides**	
AgCl	1.8×10^{-10}	$PbSO_4$	6.3×10^{-7}	$Al(OH)_3$	3.0×10^{-34}
AgBr	5.0×10^{-13}	$BaSO_4$	1.1×10^{-10}	$Zn(OH)_2$	3.0×10^{-16}
AgI	8.3×10^{-17}	$CaSO_4$	2.4×10^{-5}	$Ca(OH)_2$	6.5×10^{-6}
$PbCl_2$	1.7×10^{-5}	**Sulfides**		$Mg(OH)_2$	7.1×10^{-12}
$PbBr_2$	2.1×10^{-6}	NiS	4.0×10^{-20}	$Fe(OH)_2$	7.9×10^{-16}
PbI_2	7.9×10^{-9}	CuS	8.0×10^{-37}	**Carbonates**	
PbF_2	3.6×10^{-8}	Ag_2S	8.0×10^{-51}	$CaCO_3$	4.5×10^{-9}
CaF_2	3.9×10^{-11}	ZnS	3.0×10^{-23}	$SrCO_3$	9.3×10^{-10}
Chromates		FeS	8.0×10^{-19}	$ZnCO_3$	1.0×10^{-10}
$PbCrO_4$	1.8×10^{-14}	CdS	1.0×10^{-27}	Ag_2CO_3	8.1×10^{-12}
Ag_2CrO_4	1.2×10^{-12}	SnS	1.3×10^{-26}	$BaCO_3$	5.0×10^{-9}
		PbS	3.0×10^{-28}		

As long as some undissolved (solid) AgCl is present, the concentration of the AgCl is a constant. Thus the concentration of AgCl can be combined with the equilibrium constant to form a new constant.

$$K_{eq} \times [AgCl] = [Ag^+] \times [Cl^-] = K_{sp}$$

This new constant, called the **solubility product constant** (K_{sp}), equals the product of the concentration terms each raised to the power of the coefficient of the substance in the dissociation equation. The coefficients for the dissociation of silver chloride are each 1. The value of K_{sp} for silver chloride at 25 °C is 1.8×10^{-10}.

$$K_{sp} = [Ag^+] \times [Cl^-] = 1.8 \times 10^{-10}$$

The solubility product constants for some common sparingly soluble salts are given in **Table 21.4.** Although most compounds of barium are highly toxic, barium sulfate is so insoluble ($K_{sp} = 1.1 \times 10^{-10}$) that it is routinely used by physicians for gastrointestinal examinations. See **Figure 21.15.**

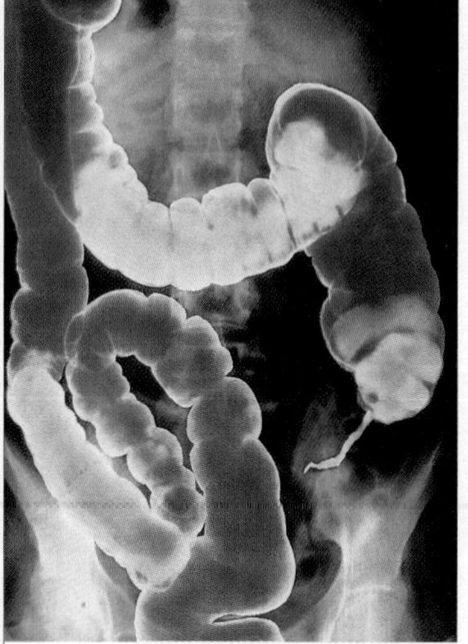

Figure 21.15
Barium sulfate is used for making x-ray images of the digestive tract. Barium sulfate absorbs the x-rays, thereby producing light areas on the developed x-ray film. Barium sulfate can be used safely for this procedure because it is insoluble in body fluids. Soluble barium salts are highly toxic.

Meeting Diverse Needs

At Risk Students who lack a strong background in mathematics will probably need help in taking square and cube roots of numbers written in scientific notation. Have them work in pairs or small groups with other students who understand the procedure until they can do it on their own.

CHEMath

ALGEBRA AND K_{sp}

The solubility product constant K_{sp} is closely related to K_{eq}, the equilibrium constant introduced in Chapter 19. This page offers a review of the math skills needed to solve problems involving K_{eq} and K_{sp}.

Consider the slightly soluble salt AB, which partially dissociates into A^+ and B^- ions.

$$AB(s) \rightleftharpoons A^+(aq) + B^-(aq)$$

The general formula for calculating K_{sp} is

$$K_{sp} = [A]^x[B^-]^y$$

where [A] and [B] are concentrations, and x and y are their respective coefficients from the dissociation equation. Thus K_{sp} for the dissociation of AB is

$$K_{sp} = [A^+][B^-]$$

When silver chloride (AgCl) dissolves in water, one Ag^+ ion forms for every Cl^- ion, thus $x = 1$ and $y = 1$.

$$AgCl(s) \rightleftharpoons Ag^+(aq) + Cl^-(aq) \qquad K_{sp} = [Ag^+][Cl^-]$$

When lead fluoride (PbF_2) dissolves in water, one Pb^{2+} ion forms for every two F^- ions, thus $x = 1$ and $y = 2$.

$$PbF_2(s) \rightleftharpoons Pb^{2+}(aq) + 2F^-(aq) \qquad K_{sp} = [Pb][F]^2$$

Without knowing the actual concentration of either ion in solution, you can use algebra to relate the relative concentrations of the ions. You can then write an equation involving a known value of K_{sp} to calculate each ion's exact concentration. This is done by substituting x for the concentration of one of the ions and using the dissociation equation to write the other ion's concentration, also in terms of x. Solving the equation then usually involves the use of square roots or cube roots.

Example

What are the concentrations of nickel and sulfide ions in a saturated solution of nickel sulfide (NiS) at 25 °C? ($K_{sp} = 4.0 \times 10^{-20}$)

First, write the dissociation equation.

$$NiS(s) \rightleftharpoons Ni^{2+}(aq) + S^{2-}(aq)$$

Note that one Ni^{2+} ion forms for every S^{2-} ion. Recall the general equations for K_{sp}.

$$K_{sp} = [A^+]^x[B^-]^y, \text{ thus } K_{sp} = [Ni^{2+}][S^{2-}]$$

Substituting the known value for K_{sp} yields,

$$4.0 \times 10^{-20} = [Ni^{2+}][S^{2-}]$$

Set $x = [Ni^{2+}]$. Because $[Ni^{2+}]$ and $[S^{2-}]$ are equal, x also equals $[S^{2-}]$. Substituting for $[Ni^{2+}]$ and $[S^{2-}]$ yields

$$4.0 \times 10^{-20} = (x)(x)$$

$$4.0 \times 10^{-20} = x^2$$

Taking the square root of both sides of the equation yields,

$$\sqrt{4.0 \times 10^{-20}} = \sqrt{x^2}$$

$$2.0 \times 10^{-10} = x$$

Going back to a relationship between x and ion concentrations yields

$$[Ni^{2+}] = [S^{2-}] = 2.0 \times 10^{-10} M$$

Note that for a dissociation that yields two A ions for every one B ion, algebra can still be used to relate the concentrations. In this case, x equals the concentration of B, $x = [B]$, and $2x$ equals the concentration of A, $2x = [A]$. Substituting these x values into the K_{sp} equation, the ion concentrations can be calculated.

Practice Problems

Prepare for upcoming problems in this chapter by using solubility products to calculate the following concentrations.

A. Write the solubility product expression for the salts CuS, $ZnCO_3$, and $PbCrO_4$.

B. What are the concentrations of silver and bromide ions in a saturated solution of silver bromide (AgBr) at 25 °C? $K_{sp} = 5 \times 10^{-13}$

C. What is the concentration of iodide ions in a saturated solution of lead iodide (PbI_2) at 25.0 °C? $K_{sp} = 7.9 \times 10^{-9}$

CHEMath

Teaching Tips

Explain that the solubility product is very closely related to the equilibrium constant that students learned about in Chapter 20. Point out similarities in the equations for these two quantities. Write the example problems on the board and carry out the solutions for the solubility product, K_{sp}. Focus on how to set up the algebraic expression.

Answers to Practice Problems

A. CuS: $[Cu^{2+}][S^{2-}]$, $ZnCO_3$: $[Zn^{2+}][CO_3^{2-}]$, $PbCrO_4$: $[Pb^{2+}][CrO_4^{2-}]$

B. $[Ag^+] = [Br^-] = 7 \times 10^{-7} M$

C. $[I^-] = 2.5 \times 10^{-3} M$

MEETING DIVERSE NEEDS

At Risk Pair students with weak math skills with those proficient at algebra and problem solving. Have these students work through the practice problems, taking turns being the lead solver. The lead student should explain the process being used to arrive at the solution.

Practice Problems Plus

Additional Practice Problem

The Chem ASAP! CD-ROM contains this problem related to Sample Problem 21-10: What is the concentration of barium and sulfate ions in a saturated barium sulfate solution at 25°C? $K_{sp} = 1.1 \times 10^{-10}$ $(1.0 \times 10^{-5} M)$

TEACHER DEMO

Prepare saturated solutions of silver chloride and silver nitrate in two test tubes. Write the solubility product constant expression for the silver chloride salt on the board.

$K_{sp} = [Ag^+] \times [Cl^-] = 1.8 \times 10^{-10}$

Have students hypothesize on what would occur if a few drops of the silver nitrate solution were added to the test tube containing the silver chloride. Add the drops of silver nitrate solution and ask the students what they observe. (A precipitate, silver chloride, comes out of solution.) Explain that the value of $[Ag^+]$ increases in the K_{sp} expression as the silver nitrate is added so that silver chloride must precipitate out until the solubility product again equals the K_{sp}.

Sample Problem 21-10

What is the concentration of silver ions and chloride ions in a saturated silver chloride solution at 25 °C? ($K_{sp} = 1.8 \times 10^{-10}$)

1. ***ANALYZE*** ***List the knowns and the unknowns.***

Knowns:
- $K_{sp} = 1.8 \times 10^{-10}$
- $K_{sp} = [Ag^+] \times [Cl^-]$
- $AgCl(s) \rightleftharpoons Ag^+(aq) + Cl^-(aq)$

Unknowns:
- $[Ag^+] = ?M$
- $[Cl^-] = ?M$

The equation shows that for each Ag^+ ion formed, one Cl^- ion is formed.

2. ***CALCULATE*** ***Solve for the unknowns.***

$$K_{sp} = [Ag^+] \times [Cl^-] = 1.8 \times 10^{-10}$$

At equilibrium $[Ag^+] = [Cl^-]$, so $[Ag^+]$ can be substituted for $[Cl^-]$ in the expression for K_{sp}. This gives an equation with one unknown.

$$K_{sp} = [Ag^+] \times [Ag^+] = 1.8 \times 10^{-10}$$

Solving for $[Ag^+]$:

$$[Ag^+]^2 = 1.8 \times 10^{-10}$$
$$[Ag^+] = 1.3 \times 10^{-5} M$$

The equilibrium concentration of Cl^- is also $1.3 \times 10^{-5} M$, because $[Ag^+] = [Cl^-]$.

3. ***EVALUATE*** ***Do the results make sense?***

The product $[Ag^+] \times [Cl^-]$ is very close to the value of K_{sp}. The answer should have two significant figures.

Practice Problems

25. Lead(II) sulfide (PbS) has a K_{sp} of 3×10^{-28}. What is the concentration of lead(II) ions in a saturated solution of PbS?

26. What is the concentration of calcium and carbonate ions in a saturated calcium carbonate solution at 25 °C? ($K_{sp} = 4.5 \times 10^{-9}$)

Chem ASAP!

Problem-Solving 26

Solve Problem 26 with the help of an interactive guided tutorial.

Sample Problem 21-11

Calcium fluoride has a K_{sp} of 3.9×10^{-11} at 25 °C. What is the fluoride-ion concentration at equilibrium?

1. ***ANALYZE*** ***List the knowns and the unknowns.***

Knowns:
- $K_{sp} = 3.9 \times 10^{-11}$
- $CaF_2 \rightleftharpoons Ca^{2+}(aq) + 2F^-(aq)$
- $K_{sp} = [Ca^{2+}] \times [F^-]^2$

Unknowns:
- $[Ca^{2+}] = ?M$
- $[F^-] = ?M$

There are twice as many F^- ions as Ca^{2+} ions formed when CaF_2 dissociates. To determine the $[F^-]$, the $[Ca^{2+}]$ must also be determined.

Sample Problem 21-11 (cont.)

2. ***CALCULATE*** ***Solve for the unknowns.***
When the $[Ca^{2+}]$ is x, the $[F^-]$ is $2x$. If these terms are substituted into the K_{sp} expression, the expression can be solved for x.

$$K_{sp} = (x)(2x)^2 = 3.9 \times 10^{-11}$$
$$4x^3 = 3.9 \times 10^{-11}$$
$$x^3 = 9.8 \times 10^{-12}$$
$$x = 2.1 \times 10^{-4}$$

The $[Ca^{2+}] = 2.1 \times 10^{-4}M$. The $[F^-]$ is twice the $[Ca^{2+}]$. Thus $[F^-] = 4.2 \times 10^{-4}M$.

3. ***EVALUATE*** ***Do the results make sense?***
The concentrations are reasonable because 10^{-4} cubed (10^{-12}) is close to the value for K_{sp}. The answers should have two significant figures.

Practice Problems

27. The K_{sp} of silver sulfide (Ag_2S) is 8×10^{-51}.
    a. What is the silver-ion concentration of a saturated solution of silver sulfide?
    b. What is the sulfide-ion concentration of the same solution?
28. Write the solubility product expression for each salt.
    a. $PbCl_2$
    b. Ag_2CrO_4

The Common Ion Effect

In a saturated solution of lead(II) chromate, an equilibrium is established between the solid lead(II) chromate and its ions.

$$PbCrO_4(s) \rightleftharpoons Pb^{2+}(aq) + CrO_4^{2-}(aq) \qquad K_{sp} = 1.8 \times 10^{-14}$$

What do you think would happen if you added some lead nitrate to this solution? Immediately after the addition, the product of $[Pb^{2+}]$ and $[CrO_4^{2-}]$ would be greater than K_{sp}. Applying Le Châtelier's principle (Section 19.2), the stress of the additional Pb^{2+} can be relieved if the reaction shifts to the left. As you can see in **Figure 21.16,** lead ions would combine with chromate ions in solution to form additional solid $PbCrO_4$. In fact, $PbCrO_4$ will precipitate until the product of $[Pb^{2+}]$ and $[CrO_4^{2-}]$ once again equals 1.8×10^{-14}.

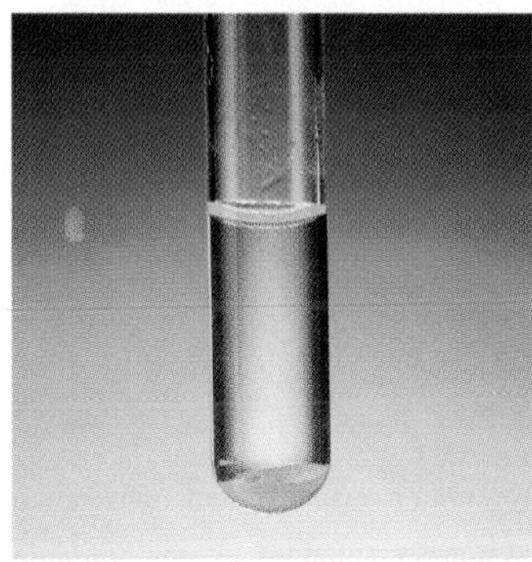

(a) $PbCrO_4(s) \rightleftharpoons Pb^{2+}(aq) + CrO_4^{2-}$

(b) $PbCrO_4(s) \rightleftharpoons Pb^{2+}(aq) + CrO_4^{2-}$ (more Pb^{2+} added)

Figure 21.16
(a) *Lead(II) chromate, a bright-yellow compound, is only slightly soluble in water. A saturated solution of this compound is very pale yellow due to the presence of* $CrO_4^{2-}(aq)$*, as shown in the equilibrium expression.* ***(b)*** *When a few drops of lead nitrate* ($Pb(NO_3)_2$) *are added to the clear, pale-yellow solution, more lead(II) chromate precipitates. The amount of lead(II) chromate* ($PbCrO_4$)(s) *increases because its solubility is lower in the presence of the common ion* (Pb^{2+}).

Use the Visual

Have students examine the photographs in Figure 21.16 and observe what happens when additional chromate ions are added to the saturated lead(II) chromate solution. To maintain the solubility product constant equality, a precipitate forms.

Discuss

Point out that the common-ion effect is not a new concept, but an example of Le Châtelier's principle at work. Explain that adding an ion to an equilibrium system containing that ion will cause the equilibrium to shift in the direction that uses up the ion.

Practice Problems Plus

Related Chapter Review Problem
Chapter Review problem 64 is related to Sample Problem 21-11.

Practice Problems Plus

Related Chapter Review Problem
Chapter Review problem 68 is related to Sample Problem 21-12.

Think Critically

Have students hypothesize how the common-ion effect could be used to identify ions in solution. Suggest that they use Table 21.3 to find a common property of salts listed in Table 21.4 (insolubility). Explain that saturated solutions of many of the salts in Table 21.4 can be used to test for the presence of different ions in solution. If a precipitate forms when a few drops of an unknown solution is added to a known saturated solution, the unknown solution must have an ion in common with the known solution. Such tests are part of a branch of chemistry called qualitative analysis. Have small groups of students research qualitative tests designed to identify specific ions. Compile a list of their results, including descriptions of the tests.

Practice Problems Plus

Additional Practice Problem
The Chem ASAP! CD-ROM contains this problem related to Sample Problem 21-13: A chemist prepares a solution by combining 0.045 mol $CaCl_2$ and 0.045 mol NaOH and adds water to make 2.5 L of solution. What is the ion product? Will a precipitate of $Ca(OH)_2$ form in this solution? (ion product = 5.8×10^{-6}; no)

In this example, the lead ion is called a common ion. A **common ion** is an ion that is common to both salts. Adding lead nitrate to a saturated solution of $PbCrO_4$ causes the solubility of $PbCrO_4$ to be decreased. The solubility of $PbCrO_4$ is less in the presence of a solution of $Pb(NO_3)_2$ than it is in pure water. The lowering of the solubility of a substance by the addition of a common ion is called the **common ion effect.**

Adding sodium chromate would also produce the common ion effect in this example. The additional chromate ion, a different common ion, would cause the reaction to shift to the left. More $PbCrO_4$ would be formed and the solubility of $PbCrO_4$ would again decrease until the K_{sp} equaled 1.8×10^{-14}. Sample Problem 21-12 illustrates how to use the common ion effect to predict the iodide-ion concentration of a solution of silver iodide when silver nitrate solution is added. In this case, silver is the common ion.

Sample Problem 21-12

The K_{sp} of silver iodide is 8.3×10^{-17}. What is the iodide-ion concentration of a 1.00-L saturated solution of AgI to which 0.020 mol $AgNO_3$ is added?

1. ***ANALYZE*** ***List the knowns and the unknown.***

   Knowns:
   - $K_{sp} = 8.3 \times 10^{-17}$
   - mol $AgNO_3$ added = 0.020 mol
   - $AgI(s) \rightleftharpoons Ag^+(aq) + I^-(aq)$
   - $K_{sp} = [Ag^+] \times [I^-]$

   Unknown:
   - $[I^-] = ?\,M$

   If the equilibrium concentration of iodide ion from the dissociation is x, then the equilibrium concentration of silver ion is $x + 0.020$.

2. ***CALCULATE*** ***Solve for the unknown.***

   Because of the small value of K_{sp}, x will be small compared with 0.020. Therefore the $[Ag^+]$ at equilibrium equals 0.020*M*. Substitute these values into the K_{sp} expression.

$$K_{sp} = [Ag^+] \times [I^-]$$
$$8.3 \times 10^{-17} = (0.020)(x)$$
$$x = \frac{8.3 \times 10^{-17}}{0.020}$$
$$x = 4.2 \times 10^{-15}$$

   The equilibrium concentration of iodide ion is $4.2 \times 10^{-15}M$.

3. ***EVALUATE*** ***Does the result make sense?***

   Because $[Ag^+]$ is so high, it is to be expected that $[I^-]$ will be very low, approaching the numerical value for K_{sp}. The answer should have two significant figures.

Practice Problems

29. What is the concentration of sulfide ion in a 1.0-L solution of iron(II) sulfide to which 0.04 mol of iron(II) nitrate has been added? The K_{sp} of FeS is 8×10^{-19}.

30. The K_{sp} of $SrSO_4$ is 3.2×10^{-7}. What is the equilibrium concentration of sulfate ion in a 1.0-L solution of strontium sulfate to which 0.10 mol $Sr(CH_3COO)_2$ has been added?

The solubility product (K_{sp}) can be used to predict whether a precipitate will form when solutions are mixed. If the ion-product concentration of two ions in the mixture is greater than the K_{sp} of the compound formed from the ions, a precipitate will form. If the ion-product concentration of two ions is less than the K_{sp}, no precipitate will form, and the solution will be unsaturated. What do you think will happen if the ion-product concentration of two ions equals K_{sp}? ❶

Sample Problem 21-13

Predict whether barium sulfate will precipitate when 0.50 L of $0.002M$ $Ba(NO_3)_2$ mixes with 0.50 L of $0.008M$ Na_2SO_4 to form 1.00 L of solution. The K_{sp} of $BaSO_4$ is 1.1×10^{-10}.

1. *ANALYZE* ***List the knowns and the unknown.***

Knowns:

- $K_{sp} = 1.1 \times 10^{-10}$
- 0.50 L of $0.002M$ $Ba(NO_3)_2$
- 0.50 L of $0.008M$ Na_2SO_4

Unknown:

- $[Ba^{2+}] \times [SO_4^{2-}] > K_{sp}$?

Precipitation will occur if the product of the concentrations of the two ions exceeds the K_{sp} of $BaSO_4$.

2. *CALCULATE* ***Solve for the unknown.***

Halve the concentration of each ion because the volume is doubled when the original solutions mix. So in the final solution, $[Ba^{2+}] = 0.001M$ and $[SO_4^{2-}] = 0.004M$. After mixing, the ion product is

$$[Ba^{2+}] \times [SO_4^{2-}] = (0.001M) \times (0.004M) = 4 \times 10^{-6}$$

This number is larger than the K_{sp}.

3. *EVALUATE* ***Does the result make sense?***

Because the ion-product concentration of the two ions exceeds K_{sp}, $BaSO_4$ will precipitate.

Practice Problems

31. Predict whether calcium carbonate will precipitate when 0.5 L of $0.001M$ $Ca(NO_3)_2$ is mixed with 0.5 L of $0.0008M$ Na_2CO_3 to form 1.0 L of solution. The K_{sp} of $CaCO_3$ is 4.5×10^{-9}.
32. A student prepares a solution by combining 0.025 mol $CaCl_2$ and 0.015 mol $Pb(NO_3)_2$ and adding water to make 1.0 L of solution. Will a precipitate of $PbCl_2$ form in this solution?

Chem ASAP!

Problem-Solving 32

Solve Problem 32 with the help of an interactive guided tutorial.

section review 21.2

33. Using equations, show what happens when acid is added to an ammonium ion–ammonia buffer. What happens when base is added?
34. What is the K_{sp} of nickel(II) sulfide if the equilibrium concentrations of a saturated solution of Ni^{2+} and S^{2-} are each $2 \times 10^{-10}M$?
35. Would precipitation occur when 500 mL of a $0.02M$ solution of $AgNO_3$ is mixed with 500 mL of a $0.001M$ solution of NaCl? Explain.

Chem ASAP! **Assessment 21.2** Check your understanding of the important ideas and concepts in Section 21.2.

21.2

3 Assess

Evaluate Understanding

- **Have students explain in their own words what determines whether a solution containing a hydrolyzing salt is acidic or basic?** (Salts formed from a strong acid and strong base are neutral. Those formed from a strong base and weak acid are basic; those formed from a strong acid and weak base are acidic.)
- **Ask students to explain how ethanoate ions can effectively remove hydrogen ions from solution.** (The ethanoate ions are a Brønsted–Lowry base that accept the hydrogen ions to form ethanoic acid.)
- **Have students reclassify the salts in Table 21.4 on the basis of solubility and explain the relationship between the solubility product constant and the relative solubility of a salt?** (the larger the solubility product constant, the more soluble the salt)

Reteach

Point out that a hydrolysis reaction always uses one part of the water molecule and leaves behind the other. Use equations to show how the salt of a weak acid always reacts with water to produce hydroxide ions and the salt of a weak base always reacts to produce hydrogen ions. Remind students that chemical buffers are equilibrium systems that tend to resist change due to external influences. Emphasize that the behavior of a buffer is an application of Le Châtelier's principle. Adding an acid or base to a buffered system is a stress. As long as the buffer capacity is not exceeded, the system will remove the excess ions and keep the pH constant.

Answers

❶ The solution will be saturated; adding additional ions will cause a precipitate to form.

SECTION REVIEW 21.2

33. $NH_3 + H^+ \rightleftharpoons NH_4^+$
    $NH_4^+ + OH^- \rightleftharpoons NH_3 + H_2O$
34. 4×10^{-20}
35. Yes; the ion product is greater than K_{sp}.

Chemistry Serving... Industry

DISCUSS

For students unfamiliar with bacteria, explain that bacteria are one-celled organisms that range in size from 1 to 10 micrometers. Bacteria are prokaryotic cells, which are cells that lack a nucleus.

To help students understand the terms *extremophile*, *alkaliphile*, and *acidophile*, point out that they all contain the suffix *-phile.* This suffix comes from the Greek *philos,* which means "fond of" or "liking" or "attracted to." An extremophile thrives in extreme conditions, an acidophile thrives in acidic conditions, and an alkaliphile thrives in basic conditions.

As an extension, you may wish to have students do research on the organisms that live in Mono Lake in California. The water in Mono Lake has a higher concentration of salts than does ocean water and is highly alkaline. Although no fish can live in the lake, other organisms, including the alkali fly, can be found in huge numbers.

CHEMISTRY IN CAREERS

Encourage students to connect to the Internet address shown on this page to find out more about a career as a microbiologist. Information may also be found on page 878 of the text.

Chemistry Serving...Industry

EXTREME ORGANISMS' VALUABLE ENZYMES

You probably would not want to jump into the pool of water shown in this photo. Even if you could survive the near-boiling temperature, the pH might begin to eat away at your skin. Although an environment like this is inhospitable to human life, it is not necessarily lifeless. Scientists have discovered that a variety of extreme environments on Earth support living organisms. These environments include boiling hot springs, frozen glaciers, lagoons much saltier than the ocean, highly alkaline soda lakes, and acidic mine pits.

The organisms that live in these extreme environments are called extremophiles. Most of these organisms are very simple, primitive bacteria. Recently, scientists have found that the chemical tricks extremophiles use to survive in their harsh habitats may have many useful applications in industry.

In particular, scientists are interested in the enzymes that extremophiles produce for carrying out their life processes. Enzymes are biological catalysts. Like the synthetic catalysts you have already learned about, enzymes greatly speed up chemical reactions without themselves being altered. Enzymes are very important in many industrial processes today. But most enzymes cannot function under extreme conditions. Heat destroys them, cold inactivates them, and extremes of pH break them down. In contrast, the enzymes produced by extremophiles—called extremozymes—are able to survive harsh conditions such as these. Therefore they have the potential for being used where typical enzymes cannot.

Two types of extremophiles of interest to scientists are acidophiles and alkaliphiles. Acidophiles live in very acidic environments ($pH < 5$), and alkaliphiles live in very basic environments ($pH > 9$). Some of the extremozymes produced by acidophiles work to break down food outside the organism so it can be absorbed or ingested. These extremozymes could be used to increase the digestibility of animal feed. When added to the feed, these enzymes would work in the acid environment of animals' stomachs to help break down inexpensive grains that would normally be only partially digested. Other acid-tolerant extremozymes could be used as catalysts for producing compounds that must be synthesized under acidic conditions.

The chemical tricks extremophiles use to survive in their harsh habitats may have many useful applications in industry.

The alkaline-tolerant enzymes produced by alkaliphiles also have many possible uses. Detergent makers, for example, are very interested in the ability of alkaliphiles to remove food stains from clothes. Certain typical enzymes work well as stain removers because they help break down the proteins and fats that make up the stains. But these enzymes are destroyed in the highly alkaline environment created by detergents. Alkaline-tolerant versions of these enzymes would solve this problem, allowing detergent makers to create powerful stain-fighting detergents that would be of value to many people.

CHEMISTRY IN CAREERS

MICROBIOLOGIST

Microbiologist needed for the development of genetically engineered bacteria.

See page 878.

CHEMICAL SPECIALIST

Environmental GC. Must know

TAKE IT TO THE NET

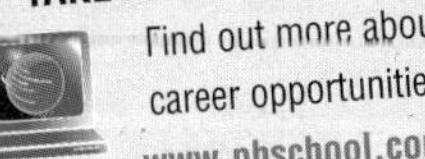

Find out more about career opportunities: **www.phschool.com**

CHEMICAL SPECIALIST

Local food service distributor se responsible self-motivated indiv

Chapter 21 STUDENT STUDY GUIDE

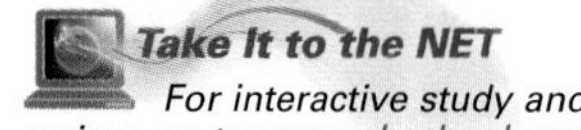

For interactive study and review, go to www.phschool.com

KEY TERMS

- buffer *p. 628*
- buffer capacity *p. 629*
- common ion *p. 636*
- common ion effect *p. 636*
- end point *p. 617*
- equivalence point *p. 622*
- equivalent (equiv) *p. 619*
- gram equivalent mass *p. 619*
- neutralization reaction *p. 614*
- normality (N) *p. 621*
- salt hydrolysis *p. 626*
- solubility product constant (K_{sp}) *p. 632*
- standard solution *p. 617*
- titration *p. 617*

KEY EQUATIONS AND RELATIONSHIPS

- Acid + base $\longrightarrow$ salt + water
- Gram equivalent mass $= \dfrac{\text{molar mass}}{\text{number of ionizable hydrogens}}$
- Normality (N) = equiv/L
- $N_1 \times V_1 = N_2 \times V_2$
- $N_A \times V_A = N_B \times V_B$
- For a slightly soluble salt (MX), $K_{sp} = [M^+] \times [X^-]$

CONCEPT SUMMARY

21.1 Neutralization Reactions

- In the reaction of an acid with a base, hydrogen ions and hydroxide ions react to produce a salt and water. This reaction, called neutralization, is usually carried out by titration.
- The end point in a titration is the point at which the indicator changes color. An indicator is chosen so the solution reaches the end point very near the point of neutralization.
- An equivalent of an acid is the mass of the acid that provides one mole of hydrogen ions in solution.
- A solution that contains one equivalent of an acid or a base in a single liter of solution is a one-normal ($1N$) solution.
- At the equivalence point of a titration, the number of equivalents of acid equals the number of equivalents of base.

21.2 Salts in Solution

- A salt consists of an anion from an acid and a cation from a base and forms when an acid is neutralized by a base.
- Salts of strong acid–strong base reactions produce neutral solutions in water.
- Salts formed from weak acids or weak bases hydrolyze water and produce basic or acidic solutions.
- Solutions that resist changes in pH are called buffer solutions. Within limits, components of a buffer can react with hydrogen and hydroxide ions and minimize changes in pH.
- The solubility product constant (K_{sp}) is the equilibrium constant for the reaction between an ionic solid and its ions in solution.
- The solubility of a salt is decreased by the addition of a common ion.

CHAPTER CONCEPT MAP

Use these terms to construct a concept map that organizes the major ideas of this chapter.

equivalence point | equivalent | neutralization
end point | titration | normality

Chem ASAP! Concept Map 21
Create your Concept Map using the computer.

Take It to the Net

At **www.phschool.com** students will find for this chapter

- an Internet activity
- links to related chemistry sites
- an interactive quiz
- career links

4 Close

Summary

Ask students the following questions, which require them to summarize chapter content.

- **What happens to the pH during a titration?** (There is a gradual change in pH, which becomes dramatic as the end point nears.)
- **Based on this chapter, define a salt.** (A salt contains an anion from an acid and a cation from a base.)
- **What do all buffers have in common?** (They are made from a weak acid or base and its salt.)
- **What effect does a common ion have on the solubility product constant of a solution?** (No effect, but the solubility of the original salt is decreased.)

Extension

Have students work in groups to prepare a chart that illustrates the relationships among acids, bases, and salts. Include topics such as titration, neutralization, equivalence point, pH, salt hydrolysis, solubility product constant, and buffers (common ion effect).

Looking Back . . . Looking Ahead . . .

The previous chapter described properties of acids and bases along with three acid–base theories. The current chapter discussed neutralization reactions and the properties of salts. The next chapter discusses oxidation–reduction reactions.

Answers

36. Acid + base → salt + water
37. a. $HNO_3 + KOH \rightarrow KNO_3 + H_2O$
b. $2HCl + Ca(OH)_2 \rightarrow CaCl_2 + 2H_2O$
c. $H_2SO_4 + 2NaOH \rightarrow Na_2SO_4 + 2H_2O$
38. Indicator changes color.
39. a. 1.40*M*
b. 2.61*M*
40. a. 56.1 g (1 equiv/mol)
b. 36.5 g (1 equiv/mol)
c. 49.0 g (2 equiv/mol)
41. a. 0.10 equiv
b. 3.86 equiv
c. 0.30 equiv
42. number of equivalents of solute in 1 L of solution
43. a. 1*N*
b. 2*N*
c. 0.2*N*
d. 0.2*N*
44. a. 1.0*N*
b. 0.13*N*
c. 0.074*N*
d. 0.105*N*
e. 0.80*N*
f. 0.234*N*
45. a. 0.0600*N*
b. 0.400*N*
c. 0.171*N*
d. 0.0752*N*
46. salts with a cation from a weak base and an anion from a strong acid, or with a cation from a strong base and an anion from a weak acid
47. $HCO_3^-(aq) + H_2O(l) \rightleftharpoons H_2CO_3(aq) + OH^-(aq)$
48. Weak-acid anions accept protons from water, increasing the pH of the solution. Weak-base cations donate protons to water, decreasing the pH.
49. a. basic
b. acidic
c. neutral
d. basic
e. neutral
f. acidic
50. Eventually the buffer capacity of the buffer is exceeded and the pH will change significantly with further addition of strong acid or base.

Chapter 21 REVIEW

CONCEPT MASTERY

36. Write a general word equation for a neutralization reaction. *21.1*

37. Identify the products and write balanced equations for each neutralization reaction. *21.1*
a. $HNO_3(aq) + KOH(aq) \longrightarrow$
b. $HCl(aq) + Ca(OH)_2(aq) \longrightarrow$
c. $H_2SO_4(aq) + NaOH(aq) \longrightarrow$

38. What is characteristic of the end point of a titration? *21.1*

39. What is the molarity of sodium hydroxide if 20.0 mL of the solution is neutralized by each of the following 1.00*M* solutions? *21.1*
a. 28.0 mL of HCl b. 17.4 mL of H_3PO_4

40. Determine the gram equivalent mass and the equivalents per mole for each compound. *21.1*
a. KOH b. HCl c. H_2SO_4

41. How many equivalents is each compound? *21.1*
a. 3.7 g $Ca(OH)_2$ c. 9.8 g H_3PO_4
b. 189 g H_2SO_4

42. How is the normality of a solution calculated? *21.1*

43. What is the normality of each solution? *21.1*
a. 1*M* NaOH c. 0.2*M* KOH
b. 2*M* HNO_3 d. 0.1*M* H_2SO_4

44. Determine the normality of each solution. *21.1*
a. 250 mL of solution containing 10 g NaOH
b. 750 mL of solution containing 4.9 g H_2SO_4
c. 270 mL of solution containing 0.74 g HCl
d. 2.80 L of solution containing 18.6 g HNO_3
e. 7.3 g HCl in 250 mL of solution
f. 18.4 g HNO_3 in 1250 mL of solution

45. A student titrated several solutions of unknown concentration with various standard solutions to the point of neutralization. The volume of each unknown solution and the volume and normality of the standard solution used are given below. Calculate the normality for each unknown. *21.1*
a. 25.0 mL H_2SO_4 required 15.0 mL of 0.100*N* NaOH
b. 10.0 mL NaOH required 20.0 mL of 0.200*N* HCl
c. 17.5 mL NaOH required 25.0 mL of 0.120*N* HNO_3
d. 50.0 mL CH_3COOH required 39.6 mL of 0.0950*N* KOH

46. What kinds of salts hydrolyze water? *21.2*

47. Write an equation showing why an aqueous solution of sodium hydrogen carbonate is basic. *21.2*

48. Explain why solutions of salts that hydrolyze water do not have a pH of 7. *21.2*

49. Predict whether an aqueous solution of each salt will be acidic, basic, or neutral. *21.2*
a. $NaHCO_3$ d. Na_2CO_3
b. NH_4NO_3 e. Na_2SO_4
c. KCl f. NH_4Cl

50. A buffered solution cannot absorb an unlimited amount of acid or base. Explain. *21.2*

51. Would a solution of HCl and NaCl be a good buffer? Explain. *21.2*

52. Write the solubility product expression for each salt. *21.2*
a. NiS b. $BaCO_3$

53. What does the solubility product constant (K_{sp}) represent? *21.2*

54. Use **Table 21.4** on page 632 to rank these salts from most soluble to least soluble. *21.2*
a. CuS c. $SrCO_3$
b. $BaSO_4$ d. AgI

55. How does the addition of a common ion affect the solubility of another substance? *21.2*

CONCEPT MASTERY

56. The graph shows the number of millimoles (mmol) of water formed by the dropwise addition of 1.0*N* HCl to a 25.0-mL sample of NaOH of unknown concentration.
a. Write an equation for the reaction.
b. Estimate the concentration of NaOH.

Millimoles of Water vs. Volume of 1.0*N* HCl

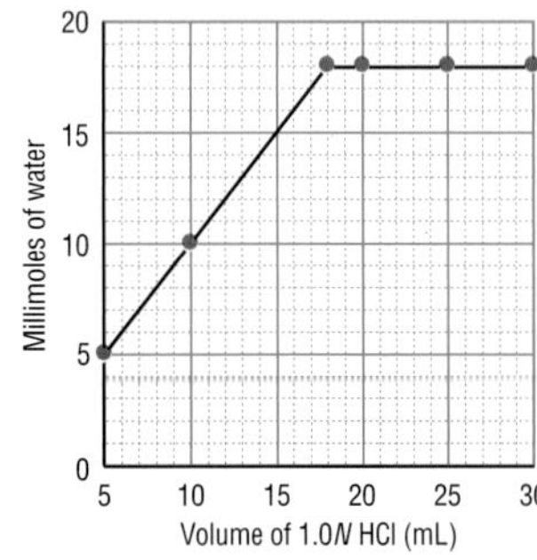

57. What must be true about the concentration of two ions if precipitation occurs when solutions of the two ions are mixed?

58. How many equivalents are in each solution?
a. 5.8 L of 0.55*N* HCl
b. 330 mL of 1.4*N* H_3PO_4
c. 0.14 L of 0.22*N* KOH

59. Use the phosphate buffer (H_2PO_4/HPO_4^{2-}) to illustrate how a buffer system works. Show, by means of equations, how the pH of a solution can be kept almost constant when small amounts of acid or base are added.

60. Find the normality of each solution.
a. 86.3 g $Mg(OH)_2$ in 2.5 L of solution
b. 5.6 g HBr in 450 mL of solution
c. 49.4 g H_2SO_3 in 1.5 L of solution

61. The following data were collected from a titration of 50.00 mL of ethanoic acid (CH_3COOH) of unknown concentration with 0.100*N* NaOH. Plot these data (pH on the *y*-axis) to obtain a titration curve.

Volume of NaOH (mL)	pH	Volume of NaOH (mL)	pH
0		50.00	8.73
10.00	4.15	50.01	8.89
25.00	4.76	51.00	11.00
40.00	5.36	60.00	11.96
49.00	6.45	75.00	12.30
49.99	8.55	100.00	12.52

a. What is the pH at the end point of this titration?
b. Use **Figure 20.8** on page 590 to identify one or more indicators that could be used to determine the end point in this titration.

62. Write an equation for the reaction of each antacid with hydrochloric acid.
a. magnesium hydroxide
b. calcium carbonate
c. aluminum hydroxide

63. How many milliliters of 2.00*N* NaOH would you need to dilute with water to make 250 mL of 0.100*N* NaOH?

64. What is the concentration of hydroxide ions in a saturated solution of each salt?
a. $Zn(OH)_2$
b. $Ca(OH)_2$
c. $Al(OH)_3$

65. How would the equilibrium between hypochlorous acid and the hypochlorite ion be affected by the addition of each?

$$HOCl(aq) + OH^-(aq) \rightleftharpoons OCl^-(aq) + H_2O(l)$$

a. HCl b. NaOH

66. Write an equation to show that an aqueous solution of sodium ethanoate will be basic.

67. Arrange these solutions in order of decreasing acidity.
a. 0.1*N* NaOH
b. 0.1*N* HCl
c. 0.1*M* ammonium chloride
d. 0.1*M* sodium ethanoate

68. What is the equilibrium concentration of barium ion in a 1.0-L saturated solution of barium carbonate to which 0.25 mol K_2CO_3 has been added?

CRITICAL THINKING

69. Choose the term that best completes the second relationship.
a. animal : dog indicator : ______
(1) pH (3) titration
(2) phenolphthalein (4) end point
b. molarity : moles normality : ______
(1) normals (3) equivalents
(2) concentration (4) ions

70. It is important for the pH of blood to be maintained in the range of 7.35 to 7.45. The hydrogen carbonate ion–carbonic acid buffer system is the most important of three buffer systems in the blood that help maintain this pH range. This system is represented by the following equation.

$$H_2O(l) + CO_2(g) \underset{\text{Lungs}}{\rightleftharpoons} H_2CO_3(aq) \underset{\text{Blood}}{\rightleftharpoons} H^+(aq) + HCO_3^-(aq)$$

Given this equation, explain how abnormal breathing patterns can lead to acid–base imbalances in the blood that are called respiratory acidosis (abnormally low pH) and respiratory alkalosis (abnormally high pH). Too-rapid and too-deep breathing, a condition called hyperventilation, leads to respiratory alkalosis. Why? Conversely, hypoventilation, the result of too-shallow breathing, can cause respiratory acidosis. Explain.

Chapter 21 REVIEW

51. No; HCl is a strong acid.
52. a. $[Ni^{2+}][S^{2-}]$
b. $[Ba^{2+}][CO_3^{2-}]$
53. the product of the ion concentrations raised to the power of their coefficients
54. c., b., d., a.
55. lowers the solubility
56. a. $NaOH + HCl \rightarrow NaCl + H_2O$
b. 0.72*N* NaOH
57. The product of the concentrations of the two ions is greater than the solubility product constant of the precipitate.
58. a. 3.2 equiv
b. 0.46 equiv
c. 0.031 equiv
59. $H_2PO_4^- + OH^- \rightleftharpoons H_2O + HPO_4^{2-}$
$HPO_4^{2-} + H^+ \rightleftharpoons H_2PO_4^-$
Added OH^- is neutralized by $H_2PO_4^-$ and added acid is neutralized by HPO_4^{2-}.
60. a. 1.2*N*
b. 0.15*N*
c. 0.80*N*
61. a. 8.73
b. phenolphthalein or thymol blue

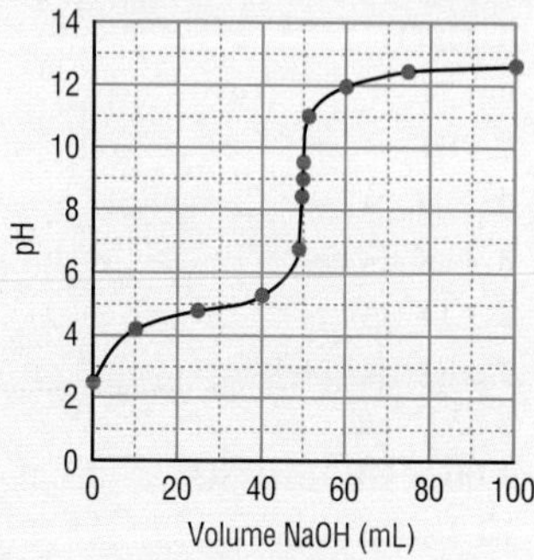

62. a. $2HCl + Mg(OH)_2 \rightarrow MgCl_2 + 2H_2O$
b. $2HCl + CaCO_3 \rightarrow H_2O + CO_2 + CaCl_2$
c. $Al(OH)_3 + 3HCl \rightarrow AlCl_3 + 3H_2O$
63. 12.5 mL
64. a. $[OH^-] = 8.4 \times 10^{-6}M$
b. $[OH^-] = 2.4 \times 10^{-2}M$
c. $[OH^-] = 5.5 \times 10^{-9}M$
65. a. shift to the left
b. shift to the right
66. $NaC_2H_3O_2 + H_2O \rightleftharpoons Na^+ + HC_2H_3O_2 + OH^-$
67. b., c., d., a.

Answers

68. $[Ba^{2+}] = 2.0 \times 10^{-8}M$
69. a. 2
 b. 3
70. Hyperventilation speeds up the removal of CO_2 from the blood. The equilibrium shifts toward CO_2, which reduces the concentration of H_2CO_3 and H^+. The result is an increase in blood pH (alkalosis). Hypoventilation allows CO_2 concentration to build up in blood. In response, the equilibrium shifts toward H_2CO_3 and H^+, causing blood pH to decrease (acidosis).
71. CO_2 concentration is higher in pure water. Less CO_2 becomes carbonate because pure water does not have the OH^- ions needed to reduce H^+ concentration.
72. a. between 6 and 8
 b. Use a pH meter.
73. a. 5.34
 b. 11.30
 c. 0.52
 d. 9.01
74. 2.25 g KCl
75. a. HSO_4^-
 b. CN^-
 c. OH^-
 d. NH_3
76. a. NaCl(*aq*)
 b. $CO_2(g)$
 c. $H_2O(l)$ at 60 °C
77. 0.5*M* NaCl
78. $1.00 \times 10^{-1}N\ H_2SO_4$
79. 74.4% $AgNO_3$
80. $HOCN + OH^- \rightleftharpoons H_2O + OCN^-$
 $OCN^- + H^+ \rightleftharpoons HOCN$

71. The following equilibria are involved in the solubility of carbon dioxide in water.

$$CO_2(g) \rightleftharpoons CO_2(aq)$$

$$CO_2(aq) + H_2O(l) \rightleftharpoons H_2CO_3(aq)$$

$$H_2CO_3(aq) \rightleftharpoons H^+(aq) + HCO_3^-(aq)$$

$$HCO_3^-(aq) \rightleftharpoons H^+(aq) + CO_3^{2-}(aq)$$

If sea water is slightly alkaline, would you expect the concentration of dissolved CO_2 to be higher or lower than in pure water? Explain.

CUMULATIVE REVIEW

72. These beakers contain the same solution. Bromocresol green has been added to the beaker on the left, and phenolphthalein has been added to the beaker on the right.

 a. What is the approximate pH of the solution?
 b. How could you determine the pH more accurately?

73. Calculate the pH of solutions with the following hydrogen-ion concentrations.
 a. $4.6 \times 10^{-6}M$
 b. $5.0 \times 10^{-12}M$
 c. $3.0 \times 10^{-1}M$
 d. $9.8 \times 10^{-10}M$
74. How many grams of potassium chloride are in 45.0 mL of a 5.00% (by mass) solution?
75. Write the formula for the conjugate base of each acid.
 a. H_2SO_4 b. HCN c. H_2O d. NH_4^+
76. In each pair, which has the higher entropy?
 a. NaCl(*s*) or NaCl(*aq*)
 b. $CO_2(s)$ or $CO_2(g)$
 c. $H_2O(l)$ at 60 °C or $H_2O(l)$ at 25 °C
77. What is the molarity of the salt in a solution that results from mixing 200 mL of 1.00*M* NaOH with 200 mL of 1.00*M* HCl?

CONCEPT CHALLENGE

78. What is the normality of an H_2SO_4 solution if 80.0 mL of the solution reacts completely with 0.424 g Na_2CO_3?

$$H_2SO_4(aq) + Na_2CO_3(aq) \longrightarrow H_2O(l) + CO_2(aq) + Na_2SO_4(aq)$$

79. An impure sample of $AgNO_3$ weighing 0.340 g was dissolved in water. After the addition of 10.0 mL of 0.200*N* HCl, 0.213 g AgCl was recovered. Calculate the percentage of $AgNO_3$ in the sample.
80. Use the cyanate buffer $HOCN/OCN^-$ to illustrate how a buffer system works. Show, by means of equations, how the pH of a solution can be kept almost constant when small amounts of acid or base are added.

Select the choice that best answers each question or completes each statement.

1. How many moles of NaOH are required to neutralize 2.4 mol H_2SO_4?
   a. 1.2 mol
   b. 2.4 mol
   c. 3.6 mol
   d. 4.8 mol
2. The net ionic equation for the neutralization reaction between solutions of potassium hydroxide and hydrochloric acid is
   a. $H^+(aq) + OH^-(aq) \longrightarrow H_2O(l)$
   b. $KOH(aq) + HCl(aq) \longrightarrow H_2O(l) + KCl(aq)$
   c. $K^+(aq) + Cl^-(aq) \longrightarrow KCl(aq)$
   d. $K^+(aq) + OH^-(aq) + H^+(aq) + Cl^-(aq) \longrightarrow KCl(aq) + H_2O(l)$
3. Calculate the molarity of an HCl solution if 25.0 mL of the solution is neutralized by 15.5 mL of 0.800*M* NaOH.
   a. 0.248*M*
   b. 0.496*M*
   c. 1.29*M*
   d. 0.645*M*
4. At 25 °C, zinc sulfide has a K_{sp} of 3.0×10^{-23}, zinc carbonate has a K_{sp} of 1.0×10^{-10}, and silver iodide has a K_{sp} of 8.3×10^{-17}. Order these salts from most soluble to least soluble.
   a. zinc carbonate, zinc sulfide, silver iodide
   b. silver iodide, zinc carbonate, zinc sulfide
   c. zinc carbonate, silver iodide, zinc sulfide
   d. zinc sulfide, silver iodide, zinc carbonate
5. Which combination of compound and ion would *not* make a useful buffer solution?
   a. ammonium ion and ammonia
   b. hydrogen carbonate ion and carbonic acid
   c. sulfate ion and sulfuric acid
   d. ethanoate ion and ethanoic acid

The lettered choices below refer to questions 6–9. A lettered choice may be used once, more than once, or not at all

(A) PQ
(B) P_2Q_3
(C) PQ_3
(D) P_3Q
(E) PQ_2

Which of the choices is the general formula for the salt formed in each of the following neutralization reactions? P is a cation; Q is an anion.

6. $HCl + Ca(OH)_2 \longrightarrow$
7. $H_3PO_4 + NaOH \longrightarrow$
8. $H_2SO_4 + Mg(OH)_2 \longrightarrow$
9. $HNO_3 + Al(OH)_3 \longrightarrow$

A student titrated three different acids with three different bases. He then used the experimental data to produce three titration curves. Use the graphs to answer questions 10–12.

a.
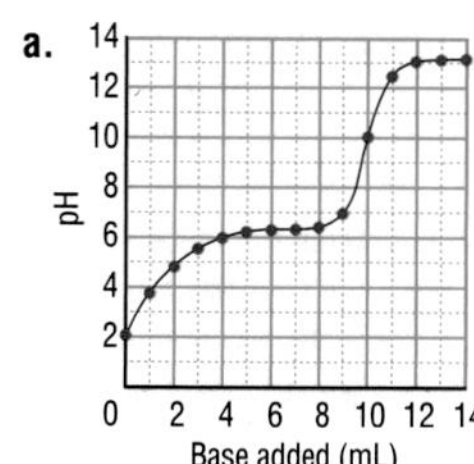

b.
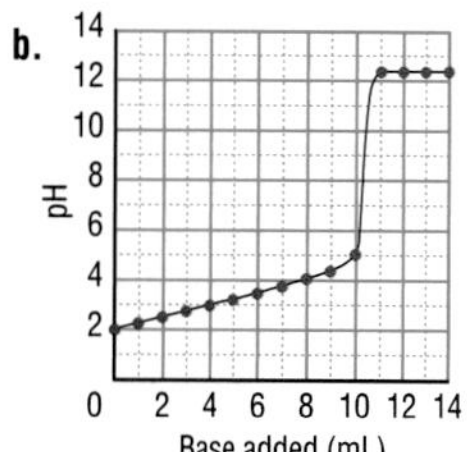

c.
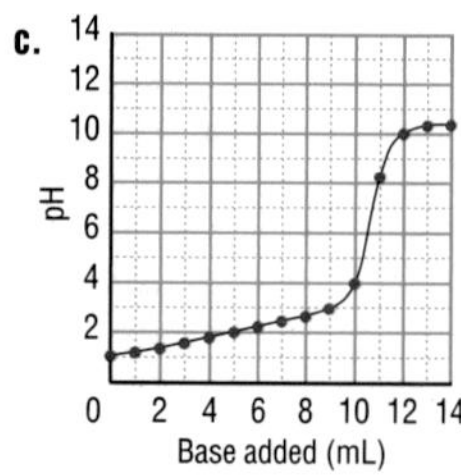

Match each graph with one acid–base pair.

10. aqueous ammonia + hydrochloric acid
11. sulfuric acid + potassium hydroxide
12. sodium hydroxide + ethanoic acid

Use the atomic windows to answer question 13.

13. Window (a) represents an aqueous solution of a weak acid, HA, and its sodium salt, Na^+A^-. For simplicity, the sodium ions (Na^+) and water molecules have been omitted. Which window shows the system after two OH^- ions are added?

a.
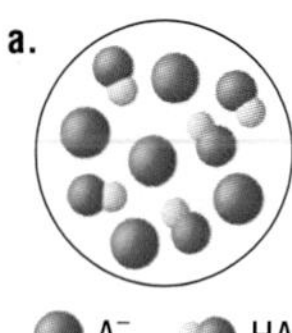

b.
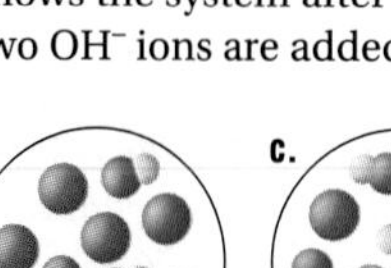

c.
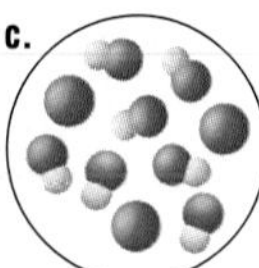

d.
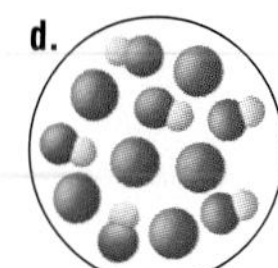

1. d
2. a
3. b
4. c
5. c
6. (E)
7. (D)
8. (A)
9. (C)
10. c
11. b
12. a
13. b

Planning Guide

SECTION OBJECTIVES	ACTIVITIES/FEATURES	MEDIA & TECHNOLOGY
22.1 The Meaning of Oxidation and Reduction ● ■ ◆ ▸ Define oxidation and reduction in terms of the loss or gain of oxygen or hydrogen and the loss or gain of electrons ▸ State the characteristics of a redox reaction, and identify the oxidizing agent and reducing agent	**SE** Discover It! *Rusting,* p. 644 **SE** **Link to Photography** *Redox in Photography,* p. 650 LM 44: *Oxidation-Reduction Reactions* SSLM 31: *Determination of an Activity Series* **TE** DEMO, p. 646	ASAP Problem Solving 1 ASAP Assessment 22.1 OT 70: *Oxidation-Reduction Reactions* CA *Elephant Toothpaste* CA *Thermite Reaction* CA *Oxidation States of Vanadium* CA *Oxidizing with Potassium Chlorate* CHM Side 8, 2: *Picture This* CHM Side 8, 10: *Space Flight* CHM Side 8, 30: *The Oxidation of Magnesium* CHM Side 8, 32: *Light Sensitivity*
22.2 Oxidation Numbers ■ ◆ ▸ Determine the oxidation number of an atom of any element in a pure substance ▸ Define oxidation and reduction in terms of a change in oxidation number, and identify atoms being oxidized or reduced in redox reactions	**TE** DEMO, p. 655	ASAP Problem Solving 11 ASAP Assessment 22.2 CA *Oxidation States of Vanadium* CA *Oxidizing with Potassium Chlorate*
22.3 Balancing Redox Equations □ ◆ ▸ Use the oxidation-number-change method to balance redox equations ▸ Break a redox equation into oxidation and reduction half-reactions, and then use the half-reaction method to balance the equation	**SE** **Link to Biology** *Bioluminescence,* p. 667 **SE** Mini Lab *Bleach It! Oxidize the Color Away,* p. 669 (LRS 22-1) PLM *Bleach It! Oxidize the Color Away* **SE** Small-Scale Lab *Half-Reactions,* p. 670 (LRS 22-2) **SE** **Chemistry Serving . . . Society** *Just Add Cold Water for a Hot Meal on the Go,* p. 671 **SE** **Chemistry in Careers** *Mechanical Engineer,* p. 671 SSLM 32: *Oxidation-Reduction Reactions* **TE** DEMO, p. 662, 665	ASAP Problem Solving 18, 20, 22 ASAP Assessment 22.3 ASAP Concept Map 22 SSV 4: *Redox Reactions* CHM Side 8, 5: *Photography and Stoichiometry* PP Chapter 22 Problems RP Lesson Plans, Resource Library AR Computer Test 22 www Activities, Self-Tests, *SCIENCE NEWS* updates TCP The Chemistry Place Web Site

KEY

- ● Conceptual (concrete concepts)
- ○ Conceptual (more abstract/math)
- ■ Standard (core content)
- □ Standard (extension topics)
- ◆ Honors (core content)
- ◇ Honors (options to accelerate)

AR	Assessment Resources
ASAP	Chem ASAP! CD-ROM
ACT	ActivChemistry CD-ROM
CHM	CHEMedia Videodiscs
CA	Chemistry Alive! Videodiscs
GCP	Graphing Calculator Problems
GRS	Guided Reading and Study Workbook
LM	Laboratory Manual
LP	Laboratory Practicals
LRS	Laboratory Recordsheets
SSLM	Small-Scale Lab Manual

PRACTICE	ASSESSMENT
SE **Sample Problem** 22-1 **GRS** Section 22.1 **RM** Practice Problems 22.1	**SE** Section Review **RM** Section Review 22.1 **RM** Chapter 22 Quiz **LP** Lab Practical 22-1
SE **Sample Problems** 22-2 to 22-4 **GRS** Section 22.2 **RM** Practice Problems 22.2	**SE** Section Review **RM** Section Review 22.2 **RM** Chapter 22 Quiz
SE **Sample Problems** 22-5 to 22-7 **GRS** Section 22.3 **RM** Practice Problems 22.3 **RM** Interpreting Graphics	**SE** Section Review **RM** Section Review 22.3 **RM** Vocabulary Review 22 **SE** Chapter Review **SM** Chapter 22 Solutions **SE** Standardized Test Prep **PHAS** Chapter 22 Test Prep **RM** Chapter 22 Quiz **RM** Chapter 22 A & B Test

OT	Overhead Transparency	**SE**	Student Edition
PHAS	PH Assessment System	**SM**	Solutions Manual
PLM	Probeware Lab Manual	**SSV**	Small-Scale Video/Videodisc
PP	Problem Pro CD-ROM	**TCP**	www.chemplace.com
RM	Review Module	**TE**	Teacher's Edition
RP	Resource Pro CD-ROM	**www**	www.phschool.com

PLANNING FOR ACTIVITIES

STUDENT EDITION

Discover It! p. 644
- iron finishing nails
- pliers
- scissors
- copper wire
- zinc strips
- fine sandpaper
- plastic wrap
- saucers
- water
- table salt
- petroleum jelly
- paper towels

Mini Lab, p. 669
- spot plates
- medicine droppers
- liquid and powder bleach, 3% H_2O_2, 1% oxalic acid, 0.2*M* $Na_2S_2O_3$
- I_2 in KI(*aq*), $KMnO_4$(*aq*), grape juice, rusty water, colored fabric, grass-stained fabric, flower petals

Small-Scale Lab, p. 670
- pencils, paper, rulers
- reaction surfaces
- 1.0*M* HCl, HNO_3, H_2SO_4
- pieces of Zn, Mg, Cu, Fe

TEACHER'S EDITION

Teacher Demo, p. 646
- 3 g copper(II) oxide (CuO)
- 3 g charcoal (C)
- crucible with cover
- Bunsen burner
- tongs
- glass plate

Teacher Demo, p. 655
- 100 mL of 0.1*M* lead(II) acetate [$Pb(CH_3COO)_2$]
- 150-mL beaker
- 1-cm × 3-cm zinc strip
- sodium sulfide
- 3*M* NaOH
- 1*M* iron(III) chloride
- stirring rod
- plastic container

Activity, p. 657
- 3 small, glass test tubes
- zinc, copper, magnesium
- 0.1*M* HCl
- medicine dropper or pipet

Teacher Demo, p. 662
- 6 crystals of potassium permanganate ($KMnO_4$)
- 540 mL of water
- large beaker
- 1 g sodium hydrogen sulfite ($NaHSO_8$)
- two 50-mL beakers
- 1 g barium chloride dihydrate ($BaCl_2{\cdot}2H_2O$)
- 3*M* H_2SO_4
- stirring rod
- plastic container

Teacher Demo, p. 665
- two 600-mL beakers
- 8 mL of 6*M* H_2SO_4
- distilled water
- 43 g of sodium pyrophosphate decahydrate ($NaP_2O_7{\cdot}10H_2O$)
- 72 mL 6*M* H_2SO_4
- 80 mL of 0.1*M* $MnSO_4$
- 40 mL of 0.25*M* potassium bromate ($KBrO_3$)
- stir bar
- magnetic stirrer
- stirring rod
- watch or clock

Activity, p. 667
- potatoes and apples
- beakers
- water
- boiled water
- lemon juice
- carbonated beverage
- sugar, salt, and vinegar solutions
- paper towels

Chapter 22

Key Terms

22.1 oxidation, reduction, oxidation-reduction reactions, redox reactions, reducing agent, oxidizing agent
22.2 oxidation number
22.3 oxidation-number-change method, half-reactions, half-reaction method

DISCOVER IT!

Students should observe that nails 1, 2, 5, and 6 show evidence of rusting whereas 3, 4, and 7 do not. Students should suggest a relationship between evidence of rusting and the presence of water, air, salt, and more-or-less active metals.

Chapter 22 OXIDATION-REDUCTION REACTIONS

Salt water increased the rate at which this ship's hull rusted.

FEATURES

DISCOVER IT! RUSTING

You need seven iron finishing nails (approximately 6 cm long), pliers, scissors, copper wire, zinc strip, fine sandpaper, plastic wrap, a saucer, water, table salt, petroleum jelly, and paper towels.

1. Use sandpaper to polish seven nails. Wipe them clean with a paper towel.
2. Place two wet paper towels on the saucer.
3. Nail 1: Using pliers, bend into a U shape. Nail 2: Wrap one end with copper wire. Nail 3: Wrap one end with a strip of zinc. Nail 4: Cover the entire nail with a thin coat of petroleum jelly. Nail 5: Moisten with water and sprinkle with salt. Nail 6: Leave untreated. Nail 7: Leave untreated.
4. Place nails 1 through 6 on the wet paper towel. Make sure the nails do not touch. Cover them with a piece of plastic wrap. Place nail 7 (the control) on top of the plastic wrap. During a 24-hour period, record your observations in a table.

Stay current with SCIENCE NEWS
Find out more about redox reactions:
www.phschool.com

22.1 ● ■ ◆
22.2 ■ ◆
22.3 ○ □ ◆

Conceptual Use Table 22.1 to summarize the processes that can occur in redox reactions. Introduce half reactions in Section 22.3 to prepare students for Chapter 23. Balancing redox reactions may be omitted.

Standard It may help to review the metals with more than one ionic charge (Table 6.3) and common polyatomic ions (Table 6.4). Allow students to refer to the rules for assigning oxidation numbers as they work through the sample problems in this chapter; they will more readily learn the rules through repeated practice. Even if students are not asked to balance redox equations, they should learn about half reactions as a prerequisite for Chapter 23.

Honors Sample Problem 22-4 can be assigned as homework if students are successful with the preceding problems. Use the Small-Scale Lab to reteach half reactions and prepare students for Chapter 23.

THE MEANING OF OXIDATION AND REDUCTION

objectives
- Define oxidation and reduction in terms of the loss or gain of oxygen or hydrogen and the loss or gain of electrons
- State the characteristics of a redox reaction, and identify the oxidizing agent and reducing agent

key terms
- oxidation
- reduction
- oxidation–reduction reactions
- redox reactions
- reducing agent
- oxidizing agent

During winter in cold climates, salt is often used on roads to prevent the buildup of slippery ice. Although salt may make driving safer, it can cause the metallic parts of cars to corrode or rust relatively quickly. This problem is often so severe that people will not drive their newer cars in winter because they fear that their car will end up looking as rusty as this ship. **What property causes metal to corrode?**

Oxygen in Redox Reactions

The combustion of gasoline in an automobile engine and the burning of wood in a fireplace produce energy. So does the metabolism of food by your body. Such reactions are among the principal sources of energy on Earth, and all involve a process called oxidation. What do you think of when you hear that term? If you answered "oxygen," you are thinking along the same lines as most early chemists.

Oxidation originally meant the combination of an element with oxygen to produce oxides. As you will soon learn, the term also has a more modern, and wider, meaning. When gasoline or wood burns in air, it oxidizes and produces carbon dioxide. So does coal, as shown in **Figure 22.1.** Methane (CH_4), a component of natural gas, also burns in air. Methane oxidizes to form oxides of carbon and hydrogen.

Not all oxidation processes involve burning, however. Bleaching is an example of oxidation that does not involve burning. Bleaches are substances that remove stains or unwanted color from fabrics and other

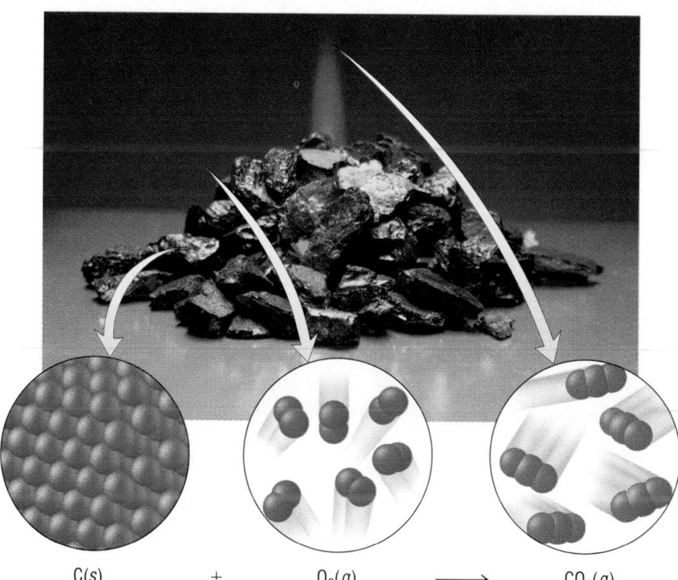

Figure 22.1
When coal, which is mostly carbon, is burned in air, carbon dioxide and heat are produced.

22.1

1 Engage

Use the Visual

Have students examine the section-opening photograph and read the text that accompanies it. Explain that metals corrode because most metals easily lose electrons to electron acceptors, such as oxygen or water. Ask students to conjecture how the presence of salt can speed up the process of corrosion based on an electron transfer mechanism. Ask students how cars might be protected from corrosion. (with a protective undercoat or plastic parts)

Check Prior Knowledge

To assess students' prior knowledge about electron transfer, ask:

- **What happens to magnesium and oxygen when they react to form magnesium oxide, MgO?** (Mg loses electrons to form Mg^{2+} and oxygen gains electrons to form O^{2-}.)
- **Compare the electron density around the carbon atoms in CH_4 and CO_2.** (Because oxygen is more electronegative than hydrogen, electron density around the carbon atom in CO_2 is less than in CH_4.)
- **Why does iron corrode easily while gold does not?** (Fe readily donates electrons; Au does not.)
- **If H_2O is converted to H_2O_2, has H_2O been oxidized?** (Yes, because oxygen has been added to form H_2O_2.)

STUDENT RESOURCES

From the Teacher's Resource Package, use:
- Section Review 22.1, Ch. 22 Practice Problems and Quizzes from the Review Module
- Laboratory Manual: Experiment 44
- Laboratory Practical 22-1
- Small-Scale Chemistry Lab Manual: Experiment 31

TECHNOLOGY RESOURCES

Relevant technology resources include:
- Chem ASAP! CD-ROM
- Resource Pro CD-ROM
- Chemistry Alive! Videodisc: *Elephant Toothpaste, Thermite Reaction, Oxidation States of Vanadium, Oxidizing with Potassium Chlorate*

2 Teach

TEACHER DEMO

The following demonstration will show students how pure metals are obtained from their oxide ores by reduction (loss of oxygen). Mix 3 g copper(II) oxide (CuO) and 3 g charcoal (C) in a crucible. Cover the crucible. Then heat strongly and evenly over a Bunsen burner for 10 minutes. Cool the crucible and remove the copper with tongs. Pass the copper around on a glass plate for the students to examine. Pass around a bit of the original CuO as well so that the students can see the change in appearance. Write the reaction that occurred on the chalkboard:

$2CuO(s) + C(s) \rightarrow 2Cu(s) + CO_2(g)$

Ask the students to identify the species that were oxidized and reduced based on loss or gain of oxygen.

Think Critically

Carbon dioxide is the carbon oxide that results from the *complete* oxidation of methane as shown in Figure 22.2. Based on what the students have learned, ask the students to infer formulas for *incompletely* oxidized compounds of methane that have plausible structural formulas. [CO (carbon monoxide), CH_2O_2 (formic acid), CH_2O (formaldehyde), CH_4O (methanol) are all possibilities.]

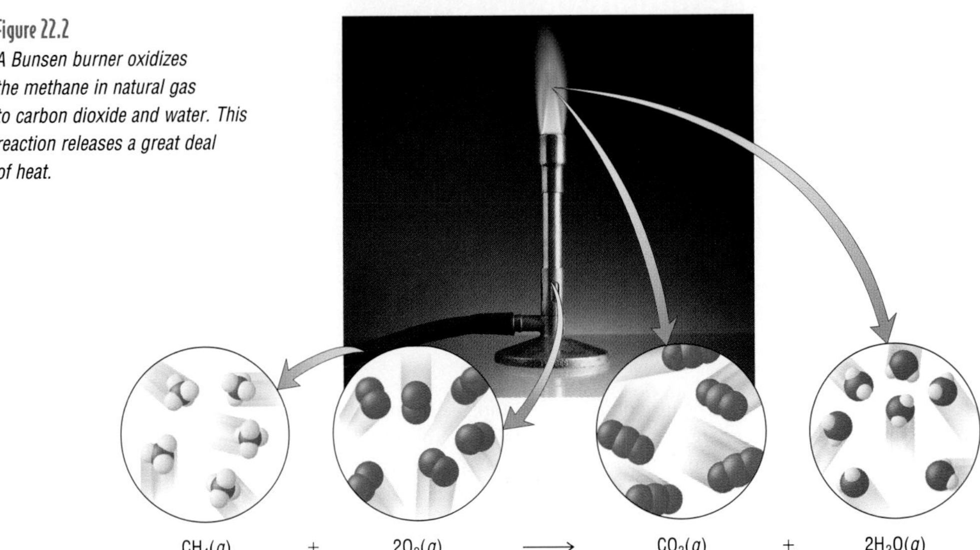

Figure 22.2
A Bunsen burner oxidizes the methane in natural gas to carbon dioxide and water. This reaction releases a great deal of heat.

materials. Common liquid household bleaches contain sodium hypochlorite (NaClO). Powder bleaches may contain calcium hypochlorite ($Ca(ClO)_2$) or sodium perborate ($NaBO_3$).

Hydrogen peroxide (H_2O_2) is another good oxidizing agent. Common household peroxide is both a bleach and a mild antiseptic that kills bacteria by oxidizing them. Notice that, in keeping with the original definition of oxidation, all these substances have one thing in common: They all contain oxygen.

Another familiar example of an oxidation process that does not involve burning is rusting. When elemental iron turns to rust, it slowly oxidizes to compounds such as iron(III) oxide (Fe_2O_3).

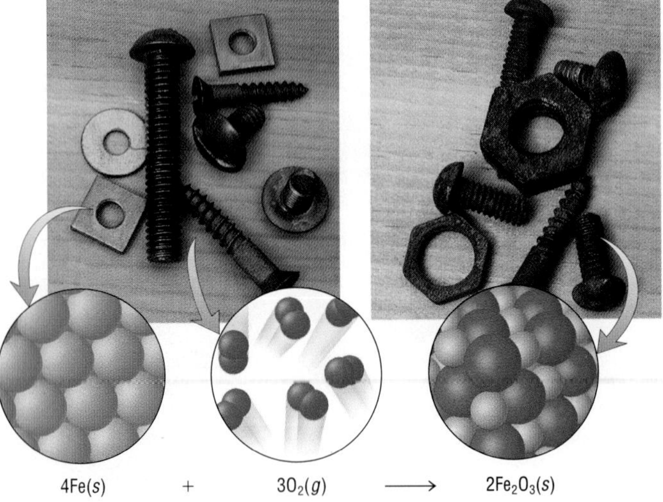

Figure 22.3
When items made of iron are exposed to moist air, the Fe atoms react with O_2 molecules. The iron rusts; it is oxidized to compounds such as iron(III) oxide (Fe_2O_3).

MEETING DIVERSE NEEDS

Gifted Have advanced students do research on redox reactions involving oxygen that are important in nature. Examples are the reduction of carbon in carbon dioxide to form carbohydrates during photosynthesis and the oxidation of glucose during cell respiration to release energy. Have students prepare a report or poster based on their research.

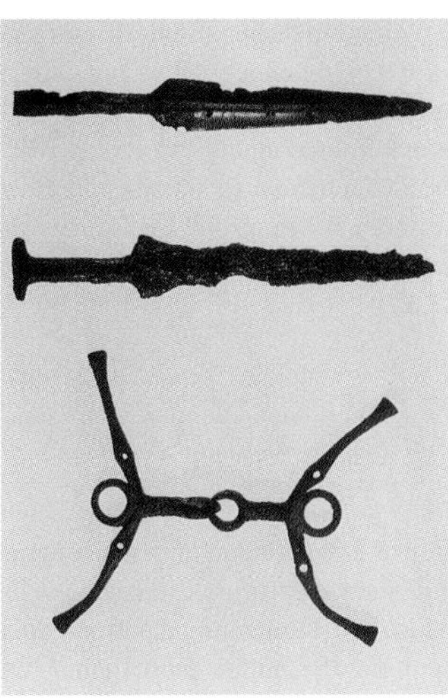

Figure 22.4
These iron objects (left) were made in ancient times. The iron was obtained by reduction of iron ore with charcoal. A similar process is carried out in a modern-day blast furnace (right).

A process called reduction is the opposite of oxidation. Originally, **reduction** meant the loss of oxygen from a compound. The reduction of iron ore to metallic iron involves the removal of oxygen from iron(III) oxide. The reduction is accomplished by heating the ore with charcoal. A large decrease in volume occurs during the reduction of a metal oxide to metal. Can you explain how reduction got its name? The equation for the reduction of iron ore is 1

$$\underset{\text{Iron(III) oxide}}{2Fe_2O_3(s)} + \underset{\text{Carbon}}{3C(s)} \longrightarrow \underset{\text{Iron}}{4Fe(s)} + \underset{\text{Carbon dioxide}}{3CO_2(g)}$$

The artifacts in **Figure 22.4** show that ancient people reduced iron ore to iron in the early Iron Age, more than 2500 years ago!

The equation for the reduction of iron also includes an oxidation process. Oxidation and reduction always occur simultaneously. As iron oxide is reduced to iron by losing oxygen, carbon is oxidized to carbon dioxide by gaining oxygen. No oxidation occurs without reduction, and no reduction occurs without oxidation. Reactions that involve these processes are therefore called **oxidation–reduction reactions.** Oxidation–reduction reactions are also known as **redox reactions.**

Electron Transfer in Redox Reactions

Today, the concepts of oxidation and reduction have been extended to include many reactions that do not even involve oxygen. Redox reactions are understood to involve a shift of electrons between reactants. **Oxidation** is redefined to mean complete or partial loss of electrons or gain of oxygen. **Reduction** is complete or partial gain of electrons or loss of oxygen.

Oxidation	Reduction
Loss of electrons	Gain of electrons
Gain of oxygen	Loss of oxygen

CHEMISTRY AND SCIENCE HISTORY

Reduction processes were used to produce some metals more than 5000 years ago. Copper oxide and sulfide ores were roasted with charcoal to form copper, carbon dioxide, and sulfur dioxide. Large amounts of copper ore were mined on the island of Cyprus, off the coast of the area now known as Turkey. The Romans, who eventually imported this valuable ore, called it *aes cyprium,* or "ore of Cyprus." The name was later shortened to *cyprium,* and then corrupted to *cuprum.* The English word "copper" derives from this corruption of the Latin word for Cyprus.

Discuss

Ask students to recall the electron structure of an oxygen atom and how oxygen atoms bond with other atoms. Be sure they identify the source of the electrons an oxygen atom gains during bonding. Point out that when a metal combines with oxygen, it loses electrons, and when oxygen is removed from the oxide of a metal, the metal gains electrons. Explain to students that this knowledge led to a broader definition of oxidation and reduction as an exchange of electrons.

Answer

1 The volume of material is greatly reduced when ore is converted to metal.

Chemistry Alive!

Elephant Toothpaste Play

Think Critically

People with anemia (low blood iron) **usually benefit from the cooking of acidic sauces, such as tomato sauce, in cast iron skillets or pots. Ask the students to conjecture why this is true based on what they have learned thus far. Iron metal can be oxidized directly to soluble, readily-ingested iron cations by acid in the sauce:**

$2Fe(s) + 6H^+(aq) \rightarrow 2Fe^{3+}(aq) + 3H_2(g)$

ACTIVITY

Show students how to construct Born-Haber cycles for the redox reactions of sodium and magnesium with oxygen by finding tabulated energy values for the heats of formation of MgO and Na_2O; heats of vaporization of Mg(*s*) and Na(*s*); heat of forming O atoms from O_2; ionization energies of gaseous Mg and Na; and the electron affinity of O. From their diagrams, the students should be able to determine the lattice energies of MgO and Na_2O. Large lattice energies indicate that these redox reactions are highly exothermic.

"*LEO* the lion goes *GER*" may help you remember the definitions of oxidation and reduction. *LEO* stands for *L*osing *E*lectrons is *O*xidation; *GER* stands for *G*aining *E*lectrons is *R*eduction.

Some examples will illustrate these redefined concepts. Consider reactions between a metal and a nonmetal. Electrons are transferred from atoms of the metal to atoms of the nonmetal. For example, the ionic compound magnesium sulfide is produced when magnesium metal is heated with the nonmetal sulfur. See **Figure 22.5.**

$$\text{Mg} + \text{S} \longrightarrow \text{Mg}^{2+} + \text{S}^{2-}$$

Magnesium atom | Sulfur atom | Magnesium ion | Sulfide ion

The result of this reaction is the transfer of two electrons from a magnesium atom to a sulfur atom. Because it loses electrons, the magnesium atom is said to be oxidized to a magnesium ion. How many electrons does
1 the magnesium atom lose? Simultaneously, the sulfur atom gains two

Figure 22.5
When magnesium and sulfur are heated together, they undergo an oxidation–reduction reaction to form magnesium sulfide. The magnesium atoms become more stable by the loss of electrons (oxidation). The sulfur atoms become more stable by the gain of electrons (reduction).

Answer

1 two

Chemistry Alive!

Thermite Reaction Play

electrons and is reduced to a sulfide ion. The overall process is represented as the two component processes below.

Oxidation: $\cdot Mg \cdot \longrightarrow Mg^{2+} + 2e^-$ (loss of electrons)

Reduction: $\cdot\ddot{S}: + 2e^- \longrightarrow :\ddot{S}:^{2-}$ (gain of electrons)

The substance in a redox reaction that loses electrons is called the **reducing agent.** By losing electrons to sulfur, magnesium reduces the sulfur. Magnesium is thus the reducing agent. The substance in a redox reaction that accepts electrons is called the **oxidizing agent.** By accepting electrons from magnesium, sulfur oxidizes the magnesium. Sulfur is thus the oxidizing agent.

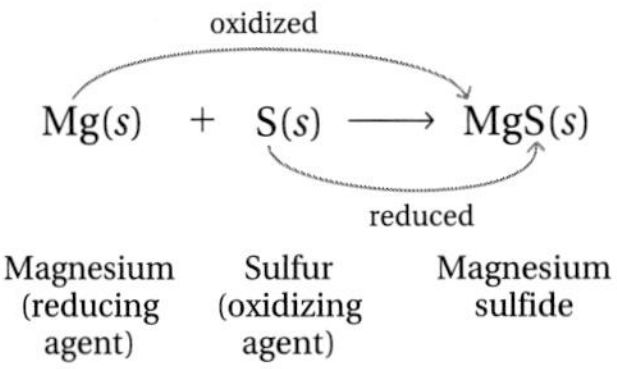

Sample Problem 22-1

What is oxidized and what is reduced in this single-replacement reaction? What is the oxidizing agent? The reducing agent?

$$2AgNO_3(aq) + Cu(s) \longrightarrow Cu(NO_3)_2(aq) + 2Ag(s)$$

1. ***ANALYZE*** ***Plan a problem-solving strategy.***
   Begin by rewriting the equation and showing the ions. Determine what loses electrons (is oxidized) and what gains electrons (is reduced).

2. ***SOLVE*** ***Apply the problem-solving strategy.***
   The rewritten equation showing the ions is:
   $$2Ag^+ + 2NO_3^- + Cu \longrightarrow Cu^{2+} + 2NO_3^- + 2Ag$$
   In this reaction, two electrons have been lost from a copper atom (Cu) because it becomes a Cu^{2+} ion. These electrons are gained by two silver ions (Ag^+), which become neutral silver atoms.
   Oxidation: $Cu \longrightarrow Cu^{2+} + 2e^-$ (loss of electrons)
   Reduction: $2Ag^+ + 2e^- \longrightarrow 2Ag$ (gain of electrons)
   The Cu is oxidized and is therefore the reducing agent. The Ag^+ is reduced and is therefore the oxidizing agent.

3. ***EVALUATE*** ***Does the result make sense?***
   Ag^+, the oxidizing agent, gained electrons and is thus reduced. Cu, the reducing agent, lost electrons and is thus oxidized. The definitions of oxidation and reduction have been correctly applied.

Practice Problems

1. Determine what is oxidized and what is reduced in each reaction. Identify the oxidizing agent and reducing agent in each case.
   a. $2Na(s) + S(s) \longrightarrow Na_2S(s)$
   b. $4Al(s) + 3O_2(g) \longrightarrow 2Al_2O_3(s)$
2. Identify these processes as either oxidation or reduction.
   a. $Li \longrightarrow Li^+ + e^-$
   b. $2I^- \longrightarrow I_2 + 2e^-$
   c. $Zn^{2+} + 2e^- \longrightarrow Zn$
   d. $Br_2 + 2e^- \longrightarrow 2Br^-$

Chem ASAP!

Problem-Solving 1

Solve Problem 1 with the help of an interactive guided tutorial.

22.1

Cooperative Learning

Separate students into groups of two or three. Have one student in each group select an active metal and a nonmetal from the periodic table. Have this student ask the other student(s) to identify the ionic compound that forms when electron transfer occurs between the metal and nonmetal. Have students write the equation for the reaction and identify the oxidizing and reducing agents.

Think Critically

Ask students why it is misleading to label the burning of methane as an oxidation reaction. (Oxidation cannot take place without a parallel reduction reaction. When methane is oxidized to form carbon dioxide, molecular oxygen is reduced to form water.)

Practice Problems Plus

Additional Practice Problem

The Chem ASAP! CD-ROM contains this problem related to Sample Problem 22-1: Determine what is oxidized and what is reduced in the following reaction. Also identify the oxidizing agent and reducing agent.

$H_2(g) + N_2(g) \rightarrow NH_3(g)$

(H is oxidized and is the reducing agent; N is reduced and is the oxidizing agent.)

Use the Visual

Have the students examine the photograph in Figure 22.6 and read the caption. On the chalkboard, write the balanced redox reaction for hydrogen burning in oxygen. Identify the reducing and oxidizing agents; explain the identities on the basis of partial loss or gain of electrons. State that another gas, acetylene (C_2H_2) is also used in torches because it produces a large amount of heat when it burns. Work with the students to write the balanced chemical equation for this redox reaction. [$2C_2H_2(g) + 5O_2(g) \rightarrow 4CO_2(g) + 2H_2O(g)$] Have the students identify the oxidizing and reducing agents; ask them to explain their answers.

Discuss

Have students read the link to photography section. Review the relationship between object colors and the amount of reflected light that enters the aperture. Explain how the diameter of the aperture and shutter speed also control the amount of light that reaches the film.

Chemistry Alive!

Oxidation States of Vanadium — Play

Oxidizing with Potassium Chlorate — Play

LINK TO PHOTOGRAPHY

Redox in Photography

Black-and-white photography involves oxidation–reduction reactions. Exposing black-and-white film to light activates very fine grains of silver bromide in the film. The film is developed by placing it in a developing solution that is actually a reducing agent. The developer, usually an organic chemical such as hydroquinone ($C_6H_4(OH)_2$), reduces the activated silver bromide to finely divided, black metallic silver. Any silver bromide that remains unactivated is removed from the film by using a solvent called a fixer. Sodium thiosulfate, commonly called hypo, is used for this purpose. The areas of the film exposed to the most light appear darkest because they have the highest concentration of metallic silver. The reversed image, called a negative, is used to produce a positive print, the black-and-white photograph, of the image.

It is easy to identify complete transfers of electrons in ionic reactions such as those just examined. But what about reactions that produce covalent compounds; that is, compounds in which complete electron transfer does not occur? Consider the reaction of hydrogen and oxygen.

$$2H_2(g) + O_2(g) \longrightarrow 2H_2O(l)$$

According to the older definition of oxidation as a combination with oxygen, it is clear that the hydrogen is oxidized to water. The newer definition involving electron transfer will also provide the same answer. Consider what happens to the bonding electrons in the formation of a water molecule. The bonding electrons in each reactant hydrogen molecule are shared equally between the hydrogen atoms. In water, however, the bonding electrons are pulled toward oxygen because it is much more electronegative than hydrogen. The result is a shift of bonding electrons away from hydrogen, even though there is not a complete transfer.

H—H electrons shared equally — H—O (with H bonded below) shift of bonding electrons away from hydrogen

Hydrogen is oxidized because it undergoes a partial loss of electrons. Thus the old gain-of-oxygen and the newer loss-of-electron definitions of oxidation agree that it is the hydrogen that is oxidized when the water forms.

What about oxygen, the other reactant? The bonding electrons are shared equally between oxygen atoms in the reactant oxygen molecule, but there is a shift of electrons toward oxygen in water. Oxygen is thus reduced because it undergoes a partial gain of electrons.

O—O electrons shared equally — H—O (with H bonded below) shift of bonding electrons toward oxygen

In every redox reaction, including those that produce covalent products, there is an oxidizing agent and a reducing agent. In the reaction of hydrogen and oxygen to produce water, hydrogen is the reducing agent because it is oxidized. Oxygen is the oxidizing agent because it is reduced. This redox reaction is highly exothermic—that is, it releases a great deal of energy, as shown in **Figure 22.6.**

Figure 22.6

This artist uses an oxyhydrogen torch to cut and weld steel to make a sculpture. When hydrogen burns in oxygen, the redox reaction generates temperatures of about 2600 °C.

In some reactions involving covalent products, the partial electron shifts are less obvious. Some general guidelines are thus helpful. For example, for carbon compounds, the addition of oxygen or the removal of hydrogen is always oxidation. **Table 22.1** lists processes that constitute oxidation and reduction. The last entry in the table identifies oxidation numbers, which are another way to describe oxidation and reduction. You will learn more about oxidation numbers in Section 22.2.

Table 22.1

Processes Leading to Oxidation and Reduction	
Oxidation	**Reduction**
Complete loss of electrons (ionic reactions)	Complete gain of electrons (ionic reactions)
Shift of electrons *away* from an atom in a covalent bond	Shift of electrons *toward* an atom in a covalent bond
Gain of oxygen	Loss of oxygen
Loss of hydrogen by a covalent compound	Gain of hydrogen by a covalent compound
Increase in oxidation number	Decrease in oxidation number

Corrosion

Billions of dollars are spent yearly to prevent and to repair damage caused by the corrosion of metals. Iron, a common construction metal often used in the form of the alloy steel, corrodes by being oxidized to ions of iron by oxygen. Water in the environment accelerates the rate of corrosion. Oxygen, the oxidizing agent, is reduced to oxide ions (in compounds such as Fe_2O_3) or to hydroxide ions. The following equations describe the corrosion of iron to iron hydroxides at moist conditions.

$$2Fe(s) + O_2(g) + 2H_2O(l) \longrightarrow 2Fe(OH)_2(s)$$
$$4Fe(OH)_2(s) + O_2(g) + 2H_2O(l) \longrightarrow 4Fe(OH)_3(s)$$

Corrosion occurs more rapidly in the presence of salts and acids. These substances produce conducting solutions that make electron transfer easier. The corrosion of some metals can be a desirable feature, as **Figure 22.7** shows.

Not all metals corrode easily. Gold and platinum are called noble metals because they are very resistant to losing their electrons by corrosion. Other metals lose electrons easily but are protected from extensive corrosion by the oxide coating formed on their surface. For example, aluminum oxidizes quickly in air to form a coating of very tightly packed aluminum oxide particles. This coating protects the aluminum object from further corrosion, as

Figure 22.7
Oxidation–reduction reactions cause corrosion. The copper on this roof reacted with water vapor, carbon dioxide, and other substances in the air to form a patina. This patina consists of a pale-green film of basic copper(II) carbonate. Because patinas enhance the surface appearance of objects, they are valued by architects and artists.

22.1

Build Writing Skills

Have students summarize in their own words the first four processes that constitute oxidation and reduction as described in Table 22.1.

Discuss

Give examples of redox reactions and identify oxidizing and reducing agents based on the processes listed in Table 22.1. Emphasize that, for a given reaction, one or more of the processes may be in effect. Tell students that the issue of oxidation numbers will be addressed in Section 22.2.

ACTIVITY

Have students try to prepare a list of metals ordered by their ease of corrosion. Have students base their lists on known behaviors of specific metals. Suggest that students consider how different metals are used as a clue to their behavior. After students compile their lists, refer them to the activity series for metals on p. 217. Have students write a paragraph discussing how closely their lists correspond to the actual activity series. Ask students to offer possible reasons for any discrepancies. (Reasons might include unavailability of data for some metals or difficulty in rating metals whose behaviors and uses are similar.)

Discuss

Lead a class discussion on metal corrosion. Ask students to identify the common chemical characteristic of all metal corrosion. (the transfer of metal electrons to oxidizing agents) **Have students cite examples of desirable and undesirable metal corrosion. Ask students to suggest ways in which the rate of corrosion could be decreased or increased.**

Think Critically

Tell students that the steel cans used for commercial canning are coated with a thin layer of tin. Ask them to infer the reason for this procedure, and to predict whether or not aluminum cans have a similar lining. Have students explain their reasoning.

Water
Aluminum oxide
Oxygen
Aluminum

(a)

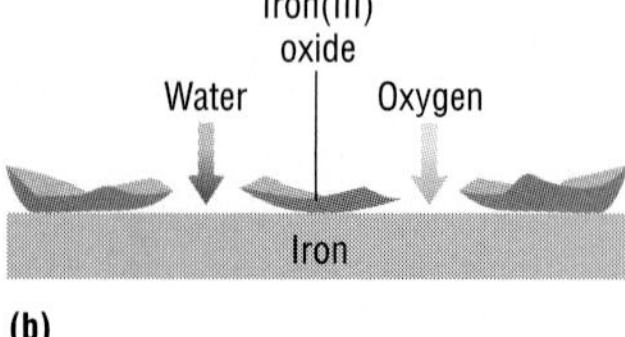

(b)

Figure 22.8
*Oxidation causes the complete corrosion of some metals. Aluminum, however, resists such corrosion because it forms a protective coating of aluminum oxide. How does the aluminum oxide on aluminum **(a)** differ from the iron(III) oxide formed on corroding iron **(b)**?* ❶

shown in **Figure 22.8.** Iron also forms a coating when it corrodes, but the coating of iron oxide that forms is not tightly packed. Water in the air can penetrate the coating and attack the iron metal below it. The corrosion continues until the iron object becomes only a pile of rust.

The corrosion of objects such as shovels or knives is a common problem but not usually a serious one. In contrast, the corrosion of a steel support pillar of a bridge or the hull of an oil tanker is much more serious and costly to repair! To prevent corrosion in such cases, the metal surface may be coated with oil, paint, plastic, or another metal, as shown in **Figure 22.9.** These coatings exclude air and water from the surface, thus preventing corrosion. If the coating is scratched or worn away, however, the exposed metal will begin to corrode.

In another method of corrosion control, one metal is used to save a second metal. For example, to protect an iron object, a piece of magnesium may be placed in electrical contact with the iron. When oxygen and water attack the iron object, the iron atoms lose electrons as the iron begins to be oxidized. However, because magnesium is a better reducing agent than iron and is more easily oxidized, the magnesium immediately transfers electrons to the iron ions, reducing them back to neutral iron atoms.

Figure 22.9
Painting a surface (left) protects it from the effects of the environment. Chromium metal also serves as a protective coating and imparts an attractive, mirrorlike finish (right). Like aluminum, chromium forms a corrosion-resistant oxide film on its surface.

Answer

❶ Unlike the iron oxide, aluminum oxide forms a tough coating on the surface of the metal. The coating protects the aluminum from further oxidation.

SECTION REVIEW 22.1

3. a–d. Oxidation: loss of electrons, loss of hydrogen, gain of oxygen, electrons shift away from atom; reduction: gain of electrons, gain of hydrogen, loss of oxygen, electrons shift toward atom.

4. In redox reactions, there is a partial or complete transfer of electrons between a reducing agent, which is oxidized, and an oxidizing agent, which is reduced.

5. a., c. oxidizing agent; **b.** reducing agent

Sacrificial zinc and magnesium blocks are sometimes attached to piers and ship hulls to prevent corrosion damage in areas submerged in water, as shown in **Figure 22.10.** Underground pipelines and storage tanks may be connected to magnesium blocks for protection. Obviously, it is easier and cheaper to replace a block of magnesium or zinc than to replace a bridge or a pipeline.

Figure 22.10
Zinc blocks, the white strips in the photograph, are attached to the steel hull of this ship to protect the steel from corrosion. The zinc blocks oxidize (corrode) and release electrons. The steel hull consumes the electrons supplied by the zinc, preventing it from corroding.

section review 22.1

3. Define oxidation and reduction in terms of
- **a.** gain or loss of electrons.
- **b.** gain or loss of hydrogen (in a covalent bond).
- **c.** gain or loss of oxygen.
- **d.** shift of electrons (in a covalent bond).

4. State the characteristics of a redox reaction, and explain how to identify the oxidizing agent and the reducing agent.

5. Which of the following would most likely be oxidizing agents and which would most likely be reducing agents? (*Hint:* Think in terms of tendencies to lose or gain electrons.)

a. Cl_2 **b.** K **c.** Ag^+

6. Refer to the electronegativity values in **Table 14.2** on page 405 to determine which reactant is oxidized and which reactant is reduced in each reaction. Also determine which reactant is the reducing agent and which is the oxidizing agent.
- **a.** $H_2(g) + Cl_2(g) \longrightarrow 2HCl(g)$
- **b.** $S(s) + Cl_2(g) \longrightarrow SCl_2(g)$
- **c.** $N_2(g) + 2O_2(g) \longrightarrow 2NO_2(g)$
- **d.** $2Li(s) + F_2(g) \longrightarrow 2LiF(s)$
- **e.** $H_2(g) + S(s) \longrightarrow H_2S(g)$

7. Use electron transfer or electron shift to identify what is oxidized and what is reduced in each reaction. Make use of electronegativity values, as needed, for molecular compounds.
- **a.** $2Na(s) + Br_2(l) \longrightarrow 2NaBr(s)$
- **b.** $N_2(g) + 3H_2(g) \longrightarrow 2NH_3(g)$
- **c.** $S(s) + O_2(g) \longrightarrow SO_2(g)$
- **d.** $Mg(s) + Cu(NO_3)_2(aq) \longrightarrow Mg(NO_3)_2(aq) + Cu(s)$

8. Why would a metal corrode more quickly in salt water than in distilled water?

Chem ASAP! **Assessment 22.1** Check your understanding of the important ideas and concepts in Section 22.1.

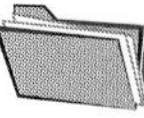

portfolio project

Shipwrecks of Spanish galleons often contain gold or silver treasures. Why is the recovered gold hardly changed while the silver has turned black? Research and write a report on how the thick layers of tarnish are removed from silver artifacts.

3 Assess

Evaluate Understanding

On the chalkboard, write several equations for redox reactions. Ask students to identify the oxidizing and reducing agents. Make sure to assess the logic students use to arrive at their answers.

Reteach

To reinforce the dual nature of redox reactions, compare redox reactions to acid–base reactions. Explain that the transfer of electrons from a reducing agent to an oxidizing agent is analogous to the transfer of protons from acids to bases. In neither case, can the donor act independent of an acceptor.

Portfolio Project

Silver is more easily oxidized than is gold. Silver tarnish is mainly silver sulfide. Tarnish can be removed if the silver is placed in a dilute NaOH bath and a weak current is passed through the solution.

6. Species oxidized is reducing agent; species reduced is oxidizing agent.
- **a.** H_2 oxidized; Cl_2 reduced
- **b.** S oxidized; Cl_2 reduced
- **c.** N_2 oxidized; O_2 reduced
- **d.** Li oxidized; F_2 reduced
- **e.** H_2 oxidized; S reduced

7.
- **a.** Na oxidized; Br_2 reduced
- **b.** H_2 oxidized; N_2 reduced
- **c.** S oxidized; O_2 reduced
- **d.** Mg oxidized; Cu^{2+} reduced

8. Ions make electron transfer easier.

22.2

1 Engage

Use the Visual

Have students examine the photograph of fireworks and read the text that accompanies it. Explain that oxidation numbers are assigned to atoms according to a set of rules. These numbers are used to keep track of electrons in oxidation-reduction reactions. Tell students that one can determine if a reaction involves oxidation/reduction by looking for changes in oxidation numbers of elements in the compounds involved.

section 22.2

objectives

- Determine the oxidation number of an atom of any element in a pure substance
- Define oxidation and reduction in terms of a change in oxidation number, and identify atoms being oxidized or reduced in redox reactions

key term

- oxidation number

OXIDATION NUMBERS

Did you know that the different colors produced by fireworks are the result of various elements and compounds being burned? Burning sodium produces yellow light. Burning barium and copper compounds produces green light and blue-green light, respectively. The fireworks shown here are called red stars—their beautiful crimson glow coming from the burning of strontium compounds. As elements burn, their oxidation state often changes. **How are oxidation numbers assigned, and how are they used to analyze redox reactions?**

Assigning Oxidation Numbers

An **oxidation number** is a positive or negative number assigned to an atom according to a set of rules. Oxidation numbers can be thought of as a chemical bookkeeping device. As you will learn in Section 22.3, complex redox equations can be balanced by the use of oxidation-number changes. As a general rule, a bonded atom's oxidation number is the charge that it would have if the electrons in the bond were assigned to the atom of the more electronegative element. In binary ionic compounds, such as NaCl and $CaCl_2$, the oxidation numbers of the atoms equal their ionic charges. The compound sodium chloride is composed of sodium ions (Na^{1+}) and chloride ions (Cl^{1-}). Thus the oxidation number of sodium is +1, and that of chlorine is −1. Notice that when oxidation numbers are written, the sign is put before the number. Sodium in NaCl has an ionic charge of 1+ and an oxidation number of +1. What are the oxidation numbers of calcium and
1 of fluorine in calcium fluoride (CaF_2)?

Because water is a molecular compound, no ionic charges are associated with its atoms. As you learned in Section 22.1, however, oxygen is reduced in the formation of water. Oxygen is more electronegative than

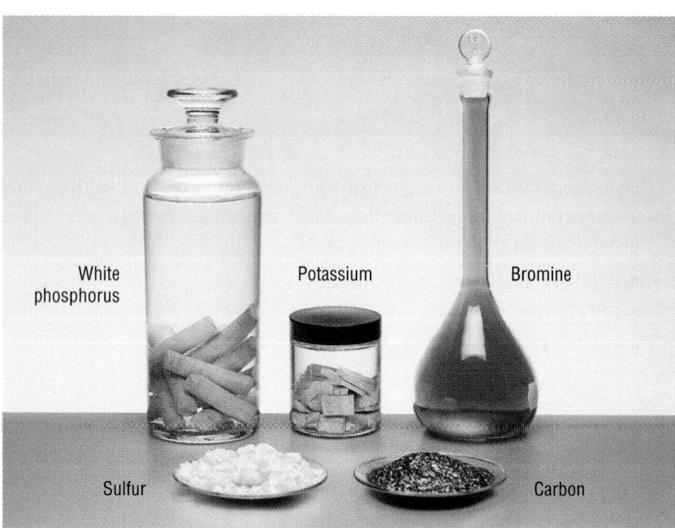

Figure 22.11
The oxidation number of any element in the free or uncombined state is zero. The elements shown here (left to right) are white phosphorus (stored under water), sulfur, potassium (stored under oil), carbon, and bromine liquid. The potassium and phosphorus are stored under a liquid to prevent them from reacting with oxygen in the air.

Student Resources

From the Teacher's Resource Package, use:

- Section Review 22.2, Ch. 22 Practice Problems and Quizzes from the Review Module (Ch. 21–24)

Technology Resources

Relevant technology resources include:

- Chem ASAP! CD-ROM
- Resource Pro CD-ROM
- Chemistry Alive! Videodisc: *Oxidation States of Vanadium, Oxidizing with Potassium Chlorate*

Figure 22.12
Yellow potassium chromate (K_2CrO_4) and orange potassium dichromate ($K_2Cr_2O_7$) each have a chromium-containing polyatomic ion. What are the formulas of the chromate ion and the dichromate ion? 3

hydrogen. In water, the two shared electrons in the H—O bond are shifted toward oxygen and away from hydrogen. Imagine that the electrons contributed by the hydrogen atoms are completely transferred to the oxygen. The charges that would result from this transfer are the oxidation numbers of the bonded elements. The oxidation number of oxygen is −2. The oxidation number of each hydrogen is +1. Oxidation numbers are often written above the chemical symbols in a formula. For example, water can be represented as

$$\overset{+1\,-2}{H_2O}$$

The following set of rules should help you determine oxidation numbers.

Rules for Assigning Oxidation Numbers

1. The oxidation number of a monatomic ion is equal in magnitude and sign to its ionic charge. For example, the oxidation number of the bromide ion (Br^{1-}) is −1; that of the Fe^{3+} ion is +3.
2. The oxidation number of hydrogen in a compound is +1, except in metal hydrides, such as NaH, where it is −1.
3. The oxidation number of oxygen in a compound is −2, except in peroxides, such as H_2O_2, where it is −1.
4. The oxidation number of an atom in uncombined (elemental) form is 0. For example, the oxidation number of the potassium atoms in potassium metal (K) or of the nitrogen atoms in nitrogen gas (N_2) is 0. See **Figure 22.11.**
5. For any neutral compound, the sum of the oxidation numbers of the atoms in the compound must equal 0.
6. For a polyatomic ion, the sum of the oxidation numbers must equal the ionic charge of the ion.

The last two rules can be used together to determine the oxidation number of atoms not covered in the first four rules. The yellow crystals and the orange crystals in **Figure 22.12** are both compounds of chromium. What is the oxidation number of chromium in each compound? 

Answers

1 Ca = +2, F = −1
2 chromate: +6; dichromate: +6
3 chromate: CrO_4^{2-}; dichromate: $Cr_2O_7^{2-}$

Meeting Diverse Needs

Gifted Have advanced students do research and report to the class why carbon monoxide is extremely poisonous to living systems whereas carbon dioxide is relatively harmless. Point out that carbon in carbon monoxide has an oxidation number that is lower than that in carbon dioxide. Therefore, one difference is that carbon monoxide is a better reducing agent than carbon dioxide.

22.2

Motivate and Relate

Explain that determining oxidation numbers of elements in compounds is a way for chemists to keep track of electron transfer during redox reactions. Ask the students to name other examples where items are numbered to keep track of movement. (The numbers on sports players' jerseys and the area codes assigned to telephone numbers in different regions are two examples.)

2 Teach

Teacher Demo

Pour 100 mL of 0.1 *M* lead(II) acetate [$Pb(CH_3COO)_2$] into a 150-mL beaker and add a 1-cm × 3-cm zinc strip. **CAUTION!** Lead(II) acetate is a suspected carcinogen and mutagen. Observe the "tree" of lead that forms over the course of a day. Dispose of the reaction mixture by adding a threefold molar excess of sodium sulfide to the contents of the beaker and stirring occasionally for one hour. Adjust the pH to neutral with 3 *M* NaOH. Filter or decant the mixture. Allow the residue to dry. Add a threefold excess of 1 *M* iron(III) chloride to the filtrate, stirring slowly. Allow the precipitate to settle; then filter or decant. Put the dry precipitate and filtered residue in a plastic container and bury in an approved landfill site. Flush the neutral filtrate down the drain with excess water.

Ask students to write a balanced equation for the redox reaction that occurred. Have them assign oxidation numbers for each element. Based on changes in oxidation number, have the students identify elements that were oxidized, reduced, and not affected.

Discuss

Lead a class discussion on determining oxidation numbers for elements in compounds by applying the rules for assigning oxidation numbers. Go over many examples until the students feel comfortable with determining these numbers. Also explain that the same element can have several oxidation numbers, depending on the compound. Derive the oxidation numbers of iodine in HIO_4, HIO_3, HIO, I_2, and HI as an example. (+7, +5, +1, 0, −1, respectively)

Cooperative Learning

Write formulas for ionic and molecular compounds on the chalkboard. Divide the class into groups of two and have the students work together to assign oxidation numbers to all of the elements in each compound. Students can use the problem-solving strategy outlined in Sample Problem 22-2.

Practice Problems Plus

Related Chapter Review Problem
Chapter Review problem 38 is related to Sample Problem 22-2.

Chemistry Alive!

Oxidation States of Vanadium — Play

Oxidizing with Potassium Chlorate — Play

Sample Problem 22-2

What is the oxidation number of each kind of atom in the following compounds?

a. SO_2 **b.** CO_3^{2-} **c.** K_2SO_4

1. *ANALYZE* *Plan a problem-solving strategy.*

a.–c. Use the set of rules you just learned to assign oxidation numbers and to calculate unknown ones.

2. *SOLVE* *Apply the problem-solving strategy.*

a. There are two oxygen atoms and the oxidation number of each oxygen is −2 (rule 3). You also know that the sum of the oxidation numbers for the neutral compound must be 0 (rule 5). Therefore the oxidation number of sulfur is +4, because $+4 + (2 \times (-2)) = 0$.

$$\overset{+4}{S}\overset{-2}{O_2}$$

b. The oxidation number of oxygen is −2 (rule 3).

$$\overset{?}{C}\overset{-2}{O_3}{}^{2-}$$

The sum of the oxidation numbers of the carbon and oxygen atoms must equal the ionic charge, 2− (rule 6). The oxidation number of carbon must be +4, because $+4 + (3 \times (-2)) = -2$.

$$\overset{+4}{C}\overset{-2}{O_3}{}^{2-}$$

c. The oxidation number of the potassium ion is the same as its ionic charge, +1 (rule 1). The oxidation number of oxygen is −2 (rule 3).

$$\overset{+1}{K_2}\overset{?}{S}\overset{-2}{O_4}$$

For the sum of the oxidation numbers in the compound to be 0 (rule 5), the oxidation number of sulfur must be +6, because $(2 \times (+1)) + (+6) + (4 \times (-2)) = 0$.

$$\overset{+1}{K_2}\overset{+6}{S}\overset{-2}{O_4}$$

3. *EVALUATE* *Do the results make sense?*

The results are consistent with the rules for determining oxidation numbers. Rule 1 was used to find the oxidation number of potassium in K_2SO_4. Rule 3 was used to find the oxidation number of oxygen in all three compounds. Rule 5 was used to find the oxidation number of sulfur in SO_2 and K_2SO_4. Rule 6 was used to find the oxidation number of carbon in CO_3^{2-}. Also, addition of the oxidation numbers correctly gives the final overall charge for the ion and the two neutral compounds.

Practice Problems

9. Determine the oxidation number of each element in these substances.

a. S_2O_3 **b.** O_2 **c.** $Al_2(SO_4)_3$ **d.** Na_2O_2

10. Find the oxidation number of each kind of atom in the following.

a. P_2O_5 **b.** NH_4^+ **c.** $Na_2Cr_2O_7$ **d.** $Ca(OH)_2$

Figure 22.13
When a copper wire is placed in a colorless silver nitrate solution (left), crystals of silver coat the wire (right). The solution slowly turns blue as a result of the formation of copper(II) nitrate. What change occurs in the oxidation number of the silver? How does the oxidation number of the copper change? ❶

Oxidation-Number Changes in Chemical Reactions

An increase in the oxidation number of an atom indicates oxidation. A decrease in the oxidation number of an atom indicates reduction. Look again at the equation in Sample Problem 22-1 on page 649 and at **Figure 22.13.** Can you identify what is being oxidized and what is being reduced on the basis of oxidation-number changes? Here is the equation with oxidation numbers added:

$$\overset{+1\ +5\ -2}{2AgNO_3(aq)} + \overset{0}{Cu(s)} \longrightarrow \overset{+2\ +5\ -2}{Cu(NO_3)_2(aq)} + \overset{0}{2Ag(s)}$$

In this reaction, the oxidation number of silver decreases from +1 to 0, which indicates reduction: Silver ions (Ag^{1+}) reduce to silver metal (Ag^0). Copper is oxidized in this reaction. Its oxidation number increases from 0 to +2 as copper metal oxidizes from Cu^0 to Cu^{2+}. Note that these results agree with those obtained by analyzing the electron transfers that occur in the reaction. **Figure 22.14** illustrates a redox reaction that shows what occurs when a shiny iron nail is dipped into a solution of copper(II) sulfate.

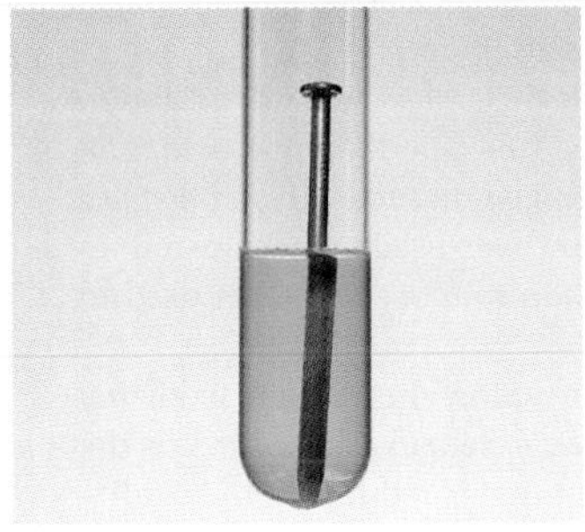

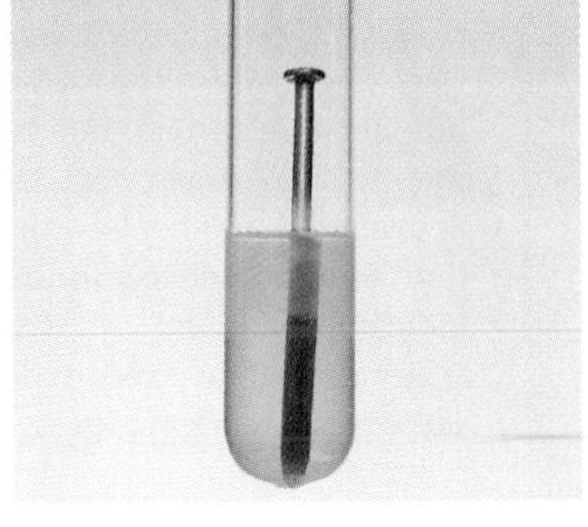

Figure 22.14
An iron nail dipped in a copper(II) sulfate solution (left) becomes coated with metallic copper (right). The iron reduces Cu^{2+} ions in solution and is simultaneously oxidized to Fe^{2+}. Write the balanced ionic equation for this redox reaction. ❷

22.2

ACTIVITY

Have students work in pairs. Give each pair a set of three small, glass test tubes containing a small portion of zinc, copper, and magnesium. Have students add approximately 10 drops of 0.1*M* HCl to the metals and note signs of reaction. They should see bubbles form in the tubes containing zinc and magnesium, but not in the tube containing copper. Ask students to guess the identity of the gas in the tubes containing zinc and magnesium [$H_2(g)$]. Have them write equations for the reactions of zinc and magnesium with HCl. Ask students if these reactions are redox reactions and to postulate what would happen if gold and sodium were exposed to aqueous HCl.

Think Critically

Have students examine Figure 22.14 and ask them to write the equation for the redox reaction. Ask students to postulate what would happen if a copper nail were added to a solution of iron(II) sulfate instead and to explain their hypothesis.

Cooperative Learning

Write the equations for many redox reactions on the chalkboard. Divide the class into groups of two and have students work together to assign oxidation numbers to all of the elements involved in each reaction whether uncombined or combined. Ask students to identify which elements are oxidized and which are reduced based on changes in oxidation numbers.

Answers

❶ The oxidation number of the silver changes from +1 to 0; the oxidation number of the copper changes from 0 to +2.
❷ $Fe(s) + Cu^{2+}(aq) \longrightarrow Fe^{2+}(aq) + Cu(s)$

Practice Problems Plus

Additional Practice Problem

The Chem ASAP! CD-ROM contains this problem related to Sample Problem 22-3: Use the changes in oxidation numbers to identify which atoms are oxidized and which are reduced in each of these reactions.

a. $C(s) + O_2(g) \rightarrow CO_2(g)$
(C is oxidized, O is reduced)

b. $Fe_2O_3(s) + 3CO(g) \rightarrow 2Fe(s) + 3CO_2(g)$
(C is oxidized, Fe is reduced)

Think Critically

Tell students that iodide anions can react with hydrogen peroxide in acidic solution to form triiodide anions and water: $3I^-(aq) + H_2O_2(aq) + 2H^+(aq) \rightarrow I_3^-(aq) + 2H_2O(l)$ Write the equation on the chalkboard and ask if the reaction is a redox reaction. Have students use changes in oxidation numbers to justify their responses. (Students should be able to determine that the oxidation number of oxygen changes from −1 to −2, which indicates a reduction of oxygen. It will be hard for students to find the corresponding oxidation reaction because it is difficult to determine the oxidation number of iodine in the triiodide ion. Thus, students will be unable to show the increase in oxidation number required for oxidation. Use this example to emphasize the limitations in the use of oxidation numbers.)

Sample Problem 22-3

Use the changes in oxidation number to identify which atoms are oxidized and which are reduced in each reaction.

a. $Cl_2(g) + 2HBr(aq) \longrightarrow 2HCl(aq) + Br_2(l)$

b. $C(s) + O_2(g) \longrightarrow CO_2(g)$

c. $Zn(s) + 2MnO_2(s) + 2NH_4Cl(aq) \longrightarrow ZnCl_2(aq) + Mn_2O_3(s) + 2NH_3(g) + H_2O(l)$

1. *ANALYZE* *Plan a problem-solving strategy.*

Use the rules to assign an oxidation number to each atom on both sides of the equation. Note the increases and decreases in oxidation numbers. On the basis of these changes, identify the atoms oxidized and those reduced. A decrease in oxidation number indicates reduction. An increase in oxidation number indicates oxidation.

2. *SOLVE* *Apply the problem-solving strategy.*

a. $\overset{0}{Cl_2}(g) + 2\overset{+1-1}{HBr}(aq) \longrightarrow 2\overset{+1-1}{HCl}(aq) + \overset{0}{Br_2}(l)$

The element chlorine is reduced because its oxidation number decreases (0 to −1). The bromide ion is oxidized because its oxidation number increases (−1 to 0).

b. $\overset{0}{C}(s) + \overset{0}{O_2}(g) \longrightarrow \overset{+4-2}{CO_2}(g)$

The element carbon is oxidized (0 to +4). The element oxygen is reduced (0 to −2).

c. $\overset{0}{Zn}(s) + 2\overset{+4\ -2}{MnO_2}(s) + 2\overset{-3+1\ -1}{NH_4Cl}(aq) \longrightarrow \overset{+2\ -1}{ZnCl_2}(aq) + \overset{+3\ -2}{Mn_2O_3}(s) + 2\overset{-3+1}{NH_3}(g) + \overset{+1\ -2}{H_2O}(l)$

The element zinc is oxidized (0 to +2). The manganese ion is reduced (+4 to +3).

3. *EVALUATE* *Do the results make sense?*

Checking the results reveals that the assignments of oxidation numbers are correct; that is, the rules for assigning oxidation numbers have been correctly applied. In each case, a decrease in oxidation number has correctly been used to indicate reduction, and an increase in oxidation number has been used to indicate oxidation.

Recall that in every redox reaction there is an oxidizing agent and a reducing agent. The element that is oxidized is the reducing agent, and the element that is reduced is the oxidizing agent. Thus once you have identified the elements that are oxidized and the elements that are reduced in a reaction, you can use this information to identify the oxidizing agent and the reducing agent. This is illustrated in Sample Problem 22-4.

Practice Problem

11. Use the changes in oxidation numbers to identify which atoms are oxidized and which are reduced in each reaction.
    a. $2H_2(g) + O_2(g) \longrightarrow 2H_2O(l)$
    b. $2KNO_3(s) \longrightarrow 2KNO_2(s) + O_2(g)$
    c. $NH_4NO_2(s) \longrightarrow N_2(g) + 2H_2O(g)$
    (*Hint:* Consider each N in NH_4NO_2 separately.)
    d. $PbO_2(aq) + 4HI(aq) \longrightarrow I_2(aq) + PbI_2(s) + 2H_2O(l)$

Chem ASAP!

Problem-Solving 11

Solve Problem 11 with the help of an interactive guided tutorial.

Sample Problem 22-4

Identify the oxidizing agent and the reducing agent in each equation in Sample Problem 22-3.

1. *ANALYZE* *Plan a problem-solving strategy.*

a.–c. Use the following facts to solve the problem: The substance oxidized in a reaction is the reducing agent, and the substance reduced is the oxidizing agent.

2. *SOLVE* *Apply the problem-solving strategy.*

a. Chlorine is reduced, so Cl_2 is the oxidizing agent. Bromide (in HBr) is oxidized, so Br^- is the reducing agent.

b. Carbon is oxidized, so C is the reducing agent. Oxygen is reduced, so O_2 is the oxidizing agent.

c. Zinc is oxidized, so Zn is the reducing agent. Manganese (in MnO_2) is reduced, so Mn^{4+} is the oxidizing agent.

3. *EVALUATE* *Do the results make sense?*

It makes sense that what is oxidized in a chemical reaction is the reducing agent because it loses electrons—it becomes the agent by which the atom that is reduced gains electrons. Conversely, it makes sense that what is reduced in a chemical reaction is the oxidizing agent because it gains electrons—it is the agent by which the atom that is oxidized loses electrons. These facts were correctly applied in solving the problem.

Practice Problem

12. Identify the oxidizing agent and the reducing agent in each equation in Practice Problem 11.

section review 22.2

13. a. What is the oxidation number of nitrogen in nitrogen gas (N_2)? Explain.

b. How would you determine the oxidation numbers of the elements in a compound?

c. How is charge used to assign oxidation numbers to the elements in a polyatomic ion?

14. How are oxidation numbers determined and used?

15. Use the changes in oxidation numbers to identify which atoms are oxidized and which are reduced in each reaction.

a. $2Na(s) + Cl_2(g) \longrightarrow 2NaCl(s)$

b. $2HNO_3(aq) + 6HI(aq) \longrightarrow 2NO(g) + 3I_2(s) + 4H_2O(l)$

c. $3H_2S(aq) + 2HNO_3(aq) \longrightarrow 3S(s) + 2NO(g) + 4H_2O(l)$

d. $2PbSO_4(s) + 2H_2O(l) \longrightarrow Pb(s) + PbO_2(s) + 2H_2SO_4(aq)$

16. Identify the oxidizing agent and the reducing agent in each reaction in Problem 15.

 Chem ASAP! **Assessment 22.2** Check your understanding of the important ideas and concepts in Section 22.2.

22.2

Practice Problems Plus

Related Chapter Review Problem
Chapter Review problem 40 is related to Sample Problem 22-4.

3 Assess

Evaluate Understanding

In order to evaluate the students' understanding of this section, ask:

- **What are the oxidation numbers of the elements in the following compounds:**
  KNO_3 (K +1, N +5, O −2)
  Na_2SO_3 (Na +1, S +4, O −2)
  P_2O_5 (P +5, O −2)
  MgH_2 (Mg +2, H −1)
  ClF_3 (Cl +3, F −1)
  HIO (H +1, I +1, O −2)
- **What does the sign and magnitude of the oxidation number of an element in a compound indicate?** (The sign indicates loss (+) or gain (−) of electrons compared to the isolated element; the magnitude indicates the degree of the gain or loss.)
- **If fluorine is converted to F^- from F_2, what is the change in oxidation number? Is this oxidation or reduction?** (change is 0 to −1; reduction)
- **If sulfur is converted to SO_2 from PbS, what is the change in oxidation number? Is this oxidation or reduction?** (change is −2 to +4; oxidation)

Reteach

Stress that oxidation numbers refer to the combining capacity of single atoms. Thus, a change in oxidation number must be calculated on a per atom basis. Use examples such as Mn_2O_3 and $Cr_2O_7^{2-}$ to explain to students that they must consider the number of atoms involved.

Answers

SECTION REVIEW 22.2

13. a. The oxidation number of an uncombined atom is zero. **b.** The sum of the oxidation numbers in a compound must equal zero. **c.** The sum of the oxidation numbers must equal the charge on the polyatomic ion.

14. Oxidation numbers are assigned by rules based on electronegativity values and used to identify oxidizing or reducing agents.

15. a. Na, oxidized; Cl_2, reduced
b. I, oxidized; N, reduced
c. S, oxidized; N, reduced
d. Pb, oxidized and reduced

16. a. Na reducing agent, Cl_2 oxidizing agent
b. N in HNO_3, oxidizing agent; I in HI, reducing agent
c. N in HNO_3, oxidizing agent; S in H_2S, reducing agent
d. Pb in $PbSO_4$, oxidizing–reducing agent

1 Engage

Use the Visual

Have students examine the section-opening photograph and read the text. Emphasize that alchemists believed lead and other metals could be changed into gold because of the change in physical appearance of gold when it reacts with *aqua regia*. Note that alchemists did not know that they were only combining elemental gold into a gold compound. Answer the section-opening question by using the half-reaction method to derive the balanced equation for this reaction:

$$Au(s) + 4H^+(aq) + 4Cl^-(aq) + NO_3^-(aq) \rightarrow AuCl_4^-(aq) + NO(g) + 2H_2O(l)$$

Check Prior Knowledge

Have students balance the following equations:

- **$CH_3OH(l) + O_2(g) \rightarrow CO_2(g) + H_2O(l)$** [$2CH_3OH(l) + 3O_2(g) \rightarrow 2CO_2(g) + 4H_2O(l)$]
- **$Cl_2(g) + KI(aq) \rightarrow KCl(aq) + I_2(aq)$** [$Cl_2(g) + 2KI(aq) \rightarrow 2KCl(aq) + I_2(aq)$]
- **$Li(s) + H_2O(l) \rightarrow LiOH(aq) + H_2(g)$** [$2Li(s) + 2H_2O(l) \rightarrow 2LiOH(aq) + H_2(g)$]

Ask:

- **What is the purpose of balancing an equation?** (A balanced equation obeys the law of conservation of mass.)

Point out that the equations listed above are all redox reactions. Explain to students that in this section they will learn more about how to balance redox reactions.

section 22.3

BALANCING REDOX EQUATIONS

objectives

- Use the oxidation-number–change method to balance redox equations
- Break a redox equation into oxidation and reduction half-reactions, and then use the half-reaction method to balance the equation

key terms

- oxidation-number–change method
- half-reactions
- half-reaction method

The unattainable, but long pursued, goal of alchemists was to change common metals, such as lead or copper, into gold. One tool that the alchemists employed in their quest was aqua regia, which means "royal water." Aqua regia is a mixture of concentrated hydrochloric (HCl) and nitric (HNO_3) acids that is capable of dissolving gold. When gold is added to aqua regia, oxidation and reduction reactions produce gaseous nitrogen monoxide (NO) and stable $AuCl_4^-$ ions.

How can a balanced chemical equation for this redox reaction be written?

Identifying Redox Reactions

In general, all chemical reactions can be assigned to one of two classes. In oxidation–reduction (redox) reactions, electrons are transferred from one reacting species to another. In all other reactions, no electron transfer occurs. For example, double-replacement reactions and acid–base reactions are not redox reactions. Many single-replacement reactions, combination reactions, decomposition reactions, and combustion reactions, however, are redox reactions. Can you write a balanced equation for each reaction shown in **Figure 22.15?** 1

(a)

(b)

Figure 22.15
(a) *Potassium metal reacts violently with water to produce hydrogen gas (which ignites) and potassium hydroxide. Is this a redox reaction? Explain.* ***(b)*** *Zinc metal reacts vigorously with hydrochloric acid to produce hydrogen gas and zinc chloride. Is this a redox reaction? Explain.* 3

You can determine whether a reaction is a redox reaction by using oxidation numbers to keep track of electrons. If the oxidation number of an element in a reacting species changes, then that element has undergone either oxidation or reduction. Therefore the reaction as a whole must be a redox reaction. Consider this example: During an electrical storm, oxygen molecules and nitrogen molecules in the air react to form nitrogen(II) oxide. The equation for the reaction is

$$N_2(g) + O_2(g) \longrightarrow 2NO(g)$$

Is this a redox reaction? How can you tell? Does the oxidation number of either reacting species change? If so, what is that change? 2

STUDENT RESOURCES

- Section Review 22.3, Ch. 22 Practice Problems, Interpreting Graphics, Vocabulary Review, Quizzes and Tests from the Review Module (Ch. 21–24)
- Laboratory Recordsheets 22-1 and 22-2
- Small-Scale Chemistry Lab Manual: Experiment 22

TECHNOLOGY RESOURCES

Relevant technology resources include:

- Chem ASAP! CD-ROM
- Resource Pro CD-ROM
- Assessment Resources CD-ROM Chapter 22 Tests

Figure 22.16
A color change can signal a redox reaction. When a colorless solution containing bromide ions is added to a solution containing permanganate ions, the distinctive purple color of the permanganate ion (MnO_4^-) is replaced by the pale brown color of bromine. What is the oxidation-number change in manganese and bromine? ❹

Various other changes often signal that an oxidation–reduction reaction has taken place. One such change is a change in color. For example, look at **Figure 22.16.** When a colorless aqueous solution of bromide ions (Br^-) is added to a purple aqueous solution of permanganate ions (MnO_4^-), the solution changes from purple to pale brown. As the following unbalanced equation shows, bromide ions are oxidized to bromine, and permanganate ions are reduced to manganese(II) ions.

$$\underset{\text{Permanganate ion (purple)}}{MnO_4^-(aq)} + \underset{\text{Bromide ion (colorless)}}{Br^-(aq)} \longrightarrow \underset{\text{Manganese(II) ion (colorless)}}{Mn^{2+}(aq)} + \underset{\text{Bromine (brown)}}{Br_2(aq)}$$

Color changes can do more than signal that a redox reaction has taken place. They can help obtain quantitive information about concentrations in redox reactions. The titration of reducing agents and oxidizing agents is similar to the titration of acids and bases, which you learned about in Section 21.1. For example, the titration of the reducing agent oxalic acid ($H_2C_2O_4$) with a solution of unknown concentration of the oxidizing agent potassium permanganate ($KMnO_4$) is illustrated in **Figure 22.17.** To obtain a quantitative result in such a titration, a balanced equation of the redox reaction is required. A known mass of colorless oxalic acid, the reducing

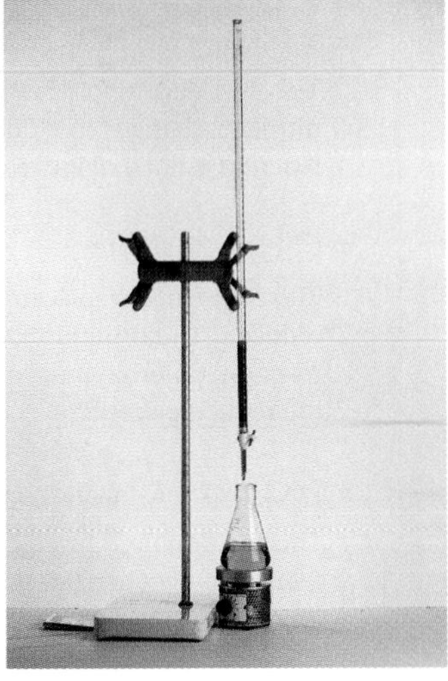

Figure 22.17
In this redox titration, the oxidizing agent in the buret (purple aqueous potassium permanganate) serves as the indicator. A known mass of the reducing agent, colorless oxalic acid, is dissolved in dilute sulfuric acid in the flask. As permanganate is added slowly to the contents of the flask, it immediately loses its purple color (left). When all the oxalic acid in the flask is oxidized, the next drop of permanganate remains unreacted, and the solution in the flask turns light purple (right). That is the end point of the redox titration.

22.3

Motivate and Relate

Remind students of the concepts that underlie the need to balance equations. For all equations, mass must be conserved by balancing the number of atoms and electrical neutrality by balancing charges. The balancing of redox reactions must also reflect the electrons gained and lost by oxidizing and reducing agents. In other words, electrons cannot be created or destroyed.

Ask students for other examples in which "balancing" is required to achieve conservation. One example, is the first law of thermodynamics, which states that energy must be conserved. To conserve energy, any energy change in a system must be balanced by an equal and opposite energy change in the surroundings. For a non-scientific example, consider a checking account. To maintain sufficient funds to cover checks written, money withdrawn would eventually need to be balanced by deposits.

2 Teach

Discuss

Ask students to recall the types of chemical reactions they studied in previous chapters. Write an example of each kind of equation on the chalkboard or overhead projector. Ask students to decide which equations represent redox reactions and which do not.

Answers

❶ $2K + 2H_2O \rightarrow 2KOH + H_2$
$Zn + 2HCl \rightarrow ZnCl_2 + H_2$
❷ Yes; nitrogen's oxidation number increases from 0 to +2; oxygen's oxidation number decreases from 0 to −2.
❸ **a.** The oxidation number of potassium goes from 0 to +1 when the metal reacts with water to form KOH. Potassium is oxidized, so this is a redox reaction.
b. The oxidation number of zinc goes from 0 to +2 when the metal reacts with HCl to form $ZnCl_2$. Zinc is oxidized, so this is a redox reaction.
❹ The oxidation number goes from +7 in MnO_4^- to +2 in Mn^{2+}; the change is −5; the oxidation number goes from −1 in Br^- to 0 in Br_2 (elemental bromine); the change is +1.

TEACHER DEMO

To demonstrate that redox reactions can sometimes be identified as a result of color changes, dissolve 6 crystals of potassium permanganate ($KMnO_4$) in 500 mL of water in a large beaker. **CAUTION!** Potassium permanganate is an irritant and suspected mutagen. Dissolve 1 g sodium hydrogen sulfite ($NaHSO_3$) in 20 mL of water in a 50-mL beaker. Add this solution to the $KMnO_4$ solution and stir. The red solution becomes clear. Dissolve 1 g barium chloride dihydrate ($BaCl_2{\cdot}2H_2O$) in 20 mL of water in another 50-mL beaker. **CAUTION!** Barium chloride is toxic. Add this solution to the large beaker and stir. The solution turns cloudy-white. Ask students to try to explain the color changes. (The first reaction involves reduction of MnO_4^- to colorless Mn^{2+} and oxidation of HSO_3^- to HSO_4^-. The second reaction is not a redox reaction; it involves the formation of insoluble $BaSO_4$.)
Dispose of the waste solution by adding 3*M* H_2SO_4 until precipitation of $BaSO_4$ appears to be complete. Let stand overnight. Filter or decant the liquid. Dry the residue, place in a plastic container, and bury in an approved landfill site. Neutralize the filtrate, and flush down the drain with excess water.

agent, is dissolved in dilute sulfuric acid in the reaction flask. The concentration of the purple potassium permanganate solution can be calculated from the volume added to reach the end point, which is reached when all the oxalic acid is oxidized and the next drop of purple $KMnO_4$ remains unreacted. The net ionic equation for the reaction can be written. From this equation, the oxidation-number changes for the oxidized and reduced species can be determined. The full balanced equation for the reaction is

$$2KMnO_4(aq) + 5H_2C_2O_4(aq) + 3H_2SO_4(aq) \longrightarrow 2MnSO_4(aq) + K_2SO_4(aq) + 10CO_2(g) + 8H_2O(l)$$

Sample Problem 22-5

Use the change in oxidation number to identify whether each reaction is a redox reaction or a reaction of some other type. If a reaction is a redox reaction, identify the element reduced, the element oxidized, the reducing agent, and the oxidizing agent.

a. $N_2O_4(g) \longrightarrow 2NO_2(g)$
b. $Cl_2(g) + 2NaBr(aq) \longrightarrow 2NaCl(aq) + Br_2(g)$
c. $PbCl_2(aq) + K_2SO_4(aq) \longrightarrow 2KCl(aq) + PbSO_4(s)$
d. $2NaOH(aq) + H_2SO_4(aq) \longrightarrow Na_2SO_4(aq) + 2H_2O(l)$

1. ***ANALYZE*** ***Plan a problem-solving strategy.***
a.–d. Use the rules to assign oxidation numbers to each element. Note whether there are any changes in oxidation number. If there are, the reaction is a redox reaction. The element with an oxidation number that increases is oxidized and is the reducing agent. The element with an oxidation number that decreases is reduced and is the oxidizing agent.

2. ***SOLVE*** ***Apply the problem-solving strategy.***

a. $\overset{+4-2}{N_2O_4}(g) \longrightarrow 2\overset{+4-2}{NO_2}(g)$

This is a decomposition reaction. Neither oxygen nor nitrogen changes in oxidation number. Therefore the reaction is not a redox reaction.

b. $\overset{0}{Cl_2}(g) + 2\overset{+1\,-1}{NaBr}(aq) \longrightarrow 2\overset{+1\,-1}{NaCl}(aq) + \overset{0}{Br_2}(g)$

This is a single-replacement reaction. The chlorine is reduced. The bromide ion is oxidized. This is a redox reaction. Chlorine is the oxidizing agent; bromide ion is the reducing agent.

c. $\overset{+2\,-1}{PbCl_2}(aq) + \overset{+1+6-2}{K_2SO_4}(aq) \longrightarrow 2\overset{+1-1}{KCl}(aq) + \overset{+2+6-2}{PbSO_4}(s)$

This is a double-replacement reaction. None of the elements changes in oxidation number. This is not a redox reaction.

Practice Problems

17. Identify which of the following are oxidation–reduction reactions. If a reaction is a redox reaction, name the element oxidized and the element reduced.

a. $BaCl_2(aq) + 2KIO_3(aq) \longrightarrow Ba(IO_3)_2(s) + 2KCl(aq)$
b. $H_2CO_3(aq) \longrightarrow H_2O(l) + CO_2(g)$
c. $Mg(s) + Br_2(l) \longrightarrow MgBr_2(s)$
d. $NH_4NO_2(s) \longrightarrow N_2(g) + 2H_2O(l)$
e. $2KClO_3(s) \longrightarrow 2KCl(s) + 3O_2(g)$

MEETING DIVERSE NEEDS

At Risk Balancing redox reactions by the half-reaction method is a lengthy process that requires good organizational skills and attention to detail. Students who are impatient or easily frustrated may have difficulty with these problems. Working in small groups can help them organize their efforts. Supervise the work closely to be sure each student is contributing.

Sample Problem 22-5 (cont.)

d. $\overset{+1\ -2+1}{2NaOH(aq)} + \overset{+1+6-2}{H_2SO_4(aq)} \longrightarrow \overset{+1\ +6-2}{Na_2SO_4(aq)} + \overset{+1\ -2}{2H_2O(l)}$

This is an acid–base (neutralization) reaction. None of the elements changes in oxidation number. This is not a redox reaction.

3. ***EVALUATE*** ***Do the results make sense?***
Only in redox reactions do the oxidation numbers of reactants change as products form. Only reaction **b.** represents a redox reaction. In **b.**, each chlorine atom (Cl) accepts an electron from a bromide ion (Br^-). The chlorine molecule is reduced to two chloride ions (Cl^-); therefore, it is the oxidizing agent in the reaction. That makes sense because chlorine is more electronegative than bromine (see **Table 14.2**) and thus more apt to gain electrons. Because the bromide ions are oxidized to molecular bromine (Br_2), the bromide ions are the reducing agent.

Practice Problems (cont.)

18. Identify which of the following are oxidation–reduction reactions. If a reaction is a redox reaction, name the element oxidized and the element reduced.

a. $CaCO_3(s) + 2HCl(aq) \longrightarrow CaCl_2(aq) + H_2O(l) + CO_2(g)$

b. $CuO(s) + H_2(g) \longrightarrow Cu(s) + H_2O(l)$

Chem ASAP!

Problem-Solving 18

Solve Problem 18 with the help of an interactive guided tutorial.

As you know, the ability to write a correctly balanced equation that accurately represents what happens in a chemical reaction is essential to chemists and chemical engineers. Many oxidation–reduction reactions are too complex to be balanced by trial and error. Fortunately, two systematic methods are available. These methods, described below, are based on the fact that the total number of electrons gained in reduction must equal the total number of electrons lost in oxidation. One method uses oxidation-number changes, and the other uses half-reactions.

Using Oxidation-Number Changes

In the **oxidation-number–change method,** a redox equation is balanced by comparing the increases and decreases in oxidation numbers. To use this method, start with a skeleton equation for the redox reaction. The reduction of iron ore is used here as an example.

$$Fe_2O_3(s) + CO(g) \longrightarrow Fe(s) + CO_2(g) \text{ (unbalanced)}$$

Step 1. Assign oxidation numbers to all the atoms in the equation. Write the numbers above the atoms.

$$\overset{+3\ -2}{Fe_2O_3(s)} + \overset{+2-2}{CO(g)} \longrightarrow \overset{0}{Fe(s)} + \overset{+4-2}{CO_2(g)}$$

Note that the oxidation number is stated as the charge per atom. So although the total positive charge of Fe ions in Fe_2O_3 is 6+, the oxidation number of each Fe ion is +3.

Step 2. Identify which atoms are oxidized and which are reduced. In this reaction, iron decreases in oxidation number from +3 to 0, a change of −3. Therefore iron is reduced. Carbon increases in oxidation number from +2 to +4, a change of +2. Thus carbon is oxidized. Oxygen does not change in oxidation number.

CHEMISTRY AND SCIENCE HISTORY

Evidence exists that the use of rockets—and of redox reactions to power them—began in the thirteenth century when Chinese armies used "arrows of flying fire" to defend their cities. The early Chinese rockets were leather cylinders that were filled with gunpowder and mounted on a stick. It is generally thought that the gunpowder mixture originated in ninth century China.

The essential components of a rocket propellant are a source of energy (the fuel) and an oxidizing agent to react with the fuel. When the mixture is ignited, the fuel is oxidized rapidly and the oxidizing agent is reduced. Gunpowder is a mixture of charcoal (carbon), sulfur, and potassium nitrate. It reacts explosively to produce potassium sulfide, gaseous carbon dioxide, and nitrogen. In the redox reaction, carbon is the fuel and other components are oxidizing agents.

Practice Problems Plus

Related Chapter Review Problems

Chapter Review problems 32 and 33 are related to Sample Problem 22-5.

Additional Practice Problems

The Chem ASAP! CD-ROM contains the following problem: Identify which of these are oxidation-reduction reactions. If a reaction is a redox reaction, name the element oxidized and reduced.

a. $3AgNO_3(aq) + FeCl_3(aq) \rightarrow 3AgCl(s) + Fe(NO_3)_3(aq)$
(not a redox reaction)

b. $C(s) + H_2O(g) \rightarrow CO(g) + H_2(g)$
(C is oxidized, H is reduced)

c. $4H_2O_2(aq) + PbS(s) \rightarrow PbSO_4(s) + 4H_2O(l)$
Note the peroxide! (S is oxidized, O is reduced.)

Think Critically

A reaction in which the same compound is both oxidized and reduced is called *disproportionation.* When NO_2 is dissolved in water, it disproportionates to form nitric acid (HNO_3) and nitric oxide (NO). Use the oxidation-number-change method to write the balanced equation for this redox reaction. Given that NO_2 is both the oxidizing and reducing agent in this reaction, how should it be treated in Step 3? (It should be the end point of two bracketing lines.) **The balanced equation for the reaction is:**

$$3NO_2(g) + H_2O(l) \rightarrow 2H^+(aq) + 2NO_3^-(aq) + NO(g)$$

Build Writing Skills

Have students describe the oxidation-number-change method of balancing redox reactions in their own words. They should use specific examples to make their descriptions clearer.

Practice Problems Plus

Related Chapter Review Problems

Chapter Review problems 30 and 41 are related to Sample Problem 22-6.

Additional Practice Problem

The Chem ASAP! CD-ROM contains this problem: Use the oxidation number change method to balance this redox equation.

$Cu(s) + HNO_3(aq) \rightarrow Cu(NO_3)_2(aq) + NO(g) + H_2O(l)$

$[3Cu(s) + 8HNO_3(aq) \rightarrow 3Cu(NO_3)_2(aq) + 2NO(g) + 4H_2O(l)]$

Figure 22.18

Potassium dichromate, shown on the left, is an orange crystalline substance. It reacts with water and sulfur to form chromium(III) oxide, the green compound shown on the right. What are the other products? 1

Step 3. Use one bracketing line to connect the atoms that undergo oxidation and another such line to connect those that undergo reduction. Write the oxidation-number change at the midpoint of each line.

$$\overset{+3\ -2}{Fe_2O_3}(s) + \overset{+2-2}{CO}(g) \longrightarrow \overset{0}{Fe}(s) + \overset{+4-2}{CO_2}(g)$$

(C: +2; Fe: −3)

Step 4. Make the total increase in oxidation number equal to the total decrease in oxidation number by using appropriate coefficients. In this example, the oxidation-number increase should be multiplied by 3 and the oxidation-number decrease should be multiplied by 2, which gives an increase of +6 and a decrease of −6. This can be done in the equation by placing the coefficient 2 in front of Fe on the right side and the coefficient 3 in front of both CO and CO_2. The formula Fe_2O_3 does not need a coefficient because the formula already indicates 2 Fe.

$$Fe_2O_3(s) + 3CO(g) \longrightarrow 2Fe(s) + 3CO_2(g)$$

($3 \times (+2) = +6$; $2 \times (-3) = -6$)

Step 5. Finally, make sure that the equation is balanced for both atoms and charge. If necessary, the remainder of the equation can be balanced by inspection.

$$Fe_2O_3(s) + 3CO(g) \longrightarrow 2Fe(s) + 3CO_2(g)$$

Sample Problem 22-6

Balance this redox equation by using the oxidation-number–change method. Samples of the initial and final chromium-containing compounds for this reaction are shown in **Figure 22.18.**

$$K_2Cr_2O_7(aq) + H_2O(l) + S(s) \longrightarrow KOH(aq) + Cr_2O_3(s) + SO_2(g)$$

1. ***ANALYZE* Plan a problem-solving strategy.**
Apply the five steps of the oxidation-number–change method to balance the equation.

2. ***SOLVE* Apply the problem-solving strategy.**
Step 1. Assign oxidation numbers.

$$\overset{+1\ +6\ -2}{K_2Cr_2O_7}(aq) + \overset{+1\ -2}{H_2O}(l) + \overset{0}{S}(s) \longrightarrow \overset{+1-2+1}{KOH}(aq) + \overset{+3\ -2}{Cr_2O_3}(s) + \overset{+4-2}{SO_2}(g)$$

Step 2. Identify which atoms are oxidized and which atoms are reduced. Cr is reduced because its oxidation number decreases, while S is oxidized because its oxidation number increases.

Practice Problems

19. Balance each redox equation using the oxidation-number–change method.
  a. $KClO_3(s) \longrightarrow KCl(s) + O_2(g)$
  b. $HNO_2(aq) + HI(aq) \longrightarrow NO(g) + I_2(s) + H_2O(l)$
  c. $As_2O_3(s) + Cl_2(g) + H_2O(l) \longrightarrow H_3AsO_4(aq) + HCl(aq)$

Answer

1 The other products are potassium hydroxide (KOH) and sulfur dioxide (SO_2).

Sample Problem 22-6 (cont.)

Step 3. Connect the atoms that change in oxidation number. Indicate the signs and magnitudes of the changes.

$$\overset{+6}{K_2Cr_2O_7}(aq) + H_2O(l) + \overset{0}{S}(s) \longrightarrow KOH(aq) + \overset{+3}{Cr_2O_3}(s) + \overset{+4}{SO_2}(g)$$

(Cr: −3; S: +4)

Step 4. Balance the increase and decrease in oxidation numbers. Four chromium atoms must be reduced ($4 \times (-3) = -12$ decrease) for each three sulfur atoms that are oxidized ($3 \times (+4) = +12$ increase). Put the coefficient 3 in front of S and SO_2 and the coefficient 2 in front of $K_2Cr_2O_7$ and Cr_2O_3,

$$2\overset{+6}{K_2Cr_2O_7}(aq) + H_2O(l) + 3\overset{0}{S}(s) \longrightarrow KOH(aq) + 2\overset{+3}{Cr_2O_3}(s) + 3\overset{+4}{SO_2}(g)$$

(Cr: (4)(−3) = −12; S: (3)(+4) = +12)

Step 5. Check the equation and finish balancing by inspection if necessary. The coefficient 4 in front of KOH balances potassium. The coefficient 2 in front of H_2O balances hydrogen and oxygen. The final equation is

$$2K_2Cr_2O_7(aq) + 2H_2O(l) + 3S(s) \longrightarrow 4KOH(aq) + 2Cr_2O_3(s) + 3SO_2(g)$$

3. *EVALUATE* *Does the result make sense?*
Inspection reveals that the equation is correctly balanced.

Practice Problems (cont.)

20. Use the oxidation-number–change method to balance each redox equation.

a. $Bi_2S_3(s) + HNO_3(aq) \longrightarrow Bi(NO_3)_3(aq) + NO(g) + S(s) + H_2O(l)$

b. $MnO_2(s) + H_2SO_4(aq) + H_2C_2O_4(aq) \longrightarrow MnSO_4(aq) + CO_2(g) + H_2O(l)$

c. $SbCl_5(aq) + KI(aq) \longrightarrow SbCl_3(aq) + KCl(aq) + I_2(s)$

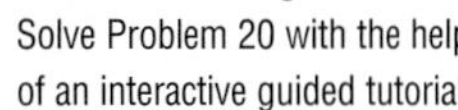

Chem ASAP!

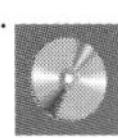

Problem-Solving 20
Solve Problem 20 with the help of an interactive guided tutorial.

Using Half-Reactions

You have just learned about the oxidation-number–change method, one of two methods often used to balance redox equations. A second method involves the use of **half-reactions,** or equations showing either the reduction or the oxidation of a species in an oxidation–reduction reaction. The **half-reaction method** is used to balance redox equations by balancing the oxidation and reduction half-reactions. The procedure is different, but the outcome is the same as with the oxidation-number–change method. The half-reaction method is particularly useful in balancing equations for ionic reactions.

The oxidation of sulfur by nitric acid in aqueous solution is one example of a redox reaction that can be balanced by the half-reaction method.

$$S(s) + HNO_3(aq) \longrightarrow SO_2(g) + NO(g) + H_2O(l) \text{ (unbalanced)}$$

The following steps show how to balance this equation using the half-reaction method.

Step 1. Write the unbalanced equation in ionic form. In this case, only HNO_3 is ionized. The products are covalent compounds.

$$S(s) + H^+(aq) + NO_3^-(aq) \longrightarrow SO_2(g) + NO(g) + H_2O(l)$$

22.3

TEACHER DEMO

In a 600-mL beaker, combine 8 mL of 6*M* H_2SO_4 and 125 mL distilled water. (**CAUTION!** Handle concentrated sulfuric acid carefully.) Add 43 g of sodium pyrophosphate decahydrate ($Na_4P_2O_7 \cdot 10H_2O$) to the beaker and stir until all the solid has dissolved. Dilute the solution to 200 mL with distilled water.
Then add 72 mL 6*M* H_2SO_4 and 80 mL of 0.1*M* $MnSO_4$.
Into another 600-mL beaker, add 40 mL of 0.25*M* potassium bromate ($KBrO_3$). Place a stir bar in the beaker. Set the beaker on the magnetic stirrer. (**CAUTION!** Potassium bromate is toxic by ingestion.)
During class, pour the solution from the first beaker into the second while the second solution is being stirred. After addition is complete, ask a student to watch the second hand of a clock. Record the exact number of seconds that pass before the colorless solution turns purple.
Explain that the appearance of the purple color is due to bromate oxidation of Mn^{2+} to Mn^{3+} with formation of the purple pyrophosphate complex of Mn^{3+}). The unbalanced half reactions are:
1) $Mn^{2+} + H_4P_2O_7 \rightarrow Mn(H_2P_2O_7)_3^{3-}$
2) $BrO_3^- \rightarrow HBrO$
On the chalkboard, show students how to use the half-reaction method to balance the overall reaction:
$4Mn^{2+}(aq) + 12H_4P_2O_7(aq) + BrO_3^-(aq) \rightarrow 4Mn(H_2P_2O_7)_3^{3-}(aq) + HBrO(aq) + 19H^+(aq) + 2H_2O(l)$.
Dispose of the waste solution by neutralization and flushing down the drain with excess water.
(*Chemical Demonstrations: A Handbook for Teachers of Chemistry,* B.Z. Shakhashiri, ed., Volume 4, 1991, p. 83)

Discuss

Lead a class discussion on using the half-reaction method to balance redox reactions. Point out that the underlying assumption when using half-reactions is that any redox reaction can always be considered as two separate processes occurring simultaneously. Therefore, the "whole" reaction can be thought of as the sum of two halves—an oxidation reaction and a reduction reaction. Go through the half-reaction method with a specific example, clearly explaining each step. Emphasize that oxidation half-reactions are always written with the electrons on the right-hand side of the equation. Reduction half-reactions are always written with the electrons on the left-hand side of the equation.

Think Critically

In Step 3, H_2O and either H^+ or OH^- can be added to each side of an equation to balance O and H atoms. Ask the students to explain why this is allowable. (The reaction itself is not changed by doing so; rather the medium of the reaction is simply being defined. In an acidic solution, H^+ ions are present in excess. In a basic solution, OH^- ions are present in excess.)

Table 22.2

Oxidation Numbers of Sulfur in Different Compounds

Compound	Oxidation Number
H_2SO_4	+6
SO_3	+6
H_2SO_3	+4
SO_2	+4
$Na_2S_2O_3$	+2
SCl_2	+2
S_2Cl_2	+1
S	0
H_2S	−2

Step 2. Write separate half-reactions for the oxidation and reduction processes. As you can see from **Table 22.2,** which shows the oxidation numbers of sulfur in various compounds, sulfur is oxidized in this reaction because its oxidation number increases from 0 to 4. Nitrogen is reduced because its oxidation number decreases from +5 to +2.

$$\text{Oxidation half-reaction: } S(s) \longrightarrow SO_2(g)$$
$$\text{Reduction half-reaction: } NO_3^-(aq) \longrightarrow NO(g)$$

Step 3. Balance the atoms in the half-reactions.

a. *Balance the oxidation half-reaction.* This reaction takes place in acid solution. In such cases, H_2O and H^+ can be used to balance oxygen and hydrogen as needed. Sulfur is already balanced in the half-reaction. Add two molecules of H_2O on the left to balance the oxygen in the half-reaction.

$$2H_2O(l) + S(s) \longrightarrow SO_2(g)$$

Oxygen is now balanced, but four hydrogen ions ($4H^+$) must be added to the right to balance the hydrogen on the left.

$$2H_2O(l) + S(s) \longrightarrow SO_2(g) + 4H^+(aq)$$

This half-reaction is now balanced in terms of atoms. Note that it is not balanced for charge. The charges will be balanced in Step 4.

b. *Balance the reduction half-reaction.* Nitrogen is already balanced. Add two molecules of H_2O on the right to balance the oxygen.

$$NO_3^-(aq) \longrightarrow NO(g) + 2H_2O(l)$$

Oxygen is balanced, but four hydrogen ions ($4H^+$) must be added to the left to balance hydrogen.

$$4H^+(aq) + NO_3^-(aq) \longrightarrow NO(g) + 2H_2O(l)$$

This half-reaction is now balanced in terms of atoms, but not for charge.

Step 4. Add sufficient electrons to one side of each half-reaction to balance the charges. Four electrons are needed on the right side in the oxidation half-reaction.

$$\text{Oxidation: } 2H_2O(l) + S(s) \longrightarrow SO_2(g) + 4H^+(aq) + 4e^-$$

Three electrons are needed on the left side in the reduction half-reaction.

$$\text{Reduction: } 4H^+(aq) + NO_3^-(aq) + 3e^- \longrightarrow NO(g) + 2H_2O(l)$$

Each half-reaction is now balanced with respect to both atoms and charge.

Step 5. Multiply each half-reaction by an appropriate number to make the numbers of electrons equal in both. In any redox reaction, the number of electrons lost in oxidation must equal the number of electrons gained in reduction. In this case, the oxidation half-reaction is multiplied by 3 and the reduction half-reaction is multiplied by 4. Therefore the number of electrons lost in oxidation and the number of electrons gained in reduction both equal 12.

$$\text{Oxidation: } 6H_2O(l) + 3S(s) \longrightarrow 3SO_2(g) + 12H^+(aq) + 12e^-$$
$$\text{Reduction: } 16H^+(aq) + 4NO_3^-(aq) + 12e^- \longrightarrow 4NO(g) + 8H_2O(l)$$

Step 6. Add the half-reactions to show an overall equation. Then subtract terms that appear on both sides of the equation. The equation is

$$6H_2O(l) + 3S(s) + 16H^+(aq) + 4NO_3^-(aq) + 12e^- \longrightarrow 3SO_2(g) + 12H^+(aq) + 12e^- + 4NO(g) + 8H_2O(l)$$

Subtracting terms that are on both the left and right produces

$$3S(s) + 4H^+(aq) + 4NO_3^-(aq) \longrightarrow 3SO_2(g) + 4NO(g) + 2H_2O(l)$$

Step 7. Add the spectator ions and balance the equation. Recall from Chapter 8 that like the spectators at an athletic event, spectator ions are present but do not participate in or change during a reaction. Because none of the ions in the reactants appear in the products, there are no spectator ions in this particular example. The balanced equation above is thus correct. However, it can be written to show the HNO_3 as un-ionized.

$$3S(s) + 4HNO_3(aq) \longrightarrow 3SO_2(g) + 4NO(g) + 2H_2O(l)$$

Sample Problem 22-7

Use the half-reaction method to balance the equation for the following redox reaction.

$$KMnO_4(aq) + HCl(aq) \longrightarrow MnCl_2(aq) + Cl_2(g) + H_2O(l) + KCl(aq)$$

1. *ANALYZE* *Plan a problem-solving strategy.*
Follow the seven steps of the half-reaction method to balance the equation.

2. *SOLVE* *Apply the problem-solving strategy.*
Step 1. Write the unbalanced equation in ionic form.

$$K^+(aq) + MnO_4^-(aq) + H^+(aq) + Cl^-(aq) \longrightarrow Mn^{2+}(aq) + 2Cl^-(aq) + Cl_2(g) + H_2O(l) + K^+(aq) + Cl^-(aq)$$

Step 2. Write separate half-reactions for the oxidation and reduction processes. Use oxidation numbers to determine the oxidation process and the reduction process.

Oxidation: $Cl^- \longrightarrow Cl_2$

Reduction: $MnO_4^- \longrightarrow Mn^{2+}$

Step 3. Balance the atoms in the half-reactions. Because the solution is acidic, use H_2O and H^+ to balance the oxygen and hydrogen if necessary. (If the solution were basic, H_2O and OH^- would be used.)

Oxidation: $2Cl^-(aq) \longrightarrow Cl_2(g)$ (atoms balanced)

Reduction: $MnO_4^-(aq) + 8H^+(aq) \longrightarrow Mn^{2+}(aq) + 4H_2O(l)$ (atoms balanced)

Step 4. Add electrons to balance the charges.

Oxidation: $2Cl^-(aq) \longrightarrow Cl_2(g) + 2e^-$ (charges balanced)

Reduction: $MnO_4^-(aq) + 8H^+(aq) + 5e^- \longrightarrow Mn^{2+}(aq) + 4H_2O(l)$ (charges balanced)

BIOLOGY

Bioluminescence

Some organisms produce light by means of oxidation–reduction reactions in a process known as bioluminescence. Light is given off when one of a class of

compounds called luciferins is oxidized by combination with oxygen. A product of these reactions is energy in the form of light. Bioluminescence serves different functions in different organisms. For example, fireflies use bioluminescence to attract mates, while some squids release a glowing cloud of "ink" to confuse and escape from their predators.

22.3

ACTIVITY

Students should be familiar with the "browning" of sliced potatoes or apples upon exposure to oxygen. The oxidation is catalyzed by enzymes in food. Explain that oxidation can be prevented or slowed down. The outer skin provides a natural physical barrier to oxygen. Chemical protection can be provided by antioxidants, which act as reducing agents or by inhibiting enzymes. Liquids such as lemon juice contain antioxidants.

Have students form groups of two or three and give each group a potato and apple. Provide beakers filled with tap water, water that has been boiled to exclude most dissolved oxygen, lemon juice, a carbonated beverage; plus sugar, salt, and vinegar solutions. Have the students quickly cut the potato and apple into slices and immediately place a slice in each beaker. Have them label locations on paper towels to match the labels on the beakers. Have students remove the slices after they have soaked for several minutes and place them on the paper towels. Have the students make periodic observations of the slices to determine the degree of "browning." On the basis of their observations, have the students decide which liquids offer the best protection. Have them try to explain why the liquids are good or bad protectants. They can present their findings in the form of a poster or written/oral report.

Practice Problems Plus

Related Chapter Review Problems
Chapter Review problems 36 and 37 are related to Sample Problem 22-7.

Additional Practice Problem
The Chem ASAP! CD-ROM contains the following problem: The following reaction takes place in basic solution. Use the half-reaction method to write a balanced ionic equation.
$Zn(s) + As_2O_3(aq) \rightarrow AsH_3(aq) + Zn^{2+}(aq)$
[$6Zn(s) + As_2O_3(aq) + 9H_2O \rightarrow + 6Zn^{2+}(aq) + 2AsH_3(aq) + 12OH^-(aq)$]

3 Assess

Evaluate Understanding

In order to evaluate the students' understanding of this section, ask:

- **Copper(II) hydroxide [$Cu(OH)_2$] is converted to copper(II) oxide (CuO) and water upon heating. Is this a redox reaction? Why?** (No, none of the elements undergoes a change in oxidation number.)
- **Copper(II) oxide (CuO) is converted to copper metal upon heating in the presence of carbon. Is this a redox reaction? Why?** (Yes, copper is reduced; its oxidation number decreases from +2 to 0. Carbon is oxidized; its oxidation number increases from 0 to +4 when it is converted to CO_2.)
- **Permanganate anion oxidizes iron(II) to iron(III) in acidic solution, and the purple color of the permanganate disappears. Write a balanced ionic equation for the reaction.** [$5Fe^{2+}(aq) + MnO_4^-(aq) + 8H^+(aq) \rightarrow 5Fe^{3+}(aq) + Mn^{2+}(aq) + 4H_2O(l)$]

Practice Problems

21. Write balanced ionic equations for the following reactions, which occur in acid solution. Use the half-reaction method.

a. $Sn^{2+}(aq) + Cr_2O_7^{2-}(aq) \longrightarrow Sn^{4+}(aq) + Cr^{3+}(aq)$

b. $CuS(s) + NO_3^-(aq) \longrightarrow Cu(NO_3)_2(aq) + NO_2(g) + SO_2(g)$

c. $I^-(aq) + NO_3^-(aq) \longrightarrow I_2(s) + NO(g)$

22. The following reactions take place in basic solution. Use the half-reaction method to write a balanced ionic equation for each.

a. $MnO_4^-(aq) + I^-(aq) \longrightarrow MnO_2(s) + I_2(s)$

b. $NiO_2(s) + S_2O_3^{2-}(aq) \longrightarrow Ni(OH)_2(s) + SO_3^{2-}(aq)$

c. $Zn(s) + NO_3^-(aq) \longrightarrow NH_3(aq) + Zn(OH)_4^{2-}(aq)$

Chem ASAP!

Problem-Solving 22
Solve Problem 22 with the help of an interactive guided tutorial.

Sample Problem 22-7 (cont.)

Step 5. Make the numbers of electrons equal. Multiply the oxidation half-reaction by 5 and the reduction half-reaction by 2, so that 10 electrons are lost in oxidation and 10 electrons are gained in reduction.

Oxidation: $10Cl^-(aq) \longrightarrow 5Cl_2(g) + 10e^-$

Reduction: $2MnO_4^-(aq) + 16H^+(aq) + 10e^- \longrightarrow 2Mn^{2+}(aq) + 8H_2O(l)$

Step 6. Add the half-reactions and subtract terms that appear on both sides of the equation.

$$10Cl^-(aq) + 2MnO_4^-(aq) + 16H^+(aq) + \cancel{10e^-} \longrightarrow 5Cl_2(g) + \cancel{10e^-} + 2Mn^{2+}(aq) + 8H_2O(l)$$

The equation, after subtracting terms that appear on both sides, is

$$10Cl^-(aq) + 2MnO_4^-(aq) + 16H^+(aq) \longrightarrow 5Cl_2(g) + 2Mn^{2+}(aq) + 8H_2O(l)$$

Step 7. Add the spectator ions and balance the equation. In the example, the permanganate ions come from $KMnO_4$. The chloride ions come from HCl. K^+ ions, which appear on both sides of the unbalanced ionic equation in Step 1, come from $KMnO_4$ and must equal the number of MnO_4^- ions (given the formula, $KMnO_4$). The K^+ ions are spectator ions as are some of the Cl^- ions from the HCl—namely, those that are still present as products combined with Mn^{2+} and K^+.

The number of spectator Cl^- ions combined with Mn^{2+} must be twice the number of Mn^{2+} ions (given the formula, $MnCl_2$). The number of spectator Cl^- ions combined with K^+ must equal the number of K^+ ions (given the formula, KCl). The spectator ions are now added to the equation and balanced. Spectator ions are shown in blue.

$$10Cl^- + 2MnO_4^- + 2K^+ + 16H^+ + 6Cl^- \longrightarrow 5Cl_2 + 2Mn^{2+} + 4Cl^- + 8H_2O + 2K^+ + 2Cl^-$$

Summing spectator and nonspectator Cl^- on each side gives

$$16Cl^-(aq) + 2MnO_4^-(aq) + 2K^+(aq) + 16H^+(aq) \longrightarrow 5Cl_2(g) + 2Mn^{2+}(aq) + 6Cl^-(aq) + 8H_2O(l) + 2K^+(aq)$$

The equation is balanced for atoms and charge. To show it balanced for the substances given in the question (rather than for ions), rewrite it as

$$2KMnO_4(aq) + 16HCl(aq) \longrightarrow 2MnCl_2(aq) + 5Cl_2(g) + 8H_2O(l) + 2KCl(aq)$$

3. *EVALUATE Does the result make sense?*
The equation is correctly balanced for atoms and charge.

MINI LAB

Bleach It! Oxidize the Color Away

PURPOSE

To test the effect of oxidizing agents on stains and dyes.

MATERIALS

- spot plate
- medicine dropper
- water
- colorimeter (optional)

Oxidizing agents

- liquid chlorine bleach (5% (m/v) sodium hypochlorite)
- powder bleach
- oxalic acid solution (1% (m/v))
- sodium thiosulfate solution (hypo) (0.2*M* $Na_2S_2O_3$)
- hydrogen peroxide (3% (v/v) H_2O_2)

Samples

- iodine solution (1% I_2 in 2% (m/v) KI)
- potassium permanganate solution (0.05*M* $KMnO_4$)
- grape juice
- rusty water
- piece of colored fabric
- colored flower petals
- grass stain on piece of white fabric

PROCEDURE

 *Sensor version available in the Probeware Lab Manual.*

1. Place samples on a spot plate. Use 4 drops of each liquid or a small piece of each solid.
2. Describe the color and appearance of each sample in Step 1.
3. Add a few drops of the first oxidizing agent to each sample.

4. Describe any immediate change in appearance and any further change after 15 minutes.
5. Repeat Steps 2–4 with each oxidizing agent.

ANALYSIS AND CONCLUSIONS

1. Make a grid and record your observations.
2. Compare the oxidizing power of the oxidizing agents.
3. How do you know that chemical changes have occurred?

section review 22.3

23. Balance each redox equation, using the oxidation-number–change method.

a. $ClO_3^-(aq) + I^-(aq) \longrightarrow Cl^-(aq) + I_2(aq)$ [acid solution]

b. $C_2O_4^{2-}(aq) + MnO_4^-(aq) \longrightarrow Mn^{2+}(aq) + CO_2(g)$ [acid solution]

c. $Br_2(l) + SO_2(g) \longrightarrow Br^-(aq) + SO_4^{2-}(aq)$ [acid solution]

24. Use the half-reaction method to write a balanced ionic equation for each reaction.

a. $MnO_4^-(aq) + ClO_2^-(aq) \longrightarrow MnO_2(s) + ClO_4^-(aq)$ [basic solution]

b. $Cr^{3+}(aq) + ClO^-(aq) \longrightarrow CrO_4^{2-}(aq) + Cl^-(aq)$ [basic solution]

c. $Mn^{3+}(aq) + I^-(aq) \longrightarrow Mn^{2+}(aq) + IO_3^-(aq)$ [basic solution]

 Chem ASAP! **Assessment 22.3** Check your understanding of the important ideas and concepts in Section 22.3.

22.3

Reteach

Point out that the distinguishing feature of redox reactions is the change in oxidation numbers that occurs. To decide whether a given reaction is a redox reaction, students need only compare the oxidation numbers of the reactants to the oxidation numbers of the products. If a change has occurred, the reaction is a redox reaction.

MINI LAB

Bleach It! Oxidize the Color Away

SAFETY

The oxidizing agents in this lab pose a slight contact hazard and should be handled carefully. Safety goggles should be worn at all times. Students should also be careful with splashing and spillage because the oxidizing solutions can bleach/destroy clothing; lab coats are recommended.

LAB PREP AND PLANNING

 *Probeware Lab Manual has Vernier, TI, and PASCO labs.*

Prepare the oxidizing agents and other solutions before class. Arrange to bring items such as grape juice and flower petals to the class on the day of the lab.

ANALYSIS AND CONCLUSIONS

1. Answers will vary.
2. Answers will vary.
3. Color changes occur.

Answers

SECTION REVIEW 22.3

23. a. $ClO_3^-(aq) + 6I^-(aq) + 6H^+(aq) \rightarrow Cl^-(aq) + 3I_2(aq) + 3H_2O(l)$

b. $5C_2O_4^{2-}(aq) + 2MnO_4^-(aq) + 16H^+(aq) \rightarrow 2Mn^{2+}(aq) + 10CO_2(g) + 8H_2O(l)$

c. $Br_2(l) + SO_2(g) + 2H_2O(l) \rightarrow 2Br^-(aq) + SO_4^{2-}(aq) + 4H^+(aq)$

24. a. $4MnO_4^-(aq) + 3ClO_2^-(aq) + 2H_2O(l) \rightarrow 4MnO_2(s) + 3ClO_4^-(aq) + 4OH^-(aq)$

b. $2Cr^{3+}(aq) + 3ClO^-(aq) + 10OH^-(aq) \rightarrow 2CrO_4^{2-}(aq) + 3Cl^-(aq) + 5H_2O(l)$

c. $6Mn^{3+}(aq) + I^-(aq) + 6OH^-(aq) \rightarrow 6Mn^{2+}(aq) + IO_3^-(aq) + 3H_2O(l)$

SMALL-SCALE LAB

HALF REACTIONS

OVERVIEW

Procedure Time: 30 minutes
Students will observe redox reactions between acids and metals and learn to write half reactions to describe what is happening.

MATERIALS

Solution	Preparation
1.0*M* HCl	82 mL of 12*M* in 1.0 L.
1.0*M* HNO_3	63 mL of 15.8*M* in 1.0 L
1.0*M* H_2SO_4	56 mL of 18*M* in 1.0 L

CAUTION! Always add acid to water carefully and slowly!

TEACHING SUGGESTIONS

Emphasize that a half reaction shows what happens to electrons in redox reactions. Use magnesium turnings or cut magnesium ribbon into 1/2-cm pieces. Use galvanized nails for zinc, staples from a stapler for iron, and pennies for copper. Recycle all the metals except the staples.

YOU'RE THE CHEMIST

1. Add a drop of any acid to the damaged part of the penny and notice that only the zinc interior reacts.
2. Many toilet-bowl cleaners and vinegar dissolve metals. Keep products containing acids away from metal pipes and fixtures.

SMALL-SCALE LAB

HALF-REACTIONS

SAFETY    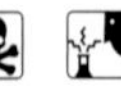

Wear safety glasses and follow the standard safety procedures outlined on page 18.

PURPOSE

To observe redox reactions and to write half-reactions that describe them.

MATERIALS

- pencil
- paper
- ruler
- reaction surface
- chemicals shown in Figure A

PROCEDURE

On separate sheets of paper, draw two grids similar to Figure A. Make each square 2 cm on each side.

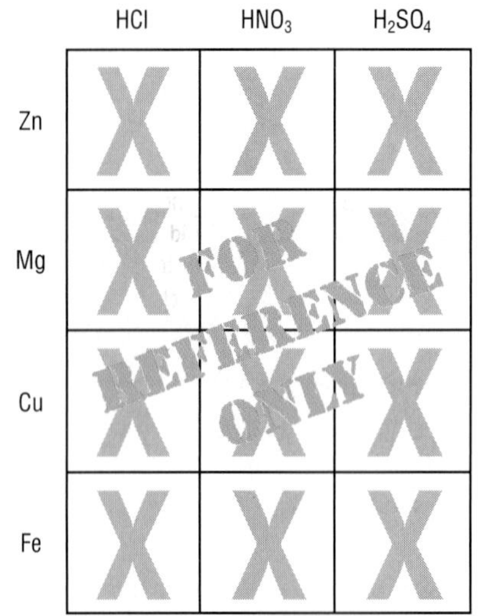

Figure A

Place a reaction surface over one of the grids and add one drop of each acid solution to one piece of each metal, as shown in Figure A. Use the second grid as a data table to record your observations for each solution.

ANALYSIS

Using your experimental data, record the answers to the following questions below your data table.

1. Which metal is the most reactive? On what observation do you base your answer? Which metal did not react with any of the acids? List the metals in order of decreasing reactivity.
2. What is the chemical formula of the gas produced in each reaction?
3. An active metal reacts with an acid to produce hydrogen gas and a salt. Write equations and net ionic equations to describe the reactions you observed. Are all of these redox reactions? Explain.
4. The half-reaction for the oxidation of Zn is
$$Zn(s) \longrightarrow Zn^{2+}(aq) + 2e^-$$
Write the oxidation half-reaction for the other metals that react.
5. The half-reaction for the reduction of H^+ from the acid is
$$2H^+ + 2e^- \longrightarrow H_2(g)$$
Notice that this half-reaction is the same for all the acids. Demonstrate how adding this half-reaction to each oxidation half-reaction results in the overall net ionic equations.

YOU'RE THE CHEMIST

The following small-scale activities allow you to develop your own procedures and analyze the results.

1. **Analyze It!** Pennies minted after 1982 are made of zinc with a thin copper coating. Find a penny that has been damaged so that a portion of the zinc shows through. Carry out experiments to compare the reactivity of the zinc and the copper toward various acids. What are your results?
2. **Design It!** Many household products, such as toilet-bowl cleaners and vinegar, contain acids. Design and carry out experiments to find out if these products also dissolve metals. When using these household products in your home, what precautions should you take?

ANALYSIS

1. Mg is most reactive because it bubbles most vigorously. Cu does not react. The order of reactivity is Mg > Zn > Fe > Cu.
2. $H_2(g)$ is produced in all the reactions.
3. $Mg(s) + 2HCl(aq) \rightarrow H_2(g) + MgCl_2(aq)$
$Mg(s) + 2H^+(aq) \rightarrow H_2(g) + Mg^{2+}(aq)$
$Fe(s) + 2HCl(aq) \rightarrow H_2(g) + FeCl_2(aq)$
$Fe(s) + 2H^+(aq) \rightarrow H_2(g) + Fe^{2+}(aq)$
All are redox reactions because the oxidation numbers of reactants change.
4. $Mg(s) \rightarrow Mg^{2+}(aq) + 2e^-$
$Fe(s) \rightarrow Fe^{2+}(aq) + 2e^-$
5. $$\begin{array}{l} 2H^+ + 2e^- \rightarrow H_2(g) \\ \quad Mg(s) \rightarrow Mg^{2+}(aq) + 2e^- \\ \hline Mg(s) + 2H^+(aq) \rightarrow H_2(g) + Mg^{2+}(aq) \end{array}$$

Chemistry Serving...Society

JUST ADD COLD WATER FOR A HOT MEAL ON THE GO

People spend a considerable amount of time and energy trying to prevent or inhibit the simple redox reaction called corrosion. Many metals oxidize when exposed to the oxygen in air. To keep them from corroding or rusting, the metals can be painted, plated with chromium, or galvanized with zinc. But corrosion is not all bad. Because the oxidation of metal is generally an exothermic reaction, it can be used to generate useful heat—as long as the oxidation takes place quickly enough to release the heat energy in a short span of time.

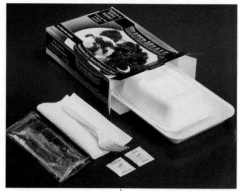

Researchers working for the U.S. Army have put corrosion to work to solve a long-standing problem: how to provide soldiers with hot meals when there is no time to cook. They developed the Flameless Ration Heater (FRH), a simple way of heating precooked, packaged meal rations. All a soldier has to do is put a meal pouch in an FRH and add water. After about 15 minutes, the food has been heated to a warm 60 °C. Already proven in the battlefield, the FRH can also be used by forest firefighters, emergency crews, and other people who need quick, warm food under demanding conditions.

Researchers working for the U.S. Army have put corrosion to work to solve a long-standing problem: how to provide soldiers with hot meals when there is no time to cook.

The Flameless Ration Heater is based on the reaction in which magnesium metal is oxidized by water to yield magnesium hydroxide, hydrogen, and heat.

$$Mg + 2H_2O \longrightarrow Mg(OH)_2 + H_2 + \text{heat}$$

Just 24 grams of magnesium (1 mol) releases 355 000 joules (85 kcal) of heat, enough energy to boil approximately one liter of water.

But developing the FRH was not as simple as packaging a few grams of magnesium in a box. When exposed to air, magnesium oxidizes to magnesium oxide (MgO), which forms a thin film on the metal's surface. This film prevents further oxidation of the metal, either by air or by water. So the researchers had to find a way of letting water get through the MgO film.

They discovered that adding common table salt (NaCl) and powdered iron to the magnesium would do the trick. When water is added to this mixture, the chloride ions (Cl^-) react with the MgO film to form MgOHCl, which eats away at the remaining MgO.

Once the MgO film is gone, the water can react directly with the magnesium metal. The powdered iron seems to help this reaction take place more rapidly than it normally would. The researchers do not understand exactly how the iron helps, but they theorize that it assists in the transfer of electrons between the water and the magnesium. In any case, with the iron present, the reaction occurs rapidly and vigorously. As a result of putting corrosion to work, the energy stored in the chemical bonds of the reactants is released in just a few minutes—and you have a hot meal!

CHEMISTRY IN CAREERS

TAKE IT TO THE NET

Find out more about career opportunities:

www.phschool.com

MECHANICAL ENGINEER

Use redox reactions to invent new electrical technology.

See page 878.

CHEMICAL SPECIALIST

Local food service distributor seeks responsible self-motivated individ-

CHEMIST NEEDED

Enviro. Lab. GC Must know EPA methods, troubleshoot-

Chemistry Serving... Society

DISCUSS

After students have read the article, ask them why water is needed to activate the flameless ration heater. (It serves as an oxidizing agent.) Have students write the two half-reactions and indicate the oxidation states of the elements before and after the reaction. Ask:

- What is the reducing agent in this reaction? (Mg)

Ask students to name other metals that might react with water to generate heat and hydrogen gas.

CHEMISTRY IN CAREERS

Encourage students to connect to the Internet address shown on this page to find out more about careers in mechanical engineering. Information may also be found on page 878 of this text.

4 Close

Summary

Ask students to define the key terms for the chapter. Have students define the terms operationally, applying the concepts to examples of redox reactions.

Extension

Despite oxygen's essential role in sustaining life, it can be destructive. It corrodes metals and causes food to become rancid. Out of control combustion can destroy both human structures and natural habitats. Scientists suspect that oxygen may contribute to the aging process. Have students research the damage caused by "free radicals," which are extremely reactive oxygen compounds found within cells. Students may also research antioxidants, compounds that protect against "free radical" damage. Students can prepare a report or poster, which can be presented to the class.

Looking Back . . . Looking Ahead . . .

The preceding chapter discussed neutralization reactions, which generally do not involve electron transfer. The current chapter discusses redox reactions, which do involve electron transfer. It describes how to determine oxidation numbers, recognize redox reactions, and balance redox equations. The next chapter addresses electrochemistry, a specific class of redox reactions.

Chapter 22 STUDENT STUDY GUIDE

For interactive study and review, go to www.phschool.com

KEY TERMS

- half-reaction *p. 665*
- half-reaction method *p. 665*
- oxidation *pp. 645, 647*
- oxidation number *p. 654*
- oxidation-number–change method *p. 663*
- oxidation–reduction reaction *p. 647*
- oxidizing agent *p. 649*
- redox reaction *p. 647*
- reducing agent *p. 649*
- reduction *p. 647*

KEY RELATIONSHIPS

X loses electron(s). X is oxidized. X is the reducing agent. X increases oxidation number.	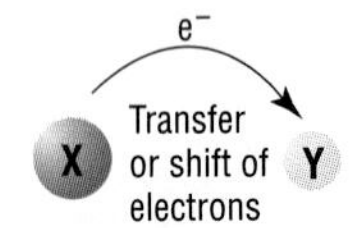	Y gains electron(s). Y is reduced. Y is the oxidizing agent. Y decreases oxidation number.

CONCEPT SUMMARY

22.1 The Meaning of Oxidation and Reduction

- Oxidation is the gain of oxygen or the loss of electrons. Reduction is the loss of oxygen or the gain of electrons.
- An oxidation process is always accompanied by a reduction process. The substance that does the oxidizing is called an oxidizing agent. The substance that does the reducing is called a reducing agent.

22.2 Oxidation Numbers

- Oxidation numbers help to keep track of electrons in redox reactions. An oxidation-number increase is oxidation. An oxidation-number decrease is reduction.
- The oxidation number of an element in an uncombined state is zero. The oxidation number of a monatomic ion is the same in magnitude and sign as its ionic charge. The sum of the oxidation numbers of the elements in a neutral compound is zero.

22.3 Balancing Redox Equations

- The oxidation-number–change method for balancing redox equations involves determining the oxidation-number change of the substances that are oxidized and reduced. Coefficients are used to make the increase in oxidation number equal to the decrease.
- In the half-reaction method, which is another way to write a balanced equation for a redox reaction, the net ionic equation is first divided into two half-reactions. One is for the oxidation and the other is for the reduction. Each half-reaction is balanced independently for atoms. H^+, OH^-, H_2O are added as needed. The charge on both sides is balanced by adding electrons. The half-reactions are then multiplied by factors to make the number of electrons lost equal to the number of electrons gained. Finally, the half-reactions are added.

CHAPTER CONCEPT MAP

Use these terms to construct a concept map that organizes the major ideas of this chapter.

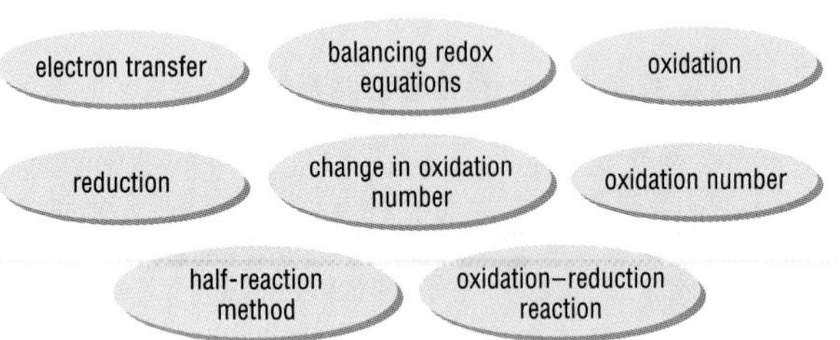

Chem ASAP! Concept Map 22
Create your Concept Map using the computer.

672 Chapter 22

Take It to the Net

At **www.phschool.com** students will find for this chapter

- an Internet activity
- links to related chemistry sites
- an interactive quiz
- career links

Chapter 22 REVIEW

CONCEPT PRACTICE

25. What chemical process must always accompany a reduction process? *22.1*

26. Balance each redox equation. *22.1*
a. $Ba(s) + O_2(g) \longrightarrow BaO(s)$
b. $CaO(s) + Al(s) \longrightarrow Al_2O_3(s) + Ca(s)$
c. $C_2H_5OH(l) + O_2(g) \longrightarrow CO_2(g) + H_2O(l)$

27. Balance each redox equation. *22.1*
a. $C_2H_4(g) + O_2(g) \longrightarrow CO_2(g) + H_2O(l)$
b. $KClO_3(s) \longrightarrow KCl(s) + O_2(g)$
c. $CuO(s) + H_2(g) \longrightarrow Cu(s) + H_2O(l)$
d. $H_2(g) + O_2(g) \longrightarrow H_2O(l)$

28. In redox reactions, a reduction is a gain. Explain. *22.2*

29. How do electrons lost and electrons gained compare in every redox reaction? *22.3*

30. Balance each redox equation. *22.3*
a. $Al(s) + Cl_2(g) \longrightarrow AlCl_3(s)$
b. $Al(s) + Fe_2O_3(s) \longrightarrow Al_2O_3(s) + Fe(s)$
c. $Cl_2(g) + KOH(aq) \longrightarrow KClO_3(aq) + KCl(aq) + H_2O(l)$
d. $HNO_3(aq) + H_2S(g) \longrightarrow S(s) + NO(g) + H_2O(l)$
e. $KIO_4(aq) + KI(aq) + HCl(aq) \longrightarrow KCl(aq) + I_2(s) + H_2O(l)$

31. Name some general types of reactions that are typically redox reactions. *22.3*

32. Identify which of these unbalanced equations represent redox reactions. *22.3*
a. $Li(s) + H_2O(l) \longrightarrow LiOH(aq) + H_2(g)$
b. $K_2Cr_2O_7(aq) + HCl(aq) \longrightarrow KCl(aq) + CrCl_3(aq) + H_2O(l) + Cl_2(g)$
c. $Al(s) + HCl(aq) \longrightarrow AlCl_3(aq) + H_2(g)$
d. $P_4(s) + S_8(s) \longrightarrow P_2S_5(s)$

33. Which of these unbalanced equations represent redox reactions? *22.3*
a. $MnO(s) + PbO_2(s) \longrightarrow MnO_4^-(aq) + Pb^{2+}(aq)$ [acidic]
b. $Cl_2(g) + H_2O(l) \longrightarrow HCl(aq) + HClO(aq)$
c. $I_2O_5(s) + CO(g) \longrightarrow I_2(s) + CO_2(g)$
d. $H_2O(l) + SO_3(g) \longrightarrow H_2SO_4(aq)$

34. For each redox equation in Problem 32, identify the oxidizing agent and reducing agent. *22.3*

35. For each redox equation in Problem 33, identify the oxidizing agent and reducing agent. *22.3*

CONCEPT MASTERY

36. Balance the equations in Problem 32 by an appropriate method.

37. Balance the equations in Problem 33 by an appropriate method.

38. Determine the oxidation number of phosphorus in each substance.
a. P_4O_8 **c.** P_2O_5 **e.** $H_2PO_4^-$
b. PO_4^{3-} **d.** P_4O_6 **f.** PO_3^{3-}

39. Sodium chlorite is a powerful bleaching agent used in the paper and textile industries. It is prepared by this reaction.

$$4NaOH(aq) + Ca(OH)_2(aq) + C(s) + 4ClO_2(g) \longrightarrow 4NaClO_2(aq) + CaCO_3(s) + 3H_2O(l)$$

a. Identify the element oxidized in this reaction.
b. What is the oxidizing agent?

40. Identify the element oxidized, the element reduced, the oxidizing agent, and the reducing agent in each unbalanced redox equation.
a. $MnO_2(s) + HCl(aq) \longrightarrow MnCl_2(aq) + Cl_2(g) + H_2O(l)$
b. $Cu(s) + HNO_3(aq) \longrightarrow Cu(NO_3)_2(aq) + NO_2(g) + H_2O(l)$
c. $P(s) + HNO_3(aq) + H_2O(l) \longrightarrow NO(g) + H_3PO_4(aq)$
d. $Bi(OH)_3(s) + Na_2SnO_2(aq) \longrightarrow Bi(s) + Na_2SnO_3(aq) + H_2O(l)$

41. Balance each redox equation in Problem 40 by using the oxidation-number–change method.

CRITICAL THINKING

42. Choose the term that best completes the second relationship.
a. oxidation : reduction
fill up : ______
(1) empty (3) dry out
(2) overflow (4) splash
b. oxidizing agent : Group 7A
reducing agent : ______
(1) Group 6A (3) Group 0
(2) Group 1A (4) Group 4A

43. Many decomposition, single-replacement, combination, and combustion reactions are also redox reactions. Why is a double-replacement reaction never a redox reaction?

Answers

25. oxidation
26. a. $2Ba(s) + O_2(g) \rightarrow 2BaO(s)$
b. $3CaO(s) + 2Al(s) \rightarrow Al_2O_3(s) + 3Ca(s)$
c. $C_2H_5OH(l) + 3O_2(g) \rightarrow 2CO_2(g) + 3H_2O(l)$
27. a. $C_2H_4(g) + 3O_2(g) \rightarrow 2CO_2(g) + 2H_2O(l)$
b. $2KClO_3(s) \rightarrow 2KCl(s) + 3O_2(g)$
c. $CuO(s) + H_2(g) \rightarrow Cu(s) + H_2O(l)$
d. $2H_2(g) + O_2(g) \rightarrow 2H_2O(l)$
28. An element gains one or more electrons.
29. must be equal
30. a. $2Al(s) + 3Cl_2(g) \rightarrow 2AlCl_3(s)$
b. $2Al(s) + Fe_2O_3(s) \rightarrow Al_2O_3(s) + 2Fe(s)$
c. $3Cl_2(g) + 6KOH(aq) \rightarrow KClO_3(aq) + 5KCl(aq) + 3H_2O(l)$
d. $2HNO_3(aq) + 3H_2S(aq) \rightarrow 3S(s) + 2NO(g) + 4H_2O(l)$
e. $KIO_4(aq) + 7KI(aq) + 8HCl(aq) \rightarrow 8KCl(aq) + 4I_2(s) + 4H_2O(l)$
31. single-replacement, combustion, combination, and decomposition

32. a., b., c., d.
33. a., b., c.
34. a. Li, reducing agent; H, oxidizing agent
b. Cr, oxidizing agent; Cl, reducing agent
c. Al, reducing agent; H, oxidizing agent
d. P, reducing agent; S, oxidizing agent
35. a. Mn, reducing agent; Pb, oxidizing agent
b. Cl, oxidizing agent; Cl, reducing agent
c. I, oxidizing agent; C, reducing agent
36. a. $2Li(s) + 2H_2O(l) \rightarrow 2LiOH(aq) + H_2(g)$
b. $K_2Cr_2O_7(aq) + 14HCl(aq) \rightarrow 2KCl(aq) + 2CrCl_3(aq) + 7H_2O(l) + 3Cl_2(g)$
c. $2Al(s) + 6HCl(aq) \rightarrow 2AlCl_3(aq) + 3H_2(g)$
d. $4P_4(s) + 5S_8(s) \rightarrow 8P_2S_5(s)$
37. a. $2MnO(s) + 5PbO_2(s) + 8H^+ \rightarrow 2MnO_4^-(aq) + 5Pb^{2+}(aq) + 4H_2O(l)$
b. $Cl_2(g) + H_2O(l) \rightarrow HCl(aq) + HClO(aq)$
c. $I_2O_5(s) + 5CO(g) \rightarrow I_2(s) + 5CO_2(g)$
d. $H_2O(l) + SO_3(g) \rightarrow H_2SO_4(aq)$
38. a. +4 **c.** +5 **e.** +5
b. +5 **d.** +3 **f.** +3
39. a. C **b.** Cl
40. a. Cl oxidized, Mn reduced, Mn oxidizing agent, Cl reducing agent
b. Cu oxidized, N reduced, N oxidizing agent, Cu reducing agent
c. P oxidized, N reduced, N oxidizing agent, P reducing agent
d. Sn oxidized, Bi reduced, Bi oxidizing agent, Sn reducing agent
41. a. $MnO_2(s) + 4HCl(aq) \rightarrow MnCl_2(aq) + Cl_2(g) + 2H_2O(l)$
b. $Cu(s) + 4HNO_3(aq) \rightarrow Cu(NO_3)_2(aq) + 2NO_2(g) + 2H_2O(l)$
c. $3P(s) + 5HNO_3(aq) + 2H_2O(l) \rightarrow 5NO(g) + 3H_3PO_4(aq)$
d. $2Bi(OH)_3(s) + 3Na_2SnO_2(aq) \rightarrow 2Bi(s) + 3Na_2SnO_3(aq) + 3H_2O(l)$

Answers

42. a. 1
b. 2

43. Double-replacement reactions never involve the transfer of electrons; instead, they involve the exchange of positive ions in aqueous solution.

44. a. 5.00
b. 10.00
c. 13.00
d. 6.52

45. a. acidic
b. basic
c. basic
d. acidic

46. a. NH_4^+ and NH_3; H_2O and H_3O^+
b. H_2SO_3 and HSO_3^-; NH_2^- and NH_3
c. HNO_3 and NO_3^-; I^- and HI

47. 56.3 mL KOH

48. a. $1.0 \times 10^{-2}M$
b. $1.0 \times 10^{-11}M$
c. $1.6 \times 10^{-9}M$

49. $1.14N\ H_3PO_4$

50. solubility $PbBr_2 = 8.1 \times 10^{-3}M$

51. Test tube B has the NaCl added to it. Due to the common-ion effect, the addition of either sulfate ion or barium ion to a saturated solution of $BaSO_4$ will cause the solubility product of $BaSO_4$ to be exceeded, and barium sulfate will precipitate, as shown in test tubes A and C.

52. Dilute 110 mL of $6.0M$ HCl to 440 mL total volume.

CUMULATIVE REVIEW

44. Calculate the pH of solutions with the following hydrogen-ion or hydroxide-ion concentrations.
a. $[H^+] = 0.000\ 010M$
b. $[OH^-] = 1.0 \times 10^{-4}M$
c. $[OH^-] = 1.0 \times 10^{-1}M$
d. $[H^+] = 3.0 \times 10^{-7}M$

45. Classify each solution in Problem 44 as acidic, basic, or neutral.

46. Identify the conjugate acid–base pairs in each equation.
a. $NH_4^+(aq) + H_2O(l) \longrightarrow NH_3(aq) + H_3O^+(aq)$
b. $H_2SO_3(aq) + NH_2^-(aq) \longrightarrow HSO_3^-(aq) + NH_3(aq)$
c. $HNO_3(aq) + I^-(aq) \longrightarrow HI(aq) + NO_3^-(aq)$

47. How many milliliters of a $4.00M$ KOH solution are needed to neutralize 45.0 mL of $2.50M$ H_2SO_4 solution?

48. What is the hydrogen-ion concentration of solutions with the following pH?
a. 2.00 **b.** 11.00 **c.** 8.80

49. What is the normality of the solution prepared by dissolving 46.4 g H_3PO_4 in enough water to make 1.25 L of solution?

50. The K_{sp} of lead(II) bromide ($PbBr_2$) at 25 °C is 2.1×10^{-6}. What is the solubility of $PbBr_2$ (in mol/L) at this temperature?

51. Bottles containing $0.1M$ solutions of Na_2SO_4, $BaCl_2$, and NaCl have had their labels accidentally switched. To discover which bottle contains the NaCl, you set up the following test. You place a clear saturated solution of $BaSO_4$ ($K_{sp} = 1.1 \times 10^{-10}$) into three test tubes. To each test tube you add a few drops of each mislabeled solution. The results are shown below. To which tube was NaCl added? Explain.

Test tube A

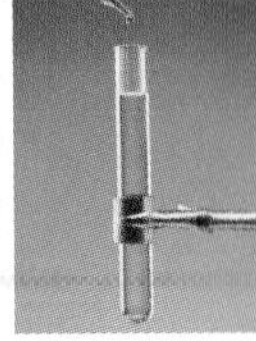
Test tube B

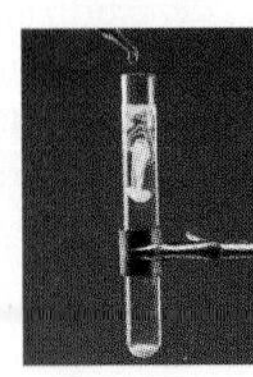
Test tube C

52. How would you make 440 mL of $1.5M$ HCl solution from a stock solution of $6.0M$ HCl?

53. The complete combustion of hydrocarbons involves the combination of both carbon and hydrogen atoms with oxygen. Carbon combines with oxygen to form carbon dioxide, and hydrogen combines with oxygen to form water. The following table lists the moles of O_2 used and the moles of CO_2 and moles of H_2O produced when a series of hydrocarbons called alkanes are burned.

Alkane burned	O_2 used (mol)	CO_2 produced (mol)	H_2O produced (mol)
CH_4	2	1	2
C_2H_6	3.5	2	3
C_3H_8	5	3	4
C_4H_{10}			
C_5H_{12}			
C_6H_{14}			

a. Complete the table.
b. Based on the data, write a balanced generalized equation for the complete combustion of any alkane. Use the following form, and write the coefficients in terms of x and y:

$$C_xH_y + \_\_\ O_2 \longrightarrow \_\_\ CO_2 + \_\_\ H_2O$$

CONCEPT CHALLENGE

54. How many grams of copper are needed to reduce completely the silver ions in 85.0 mL of $0.150M$ $AgNO_3(aq)$ solution?

55. How many milliliters of $0.280M$ $K_2Cr_2O_7(aq)$ solution are needed to oxidize 1.40 g of sulfur? First balance the equation.

$$K_2Cr_2O_7(aq) + H_2O(l) + S(s) \longrightarrow SO_2(g) + KOH(aq) + Cr_2O_3(aq)$$

56. Carbon monoxide can be removed from the air by passing it over solid diiodine pentoxide.

$$CO(g) + I_2O_5(s) \longrightarrow I_2(s) + CO_2(g)$$

a. Balance the equation.
b. Identify the element being oxidized and the element being reduced.
c. How many grams of carbon monoxide can be removed from the air by 0.55 g of diiodine pentoxide (I_2O_5)?

53. a.

Alkane burned	O_2 used (mol)	CO_2 produced (mol)	H_2O produced (mol)
C_4H_{10}	6.5	4	5
C_5H_{12}	8	5	6
C_6H_{14}	9.5	6	7

b. $C_xH_y + [x + (y/4]O_2 \rightarrow xCO_2 + (y/2)H_2O$

54. 0.406 g Cu

55. 104 mL $K_2Cr_2O_7$

56. a. $5CO + I_2O_5 \rightarrow I_2 + 5CO_2$
b. C is oxidized. I is reduced.
c. 0.22 g CO

Chapter 22 STANDARDIZED TEST PREP

Select the choice that best answers each question or completes each statement.

1. Which of these processes is *not* an oxidation?
   a. a decrease in oxidation number
   b. a complete loss of electrons
   c. a gain of oxygen
   d. a loss of hydrogen by a covalent molecule
2. In which of these pairs of nitrogen-containing ions and compounds is the oxidation number of nitrogen in the ion higher than in the nitrogen compound?
   I. N_2H_4 and NH_4^+
   II. NO_3^- and N_2O_4
   III. N_2O and NO_2^-
   a. I only
   b. I and II only
   c. I and III only
   d. II and III only
   e. I, II, and III
3. Identify the elements oxidized and reduced in this reaction.
   $ClO^- + H_2 + 2e^- \longrightarrow Cl^- + 2OH^-$
   a. Cl is oxidized; H is reduced
   b. H is oxidized; Cl is reduced
   c. Cl is oxidized; O is reduced
   d. O is oxidized; Cl is reduced
4. Which of these half-reactions represents a reduction?
   I. $Fe^{2+} \longrightarrow Fe^{3+}$
   II. $Cr_2O_7^{2-} \longrightarrow Cr^{3+}$
   III. $MnO_4^- \longrightarrow Mn^{2+}$
   a. I and II only
   b. II and III only
   c. I and III only
   d. I, II, and III
5. Which of these general types of reactions is *not* a redox reaction?
   a. single replacement
   b. double replacement
   c. combustion
   d. combination
6. What is the reducing agent in this unbalanced redox reaction?
   $MnO_4^- + SO_2 \longrightarrow Mn^{2+} + SO_4^{2-}$
   a. SO_2
   b. SO_4^{2-}
   c. Mn^{2+}
   d. MnO_4^-

Use the table to answer questions 7–11.

Arrow	Group	Metal	Metal ion	Arrow
	3	K	K^+	
	3	Ca	Ca^{2+}	
	3	Na	Na^+	
	4	Mg	Mg^{2+}	
	4	Al	Al^{3+}	
	4	Zn	Zn^{2+}	
1	4	Fe	Fe^{2+}	2
	4	Ni	Ni^{2+}	
	4	Sn	Sn^{2+}	
	4	Pb	Pb^{2+}	
	5	Cu	Cu^{2+}	
	5	Hg	Hg^{2+}	
	5	Ag	Ag^+	
	5	Au	Au^{3+}	

7. Which arrow indicates increasing ease of oxidation? Of reduction?
8. Which numbered group of metals are the strongest reducing agents?
9. Which numbered group of metals are the most difficult to oxidize?
10. Which is a stronger oxidizing agent, Na or Ni?
11. Which is a stronger reducing agent, Mg or Hg?

Use this diagram to answer questions 12 and 13. It shows the formation of an ion from an atom.

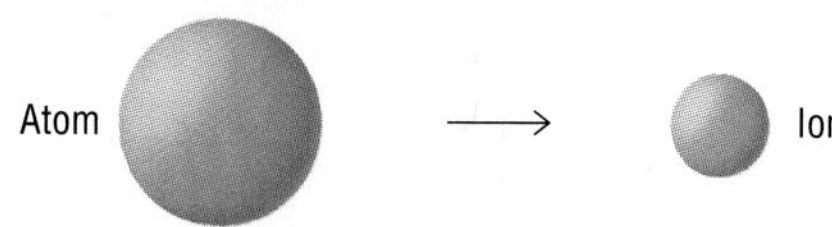

12. Does the diagram represent an oxidation or a reduction? Does the oxidation number increase or decrease when the ion forms?
13. *You Draw It!* Draw a diagram showing the formation of a sulfide ion from a sulfur atom. Make the relative sizes of the atom and ion realistic. Does your drawing represent an oxidation or a reduction?

1. a
2. d
3. b
4. b
5. b
6. a
7. arrow 1; arrow 2
8. group 3
9. group 5
10. nickel
11. magnesium
12. oxidation; oxidation number increases
13. sulfide ion should be larger than sulfur atom; reduction

Planning Guide

SECTION OBJECTIVES	ACTIVITIES/FEATURES	MEDIA & TECHNOLOGY
23.1 Electrochemical Cells ● ■ ◆ ▸ Describe how redox reactions interconvert electrical energy and chemical energy ▸ Explain the structure of a dry cell and identify the substances that are oxidized and reduced	**SE** Discover It! *A Lemon Battery,* p. 676 SSLM 33: *Small-Scale Voltaic Cells* **TE** DEMOS, pp. 677, 678, 681, 682	**ASAP** Assessment 23.1 **ACT** 17: *Electrochemistry* **SSV** 4: *Redox Reactions* **SSV** 5: *Small-Scale Voltaic Cells* **CHM** Side 8, 20: *A Closer Look at Voltaic Cells* **CHM** Side 8, 34: *Power from Voltaic Cells* **CHM** Side 8, 36: *The Dry Cell* **OT** 71: *Voltaic Cell*
23.2 Half-Cells and Cell Potentials □ ◆ ▸ Define standard cell potential and standard reduction potential ▸ Use standard reduction potentials to calculate standard cell potential	LM 45: *Corrosion* **TE** DEMO, p. 687	**ASAP** Simulation 27 **ASAP** Problem Solving 9, 10 **ASAP** Assessment 23.2 **ACT** 17: *Electrochemistry*
23.3 Electrolytic Cells ○ ■ ◆ ▸ Distinguish between electrolytic and voltaic cells, and list some possible uses of electrolytic cells ▸ Identify the products of the electrolysis of brine, molten sodium chloride, and water	**SE** **Link to Metallurgy** *Anodizing,* p. 695 **SE** Mini Lab *Tarnish Removal,* p. 697 (LRS 23-1) **SE** Small-Scale Lab *Electrolysis of Water,* p. 698 (LRS 23-2) **SE** **Chemistry Serving . . . the Environment** *How Many Kilometers Per Charge?,* p. 699 **SE** **Chemistry in Careers** *Electrochemist,* p. 699 LM 46: *Electrochemistry* **TE** DEMO, p. 693	**ASAP** Animation 27 **ASAP** Assessment 23.3 **ASAP** Concept Map 23 **SSV** 6: *Electrolysis of Solutions* **CHM** Side 8, 23: *A Closer Look at Electrochemical Cells* **CHM** Side 8, 13: *Automobile Manufacturing* **OT** 72: *Voltaic and Electrolytic Cells* **PP** Chapter 23 Problems **RP** Lesson Plans, Resource Library **AR** Computer Test 23 **www** Activities, Self-Tests, *SCIENCE NEWS* updates **TCP** The Chemistry Place Web Site

KEY

● Conceptual (concrete concepts)	**AR** Assessment Resources	**GRS** Guided Reading and Study Workbook
○ Conceptual (more abstract/math)	**ASAP** Chem ASAP! CD-ROM	LM Laboratory Manual
■ Standard (core content)	**ACT** ActivChemistry CD-ROM	**LP** Laboratory Practicals
□ Standard (extension topics)	**CHM** CHEMedia Videodiscs	LRS Laboratory Recordsheets
◆ Honors (core content)	**CA** Chemistry Alive! Videodiscs	SSLM Small-Scale Lab Manual
◇ Honors (options to accelerate)	**GCP** Graphing Calculator Problems	

PRACTICE		ASSESSMENT	
GRS	Section 23.1	SE	Section Review
RM	Section Review 23.1	RM	Section Review 23.1
		RM	Chapter 23 Quiz
SE	**Sample Problems** 23-1 to 23-2	SE	Section Review
GRS	Section 23.2	RM	Section Review 23.2
RM	Practice Problems 23.2	RM	Chapter 23 Quiz
GRS	Section 23.3	SE	Section Review
RM	Practice Problems 23.3	RM	Section Review 23.3
RM	Interpreting Graphics	LP	Lab Practical 23-1
		RM	Vocabulary Review 23
		SE	Chapter Review
		SM	Chapter 23 Solutions
		SE	Standardized Test Prep
		PHAS	Chapter 23 Test Prep
		RM	Chapter 23 Quiz
		RM	Chapter 23 A & B Test

OT	Overhead Transparency	**SE**	Student Edition
PHAS	PH Assessment System	**SM**	Solutions Manual
PLM	Probeware Lab Manual	**SSV**	Small-Scale Video/Videodisc
PP	Problem Pro CD-ROM	**TCP**	www.chemplace.com
RM	Review Module	**TE**	Teacher's Edition
RP	Resource Pro CD-ROM	**www**	www.phschool.com

PLANNING FOR ACTIVITIES

STUDENT EDITION

Discover It! p. 676
- whole lemon
- copper strip
- zinc strip
- sandpaper
- voltmeter

Mini Lab, p. 697
- baking soda
- water
- hot plate
- saucepan
- tablespoon
- shallow dish
- aluminum foil
- tarnished silver utensil
- tongs

Small-Scale Lab, p. 698
- pencil, paper, ruler
- reaction surface
- electrolysis device
- water
- Na_2SO_4, KI, NaCl, $CuSO_4$, KBr, bromthymol blue, starch

TEACHER'S EDITION

Teacher Demo, p. 677
- radio that works with cord or batteries

Teacher Demo, p. 678
- 200 mL of 0.1*M* silver nitrate ($AgNO_3$)
- 250-mL beaker
- glass stirring rod
- polished copper strip
- 50% molar excess of NaCl
- plastic container

Teacher Demo, p. 681
- dry cell
- hacksaw
- zinc electrode
- carbon electrode
- saturated ammonium chloride solution
- powdered manganese dioxide
- voltmeter

Teacher Demo, p. 682
- two lead strips
- wooden rod
- 250-mL beaker
- sulfuric acid
- 6-V DC power supply
- wire
- bell

Teacher Demo, p. 687
- pennies
- metal wires from twist ties
- table salt
- tap water
- paper towel

Teacher Demo, p. 693
- Hoffman apparatus

  or

  iron nails and two test tubes
- 6*M* NaOH solution
- beaker
- match
- insulated wire
- power supply
- glowing splint

Chapter 23

Key Terms

23.1 electrochemical process, electrochemical cell, voltaic cells, half-cell, salt bridge, electrode, anode, cathode, dry cell, battery, fuel cells
23.2 electrical potential, reduction potential, cell potential, standard cell potential, standard hydrogen electrode
23.3 electrolysis, electrolytic cell

DISCOVER IT!

Safety: Be sure the edges of the metal strips are smooth. Have students rinse and dry the strips after use to remove saliva. Students should report a tingling sensation due to a mild electric shock. (The lemon juice is a conductor, as is saliva, which completes the circuit.) There is no reaction with two copper or two zinc strips. Because an orange is a similar fruit, students should predict similar results. Theoretically any battery that can supply 12 volts should be able to start the car (but would it be rechargeable?). If possible, hook the lemon battery up to a voltmeter so students can observe the voltage produced.

Chapter 23 ELECTROCHEMISTRY

This squid emits energy in the form of light.

FEATURES

Stay current with SCIENCE NEWS
Find out more about electrochemistry:
www.phschool.com

DISCOVER IT! A LEMON BATTERY

You need a whole lemon, a copper strip (1 cm × 3 cm), a zinc strip (1 cm × 3 cm), fine sandpaper, and a voltmeter (optional).

1. Polish the copper and zinc strips with sandpaper.
2. Push the zinc strip into the lemon. Leave about 1.5 cm of the zinc strip sticking out of the lemon.
3. Push the copper strip into the lemon about 1 cm away from the zinc strip. It should not make contact with the zinc strip. Leave about 1.5 cm of the copper strip sticking out of the lemon.
4. Briefly touch both metal strips at the same time with your tongue and observe what happens.
5. If you have a voltmeter, measure the voltage of your lemon battery.

Describe the sensation you felt on your tongue. Do you think you would feel the same sensation using two copper strips or two zinc strips? Try it! What would you expect to happen if, in place of the lemon, you were to use an orange? If you connected 12 lemon batteries together in series, you would generate a voltage of about 12 V. Could you use this assembly to start a car? Explain your answer. As you read about electrochemical cells in this chapter, refer back to this activity and explain what occurred.

23.1 ● ■ ◆
23.2 □ ◆
23.3 ○ ■ ◆

Conceptual Use Figures 23.1 and 23.3 to help students visualize redox reactions in electrochemical cells. It is not necessary to cover every type of electrochemical cell or electrolytic cell; lead storage batteries, fuel cells, and electroplating may be omitted. Section 23.2 may be omitted because of its emphasis on math.

Standard Refer students back to the CHEMath feature on page 114 if they are struggling with adding and subtracting positive and negative reduction potentials in Section 23.2. Use Figure 23.11 to clear up any confusion between electrochemical cells and electrolytic cells.

Honors If students are successful with Sample Problem 23-1, assign Sample Problem 23-2 as homework. The Small-Scale Lab may be used to introduce Section 23.3.

ELECTROCHEMICAL CELLS

On a summer evening, fireflies glow to attract their mates. In the ocean depths, anglerfish emit light to attract prey. Luminous shrimp, squid, jellyfish, and even bacteria also exist. These organisms, and others, are able to give off energy in the form of light as a result of redox reactions. **How does the transfer of electrons in a redox reaction produce energy?**

objectives

- Describe how redox reactions interconvert electrical energy and chemical energy
- Explain the structure of a dry cell and identify the substances that are oxidized and reduced

key terms

- electrochemical process
- electrochemical cell
- voltaic cells
- half-cell
- salt bridge
- electrode
- anode
- cathode
- dry cell
- battery
- fuel cells

The Nature of Electrochemical Cells

As you have learned, chemical processes can either release energy or absorb energy. The energy can sometimes be in the form of electricity. Electrochemistry has many applications in the home as well as in industry. Flashlight and automobile batteries are familiar examples of devices used to generate electrochemical energy. The manufacture of sodium and aluminum metals and the silverplating of tableware involve the use of electricity. Biological systems also use electrochemistry to carry nerve impulses. In this chapter, you will learn about the relationship between redox reactions and electrochemistry.

When a strip of zinc metal is dipped into an aqueous solution of blue copper sulfate, the zinc becomes copper-plated, as shown in **Figure 23.1.** The net ionic equation involves only zinc and copper.

$$Zn(s) + Cu^{2+}(aq) \longrightarrow Zn^{2+}(aq) + Cu(s)$$

Electrons are transferred from zinc atoms to copper ions. This is a redox reaction that occurs spontaneously. As the reaction proceeds, zinc atoms lose electrons as they are oxidized to zinc ions. The zinc metal slowly dissolves. At the same time, copper ions in solution gain the electrons lost by

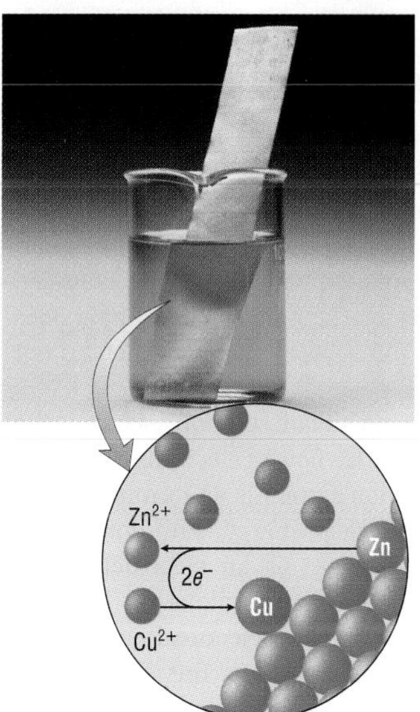

Figure 23.1
A spontaneous redox reaction occurs when a zinc strip is immersed in a solution of copper sulfate (left). As the redox reaction proceeds, copper plates out on the zinc (the copper appears black because it is finely divided), and the zinc strip is corroded. Slowly, the blue copper sulfate solution is replaced by a solution of zinc sulfate (right). The blue color of the original copper sulfate gradually diminishes, and eventually the solution becomes colorless as zinc sulfate replaces copper sulfate. At the atomic level, each Zn atom loses two electrons, which are picked up by a Cu^{2+} ion. What substance is oxidized? Reduced? (1)

23.1

1 Engage

Use the Visual

Have students examine the photograph and read the section-opening paragraph. Ask:

- **What is a redox reaction?** (Redox is a reaction that involves an oxidation and a reduction. During oxidation, an atom or ion loses electrons; during reduction, an atom or ion gains electrons.)
- **How does the transfer of electrons in a redox reaction produce energy?** (If the reduction and oxidation reactions are physically separated, the electrons released during oxidation can be forced to travel through an external electrical circuit on route to the reduction reaction. An item, such as a light bulb, can be inserted into the circuit so that the current can do useful work.)

TEACHER DEMO

Bring a radio to class that works with either a cord or with batteries. First plug in the radio to an electrical socket and turn it on. Students should be able to explain that the socket connects the radio to a supply of electrons that flow through the cord to the radio. Then shut off the radio and remove the cord. Insert the required batteries and turn on the radio. Ask students what is supplying the necessary electrons. (a chemical reaction in the batteries) Explain that students will discover that the "batteries" used in the radio do not meet the definition of a battery used by scientists.

STUDENT RESOURCES

From the Teacher's Resource Package, use:

- Section Review 23.1 and Quizzes from the Review Module (Ch. 21–24)
- Small-Scale Chemistry Lab Manual: Experiment 33

TECHNOLOGY RESOURCES

Relevant technology resources include:

- Chem ASAP! CD-ROM
- ResourcePro CD-ROM
- ActivChemistry CD-ROM: *Electrochemistry*

(1) The Zn atom is oxidized; the copper(II) ion is reduced.

2 Teach

TEACHER DEMO

Place 200 mL of 0.1 *M* silver nitrate ($AgNO_3$) in a 250-mL beaker. **CAUTION!** Silver nitrate is a suspected mutagen. Put a glass stirring rod across the top of the beaker and suspend a polished copper strip from the rod so that the strip dips into the solution. Have the students observe that the metal surface darkens and appears "fuzzy" as silver metal is deposited. Over time, a layer of silver will form on the copper. Explain that the blue color of the solution shows the formation of copper(II) ion. Write the net reaction for the electrochemical process on the board:
$2Ag^+(aq) + Cu(s) \rightarrow 2Ag(s) + Cu^{2+}(aq)$
Point out that the reaction involves a transfer of electrons from copper atoms to silver ions. The copper is oxidized and the silver ions are reduced.
Disposal: Combine liquid wastes and add a 50% molar excess of NaCl. Filter or decant and dry the AgCl residue. Put in a plastic container and bury in an approved landfill. Flush the filtrate down the drain with excess water.

Discuss

Explain that redox reactions make it possible for energy interconversion between electrical energy and chemical energy. Review oxidation and reduction half-cell reactions. Stress that the electrons must be balanced. Show how these reactions are combined to form the net ionic equation for an electrochemical process such as the one just demonstrated. The net equation summarizes the transfer of electrons from the species being oxidized to the species being reduced.

Table 23.1

Activity Series of Metals, with Half-Reactions for Oxidation Processes	Element	Oxidation half-reactions
Most active and most easily oxidized	Lithium	$Li(s) \longrightarrow Li^+(aq) + e^-$
	Potassium	$K(s) \longrightarrow K^+(aq) + e^-$
	Barium	$Ba(s) \longrightarrow Ba^{2+}(aq) + 2e^-$
Decreasing activity ↓	Calcium	$Ca(s) \longrightarrow Ca^{2+}(aq) + 2e^-$
	Sodium	$Na(s) \longrightarrow Na^+(aq) + e^-$
	Magnesium	$Mg(s) \longrightarrow Mg^{2+}(aq) + 2e^-$
	Aluminum	$Al(s) \longrightarrow Al^{3+}(aq) + 3e^-$
	Zinc	$Zn(s) \longrightarrow Zn^{2+}(aq) + 2e^-$
	Iron	$Fe(s) \longrightarrow Fe^{2+}(aq) + 2e^-$
	Nickel	$Ni(s) \longrightarrow Ni^{2+}(aq) + 2e^-$
	Tin	$Sn(s) \longrightarrow Sn^{2+}(aq) + 2e^-$
	Lead	$Pb(s) \longrightarrow Pb^{2+}(aq) + 2e^-$
	Hydrogen*	$H_2(g) \longrightarrow 2H^+(aq) + 2e^-$
	Copper	$Cu(s) \longrightarrow Cu^{2+}(aq) + 2e^-$
	Mercury	$Hg(s) \longrightarrow Hg^{2+}(aq) + 2e^-$
Least active and least easily oxidized	Silver	$Ag(s) \longrightarrow Ag^+(aq) + e^-$
	Gold	$Au(s) \longrightarrow Au^{3+}(aq) + 3e^-$

*Hydrogen is included for reference purposes.

the zinc. They are reduced to copper atoms and are deposited as metallic copper. As the copper ions in solution are gradually replaced by zinc ions, the blue color of the solution fades. Balanced half-reactions for this redox reaction can be written as follows.

$$\text{Oxidation: } Zn(s) \longrightarrow Zn^{2+}(aq) + 2e^-$$

$$\text{Reduction: } Cu^{2+}(aq) + 2e^- \longrightarrow Cu(s)$$

If you look at the activity series of metals in **Table 23.1**, you will see that zinc is higher on the list than copper. For any two metals in the table, the metal that is the higher of the two is the more readily oxidized. As **Figure 23.1** shows, zinc is more readily oxidized than copper because when dipped into a copper sulfate solution, zinc becomes plated with copper. In contrast, when a copper strip is dipped into a solution of zinc sulfate, the copper does not spontaneously become zinc-plated. This is because copper metal is not oxidized by zinc ions.

When a zinc strip is dipped into a copper sulfate solution, electrons are transferred from zinc metal to copper ions. This flow of electrons is an electric current. Thus the zinc-metal–copper-ion system is an example of the conversion of chemical energy into electrical energy. Any conversion between chemical energy and electrical energy is an **electrochemical process.** All electrochemical processes involve redox reactions.

If a redox reaction is to be used as a source of electrical energy, the two half-reactions must be physically separated. In the case of the zinc-metal–copper-ion reaction, the electrons released by zinc must

MEETING DIVERSE NEEDS

At Risk and **LEP** Have students compare the arrangement of half-reactions in Table 23.1 to the arrangement of elements in the periodic table. Remind them about what they learned in Chapter 14 about periodic trends in chemical properties. Point out that the elements on the left side of the periodic table are more easily *oxidized* than the elements on the right side of the periodic table. Group 1A and 2A metals, listed at the top of the activity series, represent some of the most reactive elements in nature. They readily form ions by giving up one or two electrons; they are easily *oxidized.* The coinage metals, Au, Ag, and Cu, represent some of the most stable elements in nature. They are not readily oxidized, and are listed at the bottom of the activity series.

pass through an external circuit to reach the copper ions if useful electrical energy is to be produced. In that situation, the system serves as an electrochemical cell. Alternatively, an electric current can be used to produce a chemical change. That system, too, serves as an electrochemical cell. An **electrochemical cell** is any device that converts chemical energy into electrical energy or electrical energy into chemical energy. Redox reactions occur in electrochemical cells.

Voltaic Cells

The Italian physicist Alessandro Volta (1745–1827) invented the first electrochemical cell. In 1800, Volta designed and built a cell that could be used to generate a direct electric current (DC). **Figure 23.2** shows a photograph of Volta's illustration of one of his early cells. **Voltaic cells** (not surprisingly named after their inventor) are electrochemical cells used to convert chemical energy into electrical energy. The energy is produced by spontaneous redox reactions within the cell. You use a voltaic cell every time you turn on a flashlight or a battery-powered calculator.

A **half-cell** is one part of a voltaic cell in which either oxidation or reduction occurs. A half-cell consists of a metal rod or strip immersed in a solution of its ions. **Figure 23.3** on the following page shows a voltaic cell that makes use of the zinc–copper reaction. In this cell, one half-cell is a zinc rod immersed in a solution of zinc sulfate. The other half-cell is a copper rod immersed in a solution of copper sulfate. The half-cells are separated by a **salt bridge**—a tube containing a strong electrolyte, often potassium sulfate (K_2SO_4). Salt bridges usually are made of agar, a gelatinous substance. A porous plate may be used instead of a salt bridge. The porous plate allows ions to pass from one half-cell to the other but prevents the solutions from mixing completely. A wire carries the electrons in the external circuit from the zinc rod to the copper rod. A voltmeter or light bulb can be connected in the circuit. The driving force of such a voltaic cell is the spontaneous redox reaction between zinc metal and copper(II) ions in solution. Is the zinc metal oxidized or reduced? 1

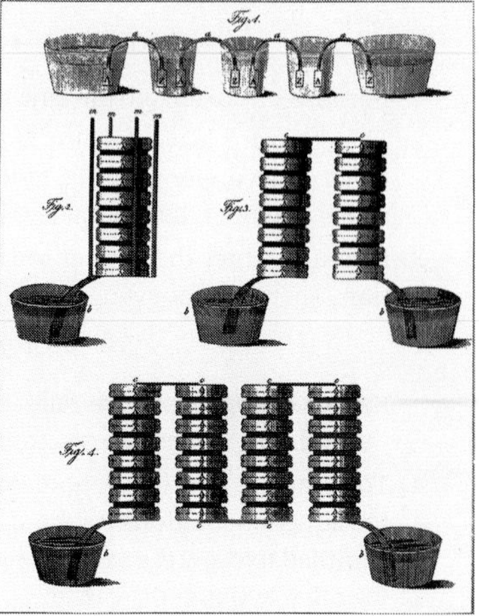

Figure 23.2
Volta built his electrochemical cell using piles of copper and zinc plates separated by cardboard soaked in salt water. He used his cell to obtain an electrical current. A working voltaic cell can be constructed using strips of copper and zinc and a lemon.

Think Critically

Challenge students to research the electrochemical nature of the nervous system. Students may want to focus on different aspects of the system and combine their findings into a visual display and/or oral report. Possible areas to be addressed include: the role of sodium and potassium ions; how signals are transmitted across synapses, what an EEG measures, and the value of squid for nervous system research.

Discuss

Remind students of the lemon battery they made in the Discover It! activity. Explain that the lemon battery is an example of a voltaic cell. Use its parts to illustrate terms such as *cathode, anode, salt bridge,* and *half-cell.* Introduce the shorthand method for representing an electrochemical cell.

Cooperative Learning

Divide the class into groups of two. Write a series of spontaneous and non-spontaneous redox equations on the board using the half-reactions from Table 23.1. Have students classify the reactions as spontaneous or nonspontaneous and justify their answers.

Answer

1 oxidized

Meeting Diverse Needs

At Risk Help the students remember the association for voltaic cells of cathode/reduction and anode/oxidation by noting that initial consonants (c and r) go together, as do initial vowels (a and o).

Use the Visual

Have the students study the voltaic cell in Figure 23.3. Remind them that voltaic cells can be used as sources of electrical energy because the two half-reactions are physically separated. Point out that oxidation (loss of electrons) occurs at the anode (− terminal) and reduction (gain of electrons) occurs at the cathode (+ terminal). The two half-reactions are connected by a salt bridge or a tube containing a conducting solution. In the voltaic cell, the electrons generated from the conversion of Zn to Zn^{2+} pass through the external wire to the copper strip where Cu^{2+} is reduced to Cu.

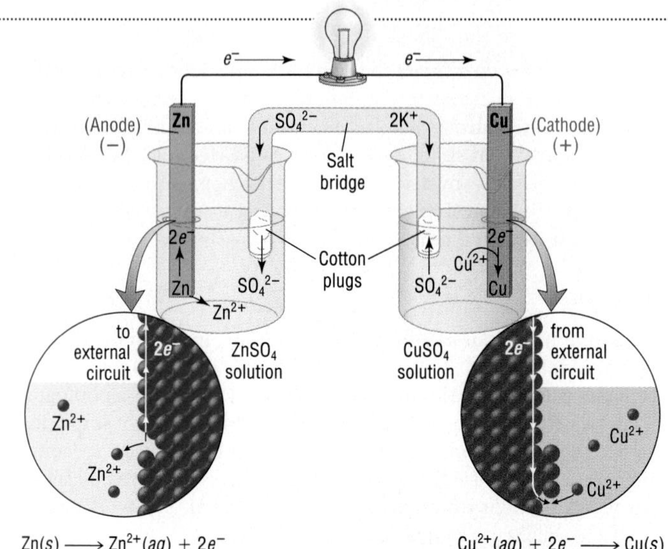

Figure 23.3
In all electrochemical cells, oxidation occurs at the anode and reduction occurs at the cathode. In this voltaic cell, the electrons generated from the oxidation of Zn to Zn^{2+} pass up the zinc strip and flow through the external circuit (the wire) into the copper strip. These electrons reduce the surrounding Cu^{2+} to Cu. As electrons flow from left to right in the external circuit, anions flow from right to left in the salt bridge to maintain neutrality in the electrolytes.

The zinc and copper rods in this voltaic cell are the electrodes. An **electrode** is a conductor in a circuit that carries electrons to or from a substance other than a metal. The reaction at the electrode determines whether the electrode is labeled as an anode or a cathode. The electrode at which oxidation occurs is called the **anode.** Because electrons are produced at the anode, it is labeled the negative electrode. The electrode at which reduction occurs is called the **cathode.** Electrons are consumed at the cathode. As a result, the cathode is labeled the positive electrode. Neither electrode is really charged, however. All parts of the voltaic cell remain balanced in terms of charge at all times. The moving electrons balance any charge that might build up as oxidation and reduction occur.

The electrochemical process that occurs in a zinc–copper voltaic cell can best be described in a number of steps. These steps actually occur at the same time.

1. Electrons are produced at the zinc rod according to the oxidation half-reaction.
$$Zn(s) \longrightarrow Zn^{2+}(aq) + 2e^-$$
Because it is oxidized, the zinc rod is the anode, or negative electrode.
2. The electrons leave the zinc anode and pass through the external circuit to the copper rod. (If a bulb is in the circuit, the electron flow will cause it to light. If a voltmeter is present, it will indicate a voltage.)
3. Electrons enter the copper rod and interact with copper ions in solution. There the following reduction half-reaction occurs.
$$Cu^{2+}(aq) + 2e^- \longrightarrow Cu(s)$$
Because copper ions are reduced at the copper rod, the copper rod is the cathode, or positive electrode, in the voltaic cell.
4. To complete the circuit, both positive and negative ions move through the aqueous solutions via the salt bridge. The two half-reactions can be summed to show the overall cell reaction. Note that the electrons in the overall reaction must cancel.

$$Zn(s) \longrightarrow Zn^{2+}(aq) + \cancel{2e^-}$$
$$Cu^{2+}(aq) + \cancel{2e^-} \longrightarrow Cu(s)$$
$$Zn(s) + Cu^{2+}(aq) \longrightarrow Zn^{2+}(aq) + Cu(s)$$

You can use a shorthand method to represent an electrochemical cell. For example, you can write the zinc–copper voltaic cell as follows.

$$Zn(s) \mid ZnSO_4(aq) \parallel CuSO_4(aq) \mid Cu(s)$$

The single vertical lines indicate boundaries of phases that are in contact. The zinc rod (Zn)(*s*) and the zinc sulfate solution ($ZnSO_4$)(*aq*), for example, are separate phases in physical contact. The double vertical lines represent the salt bridge or porous partition that separates the anode compartment from the cathode compartment. The half-cell that undergoes oxidation (the anode) is written first, to the left of the double vertical lines. When the zinc sulfate and copper(II) sulfate solutions in the voltaic half-cells are both 1.0*M*, the cell generates an electrical potential of 1.10 volts (V). If different metals are used for the electrodes in the cell, the voltage will differ. Would different solution concentrations affect the voltage as well? ❶

Dry Cells

Although the zinc–copper voltaic cell is of historical importance, it is no longer used commercially. Nevertheless, this cell is a convenient model to use when describing the production of electrical energy from a chemical change. Today, the more practical and compact dry cell is usually chosen when a portable electrical energy source is required. A **dry cell** is a voltaic cell in which the electrolyte is a paste. A type of dry cell that is very familiar to you is the common flashlight battery, which, despite the name, is not a true battery. In such a dry cell, a zinc container is filled with a thick, moist electrolyte paste of manganese(IV) oxide (MnO_2), zinc chloride ($ZnCl_2$), ammonium chloride (NH_4Cl), and water (H_2O). As shown in **Figure 23.4a,** a graphite rod is embedded in the paste. The zinc container is the anode and the graphite rod is the cathode. The thick paste and its surrounding paper liner prevent the contents of the cell from freely mixing, so a salt bridge is not needed. The half-reactions for this cell are shown on the following page.

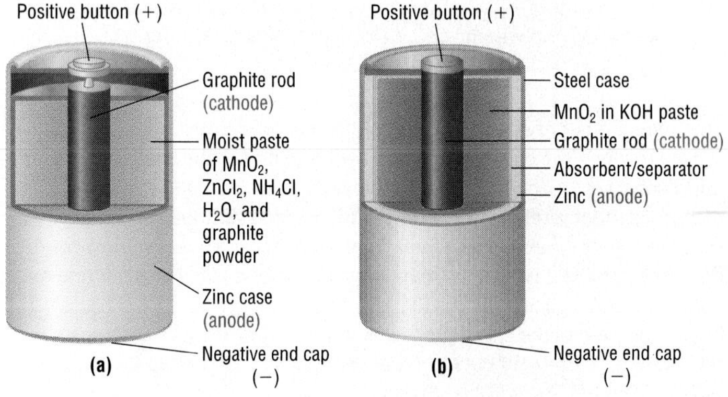

Figure 23.4
Many devices—including portable radios, toys, and flashlights—are powered by voltaic cells such as dry cells and alkaline batteries. ***(a)*** *A common dry cell is a single electrochemical cell that produces about 1.5 V. Although inexpensive, it has a short shelf-life and suffers from voltage drop when in use. What is oxidized in this cell and what is reduced?* ***(b)*** *The alkaline battery is an improved dry cell. Although more expensive than the common dry cell, it has a longer shelf-life and does not suffer from voltage drop. Both of these batteries have many uses.* ❷

Answers

❶ yes
❷ Zn is oxidized; Mn^{4+} is reduced.

TEACHER DEMO

Before class, cut through a dry cell with a hacksaw from the top down on the right side of the central carbon electrode. Have the students identify the three parts of the cell: the central carbon electrode, the zinc container that serves as the other electrode, and the manganese dioxide/ammonium chloride paste.

Construct a dry cell by placing a zinc electrode and a carbon electrode into a manganese dioxide/ammonium chloride paste. Prepare the paste by adding saturated ammonium chloride to powdered manganese dioxide until thick. Connect the cell to a voltmeter and note the potential.

Discuss

Explain to the students that a dry cell is a voltaic cell in which the electrolyte is a paste. A common flashlight battery is an example of a dry cell that has the following half-reactions:
$Zn(s) \rightarrow Zn^{2+}(aq) + 2e^-$
$2MnO_2(s) + 2NH_4^+(aq) + 2e^- \rightarrow Mn_2O_3(s) + 2NH_3(aq) + H_2O(l)$
By itself, a dry cell does not provide a complete circuit, that is, electrons cannot flow from the anode to the cathode. When devices using dry cells are turned on, an external circuit is completed, allowing the flow of electrons from the anode to the cathode.

Use the Visual

Have the students examine Figure 23.4.

- **Ask the students to describe the difference between the common dry cell and the alkaline battery.** (In the alkaline dry cell, the electrolyte is a basic KOH paste.)

Use the Visual

Have students study Figure 23.5. Explain that a battery is a group of dry cells connected in series. In an automobile, the battery consists of six 2-V cells. Point out that the total voltage from the battery is the sum of the voltages from the individual cells. The overall spontaneous reaction that occurs in the lead storage battery is:

$Pb(s) + PbO_2(s) + 2H_2SO_4(aq) \rightarrow 2PbSO_4(s) + 2H_2O(l)$

- **Write the half-cell reactions on the board and ask which substance is being reduced and which is being oxidized?** (Pb is oxidized to Pb^{2+} in $PbSO_4$. Pb^{4+} in PbO_2 is reduced to Pb^{2+} in $PbSO_4$.)

Point out that when the battery is charging, the reverse reaction occurs. This reaction is nonspontaneous.

TEACHER DEMO

To prepare a lead cell, attach two lead strips to a wooden rod so they hang vertically in a 250-mL beaker. Place the strips 4 cm apart. Pour sufficient dilute sulfuric acid into the beaker to cover two-thirds of each strip. **CAUTION!** Sulfuric acid is corrosive. Connect a 6-V DC power supply to the strips with wire, and charge for a few minutes. Connect the cell to a bell and have students observe that the bell rings. Discuss the half-cell reactions that occur.

- Ask the students to write the shorthand notation for the electrochemical cell. [$Pb(s)$ | $PbSO_4(s)$ || $PbO_2(s)$ | $PbSO_4(s)$]

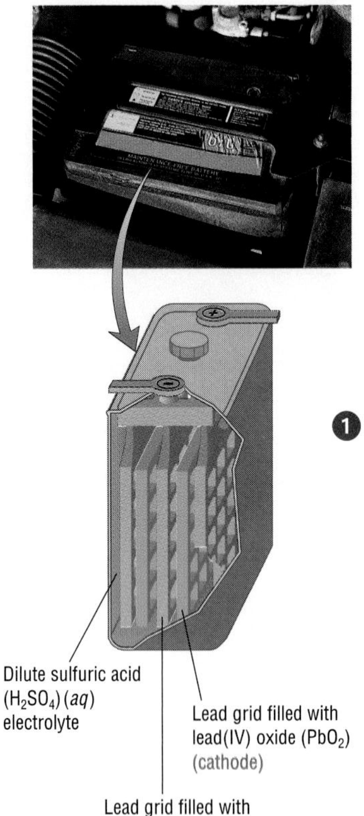

Figure 23.5
A 12-V lead storage battery, such as those used in cars, consists of six 2-V cells in series. One such cell is illustrated here. The cells do not need to be in separate compartments, but if they are, performance is improved. While the battery is producing a current, lead at the anode and lead(IV) oxide at the cathode are both converted to lead sulfate. These processes decrease the sulfuric acid concentration in the battery. Reversing the reaction recharges the battery.

Oxidation: $Zn(s) \longrightarrow Zn^{2+}(aq) + 2e^-$ (anode reaction)

Reduction: $2MnO_2(s) + 2NH_4^+(aq) + 2e^- \longrightarrow Mn_2O_3(s) + 2NH_3(aq) + H_2O(l)$ (cathode reaction)

In this cell, the graphite rod serves only as a conductor and does not undergo reduction, even though it is the cathode. The manganese in MnO_2 is the species that is actually reduced. The electrical potential of this cell starts out at 1.5 V but decreases steadily during use to about 0.8 V. Dry cells of this type are not rechargeable.

The alkaline battery, shown in **Figure 23.4b** on the previous page, is an improved dry cell used for the same purposes. In the alkaline battery, the reactions are similar to those in the common dry cell, but the electrolyte is a basic KOH paste. This change in design eliminates the buildup of ammonia gas and maintains the Zn electrode, which corrodes more slowly under alkaline conditions. What is oxidized in an alkaline battery? 1

Lead Storage Batteries

A **battery** is a group of cells connected together. People depend on lead storage batteries to start their cars. A 12-V car battery consists of six voltaic cells connected together. Each cell produces about 2 V and consists of lead grids, as shown in **Figure 23.5.** One set of grids, the anode, is packed with spongy lead. The other set, the cathode, is packed with lead(IV) oxide (PbO_2). The grids are immersed in an electrolyte of concentrated sulfuric acid. The half-reactions are as follows.

Oxidation: $Pb(s) + SO_4^{2-}(aq) \longrightarrow PbSO_4(s) + 2e^-$

Reduction: $PbO_2(s) + 4H^+(aq) + SO_4^{2-}(aq) + 2e^- \longrightarrow PbSO_4(s) + 2H_2O(l)$

When a lead storage battery discharges, it produces the electric power needed to start a car. The overall spontaneous redox reaction that occurs is the sum of the oxidation and reduction half-reactions.

$$Pb(s) + PbO_2(s) + 2H_2SO_4(aq) \longrightarrow 2PbSO_4(s) + 2H_2O(l)$$

This equation shows that during discharge, lead sulfate forms. The sulfate slowly builds up on the plates, and the concentration of sulfuric acid decreases.

The reverse reaction occurs when a lead storage battery is recharged. This reaction occurs whenever the car's generator is running.

$$2PbSO_4(s) + 2H_2O(l) \longrightarrow Pb(s) + PbO_2(s) + 2H_2SO_4(aq)$$

This reaction is nonspontaneous. To make it go, a direct current must be passed through the cell in a direction opposite that of the current flow during discharge. In theory, a lead storage battery can be discharged and recharged indefinitely, but in practice its lifespan is limited. This is because small amounts of lead sulfate fall from the electrodes and collect on the bottom of the cell. Eventually, the electrodes lose so much lead sulfate that the recharging process is ineffective or the cell is shorted out. The battery must then be replaced. The processes that occur during the discharge and recharge of a lead–acid battery are summarized in **Figure 23.6.**

Answer

1 zinc

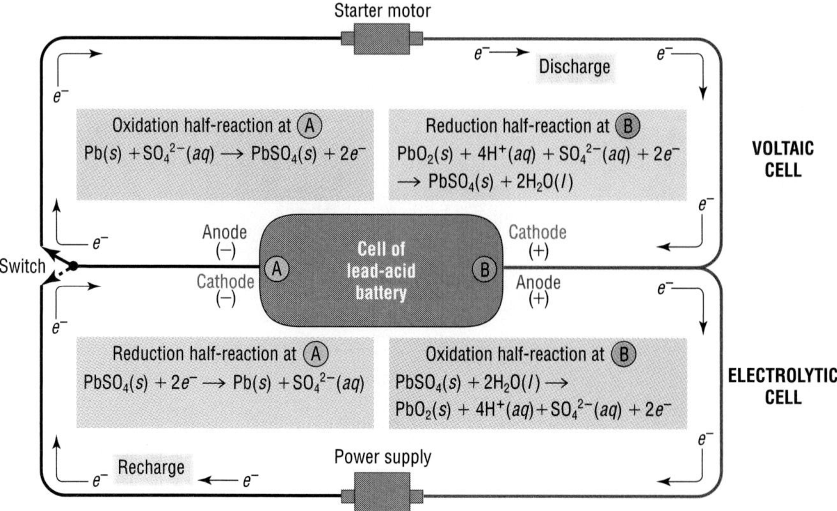

Figure 23.6

Turning a car's ignition causes energy from the battery to activate the starter motor, which starts the car's engine. During this process, the lead–acid battery is discharging and acts as a voltaic cell (top). The anode (terminal A) is negative and the cathode (terminal B) is positive. When the engine is running, some of the power it generates is used to recharge the battery. When the battery is recharging, it acts as an electrolytic cell (bottom) with a positive anode (terminal B) and a negative cathode (terminal A). You will learn more about electrolytic cells in Section 23.3.

Fuel Cells

To overcome the disadvantages associated with lead storage batteries, cells with renewable electrodes have recently been developed. Such cells, called **fuel cells,** are voltaic cells in which a fuel substance undergoes oxidation and from which electrical energy is continuously obtained. Fuel cells do not have to be recharged. They can be designed to emit no air pollutants and to operate more quietly and more cost-effectively than a conventional electrical generator.

Perhaps the simplest fuel cell is the hydrogen–oxygen fuel cell, which is shown in **Figure 23.7** on the following page. In this fuel cell, there are three compartments separated from one another by two electrodes made of porous carbon. Oxygen (the oxidizer) is fed into the cathode compartment. Hydrogen (the fuel) is fed into the anode compartment. The gases diffuse slowly through the electrodes. The electrolyte in the central compartment is a hot, concentrated solution of potassium hydroxide. Electrons from the oxidation half-reaction at the anode pass through an external circuit to enter the reduction half-reaction at the cathode.

Oxidation: $2H_2(g) + 4OH^-(aq) \longrightarrow 4H_2O(l) + 4e^-$ (anode)

Reduction: $O_2(g) + 2H_2O(l) + 4e^- \longrightarrow 4OH^-(aq)$ (cathode)

The overall reaction is the oxidation of hydrogen to form water.

$$2H_2(g) + O_2(g) \longrightarrow 2H_2O(l)$$

23.1

Use the Visual

Have students examine Figure 23.7 on page 684. Point out that the hydrogen-oxygen fuel cell is a clean, reliable source of power. Explain that the fuel in a fuel cell is the species that is oxidized to produce electrical energy continuously.

Discuss

Point out that fuel cells were developed for space travel where there is a need for lightweight, reliable power systems. The electrodes are typically porous carbon through which the gases flow into a concentrated alkaline electrolyte. Discuss the half-cell and overall reaction for the hydrogen–oxygen fuel cell. Other fuels such as methane and carbon monoxide can be used in place of hydrogen.

Cooperative Learning

Divide the students into groups of three or four. Have them research the use of fuel cells in the space shuttle and by utility companies. Ask them to prepare posters showing the different types of fuel cells and the applications for which they are used. Where possible, have students include the half-cell and overall reactions for each fuel cell. Have them describe the advantages gained by using the fuel cells over more conventional sources of electrical energy.

3 Assess

Evaluate Understanding

Have students make a sketch of a tin–lead (Sn | $SnSO_4$ || $PbSO_4$ | Pb) voltaic cell. They should label the cathode and anode, and indicate the direction of electron flow.

Have students name the substance oxidized and the substance reduced in a dry cell. (Zn is oxidized and Mn^{4+} from manganese dioxide is reduced.) **Have students explain why lead storage batteries cannot be recharged indefinitely.** (Because small amounts of lead sulfate fall from the electrodes, eventually the recharging process becomes ineffective.)

Reteach

Emphasize that a chemical reaction can produce a flow of electrons or a flow of electrons can cause a chemical reaction to occur. Note that the cathode is always the electrode where reduction occurs, and the anode is always the electrode where oxidation occurs.

Discuss the half-reactions for the charging process in a lead cell, and compare them to the half-reactions when the cell is producing electric current. Point out that the lead storage battery works because lead has three oxidation states: 0, +2, and +4.

Portfolio Project

The designs should attempt to control all variables except for brand. Challenge students to consider how the conditions under which batteries were stored before and after sale could affect the results.

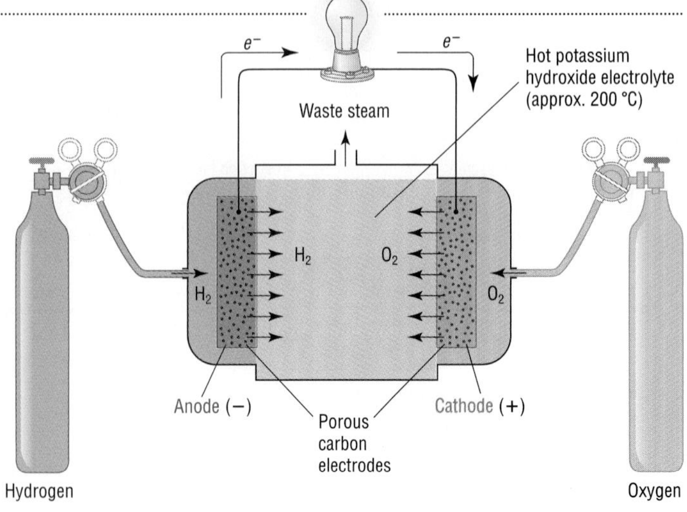

Figure 23.7
The hydrogen–oxygen fuel cell is a clean source of power. Such cells are often used in spacecraft. What waste products, if any, are produced? 2

Other fuels, such as methane (CH_4) and ammonia (NH_3), can be used in place of hydrogen. Other oxidizers, such as chlorine (Cl_2) and ozone (O_3), can be used in place of oxygen. Fuel cells are currently being built as auxiliary power sources for submarines and other military vehicles. At present, however, they are too expensive for general use. Hydrogen–oxygen fuel cells with a mass of approximately 100 kg each were used in the Apollo spacecraft missions. Can you name a benefit of producing energy with a
1 hydrogen–oxygen fuel cell on an extended space mission?

portfolio project

Design an experiment to test different brands of batteries for longevity. Should all the batteries be the same type, for example, D cells or AA cells? If your teacher approves your design, perform the experiment. Make a graph illustrating your results.

section review 23.1

1. How do redox reactions interconvert electrical energy and chemical energy?
2. What is the source of the electrical energy produced in a voltaic cell?
3. Describe the structure of a dry cell. Identify the substance oxidized and the substance reduced in this type of cell. Include a diagram in your description.
4. Represent the electrochemical reaction of a flashlight battery using the shorthand method.
5. Predict the result when a strip of copper is dipped into a solution of iron(II) sulfate.
6. Describe the composition of the anode, cathode, and electrolyte in a fully charged lead storage battery.
7. Based on the half-reactions for the hydrogen–oxygen fuel cell, would KOH be consumed in such a cell? Explain.

 Chem ASAP! **Assessment 23.1** Check your understanding of the important ideas and concepts in Section 23.1.

Answers

1 Continuous energy source, no pollutants.
2 The cells produce drinkable water.

SECTION REVIEW 23.1

1. Spontaneous redox reactions generate a flow of electrons that can be used to provide DC power to an external circuit.
2. spontaneous redox reactions
3. A dry cell is a voltaic cell in which the electrolyte is a paste. See Figure 23.4. In a flashlight battery, the Zn anode is oxidized and Mn^{4+} in MnO_2 is reduced.
4. Zn(*s*) | $ZnCl_2$(*aq*) || MnO_2(*s*) | Mn_2O_3(*s*)
5. There is no reaction.
6. The anode is spongy lead. The cathode is lead(IV) oxide. The electrolyte is sulfuric acid.
7. No; for every 4 mol OH^- used up in the oxidation half-reaction, 4 mol OH^- are produced in the reduction half-reaction.

HALF-CELLS AND CELL POTENTIALS

Batteries provide current to power lights, electronic devices, and even toys—such as the robot shown here. One type of battery most people use is a car battery. The potential difference between the negative and positive terminals of a car battery is 12 V. Other common batteries are the ones used to power flashlights, television remote controls, radios, and CD players. They are commonly known as AA, AAA, C, and D batteries. The potential difference in these batteries is 1.5 V. **How can electrical potential be calculated?**

objectives

- Define standard cell potential and standard reduction potential
- Use standard reduction potentials to calculate standard cell potential

key terms

- electrical potential
- reduction potential
- cell potential
- standard cell potential
- standard hydrogen electrode

Electrical Potential

The **electrical potential** of a voltaic cell is a measure of the cell's ability to produce an electric current. Electrical potential is usually measured in volts (V). The potential of an isolated half-cell cannot be measured. For example, you cannot measure the electrical potential of a zinc half-cell or of a copper half-cell separately. When these two half-cells are connected to form a voltaic cell, however, the difference in potential can be measured. The electrical potential of a 1*M* zinc–copper cell is +1.10 V.

The electrical potential of a cell results from a competition for electrons between the two half-cells. The half-cell that has a greater tendency to acquire electrons will be the one in which reduction occurs. Oxidation occurs in the other half-cell as there is a loss of electrons. The tendency of a given half-reaction to occur as a reduction is called the **reduction potential.** The half-cell in which reduction occurs has a greater reduction potential than the half-cell in which oxidation occurs. The difference between the reduction potentials of the two half-cells is called the **cell potential.**

$$\text{cell potential} = \begin{pmatrix}\text{reduction potential}\\ \text{of half-cell in which}\\ \text{reduction occurs}\end{pmatrix} - \begin{pmatrix}\text{reduction potential}\\ \text{of half-cell in which}\\ \text{oxidation occurs}\end{pmatrix}$$

$$\text{or} \quad E^0_{cell} = E^0_{red} - E^0_{oxid}$$

Standard Cell Potential

The **standard cell potential** (E^0_{cell}) is the measured cell potential when the ion concentrations in the half-cells are 1*M*, any gases are at a pressure of 101 kPa, and the temperature is 25 °C. The symbols E^0_{red} and E^0_{oxid} represent the standard reduction potentials for the reduction and oxidation half-cells, respectively. The relationship between these values follows the general relationship for cell potential given above.

$$E^0_{cell} = E^0_{red} - E^0_{oxid}$$

Because half-cell potentials cannot be measured, scientists have chosen an arbitrary electrode to serve as a reference. The **standard hydrogen electrode** is used with other electrodes so the reduction potentials of the other cells can be measured. The standard reduction potential of the hydrogen electrode has been assigned a value of 0.00 V. The standard hydrogen

1 Engage

Use the Visual

Have students study the photograph and read the text that opens the section. Remind students that a battery is a device that uses the energy of an oxidation-reduction reaction to produce an electrical current through an external circuit. The "force" with which electrons are driven through the external circuit is a measure of the *potential* of the cell. Ask:

- **How can electrical potential be calculated?** (The potential of an electrochemical cell depends on how readily one substance gives up electrons and how eagerly another substance accepts electrons. To calculate the cell potential, the relative activities of half-reactions occurring in separate compartments of the cell are compared on a numerical scale using units called *volts*.)

Check Prior Knowledge

Remind students that all electrochemical cells involve a redox reaction. Treating a redox reaction as two half-reactions is useful as is separating a voltaic cell into half-cells in order to understand the overall reaction. Ask:

- **How can you determine which reaction occurs at each of the electrodes?** (The oxidation reaction occurs at the anode; the reduction reaction occurs at the cathode.)

Student Resources

From the Teacher's Resource Package, use:

- Section Review 23.2, Ch. 23 Practice Problems and Quizzes from the Review Module (Ch. 21–24)
- Laboratory Manual: Experiment 45

Technology Resources

Relevant technology resources include:

- Chem ASAP! CD-ROM
- Resource Pro CD-ROM
- ActivChemistry CD-ROM: *Electrochemistry*

2 Teach

Use the Visual

Have the students examine Figure 23.8. Explain that the standard hydrogen electrode is used as a reference electrode and the reduction potential of all other substances is measured relative to the standard hydrogen electrode.

- **Ask the students to define, in their own words, the electrical potential of a voltaic cell?** (It is the ability of a cell to produce an electric current and is measured as the difference between the potentials of the two half-cells.)

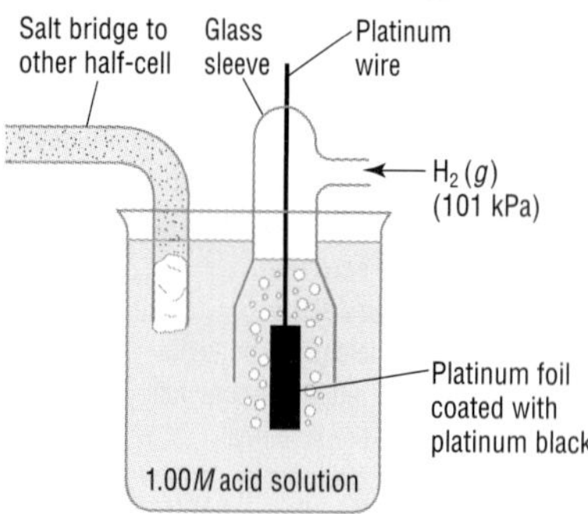

Figure 23.8
The standard hydrogen electrode is arbitrarily assigned a standard reduction potential of 0.00 V at 25 °C.

electrode, which is illustrated in **Figure 23.8,** consists of a platinum electrode immersed in a solution with a hydrogen-ion concentration of 1*M*. The solution is at 25 °C. The electrode itself is a small square of platinum foil coated with finely divided platinum, known as platinum black. Hydrogen gas at a pressure of 101 kPa is bubbled around the platinum electrode. The half-cell reaction that occurs at the platinum black surface is as follows.

$$2H^+(aq, 1M) + 2e^- \rightleftharpoons H_2(g, 101\text{ kPa}) \qquad E^0_{H^+} = 0.00\text{ V}$$

The symbol $E^0_{H^+}$ represents the standard reduction potential of H^+. The standard reduction potential of H^+ is the tendency of H^+ ions to acquire electrons and be reduced to ❶ $H_2(g)$. What is the significance of the double arrows in this reaction? Whether this half-cell reaction occurs as a reduction or as an oxidation is determined by the reduction potential of the half-cell to which the standard hydrogen electrode is connected.

Standard Reduction Potentials

A voltaic cell can be made by connecting a standard hydrogen half-cell to a standard zinc half-cell, as shown in **Figure 23.9.** To determine the overall reaction for this cell, first identify the half-cell in which reduction takes place. In all electrochemical cells, reduction takes place at the cathode and oxidation takes place at the anode. A voltmeter gives a reading of +0.76 V when the zinc electrode is connected to the negative terminal and the hydrogen electrode is connected to the positive terminal. The zinc is oxidized, which means that it is the anode. Hydrogen ions are reduced, which means that the hydrogen electrode is the cathode. You can now write the half-reactions and the overall cell reaction.

Chem ASAP!
Simulation 27
Simulate the operation of voltaic cells.

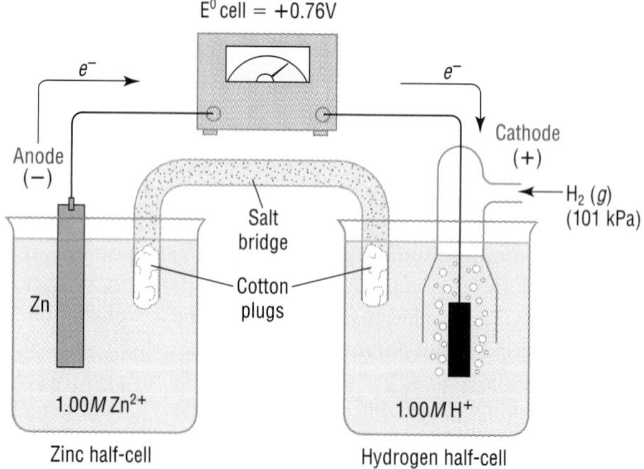

Figure 23.9
This voltaic cell consists of zinc and hydrogen half-cells. Where does reduction occur, and what is reduced in this cell? ❷

Answers

❶ The reaction is reversible.
❷ $H^+(aq)$ ions are reduced to $H_2(g)$ at the cathode.
❸ fluorine (F_2)

Oxidation: $Zn(s) \longrightarrow Zn^{2+}(aq) + 2e^-$ (at anode)

Reduction: $2H^+(aq) + 2e^- \longrightarrow H_2(g)$ (at cathode)

Cell reaction: $Zn(s) + 2H^+(aq) \longrightarrow Zn^{2+}(aq) + H_2(g)$

Use of a standard hydrogen electrode allows you to calculate the standard reduction potential for the zinc half-cell according to the equation for standard cell potential just developed.

$$E^0_{cell} = E^0_{red} - E^0_{oxid}$$

Therefore,

$$E^0 = E^0_{H^+} - E^0_{Zn^{2+}}$$

The cell potential (E^0_{cell}) was measured at +0.76 V. The reduction potential of the hydrogen half-cell is a defined standard: $E^0_{H^+}$ always equals 0.00 V. Substituting these values into the preceding equation will give the standard reduction potential for the zinc half-cell.

$$+0.76\text{ V} = 0.00\text{ V} - E^0_{Zn^{2+}}$$

$$E^0_{Zn^{2+}} = -0.76\text{ V}$$

The standard reduction potential for the zinc half-cell is −0.76 V. The value is negative because the tendency of zinc ions to be reduced to zinc metal in this cell is less than the tendency of hydrogen ions to be reduced to hydrogen gas (H_2). Consequently, the zinc ions are not reduced. Instead, the opposite occurs: Zinc metal is oxidized to zinc ions.

Many different half-cells can be paired with the hydrogen half-cell in a similar manner. Using this method, the standard reduction potential for each half-cell can be obtained. For a standard copper half-cell, for example, the measured standard cell potential is +0.34 V. Copper is the cathode (the positive electrode), and Cu^{2+} ions are reduced to Cu metal when the cell operates. The hydrogen half-cell is the anode (the negative electrode), and H_2 gas is oxidized to H^+ ions. You can calculate the standard reduction potential for copper as follows.

$$E^0_{cell} = E^0_{red} - E^0_{oxid}$$

$$E^0_{cell} = E^0_{Cu^{2+}} - E^0_{H^+}$$

$$+0.34\text{ V} = E^0_{Cu^{2+}} - 0.00$$

$$E^0_{Cu^{2+}} = +0.34\text{ V}$$

These calculations show that the standard reduction potential of copper is +0.34 V. The value is positive because the tendency of copper ions to be reduced in this cell is greater than the tendency of hydrogen ions to be reduced.

Table 23.2 on the following page lists some standard reduction potentials at 25 °C. The half-reactions are arranged in increasing order of their tendency to occur in the forward direction—that is, as a reduction. Thus the half-reactions at the top of the table have the least tendency to occur as reductions. The half-reactions at the bottom of the table have the greatest tendency to occur as reductions. For example, lithium ions (at the top) have very little tendency to be reduced to lithium metal. Which substance has the greatest tendency to be reduced? 

TEACHER DEMO

Demonstrate the galvanic corrosion of iron using pennies, metal wires from twist ties, and table salt. Moisten a paper towel with tap water and sprinkle with salt. Wrap one wire around a penny and place on the paper towel. Both metals must be clean. Add an unwrapped penny and a coil of wire as a control. Observe what happens after at least 24 hours.

- Ask the students to hypothesize about what is occurring? (In galvanic corrosion, the corrosion rate (oxidation) of one metal accelerates while that of the other is decreased. Because iron is more active than copper, its corrosion was accelerated.)

MEETING DIVERSE NEEDS

At Risk Have students look at the diagram of the voltaic cell in Figure 23.9. Pair up the students and have one of them trace the movement of the electrons from one electrode to the other. The other student should write down the half-reactions and overall reaction for the process. They should be able to explain that the electrons travel up from the zinc anode, into the external wire, through the voltmeter, and down through the wire into the hydrogen/platinum cathode.

Discuss

Explain that the two half-cells of a voltaic cell are competing for electrons and that oxidation or reduction could occur in either cell. The half-cell with the more positive reduction potential will win the competition and undergo reduction; oxidation will occur in the other half-cell. The potential produced by the electrochemical cell is the difference in the reduction potentials of the two half-cell reactions. Explain that the quantitative value of the half-cell potential is always measured against the standard hydrogen electrode. Point out that Table 23.2 contains some of the tabulated values, which can be used to calculate cell potentials.

Table 23.2

Reduction Potentials at 25 °C with 1*M* Concentrations of Aqueous Species

Electrode	Half-reaction	E^0 (V)
Li^+/Li	$Li^+ + e^- \longrightarrow Li$	−3.05
K^+/K	$K^+ + e^- \longrightarrow K$	−2.93
Ba^{2+}/Ba	$Ba^{2+} + 2e^- \longrightarrow Ba$	−2.90
Ca^{2+}/Ca	$Ca^{2+} + 2e^- \longrightarrow Ca$	−2.87
Na^+/Na	$Na^+ + e^- \longrightarrow Na$	−2.71
Mg^{2+}/Mg	$Mg^{2+} + 2e^- \longrightarrow Mg$	−2.37
Al^{3+}/Al	$Al^{3+} + 3e^- \longrightarrow Al$	−1.66
H_2O/H_2	$2H_2O + 2e^- \longrightarrow H_2 + 2OH^-$	−0.83
Zn^{2+}/Zn	$Zn^{2+} + 2e^- \longrightarrow Zn$	−0.76
Cr^{3+}/Cr	$Cr^{3+} + 3e^- \longrightarrow Cr$	−0.74
Fe^{2+}/Fe	$Fe^{2+} + 2e^- \longrightarrow Fe$	−0.44
H_2O/H_2 (pH 7)	$2H_2O + 2e^- \longrightarrow H_2 + 2OH^-$	−0.42
Cd^{2+}/Cd	$Cd^{2+} + 2e^- \longrightarrow Cd$	−0.40
$PbSO_4/Pb$	$PbSO_4 + 2e^- \longrightarrow Pb + SO_4^{2-}$	−0.36
Co^{2+}/Co	$Co^{2+} + 2e^- \longrightarrow Co$	−0.28
Ni^{2+}/Ni	$Ni^{2+} + 2e^- \longrightarrow Ni$	−0.25
Sn^{2+}/Sn	$Sn^{2+} + 2e^- \longrightarrow Sn$	−0.14
Pb^{2+}/Pb	$Pb^{2+} + 2e^- \longrightarrow Pb$	−0.13
Fe^{3+}/Fe	$Fe^{3+} + 3e^- \longrightarrow Fe$	−0.036
H^+/H_2	$2H^+ + 2e^- \longrightarrow H_2$	0.000
$AgCl/Ag$	$AgCl + e^- \longrightarrow Ag + Cl^-$	+0.22
Hg_2Cl_2/Hg	$Hg_2Cl_2 + 2e^- \longrightarrow 2Hg + 2Cl^-$	+0.27
Cu^{2+}/Cu	$Cu^{2+} + 2e^- \longrightarrow Cu$	+0.34
O_2/OH^-	$O_2 + 2H_2O + 4e^- \longrightarrow 4OH^-$	+0.40
Cu^+/Cu	$Cu^+ + e^- \longrightarrow Cu$	+0.52
I_2/I^-	$I_2 + 2e^- \longrightarrow 2I^-$	+0.54
Fe^{3+}/Fe^{2+}	$Fe^{3+} + e^- \longrightarrow Fe^{2+}$	+0.77
Hg_2^{2+}/Hg	$Hg_2^{2+} + 2e^- \longrightarrow 2Hg$	+0.79
Ag^+/Ag	$Ag^+ + e^- \longrightarrow Ag$	+0.80
O_2/H_2O (pH 7)	$O_2 + 4H^+ + 4e^- \longrightarrow 2H_2O$	+0.82
Hg^{2+}/Hg	$Hg^{2+} + 2e^- \longrightarrow Hg$	+0.85
Br_2/Br^-	$Br_2 + 2e^- \longrightarrow 2Br^-$	+1.07
O_2/H_2O	$O_2 + 4H^+ + 4e^- \longrightarrow 2H_2O$	+1.23
MnO_2/Mn^{2+}	$MnO_2 + 4H^+ + 2e^- \longrightarrow Mn^{2+} + 2H_2O$	+1.28
$Cr_2O_7^{2-}/Cr^{3+}$	$Cr_2O_7^{2-} + 14H^+ + 6e^- \longrightarrow 2Cr^{3+} + 7H_2O$	+1.33
Cl_2/Cl^-	$Cl_2 + 2e^- \longrightarrow 2Cl^-$	+1.36
PbO_2/Pb^{2+}	$PbO_2 + 4H^+ + 2e^- \longrightarrow Pb^{2+} + 2H_2O$	+1.46
MnO_4^-/Mn^{2+}	$MnO_4^- + 8H^+ + 5e^- \longrightarrow Mn^{2+} + 4H_2O$	+1.51
$PbO_2/PbSO_4$	$PbO_2 + 4H^+ + SO_4^{2-} + 2e^- \longrightarrow PbSO_4 + 2H_2O$	+1.69
F_2/F^-	$F_2 + 2e^- \longrightarrow 2F^-$	+2.87

Calculating Standard Cell Potentials

To function, a cell must be constructed of two half-cells. The half-cell reaction having the more positive (or less negative) reduction potential occurs as a reduction in the cell. With this in mind, it is possible to write cell reactions and calculate cell potentials for cells without actually assembling them. You can simply use the known standard reduction potentials for the various half-cells (from **Table 23.2**) to predict the half-cells in which reduction and oxidation will occur. This information can then be used to find the resulting E^0_{cell} value.

If the cell potential for a given redox reaction is positive, then the reaction is spontaneous. If the cell potential is negative, then the reaction is nonspontaneous. This latter reaction will be spontaneous in the reverse direction, however, and the cell potential will then have a numerically equal but positive value.

Sample Problem 23-1

Determine the cell reaction, the standard cell potential, and the half-cell that acts as the cathode for a voltaic cell composed of the following half-cells.

$Fe^{3+}(aq) + e^- \longrightarrow Fe^{2+}(aq)$ $E^0_{Fe^{3+}} = +0.77\text{ V}$

$Ni^{2+}(aq) + 2e^- \longrightarrow Ni(s)$ $E^0_{Ni^{2+}} = -0.25\text{ V}$

1. ***ANALYZE*** ***List the knowns and the unknowns.***

Knowns:

- $E^0_{Fe^{3+}} = +0.77\text{ V}$
- $E^0_{Ni^{2+}} = -0.25\text{ V}$

Unknowns:

- cell reaction = ?
- E^0_{cell} = ? V
- cathode = ?

The half-cell with the more positive reduction potential is the one in which reduction occurs (the cathode). That means the oxidation reaction occurs at the anode. Add the half-reactions, making certain that the number of electrons lost equals the number of electrons gained. Then calculate the standard cell potential using the following equation.

$E^0_{cell} = E^0_{red} - E^0_{oxid}$

2. ***CALCULATE*** ***Solve for the unknowns.***

In this cell, Fe^{3+} is reduced and Ni is oxidized. Because reduction takes place in the Fe^{3+} half-cell, this half-cell is the cathode. The half-cell reactions, written in the direction in which they actually occur, are as follows.

Oxidation: $Ni(s) \longrightarrow Ni^{2+}(aq) + 2e^-$ (at anode)

Reduction: $Fe^{3+}(aq) + e^- \longrightarrow Fe^{2+}(aq)$ (at cathode)

Practice Problems

8. A voltaic cell is constructed using the following half-reactions.

$Cu^{2+}(aq) + 2e^- \longrightarrow Cu(s)$
$E^0_{Cu^{2+}} = +0.34\text{ V}$

$Al^{3+}(aq) + 3e^- \longrightarrow Al(s)$
$E^0_{Al^{3+}} = -1.66\text{ V}$

Determine the cell reaction and calculate the standard cell potential.

Practice Problems Plus

Related Chapter Review Problem

Chapter Review problem 35 is related to Sample Problem 23-1.

Additional Practice Problem

The Chem ASAP! CD-ROM contains the following problem: A voltaic cell is constructed using the following half-reactions.

$Li^+(aq) + e^- \rightarrow Li(s)$
$E^0_{Li^+} = -3.05\text{ V}$

$Mg^{2+}(aq) + 2e^- \rightarrow Mg(s)$
$E^0_{Mg^{2+}} = -2.37\text{ V}$

Determine the standard cell reaction and the standard cell potential.

($2Li(s) + Mg^{2+}(aq) \rightarrow Mg(s) + 2Li^+(aq)$; 0.68 V)

ACTIVITY

Have the students work the sample problems on calculating standard cell potentials. Students may think that in E° calculations, as in ΔH calculations, the values must be multiplied by a coefficient in the redox equation. Point out the E° values are standard values already set in terms of a concentration of 1 *M*. Therefore, E° values should not be adjusted in such calculations.

Meeting Diverse Needs

At Risk Some students may have trouble remembering which reactions in Table 23.2 have the greatest tendency to occur as reductions. Have them compare the table with the activity series of metals in Table 23.1. Point out that the most active metals are at the top of Table 23.1; ions of these same metals are at the top of Table 23.2. Because active metals lose electrons easily, they are most easily oxidized. By the same reasoning, their ions are least likely to be reduced.

Practice Problems Plus

Related Chapter Review Problem
Chapter Review problem 34 is related to Sample Problem 23-2.

Additional Practice Problem
The Chem ASAP! CD-ROM contains the following problem: Determine whether this redox reaction will occur spontaneously by calculating the standard cell potential.
$2Al^{3+}(aq) + 3Mg(s) \rightarrow 2Al(s) + 3Mg^{2+}(aq)$
(spontaneous; $E^0_{cell} = +0.71\ V$)

Discuss

Point out that until now it has been assumed that reactions in standard half-cells always go in the spontaneous direction. Cells can be constructed in such a way that the only substance available for oxidation is silver metal and the only substance available for reduction is copper ion. In this case the reaction is nonspontaneous and the cell potential is negative.

Sample Problem 23-1 (cont.)

Before adding the half-reactions, be sure that the electrons cancel—that is, that they are present in equal number on both sides of the equation. Because Ni loses two electrons as it is oxidized, while Fe^{3+} gains only one electron as it is reduced, you must multiply the iron equation by 2.

$$Ni(s) \longrightarrow Ni^{2+}(aq) + 2e^-$$
$$2[Fe^{3+}(aq) + e^- \longrightarrow Fe^{2+}(aq)]$$
$$Ni(s) + 2Fe^{3+}(aq) \longrightarrow Ni^{2+}(aq) + 2Fe^{2+}(aq)$$

The standard cell potential can now be calculated.

$$\begin{aligned} E^0_{cell} &= E^0_{red} - E^0_{oxid} \\ &= E^0_{Fe^{3+}} - E^0_{Ni^{2+}} \\ &= +0.77\ V - (-0.25\ V) \\ &= +1.02\ V \end{aligned}$$

Note that the E^0 of a half-cell is not multiplied by any number even if one or both of the equations is multiplied by a coefficient to make the electrons cancel.

3. *EVALUATE* *Does the result make sense?*
The equations are properly set up and added, and the calculation is correct. If the reduction potential of the reduction is positive and the reduction potential of the oxidation is negative, the E^0_{cell} will always be positive. This is because subtracting a negative number is the same as adding the same positive value.

Practice Problems (cont.)

9. A voltaic cell is constructed using the following half-reactions.

$$Ag^+(aq) + e^- \longrightarrow Ag(s) \quad E^0_{Ag^+} = +0.80\ V$$
$$Cu^{2+}(aq) + 2e^- \longrightarrow Cu(s) \quad E^0_{Cu^{2+}} = +0.34\ V$$

Determine the cell reaction and the standard cell potential.

Chem ASAP!
Problem-Solving 9

Solve Problem 9 with the help of an interactive guided tutorial.

Sample Problem 23-2

Calculate E^0_{cell} to determine whether the following redox reaction is spontaneous as written.

$$Ni(s) + Fe^{2+}(aq) \longrightarrow Ni^{2+}(aq) + Fe(s)$$

1. *ANALYZE* *List the knowns and the unknowns.*

Knowns (from **Table 23.2**):
- $E^0_{Ni^{2+}} = -0.25\ V$
- $E^0_{Fe^{2+}} = -0.44\ V$

Unknowns:
- reaction spontaneous = ?
- $E^0_{cell} = ?\ V$

The redox reaction is spontaneous if the standard cell potential is positive.

Sample Problem 23-2 (cont.)

2. ***CALCULATE*** ***Solve for the unknowns.***
Write each half-reaction with its standard reduction potential. Find the standard cell potential, using $E^0_{cell} = E^0_{red} - E^0_{oxid}$. The half-reactions from the given equation are as follows.

$$\text{Oxidation: } Ni(s) \longrightarrow Ni^{2+}(aq) + 2e^-$$
$$\text{Reduction: } Fe^{2+}(aq) + 2e^- \longrightarrow Fe(s)$$

The standard reduction potentials for the half-cells, both written as reductions, are given in **Table 23.2** as

$$Ni^{2+}(aq) + 2e^- \longrightarrow Ni(s) \qquad E^0_{Ni^{2+}} = -0.25\text{ V}$$
$$Fe^{2+}(aq) + 2e^- \longrightarrow Fe(s) \qquad E^0_{Fe^{2+}} = -0.44\text{ V}$$

The equation given was written with the nickel half-cell as the oxidation half-cell and the iron half-cell as the reduction half-cell. The standard cell potential is thus the difference of the potentials of the iron half-cell and the nickel half-cell.

$$E^0_{cell} = E^0_{red} - E^0_{oxid} = E^0_{Fe^{2+}} - E^0_{Ni^{2+}}$$
$$= -0.44\text{ V} - (-0.25\text{ V}) = -0.19\text{ V}$$

Because the calculated standard cell potential is a negative number, the redox reaction is nonspontaneous as written. Energy would have to be applied to make the reaction proceed. Without such an energy input, the reverse reaction would be the one that occurs.

3. ***EVALUATE*** ***Does the result make sense?***
Because iron is above nickel in the activity series of metals, it makes sense that iron ions would not be reduced by nickel ions. Instead, iron would be oxidized in the presence of nickel ions and would reduce Ni^{2+} ions to nickel metal.

Practice Problems

10. Calculate the standard cell potential to determine whether this redox reaction will occur spontaneously.

$$3Zn^{2+}(aq) + 2Cr(s) \longrightarrow 3Zn(s) + 2Cr^{3+}(aq)$$

11. Is this redox reaction spontaneous as written?

$$Co^{2+}(aq) + Fe(s) \longrightarrow Fe^{2+}(aq) + Co(s)$$

Chem ASAP!

Problem-Solving 10
Solve Problem 10 with the help of an interactive guided tutorial.

section review 23.2

12. What is the difference between standard cell potential and standard reduction potential?

13. How can standard reduction potentials be used to determine standard cell potentials?

14. What is the standard cell potential for the zinc–copper voltaic cell?

$Zn(s) \mid ZnSO_4(aq) \parallel CuSO_4(aq) \mid Cu(s)$

15. What is the purpose of using the standard hydrogen electrode as a reference electrode?

16. The standard reduction potential for a cadmium half-cell is −0.40 V. What does this mean?

Chem ASAP! **Assessment 23.2** Check your understanding of the important ideas and concepts in Section 23.2.

23.2

3 Assess

Evaluate Understanding

Have students define standard cell potential in their own words and list the conditions under which it is measured.

- **Ask students to define the standard reduction potential of an electrode?** (It is the cell potential measured for a half-cell when it is connected to a standard hydrogen electrode, which has an assigned value of 0.00 V.)
- **Have students explain the significance of positive and negative cell potentials?** (If the cell potential is positive, the reaction is spontaneous. If the cell potential is negative, the reaction is nonspontaneous.)

Reteach

Remind students that the cell potential is measured as the difference between the reduction potentials for the two half-cells. Emphasize that a half-cell can never function by itself. As one half-cell undergoes oxidation, the other undergoes reduction. Similarly, the potential of a half-cell can never be found independently. A given half-cell must always be connected to another to produce a potential. The value of that potential is determined by the nature of the other half-cell. Standardization is necessary in order to compare half-cell potentials. Stress that all half-cell potentials are determined by connecting the given half-cell to a standard hydrogen half-cell.

Answers

SECTION REVIEW 23.2

12. The standard reduction potential of a half-cell is a measure of the tendency of a given half-reaction to occur as a reduction under standard conditions. The difference between the standard reduction potentials of the two half-cells is called the standard cell potential.

13. $E^\circ_{cell} = E^\circ_{red} - E^\circ_{oxid}$, in which E°_{red} is the standard reduction potential with the more positive (or less negative) value.

14. +1.10 V

15. A standard hydrogen electrode is used as a reference electrode with an assigned reduction potential of 0.00 V. Thus, measuring the standard cell potential will allow the standard reduction potential of the other half-cell to be determined.

16. The cadmium has a tendency to undergo oxidation.

1 Engage

Use the Visual

Have students examine the section-opening photograph and read the accompanying text. Remind students that, in a similar way, relatively inexpensive decorative objects can be produced by coating less valuable materials with gold. Note that such a plating process has widespread industrial and commercial applications. Ask:

- **How can electricity be used to deposit metal on objects?** (Electric current can be used to force a normally nonspontaneous reaction in the forward direction. If the object to be plated is placed in a solution of metal ions, those ions will be reduced and deposited on the object as metal atoms.)

Check Prior Knowledge

Ask students to recall the lead-storage batteries used in automobiles. Ask:

- **Describe the reaction that allows the battery to supply the electricity needed for ignition.** (Lead and lead(IV) oxide undergo a redox reaction in which lead(II) sulfate is formed.)
- **Explain why the storage battery is labeled rechargeable.** (Electricity from the generator reverses the redox reaction in the battery.)

section 23.3

objectives

- Distinguish between electrolytic and voltaic cells, and list some possible uses of electrolytic cells
- Identify the products of the electrolysis of brine, molten sodium chloride, and water

key terms

- electrolysis
- electrolytic cell

ELECTROLYTIC CELLS

DVDs are used to store all types of data, from multimedia computer programs to the latest popular movie. DVDs have made it possible to store large amounts of data in a very small space. To manufacture a DVD, a laser is used to transfer all the necessary data to the master DVD. Then, metal is deposited on the master using electricity. Once the metal disc is removed from the master, it is used to stamp out the duplicates sold to the public. **How can metal be deposited on objects by using electricity?**

Electrolysis of Water

In Section 23.1, you learned how a spontaneous chemical reaction can be used to generate a flow of electrons (an electric current). In this section, you will learn how an electric current can be used to make a nonspontaneous redox reaction go forward. The process in which electrical energy is used to bring about such a chemical change is called **electrolysis.** Although you may not have realized it, you are already familiar with some results of electrolysis: silverplated dishes and utensils, gold-plated jewelry, and chrome-plated automobile parts are a few examples.

The apparatus in which electrolysis is carried out is an electrolytic cell. An **electrolytic cell** is an electrochemical cell used to cause a chemical change through the application of electrical energy. An electrolytic cell uses electrical energy (DC current) to make a nonspontaneous redox reaction proceed to completion. The vast array of electrolytic cells shown in **Figure 23.10** are used for the commercial production of chlorine and sodium hydroxide from brine, a concentrated aqueous solution of sodium
1 chloride. What are some uses of chlorine and sodium hydroxide?

In both voltaic and electrolytic cells, electrons flow from the anode to the cathode in the external circuit. As shown in **Figure 23.11,** for both types of cells, the electrode at which reduction occurs is the cathode. The electrode at which oxidation occurs is the anode. The difference between the

Figure 23.10

To produce chlorine and sodium hydroxide in electrolytic cells, electricity is passed through brine, a sodium chloride solution. The products have many uses, such as disinfecting water, cleaning drains, manufacturing soap, and producing paper pulp.

Student Resources

From the Teacher's Resource Package, use:

- Section Review 23.3 and Ch. 23 Vocabulary Review, Interpreting Graphics, and Quizzes from the Review Module (Ch. 21–24)
- Laboratory Recordsheets 23-1 and 23-2
- Laboratory Manual: Experiment 46
- Laboratory Practical 23-1

Technology Resources

Relevant technology resources include:

- Chem ASAP! CD-ROM
- Resource Pro CD-ROM
- Assessment Resources CD-ROM: Chapter 23 Tests

Voltaic Cell

e^- e^-
Anode (oxidation)
Cathode (reduction)
Energy
Electrolytes
(−)
(+)
Porous plate or salt bridge

- Energy is released from a spontaneous redox reaction.
- System (cell) does work on the surroundings (light bulb).

Electrolytic Cell

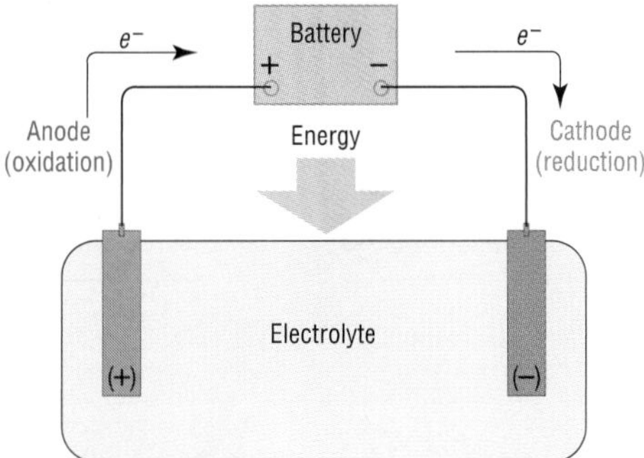

- Energy is absorbed to drive a nonspontaneous redox reaction.
- Surroundings (battery or power supply) do work on the system (cell).

Figure 23.11
Voltaic cells and electrolytic cells share some features. In both cells, reduction occurs at the cathode, and oxidation occurs at the anode. Also, in both cells, electrons flow from the anode to the cathode in the external circuit. However, the charges on the anode and cathode differ for the two kinds of cells. The redox process in the voltaic cell is spontaneous. The redox process in the electrolytic cell is nonspontaneous. What would make a nonspontaneous redox process go? 2

two cells is that in a voltaic cell the flow of electrons is the result of a spontaneous redox reaction, whereas in an electrolytic cell electrons are pushed by an outside power source such as a battery. Electrolytic and voltaic cells also differ in the assignment of charge to the electrodes. In an electrolytic cell, the cathode is considered to be the negative electrode. This is because it is connected to the negative electrode of the battery. (Remember that in a voltaic cell, the anode is the negative electrode and the cathode is the positive electrode.) The anode in the electrolytic cell is considered to be the positive electrode because it is connected to the positive electrode of the battery. It is important to remember these conventions about the two kinds of cells.

When a current is applied to two electrodes immersed in pure water, nothing happens. There is no current flow and no electrolysis. However, when an electrolyte such as H_2SO_4 or KNO_3 in low concentration is added to the pure water, the solution conducts electricity and electrolysis occurs. This process is illustrated in **Figure 23.12** on the following page. The products of the electrolysis of water are hydrogen gas and oxygen gas. Water is reduced to hydrogen at the cathode according to the following reduction half-reaction.

Reduction: $2H_2O(l) + 2e^- \longrightarrow H_2(g) + 2OH^-(aq)$ (at cathode)

Water is oxidized at the anode according to the following oxidation half-reaction.

Oxidation: $2H_2O(l) \longrightarrow O_2(g) + 4H^+(aq) + 4e^-$ (at anode)

2 Teach

TEACHER DEMO

Use a Hoffman apparatus or construct a substitute using iron nails as electrodes and two test tubes to collect the hydrogen gas (at the negative electrode) and oxygen gas (at the positive electrode). Fill the test tubes with 6*M* NaOH solution and invert in a beaker that is 1/3 filled with the NaOH solution.
CAUTION! NaOH is corrosive. Insert the iron electrodes into the lower end of the test tubes. The wire leads connected to the nails should be insulated to avoid contact with the electrolyte. Connect to the power supply. On completion, demonstrate that one test tube contains hydrogen and the other oxygen with a lit match and a glowing splint. Write the half-cell reactions for the electrolysis on the board.

Discuss

Electrolytic processes are used to produce active metals such as aluminum, magnesium, and sodium from their salts. The same process is used to recover metals from ores. Compare electrolytic cells to voltaic cells. In both, oxidation occurs at the anode and reduction at the cathode. In the voltaic cell, a spontaneous redox reaction occurs. In the electrolytic cell, external electrical energy makes a nonspontaneous redox reaction occur.

Answers

1 Answers will vary but may include the following: chlorine: bleaching and disinfecting; sodium hydroxide: soapmaking and degreasing.
2 use of a battery or other external energy source to push electrons

Meeting Diverse Needs

Gifted Explain that electrolytic cells are used in the recovery of low-grade ores in mining. Have the students select a metal and research how it is recovered and purified in the mining process.

Use the Visual

Display Figure 23.13 on an overhead projector. Point out that a spontaneous redox reaction in the battery (the voltaic cell) supplies the electrical energy needed to drive the nonspontaneous redox reaction in the electrolytic cell.
Ask:

- **What substance is being oxidized in the electrolytic cell?** (Cl^-)
- **What substance is being reduced?** (H_2O)

Use colored pens to label the cathode (+) and anode (−) on the battery, and to show the flow of electrons from the anode of the battery to the cathode of the electrolytic cell, and from the anode of the electrolytic cell to the cathode of the battery. Show students that in both the voltaic and electrolytic cells, the definitions of a cathode and an anode are the same: reduction always takes place at the cathode and oxidation always takes place at the anode. Ask:

- **Why is a DC current, and not an AC current, used to run the reaction?** (DC devices maintain a constant potential difference. The voltage potential from an AC generator alternates in sign.) **Write the overall ionic equation for the reaction on the board and ask students to use the standard reduction potentials in Table 23.2 to show why this reaction is nonspontaneous as written.**
  $Cl_2(g) + 2e^- \rightarrow 2Cl^-(aq)$
  $E° = +1.36$
  $2H_2O(l) + 2e^- \rightarrow H_2(g) + 2OH^-(aq)$
  $E° = -0.42$
  $E°_{cell} = E°_{red} - E°_{oxid}$
  $= -0.42 - 1.36$
  $= -1.78$

The potential is negative; therefore, energy is needed to make this reaction occur.

Figure 23.12
When an electric current is passed through water, the water decomposes into oxygen gas and hydrogen gas. At which electrode is hydrogen gas produced? Which gas results from the oxidation process? ❶

The region around the cathode turns basic due to the production of OH^- ions. The region around the anode turns acidic due to an increase in H^+ ions. The overall cell reaction is obtained by adding the half-reactions (after doubling the first one to balance electrons).

$$\text{Reduction: } 2[2H_2O(l) + 2e^- \longrightarrow H_2(g) + 2OH^-(aq)]$$
$$\text{Oxidation: } 2H_2O(l) \longrightarrow O_2(g) + 4H^+(aq) + 4e^-$$
$$\text{Overall cell reaction: } 6H_2O(l) \longrightarrow 2H_2(g) + O_2(g) + 4H^+(aq) + 4OH^-(aq)$$

The ions produced tend to recombine to form water.

$$4H^+(aq) + 4OH^-(aq) \longrightarrow 4H_2O(l)$$

For that reason they are not included in the net reaction.

$$2H_2O(l) \xrightarrow{\text{electrolysis}} 2H_2(g) + O_2(g)$$

Electrolysis of Brine

If the electrolyte in an aqueous solution is more easily oxidized or reduced than water, then the products of electrolysis will be substances other than hydrogen and oxygen. An example is the electrolysis of brine, a concentrated aqueous solution of sodium chloride, which simultaneously produces three important industrial chemicals: chlorine gas, hydrogen gas, and sodium hydroxide. The electrolytic cell for this process is shown in **Figure 23.13.**

During electrolysis, chloride ions are oxidized to produce chlorine gas at the anode. Water is reduced to produce hydrogen gas at the cathode. Sodium ions are not reduced to sodium metal in the process because water molecules are more easily reduced than are sodium ions. The reduction of water produces hydroxide ions as well as hydrogen gas. Thus the electrolyte in solution becomes sodium hydroxide (NaOH). The half-reactions are

$$\text{Oxidation: } 2Cl^-(aq) \longrightarrow Cl_2(g) + 2e^- \text{ (at anode)}$$
$$\text{Reduction: } 2H_2O(l) + 2e^- \longrightarrow H_2(g) + 2OH^-(aq) \text{ (at cathode)}$$

Answers

❶ Water is reduced to hydrogen at the cathode; oxygen is produced at the anode when water is oxidized.
❷ Oxidation: chlorine gas; reduction: hydrogen gas.

Battery

+ −

Cl_2 ← → H_2

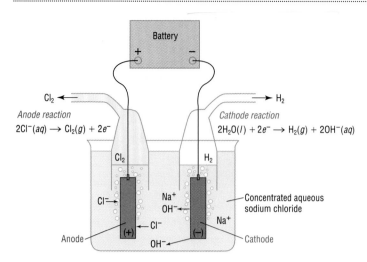

Anode reaction

$2Cl^-(aq) \longrightarrow Cl_2(g) + 2e^-$

Cathode reaction

$2H_2O(l) + 2e^- \longrightarrow H_2(g) + 2OH^-(aq)$

Cl_2 H_2

Cl^- Na^+ OH^- Cl^- OH^- Na^+

(+) (−)

Anode

Cathode

Concentrated aqueous sodium chloride

Figure 23.13

Which substances are produced by oxidation and by reduction during the electrolysis of a concentrated solution of sodium chloride (brine)? 2

The overall ionic equation is the sum of the two half-reactions.

$$2Cl^-(aq) + 2H_2O(l) \longrightarrow Cl_2(g) + H_2(g) + 2OH^-(aq)$$

The spectator ion Na^+ can be included in the equation (as part of NaCl and of NaOH) to show the formation of sodium hydroxide during the electrolytic process.

$$2NaCl(aq) + 2H_2O(l) \longrightarrow Cl_2(g) + H_2(g) + 2NaOH(aq)$$

When the sodium hydroxide solution is about 10% (m/v), it is removed from the cell and processed further.

Electrolysis of Molten Sodium Chloride

Both sodium and chlorine are commercially important. Sodium is used in sodium vapor lamps and as the coolant in some nuclear reactors. Chlorine, a toxic greenish-yellow gas, is used to sterilize drinking water and is important in the manufacture of polyvinyl chloride and various pesticides. These two elements are produced when pure molten sodium chloride, rather than an aqueous solution of NaCl, is electrolyzed. Chlorine gas comes off at the anode and molten sodium collects at the cathode. The sodium, with a melting point of 97.8 °C, remains in liquid form, floating on the more dense molten sodium chloride. The electrolytic cell in which this commercial process is carried out is called the Downs cell and is shown in **Figure 23.14** on the following page. The design of the Downs cell allows fresh sodium chloride to be added as required. The design also separates the products so they will not recombine to re-form sodium chloride. The half-reactions and overall cell reaction for this electrolysis are as follows.

$$\text{Oxidation: } 2Cl^-(l) \longrightarrow Cl_2(g) + \cancel{2e^-} \text{ (at anode)}$$

$$\text{Reduction: } 2Na^+(l) + \cancel{2e^-} \longrightarrow 2Na(l) \text{ (at cathode)}$$

$$\text{Overall cell reaction: } 2NaCl(l) \longrightarrow 2Na(l) + Cl_2(g)$$

METALLURGY

Anodizing

Anodizing is an electrolytic process for applying a corrosion-resistant oxide coating to aluminum and certain other metals. When an electric current flows through an electrolytic cell

containing an aluminum anode, a thin coat of aluminum oxide forms on the surface of the aluminum. Immersion of the aluminum object in pure boiling water permanently seals the film against corrosion. Dyeing of anodized aluminum can produce metallic red, metallic blue, and a host of other metallic colors on the metal surface. Also, the high porosity of the oxide coating permits it to absorb pigments and lubricants.

23.3

Discuss

Explain that the operational definition of cathodes and anodes is the same in voltaic and electrolytic cells (the sites of reduction and oxidation, respectively). However, the charge labels applied to the electrodes are reversed. In a voltaic cell, the anode is negative and the cathode is positive. In an electrolytic cell, the cathode is labeled negative because it is connected to the negative battery terminal (the anode); the anode is labeled positive because it is connected to the positive battery terminal (the cathode).

CHEMISTRY AND INDUSTRY

Gold and silver are the most common metals used for plating jewelry. A thin layer of the precious metal is deposited over much less expensive base materials such as copper. Nonmetallic objects may be used as the base material as long as their surface is first coated with a conductor such as graphite. Plating solutions contain a suitable electrolyte (an acid or base) to adjust pH. Small amounts of other substances may be added to vary the color of gold jewelry. Nickel, silver, copper, cobalt, or indium may be deposited along with the gold.

Use the Visual

Have students examine Figure 23.14. Ask students to explain why sodium cannot be produced from an aqueous solution of sodium chloride. (The H^+ in the aqueous solution is more easily reduced to hydrogen gas than the Na^+ is to sodium metal.)

3 Assess

Evaluate Understanding

Use several examples to point out the nature of all reactions in electrolytic cells.

- **Have students generalize about the potentials of electrolytic cells.** (Electrolytic cells have negative potentials. They are nonspontaneous redox reactions.)
- **Have students list three applications each for voltaic cells and electrolytic cells.** (electrolytic cells: plating, production of chemicals, extraction of metals from ores; voltaic cells: batteries, fuel cells, generating electricity)

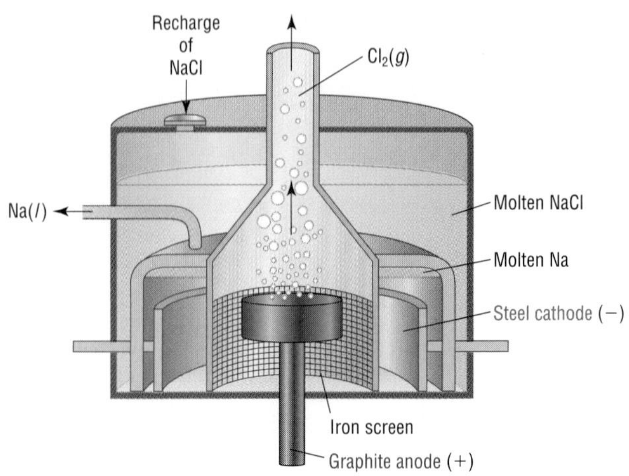

Figure 23.14
The Downs cell produces sodium metal and chlorine gas from the electrolysis of molten NaCl. The cell operates at a temperature of 801 °C to maintain the salt in a molten state. A perforated iron screen separates the circular cathode from the anode. The liquid sodium floats to the top of the molten salt and is drawn off. The chlorine gas is collected after it bubbles up and out of the molten salt.

Chem ASAP!

Animation 27
Take an atomic-level look at how electricity can bring about a chemical change.

Electroplating and Related Processes

Electroplating is the deposition of a thin layer of a metal on an object in an electrolytic cell. The process has many important applications. An object may be electroplated to protect the surface of the base metal from corrosion or to make it more attractive. The layer of the deposited metal is very thin, usually from 5×10^{-5} cm to 1×10^{-3} cm thick. Metals commonly used for electroplating include gold, silver, copper, nickel, and chromium. An object that is to be silverplated is made the cathode in an electrolytic cell. The anode is the metallic silver that is to be deposited, and the electrolyte is a solution of a silver salt, such as silver cyanide. When a DC current is applied, silver ions move from the anode to the object to be plated. What happens to the silver ions at the cathode? 1

$$\text{Reduction: } Ag^+(aq) + e^- \xrightarrow{\text{electrolysis}} Ag(s) \text{ (at cathode)}$$

The net result is that silver transfers from the silver electrode to objects being plated, like those shown in **Figure 23.15.** Controlling this reaction is a fine art. Many factors contribute to the quality of the metal coating that forms. In the plating solution, the concentration of the cations to be reduced must be carefully controlled. The solution must contain compounds to control the acidity and to increase the conductivity. Some solutions contain compounds that form complex ions with the cation to be reduced. Other compounds may be used to make the metal coating brighter or smoother. In many plating cells, the anode must be shaped like the object at the cathode in order to plate all parts of the cathode evenly.

Several useful processes involve the depositing of metal at the cathode of a cell. In electroforming, an object is reproduced by making a metal mold of it at the cathode. A phonograph record, for example, can be coated with metal so it will conduct electricity. It is then electroplated to obtain a thick coating of metal. This coating can then be stripped off and used as a mold to produce copies of the record. In another process, called electrowinning, impure metals can be purified in electrolytic cells. The cations of molten salts or aqueous solutions are reduced at the cathode to give very

Figure 23.15
Electroplating is the coating of an object with a thin layer of metal in an electrolytic cell. Pure silver was electroplated onto steel to produce this silverplated tea service.

Answer

1 The silver ions are reduced and are deposited as silver atoms (metallic silver).

Meeting Diverse Needs

At Risk Encourage students to review the chapter looking for words in bold print and other key terms. Have the students compile a glossary in which they define each term in their own words. Have them also construct a chart contrasting the properties of voltaic cells and electrolytic cells.

MINI LAB

TARNISH REMOVAL

PURPOSE

To remove tarnish from silver.

MATERIALS

- baking soda (sodium hydrogen carbonate)
- water
- hot plate
- saucepan
- tablespoon
- shallow dish (glass or plastic)
- aluminum foil
- tarnished silver item
- tongs

PROCEDURE

1. Mix 1 tablespoon of baking soda with about 1 quart of water. Heat the solution in the saucepan until it is hot but not boiling. **CAUTION:** Be careful when using the heat source and handling the hot solution.
2. Place the aluminum foil on the bottom of the shallow dish.
3. Place the tarnished item on the foil, making sure the item is in contact with the foil.
4. Carefully pour the hot solution from the saucepan into the dish so the solution covers the tarnished item.
5. After 5–10 minutes, use tongs to remove the item and rinse with water.
6. Optional: Try to retarnish the silver item by smearing egg yolk over one part of it and letting it sit for 30 minutes.

ANALYSIS AND CONCLUSIONS

1. How did the tarnished item differ in appearance before and after its treatment?
2. What do you think would have happened if the treatment solution had been cold?
3. The tarnish on silver is usually a film of the ionic compound Ag_2S. Use information from **Table 23.1** to explain how this layer of Ag_2S is removed. What is the purpose of the baking-soda solution?

pure metals. Electrowinning from molten salts is the only method for obtaining pure sodium and other very reactive metals. In the process of electrorefining, a piece of impure metal is made the anode of the cell. It is oxidized to the cation and then reduced to the pure metal at the cathode. This technique is used to obtain ultrapure silver, lead, and copper.

Other electrolytic processes are centered on the anode rather than the cathode. In electropolishing, for example, the surface of an object at the anode is dissolved selectively to give it a high polish. In electromachining, a piece of metal at the anode is partially dissolved until the remaining portion is an exact copy of the object at the cathode.

section review 23.3

17. Describe the difference between an electrolytic cell and a voltaic cell. Name several uses for electrolytic cells.
18. What are the products of the electrolysis of the following?
    a. brine b. molten sodium chloride c. water
19. What process occurs at the anode of an electrolytic cell? At the cathode?

 Chem ASAP! **Assessment 23.3** Check your understanding of the important ideas and concepts in Section 23.3.

23.3

MINI LAB

Tarnish Removal

SAFETY

Remind students to wear safety goggles.

ANALYSIS AND CONCLUSIONS

1. The black tarnish is replaced by a shiny silver surface.
2. The reaction is temperature dependent. It would have been slower at the lower temperature.
3. Aluminum is above silver in the table. This means that silver is less easily oxidized (or its cation is more easily reduced) than aluminum. Thus, when Ag^+ ions are in contact with Al atoms in the presence of an electrolyte, $NaHCO_3(aq)$, electrons are transferred from Al to Ag^+. The reaction: $Al(s) + 3Ag^+(aq) \rightarrow Al^{3+}(aq) + 3Ag(s)$

Reteach

Stress that an electrolytic cell is the reverse of a voltaic cell. This means that electrical energy is used to produce a nonspontaneous chemical change in an electrolytic cell. Also point out the similarities in both types of cells: oxidation occurs at the anode and reduction occurs at the cathode.

Answers

SECTION REVIEW 23.3

17. Voltaic cells use an electrochemical reaction to produce electrical energy; electrolytic cells use electrical energy to bring about a chemical change. Electrolytic cells are used in electroplating and extraction of metals from ores.
18. a. sodium hydroxide, chlorine gas, and hydrogen gas
    b. sodium metal and chlorine gas
    c. hydrogen gas and oxygen gas
19. oxidation at the anode; reduction at the cathode

SMALL-SCALE LAB

ELECTROLYSIS OF WATER

OVERVIEW

Procedure Time: 30 Minutes
Students will apply an electrolysis device to pure water and aqueous solutions.

MATERIALS

Solution	Preparation
0.2*M* Na_2SO_4	7.1 g in 250 mL
0.1*M* KI	4.2 g in 250 mL
0.2*M* $CuSO_4$	12.5 g $CuSO_4 \cdot 5H_2O$ in 250 mL
starch	50 mL liquid starch in 200 mL
0.04% BTB	100 mg in 16.0 mL of 0.01*M* NaOH; dilute to 250 mL
1.0*M* NaCl	14.6 g in 250 mL
0.2*M* KBr	6.0 g in 250 mL

TEACHING SUGGESTIONS

For instructions on making a set of electrolysis devices, see the *Small-Scale Chemistry Laboratory Manual.*

YOU'RE THE CHEMIST

1. Bubbles are $H_2(g)$; blue solution indicates the presence of OH^- ions; yellow solution is $I_2(aq)$, which is black in the presence of starch.
2. **NaCl**
   Cathode: $2H_2O + 2e^- \rightarrow H_2(g) + 2OH^-$ (bubbles , blue BTB)
   Anode: $2Cl^- \rightarrow Cl_2(aq) + 2e^-$ (yellow solution)
   KBr
   Cathode: $2H_2O + 2e^- \rightarrow H_2(g) + 2OH^-$ (bubbles , blue BTB)
   Anode: $2Br^- \rightarrow Br_2(aq) + 2e^-$ (yellow solution)
   $CuSO_4$
   Cathode: $Cu^{2+} + 2e^- \rightarrow H_2(g) + 2OH^-$ (Copper plates out.)
   Anode: $H_2O \rightarrow \frac{1}{2} O_2(g) + 2H^+ + 2e^-$ (bubbles , yellow BTB)
3. $H_2O \rightarrow \frac{1}{2} O_2(g) + 2H^+ + 2e^-$
   $E^o = -0.82$ V
   $2I^- \rightarrow I_2(aq) + 2e^-$; $E^o = -0.54$ V
   I^- is more likely to oxidize than H_2O because it has a more positive E^o value.

SMALL-SCALE LAB

ELECTROLYSIS OF WATER

SAFETY

Wear safety glasses and follow the standard safety procedures outlined on page 18.

PURPOSE

To electrolyze solutions and interpret your observations in terms of chemical reactions and equations.

MATERIALS

- pencil
- paper
- ruler
- reaction surface
- electrolysis device
- chemicals shown in Figure A

PROCEDURE

On separate sheets of paper, draw two grids similar to Figure A. Make each square 2 cm on each side. Place a reaction surface over one of the grids and add one drop of each solution as shown in Figure A. Apply the leads of the electrolysis device to each solution. Be sure to clean the leads between each experiment. Look carefully at the cathode (negative lead) and the anode (positive lead). Use the second grid as a data table to record your observations for each solution.

H_2O	Na_2SO_4	Na_2SO_4 + BTB
no visible reaction	bubbles at anode and cathode	anode yellow; cathode blue

Figure A

ANALYSIS

Using your experimental data, record the answers to the following questions below your data table.

1. Explain why pure water does not conduct electricity and does not undergo electrolysis.
2. Explain why water with sodium sulfate conducts electricity and undergoes electrolysis.
3. The cathode (negative lead) provides electrons to water and the following half-reaction occurs.
   $$2H_2O + 2e^- \longrightarrow H_2(g) + 2OH^-$$
   Explain how your observations correspond to the products shown in this reaction.
4. The anode (positive lead) takes away electrons from water and the following half-reaction occurs.
   $$H_2O \longrightarrow \tfrac{1}{2} O_2(g) + 2H^+ + 2e^-$$
   Explain how your observations correspond to the products shown in this reaction.
5. Add the two half-reactions to obtain the overall reaction for the electrolysis of water. Simplify the result by adding the OH^- and H^+ to get HOH, and then canceling anything that appears on both sides of the equation.

YOU'RE THE CHEMIST 

The following small-scale activities allow you to develop your own procedures and analyze the results.

1. **Analyze It!** Perform the above experiment using the chemicals shown in Figure B. Record your results. The cathode and anode reactions are
   $2H_2O + 2e^- \longrightarrow H_2(g) + 2OH^-$ (at cathode)
   $2I^- \longrightarrow I_2(aq) + 2e^-$ (at anode)
   Explain how your observations correspond to the products shown in these half-reactions.

KI	KI + starch	KI + BTB
cathode: bubbles; anode: yellow soln	cathode: bubbles; anode: black soln	cathode: bubbles and blue soln; anode: yellow soln

Figure B

2. **Design It!** Design an experiment to explore what happens when you electrolyze NaCl, KBr, and $CuSO_4$ with and without BTB. Use half-reactions to predict your results.
3. **Analyze It!** For each half-reaction listed above, look up the E^0 values in Section 23.2. Show that the E^0 values are consistent with what happens in each case.

ANALYSIS

1. Pure water has too few ions to carry an electric current.
2. Sodium sulfate dissociates into ions in solution, which carry an electric current.
3. The bubbles are $H_2(g)$, and the blue BTB solution indicates the presence of OH^- ions.
4. The bubbles are $O_2(g)$, and H^+ ions in solution impart the yellow color to the BTB solution.
5. $2H_2O + \cancel{2e^-} \rightarrow H_2(g) + 2OH^-$
   $H_2O \rightarrow \frac{1}{2} O_2(g) + 2H^+ + \cancel{2e^-}$
   $3H_2O \rightarrow H_2(g) + \frac{1}{2} O_2(g) + 2OH^- + 2H^+$
   $\cancel{3}H_2O \rightarrow H_2(g) + \frac{1}{2} O_2(g) + \cancel{2H_2O}$
   $H_2O \rightarrow H_2(g) + \frac{1}{2} O_2(g)$

Chemistry Serving...the Environment

HOW MANY KILOMETERS PER CHARGE?

Imagine this: After breakfast, you open your garage, unplug your car and drive off to school. After school, you take some friends to the football game at a school in the next town. Later you pick up your aunt at the airport. You have logged 150 kilometers on the road, but you have not burned a drop of gasoline or emitted a gram of pollutants from your tailpipe. Actually, you do not even have a tailpipe because your car runs on electricity.

This scenario is not a vision of the future: Electric cars are here! Most of the major automakers now offer an electric vehicle, or EV, for lease or sale. EVs are hard to distinguish from gasoline-powered models because they have many of the same features. The main advantage EVs have over gasoline-powered cars is that they are environmentally friendly. They do not produce air pollutants and are therefore classified as zero-emission vehicles.

Improving battery technology is the central problem for electric-car designers.

But EVs have a major drawback: Their distance range is limited. Current models can go only 150 to 200 kilometers on a single charge of their batteries, a problem that stems from the energy-storage capacity of the batteries. Improving battery technology, therefore, is the central problem for electric-car designers.

The most common type of battery used in EVs is an improved version of the lead–acid battery, which has been used in gasoline-fueled vehicles for nearly 100 years. Despite its improvements, this type of battery is a poor choice for storing energy in a vehicle. For the amount of electrical energy it can store, a lead–acid battery is very heavy. And it is an established fact that the greater the mass of a vehicle, the poorer its performance and efficiency. Consider the fact that 100 kilograms of lead–acid batteries are needed to store the energy contained in only 1 kilogram of gasoline!

Other types of more-efficient batteries are being used in EVs. These include nickel-metal hydride batteries and lithium-ion batteries, which can store more energy than lead–acid batteries. However, they are considerably more expensive because large-scale production has only just begun.

Researchers are developing other types of batteries that can store more energy in less mass. One promising technology is the sodium–sulfur battery in which sodium atoms supply electrons to the electric motor. For the battery to work, the sodium and sulfur must be molten, which requires temperatures of 300 °C or more. This type of battery has the potential to store more than four times as much energy as a lead–acid battery of the same mass. However, keeping the battery hot enough when not in use remains an unresolved problem to date.

The long-term goal for battery designers is to have batteries that are relatively inexpensive, long lasting (more than 10 years), and able to power an EV for more than 300 kilometers. Once such batteries become a reality, EVs may be on their way to replacing gasoline-powered cars. Such progress would undoubtedly benefit the environment by reducing air pollution.

CHEMISTRY IN CAREERS

TAKE IT TO THE NET
Find out more about career opportunities: www.phschool.com

ELECTROCHEMIST
Interested in pursuing a highly charged career? **See page 879.**

CHEMICAL SPECIALIST
Local food service distributor seeks responsible self-motivated individ-

CHEMIST NEEDED
Enviro. Lab. GC Must know EPA methods, troubleshoot-

Chemistry Serving... the Environment

DISCUSS

Point out that current electric cars are quiet and clean. However, they are not practical for most consumers because of their limited range. With existing lead storage batteries, a car can travel only about 60 miles before it must be recharged. Manufacturers also want batteries that would make electric cars competitive in terms of performance and cost. Fuel cells could provide performance comparable to that of today's gasoline powered automobiles. Ask:

- What would the waste products be from a car that uses a hydrogen–oxygen fuel cell? (The product of the redox reaction is water.)

Have students research the benefits and drawbacks of different alternatives to the lead storage battery. Explain that much of the motivation for the development of longer lasting batteries and fuel cells has come from NASA. The goal has been to maximize the life of the battery and minimize weight. There is also a need to minimize waste products produced during extended periods in space.

CHEMISTRY IN CAREERS

Direct students to page 879 of this text for a career feature on electrochemists. Students may wish to connect to the Internet address shown to discover more about career opportunities in this field.

4 Close

Summary

Ask questions that require students to summarize chapter content.

- **What is similar in a voltaic cell and a electrolytic cell?** (Oxidation occurs at the anode and reduction at the cathode.)
- **How do the two types of cells differ?** (In a voltaic cell, redox is spontaneous; in an electrolytic cell, it is not.)
- **How are half-cell reduction potentials determined?** (by comparison with a standard hydrogen electrode)
- **For a given redox reaction, how can you predict which half-cell undergoes reduction?** (the one with the higher reduction potential)
- **Describe some applications of electrolytic cells.** (electroplating, refining of metals, production of active chemicals)

Extension

Refer to Section 19.3 to remind students of the role enthalpy and entropy changes play in determining spontaneity. Ask students if these concepts can be applied to redox reactions.

Looking Back . . . Looking Ahead . . .

The preceding chapter introduced redox reactions and oxidation numbers. The current chapter applied the concept of redox to voltaic and electrolytic cells. The next chapter discusses the properties of metals and nonmetals in relation to their positions on the periodic table.

Chapter 23 STUDENT STUDY GUIDE

Take It to the NET For interactive study and review, go to www.phschool.com

KEY TERMS

- anode *p. 680*
- battery *p. 682*
- cathode *p. 680*
- cell potential *p. 685*
- dry cell *p. 681*
- electrical potential *p. 685*
- electrochemical cell *p. 679*
- electrochemical process *p. 678*
- electrode *p. 680*
- electrolysis *p. 692*
- electrolytic cell *p. 692*
- fuel cell *p. 683*
- half-cell *p. 679*
- reduction potential *p. 685*
- salt bridge *p. 679*
- standard cell potential *p. 685*
- standard hydrogen electrode *p. 685*
- voltaic cell *p. 679*

KEY EQUATIONS AND RELATIONSHIPS

- Representing electrochemical cells:

$Zn(s) \mid ZnSO_4(aq) \parallel CuSO_4(aq) \mid Cu(s)$

Oxidation (anode) — Reduction (cathode)

- $E^0_{cell} = E^0_{red} - E^0_{oxid}$

CONCEPT SUMMARY

23.1 Electrochemical Cells

- Spontaneous redox equations can be used to generate electrical energy in voltaic cells.
- In an electrochemical cell, oxidation occurs at the anode and reduction occurs at the cathode.
- In a voltaic cell the cathode is the positively charged electrode and the anode is the negatively charged electrode.
- The half-cells are separated by a porous plate or salt bridge that prevents the contents of the two half-cells from mixing but permits the passage of ions between the half-cells.
- The half-cell with the higher reduction potential undergoes reduction; the other half-cell undergoes oxidation.
- A dry cell is not rechargeable. The lead storage battery is rechargeable.
- Fuel cells do not have to be recharged because the fuels they consume are fed into the cell continuously.

23.2 Half-Cells and Cell Potentials

- The cell potential is the difference between the reduction potentials of the half-cells.
- The reduction potential of an isolated half-cell is determined by comparison to the standard hydrogen electrode, which is assigned a reduction potential of 0.00 V.
- Half-cells more easily reduced than the reference electrode have positive reduction potentials. Half-cells less easily reduced have negative reduction potentials.

23.3 Electrolytic Cells

- In electrolysis, reduction occurs at the cathode (negatively charged) and oxidation occurs at the anode (positively charged).
- In an electrolytic cell, electrical energy is used to bring about desired redox reactions.
- Electrolytic cells are used in electroplating, in the refining of metals, and in the production of important chemicals.

CHAPTER CONCEPT MAP

Use these terms to construct a concept map that organizes the major ideas of this chapter.

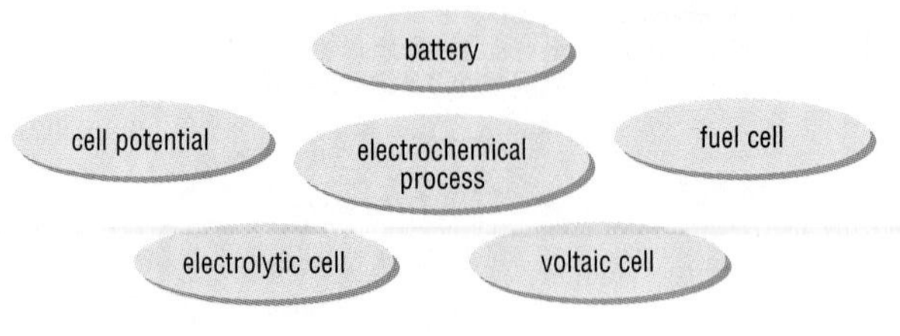

Chem ASAP! Concept Map 23
Create your Concept Map using the computer.

Take It to the Net

At **www.phschool.com** students will find for this chapter

- an Internet activity
- links to related chemistry sites
- an interactive quiz
- career links

Chapter 23 REVIEW

CONCEPT PRACTICE

20. What is meant by the term half-reaction? Write the half-reactions that occur when a strip of aluminum is dipped into a solution of copper(II) sulfate. *23.1*

21. What would you expect to happen when a strip of lead is placed in an aqueous solution of magnesium nitrate? *23.1*

22. For each pair of metals listed below, decide which metal is more readily oxidized. *23.1*

a. Hg, Cu **c.** Ni, Mg **e.** Pb, Zn
b. Ca, Al **d.** Sn, Ag **f.** Cu, Al

23. At which electrode in a voltaic cell does reduction always occur? *23.1*

24. Explain the function of the salt bridge in a voltaic cell. *23.1*

25. In a typical flashlight battery, what material is used for the anode? For the cathode? *23.1*

26. Explain why the specific gravity of the electrolyte in a lead storage battery decreases during the discharge process. *23.1*

27. Use the shorthand method to represent the electrochemical reaction in a car battery. *23.1*

28. Fuel cells can be designed to generate electrical energy while emitting no air pollutants, yet they are not widely used. Explain. *23.1*

29. List the advantages of a fuel cell over a lead storage battery. *23.1*

30. How was the standard reduction potential of the hydrogen electrode determined? *23.2*

31. What is the electric potential of a cell? *23.2*

32. How does the order of the metals in **Table 23.1** compare with the order in **Table 23.2**? Why? *23.2*

33. Explain how to determine the standard reduction potential for the aluminum half-cell. *23.2*

34. Determine whether these redox reactions will occur spontaneously. Calculate the standard cell potential in each case. *23.2*

a. $Cu(s) + 2H^+(aq) \longrightarrow Cu^{2+}(aq) + H_2(g)$
b. $2Ag(s) + Fe^{2+}(aq) \longrightarrow 2Ag^+(aq) + Fe(s)$

35. Use the information in **Table 23.2** to calculate standard cell potentials for these voltaic cells. *23.2*

a. $Ni \mid Ni^{2+} \parallel Cl_2 \mid Cl^-$
b. $Sn \mid Sn^{2+} \parallel Ag^+ \mid Ag$

36. Why is direct current, not alternating current, used in the electroplating of metals? *23.3*

37. Describe briefly how you would electroplate a teaspoon with silver. *23.3*

CONCEPT MASTERY

38. Distinguish between voltaic and electrolytic cells.

39. Why is it not possible to measure the potential of an isolated half-cell?

40. Describe the composition of the anode, cathode, and electrolyte in a fully discharged lead storage battery.

41. Predict what will happen, if anything, when an iron nail is dipped into a solution of copper sulfate. Write the oxidation and reduction half-reactions for this process and the balanced equation for the overall reaction.

42. Calculate E^0_{cell} and write the overall cell reaction for these cells.

a. $Sn \mid Sn^{2+} \parallel Pb^{2+} \mid Pb$
b. $H_2 \mid H^+ \parallel Br_2 \mid Br^-$

43. What property do lead(II) sulfate and lead dioxide have that makes salt bridges unnecessary in a lead storage battery?

44. Complete this data table for the electrolysis of water.

	H_2O used	H_2 formed	O_2 formed
a.	2.0 mol	____ mol	____ mol
b.	____ g	____ g	16.0 g
c.	____ mL	10.0 g	____ g
d.	44.4 g	____ g	____ g
e.	____ g	8.80 L (STP)	____ L (STP)
f.	66.0 mL	____ g	____ L (STP)

CRITICAL THINKING

45. Gold is not included in **Table 23.2.** In what part of the table does gold belong?

46. Lead storage batteries can be recharged. Why are dry cells not rechargeable?

47. For any voltaic cell, chemists consider the electrode that produces electrons to be negative, and they call it the anode. Most dictionaries, however, define the anode as the positively charged electrode. Explain.

Answers

CHAPTER REVIEW

20. the oxidation or reduction in a redox reaction; oxidation: $Al(s) \rightarrow Al^{3+}(aq) + 3e^-$, reduction: $Cu^{2+}(aq) + 2e^- \rightarrow Cu(s)$

21. nothing

22. **a.** Cu **c.** Mg **e.** Zn
b. Ca **d.** Sn **f.** Al

23. cathode

24. The salt bridge allows ions to pass from one half-cell to the other, but prevents the solutions from mixing.

25. anode: zinc; cathode: carbon (graphite)

26. Water is produced by the redox reaction, sulfuric acid is used up, and water has the lower density.

27. $Pb(s) \mid PbSO_4(s) \parallel PbO_2(s) \mid PbSO_4(s)$

28. Fuel cells cannot generate electricity as economically as more conventional forms of electrical generation can.

29. A fuel cell needs no recharging, and does not produce toxic wastes if the fuel is hydrogen gas.

30. It was arbitrarily set at zero.

31. the ability of a voltaic cell to produce a current

32. The relative order is the same because both tables rank the elements according to their tendency to undergo oxidation/reduction.

33. The aluminum half-cell is connected to a standard hydrogen half-cell and a voltmeter is used. The aluminum half-cell has the indicated voltage.

34. **a.** nonspontaneous, -0.34 V
b. nonspontaneous, -1.24 V

35. **a.** $+1.61$ V **b.** $+0.94$ V

36. A direct current flows in one direction only.

37. The teaspoon is the cathode in an electrolytic cell with silver cyanide as the electrolyte. When the DC current flows, the silver ions deposit on the teaspoon.

38. Voltaic cells convert chemical energy into electrical energy. Electrolytic cells use electrical energy to cause a chemical reaction.

39. Two half-cells are needed because oxidation or reduction cannot occur in isolation. One half-cell gains electrons and one loses them, producing an electric current.

40. The anode and cathode grids are both packed with $PbSO_4$. The electrolyte is very dilute sulfuric acid.

41. Some of the iron dissolves and the nail becomes coated with copper.
$Fe(s) + CuSO_4(aq) \rightarrow FeSO_4(aq) + Cu(s)$
oxidation: $Fe \rightarrow Fe^{2+} + 2e^-$
reduction: $Cu^{2+} + 2e^- \rightarrow Cu$

42. **a.** $Sn(s) + Pb^{2+}(aq) \rightarrow Sn^{2+}(aq) + Pb(s)$ $E^\circ_{cell} = +0.01$ V
b. $H_2(g) + Br_2(l) \rightarrow 2H^+(aq) + 2Br^-(aq)$ $E^\circ_{cell} = +1.07$ V

43. Lead(II) sulfate and lead dioxide are very insoluble in sulfuric acid.

Answers

44. a. 2.0, 1.0 **d.** 4.9, 39.5
b. 18.0, 2.0 **e.** 7.07, 4.40
c. 90, 80 **f.** 7.3, 41.1

45. Gold belongs near the bottom, below silver, because it is one of the least active metals.

46. The paste in a dry cell allows for the easy movement of electrons but not ions.

47. The chemists' definition focuses on the electons that are produced by oxidation at the anode of a voltaic cell; the dictionary definition is probably based on an electrolytic cell, whose electrodes are defined by the battery terminals to which they are attached.

48. a. 2 **b.** 1 **c.** 4

49. The baby shoe is being plated with copper.

50. d; The voltage falls steadily.

51. 467 mL O_2

52. a. $3H_2S(g) + 2HNO_3(aq) \rightarrow 3S(s) + 2NO(g) + 4H_2O(l)$
b. $2AgNO_3(aq) + Pb(s) \rightarrow Pb(NO_3)_2(aq) + 2Ag(s)$
c. $3Cl_2(g) + 6NaOH(aq) \rightarrow 5NaCl(aq) + NaClO_3(aq) + 3H_2O(l)$

53. Dilute 31 mL of 16*M* acid to a total volume of 500 mL.

54. a. +6 **b.** +5 **c.** +7 **d.** +3

55. a. $2AgCl + Ni \rightarrow 2Ag + NiCl_2$
$E^\circ_{cell} = +0.47\ V$
b. $3Cl_2 + 2Al \rightarrow 2AlCl_3$
$E^\circ_{cell} = +3.02\ V$

56. Oxidation: $2Cu(impure) + 2H_2SO_4 \rightarrow 2Cu^{2+} + 2H_2 + 2SO_4^{2-}$
Reduction: $2Cu^{2+} + 2SO_4^{2-} + 2H_2O \rightarrow 2Cu(pure) + 2H_2SO_4 + O_2$
$2Cu(impure) + 2H_2O \rightarrow 2Cu(pure) + 2H_2 + O_2$

57. a. The iron electrode is the anode; the nickel electrode is the cathode.
b. The anode is negative; the cathode is positive.
c. anode: $Fe(s) \rightarrow Fe^{2+}(aq) + 2e^-$
cathode: $Ni^{2+}(aq) + 2e^- \rightarrow Ni(s)$
d. $E^\circ_{cell} = 0.19\ V$

48. Choose the term that best completes the second relationship.

a. battery : cells
herd : ______
(1) birds (3) trees
(2) sheep (4) fuel cells

b. river : water
electrode : ______
(1) electrons (3) salt bridge
(2) anode (4) ions

c. anode : electrons
fire hydrant : ______
(1) fire (3) dogs
(2) firefighter (4) water

49. Describe the process that is occurring in the following illustration.

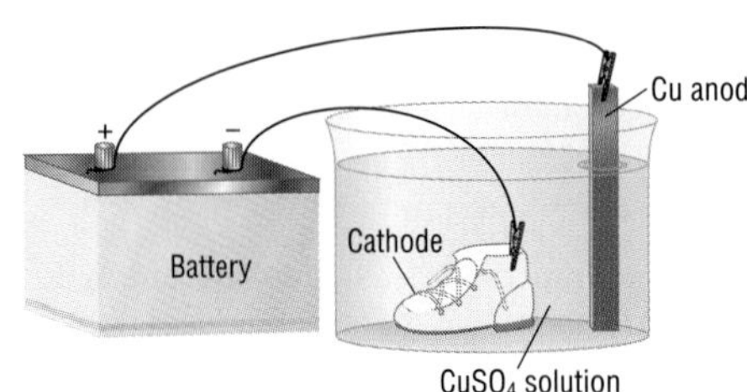

50. Which plot is characteristic of a dry cell? Explain.

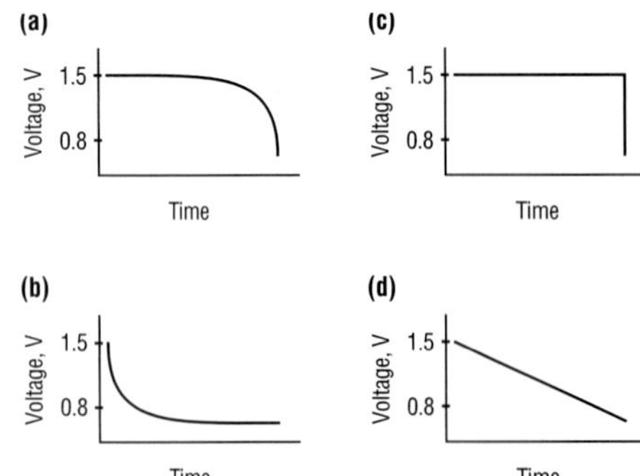

CUMULATIVE REVIEW

51. A sample of oxygen gas has a volume of 425 mL at 30 °C. What is the new volume of the gas if the temperature is raised to 60 °C while the pressure is kept constant?

52. Balance each equation.

a. $H_2S(g) + HNO_3(aq) \longrightarrow S(s) + NO(g) + H_2O(l)$

b. $AgNO_3(aq) + Pb(s) \longrightarrow Pb(NO_3)_2(aq) + Ag(s)$

c. $Cl_2(g) + NaOH(aq) \longrightarrow NaCl(aq) + NaClO_3(aq) + H_2O(l)$

53. Concentrated nitric acid is 16*M*. How would you prepare 500 mL of 1.0*M* HNO_3?

54. What is the oxidation number of the italicized element in each formula?

a. $K_2\textit{Cr}_2O_7$
b. $K\textit{I}O_3$
c. $\textit{Mn}O_4^-$
d. $\textit{Fe}Cl_3$

CONCEPT CHALLENGE

55. Write the overall cell reactions and calculate E^0_{cell} for voltaic cells composed of the following sets of half-reactions.

a. $AgCl(s) + e^- \longrightarrow Ag(s) + Cl^-(aq)$
$Ni^{2+}(aq) + 2e^- \longrightarrow Ni(s)$

b. $Al^{3+}(aq) + 3e^- \longrightarrow Al(s)$
$Cl_2(g) + 2e^- \longrightarrow 2Cl^-(aq)$

56. Impure copper is purified in an electrolytic cell. Design an electrolytic cell, with H_2SO_4 as electrolyte, that will allow you to carry out this process. Give the oxidation and reduction half-reactions and a balanced equation for the overall reaction.

57. This spontaneous redox reaction occurs in the voltaic cell illustrated below.

$$Ni^{2+}(aq) + Fe(s) \longrightarrow Ni(s) + Fe^{2+}(aq)$$

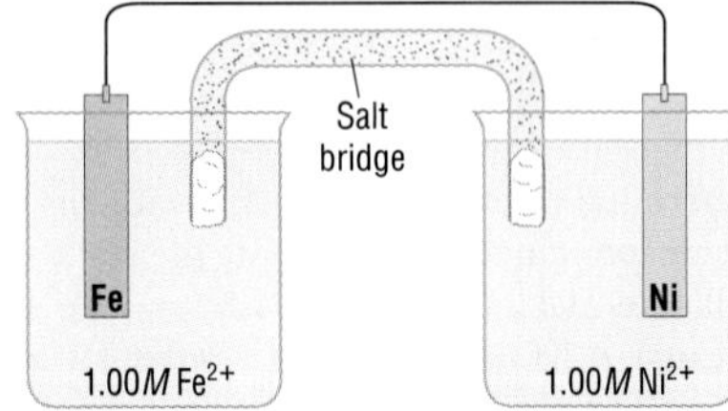

a. Identify the anode and the cathode.

b. Assign charges to the electrodes.

c. Write the half-reactions.

d. Calculate the standard cell potential when the half-cells are at standard conditions.

Chapter 23 STANDARDIZED TEST PREP

Select the choice that best answers each question or completes each statement.

1. Which statement describes electrolysis?
   a. Reduction occurs at the anode.
   b. Energy is produced.
   c. Oxidation occurs at the cathode.
   d. Positive ions move to the cathode.
2. Which of these statements about rusting is true?
   I. Iron is oxidized.
   II. Oxygen is the oxidizing agent.
   III. Iron atoms gain electrons.
   a. I and II only
   b. II and III only
   c. I only
   d. I, II, and III
3. The energy source for an ordinary flashlight is a
   a. dry cell.
   b. voltaic cell.
   c. battery.
   d. all of the above
4. Magnesium metal is prepared by the electrolysis of molten $MgCl_2$. One half-reaction is $Mg^{2+}(l) + 2e^- \longrightarrow Mg(l)$.
   a. This half-reaction occurs at the cathode.
   b. Magnesium ions are oxidized.
   c. Chloride ions are reduced at the anode.
   d. Chloride ions gain electrons during this process.
5. If the cell potential for a redox reaction is positive,
   a. the redox reaction is spontaneous.
   b. the redox reaction is not spontaneous.
   c. the reaction only occurs during electrolysis.
   d. More than one statement is correct.

Use this table to answer questions 6–12. Hydrogen is included as a reference point for the metals.

Activity Series of Selected Metals	
Element	**Oxidation half-reaction**
Lithium	$Li(s) \longrightarrow Li^+(aq) + e^-$
Potassium	$K(s) \longrightarrow K^+(aq) + e^-$
Sodium	$Na(s) \longrightarrow Na^{2+}(aq) + 2e^-$
Aluminum	$Al(s) \longrightarrow Al^{3+}(aq) + 3e^-$
Zinc	$Zn(s) \longrightarrow Zn^{2+}(aq) + 2e^-$
Iron	$Fe(s) \longrightarrow Fe^{2+}(aq) + 2e^-$
Hydrogen	$H_2(g) \longrightarrow 2H^+(aq) + 2e^-$
Copper	$Cu(s) \longrightarrow Cu^{2+}(aq) + 2e^-$

6. Which metal will more easily lose an electron, sodium or potassium?
7. Which metal is more easily oxidized, copper or aluminum?
8. What is the relationship between ease of oxidation and the activity of a metal?
9. Describe what would happen if you placed a clean strip of aluminum in a solution of copper(II) sulfate. Explain your answer.
10. Would a copper strip placed in a solution containing zinc ions react spontaneously with the zinc ions? Explain your reasoning.
11. Based on the positions of zinc and iron in the table, explain how attaching zinc blocks to a steel ship hull protects the steel from corrosion.
12. Write the half-reaction for the reduction of aluminum ions.

Use the diagram to answer questions 13–15.

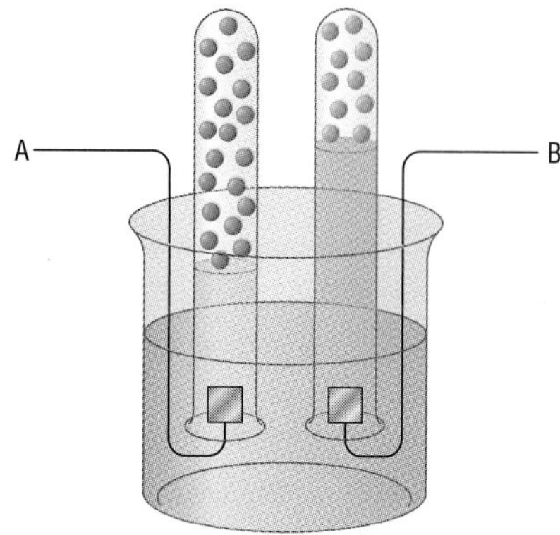

13. Write an equation for the decomposition of water by electrolysis.
14. At which electrode, A or B, is hydrogen produced?
15. The equation for the electrolysis of brine is
$$2NaCl(aq) + 2H_2O(l) \longrightarrow Cl_2(g) + H_2(g) + 2NaOH(aq)$$
How would you modify the electrolysis diagram to make it quantitatively represent the formation of hydrogen and chlorine?

Use complete sentences in response to question 16.

16. What do voltaic cells and electrolytic cells have in common? How do they differ?

Chapter 23 STANDARDIZED TEST PREP

1. d
2. a
3. d
4. a
5. a
6. potassium
7. aluminum
8. The more active the metal, the more easily it is oxidized.
9. Aluminum would be oxidized and copper ions would be reduced. Copper would plate out on the aluminum.
10. No. Because zinc is more easily oxidized than copper, copper cannot reduce the zinc ions in solution.
11. Zinc is more easily oxidized than iron. If iron atoms lose electrons through oxidation, zinc can act as a reducing agent to replace those lost electrons.
12. $Al^{3+}(aq) + 3e^- \rightarrow Al(s)$.
13. $2H_2O(l) \rightarrow O_2(g) + 2H_2(g)$
14. electrode A
15. Apparatus similar to above. The spheres representing the molecules of H_2 and Cl_2 will be in a 1:1 ratio.
16. In both cells, electrons flow from the anode to the cathode through an external circuit. Reduction occurs at the anode; oxidation occurs at the cathode. In a voltaic cell, the flow of electrons is spontaneous; an outside source of power is needed to drive the reaction in an electrolytic cell. The negative electrode is the anode in a voltaic cell; it is the cathode in an electrolytic cell.

Planning Guide

SECTION OBJECTIVES	ACTIVITIES/FEATURES	MEDIA & TECHNOLOGY
24.1 The *s*-Block Elements: Active Metals ● □ ◇ ▸ List sources, properties, and uses for the alkali metals (Group 1A) and their compounds ▸ Describe the preparation and properties of the alkaline-earth metals (Group 2A) and give uses for their compounds	**SE** **Discover It!** *Tempering of Metals,* p. 704 **SE** **CHEMath** *Classification,* p. 707 **SE** **Small-Scale Lab** *Complex Ions,* p. 711 (**LRS** 24-1) **TE** **DEMOS,** pp. 708, 710	**ASAP** Assessment 24.1 **CA** *Making Sodium Chloride* **CA** *Burning Magnesium*
24.2 The *p*-Block Elements: Metals and Non-Metals ● □ ◇ ▸ Describe properties and uses of *p*-block metals and nonmetals ▸ Explain methods for obtaining specific *p*-block metals and nonmetals from their compounds and minerals	**SE** **Link to Oceanography** *Deep-Sea Minerals,* p. 720 **LM** 48: *Allotropic Forms of Sulfur* **SSLM** 34: *Titration of Bleach* **TE** **DEMO,** p. 716	**ASAP** Assessment 24.2 **CA** *Making Sodium Chloride* **CA** *Elephant Toothpaste* **CA** *Watermelon Surprise* **CA** *Thermite Reaction* **CA** *Silicon Alien?*
24.3 The *d*- and *f*-Block Elements: Transition and Inner Transition Metals ● □ ◇ ▸ List the properties of specific transition metals ▸ Describe the chemical diversity shown by the transition metals and inner transition metals	**LM** 47: *Reactivity of Metals* **SSLM** 35: *Formation of Complex Ions with Ammonia* **TE** **DEMOS,** pp. 726, 728	**ASAP** Assessment 24.3 **CA** *Oxidation States of Vanadium* **CHM** Side 2, 10: *Neon Lights*
24.4 Hydrogen and Noble Gases ● □ ◇ ▸ Show how hydrogen is unique among the elements ▸ Explain why the noble gases, Group O, are important even though they are chemically unreactive	**SE** **Link to Environmental Awareness** *Radon Gas,* p. 734 **SE** **Mini Lab** *Decomposition of Hydrogen Peroxide,* p. 735 (**LRS** 24-2) **SE** **Chemistry Serving . . . Industry** *Diamonds are an Engineer's Best Friend,* p. 736 **SE** **Chemistry in Careers** *Gemologist,* p. 736	**ASAP** Assessment 24.4 **ASAP** Concept Map 24 **RP** Lesson Plans, Resource Library **AR** Computer Test 24 **www** Activities, Self-Tests, *SCIENCE NEWS* updates **TCP** The Chemistry Place Web Site

KEY

● Conceptual (concrete concepts)	**AR** Assessment Resources	**GRS** Guided Reading and Study Workbook
○ Conceptual (more abstract/math)	**ASAP** Chem ASAP! CD-ROM	**LM** Laboratory Manual
■ Standard (core content)	**ACT** ActivChemistry CD-ROM	**LP** Laboratory Practicals
□ Standard (extension topics)	**CHM** CHEMedia Videodiscs	**LRS** Laboratory Recordsheets
◆ Honors (core content)	**CA** Chemistry Alive! Videodiscs	**SSLM** Small-Scale Lab Manual
◇ Honors (options to accelerate)	**GCP** Graphing Calculator Problems	

PRACTICE		ASSESSMENT	
GRS	Section 24.1	**SE**	Section Review
RM	Practice Problems 24.1	**RM**	Section Review 24.1
		RM	Chapter 24 Quiz
GRS	Section 24.2	**SE**	Section Review
RM	Practice Problems 24.2	**RM**	Section Review 24.2
		RM	Chapter 24 Quiz
GRS	Section 24.3	**SE**	Section Review
RM	Practice Problems 24.3	**RM**	Section Review 24.3
		RM	Chapter 24 Quiz
		LP	Lab Practical 24-1
GRS	Section 24.4	**SE**	Section Review
RM	Practice Problems 24.4	**RM**	Section Review 24.4
RM	Interpreting Graphics	**RM**	Vocabulary Review 24
GCP	Chapter 24	**SE**	Chapter Review
		SM	Chapter 24 Solutions
		SE	Standardized Test Prep
		PHAS	Chapter 24 Test Prep
		RM	Chapter 24 Quiz
		RM	Chapter 24 A & B Test

PLANNING FOR ACTIVITIES

STUDENT EDITION

Discover It! p. 704
- hairpins
- pliers or tweezers
- glass of cold water
- heat-proof gloves
- gas burner
- heat-resistant surface

Small-Scale Lab, p. 711
- pencil
- paper
- ruler
- $CaCl_2$
- $AlCl_3$
- $FeCl_3$
- $ZnCl_2$
- KOH
- HNO_3
- NaOH
- reaction surface

Mini Lab, p. 735
- hydrogen peroxide
- manganese dioxide
- toothpicks
- candle
- matches
- test tube
- cork or plastic film wrap
- tongs

TEACHER'S EDITION

Teacher Demo, p. 708
- slides or photos of consumer products containing magnesium alloys

Teacher Demo, p. 710
- 250 mL of $0.1M$ $Ca(OH)_2$
- 250 mL seawater
- 500-mL beaker

Teacher Demo, p. 716
- safety goggles
- hood or well-ventilated room
- 250-mL beaker
- 10 mL water
- 3–4 drops of 0.1% phenolphthalein
- 30 g of ammonium chloride
- 15 g sodium hydroxide
- 500-mL Erlenmeyer flask
- one-hole stopper
- glass tube
- rubber tubing

Activity, p. 725
- product labels of at least three different brands of multivitamin supplements

Teacher Demo, p. 726
- dilute aqueous solutions (approx. $0.1M$ to $0.5M$) of different salts of transition metals such as manganese, iron, cobalt, nickel, copper and zinc

Teacher Demo, p. 728
- safety goggles
- 100 mL of concentrated sodium silicate (Na_2SiO_3)
- 200 mL water
- 600-mL beaker
- sand
- crystals of $CoCl_2$, $CuCl_2$, $PbCl_2$, $FeCl_3$, $NiCl_2$, $MnCl_2$, $CaCl_2$

Activity, p. 728
- business sections of newspapers or library references

OT	Overhead Transparency	**SE**	Student Edition
PHAS	PH Assessment System	**SM**	Solutions Manual
PLM	Probeware Lab Manual	**SSV**	Small-Scale Video/Videodisc
PP	Problem Pro CD-ROM	**TCP**	www.chemplace.com
RM	Review Module	**TE**	Teacher's Edition
RP	Resource Pro CD-ROM	**www**	www.phschool.com

Chapter 24

Key Terms

24.1 lime, slaked lime
24.3 ores, metallurgy, Monel metal, inner transition elements

DISCOVER IT!

The physical properties of carbon steel, including flexibility and hardness, depend in part on the relative amounts of iron, carbon, and iron carbide. Large amounts of iron carbide, formed by heating and rapid cooling, make steel brittle. Heating followed by slow cooling favors the formation of carbon crystals, which makes the steel more ductile. Students should observe that the tempering treatments applied in Steps 1 and 2 increase and reduce (respectively) the flexibility of the hairpins. The hairpin treated as described in Step 3 is the most flexible.

Chapter 24 THE CHEMISTRY OF METALS AND NONMETALS

FEATURES

Among the gems on this crown are rubies (Al_2O_3) and pearls ($CaCO_3$).

DISCOVER IT! TEMPERING OF METALS

You need four metal hairpins, pliers or tweezers, a glass of cold water, heatproof gloves, a heat-resistant surface, and a gas burner.

1. Wearing heatproof gloves, use the pliers to hold a hairpin by the open end. Heat the bend of the hairpin red-hot over the burner. Set the hairpin aside on the heat-resistant surface to cool.
2. Heat the bend of another hairpin red-hot, remove it from the flame, and immediately plunge it into the glass of cold water.
3. Repeat Step 1 with a third hairpin. After cooling, gently reheat it but do not let it get red-hot. Allow it to cool slowly in the air.
4. Try bending each of the three hairpins you heated, as well as a fourth hairpin that was not heated.

Tempering is the control of the hardness and flexibility of a metal by heating. Compare the flexibility of the three hairpins that were heated with the flexibility of the fourth hairpin, which serves as a control. Which has the most flexibility? The least? How does slow cooling affect the flexibility? After completing this chapter, return to this activity and explain the results.

24.1 ● ■ ◇
24.2 ● ■ ◇
24.3 ● ■ ◇
24.4 ● ■ ◇

Conceptual Because much of the content is descriptive, this chapter is particularly appropriate for conceptual students as long as they can connect the observed properties to the position in the periodic table. Use photos throughout the chapter to encourage students to observe differences and similarities in properties.

Standard The sections of this chapter are designated as extension topics because time may not permit coverage of this material. Students may retain information in this chapter more readily if they make comparisons across groups as well as within groups. For example, Table 24.1 and Table 24.2 can be used to compare the properties of alkali metals and alkaline earth metals.

Honors You may assign this chapter as enrichment reading because of its mostly descriptive content. If time permits, discuss selected groups of elements in class. You may want students to research a group cooperatively and report back to the class.

THE *s*-BLOCK ELEMENTS: ACTIVE METALS

section 24.1

The Bonneville Salt Flats, which stretch over 30 000 acres, are one of the most barren, inhospitable, and unique natural features in Utah. Part of the Great Basin, this dazzling white plain, along with the Great Salt Lake, is a remnant of ancient Lake Bonneville, which nearly 15 000 years ago covered one-third of present-day Utah and parts of neighboring states. Wind and water combine to create the flat surface of salt that has made Bonneville a world-famous destination for high-speed auto racers. Since 1896, when its potential for racing was first recognized, vehicles have hurtled along at record-breaking speeds, currently topping 600 miles per hour! The Bonneville Salt Flats are rich in salts of lithium, sodium, potassium, magnesium—metals of Groups 1A and 2A. **What are the properties of these active metals?**

objectives

- List sources, properties, and uses for the alkali metals (Group 1A) and their compounds
- Describe the preparation and properties of the alkaline-earth metals (Group 2A) and give uses for their compounds

key terms

- lime
- slaked lime

The Alkali Metals

You know from Chapter 14 that the representative elements occupy the *s* and *p* blocks in the periodic table because the outermost electrons of those elements are in the *s* and *p* sublevels. You also know that the transition elements occupy the *d* block and that the inner transition elements occupy the *f* block. About three-quarters of the more than 100 known elements are metals. The metals occupy all of the *s*, *d*, and *f* blocks and about half of the *p* block. The nonmetals occupy the remainder of the *p* block, in the upper-right corner of the periodic table. This chapter describes, block by block, some of the distinctive features of metals and nonmetals and their practical applications.

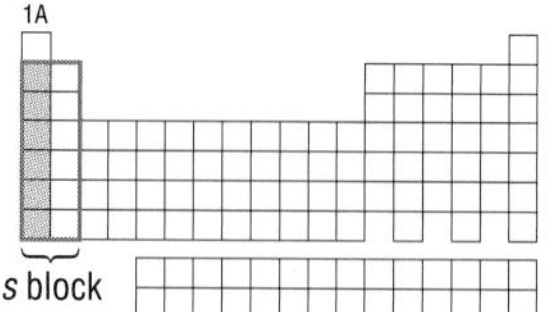

The alkali metals are the Group 1A elements. Rich deposits of alkali metal salts are found all over the world. The Bonneville Salt Flats in Utah and deposits at Searles Lake, California, contain huge quantities of sodium chloride and other alkali salts. The evaporation of ancient seas left large alkali metal salt deposits that are now underground near the Gulf Coast shores of Texas and Louisiana. Alkali metal salts are very soluble in water. Rain water leaches them out of the soil, and rivers carry them to the sea. Sea water is about 3% by mass alkali metal salts. As **Figure 24.1** shows, sea water is a source of table salt.

Figure 24.1
Large amounts of sodium chloride (table salt) are produced by the evaporation of sea water.

Student Resources

From the Teacher's Resource Package, use:

- Section Review 24.1 and Ch. 24 Quizzes from the Review Module (Ch. 21–24)
- Laboratory Recordsheet 24-1

Technology Resources

Relevant technology resources include:

- Chem ASAP! CD-ROM
- Resource Pro CD-ROM
- Chemistry Alive! Videodisc: *Making Sodium Chloride, Burning Magnesium*

24.1

1 Engage

Use the Visual

Have students study the photograph and read the text that opens the section. Have students refer to a copy of the periodic table. Remind them that the periodic table is an important tool for organizing chemical information. Ask:

- **What are the properties of the active metals?** (In general, all of the active metals in Groups 1A and 2A have low ionization energies, and low electronegativities compared with other elements in the periodic table. As a result, these metals are relatively reactive, and tend to give up their valence electrons to form cations when they react to form compounds. In this section, students will learn more about the physical and chemical properties that distinguish these elements.)

Check Prior Knowledge

To assess students' knowledge about the organization of elements in the periodic table, ask:

- **On which side of the periodic table are the metals located? Nonmetals? Metalloids?** (left side; upper right-hand corner; along the stair-step line that divides metals from nonmetals)
- **How are elements arranged in the periodic table?** (in order of increasing atomic number)
- **How can the periodic table be used to determine the number of valence electrons in an atom of an element?** (For representative elements, the group number equals the number of valence electrons.)

2 Teach

Discuss

Remind students that elements in the same group contain the same number of valence electrons. Similarities in electron configuration lead to similar chemical properties. Review the electron configurations for the Group 1A and 2A elements. Point out the noble gas cores and valence shell electrons in each case. Ask:

- **Why do alkali metals tend to be such strong reducing agents?** (Alkali metals attain noble cores when they give up a single electron.)
- **Are Group 2A metals better reducing agents or oxidizing agents?** (Reducing agents; they attain stability by giving up two electrons.)

Discuss

On the chalkboard, rewrite the chemical equation for the reaction of sodium with water. Point out that the name *alkali* refers to a chemical base. Ask students why this name makes sense in terms of the reactivity of the Group 1A metals. (Alkali metals react with water to produce an alkaline or basic solution.) **Have students write a general equation for the reaction of an alkali metal with water.** [$2X(s) + 2H_2O(l) \rightarrow 2XOH(aq) + H_2(g)$]

Chemistry Alive!

Making Sodium Chloride

Play

Table 24.1

Some Physical Properties of the Alkali Metals				
Element	Melting point (°C)	Boiling point (°C)	Density (g/cm^3)	Atomic radius (nm)
Lithium	179	1336	0.53	0.123
Sodium	98	883	0.97	0.157
Potassium	64	758	0.86	0.203
Rubidium	39	700	1.53	0.216
Cesium	28	670	1.90	0.235

The metals of Group 1A are the most reactive known. Because of their reactivity, alkali metals are not found free in nature but are typically combined with nonmetals as alkali salts. Within the group, cesium and rubidium are the most reactive. Each alkali metal reacts violently with cold water, producing hydrogen gas and a solution of the alkali metal hydroxide (simply known as an alkali or base). Sodium, for example, reacts with cold water, forming sodium hydroxide and releasing hydrogen gas.

$$2Na(s) + 2H_2O(l) \longrightarrow 2NaOH(aq) + H_2(g)$$

The reaction of alkali metals with water is so rapid and exothermic that the hydrogen often burns as it is produced. Alkali metals will also react vigorously with the moisture in skin and should never be handled with bare hands. The alkali metals are usually stored under oil or kerosene to keep them from reacting with oxygen and moisture in the air.

The alkali metals have low densities, low melting points, and high electrical conductivity. **Table 24.1** lists some physical properties of the alkali metals. If you had a 10-g sample of each alkali metal listed in the table, ❶ which sample would have the largest volume? Why? These metals have the consistency of stiff modeling clay and are soft enough to be cut with a knife, as shown in **Figure 24.2.** The freshly cut surface is shiny, with the typical silvery luster of a metal. The surface quickly dulls on exposure to air, however, due to rapid reaction with oxygen and moisture.

Figure 24.2
Sodium is shiny when it is first cut, but it quickly reacts with oxygen and moisture in the air to become dull.

Sodium is the only alkali metal manufactured on a large scale. To produce the free metal, sodium ions must be reduced. Metallic sodium is generally produced by the electrolysis of molten sodium chloride. Chlorine gas is a valuable by-product of this process.

Sodium is used as the light source in sodium vapor lamps and in the production of many chemicals. Liquid sodium is an excellent conductor of heat, and because of this it can be used in nuclear reactors to remove heat from the reactor core.

Sodium hydroxide is an ingredient in common household products used to clear clogged drains. Another compound of sodium is a popular type of bleach. An aqueous solution of sodium hypochlorite (NaClO) is an alternative to hydrogen peroxide. Do you know what important compound of sodium is present on most dining tables? What sodium compound is ❷ known commonly as baking soda?

Answers

❶ lithium, because it has the lowest density
❷ NaCl; $NaHCO_3$

Meeting Diverse Needs

Gifted Calcium is an important constituent of living systems. Bone gets its structural strength from deposits of insoluble calcium phosphate. Have students do research on the chemical composition of bone tissue and describe the formation of calcium-based minerals in bone tissue. Ask students to infer why strontium-90 is able to accumulate in bone tissue.

CHEMath

CLASSIFICATION

As you study chemistry, you will encounter more than 100 different chemical elements. In this chapter you learn how the elements are organized, or classified, into groups in the periodic table. A classification system, such as the periodic table, is a plan for organizing a large number of items into smaller groups.

Classification systems are not limited to chemistry, and they do not have to be complex. For example, do you keep your shirts, pants, and socks in different drawers? If so, you have classified your clothing.

The key to classifying anything is to divide large sets of items into smaller subsets according to similarities among items. In the case of clothes, the large set might include all of your clothes. Shirts, pants, and socks might be subsets. You can make even more subsets, such as long pants and short pants, tee shirts and sweatshirts, and so on.

Keep in mind that there is no single correct classification system. The same items can be classified in more than one way. Can you think of another way to classify your clothes? Perhaps by color, by the season in which you wear them, or by whether you wear them during the day or at night.

Biologists also use a classification system. They need to organize the more than two million different types of living things in existence. One common classification system divides living things into five major groups, called kingdoms. For example, animals are in one kingdom and plants are in another. The kingdoms are then divided into smaller groups by more specific characteristics. Animals, for example, can be divided according to whether or not they have a backbone. The more specific the characteristic, the fewer items there are in the group.

In chemistry, the elements are easier to study if they are classified into groups. Elements can be classified according to properties such as atomic size, ionic size, ionization energy, and electronegativity. Becoming familiar with how the elements are classified will help you better understand chemical concepts.

Example 1

Classify the following items according to their use.

paper clip	clothespin
pen	staple
pencil	marker

Things that hold something together:
paper clip, clothespin, staple

Things to write with:
pen, pencil, marker

Example 2

Classify the following elements according to their positions in the periodic table.

helium	beryllium
lithium	potassium
calcium	xenon

Group 1A: lithium, potassium
Group 2A: beryllium, calcium
Group 8A: helium, xenon

Practice Problems

A. Develop a classification system for all of the motor vehicles parked in the school parking lot. Compare your classification system with those of your classmates and determine the strengths and weaknesses of each.

B. Classify the items in each of the following sets and identify the characteristic you used to classify them.

1. dog, cat, shark, elephant, guppy, whale
2. book, newspaper, calculator, adding machine, letter
3. lead, tungsten, bromine, chlorine, zinc, iodine

CHEMath

Teaching Tips

For students having difficulty with the concept of classification, you may wish to challenge them to classify the letters of the alphabet. They should first classify the letters into two groups, and then into additional groups. For example, they may arrange the letters according to those that have curves in them and those that do not. They can then divide the letters without curves according to the number of lines in them. This simple exercise will help them to look for common traits. It will also help them to recognize that the groups become smaller as the classification system becomes more detailed.

For added practice, ask students to design a layout for a store. They should draw a plan, showing how the items in the store would be arranged. When they complete this activity, have them look at the periodic table. Ask them to relate their store layout to the periodic table. Help them to recognize the significance of arranging items, such as chemical elements, according to similar traits.

Answers to Practice Problems

A. Answers will vary. Check that classification systems are logical. Examples include classifying the vehicles according to the type of vehicle (car, truck, or van), manufacturer, color, or number of doors.

B.
1. lives on land: dog, cat, elephant; lives in water: shark, guppy, whale
2. related to words: book, newspaper, letter; related to numbers: calculator, adding machine
3. nonmetal: chlorine, bromine, iodine; metal: lead, tungsten, zinc

ACTIVITY

Alkaline earth metals and their compounds are used for many purposes. Ask students to keep a written record of their observations of these substances, both at home and at school, over the course of several days. Students should check the labels of foods and other products for alkaline earth metals and their compounds. In each case, they should identify the substance, describe its appearance and properties (if it is in a pure state), and describe how it is being used.

Think Critically

As shown in Figure 24.3, calcium carbonate occurs naturally in many different forms. Have students do research on the structural and chemical characteristics of limestone and marble. Ask:

- **What makes marble different from natural limestone?** (Marble is a metamorphic form of limestone. Metamorphic rocks are formed under conditions of intense heat, pressure, or both within Earth's crust.)

TEACHER DEMO

Although magnesium is a relatively reactive element, its alloys, which often contain aluminum, are strong and corrosion resistant. Bring in slides or photos of consumer products containing magnesium alloys. Examples include motorcycle, automobile, and aircraft engine parts; bicycle and backpack frames; as well as power-tool housings.

2A

s block

The Alkaline Earth Metals

The alkaline earth metals are the elements of Group 2A. Alkaline earth metal salts are less soluble in water than are the corresponding alkali metal salts. Nevertheless, the sea is a rich source of the ions of magnesium and calcium, two alkaline earth metals. Calcium ions in sea water are used by shellfish in building their calcium carbonate shells and by coral animals in forming reefs.

Some of the alkaline earth carbonates and sulfates are insoluble enough to resist weathering and the leaching action of rainwater. The most common mineral form of calcium carbonate is limestone. Other forms that occur naturally are shown in **Figure 24.3.** The White Cliffs of Dover, on the southern coast of England, are another form of calcium carbonate. There, the shells of microscopic marine animals have been pressed into a natural chalk.

Calcite

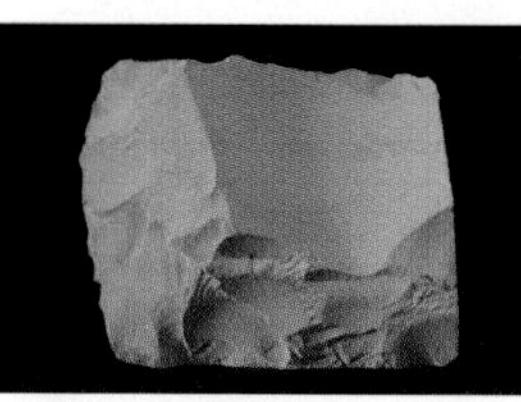

Limestone

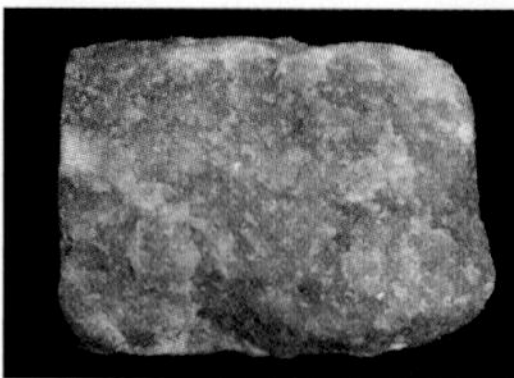

Marble

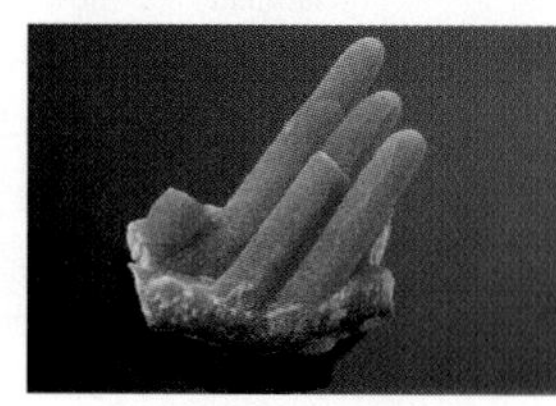

Stalactites

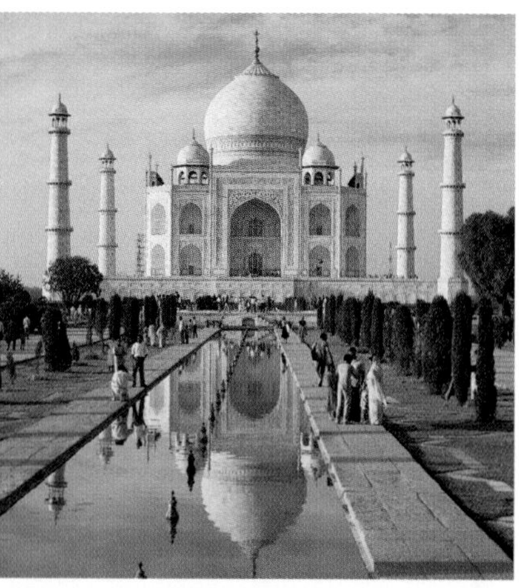

Figure 24.3
Calcium carbonate ($CaCO_3$) occurs naturally in several forms. The Taj Mahal in India is built of marble, one form of $CaCO_3$.

The alkaline earth metals react with water to produce basic, or alkaline, solutions. Their compounds are extracted from the minerals that since early times have been called "earths." The term "earths" has its origin with the alchemists of the Middle Ages who used the word to describe substances that are unchanged by fire, notably calcium oxide (CaO) and magnesium oxide (MgO). Although the alkaline earth metals are not found uncombined in nature, they are chemically less reactive than the alkali metals in Group 1A. Therefore they need not be stored under oil. Barium is the most reactive element in the group. As **Figure 24.4** shows, calcium reacts with cold water to produce hydrogen, but it does so more slowly than Group 1A metals do. Beryllium and magnesium react only with hot water or steam.

The alkaline earth metals are harder than the alkali metals. They have a gray-white luster but tarnish quickly in air to form a tough, thin oxide coating. The coating protects the metals, particularly beryllium and magnesium, from further oxidation. This property allows alloys of these metals to be used as low-density structural materials. **Table 24.2** lists some physical properties of the alkaline earth metals. What is the trend in the atomic ❶ radii of the alkaline earth metals? Why?

Answer

❶ Atomic radii increase from top to bottom within the group. Additional electrons enter larger shells.

Table 24.2

Some Physical Properties of the Alkaline Earth Metals				
Element	Melting point (°C)	Boiling point (°C)	Density (g/cm³)	Atomic radius (nm)
Beryllium	1280	1500	1.86	0.089
Magnesium	651	1107	1.75	0.136
Calcium	851	1487	1.55	0.174
Strontium	800	1366	2.60	0.191
Barium	850	1537	3.59	0.198

Figure 24.4

Alkaline earth metals are less reactive than alkali metals. Here, calcium reacts slowly with water.

Calcium and magnesium are the two most important alkaline earth metals. Calcium is produced by the electrolysis of molten calcium chloride in a reaction similar to that used to produce sodium. Magnesium is prepared from sea water, as illustrated in **Figure 24.5.** Magnesium is an important structural material and is the chief component in a number of high-tensile-strength, low-density alloys. These properties make the alloys valuable in aircraft and spacecraft construction. Magnesium is also present in asbestos, which is shown in **Figure 24.6** on the following page. Because asbestos is a

Figure 24.5

Sea water is the main source of magnesium compounds, with each ton of sea water yielding about 3 kg of magnesium. Calcium carbonate from oyster shells is converted to lime, which reacts with the Mg^{2+} in the sea water. Most commercial magnesium is prepared electrolytically from magnesium chloride, which is produced at a later stage of the process.

24.1

Use the Visual

Have students compare Tables 24.1 and 24.2. Point out how the trends in the physical properties listed correlate with the increasing atomic numbers of the elements. Ask students to describe the differences in physical properties between alkali and alkaline-earth metals. (Group 2A metals have much higher boiling and melting points and greater densities than Group 1A metals. Group 1A atoms are larger.) **Have students recall why atomic radii increase going from top to bottom within a group.** (The principle quantum number increases while the *effective* nuclear charge remains relatively constant.) **Have students name and describe other trends in properties for the alkali and alkaline-earth metals such as ionization energy and electronegativity.** (Ionization energy and electronegativity decrease from top to bottom; cesium is the least electronegative element.) **Have students look up these values and include them in expanded versions of Tables 24.1 and 24.2.**

Chemistry Alive!

Burning Magnesium

Play

Meeting Diverse Needs

At Risk Have students scan the rest of the chapter and prepare an outline of the chapter content. The outline should include section titles and subsection headings. Students can add their own notes below each subsection heading. In conjunction with the outline, students should make a copy of the periodic table from the back cover of the book and use colored pencils or highlighters to shade each block(*s*, *p*, *d*, and *f*) of the table with a different color as they study it. Encourage students to add to the periodic table, for example, by writing the outer shell electron configurations at the top of each group of elements. Students can use the outline and periodic table together to help organize the facts and concepts presented in this chapter.

TEACHER DEMO

Review the flow chart in Figure 24.5. Explain that the precipitation of $Mg(OH)_2$ is used to concentrate the magnesium. Then add 250 mL of 0.1*M* $Ca(OH)_2$ to 250 mL of seawater in a 500-mL beaker. Have students observe the precipitate formed. Point out that the precipitate can be separated from the remaining solution by decanting or filtering. Have students write the equation for the subsequent reaction of $Mg(OH)_2$ with HCl to form water and $MgCl_2$. Ask:

- What type of reaction is this? (neutralization)

3 Assess

Evaluate Understanding

Ask students to name two or more alkali and alkaline-earth metals and identify at least one important use of each metal. Have students write a general equation for the reaction of each type of metal with water.

Reteach

Help students start a table that they can use throughout their study of the elements. Include columns for *Group Number, Properties, Preparation Methods,* and *Reactions* (which will list general equations).

Portfolio Project

Carbonic acid dissolves the calcium carbonate in limestone. When groundwater containing calcium and carbonate ions drips from the roof of a cave, calcium carbonate is deposited as the icicle-shaped stalactites and round stalagmites that make underground caverns popular tourist attractions.

Figure 24.6
Magnesium is present in asbestos, which is a fibrous form of the mineral serpentine (left). Asbestos was used as an insulation material until it was discovered that inhalation of its fibers causes lung cancer. As a result, most asbestos is being removed from the sites where it was installed (right).

very poor conductor of heat, it was once commonly used as an insulator. This use has been discontinued because it is now known that asbestos fibers cause lung cancer.

Calcium oxide, commonly called **lime** or quicklime, is an important industrial chemical. Calcium oxide is produced by the high-temperature decomposition of limestone in a lime kiln.

$$\underset{\text{Limestone}}{CaCO_3(s)} \xrightarrow{900\ ^\circ C} \underset{\text{Lime}}{CaO(s)} + CO_2(g)$$

The reaction of lime with water is called slaking. The process is exothermic and the product is **slaked lime,** or calcium hydroxide.

$$\underset{\text{Lime}}{CaO(s)} + H_2O(l) \longrightarrow \underset{\text{Slaked lime}}{Ca(OH)_2(aq)} + \text{heat}$$

Slaked lime is used to make plaster and mortar and to neutralize acidic soils. The setting process for mortar involves the conversion of slaked lime back to calcium carbonate (limestone) by the action of atmospheric carbon dioxide.

$$Ca(OH)_2(aq) + CO_2(g) \longrightarrow CaCO_3(s) + H_2O(g)$$

portfolio project

Some caves are created when underground limestone deposits are dissolved. Prepare a report or poster display on the chemistry of such caves. Find out the identity and source of the acid that dissolves the limestone. Include an explanation for the formation of stalagmites and stalactites.

section review 24.1

1. List the properties of the alkali metals, their main sources, and their uses.
2. How are alkaline earth metals prepared? What are their properties and main uses?
3. Why are alkali metals stored under kerosene or mineral oil?
4. Which alkaline earth metals react with cold water and which do not? Write a typical equation that shows such a reaction.
5. Write an equation for the reaction of potassium with water.
6. Would you expect to find pure samples of alkali metals in Earth's crust? Explain.
7. How does the reactivity of the Group 2A metals compare with that of the Group 1A metals?

Chem ASAP! Assessment 24.1 Check your understanding of the important ideas and concepts in Section 24.1.

Answers

SECTION REVIEW 24.1

1. Alkali metals are the most reactive metals. (For physical properties, see Table 24.1.) They are found in underground deposits. Table salt and lye are sodium compounds.
2. Alkaline-earth metals are extracted from sea water or minerals. (For physical properties, see Table 24.2.) Magnesium alloys and calcium compounds are used in construction.
3. Alkali metals are stored under oil to prevent them from reacting with oxygen in the air.
4. Calcium, strontium, and barium react with cold water. Beryllium and magnesium react only with hot water or steam. $Ca(s) + 2H_2O(l) \rightarrow Ca(OH)_2(aq) + H_2(g)$
5. $2K(s) + 2H_2O(l) \rightarrow 2KOH(aq) + H_2(g)$
6. No; their chemical reactivity is too great; they react readily with oxygen and water.
7. Group 2A metals are less reactive than Group 1A metals.

SMALL-SCALE LAB

COMPLEX IONS

SAFETY     

Wear safety glasses and follow the standard safety procedures outlined on page 18.

PURPOSE

To observe some chemical reactions of metal ions to form complex ions.

MATERIALS

- pencil
- ruler
- paper
- reaction surface
- chemicals shown in Figure A

PROCEDURE

On separate sheets of paper, draw two grids similar to Figure A. Draw one with Xs and one without. Make each square 2 cm on a side. Place a reaction surface over the grid with Xs and add one drop of each chemical as shown in Figure A. Use the grid without Xs as a data table to record your observations for each reaction.

	KOH	KOH + HNO_3	KOH + NaOH
$CaCl_2$ (Ca^{2+})	white ppt	white ppt dissolves	white ppt no change
$AlCl_3$ (Al^{3+})	white ppt	white ppt dissolves	white ppt dissolves
$FeCl_3$ (Fe^{3+})	orange ppt	orange ppt dissolves	orange ppt no change
$ZnCl_2$ (Zn^{2+})	white ppt	white ppt dissolves	white ppt dissolves

Figure A

ANALYSIS

Using your experimental data, record the answers to the following questions below your data table.

1. An amphoteric compound is one that behaves as a base under a certain set of conditions and as an acid under a different set of conditions. An amphoteric compound reacts with both an acid and a base. Which compounds in this experiment are amphoteric?
2. Hydroxide ions commonly form precipitates with metal ions. For example, copper(II) ion reacts with hydroxide ion to produce copper(II) hydroxide.

   $$Cu^{2+}(aq) + 2OH^-(aq) \longrightarrow Cu(OH)_2(s)$$

   Write the net ionic equations that describe each precipitation reaction you observed.
3. Hydroxide precipitates dissolve readily in acid solution. For example, copper(II) hydroxide reacts with acid to form copper(II) ions and water.

   $$Cu(OH)_2(s) + 2H^+(aq) \longrightarrow Cu^{2+}(aq) + 2H_2O(l)$$

   Write the net ionic equations that describe each neutralization reaction you observed.
4. When hydrogen ions are added, do the hydroxide precipitates act as bases or acids? Explain your answer.
5. Some hydroxide precipitates dissolve in excess NaOH. For example, aluminum hydroxide reacts with excess hydroxide ions to produce an aluminum hydroxide complex ion.

   $$Al(OH)_3(s) + OH^-(aq) \longrightarrow Al(OH)_4^-(aq)$$

   Which other hydroxide precipitate(s) reacted with excess sodium hydroxide? Write the net ionic equations that describe the formation of these complex ions.
6. When a base is added in excess, do the hydroxide precipitates act as acids or bases? Explain your answer.
7. Which hydroxide precipitates in this experiment are amphoteric?

SMALL-SCALE LAB

COMPLEX IONS

OVERVIEW

Procedure Time: 40 minutes
Students test metal ion solutions to determine which form complex ions with hydroxides.

MATERIALS

Solution	Preparation
0.5*M* $CaCl_2$	13.9 g in 250 mL
0.2*M* $AlCl_3$	12.1 g $AlCl_3 \cdot 6H_2O$ in 250 mL
0.1*M* $FeCl_3$	6.8 g $FeCl_3 \cdot 6H_2O$ in 25 mL of 1.0*M* NaCl; dilute to 250 mL
0.2*M* $ZnCl_2$	6.8 g in 250 mL
0.2*M* KOH	2.8 g in 250 mL
0.5*M* NaOH	20.0 g in 1.0 L
1.0*M* HNO_3	63 mL of 15.8*M* in 1.0 L

CAUTION! Always add acid to water carefully and slowly!

TEACHING SUGGESTIONS

This lab works best if students allow the precipitates to stand for only a short time before adding excess base. Coagulation of the precipitates causes loss of surface area and prolongs the reaction with excess base. Explain that hydroxide precipitates that react with excess hydroxide ions are amphoteric.

ANALYSIS

1. $AlCl_3$ and $ZnCl_2$ are amphoteric.
2. $Ca^{2+} + 2OH^- \rightarrow Ca(OH)_2(s)$
   $Al^{3+} + 3OH^- \rightarrow Al(OH)_3(s)$
   $Fe^{3+} + 3OH^- \rightarrow Fe(OH)_3(s)$
   $Zn^{2+} + 2OH^- \rightarrow Zn(OH)_2(s)$
3. $Ca(OH)_2(s) + 2H^+ \rightarrow Ca^{2+} + 2HOH$
   $Al(OH)_3(s) + 3H^+ \rightarrow Al^{3+} + 3HOH$
   $Fe(OH)_3(s) + 3H^+ \rightarrow Fe^{3+} + 3HOH$
   $Zn(OH)_2(s) + 2H^+ \rightarrow Zn^{2+} + 2HOH$
4. Hydroxide precipitates act as bases in the presence of hydrogen ions. They are neutralized by acid.
5. $Zn(OH)_2(s) + 2OH^- \rightarrow Zn(OH)_4^-$
6. Hydroxide precipitates act as acids in the presence of excess hydroxide because they react with the hydroxide base.
7. $Al(OH)_3$ and $Zn(OH)_2$ are amphoteric because they react with both acids and bases.

1 Engage

Use the Visual

Have students study the photograph and read the text that opens the section. Ask:

- **What are the properties of the metals and nonmetals in the *p* block?** (In general, an element in the *p* block of the periodic table is characterized by the filling of its outermost *p* orbital. The *p* block contains metals, nonmetals, and metalloids. As one moves from left to right within a period in the *p* block, metallic character tends to decrease.) **Remind students about the trends in electronegativities and ionization energies observed for those elements in the *p* block.**

Check Prior Knowledge

To assess students' knowledge about properties of *p*-block elements, ask:

- **Which groups in the periodic table contain elements whose outermost atomic orbital is a partially filled *p* orbital?** (3A, 4A, 5A, 6A, and 7A)
- **Which *p*-block elements could be classified as semiconductors?** (metalloids such as Si, Ge, and As)
- **Which *p*-block group contains elements that tend to form anions with a 1− charge when they react? What name is given to this group?** (7A, halogens)
- **Classify each of the following elements as a metal, metalloid, or nonmetal: Al, C, Sb, S, and Pb.** (metals: Al, Pb; metalloid: Sb; nonmetals: C, S)

section 24.2

objectives

- Describe properties and uses of *p*-block metals and nonmetals
- Explain methods for obtaining specific *p*-block metals and nonmetals from their compounds and minerals

THE *p*-BLOCK ELEMENTS: METALS AND NONMETALS

How much do you value aluminum objects such as containers and cans? Probably not enough to consider them a treasure or to store them in a safety-deposit box. However, there are some objects made partially of aluminum that are quite valuable indeed. Some of the jewels shown in this photo actually contain compounds of aluminum. Aluminum is an element in the p block of the periodic table. **What are the properties of the metals and nonmetals that make up the groups in the *p* block?**

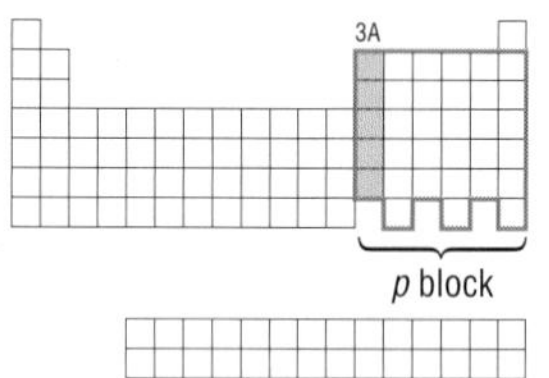

Aluminum and Group 3A

Boron is the first element in Group 3A. Boron occurs naturally in borate minerals. Although not widespread, large deposits of the borate mineral borax are found in the arid regions of the United States, particularly the Mojave Desert and Death Valley, California. Borax was once hauled from Death Valley by teams of mules, as shown in **Figure 24.7.**

Purified borax, or sodium tetraborate decahydrate ($Na_2B_4O_7 \cdot 10H_2O$), is soft, colorless, and crystalline. Borax is found in glass, in glazes used to decorate ceramics, and in fertilizers. Borax is used to soften water, to disinfect, or as a flux to help solder flow between pieces of metal.

Pure boron is black, lustrous, and extremely hard but brittle. It is also a metalloid, or semiconductor. Boron can be prepared by the reaction of its oxide with magnesium metal.

$$B_2O_3(s) + 3Mg(s) \longrightarrow 2B(s) + 3MgO(s)$$

One important compound of boron, boron carbide (B_4C), is a ceramic material approaching diamond in hardness. Boron carbide is used in the fabrication of machine tools and for other applications in which wear-resistance is important.

The elements that follow boron in Group 3A are the metals aluminum, gallium, indium, and thallium. Aluminum is the most abundant metal in Earth's crust. Although it does not exist in the uncombined state in nature, it is a major component of many rocks and minerals, especially in the form of bauxite (Al_2O_3). Aluminum is commonly found as the very hard mineral

Figure 24.7
Borax is a borate mineral. Large mule teams were once used to transport mined borax out of Death Valley, California.

STUDENT RESOURCES

From the Teacher's Resource Package, use:

- Section Review 24.2 and Ch. 24 Quizzes from the Review Module (Ch. 21–24)
- Laboratory Manual: Experiment 48
- Small-Scale Chemistry Lab Manual: Experiment 34

TECHNOLOGY RESOURCES

Relevant technology resources include:

- Chem ASAP! CD-ROM
- Resource Pro CD-ROM
- Chemistry Alive! Videodisc: *Making Sodium Chloride, Elephant Toothpaste, Watermelon Surprise, Thermite Reaction, Silicon Alien?*

Figure 24.8
Corundum is a mineral form of aluminum oxide. Rubies are corundum in which a few of the aluminum ions are replaced by chromium ions. Blue sapphires are corundum with traces of iron and titanium ions. What are the formulas for aluminum oxide and chromium(III) oxide? 2

corundum (impure aluminum oxide). Cut forms of corundum with trace amounts of other elements are the commonly known gemstones ruby and sapphire, shown in **Figure 24.8.** Corundum, in its less exotic form, is used as the abrasive in emery powder and grinding wheels.

As a pure metal, aluminum has strength and ductility in addition to low density, high electrical conductivity, and corrosion resistance. These properties make it a valuable structural and manufacturing metal. When aluminum is exposed to air, a tough film of oxide quickly forms, protecting it from further corrosion by oxygen and water.

Until the end of the nineteenth century, aluminum sold for the same price as silver. Aluminum was so expensive because there was no practical way to produce it. Paul Héroult, a Frenchman, and Charles Hall, an American, almost simultaneously invented the first commercial process for the production of aluminum. Hall's chemistry professor challenged his class to find an inexpensive way to produce aluminum. After graduation, Hall set up a laboratory in a woodshed. He found that cryolite (Na_3AlF_6) dissolves aluminum oxide, forming a low melting point solution, from which aluminum could be obtained by electrolysis. This process is essentially the method used for aluminum production today. Aluminum is widely used as a lightweight structural material in aircraft production, as shown in **Figure 24.9,** and in the manufacture of cookware. What is another everyday item made from aluminum? 1

The other Group 3A elements—gallium, indium, and thallium—are quite rare. One interesting use of gallium is in thermometers because of its extraordinarily wide liquid temperature range. Gallium has a melting point of 30 °C and a boiling point of 1980 °C. Because gallium is a semiconductor, it is also used in the manufacture of some electronic components.

Figure 24.9
Aluminum is particularly useful in the construction of aircraft because it is lightweight and very strong. It also reacts with oxygen to form a thin protective coating of aluminum oxide that will not react with water and that resists further corrosion.

2 Teach

Chemistry and Science History

Shortly after graduation from Oberlin College, Charles Martin Hall (1863-1914) devised an electrolytic process for producing aluminum. He spent several years perfecting the process and demonstrating it to potential investors. In 1888, the Pittsburgh Reduction Company, which later became the Aluminum Company of America, adopted the Hall process, and shortly thereafter made Hall the company's vice-president. By the time of his death, Hall had amassed a fortune. Hall did not forget the role that education played in his life. He bequeathed most of his money to various universities. Oberlin College alone received more than three million dollars—a stupendous amount of money in 1914.

Discuss

Although aluminum is a fairly reactive metal, it is used in mechanical parts, cooking implements, and beverage containers. Explain that the durability of aluminum-based materials derives, in part, from the way in which aluminum reacts with oxygen. Unlike the oxidation of iron, which gives rise to flakes of iron oxide that chip off, the oxidation product of aluminum, Al_2O_3, forms a thin veneer-like coating that protects the underlying aluminum from further oxidation. Beverage containers are protected in an additional way, by a plastic coating that prevents acids from reacting with aluminum. Point out that most colas have a pH much less than 7. Ask students what products are formed when Al(*s*) reacts with HCl(*aq*). [$H_2(g)$ and $AlCl_3(aq)$]

Answers

1 beverage containers
2 Al_2O_3, Cr_2O_3

24.2

Discuss

Because lead has such a low melting point, it is easily obtained from its ore, PbS. As a result, lead was one of the first metals to be purified from ore. The Romans were the first civilization to use large amounts of lead. They used lead to make plumbing, ceramic glazes, and eating utensils. Lead is very poisonous; it accumulates in brain tissue and affects mental development. Analysis of bones from the Roman era indicates that humans were exposed to significant levels of lead. Some historians theorize that lead poisoning contributed to the downfall of the Roman Empire. Today, lead is used primarily for electrodes in automobile batteries. Remind students that lead can react to form compounds in the +2 and +4 oxidation states, for example, $PbCl_2$ and PbF_4. Ask:

- **What other element in Group 4A forms more than one cation?** (Sn can react to form Sn^{2+} and Sn^{4+}.)

Chemistry Alive!

Thermite Reaction Play

Silicon Alien? Play

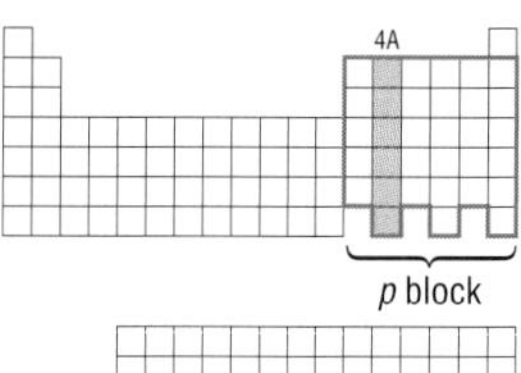

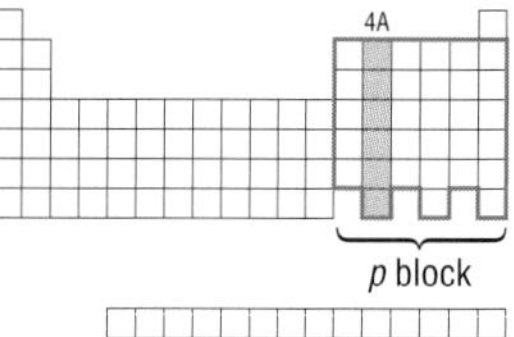

Carbon and Group 4A

Two of the most important elements on Earth, carbon and silicon, are in Group 4A. Carbon is the fundamental element in the molecules of all living things. Silicon is the fundamental element of the geologic world.

The two familiar allotropes of carbon are diamond and graphite. Diamond, the hardest material known, behaves like a typical nonmetal. It is transparent and an electric insulator—that is, a nonconductor of electricity. Graphite, however, is soft, black, and opaque and is a good conductor of electricity. In the late 1980s, a third allotrope of carbon, the soccer ball shaped buckminsterfullerene, was isolated from soot. **Figure 24.10** shows examples of the three allotropes of carbon. In general, carbon-containing compounds, which number at least 10 million, are called organic compounds. Organic compounds are the basis of all living things.

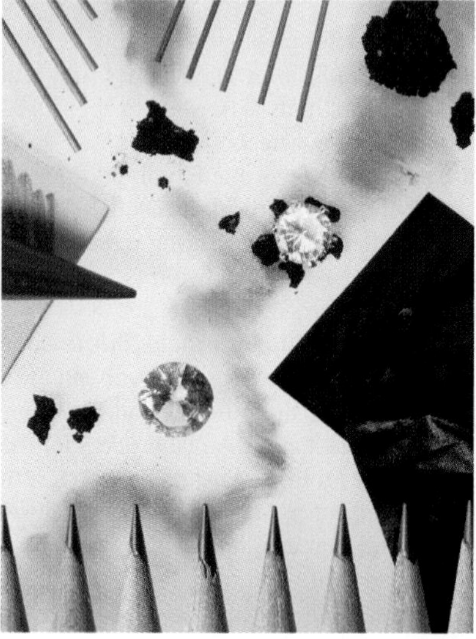

Figure 24.10 ▶
Diamond, graphite, and the recently isolated buckminsterfullerene (also called buckyballs) are three allotropes of carbon.

Silicon is the second-most-abundant element in Earth's crust. It occurs in nature in the combined state as silicon dioxide (SiO_2) in quartz, an essential component of sand, and as silicates in rocks, soils, and clays. Silicon and germanium, the third element in Group 4A, are semiconductors. At low temperatures, both substances are insulators. At high temperatures, however, they conduct electricity. These two elements, when highly purified, form the foundation of transistor technology. A silicon computer chip is shown in **Figure 24.11.** Silicon and germanium are also used in photocells for solar power units.

Figure 24.11
The size of a present-day computer chip is shown here relative to a penny. Each chip may contain millions of transistors.

Tin and lead are the remaining elements in Group 4A. Both are fairly unreactive metals. Therefore, tin and lead have many uses as free metals. Tin, a soft, silvery metal that can be rolled into thin sheets to make tin foil, is usually found in nature as its oxide (SnO_2). In tin cans, a tin coating protects the iron underneath against the corroding effect of acidic foods. Tin has been used for centuries in various alloys, such as bronze (an alloy of tin and copper) and solder (an alloy of tin and lead). One tin compound, stannous fluoride (SnF_2), is an important ingredient in some toothpastes.

Lead has a low melting point (327 °C) and is easily obtained from the sulfide mineral galena (PbS). Lead was known to the Egyptians as early as 3000 B.C., and since Roman times it has been used in plumbing. Lead's toxicity is a major health concern, and its use in certain products is now regulated. For example, plumber's solder, which was once an alloy of tin and lead, is now lead free, and the manufacture of leaded gasoline has been phased out. Currently, the largest commercial use for lead is for the electrodes in lead storage batteries.

Meeting Diverse Needs

Gifted Ask students to describe how the periodic table can be used to determine the outer-shell (valence-shell) electron configuration of an element. [Note which block (*s*, *p*, *d*, *f*) the element belongs to. Determine the principal energy level from the period number. For transition metals it is one less than the period number; for inner transition elements it is two less. Find the number of electrons in a partially filled sublevel by counting over to the element, starting from the left side of the sublevel.] Have students demonstrate their method: Point to any element in the periodic table and have them give the electron configuration. Example: Mg = [Ne]$3s^2$.

Nitrogen and Group 5A

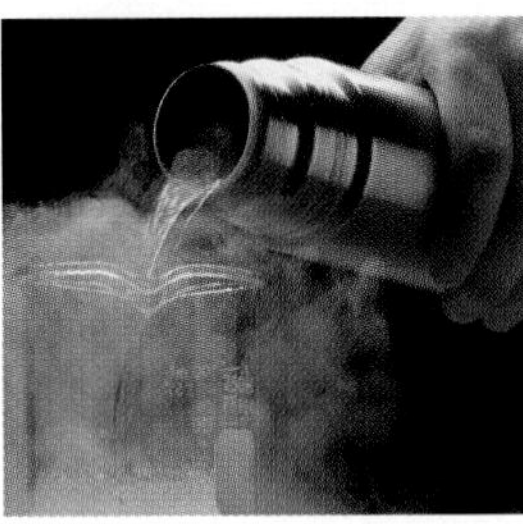

The first Group 5A element, nitrogen, is a nonmetal and is a gas at room temperature. In descending order, the next elements are phosphorus (a solid nonmetal) and arsenic and antimony, which are metalloids. The last element, bismuth, is a metal.

Nitrogen is essential to living organisms. Although 80% of the air you breathe is nitrogen, your body cannot use the element in that form. Fortunately, bacteria in the soil and in the root nodules of peas, beans, and other legumes, fix atmospheric nitrogen into a usable form. Plants can use the fixed nitrogen compounds to synthesize proteins and other biologically important nitrogen-containing compounds.

Nitrogen is isolated commercially from air by two processes. In one method, the nitrogen is fractionally distilled from liquefied air. Because liquid nitrogen, shown in **Figure 24.12,** boils at a lower temperature than liquid oxygen does, nitrogen distills off first from the mixture. In the second method of isolating nitrogen, air is moved over red-hot coke, which is almost pure carbon. The carbon combines with oxygen to form carbon dioxide while the nitrogen remains unchanged.

Elemental nitrogen is a colorless, odorless, tasteless gas, composed of diatomic molecules (N_2). It is slightly soluble in water. Its melting point is −210 °C and its boiling point is −196 °C.

The most important industrial uses of atmospheric nitrogen are in the manufacture of two compounds: ammonia (by the Haber-Bosch process, illustrated in the flowchart in **Figure 24.13)** and nitric acid (by the Ostwald method).

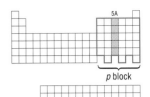

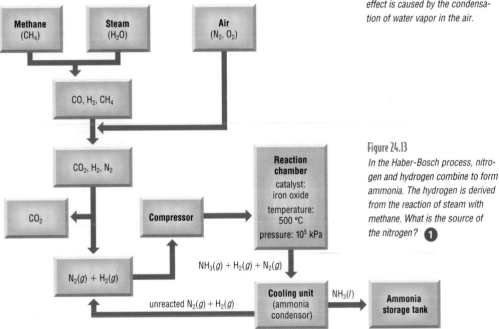

Figure 24.12
Extremely cold liquid nitrogen boils away quickly when poured from an insulated flask into a beaker at room temperature. The smokelike effect is caused by the condensation of water vapor in the air.

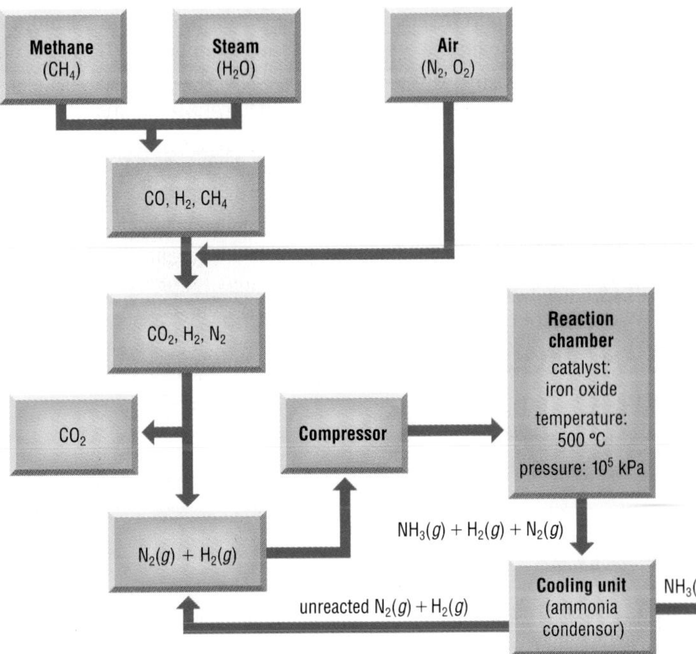

Figure 24.13
In the Haber-Bosch process, nitrogen and hydrogen combine to form ammonia. The hydrogen is derived from the reaction of steam with methane. What is the source of the nitrogen? 1

Discuss

Explain that all tissues in the body contain nitrogen. Nitrogen is an important constituent of proteins, which make up hair, skin, and muscles; and nucleic acids, which carry the genetic code in each cell. Explain to students that the nitrogen in their bodies comes from nitrogen compounds in the foods they eat, not directly from the nitrogen gas in the air they breathe.

Answer

1 N_2 in the atmosphere

TEACHER DEMO

Ammonia gas forms when ammonium chloride reacts with sodium hydroxide. **CAUTION!** Wear safety goggles. Perform the experiment in a hood or well-ventilated room. Prepare a 250-mL beaker containing 10 mL of water and 3–4 drops of 0.1% phenolphthalein. Mix 30 g of ammonium chloride with 15 g of sodium hydroxide in a 500-mL Erlenmeyer flask. Add 10 mL of water to the flask and stopper with a one-hole stopper fitted with a glass tube attached to rubber tubing. Position the open end of the rubber tubing in the beaker. Students should observe the formation of ammonia gas bubbles in the water as shown by the color change in the pH indicator. Note that ammonia is extremely soluble in water. It dissolves in the water in the beaker and lowers the pressure in the flask. Consequently, after several minutes, the water in the beaker will begin to enter the flask.

Figure 24.14
Pure liquid ammonia is called anhydrous ammonia. It is used extensively as a fertilizer and is injected directly into the ground.

In the Haber-Bosch process, nitrogen and hydrogen gases are heated to 500 °C at 6×10^4 kPa (600 atm) in the presence of an iron catalyst.

$$N_2(g) + 3H_2(g) \xrightleftharpoons{Fe} 2NH_3(g)$$

The ammonia gas produced is easily liquefied by cooling and is separated from the reactants by liquefaction.

Ammonia, which is very soluble in water, is a colorless gas with a strong, irritating odor with which you are probably familiar. Aqueous ammonia is weakly basic and is a component of many cleaning products. Liquid ammonia, aqueous ammonia, and several ammonium salts are used as fertilizers. Liquid ammonia, also called anhydrous ammonia, is sometimes applied under pressure directly to soil, as shown in **Figure 24.14.** Liquid ammonia is also used as a refrigerant, particularly in the frozen-food industry. Ammonium nitrate is the most important solid fertilizer in the world today.

About one-quarter of all ammonia produced is made into nitric acid. Nitric acid is made commercially from ammonia by the Ostwald method, a process invented by Wilhelm Ostwald (1853–1932), a German chemist. Nitric acid is used in etching processes and in the production of fertilizers and dyes. It is also an important raw material in the manufacture of explosives.

Like nitrogen, phosphorus is essential to living organisms. Phosphorus is present in the double strands of DNA, in bones, and in teeth. It is also part of ATP (adenosine triphosphate), which is the principal energy-transfer molecule in living systems. Phosphorus occurs mainly in the form of phosphate rock. As shown in **Figure 24.15,** pure phosphorus is prepared in a white form and a red form. The white form is very reactive. Red phosphorus is a less active form used in the manufacture of matches. Why do you think
1 white phosphorus is usually stored under water?

The other Group 5A elements—arsenic, antimony, and bismuth—occur in nature mainly in the form of sulfide minerals. They are not essential to living organisms. Because they expand as they solidify, alloys containing antimony and bismuth are used in making metal type characters

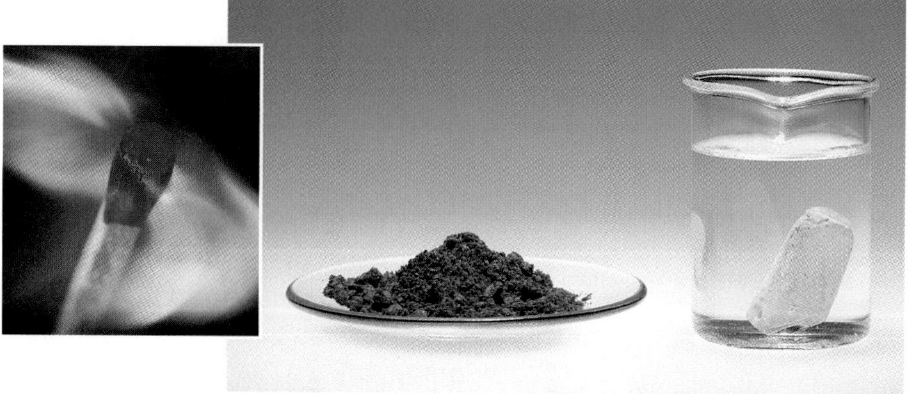

Figure 24.15
White (or yellow) phosphorus is very reactive and must be stored under water (right). Red phosphorus is much more stable, and is used to make matches (left).

Answers

1 to prevent reaction with oxygen from the air

2 A supply of oxygen is needed so the passengers and crew can breathe at the low partial pressures of atmospheric oxygen found at high altitudes.

3 paramagnetism

for traditional press printing. Because characters of type must be identical, the metals used to manufacture them must expand to fill the type molds completely.

6A

p block

Oxygen and Group 6A

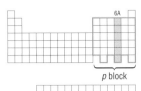

The Group 6A elements include oxygen, sulfur, selenium, tellurium, and polonium. Oxygen is a gas and a nonmetal. Sulfur is a nonmetal that occurs free in nature as a brittle yellow solid. Selenium and tellurium are both solids and metalloids, with properties intermediate to those of metals and nonmetals. Polonium is a radioactive metal that occurs only in trace quantities in radium-containing minerals.

Oxygen is Earth's most abundant element. It accounts for 50% by mass of Earth's crust, 60% by mass of the human body, and 20% by volume of the air you breathe. Most oxygen is combined in the silicate rocks of Earth's crust.

The primary source of commercial oxygen (O_2) is air. Relatively pure oxygen is obtained by liquefying air and then fractionally distilling the liquid. Oxygen distills off at −183 °C. When cooled, oxygen condenses to a clear blue liquid that freezes at −218 °C. In liquid form, oxygen can be held between the poles of a magnet, as shown in **Figure 24.16.** This property, called paramagnetism, is caused by the presence of unpaired electrons in the oxygen molecule. The major commercial use of oxygen is in the basic oxygen process for the manufacture of steel.

Medical emergency teams administer oxygen to victims of smoke inhalation, electrical shock, or drowning, as shown in **Figure 24.17.** For certain medical conditions, such as pneumonia, emphysema, or gas poisoning, a patient may need to breathe air enriched with oxygen for long periods of time. Why are tanks of oxygen carried in airplanes for use at high altitudes? 2

Ozone (O_3) is an allotrope of oxygen. Ozone is produced during electrical storms, as well as in Earth's upper atmosphere—the ozone layer—by the action of ultraviolet light on oxygen. Ozone is also produced near high-voltage generators. The ozone layer protects living things from excess ultraviolet light from the sun. In the 1970s, it was discovered that chlorofluorocarbons (CFCs), used as refrigerants and as propellants in aerosol cans, had begun to destroy the ozone layer. In 1985, a hole was found in the ozone layer over Antarctica. The nations of the world have since agreed to reduce production of CFCs and hope for 100% elimination in a few years.

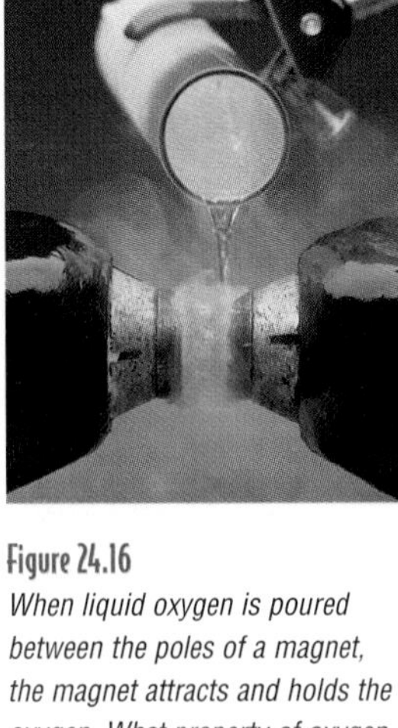

Figure 24.16
When liquid oxygen is poured between the poles of a magnet, the magnet attracts and holds the oxygen. What property of oxygen does this demonstrate? 3

Figure 24.17
Oxygen is often administered to medical patients who have difficulty breathing.

24.2

Use the Visual

Direct students' attention to Figure 24.16. Some students may think that only metals can be affected by a magnetic field. Explain that the response of an element to a magnet is related to the presence or absence of unpaired electrons. Point out that oxygen has two unpaired electrons and is paramagnetic, that is, weakly attracted to a magnet. Write the molecular orbital configuration of oxygen on the board.

$[(\sigma_{1s})^2(\sigma_{1s^*})^2(\sigma_{2s})^2(\sigma_{2s^*})^2(\pi_{2p})^4(\sigma_{2p^*})^2(\pi_{2p^*})^2]$

Point out the single electrons in the π_{2p^*} *(antibonding)* orbital. Refer back to the discussion of magnetism on p. 450.

MEETING DIVERSE NEEDS

At Risk Because nonmetals display such a wide range of properties, students may have trouble remembering which specific properties or uses are associated with a particular nonmetal. You may want students to prepare a set of bulletin board displays for elements such as carbon, silicon, oxygen, nitrogen, and sulfur. The displays could include newspaper articles, photographs, and student drawings. Students would need to prepare labels for each item displayed describing the connection between the item and the element in question.

Chemistry Alive!

Elephant Toothpaste

Play

Watermelon Surprise

Discuss

Point out that bonding in the metallic oxides is predominantly ionic and that these oxides are *basic.* Bonding in the nonmetallic oxides is predominantly covalent and these oxides are *acidic.*

Ozone is a pale-blue gas with a characteristic odor. When cooled to −112 °C, it condenses to a deep-blue, explosive liquid. Ozone is a strong oxidizing agent. It is used commercially to bleach flour, oil, and delicate fabrics and to sterilize water. Because ozone is unstable, it must be generated where it will be used. It is prepared by passing air through an electrical discharge.

When substances combine chemically with oxygen, the process is called oxidation. The product of an oxidation reaction is an oxide. Oxides of all elements, except the noble gases and a few inactive metals, can be prepared simply by heating the elements in the presence of oxygen. In general, oxides of metals are solids, and many react with water to form bases. Oxides of nonmetals may be solid, liquid, or gas, and many react with water to form acids.

Sulfur is a pale-yellow, tasteless, odorless, water-insoluble, brittle solid that has been known about since ancient times. Sulfur occurs in the elemental state in large underground deposits. Some of the world's richest deposits of elemental sulfur are on the Gulf Coast in Louisiana and Texas. These deposits are not easily mined, however, because they are buried under several hundred feet of quicksand. A German-American engineer, Herman Frasch (1851–1914), devised an ingenious method for getting this sulfur out of the ground. The method takes advantage of sulfur's low melting point (119 °C). The Frasch process produces 80% of the world's sulfur output.

In the Frasch process, wells are drilled into the sulfur bed. Then an arrangement of three concentric tubes (with 2.5-cm, 7.5-cm, and 15-cm diameters) is installed, as shown in **Figure 24.18.** Superheated water (180 °C) under pressure is pumped down the outside tube to melt the sulfur. Compressed air is pumped down the center tube. A frothy mixture of air, water, and molten sulfur rises up the third tube. The molten sulfur, 99.5% pure, is pumped into large storage vats, where it cools and solidifies into huge blocks. The blocks may be as much as 130 m long, 70 m wide, and 30 m high. For shipping, the sulfur is loosened by dynamite blasting and loaded into freight cars.

Figure 24.18
Sulfur deposits deep underground are mined by the Frasch process. The sulfur, melted by hot water and whipped into a froth by compressed air, rises to the surface. After being mined, the sulfur is dried and stored in huge blocks until it is shipped.

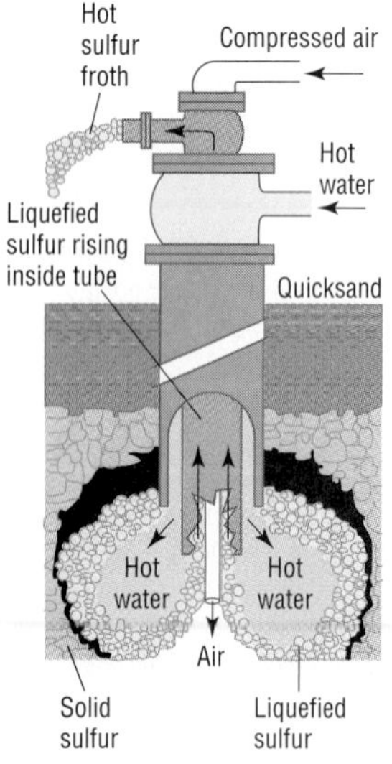

Top view Side view

(a) Crystalline sulfur

(b) Molten sulfur

(c) Monoclinic crystal (d) Rhombic crystal

Figure 24.19
*Crystalline sulfur exists as S_8 molecules **(a)**. Above the melting point, the S_8 rings of sulfur break and form chains of sulfur atoms. Near the boiling point, the chains break into smaller groups and individual atoms **(b)**. As the sulfur cools, the needlelike monoclinic crystals **(c)** gradually change to the stable rhombic form **(d)**.*

Large amounts of sulfur are also obtained from hydrogen sulfide, a poisonous gas. Hydrogen sulfide is a product of petroleum refining and can be present up to 40% in natural gas. To obtain pure sulfur, some of the hydrogen sulfide is burned in air to make sulfur dioxide. The sulfur dioxide is then treated with more hydrogen sulfide to produce elemental sulfur. This reaction can be summarized in a single equation.

$$2H_2S(g) + SO_2(g) \longrightarrow 2H_2O(l) + 3S(s)$$

Sulfur occurs in several allotropic forms. When molten sulfur cools below 119 °C, it forms crystals of monoclinic sulfur. This allotrope contains eight-membered rings of covalently bonded sulfur atoms (S_8) arranged in monoclinic crystals. Below 95.5 °C, the sulfur changes to rhombic sulfur, which is also composed of S_8 units. The crystal forms of sulfur are shown in **Figure 24.19.** A third form of sulfur is the dark-brown, rubberlike allotrope known as amorphous or plastic sulfur. It can be produced by pouring molten sulfur, near its boiling point, into cold water, as shown in **Figure 24.20.** Amorphous solids do not have a distinct crystal form. Within hours, the amorphous sulfur loses its elasticity as it is converted to rhombic crystals. Sulfur can be purified by boiling it and condensing the vapor. Tiny rhombic crystals form in flowerlike patterns on the walls of the condensation chamber. This powder is called flowers of sulfur.

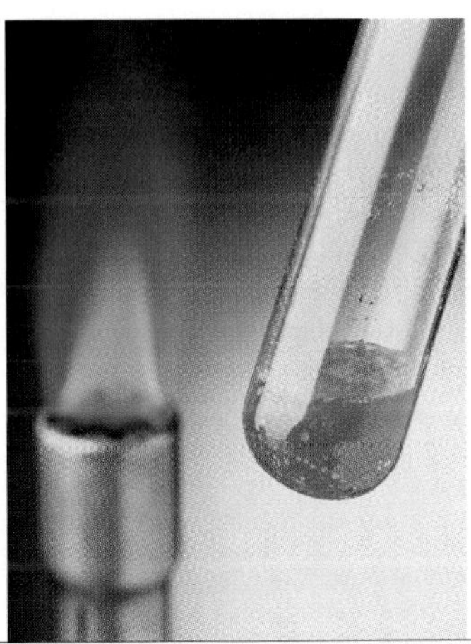

Figure 24.20
Sulfur in the liquid state after being heated over a flame (left). When the liquid sulfur is poured into water, it forms an amorphous solid (right).

Discuss

Point out that production of sulfur goes back to ancient times. It was used in religious ceremonies, for fumigating buildings, and bleaching cloth. The sulfur was generally obtained through so-called purification by fire, when the conventionally mined ores were heated to melt sulfur.

Discuss

An interesting corollary to the reactions used in the contact process has to do with air pollution. Sulfur dioxide (SO_2) is produced by the combustion of fossil fuels. For a long time, researchers were unable to determine how the conversion of SO_2 to SO_3, normally a slow process, occurred so rapidly in the air. Research has shown that dust and other particles in the air act as catalysts for the process.

Discuss

Have students read the Link to Oceanography. Explain that some of the ions in seawater, such as manganese, precipitate slowly, often growing as nodules. In some areas of the Pacific Ocean, the ocean floor at a depth of 3 to 5 km is dotted with nodules containing 32% manganese. At depths of 1 km near Western Samoa, the ocean floor crust contains up to 2% cobalt. The top meter of ocean floor sediment in the Pacific Ocean is estimated to contain 400 billion metric tons of manganese, 9 billion metric tons of copper, 16 billion metric tons of nickel, and 6 billion metric tons of cobalt.

OCEANOGRAPHY

Deep-Sea Minerals

Minerals have been found in the developing undersea ridges of the Pacific Ocean near Hawaii. Where the tectonic plates of Earth's crust are separating,

sea water comes into contact with liquid magma of the underlying volcanic rocks. This superheated water, with recorded temperatures as high as 500 °C, extracts minerals from crust material and deposits these minerals in immense underwater fields on the ocean floor. In one area of the Pacific Ocean, there is an extensive field of potato-sized manganese nodules. Rich deposits of chromium, cobalt, copper, gold, iron, nickel, and zinc have also been found.

When heated with metals (except gold and platinum), sulfur forms compounds called sulfides. Hydrogen sulfide gas, which smells like rotten eggs, is prepared by the action of an acid on a sulfide.

$$FeS(s) + 2HCl(aq) \longrightarrow FeCl_2(aq) + H_2S(g)$$

Sulfur is an extremely important raw material in the chemical industry. It is used in the preparation of paints, plastics, drugs, and dyes, but its major uses are in the vulcanization of rubber and the manufacture of sulfuric acid.

The annual production of sulfuric acid in North America is in excess of 50 million metric tons. Practically every industry makes some use of sulfuric acid. Almost half of the world's production goes into the manufacture of fertilizers, such as ammonium sulfate and superphosphate. Large amounts of sulfuric acid are used in the metal industry for pickling iron and steel. Pickling involves putting a metal in a chemical bath to remove oxides on the metal surface. Sulfuric acid is also essential to many of the processes in petroleum refining.

Today, sulfuric acid is produced mainly by the contact process. In this process, sulfur is first burned in air to produce sulfur dioxide, an irritating toxic gas. Then the sulfur dioxide is oxidized to sulfur trioxide in the presence of a vanadium(V) oxide catalyst.

$$2SO_2(g) + O_2(g) \xrightarrow{V_2O_5} 2SO_3(g)$$

The addition of sulfur trioxide to water gives sulfuric acid.

$$SO_3(g) + H_2O(l) \longrightarrow H_2SO_4(l)$$

The entire process is called the contact process because the key reaction takes place when the reactants are in contact with the surface of the solid catalyst.

Pure, anhydrous sulfuric acid is a dense, colorless, syrupy liquid. The concentrated acid generally used in laboratories is 98% H_2SO_4 and 2% H_2O. Sulfuric acid has the chemical properties of a typical acid. The dilute acid reacts with metals, oxides, hydroxides, or carbonates to form salts called sulfates. The reaction with metals also releases hydrogen gas. Sulfuric acid displaces other acids from their salts. For example, hydrogen chloride can be produced from sulfuric acid and sodium chloride according to the following equation.

$$H_2SO_4(l) + 2NaCl(s) \longrightarrow Na_2SO_4(s) + 2HCl(g)$$

When dissolved in water, the gaseous hydrogen chloride that is produced forms hydrochloric acid.

Concentrated sulfuric acid dissolves readily in water, liberating a large
1 amount of heat. Is such a process endothermic or exothermic? Its strong affinity for water makes sulfuric acid a powerful dehydrating agent. Thus it must be handled with great care. When the concentrated acid is added to sugar, a vigorous dehydration reaction takes place. A large volume of charred material is produced as the acid removes water from the sugar, leaving only carbon, as shown in **Figure 24.21.**

$$C_{12}H_{22}O_{11}(s) \xrightarrow{H_2SO_4(l)} 12C(s) + 11H_2O(g)$$

Answers

1 exothermic

2 dehydrating agent; it has a strong affinity for water

Figure 24.21
When concentrated sulfuric acid is added to sucrose (cane sugar), water is removed. The reaction releases a great deal of heat. The water turns to steam and the residual carbon expands to a porous mass. What kind of an agent is concentrated sulfuric acid in this reaction? Why? 2

Hot concentrated sulfuric acid is an oxidizing agent. For example, it reacts with certain metals to generate sulfur dioxide.

$$2H_2SO_4(aq) + Cu(s) \longrightarrow CuSO_4(aq) + SO_2(g) + 2H_2O(l)$$

Selenium, another Group 6A element, is a semiconductor. It is a poor conductor of electricity in the dark, but its conductivity increases greatly in the light. Because of that property, selenium is used in photoelectric cells, in exposure meters for cameras, and in light-sensitive switches. The xerographic process of photocopying also depends on the photoconductivity of selenium.

Tellurium is one of the rarest elements. Its compounds are toxic. The element itself plays no known role in living organisms.

The Halogens

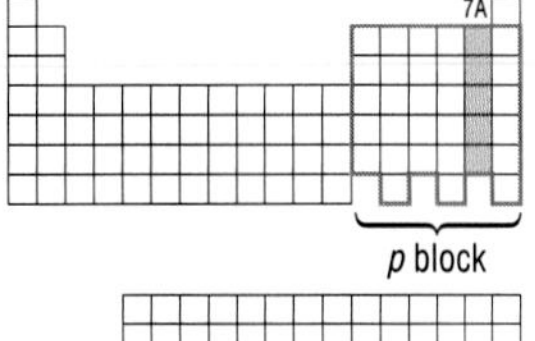

The halogens, Group 7A, do not exist in nature in the uncombined state, but their compounds are fairly abundant. The salts sodium chloride, sodium bromide, and sodium iodide are found in sea water and salt beds. Calcium fluoride is found in the form of the mineral fluorspar.

The halogens are fluorine, chlorine, bromine, iodine, and astatine. All of the halogens are nonmetals. The first two elements, fluorine and chlorine, are yellowish or yellowish-green gases at room temperature and normal atmospheric pressure. Bromine is a dark-red liquid, while iodine is a purple-black crystalline solid with a metallic sheen. The last element, astatine, is a rare radioactive solid that has not been well investigated. The colors of chlorine, bromine, and iodine vapors are shown in **Figure 24.22** on the following page. Also evident in this figure is an important property of the free halogens—free halogens are very reactive and must be handled with extreme caution. **Figure 24.22** illustrates the reactivity of chlorine gas with steel wool.

Discuss

Point out that photocopy machines operate by an electrostatic process. Because it is photoconductive, selenium will conduct an electric current when exposed to light but not when kept in the dark. The selenium in a photocopy machine is located on the surface of an aluminum drum. When a copy is requested, the drum starts rotating and is given a positive charge. Then a beam of light is passed over the document in the window above the drum. Light reflects off the places on the paper that have no printing, and this reflected light then falls on the rotating drum. At every point on the drum that is struck by light, the selenium becomes conductive and the positive charge drains through to the drum. Places where the light does not strike the drum retain their positive charge. The electrostatic image on the drum is then developed by coating the drum with a fine powder containing a carrier and a toner. Finally, a blank sheet of positively-charged paper is brought into contact with the negatively-charged toner. Ask students to find out how color photocopy machines work.

Discuss

Review the chemistry of the halogens. On the chalkboard, write a number of reactions involving the halogens, including those between Group 1A and 2A metals; and those with boron, carbon, nitrogen, and phosphorous. Remind students that Group 7A consists entirely of nonmetals. Write the halogen valence-shell electron configuration, and show that each halogen can attain a stable, noble-gas configuration by accepting one electron from another atom. As a result, the halogens are powerful oxidizing agents. Point out that fluorine is the most electronegative element in the periodic table. Molecular compounds containing fluorine are highly polarized.

Discuss

Tooth decay results when the hydroxyapatite $Ca_{10}(PO_4)_6(OH)_2$ that forms tooth enamel is eroded by reaction with hydronium ions (H_3O^+). The replacement of the hydroxide ion (OH^-) by fluoride ion (F^-) produces a stronger enamel that is less soluble in acidic solutions. Thus, fluoridized water, and fluoride-containing toothpastes help prevent tooth decay by promoting the growth of fluoroapatite crystals ($Ca_{10}(PO_4)_6F_2$).

Figure 24.22
(a) *Iodine, chlorine, and bromine (left to right) are very different in color and in physical state.*
(b) *A heated piece of steel wool reacts vigorously with chlorine gas. The smokelike cloud that appears is actually made up of fine particles of iron(III) chloride. Would the reaction be more or less vigorous in bromine? In fluorine?*
2

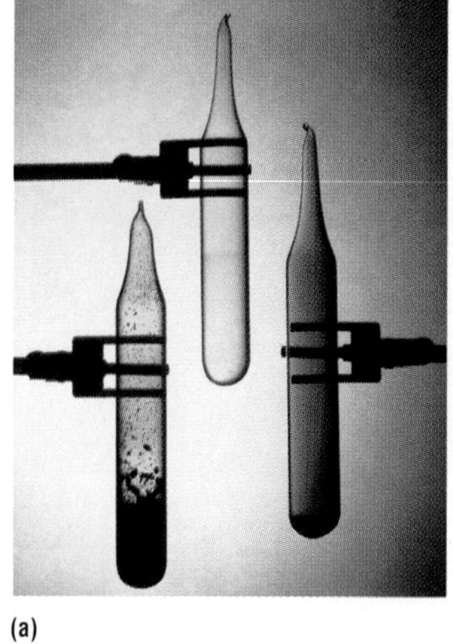
(a)

(b)

Compounds of fluorine, chlorine, and iodine are essential to your well-being and must be included in your diet. Why is fluoride ion added to many
1 municipal water supplies? Chlorine, as chloride ions, is an important component of the blood and other body fluids. Iodine, as iodide ions, is necessary to prevent goiter, an enlargement of the thyroid gland. For this reason sodium iodide is commonly added to table salt. Thus the iodized table salt you eat contains a small amount of NaI.

The halogens and their compounds have many other uses. A dilute solution of chlorine is used as a bleaching and disinfecting agent. Light-sensitive silver chloride and silver bromide are used to make photographic film. Fluorine is used in the manufacture of nonstick PTFE (*p*oly*t*etra*f*luoro*e*thylene) coatings, which are applied to frying pans and a variety of other cookware, as shown in **Figure 24.23.**

Most of the compounds of halogens are soluble in water. Halide ions are abundant in sea water and in salt beds formed by the evaporation of salt water. **Table 24.3** lists the typical concentrations of halide ions found in sea water.

Fluorine was known in compounds long before it was isolated as an element. In 1886, French chemist Henri Moissan made pure fluorine by electrolyzing an ice-cold solution of potassium fluoride in hydrogen fluoride. Fluorine is still made this way today.

Chlorine gas is made commercially by the electrolysis of brine, a concentrated solution of sodium chloride. In a separate process, the hydrogen also produced in the electrolysis is burned in the chlorine to make hydrogen chloride gas—an important source of hydrochloric acid.

Bromine is obtained commercially from sea water and salt-well brines. Sodium chloride in the water is allowed to crystallize, leaving a solution containing the more soluble bromides. Chlorine gas, which is more electronegative than bromine, is then used to displace bromide ions from the solution as free bromine.

$$2NaBr(aq) + Cl_2(g) \longrightarrow 2NaCl(aq) + Br_2(l)$$

Figure 24.23
One nonstick coating commonly used is a heat-resistant polymer composed of fluorine and carbon.

Answers

1 to strengthen tooth enamel against decay
2 less vigorous in bromine; more vigorous in fluorine
3 Fluorine is the most electronegative element.

Chemistry Alive!

Making Sodium Chloride

Play

At one time, iodine was extracted from the ashes of certain seaweeds that concentrate iodine from sea water. Now it is produced commercially from sodium iodate ($NaIO_3$), which occurs as an impurity in deposits of sodium nitrite.

$$2NaIO_3(aq) + 5NaHSO_3(aq) \longrightarrow 2Na_2SO_4(aq) + 3NaHSO_4(aq) + H_2O(l) + I_2(s)$$

Fluorine is the most chemically reactive of all the nonmetals. Where do you think fluorine ranks on the electronegativity scale? Fluorine, the strongest elemental oxidizing agent known, forms compounds with all elements except helium, neon, and argon. The reactivity of the other halogens decreases as their atomic size and atomic mass increase. Thus iodine is the least reactive of the common halogens. 3

With the exception of hydrogen fluoride, the hydrogen halides are highly ionized in water, forming strong acids. Hydrofluoric acid is a weak acid; that is, it is weakly ionized in water. Hydrogen fluoride molecules are very polar and are strongly hydrogen-bonded to one another. The other hydrogen halides exhibit this effect to a much lesser degree. Hydrofluoric acid is used to etch designs in glass and to frost light bulbs. For many years, hydrofluoric acid had to be stored in wax containers. Today, plastic bottles are used instead.

The enrichment of uranium is another important use of fluorine. In the process for separating the isotopes of uranium, the metal is converted to uranium hexafluoride (UF_6), a gas. The diffusion of this gas allows the separation of uranium-235 (the fissionable isotope) from uranium-238 (the nonfissionable isotope).

Large amounts of chlorine are used in the purification of city water supplies, swimming pools, and sewage. Because chlorine in solution is a powerful oxidizing agent, it kills disease-causing bacteria. Chlorine is also used in the synthesis of vinyl chloride ($CH_2{=}CHCl$), which reacts to form polyvinyl chloride (PVC). PVC, a plastic, is used for floor coverings and other vinyl products.

Table 24.3

Concentration of Halide Ions in Sea Water

Ion	g/L
F^-	1.3×10^{-3}
Cl^-	1.9×10^{1}
Br^-	6.5×10^{-2}
I^-	5.0×10^{-5}

section review 24.2

8. Describe some properties of *p*-block elements from the various groups of the periodic table.
9. Describe how several *p*-block elements are obtained from their compounds or minerals.
10. Why is aluminum so resistant to corrosion?
11. Aluminum is produced by electrolysis. Write the electrode reactions for the process.
12. Which of the Group 4A elements are used in electronic equipment?
13. In what form is nitrogen present in Earth's atmosphere?
14. What are some important uses of sulfuric acid?
15. What are the halogens? Describe some properties and uses of each.

Chem ASAP! **Assessment 24.2** Check your understanding of the important ideas and concepts in Section 24.2.

portfolio project

Both Joseph Priestly, the discoverer of oxygen, and Antoine Lavoisier, who gave oxygen its name, were persecuted for their political beliefs. Prepare a written report describing their scientific accomplishments and their political fates.

Answers

SECTION REVIEW 24.2

8. Answers will vary.
9. Answers will vary.
10. Aluminum reacts with oxygen in the air to form a corrosion-resistant protective layer of aluminum oxide.
11. cathode reaction: $Al^{3+} + 3e^- \rightarrow Al(s)$
anode reaction: $2O^{2-} \rightarrow O_2(g) + 4e^-$
12. Silicon and germanium are used in electronic equipment.
13. Nitrogen is found as N_2 gas in air.
14. Sulfuric acid is used in the manufacture of fertilizers, in petroleum refining, and as an oxidizing agent.
15. The halogens are fluorine, chlorine, bromine, iodine, and astatine—Group 7A. Answers will vary.

3 Assess

Evaluate Understanding

Point to a group in the *p*-block of the periodic table and have students name the properties of the elements in that group. Assign students one element from Groups 3A through 7A and ask them to describe the method used for its preparation and its major commercial uses.

Reteach

If circumstances permit, bring in samples of selected elements from each of the groups discussed in this section for students to view while you review physical and chemical properties. Ask questions such as:

- **Which are the most reactive metals shown? The most reactive nonmetals?** (Answers will depend on elements chosen for display.)
- **Which two elements can form chains of atoms?** (carbon, silicon)
- **Which Group 5A element is used to produce ammonia?** (nitrogen)
- **What are some commercial uses of chlorine?** (purification of water, synthesis of plastics, production of hydrochloric acid, bleaching)

Portfolio Project

Because Priestly (1733–1804) was a Congregational minister, a mob destroyed his chapel, house, and laboratory in 1791. Lavoiser (1743–1794) was executed during the French Revolution along with other officers of a tax-collection organization. The mathematician Joseph Lagrange said "It required only a moment to sever that head, and perhaps a century will not be sufficient to produce another like it."

24.3

1 Engage

Use the Visual

Have students study the photograph and read the text that opens the section. Ask:

- **What is this versatile metal?** (titanium; Point out the location of Ti in the periodic table. Explain that titanium is a low-density, high-strength metal that is resistant to chemical attack.)

If possible, locate and display a photograph of an artificial titanium hip joint.

Check Prior Knowledge

To assess students' knowledge about metals, have students:

- **Describe what is meant by the terms *ductile* and *malleable.*** (ductile: can be drawn into wires; malleable: can be hammered into different shapes)
- **Why is it possible to bend metals but not ionic crystals?** (Under stress, the cations in a metal slide past each other because their charges are effectively screened by a "sea" of free-floating electrons. In an ionic crystal, valence electrons are not free-floating. When ions of like charge are forced together under stress, the ions repel one another, and the crystal shatters.)
- **Explain briefly why metals are good conductors of electricity.** (To conduct electricity a material must contain mobile units of charge. The mobile valence electrons carry the electrical current.)

section 24.3

objectives

- List the properties of specific transition metals
- Describe the chemical diversity shown by the transition metals and inner transition metals

key terms

- ores
- metallurgy
- Monel metal
- inner transition elements

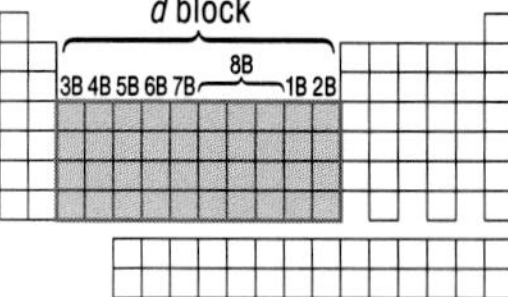

THE *d*- AND *f*-BLOCK ELEMENTS: TRANSITION AND INNER TRANSITION METALS

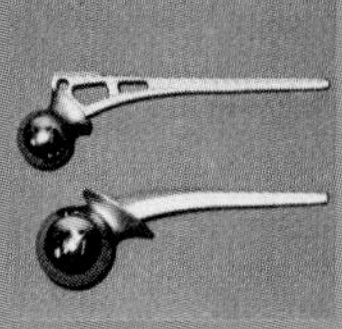

What do broken bones and aircraft engines have in common? The same metal that is strong enough to withstand the forces inside a jet engine is also lightweight and corrosion-resistant enough to be used to repair broken bones and replace joints. **What is this versatile metal?**

Overview

Most metals come from mineral deposits in Earth. Minerals used for the commercial production of metals are called **ores. Metallurgy** is the use of various procedures to separate metals from their ores. Many of these techniques were developed over the centuries by trial and error. Three basic steps are involved: concentrating the ore, chemically reducing the ore to the metal, and refining and purifying the metal.

Figure 24.24 shows aqueous solutions of some transition-metal compounds. You can see from the periodic table that these metals make up ten columns, extending from Group 3B on the left, through 7B and the three columns that together make up Group 8B, and finally to Groups 1B and 2B. Among the transition metals, as atomic number increases, there is an increase in the number of electrons in the second-to-highest energy level. The transition metals exhibit typical metallic properties. In general, they are ductile, malleable, and good conductors of heat and electricity. With the exception of copper and gold, they have a silvery luster.

Table 24.4 lists some important uses of the transition metals. Tungsten, a hard, brittle solid with a melting point of 3400 °C, is used in light-bulb filaments. At the other end of the scale is mercury, with a melting point of −38 °C. Mercury is used in making thermometers. The production of copper wire in enormous quantities attests to the high electrical conductivity of copper. Adding small amounts of cobalt, copper, chromium, nickel, or vanadium to iron makes alloy steels with widely different characteristics.

Figure 24.24

These aqueous solutions contain cations of transition metals. Note that many of the solutions have a distinctive color.

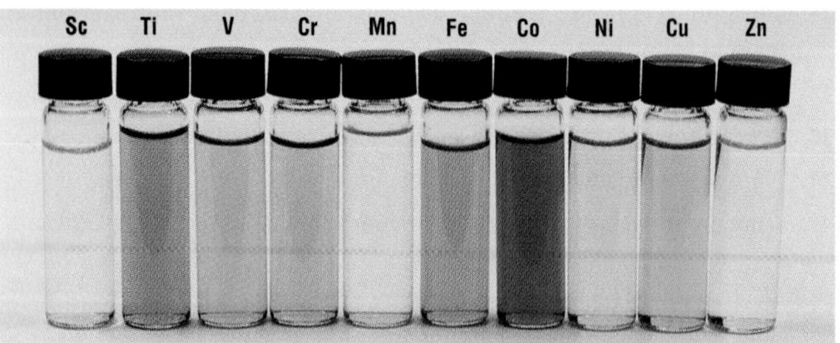

STUDENT RESOURCES

From the Teacher's Resource Package, use:

- Section Review 24.3 and Ch. 24 Quizzes from the Review Module (Ch. 21–24)
- Laboratory Manual: Experiment 47
- Laboratory Practical 24-1
- Small-Scale Chemistry Lab Manual: Experiment 35

TECHNOLOGY RESOURCES

Relevant technology resources include:

- Chem ASAP! CD-ROM
- Resource Pro CD-ROM
- Chemistry Alive! Videodisc: *Oxidation States of Vanadium*

Table 24.4

Common Transition Metals and Their Uses	
Metal	**Uses**
Cadmium (Cd)	Batteries; control rods for nuclear reactors
Chromium (Cr)	Plating; making stainless steel
Cobalt (Co)	Alloys; treatment of cancer (Cobalt-60 only)
Copper (Cu)	Electrical wiring; plumbing; coinage
Gold (Au)	Jewelry; ornaments; standard of wealth
Iron (Fe)	Steel; magnets
Manganese (Mn)	Steel; nonferrous alloys
Mercury (Hg)	Lamps; switches; thermometers; barometers
Nickel (Ni)	Hardening steel; plating; catalyst
Platinum (Pt)	Catalyst; electronics; lab-ware; jewelry
Silver (Ag)	Mirrors; jewelry; photography; coins
Tantalum (Ta)	Surgery; corrosion-resistant equipment
Titanium (Ti)	Combustion chambers for rockets and jets
Tungsten (W)	Filaments for light bulbs; alloys
Vanadium (V)	Shock-resistant steel alloys; catalyst
Zinc (Zn)	Galvanized iron; brass; dry cell batteries

Figure 24.25
Most metals are refined from ores. Those shown here are (clockwise from the top): vanadinite (vanadium), hematite (iron), wulfenite (lead, molybdenum), sphalerite (zinc), galena (lead), and malachite with azurite (copper) in the center.

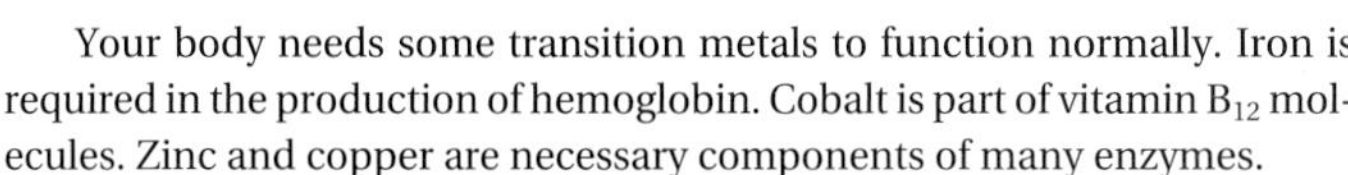

Your body needs some transition metals to function normally. Iron is required in the production of hemoglobin. Cobalt is part of vitamin B_{12} molecules. Zinc and copper are necessary components of many enzymes.

The transition metals vary greatly in their chemical reactivity. The elements scandium, yttrium, and lanthanum are similar to the Group 1A and 2A metals in that they are easily oxidized on exposure to air and react with water to liberate hydrogen. In contrast, platinum and gold are extremely unreactive and resist oxidation.

Figure 24.26
The transition-metal oxyanions that produced the colored solutions (left) are VO_4^{3-}, $Cr_2O_7^{2-}$, and MnO_4^-, respectively. Because the solid compounds (right) contain transition metal ions, they display a wide range of characteristic colors.

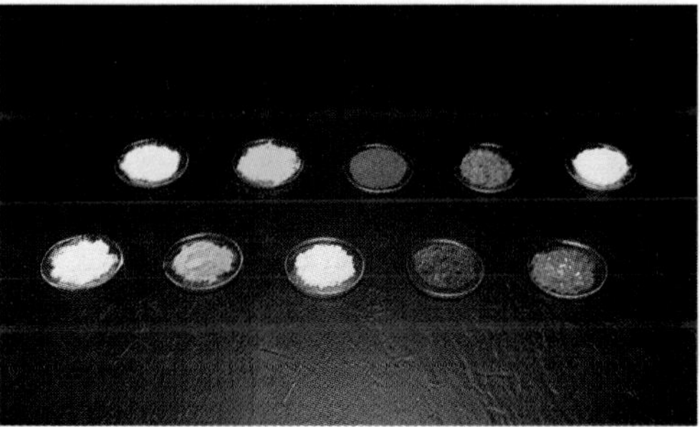

24.3

Discuss

Point out that the profitability of mining a particular ore depends on a number of factors. The ore must contain enough of the desirable metal so that the market value of the metal offsets the cost of finding, removing, and refining the ore. For abundant metals such as aluminum, the ore must contain at least 30% metal. For very rare and expensive metals such as gold, it may be profitable to mine ore containing less than 1% of the metal.

ACTIVITY

All living organisms need small amounts of a number of metals to remain healthy. These trace nutrients are usually called "minerals." Have students examine the product labels for at least three different brands of multivitamin supplements. Have them compare the kinds and amounts of metals listed on each of the labels. Ask:

- Which trace metals are important nutrients? (iron, zinc, copper, manganese, cobalt, molybdenum, and chromium)

Meeting Diverse Needs

At Risk Have students choose ten metals from the *d* and *f* blocks (other than the ones listed in Figure 24.26). Tell them to look each one up in an encyclopedia to learn which ores it comes from and what its most important uses are. Have them present their findings in a table.

TEACHER DEMO

Many, but not all, transition-metal compounds are intensely colored. The colors depend on the particular metal and on the oxidation state of the metal. Metal ions that have a partially filled *d* sublevel are usually colored, whereas metal ions that have a completely filled or completely empty *d* sublevel are colorless. Prepare dilute aqueous solutions (approximately 0.1 *M* to 0.5 *M*) of different salts of transition metals such as manganese, iron, cobalt, nickel, copper, and zinc. Have students compare their colors and write the electron configurations for each of the metal ions. Discuss the correlation of color to the electronic properties of the ion.

One of the most characteristic chemical properties of the transition metals is the occurrence of multiple oxidation states. **Figure 24.26** on the previous page shows the colorful aqueous solutions of oxyanions of vanadium, chromium, and manganese in their highest oxidation states. In addition to the characteristically colored aqueous solutions, solid compounds of the transition metals also produce a striking range of colors.

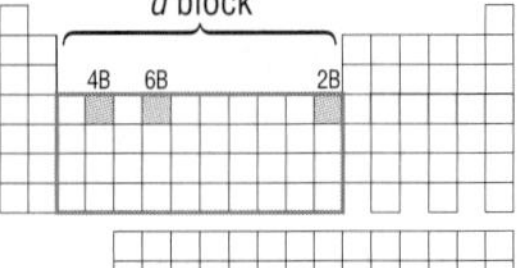

Titanium, Chromium, and Zinc

Figure 24.27 shows some uses of the metals titanium, chromium, and zinc. Titanium is an important structural metal. It is strong and corrosion-resistant. Moreover, its density is only about half that of iron. Titanium alloys are used in high-performance aircraft engines and missiles. Each engine of a 747 jetliner contains almost 1500 kg of titanium alloys. Titanium alloys are also used for joint replacements, such as hips and knees, and as screws to repair broken bones. What is an everyday application of
1 titanium? The most familiar and important compound of titanium is its dioxide (TiO_2). Titanium dioxide, or, more correctly, titanium(IV) oxide, is an opaque, white compound used as a pigment in white paint, paper, and other consumer items, such as sunscreen. Titanium is widely distributed in Earth's crust; the major ore is rutile, which is impure TiO_2.

Although relatively rare, chromium has many important uses. It is a hard, brittle metal with a luster. Its extreme resistance to corrosion makes it an ideal coating for iron and steel objects. It retains its bright surface by developing an invisible oxide coating. Chromium is an important ingredient in stainless steel. Typical stainless steel contains about 18% chromium and 8% nickel, with trace amounts of carbon, manganese, and phosphorus. The remaining percentage consists of iron.

Figure 24.27
The transition metals titanium, chromium, and zinc have many important uses. For example, titanium is used in paints and pigments. Chromium is used to produce a shiny, corrosion-resistant coating. Zinc is used to galvanize steel to help prevent rusting.

Answers

1 Titanium is used to make strong, flexible frames for eyeglasses. It is also used to make sporting equipment such as golf clubs, bicycles, and tennis rackets.

2 Fe^{3+} is reduced to Fe.

Chemistry Alive!

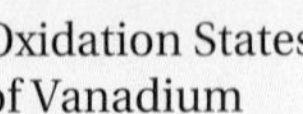
Oxidation States of Vanadium

Play

Like most transition metals, chromium has more than one oxidation state. The most common oxidation states of chromium are +2, +3, and +6. Chromium has an oxidation number of +6 in two polyatomic ions: chromate (CrO_4^{2-}) and dichromate ($Cr_2O_7^{2-}$). All compounds in which chromium is in the +6 oxidation state are powerful oxidizing agents.

Zinc is one of industry's most useful metals. Zinc is obtained from the sulfide ore zinc blend (sphalerite), which is shown in **Figure 24.25** on page 725. Large amounts of zinc are used to produce galvanized steel, which is iron coated with a thin layer of zinc to protect the iron from rusting. Zinc forms the case in dry-cell batteries and is alloyed with copper to form brass.

Iron, Cobalt, and Nickel

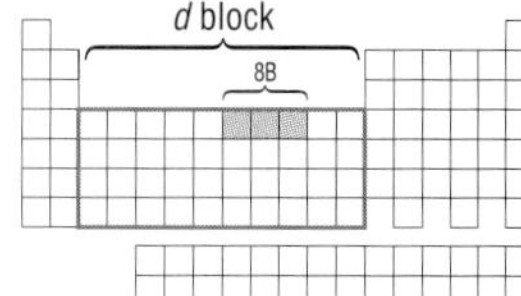

Iron, cobalt, and nickel share a number of common properties. Their densities are similar, as are their melting and boiling points. Their most striking physical property, however, is magnetism. Iron is the most strongly magnetic and nickel is the least magnetic of the three metals.

Making up about 5% of Earth's crust, iron is the second-most-abundant metal on Earth. (Aluminum is the most abundant.) Iron is the cheapest metal and, in the form of steel, the most useful. Common iron minerals are Fe_2O_3 (the red oxide hematite) and Fe_3O_4 (magnetite). Millions of tons of iron are produced annually in the United States by the reaction of Fe_2O_3 with coke (pure carbon). This reaction is carried out in a blast furnace, similar to the one shown in **Figure 24.28.** A modern blast furnace produces about 5000 tons of iron daily.

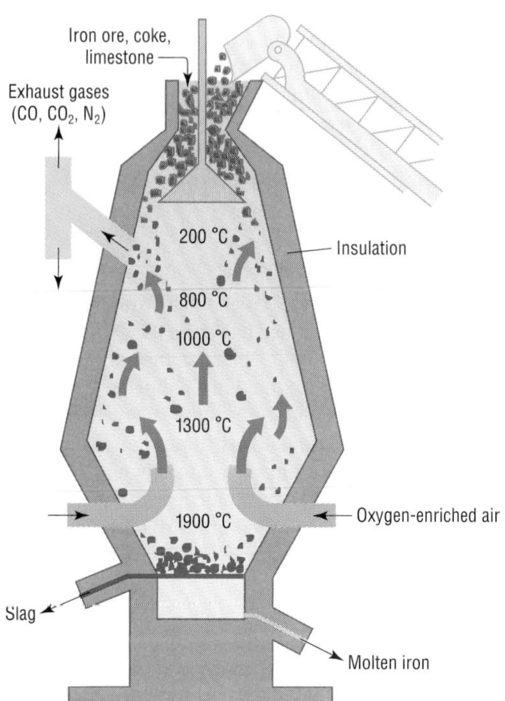

Figure 24.28
A blast furnace is about 30 m high and 10 m in diameter. Iron ore, limestone, and coke (a form of carbon) are added at the top. During the process, carbon, the reducing agent, is oxidized to CO_2. The limestone converts sand in the ore to silicate slag. What is reduced in the blast furnace? 2

24.3

Discuss

Point out that the use of zinc to galvanize iron pipes is like the use of paint to protect wood surfaces from weathering. However, unlike the paint, zinc does not simply serve as a coating that separates the material beneath it from the environment. Because it is a reactive metal, zinc also prevents corrosion of underlying materials by preferentially reacting with oxidizing agents in the environment.

CHEMISTRY AND SCIENCE HISTORY

Steel-making techniques were developed by people of various cultures in ancient times. The manufacturing of carbon steel by the Chinese has been dated to 500 B.C. Recent archaeological evidence of forced-draft furnaces on the shores of Lake Victoria and traditional oral histories indicate that Africans were also producing steel around 2000 years ago.

MEETING DIVERSE NEEDS

At Risk and LEP Have students work with a partner to create a graphic that identifies essential minerals, their functions, and the foods in which they are found. Students can make cartoons, collages, or concept maps.

TEACHER DEMO

Demonstrate the formation of silicate crystals of transition metals. **CAUTION!** Wear your safety goggles. Do not handle the chemicals with your bare hands. Add 100 mL of waterglass, a concentrated solution of sodium silicate (Na_2SiO_3), to 200 mL of water in a 600-mL beaker that contains a thick layer of sand. Into the beaker, spread five crystals each of cobalt(II) chloride, copper(II) chloride, nickel(II) chloride, manganese(II) chloride, lead(II) chloride, iron(III) chloride, and calcium chloride. (For some, you will need to use a hydrate.) After several days, have students observe the colors of the silicate crystals that form.

ACTIVITY

Have students look in the business sections of newspapers or in other library references to find the types of metals that are traded on the stock exchange. Ask:

- Which metal currently has the highest unit value?
- Which metal has the lowest unit value?

Have students list the metals traded in order of least expensive to most expensive. (Many students may be surprised to learn where gold ranks in this list.) Have students provide reasons for the ranking based on what they have learned in this chapter.

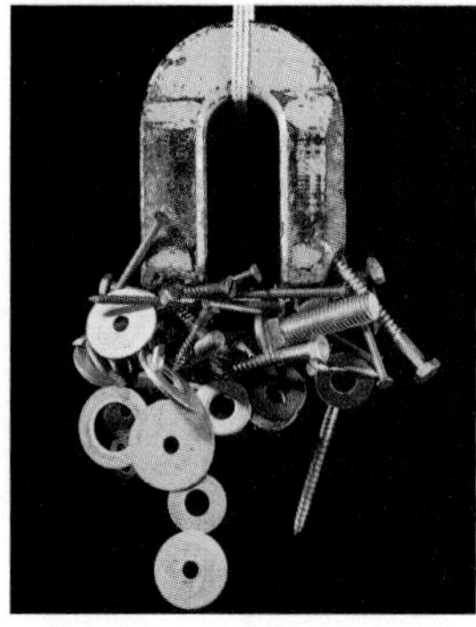

Figure 24.29
The transition metal iron, like cobalt and nickel, is attracted by magnets.

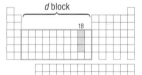

There are two types of steels: carbon steels and alloy steels. Both types actually contain carbon, but carbon steels contain no metal other than iron. About 90% of all steels produced are carbon steels. Carbon steel with less than 0.2% carbon is called mild steel. Mild steel, which is malleable and ductile, is used when load-bearing ability is not important. Medium steels, which contain from 0.2% to 0.6% carbon, are used for structural materials, such as beams and girders. High-carbon steels contain between 0.8% and 1.5% carbon and are used to make drill bits, knives, and other items for which hardness is important.

Alloy steels contain other transition metals in small amounts. Different metals give different properties to the alloy steels. Chromium improves hardness and resistance to corrosion, manganese improves resistance to wear, molybdenum and tungsten increase resistance to heat, nickel adds toughness, and vanadium adds springiness. The wide ranging properties of alloy steels make them useful in many industrial applications.

The alloy steels called stainless steels contain high percentages of chromium and nickel. Stainless steels resist corrosion. What are some uses ❶ of stainless steel? The most common stainless steel contains 18% chromium and 8% nickel.

The world's largest nickel and cobalt deposits are in Ontario, Canada. The two metals occur as sulfide ores, together with copper, silver, iron, platinum, palladium, and iridium. Most of the nickel and cobalt produced today is used in the manufacture of steels. Nickel makes steel more ductile and corrosion resistant and adds toughness. Why are these properties essential ❷ to steel? One isotope of cobalt, cobalt-60, is radioactive and is used in the treatment of certain cancers.

Unlike iron, nickel and cobalt are resistant to atmospheric corrosion. Nickel is often used to electroplate iron and steel objects before they are plated with chromium. **Monel metal,** a strong, corrosion-resistant alloy of nickel and copper, is used to make the propeller shafts of seagoing vessels. Both cobalt and nickel are used as catalysts in industrial processes. For example, finely divided nickel catalyzes the hydrogenation of vegetable oils in the manufacture of butter substitutes.

d block

1B

Copper, Silver, and Gold

Copper, silver, and gold have low chemical reactivity and often occur naturally in the free state, as shown in **Figure 24.30.** These elements are called the coinage metals and were the first metals collected and worked by humans. In areas of the Middle East, copper and gold artifacts dating to about 7000 B.C. have been found.

Copper occurs occasionally as the free metal and was probably the first metal to be reduced from its naturally occurring ores. Some of the most important copper ores are chalcopyrite ($CuFeS_2$), chalcocite (Cu_2S), and cuprite (Cu_2O). The crushed ores are processed into copper of high purity by reduction and electrolytic refining. Electrolytic refining produces 99.99% pure copper.

Copper is the most widely used metal for electrical wiring. Only silver is a better conductor of electricity. Copper is also widely used in plumbing for pipes that carry hot or cold water. **Figure 24.31** illustrates some applications of copper.

Answers

❶ sinks, cutlery, knives and other tools
❷ Steel oxidizes easily and can be brittle. Nickel makes steel more useful.
❸ Answers will vary, but students may suggest jewelry, silverware, and tea sets.

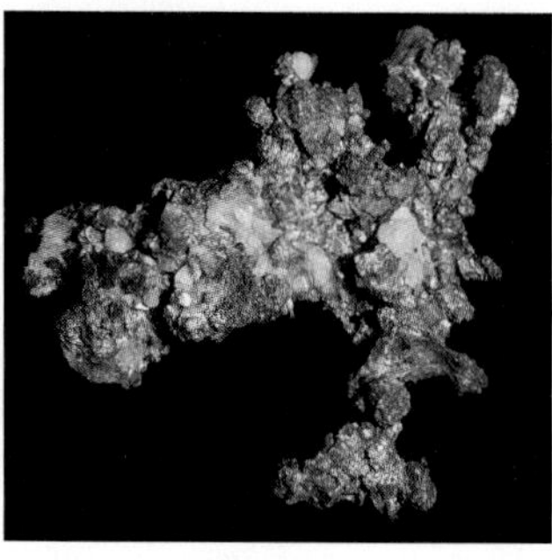

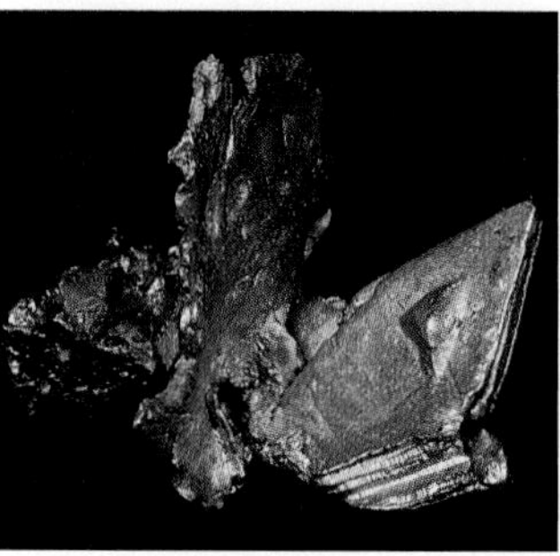

Copper (Cu) Silver (Ag) Gold (Au)

Figure 24.30
Shown above are samples of native copper, silver, and gold.

Brass is an alloy of copper with zinc, and bronze is an alloy of copper with tin. Brass and bronze are among the earliest known alloys. Brass usually contains about 35% zinc and has high ductility and strength. It is used for piping, nozzles, marine equipment, ornaments and jewelry. Bronze usually contains about 5% to 10% tin and is very corrosion resistant. It is used for casting, to make marine equipment and spark-resistant tools, and in fine-arts work.

Silver occurs as the free metal and as the sulfide ore argentite (Ag_2S). Most commercial silver is produced as a by-product in the processing of other ores such as copper, lead, and zinc. Silver is recovered from the muddy layers that collect on the bottom of the electrolysis tanks used in copper refining. Because of its high luster, silver reflects light extremely well and has long been used to coat the backs of mirrors. Silver is an excellent conductor of both electricity and heat. Perhaps you have had the painful experience of using a silver spoon to eat hot soup! What are some uses of silver that are familiar to you? The light-sensitive silver halides are also used in photographic processes. 

Figure 24.31
Copper is still a common coinage metal, though modern pennies are only copper plated. Copper is also used for crush washers or gaskets that deform under large forces to create tight seals. The colorful patina that forms when copper is exposed to weather makes it useful to artists, architects, and designers.

24.3

Discuss

Explain that black-and-white photographic film is coated with an emulsion of solid silver bromide. When exposed to light, Ag^+ ion is reduced to silver metal. The number of silver atoms formed depends on the amount of light that strikes the paper. Those areas receiving the most light become darkest. Once exposed, the film paper is "developed," with a reducing agent that enhances the intensity of the image, and "fixed" with sodium thiosulfate, to remove any unreacted silver bromide. The resulting "negative" image can be converted to a "positive" image by using it as a template through which light is shined onto a new sheet of photo-sensitive paper. The development and fixation steps are then repeated to produce the final photograph.

Meeting Diverse Needs

Gifted Silver is an essential component of color-sensitive films as well as black-and-white films. Have students prepare a report on the chemistry of color photography. Have them find out what light-sensitive reactions are necessary for film to record colors.

Discuss

Point out that the purity of gold is expressed in karats. A karat is equal to 1/24 part. Ten-karat gold is thus 10/24, or roughly 42%, gold; 14-karat gold is roughly 58% gold; and 24-karat gold is 100% gold. The price of gold rises with the number of karats. Pieces with more than 18 karats are generally not used for jewelry because the high gold content makes the alloy soft enough to be easily dented. Various metals may be added to gold to form alloys. The most common is copper, which preserves the yellow color. Metals such as zinc and nickel can be used to produce white gold. Other substances are used to give a rosy or greenish hue. The choice of alloy and color depends on the consumer's taste.

Figure 24.32
Because of its high luster, metallic silver is used to plate mirrors, such as the ones shown here.

Unfortunately, silver is a metal that tarnishes easily. You have probably seen tarnish and possibly even cleaned it off a variety of silver-plated objects such as knives, forks, spoons, jewelry, and picture frames. Tarnish is black silver sulfide (Ag_2S) that forms on the shiny surface of a silver object as a result of a redox reaction that is a form of corrosion. The reaction involves silver combining with hydrogen sulfide (H_2S) in the air. Is silver oxidized or reduced in the reaction? ❶

Gold has been used for thousands of years to make jewelry and ornate artifacts. The gold mask of the Egyptian King Tutankhamen, shown in **Figure 24.33,** was fashioned in about 1350 B.C. Gold occurs chiefly as small particles of the free metal in veins of quartz. Typically, about 5 grams of gold is produced per metric ton of gold-bearing rock. Gold is also recovered as a by-product of copper refining.

Gold is the most malleable and ductile of all metals and can be pounded into sheets so thin that they transmit light. These thin sheets, called gold leaf, are used in decorative lettering and other forms of ornamentation. Corrosion resistance coupled with high electrical and thermal conductivity make gold valuable in high-technology industries. Gold is used to plate the contacts in microcircuits, also shown in **Figure 24.33,** and to cover the external surfaces of satellite components. Gold alloys are used extensively in dentistry, and certain compounds of gold have useful therapeutic applications; for example, salts of Au^+ are used to treat certain kinds of rheumatoid arthritis. Pure gold is soft and is often alloyed to make it harder. A karat system, based on 24 karat = 100%, is used to express the purity of gold. Pure gold is 24 karat. Coinage gold is 22 karat, or $22/24 \times 100 = 92\%$ gold. White gold, which is used in jewelry, is an alloy of gold, nickel, and other metals. ❷ What percentage of gold does 14-karat gold contain?

Figure 24.33
In the past, gold was often used to adorn the burial artifacts of the dead. Gold is now commonly used for the contacts on computer chips and other electronic devices. What properties make gold well-suited for this task? ❸

Answers

❶ oxidized
❷ 58.3% gold
❸ corrosion resistance and high electrical conductivity

Lanthanides and Actinides

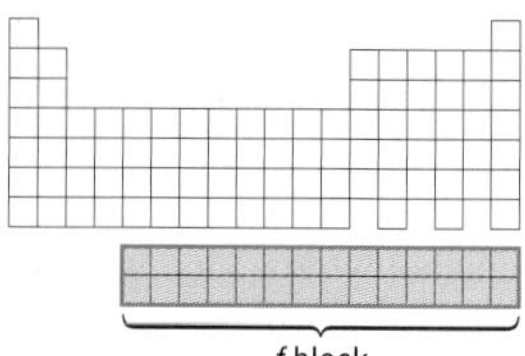

Below the main body of the periodic table is a row of 14 lanthanides, ranging from cerium (Ce, atomic number 58) through lutetium (Lu, atomic number 71). Below that row is another row consisting of 14 actinides, ranging from thorium (Th, atomic number 90) through lawrencium (Lw, atomic number 103). The lanthanides and actinides are called the **inner transition elements** because their seven inner 4*f* and 5*f* orbitals, respectively, are being filled.

The lanthanides are often called the rare-earth elements because it was once believed that they existed only in small quantities. Actually, they are not rare. For example, the lanthanide cerium (Ce) is about five times more abundant than lead. All the lanthanides are silvery, high-melting (800 °C to 1600 °C) metals. The chemical properties of the inner transition elements are so similar that nineteenth-century chemists had great difficulty in separating the metals individually from their ores. Several of the lanthanide oxides are used for tinting sunglasses and welder's goggles and in the manufacture of high-quality camera lenses.

The actinides are radioactive with similar chemical properties. Except for technetium (Tc, 43) and promethium (Pr, 61), elements up to and including uranium (atomic number 92) occur in nature. Elements with atomic numbers greater than 92 are called transuranium elements. These elements are synthesized in particle accelerators, but usually in very small amounts—in some cases, only a few atoms!

Figure 24.34
Lanthanide oxides are sometimes used to make tinted sunglasses, such as these, as well as certain camera lenses. Compounds of europium and ytterbium are used in the picture tubes of color televisions. High-power lasers use neodymium.

section review 24.3

16. Name five transition metals and describe their properties.
17. Describe the range of chemical diversity shown by the transition metals and inner transition metals.
18. Suggest reasons why copper, silver, and gold are valued for the manufacture of electronic devices. Why is copper preferred to silver and gold for home wiring?
19. What is an ore? What are hematite and magnetite?
20. What are the properties and uses of nickel? Of titanium?

 **Chem ASAP! Assessment 24.3** Check your understanding of the important ideas and concepts in Section 24.3.

24.3

3 Assess

Evaluate Understanding

Have students write a brief biography of a transition element. Students should include when the element was discovered, the ore it is refined from, the main method used for its preparation, and its major commercial use. Encourage students to write about those distinctive physical or chemical properties that distinguish the element from other elements.

Alternatively, have students write an autobiography of an element, describing the above mentioned aspects and properties without actually naming the element. Have students trade autobiographies and try to identify the elements based on their descriptions.

Reteach

Prepare a display of selected elements and ores to represent as many *d* and *f* block groups as possible. Also show typical compounds of these elements. Have students view the display and list the obvious properties of the substances and minerals. Discuss the methods used to refine the elements from their ores.

Answers

SECTION REVIEW 24.3

16. Answers will vary.
17. Answers will vary.
18. Copper, silver, and gold are the best electrical conductors. Copper is the least expensive.
19. An ore is a mineral used for the commercial production of a metal. Hematite and magnetite are the most common minerals of iron.
20. Nickel is corrosion resistant and can be magnetized. It is used in alloys and as an industrial catalyst. Titanium is half as dense as iron, strong, and corrosion resistant. The most important compound of titanium is titanium dioxide (TiO_2).

24.4

1 Engage

Use the Visual

Have students study the photograph and read the text that opens the section. Ask:

- **What properties of hydrogen make it so useful as a rocket fuel?** (Liquid hydrogen is lightweight and releases a large amount of energy when burned.)

Have students refer to Table 11.4. Ask:

- **Which produces the most energy when burned: a gram of hydrogen, a gram of graphite, or a gram of sucrose?** (H_2: 141 kJ/g; C: 32.8 kJ/g; $C_{12}H_{22}O_{11}$: 17.0 kJ/g) **Of the fuels listed, hydrogen provides the most energy per gram. This property makes hydrogen one of the most efficient fuels to carry and utilize.**

Check Prior Knowledge

To assess students' knowledge about the properties of hydrogen, ask:

- **How many neutrons, protons, and electrons are in most hydrogen atoms?** (0 neutrons, 1 proton, and 1 electron)
- **How are the isotopes hydrogen-1, hydrogen-2, and hydrogen-3 alike? How are they different? Write the chemical symbol, including the atomic number and mass number, for each.** [Each has 1 proton and 1 electron, so their chemical properties are the same; they have different numbers of neutrons: $^{1}_{1}H$ (0 neutrons), $^{2}_{1}H$ (1 neutron), $^{3}_{1}H$ (2 neutrons).]

section 24.4

objectives

- *Show how hydrogen is unique among the elements*
- *Explain why the noble gases, Group 0, are important even though they are chemically unreactive*

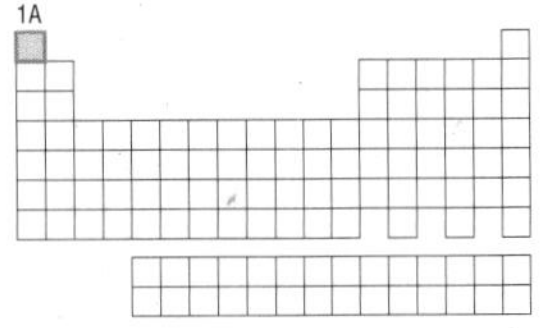

HYDROGEN AND NOBLE GASES

You might expect the fuel used in the main engines of a space shuttle to have a complicated molecular formula that only a rocket scientist would understand. In fact, these engines burn a mixture of hydrogen (the simplest element) and oxygen. **What properties of hydrogen make it so useful as a rocket fuel?**

Hydrogen—A Group by Itself

Hydrogen is the most abundant element in the universe. Free (or elemental) hydrogen is very rare on Earth, but compounds of hydrogen are common. They account for about 1% of Earth's crust. Water is the most abundant hydrogen-containing compound on Earth. Combined with carbon and oxygen, hydrogen is present in all sugars, starches, fats, and proteins. Proteins also contain nitrogen. These complex compounds are abundant in living tissues. Coal, natural gas, and petroleum products, such as gasoline, kerosene, and lubricating oils, also contain hydrogen. **Figure 24.35** shows a wide range of hydrogen-containing items.

Naturally occurring hydrogen and its compounds are composed of three isotopes. The most abundant of the isotopes is protium, commonly referred to simply as hydrogen. Protium (hydrogen-1) and deuterium (hydrogen-2) account for approximately 99.98% and 0.02% of naturally occurring hydrogen samples, respectively. Tritium (hydrogen-3), an unstable (radioactive) form of hydrogen, is present in extremely small amounts.

Hydrogen usually appears at the top of Group 1A in the periodic table. Unlike the true Group 1A elements, however, hydrogen is not a metal nor is it a good conductor of heat or electricity. Like the alkali metals, however, hydrogen does react with the halogens. In some periodic tables, hydrogen appears at the top of Group 7A as well as at the top of Group 1A. There is some justification for placing it in Group 7A because, like the halogens, hydrogen has one fewer valence electron than the noble gas it precedes. Like the halogens, it reacts with the alkali metals and takes on an oxidation number of −1. Thus hydrogen is unique in this regard, behaving like both an alkali metal and a halogen.

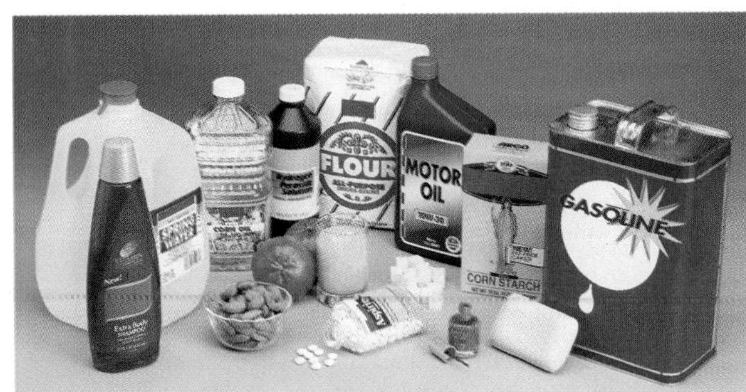

Figure 24.35
Hydrogen is seldom found free in nature but is present in a large number of compounds. All of these items contain compounds of hydrogen.

Student Resources

From the Teacher's Resource Package, use:

- Section Review 24.4 and Ch. 24 Vocabulary Review, Interpreting Graphics, Quizzes, and Tests from the Review Module (Ch. 21–24)
- Laboratory Recordsheet 24-2

Technology Resources

Relevant technology resources include:

- Chem ASAP! CD-ROM
- Resource Pro CD-ROM
- Assessment Resources CD-ROM: Chapter 24 Tests

Figure 24.36
Mixtures of hydrogen and oxygen explode violently when ignited by a spark. The Hindenburg, shown here, was one of 73 hydrogen-filled dirigibles that exploded. The nonflammable gas helium is now used to inflate balloons and blimps.

Hydrogen combines directly with a number of metallic and nonmetallic elements. With chlorine, it forms hydrogen chloride (HCl); with calcium, it forms calcium hydride (CaH_2); with nitrogen, it forms ammonia (NH_3). Hydrogen loses its single electron easily. Does this make it a good oxidizing agent or a good reducing agent? 1

The major use of hydrogen is in the manufacture of ammonia. Large volumes of hydrogen are also used in the conversion of vegetable oils, such as peanut and coconut oil, into solid fats. This process, called hydrogenation, involves treating the oils with hydrogen at high temperature and pressure in the presence of a catalyst. Finely divided nickel or platinum is commonly used. Solid shortenings and margarine are produced in this way. Liquid hydrogen is an important rocket fuel because of its high chemical energy and the fact that it is lightweight. Hydrogen gas was once used to fill lighter-than-air balloons and dirigibles, sometimes with disastrous results, as you can see in **Figure 24.36.** Helium, a nonflammable gas, is now used instead.

The electrolysis of water can produce very pure hydrogen. This method is practical, however, only in small-scale operations. Although the raw material, water, is cheap and plentiful, the cost of the electricity makes the process expensive.

Hydrogen can be prepared commercially from water by the Bosch process. In this method, steam is passed over red-hot iron filings. The iron combines with the oxygen in the steam, and hydrogen gas is liberated.

$$3Fe(s) + 4H_2O(g) \longrightarrow Fe_3O_4(s) + 4H_2(g)$$

More than half of the hydrogen produced in North America is produced by the reaction between steam and natural gas (methane), in a process called steam reforming. Methane reacts with the steam over a finely divided nickel catalyst at 700 °C to 1000 °C.

$$CH_4(g) + H_2O(g) \xrightarrow{\text{catalyst}} CO(g) + 3H_2(g)$$

The carbon monoxide that forms is removed by cooling and compressing the mixture of gases. Hydrogen remains a gas after the carbon monoxide liquefies out of the mixture.

24.4

Discuss

Explain that, chemically, hydrogen can behave as a metal or nonmetal. As a nonmetal, it can react with Group 1A and 2A metals to form *ionic hydrides;* as a metal it can react with a halogen, carbon, nitrogen, or oxygen to form *covalent hydrides.* Ask students to write the formulas of some ionic hydrides (NaH, CaH_2), and some covalent hydrides (CH_4, NH_3, H_2O, HCl). Make sure students understand that the bonding in NaH is ionic and the bonding in HCl is covalent, although each is a strong electrolyte. Compare the dissociation of each of these compounds in water. Write the complete ionic equations on the board. Point out that in one case hydrogen acts as a base, in the other as an acid. Ask:

- **Is the hydride ion (H^-) an acid or a base?** (In fact, H^- is one of the strongest bases known.)
- **What is the conjugate acid of H^-? [$H_2(g)$]**

Answer

1 reducing agent

Meeting Diverse Needs

At Risk When certain transition metals are exposed to hydrogen gas under pressure, hydrogen will diffuse into the metal forming what is sometimes referred to as an *interstitial hydride.* Have students do research to find out how these hydrides are formed and how they could be used for the safe transport and utilization of hydrogen fuel.

Discuss

Point out that the noble gases that form compounds most readily have the larger atoms in which many electron shells shield the nucleus. Because the outer electrons in these atoms are weakly held, they can be "borrowed" more easily by an active element such as fluorine. Have students read the Link to Environmental Awareness. Ask:

- **What properties of radon make it harmful to humans?**
- **What regions of the country experience the highest radon levels? Why?**
- **What methods are used to test for radon gas? Are home-test kits available commercially?**

3 Assess

Evaluate Understanding

Ask:

- **In what sense does hydrogen behave like both a metal and a nonmetal?** (In the chemical sense, hydrogen behaves as a metal when it reacts with a halogen to form compounds in which it exists in the +1 oxidation state, e.g., HCl; hydrogen behaves as a nonmetal when it reacts with an alkali metal to form a compound in which it exists as an anion with a 1− charge.)
- **Name one method for the preparation of hydrogen. What is the hydrogen source used in any method?** (Correct answers include: electrolysis of water, Bosch process, or steam reforming. In each case water serves as the source of hydrogen.)
- **Name and write the chemical symbol for the least massive noble gas.** (He)

ENVIRONMENTAL AWARENESS

Radon Gas

In the 1970s, the noble gas radon generated concern as an environmental hazard. At first, radon was found in homes built on the tailings of uranium and phosphate mines. Further research has shown that there are measurable amounts of radon in most regions of the United States. The highest radon levels are recorded in regions where the bedrock is granite. Radon is a natural product of the decay of radioactive minerals that contain uranium-238. It seeps through cracks in basement floors, and because it is denser than air, it tends to collect in the lower levels of a house. The danger of radon is that high levels in the lungs may increase the risk of lung cancer. Testing homes for radon can be done easily. High levels of radon are reduced by improved ventilation in basements and in closed areas where it collects.

The Noble Gases

When compared with all other elements, the Group 0 elements helium, neon, argon, krypton, xenon, and radon are extremely unreactive. Therefore, these elements are now called noble gases. This name emphasizes their tendency to exist as separate atoms rather than in combination with other atoms.

The name rare gases was originally used to describe the Group 0 elements because they occur in the atmosphere in only very small amounts. Early chemists also called these elements inert gases because they were thought to be completely unable to combine with other elements. In 1962, however, a Canadian chemist named Neil Bartlett prepared xenon tetrafluoride (XeF_4), a compound of xenon. Since that time, compounds of krypton, radon, and argon have also been prepared, but scientists have been unable to react helium and neon.

Despite their low reactivity, the noble gases have many uses. Helium is used to fill weather balloons. Both helium and neon are mixed with oxygen for use in artificial atmospheres, such as those required in deep-sea diving. An artificial atmosphere is preferred to air because decompression sickness, or the bends, is less likely to occur than when nitrogen (a more soluble gas) is present. Argon, krypton, and xenon are used to produce the inert atmospheres needed for photographic flashbulbs and aluminum welding. The noble gases are also used to fill gas discharge tubes for neon signs.

Figure 24.37

Underwater research facilities commonly have an artificial atmosphere consisting primarily of oxygen and a noble gas. Noble gases are also used in advertising signs and light bulbs. Why are light bulbs filled with noble gases rather than with air? ❶

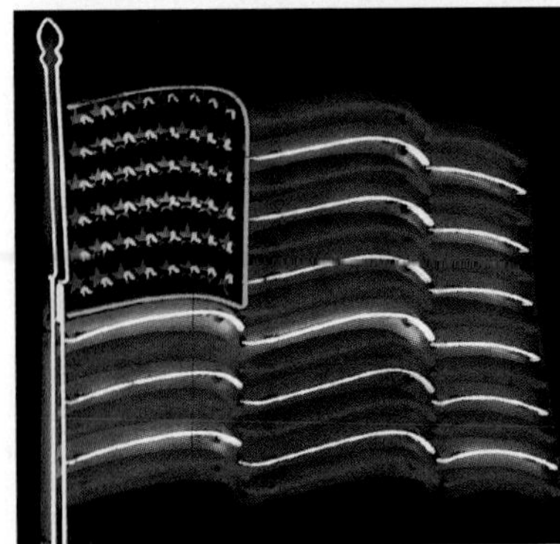

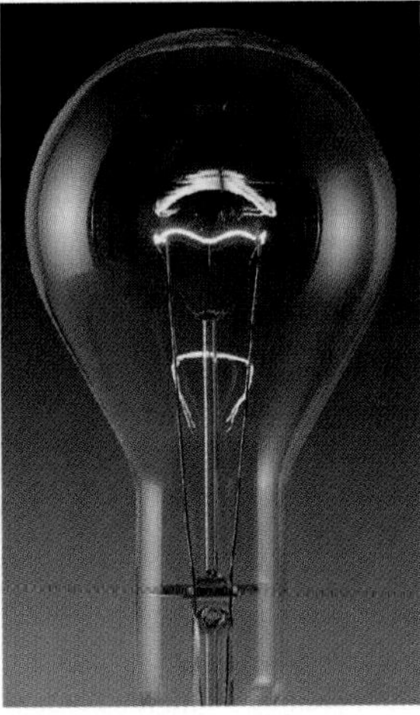

Answers

❶ Unlike oxygen, noble gases will not react with the filaments.

SECTION REVIEW 24.4

21. It behaves somewhat like an alkali metal as well as somewhat like a halogen.
22. They have many uses in areas where unreactive gases are needed.

MINI LAB

Decomposition of Hydrogen Peroxide

PURPOSE

To prepare and identify a gas evolved during the decomposition of hydrogen peroxide.

MATERIALS

- hydrogen peroxide (3% solution from drugstore)
- manganese dioxide
- toothpicks
- candle
- matches
- test tube
- cork to fit test tube (or plastic film wrap)
- tongs

PROCEDURE

1. Fill a test tube about one-third full with the hydrogen peroxide solution.
2. Using the flat end of a toothpick as a spatula, add a small portion of manganese dioxide to the hydrogen peroxide solution.
3. When the bubbling and foaming have stopped, cork the test tube.
4. Hold a toothpick with tongs and light it from a candle flame. When the toothpick is burning well, blow it out, uncork the test tube, and plunge the glowing end of the toothpick into the gas above the liquid in the test tube.

ANALYSIS AND CONCLUSIONS

1. What happened to the glowing toothpick when it was plunged into the gas above the liquid in the test tube?
2. What must be the gas that is released from the hydrogen peroxide solution? Write an equation for this decomposition of hydrogen peroxide.
3. Does the manganese dioxide undergo any apparent change during the reaction?
4. What is the role of manganese dioxide in the decomposition reaction?

section review 24.4

21. In what sense is hydrogen unique among the elements?
22. Why are the noble gases important to industry despite their chemical unreactivity?
23. What are the names, atomic symbols, and mass numbers of the isotopes of hydrogen?
24. The elements in Group 0 have been called rare, inert, and noble. Explain the origin of each of these terms. Are any of these terms misleading based on what you know?
25. What is the largest industrial use of hydrogen? Write an equation for the preparation of hydrogen by the Bosch process.
26. Describe what happens to hydrogen when it combines with chlorine. With calcium. With nitrogen. What common compounds are formed in these reactions?
27. Name an advantage and a disadvantage of producing hydrogen through electrolysis of water.

 Chem ASAP! **Assessment 24.4** Check your understanding of the important ideas and concepts in Section 24.4.

24.4

MINI LAB

Decomposition of Hydrogen Peroxide

SAFETY

Always wear safety goggles. Avoid skin contact with hydrogen peroxide.

ANALYSIS AND CONCLUSIONS

1. The glowing toothpick burns brightly because oxygen fuels the combustion reaction.
2. oxygen gas; $2H_2O_2(aq) \longrightarrow 2H_2O(l) + O_2(g)$
3. No; it should appear the same since it is not used up in the reaction.
4. It catalyzes the decomposition of hydrogen peroxide.

Reteach

Write the chemical equations for the three processes that can be used to produce hydrogen. If possible, show the heats of reaction in each case, or have students calculate these using standard heats of formation. Discuss the nature of the reactions (oxidation–reduction) and show students that a source of electrons is needed to reduce hydrogen in water to hydrogen gas. Compare the energy consumed to produce one mole of hydrogen to the energy released by the combustion of one mole of hydrogen. Ask students to discuss the practical significance of these values.

SECTION REVIEW 24.4 CONTINUED

23. protium (hydrogen-1), ^1_1H, mass number = 1; deuterium (hydrogen-2), ^2_1H, mass number = 2; tritium (hydrogen-3), ^3_1H, mass number = 3
24. Rare described their abundance; inert, their inability to react with other elements; and noble, their natural existence as separate atoms. Helium is the second most abundant element in the universe; noble gases can react to form compounds.
25. the manufacture of ammonia; $3Fe(s) + 4H_2O(g) \rightarrow Fe_3O_4(s) + 4H_2(g)$
26. Hydrogen loses its electron easily. Hydrogen combines with chlorine to form hydrogen chloride. With calcium, hydrogen forms calcium hydride. Ammonia is formed when hydrogen combines with nitrogen.
27. Advantages include an inexpensive and plentiful raw material and the purity of the hydrogen obtained. A disadvantage is the high cost of the electricity needed.

Chemistry Serving... Industry

DISCUSS

After students have read the article, explain that there are many current and potential applications of synthetic diamond films. For example, because a diamond film is "slippery," it could be used as a longer-lasting alternative to the current materials used to create non-stick surfaces on kitchen cookware. In addition, it has superb optical transparency from the infrared through the ultraviolet range of the electromagnetic spectrum. A diamond film is very resistant to scratches and ionizing radiation, making it an excellent protective coating for optical instruments, such as space telescopes, which are exposed to large doses of radiation and collisions with particulate debris. Have students do library research and report on one new discovery made in the field of growing diamonds from vapor since the development of this process in 1987. The written report should include information about new applications that could result from this research.

CHEMISTRY IN CAREERS

Encourage students to connect to the Internet address shown on this page to find out more about careers in gemology. See also the information on page 879 of this text.

Chemistry Serving...Industry

DIAMONDS ARE AN ENGINEER'S BEST FRIEND

Diamond—one of the three allotropes of carbon—is the hardest material known. This property makes it extremely valuable for certain industrial applications, such as cutting, grinding, or shaping other materials. Naturally occurring diamond, however, is rare and very costly. It is formed hundreds of kilometers beneath Earth's surface under conditions of extreme temperature and pressure. Only chance geologic events bring it close enough to the surface to be mined.

Fortunately, scientists have discovered ways of producing synthetic diamond. This material, chemically identical to natural diamond, is used in industry more extensively than natural diamond is.

In one sense, creating synthetic diamond is simple because diamond is made of only carbon, and carbon is extremely easy to obtain. In practice, however, getting carbon atoms to arrange themselves in the tightly packed tetrahedral arrangement of diamond is very difficult.

Diamond's hardness makes it extremely valuable for certain industrial applications, such as cutting, grinding, or shaping other materials.

Thermodynamically, graphite is favored over diamond. If diamond is heated in the absence of air to 1200 °C, its very stable sp^3 bonds will be broken and rearranged into the sp^2 bonds of graphite. To reverse this reaction—or to make diamond from graphite—even higher temperatures and very high pressures are required.

The first attempt to create the necessary conditions for forming diamond from graphite was a failure. Researchers created a chamber that could maintain a pressure of 1.5 million pounds per square inch and a temperature of 2500 °C. This temperature, however, was not high enough to melt the graphite.

Then the researchers tried another approach: They added small amounts of a metal with a melting point below 2500 °C. The graphite dissolved in the molten metal and then crystallized out as diamond. The diamond crystals were only about 1 mm in size, but they were as hard as natural diamonds. Today, this high-pressure, high-temperature (HPHT) method is still used to make large quantities of synthetic diamond, mostly for use as industrial abrasives.

After the first synthetic diamonds were created, researchers explored other ways of growing diamonds. One method that has become very important in industry involves using carbon vapor to grow films of diamond. The carbon can come from various sources, including methane (CH_4) or graphite. The carbon is vaporized with intense heat or with the energy from microwaves, laser beams, or electric discharges. The vaporized carbon is then deposited as an ice-like film on diamond "seeds" or on other substances such as quartz or silicon.

Diamond films have many uses. A diamond coating can make a lens durable and scratch-proof. Cutting tools coated with diamond never need sharpening. Diamond's unmatched ability to conduct heat also makes diamond films useful in many other applications, such as in the manufacture of extremely fast computer chips.

Natural diamonds may always be consumers' choice for adornment, but synthetic diamonds are the preference of industry.

CHEMISTRY IN CAREERS

GEMOLOGIST

Grade the quality of valuable gems using your knowledge of crystal structure.

See page 879.

CHEMIST NEEDED

Enviro. Lab. GC Must know

TAKE IT TO THE NET

Find out more about career opportunities. **www.phschool.com**

CHEMICAL SPECIALIST

Local food service distributor see responsible self-motivated indiv

STUDENT STUDY GUIDE

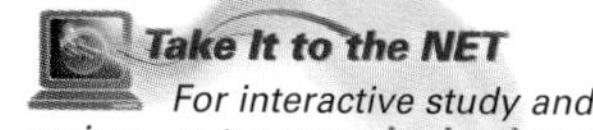

For interactive study and review, go to **www.phschool.com**

KEY TERMS

- inner transition element *p. 731*
- lime *p. 710*
- metallurgy *p. 724*
- Monel metal *p. 728*
- ore *p. 724*
- slaked lime *p. 710*

KEY EQUATIONS

- Preparation of sulfuric acid:
  $2SO_2(g) + O_2(g) \xrightleftharpoons{\text{catalyst}} 2SO_3(g)$
  $SO_3(g) + H_2O(l) \longrightarrow H_2SO_4(aq)$
- Bosch process:
  $3Fe(s) + 4H_2O(g) \longrightarrow Fe_3O_4(s) + 4H_2(g)$
- Haber-Bosch process:
  $N_2(g) + 3H_2(g) \xrightleftharpoons{\text{Fe}} 2NH_3(g)$

CONCEPT SUMMARY

24.1 The *s*-Block Elements: Active Metals

- The alkali and alkaline earth metals have low ionization energies and high reactivities.
- Metallic sodium is produced by electrolysis. It is used in sodium vapor lamps and in the production of many chemicals.
- Calcium is produced by electrolysis. Calcium oxide, or lime, is an industrial chemical.
- Magnesium is produced from sea water. It is an important structural material.

24.2 The *p*-Block Elements: Metals and Nonmetals

- Aluminum is the most abundant metal in Earth's crust. It is not found as a free metal. Its value in industry derives from its corrosion resistance and high electrical conductivity.
- Carbon and silicon are the basic elements of life and minerals, respectively. Tin and lead have numerous uses as both free metals and in alloys such as bronze and solder.
- Nitrogen exists as N_2. It makes up 80% of the air by volume. Two important nitrogen compounds are ammonia and nitric acid.
- Oxygen is the most abundant element in Earth's crust. It exists in the atmosphere mainly as O_2.
- Sulfur readily forms compounds with most metals and nonmetals. Sulfuric acid is an important industrial chemical.
- The halogens exist in nature as halide ions found in sea water and salt beds.

24.3 The *d*- and *f*-Block Elements: Transition and Inner Transition Metals

- Many transition metals are used to make alloys. Monel metal is a corrosion-resistant alloy of copper and nickel. Copper, silver, and gold are excellent conductors.
- The lanthanides and actinides are the inner transition metals. The lanthanides are silvery, high-melting metals. The actinides are radioactive.

24.4 Hydrogen and Noble Gases

- Free hydrogen is rare on Earth. The three isotopes of hydrogen are protium, deuterium, and tritium. Hydrogen is used as a reducing agent, in ammonia production, and in the hydrogenation of vegetable oils.
- The noble gases are chemically unreactive. They are used in weather balloons, artificial atmospheres, and gas discharge tubes.

CHAPTER CONCEPT MAP

Use these terms to construct a concept map that organizes the major ideas of this chapter.

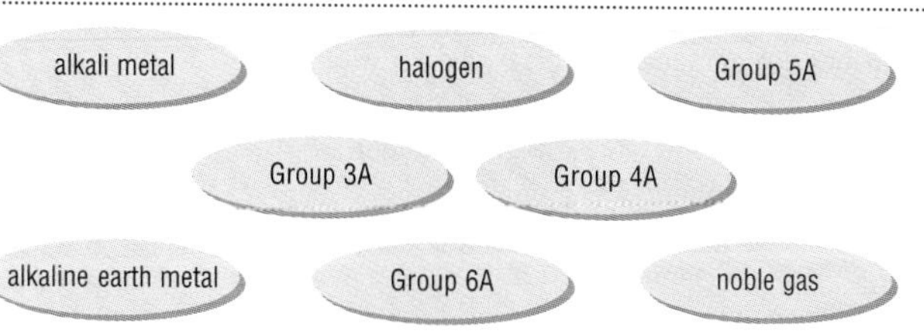

 **Chem ASAP! Concept Map 24**
Create your Concept Map using the computer.

Take It to the Net

At **www.phschool.com** students will find for this chapter

- an Internet activity
- links to related chemistry sites
- an interactive quiz
- career links

4 Close

Summary

Assign students to teams to construct a concept map using the terms shown on the bottom of this page. Have teams amend their maps to include materials discussed in Section 24.3. Then have each team decide on an element to represent each of the eight categories shown. Have each team present its choices and describe some of the important properties and uses of each element chosen.

Extension

Have students choose an essential trace element and conduct library research on its biological role. Students should write a brief report outlining its overall prevalence in the body in relation to other trace elements, important metabolic functions, and possible explanations as to why that particular element is well suited for those functions. Examples for study include zinc, cobalt, and copper.

Looking Back . . . Looking Ahead . . .

The previous chapter discussed the interconversion between electrical energy and chemical energy. This chapter presents an overview of the properties of elements from each block of the periodic table. In the next chapter, students are introduced to organic chemistry. They begin by studying the structure of hydrocarbons.

Answers

28. $2Na(s) + O_2(g) \rightarrow Na_2O_2(s)$
$2Na(s) + 2H_2O(l) \rightarrow$
$2NaOH(aq) + H_2(g)$

29. Group 1A: alkali metals; Group 2A: alkaline-earth metals

30. calcium oxide (CaO); in a kiln at 900 °C

31. Their atomic diameters are smaller.

32. $Ca(s) + 2H_2O(l) \rightarrow Ca(OH)_2(s) + H_2(g)$;
$CaO(s) + H_2O(l) \rightarrow Ca(OH)_2(s)$

33. bauxite

34. Aluminum has strength and ductility, low density, high electrical conductivity, and high corrosion resistance.

35. diamond, graphite, and buckminsterfullerene

36. colorless gas, strong odor, relatively high boiling and melting point, high heat of vaporization, very soluble in water

37. Nitric acid is used in dye and fertilizer manufacture, etching, and production of explosives.

38. Phosphorus occurs in white and red forms; the white form is more reactive.

39. Plants cannot directly incorporate atmospheric nitrogen (N_2) into their tissues. Nitrogen-fixing bacteria must transform it into nitrogen compounds.

40. Ammonia is used to manufacture products such as fertilizers, cleaning products, and nitric acid. It is also a refrigerant.

41. The three conditions are a temperature of 500 °C, a pressure of 10^5 kPa, and an iron oxide catalyst.

42. Ammonia (NH_3), ammonium sulfate ($(NH_4)_2SO_4$), and ammonium nitrate (NH_4NO_3) are used as fertilizers.

43. **a.** manufacturing of steel
b. to produce sulfuric acid
c. to produce photoelectric cells and other light-sensitive instruments

44. **a.** colorless, odorless, gaseous
b. pale-blue, odorous, gaseous
c. pale-yellow, tasteless, odorless, brittle, solid

45. Hydrogen peroxide is used as a bleach and an antiseptic.

46. Oxygen is found as an atmospheric gas, in water molecules, and in compounds found in rocks and soils.

47. 20%

48. Paramagnetism is the magnetic attraction caused by unpaired electrons. Molecules of liquid oxygen appear to have two unpaired electrons, one on each oxygen atom, and are weakly attracted to a magnet.

49. the manufacture of steel

50. Ozone is produced in Earth's upper atmosphere.

51. **a.** $2Mg(s) + O_2(g) \rightarrow 2MgO(s)$
b. $2H_2(g) + O_2(g) \rightarrow 2H_2O(g)$
c. $S(s) + O_2(g) \rightarrow SO_2(g)$

52. Stable rhombic crystals contain S_8 rings.

53. Sulfur is used mostly for the production of sulfuric acid, which is used in manufacturing fertilizers, in the steel industry, in petroleum

Chapter 24 REVIEW

CONCEPT PRACTICE

28. When sodium is exposed to oxygen, it forms a peroxide. A solution of sodium hydroxide can be obtained by adding sodium to water. Write balanced equations for these two reactions. *24.1*

29. What are the group names for the elements in Group 1A and Group 2A? *24.1*

30. What is lime? How is lime made from calcium carbonate ($CaCO_3$)? *24.1*

31. Why do the alkaline-earth metals as a group have higher densities than the alkali metals? *24.1*

32. Give the equations for two reactions by which $Ca(OH)_2$ can be made. *24.1*

33. Name an aluminum ore rich in Al_2O_3. *24.2*

34. Describe the physical properties of aluminum that make it a commercially valuable metal. *24.2*

35. Name the three allotropes of carbon. *24.2*

36. Give four physical properties of ammonia. *24.2*

37. What are three uses of nitric acid? *24.2*

38. What are two forms of elemental phosphorus? How do they differ in their reactivity with oxygen? *24.2*

39. Why is atmospheric nitrogen not directly useful to plants? What makes it possible for nitrogen to be used by plants? *24.2*

40. List major uses of ammonia in industry. *24.2*

41. What are the three conditions of the Haber-Bosch process that make the commercial production of ammonia possible? *24.2*

42. Give the names and formulas of two nitrogen-containing substances used as fertilizers. *24.2*

43. Name at least one industrial use of each. *24.2*
a. oxygen **b.** sulfur **c.** selenium

44. Give three physical properties of each of the following substances. *24.2*
a. oxygen **b.** ozone **c.** sulfur

45. List some uses of hydrogen peroxide. *24.2*

46. In what three major forms is oxygen present on Earth? *24.2*

47. What is the volume percentage of oxygen in Earth's atmosphere? *24.2*

48. Explain paramagnetism. Use liquid oxygen as your example. *24.2*

49. What is the largest industrial use of oxygen? *24.2*

50. Where is ozone produced naturally? *24.2*

51. Complete and balance these equations. *24.2*
a. $Mg(s) + O_2(g) \longrightarrow$
b. $H_2(g) + O_2(g) \longrightarrow$
c. $S(s) + O_2(g) \longrightarrow$

52. Describe the common allotrope of sulfur that is stable at room temperature. *24.2*

53. What are some uses of sulfur? *24.2*

54. Give the names and molecular formulas of the halogens. *24.2*

55. Why is chlorine added to drinking water and swimming pools? *24.2*

56. How is chlorine gas normally prepared for commercial use? *24.2*

57. Describe how iodine is obtained commercially. *24.2*

58. What process is used to make fluorine? *24.2*

59. Give colors and physical states at STP of chlorine, bromine, and iodine. *24.2*

60. Write formulas for the oxides of iron. *24.3*

61. Nickel is an important metal. Why? What is Monel metal? *24.3*

62. Copper, silver, and gold are among the few metals that have been used for thousands of years. Explain why. *24.3*

63. Reduction—not oxidation—is required to convert the metal in a compound to the free metal. Why? *24.3*

64. Name three practical uses of various noble gases. *24.4*

65. Explain how superheated water and compressed air are used in sulfur recovery by the Frasch process. What conditions prevent sulfur from being mined in the same way coal is? *24.4*

66. Large volumes of hydrogen are used in the hydrogenation of vegetable oils. Describe the process, and comment on its commercial value. *24.4*

CONCEPT MASTERY

67. Calculate the mass percent of aluminum in cryolite (Na_3AlF_6).

68. At which electrode is the free metal produced in the electrolysis of a metal compound? Explain your answer.

69. The following data table shows the quantities of iron ore (Fe_2O_3) and coke (C) that react in a blast furnace to produce pig iron.

Mass of Fe_2O_3 (kg)	Amount of Fe_2O_3 (mol)	Mass of C (kg)	Amount of C (mol)
8.65×10^4		1.95×10^4	
1.26×10^5		2.84×10^4	
2.01×10^5		4.54×10^4	
6.56×10^5		1.48×10^5	
9.61×10^5		2.17×10^5	

a. Complete the table.
b. Plot moles of iron ore (y-axis) versus moles of carbon (x-axis).
c. From the slope of the line, determine the mole ratio by which iron reacts with carbon in a blast furnace.

70. A commercially important source of titanium is ilmenite ($FeTiO_3$). Calculate the percent by mass of titanium in ilmenite.

71. Calculate the percent by mass of cobalt in cobaltite (CoAsS). How many kilograms of cobalt would be contained in 1.00×10^3 kg of an ore containing 72.6% CoAsS by mass.

72. Describe the process used to obtain aluminum from its natural source.

73. Name the two allotropic forms of the element oxygen. Describe one method of preparation for each.

74. Distinguish between oxygen and ozone.

75. Write a balanced equation for the reaction of hydrogen with these elements.
a. nitrogen **b.** chlorine **c.** calcium

76. List the halogens in order of increasing electronegativity.

77. List some of the major uses of sulfuric acid.

78. The following data table lists the melting and boiling points of the halogens.

Element	Melting point (°C)	Boiling point (°C)
F	−219	−188
Cl	−107	−34
Br	−7	58
I	113	184

a. Describe the trends you observe.
b. Explain these trends. Are they likely to occur with other groups in the periodic table?

79. Explain why ammonia exhibits many waterlike properties.

80. Name a nonmetallic oxide that reacts with water to form a base and a metallic oxide that reacts with water to form an acid. Write a balanced equation for each reaction.

CRITICAL THINKING

81. Choose the term that best completes the second relationship.
a. sister : brother
beryllium : _______
(1) cesium (3) barium
(2) lead (4) sodium
b. protium : deuterium
nitrogen-14 : _______
(1) isotope (3) carbon-14
(2) tritium (4) nitrogen-15

82. Hydrogen and helium are by far the most abundant elements in the universe. Why are helium and the free, or elemental, form of hydrogen very rare on Earth?

83. Compare the methods used to produce hydrogen. What are the disadvantages of each for producing hydrogen as a fuel?

84. What advantage does the recycling of aluminum have over the production of aluminum from its minerals?

85. The period of human history called the Bronze Age began about 3000 B.C. and was followed by the Iron Age. Why was bronze widely used for tools and weapons before iron?

86. The histogram below classifies chlorides of Pb, Cu, Co, Ni, Mn, Ca, and Fe by percent chloride composition. Give the chemical formula for each compound shown.

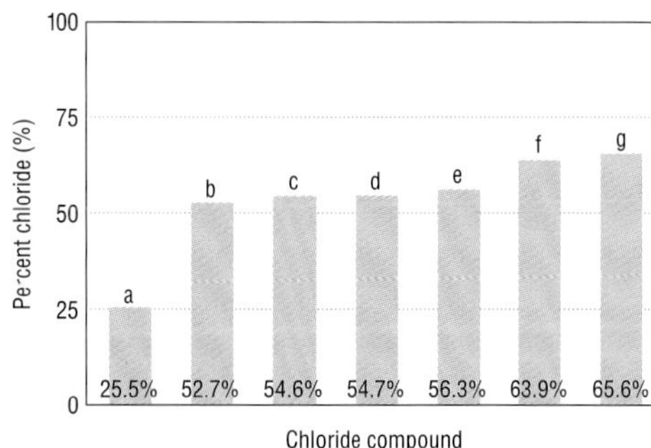

refining, and in many other industries.
54. fluorine (F_2); chlorine (Cl_2); bromine (Br_2); iodine (I_2); astatine (At_2)
55. Chlorine in solution is a strong oxidizing agent that kills disease-causing bacteria.
56. Chlorine is produced by the electrolysis of aqueous or molten sodium chloride.
57. Iodine is obtained by reacting sodium iodate with sodium hydrogen sulfite solution to precipitate iodine crystals.
58. Fluorine is obtained by the electrolysis of an ice-cold solution of potassium fluoride dissolved in hydrogen fluoride.
59. Cl_2: yellow-green gas; Br_2: dark-red liquid; I_2: purple-black crystalline solid
60. Fe_2O_3 (hematite); Fe_3O_4 (magnetite)
61. Nickel is used to electroplate iron and steel and to catalyze industrial reactions. Monel metal is a strong, corrosion-resistant alloy of nickel and copper.
62. They occur in the free state and were collected by people in ancient times.
63. Reduction converts positively charged metal ions in a compound to the free metal.
64. lighter-than-air balloons, artificial and inert atmospheres, and gas discharge tubes
65. Superheated water and compressed air are sent down two tubes into the sulfur beds. A third tube brings sulfur froth to the surface. Sulfur is not mined because it is found under quicksand.
66. In hydrogenation, oil is treated with hydrogen gas at a high temperature and pressure in the presence of a catalyst. This process converts liquid oils into solid fats. Hydrogenated vegetable oils are found in many foods and constitute a large market.
67. 12.9% Al
68. When a free metal is made from its ion, reduction occurs at the cathode.
69. a. Amounts of Fe_2O_3: 5.42×10^5, 7.89×10^5, 1.26×10^6, 4.11×10^6, 6.02×10^6
Amounts of C: 1.63×10^6, 2.37×10^6, 3.78×10^6, 1.23×10^7, 1.81×10^7
b. Coordinates on students' graphs should be plotted to reflect answers in (a); line should be straight.
c. slope = 1 mol Fe_2O_3/3 mol C
70. 31.6% Ti
71. 35.5% Co in CoAsS; 2.58×10^2 kg Co
72. Aluminum is refined by electrolysis of a solution of Al_2O_3 in cryolite.
73. O_2 and O_3; oxygen gas (O_2) is obtained from liquid air by fractional distillation. Ozone (O_3) is made by passing an electrical discharge through oxygen gas.
74. Oxygen (O_2) is stable, odorless, and colorless. Ozone (O_3) is unstable, has a sharp odor, and is pale blue.

CUMULATIVE REVIEW

87. Write the formula of each compound.
a. calcium oxide **d.** aluminum oxide
b. mercury(II) oxide **e.** sulfur dioxide
c. carbon monoxide **f.** sodium peroxide

88. Calculate the mass percent of iron in each mineral compound.
a. Fe_3O_4 **b.** Fe_2O_3

89. Write complete electron configurations for each.
a. Fe **c.** Ag **e.** Cu **g.** Zn
b. V **d.** Fe^{3+} **f.** Ni^{2+} **h.** Ag^+

90. A sample of spring water contains 46.0 mg of magnesium ions per liter.
a. What is the molarity of Mg^{2+} in the sample?
b. What volume of this water contains 1.00 mol Mg^{2+}?

91. What ions are produced when each substance is dissolved in water?
a. LiBr **b.** $Ca(OH)_2$ **c.** KI **d.** $CdBr_2$

92. Classify each compound as a nonelectrolyte, a weak electrolyte, or a strong electrolyte.
a. CH_3COCH_3 (acetone) **d.** KNO_3
b. C_2H_5OH (ethyl alcohol) **e.** Na_2SO_4
c. CH_3COOH (acetic acid) **f.** $CaCl_2$

93. Determine the oxidation number of nitrogen in each.
a. N_2 **b.** NO_3^- **c.** NO_2 **d.** N_2O **e.** NO_2^-

94. Identify the oxidizing agent and the reducing agent in the following reaction.

$$I_2O_5(s) + 5CO(g) \longrightarrow 5CO_2(g) + I_2(s)$$

95. Calcium carbonate is used as an antacid to neutralize HCl in the stomach. Write the equation for the neutralization reaction.

CONCEPT CHALLENGE

96. Calculate the mass, in kilograms, of sulfur dioxide produced in the roasting of 1000 kg of chalcocite ore containing 7.2% Cu_2S, 0.6% Ag_2S, and no other sulfur compounds.

97. A 0.50-g sample of metallic sodium is converted to NaCl by reaction with chlorine gas. What volume of Cl_2 at 20 °C and 98.6 kPa will be required to react completely with the sodium?

98. Azurite, a dark-blue copper mineral, has the following composition: Cu 55.3%, C 6.97%, O 37.1%, H 0.585%. Determine the simplest formula of this mineral.

99. Use this equation to answer the questions.

$$Mg(OH)_2 + 2HCl \longrightarrow MgCl_2 + 2H_2O$$

a. How many moles of H^+ are required to react with 3.20 g $Mg(OH)_2$?
b. What volume of 3.00*M* HCl is required to react with 3.20 g $Mg(OH)_2$?

100. Barium peroxide decomposes above 700 °C, according to this equation.

$$2BaO_2(s) \longrightarrow 2BaO(s) + O_2(g)$$

The oxygen liberated when 20.0 g of barium peroxide is heated is collected in a 1-L flask at 25 °C. Calculate the pressure in the flask.

101. How many kilograms of Br^- are in a cubic mile of sea water?

102. What mass of sulfur dioxide is produced when 2.00×10^2 g of sulfur is burned in oxygen?

103. Hydrogen peroxide decomposes according to the equation:

$$2H_2O_2(l) \longrightarrow 2H_2O(l) + O_2(g)$$

A solution is 3.00% H_2O_2 by volume. The density of H_2O_2 is 1.44 g/cm³. How many liters of oxygen gas, measured at STP, are produced when 1 L of 3.00% H_2O_2 is decomposed?

104. In the following graph, the SO_2 emissions of coal-burning electric plants in a recent year are shown to vary according to generating capacity and plant age. In the year shown, all coal-burning electric plants emitted approximately 13 million tons of SO_2.

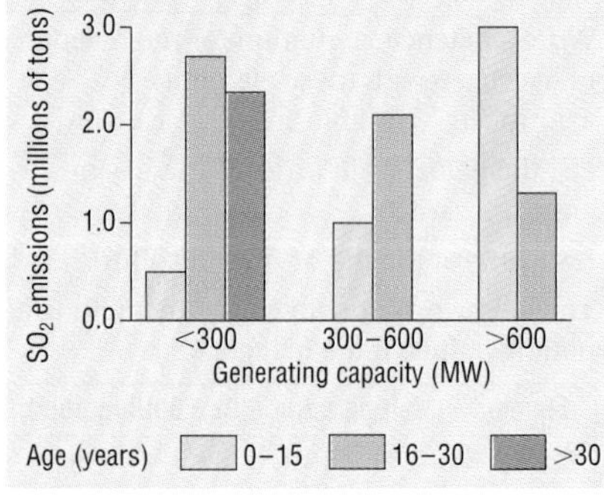

a. What are the three types of plants with the largest SO_2 emissions?
b. Approximately what percent of the year's emissions were due to the three types of plants identified in question **a**?

Answers

75. a. $3H_2(g) + N_2(g) \rightarrow 2NH_3(g)$
b. $H_2(g) + Cl_2(g) \rightarrow 2HCl(g)$
c. $H_2(g) + Ca(s) \rightarrow CaH_2(s)$

76. I, Br, Cl, F

77. Sulfuric acid is used to prepare fertilizers and to pickle iron and steel. It is also used in petroleum refining and in many other industries.

78. a. Both melting points and boiling points increase with increasing atomic mass.
b. The larger number of electrons in larger atoms results in stronger molecular interactions between molecules. Therefore, melting and boiling points should increase in moving down any group on the periodic table.

79. Ammonia is a small, polar molecule with hydrogen bonding. It is covalently bonded, with bond angles similar to those in water.

80. Answers will vary.
nitrogen dioxide: $3NO_2(g) + H_2O(l) \rightarrow 2HNO_3(aq) + NO(g)$
calcium oxide: $CaO(s) + H_2O(l) \rightarrow Ca(OH)_2(s)$

81. a. 3 **b.** 4

82. Earth is not massive enough to hold these lightweight elements, except as compounds in the case of hydrogen. The stars, however, are made up mostly of hydrogen and helium and can hold on to them because of their much larger mass and stronger gravitational pull.

83. All the processes require an input of energy, either electricity or heat. Steam reforming produces poisonous carbon monoxide and requires the use of a nonrenewable resource, natural gas. The method using steam and white-hot coke produces carbon dioxide, which contributes to global warming.

84. The production of aluminum from it minerals requires enormous amounts of electrical energy. Recycling aluminum metal would probably be less expensive. The environmental pollution caused by generating electricity would be reduced.

85. Bronze is an alloy mostly of copper and tin. The technology for producing copper and tin was developed relatively easily. The separation of iron from its ore, however, requires much greater heat and more-advanced technology.

86. a. $PbCl_2$ **d.** $NiCl_2$ **g.** $FeCl_3$
b. $CuCl_2$ **e.** $MnCl_2$
c. $CoCl_2$ **f.** $CaCl_2$

87. a. CaO **c.** CO **e.** SO_2
b. HgO **d.** Al_2O_3 **f.** Na_2O_2

88. a. 72.4% Fe **b.** 69.9% Fe

89. a. $[Ar]3d^64s^2$ **e.** $[Ar]3d^{10}4s^1$
b. $[Ar]3d^34s^2$ **f.** $[Ar]3d^8$
c. $[Kr]4d^{10}5s^1$ **g.** $[Ar]3d^{10}4s^2$
d. $[Ar]3d^5$ **h.** $[Kr]4d^{10}$

STANDARDIZED TEST PREP

Select the choice that best answers each question or completes each statement.

1. The *s*-block elements are the
   a. halogens and noble gases.
   b. alkali metals and alkaline earth metals.
   c. alkaline earth metals and transition metals.
   d. transition metals and inner transition metals.
2. Which of these metals would react most vigorously with water?
   a. aluminum c. cesium
   b. zirconium d. magnesium
3. The most reactive nonmetallic element is
   a. helium. c. chlorine.
   b. oxygen. d. fluorine.
4. Variations in ionic charge are found in the
   a. noble gases.
   b. transition metals.
   c. alkali metals.
   d. alkaline earth metals.
5. The elements known for their limited chemical reactivity are the
   a. inner transition metals.
   b. halogens.
   c. noble gases.
   d. *d*-block elements.
6. An allotrope of which element protects Earth from ultraviolet radiation?
   a. oxygen c. sulfur
   b. carbon d. helium

Questions 7–9 describe three different elements labeled X, Y, and Z, respectively. Use the descriptions to identify the elements.

7. Element X has several allotropes. Herman Frasch devised a method for mining underground deposits of element X. An oxide of element X reacts to produce an acid used in almost every industry.
8. Element Y often is found uncombined in nature. It is an excellent conductor of both electricity and heat. Element Y darkens when it reacts with hydrogen sulfide gas. Light-sensitive halides of element Y are used to coat photographic film.
9. Element Z is the second most abundant element in Earth's crust. It is an insulator at low temperatures and a conductor at high temperatures. A highly pure form of element Z is used to make computer chips.

The lettered choices below refer to questions 10–13. A lettered choice may be used once, more than once, or not at all.

(A) alkaline earth metal
(B) noble gas
(C) halogen
(D) transition metal
(E) alkali metal

Which of the above element types describes each of the following elements?

10. potassium
11. iodine
12. xenon
13. molybdenum

Use the atomic windows to answer question 14.

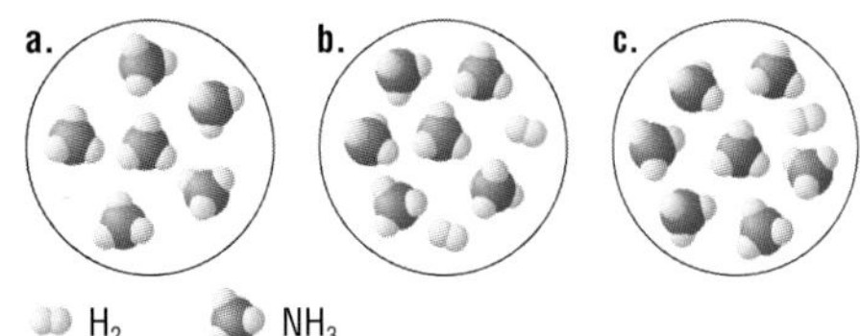

14. Ammonia is produced by the Haber-Bosch process. The overall equation is
    $N_2(g) + 3H_2(g) \longrightarrow 2NH_3(g)$
    Which of the windows shows the correct result when 3 molecules of nitrogen react with 11 molecules of hydrogen?

Methane gas (CH_4) reacts with steam at a high temperature to produce carbon monoxide gas and hydrogen gas. Use this information to answer questions 15 and 16.

15. Write a balanced equation for the reaction.
16. *You Draw It!* Using the equation from question 15 and the symbols provided below, draw the atomic window that shows the result of reacting 2 molecules of methane with 3 molecules of steam.

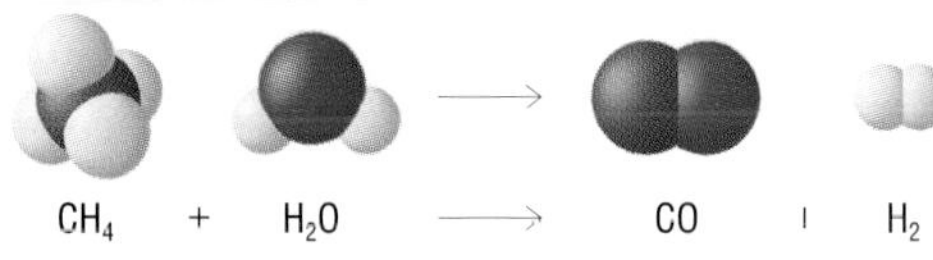

CHAPTER 24 STANDARDIZED TEST PREP

1. b
2. c
3. d
4. b
5. c
6. a
7. sulfur
8. silver
9. silicon
10. (E)
11. (C)
12. (B)
13. (D)
14. b
15. $CH_4(g) + H_2O(g) \rightarrow CO(g) + 3H_2(g)$
16. The window would contain two carbon monoxide molecules, six hydrogen molecules, and one water molecule.

90. **a.** $1.89 \times 10^{-3} M\ Mg^{2+}$ **b.** 5.29×10^2 L
91. **a.** Li^+, Br^- **b.** Ca^{2+}, OH^- **c.** K^+, I^- **d.** Cd^{2+}, Br^-
92. **a., b.:** nonelectrolyte; **c.** weak electrolyte **d., e., f.:** strong electrolyte
93. **a.** 0 **b.** +5 **c.** +4 **d.** +1 **e.** +3
94. I_2O_5, oxidizing agent; CO, reducing agent
95. $CaCO_3(s) + 2HCl(aq) \rightarrow CaCl_2(aq) + H_2O(l) + CO_2(g)$
96. 3×10^1 kg SO_2
97. 2.7×10^{-1} L Cl_2
98. $Cu_3C_2O_8H_2$
99. **a.** 1.10×10^{-1} mol H^+ **b.** 3.67×10^1 mL
100. 1.46×10^2 kPa
101. 2.8×10^8 kg Br^-
102. 4.00×10^2 g SO_2
103. 1.42×10^1 L O_2
104. **a.** greater than 600 MW, less than 15 years old; less than 300 MW, 16–30 years old; less than 300 MW; greater than 30 years old **b.** about 65%

Planning Guide

SECTION OBJECTIVES	ACTIVITIES/FEATURES	MEDIA & TECHNOLOGY
25.1 Hydrocarbons ● ■ ◆ ▸ Describe the bonding in hydrocarbons ▸ Distinguish between straight-chain and branched-chain alkanes	SE **Discover It!** *What Dissolves What?*, p. 742 LM 49: *Hydrocarbons: A Structural Study* SSLM 36: *Molecular Structure of Hydrocarbons* TE **DEMOS,** pp.744, 745, 746, 750	ASAP Animation 28 ASAP Problem Solving 4, 6 ASAP Assessment 25.1 CA *Methane Bubbles*
25.2 Unsaturated Hydrocarbons ○ ■ ◆ ▸ Explain the difference between unsaturated and saturated hydrocarbons ▸ Differentiate between the structures of alkenes and alkynes	LM 49: *Hydrocarbons: A Structural Study*	ASAP Assessment 25.2 CHM Side 10, 36: *Unsaturated Hydrocarbons*
25.3 Isomerism □ ◆ ▸ Distinguish among structural, geometric, and stereoisomers ▸ Identify the asymmetric carbon or carbons in stereoisomers	SE **Mini Lab** *Structural Isomers of Heptane*, p. 757 (**LRS** 25-1) SE **Small-Scale Lab** *Hydrocarbon Isomers*, p. 758 (**LRS** 25-2)	ASAP Simulation 28 ASAP Assessment 25.3 CA *Carbide Cannon* OT 73: *Isomerism*
25.4 Hydrocarbon Rings ○ □ ◆ ▸ Identify common cyclic ring structures ▸ Explain resonance in terms of the aromatic ring of benzene		ASAP Assessment 25.4
25.5 Hydrocarbons from the Earth ● ■ ◇ ▸ Identify three important fossil fuels and describe their origins ▸ Name some products obtained from natural gas, petroleum, and coal	SE **Link to Ecology** *Oil Spills*, p. 764 SE **Chemistry Serving . . . the Consumer** *A Number You Can't Knock*, p. 766 SE **Chemistry in Careers** *Organic Chemist*, p. 766	ASAP Assessment 25.5 ASAP Concept Map 25 CHM Side 10, 10: *Petroleum* OT 74: *Petroleum Distillation* PP Chapter 25 Problems RP Lesson Plans, Resource Library AR Computer Test 25 www Activities, Self-Tests, *SCIENCE NEWS* updates TCP The Chemistry Place Web Site

KEY

● Conceptual (concrete concepts)	AR Assessment Resources	GRS Guided Reading and Study Workbook
○ Conceptual (more abstract/math)	ASAP Chem ASAP! CD-ROM	LM Laboratory Manual
■ Standard (core content)	ACT ActivChemistry CD-ROM	LP Laboratory Practicals
□ Standard (extension topics)	CHM CHEMedia Videodiscs	LRS Laboratory Recordsheets
◆ Honors (core content)	CA Chemistry Alive! Videodiscs	SSLM Small-Scale Lab Manual
◇ Honors (options to accelerate)	GCP Graphing Calculator Problems	

PRACTICE		ASSESSMENT	
SE	**Sample Problems** 25-1 to 25-3	SE	Section Review
GRS	Section 25.1	RM	Section Review 25.1
RM	Practice Problems 25.1	RM	Chapter 25 Quiz
		LP	Lab Practical 25-1
GRS	Section 25.2	SE	Section Review
RM	Practice Problems 25.2	RM	Section Review 25.2
		RM	Chapter 25 Quiz
		LP	Lab Practical 25-1
SE	**Sample Problem** 25-4	SE	Section Review
GRS	Section 25.3	RM	Section Review 25.3
RM	Practice Problems 25.3	RM	Chapter 25 Quiz
GRS	Section 25.4	SE	Section Review
RM	Practice Problems 25.4	RM	Section Review 25.4
RM	Interpreting Graphics	RM	Chapter 25 Quiz
GRS	Section 25.5	SE	Section Review
RM	Practice Problems 25.5	RM	Section Review 25.5
GCP	Chapter 25	RM	Vocabulary Review 25
		SE	Chapter Review
		SM	Chapter 25 Solutions
		SE	Standardized Test Prep
		PHAS	Chapter 25 Test Prep
		RM	Chapter 25 Quiz
		RM	Chapter 25 A & B Test

PLANNING FOR ACTIVITIES

STUDENT EDITION

Discover It! p. 742
- candle wax
- vegetable oil
- petroleum jelly
- rubbing alcohol
- mineral oil
- water
- saucers
- toothpicks

Mini Lab, p. 757
- ball-and-stick molecular model kits

Small-Scale Lab, p. 758
- pencils
- paper
- toothpicks
- modeling clay

TEACHER'S EDITION

Teacher Demo, p. 744
- samples of methane, propane, butane

Teacher Demos, pp. 744 and 745
- Styrofoam balls and applicator sticks

  or
- molecular model sets

Activity, p. 745
- Styrofoam balls and applicator sticks

  or
- molecular model sets

Teacher Demo, p. 746
- samples of gaseous (methane), liquid (hexane), and solid (candle wax) alkanes
- Table 25.1

Teacher Demo, p. 750
- large test tube with stopper
- water
- mineral or baby oil
- food coloring

OT	Overhead Transparency	SE	Student Edition
PHAS	PH Assessment System	SM	Solutions Manual
PLM	Probeware Lab Manual	SSV	Small-Scale Video/Videodisc
PP	Problem Pro CD-ROM	TCP	www.chemplace.com
RM	Review Module	TE	Teacher's Edition
RP	Resource Pro CD-ROM	www	www.phschool.com

Chapter 25

Key Terms

25.1 hydrocarbons, alkanes, straight-chain alkanes, homologous series, condensed structural formulas, substituent, alkyl group, branched-chain alkanes
25.2 alkenes, unsaturated compounds, saturated compounds, alkynes
25.3 structural isomers, *trans* configuration, *cis* configuration, geometric isomers, asymmetric carbons, stereoisomers
25.4 cyclic hydrocarbons, aliphatic compounds, arenes, aromatic compounds
25.5 cracking

DISCOVER IT!

Students should observe that petroleum jelly and vegetable oil are soluble in mineral oil. Candle wax does not dissolve in mineral oil. None of the materials is soluble in water or rubbing alcohol. Students should suggest that substances that dissolve in mineral oil will probably not dissolve in water or rubbing alcohol. The properties of rubbing alcohol are closer to water than to mineral oil.

Chapter 25 HYDROCARBON COMPOUNDS

This painting shows Saturn's moon, Titan, as a place to refuel in space.

FEATURES

Stay current with SCIENCE NEWS
Find out more about hydrocarbons:
www.phschool.com

DISCOVER IT! WHAT DISSOLVES WHAT?

You need candle wax, vegetable oil, petroleum jelly, rubbing alcohol, mineral oil, water, three saucers, and nine toothpicks.

1. Place 3 pea-sized portions of petroleum jelly 10 cm apart on a saucer.
2. Add 5 drops of rubbing alcohol to one portion of the petroleum jelly, 5 drops of mineral oil to another, and 5 drops of water to the third.
3. Use a clean toothpick to mix each as completely as possible. Observe whether the substances appear to dissolve.
4. On a second saucer, repeat steps 1 through 3 using pea-sized portions of candle wax.
5. On a third saucer, repeat steps 1 through 3 using pea-sized portions of vegetable oil.

How did the solubility of these three materials (petroleum jelly, candle wax, and vegetable oil) in water compare with their solubility in mineral oil? Was the solubility of these substances in rubbing alcohol more like that in water or in mineral oil? Use your observations to develop a statement about solubility trends (similarities and differences). When you have completed the chapter, return to your statement to see if it is correct.

25.1 ● ■ ◆
25.2 ○ ■ ◆
25.3 □ ◆
25.4 ○ □ ◆
25.5 ● ■ ◇

Conceptual Allow students to refer to Table 25.1 and the rules for naming alkanes on pages 748–749 as they work through the sample problems in Section 25.1. In Section 25.2, direct students to check that each carbon always has four bonds. Students may have difficulty with the concept of resonance in Section 25.4. The discussion of fuels in Section 25.5 should help students appreciate the importance of hydrocarbons.

Standard In Section 25.3, the Small-Scale Lab will give students an opportunity to create molecular models and improve their ability to visualize geometric isomers. Counting the bonds for each carbon will help students distinguish cyclohexane from benzene.

Honors Students should be able to do the sample problems in this chapter as homework. You may assign Section 25.5 for self-study because of its mostly descriptive content.

HYDROCARBONS

People use fuels such as gasoline and diesel fuel every day—in cars and trucks, trains, airplanes and lawnmowers, and many other machines. These fuels and others contain a mixture of many compounds called hydrocarbons. **What are hydrocarbons, and why are there so many compounds?**

section 25.1

objectives

- Describe the bonding in hydrocarbons
- Distinguish between straight-chain and branched-chain alkanes

key terms

- hydrocarbons
- alkanes
- straight-chain alkanes
- homologous series
- condensed structural formulas
- substituent
- alkyl group
- branched-chain alkane

Organic Chemistry and Hydrocarbons

Fewer than 200 years ago, it was thought that the carbon compounds found in nature could be synthesized, or built, only by living organisms. From the word *organism* came the classification of these carbon compounds as organic and the chemistry of carbon compounds as organic chemistry. The prevailing theory held that the formation of carbon compounds was directed by a mysterious vital force. The German chemist Friedrich Wöhler (1800–1882) refuted this theory in 1828 when he synthesized urea from inorganic materials. Urea is a carbon-containing compound found in urine.

Today, organic chemistry includes the chemistry of virtually all carbon compounds, regardless of their origin. Carbon compounds number more than a million, with a dazzling array of valuable properties. One of the reasons carbon compounds are so numerous is carbon's unique bonding ability. Your study of organic chemistry starts with the simplest organic compounds, the hydrocarbons.

Organic compounds that contain only carbon and hydrogen are called **hydrocarbons.** The simplest hydrocarbons are the **alkanes,** which contain only single covalent bonds. Methane is the simplest alkane. It is the major component of natural gas. Methane is sometimes called marsh gas because it is formed by the action of bacteria on decaying vegetation in swamps and other marshy areas. Livestock and termites also emit substantial quantities of methane.

Chem ASAP!

Animation 28

Get a glimpse of the staggering variety of hydrocarbon compounds.

Figure 25.1

Bacteria feeding on decaying vegetation produce methane. Termites produce methane as a byproduct of digestion. Methane is the main component of natural gas, used for cooking and heating.

STUDENT RESOURCES

- Section Review 25.1, Ch. 25 Practice Problems and Quizzes from the Review Module (Ch. 25–28)
- Laboratory Manual: Experiment 49
- Laboratory Practical 25-1
- Small-Scale Chemistry Lab Manual: Experiment 36

TECHNOLOGY RESOURCES

Relevant technology resources include:

- Chem ASAP! CD-ROM
- Resource Pro CD-ROM
- Chemistry Alive! Videodisc: *Methane Bubbles*

25.1

1 Engage

Use the Visual

Have students examine the photo and read the text that opens the section. Explain that gasoline and diesel fuels are each mixtures of hydrocarbons obtained by refining petroleum, or crude oil. Ask:

- **What are hydrocarbons, and why are there so many compounds?** (Hydrocarbons are organic compounds that contain only carbon and hydrogen. Because carbon is tetravalent and because it can form bonds to itself and to other elements—such as O, N, S, P, and the halogens—millions of carbon-containing compounds are possible.)

Check Prior Knowledge

To assess students' prior knowledge about carbon chemistry, ask them to:

- **Explain the differences between ionic and covalent bonds.** (Electrons are transferred between atoms to form an ionic bond. Electrons are shared between atoms to form a covalent bond.)
- **Explain which type of bonding carbon uses.** (Covalent bonding)
- **Describe valence electrons.** (Valence electrons are those electrons used in the formation of chemical bonds; for representative elements, the number of valence electrons is equal to the group number in the periodic table.)

TEACHER DEMO

Burn samples of methane (natural gas, Bunsen burner), propane (torch), and butane (Bic® lighter). Ask students to compare and contrast these three substances by naming similarities and differences, for example:

- All three can be used as fuels; they react with oxygen to produce heat and light.
- All three are hydrocarbons (CH_4, C_3H_8, C_4H_{10}).

The discussion should include terms and concepts such as *exothermic, heat of combustion* (ΔH), *boiling point, molecular formula,* and *molar mass.*

2 Teach

TEACHER DEMO

Use both ball-and-stick and space filling models to illustrate the three-dimensional shape of the simplest hydrocarbon, methane. (Colored styrofoam balls and applicator sticks can be used to construct models that are large enough for all the students in the classroom to see.) Discuss the tetravalency of carbon as well as the octet rule. To assess students' knowledge about molecular shape, ask:

- Why does methane have a tetrahedral shape? (VSEPR theory states that because electron pairs repel each other, molecules adjust their shapes so that the valence-electron pairs are as far apart as possible.)

Explain that one of the key ideas of organic chemistry is "structure dictates function." The arrangement of atoms is as important as the number and type of atoms.

The methane molecule, which consists of four hydrogen atoms and one carbon atom, is a good example of carbon–hydrogen bonding. As you will recall (or a glance at the periodic table will remind you), a carbon atom has four valence electrons. Four hydrogen atoms, each with one valence electron, form four covalent carbon–hydrogen bonds. This combination is a molecule of methane.

$$\cdot\dot{\underset{\cdot}{C}}\cdot \;+\; 4H\cdot \longrightarrow H{:}\overset{H}{\underset{H}{\ddot{\underset{..}{C}}}}{:}H$$

Carbon atom | Hydrogen atoms | Methane molecule

The carbon–hydrogen bonding in methane illustrates an important principle: Because a carbon atom contains four valence electrons, it always forms four covalent bonds. Remembering this principle will help you to write complete and correct structures for organic molecules.

For simplicity, the structural formulas you learned to write in Chapter 16 will be used here to represent organic molecules. Remember, the line between the atomic symbols represents two bonding electrons.

$$H-\overset{\displaystyle H}{\underset{\displaystyle H}{\overset{|}{\underset{|}{C}}}}-H$$

Line represents shared electron pair

Methane molecule

Structural formulas are convenient to write on a page, but keep in mind that they are only two-dimensional representations of three-dimensional molecules. Molecular models represent the shapes of molecules more accurately. Throughout this chapter and the next, ball-and-stick molecular models and space-filling molecular models will be used with structural formulas to represent the shapes of organic molecules more accurately. These shapes are predicted by VSEPR theory and hybrid orbital theory. For example, methane has a tetrahedral shape shown in **Figure 25.2.**

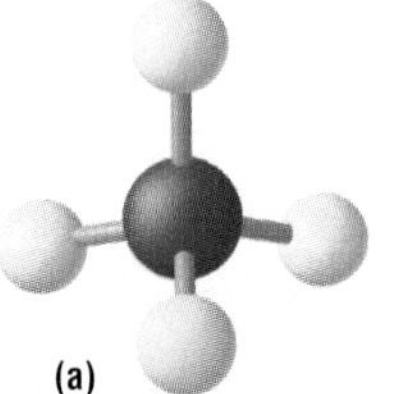

(a)

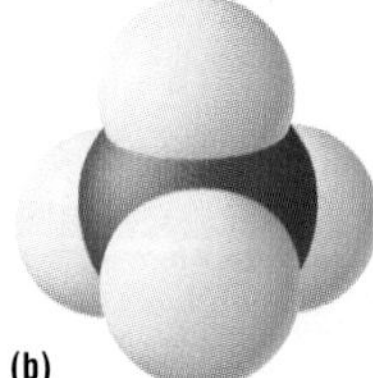

(b)

Figure 25.2
*The tetrahedral shape of the methane molecule is illustrated in the ball-and-stick model **(a)** and the space-filling model **(b)**.*

Carbon has the unique ability to make stable carbon–carbon bonds and to form chains. This is the major reason for the vast number of organic molecules. Silicon also forms chains, but they are unstable in an oxygen environment.

Ethane (C_2H_6) is the simplest alkane containing a carbon–carbon bond. Similar to methane, ethane is a gas at standard temperature and pressure. When ethane is formed from carbon and hydrogen, two carbon atoms share a pair of electrons. A carbon–carbon covalent bond is formed. The

MEETING DIVERSE NEEDS

LEP Encourage LEP students to preview the chapter looking for words in bold print and other key terms. Have students compile a glossary in which they define each term in English and in their native language. Suggest that students also include illustrations when appropriate.

At Risk Have students list the section objectives in their notebooks, leaving space after each objective. As they read each section, they should list terms and phrases that provide the information requested for each objective. Check their entries for accuracy.

remaining six valence electrons form bonding pairs with the electrons from six hydrogen atoms.

$$2\dot{\underset{\cdot}{C}}\cdot + 6H\cdot \longrightarrow \cdot\underset{\cdot}{\dot{C}}-\underset{\cdot}{\dot{C}}\cdot + 6H\cdot \longrightarrow H-\underset{H}{\overset{H}{C}}-\underset{H}{\overset{H}{C}}-H$$

Carbon atoms, Hydrogen atoms, Ethane molecule

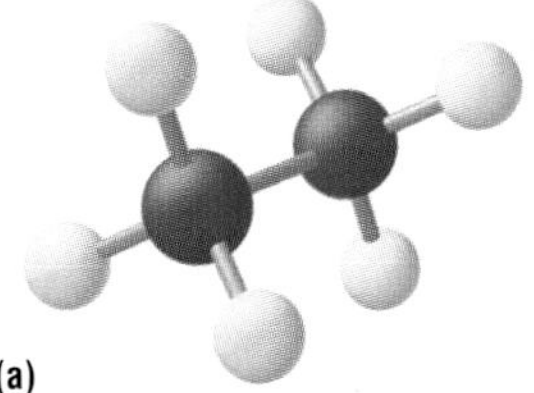

(a)

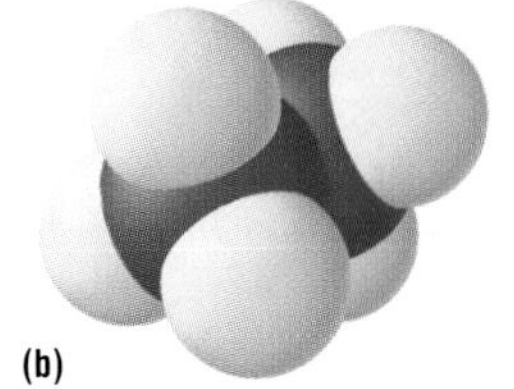

(b)

Figure 25.3
*A ball-and-stick model **(a)** and a space-filling model **(b)** of ethane.*

Straight-Chain Alkanes

Straight-chain alkanes contain any number of carbon atoms, one after the other, in a chain. To draw a structural formula for a straight-chain alkane, write the symbol for carbon as many times as necessary to get the proper chain length. Then fill in with hydrogens and lines representing covalent bonds. Remember that carbon forms four covalent bonds. **Table 25.1** shows the straight-chain alkanes containing up to ten carbons. Note that the names of alkanes always end with *-ane.* According to **Table 25.1**, what is the name of the straight-chain alkane with six carbons? Eight carbons? How many carbons does butane have? 1

The straight-chain alkanes are an example of a homologous series. A group of compounds forms a **homologous series** if there is a constant increment of change in molecular structure from one compound in the series to the next. The $—CH_2—$ group is the increment of change in straight-chain alkanes. This change can be traced in the structural formulas in **Table 25.1.**

Table 25.1

The First Ten Straight-Chain Alkanes

Name	Molecular formula	Structural formula	Boiling point (°C)
Methane	CH_4	CH_4	−161.0
Ethane	C_2H_6	CH_3CH_3	−88.5
Propane	C_3H_8	$CH_3CH_2CH_3$	−42.0
Butane	C_4H_{10}	$CH_3CH_2CH_2CH_3$	0.5
Pentane	C_5H_{12}	$CH_3CH_2CH_2CH_2CH_3$	36.0
Hexane	C_6H_{14}	$CH_3CH_2CH_2CH_2CH_2CH_3$	68.7
Heptane	C_7H_{16}	$CH_3CH_2CH_2CH_2CH_2CH_2CH_3$	98.5
Octane	C_8H_{18}	$CH_3CH_2CH_2CH_2CH_2CH_2CH_2CH_3$	125.6
Nonane	C_9H_{20}	$CH_3CH_2CH_2CH_2CH_2CH_2CH_2CH_2CH_3$	150.7
Decane	$C_{10}H_{22}$	$CH_3CH_2CH_2CH_2CH_2CH_2CH_2CH_2CH_2CH_3$	174.1

Answer

1 hexane; octane; 4

ACTIVITY

Using molecular model sets, or styrofoam balls and applicator sticks, have students construct a ball-and-stick model of ethane. Check the students' models for tetrahedral geometry at each carbon. Ask:

- How many covalent bonds are there in ethane? (7)
- Why do three-dimensional models represent organic molecules more accurately than two-dimensional models? (Three-dimensional models convey the spatial relationships among the various atoms.)

TEACHER DEMO

Write the complete structural formulas for methane, ethane, propane, and butane on the board or overhead projector. Ask students how the structures are similar and how they are different. Show students molecular models for each of the compounds. Pull apart an ethane model and show how a $-CH_2-$ fragment can be inserted between the CH_3- (methyl) sections to make larger alkanes. Use the models of propane and butane to show how the tetrahedral arrangement of the atoms around each carbon results in a "zig-zag" shape for the carbon backbone. Illustrate the rotational freedom of the atoms in a saturated hydrocarbon by twisting the models of propane and butane about the carbon-carbon axis. Once students understand how the atoms are joined together in the larger alkanes, write the condensed formulas under the complete structural formulas for ethane, propane and butane. Explain that the condensed formulas are a convenient shorthand notation but do not represent the true structure.

Use The Visual

Use an overhead transparency to display Table 25.1 on page 745. Explain the IUPAC system for naming alkanes. Suggest that students memorize the names of the first 10 alkanes. This will facilitate learning any nomenclature that is encountered later. Ask:

- **Why are all of these compounds considered straight-chain alkanes?** (All of the carbon atoms can be connected with an unbroken line. Each carbon atom within the chain is bonded to no more than two other carbon atoms.)

TEACHER DEMO

If possible, bring samples of gaseous (methane), liquid (hexane), and solid (candle wax) alkanes to class. With the condensed formulas on the board and Table 25.1 on the overhead projector, ask:

- What trend do you see regarding the number of carbon atoms and the boiling point of the alkanes? (the longer the carbon atom chain, the greater the boiling point)
- What can you conclude about the attraction between molecules as the number of carbon atoms increases? (The sum total of the intermolecular forces holding molecules together increases.)
- Given the information in Table 25.1, would it be possible to predict the boiling point of $C_{11}H_{24}$? (Yes; Construct a graph. Molecular formula is the independent variable and boiling point is the dependent variable. Extrapolate to find the boiling point of $C_{11}H_{24}$.)

Figure 25.4
Pressurized tanks of propane are used to fuel the burners in hot-air balloons. Butane serves as the fuel for many lighters.

Figure 25.4 shows some common uses for these two hydrocarbons. As the number of carbons in the straight-chain alkanes increases, so does the boiling point. This is also true of the melting point.

Complete structural formulas show all the atoms and bonds in a molecule. Sometimes, however, shorthand or condensed structural formulas work just as well. **Condensed structural formulas** leave out some bonds and/or atoms from the structural formula. Although they do not appear, you must understand that these bonds and atoms are there. The following illustrates several ways to draw condensed structural formulas for butane.

C_4H_{10}	Molecular formula
H–C(H)(H)–C(H)(H)–C(H)(H)–C(H)(H)–H	Complete structural formula
$CH_3-CH_2-CH_2-CH_3$	Condensed structural formula; C—H bonds understood
$CH_3CH_2CH_2CH_3$	Condensed structural formula; C—H and —C—C— bonds understood
$CH_3(CH_2)_2CH_3$ (Methylene units; Subscript)	Condensed structural formula; all bonds understood; parentheses indicate CH_2 units are linked together in a continuous chain (the $-CH_2-$ unit is called a methylene group); subscript 2 to the right of parenthesis indicates there are two methylene units linked together
C—C—C—C	Carbon skeleton; all hydrogens and C—H bonds understood

The names of the straight-chain alkanes listed in **Table 25.1** are recommended by the International Union of Pure and Applied Chemistry (IUPAC). As you can see, all the names end with the suffix *-ane*. The root part of an

Sample Problem 25-1

Draw complete structural formulas for the straight-chain alkanes that have three and four carbons.

1. ***ANALYZE*** ***Plan a problem-solving strategy.***
   Apply the rules for carbon–hydrogen and carbon–carbon bonding. Because these are straight-chain alkanes, the appropriate number of carbons will be written in a straight line and then connected to each other by single bonds. The chain will then be filled in with hydrogen atoms so that each carbon forms four covalent bonds.

2. ***SOLVE*** ***Apply the problem-solving strategy.***
   A three-carbon straight-chain alkane has three carbon atoms connected to each other in a straight line. This gives the center carbon two covalent bonds, so it needs two hydrogen atoms bonded to it. The two end carbons each need three hydrogen atoms bonded to them. A four-carbon straight-chain alkane has four carbon atoms connected to each other in a straight line. The two center carbons each need two hydrogen atoms, and the two end carbons each need three hydrogen atoms to give each carbon atom four single covalent bonds. The correct structural formulas are shown below.

```
    H  H  H           H  H  H  H
    |  |  |           |  |  |  |
 H—C—C—C—H       H—C—C—C—C—H
    |  |  |           |  |  |  |
    H  H  H           H  H  H  H
```

3. ***EVALUATE*** ***Do the results make sense?***
   The carbon chains have three and four carbons. Each carbon atom forms a total of four single covalent bonds with carbon and hydrogen atoms.

Practice Problems

1. Draw complete structural formulas for the straight-chain alkanes with five and six carbons.
2. How many single bonds are in a propane molecule?

alkane's name indicates how many carbon atoms it contains. For example, *meth-* means one carbon atom, *prop-* means three carbon atoms, *pent-* means five carbon atoms, and *dec-* means ten carbon atoms. You may find it helpful to learn these names now. They are the basis of a precise, internationally accepted system of naming organic compounds called the IUPAC system. As precise as the IUPAC system is, however, organic chemists still rely on a combination of systematic, semisystematic, and common names to identify organic compounds.

Many organic compounds are best known by their common names. Although these names bear no relation to IUPAC names or to the molecular structure of the compounds, they have been used for a long time. In most instances, common names are simpler than IUPAC names, which can be long and cumbersome. Like any living language, the language of science is constantly changing. IUPAC, semisystematic, and common names are used in this textbook.

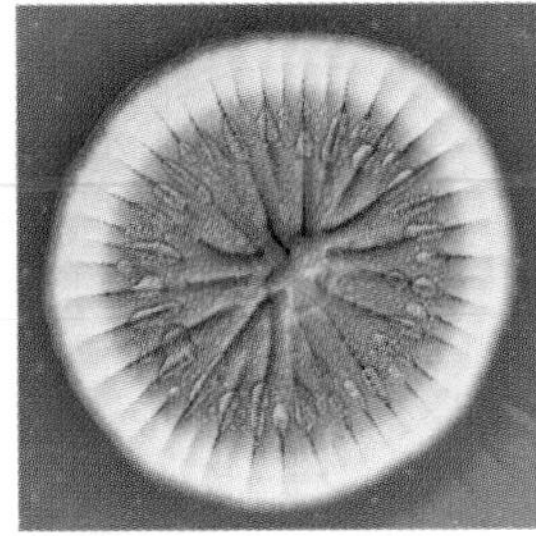

Figure 25.5
Many organic compounds isolated from nature have common names that reflect their origins. For example, penicillin, a common antibiotic, is named for the mold it is derived from, Penicillium notatum (shown here).

25.1

Practice Problems Plus

Related Chapter Review Problem
Chapter Review problem 30 is related to Sample Problem 25-1.

Think Critically

Discuss sources of methane and methane's relationship to the greenhouse effect and global warming. Ask questions such as:

- **What is the greenhouse effect, and how does methane contribute to it?**
- **What are the major sources of the methane in Earth's atmosphere?**
- **Is the methane content of Earth's atmosphere increasing or decreasing?**

Chemistry Alive!

Methane Bubbles

Play

Discuss

Discuss branched-chain alkanes and how they differ from straight-chain alkanes. Using ball-and-stick models, show how straight-chain butane (C_4H_{10}) can be arranged as a branched-chain structure (2-methylpropane) with the same molecular formula. Stress that each arrangement of atoms represents a different molecule with different physical and chemical properties. Show, with models, several other branched-chain alkanes. Use an overhead transparency to display the IUPAC rules for naming branched-chain alkanes, and use them to name the branched-chain structures you have built. Practice additional examples on the board. To help the students identify the parent alkane, use colored chalk to circle the longest continuous chain. Point out that the longest chain of carbon atoms is not always a straight line. Discuss the naming of alkyl groups (substituents), showing how they are formed from alkanes.

Branched-Chain Alkanes

Hydrogen atoms are not the only atoms that can bond to the carbon atoms in a hydrocarbon. The halogens and groups of atoms including carbon, hydrogen, oxygen, nitrogen, sulfur, or phosphorus may take the place of a hydrogen atom. An atom or group of atoms that can take the place of a hydrogen atom on a parent hydrocarbon molecule is called a **substituent.**

```
    Substituent                      Substituents
     C                         C   C   C
     |                         |   |   |
  C—C—C                    C—C—C—C—C—C
Parent alkane (propane)    Parent alkane (hexane)
```

A hydrocarbon substituent is called an **alkyl group.** An alkyl group can be one carbon or several carbons long. Three common alkyl groups are the methyl group (CH_3—), the ethyl group (CH_3CH_2—), and the propyl group ($CH_3CH_2CH_2$—). As you can see, an alkyl group consists of an alkane with one hydrogen removed. Alkyl groups are sometimes referred to as radicals. They are named by removing the *-ane* ending from the parent hydrocarbon name and adding *-yl.* What is the name of the alkyl group with the formula
1 $CH_3CH_2CH_2CH_2CH_2$—?

When a substituent alkyl group is added to a straight-chain hydrocarbon, branches are formed. An alkane with one or more alkyl groups is called a **branched-chain alkane.** The IUPAC rules for naming branched-chain alkanes are quite straightforward. The following compound can be used as an example.

```
 7     6     5     4    3    2    1
CH3—CH2—CH2—CH—CH—CH—CH3
                   |    |    |
                  CH2  CH3  CH3
                   |
                  CH3
```

1. Find the longest chain of carbons in the molecule. This chain is considered the parent structure. In the example, the longest chain contains seven carbon atoms. Therefore, the parent hydrocarbon structure is heptane.
2. Number the carbons in the main chain in sequence. To do this, start at the end that will give the groups attached to the chain the smallest numbers. This has already been done in the example. As you can see, the numbers go from right to left, which places the substituent groups at carbon atoms 2, 3, and 4. If the chain were numbered from left to right, the groups would be at positions 4, 5, and 6. These are higher numbers and therefore violate the rule.
3. Add numbers to the names of the substituent groups to identify their positions on the chain. These numbers become prefixes to the name of the parent alkane. In this example the substituents and positions are 2-methyl, 3-methyl, and 4-ethyl.
4. Use prefixes to indicate the appearance of a group more than once in the structure. Common prefixes are *di-* (twice), *tri-* (three times), *tetra-*

Answer

1 pentyl

(four times), and *penta-* (five times). This example has two methyl substituents. Thus the word *dimethyl* will be part of the complete name.

5. List the names of alkyl substituents in alphabetical order. For purposes of alphabetizing, ignore the prefixes *di-*, *tri-*, and so on. In this example, the 4-ethyl group is listed before the 2-methyl and 3-methyl groups (which are combined as 2,3-dimethyl in the name).
6. Use proper punctuation. This is very important in writing the names of organic compounds in the IUPAC system. Commas are used to separate numbers. Hyphens are used to separate numbers and words. The entire name is written without any spaces.

According to the IUPAC rules, the name of this compound is 4-ethyl-2,3-dimethylheptane. Note that the name of the parent alkane, heptane, follows directly after the final prefix, dimethyl. (4-ethyl-2,3-dimethyl heptane would be incorrect.) A ball-and-stick model and a space-filling model of this compound are shown in **Figure 25.6.**

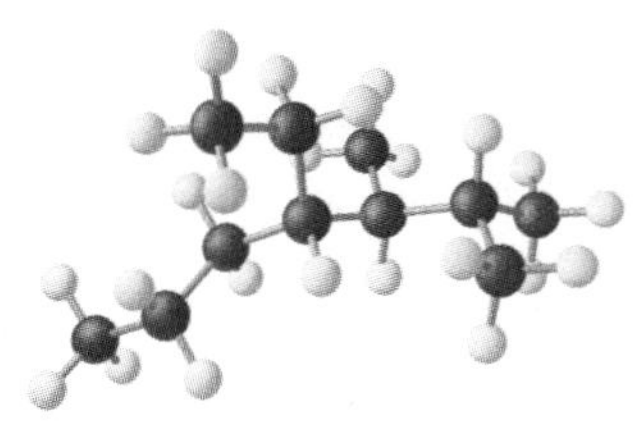
(a)

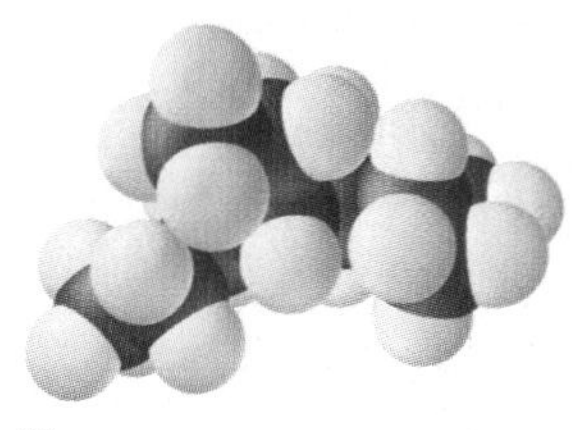
(b)

Figure 25.6
A ball-and-stick model ***(a)*** *and a space-filling model* ***(b)*** *of 4-ethyl-2,3-dimethylheptane.*

Sample Problem 25-2

Name these compounds using the IUPAC system. Notice that the longest chain is not written in a straight line in molecule **a**.

a.

```
                 CH3
                 |
CH3 — CH2 — C — CH3
                 |
                 CH2
                 |
                 CH2
                 |
                 CH3
```

b.

```
        CH3          CH3
        |            |
CH3 — C — CH2 — C — CH3
        |            |
        CH3          CH3
```

1. ***ANALYZE*** ***Plan a problem-solving strategy.***
For each compound, follow the IUPAC rules. Find the longest chain of carbon atoms, or the parent structure, and name it. Number the carbons in the main chain in sequence, making sure that the carbons to which the alkyl substituents are attached have the lowest numbers possible. These location numbers become part of the name as prefixes. If a substituent appears more than once in a structure, another appropriate prefix must be used. List the names of the alkyl substituents in alphabetical order. Separate numbers with commas; separate numbers and words with hyphens.

Practice Problems

3. Name the following alkanes.

a.

```
    H  H  H
    |  |  |
H — C — C — C — H
    |  |  |
    H  H  H
```

b.

```
    H  H  H  H  H
    |  |  |  |  |
H — C — C — C — C — C — H
    |  |  |  |  |
    H  H  H  H  H
```

c.

```
    H  H  H  H  H  H
    |  |  |  |  |  |
H — C — C — C — C — C — C — H
    |  |  |  |  |  |
    H  H  H  H  H  H
```


CHEMISTRY AND SCIENCE HISTORY

George Olah won the 1994 Nobel Prize in Chemistry for his work with hydrocarbons. Olah showed that chemical reactions involving hydrocarbons proceed through short-lived, highly reactive, intermediate compounds which he called "carbocations." Olah was able to produce, isolate, and preserve carbocations so that they could be studied in detail.

Practice Problems Plus

Related Chapter Review Problems
Chapter Review problems 29, 31, and 32 are related to Sample Problem 25-2.

Additional Practice Problem
The Chem ASAP! CD-ROM contains the following problem: Name the following compounds according to the IUPAC system.

a. $CH_3-CH(CH_3)-CH_2-CH_2-CH_3$

b. $CH_2(CH_3)-CH_2-CH_2-CH(CH_2CH_3)-CH_2(CH_3)$

(**a.** 2-methylpentane, **b.** 3-ethylheptane)

TEACHER DEMO

To illustrate the inability of water and oil to mix, prepare a large test tube with equal volumes of water and either mineral oil or baby oil. A small amount of food coloring can be added to the water to more clearly demarcate the polar and nonpolar phases. Shake the test tube vigorously and allow the phases to separate. Discuss the rule "like dissolves like." Ask:

- Why don't the oil and water form a solution? (Water is a polar compound, oil is a nonpolar compound.)

Review concepts such as polar and nonpolar bonds, dipole-dipole interactions, and van der Waals forces. Review the role that molecular shape plays in determining if a molecule is polar or nonpolar.

Sample Problem 25-2 (cont.)

2. *SOLVE* ***Apply the problem-solving strategy.***

a. The longest carbon chain in the molecule is hexane (six carbons). There are two methyl substituents on carbon 3, so the prefix is 3,3-dimethyl. The correct IUPAC name is 3,3-dimethylhexane.

b. The longest carbon chain in the molecule is pentane (five carbons). There are two methyl substituents each on carbon 2 and on carbon 4. The prefix is 2,2,4,4-tetramethyl, and the correct IUPAC name is 2,2,4,4-tetramethylpentane.

3. *EVALUATE* ***Do the results make sense?***
The names of the parent structures conform to the longest chains. The substituents are given the lowest numbers and the correct names. Prefixes are used to indicate the appearance of the substituents, if they appear more than once. Numbers are separated by commas, and numbers are separated from words by hyphens.

Practice Problems (cont.)

4. Name these compounds according to the IUPAC system.

a. $CH_3-CH_2-CH(CH_3)-CH_3$

b. $CH_2(CH_3)-CH_2-CH(CH_2CH_3)-CH_2-CH_3$

Chem ASAP!
Problem-Solving 4
Solve Problem 4 with the help of an interactive guided tutorial.

With an alkane name and knowledge of the IUPAC rules, it is easy to reconstruct the structural formula.

1. Find the root word (ending in *-ane*) in the hydrocarbon name. Then write the longest carbon chain to create the parent structure.
2. Number the carbons on this parent carbon chain.
3. Identify the substituent groups. Attach the substituents to the numbered parent chain at the proper positions.
4. Add hydrogens as needed.

Sample Problem 25-3

Draw complete structural formulas for each of the following compounds.

a. 3-ethylhexane
b. 2,2,4-trimethylpentane

1. *ANALYZE* ***Plan a problem-solving strategy.***
Work backward from each name to arrive at the structure. The prefix indicates the types of alkyl substituents, the number of times each appears, and the carbon to which each is attached. The root word in the hydrocarbon name, the word that ends in *-ane,* is the parent structure. The carbons on the parent chain are numbered to give the carbons bonded to the alkyl substituents the lowest numbers. Hydrogens are added as needed.

Practice Problems

5. Draw the structural formula for the compound 2,3-dimethylhexane.

Sample Problem 25-3 (cont.)

2. ***SOLVE*** ***Apply the problem-solving strategy.***
   a. The parent structure is a straight chain of six carbon atoms (hexane). An ethyl substituent is found on carbon 3. There are a total of 13 additional hydrogens added to the parent chain.

```
                CH3
                |
                CH2
                |
CH3—CH2—CH—CH2—CH2—CH3
 1    2    3    4    5    6
```

   b. The parent structure is a five-carbon straight chain (pentane) with three methyl substituents—two on carbon 2 and one on carbon 4. A total of nine hydrogens are added to the parent chain.

```
        CH3        CH3
        |          |
CH3—C—CH2—CH—CH3
 1    2|   3    4    5
        CH3
```

3. ***EVALUATE*** ***Do the results make sense?***
   Chains of the proper length are written and numbered. Substituents are placed on the carbons indicated by the names. The correct number of hydrogens is present.

Practice Problems (cont.)

6. Draw structural formulas for these compounds.
   a. 3-ethylpentane
   b. 4-ethyl-2,3,4-trimethyloctane

Chem ASAP!

Problem-Solving 6
Solve Problem 6 with the help of an interactive guided tutorial.

Properties of Alkanes

The electron pair in a carbon–hydrogen or a carbon–carbon bond is shared about equally by the nuclei of the atoms involved. Therefore, hydrocarbon molecules such as alkanes are nonpolar. The nonpolar attractions between hydrocarbon molecules are the very weak van der Waals forces, which you learned about in Chapter 16. Thus hydrocarbons of low molar mass tend to be gases or low-boiling liquids.

Nonpolar organic molecules such as hydrocarbons are not attracted to water. For example, oil (a hydrocarbon) and water do not mix. A good rule of thumb is that "like dissolves like." This means that two nonpolar compounds will form a solution as will two polar compounds. But a nonpolar compound and a polar compound will not form a solution.

Figure 25.7
The nonpolar molecules in this oil spill are not attracted to the polar water molecules in the ocean. The insoluble oil from an oil spill remains on the surface of the water because oil has a lower density than water.

section review 25.1

7. Describe the bonding in hydrocarbons.
8. Differentiate between a straight-chain and a branched-chain alkane.
9. What is a homologous series? Why are the alkanes such a series?
10. Explain why alkane molecules are nonpolar.

 **Chem ASAP!** **Assessment 25.1** Check your understanding of the important ideas and concepts in Section 25.1.

25.1

Practice Problems Plus

Additional Practice Problem
The Chem ASAP! CD-ROM contains this problem related to Sample Problem 25-3: Draw structural formulas for these compounds.
a. 3-methylhexane
b. 3-ethyl-3, 4-dimethyl-octane

```
a. CH3—CH—CH2—CH2
          |          |
          CH2        CH3
          |
          CH3

b.              CH2—CH3
                |
CH3—CH2—C—CH—CH2—CH2—CH2—CH3
                |    |
                CH3 CH3
```

3 Assess

Evaluate Understanding

To assess students' knowledge of alkanes, ask them to:

- **Explain why straight-chain alkanes are considered a homologous series.**
- **Discuss the differences between straight-chain and branched-chain alkanes.**
- **Explain the observation that alkanes do not mix with water.**

Reteach

Quiz students orally to ensure memorization of the first ten prefixes used in hydrocarbon nomenclature. Then write the following names on the board:
3-ethylpentane
2,3,4-trimethylhexane
2,2-dimethylbutane
Have students write structural formulas for these compounds. Remind students to (1) write the alkane carbon skeleton, (2) attach the substituent groups, and (3) add hydrogen atoms to all the unfilled bonding sites.

Answers

SECTION REVIEW 25.1

7. Atoms of hydrogen and atoms of carbon are covalently bonded to each other in hydrocarbon compounds.
8. In a straight-chain alkane, there are at least two hydrogens attached to each carbon. In a branched-chain alkane, substituents such as alkyl groups replace some hydrogens.
9. A homologous series is a set of compounds that differ by a constant increment of change in molecular structure. Alkanes form a homologous series because they differ from each other only by the number of —CH_2— groups they contain.
10. The bonds between carbons are nonpolar and carbon–hydrogen bonds are very weakly polar; thus alkanes are nonpolar.

25.2

1 Engage

Use the Visual

Have students study the photograph and read the text that opens the section. Explain that many naturally occurring biologically active chemicals, such as pheromones, hormones, and vitamins, are derived from unsaturated hydrocarbons. Ask:

- **How do unsaturated hydrocarbons differ from alkanes?** (Alkanes are saturated hydrocarbons, which contain only single bonds. Unsaturated hydrocarbons contain one or more carbon–carbon double or triple bonds. Unsaturated hydrocarbons contain less than the maximum number of hydrogens in their structure.)

Check Prior Knowledge

To assess students' knowledge about multiple covalent bonds, ask:

- **What kind of bonding occurs between carbon and oxygen atoms in a molecule of carbon dioxide, CO_2 and between nitrogen atoms in a molecule of nitrogen gas, N_2?** (A double bond, as occurs in CO_2, is a covalent bond in which two pairs of electrons are shared between two atoms. A triple bond, as occurs in N_2, is a covalent bond in which three pairs of electrons are shared between two atoms.)
- **Why do some atoms form double and triple bonds?** (to satisfy the octet rule: to obtain an electron configuration similar to that of one of the noble gases)

section 25.2

objectives

- Explain the difference between unsaturated and saturated hydrocarbons
- Differentiate between the structures of alkenes and alkynes

key terms

- alkenes
- unsaturated compounds
- saturated compounds
- alkynes

UNSATURATED HYDROCARBONS

Roach traps are filled with pheromones, chemicals which are given off by many insects to attract mates. When a roach senses the pheromones inside the trap, it crawls in and is stuck on sticky paper. Unsaturated hydrocarbons form the basis for the molecular structures of many pheromones.

How do unsaturated hydrocarbons differ from alkanes?

Alkenes

As you just learned, the carbon–carbon bonds in alkanes are single covalent bonds. But single covalent bonds are not the only kind of carbon–carbon bond to be found in hydrocarbons. Multiple bonds between carbons also exist. Hydrocarbons containing carbon–carbon double covalent bonds are called **alkenes.** This is the carbon–carbon double bond found in alkenes.

$$\gt C=C \lt$$

Organic compounds that contain double or triple carbon–carbon bonds are called **unsaturated compounds.** The compounds have this name because they contain fewer than the maximum number of hydrogens in their structure. The alkanes, which contain the maximum number of hydrogens, are called **saturated compounds.**

To name an alkene by the IUPAC system, find the longest chain in the molecule that contains the double bond. This chain is the parent alkene. It gets the root name of the alkane with the same number of carbons plus the ending *-ene.* The chain is numbered so that the carbon atoms of the double bond get the lowest possible numbers. Substituents on the chain are named and numbered in the same way they are for the alkanes. Ethene and propene are the simplest alkenes. They are often called by the common names ethylene and propylene. Some examples of the structures and IUPAC names of simple alkenes are shown below.

$H_2C=CH_2$	$CH_3-CH=CH-H$	$CH_2=CH-CH_2-CH_3$
Ethene (ethylene, the simplest alkene)	Propene (propylene)	1-butene

Figure 25.8
Ethene, the simplest alkene, stimulates the growth of plants and the ripening of their fruits.

Figure 25.9 shows the ball-and-stick and space-filling models of ethene. Notice that the four hydrogens that project from the double-bonded carbons lie in a plane and are 120° apart. Try to imagine freely rotating each CH_2 group around the double bond. It is impossible. The same is true of the double bond in real alkene molecules. No rotation occurs about a carbon–carbon double bond.

STUDENT RESOURCES

From the Teacher's Resource Package, use:

- Section Review 25.2, Ch. 25 Practice Problems and Quizzes from the Review Module (Ch. 25–28)
- Laboratory Manual: Experiment 49
- Laboratory Practical 25-1

TECHNOLOGY RESOURCES

Relevant technology resources include:

- Chem ASAP! CD-ROM
- Resource Pro CD-ROM

❶ No; for rotation to occur, the pi bond, which is formed by the overlap of two *p*-orbital electrons, must be broken, a process that requires a large amount of energy.

Alkynes

Hydrocarbons containing carbon–carbon triple covalent bonds are called **alkynes.** Like alkenes, alkynes are unsaturated compounds. This is the carbon–carbon triple bond found in alkynes.

$$-C \equiv C-$$

Alkynes are not plentiful in nature. The simplest alkyne is the gas ethyne (C_2H_2). The common name for ethyne is acetylene. Acetylene is the fuel burned in oxyacetylene torches used in welding. **Figure 25.10** shows that the single bonds that extend from the carbons in the carbon–carbon triple bond of ethyne are separated by the maximum angle of 180°. This makes ethyne a linear molecule.

The major attractions between alkane molecules, alkene molecules, or alkyne molecules are weak van der Waals forces. As a result, the introduction of a double or triple bond into a hydrocarbon does not have a dramatic effect on physical properties such as boiling point. See **Table 25.2.**

Figure 25.9
The six atoms of ethene lie in one plane. Can rotation occur about the double bond? 1

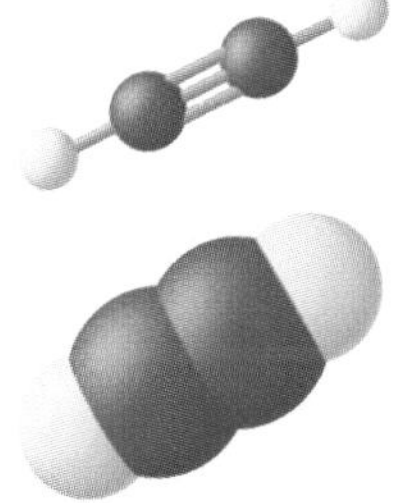

Figure 25.10
The triple bond in ethyne is rigid, so rotation about this bond does not occur. Ethyne is the simplest alkyne. The common name for ethyne is acetylene.

Table 25.2

Boiling Points of Homologous Alkanes, Alkenes, and Alkynes

Name	Molecular structure	Boiling point (°C)
C_2		
Ethane	$CH_3{-}CH_3$	−88.5
Ethene	$CH_2{=}CH_2$	−103.9
Ethyne	$CH{\equiv}CH$	−81.8
C_3		
Propane	$CH_3CH_2CH_3$	−42.0
Propene	$CH_3CH{=}CH_2$	−47.0
Propyne	$CH_3C{\equiv}CH$	−23.3
C_4		
Butane	$CH_3CH_2CH_2CH_3$	−0.5
1-Butene	$CH_3CH_2CH{=}CH_2$	−6.3
1-Butyne	$CH_3CH_2C{\equiv}CH$	8.6
C_5		
Pentane	$CH_3CH_2CH_2CH_2CH_3$	36.0
1-Pentene	$CH_3CH_2CH_2CH{=}CH_2$	30.0
1-Pentyne	$CH_3CH_2CH_2C{\equiv}CH$	40.0

section review 25.2

11. Differentiate between unsaturated and saturated hydrocarbons.

12. Write electron dot structures for ethene and ethyne. Describe the shape of each.

13. Draw and name all the alkenes with the molecular formula C_4H_8.

Chem ASAP! **Assessment 25.2** Check your understanding of the important ideas and concepts in Section 25.2.

2 Teach

ACTIVITY

Have students use their hands to model the rigidity of a double bond. Begin by having them touch just the tips of the first fingers of both hands. With the fingers still touching, have them rotate their hands in opposite directions and observe the position of their thumbs. Then have them repeat the procedure, but with two fingers of each hand touching. They will find that rotation about the "double bond" is impossible. For example, if both thumbs are in the "up" position, they must stay there.

3 Assess

Evaluate Understanding

Have students explain why compounds with double or triple bonds are called unsaturated compounds.

Reteach

Using molecular model sets, have students construct ball-and-stick models of ethane, ethene, and ethyne. Ask students to write the complete structural formula for each molecule. Point out the lack of rotation about the carbon-carbon double and triple bond. Have students describe the molecular geometry of each molecule. (tetrahedral, planar, and linear, respectively) **Ask:**

- **How many covalent bonds are there in ethane?** (7) **In ethene?** (6; 5 sigma bonds and 1 pi bond) **In ethyne?** (5; 3 sigma bonds and 2 pi bonds)
- **How many hydrogen atoms are in ethane?** (6) **In ethene?** (4) **In ethyne?** (2)

Answers

SECTION REVIEW 25.2

11. Saturated hydrocarbons contain only carbon-hydrogen and carbon-carbon single bonds; unsaturated hydrocarbons contain at least one carbon-carbon double or triple bond.

12. H:C::C:H (with H above each C and dots) — ethene; H:C:::C:H — ethyne

Ethene, the simplest alkene, is a planar molecule; ethyne, the simplest alkyne, is a linear molecule.

13. $CH_2{=}CH{-}CH_2{-}CH_3$ 1-butene

$CH_3{-}CH{=}CH{-}CH_3$ (H on each double-bonded C) 2-butene

$CH_2{=}C(CH_3){-}CH_3$ methylpropene

1 Engage

Use the Visual

Have students study the photograph and read the text that opens the section. Explain that the structure of a molecule greatly influences its chemical behavior. Many biological processes depend on structural rearrangements in molecules. Ask:

- **What is an isomer and what types of isomerization are possible?** (Isomers are compounds that have the same molecular formula but different molecular structures. Use butane and 2-methylpropane, and 1-pentene and 2-pentene, as examples of two possible types of isomerization.)

Check Prior Knowledge

To assess students' understanding of the term *isomer*, explain that it derives from Greek terms meaning "same parts." Ask:

- **Identify and define another chemical term that begins with the prefix *iso-*.** (Students should recall the term *isotope* from the chapter on atomic structure. Isotopes have the same number of protons, but different number of neutrons. Isotopes are atoms of the same elements with different masses.)

section 25.3

objectives

- Distinguish among structural, geometric, and stereoisomers
- Identify the asymmetric carbon or carbons in stereoisomers

key terms

- structural isomers
- *trans* configuration
- *cis* configuration
- geometric isomers
- asymmetric carbon
- stereoisomers

ISOMERISM

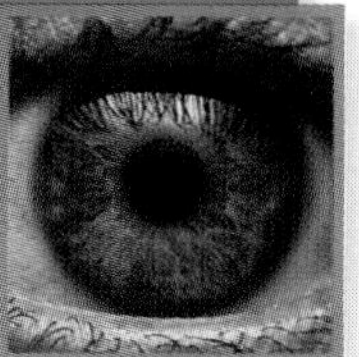

Retinal is a molecule in your eyes that has a hydrocarbon skeleton. The first step in the process of vision occurs when a ray of light enters your eye and strikes retinal, causing its hydrocarbon skeleton to rearrange to a different form called an isomer. **What is an isomer and what types of isomerization are possible?**

Structural Isomers

You may have noticed that the structures of some hydrocarbons are similar, differing only in the positions of the substituent groups or of the multiple bonds in their molecules. Compounds that have the same molecular formula but different molecular structures are called **structural isomers.** For example, two different molecules having the formula C_4H_{10} are shown below. One molecule is butane; the other molecule is 2-methylpropane. They are structural isomers of each other.

$$CH_3-CH_2-CH_2-CH_3$$

Butane (C_4H_{10})
(bp −0.5 °C)

$$CH_3-\underset{}{\overset{CH_3}{\overset{|}{C}}}H-CH_3$$

2-Methylpropane (C_4H_{10})
(bp −10.2 °C)

Structural isomers differ in physical properties such as boiling point and melting point. They also have different chemical reactivity. In general, the more highly branched the hydrocarbon structure, the lower its boiling point compared with its other structural isomers. For example, 2-methylpropane has a lower boiling point than butane has.

Geometric Isomers

The lack of rotation around carbon–carbon double bonds has an important structural implication. Look at the structure of 2-butene in **Figure 25.11.** Two arrangements are possible for the methyl groups with respect to the rigid double bond. In the ***trans* configuration,** the substituted groups are on opposite sides of the double bond. In the ***cis* configuration,** the substituted groups are on the same side of the double bond. *Trans*-2-butene and *cis*-2-butene are geometric isomers. **Geometric isomers** differ only in the geometry of their substituted groups. Like other structural isomers, isomeric 2-butenes are distinguishable by their physical and chemical

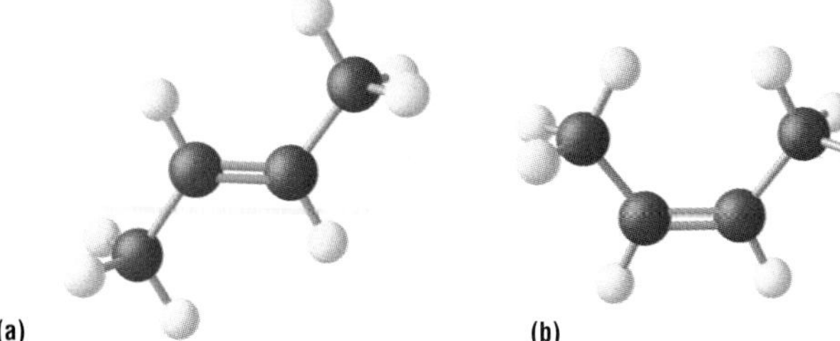

Figure 25.11
*These are the two geometric isomers of 2-butene. How does the trans configuration **(a)** differ from the cis configuration **(b)**?* 1

STUDENT RESOURCES

From the Teacher's Resource Package, use:

- Section Review 25.3, Ch. 25 Practice Problems and Quizzes from the Review Module (Ch. 25–28)
- Laboratory Recordsheets 25-1 and 25-2

TECHNOLOGY RESOURCES

Relevant technology resources include:

- Chem ASAP! CD-ROM
- Resource Pro CD-ROM
- Chem Alive! Videodisc: *Carbide Cannon*

properties. The groups on the carbons of the double bond do not need to be the same. Geometric isomerism is possible whenever each carbon of the double bond has at least one substituent. Why does 2-methyl-1-butene have no *cis, trans* isomers? **2**

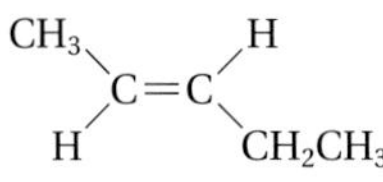

trans-2-Pentene

CH_3 CH_2CH_3
C=C
H H

cis-2-Pentene

CH_3 H
C=C
CH_3CH_2 H

2-Methyl-1-butene (no *cis, trans* isomers)

Stereoisomers

The structures that you have drawn up until now have ignored the fact that molecules are really three-dimensional. Could two molecules that look the same on paper differ in the third dimension—the positions of their atoms in space? They can, if the molecules are nonsuperimposable mirror images of each other. Placing an object in front of a mirror can give two different results. If the object is symmetrical, like a ball, then its mirror image is superimposable. That is, the appearance of the ball and its reflection are indistinguishable. By contrast, a pair of hands is distinguishable even though the hands consist of identical parts. The right hand reflects as a left hand and the left hand reflects as a right hand. Your hands are examples of nonsuperimposable mirror images. They are mirror images that cannot be placed on top of each other to obtain a match. Many pairs of ordinary objects, such as ears, feet, shoes, and bird wings, are similarly related.

You learned earlier that four groups attached to a carbon by single covalent bonds form a tetrahedron with the carbon at the center. In **Figure 25.12**, the carbon atom has four different groups attached: F, H, Cl, and Br. A carbon with four different groups attached is called an **asymmetric carbon.** Compounds with molecules that contain an asymmetric carbon have handedness. For these compounds, two kinds of molecules exist that are related to each other in much the same way as a pair of hands. **Figure 25.12** shows the mirror images of CHFClBr. Like hands, these mirror images are nonsuperimposable. Unless bonds are broken, these molecules cannot be superimposed. The four atoms attached to the carbon will not all match at once. **Stereoisomers** are molecules of the same molecular structure that differ only in the arrangement of the atoms in space. The mirror images of CHFClBr are stereoisomers.

Chem ASAP!

Simulation 28

Play the isomer game of "Pick the Pairs."

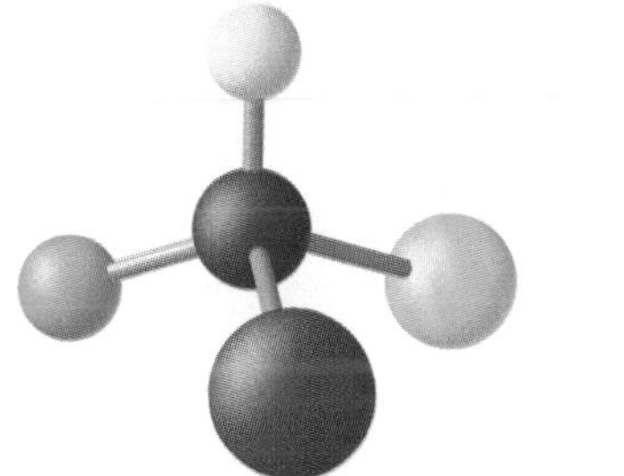

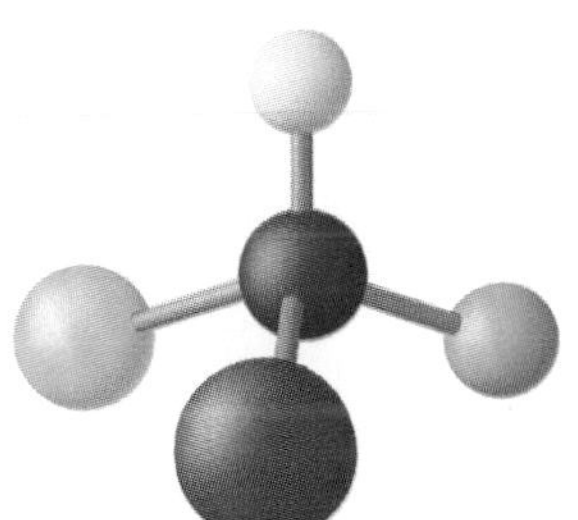

Figure 25.12

These ball-and-stick models illustrate the stereoisomers of CHFClBr. Stereoisomers are mirror-image molecules that cannot be superimposed on each other. Build models from toothpicks and clay to prove to yourself that these two isomers cannot be superimposed.

25.3

2 Teach

Discuss

Point out that the number of possible structural isomers for an alkane increases dramatically with increasing numbers of carbon atoms. With more carbon atoms there are more ways to arrange the atoms in space. Remind students that each isomer has a unique set of physical and chemical properties. Show the three isomers of pentane and name each. Have students work out the five isomers of hexane on their own. Point out that structural isomerism is one of the major reasons that so many different organic compounds exist.

Use the Visual

Have students study the models in Figure 25.12. Tell students the key to the existence of stereoisomers is the presence of a carbon atom with four different groups attached to it. Such an atom is called *asymmetric*. If a carbon atom is asymmetric, stereoisomers must exist.

Answers

1 In the *trans* configuration, the substituted groups are on opposite sides of the double bond. In the *cis* configuration, the substituted groups are on the same side of the double bond.

2 *Cis-trans* isomerism is not possible when one of the double-bonded carbons bears two identical substituents.

Chemistry Alive!

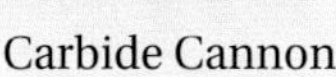

Carbide Cannon Play

CHEMISTRY AND BIOLOGY

Tell students that mirror images exist in nature. For example, 99% of the shells of certain snails are right-handed spirals and 1% are left-handed spirals. The right- and left-handed spirals are mirror images. Other species have shells that are normally left-handed.

Discuss

Point out that alkenes, like alkanes, can exist in different structural forms. In fact, the number of possible isomers for an alkene can be much greater than for an alkane. Draw the complete structural formulas of 1-butene, *cis*-2-butene, and *trans*-2-butene on the board to show the three possible isomers for butene (C_4H_8). Note that structural isomers for alkenes result from different locations of the double bond and from different spatial orientations of the atoms about the carbon–carbon double bond. Introduce the term *geometric isomers* and explain how *cis*-2-butene and *trans*-2-butene are different molecules due to the lack of free rotation about the carbon–carbon double bond.

Practice Problems Plus

Related Chapter Review Problems
Chapter Review problems 41 and 42 are related to Sample Problem 25-4.

Sample Problem 25-4

Which of the following compounds have an asymmetric carbon?

a. CH_3CHCH_3 (with OH on the central carbon)

b. $CH_3CHCH_2CH_3$ (with OH on the second carbon)

c. CH_3CHCHO (with OH on the second carbon)

1. *ANALYZE* *Plan a problem-solving strategy.*
Use the definition of an asymmetric carbon to analyze the structures: An asymmetric carbon has four different groups attached.

2. *SOLVE* *Apply the problem-solving strategy.*
Draw the structures in a more complete form to determine the presence of a carbon atom with four different groups attached.

a. The central carbon has one H, one OH, and two CH_3 groups attached. It is not asymmetric.

```
      H
      |
CH3—C—CH3
      |
      OH
```

b. The central carbon has one H, one OH, one CH_3, and one CH_2CH_3 group attached. Because these four groups are different, the central carbon is asymmetric. It can be marked with an asterisk. None of the other carbons in this molecule is asymmetric.

```
      H
      |
CH3—*C—CH2CH3
      |
      OH
```

c. This molecule also has an asymmetric carbon because the central carbon has four different groups attached: H, OH, CH_3, CHO. The asymmetric carbon is marked with an asterisk.

```
      H
      |
CH3—*C—CHO
      |
      OH
```

3. *EVALUATE* *Do the results make sense?*
Only the two carbons with four different groups attached have been correctly identified as asymmetric. All other carbons shown have three or fewer different groups attached. The molecule that does not include an asymmetric carbon has a superimposable image.

Practice Problems

14. Identify the asymmetric carbon, if any, in each of the following structures.

a. CH_3CH_2CHO

b. CH_3CHCHO (with Cl on the second carbon)

c. CH_3CHOH (with CH_3 on the second carbon)

15. For which of the following molecular formulas can mirror image molecules be drawn? Why?

a. CH_2Cl_2

b. HS—C—OH (with F above and F below the carbon)

c. $CH_3CF_2CH_3$

d. CH_3CH_2—C—Br (with CH_3 above and F below the carbon)

Answers

SECTION REVIEW 25.3

16. Structural isomers have the same molecular formula but different molecular structures. Geometric isomers differ in the positions of substituent groups relative to a carbon-carbon double bond.

17. Geometric isomers differ in the positions of substituent groups on the same side or on opposite sides of the carbon-carbon double bond. Stereoisomers are mirror images of the same molecule.

18. a.

```
H         CH2CH2CH3
  \      /
   C=C
  /      \
H         H
```

1-pentene
(no geometric isomers)

b.

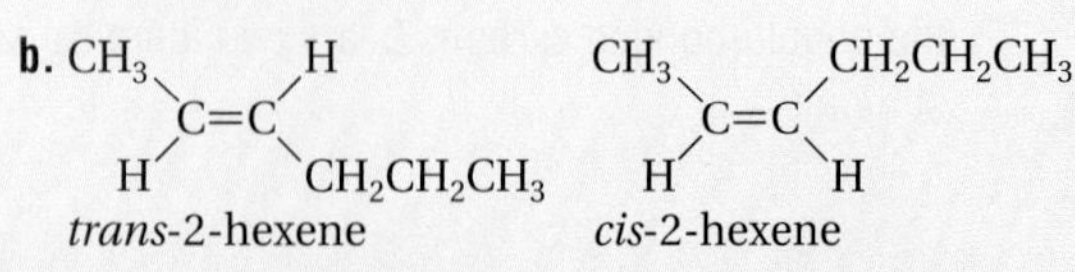

MINI LAB

Structural Isomers of Heptane

PURPOSE

To build ball-and-stick models and name the nine structural isomers of heptane (C_7H_{16}).

MATERIALS

- ball-and-stick molecular model kit

PROCEDURE

1. Build a model for the straight-chain molecule of C_7H_{16}. Draw the structural formula and write the IUPAC name for this compound.
2. Move one carbon atom from the end of the chain and use it as a methyl substituent, making a new compound. Draw the structural formula and name this structural isomer.
3. Move the methyl group to a new position on the chain and draw and name this third structural isomer. Is there a third position that this methyl group can be moved to on the chain of six carbons, giving yet another compound?
4. Make other structural isomers by shortening the longest continuous chain and using the removed carbons as substituents. Draw the structural formulas and name each compound.

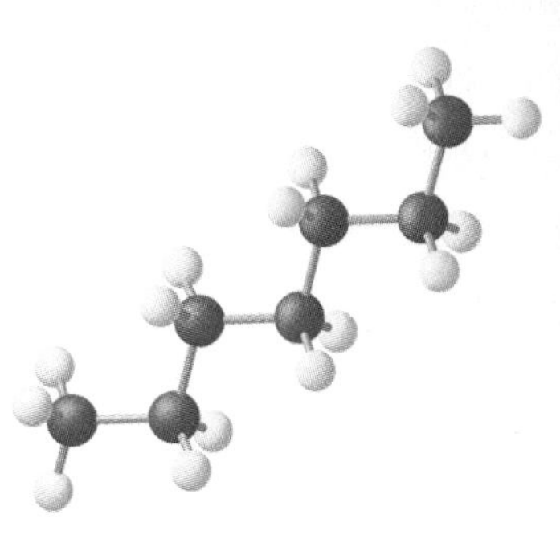

ANALYSIS AND CONCLUSIONS

1. What are structural isomers?
2. What are the names of the nine structural isomers of C_7H_{16}?
3. What was the shortest continuous chain?
4. Did each structural isomer have its own unique name? Should it?

section review 25.3

16. How do geometric isomers differ from structural isomers?
17. How do geometric isomers differ from stereoisomers?
18. Draw structural formulas for the following alkenes. If a compound has geometric isomers, draw both the *cis* and *trans* forms.
    a. 1-pentene
    b. 2-hexene
    c. 2-methyl-2-hexene
    d. 2,3-dimethyl-2-butene
19. How can an asymmetric carbon be identified? Why does the presence of an asymmetric carbon produce stereoisomers?

Chem ASAP! Assessment 25.3 Check your understanding of the important ideas and concepts in Section 25.3.

SECTION REVIEW 25.3 CONTINUED

18. c. CH_3 / CH_3 C=C H / $CH_2CH_2CH_3$

2-methyl-2-hexene
(no geometric isomers)

d.

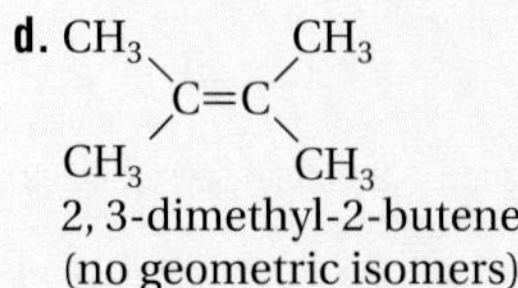

CH_3 / CH_3 C=C CH_3 / CH_3

2, 3-dimethyl-2-butene
(no geometric isomers)

19. An asymmetric carbon atom can be identified by four different substituents bonded to it. Molecules containing asymmetric carbons have handedness, and exist as stereoisomers.

25.3

3 Assess

Evaluate Understanding

Have students explain what the term *isomer* means and then describe the differences between structural isomers, geometric isomers, and stereo-isomers. Have them draw examples of each kind of iso-mer.

Reteach

Have students model *cis* and *trans* geometric isomers with their hands. Tell them their thumbs represent substituted groups on the carbons of a double bond represented by touching two fingers of each hand together. If both thumbs are on the same side of the bond, the molecule is the *cis*-configuration; if the thumbs are on opposite sides of the bond, it is the *trans*-configuration.

MINI LAB

Structural Isomers of Heptane

ANALYSIS AND CONCLUSIONS

1. Compounds that have the same molecular formula but different molecular structures.
2. heptane, 2-methylhexane, 3-methylhexane, 2,3-dimethylpentane, 2,4-dimethylpentane, 2,2-dimethylpentane, 3,3-dimethylpentane, 3-ethylpentane, 2,2,3-trimethylbutane
3. 4
4. Yes; they should each have a unique name because each is a unique structure.

SMALL-SCALE LAB

HYDROCARBON ISOMERS

OVERVIEW

Procedure Time: 20 minutes
Students learn to write line-angle formulas for hydrocarbons.

ANALYSIS

1. Students should draw formulas for pentane, 2-methyl butane, and 2,2-dimethyl propane.
2. Subtract from four the number of lines drawn to any point.
3. 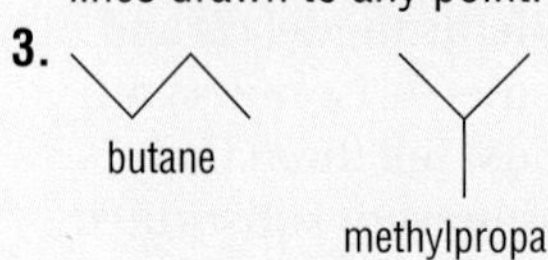

YOU'RE THE CHEMIST

1. 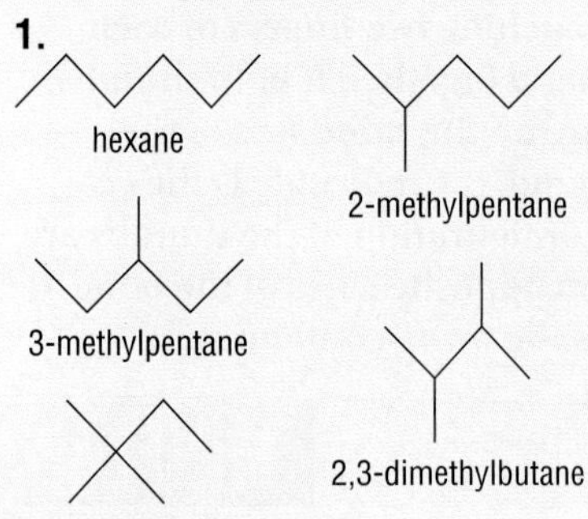

2. 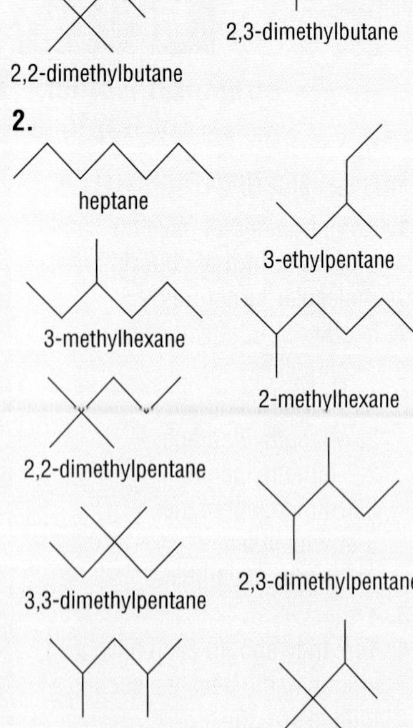

SMALL-SCALE LAB

HYDROCARBON ISOMERS

SAFETY

Use safe and proper laboratory procedures.

PURPOSE

To draw line-angle formulas and name some of the isomers in gasoline.

MATERIALS

- pencil
- paper
- toothpicks
- modeling clay

PROCEDURE

Gasoline is a complex mixture of hydrocarbon molecules. Each molecule contains between five and nine carbon atoms. Study the formulas and names of the isomers of C_5H_{12} in Figure A. Make a model of each isomer using toothpicks and modeling clay. Compare the models and make accurate drawings on your paper.

Various Formulas Representing Isomers of C_5H_{12}

Condensed	Line-angle	Space-filling
$CH_3CH_2CH_2CH_2CH_3$ pentane		
$CH_3CHCH_2CH_3$ with CH_3 on C-2; 2-methylbutane		
CH_3CCH_3 with CH_3 above and below; 2, 2-dimethylpropane		

Figure A

ANALYSIS

Using your experimental data, record the answers to the following questions below your drawings.

1. Expanded structural formulas represent all chemical bonds in a molecule as short lines. Draw the expanded structural formulas for each isomer of C_5H_{12} in Figure A.
2. In a line-angle formula, each line represents a C—C bond. Each end of a line, as well as the intersection of lines, represents a carbon atom. All hydrogen atoms are implied. Knowing that carbon always forms four bonds in organic compounds, explain how to determine the number of hydrogen atoms bonded to each carbon in a line-angle formula.
3. Because of its volatility, butane is used in the formulations of gasolines in cold climates during winter. Draw the condensed structural formulas, the line-angle formulas, and the space-filling formulas for the two isomers of butane (C_4H_{10}). Make models of each isomer.

YOU'RE THE CHEMIST

The following small-scale activities allow you to develop your own procedures and analyze the results.

1. **Analyze It!** Gasoline contains isomers of hexane. Draw the line-angle formulas and name the five isomers of C_6H_{14}. Make a model of one isomer and convert that model into the other four.
2. **Analyze It!** There are nine isomers of C_7H_{16}. Draw their line-angle formulas and name them.
3. **Analyze It!** There are eighteen isomers of C_8H_{18}. Draw their line-angle formulas and name them.
4. **Analyze It!** See how many isomers of C_9H_{20} you can draw and name.
5. **Analyze It!** Gasoline also contains cyclic alkanes. Draw the line-angle formulas and name several isomers of C_5H_{10} and C_6H_{12}.

3. The isomers increase in complexity from octane to 2,2,3,3-tetramethylbutane.

methylcyclopentane

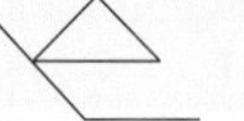

4. There are 35 isomers of C_9H_{20}
5. The 15 possible answers include

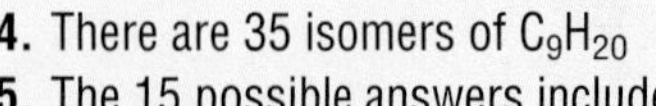

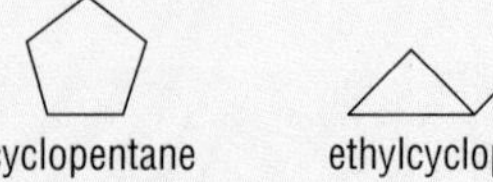

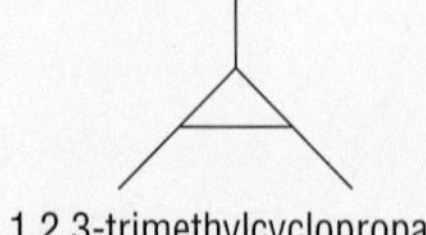

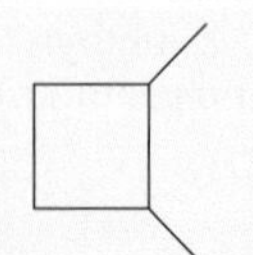

HYDROCARBON RINGS

section 25.4

Not everyone loves carrots, but you have probably heard about how good they are for you. One reason carrots are a healthful food choice is that they contain betacarotene, an important nutrient for vision. Betacarotene also gives carrots their orange color. The hydrocarbon skeleton of betacarotene contains hydrocarbon rings. **What is a hydrocarbon ring?**

objectives

- Identify common cyclic ring structures
- Explain resonance in terms of the aromatic ring of benzene

key terms

- cyclic hydrocarbons
- aliphatic compounds
- arenes
- aromatic compound

Cyclic Hydrocarbons

So far, you have seen a wide variety of short and long carbon chains; chains containing single, double, and triple covalent bonds between carbon atoms; and chains containing a number of different substituents. What if the ends of these chains linked up to form rings?

In some hydrocarbon compounds, the two ends of a carbon chain are attached to form a ring. Compounds that contain a hydrocarbon ring are called **cyclic hydrocarbons.** The structures of some cyclic hydrocarbons are shown in **Figure 25.13.** Rings containing from 3 to 20 carbons are found in nature. Five- and six-membered rings are the most abundant. Hydrocarbon compounds that do not contain rings are known as **aliphatic compounds.** They include compounds with both short and long carbon chains. The hydrocarbon compounds you have learned about in the previous sections are aliphatic compounds.

Aromatic Hydrocarbons

A special group of unsaturated cyclic hydrocarbons are known as **arenes.** These compounds contain single rings or groups of rings. The arenes were originally called aromatic compounds because many of them have pleasant odors. Benzene (C_6H_6) is the simplest arene. Today the term **aromatic compound** is applied to any substance in which the bonding is like that of benzene.

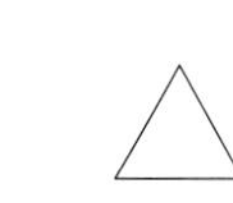

Cyclopropane (bp −34.4 °C)

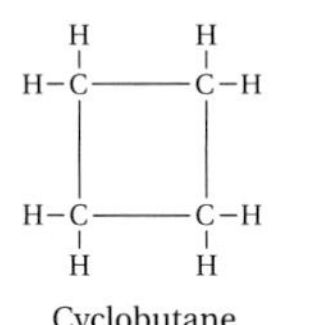

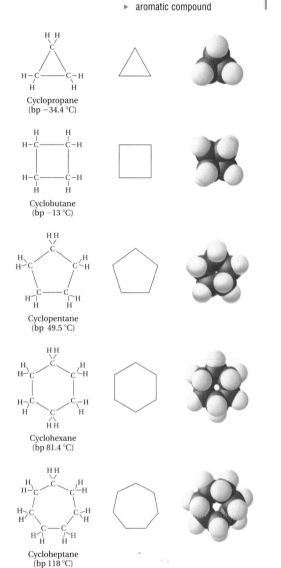

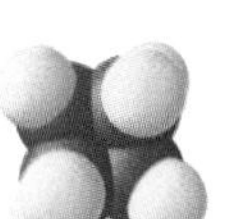

Cyclobutane (bp −13 °C)

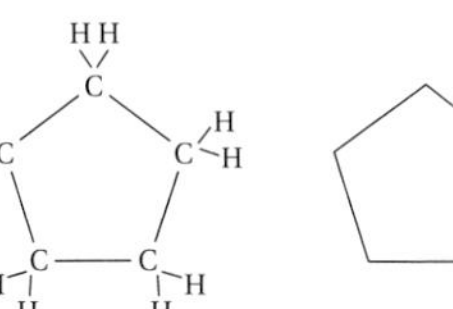

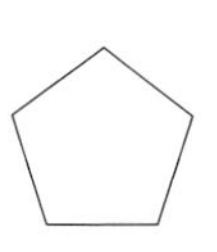

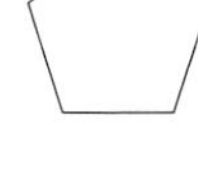

Cyclopentane (bp 49.5 °C)

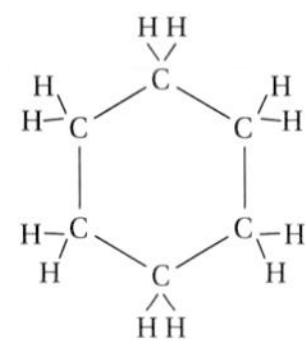

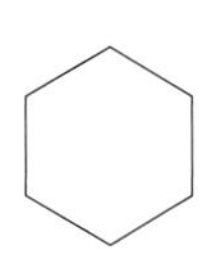

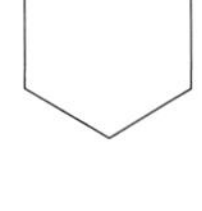

Cyclohexane (bp 81.4 °C)

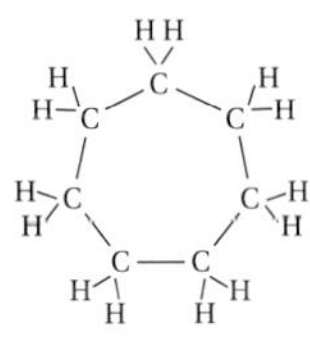

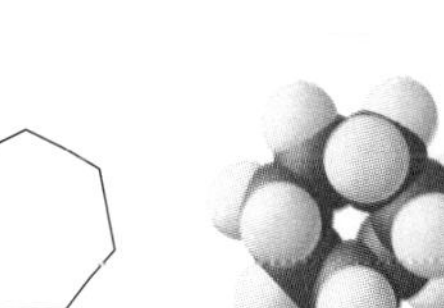

Cycloheptane (bp 118 °C)

Figure 25.13 ▶

Molecular structures for a series of cycloalkanes are shown in the left column. The middle column shows a convenient way to represent cyclic rings. In these minimal structures, each intersection of lines represents a carbon atom and the hydrogens bonded to it. The space-filling models in the right column show the shapes of the rings. Cycloalkenes can be drawn similarly but are not shown here.

25.4

1 Engage

Use the Visual

Have students study the photograph and read the text that opens the section. Write the structure for β-carotene, which yields Vitamin A, on the board.

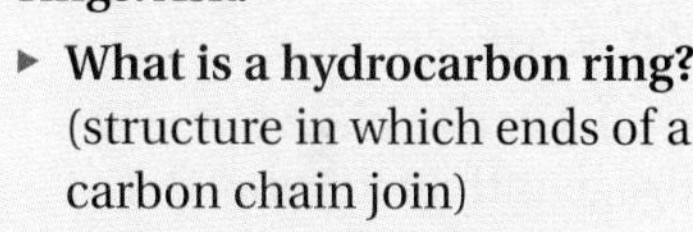

Explain that the flexibility of a carbon chain makes it possible for the carbon atoms at the ends of a chain to meet, attach, and form a hydrocarbon ring. Many naturally occurring hydrocarbon compounds contain rings. Ask:

- **What is a hydrocarbon ring?** (structure in which ends of a carbon chain join)

Check Prior Knowledge

To assess students' prior knowledge about saturated and unsaturated hydrocarbons, write the formulas for hexene and hexane on the board. Ask:

- **Why does hexene have two fewer hydrogen atoms than hexane?** (Students should explain that the presence of a carbon–carbon double bond reduces the number of electrons used for forming covalent bonds with other atoms, such as hydrogen.)
- **Is the classification of a compound as unsaturated determined by the number of hydrogens in one of its molecules or by the types of bonds?** (The types of bonds; to be unsaturated, the molecule must have at least one double or triple bond.)

STUDENT RESOURCES

From the Teacher's Resource Package, use:

- Section Review 25.4, Ch. 25 Practice Problems, Interpreting Graphics, and Quizzes from the Review Module (Ch. 25–28)

TECHNOLOGY RESOURCES

Relevant technology resources include:

- Chem ASAP! CD-ROM
- Resource Pro CD-ROM

2 Teach

Discuss

Students should be made aware that the formation of a ring causes bond strain in all cycloalkanes with rings that contain fewer than six carbons. The bond angle in these cycloalkanes tends to be less than the normal carbon–carbon bond angle of 109.5°. This bonding strain explains the increased chemical reactivity of small–ring cycloalkanes. Discuss the convenient minimal structural formulas for cycloalkanes in which carbon atoms are represented by the intersection of lines. Ask:

- **Cyclohexane has 2 fewer hydrogens than hexane. Does this mean it is unsaturated?** (No; it does not have any double or triple bonds.)

CHEMISTRY AND SCIENCE HISTORY

August Kekulé (1829–1896) proposed that carbon atoms link together to form chains. Later he pondered the problem of benzene, which seemed to have too few hydrogen atoms to exist in chain form. After working on the benzene problem one evening, Kekulé fell asleep in front of his fireplace and dreamed of snakes twisting and turning. One of the snakes grasped its tail in its mouth and formed itself into a ring. Kekulé awoke at once and worked out his ring structure for benzene that night. Afterwards, when other scientists were confronted with difficult problems, Kekulé told them: "Let us learn to dream, gentlemen, and then perhaps we shall learn the truth."

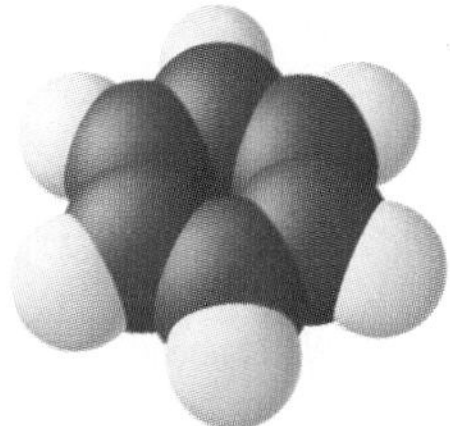

Figure 25.14
A space-filling model of benzene.

The benzene molecule has a six-membered carbon ring with one hydrogen attached to each carbon. This leaves one electron from each carbon free to participate in a double bond. Two different structures with double bonds can be written for benzene.

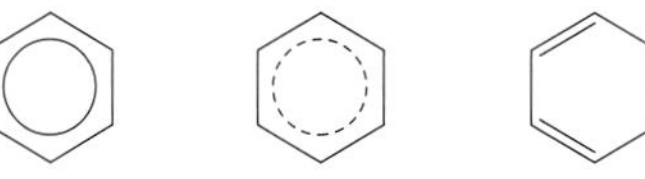

These structural formulas show only the extremes in electron sharing between any two adjacent carbons in benzene. One extreme is a normal single bond. The other extreme is a normal double bond. When two or more equally valid structures can be drawn for a molecule, resonance occurs. The benzene molecule exhibits resonance. Benzene and other molecules that exhibit resonance are more stable than similar molecules that do not exhibit resonance. Thus benzene is not as reactive as six-carbon alkenes.

The inscribed circle is a good way to represent the nature of resonance bonding. However, it does not show the number of electrons involved. For this reason, the traditional structure, which is the structure shown on the right, is used in this textbook.

Compounds containing substituents attached to the benzene ring are named as derivatives of benzene. Sometimes the benzene ring is named as a substituent on an alkane. In such instances the C_6H_5— group is called a phenyl group.

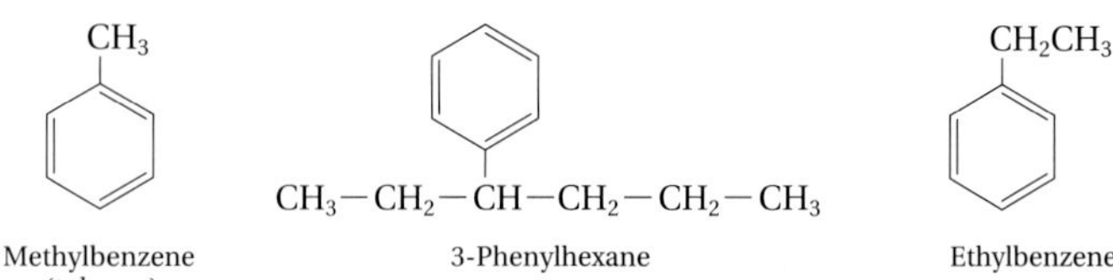

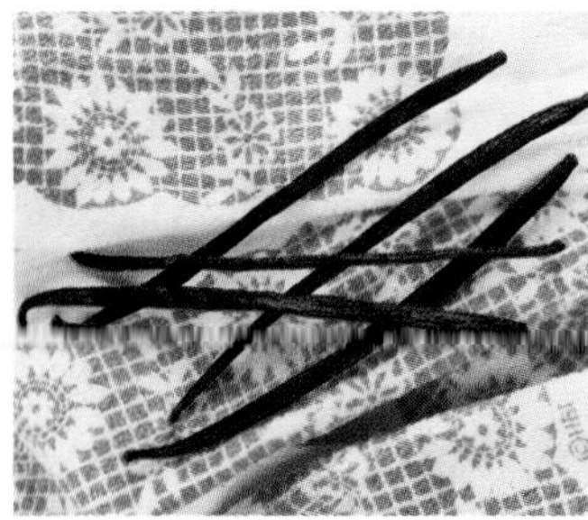

Figure 25.15
Part of the molecule that produces the distinctive aroma of vanilla beans is derived from aromatic hydrocarbons. Portions of many dyes are also derived from aromatic hydrocarbons.

Some derivatives of benzene have two substituents. These derivatives are called disubstituted benzenes. Three different structural isomers occur for the liquid aromatic compound dimethylbenzene ($C_6H_4(CH_3)_2$). Once again, the physical properties of structural isomers are different, as indicated by their boiling points.

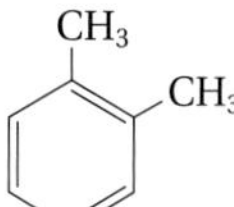

1,2-dimethylbenzene
(*o*-dimethylbenzene,
o-xylene)
(bp 144 °C)

CH_3 / CH_3

1,3-dimethylbenzene
(*m*-dimethylbenzene,
m-xylene)
(bp 139 °C)

CH_3 / CH_3

1,4-dimethylbenzene
(*p*-dimethylbenzene,
p-xylene)
(bp 138 °C)

In the IUPAC naming system, the possible positions of two substituents in disubstituted benzene are designated as 1,2; 1,3; or 1,4. Common names for disubstituted benzenes use the terms *ortho, meta,* and *para* (abbreviated *o, m,* and *p*) in place of numbers. The dimethylbenzenes are also called xylenes.

section review 25.4

20. Name the following compounds.

a. H_2C, $C H_2$, CH_2, H_2C, CH_2, $C H_2$

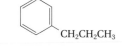

c.
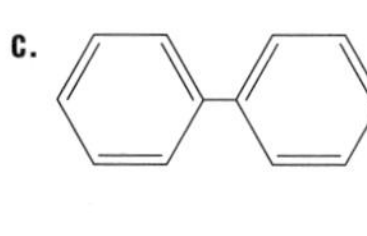

b. CH_2CH_3, $CH_2CH_2CH_3$

d. CH_2CH_3
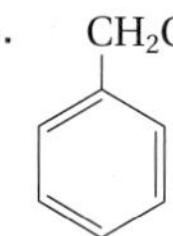

21. Using structures, explain why benzene exhibits resonance. How does resonance affect stability?

22. The molecules of both cyclohexane and benzene have six carbon atoms bonded in a ring. What are some differences between these two compounds?

23. Name the following compounds.

a.
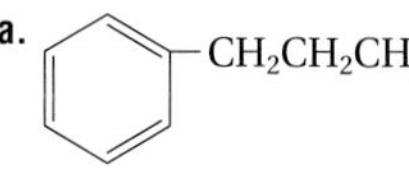

b. $CH_3—CH—CH_3$
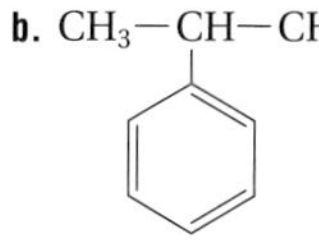

Chem ASAP! Assessment 25.4 Check your understanding of the important ideas and concepts in Section 25.4.

portfolio project

Research the origin of the International Union of Pure and Applied Chemistry (IUPAC). Find out when, where, and why the first international congress was held. Present your findings in a written report.

25.4

Discuss

Using the diagrams at the bottom of page 760 and on the top of page 761, discuss the nomenclature of aromatic compounds. Discuss additional examples such as *para*-dichlorobenzene (mothballs) or DDT, dichlorodiphenyltrichloroethane.

3 Assess

Evaluate Understanding

Ask students to compare and contrast the following terms as they relate to hydrocarbons.

1. *aromatic* and *aliphatic*
2. *chain* and *cyclic*
3. *saturated* and *unsaturated*

Reteach

Students are being introduced to many new terms in this chapter. Before they learn any more terms, you may wish to review *alkane, alkene, alkyne, cycloalkane, aliphatic, arene,* and *aromatic compounds.* Note that both groups of unsaturated hydrocarbons have a name ending in -ene. Explain that it is not necessary for a compound to have a distinctive odor for it to be an aromatic compound. *Aromatic* refers strictly to the presence of a benzene ring in the structure.

Portfolio Project

The first IUPAC international congress was held in Geneva, Switzerland, in 1892. It was organized by scientists who wanted to standardize the naming of organic compounds.

Answers

SECTION REVIEW 25.4

20. a. cyclohexane
b. 1-ethyl-3-propylbenzene
c. phenylbenzene (biphenyl)
d. ethylbenzene

21. Molecules that exhibit resonance are more stable than similar molecules that do not exhibit resonance.

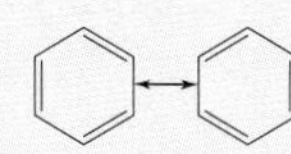

22. Cyclohexane is a saturated cyclic hydrocarbon; it behaves chemically like other saturated hydrocarbons. Benzene is an unsaturated aromatic ring, and its apparent double bonds are stabilized by resonance; compared with alkenes benzene is relatively unreactive.

23. a. 1-phenylpropane
b. 2-phenylpropane

1 Engage

Use the Visual

Have students study the photograph and read the text that opens the section. Explain that Earth's atmosphere does not contain methane or ethane to any appreciable extent. Ask:

- **What are the gases in Earth's atmosphere?** (N_2 (78%), O_2 (21%), Ar (1%), and CO_2 (0.03%) are the main components. Water vapor, H_2O, is present in variable amounts.)
- **What are the sources of hydrocarbons on Earth and how do they form?** (Ultimately, all hydrocarbon compounds are directly or indirectly produced by photosynthetic organisms, which use solar energy to fix the carbon in CO_2 into hydrocarbons and hydrocarbon derivatives. Photosynthetic organisms that lived and died millions of years ago, together with the marine and land animals that fed on them, are the primary sources of the hydrocarbons used on Earth today.)

Check Prior Knowledge

To assess students' prior knowledge of fossil fuels, ask:

- **When did people first begin to utilize petroleum and natural gas as energy sources?** (It was not until the invention of the kerosene lamp that people valued petroleum. By the time electric lights reduced the need for kerosene, automobiles required gasoline, which previously had been discarded as a useless by-product of kerosene production)
- **Before the petroleum era began, what types of energy sources were used?** (fats from domesticated animals, whale oil, wood, and coal, which is also a fossil fuel)

section 25.5

objectives

- Identify three important fossil fuels and describe their origins
- Name some products obtained from natural gas, petroleum, and coal

key term

- cracking

HYDROCARBONS FROM THE EARTH

Need a fill-up? If you were traveling near Saturn's moon, Titan, you might easily fill up your tank with some of the many hydrocarbons in Titan's atmosphere. But here on Earth, hydrocarbon fuels are more difficult to obtain.

What are the sources of hydrocarbons on Earth, and how do they form?

Natural Gas

You have learned quite a lot about hydrocarbons. But one question that has yet to be answered is where hydrocarbons come from. To explore the origins of hydrocarbons, you will need to look back into Earth's past, back before the age of dinosaurs.

Much of the world's energy is supplied by burning fossil fuels. Fossil fuels are organic because they are derived from the decay of organisms in geologic history. Millions of years ago, marine life settled on the ocean floors and became buried in ocean sediments. Heat, pressure, and the action of bacteria changed this residue into petroleum and natural gas, which are two important fossil fuels. Petroleum and natural gas contain mostly aliphatic, or straight-chain, hydrocarbons. **Figure 25.16** illustrates how natural gas is often found overlying oil deposits or in separate pockets in rock.

Natural gas is an important source of alkanes of low molar mass. Typically, natural gas is composed of about 80% methane, 10% ethane, 4%

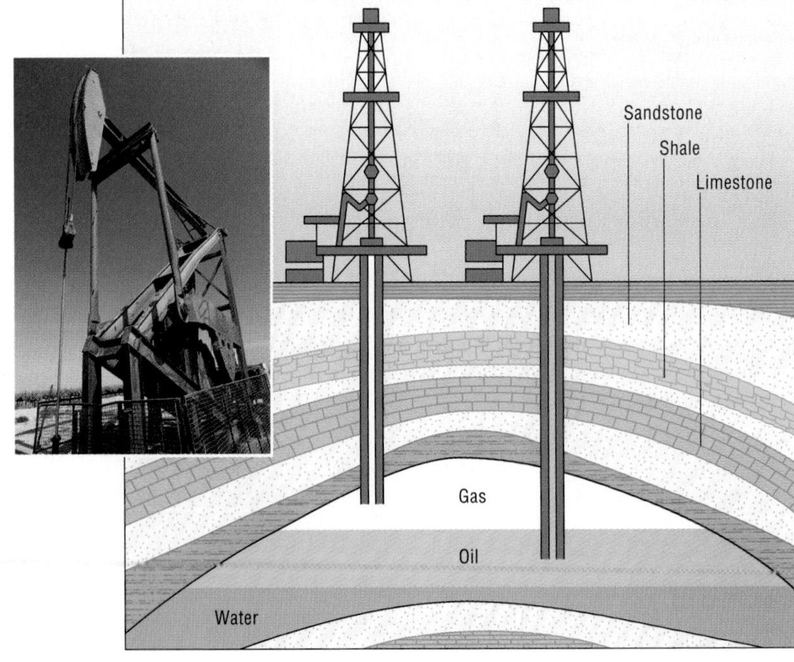

Figure 25.16
Natural gas and oil are typically found in dome-shaped geological formations. Petroleum prospectors drill oil wells to tap into the gas and oil. Sometimes the gas is under pressure and will force the oil up the well pipe, but pumping is usually required.

STUDENT RESOURCES

From the Teacher's Resource Package, use:

- Section Review 25.5, Ch. 25 Practice Problems, Vocabulary Review, Quizzes, and Tests from the Review Module (Ch. 25–28)
- Solutions Manual for Chapter Reviews
- Graphing Calculator Problems

TECHNOLOGY RESOURCES

Relevant technology resources include:

- Chem ASAP! CD-ROM
- ResourcePro CD-ROM
- Assessment Resources CD-ROM: Ch. 25 Tests
- Chemistry Alive! Videodisc: *Methane Bubbles*

propane, and 2% butane. The remaining 4% consists of nitrogen and higher molar mass hydrocarbons. Natural gas also contains a small amount of the noble gas helium and is one of its major sources. Methane, the major constituent of natural gas, is especially prized for combustion because it burns with a hot, clean flame.

$$CH_4(g) + 2O_2(g) \longrightarrow CO_2(g) + 2H_2O(g) + \text{heat}$$

Propane and butane, which are separated from the other gases by liquefaction, are also good heating fuels. They are sold in liquid form in pressurized tanks as liquid petroleum gas (LPG).

A sufficient supply of oxygen is necessary to oxidize a hydrocarbon fuel completely and to obtain the greatest amount of heat. Complete combustion of a hydrocarbon gives a blue flame. Incomplete combustion gives a yellow flame. This is due to the formation of small, glowing carbon particles that are deposited as soot when they cool. Carbon monoxide, a toxic gas, is also formed along with carbon dioxide and water during incomplete combustion.

Petroleum

The organic compounds found in petroleum, or crude oil, are more complex than those in natural gas. Most of the hydrocarbons in petroleum are straight-chain and branched-chain alkanes. Petroleum also contains small amounts of aromatic compounds and sulfur-, oxygen-, and nitrogen-containing organic compounds.

Humans have known about petroleum for centuries; ancient peoples found it seeping from the ground in certain areas. In the late 1850s, a vast deposit of petroleum was discovered in Pennsylvania when a well was drilled to obtain petroleum for use as a fuel. Within decades, petroleum deposits had also been found in the Middle East, Europe, and the East Indies. Petroleum has since been found in other parts of the world as well.

Crude oil must be refined before it is commercially useful. The refining process consists of distilling crude oil to divide it into fractions according to boiling point. A schematic of a petroleum refining distillation tower is shown in **Figure 25.17.** Each distillation fraction contains several different hydrocarbons. The fractions, which have a variety of uses, are listed in **Table 25.3.**

Fractionating Column

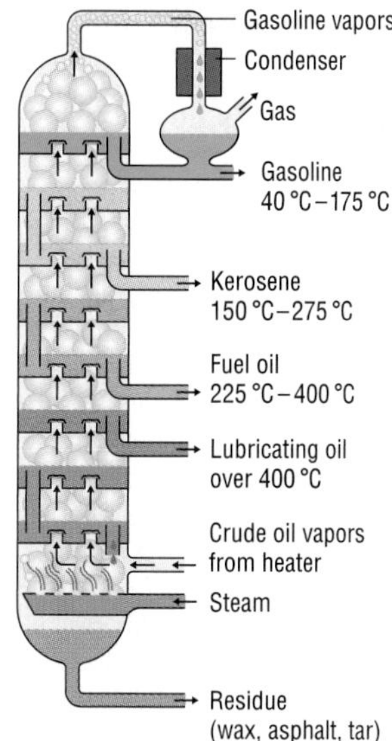

Figure 25.17

Petroleum is refined by fractional distillation. The crude oil is heated so that it vaporizes and rises through the fractionating column. Compounds with the highest boiling points condense first near the bottom. Compounds with the lowest boiling points condense last near the top.

Table 25.3

Fractions Obtained from Crude Oils

Fraction	Composition of carbon chains	Boiling range (°C)	Percent of crude oil
Natural gas	C_1 to C_4	Below 20	
Petroleum ether (solvent)	C_5 to C_6	30 to 60	10%
Naphtha (solvent)	C_7 to C_8	60 to 90	
Gasoline	C_6 to C_{12}	40 to 175	40%
Kerosene	C_{12} to C_{15}	150 to 275	10%
Fuel oils, mineral oil	C_{15} to C_{18}	225 to 400	30%
Lubricating oil, petroleum jelly, greases, paraffin wax, asphalt	C_{16} to C_{24}	Over 400	10%

25.5

2 Teach

Discuss

Ask students why burning natural gas, rather than coal or oil, is such a popular method for heating homes, drying clothes, and cooking food. (Included in their answers should be things like ease of transportation, ease of regulation, burns clean, high heat of combustion, and relatively low cost.) **Review the balanced equation for combustion of methane and explain the conditions that lead to the formation of CO.** (When oxygen is in limited supply, incomplete combustion occurs.)

Cooperative Learning

Divide the class into five groups. Have each group create a display on one of the fractions obtained from crude oil (see Table 25.3). Each group should make a pie graph of the percents of the fractions in crude oil, highlighting the slice that represents their fraction. Displays should also include information on the composition of the fraction and uses of the products derived from that fraction. Students can illustrate their displays with original art or magazine photographs.

Meeting Diverse Needs

Gifted Have students research the energy sources used in the United States. Have them summarize their findings in a bar chart showing to what extent the United States depends on each type of energy source. The bar chart should profile the energy sources for the years 1850, 1900, 1950, 1990, and prediction for the year 2000.

Gifted Have gifted students research one or more specific examples in which bacteria have been used to clean up an oil spill. Was the method successful? Why or why not? Challenge the students to explain how such organisms are able to metabolize hydrocarbons. What are the products of the bacterial digestion of oil?

Discuss

Have students study and read the caption to Figure 25.18. If possible, show the students samples of peat, lignite, bituminous, and anthracite. Explain that coal can be converted to methane by a process called coal gasification in which coal and hydrogen react under high temperature and pressure. Synthetic crude oil can be made by a process called coal liquefaction in which coal is dissolved in a solvent at high pressure and temperature. The product, called solvent refined coal, is then purified and heated. The hard, brittle residue that remains can be melted and used as a liquid fuel.

Build Writing Skills

Provide students with an opportunity to research some aspect of coal usage, either as a source of energy or of other chemicals. Reports should include information about economic value, environmental concerns, and/or energy concerns.

Chemistry Alive!

Methane Bubbles

Play

ECOLOGY

Oil Spills

Spilled oil from oil tankers can destroy fragile marine ecosystems. Because cleaning up oil spills is difficult and expensive, researchers are developing new cleanup technologies and safer transportation methods. Most current cleanup methods involve containing the oil and either pumping it away or soaking it up with absorbent materials. Some of the new developments include the genetic engineering of bacteria to "eat" up the oil spill and the design of double-hulled tankers that are less likely to spill their oil in a collision.

The fractions containing compounds of higher molar mass can be "cracked" to produce the more useful short-chain components of gasoline and kerosene. **Cracking** is a controlled process by which hydrocarbons are broken down or rearranged into smaller, more useful molecules. Hydrocarbons are cracked with the aid of a catalyst and with heat. By this process, petroleum is the principal source of raw materials for the organic chemicals industry. For example, low-molar-mass alkanes are starting materials for the manufacture of paints and plastics.

Coal

Geologists believe that coal had its origin some 300 million years ago when huge tree ferns and mosses grew abundantly in swampy tropical regions. When the plants died, they formed thick layers of decaying vegetation. Layer after layer of soil and rock eventually covered the decaying vegetation, which caused a buildup of intense pressure. This pressure, together with heat from Earth's interior, slowly turned the plant remains into coal.

The first stage in the formation of coal is an intermediate material known as peat. Peat is a soft, brown, spongy, fibrous material rather like decayed and compressed garden refuse. When first dug out of a peat bog, it has a very high water content. After it has been allowed to dry, it produces a low-cost but smoky fuel. If peat is left in the ground, it continues to change. After a long period of time, it loses most of its fibrous texture and becomes lignite, or brown coal. Lignite is much harder than peat and has a higher carbon content (about 50%). The water content, however, is still high. Continued pressure and heat slowly change lignite to bituminous, or soft coal. Bituminous coal has a lower water content and higher carbon content

Figure 25.18
Coal formed when plants similar to these died and formed layers of decaying organic material. The material was compressed over millions of years between layers of soil and rock. The first stage in coal formation is peat. Continued pressure and heat transform peat into forms of coal—lignite, bituminous, and anthracite.

Figure 25.19
Coal is mined from both surface mines, shown here, and underground mines. Underground coal seams may be thin and wandering, making mining difficult and hazardous. Modern machinery is reducing the number of miners who have to work underground.

(70–80%) than lignite. In some regions of Earth's crust, even greater pressures have been exerted. In those places, such as eastern Pennsylvania, soft coal has been changed into anthracite, or hard coal. Anthracite has a carbon content that exceeds 80%, making it an excellent fuel source.

Coal, which is usually found in seams from 1 to 3 meters thick, is obtained from both underground and surface mines. In North America, coal mines are usually less than 100 meters underground. Much of the coal is so close to the surface that it is strip-mined. By contrast, many coal mines in Europe and other parts of the world extend 1000 to 1500 meters below Earth's surface.

Coal consists largely of condensed ring compounds of extremely high molar mass. These compounds have a very high proportion of carbon compared with hydrogen. Due to the high proportion of these aromatic compounds, coal leaves more soot upon burning than do the more aliphatic fuels obtained from petroleum. The majority of the coal once burned in North America contained about 7% sulfur, which burns to form the major air pollutants SO_2 and SO_3.

Coal may be distilled to obtain a variety of products: coal gas, coal tar, coke, and ammonia. Coal gas consists mainly of hydrogen, methane, and carbon monoxide, all of which are flammable. Coal tar can be distilled further into benzene, toluene, naphthalene, phenol, and pitch. Coke is used as a fuel in many industrial processes. Because it is almost pure carbon, coke produces intense heat and little or no smoke when it burns. The ammonia from distilled coal is converted to ammonium sulfate for use as a fertilizer.

section review 25.5

24. What are the three major fossil fuels? How did each form?
25. What is the source of methane? Of aliphatic hydrocarbons? Of aromatic hydrocarbons?
26. Write a balanced equation for the complete combustion of pentane. What other products would be formed if combustion were incomplete?
27. Define petroleum refining and cracking.

Chem ASAP! **Assessment 25.5** Check your understanding of the important ideas and concepts in Section 25.5.

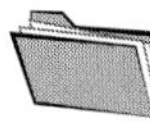

portfolio project

Research the structures that petroleum geologists use to locate oil. Prepare a poster with labeled diagrams showing each type of geologic structure. Explain what the structures have in common that makes them good indicators for oil deposits.

25.5

3 Assess

Evaluate Understanding

Have students compare the formation of coal, natural gas, and petroleum. (Petroleum and natural gas are derived from marine organisms buried under ocean sediments; coal, from buried wetland plants. Natural gas and petroleum contain mostly aliphatic compounds; coal contains a large percentage of aromatic compounds.) **Have students name some products made from natural gas, petroleum, and coal.** (Answers will vary.) **Ask students to put the forms of coal** (bituminous, anthracite, peat, and lignite) **in order according to hardness and potential energy.** (peat, lignite, bituminous, and anthracite. The harder the coal, the more energy it provides.)

Reteach

Display Figure 25.17 on an overhead projector. If possible, bring a sample of crude oil to class so that students can observe its color and odor. Explain that petroleum is a complex mixture of hydrocarbons whose molecules range in size from 5 to 30 carbon atoms. To be used effectively, the crude oil must be refined by fractional distillation, which separates compounds on the basis of their boiling points. Explain that in general, low molar mass compounds have low boiling points and high molar mass compounds have high boiling points; smaller hydrocarbons boil off first.

Portfolio Project

Petroleum geologists look for sedimentary rock structures that can trap oil and allow it to collect in large quantities. For example, oil often is found near salt domes and faults.

Answers

SECTION REVIEW 25.5

24. coal, petroleum, and natural gas; Petroleum and natural gas were formed from marine life buried in the sediments of the oceans. Coal was formed from decayed vegetation buried under rocks and soil.
25. Methane, the major constituent of natural gas, is the by-product of the transformation of marine life into petroleum. Aliphatic hydrocarbons come from petroleum; aromatic hydrocarbons come from coal.
26. $C_5H_{12} + 8O_2 \rightarrow 5CO_2 + 6H_2O$; CO and C
27. Petroleum refining is the distillation of crude oil to separate it into fractions. Cracking is the breaking down and rearranging of hydrocarbon chains into smaller molecules.

Chemistry Serving... the Consumer

DISCUSS

Review the refining processes used to convert large molar mass hydrocarbons to smaller branched-chain hydrocarbons. Point out that the petroleum industry uses *cracking, reforming,* and *isomerization* to produce gasolines with the desired octane rating. Cracking is the process by which large hydrocarbons are broken down into smaller molecules:

$$C_{16}H_{34} + H_2 \xrightarrow{\text{catalyst, heat}} 2\ C_8H_{18}$$

Reforming and isomerization are the processes used to convert straight-chain alkanes to branched-chain alkanes:

$$C_8H_{18} \xrightarrow{\text{catalyst}} CH_3C(CH_3)_2CH_2CH(CH_3)_2$$

(Draw the structural formula for 2,2,4-trimethylpentane.) Ask:

- Why is this reaction called an isomerization? (The reactant and product are structural isomers.)
- What are the names of the reactant and product of this reaction? (*n*-octane; 2,2,4-trimethylpentane) Have students identify other branched chain isomers of octane that could be produced by the isomerization reaction. (Examples include: 2,5-dimethylhexane, 2,4-dimethylhexane, and 2,3,4-trimethylpentane.)

CHEMISTRY IN CAREERS

Encourage students to connect to the Internet address shown on this page to find out more about careers in organic chemistry. Also see the information on page 880 of this text.

Chemistry Serving...the Consumer

A NUMBER YOU CAN'T KNOCK

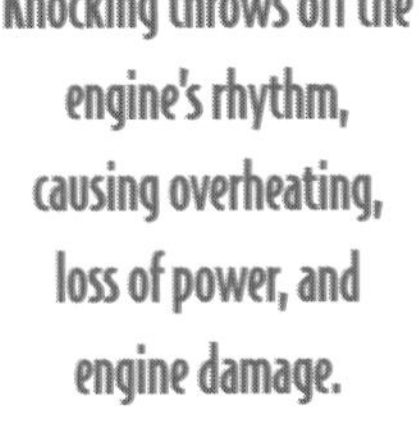

When you buy gas for your car, you have a choice of fuels with different octane numbers, usually from 87 to 93. What do these numbers mean? Because higher octane gasoline costs more, many people believe that octane is an indicator of fuel quality—that higher octane fuel will give them better performance. But this belief is not quite accurate. The octane number is actually a measurement of the fuel's ability to resist engine "knock", which occurs when the fuel ignites by itself in the engine cylinder.

Engine knock has been a problem in internal combustion gasoline engines since they began to be widely used at the beginning of the twentieth century. Knocking throws off the engine's rhythm, causing overheating, loss of power, and engine damage. It is influenced by two main variables.

Knocking throws off the engine's rhythm, causing overheating, loss of power, and engine damage.

The first variable is engine compression—the degree to which the gas-air mixture in the cylinder is compressed prior to ignition by the spark plug. The higher the compression, the more power delivered by the ignition of the fuel. But higher compression also means greater likelihood of engine knock.

The second variable is the chemical structure of the hydrocarbons that make up the fuel. Straight-chain alkanes are more prone to knocking than are branched-chain alkanes, cycloalkanes, and aromatic hydrocarbons.

In practice, these two variables mean that engine compression—and thus performance—is limited by the composition of the gasoline used. In the early part of the twentieth century, this was a problem for automobile engineers. They could design an engine with any compression ratio, but they could not be sure of the quality of the fuel the engine would burn. They needed a method for measuring each fuel's resistance to knocking. The octane rating system was born out of this need.

The designers of the octane system developed a way of testing fuels in a special, one-cylinder engine with a variable compression ratio. For each fuel tested, they could increase the compression until the engine began to knock. Then they chose two pure fuels as standards against which to measure others: isooctane (2,2,4-trimethylpentane) and heptane. These two hydrocarbons had similar physical properties, but isooctane was very resistant to knocking, and heptane was very prone to it. The designers arbitrarily assigned the heptane an octane rating of 0 and the isooctane a rating of 100. Every fuel tested could then be assigned an octane number relative to these. An octane rating of 89, for example, meant that a fuel performed the same in the special test engine as a mixture that was 89 parts isooctane and 11 parts heptane.

This fuel rating system is still in use today. The major modification is that two somewhat different testing methods are used. One, called the motor octane rating (M), tests the fuel under conditions that approximate high-speed driving with heavy loads. The other, called the research octane rating (R), uses conditions that simulate more normal driving. Usually, the octane number assigned to gasolines is an average of the numbers given by these two methods. You can see this indicated on gasoline pumps as (R + M) / 2.

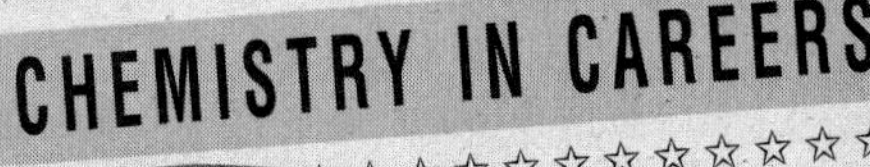

ORGANIC CHEMIST

Does synthesizing organic compounds match your career interests?

See page 880.

CHEMIST NEEDED

Enviro. Lab. GC Must know

TAKE IT TO THE NET

Find out more about career opportunities: www.phschool.com

CHEMICAL SPECIALIST

Local food service distributor seeks responsible self-motivated indivi...

Chapter 25 STUDENT STUDY GUIDE

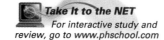

Take It to the NET
For interactive study and review, go to www.phschool.com

KEY TERMS

- aliphatic compound *p. 759*
- alkane *p. 743*
- alkene *p. 752*
- alkyl group *p. 748*
- alkyne *p. 753*
- arene *p. 759*
- aromatic compound *p. 759*
- asymmetric carbon *p. 755*
- branched-chain alkane *p. 748*
- *cis* configuration *p. 754*
- condensed structural formula *p. 746*
- cracking *p. 764*
- cyclic hydrocarbon *p. 759*
- geometric isomer *p. 754*
- homologous series *p. 745*
- hydrocarbons *p. 743*
- saturated compound *p. 752*
- stereoisomer *p. 755*
- straight-chain alkane *p. 745*
- structural isomer *p. 754*
- substituent *p. 748*
- *trans* configuration *p. 754*
- unsaturated compound *p. 752*

CONCEPT SUMMARY

25.1 Hydrocarbons

- Carbon makes stable covalent bonds with other carbons to form chain and ring compounds.
- Hydrocarbons are compounds containing only carbon and hydrogen. Alkanes contain only carbon–carbon single bonds.
- Straight-chain hydrocarbons contain carbon atoms in a chain, one after another. Branched-chain hydrocarbons are formed when one or more substituent alkyl groups are added to a straight-chain hydrocarbon.

25.2 Unsaturated Hydrocarbons

- Alkenes are unsaturated hydrocarbons containing at least one carbon–carbon double bond.
- Alkynes are unsaturated hydrocarbons containing at least one carbon–carbon triple bond.

25.3 Isomerism

- Structural isomers have the same molecular formula but different molecular structures.
- Alkenes may exhibit geometric isomerism according to whether substituents are on the same side or on opposite sides of the double bond.
- Stereoisomers can occur if four different groups are attached to a single carbon atom.

25.4 Hydrocarbon Rings

- Aromatic hydrocarbons, or arenes, are related to the hydrocarbon benzene.
- As a result of resonance, the interior bonds of the benzene ring are somewhere between ordinary single bonds and double bonds. Benzene is less reactive than ordinary alkenes because of this unusual bonding.

25.5 Hydrocarbons from the Earth

- The fossil fuels—coal, petroleum, and natural gas—had their origin in marine life buried in ocean sediments millions of years ago and subjected to the effects of heat, pressure, and bacteria.
- Aliphatic hydrocarbons come mainly from natural gas and petroleum. Aromatic hydrocarbons come mainly from coal.
- The molecular structures of the hydrocarbons present in natural petroleum can be broken down into other useful products by cracking.

CHAPTER CONCEPT MAP

Use these terms to construct a concept map that organizes the major ideas of this chapter.

ChemASAP! Concept Map 25
Create your Concept Map using the computer.

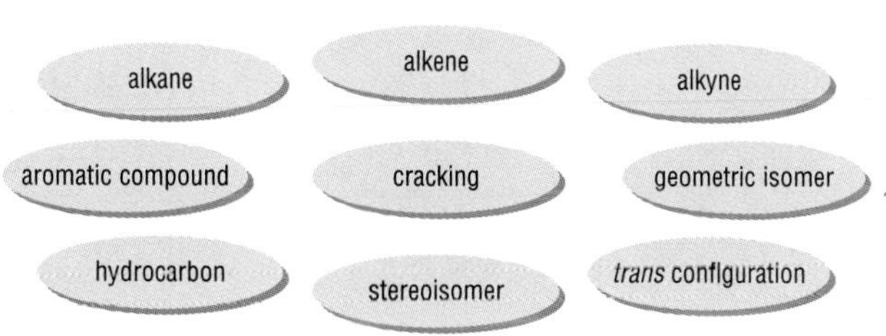

Take It to the Net

At **www.phschool.com** students will find for this chapter

- an Internet activity
- links to related chemistry sites
- an interactive quiz
- career links

4 Close

Summary

Ask the following question, which requires students to summarize information contained in the chapter.

- **What are the differences among saturated, unsaturated, and aromatic hydrocarbons? Give an example of each. List your examples in order of increasing chemical stability and in order of increasing boiling point. Explain the trend in boiling points in terms of the kinds of attractive intermolecular forces that hold hydrocarbon molecules together in the liquid state.**

Extension

In the carbon cycle, carbon atoms move back and forth between atmospheric carbon dioxide and solid nutrients. Have students research the carbon cycle and create a diagram showing the movement of carbon between marine and land ecosystems. Ask them to describe the impact that human activities have had on the carbon cycle.

Looking Back . . . Looking Ahead . . .

The preceding chapter presented an overview of the general chemistry of metallic and nonmetallic elements, including their sources, properties, and uses. This chapter discusses an important nonmetal, carbon: its chemistry and compounds. The next chapter will discuss hydrocarbon derivatives.

Answers

28. $CH_3CH_2CH_2CH_2CH_3$
pentane
$CH_3CH_2CH_2CH_2CH_2CH_3$
hexane

29. **a.** propane **c.** pentane
b. octane

30.

```
  H        H  H  H
  |        |  |  |
H-C-     H-C--C--C-
  |        |  |  |
  H        H  H  H
methyl      propyl

  H  H
  |  |
H-C--C-
  |  |
  H  H
  ethyl
```

31. **a.** The prefix *di-* means two methyl groups are bonded to the parent chain. Locations of both groups must be identified by a number.
b. pentane; The longest carbon chain is misidentified.
c. 2-methylbutane; Lowest numbering for substituent is not followed.
d. 3-methylpentane; The longest carbon chain is misidentified.

32. **a.** 2-methylbutane
b. 2,3-dimethylbutane
c. 3-methylhexane

33. The bonds between carbons are nonpolar and carbon-hydrogen bonds are very weakly polar; thus alkanes are nonpolar.

34. The prefix *meth-* implies a single carbon atom and the suffix *-ene* implies the presence of a double bond. Alkenes have a double bond between two carbon atoms.

35. **a.** propene
b. *trans*-2-pentene
c. 4-methyl-1-pentene
d. 3-ethyl-2-methyl-2-pentene
e. 1-hexene

36. $CH_2{=}CHCH_2CH_2CH_3$
1-pentene
$CH_3CH{=}CHCH_2CH_3$
2-pentene

```
     CH3
     |
CH3-C=CH-CH3
```
2-methyl-2-butene

Chapter 25 REVIEW

CONCEPT PRACTICE

28. Draw condensed structural formulas for pentane and hexane. Assume that the C—H and C—C bonds are understood. *25.1*

29. Name the alkanes that have the following molecular or structural formulas. *25.1*

a. $CH_3CH_2CH_3$

b. $CH_3(CH_2)_6CH_3$

c.
```
  H  H  H  H  H
  |  |  |  |  |
H-C--C--C--C--C-H
  |  |  |  |  |
  H  H  H  H  H
```

30. Draw structures for the alkyl groups derived from methane, ethane, and propane. *25.1*

31. Why are the following names incorrect? What are the correct names? *25.1*
a. 2-dimethylpentane
b. 1,3-dimethylpropane
c. 3-methylbutane
d. 3,4-dimethylbutane

32. Give the IUPAC name for each compound. *25.1*

a.
```
CH3-CH-CH2
    |   |
   CH3 CH3
```

b.
```
CH3-CH-CH-CH3
    |   |
   CH3 CH3
```

c.
```
CH3-CH-CH2-CH2
    |       |
    CH2     CH3
    |
    CH3
```

33. Why are alkane molecules nonpolar? *25.1*

34. You cannot draw a structural formula for methene. Explain. *25.2*

35. Give a systematic name for these alkenes. *25.2*

a. $CH_3CH{=}CH_2$

b.
```
CH3       H
   \     /
    C=C
   /     \
  H       CH2CH3
```

c.
```
CH3CHCH2CH=CH2
    |
    CH3
```

d.
```
CH3       CH2CH3
   \     /
    C=C
   /     \
CH3       CH2CH3
```

e.
```
CH2=CHCH2CH2
          |
          CH2CH3
```

36. Draw a structural formula for each alkene with the molecular formula C_5H_{10}. Name each compound. *25.2*

37. Draw all the structural isomers with the molecular formula C_6H_{14}. Name each one. (For convenience, you may wish to draw only the carbon skeleton for each structure.) *25.3*

38. Draw one structural isomer of each compound. *25.3*

a.
```
     CH3
     |
CH3-CH-CH3
     |
     CH3
```

b.
```
     CH3
     |
CH3-CH-CH-CH3
        |
        CH2
        |
        CH3
```

c.
```
     CH3         CH3
     |           |
CH3-CH-CH2-CH-CH
            |   |
            CH2 CH2
            |   |
            CH3 CH3
```

39. Draw a structural formula or carbon skeleton for each of the following alkenes. Include both *cis* and *trans* forms if the compound has geometric isomers. *25.3*
a. 2-pentene
b. 2-methyl-2-pentene
c. 3-ethyl-2-pentene

40. Show how lack of rotation about a carbon–carbon double bond leads to geometric isomerism. Use the isomers of 2-pentene to illustrate your answer. *25.3*

41. Do all molecules have stereoisomers? Explain. *25.3*

37. Five structural isomers of molecular formula C_6H_{14} exist.

```
C-C-C-C-C-C
```
hexane

```
    C
    |
C-C-C-C
    |
    C
```
2, 2-dimethylbutane

```
  C
  |
C-C-C-C-C
```
2-methylpentane

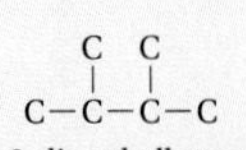
2, 3-dimethylbutane

```
    C
    |
C-C-C-C-C
```
3-methylpentane

38. Accept any isomer
a. with 5 carbons and 12 hydrogens.
b. with 7 carbons and 16 hydrogens.
c. with 11 carbons and 24 hydrogens.

39. **a.**
```
CH3       H
   \     /
    C=C
   /     \
  H       CH2CH3
```
trans-2-pentene

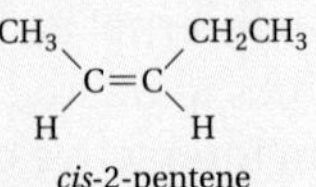
cis-2-pentene

b.
```
CH3       H
   \     /
    C=C
   /     \
CH3       CH2CH3
```
2-methyl-2-pentene

c.
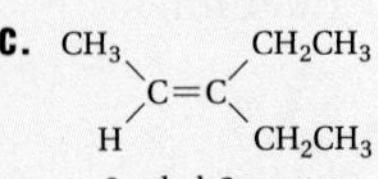
3-ethyl-2-pentene

42. Can you draw a structural isomer of hexane (C_6H_{14}) that has an asymmetric carbon? *25.3*

43. Draw a structure for each compound. *25.4*
   a. *p*-diethylbenzene
   b. 2-methyl-3-phenylpentane
   c. *p*-xylene
   d. toluene

44. Explain why both of these structures represent 1,2-diethylbenzene. *25.4*

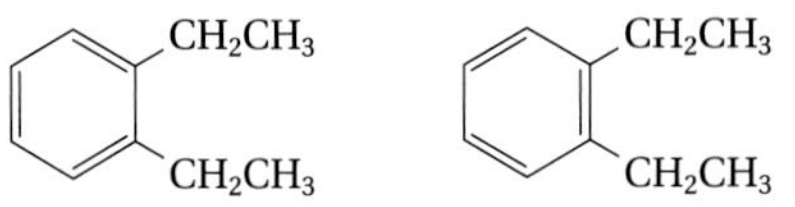

45. Write an equation for the complete combustion of octane (C_8H_{18}). *25.5*

46. Rank these materials in order of increasing hardness: bituminous coal, peat, lignite, and anthracite coal. *25.5*

47. Discuss the chemical structure of coal. *25.5*

48. How does the amount of sulfur in coal affect its potential use? *25.5*

CONCEPT MASTERY

49. For each hydrocarbon pictured below, identify the type of bonding and name the compound.

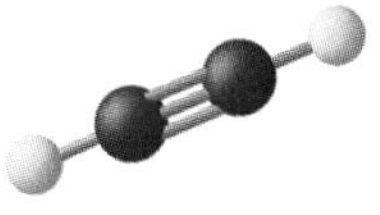

(a)

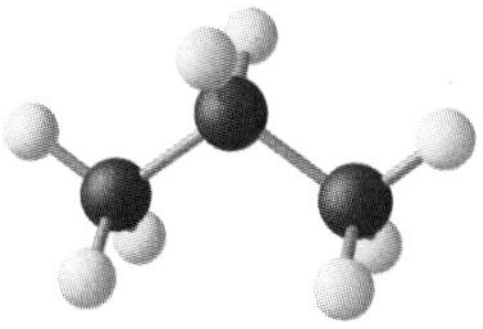

(b)

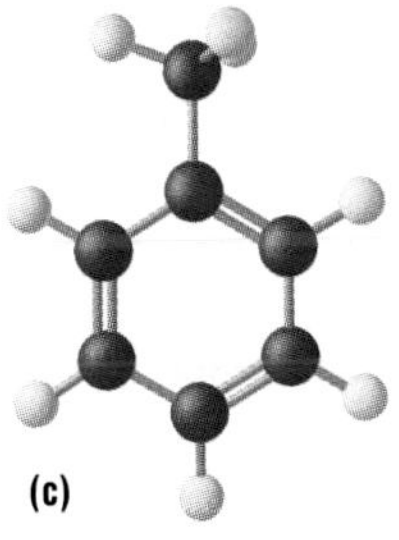

(c)

50. Write structural formulas for the following compounds.
   a. propyne
   b. cyclohexane
   c. 2-phenylpropane
   d. 2,2,4-trimethylpentane
   e. 2,3-dimethylpentane
   f. 1,1-diphenylhexane

51. Name the next three higher homologs of ethane.

52. Draw electron dot structures for each compound.
   a. ethene **c.** ethyne
   b. propane **d.** cyclobutane

53. Compare these three molecular structures. Which would you expect to be most stable? Why?

54. The seven organic chemicals produced in the largest amounts in the United States in a recent year are listed in the table below. Answer the following questions based on the data given.

Chemical	Amount produced (billions of kg)
Ethylene	15.9
Propylene	8.4
Urea	6.8
Ethylene dichloride	6.3
Benzene	5.3
Ethyl benzene	4.3
Vinyl chloride	3.7

   a. How many billion kilograms of aromatic compounds were produced?
   b. Of the total mass of all seven compounds produced, what percent by mass was made up of aliphatic compounds?

55. Are these two structures geometric isomers? Explain.

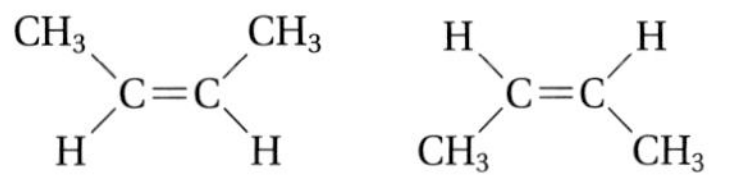

40. Because no rotation can occur around the double bond, the substituents on carbon 3 cannot be rotated from the *trans* configuration into the *cis* configuration.

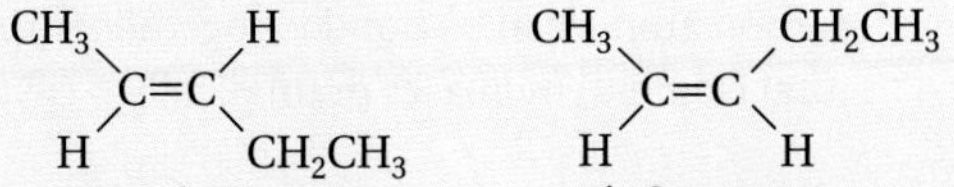

trans-2-pentene *cis*-2-pentene

41. No; only molecules with at least one asymmetric carbon have stereoisomers.

42. No

43. a.

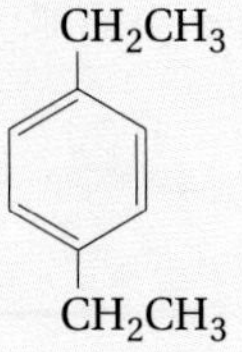

b.

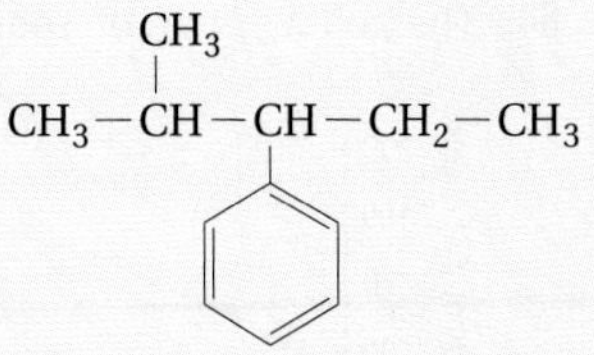

c.

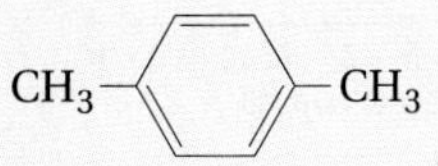

d.

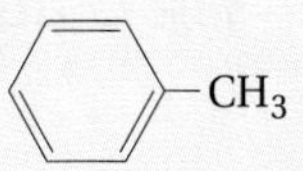

44. Two different structural formulas are possible because this compound exhibits resonance.

45. $2C_8H_{18} + 25O_2 \rightarrow 16CO_2 + 18H_2O$

46. peat, lignite, bituminous coal, anthracite coal

47. Coal is mostly large aromatic rings.

48. The cost of pollution control must be factored into the cost of burning high-sulfur-containing coal.

49. a. single bonds and triple bond; ethyne (acetylene)
b. single bonds; propane
c. single bonding within substituents and between substituents and ring; aromatic bonding (alternating single and doublebonds) within the ring; methylbenzene

50. a. $CH \equiv C-CH_3$

b.

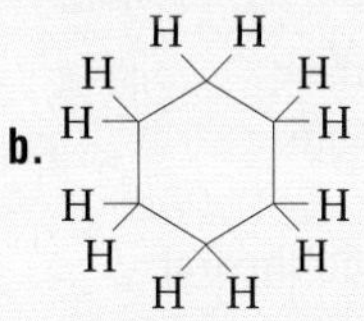

c.

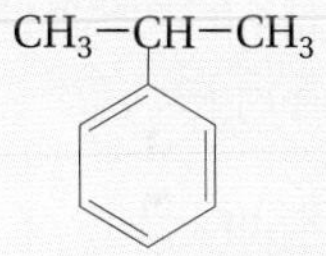

d.

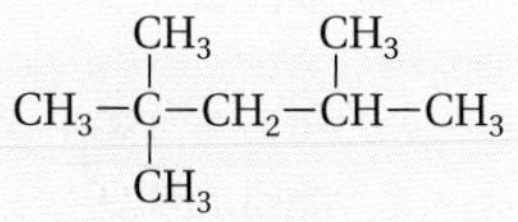

e.

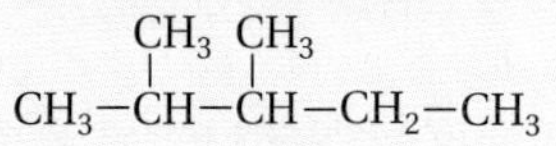

f.

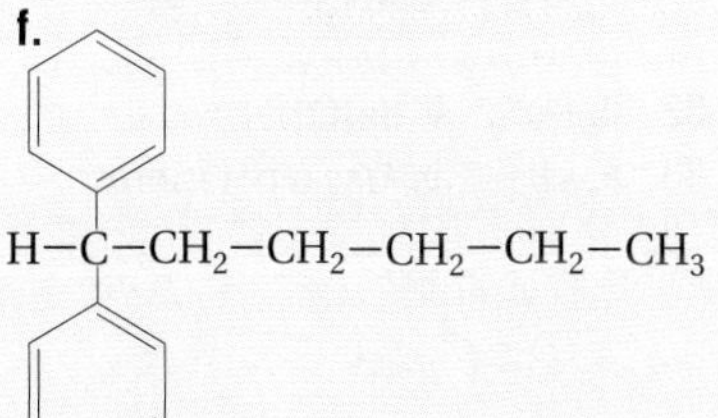

51. propane, butane, pentane

52. a.

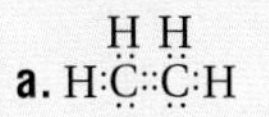

c. H:C:::C:H

b.

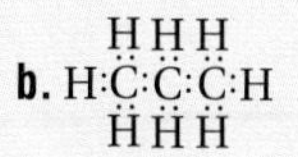

d.

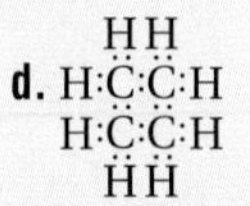

53. The middle structure is most stable due to resonance within its ring.

Answers

54. a. 9.6 billion
b. 81%

55. No; The structures are identical; one has been flipped over.

56. a. (1) alkyl substituent
b. (4) line
c. (2) six

57. Cyclic hydrocarbons are more structurally rigid, so their van der Waals forces are stronger. Thus, it takes more energy (a higher temperature) to liberate these molecules from the liquid state.

58. a. 10.00 **b.** 7.59 **c.** 12.00
d. 11.70

59. a. Ca, +2; C, +4; O, −2
b. Cl, 0
c. Li, +1; I, +5; O, −2
d. Na, +1; S, +4; O, −2

60. a. H:P̤̈:H pyramidal (with H below P)
b. :C:::O: linear
c. :S̈::C::S̈: linear
d. :F̤̈:C:F̤̈: tetrahedral (with :F: above and below C)

61. a. $K_{eq} = \frac{[ICl]^2}{[I_2][Cl_2]}$

b. $K_{eq} = \frac{[H_2][Br_2]}{[HBr]^2}$

c. $K_{eq} = \frac{[HCl]^4[S]^3[SO_2]}{[S_2Cl_2]^2[H_2O]^2}$

d. $K_{eq} = \frac{[NH_3]^2}{[N_2][H_2]^3}$

62. d. CaS **f.** $Ba(OH)_2$

63. a. ~8 **b.** Use a pH meter.

CRITICAL THINKING

56. Choose the term that best completes the second relationship.

a. tree : branch
hydrocarbon chain: _______
(1) alkyl substituent (3) isomer
(2) methane (4) double bond

b. cyclic : aliphatic
circle: _______
(1) oval (3) square
(2) parallelogram (4) line

c. propane : three
hexene: _______
(1) seven (3) five
(2) six (4) four

57. Why do most cyclic hydrocarbons have higher boiling points than alkanes with the same number of carbons?

CUMULATIVE REVIEW

58. What are the pH values for aqueous solutions containing each of the following hydroxide-ion concentrations?
a. $1.0 \times 10^{-4}M$
b. $3.9 \times 10^{-7}M$
c. $0.010M$
d. $0.0050M$

59. Give the oxidation number of each element in the following compounds.
a. $CaCO_3$ **c.** $LiIO_3$
b. Cl_2 **d.** Na_2SO_3

60. Draw electron dot structures and predict the shapes of the following molecules.
a. PH_3 **b.** CO **c.** CS_2 **d.** CF_4

61. Write equilibrium constant expressions for the following reactions.
a. $Cl_2(g) + I_2(g) \rightleftharpoons 2ICl(g)$
b. $2HBr(g) \rightleftharpoons H_2(g) + Br_2(g)$
c. $2S_2Cl_2(g) + 2H_2O(g) \rightleftharpoons 4HCl(g) + 3S(g) + SO_2(g)$
d. $N_2(g) + 3H_2(g) \rightleftharpoons 2NH_3(g)$

62. Identify any incorrect formulas among the following compounds.
a. H_2O_2 **c.** SrO **e.** $CaHPO_4$
b. $NaIO_4$ **d.** CaS_2 **f.** BaOH

63. A colorless solution of unknown pH turns blue when tested with the acid-base indicator bromthymol blue. It remains colorless when tested with phenolphthalein.
a. What is the approximate pH of the solution?
b. How could you determine the pH more accurately?

CONCEPT CHALLENGE

64. Use the data in **Table 25.1** to make a graph of boiling point versus number of carbons for the first ten straight-chain alkanes. Does the graph describe a straight line? On the basis of this graph, what would you predict to be the boiling point of undecane, the straight-chain alkane containing eleven carbons? Use a chemistry handbook to find the actual boiling point of undecane. Compare this boiling point with your prediction.

65. Fossil fuels such as oil and natural gas are the raw materials for many consumer products. Do you believe it is wise to continue to use the limited supplies of fossil fuels as an energy source?

66. The graph shows the number of structural isomers for alkanes with three through ten carbon atoms.

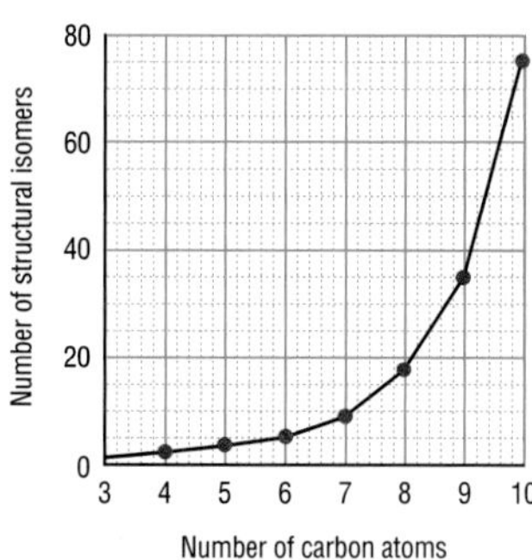

a. According to the graph, how many structural isomers are there for the C_6, C_7, C_8, C_9, and C_{10} alkanes?
b. The difference between the number of isomers for C_7 and C_8 is 9. The difference between the number of isomers for C_9 and C_{10} is 40. In each case, one additional carbon atom is added to the molecule. Why is the change in the number of isomers so different?

64. The boiling point of undecane is 196°C.

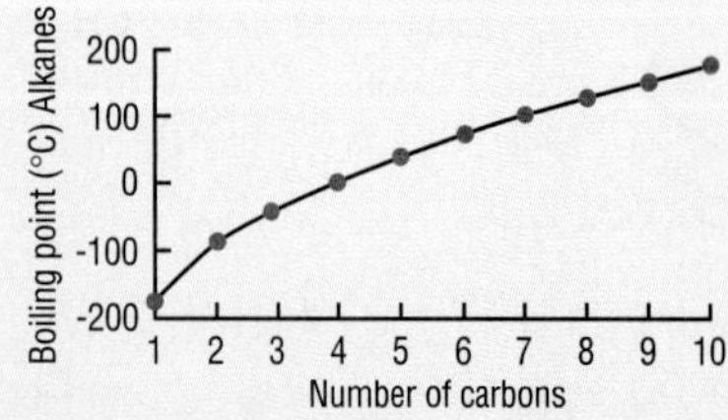

65. Answers will vary.

66. a. $C_6 = 5$, $C_7 = 9$, $C_8 = 18$, $C_9 = 35$, $C_{10} = 75$
b. As the size of the alkane molecule gets larger, the number of different ways that the carbon atoms can be bonded together (structural isomers) increases dramatically.

Chapter 25 STANDARDIZED TEST PREP

Select the choice that best answers each question or completes each statement.

1. What is the name of the compound with the following structural formula?

```
      CH3 H  CH3
       |  |  |
CH3 —  C— C— C — CH3
       |  |  |
       H  H  H
```

a. 1,2,3,3-tetramethylpropane
b. heptane
c. 2,4-dimethylpentane
d. 1,5-demethylbutane

2. Which of these are characteristic of all alkenes?
I. unsaturated
II. carbon-carbon double bond
III. stereoisomers
a. I and II only
b. II and III only
c. I and III only
d. I, II, and III

3. How many carbon atoms are in a molecule of 4,5-diethyloctane?
a. 10 **c.** 14
b. 12 **d.** 16

4. *Cis-trans* geometric isomerism is possible in
a. 2-pentene. **c.** propyne.
b. 2-butane. **d.** benzene.

5. A structural isomer of heptane is
a. methylbenzene.
b. 3,3-dimethylpentane.
c. cycloheptane.
d. 3-methylhexene.

6. Which molecule has stereoisomers?
a. CH_4 **c.** CFClBrI
b. CF_2H_2 **d.** CF_2Cl

Use the space-filling models of pentane isomers to answer questions 7 and 8.

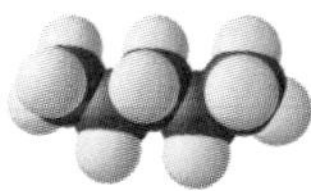

7. Write structural formulas for three structural isomers of pentane, C_5H_{12}. Name each isomer.

8. *You Draw It!* Write structural formulas for the four structural isomers of cyclopentane. Draw ball-and-stick models of the isomers.

The lettered choices below refer to questions 9–12. A lettered choice may be used once, more than once, or not at all.

(A) alkene
(B) arene
(C) alkyne
(D) alkane

In which of the above classes of hydrocarbons does each of the following compounds belong?

9. C_7H_{16}
10. C_5H_8
11. C_6H_6
12. C_8H_{16}

Use the molecular structures below to answer questions 13–16. A molecular structure may be used once, more than once, or not at all.

(A)
```
CH3     H
   \   /
    C=C
   /   \
  H     CH3
```

(B) $CH_2{=}CH{-}CH{=}CH_2$

(C)
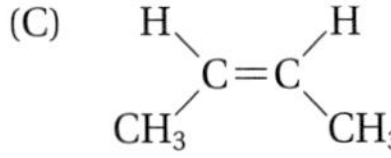

(D)
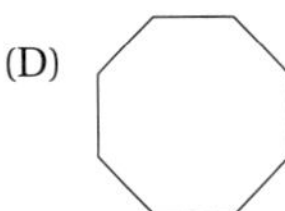

13. Which structure is a cycloalkane?
14. Which structure is a saturated hydrocarbon?
15. Which structure is a *cis*-isomer?
16. Which structure is a *trans*-isomer?

Use the labeled features in the molecular structure below to answer questions 17–21.

```
                (2)
   (1)      CH3CH2CH2
 C6H5 \        |      H
        [benzene ring]
                         (3)
(5) CH3 /             \ CH=CH2
               |
          Cl — C — Br
               |
               F
              (4)
```

17. Which label identifies a double bond?
18. Which label identifies a phenyl group?
19. Which label identifies a methyl group?
20. Which label identifies an asymmetric carbon?
21. Which label identifies a propyl group?

Chapter 25
STANDARDIZED TEST PREP

1. c
2. a
3. b
4. a
5. b
6. c
7.

```
   H  H  H  H  H
   |  |  |  |  |
H— C— C— C— C— C—H
   |  |  |  |  |
   H  H  H  H  H
```
pentane

```
   H  H  H  H
   |  |  |  |
H— C— C— C— C—H
   |  |  |  |
   H  |  H  H
   H— C—H
      |
      H
```
2-methylbutane

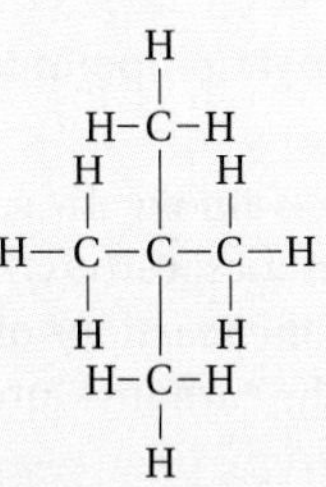

2,2-dimethylpropane

8. See structural formulas and ball-and-stick models below. In the models, the black spheres represent carbon and the white spheres represent hydrogen.
9. (D)
10. (C)
11. (B)
12. (A)
13. d
14. d
15. c
16. a
17. (3)
18. (1)
19. (5)
20. (4)
21. (2)

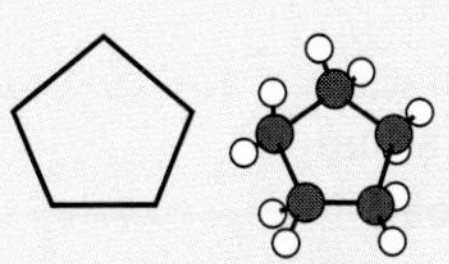
cyclopentane

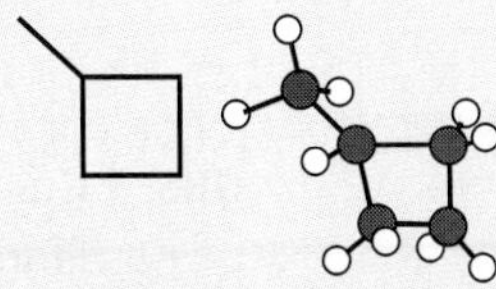
methylcyclobutane

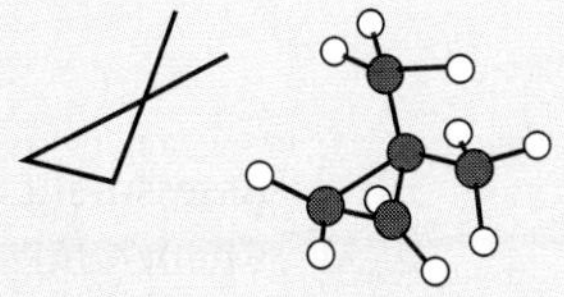
1,1-dimethylcyclopropane

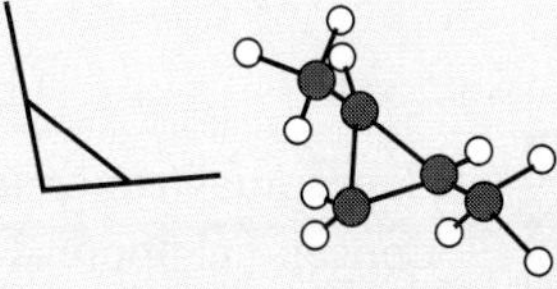
1,2-dimethylcyclopropane

Planning Guide

SECTION OBJECTIVES	ACTIVITIES/FEATURES	MEDIA & TECHNOLOGY
26.1 Introduction to Functional Groups ○ ■ ◆ ▸ Define a functional group and give several examples ▸ Describe halocarbons and the substitution reactions they undergo	**SE** Discover It! *Making the Slimiest Polymer,* p. 772 **TE** DEMOS, pp. 774, 776	**ASAP** Assessment 26.1 **OT** 75: *Functional Groups*
26.2 Alcohols and Ethers □ ◆ ▸ Describe the structures and naming of alcohols and ethers ▸ Define an addition reaction and give several examples ▸ Compare the properties of alcohols and ethers	**SE** **Link to Pollution** *Alternative Fuels,* p. 784 **TE** DEMOS, pp. 780, 781, 782	**ASAP** Assessment 26.2
26.3 Carbonyl Compounds □ ◆ ▸ Distinguish among the carbonyl groups of aldehydes, ketones, carboxylic acids, and esters ▸ Describe the reactions of compounds that contain the carbonyl functional group	**SE** **Link to Genetics** *Scientists Growing Flowers,* p. 790 **SE** Mini Lab *Testing for an Aldehyde,* p. 794 (LRS 26-1) LM 50: *Esters of Carboxylic Acids* SSLM 37: *Vitamin C in Tablets* SSLM 38: *Vitamin C in Drinks* **TE** DEMOS, pp. 790, 793	**ASAP** Assessment 26.3
26.4 Polymerization □ ◆ ▸ Define polymer and monomer ▸ Name and describe the uses of some important addition and condensation polymers	**SE** Small-Scale Lab *Polymers,* p. 801 (LRS 26-2) **SE** **Chemistry Serving . . . the Consumer** *The Sweet Smell of Functional Groups,* p. 802 **SE** **Chemistry in Careers** *Flavor Developer,* p. 802 **TE** DEMOS, pp. 796, 797	**ASAP** Assessment 26.4 **ASAP** Concept Map 26 **CHM** Side 10, 23: *Polymers* **CHM** Side 10, 30: *Condensation Polymers* **CHM** Side 10, 32: *Addition Polymers* **RP** Lesson Plans, Resource Library **AR** Computer Test 26 **www** Activities, Self-Tests, *SCIENCE NEWS* updates **TCP** The Chemistry Place Web Site

KEY

●	Conceptual (concrete concepts)	**AR**	Assessment Resources	**GRS**	Guided Reading and Study Workbook
○	Conceptual (more abstract/math)	**ASAP**	Chem ASAP! CD-ROM		
■	Standard (core content)	**ACT**	ActivChemistry CD-ROM	LM	Laboratory Manual
□	Standard (extension topics)	**CHM**	CHEMedia Videodiscs	**LP**	Laboratory Practicals
◆	Honors (core content)	**CA**	Chemistry Alive! Videodiscs	LRS	Laboratory Recordsheets
◇	Honors (options to accelerate)	**GCP**	Graphing Calculator Problems	SSLM	Small-Scale Lab Manual

PRACTICE / ASSESSMENT

PRACTICE		ASSESSMENT	
GRS	Section 26.1	**SE**	Section Review
RM	Practice Problems 26.1	**RM**	Section Review 26.1
RM	Chapter 26 Quiz	**RM**	Chapter 26 Quiz
GRS	Section 26.2	**SE**	Section Review
RM	Practice Problems 26.2	**RM**	Section Review 26.2
RM	Chapter 26 Quiz	**RM**	Chapter 26 Quiz
GRS	Section 26.3	**SE**	Section Review
RM	Practice Problems 26.3	**RM**	Section Review 26.3
RM	Chapter 26 Quiz	**RM**	Chapter 26 Quiz
		LP	Lab Practical 26-1
GRS	Section 26.4	**SE**	Section Review
RM	Practice Problems 26.4	**RM**	Section Review 26.4
RM	Interpreting Graphics	**RM**	Vocabulary Review 26
		SE	Chapter Review
		SM	Chapter 26 Solutions
		SE	Standardized Test Prep
		PHAS	Chapter 26 Test Prep
		RM	Chapter 26 Quiz
		RM	Chapter 26 A & B Test

OT	Overhead Transparency	**SE**	Student Edition
PHAS	PH Assessment System	**SM**	Solutions Manual
PLM	Probeware Lab Manual	**SSV**	Small-Scale Video/Videodisc
PP	Problem Pro CD-ROM	**TCP**	www.chemplace.com
RM	Review Module	**TE**	Teacher's Edition
RP	Resource Pro CD-ROM	**www**	www.phschool.com

PLANNING FOR ACTIVITIES

STUDENT EDITION

Discover It! p. 772
- white paper glue
- borax
- large glasses
- paper or foam cups
- spoons
- 10-mL and 100-mL graduated cylinders

Mini Lab, p. 794
- 1*M* sodium hydroxide
- 6*M* aqueous ammonia
- 5% silver nitrate
- small test tubes
- test tube racks
- medicine droppers
- methanal
- propanone
- ethanol

Small-Scale Lab, p. 801
- $3\frac{1}{2}$ oz plastic cups
- plastic spoons
- soda straws
- 5% borax solution
- powdered guar gum
- pipets

TEACHER'S EDITION

Teacher Demo, p. 774
- organic alcohols, acids, aldehydes, esters and amines

Teacher Demo, p. 776
- piece of copper wire
- Bunsen burner
- bromobenzene
- benzoic acid

Teacher Demo, p. 781
- common household products containing alcohols

Activity, p. 781
- test tubes with stoppers
- distilled water
- ethanol, 1-butanol, 1-hexanol, 1-octanol, hexanediol
- litmus paper

Teacher Demo, p. 781
- litmus paper
- several alcohols

Teacher Demo, p. 782
- 0.1 g of $KMnO_4$
- 15 g of NaOH
- 100 mL of distilled water
- 12 mL of cyclohexane and cyclohexene
- 2 test tubes with stoppers

Teacher Demo, p. 790
- warm water bath
- hot plate
- variety of fragrant esters
- 1–2 mL (or 0.2–0.6 g if the acid is solid) of each carboxylic acid
- 3–5 mL of corresponding alcohol
- 1–2 drops of concentrated sulfuric acid
- filter paper
- test tubes
- clean watch glass

Teacher Demo, p. 793
- 0.25 g of CrO_3
- 0.75 mL water
- 0.25 mL of concentrated sulfuric acid
- 3 test tubes
- 3 mL of acetone
- 5 drops each of ethanol, isopropanol, and *t*-butyl alcohol

Teacher Demo, p. 796
- metal link chain
- paper clips

Teacher Demo, p. 797
- 5 mL of a 5% aqueous solution of 1,6,-hexanediamine
- small beaker
- 5 drops of 20% sodium hydroxide
- 5 mL of a 5% solution of adipoyl chloride in cyclohexane
- forceps or tongs
- paper towels

Chapter 26

Key Terms

26.1 functional group, halocarbons, alkyl halides, aryl halides, substitution reaction
26.2 alcohols, hydroxyl group, fermentation, denatured alcohol, addition reaction, hydration reaction, hydrogenation reaction, ethers
26.3 carbonyl group, aldehydes, ketones, carboxylic acids, carboxyl group, fatty acids, esters, dehydrogenation reaction, Benedict's test, Fehling's test
26.4 polymer, monomers

DISCOVER IT!

Have students work in pairs. They should summarize their observations in a table. Ask the students to compare the properties of the product to those of the materials with which they started. Have them propose a hypothesis for what may have happened at the molecular level. (The properties of the product are those of a deformable solid rather than those of the liquids from which it was made. The borax causes interlinking between the polymers in the glue.) CAUTION: Note that borax is toxic if ingested and can irritate the skin if left in contact for long periods. Have students wash their hands thoroughly after performing this activity.

Chapter 26 FUNCTIONAL GROUPS AND ORGANIC REACTIONS

Many desserts owe their flavors to organic aldehydes and esters.

FEATURES

DISCOVER IT!
Making the Slimiest Polymer

SMALL-SCALE LAB
Polymers

MINI LAB
Testing for an Aldehyde

CHEMISTRY SERVING ... THE CONSUMER
The Sweet Smell of Functional Groups

CHEMISTRY IN CAREERS
Flavor Developer

LINK TO POLLUTION
Alternative Fuels

LINK TO GENETICS
Scientists Growing Flavors

Stay current with SCIENCE NEWS
Find out more about organic reactions:
www.phschool.com

DISCOVER IT! MAKING THE SLIMIEST POLYMER

You need white paper glue, borax, two large glasses, four paper or foam cups, a spoon, a 10-mL graduated cylinder, and a 100-mL graduated cylinder.

1. Mix 60 mL (about 2 oz) of white paper glue with 60 mL of water in a glass.
2. Pour 30 mL of this mixture into each of the four cups.
3. Make a borax solution by stirring 5 g of borax into 100 mL of water in the other glass.
4. Add 5 mL of borax solution to the first cup of glue, 10 mL to the second, 15 mL to the third, and 20 mL to the fourth. Stir the mixture in each cup and knead it with your fingers until it is smooth.
5. Take the contents of the cup with 5 mL of borax and slowly stretch it into a long strand. Repeat for the other three cups.

Which amount of borax allowed you to pull the longest strand? Do you think adding less than 5 mL or more than 20 mL of the borax solution would allow you to pull a longer strand? Would the results be different if you did not add water to the glue initially? Try it! After you complete this chapter, try to determine what polymer you made.

26.1 ○ ■ ◆
26.2 □ ◆
26.3 □ ◆
26.4 □ ◆

Conceptual Use Section 26.1 to give students a feel for functional groups; emphasize that they do not need to memorize details in this section. Table 26.3 will help students see the connection between intermolecular forces and halocarbon properties.

Standard Use the photos and illustrations in Sections 26.2 through 26.4 to give an overview of alcohols, ethers, carbonyl compounds, and polymers. If time permits, explore one or more classes of compounds in more detail. The topic of polymers is a prerequisite if carbohydrates, proteins, or nucleic acids will be studied in Chapter 27.

Honors Encourage students to use Tables 26.3, 26.5, and 26.6 to analyze trends in properties based on differences in bonding. If students find the content load in this chapter overwhelming, they might research one or two classes of compounds and report back to the class. The Small-Scale Lab may be used to introduce the topic of polymers.

INTRODUCTION TO FUNCTIONAL GROUPS

section 26.1

The lights dim and the audience chatter subsides as the musicians enter the hall. All the musicians wear black, and from a distance they look nearly identical. But their differences become apparent when they pick up their instruments. As the conductor signals the beginning of the piece, a single flute is heard. The warm sounds of the stringed instruments join in, and soon each musician contributes a unique sound to the music. In a similar way, one hydrocarbon is nearly identical to another until it picks up a functional group. **How do functional groups determine the character of organic compounds?**

objectives
- Define a functional group and give several examples
- Describe halocarbons and the substitution reactions they undergo

key terms
- functional group
- halocarbons
- alkyl halides
- aryl halides
- substitution reaction

Functional Groups

In Chapter 25, you learned about hydrocarbon chains and rings—the essential components of every organic compound. Yet in most chemical reactions involving organic molecules, the saturated hydrocarbon skeletons of the molecules are chemically inert. How, then, can there be hundreds of different kinds of organic reactions?

Most organic chemistry involves substituents, which are groups attached to hydrocarbon chains. The substituents of organic molecules often contain oxygen, nitrogen, sulfur, or phosphorus. They are called functional groups because they are the chemically functional parts of the molecules. A **functional group** is a specific arrangement of atoms in an organic compound that is capable of characteristic chemical reactions. Most organic chemistry is functional-group chemistry. Organic compounds can be classified according to their functional groups. The symbol R represents any carbon chains or rings attached to the functional group. Because the double and triple bonds of alkenes and alkynes are chemically reactive, they are considered functional groups. **Table 26.1** on the following page lists the other functional groups that you will learn about in this chapter and in Chapter 27. What do most of the functional groups in **Table 26.1** have in common? You will find it helpful to refer to this table throughout the chapter. 1

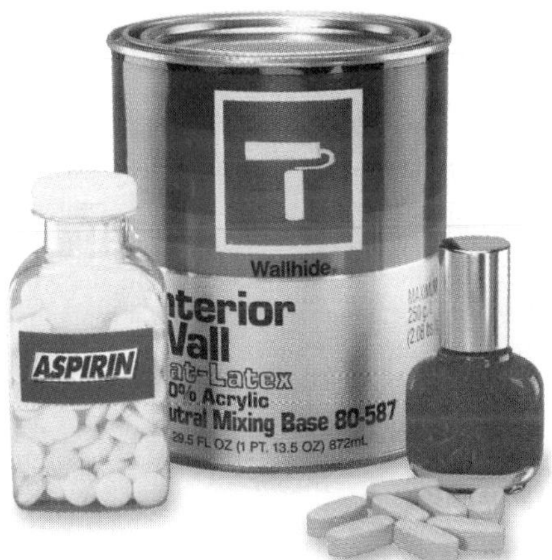

Figure 26.1
A wide array of consumer products contains hydrocarbon derivatives. The hydrocarbon skeletons in these products are chemically similar. Functional groups give each product unique properties and uses.

26.1

1 Engage

Use the Visual

Have students study the photograph and read the text that opens the section. Stress the role that structure plays in the chemistry of organic molecules. Molecular shape is dictated by the types of functional groups that are present. Ask students to hypothesize about ways in which the structure of a molecule that is eaten or inhaled is related to its particular taste or odor. (Both senses rely, in part, on the ability of chemoreceptor molecules on cell surfaces to recognize a molecule by its shape.) **Ask:**

- **How do functional groups determine the character of organic compounds?** (Each functional group has certain chemical properties that affect the behavior of any compound that contains the functional group. Chemists can predict the behavior of large classes of organic compounds based on functional groups and use this behavior to identify samples of unknown compounds.)

Check Prior Knowledge

To assess students' knowledge about the attractive forces that exist between neutral molecules, ask:

- **Why are some molecular compounds gases at room temperature while others are liquids or solids?** (Intermolecular forces of attraction exist between molecules. The stronger the attractive forces, the higher the boiling point and the more likely the substance will be a liquid or solid at room temperature.) **Challenge students to name the three types of intermolecular forces.** (dipole, dispersion, and hydrogen bonding)

STUDENT RESOURCES

From the Teacher's Resource Package, use:
- Section Review 26.1, Ch. 26 Practice Problems and Quizzes from the Review Module (Ch. 25–28)

TECHNOLOGY RESOURCES

Relevant technology resources include:
- Chem ASAP! CD-ROM
- Resource Pro CD-ROM

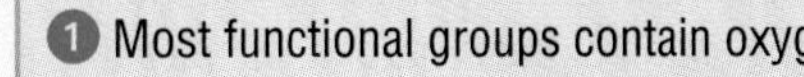

1 Most functional groups contain oxygen.

2 Teach

Use the Visual

Discuss Table 26.1. Encourage students to memorize the general compound structure for each functional group. It may be helpful to list specific examples of compounds for each functional group, so that students become more familiar with these groups in larger molecules. Include examples of alkenes and alkynes and point out that the carbon–carbon double and triple bonds are considered to be functional groups.

TEACHER DEMO

Chemists have learned that compounds that contain the same functional group have similar chemical and physical properties. The odor associated with a functional group is one example of a physical property.

Bring samples of organic compounds to class. Allow students to observe the characteristic odors associated with alcohols (2-propanol), acids (propanoic and butanoic acid), amines (butyl amine), aldehydes (vanillin and cinnamaldehyde), and esters (the orange odor of octyl acetate and the pineapple odor of ethyl butyrate).

CAUTION! Remind students that normally it is not safe to smell chemicals. An exception is being made for these known compounds Have students record their observations; make a summary table on the board. Ask students to consider what other properties molecules with the same functional groups might have in common.

Table 26.1

Organic Compounds Classified by Functional Group		
Compound type	**Compound structure**	**Functional group**
Halocarbon	R—X (X = F, Cl, Br, I)	Halogen
Alcohol	R—OH	Hydroxyl
Ether	R—O—R	Ether
Aldehyde	R—C(=O)—H	Carbonyl
Ketone	R—C(=O)—R	Carbonyl
Carboxylic acid	R—C(=O)—OH	Carboxyl
Ester	R—C(=O)—O—R	Ester
Amine	R—NH_2	Amino
Amide	R—C(=O)—N(H)—R	Amido

Halogen Substituents

Halocarbons are a class of organic compounds containing covalently bonded fluorine, chlorine, bromine, or iodine. The IUPAC rules for naming halocarbons are based on the name of the parent hydrocarbon. The halogen groups are added as substituents. Examples of IUPAC names for several simple halocarbons along with their structural formulas are given below. (Common names are in parentheses.)

CH_3—Cl — Chloromethane (methyl chloride)

H(H)C=C(H)Cl — Chloroethene (vinyl chloride)

C_6H_5—Cl — Chlorobenzene (phenyl chloride)

(a)

(b)

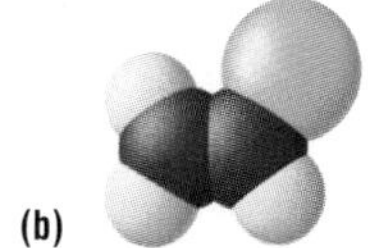

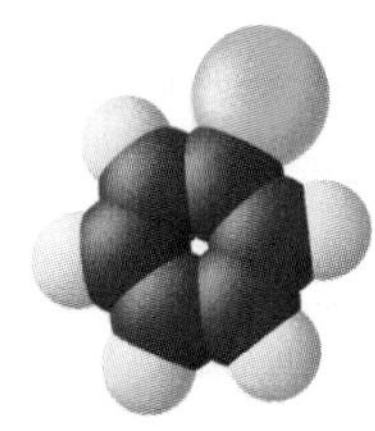

(c)

Figure 26.2
*Space-filling models of chloromethane **(a)**, chloroethene **(b)**, and chlorobenzene **(c)**.*

Common names of halocarbons consist of two parts. The first part names the hydrocarbon portion of the molecule as an alkyl group, such as methyl, ethyl, butyl, or phenyl. The second part gives the halogen with an *-ide* ending. Methyl chloride (CH_3Cl) is an example. Remember, however, that the bonding in a halocarbon is covalent, not ionic.

On the basis of their common names, halocarbons in which a halogen attaches to a carbon of an aliphatic chain are called **alkyl halides. Table 26.2** lists the names of alkyl groups other than methyl, ethyl, and propyl. Halocarbons in which a halogen attaches to a carbon of an arene ring are called **aryl halides.**

Meeting Diverse Needs

Gifted Have gifted students research other examples in which the structure of a chemical plays an important role in a biological process. Suggestions for research include neurotransmitters such as acetylcholine and hormones such as estradiol and testosterone. Have students research how environmental pollutants can interfere with the action of these natural chemicals.

Table 26.2

Names of Some Common Alkyl Groups		
Name	**Alkyl group**	**Remarks**
Isopropyl	$CH_3-CH(CH_3)-$ (with H on the central C)	The prefix *iso-* is used when there is a methyl group on the carbon second from the unsubstituted end of the longest chain.
Isobutyl	$CH_3-CH(CH_3)-CH_2-$ (primary carbon)	The carbon joining this alkyl group to another group is bonded to one other carbon; it is a primary carbon.
Secondary butyl (*sec*-butyl)	$CH_3-CH_2-CH(-)-CH_3$ (secondary carbon)	The carbon joining this alkyl group to another group is bonded to two other carbons; it is a secondary carbon.
Tertiary butyl (*tert*-butyl)	$CH_3-C(CH_3)_2-$ (tertiary carbon)	The carbon joining this alkyl group to another group is bonded to three carbons; it is a tertiary carbon.
Vinyl	$H_2C=CH-$	When used as an alkyl group in giving compounds common names, this group is called vinyl.
Phenyl	C_6H_5-	Phenyl is derived from benzene.

The attractions between halocarbon molecules are primarily the result of the weak van der Waals interactions called dispersion forces. These attractions increase with the degree of halogen substitution. Thus more highly halogenated organic compounds have higher boiling points, as illustrated in **Table 26.3.**

Table 26.3

Molecular Masses and Boiling Points of the Chloromethanes			
Molecular structure	**Name**	**Molar mass**	**Boiling point (°C)**
CH_4*	Methane	16	−161
CH_3Cl	Chloromethane (methyl chloride)	50.5	−24
CH_2Cl_2	Dichloromethane (methylene chloride)	85.0	40
$CHCl_3$	Trichloromethane (chloroform)	119.5	61
CCl_4	Tetrachloromethane (carbon tetrachloride)	154	74

* Included for purposes of comparison.

Meeting Diverse Needs

LEP Encourage LEP students to practice writing and then naming the structures of 1, 2, and 3 carbon alkyl halides. Have them build a model of one of these structures and discuss how it is produced from the corresponding alkane.

Discuss

Display Table 26.2 using the overhead projector and discuss the rules for naming alkyl halides. Write examples of primary, secondary, tertiary, vinyl, and aryl halides on the board. Help students to recognize the alkyl portion of the molecule by circling it in colored chalk. Write the structure of the original parent hydrocarbon next to the alkyl halide. Once students learn the meaning of the terms primary, secondary, and tertiary, point out the derivation of the common names: *sec*-butylbromide and *tert*-butylbromide.

Cooperative Learning

Divide the class into groups of two or three students. Have each group practice the naming of halocarbons from structural formulas and the writing of structural formulas from the names. Students in each group should take turns devising questions and giving answers.

Discuss

Use examples from this section to show how a chemist can start with a simple alkane, change it to an alkyl halide, then change the halide to an alcohol. Point out that this is one way organic chemists synthesize molecules.

Think Critically

Have students study Table 26.3. Point out the trend in boiling-point temperatures. Have students:

- **Infer the cause of this pattern.**
- **List the bromoethanes in order of increasing boiling point.** (Students can verify their lists by looking up the actual boiling points in a chemical handbook. Determine the numerical change in boiling-point temperature as a function of the degree of halogen substitution.)

TEACHER DEMO

The presence of halogen functionalities in organic compounds can be determined using the Beilstein test. Point out that the identification and characterization of unknown organic compounds is an important part of organic chemistry. Organic chemists often perform qualitative chemical tests to identify unknown organic compounds. Prepare a piece of copper wire with a small loop in one end. Heat the looped end of the wire in a Bunsen burner flame and allow the wire to cool. Next, dip the loop into a sample of aryl halide (bromobenzene). Heat the wire in the Bunsen burner flame again. A green flame will be produced, indicating the presence of a halogen. Repeat the test with a negative control, benzoic acid.

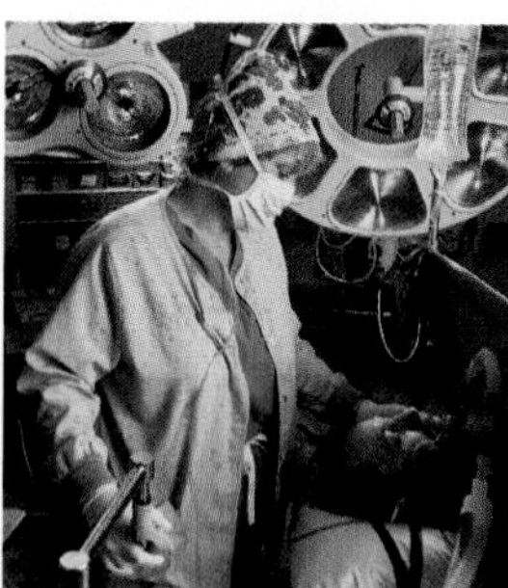

Figure 26.3
Halothane (2-bromo-2-chloro-1,1,1-trifluoroethane) is used as an anesthetic. Is this molecule fully halogenated? ❶

Very few halocarbons are found in nature, but they are readily prepared and used for many purposes. For example, halothane is used as an anesthetic. Hydrofluorocarbons are used as refrigerants in automobile air-conditioning systems.

Substitution Reactions

Organic reactions often proceed more slowly than the reactions of inorganic molecules and ions. This is because reactions of organic molecules commonly involve the breaking of relatively strong covalent bonds. Chemists are therefore constantly seeking new catalysts and improved procedures for conducting organic reactions. Many organic reactions are complex, often producing a mixture of products. The desired product must then be separated by distillation, crystallization, or other means. A common type of organic reaction is a **substitution reaction,** in which an atom or group of atoms replaces another atom or group of atoms.

A halogen can replace a hydrogen on an alkane to produce a halo carbon. The symbol X stands for a halogen in this generalized equation.

$$\underset{\text{Alkane}}{R{-}H} + \underset{\text{Halogen}}{X_2} \longrightarrow \underset{\text{Halocarbon}}{R{-}X} + \underset{\text{Hydrogen halide}}{HX}$$

Sunlight or another source of ultraviolet radiation is usually a sufficient catalyst for this reaction. From the generalized equation, you can write a specific one.

$$\underset{\text{Methane}}{CH_4} + \underset{\text{Chlorine}}{Cl_2} \xrightarrow[\text{light}]{\text{UV}} \underset{\text{Chloromethane}}{CH_3Cl} + \underset{\text{Hydrogen chloride}}{HCl}$$

Even under controlled conditions, this simple halogenation reaction produces a mixture of mono-, di-, tri-, and tetrachloromethanes.

Treating benzene with a halogen in the presence of a catalyst causes the substitution of a ring hydrogen. Iron compounds are often used as catalysts for aromatic substitution reactions. Even a rusty nail dropped in the reaction flask acts as a catalyst.

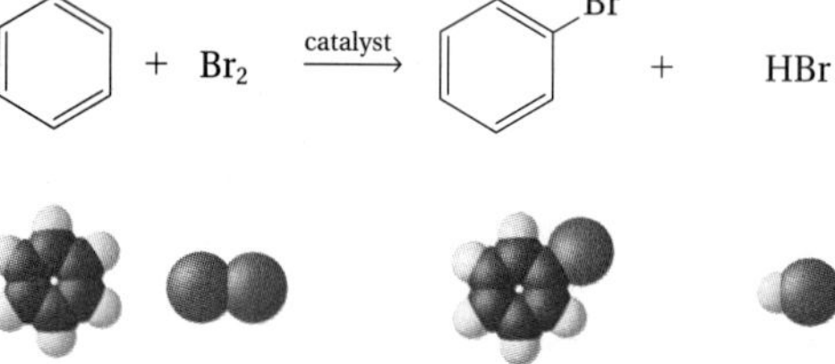

Figure 26.4
Space-filling models can be used to represent this substitution reaction visually.

Halogens on carbon chains are readily displaced by hydroxide ions to produce an alcohol and a salt. The general reaction is as follows.

$$\underset{\text{Halocarbon}}{R{-}X} + \underset{\text{Hydroxide ion}}{OH^-} \xrightarrow[100\,^\circ C]{H_2O} \underset{\text{Alcohol}}{R{-}OH} + \underset{\text{Halide ion}}{X^-}$$

Answers

❶ No, it contains one hydrogen atom attached to C-2.

❷ Each compound is a strong electrolyte, and completely dissociates to form ions when dissolved in aqueous solution.

SECTION REVIEW 26.1

1. A functional group is a specific arrangement of atoms in an organic compound that is capable of characteristic chemical reactions.
2. **a.** —OH, hydroxyl
   b. —NH_2, amino
   c. —COOH, carboxyl
   d. —Br, halogen
   e. —C—O—C—, ether
3. A halocarbon is an organic compound containing covalently bonded fluorine, chlorine, bromine, or iodine.

Chemists usually use aqueous solutions of sodium or potassium hydroxide as the source of hydroxide ions. The chemical equations for two specific examples are shown below.

$$CH_3-I(l) + KOH(aq) \xrightarrow[100\,°C]{H_2O} CH_3-OH(l) + KI(aq)$$

Iodomethane (methyl iodide) — Potassium hydroxide — Methanol — Potassium iodide

$$CH_3CH_2Br(l) + NaOH(aq) \xrightarrow[100\,°C]{H_2O} CH_3CH_2OH(l) + NaBr(aq)$$

Bromoethane (ethyl bromide) — Sodium hydroxide — Ethanol — Sodium bromide

How do KOH, KI, NaOH, and NaBr occur in aqueous solution? Fluoro groups are not easily displaced. Thus fluorocarbons are seldom, if ever, used to make alcohols. 2

section review 26.1

1. What is a functional group?
2. Identify the functional group in each structure.
   a. CH_3-OH
   b. $CH_3-CH_2-NH_2$
   c. 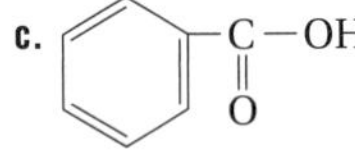
   d. $CH_3-CH_2-CH_2-Br$
   e. $CH_3-CH_2-O-CH_2-CH_3$
3. What is a halocarbon? Write IUPAC names for each halocarbon.
   a. 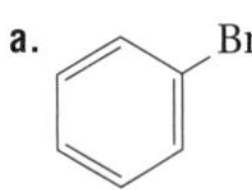
   b. CH_3CH_2Cl
   c. $CH_3CH(Cl)CH{=}CH_2$
4. Give the structural formula for each compound.
   a. isopropyl chloride
   b. 1-iodo-2,2-dimethylpentane
   c. *p*-bromotoluene
5. What is a substitution reaction? Give the general equations for the substitution of an alkane to form a halocarbon and the substitution of a halocarbon to form an alcohol.
6. Write the names of all possible dichloropropanes that could form from the chlorination of propane.

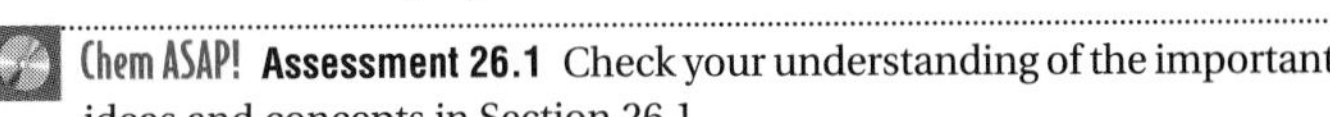

Chem ASAP! Assessment 26.1 Check your understanding of the important ideas and concepts in Section 26.1.

26.1

3 Assess

Evaluate Understanding

Write the structural formulas for ethene, ethyl chloride, and ethanol on the board. Ask students to name the compounds and identify their functional groups. (double bond, halogen, hydroxyl group) **Ask students to choose two of the compounds to illustrate a substitution reaction.** (hydroxide ions can replace the chlorine in ethyl chloride to form ethanol) **Have students explain the terms *primary, secondary,* and *tertiary*.** (used to characterize the carbon atom that joins an alkyl group to other groups)

Reteach

Work with students to construct a concept map showing the connection between alkanes and their synthetic derivatives: alkenes, alkynes, halocarbons, and alcohols. Use propane as a specific example and start with the heading alkanes. Alcohols should be shown as a branch of halocarbons to indicate the use of hydrolysis to prepare alcohols from alkyl halides. Write the structural formula for the propyl derivative in each case.

SECTION REVIEW 26.1 CONTINUED

a. bromobenzene
b. chloroethane
c. 3-chloro-1-butene

4. a. 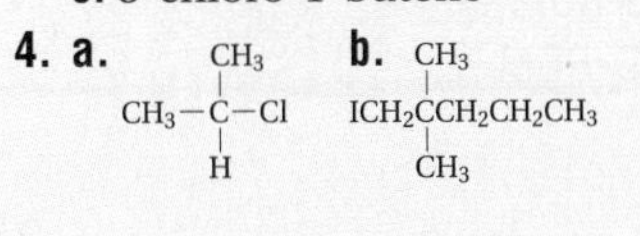 b. $ICH_2C(CH_3)_2CH_2CH_2CH_3$ c. 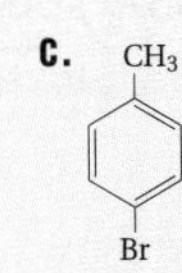

5. A substitution reaction is a reaction in which an atom or group of atoms is replaced by another atom or group of atoms.

$$R-H + X_2 \rightarrow R-X + HX$$

Alkane — Halogen — Halocarbon — Hydrogen halide

$$R-X + OH^- \xrightarrow[100\,°C]{H_2O} R-OH + X^-$$

Halocarbon — Hydroxide ion — Alcohol — Halide ion

6. 1,1-dichloropropane
   1,2-dichloropropane
   2,2-dichloropropane
   1,3-dichloropropane

1 Engage

Use the Visual

Have students study the photograph and read the text that opens the section. Explain that many biochemically active molecules contain carbon atoms bonded to oxygen atoms. Point out that the presence of a single oxygen atom significantly alters the chemical and physical properties of a hydrocarbon. Explain that alcohols and ethers are two important classes of organic compounds that contain oxygen. Ask:

- **What are the chemical characteristics of ethers?** (Ethers are compounds in which oxygen is covalently bonded to two carbon groups.)
- **Ask students to consider why ethyl ether is no longer used as an anesthetic.** (causes nausea; highly flammable; difficult to control dosage of inhaled vapor)

Ethers are used as nonpolar solvents to extract natural products such as lipids. Some low molar mass ethers are water soluble due to the ability of the oxygens in the ether functional groups to form hydrogen bonds. Solubility decreases as the mass of the carbon groups increases.

section 26.2

objectives

- Describe the structures and naming of alcohols and ethers
- Define an addition reaction and give several examples
- Compare the properties of alcohols and ethers

key terms

- alcohols
- hydroxyl group
- fermentation
- denatured alcohol
- addition reaction
- hydration reaction
- hydrogenation reaction
- ethers

ALCOHOLS AND ETHERS

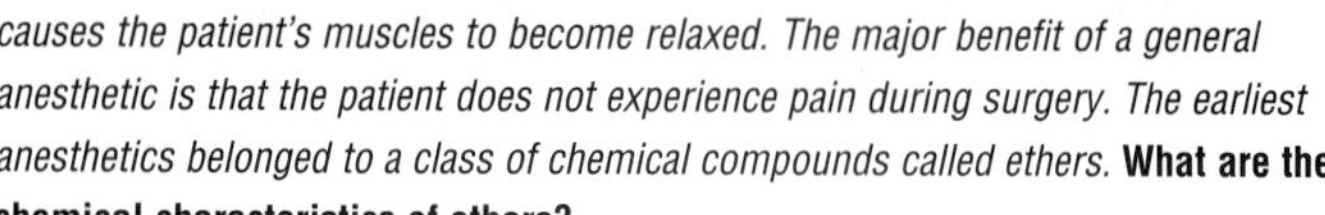

Prior to the 1840s, patients had to endure surgery while they were fully conscious. Today, when having major surgery, a patient often receives a general anesthetic. In addition to causing the patient to lose consciousness, a general anesthetic also causes the patient's muscles to become relaxed. The major benefit of a general anesthetic is that the patient does not experience pain during surgery. The earliest anesthetics belonged to a class of chemical compounds called ethers. **What are the chemical characteristics of ethers?**

Alcohols

When you hear the word alcohol, what do you think of? Probably not a general class of thousands of organic compounds! **Alcohols** are organic compounds with an —OH group.

$$\mathrm{R{-}\ddot{\underset{..}{O}}{-}H}$$

Alcohol molecule

The —OH functional group in alcohols is called a **hydroxyl group** or
1 hydroxy function. How does a hydroxyl group differ from a hydroxide ion? (If you need to review the structure of a hydroxide ion, refer back to Section 20.2.) Aliphatic alcohols can be arranged into structural categories according to the number of R groups attached to the carbon with the hydroxyl group. If one R group is attached, the alcohol is a primary alcohol; if two R groups, a secondary alcohol; if three R groups, a tertiary alcohol. This nomenclature is summarized below.

Primary alcohol	$R{-}CH_2{-}OH$	Only one R group is attached to C—OH of a primary (abbreviated 1°) alcohol.
Secondary alcohol	$R{-}\overset{\large R}{\overset{\vert}{C}}H{-}OH$	Two R groups are attached to C—OH of a secondary (2°) alcohol.
Tertiary alcohol	$R{-}\underset{\underset{\large R}{\vert}}{\overset{\overset{\large R}{\vert}}{C}}{-}OH$	Three R groups are attached to C—OH of a tertiary (3°) alcohol.

Both IUPAC and common names are used for alcohols. When using the IUPAC system to name continuous-chain and substituted alcohols, drop the *-e* ending of the parent alkane name and add the ending *-ol.* The parent alkane is the longest continuous chain of carbons that includes the carbon attached to the hydroxyl group. In numbering the longest continuous chain, the position of the hydroxyl group is given the lowest possible number. Alcohols containing two, three, and four —OH substituents are named diols, triols, and tetrols, respectively.

STUDENT RESOURCES

From the Teacher's Resource Package, use:

- Section Review 26.2, Ch. 26 Practice Problems and Quizzes from the Review Module (Ch. 25–28)

TECHNOLOGY RESOURCES

Relevant technology resources include:

- Chem ASAP! CD-ROM
- Resource Pro CD-ROM

Common names of aliphatic alcohols are written in the same way as those of the halocarbons. The alkyl group methyl, for example, is named and followed by the word alcohol, as in methyl alcohol. Compounds with more than one —OH substituent are called glycols. The structural formulas for some simple aliphatic alcohols along with their IUPAC and common names are shown below.

$$CH_3-OH$$
Methanol (methyl alcohol)

$$CH_3-CH_2-OH$$
Ethanol (ethyl alcohol)

$$CH_3-CH_2-CH_2-OH$$
1-propanol (propyl alcohol)

$$CH_3-\overset{\displaystyle OH}{\overset{|}{CH}}-CH_3$$
2-propanol (isopropyl alcohol)

$$CH_3-\overset{\displaystyle CH_3}{\overset{|}{CH}}-CH_2-OH$$
2-methyl-1-propanol (isobutyl alcohol)

$$CH_3-CH_2-\underset{\displaystyle OH}{\underset{|}{CH}}-CH_3$$
2-butanol (*sec*-butyl alcohol)

$$CH_3-\overset{\displaystyle CH_3}{\overset{|}{\underset{\displaystyle OH}{\underset{|}{C}}}}-CH_3$$
2-methyl-2-propanol (*tert*-butyl alcohol)

$$\underset{\displaystyle OH}{\underset{|}{CH_2}}-\underset{\displaystyle OH}{\underset{|}{CH_2}}$$
1,2-ethanediol (ethylene glycol)

$$CH_3-\underset{\displaystyle OH}{\underset{|}{CH}}-\underset{\displaystyle OH}{\underset{|}{CH_2}}$$
1,2-propanediol (propylene glycol)

$$\underset{\displaystyle OH}{\underset{|}{CH_2}}-\underset{\displaystyle OH}{\underset{|}{CH}}-\underset{\displaystyle OH}{\underset{|}{CH_2}}$$
1,2,3-propanetriol (glycerol)

Phenols are compounds in which a hydroxyl group is attached directly to an aromatic ring. Phenol is the parent compound. Cresol is the common name for the *o, m,* and *p* structural isomers of methylphenol.

Properties of Alcohols

Like water, alcohols are capable of intermolecular hydrogen bonding. Alcohols, therefore, boil at a higher temperature than do alkanes and halocarbons containing comparable numbers of atoms.

Because alcohols are derivatives of water (the hydroxyl group is part of a water molecule), you might expect them to be soluble in water. To a point, you would be correct. Alcohols of up to four carbons are soluble in water in all proportions. The solubility of alcohols with four or more carbons in the chain is usually much lower. The explanation for this solubility difference is that alcohols consist of two parts: the carbon chain and the hydroxyl group. The carbon chain is nonpolar and is not attracted to water. The hydroxyl group is polar and strongly interacts with water through hydrogen bonding.

Figure 26.5
Ethanol is a common component of household products.

26.2

Motivate and Relate

Write the structures of methanol, ethanol, and isopropyl alcohol on the board and point out to students that they already know a lot about the uses and properties of alcohols. Help students name the structures shown and challenge students to cite everyday examples where these alcohols are used. (Methanol is a good fuel for internal combustion engines; it is sometimes used for race cars. Methanol is also an excellent solvent, and is used for paints, shellacs, and varnishes. Ethanol is found in alcoholic beverages and is sometimes used as a gasoline additive to produce cleaner burning fuels. Isopropyl alcohol is used as an antiseptic—rubbing alcohol.)

Answer

❶ The bonding between carbon and the hydroxyl group in alcohols is covalent. The bonding between a metal and the hydroxide ion in bases is ionic.

2 Teach

TEACHER DEMO

Write the structure of water and the general structures of alcohols and ethers on the board. Point out that alcohols and ethers can be viewed as derivatives of water. Allow students to handle ball-and-stick and space-filling models of water, ethanol, and diethyl ether. Write the condensed structural formulas for each on the board. Ask students to describe the structural similarities between all three compounds. (Students should identify the bent geometry around the oxygen atom.) Ask:

- What significance does this bent shape have for the overall polarity of the molecules? (Each O—H and O—C bond is polar. Because the molecules are bent, the bond polarities do not cancel, and the molecules as a whole are polar. The region around the oxygen atom is slightly negatively charged.) Point out that the polar character of alcohols and ethers is tempered by the alkyl groups which are nonpolar. Challenge students to name the important intermolecular forces holding like molecules together in each case. Ask students to list water, ethanol, and diethyl ether in order of increasing boiling point and to explain their reasoning.

Discuss

Discuss the IUPAC conventions for naming aliphatic alcohols. Write the condensed structural formulas of several primary, secondary, and tertiary alcohols on the board, and help students to name them. Remind students how to distinguish primary, secondary, and tertiary carbons.

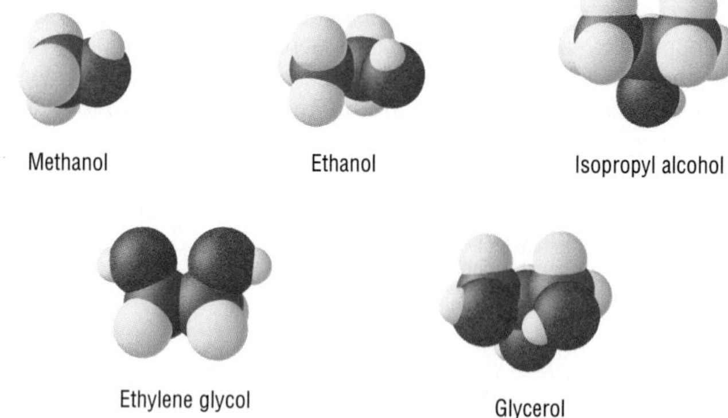

Figure 26.6
The alcohols contain one, two, or three hydroxyl groups. Which space-filling model depicts a diol? A triol? ❶

For alcohols of up to four carbons, the polarity of the hydroxyl group is more significant than the nonpolarity of the carbon chain. Thus these alcohols are soluble in water. As the number of carbon atoms increases above four, however, the nonpolarity of the chain becomes more significant and the solubility decreases.

Many aliphatic alcohols are used in laboratories, clinics, and industry. Isopropyl alcohol (IUPAC: 2-propanol), which is more familiarly known as rubbing alcohol, is a colorless, nearly odorless liquid (bp 82 °C) used as an antiseptic. It is also used as a base for perfumes, creams, lotions, and other cosmetics. Ethylene glycol (IUPAC: 1,2-ethanediol) is the principal ingredient of certain antifreezes. Its boiling point is 197 °C. Its advantages over other liquids with high boiling points are its solubility in water and a freezing point of −17.4 °C. If water is added to ethylene glycol, the mixture freezes at an even lower temperature: A 50% (v/v) aqueous solution of ethylene glycol freezes at −36 °C. Glycerol (IUPAC: 1,2,3-propanetriol) is a viscous, sweet-tasting, water-soluble liquid used as a moistening agent in cosmetics, foods, and drugs. Glycerol is also an important component of fats and oils, which you will learn about in Chapter 27.

Figure 26.7
Many cosmetic products contain glycerol (left). Isopropyl alcohol is an effective antiseptic (center). Ethylene glycol is the main ingredient in antifreeze (right). How does this poisonous substance keep the water in a car's radiator from freezing at 0 °C and boiling at 100 °C? ❷

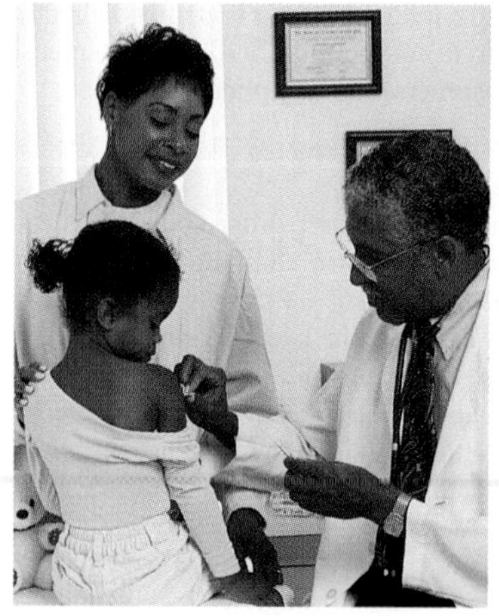

Answers

❶ Ethylene glycol represents a diol. Glycerol represents a triol.
❷ Addition of a nonvolatile solute to a liquid solvent depresses the freezing point and elevates the boiling point of the solvent. These are examples of colligative properties of solution.
❸ It evaporates.

Ethyl alcohol (IUPAC: ethanol), which has a boiling point of 78.5 °C, is also called grain alcohol. It is an important industrial chemical. Some ethanol is still produced by yeast fermentation of sugar. **Fermentation** is the production of ethanol from sugars by the action of yeast or bacteria. The enzymes of the yeast or bacteria serve as catalysts for the transformation. The breakdown of the sugar glucose ($C_6H_{12}O_6$) is an important fermentation reaction.

$$\underset{\text{Glucose}}{C_6H_{12}O_6(aq)} \longrightarrow \underset{\text{Ethanol}}{2CH_3CH_2OH(aq)} + \underset{\text{Carbon dioxide}}{2CO_2(g)}$$

Ethanol is the intoxicating substance in alcoholic beverages. It is a depressant that can be fatal if taken in large doses at once. Over time, continuous abuse of alcoholic beverages can damage the liver, which can lead to death.

The ethanol used in industrial applications is denatured. **Denatured alcohol** is ethanol with an added substance to make it toxic (poisonous). That added substance, or denaturant, is often methyl alcohol (IUPAC: methanol). Methyl alcohol is sometimes called wood alcohol because prior to 1925 it was prepared by the distillation of wood. Wood alcohol is extremely toxic: As little as 10 mL has been reported to cause permanent blindness and as little as 30 mL has been known to cause death.

Figure 26.8
Cracker and bread dough alike are made primarily of flour and water, but bread dough also contains yeast. Carbon dioxide, a product of sugar fermentation by yeast, causes bread to rise. What happens to the ethanol that is also produced? 3

Addition Reactions

As you already know, carbon–carbon single bonds are not easy to break. However, one of the bonds in an alkene double bond is somewhat weaker and thus easier to break than a carbon–carbon single bond. So it is sometimes possible for a compound of general structure X—Y to add to a double bond. In an **addition reaction,** a substance is added at the double or triple bond of an alkene or alkyne. Addition reactions are an important method of introducing new functional groups into organic molecules. In the general reaction shown below, X and Y stand for the two parts of the compound that are added.

$$>C=C< + X-Y \longrightarrow -\overset{X}{\underset{|}{\overset{|}{C}}}-\overset{Y}{\underset{|}{\overset{|}{C}}}-$$

The addition of water to an alkene is a **hydration reaction.** Hydration reactions usually occur when the alkene and water are heated to about 100 °C in the presence of a trace of strong acid. The acid, usually hydrochloric acid or sulfuric acid, serves as a catalyst for the reaction. The addition of water to ethene, the equation for which is shown below, is a typical hydration reaction. The parts of ethanol that come from the addition of water are shown in blue.

$$\underset{\text{Ethene}}{H_2C=CH_2} + \underset{\text{Water}}{H-OH} \xrightarrow[100°C]{H^+} \underset{\text{Ethanol}}{H-\overset{H}{\underset{H}{C}}-\overset{OH}{\underset{H}{C}}-H}$$

MEETING DIVERSE NEEDS

Gifted Have students research (1) the role of microorganisms in fermentation processes and (2) fermentation as a cellular process that extracts energy from glucose in the absence of oxygen. Have students design a simple experiment to show that fermentation can occur when the starting ingredient is table sugar (sucrose) rather than glucose.

At Risk Students may not understand why breaking *one* of the bonds in a double bond opens up *two* bonding sites. Have them count the number of bonds each carbon atom can form after the double bond has been broken. Remind them that each carbon atom can form single bonds with four other atoms. Therefore, one atom is added to each carbon involved.

TEACHER DEMO

Bring some common household products that contain alcohols to class. Read the name of the alcohol given on each label, and help students to write the structure for each alcohol. Explain why alcohols are used in so many commercial products. (Because alcohols have both polar and nonpolar characteristics, and are capable of forming hydrogen bonds, they make excellent solvents for a wide variety of polar and nonpolar compounds.)

ACTIVITY

Have students observe the aqueous solubilities of alcohols and glycols. Prepare two or three sets of 5 test tubes, labeled 1–5, each containing a small volume of distilled water and an equal volume of ethanol, 1-butanol, 1-hexanol, 1-octanol, and a hexanediol. Stopper the tubes securely. Instruct students to mix the contents of each tube and then allow the mixture to settle. Have the students record their observations and describe any trends they observe. Students should generalize about the aqueous solubility of straight chain alcohols and about the role that –OH groups play in this solubility.

TEACHER DEMO

Point out that, although they contain an –OH group, alcohols are not hydroxides. Alcohols are not ionic compounds; the –OH group is covalently bonded to carbon. Test several alcohols with litmus paper to show that they have no basic properties.

Use the Visual

Direct students to study Figure 26.9. Remind students that a double or triple bond is more reactive than a single covalent bond. Write the equation for the addition of Br_2 to 2-butene on the board. Explain that when a molecule containing a double bond reacts, one of the bonds in the double bond is broken. This reaction opens up two bonding sites, one on each of the carbon atoms that previously shared the double bond. Explain that the addition of molecules to alkenes makes it possible for chemists to synthesize a large variety of compounds. In Section 26.4, students will learn how plastics such as polyethylene are made by the addition polymerization of alkenes.

TEACHER DEMO

Explain that alkenes and alkynes react with oxidizing agents such as potassium permanganate and that this reaction can be used to test for unsaturated compounds. Prepare an alkaline solution of potassium permanganate by dissolving 0.1 g of $KMnO_4$ and 15 g of NaOH in 100 mL of distilled water. Put 12 mL of cyclohexane and cyclohexene in two separate test tubes. Add 1 mL of the alkaline permanganate solution to each. Stopper and shake. The general equation for the reaction of permanganate with an unsaturated hydrocarbon is

$$R{-}RCH{=}CH{-}R + MnO_4^- \rightarrow R{-}CHOH{-}CHOH{-}R + MnO_2$$

The decolorization indicates the presence of a double or triple bond.

Figure 26.9

When a few drops of orange bromine solution (top) are added to an unsaturated organic compound (center), the bromine reacts to form a colorless halocarbon (bottom). Bromine can be used to identify unsaturated compounds.

When the reagent X—Y is a halogen molecule such as chlorine or bromine, the product of the reaction is a disubstituted halocarbon. The addition of bromine to ethene to form the disubstituted halocarbon 1,2-dibromoethane is an example.

$$H_2C{=}CH_2 + Br{-}Br \longrightarrow H{-}CHBr{-}CHBr{-}H$$

Ethene (colorless) + Bromine (brownish orange) → 1,2-dibromoethane (colorless)

The addition of bromine to carbon–carbon multiple bonds is often used as a chemical test for unsaturation in an organic molecule. Bromine has a brownish-orange color, but most organic compounds of bromine are colorless. The test for unsaturation is performed by adding a few drops of a 1% solution of bromine in carbon tetrachloride (CCl_4) to the suspected alkene. As **Figure 26.9** shows, the loss of the orange color is a positive test for unsaturation. If the orange color remains, what would you conclude about 1 the sample?

Hydrogen halides, such as HBr or HCl, can also add to a double bond. Because the product contains only one substituent, it is called a monosubstituted halocarbon. The addition of hydrogen chloride to ethene is an example.

$$H_2C{=}CH_2 + H{-}Cl \longrightarrow H{-}CH_2{-}CHCl{-}H$$

Ethene (ethylene) + Hydrogen chloride → Chloroethane (ethyl chloride)

The addition of hydrogen to a carbon–carbon double bond to produce an alkane is a **hydrogenation reaction.** A hydrogenation reaction usually requires a catalyst. Finely divided platinum (Pt) or palladium (Pd) is often used. A common application of a hydrogenation reaction is the manufacture of margarine from unsaturated vegetable oils.

$$H_2C{=}CH_2 + H{-}H \xrightarrow{Pt} H{-}CH_2{-}CH_2{-}H$$

Ethene (ethylene) + Hydrogen → Ethane

$$C_6H_{10} + H{-}H \xrightarrow{Pt} C_6H_{12}$$

Cyclohexene + Hydrogen → Cyclohexane

Answers

1 The sample is saturated.

2 Due to resonance stabilization, benzene is more stable than alkenes and, thus, less reactive than alkenes.

The hydrogenation of a double bond is a reduction reaction. In the examples on the previous page, ethene is reduced to ethane, and cyclohexene is reduced to cyclohexane.

Under normal conditions, benzene resists hydrogenation. It also resists the addition of a halogen or a hydrogen halide. Under conditions of high temperatures and high pressures of hydrogen, however, three molecules of hydrogen reduce one molecule of benzene to cyclohexane. What property of benzene distinguishes it from alkenes and causes it to be less reactive? 

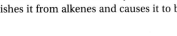

$$C_6H_6 + 3H{-}H \xrightarrow{Pt} C_6H_{12}$$

Benzene | Hydrogen | Cyclohexane

Ethers

Another class of organic compounds may sound familiar to you—ethers. **Ethers** are compounds in which oxygen is bonded to two carbon groups. The general structure of an ether is R—O—R. Ethers are easy to name. The alkyl groups attached to the ether linkage are named in alphabetical order and are followed by the word ether.

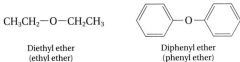

R—Ö—R (Ether molecule) | $CH_3CH_2{-}O{-}CH_3$ (Ethylmethyl ether) | $CH_3{-}O{-}C_6H_5$ (Methylphenyl ether (anisole))

Some ethers, such as ethylmethyl ether and methylphenyl ether, are nonsymmetric. This is because the R groups attached to the ether oxygen are different. When both R groups are the same, the ether is symmetric. Symmetric ethers are named by using the prefix *di-*. Sometimes, however, the prefix *di-* is dropped and a compound such as diethyl ether is simply called ethyl ether.

$CH_3CH_2{-}O{-}CH_2CH_3$ (Diethyl ether (ethyl ether)) | $C_6H_5{-}O{-}C_6H_5$ (Diphenyl ether (phenyl ether))

Diethyl ether, a volatile liquid (bp 35 °C), was the first reliable general anesthetic. (An anesthetic is a substance that produces a loss of sensation.) Originally reported in 1842 by Crawford W. Long, an American physician, diethyl ether was used by doctors for more than a century. It has been replaced by other anesthetics because it is highly flammable and often causes nausea. An anesthetic used today is halothane, a halocarbon you read about in Section 26.1.

(a)

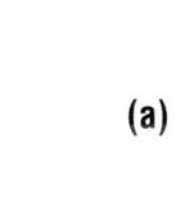

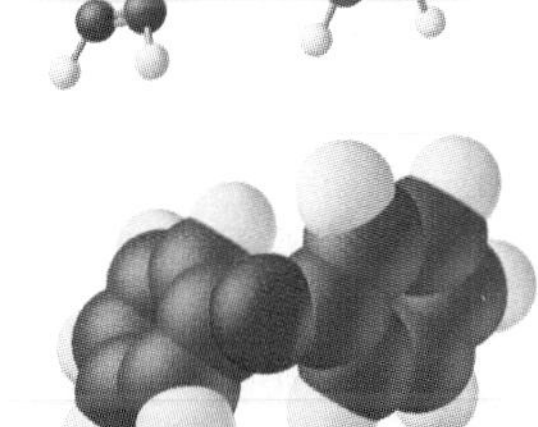

(b)

Figure 26.10
*Ethylmethyl ether **(a)** is an example of a nonsymmetric ether, and diphenyl ether **(b)** is an example of a symmetric ether.*

26.2

Discuss

Write several examples of addition reactions on the board and help students predict the structure of the expected product in each case.

Use the Visual

Have students study the models in Figure 26.10. Write the structures for these molecules on the board and describe the molecular structure of ethers. Point out that ethers are slightly polar due to the bent geometry around the oxygen atom. Ask:

- **Are ethers more or less soluble in water than alkanes of similar size?** (Ethers are generally more soluble due to their ability to participate in dipole-dipole interactions with water.)

Discuss the naming of ethers using a number of specific examples. Explain that because ethers contain nonpolar and polar regions, and because they are very inert to most chemical reagents, they make excellent solvents for reactions involving nonpolar and polar organic compounds.

Answers

SECTION REVIEW 26.2

7. R—OH; R—O—R; Alcohol: Replace the *-e* in an alkane name with *-ol*. Ether: List the alkyl groups attached to the ether linkage in alphabetical order and add *ether*.
8. Addition: substance is added at the double or triple bond of a hydrocarbon. Hydrogenation: addition of H_2 to form an alkane; hydration: addition of H_2O to form an alcohol.

26.2

Think Critically

Ask students whether it is possible to have a compound that contains both ether and alcohol functional groups. (Example: 4-methoxy-2-butanol)

3 Assess

Evaluate Understanding

Have students draw the structural formulas for 3-pentanol, 2-methyl-2-butanol, 1-heptanol, ethylpropyl ether, and diphenyl ether. Have students classify each alcohol as primary, secondary, or tertiary. Have students name and describe a chemical test for unsaturation in an organic molecule. Have them write the chemical equation for the reaction. Write the structure for 2-butanol on the board and ask:

- **What reactants could a chemist use to prepare this compound?** (2-butene and aqueous acid)

Reteach

Using the reactions given in this section, point out that the two atoms being added can be identical or they can be different. In hydrogenation, the atoms added are the same. When HCl reacts with an alkene to produce an alkyl chloride or when water and an alkene react to produce alcohol, the atoms added are different.

Portfolio Project

Engines knock when gasoline burns too soon or too quickly. MTBE replaced tetraethyl lead as an anti-knocking additive. MTBE from leaking storage tanks and fuel spills contaminates groundwater.

POLLUTION

Alternative Fuels

The 1970 Clean Air Act required vehicles to use cleaner fuels. Auto makers now produce vehicles that run on gasoline or mixtures of gasoline and ethanol—E10 (10% ethanol) or E85 (85% ethanol). Because E10 and E85 burn more completely than regular gasoline, they produce fewer pollutants. In fact, much of the gasoline sold today is E10. The use of E85 is limited because it is hard to find and few owners realize that their cars can burn it. Another new fuel, biodiesel, is made from soybean oil. Regular diesel fuel produces an irritating, black exhaust. Colorless biodiesel exhaust smells like French fries!

Ethers usually have lower boiling points than alcohols of comparable molar mass. They have higher boiling points than comparable hydrocarbons and halocarbons. Ethers are more soluble in water than hydrocarbons and halocarbons, but less soluble than alcohols. The reason for this is that the oxygens in ethers are hydrogen acceptors. However, ethers have no hydroxyl hydrogens to donate in hydrogen bonding. Ethers, therefore, form more hydrogen bonds than hydrocarbons and halocarbons but fewer hydrogen bonds than alcohols.

section review 26.2

7. Write general structural formulas for an alcohol and an ether. How would you name these molecules?
8. What is an addition reaction? Give examples of a hydration reaction and a hydrogenation reaction.
9. How does the solubility of alcohols compare with that of ethers? The boiling points? Explain your answers.
10. Give the following alcohols IUPAC names and classify each as primary, secondary, or tertiary.
    - a. $CH_3CH_2CH_2CH_2OH$
    - b. $CH_3CH(CH_3)OH$ (CH₃ branch on the CH carbon)
    - c. $CH_3CH_2CH(CH_3)CH_2OH$
11. Write the common names for the following compounds.
    - a. $CH_3CH_2CH(OH)CH_3$
    - b. $CH_3CH(CH_3)CH_2OH$
    - c. $CH_3CH_2OCH_2CH_3$
    - d. $CH_3CH_2CH_2OCH_2CH_2CH_2CH_3$
12. Give the structure for the expected organic product from each of the following reactions.
    - a. $(CH_3)(H)C{=}C(H)(CH_3) + HBr \longrightarrow$
    - b. benzene $+ Cl_2 \xrightarrow{\text{catalyst}}$
    - c. benzene $+ 3H_2 \xrightarrow[\text{pressure}]{\text{catalyst}}$

portfolio project

Research and write a report on methyl-*tert*-butyl ether (MTBE), which is a gasoline additive. Find out why MTBE was introduced and why there are efforts to limit its use.

Chem ASAP! Assessment 26.2 Check your understanding of the important ideas and concepts in Section 26.2.

SECTION REVIEW 26.2 CONTINUED

9. Alcohols are more soluble in water than ethers because the hydroxyl group of alcohols interacts with water through hydrogen bonding. Ethers have lower boiling points than comparable alcohols because ethers do not undergo intermolecular hydrogen bonding.
10. **a.** 1-butanol, primary
    b. 2-propanol, secondary
    c. 2-methyl-1-butanol, primary
11. **a.** *sec*-butyl alcohol
    b. isobutyl alcohol
    c. diethyl ether (ethyl ether)
    d. butylpropyl ether
12. **a.** $H{-}C(CH_3)(H){-}C(CH_3)(Br){-}H$ **b.** 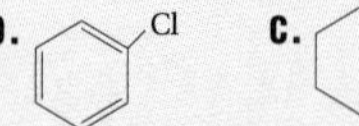 **c.** cyclohexane

CARBONYL COMPOUNDS

***W**hat is your favorite flavor, benzaldehyde or vanillin? Although these flavors might sound like they belong in the foods of a science-fiction movie, the chances are good that you have eaten these organic molecules, called aldehydes, on many occasions. Aldehydes and many other carbonyl compounds have important everyday uses—even though they may have unfamiliar names.* **What properties are associated with carbonyl compounds?**

objectives

- Distinguish among the carbonyl groups of aldehydes, ketones, carboxylic acids, and esters
- Describe the reactions of compounds that contain the carbonyl functional group

key terms

- carbonyl group
- aldehydes
- ketones
- carboxylic acids
- carboxyl group
- fatty acids
- esters
- dehydrogenation reaction
- Benedict's test
- Fehling's test

Aldehydes and Ketones

You have now learned that in an alcohol, an oxygen atom is bonded to a carbon group and a hydrogen atom. In an ether, an oxygen atom is bonded to two carbon groups. An oxygen atom can also be bonded to a single carbon atom by a double covalent bond. Such an arrangement is called a carbonyl group. A **carbonyl group** consists of a carbon atom and an oxygen atom joined by a double bond. It is the functional group in aldehydes and ketones.

Aldehydes are organic compounds in which the carbon of the carbonyl group is always joined to at least one hydrogen. The general formula for an aldehyde is RCHO.

Ketones are organic compounds in which the carbon of the carbonyl group is joined to two other carbons. The general formula for a ketone is RCOR.

$$\underset{\text{Aldehyde}}{R-\overset{\overset{\displaystyle O}{\|}}{C}-H} \quad \leftarrow \text{Carbonyl group} \qquad \underset{\text{Ketone}}{R-\overset{\overset{\displaystyle O}{\|}}{C}-R} \quad \leftarrow \text{Carbonyl group}$$

The IUPAC system may be used for naming aldehydes and ketones. For either class of compounds, you must first identify the longest hydrocarbon chain that contains the carbonyl group. The *-e* ending of the hydrocarbon is replaced by *-al* to designate an aldehyde. In the IUPAC system, the continuous-chain aldehydes are named methanal, ethanal, propanal, butanal, and so forth.

Ketones are named by changing the ending of the longest continuous carbon chain that contains the carbonyl group from *-e* to *-one*. **Table 26.4** on the following page illustrates the naming of some common aldehydes and ketones. If the carbonyl group of a ketone could occur at several places on the chain, then its position is designated by the lowest possible number. Why is it unnecessary to use a number prefix when naming an aldehyde? 1

Aldehydes and ketones can form weak hydrogen bonds between the carbonyl oxygen and the hydrogens of water. The lower members of the series—methanal (formaldehyde), ethanal (acetaldehyde), and propanone (acetone)—are soluble in water in all proportions. As the length of the hydrocarbon chain increases, however, water solubility decreases. When the carbon chain exceeds five or six carbons, solubility of both aldehydes and ketones is very low. As might be expected, all aldehydes and ketones are soluble in nonpolar solvents.

(a)
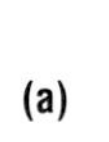

(b)

(c)
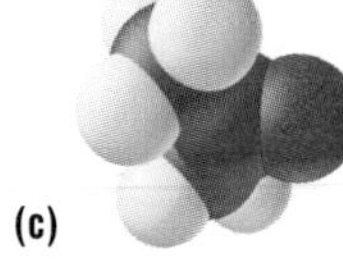

Figure 26.11
*The low-molar-mass carbonyl-group compounds methanal **(a)**, ethanal **(b)**, and propanone **(c)** are completely soluble in water.*

STUDENT RESOURCES

- Section Review 26.3, Ch. 26 Practice Problems and Quizzes from the Review Module (Ch. 25–28)
- Laboratory Recordsheet 26-1
- Laboratory Manual: Experiment 50
- Laboratory Practical 26-1
- Small-Scale Chemistry Lab Manual: Experiments 37 and 38

TECHNOLOGY RESOURCES

Relevant technology resources include:

- Chem ASAP! CD-ROM
- Resource Pro CD-ROM

1 The carbonyl carbon of an aldehyde is, by definition, joined to at least one hydrogen atom. This can only occur if the carbonyl carbon is at the end of a hydrocarbon chain.

26.3

1 Engage

Use the Visual

Have students identify foods in the section-opening photograph while gently sniffing the contents of labeled test tubes containing small amounts of vanillin, isoamyl acetate, and benzaldehyde. Ask:

- **Which odor is that of vanilla ice cream?** (vanillin) **Bananas?** (isoamyl acetate) **Maraschino cherries?** (benzaldehyde)

Write structures for the compounds on the board. Refer to Table 26.4 for benzaldehyde and vanillin; isoamyl acetate is $CH_3COOCH_2CH_2CH(CH_3)_2$. Explain that compounds with carbonyl groups account for the odor/taste of many foods.

Check Prior Knowledge

Many of the reactions involving carbonyl groups are reversible. The results of these reactions depend on the equilibrium position. To assess students' knowledge about chemical equilibrium, ask:

- **What is the significance of a double arrow in a chemical equation?** (It shows that the reaction is reversible.)
- **When has a reaction reached chemical equilibrium?** (when the forward and reverse reactions are taking place at the same rate)
- **How do the amounts of reactants and products change once a reaction has achieved chemical equilibrium?** (At equilibrium, the concentrations of reactants and products do not change.)
- **What factors will affect the equilibrium position of a reaction?** (changes in concentration of reactants or products, changes in temperature, changes in pressure)

2 Teach

Use the Visual

Write several examples of aldehydes and ketones on the board and discuss the naming of these compounds. Explain to students that the difference between an aldehyde and a ketone lies in the position of the carbonyl group. Display Table 26.4 using an overhead projector and point out that in ketones, as in ethers, the functional group lies between two R groups. Except for methanal, the carbonyl group in aldehydes is attached to one R group and a hydrogen atom.

Table 26.4

Condensed formula	Structural formula	IUPAC name	Common name
Aldehydes			
HCHO	H—C(=O)—H	Methanal	Formaldehyde
CH_3CHO	CH_3—C(=O)—H	Ethanal	Acetaldehyde
C_6H_5CHO	C_6H_5—C(=O)—H	Benzaldehyde	Benzaldehyde
$C_6H_5CH{=}CHCHO$	C_6H_5—CH=CH—C(=O)—H	3-phenyl-2-propenal	Cinnamaldehyde
$CH_3O(OH)C_6H_3CHO$	H—C=O on ring with HO and OCH_3	4-hydroxy-3-methoxybenzaldehyde	Vanillin
Ketones			
CH_3COCH_3	CH_3—C(=O)—CH_3	Propanone	Acetone (dimethyl ketone)
$C_6H_5COC_6H_5$	C_6H_5—C(=O)—C_6H_5	Diphenylmethanone	Benzophenone (diphenyl ketone)

Some Common Aldehydes and Ketones

Aldehydes and ketones cannot form intermolecular hydrogen bonds because they lack hydroxyl (—OH) groups. Consequently, they have boiling points lower than those of the corresponding alcohols. Aldehydes and ketones can attract each other, however, through polar–polar interactions of their carbonyl groups. As a result, their boiling points are higher than those of the corresponding alkanes. These attractive forces account for the fact that nearly all aldehydes and ketones are either liquids or solids at room temperature. The exception is methanal, which is an irritating, pungent gas. **Table 26.5** compares the boiling points of alkanes, aldehydes, and alcohols of similar molar mass.

A wide variety of aldehydes and ketones have been isolated from plants and animals. Many of them, particularly those with high molar masses, have fragrant or penetrating odors. They are usually known by their common names, which can indicate their natural sources or perhaps a characteristic property. Aromatic aldehydes are often used as flavoring

Meeting Diverse Needs

Gifted Have gifted students study the structure of cinnamaldehyde and other carbonyl-containing compounds that contain carbon-carbon bonds conjugated with a carbon-oxygen double bond. Have students consider how the location of a carbonyl group next to an alkene group affects the stability of the molecule through resonance.

Table 26.5

Compound	Formula	Molar mass	Boiling point (°C)	Comments
One carbon				
Methane	CH_4	16	−161	No hydrogen-bonding or polar-polar interactions
Methanal	HCHO	30	−21	Polar-polar interactions
Methanol	CH_3OH	32	65	Hydrogen bonding
Two carbons				
Ethane	C_2H_6	30	−89	No hydrogen-bonding or polar-polar interactions
Ethanal	CH_3CHO	44	20	Polar-polar interactions
Ethanol	CH_3CH_2OH	46	78	Hydrogen bonding

Caption: Boiling Points of Some Compounds with One and Two Carbons

agents. Benzaldehyde is the simplest aromatic aldehyde. Also known as oil of bitter almond, benzaldehyde is a constituent of almonds. It is a colorless liquid with a pleasant almond odor. Cinnamaldehyde imparts the characteristic odor of oil of cinnamon. Vanillin, which is responsible for the popular vanilla flavor, was at one time obtainable only from the podlike capsules of certain climbing orchids, such as those shown in **Figure 26.12.** Today much vanillin is synthetically produced.

The simplest aldehyde is methanal, which is also called formaldehyde. Methanal is very important industrially, having its greatest use in the manufacture of synthetic resins, but it is inconvenient to handle in the gaseous state. Methanal is usually available as a 40% aqueous solution, known as formalin. Formalin is used to preserve biological specimens. The methanal in solution combines with protein in tissues to make the tissues hard and insoluble in water. This process prevents the specimen from decaying. Today, formalin is used infrequently as a preservative because it is a carcinogen, or cancer-causing agent.

The most important industrial ketone is propanone, also called acetone. It is a colorless, volatile liquid that boils at 56 °C. Propanone is used as a solvent for resins, plastics, and varnishes and is often found in nail-polish removers. Propanone is miscible with water in all proportions. Can you suggest a reason why propanone is used in nail-polish removers? 

Figure 26.12
Vanilla beans, the seed pods from this orchid, are the natural source of vanilla flavor for ice cream and other foods. What is the IUPAC name for vanillin? ❷

26.3

ACTIVITY

Have students compare and contrast the structures of alcohols, aldehydes, and ketones. Ask students to think of some examples of simple organic molecules that contain each type of functional group. Have them write the structures and build models of each molecule. Ask:

- What are some of the structural features of carbonyl-containing molecules? (Carbonyl groups display trigonal planar geometry.)

Point out that a carbonyl carbon represents the positive center of aldehydes and ketones and that an oxygen atom represents the negative center. Ask students to discuss the types of intermolecular forces operating between the functional groups in aldehydes and ketones. How does the shape of a molecule affect the strength of intermolecular attractions? How do the aqueous solubilities of aldehydes and ketones change as the hydrocarbon portions get larger?

Answers

❶ Propanone is a powerful solvent which is able to dissolve a wide variety of polar and nonpolar compounds, including nail polish as well as many types of plastics.
❷ 4-hydroxy-3-methoxybenzaldehyde

Discuss

Write the structural formula for a carboxylic acid on the chalkboard or overhead projector. Ask:

- **What functional groups are present and what characteristics would you expect the molecule to have?** (Students should recognize the hydroxyl and carbonyl groups, and may suggest that the molecule would have properties of both an alcohol and an aldehyde.)
- **Why are carboxylic acids considered to be acids?** (They contain an ionizable proton.)

Explain that the presence of the double-bonded oxygen weakens the bond between the other oxygen atom and the hydrogen atom attached to it, so that the hydrogen atom can ionize easily. The overall functional group is called a carboxyl group and the molecule containing it is a carboxylic acid.

Use the Visual

Have students study the photograph in Figure 26.13. Explain that carboxylic acids are found in many natural products. Ask students to name the acids in the food products shown. Write the structural formulas for the acids on the board. Point out the carboxyl functional group in each. Have students name other examples of food products that contain acids and describe the tastes and odors associated with carboxylic acids.

Carboxylic Acids

Carboxylic acids are compounds with a carboxyl group. A **carboxyl group** consists of a carbonyl group attached to a hydroxyl group.

$$\mathrm{R-\overset{\overset{\displaystyle O}{\|}}{C}-OH}$$

Carbonyl group (C=O); Hydroxyl group (OH)

Carboxyl group
(also written $-CO_2H$ or $-COOH$)

The general formula for a carboxylic acid is RCOOH. Carboxylic acids are weak acids because they ionize only slightly in solution to give a carboxylate ion and a hydrogen ion.

$$\underset{\text{Carboxylic acid}}{\mathrm{R-\overset{\overset{\displaystyle O}{\|}}{C}-OH}} \rightleftharpoons \underset{\text{Carboxylate ion}}{\mathrm{R-\overset{\overset{\displaystyle O}{\|}}{C}-O^-}} + \underset{\text{Hydrogen ion (proton)}}{\mathrm{H^+}}$$

In the IUPAC system, carboxylic acids are named by replacing the *-e* ending of the parent alkane with the ending *-oic acid.* Remember, the parent alkane is the hydrocarbon with the longest continuous carbon chain containing the carboxyl group.

Carboxylic acids are abundant and widely distributed in nature. Many have common names derived from a Greek or Latin word describing their natural sources. For example, the common name for ethanoic acid is acetic acid, which comes from the Latin word *acetum,* meaning vinegar. Acetic acid is produced when wine turns sour and becomes vinegar. The pungent aroma of vinegar comes from acetic acid. Common household vinegar contains about 5% (v/v) acetic acid. Many continuous-chain carboxylic acids were first isolated from fats and are called **fatty acids.** Propionic acid, the three-carbon acid, literally means first fatty acid. Common names are used more often than IUPAC names for carboxylic acids. **Table 26.6** lists the names and formulas of some common aliphatic carboxylic acids. What relationship do you notice between the number of carbon atoms and the
1 melting point for the fatty acids?

Figure 26.13
Carboxylic acids give a variety of foods—spoiled as well as fresh—a distinctive sour taste.

The low-molar-mass members of the aliphatic carboxylic acid series are colorless, volatile liquids. They have sharp, unpleasant odors. The smells of rancid butter and dirty feet are due in part to butyric acid, the four-carbon acid. The higher members of the series are nonvolatile, waxy solids with low melting points. Stearic acid, an 18-carbon acid obtained from beef fat, is used to make inexpensive wax candles. Stearic acid and other long-chain fatty acids have very little odor.

Like alcohols, carboxylic acids can form intermolecular hydrogen bonds. As a result, carboxylic acids have higher boiling points and higher melting points than other compounds of similar molar mass. All aromatic carboxylic acids are crystalline solids at room temperature.

The carboxyl group in carboxylic acids is polar and readily forms hydrogen bonds with water molecules. This should indicate an important

Answers

1 As the length of the carbon chain increases, the melting point increases.

2 The carboxyl group is polar and can form hydrogen bonds with water molecules. The combined van der Waals forces between the single methyl groups of acetic acid are not great enough to prevent water miscibility.

Meeting Diverse Needs

At Risk Write the general structures of an aldehyde, ketone, carboxylic acid, and an ester on the board. Have students copy the structures in their notebooks. Have them label the carbonyl groups, hydroxyl groups, aliphatic, and aryl groups. Have them indicate whether each structure is an aldehyde, ketone, ester, or carboxylic acid.

Table 26.6

Saturated Aliphatic Carboxylic Acids				
Formula	Carbon atoms	IUPAC name	Common name	Melting point (°C)
$HCOOH$	1	Methanoic acid	Formic acid	8
CH_3COOH	2	Ethanoic acid	Acetic acid	17
CH_3CH_2COOH	3	Propanoic acid	Propionic acid	−22
$CH_3(CH_2)_2COOH$	4	Butanoic acid	Butyric acid	−6
$CH_3(CH_2)_4COOH$	6	Hexanoic acid	Caproic acid	−3
$CH_3(CH_2)_6COOH$	8	Octanoic acid	Caprylic acid	16
$CH_3(CH_2)_8COOH$	10	Decanoic acid	Capric acid	31
$CH_3(CH_2)_{10}COOH$	12	Dodecanoic acid	Lauric acid	44
$CH_3(CH_2)_{12}COOH$	14	Tetradecanoic acid	Myristic acid	58
$CH_3(CH_2)_{14}COOH$	16	Hexadecanoic acid	Palmitic acid	63
$CH_3(CH_2)_{16}COOH$	18	Octadecanoic acid	Stearic acid	70

property of some carboxylic acids. Methanoic, ethanoic, propanoic, and butanoic acids are completely miscible with water. The solubility of carboxylic acids of higher molar mass drops sharply, however. Most carboxylic acids dissolve in organic solvents such as ethanol or propanone.

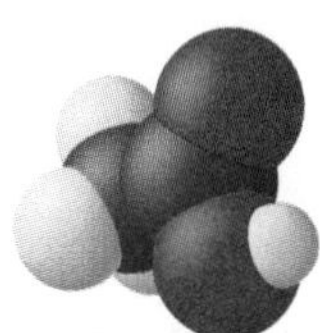

Figure 26.14
Ethanoic acid, shown in this space-filling model, is a colorless, volatile liquid. Is it soluble in water? Why? 2

Esters

Esters are probably the most pleasant and delicious organic compounds to study. **Esters** are derivatives of carboxylic acids in which the —OH of the carboxyl group has been replaced by an —OR from an alcohol. Thus esters contain a carbonyl group and an ether link to the carbonyl carbon. Many esters have pleasant, fruity odors. Esters give blueberries, pineapples, apples, pears, bananas, and many other fruits their characteristic odors. They are also responsible for giving some perfumes their fragrances.

Carbonyl group (from the acid)

$$R-C(=O)-O-R$$

Alkyl or aryl group (from the alcohol)

The general formula for a carboxylate ester is RCOOR. The R group can be a short-chain or long-chain aliphatic (alkyl) or aromatic (aryl) group, saturated or unsaturated.

Simple esters are neutral substances. Although the molecules are polar, they cannot form hydrogen bonds with one another because they do not contain hydrogen attached to oxygen or another electronegative atom. As a result, only weak attractions hold ester molecules to one another. As you might expect, esters have much lower boiling points than the strongly hydrogen-bonded carboxylic acids from which they are derived. The low-formula-mass esters are somewhat soluble in water, but esters containing more than four or five carbons have very limited solubility.

26.3

Use the Visual

Use an overhead projector to display Table 26.6. Discuss the IUPAC rules for naming carboxylic acids. Remind students that the systematic names are derived from the parent alkane. Discuss the basis for the observed trends in physical properties for the homologous series of aliphatic carboxylic acids. Ask:

- **What is the name of an aliphatic carboxylic acid that is a solid at room temperature?** (Answers will vary, but should be restricted to those acids with melting points higher than 25°C.)

Think Critically

Ask students why carboxylic acids are considered weak acids. Write the chemical equation for the ionization of dilute aqueous ethanoic acid on the board. Point out that the acid is only partly ionized at equilibrium. Ask students to consider how resonance stabilization plays a role in the position of the equilibrium. Challenge students to write the equilibrium constant expression for the reaction shown. Have students write the chemical equation for the ionization of propanoic acid.

26.3

TEACHER DEMO

Explain that esters are derivatives of carboxylic acids in which the —OH group of the carboxyl group has been replaced by an —OR group from an alcohol. The synthesis of an ester from an acid and an alcohol is called *esterification.* Synthesize a variety of fragrant esters in class and give students a chance to smell the products. Set up a water bath on a hot plate to maintain the temperature at about 80°C. To synthesize the ester, mix 1–2 mL (or 0.2-0.6 g if the acid is a solid) of each carboxylic acid with 3–5 mL of the corresponding alcohol in a test tube. Add 1–2 drops of concentrated sulfuric acid and heat for 15 minutes in the water bath. Dip a piece of filter paper into the test tube and put it on a clean watch glass. Have students record the esters you prepare and their associated odors. Ask them to write the chemical equation for each reaction.

acid	*alcohol*	*odor of ester*
acetic	ethyl	apple
acetic	isoamyl	banana
acetic	amyl	apricot
acetic	octyl	orange
butyric	ethyl	pineapple
salicylic	methyl	wintergreen

Point out that the only difference in these reactions is in the "R" group attached to the carboxyl and alcohol functional groups. Neutralize the combined reaction mixtures and flush them down the drain.

Figure 26.15
Esters impart the characteristic odors and flavors of many flowers and fruits.

If an ester is heated with water for several hours, usually very little happens. In strong acid or base solutions, however, the ester breaks down. An ester is hydrolyzed by the addition of water to produce a carboxylic acid and an alcohol. The reaction is rapid in acidic solution. How might you explain this? 1

$$CH_3-\overset{\overset{\displaystyle O}{\|}}{C}-OCH_2CH_3 + H-OH \xrightleftharpoons{H^+} CH_3-\overset{\overset{\displaystyle O}{\|}}{C}-OH + HOCH_2CH_3$$

Ethyl ethanoate → Ethanoic acid + Ethanol

GENETICS

Scientists Growing Flavors
Scientists are learning how to grow flavor molecules and other naturally occurring compounds in plant cell cultures and genetically altered yeast cells. For example, Japanese scientists are using a tissue culture technique called root culture to produce ginseng for teas and medicines and for a red pigment used in cosmetics. In the United States, scientists are obtaining vanilla flavor from cells surgically removed from a vanilla orchid and cultured in a glass bioreactor. Prior to any commercial production, a patent application is usually filed, and manufacturers must seek approval from the Food and Drug Administration to use the prepared flavor molecules in food products.

Hydroxide ions also promote this reaction. The usual agent for ester hydrolysis is an aqueous solution of sodium hydroxide or potassium hydroxide. Because many esters do not dissolve in water, a solvent such as ethanol is added to make the solution homogeneous. The reaction mixture is usually heated. All of the ester is converted to products. The carboxylic acid product is in solution as its sodium or potassium salt.

$$CH_3-\overset{\overset{\displaystyle O}{\|}}{C}-OCH_2CH_3 + NaOH \longrightarrow CH_3-\overset{\overset{\displaystyle O}{\|}}{C}-O^-Na^+ + HOCH_2CH_3$$

Ethyl ethanoate → Sodium ethanoate + Ethanol

If the reaction mixture is acidified, the carboxylic acid forms.

$$CH_3-\overset{\overset{\displaystyle O}{\|}}{C}-O^-Na^+ + HCl \longrightarrow CH_3-\overset{\overset{\displaystyle O}{\|}}{C}-OH + NaCl$$

Sodium ethanoate → Ethanoic acid

Esters may be prepared from an acid and an alcohol. Esterification is the formation of an ester from a carboxylic acid and an alcohol. The reactants, usually a carboxylic acid and a primary or secondary alcohol, are heated with a trace of mineral acid as a catalyst. The reaction is reversible.

$$R-C\begin{matrix}\nearrow O \\ \searrow OH\end{matrix} + RO-H \xrightleftharpoons{H^+} R-C\begin{matrix}\nearrow O \\ \searrow OR\end{matrix} + H-OH$$

Carboxylic acid + Alcohol → Carboxylate ester + Water

Answer

1 The acid acts as a catalyst, decreasing the time it takes to reach equilibrium.

The formation of the ester ethyl ethanoate from ethanoic acid and ethanol is an example of esterification.

$$CH_3-C(=O)-OH + CH_3CH_2O-H \overset{H^+}{\rightleftharpoons} CH_3-C(=O)-OCH_2CH_3 + H_2O$$

Ethanoic acid — Ethanol — Ethyl ethanoate — Water

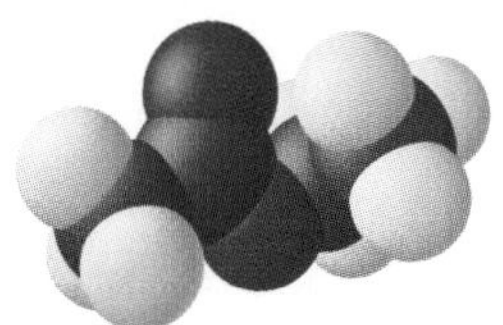

Figure 26.16
Ethyl ethanoate, shown in this space-filling model, is a low-molar-mass ester.

Oxidation–Reduction Reactions

All the classes of organic compounds you have just studied are related by oxidation and reduction reactions. Before you explore this concept further, it would probably be helpful to review the definitions of these reactions. As you have learned, oxidation is the gain of oxygen, loss of hydrogen, or loss of electrons. Reduction is the reverse: loss of oxygen, gain of hydrogen, or gain of electrons. Oxidation and reduction reactions are coupled: One does not occur without the other.

In organic chemistry, the number of oxygens and hydrogens attached to carbon indicates the degree of oxidation of a compound. The fewer the number of hydrogens on a carbon–carbon bond, the more oxidized the bond. Thus a triple bond (characteristic of an alkyne) is more oxidized than a double bond (an alkene) and a single bond (an alkane). Expressed another way, an alkane can be oxidized to an alkene and then to an alkyne. For example, ethane (an alkane) can be oxidized to ethene (an alkene) and then to ethyne (an alkyne).

The loss of hydrogen is a **dehydrogenation reaction.** Strong heating and a catalyst are usually necessary to make dehydrogenation reactions occur. The loss of each molecule of hydrogen involves the loss of two electrons from the organic molecule. The remaining carbon electrons pair to make a second or third bond, as shown below.

$$H_3C-CH_3 \xrightarrow[\text{oxidation}]{\text{loss of hydrogen (dehydrogenation)}} H_2C=CH_2 \xrightarrow[\text{oxidation}]{\text{loss of hydrogen (dehydrogenation)}} H-C\equiv C-H$$

Least oxidized (most reduced) — Most oxidized (least reduced)

These reactions are reversible. Alkynes can be reduced to alkenes, and alkenes can be reduced to alkanes by the addition of hydrogen to a double bond.

$$H-C\equiv C-H \xrightarrow[\text{reduction}]{\text{gain of hydrogen}} H_2C=CH_2 \xrightarrow[\text{reduction}]{\text{gain of hydrogen}} H_3C-CH_3$$

Most oxidized (least reduced) — Most reduced (least oxidized)

Oxidation in organic chemistry also involves the number and degree of oxidation of oxygens attached to carbon. For example, methane, a saturated hydrocarbon, can be oxidized in steps to carbon dioxide. This occurs

26.3

Think Critically

Explain that the formation of esters from a carboxylic acid and an alcohol is an excellent example of dynamic equilibrium. Under the proper conditions, the equilibrium can be shifted greatly in favor of the products. Other conditions favor the decomposition of the ester into the carboxylic acid and alcohol. Have students review Le Châtelier's principle, and use it to predict conditions that would favor the reactants and conditions that would favor the products. Ask:

- **How can a chemist improve the yield of an ester from an esterification reaction?** (Use an excess of one of the reactants, or remove the water or ester, by distillation, as it is formed.)

Meeting Diverse Needs

Gifted Have students assign oxidation numbers to carbon for each step in the sequence of oxidation reactions shown in the text in which methane is converted to carbon dioxide. Ask students how many molecules of hydrogen are given up by methane. How many electrons are lost? How many molecules of H_2 would ethane yield?

LEP and At Risk Students may find it difficult to understand that oxidation can be defined in terms of the gain or loss of hydrogen. Pair up the students who are having trouble with this concept. Have each student write out each series of oxidation-reduction reactions, noting the number of hydrogen and oxygen atoms attached to carbon at each step.

Discuss

Discuss the base-catalyzed hydrolysis of an ester. Write a number of examples of reactions on the chalkboard or overhead projector. Explain which part of the ester yields the alcohol and which part yields carboxylic acid products. Ask students about the difference in volatility of the products and how a chemist might separate them once they are formed. Write the molecular structure for propyl ethanoate on the board. Ask:

- **What products would be formed from the base-catalyzed hydrolysis of propyl ethanoate?** (propyl alcohol and acetic acid)

Discuss

Go through the sequence of oxidations that occurs for alkanes. Explain that carbon dioxide is the most oxidized form of carbon, and an alkane is the most reduced form of carbon. Discuss each step as being an example of an oxidation or reduction. Remind students that the loss of hydrogen is oxidation, and the gain of hydrogen is reduction. Therefore, in any series of oxidations or reductions involving hydrocarbons, an alkane is the least oxidized and an alkyne is the most oxidized.

if it alternately gains oxygens and loses hydrogens. Methane is oxidized to methanol, then to methanal, then to methanoic acid, and finally to carbon dioxide. The same sequence of oxidations occurs for other alkanes. Each series consists of an alkane, alcohol, aldehyde (or ketone), carboxylic acid, and carbon dioxide. The carbon dioxide is most oxidized or least reduced, and the alkane is least oxidized or most reduced.

$$\underset{\substack{\text{Methane}\\\text{(most energetic}\\\text{molecule)}}}{H{-}\overset{H}{\underset{H}{C}}{-}H} \xrightarrow[\text{oxidation}]{\substack{\text{gain of}\\\text{oxygen}}} \underset{\substack{\text{Methanol}\\\text{(methyl alcohol)}}}{H{-}\overset{OH}{\underset{H}{C}}{-}H} \xrightarrow[\text{oxidation}]{\substack{\text{loss of}\\\text{hydrogen}}} \underset{\substack{\text{Methanal}\\\text{(formaldehyde)}}}{H{-}\overset{O}{\overset{\|}{C}}{-}H} \xrightarrow[\text{oxidation}]{\substack{\text{gain of}\\\text{oxygen}}} \underset{\substack{\text{Methanoic acid}\\\text{(formic acid)}}}{H{-}\overset{O}{\overset{\|}{C}}{-}OH} \xrightarrow[\text{oxidation}]{\substack{\text{loss of}\\\text{hydrogen}}} \underset{\substack{\text{Carbon dioxide}\\\text{(least energetic}\\\text{molecule)}}}{O{=}C{=}O}$$

The more reduced a carbon compound, the more energy it can release upon its complete oxidation to carbon dioxide. Are oxidation reactions (1) endothermic or exothermic? The energy-releasing properties of oxidation reactions are extremely important for energy production in living systems. They also explain why the combustion of hydrocarbons such as methane is a good source of energy. **Figure 26.17** shows examples of energy production by oxidation reactions.

Primary alcohols can be oxidized to aldehydes, and secondary alcohols can be oxidized to ketones.

$$\underset{\text{Primary alcohol}}{R{-}\overset{OH}{\underset{H}{C}}{-}H} \xrightarrow[-2H]{\text{oxidation}} \underset{\text{Aldehyde}}{R{-}\overset{O}{\overset{\|}{C}}{-}H} \qquad \underset{\text{Secondary alcohol}}{R{-}\overset{OH}{\underset{H}{C}}{-}R} \xrightarrow[-2H]{\text{oxidation}} \underset{\text{Ketone}}{R{-}\overset{O}{\overset{\|}{C}}{-}R}$$

Tertiary alcohols, however, cannot be oxidized because there is no hydrogen present on the carbon bearing the hydroxyl group. A comparison of the structure of primary, secondary, and tertiary alcohols is shown below.

$$\underset{\text{Primary alcohol}}{H{-}\overset{H}{\underset{H}{C}}{-}\overset{H}{\underset{H}{C}}{-}\overset{H}{\underset{H}{C}}{-}OH} \qquad \underset{\text{Secondary alcohol}}{H{-}\overset{H}{\underset{H}{C}}{-}\overset{H}{\underset{OH}{C}}{-}\overset{H}{\underset{H}{C}}{-}H} \qquad \underset{\text{Tertiary alcohol}}{H{-}\overset{H}{\underset{H}{C}}{-}\overset{CH_3}{\underset{OH}{C}}{-}\overset{H}{\underset{H}{C}}{-}H}$$

Figure 26.17

These soccer players are energized by oxidation reactions. Much of the world relies on hydrocarbon combustion as a source of energy.

Answers

(1) The oxidation of organic compounds is exothermic.

(2) The product can be distilled from the reaction mixture as it is formed.

(3) +1

The primary alcohols methanol and ethanol can be oxidized to aldehydes by warming them at about 50 °C with acidified potassium dichromate ($K_2Cr_2O_7$). In these reactions, methanol produces formaldehyde, and ethanol produces acetaldehyde.

$$\text{H}-\underset{\text{H}}{\underset{|}{\overset{\text{OH}}{\overset{|}{\text{C}}}}}-\text{H} \xrightarrow[H_2SO_4]{K_2Cr_2O_7} \text{H}-\overset{\text{O}}{\overset{\|}{\text{C}}}-\text{H} \qquad \text{CH}_3-\underset{\text{H}}{\underset{|}{\overset{\text{OH}}{\overset{|}{\text{C}}}}}-\text{H} \xrightarrow[H_2SO_4]{K_2Cr_2O_7} \text{CH}_3-\overset{\text{O}}{\overset{\|}{\text{C}}}-\text{H}$$

Methanol (methyl alcohol) (bp 65 °C) — Methanal (formaldehyde) (bp -21 °C) — Ethanol (ethyl alcohol) (bp 78 °C) — Ethanal (acetaldehyde) (bp 21 °C)

Preparing an aldehyde by this method is often a problem because aldehydes are easily oxidized to carboxylic acids.

$$\text{R}-\overset{\text{O}}{\overset{\|}{\text{C}}}-\text{H} \xrightarrow[H_2SO_4]{K_2Cr_2O_7} \text{R}-\overset{\text{O}}{\overset{\|}{\text{C}}}-\text{OH}$$

Aldehyde — Carboxylic acid

However, further oxidation of aldehydes that have low boiling points, such as acetaldehyde, is not a problem. Why is this? 2

Oxidation of the secondary alcohol 2-propanol by warming it with acidified potassium dichromate produces acetone.

$$\text{CH}_3-\underset{\text{H}}{\underset{|}{\overset{\text{OH}}{\overset{|}{\text{C}}}}}-\text{CH}_3 \xrightarrow[H_2SO_4]{K_2Cr_2O_7} \text{CH}_3-\overset{\text{O}}{\overset{\|}{\text{C}}}-\text{CH}_3$$

2-Propanol (isopropyl alcohol) — Propanone (acetone)

Unlike aldehydes, ketones are resistant to further oxidation, so there is no need to remove them from the reaction mixture during the course of the reaction.

Chemists have taken advantage of the ease with which an aldehyde can be oxidized to develop several tests for their detection. **Benedict's test** and **Fehling's test** are commonly used for aldehyde detection. Benedict's and Fehling's reagents are deep-blue alkaline solutions of copper(II) sulfate of slightly differing composition. **Figure 26.18** illustrates Fehling's test for an aldehyde. When an aldehyde is oxidized with Benedict's or Fehling's reagent, a red precipitate of copper(I) oxide (Cu_2O) is obtained. The aldehyde is oxidized to its acid, and copper(II) ions (Cu^{2+}) are reduced to copper(I) ions (Cu^+).

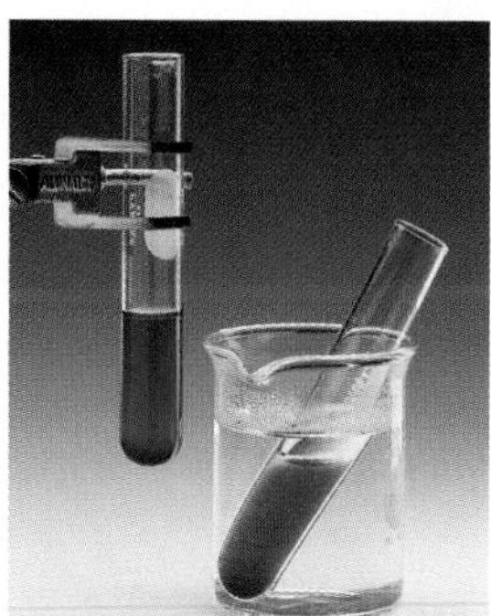

Figure 26.18
When an aldehyde is mixed with Fehling's reagent (left) and heated, the blue copper(II) ions in Fehling's reagent are reduced to form Cu_2O, a red precipitate (right). What is the oxidation state of copper in the product? 3

$$\text{CH}_3-\overset{\text{O}}{\overset{\|}{\text{C}}}-\text{H}(aq) + 2\text{Cu}^{2+}(aq) + 5\text{OH}^-(aq) \longrightarrow \text{CH}_3-\overset{\text{O}}{\overset{\|}{\text{C}}}-\text{O}^-(aq) + \text{Cu}_2\text{O}(s) + 3\text{H}_2\text{O}(l)$$

Ethanal — Copper(II) ion complex (blue solution) — Ethanoic acid (as ethanoate ion) — Copper(I) oxide (red precipitate)

26.3

TEACHER DEMO

Illustrate the oxidation of primary and secondary alcohols to carboxylic acids and ketones respectively by chromic acid. Explain how the loss of a molecule of hydrogen by an alcohol can lead to the formation of a carbonyl group. Prepare an acidic solution of CrO_3 by dissolving 0.25 g of CrO_3 in 0.25 mL of concentrated sulfuric acid. Add 0.75 mL of water. A clear orange solution should be obtained. Set up three test tubes each containing 1 mL of acetone. Dissolve about 5 drops of ethanol, isopropanol, and *t*-butyl alcohol in separate tubes. Add 1–2 drops of chromic acid reagent to each tube and gently tap the tubes to mix. Note any changes that occur. A positive test for a primary or secondary alcohol is indicated by a change in color of the solution from orange to blue-green. No change should occur in the case of *t*-butyl alcohol. Have students record their observations and write the chemical equations for any reactions they observed. Ask:

- Why was no reaction observed for the test tube containing *t*-butyl alcohol? (There is no hydrogen on the carbon bearing the hydroxyl group.)

MINI LAB

Testing for an Aldehyde

SAFETY

Do not store Tollens' reagent for extended periods of time. It decomposes on standing and yields an explosive mixture. Avoid skin contact with these chemicals.

LAB TIPS

The deposition of a silver mirror on the inner walls of the test tube is a positive indicator for the presence of an aldehyde functional group. In some cases it may be necessary to warm the test tube in a bath of warm water.

ANALYSIS AND CONCLUSIONS

1. A layer of reflective, silver metal is deposited on the inner walls of test tube 1. No reaction is observed for test tubes 2 and 3.
2. In an alkaline solution of silver nitrate, the aldehyde is oxidized to a carboxylic acid according to the following reaction:
   $RCHO(aq) + 2Ag(NH_3)_2OH(aq) \rightarrow 2Ag(s) + RCOO^-(aq) + NH_4^+(aq) + H_2O(l) + 3NH_3(aq)$
3. The test is useful for distinguishing aldehydes from alcohols and ketones.

3 Assess

Evaluate Understanding

Write the molecular structures of vanillin, 3-heptanone, ethyl butanoate, and pentanoic acid on the board. Have students classify each compound as an aldehyde, ketone, ester, or carboxylic acid.

Reteach

Review the oxidation states of carbon in organic compounds. Remind students that the oxidation state of carbon can cycle from −4 to +4.

MINI LAB

Testing for an Aldehyde

PURPOSE

To distinguish an aldehyde from an alcohol or a ketone using Tollens's reagent.

MATERIALS

- 1*M* sodium hydroxide
- 6*M* aqueous ammonia
- 5% silver nitrate
- 4 small test tubes
- test tube rack
- plastic eyedroppers
- methanal
- propanone
- ethanol

PROCEDURE

1. Add 1 drop of 1*M* sodium hydroxide to 2 mL of 5% silver nitrate in a test tube. Add 6*M* aqueous ammonia drop by drop, shaking after each addition until the brownish precipitate dissolves. This will be your Tollens's reagent.
2. Place 10 drops of Tollens's reagent in each of three clean, labeled test tubes.
3. To test tube 1, add 2 drops of methanal solution. To test tube 2, add 2 drops of propanone. To test tube 3, add 2 drops of ethanol. Shake the test tubes to mix the contents.
4. Observe the test tubes, leaving them undisturbed for at least 5 minutes.

ANALYSIS AND CONCLUSIONS

1. What evidence of a chemical reaction did you observe in test tube 1? In test tube 2? In test tube 3?
2. Write the equation for any chemical reaction you observed.
3. If you observed a chemical reaction in one or more of the test tubes, what practical uses might the reaction have?

section review 26.3

13. What is a carbonyl group? Describe the carbonyl groups that are characteristic of aldehydes, ketones, carboxylic acids, and esters.
14. What reactants are needed to make the ester propyl ethanoate?
15. What is dehydrogenation? What products are expected when the following compounds are oxidized?
    a. $CH_3CH_2CH_2CH_2OH$ **b.** $CH_3CH_2C(OH)(CH_3)CH_3$ **c.** $CH_3CH_2CH(OH)CH_3$
16. Give the IUPAC name for aldehyde and ketone.
    a. CH_3CH_2CHO **b.** $CH_3CH_2CH_2C(=O)CH_2CH_3$ **c.** $CH_3CH_2CH(CH_3)CH_2CHO$ **d.** $CH_3C(=O)CH_2CH_3$

Chem ASAP! **Assessment 26.3** Check your understanding of the important ideas and concepts in Section 26.3.

Answers

SECTION REVIEW 26.3

13. A carbonyl group is a carbon atom double bonded to an oxygen atom. Aldehyde—carbon of the carbonyl group is always joined to at least one hydrogen (RCHO). Ketone—carbon is joined to two other carbons (RCOR). Carboxylic acid—carbon is attached to a hydroxyl group (RCOOH). An ester is a derivative of a carboxylic acid in which the —OH of the carboxyl group has been replaced by an —OR from an alcohol (RCOOR).
14. 1-propanol and ethanoic acid
15. Dehydrogenation is the loss of hydrogen.
    a. $CH_3CH_2CH_2CHO$ **b.** no reaction
    c. $CH_3CH_2C(=O)CH_3$
16. **a.** propanal **b.** 3-hexanone
    c. 3-methylpentanal **d.** butanone

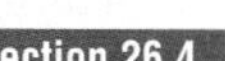

POLYMERIZATION

The base of the Tower of the Americas in San Antonio, Texas, is actually hundreds of feet tall! The restaurant at the top was built and then elevated as the section below it was constructed. This section was then pushed upward as the next section was put into place. In this fashion, the entire tower was established by piecing together unit after unit until the original base ended up at the top. Chemical compounds called polymers are very much like this tower. **What are some characteristics of polymers and monomers?**

objectives

- Define polymer and monomer
- Name and describe the uses of some important addition and condensation polymers

key terms

- polymer
- monomers

Addition Polymers

Most of the reactions that you have learned about so far involve reactants and products of low molar mass. Some of the most important organic compounds that exist, however, are giant molecules called polymers. Each day, you see many different polymers. For example, the materials you know as plastics are polymers. The kinds and uses of plastics are numerous indeed!

A **polymer** is a large molecule formed by the covalent bonding of repeating smaller molecules. The smaller molecules that combine to form a polymer are called **monomers.** Some polymers contain only one type of monomer. Others contain two or more types of monomers. Polymerization is the reaction that joins monomers to form a polymer. Most polymerization reactions require a catalyst.

An addition polymer forms when unsaturated monomers react to form a polymer. Ethene undergoes addition polymerization. The ethene molecules bond to one another to form the long-chain polymer polyethylene.

$$x\,H_2C{=}CH_2 \longrightarrow H{-}(CH_2{-}CH_2)_x{-}H$$

Ethene (ethylene) → Polyethylene

x is number of ethylene units that combine to form long chain

x is number of repeating $-CH_2-CH_2-$ units in polymer; parentheses identify the repeating unit

Polyethylene, which is chemically resistant and easy to clean, is an important industrial product. It is used to make refrigerator dishes, plastic milk bottles, plastic wrappings, and many other familiar items found in the home. The physical properties of polyethylene can be controlled by shortening or lengthening the carbon chains. Polyethylene containing relatively short chains ($x = 100$) has the consistency of paraffin wax. Polyethylene with long chains ($x = 1000$) is harder and more rigid.

Polymers of substituted ethenes are also useful. Polypropylene is prepared by the polymerization of propene.

$$x\,CH_2{=}CH(CH_3) \longrightarrow (CH_2{-}CH(CH_3))_x$$

Propene (propylene) → Polypropylene

26.4

1 Engage

Use the Visual

Have students study the photograph and read the text that opens the section. Explain that *polymer* derives from the Greek words *poly* (many) and *meros* (parts). In organic chemistry the term *polymer* refers to large molecules formed by linking together several thousand small molecules, or monomers, to form a continuous chain of repeating units. Relate the construction of the Tower of the Americas to the formation of a polymer. Ask:

- **What is the repeating unit, or monomer, in the Tower of the Americas?** (each section of the tower)
- **What do you think are some of the characteristics of polymers and monomers?** (Answers will vary.)

Motivate and Relate

Discuss the similarities between the addition of hydrogen and halogens to alkenes and the formation of addition polymers. Explain that chemists have also found that when certain catalysts are used, alkenes will add to one another to form a long chain of hydrocarbon units. Ask:

- **What is a catalyst?** (A catalyst is a substance that increases the rate of a reaction without being used up itself in the reaction. Catalysts lower the activation energy barrier of a reaction permitting more reactants to form products within a given time.)

STUDENT RESOURCES

From the Teacher's Resource Package, use:

- Section Review 26.4, Ch. 26 Practice Problems, Interpreting Graphics, Vocabulary Review, Quizzes, and Tests from the Review Module (Ch. 25–28)
- Laboratory Recordsheet 26–2
- Solutions Manual for Chapter Reviews

TECHNOLOGY RESOURCES

Relevant technology resources include:

- Chem ASAP! CD-ROM
- Resource Pro CD-ROM
- Assessment Resources CD-ROM: Ch. 26 Tests

2 Teach

TEACHER DEMO

Bring a length of metal link chain to class. Use the metal chain as a model for a polymer. Explain that a polymerization reaction is a reaction that joins monomers to form a polymer. Point out the features in the chain that are similar to a polymer molecule. In addition, model a polymerization reaction using paperclips. Each paperclip represents a monomer. Explain that each end of the paperclip represents a reaction site. Build a chain by successively connecting one paperclip to another. The chain formed is analogous to a polymer. Point out that the main difference between polymers in organic chemistry is the type of monomer used to form the chains.

Discuss

Explain that the key to polymerization is the ability of molecules to form repeating chains. For this to happen, each reacting molecule must have at least *two* reactive sites. In addition polymerization, the two active sites become available as double bonds are opened up. In condensation polymerization, the sites become available as two groups (such as —H and —OH) are split off of the molecules.

Polypropylene, a stiffer polymer than polyethylene, is used extensively in utensils and containers. Polystyrene is prepared by the polymerization of styrene. Polystyrene in the form of a rigid foam is a poor heat conductor, which makes it useful for insulating homes and for manufacturing molded items such as coffee cups and picnic coolers. What brand name is
1 commonly used to describe this substance?

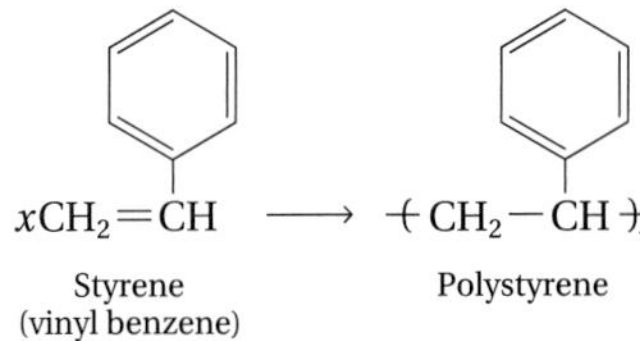

Many halocarbon polymers, including polyvinyl chloride, have useful properties. Vinyl chloride is the monomer of polyvinyl chloride.

$$x\text{CH}_2\text{=CH(Cl)} \longrightarrow \text{+CH}_2\text{—CH(Cl)+}_x$$

Chloroethene (vinyl chloride) — Polyvinyl chloride (PVC)

Polyvinyl chloride (PVC) is used for pipes in plumbing. It is also produced in sheets, sometimes with a fabric backing, for use as a tough plastic upholstery covering. Polytetrafluoroethene (Teflon™ or PTFE) is the product of

(a)

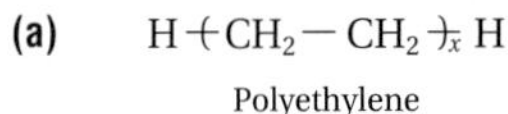

(b)

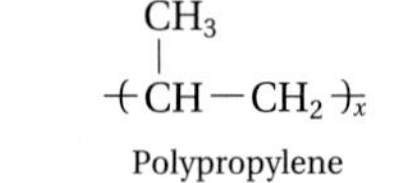

(c)

$\text{+CH}_2\text{—CH(Cl)+}_x$

Polyvinyl chloride (PVC)

Figure 26.19

*These common polymers are used for various applications: **(a)** Polyethylene is shown here as it is formed into a film. This film is commonly used to wrap food. **(b)** Polypropylene is used in the manufacture of a variety of items, such as these whistles. **(c)** Polyvinyl chloride is used for pipes in plumbing.*

MEETING DIVERSE NEEDS

Gifted Have students do research on the pioneering work of Elias James Corey, a professor at Harvard University, who received the Nobel Prize in Chemistry for his work in synthetic organic chemistry. Tell students that his work led to the synthesis of more than 100 important drugs and other naturally occurring compounds. Ask students to write a short report.

At Risk Have students collect several household items composed of different polymer materials. Ask the students to create a table listing the items and the Plastic Container Codes found on the bottom of each item. Have students do research to find out how the codes are involved in recycling.

Figure 26.20 ▶
PTFE is used to coat cookware and to insulate wires, cables, motors, and generators. It is also suspended in motor oils as a friction-reducing agent.

the polymerization of tetrafluoroethene monomers. PTFE is very resistant to heat and chemical corrosion. You are probably familiar with this polymer as the coating on nonstick cookware. Because PTFE is very durable and slick, it is formed into bearings and bushings used in chemical reactors.

$$\underset{\text{Tetrafluoroethene}}{xCF_2{=}CF_2} \longrightarrow \underset{\text{Teflon (PTFE)}}{\left(CF_2 - CF_2\right)_x}$$

Polyisoprene is the polymer that constitutes natural rubber. It is used in the manufacture of boots, tires, and rubber tubing.

$$\underset{\text{Isoprene}}{xCH_2{=}C(CH_3)CH{=}CH_2} \longrightarrow \underset{\text{Polyisoprene}}{\left(CH_2(CH_3)C{=}CH\,CH_2\right)_x}$$

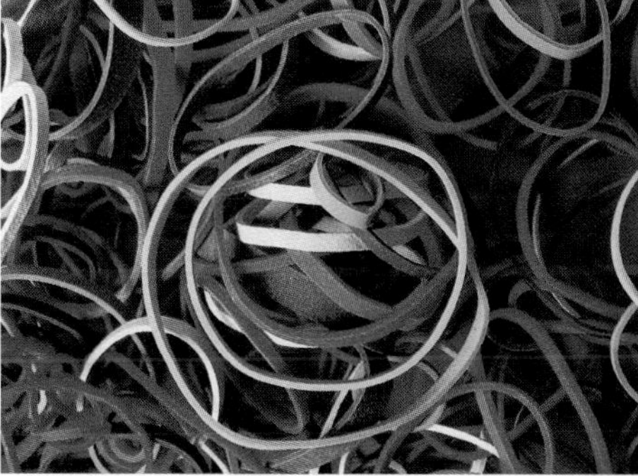

Figure 26.21
Natural rubber is used in a variety of products. What is the source of this polyisoprene? ②

Answers

① styrofoam
② tropical plants such as the rubber tree

TEACHER DEMO

Demonstrate the synthesis of nylon. Pour 5 mL of a 5% aqueous solution of 1,6-hexanediamine into a small beaker. Add 5 drops of 20% sodium hydroxide. Slowly add 5 mL of a 5% solution of adipoyl chloride in cyclohexane by pouring it down the wall of the beaker. A two-phase system will form. Note the formation of a polymer film at the interface. Use forceps or tongs and slowly draw out a long strand of nylon. Rinse the thread of nylon thoroughly with water and set it on a paper towel to dry. Have students examine the color, hardness, and strength. Ask students to consider how many molecules are linked together to form a strand of nylon that long. Discuss some of the uses of nylon, and explain that nylon is a polyamide and an example of a condensation polymer. Note that millions of tons of nylon are produced annually in the United States.

Discuss

Have students study the chemical equations for the formation of the addition polymers discussed in this section. Write the equations on the board and help students identify the repeating units. Discuss the features of an addition polymerization and how the name of a polymer is derived from the monomer units. Bring samples of plastic items to class that are composed of addition polymers. Have students describe and compare their physical characteristics. Ask:

- **What are some of the advantages of plastics over metals and other natural materials?** (Plastics are moldable, lightweight, chemically resistant, and easy to clean.)

ACTIVITY

Write the formula for the simplest repeating unit of several polymers on the board. Ask students to identify the polymer and to write out the structure for each polymer showing how 3 successive monomers are linked together. Ask:

- Why can't an exact formula for a polymer be written? (The length of the carbon chain varies from polymer molecule to polymer molecule. The best that can be done is to write the formula for the simplest repeating unit, the monomer, in parentheses. The subscript *x* indicates a very large number.)

Condensation Polymers

Condensation polymers are formed by the head-to-tail joining of monomer units. This is usually accompanied by the loss of water from the reacting monomers and the formation of water as a reaction product; thus the name condensation polymers. What is the opposite of a condensation
1 reaction? The formation of polyesters is an example of condensation polymerization. Polyesters are polymers consisting of many repeating units of dicarboxylic acids and dihydroxy alcohols joined by ester bonds. The formation of a polyester is represented by a block diagram, which shows only the functional groups involved in the polymerization reaction. The squares and circles represent unreactive parts of the organic molecules.

Condensation polymerization requires that there be two functional groups on each monomer molecule.

$$x\text{HO}-\overset{\text{O}}{\overset{\|}{\text{C}}}-\square-\overset{\text{O}}{\overset{\|}{\text{C}}}-\text{OH} + x\text{HO}-\bigcirc-\text{OH} \longrightarrow \left[\overset{\text{O}}{\overset{\|}{\text{C}}}-\square-\overset{\text{O}}{\overset{\|}{\text{C}}}-\text{O}-\bigcirc-\text{O}\right]_x + 2x\text{H}_2\text{O}$$

Dicarboxylic acid | Dihydroxy alcohol | Representative polymer unit of a polyester

The polyester polyethylene terephthalate (PET) is formed from terephthalic acid and ethylene glycol.

$$\text{HO}-\overset{\text{O}}{\overset{\|}{\text{C}}}-\text{C}_6\text{H}_4-\overset{\text{O}}{\overset{\|}{\text{C}}}-\text{OH} + \text{HO}-\text{CH}_2\text{CH}_2-\text{OH} \longrightarrow \left[\overset{\text{O}}{\overset{\|}{\text{C}}}-\text{C}_6\text{H}_4-\overset{\text{O}}{\overset{\|}{\text{C}}}-\text{O}-\text{CH}_2\text{CH}_2-\text{O}\right]_x + 2x\text{H}_2\text{O}$$

Terephthalic acid | Ethylene glycol | Representative polymer unit of polyethylene terephthalate (PET)

PET fibers form when the compound is melted and forced through tiny holes in devices called spinnerettes. The fibers, sold as Dacron™, Fortrel®, or Terylene (depending on the manufacturer), are used for tire cord and permanent-press clothing. The fibers are often blended with cotton to make garments that are more comfortable on hot, humid days than those containing 100% polymer. These garments retain the wrinkle resistance of 100% polyester. PET melts may be forced through a narrow slit to produce sheets of Mylar™, a polymer used extensively as magnetic tape for tape
2 recorders and computers. Can you think of another common use of Mylar?

Figure 26.22
Woven Dacron tubing can be used to replace major blood vessels.

Answers

1 hydrolysis
2 balloons
3 rope

Many important polymers are formed by the reaction of carboxylic acids and amines. The amines used to make polymers generally contain the amino functional group $-NH_2$. The condensation of a carboxylic acid and an amine produces an amide.

$$\underset{\text{Carboxylic acid}}{R-\overset{\overset{\large O}{\|}}{C}-OH} + \underset{\text{Amine}}{H-\overset{\overset{\large H}{|}}{N}-R} \longrightarrow \underset{\text{Amide}}{R-\overset{\overset{\large O}{\|}}{C}-\overset{\overset{\large H}{|}}{N}-R} + \underset{\text{Water}}{HO-H}$$

Polyamides are polymers in which the carboxylic acid and amine monomer units are linked by amide bonds. Various types of nylon are polyamides. You are probably familiar with a range of nylon products. Nylon polymer has a molar mass of about 1×10^4 g/mol and a melting point of 250 °C. The representative polymer unit of nylon is derived from 6-aminohexanoic acid, a compound that contains carboxylic acid and amino functional groups. The long polymer chain is formed by the successive attachment of the carboxyl group of one molecule of the acid to the amino group of the next by the formation of an amide bond.

$$\underset{\text{6-Aminohexanoic acid}}{xH_2N-CH_2(CH_2)_4\overset{\overset{\large O}{\|}}{C}-OH} \xrightarrow{\text{heat}} \underset{\text{Representative polymer unit of nylon}}{\left(CH_2(CH_2)_4\overset{\overset{\large O}{\|}}{C}-\overset{\overset{\large H}{|}}{N}\right)_x} + xH_2O$$

The melted polymer can be spun into very fine, yet very strong fibers. Nylon fibers are used for carpeting, tire cord, fishing lines, and textiles. They have replaced natural silk in sheer hosiery. Nylon is also molded into gears, bearings, and zippers. Can you think of additional uses of nylon? 3

Figure 26.23

Nylon fishing lines are lightweight, yet very strong. Flame-resistant clothing is made of Nomex™, a polyamide with aromatic rings.

26.4

Discuss

Discuss the features of a condensation polymerization. Compare and contrast condensation reactions with addition reactions. Write a number of specific examples of condensation polymers on the board. Relate the structure of polyamides to the structure of a protein. Point out that cells use enzymes to catalyze condensation reactions between amino acids to form a polymer of amino acids called a protein. Ask students to name some other kinds of naturally occurring polymers. (polysaccharides such as cellulose, starch, and chitin; nucleic acids such as DNA; proteins)

3 Assess

Evaluate Understanding

Have students write sentences in their own words describing what takes place during the formation of a polymer. Ask:

- **What is the difference between an addition polymer and a condensation polymer? Give examples of each.** (An addition polymer is formed when unsaturated monomers react to form a polymer. One example is the reaction of ethene to form polyethylene. Condensation polymers are formed by the head-to-tail joining of monomer units. The other product of the condensation reation is a small molecule such as water. For example, a dicarboxylic acid and a dihydroxy alcohol can join to form a polyester and water.)

Have students describe the uses of some addition and condensation polymers.

Reteach

Remind students that the area of chemistry devoted to the study of polymers involves more than just plastics. Synthetic fibers are used to create clothes and automobile tires, as well as materials used in the medical and dental fields. Very few elements contribute to the large number of synthetic polymers; their abundance can be traced to the unique bonding ability of carbon. To show how small changes in the structures of monomers can lead to dramatic differences in physical properties, compare polymers based on ethene and vinyl chloride.

Polyamides that contain aromatic rings are extremely tough and flame resistant. The aromatic rings make the resulting fiber stiffer and tougher. Nomex™ is a polyamide with a carbon skeleton consisting of aromatic rings derived from isophthalic acid and *m*-phenylenediamine.

$$\text{HO}-\overset{\text{O}}{\overset{\|}{\text{C}}}-\text{C}_6\text{H}_4-\overset{\text{O}}{\overset{\|}{\text{C}}}-\text{OH} + \text{H}_2\text{N}-\text{C}_6\text{H}_4-\text{NH}_2 \longrightarrow \left[\overset{\text{O}}{\overset{\|}{\text{C}}}-\text{C}_6\text{H}_4-\overset{\text{O}}{\overset{\|}{\text{C}}}-\underset{\text{H}}{\underset{|}{\text{N}}}-\text{C}_6\text{H}_4-\underset{\text{H}}{\underset{|}{\text{N}}}\right]_x + 2x\text{H}_2\text{O}$$

Isophthalic acid — *m*-Phenylenediamine — Representative unit of Nomex

Like nylon, Nomex is a poor conductor of electricity. Because it is more rigid than nylon, it is used to make parts for electrical fixtures. Nomex is also used in the manufacture of flame-resistant clothing for race-car drivers and firefighters and in the fabrication of flame-resistant building materials.

Kevlar™ has a structure similar to that of Nomex. It is a polyamide made from terephthalic acid and *p*-phenylenediamine. What type of reaction do these components undergo to form Kevlar? ❶

$$\text{HO}-\overset{\text{O}}{\overset{\|}{\text{C}}}-\text{C}_6\text{H}_4-\overset{\text{O}}{\overset{\|}{\text{C}}}-\text{OH} + \text{H}_2\text{N}-\text{C}_6\text{H}_4-\text{NH}_2 \longrightarrow \left[\overset{\text{O}}{\overset{\|}{\text{C}}}-\text{C}_6\text{H}_4-\overset{\text{O}}{\overset{\|}{\text{C}}}-\overset{\text{H}}{\overset{|}{\text{N}}}-\text{C}_6\text{H}_4-\overset{\text{H}}{\overset{|}{\text{N}}}\right]_x + 2x\text{H}_2\text{O}$$

Terephthalic acid — *p*-Phenylenediamine — Representative unit of Kevlar

Figure 26.24 *Bulletproof vests are made of Kevlar.*

Kevlar is used extensively where strength and flame resistance are needed. You can see one application of Kevlar in **Figure 26.24.** A properly constructed vest made of Kevlar is strong enough to stop high-speed bullets, yet light and flexible enough to be worn under normal clothing.

Proteins, which are polyamides of naturally occurring amino acids, rank among the most important of all biological molecules. You will learn about these essential polymers in more detail in Chapter 27.

section review 26.4

17. What is a polymer? A monomer?
18. Describe addition and condensation polymerization. Give an example of each type of polymer.
19. Give names and uses for three types of polymers.

Chem ASAP! Assessment 26.4 Check your understanding of the important ideas and concepts in Section 26.4.

Answers

❶ condensation

SECTION REVIEW 26.4

17. A polymer is a large molecule formed by the covalent bonding of repeating small molecules. The smaller molecules that combine to form the repeating unit of a polymer are monomers.
18. Addition polymers are formed by the addition of one monomer unit to another; polyethylene is an addition polymer. Condensation polymers are formed by the head-to-tail joining of monomer units, usually with the loss of water; polyethylene terephthalate is a condensation polymer.
19. Answers will vary. Polyethylene is used for milk bottles and other containers. Polyester is used as a textile for permanent-press clothing, and Kevlar is used in bullet-resistant vests.

SMALL-SCALE LAB

POLYMERS

SAFETY  

Wear safety glasses and follow the standard safety proceedures outlined on page 18.

PURPOSE

To cross-link some polymers and examine their properties.

MATERIALS

- $3\frac{1}{2}$-oz plastic cup
- plastic spoon
- soda straw
- 4% borax solution
- powdered guar gum
- pipet

PROCEDURE

1. Half fill a $3\frac{1}{2}$-oz cup with water.
2. Use a soda straw as a measuring scoop to obtain approximately 2 cm of powdered guar gum. Gently sprinkle the guar gum powder into the water while stirring with a plastic spoon. Add the guar gum powder slowly to prevent it from clumping. Stir the mixture well.
3. While stirring, add one full pipet (about 4 mL) of borax solution. Continue to stir until a change occurs.

ANALYSIS

Using your experimental data, record the answers to the following questions below your observations.

1. Describe the polymer you just made. Is it a liquid or a solid? What special characteristics does it have?
2. Guar gum is a carbohydrate, or a polymer with many repeating alcohol functional groups (—OH). Draw a zig-zag line to represent a crude polymer chain.
3. Add —OH groups along the chain to represent the alcohol functional groups.
4. Borate ions combine with alcohol to form water and borate complexes of the alcohol.

HO, OH, $^{-}$O, OH on B $+$ R—OH $\longrightarrow$ HO, OH, $^{-}$O, OR on B $+$ HOH

Write a similar equation that replaces all of the —OH groups on the borate with —OR groups.

5. If two polymer chains each contain two nearby —OH groups, borate will cross-link the polymer chains by forming a complex with two alcohols on each chain. Draw a structure similar to the one you drew for Question 4, but replace your four R groups with two polymer chains.

YOU'RE THE CHEMIST

The following small-scale activities allow you to develop your own procedures and analyze the results.

1. **Design It!** Try using borax to cross-link other common carbohydrate polymers, such as corn starch or liquid laundry starch. For each polymer, half fill a $3\frac{1}{2}$-oz cup with the chosen carbohydrate polymer and add enough water to bring the liquid to within about 1 cm of the rim. Stir carefully and thoroughly. Add one full pipet (about 4 mL) of borax solution while stirring. Describe the similarities and differences between this cross-linked polymer and the polymer you made previously. Compare the properties of these polymers.
2. **Analyze It!** Cut a 1 cm × 15 cm strip of paper and use a drop of glue or a stapler to fasten one end of the strip of paper to the other end to form a ring. Now cut out some identical-sized strips of paper and glue or staple them together into an interlocking chain of paper rings. Explain how this chain is like a polymer. Now make another paper ring that links your chain to one of your classmate's chains. Explain how these linked chains are like cross-linked polymers.

SMALL-SCALE LAB

POLYMERS

OVERVIEW

Procedure Time: 30 minutes

Students will use a borax solution to cross link carbohydrate polymers. They will write structural formulas for cross-linked polymers.

MATERIALS

Solution	Preparation
4% borax	10 g $Na_2B_4O_7 \cdot 10H_2O$ in 250 mL
liquid starch	Use laundry starch.
white glue	Use white school glue.

TEACHING SUGGESTIONS

The school glue tends to adhere to the stirring spoon as it cross links. Decant any excess liquid that does not adhere. Wooden craft sticks make great stirring sticks.

SAFETY AND DISPOSAL

Students should not take home the polymers as the mixtures tend to develop mold in a few days. Dispose of these polymers in the waste basket, not down the sink. While the cross linked polymers are non-toxic, borax is toxic if ingested. Remind students to wash their hands thoroughly when finished.

YOU'RE THE CHEMIST

1. Through experimentation, students are able to produce an amazing variety of polymers with different properties.
2. The chain is similar to a polymer because it contains many repeating units linked end to end. The rings that link two chains together are like the borate ion that cross-links polymer chains.

ANALYSIS

1. a viscous gel that oozes

2–3.

OH OH OH OH

4. HO, OH, HO, OH on B $+$ 4R—OH $\longrightarrow$ RO, OR, RO, OR on B $+$ 4HOH

5.

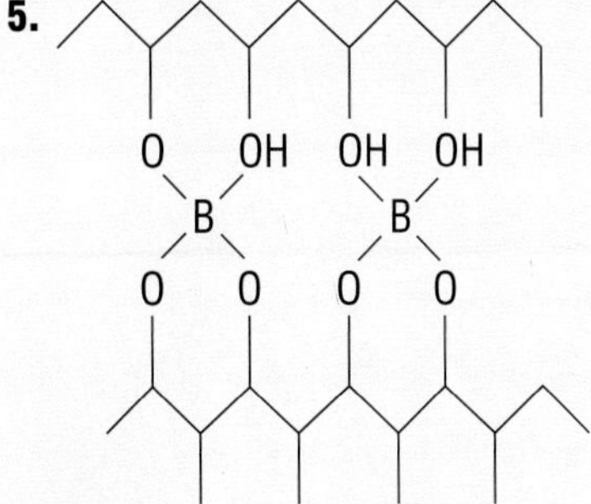

Chemistry Serving... the Consumer

DISCUSS

Have students study the text. Explain that many familiar smells and flavors come from organic compounds. For example, many ketones and aldehydes, particularly those with high molar masses, have fragrant odors. Ask:

- What are some other functional groups in molecules that can have an effect on the particular odor of an organic compound? (alcohols, esters, amines)

Have students name some of their favorite scents. Ask:

- What are the natural sources of these perfumes?

Explain that perfume makers usually combine many different compounds to create a complex scent. These compounds come from animal oils and plant extracts, or they can be synthesized. Write the structures for citral (roses, citrus), and civetone (musk) on the board and point out the functional groups in each.

$$CH_3-\overset{\overset{CH_3}{|}}{C}=CHCH_2CH_2\overset{\overset{CH_3}{|}}{C}=CH\overset{\overset{O}{\|}}{C}-H$$

citral

$$\begin{array}{c} O \\ \| \\ C \\ (CH_2)_7 \quad\quad (CH_2)_7 \\ C=C \\ |\quad\;| \\ H\quad H \end{array}$$

civetone

CHEMISTRY IN CAREERS

Have students read about the field of fragrance development on page 880. If you have access to the Internet in your classroom or school, you may wish to have students connect to the address shown. Have students do research to find out how perfumes and colognes are made.

Chemistry Serving...the Consumer

THE SWEET SMELL OF FUNCTIONAL GROUPS

What happens when you smell good food cooking? The scent of a freshly mowed lawn? The salt air of the ocean? If you are like most people, smells have the power to change the way you feel and to evoke memories of the past.

The power of smell has always led human beings to find ways to capture and produce certain scents that are particularly pleasing or meaningful. Long ago, people burned the wood and leaves of aromatic plants to release their scents, often as part of religious practices. Oils from the scent glands of certain animals, such as the musk deer, were also used for their compelling odors. Applied to the body, these oils became the first perfumes.

Attempts to capture a wider variety of odors in a concentrated, storable form led to the beginnings of perfumery. From its earliest days up to the present, perfumery has been an art—but it has also always involved quite a bit of chemistry.

Most of the fragrances that have enchanted people throughout the ages, such as jasmine, patchouli, lavender, and myrrh, come from the essential oils found in plants. These oils, containing the organic compounds that give each plant its characteristic scent, are found in a variety of plant parts, including flowers, bark, roots, seeds, and fruits. The oils must be extracted from the plant parts to be used in perfume.

The ancient Egyptians were perhaps the first to extract the essential oils of plant parts. They soaked flower petals in liquid fat, which dissolved the essential oils. Then they mixed the fat with ethanol. The essential oils were more soluble in ethanol than in fat, and dissolved in the ethanol, resulting in a strongly scented solution that could be stored and applied to the body.

> From its earliest days up to the present, perfumery has been an art—but it has also always involved quite a bit of chemistry.

Since the time of the Egyptians, various other extraction methods have been developed. Some are similar to the Egyptians' method but employ a variety of different solvents. Other methods take advantage of differences in boiling point between essential oils and solvents by using distillation to concentrate the oils.

During the past century, advances in chemistry techniques have revolutionized the perfume industry. Natural essential oils are still extracted from plants, but most perfumes are made mostly from synthetic compounds.

The synthetic compounds used in perfumes are either identical or similar to compounds in natural essential oils. To make such a compound, chemists first have to identify the active compound in the essential oil and learn its structure. Then they can find ways of synthesizing it from more common substances.

Analysis of essential oils has revealed that many of the scents that please human noses are produced by alcohols, aldehydes, and esters. And in many, the carbon backbone of the molecule contains a benzene ring or one or more carbon–carbon double bonds. Scientists believe that molecules such as these have their effect on our noses because of their shapes and the functional groups they contain.

CHEMISTRY IN CAREERS

FLAVOR DEVELOPER

Put your money where your mouth is! Learn the art of making new flavors and scents for various industries.

See page 880.

TAKE IT TO THE NET

Find out more about career opportunities: www.phschool.com

CHEMICAL SPECIALIST

Local food service distributor seek responsible self-motivated individ

Chapter 26 STUDENT STUDY GUIDE

Take It to the NET
For interactive study and review, go to www.phschool.com

KEY TERMS

- addition reaction *p. 781*
- alcohol *p. 778*
- aldehyde *p. 785*
- alkyl halide *p. 774*
- aryl halide *p. 774*
- Benedict's test *p. 793*
- carbonyl group *p. 785*
- carboxyl group *p. 788*
- carboxylic acid *p. 788*
- dehydrogenation reaction *p. 791*
- denatured alcohol *p. 781*
- ester *p. 789*
- ether *p. 783*
- fatty acid *p. 788*
- Fehling's test *p. 793*
- fermentation *p. 781*
- functional group *p. 773*
- halocarbon *p. 774*
- hydration reaction *p. 781*
- hydrogenation reaction *p. 782*
- hydroxyl group *p. 778*
- ketone *p. 785*
- monomer *p. 795*
- polymer *p. 795*
- substitution reaction *p. 776*

CONCEPT SUMMARY

26.1 Introduction to Functional Groups

- Saturated hydrocarbons undergo few useful reactions. The chemical reactions of most organic compounds involve functional groups.
- A functional group is a specific arrangement of atoms in an organic compound capable of characteristic chemical reactions.
- A substitution reaction occurs when an atom or group of atoms replaces another atom or group of atoms.

26.2 Alcohols and Ethers

- An alcohol is an organic compound with a hydroxyl group. An ether is a compound in which oxygen is bonded to two carbon groups.
- A common functional group is the carbon–carbon double bond of an alkene. Alkenes undergo addition of water to form alcohols. Alkenes can add a hydrogen halide or a halogen to form halocarbons. They can add hydrogen to form alkanes.
- A benzene ring usually undergoes a substitution reaction rather than addition.

26.3 Carbonyl Compounds

- Common functional groups containing oxygen include alcohols, ethers, aldehydes, ketones, carboxylic acids, and esters. The latter four types of functional groups contain a carbon–oxygen double bond, or carbonyl group.
- Alcohols may be oxidized to aldehydes or ketones. Aldehydes may be oxidized to carboxylic acids. Fehling's and Benedict's solutions can be used to detect the presence of aldehydes.
- Esters result from the combination of carboxylic acids and alcohols. Hydrolysis of an ester produces these components.

26.4 Polymerization

- Alkenes can undergo addition polymerization to form useful materials such as the long-chain polymer polyethylene.
- Condensation polymerization of monomer units containing hydroxyl and carboxyl groups in the same molecule produces polyesters.

CHAPTER CONCEPT MAP

Use these terms to construct a concept map that organizes the major ideas of this chapter.

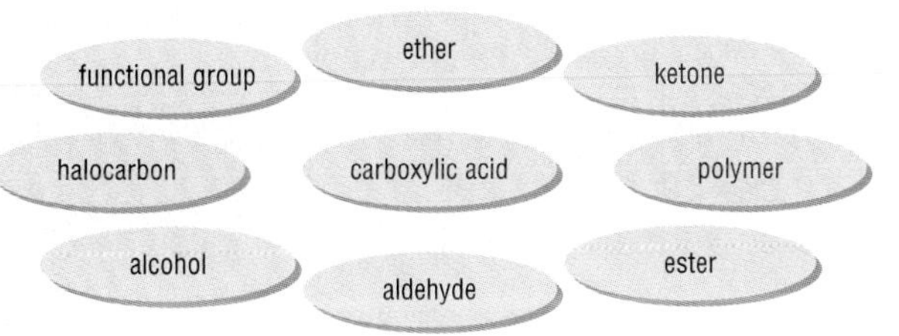

Chem ASAP! **Concept Map 26**
Create your Concept Map using the computer.

Take It to the Net

At **www.phschool.com** students will find for this chapter

- an Internet activity
- links to related chemistry sites
- an interactive quiz
- career links

Chapter 26
STUDENT STUDY GUIDE

4 Close

Summary

Ask the following questions, which require students to summarize the information contained in the chapter.

- **What is meant by the term *functional group*?** (a specific arrangement of atoms in a molecule that undergoes predictable reactions)
- **What series of reactions could be applied to synthesize 2-butanone from 1-butene?** (Addition: 1-butene + HBr to produce 2-bromobutane; Substitution: 2-bromobutane + NaOH to produce 2-butanol; Oxidation: 2-butanol + acidified potassium dichromate to produce 2-butanone)

Write structures for organic molecules. Have students classify each molecule according to the functional groups it contains and name some physical and chemical properties associated with each functional group.

Extension

Have students research the use of synthetic polymers in medicine. Some applications include prosthetics, skin substitutes, artificial artery grafts, and heart valve replacements.

Looking Back . . . Looking Ahead . . .

The preceding chapter used the structure and physical properties of hydrocarbons as an introduction to organic chemistry. This chapter discussed the role functional groups play in determining the behavior of organic compounds. The next chapter will introduce the main categories of biological molecules: carbohydrates, lipids, proteins, and nucleic acids.

Answers

20. a carbon chain or ring attached to the functional group

21. a. $ClCH_2CCl_2CH_3$ (Cl above and below the central C)

c. cyclohexane ring with Cl on adjacent carbons

b. benzene ring with Br, Br, Br (1,3,5)

22. a. 3-chloropropene
b. 1,2-dichloro-4-methylpentane
c. 1,3-dibromobenzene

23. a. $CH(Cl)_2-CH_2-CH_3$
1, 1-dichloropropane

$CH_2(Cl)-CH(Cl)-CH_3$
1, 2-dichloropropane

$CH_2(Cl)-CH_2-CH_2(Cl)$
1, 3-dichloropropane

$CH_3-C(Cl)_2-CH_3$
2, 2-dichloropropane

b. $CH_3-CH_2-CH_2-CH_2Br$
1-bromobutane

$CH_3-CH_2-CH(Br)-CH_3$
2-bromobutane

$CH_3-CH(CH_3)-CH_2Br$
1-bromo-2-methylpropane

$CH_3-C(CH_3)(Br)-CH_3$
2-bromo-2-methylpropane

24. a. benzene ring–OH b. cyclohexane ring–OH

c. $CH_3-CH(CH_3)-OH$ d. benzene ring–Br

25. a. $CH_2(Br)CH(Br)CH_2CH_3$ c. $CH_3CH(H)CH(H)CH_3$

b. $CH_3CH(I)CH(I)CH_3$ d. cyclohexane ring with Cl on adjacent carbons

26. a. $CH_2(H)-CH_2(Br)$
bromoethane

d. $CH_2(H)-CH_2(H)$
ethane

b. $CH_2(Cl)-CH_2(Cl)$
1, 2-dichloroethane

e. $CH_2(H)-CH_2(Cl)$
chloroethane

c. $CH_2(H)-CH_2(OH)$
ethanol

Chapter 26 REVIEW

CONCEPT PRACTICE

20. What does R in the formula $R-CH_2Cl$ represent? *26.1*

21. Write a structural formula for each compound. *26.1*
a. 1,2,2-trichlorobutane
b. 1,3,5-tribromobenzene
c. 1,2-dichlorocyclohexane

22. Name the following halocarbons. *26.1*
a. $CH_2=CHCH_2Cl$
b. $CH_3CH(CH_3)CH_2CH(Cl)CH_2Cl$
c. benzene ring with Br at positions 1 and 3

23. Write structural formulas and give IUPAC names for all the isomers of the following compounds. *26.1*
a. $C_3H_6Cl_2$
b. C_4H_9Br

24. What organic products are formed in the following reactions? *26.1*
a. $C_6H_5-Br + NaOH \xrightarrow{heat}$ ____ $+ NaBr$
b. $C_6H_{11}-Cl + NaOH \longrightarrow$ ____ $+ NaCl$
c. $CH_3CH(CH_3)Cl + NaOH \longrightarrow$ ____ $+ NaCl$
d. $C_6H_6 + Br_2 \xrightarrow{catalyst}$ ____ $+ HBr$

25. Write the structure for the expected product from each reaction. *26.2*
a. $CH_2=CHCH_2CH_3 + Br_2 \longrightarrow$
b. $CH_3CH=CHCH_3 + I_2 \longrightarrow$
c. $CH_3CH=CHCH_3 + H_2 \longrightarrow$
d. cyclohexene $+ Cl_2 \longrightarrow$

26. Write structures and names of the products obtained upon addition of each of the following reagents to ethene. *26.2*
a. HBr
b. Cl_2
c. H_2O
d. H_2
e. HCl

27. Name the following ethers. *26.2*
a. $CH_3OCH_2CH_3$
b. $C_6H_5-O-CH_2CH_3$
c. $CH_2=CHOCH=CH_2$
d. $CH_3CH(CH_3)OCH(CH_3)CH_3$

28. Explain why diethyl ether is more soluble in water than dihexyl ether is. Would you expect propane or diethyl ether to be more soluble in water? Why? *26.2*

29. Explain why 1-butanol has a higher boiling point than diethyl ether has. Which compound would you expect to be more soluble in water? Why? *26.2*

30. Name these aldehydes and ketones. *26.3*
a. $CH_3C(=O)CH_3$
b. $CH_3CH(CH_3)CH_2CHO$
c. $C_6H_5-CH_2CHO$
d. $C_6H_5-C(=O)-C_6H_5$
e. CH_3CHO
f. $CH_3CH_2C(=O)CH_2CH_2CH_3$

31. How would you expect the boiling points of propane, propanol, and propanal to compare? *26.3*

32. Propane ($CH_3CH_2CH_3$) and acetaldehyde (CH_3CHO) have the same molar mass, but propane boils at −42 °C and acetaldehyde boils at 20 °C. Account for this difference. *26.3*

27. a. ethylmethyl ether
b. ethylphenyl ether
c. divinyl ether or vinyl ether
d. diisopropyl ether or isopropyl ether

28. The oxygen atom in diethyl ether polarizes the small molecule. This enables diethyl ether to dissolve in water, which is also polar. The large dihexyl ether molecule has large nonpolar parts and does not dissolve. Propane is less soluble in water than is diethyl ether because propane is nonpolar.

29. The alcohol molecules form hydrogen bonds with one another, resulting in a higher boiling point. They also form hydrogen bonds with water molecules, causing 1-butanol to be more soluble than diethyl ether. (Although diethyl ether is polar, 1-butanol has greater polarity.)

30. a. propanone or acetone
b. 3-methylbutanal
c. 2-phenylethanal
d. diphenylmethanone or diphenyl ketone or benzophenone

33. How would you expect the water solubility of ethanoic and decanoic acids to compare? *26.3*

34. What are the products of each reaction? *26.3*
 a. $HCOOH + KOH \longrightarrow$
 b. $CH_3CH_2COOH + NaOH \longrightarrow$
 c. $CH_3COOH + NaOH \longrightarrow$

35. Give common names for each carboxylic acid. *26.3*
 a. $HCOOH$
 b. CH_3COOH
 c. CH_3CH_2COOH
 d. $CH_3(CH_2)_{16}COOH$

36. Give the name and structure of the alcohol that must be oxidized to make the following compounds. *26.3*
 a. CH_3CH_2CHO
 b. $CH_3CH_2CCH_3$ (with C=O)
 c. CH_3CH_2CHCHO (with CH_3 on CH)

37. Complete the reactions by writing the structure of the expected products. *26.3*
 a. $CH_3CH_2OH \xrightarrow{K_2Cr_2O_7}$
 b. $CH_3CH_2CHO \xrightarrow{K_2Cr_2O_7}$
 c. $CH_3CH_2CHOH \xrightarrow{K_2Cr_2O_7}$ (with CH_3 on CH)
 d. 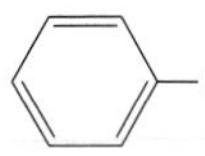 $C_6H_5-CH_2CHO \xrightarrow{K_2Cr_2O_7}$
 e. $CH_3CH_2CCH_3$ (with C=O) $\xrightarrow{K_2Cr_2O_7}$

38. Write the name and structure for the alcohol that must be oxidized to make each carbonyl compound. *26.3*
 a. $HCHO$
 b. CH_3CCH_3 (with C=O)
 c. CH_3CHCHO (with CH_3 on CH)
 d. 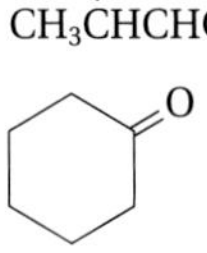 (cyclohexanone: ring with =O)

39. Write the structures of the expected products for these reactions. *26.3*
 a. $CH_3CH_2COOCH_2CH_3 \xrightarrow{NaOH}$
 b. $CH_3COO-C_6H_5 \xrightarrow{KOH}$
 c. $CH_3CH_2COOCH_2CHCH_3 \xrightarrow{HCl}$ (with CH_3 on CH)

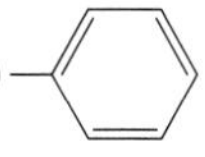

40. Complete the following reactions by writing the structures of the expected products and by naming the reactants and products. *26.3*
 a. $CH_3COOCH_3 + H_2O \xrightarrow{HCl}$
 b. $CH_3CH_2CH_2COOCH_2CH_2CH_3 + H_2O \xrightarrow{NaOH}$
 c. $HCOOCH_2CH_3 + H_2O \xrightarrow{KOH}$

41. Write the structure and name of the ester that could be produced from each reaction. *26.3*
 a. formic acid + methanol $\longrightarrow$
 b. butyric acid + ethanol $\longrightarrow$
 c. acetic acid + propanol $\longrightarrow$

42. Different samples of a polymer such as polyethylene can have different properties. Explain. *26.4*

43. What is the structure of the repeating units in a polymer that has the following monomers? *26.4*
 a. 1-butene
 b. 1,2-dichloroethene

CONCEPT MASTERY

44. Write a general structure for each type of compound.
 a. halocarbon c. ester
 b. ketone d. amide

45. Predict which compound has the highest boiling point. Molar masses are given in parentheses.
 a. CH_3CHO (44)
 b. CH_3CH_2OH (46)
 c. $CH_3CH_2CH_3$ (44)

46. Write the structure and name of the expected products for each reaction.
 a. $CH_3COOH + CH_3OH \xrightarrow{H^+}$
 b. $CH_3CH_2CH_2COOCH_2CH_3 + H_2O \xrightarrow{NaOH}$
 c. $CH_3CH_2OH \xrightarrow{K_2Cr_2O_7}$

47. Explain why a carbon–carbon double bond is nonpolar, but a carbon–oxygen double bond is very polar.

e. ethanal or acetaldehyde
f. 3-hexanone or ethylpropyl ketone

31. in order of increasing boiling point: propane, propanal, and propanol

32. Acetaldehyde is polarized by its carbonyl oxygen, forming stronger intermolecular attractions. Nonpolar propane has weak intermolecular attractions. Thus propane molecules are more easily liberated from the liquid state.

33. The short-chain ethanoic acid has a higher water solubility.

34. a. $HCOOH + KOH \rightarrow HCOO^-K^+ + H_2O$
 b. $CH_3CH_2COOH + NaOH \rightarrow CH_3CH_2COO^-Na^+ + H_2O$
 c. $CH_3COOH + NaOH \rightarrow CH_3COO^-Na^+ + H_2O$

35. a. formic acid c. propionic acid
 b. acetic acid d. stearic acid

36. a. 1-propanol, $CH_3CH_2CH_2OH$ b. 2-butanol, $CH_3CH_2CHCH_3$ (with OH on CH)
 c. 2-methyl-1-butanol, $CH_3CH_2CHCH_2OH$ (with CH_3 on CH)

37. a. $CH_3CH_2OH \xrightarrow{K_2Cr_2O_7} CH_3-C(=O)-H$
 b. $CH_3CH_2CHO \xrightarrow{K_2Cr_2O_7} CH_3-CH_2-C(=O)-OH$
 c. CH_3CH_2CHOH (with CH_3) $\xrightarrow{K_2Cr_2O_7} CH_3-CH_2-C(=O)-CH_3$
 d. 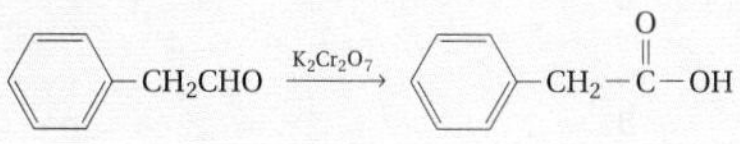 $C_6H_5-CH_2CHO \xrightarrow{K_2Cr_2O_7} C_6H_5-CH_2-C(=O)-OH$
 e. $CH_3CH_2C(=O)CH_3 \xrightarrow{K_2Cr_2O_7}$ no reaction

38. a. CH_3OH, methanol
 b. $CH_3-CH(OH)-CH_3$, 2-propanol
 c. $CH_3-CH(CH_3)-CH_2-OH$, 2-methyl-1-propanol
 d. 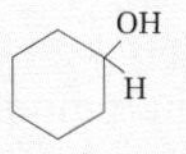 cyclohexanol (ring with OH and H on the same carbon)

39. a. $CH_3CH_2COO^-Na^+$, CH_3CH_2OH
 b. $CH_3COO^-K^+$, C_6H_5-OH
 c. CH_3CH_2COOH, CH_3CHCH_2OH (with CH_3 on CH)

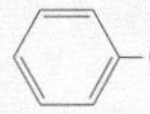

40. a. $CH_3COOCH_3 + H_2O \xrightarrow{HCl} CH_3-C(=O)-OH + CH_3OH$
 methyl ethanoate (methyl acetate), water, ethanoic acid (acetic acid), methanol
 b. $CH_3CH_2CH_2COOCH_2CH_2CH_3 + H_2O \xrightarrow{NaOH}$
 propyl butanoate (propyl butyrate), water
 $CH_3CH_2CH_2-C(=O)-O^-Na + CH_3CH_2CH_2OH$
 sodium butanoate (sodium butyrate), 1-propanol
 c. $HCOOCH_2CH_3 + H_2O \xrightarrow{KOH}$
 ethyl methanoate (ethyl formate), water
 $H-C(=O)-O^-K + CH_3CH_2OH$
 potassium methanoate (potassium formate), ethanol

41. a. $H-C(=O)-O-CH_3$
 methyl methanoate (methyl formate)
 b. $CH_3CH_2CH_2-C(=O)-O-CH_2CH_3$
 ethyl butanoate (ethyl butyrate)
 c. $CH_3-C(=O)-O-CH_2CH_2CH_3$
 propyl ethanoate (propyl acetate)

42. The properties of polyethylene vary with the length of the chains.

43. a. $-CH_2-CH(CH_2CH_3)-$ b. $-CH(Cl)-CH(Cl)-$

Chapter 26 REVIEW

Answers

44. a. R–X **c.** $R-\overset{O}{\overset{\|}{C}}-O-R$

b. $R-\overset{O}{\overset{\|}{C}}-R$ **d.** $R-\overset{O}{\overset{\|}{C}}-\overset{H}{\overset{|}{N}}-R$

45. b. CH_3CH_2OH (46) has the highest boiling point.

46. a. $CH_3-\overset{O}{\overset{\|}{C}}-OCH_3 + H_2O$

b. $CH_3CH_2CH_2-\overset{O}{\overset{\|}{C}}-O^-Na^+ + CH_3CH_2OH$

c. $CH_3-\overset{O}{\overset{\|}{C}}-H$

47. Both atoms in a carbon-carbon double bond have the same electronegativity, so the bond is nonpolar. Because oxygen is more electronegative than carbon, a carbon-oxygen double bond is very polar.

48. a. phenol
b. ether
c. alcohol
d. phenol
e. alcohol

49. a. $CH_3CH_2\overset{Cl}{\overset{|}{C}H}-\overset{Cl}{\overset{|}{C}H_2}$

b. $CH_3CH_2\overset{Br}{\overset{|}{C}H}-\overset{Br}{\overset{|}{C}H_2}$

c. (cyclohexane ring with H and Br on one carbon)

50. a. carboxylic acid, ethanoic acid (acetic acid)
b. ketone, propanone (acetone)
c. ether, diethyl ether
d. alcohol, ethanol (ethyl alcohol)

51. The chemical properties (and toxicity) of organic compounds are determined by the compound as a whole. As a substituent in a molecule, a phenyl group ring does not have the same properties as benzene.

52. $0.117M$ $Ca(NO_3)_2$

53. At any given moment, the rate of dissolving of solute is equal to the rate of precipitation of solute. As a result, the concentration of the solution remains constant.

54. b. 3

55. a. $:\ddot{F}\cdot + e^- \longrightarrow :\ddot{F}:^-$

b. $H\cdot + :\ddot{O}: + e^- \longrightarrow (:\ddot{O}:H)^-$

56. 2.86 g SO_2

48. Classify each compound as an alcohol, a phenol, or an ether.

a. (naphthalene ring with OH)

b. (phenyl–O–phenyl)

c. (phenyl–CH_2OH)

d. (benzene ring with OH)

e. $CH_3CH_2\underset{CH_3}{\underset{|}{C}}HOH$

49. Write the structural formulas for the products of these reactions.

a. $CH_3CH_2CH{=}CH_2 + Cl_2 \longrightarrow$

b. $CH_3CH_2CH{=}CH_2 + Br_2 \longrightarrow$

c. (cyclohexene) $+ HBr \longrightarrow$

50. For each compound pictured, identify the functional group and name the compound. The red atoms represent oxygen.

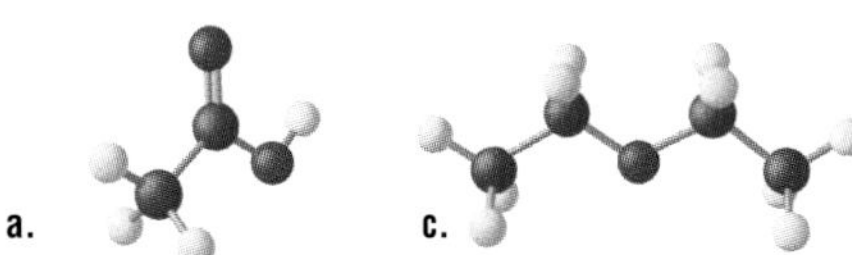

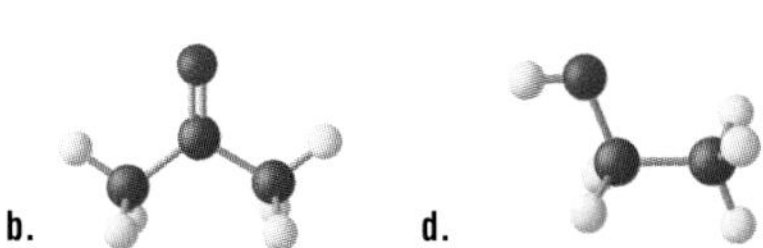

CRITICAL THINKING

51. Benzene is poisonous and a proven carcinogen. Yet many compounds containing benzene rings, such as benzaldehyde, are common in the foods you eat. Why are some organic compounds with phenyl groups safe to eat?

CUMULATIVE REVIEW

52. A solution is made by diluting 250 mL of $0.210M$ $Ca(NO_3)_2$ solution with water to a final volume of 450 mL. Calculate the molarity of $Ca(NO_3)_2$ in the diluted solution.

53. In a saturated solution containing undissolved solute, the solute is continually dissolving, but the solution concentration remains constant. Explain.

54. What is the maximum number of orbitals in the p sublevel of an atom?
a. 1
b. 3
c. 5
d. 9

55. Using electron dot structures, illustrate the formation of F^- from a fluorine atom and a hydroxide ion from atoms of hydrogen and oxygen.

56. Calculate the mass, in grams, of one liter of SO_2 at standard temperature and pressure.

CONCEPT CHALLENGE

57. Cholesterol is a compound in your diet and is also synthesized in the liver. Sometimes it is deposited on the inner walls of blood vessels, causing hardening of the arteries. Describe the structural features and functional groups of this important molecule.

(cholesterol structure: HO; CH_3; CH_3; CH_3–$CHCH_2CH_2CH$–CH_3, CH_3)

58. Hydrocarbons from petroleum are an important source of raw material for the chemical industry. Using reactions covered in this chapter and any required inorganic chemicals, propose a scheme for the manufacture of ethylene glycol, a major component of antifreeze, from petrochemical ethene.

57. Cholesterol is an alcohol with a hydroxyl group on a cycloalkane. It has four nonaromatic rings. It has a double bond on one of its rings, as well as a large alkyl group, making it nonpolar.

58. $CH_2CH_2(g) + Br_2(l) \rightarrow CH_2BrCH_2Br(l)$
$CH_2BrCH_2Br(l) + 2NaOH(aq) \rightarrow$
$CH_2OHCH_2OH(l) + 2NaBr(aq)$

Chapter 26 STANDARDIZED TEST PREP

Select the choice that best answers each question or completes each statement.

1. Which type of organic compound is associated with the odors of fruits?
   a. alcohols c. esters
   b. amines d. ethers
2. The acid-catalyzed hydrolysis of an ester gives a carboxylic acid and
   a. an amine. c. an alcohol.
   b. an ether. d. an alkene.
3. Ethane, methanal, and methanol have similar molar masses. Which series lists the compounds in order of increasing boiling point?
   a. ethane, methanal, methanol
   b. methanal, methanol, ethane
   c. methanol, methanal, ethane
   d. ethane, methanol, methanal
4. A carbonyl group is characterized by a
   a. carbon-carbon double bond.
   b. carbon-oxygen double bond.
   c. carbon-nitrogen single bond.
   d. carbon-oxygen single bond.

The lettered choices below refer to questions 5–8. A lettered choice may be used once, more than once, or not at all.

(A) alcohol
(B) ketone
(C) carboxylic acid
(D) ether
(E) aldehyde

To which class of organic compounds does each of the following compounds belong?

5. CH_3CH_2COOH
6. $CH_3CH_2OCH_3$
7. $CH_3CH_2CH_2OH$
8. CH_3COCH_3

Use the space-filling models with questions 9 and 10.

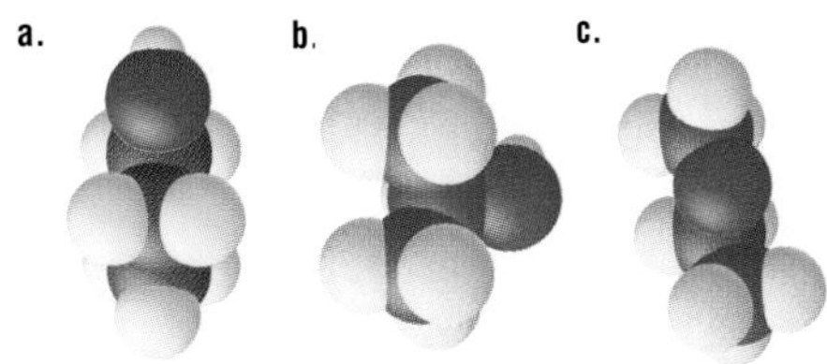

9. Using the space-filling models as a guide, draw ball-and-stick models of the three compounds that have the formula C_3H_8O. Name each compound.
10. *You Draw It!* There are two compounds with a carbonyl group that have the molecular formula C_3H_6O. Write a complete structural formula and draw a ball-and-stick model for each compound. Name each compound.

Characterize the reactions in questions 11–15 as addition, esterification, oxidation, polymerization, or substitution reactions.

11. $CH_3CHO \xrightarrow[H_2SO_4]{K_2Cr_2O_7} CH_3COOH$
12. $CH_2{=}CH_2 + HCl \longrightarrow CH_3CH_2Cl$
13. $CH_3CO_2H + CH_3CH_2OH \xrightarrow{H^+} CH_3COOCH_2CH_3 + H_2O$
14. $xCH_2{=}CH_2 \longrightarrow H{+}CH_2{-}CH_2{+}_xH$
15. (benzene) $+ Br_2 \xrightarrow{catalyst}$ (benzene ring with Br) $+ HBr$

For each question there are two statements. Decide whether each statement is true or false. Then decide whether Statement II is a correct explanation for Statement I.

	Statement I		Statement II
16.	The addition of hydrogen to an alkene is a reduction reaction.	BECAUSE	The addition of hydrogen to any molecule is a reduction reaction.
17.	Aldehydes react readily with oxidizing agents.	BECAUSE	Oxidation of aldehydes produces alcohols.
18.	Ethanol (CH_3CH_2OH) is immiscible in water in all proportions.	BECAUSE	Ethanol molecules can form hydrogen bonds with other ethanol molecules.
19.	The hydrogenation of benzene is easier than the hydrogenation of ethene.	BECAUSE	The benzene ring is stabilized by resonance.

Chapter 26 STANDARDIZED TEST PREP

1. c
2. c
3. a
4. b
5. (C)
6. (D)
7. (A)
8. (B)

9. a.

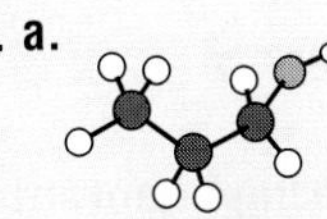

1-propanol

b.

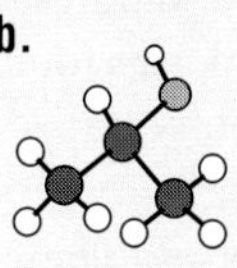

2-propanol

c.

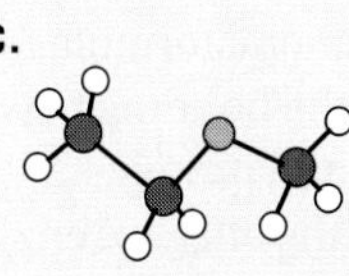

ethylmethyl ether

10. See structural formulas and ball-and-stick models for propanal and propanone (also known as dimethyl ketone and acetone) below. In the models, black spheres represent carbon, gray spheres represent oxygen, and white spheres represent hydrogen.
11. oxidation
12. addition
13. esterification
14. polymerization
15. substitution
16. True, True, correct explanation
17. True, False
18. False, True
19. False, True

```
   H  H  H
   |  |  |
H—C—C—C=O
   |  |
   H  H
```

propanal

```
   H  O  H
   |  ‖  |
H—C—C—C—H
   |     |
   H     H
```

propanone

Planning Guide

SECTION OBJECTIVES	ACTIVITIES/FEATURES	MEDIA & TECHNOLOGY
27.1 A Strategy for Life ● ■ ◇ ▸ Describe the structure of a typical eukaryotic cell ▸ Explain the relationship between photosynthesis and all life on Earth	**SE** **Discover It!** *Biological Catalysis,* p. 808	**ASAP** Animation 29 **ASAP** Assessment 27.1
27.2 Carbohydrates □ ◆ ▸ Describe the important structural characteristics of monosaccharides, disaccharides, and polysaccharides ▸ List sources and uses for a number of important carbohydrates	**TE** **DEMO,** p. 814	**ASAP** Assessment 27.2 **CHM** Side 10, 34: *Carbohydrates*
27.3 Amino Acids and Their Polymers □ ◆ ▸ Write a general formula for an amino acid, and describe the bonding between amino acids in peptides and proteins ▸ Describe the effect of enzymes on biochemical reactions	**SE** **Link to Biology** *Ethnobiology,* p. 816 **SE** **Small-Scale Lab** *The Egg: A Biochemical Storehouse,* p. 820 (LRS 27-1) **SSLM** 39: *Reaction of Biomolecules* **TE** **DEMOS,** pp. 816, 817	**ASAP** Assessment 27.3
27.4 Lipids □ ◆ ▸ Characterize the molecular structures of triglycerides, phospholipids, and waxes ▸ Describe the functions of phospholipids and proteins in cell membranes	**LM** 51: *Preparation of Soap*	**ASAP** Assessment 27.4 **CHM** Side 10, 20: *Soaps and Detergents*
27.5 Nucleic Acids □ ◆ ▸ Describe the structural components of nucleotides and nucleic acids, including DNA ▸ Give simple examples of genetic mutations ▸ Explain what is meant by recombinant DNA technology	**SE** **Mini Lab** *A Model of DNA,* p. 826 (LRS 27-2)	**ASAP** Simulation 29 **ASAP** Assessment 27.5
27.6 Metabolism □ ◆ ▸ Describe the role of ATP in energy production and energy use in the cell ▸ Define metabolism and explain the relationship between catabolism and anabolism	**SE** **Link to Exercise Physiology** *Training Energy Pathways,* p. 833 **SE** **Chemistry Serving . . . the Consumer** *Fake Fat,* p. 835 **SE** **Chemistry in Careers** *Biochemist,* p. 835	**ASAP** Assessment 27.6 **ASAP** Concept Map 27 **OT** 76: *Metabolism* **PP** Chapter 27 Problems **RP** Lesson Plans, Resource Library **AR** Computer Test 27 **www** Activities, Self-Tests **TCP** The Chemistry Place Web Site

KEY

● Conceptual (concrete concepts)	**AR** Assessment Resources	**GRS** Guided Reading and Study Workbook
○ Conceptual (more abstract/math)	**ASAP** Chem ASAP! CD-ROM	**LM** Laboratory Manual
■ Standard (core content)	**ACT** ActivChemistry CD-ROM	**LP** Laboratory Practicals
□ Standard (extension topics)	**CHM** CHEMedia Videodiscs	**LRS** Laboratory Recordsheets
◆ Honors (core content)	**CA** Chemistry Alive! Videodiscs	**SSLM** Small-Scale Lab Manual
◇ Honors (options to accelerate)	**GCP** Graphing Calculator Problems	

PRACTICE		ASSESSMENT	
GRS	Section 27.1	**SE**	Section Review
RM	Practice Problems 27.1	**RM**	Section Review 27.1
		RM	Chapter 27 Quiz
GRS	Section 27.2	**SE**	Section Review
RM	Practice Problems 27.2	**RM**	Section Review 27.2
		RM	Chapter 27 Quiz
GRS	Section 27.3	**SE**	Section Review
RM	Practice Problems 27.3	**RM**	Section Review 27.3
		RM	Chapter 27 Quiz
GRS	Section 27.4	**SE**	Section Review
RM	Practice Problems 27.4	**RM**	Section Review 27.4
		RM	Chapter 27 Quiz
GRS	Section 27.5	**SE**	Section Review
RM	Practice Problems 27.5	**RM**	Section Review 27.5
RM	Interpreting Graphics	**RM**	Chapter 27 Quiz
GRS	Section 27.6	**SE**	Section Review
RM	Practice Problems 27.6	**RM**	Section Review 27.6
		RM	Vocabulary Review 27
		SE	Chapter Review
		SM	Chapter 27 Solutions
		PHAS	Chapter 27 Test Prep
		RM	Chapter 27 Quiz & Test

OT	Overhead Transparency	**SE**	Student Edition
PHAS	PH Assessment System	**SM**	Solutions Manual
PLM	Probeware Lab Manual	**SSV**	Small-Scale Video/Videodisc
PP	Problem Pro CD-ROM	**TCP**	www.chemplace.com
RM	Review Module	**TE**	Teacher's Edition
RP	Resource Pro CD-ROM	**www**	www.phschool.com

PLANNING FOR ACTIVITIES

STUDENT EDITION

Discover It! p. 808
- raw and cooked meat
- 3% hydrogen peroxide
- medicine dropper
- plates

Small-Scale Lab, p. 820
- pencil, paper, ruler
- egg
- powdered milk
- balance
- 1*M* HCl, 0.2*M* $CuSO_4$, 0.5*M* NaOH

Mini Lab, p. 826
- cardboard tube
- 10 toothpicks
- felt-tip markers
- thumbtack
- metric ruler

TEACHER'S EDITION

Teacher Demo, p. 814
- hot water bath
- 2–5 mL of 1% carbohydrate solution (corn syrup)
- 2 test tubes
- distilled water
- 2–3 mL of Benedict's reagent

Teacher Demo, p. 816
- amino acid models

Teacher Demo, p. 817
- thighbone from cooked poultry
- 1*M* HCl
- water

Activity, p. 822
- butter
- regular margarine
- low-calorie margarine
- 3 test tubes
- warm water

Activity, p. 828
- spectrometer
- purified DNA
- distilled water
- quartz cuvettes
- thermometer
- 0.2*M* HCl
- tris-buffer (pH 8.0)
- 0.2*M* NaOH

Chapter 27

Key Terms

27.1 photosynthesis
27.2 carbohydrates, monosaccharides, disaccharides, polysaccharides
27.3 amino acid, peptide, peptide bond, amino acid sequence, polypeptide, protein, enzymes, substrates, active site, enzyme–substrate complex, coenzymes
27.4 lipids, triglycerides, saponification, phospholipids, waxes
27.5 nucleic acids, nucleotides, gene, genetic code
27.6 adenosine triphosphate, metabolism, catabolism, anabolism

DISCOVER IT!

Students should observe bubbles of gas when H_2O_2 is added to the raw meat, but no bubbles when H_2O_2 is added to the cooked meat. Students may recall from earlier chapters that the rate of decomposition of H_2O_2 into O_2 and H_2O increases greatly in the presence of a catalyst. Thus, they may infer that a catalyst exists in raw meat that is destroyed when the meat is cooked. (The catalyst is called catalase.)

Chapter 27 THE CHEMISTRY OF LIFE

FEATURES

Stay current with SCIENCE NEWS
Find out more about biochemistry:
www.phschool.com

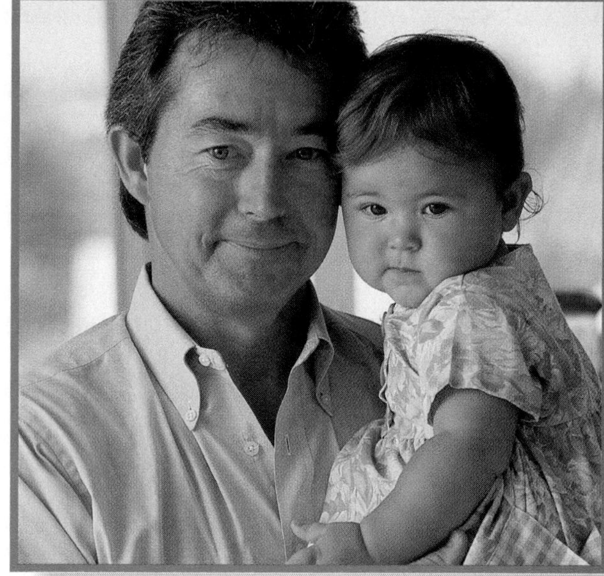

Children resemble their biological parents because of inherited DNA.

DISCOVER IT! BIOLOGICAL CATALYSIS

You need a small piece of raw red meat, such as ground beef, a similar-size piece of the same kind of meat that has been cooked, two plates, a small amount of 3% hydrogen peroxide, and a medicine dropper.

1. Place the two meat samples on separate plates. Add a few drops of hydrogen peroxide solution to the raw and the cooked meat samples. Both samples should be at room temperature.
2. Observe the samples for about five minutes.

Do you observe any difference in the appearance of the hydrogen peroxide placed on the raw meat and on the cooked meat? What do you think causes the difference in the appearance of the hydrogen peroxide on the raw and cooked meats? What can you conclude from this experiment? Return to your results as you read about chemical processes in this chapter and try to explain what you have observed.

27.1 ● ■ ◇
27.2 □ ◆
27.3 □ ◆
27.4 □ ◆
27.5 □ ◆
27.6 □ ◆

Conceptual Section 27.1 can be used to give students an overview of biochemistry. The remainder of the chapter may be omitted due to the breadth of topics and the complexity of the molecules.

Standard Use the illustrations in this chapter to provide students with a visual overview of those compounds that are important in biochemistry. Section 25.5, Nucleic Acids, is an important section for building students' science literacy.

Honors Section 27.1 may be accelerated because students should have encountered this material in a biology course. Use a theme of polymers to help students organize this chapter. Figure 27.27 will help students to summarize chapter concepts in the context of metabolism.

A STRATEGY FOR LIFE

section 27.1

The air you breathe is composed mainly of oxygen (O_2) and nitrogen (N_2). A hypothesis regarding Earth's early atmosphere suggests, however, that the air was very different and most inhospitable to life. It is thought that the atmosphere changed over time as a result of photosynthesis, a process carried out by certain living organisms. One of the products of photosynthesis is oxygen, which began to accumulate in the atmosphere, changing its composition. **What are the reactants and products of photosynthesis, and in what organisms does the process occur?**

objectives
- Describe the structure of a typical eukaryotic cell
- Explain the relationship between photosynthesis and all life on Earth

key term
- photosynthesis

The Structure of Cells

Life! You are certainly familiar with it, but what does it really mean? Until recently, life has been defined as the ability of an organism to grow and to reproduce its own kind. However, recent discoveries made at the fringes of life seem to blur this simple definition. As difficult as it is to define life, you can generally regard tiny structures called cells as the fundamental units of life. In this section, you will learn some basic ideas about the construction of cells and about the chemistry that cells use to obtain the energy they need to sustain life.

Organisms are composed of as few as one cell or as many as billions of cells. Two major cell designs occur in nature. Prokaryotic cells are the cells of bacteria. The cells of all other organisms, including green plants and humans, are called eukaryotic cells. The prokaryotic cell is the more ancient of the two. Microscopic examination of fossilized remains shows that prokaryotic cells were present on Earth at least 3 billion years ago. Eukaryotic cells did not appear until about 1 billion years ago. **Figure 27.1** shows both types of cells.

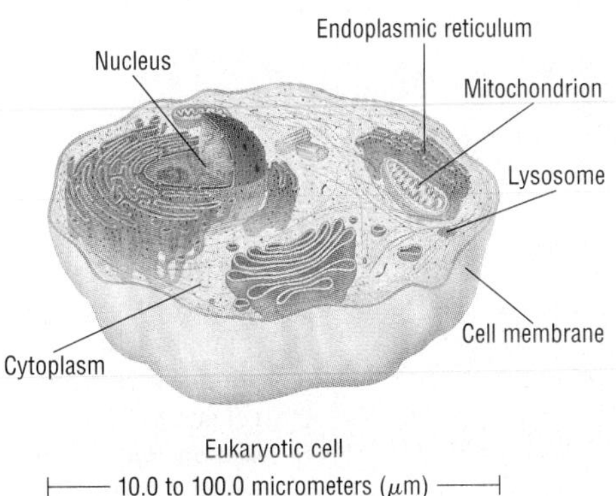

Figure 27.1
The structures of a typical prokaryotic cell and a typical eukaryotic cell are shown here. Note that only the eukaryotic cell has a nucleus.

27.1

1 Engage

Use the Visual

Have students study the photograph and read the text that opens the section. Explain that scientists can study gases emitted from volcanoes and the atmospheres of other planets to infer what Earth's early atmosphere was like. Ask students to speculate about what types of gases may have dominated before the appearance of photosynthetic organisms. (Studies suggest the following: CO_2, CH_4, H_2, NH_3, N_2, and H_2O. Photosynthesis, directly and indirectly, relies on the presence of these simple compounds.) **Ask:**

- **What are the reactants and products of photosynthesis, and in which type of organisms does the process occur?** (In photosynthesis, solar energy is used to drive the production of an organic compound, glucose, from simple inorganic compounds—CO_2 and H_2O. A necessary byproduct of this reaction is O_2. These reactions occur in many types of bacteria and in all plant species.)

STUDENT RESOURCES

From the Teacher's Resource Package, use:
- Section Review 27.1, Ch. 27 Quizzes from the Review Module (Ch. 25–28)

TECHNOLOGY RESOURCES

Relevant technology resources include:
- Chem ASAP! CD-ROM
- ResourcePro CD-ROM

Motivate and Relate

To help students begin to focus on biochemical processes, have them work in teams to create a list of characteristics that distinguish living from nonliving systems. (Living things have one or more cells; they obtain and consume energy; they respond to changing conditions; they grow and develop, and they can reproduce.) **Once students have finished, present a series of photographs of living and nonliving things and help students use their list to classify each subject as living or nonliving.**

2 Teach

Use the Visual

Display Figure 27.1 on an overhead projector. Point out that a cell's size and shape can vary and are related to its function. For example, compare the functions of nerve cells and blood cells. Emphasize that many metabolic reactions in eukaryotic cells are compartmentalized. Ask:

- **What possible advantages would organelles provide for a cell?** (Many of the chemical activities that occur in cells involve similar reactants and products. By nature, they are incompatible. If these processes are physically separated, they can proceed concurrently without conflict.)

Both types of cells include the chemicals necessary for life, encased in a cell membrane. The cell membrane is a sack that holds the contents of a cell and that acts as a selective barrier for the passage of substances into and out of the cell. Eukaryotic cells are considerably larger and more complicated than prokaryotic cells, but the chemical processes carried out by both types of cells are very similar, and both are extremely efficient chemical factories. One major feature that distinguishes eukaryotic cells is the presence of certain organelles, which means little organs. Organelles are small, usually membrane-enclosed, structures suspended in the interior cellular fluid, or cytoplasm. The organelles are the sites of many specialized functions in eukaryotic cells.

The nucleus, one structure that is important in eukaryotic cell reproduction, is not present in prokaryotic cells. Mitochondria (singular: mitochondrion), are the source of cellular energy in eukaryotic cells that use oxygen. Mitochondria are often referred to as the powerhouses of the cell. Lysosomes are the sites for the digestion of substances taken into a cell. Yet another membrane-encased structure in eukaryotic cells is the highly folded, netlike endoplasmic reticulum (ER). Among its various functions, the ER serves as an attachment site for ribosomes. The ribosomes, small organelles that are not membrane-encased, are the sites where essential substances called proteins are made.

Energy and Carbon Cycle

Organisms must have energy to survive. Sunlight is the source of all energy obtained by organisms. **Photosynthesis** is the process by which cells directly capture and use solar energy to make food. Chloroplasts, which are specialized organelles in the cells of green plants, and some algae contain a light-capturing system that converts light energy into chemical energy. Photosynthetic organisms also absorb carbon dioxide from the atmosphere. The energy obtained from sunlight is used to reduce the carbon dioxide to compounds that contain C—H bonds, mainly in the form of the sugar glucose ($C_6H_{12}O_6$).

$$\underset{\text{Carbon dioxide (carbon in more oxidized state)}}{6CO_2} + \underset{\text{Water}}{6H_2O} + \underset{\text{from sunlight}}{\text{Energy}} \longrightarrow \underset{\text{Glucose (carbon in more reduced state)}}{C_6H_{12}O_6} + \underset{\text{Oxygen}}{6O_2}$$

Figure 27.2 illustrates the relationship between photosynthesis and the carbon compounds used by all organisms. In the energy and carbon cycle, photosynthetic organisms produce necessary carbon compounds. Animals, which do not carry out photosynthesis, get those carbon compounds by eating plants or by eating animals that feed on plants. Animals get energy by unleashing the energy stored in the chemical bonds of these carbon compounds. The nutrients are oxidized back to carbon dioxide and water in the process.

$$\underset{\text{Glucose (carbon in more reduced state)}}{C_6H_{12}O_6} + \underset{\text{Oxygen}}{6O_2} \longrightarrow \underset{\text{Carbon dioxide (carbon in more oxidized state)}}{6CO_2} + \underset{\text{Water}}{6H_2O} + \text{Energy}$$

Meeting Diverse Needs

LEP Have students draw a cell showing the organelles. Ask them to label each one and write a short description of its function.

Gifted Have students write an essay about a scientist who made a major contribution to one principle of photosynthesis. Volunteers can share their essays with the class. Scientists who devoted their professional careers to the study of photosynthesis include Robert Emerson, Cornelis Van Niel, Robert Hill, and Melvin Calvin.

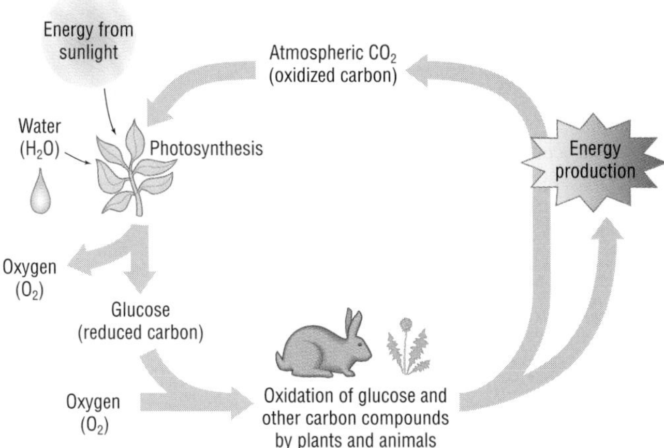

Figure 27.2
In the energy and carbon cycle, the processes of photosynthesis and the oxidation of glucose are responsible for the major transformations and movements of carbon.

Thus, although plant life could survive without animals, animal life could never survive without plants. Without photosynthesis the supply of carbon compounds needed by animals for energy production would not exist.

Oxygen is also an important product of photosynthesis. Photosynthetic organisms produce the oxygen found in Earth's atmosphere. The oxygen is necessary for most organisms to sustain life. The importance of photosynthetic organisms in producing carbon compounds and oxygen is a major reason for the concern about the loss of such organisms through the destruction of rain forests, shown in **Figure 27.3,** and the pollution of the oceans.

All biological processes, including photosynthesis, are based on certain essential kinds of chemical substances. Surprisingly, the great complexity of life arises from just a few types of biological molecules. In the remainder of this chapter, you will learn about the molecular structures of the great classes of biological molecules. You will also learn about the roles they play.

Chem ASAP!
Animation 29
Explore the complexity of chemicals essential to all life.

Figure 27.3
Plants, such as these rain forest trees, release essential oxygen into the atmosphere through photosynthesis. The destruction of these trees raises serious concerns about the future of Earth's atmosphere.

section review 27.1

1. Describe the structure of a eukaryotic cell.
2. Explain how the existence of life on Earth depends on photosynthesis.
3. Write an equation that summarizes the process of photosynthesis.

 Chem ASAP! **Assessment 27.1** Check your understanding of the important ideas and concepts in Section 27.1.

Answers

SECTION REVIEW 27.1

1. A eukaryotic cell has a cell membrane, a nucleus, and various organelles, which include mitochondria, lysosomes, an endoplasmic reticulum, and ribosomes.
2. The existence of life depends on photosynthesis because its principal product, glucose, captures and stores the energy needed by both plants and animals. The oxygen produced is required to release the stored energy.
3. $6CO_2 + 6H_2O + \text{Sunlight} \rightarrow C_6H_{12}O_6 + 6O_2$

27.1

3 Assess

Evaluate Understanding

Write the equation for photosynthesis on the chalkboard. Ask:

- **Should this reaction be classified as acid–base, redox, or combination?** [redox; Note that carbon dioxide forms during the combustion (oxidation) of glucose. In the reverse reaction, photosynthesis, CO_2 is reduced and H_2O is oxidized.]
- **Is photosynthesis endothermic or exothermic?** (endothermic)
- **Which part of the plant cell specializes in the light-harvesting reactions of photosynthesis?** (the chloroplast)

Reteach

Have students examine Figure 27.2. Explain that the diagram represents the flow of energy and nutrients through Earth's biosphere, the regions on Earth where living organisms exist. Point out that energy is required to maintain the highly organized chemical environments found in living cells. Point out that, while nutrients continuously cycle between living organisms and the environment, the flow of energy is unidirectional; thus, life on Earth depends on a continuous supply of energy from the sun. Ask students why plants and other photosynthetic organisms are called "producers" and animals are called "consumers." (Photosynthetic organisms produce high energy forms of carbon compounds that are consumed by animals.)

1 Engage

Use the Visual

Have students study the photograph and read the text that opens the section. Point out that carbohydrates represent the most abundant class of biological compounds on Earth. Plants, for example, are mostly composed of a carbohydrate known as cellulose. Ask:

- **What are carbohydrates, and what is the general formula of these compounds?** (Carbohydrates are a group of organic compounds that include simple sugars, such as glucose, and more complex sugars such as starches and cellulose. Plants and animals use different carbohydrates to store energy and to provide structural support, such as chitin used by arthropods to build exoskeletons. Most carbohydrates have the general formula $C_nH_{2n}O_n$.)

Check Prior Knowledge

To assess students' knowledge about polymers, ask:

- **What is a polymer?** (A high molar mass molecule formed by the covalent bonding of repeating smaller units, called monomers.)
- **Explain why different samples of the same polymer can have different physical properties.** (The molar mass, shape, aqueous solubility, melting point, and flexibility of a particular polymer will depend on the length of the polymer chains, the amount of branching, and the amount of cross-linking between chains.)

section 27.2

CARBOHYDRATES

objectives

- Describe the important structural characteristics of monosaccharides, disaccharides, and polysaccharides
- List sources and uses for a number of important carbohydrates

key terms

- carbohydrates
- monosaccharides
- disaccharides
- polysaccharides

What's wrong with this cicada? Nothing! This cicada is molting—shedding its old exoskeleton and forming a new one. An exoskeleton is a hard protective covering characteristic of insects, lobsters, and other arthropods. As an arthropod grows, it must molt to make room for a bigger body. An arthropod's exoskeleton is made of a polymer called chitin, which belongs to a class of organic molecules called carbohydrates. **What are carbohydrates, and what is the general formula of these compounds?**

Monosaccharides

Long-distance runners often prepare for a big race by eating a great deal of food such as bread and pasta, a process known as carbohydrate loading. Breads and pastas are excellent sources of the family of important molecules called carbohydrates. **Carbohydrates** are monomers and polymers of aldehydes and ketones that have numerous hydroxy groups attached; they are made up of carbon, hydrogen, and oxygen. The name carbohydrate comes from the early observation that many of these compounds have the general formula $C_n(H_2O)_n$ and, as a result, appear to be hydrates of carbon. (They are not, in fact, true hydrates.)

You would probably call a diet based mainly on bread and pasta a starchy diet. You would be correct. Such foods contain carbohydrates called starches. In this section, you will learn about the similarities and differences among some well-known types of carbohydrates.

Carbohydrates are present in most foods, as **Figure 27.4** shows. The simplest carbohydrate molecules are called simple sugars, or **monosaccharides.** Examples of simple sugars are glucose and fructose, which are also structural isomers, because they both have the molecular formula $C_6H_{12}O_6$. However, glucose has an aldehyde functional group, whereas fructose has a ketone functional group. They both undergo many of the same reactions as ordinary aldehydes and ketones.

Figure 27.4
Carbohydrates are the most abundant sources of energy in food. The foods in the photograph on the left contain simple carbohydrates called sugars. The foods in the photograph on the right are good sources of complex carbohydrates called starches.

STUDENT RESOURCES

From the Teacher's Resource Package, use:

- Section Review 27.2, Ch. 27 Quizzes from the Review Module (Ch. 25–28)

TECHNOLOGY RESOURCES

Relevant technology resources include:

- Chem ASAP! CD-ROM
- Resource Pro CD-ROM

In aqueous solution, simple sugars such as glucose and fructose exist in dynamic equilibrium in straight-chain and cyclic forms. The cyclic form predominates. Glucose is abundant in plants and animals. Depending on the source, glucose has also been called corn sugar, grape sugar, or blood sugar. Fructose occurs in a large number of fruits and in honey. Here are the structures for each sugar in its two forms:

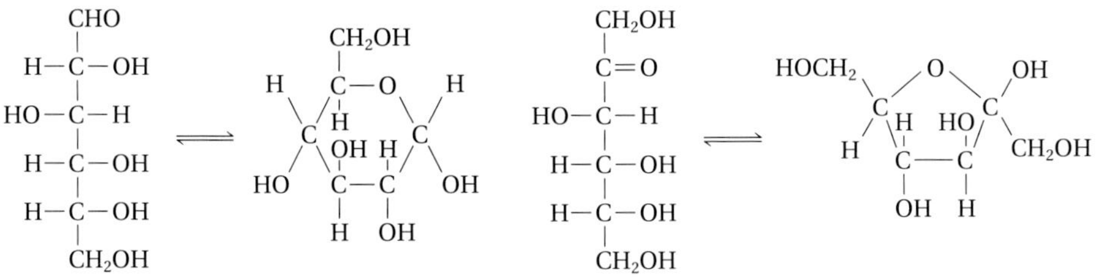

Straight-chain and cyclic forms of glucose

Straight-chain and cyclic forms of fructose

Note the aldehyde functional group (—CHO) on the straight-chain glucose, and the ketone functional group ($-\overset{O}{\overset{\|}{C}}-$) on the straight-chain fructose. What difference do you notice between the cyclic form of glucose and the cyclic form of fructose? ❶

Disaccharides and Polysaccharides

The cyclic forms of two simple sugars can be linked with the loss of water. The linking of glucose and fructose by means of such a condensation reaction gives sucrose, or common table sugar. Sugars that form from two monosaccharides in this way are known as **disaccharides.** Sucrose is thus an example of a disaccharide. The reaction by which it forms is as follows.

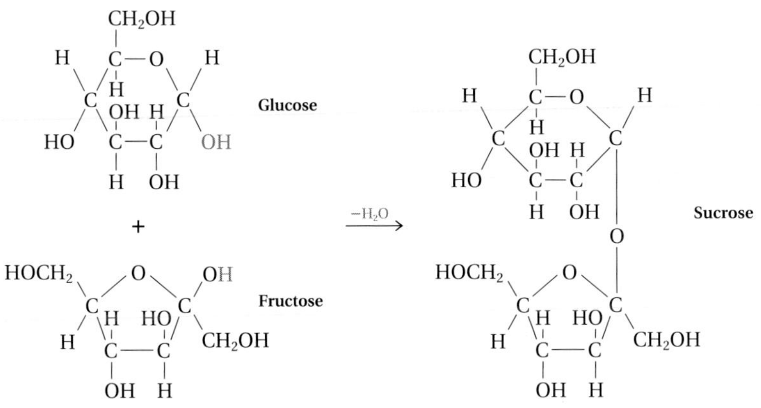

Glucose + Fructose $\xrightarrow{-H_2O}$ Sucrose

Sucrose is obtained commercially mainly from the juice of sugarcane and sugar beets. The world's production from these sources exceeds 7×10^9 metric tons per year.

2 Teach

Discuss

Construct ball-and-stick models of glucose and fructose for students to examine while studying the molecular structures shown on this page. Point out the aldehyde, ketone, and hydroxyl functional groups on both molecules. Use the models to demonstrate the cyclization of the straight-chain forms. Remind students about ring strain and the consequent stability of six- and five-membered rings. Connect the two cyclic forms together to form a disaccharide and note the loss of a water molecule. Point out that if this condensation reaction is allowed to continue, many monosaccharide units could join to form a polysaccharide such as starch or cellulose.

Discuss

Point out that the synthesis of polymers very often requires a catalyst. In a living cell, the polymerization of monosaccharides is catalyzed by specialized proteins called enzymes. One way enzymes facilitate the linkage of sugars is by providing a type of docking surface on which sugars are aligned in the proper orientation for chemical bonding. Tell students that they will learn about enzymes in Section 27.3.

Meeting Diverse Needs

At Risk People use carbohydrates as a source of food, clothing, and shelter. Divide the class into groups of four or five and have each group make a poster showing how people use carbohydrates. Groups can use original art, pictures from magazines, or other sources to illustrate the usefulness of carbohydrates in daily life.

Answer

❶ Glucose has five carbons and one oxygen in its ring. Fructose has four carbons and one oxygen in its ring.

TEACHER DEMO

The ability to detect mono- and polysaccharides is an important function of clinical laboratories and at-home tests for those afflicted with diabetes mellitus. Simple sugars based on aldehydes are known as reducing sugars; they can be detected with Benedict's reagent, an alkaline solution of copper(II) sulfate. The Cu^{2+} ions react with reducing sugars to form a brick-red cuprous oxide (Cu_2O) precipitate. The amount of precipitate formed depends on the amount of sugar present. To demonstrate the Benedict's test, prepare a hot-water bath. Place 2–5 mL of 1% carbohydrate solution (corn syrup) in one test tube; place distilled water in another test tube as a negative control. Add 2–3 mL of Benedict's reagent to each tube and heat for a few minutes. Have students observe the formation of a red precipitate. Test a number of mono- and polysaccharides and have students classify them as reducing or nonreducing.

3 Assess

Evaluate Understanding

Have students describe the structure of mono-, di-, and polysaccharides, and give a common example of each. Ask students to specify the types of elements, functional groups, and linkages found in carbohydrates.

Reteach

Point out that polysaccharides are polymers with more than nine units. Glucose is the basic unit in starch; glycogen, and cellulose, are the most important natural polysaccharides. Explain that plants are sources of cellulose and starch. Animals store polysaccharides in the form of glycogen.

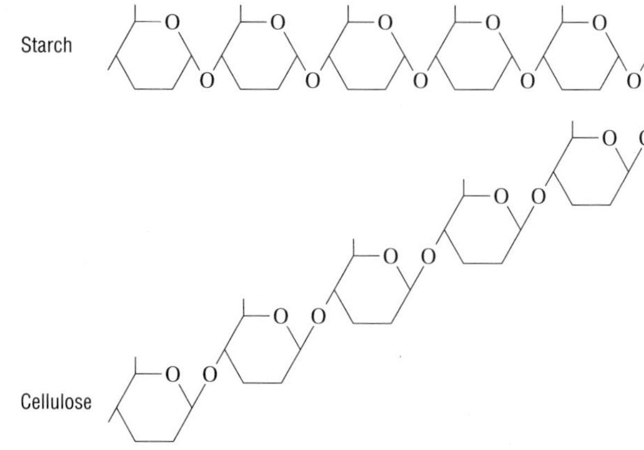

Figure 27.5
Starch and cellulose are similar polymers made up of hundreds of glucose monomers. They differ in the orientation of the bond between the glucose units. Because of this difference, starch is readily digestible, but cellulose is indigestible by most organisms. What organisms can digest cellulose? ❶

The formation of a disaccharide is sometimes the first step in a condensation polymerization that produces extremely large molecules. The polymers produced by the linkage of many monosaccharide monomers are called **polysaccharides.** Starches, the major storage form of glucose in plants, are polysaccharide polymers consisting of glucose monomers. **Figure 27.5a** shows a portion of a starch molecule. A typical linear starch molecule contains hundreds of glucose monomers. Some starches are branched molecules, each branch containing about a dozen glucose units. Glycogen, the starchlike polysaccharide produced by animals, is a source of stored energy for animals and is more highly branched than plant starches. Molecules of glycogen are stored in the liver and muscles of animals.

Cellulose is probably the most abundant biological molecule. As **Figure 27.5b** shows, cellulose is also a polymer of glucose. The orientation of the bond that links the glucose monomers in cellulose is different, however, from the orientation in starch. Starch is digestible by most organisms and is partially soluble in water. Cellulose, however, can be digested by only a few kinds of microorganisms, such as those that live in the digestive tracts of cattle and termites. Cellulose is insoluble in water and is an important structural polysaccharide that provides form, hardness, and rigidity in plants. Plant cell walls, such as those in wood, are made of cellulose. Cotton is about 80% cellulose.

section review 27.2

4. Describe the main characteristics of monosaccharides, disaccharides, and polysaccharides.
5. Give a source for each.
   a. starch b. cellulose c. glycogen
6. Glycogen and cellulose have different properties, but both are composed of glucose units. Explain what makes them different.
7. Distinguish between the important structural features of sucrose, glucose, and fructose.

 Chem ASAP! **Assessment 27.2** Check your understanding of the important ideas and concepts in Section 27.2.

Answers

❶ Organisms in the digestive tracts of termites and cattle (or other ruminants) can digest cellulose.

SECTION REVIEW 27.2

4. Monosaccharides (simple sugars) can exist in straight-chain or cyclic form. Disaccharides form when two simple sugars condense. Polysaccharides are polymers of monosaccharides.
5. **a.** Plants store glucose in starches.
   b. Plant cell walls contain cellulose.
   c. Glycogen is produced by animals; it is stored in the liver and muscles.
6. The orientation of the bonds that hold glucose units together is different.
7. Sucrose is a disaccharide composed of glucose and fructose; glucose is a 6-carbon monosaccharide with an aldehyde group; fructose is a 6-carbon monosaccharide with a ketone group.

AMINO ACIDS AND THEIR POLYMERS

Many people are lactose intolerant, meaning they cannot digest milk or products containing milk. These people cannot digest milk products because their bodies do not produce enough of the enzyme lactase to digest lactose, the sugar found in milk. The undigested lactose causes bloating and stomach upset. Some people with lactose intolerance can enjoy milk products if they take a pill containing the enzyme lactase before eating. **What are enzymes, and what function do they serve in the body?**

objectives

- Write a general formula for an amino acid, and describe the bonding between amino acids in peptides and proteins
- Describe the effect of enzymes on biochemical reactions

key terms

- amino acid
- peptide
- peptide bond
- amino acid sequence
- polypeptide
- protein
- enzymes
- substrates
- active site
- enzyme–substrate complex
- coenzymes

Amino Acids

Many biological compounds contain nitrogen in addition to carbon, oxygen, and hydrogen. Plants obtain nitrogen through nitrogen fixation, a process in which atmospheric nitrogen (N_2) is reduced to ammonia (NH_3). Legumes, a family of plants that includes beans, peas, and alfalfa, carry out most nitrogen fixation. The ammonia produced by nitrogen fixation is incorporated into biological molecules. This section will introduce you to the amino acids, a very important class of nitrogen-containing molecules. In fact, the polymers of amino acids, which you will also learn about, make up more than one half of the dry weight of your body.

An **amino acid** is any compound that contains amino ($-NH_2$) and carboxylic acid ($-COOH$) groups in the same molecule. For chemists and biochemists, however, the term is usually reserved for amino acids that are formed and used in living organisms. All 20 amino acids common in nature have a skeleton consisting of a carboxylic acid group and an amino group, both of which are covalently bonded to a central carbon atom. The remaining two groups on the central carbon atom are hydrogen and an R group that constitutes the amino acid side chain.

$$H_2N-\underset{\displaystyle H}{\overset{\displaystyle R}{\overset{|}{\underset{|}{C}}}}-COOH$$

Side chain
Central carbon atom
Amino group
Carboxylic acid group

Because the central carbon of the amino acids is asymmetric, these compounds can exist as stereoisomers. As you may recall from Section 25.3, stereoisomers may be right- or left-handed. Nearly all the amino acids found in nature are of the left-handed, or L form. What is the major difference in the composition of amino acids and carbohydrates? 2

The chemical nature of the side-chain group accounts for the differences in properties among the 20 amino acids. In some cases, the side chains are nonpolar aliphatic or aromatic hydrocarbons. In other cases, the side chains are neutral but polar. In still other cases, the side chains are acidic or basic. **Table 27.1** gives the names of the amino acids with their three-letter abbreviations. What is the three-letter abbreviation for the amino acid phenylalanine? For aspartic acid? 3

Table 27.1

Abbreviations for Amino Acids

Amino acid	Abbreviation
Alanine	Ala
Arginine	Arg
Asparagine	Asn
Aspartic acid	Asp
Cysteine	Cys
Glutamine	Gln
Glutamic acid	Glu
Glycine	Gly
Histidine	His
Isoleucine	Ile
Leucine	Leu
Lysine	Lys
Methionine	Met
Phenylalanine	Phe
Proline	Pro
Serine	Ser
Threonine	Thr
Tryptophan	Trp
Tyrosine	Tyr
Valine	Val

27.3

1 Engage

Use the Visual

Have students study the photograph and read the text that opens the section. Explain that lactose, a disaccharide made from glucose and galactose, is found in all dairy products. Almost all human infants can digest lactose. However, some lose this ability as they mature because they stop producing the enzyme needed to metabolize lactose. Ask:

- **What are enzymes, and what function do they serve in the body?** (Enzymes are proteins that catalyze metabolic reactions. Virtually every metabolic reaction is catalyzed by a specific enzyme.)

Check Prior Knowledge

Have students use energy profile curves for catalyzed and uncatalyzed reactions to explain how catalysts affect the rate of reaction. (Catalysts increase the rate of reaction by lowering the activation energy barrier.) **Remind students that catalysts have an equal effect on both the forward and reverse reactions. Thus, catalysts reduce the amount of time required to reach equilibrium, but do not influence the equilibrium position.**

STUDENT RESOURCES

From the Teacher's Resource Package, use:

- Section Review 27.3, Ch. 27 Quizzes from the Review Module (Ch. 25–28)
- Laboratory Recordsheet 27-1
- Small-Scale Chemistry Lab Manual Experiment 39

TECHNOLOGY RESOURCES

Relevant technology resources include:

- Chem ASAP! CD-ROM
- ResourcePro CD-ROM

2 Amino acids contain nitrogen.
3 Phe, Asp

2 Teach

TEACHER DEMO

If possible, use models to demonstrate how amino acids link through amide bonds to form peptides. Point out that peptide formation is another example of a condensation polymerization. Have students imagine how continuing the polymerization process produces polypeptides and protein molecules.

Discuss

Point out that the type of amino acids in proteins have a large influence on the physical and chemical properties of the peptide. The twenty common amino acids can be grouped into two major categories: polar or nonpolar. The ratio of polar to nonpolar amino acids profoundly affects the shape of a protein. When a protein folds, nonpolar amino acids are sequestered to the interior of the protein, away from the aqueous environment. Polar amino acid residues often predominate on the surface of proteins.

Peptides

A **peptide** is any combination of amino acids in which the amino group of one acid is united with the carboxylic acid group of another. The amide bond between the carbonyl group of one amino acid and the nitrogen of the next amino acid in the peptide chain is called a **peptide bond.** The bonds between the amino acids always involve the central amino and central carboxylic acid groups. The side chains are not involved in the bonding.

$$H_2N-\underset{\displaystyle H}{\overset{\displaystyle R}{C}}-\overset{\displaystyle O}{\overset{\|}{C}}-OH + H-\underset{\displaystyle H}{N}-\underset{\displaystyle H}{\overset{\displaystyle R}{C}}-\overset{\displaystyle O}{\overset{\|}{C}}-OH \longrightarrow H_2N-\underset{\displaystyle H}{\overset{\displaystyle R}{C}}-\overset{\displaystyle O}{\overset{\|}{C}}-\underset{\displaystyle H}{N}-\underset{\displaystyle H}{\overset{\displaystyle R}{C}}-\overset{\displaystyle O}{\overset{\|}{C}}-OH + H_2O$$

Amino acid — Amino acid — Peptide (Peptide bond)

BIOLOGY

Ethnobiology

Often exploring the remotest jungles and forests, ethnobiologists search for information about plants, fungi, and even insects that people use as medicine.

These scientists carefully interview local specialists, who administer or use the remedies; collect samples of the remedies; and send these samples to laboratories for analysis of their chemical components. Ethnobiologists carry out an important task—fewer than 3% of the flowering plants on Earth have been examined for medicinal value, but about 40% of medicines in use today come from plant sources.

Note that a free amino group remains at one end of the resulting peptide. The convention is to write the formula of the peptide so that this free amino group is at the left end. There is also a free carboxylic acid group, which appears at the right end.

More amino acids may be added to the peptide in the same fashion to form long chains by condensation polymerization. The order in which the amino acids of a peptide molecule are linked is called the **amino acid sequence** of that molecule. The amino acid sequence of a peptide is conveniently expressed using the three-letter abbreviations for the amino acids. For example, Asp—Glu—Gly represents a peptide containing three amino acids. This tripeptide contains aspartic acid, glutamic acid, and glycine, in that order, with the free amino group assumed to be on the left end (on the Asp) and the free carboxyl group on the right end (on the Gly). Note that Asp—Glu—Gly is a different peptide from Gly—Glu—Asp, because the order of amino acids is reversed, and thus the free amino group and free carboxyl group are on different amino acids.

Proteins

In theory, the process of adding amino acids to a peptide chain may continue indefinitely. Any peptide with more than ten amino acids is called a **polypeptide.** A peptide with more than about 100 amino acids is called a **protein.** On average, a molecule of 100 amino acids has a molar mass of about 10 000 amu. Proteins are an important class of biomolecules. Your skin, hair, nails, muscles, and the hemoglobin molecules in your blood are
1 made of protein. Why is it so important to include protein in your diet?

Differences in the chemical and physiological properties of peptides and proteins result from differences in the amino acid sequence. The number of ways in which 20 amino acids can be linked in a protein molecule is enormous. For example, as many as 20^{100} amino acid sequences are possible for a protein of 100 amino acids containing a combination of the 20 different amino acids.

The long peptide chains of proteins, shown in **Figure 27.6a,** can fold into relatively stable shapes. **Figure 27.6b** shows how sections of peptide chain may coil into a regular spiral, known as a helix. Peptide chains may also be arranged side by side to form a wavy sheet, as shown in **Figure 27.6c.**

Answer

1 Proteins provide the amino acids necessary to build many important biomolecules.

(a)

(b)

(c)

Figure 27.6
(a) *This is a representation of amino acids in a peptide chain.* ***(b)*** *The chain may fold into a helix.* ***(c)*** *Two peptide chains may become arranged in a sheetlike structure.*

Irregular folding of the chains can also occur. Three-dimensional protein shape is determined by interactions between the amino acids of its peptide chains. Protein shape is partly maintained by hydrogen bonds between adjacent parts of the folded chains. Covalent bonds also form between sulfur atoms in the side-chain groups of cysteine monomers that are folded near each other. In that way, separate polypeptide chains may be joined into a single protein. **Figure 27.7** traces the shape of myoglobin, a protein that stores oxygen in muscle cells. As is shown in the figure, the peptide chains of most of the myoglobin molecule are twisted into helixes.

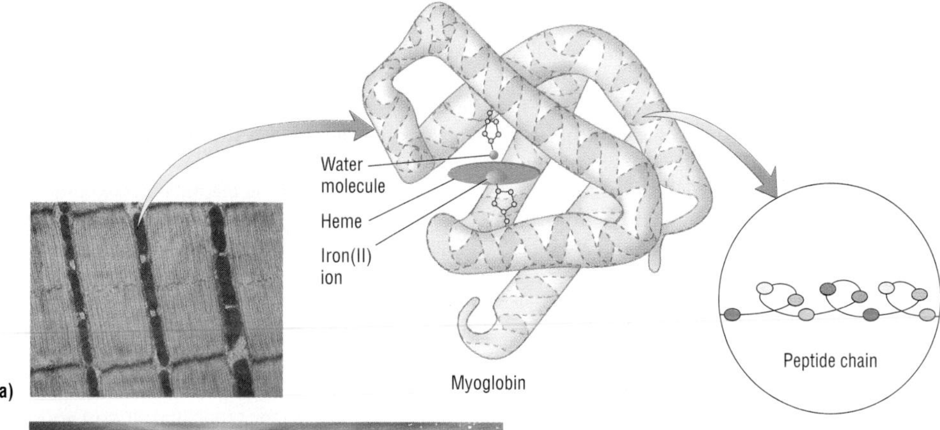

Figure 27.7
(a) *The three-dimensional structure of myoglobin, the oxygen storage protein of muscle, is shown here. Most of the peptide chain of myoglobin is wound into helixes. Myoglobin also contains a nonprotein structure called heme. The heme group is shown as a disk in the myoglobin structure. Heme contains four linked rings with an iron(II) ion (Fe^{2+}) at the center. Molecular oxygen binds to the heme iron.* ***(b)*** *Dolphins and other marine animals have a large concentration of myoglobin in their muscles. This allows them to store oxygen during long dives.*

Meeting Diverse Needs

Gifted Scientists have long sought a reliable, and general method for predicting the three-dimensional shape of any protein based on its amino acid sequence. Because a protein's shape is so closely associated with its substrate binding properties, a code for predicting protein structure would allow scientists to design new proteins and modify existing ones for a variety of purposes. Have students do research about how the amino acid sequence of a protein dictates its secondary and tertiary structure.

Think Critically

Point out that the number of possible ways for an organism to assemble amino acids into a protein is enormous. Ask:

- **If an amino acid can be used once, multiple times, or not at all, how many different ways are there are to arrange 20 amino acids in a protein that is 20 amino acids long?** (20^{20} possible arrangements)
- **How many combinations are possible in a protein 40 amino acids long?** (20^{40})

Point out to students that an average protein usually contains more than 100 amino acids!

Teacher Demo

Not all proteins produced in the body are used to catalyze metabolic reactions. Some serve purely structural roles in soft and hard tissues alike. Students may be surprised to learn that protein is a major structural constituent of bone. Illustrate this fact by soaking a thigh bone from cooked poultry in 1*M* HCl overnight. The next day, rinse the bone with water. Upon visual inspection it may appear as if no reaction has occurred at all; demonstrate that the bone can be bent and twisted quite easily. Point out that the mineral salts of the bone were dissolved in the acid. The acid-insoluble portion remaining consists of a special protein called collagen. Ask students to propose possible roles for collagen in the formation and maintenance of healthy bone tissue. (Collagen provides a framework for the deposition of calcium phosphate, the mineral that makes up bone, and for the growth of bone marrow cells.)

Discuss

Almost nowhere else in chemistry is the phrase "structure dictates function" so elegantly illustrated as in the three-dimensional conformation of a protein. The ability of an enzyme to recognize a particular substrate and, therefore, to catalyze a reaction involving the substrate, depends on the conformation, or shape, of the protein. The amino acid sequence of a protein directs how a protein will fold, and what shape its active site will ultimately assume. Because different enzymes have different amino acid sequences, their active sites have different shapes and bind different substrates.

Use the Visual

Have students study Figure 27.8. Compare the binding of a substrate and an enzyme to the way in which a key fits into a lock. To fit into the active site, a substrate must have a matching, or complementary, shape. Point out, however, that an enzyme often changes shape slightly when a substrate is bound, trapping the substrate in place.

For a dramatic illustration of the rate-enhancing ability of an enzyme, use lactase and fresh milk. Treat one container of milk with Lactaid® and one with an equal number of water drops. Store the milk cartons in the refrigerator for 1–2 days. Then have student volunteers do a comparative taste test.

Enzymes

Enzymes are proteins that act as biological catalysts. Enzymes increase the rates of chemical reactions in living things. In 1926, the American chemist James B. Sumner reported the first isolation and crystallization of an enzyme. The enzyme he isolated was urease, which is able to hydrolyze urea, a constituent of urine, into ammonia and carbon dioxide. The strong ammonia smell of dirty diapers allowed to stand for a time is the result of the action of bacteria that contain this enzyme. The reaction is shown here.

$$\underset{\text{Urea}}{H_2N-\overset{\overset{\displaystyle O}{\|}}{C}-NH_2(aq)} + \underset{\text{Water}}{H_2O(l)} \xrightarrow{\text{urease}} \underset{\text{Ammonia}}{2NH_3(g)} + \underset{\text{Carbon dioxide}}{CO_2(g)}$$

Since the discovery of urease, hundreds of enzymes have been isolated and structurally characterized as proteins.

In addition to being able to promote reactions, enzymes have two other properties of true catalysts. First, they are unchanged by the reaction they catalyze. Second, they do not change the normal equilibrium position of a chemical system. The same amount of product is eventually formed whether or not an enzyme is present. Few reactions in cells ever reach equilibrium, however. The products tend to rapidly convert to another substance in a subsequent enzyme-catalyzed reaction. According to Le Châtelier's principle, such removal of a product pulls the reaction toward completion.

Enzymes catalyze most of the chemical changes that occur in the cell. **Substrates** are the molecules on which an enzyme acts. **Figure 27.8** shows how substrates are transformed into products by bond-making and bond-breaking processes. In an enzymatic reaction, the substrate interacts with side chains of the amino acids on the enzyme. These interactions cause the making and breaking of bonds. A substrate molecule must make contact with and bind to an enzyme molecule before the substrate can be transformed into product. The **active site** is the place on an enzyme where processes that convert substrates to products occur. The active site is usually a pocket or crevice formed by folds in the peptide chains of the enzyme protein. The peptide chain of an enzyme is folded in a unique way to accommodate the substrate at the active site. **Figure 27.9** shows this folding for an enzyme called the HIV protease. The HIV protease, which is produced by the virus that causes AIDS, allows the replicated virus to package itself for release from an infected cell.

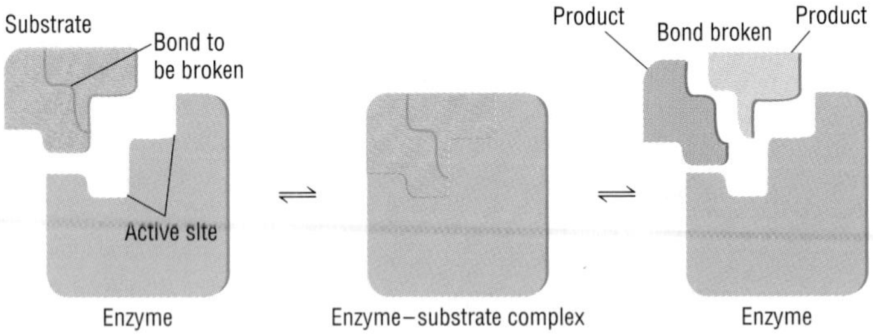

Figure 27.8
A substrate fits into a distinctively shaped active site on an enzyme. Bond-breaking occurs at the active site to produce the products of the reaction. What would happen if access to the active site were blocked by another molecule? (1)

Meeting Diverse Needs

At Risk Many laundry detergents contain enzymes that are purported to help remove stains. Ask students to read the labels of a number of laundry detergents and list the types of enzymes they contain. Ask them to hypothesize how enzymes improve the cleaning performance of a detergent.

Answer

(1) Transformation of the substrate into product could not occur.

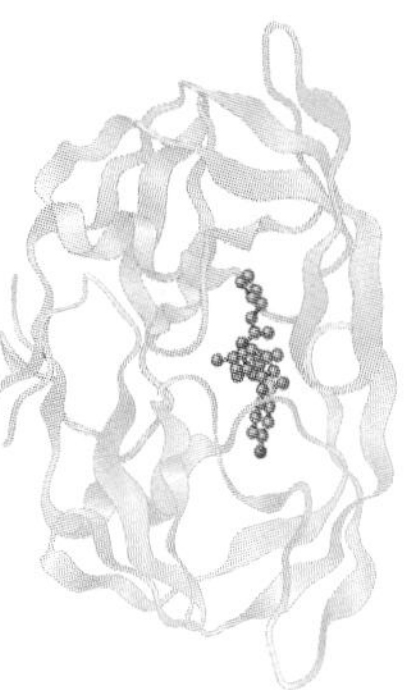

Figure 27.9

The scanning electron micrograph shows HIV infecting a human white blood cell. The illustration models the enzyme HIV protease. The ribbons trace the two peptide chains of the enzyme. A substrate molecule is embedded in the active site formed between the two peptide chains.

The active site of each enzyme has a distinctive shape. This allows only one specific substrate molecule to fit into the enzyme, much as only one key will fit into a certain lock. Each enzyme catalyzes only one chemical reaction, with only one substrate. An **enzyme–substrate complex** is formed when an enzyme molecule and a substrate molecule are joined.

To get an idea of the tremendous efficiency of enzymes, consider the action of an enzyme called carbonic anhydrase. It catalyzes only one reaction, the reversible breakdown of carbonic acid to water and carbon dioxide.

$$\underset{\text{Carbonic acid}}{H_2CO_3(aq)} \xrightleftharpoons{\text{carbonic anhydrase}} \underset{\text{Carbon dioxide}}{CO_2(g)} + \underset{\text{Water}}{H_2O(l)}$$

A single molecule of carbonic anhydrase can catalyze the breakdown of about 36 million molecules of carbonic acid in one minute!

Some enzymes can directly catalyze the transformation of biological substrates without assistance from other substances. Other enzymes need nonprotein coenzymes, also called cofactors, to assist the transformation. **Coenzymes** are metal ions or small organic molecules that must be present for an enzyme-catalyzed reaction to occur. Many water-soluble vitamins, such as B vitamins, are coenzymes. Metal ions that act as cofactors include the cations of magnesium, potassium, iron, and zinc. The enzyme catalase, for example, includes an iron(III) ion in its molecular structure. Catalase catalyzes the decomposition of hydrogen peroxide to water and oxygen.

$$2H_2O_2(aq) \xrightarrow{\text{catalase}} 2H_2O(l) + O_2(g)$$

section review 27.3

8. Write the general formula for an amino acid, and describe the bonding between such acids in peptides and proteins.
9. Describe how enzymes affect biochemical reactions.
10. What is meant by the amino acid sequence of a protein?
11. Describe three properties of enzymes.

 Chem ASAP! **Assessment 27.3** Check your understanding of the important ideas and concepts in Section 27.3.

Discuss

Explain that coenzymes are nonprotein organic molecules or metal ions whose presence is required for the activity of many enzymes. Coenzymes are often vitamins. For example, B vitamins help release energy from food. In general, coenzymes accept electrons or functional groups that are removed from organic substrates.

3 Assess

Evaluate Understanding

Have students sketch the general structure of a tripeptide—such as gly-val-trp—using letter abbreviations in place of the R group. Ask them to label the peptide bond, and the carboxyl and amino ends. Ask:

- **Which element do proteins contain that carbohydrates do not?** (nitrogen)
- **What is an enzyme? Describe one enzyme discussed and its function.** (Enzymes are proteins that function as biological catalysts. One example is urease, which catalyzes the hydrolysis of urea.)

Reteach

Use dryer duct tubing or slinky toys to help illustrate the secondary and tertiary structure of proteins. Point out that many of the amino acids in an enzyme are not directly involved in catalysis because they lie far from the active site. However, they do participate in intermolecular forces, which maintain the folded shape of the protein, a prerequisite for its biological activity.

Answers

SECTION REVIEW 27.3

8. $CH(R)(NH_2)COOH$; A peptide is formed when the amino group of one amino acid combines with the carboxylic acid of another. A protein is a polypeptide that contains more than 100 amino acids.
9. Enzymes act as catalysts; they speed up the rate of reaction.
10. The amino acid sequence of a protein is the order of amino acids in the protein molecule.
11. Enzymes are catalysts (they remain unchanged in a reaction); they are proteins; and they are specific to the reactions they catalyze.

SMALL-SCALE LAB

THE EGG: A BIOCHEMICAL STOREHOUSE

OVERVIEW

Procedure Time: 20 minutes

Students examine the physical and chemical properties of a chicken egg.

MATERIALS

Solution	Preparation
0.5*M* NaOH	20.0 g in 1.0 L
0.2*M* $CuSO_4$	12.5 g $CuSO_4 \cdot 5H_2O$ in 250 mL
1.0*M* HCl	82 mL of 12*M* in 1.0 L

CAUTION! Always add acid to water carefully and slowly!

TEACHING SUGGESTIONS

For each class, obtain at least a dozen eggs of two or three different sizes: medium, large, and extra large. Some markets also sell small or jumbo eggs. This approach will allow students to obtain data from classmates about eggs that are a different size than their own. Have students write their names on their eggs so they can identify them the next day. Store some eggs in a refrigerator and some at room temperature if students are going to do activities 1, 2, 3, and 8, in the YOU'RE THE CHEMIST section.

ANALYSIS

Sample data: length = 5.90 cm; width = 4.55 cm; mass = 62.42 g

1. shape index = $\frac{4.55 \text{ cm}}{5.90 \text{ cm}} \times 100\% = 77.1$
2. $V = (0.5236)(5.90 \text{ cm})(4.55 \text{ cm})^2 = 64.0 \text{ cm}^3$
   $M = (0.5632)(5.90 \text{ cm})(4.55 \text{ cm})^2 = 68.8 \text{ g}$
   $A = 3.138[(5.90 \text{ cm})(4.55 \text{ cm})^2]^{2/3} = 77.3 \text{ cm}$
3. The measured mass of 62.42 g is less than the calculated mass of 68.8 g by 6.4 g. The egg may have lost water.
4. $d = \frac{62.42 \text{ g}}{64.0 \text{ cm}^3} = 0.975 \text{ g/cm}^3$
   This is less than the density of a freshly laid egg.

SMALL-SCALE LAB

THE EGG: A BIOCHEMICAL STOREHOUSE

SAFETY

Use safe and proper laboratory procedures.

PURPOSE

To explore some physical and chemical properties of a chicken egg.

MATERIALS

- chicken egg
- pencil
- paper
- ruler
- balance

PROCEDURE

Obtain a chicken egg. Examine the egg's shape and measure its length and width in centimeters. Measure the mass of the egg. Make an accurate, life-size sketch of your egg and record all your data on the sketch.

ANALYSIS

Using your experimental data, record the answers to the following questions below your drawing.

1. A common way to compare the shapes of eggs is by using the shape index. The shape index is the width of an egg expressed as a percentage of its length.
   $$\text{Shape index} = \frac{\text{width}}{\text{length}} \times 100\%$$
   Calculate the shape index of your egg.
2. The volume, original mass (when freshly laid), and surface area of an egg can easily be estimated by using the following equations.
   $V = (0.5236)(lw^2)$ $M = (0.5632)(lw^2)$
   $A = (3.138)(lw^2)^{2/3}$
   V = volume M = original mass
   A = surface area l = length w = width
   Use your data to calculate the volume, original mass, and surface area of your egg. Show your work and record your results.
3. Compare your measured mass of the egg with your calculated mass. Which is greater? Suggest why the mass of an egg might change over time.
4. Using your measured mass and your calculated volume, calculate the density of your egg. Compare this value with the density of a freshly laid egg ($d = 1.075 \text{ g/cm}^3$).

YOU'RE THE CHEMIST

The following small-scale activities allow you to develop your own procedures and analyze the results.

1. **Design It!** Design an experiment to answer the following question: Does the mass of an egg change over time?
2. **Analyze It!** Using your measured mass, your calculated original mass, and your experiments on the mass loss of an egg over time, estimate the age of your egg. What assumptions must you make?
3. **Design It!** Design and carry out an experiment to measure the volume of your egg. Write down what you did and what you found.
4. **Design It!** Carry out a series of experiments, or consult with your classmates and use their data, to determine if and how the shape index varies with the size of the egg (small, medium, large, extra large, jumbo).
5. **Analyze It!** Determine how the mass of an egg varies with its size (small, medium, large, extra large, jumbo).
6. **Analyze It!** An eggshell contains a calcium carbonate matrix with a protein cuticle. Place one drop of HCl on an eggshell and observe what happens. Write a chemical equation for this reaction.
7. **Analyze It!** Proteins can be detected by adding aqueous solutions of copper(II) sulfate and sodium hydroxide to a sample. A violet color indicates the presence of protein. Test powdered milk and an eggshell for protein. What are your results?
8. **Design It!** Design and carry out an experiment to answer the following question. Does temperature affect the mass of an egg over time?

YOU'RE THE CHEMIST

1. Measure the mass each day for two or three days. Eggs lose 0.2 to 0.5 grams per day.
2. Assuming the egg loses 0.20 g per day, the age of the egg is
   $(68.8 \text{ g} - 62.42 \text{ g}) \times \frac{1 \text{ day}}{0.20 \text{ g}} = 32$ days old.
3. Measure the volume by water displacement
4. Larger eggs have smaller shape indexes.
5. Extra large eggs are usually more than 70 g, and medium eggs are less than 50 g.
6. $2HCl + CaCO_3 \rightarrow CO_2(g) + H_2O + CaCl_2$
7. Both powdered milk and the eggshell produce a violet color
8. Measure the mass of a refrigerated egg for three days. Measure its mass for three more days while storing it at room temperature. The warmer the temperature, the greater the mass loss.

LIPIDS

Candles were an early invention of ancient civilizations. Candlesticks that date to at least 5000 years ago have been found with other artifacts from the civilizations of ancient Egypt and Crete. Before the invention of electric lighting, candles were the major source of illumination in homes. **What type of molecules make up waxes such as candle wax?**

section 27.4

objectives

- Characterize the molecular structures of triglycerides, phospholipids, and waxes
- Describe the functions of phospholipids and proteins in cell membranes

key terms

- lipids
- triglycerides
- saponification
- phospholipids
- waxes

Triglycerides

Experts recommend that your fat intake make up less than 30% of your daily caloric intake of food. Although excessive dietary fat is harmful, you do need some fat in your diet to stay healthy. Fat provides an efficient way for your body to store energy. It is also needed to keep your cell membranes healthy. Fats, oils, and other water-insoluble compounds are called **lipids.** In this section, you will learn about the chemical composition and biological uses of lipids.

Lipids tend to dissolve readily in organic solvents, such as ether and chloroform, rather than in highly polar solvents such as water. This property sets them apart from biological substances such as carbohydrates and proteins. Natural **triglycerides** are fats and oils that are triesters of glycerol with fatty acids, which are long-chain carboxylic acids (C_{12} through C_{24}). Triglycerides are important as the long-term storage form of energy in the human body. The following equation shows the general reaction for the formation of triglycerides.

$$\begin{array}{ccccccc} CH_2OH & & HO-\overset{\overset{\large O}{\|}}{C}-R & & CH_2-O-\overset{\overset{\large O}{\|}}{C}-R & & \\ | & & & & | & & \\ CHOH & + & HO-\overset{\overset{\large O}{\|}}{C}-R & \longrightarrow & CH-O-\overset{\overset{\large O}{\|}}{C}-R & + & 3H_2O \\ | & & & & | & & \\ CH_2OH & & HO-\overset{\overset{\large O}{\|}}{C}-R & & CH_2-O-\overset{\overset{\large O}{\|}}{C}-R & & \\ \text{Glycerol} & & \text{3 Fatty acid molecules} & & \text{Triglyceride (triester of glycerol)} & & \text{Water} \end{array}$$

The difference between fats and oils is simply that fats are solid at room temperature and oils are liquid. Most fats, such as lard, are obtained from animals. The fats from palm kernels and coconuts, however, are exceptions. Most oils, such as olive oil, are plant products. From the organic chemist's point of view, however, fats and oils are both esters. Like other esters, they are easily hydrolyzed in the presence of acids and bases. The hydrolysis of oils or fats by boiling with an aqueous solution of an alkali-metal hydroxide is called **saponification.** Saponification is a process used to make soap. Soaps are thus the alkali metal (Na, K, or Li) salts of fatty acids. A typical saponification reaction is shown at the top of the following page. Which alkali metal hydroxide is used in this saponification reaction? 1

Figure 27.10
Moderate levels of dietary fats and oils are essential to health.

1 Engage

Use the Visual

Have students study the photograph and read the text that opens the section. Ask students to recall the general chemical structure and properties of carboxylic acids (presented in Chapter 26). Remind them that many long-chain carboxylic acids were first isolated from fats and are sometimes called "fatty acids." Fatty acids belong to a class of compounds known as *lipids.* Bring to class a bottle of vegetable oil and a candle; have students compare and contrast the physical properties of each substance. Ask:

- **What type of molecules make up waxes such as candle wax?** (Students may not know, but guide them to understand that an important difference between waxes and oils is the length of the hydrocarbon backbone. Waxes are derivatives of high molar mass carboxylic acids. Remind students that carboxylic acids react with alcohols to form esters. Waxes are essentially high molar mass esterification products of carboxylic acids.)

STUDENT RESOURCES

From the Teacher's Resource Package, use:

- Section Review 27.4 and Ch. 27 Quizzes from the Review Module (Ch. 25–28)
- Laboratory Manual: Experiment 51

TECHNOLOGY RESOURCES

Relevant technology resources include:

- Chem ASAP! CD-ROM
- Resource Pro CD-ROM

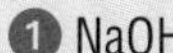 1 NaOH

2 Teach

Discuss

Explain that lipids do contain identifiable functional groups, but their most distinguishing property is their insolubility in water. Lipids are nonpolar compounds that must be extracted from animal and plant sources using nonpolar solvents such as hexane or benzene.

ACTIVITY

Have students compare the composition of butter, regular margarine, and low-calorie margarine by melting the same mass of each kind of spread in separate test tubes. Put the test tubes in warm water to liquefy the samples. Then allow the samples to settle. The liquid will form two layers: the top layer is a lipid (butterfat or partially hydrogenated vegetable oil) and the bottom layer is water. Students can compare the samples quantitatively and qualitatively, and calculate the percent of water in each.

Discuss

Lead a discussion on the uses of fats, oils, and soaps. You may wish to use the preceding activity to initiate a discussion on the fat content of various foods and the recent introduction of fat substitutes. Have students debate the pros and cons of fat substitutes

Think Critically

Have students discuss what would happen to the cell membrane if it were exposed to a nonpolar solvent.

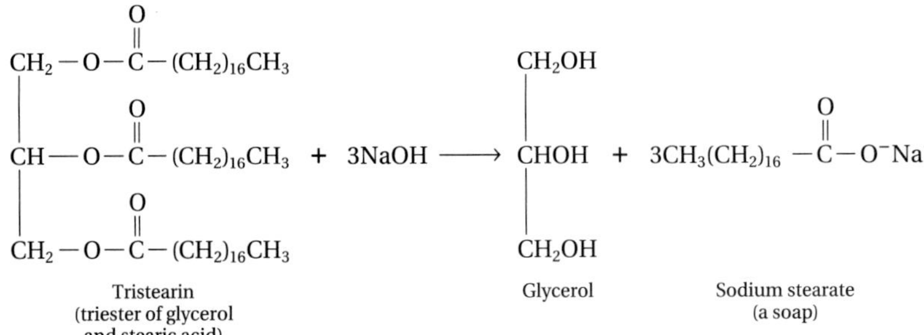

As shown in **Figure 27.11**, soap can be made by heating a fat, such as beef tallow or coconut oil, with an excess of sodium hydroxide. When sodium chloride is added to the saponified mixture, the sodium salts of the fatty acids separate as a thick curd of crude soap. Glycerol is an important by-product of saponification reactions. It is recovered by evaporating the water layer. The crude soap is then purified. Coloring agents and perfumes are added according to market demands.

Figure 27.11
These photographs illustrate soapmaking. Once the soap is formed, it is poured into molds. Later it may be milled, or shredded, with scent or color added, and then remolded to produce a finished product.

Phospholipids

Phospholipids, or lipids that contain phosphate groups, are abundant in cells. **Figure 27.12** shows a typical phospholipid lecithin molecule. The lecithin molecule has a hydrophilic (water-loving) ionic head and an oily or hydrophobic (water-hating) hydrocarbon tail. You know the solubility rule, "Like dissolves like." In water, lecithin, which is only partly hydrophobic and only partly hydrophilic, behaves like both an insoluble hydrocarbon and a soluble ionic compound. In water, the hydrophobic carbon chains aggregate to exclude water. The hydrophilic part is drawn to water, which can solvate it. This results in the spontaneous formation of a spherical double layer of phospholipid called a lipid bilayer. **Figure 27.13** shows a lipid
1 bilayer. Where are the hydrophobic tails located in a lipid bilayer?

Cell membranes consist primarily of lipid bilayers. The lipid bilayer of a cell membrane is a barrier against the passage of molecules and ions into and out of the cell. However, cells do need to take in certain ions and molecules, such as nutrients, while excluding other materials. Selective absorption is accomplished by the protrusion of protein molecules through the lipid bilayer. These proteins form channels through which specific ions

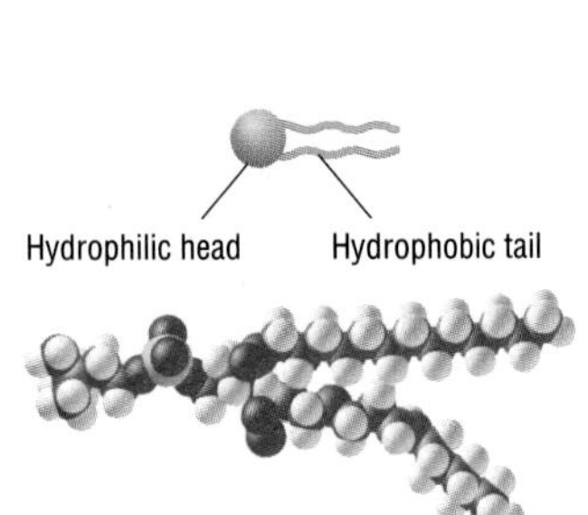

Figure 27.12
A space-filling model of lecithin is shown here. In the simplified representation, the hydrophilic head is shown as a sphere and the hydrophobic tails as wavy lines.

MEETING DIVERSE NEEDS

Sight Impaired Create a model of a cell membrane by floating a layer of table-tennis balls in a tub of water. Add a few tennis balls. Explain that the water underneath the balls represents the cytoplasm of the cell, the table-tennis balls represent the phospholipids, and the tennis balls, the proteins.

Gifted Although soybeans are used primarily for human food products and livestock feeds, hundreds of other uses for soybean derivatives have been developed. Have students investigate nonedible commercial applications of soybean products. (Ink is a good example. In 1989, soy ink had replaced petroleum-based ink for about half of the daily newspapers in the United States.)

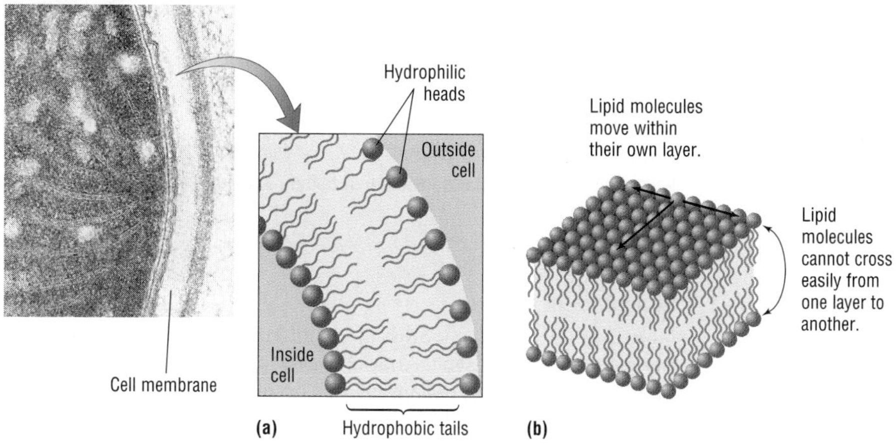

and molecules can selectively pass. Not all membrane proteins extend all the way through the membrane. Proteins, such as enzymes, may be bound to the interior surface of the membrane. Many membrane proteins have attached carbohydrate molecules. The carbohydrate portion is on the exterior of the lipid bilayer, where it can hydrogen-bond with water. The protein portion is on the interior of the lipid bilayer, so it does not contact the water.

Figure 27.13
(a) *A cell membrane typically includes a lipid bilayer.* ***(b)*** *The lipid molecules move easily within their own layer but do not readily move to the other layer. In such a lipid bilayer, the hydrophilic heads are in contact with water, but the hydrophobic tails are not.*

Waxes

Waxes are also part of the lipid family. **Waxes** are esters of long-chain fatty acids and long-chain alcohols. The hydrocarbon chains for both the acid and the alcohol usually contain from 10 to 30 carbon atoms. Waxes are low-melting, stable solids that occur in both plants and animals. In many plants, a wax coat protects the surfaces of leaves from water loss and attack by microorganisms. For example, carnauba wax, a major ingredient in car wax and floor polish, is found on the leaves of a South American palm tree. In many animals, waxes coat the skin, hair, and feathers and are used to help keep these structures pliable and waterproof. Beeswax, shown in **Figure 27.14**, is largely myricyl palmitate, the ester of myricyl alcohol and palmitic acid.

$$CH_3(CH_2)_{14}-\overset{\displaystyle O}{\overset{\|}{C}}-O(CH_2)_{29}CH_3$$

Myricyl palmitate

Figure 27.14
Bees construct their honeycomb from beeswax. Waxes are esters of long-chain fatty acids and long-chain alcohols.

section review 27.4

12. Describe the molecular structures of the three main types of lipids.

13. What role do phospholipids and proteins play in cell membranes?

14. Name the two classes of organic compounds produced when waxes are hydrolyzed.

Chem ASAP! Assessment 27.4 Check your understanding of the important ideas and concepts in Section 27.4.

27.4

Use the Visual

Have students refer to Figure 27.13. Ask:

- **Why is the cell membrane called a lipid bilayer?** (It contains two layers of phospholipid molecules.)
- **Why are the phospholipid tails located facing each other?** (Because they are hydrophobic and the membrane is surrounded by water.)

Point out that cell membranes also contain proteins. Unlike proteins in the cell cytoplasm, membrane proteins fold so that their nonpolar amino acid side chains are exposed on the surface of the protein. This arrangement enhances their solubility in the hydrophobic interior of the cell membrane. Membrane proteins often function as receptors for cell-to-cell signaling.

3 Assess

Evaluate Understanding

Have students describe the physical properties of lipids and name a naturally occurring lipid. (aqueous insoluble, high molar mass; triglycerides)

- **What are soaps and how are they formed?** (Soaps are alkali metal salts of fatty acids; they are formed by the hydrolysis of fats and oils by strong bases.)

Reteach

Point out that fats and oils hydrolyze in the presence of acids or bases because lipids are esters. Saponification is hydrolysis in which an ester reacts with an alkali hydroxide to form glycerol and an alkali metal salt of the fatty acid.

Answers

1 They are located between the outside and inside surfaces of the lipid bilayer.

SECTION REVIEW 27.4

12. Triglycerides are triesters of glycerol with long chains of carboxylic (fatty) acids attached. Phospholipids are lipids that contain phosphate groups; they have a hydrophobic and hydrophilic end. Waxes are esters of long-chain fatty acids and long-chain alcohols.

13. Phospholipids make up the lipid bilayer of the cell membrane. Protein molecules protrude from this layer, forming channels through which nutrients can pass.

14. Long-chain carboxylic acids (fatty acids) and long-chain alcohols are produced when waxes are hydrolyzed.

1 Engage

Use the Visual

Have students study the photograph and read the text that opens the section. Begin the discussion by considering blueprints in general. Ask:

- **Why do all the restaurants of a fast-food chain look alike?** (To assure market identity, the parent company supplies a blueprint to each franchise. The construction blueprint contains a set of detailed instructions that specify how and in what order components should be assembled.)
- **Which molecules in the body contain the instructions for building proteins?** (DNA or deoxyribonucleic acid)
- **What are these instructions called, and how do they code for proteins?** (They are called the *genetic code.* Sections of nucleic acid, called *genes,* specify the sequence of amino acids required to synthesize individual proteins.)

section 27.5

objectives

- Describe the structural components of nucleotides and nucleic acids, including DNA
- Give simple examples of genetic mutations
- Explain what is meant by recombinant DNA technology

key terms

- nucleic acids
- nucleotides
- gene
- genetic code

NUCLEIC ACIDS

Maybe people have told you that you have your mother's eyes or your father's nose. Although this is not literally true (your eyes and nose are your own), you do inherit the instructions for assembling the proteins of your body from your parents. **What are these instructions called, and how do they code for proteins?**

DNA and RNA

More than 100 years ago, a Swiss biochemist discovered a class of nitrogen-containing compounds in the nuclei of cells. The nuclei were first obtained from dead white blood cells in the pus of infected wounds. The eventual understanding of the biological role of the compounds has led to a revolution in biochemistry.

These nitrogen-containing compounds, called **nucleic acids,** are polymers found primarily in cell nuclei. They are indispensable components of every living thing. Two kinds of nucleic acids are found in cells—*d*eoxyribo*n*ucleic *a*cid (DNA) and *r*ibo*n*ucleic *a*cid (RNA). DNA stores the information needed to make proteins and governs the reproduction and growth of cells and new organisms. RNA has a key role in the transmission of the information stored in DNA.

The monomers that make up the DNA and RNA polymers are called **nucleotides.** Nucleic acids are therefore polynucleotides. As shown below, each nucleotide consists of a phosphate group, a five-carbon sugar, and a nitrogen-containing unit called a nitrogen base.

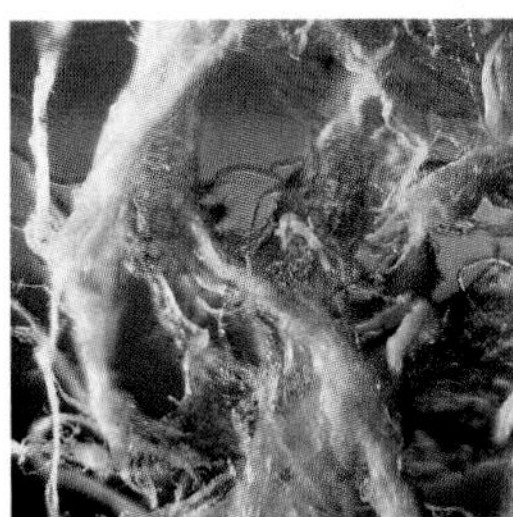

Figure 27.15
This photo shows strands of DNA extracted from cellular material.

The sugar unit in the nucleotides of DNA is the five-carbon monosaccharide known as deoxyribose. The nitrogen bases may be any one of four compounds—adenine, guanine, thymine, and cytosine. Adenine and guanine are each composed of a double ring. Thymine and cytosine are each composed of a single ring. These four bases are abbreviated A, G, T, and C, respectively, and are shown in a short segment of a DNA molecule in **Figure 27.16.** Ribose, which has one more oxygen than deoxyribose, is the sugar in the nucleotide monomers of RNA. The base thymine is never found in RNA. Instead, it is replaced by a fifth nitrogen base, uracil, abbreviated U.

Chemists studying nucleic acids discovered that the amount of adenine in DNA always equals the amount of thymine (A = T). Similarly, the amount of guanine always equals the amount of cytosine (G = C). The significance of this fact was not apparent until 1953. At that time, two scientists—James Watson and Francis Crick—proposed that DNA molecules consist of two polynucleotide chains wrapped into a spiral shape, as shown in **Figure 27.17.** This is the famous double helix of DNA. For the nitrogen bases to fit neatly into the double helix, every double-ringed base

Student Resources

From the Teacher's Resource Package, use:

- Section Review 27.5 and Ch. 27 Interpreting Graphics and Quizzes from the Review Module (Ch. 25–28)
- Laboratory Recordsheet 27-2

Technology Resources

Relevant technology resources include:

- Chem ASAP! CD-ROM
- Resource Pro CD-ROM

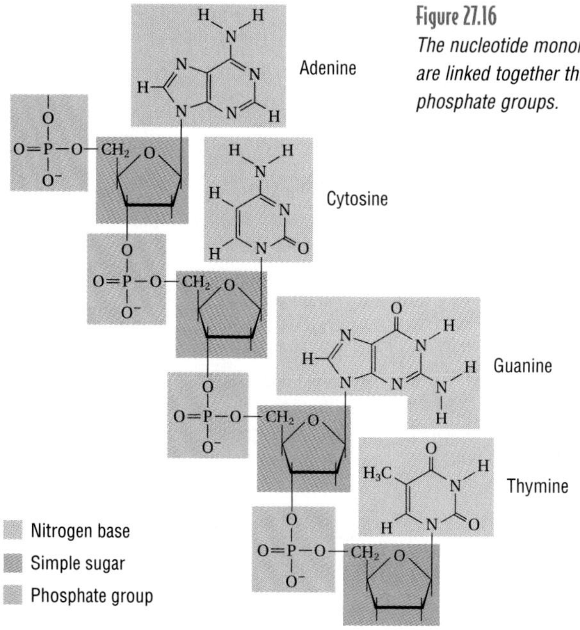

Adenine

Cytosine

Guanine

Thymine

Nitrogen base

Simple sugar

Phosphate group

Figure 27.16
The nucleotide monomers of DNA are linked together through their phosphate groups.

Figure 27.17
This is a space-filling model of a segment of the DNA double helix.

on one strand must be matched with a single-ringed base on the opposing strand. The pairing of A with T and G with C not only provides the best possible fit, it also gives the maximum number of hydrogen bonds between the opposing bases, as **Figure 27.18** shows. Thus the pairing of A with T (two hydrogen bonds between the opposing bases) and of G with C (three hydrogen bonds) makes for the most stable arrangement in the double helix.

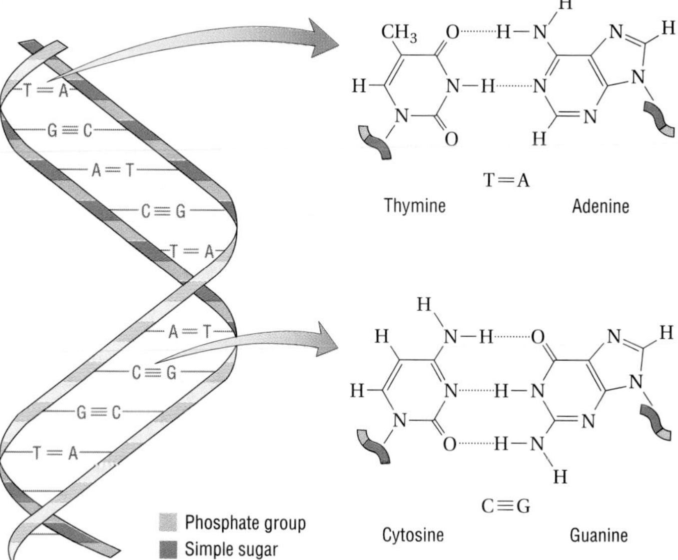

Figure 27.18
The two DNA strands in a double helix are held together by many hydrogen bonds; two hydrogen bonds between each thymine (T) and adenine (A) and three hydrogen bonds between each cytosine (C) and guanine (G).

Chem ASAP!

Simulation 29
Construct a portion of a DNA molecule.

MEETING DIVERSE NEEDS

Gifted Many people have noted various health risks associated with the consumption of caffeinated beverages. One concern raised is the potential mutagenic effects due to the misincorporation of caffeine into nucleic acids. Have students look up the structure of caffeine and suggest reasons why these concerns have been raised. (Caffeine is a nitrogen compound that is structurally related to purine bases, especially guanine. Its resemblance may allow it to bind to DNA-synthesizing enzymes, which would incorporate it into a polynucleotide. However, caffeine lacks hydrogens for proper base-pairing which would destabilize the DNA helix.)

27.5

Check Prior Knowledge

Intermolecular forces play significant roles in the functions of all biological molecules, especially nucleic acids. Ask:

- **What are intermolecular forces? Name the three main types of intermolecular forces.** (Intermoleculor forces are attractive forces that operate between separate but adjacent molecules; dipole–dipole, hydrogen bonding, and dispersion forces. Point out how these same attractive forces can also act within a molecule between two parts of a long coil, such as a polypeptide or polynucleotide.)
- **What is a hydrogen bond?** (an electrostatic interaction between a hydrogen atom covalently bonded to a very electronegative atom—F, O, N—and a lone pair of electrons on another very electronegative atom)

2 Teach

Use the Visuals

Have students study Figures 27.16 and 27.18 Point out the sequence *complementarity* between strands. Ask:

- **How does the experimental data that moles A = moles T, and moles G = moles C support the concept of "base-pairing?"** (A and G are always hydrogen bonded to T and C respectively in the opposite strand. Thus, there is a one-to-one mole correspondence.)
- **Why are adenine, cytosine, guanine, and thymine called nitrogen bases?** (They contain one or more nitrogen atoms. Nitrogen behaves as a weak base.)

MINI LAB

A Model of DNA

LAB TIPS

To get parallel strands, have students cut a thin strip of paper that just wraps around the tube (following the spiral seam); then measure the length of the paper, marking points along it every 5 cm. Students can use the paper to mark points on the tube and connect points to draw the parallel strand.

ANALYSIS AND CONCLUSIONS

1. Base pairs are neatly stacked inside the double helix.
2. Phosphate groups are on the exterior of the double helix.
3. Sugar units are on the exterior of the double helix.

Think Critically

Have students critically examine the double helical structure by referring to the model built in the MINI LAB and the diagrams on page 825. Ask:

- **What forces hold the two strands of DNA together?** (primarily hydrogen bonding; dispersion forces also act between stacked bases)
- **What forces the two strands apart?** (repulsion of negative charges on phosphate groups; two strands have greater entropy than a single helical structure)

Point out that DNA is held together by many reinforcing bonds. Ask students to speculate about the influence of pH on the stability of DNA? (Extremes in pH could disrupt the hydrogen bonding between bases in opposite strands due to ionization of nitrogen.)

MINI LAB

A Model of DNA

PURPOSE

To construct a model of double-stranded DNA.

MATERIALS

- cardboard tube from paper-towel roll
- 10 toothpicks
- felt-tip markers (two colors)
- thumbtack
- metric ruler

PROCEDURE

1. The typical tube has a seam that when viewed from one end describes a spiral that moves away from the observer. This spiral is a helix. Outline the spiral seam with a colored marker.
2. Using a different-colored marker, draw a second spiral midway between the lines of the first. These two spirals represent the two strands of double-stranded DNA.
3. Measure along the tube and mark a dot on each spiral every 5 cm. Label each dot with the letter S to indicate a sugar unit. Make a hole in the spirals at each S mark with the thumbtack. Move down each spiral and mark a letter P to indicate a phosphate group halfway between each two S dots.
4. Color each toothpick along half its length with a marker. A toothpick represents a base pair in the DNA molecule.
5. Starting at the top of the tube, insert a toothpick in one hole at an S label and guide it so it emerges through the hole in the S on the opposite side of the tube. Repeat the process for the other holes.

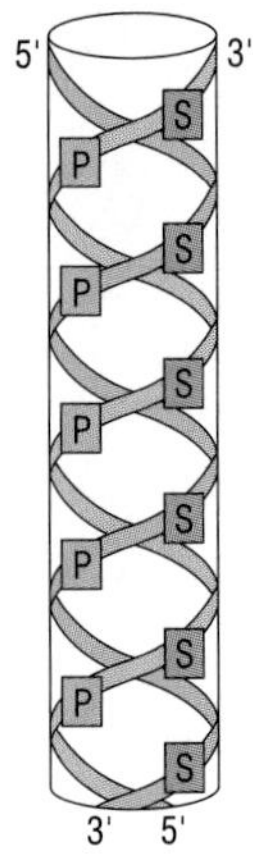

ANALYSIS AND CONCLUSIONS

1. Are the bases on the interior or the exterior of the double helix? Are they randomly arranged or neatly stacked?
2. Are the phosphate groups on the exterior or the interior of the DNA structure?
3. Are the sugar groups on the interior or the exterior of the DNA molecule?

The Genetic Code

An organism contains many peptides and proteins characteristic of that organism. The proteins of cats are different from the proteins of pine trees, which are different from the proteins of people. How do cells in a given kind of organism know which proteins to make? The cells use the instructions contained in DNA. A **gene** is a segment of DNA that codes for one kind of protein. Thus the products of genes are the peptides and proteins found in an organism.

You can think of DNA as a reference manual that stores the instructions for building proteins. The instructions are written in a simple language that has 4 "letters"— the bases A, T, G, and C. Experimental data shows that each "word" in a DNA manual is exactly three letters in length. The three-letter base sequences, or triplets, are the code words for 20 common amino acids. Some triplets are synonyms. The arrrangement of the code words in DNA is called the **genetic code.** The code words are strung together to form segments of DNA called genes, which specify proteins. Three bases of

Meeting Diverse Needs

Gifted Challenge gifted students to write a computer program that converts any sequence of DNA bases into an amino acid sequence using the three-letter code words listed in Table 27.2.

DNA arranged in a specific sequence are required to specify one amino acid in a peptide or protein chain.

Table 27.2 provides the DNA code words for the 20 common amino acids. For example, you can see that the DNA code word for the amino acid phenylalanine is AAA and that the DNA code word for the amino acid alanine is CGA. An amino acid may be specified by more than one code word, but a code word never specifies more than one amino acid. With DNA code words of three letters, 900 bases arranged in a specific sequence would be required to code for a protein made up of 300 amino acids arranged in a specific sequence.

Three code words (ATT, ATC, and ACT) are reserved as end, or termination, code words. The translation of a gene sequence of DNA to the amino acid sequence of a protein begins after one termination code word and runs continuously until another termination code word is reached. The termination code word signals a stop to the addition of amino acids to the protein, which is thus completed. You can think of a termination code as being similar to the period at the end of a sentence.

The molar masses of DNA molecules reach into the millions and possibly billions. Even with only four bases, the number of possible sequences of nucleotides in a DNA chain is enormous. The sequence of the nitrogen bases A, T, G, and C in the DNA of an organism constitutes the genetic plan, or blueprint, for that organism. This genetic plan is inherited from parents and passed to offspring. Differences in the number and order of the bases in DNA ultimately are responsible for the great diversity of living creatures found on Earth.

Table 27.2

Three-Letter DNA Code Words for the Amino Acids

First Letter in Code Word	Second Letter in Code Word: A	G	T	C	Third Letter in Code Word
A	AAA Phe	AGA Ser	ATA Tyr	ACA Cys	A
A	AAG Phe	AGG Ser	ATG Tyr	ACG Cys	G
A	AAT Leu	AGT Ser	ATT End	ACT End	T
A	AAC Leu	AGC Ser	ATC End	ACC Trp	C
G	GAA Leu	GGA Pro	GTA His	GCA Arg	A
G	GAG Leu	GGG Pro	GTG His	GCG Arg	G
G	GAT Leu	GGT Pro	GTT Gln	GCT Arg	T
G	GAC Leu	GGC Pro	GTC Gln	GCC Arg	C
T	TAA Ile	TGA Thr	TTA Asn	TCA Ser	A
T	TAG Ile	TGG Thr	TTG Asn	TCG Ser	G
T	TAT Ile	TGT Thr	TTT Lys	TCT Arg	T
T	TAC Met	TGC Thr	TTC Lys	TCC Arg	C
C	CAA Val	CGA Ala	CTA Asp	CCA Gly	A
C	CAG Val	CGG Ala	CTG Asp	CCG Gly	G
C	CAT Val	CGT Ala	CTT Glu	CCT Gly	T
C	CAC Val	CGC Ala	CTC Glu	CCC Gly	C

27.5

Discuss

The fact that millions of different sequences are possible from the combination of only four nucleotides can be compared to the fact that it is possible to form all the words of the English language from only two symbols—the dot and dash of Morse code. Moreover, both systems use patterns to communicate information. Point out, however, that Morse code is a relatively simple human invention that uses arbitrary combinations of dots and dashes to represent letters of the alphabet. The genetic code is a complex natural communication system that directs and controls all life processes.

Discuss

Cracking the genetic code is one of the greatest achievements of modern science. Point out that the genetic code is (nearly) *universal.* The three-letter codes used to translate DNA to protein are the same in humans and in bacteria. The fact that the genetic code is pervasive through the plant and animal kingdoms supports the argument that life on Earth evolved only once.

Meeting Diverse Needs

Gifted Have students read the autobiography of the Nobel prize-winning chemist, Arthur Kornberg, *For the Love of Enzymes: The Odyssey of a Biochemist.* Dr. Kornberg, a central player during the "Golden Age" of genetics research, provides an exceptionally readable, comprehensive historical account of the discovery and characterization of DNA and nucleic acids in general. This book documents the drama and excitement that surrounded many of the important biological discoveries made in the 20th century. As they read the book, encourage students to keep a time-line, "A Short History of DNA."

ACTIVITY

If you have access to a spectrometer that is able to accurately measure absorbencies in the ultraviolet region, have students investigate the physical properties of DNA. Factors such as temperature, pH, and ionic strength can be studied for their effects on the reversible denaturation of the DNA helix. Purified DNA (salmon sperm, calf thymus, sheared or unsheared) is available, at low cost, from chemical and biological supply houses. A 1 mg/mL solution of DNA in distilled water can be used for the studies. Use quartz cuvettes to measure the absorbance of DNA at 260 nm. Have students measure the absorbance versus temperature at 4 °C, 25 °C, 50 °C, and 100 °C. To study the affect of pH on absorbance, provide students with DNA in 0.2*M* HCl, tris-buffer (pH 8.0), and in 0.2*M* NaOH.

If time permits, have students repeat the measurements at different ionic strengths—0.5 mM, 2mM and 10mM $MgCl_2$. Explain that the absorbance of UV light by DNA increases when the DNA strands fall apart. If students follow the absorbance change of a solution initially at 100 °C they will find that the denaturation process is reversible. Point out that specialized enzymes in the cell nucleus make it possible for DNA to open and close at much lower temperatures. Ask:

- Why would it be important for the DNA strands to separate at certain times in the cell nucleus? (In order to translate the genetic code into protein, cellular enzymes must have access to the bases, which are stacked inside the helix.)

Gene Mutations

Substitutions, additions, or deletions of one or more nucleotides in the DNA molecule are called gene mutations. The effect of the deletion of a single base from a gene can be illustrated by starting with a string of letters of the alphabet. Suppose the string goes as follows.

PATTHEBADCAT

The letters may not make sense at first glance. However, if you separate them into three-letter words, they form a perfectly sensible statement.

PAT THE BAD CAT

Now remove the first letter, and again separate the string into three-letter segments.

ATT HEB ADC AT

This last sequence is nonsensical. Similarly, the deletion of a base in the DNA genetic code can turn the code into nonsense. A sequence that once may have coded for the proper sequence of amino acids in a necessary protein may be replaced by a code that produces a useless or damaging amino acid sequence. The same sort of harmful effect may be obtained by mutations involving substitutions or additions of nucleotides.

Such mutations might result in the production of a faulty protein or of no protein at all. Diseases that result from gene mutations are called inborn errors or molecular diseases. More than 200 such molecular diseases are known. For example, sickle cell anemia, which affects mainly people of African descent, is due to a mutation in the peptide chain of hemoglobin, the oxygen-carrying protein of blood. **Figure 27.19** shows the three-dimensional structure of hemoglobin. The defective hemoglobin reduces the oxygen supply to the tissues, triggering the painful episodes of the disease. In methemoglobinemia, another molecular disease, iron in some of the hemoglobin is in the oxidized iron(III) state rather than the normal reduced iron(II) state. A faulty hemoglobin chain permits the undesirable oxidation of the iron(II) to occur.

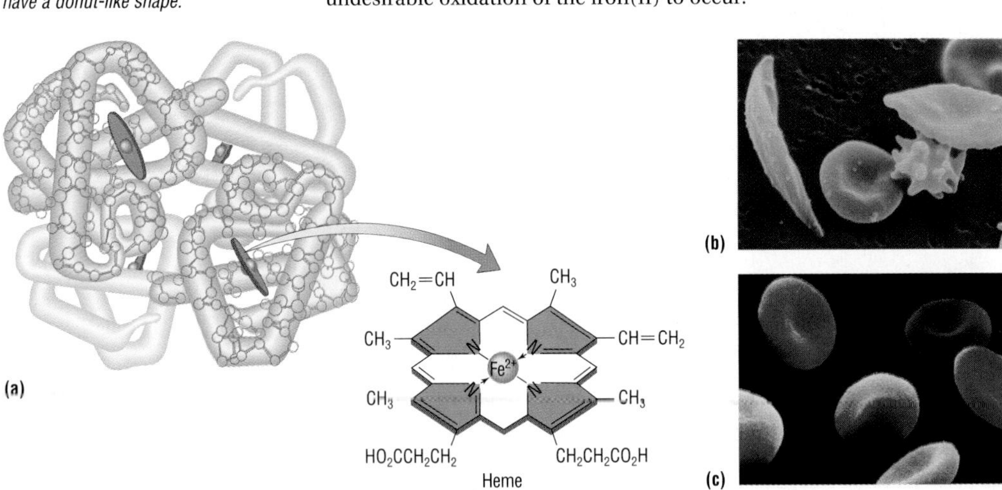

Figure 27.19
(a) *The abbreviated molecular structure of hemoglobin, the oxygen-transporting protein of blood, is shown here. The hemoglobin molecule consists of four peptide chains, each containing a heme group.* ***(b)*** *In the molecular disease called sickle cell anemia, defective hemoglobin causes red blood cells to take on a distorted shape.* ***(c)*** *Normal red blood cells have a donut-like shape.*

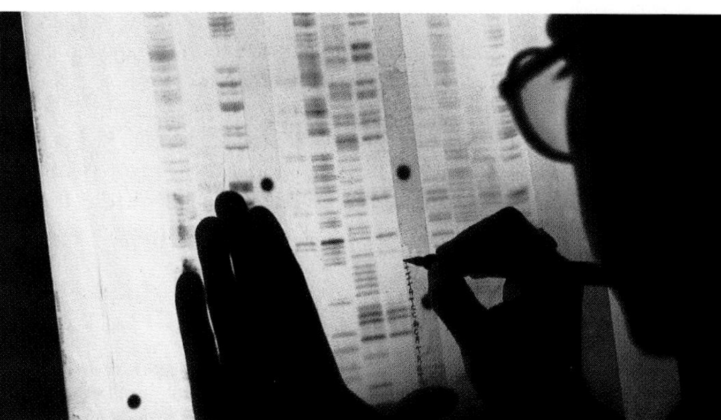

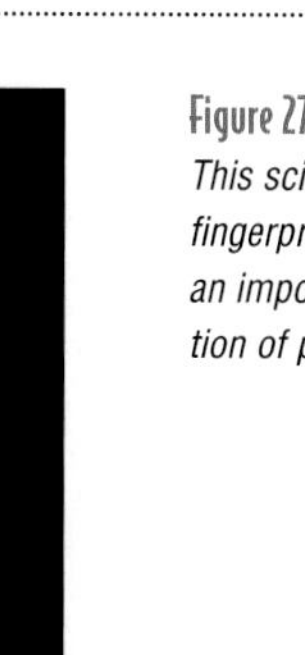

Figure 27.20
This scientist is comparing DNA fingerprints. DNA fingerprinting is an important tool in the identification of people.

Not all gene mutations are harmful, however. Some can result in the synthesis of a protein that can perform its function better than the version that previously existed. Such a mutation could thus be beneficial to the survival of an organism.

DNA Fingerprinting

Only about 5% of the DNA of human beings is used for coding information for the synthesis of proteins. The remaining 95% of DNA consists of repeating, noncoding base sequences that separate or sometimes interrupt gene coding sequences. The role of these stretches of noncoding DNA is unclear. The noncoding sequences are similar for members of the same family but are slightly different for almost every individual. Differences also exist in the coding portions of DNA. The base sequences of DNA are different for different individuals, except for identical twins.

The variation in the DNA of individuals forms the basis of a method for identifying a person from samples of their hair, skin cells, or body fluid. Because DNA sequences, like fingerprints, are unique for each individual, the DNA method is called DNA fingerprinting. To construct a DNA fingerprint, scientists first isolate the DNA in the sample. Only a tiny sample is needed because a method is available to amplify the amount of DNA in a sample by many millions of times. Enzymes that cut DNA chains between specific base sequences are used to cleave the DNA chains in the sample, thus providing a large number of DNA fragments. The fragments are of different lengths and base compositions for different individuals and can be separated and visualized. The pattern, or DNA fingerprint, that emerges can then be compared with the DNA of a sample from a known individual, as **Figure 27.20** shows.

If the DNA fingerprints are identical, it can be stated with a high degree of certainty that the DNA in the unknown sample is from the known individual. The chance in favor of a DNA fingerprint being unique to one individual can be as high as 1×10^{19} to 1—a high degree of certainty, given that the world's population is only about 6×10^{9} people. In practice, however, the chance of positively identifying an individual from a DNA sample is generally lower and is often subject to interpretation. For example, a person claiming to be the long-lost heir to a family fortune would have a DNA fingerprint that

ACTIVITY

Have students practice "reading" the genetic code. Write a number of short DNA sequences on the board and ask students to use Tables 27.1 and 27.2 to translate the genetic information into an amino acid sequence. Example: ACGGCAAGAGAATGGACT = (Cys-Arg-Ser-Leu-Thr-End)
Next, ask students to interpret the effects of *substitutions, additions,* and *deletions* on the translation of the DNA sequence. For example, ask:

- How would the sequence be affected if a thymine were substituted for the first guanine? (This mutation would insert a "stop" codon. The protein synthesis would prematurely terminate, and there would most likely be no biological activity.)

Students may also enjoy writing the DNA sequence that corresponds to a short polypeptide sequence. Point out that, using these same principles, scientists can use the amino acid sequence of a protein to search for the corresponding gene in the nuclear DNA.

Build Writing Skills

Have students do library research on DNA fingerprinting and write a short paper explaining the technique and its applications.

Discuss

Explain that recombinant DNA technology makes use of small circular DNA molecules called *plasmids* that occur naturally in bacteria and in some yeast. Once a human gene has been combined with the plasmid DNA, the recombinant molecule is reintroduced into bacteria. Because human DNA and bacterial DNA use the same bases, and because the genetic code is universal, the bacteria translate the human gene in the recombinant DNA into protein exactly as they would their own genes. When grown in large incubators, the bacteria are capable of producing massive quantities of the recombinant protein.

Think Critically

Point out to students that the treatment of human diseases with recombinant proteins, though effective, is still only a replacement therapy: patients must have a continuous supply of the protein to remain asymptomatic. Ultimately, genetic engineers would like to develop methods for treating the *cause* of a genetically linked disorder by providing the patient's own cells with the ability to produce a missing or faulty protein, an area of research referred to as *gene therapy.* Have students research this approach and report on a successful application of gene therapy. (Suggestions include cystic fibrosis and hemophilia.)

contains DNA fragments that are characteristic of his or her father and mother but that would not be identical to that of the father or the mother.

Questions about the certainty of identification by means of DNA fingerprints are often at issue in criminal cases. DNA fingerprints can be obtained from evidence left at the scene of a crime, such as blood, but there is no standard for absolute determination of innocence or guilt from DNA fingerprinting, nor is there likely to be. Juries must decide whether the evidence is convincing. Evidence based on DNA fingerprints is permitted in more than half of the states in the U.S.

Recombinant DNA Technology

Scientists have learned to manipulate genes by various methods. Recombinant DNA technology consists of methods for cleaving a DNA chain, inserting a new piece of DNA into the gap created by the cleavage, and resealing the chain. **Figure 27.21** illustrates such a method.

The first practical application of recombinant DNA technology was to insert the gene for making human insulin into bacteria. Most people naturally make insulin, a polypeptide that controls levels of blood sugar, but insufficient insulin production results in diabetes, a potentially deadly disease. The symptoms of diabetes can often be controlled by insulin injections. In the past, human insulin was not available for this purpose. Pig insulin, which is quite similar in structure to human insulin, was used as a substitute. Some patients, however, were allergic to pig insulin. Today, diabetic patients use the human form of insulin produced by the bacteria altered by recombinant DNA technology. Such bacteria can produce large amounts of insulin in very pure form. Use of this insulin removes the need for the potentially dangerous use of pig insulin.

Other proteins produced by recombinant DNA technology are used as medicinal drugs. For example, an enzyme called tissue plasminogen activator is used to dissolve blood clots in patients who have suffered heart attacks. Another protein, interferon, is thought to alleviate or delay some of the debilitating effects of multiple sclerosis. Many other proteins produced by recombinant DNA technology are in use, under investigation, or on

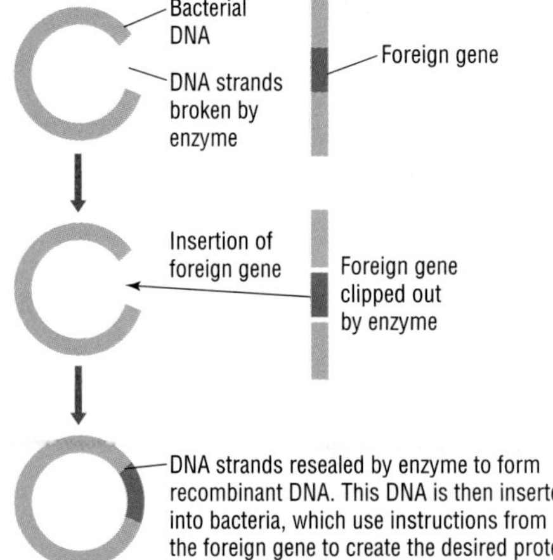
Bacterial DNA
DNA strands broken by enzyme
Foreign gene
Insertion of foreign gene
Foreign gene clipped out by enzyme
DNA strands resealed by enzyme to form recombinant DNA. This DNA is then inserted into bacteria, which use instructions from the foreign gene to create the desired protein.

Figure 27.21
Included here are the elements of an experiment involving recombinant DNA. In this experiment, DNA from one organism is inserted into the DNA of a different organism.

Figure 27.22
These genetically engineered tomatoes were altered to increase the period it takes them to ripen.

their way to eventual approval for human use. Recombinant DNA technology is also being applied to the cure of molecular diseases.

In agriculture, new recombinant DNA techniques can make plants resistant to pests and weed killers and produce fruits and vegetables that are better suited for shipping and storage. **Figure 27.22** shows genetically altered tomatoes that have a longer ripening period, which prevents spoilage and allows for their year-round availability. Genetically altered organisms have many potential benefits, but some people have concerns about the possibility of mistakes or misuse.

Ethical concerns were raised recently when Scottish scientists announced the birth of a lamb named Dolly, shown in **Figure 27.23.** In normal animal reproduction, an offspring is a genetic mixture of the characteristics of both parents; however, Dolly was a clone, an offspring of a single individual. A clone is an exact genetic copy of its parent because it is formed using the DNA of only that parent. The birth of Dolly has raised the question of whether human beings might eventually be cloned. Many people are horrified by some of the possible outcomes of cloning identical individuals. These reactions are one aspect of more general concerns about the uniqueness of life. Human cloning experiments have been outlawed in several countries for such reasons.

Figure 27.23
The sheep Dolly has no father. She was cloned from a single cell taken from her mother.

section review 27.5

15. Describe the structure of a nucleotide and of the nucleic acid DNA.
16. What are some types of gene mutations?
17. What is recombinant DNA technology?
18. Why do you think recombinant DNA research is controversial? What are your personal thoughts on this matter?

 Chem ASAP! **Assessment 27.5** Check your understanding of the important ideas and concepts in Section 27.5.

Answers

SECTION REVIEW 27.5

15. A nucleotide is composed of covalently bonded units: phosphate, a sugar, and a nitrogen base. A nucleic acid is a condensation polymer of nucleotide units. The sugar in DNA is deoxyribose; there are four possible nitrogen bases.
16. Gene mutations occur by the addition, deletion, or substitution of bases in a segment of DNA that constitutes a gene.
17. Recombinant DNA technology inserts a new gene into an existing DNA sequence.
18. People have different opinions about issues such as the dignity of human life and interfering with nature.

3 Assess

Evaluate Understanding

Ask:

- **What is the fundamental chemical unit of DNA and what is its general structure?** (the nucleotide, which consists of a phosphate linked to a sugar linked to a nitrogen base)
- **What nitrogen bases are found in DNA?** (adenine, guanine, thymine, cytosine)
- **Describe the structure of DNA. What type of bonding occurs between DNA strands?** (DNA has a double helical structure; two polynucleotide strands, running in opposite directions, wind around each other. They are held together primarily through hydrogen bonds between bases.)
- **How is the chemical data in DNA used by the cell to direct the synthesis of a protein?** (The cell's protein-synthesis system translates the sequence of DNA bases into an arrangement of amino acids that matches a specific protein.)
- **Bacteria and humans are very different organisms. How is it possible to combine their DNA to make a human protein?** (Although they are very different, human and bacterial DNA are made from the same bases and use the same genetic code.)

Reteach

Display Figure 27.18 on an overhead projector. Use it to discuss the chemical and functional differences between DNA and RNA. (RNA contains a different sugar and one different base. RNA transmits genetic information from the nucleus to the cytoplasm in a eukaryotic cell.)

27.6

1 Engage

Use the Visual

Have students study the photograph and read the text that opens the section. Explain that a hormone is a signaling molecule released by a cell and distributed by the bloodstream to other target cells in the body. Insulin is another example of a hormone. Ask students to recall what effect insulin has on target cells. (It stimulates the uptake of glucose from the blood.) **Thyroxine stimulates energy production in cells. Animal cells harvest energy by breaking down larger, energy rich molecules into smaller molecules. (Compare this process to that of photosynthesis.) Because each molecule of thyroxine contains four atoms of iodine, the absence of iodine in the diet prevents the synthesis of thyroxine, and results in lower metabolic activity. Ask students to name one of the primary dietary sources of iodine.** (iodized salt)

- **What kind of metabolic reactions break down biomolecules in a person's diet?** (catabolic reactions)

Check Prior Knowledge

Have students compare the free energy changes associated with spontaneous and nonspontaneous processes. Energy released by a reaction that is available to do work is called *free energy.* Spontaneous processes, which favor formation of products, release free energy.

section 27.6

objectives

- Describe the role of ATP in energy production and energy use in the cell
- Define metabolism and explain the relationship between catabolism and anabolism

key terms

- adenosine triphosphate
- metabolism
- catabolism
- anabolism

METABOLISM

The body's metabolic rate is controlled by a hormone called thyroxine that is released by the thyroid gland. When a person's diet does not contain adequate iodine, the thyroid gland is unable to produce and release thyroxine, and the gland begins to swell, forming a goiter. **What kind of metabolic reactions break down biomolecules in a person's diet?**

ATP

Adenosine triphosphate (ATP), shown in **Figure 27.24,** is the molecule that transmits the energy needed by cells of all living things. The function of ATP can be compared with a belt connecting an electric motor to a pump. The motor generates energy capable of operating the pump. But if a belt does not connect the motor to the pump, the energy produced by the motor is wasted. You can think of ATP as the belt that couples the production and use of energy by cells as shown in **Figure 27.25.**

Recall that oxidation reactions, such as the combustion of methane in a furnace or the oxidation of glucose in a living cell, produce energy. Energy is also captured when adenosine diphosphate (ADP) condenses with a molecule of phosphoric acid and becomes ATP. The addition of a phosphate group, called phosphorylation, occurs during certain biochemical oxidation reactions.

$$\underset{\text{Adenosine diphosphate (ADP)}}{\text{Adenosine}-\underset{\text{OH}}{\overset{\text{O}}{\overset{\|}{\text{P}}}}-\text{O}-\underset{\text{OH}}{\overset{\text{O}}{\overset{\|}{\text{P}}}}-\text{OH}} + \underset{\text{Inorganic phosphate } (P_i)}{\text{HO}-\underset{\text{OH}}{\overset{\text{O}}{\overset{\|}{\text{P}}}}-\text{OH}} \longrightarrow \underset{\text{Adenosine triphosphate (ATP)}}{\text{Adenosine}-\underset{\text{OH}}{\overset{\text{O}}{\overset{\|}{\text{P}}}}-\text{O}-\underset{\text{OH}}{\overset{\text{O}}{\overset{\|}{\text{P}}}}-\text{O}-\underset{\text{OH}}{\overset{\text{O}}{\overset{\|}{\text{P}}}}-\text{OH}} + \underset{\text{Water}}{H_2O}$$

The formation of ATP efficiently captures energy produced by the oxidation reactions in living cells. Every mole of ATP produced by the phosphorylation of ADP stores about 30.5 kJ of energy. The reverse happens when ATP is hydrolyzed back to ADP: Every mole of ATP hydrolyzed back to ADP releases about 30.5 kJ of energy. Cells use this released energy to drive processes that would ordinarily be nonspontaneous. Because of its ability to capture energy from one process and transmit it to another,

Figure 27.24
ATP is ADP that has been phosphorylated. ATP is important because it occupies an intermediate position in the energetics of the cell.

STUDENT RESOURCES

From the Teacher's Resource Package, use:

- Section Review 27.6, Ch. 27 Vocabulary Review, Quizzes and Tests from the Review Module (Ch. 25–28)

TECHNOLOGY RESOURCES

Relevant technology resources include:

- Chem ASAP! CD-ROM
- ResourcePro CD-ROM
- Assessment Resources CD-ROM Chapter 27 Tests

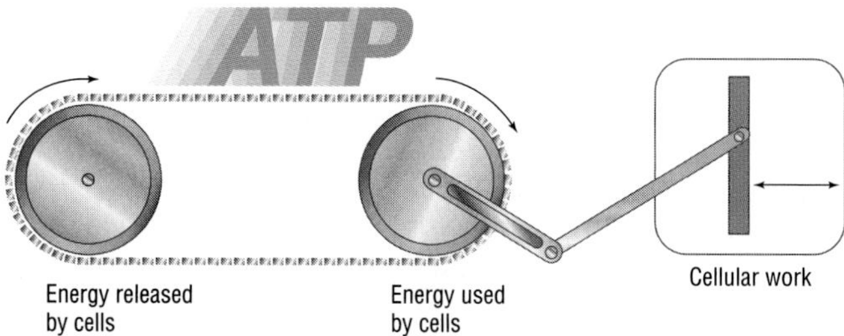

Figure 27.25
ATP is the energy carrier between the metabolic reactions that release energy and those that use energy.

ATP is sometimes referred to as a high-energy compound; however, the energy produced by the breakdown of ATP to ADP is not particularly high for the breaking of a covalent bond. ATP is important because it occupies an intermediate position in the energetics of the cell. It can be formed by using the energy obtained from a few higher-energy oxidation reactions. The energy conserved in ATP is then used to drive other cellular processes.

Catabolism

Metabolism is the entire set of chemical reactions carried out by an organism. In metabolism, food and unneeded cellular components are degraded to simpler compounds by chemical reactions collectively called **catabolism.** The degradation of complex biological molecules (carbohydrates, lipids, proteins, and nucleic acids) during catabolism provides building blocks for the construction of new compounds needed by the cell.

Through the formation of ATP, catabolic reactions provide the energy for such needs as body motion and the transport of nutrients to cells where they are required. The oxidation reactions of catabolism also provide energy in the form of heat. These reactions help keep your body temperature constant at 37 °C.

The complete oxidation of glucose to carbon dioxide and water is one of the most important energy-yielding processes of catabolism. Study **Figure 27.26,** which summarizes the major steps in the degradation of glucose to six molecules of carbon dioxide. The complete oxidation actually involves many reactions, which are not shown. As you can see, the major carbon-containing reactants and products are named, and they are also referred to according to the number of carbons they contain.

The combustion of 1 mol of glucose to carbon dioxide and water, whether by fire or by oxidation by a living organism, produces 2.82×10^3 kJ of energy.

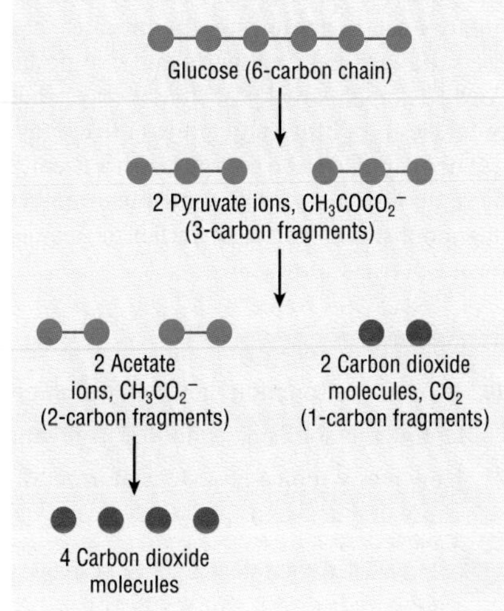

Figure 27.26
The oxidation of glucose to carbon dioxide and water is one of the most important energy-yielding processes of catabolism.

EXERCISE PHYSIOLOGY

Training Energy Pathways

In the human body, glycogen is stored in skeletal muscles and the liver. Liver glycogen is used to maintain a normal blood sugar level. Muscle glycogen, however, is used solely for muscle work. Large amounts of carbohydrates, including glycogen, are used by the body during athletic events that require power or speed. The carbohydrates provide the necessary energy. Interval training, a technique that alternates rest and work phases, is designed as much to train biochemical energy pathways as to train the muscles used in an athletic activity. Proper training and diet can actually increase the amount of glycogen stored after a training session by more than a thousandfold.

2 Teach

Discuss

Draw a schematic diagram on the chalkboard that illustrates the exchange of energy between anabolic and catabolic reactions. Connect the terms *ATP* and *ADP* by upper and lower arrows. Label the upper arrow *anabolism;* in anabolic reactions, small molecules are put together to form larger ones. These nonspontaneous processes must be coupled to the free energy of ATP hydrolysis; thus, ADP is produced during anabolic reactions. Label the lower arrow *catabolism;* in catabolic reactions, large molecules are broken down. Catabolic reactions most often release energy that can be used to drive the synthesis of ATP.

Remind students that all living things require a continuous input of free energy for muscle contraction, transport of nutrients, and synthesis of biomolecules. More than 50% of the free energy released by cellular reactions is actually used to cause other chemical changes. The rest is lost as heat. Compare this efficiency to that of the internal combustion engine. At most, 30% of the free energy of combustion is used to propel a car.

27.6

Discuss

Use Hess's law of heat summation to show how the free energy of ATP hydrolysis can be harnessed by the cell to drive a nonspontaneous reaction. Consider the reversible, nonspontaneous reaction:

$X \rightleftharpoons Y \qquad \Delta G° = +21\ kJ/mol$

Normally, K_{eq} for this reaction is much less than 1, but if the reaction is coupled to the hydrolysis of ATP:

$ATP + H_2O \rightarrow ADP + P + H^+ \qquad \Delta G° = -30.51\ kJ/mol$

$X + ATP + H_2O \rightarrow YP + ADP + H^+ \qquad \Delta G° = -9.5\ kJ/mol$

Conversion of X to Y is driven forward by high intracellular concentrations of ATP.

3 Assess

Evaluate Understanding

Ask the students to compare photosynthesis and glucose degradation. (Photosynthesis is an anabolic process fueled by solar energy; glucose degradation is a catabolic process that yields energy through ATP formation.)

Reteach

Point out that it is the phosphate–oxygen anhydride bonds in ATP that store the chemical energy in ATP. When broken by water, these bonds release energy that is used to form new types of chemical bonds in other molecules.

Portfolio Project

The basal metabolic rate is the rate at which the energy from food is released when a person is resting and fasting. This energy supports functions such as breathing, circulation, and maintaining body temperature. The basal metabolic rate varies with age, size, body type, and degree of physical activity.

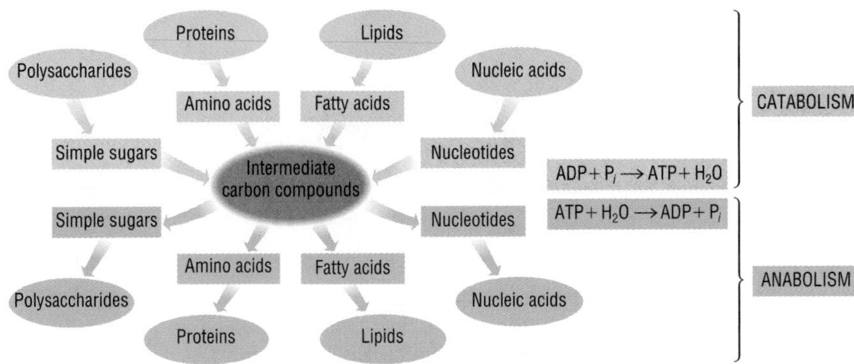

Figure 27.27
Simple compounds produced by catabolism are used in the synthesis reactions of anabolism. Catabolism produces energy, whereas anabolism uses energy.

Cells that use oxygen may produce up to 38 mol of ATP by capturing the energy produced by the complete oxidation of a single mole of glucose! (1) How many kilojoules of energy are produced per mole of ATP produced? The large amount of ATP produced from the oxidation of glucose makes it the likeliest mode of energy production for most kinds of cells. In fact, if glucose is available, brain cells use no other source of carbon compounds for energy production.

Anabolism

Some of the simple compounds produced by catabolism are used to synthesize more-complex biological molecules (carbohydrates, lipids, proteins, and nucleic acids) necessary for the health and growth of an organism. The synthesis reactions of metabolism are called **anabolism.** Unlike catabolism, which produces energy, anabolism uses energy.

Figure 27.27 gives an overview of the relationship between catabolism and anabolism. As you have learned, nutrients and unneeded cell components are degraded to simpler components by the reactions of catabolism. The oxidative reactions of catabolism yield energy captured in the formation of ATP. In anabolism, the products and the energy of catabolism are used to make new cell parts and compounds needed for cellular life and growth. You already know that energy produced by physical and chemical processes is of little value unless the energy can be captured to do work. If it is not captured, the energy is lost as heat. The chemical energy produced by catabolism must have some means of being used for the chemical work of anabolism. As you read earlier in this section, the ATP molecule is that means of transmitting energy.

portfolio project

Prepare a written report on human basal metabolism. Define basal metabolism and discuss how it varies with the age, weight, and gender of an individual. Include the units of basal metabolism.

section review 27.6

19. Discuss the function of ATP in cellular metabolism.
20. Define metabolism, catabolism, and anabolism.
21. How many moles of ATP are formed from the complete oxidation of 1 mol of glucose in a cell that uses oxygen?

 Chem ASAP! **Assessment 27.6** Check your understanding of the important ideas and concepts in Section 27.6.

Answers

(1) 74 kJ of energy per mole of ATP is produced.

SECTION REVIEW 27.6

19. ATP stores and transmits energy produced during metabolic processes.
20. Metabolism is the entire set of reactions carried out by an organism. Catabolism is the subset of metabolic reactions that involve degradation of complex molecules. Anabolism is the subset of metabolic reactions that involve synthesis.
21. 38 mol

Chemistry Serving...the Consumer

FAKE FAT

What are your favorite foods? If you are like most people, the list probably includes a number of foods that are high in fat, such as cheeseburgers, pizza, french fries, and chocolate. Although they taste good, foods like these may not be good for your health if you eat them often. Medical research shows that a diet high in fat greatly increases the risk of heart disease, cancer, diabetes, and other health problems. For these reasons, the Food and Drug Administration (FDA) recommends that no more than 30% of your daily caloric intake come from fat.

But limiting your intake of fat can be very difficult. One reason is that much of the food available is either cooked in fat or has fat added to it. Try looking at the ingredients list next time you eat potato chips, a candy bar, or crackers—you will find that some kind of vegetable oil or animal fat is among the first ingredients listed. Another reason it is hard to cut down on fat is that it tastes good.

Now, thanks to biochemical research, fatty-food lovers have another option other than avoiding their favorite foods. They can eat foods containing Olestra™, a fat substitute with none of the drawbacks of real fat (but perhaps some of its own). Olestra has the same properties as fats and oils, but because of its chemical structure, it is not digested or absorbed by the body.

Biochemists spent many years designing the Olestra molecule. It is made up of six to eight fatty acid groups connected to a sucrose base. A normal fat molecule, in contrast, is a triglyceride, which is an organic compound made up of three fatty acid groups attached to a glycerol base. Olestra is similar enough to fat to share its properties, but the substitution of the sucrose for the glycerol base fools the body's digestive enzymes. When you ingest real fat molecules, digestive enzymes bind to them and act as catalysts in the hydrolysis of the bonds between the glycerol and the fatty acids. These same enzymes, however, are unable to bind to Olestra molecules. Olestra, therefore, passes through the digestive system unchanged.

Eating a 1-ounce serving of Olestra potato chips instead of regular potato chips will cut about 11 grams of fat out of your diet.

But Olestra has drawbacks that will likely restrict its total replacement of fat in foods. It can tie up fat-soluble vitamins, which include vitamins A, D, E, and K. If foods containing these vitamins are cooked with Olestra, the vitamins will dissolve in the Olestra just as they would in a triglyceride fat. Because the body cannot digest the Olestra, it cannot absorb the vitamins dissolved in the Olestra. Olestra can also cause digestive problems in certain people.

In January 1998, the FDA approved Olestra for use in snack foods such as chips and crackers. Snack foods are an ideal candidate for Olestra because they are not an important source of fat-soluble vitamins. At the same time they are often very high in fat. Eating a 1-ounce serving of Olestra potato chips instead of regular potato chips will cut about 11 grams of fat out of your diet.

Olestra will never completely replace fat, but eating high-fat foods may become a little more guilt free.

CHEMISTRY IN CAREERS

TAKE IT TO THE NET

Find out more about career opportunities:

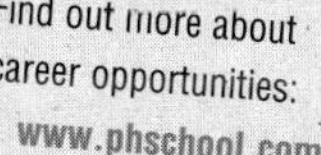

www.phschool.com

CHEMICAL SPECIALIST

Local food service distributor seeks responsible self-motivated individ-

BIOCHEMIST

Study and manipulate the molecules of living organisms. Identify enzymes and biochemical pathways.

See page 881.

CHEMIST NEEDED

Chemistry Serving... the Consumer

DISCUSS

After students have read the article, ask students about the meaning and significance of food labels that say "Low Fat," "Low in Unsaturated Fat," or "No tropical Oils." Have them recall that the word *fat* refers to a large class of related, water-insoluble molecules. Ask students to describe the differences between fats and oils. (length and degree of unsaturation of the fatty acid chain) Saturated fats tend to be found in tropical oils and animal fats; unsaturated fats are found in other plant oils.

Have students discuss the pros and cons of fat substitutes using the issues presented in the article. Ask students to propose an explanation about why Olestra is not metabolized by the body. (Human enzymes do not recognize or bind the molecule; thus, they cannot catalyze its breakdown.)

CHEMISTRY IN CAREERS

Encourage students to connect to the Internet address shown on this page to find out more about careers in biochemistry. See page 881 for additional information.

4 Close

Summary

Have students name the major classes of biological compounds and briefly describe their physical and chemical features. Display photos of foods and ask students to sort them according to the nutrients they supply. Ask:

- **How is genetic information stored?** (in DNA through nucleotide sequences)
- **How do proteins function in the cell?** [as structural units and catalysts (enzymes)]

Extension

Refer students to Table 27.2. Ask:

- **How many three-letter codes are there? Are all possible combinations of bases used?** (64; yes, $4^3 = 64$)
- **Why are there more codes than amino acids?** (The code is *degenerate;* some amino acids have more than one code.)
- **Is there an advantage to redundant codes?** (might protect against harmful mutations)

Looking Back . . . Looking Ahead . . .

The preceding chapter introduced the concept of functional groups. This chapter discussed the major classes of biomolecules: carbohydrates, lipids, proteins, and nucleic acids. The next chapter focuses on nuclear chemistry.

Chapter 27 STUDENT STUDY GUIDE

Take It to the NET
For interactive study and review, go to www.phschool.com

KEY TERMS

- active site *p. 818*
- adenosine triphosphate (ATP) *p. 832*
- amino acid *p. 815*
- amino acid sequence *p. 816*
- anabolism *p. 834*
- carbohydrate *p. 812*
- catabolism *p. 833*
- coenzyme *p. 819*
- disaccharide *p. 813*
- enzyme *p. 818*
- enzyme–substrate complex *p. 819*
- gene *p. 826*
- genetic code *p. 826*
- lipid *p. 821*
- metabolism *p. 833*
- monosaccharide *p. 812*
- nucleic acid *p. 824*
- nucleotide *p. 824*
- peptide *p. 816*
- peptide bond *p. 816*
- phospholipid *p. 822*
- photosynthesis *p. 810*
- polypeptide *p. 816*
- polysaccharide *p. 814*
- protein *p. 816*
- saponification *p. 821*
- substrate *p. 818*
- triglyceride *p. 821*
- wax *p. 823*

KEY EQUATIONS

- Photosynthesis:

$$6CO_2 + 6H_2O + \text{Energy} \longrightarrow C_6H_{12}O_6 + 6O_2$$

- Oxidation of glucose:

$$C_6H_{12}O_6 + 6O_2 \longrightarrow 6CO_2 + 6H_2O + \text{Energy}$$

CONCEPT SUMMARY

27.1 A Strategy for Life
- Photosynthetic organisms use the sun's energy during photosynthesis to reduce carbon dioxide to glucose. Plants and animals use glucose as an energy source.

27.2 Carbohydrates
- Carbohydrates are made up of carbon, hydrogen, and oxygen and are often found in nature as simple molecules containing one or a few sugar units called monosaccharides.
- Condensation polymers called polysaccharides contain hundreds of thousands of sugar units.

27.3 Amino Acids and Their Polymers
- Peptides and proteins are condensation polymers of amino acids.
- Enzymes, which are proteins, serve as catalysts of biological reactions.

27.4 Lipids
- Lipids, which include fats and oils, are water-insoluble substances.
- Phospholipids, lipids that contain phosphate groups, make up cell membranes.

27.5 Nucleic Acids
- Nucleic acids are polymers of nucleotides.
- Alterations in base sequences are gene mutations, which can cause errors in protein production.

27.6 Metabolism
- ATP is the molecule that stores and transmits the energy needed by cells of all living things.
- Metabolism is the sum of all the reactions carried out by an organism.
- The energy and compounds generated by catabolism are used for anabolism, the nonspontaneous synthesis reactions.

CHAPTER CONCEPT MAP

Use these terms to construct a concept map that organizes the major ideas of this chapter.

Chem ASAP! Concept Map 27
Create your Concept Map using the computer.

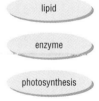

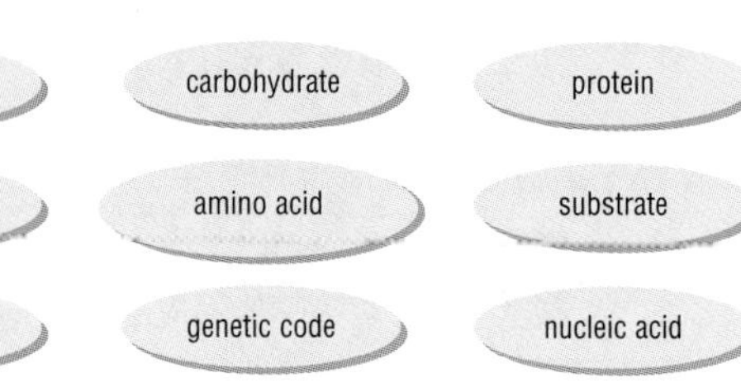

Take It to the Net

At **www.phschool.com** students will find for this chapter

- an Internet activity
- links to related chemistry sites
- an interactive quiz
- career links

Chapter 27 REVIEW

CONCEPT PRACTICE

22. Explain what happens in photosynthesis. *27.1*

23. Where do biological molecules originate? *27.1*

24. Write a balanced equation for the complete oxidation of glucose. *27.2*

25. Name two important monosaccharides. *27.2*

26. Give sources for glucose and fructose. *27.2*

27. How does the carbonyl functional group differ in glucose and fructose? *27.2*

28. Which monosaccharides are combined to form the disaccharide sucrose? *27.2*

29. What is the product of the complete hydrolysis of each polysaccharide? *27.2*
   a. starch
   b. glycogen

30. What is the name given to the bond connecting two amino acids in a peptide chain? *27.3*

31. Consider the tripeptide Ser—Gly—Phe. How many peptide bonds does this molecule have? *27.3*

32. Describe what an enzyme does. *27.3*

33. Describe two common patterns found in the folding of protein chains. *27.3*

34. What is the meaning of each abbreviation? Are the structures of these tripeptides the same? Explain. *27.3*
   a. Ala—Ser—Gly
   b. Gly—Ser—Ala

35. Distinguish between a fat and an oil. *27.4*

36. What is a triglyceride? *27.4*

37. What is a soap? *27.4*

38. Draw structural formulas for the products of complete hydrolysis of tristearin. *27.4*

39. Draw a simple representation of a lipid bilayer. *27.4*

40. Name two functions of membrane proteins. *27.4*

41. Cells contain two types of nucleic acids. What are they called? *27.5*

42. What are the components of a nucleotide? *27.5*

43. What is the structural difference between the sugar unit in RNA and the sugar unit in DNA? *27.5*

44. What type of bonding helps hold a DNA double helix together? *27.5*

45. Does every code word in DNA specify an amino acid in protein synthesis? Explain. *27.5*

46. Which of the following base pairs are found in a DNA molecule? *27.5*
   a. A—A **c.** C—G **e.** A—U
   b. A—T **d.** G—A **f.** T—U

47. What causes a gene mutation? *27.5*

48. An average adult expends about 8400 kJ of energy every day. How many moles of ATP must be converted to ADP to provide this amount of energy? *27.6*

49. Where do the complex biomolecules your body degrades during catabolism come from? *27.6*

50. Write an abbreviated, balanced equation for the hydrolysis of ATP to ADP. *27.6*

51. Compare and contrast catabolism and anabolism. *27.6*

52. Some experts recommend that complex carbohydrates provide 50–55% of dietary calories. What is the relationship between energy production and complex carbohydrates? *27.6*

CONCEPT MASTERY

53. What is glycogen? How does it differ from starch?

54. What is cellulose and where is it found?

55. Give the names and structures of the compounds produced by the hydrolysis of sucrose.

56. What is an enzyme–substrate complex?

57. What is the active site of an enzyme?

58. What role do coenzymes play?

59. The formula for palmitic acid is $CH_3(CH_2)_{14}CO_2H$. A popular soap is mostly sodium palmitate. Write the structure of sodium palmitate.

60. Which part of a phospholipid molecule can interact favorably with water?

61. Consider the following segment of DNA.
G—C—C—C—C—A—A—C—G—T—T—A
   a. Using abbreviations for the amino acids from **Table 27.2**, write the amino acid sequence formed by translation of this information into a peptide.
   b. What amino acid sequence would result from the substitution of adenine (A) for the second cytosine (C)?

CHAPTER REVIEW

22. Photosynthetic organisms capture and use energy from sunlight to synthesize glucose.

23. Glucose fuels their synthesis.

24. $C_6H_{12}O_6 + 6O_2 \rightarrow 6CO_2 + 6H_2O + \text{energy}$

25. glucose and fructose

26. **a.** Glucose is found in blood, corn, and grapes. **b.** Fructose is found in honey and many fruits.

27. Glucose is an aldehyde; fructose is a ketone.

28. glucose and fructose

29. **a.** glucose **b.** glucose

30. peptide bond

31. two

32. Enzymes catalyze biological reactions.

33. Peptide chains fold into spiral helixes or into sheets in which peptide chains lie side by side.

34. Both tripeptides contain alanine, serine, and glycine, but the sequences are different.

35. At room temperature, animal fats are solid; plant oils are liquid.

Chapter 27 REVIEW

36. triester of glycerol and fatty acids

37. alkali metal salt of a fatty acid

38.

$$\begin{array}{l} CH_2OH \\ | \\ CHOH + 3CH_3(CH_2)_{16}-\overset{\overset{\Large O}{\|}}{C}-O^-Na^+ \\ | \\ CH_2OH \end{array}$$

Glycerol — Sodium stearate

39. Hydrophilic head faces water; Hydrophobic tail protected from water; Lipid bilayer

40. channels for the transport of ions and molecules; enzymes

41. deoxyribonucleic acid (DNA) and ribonucleic acid (RNA)

42. a phosphate group, a 5-carbon sugar unit, and a nitrogen base

43. Ribose has one more oxygen than deoxyribose.

44. hydrogen bonding

45. No; termination codes signal the end of a peptide chain.

46. **b.** A—T **c.** C—G

47. Addition, deletion, or substitution of one or more DNA bases

48. 2.8×10^2 mol of ATP

49. foods or existing body tissues

50.

$$\text{adenosine}-\overset{\overset{O}{\|}}{\underset{\underset{OH}{|}}{P}}-O-\overset{\overset{O}{\|}}{\underset{\underset{OH}{|}}{P}}-O-\overset{\overset{O}{\|}}{\underset{\underset{OH}{|}}{P}}-OH$$

ATP

$$+ H_2O \rightarrow \text{Adenosine}-\overset{\overset{O}{\|}}{\underset{\underset{OH}{|}}{P}}-O-\overset{\overset{O}{\|}}{\underset{\underset{OH}{|}}{P}}-OH$$

ADP

$$+ HO-\overset{\overset{O}{\|}}{\underset{\underset{OH}{|}}{P}}-OH + 30.5\text{kJ}$$

inorganic phosphate

51. In catabolism, large biomolecules are broken down and energy is captured through the production of ATP. In anabolism, the products and energy from catabolism are used to make biomolecules.

REVIEW

52. Catabolism of one mol glucose yields 2.82×10^3 kJ of energy.

53. Glycogen, a polysaccharide produced by animals, has more branches than do plant starches.

54. Cellulose, a polysaccharide of glucose, is found in plant cell walls.

55.

```
     CHO              CH2OH
      |                 |
   H—C—OH              C=O
      |                 |
  HO—C—H           HO—C—H
      |                 |
   H—C—OH           H—C—OH
      |                 |
   H—C—OH           H—C—OH
      |                 |
    CH2OH             CH2OH
   glucose           fructose
```

56. a substrate molecule bonded to the active site of an enzyme

57. location where a substrate binds and is converted to products

58. Some catalysts require coenzymes to function.

59. $CH_3(CH_2)_{14}-\overset{O}{\overset{\|}{C}}-O^-Na^+$

sodium palmitate

60. the hydrophilic head

61. **a.** Arg—Gly—Cys—Asn
b. Arg—Gly—Cys—Asn

62. One possible combination is C—G—A—C—C—A—A—G—A Other combinations are possible.

63. Met—Arg—Val—Tyr

64. G—C—T—A—G—G—T

65. Diseases such as sickle cell anemia result from faulty genes.

66. medicines, control of molecular diseases, disease-resistant plants

67. Transplant normal gene into individual with faulty gene.

68. **a.** Fe(II) **b.** Fe(III)

69. beneficial, harmful, or neutral mutations

70. Brain cells use only ATP produced from glucose.

71. More oxygen is needed for energy-producing processes; rids the body of excess CO_2.

72. $30.5\ kJ/mol \times 38\ mol \div 2.82 \times 10^3\ kJ/mol \times 100\% = 41.1\%$

73. **a.** (1) **b.** (4)

74. CO_2 cannot produce energy through oxidation. Glucose can be oxidized to CO_2.

75. Inside the helix; strands must unwind.

62. Use **Table 27.2** to write a base sequence for DNA that codes for the tripeptide Ala—Gly—Ser.

63. Determine the amino acid sequence of a peptide if the corresponding base sequence on DNA is

T—A—C—G—C—A—C—A—T—A—T—A

64. A segment of a DNA strand has the following base sequence.

C—G—A—T—C—C—A

Write the base sequence that appears on its opposing strand in the double helix.

65. What is a molecular disease? Name two such diseases.

66. What are some outcomes of recombinant DNA research?

67. Describe a possible use of gene therapy.

68. What is the oxidation state of iron in the following?
a. normal hemoglobin
b. methemoglobinemia

69. What are the possible consequences of an error in a DNA sequence?

70. How is the catabolism carried out in brain cells different from that in other body cells?

71. Why is it necessary to breathe deeply during vigorous exercise?

72. The complete oxidation of glucose releases 2.82×10^3 kJ/mol of energy, and the formation of ATP from ADP requires 30.5 kJ/mol. What percent of the energy released in the complete oxidation of glucose is captured in the formation of ATP?

CRITICAL THINKING

73. Choose the term that best completes the second relationship.

a. animals : carbon dioxide   plants : ______
(1) oxygen   (3) methanol
(2) ethanol   (4) sugar

b. polymer : monomer   protein : ______
(1) monosaccharide   (3) fatty acid
(2) nucleic acid   (4) amino acid

74. Interpret this statement: "Carbon dioxide is an energy-poor molecule, but glucose is an energy-rich molecule."

75. In the DNA double helix, are the base pairs part of the backbone structure, inside the double helix, or outside the double helix? What must happen before the protein-making machinery of the cell can recognize the DNA bases?

CUMULATIVE REVIEW

76. Write a molecular formula for each compound.
a. methyl acetate   **c.** 5-methyl-3-hexene
b. heptane   **d.** butanal

77. What must be true at the end point of an acid–base titration?

78. How many carbon atoms are in 2.00 mol of ethanol?

79. Name the four structural isomers of C_4H_9Cl.

80. How many electrons are in each ion?
a. Fe^{3+} **b.** S^{2-} **c.** As^{3-} **d.** V^{5+}

81. Identify the oxidizing agent in each reaction.
a. xenon + fluorine $\longrightarrow$ xenon tetrafluoride
b. sulfur + oxygen $\longrightarrow$ sulfur trioxide
c. gaseous chlorine + aqueous sodium bromide $\longrightarrow$ aqueous bromine + aqueous sodium chloride

82. Calculate the pH of each solution.
a. $[H^+] = 7.0 \times 10^{-5}M$
b. $[OH^-] = 1.8 \times 10^{-9}M$
c. $[OH^-] = 6.1 \times 10^{-2}M$
d. $[H^+] = 4.4 \times 10^{-11}M$

83. What is a necessary condition for a molecule to have stereoisomers?

CONCEPT CHALLENGE

84. Explain why a lack of nutrients stops cell growth.

85. The following compound is hydrolyzed by boiling with sodium hydroxide.

```
              O
              ‖
CH2—O—C—(CH2)14CH3
 |
 |            O
 |            ‖
CH —O—C—(CH2)12CH3
 |
 |            O
 |            ‖
CH2—O—C—(CH2)16CH3
```

What are the saponification products?

86. What causes the spontaneous formation of lipid bilayers?

76. **a.** $C_3H_6O_2$ **b.** C_7H_{16} **c.** C_7H_{14} **d.** C_4H_8O

77. Equivalents of acid and base must be equal.

78. 2.41×10^{24}

79. 1-chlorobutane, 2-chlorobutane, 1-chloro-2-methylpropane, 2-chloro-2-methylpropane

80. **a.** 23 **b.** 18 **c.** 36 **d.** 18

81. **a.** fluorine **b.** oxygen **c.** chlorine

82. **a.** 4.15 **b.** 5.26 **c.** 12.79 **d.** 10.36

83. at least one asymmetric carbon

84. Nutrients supply carbon compounds, ions, and energy needed for growth.

85.

```
                            O
                            ‖
                 CH3(CH2)14—C—O⁻Na⁺
 CH2OH                      O
  |                         ‖
 CHOH      +     CH3(CH2)12—C—O⁻Na⁺
  |                         O
  |                         ‖
 CH2OH           CH3(CH2)16—C—O⁻Na⁺
glycerol           salts of fatty acids
```

86. Phospholipids have hydrophobic tails that exclude water and polar heads that dissolve in water.

Chapter 27 STANDARDIZED TEST PREP

Select the choice that best answers each question or completes each statement.

1. What phrase best describes ATP?
   a. energy producer
   b. energy consumer
   c. energy transmitter
   d. energy pump
2. Which element is *not* found in amino acids?
   a. phosphorus
   b. nitrogen
   c. carbon
   d. oxygen
   e. hydrogen
3. For any enzyme to function, the substrate must bind to the
   a. product.
   b. cofactor.
   c. active site.
   d. peptide.

Use the paragraph to answer questions 4–6.

Because an amino acid contains both a carboxylic acid group and an amino group, it is amphoteric, that is, it can act as either an acid or a base. Crystalline amino acids have some properties—relatively high melting points and high water solubilities—that are more characteristic of ionic substances than molecular substances.

4. Write an equation showing glycine acting as an acid in a reaction with water.
5. Write an equation showing glycine acting as a base in a reaction with water.
6. It is possible for glycine to undergo an internal Brønsted-Lowry acid–base neutralization reaction. Write the resulting structural formula. Explain how this reaction would account for the ionic properties of glycine.

Use the photo to answer question 7.

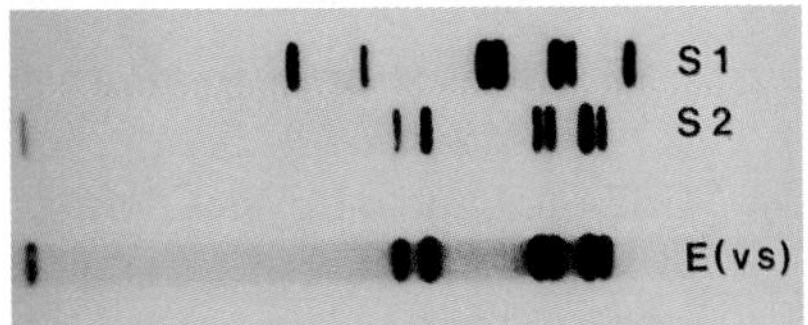

7. The photo shows three DNA fingerprints. The DNA in sample E was taken from a crime scene. The DNA in samples S1 and S2 were taken from suspects. Based on the DNA evidence, which suspect did not commit the crime?

For questions 8–11, name the category of organic compounds most closely identified with each biological molecule.

8. proteins
9. nucleic acids
10. lipids
11. carbohydrates

Use these models to answer question 12.

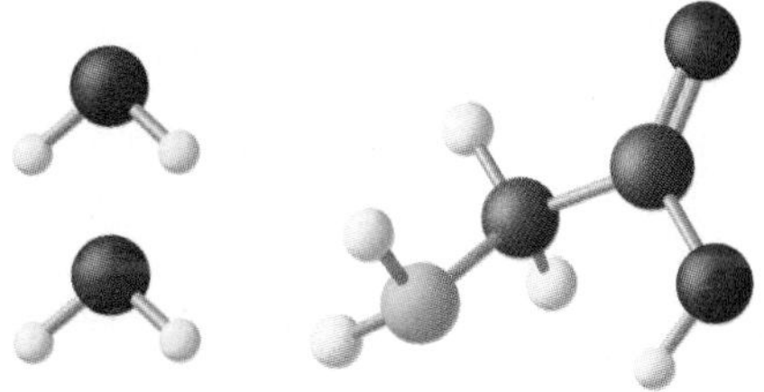

12. *You Draw It!* Glycine, which is the simplest amino acid, can form hydrogen bonds with water. Draw a glycine molecule with at least two water molecules bonded to each end of the glycine molecule.

For each question there are two statements. Decide whether each statement is true or false. Then decide whether Statement II is a correct explanation for Statement I.

Statement I		Statement II
13. Lipids tend to be insoluble in water.	BECAUSE	Lipid molecules have mainly nonpolar bonds.
14. Starch and cellulose are both digestible by most organisms.	BECAUSE	Glucose is the monomer in both starch and cellulose.
15. Many of the reactions in catabolism are oxidation reactions.	BECAUSE	Oxidation reactions tend to be energy-producing reactions.
16. The sequence of bases in DNA contains the code for making proteins.	BECAUSE	Each pair of bases in DNA codes for a specific amino acid.

Chapter 27 STANDARDIZED TEST PREP

1. c
2. a
3. c
4. $H_2NCH_2COOH(aq) + H_2O(l) \rightarrow H_2NCH_2COO^-(aq) + H_3O^+(aq)$
5. $H_2NCH_2COOH(aq) + H_2O(l) \rightarrow H_3N^+CH_2COOH(aq) + OH^-(aq)$
6. The carboxylic acid group can donate a proton to the amino group to form $H_3N^+CH_2COO^-$.
7. suspect S1
8. amino acids
9. nucleotides
10. fatty acids
11. monosaccharides
12. See the drawing below. In the drawing, black spheres represent carbon, gray spheres represent oxygen, and white spheres represent hydrogen.
13. True, True, correct explanation
14. False, True
15. True, True, correct explanation
16. True, False

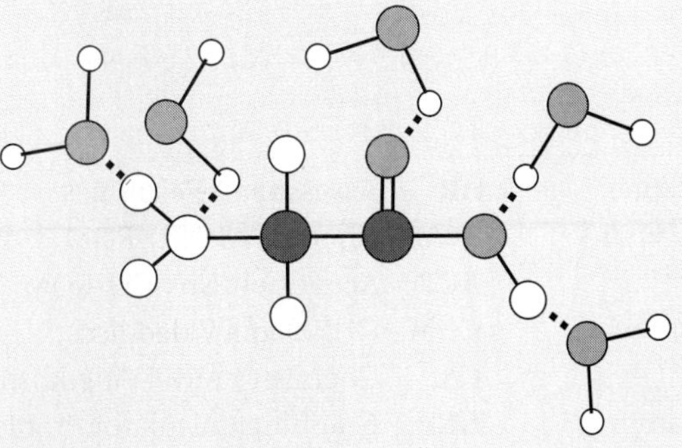

Planning Guide

SECTION OBJECTIVES	ACTIVITIES/FEATURES	MEDIA & TECHNOLOGY
28.1 Nuclear Radiation ● ■ ◆ ▸ Discuss the processes of radioactivity and radioactive decay ▸ Characterize alpha, beta, and gamma radiation in terms of composition and penetrating power	**SE** **Discover It!** *Simulating Radioactive Decay,* p. 840 **SE** **Link to Physics** *Particle Accelerators,* p. 844 **LM** 52: *Radioactivity and Radiation* **TE** **DEMO,** p. 842	**ASAP** Assessment 28.1 **OT** 77: *Types of Radiation* **OT** 78: *Radioactive Decay*
28.2 Nuclear Transformations ○ ■ ◆ ▸ Use half-life information to determine the amount of a radioisotope remaining at a given time ▸ Give examples of equations for the synthesis of transuranium elements by transmutation	**SE** **Link to Archeology** *Carbon Dating,* p. 847 **SE** **CHEMath** *Calculating Half-Life,* p. 848 **SE** **Small-Scale Lab** *Radioactivity and Half-Lives,* p. 852 (**LRS** 28-1) **SSLM** 40: *Half-Lives and Reaction Rates* **TE** **DEMO,** p. 847	**ASAP** Simulation 30 **ASAP** Problem Solving 4 **ASAP** Assessment 28.2 **CHM** Side 10, 2: *Ancient Cultures, Modern Chemistry* **OT** 78: *Radioactive Decay* **OT** 79: *Uranium-238 Decay Series*
28.3 Fission and Fusion of Atomic Nuclei ● ■ ◆ ▸ Compare nuclear fission and nuclear fusion, and comment on their potential as sources of energy ▸ Describe the methods used in nuclear power plants to produce and control fission reactions ▸ Explain the issues involved in storage, containment, and disposal of nuclear waste	**TE** **DEMO,** p. 854	**ASAP** Animation 30 **ASAP** Assessment 28.3 **OT** 80: *Nuclear Fission and Fusion*
28.4 Radiation in Your Life ● ■ ◇ ▸ Describe three methods of detecting radiation ▸ List some applications of radioisotopes in research and medicine	**SE** **Mini Lab** *Studying Inverse-Square Relationships,* p. 858 (**LRS** 28-2) **PLM** *Studying Inverse-Square Relationships* **SE** **Chemistry Serving . . . the Environment** *Nuclear Waste: Storage, Disposal, and Containment,* p. 862 **SE** **Chemistry in Careers** *Nuclear Physician,* p. 862 **TE** **DEMO,** p. 858	**ASAP** Assessment 28.4 **ASAP** Concept Map 28 **CHM** Side 2, 32: *Hot Pocket Change* **CA** Side 10, 13: *PET Scan* **PP** Chapter 28 Problems **RP** Lesson Plans, Resource Library **AR** Computer Test 28 **www** Activities, Self-Tests, *SCIENCE NEWS* updates **TCP** The Chemistry Place Web Site

KEY

● Conceptual (concrete concepts)	**AR** Assessment Resources	**GRS** Guided Reading and Study Workbook
○ Conceptual (more abstract/math)	**ASAP** Chem ASAP! CD-ROM	**LM** Laboratory Manual
■ Standard (core content)	**ACT** ActivChemistry CD-ROM	**LP** Laboratory Practicals
□ Standard (extension topics)	**CHM** CHEMedia Videodiscs	**LRS** Laboratory Recordsheets
◆ Honors (core content)	**CA** Chemistry Alive! Videodiscs	**SSLM** Small-Scale Lab Manual
◇ Honors (options to accelerate)	**GCP** Graphing Calculator Problems	

PRACTICE	ASSESSMENT
GRS Section 28.1 **RM** Practice Problems 28.1	**SE** Section Review **RM** Section Review 28.1 **RM** Chapter 28 Quiz **LP** Lab Practical 28-1
SE **Sample Problem** 28-1 **GRS** Section 28.2 **RM** Practice Problems 28.2 **RM** Interpreting Graphics	**SE** Section Review **RM** Section Review 28.2 **RM** Chapter 28 Quiz **LP** Lab Practical 28-2
GRS Section 28.3 **RM** Practice Problems 28.3	**SE** Section Review **RM** Section Review 28.3 **RM** Chapter 28 Quiz
GRS Section 28.4 **RM** Practice Problems 28.4 **GCP** Chapter 28	**SE** Section Review **RM** Section Review 28.4 **RM** Vocabulary Review 28 **SE** Chapter Review **SM** Chapter 28 Solutions **SE** Standardized Test Prep **PHAS** Chapter 28 Test Prep **RM** Chapter 28 Quiz **RM** Chapter 28 A & B Test

OT	Overhead Transparency	**SE**	Student Edition
PHAS	PH Assessment System	**SM**	Solutions Manual
PLM	Probeware Lab Manual	**SSV**	Small-Scale Video/Videodisc
PP	Problem Pro CD-ROM	**TCP**	www.chemplace.com
RM	Review Module	**TE**	Teacher's Edition
RP	Resource Pro CD-ROM	**www**	www.phschool.com

PLANNING FOR ACTIVITIES

STUDENT EDITION

Discover It! p. 840
- small containers with lids
- pennies
- flat table or desk surface
- scientific calculators
- loose-leaf paper
- graph paper
- pens or pencils

Small-Scale Lab, p. 852
- pencils, paper, rulers
- graph paper
- pennies
- dice

Mini Lab, p. 858
- flashlights
- strips of duct tape
- scissors
- poster board, white (50 cm × 50 cm)
- meter rulers or tape measures
- flat surface, long enough to hold a meter ruler
- loose-leaf paper
- graph paper
- pens or pencils

TEACHER'S EDITION

Teacher Demo, p. 842
- Geiger counter equipped with a rate meter and a loudspeaker
- objects containing radioactive elements such as the luminous dial of a watch, smoke detector, or electronic stud finder

Teacher Demo, p. 847
- 16 pennies

Teacher Demo, p. 854
- wooden matchsticks
- fire extinguisher on hand

Teacher Demo, p. 858
- metal key
- unexposed photographic film wrapped in black paper
- radiation source

Chapter 28

Key Terms

28.1 radioisotopes, radioactivity, radiation, radioactive decay, alpha radiation, alpha particles, beta radiation, beta particles, gamma radiation
28.2 band of stability, positron, half-life, transmutation, transuranium elements
28.3 fission, neutron moderation, neutron absorption, fusion
28.4 ionizing radiation, Geiger counter, scintillation counter, film badge

DISCOVER IT!

This activity provides a model for radioactive decay. After studying Section 28.2, students should realize that the graphs they prepared by plotting "number of heads" versus "trial number" resemble exponential decay curves. When students plot "log of number of heads" versus "trial number," the graph is linear. With careful study of the data, some students may realize that the number of heads appearing after *n* trials is approximately $128 \times (1/2)^n$.

Chapter 28 NUCLEAR CHEMISTRY

A bubble chamber can detect ionizing radiation.

FEATURES

Stay current with SCIENCE NEWS
Find out more about nuclear chemistry:
www.phschool.com

DISCOVER IT! SIMULATING RADIOACTIVE DECAY

You need a small container with a lid, 128 pennies, a flat table or desk surface, a scientific calculator, loose-leaf paper, graph paper, and a pen or pencil.

1. Make a two-column data table with the headings "Trial" and "Number of heads."
2. Put 128 pennies in the container and shake them.
3. Carefully pour the pennies onto a flat surface.
4. Pick up, count, and set aside those pennies that are turned heads up. In the table, enter the number of heads for Trial 1.
5. Return the remaining pennies (those that were tails) to the container and repeat Steps 3–5 for a total of five trials.
6. If time permits, repeat the entire experiment four more times and average your data or the data collected by the class.

Prepare a graph of the number of heads (*y*-axis) versus the trial number (*x*-axis). Is the result a straight line or a curve? Add a third column titled "Log of number of heads" to the data table, and use your calculator to complete the table. Prepare a graph of the data in Column 3 (*y*-axis) versus trial number (*x*-axis). Is the result a straight line or a curve? After you have read this chapter, relate the results of this activity to radioactive decay and half-life.

28.1 ● ■ ◆
28.2 ○ ■ ◆
28.3 ● ■ ◆
28.4 ● ■ ◇

Conceptual Use Table 28.1 to summarize properties of radiation. In Section 28.2, use Figure 28.7 to help students understand radioactive decay at the particle level. The Small-Scale Lab provides a hands-on opportunity to make half-lives less abstract for students. Sample Problem 28-1 may be omitted.

Standard Use Table 28.1 and Figure 28.2 to compare and contrast several types of radiation. The CHEMath feature will help students who have difficulty with Sample Problem 28-1 and Section Review question 10. Section 28-3 is an important section for building students' scientific literacy.

Honors Section 28.4 contains descriptive content that is suitable for accelerated study. The Mini Lab in this section could also be assigned as a homework project.

NUCLEAR RADIATION

section 28.1

Marie Curie was a Polish scientist whose research led to many discoveries about radiation and radioactive elements. In 1903 she and her husband Pierre won the Nobel Prize for physics. She was also awarded the Nobel Prize for chemistry in 1911 for her research on radioactive elements. Marie Curie's research was invaluable to the understanding and use of newly discovered radioactive elements. In 1934 she died from leukemia caused by her long-term exposure to radiation. **What types of radiation are there, and how harmful are they?**

objectives

- Discuss the processes of radioactivity and radioactive decay
- Characterize alpha, beta, and gamma radiation in terms of composition and penetrating power

key terms

- radioisotopes
- radioactivity
- radiation
- radioactive decay
- alpha radiation
- alpha particles
- beta radiation
- beta particles
- gamma radiation

Radioactivity

In the preceding chapters, you read about chemical reactions. In such reactions, atoms tend to attain stable electron configurations. This chapter deals with nuclear reactions, reactions in which the nuclei of unstable isotopes, called **radioisotopes,** gain stability by undergoing changes. These changes are always accompanied by the emission of large amounts of energy. Unlike chemical reactions, nuclear reactions are not affected by changes in temperature, pressure, or the presence of catalysts. They are also unaffected by the compounds in which the unstable isotopes are present, and they cannot be slowed down, speeded up, or turned off.

In 1896, the French chemist Antoine Henri Becquerel (1852–1908) made an interesting accidental discovery. He was studying the ability of uranium salts that had been exposed to sunlight to fog photographic film plates. During a period of bad weather in Paris, Becquerel realized that even uranium salts not exposed to the sun caused the same result in the film, as shown in **Figure 28.1.** At that time, two of Becquerel's associates were Marie Curie (1867–1934) and Pierre Curie (1859–1906). The Curies were able to show that the fogging of the plates was caused by rays emitted by the uranium atoms in the ore. Marie Curie named the process by which materials give off such rays **radioactivity.** The penetrating rays and particles emitted by a radioactive source are called **radiation.** Pierre Curie assisted his wife in the isolation of several radioactive elements. Together with Becquerel, they won the Nobel Prize in physics in 1903 for their work.

Figure 28.1
Uranium salts, which are radioactive, can fog a photographic plate like this one. This effect was discovered by the French chemist Antoine Henri Becquerel.

28.1

1 Engage

Use the Visual

Have students examine the photograph and read the text that opens the section. Explain that the nuclei of a radioactive element spontaneously decompose. Nuclear chemistry is the study of changes in matter that originate in atomic nuclei. Ask:

- **What types of radiation exist, and how harmful are they?** (The three most common types of radiation emitted by unstable nuclei are alpha (α), beta (β), and gamma (γ) radiation. Although all forms of radiation are somewhat harmful, gamma rays are particularly dangerous because they penetrate body tissues.)

Check Prior Knowledge

Ask students what they recall about atomic structure. Draw a model that reflects their understanding. Ask:

- **What particles in an atom are involved in chemical reactions?** (valence electrons)
- **What is the difference between a chemical reaction and a nuclear reaction?** (Chemical reactions involve the transfer or sharing of electrons; the nucleus remains unchanged. Nuclear reactions, by definition, involve changes in the nuclei of atoms, a situation students have not had to consider up to this point.)

STUDENT RESOURCES

From the Teacher's Resource Package, use:

- Section Review 28.1, Ch. 28 Practice Problems and Quizzes from the Review Module (Ch. 25–28)
- Laboratory Manual: Experiment 52
- Laboratory Practical 28-1

TECHNOLOGY RESOURCES

Relevant technology resources include:

- Chem ASAP! CD-ROM
- Resource Pro CD-ROM

2 Teach

TEACHER DEMO

Using a Geiger counter equipped with a rate meter and a loudspeaker, have students listen to the random clicking noise. Be sure that there are no radioactive sources nearby. Point out that the counter is recording background radiation, which is always present. Ask students about the possible sources of background radiation. (cosmic, soil, and interior of Earth) Explain that background radiation varies from location to location. To make an accurate measurement of radiation from a given source, you must take local background radiation into account. Next place the Geiger counter near some objects containing radioactive elements such as the luminous dial of a watch or a smoke detector.

Use the Visual

Have students examine Table 28.1. Encourage students to memorize the names and symbols used to represent each type of radiation. Point out the differences in charge and mass. Students may wish to recreate Table 28.1 on an index card and use the information to help them write and balance nuclear equations. Point out the relative penetrating power of each type of radiation and the precautions necessary to effectively block different types of radiation.

Table 28.1

Characteristics of Some Ionizing Radiations			
Property	**Alpha radiation**	**Beta radiation**	**Gamma radiation**
Composition	Alpha particle (helium nucleus)	Beta particle (electron)	High-energy electromagnetic radiation
Symbol	α, ${}^{4}_{2}\text{He}$	β, ${}^{\ 0}_{-1}\text{e}$	γ
Charge	2+	1−	0
Mass (amu)	4	1/1837	0
Common source	Radium-226	Carbon-14	Cobalt-60
Approximate energy	5 MeV*	0.05 to 1 MeV	1 MeV
Penetrating power	Low (0.05 mm body tissue)	Moderate (4 mm body tissue)	Very high (penetrates body easily)
Shielding	Paper, clothing	Metal foil	Lead, concrete (incompletely shields)

*(1 MeV = 1.60×10^{-13} J)

The discovery of radioactivity dealt a deathblow to Dalton's theory of indivisible atoms. A radioactive atom, or radioisotope, undergoes drastic changes as it emits radiation. Such radioisotopes have unstable nuclei. The stability of a nucleus depends on the relative proportion of neutrons to protons in the nucleus as well as on the overall size of the nucleus. The presence of too many or too few neutrons, relative to the number of protons, leads to an unstable nucleus. An unstable nucleus loses energy by emitting radiation during the process of **radioactive decay.** Eventually, unstable radioisotopes of one element are transformed into stable (nonradioactive) isotopes of a different element. Radioactive decay is spontaneous and does not require any input of energy.

Types of Radiation

Several types of radiation can be emitted during radioactive decay. **Table 28.1** summarizes the characteristics of three of these types of radiation. The different types of radiation from a radioactive source can be separated by an electric or magnetic field, as shown in **Figure 28.2**.

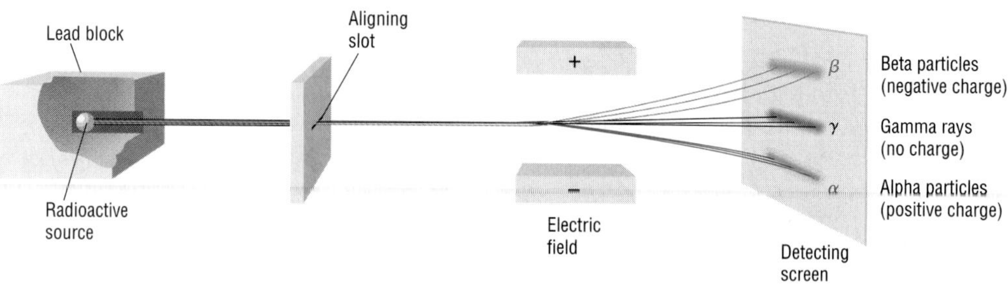

Figure 28.2
An electric field has different effects on these three types of radiation. Alpha particles and beta particles are deflected in opposite directions—alpha particles toward the negative plate and beta particles toward the positive plate. Gamma rays are undeflected.

A type of radiation called **alpha radiation** consists of helium nuclei that have been emitted from a radioactive source. These emitted particles, called **alpha particles,** contain two protons and two neutrons and have a double positive charge. In nuclear equations, an alpha particle is written ${}^{4}_{2}\text{He}$ or α. The electric charge is generally omitted.

The radioisotope uranium-238 releases alpha radiation and is transformed into another radioisotope, thorium-234. **Figure 28.3** illustrates this transformation.

$$\underset{\text{Uranium-238}}{{}^{238}_{92}\text{U}} \xrightarrow[\text{decay}]{\text{Radioactive}} \underset{\text{Thorium-234}}{{}^{234}_{90}\text{Th}} + \underset{\text{Alpha particle}}{{}^{4}_{2}\text{He}} \ (\alpha \text{ emission})$$

Note that this nuclear equation is balanced. The sum of the mass numbers (superscripts) on the right equals the sum on the left. How does the sum of the atomic numbers (subscripts) on the right compare with the sum on the left? ❶

When an atom loses an alpha particle, the atomic number of the product atom is lower by two and its mass number is lower by four. Because of their large mass and charge, alpha particles do not tend to travel very far and are not very penetrating. They are easily stopped by a sheet of paper or by the surface of your skin. However, radioisotopes that emit alpha particles are dangerous when ingested. The particles do not have to travel far to penetrate soft tissue and cause damage.

Beta radiation consists of fast-moving electrons formed by the decomposition of a neutron in an atom. The neutron decomposes into a proton, which remains in the nucleus, and an electron, which is released.

$$\underset{\text{Neutron}}{{}^{1}_{0}\text{n}} \longrightarrow \underset{\text{Proton}}{{}^{1}_{1}\text{H}} + \underset{\text{Electron (beta particle)}}{{}^{0}_{-1}\text{e}}$$

The fast-moving electrons released by a nucleus are called **beta particles.** The fact that they are negatively charged is reflected in the subscript -1 where atomic number is generally written. Their virtual lack of mass is represented by the superscript 0, corresponding to mass number.

Carbon-14 emits a beta particle as it undergoes radioactive decay to form nitrogen-14. **Figure 28.4** shows this beta emission.

$$\underset{\text{Carbon-14 (radioactive)}}{{}^{14}_{6}\text{C}} \longrightarrow \underset{\text{Nitrogen-14 (stable)}}{{}^{14}_{7}\text{N}} + \underset{\text{Beta particle}}{{}^{0}_{-1}\text{e}} \ (\beta \text{ emission})$$

The mass number of the nitrogen-14 atom produced is the same as that of carbon-14. Its atomic number has increased by 1. The nucleus now contains an additional proton and one fewer neutron. Note that the nuclear equation is balanced. The sum of the atomic numbers on the right $(7 + (-1))$ equals the sum of the atomic numbers on the left (6). The mass numbers are also balanced (14 on each side). Beta particles have less charge and much less mass than alpha particles. Consequently, they are more penetrating. Beta particles are stopped by aluminum foil or thin pieces of wood.

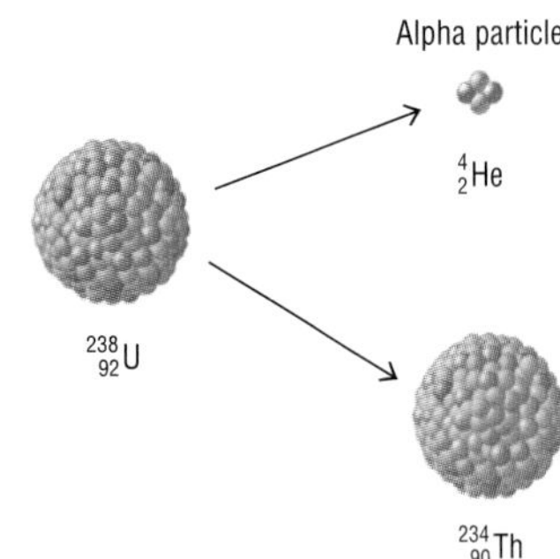

Figure 28.3
Uranium-238 is shown undergoing alpha decay to form thorium-234. What particle is emitted during this decay process? ❷

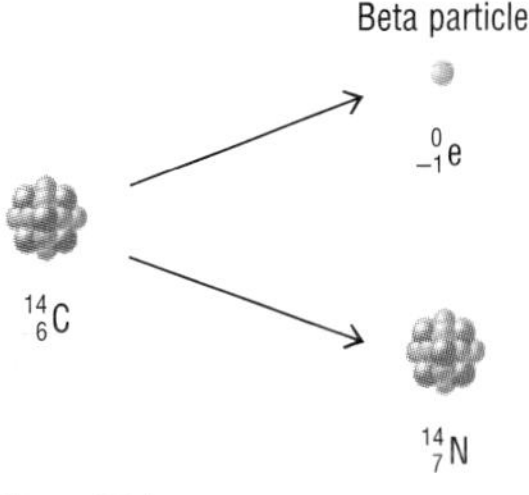

Figure 28.4
Carbon-14 is shown undergoing beta decay to form nitrogen-14. What particle is emitted during this decay process? ❸

28.1

Discuss

Explain that, nuclear decay, unlike chemical reactions, is unaffected by temperature, pressure, or the presence of catalysts. Spontaneous nuclear decay cannot be slowed down, speeded up, or stopped. Nuclear decay illustrates the interconnected nature of matter and energy; matter is related to energy according to the equation $E = mc^2$, where c is a constant equal to the speed of light.

Discuss

Explain that radioactivity reflects the tendency of atomic nuclei to achieve stability. Ask:

- **What makes a nucleus unstable?** (too many or too few neutrons relative to the number of protons)

Note that the energy produced during nuclear decay can be released as electromagnetic radiation (e.g., gamma rays) or as kinetic energy, which is carried by particles emitted from the nucleus (e.g., alpha particles).

Think Critically

Present students with the partial equation for the decay of chlorine-32 into sulfur-32. Ask:

- **Which kind of particle would need to be emitted to balance the equation?** (a positron, a positively charged particle with the same mass as an electron)

Explain that during positron emission, a proton is converted into a neutron. Ask:

- **What kind of unstable nuclei would be likely to emit positrons?** (one with too many protons for the number of neutrons)

Answers

❶ They are equal.
❷ alpha particle
❸ beta particle (electron)

Discuss

Point out that radio waves, microwaves, visible light, ultraviolet light, and X-rays are all forms of electromagnetic radiation. However, they are not directly produced by the radioactive decay of a nucleus.

Discuss

Have students reexamine the photograph on page 840 after reading the Link to Physics on this page. Ask:

- **What is smallest particle of matter? In other words, "What is the fundamental nature of matter?"** (This intriguing question is currently unanswerable, and remains one of the greatest challenges in nuclear physics.)

3 Assess

Evaluate Understanding

Write several partial equations for nuclear decay involving alpha and beta particles. Have students compare the parent and daughter nuclei to identify the type of particle emitted. Ask students to justify their conclusions. Then ask what other type of emission is possible during nuclear decay. (emission of gamma radiation)

Reteach

Remind students that when writing a nuclear equation, the sums of the mass numbers and atomic numbers of the reactants must equal the sums of the mass and atomic numbers of the products. Students should check their equations by comparing the sums of the superscripts and subscripts on each side of the equation.

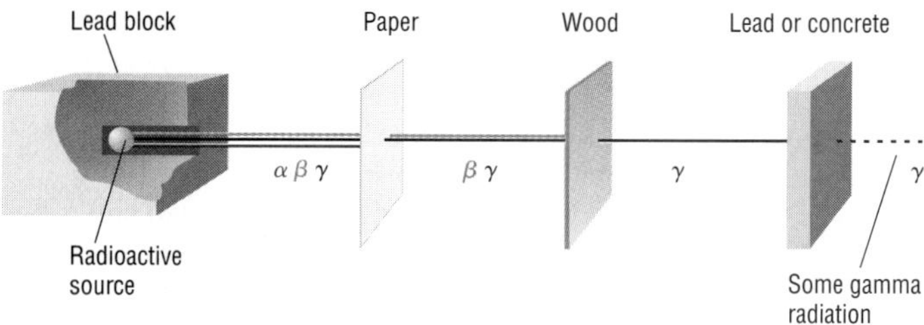

Figure 28.5
Because of their large mass and charge, alpha particles (red) have the least penetrating power of the three main types of radiation. Gamma rays (black) have no mass or charge and are the most penetrating.

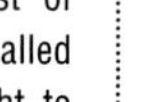
LINK TO PHYSICS

Particle Accelerators
Since the 1930s, physicists have used particle accelerators to probe the mysteries of the atom. In a particle accelerator, a beam of subatomic particles traveling near the speed of light collides with a nucleus, smashing it into fragments. Using a detection device called a bubble chamber, the path that these fragments take may be traced. Researchers have discovered evidence that suggests that protons and neutrons consist of even smaller particles called quarks. Quarks are thought to be held together by other particles, called gluons.

Gamma radiation is high-energy electromagnetic radiation given off by a radioisotope. Visible light, or the light you see, is also electromagnetic radiation, but of much lower energy. Gamma rays are often emitted along with alpha or beta radiation by the nuclei of disintegrating radioactive atoms. The following examples demonstrate this process.

$$^{230}_{90}\text{Th} \longrightarrow {}^{226}_{88}\text{Ra} + {}^{4}_{2}\text{He} + \gamma$$

Thorium-230 → Radon-226 + Alpha particle + Gamma ray

$$^{234}_{90}\text{Th} \longrightarrow {}^{234}_{91}\text{Pa} + {}^{0}_{-1}\text{e} + \gamma$$

Thorium-234 → Protactinium-234 + Beta particle + Gamma ray

Gamma rays have no mass and no electrical charge. Thus the emission of gamma radiation in itself does not alter the atomic number or mass number of an atom.

X-radiation, or x-rays, behave essentially the same as gamma rays, but their origin is different. X-rays are not emitted during radioactive decay. They are produced as excited electrons in certain metals lose their energy. X-rays are extremely penetrating and potentially very dangerous. Both gamma rays and x-rays pass easily through paper, wood, and the human body. They can be stopped, although not completely, by several meters of concrete or several centimeters of lead, as shown in **Figure 28.5.** Why does ❶ your dentist place a heavy apron over your torso before taking an x-ray?

section review 28.1

1. Explain what is meant by radioactivity and radioactive decay.
2. Distinguish among alpha, beta, and gamma radiation on the basis of the following.
   a. mass **b.** charge **c.** penetrating power
3. What part of an atom undergoes change during radioactive decay?

 Chem ASAP! **Assessment 28.1** Check your understanding of the important ideas and concepts in Section 28.1.

Answers

❶ The lead apron acts as a shield against the penetrating X-rays.

SECTION REVIEW 28.1

1. Radioactivity is the process by which an atomic nucleus gives off radiation. Radioactive decay is the process in which an unstable nucleus disintegrates.
2.

	Alpha	Beta	Gamma
a. Mass	4 amu	1/1837 amu	0
b. Charge	2+	1−	0
c. Penetrating power	low	moderate	high

3. The nucleus undergoes change.

NUCLEAR TRANSFORMATIONS

Although weather stripping and insulation help conserve energy, thus lowering heating and cooling bills, such efforts can prevent an ample exchange of fresh air. As a result, radioactive substances such as radon gas can accumulate and pose a health risk. Radon-222 is a radioactive isotope that is present naturally in the soil in some areas. It has a constant rate of decay. **How is the decay rate of a radioactive substance expressed?**

objectives

- Use half-life information to determine the amount of a radioisotope remaining at a given time
- Give examples of equations for the synthesis of transuranium elements by transmutation

key terms

- band of stability
- positron
- half-life
- transmutation
- transuranium elements

Nuclear Stability and Decay

About 1500 different nuclei are known. Of those, only 264 are stable and do not decay or change with time. As mentioned earlier, the stability of a nucleus depends on its neutron-to-proton ratio. For elements of low atomic number (below about 20), the ratio for stability is about 1. That means the stable nuclei have roughly equal numbers of neutrons and protons. For example, the isotopes ${}^{12}_{6}C$, ${}^{14}_{7}N$, and ${}^{16}_{8}O$ are stable. Above atomic number 20, stable nuclei have more neutrons than protons. The neutron-to-proton ratio reaches about 1.5 for heavy elements. The lead isotope ${}^{206}_{82}Pb$, for example, with 124 neutrons and 82 protons, is stable. Its ratio is $\frac{124}{82} \approx 1.5$.

Figure 28.6 shows a plot of the number of neutrons versus the number of protons for all the known stable nuclei. The stable nuclei on a neutron-versus-proton plot are located in a region called the **band of stability.** Unstable nuclei undergo spontaneous radioactive decay. The type of decay that occurs depends on the neutron-to-proton ratio of the unstable nucleus.

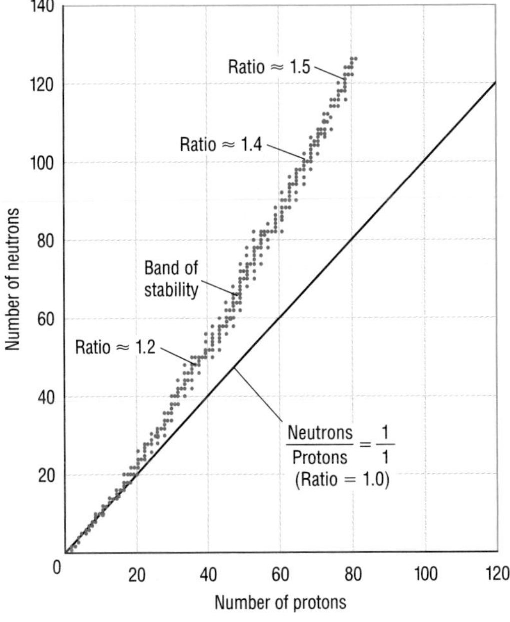

Figure 28.6
A neutron-versus-proton plot of all the known stable nuclei forms a pattern called the band of stability (shown in red). For isotopes of low atomic number, the stability ratio is about 1:1; for the heavier isotopes, the ratio increases to about 1.5:1.

28.2

1 Engage

Use the Visual

Have students study the photograph and read the text that opens the section. Ask students to locate radon on a periodic table and to write the chemical symbol for radon-222 (${}^{222}_{86}Rn$) Point out that radon is a relatively unreactive noble gas. Thus it is able to percolate up through the ground without being trapped chemically. Radon that is inhaled can deposit solid polonium in the lungs. Write the partial nuclear equation for the decomposition of radon.

${}^{222}_{86}Rn \rightarrow {}^{218}_{84}Po$

Ask:

- **Which type of radiation does radon emit?** (alpha particles; Polonium also undergoes alpha decay. Ask students to write the nuclear equation for the decay of polonium.)
- **How do scientists express the decay rates of radioactive isotopes?** (as a half-life—the time it takes for half of the nuclei in the sample to decay)

Motivate and Relate

Point out that nearly 85% of all known nuclei are unstable. (This does not mean that 85% of all matter is unstable; there isn't an even distribution of stable and unstable isotopes.) **Because neutrons help bind protons together in the nucleus, the ratio of protons to neutrons is a major factor in determining stability.**

STUDENT RESOURCES

- Section Review 28.2, Ch. 28 Practice Problems, Interpreting Graphics, and Quizzes from the Review Module (Ch. 25–28)
- Laboratory Recordsheet 28-1
- Small-Scale Chemistry Lab Manual: Experiment 40
- Laboratory Practical 28-2

TECHNOLOGY RESOURCES

Relevant technology resources include:

- Chem ASAP! CD-ROM
- Resource Pro CD-ROM

2 Teach

Discuss

For each element, there exists only a small range of neutron-to-proton ratios that produce stable nuclei. If a nucleus does not reflect a stable ratio, it spontaneously decays until a stable ratio of neutrons to protons results.

Discuss

Explain that the nuclear stability that results from a proper ratio of neutrons to protons in an atom is like the structural stability that results from a proper ratio of mortar to bricks in a building. In the building, gravitational and adhesive forces are balanced; in the nucleus, forces of repulsion and attraction are balanced.

CHEMISTRY AND SCIENCE HISTORY

Have students read the Link to Archaeology. Explain that the technique of carbon dating was developed in the 1940s by U.S. chemist Willard Libby. He received the 1960 Nobel Prize in Chemistry for his work. Current methods for carbon-14 dating involve ionizing the sample and passing the ions through a mass spectroscope.

A nucleus may be unstable for several reasons. Some nuclei have too many neutrons relative to the number of protons. These nuclei decay by turning a neutron into a proton and by emitting a beta particle (an electron) from the nucleus.

$$^{1}_{0}n \longrightarrow {}^{1}_{1}H + {}^{0}_{-1}e$$

This process is known as beta decay or beta emission. It produces a simultaneous increase in the number of protons and a decrease in the number of neutrons. Here are the nuclear equations for several isotopes that undergo beta emission.

$$^{66}_{29}Cu \longrightarrow {}^{66}_{30}Zn + {}^{0}_{-1}e$$

$$^{14}_{6}C \longrightarrow {}^{14}_{7}N + {}^{0}_{-1}e$$

Other nuclei are unstable because they have too few neutrons relative to the number of protons. These nuclei increase their stability by converting a proton to a neutron. As seen in the following examples, an electron is captured by a nucleus during this process.

$$^{59}_{28}Ni + {}^{0}_{-1}e \longrightarrow {}^{59}_{27}Co$$

$$^{37}_{18}Ar + {}^{0}_{-1}e \longrightarrow {}^{37}_{17}Cl$$

A **positron** is a particle with the mass of an electron but a positive charge. Its symbol is ${}^{0}_{+1}e$. A positron may be emitted as a proton changes to a neutron, as in the following cases.

$$^{8}_{5}B \longrightarrow {}^{8}_{4}Be + {}^{0}_{+1}e$$

$$^{15}_{8}O \longrightarrow {}^{15}_{7}N + {}^{0}_{+1}e$$

When a proton is converted to a neutron, the atomic number decreases by 1 and the number of neutrons increases by 1. How does this change affect
1 the mass number?

All nuclei with atomic number greater than 83 are radioactive. These nuclei lie in the upper end of the band of stability, and are especially heavy. They have both too many neutrons and too many protons to be stable. Therefore they undergo decay. Most of them emit alpha particles. Alpha emission results in an increase in the neutron-to-proton ratio, which tends to increase the stability of the nucleus.

$$^{204}_{82}Pb \longrightarrow {}^{200}_{80}Hg + {}^{4}_{2}He$$

$$^{226}_{88}Ra \longrightarrow {}^{222}_{86}Rn + {}^{4}_{2}He$$

$$^{232}_{90}Th \longrightarrow {}^{228}_{88}Ra + {}^{4}_{2}He$$

In alpha emission, the mass number decreases by four and the atomic number decreases by two. If the masses of the reactants and products of a nuclear reaction could be determined with a sufficiently sensitive balance, you would find that mass is not conserved! In fact, an infinitesimally small quantity of mass is lost. The lost mass is converted to energy associated with radiation.

Answer

1 The mass number is unchanged.

Meeting Diverse Needs

Gifted After students examine the graph in Figure 28.6, have them consider the role neutrons play in stabilizing the nuclei of atoms. Encourage students to do research on the *nuclear force,* the strong force of attraction that holds nucleons together in a stable nucleus.

Initial amount of radioisotope

After 1 half-life

After 2 half-lives

After 3 half-lives

Radioisotope remaining (%): 100, 50, 25, 12.5, 0

$t_{1/2}$ $t_{1/2}$ $t_{1/2}$

Number of half-lives: 1, 2, 3, 4

Figure 28.7
This decay curve for a radioactive element shows that by the end of each half-life, half of the remaining original radioactive atoms have decayed into atoms of another element.

Half-Life

Every radioisotope has a characteristic rate of decay measured by its half-life. A **half-life** ($t_{1/2}$) is the time required for one-half of the nuclei of a radioisotope sample to decay to products. After one half-life, half of the original radioactive atoms have decayed into atoms of a new element. The other half are still unchanged at that point, as demonstrated in **Figure 28.7.** After a second half-life, only one-quarter of the original radioactive atoms remain.

Half-lives may be as short as a fraction of a second or as long as billions of years. **Table 28.2** shows the half-lives of some naturally occurring radioisotopes. Scientists use the half-lives of some naturally occurring radioisotopes to determine the age of ancient artifacts. Many artificially produced radioisotopes have very short half-lives, a feature that is a great advantage in nuclear medicine. The rapidly decaying isotopes do not pose long-term biological radiation hazards to the patient.

Table 28.2

Half-Lives and Radiation of Some Naturally Occurring Radioisotopes

Isotope	Half-life	Radiation emitted
Carbon-14	5.73×10^3 years	β
Potassium-40	1.25×10^9 years	β, γ
Radon-222	3.8 days	α
Radium-226	1.6×10^3 years	α, γ
Thorium-230	7.54×10^4 years	α, γ
Thorium-234	24.1 days	β, γ
Uranium-235	7.0×10^8 years	α, γ
Uranium-238	4.46×10^9 years	α

ARCHAEOLOGY

Carbon Dating

Carbon-14, which has a half-life of 5730 years, is used extensively to date artifacts that are made of organic material. All living organisms contain carbon-12

and carbon-14 in a fixed ratio. After an organism dies, however, the ratio of carbon-12 to carbon-14 changes as the carbon-14 decays to nitrogen-14. An object that was made of plant fibers might now have a carbon-14 concentration of one-half its original value. This object must therefore be about 5730 years old (one half-life). Carbon-14 dating can be used to determine the ages of objects between 200 and 50 000 years old. The Dead Sea Scrolls, for example, were determined to be 1940 ± 70 years old.

Use the Visual

Refer students to Table 28.2 and have them note the range of half-life values. Remind them that these values do not indicate how long a given atom of an isotope will exist, but only how long it takes for half of the atoms in a sample to undergo radioactive decay.

TEACHER DEMO

Place 16 pennies, heads up, together on a table. Tell students that each represents an atom of a radioactive isotope. At the end of a period that you define as one half-life, turn half the coins over signifying the decay of some radioactive atoms to nonradioactive ones. Then, after the same amount of time passes again, turn over half the remaining coins. The fact that some coins still remain heads up demonstrates that atoms of the original isotope will still exist after a period of two half-lives. If you wish, continue the process for two more half-lives.

Represent this process algebraically by writing $16 \times 1/2 \times 1/2 \times 1/2 \times 1/2 \ldots$ Point out that, theoretically, in terms of actual atomic nuclei, there will always be some atoms of radioisotope remaining. However, the amount may be too small to detect.

MEETING DIVERSE NEEDS

At Risk To reinforce the concept of half-life, have students consider the following problem: Imagine winning a $1000 prize but the conditions of the award require that half of the remainder of the prize is spent each month. After how many months would you be left with less than $1? What is the half-life of this prize? (9 months; 1 month) As an extension, have students plot a decay curve, like the one in Figure 28.7, except the *x*-axis should represent time (days or years), for one of the elements in Table 28.2. Tell students that the initial amount of radioisotope is 100 g.

CHEMath

Teaching Tips

Review exponential notation and rules of exponents as they apply to logarithms. Write numbers in longhand form and have students write the number in scientific notation. For example write:
$3.5 \times 10 \times 10 \times 10 \times 10$
or 35 000 and have students write the number as 3.5×10^4.
Have students simplify powers and roots of numbers and variables such as $(2a)^3$, $(1/2)^5$, and $\sqrt{(8a)^2}$. Review the rules for manipulating logs such as $\log xy = \log x + \log y$ and $\log x^a = a\log x$. Assign practice problems that require students to apply these rules.

Answers to Practice Problems

A. $A = 64\left(\frac{1}{2}\right)^{t/18}$

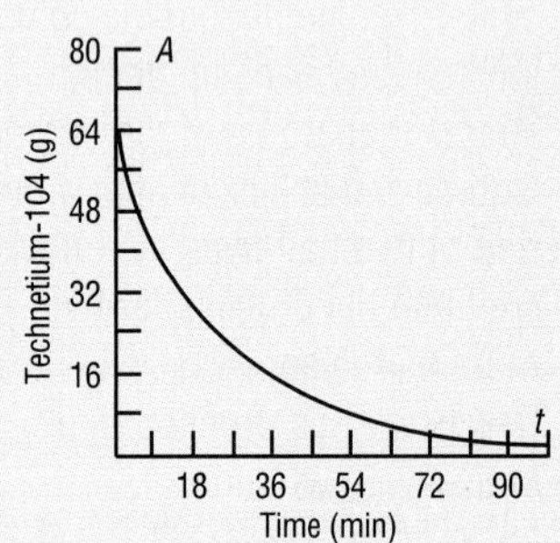

B. $A = 248\left(\frac{1}{2}\right)^{t/29}$; 72 g

C. 5.25 years; 17.4 g

CHEMath

CALCULATING HALF-LIFE

This CHEMath feature provides more detailed information about half-life. With this information, you will have the skills needed to write the exponential decay function for a radioactive element. The table shows the amount of radioactive isotope remaining after 0, 1, and 2 half-lives if the initial amount is A_0.

Do you see the pattern? After n half-lives, the amount remaining is $A = A_0 \times \left(\frac{1}{2}\right)^n$. Since n is the time elapsed divided by the half-life, this equation can be written as $A = A_0 \times \left(\frac{1}{2}\right)^{t/T}$, where t is time and T is the half-life (using the same units). This type of equation is known as an exponential decay function.

Number of Half-Lives	Amount Remaining	Exponential Form
0	A_0	$A_0 \times \left(\frac{1}{2}\right)^0$
1	$A_0 \times \frac{1}{2}$	$A_0 \times \left(\frac{1}{2}\right)^1$
2	$A_0 \times \frac{1}{2} \times \frac{1}{2}$	$A_0 \times \left(\frac{1}{2}\right)^2$

Logarithms can be used to algebraically calculate half-lives. You will use the formula $\log (a^b) = b \times \log a$, and you will want to use a calculator. See Example 2.

Example 1

Iodine-131 has a half-life of 8 days. Write and graph an equation for the amount (A) of iodine-131 after t days if the sample initially contains 80 g iodine-131.
Using $A = A_0 \times \left(\frac{1}{2}\right)^{t/T}$, the equation is $A = 80 \times \left(\frac{1}{2}\right)^{t/8}$.

The graph includes the points listed below.

t (days)	0	8	16	24	32
A (grams)	80	40	20	10	5

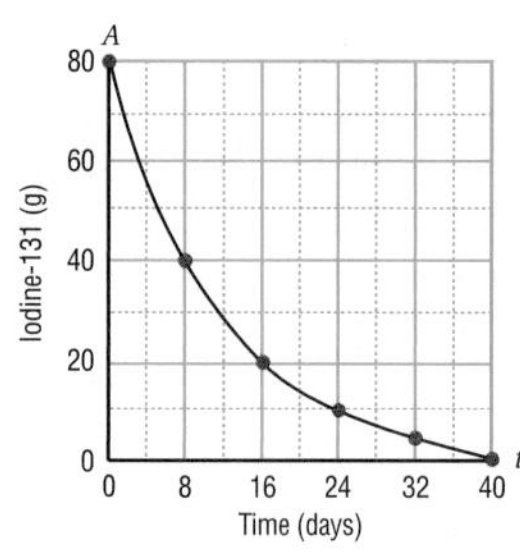

Example 2

A sample initially contains 100.0 g of thorium-234. After 16.0 days, the sample contains 63.1 g of thorium-234. Calculate the half-life.

Use the formula $A = A_0 \times \left(\frac{1}{2}\right)^{t/T}$.

$$63.1 \text{ g} = (100 \text{ g}) \times \left(\frac{1}{2}\right)^{(16.0 \text{ days})/T}$$

$$\frac{63.1}{100} = \left(\frac{1}{2}\right)^{(16.0 \text{ days})/T}$$

$$0.631 = 0.5^{(16.0 \text{ days})/T}$$

$$\log 0.631 = \log \left(0.5^{(16.0 \text{ days})/T}\right)$$

$$\log 0.631 = \frac{(16.0 \text{ days})}{T} \times \log 0.5$$

$$T = (16.0 \text{ days}) \times \frac{\log 0.5}{\log 0.631} = 24.1 \text{ days}$$

The half-life of thorium-234 is 24.1 days.

Practice Problems

Practice writing and using exponential decay functions by solving each of the following.

A. Technetium-104 has a half-life of 18 minutes. Write and graph an equation for the amount A of technetium-104 after t minutes if the sample initially contains 64 g of technetium-104.

B. A sample initially contains 248 g of strontium-90, which has a half-life of 29 years. Write an equation for the amount A of strontium-90 after t years. What is the amount of strontium-90 at $t = 52$ years?

C. A sample initially contains 50.0 g of cobalt-60. After 2.00 years, the sample contains 38.4 g of cobalt-60. Calculate the half-life of cobalt-60. What is the amount of cobalt-60 at $t = 8.00$ years?

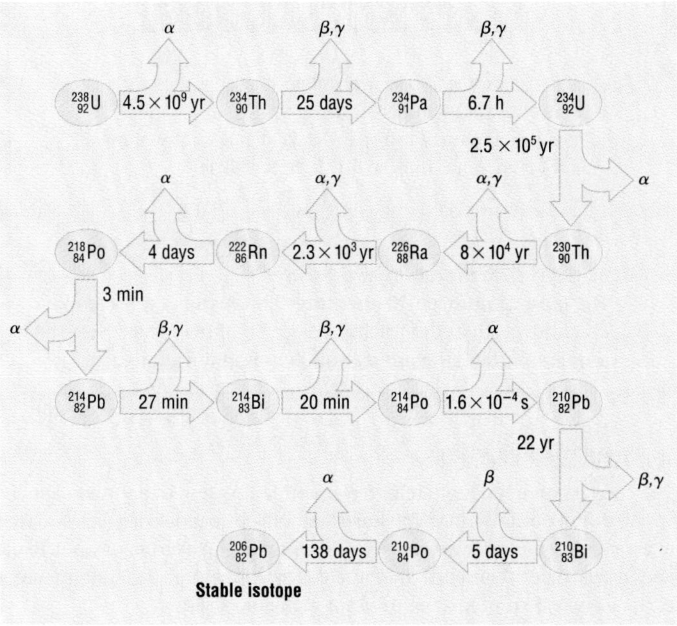

$^{238}_{92}U$ 4.5 × 10^9 yr (α) $^{234}_{90}Th$ 25 days (β,γ) $^{234}_{91}Pa$ 6.7 h (β,γ) $^{234}_{92}U$

2.5 × 10^5 yr (α)

$^{230}_{90}Th$ 8 × 10^4 yr (α,γ) $^{226}_{88}Ra$ 2.3 × 10^3 yr (α,γ) $^{222}_{86}Rn$ 4 days (α) $^{218}_{84}Po$

3 min (α)

$^{214}_{82}Pb$ 27 min (β,γ) $^{214}_{83}Bi$ 20 min (β,γ) $^{214}_{84}Po$ 1.6 × 10^{-4} s (α) $^{210}_{82}Pb$

22 yr (β,γ)

$^{210}_{83}Bi$ 5 days (β) $^{210}_{84}Po$ 138 days (α) $^{206}_{82}Pb$

Stable isotope

Figure 28.8
Uranium-238 decays through a complex series of radioactive intermediates, including radon gas (Rn). What is the stable end-product of this series? ❶

Chem ASAP!
Simulation 30
Simulate the decay of several radioisotopes.

One isotope that has a long half-life is uranium-238, which decays through a complex series of radioactive intermediates to the stable isotope lead-206. **Figure 28.8** illustrates this process. The age of uranium-containing minerals can be estimated by measuring the ratio of uranium-238 to lead-206. Because the half-life of uranium-238 is 4.5 × 10^9 years, it is possible to use this method to date rocks nearly as old as the solar system.

Sample Problem 28-1

Nitrogen-13 emits beta radiation and decays to carbon-13 with a half-life ($t_{1/2}$) of 10 min. Assume a starting mass of 2.00 g of nitrogen-13.

a. How long is three half-lives?
b. How many grams of the isotope will still be present at the end of three half-lives?

1. ***ANALYZE*** ***List the knowns and the unknowns.***

Knowns:	*Unknowns:*
• ($t_{1/2}$) = 10 min	• 3 half-lives = ? min
• initial mass = 2.00 g	• mass after 3 half-lives = ? g
• number of half-lives = 3	

Calculate the time required for three half-lives by multiplying the length of each half-life by the number of half-lives (3). Determine the mass remaining by multiplying the original mass by $\frac{1}{2}$ for each half-life.

Practice Problems

4. Manganese-56 is a beta emitter with a half-life of 2.6 h. What is the mass of manganese-56 in a 1.0-mg sample of the isotope at the end of 10.4 h?

28.2

Discuss

Emphasize that the rate of disintegration of the nuclei of an isotope is constant. The rate is unaffected by factors such as the number of nuclei present, heat, pressure, or chemical reactions. Note that the half-life for a particular radioactive isotope is unique to that isotope. No two isotopes have exactly the same half-life.

Think Critically

Have students compare and contrast the change in radioactivity levels over time for one mole samples of the isotopes potassium-40 and thorium-234.

Use the Visual

Have students study Figure 28.8, which charts the radioactive decay of uranium-238. Explain that many radioisotopes go through a complex series of nuclear reactions before a stable product is formed.

Practice Problems Plus

Related Chapter Review Problems
Chapter Review problems 36 and 49 are related to Sample Problem 28-1.

Additional Practice Problem
The Chem ASAP! CD-ROM contains the following problem: Strontium-90 is a beta emitter with a half-life of 29 years. What is the mass of strontium-90 in a 5.0-g sample of the isotope at the end of 87 years? (0.63 g)

Answer

❶ lead-206

Discuss

Explain that one of the main goals of medieval alchemists was the conversion of common metals to precious metals. No chemical reaction can achieve their goal. However, through transmutations modern chemists can change one element into another. Point out that transmutation reactions also allow chemists to produce elements that do not occur naturally, such as the transuranium elements with atomic numbers 93 through 112.

ACTIVITY

Write equations for a number of transmutation reactions on the board. Point out that the atomic masses and atomic numbers for reactants and products in transmutation reactions are balanced just as in nuclear decay reactions. Write partial equations such as

$^{239}_{94}Pu + ? \rightarrow ^{242}_{96}? + ^{1}_{0}n$

$^{238}_{92}U + ^{?}_{?}C \rightarrow ^{246}_{98}Cf + 4^{1}_{0}n$

on the chalkboard. Then have students supply the missing information.

Cooperative Learning

Particle accelerators such as linear accelerators, cyclotrons, and synchrotrons are used in transmutation experiments. Divide the class into groups of four to five students. Have each group choose one type of accelerator and conduct library research on its design and function. The group should create a model of the accelerator, which they can display and explain in class.

Practice Problems (cont.)

5. A sample of thorium-234 has a half-life of 25 days. Will all the thorium undergo radioactive decay in 50 days? Explain.

Chem ASAP!

Problem-Solving 4

Solve Problem 4 with the help of an interactive guided tutorial.

Sample Problem 28-1 (cont.)

2. *CALCULATE* ***Solve for the unknowns.***
   a. $3 \times 10\ \text{min} = 30\ \text{min}$
   b. The initial mass of nitrogen-13, 2.00 g, is cut by one-half for each half-life, so for three half-lives

$$2.00\ \text{g} \times \frac{1}{2} \times \frac{1}{2} \times \frac{1}{2} = 0.250\ \text{g}$$

3. *EVALUATE* ***Do the results make sense?***
   The mass of nitrogen-13 after three half-lives is expected to be much lower than the original mass. The final answer has the proper units and the proper number of significant figures.

Transmutation Reactions

The conversion of an atom of one element to an atom of another element is called **transmutation.** As you have learned, radioactive decay is one way in which transmutation occurs. A transmutation can also occur when high-energy particles bombard the nucleus of an atom. The high-energy particles may be protons, neutrons, or alpha particles.

Many transmutations occur in nature. The production of carbon-14 from naturally occurring nitrogen-14, for example, takes place in the upper atmosphere. Another naturally occurring isotope, uranium-238, undergoes 14 transmutations before reaching a stable isotope, as was shown in **Figure 28.8.** Many other transmutations are done in laboratories or in nuclear reactors. The earliest artificial transmutation was performed in 1919 by Ernest Rutherford (1871–1937). He bombarded nitrogen gas with alpha particles to produce an unstable isotope of fluorine. The results of this reaction are shown in **Figure 28.9.** The reaction is

$$\underset{\text{Nitrogen-14}}{^{14}_{7}N} + \underset{\text{Alpha particle}}{^{4}_{2}He} \longrightarrow \underset{\text{Fluorine-18}}{^{18}_{9}F}$$

The fluorine isotope quickly decomposes to a stable isotope of oxygen and a proton.

$$\underset{\text{Fluorine-18}}{^{18}_{9}F} \longrightarrow \underset{\text{Oxygen-17}}{^{17}_{8}O} + \underset{\text{Proton}}{^{1}_{1}H}$$

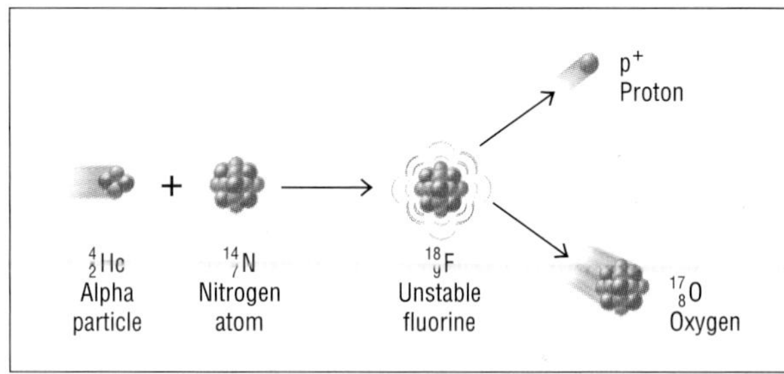

Figure 28.9
In 1919, Ernest Rutherford carried out the first artificial transmutation when he bombarded nitrogen gas with alpha particles. What particles were formed? ❶

Answers

❶ ultimately, oxygen-17 atoms and protons

❷ beta particle

Rutherford's experiment eventually helped lead to the discovery of the proton. James Chadwick's discovery of the neutron in 1932 also involved a transmutation experiment. Neutrons were produced when beryllium-9 was bombarded with alpha particles.

$$ {}^{9}_{4}Be \text{ (Beryllium-9)} + {}^{4}_{2}He \text{ (Alpha particle)} \longrightarrow {}^{12}_{6}C \text{ (Carbon-12)} + {}^{1}_{0}n \text{ (Neutron)} $$

The elements in the periodic table with atomic numbers above 92, called the **transuranium elements,** all undergo transmutation. None of them occurs in nature, and all of them are radioactive. These elements have been synthesized in nuclear reactors and nuclear accelerators, one of which is shown in **Figure 28.10.** Such devices accelerate bombarding particles to very high speeds. When uranium-238 is bombarded with relatively slow neutrons from a nuclear reactor, some uranium nuclei capture neutrons to produce uranium-239.

$$ {}^{238}_{92}U + {}^{1}_{0}n \longrightarrow {}^{239}_{92}U $$

Uranium-239 is radioactive and emits a beta particle. The product is an isotope of the artificial radioactive element neptunium (atomic number 93).

$$ {}^{239}_{92}U \longrightarrow {}^{239}_{93}Np + {}^{0}_{-1}e $$

Neptunium is unstable and decays to produce a second artificial element, plutonium (atomic number 94). What particle is emitted in this reaction? 2

$$ {}^{239}_{93}Np \longrightarrow {}^{239}_{94}Pu + {}^{0}_{-1}e $$

Plutonium and neptunium are both transuranium elements and do not occur naturally. These first artificial elements were synthesized in 1940 by scientists in Berkeley, California. Since that time, more than 20 additional transuranium elements have been synthesized.

Figure 28.10
The Stanford Linear Accelerator Center is operated by Stanford University in California. This research facility houses a linear accelerator that is two miles long.

section review 28.2

6. How is half-life used to determine the amount of a radioisotope remaining at a given time?

7. Give two examples of equations for the synthesis of transuranium elements by transmutation.

8. Complete and balance the equations for the following nuclear reactions.

a. ${}^{27}_{13}Al + {}^{4}_{2}He \longrightarrow {}^{30}_{14}Si + ?$

b. ${}^{214}_{83}Bi \longrightarrow {}^{4}_{2}He + ?$

c. ${}^{27}_{14}Si \longrightarrow {}^{0}_{-1}e + ?$

d. ${}^{66}_{29}Cu \longrightarrow {}^{66}_{30}Zn + ?$

9. Explain the process of transmutation. Write at least three nuclear equations to illustrate your answer.

10. The mass of cobalt-60 in a sample is found to have decreased from 0.800 g to 0.200 g in a period of 10.5 years. From this information, calculate the half-life of cobalt-60.

Chem ASAP! **Assessment 28.2** Check your understanding of the important ideas and concepts in Section 28.2.

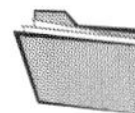

portfolio project

Research the methods used by archeologists to date materials such as pottery, coral, and stone. Prepare a poster display on the radioisotopes used, their half-lives, and their limitations.

28.2

3 Assess

Evaluate Understanding

Ask students how the ratio of neutrons to protons changes in nuclei that undergo beta particle, positron, and alpha particle emission. Ask students to explain the concept of half-life using a numerical example. (For example: original state = 100 atoms; one half-life = 50 atoms; two half-lives = 25 atoms, and so on.)
Write several balanced and unbalanced nuclear equations on the board and have students identify and correct the unbalanced equations.

Reteach

Point out the three natural processes that can result in a nucleus attaining a stable ratio of neutrons to protons. A beta emission increases the number of protons and decreases the number of neutrons. A positron emission decreases the number of protons and increases the number of neutrons. In heavy elements, the nucleus may emit an alpha particle, which decreases the numbers of both neutrons and protons.

Portfolio Project

C-14 is used to date once-living artifacts such as bones. Its range is limited to 50 000 years by its relatively short half-life. The oldest rocks are dated with U-238. Its half-life is about 4.5 billion years. The decay product is Pb-207. K-40 has a half-life of 1.3×10^9 years. It decays to Ar-40. Items made of clay, such as pottery and hearths, can be dated using a property called thermoluminescence.

Answers

SECTION REVIEW 28.2

6. A half-life is the time required for one-half of the nuclei of a radioisotope to decay to products. Use x to represent the number of half-lives. Then $1/2^x$ will represent the fraction of original isotope that remains.

7. Answers will vary.

8. a. ${}^{27}_{13}Al + {}^{4}_{2}He \rightarrow {}^{30}_{14}Si + {}^{1}_{1}H$

b. ${}^{214}_{83}Bi \rightarrow {}^{4}_{2}He + {}^{210}_{81}Tl$

c. ${}^{27}_{14}Si \rightarrow {}^{0}_{-1}e + {}^{27}_{15}P$

d. ${}^{66}_{29}Cu \rightarrow {}^{66}_{30}Zn + {}^{0}_{-1}e$

9. Transmutation is the conversion of an atom of one element to an atom of another element. Examples are

${}^{238}_{92}U \rightarrow {}^{234}_{90}Th + {}^{4}_{2}He$ (alpha emission)
${}^{14}_{7}N + {}^{4}_{2}He \rightarrow {}^{18}_{9}F$ (alpha bombardment)
${}^{239}_{92}Np \rightarrow {}^{239}_{94}Pu + {}^{0}_{-1}e$ (beta emission)

10. 5.25 years

SMALL-SCALE LAB

RADIOACTIVITY AND HALF-LIVES

OVERVIEW

Procedure Time: 20 minutes
Students will flip a coin and plot the disappearance of heads over time. They will graph their data and relate it to half-lives and radioactive decay.

TEACHING SUGGESTIONS

Emphasize that the trials involving flipping a coin simulate radioactivity. The production and removal of a "head" is analogous to the decay of an unstable nuclei. The rate of removal is analogous to the half-life of a radioactive isotope—around 50%. Although the number of heads decreases over time, the percent of heads produced remains steady until the sample becomes too small.

Note that in a sample as small as 100, the likelihood of producing a number of heads other than 50 is high. If the sample size were increased, the relative error would decrease. As the sample size approaches zero, the results are no longer statistically reliable.

Remind students that the probabilities relate to the overall sample, not to any individual coin or atom. Predictions for a mole of atoms, which provides a very large sample, can be quite accurate.

ANALYSIS

1\.

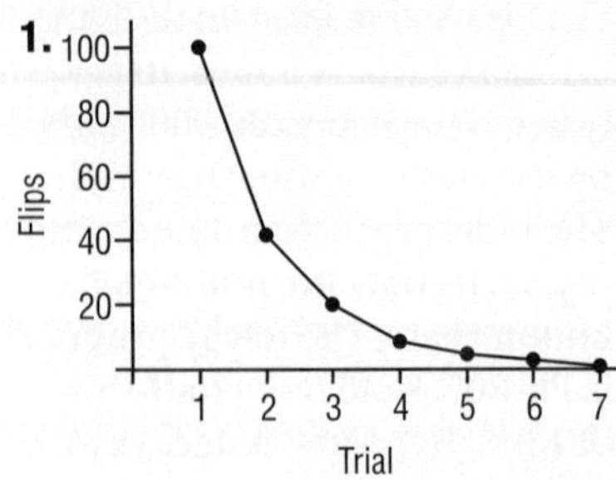

2\. Non-linear; the rate decreases over time.
3\. For each flip the probability of a head is 0.50.
4\. one trial

SMALL-SCALE LAB

RADIOACTIVITY AND HALF-LIVES

SAFETY

Use safe and proper laboratory procedures.

PURPOSE

To simulate the chemical conversion of a reactant over time and to graph the data and relate it to radioactive decay and half-lives.

MATERIALS

- pencil
- paper
- ruler
- graph paper
- penny

PROCEDURE

On a sheet of paper, draw a grid similar to Figure A. Flip a penny 100 times and, in your grid, record the total number of heads that result. Now flip the penny the same number of times as the number of heads that you obtained in the first 100 flips. Record the total number of flips and the number of heads that result. Continue this procedure until you obtain no more heads. Record all your data in Figure A.

Trial #	Number of flips	Number of heads
1	100	42
2	42	20
3	20	9
4	9	5
5	5	3
6	3	1
7	1	0

Figure A

ANALYSIS

Using your experimental data, record the answers to the following questions below your data table.

1. Use graph paper to plot the number of flips (y-axis) versus the trial number (x-axis). Draw a smooth line through the points.
2. Examine your graph. Is the rate of the number of heads produced over time linear or nonlinear? Is the rate constant over time or does it change?
3. Why does each trial reduce the number of heads by approximately one-half?
4. A half-life is the time required for one-half of the atoms of a radioisotope to emit radiation and to decay to products. What value represents one half-life for the process of flipping coins?

YOU'RE THE CHEMIST

The following small-scale activities allow you to develop your own procedures and analyze the results.

1. **Design It!** Design and carry out an experiment using a single die to model radioactive decay. Plot your data.
2. **Analyze It!** Many radioisotopes undergo alpha decay. They emit an alpha particle (helium nucleus ${}^{4}_{2}He$). For example,

$$ {}^{222}_{86}Rn \longrightarrow {}^{218}_{84}Po + {}^{4}_{2}He $$

Write similar balanced nuclear equations for the alpha decay of each of the following.

a. Pa-231 **b.** Am-241 **c.** Ra-226 **d.** Es-252

3. **Analyze It!** Other radioisotopes undergo beta decay, emitting a beta particle (electron ${}^{0}_{-1}e$). For example,

$$ {}^{14}_{6}C \longrightarrow {}^{14}_{7}N + {}^{0}_{-1}e $$

Write similar balanced nuclear equations for the beta decay of each of the following.

a. H-3 **b.** Mg-28 **c.** I-131 **d.** Se-75

YOU'RE THE CHEMIST

1\. Count the total number of even numbers that result in 100 rolls of the die. Roll the die again a number of times equal to the number obtained on the first trial. Do trials until the number of evens equals zero. Plot number of evens vs. trial.

2\. **a.** ${}^{231}_{91}Pa \rightarrow {}^{227}_{89}Ac + {}^{4}_{2}He$

b. ${}^{241}_{95}Am \rightarrow {}^{237}_{93}Np + {}^{4}_{2}He$

c. ${}^{226}_{88}Ra \rightarrow {}^{222}_{86}Rn + {}^{4}_{2}He$

d. ${}^{252}_{99}Es \rightarrow {}^{248}_{97}Bk + {}^{4}_{2}He$

3\. **a.** ${}^{3}_{1}H \rightarrow {}^{3}_{2}He + {}^{0}_{-1}e$

b. ${}^{28}_{12}Mg \rightarrow {}^{28}_{13}Al + {}^{0}_{-1}e$

c. ${}^{131}_{53}I \rightarrow {}^{131}_{54}Xe + {}^{0}_{-1}e$

d. ${}^{75}_{34}Se \rightarrow {}^{75}_{35}Br + {}^{0}_{-1}e$

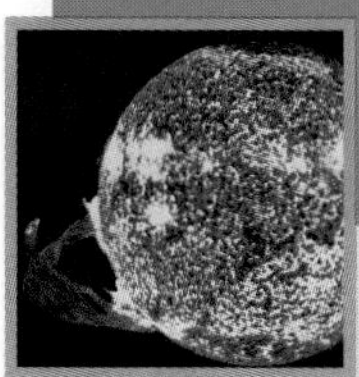

FISSION AND FUSION OF ATOMIC NUCLEI

The sun appears as a fiery ball in the sky—so bright it should never be looked at with unprotected eyes. Although its surface temperature is about 5800 K, the sun is not actually burning. If the energy given off by the sun were the product of a combustion reaction, the sun would have burned out approximately 2000 years after it was formed, long before today. **How is the energy given off by the sun produced?**

Nuclear Fission

When the nuclei of certain isotopes are bombarded with neutrons, they undergo **fission,** the splitting of a nucleus into smaller fragments. Uranium-235 and plutonium-239 are fissionable materials. As shown in **Figure 28.11,** a fissionable atom, such as uranium-235, breaks into two fragments of roughly the same size when struck by a slow-moving neutron. At the same time, more neutrons are released by the fission. These neutrons strike the nuclei of other uranium-235 atoms, continuing the fission by a chain reaction. In a chain reaction, some of the neutrons produced react with other fissionable atoms, producing more neutrons, which react with still more fissionable atoms. This process is similar to the toppling of dominoes shown in **Figure 28.11**.

Nuclear fission can unleash enormous amounts of energy. The fission of 1 kg of uranium-235, for example, releases an amount of energy equal to that generated in the explosion of 20 000 tons of dynamite. In an uncontrolled nuclear chain reaction, the total energy release is nearly instantaneous. The entire reaction takes only fractions of a second. Atomic bombs are devices that start uncontrolled nuclear chain reactions.

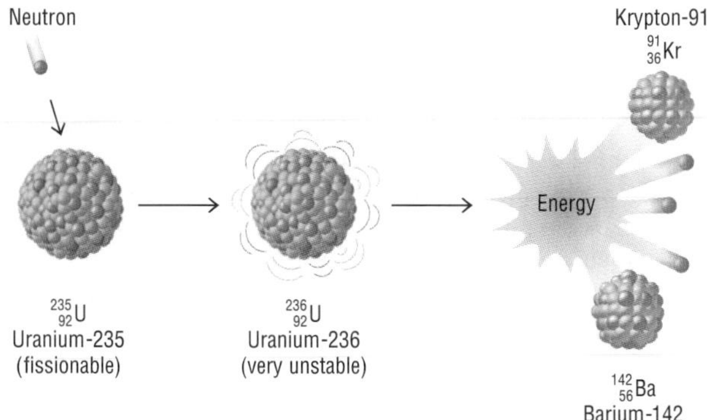

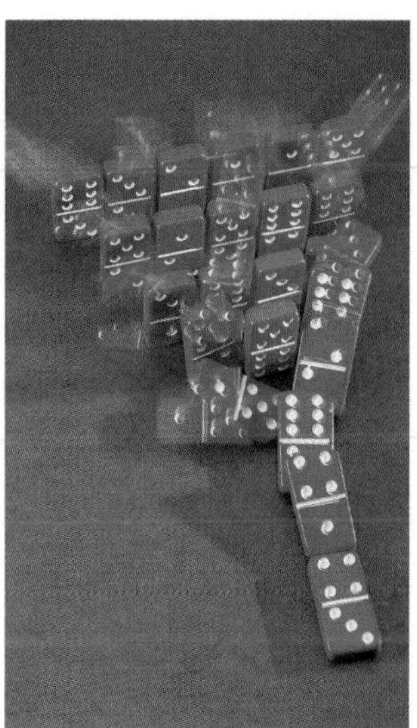

Figure 28.11
In nuclear fission, uranium-235 breaks into two fragments. What is produced? The released neutrons can split other uranium-235 atoms, creating a chain reaction. The toppling dominoes shown in the photograph illustrate the idea of a chain reaction. ❶

section 28.3

objectives

- Compare nuclear fission and nuclear fusion, and comment on their potential as sources of energy
- Describe the methods used in nuclear power plants to produce and control fission reactions
- Explain the issues involved in storage, containment, and disposal of nuclear waste.

key terms

- fission
- neutron moderation
- neutron absorption
- fusion

Chem ASAP!

Animation 30
Take a close look at a nuclear fission chain reaction.

28.3

1 Engage

Use the Visual

Have students study the photograph and read the text that opens the section. Explain that the sun contains hydrogen nuclei. Ask:

- **In which state of matter do these hydrogen nuclei exist?** (as plasma)
- **Describe a hydrogen nucleus.** (a proton)

Explain that protons can combine with electrons to produce helium nuclei.

- **How is the energy of the sun produced?** (Students should infer that energy is released during the formation of helium nuclei.)

Check Prior Knowledge

Ask students to provide general definitions for the terms *fission* and *fusion*. (During fission an object splits into smaller parts. During fusion objects combine into a larger whole.)
Explain that these terms have more precise meanings when applied to nuclear reactions, but still include the concepts of fragmentation and merger.

STUDENT RESOURCES

From the Teacher's Resource Package, use:

- Section Review 28.3, Ch. 28 Practice Problems and Quizzes from the Review Module (Ch. 25–28)

TECHNOLOGY RESOURCES

Relevant technology resources include:

- Chem ASAP! CD-ROM
- Resource Pro CD-ROM

❶ krypton-91, barium-142, energy, and additional neutrons

2 Teach

TEACHER DEMO

To demonstrate a chain reaction, cut some wooden matchsticks in half and arrange the heads in branching chains on a noncombustible surface. Construct the chain so that the end of each matchstick is touching the heads of two other matchsticks, forming a "Y". **CAUTION!** Keep all flammable materials away from the demonstration. Have a fire extinguisher ready. Ignite the first matchstick head. Have students observe the matches as they ignite in a chain-reaction.

ACTIVITY

Display Figure 28.11 on an overhead projector and explain that uranium-235 does not spontaneously fission. Uranium-235 is called a *fissionable material* because its nucleus becomes unstable when it is struck by a neutron. Note that two additional neutrons are produced by each fission event. Fission reactions also produce radioactive waste products such as krypton-91 and barium-142. Ask:

- Which types of radiation do these two products emit? (Barium-142 and krypton-91 both emit beta particles.)

Have students draw a diagram showing how the neutrons from the single fission process in Figure 28.11 can be used to induce subsequent fissions in other uranium-235 nuclei. The diagrams should illustrate the exponential growth of fission processes. Have students use their diagrams to discuss why fission reactors must be carefully monitored and controlled.

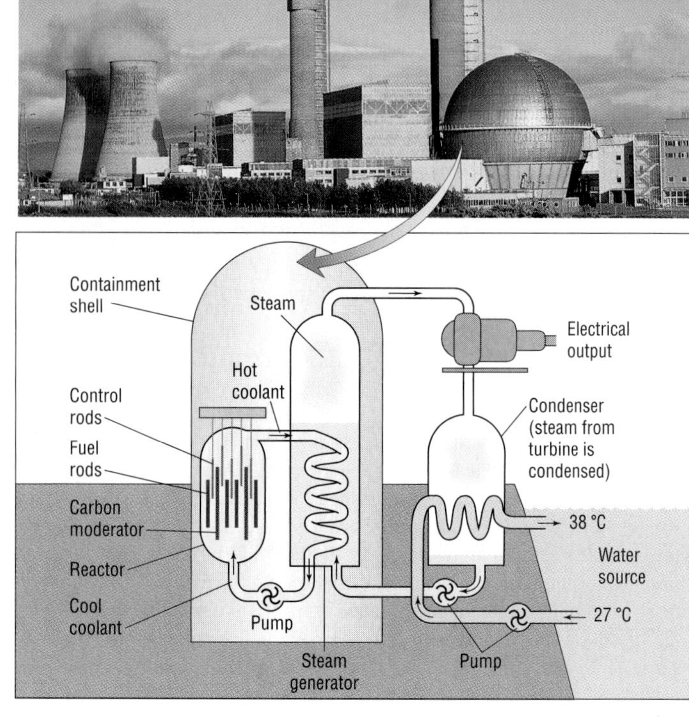

Figure 28.12
The illustration shows the basic components of a nuclear reactor. Heat produced in the reactor by the fission process is removed by circulating coolant. The removed heat is used to power a steam-driven turbine, which generates electricity.

Fission can be controlled so energy is released more slowly. Nuclear reactors, such as the one diagrammed in **Figure 28.12,** use controlled fission to produce useful energy. In the controlled fission reaction within a nuclear reactor, much of the energy generated is in the form of heat. A suitable coolant fluid, usually liquid sodium or water, removes the heat from the reactor core. The heat is used to generate steam, which drives a turbine that in turn generates electricity. The control of fission in a nuclear reactor involves two steps.

1. **Neutron moderation** is a process that reduces the speed of neutrons so they can be captured by the reactor fuel (usually uranium-235 or plutonium-239) in order to continue the chain reaction. Moderation is necessary because most of the neutrons produced move so fast that they will pass right through a nucleus without being absorbed. Water and carbon are good moderators because they slow the neutrons so the chain reaction can be sustained.
2. **Neutron absorption** is a process that decreases the number of slow moving neutrons. To prevent the chain reaction from going too fast, some of the slowed neutrons must be trapped before they hit fissionable atoms. Neutron absorption is carried out by control rods made of a material, such as cadmium, that absorbs neutrons. When the control rods extend almost all the way into the reactor core, they absorb many neutrons, and fission occurs slowly. As the rods are pulled out, they absorb fewer neutrons and the fission process speeds up. If the chain reaction were to go too fast, heat might be produced faster than

MEETING DIVERSE NEEDS

LEP Have students write the definitions of fission and fusion in their notebooks in their native language and in English. Have them prepare a table, complete with sample reactions, describing the characteristics of fission and fusion processes in nuclear chemistry.

it could be removed by the coolant. In this case, the reactor core would overheat, which could lead to mechanical failure and release of radioactive materials into the atmosphere. Ultimately, a meltdown of the reactor core might occur.

Despite other dangers, a nuclear reactor cannot produce a nuclear explosion. The fuel elements are widely separated and cannot physically connect to produce the critical mass required. Once a nuclear reactor is started, however, it remains highly radioactive for many generations. Shields protect the reactor structure from radiation damage. Walls of high-density concrete are also designed to protect the operating personnel.

Nuclear Waste

Fuel rods from nuclear power plants are one major source of nuclear waste. The fuel rods are made from a fissionable isotope, either uranium-235 or plutonium-239. The fuel rods are long and narrow—typically 3 meters long with a 0.5-cm diameter. Three hundred fuel rods are bundled together to form an assembly and one hundred assemblies are arranged to form the reactor core. During fission, the amount of fissionable isotope in each fuel rod decreases. There comes a time when there is no longer enough fuel in the rods to ensure that the output of the power station remains constant. The isotope-depleted, or spent, fuel rods must be removed and replaced with new fuel rods.

Spent fuel rods are classified as high-level nuclear waste. They contain the remainder of the fissionable isotope along with the fission products, a complex mixture of highly radioactive isotopes. Some of these fission products have very short half-lives, on the order of fractions of seconds. Others have half-lives of hundreds or thousands of years. All nuclear power plants have holding tanks, or swimming pools, for spent fuel rods. These pools, which are typically 12 meters deep, are filled with water as shown in **Figure 28.13.** Storage racks at the bottom of these pools are designed to hold the spent fuel assemblies. There are two reasons fuel assemblies are stored in water. Water cools the rods, which continue to produce heat for years after their removal from the core. Water also acts as a radiation shield to reduce the radiation levels from the spent fuel rods.

The assemblies of spent fuel rods may spend a decade or more in a holding pool. Plant operators expected used fuel rods to be reprocessed to recover the remaining fissionable isotope, which would be recycled in the manufacture of new fuel rods. However, with large deposits of uranium ore available—many in the United States—it is less expensive to mine new fuel than to reprocess depleted fuel. At some nuclear plants, there is no space left in the storage pool. In order to keep these plants open, their fuel rods must be moved to off-site storage facilities.

The number of years a nuclear plant can operate is limited. Eventually the plant must be decommissioned and dismantled. Because the plant is usually contaminated with radioactive materials, dismantling produces thousands of tons of low-level nuclear waste. Low-level waste is taken to licensed burial sites that are monitored and controlled by the Department of Transportation (DOT) and the Nuclear Regulatory Commission (NRC). For more information on nuclear waste, see the feature on page 862.

Figure 28.13
The racks at the bottom of this pool contain spent fuel rods.

Discuss

Explain that the wastes produced in fission reactors contain isotopes with half-lives measured in the thousands or hundreds of thousands of years. Many plans to store or dispose of these wastes involve methods and materials that may be highly unsuitable. For example, the placement of wastes in thick drums that would then be sunk in the oceans—one suggested method—may contain the waste for only decades or a few centuries. The containers would begin to leak long before the contents were safe. The disposal problem has led some people to propose permanent abandonment of nuclear reactors as sources of energy. After students read the text and the feature on p. 862, lead a discussion on the pros and cons of nuclear energy.

Build Writing Skills

Have students compare and contrast nuclear fission to nuclear fusion. Ask students to describe the advantages and disadvantages of each type of process as a means to meet future energy needs. Ask:

- **What types of technical hurdles remain to be solved in order to utilize the energy produced by nuclear fusion?**

3 Assess

Evaluate Understanding

Ask students to write examples of nuclear equations describing nuclear fission and nuclear fusion. Have students hypothesize about how a meltdown of a nuclear reactor might occur. (Rapid removal of the control rods could allow too many neutrons to react with fissionable nuclei.) **Have students describe the conditions under which nuclear fusion will occur.** (Temperatures in excess of 40 000 000 °C are required, as are methods to confine and control the dense plasmas.)

Reteach

Explain that both nuclear fission and nuclear fusion produce energy by the conversion of matter to energy. In fission, very heavy nuclei are split into lighter nuclei. In fusion, very light nuclei combine to form heavier nuclei. Fission reactions are relatively easy to control but produce radioactive wastes. Fusion reactions are difficult to initiate and control but produce little radioactive waste.

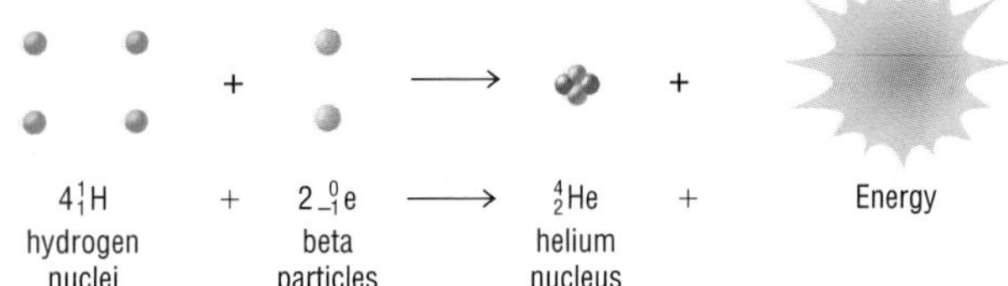

Figure 28.14
Thermonuclear fusion reactions occurring in the sun have provided Earth with energy for billions of years.

Nuclear Fusion

The sun is an extraordinary energy source. The energy released by the sun results from a thermonuclear reaction, or nuclear fusion. **Fusion** occurs when nuclei combine to produce a nucleus of greater mass. In solar fusion, hydrogen nuclei (protons) fuse to make helium nuclei. **Figure 28.14** shows that the reaction also requires two beta particles. Fusion reactions tend to release more energy than do fusion reactions. However, fusion reactions occur only at very high temperatures—in excess of 40 000 000 °C.

The use of controlled nuclear fusion as an energy source on Earth is appealing. The potential fuels are inexpensive and readily available, and the fusion products are usually not radioactive. One reaction that scientists are studying is the combination of a deuterium (hydrogen-2) nucleus and a tritium (hydrogen-3) nucleus to form a helium nucleus.

$$^2_1H + ^3_1H \longrightarrow ^4_2He + ^1_0n + \text{energy}$$

The problems with fusion lie in achieving the high temperatures necessary to start the reaction, and in containing the reaction once it has started. The high temperatures required to initiate fusion reactions have been achieved by using a fission bomb. Such a bomb is the triggering device used for setting off a hydrogen bomb, which is an uncontrolled-fusion device. Such a process is clearly of no use, however, as a controlled generator of power.

At the very high temperatures involved in fusion, matter exists as a plasma, a high-energy state in which ions exist in a gaslike form. To contain a plasma is a formidable task. No known structural material can withstand the hot, corrosive plasma. Scientists are experimenting with magnetic fields to contain plasma.

section review 28.3

11. Compare nuclear fission and fusion. Evaluate the reliability of each as a source of energy.
12. Describe how a nuclear fission power plant operates.
13. Explain what happens in a nuclear chain reaction.
14. Identify two types of nuclear waste produced by nuclear power plants.
15. Assuming technical problems could be overcome, what are some advantages to producing electricity in a fusion reactor?

Chem ASAP! Assessment 28.3 Check your understanding of the important ideas and concepts in Section 28.3.

Answers

SECTION REVIEW 28.3

11. Fission involves splitting nuclei into smaller fragments. This is a reliable, controllable source of energy but poses operational dangers and produces radioactive wastes. Fusion occurs when two nuclei combine to produce a nucleus of greater mass. Fusion reactions require very high temperatures and are difficult to contain.
12. The heat produced by fission is removed from the reactor core and used to generate steam to drive a turbine, which generates electricity.
13. A nuclear chain reaction involves the splitting of atomic nuclei that release energetic neutrons that split more nuclei.
14. high-level waste in spent fuel rods and low-level waste in contaminated structures that must be disposed of when the plant is dismantled
15. abundant fuels with no radioactive waste

RADIATION IN YOUR LIFE

The Scottish physicist C.T.R Wilson wanted to study the formation of clouds, so he built a chamber in which he could duplicate cloud formation. In the chamber, water in the air condensed on dust particles, forming clouds. When Wilson used dust-free air, clouds continued to condense, provided the air was saturated enough. He hypothesized that clouds condensed on ions in the air. To test his hypothesis, Wilson exposed his chamber to ionizing radiation. The radiation left trails of condensed water droplets. When Wilson finished his work in 1912, his chamber had become a radiation detector used by many scientists of the day. **What are some other devices used to detect radiation, and how do they function?**

objectives
- Describe three methods of detecting radiation
- List some applications of radioisotopes in research and medicine

key terms
- ionizing radiation
- Geiger counter
- scintillation counter
- film badge

Detecting Radiation

X-rays and the radiation emitted by radioisotopes are called ionizing radiation. **Ionizing radiation** is radiation with enough energy to knock electrons off some atoms of the bombarded substance to produce ions. You cannot detect ionizing radiation with any of your senses. Instead, various instruments and monitoring devices are used for this purpose. One such device, called a **Geiger counter,** uses a gas-filled metal tube to detect radiation. **Figure 28.15** shows the construction of a Geiger counter. The tube has a central wire electrode that is connected to a power supply. When ionizing radiation penetrates a thin window at one end of tube, the gas inside the tube becomes ionized. Because of the ions and free electrons produced, the gas becomes an electrical conductor. Each time a Geiger tube is exposed to radiation, current flows. The bursts of current drive electronic counters or cause audible clicks from a built-in speaker. Geiger counters are used primarily to detect beta radiation. Alpha particles cannot pass through the end window. Most gamma rays and x-rays pass directly through the gas, causing few ionizations.

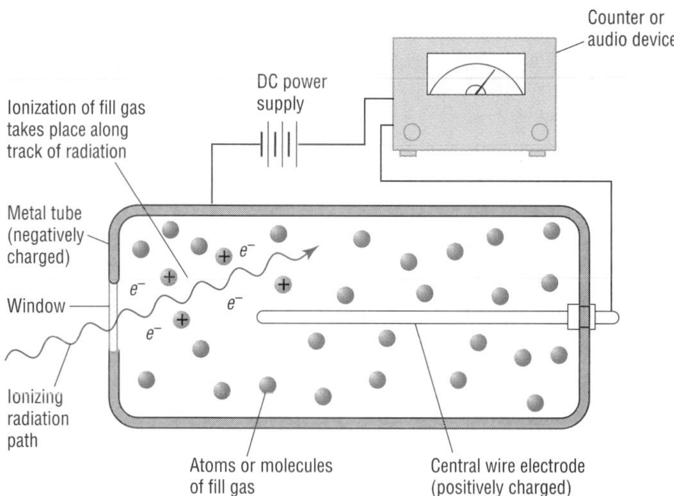

Figure 28.15
Radiation cannot be seen, heard, felt, or smelled. Thus warning signs and radiation-detection instruments must be used to alert people to the presence of radiation and to monitor its level. The Geiger counter is one such instrument that is widely used. What type of radiation does a Geiger counter primarily detect? ❶

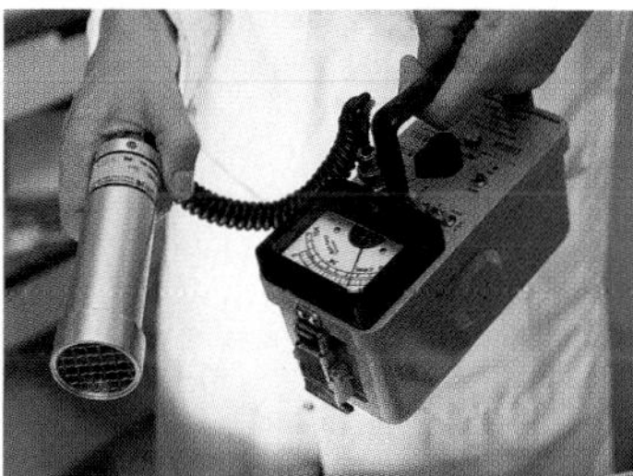

28.4

1 Engage

Use the Visual

Have students study the photograph and read the text that opens the section. Explain that a cloud chamber is a device used to detect *ionizing* radiation—radiation that is able to remove electrons from atoms. Ask:

- **What are some other devices used to detect radiation and how do they function?** (Radiation can expose a photographic plate; cause the sealed gas in a Geiger counter to be ionized so that it conducts electricity; or produce a flash of light on the phosphor-coated surface of a scintillation counter.)

Motivate and Relate

Hans Geiger (1882–1945) was a student of Ernest Rutherford. It was Geiger, along with another student named Marsden, who carried out the gold-foil, alpha-scattering experiment at Rutherford's direction. Geiger gained experience with the problem of detecting radiation, which ultimately led to his invention of the Geiger counter. Ask students to review the gold-foil experiment and describe the detection method used to measure the scattering of alpha particles. (a precursor of the scintillation counter)

STUDENT RESOURCES

From the Teacher's Resource Package, use:
- Section Review 28.4, Ch. 28 Practice Problems, Vocabulary Review, Tests, and Quizzes from the Review Module (Ch. 25–28)
- Laboratory Recordsheet 28-2

TECHNOLOGY RESOURCES

Relevant technology resources include:
- Chem ASAP! CD-ROM
- Resource Pro CD-ROM
- Assessment Resources CD-ROM: Chapter 28 Tests

❶ beta radiation

2 Teach

MINI LAB

Studying Inverse-Square Relationships

LAB TIPS

Probeware Lab Manual has Vernier, TI, and PASCO labs.

The brightness of any part of an illuminated surface indicates the intensity of the radiant energy being reflected. In this activity, it is assumed that the screen reflects the same amount of energy at all locations. However, because the illuminated area increases with distance from the source, the intensity decreases and the screen appears dimmer.

ANALYSIS AND CONCLUSIONS

1. The intensity decreases.
2. The area is quadrupled. The area increases nine-fold.

TEACHER DEMO

Explain that radioactivity reduces silver in photographic film. This property led to the discovery of radioactivity by Becquerel. Place a metal key on top of unexposed photographic film that has been wrapped in black paper. Place a radiation source on top of the key. **CAUTION!** Use only safely packaged radiation sources. Do not directly touch any radioactive material. After one week, develop the film. Have students note that the area of the film that was beneath the key is unexposed, unlike the rest of the film. Point out that the film badges worn by persons working with radioactivity operate by the same mechanism. They serve as important radiation detectors, safeguarding against over-exposure to ionizing radiation.

MINI LAB

Studying Inverse-Square Relationships

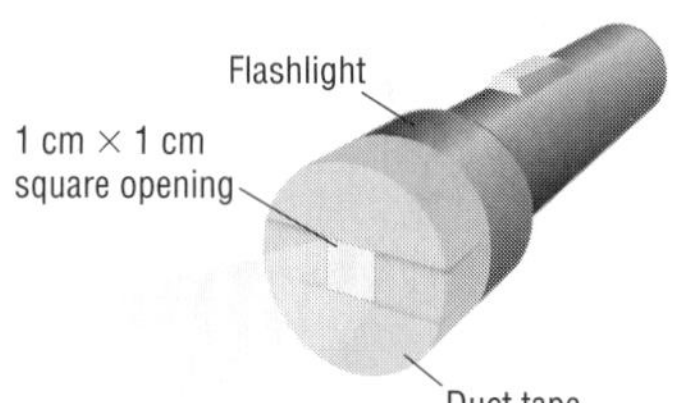

PURPOSE

To demonstrate the relationship between radiation intensity and the distance from the radiation source.

MATERIALS

- flashlight
- strips of duct tape
- scissors
- poster board, white (50 cm × 50 cm)
- meter ruler or tape measure
- flat surface, long enough to hold the meter ruler
- loose-leaf paper
- graph paper
- pen or pencil
- light sensor (optional)

PROCEDURE

Sensor version available in the Probeware Lab Manual.

1. Measure and record the distance (A) from the bulb filament to the front surface of the flashlight.
2. Cover the end of a flashlight with tape. Leave a 1-cm × 1-cm square hole in the center for light to pass through.
3. Place the flashlight on its side on a flat, horizontal surface. Turn on the flashlight. Darken the room.
4. Mount a large piece of white poster board directly in front of the flashlight, perpendicular to the horizontal surface.
5. Move the flashlight backward from the vertical board in short increments. At each position you select, record the distance (B) from the flashlight to the vertical board and the length (L) of one side of the square image projected on the board.
6. On a sheet of graph paper, plot L on the y-axis versus $A + B$ on the x-axis. On another sheet, plot L^2 on the y-axis versus $A + B$ on the x-axis.

ANALYSIS AND CONCLUSIONS

1. As the flashlight is moved away from the vertical board, what do you notice about the intensity of the light in the illuminated square? Use your graphs to demonstrate the relationship between intensity and distance.
2. When the distance of the flashlight from the board is doubled and tripled, what can you say about the areas and intensities of the illuminated squares?

A **scintillation counter** is a device that uses a specially coated phosphor surface to detect radiation. Ionizing radiation striking the phosphor surface produces bright flashes of light, or scintillations. The number of flashes and their respective energies are detected electronically. The information is then converted into electronic pulses, which are measured and recorded. Scintillation counters have been designed to detect all types of ionizing radiation. Such devices are similar to television screens coated with zinc sulfide (ZnS) as the phosphor. Inside a television, electrons are shot at a phosphor screen, producing scintillations. The pattern of these scintillations produces the television picture.

Film badges are important radiation detectors for persons who work near radiation sources. A **film badge,** shown in **Figure 28.16,** consists of several layers of photographic film covered with black lightproof paper, all encased in a plastic or metal holder. The badge is worn the entire time the person is at work. At specific intervals, with the frequency depending on the type of work involved, the film is removed and developed. The strength and type of radiation exposure are determined from the darkening of the film.

Filter
Filter
Film
Case (plastic)

Figure 28.16
People who work with ionizing radiation wear film badge detectors. The film inside such a badge is exposed in proportion to the amount and type of radiation it receives.

Records are kept of the results. Film badges do not protect a person from radiation exposure. They merely serve as precautionary monitoring devices. Protection against radiation is achieved by keeping a safe distance from the source and by using adequate shielding.

Using Radiation

Although radiation can be harmful and should always be handled with care, it can be used safely and is important in many scientific procedures. Neutron activation analysis is a procedure used to detect trace amounts of elements in samples. In this procedure, a sample of interest is bombarded with neutrons from a radioactive source, which causes some atoms in the sample to become radioactive. The half-life and type of radiation emitted by the radioisotopes are detected, and this information is processed by a computer. Because this information is characteristic for each element, scientists can determine what radioisotopes were produced and what elements were originally present in the sample. This is a sensitive technique used to detect trace amounts of elements. It is capable of measuring 10^{-9} g of an element in a sample. Neutron activation analysis is used by museums to detect art forgeries, and by crime laboratories to analyze gunpowder residues.

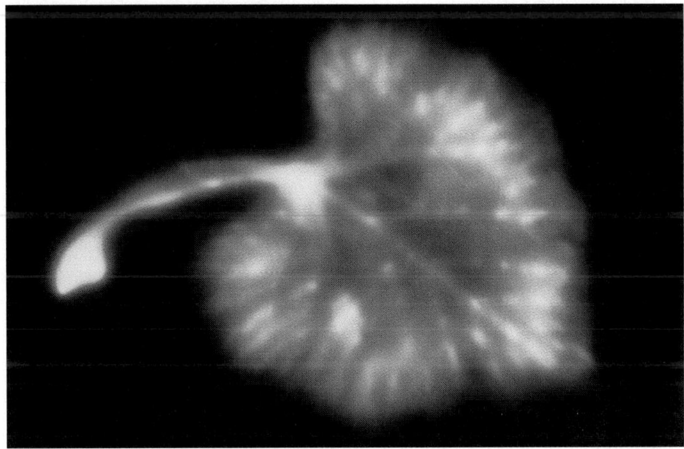

Figure 28.17
A radioactive tracer is used to determine where an absorbed pesticide or fertilizer is taken up by a plant.

Discuss

Explain that the most commonly used radiation detector, the Geiger counter, is primarily used for the detection of beta radiation, because alpha particles cannot penetrate the tube and most gamma and X-rays pass through the gas without causing many ionizations. Scintillation counters, though not generally portable, are designed to detect all types of radiation. They rely on a phosphor surface, which produces a bright flash when struck by ionizing radiation. The flashes and their energies can be detected electronically. The process is analogous to electrons striking the phosphor screen inside a television set. Explain that the third type of radiation detector, the film badge, is a modern day version of Becquerel's photographic plate.

Discuss

Explain that many of the radioactive isotopes used in medicine and agriculture are not naturally occurring. They are produced artificially by methods such as neutron bombardment. Radioisotopes are used by physicians to diagnose and treat some forms of cancer. Actively dividing cells are usually more sensitive to radiation than are normal cells.

Meeting Diverse Needs

Gifted Have students do research on the development and use of *radioimmunoassay* (RIA) to detect minute traces of substances in the body. This technique was developed by Rosalyn Yalow, who received the 1977 Nobel Prize in Medicine for her contributions. Today, RIA is used to assay for hundreds of hormones, drugs, enzymes, and viruses. Because RIA is so sensitive, the medical establishment is able to diagnose and treat diseases at very early stages.

CHEMISTRY AND SCIENCE HISTORY

Explain that a number of women have done pioneering work in nuclear chemistry and physics and have been awarded the Nobel Prize. Marie Curie was awarded the physics prize in 1903 for her work on natural radioactivity, and the chemistry prize in 1911 for discovering the elements polonium and radium. Irene Joliot-Curie, Marie's daughter, was awarded the chemistry prize in 1935 for her contributions to the discovery of artificial radioactivity. Marie Goeppert Mayer's theoretical work describing the shell model of stable nuclei was recognized with a Nobel Prize for Physics in 1963.

Radioisotopes called tracers are used to study chemical reactions and molecular structures. During this procedure, one of the reactants is labeled with a radioisotope and added to the reaction mixture. After the reaction is complete, the radiation of the product is measured to determine the uptake of the tracer. By comparing this amount with the amount originally added, much can be learned about the reaction mechanism. Reactions with many steps can be studied using this method.

Radioisotope tracers are also used in agricultural research to test the effects of herbicides, pesticides, and fertilizers. During this procedure, the tracer is first introduced into the substance being tested to make the substance radioactive. Next, plants are treated with the radioactively labeled substance. Then, the radioactivity of the plants is measured to determine the location of the substance, as shown in **Figure 28.17** on the previous page. Often, the tracer is also monitored in animals that consume the plants, in water, and in soil. This information helps scientists determine the effects of using the substance.

Radioisotopes can even be used to diagnose some diseases. Iodine-131, for example, is used to detect thyroid problems. The thyroid gland extracts iodide ions from the bloodstream and uses them to make the hormone thyroxine. To diagnose thyroid disease, the patient is given a drink containing a small amount of the radioisotope iodine-131. After about two hours, the amount of iodide uptake is measured by scanning the patient's throat with a radiation detector. **Figure 28.18** shows the results of such a scan. In a similar way, the radioisotope technetium-99m is used to detect brain tumors and liver disorders. Phosphorus-32 is used to detect skin cancer.

Radiation has become a routine part of the treatment of some cancers. Cancer is a disease in which abnormal cells in the body are produced at a rate far beyond the rate for normal cells. The mass of cancerous tissue resulting from this runaway growth is called a tumor. Radiation therapy is often used to treat cancer because the fast-growing cancer cells are more susceptible to damage by high-energy radiation such as gamma rays than are the healthy cells. The cancerous area can be treated with radiation to kill the cancer cells. Some normal cells are also killed, however, and cancer cells at the center of the tumor may be resistant to the radiation. Therefore, the benefits of the treatment and the risks to the patient must be carefully evaluated before radiation treatment begins.

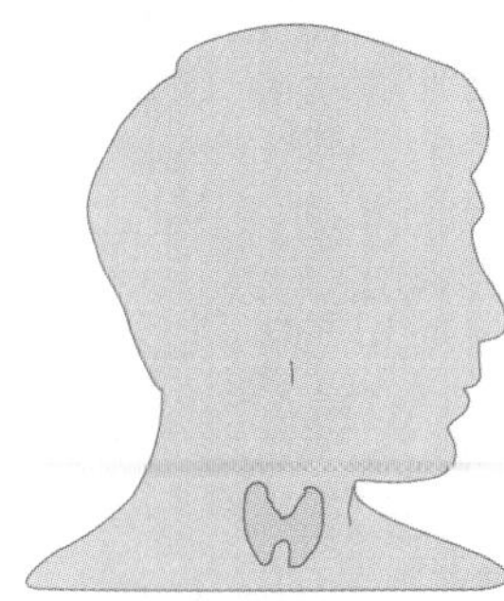

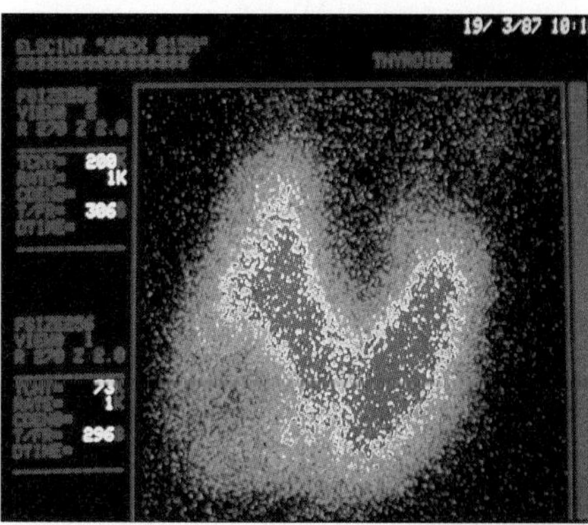

Figure 28.18

This scanned image of a thyroid gland shows where radioactive iodine-131 has been absorbed. Doctors use these images to identify thyroid disorders.

Answers

1. Cancer cells divide rapidly.

SECTION REVIEW 28.4

16. Geiger counters detect beta radiation that ionizes gas in the counter's tube, producing a current and an audible or visible signal. Scintillation counters use phosphors to convert a portion of the energy from ionizing radiation into easily detectable signals. Scintillation counters detect all types of ionizing radiation. Film badges are enclosed layers of photographic film that are developed to reveal the strength and type of radiation exposure.
17. Radioisotopes can be used to analyze samples for age and content; study chemical reactions, molecular structure, and agricultural assimilation of herbicides, pesticides, and fertilizers; and diagnose and treat diseases.

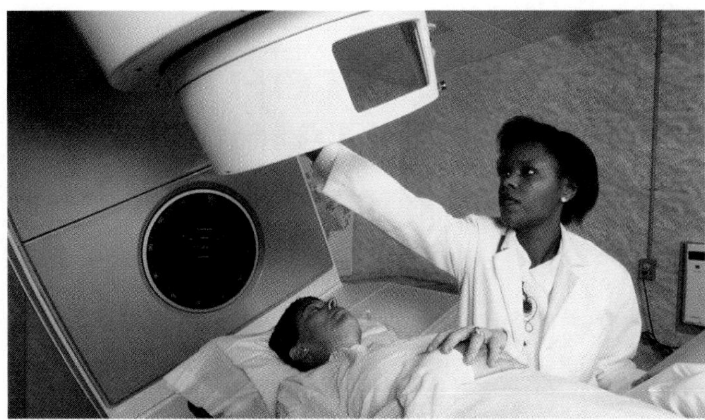

Figure 28.19
Radiation therapy is commonly used to treat cancer. The unit shown here emits a narrow, intense beam of radiation that destroys the ability of cells to reproduce. Cells are most vulnerable to the radiation when they are dividing. Why are cancer cells generally more sensitive to radiation than are normal cells? 1

In a technique called teletherapy, a narrow beam of high-intensity gamma radiation is directed at cancerous tissue. A radiation therapy unit used for teletherapy is shown in **Figure 28.19.** Cobalt-60 and cesium-137 are commonly used as the radiation sources. To minimize damage to healthy tissue, the patient is positioned so that only the cancerous region is within the radiation beam at all times. The unit rotates so that the radiation dose to the skin and surrounding normal tissue is minimized and distributed in a belt that extends all the way around the patient.

Salts of radioisotopes can also be sealed in gold tubes and directly implanted in tumors. These seeds emit beta and gamma rays that kill the surrounding cancer cells. Because the radioisotope is in a sealed container, it is prevented from traveling throughout the body.

Pharmaceuticals containing radioisotopes of gold, iodine, or phosphorus are sometimes given in radiation therapy. For example, a dose of iodine-131 larger than that used simply to detect thyroid diseases can be given to treat the diseased thyroid. The radioactive iodine accumulates in the thyroid and emits beta and gamma rays to provide therapy.

section review 28.4

16. Describe several methods of detecting radiation.
17. Describe several applications of radioisotopes in scientific research and in medicine.
18. Name an advantage of a scintillation counter over a Geiger counter.
19. Suppose that a radioactive solution containing an alpha emitter accidentally gets on your hands. You wash your hands with soap and water and then check them with a Geiger counter for residual radioactive contamination. The Geiger counter does not register any radioactivity. Are your hands definitely free from radioactive contamination? Explain.
20. What are some uses of radioactive tracers?
21. What is an advantage of using a radioactive seed to treat a cancerous tumor?

 Chem ASAP! **Assessment 28.4** Check your understanding of the important ideas and concepts in Section 28.4.

portfolio project

Research and report on the use of technetium-99m to diagnose diseases and disorders of the liver or gallbladder. Explain how doctors use this technique to pinpoint the source of a problem without surgery.

SECTION REVIEW 28.4 CONTINUED

18. A Geiger counter only detects beta radiation; a scintillation counter can detect all types of ionizing radiation.
19. No, because a Geiger counter cannot detect alpha radiation.
20. Tracers are used to study reaction mechanisms and the uptake of substances by organisms.
21. The seed can be placed at the location of the tumor, which minimizes the effect on normal cells, and there are no radioactive waste products to be disposed of.

28.4

3 Assess

Evaluate Understanding

Have students compare and contrast detection of radiation by a film badge and detection of radiation by a scintillation counter. Have students describe two medical uses of radioactive isotopes. (as a tracer in disease diagnosis and in the treatment of cancer)

Reteach

Explain that the most common home smoke detector is based on the same principles as a radiation detector. The smoke detector contains a radioisotope that ionizes the air in a small chamber. The ionized air allows a current to flow in the chamber. When smoke particles enter the chamber, the ionization decreases, which decreases the current. The decreased current is detected by a secondary circuit, which triggers an alarm. Point out that radioisotopes of an element are not chemically different from other isotopes of that element. Chemical reactivity depends on the electron structure of the element, not its nuclear makeup.

Portfolio Project

The liver and gallbladder are an organ system that secretes bile and helps digest fats. Cirrhosis of the liver, liver cancer, and gallstones cause the system to malfunction. To pinpoint the problem, doctors inject a mixture that should be absorbed by the liver. If gamma radiation emitted by the technetium-99m atoms in the mixture is detected in the kidneys, there is likely a problem with the liver. If the mixture reaches the liver but does not reach the gallbladder, the duct connecting the organs is probably blocked.

Chemistry Serving... the Environment

DISCUSS

Explain that the concerns about the disposal of nuclear wastes parallel concerns communities have about chemical waste disposal. However, concerns about nuclear wastes are magnified by the timeframe during which the wastes will be hazardous.

As of 1997, 440 nuclear power plants were in operation worldwide and about 30 countries got some of their electricity from nuclear power stations. In France, for example, 75% of the electricity is generated by nuclear power. By contrast, only about 20% of electricity generated in the United States comes from nuclear power plants.

The DOE is responsible for cleaning up 130 nuclear sites and safely managing their waste. The sites include locations where uranium was milled, research labs, and former nuclear weapons production facilities. Have students research how the cleanups are progressing.

CHEMISTRY IN CAREERS

Have students read about the duties and responsibilities of a nuclear physician on page 881. Also encourage them to connect to the Internet address shown.

Chemistry Serving...the Environment

NUCLEAR WASTE: STORAGE, DISPOSAL, AND CONTAINMENT

The proper disposal of chemical wastes, which can pollute air, land, and water, is an ongoing issue for society. Nuclear waste offers an even greater challenge because the ionizing radiation it emits can harm both existing organisms and future generations. Nuclear waste began accumulating in the 1940s at nuclear weapons facilities, which no longer operate. Today, nuclear waste is created by power plants and sectors of society such as hospitals that use radioisotopes. The nuclear waste accumulated in the last fifty years has been in temporary storage awaiting sites and methods for permanent disposal.

Transuranic waste is low- to mid-level waste contaminated with alpha-emitting radioisotopes with half-lives greater than 20 years and atomic numbers greater than 92. This waste consists of disposable items such as tools, clothing, and debris.

The photograph shows the interior of the Waste Isolation Pilot Plant (WIPP) near Carlsbad, New Mexico. The plant is located on unpopulated, federally owned land near the site where more than 900 nuclear tests were conducted. WIPP is a network of underground excavations in an ancient bed of rock salt about 660 meters beneath the desert. WIPP is an example of a geologic repository, which is a storage facility located in rock deep underground. WIPP is the world's first geologic repository for the permanent disposal of defense-generated transuranic waste. It is vital to the Department of Energy's (DOE) long-range environmental cleanup plan.

WIPP is designed to store thousands of tons of transuranic waste. The plant is expected to receive more than 35 000 loads of transuranic waste from 23 DOE sites for a period of about 35 years. The waste must be contained before shipment. Liquid waste will be solidified. All waste will be packed in 55-gallon drums.

In 1982, the United States Congress passed the Nuclear Waste Policy Act, which set policy guidelines for the disposal of all high-level nuclear waste. This waste consists of spent fuel rods from commercial nuclear reactors and high-level waste from the production of nuclear weapons.

Nuclear waste has been in temporary storage awaiting sites and methods for permanent disposal.

The high-level waste will be vitrified and packed in corrosion-resistant containers. Vitrification is the technical term for the production of glass. The contained waste will then be transported for permanent burial in a geologic repository.

Choosing a site for the repository is difficult because it takes more than 600 years for the fission product strontium-90 and more than 20 000 years for plutonium-239 to decay to safe levels. The site must meet health and safety standards over thousands of years. For more than a decade, studies have been done at Yucca Mountain, Nevada. Approval by the President and Congress, licensing by the Nuclear Regulatory Commission, and construction by the Department of Energy could take another decade. Meanwhile spent fuel rods will remain in their on-site or off-site temporary storage facilities.

CHEMISTRY IN CAREERS

NUCLEAR PHYSICIAN

Use radioactive isotopes to diagnose, manage, treat, and prevent diseases and disorders.

See page 881.

CHEMIST NEEDED

Enviro. Lab. GC Must know

TAKE IT TO THE NET

Find out more about career opportunities: www.phschool.com

CHEMICAL SPECIALIST

Local food service distributor seeks responsible self-motivated, individ- ... established San Jose

Chapter 28 STUDENT STUDY GUIDE

Take It to the NET For interactive study and review, go to www.phschool.com

KEY TERMS

- alpha particle *p. 843*
- alpha radiation *p. 843*
- band of stability *p. 845*
- beta particle *p. 843*
- beta radiation *p. 843*
- film badge *p. 858*
- fission *p. 853*
- fusion *p. 855*
- gamma radiation *p. 844*
- Geiger counter *p. 857*
- half-life *p. 847*
- ionizing radiation *p. 857*
- neutron absorption *p. 854*
- neutron moderation *p. 854*
- positron *p. 846*
- radiation *p. 841*
- radioactive decay *p. 842*
- radioactivity *p. 841*
- radioisotope *p. 841*
- scintillation counter *p. 858*
- transmutation *p. 850*
- transuranium elements *p. 851*

KEY EQUATIONS

- $^{238}_{92}\text{U} \xrightarrow[\text{decay}]{\text{Radioactive}} {}^{234}_{90}\text{Th} + {}^{4}_{2}\text{He}$
- $^{14}_{6}\text{C} \longrightarrow {}^{14}_{7}\text{N} + {}^{0}_{-1}\text{e}$
- $^{230}_{90}\text{Th} \longrightarrow {}^{226}_{88}\text{Ra} + {}^{4}_{2}\text{He} + \gamma$

CONCEPT SUMMARY

28.1 Nuclear Radiation

- Isotopes with unstable nuclei are radioactive and are called radioisotopes. The nuclei of radioisotopes emit radiation as they decay to stable nuclei.
- The radiation may be alpha (positively charged helium nuclei), beta (electrons), or gamma (electromagnetic radiation).

28.2 Nuclear Transformations

- When a radioactive nucleus emits an alpha particle, its atomic number decreases by 2 and its mass number decreases by 4. For beta emission, the atomic number increases by 1 and the mass number stays the same. The atomic number and the mass number stay the same for gamma emission.
- Every radioisotope decays at a characteristic rate. A half-life is the time required for one-half of the nuclei in a radioisotope to decay.

28.3 Fission and Fusion of Atomic Nuclei

- In nuclear fission, fissionable isotopes split when bombarded with neutrons. The isotopes release neutrons that cause a chain reaction.
- Nuclear power plants control fission through neutron moderation and neutron absorption. These plants generate electricity and high-level nuclear waste. Spent fuel rods removed from nuclear cores are stored in on-site pools.
- In nuclear fusion, light nuclei fuse to make more massive nuclei.

28.4 Radiation in Your Life

- Radiation may be detected with a Geiger counter or a scintillation counter. A film badge monitors radiation exposure of individuals who work with radioactive materials.
- Radioactivity and radiation are used extensively in agricultural research, in medical diagnosis, and in the treatment of some diseases.

CHAPTER CONCEPT MAP

Use these terms to construct a concept map that organizes the major ideas of this chapter.

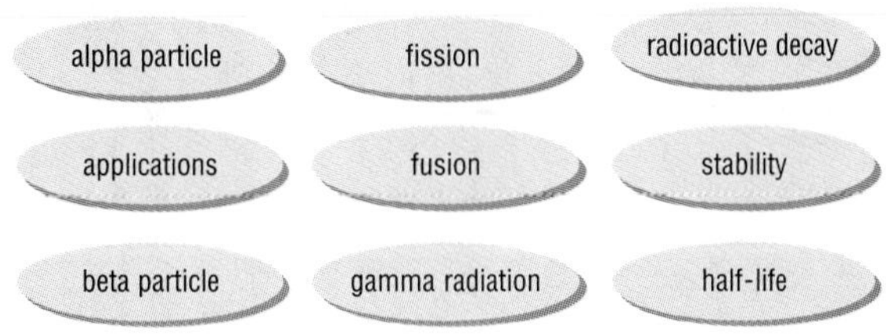

Chem ASAP! Concept Map 28
Create your Concept Map using the computer.

Take It to the Net

At **www.phschool.com** students will find for this chapter

- an Internet activity
- links to related chemistry sites
- an interactive quiz
- career links

Chapter 28
STUDENT STUDY GUIDE

4 Close

Summary

Ask these questions to summarize chapter content.

- **Name three types of radiation that are emitted by unstable nuclei.** (alpha particles, beta particles, or gamma radiation)
- **What is a transmutation reaction and how does it occur?** (The conversion of an atom of one element into an atom of another element. It can occur when a nucleus spontaneously decays or when a nucleus is bombarded by high-energy particles.)
- **What is half-life?** (the time required for 1/2 of the nuclei in a given sample to decay)
- **Have students contrast nuclear fission and fusion.** (Fission occurs when isotopes bombarded by neutrons split into fragments. In fusion, light nuclei fuse to make heavier nuclei at extremely high temperatures.)

Extension

Ask students to research some aspect of cancer treatment that involves radiation. Have them write a report that includes a description of the treatment and the theory behind it.

Looking Back . . . Looking Ahead . . .

The preceding chapter introduced classes of biological molecules and their functions. This chapter introduced nuclear chemistry with emphasis on the nature and rate of decay; fission and fusion; detection; and practical applications of radiation.

Answers

22. Each isotope of an element has the same atomic number but a different atomic mass. A radioisotope is an isotope that is radioactive.

23. $^{226}_{88}Ra \rightarrow ^{222}_{86}Rn + ^{4}_{2}He$

24. $^{210}_{82}Pb \rightarrow ^{210}_{83}Bi + ^{0}_{-1}e$

25. **a.** $\alpha, 2+$ **b.** $\beta, 1-$ **c.** $\gamma, 0$

26. **a.** $^{238}_{92}U \rightarrow ^{234}_{90}Th + ^{4}_{2}He$; thorium-234
b. $^{230}_{90}Th \rightarrow ^{226}_{88}Ra + ^{4}_{2}He$; radium-226
c. $^{235}_{92}U \rightarrow ^{231}_{90}Th + ^{4}_{2}He$; thorium-231
d. $^{222}_{86}Rn \rightarrow ^{218}_{84}Po + ^{4}_{2}He$; polonium-218

27. **a.** $^{14}_{6}C \rightarrow ^{14}_{7}N + ^{0}_{-1}e$
b. $^{90}_{38}Sr \rightarrow ^{90}_{39}Y + ^{0}_{-1}e$
c. $^{40}_{19}K \rightarrow ^{40}_{20}Ca + ^{0}_{-1}e$
d. $^{13}_{7}N \rightarrow ^{13}_{8}O + ^{0}_{-1}e$

28. **a.** mass number: unchanged; atomic number: increases by 1
b. mass number: decreases by 4; atomic number: decreases by 2
c. mass number and atomic number: unchanged

29. **a.** $^{234}_{92}U$ **c.** $^{206}_{82}Pb$
b. $^{206}_{81}Tl$ **d.** $^{226}_{88}Ra$

30. It undergoes radioactive decay.

31. $^{17}_{9}F \rightarrow ^{17}_{8}O + ^{0}_{+1}e$

32. **a.** $^{13}_{6}C$ **c.** $^{16}_{8}O$
b. $^{1}_{1}H$ **d.** $^{14}_{7}N$

33. **a.** platinum **e.** xenon
b. thorium **f.** californium
c. francium **g.** vanadium
d. technetium **h.** palladium
Francium **(c)**, technetium **(d)**, and californium **(f)** have no stable isotopes.

34. so exposure to the radioactivity is limited in time

35. One half-life is the time required for one-half of the atoms of a radioisotope to emit radiation and decay.

36. 6.3×10^{-1} mg

37. Natural radioactivity comes from elements in nature. Artificial radioactivity comes from elements created in nuclear reactors and accelerators.

Chapter 28 REVIEW

CONCEPT PRACTICE

22. Explain the difference between an isotope and a radioisotope. *28.1*

23. The disintegration of the radioisotope radium-226 produces an isotope of the element radon and alpha radiation. The atomic number of radium (Ra) is 88; the atomic number of radon (Rn) is 86. Write a balanced equation for this transformation. *28.1*

24. A radioisotope of the element lead (Pb) decays to an isotope of the element bismuth (Bi) by emission of a beta particle. Complete the equation for the decay process by supplying the missing atomic number and mass number. *28.1*

$$^{210}Pb \longrightarrow {}_{83}Bi + ^{0}_{-1}e$$

25. Write the symbol and charge for each. *28.1*
a. alpha particle
b. beta particle
c. gamma ray

26. Alpha radiation is emitted during the disintegration of the following isotopes. Write balanced nuclear equations for their decay processes. Name the element produced in each case. *28.1*
a. uranium-238 ($^{238}_{92}U$)
b. thorium-230 ($^{230}_{90}Th$)
c. uranium-235 ($^{235}_{92}U$)
d. radon-222 ($^{222}_{86}Rn$)

27. The following radioisotopes are beta emitters. Write balanced nuclear equations for their decay processes. *28.1*
a. carbon-14 ($^{14}_{6}C$)
b. strontium-90 ($^{90}_{38}Sr$)
c. potassium-40 ($^{40}_{19}K$)
d. nitrogen-13 ($^{13}_{7}N$)

28. How are the mass number and atomic number of a nucleus affected by the loss of the following? *28.1*
a. beta particle
b. alpha particle
c. gamma ray

29. The following radioactive nuclei decay by emitting alpha particles. Write the product of the decay process for each. *28.1*
a. $^{238}_{94}Pu$ **c.** $^{210}_{84}Po$
b. $^{210}_{83}Bi$ **d.** $^{230}_{90}Th$

30. What happens to an atom with a nucleus that falls outside the band of stability? *28.2*

31. Write an equation for the radioactive decay of fluorine-17 by positron emission. *28.2*

32. Identify the more stable isotope in each pair. *28.2*
a. $^{14}_{6}C$, $^{13}_{6}C$
b. $^{3}_{1}H$, $^{1}_{1}H$
c. $^{16}_{8}O$, $^{18}_{8}O$
d. $^{14}_{7}N$, $^{15}_{7}N$

33. Name the elements represented by the following symbols and indicate which of them would have no stable isotopes. *28.2*
a. Pt **e.** Xe
b. Th **f.** Cf
c. Fr **g.** V
d. Tc **h.** Pd

34. Why is it important that radioactive isotopes used internally for diagnosis or treatment have relatively short half-lives? *28.2*

35. Explain half-life. *28.2*

36. A patient is administered 20 mg of iodine-131. How much of this isotope will remain in the body after 40 days if the half-life for iodine-131 is 8 days? *28.2*

37. What is the difference between natural and artificial radioactivity? *28.2*

38. What are the transuranium elements? Why are they unusual? *28.2*

39. Describe the process of nuclear fission and define a nuclear chain reaction. *28.3*

40. Why are spent fuel rods removed from a reactor core? What do they contain? What happens to them after they are removed? *28.3*

41. Fusion reactions produce enormous amounts of energy. Why is fusion not used to generate electrical power? *28.3*

42. Why are x-rays and the radiation emitted by radioisotopes called ionizing radiation? *28.4*

43. What is the purpose of wearing a film badge when working with ionizing radiation sources? *28.4*

44. Explain how iodine-131 is used in both the diagnosis and treatment of thyroid disease. *28.4*

38. The elements with atomic number greater than 92; none occurs in nature and all are radioactive.

39. The nuclei of certain isotopes are bombarded with neutrons. The nuclei break into two fragments and release more neutrons. Released neutrons hit other nuclei to start a chain reaction that releases large amounts of energy.

40. A power plant cannot maintain a constant output of electricity with spent fuel rods, which contain depleted fissionable isotopes and fission products. The rods are stored in pools.

41. Fusion requires extremely high temperatures, which would destroy any container.

42. Ionizing radiation, such as x-rays and gamma radiation, has sufficient energy to remove electrons from the atoms it hits.

43. The film badge measures radiation exposure; an exposed film badge indicates how much radiation a worker has received.

44. In diagnosis, the amount of iodine uptake in

CONCEPT MASTERY

45. Write nuclear equations for these conversions.
    a. $^{30}_{15}P$ to $^{30}_{14}Si$
    b. $^{13}_{6}C$ to $^{14}_{6}C$
    c. $^{131}_{53}I$ to $^{131}_{54}Xe$

46. What is the difference between the nuclear reactions taking place in the sun and the nuclear reactions taking place in a nuclear reactor?

47. Complete these nuclear equations.
    a. $^{32}_{15}P \longrightarrow \_\_\_\_ + {}^{0}_{-1}e$
    b. $\_\_\_\_ \longrightarrow {}^{14}_{7}N + {}^{0}_{-1}e$
    c. $^{238}_{92}U \longrightarrow {}^{234}_{90}Th + \_\_\_\_$
    d. $^{141}_{56}Ba \longrightarrow \_\_\_\_ + {}^{0}_{-1}e$
    e. $\_\_\_\_ \longrightarrow {}^{181}_{77}Ir + {}^{4}_{2}He$

48. Write nuclear equations for the beta decay of the following isotopes.
    a. $^{90}_{38}Sr$
    b. $^{14}_{6}C$
    c. $^{137}_{55}Cs$
    d. $^{239}_{93}Np$
    e. $^{50}_{22}Ti$

49. The following graph shows the radioactive decay curve for thorium-234. Use the graph to answer the questions.

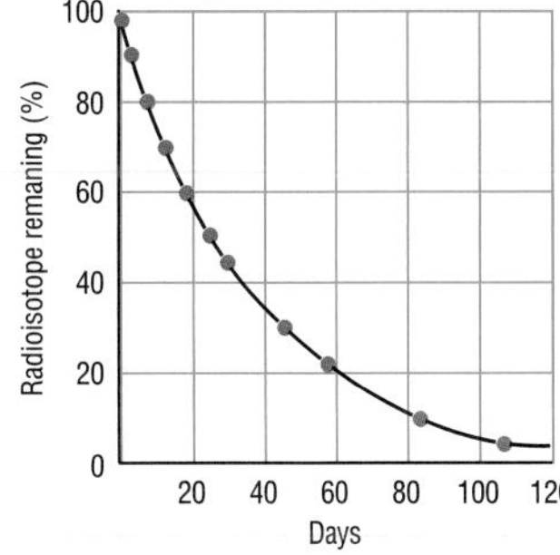

    a. What percent of the isotope remains after 60 days?
    b. How many grams of a 250-g sample of thorium-234 would remain after 40 days had passed?
    c. How many days would pass while 44 g of thorium-234 decayed to 4.4 g of thorium-234?
    d. What is the half-life of thorium-234?

50. Write a nuclear equation for each word equation.
    a. Radon-222 emits an alpha particle to form polonium-218.
    b. Radium-230 is produced when thorium-234 emits an alpha particle.
    c. When polonium-210 emits an alpha particle, the product is lead-206.

51. Describe the various contributions the following people made to the fields of nuclear and radiation chemistry.
    a. Marie Curie
    b. Antoine Henri Becquerel
    c. James Chadwick
    d. Ernest Rutherford

CRITICAL THINKING

52. Choose the term that best completes the second relationship.
    a. decomposition : combination
       fission : _______
       (1) energy (3) radioactivity
       (2) nuclei (4) fusion
    b. anion : cation
       electron : _______
       (1) positron (3) beta particle
       (2) neutron (4) alpha particle
    c. umbrella : rain
       wood : _______
       (1) radioactive decay (3) beta radiation
       (2) gamma radiation (4) x-radiation

53. Compare the half-life of an element to a single-elimination sports tournament.

54. Why does the relatively large mass and charge of an alpha particle limit its penetrating power?

55. Why might radioisotopes of C, N, and O be especially harmful to living creatures?

CUMULATIVE REVIEW

56. What is the Pauli exclusion principle? What is Hund's rule?

57. Balance the following equations.
    a. $Ca(OH)_2 + HCl \longrightarrow CaCl_2 + H_2O$
    b. $Fe_2O_3 + H_2 \longrightarrow Fe + H_2O$
    c. $NaHCO_3 + H_2SO_4 \longrightarrow Na_2SO_4 + CO_2 + H_2O$
    d. $C_2H_6 + O_2 \longrightarrow CO_2 + H_2O$

the thyroid is measured; in treatment, the radioactive iodine-131 is concentrated in and by the thyroid.

45. a. $^{30}_{15}P + {}^{0}_{-1}e \rightarrow {}^{30}_{14}Si$
    b. $^{13}_{6}C + {}^{1}_{0}n \rightarrow {}^{14}_{6}C$
    c. $^{131}_{53}I \rightarrow {}^{131}_{54}Xe + {}^{0}_{-1}e$

46. Nuclear fusion takes place in the sun. A nuclear reactor utilizes nuclear fission.

47. a. $^{32}_{16}S$ c. $^{4}_{2}He$ e. $^{185}_{79}Au$
    b. $^{14}_{6}C$ d. $^{141}_{57}La$

48. a. $^{90}_{38}Sr \rightarrow {}^{90}_{39}Y + {}^{0}_{-1}e$
    b. $^{14}_{6}C \rightarrow {}^{14}_{7}N + {}^{0}_{-1}e$
    c. $^{137}_{55}Cs \rightarrow {}^{137}_{56}Ba + {}^{0}_{-1}e$
    d. $^{239}_{93}Np \rightarrow {}^{239}_{94}Pu + {}^{0}_{-1}e$
    e. $^{50}_{22}Ti \rightarrow {}^{50}_{23}V + {}^{0}_{-1}e$

49. a. about 20% c. about 83 days
    b. about 85 g d. 25 days

50. a. $^{222}_{86}Rn \rightarrow {}^{218}_{84}Po + {}^{4}_{2}He$
    b. $^{234}_{90}Th \rightarrow {}^{230}_{88}Ra + {}^{4}_{2}He$
    c. $^{210}_{84}Po \rightarrow {}^{206}_{82}Pb + {}^{4}_{2}He$

51. a. Named radioactivity and discovered several radioactive elements.
    b. Discovered natural radioactivity from uranium ores.
    c. Discovered the neutron.
    d. Transmuted elements.

52. a. 4 b. 1 c. 3

53. In every round of the tournament, one-half the teams are eliminated; in every half-life, one-half of the substance decays. In the tournament a single team eventually emerges as the winner and the tournament stops, but radioactive decay continues (almost) indefinitely.

54. An alpha particle is much more likely than other kinds of radiation to collide with another particle and be stopped. At the atomic level, the larger the size of a particle, the greater is the chance of its striking another particle. The greater the magnitude of a particle's charge, the more strongly it will be attracted to particles of opposite charge.

55. Radioactive isotopes of these elements can be incorporated into the body tissues of organisms. When the isotopes decay, they can damage tissues very easily.

56. The Pauli exclusion principle states that no two electrons in an atom can have the same quantum numbers. Hund's rule states that electrons occupying orbitals of equal energy are distributed with unpaired spins as much as possible.

57. a. $Ca(OH)_2 + 2HCl \rightarrow CaCl_2 + 2H_2O$
    b. $Fe_2O_3 + 3H_2 \rightarrow 2Fe + 3H_2O$
    c. $2NaHCO_3 + H_2SO_4 \rightarrow Na_2SO_4 + 2CO_2 + 2H_2O$
    d. $2C_2H_6 + 7O_2 \rightarrow 4CO_2 + 6H_2O$

Answers

58. 6.7 mL

59. **a.** 26 p^+, 26 e^-, 33 n
b. 92 p^+, 92 e^-, 143 n
c. 24 p^+, 24 e^-, 28 n

60. **a.** covalent **c.** covalent
b. ionic **d.** ionic

61. 9216 cm^3 H_2; 0.4114 mol H_2

62. 2135

63. This graph shows the radioactive decay of carbon-14, along with the increase of the nitrogen product.

64. $^{211}_{83}Bi \rightarrow ^{207}_{81}Tl + ^{4}_{2}He$; thallium-207
$^{207}_{81}Tl \rightarrow ^{207}_{82}Pb + ^{0}_{-1}e$; lead-207

65. Bismuth-214 remains.

66. The reasoning is not sound. Cells other than cancer cells may be fast-growing and therefore killed by radiation as well.

58. You have a 0.30*M* solution of sodium sulfate. What volume (in mL) must be measured to give 0.0020 mol of sodium sulfate?

59. How many protons, neutrons, and electrons are in an atom of each isotope?
a. iron-59
b. uranium-235
c. chromium-52

60. Identify the bonds between each pair of atoms as ionic or covalent.
a. carbon and silicon
b. calcium and fluorine
c. sulfur and nitrogen
d. bromine and cesium

61. How many cubic centimeters of hydrogen gas (at STP) will be produced when 10.00 g of magnesium metal reacts with an excess of sulfuric acid? How many moles is this?

CONCEPT CHALLENGE

62. The radioisotope cesium-137 has a half-life of 30 years. A sample decayed at the rate of 544 counts per minute (cpm) in the year 1985. In what year will the decay rate be 17 cpm?

63. Describe the process depicted in the following graph.

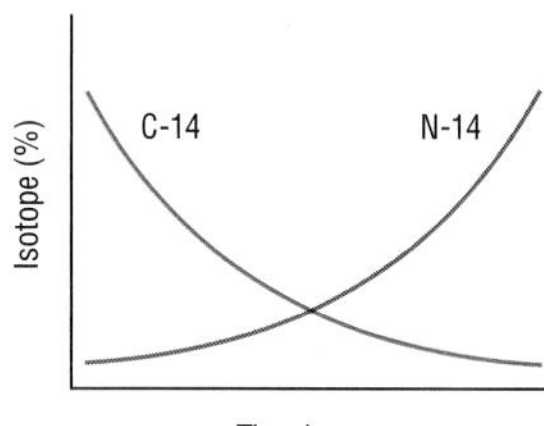

64. Bismuth-211 is a radioisotope. It decays by alpha emission to yield another radioisotope, which emits beta radiation as it decays to a stable isotope. Write equations for the nuclear reactions and name the decay products.

65. What isotope remains after three beta particles and five alpha particles are lost from a thorium-234 isotope? (Refer to the uranium-238 decay series to check your answer.)

66. Write a paragraph analyzing the overall logic of the reasoning in the following argument.
(1) Radiation kills fast-growing cells.
(2) Cancer cells are fast-growing.
(3) Therefore, radiation kills only cancer cells.

Chapter 28 STANDARDIZED TEST PREP

Select the choice that best answers each question or completes each statement.

1. If a radioisotope undergoes beta emission,
a. the atomic number changes.
b. the number of neutrons remains constant.
c. the isotope loses a proton.
d. the mass number changes.

2. The radioisotope radon-222 has a half-life of 3.8 days. How much of an initial 20.0-g sample of radon-222 would remain after 15.2 days?
a. 5.00 g **c.** 1.25 g
b. 12.5 g **d.** 2.50 g

3. Spent fuel rods
a. are no longer radioactive.
b. are stored under water for at least a decade.
c. contain only one isotope of uranium, ^{238}U.
d. remain radioactive for less than 100 years.

4. What particle is needed to balance this equation?

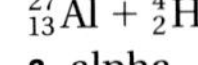

$^{27}_{13}Al + ^{4}_{2}He \longrightarrow$ ________ $+ ^{30}_{15}P$

a. alpha **c.** proton
b. electron **d.** neutron

For each nuclear equation in questions 5–8, name the particle that is being emitted or captured.

5. $^{59}_{26}Fe \longrightarrow ^{59}_{27}Co + ^{0}_{-1}e$

6. $^{185}_{79}Au \longrightarrow ^{181}_{77}Ir + ^{4}_{2}He$

7. $^{59}_{27}Co + ^{1}_{0}n \longrightarrow ^{60}_{27}Co$

8. $^{118}_{54}Xe \longrightarrow ^{118}_{53}I + ^{0}_{+1}e$

Use the graph to answer question 9.

9. Use the graph to determine whether neon-21, zirconium-90, and neodymium-130 have nuclei that are stable.

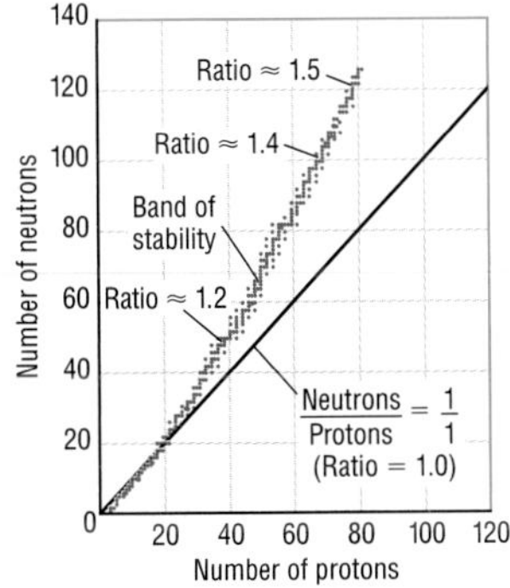

Use the graph to answer questions 10–12.

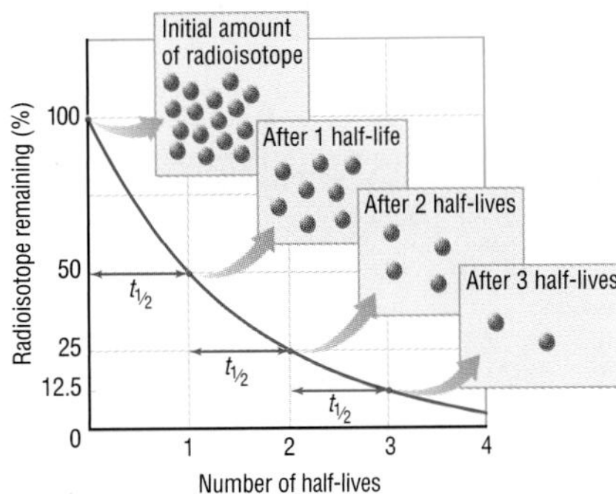

Estimate the percent remaining of the radioisotope after the given number of half-lives.

10. 0.5 $t_{1/2}$

11. 1.25 $t_{1/2}$

12. 3.75 $t_{1/2}$

Use the drawings of atomic nuclei to answer questions 13 and 14.

a.  **b.** 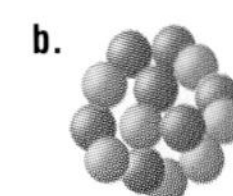 **c.** 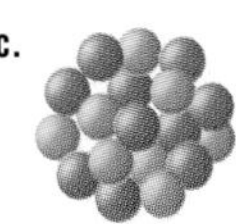

13. Write the name and symbol for each isotope.

14. Which isotope is radioactive?

The lettered choices below refer to questions 15–19. A lettered choice may be used once, more than once, or not at all.

(A) film badge
(B) radioactive tracer
(C) radiation therapy
(D) neutron activation analysis
(E) Geiger counter

Which of the above items or processes is best described by each of the following applications?

15. treating some cancers

16. detect ionizing radiation

17. monitoring exposure to radiation

18. diagnosing some diseases

19. detecting art forgeries

Chapter 28 STANDARDIZED TEST PREP

1. a
2. c
3. b
4. d
5. β emission
6. α emission
7. neutron capture
8. positron emission
9. Zirconium-90 and neon-21 nuclei are stable; neodymium-130 nuclei are unstable.
10. 71%
11. 42%
12. 7.5%
13. (a) is carbon-14, $^{14}_{6}C$; (b) is nitrogen-14, $^{14}_{7}N$; (c) is oxygen-18, $^{18}_{8}O$.
14. Carbon-14 is radioactive.
15. (C)
16. (E)
17. (A)
18. (B)
19. (D)

CHEMISTRY IN CAREERS

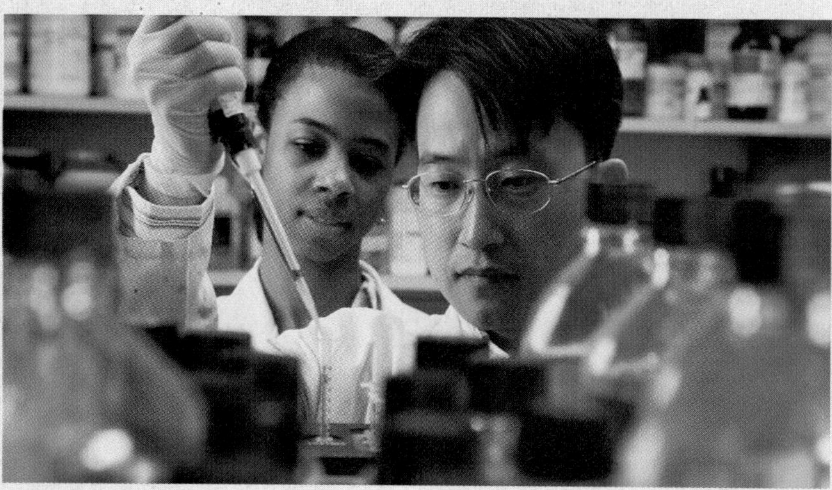

CHEMIST

The job descriptions of a chemist are very diverse. Some chemists solve problems that help make industrial processes more efficient. These industrial chemists perform quality control to ensure that the highest production standards are met. Other chemists conduct research in which they isolate new chemicals from a dazzling array of sources. Research chemists may invent processes that earn patents, or they may discover new chemicals that they get to name. Other chemists measure levels of chemicals in the environment, solve crimes, analyze the nutritional content of food, or invent new flavors.

Most chemists begin their careers by earning an undergraduate degree in chemistry. Many go on to get advanced degrees in fields such as analytical chemistry, forensic chemistry, biochemistry, organic chemistry, or nuclear chemistry. Throughout this textbook, you will have the opportunity to explore many of the career possibilities in chemistry.

MATERIALS SCIENTIST

Some scientists work solely with solid state materials. Some examples of solid state materials are safety glass, synthetic gems, superconductors, and ceramics. Materials scientists study properties such as heat conduction, strength-to-density ratios, electrical resistance, crystalline structure, and chemical reactivity. The fundamental knowledge gained by materials scientists is used to develop materials for specific applications.

A scientist who develops and describes solid state materials must be a physicist as well as a chemist. These scientists must know how a solid's crystal structure is formed and what it is made of. Such information is important if the scientist wants to describe how and where the different atoms in the solid are located. One way to map the crystalline structure of a solid is by using x-rays to probe a sample of the substance. This technique is called x-ray diffraction (XRD). When a sample is bombarded with x-rays, the x-rays scatter in a characteristic pattern, providing information on how the sample's atoms are arranged.

In addition to courses in general chemistry and physics, courses in the electronic structure of solids, the movement of heat, and applied mathematics are also recommended. A four- year degree can cover the basics of materials science if your courses are chosen with care. You also may pursue graduate studies in specialty

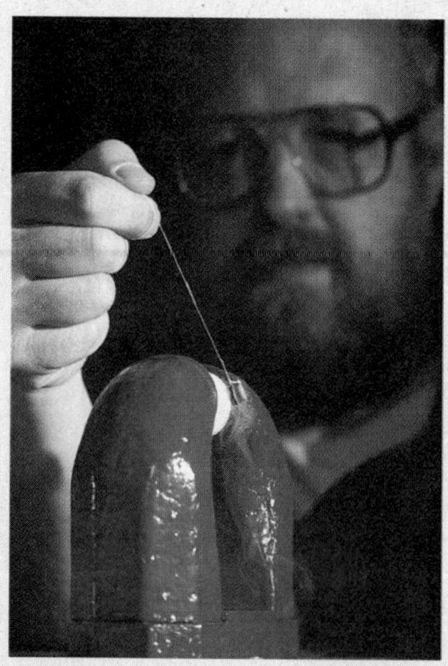

areas such as superconductivity, ceramics, corrosion resistance, and composite materials.

ANALYTICAL CHEMIST

Analytical chemists focus on making quantitative measurements. They must be familiar with many analytical techniques to work successfully on a wide variety of tasks. As an analytical chemist, you would spend your time making measurements and calculations to solve laboratory and math-based research problems. You could, for example, be involved in analyzing the composition of biomolecules. Pharmaceutical companies need people to analyze the composition of medicines and research new combinations of compounds to use as drugs. As an analytical chemist, you must be able to think creatively and develop new means for finding solutions.

Many exciting new fields, such as biomedicine and biochemistry, are now hiring analytical chemists. More traditional areas, including industrial manufacturers, also employ analytical chemists. The educational background you need to enter this field is quite extensive. You would need advanced chemical training, including organic chemistry and quantitative chemistry, as well as some training in molecular biology and computer operation. A master's degree in chemistry may be required, and certain positions require a Ph.D.

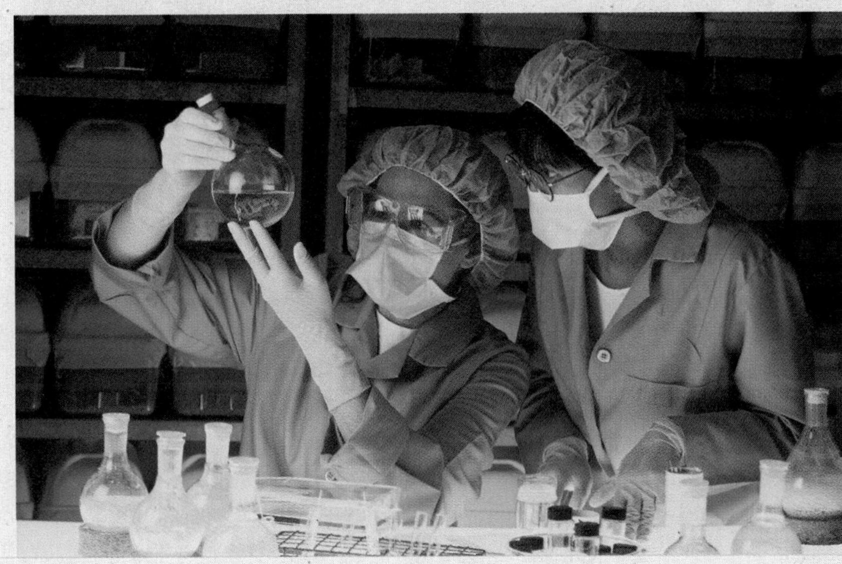

MEDICAL LABORATORY TECHNICIAN

When you go to a doctor's office for a physical examination, you typically provide a urine sample and have some blood drawn by a medical laboratory technician (MLT). Medical laboratory technicians are part medical aide and part chemist. These health care professionals perform standard laboratory tests that are used in the detection and treatment of human diseases and disorders. Typical tasks of an MLT may include collecting blood samples, classifying blood by type, and counting the different types of cells in a blood sample. They also measure the levels of enzymes and ions in the blood to detect abnormalities. MLTs prepare slides of tissues and fluids and examine them under a microscope. They isolate and identify microorganisms from samples such as urine and throat swabs. Working as part of a team with doctors, nurses, and other health care workers, MLTs help to maintain their patients' health. And knowledge of chemistry can go a long way in this endeavor!

To become an MLT, you must complete an accredited medical laboratory technician program. Such programs are offered by many community colleges and some hospitals. High school courses in biology, chemistry, computer science, and mathematics are recommended. MLTs are employed by hospitals, clinics, and public and private laboratories.

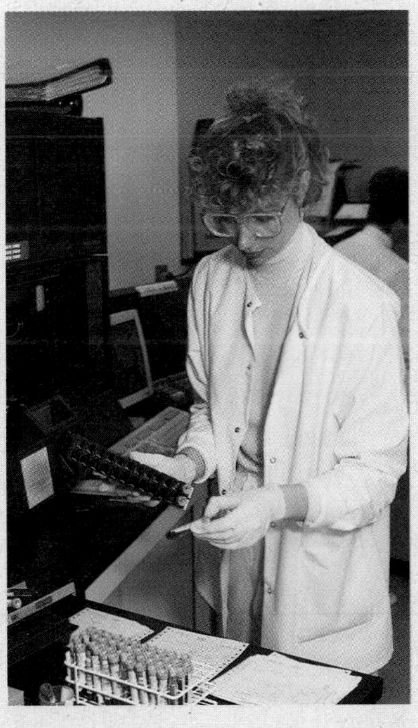

CHEMISTRY IN CAREERS

ARCHAEOLOGIST

Archaeologists are detectives of the past, sifting for clues that uncover secrets of past civilizations. Archaeologists excavate ancient cities and dwellings looking for artifacts that indicate what kinds of foods ancient people ate, how they built their homes, what types of tools they used, and how they interacted with one another as a society. Often, archaeologists must draw conclusions based on indirect evidence.

Knowing when an event occurred or when an artifact was made can provide important information. To obtain such information, archaeologists use radiometric-dating techniques such as carbon dating. This method tells them the age of a sample within a certain range, and is used with the greatest accuracy for samples that are not more than 10 000 years old. Carbon dating is often used in conjunction with other kinds of evidence, such as the types of tools or pottery present.

Archaeologists also perform chemical tests on samples of artifacts to determine their composition. For example, archaeologists might analyze the glazes used on pottery, or the dyes and fabrics used for clothing. Archaeologists may also use chemicals to preserve artifacts that have been unearthed so that the artifacts can be handled and examined without being damaged.

Archaeology requires a background in both history and science. Archaeologists often spend as much time in the laboratory studying their finds as they do out in the field excavating sites. Archaeologists take courses in archaeological techniques, biology, anatomy, chemistry, math, and history.

PHARMACIST

Pharmacists work in pharmacies where they dispense prescription drugs to customers. Pharmacists also advise people on how and when they should take medicine, describe any side effects they might experience, and answer any other questions the customers might have. Pharmacists also make sure that physicians have not prescribed a medicine or dosage that could harm the patient. Although pharmacists are trained to mix chemicals to prepare drugs, they usually dispense medicines prepared by large pharmaceutical companies.

Many pharmacists obtain a degree in pharmacology from college. This degree usually takes about five years to complete. Pharmacists must study chemistry, biology, mathematics, and statistics. They also take courses about the biological effects of drugs and drug interactions. After obtaining a degree, pharmacists have to pass a state test to obtain a license to dispense drugs. To complete the licensing procedure, they must work for a specified period of time under the supervision of another pharmacist.

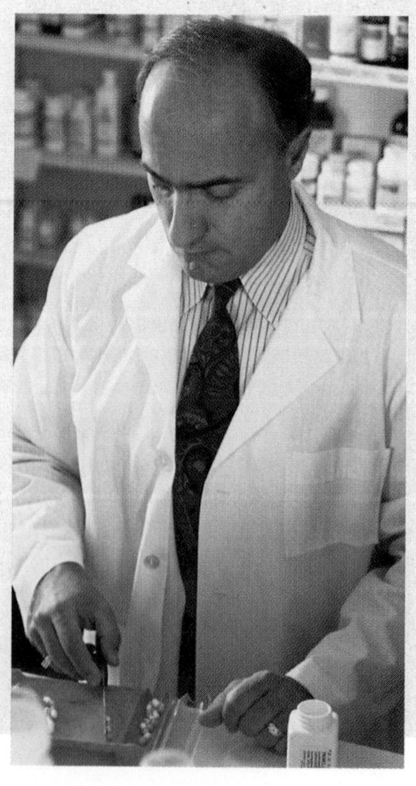

ECOLOGIST

Ecologists study relationships between the living and nonliving aspects of ecosystems. By experiment and observation, ecologists can determine how healthy an ecosystem is and pinpoint potential problems facing the system. As a water ecologist, you would work with water ecosystems such as ponds, rivers, and streams.

Water ecologists use chemistry as they study biology. One of the most important tools water ecologists use is called a dissolved gas test. Dissolved gas tests can be used to check the amounts of dissolved carbon dioxide and dissolved oxygen in different locations of a body of water, such as pools, riffles (areas of moving water), shaded regions, and full-sun regions. The amount of gases found in these areas helps the ecologist understand where life can exist and whether this life is more likely to be plant or animal. Water ecologists may dredge up the bottom of a body of water to uncover specimens of invertebrates and vertebrates living there. They may also map the land around bodies of water, noting plant and animal species as well as possible sources of pollution or human disturbance to the water.

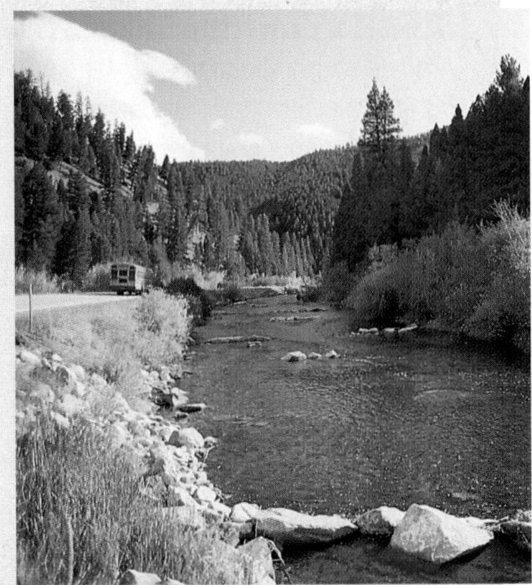

As a water ecologist, you must be skilled in designing and carrying out experiments, working in the field, analyzing data, recognizing different species of aquatic life, and collecting specimens. A bachelor's degree in biology or chemistry is required as well as training in water chemistry and in invertebrate, vertebrate, and plant classification.

FIREFIGHTER

If you are courageous, mentally alert, team-oriented, and have a strong sense of public service, you may wish to consider a career as a firefighter. Firefighters respond to fires, medical emergencies, and hazardous chemical spills. In responding to a fire, firefighters must know what types of chemicals can be used on different classes of fires. Having knowledge of the requirements of different chemical fires and spills enables firefighters to put out, rather than feed fires. In addition to these tasks, firefighters clean and maintain equipment, write reports, practice drills, and conduct fire inspections. Firefighters also engage in physical fitness activities to build stamina and improve agility and coordination. Regular reviews of fire science literature keep firefighters up to date on current technological developments and policy changes.

Although the job is hazardous and includes unpredictable hours, competition for jobs is relatively high. To become a firefighter, you must be at least 18 years of age with a high school education or equivalent. Selection for positions in fire departments often depends on passing a written test as well as a medical examination that includes drug screening and a physical fitness test. Once hired, training consists of instruction in firefighting and rescue techniques, the use of fire and rescue tools, emergency medical procedures, basic chemistry of fires and firefighting, and instruction in local building codes and fire prevention methods. Promotion within the department often depends on further training and education, such as college-level courses in fire engineering and fire science. Higher positions may require a master's degree in administration.

CHEMISTRY IN CAREERS

QUALITY CONTROL CHEMIST

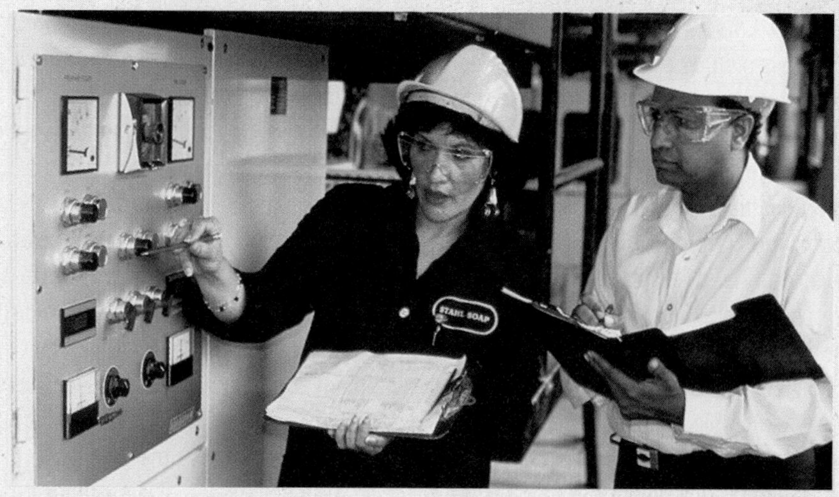

A quality control chemist is responsible for seeing that products meet manufacturing standards. The consistent quality of many products you use is due to the work of a quality control chemist. They work in a wide variety of industries, including those that manufacture airbags for automobiles. They are involved in every aspect of producing the product. They contribute to the original research and development of the product, and help refine its manufacturing process. They then develop methods to test the product, or to determine why it has failed to perform for the consumer.

Quality control chemists must know how to make and interpret measurements. Experience in analytical chemistry and the ability to develop new analytical methods are particularly important. The specific requirements for the job depend on the industry and company that employs the quality control chemist. A good quality control chemist assures that his or her company's customers receive the highest-quality products.

CLIMATOLOGIST

If you want to know what tomorrow's weather will be, you may watch a television weather forecast. What if you wanted to know what the weather will be like fifty years from now? A climatologist is a scientist who studies patterns of weather change and uses them to make such predictions. A climatologist might use complex computer programs, called forecast models, that take into account many atmospheric variables, such as temperature, wind speed, and pressure. These programs help climatologists make long-term predictions. For example, a forecast model might indicate that an increase in the level of greenhouse gases in the atmosphere will result in an increase in the level of Earth's oceans. A climatologist would use this forecast model to help governments develop plans to minimize the adverse effects that a rise in greenhouse gases would have on such areas as biodiversity, agriculture, and health.

Climatologists also study what Earth's climate was like thousands, or even millions, of years ago. They can use such data to learn how weather has changed over time and predict how it may change in the future. For example, climatologists study ancient ice core samples. The chemical composition of the gases trapped in the ice can be determined, indicating the makeup of the atmosphere at the time the ice formed.

A climatologist needs training in many scientific and mathematical areas, including physical chemistry, meteorology, physics, statistics, and ecology. A bachelor's or master's degree in meteorology or a related science is required. A climatologist may work in field areas, in a laboratory, or at a computer terminal. Research and computer skills are required.

SOLAR ENGINEER

A solar engineer is responsible for evaluating the performance of photovoltaic cells and modules made of different types, and percentages of, semiconducting materials. Photovoltaic cells are devices that convert the sun's energy into electricity. A solar engineer works with the consumer in mind, seeking to increase the safety, performance levels, and reliability of photovoltaic devices while keeping costs low. The engineer seeks to improve the function of the photovoltaic mechanism: converting sunlight directly into electricity.

Photovoltaic cells and modules are used for a wide variety of tasks. They power the emergency call boxes that are seen along most interstate highways, as well as street lights, solar calculators, and some traffic signals. A solar engineer is responsible for designing and conducting experiments in the field and in the laboratory. The engineer must be able to analyze and draw conclusions from experimental data. He or she must also understand the basics of solar energy, and have a working knowledge of designing and building photovoltaic devices and systems. A bachelor's degree in physics or a related field is usually required, as well as experience with photovoltaics.

COMMERCIAL DIVER

Commercial divers work under tremendous pressure: pressure not only from needing to do a difficult job correctly, but also from working under water. Underwater pressures can exceed ten atmospheres. Commercial divers work as welders and general construction workers. They also perform underwater inspections of dams, pilings, cables and pipelines, bridge and pier foundations, and ship hulls. Sometimes they have to spend many hours undergoing decompression for doing just five minutes of work.

Commercial divers work in a wide variety of jobs. They undergo specialized training in the underwater techniques of construction, welding, and demolition. They take courses that train them to use air and other specialized gases for underwater breathing purposes. It is important for divers to understand the chemistry of gases, as proper blood levels of the dissolved gases oxygen and nitrogen are critical to the divers' safety. In addition, divers must be aware of possible hazardous reactions from materials they use to perform various underwater repair and demolition tasks. Commercial divers need to understand important principles of mathematics, physics, and chemistry to perform their jobs.

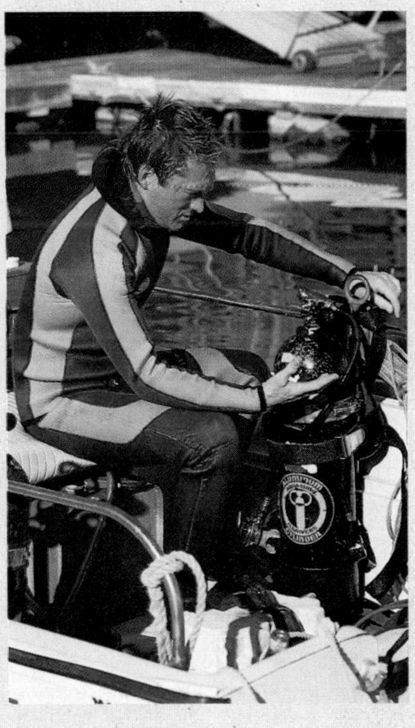

CHEMISTRY IN CAREERS

LASER TECHNICIAN

As lasers become a more commonly used tool in many industries, there will be a greater demand for people who know how to operate and maintain them. These jobs often belong to laser technicians. One type of laser technician, a medical laser specialist (MLS), is responsible for maintaining medical laser equipment and for supervising the operation of the lasers. The MLS prepares the laser equipment for use by a surgeon, makes suggestions about settings and adjustments, and operates the controls while the surgeon performs a medical procedure. The MLS is also responsible for the safety of the staff and patients while the laser is in use, and must keep surgeons up to date on new equipment and procedures.

Another type of laser technician is an industrial laser technician (ILT). An ILT is responsible for setting up and maintaining industrial laser equipment, such as laser cutting and welding devices. The ILT must be able to program laser equipment to perform specific tasks and to train other personnel in the proper use of the laser. ILTs are often responsible for the maintenance and repair of the laser equipment, as well as maintaining laser safety in the workplace.

Laser technicians usually have a two-year Associate of Applied Science degree in laser technology from a community or technical college. Laser technicians must complete courses in math and science to understand how lasers work and how to operate them safely.

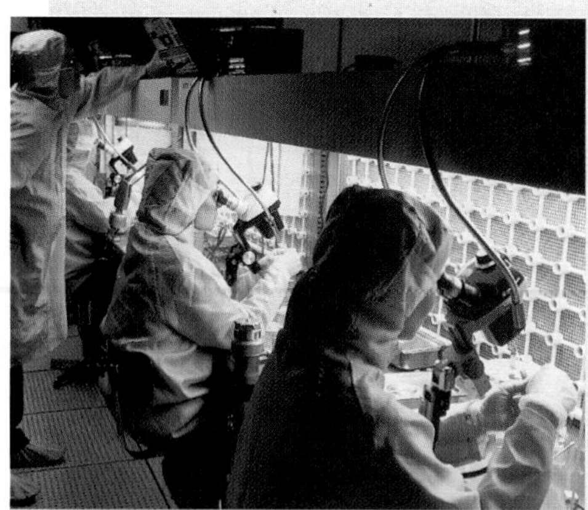

SOLID STATE CHEMIST

By now you are familiar with chemical reactions in aqueous solutions and the gaseous state. But did you know that chemical reactions can also take place in solids, without the addition of a liquid or a gas? The process of doping semiconductors, for example, involves reactions that take place in the solid state. Manufacturing semiconductors is not the only application for solid state chemistry, however. The lighter on a gas grill, for example, is made of a type of ceramic called a piezoelectric. Devices that operate on the piezoelectric effect produce a voltage when they are subjected to mechanical pressure. Solid photopolymers are now used in place of gelatin in some photographic processes. Even the darkening process used in photosensitive lenses is a solid state chemical reaction.

Solid state chemists are interested in the study of crystal structures, superconductors, semiconductors, and electrical, magnetic, and optical properties of solid compounds. They often work at major universities, teaching and performing basic research, or for corporations, doing research or developing applications for new discoveries in this field. Companies in the petroleum, electronic, and computer industries are just a few of the places a solid state chemist might be employed. To work in this field usually requires an undergraduate or graduate degree in chemistry or chemical engineering. Courses studied include chemistry, physics, and mathematics.

WASTEWATER ENGINEER

Wastewater must be physically and chemically treated before it is returned to the environment or recycled for human use. A wastewater engineer is responsible for monitoring this process. These engineers control the amount of water treated, the level of treatment, and the quality of water produced. When water is received at a treatment plant, a wastewater engineer oversees primary treatment that involves filtering the water for solids and debris. The engineer then adds microorganisms to the water that convert the remaining dissolved organic matter into solids that can be removed as sludge. If this water is to be released into a river or stream, the engineer then tests the water for levels of remaining organic compounds. If amounts of these compounds fall below recommended levels, the water is safe to be released. Production of drinking water often requires additional steps. The engineer must test the water for nitrogen and phosphorus compounds that would need to be removed. The engineer then disinfects the water using chlorine, ozone, or other disinfectants, and might possibly add fluoride to strengthen consumers' teeth.

A wastewater engineer must be able to test water for very small amounts of ions, disease-causing organisms, and other unwanted chemicals or microbes. Interpreting the results of such tests requires experience and an understanding of the entire chemical, and often biological, environment. Wastewater engineers usually have a degree in engineering. They must take courses in water toxicology, organic chemistry, and environmental biology.

ONCOLOGIST

If helping others and studying cells interests you, a career as a cancer specialist, or oncologist, may be for you. An oncologist diagnoses the various forms and stages of cancer and determines the appropriate treatment. Perhaps you will surgically remove abnormal cancerous skin cells or give your patients the latest anticancer drugs developed by biotechnology. You may have to combine surgery and chemotherapy to maximize the destruction of the cancerous cells. Radiation therapy is another often-used treatment. Application of radiation to the affected area kills cancer cells and shrinks tumors by breaking bonds and damaging cell structures. Unfortunately, radiation also damages healthy cells, so you must carefully balance the benefits of radiation treatment with the damaging side effects. As a cancer specialist, you must be aware of the latest medical techniques and drugs, and the potential side effects of treatments. You must also have a strong knowledge of patient care and disease statistics.

As there are many new cancer cases diagnosed every day in the United States, being an oncologist is a valuable profession. Hospitals, medical research centers, and pharmaceutical and biotechnology companies employ oncologists. Becoming an oncologist requires an undergraduate background in chemistry and biology, a four-year medical degree, and several additional years of specialized training.

CHEMISTRY IN CAREERS

OCEANOGRAPHER

Oceanography is the study of oceans. Because oceans are complex systems, oceanography is broken down into several interrelated subfields.

Physical oceanographers study the movement of seawater by waves, currents, and tides, and the interactions between the ocean and the atmosphere. Chemical oceanographers are concerned with the composition of seawater and the interactions between seawater and the organic and inorganic material in it. Geological oceanographers study the structure, features, and evolution of the ocean floor. Biological oceanographers study the diverse plants and animals of the sea and their relationship to their environment.

Oceanographers use many different tools to explore the oceans. Some use submersible vehicles to observe conditions in and take samples from the deepest parts of the ocean. Others wear scuba gear and visit the ocean floor. Oceanographers also use aircraft and satellites to study water temperatures and wave heights. When they are not in the field, oceanographers use computers to analyze data and create theoretical models.

Most oceanographers have advanced degrees. They begin their college education by obtaining a bachelor's degree in biology, marine biology, chemistry, or geology. They then obtain a master's degree or doctorate in a more specialized field, such as marine geology, meteorology, marine biology, or aquatic chemistry.

NEPHROLOGY NURSE

Because renal disease (disease of the kidneys) can be so debilitating to patients who have it, caring for these patients is an important task. Nephrology nurses are health-care professionals who specialize in the treatment and care of renal disease patients. The word nephrology comes from nephron, the name given to each of the more than one million tiny blood-processing units contained in each kidney. Once a patient has been diagnosed with renal disease, a nephrology nurse will assist the patient's doctor in assessing the disease's impact on the patient. Nephrology nurses are often the health-care professionals who help patients during hemodialysis. A major goal of nephrology nurses is to help patients regain the highest level of functioning while undergoing lengthy dialysis treatments. Duties include connecting patients to the dialysis machine, answering questions about treatment and progress, and providing moral support to patients who must endure hours of treatment. Nephrology nurses also assist in renal surgery such as kidney transplants.

Nephrology nurses are usually registered nurses (RNs) who specialize in nephrology. This level of certification requires an undergraduate degree in nursing plus special training in nephrology. Many vocational nurses are also involved in nephrology, and must undergo several years of training after high school.

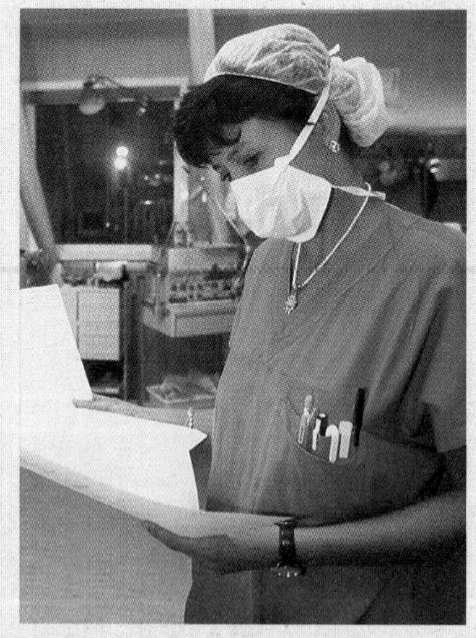

FDA INSPECTOR

The Food and Drug Administration (FDA) touches the lives of most Americans every day. It is the FDA's job to see that the food we eat is safe and wholesome, that the cosmetics we use will not harm us, and that the medicines we take are safe and effective. The FDA also ensures that these products are labeled accurately with the information that people need to use them properly.

FDA is a public health agency with approximately 1100 inspectors who visit more than 15 000 facilities every year. FDA inspectors are responsible for conducting inspections of the procedures and products related to the manufacture of cosmetics, dietary supplements, and food additives. They also do investigative work related to biotechnology and to food-borne illnesses. Inspecting and regulating food labeling and nutrition information, as well as the import and export of foods, are other responsibilities of FDA inspectors. FDA inspectors also test food for pesticides and chemical contaminants. Because of the extensive range of job tasks, FDA inspectors tend to specialize in only one area.

FDA inspectors usually have college degrees in biology, microbiology, chemistry, or food and nutrition science. They also take courses in law, economics, mathematics, and statistics. FDA inspectors have years of on-the-job training, and they take annual courses in regulatory changes.

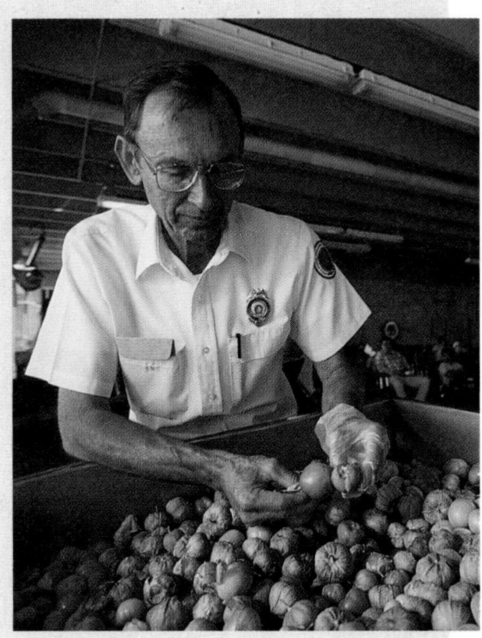

STONE CONSERVATOR

As more is learned about the damaging effects of airborne pollutants, such as acid rain, there is a growing concern to preserve historic buildings and pieces of sculpture from damage. Stone conservators work to both prevent and repair the damage to stone used in buildings and sculpture.

Stone conservators must clean statues properly to remove pollution deposits. One method they use is to apply a thin, clay mudpack to the stone's surface to pull out the deposits. Lasers are also used to remove pollution from stone. When making repairs, stone conservators sometimes use surgical microscopes to examine the surfaces of statues. If the stone was originally painted, the original paint is chemically analyzed to determine its composition. New paint is then produced that closely matches the original paint's composition. In addition, conservators research and apply methods to preserve the stone once it has been repaired. One preservation technique is to seal the stone to prevent water, which carries dissolved gases and salts, from seeping into the pores of the stone.

Stone conservators often work for museums, universities, or private companies that specialize in the cleaning and restoration of stone objects. They must have a knowledge of both chemistry and art because they often work closely with both chemists and museum personnel. A stone conservator's level of education may range from a bachelor's degree to a doctorate.

CHEMISTRY IN CAREERS

MICROBIOLOGIST

Microbiologists generally work with microscopic organisms, or organisms that are too small to be seen with the unaided eye. These organisms include bacteria, molds, protozoans, and algae. Microbiologists study the structure and function of these organisms, looking for ways in which these tiny life forms can benefit humans in fields such as industry and medicine.

For example, in the mining industry, microbiologists have discovered several different species of bacteria that can be used to retrieve specific metals from ores. These types of bacteria usually convert a chemical compound in the ore into a different compound, which is then extracted from the bacteria. The bacteria-formed compound then undergoes a third reaction to yield the pure metal. Currently, bacteria are used in the mining of copper, and research is being done on the use of bacteria in the mining of zinc and lead.

In medicine, microbiologists have developed ways to use bacteria to produce human proteins. This is done by inserting a human gene with the genetic code for a specific protein into the DNA of a bacterium. The bacterium will then produce the desired protein. Using this technique, genetically engineered bacteria are used to produced insulin, an important human hormone. People with diabetes do not produce enough insulin in their own bodies and must rely on daily insulin injections to survive.

Microbiologists usually have at least a bachelor's degree in microbiology, which requires some courses in chemistry and mathematics. Jobs involving higher-level research generally require a master's or doctorate degree.

MECHANICAL ENGINEER

Redox reactions can be used to produce heat, such as in the Flameless Ration Heater; they can also be used in fuel cells to create electricity. Fuel cells may be supplying your electrical needs in the future, but proving that a concept is theoretically workable is a far cry from putting it to practical use in people's homes.

Mechanical engineers work as part of a large team of scientists and other engineers who design not only fuel cells that can be produced easily and affordably, but also the components that complement such technology. For example, engineers must also design power grids to distribute the electricity produced by fuel cells.

Mechanical engineering is certainly not limited to the field of fuel-cell technology. Because mechanical engineers help to design all types of machines, they are employed in any industry that involves the manufacture or use of machinery. To be successful in their career, mechanical engineers must be knowledgeable about the interactions of different types of materials: Strength, corrosiveness, and a variety of other properties must be considered in the design.

A mechanical engineer typically has a bachelor's degree with college courses in mathematics, chemistry, physics, and computer science. Good communication skills are also important to a mechanical engineer because oral or written expression of ideas is an essential element of the job.

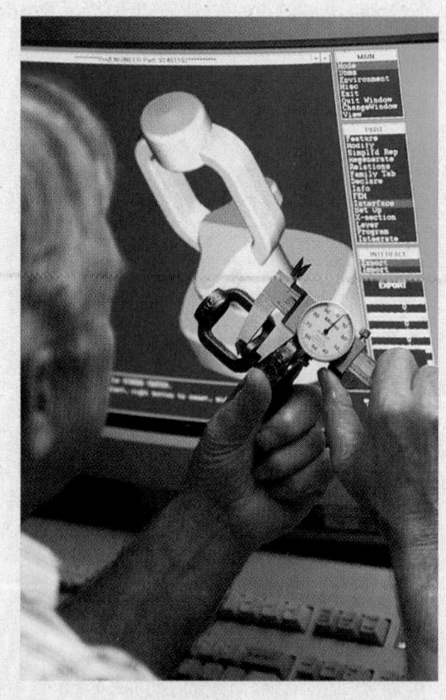

ELECTROCHEMIST

Electrochemistry provides a variety of valuable applications. Some electrochemists are interested in developing high-energy batteries that convert chemical energy to electrical power. These batteries can be used to power electric vehicles. The goal is to make batteries that can hold a charge longer and still be small enough to use in a car. Along with batteries, electrochemists investigate the function and design of high-performance fuel cells. Among other things, they are interested in the effects of temperature variation on the performance of these cells, because of their use in space and other hostile environments.

Electrochemists are also interested in designing instruments that will detect minute levels of environmental pollutants, such as carcinogens. Other electrochemists study corrosion. They want to know how to slow down or completely stop the oxidation of metal used to build bridges and roadways, the hulls of ships, and the components of space vehicles. Metal refining and purification is another area of interest to electrochemists.

Electrochemists are employed by universities to teach and work in research laboratories. They can also find work in the computer and electronics industries. Steel companies and the space program also hire electrochemists. Degrees in electrochemistry can be obtained at all levels, from a bachelor's degree to a doctorate. Electrochemists can also have degrees in materials science and chemical engineering. During the course of study, a prospective electrochemist may be required to take classes in physics and math, in addition to chemistry.

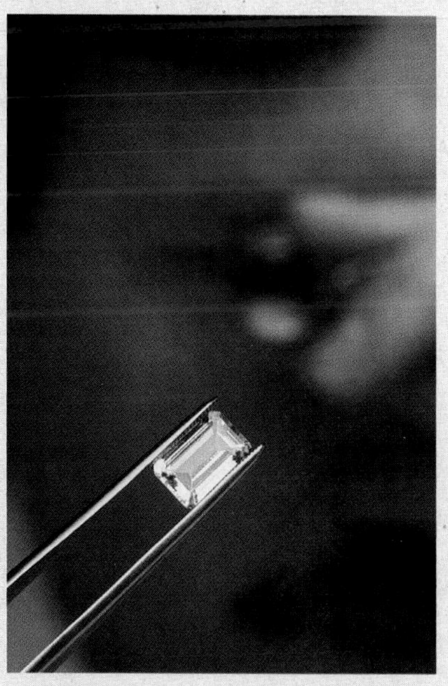

GEMOLOGIST

A gemologist grades gemstones on the basis of their color, clarity, and the presence or absence of flaws. Some gemologists specialize in cutting gemstones, which is an art that requires years of practice to perfect. Cutting gemstones brings out the crystal characteristics that make the gemstones sparkle. Other gemologists specialize in matching gemstones with jewelry settings.

Becoming a gemologist requires years of specialized training. Gemologists learn about different gemstones and their chemical composition. They must study crystal structure and understand how it enhances a gem. They must be familiar with compounds of transition elements that produce the different colors and shades of the gems. And they must know how these chemical features relate to value.

Gemologists undergo training in gem appraisal and cutting. This training involves examining many different gemstones to become familiar with identifying the unique characteristics of each gem. Gemologists usually specialize in one or a few types of gemstones. Gemologists are often employed in jewelry stores, insurance companies, and gem mines.

CHEMISTRY IN CAREERS

ORGANIC CHEMIST

Organic chemists work with carbon-containing compounds. Most organic chemists create and analyze new chemicals made from these carbon-containing building blocks. You use these compounds in just about every part of your life, from medicine to cosmetics, to food. An organic chemist might be best described as part inventor and part detective. Organic chemists make use of both scientific theories and techniques to build specific molecules that react in certain ways. To confirm the identity of these new molecules, organic chemists use many advanced technologies. Infrared (IR) spectrophotometry and nuclear magnetic resonance (NMR) spectrometry are two of the most common techniques for characterizing a chemical sample.

To become an organic chemist, you will need at least an undergraduate education in chemistry. You must have a firm background in physical, inorganic, and analytical chemistry in addition to courses in organic chemistry. If you are interested in the computer-aided design of bioactive molecules, such as new drugs, biochemistry and computer training are also important. An organic chemist must know how to use laboratory equipment and must be aware of safety guidelines for handling dangerous substances.

FLAVOR DEVELOPER

Do you love ice cream? Could you create flavor after flavor after flavor? If so, then the job of flavor developer could be for you. But don't be fooled! Flavor chemistry is not as easy as it sounds. A flavor developer needs to be familiar with the tools of instrumental and sensory analysis. You will also need to know the basics of biochemistry, chromatography, flavor compound isolation and characterization, and flavor model design. Knowledge of organometallics, a class of compounds used in flavor chemistry, would also be beneficial.

As a flavor chemist, you may find yourself working on a variety of projects beyond basic flavor development. For example, you may face the task of reducing the bitterness of a compound, or of replacing a fatty acid with a synthetic component. You may be involved in researching new ways to make flavor compounds, or ways to use esters or aromatic aldehydes to enhance a product's fragrance. You may even be hired to use the 9000 taste buds in your mouth to check ice cream (or other foods) for taste quality and consistency.

To get a job in flavor chemistry, you will need either a bachelor's degree or a master's degree in food science, biochemistry, or chemistry.

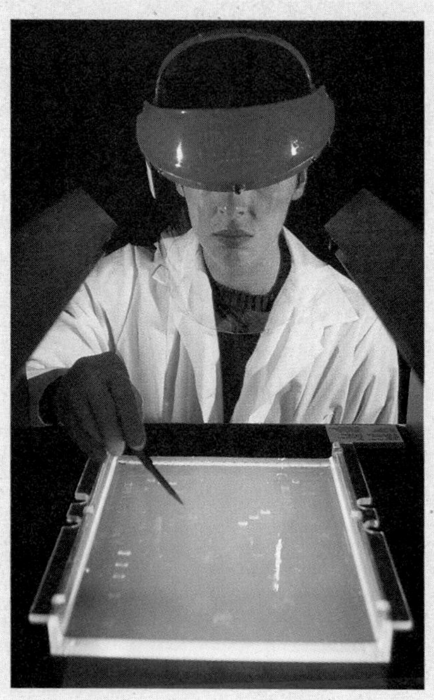

BIOCHEMIST

Biochemists study the molecules that make up organisms and that participate in biological processes. Biochemists identify enzymes and determine reaction mechanisms for biochemical reactions. Some biochemists specialize in food chemistry and digestion processes. They develop new foods, flavors, and preservatives, and study how vitamins and minerals are used in the body. Other biochemists work with DNA and RNA to identify organisms and their evolutionary relationships. Biochemists also work for pharmaceutical companies, where they develop new drugs and study their effects.

Biochemists usually earn either a master's degree or a doctorate after obtaining a bachelor's degree in either biology or chemistry. During their studies, biochemists take courses in mathematics, statistics, genetics, and physics in addition to courses in biology and chemistry. When obtaining an advanced degree, they usually specialize in one area, such as the biochemistry of nutrients, cellular processes, or genetics.

NUCLEAR PHYSICIAN

Nuclear physicians are medical doctors who specialize in the use of radiation and radioisotopes to diagnose, manage, treat, and prevent serious diseases. Nuclear physicians are often involved in nuclear imaging of the body, which is a safe and highly effective method of studying the structure and function of organs.

One use of isotopes is to inject them into the body. Different organs take up the isotope in different amounts. By recording the amount of radiation emitted and the concentration of the isotope at different points in the body, nuclear physicians can determine the size and shape of irregularities present in an organ. Nuclear physicians have found that different isotopes have a tendency to localize in certain organs. For example, iodine-131 concentrates in the thyroid and, therefore, can be useful in diagnosing defects in that organ. Also, carbon-14 can be used to study diabetes. Nuclear physicians also administer small and safe doses of radioisotopes to treat certain diseases.

A nuclear physician must first earn a medical degree and specialize in a specific field; then at least three years of postdoctoral training is necessary to become a licensed nuclear physician. Nuclear physicians must take courses in biology, chemistry, mathematics, physics, and computer science before entering a medical school program. Entrance into medical school is highly competitive, so care must be taken to maintain a high grade-point average.

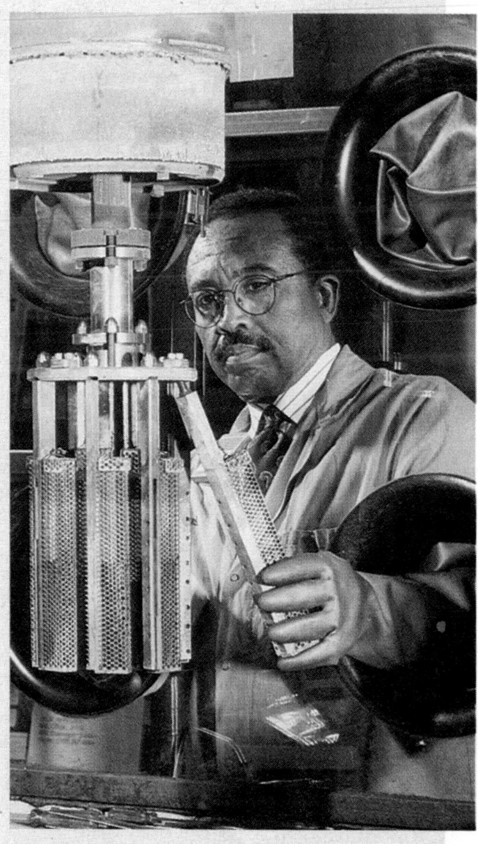

APPENDIX A REFERENCE TABLES

Table A.1 Some Properties of the Elements

Element	Symbol	Atomic number	Atomic mass	Melting point (°C)	Boiling point (°C)	Density (g/cm³) (gases at STP)	Major oxidation states
Actinium	Ac	89	(227)	1050	3200	10.07	+3
Aluminum	Al	13	26.98154	660.37	2467	2.6989	+3
Americium	Am	95	243	994	2607	13.67	+3, +4, +5, +6
Antimony	Sb	51	121.75	630.74	1950	6.691	−3, +3, +5
Argon	Ar	18	39.948	−189.2	−185.7	0.0017837	
Arsenic	As	33	74.9216	817	613	5.73	−3, +3, +5
Astatine	At	85	(210)	302	337	—	
Barium	Ba	56	137.33	725	1640	3.5	+2
Berkelium	Bk	97	(247)	986	—	14.78	
Beryllium	Be	4	9.01218	1278	2970	1.848	+2
Bismuth	Bi	83	208.9804	271.3	1560	9.747	+3, +5
Bohrium	Bh	107	(264)	—	—	—	
Boron	B	5	10.81	2079	3675	2.34	+3
Bromine	Br	35	79.904	−7.2	58.78	3.12	−1, +1, +5
Cadmium	Cd	48	112.41	320.9	765	8.65	+2
Calcium	Ca	20	40.08	839	1484	1.55	+2
Californium	Cf	98	(251)	900	—	14	
Carbon	C	6	12.011	3550	4827	2.267	−4, +2, +4
Cerium	Ce	58	140.12	799	3426	6.657	+3, +4
Cesium	Cs	55	132.9054	28.40	669.3	1.873	+1
Chlorine	Cl	17	35.453	−100.98	−34.6	0.003214	−1, +1, +5, +7
Chromium	Cr	24	51.996	1857	2672	7.18	+2, +3, +6
Cobalt	Co	27	58.9332	1495	2870	8.9	+2, +3
Copper	Cu	29	63.546	1083.4	2567	8.96	+1, +2
Curium	Cm	96	(247)	1340	—	13.51	+3
Dubnium	Db	105	(262)	—	—	—	
Dysprosium	Dy	66	162.50	1412	2562	8.550	+3
Einsteinium	Es	99	(252)	—	—	—	
Erbium	Er	68	167.26	159	2863	9.066	+3
Europium	Eu	63	151.96	822	1597	5.243	+2, +3
Fermium	Fm	100	(257)	—	—	—	
Fluorine	F	9	18.998403	−219.62	−188.54	0.001696	−1
Francium	Fr	87	(223)	27	677	—	+1
Gadolinium	Gd	64	157.25	1313	3266	7.9004	+3
Gallium	Ga	31	69.72	29.78	2403	5.904	+3
Germanium	Ge	32	72.59	937.4	2830	5.323	+2, +4
Gold	Au	79	196.9665	1064.43	3080	19.3	+1, +3
Hafnium	Hf	72	178.49	2227	4602	13.31	+4
Hassium	Hs	108	(265)	—	—	—	
Helium	He	2	4.00260	−272.2	−268.934	0.001785	
Holmium	Ho	67	164.9304	1474	2695	8.795	+3
Hydrogen	H	1	1.00794	−259.14	−252.87	0.00008988	+1
Indium	In	49	114.82	156.61	2080	7.31	+1, +3
Iodine	I	53	126.9045	113.5	184.35	4.93	−1, +1, +5, +7
Iridium	Ir	77	192.22	2410	4130	22.42	+3, +4
Iron	Fe	26	55.847	1535	2750	7.874	+2, +3
Krypton	Kr	36	83.80	−156.6	−152.30	0.003733	
Lanthanum	La	57	138.9055	921	3457	6.145	+3
Lawrencium	Lr	103	(262)	—	—	—	+3
Lead	Pb	82	207.2	327.502	1740	11.35	+2, +4
Lithium	Li	3	6.941	180.54	1342	0.534	+1
Lutetium	Lu	71	174.967	1663	3395	9.840	+3
Magnesium	Mg	12	24.305	648.8	1090	1.738	+2
Manganese	Mn	25	54.9380	1244	1962	7.32	+2, +3, +4, +7
Meitnerium	Mt	109	(268)	—	—	—	

Table A.1 Some Properties of the Elements (cont.)

Element	Symbol	Atomic number	Atomic mass	Melting point (°C)	Boiling point (°C)	Density (g/cm³) (gases at STP)	Major oxidation states
Mendelevium	Md	101	257	—	—	—	+2, +3
Mercury	Hg	80	200.59	−38.842	356.58	13.55	+1, +2
Molybdenum	Mo	42	95.94	2617	4612	10.22	+6
Neodymium	Nd	60	144.24	1021	3068	6.90	+3
Neon	Ne	10	20.179	−248.67	−246.048	.0008999	
Neptunium	Np	93	(237)	640	3902	20.25	+3, +4, +5, +6
Nickel	Ni	28	58.69	1453	2732	8.902	+2, +3
Niobium	Nb	41	92.9064	2468	4742	8.57	+3, +5
Nitrogen	N	7	14.0067	−209.86	−195.8	.0012506	−3, +3, +5
Nobelium	No	102	(259)	—	—	—	+2, +3
Osmium	Os	76	190.2	3045	5027	22.57	+3, +4
Oxygen	O	8	15.9994	−218.4	−182.962	.001429	−2
Palladium	Pd	46	106.42	1554	2970	12.02	+2, +4
Phosphorus	P	15	30.97376	44.1	280	1.82	−3, +3, +5
Platinum	Pt	78	195.08	1772	3827	21.45	+2, +4
Plutonium	Pu	94	(244)	641	3232	19.84	+3, +4, +5, +6
Polonium	Po	84	(209)	254	962	9.32	+2, +4
Potassium	K	19	39.0982	63.25	760	.862	+1
Praseodymium	Pr	59	140.9077	931	3512	6.64	+3
Promethium	Pm	61	(145)	1168	2460	7.22	+3
Protactinium	Pa	91	231.0359	1560	4027	15.37	+4, +5
Radium	Ra	88	(226)	700	1140	5.5	+2
Radon	Rn	86	(222)	−71	−61.8	.00973	
Rhenium	Re	75	186.207	3180	5627	21.02	+4, +6, +7
Rhodium	Rh	45	102.9055	1966	3727	12.41	+3
Rubidium	Rb	37	85.4678	38.89	686	1.532	+1
Ruthenium	Ru	44	101.07	2310	3900	12.41	+3
Rutherfordium	Rf	104	(261)	—	—	—	
Samarium	Sm	62	150.36	1077	1791	7.520	+2, +3
Scandium	Sc	21	44.9559	1541	2831	2.989	+3
Seaborgium	Sg	106	(263)	—	—	—	
Selenium	Se	34	78.96	217	684.9	4.79	−2, +4, +6
Silicon	Si	14	28.0855	1410	2355	2.33	−4, +2, +4
Silver	Ag	47	107.8682	961.93	2212	10.50	+1
Sodium	Na	11	22.98977	97.81	882.9	0.971	+1
Strontium	Sr	38	87.62	769	1384	2.54	+2
Sulfur	S	16	32.06	112.8	444.7	2.07	−2, +4, +6
Tantalum	Ta	73	180.9479	2996	5425	16.654	+5
Technetium	Tc	43	(98)	2172	4877	11.50	+4, +6, +7
Tellurium	Te	52	127.60	449.5	989.8	6.24	−2, +4, +6
Terbium	Tb	65	158.9254	1356	3123	8.229	+3
Thallium	Tl	81	204.383	303.5	1457	11.85	+1, +3
Thorium	Th	90	232.0381	1750	4790	11.72	+4
Thulium	Tm	69	168.9342	1545	1947	9.321	+3
Tin	Sn	50	118.69	231.968	2270	7.31	+2, +4
Titanium	Ti	22	47.88	1660	3287	4.54	+2, +3, +4
Tungsten	W	74	183.85	3410	5660	19.3	+6
Ununbium	Uub	112	(277)	—	—	—	
Ununhexium	Uuh	116	—	—	—	—	
Ununnilium	Uun	110	(269)	—	—	—	
Ununoctium	Uuo	118	—	—	—	—	
Ununquadium	Uuq	114	—	—	—	—	
Unununium	Uuu	111	(272)	—	—	—	
Uranium	U	92	238.0289	1132.3	3818	18.95	+3, +4, +5, +6
Vanadium	V	23	50.9415	1890	3380	6.11	+2, +3, +4, +5
Xenon	Xe	54	131.29	−111.9	−107.1	.005887	
Ytterbium	Yb	70	173.04	819	1194	6.965	+2, +3
Yttrium	Y	39	88.9059	1522	3338	4.469	+3
Zinc	Zn	30	65.38	419.58	907	7.133	+2
Zirconium	Zr	40	91.22	1852	4377	6.506	+4

Table A.2 Electron Configurations of the Elements

	Elements	Sublevels																		
		1*s*	2*s*	2*p*	3*s*	3*p*	3*d*	4*s*	4*p*	4*d*	4*f*	5*s*	5*p*	5*d*	5*f*	6*s*	6*p*	6*d*	7*s*	7*p*
1	hydrogen	1																		
2	helium	2																		
3	lithium	2	1																	
4	beryllium	2	2																	
5	boron	2	2	1																
6	carbon	2	2	2																
7	nitrogen	2	2	3																
8	oxygen	2	2	4																
9	fluorine	2	2	5																
10	neon	2	2	6																
11	sodium	2	2	6	1															
12	magnesium	2	2	6	2															
13	aluminum	2	2	6	2	1														
14	silicon	2	2	6	2	2														
15	phosphorus	2	2	6	2	3														
16	sulfur	2	2	6	2	4														
17	chlorine	2	2	6	2	5														
18	argon	2	2	6	2	6														
19	potassium	2	2	6	2	6		1												
20	calcium	2	2	6	2	6		2												
21	scandium	2	2	6	2	6	1	2												
22	titanium	2	2	6	2	6	2	2												
23	vanadium	2	2	6	2	6	3	2												
24	chromium	2	2	6	2	6	5	1												
25	manganese	2	2	6	2	6	5	2												
26	iron	2	2	6	2	6	6	2												
27	cobalt	2	2	6	2	6	7	2												
28	nickel	2	2	6	2	6	8	2												
29	copper	2	2	6	2	6	10	1												
30	zinc	2	2	6	2	6	10	2												
31	gallium	2	2	6	2	6	10	2	1											
32	germanium	2	2	6	2	6	10	2	2											
33	arsenic	2	2	6	2	6	10	2	3											
34	selenium	2	2	6	2	6	10	2	4											
35	bromine	2	2	6	2	6	10	2	5											
36	krypton	2	2	6	2	6	10	2	6											
37	rubidium	2	2	6	2	6	10	2	6			1								
38	strontium	2	2	6	2	6	10	2	6			2								
39	yttrium	2	2	6	2	6	10	2	6	1		2								
40	zirconium	2	2	6	2	6	10	2	6	2		2								
41	niobium	2	2	6	2	6	10	2	6	4		1								
42	molybdenum	2	2	6	2	6	10	2	6	5		1								
43	technetium	2	2	6	2	6	10	2	6	5		2								
44	ruthenium	2	2	6	2	6	10	2	6	7		1								
45	rhodium	2	2	6	2	6	10	2	6	8		1								
46	palladium	2	2	6	2	6	10	2	6	10										
47	silver	2	2	6	2	6	10	2	6	10		1								
48	cadmium	2	2	6	2	6	10	2	6	10		2								
49	indium	2	2	6	2	6	10	2	6	10		2	1							
50	tin	2	2	6	2	6	10	2	6	10		2	2							
51	antimony	2	2	6	2	6	10	2	6	10		2	3							
52	tellurium	2	2	6	2	6	10	2	6	10		2	4							
53	iodine	2	2	6	2	6	10	2	6	10		2	5							
54	xenon	2	2	6	2	6	10	2	6	10		2	6							
55	cesium	2	2	6	2	6	10	2	6	10		2	6			1				
56	barium	2	2	6	2	6	10	2	6	10		2	6			2				
57	lanthanum	2	2	6	2	6	10	2	6	10		2	6	1		2				

Table A.2 Electron Configurations of the Elements (cont.)

	Elements	Sublevels																		
		1*s*	2*s*	2*p*	3*s*	3*p*	3*d*	4*s*	4*p*	4*d*	4*f*	5*s*	5*p*	5*d*	5*f*	6*s*	6*p*	6*d*	7*s*	7*p*
58	cerium	2	2	6	2	6	10	2	6	10	2	2	6			2				
59	praseodymium	2	2	6	2	6	10	2	6	10	3	2	6			2				
60	neodymium	2	2	6	2	6	10	2	6	10	4	2	6			2				
61	promethium	2	2	6	2	6	10	2	6	10	5	2	6			2				
62	samarium	2	2	6	2	6	10	2	6	10	6	2	6			2				
63	europium	2	2	6	2	6	10	2	6	10	7	2	6			2				
64	gadolinium	2	2	6	2	6	10	2	6	10	7	2	6	1		2				
65	terbium	2	2	6	2	6	10	2	6	10	9	2	6			2				
66	dysprosium	2	2	6	2	6	10	2	6	10	10	2	6			2				
67	holmium	2	2	6	2	6	10	2	6	10	11	2	6			2				
68	erbium	2	2	6	2	6	10	2	6	10	12	2	6			2				
69	thulium	2	2	6	2	6	10	2	6	10	13	2	6			2				
70	ytterbium	2	2	6	2	6	10	2	6	10	14	2	6			2				
71	lutetium	2	2	6	2	6	10	2	6	10	14	2	6	1		2				
72	hafnium	2	2	6	2	6	10	2	6	10	14	2	6	2		2				
73	tantalum	2	2	6	2	6	10	2	6	10	14	2	6	3		2				
74	tungsten	2	2	6	2	6	10	2	6	10	14	2	6	4		2				
75	rhenium	2	2	6	2	6	10	2	6	10	14	2	6	5		2				
76	osmium	2	2	6	2	6	10	2	6	10	14	2	6	6		2				
77	iridium	2	2	6	2	6	10	2	6	10	14	2	6	7		2				
78	platinum	2	2	6	2	6	10	2	6	10	14	2	6	9		1				
79	gold	2	2	6	2	6	10	2	6	10	14	2	6	10		1				
80	mercury	2	2	6	2	6	10	2	6	10	14	2	6	10		2				
81	thallium	2	2	6	2	6	10	2	6	10	14	2	6	10		2	1			
82	lead	2	2	6	2	6	10	2	6	10	14	2	6	10		2	2			
83	bismuth	2	2	6	2	6	10	2	6	10	14	2	6	10		2	3			
84	polonium	2	2	6	2	6	10	2	6	10	14	2	6	10		2	4			
85	astatine	2	2	6	2	6	10	2	6	10	14	2	6	10		2	5			
86	radon	2	2	6	2	6	10	2	6	10	14	2	6	10		2	6			
87	francium	2	2	6	2	6	10	2	6	10	14	2	6	10		2	6		1	
88	radium	2	2	6	2	6	10	2	6	10	14	2	6	10		2	6		2	
89	actinium	2	2	6	2	6	10	2	6	10	14	2	6	10		2	6	1	2	
90	thorium	2	2	6	2	6	10	2	6	10	14	2	6	10		2	6	2	2	
91	protactinium	2	2	6	2	6	10	2	6	10	14	2	6	10	2	2	6	1	2	
92	uranium	2	2	6	2	6	10	2	6	10	14	2	6	10	3	2	6	1	2	
93	neptunium	2	2	6	2	6	10	2	6	10	14	2	6	10	4	2	6	1	2	
94	plutonium	2	2	6	2	6	10	2	6	10	14	2	6	10	6	2	6		2	
95	americium	2	2	6	2	6	10	2	6	10	14	2	6	10	7	2	6		2	
96	curium	2	2	6	2	6	10	2	6	10	14	2	6	10	7	2	6	1	2	
97	berkelium	2	2	6	2	6	10	2	6	10	14	2	6	10	9	2	6		2	
98	californium	2	2	6	2	6	10	2	6	10	14	2	6	10	10	2	6		2	
99	einsteinium	2	2	6	2	6	10	2	6	10	14	2	6	10	11	2	6		2	
100	fermium	2	2	6	2	6	10	2	6	10	14	2	6	10	12	2	6		2	
101	mendelevium	2	2	6	2	6	10	2	6	10	14	2	6	10	13	2	6		2	
102	nobelium	2	2	6	2	6	10	2	6	10	14	2	6	10	14	2	6		2	
103	lawrencium	2	2	6	2	6	10	2	6	10	14	2	6	10	14	2	6	1	2	
104	rutherfordium	2	2	6	2	6	10	2	6	10	14	2	6	10	14	2	6	2	2	
105	dubnium	2	2	6	2	6	10	2	6	10	14	2	6	10	14	2	6	3	2	
106	seaborgium	2	2	6	2	6	10	2	6	10	14	2	6	10	14	2	6	4	2	
107	bohrium	2	2	6	2	6	10	2	6	10	14	2	6	10	14	2	6	5	2	
108	hassium	2	2	6	2	6	10	2	6	10	14	2	6	10	14	2	6	6	2	
109	meitnerium	2	2	6	2	6	10	2	6	10	14	2	6	10	14	2	6	7	2	
110	ununnilium	2	2	6	2	6	10	2	6	10	14	2	6	10	14	2	6	9	1	
111	unununium	2	2	6	2	6	10	2	6	10	14	2	6	10	14	2	6	10	1	
112	ununbium	2	2	6	2	6	10	2	6	10	14	2	6	10	14	2	6	10	2	
114	ununquadium	2	2	6	2	6	10	2	6	10	14	2	6	10	14	2	6	10	2	2
116	ununhexium	2	2	6	2	6	10	2	6	10	14	2	6	10	14	2	6	10	2	4
118	ununoctium	2	2	6	2	6	10	2	6	10	14	2	6	10	14	2	6	10	2	6

Table A.3 Symbols of Common Elements

Symbol	Element	Symbol	Element	Symbol	Element
Ag	silver	Cu	copper	O	oxygen
Al	aluminum	F	fluorine	P	phosphorus
As	arsenic	Fe	iron	Pb	lead
Au	gold	H	hydrogen	Pt	platinum
Ba	barium	Hg	mercury	S	sulfur
Bi	bismuth	I	iodine	Sb	antimony
Br	bromine	K	potassium	Sn	tin
C	carbon	Mg	magnesium	Sr	strontium
Ca	calcium	Mn	manganese	Ti	titanium
Cl	chlorine	N	nitrogen	U	uranium
Co	cobalt	Na	sodium	W	tungsten
Cr	chromium	Ni	nickel	Zn	zinc

Table A.4 Symbols of Common Polyatomic Ions

Symbol	Ion	Symbol	Ion	Symbol	Ion
$C_2H_3O_2^-$	acetate	$Cr_2O_7^{2-}$	dichromate	NO_3^-	nitrate
ClO^-	hypochlorite	HCO_3^-	hydrogen carbonate	NO_2^-	nitrite
ClO_2^-	chlorite	H_3O^+	hydronium	O_2^{2-}	peroxide
ClO_3^-	chlorate	HPO_4^{2-}	hydrogen phosphate	OH^-	hydroxide
ClO_4^-	perchlorate	HSO_3^-	hydrogen sulfite	PO_4^{3-}	phosphate
CN^-	cyanide	HSO_4^-	hydrogen sulfate	SiO_3^{2-}	silicate
CO_3^{2-}	carbonate	MnO_4^-	permanganate	SO_3^{2-}	sulfite
CrO_4^{2-}	chromate	NH_4^+	ammonium	SO_4^{2-}	sulfate

Table A.5 Other Symbols and Abbreviations

Symbol	Meaning	Symbol	Meaning	Symbol	Meaning
α	alpha rays	gmm	gram molecular mass	m	mass
β	beta rays	H	enthalpy	m	molality
γ	gamma rays	H_f	heat of formation	mL	milliliter (*volume*)
Δ	change in	h	hour	mm	millimeter (*length*)
$\delta+, \delta-$	partial ionic charge	h	Planck's constant	mol	mole (*amount*)
λ	wavelength	Hz	hertz (*frequency*)	mp	melting point
π	pi bond	J	joule (*energy*)	N	normality
σ	sigma bond	K	kelvin (*temperature*)	n^0	neutron
ν	frequency	K_a	acid dissociation constant	n	number of moles
amu	atomic mass unit	K_b	base dissociation constant	n	principal quantum number
(*aq*)	aqueous solution	K_b	molal boiling point elevation constant	P	pressure
atm	atmosphere (*pressure*)	K_{eq}	equilibrium constant	p^+	proton
bp	boiling point	K_f	molal freezing point depression constant	Pa	pascal (*pressure*)
°C	degree Celsius (*temperature*)	K_w	ion product constant for water	R	ideal gas constant
c	speed of light in a vacuum	K_{sp}	solubility product constant	S	entropy
cm	centimeter (*length*)	kcal	kilocalorie (*energy*)	s	second
D	density	kg	kilogram (*mass*)	(*s*)	solid
E	energy	kPa	kilopascal (*pressure*)	SI	International System of Units
e^-	electron	L	liter (*volume*)	STP	standard temperature and pressure
fp	freezing point	(*l*)	liquid	T	temperature
G	Gibb's free energy	M	molarity	$t_{\frac{1}{2}}$	half-life
g	gram (*mass*)	m	meter (*length*)	V	volume
(*g*)	gas			v	velocity, speed
gam	gram atomic mass				
gfm	gram formula mass				

Table A.6 Physical Constants

Atomic mass unit	1 amu $= 1.6605 \times 10^{-24}$ g
Avogadro's number	$N = 6.0221 \times 10^{23} \dfrac{\text{particles}}{\text{mole}}$
Gas constant	$R = 8.31 \dfrac{\text{L} \times \text{kPa}}{\text{K} \times \text{mol}}$
Ideal gas molar volume	$V_m = 22.414 \dfrac{\text{L}}{\text{mol}}$
Masses of fundamental particles	
Electron (e^-)	$m_e = 0.0005486$ amu $= 9.1096 \times 10^{-28}$ g
Proton (p^+)	$m_p = 1.007277$ amu $= 1.67261 \times 10^{-24}$ g
Neutron (n^0)	$m_n = 1.008665$ amu $= 1.67492 \times 10^{-24}$ g
Speed of light (in vacuum)	$c = 2.997925 \times 10^{8} \dfrac{\text{m}}{\text{s}}$

Table A.7 Solubilities of Compounds at 25 °C and 101.3 kPa

	acetate	bromide	carbonate	chlorate	chloride	hydroxide	iodide	nitrate	oxide	perchlorate	phosphate	sulfate	sulfide
aluminum	S	S	—	S	S	I	S	S	I	S	I	S	d
ammonium	S	S	S	S	S	S	S	S	—	S	S	S	S
barium	S	S	I	S	S	S	S	S	sS	S	I	I	d
calcium	S	S	I	S	S	S	S	S	sS	S	I	sS	I
copper(II)	S	S	—	S	S	I	S	S	I	S	I	S	I
iron(II)	S	S	I	S	S	I	S	S	I	S	I	S	I
iron(III)	S	S	—	S	S	I	S	S	I	S	I	sS	d
lithium	S	S	sS	S	S	S	S	S	S	S	sS	S	S
magnesium	S	S	I	S	S	I	S	S	I	S	I	S	d
potassium	S	S	S	S	S	S	S	S	S	S	S	S	S
silver	sS	I	I	S	I	—	I	S	I	S	I	sS	I
sodium	S	S	S	S	S	S	S	S	S	S	S	S	S
strontium	S	S	I	S	S	S	S	S	S	S	I	I	I
zinc	S	S	I	S	S	I	S	S	I	S	I	S	I

Key:

S = soluble
sS = slightly soluble
I = insoluble
d = decomposes in water
— = no such compound

APPENDIX B MATH REVIEW

B.1 Graphing

The relationship between two variables in an experiment is often determined by graphing the experimental data. A graph is a "picture" of the data. Once a graph is constructed, additional information can be derived about the variables.

In constructing a graph, you must first label the axes. The independent variable is plotted on the *x-axis (abscissa)*. This is the horizontal axis. The independent variable is generally controlled by the experimenter. When the independent variable is changed, a corresponding change in the dependent variable is measured. The dependent variable is plotted on the *y-axis (ordinate)*. This is the vertical axis. The label on each axis should include the unit of the quantity being graphed.

Before data can be plotted on a graph, each axis must be scaled. The scale must take into consideration the smallest and largest values of each quantity. Each interval (square on the graph paper) on the scale must represent the same amount. To make it easy to find numbers along the scale, the interval chosen is usually a multiple of 1, 2, 5, or 10. Although each scale can start at zero, this is not always practical.

Data are plotted by putting a point at the intersection of corresponding values of each pair of measurement. Once the data have been plotted, the points are connected by a smooth curve. This is not the same as "connecting-the-dots," which is an incorrect approach to drawing a line. A smooth curve comes as close as possible to all the plotted points. It may, in fact, not touch any of them.

Depending on the relationship between two variables, the curve may or may not be a straight line. Two common curves are shown in Graphs A and B.

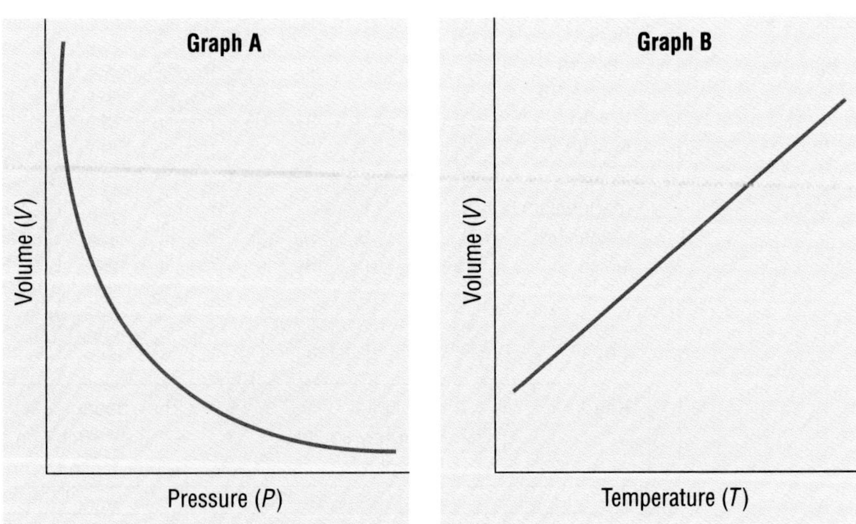

Graphs similar to these are found in Chapter 12 of this textbook. Graph A is typical of an inverse proportionality. As the independent variable (P) increases, the dependent variable (V) decreases. The product of the two variables at any point on the curve of an inverse proportionality is a constant. For graph A, $V \times P$ = constant.

The straight line in graph B is typical of a direct proportionality. As the independent variable (T) increases, there is a corresponding increase in the dependent variable (V). A straight line is represented by this general equation.

$$y = mx + b$$

Here y and x are the variables plotted on the vertical and horizontal axes, respectively; m is the slope of the line; and b is the intercept on the y-axis.

The y intercept, b, is the value of y when x is zero. The slope, m, is the ratio of the change in y (Δy) for a corresponding change in x (Δx). This relationship is often symbolized as follows:

$$m = \frac{\Delta y}{\Delta x}$$

As an example, consider these data about a bicyclist's trip. Assume that the bicyclist rode at a constant speed.

Distance from home (km)	15	25	35	50	75
Time (h)	1	2	3	4.5	7

Graph these data using time as the independent variable, and then use the graph to answer the following questions.

a. How far from home was the bicyclist at the start of the trip?

b. How many hours did it take the bicyclist to get 40 km from home?

c. What was the bicyclist's average speed, in kilometers per hour, on the trip?

The plotted points are shown in Figure 1. Each point was plotted by finding the value of time on the x-axis, then moving up vertically to the value of the other variable (distance). A smooth curve, which in this case happens to be a straight line, has been drawn through the points.

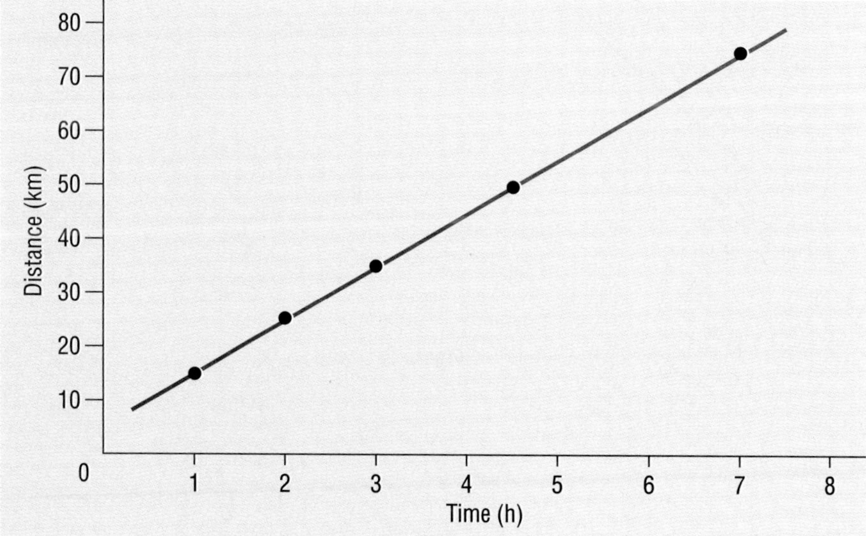

Figure 1

a. The graph in Figure 2 shows that the bicyclist started the trip 5 km from home. This is the value of the vertical axis (distance) when the time elapsed is zero (point *a* on the graph).

b. Find the given value, 40 km, on the vertical axis in Figure 2. Move to the right (horizontally) in the graph until you reach the line. Drop down vertically and read the value of time at this point (point *b*). It takes the bicyclist 3.5 h to get 40 km from home. As another, similar example, how far is the bicyclist from home after riding 5 h? Using the graph in Figure 2, start at 5 h and go up to the line. Then move horizontally to the left. The distance, point *c*, is 55 km.

c. Speed is distance/time. The average speed of the bicyclist is the slope of the line. Calculate the slope using the values for time and distance discussed in the previous paragraph **(b)**.

$$m = \frac{\Delta y}{\Delta x} = \frac{55\text{ km} - 40\text{ km}}{5\text{ h} - 3.5\text{ h}} = \frac{15\text{ km}}{1.5\text{ h}} = 10\text{ km/h}$$

The equation for this line shows the relationship between time and distance traveled by the bicyclist.

$$\text{Distance} = (10\text{ km/h})(\text{time}) + 5\text{ km}$$

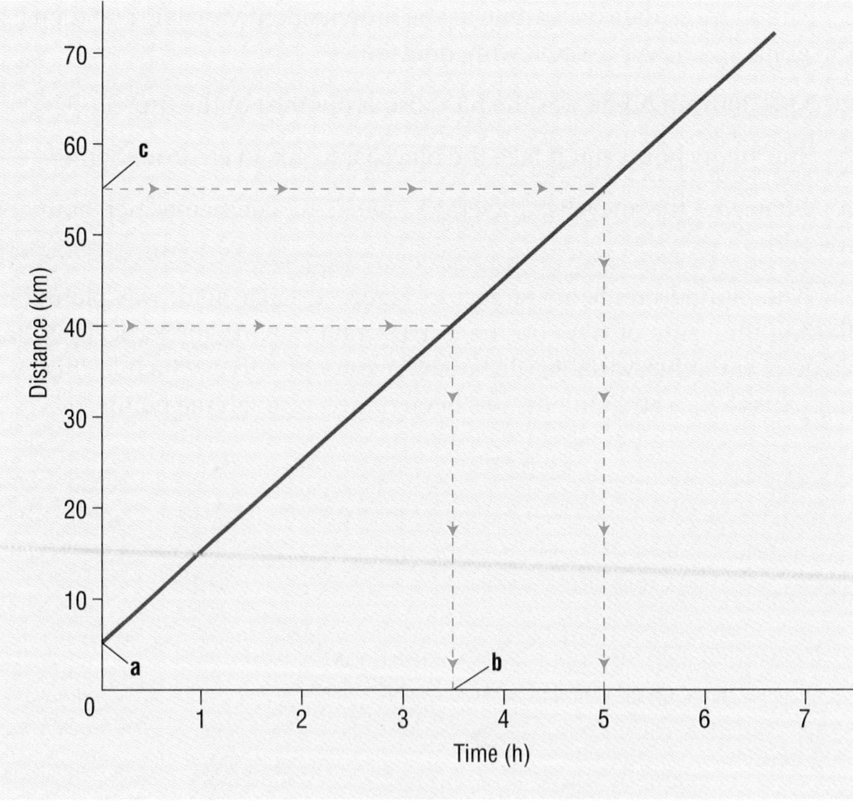

Figure 2

B.2 Logarithms

A logarithm is the exponent to which a fixed number (base) must be raised in order to produce a given number. Common logarithms use 10 as the base. A logarithm to the base 10 of a number is the exponent to which 10 must be raised to obtain the number. If $x = 10^y$, then $\log x = y$. Here x is the number and y is the logarithm of x to the base 10. Common logarithms must not be confused with natural logarithms, which use base e, where $e = 2.71828$. When base 10 is used, logarithm is abbreviated $\log_{10}$, or often simply log. When the base e is used, logarithm is abbreviated ln.

The logarithm of a number has two parts, the characteristic, or whole-number part, and the mantissa, or decimal part. In the example $\log 421.6 = 2.6249$, the characteristic is 2 and the mantissa is 0.6249. For numbers written in standard exponential form, the characteristic corresponds to the exponent of 10.

$$\log 10^1 = 1,\ \log 10^2 = 2,\ \log 10^3 = 3,\ \log 10^{-2} = -2$$

The mantissa is the decimal part of a logarithm and can be looked up in a log table. The number of significant figures in a logarithm is the number of figures in the mantissa. The logarithm of 176 is found as follows.

1. Write 176 in exponential notation as 1.76×10^2.
2. Locate the number 1.7 in the column labeled N; then under the column headed 6 read the mantissa as 0.2455.
3. Because 1.76×10^2 contains only three significant figures, the mantissa is rounded to three digits. Thus the mantissa is 0.246.
4. The characteristic is the exponent, 2.
5. The logarithm of 176 is 2.246 or $\log 176 = 2.246$.

Logarithms can also be found using the log key of a calculator. Simply enter the number and press the log key.

The reverse process of converting a logarithm into a number is referred to as obtaining the antilogarithm. The antilogarithm of the logarithm of a number is the number itself. Here x represents any number.

$$\text{antilog}\ (\log x) = x$$

Antilogarithms can be obtained from a table of common logarithms. They also can be obtained with a calculator by using the antilog key, the 10^x key, or the inverse key in conjunction with the log key. For example, what is the antilog of the logarithm 4.618? Look in the body of the log table for the mantissa, 0.618. This value is found in the row with an N of 4.1 and in the column labeled 5. Thus the antilog of 0.618 is 4.15. The characteristic is 4. This corresponds to an exponential term of 10^4. The number whose log is 4.618 is 4.15×10^4.

There are rules based on the laws of exponents that must be followed when using logarithms for calculations. The logarithm of the product of two numbers is the sum of their logs.

$$\log(a \times b) = \log a + \log b$$

The log of the ratio of two numbers is the log of the numerator minus the log of the denominator.

$$\log(a/b) = \log a - \log b$$

Table B.1 Common Logarithms

x	0	1	2	3	4	5	6	7	8	9
1.0	.0000	.0043	.0086	.0128	.0170	.0212	.0253	.0294	.0334	.0374
1.1	.0414	.0453	.0492	.0531	.0569	.0607	.0645	.0682	.0719	.0755
1.2	.0792	.0828	.0864	.0899	.0934	.0969	.1004	.1038	.1072	.1106
1.3	.1139	.1173	.1206	.1239	.1271	.1303	.1335	.1367	.1399	.1430
1.4	.1461	.1492	.1523	.1553	.1584	.1614	.1644	.1673	.1703	.1732
1.5	.1761	.1790	.1818	.1847	.1875	.1903	.1931	.1959	.1987	.2014
1.6	.2041	.2068	.2095	.2122	.2148	.2175	.2201	.2227	.2253	.2279
1.7	.2304	.2330	.2355	.2380	.2405	.2430	.2455	.2480	.2504	.2529
1.8	.2553	.2577	.2601	.2625	.2648	.2672	.2695	.2718	.2742	.2765
1.9	.2788	.2810	.2833	.2856	.2878	.2900	.2923	.2945	.2967	.2989
2.0	.3010	.3032	.3054	.3075	.3096	.3118	.3139	.3160	.3181	.3201
2.1	.3222	.3243	.3263	.3284	.3304	.3324	.3345	.3365	.3385	.3404
2.2	.3424	.3444	.3464	.3483	.3502	.3522	.3541	.3560	.3579	.3598
2.3	.3617	.3636	.3655	.3674	.3692	.3711	.3729	.3747	.3766	.3784
2.4	.3802	.3820	.3838	.3856	.3874	.3892	.3909	.3927	.3945	.3962
2.5	.3979	.3997	.4014	.4031	.4048	.4065	.4082	.4099	.4116	.4133
2.6	.4150	.4166	.4183	.4200	.4216	.4232	.4249	.4265	.4281	.4298
2.7	.4314	.4330	.4346	.4362	.4378	.4393	.4409	.4425	.4440	.4456
2.8	.4472	.4487	.4502	.4518	.4533	.4548	.4564	.4579	.4594	.4609
2.9	.4624	.4639	.4654	.4669	.4683	.4698	.4713	.4728	.4742	.4757
3.0	.4771	.4786	.4800	.4814	.4829	.4843	.4857	.4871	.4886	.4900
3.1	.4914	.4928	.4942	.4955	.4969	.4983	.4997	.5011	.5024	.5038
3.2	.5051	.5065	.5079	.5092	.5105	.5119	.5132	.5145	.5159	.5172
3.3	.5185	.5198	.5211	.5224	.5237	.5250	.5263	.5276	.5289	.5307
3.4	.5315	.5328	.5340	.5353	.5366	.5378	.5391	.5403	.5416	.5428
3.5	.5441	.5453	.5465	.5478	.5490	.5502	.5514	.5527	.5539	.5551
3.6	.5563	.5575	.5587	.5599	.5611	.5623	.5635	.5647	.5658	.5670
3.7	.5682	.5694	.5705	.5717	.5729	.5740	.5752	.5763	.5775	.5786
3.8	.5798	.5809	.5821	.5832	.5843	.5855	.5866	.5877	.5888	.5899
3.9	.5911	.5922	.5933	.5944	.5955	.5966	.5977	.5988	.5999	.6010
4.0	.6021	.6031	.6042	.6053	.6064	.6075	.6085	.6096	.6107	.6117
4.1	.6128	.6138	.6149	.6160	.6170	.6180	.6191	.6201	.6212	.6222
4.2	.6232	.6243	.6253	.6263	.6274	.6284	.6294	.6304	.6314	.6325
4.3	.6335	.6345	.6355	.6365	.6375	.6385	.6395	.6405	.6415	.6425
4.4	.6435	.6444	.6454	.6464	.6474	.6484	.6493	.6503	.6513	.6522
4.5	.6532	.6542	.6551	.6561	.6571	.6580	.6590	.6599	.6609	.6618
4.6	.6628	.6637	.6646	.6656	.6665	.6675	.6684	.6693	.6702	.6712
4.7	.6721	.6730	.6739	.6749	.6758	.6767	.6776	.6785	.6794	.6803
4.8	.6812	.6821	.6830	.6839	.6848	.6857	.6866	.6875	.6884	.6893
4.9	.6902	.6911	.6920	.6928	.6937	.6946	.6955	.6964	.6972	.6981
5.0	.6990	.6998	.7007	.7016	.7024	.7033	.7042	.7050	.7059	.7067
5.1	.7076	.7084	.7093	.7101	.7110	.7118	.7126	.7135	.7143	.7152
5.2	.7160	.7168	.7177	.7185	.7193	.7202	.7210	.7218	.7226	.7235
5.3	.7243	.7251	.7259	.7267	.7275	.7284	.7292	.7300	.7308	.7316
5.4	.7324	.7332	.7340	.7348	.7356	.7364	.7372	.7380	.7388	.7396

Table B.1 Common Logarithms (cont.)

x	0	1	2	3	4	5	6	7	8	9
5.5	.7404	.7412	.7419	.7427	.7435	.7443	.7451	.7459	.7466	.7474
5.6	.7482	.7490	.7497	.7505	.7513	.7520	.7528	.7536	.7543	.7551
5.7	.7559	.7566	.7574	.7582	.7589	.7597	.7604	.7612	.7619	.7627
5.8	.7634	.7642	.7649	.7657	.7664	.7672	.7679	.7686	.7694	.7701
5.9	.7709	.7716	.7723	.7731	.7738	.7745	.7752	.7760	.7767	.7774
6.0	.7782	.7789	.7796	.7803	.7810	.7818	.7825	.7832	.7839	.7846
6.1	.7853	.7860	.7868	.7875	.7882	.7889	.7896	.7903	.7910	.7917
6.2	.7924	.7931	.7938	.7945	.7952	.7959	.7966	.7973	.7980	.7987
6.3	.7993	.8000	.8007	.8014	.8021	.8028	.8035	.8041	.8048	.8055
6.4	.8062	.8069	.8075	.8082	.8089	.8096	.8102	.8109	.8116	.8122
6.5	.8129	.8136	.8142	.8149	.8156	.8162	.8169	.8176	.8182	.8189
6.6	.8195	.8202	.8209	.8215	.8222	.8228	.8235	.8241	.8248	.8254
6.7	.8261	.8267	.8274	.8280	.8287	.8293	.8299	.8306	.8312	.8319
6.8	.8325	.8331	.8338	.8344	.8351	.8357	.8363	.8370	.8376	.8382
6.9	.8388	.8395	.8401	.8407	.8414	.8420	.8426	.8432	.8439	.8445
7.0	.8451	.8457	.8463	.8470	.8476	.8482	.8488	.8494	.8500	.8506
7.1	.8513	.8519	.8525	.8531	.8537	.8543	.8549	.8555	.8561	.8567
7.2	.8573	.8579	.8585	.8591	.8597	.8603	.8609	.8615	.8621	.8627
7.3	.8633	.8639	.8645	.8651	.8657	.8663	.8669	.8675	.8681	.8686
7.4	.8692	.8698	.8704	.8710	.8716	.8722	.8727	.8733	.8739	.8745
7.5	.8751	.8756	.8762	.8768	.8774	.8779	.5785	.8791	.8797	.8802
7.6	.8808	.8814	.8820	.8825	.8831	.8837	.8842	.8848	.8854	.8859
7.7	.8865	.8871	.8876	.8882	.8887	.8893	.8899	.8904	.8910	.8915
7.8	.8921	.8927	.8932	.8938	.8943	.8949	.8954	.8960	.8965	.8971
7.9	.8976	.8982	.8987	.8993	.8998	.9004	.9009	.9015	.9020	.9025
8.0	.9031	.9036	.9042	.9047	.9053	.9058	.9063	.9069	.9074	.9079
8.1	.9085	.9090	.9096	.9101	.9106	.9112	.9117	.9122	.9128	.9133
8.2	.9138	.9143	.9149	.9154	.9159	.9165	.9170	.9175	.9180	.9186
8.3	.9191	.9196	.9201	.9206	.9212	.9217	.9222	.9227	.9232	.9238
8.4	.9243	.9248	.9253	.9258	.9263	.9269	.9274	.9279	.9284	.9289
8.5	.9294	.9299	.9304	.9309	.9315	.9320	.9325	.9330	.9335	.9340
8.6	.9345	.9350	.9355	.9360	.9365	.9370	.9375	.9380	.9385	.9390
8.7	.9395	.9400	.9405	.9410	.9415	.9420	.9425	.9430	.9435	.9440
8.8	.9445	.9450	.9455	.9460	.9465	.9469	.9474	.9479	.9484	.9489
8.9	.9494	.9499	.9504	.9509	.9513	.9518	.9523	.9528	.9533	.9538
9.0	.9542	.9547	.9552	.9557	.9562	.9566	.9571	.9576	.9581	.9586
9.1	.9590	.9595	.9600	.9605	.9609	.9614	.9619	.9624	.9628	.9633
9.2	.9638	.9643	.9647	.9652	.9657	.9661	.9666	.9671	.9675	.9680
9.3	.9685	.9689	.9694	.9699	.9703	.9708	.9713	.9717	.9722	.9727
9.4	.9731	.9736	.9741	.9745	.9750	.9754	.9759	.9763	.9768	.9773
9.5	.9777	.9782	.9786	.9791	.9795	.9800	.9805	.9809	.9814	.9818
9.6	.9823	.9827	.9832	.9836	.9841	.9845	.9850	.9854	.9859	.9863
9.7	.9868	.9872	.9877	.9881	.9886	.9890	.9894	.9899	.9903	.9908
9.8	.9912	.9917	.9921	.9926	.9930	.9934	.9939	.9943	.9948	.9952
9.9	.9956	.9961	.9965	.9969	.9974	.9978	.9983	.9987	.9991	.9996

APPENDIX C ANSWERS TO SELECTED PROBLEMS

CHAPTER 1

17. Chemistry is the study of matter and its changes for the sake of understanding them; chemical technology is the application of this knowledge to attain specific goals.

19. **a.** 2 **b.** 5 **c.** 1 **d.** 6 **e.** 3 **f.** 4

21. The ozone layer protects Earth from harmful ultraviolet rays.

23. Experiments are used to test hypotheses.

25. If you are certain that your experiment is not flawed, you must revise your hypothesis.

27. A language uses its vocabulary to express thoughts about anything—food, politics, religion, and so forth. The vocabulary of the sciences is used to express concepts and facts about the workings of the material world.

29. Students' diagrams should show one string that is threaded through both holes at A and C. The string at hole B is a separate thread from the string passing through holes A and C.

31. The experiment may be flawed; if not, the results are evidence that the theory may need to be revised.

33. **a.** 1 **b.** 3 **c.** 1

35. No; for any discovery to take place, the discoverer must have the knowledge to recognize the significance of the observation or data.

37. **a.** The textbook contains all of the major content and concepts covered in the course. The visuals, reading/studying guides, and practice/review features are essential elements in learning and understanding chemistry.
b. (false statement) Chemistry concepts are interwoven and are best understood as parts of a big picture. Seeing connections and building a framework imply understanding, not memorization.
c. Learning in small chunks is more effective. Studying on a regular basis helps build a big picture of chemistry into which individual facts can be placed. It assures understanding rather than memorization.
d. An exchange of ideas and perspectives is valuable.
e. Talking about chemistry aids understanding and retention.

CHAPTER 2

5. the attraction of iron, but not salt, to a magnet; the solubility of salt, but not iron, in water

6. **a.** heterogeneous **b.** homogeneous **c.** heterogeneous **d.** homogeneous **e.** homogeneous

13. The liquid was not an element because a solid was left when the liquid evaporated. Evaporation is a way to physically separate a mixture. Compounds cannot be separated by physical means such as evaporation.

25. **a.** solid **b.** liquid **c.** gas **d.** solid **e.** liquid **f.** liquid

27. a vapor; The term "vapor" refers to the gaseous state of a substance that normally exists as a liquid or solid at room temperature.

29. chlorine, mercury, bromine, and water; Chlorine condenses, and mercury, bromine and water all freeze when the temperature drops within the stated range.

31. one; A solution is a system with uniform composition and properties. Solutions are homogeneous mixtures, consisting of a single phase.

33. **a.** nitrogen, hydrogen, chlorine
b. potassium, manganese, oxygen
c. carbon, hydrogen, oxygen
d. calcium, iodine

35. **a.** physical **b.** chemical **c.** chemical **d.** physical

37. As the wax burns, the chemical composition of the wax changes, producing the products water and carbon dioxide, which are released into the surrounding air.

39. Mixtures in both cases.

41. **a.** homogeneous mixture
b. homogeneous mixture
c. heterogeneous mixture
d. homogeneous mixture
e. heterogeneous mixture
f. compound
g. homogeneous mixture
h. heterogeneous mixture

43. **a.** color and odor change, irreversible
b. gas is produced
c. formation of a precipitate (solid), not easily reversed
d. color change, not easily reversed, change in texture
e. release of energy; change in odor and color; irreversible

45. In gases, particles are far apart; in liquids, particles are in contact; in solids, particles are tightly packed.

47. **a.** Yes; because the graph is a straight line, the proportion of iron to oxygen is a constant, which is true for a compound.
b. No; plotting these values on the graph would not give a point on the line indicating that the mass ratio of iron to oxygen is different from the other four samples.

49. **a.** two, mercury and sulfur
b. Sulfur melts at 113 °C and boils at 445 °C. Between 113 °C and 445 °C, it exists as a liquid. Mercury melts at −39 °C, and boils at 357 °C. In between these temperatures, it exists as a liquid.
c. Possibilities include: by color, alphabetically, by increasing boiling point, by increasing density.

CHAPTER 3

5. **a.** 4 **b.** 4 **c.** 2 **d.** 5

6. **a.** 3 **b.** 2 **c.** 4 **d.** 4

7. **a.** 8.71×10^1 m **b.** 4.36×10^8 m **c.** 1.55×10^{-2} m **d.** 9.01×10^3 m **e.** 1.78×10^{-3} m **f.** 6.30×10^2 m

8. **a.** 9×10^1 m **b.** 4×10^8 m **c.** 2×10^{-2} m **d.** 9×10^3 m **e.** 2×10^{-3} m **f.** 6×10^2 m

9. **a.** 79.2 m **b.** 7.33 m **c.** 11.53 m **d.** 17.3 m

10. 23.8 grams

11. **a.** 1.8×10^1 m^2 **b.** 6.75×10^2 m **c.** 5.87×10^{-1} min

12. 1.3×10^3 m^3

23. 2.50 g/cm^3; no

24. 6.5 cm^3

30. –196 °C

31. melting point: 1234 K; boiling point: 2485 K

37. **a.** precision **b.** accuracy **c.** precision **d.** precision **e.** accuracy **f.** accuracy

39. Lissa: inaccurate and imprecise
Lamont: accurate and precise
Leigh Anne: inaccurate but precise

41. **a.** infinite **b.** infinite **c.** infinite **d.** 4 **e.** infinite **f.** 3

43. **a.** 9.85×10^1 L **b.** 7.63×10^{-4} cg **c.** 5.70×10^1 m **d.** 1.22×10^1 °C **e.** 7.50×10^1 mm **f.** 1.76×10^3 mL

45. **a.** 4.3×10^1 g **b.** 7.3×10^0 cm^2 **c.** 2.258×10^2 L **d.** 9.20×10^1 kg **e.** 3.24×10^1 m^3 **f.** 1.04×10^2 m^3

47. The absolute value of the error is used.

49. 23.9 g

51. **a.** second, s **b.** meter, m **c.** kelvin, K **d.** kilogram, kg

53. **a.** 2.4 mm **b.** 17.6 cm **c.** 27.6 cm

55. 7.9×10^2 cm^3

57. **a.** 0.01 g **b.** 0.000 001 g **c.** 1000 g **d.** 0.001 g

59. **a.** 1 **b.** 4 **c.** 2 **d.** 3

61. No; the density of the metal bar is 12 g/cm^3, but the density of gold is 19 g/cm^3.

63. **a.** 2.70 **b.** 13.6 **c.** 0.917

65. germanium

67. **e.**, **d.**, **c.**, **f.**, **a.**, **b.**

69. °F = 1.8 °C + 32

71. The digit to the right of the last significant figure is dropped if it is less than five.

73. Yes; an object can change position such that the force of gravity acting on it changes, causing its weight to change.

75. **a.** 2 **b.** 1

77. You do not change your estimate. The extra 15 ducks are negligible compared to 850 000 ducks.

79. to understand the world around us

81. **a.** chemical **b.** physical **c.** physical **d.** chemical **e.** chemical **f.** physical

83. The components of a homogeneous mixture can be separated by physical means; a compound must be chemically broken apart into its components. A homogeneous mixture has a variable composition; a compound has a definite composition.

85. volume of iron = 45.1 cm^3; mass of lead = 514 g Pb

87. The slope of the line yields: density of sulfur = 2.1 g/cm^3

CHAPTER 4

1. 27.0 cm^3

2. 269 °C

9. 67 students

10. 86.4 °F

11. **a.** 44 m **b.** 4.6×10^{-3} g **c.** 89 dm **d.** 10.7 cg

12. **a.** 1.5×10^{-2} L **b.** 7.38×10^{-3} kg **c.** 6.7×10^2 ms **d.** 9.45×10^7 µg

13. **a.** 6.32 cm^3 **b.** 5.0 g **c.** 0.342 cm^3

14. See answers for problem 13.

20. 2.27×10^{-8} cm

21. 1.3×10^8 dm

22. 10 080 min

23. 144 000 s

24. 1.93×10^4 kg/m^3

25. 7.0×10^{12} RBC/L

33. 155.3 g

35. one

37. They must equal one another.

39. The unit of the conversion factor in the denominator must be identical to the unit in the given measurement.

41. Estimate an answer to see if your calculator answer makes sense and round the answer off to the correct number of significant figures.

43. **a.** 1.57 s **b.** 4.27×10^4 mL **c.** 2.61×10^{-4} mm **d.** 6.5×10^2 dm **e.** 6.42×10^{-3} kg **f.** 8.25×10^9 µg

45. 10^6 mL

47. 31.1 m/s

49. 3.6 min lost

51. **a.** $C_2 = -90$ °C, $C_4 = 0$ °C, $C_6 = 70$ °C, $C_8 = 125$ °C
b. C_1 through C_4
c. three
d. Over the range C_1 through C_9, the increase is about 38 °C/additional carbon. Over the range C_3 through C_9, the increase is about 32 °C/additional carbon.

53. 73 g

55. 24.0 kg of water

57. Answers will vary.

59. The mass of fuel plus the mass of oxygen must equal the mass of ash plus gaseous products.

61. a. mixture
b. mixture
c. compound
d. element

63. a. centigram
b. kiloliter
c. millisecond
d. cubic decimeter
e. micrometer

65. a. 3 b. 4 c. 2 d. 4 e. 5 f. 4

67. 8.0 g Sr

69. 32 cm

71. a. 85 g b. 1.3 g/mL

CHAPTER 5

7. a. 9,9 b. 13, 13 c. 20, 20

8. K: 19; B: 5, 5; S:16, 16; V: 23, 23

9. a. 8 b. 16 c. 61 d. 45 e. 125

10. a. ${}^{12}_{6}C$ b. ${}^{19}_{9}F$ c. ${}^{9}_{4}Be$

11. a. 8 b. 16 c. 47 d. 35 e. 82

12. ${}^{16}_{8}O$, ${}^{17}_{8}O$, ${}^{18}_{8}O$

13. 26, 28, 29

14. boron–11

15. Silicon–28 must be by far the most abundant. The others must be present in only small amounts.

16. 63.6 amu

17. 79.90 amu

33. Dalton would agree with all four statements because they all fit his atomic theory.

35. repel

37. Every atomic nucleus is positively charged.

39. number of protons in the nucleus

41. The atomic number is the number of protons. The mass number is the sum of the number of protons and number of neutrons.

43. mass numbers, atomic masses, number of neutrons, relative abundance

45. which isotopes exist, their masses, and their natural percent abundance

47. Answers will vary.

49. a. C
b. La
c. B, Ne, P, Br
d. Hg, Br
e. Bi

51. five protons and six neutrons in the nucleus; five electrons outside the nucleus

53. 207 amu

55. Atoms are the smallest particle of an element that retains the properties of that element.

57. a. 3 b. 1 c. 4

59. The following are reasonable hypotheses: The space in an individual atom is large relative to the volume of the atom, but very small relative to an object the size of a hand. There are many layers of atoms in a wall or a desk. The space that exists is distributed evenly throughout the solid, similar to the distribution of air pockets in foam insulation.

61. Answers will vary.

63. a. 4 b. 3 c. 3 d. 4

65. 4.84×10^5 cm^3

67. a. element
b. mixture
c. mixture
d. mixture
e. mixture
f. mixture

69. 92.5%

CHAPTER 6

1. a. sulfide ion (S^{2-})
b. aluminum ion (Al^{3+})
c. calcium ion (Ca^{2+})

2. a. 2 lost b. 3 gained c. 2 lost

10. 2:1

11. No; the ratio in the compound is 1.62:1 rather than 1.75:1, or 7:4.

16. a. 2– b. 2+ c. 1+ d. 3–

17. a. 3 lost b. 2 gained c. 1 lost d. 2 lost

18. a. selenide ion, anion
b. barium ion, cation
c. cesium ion, cation
d. phosphide ion, anion

19. a. iron(III) or ferric ion
b. oxide ion
c. copper(I) or cuprous ion
d. cadmium ion

24. a. BaS b. Li_2O c. Ca_3N_2 d. CuI_2

25. a. NaI b. $SnCl_2$ c. K_2S d. CaI_2

26. a. zinc sulfide
b. potassium chloride
c. barium oxide
d. copper(II) or cupric bromide

27. a. calcium oxide
b. copper(I) or cuprous selenide
c. iron(II) or ferrous sulfide
d. aluminum fluoride

28. a. $(NH_4)_2SO_3$
b. $Ca_3(PO_4)_2$
c. $Al(NO_3)_3$
d. K_2CrO_4

29. a. $LiHSO_4$
b. $Cr(NO_2)_3$
c. $HgBr_2$
d. $(NH_4)_2Cr_2O_7$

30. a. calcium oxalate
b. potassium hypochlorite
c. potassium permanganate
d. lithium sulfite

31. a. aluminum hydroxide
b. sodium chlorate
c. tin(II) or stannous phosphate
d. sodium chromate

37. **a.** oxygen difluoride
b. dichlorine octoxide
c. sulfur trioxide

38. **a.** NF_3 **b.** S_2Cl_2 **c.** N_2O_4

45. Ions are formed when atoms lose electrons or gain electrons.

47. **a.** bromide ion, anion **d.** calcium ion, cation
b. sodium ion, cation **e.** copper(I) ion, cation
c. arsenide ion, anion **f.** hydride ion, anion

49. **a.** ionic **c.** ionic **e.** ionic
b. molecular **d.** molecular **f.** molecular

51. **a.** carbon 6, hydrogen 8, oxygen 6
b. carbon 5, hydrogen 8, oxygen 4, sodium 1
c. carbon 12, hydrogen 22, oxygen 11
d. carbon 7, hydrogen 5, nitrogen 3, oxygen 6
e. nitrogen 2, hydrogen 4, oxygen 3

53. **a.** O^{2-} **c.** Li^{+} **e.** Cu^{2+}
b. Pb^{2+} **d.** N^{3-} **f.** F^{-}

55. cyanide (CN^{-}) and hydroxide (OH^{-})

57. The net ionic charge is zero. Ionic compounds are electrically neutral and thus have no net ionic charge.

59. Knowing the number of each ion in the formula and the ionic charge of the anion, the charge of the cation must be such that the net ionic charge is zero.

61. **a.** $CaCO_3$ **c.** $LiClO$
b. $Ba(HCO_3)_2$ **d.** $Sn(Cr_2O_7)_2$

63. two nonmetals

65. **a.** hydrochloric acid **c.** H_2SO_4
b. nitric acid **d.** $HC_2H_3O_2$

67. **a.** sodium chlorate **e.** tin(IV) oxide
b. mercury(I) bromide **f.** iron(III) acetate
c. potassium chromate **g.** potassium hydrogen sulfate
d. aluminum iodide **h.** calcium hydride

69. **a.** lithium perchlorate **e.** barium phosphate
b. dichlorine monoxide **f.** iodine
c. mercury(II) fluoride **g.** strontium sulfate
d. calcium oxide **h.** copper(I) acetate

71. **a.** magnesium permanganate **e.** lithium hydroxide
b. beryllium nitrate **f.** barium fluoride
c. potassium carbonate **g.** phosphorus triiodide
d. dinitrogen tetrahydride **h.** zinc oxide

73. **a.** $CaBr_2$ **d.** NO_2 **g.** $Sr(C_2H_3O_2)_2$
b. $AgCl$ **e.** $Sn(CN)_4$ **h.** Na_2SiO_3
c. Al_4C_3 **f.** LiH

75. A molecular formula shows the number of each kind of atom in a molecule of the compound. The formula unit shows the lowest whole-number ratio of ions in a compound.

77. Common names would vary in different languages and might even vary in the same language.

79. Possible answers include: cations always come before anions; when a cation has more than one ionic charge, the charge is indicated by a Roman numeral; monatomic anions use an *-ide* ending.

81. **a.** 12 protons, 10 electrons
b. 35 protons, 36 electrons
c. 38 protons, 36 electrons
d. 16 protons, 18 electrons

83. 0.538 g/cm^3

85. **a.** Potassium carbonate has a much greater water solubility than $CaCO_3$.
b. Copper compound is blue; iron compound is white.
c. Because of differences in solubilities, water could be added to dissolve the NH_4Cl. The resulting solution could be filtered leaving the insoluble $BaSO_4$.
d. chlorine (nonmetal), sulfur (nonmetal), bromine (nonmetal), barium (metal), iodine (nonmetal), mercury (metal)
e. barium sulfate, calcium carbonate, potassium carbonate, copper(II) sulfate pentahydrate, iron(II) sulfate pentahydrate, ammonium chloride
f. 639 g
g. 7.54 cm^3
h. Color, density, melting point, and boiling point could all be used to distinguish among the halogens.

CHAPTER 7

1. 5.0 kg

2. 672 seeds

3. 4.65 mol Si

4. 2.17×10^{23} molecules H_2O

5. 2.75×10^{24} atoms

6. 7.72 mol NO_2

7. **a.** 30.0 g **b.** 137.5 g **c.** 60.0g **d.** 108.0 g

8. **a.** 71.0 g **b.** 46.0 g **c.** 331.6 g **d.** 60.1 g

9. **a.** 94.2 g **b.** 136.2 g **c.** 317.3 g

10. **a.** 175.3 g **c.** 84.0 g
b. 139.6 g **d.** 294.3 g

16. **a.** 1.30×10^2 g **b.** 1.27 g **c.** 1.55 g

17. **a.** 355 g **b.** 225 g

18. **a.** 3.43×10^{-2} mol B
b. 0.343 mol TiO_2
c. 8.82 mol $(NH_4)_2CO_3$

19. **a.** 0.987 mol N_2O_3
b. 2.68 mol N_2
c. 1.21 mol Na_2O

20. **a.** 7.17×10^{-2}L **b.** 21.5 L **c.** 82.9 L

21. **a.** 3.00 mol SO_2
b. 3.93×10^{-2} mol He
c. 44.6 mol C_2H_6

22. 80.2 g/mol

23. 3.74 g/L

29. **a.** 72.2% Mg, 27.8% N **b.** 87.1% Ag, 12.9% S

30. 93.0% Hg, 7.0% O

31. **a.** 80.0% C, 20% H
b. 19.2% Na, 0.83% H, 26.7% S, 53.3% O
c. 26.2% N, 7.5 % H, 66.4% Cl

32. **a.** 46.7% N **b.** 82.4% N **c.** 35.0% N

33. **a.** 70 g H **b.** 0.17 g H **c.** 0.16 g H

34. **a.** 58.4 g N **b.** 103 g N **c.** 43.8 g N

35. **a.** OH **b.** CH_3 **c.** $HgSO_4$ **d.** C_2HNO_3

36. C_3H_8N

37. a. $C_2H_6O_2$ b. $C_6H_4Cl_2$

38. a

45. a. molecule b. formula unit c. molecule d. atom

47. All contain 6.02×10^{23} molecules.

49. a. 1.81×10^{24} atoms Sn
b. 2.41×10^{23} formula units KCl
c. 4.52×10^{24} molecules SO_2
d. 2.89×10^{21} formula units NaI

51. a. 60.1 g b. 28.0 g c. 106.8 g d. 63.5 g

53. 71.0 g Cl_2

55. a. 0.258 mol SiO_2 d. 0.106 mol KOH
b. 4.80×10^{-4} mol AgCl e. 5.93 mol $Ca(C_2H_3O_2)_2$
c. 1.12 mol Cl_2 f. 2.00×10^{-2} mol Ca

57. a. 1.7×10^2 L Ar b. 9.9 L C_2H_6 c. 26.9 L O_2

59. a. 234 L SO_3
b. 2.99×10^{-22} g $C_9H_8O_4$
c. 3.13×10^{25} atoms

61. a. 3.33 g S b. 5.65 g N c. 40.6 g Mg d. 152 g P

63. H_2O_2

65. a. $C_3H_6O_3$ b. Hg_2Cl_2 c. $C_6H_{10}O_4$

67. You can measure the mass of 22.4 L of the compound at STP; this is the molar volume of the gas. The mass of the molar volume is the molar mass.

69. a. A molecule is composed of two or more atoms.
b. A compound has a gram molecular mass; not a gram atomic mass.
c. 1 mol CO_2 has 3 times Avogadro's number of atoms.

71. 24.5 g

73. a. 27 amu b. aluminum

75. a. $CuBr_2$ b. CH_3

77. A molecular formula is a whole number multiple of its empirical formula.

79. Gas molecules are separated by so much empty space, their own volumes are insignificant when considering how much space a certain quantity of gas occupies.

81. 1.59 mol Pt

83. a. 4.72×10^3 mg b. 97 km/hr c. 4.4×10^{-2} dm

85. a. KNO_3 b. CuO c. Mg_3N_2 d. AgF

87. 21.9 cm^3

89. a.

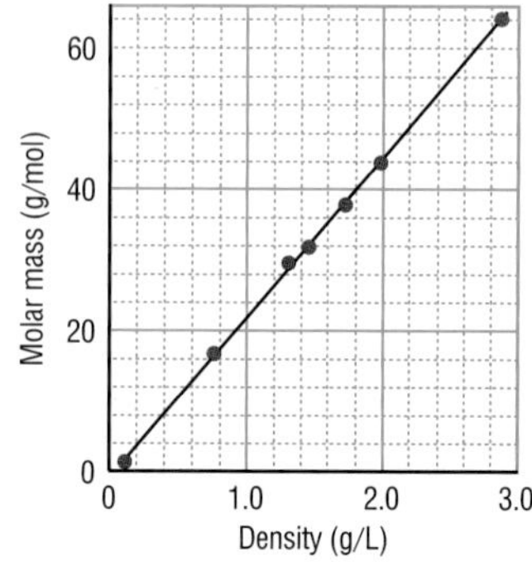

b. 22.4 L/mol c. 24.6 g/mol d. 2.5 g/L

91. 6.025×10^{23} formula units/mol

CHAPTER 8

1. a. $S(s) + O_2(g) \longrightarrow SO_2(g)$
b. $KClO_3(s) \xrightarrow{MnO_2, \Delta} KCl(s) + O_2(g)$

2. a. Mixing aqueous potassium hydroxide and aqueous sulfuric acid produces water and aqueous potassium sulfate.
b. When solid sodium is added to water, aqueous sodium hydroxide and hydrogen gas are produced.

3. a. $2AgNO_3 + H_2S \longrightarrow Ag_2S + 2HNO_3$
b. $MnO_2 + 4HCl \longrightarrow MnCl_2 + 2H_2O + Cl_2$
c. $3Zn(OH)_2 + 2H_3PO_4 \longrightarrow Zn_3(PO_4)_2 + 6H_2O$

4. a. $H_2 + S \longrightarrow H_2S$
b. $2FeCl_3 + 3Ca(OH)_2 \longrightarrow 2Fe(OH)_3 + 3CaCl_2$

5. $3CO + Fe_2O_3 \longrightarrow 2Fe + 3CO_2$

6. $2C + O_2 \longrightarrow 2CO$

7. a. $FeCl_3 + 3NaOH \longrightarrow Fe(OH)_3 + 3NaCl$
b. $CS_2 + 3Cl_2 \longrightarrow CCl_4 + S_2Cl_2$
c. $CH_4 + Br_2 \longrightarrow CH_3Br + HBr$

8. a. $2Na + 2H_2O \longrightarrow 2NaOH + H_2$
b. $Ca(OH)_2 + H_2SO_4 \longrightarrow CaSO_4 + 2H_2O$

13. a. $2Be + O_2 \longrightarrow 2BeO$
b. $SO_2 + H_2O \longrightarrow H_2SO_3$

14. a. $Sr + I_2 \longrightarrow SrI_2$
b. $3Mg + N_2 \longrightarrow Mg_3N_2$

15. a. $2HI \longrightarrow H_2 + I_2$
b. $Mg(ClO_3)_2 \longrightarrow MgCl_2 + 3O_2$

16. a. HBr b. NaCl

17. a. $Fe(s) + Pb(NO_3)_2(aq) \longrightarrow Fe(NO_3)_2(aq) + Pb(s)$
b. $Cl_2(g) + 2NaI(aq) \longrightarrow 2NaCl(aq) + I_2(s)$
c. $Ca(s) + 2H_2O(l) \longrightarrow Ca(OH)_2(aq) + H_2(g)$

18. a. $3NaOH + Fe(NO_3)_3 \longrightarrow Fe(OH)_3 + 3NaNO_3$
b. $3Ba(NO_3)_2 + 2H_3PO_4 \longrightarrow Ba_3(PO_4)_2 + 6HNO_3$

19. a. $3KOH + H_3PO_4 \longrightarrow K_3PO_4 + 3H_2O$
b. $3H_2SO_4 + 2Al(OH)_3 \longrightarrow Al_2(SO_4)_3 + 6H_2O$

20. a. $2HCOOH + O_2 \longrightarrow 2CO_2 + 2H_2O$
b. $C_7H_{16} + 11O_2 \longrightarrow 7CO_2 + 8H_2O$

21. $C_6H_{12}O_6 + 6O_2 \longrightarrow 6CO_2 + 6H_2O$

25. a. $PB_2^+(aq) + 2I^-(aq) \longrightarrow PbI_2(s)$
b. $Zn(s) + 2H^+(aq) \longrightarrow Zn_2^+(aq) + H_2(g)$
c. $H^+(aq) + OH^-(aq) \longrightarrow H_2O(l)$

26. $Pb^{2+}(aq) + 2Cl^-(aq)\ PbCl_2(s)$

27. $Fe^{3+}(aq) + 3NO_3^-(aq) + 3Na^+(aq) + 3OH^-(aq) \longrightarrow Fe(OH)_3(s) + 3Na^+(aq) + 3NO_3^-(aq)$
$Fe^{3+}(aq) + 3OH^-(aq) \longrightarrow Fe(OH)_3(s)$

33. Dalton said that the atoms of reactants are rearranged to form new substances as products.

35. a. Gaseous ammonia and oxygen react in the presence of platinum to produce nitrogen monoxide gas and water vapor.
b. Aqueous solutions of sulfuric acid and barium chloride are mixed to produce a precipitate of barium sulfate and aqueous hydrochloric acid.
c. The gas dinitrogen trioxide reacts with water to produce an aqueous solution of nitrous acid.

37. a. $C + 2F + 2G \longrightarrow CF_2G_2$
b. $F + 3W + S + 2P \longrightarrow FW_3SP_2$

39. a. $2PbO_2 \longrightarrow 2PbO + O_2$
b. $2Fe(OH)_3 \longrightarrow Fe_2O_3 + 3H_2O$
c. $(NH_4)_2CO_3 \longrightarrow 2NH_3 + H_2O + CO_2$
d. $2NaCl + H_2SO_4 \longrightarrow Na_2SO_4 + 2HCl$
e. $4H_2 + Fe_3O_4 \longrightarrow 3Fe + 4H_2O$
f. $2Al + 3CuSO_4 \longrightarrow Al_2(SO_4)_3 + 3Cu$

41. Use ionic charges to write an electrically neutral formula.

43. a. $2Mg + O_2 \longrightarrow 2MgO$
b. $4P + 5O_2 \longrightarrow 2P_2O_5$
c. $Ca + S \longrightarrow CaS$
d. $2Fe + O_2 \longrightarrow 2FeO$
e. $N_2O_5 + H_2O \longrightarrow 2HNO_3$

45. a single reactant

47. a. no reaction
b. $Zn(s) + 2AgNO_3(aq) \longrightarrow Zn(NO_3)_2(aq) + 2Ag(s)$
c. $2Al(s) + 3H_2SO_4(aq) \longrightarrow Al_2(SO_4)_3(aq) + 3H_2(g)$
d. no reaction
e. $2Al(s) + 3CuSO_4(aq) \longrightarrow Al_2(SO_4)_3(aq) + 3Cu(s)$

49. oxygen

51. a. $3Hf + 2N_2 \longrightarrow Hf_3N_4$; combination
b. $Mg + H_2SO_4 \longrightarrow MgSO_4 + H_2$; single-replacement
c. $2C_2H_6 + 7O_2 \longrightarrow 4CO_2 + 6H_2O$; combustion
d. $Pb(NO_3)_2 + 2NaI \longrightarrow PbI_2 + 2NaNO_3$; double-replacement
e. $3Fe + 2O_2 \longrightarrow Fe_3O_4$; combination
f. $2Pb(NO_3)_2 \longrightarrow 2PbO + 4NO_2 + O_2$; decomposition
g. $Hg(NO_3)_2 + 2NH_4SCN \longrightarrow Hg(SCN)_2 + 2NH_4NO_3$; double-replacement
h. $(NH_4)_2SO_4 + 2NaOH \longrightarrow 2NH_3 + 2H_2O + Na_2SO_4$; double-replacement, then decomposition

53. an ion that does not participate in the reaction

55. a. $Cl_2(g) + 2KI(aq) \longrightarrow I_2(s) + 2KCl(aq)$
b. $2Fe(s) + 6HCl(aq) \longrightarrow 2FeCl_3(aq) + 3H_2(g)$
c. $P_4O_{10}(s) + 6H_2O(l) \longrightarrow 4H_3PO_4(aq)$
d. $2Ag_2O(s) \xrightarrow{\Delta} 4Ag(s) + O_2(g)$
e. $I_2(s) + 3Cl_2(g) \longrightarrow 2ICl_3(s)$
f. $4HgS(s) + 4CaO(s) \xrightarrow{\Delta} 4Hg(l) + 3CaS(s) + CaSO_4(s)$

57. a. $Na_2O(s) + H_2O(l) \longrightarrow 2NaOH(aq)$
b. $H_2(g) + Br_2(g) \longrightarrow 2HBr(g)$
c. $Cl_2O_7(l) + H_2O(l) \longrightarrow 2HClO_4(aq)$

59. a. A
b. $2Na(s) + 2H_2O(l) \longrightarrow 2NaOH(aq) + H_2(g)$ single-replacement

61. a. $2Al_2O_3 \xrightarrow{\text{energy}} 4Al + 3O_2$
b. $Sn(OH)_4 \xrightarrow{\Delta} SnO_2 + 2H_2O$
c. $Ag_2CO_3 \xrightarrow{\Delta} Ag_2O + CO_2$

63. a. 4 b. 2 c. 1

65. a. $C_5H_{12} + 8O_2 \longrightarrow 5CO_2 + 6H_2O$
$C_9H_{20} + 14O_2 \longrightarrow 9CO_2 + 10H_2O$
b. $2C_{12}H_{26} + 37O_2 \longrightarrow 24CO_2 + 26H_2O$
$C_{17}H_{36} + 26O_2 \longrightarrow 17CO_2 + 18H_2O$
c. $n = CO_2$; $(n + 1) = H_2O$

67. 425 g Au

69. a.

$CaCl_2$ (mol)	H_2O (mol)
0.156	0.312
0.439	0.878
1.12	2.24
3.03	6.06

b.

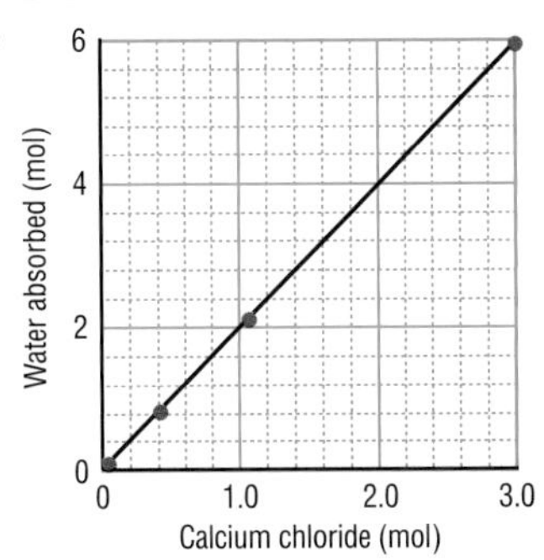

c. Two molecules of water are absorbed by each formula unit of $CaCl_2$.

71. a. 1.01 g/mol protons, 5.48×10^{-4} g/mol electrons
b. 1.84×10^3 electrons/protons

CHAPTER 9

1. 288 seats, 864 wheels, 576 pedals

2. Answers will vary but should include the correct number of "parts" to make the product.

3. 2 molecules H_2 + 1 molecule $O_2 \longrightarrow$ 2 molecules H_2O
2 mol H_2 + 1 mol $O_2 \longrightarrow$ 2 mol H_2O
44.8 L H_2 + 22.4 L $O_2 \longrightarrow$ 44.8 L H_2O

4. 2 mol C_2H_2 + 5 mol $O_2 \longrightarrow$ 4 mol CO_2 + 2 mol H_2O
44.8 L C_2H_2 + 112 L $O_2 \longrightarrow$ 89.6 L CO_2
212 g reactants $\longrightarrow$ 212 g products

9. a. $\frac{4\text{ mol Al}}{3\text{ mol }O_2}$ $\frac{3\text{ mol }O_2}{4\text{ mol Al}}$ $\frac{4\text{ mol Al}}{2\text{ mol }Al_2O_3}$
$\frac{2\text{ mol }Al_2O_3}{4\text{ mol Al}}$ $\frac{2\text{ mol }Al_2O_3}{3\text{ mol }O_2}$ $\frac{3\text{ mol }O_2}{2\text{ mol }Al_2O_3}$
b. 7.4 mol

10. a. 11.1 mol b. 0.52 mol

11. 2.03 g C_2H_2

12. 1.36 mol CaC_2

13. 4.82×10^{22} molecules O_2

14. 11.5 g NO_2

15. 1.93 L O_2

16. 0.28 L PH_3

17. 18.6 mL SO_2

18. 1.9 dL CO_2

23. a. 8.10 mol O_2 required; O_2 is the limiting reagent.
b. 4.20 mol H_2O

24. a. 5.40 mol O_2 required; C_2H_4 is the limiting reagent.
b. 5.40 mol H_2O

25. a. HCl b. 0.16 g H_2

26. 43.2 g H_2O

27. 91.6%

28. 83.5%

33. a. Two formula units $KClO_3$ decompose to form 2 formula units KCl and 3 molecules O_2.
b. Four molecules NH_3 react with 6 molecules NO to form 5 molecules N_2 and 6 molecules H_2O.
c. Four atoms K react with 1 molecule O_2 to form 2 formula units K_2O.

35. a. 245.2 g b. 248.0 g c. 188.4 g
All obey the law of conservation of mass.

37. a. 0.54 mol c. 0.984 mol
b. 13.6 mol d. 236 mol

39. a. 372 g F_2 b. 1.32 g NH_3 c. 123 g N_2F_4

41. a. 51.2 g H_2O
b. 5.71×10^{23} molecules NH_3
c. 23.2 g Li_3N

43. To identify the limiting reagent (the reagent that determines how much product is formed), express quantities of reactants as moles and compare to the mole ratios from the balanced equation.

45. a. 3.5 mol $AlCl_3$ c. 0.96 mol H_3PO_4
b. 6.4 mol H_2O d. 7.20 mol P_2O_5

47. 91.5%

49. The efficiency of a reaction; the actual yield/theoretical yield expressed as a percent.

51. a. 2.36 g H_3PO_4 b. 1.89 g CO_2

53. a. 7.0×10^2 L N_2 b. no reagent in excess

55. 10.7 kg $CaSO_4$

57. a. 2 b. 4

59. Yes; a net ionic equation is balanced and thus obeys the law of conservation of mass.

61. a. $2Pb(NO_3)_2 \longrightarrow 2PbO + 4NO_2 + O_2$
b. $2C_3H_7OH + 9O_2 \longrightarrow 6CO_2 + 8H_2O$
c. $2Al + 3FeO \longrightarrow 3Fe + Al_2O_3$

63. a. 1, 1, 1, 2 b. 1, 3, 3, 1 c. 1, 1, 1, 2

65. a. sodium ion and nitrate ion
b. aluminum ion and nitrate ion
c. magnesium ion and sulfate ion

67. 1.30×10^{-22} g C_6H_6

69. $C_2H_2O_4$

71. 13 days

73. 87.4% $CaCO_3$

CHAPTER 10

1. 51.3 kPa, 0.507 atm

2. 33.7 kPa is greater than 0.25 atm.

21. pascal (Pa), millimeter of mercury (mm Hg), atmosphere (atm); Pa is the SI unit.

23. a. 25 kPa b. 0.25 atm

25. The kinetic energy of the particles and the absolute temperature are directly proportional.

27. standard temperature and pressure, 0 °C and 1 atm

29. The average kinetic energy of particles at absolute zero is zero.

31. A liquid is condensed in the sense that its volume is minimally affected by an increase in pressure.

33. In both cases, particles with sufficient kinetic energy move from the liquid to the vapor phase. A dynamic equilibrium is set up between a contained liquid and its vapor; this is not possible with an uncontained liquid.

35. A larger percentage of molecules have enough energy to escape attractions within the liquid.

37. It increases the kinetic energy, which increases the vapor pressure.

39. a. ~55 mm Hg
b. T~93 °C
c. The vapor pressure is 760 mm Hg, which is standard pressure.

41. about 35 kPa

43. Answers will vary, but should include the fact that ionic compounds have crystalline structures with relatively high melting points.

45. The forces between the molecules of molecular solids are weaker.

47. The molecules escaping the liquid remove the added heat. The average kinetic energy of the liquid remains the same.

49. 77 °C

51. There is a continual exchange of particles, though net amounts of vapor molecules and liquid molecules remain constant.

53. It increases.

55. The substance has sufficient average kinetic energy so that the forces holding the solid crystal together are disrupted.

57. Possible answers include: odors will travel through a room; ink will move throughout a beaker of water.

59. No; collisions between objects other than at the atomic level involve the transformation of kinetic energy into heat energy.

61. Some compounds have stronger intermolecular forces than others.

63. 3.74 g/L = 0.003 74 g/cm^3

65. a. 13.9 mol SO_2
b. 0.0472 mol NH_3
c. 0.021 mol CO_2

67. Vapor pressure depends only on the kinetic energy of the escaping molecules.

69. a. orthorhombic
b. rhombohedral
c. tetragonal
d. triclinic
e. cubic

CHAPTER 11

1. 2.0 J/(g × °C)
2. 0.511 J/(g × °C)
3. 1.8 kJ
11. 1.5 kJ
12. 150 J
13. 6.63 kJ
14. 89.4 kJ
20. 1.20 g
21. 3.34 kJ
22. 144 kJ
23. 0.19 kJ
24. 301 kJ
25. 3.42 mol $NH_4NO_3(s)$
30. **a.** −30.91 kJ **b.** 178.4 kJ **c.** −113.0 kJ
31. CO is a compound.
37. Heat flows from the object at the higher temperature to the object at the lower temperature.
39. the chemical composition of the substance and its mass
41. **a.** 8.50×10^{-1} Calorie
    b. 1.86×10^3 J
    c. 1.8×10^3 J
    d. 1.1×10^2 cal
43. A negative sign is given to heat flow from the system to the surroundings. A positive sign is given to heat flow to the system from the surroundings.
45. **a.** exothermic **c.** exothermic
    b. endothermic **d.** endothermic
47. A calorimeter is an instrument used to measure heat changes in physical or chemical processes.
49. bomb calorimeter
51. amount of heat released or absorbed in the chemical change at constant pressure
53. **a.** -2.10×10^1 kJ
    b. -1.8×10^1 kJ
    c. -5.56×10^2 kJ
    d. 6.5 kJ
55. It allows the calculation of the enthalpy of a reaction from the known enthalpies of two or more other reactions.
57. 3.02×10^1 kJ
59. -1.259×10^2 kJ
61. This statement is true, because stability implies lower energy. The greater the release of heat, the more stable is the compound relative to its elements (all of which have $\Delta H_f^0 = 0$).
63. ΔH_f^0 for the formation of one mole of a compound from its elements
65. 4.00×10^1 g water; 9.60×10^2 g ice
67. 2.44×10^4 cal; 1.02×10^5 J
69. **a.** -2.21×10^2 kJ
    b. -1.96×10^2 kJ
    c. -9.046×10^2 kJ
71. 2.36×10^1 kJ
73. 2.38×10^2 kJ
75. 6.72×10^1 kJ
77. The region denoted by ΔH_{fus} represents the coexistence of solid and liquid. The region denoted by ΔH_{vap} represents the coexistence of liquid and vapor.
79. 1.20×10^{24} H_2 molecules
81. 1.18×10^1 g O_2
83. -1.37×10^2 kJ
85. 45.4 °C

CHAPTER 12

10. 6.48 L
11. 6.83×10 kPa
12. 3.39 L
13. 8.36 L
14. 2.58 kPa
15. 341 K, 68 °C
16. 3.42×10^{-1} L
17. 1.29×10^2 kPa
22. 2.51×10^2 mol He(*g*)
23. 1.71×10^3 kPa
24. 2.5 g air
25. 1.76×10^1 L $O_2(g)$
31. 5.60 L
32. 1.66×10^1 L
33. 1.38×10^{23} nitrogen molecules
34. 1.50 L He(*g*)
35. 4.48 L
36. 7.67×10^1 L
37. 9.34×10^1 kPa
38. 3.3 kPa
45. The increased kinetic energy of the gas particles causes collisions to occur with more force.
47. The pressure doubles.
49. Temperatures measured on the Kelvin scale are always positive and directly proportional to the average kinetic energy of the gaseous particles.
51. 1.00×10^2 kPa
53. 1.8×10^1 L
55. High temperatures can sufficiently increase the pressure of the gas remaining in the container to cause it to explode.

57. 1.10×10^3 kPa

59. Gas particles have a finite volume and are attracted to one another, especially at low temperatures.

61. At lower temperatures, gas particles are attracted to one another; the finite volume of gas particles is significant at high pressures.

63. **a.** 5.6×10^1 L **b.** 6.7 L **c.** 7.84 L

65. At any temperature, hydrogen gas diffuses faster than chlorine gas by an approximate factor of six.

67. 2.25:1

69. The variables are directly proportional.

71. 165 °C

73. **a.** (4) **b.** (1) **c.** (1) **d.** (3) **e.** (2)

75. A vacuum contains no matter to allow the transfer of kinetic energy between molecules.

77. **a.** 1.58×10^2 g **c.** 3.42×10^2 g
b. 9.80×10^1 g **d.** 3.31×10^2 g

79. **a.** 2.06×10^2 g
b. 82 protons, 82 electrons, 124 neutrons

81. **a.** $Ca + 2H_2O \longrightarrow Ca(OH)_2 + H_2$
b. $P_4O_{10} + 6H_2O \longrightarrow 4H_3PO_4$
c. $2HgO \longrightarrow 2Hg + O_2$
d. $Al_2S_3 + 6H_2O \longrightarrow 2Al(OH)_3 + 3H_2S$

83. **a.** $C_4H_8O_2$ **b.** C_8H_8 **c.** $C_2H_2O_4$

85. 60.0% C, 13.3% H, 26.7% O

87. 2 mol KNO_3 for each 1 mol O_2

89. 46% CH_4

91. Because attractions between molecules in gases such as nitrogen and oxygen are insignificant, these gases have the molar volume of an ideal gas—22.41 L at STP. Based on their molar volumes at STP, attractions between molecules increase in strength and effect from methane to carbon dioxide to ammonia.

CHAPTER 13

5. **a.** $1s^22s^22p^2$ **b.** $1s^22s^22p^63s^23p^6$

6. **a.** $1s^22s^22p^1$; 1 unpaired electron
b. $1s^22s^22p^63s^23p^2$; 2 unpaired electrons

11. 2.00×10^{-3} cm; longer wavelength than red light

12. $6.00 \times 10^{15}\ s^{-1}$; ultraviolet

13. 2.12×10^{-22} J

14. 7.6×10^{-17} J

21. Electrons have fixed energies. To move from one energy level to another, they must emit or absorb specific amounts of energy, called quanta.

23. 90% of the time an electron is found inside this boundary.

25. The 1*s* orbital is spherical. The 2*s* orbital is spherical with a diameter larger than that of the 1*s* orbital. The 2*p* orbital is dumbbell shaped and reaches beyond the diameter of the 2*s* orbital.

27. The Aufbau principle states that electrons occupy the lowest possible energy levels. The Pauli exclusion principle states that an atomic orbital can hold at most two electrons. Hund's rule states that before any pairing of electrons occurs, one electron occupies each of a set of orbitals with equal energies.

29. The *p* orbitals in the third quantum level have three electrons.

31. **a.** 2 **b.** 6 **c.** 2 **d.** 10 **e.** 6 **f.** 2 **g.** 14 **h.** 6

33. **a.** $1s^22s^22p^63s^23p^63d^{10}4s^24p^4$ **c.** $1s^22s^22p^63s^23p^63d^84s^2$
b. $1s^22s^22p^63s^23p^63d^34s^2$ **d.** $1s^22s^22p^63s^23p^64s^2$

35. Frequency is the number of wave cycles that pass a given point per unit time. Frequency units are cycles, reciprocal seconds (s^{-1}), or hertz. Frequency and wavelength are inversely related; the speed of light divided by the wavelength of the light equals the frequency.

37. Classical physics viewed energy changes as continuous and occurring in any quantity. In the quantum concept, energy changes occur in tiny discrete units called quanta.

39. Planck showed that radiation is emitted and absorbed in quanta, dependent on the frequency of radiation.

41. A photon is a quantum of light energy. A quantum is a discrete amount of energy.

43. The outermost electron of sodium absorbs photons of wavelength 589 nm as it jumps to a higher orbital.

45. **a.** Ar **b.** Ru **c.** Gd

47. **a.** 2 **b.** 4 **c.** 10 **d.** 6

49. 2.61×10^4 cm

51. 3.08×10^{-19} J; lower energy

53. **a.** 1 **b.** 4 **c.** 1

55. The model of the atom uses the abstract idea of probability; light is considered a particle and a wave at the same time. Atoms and light cannot be compared with familiar objects or observations because humans cannot experience atoms or photons directly and because matter and energy behave differently at the atomic level than at the level humans can observe directly.

57. **a.** 2, 1, 1, 2 **b.** 1, 1, 2, 1 **c.** 2, 1, 2 **d.** 1, 2, 1, 2

59. **a.** 46.8% Si, 53.2% O **c.** 11.1% H, 88.9% O
b. 34.4% Fe, 65.6% Cl **d.** 2.1% H, 32.7% S, 65.2% O
e. 40% Ca, 12% C, 48% O

61. **a.** 55 p^+, 55 e^- **c.** 48 p^+, 46 e^-
b. 47 p^+, 46 e^- **d.** 34 p^+, 36 e^-

63. 154 g; 0.154 Kg

65. 6.93×10^2 s

67. H: 1312 kJ/mol ($n = 1$); 328 kJ ($n = 2$);
Li^{2+}: 1.18×10^4 kJ ($n = 1$).

CHAPTER 14

1. **a.** $1s^22s^22p^2$
b. $1s^22s^22p^63s^23p^63d^34s^2$
c. $1s^22s^22p^63s^23p^63d^{10}4s^24p^65s^2$

2. **a.** He, Be, Mg, Ca, Sr, Ba, Ra
b. F, Cl, Br, I, At
c. Ti, Zr, Hf, Rf

11. **a.** $1s^2 2s^2 2p^1$
b. $1s^2 2s^2 2p^6 3s^2$
c. $1s^2 2s^2 2p^6 3s^2 3p^6 3d^{10} 4s^2 4p^3$

13. **a.** Ar: $1s^2 2s^2 2p^6 3s^2 3p^6$
b. Ge: $1s^2 2s^2 2p^6 3s^2 3p^6 3d^{10} 4s^2 4p^2$
c. Ba: $1s^2 2s^2 2p^6 3s^2 3p^6 3d^{10} 4s^2 4p^6 4d^{10} 5s^2 5p^6 6s^2$

15. **a.** $1s^2 2s^2 2p^5$
b. $1s^2 2s^2 2p^6 3s^2 3p^6 3d^{10} 4s^2$
c. $1s^2 2s^2 2p^6 3s^2 3p^1$
d. $1s^2 2s^2 2p^6 3s^2 3p^6 3d^{10} 4s^2 4p^6 4d^{10} 5s^2 5p^2$

17. Fluorine has a smaller atomic radius than oxygen because fluorine has one more nuclear charge. Fluorine has a smaller radius than chlorine because fluorine has eight fewer electrons.

19. The first ionization energy is the energy needed to remove the outermost electron. The second ionization energy is the energy needed to remove the second-outermost electron.

21. nonmetals; The nuclear charge increases, but the shielding effect is the same, thus creating greater electron attraction.

23. An atom of an alkali metal becomes stable by losing one electron. Removing the second electron involves removing an electron from an ion that has a stable noble-gas configuration. This requires much more energy.

25. The radius of a cation is smaller than the atom from which it forms.

27. **a.** F **b.** N **c.** Mg **d.** As

29. Zinc has more protons than calcium, thus attracting the $4s$ electrons more.

31. **a.** potassium, K, $[Ar]4s^1$
b. aluminum, Al, $[Ne]3s^2 3p^1$
c. sulfur, S, $[Ne]3s^2 3p^4$
d. barium, Ba, $[Kr]6s^2$

33. scandium, Sc, $[Ar]3d^1 4s^2$
titanium, Ti, $[Ar]3d^2 4s^2$
vanadium, V, $[Ar]3d^3 4s^2$
chromium, Cr, $[Ar]3d^5 4s^1$
manganese, Mn, $[Ar]3d^5 4s^2$
iron, Fe, $[Ar]3d^6 4s^2$
cobalt, Co, $[Ar]3d^7 4s^2$
nickel, Ni, $[Ar]3d^8 4s^2$
copper, Cu, $[Ar]3d^{10} 4s^1$
zinc, Zn, $[Ar]3d^{10} 4s^2$

35. Magnesium achieves a stable electron configuration by losing two electrons; aluminum achieves a stable electron configuration by losing three electrons.

37. **a.** Possible cations are Rb^+ and Sr^{2+}; possible anions are Br^-, Se^{2-}, and As^{3-}.
b. No; a cation is isoelectronic with the noble gas in the preceding period; an anion is isoelectronic with the noble gas in the same period.

39. **a.** $2Ag + S \longrightarrow Ag_2S$
b. $Na_2SO_4 + Ba(OH)_2 \longrightarrow BaSO_4 + 2NaOH$
c. $Zn + 2HNO_3 \longrightarrow Zn(NO)_2 + H_2$
d. $2H_2O + 2SO_2 + O_2 \longrightarrow 2H_2SO_4$

41. **a.** Li_2SO_4 **b.** $Zn_3(PO_4)_2$ **c.** $KMnO_4$ **d.** $SrCO_3$

43. The ionic radii decrease from S^{2-}, Cl^-, Ar, K^+, Ca^{2+}, to Sc^{3+} as the number of protons increases. The radii decrease from O^{2-}, F^-, Ne, Na^+, Mg^{2+}, to Al^{3+} for the same reason.

CHAPTER 15

7. **a.** KI **b.** Al_2O_3

8. **a.** potassium iodide **b.** aluminum oxide

21. fluorine (F), chlorine (Cl), bromine (Br), iodine (I); Group 7A; seven valence electrons

23. **a.** :C̈l· **b.** :S̤̈· **c.** ·Al̇· **d.** Li·

25. **a.** 2 **b.** 3 **c.** 1 **d.** 2

27. **a.** $1s^2 2s^2 2p^6 3s^2 3p^6 3d^3$
b. $1s^2 2s^2 2p^6 3s^2 3p^6 3d^4$
c. $1s^2 2s^2 2p^6 3s^2 3p^6 3d^5$

29. **a.** S^{2-} **b.** Na^+ **c.** F^- **d.** P^{3-}

31. **a.** Br^- **b.** H^- **c.** As^{3-} **d.** Se^{2-}

33. The positive charges of the cations equal the negative charges of the anions.

35. **a.** K^+Cl^-
b. $Ba^{2+}SO_4^{2-}$
c. $Mg^{2+}Br^-$
d. $Li^+CO_3^{2-}$

37. Their network of electrostatic attractions and repulsions forms a rigid structure.

39. They have many mobile valence electrons. Electrons in the current replace the electrons leaving the metal.

41. Answers will vary and could include tableware, steel in cars and buses, high-speed dental drill bits, solder in stereos and televisions, and structural steel in buildings.

43. Brass is a mixture of copper and zinc. The properties of a particular sample of brass will vary with the relative proportions of the two metals.

45.

Group number	Valence electrons lost or gained	Ion formula
1A	1	Na^+
2A	2	Ca^{2+}
3A	3	Al^{3+}
5A	3	N^{3-}
6A	2	S^{2-}
7A	1	Br^-

47. For the representative elements, the number of electrons in the electron dot structure is the group number.

49. It has gained valence electrons.

51. **a.** $1s^2 2s^2 2p^6 3s^2 3p^6 3d^6$
b. $1s^2 2s^2 2p^6 3s^2 3p^6 3d^7$
c. $1s^2 2s^2 2p^6 3s^2 3p^6 3d^8$

53. Hexagonal closest-packed unit cells have 12 neighbors for every atom or ion. Face-centered cubic unit cells also have 12 neighbors for every atom or ion, with an atom or ion in the center of each face. Body-centered cubic units cells have 8 neighbors for every atom or ion, with an atom or ion at the center of each cube.

55. **a.** $1s^2 2s^2 2p^6 3s^2 3p^6$ **b.** $1s^2 2s^2 2p^6$

Each has a noble-gas electron configuration.

57. Both are composed of ions and are held together by electrostatic bonds. Metals always conduct electricity, but ionic compounds conduct only when melted or in aqueous solution. Ionic compounds are composed of cations and anions, but metals are composed of cations and free-floating valence electrons. Metals are malleable, but ionic compounds are brittle.

59. Gases and liquids assume the shapes of their containers. Solids have a definite shape. Liquids and solids have a definite volume; gases do not. Gases have a low density; liquids have an intermediate density; solids have a high density. The molecules in a gas move freely and randomly. The molecules of a liquid flow. The molecules of a solid vibrate and rotate around a fixed position.

61. The average kinetic energy and the speed of the gas molecules increase.

63. **a.** At 40 °C more water molecules have energies that allow them to overcome intermolecular attractions than at 20 °C.
b. Steam at 100 °C has more potential energy.
c. The molecules' kinetic energy has increased, creating a greater vapor pressure.
d. Because water molecules require more energy to evaporate than do ether molecules, the attractions between water molecules must be greater than those between ether molecules.
e. Ice absorbs energy from the tea. The kinetic energy of the tea decreases.
f. The goal is to change liquid water to water vapor without destroying chemical bonds by heating; lowering the external pressure lowers the temperature at which liquid water molecules have enough energy to vaporize.

65. 1.36×10^2 kPa

67. **a.** 0.0147
0.0373
0.0639
0.0879

b.

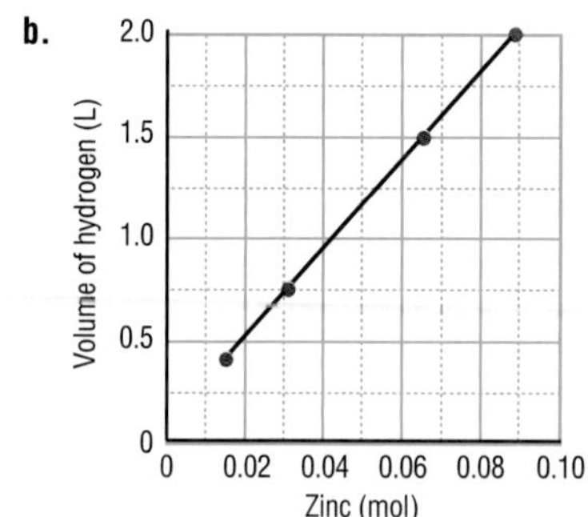

c. 22.4 L/mol
d. 1:1 ratio

69. 0.1445 nm

71. **a.** iron(II); Fe^{2+}
b. iron(III); Fe^{3+}
c. copper(II); Cu^{2+}
d. copper(I); Cu^{+}
e. cobalt(III); Co^{3+}

CHAPTER 16

1. **a.** :C̈l:C̈l: **b.** :B̈r:B̈r: **c.** :Ï:Ï:

2. **a.** H:Ö:Ö:H **b.** :C̈l:P̈:C̈l: / :C̈l:

3. [H:Ö:]$^-$

4. [F:B:F with F above and below]$^-$

5. [:Ö:S:Ö: with :Ö: above and below]$^{2-}$ [:Ö:C:Ö: with :Ö: above]$^{2-}$

6. [H:Ö:C:Ö with :Ö: below]$^-$

19. **a.** moderately polar covalent
b. ionic
c. moderately to very polar covalent
d. moderately to very polar covalent
e. ionic
f. nonpolar covalent

20. In order of increasing polarity: e, c, d, b, a

27. Neon has an octet of valence electrons. A chlorine atom achieves an octet by sharing an electron with another chlorine atom.

29. Ionic bonds depend on electrostatic attraction between ions. Covalent bonds depend on electrostatic attraction between shared electrons and nuclei of combining atoms.

31. The element of which there is only one atom is the central atom of the molecule.

33. **a.** :F̈:F̈: **b.** H:C̈l: **c.** H:C::C:H **d.** H:C:::N:

35. An unshared pair of electrons is needed for a coordinate covalent bond. There are no unshared pairs in C—H or C—C bonds.

37. [Ö::C:Ö: / :Ö:]$^{2-}$ ⟷ [:Ö:C::Ö: / :Ö:]$^{2-}$ ⟷ [:Ö:C:Ö: / :Ö:]$^{2-}$

39. The measured mass of a paramagnetic substance appears greater when measured in the presence of a magnetic field than when measured in the absence of a magnetic field.

41. (b) and (c); assuming only single bonds the P and S atoms each have 10 valence electrons.

43. Increasing bond dissociation energy is linked to lower chemical reactivity.

45.

H:C:::C:H		1694 kJ/mol
2H:C	2 × 393 kJ/mol =	786 kJ/mol
C:::C		908 kJ/mol

47. **a.** 2 sigma, 2 pi **b.** 6 sigma, 2 pi

49. The 2*s* and the 2*p* orbitals form two sp^2 hybrid orbitals in the carbon atom. One sp^2 hybrid orbital from each oxygen atom forms a sigma bond with the carbon atom. Pi bonds between each oxygen atom and the carbon are formed by the unhybridized 2*p* orbitals.

51. The two atoms involved must have different electronegativities.

53. b., d., e., a., c., f.

55.

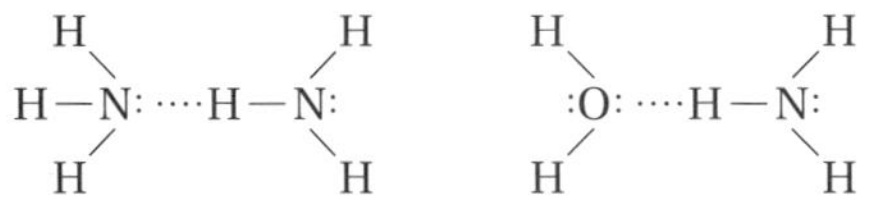

```
  H              H        H              H
   \            /          \            /
  H—N: ····H—N:            :O: ····H—N:
   /            \          /            \
  H              H        H              H
```

57. A hydrogen bond is formed by an electrostatic attraction between a hydrogen atom covalently bonded to a more electronegative atom and an unshared electron pair of a nearby atom.

59. The 3s and three 3p orbitals of phosphorus hybridize to form four sp^3 hybrid orbitals. The resulting shape is pyramidal with a bond angle of 107° between the sigma bonds.

61.

```
:C̈l:S̈:C̈l:
    :Ö:
```

63. **a.** tetrahedral, 109.5° **b.** trigonal planar, 120° **c.** tetrahedral, 109.5° **d.** bent, 105°

65. $\Delta H = -55$ kJ/mol

67. An antibonding orbital has a higher energy because the atomic orbitals overlap on opposite sides of the nuclei, not between them.

69. C, O, H, S, N, F, Cl, I, Br; These elements are all nonmetals.

71. a., b., c., d.

73. **a.** $1s^22s^22p^63s^1$ **b.** $1s^22s^22p^63s^23p^4$ **c.** $1s^22s^22p^63s^23p^3$ **d.** $1s^22s^22p^3$

75. **a.** 1.20×10^{22} **b.** 9.27×10^{20} **c.** 5.26×10^{15} **d.** 1.81×10^{18}

77.

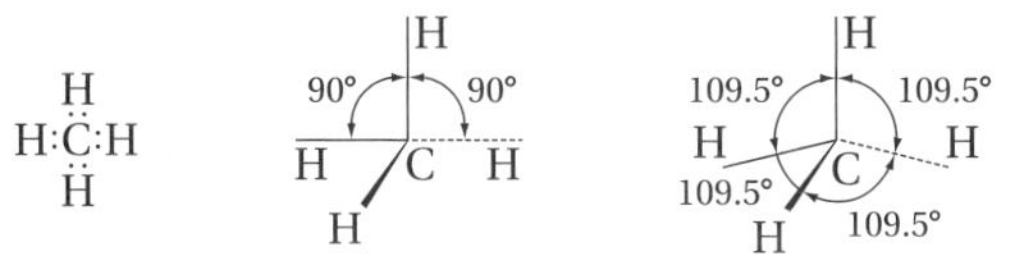

```
   H                    |H                          |H
 H:C̤:H          90°     |     90°         109.5°    |     109.5°
   H             H     /C      H          H        (C)        H
                      H                   109.5°         109.5°
                                                H
```

The first sketch is tetrahedral. The second sketch is a tetrahedron. The bond angles in the first sketch are not all the same, with some being 90°. The bond angles in the second sketch are all 109.5°. The second sketch is correct. (Note: The wedge-shaped lines come out of the page; the dotted lines recede into the page.)

CHAPTER 17

8. 36.1%

9. **a.** $CaCl_2 + 6H_2O \rightarrow CaCl_2{\bullet}6H_2O$
b. 49.3%

19. Oxygen is more electronegative than hydrogen. Because the water molecules are bent, the bond polarities do not cancel.

21. Some water molecules have sufficient energy to leave the surface of the liquid. Vapor pressure is the pressure of the water vapor above the liquid.

23. It physically interferes with hydrogen bonding and reduces surface tension.

25. water, 2.0×10^4 cal; iron, 2.2×10^3 cal

27. Ammonia molecules form hydrogen bonds, whereas methane molecules do not. Substances whose molecules form hydrogen bonds tend to have higher boiling points.

29. Ice floats in liquid water. Solids generally sink in their liquids because a substance is generally more dense in the solid state.

31. Bodies of water would freeze from the bottom up. This would kill many forms of aquatic life.

33. The polar water molecules electrostatically attract ions and polar covalent molecules. Polar compounds will dissolve but nonpolar compounds are unaffected because they have no charges.

35. The positive ions are attracted by the negatively charged end of the polar water molecule; the negative ions are attracted by the positively charged end. As the ions are pulled away from the crystal, they are surrounded by the water molecules.

37. Its ions are free to move toward an electrode.

39. water in the crystal structure of a substance

41. **a.** tin(IV) chloride pentahydrate
b. iron(II) sulfate heptahydrate
c. barium bromide tetrahydrate
d. iron(III) phosphate tetrahydrate

43. A suspension is a mixture with large particles that settle on standing. A colloid is a mixture with particles of intermediate size that do not settle.

45. the random motion of the dispersion medium molecules

47. Water molecules at 4 °C are tightly packed and have maximum density. Below 4 °C the water molecules arrange in a regular network because of the attractions between them. As a result, ice has a lower density than water and floats.

49. **a.** sodium carbonate monohydrate, 14.5% H_2O
b. magnesium sulfate heptahydrate, 51.1% H_2O

51. Ions in solution are surrounded by water molecules. Negative ions are attracted to the hydrogen atoms and positive ions are attracted to the oxygen atoms.

53. **a.** no
b. drying to examine the crystals, testing for electrical conductivity, doing a flame test

55. **1.** a, b **3.** a **5.** b, c **7.** b **9.** b
2. b, c **4.** c **6.** a, b **8.** a

57. Ethanol has a polar hydroxyl end (—OH) that dissolves in water, and a nonpolar hydrocarbon end (C_2H_5—) that dissolves in gasoline.

59. No; nonpolar molecules do not dissolve in polar molecules.

61. **a.** 1.00 g/mL **b.** 4 °C
c. No; the density of ice is 0.917 g/mL at 0 °C. There would be a break in the curve at 0 °C as liquid water at 0 °C changes to solid water (ice) at 0 °C.

63. The blood and all of the body's cells contain large amounts of water. Most of the important chemical reactions of life take place in aqueous solution inside cells.

65. **a.** 195 g **b.** 1.95×10^5 mg **c.** 0.195 kg

67. **a.** raises the boiling point
b. lowers the boiling point

69. 9.00 g H_2, 72.0 g O_2

71. 636 g C_2H_4O

73. **a.** As the molar mass increases, the boiling point increases.
b. The boiling point more than doubles.

75. **a.** hydrogen **c.** oxygen
b. 4.9×10^{-2} g H_2O **d.** 10 cm^3 O_2

CHAPTER 18

1. 0.44 g/L

2. 2.6 atm

8. 0.10*M*

9. 2.8*M*

10. 0.142 mol

11. 0.50 mol $CaCl_2$; 56 g $CaCl_2$

12. 47.5 mL

13. Use a pipet to transfer 50 mL of the 1.0*M* solution to a 250-mL volumetric flask. Then add distilled water up to the 250-mL mark.

14. 5% (v/v)

15. 12 mL

16. 0.25 g $MgSO_4$

17. 3.6% $CuSO_4$ (m/v)

28. 12.6 g

29. 0.285*m* NaCl

30. $X_{C_2H_5OH} = 0.190; X_{H_2O} = 0.810$

31. $X_{CCl_4} = 0.437; X_{CHCl_3} = 0.563$

32. 101.37 °C

33. 114 g NaCl

34. 39.2 g/mol

35. 169 g/mol

41. Random collisions of the solvent molecules with the solute molecules or ions provide enough force to overcome gravity.

43. Solubility is the amount of solute dissolved in a given amount of solvent to make a saturated solution at a given temperature. A saturated solution has the maximum amount of solute that can be dissolved in a given amount of solvent at a given temperature. An unsaturated solution has less dissolved solute than a saturated solution.

45. Particles of solute crystallize.

47. Increasing pressure increases the solubility.

49. Molarity gives you the exact number of moles of solute per liter of solution. Dilute and concentrated are relative terms and are not quantitative.

51. the number of moles of solute dissolved in one liter of solution

53. **a.** 23 g NaCl **b.** 2.0 g $MgCl_2$

55. **a.** 100.26 °C **b.** 101.54 °C

57. A 1*M* solution has 1 mol of solute dissolved in 1 L of solution. A 1*m* solution has 1 mol of solute dissolved in 1000 g of solvent.

59. **a.** −1.1 °C **b.** −0.74 °C **c.** −1.5 °C

61. **a.** Because the freezing-point depression is twice as great for solute B compared with solute A, solute B must give twice as many particles in solution compared with solute A.
b. Solute A forms a saturated solution in this quantity of water when 2.4 mol of A is added.

63. Each gram of acetone requires 0.93 g of water.

65. The mole fraction of NaCl is 0.00269; the mole fraction of H_2O is 0.997.

67. **a.** about 1.14 **b.** about −7.2 °C **c.** about −9.5 °C

69. $X_{C_2H_5OH} = 0.20; X_{H_2O} = 0.80$

71.

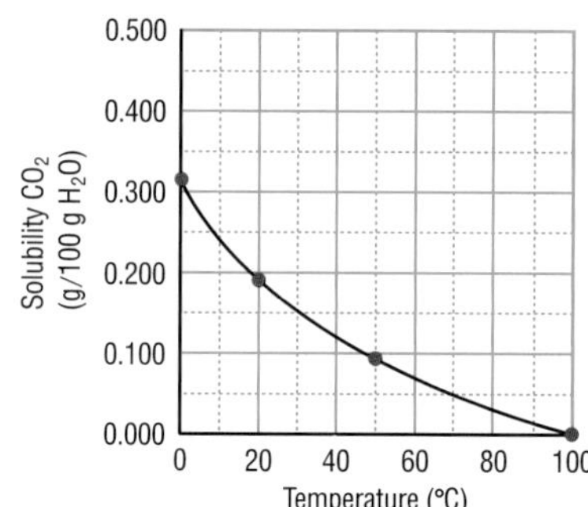

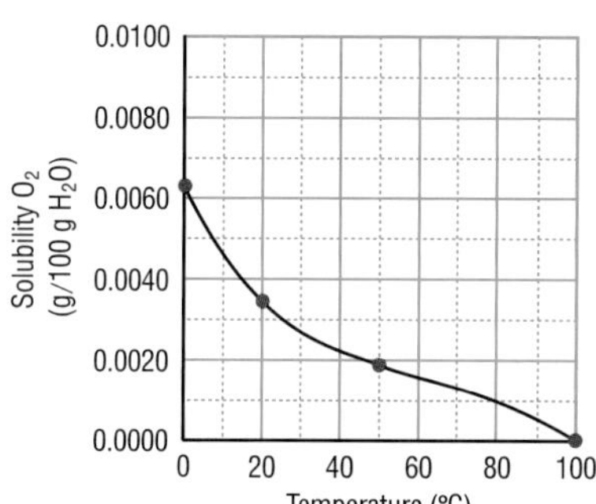

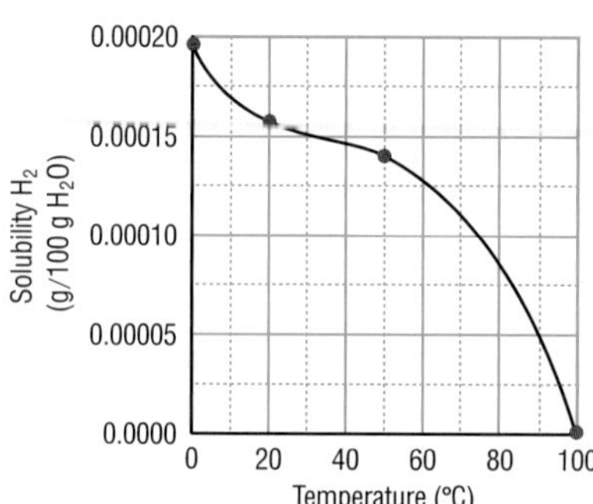

73. **a.** 0.30 mol **b.** 0.40 mol **c.** 0.50 mol **d.** 0.20 mol

75. unsaturated

77. **a.** 7.5 g H_2O_2 **b.** 0.88*M*

79. 85.5 g/mol

81. **a.** 4 **b.** 1

83. 1 mol of calcium chloride dissolves into 3 mol of particles, and 1 mol of sodium chloride dissolves into 2 mol of particles; freezing-point depression is proportional to the number of particles of solute.

85. The stronger the intermolecular attractions in a liquid, the greater is the surface tension.

87. raises the boiling points

89. a. the substance that is dissolved
b. the substance used for dissolving
c. No additional solute can be dissolved.
d. More solute can be dissolved.
e. The solution contains more solute than a saturated solution.
f. a solute that cannot conduct electricity
g. a solute that conducts a small amount of electricity
h. a solute that can conduct electricity well

91. 1.10×10^2 mL HNO_3

93. $0.090M$ Na_2SO_4

CHAPTER 19

6. a. reactants favored
b. reactants favored
c. products favored
d. reactants favored

7. a. favors reactants
b. favors products
c. no effect
d. favors products

8. $K_{eq} = 12$

9. a. $K_{eq} = 0.084$ or 8.4×10^{-2}
b. One is the inverse of the other.

10. $K_{eq} = 1.6 \times 10^{-3}$

11. $K_{eq} = 0.79$

12. 3.5 mol HI and 0.50 mol I_2

13. a. $K_{eq} = \frac{[H_2] \times [Br_2]}{[HBr]^2}$
b. $K_{eq} = \frac{[SO_2]^2 \times [O_2]}{[SO_3]^2}$
c. $K_{eq} = \frac{[CO] \times [H_2O]}{[CO_2] \times [H_2]}$
d. $K_{eq} = \frac{[H_2O]^6 \times [NO]^4}{[NH_3]^4 \times [O_2]^5}$

25. 254.7 J/K•mol

26. −0.7 J/K•mol

27. −46.2 kJ/mol

28. a. 38.1 kJ; nonspontaneous
b. −1097.7 kJ; spontaneous

29. a. 514.2 kJ; nonspontaneous
b. 33.02 kJ; nonspontaneous

33. Rate = k[A]; rate is moles per liter per second. [A] is moles per liter. k = rate/[A]. $k = 1/s = s^{-1}$

34. 0.25 mol/L•s; 0.125 mol/L•s

39. Reactant particles must have a certain minimum amount of energy to react to form product, just as it takes a certain minimum amount of energy to climb over a wall or barrier.

41. A catalyst increases the rate of reactions by providing an alternative reaction mechanism with a lower activation energy.

43. Gas molecules and oxygen molecules mix readily but do not have enough energy to react at room temperature. The flame raises the temperature and the energy of collisions, so the reaction rate is increased. The heat released by the reaction maintains the high temperature, and the reaction continues spontaneously.

45. The rate of formation of products from reactants and the rate of formation of reactants from products are equal.

47. A system in dynamic equilibrium changes to relieve the stress applied to it. Carbonated drinks in closed containers have achieved a state of dynamic equilibrium between the carbon dioxide in the liquid and gas states. When the containers are opened, carbon dioxide gas escapes. Carbon dioxide from the liquid goes into the gas state in an attempt to re-establish equilibrium.

49. a. $K_{eq} = \frac{[H_2S]^2 \times [CH_4]}{[H_2]^4 \times [CS_2]}$
b. $K_{eq} = \frac{[PCl_3] \times [Cl_2]}{[PCl_5]}$
c. $K_{eq} = \frac{[NO_2]^2}{[NO]^2 \times [O_2]}$
d. $K_{eq} = \frac{[H_2] \times [CO_2]}{[CO] \times [H_2O]}$

51. the energy in a system available to do work; a spontaneous reaction has a negative free energy

53. unfavorable

55. a. completed jigsaw puzzle
b. 50 mL of ice
c. sodium chloride crystals

57. No; some endothermic processes are spontaneous because of their favorable change in entropy.

59. In combination, the change in heat content (enthalpy) and the change in entropy determine whether a process is spontaneous.

61. a. the proportionality constant relating the concentrations of the reactants to the reaction rate
b. a chemical reaction with a rate proportional to the concentration of one reactant
c. an equation relating the reaction rate to the concentrations of the reactants

63. 100 minutes

65. a. 1.22×10^{-4} mol/(L · s) b. 1.94×10^{-4} mol/(L · s)

67. 25 kJ

69. a. yes b. yes c. no d. yes

71. $K_{eq} = 0.659$

73. Increasing pressure tends to reduce volume and increase density, so the system responds by favoring production of the liquid, which has the greater density.

75. The reaction is slow (activation energy is high).

77. −89.8 J/K

79. A catalyst increases the efficiency of the collisions; a greater number of collisions results in the formation of the product.

81. a. fluoride ion (anion)
b. copper(II) ion (cation)
c. phosphide ion (anion)
d. hydrogen ion (cation)
e. sodium ion (cation)
f. iodide ion (anion)
g. oxide ion (anion)
h. magnesium ion (cation)

83. Solid potassium chloride is an ionic compound of K^+ and Cl^-, not of KCl molecules. Each ion is surrounded by six ions of opposite charge in a simple cubic unit cell crystal.

85. crystalline solids, high melting points; insulators when solid but conductors when molten; not malleable or ductile; often soluble in water

87. **a.** $Li^+ :\ddot{Br}:^-$ **c.** $^-:\ddot{F}:Mg^{2+}:\ddot{F}:^-$

b. $^-:\ddot{Cl}:\ddot{Al}^{3+}:\ddot{Cl}:^-$ (with $:\ddot{Cl}:^-$ above Al) **d.** $Na^+ :\ddot{S}:^{2-} Na^+$

89. positive ions: **b.**, **e.**, **f.**, **g.**, **h.**, and **j.**; negative ions: **a.**, **c.**, **d.**, and **i.**

91. **a.**

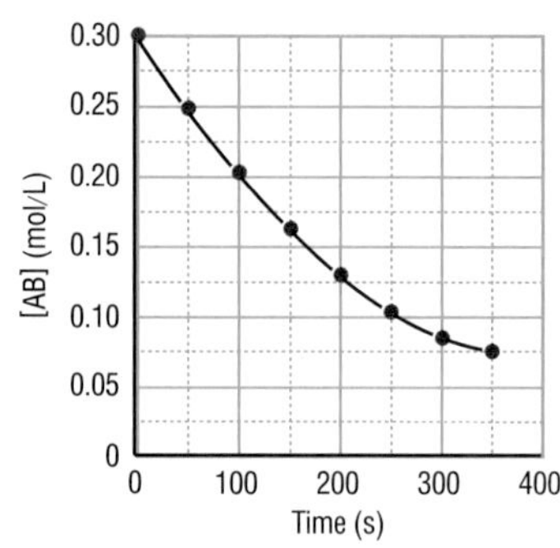

b. The rate when $t = 100$ is 8×10^{-4} mol/L•*s*. The rate when $t = 250$ is 4×10^{-4} mol/L•*s*.

CHAPTER 20

1. **a.** hydrofluoric acid **b.** nitric acid **c.** potassium hydroxide **d.** sulfuric acid

2. **a.** H_2CrO_4 **b.** $Fe(OH)_2$ **c.** HI **d.** LiOH

6. $1.0 \times 10^{-11}M$; basic

7. **a.** basic **b.** basic **c.** acidic **d.** neutral

8. **a.** 4.00 **b.** 3.00 **c.** 9.00

9. **a.** 12.00 **b.** 2.00 **c.** 4.00

10. **a.** $1.0 \times 10^{-5}M$ **b.** $1.0 \times 10^{-7}M$ **c.** $1.0 \times 10^{-6}M$ **d.** $1.0 \times 10^{-3}M$

11. **a.** $1.0 \times 10^{-4}M$ **b.** $1.0 \times 10^{-11}M$ **c.** $1.0 \times 10^{-8}M$

12. **a.** 5.30 **b.** 9.08 **c.** 6.57

13. **a.** 9.63 **b.** 9.30 **c.** 3.65

14. **a.** $5.0 \times 10^{-8}M$ **b.** $1.6 \times 10^{-2}M$ **c.** $8.9 \times 10^{-8}M$ **d.** $2.0 \times 10^{-7}M$

15. **a.** $2.0 \times 10^{-7}M$ **b.** $6.3 \times 10^{-13}M$ **c.** $1.1 \times 10^{-7}M$ **d.** $5.0 \times 10^{-8}M$

19. Lewis base; it has a nonbonding pair of electrons, which it can donate.

20. **a.** H^+ is the Lewis acid; H_2O is the Lewis base.

b. $AlCl_3$ is the Lewis acid; Cl^- is the Lewis base.

26. $K_a = 1.8 \times 10^{-4}$

27. **a.** pH = 3.006 **b.** $K_a = 4.89 \times 10^{-6}$

35. $H_2O \rightleftharpoons H^+ + OH^-$

37. the negative logarithm of the hydrogen-ion concentration

39. **a.** pH = 2.00, acidic **b.** pH = 12.00, basic **c.** pH = 6.00, acidic **d.** pH = 6.00, acidic

41. **a.** 5.62 **b.** 8.04 **c.** $6.3 \times 10^{-14}M$ **d.** $2.0 \times 10^{-7}M$

43. **a.** $KOH \longrightarrow K^+ + OH^-$
b. $Mg(OH)_2 \longrightarrow Mg^{2+} + 2OH^-$

45. **a.** base **b.** acid **c.** base **d.** acid **e.** acid **f.** acid

47. **a.** $2Li + 2H_2O \longrightarrow 2LiOH + H_2$
b. $Ba + 2H_2O \longrightarrow Ba(OH)_2 + H_2$

49. **a.** HNO_3 with NO_3^-, H_2O with H_3O^+
b. CH_3COOH with CH_3COO^-, H_2O with H_3O^+
c. H_2O with OH^-, NH_3 with NH_4^+
d. H_2O with OH^-, CH_3COO^- with CH_3COOH

51. A Lewis acid accepts a pair of electrons to form a covalent bond. A Lewis base donates a pair of electrons to form a covalent bond. The Lewis theory includes as acids or bases many substances that do not donate or accept hydrogen ions.

53. A strong acid is completely dissociated; K_a must be large.

55. **a.** $K_a = \frac{[H^+][F^-]}{[HF]}$ **b.** $K_a = \frac{[H^+][HCO_3^-]}{[H_2CO_3]}$

57. Yes; acids such as acetic acid dissolve well but ionize poorly.

59. $4.5 \times 10^{-4}M$

61. **a.** $HClO_2$, chlorous acid **b.** H_3PO_4, phosphoric acid **c.** H_3O^+, hydronium ion **d.** NH_4^+, ammonium ion

63. **a.** $[OH^-] = 4.0 \times 10^{-10}M$ **b.** $[OH^-] = 2.0 \times 10^{-5}M$ **c.** pH = 12.26 **d.** pH = 5.86

65. **a.** KOH is the base; HBr is the acid.
b. HCl is the acid; H_2O is the base.

67. Answers will vary; students are likely to consider the Arrhenius theory the easiest to understand. All three theories provide definitions and describe accepted behavior of a certain group of compounds. Because it is more general, the Brønsted–Lowry theory includes a greater number of compounds than the Arrhenius theory. The Lewis theory includes the greatest number of compounds because it is the most general. Each theory has advantages in certain circumstances.

69. The *y*-axis might correspond to $[H^+]$ or pOH because HCl is a strong acid.

71. 0.47 L

73. **a.** $K_{eq} = \frac{[CO]^2[O_2]}{[CO_2]^2}$ **b.** $K_{eq} = \frac{[NH_3]^2}{[H_2]^3[N_2]}$

75. $K_w = K_aK_b = \frac{[H^+][\cancel{A^-}]}{[\cancel{HA}]} \times \frac{[\cancel{HA}][OH^-]}{[\cancel{A^-}]} = [H^+][OH^-]$

77. 50.0 mL; The pH = 7 when $[H^+] = [OH^-]$. Because HCl is a strong acid that supplies one hydrogen ion per formula unit and NaOH is a strong base that supplies one hydroxide ion per formula unit, $[H^+] = [OH^-]$ when equal volumes of solutions of the same molarity are combined.

CHAPTER 21

1. 4.68 mol KOH
2. 0.20 mol NaOH
3. 56 mL HCl
4. 0.128*M*
5. **a.** 32.7 g **b.** 40.0 g
6. **a.** 29.2 g/equiv **b.** 60.0 g/equiv
7. **a.** 0.50 equiv **b.** 0.20 equiv
8. **a.** 0.30 equiv **b.** 0.40 equiv
9. **a.** 0.99 equiv **b.** 0.80 equiv **c.** 0.070 equiv
10. **a.** 0.50*N* **b.** 0.20*N* **c.** 1.03*N* **d.** 1.79*N*
11. 58 mL of 3.0*N* KOH
12. Dilute 25 mL of 4.0*N* H_2SO_4 to 500 mL.
13. 0.260*N* base
14. 1.2*N* base
15. 14.1 mL
16. 29.5 mL

23. **a.** $HPO_4^{2-} + H^+ \rightleftharpoons H_2PO_4^-$
    b. $H_2PO_4^- + OH^- \rightleftharpoons HPO_4^{2-} + H_2O$
24. $CH_3COO^- + H^+ \rightleftharpoons CH_3COOH$
25. $2 \times 10^{-14}M$
26. $6.7 \times 10^{-5}M$
27. **a.** $2 \times 10^{-17}M$ **b.** $1 \times 10^{-17}M$
28. **a.** $[Pb^{2+}][Cl^-]^2$ **b.** $[Ag^+]^2[CrO_4^{2-}]$
29. $2 \times 10^{-17}M$
30. $3.2 \times 10^{-6}M$
31. It will.
32. Yes

37. **a.** $HNO_3 + KOH \longrightarrow KNO_3 + H_2O$
    b. $2HCl + Ca(OH)_2 \longrightarrow CaCl_2 + 2H_2O$
    c. $H_2SO_4 + 2NaOH \longrightarrow Na_2SO_4 + 2H_2O$
39. **a.** 1.40*M* **b.** 2.61*M*
41. **a.** 0.10 equiv **b.** 3.86 equiv **c.** 0.30 equiv
43. **a.** 1*N* **b.** 2*N* **c.** 0.2*N* **d.** 0.2*N*
45. **a.** 0.0600*N* **c.** 0.171*N*
    b. 0.400*N* **d.** 0.0752*N*
47. $HCO_3^-(aq) + H_2O(l) \rightleftharpoons H_2CO_3(aq) + OH^-(aq)$
49. **a.** basic **c.** neutral **e.** neutral
    b. acidic **d.** basic **f.** acidic
51. No; HCl is a strong acid.
53. the product of the ion concentrations raised to the power of their coefficients
55. lowers the solubility
57. The product of the concentrations of the two ions is greater than the solubility product constant of the precipitate.
59. $H_2PO_4^- + OH^- \rightleftharpoons H_2O + HPO_4^{2-}$
    $HPO_4^{2-} + H^+ \rightleftharpoons H_2PO_4^-$
    Added OH^- is neutralized by $H_2PO_4^-$ and added acid is neutralized by HPO_4^{2-}.
61. **a.** 8.73
    b. phenolphthalein or thymol blue

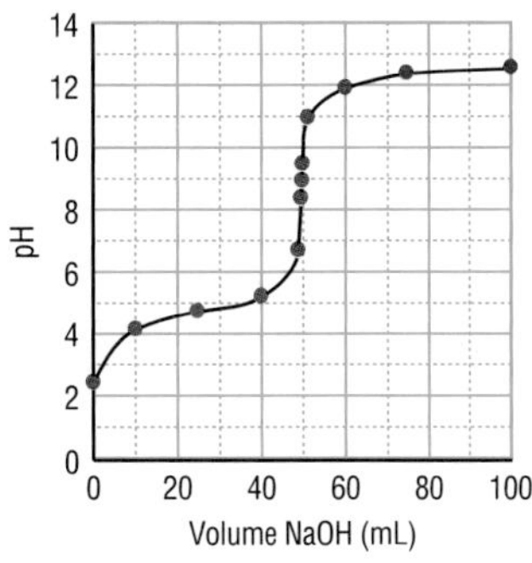

63. 12.5 mL
65. **a.** shift to the left **b.** shift to the right
67. **b.**, **c.**, **d.**, **a.**
69. **a.** 2 **b.** 3
71. CO_2 concentration is higher in pure water. Less CO_2 becomes carbonate because pure water does not have the OH^- ions needed to reduce H^+ concentration.
73. **a.** 5.34 **b.** 11.30 **c.** 0.52 **d.** 9.01
75. **a.** HSO_4^- **b.** CN^- **c.** OH^- **d.** NH_3
77. 0.5*M* NaCl
79. 74.4% $AgNO_3$

CHAPTER 22

1. **a.** Na: oxidized (reducing agent);
   S: reduced (oxidizing agent)
   b. Al: oxidized (reducing agent);
   O_2: reduced (oxidizing agent)
2. **a.** oxidation **c.** reduction
   b. oxidation **d.** reduction
9. **a.** S: +3; O: −2 **c.** Al: +3; S: +6; O: −2
   b. O: 0 **d.** Na: +1; O: −1
10. **a.** P: +5; O: −2 **c.** Na: +1; Cr: +6; O: −2
    b. N: −3; H: +1 **d.** Ca: +2; O: −2; H: +1
11. **a.** H_2 oxidized, O_2 reduced
    b. K unchanged, N reduced, O oxidized
    c. N in NH_4^+ oxidized, H unchanged,
    N in NO_2^- reduced, O unchanged
    d. Pb reduced, O unchanged, H unchanged, I oxidized
12. **a.** H_2, reducing agent; O_2, oxidizing agent
    b. N in NO_3^-, oxidizing agent;
    O in NO_3^-, reducing agent
    c. N in NH_4^+, reducing agent;
    N in NO_2^-, oxidizing agent
    d. Pb, oxidizing agent; I, reducing agent

17. The following are redox reactions: **c.** Mg is oxidized; Br_2 is reduced. **d.** Nitrogen in NO_2^- is oxidized; nitrogen in NH_4^+ is reduced. **e.** Oxygen is oxidized; chlorine is reduced.

18. **b.** is an oxidation-reduction reaction; Cu is reduced; H_2 is oxidized.

19. **a.** 2, 2, 3
b. 2, 2, 2, 1, 2
c. 1, 2, 5, 2, 4

20. **a.** 1, 8, 2, 2, 3, 4
b. 1, 1, 1, 1, 2, 2
c. 1, 2, 1, 2, 1

21. **a.** $3Sn^{2+} + 14H^+ + Cr_2O_7^{2-} \longrightarrow 3Sn^{4+} + 2Cr^{3+} + 7H_2O$
b. $CuS + 8NO_3^- + 8H^+ \longrightarrow Cu(NO_3)_2 + SO_2 + 6NO_2 + 4H_2O$
c. $6I^- + 2NO_3^- + 8H^+ \longrightarrow 3I_2 + 2NO + 4H_2O$

22. **a.** $6I^- + 2MnO_4^- + 4H_2O \longrightarrow 3I_2 + 2MnO_2 + 8OH^-$
b. $S_2O_3^{2-} + 2NiO_2 + 2OH^- + H_2O \longrightarrow 2SO_3^{2-} + 2Ni(OH)_2$
c. $4Zn + NO_3^- + 6H_2O + 7OH^- \longrightarrow 4Zn(OH)_4^{2-} + NH_3$

25. oxidation

27. **a.** $C_2H_4(g) + 3O_2(g) \longrightarrow 2CO_2(g) + 2H_2O(l)$
b. $2KClO_3(s) \longrightarrow 2KCl(s) + 3O_2(g)$
c. $CuO(s) + H_2(g) \longrightarrow Cu(s) + H_2O(l)$
d. $2H_2(g) + O_2(g) \longrightarrow 2H_2O(l)$

29. Must be equal.

31. single-replacement, combustion, combination, and decomposition

33. **a.**, **b.**, **c.**

35. **a.** Mn, reducing agent; Pb, oxidizing agent
b. Cl, oxidizing agent; Cl, reducing agent
c. I, oxidizing agent; C, reducing agent

37. **a.** $2MnO(s) + 5PbO_2(s) + 8H^+ \longrightarrow 2MnO_4^-(aq) + 5Pb^{2+}(aq) + 4H_2O(l)$
b. $Cl_2(g) + H_2O(l) \longrightarrow HCl(aq) + HClO(aq)$
c. $I_2O_5(s) + 5CO(g) \longrightarrow I_2(s) + 5CO_2(g)$
d. $H_2O(l) + SO_3(g) \longrightarrow H_2SO_4(aq)$

39. **a.** C **b.** Cl

41. **a.** $MnO_2(s) + 4HCl(aq) \longrightarrow MnCl_2(aq) + Cl_2(g) + 2H_2O(l)$
b. $Cu(s) + 4HNO_3(aq) \longrightarrow Cu(NO_3)_2(aq) + 2NO_2(g) + 2H_2O(l)$
c. $3P(s) + 5HNO_3(aq) + 2H_2O(l) \longrightarrow 5NO(g) + 3H_3PO_4(aq)$
d. $2Bi(OH)_3(s) + 3Na_2SnO_2(aq) \longrightarrow 2Bi(s) + 3Na_2SnO_3(aq) + 3H_2O(l)$

43. Double-replacement reactions never involve the transfer of electrons; instead, they involve the exchange of positive ions in aqueous solution.

45. **a.** acidic **b.** basic **c.** basic **d.** acidic

47. 56.3 mL KOH

49. $1.14N$ H_3PO_4

51. Test tube B has the NaCl added to it. Due to the common-ion effect, the addition of either sulfate ion or barium ion to a saturated solution of $BaSO_4$ will cause the solubility product of $BaSO_4$ to be exceeded, and barium sulfate will precipitate, as shown in test tubes A and C.

53. **a.**

Alkane burned	O_2 used (mol)	CO_2 produced (mol)	H_2O produced (mol)
C_4H_{10}	6.5	4	5
C_5H_{12}	8	5	6
C_6H_{14}	9.5	6	7

b. $C_xH_y + \left[x + \left(\frac{y}{4}\right)\right]O_2 \longrightarrow xCO_2 + \left(\frac{y}{2}\right)H_2O$

55. 104 mL $K_2Cr_2O_7$

CHAPTER 23

8. $2Al(s) + 3Cu^{2+}(aq) \longrightarrow 2Al^{3+}(aq) + 3Cu(s)$; $E^0_{cell} = +2.00$ V

9. $E^0_{cell} = +0.46$ V

10. nonspontaneous: $E^0_{cell} = -0.02$ V

11. Yes; $E^0_{cell} = +0.16$ V

21. nothing

23. cathode

25. anode: zinc; cathode: carbon (graphite)

27. $Pb(s) \mid PbSO_4(s) \parallel PbO_2(s) \mid PbSO_4(s)$

29. A fuel cell needs no recharging and does not produce toxic wastes if the fuel is hydrogen gas.

31. the ability of a voltaic cell to produce a current

33. The aluminum half-cell is connected to a standard hydrogen half-cell and a voltmeter is used. The aluminum half-cell has the indicated voltage.

35. **a.** +1.61 V **b.** +0.94 V

37. The teaspoon is the cathode in an electrolytic cell with silver cyanide as the electrolyte. When the DC current flows, the silver ions deposit on the teaspoon.

39. Two half-cells are needed because oxidation or reduction cannot occur in isolation. One half-cell gains electrons and one loses them, producing an electric current.

41. Some of the iron dissolves and the nail becomes coated with copper.

$$Fe(s) + CuSO_4(aq) \longrightarrow FeSO_4(aq) + Cu(s)$$

oxidation: $Fe \longrightarrow Fe^{2+} + 2e^-$

reduction: $Cu^{2+} + 2e^- \longrightarrow Cu$

43. Lead(II) sulfate and lead dioxide are very insoluble in sulfuric acid.

45. Gold belongs near the bottom, below silver, because it is one of the least active metals.

47. The chemists' definition focuses on the electrons that are produced by oxidation at the anode of a voltaic cell; the dictionary definition is probably based on an electrolytic cell, whose electrodes are defined by the battery terminals to which they are attached.

49. The baby shoe is being plated with copper.

51. 467 mL O_2

53. Dilute 31 mL of 16*M* acid to a total volume of 500 mL.

55. a. $2AgCl + Ni \longrightarrow 2Ag + NiCl_2$
$E^0_{cell} = +0.47\ V$
b. $3Cl_2 + 2Al \longrightarrow 2AlCl_3$
$E^0_{cell} = +3.02\ V$

57. a. The iron electrode is the anode; the nickel electrode is the cathode.
b. The anode is negative; the cathode is positive.
c. anode: $Fe(s) \longrightarrow Fe^{2+}(aq) + 2e^-$
cathode: $Ni^{2+}(aq) + 2e^- \longrightarrow Ni(s)$
d. $E^0_{cell} = +0.19\ V$

CHAPTER 24

29. Group 1A: alkali metals; Group 2A: alkaline-earth metals

31. Their atomic diameters are smaller.

33. bauxite

35. diamond, graphite, and buckminsterfullerene

37. Nitric acid is used in dye and fertilizer manufacture, etching, and the production of explosives.

39. Plants cannot directly incorporate atmospheric nitrogen (N_2) into their tissues. Nitrogen-fixing bacteria must first transform it into nitrogen compounds.

41. The three conditions are a temperature of 500°C, a pressure of 10^5 kPa, and an iron oxide catalyst.

43. a. manufacturing of steel
b. to produce sulfuric acid
c. to produce photoelectric cells and other light-sensitive instruments

45. Hydrogen peroxide is used as a bleach and an antiseptic.

47. 20%

49. the manufacture of steel

51. a. $2Mg(s) + O_2(g) \longrightarrow 2MgO(s)$
b. $2H_2(g) + O_2(g) \longrightarrow 2H_2O(g)$
c. $S(s) + O_2(g) \longrightarrow SO_2(g)$

53. Sulfur is used mostly for the production of sulfuric acid, which is used extensively in manufacturing fertilizers, in the steel industry, in petroleum refining, and in many other industries.

55. Chlorine in solution is a strong oxidizing agent that kills disease-causing bacteria.

57. Iodine is obtained by reacting sodium iodate with sodium hydrogen sulfite solution to precipitate iodine crystals.

59. Cl_2: yellow-green gas; Br_2: dark-red liquid; I_2: purple-black crystalline solid

61. Nickel is used to electroplate iron and steel and to catalyze industrial reactions. Monel metal is a strong, corrosion-resistant alloy of nickel and copper.

63. Reduction converts positively charged metal ions in a compound to the free metal.

65. Superheated water and compressed air are sent down two tubes into the sulfur beds. A third tube brings sulfur froth to the surface. Sulfur is not mined because it is found under quicksand.

67. 12.9% Al

69. a. Amounts of Fe_2O_3: 5.42×10^5, 7.89×10^5, 1.26×10^6, 4.11×10^6, 6.02×10^6
Amounts of C: 1.63×10^6, 2.36×10^6, 3.78×10^6, 1.23×10^7, 1.81×10^7
b.

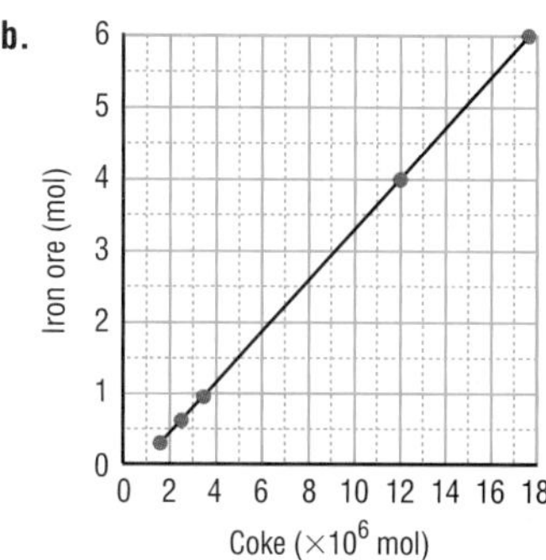

c. slope = 1 mol Fe_2O_3/3 mol C

71. 35.5% Co in CoAsS; 2.58×10^2 kg Co

73. O_2 and O_3; oxygen gas (O_2) is obtained from liquid air by fractional distillation. Ozone (O_3) is made by passing an electrical discharge through oxygen gas.

75. a. $3H_2(g) + N_2(g) \longrightarrow 2NH_3(g)$
b. $H_2(g) + Cl_2(g) \longrightarrow 2HCl(g)$
c. $H_2(g) + Ca(s) \longrightarrow CaH_2(s)$

77. Sulfuric acid is used to prepare fertilizers and to pickle iron and steel. It is also used in petroleum refining and in many other industries.

79. Ammonia is a small, polar molecule with hydrogen bonding. It is covalently bonded, with bond angles similar to those in water.

81. a. 3 b. 4

83. All the processes require an input of energy, either electricity or heat. Steam reforming produces poisonous carbon monoxide and requires the use of a nonrenewable resource, natural gas. The method using steam and white-hot coke produces carbon dioxide, which contributes to global warming.

85. Bronze is an alloy mostly of copper and tin. The technology for producing copper and tin was developed relatively easily. The separation of iron from its ore, however, requires much greater heat and more-advanced technology.

87. a. CaO c. CO e. SO_2
b. HgO d. Al_2O_3 f. Na_2O_2

89. a. $[Ar]3d^64s^2$ d. $[Ar]3d^54s^2$ g. $[Ar]3d^{10}4s^2$
b. $[Ar]3d^34s^2$ e. $[Ar]3d^{10}4s^1$ h. $[Kr]4d^{10}$
c. $[Kr]4d^{10}5s^1$ f. $[Ar]3d^84s^2$

91. a. Li^+, Br^- b. Ca^{2+}, OH^- c. K^+, I^- d. Cd^{2+}, Br^-

93. a. 0 b. +5 c. +4 d. +1 e. +3

95. $CaCO_3(s) + 2HCl(aq) \longrightarrow CaCl_2(aq) + H_2O(l) + CO_2(g)$

97. 2.7×10^{-1} L Cl_2

99. a. 1.10×10^{-1} mol H^+ b. 3.67×10^1 mL

101. 2.8×10^8 kg Br^-

103. 1.42×10^1 L O_2

CHAPTER 25

1.

```
    H   H   H   H   H            H   H   H   H   H   H
    |   |   |   |   |            |   |   |   |   |   |
H - C - C - C - C - C - H    H - C - C - C - C - C - C - H
    |   |   |   |   |            |   |   |   |   |   |
    H   H   H   H   H            H   H   H   H   H   H
```

2. 10

3. **a.** propane **b.** pentane **c.** hexane

4. **a.** 2-methylbutane **b.** 3-ethylhexane

5.

```
   CH3
   |
CH3CHCHCH2CH2CH3
      |
      CH3
```

6. **a.**

```
                 CH3
                 |
                 CH2
                 |
CH3 - CH2 - CH - CH2 - CH3
```

b.

```
      CH3  CH3  CH3
      |    |    |
CH3 - CH - CH - C - CH2 - CH2 - CH2 - CH3
                |
                CH2
                |
                CH3
```

14. Molecule **b** has an asymmetric carbon, which is marked with an asterisk in the answer.

```
        H
        |
CH3 - *C - CHO
        |
        Cl
```

15. Molecule **d** has an asymmetric carbon; molecules **a, b,** and **c** do not have nonsuperimposable, mirror images because they have no asymmetric carbons.

29. **a.** propane **b.** octane **c.** pentane

31. **a.** The prefix *di-* means two methyl groups are bonded to the parent chain. Locations of both groups must be identified by a number.
b. pentane; The longest carbon chain is misidentified.
c. 2-methylbutane; Lowest numbering for substituent is not followed.
d. 3-methylpentane; The longest carbon chain is misidentified.

33. The bonds between carbons are nonpolar and carbon-hydrogen bonds are very weakly polar; thus alkanes are nonpolar.

35. **a.** propene (or 1-propene)
b. *trans*-2-pentene
c. 4-methyl-1-pentene
d. 3-ethyl-2-methyl-2-pentene
e. 1-hexene

37. Five structural isomers of molecular formula C_6H_{14} exist.

```
C-C-C-C-C-C          C
  hexane             |
                   C-C-C-C
                     |
                     C
               2, 2-dimethylbutane

  C                  C C
  |                  | |
C-C-C-C-C          C-C-C-C
2-methylpentane    2, 3-dimethylbutane

      C
      |
  C-C-C-C-C
  3-methylpentane
```

39. **a.**

```
CH3       H            CH3       CH2CH3
   \     /                \     /
    C=C                    C=C
   /     \                /     \
  H       CH2CH3         H       H
 trans-2-pentene         cis-2-pentene
```

b.

```
CH3       H
   \     /
    C=C
   /     \
CH3       CH2CH3
2-methyl-2-pentene
```

c.

```
CH3       CH2CH3
   \     /
    C=C
   /     \
  H       CH2CH3
3-ethyl-2-pentene
```

41. No; only molecules with a least one asymmetric carbon have stereoisomers.

43. **a.**

```
  CH2CH3
    |
 [benzene ring]
    |
  CH2CH3
```

b.

```
       CH3
       |
CH3 - CH - CH - CH2 - CH3
            |
      [benzene ring]
```

c.

```
CH3 - [benzene ring] - CH3
```

d.

```
[benzene ring] - CH3
```

45. $2C_8H_{18} + 25O_2 \longrightarrow 16CO_2 + 18H_2O$

47. Coal is mostly large aromatic rings.

49. **a.** single bonds and triple bond; ethyne (acetylene)
b. single bonds; propane
c. single bonding within substituents and between substituents and ring; aromatic bonding (alternating single and double bonds) within the ring; methylbenzene

51. propane, butane, pentane

53. The middle structure is most stable due to resonance within its ring.

55. No; The structures are identical; one has been flipped over.

57. Cyclic hydrocarbons are more structurally rigid, so their van der Waals forces are stronger. Thus, it takes more energy (a higher temperature) to liberate these molecules from the liquid state.

59. **a.** Ca, +2; C, +4; O, −2 **b.** Cl, 0 **c.** Li, +1; I, +5; O, −2 **d.** Na, +1; S, +4; O, −2

61. **a.** $K_{eq} = \frac{[ICl]^2}{[I_2][Cl_2]}$ **b.** $K_{eq} = \frac{[H_2][Br_2]}{[HBr]^2}$ **c.** $K_{eq} = \frac{[HCl]^4[S]^3[SO_2]}{[S_2Cl_2]^2[H_2O]^2}$ **d.** $K_{eq} = \frac{[NH_3]^2}{[N_2][H_2]^3}$

63. **a.** ~8 **b.** Use a pH meter.

65. Answers will vary.

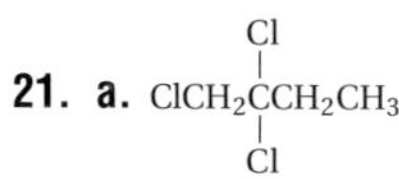

21. a. $ClCH_2CCl_2CH_2CH_3$ (Cl above and below the second carbon) b. (benzene ring with Br, Br, Br) c. (cyclohexane ring with Cl, Cl)

23. a. $CH(Cl)_2-CH_2-CH_3$
1, 1-dichloropropane

$CH_2(Cl)-CH(Cl)-CH_3$
1, 2-dichloropropane

$CH_2(Cl)-CH_2-CH_2(Cl)$
1, 3-dichloropropane

$CH_3-C(Cl)_2-CH_3$
2, 2-dichloropropane

b. $CH_3-CH_2-CH_2-CH_2Br$
1-bromobutane

$CH_3-CH_2-CH(Br)-CH_3$
2-bromobutane

$CH_3-CH(CH_3)-CH_2Br$
1-bromo-2-methylpropane

$CH_3-C(CH_3)(Br)-CH_3$
2-bromo-2-methylpropane

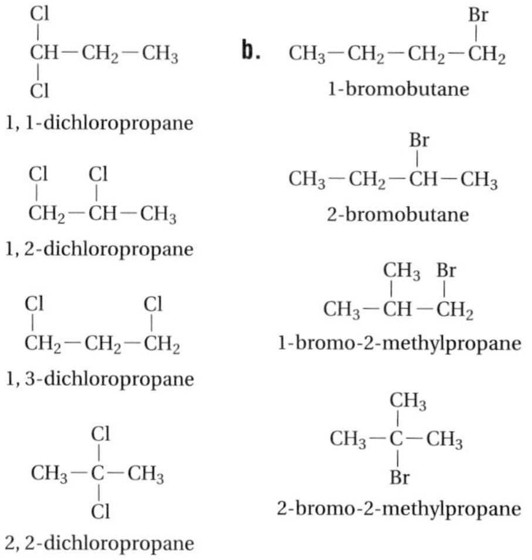

25. a. $CH_2(Br)CH(Br)CH_2CH_3$ c. $CH_3CH(H)CH(H)CH_3$
b. $CH_3CH(I)CH(I)CH_3$ d. (cyclohexane ring with Cl, Cl)

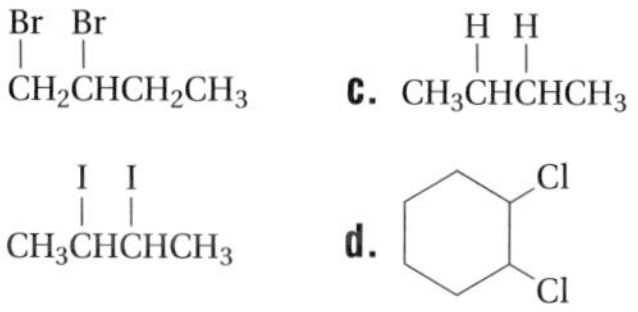

27. a. ethylmethyl ether
b. ethylphenyl ether
c. divinyl ether or vinyl ether
d. diisopropyl ether or isopropyl ether

29. The alcohol molecules form hydrogen bonds with one another, resulting in a higher boiling point. They also form hydrogen bonds with water molecules, causing 1-butanol to be more soluble than diethyl ether. (Although diethyl ether is polar, l-butanol has greater polarity.)

31. in order of increasing boiling point: propane, propanal, and propanol

33. The short-chain ethanoic acid has a higher water solubility.

35. a. formic acid c. propionic acid
b. acetic acid d. stearic acid

37. a. $CH_3CH_2OH \xrightarrow{K_2Cr_2O_7} CH_3-C(=O)-H$

b. $CH_3CH_2CHO \xrightarrow{K_2Cr_2O_7} CH_3-CH_2-C(=O)-OH$

c. $CH_3CH_2CH(CH_3)OH \xrightarrow{K_2Cr_2O_7} CH_3-CH_2-C(=O)-CH_3$

d. $C_6H_5-CH_2CHO \xrightarrow{K_2Cr_2O_7} C_6H_5-CH_2-C(=O)-OH$

e. $CH_3CH_2C(=O)CH_3 \xrightarrow{K_2Cr_2O_7}$ no reaction

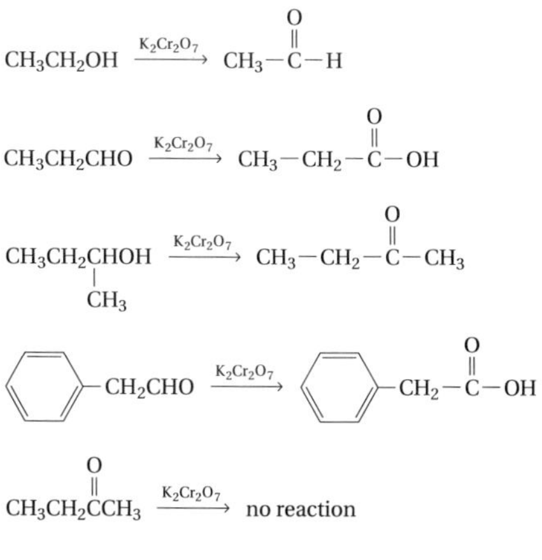

39. a. $CH_3CH_2COO^-Na^+$, CH_3CH_2OH

b. $CH_3COO^-K^+$, C_6H_5-OH

c. CH_3CH_2COOH, $CH_3CH(CH_3)CH_2OH$

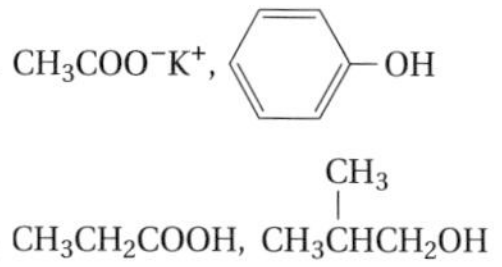

41. a. $H-C(=O)-O-CH_3$
methyl methanoate
(methyl formate)

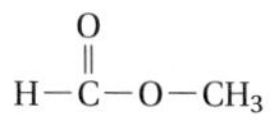

b. $CH_3CH_2CH_2-C(=O)-O-CH_2CH_3$
ethyl butanoate
(ethyl butyrate)

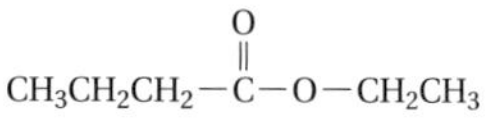

c. $CH_3-C(=O)-O-CH_2CH_2CH_3$
propyl ethanoate
(propyl acetate)

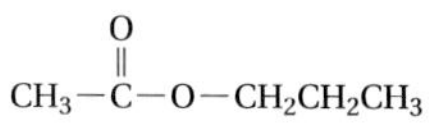

43. a. $-CH_2-CH(CH_2CH_3)-$ b. $-CH(Cl)-CH(Cl)-$

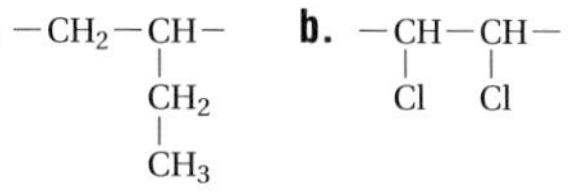

45. b. CH_3CH_2OH (46) has the highest boiling point.

47. Both atoms in a carbon–carbon double bond have the same electronegativity, so the bond is nonpolar. Because oxygen is more electronegative than carbon, a carbon–oxygen double bond is very polar.

49. a. $CH_3CH_2CH(Cl)-CH_2(Cl)$ b. $CH_3CH_2CH(Br)-CH_2(Br)$ c. (cyclohexane ring with H, Br)

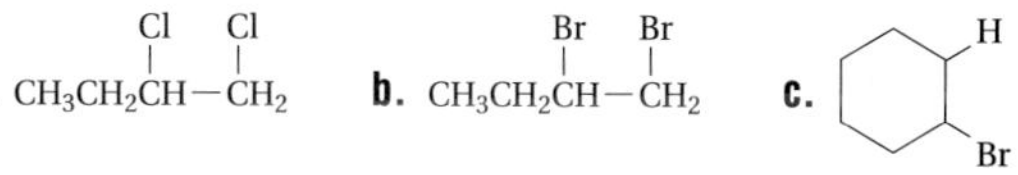

51. The chemical properties (and toxicity) of organic compounds are determined by the compound as a whole. As a substituent in a molecule, a phenyl group ring does not have the same properties as benzene.

53. At any given moment, the rate of dissolving of solute is equal to the rate of precipitation of solute. As a result, the concentration of the solution remains constant.

55. a. $:\ddot{F}\cdot + e^- \longrightarrow :\ddot{F}:^-$ b. $H\cdot + :\ddot{O}: + e^- \longrightarrow (:\ddot{O}:H)^-$

57. Cholesterol is an alcohol with a hydroxyl group on a cycloalkane. It has four nonaromatic rings. It has a double bond on one of its rings, as well as a large alkyl group, making it nonpolar.

CHAPTER 27

23. The glucose produced during photosynthesis is used by both plants and animals to fuel the synthesis of more complex biomolecules.

25. Glucose and fructose are two important monosaccharides.

27. Glucose contains an aldehyde functional group; fructose contains a ketone functional group.

29. **a.** glucose **b.** glucose

31. Ser—Gly—Phe contains two peptide bonds.

33. Peptide chains commonly fold into spiral shapes called helixes and into sheets in which peptide chains lie side by side.

35. Fats are solid at room temperature and oils are liquid; most fats are animal products, and most oils are plant products.

37. A soap is the alkali metal salt of a fatty acid.

39.

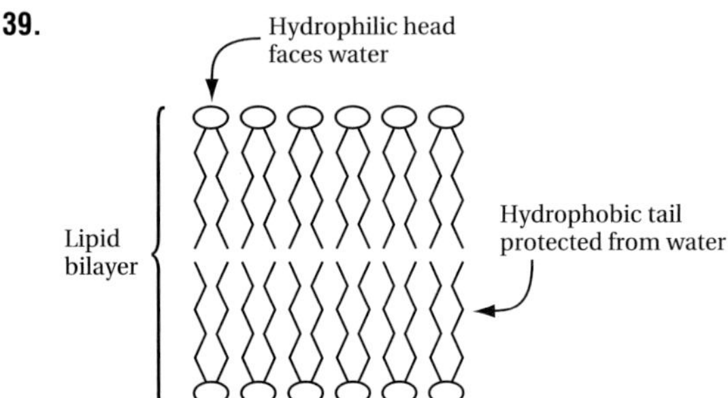

41. deoxyribonucleic acid (DNA) and ribonucleic acid (RNA)

43. Ribose, the sugar unit in RNA, has one more oxygen than deoxyribose, the sugar in DNA.

45. No; there are termination code words that signal the end of a peptide chain in protein synthesis; they do not specify an amino acid.

47. The addition, deletion, or substitution of one or more bases of DNA causes a gene mutation.

49. These complex biomolecules are either the carbohydrates, lipids, and proteins in your diet, or they are the carbohydrates, lipids, and proteins of your own body.

51. Catabolism and anabolism are complementary sets of reactions of metabolism. In catabolism, large biomolecules are broken down and energy is captured in the production of ATP. In anabolism, the products and energy from catabolism are used to make cell parts and other compounds.

53. Glycogen, a polysaccharide produced by animals, is more highly branched than are plant starches.

55. Sucrose gives glucose and fructose when hydrolyzed. The structures are

Glucose	Fructose
CHO	CH_2OH
$H-C-OH$	$C{=}O$
$HO-C-H$	$HO-C-H$
$H-C-OH$	$H-C-OH$
$H-C-OH$	$H-C-OH$
CH_2OH	CH_2OH

57. The active site is the physical location on an enzyme molecule where the substrate binds and is converted to products.

59.

$$CH_3(CH_2)_{14}-\overset{\displaystyle O}{\overset{\|}{C}}-O^-Na^+$$

Sodium palmitate

61. **a.** Arg—Gly—Cys—Asn **b.** Arg—Gly—Cys—Asn

63. The peptide is Met—Arg—Val—Tyr.

65. Molecular diseases result from faulty genes. Sickle cell anemia, methemoglobinemia, and alcaptonuria are examples of molecular diseases.

67. In theory, gene therapy could be used to transplant a normal gene into an individual who has a faulty version of that gene, thereby curing a molecular disease.

69. Gene mutations will occur. The mutations may be beneficial, harmful, or neutral to the individual and to the species.

71. Deep breathing provides a large supply of oxygen for energy-producing cell processes and rids the body of excess carbon dioxide.

73. **a.** (1) oxygen **b.** (4) amino acid

75. Base pairs are inside the double helix; strands must unwind for the bases to be recognized.

77. The equivalents of acid must equal the equivalents of base.

79. 1-chlorobutane, 2-chlorobutane, 1-chloro-2-methylpropane, 2-chloro-2-methylpropane

81. **a.** fluorine **b.** oxygen **c.** chlorine

83. The molecule must have at least one asymmetric carbon.

85.

$$\begin{array}{lcl} CH_2OH & & CH_3(CH_2)_{14}-\overset{\displaystyle O}{\overset{\|}{C}}-O^-Na^+ \\ | & & \\ CHOH & + & CH_3(CH_2)_{12}-\overset{\displaystyle O}{\overset{\|}{C}}-O^-Na^+ \\ | & & \\ CH_2OH & & CH_3(CH_2)_{16}-\overset{\displaystyle O}{\overset{\|}{C}}-O^-Na^+ \end{array}$$

Glycerol Sodium salts of three fatty acid molecules

CHAPTER 28

4. 0.063 mg Mn-56

5. No; one fourth of the sample will remain.

23. $^{226}_{88}\text{Ra} \longrightarrow ^{222}_{86}\text{Rn} + ^{4}_{2}\text{He}$

25. **a.** α, 2+ **b.** β, 1− **c.** γ, 0

27. **a.** $^{14}_{6}\text{C} \longrightarrow ^{14}_{7}\text{N} + ^{0}_{-1}\text{e}$
    b. $^{90}_{38}\text{Sr} \longrightarrow ^{90}_{39}\text{Y} + ^{0}_{-1}\text{e}$
    c. $^{40}_{19}\text{K} \longrightarrow ^{40}_{20}\text{Ca} + ^{0}_{-1}\text{e}$
    d. $^{13}_{7}\text{N} \longrightarrow ^{13}_{8}\text{O} + ^{0}_{-1}\text{e}$

29. **a.** $^{234}_{92}\text{U}$ **b.** $^{206}_{81}\text{Tl}$ **c.** $^{206}_{82}\text{Pb}$ **d.** $^{226}_{88}\text{Ra}$

31. $^{17}_{9}\text{F} \longrightarrow ^{17}_{8}\text{O} + ^{0}_{+1}\text{e}$

33. **a.** platinum **d.** technetium **g.** vanadium
    b. thorium **e.** xenon **h.** palladium
    c. francium **f.** californium

    Francium **(c)**, technetium **(d)**, and californium **(f)** have no stable isotopes.

35. One half-life is the time required for one-half of the atoms of a radioisotope to emit radiation and decay.

37. Natural radioactivity comes from radioactive elements present in nature. Artificial radioactivity comes from elements created in nuclear reactors and accelerators.

39. The nuclei of certain isotopes are bombarded with neutrons. The nuclei break into two fragments and release more neutrons. The released neutrons hit other nuclei to start a chain reaction that releases large amounts of energy.

41. Fusion requires extremely high temperatures, which would destroy any container.

43. The film badge measures radiation exposure; an exposed film badge indicates how much radiation a worker has received.

45. **a.** $^{30}_{15}\text{P} + ^{0}_{-1}\text{e} \longrightarrow ^{30}_{14}\text{Si}$
    b. $^{13}_{6}\text{C} + ^{1}_{0}\text{n} \longrightarrow ^{14}_{6}\text{C}$
    c. $^{131}_{53}\text{I} \longrightarrow ^{131}_{54}\text{Xe} + ^{0}_{-1}\text{e}$

47. **a.** $^{32}_{16}\text{S}$ **b.** $^{14}_{6}\text{C}$ **c.** $^{4}_{2}\text{He}$ **d.** $^{141}_{57}\text{La}$ **e.** $^{185}_{79}\text{Au}$

49. **a.** about 20%
    b. about 85 g
    c. about 83 days
    d. 25 days

51. **a.** Named radioactivity and discovered several radioactive elements.
    b. Discovered natural radioactivity from uranium ores.
    c. Discovered the neutron.
    d. Transmuted elements.

53. In every round of the tournament, one-half the teams are eliminated; in every half-life, one-half of the substance decays. In the tournament a single team eventually emerges as the winner and the tournament stops, but radioactive decay continues (almost) indefinitely.

55. Radioactive isotopes of these elements can be incorporated into the body tissues of organisms. When the isotopes decay, they can damage tissues very easily.

57. **a.** $Ca(OH)_2 + 2HCl \longrightarrow CaCl_2 + 2H_2O$
    b. $Fe_2O_3 + 3H_2 \longrightarrow 2Fe + 3H_2O$
    c. $2NaHCO_3 + H_2SO_4 \longrightarrow Na_2SO_4 + 2CO_2 + 2H_2O$
    d. $2C_2H_6 + 7O_2 \longrightarrow 4CO_2 + 6H_2O$

59. **a.** 26 p^+, 26 e^-, 33 n
    b. 92 p^+, 92 e^-, 143 n
    c. 24 p^+, 24 e^-, 28 n

61. 9216 cm^3 H_2; 0.4114 mol H_2

63. This graph shows the radioactive decay of carbon-14, along with the increase of the nitrogen product.

65. Bismuth-214 remains.

GLOSSARY

A

absolute zero the zero point on the Kelvin temperature scale, equivalent to -273.15 °C; all molecular motion theoretically stops at this temperature. *3.5*

accepted value a quantity used by general agreement of the scientific community. *3.2*

accuracy the closeness of a measurement to the true value of what is being measured. *3.2*

acid a compound that produces hydrogen ions in solution, is a hydrogen-ion donor, or an electron-pair acceptor. *20.1*

acid dissociation constant (K_a) the ratio of the concentration of the dissociated form of an acid to the undissociated form; stronger acids have larger K_a values than weaker acids. *20.4*

acidic solution any solution in which the hydrogen-ion concentration is greater than the hydroxide-ion concentration. *20.2*

activated complex an unstable arrangement of atoms that exists momentarily at the peak of the activation-energy barrier; an intermediate or transitional structure formed during the course of a reaction. *19.1*

activation energy the minimum energy colliding particles must have in order to react. *19.1*

active site a groove or pocket in an enzyme molecule into which the substrate (reactant molecule) fits; where the substrate is converted to products. *27.3*

activity series of metals a table listing metals in order of decreasing activity. *8.2*

actual yield the amount of product that forms when a reaction is carried out in the laboratory. *9.3*

addition reaction a reaction in which a substance is added at the double bond of an alkene or at the triple bond of an alkyne. *26.2*

adenosine triphosphate (ATP) a molecule that transmits the energy needed by cells of all living things. *27.6*

alcohol an organic compound having an —OH (hydroxyl) group; the general structure is R—OH. *26.2*

aldehyde an organic compound in which the carbon of the carbonyl group is joined to at least one hydrogen; the general formula is RCHO. *26.3*

aliphatic compound a hydrocarbon compound that does not contain a ring structure. *25.4*

alkali metal any metal in Group 1A of the periodic table. *5.4*

alkaline earth metal any metal in Group 2A of the periodic table. *5.4*

alkaline solution a basic solution. *20.2*

alkane a hydrocarbon containing only single covalent bonds; alkanes are saturated hydrocarbons. *25.1*

alkene a hydrocarbon containing one or more carbon–carbon double bonds; alkenes are unsaturated hydrocarbons. *25.2*

alkyl group a hydrocarbon substituent; the methyl group ($—CH_3$) is an alkyl group. *25.1*

alkyl halide a halocarbon in which one or more halogen atoms are attached to the carbon atoms of an aliphatic chain. *26.1*

alkyne a hydrocarbon containing a carbon–carbon triple bond; alkynes are unsaturated hydrocarbons. *25.2*

allotrope one of two or more different molecular forms of an element in the same physical state; oxygen (O_2) and ozone (O_3) are allotropes of the element oxygen. *10.3*

alpha particle a positively charged particle emitted from certain radioactive nuclei; it consists of two protons and two neutrons and is identical to the nucleus of a helium atom. *28.1*

alpha radiation alpha particles emitted from a radioactive source. *28.1*

amino acid an organic compound having amino ($—NH_2$) and carboxylic acid (—COOH) groups in the same molecule; proteins are made from the 20 naturally occurring amino acids. *27.3*

amino acid sequence the order in which amino acids are linked in a peptide or protein molecule. *27.3*

amorphous solid a term used to describe a solid that lacks an ordered internal structure; denotes a random arrangement of atoms. *10.3*

amphoteric a substance that can act as both an acid and a base; water is amphoteric. *20.3*

amplitude the height of a wave from the origin to the crest. *13.3*

anabolism synthetic processes in the metabolism of cells; which usually require the expenditure of energy. *27.6*

analytical chemistry the study of the composition of substances. *1.1*

anion any atom or group of atoms with a negative charge. *6.1*

anode the electrode at which oxidation occurs. *23.1*

antibonding orbital a molecular orbital with an energy that is higher than that of the atomic orbitals from which it is formed. *16.2*

aqueous solution *(aq)* a solution in which the solvent is water. *17.3*

arene any member of a special group of unsaturated cyclic hydrocarbons. *25.4*

aromatic compound a name originally given to the arenes because many of them have pleasant odors; any compound with bonding like that of benzene. *25.4*

aryl halide a halocarbon in which one or more halogens are attached to the carbon atoms of an arene ring. *26.1*

asymmetric carbon a carbon atom that has four different groups attached. *25.3*

atmospheric pressure the pressure exerted by air molecules in the atmosphere surrounding Earth, resulting from collisions of air molecules with objects. *10.1*

atom the smallest particle of an element that retains the properties of that element. *5.1*

atomic emission spectrum the pattern of frequencies obtained by passing light emitted by atoms of an element in the gaseous state through a prism; the emission spectrum of each element is unique to that element. *13.3*

atomic mass the weighted average of the masses of the isotopes of an element. *5.3*

atomic mass unit (amu) a unit of mass equal to one-twelfth the mass of a carbon-12 atom. *5.3*

atomic number the number of protons in the nucleus of an atom of an element. *5.3*

atomic orbital a region in space around the nucleus of an atom where there is a high probability of finding an electron. *13.1*

atomic radius one-half the distance between the nuclei in a molecule consisting of identical atoms. *14.2*

Aufbau principle electrons enter orbitals of lowest energy first. *13.2*

Avogadro's hypothesis equal volumes of gases at the same temperature and pressure contain equal numbers of particles. *12.5*

Avogadro's number the number of representative particles contained in one mole of a substance; equal to 6.02×10^{23} particles. *7.1*

B

balanced equation a chemical equation in which mass is conserved; each side of the equation has the same number of atoms of each element. *8.1*

band of stability the location of stable nuclei on a neutron-vs-proton plot. *28.2*

barometer an instrument used to measure atmospheric pressure. *10.1*

base a compound that produces hydroxide ions in solution, is a hydrogen-ion acceptor, or an electron-pair donor. *20.1*

base dissociation constant (K_b) the ratio of the concentration of the dissociated form of a base to the undissociated form. *20.4*

basic solution any solution in which the hydroxide-ion concentration is greater than the hydrogen-ion concentration. *20.2*

battery a group of voltaic cells that are connected to one another. *23.1*

Benedict's test a test commonly used to detect the presence of aldehydes. *26.3*

beta particle a fast-moving electron emitted from certain radioactive nuclei; it is formed when a neutron decomposes into a proton and an electron. *28.1*

beta radiation fast-moving electrons (beta particles) emitted from a radioactive source. *28.1*

binary compound a compound composed of two elements; NaCl and Al_2O_3 are binary compounds. *6.4*

biochemistry the study of the composition and behavior of substances in living organisms. *1.1*

boiling point (bp) the temperature at which the vapor pressure of a liquid is just equal to the external pressure on the liquid. *10.2*

boiling-point elevation the difference in temperature between the boiling point of a solution and of the pure solvent. *18.3*

bond dissociation energy the amount of energy required to break a covalent bond between atoms; this value is usually expressed in kJ per mol of substance. *16.1*

bonding orbital a molecular orbital whose energy is lower than that of the atomic orbitals from which it is formed. *16.2*

Boyle's law for a fixed mass of gas at constant temperature, the volume of the gas varies inversely with pressure. *12.3*

branched-chain alkane an alkane with one or more alkyl groups attached. *25.1*

Brownian motion the chaotic movement of colloidal particles, caused by collision with particles of the solvent in which they are dispersed. *17.4*

buffer a solution in which the pH remains relatively constant when small amounts of acid or base are added; a buffer can be either a solution of a weak acid and the salt of a weak acid or a solution of a weak base with the salt of a weak base. *21.2*

buffer capacity a measure of the amount of acid or base that may be added to a buffer solution before a significant change in pH occurs. *21.2*

C

calorie (cal) the quantity of heat that raises the temperature of 1 g of pure water 1 °C. *11.1*

calorimeter a device for measuring heat changes. *11.2*

calorimetry the measurement of heat changes for physical and chemical processes. *11.2*

carbohydrate the name given to monomers and polymers of aldehydes and ketones that have numerous hydroxyl groups; sugars and starches are carbohydrates. *27.2*

carbonyl group a functional group having a carbon atom and an oxygen atom joined by a double bond;

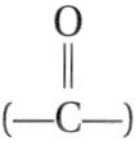

it is found in aldehydes, ketones, esters, and amides. *26.3*

carboxyl group a functional group consisting of a carbonyl group attached to a hydroxyl group;

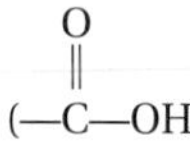

it is found in carboxylic acids. *26.3*

carboxylic acid an organic acid containing a carboxyl group; the general formula is RCOOH. *26.3*

catabolism the reactions in living cells in which substances are broken down and energy is produced. *27.6*

catalyst a substance that increases the rate of reaction by lowering the activation-energy barrier; the catalyst is not used up. *8.1, 19.1*

cathode the electrode at which reduction occurs. *23.1*

cathode ray a stream of electrons produced at the negative electrode (cathode) of a tube containing a gas at low pressure. *5.2*

cation any atom or group of atoms with a positive charge. *6.1*

cell membrane the partition made of phospholipids and proteins that forms the outermost edge of the cell; it is the barrier between the cell and its environment. *27.1*

cell potential the difference between the reduction potentials of two half-cells. *23.2*

Celsius scale the temperature scale on which the freezing point of water is 0 °C and the boiling point is 100 °C. *3.5*

Charles's law the volume of a fixed mass of gas is directly proportional to its Kelvin temperature if the pressure is kept constant. *12.3*

chemical equation an expression representing a chemical reaction; the formulas of the reactants (on the left) are connected by an arrow with the formulas for the products (on the right). *8.1*

chemical equilibrium a state of balance in which forward and reverse reactions are taking place at the same rate; no net change in the amounts of reactants and products occurs in the chemical system. *19.2*

chemical formula shows the number and type of atoms present in the smallest representative unit of a substance; the chemical formula of ammonia, with one nitrogen and three hydrogens, is NH_3. *6.2*

chemical potential energy energy stored within the structural units of chemical substances. *11.1*

chemical property the ability of a substance to undergo chemical reactions and to form new substances. *2.4*

chemical reaction the changing of substances to other substances by the breaking of bonds in reactants and the formation of bonds in products. *2.4*

chemical symbol a one or two letter representation of an element. *2.3*

chemistry the study of the structure, properties, and composition of substances, and the changes that substances undergo. *1.1*

***cis* configuration** a term applied to geometric isomers; it denotes an arrangement in which the substituted groups are on the same side of the double bond. *25.3*

coefficient a small whole number that appears in front of a formula in a balanced chemical equation. *8.1*

coenzyme a small organic molecule or metal ion necessary for an enzyme's biological activity. *27.3*

colligative properties a property of a solution that depends only on the number of the solute particles; boiling-point elevation, freezing-point depression, and vapor-pressure lowering are colligative properties. *18.3*

collision theory atoms, ions, and molecules can react to form products when they collide, provided that the particles have enough kinetic energy. *19.1*

colloid a mixture whose particles are intermediate in size between those of a suspension and a solute solution. *17.4*

combination reaction a chemical change in which two or more substances react to form a single new substance; also called a synthesis reaction. *8.2*

combined gas law a relationship describing the behavior of gases that combines Boyle's law, Charles's law, and Gay-Lussac's law. *12.3*

combustion reaction a chemical change in which oxygen reacts with another substance, often producing energy in the form of heat and light. *8.2*

common ion an ion that is common to both salts in a solution; in a solution of silver nitrate and silver chloride, Ag^+ would be a common ion. *21.2*

common ion effect a decrease in the solubility of a substance caused by the addition of a common ion. *21.2*

complete ionic equation an equation for a reaction in solution showing all strong electrolytes as ions. *8.3*

compound a substance that can be separated into simpler substances (elements or other compounds) only by chemical reactions. *2.3*

compressibility a measure of how much the volume of matter decreases under pressure. *12.1*

concentrated solution a solution containing a large amount of solute. *18.2*

concentration a measurement of the amount of solute that is dissolved in a given quantity of solvent; usually expressed as mol/L. *18.2*

condensed structural formula a structural formula that leaves out some bonds and/or atoms; the presence of these atoms or bonds is understood. *25.1*

conjugate acid the particle formed when a base gains a hydrogen ion; NH_4^+ is the conjugate acid of the base NH_3. *20.3*

conjugate acid-base pair two substances that are related by the loss or gain of a single hydrogen ion. Ammonia (NH_3) and the ammonium ion (NH_4^+) are a conjugate acid-base pair. *20.3*

conjugate base the particle that remains when an acid has donated a hydrogen ion; OH^- is the conjugate base of the acid water. *20.3*

conversion factor a ratio of equivalent measurements used to convert a quantity from one unit to another. *4.2*

coordinate covalent bond a covalent bond formed when one atom contributes both bonding electrons. *16.1*

coordination number the number of ions of opposite charge that surround each ion in a crystal. *15.2*

cracking the controlled process by which hydrocarbons are broken down or rearranged into smaller, more useful molecules. *25.5*

crystal a substance in which the atoms, ions, or molecules are arranged in an orderly, repeating, three-dimensional pattern called a crystal lattice. *10.3*

cyclic hydrocarbon an organic compound that contains a hydrocarbon ring. *25.4*

D

Dalton's atomic theory the first theory to relate chemical changes to events at the atomic level. *5.1*

Dalton's law of partial pressures at constant volume and temperature, the total pressure of a mixture of gases is the sum of the partial pressures of all the gases present. *12.5*

de Broglie's equation an equation that describes the wavelength of a moving particle; it predicts that all matter exhibits wavelike motions. *13.3*

decomposition reaction a chemical change in which a single compound is broken down into two or more simpler products. *8.2*

dehydrogenation reaction a reaction in which hydrogen is lost. *26.3*

deliquescent a term describing a substance that removes sufficient water from the air to form a solution; the solution formed has a lower vapor pressure than that of the water in the air. *17.3*

denatured alcohol ethanol to which a poisonous substance has been added to make it unfit to drink. *26.2*

density the ratio of the mass of an object to its volume. *3.4*

desiccant a hygroscopic substance used as a drying agent. *17.3*

diamagnetic a substance that is weakly repelled by a magnetic field. *16.1*

diffusion the tendency of molecules and ions to move toward areas of lower concentration until the concentration is uniform throughout the system. *12.5*

dilute solution a solution that contains a small amount of solute. *18.2*

dimensional analysis a technique of problem-solving that uses the units that are part of a measurement to help solve the problem. *4.2*

dipole a molecule that has two electrically charged regions, or poles. *16.3*

dipole interaction a weak intermolecular force resulting from the attraction of oppositely charged regions of polar molecules. *16.3*

diprotic acid any acid that contains two ionizable protons (hydrogen ions); sulfuric acid (H_2SO_4) is a diprotic acid. *20.3*

disaccharide a carbohydrate formed from two monosaccharide units; common table sugar (sucrose) is a disaccharide. *27.2*

dispersion force the weakest kind of intermolecular attraction; this attraction is thought to be caused by the motion of electrons. *16.3*

distillation a purification process in which a liquid is evaporated and then condensed again to a liquid; used to separate dissolved solids from liquids or liquids from liquids according to boiling point. *2.2*

double covalent bond a covalent bond in which two pairs of electrons are shared by two atoms. *16.1*

double-replacement reaction a chemical change that involves an exchange of positive ions between two compounds. *8.2*

dry cell a commercial voltaic cell in which the electrolyte is a moist paste; despite their name, the compact, portable batteries used in flashlights are dry cells. *23.1*

E

effloresce to lose water of hydration; the process occurs when the hydrate has a vapor pressure higher than that of water vapor in the air. *17.3*

effusion a process that occurs when a gas escapes through a tiny hole in its container. *12.5*

electrical potential the ability of a voltaic cell to produce an electric current. *23.2*

electrochemical cell any device that converts chemical energy into electrical energy or electrical energy into chemical energy. *23.1*

electrochemical process the conversion of chemical energy into electrical energy or electrical energy into chemical energy; all electrochemical processes involve redox reactions. *23.1*

electrode a conductor in a circuit that carries electrons to or from a substance other than a metal. *23.1*

electrolysis a process in which electrical energy is used to bring about a chemical change; the electrolysis of water produces hydrogen and oxygen. *23.3*

electrolyte a compound that conducts an electric current in aqueous solution or in the molten state; all ionic compounds are electrolytes, but most covalent compounds are not. *17.3*

electrolytic cell an electrochemical cell used to cause a chemical change through the application of electrical energy. *23.3*

electromagnetic radiation a series of energy waves that travel in a vacuum at 3.0×10^{10} cm/s; includes radio waves, microwaves, visible light, infrared and ultraviolet light, x-rays, and gamma rays. *13.3*

electron a negatively charged subatomic particle. *5.2*

electron configuration the arrangement of electrons around the nucleus of an atom in its ground state. *13.2*

electron dot structure a notation that depicts valence electrons as dots around the atomic symbol of the element; the symbol represents the inner electrons and atomic nucleus; also called Lewis dot structures. *15.1*

electronegativity the tendency for an atom to attract electrons to itself when it is chemically combined with another element. *14.2*

element a substance that cannot be changed into simpler substances under normal laboratory conditions. *2.3*

elementary reaction a reaction in which reactants are converted to products in a single step. *19.5*

empirical formula a formula with the lowest whole-number ratio of elements in a compound; the empirical formula of hydrogen peroxide (H_2O_2) is HO. *7.3*

emulsion the colloidal dispersion of one liquid in another. *17.4*

endothermic process a heat-absorbing process. *11.1*

end point the point in a titration at which neutralization is achieved. *21.1*

energy the capacity for doing work; it exists in several forms, including chemical, nuclear, electrical, radiant, mechanical, and thermal energies. *11.1*

energy level a region around the nucleus of an atom where an electron is likely to be moving. *13.1*

enthalpy (H) the heat content of a system at constant pressure. *11.2*

entropy (S) a measure of the disorder of a system; systems tend to go from a state of order (low entropy) to a state of maximum disorder (high entropy). *19.3*

enzyme proteins that act as biological catalysts; most of the reactions that occur in cells require catalysts. *27.3*

enzyme–substrate complex the structure formed when a substrate molecule joins an enzyme at its active site. *27.3*

equilibrium constant (K_{eq}) the ratio of product concentrations to reactant concentrations at equilibrium with each concentration raised to a power equal to the number of moles of that substance in the balanced chemical equation. *19.2*

equilibrium position the relative concentrations of reactants and products of a reaction that has reached equilibrium; indicates whether the reactants or products are favored in the reversible reaction. *19.2*

equivalence point the point in a titration at which the number of equivalents of acid and base are equal. *21.1*

equivalent (equiv) one equivalent is the amount of an acid (or base) that can give one mole of hydrogen (or hydroxide) ions. *21.1*

error the difference between the accepted value and the experimental value. *3.2*

ester a derivative of a carboxylic acid in which the —OH of the carbonyl group has been replaced by the —OR from an alcohol; the general formula is RCOOR. *26.3*

ether an organic compound in which oxygen is bonded to two carbon groups; the general formula is R—O—R. *26.2*

evaporation vaporization that occurs at the surface of a liquid that is not boiling. *10.2*

excess reagent a reagent present in a quantity that is more than sufficient to react with a limiting reagent; any reactant that remains after the limiting reagent is used up in a chemical reaction. *9.3*

exothermic process a heat-dissipating process. *11.1*

experiment a carefully controlled, repeatable procedure for gathering data to test a hypothesis. *1.3*

experimental value a quantitative value measured during an experiment. *3.2*

F

fatty acid the name given to continuous-chain carboxylic acids that were first isolated from fats. *26.3*

Fehling's test a test used to detect aldehydes. *26.3*

fermentation the production of ethanol from sugars by the action of yeast or bacteria. *26.2*

film badge a small radiation detector worn by persons who work near radiation sources; it consists of several layers of photographic film covered with black lightproof paper encased in a plastic or metal holder. *28.4*

first-order reaction a reaction in which the reaction rate is proportional to the concentration of only one reactant. *19.5*

fission the splitting of a nucleus into smaller fragments, accompanied by the release of neutrons and a large amount of energy. *28.3*

formula unit the lowest whole-number ratio of ions in an ionic compound; in magnesium chloride, the ratio of magnesium ions to chloride ions is 1:2 and the formula unit is $MgCl_2$. *6.2*

free energy the energy available to do work. *19.3*

freezing-point depression the difference in temperature between the freezing point of a solution and of the pure solvent. *18.3*

frequency (υ) the number of wave cycles that pass a given point per unit of time; there is an inverse relationship between the frequency and wavelength of a wave. *13.3*

fuel cell a voltaic cell that does not need to be recharged; the fuel is oxidized to produce a continuous supply of electrical energy. *23.1*

functional group a specific arrangement of atoms in an organic compound that is capable of characteristic chemical reactions; the chemistry of an organic compound is determined by its functional groups. *26.1*

fusion a reaction in which two light nuclei combine to produce a nucleus of heavier mass, accompanied by the release of a large amount of energy. *28.3*

G

gamma radiation high energy electromagnetic radiation emitted by certain radioactive nuclei; gamma rays have no mass or electrical charge. *28.1*

gas matter that has no definite shape or volume; it adopts the shape and volume of its container. *2.1*

gas pressure the force exerted by a gas per unit surface area of an object; due to collisions of gas particles with the object. *10.1*

Gay-Lussac's law the pressure and Kelvin temperature of a gas are directly proportional at constant volume. *12.3*

Geiger counter a gas-filled metal tube used to detect the presence of beta radiation. *28.4*

gene a segment of DNA that codes for a single protein; any alteration in the base sequence (a mutation) affects protein synthesis. *27.5*

genetic code the arrangement of code words in DNA; there are 61 triplet code words used to specify the amino acids in protein synthesis plus 3 triplet code words that signal "stop" after the protein synthesis is complete. *27.5*

geometric isomer an organic compound that differs from another compound only in the geometry of their substituted groups. *25.3*

Gibbs free energy change (ΔG) the maximum amount of energy that can be coupled to another process to do useful work. *19.4*

glass the optically transparent fusion product of inorganic materials that have cooled to a rigid state without crystallizing. *10.3*

Graham's laws of effusion the rate of effusion of a gas is inversely proportional to the square root of its molar mass; this relationship is also true for the diffusion of gases. *12.5*

gram (g) a metric mass unit equal to the mass of 1 cm^3 of water at 4 °C. *3.3*

gram atomic mass (gam) the mass, in grams, of one mole of atoms in a monatomic element; it is numerically equal to the atomic mass in amu. *7.1*

gram equivalent mass the mass of one equivalent of a substance. *21.1*

gram formula mass (gfm) the mass of one mole of an ionic compound. *7.1*

gram molecular mass (gmm) the mass, in grams, of one mole of a molecular substance. *7.1*

ground state the lowest energy level occupied by an electron when an atom is in its most stable energy state. *13.3*

group a vertical column of elements in the periodic table; the constituent elements of a group have similar chemical and physical properties. *5.4*

H

half-cell the part of a voltaic cell in which either oxidation or reduction occurs; it consists of a single electrode immersed in a solution of its ions. *23.1*

half-life ($t_{\frac{1}{2}}$) the time required for one-half of the atoms of a radioisotope to emit radiation and decay to products. *28.2*

half-reaction an equation showing either the reduction or the oxidation of a species in an oxidation–reduction reaction. *22.3*

half-reaction method a method of balancing a redox equation by balancing the oxidation and reduction half-reactions. *22.3*

halide ion a negative ion formed when a halogen atom gains an electron. *15.1*

halocarbon any member of a class of organic compounds containing covalently bonded fluorine, chlorine, bromine, or iodine. *26.1*

halogen any member of the nonmetallic elements in Group 7A of the periodic table. *5.4*

heat (q) the energy that is transferred from one body to another because of a temperature difference. *11.1*

heat capacity the quantity of heat required to change an object's temperature by exactly 1 °C. *11.1*

heat of combustion the heat released during a chemical reaction in which one mole of a substance is completely burned. *11.2*

heat of reaction the heat released or absorbed during a chemical reaction; equivalent to h, the change in enthalpy. *11.2*

Heisenberg uncertainty principle it is impossible to know both the velocity and the position of a particle at the same time. *13.3*

Henry's law at a given temperature the solubility of a gas in a liquid is directly proportional to the pressure of the gas above the liquid. *18.1*

hertz (Hz) the SI unit of frequency, equal to one cycle per second. *13.3*

Hess's law of heat summation in going from a particular set of reactants to a particular set of products, the enthalpy change is the same whether the reaction takes place in one step or in a series of steps. *11.4*

heterogeneous mixture a mixture that is not uniform in composition; its components are readily distinguished. *2.2*

homogeneous mixture a mixture that is completely uniform in composition; its components are not readily distinguished. *2.2*

homologous series a group of compounds in which there is a constant increment of change in molecular structure from one compound in the series to the next. *25.1*

Hund's rule when electrons occupy orbitals of equal energy, one electron enters each orbital until all orbitals contain one electron with their spins parallel. *13.2*

hybridization a process in which several atomic orbitals (such as *s* and *p* orbitals) mix to form the same number of equivalent hybrid orbitals. *16.2*

hydration reaction a reaction in which water is added to an alkene. *26.2*

hydrocarbon an organic compound that contains only carbon and hydrogen. *25.1*

hydrogenation reaction a reaction in which hydrogen is added to a carbon–carbon double bond to give an alkane. *26.2*

hydrogen bond a relatively strong intermolecular force in which a hydrogen atom that is covalently bonded to a very electronegative atom is also weakly bonded to an unshared electron pair of another electronegative atom in the same molecule or one nearby. *16.3*

hydrogen-ion acceptor a base, according to the Brønsted-Lowry theory; ammonia acts as a base when it accepts hydrogen ions from water. *20.3*

hydrogen-ion donor an acid, according to the Brønsted-Lowry theory. *20.3*

hydrometer a device used to measure the specific gravity of a liquid. *3.4*

hydronium ion (H_3O^+) the positive ion formed when a water molecule gains a hydrogen ion; all hydrogen ions in aqueous solution are present as hydronium ions. *20.2*

hydroxide ion (OH^-) the negative ion formed when a water molecule loses a hydrogen ion. *20.2*

hydroxyl group the —OH functional group present in alcohols. *26.2*

hygroscopic a term describing salts and other compounds that remove moisture from the air. *17.3*

hypothesis a proposed explanation for observations. *1.3*

I

ideal gas constant (*R*) a term in the ideal gas law, which has the value 8.31 (L × kPa)/(K × mol). *12.4*

ideal gas law the relationship $P \times V = n \times R \times T$, which describes the behavior of an ideal gas. *12.4*

immiscible describes liquids that are insoluble in one another; oil and water are immiscible. *18.1*

inhibitor a substance that interferes with the action of a catalyst. *19.1*

inner transition element *see* inner transition metal. *24.3*

inner transition metal an element in the lanthanide and actinide series; characterized by addition of electrons to *f* orbitals. *5.4, 14.1*

inorganic chemistry primarily the study of substances that do not contain carbon. *1.1*

intermediate a product of a reaction that immediately becomes a reactant of another reaction. *19.5*

International System of Units (SI) the revised version of the metric system, adopted by international agreement in 1960. *3.3*

ion an atom or group of atoms that has a positive or negative charge; cations are ions with a positive charge, and anions are ions with a negative charge. *6.1*

ionic bond the electrostatic attraction that binds oppositely charged ions together. *15.2*

ionic compound a compound composed of positive and negative ions. *6.1*

ionization energy the energy required to remove an electron from a gaseous atom. *14.2*

ionizing radiation radiation with enough energy to produce ions by knocking electrons off some atoms of a bombarded substance. *28.4*

ion-product constant for water (K_w) the product of the hydrogen ion and hydroxide ion concentrations in water; it is 1×10^{-14} at 25 °C. *20.2*

isotopes atoms of the same element that have the same atomic number but different atomic masses due to a different number of neutrons. *5.3*

J

joule (j) the SI unit of energy; 4.184 J equal one calorie. *11.1*

K

Kelvin scale the temperature scale in which the freezing point of water is 273 K and the boiling point is 373 K; 0 K is absolute zero. *3.5*

ketone an organic compound in which the carbon of the carbonyl group is joined to two other carbons; the general formula is RCOR. *26.3*

kilogram (kg) the mass of 1 L of water at 4 °C; it is the base unit of mass in SI. *3.3*

kinetic energy the energy an object has because of its motion. *10.1*

kinetic theory a theory explaining the states of matter, based on the concept that the particles in all forms of matter are in constant motion. *10.1*

L

law of conservation of energy energy is neither created nor destroyed in an ordinary chemical or physical process. *11.1*

law of conservation of mass mass can be neither created nor destroyed in an ordinary chemical or physical process. *2.4*

law of definite proportions in any sample of a chemical compound, the elements are always combined in the same proportion by mass. *6.2*

law of disorder it is a natural tendency of systems to move in the direction of maximum chaos or disorder. *19.3*

law of multiple proportions whenever two elements form more than one compound, the different masses of one element that combine with the same mass of the other element are in the ratio of small whole numbers. *6.2*

Le Châtelier's principle when stress is applied to a system at equilibrium, the system changes to relieve the stress. *19.2*

Lewis acid any substance that can accept a pair of electrons to form a covalent bond. *20.3*

Lewis base any substance that can donate a pair of electrons to form a covalent bond. *20.3*

lime calcium oxide (CaO); it is sometimes called quicklime. *24.1*

limiting reagent any reactant that is used up first in a chemical reaction; it determines the amount of product that can be formed in the reaction. *9.3*

lipid a member of a large class of relatively water-insoluble organic compounds; fats, oils, and waxes are lipids. *27.4*

liquid a form of matter that flows, has a fixed volume, and takes the shape of its container. *2.1*

liter (L) the volume of a cube measuring 10 centimeters on each edge (1000 cm^3); it is the common unprefixed unit of volume in the metric system. *3.3*

M

mass the amount of matter that an object contains; the SI base unit of mass is the kilogram. *2.1*

mass number the total number of protons and neutrons in the nucleus of an atom. *5.3*

matter anything that takes up space and has mass. *2.1*

melting point (mp) the temperature at which a substance changes from solid to a liquid; the melting point of water is 0 °C. *10.3*

metabolism all the chemical reactions carried out by an organism; includes energy-producing (catabolism) reactions and energy-absorbing (anabolism) reactions. *27.6*

metal one of a class of elements that includes a large majority of the known elements; metals are characteristically lustrous, malleable, ductile, and good conductors of heat and electricity. *5.4*

metallic bond the force of attraction that holds metals together; it consists of the attraction of free-floating valence electrons for positively charged metal ions. *15.3*

metalloid one of a class of elements having properties intermediate to metals and nonmetals. *5.4*

metallurgy the various procedures used to separate metals from their ores. *24.3*

meter (m) the base unit of length in SI. *3.3*

miscible describes liquids that dissolve in each other. *18.1*

mixture a physical blend of two or more substances that are not chemically combined. *2.2*

molal boiling-point elevation constant (K_b) the change in boiling point for a 1-molal solution of a nonvolatile molecular solute. *18.4*

molal freezing-point depression constant (K_f) the change in freezing point for a 1-molal solution of a nonvolatile molecular solute. *18.4*

molality (*m*) the concentration of solute in a solution expressed as the number of moles of solute dissolved in 1 kilogram (1000 g) of solvent. *18.4*

molar heat of condensation the heat, in joules or calories, released by 1 mole of a substance as it changes from a vapor to a liquid at the liquid's boiling point. *11.3*

molar heat of fusion the energy, in kilojoules, required to melt 1 mole of a solid. *11.3*

molar heat of solidification the heat, in joules or calories, released by 1 mole of a substance as it changes from a liquid to a solid at the solid's melting point. *11.3*

molar heat of solution the heat, in joules or calories, released or absorbed by 1 mole of a substance as it dissolves in water to produce 1 liter of a 1 molar solution. *11.3*

molar heat of vaporization the energy, in kilojoules, required to vaporize 1 mole of a liquid. *11.3*

molarity (*M*) the concentration of solute in a solution expressed as the number of moles of solute dissolved in 1 liter of solution. *18.2*

molar mass a general expression used to refer to the mass of a mole of any substance unless a more specific term is needed to avoid confusion. *7.2*

molar volume the volume occupied by 1 mole of a gas at standard temperature and pressure (STP); 22.4 L. *7.2*

mole (mol) the amount of a substance that contains 6.02×10^{23} representative particles of that substance. *7.1*

molecular compound a compound that is composed of molecules. *6.1*

molecular formula a chemical formula that shows the actual number and kinds of atoms present in a molecule of a compound. *6.2*

molecular orbital an orbital resulting from the overlapping of atomic orbitals when two atoms combine. *16.2*

molecule a neutral chemically bonded group of atoms that act as a unit. *6.1*

mole fraction the ratio of the moles of solute in solution to the total number of moles of both solvent and solute. *18.4*

monatomic ion a single atom with a positive or negative charge as a result of losing or gaining valence electrons. *6.3*

Monel metal a strong, corrosion resistant alloy of nickel and copper. *24.3*

monomer a simple molecule that repeatedly combines to form a polymer. *26.4*

monoprotic acid any acid that contains one ionizable proton (hydrogen ion); nitric acid (HNO_3) is a monoprotic acid. *20.3*

monosaccharide a carbohydrate consisting of one sugar unit; it is also called a simple sugar. *27.2*

N

net ionic equation an equation for a reaction in solution showing only those particles that are directly involved in the chemical change. *8.3*

network solid a substance in which all of the atoms are covalently bonded to each other. *16.3*

neutralization reaction a reaction in which an acid and a base react in an aqueous solution to produce a salt and water. *21.1*

neutral solution an aqueous solution in which the concentrations of hydrogen and hydroxide ions are equal, that is, 1.0×10^{-7} mol/L; it has a pH of 7.0. *20.2*

neutron a subatomic particle with no charge and a mass of 1 amu; found in the nucleus of the atom. *5.2*

neutron absorption a process used in a nuclear reactor to slow the chain reaction by decreasing the number of moving neutrons; this is done with control rods made of a material such as cadmium that absorbs neutrons. *28.3*

neutron moderation a process used in a nuclear reactor to slow the neutrons so they can be captured by the reactor fuel to continue the chain reaction; water and carbon are good moderators. *28.3*

noble gas any member of a group of gaseous elements in Group 0 of the periodic table; the *s* and *p* sublevels of their outermost energy level are filled. *5.4, 14.1*

nonelectrolyte a compound that does not conduct an electric current in aqueous solution or in the molten state. *17.3*

nonmetal one of a class of elements that are not lustrous and are generally poor conductors of heat and electricity; nonmetals are grouped on the right side of the periodic table. *5.4*

nonpolar covalent bond a bond formed when the atoms in a molecule are alike and the bonding electrons are shared equally. *16.3*

nonspontaneous reaction a reaction that does not favor the formation of products at the specified conditions. *19.3*

normal boiling point the boiling point of a liquid at a pressure of 1 atmosphere. *10.2*

normality (*N*) the concentration of a solution expressed as the number of equivalents of solute in 1 liter of solution. *21.1*

nucleic acid a polymer of ribonucleotides (RNA) or deoxyribonucleotides (DNA) found primarily in cell nuclei; nucleic acids play an important role in the transmission of hereditary characteristics, protein synthesis, and the control of cell activities. *27.5*

nucleotide one of the monomers that make up DNA and RNA; it consists of a nitrogen-containing base (a purine or pyrimidine), a sugar (ribose or deoxyribose), and a phosphate group. *27.5*

nucleus the dense central portion of an atom, composed of protons and neutrons. *5.2*

O

observation information obtained through the senses; observation in science often involves a measurement. *1.3*

octet rule atoms react by gaining or losing electrons so as to acquire the stable electron structure of a noble gas, usually eight valence electrons. *15.1*

ore a mineral used for commercial production of a metal. *24.3*

organic chemistry the study of compounds that contain the element carbon. *1.1*

oxidation a process that involves complete or partial loss of electrons or a gain of oxygen; it results in an increase in the oxidation number of an atom. *22.1*

oxidation number a positive or negative number assigned to a combined atom according to a set of arbitrary rules; generally it is the charge an atom would have if the electrons in each bond were assigned to the atoms of the more electronegative element. *22.2*

oxidation-number-change method a method of balancing a redox equation by comparing the increases and decreases in oxidation numbers. *22.3*

oxidation–reduction reaction a reaction that involves the transfer of electrons between reactants. *22.1*

oxidizing agent the substance in a redox reaction that accepts electrons; in the reaction, the oxidizing agent is reduced. *22.1*

P

paramagnetic a term used to describe a substance that shows a relatively strong attraction to an external magnetic field; these substances have molecules containing one or more unpaired electrons. *16.1*

partial pressure the pressure exerted by each gas in a gaseous mixture. *12.5*

pascal (Pa) the SI unit of pressure. *10.1*

Pauli exclusion principle no more than two electrons can occupy an atomic orbital; these electrons must have opposite spins. *13.2*

peptide an organic compound formed by a combination of amino acids in which the amino group of one acid is united with the carboxylic group of another through an amide bond. *27.3*

peptide bond the bond between the carbonyl group of one amino acid and the nitrogen of the next amino acid in the peptide chain; the structure is

$$\begin{array}{cc} \mathrm{O} & \mathrm{H} \\ \| & | \\ \text{—C} & \text{—N—} \end{array}$$

27.3

percent composition the percent by mass of each element in a compound. *7.3*

percent error the percent that a measured value differs from the accepted value. *3.2*

percent yield the ratio of the actual yield to the theoretical yield for a chemical reaction expressed as a percentage; a measure of the efficiency of a reaction. *9.3*

period a horizontal row of elements in the periodic table. *5.4*

periodic law when the elements are arranged in order of increasing atomic number, there is a periodic repetition of their physical and chemical properties. *5.4*

periodic table an arrangement of elements into rows and columns according to similarities in their properties. *5.4*

pH a number used to denote the hydrogen-ion concentration, or acidity, of a solution; it is the negative logarithm of the hydrogen-ion concentration of a solution. *20.2*

phase any part of a system with uniform composition and properties. *2.2*

phase diagram a diagram showing the conditions at which a substance exists as a solid, liquid, or vapor. *10.4*

phospholipid a lipid that contains a phosphate group; because phospholids have hydrophilic heads and hydrophobic tails, they form the lipid bilayers found in cell membranes. *27.4*

photoelectric effect electrons are ejected by certain metals when they absorb light with a frequency above a threshold frequency. *13.3*

photon a quantum of light; a discrete bundle of electromagnetic energy that behaves as a particle. *13.3*

photosynthesis the process by which green plants and algae use radiant energy from the sun to synthesize glucose from carbon dioxide and water. *27.1*

physical change an alteration of a substance that does not affect its chemical composition. *2.1*

physical chemistry the study of the theoretical basis of chemical behavior, relying on mathematics and physics. *1.1*

physical property a quality of a substance that can be observed or measured without changing the substance's chemical composition. *2.1*

pi bond (π bond) a bond in which the bonding electrons are most likely to be found in the sausage-shaped regions above and below the nuclei of the bonded atoms. *16.2*

Planck's constant (h) a number used to calculate the radiant energy (E) absorbed or emitted by a body based on the frequency of radiation. *13.3*

polar bond a bond formed when two different atoms are joined by a covalent bond and the bonding electrons are shared unequally. *16.3*

polar covalent bond *see* polar bond. *16.3*

polar molecule a molecule, like water, in which one or more atoms is slightly negative and one or more is slightly positive, unless molecular geometry causes the polarities to cancel each other out. *16.3*

polyatomic ion a tightly bound group of atoms that behaves as a unit and carries a charge. *6.3*

polymer a very large molecule formed by the covalent bonding of repeating small molecules, known as monomers. *26.4*

polypeptide a peptide with more than 10 amino acids. *27.3*

polysaccharide a complex carbohydrate polymer formed by the linkage of many monosaccharide monomers; starch, glycogen, and cellulose are polysaccharides. *27.2*

positron a particle that has the same mass as an electron but that has a positive charge. *28.2*

precision describes the closeness, or reproducibility, of a set of measurements taken under the same conditions. *3.2*

product a substance formed in a chemical reaction. *2.4*

protein any peptide with more than 100 amino acids. *27.3*

proton a positively charged subatomic particle found in the nucleus of an atom. *5.2*

Q

qualitative measurement a measurement that gives descriptive, nonnumeric results. *3.1*

quantitative measurement a measurement that gives definite, usually numeric results. *3.1*

quantum the amount of energy needed to move an electron from its present energy level to the next higher one. *13.1*

quantum mechanical model the modern description, primarily mathematical, of the behavior of electrons in atoms. *13.1*

R

radiation the penetrating rays and particles emitted by a radioactive source. *28.1*

radioactive decay the spontaneous emission of radiation by an unstable nucleus; the rate of decay is unaffected by temperature, pressure, or catalysts. *28.1*

radioactivity the processes by which unstable atomic nuclei achieve stability. *28.1*

radioisotope an isotope that has an unstable nucleus and undergoes radioactive decay. *28.1*

rate describes the speed of change over an interval of time. *19.1*

rate law an expression relating the rate of a reaction to the concentration of the reactants. *19.5*

reactant a starting substance in a chemical reaction. *2.4*

reaction mechanism a series of elementary reactions that take place during the course of a complex reaction. *19.5*

redox reaction another name for an oxidation–reduction reaction. *22.1*

reducing agent the substance in a redox reaction that donates electrons; in the reaction, the reducing agent is oxidized. *22.1*

reduction a process that involves a complete or partial gain of electrons or the loss of oxygen; it results in a decrease in the oxidation number of an atom. *22.1*

reduction potential a measure of the tendency of a given half-reaction to occur as a reduction (gain of electrons) in an electrochemical cell. *23.2*

representative element Group A element on the periodic table; together, these elements, which have only partially filled outermost *s* and *p* sublevels, illustrate the entire range of chemical properties. *5.4, 14.1*

representative particle the smallest unit into which a substance can be broken down without a change in composition; the term refers to whether a substance commonly exists as atoms, ions, or molecules. *7.1*

resonance structure one of two or more equally valid electron dot structures for a molecule or polyatomic ion; the actual bonding is a hybrid, or mixture, of the resonance structures. *16.1*

reversible reaction a reaction in which the conversion of reactants into products and the conversion of products into reactants occur simultaneously. *19.2*

S

salt bridge a tube containing a conducting solution used to connect half-cells in a voltaic cell; it allows the passage of ions from one compartment to another but prevents the solutions from mixing completely. *23.1*

salt hydrolysis a process in which the cations or anions of a dissociated salt accept hydrogen ions from water or donate hydrogen ions to water; solutions containing hydrolyzed salts may be either acidic or basic. *21.2*

saponification the process used to make soap; it involves the hydrolysis of fats or oils by a hot aqueous alkali-metal hydroxide. *27.4*

saturated compound an organic compound in which all carbon atoms are joined by single covalent bonds; it contains the maximum number of hydrogen atoms. *25.2*

saturated solution a solution containing the maximum amount of solute for a given amount of solvent at a constant temperature and pressure; an equilibrium exists between undissolved solute and ions in solution. *18.1*

scientific law a concise statement that summarizes the results of many observations and experiments. *1.3*

scientific method a method of inquiry involving observation, experiments, hypotheses, and broad explanations called theories. *1.3*

scientific notation expression of numbers in the form $n \times 10^n$ where n is equal to or greater than 1 and less than 10 and n is an integer. *3.1*

scintillation counter a device that uses a surface coated with a phosphor to detect radiation; ionizing radiation striking the phosphor surface produces bright flashes of light (scintillations). *28.4*

self-ionization a term describing the reaction in which two water molecules react to produce ions. *20.2*

sigma bond (σ bond) a bond formed when two atomic orbitals combine to form a molecular orbital that is symmetrical along the axis connecting the two atomic nuclei. *16.2*

significant figures all the digits that can be known precisely in a measurement, plus a last estimated digit. *3.2*

single covalent bond a bond formed when a pair of electrons is shared between two atoms. *16.1*

single-replacement reaction a chemical change in which one element replaces a second element in a compound; also called a displacement reaction. *8.2*

skeleton equation a chemical equation that does not indicate the relative amounts of reactants and products. *8.1*

slaked lime calcium hydroxide ($Ca(OH)_2$). *24.1*

solid matter that has a definite shape and volume. *2.1*

solubility the amount of a substance that dissolves in a given quantity of solvent at specified conditions of temperature and pressure to produce a saturated solution. *18.1*

solubility product constant (K_{sp}) an equilibrium constant that can be applied to the solubility of electrolytes; it is equal to the product of the concentration terms each raised to the power of the coefficient of the substance in the dissociation equation. *21.2*

solute dissolved particles in a solution. *17.3*

solution a homogeneous mixture. *2.2*

solvation a process that occurs when an ionic solute dissolves; in solution, the ions are surrounded by solvent molecules. *17.3*

solvent the dissolving medium in a solution. *17.3*

specific gravity the ratio of the density of a substance to that of a standard substance (usually water). *3.4*

specific heat (c) the quantity of heat, in joules or calories, required to raise the temperature of 1 g of a substance 1 °C. *11.1*

specific heat capacity *see* specific heat. *11.1*

specific rate constant a proportionality constant relating the concentrations of reactants to the rate of the reaction. *19.5*

spectator ion an ion that is not directly involved in a chemical reaction; an ion that does not change oxidation number or composition during a reaction. *8.3*

spectrum range of wavelengths of electromagnetic radiation; wavelengths of visible light are separated when a beam of white light passes through a prism. *13.3*

spontaneous reaction a reaction that favors the formation of products at the specified conditions; spontaneity depends on enthalpy and entropy changes. *19.3*

standard atmosphere (atm) a unit of pressure; it is the pressure required to support 760 mm of mercury in a mercury barometer at 25 °C; this is the average atmospheric pressure at sea level. *10.1*

standard cell potential the measured cell potential when the ion concentration in the half-cells are 1.00 M at 1 atm of pressure and 25 °C. *23.2*

standard entropy (S^0) the entropy of a substance in its stable state at 25 °C and 1 atm. *19.4*

standard heat of formation (H_f^0) the change in enthalpy for a reaction in which one mole of a compound is formed from its constituent elements. *11.4*

standard hydrogen electrode an arbitrary reference electrode (half-cell) used with another electrode (half-cell) to measure the standard reduction potential of that cell; the standard reduction potential of the hydrogen electrode is assigned a value of 0.00 V. *23.2*

standard solution a solution of known concentration used in carrying out a titration. *21.1*

standard temperature and pressure (STP) the conditions under which the volume of a gas is usually measured; standard temperature is 0 °C and standard pressure is 1 atmosphere (atm). *7.2*

stereoisomer an organic molecule having the same molecular structure as another molecule, but differing in the arrangement of atoms in the space. *25.3*

stoichiometry that portion of chemistry dealing with numerical relationships in chemical reactions; the calculation of quantities of substances involved in chemical equations. *9.1*

straight-chain alkane a saturated open-chain hydrocarbon in which all carbons are arranged consecutively; that is, there is no branching. *25.1*

strong acid an acid that is completely (or almost completely) ionized in aqueous solution. *20.4*

strong base a base that completely dissociates into metal ions and hydroxide ions in aqueous solution. *20.4*

strong electrolyte a solution in which a large portion of the solute exists as ions. *17.3*

structural formula a chemical formula that shows the arrangement of atoms in a molecule or a polyatomic ion; each dash between two atoms indicates a pair of shared electrons. *16.1*

structural isomer a compound that has the same molecular formula as another compound but that has a different molecular structure. *25.3*

sublimation the conversion of a solid to a gas or vapor without passing through the liquid state. *10.4*

substance a sample of matter having a uniform and definite composition; it can be either an element or a compound. *2.1*

substituent an atom or group of atoms that can take the place of a hydrogen atom on a parent hydrocarbon molecule. *25.1*

substitution reaction common type of organic reaction; involves the replacement of an atom or group of atoms by another atom or group of atoms. *26.1*

substrate a molecule on which an enzyme acts. *27.3*

supersaturated solution a solution that contains more solute than it can theoretically hold at a given temperature; excess solute precipitates if a seed crystal is added. *18.1*

surface tension an inward force that tends to minimize the surface area of a liquid; it causes the surface to behave as if it were a thin skin. *17.1*

surfactant a surface active agent; any substance with molecules that interfere with the hydrogen bonding between water molecules, reducing surface tension; soaps and detergents are surfactants. *17.1*

surroundings the remainder of the universe that is outside the system. *11.1*

suspension a mixture from which some of the particles settle out slowly upon standing. *17.4*

system any part of the universe upon which attention is focused. *11.1*

T

temperature a measure of the average kinetic energy of particles in matter; temperature determines the direction of heat transfer. *3.5*

ternary compound a compound containing atoms of three different elements, usually containing at least one polyatomic ion; Na_2CO_3 and $Mg(OH)_2$ are ternary ionic compounds. *6.4*

tetrahedral angle a bond angle of 109.5° created when a central atom forms four bonds directed toward the corners of a regular tetrahedron. *16.2*

theoretical yield the amount of product that could form during a reaction calculated from a balanced chemical equation; it represents the maximum amount of product that could be formed from a given amount of reactant. *9.3*

theory a thoroughly tested model that explains why experiments give certain results. *1.3*

thermochemical equation a chemical equation that includes the amount of heat produced or absorbed during the reaction. *11.2*

thermochemistry the study of heat changes in chemical reactions. *11.1*

titration method used to determine the concentration of a solution (often an acid or base); a solution of known concentration (the standard) is added to a measured amount of the solution of unknown concentration until an indicator signals the end point. *21.1*

***trans* configuration** a term applied to geometric isomers; it denotes an arrangement in which the substituted groups are on opposite sides of the double bond. *25.3*

transition metal Group B element characterized by addition of electrons to *d* suborbitals. *5.4, 14.1*

transition state a term sometimes used to refer to the activated complex. *19.1*

transmutation natural or artificial conversion of an atom of one element into an atom of another element by the emission of radiation from an unstable nucleus. *28.2*

transuranium element an element in the periodic table with an atomic number that is greater than 92. *28.2*

triglyceride an ester in which all three hydroxyl groups on a glycogen molecule have been replaced by long-chain fatty acids; fats are triglycerides. *27.4*

triple covalent bond a covalent bond in which three pairs of electrons are shared by two atoms. *16.1*

triple point the point on a phase diagram that represents the only set of conditions at which all three phases exist in equilibrium with one another. *10.4*

triprotic acid any acid that contains three ionizable protons (hydrogen ions); phosphoric acid (H_3PO_4) is a triprotic acid. *20.3*

Tyndall effect scattering of light by particles in a colloid or suspension, which causes a beam of light to become visible. *17.4*

U

unit cell the smallest group of particles within a crystal that retains the geometric shape of the crystal. *10.3*

universe the totality of all existing things. *11.1*

unsaturated a solution that contains less solute than a saturated solution at a given temperature and pressure. *18.1*

unsaturated compound an organic compound with one or more double or triple carbon–carbon bonds. *25.2*

unshared pair a pair of valence electrons that is not involved in bonding. *16.1*

V

vacuum a space where no particles of matter exist. *10.1*

valence electron an electron in the highest occupied energy level of an atom. *15.1*

van der Waals force a term used to describe the weakest intermolecular attractions; these include dispersion forces and dipole interactions. *16.3*

vapor a substance in the gaseous state that is ordinarily a liquid or solid at room temperature. *2.1*

vaporization the conversion of a liquid to a gas or a vapor. *10.2*

vapor pressure the pressure produced when vapor particles above a liquid in a sealed container collide with the container walls; a dynamic equilibrium exists between the vapor and the liquid. *10.2*

voltaic cell an electrochemical cell used to convert chemical energy into electrical energy; the energy is produced by a spontaneous redox reaction. *23.1*

volume the space occupied by a sample of matter. *3.3*

VSEPR theory valence-shell electron-pair repulsion theory; because electron pairs repel, molecules adjust their shapes so that valence-electron pairs are as far apart as possible. *16.2*

W

water of hydration water molecules that are an integral part of a crystal structure. *17.3*

wavelength (λ) the distance between two adjacent crests of a wave. *13.3*

wax an ester of a long-chain fatty acid and a long-chain alcohol. *27.4*

weak acid an acid that is only slightly ionized in aqueous solution. *20.4*

weak base a base that does not dissociate completely in aqueous solution. *20.4*

weak electrolyte a solution in which only a fraction of the solute exists as ions. *17.3*

weight force that measures the pull of gravity on a given mass. *3.3*

INDEX

The page on which a term is defined is indicated in **boldface** type.

A

D

H

I

J

K

L

M

N

O

P

Q

R

S

T

U

V

W

X

Y

Z

CREDITS

CP-Color-Pic, Inc., **FP**-Fundamental Photographs, **OPC**-Omni-Photo Communications, **PA**-Peter Arnold, Inc., **PR**-Photo Researchers, Inc., **SM**-The Stock Market

Photographs

Table of Contents iiiL Richard Megna/FP; iiiC Telegraph Colour Library/FPG International; iiiR Stephen Frisch*; ivLC Robert E. Daemmrich/Tony Stone Images; ivTR Coco McCoy/Rainbow; ivBR, Elisa Leonelli/TonyStone Images; vT Richard Megna/FP; vB Phil Degginger/CP; viL Wayne Estep/Tony Stone Images; viBL Richard Megna/FP; viBR Richard Megna/FP; viiT The Granger Collection; viiC Scott Smith/Earth Scene; viiB Hermann Lustbader/PR; viiiTR Richard Megna/FP; viiiCL, Molkenthin Studio/SM; viiiB Chris Cole/Tony Stone Images; ixT Richard Megna/FP; ixC Metcalfe/Thatcher/Tony Stone Images; ixB Ray Massey/Tony Stone Images; xTL Alvis Upitis/The Image Bank; xTR Carolyn Iverson; xBL Norbert Schafer/SM; xCR Reuters/Ho/Archive Photos; xBR E.R. Degginger/FPSA; xiT Paul Silverman/FP; xiB Richard Hutchings*; xiii R. Ian Lloyd/Westlight

Chapter 1 2, Courtesy NASA; 3T Courtesy NASA; 3B Mark A. Johnson/SM; 4TL Erik A. Svensson/SM; 4TC John Lamb/Tony Stone Images; 4TR Don Mason/SM; 4B Gregory Ochocki/PR; 5TL Telegraph Colour Library/FPG International; 5TC Courtesy NASA/PR; 5TR Thomas Braise/Tony Stone Images; 5CL Scott Camazine/PR; 5C Joe Bator/SM; 5BL Robert E. Daemmrich/Tony Stone Images; 5BC FOTOPIC/OPC; 6 Peter Johansky/FPG International; 7T Richard Laird/FPG International; 7BL BPL; 7BC James L. Amos/PA; 7BR Briolle Rapho/PR; 8L Kunio Owaki/SM; 8C Lowell Georgia/Science Source/PR; 8R Mark Lewis/Tony Stone Images; 9TL Bohdan Hrynewych/Stock, Boston; 9CR Fran Heyl Associates; 9BL Chromosohm/Sohm/PR; 9BC Ken Karp*; 9BR Ken Karp*; 10L Phil Degginger/CP; 10R Norbert Schafer/SM; 11T Gary Farber/The Image Bank; 11C Prof. K. Seddon & Dr. T. Evans, Queens Univ. Belfast Science Photo Library/PR; 11B Roslin Institute/Phototake NYC; 12TL G. David Lewis/CP; 12TC E.R. Degginger/CP; 12TR Nigel Cattlin/PR; 12B Ed Pritchard/Tony Stone Images; 13L NYC Parks Photo Archive/FP; 13R Kristen Brochmann/FP; 14TL Corbis-Bettmann; 14TR Coursesy NASA; 14CL Agence France Presse/Corbis-Bettmann; 15 Runk/Schoenberger/Grant Heilman Photography; 16TC Richard Megna/FP; 16TR Ken Karp*; 16BC Ken Karp*; 20TL Amanda Kavanaugh; 20BL Rosemary Weller/Tony Stone Images; 20BC Dr. Dennis Kunkel/Phototake; 20BR Richard Megna/FP; 21 Ken Karp*; 22 J. Nourok/PhotoEdit; 23 Ted Horowitz/SM; 25TL Telegraph Colour Library/FPG International; 25BL Telegraph Colour Library/FPG International; 25TR William Johnson/Stock, Boston; 25CRT Corbis; 25CRB Bruce Iverson; 25BR Phil Degginger/CP

Chapter 2 28 Dick Durrance II/SM; 29 Dick Durrance II/SM; 30 Geoffrey Nilsen Photography*; 31 Richard Megna/FP; 32T Neal & Mary Mishler/Tony Stone Images; 32B Ken Karp*; 33 David Lawrence*; 36T Lynn St. John/International Stock; 36B David Lawrence*; 37CL Yoav Levy/Phototake; 37C Art Stein/PR; 37CR David Lawrence*; 37BL Mike Hewitt/Tony Stone Images; 37BC Rich Treptow/PR; 37BR David Lawrence*; 38BL E. R. Degginger/CP; 38BCL Coco McCoy/Rainbow; 38BCR E.R. Degginger/CP; 38BR Monkmeyer/Conklin/Monkmeyer Press Photo; 41T Cameron Davidson/Tony Stone Images; 41BL Laurence Dutton/Tony Stone Images; 41BC Ed Pritchard/Tony Stone Images; 41BR Kim Heacox/Tony Stone Images; 42 Richard Megna/FP; 43 Richard Megna/FP; 45 James Stoots Lawrence/Livermore National Laboratory

Chapter 3 50 Dick Hanley/PR; 51T Jerry Cooke/PR; 51BL Leonard Lessin/FBPA; 51BC Comstock; 51BR Leonard Lessin/PA; 52T PA; 52B Fran Heyl Associates; 54 SYGMA; 55 Richard Hutchings*; 57 David Lawrence*; 58 Ken Karp*; 61 David Lawrence*; 63 Erich Lessing/Art Resource, NY; 64 David Lawrence*; 65TL Ken Karp*; 65TC David Lawrence*; 65TR David Lawrence*; 65CR David Lawrence*; 66TL Richard Hutchings*; 66TR Ken Karp*; 66BL Courtesy NASA; 67 Richard Hutchings*; 68 Dick Hanley/PR; 69 Richard Megna/FP; 71 Ken Karp*; 72L Phil Degginger/CP; 72R Phil Degginger/CP; 74TR Runk Schoenberger/Grant Heilman Photography; 74TL PA; 74CL Gordon Wiltsie/PA; 74BL Alan Schein/SM; 76 Bureau International des Poids & Measures; 80 Geoffrey Nilsen Photography*

Chapter 4 82 Pat Kepic*; 83T Pat Kepic*; 83B Ken Karp*; 86 Richard Hutchings*; 89 Ken Reid/FPG International; 91 Richard Hutchings*; 92 P. Young-Wolff/PhotoEdit; 93 Ken Karp*; 97 Ana Nance; 98 SuperStock; 99 Gary M. Prior/Allsport; 101 Grant Heilman Photography; 104 Roy Schneider/SM

Chapter 5 106 Physics Department, Imperial College/Science Photo Library/PR; 107 Physics Department, Imperial College/Science Photo Library/PR; 108 Fran Heyl Associates; 109T Science Photo Library/PR; 109B Stephen G. St. John/National Geographic Society; 110 Richard Megna/FP; 112 Steve Bronstein/The Image Bank; 113 Richard Shiell/Earth Scenes; 116 PA; 117 Elisa Leonelli/Tony Stone Images; 118 Steven Sutton/Duomo; 123T Chuck Keeler/Tony Stone Images; 123B The Granger Collection; 125TL Ken Karp*; 125TC Ken Karp*; 125TR Ralph Reinhold/Earth Scenes; 125BL Richard Megna/FP; 125BR Kip Peticolas/FP; 126BL George Whiteley/PR; 126T Stephen Marks/The Image Bank; 126BR Dawn Goss/Phototake; 127 PR

Chapter 6 132 Dr. Pepper Museum; 133T Rob Matheson/SM; 133BL Ken Karp*; 133BR Ginger Chih/PA; 134L Alese/Mortpechter/SM; 134R Kent Wood/PR; 138 Dick Luria/PG International; 139 David Young-Wolff/PhotoEdit; 140 Bruce Iverson; 141 Ken Karp*; 143 John Michael/International Stock; 144 E.R.Degginger/CP; 145L Ken Karp*; 145R Alan Oddie/PhotoEdit; 148L Paul Silverman/FP; 148C E.R. Degginger/CP; 148R Charles D. Winters/PR; 149 Dr. Pepper Museum; 150 David Barber/PhotoEdit; 153L Ken Karp*; 153R Haroldo De Faria Castro/FPG International; 154L SuperStock; 154C Paul Silverman/FP; 154R Randy Duchaine; 158T T. Kevin Smyth/SM; 158BL Tim Davis/PR; 158BR Spencer Grant/PhotoEdit; 160 Ken Karp*; 161 Fran Heyl Associates; 162 Jerry L. Ferrara/PR; 164 Ken Karp*

Chapter 7 170 Jim Zuckerman/Westlight; 171T Jim Zuckerman/Westlight; 171BL Richard Hutchings*; 171BR David Lawrence*; 172 Ken Karp*; 173 The Granger Collection; 174TL Richard Hutchings*; 174CL David Lawrence*; 174BL David Lawrence*; 175 Ken Karp*; 176 ZEFA-Lange/SM; 177 Geoffrey Nilsen Photography*; 179 Geoffrey Nilsen Photography*; 180 David Lawrence*; 182 Corbis; 184 David Lawrence*; 185 Ken Karp*; 188TR Alice Gurlich-Jones/OPC; 188CL David Lawrence*; 189BL Ewing Galloway/American Stock Photography; 189BR Rod Planck/Tony Stone Images; 193T Phil Degginger/CP; 193B Felicia Martinez/PhotoEdit; 196 Roy Morsch/SM

Chapter 8 202 Will and Deni McIntyre/PR; 203T Archive Photos; 203BL David Lawrence*; 203BC David Lawrence*; 203BR SuperStock; 204L Jacques M. Chenet/Woodfin Camp & Associates; 204C Oliver Meckes/PR; 204R SM; 205L Ken Karp*; 205R Ken Karp*; 210 David Lawrence*; 212 Will and Deni McIntyre/PR; 213L David Lawrence*; 213C David Lawrence*; 213R Richard Megna/FP; 215TL Ken Karp*; 215TC Ken Karp*; 215TR Ken Karp*; 215B Ron Watts/Westlight; 216 David Lawrence*; 217L E.R. Degginger/CP; 217R Charles D. Winters/PR; 219 Richard Megna/FP; Richard Megna/FP; 222TL David Lawrence*; 222TR David Lawrence*; 222BL Ken Karp*; 222BC Ken Karp*; 222BR Ken Karp*; 223T David Lawrence*; 223CL Richard Megna/FP; 223C Richard Megna/FP; 223CR Richard Megna/FP; 223B Richard Megna/FP; 225 Phil Schermeister/Tony Stone Images; 227 David Lawrence*; 230 Superstock; 235TL HMS Images; 235TR HMS Images; 235BL Jody Dole; 235BC Ken Karp*; 235BR Richard Megna/FP

Chapter 9 236 David Woods/SM; 237T Ken Straiton/SM; 237B Gianni Cigolini/The Image Bank; 239 SuperStock; 240 E.R. Degginger/CP; 242T David Woods/SM; 242L&R Grant Heilman Photography; 244 Richard Hutchings*; 246TR Courtesy NASA; 246L Runk/Schoenberger/Grant Heilman Photography; 248 Runk/Schoenberger/Grant Heilman; 249 Gerard Fritz/Tony Stone Images; 252T Tom Stewart/SM; 252B Ken Karp*; 254 E.R. Degginger/CP; 255 Ken Karp*; 256 Uniphoto; 257T Ken Karp*; 257B Richard Hutchings*; 260 David Young-Wolff/PhotoEdit; 265 SuperStock

Chapter 10 266 Ed Reschke/PA; 267 G.C. Kelley/PR; 268 Ken Karp*; 269 Runk/Schoenberger/Grant Heilman Photography; 274T OPC; 274B Phil Degginger/CP; 275 Richard Megna/FP; 276 Ken Karp*; 280 Ken Eward/PR; 281T Bruce Iverson; 281CL Charles D. Winters/PR; 281C BioPhoto Associates/PR; 281CR BioPhoto Associates/PR; 281BL E.R. Degginger/PR; 281BCL Gary Rutherford/PR; 281BCR Arthur Singer/Phototake NYC; 281BR George Roos/PA; 283 James L. Amos/PA; 284T Ed Reschke; 284B Ken Karp*; 287 Sydney Thomson/Earth Science; 290 Richard Megna/FP

Chapter 11 292 Lee Snyder/PR.; 293T Richard A. Cooke III/Tony Stone Images; 293BL D.R. Stoecklein/SM; 293BR Brian Czobat/International Stock; 295L Richard Hutchings*; 295R William Sallaz/Duomo; 296 David Lawrence*; 297T Accuweather; 297B Wayne Eastep/Tony Stone Images; 300 Lee Snyder/PR.; 303 Richard Megna/FP; 306L Kennan Ward/SM; 306R N.R. Rowan/PR; 307 Craig Hammell/SM; 309 Bob Abraham/SM; 312L Geoffrey Nilsen Photography*; 312R Richard Hutchings*; 314 Andy Caulfield/The Image Bank; 320 John Roberts/SM

Chapter 12 326 Alese/Mort Pechter/SM; 327T W. Geiersperger/SM; 327B Romilly Lockyer/The Image Bank; 328 Ken Karp*; 330T S.L. Craig Jr./Bruce Coleman Inc.; 330C Phil Degginger/CP; 331 Ken Karp*; 333 Alese/Mort Pechter/SM; 337 Ken Karp*; 341 Roger Tully/Tony Stone Images; 342 Erik Leigh Simmons/The Image Bank; 344 Barry L. Runk/Grant Heilman Photography; 345 Diane Hirsch/FP; 347 Butch Adams/The Image Bank; 351L Joe Towers/SM; 351R Keren Su/Stock, Boston; 352 Ken Karp*; 354 David & Doris Krumholz/FP; 357 Diane Hirsch/FP

Chapter 13 360 Richard Pasley/Stock, Boston; 361T Embry Riddle Aeronautical Univ.; 361BL Science Photo Library/PR; 361BCL Science Photo Library/PR; 361BCR Corbis-Bettman; 361BR Science Photo Library/PR; 367 Allan Morgan/PA; 370TL Doug Martin/PR; 370TR Guy Gillette/PR; 372 Richard Pasley/Stock, Boston; 374BL Kunio Owaki/SM; 374BC Richard Megna/FP; 374BR Richard Megna/FP; 374 Wabash Instrument Corp./FP; 375L Ray Pfortner/PA; 375R Blair Seitz/PR; 376T European Space Agency/PR; 376B Phil Degginger/CP; 377 Chromosohm/Sohm/SM; 381 Martin Dohrn/PR; 384 David Parker/PR; 387 Richard Megna/FP

Chapter 14 390 The Granger Collection; 391 The Granger Collection; 394L John Michael/International Stock; 394R E.R. Degginger/CP; 395TL David Sailors/SM; 395TR Charles O'Rear/Westlight; 395BL Stephen Frisch/Stock, Boston; 395BR E.R. Degginger/CP; 398 SuperStock; 402 Phil Degginger/CP; 407 Dr. Jeremy Burgess/Science Photo Library/PR

Chapter 15 412 Pat Lynch/PR; 413T Pat Lynch/PR; 413BL Amos Zezmer/OPC; 413BR Pawel Kumelowski/OPC; 414T Yoav Levy/Phototake; 414B Joel Greenstein/OPC; 416L Erich Lessing/Art Resource; 416R Nathan Beck/OPC; 419T Gary Retherford/PR; 419R Ken Karp*; 422TL Breck P. Kent; 422TR J&L Weber/PA; 422CL J&L Weber/PA; 422CC Breck P. Kent/Earth Scenes; 422CR Grace Davies/OPC; 422BL George Whiteley/PR; 422BC Tomisch/PR; 422BR Paul Silverman/FP; 423T Omikron/Science Source/PR; 423B Tomisch/PR; 424 Phil Degginger/Earth Scenes; 427 Bruce Frisch/PR; 428T Steve Elmore/SM; 428B Fran Heyl Associates; 429L Scott Smith/Earth Scenes; 429R Kourosh Behbahani/Tony Stone Images; 430 Barry L. Runk/Grant Heilman Photography

Chapter 16 436 Gerben Oppermans/Tony Stone Images; 437 Courtesy NASA; 438L Robert Frerck/Woodfin Camp and Assoc.; 438R Garry Hunter/Tony Stone Images; 442 Hugh Sitton/Tony Stone Images; 443 Susan M. Klemens/Stock, Boston; 444 Richard Hutchings*; 449 A. Ramey/Woodfin Camp and Assoc.; 452 James Schnepf/Liaison International; 453 United States Dept. of Interior Geological Survey; 460T Gerben Oppermans/Tony Stone Images; 460B George Disario/SM; 464 Dr. Arthur Lesk, Laboratory for Molecular Biology/SPL/PR; 465 Craig Tuttle/SM; 468 Allen Penn*

Chapter 17 474 Frank Rossotto/SM; 475T Frank Rossotto/SM; 475R William and Marcia Levy/PR; 476 Hermann Eisenbeiss/PR; 477L Michael Lustbader/PR; 477R Ken Karp*; 478L Richard Hutchings*; 478R Four by Five; 479T Critical Position of H.M.S. Investigator on the North Coast of Baring Island, August 20th 1851, drawn by Lieu. S. Gurney Cresswell, pub. 1854 by Day & Son and Ackermann & Co.; Royal Geographical Society, London, UK/Bridgeman Art Library, London/New York; 479BL Thomas R. Fletcher/Stock, Boston; 479BC E.R. Degginger/CP; 479BR SuperStock; 481T Nuridsany et Perennou/Science Source PR; 481B Richard Hutchings*; 482 Russ Lappa*; 483 Cary Wolinsky/Stock, Boston; 484 FP; 486T Ken Karp*; 486L Kristen Brochman/FP*; 487 Ken Karp*; 490 Russ Lappa*; 491 Ken Karp*; 491(2nd from bottom) Johnathan T. Wright/Bruce Coleman, Inc.; 492 Robert Essel/SM; 493 Ken Karp*; 494 SuperStock

Chapter 18 500 St. Petersburg Times/Liason International; 501T St. Petersburg Times/Liason International; 501BR Richard Megna/FP; 501B Ken Karp*; 503BL Phil Degginger/Tony Stone Images; 503BR Jim Olive/PA; 505 Runk/Schoenberger/Grant Heilman Photography; 506T 1990 Richard Megna/FP; 506B John Lamb/Tony Stone Images; 507 Jim Brandenburg; 508 Richard Megna/FP*; 509T Tony Stone Images; 509B Ken Karp*; 510L Tom McCarthy/Rainbow; 510R Richard Hutchings*; 512T Ken Karp*; 512B Richard Hutchings*; 514 Ken Karp*; 517 John M Burnley/PR; 518 Ken Karp*; 519 Ken Karp*; 520 Ken Karp*; 522 Phil Degginger/CP; 526 Dan McCoy/Rainbow

Chapter 19 532 Associated Press; 533T J. Langevin/Sygma; 533B Gray Mortimore/Tony Stone Images; 534TL Thomas Brase/Tony Stone Images; 534TR Nathan Beck/OPC; 534BL Erika Stone/PA; 534BR Werner H. Muller/PA; 536 Ken Karp*; 537 The Bettman Archives; 538 Richard Hutchings*; 539 Chuck Kuhn/The Image Bank; 540 Martin Rogers/Stock, Boston; 542 Chris Cole/Tony Stone Images; 545 Geoffrey Nilsen Photography*; 549T Associated Press; 549B Chris Hamilton/SM; 550 Ken Karp*; 551 John Kaprielian/PR; 552 Richard Hutchings*; 558 Frank Siteman/OPC; 561 AP/Wide World Photos; 562 Fotopic/OPC; 566 Reuters/Corbis-Bettman; 570 Ken Karp*

Chapter 20 576 Ken Karp*; 577T J. H. Robinson/PR; 577TR Richard Megna/FP; 577C Bill Bachman/PR; 577L Don Mason/SM; 577BR Russ Lappa*; 578T Molkenthin Studio/SM; 578B Tom Hollyman/PR; 580 Martyn F. Chillmaid/Science Photo Library/PR; 581L Michael Dalton/FP; 581C E.R. Degginger/CP; 581R Ken Karp*; 589 Richard Megna/FP*; 591 Richard Megna/FP*; 592T E.R. Degginger/CP; 592BL Richard Hutchings*; 592BR Matt Meadows/PA; 594T Stephen Kraseman/Nature Conservancy/PR; 594BL Mike Neumann/PR; 594, John Maher/Stock Boston/Picture Quest; 594BR Bob Daemmrich/ Stock, Boston; 600 Ken Karp*; 607 Simon Fraser/Science Photo Library/PR

Chapter 21 612 Paul Silverman/FP*; 613T Paul Silverman/FP*; 613BL E.R. Degginger/CP; 613BR Roy Morsch/SM; 616 Ken Karp*; 617T Richard Megna/FP*; 617B Ken Karp*; 626 Bill Longcore/PR; 627T Richard Megna/FP; 627B Ken Karp*; 628 Ken Karp*; 629 No credit; 631 Ken Karp*; 632 CNRI/Science Photo Library/PR; 635 Ken Karp*; 638 Pete Saloutos/SM; 642 Ken Karp*

Chapter 22 644 Bill Stormont/SM; 645T Bill Stormont/SM; 645B Ken Karp*; 646T Richard Megna/FP*; 646B E.R. Degginger/CP*; 647L The Granger Collection; 647R Crown Copyright/Health & Safety Laboratory/Science Photo Library/PR; 648L Paul Silverman/FP*; 648C Ken Karp*; 648R Ken Karp*; 650 Jens J. Jensen/World Images, Inc./SM; 651 Jan Halaska/PR; 652L Michael Newman/Photo Edit; 652R Julie Houck/Stock, Boston; 653T No credit; 653B, Jonathan Blair/Corbis; 654T Werner H. Muller/PA; 654B Ken Karp*; 655 Ken Karp*; 657 Ken Karp*; 660T The Granger Collection-New York; 660B Richard Megna/FP*; 661T E.R. Degginger/CP; 661B Ken Karp*; 664 Ken Karp*; 667 Ivan Polunin/Bruce Coleman Inc.; 669 Leonard Lessin/FBPA; 671 Photri-Microstock/OPC; 674 Ken Karp*

Chapter 23 676 Clay Wiseman/Animals Animals; 677T Clay Wiseman/Animals Animals; 677B Richard Megna/FP*; 679L Science Photo/PR; 679C The Granger Collection; 679R Paul Silverman/FP; 682 Leonard Lessin/FBPA; 684 Frank Rosotto/SM; 685 D. Young-Wolff/Photo Edit; 692T Robert Rathe/Stock, Boston; 692TL John Clark/SM; 692BL Stephen Frisch/Stock, Boston; 692BR Occidental Chemical Corp.; 694 Charles D. Winters/PR; 695 E.R. Degginger/CP; 696 Felicia Martinez/Photo Edit; 699 Will and Deni McIntyre/PR

Chapter 24 704 Erich Lessing/Art Resource; 705T Corbis/Bettman; 705B Jim Wark/PA; 706 Richard Megna/FP; 708TL Charles D. Winters/PR; 708TC E.R. Degginger/CP; 708BL E.R. Degginger/CP; 708BC Paul Silverman/FP; 708R Agra Uttar Pradesh/SuperStock; 709T E.R. Degginger/CP*; 709C E.R. Degginger/PR; 709B Ken Karp*;710L Paul Silverman/FP; 710R Jim Olive/Uni Photo Picture Agency; 712T Erich Lessing/Art Resource; 712BL S.J. Krasemann/PA; 712BR American Stock; 713T Paul Silverman/FP; 713B Matthew McVay/All Stock; 714L C. Falco/PR; 714R Kristen Brochmann/FP; 715 Richard Megna/FP; 716T Grant Heilman Photography; 716BL Metcalfe/Thatcher/Tony Stone Images; 716BR Geoffrey Nilsen Photography*; 717L Barrie Fanton/OPC; 717R Richard Megna/FP; 718 Phil Degginger/CP; 719L Richard Megna/FP; 719R Charles D. Winters/PR; 720 Dr. Ken McDonald/Science Photo Library; 721 Ken Karp*; 722TL Stephen Frisch*; 722TR Richard Megna/FP; 722B Geoffrey Nilsen Photography*; 724T Mauritius-Wesche/The Stock Shop Inc.; 724B E.R. Degginger/CP; 725T Geoffrey Nilsen Photography*; 725BL Stephen Frisch; 726L Thomas Brase/Tony Stone Images; 726C Ray Massey/Tony Stone Images; 726R Ken Karp*; 728 Runk/Scheonberger/Grant Heilman Photography; 729TL Paul Silverman/FP; 729TC John Cancalosi/PA; 729TR E.R. Degginger/CP; 729B Paul Silverman/FP; 730T Ken Karp*; 730BL Egyptian Museum, Cairo, Egypt/Giraudon/SuperStock; 730BR Sepp Seitz/Woodfin Camp & Associates; 731L Alberto Incrocci/The Image Bank; 731C Derek Berwin/The Image Bank; 731R Maximilian Stock Ltd/Science Photo Library/PR; 732T Courtesy NASA; 732B Ken Karp*; 733 Mary Evans/PR; 734T E.R. Degginger; 734B Cralle/The Image Bank; 734R E.R. Degginger; 736 Peter Lamberti/Tony Stone Images

Chapter 25 742 Jet Propulsion Lab; 743TL Richard A. Cooke III/Tony Stone Images; 743BL Michael P. Gadomski/PR; 743CR James H. Robinson/PR; 743BR David R. Frazier/PR; 746L Jacques Cochin/Agence Vandystadt/PR; 746R Leonard Lessin/FBPA; 747 Andrew McClenaghan/Science Photo Library/PR; 751 Edward Parker/PA; 752TR Nigel Cattlin/PR; 752CL Alan & Linda Detrick/PR; 752BL Ross M. Horowitz/The Image Bank; 754 Martin Dohrn/Science Photo Library/PR; 759 Miwako Ikeda/International Stock; 760L Barry L. Runk/Grant Heilman Photography; 760R Malcolm Fielding, Johnson Matthey/Science Photo Library/PR; 762T Jet Propulsion Lab; 762B Joseph Sohm/Chromosohm/PR; 764C Earth History Hall, American Museum of Natural History; 764CL Bob O'Shaughnessy/SM; 764BC E.R. Degginger/CP; 764BR Carolina Biological Supply Company/Phototake; 764CR E.R. Degginger/CP; 765 William Felger/Grant Heilman Photography; 766 Jeffrey Mark Dunn/Stock, Boston

Chapter 26 772 Ken Karp*; 773T Robert Fried/Stock, Boston; 773B Michael Groen*; 776 Alan Levenson/Tony Stone Images; 778 Brown Brothers; 779 Michael Groen*; 780L Michael Groen*; 780C Michael A. Keller/SM; 780R Phil Degginger/CP; 781T Ken Karp*; 781B Ken Karp*; 782 Richard Megna/FP; 785 Ken Karp*; 787 Jane Grushow/Grant Heilman Photography; 788 Michael Groen*; 790L Steve Solum/Bruce Coleman Inc.; 790C Danny Eilers/Grant Heilman Photography; 790R Barry L. Runk/Grant Heilman Photography; 792L Tony Freeman/PhotoEdit; 792R Berenholtz/SM; 793 Richard Megna/FP; 795 Andrea Pistolesi/The Image Bank; 796L Bob Masini/Phototake; 796C Alan Schein/SM; 796R Grant Heilman Photography; 797TR Leonard Lessin/FBPA; 797CL Bill Stanton/International Stock; 797CR Rosemary Weller/Tony Stone Images; 797BL Chris Sorensen/SM; 797BR Garry Gay/The Image Bank; 798L Michael English/Custom Medical Stock Photo; 798R Michelle Del Guercio/Custom Medical Stock Photo; 799L Zephyr/Camerique; 799R Alvis Upitis/The Image Bank; 800 Bob Daemmrich/Stock, Boston; 802 Michael Busselle/Tony Stone Images; 802Inset Michael Groen*

Chapter 27 808 George W. Disario/SM; 809T Randy Wells/Tony Stone Images; 809B Carlyn Iverson; 811 Kunio Owaki/SM; 812T Paul Skelcher/Rainbow; 812B Ken Karp*; 815 J. Barry O'Rourke/SM; 816 John Mitchell/PR; 817T A.J. Wasserman/Visuals Unlimited; 817B E.R. Degginger/CP; 819 NIBSC/Science Photo Library/PR; 821T Dennis Blachut/SM; 821B Richard Hutchings*; 822 Richard Hutchings*; 823T Dr. Dennis Kunkel; 823B Hans Pfletschinger/PA; 824T George W. Disario/SM; 824B Phil A. Harrington/PA; 825 Richard Megna/FP; 828T Jackie Lewin, Royal Free Hospital-Science Photo Library/PR; 828B Ken Eward/Science Source/PR; 829 Dan McCoy/Rainbow; 831T Caloene/PA; 831B Reuters/Ho/Archive Photos; 832 John Paulkay/PA; 835 Michael Pohuski/Envision; 839 Leonard Lessin/PA

Chapter 28 840 Lawrence Radiation/PR; 841T Novosti/Science Photo Library/PR; 841BL The Granger Collection; 841BR Science Photo Library/PR; 845 Richard Megna/FP*; 847 Archive Photos; 851 Robert Isaacs/Science Source PR; 853T Courtesy NASA; 853B Ken Karp*; 854R Dr. Jeremy Burgess/PR; 855 Daniel MacDonald/Stock Boston/Pictue Quest; 857T Lawrence Radiation Lab/PR; 857B Hank Morgan/PR; 859T Yoav Levy/Phototake NYC; 859B Runk/Schoen Erger/Grant Heilman Photography; 860 CNRI/Phototake NYC; 861 Larry Mulvehill/PR; 862 Alan Rostro/Waste Isolation Pilot Plant (WIPP)

Chemistry In Careers 868T Bill Varie/Westlight; 868B Iowa State University/Science Photo Library/PR; 869T Steve Chenn Photography/Westlight; 869B Mac Donald Photography/The Stockhouse, Inc.; 870T Wesley Bocxe/PR; 870B SuperStock; 871T Peter French/Bruce Coleman, Inc.; 871B Alvis Upitis/The Image Bank; 872T Jeff Greenberg/Unicorn Photos; 872B Michael Sewell/PA; 873T US Department of Energy/Mark Marten/PR; 873B Lawrence Migdale/Stock, Boston; 874T Firefly Productions/Firefly; 874B George Haling/PR; 875T Tom Hollyman/PR; 875B Michael A. Keller Studios LTD./SM; 876T Brian Smith/Stock, Boston; 876B Giancarlo DeBellis/OPC; 877T C.E. Mitchell/Black Star; 877B Y. Arthus Bertrand/PA; 878T Stephen Derr/The Image Bank; 878B Chris Jones/SM; 879T Randy Duchaine/SM; 879B Andy Duchaine; 880T Matt Meadows/PA; 880B Russ Lappa*; 881T Hank Morgan/Rainbow; 881B Archive Photos

Chem Serving Borders SOCIETY Nick Wheeler/Westlight; THE ENVIRONMENT, INDUSTRY C. Murray/Westlight; THE CONSUMER R. Ian Lloyd/Westlight

*Photographs taken expressly for Addison-Wesley Chemistry © 2000 and © 2002.

Special thanks to Kris Wynne-Jones

Illustrations

Academy Artworks iv, vii, ix, xi, 21, 117, 118, 129, 179, 192, 213, 215, 216, 219, 221, 239, 248, 253, 362, 382, 389, 423, 437, 438, 439, 442, 443, 444, 449, 451, 453, 454, 455, 456, 457, 458, 459, 461, 463, 464, 465, 466, 563, 694, 719, 745, 746, 747, 749, 753, 754, 755, 757, 758, 759, 760, 769, 771, 774, 776, 780, 783, 785, 789, 791, 806, 807, 839

Jim DeLapine 295, 303, 307, 310, 315, 317, 318

Guilbert Gates 4, 16, 21, 39, 43, 52, 54, 58, 62, 75, 84, 89, 99, 169, 177, 192, 239, 247, 253, 263

Network Graphics iii, viTL, 19, 27, 49, 81, 105, 169, 178, 201, 208, 209, 235, 265, 268, 274, 281, 283, 285, 291, 325, 359, 389, 411, 435, 473, 475, 476, 481, 484, 499, 518, 531, 539, 553, 575, 580, 595, 597, 598, 611, 614, 626, 631, 643, 645, 646, 648, 675, 677, 680, 703, 741, 771, 807, 822, 839, 843, 850, 853, 856, 867

J/B Woolsey Associates v, viCR, vii, 30, 34, 49, 56, 68, 107, 108, 109, 110, 111, 124, 131, 169, 186, 199, 235, 259, 268, 269, 275, 277, 278, 280, 281, 282, 286, 291, 294, 328, 330, 331, 332, 333, 336, 338, 345, 347, 350, 363, 373, 374, 377, 380, 392, 394, 395, 398, 399, 400, 401, 403, 404, 404, 406, 410, 411, 412, 419, 424, 425, 427, 433, 460, 467, 477, 492, 493, 503, 505, 512, 517, 520, 522, 533, 534, 535, 543, 545, 552, 553, 573, 593, 621, 622, 652, 672, 680, 681, 682, 683, 684, 686, 693, 695, 696, 702, 718, 727, 817, 818, 823, 825, 826, 830, 833, 842, 844, 845, 847, 849, 854, 857, 858, 859, 860, 867

Precision Graphics 811, 819, 822, 828

Molly K. Scanlon 273, 629

Michael Sloan 15, 16

Steven Stankiewicz 9, 13, 16, 26, 38, 48, 105, 188, 272, 279, 284, 291, 335, 336, 344, 372, 373, 504, 505, 523, 538, 583, 584, 584, 590, 709, 715, 739, 740, 833, 834

Nina Wallace 35, 72, 207, 238, 300, 301, 427, 762, 763